IFMBE Proceedings

Volume 76

The IFMBE Proceedings Book Series is an official publication of *the International Federation for Medical and Biological Engineering* (IFMBE). The series gathers the proceedings of various international conferences, which are either organized or endorsed by the Federation. Books published in this series report on cutting-edge findings and provide an informative survey on the most challenging topics and advances in the fields of medicine, biology, clinical engineering, and biophysics.

The series aims at disseminating high quality scientific information, encouraging both basic and applied research, and promoting world-wide collaboration between researchers and practitioners in the field of Medical and Biological Engineering.

Topics include, but are not limited to:

- Diagnostic Imaging, Image Processing, Biomedical Signal Processing
- Modeling and Simulation, Biomechanics
- Biomaterials, Cellular and Tissue Engineering
- Information and Communication in Medicine, Telemedicine and e-Health
- Instrumentation and Clinical Engineering
- Surgery, Minimal Invasive Interventions, Endoscopy and Image Guided Therapy
- Audiology, Ophthalmology, Emergency and Dental Medicine Applications
- Radiology, Radiation Oncology and Biological Effects of Radiation

IFMBE proceedings are indexed by SCOPUS and EI Compendex. They are also submitted for ISI proceedings indexing.

Proposals can be submitted by contacting the Springer responsible editor shown on the series webpage (see "Contacts"), or by getting in touch with the series editor Ratko Magjarevic.

More information about this series at http://www.springer.com/series/7403

Jorge Henriques · Nuno Neves ·
Paulo de Carvalho
Editors

XV Mediterranean Conference on Medical and Biological Engineering and Computing – MEDICON 2019

Proceedings of MEDICON 2019, September 26–28, 2019, Coimbra, Portugal

Set 2

Springer

Editors
Jorge Henriques
University of Coimbra
Coimbra, Portugal

Nuno Neves
Universidade do Minho
Braga, Portugal

Paulo de Carvalho
University of Coimbra
Coimbra, Portugal

ISSN 1680-0737 ISSN 1433-9277 (electronic)
IFMBE Proceedings
ISBN 978-3-030-31634-1 ISBN 978-3-030-31635-8 (eBook)
https://doi.org/10.1007/978-3-030-31635-8

This Springer imprint is published by the registered company Springer Nature Switzerland AG
The registered company address is: Gewerbestrasse 11, 6330 Cham, Switzerland

Preface

MEDICON 2019 is the XV in the series of regional meetings of the International Federation of Medical and Biological Engineering (IFMBE) in the Mediterranean. MEDICON 2019 will be organized for the first time in Portugal and will be hosted by the UNESCO World Heritage University, the University of Coimbra.

The goal of MEDICON 2019 is to provide updated information on the state of the art on Medical and Biological Engineering and Computing under the main theme "**Patient empowerment**". Patient empowerment has emerged as a new paradigm that positions the patients at the heart of the health system and encourages them to be actively involved in managing their own healthcare needs. Effective patient empowerment requires a holistic approach, combining multiple dimensions of needs and patient contexts. Medical and Biological Engineering and Computing is a discipline at the heart of patient empowerment. Research and development in these areas are impacting the science and technology by advancing fundamental concepts in translational medicine and understanding in human physiology, function and behaviour at multiple levels. This is leading to improved tools and techniques for the detection, prevention, treatment and management of diseases. MEDICON 2019 provides a common platform for the cross fertilization of ideas and to help shape knowledge and scientific achievements by bridging complementary disciplines into an interactive and attractive forum under the special theme of the conference that is "improving health care through holistic patient empowerment".

The programme consists of some approximately 250 invited and submitted papers on new developments around the Conference theme, presented in 8 plenary sessions, 18 parallel scientific sessions and 19 special sessions and also includes a set of competitions and awards.

More specifically, the parallel scientific sessions cover the topics of biomedical signal processing; biomedical imaging and image processing; bio-instrumentation, bio-senso and bio-micro/nanotechnologies; bioinformatics, computational biology and systems biology; biomechanics, robotics and rehabilitation; therapeutic and diagnostic systems, devices and technologies and clinical engineering; information technology in health systems; assistive technologies; technologies for active ageing;

biomedical engineering education and society; clinical engineering and health technology assessment; neuroengineering, neurosystems; technologies for preventive health care; biomedical technologies for developing countries; standardization of open data; biomaterials and tissue engineering.

The special sessions include the topics of optimization in medicine and biology; ontological engineering in biomedical informatics; electronics and smart algorithms for the effective lung monitoring and COPD management; non-invasive temperature assessment using ultrasound; computational biology and medical applications; smartphone-based, patient-centred technologies; computational and experimental modelling for designing bone-implant systems; artificial organs: extracorporeal blood circulation medical devices; diabetes and cardiovascular diseases: Ibero-American trends; smart robotic assistant for minimally invasive surgery: the SMARTsurg project experience; intelligent computational systems in biomedical engineering; INT4DAT - Intelligent systems and technologies for diagnostic, assistance and therapeutics; upper limb exoskeletons for a better quality of life: what is real, what is useful, and what is next?; neurosystems and connectivity; therapeutic applications of imaging and neurostimulation; value-based health technology assessment; international collaborative on medical device assessment; ocular imaging; assessing human error in cognitive/intellectual demanding tasks: case study on software engineering.

The conference programme also includes three competitions and two awards: IFMBE Scientific Challenge competition; Fraunhofer Best Portuguese PhD thesis competition in Biomedical Engineering; Fraunhofer Best Portuguese MSc thesis competition in Biomedical Engineering; Best Student Paper Award; Young Investigator Competition Award.

Furthermore, the conference programme is highlighted by eight plenary sessions: Digitally empowered patients, presented by Aart van Halteren; Cardiovascular modelling and simulations—applications to some clinical studies, presented by Adélia Sequeira; Biomedical signals and images processing: towards innovative paradigms of information integration in the era of precision medicine and big data in health, presented by Sergio Cerutti; Multilingual dictionary of medical physics terms—update and relevance for clinical engineering, presented by Slavik Tabakov; Prevention Engineering: evolving challenges for biomedical and clinical engineering, presented by Luis Kun; Optical coherence tomography: a window into the mechanism of neurodegenerative disorders, presented by Rui Bernardes; Towards a value based healthcare system supported by process mining techniques, presented by Vicente Traver; In silico clinical trials: towards transforming the biomedical industry and the healthcare delivery, presented by Dimitrios Fotiadis.

Particular thanks are expressed to the kind support and effort of a number of external sponsors to which we would like to express our appreciation. Finally, a heartfelt thanks to all of you, the participants for your paper contributions, wishing you every success in your work at the conference. We hope that MEDICON 2019 will offer opportunities for professional growth and establishing new contacts with fellow colleagues. Our intention is to do all we can to make your participation in MEDICON 2019 worthwhile and your stay in Coimbra enjoyable and memorable.

We hope you will appreciate this proceedings volume as much as we are proud of it!

September 2019

Paulo de Carvalho
General Co-chair

Jorge Henriques
Nuno Neves
Program Committee Co-chairs

Organization

Editors

Jorge Henriques	University of Coimbra, Portugal
Nuno Neves	University of Minho, Portugal
Paulo de Carvalho	University of Coimbra, Portugal

Organizers

Thematic Tracks
T1: Biomedical Signal Processing

Anna Bianchi	Politecnico di Milano, Italy
Ana Paula Rocha	University of Porto, Portugal

T2: Biomedical Imaging and Image Processing

Ana Mendonça	University of Porto, Portugal
João Sanches	University of Lisbon, Portugal

T3: Bio-instrumentation, Biosensors and Bio-micro/nanotechnologies

Pedro Vieira	University of Lisbon, Portugal
João Paulo Cunha	University of Porto, Portugal

T4: Bioinformatics, Computational Biology and Systems Biology

Paula Oliveira	University of Coimbra, Portugal
Joel Arrais	University of Coimbra, Portugal

T5: Biomechanics, Robotics and Rehabilitation

Renato Natal Jorge	University of Porto, Portugal
Urbano Nunes	University of Coimbra, Portugal

T6: Therapeutic and Diagnostic Systems, Devices and Technologies and Clinical Engineering

Altamiro Pereira	University of Porto, Portugal
Jens Muehlsteff	Philips Research, Netherlands

T7: Information Technology in Health Systems

Nicos Maglaveras	Northwestern University, USA

T8: Assistive Technologies

Vicente Traver	Universitat Politècnica de València, Spain
Antonio Martinez Millana	Universitat Politècnica de València, Spain

T9: Technologies for Active Ageing

João Malva	University of Coimbra, Portugal
Maria Arredondo	Universidad Politècnica de Madrid, Spain

T10: Biomedical Engineering Education and Society

Shankar Krishnan	Wentworth Institute of Technology, USA
Monique Frize	University of Ottawa, Canada

T11: Clinical Engineering and Health Technology Assessment

Leandro Pecchia	University of Warwick, UK
Ernesto Iadanza	University of Florence, Italy

T12: Neuroengineering, Neurosystems

Miguel Castelo-Branco	University of Coimbra, Portugal
Rita Nunes	University of Lisbon, Portugal

T13: Technologies for Preventive Health Care

Miguel Coimbra	University of Porto, Portugal
Ioanna Chouvarda	Aristotle University of Thessaloniki, Greece

T14: Biomedical Technologies for Developing Countries

Graça Ruano	University of Algarve, Portugal
Martha Zequera	Pontificia Universidad Javeriana, Colombia

T15: Standardization of Open Data

Ratko Magjarevic	University of Zagreb, Croatia

T16: Biomaterials and Tissue Engineering

Birgit Glasmacher	Leibniz Universität Hannover, Germany

Special Sessions

SS01: Optimization in Medicine and Biology

Joana Matos Dias	University of Coimbra, Portugal
Humberto Rocha	University of Coimbra, Portugal

SS02: Electronics and Smart Algorithms for the Effective Lung Monitoring and COPD Management

Nicos Maglaveras	Aristotle University of Thessaloniki, Greece
Andreas Raptopoulos	EXODUS Innovation, Greece
Rui Paiva	University of Coimbra, Portugal

SS03: Non-invasive Temperature Assessment Using Ultrasound

César Teixeira	University of Coimbra, Portugal
André Alvarenga	INMETRO, Rio de Janeiro, Brazil
Wagner Pereira	Federal University of Rio de Janeiro, Brazil

SS04: Computational Biology and Medical Applications

José Ferreira	University of Coimbra, Portugal

SS05: Smartphone-Based, Patient-Centred Technologies

Rute Almeida	CINTESIS, Porto, Portugal
Ana Ferreira	CINTESIS, Porto, Portugal
João Fonseca	University Medical School of Porto, Portugal

SS06: Computational and Experimental Modelling for Designing Bone-Implant Systems

Ana Amaro (Organizer)	University of Coimbra, Portugal

SS07: Artificial Organs: Extracorporeal Blood Circulation Medical Devices

Maria Norberto Pinho	University of Lisbon, Portugal
Mónica Faria	University of Lisbon, Portugal

SS08: Diabetes and Cardiovascular Diseases: Ibero-American Trends

M. Graça Ruano	University of Algarve, Portugal

SS9: Smart Robotic Assistant for Minimally Invasive Surgery: The Smartsurg Project Experience

Sanja Dogramadzi,	University of the West of England, UK
Elana De Momi	Politecnico di Milano, Italy

SS10: Intelligent Computational Systems in Biomedical Engineering

Jan Kubicek	Technical University of Ostrava, Czech Republic
Marek Penhaker	Technical University of Ostrava, Czech Republic
Martin Cerny	Technical University of Ostrava, Czech Republic
Martin Augustynek	Technical University of Ostrava, Czech Republic

SS11: INT4DAT – Intelligent Systems and Technologies for Diagnostic, Assistance and Therapeutics

Joao Ruivo Paulo	University of Coimbra, Portugal
Jerome Mendes	Spanish National Research Council, Spain
Teresa Sousa	University of Coimbra, Coimbra
Eduardo Rocon	University of Coimbra, Portugal

SS12: Upper Limb Exoskeletons for a Better Quality of Life: What Is Real, What is Useful, and What is Next?

Alessandra Pedrocchi	Politecnico di Milano, Italy
Emilia Ambrosini	Politecnico di Milano, Italy
Marta Gandolla	Politecnico di Milano, Italy

SS13: Neurosystems and Connectivity

Miguel Castelo-Branco	University of Coimbra, Portugal
Rita Nunes	University of Lisbon, Portugal

SS14: Therapeutic Applications of Imaging and Neurostimulation

Miguel Castelo-Branco	University of Coimbra, Portugal
Rita Nunes	University of Lisbon, Portugal

SS15: Value-Based Health Technology Assessment

Dan Clark	University of Nottingham, UK
Giuditta Callea	Bocconi University, Italy
Marjan Hummel	Philips Research, Eindhoven, The Netherlands
Martina Andellini	Bambino Gesù Children's Hospital, Italy
Ernesto Iadanza	University of Florence, Italy

SS16: International Collaborative on Medical Devices Assessment

Ernesto Iadanza	University of Florence, Italy
Julie Polisena	Canadian Agency for Drugs and Technologies in Health, Canada
Leandro Pecchia	University of Warwick, UK

SS17: Ocular Imaging

Miguel Morgado	University of Coimbra, Portugal
Rui Bernardes	University of Coimbra, Portugal

SS18: Assessing Human Error in Cognitive/Intellectual Demanding Tasks: Case Study on Software Engineering

Ricardo Couceiro	University of Coimbra, Portugal
Henrique Madeira	University of Coimbra, Portugal

Challenges

IFMBE Scientific Challenge

Marco Simões	University of Coimbra, Portugal
Jorge Henriques	University of Coimbra, Portugal

Fraunhofer Best Portuguese PhD Thesis Competition in Biomedical Engineering

Aurélio Campilho	University of Porto, Portugal

Fraunhofer Best Portuguese MsC Thesis Competition in Biomedical Engineering

Aurélio Campilho	University of Porto, Portugal

Student Competition

Anna Bianchi	Politecnico di Milano, Italy

Young Investigator Competition

Anna Bianchi	Politecnico di Milano, Italy

Organizing Committees

General Co-chairs

Mário Secca	New University of Lisbon, Portugal
Paulo de Carvalho	University of Coimbra, Portugal

Programme Committee

Jorge Henriques	University of Coimbra, Portugal
Nuno Neves	University of Minho, Portugal

Financial Chair

César Teixeira	University of Minho, Portugal

Industrial Track Co-chairs

Miguel Morgado	University of Coimbra, Portugal
Hugo Gamboa	New University of Lisbon, Portugal

Publicity Co-chairs

Paula Oliveira	University of Coimbra, Portugal
Rui Pedro Paiva	University of Coimbra, Portugal

Local Arrangement

César Teixeira	University of Coimbra, Portugal
Jorge Henriques	University of Coimbra, Portugal
Márcia Santos	University of Coimbra, Portugal
Miguel Castelo-Branco	University of Coimbra, Portugal
Miguel Morgado	University of Coimbra, Portugal
Paula Oliveira	University of Coimbra, Portugal
Ricardo Couceiro	University of Coimbra, Portugal
Rui Pedro Paiva	University of Coimbra, Portugal

International Advisory Board

Aurélio Campilho	University of Porto, Portugal
Constantinos S. Pattichis	University of Cyprus, Cyprus
Damijan Miklavčič	University of Ljubljana, Slovenia
Dimitrios Fotiadis	University of Ioannina, Greece
James Goh	National University of Singapore, Singapore
José Princípe	University of Florida, USA
Laura Roa	University of Seville, Spain
Lenka Lhotská	Czech Technical University in Prague, Czech Republic
Luis Kun	National Security at CHDS at NDU, USA
Martha Zequera	Pontificia Universidad Javeriana, Colombia
Nitish Thakor	Johns Hopkins University, USA
Pablo Laguna	Universidad de Zaragoza, Spain
Ratko Magjarevic	University of Zagreb, Croatia
Shankar Krishnan	Wentworth Institute of Technology, USA
Timo Jämsä	University of Oulu, Finland
YT Zhang	Chinese University of Hong Kong, Hong Kong

International Programme Committee

Aurélio Campilho	University of Porto, Portugal
Adam Idzkowski	Bialystok University of Technology, Poland
Ákos Jobbágy	Budapest University of Technology and Economics, Hungary
Alan Murray	Newcastle University, UK
Ana Castro	University of Porto, Portugal
Ana Mendonça	University of Porto, Portugal
Ana Paula Rocha	University of Aveiro, Portugal
Ana Paula Rocha	University of Porto, Portugal
Andres Santos	Universidad Politécnica de Madrid, Spain
Andriana Prentza	University of Piraeus, Greece

Antonio Azevedo	University of Porto, Portugal
Antonio Dourado	University of Coimbra, Portugal
Antonio Miguel Morgado	University of Coimbra, Portugal
Antonis Billis	Aristotle University of Thessaloniki, Greece
Argentina Leite	University of Trás-os-Montes e Alto Douro, Portugal
Aristides Vagelatos	Computer Technology Institute and Press, Greece
Aristotelis Chatziioannou	University of Cyprus, Cyprus
Branko Babušiak	University of Žilina, Slovakia
Carlos Ferreira	INESC TEC, Portugal
Carlos Silva	University of Minho, Portugal
Catarina Dias	University of Porto, Portugal
Catarina Nunes	Universidade Aberta, Portugal
Cesar Teixeira	University of Coimbra, Portugal
Christos Frantzidis	Aristotle University of Thessaloniki, Greece
Christos Schizas	University of Cyprus, Cyprus
Constantinos Pattichis	University of Cyprus, Cyprus
Cristina Santos	University of Minho, Portugal
Damijan Miklavcic	University of Ljubljana, Slovenia
Dan Clark	Nottingham University Hospitals NHS Trust, UK
Daniela Giordano	University of Catania, Italy
Diana Mendes	University of Coimbra, Portugal
Dimitrios I Fotiadis	University of Ioannina, Greece
Dimitris Iakovidis	University of Thessaly, Greece
Dimitris Kaolis	Ministry of Health, Cyprus
Dinesh Kumar	University of Coimbra, Portugal
Eduardo Castro	INESC TEC, Portugal
Eduardo Rocon	University of Coimbra, Portugal
Efi Psarouli	Aristotle University of Thessaloniki, Greece
Efthyvoulos Kyriacou	Frederick University, Cyprus
Eftychios Christoforou	University of Cyprus, Cyprus
Eleni Dafli	Aristotle University of Thessaloniki, Greece
Eleni Kaldoudi	Democritus University of Thrace, Greece
Emil Valchinov	University of Patras, Greece
Ernesto Iadanza	University of Florence, Italy
Estela Bicho	University of Minho, Portugal
Evdokimos Konstantinidis	Aristotle University of Thessaloniki, Greece
Gabriel Pires	University of Coimbra, Portugal
George Eleftherakis	The University of Sheffield, UK
George Hadjichristofi	Frederick University, Cyprus
Georgios Matis	Uniklinik Köln, Germany
Giandomenico Nollo	University of Trento, Italy
Gil Goncalves	University of Porto, Portugal
Giuditta Callea	SDA Bocconi School of Management, Italy
Giulia Matrone	University of Pavia, Italy

Hernâni Gonçalves	University of Porto, Portugal
Hugo Gamboa	New University of Lisbon, Portugal
Hugo Silva	New University of Lisbon, Portugal
Humberto Rocha	University of Coimbra, Portugal
Huseyin Seker	The University of Northumbria at Newcastle, UK
Igor Lackovic	University of Zagreb, Croatia
Ilias Maglogiannis	University of Piraeus, Greece
Ioanna Chouvarda	Aristotle University of Thessaloniki, Greece
James Goh	National University of Singapore, Singapore
Jan Havlik	Czech Technical University in Prague, Czech Republic
Jens Haueisen	Technical University Ilmenau, Germany
Jerome Mendes	University of Coimbra, Portugal
Jiri Holcik	Masaryk University, Brno, Czech Republic
Joana Dias	University of Coimbra, Portugal
Joana Paiva	INESC TEC, Portugal
Joao Carvalho	University of Lisbon, Portugal
João Ruivo Paulo	University of Coimbra, Portugal
João Ribeiro Pinto	University of Porto, Portugal
Joe Barbenel	University of Strathclyde, UK
Joel Arrais	University of Coimbra, Portugal
Joel Rodrigues	National Institute of Telecommunications (INATEL), Brazil
John Munoz Cardona	University of Waterloo, Canada
Jorge Dias	University of Coimbra, Portugal
Jorge Henriques	University of Coimbra, Portugal
José Luis Oliveira	University of Aveiro, Portugal
Joseph Mizrahi	Israel Institute of Technology, Israel
Jozef Wiora	Silesian University of Technology, Poland
Julie Polisena	Canadian Agency for Drugs and Technologies in Health, Canada
Konstantinos Delibasis	University of Thessaly, Greece
Konstantinos Karpouzis	National Technical University of Athens, Greece
Kristina Bliznakova	Technical University of Varna, Bulgaria
Leandro Pecchia	University of Warwick, Italy
Lino Ferreira	University of Coimbra, Portugal
Luca Faes	University of Palermo, Italy
Manousos Klados	Aston University, UK
Marek Penhaker	VSB - Technical University of Ostrava, Czech Republic
Maria Beatriz Carmo	University of Lisbon, Portugal
Maria Ruano	University of Algarve, Portugal
Mario Forjaz Secca	New University of Lisbon, Portugal
Mario Medvedec	University Hospital Centre Zagreb, Croatia
Mario Sansone	University “Federico II” of Napoli, Italy

Marjan Hummel	University of Twente, Netherlands
Martin Cerny	VSB - Technical University of Ostrava, Czech Republic
Martina Andellini	Bambino Gesù Children's Hospital, Italy
Maurizio Schmid	Roma Tre University, Italy
Michal Gála	University of Žilina, Slovakia
Michela Comune	Tel Aviv University, Israel
Miguel Amador	University of Lisbon, Portugal
Miguel Caixinha	University of Beira Interior, Portugal
Miguel Coimbra	University of Porto, Portugal
Monique Frize	Carleton University, Canada
Ofer Barnea	Tel Aviv University, Israel
Olof Lindahl	Umeå University and Luleå University of Technology, Sweden
Panagiotis Bamidis	Aristotle University of Thessaloniki, Greece
Panayiotis Kyriacou	City University London, UK
Paulo Crespo	University of Coimbra, Portugal
Paulo de Carvalho	University of Coimbra, Portugal
Paulo Maia	INESC TEC, Portugal
Paulo Mendes	University of Minho, Portugal
Robert Allen	University of Southampton, UK
Romuald Jolivot	Bangkok University, Thailand
Rui Bastos	University of Minho, Portugal
Rui Bernardes	University of Coimbra, Portugal
Samuel Silva	University of Aveiro, Portugal
Selma Supek	University of Zagreb, Croatia
Simão Paredes	Polytechnic Institute of Coimbra, Portugal
Sofia Rita Fernandes	University of Lisbon, Portugal
Sotiris Pavlopoulos	National Technical University of Athens, Greece
Spyros Kitsiou	University of Illinois at Chicago, USA
Stathis Konstantinidis	University of Nottingham, UK
Stavros Karkanis	University of Thessaly, Greece
Stavroula Mougiakakou	University of Bern, Germany
Stergiani Spyrou	Aristotle University of Thessaloniki, Greece
Styliani Petroudi	University of Cyprus, Cyprus
Stylianos Hatzipanagos	University of West London, UK
Susana Brás	University de Aveiro, Portugal
Susana Catarino	University of Minho, Portugal
Telemachos Stamkopoulos	University Ecclesiastical Academy of Thessaloniki, Greece
Teresa Rocha	Polytechnic Institute of Coimbra, Portugal
Teresa Sousa	University of Coimbra, Portugal
Themis Exarchos	University of Ioannina, Greece

Contents

Regular Sessions: Information Technology in Health Systems

Challenges: IFMBE Scientific Challenge Competition

Regular Sessions: Technologies for Active Ageing

"Patient Station" – Telerehabilitation System for People with Parkinson's Disease

Marek Żyliński[1], Wiktor Niewiadomski[2(✉)], Aleksandra Wacławek[3], Aleksandra Budzyńska[3], Anna Gąsiorowska[2,4], Anna Stępniewska[2], Adam Becmer[5], Maciej Jagielski[3], and Gerard Cybulski[1]

[1] Institute of Metrology and Biomedical Engineering, Faculty of Mechatronics, Warsaw University of Technology, Warsaw, Poland
[2] Mossakowski Medical Research Centre, Polish Academy of Sciences, Warsaw, Poland
wniewiadomski@imdik.pan.pl
[3] Faculty of Electronics and Information Technology, Warsaw University of Technology, Warsaw, Poland
[4] Neurobiology Center, Nencki Institute of Experimental Biology, Warsaw, Poland
[5] Riget Software Sp. z o.o., Warsaw, Poland

Abstract. Parkinson's disease is incurable and progressive impairment of movement. The physical activity combined with adequate exercises may improve motor capacity and delay the onset of activity limitation.

European Physiotherapy Guideline for Parkinson's Disease recommend strategies aimed to change behavior by facilitating intrinsic motivation. Telerehabilitation system which combines self-management of people with Parkinson's disease (PWP) with ongoing supervision by a physiotherapist and/or neurologist can improve the quality of treatment of PWP and promote the individual approach of the therapist to the patient.

In this paper, we describe the idea and realization of the core of such system, termed "Patient Station", i.e. the platform to communicate between PWP and physiotherapist and/or neurologist. The system allows: (a) the assessment of the patient's health state, (b) scheduling exercises, (c) assisting the patient when performing prescribed exercises, (d) controlling the execution of the schedule by PWP, (e) assessment the effectiveness of the rehabilitation, (f) supplying the documentation.

The patient's side of the system is based on smartphone application and inertial measurement units, communicating with a smartphone via Bluetooth connected with the therapist's side by the internet-based database.

Keywords: Tele-rehabilitation system · Parkinson's Disease · Self-management · Supervision

J. Henriques et al. (Eds.): MEDICON 2019, IFMBE Proceedings 76, pp. 965–970, 2020.
https://doi.org/10.1007/978-3-030-31635-8_117

1 Introduction

People with Parkinson's disease (PWP) are less active compared to their healthy contemporaries, what is associated with disease severity and specifically with gait impairments [1]. Inactivity may be part of a compensatory strategy to prevent falls. Fear of falling, common in PWP, may result in a reduction of physical activities [2]. Inactivity reduces muscle strength and length, especially in older people [3]. Reduced muscle power is related to the reduced performance of balance and functional mobility [4] which can further increase the risk and fear of falling.

PWP are encouraged to strive for the WHO recommendations for physical activity, that include at least 150 min per weak of aerobic physical activity at moderate intensity [5] like walking, bicycling, swimming or Tai Chi. Dance can also improve motor and cognitive function in PWP [6].

European Physiotherapy Guideline for Parkinson's Disease [7] recommend strategies aiming to change behavior by facilitating intrinsic motivation. These strategies require to provide PWP a knowledge about his or her problem and benefits of interventions, to make PWP feels sufficiently competent to perform the required, new behavior and considers it's worth the effort.

A patient should be in center physiotherapy interventions. They must fit with the needs, motivation, and abilities of the PWP. Increased self-control growths motivation, self-management can also be promoted [8]. Access to information technology can facilitate such an approach.

Advances in the technology allow the development of systems for monitoring of self-rehabilitation in PWP [9]. The smartphones are widespread, provide a friendly user interface and support connection to the internet. Monitoring physical activity of PWP can be performed by the built-in device sensor or by a dedicated whole-body multi-sensor kinematic recordings system [10]. Telerehabilitation can improve the quality of life of PWP and give them additional motivation and information on accomplished goals.

Tele-rehabilitation system can improve the quality of treatments of PWP. They can promote self-management of PWP and sustain their autonomy. In this paper, we describe the assumption and construction of our system.

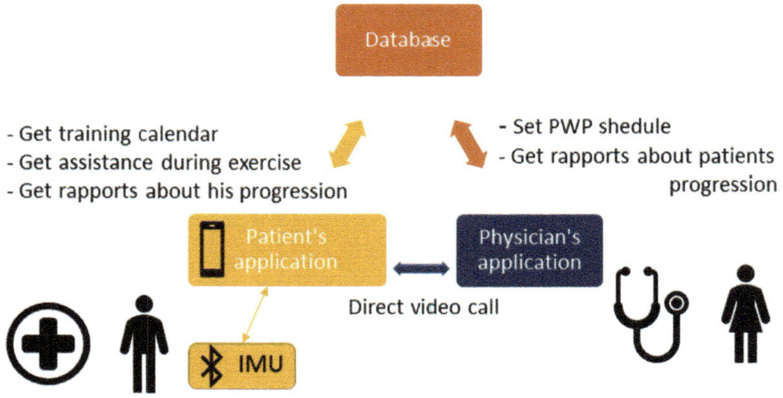

Fig. 1. Scheme of the "Patient Station" system organization.

2 System Organization

The main goal of "Patient Station" project is to develop a platform to communicate between PWP and neurologist. The system should allow: (a) the assessment of the patient's health state, (b) scheduling exercises, (c) assisting the patient when performing prescribed exercises, (d) controlling the execution of the schedule by PWP, (e) assessment the effectiveness of the rehabilitation, (f) supplying the documentation. The system should encourage PWP to self-rehabilitation and allow individual approach of the neurologist to the patient.

The system is divided into two applications: first dedicated for physician and second for PWP (Fig. 1). Activities of the patient are recorded by inertial measurement units (IMU) and send to the database. The physician can monitor the activity of patients and get reports about patients' progression. On-demand direct video call between physician and patient is possible.

During monitored session patients wear three IMU: one on the thorax and on both ankles. We use MPU-9250 sensor (InvenSense Inc., United States of America), the device contains three-axis accelerometer, gyroscope, and magnetometer. Data are sent to a smartphone by Bluetooth 4.0 interface (NRF-51822, Nordic Semiconductor, Norway).

To store the data, we use Firebase (Google LLC, United States of America). Firebase is a comprehensive mobile development platform. Information about scheduled exercises and data from IMUs are stored in cloud NoSQL database (Cloud Firestore service). Service is free for limited capacity (1 GB) and a limited number of writes operation (20K/day). At present, it is enough for our small test system. Firebase provides also a secure user authentication service, which can be used to identify the patient.

3 Patient's Application

Today smartphones are a primary communication device for most people. The touch screen is one of the most intuitive user interfaces. Capabilities of the smartphone can be extended through an external application. The well-designed application can be easily used by PWP.

The crucial part, which functionalities decide about PWP acceptance of the system, is the patient's application. The friendly application can support self-rehabilitation, disagreeable and "bossy" may discourage PWP.

The main screen of the patient application is shown in Fig. 2. The appearance of our application is as simple as it is reasonably possible. All buttons are large, and their function is described both by text and icon. Patient's application provides PWP the ability to:

- see his or her training schedule,
- assist performance of a prescribed set of exercises,
- see reports about his or her progression,

- send feedback about their state by the survey, the physician also can define own form and send them to select patients.

Exercise can be performed under or without monitoring of movement by IMU. In some cases, eq. swimming, wearing a smartphone and sensors are not recommended. Generally, monitoring of activity is suggested, collected data can provide additional information about the patient's health eq. monitoring of walking can provide information about gait and balance. In the next section, we describe a standardized set of exercise used to assess patient progression, where use of IMUs is indispensable.

Fig. 2. The main screen of the patient's application. The accessible user interface allows fast navigation between main application features (schedule, survey, statistic and perform exercise). Application was written in JS (React Native) and use Google Firebase.

4 Assessment of PWP Progression

European Physiotherapy Guideline for Parkinson's Disease [7] suggests many different tests delivering quantitative readouts, like timed up & go; five times sit to stand, 10-m walk; six-minute walk distance, rapid turns etc. Different tests address different core areas addressed by physiotherapy: balance, gait, transfer, dexterity and physical capacity. Not every test can be performed easily and safely by PWP in the home environment.

As proposed by Giuberti et al. [13], we chose three tests: leg agility task where patient rhythmically stamps his legs, sit-to-stand test and if possible, gait test where patient walk 10 steps, turns around, returns to a chair and sits. The result of these tests can be used to estimate UPDRS score [14].

Each measurement is associated with error. Therefore, in order to detect the real progression of the patients, the difference between scores of two measurement points should be larger than the measurement error. This error may be minimized by following precise instruction issued by our system, which allows to perform these tests in a standardized way.

5 Conclusion

We develop a "Patient Station" platform to communicate between PWP and neurologist, which allows the assessment and documentation of the patient's health state and support monitoring of the effectiveness of the therapy. The system should encourage PWP to self-rehabilitation and allow individual approach of the neurologist to the patient.

Reliable assessment of the effects of self-rehabilitation is a challenge, therefore we implemented the set of standardized exercises (sit-to-stand and leg agility tasks) to evaluate patient's progression.

References

1. van Nimwegen, M., Speelman, A.D., Hofman-van Rossum, E.J., Overeem, S., Deeg, D.J., Borm, G.F., et al.: Physical inactivity in Parkinson's disease. J. Neurol. **258**(12), 2214–2221 (2011)
2. Nilsson, M.H., Drake, A.M., Hagell, P.: Assessment of fall-related self-efficacy and activity avoidance in people with Parkinson's disease. BMC Geriatr. **10**(1), 78 (2010)
3. Manini, T.M., Clark, B.C.: Dynapenia and aging: an update. J. Gerontol. Ser. A Biomed. Sci. Med. Sci. **67**(1), 28–40 (2011)
4. Allen, N.E., Sherrington, C., Canning, C.G., Fung, V.S.C.: Reduced muscle power is associated with slower walking velocity and falls in people with Parkinson's disease. Parkinsonism Relat. Disord. **16**(4), 261–264 (2010)
5. https://www.who.int/dietphysicalactivity/factsheet_olderadults/en/
6. de Natale, E.R., et al.: Dance therapy improves motor and cognitive functions in patients with Parkinson's disease. NeuroRehabilitation **40**(1), 141–144 (2017)
7. Keus, S., Munneke, M., Graziano, M., Paltamaa, J., Pelosin, E., Domingos, J., et al.: European physiotherapy guideline for Parkinson's disease. The Netherlands, KNGF/ParkinsonNet (2014)
8. Tickle-Degnen, L., Ellis, T., Saint-Hilaire, M.H., Thomas, C.A., Wagenaar, R.C.: Self-management rehabilitation and health-related quality of life in Parkinson's disease: a randomized controlled trial. Mov. Disord. **25**(2), 194–204 (2010)
9. Rocchi, L., Farella, E., Greenlaw, R., Chiari, L.: Tele-rehabilitation system based on augmented feedback for people with Parkinson's disease: design principles. In: Hu, F. (ed.) Telehealthcare Computing and Engineering: Principles and Design. CRC Press, Boca raton (2013). p. 99
10. Rahimi, F., Duval, C., Jog, M., Bee, C., South, A., Jog, M., et al.: Capturing whole-body mobility of patients with Parkinson disease using inertial motion sensors: expected challenges and rewards. In: 2011 Annual International Conference of the IEEE Engineering in Medicine and Biology Society, pp. 5833–5838. IEEE, August 2011
11. Amatya, B., et al.: Effectiveness of telerehabilitation interventions in persons with multiple sclerosis: a systematic review. Mult. Scler. Relat. Disord. **4**(4), 358–369 (2015)

12. Huang, S.L., Hsieh, C.L., Wu, R.M., Tai, C.H., Lin, C.H., Lu, W.S.: Minimal detectable change of the timed "up & go" test and the dynamic gait index in people with Parkinson disease. Phys. Ther. **91**(1), 114–121 (2011)
13. Contin, L., Cimolin, V., Azzaro, C., Albani, G., Mauro, A.: BSN-based Kinematic Characterization and Comparative Outlook of UPDRS Scoring in Leg Agility, Sit-to-Stand, and Gait Tasks for Parkinsonians (2019)
14. Giuberti, M., Ferrari, G., Contin, L., Cimolin, V., Azzaro, C., Albani, G., Mauro, A.: Automatic UPDRS evaluation in the sit-to-stand task of Parkinsonians: kinematic analysis and comparative outlook on the leg agility task. IEEE J. Biomed. Health Inf. **19**(3), 803–814 (2015)

An Overview of Assistive Robotics and Technologies for Elderly Care

Eftychios G. Christoforou[1](✉), Andreas S. Panayides[2], Sotiris Avgousti[3], Panicos Masouras[3], and Constantinos S. Pattichis[2,4]

[1] Department of Mechanical and Manufacturing Engineering, University of Cyprus, Nicosia, Cyprus
e.christoforou@ucy.ac.cy

[2] Department of Computer Science, University of Cyprus, Nicosia, Cyprus

[3] Nursing Department, School of Health and Science, Cyprus University of Technology, Limassol, Cyprus

[4] Research Centre on Interactive Media, Smart Systems and Emerging Technologies (RISE CoE), Nicosia, Cyprus

Abstract. Population ageing is already having an impact on societies. This study briefly reviews associated technologies employed within the framework of elderly care, namely: robotic nursing, ambient assisted living and assistive robotics. Their current status is considered together with their potential and the associated implementation challenges.

Keywords: Assistive robotics · Elderly-care technologies · Nursing robots · Telerobotics · Human factors design

1 Introduction

Population aging is becoming a major concern internationally. An ageing population presents many challenges to labor markets, government tax, government spending and the wider economy. At a personal level, elderly individuals are challenged in different aspects [1]: social (neglect, isolation, fear, loneliness, boredom), financial (low income, fear of being a burden, lack of insurance), psychological (depression, poor memory, dementia, insomnia), and physiological (declining mental abilities, less efficient reflexes, muscle weakness, gait abnormality, fragile bones). Today, the prominent elderly care model involves ageing-in-place and the mission for robotics and modern technologies is to support the elderly both physically and emotionally. Relevant technologies include: robotic nursing, ambient assisted living, and assistive robotics. These technologies are briefly reviewed within the general framework of elderly care, while the primary challenges associated to design and development are discussed.

J. Henriques et al. (Eds.): MEDICON 2019, IFMBE Proceedings 76, pp. 971–976, 2020.
https://doi.org/10.1007/978-3-030-31635-8_118

2 Nursing Robots

Within hospitals robots may serve as supplemental healthcare workers to ease the burden on nursing staff. They have been considered for distributing food trays, medicines, and laboratory specimens. Robots can be integrated with other hospital technologies effectively supporting continuity of care. It is possible to carry-out laborious physical tasks like patient transfers and lifting to reduce physical stress on nurses, allowing them to focus on their primary duties. Such a commercial system called Robear was developed to lift patients from a bed into a wheelchair (RIKEN and Sumitomo Riko Co. Ltd., Japan). Telehealth robotic service is also among the available options that can be extended to home care. In the latter case, robotic nurses may serve as interfaces for doctors to communicate with elders over distance and facilitate medical routine tasks.

3 Smart Homes and Ambient Assisted Living

Ambient assisted living refers to the use of information and communication technologies (ICT) in the elderly daily living to enable them to stay active, socially connected, and live independently. Implementations typically involve wireless sensor networks for applications including health status monitoring. Sensors used for elderly-care applications can be: (i) wearable, (ii) non-wearable stationary, and (iii) mobile robot-mounted. Restrictions regarding wearable sensors may discourage the elderly people from adopting them. On the other hand, ambient sensors like cameras may be perceived as intrusive and thus rejected. The collection of personal data involved in smart home applications raises a range of privacy management issues that need to be carefully addressed.

Sensors can also be used to monitor everyday living activities including the use of tap water and visits to toilet. The information can be exploited in two ways [2]: (i) Identify emergencies and generate alerts, (ii) Identify long-term variations in health status by analyzing behavioral patterns in daily activities. Enhancements to smart home implementations can be accommodated via internet-of-things (IOT) technologies [1], also leveraging the integration of robots within the setup. Mobile robots may overcome limitations of embedded assistive technologies by ensuring an effective coverage, as well as capture of dynamic activities. Their physical appearance and explicit presence may be easier to accept, while providing for interactions that are more meaningful.

4 Assistive Robotics

The ultimate goal of assistive robotics is to enable elderly/disabled people to pursue healthy, independent, and productive lives. Depending on their main function, assistive robots can be grouped as: (i) Socially assistive, and (ii) Physically assistive.

4.1 Socially Assistive Robots

These robots provide assistance to end-users through social interaction. An example of socially assistive robots is NAO, which is a small-size anthropomorphic robot. Its interactions with autistic children are reported in [3]. Human perceptions towards robots constitute socially assistive robots more effective than a computer program. Companion robots may act as interfaces that enable the elderly to enrich their social lives and connect with their families and friends. Potential uses of robots are also seen in the care of individuals during physical recovery, rehabilitation, and to satisfy their training needs. In the case of post-stroke rehabilitation, the purpose of the robot is to motivate, facilitate, and monitor the process. A movement therapy robot may provide a diagnostic (measurement and assessment) or therapeutic (improvement of function) benefit.

Supporting adults with dementia is another potential role [4]. Symptoms of dementia cause particular stress and pressure on caregivers/families and often become a reason for institutionalization. Various pet-like robots have been developed to stimulate senses and provide motivation through interaction. An example is Paro, which has the appearance of a baby harp seal and interactive behavior [5]. AIBO (Sony Corporation, Japan) is a robotic dog that has been considered for the same purpose [6].

Various robotic systems have also been considered to systematically engage elderly users in physical exercise [7]. A robot may conduct customized exercise sessions, evaluate user performance, and give the user real-time feedback completely autonomously. The first challenge related to this approach is to analyze the coach's gestures automatically, as required for being able to reproduce them adequately. The second challenge is due to the fact that the robot possesses a different physical embodiment than the coach, and hence is inherently an imperfect intermediary interface [8].

4.2 Physically Assistive Robots

A key element of life quality for the ageing population is preservation of mobility. A relevant commercial system is LEA - Lean Empowering Assistant (Robot Care Systems, The Netherlands), which provides autonomous functions to help the elderly when getting out of bed or chair and when walking. Robotic wheelchairs also facilitate mobility and they can be endowed with obstacle avoidance capabilities. Autonomous vehicles is an emerging field within the scope of robotics that is also expected to play a role in elderly transport. Future developments in autonomous vehicles should be appealing and consider the accessibility requirements of the elderly and disabled users.

Specially designed assistive robotic manipulators/devices can support people with motor impairments and help them carry out tasks including eating and drinking, personal care, object manipulation, work and leisure. The manipulators can be divided into: Task-specific and general-purpose. Single-task systems include commercially available robotic feeding devices for people with disabilities [9]. General-purpose assistive robotic manipulators include MANUS and its upgraded version called iARM [9]. Wheelchair mounted arms have the advantage to always accompany the user [10].

A special category of sensory assistive devices is dedicated to the support of mobility and independent travel for blind and partially sighted/older people. Robotic guides exhibit some of the characteristics of a human guide or guide dog (while avoiding the animal care responsibilities). Examples include Guido and GuideCane [9]. Devices may operate in automatic mode (full capability to avoid obstacles) or semi-autonomously.

On a different note, commercially available domestic work robots are often encountered in many modern households and public areas. Indoor uses include robotic vacuum cleaners and floor-washing robots. Outdoor systems include robotic lawn mowers and swimming pool cleaners. Assigning day-to-day needs to technology can effectively increase the quality time the elderly can spend with family and friends.

5 Telemedicine/Telerobotics and Elderly-Care

Teleoperated robots are currently used in medicine [11]. A hospital scenario involves the physician sending a mobile robot out to find the patient in the clinic. The doctor's face comes on the screen of the robot. Using the camera, the doctor can proceed with a diagnosis and also connect to vital signs. An analogous home scenario can be implemented [12] to facilitate social interaction from a distance, support the elderly to remain socially engaged, and allow relatives to make virtual visits. Likewise, contact with doctors and nurses may also be enabled. Compared to video calls, a telepresence robot enhances interaction to a more natural level. Wired and/or wireless communication networks and video compression technologies are involved in such implementations.

6 Ethical, Safety and Legal Issues

Legal and liability issues are associated with the use of the aforementioned elderly-care technologies. They include user safety, and the risk of unauthorized access to health-care databases and sensitive private information. Privacy concerns arise from the use of pervasive technologies for continuous monitoring. In fact, robots may increase the level of privacy by avoiding the need for human assistance for tasks which are potentially embarrassing or private. Although robots are not human, programmed reactions may affect users' sense of privacy. Also, prolonged human-robot interactions may have other psychological effects including the attachment to the robot and deception about its abilities. This may potentially influence interpersonal relationships and raises ethical issues [13].

7 Assistive Robotics and Human Factors Design

Human factors engineering is the study and practice of designing equipment, machines, and environments to accommodate human users. It maximizes performance, safety, comfort, and user satisfaction while minimizing the likelihood of errors, inefficiencies, injuries, and fatigue. User-centered design requires the engagement of end-users as well as the rest of stakeholders (families, physicians, nurses, therapists, psychologists, etc.). The special characteristics of the elderly users can be grouped into: (i) Physical impairments, (ii) Perceptual impairments (sensory limitations), and (iii) Cognitive limitations.

The appearance of assistive robots is important to users [9], even though it doesn't seem to exist a consensus on preferred appearances. Physical appearance also leads to social expectations. For example, a close resemblance to human appearance can be deceiving and may lead to unrealistic expectations about robot's capabilities. Ideally, the system should be able to establish an effective relationship with the user that does not go beyond its intended purpose. Note that considerable design constraints are often imposed on appearance by the required functionality.

8 Conclusions

Elderly care technologies have attracted extensive research interest but few solutions are in common use. This is partly due to high costs, legal and safety issues that need to be resolved. Elderly are also concerned that they will eventually lose personal assistance and companionship. Technologies should be implemented so that the users do not feel that their privacy is violated and retain control, without compromising effectiveness. Human factors are critical towards that direction. Elderly-care technologies are expected to play a key role in future societies. Their full potential is yet to be unleashed!

Acknowledgements. This work was partly funded by the European Union's H2020-MSCA-RISE scheme under grant agreement ENDORSE 823887.

References

1. Sathish, K., Sachin, P., Vignesh, P., Ahmed, M.R.: Architecture for IOT based geriatric care. In: International Conference on Intelligent Computing and Control Systems, pp. 1099–1104 (2017)
2. Tsukiyama, T.: In-home health monitoring system for solitary elderly. Procedia Comput. Sci. **63**, 229–235 (2015)
3. Shamsuddin, S., et al.: Initial response of autistic children in human-robot interaction therapy with humanoid robot NAO. In: 2012 IEEE 8th International Colloquium on Signal Processing and Its Applications, pp. 188–193 (2012)
4. Begum, M., Wang, R., Huq, R., Mihailidis, A.: Performance of daily activities by older adults with dementia: the role of an assistive robot. In: IEEE International Conference on Rehabilitation Robotics, pp. 1–8 (2013)
5. Shibata, T.: Therapeutic seal robot as biofeedback medical device: qualitative and quantitative evaluations of robot therapy in dementia care. Proc. IEEE **100**(8), 2527–2538 (2012)
6. Kramer, S.C., Friedmann, E., Bernstein, P.L.: Comparison of the effect of human interaction, animal-assisted therapy, and AIBO-assisted therapy on long-term care residents with dementia. Anthrozoos **22**(1), 43–57 (2009)

7. Fasola, J., Matarić, M.J.: Socially assistive robot exercise coach: motivating older adults to engage in physical exercise, pp. 463–479 (2013)
8. Görer, B., Salah, A.A., Akin, H.L.: A robotic fitness coach for the elderly. In: Lecture Notes in Computer Science, LNCS, vol. 8309, pp. 124–139 (2013)
9. Hersh, M.: Overcoming barriers and increasing independence–service robots for elderly and disabled people. Int. J. Adv. Robot. Syst. **12**(8), 114 (2015)
10. Ktistakis I.P., Bourbakis, N.G.: A survey on robotic wheelchairs mounted with robotic arms. In: 2015 National Aerospace and Electronics Conference, pp. 258–262 (2015)
11. Avgousti, S., Christoforou, E.G., Panayides, A.S., Voskarides, S., Novales, C., Nouaille, L., Pattichis, C.S., Vieyres, P.: Medical telerobotic systems: current status and future trends. Biomed. Eng. Online **15**(96), 1–44 (2016)
12. Reis, A., Xavier, R., Barroso, I., Monteiro, M.J., Paredes, H., Barroso, J.: The usage of telepresence robots to support the elderly. In: 2018 2nd International Conference on Technology and Innovation in Sports, Health and Wellbeing, pp. 1–6 (2018)
13. Feil-Seifer, D., Mataric, M.: Socially assistive robotics. IEEE Robot. Autom. Mag. **18**(1), 24–31 (2011)

Artificial Intelligence Gamified AAL Solution

Marta Pinto[1](✉), Mário Pereira[1], Diana Raposo[1], Marco Simões[2], and Miguel Castelo-Branco[3]

[1] Intellicare, Coimbra, Portugal
info@intellicare.pt
[2] CIBIT, FMUC and CISUC, Coimbra, Portugal
msimoes@dei.uc.pt
[3] CIBIT, FMUC and ICNAS, Coimbra, Portugal
mcbranco@fmed.uc.pt

Abstract. GameAAL - Gamification Supporting Active and Assisted Living is an innovative system to monitor elderly daily life activities and provide neurocognitive stimulation games. The solution is focused at filling a gap on existing Ambient Assisted Living solutions, promoting social interaction and mobility of end-users using gamification and machine learning techniques. For each goal achieved, the system unlocks a reward providing the user with the chance of winning free admissions in cultural events supplied by community partners. A Proof of Concept of AAL ecosystem and TV interface is ongoing with 26 real end-users, reporting good acceptability and willingness to use the system.

Keywords: AAL · ADL · Pattern recognition · Gamification · Cognitive training

1 Introduction

European population aged over 80 years old are expected to double by 2080, accounting 13% of the whole population [1, 7]. Overloaded healthcare and social systems struggle to supply the increasing demand for professional caregiving by implementing traditional care models comprising high operational costs and unable to be scaled for larger population segments [3, 5].

Current market products on healthy ageing offer ICT-based solutions to fulfill these challenges on elderly welfare through gamification [5], social network applications [7] and remote monitoring for ambient, health and bio-medical parameters. These functionalities, however, are unintegrated and unable to produce crucial correlated information required for a personalized and comprehensive care optimization [3].

On this subject, GameAAL is an ambient assisted living (AAL) system for ageing in place, at home and in the community, integrating gamification techniques to monitor activity and motivate senior citizens to have healthy behaviors [5].

J. Henriques et al. (Eds.): MEDICON 2019, IFMBE Proceedings 76, pp. 977–982, 2020.
https://doi.org/10.1007/978-3-030-31635-8_119

2 Activities of Daily Life

Activities of daily life (ADL) embody a method to assess functional aptitude of healthy adults and in elderly individuals used as forecaster to the need for comprehensive care assistance. ADL refers to self-care, comprising activities/tasks a person needs to perform reliably to be able to live autonomously [3].

ADL information gives a behavioral portrait of the individuals. Decline in ADL performance is an adverse outcome of frailty, being so the main reason for institutionalization. Recognizing systematic behavioral patterns on routine activities aids early detection of anomalous traits as well as changes in health status, providing care optimization towards individual's needs, relieving care costs and the consume of resources [3].

Sensors measuring ADL follow 2 approaches on data assortment: wearable sensors collect data directly from subject; ambient sensors gather data from the environment surrounding the individual, deriving behavioral information and ADL performance.

3 Methods

GameAAL follows an User Centric Design methodology with permanent iteration with its stakeholders and partners, mitigating risks related to acceptance and receptivity by end-users and caregivers. Meetings and Focus Groups were conducted with relevant institutions in the elderly area to obtain feedback from end-users and caregivers concerning expectations from GameAAL and methods to address their daily life needs.

In addition, GameAAL counts with the partnership of event promoters, whose core business targets senior population. By being partners, event promoters disclose their events while giving some free entrances. Currently, GameAAL counts with a museum, a theater, a concert room and a newspaper as its main event promoters.

4 Design and Implementation

4.1 General Architecture

GameAAL tracks daily activity in a non-intrusive way, transforming raw data into valuable information on activity and behavior using big data and machine learning techniques. GameAAL system makes use of ambient sensors to detect and quantify ADL (Fig. 1).

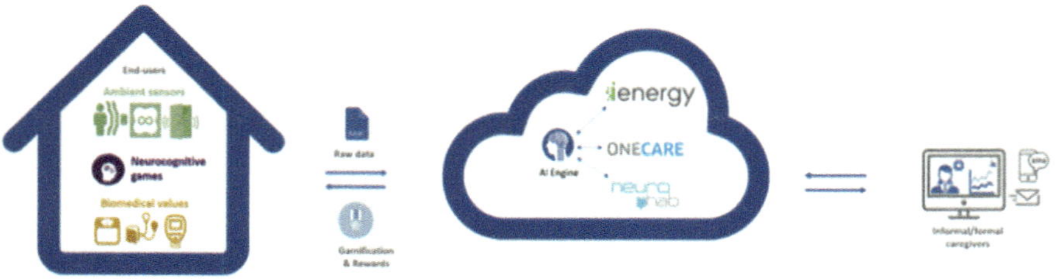

Fig. 1. GameAAL general architecture scheme

Smart plugs measure key electrical appliances used by the senior: to make breakfast and cook; the last appliance used before sleep; the TV. Acquired data is provided to the Artificial Intelligent Engine (AI Engine) to infer routines related to cook, sleep and leisure. Biomedical data is also collected by medical wireless devices (scale, blood pressure monitor and glucometer), and then processed on AI Engine which evaluates if the health parameters are normal, according to the senior's pathologies and clinical history. Caregivers receive notifications when it is detected variations of normal pattern are detected. NeuroHab platform [4] addresses distinct domains of neurocognitive impairment by interacting with the information extracted from the sensors in order to create hybrid trainings with virtual and real activities, providing metrics of neurocognitive training games played by the senior, later used by AI Engine to adapt cognitive training sessions according to the performance of the senior. A web-based platform is available to caregivers for managing senior daily tasks/goals and rewards. Furthermore, it provides caregivers an overview of the senior status, containing detailed ADL, health and cognitive gaming performance information.

4.2 Pattern Recognition Contextualization

GameAAL uses pattern recognition methods to evaluate ADL performance, computing metrics/indicators to assess individual functional capability and detect deviation from normal behavior during ADL routines. As previously mentioned, cook, sleep, sedentary level and leisure activities are the scope of GameAAL study:

Cook - Elderly people are often left alone to fend for themselves to maintain their health, which may interfere with the maintenance of a good nutritional status. Increasing prevalence of malnutrition in this population is associated with the decline of functional status [2]. **Sleep** - Sleep foundation recommends people above 65 years old to have sleep range of 7–8 h. Sleep routine has impact on several health indicators and sleep deprivation is an often overlooked but is a significant risk factor for short and long-term memory, type 2 diabetes, high blood pressure, heart disease overweight and weakened immunity [8]. **TV** - Sedentary lifestyle is a substantial risk factor for overweight and associated metabolic diseases. Despite this well-known negative impact, data from 2018 reveal the majority of older generation spend an average of 5 h and 30 min per day watching TV as privileged leisure activity at home [6].

4.3 Pattern Recognition Implementation

GameAAL pattern recognition models development followed two approaches: Monitoring daily routines by creating a model using data derived from ambient sensors; monitoring health and mental condition by creating a model using biomedical parameter values and metrics from neurocognitive games.

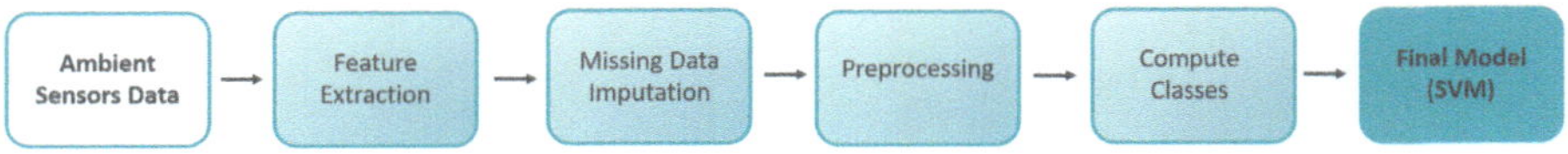

Fig. 2. Steps implemented on Final Model

In this paper, we detail the daily routines monitoring model which were first implemented, being tested on 11 real end-users. Development of the second part is currently undergoing its final steps and is estimated to be ready for validation on May 2019. Implementation of ADL monitoring model with real data revealed challenging. First difficulty was the inexistence of labeled data. Labeling the data implies a great time-consuming effort, involving observations at seniors' homes. The Final Model was trained using 3 months of real end-user data, providing enough quantity and quality. Decision about development technology fell on Scikit-learn in Pyhton and a set of auxiliary libraries. Final version of the model includes the steps on the scheme of Fig. 2.

Feature Extraction. Performing feature extraction, one has to deal with missing, unstructured or unlabeled data. For each routine – cook, sleep and tv – was computed event frequency and its duration. For instance, how many times and how long a senior gets to cook, allowing model to obtain sleep, cook, watch tv and breakfast patterns. **Missing Data Imputation.** Despite had enough quality samples of training data, there was missing data on dataset after feature extraction with a percentage between 7.6% to 57.6%. Mean Substitution Method was used to handle missing data, not only for training model, but also to prevent future problems. **Preprocessing.** Input dataset does not have categorical, nor binary variables. All seven variables are numeric, thus basic preprocessing procedures were applied: normalization to scale all data to same range and Principal Component Analysis to get 2-D input dataset with most variance. **Label data.** To determine number of clusters (classes), DBScan method was applied to data with 2 clusters found to be the best. Then, clustering method K-Means was used for each individual. Elderly have creased routines, thus cluster with more samples (days) has been labeled data as "Normal day" and remaining as "Deviation day". **Classification.** After testing some supervised learning algorithms, Final Model was obtained using Support Vector Machines with a linear kernel, good for unbalanced datasets, which is the case (i.e. more *Normal Days* than Deviation Days). Mean train accuracy was 98%. Model validation with 42 days of unseen data resulted in 93% of accuracy, certifying the model robustness and efficiency. **Update Final Model.** Final model is updated weekly, assuring assumptions are updated according to actual behavior and routines of end-users.

5 Proof of Concept

A solution like GameAAL needs to be accepted by elderly population, so it demands an optimized user experience and interaction with its interface to be successful and preserve end-user engagement. End-users use their own remote control to interact with the interface for Smart TVs. The interface has a simple and clear language, graphical elements with increased sizes and audio instructions for easing comprehension of the games' objectives and working mode. Main screen displays a to-do list of the daily tasks that are important to this population. Gamification component presents daily goals status and educational information related to healthy lifestyle, promoting health literacy. Periodically, quizzes are shown to seniors allowing them to earn points, counting towards daily goals. Users can access on the interface: cognitive sessions; standalone neurocognitive games; news; events; physiological data and routines information; points and current progress to unlock rewards and merit; history on achieved rewards.

A Proof of Concept is going on with 15 participants to evaluate a set of cognitive training games and validate the interface. Users shown great acceptability and provided important feedback to adapt the games to their needs and gave valuable opinion concerning system's usability and suggestions who to improve it.

6 Discussion and Future Work

GameAAL revealed to be an added-value solution to all stakeholders involved. It provides end-users an ecosystem that reduces the negative feelings related to ageing as it promotes social interaction and mobility. It impacts end-users as they are always updated from cultural agenda and other ongoing initiatives, according to their preferences and interests and be always in touch with relatives and friends. Within home, senior citizens prolong independent living, keep social interactions and continue having an active role on society. Moreover, it delivers caregivers insights about the care status, progress on health indicators and helps families to delay seniors' institutionalization.

Undergoing developments and tests intend to deliver a consistent and robust final prototype, mainly on algorithms that dynamically adapt neurocognitive planning sessions according to stimulation needs. It is also necessary to technically validate GameAAL as a feasible and reliable system that prolongs independent living, capable to be scaled to a large population.

Acknowledgments. GameAAL project is co-funded by Centro Portugal Regional Operational Programme (CENTRO 2020), PORTUGAL 2020 and the European Union, through the ERDF and results from experience and knowledge on health remote monitoring of Intellicare together with the scientific knowledge of University of Coimbra research group CIBIT.

References

1. Eurostat – statistics explained Webpage. ec.europa.eu/eurostat/statistics-explained/index.php?title=Population_structure_and_ageing. Accessed 19 Apr 2019
2. Amarya, S., Singh, K., et al.: Changes during aging and their association with malnutrition. J. Clin. Gerontol. Geriatr. **6**, 78–84 (2015)
3. Urwyler, P., Stucki, R., et al.: Cognitive impairment categorized in community-dwelling older adults with and without dementia using in-home sensors that recognise activities of daily living. Sci. Rep. **7**, 42084 (2017)

4. Simões, M., Mouga, S., Castelo-Branco, M., et al.: Neurohab: a platform for virtual training of daily living skills in autism spectrum disorder. Procedia Technol. **16**, 1417–1423 (2014). Conference 2014, HCIST
5. Kattimeri, C., Qvist, P., et al.: Gamified solutions in healthcare - testing rehabilitation games in Finland and Asia. Acta Tech. Jaur. **10**(1), 35–49 (2017)
6. Harvey, J., Chastin, S., et al.: Prevalence of sedentary behavior in older adults: a systematic review. Int. J. Environ. Res. Public Health **10**(12), 6645–6661 (2013)
7. Coto, M., Lizano, F., et al.: Social media and elderly people: research trends. In: Social Computing and Social Media. Applications and Analytics: 9th International Conference, SCSM 2017. LNCS, Canada, pp. 65–81 (2017)
8. National Sleep Foundation Homepage. www.sleepfoundation.org. Accessed 19 Apr 2019

Empowering Community Dwelling Older Citizens to Improve Their Balance with a Novel Technology Platform

Dimitrios Gatsios[1,6], Doris Eva Bamiou[2], Sergi Costafreda[2], Eleni I. Georga[1], Konstantina K. Kourou[1], Themis Exarchos[3], Kostas M. Tsiouris[1,4], and Dimitrios I. Fotiadis[1,5](✉)

[1] Unit of Medical Technology and Intelligent Information Systems, Department of Materials Science and Engineering, University of Ioannina, 45110 Ioannina, Greece
fotiadis@cc.uoi.gr

[2] UCL Ear Institute & University College London Hospitals Biomedical Research Centre Hearing & Deafness, London WC1X 8EE, UK

[3] Department of Informatics, Ionian University, 49100 Corfu, Greece

[4] School of Electrical and Computer Engineering, National Technical University of Athens, 15773 Athens, Greece

[5] Department of Biomedical Research, IMBB-FORTH, 45110 Ioannina, Greece

[6] Department of Neurology, School of Medicine, University of Ioannina, 45110 Ioannina, Greece

Abstract. The prevalence of balance deficits increases as the population is ageing. Such deficits are associated with the increased incidence of falls which in turn is linked with substantial limited functionality and morbidity. Vestibular rehabilitation therapy (VRT) as a component of the treatment has been shown to be effective in reducing symptoms and improving balance. HOLOBALANCE is an intervention based on a novel technology platform for providing VRT unsupervised, at home which means that motivating citizens to be compliant and promoting empowerment are the cornerstones for its wide adoption. Here we present how citizens empowerment is being addressed in HOLOBALANCE.

Keywords: Empowerment · Vestibular rehabilitation therapy · Human machine interaction

1 Introduction

As population ages, an increasing number of community dwelling citizens face permanent balance problems that lead to limited functionality. For these citizens, dizziness, which is a significant risk factor for falls [1], and/or imbalance are the most frequent reason for visiting a physician [2]. Falls are a leading cause of serious injury. The overall rate of older adult deaths from falls increased 31% from 2007 to 2016 (3.0% per year) [3]. In a recent study [4] the estimated medical costs attributable to fatal and nonfatal falls was approximately $50.0 billion and the overall medical spending for fatal falls was estimated to be $754 million. It is estimated that peripheral and central vestibular deficits cause around 50% of the cases of dizziness.

J. Henriques et al. (Eds.): MEDICON 2019, IFMBE Proceedings 76, pp. 983–988, 2020.
https://doi.org/10.1007/978-3-030-31635-8_120

Vestibular rehabilitation therapy (VRT) is a form of physiotherapy that aims to retrain postural, movement and sensory strategies and to adapt these to changing contexts by means of a range of gaze, standing, and gait exercises [5]. In recent trials [6] and studies [7] VRT has been shown to be more effective than generic exercises in resolving symptoms after vestibular impairment [8]. VRT is predominantly delivered at home, daily, without supervision. It is personalized, defined and directed by expert physiotherapists and it is largely dependent on patient engagement and motivation, since the number of clinical visits is very limited and varies among settings. The exercises are progressive with increasing difficulty as physiological compensation (i.e. the recovery process that harnesses the adaptive, plastic properties of the posture and balance control system) and symptom habituation occur. Due to the nature of the intervention and as literature confirms [9–11] empowerment is of outmost importance for keeping the elderly individuals engaged. Citizen and patient empowerment is one of the strategic directions and priority areas for European policy action for health and well-being up to 2020 according to WHO [12] which actively and continuously promotes it.

2 The Concept of Empowerment in HOLOBALANCE

Empowerment is a key concept when developing health promotion technological solutions and strategies for community dwelling citizens. In the context of balance disorders, empowerment can be achieved by providing expert guided vestibular rehabilitation at home, that is to a large extent informed by citizen choice (in terms of rehabilitation goals, types of exercise and activities, timing of these etc.) and different tools to support adoption of a more active lifestyle. This is made possible and provided via a novel technological platform [13].

The HOLOBALANCE platform among other functionalities, offers a range of choices to the patient and enables clinicians to monitor their patients and their behaviors during the different interventions employed in it in order to understand their patients' engagement and willingness for change. As evidence suggests, shared decision-making [14–16], rewarding [17], goal setting [18], action planning [19] and self-engagement [20], are the other core elements for achieving empowerment. All of these can be facilitated by technology.

The design of the system is user-centric and based on direct feedback from older users [21] and the software development approach is agile in order to iteratively include that feedback. The HOLOBALANCE platform makes use of the Internet of Things (IoT) and edge computing for monitoring and evaluating in real time the performance of the users during physiotherapy exercises, exergames and cognitive training tasks. Personalized and adaptive coaching is provided to the users during the daily sessions, by employing progressive and reinforcement learning methods. Data are collected in the cloud backend for: (a) exercise performance and adherence using processing of sensor data and analytic methods, (b) stress and frustration using machine learning algorithms, (c) active and sedentary periods and steps, distance etc. using the Fitbit backend, and (d) symptoms related to the exercise performance during the rehabilitation regime. These data are made available through a Dashboard to the physiotherapists, thus facilitating the remote monitoring of patients' progress in the rehabilitation program and supporting the physiotherapist decision making for adjustments and progression in the management plan.

The novel human machine interaction (HMI) consists of a hologram acting as a physiotherapist surrogate, who guides older users with balance disorders through the whole multisensory balance rehabilitation regime. Holograms aim to take HMI a step beyond the current state of the art [22]. Augmented reality exergames consist of virtual objects and tasks presented in a mixed reality environment to provide stimuli to the users. Such interventions are gaining interest and are effective [23] in reducing falls. Both holograms and exergames are projected through a head mounted device with a smartphone embedded on it, which is a solution with reduced costs, compared to similar mixed reality headsets, and increased accessibility and usability. A mobile app is also developed for remote auditory training aiming to improve speech in noise perception and auditory memory in accord with recent studies [24] (Fig. 1).

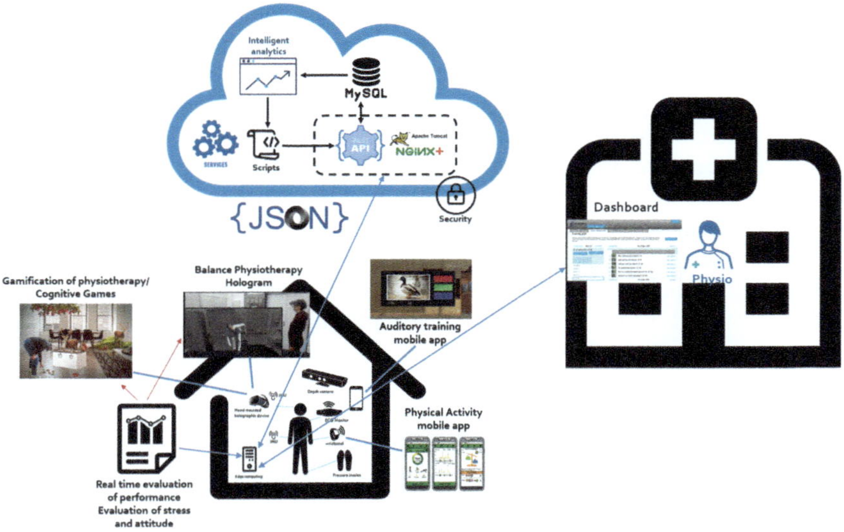

Fig. 1. The HOLOBALANCE conceptual architecture.

HOLOBALANCE adopts a holistic approach that aims to empower citizens. In addition to the intervention related empowering features, it provides citizens with access and control over their data, since this is one of the key elements of empowerment, as also emphasized in GDPR too [25]. Data download on demand, tagging and tracking personal data, transparency over data usage and permissions including the right to revoke are all included in this regulation.

HOLOBALANCE is personalized, in that the vestibular rehabilitation therapy plan and the activity goals will be co-decided by the patients and the physiotherapists at baseline and will be revised in follow up visits every 4 weeks. In between visits the physiotherapist will revise the weekly plan based on the data from the platform about performance of exercises, stress levels, compliance, progress in activity goals etc. against automated suggestions coming from reinforcement learning methods.

The HOLOBALANCE approach is based on the Capability, Opportunity and Motivation (COM) and Behavior (B) theoretical model in order to systematically identify the sources of the behavior that should be targeted for intervention and propose appropriate and specific Information Technology components (holograms, exergames, cognitive games, mobile apps) that provide the different intervention strands (balance physiotherapy, physical activity, cognitive and auditory training) to the users in order to achieve motivation and compliance to the home based rehabilitation.

HOLOBALANCE further aims to educate the patient by providing knowledge and information thus promoting health literacy, i.e. one's knowledge about his own health status and the available options to manage, since these also are core elements of empowerment. This element is achieved by educating the patient during the baseline and follow up visits, and with phone calls whenever it is needed, e.g. if the patient is not adherent to the program. Additionally, knowledge about balance disorders as well as about treatment options will be available in the patients' mobile app (videos, written materials, graphs) that will provide a continuous and reliable source of information. Finally, this will be achieved within the patient e-forum and virtual community, since patients tend to exchange information and tips in these.

The platform will promote engagement and enjoyment with feedback, challenge and rewarding mechanisms that will be embedded in all technological components. The hologram instructs the user, shows him the correct performance of the exercises and provides feedback for correcting him as well as rewards dependent on the performance to motivate him. The exergames provide feedback and rewards in a similar to the hologram way. Progress of auditory and cognitive training is also monitored and evaluated for the provision of personalized feedback. Self-monitoring is enabled through the mobile app in which all rewards and achievements are recorded and available. It also enables action planning and goal setting. The users can set their specific activity goals, mainly for steps and distance covered, as well as goals co-decided with their clinicians. Such shared decision making is nowadays central to evidence-based medicine and it is core component of patient-centered care. Patient empowerment is key to shared decision-making which can only happen when a patient admits responsibility for his health.

3 Conclusions

Current theoretical models and scientific evidence indicate that patient empowerment for patients with balance disorders at risk of falls can be achieved by adopting a holistic approach, to provide personalized home based, unsupervised interventions, by employing advanced technology and novel human machine interactions. Health literacy is the basis of such intervention since it enables citizens to use their own care information and information about balance problems in order to motivate them to engage with and complete their treatment plan. Feedback, goal setting, rewarding and action planning are other core elements of this intervention. Measuring empowerment within HOLOBALANCE, which is a challenging task [26], will add evidence to the field and will be one of the outcomes of the multicenter, randomized controlled trial that will be conducted towards the end of this project to assess the feasibility and estimate the efficacy of the whole concept.

Acknowledgment. This work has received funding from the European Union's Horizon 2020 research and innovation programme under grant agreement No 769574.

Conflict of Interest. The authors declare that they have no conflict of interest.

References

1. Agrawal, Y., Carey, J.P., Della Santina, C.C., Schubert, M.C., Minor, L.B.: Disorders of balance and vestibular function in US adults: data from the National Health and Nutrition Examination Survey, 2001–2004. Arch. Intern. Med. **169**(10), 938–944 (2009)
2. Neuhauser, H.K., Radtke, A., Von Brevern, M., Feldmann, M., Lezius, F., Ziese, T., Lempert, T.: Migrainous vertigo: prevalence and impact on quality of life. Neurology **67**(6), 1028–1033 (2006)
3. Burns, E., Kakara, R.: Deaths from falls among persons aged ≥ 65 years—United States, 2007–2016. Morb. Mortal. Wkly Rep. **67**(18), 509 (2018)
4. Florence, C.S., Bergen, G., Atherly, A., Burns, E., Stevens, J., Drake, C.: Medical costs of fatal and nonfatal falls in older adults. J. Am. Geriatr. Soc. **66**(4), 693–698 (2018)
5. Wrisley, D.M., Pavlou, M.: Physical therapy for balance disorders. Neurol. Clin. **23**(3), 855–874 (2005)
6. Rossi-Izquierdo, M., Gayoso-Diz, P., Santos-Pérez, S., Del-Río-Valeiras, M., Faraldo-García, A., Vaamonde-Sánchez-Andrade, I., et al.: Vestibular rehabilitation in elderly patients with postural instability: reducing the number of falls—a randomized clinical trial. Aging Clin. Exp. Res. **30**(11), 1353–1361 (2018)
7. Kundakci, B., Sultana, A., Taylor, A.J., Alshehri, M.A.: The effectiveness of exercise-based vestibular rehabilitation in adult patients with chronic dizziness: a systematic review. F1000Research **7**, 276 (2018). https://doi.org/10.12688/f1000research.14089.1
8. McDonnell, M.N., Hillier, S.L.: Vestibular rehabilitation for unilateral peripheral vestibular dysfunction. Cochrane Database Syst. Rev. (1) (2015). Article no. CD005397. https://doi.org/10.1002/14651858.CD005397.pub4
9. Madsen, W., Ambrens, M., Ohl, M.: Enhancing resilience in community-dwelling older adults: a rapid review of the evidence and implications for public health practitioners. Front. Public Health **7**(7), 14 (2019)
10. Seah, B., Kowitlawakul, Y., Jiang, Y., Ang, E., Chokkanathan, S., Wang, W.: A review on healthy ageing interventions addressing physical, mental and social health of independent community-dwelling older adults. Geriatr. Nurs. **40**(1), 37–50 (2018)
11. Werbrouck, A., Swinnen, E., Kerckhofs, E., Buyl, R., Beckwée, D., De Wit, L.: How to empower patients? A systematic review and meta-analysis. Transl. Behav. Med. **8**(5), 660–674 (2018)
12. World Health Organization: Regional Office for Europe. Health 2020 policy framework and strategy document [Internet]. WHO, Geneva (2012). Accessed 28 Feb 2015
13. Kouris, I., Sarafidis, M., Androutsou, T., Koutsouris, D.: HOLOBALANCE: an augmented reality virtual trainer solution for balance training and fall prevention. In: 2018 40th Annual International Conference of the IEEE Engineering in Medicine and Biology Society (EMBC), Honolulu, Hawaii, pp. 4233–4236. IEEE (2018)
14. Small, N., Bower, P., Chew-Graham, C.A., Whalley, D., Protheroe, J.: Patient empowerment in long-term conditions: development and preliminary testing of a new measure. BMC Health Serv. Res. **13**(1), 263 (2013)

15. Fotoukian, Z., Shahboulaghi, F.M., Khoshknab, M.F., Mohammadi, E.: Concept analysis of empowerment in old people with chronic diseases using a hybrid model. Asian Nurs. Res. **8** (2), 118–127 (2014)
16. Salmon, P., Hall, G.M.: Patient empowerment or the emperor's new clothes. J. R. Soc. Med. **97**(2), 53–56 (2004)
17. Allam, A., Kostova, Z., Nakamoto, K., Schulz, P.J.: The effect of social support features and gamification on a web-based intervention for rheumatoid arthritis patients: randomized controlled trial. J. Med. Internet Res. **17**(1), e14 (2015). https://doi.org/10.2196/jmir.3510
18. Young, H., Miyamoto, S., Ward, D., Dharmar, M., Tang-Feldman, Y., Berglund, L.: Sustained effects of a nurse coaching intervention via telehealth to improve health behavior change in diabetes. Telemed. e-Health **20**(9), 828–834 (2014)
19. Riva, S., Camerini, A.L., Allam, A., Schulz, P.J.: Interactive sections of an internet-based intervention increase empowerment of chronic back pain patients: randomized controlled trial. J. Med. Internet Res. **16**(8), e180 (2014). https://doi.org/10.2196/jmir.3474
20. McCarrier, K.P., Ralston, J.D., Hirsch, I.B., Lewis, G., Martin, D.P., Zimmerman, F.J., Goldberg, H.I.: Web-based collaborative care for type 1 diabetes: a pilot randomized trial. Diabetes Technol. Ther. **11**(4), 211–217 (2009)
21. Mostajeran, F., Katzakis, N., Ariza, O., Freiwald, J.P., Steinicke, F.: Welcoming a holographic virtual coach for balance training at home: two focus groups with older adults. In: Proceedings of 26th IEEE Conference on Virtual Reality and 3D User Interfaces, Osaka. IEEE (2019, ahead of publish)
22. de Amorim, J.S.C., Leite, R.C., Brizola, R., Yonamine, C.Y.: Virtual reality therapy for rehabilitation of balance in the elderly: a systematic review and META-analysis. Adv. Rheumatol. **58**(1), 18 (2018)
23. Stanmore, E.K., Mavroeidi, A., de Jong, L.D., Skelton, D.A., Sutton, C.J., Benedetto, V., Munford, L.A., Meekes, W., Bell, V., Todd, C.: The effectiveness and cost-effectiveness of strength and balance exergames to reduce falls risk for people aged 55 years and older in UK assisted living facilities: a multi-centre, cluster randomised controlled trial. BMC Med. **17** (1), 49 (2019)
24. Abrams, H.B., Bock, K., Irey, R.L.: Can a remotely delivered auditory training program improve speech-in-noise understanding? Am. J. Audiol. **24**(3), 333–337 (2015)
25. De Hert, P., Papakonstantinou, V., Malgieri, G., Beslay, L., Sanchez, I.: The right to data portability in the GDPR: towards user-centric interoperability of digital services. Comput. Law Secur. Rev. **34**(2), 193–203 (2018)
26. Barr, P.J., Scholl, I., Bravo, P., Faber, M.J., Elwyn, G., McAllister, M.: Assessment of patient empowerment–a systematic review of measures. PloS One **10**(5), e0126553 (2015). https://doi.org/10.1371/journal.pone.0126553

Assessment of Tripping Hazards by a Single Step Evaluated by Principal Component Analysis of Pedestrian Feet Movements and Eye Behaviours

Tatsuto Suzuki[1(✉)], I. Wa Liu[2], Nikolaos Papadosifos[1], Derrick Boampong[1], Pak Sum Fung[2], and Nick Tyler[2]

[1] Pedestrian Accessibility Movement Environment Laboratory, Department of Civil, Environment, Geomatic Engineering, University College London, London N19 5UN, UK
t.suzuki@ucl.ac.uk

[2] Department of Civil, Environment, Geomatic Engineering, University College London, London WC1E 6BT, UK

Abstract. The aims in this study was to investigate feet movements and eye behaviours of young and older people when stepping over a small single step up to 25 mm height while walking. In addition, this study investigated what factors in feet movements and eye behaviours can be used to show effectiveness of the preventive measures of tripping. Healthy 12 young and 5 older participants joined this study. A walkway (9.6 m length × 1.2 m width) with a single step chosen from 0, 10, and 25 mm in one of three positions 4.8, 6, and 7.2 m from the start line, was used for measurement of feet movements and eye behaviours in stepping over. Principal component analysis demonstrated to evaluate pedestrian capability of stepping over the single step and showed differences between young and older groups. First principal component might show different walking style, second principal component might show foot strategy of stepping over the single step, and third principal component might show that awareness of the single step. The awareness level of 25 mm step by older group was similar level of the awareness of 10 mm by young group, and older group was struggle to aware 10 mm step, compared with 0 mm case. These findings might be useful to consider how to improve outdoor and indoor environments as well as to assess how physical training, preventive measures, and assistive technologies to reduce the risk of tripping in stepping over the single step.

Keywords: Tripping · Single step · Feet movements · Eye behaviours · Principal component analysis

1 Introduction

Small single steps can be seen in outside environments in daily life. Pedestrians walking in outside meet the edge of lifted block pavements caused by the raised ground, the edge of dropped pedestrian kerb, the edge of a train floor on the platform in level access, for example. UK footway maintenance policy basically guides to repair over 20 mm height

J. Henriques et al. (Eds.): MEDICON 2019, IFMBE Proceedings 76, pp. 989–995, 2020.
https://doi.org/10.1007/978-3-030-31635-8_121

level uneven part on footways, and leaves under 20 mm level untreated [1]. Pedestrians can detect such a small single step before approaching, and then step over it naturally, however, sometimes tripping happens for different reasons [1].

Tripping while walking is a major cause of falling [2], which sometimes leads serious injuries in people aged 65 years and over [3] because ageing slows counter actions of legs to recovery walking balance and of hands to protect a body from crushing to the ground. In addition, ageing slows cognitive process in brain. Tripping occurs when natural movement of a swing foot in walking was forced to limit by a contact with an object, the ground, or the single steps [4].

Previous studies with 3D marker measurement systems focused on minimum foot clearance in walking on flat surface [5] as well as feet movement when stepping over an obstacle [6, 7], in order to investigate how pedestrians naturally prevent their feet movements from tripping on the ground or obstacles higher than 25 mm. Recent study focused on cognitive process how pedestrians can detect an object or a step surface while walking [8, 9]. Previous findings partly provide evidences how pedestrians naturally walk without tripping, however, no investigation with feet movements and eye behaviours together when stepping over a step has been done.

Meanwhile, coloured edge of steps, warning signs, reminders in the railway platform, increased floor illuminance, etc. are used to prevent pedestrians from tripping. These preventive measures would reduce tripping hazard, however, it's difficult to expect how these measures are effective before implementation.

The aims in this study was to investigate feet movements and eye behaviours when stepping over a small single step up to 25 mm height while walking, in order to explore in more details how young and older pedestrians prevent from tripping. In addition, this study investigated what factors in feet movements and eye behaviours can be used to show effectiveness of the preventive measures of tripping.

2 Methods

2.1 Experiment Setup

A walk way for this study in Fig. 1 was situated in the middle of the platform in the Pedestrian Accessibility Movement Laboratory (PAMELA), University College London. The total distance of walkway was 9.6 m from a start line to an end line and its width was 1.2 m. The size of the PAMELA platform was 12 × 7.2 m and consisted of 60 modules having a 1.2 × 1.2 m concrete surface on the top of each module. Each module could control height of the concrete surface by four actuators supporting each corner. The concrete surface of each module had nine 400 × 400 mm square pavers. Both the start and end lines were marked by yellow black stripe 1.2 × 0.05 m. The height of a single step was chosen from 0, 10, and 25 mm, and was created at one of three positions 4.8, 6, and 7.2 m from the start line. Only one single step was faced in each trial and the height between the single step and the end line was adjusted to the same level. The height and position of the single step was randomised for trials so that participants could not expect the position and height of the single step. Floor illuminance on the walkway was calibrated to 256 lx by a LED lamp system fixed at 3 m height from the walkway.

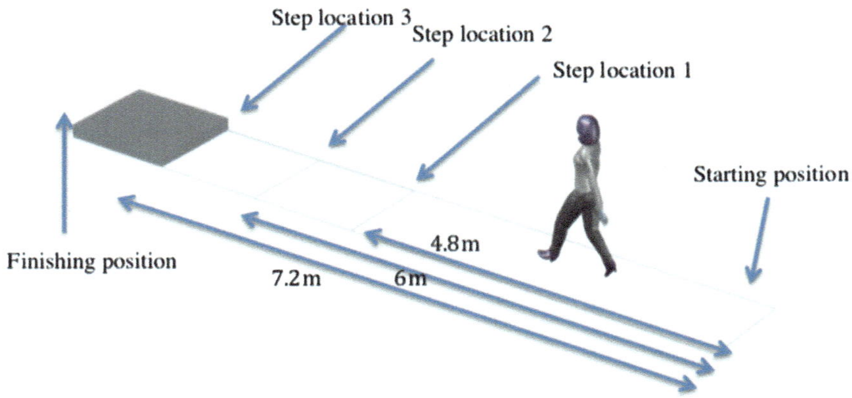

Fig. 1. Walkway with three different positions of a single step with height 0, 10, or 25 mm. Only one single step was faced in each trial and the height between the single step and the end line was adjusted to the same level.

2.2 Measurement System

Active marker 3D measurement system by infrared LED (CODA system by CODAmotion - Charnwood Dynamics Ltd.), which can measure marker 3D positions in 0.3 mm accuracy, was used to capture 3D positions of participant's feet and single step on the walkway. Two CODA cameras were placed after confirming enough area coverage to capture feet movements over three single step positions and were calibrated before trials. A pair of CODA makers were placed on both sides of the edge of each single step position on the walkway and four shoe positions (Toe, left, right, and heel) of both feet. Sampling frequency of CODA measurements were 100 Hz. Participants wore mobile eye tracking glasses (SensoMotoric Instruments GmbH), which measured the position and duration of fixations at 30 Hz sampling.

2.3 Experiment Protocol

Before trials, an examiner had asked a participant to naturally walk to the end line of the walkway from the start line. Participants allowed free fixation and head movement while walking. The single step had been created by remote control before each trial while the participant waited with sitting on a chair behind of a screen, which blocked visual view of the participants to see the walkway. After finishing set up of the single step, the participant was asked to stand at the start line, and then, started walking to the end line after the examiner moved out the screen. When the participant reached the end line, each trial ended, and the participant was asked to come back and to wait for next trial with sitting on the chair in the back of the screen.

2.4 Participants

Total 17 healthy participants without any visual, mobility and cognitive disorders, joined this study after providing written informed consent on the experiment protocol approved by the University College London Research Ethics Committee. Young group was formed by 12 participants (age = 25.9 ± 7.6 years, height = 1.66 ± 0.07 m, and body weight = 56.1 ± 5.3 kg), and older group was formed by five participants (age = 69.3 ± 8.6 years, height = 1.56 ± 0.05 m, and body weight = 46.1 ± 3.1 kg).

2.5 Data Analyses

With CODA measurements of both feet movements, a swing foot (SWF) stepping over the single step was identified in the beginning of analysis, and then stance foot (STF) was determined. Extracted parameters from SWF movement were vertical toe clearance from the edge of the single step, vertical maximum toe clearance from the ground, horizontal heel position from the edge of the single step after heal contact of SWF, mean SWF speed, and SWF step length. From STF movement, horizontal toe position from the edge of the single step was extracted. Eye tracking data were used for extracting total duration of fixations at the step, ±1 m step area, and travel path at each trial, and these fixation durations were normalised by travel time. All extracted parameters were analysed with principal component analysis [10].

3 Results

Total variance explained of data difference by from first principal component (PC1) to third one (PC3) was 75.34%. The Loadings of all parameters in the PC1-3 were shown in Fig. 2. PC1's main parameters were SWF step length (−0.48), SWF mean speed (−0.48), and travel time (0.45). PC2's main parameters were SWF maximum toe clearance (0.58), SWF toe clearance (−0.56), and STF toe position (0.55). Main parameters of PC3 were duration of fixation at the single step (0.78), and SWF heel position (0.57). Loading of parameters provides how parameters individually contribute to PC scores, and the range of loadings is from −1 to 1.

Mean PC scores grouped by groups and single step heights are shown with standard error of means in Fig. 3. PC1 showed large difference between young and old groups. Young group in three step height cases showed similar PC1 scores −0.67[0 mm], −0.68[10 mm], −0.75[25 mm], and Older group showed larger PC1 scores 0.80 [0 mm], 1.26[10 mm], and 1.32[25 mm]. PC2 showed similarity between young and old with around zero, except 0.36[25 mm] by young group and −0.61[10 mm] by older group. PC3 scores of both groups had a trend to increase by step height from similar base level at 0 mm step height. PC3 scores of young group were −0.58[0 mm], −0.05 [10 mm], and 0.57[25 mm]. Older group PC3 scores were −0.68[0 mm], −0.40 [10 mm], and 0.04[25 mm].

All data distribution PC3 by PC1 is shown in Fig. 4. Young group's PC scores were placed in left side plane, and young 25 mm cases moved to left upper area from left lower area. Older group PC scores were placed in right side place, and older 25 mm cases spread to left side area from lower middle area.

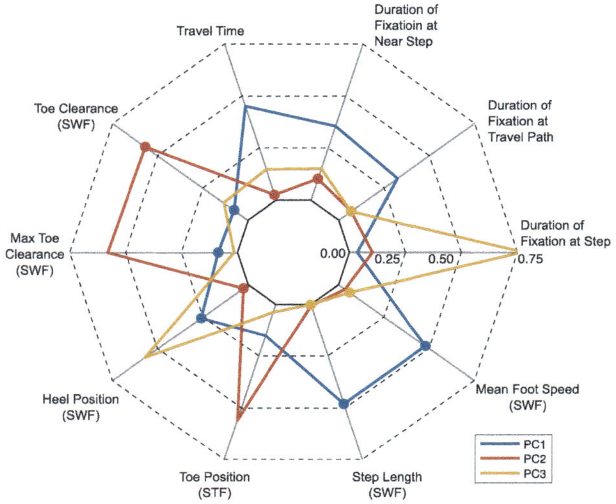

Fig. 2. Loading of all parameters of PC1–PC3. Negative loadings are shown with a circle.

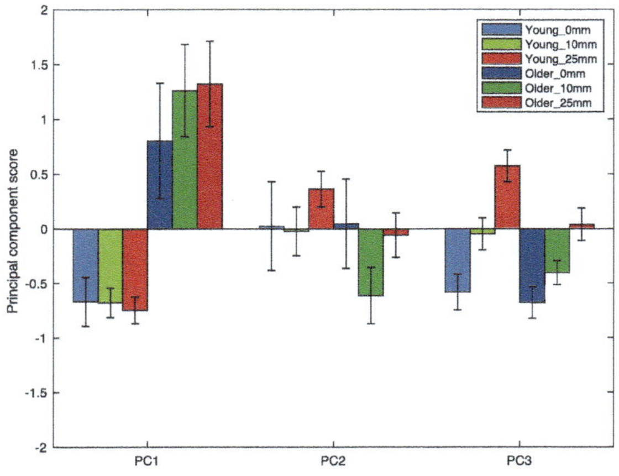

Fig. 3. Mean scores of PC1 to PC3 with standard error of means

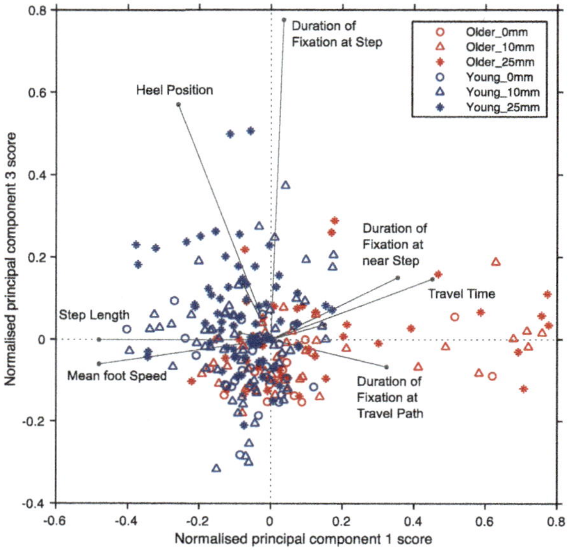

Fig. 4. Biplot of PC1 and PC3

4 Discussions

PC1 might mean individual walking style of young or older. PC1 score in Fig. 3 clearly shows difference between young and older groups. Older group took longer travel time caused by shorter step length and lower mean SWF speed with longer duration of fixation on near step and travel path. In opposite, younger group walking fast without looking on the ground. This might be because young group might detect the single step by peripheral vision. The step height didn't affect walking style of young group, but older group became more careful of walking on 25 mm case than 0 mm.

PC2 might mean general foot strategy of stepping over the single step. SWF's Higher maximum toe clearance and lower toe clearance happened when horizontal STF toe position was near to the edge of the single step. A possible reason was that STF was placed near the single step in order to keep flexibility of SWF movement and SWF started moving near the edge of the step, so toe clearance was lowered and participants intentionally raised toe to avoid tripping. PC2 scores of both groups in Fig. 3 were similar, however, young group at 25 mm case used this strategy clearly compared with other cases, and older group didn't take this strategy in 10 mm cases.

PC3 might mean that awareness of the single step. Longer duration of fixation at the single step contributes further horizontal SWF heel position after heel contact. The trend of PC3 by the step height in Fig. 3 possibly suggest that younger group showed increased awareness of the step by the increase of the step height proportionally, and older group's awareness of the single step 10 mm were slightly better than 0 mm, but not a similar difference between 0 and 10 mm cases in young group. The results might say the awareness level of 25 mm step by older group was similar level of the awareness of 10 mm by young group, and older group struggled to aware 10 mm step, compared with 0 mm case.

In the biplot PC1-PC3 in Fig. 4, it might be possible to spot individual problem. These three PC scores showed typical manners and capability of stepping over a single step by young and older groups and can be used to evaluate tripping hazard and how preventive measures improve older people's manner and capability.

This study was carried out with limitations; (1) A black battery boxes 33 × 51 × 14 mm with black active markers 12 × 8 × 5 mm were visible on three single step positions on the walkway, (2) The edge of the single step created by the PAMELA platform was chamfered, and (3) The walkway had typical concrete grey colour.

5 Conclusions

Principal component analysis in this study demonstrated to evaluate pedestrian capability how to manage stepping over the single step and showed differences between young and older groups. These findings might be useful to consider how to improve outdoor and indoor environments as well as to assess how physical training, preventive measures, and assistive technologies to reduce the risk of tripping in stepping over the single step.

References

1. Bird, S.: The risk of tripping accidents on public footways. In: Proceedings of the Institution of Civil Engineers - Municipal Engineer, vol. 161, pp. 129–136 (2008)
2. Berg, W.P., Alessio, H.M., Mills, E.M., Tong, C.: Circumstances and consequences of falls in independent community-dwelling older adults. Age Ageing **26**, 261–268 (1997)
3. Mindell, J.S., Leslie, D., Wardlaw, M.: Exposure-based, "like-for-like" assessment of road safety by travel mode using routine health data. PLoS ONE **7**(12), e50606 (2012)
4. Tinetti, M.E., Speechley, M., Ginter, S.F.: Risk factors for falls among elderly persons living in the community. N. Engl. J. Med. **319**(26), 1701–1707 (1988)
5. Mills, P.M., Barrett, R.S., Morrison, S.: Toe clearance variability during walking in young and elderly men. Gait & Posture **28**, 101–107 (2008)
6. Chen, H.-C., Ashton-Miller, J.A., Alexander, N.B., Schultz, A.B.: Stepping over obstacles: gait patterns of healthy young and old adults. J. Gerontol. **46**(6), M196–203 (1991)
7. Loverro, K.L., Mueske, N.M., Hamel, K.A.: Location of minimum foot clearance on the shoe and with respect to the obstacle changes with locomotor task. J. Biomech. **46**, 1842–1850 (2013)
8. Uttley, J., Fotios, S., Cheal, C.: Effect of illuminance and spectrum on peripheral obstacle detection by pedestrians. Lighting Res. Technol. **49**, 211–227 (2017)
9. Cheng, T.J., Yang, B., Holloway, C., Tyler, N.: Effect of environmental factors on how older pedestrians detect an upcoming step. Lighting Res. Technol. **50**, 405–415 (2018)
10. Pearson, K.: On lines and planes of closest fit to systems of points in space. Phil. Mag. **2**(11), 559–572 (1901)

Regular Sessions: Biomedical Engineering Education and Society

Automatic Lung Reference Model

Marlene Machado[1], Carlos A. Ferreira[2], João Pedrosa[2], Eduardo Negrão[3], João Rebelo[3], Patrícia Leitão[3], André S. Carvalho[3], Márcio C. Rodrigues[3], Isabel Ramos[3], António Cunha[2,4], and Aurélio Campilho[2,5](✉)

[1] Signinum, Braga, Portugal
[2] INESC-TEC - Institute for Systems and Computer Engineering, Technology and Science, Porto, Portugal
aurelio.campilho@inesctec.pt
[3] Department of Radiology, Hospital de São João, Porto, Portugal
[4] University of Trás-os-Montes e Alto Douro, Vila Real, Portugal
[5] Faculty of Engineering, University of Porto, Porto, Portugal

Abstract. The lung cancer diagnosis is based on the search of lung nodules. Besides its characterization, it is also common to register the anatomical position of these findings. Even though computed-aided diagnosis systems tend to help in these tasks, there is still lacking a complete system that can qualitatively label the nodules in lung regions. In this way, this paper proposes an automatic lung reference model to facilitate the report of nodules between computed-aided diagnosis systems and the radiologist, and among radiologists. The model was applied to 115 computed tomography scans with manually and automatically segmented lobes, and the obtained sectors' variability was evaluated. As the sectors average variability within lobes is less or equal to 0.14, the model can be a good way to promote the report of lung nodules.

Keywords: Lung reference model · Lung division · Nodule localization

1 Introduction

Computer tomography (CT) is the main imaging modality for lung cancer diagnosis. During the diagnosis process, radiologists and physicians report the location of the finding through anatomic references, such as the lung lobes and pulmonary segments. The lungs are composed of three lobes in the right lung (the upper, middle and lower) and two lobes in the left lung (the upper and lower), physically separated by fissures. Each lobe is subdivided into several pulmonary segments. There are ten segments on right lobes and eight segments on left lobes [7], as can be seen in Fig. 1. The segments are portions of the lung supplied by specific bronchus and arteries. A lesion may be regional, and affect only one segment, or heterogeneous, and affect several segments, the entire lobe or more than one lobe. In this sense, if the exact location of the nodules is known, it is possible to quantify the spread of the disease, guiding surgical interventions with more precision, simplifying treatment and monitoring [11].

J. Henriques et al. (Eds.): MEDICON 2019, IFMBE Proceedings 76, pp. 999–1008, 2020.
https://doi.org/10.1007/978-3-030-31635-8_122

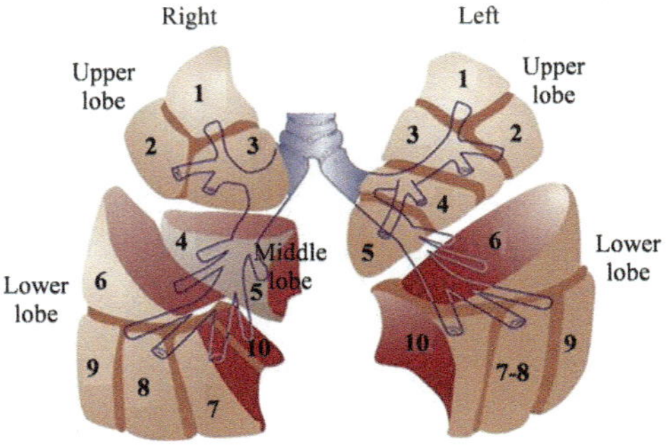

Fig. 1. Scheme of the pulmonary segments.

As the assessment of lung regions allows a localised diagnosis, a computed-aided diagnosis (CAD) module that reports the nodules in a standard way to radiologists should be a valuable contribution for lung cancer diagnosis. However, this is not trivial due to the difficulty in segmentating lobes and segments. Lately promising results have been achieved with supervised machine learning approaches, but the ground truth labeling continues to be a hard task to work around - manual delineation of fissures and lobes in several hundreds of CT slices is time-consuming and unthinkable in clinical routine. Other problems in automatic segmentation are the visibility of the fissures, the incomplete fissures [10], the physical limits of the segments, and the thin membranes of connective tissue, not recognisable on CTs [1,9]. Then, only few approaches for the pulmonary segments segmentation are reported in the literature [1,8,9], and when lobe segmentation is possible, the nodule localization in a particular lobe is very imprecise because of the large dimension of the lobes.

In this study, we describe an automatic lung reference model (LRM) that is capable to overcome the segmentation challenge of pulmonary segments. It splits the lung into sectors and name them according to their anatomic localisation for ease their interpretation by the radiologists. The LRM was thought to facilitate the communication of findings between CADs and the radiologist, and between the radiologists themselves. Furthermore, the model could be used to quickly search nodules in regions of high probability of occurrence, as top regions, and for statistical purposes as already used in [3].

2 Material

115 CTs scans were used: 17 from the dataset published by VESSEL12 Challenge [6] and 98 from a private database. The 17 scans from VESSEL12 had between 355 and 543 transversal slices and were reconstructed with slice-thickness of 0.7 mm and 1 mm. On the other hand, the 87 scans from the private dataset had between 238 and 423 transversal slices and were reconstructed with 1 mm. The resolution was, in both cases, 512×512 pixels. The lobes in VESSEL12 scans were manually

segmented by a radiologist and the lobes segmentations from the private database were obtained automatically using the algorithm proposed in [2].

3 Methodology

The LRM is built in 2 steps according to the diagram shown in Fig. 2: the "Pre-processing" step, which segments the lungs in the CT volume, and the "Automatic lungs division" step, which divides each lung into the LRM sectors.

3.1 Pre-processing

A CT scan is the input to the "pre-processing step", where the segmentation of the lungs parenchyma (I_S) is performed using the technique described in [5], followed by the division in left (I_L) and right lung (I_R), based on the minimum sum of column pixels per slice.

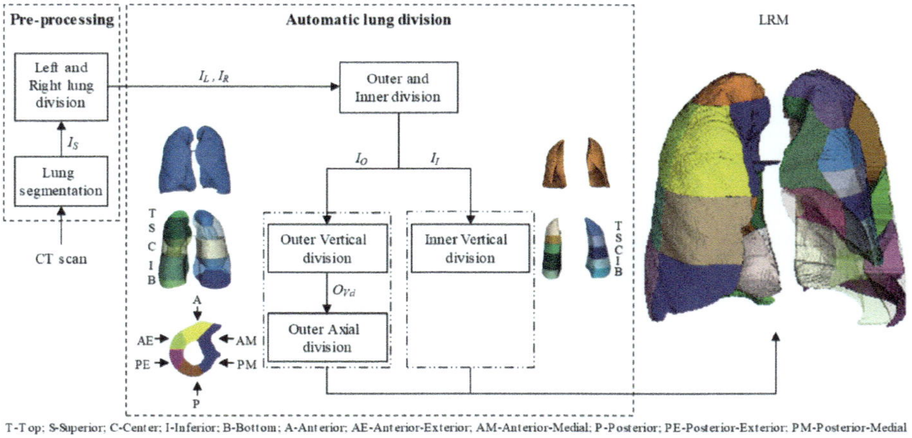

Fig. 2. Methodology for automatic design of the LRM.

As the average axes variations of (I_L) and (I_R) volumes were computed and were less than 1°, the CT volumes have similar orientations.

3.2 Automatic Lung Division

The method for division in sectors was made following 3 premises: (i) the sectors should have a visual position and dimension close to pulmonary segments; (ii) a unique reference model should be equal in both lungs; and (iii) the sectors' name should allow an intuitive understanding of the lung anatomic location.

In this way, the lungs were divided into two regions and each region into sectors. The LRM has an outer region (I_O) and an inner region (I_I): at the I_O, it is expected to find peripheral and juxta-pleural nodules; and at the I_I, it is probable to find hilar and juxta-vascular nodules. As the I_O and I_I regions are quite large, the subdivision into sectors with dimension closer to the pulmonary segments is made: the I_O is subdivided into twenty sectors, and the I_I is subdivided into five sectors. In the next sections (from 3.2.1 until 3.2.4) are described these remaining lung divisions represented in Fig. 2. The chosen syntax to the sectors names' is described in Sect. 3.2.5.

3.2.1 Outer and Inner Regions Division

The division of each lung volumes I_L and I_R into I_O and I_I is the first processing step. A 3D distance map (*dmap*) is calculated for each volume I_L and I_R using the Euclidean distance transform (DT) [4]. Then, the I_O and I_I are obtained, applying a threshold (*th*) according to (1):

$$\begin{cases} dmap < th, I_O \\ otherwise, I_I \end{cases} \tag{1}$$

The *th* is iteratively computed for each lung until the $^{20}/_{25}$ volume proportion is achieved, guaranteeing a balanced division between I_O (20 sectors) and I_I (5 sectors). A 3D representation of the resulted volumes is presented at blue and orange in Fig. 2.

3.2.2 Outer Vertical Division

The volume I_O is split into five vertical divisions (O_{Vd}) designated as Top (T), Superior (S), Centre (C), Inferior (I) and Bottom (B). To prevent a rectangular division and resemble the shape of the pulmonary segments, the DT is iteratively applied until the sectors Top and Bottom have, approximately, a proportion $^{1}/_{25}$ of the volume of the lung and the sectors Superior, Centre and Inferior to have, approximately, a proportion $^{6}/_{25}$ of the volume of the lung. A 3D representation of this division, for both lungs, is shown in Fig. 2 at the left of "Outer vertical division" box.

3.2.3 Outer Axial Division

The Superior, Centre and Inferior (O_{Vd}) divisions are angularly divided into six axial sectors anatomically positioned, following the usual terminology used by radiologists, namely: Anterior, Anterior-Exterior, Anterior-Medial, Posterior, Posterior-Exterior, and Posterior-Medial. These anatomical positions were defined empirically through 6 fix angles (Fig. 3). For each axial slice is computed the corresponding centre of mass of the I_L and I_R in axial plane before computing the divisions. As the shape of each lung in each slice of a CT scan is not a circle, the regions were not delimited by the same angle. The axial sectors were designed to follow the anatomic position, their volumes are different. Also, as the left lung is smaller than the right lung, the left sectors are generally slightly lower. An example of axial sectors is show in Fig. 2, at the left of the "Outer Axial division" box.

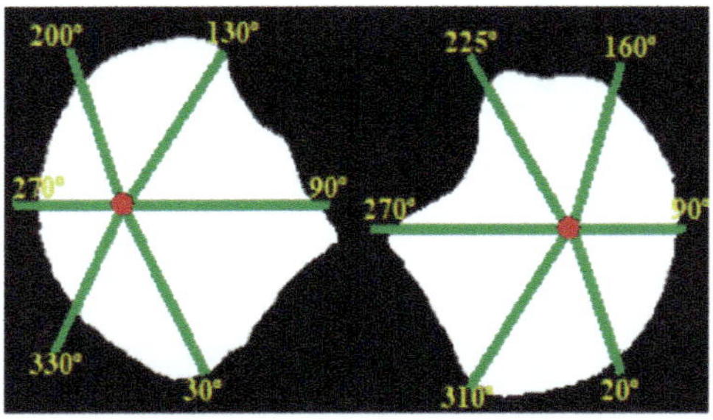

Fig. 3. Outer axial divisions angles.

3.2.4 Inner Vertical Division

As in the "Outer Axial division", the I_I volume is subdivided into five vertical divisions (I_{Vd}) with proportion $1/25$ of the volume of the lung, and are similarly named. A 3D representation of this division can be seen in Fig. 2, at the right of the "Inner Vertical division" box.

3.2.5 Sectors Name

The syntax used to nominate the sectors allows a direct understanding of the respective lung anatomic location. It is composed by: (i) the location in the left or right lung; (ii) the location in I_O sectors. For example: *Right/Left: Outer/Inner{Vertical divisions[Axial divisions;...];...}*.

4 Results

The proposed LRM generates automatically a set of lung sectors based on the lung anatomy. For the evaluation of the sectors in the lobes, the LRM was applied to the CTs dataset previously presented in Sect. 2. The variability of the sectors regarding the lobes was estimated. Lower variability of the sectors in the lobes leads to a better model. The obtained results are presented and discussed in Sect. 4.1.

Table 1. Compiled results - average percentage of vertical sectors (right) and each individual sector (left) per lobe, with the corresponding standard deviations.

	Mean % of the vertical sectors in lobe						
Lung	Vertical Sectors		Upper	Middle	Lower	**Mean % of the sectors in lobes (≥0.15)**	*Std*
Right	Outer	Top	1.00	0.00	0.00	- Upper: Outer{T(1.00±0.00); S[AM (0.97±0.04); AE (0.96±0.06); PE (0.96±0.06); A (0.94±0.07)}; Inner{T (0.99±0.03)} - Upper: Outer{S[P (0.71±0.16); PM (0.70±0.15)]}; Inner{C (0.79±0.23)} - Upper: Outer{C[PE (0.24±0.16); AM (0.22±0.13); AE (0.21±0.15); A (0.17±0.13)]}; Inner{I (0.27±0.23)}	0.10
		Superior	0.87	0.02	0.10		
		Center	0.15	0.31	0.53		
		Inferior	0.00	0.15	0.86		
		Bottom	0.00	0.01	0.99	- Middle: Outer{C[A (0.79±0.12); AE (0.56±0.15); AM (0.43±0.18)]; I[A (0.60±0.16)]} - Middle: Outer{I[AE (0.17±0.14)]}; Inner{I (0.30±0.15); B (0.30±0.11)}	0.14
	Inner	Top	0.99	0.00	0.01		
		Superior	0.00	0.04	0.96	- Lower: Outer{B (0.99±0.04); C[P (0.96±0.07); PM (0.95±0.06)]; I[P (1.00±0.01); PM (1.0±0.00); PE (0.99±0.02)]} - Lower: Outer{I[AM (0.94±0.09); AE (0.83±0.14)]}; Inner{S (0.96±0.06)} - Lower: Outer{C[PE (0.68±0.21)]}; Inner{B (0.69±0.13)} - Lower: Outer{I[A (0.40±0.16)]; C[AM (0.34±0.20); AE (0.23±0.16)]; S[P (0.29±0.16); PM (0.29±0.15)]}; Inner{I (0.43±0.15); C(0.16±0.13)}	0.11
		Center	0.79	0.04	0.16		
		Inferior	0.27	0.30	0.43		
		Bottom	0.02	0.30	0.69		
Left	Outer	Top	1.00	-	0.00	- Upper: Outer{T(1.00±0.02); S[A (1.00±0.01); AE (0.99±0.03); AM (0.99±0.04); PE (0.92±0.13)]; C[A (0.94±006)]}; Inner{T (0.96±0.07)} - Upper: Outer{S[P (0.75±0.19); C[AE (0.72±0.18); AM (0.65±0.20)]}; Inner{C (0.83±0.14)} - Upper: Outer{I[A (0.52±0.19)]; S[P (0.37±0.21)]; C[PE (0.33±0.24)]}; Inner{I (0.61±0.22)} - Upper: Outer{I[AE (0.20±0.18)]}; Inner{B (0.20±0.16)}	0.13
		Superior	0.84	-	0.16		
		Center	0.46	-	0.54		
		Inferior	0.12	-	0.87		
		Bottom	0.02	-	0.98		
	Inner	Top	0.96	-	0.04	- Lower: Outer{I[PM (1.00±0.01); P(1.00±0.01); PE (0.99±0.03); AM (0.97±0.06)]; B (0.98±0.07)]} - Lower: Outer{C[P (0.98±0.05); PM (0.88±0.11)]}; Inner{S (0.99±0.04)} - Lower: Outer{I[AE (0.80±0.18); A (0.48±0.19)]; S[P (0.63±0.21)]; C[PE (0.67±0.24)]}; Inner{B (0.79±0.18); I (0.39±0.22)} - Lower: Outer{S[PM (0.24±0.19)]; C[AM (0.17±0.14)]} - Lower: Outer{C[AE (0.28±0.18)]}; Inner{C (0.17±0.14)}	0.12
		Superior	0.01	-	0.99		
		Center	0.83	-	0.17		
		Inferior	0.61	-	0.39		
		Bottom	0.20	-	0.79		

T-Top; S-Superior; C-Center; I-Inferior; B-Bottom; A-Anterior; AE-Anterior-Exterior; AM-Anterior-Medial; P-Posterior; PE-Posterior-Exterior; PM-Posterior-Medial

4.1 LRM Evaluation

The variation of the sectors in the lobes was calculated for each lung. The variation of the total sectors' voxels is estimated in each lobe through: (i) the average percentage of each vertical sector (O_{Vd} and I_{Vd}) per lobe; (ii) the average percentage ($\geq$0.15) of each individual sector per lobe; and (iii) the average standard deviation (*std*) of the sectors in each lobe.

The quantitative results are presented in Table 1 and discussed in the following 2 subsections. A CT example with qualitative results of the sectors that belong to each lobe is represented in Fig. 4.

4.2 Right Lung

Right lung is composed by 3 lobes. The vertical sectors that belong to these lobes are:

- Upper lobe: *Outer* {*Top, Superior and Center*} and *Inner* {Top, Center and Inferior};
- Middle lobe: *Outer* {*Center and Inferior*} and *Inner* {Inferior and Bottom};
- Lower lobe: *Outer* {*Center, Inferior and Bottom*} and *Inner* {Superior, Center, Inferior and Bottom};

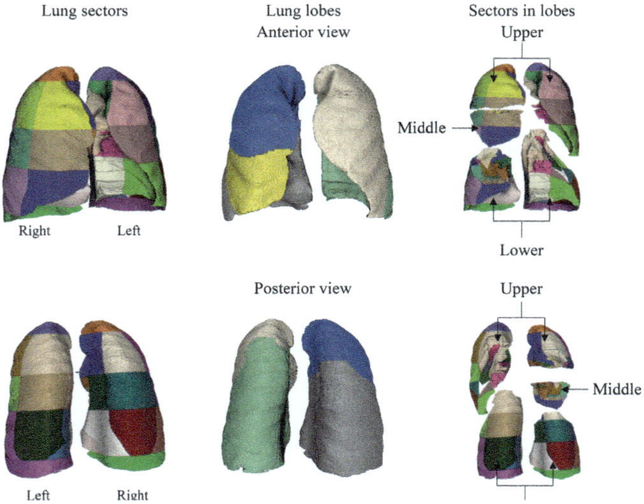

Fig. 4. CT example with the position of the sectors in lobes.

Some sectors integrate one lobe entirely, as is the case of the O_{Vd} *Outer Top* (1.0 ± 0.00) and I_{Vd} *Inner Top* (0.99 ± 0.03) in upper lobe; and the O_{Vd} *Outer Bottom* (0.99 ± 0.04) in lower lobe. These sectors represent 12%, 10% and 9% of the respective lobe and their position is very reliable because they show little variations.

The sectors predominantly (≥94%) belonging to a lobe are *Outer* {*Superior [Anterior-Medial* (0.97 ± 0.04), *Anterior-Exterior* (0.96 ≥ 0.06), *Posterior-Exterior* (0.96 ≥ 0.06), and *Anterior* (0.94 ≥ 0.07)] in upper lobe, and *Outer* {*Center* [*Posterior* (0.96 ≥ 0.07), *Posterior-Medial* (0.95 ≥ 0.06)], *Inferior* [Posterior (1.0 ≥ 0.01), *Posterior-Medial* (1.0 ≥ 0.00), *Posterior-Exterior* (0.99 ≥ 0.02), and *Anterior-Medial* (0.94 ≥ 0.09)/} in lower lobe. As previous sectors, these sectors have a consistent anatomic position with tiny variations.

The others sectors are spread by 2 or by the 3 lobes, as the sectors *Outer* {*Superior [Posterior*, *Posterior-Medial]*, *Center Posterior-Exterior*} and *Inner Center* in upper and lower lobes; *Outer* {*Inferior [Anterior*, *Anterior-Exterior]*} and *Inner Bottom* in middle and lower lobes; and *Outer Center* [all *Anterior*] and *Inner Inferior* that is part of the 3 lobes.

In the upper lobe, the sectors belonging to other lobes have higher *std* variation, being the variation of (0.15 ≤ *std* ≤ 0.18) and (*std* = 0.13) in the sectors inherent to upper and lower lobe, and upper and middle lobe, respectively, and of (0.13 ≤ *std* ≤ 0.23) in the sectors distributed by 3 lobes. The average *std* of all sectors that belong to the lobe is 0.10.

The middle lobe is the smallest lobe, having the higher average *std* (0.14) of both lungs because its sectors are all shared by the other 2 lobes being the variation between ($0.11 \leq std \leq 0.18$).

In the lower lobe, the sectors variation is also higher in the sectors that are spread by other lobes, being between ($0.13 \leq std \leq 0.21$) and the average *std* of all sectors is 0.11.

The sectors spread by more than one lobe have a more variable position in lobes due to lung anatomy differences in each person. For example, the sector *Outer Superior Posterior-Medial* has a relative average percentage of (0.70 ± 0.15) in upper lobe and an average percentage of (0.29 ± 0.15) in lower lobe, being hard to correlate the nodule location to one of lobes.

4.3 Left Lung

As in the right lung, the left lung has exclusive sectors of the two existing lobes and others that belong to both lobes. The vertical sectors belonging to 2 lobes are:

- Upper lobe: *Outer* {*Top, Superior, Center and Inferior*} and *Inner* {Top, Center, Inferior and Bottom};
- Lower lobe: *Outer* {*Superior, Center, Inferior and Bottom*} and *Inner* {Superior, Center, Inferior and Bottom}.

The sectors belonging integrally or mostly (≥94%) to a lobe are *Outer* {*Top* (1.0 ± 0.02), *Superior [Anterior* (1.0 ± 0.01), *Anterior-Exterior* (0.99 ± 0.03), *Anterior-Medial* (0.99 ± 0.04)], *Center Anterior* (0.94 ± 0.06) and *Inner Top* (0.96 ± 0.07)} that are always in upper lobe; and sectors *Outer* {*Inferior [Posterior* (1.0 ± 0.01), *Posterior-Medial* (1.0 ± 0.01), *Posterior-Exterior* (0.99 ± 0.03), *Anterior-Medial* (0.97 ± 0.06)], *Bottom* (0.98 ± 0.07)} and *Inner Superior* (0.99 ± 0.04) in lower lobe. In this lung side, the *Outer Top* and *Inner Top* sectors represent both 8% of the upper lobe, and *Outer Bottom* and *Inner Superior* represent 8% and 9%, respectively, of the lower lobe. The position of these sectors is very constant and reliable as happen in the right lung.

The sectors *Outer* {*Superior [Posterior, Posterior-Medial], Center [Anterior-Exterior, Anterior-Medial, Posterior-Medial, Posterior-Exterior], Inferior [Anterior, Anterior-Exterior]*}, and *Inner* {*Center, Inferior and Bottom*} are in both lobes.

In the left lung occurs the higher sectors variation ($std = 0.24$) being the sectors variation between ($0.11 \leq \text{std} \leq 0.24$) those that do part of both lobes. The average std of all sectors is 0.13 in upper lobe and 0.12 in lower lobe.

4.4 Sectors Visual Perception

A visual perception allows to do a rough correspondence between sectors and lung segments. In general, a segment is constituted by several sectors, but, in some cases, the entire sector is within a segment. For example, *Outer Top* sector correspond to Apical and Apico-posterior segments in the right and left lung respectively; *Outer Superior [Anterior and Anterior-Medial]* and *Inner Top* are inserted in Anterior segment in right lung; and *Outer Bottom* is placed in Anterior basal, Lateral basal, Posterior basal, and Anterior-medial segments of both lungs.

Other landmarks observed in the LRM are related to the airways, with carina lying at the level of sector *Outer Superior-Medial* and the main bronchi at the level of the sectors *Outer Center Posterior-Medial.*

5 Conclusions

In this study, an automatic pulmonary reference model was proposed to facilitate the communication of findings between CADs and radiologists, and between radiologists themselves. To the best of our knowledge, there is no method that allows the labelling of nodules in standard sectors. In this sense, this method can be a valuable contribution for the community allowing to save time to the radiologists and adding information to the cancer diagnosis. As results, it was verified that the average variability of the sectors within the lobes is less than or equal to 0.14, allowing an accurate communication of the findings in the lungs and validating the proposed reference model. In our point of view, the fact of some sectors are sharing different lobes don't compromise the quantitative analysis allowed by this method.

Acknowledgments. This study is associated with LNDetector, which is financed by the ERDF - European Regional Development Fund through the Operational Programme for Competitiveness - COMPETE 2020 Programme and by the National Fundus through the Portuguese funding agency, FCT - Fundação para a Ciência e Tecnologia within project POCI-01-0145-FEDER-016673.

References

1. Busayarat, S., Zrimec, T.: Detection of bronchopulmonary segments on high-resolution CT–preliminary results. In: Twentieth IEEE International Symposium on Computer-Based Medical Systems (CBMS 2007), pp. 199–204. IEEE (2007)
2. Ferreira, F.T., Sousa, P., Galdran, A., Sousa, M.R., Campilho, A.: End-to-end supervised lung lobe segmentation. In: 2018 International Joint Conference on Neural Networks (IJCNN), pp. 1–8. IEEE (2018)
3. Machado, M., Aresta, G., Leitão, P., Carvalho, A.S., Rodrigues, M., Ramos, I., Cunha, A., Campilho, A.: Radiologists' gaze characterization during lung nodule search in thoracic CT. In: 2018 International Conference on Graphics and Interaction (ICGI), pp. 1–7. IEEE (2018)
4. Maurer, C.R., Qi, R., Raghavan, V.: A linear time algorithm for computing exact euclidean distance transforms of binary images in arbitrary dimensions. IEEE Trans. Pattern Anal. Mach. Intell. **25**(2), 265–270 (2003)
5. Novo, J., Rouco, J., Mendonça, A., Campilho, A.: Reliable lung segmentation methodology by including juxtapleural nodules. In: International Conference Image Analysis and Recognition, pp. 227–235. Springer (2014)

6. Rudyanto, R.D., Kerkstra, S., Van Rikxoort, E.M., Fetita, C., Brillet, P.Y., Lefevre, C., Xue, W., Zhu, X., Liang, J., Öksüz, İ., et al.: Comparing algorithms for automated vessel segmentation in computed tomography scans of the lung: the vessel12 study. Med. Image Anal. **18**(7), 1217–1232 (2014)
7. Shields, T.W., LoCicero, J., Reed, C.E., Feins, H.R.: General Thoracic Surgery, 7th edn. Lippincott Williams & Wilkins, Philadelphia (2009)
8. Stoecker, C., Welter, S., Moltz, J.H., Lassen, B., Kuhnigk, J.M., Krass, S., Peitgen, H.O.: Determination of lung segments in computed tomography images using the Euclidean distance to the pulmonary artery. Med. Phys. **40**(9), 1–10 (2013)
9. Van Rikxoort, E.M., de Hoop, B., van de Vorst, S., Prokop, M., van Ginneken, B.: Automatic segmentation of pulmonary segments from volumetric chest CT scans. IEEE Trans. Med. Imaging **28**(4), 621–630 (2009)
10. Van Rikxoort, E.M., Prokop, M., de Hoop, B., Viergever, M.A., Pluim, J.P., van Ginneken, B.: Automatic segmentation of pulmonary lobes robust against incomplete fissures. IEEE Trans. Med. Imaging **29**(6), 1286–1296 (2010)
11. Van Rikxoort, E.M., Van Ginneken, B.: Automated segmentation of pulmonary structures in thoracic computed tomography scans: a review. Phys. Med. Biol. **58**(17), R187 (2013)

Preliminary Validation of an Editable Virtual Reality Simulator for Minimally Invasive Surgical Training

M. Rodríguez[1](✉), D. Camba-Lamas[1], I. Oropesa[1], K. Juhos[2], L. Wauben[3], J. Dankelman[3], F. W. Jansen[4], G. Weber[2], E. J. Gómez[1,5], and P. Sánchez-González[1,5]

[1] Biomedical Engineering and Telemedicine Centre, ETSI Telecomunicación, Center for Biomedical Technology, Universidad Politécnica de Madrid, Madrid, Spain
manuel.rodriguez.matesanz@upm.es
[2] Department of Surgical Research and Techniques, Semmelweis University, Budapest, Hungary
[3] Department of Biomechanical Engineering, Faculty of Mechanical, Maritime and Materials Engineering, Delft University of Technology, Delft, The Netherlands
[4] Department of Gynecology, Leiden University Medical Center, Leiden, The Netherlands
[5] Centro de Investigación Biomédica en Red, Biomateriales y Nanomedicina (CIBER-BBN), Madrid, Spain

Abstract. MIS-SIM is a virtual reality (VR) environment designed and developed for the creation of virtual scenarios that can be used to train and acquire basic and advance laparoscopic skills. The environment is composed by a task editor where a content creator design and develop tasks for the simulator to play. Once they are completed, objective metrics are automatically stored and examined in MIS-SIM's server so they can be displayed by an online platform. The project was validated in Semmelweis University in Budapest, Hungary where an experienced professor designed tasks for 16 young surgeons (PGY 3-4-5) from different surgical fields (gynaecology, general-, plastic-, vascular-, thoracic-, neurosurgery, etc.) with different experiences in laparoscopy. Each participant fulfilled each task as if they were completing them on physical simulator.

Keywords: Virtual reality · Surgical training · MIS-SIM · Personalization

1 Introduction

Virtual reality (VR) is nowadays established as a useful and valid tool to train minimally invasive surgical technical skills [1–3]. Like box trainers, they allow for deliberate practice of technical skills in patient-free environments, with the added value of task repeatability and capability of representing complex scenarios. In addition, computer simulations can determine the level of competence of a surgeon before operating a patient by analysing objective data generated during the simulation.

J. Henriques et al. (Eds.): MEDICON 2019, IFMBE Proceedings 76, pp. 1009–1017, 2020.
https://doi.org/10.1007/978-3-030-31635-8_123

If the actions and behaviours the user need to do in virtual environments are implemented correctly, VR can reinforce 3D-orientation, hand-eye coordination, instrument handling or improving the perception of stiffness in surgery for avoiding the fulcrum effect [6]. However, current systems restrict usage to tasks and modules offered, without possibility of personalization. When new procedures appear for training a determined skill, new simulators must be created for them. There are no simulator editors where not only tasks can be created but the behaviours of each object in those tasks be defined.

In this study we present the preliminary validation of the new virtual reality paradigm for minimally invasive surgical training: Minimally Invasive Surgery Simulator (MIS-SIM) scenario editor. MIS-SIM is designed and developed to fill an existing gap in the market: an editor can be used to create, edit and train with personalized virtual tasks and scenarios designed for surgical training.

2 Materials and Methods

2.1 MIS-SIM Environment

MIS-SIM is a multiplatform environment developed to facilitate the creation of virtual tasks for laparoscopic skills training with a custom built-in editor [2]. MIS-SIM follows the logic of physical box trainers where a collection of simple actions is used for training different skills used in laparoscopic surgery [2]. If box trainer's exercises are analysed, each action is done by an object in relation to other (Fig. 1). A task is understood as a sequence of actions between objects. Any object that makes an action in relation to other is called active. The objects that receive the action performed by the active object are called passive. If they do not perform any action on relation to other object, they are considered as regular objects.

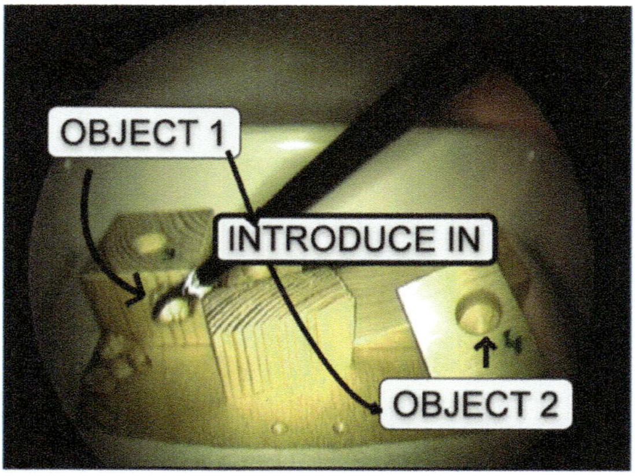

Fig. 1. Box trainer exercise where the user must introduce a ball in a specific hole. Object 1 is active. Object 2 is passive. "Introduce In" is the action.

MIS-SIM's editor displays a 3D scene where regular, active and passive objects are placed by clicking in the resource's browser. Content creators can include as many objects as needed and relate them with the desired actions. Objects have physics simulation based on AABB colliders and rigid bodies belonging to Nvidia Physics Engine [7]. Depending on the object, the collider could be a box, sphere, capsule or a collider with the form of the object's mesh. The objects in the scene can be movable (dynamic) or not (static) depending on the task and the defined behaviours. A dynamic object is an object that can be movable in a 3D scene. On the other hand, a static object is fixed to a determined position and rotation. For example, if the task consists in introducing a ball in a bottle, the bottle should be static as its movement is not needed.

MIS-SIM owns a custom server where tasks can be uploaded and downloaded if the user has internet connection. This allow users access requiring only the software itself and an internet connection to start any simulation the trainer has prepared. Likewise, content creators and teachers may access the editor and create/upload tasks anywhere, for example, from their own home. If a task is not behaving as desired, it can be modified anytime, anywhere.

Once a task is created, each object's behaviours (movement, dragging, physics…) are simulated in play-mode. The simulation can be performed by just clicking the Play button and connecting the input device to the computer. As MIS-SIM is created with a multiplatform engine, it can use any C# compatible haptic device as input.

After performance of each task, metric objective data are collected and automatically uploaded to MIS-SIM's server. These data can be used for exhaustive monitoring of trainees' laparoscopic skills without the need of the supervision of a trained surgeon.

2.2 Preliminary Validation

A testing course was given at Universidad Politécnica de Madrid, Spain in September 2018, where a group of experimented surgeons from Semmelweis University (Hungary) and Leiden University Medical Center (The Netherlands) could test MIS-SIM via streaming. They were able to create and modify tasks. Later, they could test them using a haptic device. Thus, they were able to create and modify tasks. Later, they could test them using their own haptic device.

In February 2019, a preliminary course was carried out as part of a three-hour preparatory course at the Department of Surgical Research and Techniques of Semmelweis University in Budapest, Hungary, to validate if MIS-SIM could be used for improving students' laparoscopic skills. The course was imparted by an experienced professor of Semmelweis University to a small group of residents (PGY 3-4-5) from diverse surgical fields (gynaecology, general-, plastic-, vascular-, thoracic-, neurosurgery, etc.) with different degrees of experience in laparoscopy. The students (participants) were used to box trainer exercises, but they had never used VR simulators. Tests were performed on a PC linked to the haptic device.

Prior to the course, the professor replicated 4 box trainer exercises [5] with the built-in editor based on the type of actions allowed in those exercises (Table 1). At the time of the validation, only 4 type of tasks were allowed: Touch, Introduce In, Pass Through and Join. Each task belongs to a group of one of these actions or a combination of some of them.

Table 1. Exercises based on actions

Task type by action	Descriptions	Objective	Metrics
Touch (Fig. 1A)	Touch tasks are created under the premise that an object with physics colliders enters or at least touches another object's hit collider	– Touch objects in relation to other – Move an object to a determined place	– 2 min as max. time to solve the task – 30 s as objective time to solve the task – 1–6 objectives per task – Floor and walls were not considered as an error – Dragging Time, average velocity and distance – No shadows – Number of errors – Description of how to solve them
Introduce In (Fig. 1B)	Introduce-In tasks are created based on regular box trainer simulators where an object like a ball or a tube is introduced in a bar or a bottle. The passive objects (where other objects are introduced) are static for better performance and more accuracy. If the task has more than one object, the student must introduce it in one place and then change it to another	– Introduce an object inside another	– 4 min as max. time to solve the task – 2 min as objective time to solve the task – 3 objectives per task – Floor and walls were not considered as an error – Dragging Time, average velocity and distance – No shadows – Number of errors – Description of how to solve them

(continued)

Table 1. (*continued*)

Task type by action	Descriptions	Objective	Metrics
Pass Through (Fig. 1C)	Pass-Through tasks are created with a defined path that the student must follow in order without touching the loops or letting any of the objects fall to the ground	– Pass an object through predefined objects	– 10 min as max. time to solve the task – 5 min as objective time to solve the task – 3 objectives per task – Floor and walls were not considered as an error – Dragging Time, average velocity and distance – No shadows – Number of errors – Description of how to solve them
Join (Fig. 1D)	Join tasks are similar to Introduce-in tasks, but in this case one rubber object had to connect two passive objects	– Join 2 parts of an object to two different objects, joining those two by the active object	– 4 min as max. time to solve the task – 2 min as objective time to solve the task – 3 objectives per task – Floor and walls were not considered as an error – Dragging Time, average velocity and distance – No shadows – Number of errors – Description of how to solve them

Every participant had to complete all tasks once. They did not practice before performing any of the tasks, but nevertheless they could see how the other residents completed the tasks. After each realization MIS-SIM automatically calculated objective metrics on performance (Table 2). Metric data were stored in an encrypted JSON file that was stored in the performance computer locally.

Once all the participants completed each task, they could create new tasks and modify the previous ones so that other participants could solve them one by one in the same way the previous tasks were performed. The students defined the new description, metrics, and settings about how to solve their versions of the tasks based on the

Table 2. Metrics defined in the actions within the validation

Metric name	Description
Description	Detailed description about how to solve a specific task
Max. time	Maximum time to complete the task
Time	Time spent doing the task
Errors	Mistakes produced by the participants
Grab time	Time spent grabbing the different objects
Collisions	Collision detection. How to
Dragging settings	Dragging time, velocity and distance of each object
Sensibility	Sensibility
Depth	Camera rendering shadows for better depth perception

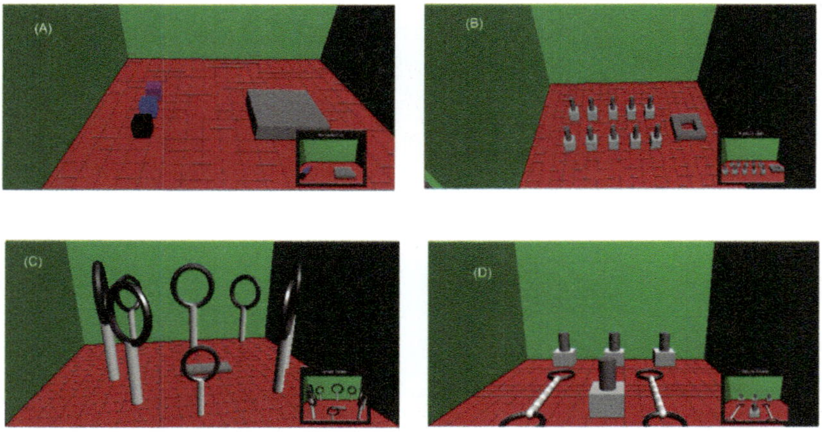

Fig. 2. Exercises based on the actions. **(A):** In this specific task the student had to hold the blue cube with one grasper while moving the other two to the top of the grey flat cube. **(B):** In this specific task the student had to introduce the prism with a hole in some poles avoiding the others. **(C):** In this specific task the student had to follow the path of loops, passing through them with the prism of the centre. The starting loop did not matter but all of them needed to be completed clockwise. **(D):** In this specific task the student had to connect each loop of the elastic bands to their corresponding poles in specific order.

different actions allowed in the task. Four of them were action-only (just one action) but with more than one objective with the same action (Fig. 2). The students explained to the others what they should do and then saved those tasks in the computer and then run them so that other participants could test them with the haptic device in play-mode. At the end, an informal interview was carried out with them to gather feedback and impressions on its usefulness and usability.

3 Results

The professor first tested in vivo the editor and the play-mode to show how easy could be to configure a new task and how to test them in front of the participants (Fig. 3). When completing the created task, he explained how to solve it to reinforce the already given instructions.

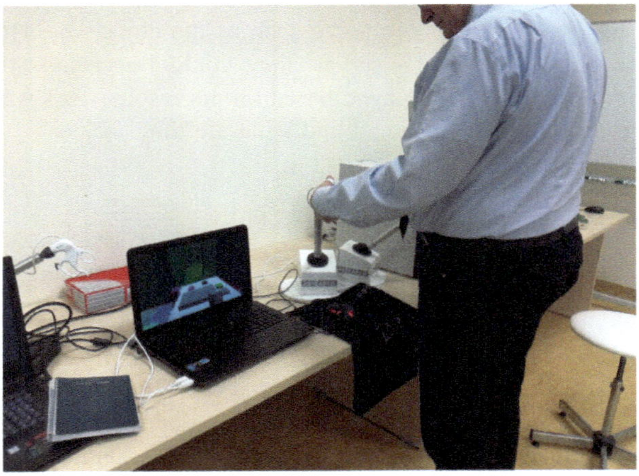

Fig. 3. Professor showing how to use MIS-SIM's environment.

Sixteen participants of the Budapest course tested the MIS-SIM environment. None of the participants had troubles completing the tasks with two hands. However, as they had to complete the tasks one by one watching how other students completed them and following the professor's instructions. Then, all of them were interviewed about the usability and what they thought that could be improvable for a surgical trainee.

In the interviews, the professor described MIS-SIM as powerful, but found difficultness in its use given the high number of more than required customization options. On the other hand, the students, found MIS-SIM professional, easy to use and enjoyable. The built-in editor allows the user to modify all the needed properties for most of the possibilities, but it demands them to set those correctly. Some of the participants, and even the professor first forgot to setup correctly the metrics and non-desired behaviours appeared when performing some of the tasks. They also did not set-up correctly the camera in some of the introduce-in tasks and when trying to perform them, some of the objects were far enough to not reach them with the grasper. To solve this, they went back to the editor, modify those tasks and retry the performance again. They found this kind of trial and error annoying but regular in similar 3D Editors like 3Ds Max or Blender.

Some of the obtained feedback was about the feeling when grabbing object with the graspers and how their meshes are adapted to the different objects placed in the 3D space when grabbing them. As there was no force feedback returned by the haptic device, they only had visual feedback given when performing the grabbing. At first, the virtual graspers adapted to the form of the draggable meshes when grabbing them, so a more realistic visual effect was achieved, but the participants complained about being able to close completely the haptic scissor without visual feedback in the virtual world. Thus, this feature was removed and when closing the scissor, the virtual grasper closed completely passing through draggable objects. All the participants preferred this but complained about not having force-feedback with the haptic when grabbing objects. This, however, did not affect task completion.

The generated data about the realization of each task was saved locally in the computer used by the participants to solve each of the objectives, However, if those data were uploaded to MIS-SIM's server, they could be visualized in MIS-SIM's online platform where a graph displayed based on the results. This way, the professor can follow the progress of each student and what are their lacks.

4 Conclusions

MIS-SIM aspires to break the barrier between VR and medical education by empowering users to create their own tasks. Thus, MIS-SIM improves creativity and generates objective data that can be followed for testing new procedures and virtual technology in virtual reality environments designed for improving minimally invasive skills. For most of the participants, laparoscopy is an important part of their speciality, thus the acquired skills during the course are going to be essential in their future work.

Feedback received by users within the preliminary validation shows the potential of MIS-SIM to (1) create personalised medical learning contents tailored to preferred learning styles, allowing the creation of individualize learning paths and (2) improve the efficiency of training by focusing on the training needs of the learners.

The challenges we face in the next months include the streamlining the user interface and functionalities to make the environment accessible and user friendly to teachers and content creators. Once these improvements are completed, we are confident of the benefits of this new paradigm based on personalized VR training. Moreover, we believe that this could be exported to other medical specialities.

Acknowledgements. This work has been partially funded by EIT Health. EIT Health is supported by the EIT, a body of the European Union.

References

1. Lamata de la Orden, P.: Metodología de análisis, diseño y evaluación de simuladores virtuales laparoscópicos. Ph.D. thesis, Universidad Politécnica de Madrid (2006)
2. Våpenstad, C., Buzink, S.N.: Procedural virtual reality simulation in minimally invasive surgery. Surg. Endosc. **27**, 364–377 (2013)
3. Feldman, L.S., Sherman, V., Fried, G.M.: Using simulators to assess laparoscopic competence: ready for widespread use? Surgery **135**, 28–42 (2004)
4. Sánchez-González, P., Oropesa, I., Davis, M., Rodríguez, M., Camba, D., Gómez, E.J.: A new virtual reality-based environment for surgical training. In: XXXVI Congreso Anual de la Sociedad Española de Ingeniería Biomédica, pp. 235–238. Ciudad Real (2018)
5. Oropesa, I., Chmarra, M.K., Gutiérrez, D., Sánchez-González, P., Guzmán-García, C., Sánchez-Peralta, L.F., Juhos, K., Negoita, A., Wéber, G., Tiu, C., Sánchez-Margallo, F.M., Dankelman, J., Gómez, E.J.: In: XXXVI Congreso Anual de la Sociedad Española de Ingeniería Biomédica, pp. 175–178. Ciudad Real (2018)
6. Niskya, I., Huangb, F., Milsteina, A., Pughc, C.M., Mussa-ivaldib, F.A., Karniela, A.: Perception of stiffness in laparoscopy – the fulcrum effect. Stud. Health Technol. Inform. **173**, 313–319 (2012)
7. http://gameworksdocs.nvidia.com/simulation.html. Accessed 28 Mar 2019
8. Macklin, M., Müller, M., Chentanez, N., Kim, T.Y.: Unified particle physics for real-time applications. ACM Trans. Graph. (SIGGRAPH) **33**(4), 153 (2014)
9. Guerrero-Hernández, A.J., Palacios-Zertuche, J.T., Reyna-Sepúlveda, F.J., Muñoz-Maldonado, G.E.: Laparoscopic training by use of a physical simulator and its application in the general surgery residency. Medicina Universitaria **18**(73), 189–193 (2017)
10. Gurusamy, K.S., Nagendran, M., Toon, C.D., Davidson, B.R.: Laparoscopic surgical box model training for surgical trainees. Cochrane Database Syst. Rev. **3** (2014)
11. Oropesa, I., Lamata, P., Sánchez-González, P., Pagador, J.B., García, M.E., Sánchez-Margallo, F.M., Gómez. E.J.: Virtual reality simulators for objective evaluation on laparoscopic surgery: current trends and benefits. In: Virtual Reality. Intech. Spain (2010)
12. Satava, R.M.: Medical applications of virtual reality. J. Med. Syst. **19**(3), 275–280 (1995)
13. Gallagher, A.G., Lederman, A.B., McGlade, K., Satava, R.M., Smith, C.D.: Discriminative validity of the minimally invasive surgical trainer in virtual reality (MIST-VR) using criteria levels based on expert performance. Surg. Endosc. **18**, 660–665 (2004)
14. Gallagher, A.G., McClure, N., McGuigan, J., Crothers, I., Browning, J.: Virtual reality training in laparoscopic surgery: a preliminary assessment of minimally invasive surgical trainer virtual reality (MIST VR), **31**(4), 310–313 (1999)

Regular Sessions: Clinical Engineering and Health Technology Assessment

Regulation and Approval of Continuous Non-invasive Blood-Pressure Monitoring Devices

Toshiyo Tamura(✉)

Waseda University, Tokyo, Japan
t.tamura3@kurenai.waseda.jp

Abstract. New-generation devices that enable continuous non-invasive blood-pressure (cNIBP) monitoring as an alternative to invasive arterial blood-pressure monitoring are being developed. These include a volume-clamp method using transmural pressure control, a tonometer based on the principles of tonometry, and photoplethysmographic sensing calculated using either pulse transit time (PTT) or the second derivative signal. However, meta-analyses of cNIBPs indicate that they are not as accurate as intermittent cuff-based sphygmomanometer measurements. While most of the cNIBP devices have been approved by medical device authorities, regulatory bodies are only starting to discuss standards for them. This paper reviews the benefits and limitations of cNIBP devices and current trends in their regulation and approval.

Keywords: Cuffless blood-pressure monitor · Standard · Regulation · ISO · IEEE

1 Introduction

Cuff-based automatic sphygmomanometers are commonly used to measure systolic and diastolic blood pressures (BPs) in both clinical practice and at home. These devices use a simple methodology based on cuff pressure change. Recently, continuous non-invasive BP monitoring has been attempted. Although invasive BP sensors that measure intra-arterial BP have been used in intensive care units, during surgery, and for emergency care, continuous non-invasive BP (cNIBP) monitoring is being developed as an alternative. While most commercial cNIBP monitors have been approved by the appropriate healthcare authorities, the only standard governing them is the Institute of Electrical and Electronics Engineers (IEEE) standard P1708-2014.

This review discusses current trends in the regulation and approval of cNIBP monitors. The volume-clamp type has been developed to replace direct measurements made by cannulating arteries or central veins. The device is placed on a finger and an algorithm based on the cardiac transfer function is used to convert the radial artery BP into the brachial BP.

J. Henriques et al. (Eds.): MEDICON 2019, IFMBE Proceedings 76, pp. 1021–1027, 2020.
https://doi.org/10.1007/978-3-030-31635-8_124

2 New BP Monitors

2.1 Continuous Non-invasive BP Monitors

There are three types of commercial cNIBPs: volume-clamp, tonometer, and cuffless photoplethysmograph (PPG) monitors. cNIBP monitors are designed to measure trends in BP, beat-to-beat rhythms, and waveforms. The volume-clamp method consists of a flexible finger cuff that is used with a photoplethysmograph (PPG), detector, and feedback control system. The finger cuff keeps the finger blood volume constant during each pulse wave by using the photoplethysmographic absorbance to adjust the cuff pressure in real time. Hence, the finger cuff inflates during systole and deflates during diastole using a fast-electronic control loop to keep the photoplethysmographic signal constant. The brachial BP is then reconstructed mathematically or calibrated against the arm oscillometric NIBP. Volume-clamp devices [1, 2] currently on the market include the Nexfin/ClearSight (Edwards Lifesciences, CA, USA) [3, 4], CNAP (CNSystems, Graz, Austria) [5, 6], and Finapres/Finometer (Finapres Medical Systems, Enschede, Netherlands) systems [7].

Arterial applanation tonometry involves placing a pressure transducer over the skin that gently compresses (i.e., applanates) the underlying artery. A proprietary algorithm considers the soft-tissue-related signal loss. Consequently, the tonometer estimates the arterial wall tension and quantifies the arterial pulse, which physicians otherwise assess subjectively through radial palpation. The T-Line device (Tensys Medical, San Diego CA, USA) is available [8–11].

The third type of device uses either a photoplethysmography device and the electrocardiogram (ECG) R wave [12, 13], or one PPG waveform [14]. An alternative approach to continuous non-invasive indirect BP measurement is based on changes in the pulse wave velocity (PWV), which is the speed of a pressure pulse propagating along the arterial wall. This can be calculated from the pulse transit time (PTT), which is the time between two pulse waves propagating during the same cardiac cycle from two separate arterial sites. The pulse arrival time (PAT), which is the interval between the ECG R wave and the sum of the PTT and the pre-ejection period, is gaining popularity for tracking BP because it is simple to estimate. The estimating equations can be found Table 1 of reference [13].

An important issue is the outcome of cNIBP monitoring, especially for wearable NIBP monitors that are used mainly for home healthcare. Research has demonstrated that the accuracy and repeatability of these devices in the resting state are sufficient for use in home healthcare. However, no study has examined their use in monitoring antihypertensive drug treatment.

2.2 Evaluation of cNIBP Monitoring in Clinical Practice

Continuous non-invasive arterial pressure monitoring devices are available for bedside use, and the accuracy and precision of these devices have been evaluated in a systematic review and meta-analysis [15–23]. The meta-analysis found that the accuracy and precision of cNIBP monitoring devices were unacceptable for clinical practice. This finding may have implications for clinical situations where cNIBP is being used for decisions about patient care.

Current technologies, although not perfect, enable much wider application of continuous monitoring. Hopefully, this will lead to improved treatment and reduce undesirable BP fluctuations and hypotensive periods. Sophisticated analyses of the arterial pressure curve make it possible to monitor not only BP but also blood flow and its variation. These new monitoring tools may significantly influence perioperative care, especially for intermediate-risk patients. However, more research is needed to determine the extent to which improvements in macrohemodynamic parameters affect postoperative outcomes in this patient population.

In clinical practice, continuous BP monitoring is important, especially at night. Many studies have measured nighttime BP. Currently, continuous BP measurements are collected using ambulatory cuff-based monitors. However, the measurement interval is at least 15 min because the cuff must be inflated. Combined pulse oximetry and cuff-based sphygmomanometers can increase precision and reduce disturbance to the patient during nocturnal BP monitoring. Changes in oxygen saturation (OS) cause BP changes. An OS threshold value is determined in advance. If the OS change exceeds this value, the cuff-based BP monitor is activated [24]. Alternatively, BP can be monitored using unloaded tonometric and photoplethysmographic methods, but these are not acceptable medically. If intermittent, long-term stabilized NIBP is adopted, personal health management should become more popular.

3 Continuous BP Monitors: Regulatory Bodies

cNIBP monitors are medical devices and two paths to their approval are available. One involves approval by agencies such as the US Food and Drug Administration (FDA). This depends on its risk classification. Each device is placed in regulatory Class I, II, or III based on the level of control necessary to provide reasonable assurance of its safety and effectiveness. BP devices are classified into Class II. FDA guidelines led to the American National Standards Institute (ANSI)/Association for the Advancement of Medical Instrumentation (AAMI) standard SP-10-1992 for electronic or automated sphygmomanometers. All tests must follow the SP-10 protocol, and the manufacturer must demonstrate the effectiveness of any new device based on clinical evidence. The above-mentioned volume-clamp and tonometric devices have already been approved by the FDA. Currently, the FDA requires that cuffless BP devices be calibrated using a cuff-based sphygmomanometer before and after measurements. Figure 1(a) presents an FDA-approved vital sensor [25]. Figure 1(b) presents another approved device based on the pulse decomposition principle [26, 27].

The second approval pathway involves the creation of a new standard with proven accuracy, which is implemented by medical approval agencies.

Worldwide, there are three main regulatory bodies. The International Standards Organization (ISO) formalizes specifications, the construction materials to be used, product and process features, and the services to be offered. The IEEE is the world's largest technical body, dealing principally with electronic communications; it produces many standards jointly with the ANSI. The International Electrotechnical Commission (IEC) is similar to the IEEE, and IEC and IEEE standards are nearly identical.

The three bodies work together to develop mutually acceptable standards. The IEEE considers regulations after accepting a new technology, while the ISO considers regulations before such acceptance, thereby avoiding confusion and the licensing of poorly performing devices.

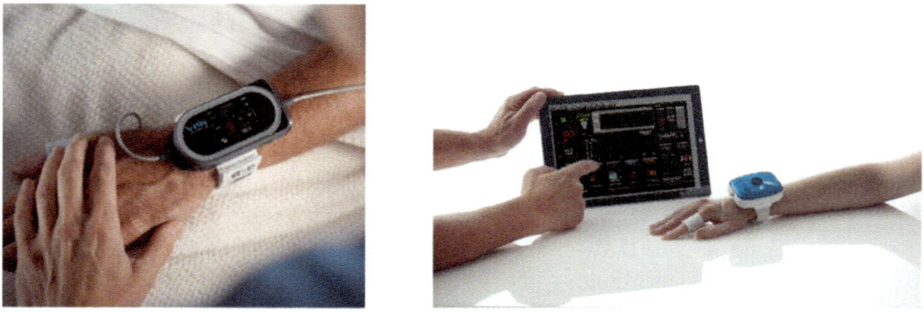

Fig. 1. Examples of FDA-approved devices: (a) the Visi mobile system (Sotera Wireless, San Diego CA, USA); and (b) the Caretaker (Caretaker Medical, Charlottesville, VA, USA).

4 A New Standard for Cuffless BP Monitors

4.1 The Current Regulations

IEEE standard 1708-2014 is relatively recent and deals with wearable cuffless BP monitors. The accuracy and validation rules reflect the established clinical guidelines for cuff-based BP monitors issued by the ANSI/AAMI/ISO in document 81060-2; these are the standards currently used by the US FDA and the European Union to evaluate cuff-based BP monitors. The differences lie in the sample size required for evaluation and the accuracy of dynamic changes. Furthermore, while the terms “wearable” and “embedded” are defined, some definitions are unclear. For example, BP is measured as the average of 20-s beat-to-beat recordings, but the start points are not defined. Additionally, “long-term stability” and “validation” are not defined. Currently, the working group is revising the 2014 version.

4.2 Upcoming Regulations

ISO81060-3 deals with non-invasive, continuous BP monitoring during clinical investigations, and was formulated by an ISO/Technical Committee (TC) 121/Subcommittee (SC) 3/Joint working group (JWG) 7. The performance of an automated non-invasive continuous sphygmomanometer is similar to that of an automated non-invasive intermittent sphygmomanometer, except in terms of BP changes and long-term stability. The standard focuses on existing automated non-invasive continuous sphygmomanometers that use either unloaded or tonometric methods in clinical practice. The discussion is mainly about the accuracy of BP determination. We propose both invasive and auscultatory references. An invasive reference can be compared with the device under test (DUT) from beat to beat. An auscultatory reference is rather complex. The average beat-to-beat BP values of the DUT are compared with an auscultatory reference. The total measurement time and start and end times of the continuous wave must be defined. Currently, ISO81060-3 is in the Committee Draft stage and will be published in 2020. The JWG7 will not presently discuss either a new standard or an amendment of ISO81060-3 for cuffless BP monitors.

4.3 Alternative Standard

Many manufacturers, especially new medical-device companies, claim that it is very difficult to obtain approval for medical devices. Although heart-rate monitors are used in clinical practice, these are considered healthcare devices rather than clinical devices and no medical approval is required. However, it is illegal to sell a cuffless BP monitor as a healthcare device; such a monitor is defined as a medical device. One solution would be to present BP values as, for example, percentage increases/decreases from earlier values (i.e., not in mmHg); this would help identify or predict disease. The IEC sets the standards for healthcare devices. We are assembling a consortium that will submit a New Work Item Proposal to either TC124 (the TC responsible for wearable electronic devices/technologies) or TC100 (which handles audio, video, and multimedia systems and equipment).

5 Conclusion

In conclusion, there is a growing clinical requirement for non-invasive monitoring of continuous BP, fluid, and stroke volume. Such devices are now available clinically and are being used in increasing numbers. Therefore, it is very important to discuss acceptable and reasonable cNIBP evaluation standards. The accuracy and validation of cuffless BP monitors are of paramount importance in terms of product quality.

We must also find evidence of the importance of changes in continuous BP values during daily life and while sleeping. To predict hypertension, an early diagnosis requires long-term BP monitoring. Urgent action is required, given the popularity of such systems and devices. A new evaluation standard that considers the inherent beat-to-beat nature and trending capability of these new devices is required.

Acknowledgments. This work was supported in part by grants-in-aid from the Strategic Information and Communications R&D Promotion Program in 2014–2016, the Japanese Ministry of Internal Affairs, and Communication and Scientific Research (C) (Kakenhi) (#17K01440), the Japan Medical Research and Development Organization's Integrated Research and Development Project for Persons with Disabilities (JP18dk0310077), and a Ministry of Economy, Trade and Industry project promoting a new ISO/IEC proposal.

References

1. Peňáz, J.: Photoelectric measurement of blood pressure, volume and flow in the finger. In: Digest of the 10th International Conference on Medical and Biological Engineering, Dresden (1973)
2. Wesseling, K.H.: Finapres, continuous noninvasive finger arterial pressure based on the method of Peňáz. In: Meyer-Sabellek, W., Gotzen, R., Anlauf, M., Steinfeld, L. (eds.) Blood Pressure Measurements. Steinkopff Blood Pressure Measurements, pp. 161–172 (1990)
3. Martina, J.R., Westerhof, B.E., van Goudoever, J., De Jonge, N., Van Lieshout, J.J., Lahpor, J.R., De Mol, B.A.: Noninvasive blood pressure measurement by the Nexfin monitor during reduced arterial pulsatility: a feasibility study. ASAIO J. **56**(3), 221–227 (2010). https://doi.org/10.1097/mat.0b013e3181d70227
4. Martina, J.R., Westerhof, B.E., van Goudoever, J., de Beaumont, E.M.F.H., Truijen, J., Kim, Y.-S., Immink, R.V., Jöbsis, D.A., Hollmann, M.W., Lahpor, J.R., de Mol, B.A.J.M., van Lieshout, J.J.: Noninvasive continuous arterial blood pressure monitoring with Nexfin®. Anesthesiology **116**(5), 1092–1103 (2012). https://doi.org/10.1097/aln.0b013e31824f94ed
5. Jeleazcov, C., Krajinovic, L., Münster, T., Birkholz, T., Fried, R., Schüttler, J., Fechner, J.: Precision and accuracy of a new device (CNAP™) for continuous non-invasive arterial pressure monitoring: assessment during general anaesthesia. Br. J. Anaesth. **105**(3), 264–272 (2010). https://doi.org/10.1093/bja/aeq143
6. Hahn, R., Rinösl, H., Neuner, M., Kettner, S.C.: Clinical validation of a continuous non-invasive haemodynamic monitor (CNAP™ 500) during general anaesthesia. BJA: Br. J. Anaesth. **108**(4), 581–585 (2012). https://doi.org/10.1093/bja/aer499
7. Stroke, D.N., Clutton-brock, T., Patil, C., Thompson, J.M., Hutton, P.: Comparison of invasive and non-invasive measurement of continuous arterial pressure using the Finapres. Br. J. Anaesth. **67**(1), 26–35 (1991). https://doi.org/10.1093/bja/67.1.26
8. Gregory, J.M., Nikolaus, G.: Anesthesia & Analgesia: an accuracy evaluation of the T-Line® Tensymeter (Continuous noninvasive blood pressure management device) versus conventional invasive radial artery monitoring in surgical patients **102**(2), 484–490 (2006). https://doi.org/10.1213/01.ane.0000194873.52453.bd
9. Szmuk, P., Pivalizza, E., Warters, R.D., Ezri, T., Gebhard, R.: An evaluation of the T-Line® Tensymeter continuous noninvasive blood pressure device during induced hypotension. **63** (3), pp. 307–312 (2006). https://doi.org/10.1111/j.1365-2044.2007.05369.x
10. Saugel, B., Fassio, F., Hapfelmeier, A., Meidert, A.S., Schmid, R.M., Huber, W.: The T-Line TL-200 system for continuous non-invasive blood pressure measurement in medical intensive care unit patients. Intensive Care Med. **38**, 1471 (2012). https://doi.org/10.1007/s00134-012-2617-x
11. Meidert, A.S., Huber, W., Müller, J.N., Schöfthaler, M., Hapfelmeier, A., Langwieser, N., Wagner, J.Y., Eyer, F., Schmid, R.M., Saugel, B: Radial artery applanation tonometry for continuous non-invasive arterial pressure monitoring in intensive care unit patients: comparison with invasively assessed radial arterial pressure. Br. J. Anaesth. **112**(3), 521 (2014). https://doi.org/10.1093/bja/aet400
12. Ding, X.-R., Zhang, Y.-T., Liu, J., Dai, W.-X., Tsang, H.K.: Continuous cuffless blood pressure estimation using pulse transit time and photoplethysmogram intensity ratio. IEEE Trans. Biomed. Eng. **63**, 964–972 (2016)
13. Ding, X., Yan, B., Zhang, Y., Liu, J., Ni, Zhao, Tsang, H.: Pulse transit time based continuous cuffless blood pressure estimation: a new extension and a comprehensive evaluation. Sci. Rep. **7**, 11554 (2017)

14. Baruch, M.C., Kalantari, K., Gerdt, D.W., Adkins, C.M.: Validation of the pulse decomposition analysis algorithm using central arterial blood pressure. BioMed. Eng. OnLine **13**, 96 (2014)
15. Chung, E., Chen, G., Alexander, B., Cannesson, M.: Non-invasive continuous blood pressure monitoring: a review of current applications. Front. Med. **7**(1), 91–101 (2013)
16. Kim, S.H., Lilot, M., Sidhu, K.S., Rinehart, J., Yu, Z., Canales, C., Cannesson, M.: Accuracy and precision of continuous noninvasive arterial pressure monitoring compared with invasive arterial pressure: a systematic review and meta-analysis. Anesthesiology **120**, 1080–1097 (2014)
17. De Hert, S.: Noninvasive hemodynamic monitoring devices: new tools or just another toy? Anesthesiology **120**(5), 1065–1066 (2014)
18. Fortin, J., Lerche, K., Flot-Zinger, D., O'Brien, T.: Is the standard supplied by the association for the advancement of medical instrumentation the measure of all things for noninvasive continuous hemodynamic devices? Anesthesiology **122**(1), 208–209 (2015)
19. Sharma, M., Barbosa, K., Ho, V., Griggs, D., Ghirmai, T., Krishnan, S.K., Hsiai, T.K., Chiao, J.-C., Cao, H.: Cuff-less and continuous blood pressure monitoring: a methodological review. Technologies **5**, 21 (2017)
20. Cannesson, M., Joseph Rinehart, J., Kim, S.H.: In reply. Anesthesiology **122**(1), 209–210 (2015)
21. Stenglova, A., Benes, J.: Continuous non-invasive arterial pressure assessment during surgery to improve outcome. Front. Med. (2017). 4 article #202
22. Lakhal, K., Ehrmann, S., Boulain, T.: Noninvasive BP monitoring in the critically ill: time to abandon the arterial catheter? Chest **153**(4), 1023–1039 (2018)
23. Meidert, A.S., Saugel, B.: Techniques for non-invasive monitoring of arterial blood pressure. Front. Med. (2018). 4 article #231
24. Kario, K.: Essential Manual of 24-Hour Blood Pressure Management: From Morning to Nocturnal Hypertension. Wiley, London (2015)
25. Sotera. Wireless "ViSi Mobile|A Product by Sotera Wireless". http://www.visimobile.com/. Accessed 19 June 2019
26. Baruch, M.C., Darren, E.R., Warburton, D., Bredin, S, Cote, A., Gerdt, D.W., Adkins, C.M.: Pulse decomposition analysis of the digital arterial pulse during hemorrhage simulation. Nonlinear Biomed. Phys. **5**, 1 (2011). https://doi.org/10.1186/1753-4631-5-1
27. Gratz, I., Deal, E., Spitz, F., Baruch, M., Elaine Allen, I., Seaman, J.E., Pukenas, E., Jean, S.: Continuous non-invasive finger cuff CareTaker® comparable to invasive intra-arterial pressure in patients undergoing major intra-abdominal surgery. BMC Anesth. BMC series – open, inclusive and trusted **17**, 48 (2017). https://doi.org/10.1186/s12871-017-0337-z

Evaluation of a New Endobronchial Double Lumen Tube with Integrated Camera: A Hospital Based HTA Experience

Michela D'Antò[1(✉)], Carlo Cosentino[2], Arturo Cuomo[1], Rossana Accardo[1], Paolo Bifulco[3], Leandro Donisi[4,5], and Maria Romano[2,5]

[1] National Cancer Institute "G. Pascale", Naples, Italy
micheladanto@hotmail.it
[2] Department of Experimental and Clinical Medicine (DMSC), University "Magna Graecia" of Catanzaro, Catanzaro, Italy
mariar@unicz.it
[3] Department of Electrical Engineering and Information Technology, DIETI, University of Naples Federico II, Naples, Italy
[4] Dipartimento di Scienze Biomediche Avanzate, University of Naples Federico II, Naples, Italy
[5] Istituti Clinici Scientifici Maugeri, Telese Terme, Benevento, Italy

Abstract. Over the last few decades, the developed countries have witnessed a strong increase in healthcare spending, essentially due to the emergence of new diseases, the aging of the population, and the development of new and costly technologies. The Health Technology Assessment (HTA) is a powerful tool for connecting the technical-scientific and decision-making world, which helps to prevent the introduction of inappropriate, ineffective, or superfluous technologies within the health system, thus limiting the expense, and improving the overall quality of medical care. This paper describes a hospital-based HTA procedure (costs-efficacy evaluation), which has been exploited to compare an innovative device, with integrated camera, for endobronchial intubation in thoracic surgery, with the instrumentation currently in use at the National Cancer Institute "G. Pascale" of Naples, which is the largest Institute of Hospitalization, Scientific Research and Care for the Oncology in Southern Italy. In the case under study, by introducing the new technology, direct variable costs could result in an actual significant spending increase for the "G. Pascale" Institute; however, despite that, the results of the HTA procedure and of the Analysis of Costs Minimization show that the introduction of the innovative device could allow to significantly decrease the intubation time, the overall surgery duration and the costs incurred by the Institute. These results confirm also the usefulness of the hospital based HTA procedures.

Keywords: Hospital based HTA · Endobronchial tubes · Health management

J. Henriques et al. (Eds.): MEDICON 2019, IFMBE Proceedings 76, pp. 1028–1037, 2020.
https://doi.org/10.1007/978-3-030-31635-8_125

1 Introduction

1.1 Background

Nowadays, in fact, technology plays an increasingly central role in modern medicine. Innovative technologies have helped to improve the efficacy, safety and quality of diagnosis and therapies; therefore, they have significantly improved health and quality of life, but, at the same time, have led themselves to an increase of health expenditure. Technological improvement, in fact, allows dealing with previously untreatable conditions and/or patients, and it can also happen that there is no complete replacement of previous technologies (overlapping).

To deal with this situation, during the 1980s and 1990s, almost all industrialized countries aimed at rationalizing the resources. This process clearly needs to evaluate the effectiveness of the different diagnostic-therapeutic protocols and technologies with respect to the population concerned, evaluating, at the same time, the service costs. To meet this need, an international multidisciplinary field research, known as "Health Technology Assessment (HTA)" [1, 2], has developed internationally.

The HTA is a powerful tool for connecting the technical-scientific and decision-making world: it allows to determine, objectively, if the introduction of a new technology is economically sustainable, to assess its appropriateness and safety etc. Therefore, HTA helps to limit the expense and to improve the overall quality of medical care [3, 4].

The evaluation carried on by HTA covers several aspects, including for example [5]:

- technical characteristics of the new technology;
- concerned health problem and technology currently in use;
- performance; which may include: sensitivity and specificity of some diagnostic tests, ease of use and maintenance;
- clinical safety;
- economic impact; for the estimation of this aspect different evaluation techniques can be employed, for example the cost-effectiveness analysis or the break-even analysis;
- organizational aspects;
- It is important to put in evidence that for the evaluation of many of these aspects, we need a phase of research and analysis of objective "evidence" carried out through experimental clinical research.

1.2 Aim of the Work

HTA makes it possible to determine, objectively, the opportunity of introducing a new technology, evaluating several aspects at the same time.

The present work, framed in this context, describes an evaluation of Hospital Based HTA ex post, concerning the medical devices for endobronchial intubation in thoracic surgery, for the National Cancer Institute "G. Pascale" of Naples (Italy). The National Cancer Institute "G. Pascale" of Naples (in the following simply called

"G. Pascale" Institute) is the largest Institute of Hospitalization, Scientific Research and Care for the Oncology of Southern Italy and it is a Reference Center for the national and regional oncology network. Of great importance within the Institute is the complex structure of Oncological Thoracic Surgery which, in July 2017, obtained European accreditation from the European Society of Thoracic Surgeons (ESTS), bearing witness to the respect of the international parameters of performance in the execution of interventions for lung cancer diseases and other large intrathoracic formations.

Commonly, in thoracic surgery a balanced general anesthesia is performed and the monopolmonary ventilation is applied which allows to exclude the ventilation of one of the two lungs while maintaining the contralateral one. This selectivity, with consequent atelectasis of the diseased lung, allows a perfect surgical exposure. There are several systems for the monopolmonary ventilation [6]; but, regardless of the kind of method, the control of the tube's exact positioning is of crucial importance, since the correct establishment of the monopolmonary ventilation depends on this. The results of the studies available in the literature indicate that the use of a flexible bronchoscope to ensure the correct positioning of an endobronchial tube should be considered the best practice [7]. However, the bronchoscopes are subject to repeated cycles of disinfection and accidental breakage; moreover, incorrect disinfection processes can be a source of infection transmission for the patient. Furthermore, for difficult intubations, it is necessary, for practice, the endoscopist's control. He is not always available, so that his waiting lengthens the surgery time. The problem is even more felt when the endoscopy department is far from the operating room, as in the case under study.

An innovative "Vivasight-DL" device with integrated camera [8], which allows to avoid the use of a fibrobroncoscope, was recently introduced on the market; it was the subject of the HTA procedure (costs-efficacy evaluation) here presented.

2 Material and Methods

2.1 Instrumentation

The Current Instrumentation

For monopolmonary ventilation, endobronchial tubes, right and left, double (DLT) and single (SLT) lumen, in PVC, and bronchial blockers associated with single-lumen endotracheal tubes are normally used. However, DLT, employed at the "G. Pascale" Institute, require less time and simpler maneuvers to be positioned than bronchial blockers; moreover, they allow a faster collapse of the lung, and guarantee a more efficient suction of secretions. For the direct endoscopic visual examination of the airways, a flexible fiberoptic bronchoscope, also called fiberbronchoscope, is used (see Fig. 1); it can be pushed directly into the bronchi to be examined. At the ends of the tube there is a mini-camera that records a series of images, then projected on a monitor.

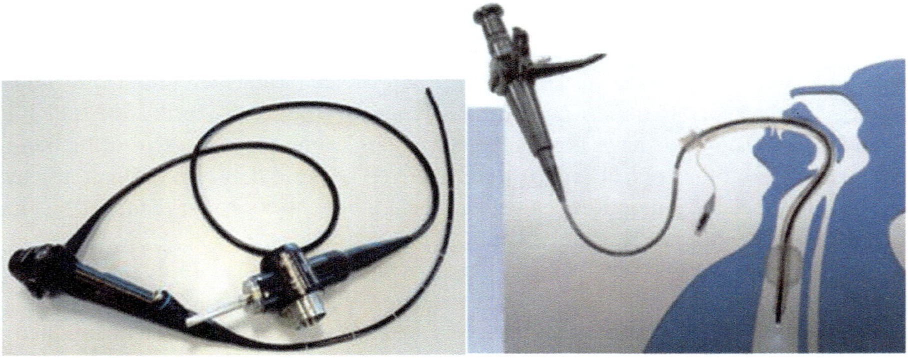

Fig. 1. On the left, image of a flexible bronchoscope; on the right, intubation by a fiberbronchoscope.

The Innovative Instrumentation

The new device, Vivasight-DL (see Fig. 2), is a double lumen endobronchial left tube, sterile, single use, in PVC, with high resolution camera integrated in the apex, to be used for the continuous visual control of anatomical structures during intubation and ventilation maneuvers.

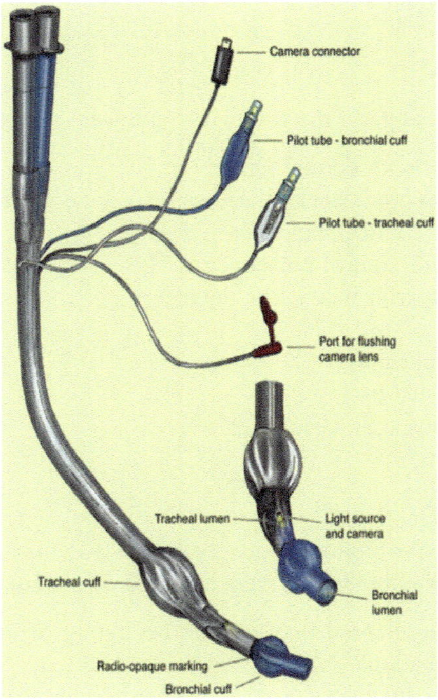

Fig. 2. Left endobronchial tube VivaSight-DL.

2.2 Data

Firstly, the analysis involved 18 patients in the study group (treated with the new technology) and 20 patients in the control group (treated with the current instrumentation). All the patients underwent to thoracic surgery between April 2017 and July 2018. Then, in order to have homogeneous groups, some patients (2 from the study group and 2 from the control group) were removed because they had difficult airways and therefore needed a mono-lumen endotracheal tube with a smaller diameter blocker than double-lumen endobronchial tubes.

The patients were grouped on the basis of the surgery kind (Table 1), for each patient were gathered (data not shown for sake of brevity): pre-surgery data (such as personal data, ASA score, etc.), intra-surgery data (intubation duration and simplicity, collapsed lung) and post-surgery data (airway injuries and disorders such as sore throat, hoarseness, cough and their intensity).

Table 1. Classification of eligible cases for surgery kind (VAT = videothoracoscopy).

Surgery	Number	Percentage %
Lobectomy	10	29,4
Wedge	6	17,6
VATs	9	26,6
Thoracotomy	3	8,8
Pneumonectomy	3	8,8
Others	3	8,8

Moreover, for each surgery, the following data were stored.

- In case of intubation with current DLT:
  - Number of endoscopic checks (necessary for tube positioning control)
  - Waiting time for control maneuver
  - Necessary time for control maneuver
- In case of intubation with innovative device:
  - Image quality
  - Number of lens cleaning
  - Simplicity in device use
  - Necessity of endoscopic check.

2.3 Efficacy Analysis

On the basis of an extensive bibliographic research [9–13], the following outcomes were selected, and different possible values were assigned on a qualitative-quantitative basis.

- Intubation time (time elapsed between the beginning of the image display to the certainty of the intubation)
- Intubation simplicity, evaluated by the anesthesiologist: from 1, very simple, to 5, impossible

- Quality of lung collapse, evaluated by the thoracic surgeon: from 1, excellent, to 3, poor
- Post-surgery disorders, evaluated by the medical staff at 4–5 h after the surgery end.

Statistical Analysis

Due to the low samples size, non-parametric tests were carried out. In particular, the statistical significance between the eventual differences in intubation times with the two devices (current and innovative) was assessed by means of a Mann-Whitney U test. The other parameters underwent to a Fisher test. All results were considered statistically significant if $p < 0.05$.

2.4 Costs Analysis

An Analysis of Costs Minimization (ACM) was considered for the evaluation, an approach used in business decision making to evaluate the opportunity to introduce a new technology when, after studying the clinical results of alternative technologies, they resulted equivalent in the relevant aspects.

In particular, costs involved were the following:

- Consumable costs
  - Robertshaw double-lumen tube = 70 € (IVA included)
  - VivaSight DLT = 230 € (IVA included)
- Team costs
  - Surgeon = 1.67 €/minute (per 4 surgeons)
  - Anesthetist = 1.67 €/minute
  - Professional nurse = 0.73 €/minute (per 2 nurses)
  - Endoscopist = 1.67 €/minute (in case of fiberbronchoscopic control)
  - Endoscopist technician = 0.73 €/minute (in case of fiberbronchoscopic control)
- Operating room = 4.3 €/minute
- Fiberbronchoscopic sterilization = 21.5 €

Costs were evaluated considering both "G. Pascale" Institute data and literature [14].

In order to carried out the economical evaluation, it has been necessary to identify the operational times to which direct variable costs are associated:

- T1 = intubation time; during this interval time all the operating team is present
- $T2_I$ = waiting time for fiberbronchoscopic control; all the operating team is present
- $T3_I$ = time of maneuver (fiberbroncoscopic to control the correct tube positioning); all the operating team plus endoscopist and endoscopist technician are present
- $T2_{II}$ = waiting time for fiberbronchoscopic control, in case it is again necessary during the surgeon; all the operating team is present
- $T3_{II}$ = time of maneuver; all the operating team plus endoscopist and endoscopist technician are present
- T4 = time necessary to move from endoscopic ward to surgeon ward and back; this time concerns endoscopist and endoscopist technician
- T5 = time taken by the professional nurse for the fiberendoscopic reprocessing.

Obviously, after computing average values of these times, the costs of each activity were calculated by multiplying these time values for the number of people involved (for each professional figure) and for the cost per minute, and then adding the other costs.

Finally, different scenarios were considered on the basis of the number of the needed fiberendoscopic controls; the first couple concerns the use of the current endotracheal tube, while the second couple concerns the use of the innovative system:

- Case 1: intubation with current DLT and 1 fiberendoscopic control (75%)
- Case 2: intubation with current DLT and 2 fiberendoscopic controls (25%)
- Case 3: intubation with innovative VivaSight-DL without fiberendoscopic control (95%)
- Case 4: intubation with innovative VivaSight-DL with 1 fiberendoscopic control (5%).

3 Results

3.1 Efficacy Analysis

Results of the statistical analysis concerning the outcomes selected for the efficacy analysis are shown in Table 2.

Table 2. Outcome and corresponding p-values (n.s. = not shown, for sake of brevity).

	Current DLT	VivaSight-DL	p-value
Intubation time [s]	78,6 ± 28	50,8 ± 16	<0,05
Intubation simplicity 1	1 (6,25%)	1 (5,6%)	>0.99
Intubation simplicity 2	7 (43,75%)	7 (38,9%)	
Intubation simplicity 3	7 (43,75%)	8 (44,4%)	
Intubation simplicity 4	1 (6,25%)	2 (11,1%)	
Intubation simplicity 5	0	0	
Collapse 1	11 (69%)	13 (87%)	0.45
Collapse 2	5 (31%)	3 (13%)	
Collapse 3	0	0	
Post-surgery disorders	n.s.	n.s.	>0.99

It is worth highlighting that the innovative system, in addition to significantly decrease the intubation time, required the fibroendoscopic control only once, therefore it allows reducing the overall surgery duration, improving the turnover of the operating room and reducing costs.

With regard to the post-surgery disorders, the two technologies resulted comparable, in accordance with the literature [10, 11, 13].

Finally, about the VivaSight-DL, the image quality was very good in almost all cases (16/18) and the lens cleaning very rarely necessary.

3.2 Costs Analysis

As said in the previous section, the innovative system required the fiberendoscopic control during surgeon only once, hence $T2_{II}$ and $T3_{II}$ had not statistical significance and for the corresponding costs were considered the values computed for the current technique.

Obtained results are shown in Table 3.

Table 3. Variable direct costs for each considered scenario. All costs are in €.

	Case 1	Case 2	Case 3	Case 4
Cost T1	18,48	18,48	11,99	11,99
Cost $T2_I$	126,99	126,99	0,00	0,00
Cost $T2_{II}$	0,00	115,28	0,00	115,28
Cost $T3_I$	82,38	82,38	0,00	0,00
Cost $T3_{II}$	0,00	40,12	0,00	40,12
Cost T4	36,00	36,00	0,00	36,00
Cost T5	7,30	7,30	0,00	7,30
Cost consumable	70,00	70,00	230,00	230,00
Cost sterilization	21,50	21,50	0,00	21,50
Total cost	**362,65**	**518,05**	**241,99**	**462,19**

As confirmed by the percentage of occurrence, the Case 4 is not common, therefore an evaluation of spending savings was carried out by comparing Case 3 with Case 1 and Case 2, obtaining spending decreases of about 33% and more than 50% respectively.

4 Discussion

As it is known, HTA is the systematic evaluation of technologies adopted in health care, with the final aim of providing all the necessary operational suggestions for the health system. The objective of this paper has been to adopt a hospital based HTA approach for the comparison between two different technologies used for intubation during thoracic surgery at the “G. Pascale” Institute, the largest and most important Institute of Hospitalization, Scientific Research and Care for the Oncology of Southern Italy.

It was considered relevant to carried out an evaluation based on clinical parameters of efficacy and related costs. Data were collected from two groups of patients which are significant for the surgery kind since in clinical practice, and also at the “G. Pascale” Institute, double-lumen endobronchial tubes are used more frequently, because, as said, they allow a more rapid lung collapse, an easier blocker positioning and ensure a more effective secretions suction [6, 15].

With reference to the statistical analysis, it appears that the ease of intubation and the quality of the pulmonary collapse are comparable in the two groups (see Table 2). However, the results of the ACM show how, despite the cost of the innovative device consumables is higher than that of the current DLT, the overall saving that on average affects each intervention is very high. Therefore, the economic evaluation suggests, as best choice, the VivaSight-DL. Moreover, the use of the VivaSight-DL yields to a time reduction of the surgery of at least 14 min (data not shown).

As a secondary but important effect, it should be emphasized that the continuous visualization of the position of the endobronchial tube appears to provide a clinical benefit, increasing the safety perceived by the patient and by the operators.

To confirm the obtained results, which encourage the introduction of the new technology at the "G. Pascale" Institute, a Budget Impact Analysis has been carried out; details and obtained results, in line with conclusions here discussed, are not shown for sake of brevity.

Conflict of Interest Declaration. The authors declare they have no conflict of interests.

References

1. D'Antò, M.: Hospital based HTA of the vascular synthesis means in thyroidectomy. Master Thesis, University of Trieste. Higher Education in Clinical Engineering (2014)
2. Leogrande, L.: L'innovazione tecnologica in sanità: il ruolo dell'Ingegnere Clinico. Seminary in Clinical Engineering, University of Naples Federico II (2015)
3. Favaretti, C., Torri, B.: Che cos'è l'Health technology assessment, RIMeL/IJLaM **3**(Suppl) (2007)
4. Improta, G.: HTA for the management of biomedical equipment in a hospital: reallocation, donation to developing countries, design and planning of a virtual community. Ph.D. Thesis, University of Naples Federico II (2010)
5. Cerbo, M., Amicosante, A.V., Cavallo, A., Chiarolla, E., Corio, M., Gillespie, F., Jefferson, T.O., Lo Scalzo, A., Migliore, A., Paone, S., Perrini, M.R., Vicari, N.: Manuale delle procedure HTA (2014)
6. Merli, G., Guarino, A., Della Rocca, G., Frova, G., Petrini, F., Sorbello, M., Coccia, C.: In cooperation with SIAARTI Studying Group on Difficult Airway: Recommendations for airway control and difficult airway management in thoracic anesthesia and lung separation procedures. Minerva Anestesiol (2009)
7. De Bellis, M., Accardo, R., Di Maio, M., La Manna, C., Rossi, G.B., Pace, M.C., Romano, V., Rocco, G.: Is flexible bronchoscopy necessary to confirm the position of double-lumen tubes before thoracic surgery? Eur. J. Cardiothorac. Surg. **40**, 912–916 (2011)
8. Dean, C., Dragnea, D., Anwar, S., Ong, C.: The VivaSight-DL double-lumen tube with integrated camera. Eur. J. Anaesthes. **33**(4), 305–308 (2016)
9. Tvede, M.F., Kristensen, M.S., Nyhus-Andreasen, M.: A cost analysis of reusable and disposable flexible optical scopes for intubation. Acta Anaesthesiol. Scand. **56**, 577–584 (2012)
10. Levy-Faber, D., Malyanker, Y., Nir, R.R., Best, L.A., Barak, M.: Comparison of VivaSight double-lumen tube with a conventional double-lumen tube in adult patients undergoing video-assisted thoracoscopic surgery. Anaesthesia **70**, 1259–1263 (2015)

11. Heir, J.S., Purugganan, R., Jackson, T.A., Norman, P.H., Cata, J.P., Kosturakis, A., Thakar, D.: A retrospective evaluation of the use of video-capable double lumen endotracheal tubes in thoracic surgery. J. Cardiothorac. Vasc. Anesth. **28**, 870–872 (2014)
12. Koopman, E.M., Barak, M., Weber, E., Valk, M.J.A., de Schepper, R.T.I., Bouwman, R.A., Huitink, J.M.: Evaluation of a new double-lumen endobronchial tube with an integrated camera (VivaSight-DL™): a prospective multicenter observational study. Anaesthesia **70**, 962–968 (2015)
13. Saracoglu, A., Saracoglu, K.T.: VivaSight: a new era in the evolution of tracheal tubes. J. Clin. Anesth. **33**, 442–449 (2016)
14. Frazzetta, M., Furgiuele, G., Raimondo, D., Sammartano, A., Mannino, E., De Simone, G., Vetri, G., Bonventre, S., Di Gesù, G.: Tiroidectomia totale con impiego del dissettore ad ultrasuoni: risultati di uno studio prospettico randomizzato. Il giornale di chirurgia (2005)
15. Lavini, C., Ruggiero, C., Morandi, U.: Chirurgia toracica videoassistita. Editore Springer Science & Business Media (2008)

Design of an Evaluation Tool to Assess IoT Solutions for Active and Healthy Aging

Gloria Cea, Alba Gallego, Maria Teresa Arredondo, and Giuseppe Fico(✉)

Universidad Politécnica de Madrid, Avenida Complutense, 30, 28040 Madrid, Spain
gfico@lst.tfo.upm.es

Abstract. Europe is currently the oldest continent having the "old-age dependency ratio" - persons over 65 compared to the active population. To cope with this situation, the deployment of Active and Healthy Aging (AHA) services through IoT is becoming increasingly relevant, to make this population as independent as possible within their environment. This study presents the whole process to design and implement the "Public Evaluation Website", a tool that will present the results and achievements generated in ACTIVAGE, a Large-Scale Pilot that implements Internet of Things solutions for AHA in 9 European regions and cities.

Keywords: Active and Healthy Aging · Internet of Things · Large Scale Pilot

1 Introduction

In the last years, the mortality rate has significantly decreased, and life expectancy is increasing throughout Europe, however, unhealthy life-years make up around 20% of a person's life [1]. Living longer does not necessarily mean living a healthier, more active and independent life. This fact is due to advances in public health systems and the evolution in socioeconomic conditions. Furthermore, according to the 2018 Ageing Report [2], people aged 60 are now expected to survive an additional 18.5 to 21.6 years and soon the world will have a higher number of older adults than children. This transformation is expected to continue, with the +65-age group growing from 20.0% in 2016 to 29.0% of the EU population by 2070.

The Active & Healthy Ageing (AHA) community is wide and heterogeneous in terms of needs, demands and living environments. AHA services based on the Internet of Things (IoT) are a strategic component to support the creation of ecosystem able to dynamically answer and prevent the challenges faced by the health and social care systems [3]. In this context, the European Commission is financing ACTIVAGE [4], a flagship project funded by the European Commission to demonstrate, in a Large Scale Pilot (LSP), the value of the IoT paradigm to enable the creation of smart living environments supporting AHA of European citizens. The pilot is deploying and demonstrating solutions in nine Deployment Sites located in seven European countries involving about 7400 end-users.

J. Henriques et al. (Eds.): MEDICON 2019, IFMBE Proceedings 76, pp. 1038–1046, 2020.
https://doi.org/10.1007/978-3-030-31635-8_126

Global and local socio-economic impact assessment and evaluation framework has been tackled with the overarching guidance of the triple win approach pursued by the EIP-AHA [5]: impacting in health status and quality of life (QoL) of elderly and care network; improving the efficiency and sustainability of health systems; and fostering innovation and growth in the IoT ecosystems.

In this research work we present the design and implementation process of the *Public Evidence Website*, for evidence gathering and data interpretation across the different Deployment Sites and for the ACTIVAGE LSP, in order to: (1) check the effectiveness of the solutions and services based on the IoT paradigm in the AHA societal challenge; (2) evaluate the KPIs and study the validity of the collected data; (3) provide a transparent user interface with objective data and easy access for users; (4) share the information on the novelties of the project and the current situation about solutions for the aging population; (5) contribute to the creation of a sustainable European IoT ecosystem focused on AHA solutions.

2 Methods

The CeHres Roadmap methodology has been applied to plan, coordinate and execute the participatory process of the tool development following a User Center Design approach [10], in which potential end-users in the AHA-IoT domain such as business companies, evaluation experts, Deployment Sites and service providers are involved in the design process to create highly usable and accessible products. We started with the definition of the objectives to be achieved and the problems that we aimed to solve in order to better focus on the design of the expected solution.

The full sequence of steps followed to perform this action are shown in Fig. 1 in which the phases of the ceHres roadmap methodology are merged.

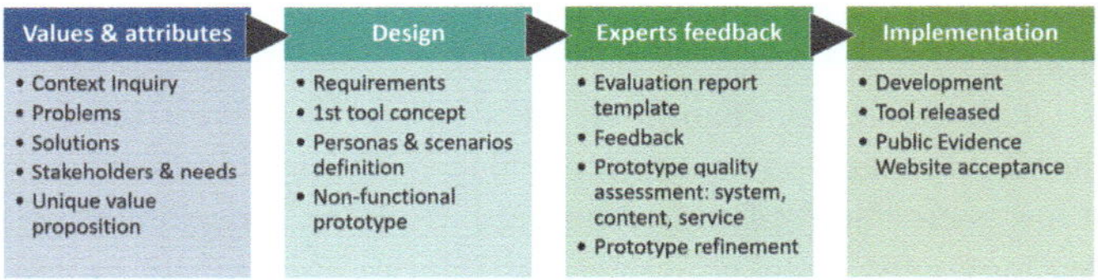

Fig. 1. Sequence of the results steps diagram

2.1 ceHres Roadmap

The ceHres (center for eHealth research) roadmap for the development of eHealth technologies [6] consists of two interdependent strategies: Human-Centered Design (HCD) and Business Modelling (BM). It is based on a holistic approach organized into five co-creation phases. Between them, the formative evaluation process is carried out, which evaluates each step of the process to obtain information that helps to improve the product.

Phase I. Context Inquiry: it consists of studying and analysing the context that frames the project we are developing from different perspectives: social, economic, political, health and technological. The Lean Canvas model, which is based on the Canvas Business Model [7], is suggested for phase I. Next step is to identify the stakeholders who are invited to participate in co-creation sessions (workshops, interviews, focus groups) in which the problems are identified and how to solve them through scenarios. Phase II. Value Specification: it aims to determine the expected improvements, as well as the possibilities and limitations to achieve values. Values refer to the improvements expected by stakeholders, which can be classified into (1) improvement in health and well-being in the living environments; (2) improvement in socio-health services for the elderly. Based on the values, attributes and how they are intended to be achieved are then identified by stakeholders. Values are related to 'The 10 "E's" of Eysenbach' [8].

Phase III. Design: it is based on the user requirements and the values specified, using persuasive theories to develop technologies that make sense to their users. The definition of problems, attributes and values will lead to determining the requirements, both technical and design. Based on these requirements the 'persona' is outlined, who is a user model represented as individual human beings that is based on the characteristics of real people in which their personification and empathy are key, identifying their roles, needs, activities, attitudes, aptitudes, motivations, abilities and behaviours. As a complement, the scenarios are described and reflect how the 'persona' interacts with the product. After this process, a first non-functional prototype of the tool is created by means of mock-ups that will be evaluated with the design team and modified until the system meets the requirements of the users.

Phase IV. Operationalization: it refers to planning and actions that lead to the dissemination, adoption, and incorporation of technology. If these conditions are ignored, the usefulness of the technology can be limited, and decision making can be delayed. During the development of the process, the key stakeholders decide the strategies and activities for the development of a business model to implement this eHealth technology.

Phase V. Summative Evaluation: it is a research activity aimed at measuring whether technology makes sense and has value for all stakeholders as well as determining what has been achieved at a given time. Both the acceptance and the impact of eHealth technology must be taken into consideration. This impact should be measured to assess if the objectives that were marked in phase II are being achieved.

2.2 Power BI

The user interface has been implemented using the Power BI platform [9], which is a business analytics solution allowing to visualize data embedded in an application or website, what has facilitated adding analytical and visualization functionalities in am effective way. It provides interactive visualizations with self-service business intelligence capabilities, where end users can create reports and dashboards by themselves, without having to depend on information technology staff or database administrators.

3 Results

3.1 Values and Attributes

First step was to analyse the context of the project. The result is summarized in Table 1. Public Evidence Website collects the information and results from the Deployment Sites, whose data should be integrated and unified to demonstrate the effects of IoT in the AHA domain.

Table 1. Problems, solutions & unique value proposition.

Problems	Solution	Unique value proposition
Lack of knowledge and inaccessibility regarding the project	Share information in a public and accessible interface	Public, multidimensional, innovative and accessible tool that undertake the promotion of the implementation of an IoT ecosystem in Europe and evaluate its results, for Active and Healthy Aging societal challenge. It shows high readability and facilitates understanding, so that all types of users can understand it, taking into account their background and experience
Non-homogeneous data from different Deployment Sites	Integration of information in a common structure of the open database showing the effects of technologies	
Uncertainty about the effectiveness and efficiency of the technologies implemented	Indicators assessment and validation of the results	

Then, the involved stakeholders have been identified along with their functions, motivations, benefits, responsibilities and objectives as shows Table 2. They can be anyone interested in using the tool or that influence the development and deployment of the project. In all categories both public and private companies/entities could be found.

Table 2. Stakeholders and needs

Stakeholders	Needs
Target user	Make use of the tool. They can be elderly, patient, end-user of an implemented AHA service, formal or informal caregivers, healthcare professionals, etc
Authorities & governments	Support and regulate the plan. They take care of the socioeconomic and legal context, thus having a role in the dissemination and acceptance of the project
Technological developers	Hardware and software design, development and testing to implement the tool
Suppliers	Technical services and technological solutions
Investors	Financing

Next step was the values definition and to discuss how to be achieved through attributes taking into account that each value is related to their respective attributes and related to one or several of the terms that make up the "Eysenbach 10 e's" (see Table 3).

(a) Improvement in health and well-being in the living environments: the tool participates in improving the quality of life, lifestyle and changes in attitude, intention and greater knowledge in the self-care, due the fact that it is an essential tool for the development of solutions that promotes AHA.
(b) Improvement in socio-health services for the elderly: the project promotes continuity, interaction and access, guarantees transparency and participates in the decision-making support since it is an evaluation tool.

Table 3. Relation between values, Eysenbach 10 e's in "e-health" and value attributes

Value(s)	Eysenbach 10 e's	Attribute(s)
Quality of Life	Enhancing quality of care	Inform and promote the development of AHA solutions
Lifestyle	Empowerment of consumers	
Self-care	Education	
Access	Equity	Public and always enabled
Transparency	Ethics	Use of standards and protocols
Continuity	Enabling information exchange	More continuous feedback Improve in communication
Interaction	Enabling information exchange	Keep up with the evidences and novelties
Decision support	Evidence based	Decision-making based on the results of the evaluation tool

3.2 Design

First paper-mockup design started after a brainstorming session to determine how we expect the tool should be. Search filters were considered, including some information fields such as Deployment Sites, IoT solutions, users characteristics, use cases, KPIs, innovation phases. Afterwards, the goal was to extract the design requirements by defining the 'persona'. Three profiles have been introduced with their respective scenarios, which allowed us to create a simple, intuitive, attractive and usable user interface:

- David, 66 years old, basic technology level. He wants to learn about IoT solutions that best suit his circumstances. He accesses to Public Evidence Website to notice the ACTIVAGE progress, its current status and its prospects.
- Lisa, 35 years old, high technology level. Her ICT company is ACTIVAGE partner and access the Public Evidence Website to gather information on the phases of the projects that are being carried out in other Deployment Sites and check the KPIs evolution.

- Harry, 48 years old, medium technology level. He is a German businessman linked to the world of technology who makes the decision to participate and invest in ACTIVAGE. He accesses the Public Evidence Website to be informed about the project and analyse what his company's contribution might be.

After an iterative process in which several mock-ups were created and assessed, useful feedbacks were received that helped to refine it such as: improve the search filtering functionality, a more professional design was needed, improve the content and structure of the main page, stronger contrasts between backgrounds and texts, etc. Considering all the proposals and refining the emerging ideas, the non-functional prototype was designed, very close to the initial expectations whose result can be seen in Fig. 2.

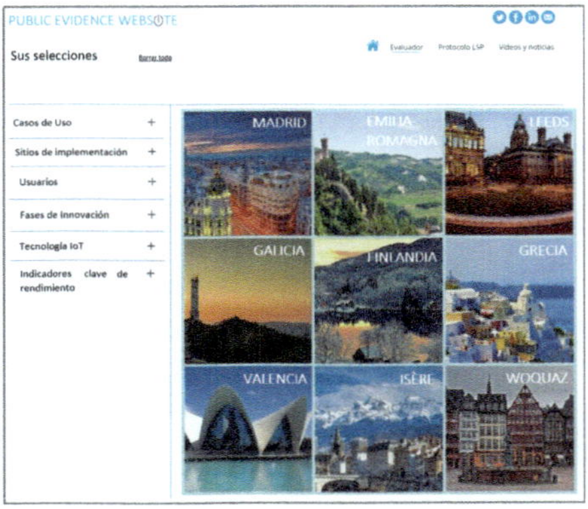

Fig. 2. Public evidence website non-functional prototype

3.3 Experts Feedback

Once the prototype design process was finished, a quality assessment report template was generated in order to explain the main objectives that the tool aims to cover and its functionalities. In addition, a set of questions was raised, which evaluate different aspects of the Public Evidence Website, pointed to check whether it meets the requirements of the User-Center Design, the objectives set and the design characteristics suitable for the users.

The experts that evaluated the tool belong to a strategic consultancy that has a wealth of experience in the management and development of innovative projects in the ICT domain, connected health and active aging. Their feedback was crucial, they completed the quality assessment report regarding the system, the contents and the services. They suggested significant improvements and identified tool strengths, which highly contribute to the refinement of the final prototype.

3.4 Implementation

The result of this process was the development of the Public Evidence Website, a public user interface, accessed openly by any person, professional, institution, company that may have an interest in be informed about the results and evidence that ACTIVAGE is generating from diverse perspectives: deployment sites, user acceptance & co-creation, socio-economic, standardization, dissemination business & exploitation, and IoT. The actors involved in the evaluation activities are diverse, each of them has a different outlook and this heterogeneous cluster enriches the framework. Representatives of the demand side such as user associations, government organisms, regulatory organisations or new Deployment Sites, are interested to verify whether the indicators are prospectively relevant to make a real and sustainable impact.

Results of the LSP and its data generated can be visualized and explored according to different categories and stratification levels that correspond directly with the six cards of the home page (see Fig. 3): Deployment Sites, ACTIVAGE overview, Operational status, Socio-economic KPIs, IoT KPIs and Impact Attainment Strategy (IAS) KPIs.

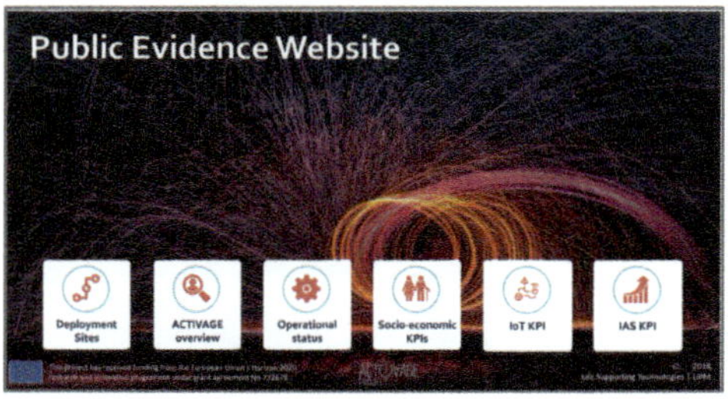

Fig. 3. Public evidence website home page

- The *Deployment Sites card* (Fig. 4) presents a brief description including a picture and the socio-demographic data related to each DS. These data are: gender, the gender percentage participating in the project, the living characteristics, the technology level, the educational level, the number of involved users classified in elderly, informal or formal carer.

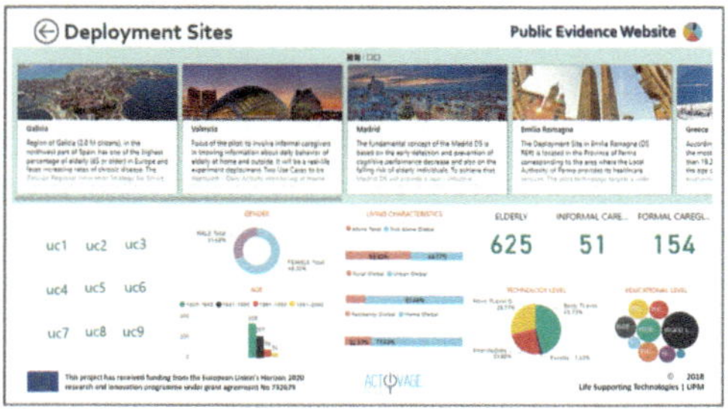

Fig. 4. Deployment sites card

- The *ACTIVAGE overview card* is composed by a map with the nine Deployment sites and shows the progress aligned with to the project advancements.
- The *Operational status card* represents the evolution during the experiment deployment and running through nine charts which can be filtered by Deployment Site. The parameters shown are the number of users, number of installations, percentage of signed consent forms, use of equipment or services, types of devices (alarm, gateway, wearable, environmental sensors, UI, communication).
- The *Socio-economic KPIs card* (Fig. 5) is composed by charts representing the number of users evaluated (elderly, formal caregivers, informal caregivers) and results of questionnaires about QoL, social isolation, physical wellbeing and self-perception parameters.

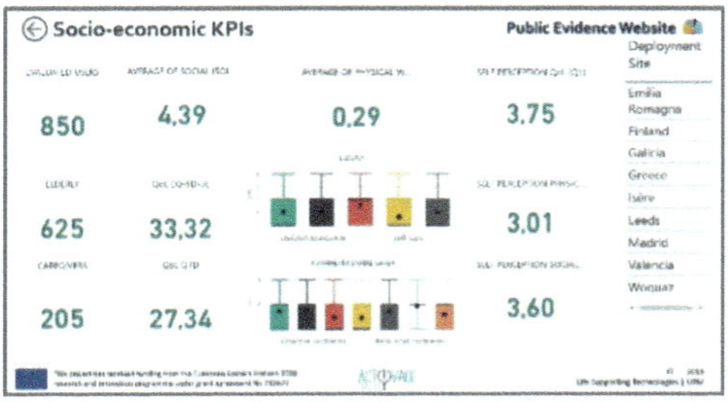

Fig. 5. Socio-economic KPIs card

- The *IoT KPIs card* shows the current status of the project and if it fulfils objectives from a technical perspective. Four KPIs are represented: application support tools and service layer components, secure interoperability layer software components, formal description of the architecture, privacy and security.
- The *IAS KPIs card* depicts the indicators from a business perspective through its goal and its current state in the areas: business & exploitation, standardization, dissemination, user acceptance & co-creation.

Additionally, search filters and visualization criteria allow to show information according to different perspectives: AHA services, deployment sites, KPIs, sociodemographic data, IoT elements, innovation phases and use cases. Information is shown according to the relevance, following common sense and high readability, to be understood from different profiles in terms of expertise and background.

4 Conclusions

Due to the aging population in Europe, innovative solutions are emerging in the market. Concretely, AHA services based on IoT technologies are being incorporated into the living environments of the elderly people and their relatives. Given the need to assess the effects of these solutions, Public Evidence Website is providing information on the evidence and impact that the ACTIVAGE is bringing to end-users and stakeholders of the AHA-IoT ecosystem towards the generation of evidence and value. The tool allows to aggregate information from each deployment site and to collect and process the KPIs. The tool has been evaluated by experts in the field to verify that it meets all the objectives established and noting that it is a tool with an intuitive design, simple to use and of great impact for the project.

Acknowledgements. The authors wish to acknowledge the partners of the ACTIVAGE project (http://www.activageproject.eu/). This project receives funding from the European Union's Horizon 2020, under grant agreement No. 732679.

References

1. European Innovation Partnership on Active and Healthy Ageing. EIP-AHA. https://ec.europa.eu/eip/ageing/about-the-partnership_en
2. European Commission: The 2018 ageing report. Economic & Budgetary Projections for the 28 EU Member States (2016–2070), underlying assumptions and projection, methodologies. Publications Office of the European Union, Luxembourg (2018). https://doi.org/10.2765/61563, https://ec.europa.eu/info/publications/economic-and-financial-affairs-publications_en. ISBN 978-92-79-77460-7, ISSN 2443-8014
3. United Nations, Department of Economic and Social Affairs, Population Division (2017). World Population Ageing 2017 - Highlights (ST/ESA/SER.A/397). ISBN 978-92-1-151551-0
4. The ACTIVAGE Project http://www.activageproject.eu/. This project has received funding from the European Union's Horizon 2020 research and innovation programme under grant agreement No 732679
5. EIP-AHA. Action Plan on 'Innovation for Age-friendly buildings, cities & environments'
6. Nijland, N.: Grounding eHealth: towards a holistic framework for sustainable eHealth technologies (2011)
7. Osterwalder, A., Pigneur, Y.: Business Model Generation: A Handbook for Visionaries, Game Changers, and Challengers. Wiley, Hoboken (2010)
8. Eysenbach, G.: What is e-Health? (2001). https://www.ncbi.nlm.nih.gov/pmc/articles/PMC1761894/
9. Bring your data to life with Microsoft Power BI. Microsoft.com. Microsoft. https://powerbi.microsoft.com/en-us/
10. The Interaction Design Foundation. What is User Centered Design? https://www.interaction-design.org/literature/topics/user-centered-design

Practical Use of Early Stage Health Technology Assessment of Medical Devices: Systematic Literature Review

Mariia Simonova(✉), Vladimír Rogalewicz, Gleb Donin, and Peter Kneppo

Department of Biomedical Technology, Faculty of Biomedical Engineering, Czech Technical University, Prague, Kladno, Czech Republic
mariia.simonova@fbmi.cvut.cz

Abstract. The early stage HTA (eHTA) is a commonly used tool for identifying economic indicators of health technology in early phases of research and development. It provides powerful influence instruments to manage the manufacturer's investment risks and to improve the device design.

There is no uniform generally accepted standard of eHTA implementation, but an analysis of several large-scale theoretical studies showed that, in general, the eHTA process for medical devices consists of three stages: conceptual modelling of the new medical technology, determination of new technology parameters by expert elicitation methods, and simulation allowing for an analysis of commercial options.

The aim of this study is a systematic review of studies published from 2014 till 2019 and an analysis of selected articles in terms of the methods used in each step of the early stage health technology assessment.

Most of the analyzed case studies deal only with a part of the early stage medical device assessment, probably because of its confidential character. But we have found the popularity of decision-making models, the applicability of AHP (the analytic hierarchy process) methods to the elicitation process, and information on the use of CEA (cost-effectiveness analysis) methods and headroom methods as economic analysis tools.

Keywords: Early stage HTA · eHTA · Health technology assessment · Medical device assessment · Medical device development

1 Introduction

Advanced society is based on knowledge and modern technologies [1]. Technology has a critical role, especially in connection with the acquisition, processing and application of information [2]. At present, monitoring, evaluating, and detecting new trends in various medical devices is an exclusively interdisciplinary issue [3]. The interdisciplinary approach is one of the ways to cope with the growing complexity, dynamism and variability of the society [4, 5].

J. Henriques et al. (Eds.): MEDICON 2019, IFMBE Proceedings 76, pp. 1047–1056, 2020.
https://doi.org/10.1007/978-3-030-31635-8_127

The emerging demand for health technology assessment at the early stage of development is caused by several factors. The introduction of the new Medical Devices European Regulation EU 2017/745 belongs to the most important ones. Under the new regulation, medical devices manufacturers have to focus more on clinical trials. The clinical trial may be based on the device clinical data only if the technical, biological and clinical equivalence with the device concerned can be demonstrated (Section 3, part A, Annex XIV, MDR 2017/745). In addition, in the case of implantable devices and Class III devices, the clinical trial must be definitely carried out, unless the device falls under one of the exceptional rules (Section 4, Article 61, Chapter 4, MDR 2017/745) [6]. This fact means that most new medical devices and new health technologies will have to go through a clinical trial. For the manufacturers, clinical trials always mean a huge expense, especially in case of implantable medical devices and medical devices with a metabolic effect. In this situation, there is a growing need for a full-scale implementation of a methodological process that would allow for health technology assessment at an early stage of development, before clinical testing, and even before design finalization. Such a solution is provided by the early stage HTA that possesses powerful tools to manage the manufacturer's investment risks and to improve the device design.

The early stage HTA (eHTA) is an assessment of health technologies at the stage of research and development. The eHTA is most commonly used as a tool for identifying economic indicators of health technologies. Early stage models can be used to gain information for the design and management stage of new health technologies to mitigate the risks associated with placing the technology on the market and its inclusion into the public insurance reimbursement systems [7].

eHTA, combined with early health economics modelling, is being increasingly used by manufacturers as an approach to identifying the added value of new health technologies. The results of such a process are useful for:

- deciding on the next direction of development,
- setting minimum efficiency limits for a new technology compared to currently available comparators,
- providing support for pricing and payment settings [8].

An analysis of several large-scale theoretical studies showed that, in general, the eHTA process for medical devices consists of three stages as shown in Fig. 1.

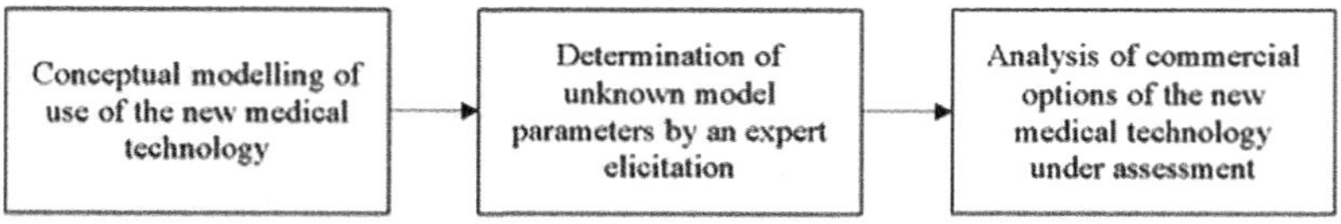

Fig. 1. eHTA steps. (Source: own analysis).

The first step is detailed conceptual modelling of use of the medical technology. It is important to consider all possible situations and conditions. Subsequently, the model parameters are determined. In the case that these parameters are not generally known, expert elicitation methods are mostly used in eHTA. Finally, statistical evaluation and introduction of unknown parameters into the model, simulation, and an analysis of commercial options are carried out after the elicitation.

There is no uniform generally accepted standard of eHTA implementation. That´s why the aim of this study is to provide a review of the most popular methodologies used for a practical implementation of early assessment of medical devices in the last five years.

2 Methodology

The study is based on a systematic literature review of published papers on eHTA of medical devices. One of the aims of this study is to provide topical information on the current state of development and use of early stage HTA methodologies. It will show the real range and specificities of eHTA usage.

A citation analysis of the early HTA was performed using keywords and subsequent manual search in Scopus. The period was selected from 2014 to February 2019. The following keywords and their various combinations were used for the search: Health Technology Assessment, Biomedical Technology Assessment, Medical Devices, Early Health Technology Assessment. Results from pharmacology, toxicology and pharmacy were excluded. A total of 118 results were found.

Subsequently, a review of the title and abstract of all identified studies was conducted, and finally 37 studies were selected for a detailed analysis. After reading full texts of all 37 documents, another choice was made. Selected articles should include at least a partial description of the methodological procedure or a case study of health technology assessment at an early stage of design and development (see Fig. 2).

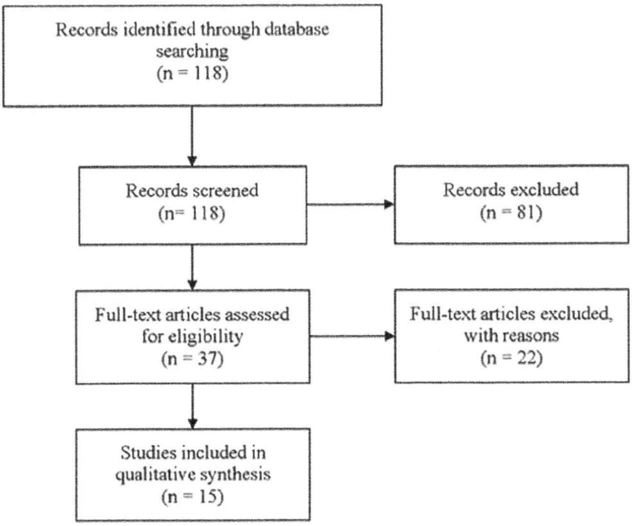

Fig. 2. The PRISMA flow chart of the systematic literature review (Source: own analysis).

In the end, 15 articles published from 2014 to February 2019 that met the requirements were included in the study. The following criteria were analyzed in the selected papers: the kind of health technology, the modelling method, the way of data receiving/elicitation, and the methods of economic analysis.

3 Results

eHTA is still a relatively new area, and it is quite difficult to find published studies. Neither terminology nor keywords are properly unified. Some ambiguities also arise from confusing translations of terms.

Table 1 shows methods used in each step of early stage health technology assessment at each of the selected studies.

Table 1. Methods used in each step of early stage health technology assessment

Author	Kind of health technology	Modelling method	Data receiving/elicitation	Method of economic analysis
Chapman [9]	Stapled Hemorrhoidectomy	N/A	Statistic data	The headroom method
Cancela [10]	Telehealth system for parkinson's disease management	N/A	AHP focus groups	N/A
Kolominsky-Rabas [11]	Mobile Stroke Units	Hybrid simulation consisting of system dynamics models for macro-simulation and agent-based models for micro-simulation	Statistic data	N/A
Buisman [12]	Diagnostic test to detect early rheumatoid arthritis (RA) and Prognostic test to assess the risk of a recurrent ischemic stroke	N/A	N/A	Early-CEA
Girling [13]	Implantable Heart Device	N/A	N/A	The headroom method

(*continued*)

Table 1. (*continued*)

Author	Kind of health technology	Modelling method	Data receiving/elicitation	Method of economic analysis
Markiewicz [14]	The Injectable healing plasters (IHP), A portable point-of-care device (POC-CKD), Continuous blood pressure measurement device (CBPM), A computed tomography photo-acoustic instrument for imaging inter-phalangeal joints in the hand (RAPAI), The flexible endoscopy robot (FLEX), Home Brain Monitoring device (HBM)	N/A	Elicitation and literary data without specifying of methods	Headroom method combined with ROI-analysis
Van Nimwegen [15]	Complex pediatric neurology	N/A	N/A	The headroom method
Miquel-Cases [16]	Predictive biomarkers in breast cancer	Decision model	AHP	The headroom method and Early-CEA
Degeling [17]	Biomarker-Based Treatment	Discrete event simulation and timed automata models	N/A	CEA
Hummelin [18]	Bilateral DIEP flap breast reconstruction	Decision model	Elicitation and literary data without specifying of methods	The headroom method
Khoudigian-Sinani [19]	Predictive technology "StraticyteTM" for the early detection of oral cancer	Decision-analytic model	Belief elicitation method, systematic review and meta-analysis	Early-CEA

(*continued*)

Table 1. *(continued)*

Author	Kind of health technology	Modelling method	Data receiving/elicitation	Method of economic analysis
Kip [20]	New triple biomarker test	Decision tree model	Elicitation without specifying of methods. Answer options to questions consisted of percentages with a range of 25%, and including 0% and 100%.	CEA
De Graaf [21]	Biomarkers	N/A	N/A	The headroom method
Brandes [22]	Hypothetical vascular closure device	Decision tree model	Meta-analysis	CEA
Grigore [23]	Alternative treatments for prostate cancer	Decision model	The histogram and hybrid elicitation methods	CEA

(Source: own processing)
AHP the analytic hierarchy process, *CEA* cost-effectiveness analysis, *N/A* information not available, *ROI* return on investment analysis

In earlier studies dealing with similar literary reviews and frequently used eHTA methods, the popularity of Markov's models [8, 24, 25], simple stochastic models assuming that future states depend solely on the current state and not the sequence of past events (states), has been widely discussed. However, it is clear that decision-making models are more commonly used in recent years for an assessment of medical devices and technologies at an early stage of development.

Some of the studies that dealt with a comparison of Markov models and decision models have concluded that the algorithms behave in a similar way and that the differences are very small. The primary difference between Markov models and decision models is that the first one models the risks of recurring events in a straightforward manner. It is argued that the decision model is more appropriate for generalizing samples in a particular context, while the Markov model is more appropriate when an exact sample has to be derived [26, 27].

The use of cost-benefit analysis (CEA) in assessing the costs and health impacts of new health technologies comparing to current practice. The estimated cost-effectiveness of the proposed new intervention is compared either with the cost-effectiveness of a set of existing interventions or with a fixed price threshold (cut-off point), which represents the expected social willingness to pay for an additional unit of health [28].

The headroom method of economic evaluation has also been very popular. It can offer a practical way to manage investments, justifying the choice to develop the potential medical product [9, 13–15].

The least described and discussed parts of eHTA studies are methods of obtaining unknown model parameters. The classical concept of health technology assessment at the early stage of development assumes mainly an application of elicitation methods. Some of the analyzed studies describe benefits of AHP-based methodologies within expert elicitation and evaluating the responses. The method of analyzing the hierarchy process has been used for a procedure for the synthesis of priorities calculated on the basis of subjective judgments of experts. However, the details of the use of elicitation methods were only briefly mentioned in one study analyzed (Kip et al. [20]). The authors do not even mention the name of the method, but according to the description it can be assumed that the fixed interval method of expert elicitation was used [23].

4 Discussion and Conclusion

The early stage medical device assessment can be divided into three steps, and this paper is focused on an overview of the practical methods used in the eHTA process. during the last five years.

According to our results decision-making models have been more commonly used in recent years for conceptual modeling of the use of medical technology within the framework of eHTA. Expert elicitation is still the least described process of early assessment of medical devices, but in many studies the authors mention AHP methods as useful tool for objective evaluation of responses. For the methods of economic analysis in recent years, CEA methods and headroom methods are more typical, sometimes in their combination.

Overall, it can be said that the eHTA process is still insufficiently described in published papers. What has been represented especially scarcely are studies about practical usage of medical devices assessment at the stage of research and development. Most of the analyzed case studies published from 2014 to February 2019 deal only with a part of the early stage medical device assessment. Probably because of the confidential character of the studies, medical devices and technologies and their characteristics are not usually described in detail. This is understandable, competitive relationships and industrial espionage force manufacturers to protect their know-how as much as possible until they enter the market. At the same time, it can reasonably and logically be assumed that, based on the findings of the eHTA, the manufacturer can influence and change the final characteristics of the final device or technology model provided the evaluation was carried out in the development phase.

We also assume that the practical methodology of eHTA is used more than it seems to be according to the number and content of published studies. In some European countries, special agencies and organizations are involved in the assessment of medical devices and technologies in the early stage of development. However, there are no European directives, recommendations or approved methodological guidelines. At the same time, different healthcare systems, pricing and reimbursement settings in different countries need to be taken into account. Therefore, the question of the possibility and, above all, the usefulness of introducing a unified system for eHTA is justified.

Acknowledgment. This research was supported by the Grant Agency of the Czech Technical University in Prague, grant No. SGS19/206/OHK5/3T/17.

References

1. Hospodková, P., Barták, M., Vacíková, K., Urbánková, E.: Global centers of medical device technology: United States, Europe and China. Lékař a technika – Clin. Technol. **48**(4), 136–144 (2018). ISSN 0301-5491
2. Donin, G., Barták, M., Kneppo, P.: Estimation of medical equipment prices–a case study of tomotherapy equipment in the Czech Republic. J. Bus. Econ. Manag. **18**(6), 1193–1211 (2017). ISSN 2029-4433
3. Gavurová, B., Kočišová, K., Sopko, J.: An analysis of health spending: trends in selected countries. Int. J. Appl. Bus. Econ. Res. **15**(9), 1–12 (2017). ISSN 0972-7302
4. Rosina, J., Rogalewicz, V., Ivlev, I., Juřičková, I., Donin, G., Jantosová, N., Vacek, J., Otawová, R., et al.: Health technology assessment for medical devices. Lékař a technika – Clin. Technol. **44**(3), 23–36 (2014). ISSN 0301-5491
5. Rogalewicz, V.: Health technology assessment as a tool for medical devices management in hospitals. In: The 5th IEEE International Conference on E-Health and Bioengineering. Gr. T. Popa University of Medicine and Pharmacy, Iasi (2015). ISBN 978-1-4673-7545-0
6. European Parliament. Regulation (EU) 2017/745 of the European Parliament and of the Council of 5 April 2017 on medical devices, amending Directive 2001/83/EC, Regulation (EC) No 178/2002 and Regulation (EC) No 1223/2009 and repealing Council Directives 90/385/EEC and 93/42/EEC. Off J Europ Uni. L 117, 5 May 2017
7. Steuten, L.M.G.: Early stage health technology assessment for precision biomarkers in oral health and systems medicine. OMICS: J. Integr. Biol. **20**(1), 30–35 (2016). https://doi.org/10.1089/omi.2015.0174
8. IJzerman, M.J., Koffijberg, H., Fenwick, E., Krahn, M.: Emerging use of early health technology assessment in medical product development: a scoping review of the literature. PharmacoEconomics **35**(7), 727–740 (2017). https://doi.org/10.1007/s40273-017-0509-1
9. Chapman, A.M., Taylor, C.A., Girling, A.J.: Early HTA to inform medical device development decisions - the headroom method. In: XIII Mediterranean Conference on Medical and Biological Engineering and Computing 2013, pp. 1151–1154 (2014). https://doi.org/10.1007/978-3-319-00846-2_285
10. Cancela, J., Fico, G., Pastorino, M., Arredondo, M.T.: Hierarchy definition for the evaluation of a telehealth system for parkinson's disease management. In: 6th European Conference of the International Federation for Medical and Biological Engineering, pp. 1000–1003 (2015). https://doi.org/10.1007/978-3-319-11128-5_250
11. Kolominsky-Rabas, P.L., Djanatliev, A., Wahlster, P., Gantner-Bär, M., Hofmann, B., German, R., Kriza, C.: Technology foresight for medical device development through hybrid simulation: the ProHTA Project. Technol. Forecast. Soc. Chang. **97**, 105–114 (2015). https://doi.org/10.1016/j.techfore.2013.12.005
12. Buisman, L.R., Rutten-van Mölken, M.P.M.H., Postmus, D., Luime, J.J., Uyl-de Groot, C. A., Redekop, W.K.: The early bird catches the worm: early costeffectiveness analysis of new medical tests. Int. J. Technol. Assess. Health Care **32**(1–2), 46–53 (2016). https://doi.org/10.1017/s0266462316000064

13. Girling, A., Lilford, R., Cole, A., Young, T.: Headroom approach to device development: current and future directions. Int. J. Technol. Assess. Health Care **31**(05), 331–338 (2015). https://doi.org/10.1017/s0266462315000501
14. Markiewicz, K., van Til, J.A., Steuten, L.M.G., IJzerman, M.J.: Commercial viability of medical devices using headroom and return on investment calculation. Technol. Forecast. Soc. Chang. **112**, 338–346 (2016). https://doi.org/10.1016/j.techfore.2016.07.041
15. Van Nimwegen, K.J.M., Lilford, R.J., van der Wilt, G.J., Grutters, J.P.C.: Headroom beyond the quality-adjusted life-year: the case of complex pediatric neurology. Int. J. Technol. Assess. Health Care **33**(01), 5–10 (2017). https://doi.org/10.1017/s0266462317000046
16. Miquel-Cases, A., Schouten, P.C., Steuten, L.M.G., Retèl, V.P., Linn, S.C., van Harten, W. H.: (Very) Early technology assessment and translation of predictive biomarkers in breast cancer. Cancer Treat. Rev. **52**, 117–127 (2017). https://doi.org/10.1016/j.ctrv.2016.11.008
17. Degeling, K., Schivo, S., Mehra, N., Koffijberg, H., Langerak, R., de Bono, J.S., IJzerman, M.J.: Comparison of timed automata with discrete event simulation for modeling of biomarker-based treatment decisions: an illustration for metastatic castration-resistant prostate cancer. Value in Health **20**(10), 1411–1419 (2017). https://doi.org/10.1016/j.jval.2017.05.024
18. Hummelink, S., Gerrits, J.G.W., Schultze Kool, L.J., Ulrich, D.J.O., Rovers, M.M., Grutters, J.P.C.: The merits of decision modelling in the earliest stages of the IDEAL framework: an innovative case in DIEP flap breast reconstructions. J. Plast. Reconstr. Aesthetic Surg. **70** (12), 1696–1701 (2017). https://doi.org/10.1016/j.bjps.2017.07.011
19. Khoudigian-Sinani, S., Blackhouse, G., Levine, M., Thabane, L., O'Reilly, D.: The premarket assessment of the cost-effectiveness of a predictive technology "Straticyte™" for the early detection of oral cancer: a decision analytic model. Health Econ. Rev. **7**(1) (2017). https://doi.org/10.1186/s13561-017-0170-6
20. Kip, M.M., Steuten, L.M., Koffijberg, H., IJzerman, M.J., Kusters, R.: Using expert elicitation to estimate the potential impact of improved diagnostic performance of laboratory tests: a case study on rapid discharge of suspected non-ST elevation myocardial infarction patients. J. Eval. Clin. Pract. **24**(1), 31–41 (2016). https://doi.org/10.1111/jep.12626
21. De Graaf, G., Postmus, D., Westerink, J., Buskens, E.: The early economic evaluation of novel biomarkers to accelerate their translation into clinical applications. Cost Effectiveness and Resource Allocation, **16**(1) (2018). https://doi.org/10.1186/s12962-018-0105-z
22. Brandes, A., Sinner, M.F., Kääb, S., Rogowski, W.H.: Early decision-analytic modeling – a case study on vascular closure devices. BMC Health Serv. Res. **15**(1) (2015). https://doi.org/10.1186/s12913-015-1118-3
23. Grigore, B., Peters, J., Hyde, C., Stein, K.: A comparison of two methods for expert elicitation in health technology assessments. BMC Med. Res. Methodol. **16**(1) (2016). https://doi.org/10.1186/s12874-016-0186-3
24. Pecchia, L., Craven, M.P.: Early stage Health Technology Assessment (HTA) of biomedical devices. The MATCH experience. In: World Congress on Medical Physics and Biomedical Engineering, 26–31 May 2012, Beijing, China, pp. 1525–1528 (2013). https://doi.org/10.1007/978-3-642-29305-4_401
25. Cao, Q., Postmus, D., Hillege, H.L., Buskens, E.: Probability elicitation to inform early health economic evaluations of new medical technologies: a case study in heart failure disease management. Value in Health **16**(4), 529–535 (2013). https://doi.org/10.1016/j.jval.2013.02.008

26. Boeck, J.D., Verpoorten, K., Luyten, K., Coninx, K.: A Comparison between decision trees and markov models to support proactive interfaces. In: 18th International Conference on Database and Expert Systems Applications (DEXA 2007) (2007). https://doi.org/10.1109/dexa.2007.4312864
27. Kuntz, K., Sainfort, F., Taylor, B.C., et al.: Decision and Simulation Modeling in Systematic Reviews. Agency for Healthcare Research and Quality: Methods Research Report. (Report number: 11(13)-EHC037-EF) (2013). https://doi.org/10.13140/2.1.1205.5842
28. Edejer Tan-Torres, T.: Making choices in health: WHO guide to cost-effectiveness analysis. World Health Organization, Geneva (2003). ISBN 92-415-4601-8

Usefulness of the Blink Reflex to Assess the Effect of Propofol During Induction of Anesthesia in Surgical Patients

Ana Leitão Ferreira[1,2(✉)], Catarina S. Nunes[2,3], Joaquim Gabriel Mendes[1], and Pedro Amorim[2]

[1] LAETA, INEGI, Faculdade de Engenharia, Universidade do Porto, Porto, Portugal
ana.leitao.ferreira@gmail.com
[2] Centro de Investigação Clínica em Anestesiologia, Serviço de Anestesiologia, Centro Hospitalar do Porto, Porto, Portugal
[3] DceT, Delegação do Porto, Universidade Aberta, Lisbon, Portugal

Abstract. The aim of this study was to investigate the relation between the blink reflex evoked by an electrical stimulus and the depth of anesthesia induced with intravenous anesthetic drug propofol. The blink reflex was stimulated before the propofol infusion started (baseline) and after, every 6 s. The electromyographic responses and the level of sedation/anesthesia scores as well as the estimated effect-site concentration of propofol were recorded in 11 patients. The blink reflex responses were abolished when patients were still conscious. The clinical scale of anesthesia increased with increasing concentrations of propofol. To predict the level of sedation/anesthesia a multinomial logistic regression was performed using blink reflex extracted features at the frequency domain. Several features proved to be good predictor estimates and the model showed to be useful. This information could be helpful to assess the moment of loss of consciousness and thus personalize anesthesia.

Keywords: Propofol · Anesthesia · Loss of consciousness · Blink reflex · Electromyography · Multinomial logistic regression · Frequency analysis

1 Introduction

In clinical practice, anesthesiologists use a combination of anesthetic drugs for surgery, including the most commonly used, the intravenous drug propofol, to render the patient unconsciousness. In a previously published work [1, 2], we determined that the amount of propofol required to induce loss of consciousness (LOC) varied widely between patients (≈300%) and that more than two-thirds of the patients required less than the initial dose recommended by the drug package inserts.

Precisely identification of the moment of LOC during the induction phase of general anesthesia is of extreme importance for determining the amount of propofol each patient requires. Using that information may serve to guide the drug infusion rate required to maintain an adequate level of anesthesia throughout surgery.

J. Henriques et al. (Eds.): MEDICON 2019, IFMBE Proceedings 76, pp. 1057–1063, 2020.
https://doi.org/10.1007/978-3-030-31635-8_128

Currently, methods to decide the exact moment of LOC are still based on subjective analysis [3]. Objectively, there are depth of anesthesia monitors, such as the Bispectral Index (BIS) (Aspect Medical System, Newton, MA, USA), which allow the maintenance of a steady state during surgery, but it does not allow the determination of the instant of LOC, which remains as open issue. BIS evaluation after acquisition of new dataset is characterized by a delay, which can even reach one minute [4]. The anesthesiologists should be aware of situations that cause false BIS readings to avoid complications, such as awareness [5, 6] or excessive anesthesia.

In a previous study, Mourisse et al. [7] showed that the components of the blink reflex are attenuated and abolished with increasing concentrations of propofol. They also showed that the blink reflex was more sensitive than BIS [8]. Their results suggested that the differential sensitivity of the components of the blink reflex could be useful to monitor depth of sedation/anesthesia. However, Mourisses's results were demonstrated with 10 min stepwise increases in propofol, which is not compatible with anesthesia induction in a surgical setup.

We hypothesized that the blink reflex could be used to assess LOC for induction of anesthesia with propofol. Thus, based on a constant infusion of propofol, with the present study we intend to investigate the relation between the electrically evoked blink reflex and the level of sedation/anesthesia.

2 Methods

Under Hospital Ethical Committee approval and with written, informed consent, eleven patients scheduled for neurosurgical procedures, subjected to general anesthesia with propofol were studied.

The analgesic drug remifentanil infusion starts with a Target Controlled Infusion (TCI) effect-site (Ce) concentration target of 2.5 ηg/mL. After the remifentanil Ce target was achieved, 1% propofol constant infusion started at 3.3 mL/kg/h. The level of consciousness was assessed using a Modified version of the Richmond Agitation-Sedation Scale (MRASS) [7], based on the evaluation of the patient eye opening where 0 corresponds to the patient fully awake and alert, 1 not fully awake but has sustained awakening to voice, 2 open the eyes to name calling, 3 does not open the eyes to name calling, 4 does not open the eyes to painful stimulation but has corneal reflex and 5 lost corneal reflex (evaluated by using a drop of sterile water to touch the cornea). At LOC, defined as a scale of 5 (loss of corneal reflex) the propofol infusion was stopped and the system was switched to effect-site TCI mode, with a Ce set to a target 25% of the Ce of LOC. Drugs were then titrated to maintain a BIS between 40 and 60.

Electromyographic (EMG) signals were recorded from both orbicularis oculi muscles through surface electrodes. The positioning of the stimulating and recording electrodes for recording of electrically evoked blink reflexes is presented in Fig. 1. Prior to electrode application, all skin surfaces were cleaned with an exfoliating agent and the electrodes coated with conductive paste (electrode impedance < 8 kΩ). The supraorbital nerve was electrically stimulated with a duration of 0.1 ms at 1/6 Hz. Recordings and stimulations were conducted with the VikingQuestTM neurophysiological device (VikingQuest, Nicolet, WI, USA) at 200 ms total sweep time. Band pass filters were set at 20 Hz and 10 kHz. The EMG records have at least two components: R_1, after a latency of 10 ms and R_2, after a latency of 30 ms [9].

Baseline blink reflexes and MRASS were recorded before and after propofol was administered, every six seconds. The estimated Ce of propofol was also noted and recorded every six seconds.

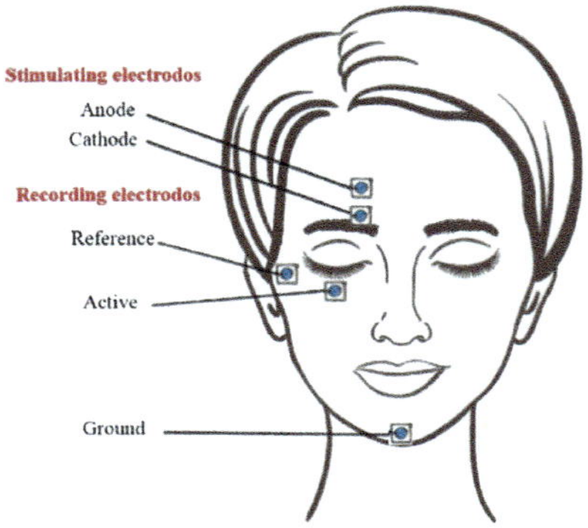

Fig. 1. Diagram showing positions of the stimulating and recording electrodes for recording of electrically evoked blink reflexes in the orbicularis oculi muscle. The active electrode is placed in the middle of the inferior rim of the orbit and the reference electrode halfway on the eye-ear line. A ground electrode is placed on the cheek or on the shoulder. One of the supraorbital nerves was stimulated transcutaneously with a bipolar block electrode (electrode surface 1.4 cm^2 and interelectrode distance 2 cm). The cathode was placed beneath the eyebrow over the supraorbital notch and the anode placed above the eyebrow.

Blink reflex responses were used to calculate the values of the following variables: latency, duration, amplitude and area under the curves R_1 and R_2. To extract features from the frequency domain, a non-parametric Fast Fourier Transform (FFT) using Welch's modified periodogram was used. Each blink response was windowed with a Hamming window and the modified periodogram was calculated for each segment. Power Spectrum Density (PSD) was finally estimated by averaging over all resulting periodograms. For this purpose, Matlab 2018a (MathWorks, USA) was used. The following features were extracted: mean power (dB) - the average power of the PSD within the epoch; mean frequency (Hz) - the average frequency of the PSD within the epoch; median frequency (Hz) - frequency at which 50% of the PSD is reached; band power (dB) - power in the frequency range 30–120 Hz of the PSD; peak frequency (Hz) - frequency at which the maximum PSD occurs; total harmonic distortion (THD) (dB) - the ratio of the PSD of all harmonics with the power in the fundamental frequency; signal to noise ratio (SNR) (dB) - the ratio of signal PSD to the noise PSD; total power (dB) - the sum of the PSD at all frequencies of the PSD spectrum; power bandwidth (dB) - the 3-dB (half-power) bandwidth of the PSD within the resolution bandwidth 30–120 Hz; spectral entropy - irregularity, complexity or amount of PSD disorders.

It was considered that LOR_1 (Loss of R_1 component) and LOR_2 (Loss of R_2 component) as the moment when R_1 and R_2 components were last seen in the EMG signal after an electrically evoked blink reflex. LOC (Loss Of Consciousness) was defined as the moment at which patient loses consciousness stated by the anesthesiologist as lack of corneal reflex.

The Shapiro-Wilk test was used to determine whether datasets were normally distributed. Correlation analysis was performed using the Pearson r test for normally distributed data or the Spearman rank r, when data were not normally distributed. Statistical Package for Social Sciences was used (IBM SPSS version 24.0 Il, USA) for all statistical analysis. A p-value inferior to 0.05 was considered statistically significant.

3 Results

In the present study 11 patients aged 47 (16) years old, weighing 73.73 (10.99) kg, heighten 165.82 (11.48) cm, 1 ASA I, 8 ASA II, 2 ASA III, 3 males and 8 females were considered. The supraorbital nerve was stimulated with 25.7 (8) mA. All patients reached the end of the depth of anesthesia scale. No patient had hemodynamic or respiratory problems.

The baseline values of the latency, duration, area and amplitude of R_1 and R_2 components are presented in Table 1.

Table 1. Baseline values of the latency, duration, amplitude and area of the blink reflex components. Baselines were recorded before propofol was administered.

	Blink reflex R_1 component	Blink reflex R2 component
Latency (ms)	10.81 (0.68)	39.31 (5.36)
Duration (ms)	9.53 (1.37)	31.79 (10.11)
Amplitude (mV)	418.80 (292.68)	485.15 (157.80)
Area (mV.ms)	961.63 (614.41)	2796.73 (1369.62)

Estimated Ce concentrations of propofol, MRASS values and times at which LOR_2, LOR_1 and LOC occurred are presented in Table 2.

Table 2. Estimated Ce concentrations of propofol, MRASS values and times at which LOR_2, LOR_1 and LOC occurred. Results are mean (standard deviation) or median [minimum, maximum].

	LOR_2	LOR_1	LOC
Time (s)	54.78 (16.18)	118.45 (33.13)	143.27 (35.87)
Propofol Ce (μg/ml)	1.11 (0.47)	3.37 (1.23)	4.10 (1.26)
MRASS	0	2 [0, 4]	5

R_2 was the first component to be abolished, followed by the R_1 component. All patients were conscious (MRASS < 5) when the R_2 and R_1 responses were last seen. There was a statistically significant difference between the estimated Ce concentration of propofol at LOC and at LOR_1 ($t(10) = -3.573$, $p = 0.005$, Paired Samples Test).

Because R_2 and R_1 responses were abolished before LOC, latency and duration cannot be measured and the amplitude and area are zero, resulting in an insufficient number of data points, a frequency domain analysis was performed.

The correlations between the extracted features from the EMG signal in the frequency domain and the estimated Ce concentrations of propofol and between the features and MRASS score are shown in Table 3.

Table 3. Correlation coefficients (Spearman r) between the estimated effect-site (Ce) concentrations of propofol and the extracted features from the EMG signal in the frequency domain. Correlation coefficients (Spearman r) between the MRASS scores and the extracted features from the EMG signal. The extracted features are: mean power (dB), mean frequency (Hz), median frequency (Hz), power within the band 30–120 Hz (dB), peak frequency (Hz), total harmonic distortion (THD) (dB), signal to noise ratio (SNR) (dB), total power bandwidth (dB), bandwidth within the band 30–120 Hz (dB) and spectral entropy. Spearman rank coefficient from pooled data of all patients (n = 11) is shown.

Feature	Spearman *r* with propofol Ce	Spearman *r* with RASS
MRASS	0,905*	1.000
Mean power (dB)	−0.561*	−0,409
Mean frequency (Hz)	−0.690*	−0,653*
Median frequency (Hz)	−0.573*	−0,538*
Band power (30–120 Hz) (dB)	−0.554*	−0,427
Peak frequency (Hz)	−0,471	−0,304
THD (dB)	Ns	Ns
SNR (dB)	0,365	Ns
Total power bandwidth (dB)	Ns	Ns
Power bandwidth (30–120 Hz) (dB)	Ns	Ns
Spectral entropy	0.474	Ns

Ns – not statistically significant at the 0.05 level.
*Means correlation was significant at the level $p < 0.001$.

The clinical scale of anesthesia MRASS increased monotonically with increasing propofol Ce concentration, revealing an increasing deepening of anesthesia.

Mean power, mean and median frequency and band power (30–120 Hz) features showed a negative correlation with estimated Ce concentration of propofol. Mean and median frequency features also showed a negative correlation with MRASS score. Mean power was the most consistent feature.

The capacity of all features to predict the clinical depth of anesthesia was evaluated by the multinomial logistic regression analysis, applied to the entire dataset. Only the five features with significant correlations with the estimated Ce concentrations of propofol and MRASS were used (mean power, mean frequency, median frequency, band power (30–120 Hz) and peak frequency). The maximum likelihood method was used to calculate the parameter coefficients of the model.

The overall effectiveness of the model was assessed using the Chi-squared statistic. The Chi-square value of 245.576 and its respective p-value < 0.001 indicates a significant relationship between the depth of sedation/anesthesia scale and the set of parameters in the final model. The Cox and Snell pseudo R-square was 0.576 and the Nagelkerte pseudo R-square was 0.611, suggesting that 60% of the variability is explained by the measured features. The overall predictive accuracy for the present model is 64%. In terms of groups wise, 62.9% of the cases were correctly predicted into awake patients and 36.7% of the cases were correctly predicted into unconsciousness patients. Likelihood ratio test shows the contribution of each variable to model (Table 4). All variables are significant independent variables related to the depth of sedation/anesthesia scale.

Table 4. Likelihood ratio tests.

	-2log likelihood	Chi-square	df	p-value
Intercept	633.821	60.145	5	<0.001
Mean Power (dB)	600.848	27.172	5	<0.001
Mean Frequency (Hz)	619.434	45.758	5	<0.001
Median Frequency (Hz)	599.425	25.749	5	<0.001
Band Power (30–120 Hz) (dB)	588.896	15.220	5	0.009
Peak Frequency (Hz)	600.367	26.691	5	<0.001

4 Discussion

Baselines of the blink reflex components were similar to those in the literature [8, 10]. Areas and amplitudes of the R_1 component of the blink reflex decrease with increasing depth of anesthesia. Loss of R_1 component occurred earlier than the loss of corneal reflex. Latency and duration of the R_1 component cannot be measured from the moment it was abolished, and the amplitude and area of the R_1 components were zero from the moment R_1 component was abolished, which resulted in an insufficient number of data points to continue analyzing the data in the time domain. EMG data was then analyzed in the frequency domain. Several features were extracted and their capacity of predicting the clinical depth of sedation/anesthesia was evaluated by a multinomial logistic regression. As multinomial regression analysis was used to assess the significant independent factors associated with the level of sedation/anesthesia during the induction phase of anesthesia with propofol. The overall predictive accuracy for the model was 64%, suggesting that the model was useful. This accuracy can be improved by using a larger sample size or maybe using a different predictive algorithm, such as Support-Vector Machines. Factors like mean power, mean frequency, median frequency, band power and peak frequency were significant predictors of depth of sedation/anesthesia.

5 Conclusions and Future Work

From a clinical practice point of view, it would be very helpful if one could predict the moment of loss of consciousness. The possibility of prediction for the total amount of propofol required for LOC during general anesthesia may allow to individualize the infusion rate of propofol required to maintain an adequate level of anesthesia during surgery. Loss of R_1 component of the electrically evoked blink reflex showed to have a strong correlation with the estimated effect-site concentrations of propofol. Several features extracted from the blink reflex showed to be good predictor estimates for the level of sedation/anesthesia during the induction phase of anesthesia with propofol.

In the future we intend to add more variables to the model and combinate the existent extracted features with each other. In addition, since a unique, simple stimuli was used to evoke the blink reflex, we intend to stimulate and record four successive blink reflexes within a short period of time and use the mean of four successive signals to perform the analysis in the time and in the frequency domain.

References

1. Ferreira, A., Nunes, C., Ferreira, A.L., Tedim, R., Amorim, P.: Inter-patient variability and predictive factors of propofol requirements and estimated concentrations for loss of consciousness and recovery. J. Neurosurg. Anesthesiol. **27**, 260–261 (2015)
2. Ferreira, A.C., Nunes, C.S., Castro, A.I., Ferreira, A.L., Amorim, P.: Propofol requirements for anesthesia induction show wide individual variability, independently of age, gender, weight and height. In: Proceedings from the American Society of Anesthesiologists (2015)
3. Ferreira, A.L., Nunes, C., Mendes, J.G., Amorim, P.: ¿Contamos actualmente con un método fiable para detectar el momento de pérdida de consciencia durante la inducción de la anestesia? Rev. Esp. Anestesiol. Reanim. **66**, 93–103 (2019)
4. Ferreira, A.L., Nunes, C.S., Gabriel, J., Amorim, P.: The influence of two different drug infusion profiles on the pharmacodynamics model performance. In: Jaffray, D.A. (ed.) World Congress on Medical Physics and Biomedical Engineering, Toronto, Canada, pp. 874–880. Springer, Cham, 7–12 June 2015
5. Ghoneim, M.M.: Incidence of and risk factors for awareness during anaesthesia. Best Pract. Res. Clin. Anaesthesiol. **21**, 327–343 (2007)
6. Mashour, G.A., Shanks, A., Tremper, K.K., Kheterpal, S., Turner, C.R., Ramachandran, S. K., Picton, P., Schueller, C., Morris, M., Vandervest, J.C., Lin, N., Avidan, M.S.: Prevention of intraoperative awareness with explicit recall in an unselected surgical population: a randomized comparative effectiveness trial. Anesthesiology **117**, 717–725 (2012)
7. Mourisse, J., Gerrits, W., Lerou, J., van Egmond, J., Zwarts, M.J., Booij, L.: Electromyographic assessment of blink and corneal reflexes during midazolam administration: useful methods for assessing depth of anesthesia? Acta Anaesthesiol. Scand. **47**, 593–600 (2003)
8. Mourisse, J., Lerou, J., Struys, M., Zwarts, M., Booij, L.: Multi-level approach to anaesthetic effects produced by sevoflurane or propofol in humans: 1. BIS and blink reflex. Br. J. Anaesth. **98**, 737–745 (2007)
9. Preston, D.C., Shapiro, B.E.: Electromyography and neuromuscular disorders : clinical-electrophysiological correlations. Elsevier Saunders (2013)
10. Mourisse, J., Lerou, J., Zwarts, M., Booij, L.: Electromyographic assessment of blink reflexes correlates with a clinical scale of depth of sedation/anaesthesia and BIS during propofol administration. Acta Anaesthesiol. Scand. **48**, 1174–1179 (2004)

A Novel Technique to Trigger High Beta and Low Gamma Activity in Patients with Schizophrenia

Eysteinn Ívarsson[1(✉)], Alec Shaw[2], Aníta Ósk Georgsdóttir[3], Brynja B. Magnúsdóttir[3], Aron D. Jónasson[1], Eric Wassermann[4], Paolo Gargiulo[2], Sigurjón B. Stefansson[1], and Ovidiu C. Banea[1,2]

[1] Clinical Neurophysiology Unit, National University Hospital of Iceland, Reykjavík, Iceland
eysteinn.ivarsson@gmail.com

[2] Reykjavik University/School of Science and Engineering, Biomedical Engineering, Reykjavik, Iceland

[3] Department of Psychology, Reykjavik University/School of Business, Reykjavik, Iceland

[4] Behavioral Neurology Unit, National Institute of Neurological Disorders and Stroke, National Institutes of Health, Bethesda, MD, USA

Abstract. The main purpose of this study was to investigate the relationship between resting state and auditory-motor task electroencephalogram beta and gamma distribution in healthy subjects and subjects with schizophrenia. First, we looked to changes in the resting state EEG distribution in three healthy subjects and in three patients with schizophrenia. We also analyzed high-beta and gamma cortical activity from high-density EEG during a cognitively-driven auditory-motor task in two participants, one normal subject and one diagnosed with schizophrenia. For the auditory-motor task, we asked the participants to press a button using the thumb of both hands independently, during two three-minute sessions. Resting state EEG showed more fragmentation in schizophrenia patients when compared with the healthy participants. During the auditory-motor task, we observed increased cortical fragmentation and clustering during the hand response in the patient and healthy control compared with a resting state EEG. The fragmentation remains stable during both hand response time and reference period located between the command "press" and "do not press" suggesting that this task activates a network which remains similar during the entire task.

Keywords: Schizophrenia · Beta activity · Gamma activity · Auditory-motor task

1 Introduction

Neural oscillations are a fundamental mechanism for enabling coordinated activity during normal brain function and represent a crucial target for schizophrenia research [1]. Sensory and cognitive events evoke superimposed and/or parallel oscillations

J. Henriques et al. (Eds.): MEDICON 2019, IFMBE Proceedings 76, pp. 1064–1070, 2020.
https://doi.org/10.1007/978-3-030-31635-8_129

(multiple oscillations) that are transferred to distributed structures with various degrees of intensity, synchronization, duration and delay [2]. In a study with intracranial recordings, auditory-verbal stimuli elicited augmented gamma-oscillations in the posterior portion of the superior temporal gyrus, whereas hand-motor responses elicited gamma-augmentation in the pre- and postcentral gyri [3]. Gamma activity has been investigated in one of the following paradigms: (1) at rest, (2) during "bottom-up" sensory stimulation, or (3) "top-down" cognitively driven tasks [4]. Based on a comprehensive review of the literature, it was concluded that at rest gamma activity is elevated and that task-driven 'evoked' gamma-band responses are reduced in schizophrenia [4].

The aim of our study is to investigate the resting state EEG cortical beta and gamma distribution and if the auditory-motor (AM) task modulates beta and gamma bands activity in healthy subjects (HS) and in patients with auditory verbal hallucinations (AVH). It has been suggested that schizophrenic symptoms can be explained by over-arousal causing depression in neural activity or inverted-U relationship between performance and arousal, also called Yerkes-Dodson Law [5, 6]. Our hypothesis is based on this suggestion and we expect that in schizophrenia the high-beta and gamma activity will break up in clusters and decrease during a task requiring attention, whereas in normal subjects these bands activity will increase.

2 Methods

2.1 Participants

Six participants were selectively recruited through convenience sampling. Three participants (2 male; mean age = 36; range = 30–48) were patients diagnosed with schizophrenia, while the remaining three were healthy controls (2 male; mean age = 26; range = 25–28). Patients were recruited through the National Hospital database of diagnosed schizophrenia patients, following the ICD-10 schizophrenia classification (F20). Only those still experiencing persistent auditory verbal hallucinations after finishing at least two 8 week drug prescription treatments were selected.

2.2 Design and Procedure

An EEG recording was acquired while subjects completed a resting baseline measurement and an auditory-motor (AM) task. The baseline measurement lasted a total of 4 min in two parts: 1 min with eyes closed, and 3 min with eyes opened. A recording was collected for two AM tasks, each lasting three minutes. The participants were awake and comfortably seated in front of a computer screen and were asked to held a button-controller in one hand, with the hand resting on their thigh. During the task, verbal commands were presented, adjusted to a comfortable volume (Nagasawa et al. 2010). Each subject was instructed to press the button using the thumb when a pre-recorded verbal command saying "press" (*ýttu*) was given, and to not press the button when a verbal command saying "do not press" (*ekki ýta*) was given. Each subject completed the AM task for both hands. The task contained 40 trials; 20 auditory-verbal commands saying "press" and 20 commands saying "do not press", presented in a pseudorandom order, following the latin square method.

EEG Recording and Preprocessing. The EEG was recorded using high-density 256 channel ANT Neuro eego sports system with an electrooculogram (EOG) electrode placed below the right eye, and a ground electrode placed on the left side of the neck. The EEG was recorded at a sampling rate of 1024 Hz. The raw EEG was exported and analyzed using the EEGLAB toolbox (v14.2.2) in MATLAB 2018b. EEG signals were first notch filtered at 50 Hz, and a later band-pass filter was applied between 0.1 and 80 Hz [7]. Eye movements and muscle artifacts were removed by visual inspection. Bad channels were identified by automatic detection using kurtosis, or spectrum measures and later interpolated using the spherical method. The length of the resulting resting state pre-processed data was 40 s $\pm$ 10 s for each participant. Finally, the remaining artifact-free continuous data was analyzed.

Data Analysis. First, we analyzed 40 $\pm$ 10 s resting state EEG with eyes closed for three patients with schizophrenia and we compared it with the EEG of three healthy control participants. For the auditory motor task, epochs related to the cognitively driven "motor cortical activation" (MCA) were segmented between −500 ms to +500 ms (1000 ms) relative to the onset of the button code (hand reaction) seen in the EEG trace. The reference period was set between +1500 ms to +2500 ms post verbal command or auditory stimulus onset (1000 ms). Finally, the remaining artifact-free trials were analyzed. Here we show preliminary results of motor cortical activation (MCA) for one healthy subject and one patient with auditory verbal hallucinations (AVH), using power spectral density (PSD) and topographical frequency maps including both hands sensory-motor regions. We looked at specific frequencies related to motor reaction: low beta (13–20 Hz), high beta (20–30 Hz), low gamma (30–45 Hz) and high gamma (45–80 Hz). The brain maps showed topographical changes on specific frequencies: 16, 25, 35 and 65 Hz. We compared data with the reference period and with the resting state EEG.

3 Results

Resting state EEG recorded with eyes closed showed topographical and clustering differences of beta and gamma activity between healthy (Fig. 1) and schizophrenia participants (Fig. 2.). In patients with schizophrenia it seems that the EEG is more fragmented. During AM task beta and gamma activity is more visible (higher Z-score)

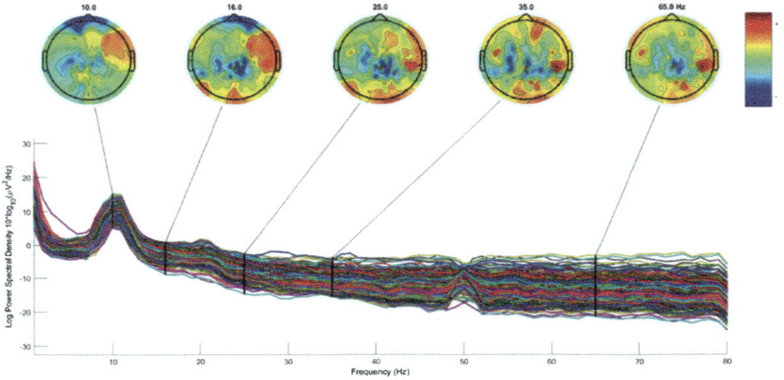

Fig. 1. Power spectral density and specific topographical maps of ***resting state EEG*** averaged for three healthy subjects. 256 channels EEG 40–50 s records with eyes closed.

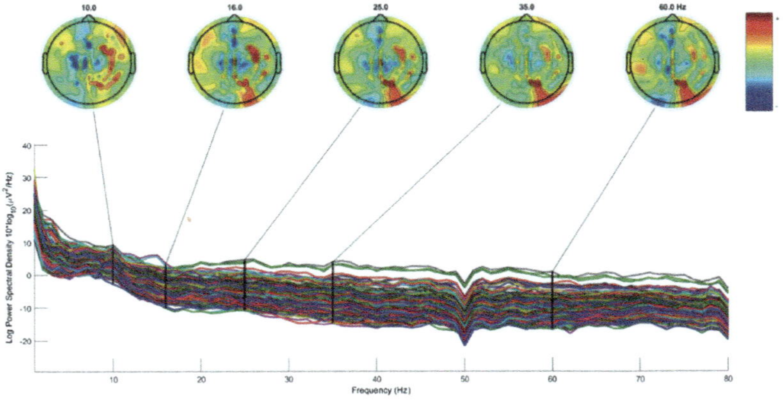

Fig. 2. Power spectral density and specific topographical maps of ***resting state EEG*** averaged for three patients with schizophrenia. 256 channels EEG 40–50 s records with eyes closed.

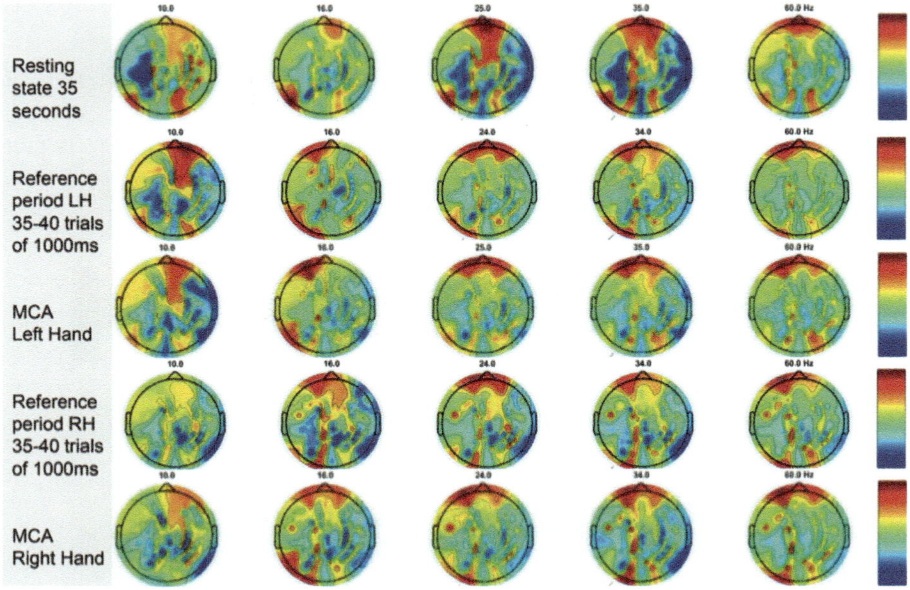

Fig. 3. Auditory-motor task in a healthy participant. Cortical EEG fragmentation is showing similarities between the reference period and hand response during both lateralized tasks. Overall, high beta and gamma activity is reduced during the auditory-motor task.

and more clustered than during resting state EEG for the analyzed healthy participant and this activity remained stable for both bands even during the reference period (Fig. 3). The schizophrenia participant showed reduction of high beta and gamma activity during the AM task, primarily when the task was performed with the dominant hand. The laterality of the cortical activity during left or right hand reaction was not visible (Fig. 4).

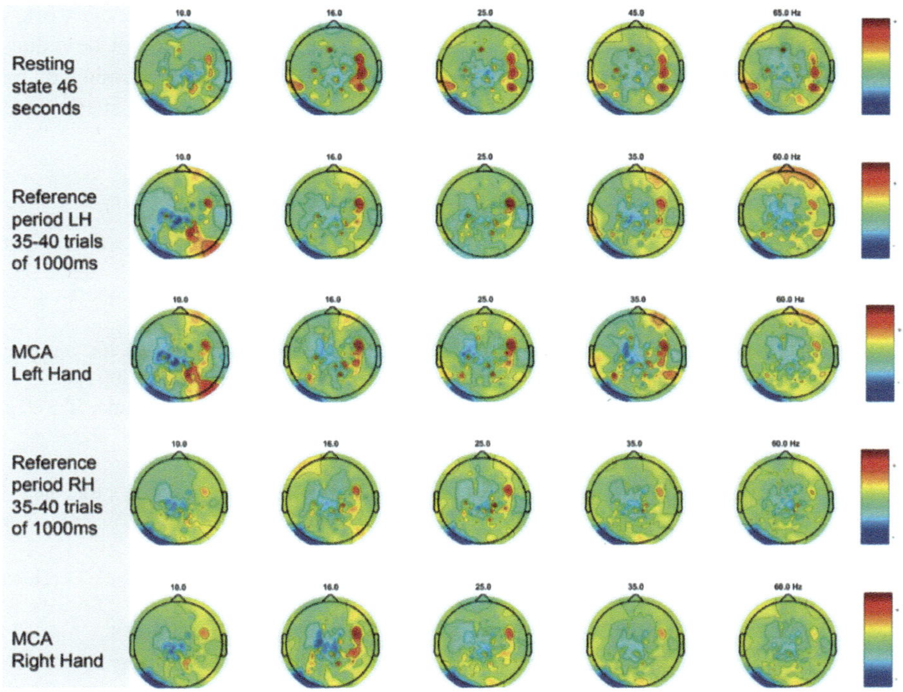

Fig. 4. Auditory-motor task in patient with schizophrenia and auditory verbal hallucinations. Cortical EEG fragmentation is showing similarities between the reference period and hand reaction time during both lateralized tasks. Overall, high beta and gamma activity is reduced during the auditory-motor task.

4 Discussions

The aim of this experiment was to investigate whether there are differences in cortical activity between healthy subjects and patients with schizophrenia during resting state and during an auditory-motor task. Furthermore, we wanted to investigate if the auditory-motor task would modulate beta and gamma activity. Our hypotheses were that: (1) The resting state EEG shows differences of cortical activity in beta and gamma frequencies between healthy control participants and patients with schizophrenia, being more clustered in schizophrenia patients and (2) The auditory-motor task would reduce beta and gamma cortical distribution, compared to the resting state in schizophrenia patient when compared with the healthy control participant. The results are congruent with our hypotheses, showing more clustered and divided cortical distribution in patients with schizophrenia and a reduced activity during the auditory-motor task, even during the reference period between the auditory commands. The activity is stable during the task, with little lateralization effect of the hand responses.

These results indicate that the auditory-motor task triggers a stable network of attention that is held active throughout the task. Cortical activity seemed to be reduced in the schizophrenia participant during the AM task. However, larger number of participants and further analysis with different quantification and clustering methods are needed to test our hypotheses.

References

1. Uhlhaas, P.J., Singer, W.: Abnormal neural oscillations and synchrony in schizophrenia. Nat. Rev. Neurosci. **11**, 100–113 (2010)
2. Başar, E., Başar-Eroglu, C., Karakaş, S., Schürmann, M.: Gamma, alpha, delta, and theta oscillations govern cognitive processes. Int. J. Psychophysiol. **39**, 241–248 (2001)
3. Nagasawa, T., Rothermel, R., Juhász, C., Fukuda, M., Nishida, M., Akiyama, T., et al.: Cortical gamma-oscillations modulated by auditory-motor tasks-intracranial recording in patients with epilepsy. Hum. Brain Mapp. **31**, 1627–1642 (2010)
4. Gandal, M.J., Edgar, J.C., Klook, K., Siegel, S.J.: Gamma synchrony: towards a translational biomarker for the treatment-resistant symptoms of schizophrenia. Neuropharmacology **62**, 1504–1518 (2012)
5. Grossberg, S.: The imbalanced brain: from normal behavior to schizophrenia. Biol. Psychiatry pp. 81–98 (2000). https://doi.org/10.1016/s0006-3223(00)00903-3
6. Yerkes, R.M., Dodson, J.D.: The relation of strength of stimulus to rapidity of habit-formation. J. Comp. Neurol. Psychol. pp. 459–482 (1908). https://doi.org/10.1002/cne.920180503
7. Hall, M.-H., Taylor, G., Salisbury, D.F., Levy, D.L.: Sensory gating event–related potentials and oscillations in schizophrenia patients and their unaffected relatives. Schizophr. Bull. pp. 1187–1199 (2011). https://doi.org/10.1093/schbul/sbq027

P50 and P300 Event Related Potentials in Patients with Schizophrenia Recorded from High-Density EEG

Ovidiu C. Banea[1,2](✉), Elena Pegolo[2], Sara Marcu[2], Rún Friðriksdóttir[2], Eysteinn Ívarsson[1], Aron D. Jónasson[1], Viktor D. Jónasson[3], Brynja B. Magnusdóttir[3], Magnús Haraldsson[4], Eric Wassermann[5], and Paolo Gargiulo[2]

[1] National University Hospital of Iceland, Clinical Neurophysiology Unit, Reykjavík, Iceland
ovidiubanea@gmail.com
[2] School of Science and Engineering, Biomedical Engineering, Reykjavik University, Reykjavik, Iceland
[3] School of Business, Department of Psychology, Reykjavik University, Reykjavik, Iceland
[4] Icelandic Psychiatric Hospital Kleppur, National University Hospital of Iceland, Reykjavík, Iceland
[5] National Institute of Neurological, Disorders and Stroke, National Institutes of Health/Behavioral Neurology Unit, Bethesda, MD, USA

Abstract. The main objectives of this study were to describe P50 and P300 cortical topography in patients with schizophrenia and to define a robust methodology of signal quantification using high density EEG. Within a clinical trial in which patients with schizophrenia and auditory verbal hallucinations were submitted to 10 days rTMS treatment, P50 and P300 were investigated as possible neurophysiological measurements to assess treatment effectiveness together with psychometric scales like Psychotic symptom rating scale (PSY-RATS), depression, anxiety and stress scale (DASS) and Quality of Life scale (QoL). Here we describe the technique of collecting P50 and P300 using high-density EEG and we reproduce the preliminary data of P300 in one healthy subject and P50 in two patients and two healthy subjects.

Keywords: P50 · P300 · Schizophrenia · High-density EEG

1 Introduction

Event-related potentials (ERP) provide safe and noninvasive approach to study psychophysiological correlates of mental processes [1]. A meta-analysis confirmed the existence of P300 and P50 deficits in schizophrenia of similar magnitude as findings reported in neuroimaging and neuropsychology [2]. The auditory P300 is a time-locked ERP component indexing attentional resources allocated to the target stimuli and/or the context updating. The P300 peak is defined as the largest positive deflection in the time range from 270 to 470 ms [3]. P50 is a component at the most positive peak between 30 and 70 ms post-stimulus (auditory) onset [4].

J. Henriques et al. (Eds.): MEDICON 2019, IFMBE Proceedings 76, pp. 1071–1077, 2020.
https://doi.org/10.1007/978-3-030-31635-8_130

While the examination of P300 topography in chronically ill schizophrenic and psychotic bipolar patients showed that schizophrenia was associated with a specific left-lateralized posterior abnormality, suggesting underlying posterior temporal lobe pathology with abnormalities of a generator located in the left superior temporal gyrus (STG) [5, 6], the cerebral topography of P50 remains largely unknown [7]. The information processing dysfunction in schizophrenia patients has been associated with a morphological abnormality in the brain [8]. EEG source analysis and the standard paradigm to clarify the neural structures associated with P50 have been used to assess the previously suggested generators of sensory gating. These were found to be STG, hippocampus, dorsolateral prefrontal cortex (DLPFC) and thalamus [9]. In a double auditory stimulus paradigm, the P50 response amplitude of the second (S2) or test stimulus [to that of the first (S1) or conditioned stimulus] demonstrates sensory gating. Deficits in sensory gating are an important endophenotype for schizophrenia [4]. We aimed to describe P50 and P300 in patients with schizophrenia and auditory verbal hallucinations using 256-channel "high-density" EEG. In this work we present data on P300 and P50 topography with signal processing of P300 in one healthy subject and with changes of P50 in baseline conditions from two healthy subjects and two patients with schizophrenia.

2 Materials and Methods

2.1 AVH TMS Clinical Trial in Iceland

The research is part of the ongoing AVH TMS Icelandic project. The goal of this clinical applied research is to assess the effectiveness of repetitive transcranial magnetic stimulation (rTMS) as a treatment of schizophrenic patients with persistent auditory verbal hallucinations (AVH). Sensory gating and cognitive functions using P50 and P300 waveforms are calculated before and after the rTMS treatment. Here we present the preliminary results of P300 topography in one patient and P50 in two patients. Difference of P50 amplitude (S1-S2) and changes of P50 ratio (S2/S1) in baseline conditions of two patients were compared with the results obtained from two healthy control participants.

A total of eighteen patients (range: 18–60 years) will be recruited from the psychiatric wards and outpatient clinics through the National Hospital database of diagnosed schizophrenia patients, following the ICD-10 schizophrenia classification (F20). Only those still experiencing persistent auditory verbal hallucinations after finishing at least two 6–8 week drug prescription treatments will be selected. Permission from the Health Research Ethics Committee at the University Hospital of Iceland was obtained (approval no. 21.2018). P50 and P300 topography in baseline condition for all eighteen patients will be further investigated in two different groups of 9 patients. One group will receive 10 sessions rTMS sham treatment over vertex and the second group will receive treatment with rTMS over left T3P3 scalp location. At the end, baseline conditions will be compared with nine healthy controls.

2.2 EEG Recording and Analysis

The ERP signals were obtained from 256-channel EEG waveguard cap using eego mylab system (ANT Neuro, Netherlands). We grouped the electrodes in seven regions of interest (ROI) with 15 electrodes in each region:

(1) Left anterior L1E-L5E, L1D-L5D, L1C-L5C or 88, 89, 90, 91, 92, 79, 80, 81, 82, 83, 71, 72, 73, 74, 75;
(2) Left posterior L5F-L8F, L6E-L10E, L6D-L9D, L6B-L7B or 102, 103, 104, 105, 93, 94, 95, 96, 97, 84, 85, 86, 87, 69, 70;
(3) Medial anterior L3L-L7L, Z3Z-Z7Z, R3R-R7R or 42, 43 44, 45, 46, 120 121, 122 123, 124, 169, 170, 171, 172, 173;
(4) Medial central L8L-L12L, Z8Z-Z12Z, R8R-R12R or 47, 48, 49, 50, 51, 125, 126, 127, 245, 246, 174, 175, 176, 177, 178;
(5) Medial posterior L13L-L17L, Z13Z-Z17Z, R13R-R17R or 52, 53, 54, 55, 56, 247, 248, 249, 250, 251, 179, 180, 181, 182, 183;
(6) Right anterior R1E-R5E, R1D-R5D, R1C-R5C or 215, 216, 217, 218, 219, 206, 207, 208, 209, 210, 198, 199, 200, 201, 202;
(7) Right posterior R5F-R8F, R6E-R10E, R6D-R9D, R6B-R7B or 229, 230, 231, 232, 220, 221, 222, 223, 224, 211, 212, 213, 214, 196, 197.

The EEG was recorded at a sampling rate of 1024 Hz with an electrooculogram (EOG) electrode placed below the right eye, and a ground electrode placed on the left side of the neck. Electrode resistances were set less than 10 kΩ and referenced to the average of left and right mastoid electrodes (R19R, L19L). The raw EEG was exported and analyzed using Brainstorm [10] and MATLAB 2018b. EEG signals were first notch filtered at 50 Hz, and a later band-pass filter was applied between 0.1 and 80 Hz [11].

P300. In our study P300 response was measured with an auditory oddball paradigm attention task. The recordings were carried out between 11AM and 14 PM. The subjects were sitting in a comfortable chair with their eyes closed. The frequent (F) and the rare (R) auditory stimuli were presented binaurally through headphones at an inter-stimulus interval between tones of constant 1.1 s. For each subject there was 1 trial of 160 tones which will occurred randomly with a probability of 0.2 [12]. We instructed the participants to pay attention at the rare stimuli without counting or moving a finger. The task was performed with eyes closed. The signals were digitized for an epoch of 1000 ms starting 100 ms prior to the presentation of each auditory stimulus (−100 ms to +900 ms). Data analysis was performed with Brainstorm [10] and EEGLAB (v14.1.2b) toolbox of MATLAB 2018b. In this work we present N100-P300 complex signals over the seven scalp ROI of a healthy control participant.

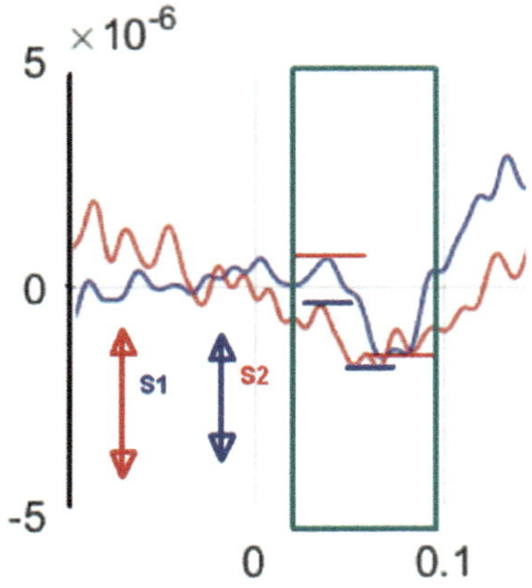

Fig. 1. P50 analysis. Testing stimulus response amplitude S2 (red trace) and conditioning stimulus potential amplitude S1 (blue trace) in 30–70 ms time range are used to determine sensory gating.

P50. The paired-click paradigm was performed to elicit the P50 component. A pure tone (1500 Hz, 6-ms duration) was used as the click sound and presented during a 500-ms interval through headphones. We presented 150 paired stimuli in 5 blocks with interstimulus interval of 10 s, which provided 25 min of EEG measurement. In consideration of participant load and ear comfort that could influence EEG measurement, we instructed participants to watch a silent film. The signals were digitized for an epoch of 500 ms starting 100 ms prior to the presentation of each auditory stimulus (−100 ms to +400 ms).

Individual trials were rejected when EEG voltage was greater than ±80 μV, indicative of excessive muscle activity, eye movements, or other artifacts. We measured the peak-to-peak P50 amplitude from a preceding negative trough to the positive peak at 30–70 ms range from the stimulus onset (Fig. 1). Sensory gating was measured with the difference between testing stimulus (S2) and conditioning stimulus (S1) and the P50 ratio was calculated as the test stimulus response divided by the conditioning stimulus response. For this current study we present data of four participants, two patients diagnosed with schizophrenia and two healthy controls.

3 Results

3.1 P300

The projection of ERP over 256 channel EEG is difficult to quantify and this is one of the limits to measure the rTMS changes on ERPs. To correct this limitation we grouped the scalp electrodes into seven ROI, each region containing 15 electrodes (Fig. 2).

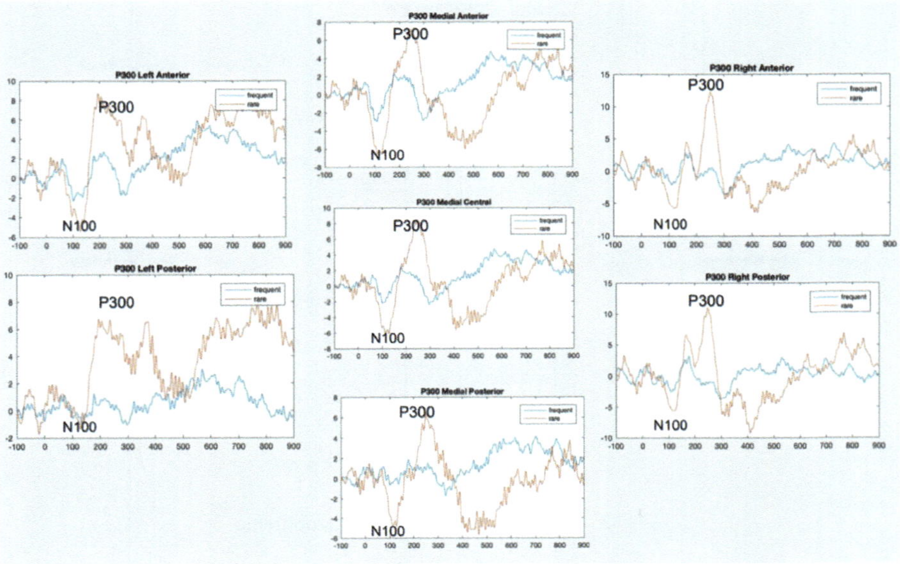

Fig. 2. P300 signals related to the rare auditory stimulus in a healthy control participant (brown). Each N100-P300 complex represents the averaged signal of 15 electrodes. The reorganization of 256 sensors signal into seven regions representing a total of 105 channels might represent an easier tool for P300 amplitude and latency quantification.

3.2 P50

S1 and S2 signals were averaged for the 7 ROI (Fig. 3) in two patients and two healthy controls. S1, S2 amplitude differences S1–S2 (Fig. 4) and S2:S1 ratios (Fig. 5) were calculated. Higher difference of S1–S2 with S1 larger than S2 and reduced S2:S1 ratio explained by reduced S2 amplitude would be correlated with sensory gating.

Patients showed reduced gating in comparison with healthy controls on left posterior, medial anterior and right posterior regions. Left anterior and left posterior regions showed the highest ratio (less sensory gating) in schizophrenia patients in comparison with the healthy controls. Healthy participants showed better response over right posterior or temporo - parietal region (lower value of S2:S1 ratio).

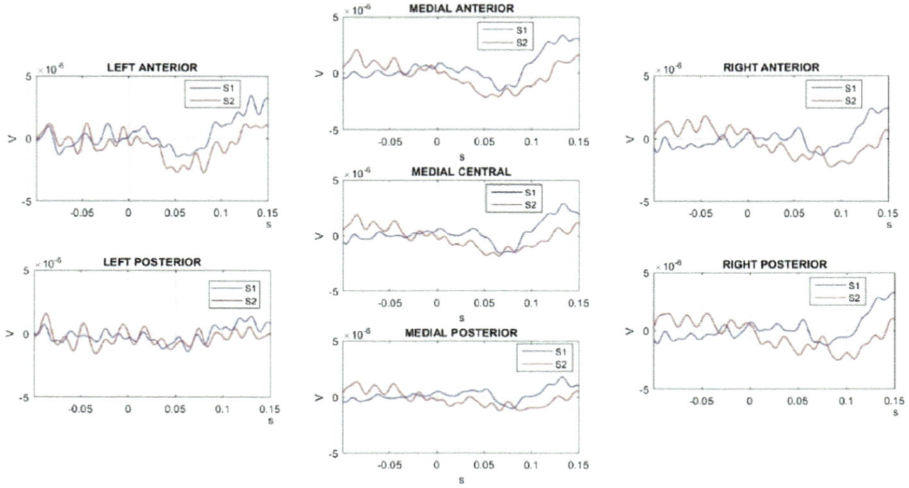

Fig. 3. P50 in a patient with schizophrenia.

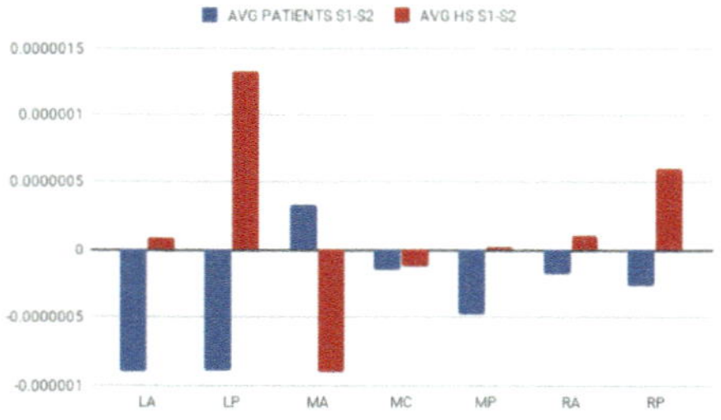

Fig. 4. P50 calculated with amplitude difference S1-S2.

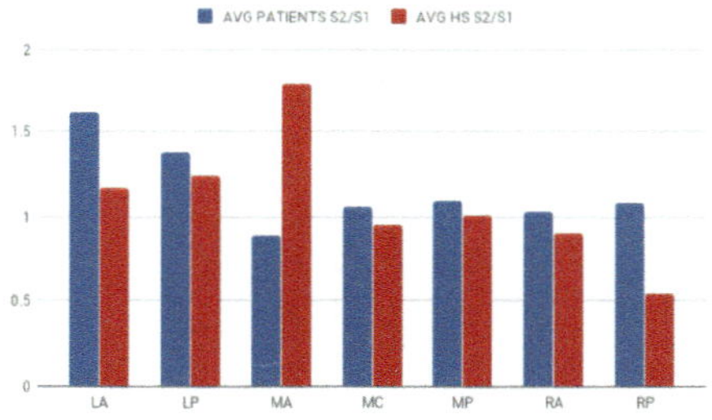

Fig. 5. P50 calculated with S2:S1 ratio.

4 Conclusions

P50 and P300 recorded at scalp level with high density EEG showed high inter-individual and intra-individual variability of signal amplitude, latencies and sensors artifacts. Topographical distribution and best signal localization are serious limits if these waveforms are used for large numbers of participants. Modern quantitative EEG and digitization of signal with large data processing and group analysis might be of interest but component analysis techniques and mixing data of several participants should be carefully considered in order to process all necessary data. To correct these limits we suggested to divide the EEG data in seven scalp regions of interest and we averaged the waves of 15 electrodes for each region. P300 in a healthy control participant showed clear N100-P300 complexes in all seven regions. Preliminary data on P50 analyzed for two patients with schizophrenia showed higher S2:S1 ratio over left anterior and left posterior regions in comparison with the healthy controls suggesting that these regions might be functionally affected.

Greater number of patients and healthy controls will be analyzed following the presented methodology in order to test these assumptions and preliminary results. Further, N100-P300 complex amplitude, P300 latency and P50 S2:S1 ratio will be used as neurophysiological markers to assess rTMS treatment effectiveness in patients with schizophrenia and persisting auditory verbal hallucinations.

References

1. Sur, S., Sinha, V.K.: Event-related potential: an overview. Ind. Psychiatry J. **18**, 70–73 (2009)
2. Bramon, E., Rabe-Hesketh, S., Sham, P., Murray, R.M., Frangou, S.: Meta-analysis of the P300 and P50 waveforms in schizophrenia. Schizophr. Res. **70**, 315–329 (2004)
3. Strik, W.K., Dierks, T., Franzek, E., Stöber, G., Maurer, K.: P300 in schizophrenia: interactions between amplitudes and topography. Biol. Psychiatry **35**, 850–856 (1994)
4. Toyomaki, A., Hashimoto, N., Kako, Y., Tomimatsu, Y., Koyama, T., Kusumi, I.: Different P50 sensory gating measures reflect different cognitive dysfunctions in schizophrenia. Schizophr. Res. Cogn. **2**, 166–169 (2015)
5. Morstyn, R.: Altered P300 topography in schizophrenia. Arch. Gen. Psychiatry, 729 p. (1983). https://doi.org/10.1001/archpsyc.1983.01790060027003
6. Salisbury, D.F., Shenton, M.E., McCarley, R.W., Yurgelun-Todd, D.A., Tohen, M.: Asymmetrical P3 topography in first episode schizophrenia-like psychosis. Schizophr. Res., 184–185 (1995). https://doi.org/10.1016/0920-9964(95)95568-t
7. Kurthen, M., Trautner, P., Rosburg, T., Grunwald, T., Dietl, T., Kühn, K.-U., et al.: Towards a functional topography of sensory gating areas: invasive P50 recording and electrical stimulation mapping in epilepsy surgery candidates. Psychiatry Res. **155**, 121–133 (2007)
8. Kawasaki, Y., Maeda, Y., Higashima, M., Nagasawa, T., Koshino, Y., Suzuki, M., et al.: Reduced auditory P300 amplitude, medial temporal volume reduction and psychopathology in schizophrenia. Schizophr. Res. **26**, 107–115 (1997)
9. Williams, T.J., Nuechterlein, K.H., Subotnik, K.L., Yee, C.M.: Distinct neural generators of sensory gating in schizophrenia. Psychophysiology **48**, 470–478 (2011)
10. Tadel, F., Baillet, S., Mosher, J.C., Pantazis, D., Leahy, R.M.: Brainstorm: a user-friendly application for MEG/EEG analysis. Comput. Intell. Neurosci. **2011**, 879716 (2011)
11. Hall, M.-H., Taylor, G., Salisbury, D.F., Levy, D.L.: Sensory gating event–related potentials and oscillations in schizophrenia patients and their unaffected relatives. Schizophr. Bull., 1187–1199 (2011). https://doi.org/10.1093/schbul/sbq027
12. Stefánsson, S.B., Jónsdóttir, T.J.: Auditory event-related potentials, auditory digit span, and clinical symptoms in chronic schizophrenic men on neuroleptic medication. Biol. Psychiatry **40**, 19–27 (1996)

Total Cost of Ownership as a Management Tool for Medical Devices Planning: A Case Study of a ST-Analyzer in Perinatology

Petra Hospodková(✉), Petr Kudrna, and Vladimír Rogalewicz

CzechHTA, Department of Biomedical Engineering, Czech Technical University in Prague, Nám. Sítná 3105, 272 01 Kladno, Czech Republic
petra.hospodkova@fbmi.cvut.cz

Abstract. Total cost of ownership is studied as a part of Hospital-based HTA analyses for decision-making in the area of purchasing, implementation and/or disinvestment of technologies or interventions in hospitals. An estimation of Total cost of ownership for a D41 ST-analyzer for a neonatal department of a university hospital is presented as a case study. The choice of the time horizon and cost items to be involved is discussed. The necessity of discounting all cost items is accented. The method is fitting for estimation of life-cycle resource requirements in case of asset acquisition through donation or public subsidy.

Keywords: Total cost of ownership · TCO · ST-Analyser · HB-HTA

1 Introduction

Recently, a growing interest in hospital-based HTA (HB-HTA) methods has been observed, since it can provide an important support for decision-making in the area of purchasing, implementation and/or disinvestment of technologies or interventions in hospitals [1, 2]. HB-HTA utilizes a large selection of diverse methods from multiple-criteria decision analysis and expert group opinion searching over mathematical modelling [3] to different types of cost analyses (CEA, CUA, CBA, CCA) [4]. It is desirable to introduce further methods that would support HB-HTA applications. Our paper shows the potential of the Total Cost of Ownership (TCO) calculation method for HB-HTA studies.

TCO is based on calculation/estimation of total costs connected with an investment during the whole life of the procured item. It was shown that the purchase price is not the decisive factor for the purchase, it is usually just a small part of the total cost, about 20–25% [5, 6]. Purchase of the most expensive variant does not guarantee the lowest operational costs [7, 8]. The method is suitable also in the case of considering procurement of new modern technologies to a health facility through a donorship [9].

The objective of this paper is an estimation of TCO for a cardiotocograph (CTG) with the addition of a ST segment changes analysis of the fetal ECG, particularly the STAN S41 device produced by Neoventa Medical (Sweden). The TCO analysis was calculated for a perinatology unit of a Czech university hospital. That technology allows for an early detection and warning in case of imminent fetus

J. Henriques et al. (Eds.): MEDICON 2019, IFMBE Proceedings 76, pp. 1078–1084, 2020.
https://doi.org/10.1007/978-3-030-31635-8_131

hypoxia. The apparatus is predestined for round-the-clock service for both planned and acute examinations.

2 Methods

The crucial question before any TCO calculation is the choice of the time horizon. The first option is the depreciable life representing the time of asset depreciation, while this time parameter depends on the respective national law. Furthermore, we can consider the economic life, in fact the number of years when the asset yield surpasses the costs of its operation and maintenance. The service life represents number of years that the asset is actually in service. In some cases, the life horizon is fixed to be a predefined number of years, e.g. 3, 5 or 10 years; e.g. if the actual economic life and/or service life of the acquisition are uncertain [5]. All calculations in this case study are done for the span of 5 years (this time horizon was chosen by a managerial decision).

The process of the TCO analysis can be summarized in five basic steps (sse Fig. 1).

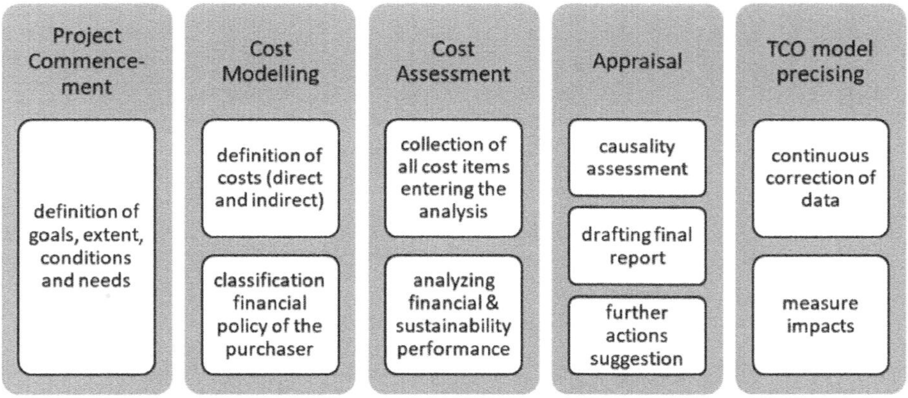

Fig. 1. Main steps of the TCO analysis [10, 11]

The Total Cost of Ownership over the investment's life time can be calculated using the following formula [12]:

$$TCO = Ca + Cc + Co + Cm + Cp + Cd, \tag{1}$$

where *TCO* means the Total Cost of Ownership, *Ca* the costs of acquisition, *Cc* the costs of commissioning, *Co* the costs of operation, *Cm* the costs of maintenance, *Cp* the costs of production[1], and *Cd* the removal and disposal costs. Selected cost items within the TCO model are shown in Table 1 [13, 14].

[1] *Cp* = costs of production—includes production losses, costs of quality management, environmental costs and costs of redundancy. This item is not considered in our research.

Although the TCO model analyses costs in the course of time, incorporation of the time factor in the form of a discount rate has been rather isolated in specialized literature [11, 12]. For this study, a 4% real financial discount rate was applied, which is recommended by the European Commission for public investment projects co-funded from European funds [13]. The calculation of the discounted TCO proceeds according to the formula (2):

$$dTCO = \sum_{i=0}^{t} \frac{(CO)_i}{(1+r)^t}, \qquad (2)$$

where *dTCO* means Discounted Total Cost of Ownership, *CO* are costs of ownership in individual years, *i* is the *i*-th year of life of the asset, *t* is the life span of the investment, and *r* is the discount rate (expressed as a fraction).

Table 1. Cost categories for TCO calculation

Cost category	Ca	Cc	Co	Cm	Cd
Selected items within the category	Purchase price Costs of acquisition, inspection and documentation	Costs of installation Costs of training Costs of IT Costs of pre-installation, refurbishing Costs of testing	Operating personnel Facility-related costs (power supply, rent, etc.) Consumables and pharma-ceuticals costs Cleaning and waste management costs Depreciations	Service contract costs Costs of service and repairs not included in the service contract Software updates costs Costs of spare parts and attachments	Costs of machine disposal Costs of environmentally-friendly disposal

3 Results

The average time necessary for one STAN examination was estimated at 40 min. This includes preparation of the apparatus for examination, input of patient data, application of electrodes and probes, the examination itself, and then removal of the electrodes, and finally switching off the apparatus. The expected number of required examinations on the ST-analyser in this department was estimated at 2,000 per year.

The costs of acquisition consists of the purchase price of the device that covers the price of the basic unit, colour LCD display, simultaneous digital and graphic imaging, built-in CTG analysis module, fetal movement monitoring, maternal parameters monitoring, thermal printer, ST analysis software, carriage with attachments, US and TOCO probes, belt, and NIBP cuff in two sizes. The purchase price of the apparatus covers also the costs of its transportation. The costs of commissioning includes the proper installation and staff training. This category normally covers also building adaptations; however, the STAN S41 apparatus is mobile requiing no adaptations of the rooms. Costs of connecting the apparatus into the CIS are also included. The costs of maintenance cover e.g. regular safety inspections carried out according to the producer´s recommendation once a year. No software updates are necessary in this apparatus. Thus, costs related to service contracts are not taken into account due to a negligible price of spare parts. The costs of operation cover human resources[2], power supply costs[3], consumables[4], and costs of cleaning services and waste disposal. Removal and disposal costs include just costs of the disposal, as there is no need for dismantling the apparatus by special personnel due to its mobile character.

A decomposition of cost items for the calculation of TCO in the case of the STAN S41 analyser is shown in Table 2. The discounted value dTCO was figured out to be CZK 8,778,753 (approx. EUR 342,345).

Table 2. Cost items for STAN S41 in a 5-year horizon (in CZK) (The 2018 average exchange rate was EUR 1=CZK 25.643 (http://www.cnb.cz/en)).

Cost category	Year					
	1	2	3	4	5	Total
Costs of acquisition (including VAT)	**1 317 127**					**1 317 127**
Cost of user training	12 000	6 000	6 000	6 000	6 000	36 000
Costs of commissioning	**12 000**	**6 000**	**6 000**	**6 000**	**6 000**	**36 000**
HR costs	1 262 209	1 262 209	1 262 209	1 262 209	1 262 209	6 311 045
Power supply costs	150	150	150	150	150	750
Consumables costs	830 000	830 000	830 000	830 000	830 000	4 150 000
Depreciations	−144 884	−293 061	−293 061	−293 061	−293 061	−1 317 128
Costs of cleaning and waste management	12 000	12 000	12 000	12 000	12 000	60 000
Costs of operation	**1 959 475**	**1 811 298**	**1 811 298**	**1 811 298**	**1 811 298**	**9 204 667**
Costs of service and repairs not included in the service contract	10 285	10 285	10 285	10 285	10 285	51 425
Costs of spare parts and attachments			33 238		33 238	66 476
Costs of maintenance	**10 285**	**10 285**	**43 523**	**10 285**	**43 523**	**117 901**
Removal and disposal costs					**5 000**	**5 000**
Total TCO	**3 298 887**	**1 827 583**	**1 860 821**	**1 827 583**	**1 865 821**	**10 680 695**

[2] Costs of HR were calculated according to the 2018 salary tables, the estimated number of examinations per year (2,000), and the average time of one examination (40 min).

[3] The working time of the apparatus is 1,333 h, its power input is 46 W, and the price of electricity of the General University Hospital in Prague that was, according to the 2016 Annual Report, equal to CZK 2.3 for 1 kWh incl. VAT.

[4] The costs take into account the price of recording sheets and colours for the printer, ultrasound gels, electrodes, disinfection degreasing agents, etc.

4 Discussion and Conclusion

Hospital-based HTA (HB-HTA) consists in implementation of assessment activities "in" or "for" hospitals; hence, it covers processes and methods supporting organization and execution of health technology assessment (HTA) at the level of individual hospitals. This process is multidisciplinary, systematic and evidence-based. It is aimed at supporting hospital management in making strategic decisions, above all about expensive medical devices [2, 15, 16]. HB-HTA objectives and methods thus differ from the classic utilization of HTA at the national regulator level. Nonetheless, the core of each HB-HTA study is a cost analysis.

The TCO approach proved to belong to useful managerial tools in the fields of budgeting and planning, asset life cycle management, prioritizing capital purchase proposals, vendor selection etc. [5]. Its combination with methods the multiple-criteria decision-making is promising, particularly for HB-HTA. Nevertheless, application possibilities of the TCO have not been fully utilized yet. As concerns the field of healthcare, the TCO method was applied there by Weber et al. [17], who combined TCO with Activity Based Costing, and implied the resulting method in production of state-of-the-art medical devices.

The Reactiondata company has focused on the development of a standardized TCO model to be used hospital financial and/or operation directors by for planning of investments in electronic health record systems [18]. Chakravarty et al. [13] published a TCO analysis comparing a purchase of a new 3T MR system and an upgrade of the existing 1.5T MR, concluding that the TCO is a useful method for assessing alternative options and decision-making in the field of capital investments in healthcare.

According to the relevant literature, the breakdown of cost items for the purpose of the TCO calculations differs case to case. One possibility is the division to obvious costs (the actual price and maintance costs) and hidden costs (acquisition costs, upgrade, refurbishing costs, operating costs, change management costs, infrastucture support costs, environmental impact costs, insurance, security costs, setup and deployment costs, financing costs and depreciation expences)[5] [5]. Morfonios et al. [14] and Chakravarty et al. [13] usually apply the sum of acquisition costs, maintenance and repair costs, and disposal costs. Martinko [19] analyses the TCO through a division to process factors (costs related to managing and administering equipment & instrument ownership processes), product factors (costs related to owning instruments & equipment), and productivity factors (costs related to benchmarking and measuring Key Performance Metrics). Figure 2 shows the cost item breakdown in the case of the STAN S41 study.

[5] Negative costs.

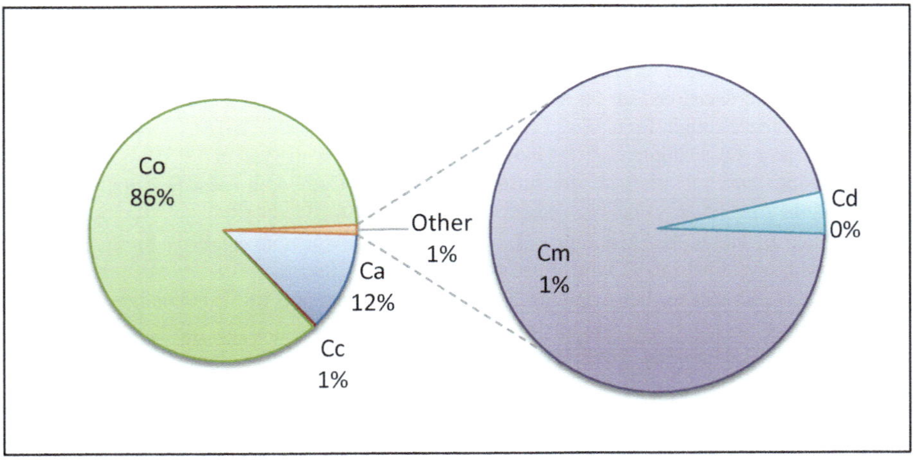

Fig. 2. Breakdown of cost items in the TCO calculation for STAN S41

The TCO approach is beneficial above all in the case of strategic decision-making concerning acquisition of specialized highly unique technologies, where sustainability is the principle decisive factor. Especially if the technology comes on the basis of a subsidy or donation, TCO can indicate the likely financial burden linked with its operation and maintenance during its lifetime.

References

1. Gagnon, M.: Hospital-based health technology assessment: developments to date. Pharmacoeconomics **32**, 819–824 (2014)
2. Gagnon, M.P., Desmartis, M., Poder, T., Witteman, W.: Effects and repercussions of local/hospital-based health technology assessment (HTA): a systematic review. Syst. Rev. **3**, 129 (2014)
3. Avdeyev, A., Tabarov, A., Akhetov, A., Shanazarov, N., Kaptagayeva, A., Hailey, D.: OP102 multiple criteria decision analysis in the field of hospital-based health technology assessment. Int. J. Tech. Assess. Health Care **34**, 38 (2018)
4. Kocaman, M., Turgut, C.G., Ozer, E., Cakmak, H.S., Baysu, F., Kahveci, R.: The feasibility of applying the multiple criteria decision analysis for the evaluation of medical devices in the context of hospital-based health technology assessment in Turkey. Value Health **21**, S266 (2018)
5. Schmidt, M.: Total cost of ownership TCO for Assets and Acquisitions. How to Uncover the Hidden Costs of Ownership. Solution Matrix Ltd, Boston (2019). Business Encyclopedia at https://www.business-case-analysis.com/total-cost-of-ownership.html
6. Hospodkova, P., Vochyanova, A.: The application of the total cost of ownership approach to medical equipment-case study in the Czech Republic. In: IUPESM World Congress on Medical Physics and Biomedical Engineering, pp. 361–366 (2016)

7. Saccani, N., Perona, M., Bacchetti, A.: The total cost of ownership of durable consumer goods: a conceptual model and an empirical application. Int. J. Prod. Econ. **183**, 1–13 (2017)
8. Corciova, C., Andritoi, D., Luca, C., Ciorap, R.: Prioritization of medical devices for maintenance decisions. In: 5th International Conference on Advancements of Medicine and Health Care through Technology (MEDITECH), pp. 323–326 (2016)
9. Emmerling, D., Dahinten, A., Malkin, R.A.: Problems with systems of medical equipment provision: an evaluation in Honduras, Rwanda and Cambodia identifies opportunities to strengthen healthcare systems. Health Technol. **8**, 129–135 (2018)
10. Heilala, J., Helin, K., Montonen, J.: Total cost of ownership analysis for modular final assembly systems. Int. J. Prod. Res. **44**, 3967–3988 (2006)
11. Reeve, T., Everdene, B.: Applying Total Cost of Ownership to Sustainability Purchaising. Workbook. Version 1.0. Sustainability Purchasing Network
12. Dabbs, T.: Optimizing Total Cost of Ownership (TCO). White Paper. ITT Corporation, White Plains, N.Y
13. Chakravarty, A., Debnath, J.: Life cycle costing as a decision making tool for technology acquisition in radio-diagnosis. Med. J. Armed Forces India **71**, 38–42 (2015)
14. Morfonios, A., Daphne, K., Filntisis, G., Baltopoulos, G., Pavlos, M.: Economic evaluation of multislice computed tomography scanners through a life cycle cost analysis. Indian J. Appl. Res. **4**, 158–161 (2011)
15. Cicchetti, A., Marchetti, M., Dibidino, R., Corio, M.: Hospital based health technology assessment world-wide survey. Health Technol. Assess. Int., Edmonton (2008)
16. Rogalewicz, V.: Health technology assessment as a tool for medical devices management in hospitals. In: 2015 E-Health and Bioengineering Conference (EHB) (2015)
17. Weber, M., Hiete, M., Lauer, L., Rentz, O.: Low cost country sourcing and its effects on the total cost of ownership structure for a medical devices manufacturer. J. Purch. Supply Manag. **16**, 4–16 (2010)
18. WHITE PAPER: The Total Cost of Ownership of Electronic Health Record Systems. Reaction (2015)
19. Martinko, K.: How Understanding the Total Cost of Ownership of Your Equipment or Instrumentation Can Reduce Costs, Increase Performance and Improve Workforce Productivity (2010)

Cost-Effectiveness Analysis of Selected Methods of Haemostatis Evaluation

Martin Zavadil[1,2(✉)], Michaela Blahýnková[2], Miroslav Selčan[1], and Vladimír Rogalewicz[1]

[1] Department of Biomedical Technology, Faculty of Biomedical Engineering, Czech Technical University in Prague, Kladno, Czech Republic
[2] University Hospital Motol, Prague, Czech Republic
Martin.Zavadil@fnmotol.cz

Abstract. Viscoelastometric tests have a specific position in haemocoagulation techniques. Representatives of these methods are thromboelastography (TEG), and/or rotational thrombo-elastometry (ROTEM). These are fast bedside (POCT) methods based on a measurement of viscoelastic whole blood properties during the formation, stabilization and lysis of a blood clot. It was suggested that these methods could replace standard laboratory tests used up to now. The goal of this paper was to calculate and assess the cost-effectiveness in order to compare the viscoelastometric tests represented by the ROTEM delta device and standard laboratory tests (considered for a comparator) from the point of view of a healthcare provider. For this purpose, the purchase, maintenance and operation costs of both technologies were estimated, and the effects of both technologies were assessed using multiple-criteria decision-making methods. As a result, the use of viscoelastic tests proved to be the more cost-effective option.

Keywords: Haemostasis · Coagulation tests · Viscoelastic methods · Cost-effectiveness

1 Introduction

Treatment of thromboembolic disease carries the risk of unwanted bleeding complications. Therefore, dosing of anticoagulants should be corrected by repeated blood tests. Haemostasis and haemocoagulation tests are widely used in patients after a history of a heart attack or stroke, in patients with sepsis, disseminated intravascular coagulopathy, systemic inflammatory response syndrome, and/or multi-organ dysfunction syndrome, in patients undergoing severe surgical procedures, or in women in the peripartum period, and in patients with a polytrauma accompanied by life-threatening bleeding.

Over time, there has been a change in the view of the blood clotting process, retreating from the classical theory of coagulation cascades to the so-called cell haemostasis model. When responding to this change, the development of new health technologies was essential. Although the laboratory diagnostic methods nowadays allow us to effectively search for rare congenital coagulopathies (e.g. haemophilia or von Willebrand's disease), they still cannot cover all the components involved in maintaining haemostasis. Hence, other techniques will need to be developed for a

J. Henriques et al. (Eds.): MEDICON 2019, IFMBE Proceedings 76, pp. 1085–1090, 2020.
https://doi.org/10.1007/978-3-030-31635-8_132

careful analysis. All the above-described diagnoses often have very negative consequences and are the most common causes of mortality or morbidity. This creates a significant global socio-economic problem, and for these reasons, haemostasis assessment certainly deserves attention.

Viscoelastometric tests have a specific position in haemocoagulation techniques. Representatives of these methods are thromboelastography (TEG), and/or rotational thrombo-elastometry (ROTEM). These are fast bedside methods based on a measurement of viscoelastic whole blood properties during the formation, stabilization and lysis of a blood clot. The measurement provides a graphical record (thromboelastrogram) and numerical values of the measured parameters. The basic form is rather of a tentative broad character. However, in some cases, the test can be focused on detecting a specific value [1].

The only current study of viscoelastic methods conducted in the Czech Republic, which includes some cost analysis elements, is the research that changed the view of using ROTEM and introduced new paradigms for evaluating haemostasis in patient care at departments of anaesthesiology and critical care [2]. A team of researchers studied the application of standard laboratory tests (SLT) and viscoelastic methods (VEM) coagulation profiles before tracheostomy. In addition to the occurrence of bleeding complications, they compared the number of freshly frozen plasma units consumed before (2009–2012) and after (2013–2015) implementation of ROTEM in routine practice [2].

The goal of this paper was to calculate and assess the cost-effectiveness in order to compare SLT and VEM tests. VEM tests are still a topic of intense research [3, 4], and their cost-effectiveness has not been decided yet [5].

2 Methods

First, it was necessary to choose the appropriate perspective that we would analyse the cost-effectiveness from. As the most appropriate viewpoint, we chose the healthcare provider's perspective, as hospitals have to decide on which method to use and under what circumstances.

Data collection was arranged in cooperation with the Department of Anaesthesiology and Critical Care (DACC) of the Motol University Hospital in Prague (MUH). The department has 22 beds, being the largest of its kind in the Czech Republic. Due to a merger of two original parts into the current department, we had to exclude the option of a retrospective information retrieval due to a possible hazard of an inconsistent data set. Out of the possible alternatives of a prospective clinical data collection, we decided to compare two groups of patients. One examined with SLT, and the other with VEM.

The sample was defined by the following inclusion/exclusion criteria:

- Patients requiring intensive and resuscitation care at DACC MUH in Prague.
- Patients over 18 years of age.
- Patients scheduled to be admitted from operating theatres (except patients after transplantation or on a waiting list).
- Patients without evidence of congenital coagulation disorder (e.g. von Willebrand's disease)

The patients were randomly distributed into two test arms (A and B). A complete set of standard coagulation laboratory tests was carried out in A-group patients at the time of admission to the department. If the INR value exceeded 1.5, the ROTEM test was also performed. Only ROTEM tests (most often EXTEM and INTEM) were performed in B-group patients on admission. Depending on the results, other types of ROTEM tests could be carried out as well. (Regular morning sample collections were not included in this study.)

The data were collected in spring 2018. Each arm consisted of 30 patients; their demographic characteristics proved to be similar (see Table 1). At the time of admission, most patients were 61 to 80 years old. In both arms, men were predominant. The distribution of operating theatres (departments) from which patients were transferred is also almost identical.

The cost-effectiveness ratio (C/E) was calculated using multiple-criteria decision-making methods (MCDA) [6–8]. The costs were divided into two parts. Costs directly related to the analysis type and the analyser (purchase price of the device, device maintenance costs and device operation costs) were calculated on the "costs" side. The other costs, i.e. those related to the care of the patient, were included in the "effect" side. The effects were evaluated using the MCDA. Together with hospital specialists covering all related professions, 41 effects (20 clinical, 14 economic, 7 technical) were chosen and weights were attached to them according to their relative importance. Then the same specialists evaluated both arms according to these effects. The TOPSIS method was used to calculate the resulting effect values.

The study was approved by the ethical board of the Motol University Hospital. Details about the study design and C/E calculation can be found in [9].

Table 1. Demographic characteristics of the patients.

Age	Arm A	Arm B	Admission from	Arm A	Arm B
0–20	1	0	Otorhinolaryngology	4	4
21–40	4	2	Orthopaedics	3	2
41–60	5	9	General surgery	11	12
61–80	18	17	Urology	6	7
81–100	2	2	Gynaecology	1	0
Sex	*Arm A*	*Arm B*	Spine surgery	4	2
Male	21	19	Traumatology	1	3
Female	9	11			

Source: own processing

3 Results

3.1 Cost-Effectiveness Analysis

To calculate the cost-effectiveness of selected methods of haemostasis evaluation, the results of multiple-criterial decision-making using the TOPSIS method were used as the final effects. The resulting values did not differ much (0.58 for SLT, and 0.69 for VET). The differences were found more in clinical effects than in economic and technical ones. The effects calculated by MCDA have no practical interpretation; opposite to QALYs, they only show the relative relation between outcomes of the compared technologies. On the other hand, the QALY concept has been criticised recently showing that it is neither robust nor stable [10, 11]. The costs were calculated on a one-year basis. The purchase costs were divided by the expected length of life of the device (six years), maintenance cost were calculated for a one-year period, and operation costs were based on the estimation of 400 patients per year (half of the patients hospitalized at DACC MUH). Resulting cost estimates are CZK 1,157,750 for SLT, and CZK 763,876 for VEM (the average 2018 exchange rate was EUR 1 = CZK 25.643 [12]).

These effects and costs were used for C/E determination. The results are shown in Table 2. According to our results, VEM proved to be evidently more cost-effective than the standard laboratory tests (Table. 3).

Table 2. One-year costs

	SLT	VEM
Purchase price (one year depreciations)	185 750	90 876
Maintenance and service	810 000	491 000
Cost of tests (400 patients per year)	162 000	182 000
TOTAL	**1 157 750**	**763 876**

Source: own processing

Table 3. Results of the cost-effectivness analysis

	SLT	VEM
Effect	0,57	0,69
Costs (CZK)	1 157 750	763 876
C/A (per year)	2 031 140	1 107 067
C/A (per patient)	5 077.85	2 767.67

Source: own processing

4 Discussion and Conclusions

The compared alternatives of the solution were complete coagulation runs and two ROTEM tests. ROTEM is an example of point-of-care testing (POCT), a modern solution that has found numerous applications in the last years as a tool providing rapid diagnostics. The health care provider's perspective was chosen, and the evaluation criteria (clinical, economic and technical) were discussed with representatives of individual professions concerned with the given category of effects.

The resulting finding of cost-effectiveness analysis was that the use of viscoelastic methods in this case is nearly twice as effective as conventional methods. From a clinical point of view, the greatest differences in length of examination, blood loss, number of fresh frozen plasma (FFP) and resuspended red blood cells and related complications were noted. The subject of further discussion may be the question of why patients have lower blood loss when using VEM. A possible explanation could be the effect of administration of freshly frozen plasma and other derivatives on INR values. Certainly, it would be appropriate to conduct a study of this phenomenon. If FFP administration were shown to increase patient bleeding, it would also explain why there is a greater need for transfusions, acute curbside consultation, and the like, when using SLT.

The division of the costs into two parts – the costs related to the device were included in the "cost" part of the C/E ratio, the other costs (above all care of the patient) in the "effect" part – in this case is rather non-standard. This was enabled by using the MCDA. Typical recommendations are of two kinds: (i) to include all costs in the "cost" part and to use the MCDA only for clinical and technical effects, or (ii) to replace the whole C/E ratio by the MCDA (meaning to include all costs in the MCDA). However, sometimes it can be functional to divide costs like in this study. This approach was suggested by Ivana Jurickova-Kubatova in her PhD Dissertation in 2015 [13] and used in several HTA studies on devices, e.g. [14]. However, when dividing the costs into two parts, one has to be extremely cautious to avoid any sign of bias.

Based on an overall assessment of all analyses performed, it can be argued that the viscoelastic tests represented by the ROTEM delta are more clinically economically advantageous than the current methods of haemostasis testing in patients hospitalized at DACC MUH. Our result coincides with information from literature that inclines to the hypothesis that VEM is more cost-effective [15, 16]. However, recent systematic reviews call for more evidence to make a final decision [5, 17, 18]. A further study with a larger number of patients could help to achieve stronger results.

References

1. Salaj, P.: Poruchy hemostázy [Hemostasis disorders], Maxdorf, Praha (2017)
2. Durila, M., Lukas, P., Astraverkhava, M., Berousek, J., Zabrodsky, M., Vymazal, T.: Tracheostomy in intensive care unit patients can be performed without bleeding complications in case of normal thromboelastometry results (EXTEM CT) despite increased PT-INR: a prospective pilot study. BMC Anesthesiol. **15**, 89 (2015)
3. Trevisan, D., Zavatti, L., Gabbieri, D., Pedulli, M., Giordano, G., Meli, M.: Point-of-care-based protocol with first-line therapy with coagulation factor concentrates is associated with decrease allogenic blood transfusion and costs in cardiovascular surgery: an Italian single-center experience. Minerva Anestesiol. **82**, 1077–1088 (2016)
4. Si, M.S.: Rotational thromboelastometry and aortic surgery. J. Thorac. Cardiovasc. Surg. **157**, 1058 (2019)
5. MacDonald, E., Severn, M.: Thromboelastography or Rotational Thrombo-elasto-graphy for Trauma: A Review of the Clinical and Cost-Effectiveness and Guidelines. CADTH (2017)

6. Rogalewicz, V., Jurickova, I.: Multiple-criteria decision making: application to medical devices. In: Proceedings IWBBIO2014: International Work-Conference on Bioinformatics and Biomedical Engineering, vol. 1 and 2, pp. 1359–1372 (2014)
7. Rosina, J., Rogalewicz, V., Ivlev, I., Juřičková, I., Donin, G., Jantosová, N., Vacek, J., Otawová, R., Kneppo, P.: Health technology assessment for medical devices. Lekar a technika - Clin. Technol. **44**, 23–36 (2014)
8. Selcan, M., Kubatova, I.: Application of HTA in optometry. In: IUPESM 2018 World Congress on Medical Physics and Biomedical Engineering, Prague, pp. 87–93 (2019)
9. Blahýnková, M.: Analýza nákladové efektivity vybraných metod hodnocení hemostázy [Cost-effectiveness analysis of selective methods of haemostatis evalu-at-ion]. Master Thesis, Czech Technical University in Prague, Kladno, p. 88 (2018)
10. Beresniak, A., Medina-Lara, A., Auray, J., De Wever, A., Praet, J., Tarricone, R., Torbica, A., Dupont, D., Lamure, M., Duru, G.: Validation of the underlying assumptions of the quality-adjusted life-years outcome: results from the ECHOUTCOME European project. Pharmacoeconomics **33**, 61–69 (2015)
11. Rogalewicz, V., Bartak, M.: Kontroverze okolo QALY [Controversies around QALYs]. Vnitřní lékařství **63**, 242–248 (2017)
12. CNB: Czech National Bank. http://www.cnb.cz/en
13. Kubátová, I.: Application of value engineering and multiple-criteria decision making in medical devices assessment [Využití hodnotového inženýrství a multikriteriálního rozhodování při hodnocení zdravotnické techniky]. PhD Dis-sertation, Faculty of Biomedical Engineering, Czech Technical University in Prague, Kladno, p. 120 (2015)
14. Jurickova, I., Kraina, A.: Case study: Mobile X-ray equipment selection for a traumatology department using value engineering and multi-criteria decision methods. In: Proceedings Iwbbio 2014: International Work-Conference on Bioinformatics and Biomedical Engineering, vol. 1 and 2, pp. 1389–1402 (2014)
15. Mallett, S.V., Chowdary, P., Burroughs, A.K.: Clinical utility of viscoelastic tests of coagulation in patients with liver disease. Liver Int. **33**, 961–974 (2013)
16. Craig, J., Aguiar-Ibanez, R., Bhattacharya, S., Downie, S., Duffy, S., Kohli, H., Nimmo, A., Trueman, P., Wilson, S., Yunni, Y.: Health technology assessment report 11: the clinical and cost effectiveness of thromboelastography/thrombo-elasto-metry. NHS Quality Improvement Scotland, Glasgow, Edinburgh (2008)
17. Da Luz, L.T., Nascimento, B., Shankarakutty, A.K., Rizoli, S., Adhikari, N.K.J.: Effect of thromboelastography (TEG (R)) and rotational thromboelastometry (ROTEM (R)) on diagnosis of coagulopathy, transfusion guidance and mortality in trauma: descriptive systematic review. Crit. Care **18**(5), 518 (2014)
18. Fahrendorff, M., Oliveri, R.S., Johansson, P.I.: The use of viscoelastic haemostatic assays in goal-directing treatment with allogeneic blood products - a systematic review and meta-analysis. Scandinavian Journal of Trauma Resuscitation & Emergency Medicine **25**(1), 39 (2017)

Regular Sessions: Neuro Engineering, Neuro Systems

Multimodal Approach for Epileptic Seizure Detection in Epilepsy Monitoring Units

Paulo Maia[1(✉)], Elodie Lopes[1], Elisabeth Hartl[2], Christian Vollmar[2], Soheyl Noachtar[2], and Joao Paulo Silva Cunha[1]

[1] INESC-TEC, Faculty of Engineering of University of Porto, Porto, Portugal
paulo_maia20@live.com
[2] Ludwig Maximillian University of Munich, Munich, Germany

Abstract. Epilepsy is one of the most common neurological disorders, affecting up to 1% of the World population. Patients with epilepsy may suffer from severe consequences from seizures (e.g. injuries) when not monitored. Automatic seizure detection systems could mitigate this problem, improving seizure tracking and alerting a caregiver during a seizure. Existing unimodal solutions for seizure detection, based on electroencephalogram (EEG) and electrocardiogram (ECG) still have an unacceptable level of false positives, which can be reduced by combining these two biosignals. In this paper, EEG and ECG data from 7 epileptic patients with diverse recording length and seizure types were used for analyzing the importance of multimodal seizure detection, at a total of around 110 h 2 m. A leave one seizure out cross validation was selected, grouping data containing the period before a seizure and the seizure period. A proof of concept of multimodal seizure detection which uses a deep learning architecture directly on raw data is performed - a Fully Convolutional Neural Network and an architecture based on LSTM were tested. The network based on LSTM achieved better performance - using the best of one or a combination of both signals, all patients had above 91% detected seizures, a specificity per epoch above 0.96 ± 0.06 and a detection delay below 8.5 ± 12 s. These results show potential for developing a patient-specific approach for seizure detection that can be transferred to the ambulatory.

Keywords: Epilepsy · Multimodal seizure detection

1 Introduction

1.1 Background

Epilepsy is one of the most common neurological disorders, affecting from 0.5 to 1% of the world population [1]. It is characterized by a predisposition to generate epileptic seizures, a transient and temporary occurrence of signs/symptoms which are consequence of an abnormal neuronal activity in the brain [2].

J. Henriques et al. (Eds.): MEDICON 2019, IFMBE Proceedings 76, pp. 1093–1104, 2020.
https://doi.org/10.1007/978-3-030-31635-8_133

Currently, for a good diagnosis and therapy planning, physicians need to admit patients into specialized centres with an Epilepsy Monitoring Unit (EMU) where continuous video-electroencephalography (vEEG) is carried out [3].

The required treatment for epilepsy is dependent on seizure type and the diagnosed epilepsy syndrome. As such, the available treatments focus on the suppression of the seizures [4], which is mainly done using anti epileptic drugs. Around 60% of the patients become seizure-free with up to three anti-seizure medication trials [5]. However, about 25% of the patients suffering from epilepsy continue having more than one seizure per month [6], 12% and 8% have weekly and daily seizure frequency, respectively. Besides, the mortality rate of patients with active epilepsy is 5 times higher than those which are seizure free, number which has remained stable despite the introduction of new antiepileptic drugs [7,8].

Outside EMUs, there may be severe consequences associated with seizures, such as status epilepticus (seizures with a duration over 5 min), injury or even death from falls or other accidents [9].

In EMUs, patient safety can be ensured by continuous observation through trained personnel. However, only about 60% of EMUs can provide this continuous patient surveillance, and around 17% of EMUs use online seizure detection and warning systems. Detection systems could optimize data review and make personnel assignment more efficient [10].

Moreover, seizure tracking, which is relevant for studying the effectiveness of an antiepileptic drug, is dependent on a subjective patient and family recall and may be influenced by the capacity of remembering details post-seizure, by the level of awareness during the seizure and by the ability to identify seizures [11]. Accuracy of seizure detection, if based on videoEEG, can be improved by 29-fold and 7-fold, respectively, if observed by families or nurses [12].

Based on these challenges, a wearable device for automatic seizure detection which can be used in both EMUs and ambulatory is of utmost interest. The best solution for seizure detection should include EEG and other physiological signals, depending on a patient's seizure type.

1.2 State of the Art and Proposal

For seizure detection, depending on the actual state of the epoch being classified, one out of four outcomes are possible: a true-seizure period is missed (false negative - FN), correctly detected (true positive - TP), wrongly flagged (false positive - FP) or rejected if it is not a seizure period (true negative - TN). Three metrics are typically calculated for the first phase of validation of seizure detection algorithms: detection latency, sensitivity and false positive per hour [13].

The detection latency (DL) is the time interval from seizure onset to actual detection, and as such, is directly related to detection speed.

Sensitivity (Eq. 1) expresses the likelihood that a true seizure is detected. Sensitivity can be calculated either by seizure epoch or by seizure. When calculated by seizure, it can be done using a test set or with a grouped cross validation approach.

$$Sens = \frac{TP}{TP + FN} \tag{1}$$

The False Positive Rate (FPR) is defined as the number of false positives occurring per unit time (typically, by hour or by 24 h). Specificity (Eq. 2) may be calculated instead, for estimating the potential of having false positives, for shorter recordings, since it is a normalized value.

$$Spec = \frac{TN}{FP + TN} \tag{2}$$

Many solutions for seizure detection are based on a single sensor (unimodal). However, unimodal seizure detectors still have an unacceptable amount of false alarms, and are not able to detect all types of seizures, as different seizures trigger different physiological responses [14]. Multimodal algorithms might help lower this false alarm rate - for instance, a non-motor seizure shows no motor changes, and as such, a seizure detector based on electromiography (EMG) or accelerometer would not work for these types of seizures. In fact, a 30–57% reduction of the false alarm rate was observed when comparing multimodal and unimodal approaches [15].

On PubMed, a search for at least one of the keywords 'seizure detection', 'seizure prediction', 'EEG seizure detection and prediction', 'EMG seizure detection', 'ECG seizure detection and prediction' returns a total of 4177 results. Of those, Table 1 shows the most relevant examples of multimodal seizure detection algorithms. The major difference between multimodal methods is the way the biosignals are integrated with each other for achieving the final result, such as late fusion, which uses one classifier for each signal [16] and fuses the result using logical operations, or early fusion [17], in which the features from each signal are concatenated.

Machine learning approaches are very target-directed, and there is no way of knowing *a priori* which feature extraction algorithm is best for a given problem. As such, it requires a long time to optimize the best features for a given problem [20]. In the past few years, deep learning has achieved very good results in challenges that conventional machine learning could not solve. This is due to an increase in computing power and the elimination of manual feature extraction and selection. Besides, deep learning algorithms learn the optimal features intrinsically, considering all the available evidence, leading to a more natural and unbiased representation [20].

Although deep learning methodologies are commonly used in biosignals other than EEG [20], there are no studies regarding its performance in epileptic seizure detection nor in multimodal approaches. This work shows a proof of concept on 7 patients for a personalized algorithm which uses electrocardiogram (ECG) and EEG for epileptic seizure detection. To the best of our knowledge, this the first deep learning approach which combines these two biosignals for seizure detection.

Table 1. State of the art on multimodal seizure detection algorithms.

Reference	Sensor(s)	Algorithm	Population & Data	Avg performance	Innovation
[18]	EEG EMG ECG	*EEG:* Rhythmic pattern detection compared to average spectrum; *EMG:* Line Length compared to baseline; *ECG:* CSI100 over 100 beats; Threshold; Late fusion (*OR*)	92 pt 494sz/11 978h	Sens: 86% FAR: 0.7/h	High sensitivity using reduced electrode montages
[17]	EDA Accelerometer	*Acc:* time/frequency domain features non-linear features; *EDA:* time domain features; SVM classifier (early fusion)	80 pt (7 with sz.) 16sz/4,213h	FAR: 0.03/h Sens: 94%	Novel methodology for generalized tonic clonic sz. detection
[19]	*First stage:* HR, SpO2 and EDA *Second stage:* HR, SpO2, EDA, ECG *Third stage:* EEG	*First stage:* Search for increases in HR and EDA and decreases in SpO2 *Second stage:* Personalized parameters for each patient *Third stage (for some subjects):* EEG-feature extraction; 3-Channel selection based on best features; kNN classifier	10 pt 26sz/339h	*First stage:* Sens: 100% (7/10 pat., 10 sz) *Second stage:* Sens: 100% (6/10 pat., 11 sz) *Third stage:* Sens: 100% (2/3 pat)	Wearable approach concerning a patient-specific biosignal selection
[15]	Accelerometer EMG ECG	*ECG:* time domain features; *Acc.:* frequency and time domain features; *EMG:* time domain features; SVM classifier; Logical OR on both alarms, in a certain time window	7 pt 22 sz/224h	ECG+Acc. (best) Sens: 90,9% FAR: 0.08/h	Multimodal sensors lead to 75% less false alarms; ACM changes are slower than EMG

Abbreviations: SpO2: Oxygen Saturation; sz: seizure; Acc: Accelerometer; SVM: Support Vector Machine; CSI100: Cardiac Sympathetic Index; pt: patient; sz: seizure; FAR: False Alarm Rate; Sens: Sensitivity; Spec: Specificity

2 Methods

2.1 Dataset

In a joint research line between University of Porto and the Ludwig Maximillian University of Munich, we have been working on 3D Video-EEG, and have been building the largest multimedia database of this kind. Currently, the database holds more than 300 seizures from over 100 patients [21]. Besides RGB video and 3D infrared radar, we store long term surface EEG and ECG signals synchronously recorded at 256 Hz and stored. Three minutes post seizures were removed from this dataset as this is not representative of natural monitoring conditions.

The rest of the dataset was taken from Temple University Hospital [22] (TUH-EEG Corpus), is an ongoing data collection effort that has recently released a training set containing 1327 seizures from 130 patients, of which the irrelevant signal parts were manually removed by healthcare professionals.

A subset of 7 patients from both Munich and TUH-EEG was chosen to represent different seizure types, lengths and recording durations, further detailed in Table 2.

2.2 Dataset Pre-processing

EEG and ECG data (256 Hz) was split in three seconds segments. Each segment was labeled as ictal (seizure period) or interictal (non-seizure period) depending on the majority of samples of each class.

Channel selection was performed for EEG, as it can significantly reduce the computational cost of the algorithm while also decreasing the amount of irrelevant information the classifier needs to learn. In EEG recordings, if no information regarding seizure location was available, the four channels with highest variance during the ictal period were selected, as previously recommended in the literature [23]. However, if healthcare professionals had previously done the

Table 2. Detailed information of the 7 patients used for this study.

Patient ID	Sz. amount	Duration	Interictal/ictal ratio	Syndrome	Seizure type(s)
MUN1	19	37 h	182	Temporal Lobe Epilepsy	Hypermotor to generalized tonic-clonic Bilateral tonic to automotor
MUN2	11	24 h 04 m	759	Paracentral Epilepsy	Aura to Tonic Clonic Seizure
MUN3	6	47 h 40 m	1037	Frontal Lobe Epilepsy	Aura to Complex Motor
TUH1	8	21 m 41 s	1.6	Unknown	Tonic-clonic seizures
TUH2	11	22 m 43 s	19.7	Frontal Lobe Epilepsy	Absence seizures
TUH3	17	24 m 54 s	2.17	Occipital Lobe Epilepsy	Complex partial seizures
TUH4	12	12 m 25 s	0.58	Temporal Lobe Epilepsy	Generalized seizures

task of labeling the channels involved in seizures (or the epilepsy syndrome), manually selected channels from that region were used, to decrease the source of errors. As such, in all cases except TUH1, channels were manually selected.

2.3 Classification

Two different architectures were tested: Fully Convolutional Neural Networks (FCNs) and a network based on Long Short-Term Memory Networks (LSTM).

FCNs were recently proposed for classifying univariate time series [24], and were chosen here as a classifier due to their efficacy for time series classification in several benchmark datasets.

The used FCN architecture is built by three convolutional blocks, each with three operations: convolution, batch normalization and ReLu activation. Then, a global average pooling (GAP) is applied, averaging the results of the last convolutional block. A sigmoid function is then fully connected to GAP's output. Regarding filter size, the convolution's stride is equal to 1, with zero padding applied. The convolutions contain 128, 256 and 128 filters, respectively, with filter lengths equal to 8, 5 and 3. Figure 1 shows a representation of the FCN network used.

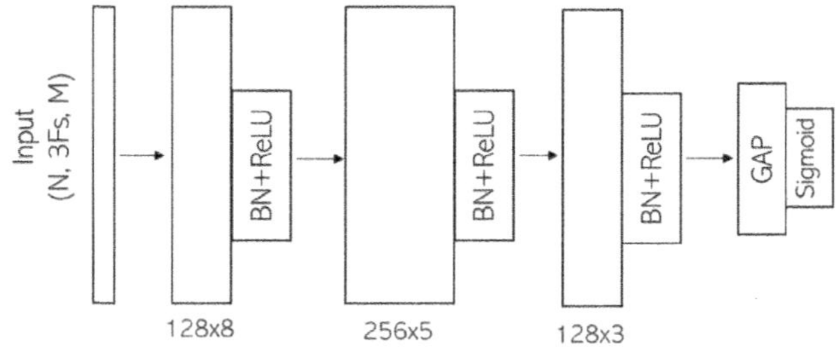

Fig. 1. Architecture of the Fully Convolutional Neural Network, extracted from [24]. In the input layer, N is the number of epochs used for training, 3Fs is the sampling frequency multiplied by three seconds and M is the number of channels (1 for ECG and 4 for EEG).

The LSTM architecture was previously proposed for seizure detection in an EEG dataset [25]. The data samples are input into a fully connected LSTM layer of 100 neurons, as a way of learning the short and long term dependencies between segments in each signal. The interest of such an architecture is harnessing the LSTMs' ability of learning and extracting features across the time steps in a window of 3 s, and not across windows. Afterwards, a time-distributed Dense layer of 50 units is applied, to calculate the cost function on all timesteps. Finally, the output of the dense layer is passed through a 1D average pooling layer, which is then input into a final sigmoid layer.

During training, a grouped leave one seizure out cross validation was performed. Epochs were grouped from the end of a seizure to the end of the following seizure (Fig. 2). As such, in cross validation it is guaranteed that temporally close epochs from the same seizure are not in the train and test folds, which could give an inflated estimate of model performance.

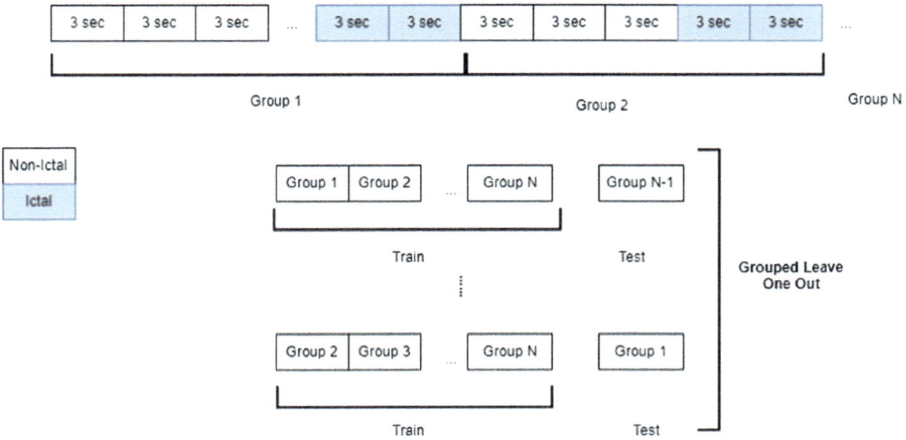

Fig. 2. Cross validation strategy for seizure detection and prediction.

In each fold of the training set, the classifier's weights were set to random and the interictal period was downsampled to be eight times the ictal period, decreasing the effect of class imbalance. In the cases where the interictal to ictal ratio was not higher than 8 (see Table 2), downsampling was performed so that the ratio between both classes was 1. These values were determined iteratively, by cross validation. In the validation set, the original size segment was used. This realistically simulates a production environment, as the model is intended to be trained in the Epilepsy Monitoring Unit on available seizure data, and should have similar performance for detecting other seizures in a longer period, where there's a higher percentage of interictal time.

Data augmentation was applied to each training epoch as a way to increase training data - random noise with a standard deviation of 0.05 was added, and a random scaling with a normal distribution having standard deviation of 0.01 was applied. These values were selected as they did not lead to significant changes in the signal's shape (by visual analysis) and led to a better model convergence and lower false positive rate during cross-validation.

The network was trained for a maximum of 100 epochs or after the loss stopped decreasing by a factor of 0.05 for 40 epochs. Adam optimizer [26] was used with an initial learning rate of 0.001 for FCN and 0.01 for LSTM. The learning rate was decreased by a factor of 0.5 when the loss stopped decreasing for 5 epochs, up to a minimum of 0.0001. The model with the highest geometric mean between specificity and sensitivity in the validation set was then chosen as the best model. Due to class imbalance, classes were weighted accordingly to the ratio of interictal to ictal samples, and the binary crossentropy loss was used.

The output of the classifier was post-processed, and considered to be ictal if N consecutive epochs were classified as ictal, where N is patient-dependent, as different seizure types have different lengths and patterns. Depending on how much of the signal we need to see before making the optimal decision, the detection latency varies. As N is picked according to the cross-validated output, this decreases the additional bias of the classifier.

To obtain a multimodal approach, the logical AND operator was used in the output of all classifiers before post-processing.

3 Results

Tables 3 and 4 summarize the obtained results for each patient, for the FCN and the LSTM architecture, respectively. As the seizure group duration is very variable, specificity was used instead of false positive rate, since it is a relative value. The sensitivity per epoch and per seizure, the detection latency and the number of consecutive epochs required to trigger a seizure alarm are also reported.

4 Discussion

This work is a proof of concept of multimodal seizure detection which uses a deep learning architecture directly on raw data, with no pre processing besides channel selection. A low number of patients was used, although it is in line with the literature, as studies with similar goals presented in Table 1 used less than 10 patients. Moreover, it respects the number of patients needed for initial validation of a seizure detection device (1–10) according to previously defined standards [13].

On average, high sensitivity values, similar to the state of the art, are obtained for detecting seizures. For classification per epoch, sensitivity is more variable. However, what is most important is detecting a seizure and triggering an alarm, which may prevent life-threatening conditions, and not estimating its full duration. Specificity shows an increase in multimodal approaches, on average, but with a decrease in sensitivity, and a higher variability in sensitivity per epoch, in some cases. It is preferable to have high sensitivity in difficult cases, with motor symptoms which may lead to injuries, such as tonic clonic seizures.

The architecture based on LSTM shows more potential for detecting seizures, as all patients have sensitivities per seizure above 80%, while for FCN the minimum value is 33%. As such, LSTM was chosen to be the best classifier for this subset.

For patient TUH1, both EEG and ECG approaches showed relevant results as unimodal detectors. Multimodal detection did not lead to significant improvements. These results are to be expected since the patient has violent tonic-clonic seizures, easily detected using either EEG or ECG. This patient could benefit

Table 3. Summary of the results for EEG, ECG and multimodal fusion (FCN).

Patient	EEG	ECG	MM	Alarm size
MUN1	**Sens/sz:** 1.00 **Sens/epoch:** 0.78 ± 0.15 **Spec/epoch:** 0.99 ± 0.04 **DL (s):** 3.6 ± 4.1	**Sens/sz:** 0.79 **Sens/epoch:** 0.46 ± 0.29 **Spec/epoch:** 0.99± 0.01 **DL (s):** 4.8 ± 6.1	**Sens/sz:** 0.79 **Sens/epoch:** 0.4 ± 0.28 **Spec/epoch:** 1.00 ± 0.0 **DL (s):** 7.4 ± 9.2	2
MUN2	**Sens/sz:** 0 **Sens/epoch:** 1.00 ± 0 **Spec/epoch:** 1.00 ± 0 **DL (s):** -	**Sens/sz:** 0.63 **Sens/epoch:** 0.37 ± 0.32 **Spec/epoch:** 0.98 ± 0.04 **DL (s):** 3.9 ± 3.09	**Sens/sz:** 0 **Sens/epoch:** 0 **Spec/epoch:** 1 ± 0 **DL (s):** -	1
MUN3	**Sens/sz:** 0.33 **Sens/epoch:** 0.174 ± 0.33 **Spec/epoch:** 1.00 ± 0.001 **DL (s):** 7.5 ± 1.5	**Sens/sz:** 0.67 **Sens/epoch:** 0.43 ± 0.36 **Spec/epoch:** 0.96 ± 0.03 **DL (s):** 1.5 ± 1.5	**Sens/sz:** 0.17 **Sens/epoch:** 0.05 ± 0.11 **Spec/epoch:** 1 ± 0 **DL (s):** 12 ± 0	1
TUH1	**Sens/sz:** 1.00 **Sens/epoch:** 0.95 ± 0.03 **Spec/epoch:** 0.98 ± 0.05 **DL (s):** 1.5 ± 1.5	**Sens/sz:** 1.00 **Sens/epoch:** 0.94 ± 0.03 **Spec/epoch:** 0.99 ± 0.03 **DL (s):** 4.25 ± 2.2	**Sens/sz:** 1.00 **Sens/epoch:** 0.91 ± 0.03 **Spec/epoch:** 1.0 ± 0.00 **DL (s):** 4.25 ± 2.2	2
TUH2	**Sens/sz:** 0.9 **Sens/epoch:** 0.9 ± 0.3 **Spec/epoch:** 0.84 ± 0.29 **DL (s):** 0	**Sens/sz:** 0.9 **Sens/epoch:** 0.77 ± 0.33 **Spec/epoch:** 0.89 ± 0.29 **DL (s):** 0.33 ± 0.94	**Sens/sz:** 0.91 **Sens/epoch:** 0.77 ± 0.33 **Spec/epoch:** 0.9 ± 0.3 **DL (s):** 0.33 ± 0.94	1
TUH3	**Sens/sz:** 0.94 **Sens/epoch:** 0.45 ± 0.23 **Spec/epoch:** 0.94 ± 0.11 **DL (s):** 2.65 ± 2.34	**Sens/sz:** 0.71 **Sens/epoch:** 0.48 ± 0.41 **Spec/epoch:** 0.65 ± 0.42 **DL (s):** 6.5 ± 12.84	**Sens/sz:** 0.65 **Sens/epoch:** 0.26 ± 0.24 **Spec/epoch:** 0.96 ± 0.11 **DL (s):** 3.6 ± 2.81	1
TUH4	**Sens/sz:** 1.00 **Sens/epoch:** 0.58 ± 0.22 **Spec/epoch:** 0.96 ± 0.23 **DL (s):** 4.2 ± 1.5	**Sens/sz:** 1.00 **Sens/epoch:** 0.60 ± 0.3 **Spec/epoch:** 1 **DL (s):** 3 ± 2.3	**Sens/sz:** 1.00 **Sens/epoch:** 0.44 ± 0.27 **Spec/epoch:** 1 **DL (s):** 4.2 ± 1.5	2

from a single wearable, such as a non intrusive ECG device, which has very high sensitivity and low detection delay, showing potential for injury prevention. Patient TUH2 shows the best sensitivity with EEG analysis, although ECG has similar results with reduced false positives, and is a less intrusive sensor and with less noise on ambulatory. Patient TUH3 has very high sensitivity per seizure in ECG, and an increased specificity in the multimodal approach, although with this data fusion methodology the sensitivity slightly decreases. As such, this patient would benefit from having a wearable with few EEG electrodes. TUH4 benefits from having a multimodal approach in terms of decrease of false positives, sacrificing sensitivity per epoch.

MUN1 shows similar behavior to TUH4. In MUN2, there is a decrease in the sensitivity per seizure in the multimodal approach, with EEG being the optimal sensor. Finally, in MUN3 the multimodal approach shows higher specificity but lower sensitivity, and also a lower delay when compared to EEG, the best biosignal for this patient.

For both MUN1 and TUH4, a high sensitivity for seizure detection using ECG is likely to happen, as temporal lobe seizures, as in 82% of the patients with epilepsy, seizures are accompanied by ictal sinus tachycardia [27], due to

Table 4. Summary of the results for EEG, ECG and multimodal fusion (LSTM).

Patient	EEG	ECG	MM	Alarm size
MUN1	**Sens/sz:** 1.00 **Sens/epoch:** 0.85 ± 0.12 **Spec/epoch:** 0.99 ± 0.01 **DL (s):** 2.5 ± 3.8	**Sens/sz:** 1.00 **Sens/epoch:** 0.55 ± 0.2 **Spec/epoch:** 0.96 ± 0.05 **DL (s):** 5.55 ± 6.1	**Sens/sz:** 1.00 **Sens/epoch:**0.5 ± 0.19 **Spec/epoch:** 0.99 ± 0.01 **DL (s):**6.3 ± 5.9	2
MUN2	**Sens/sz:** 1.00 **Sens/epoch:** 0.79 ± 0.18 **Spec/epoch:** 0.96 ± 0.01 **DL (s):** 1.6 ± 1.5	**Sens/sz:** 1.00 **Sens/epoch:**0.65 ± 0.19 **Spec/epoch:** 0.92 ± 0.08 **DL (s):** 3 ± 2.56	**Sens/sz:** 0.91 **Sens/epoch:** 0.55 ± 0.27 **Spec/epoch:** 0.99 ± 0.01 **DL (s):** 3.6 ± 2.3	1
MUN3	**Sens/sz:** 1.00 **Sens/epoch:** 0.57 ± 0.29 **Spec/epoch:** 0.97 ± 0.02 **DL (s):** 9 ± 13	**Sens/sz:** 0.83 **Sens/epoch:** 0.6 ± 0.32 **Spec/epoch:** 0.97 ± 0.02 **DL (s):** 2.4 ± 2.25	**Sens/sz:** 0.83 **Sens/epoch:** 0.44 ± 0.26 **Spec/epoch:** 0.99 ± 0.01 **DL (s):** 4.8 ± 5.6	2
TUH1	**Sens/sz:** 1.00 **Sens/epoch:** 0.93 ± 0.08 **Spec/epoch:** 0.97 ± 0.07 **DL (s):** 3 ± 1.7	**Sens/sz:** 1.00 **Sens/epoch:** 0.91 ± 0.08 **Spec/epoch:** 1 **DL (s):** 5 ± 4.1	**Sens/sz:** 1.00 **Sens/epoch:** 0.88 ± 0.12 **Spec/epoch:** 1.0 ± 0.00 **DL (s):** 5.8 ± 3.9	2
TUH2	**Sens/sz:** 0.91 **Sens/epoch:** 0.88 ± 0.29 **Spec/epoch:** 0.87 ± 0.28 **DL (s):** 0.3 ± 0.9	**Sens/sz:** 0.91 **Sens/epoch:** 0.83 ± 0.31 **Spec/epoch:** 0.96 ± 0.06 **DL (s):** 0.3 ± 0.9	**Sens/sz:** 0.91 **Sens/epoch:** 0.83 ± 0.31 **Spec/epoch:** 0.91 ± 0.29 **DL (s):** 0.3 ± 0.9	1
TUH3	**Sens/sz:** 1.00 **Sens/epoch:** 0.5 ± 0.23 **Spec/epoch:** 0.96 ± 0.06 **DL (s):** 8.5 ± 12	**Sens/sz:** 0.76 **Sens/epoch:** 0.3 ± 0.25 **Spec/epoch:** 0.93 ± 0.12 **DL (s):** 9.2 ± 5.3	**Sens/sz:** 0.65 **Sens/epoch:** 0.2 ± 0.2 **Spec/epoch:** 0.9 ± 0.02 **DL (s):** 9.5 ± 4.09	2
TUH4	**Sens/sz:** 1.00 **Sens/epoch:** 0.73 ± 0.26 **Spec/epoch:** 0.92 ± 0.15 **DL (s):** 1.2 ± 2	**Sens/sz:** 1.00 **Sens/epoch:** 0.68 ± 0.28 **Spec/epoch:** 0.79 ± 0.3 **DL (s):** 2.7 ± 4.5	**Sens/sz:** 1.00 **Sens/epoch:** 0.46 ± 0.24 **Spec/epoch:** 0.99 ± 0.04 **DL (s):** 3.6 ± 4.4	1

the activation of the central autonomic network caused by epileptic discharges. This occurs especially in patients with generalized tonic-clonic seizures and focal impaired awareness seizures, originating from the temporal lobe [28]. However, it is not clear if the seizures were detected due to tachycardia or to other factors, since this deep learning based approach has poor feature interpretability.

Data from University of Munich is more representative of realistic conditions, regarding the fact that it is composed of long term recordings, in which patients do their daily routine while in bed. On the contrary, TUH data has removed irrelevant signal segments that could be important for validating an alarm system, which is why the interictal to ictal ratio is lower than data from Munich.

In seizure detectors, the false alarm rate is dependent on the number of seizures the patient has per day. With the dataset we are developing, containing long term recordings, we can calculate newer metrics, such as the sensitivity per hour normalized by the average distance per seizures.

In the future, more patients with different conditions and sensors will be used. Moreover, complex alternatives for fusing the results from different unimodal sensors will be studied, which may help increase the performance of the classifier. These results show that logically combining the output of two signals with an "AND" operator is most beneficial when both signals give good results. If one seizure is not detected with one of the classifiers, the multimodal approach's sensitivity also decreases. As such, for some cases with good specificity per each biosignal, an approach based on the "OR" operator could be beneficial, for instance.

References

1. Epilepsy Foundation (2018). https://www.epilepsy.com/learn/
2. Fisher, R.S., et al.: Epileptic seizures and epilepsy: definitions proposed by the international league against epilepsy (ILAE) and the international bureau for epilepsy (IBE). Epilepsia **46**(4), 470–472 (2005)
3. Rosenow, F., Luders, H.: Presurgical evaluation of epilepsy. Brain **124**(9), 1683–1700 (2001)
4. Carrette, E., Vonck, K., Boon, P.: The management of pharmacologically refractory epilepsy. Curr. Med. Lit. Neurol. **26**(4), 104–121 (2011)
5. Kwan, P., Brodie, M.J.: Early identification of refractory epilepsy. New Engl. J. Med. **342**(5), 314–319 (2000)
6. Forsgren, L., et al.: The epidemiology of epilepsy in Europe - A systematic review (2005)
7. Neligan, A., et al.: Temporal trends in the mortality of people with epilepsy: a review. Epilepsia **51**(11), 2241–2246 (2010)
8. Nashef, L., et al.: Incidence of sudden unexpected death in an adult outpatient cohort with epilepsy at a tertiary referral centre. J. Neurol. Neuro - Surg. Psychiatry (1995)
9. Vel, V.A., et al.: Non-EEG seizure detection systems and potential SUDEP prevention: state of the art: review and update. Seizure **41**, 141–153 (2016)
10. Baumgartner, C., Koren, J.P., Rothmayer, M.: Automatic computer-based detection of epileptic seizures. Front. Neurol. **9**, 639 (2018)
11. Ulate-Campos, A., et al.: Automated seizure detection systems and their effectiveness for each type of seizure. Seizure **40**, 88–101 (2016)
12. Nijsen, T.M., et al.: The potential value of three-dimensional accelerometry for detection of motor seizures in severe epilepsy. Epilepsy Behav. **7**(1), 74–84 (2005)
13. Beniczky, S., Ryvlin, P.: Standards for testing and clinical validation of seizure detection devices. Epilepsia **59**, 9–13 (2018)
14. van Westrhenen, A., De Cooman, T., Lazeron, R.H., Van Huffel, S., Thijs, R.D.: Ictal autonomic changes as a tool for seizure detection: a systematic review. Clin. Auton. Res. **29**(2), 161–181 (2019)
15. De Cooman, T., et al.: Comparison and combination of electrocardiogram, electromyogram and accelerometry for tonic-clonic seizure detection in children. In: 2018 IEEE EMBS International Conference on Biomedical and Health Informatics, BHI 2018 (2018)
16. Qaraqe, M., et al.: Epileptic seizure onset detection based on EEG and ECG data fusion. Epilepsy Behav. **58**, 48–60 (2016)
17. Poh, M.Z., et al.: Convulsive seizure detection using a wrist-worn electrodermal activity and accelerometry biosensor. Epilepsia **53**(5), e93–e97 (2012)
18. Fürbass, F., et al.: Automatic multimodal detection for long-term seizure documentation in epilepsy. Clin. Neurophysiol. **128**(8), 1466–1472 (2017)

19. Cogan, D., et al.: Multi-biosignal analysis for epileptic seizure monitoring. Int. J. Neural Syst. **27**(01), 1650031 (2017)
20. Faust, O., et al.: Deep learning for healthcare applications based on physiological signals: a review. Comput. Methods Programs Biomed. **161**, 1–13 (2018)
21. Pereira Choupina, H.M., et al.: NeuroKinect 3.0: multi-bed 3Dvideo-EEG system for epilepsy clinical motion monitoring. Stud. Health Technol. Inform. **247**, 46–50 (2018)
22. Obeid, I., Picone, J.: The temple university hospital EEG data corpus. Front. Neurosci. **10**, 196 (2016)
23. Duun-Henriksen, J., et al.: Channel selection for automatic seizure detection. Clin. Neurophysiol. **123**(1), 84–92 (2012)
24. Wang, Z., Yan, W., Oates, T.: Time series classification from scratch with deep neural networks: a strong baseline. In: Proceedings of the International Joint Conference on Neural Networks (2017)
25. Hussein, R., et al.: Epileptic Seizure Detection: A Deep Learning Approach, arXiv preprint arXiv:1803.09848 (2018)
26. Konur, O.: Adam Optimizer. Energy Education Science and Technology Part B: Social and Educational Studies (2013)
27. Eggleston, K.S., Olin, B.D., Fisher, R.S.: Ictal tachycardia: the head-heart connection. Seizure **23**(7), 496–505 (2014)
28. Sevcencu, C., Struijk, J.J.: Autonomic alterations and cardiac changes in epilepsy. Epilepsia **51**(5), 725–737 (2010)

Modulation of EEG Theta and Alpha Power by an Internal Attention Task with and Without Visual Distractors

Elisa Magosso(✉), Giulia Ricci, and Mauro Ursino

Department of Electrical, Electronic and Information Engineering, "Guglielmo Marconi", University of Bologna, Campus of Cesena, Via dell'Università 50, 47522 Cesena, Italy
elisa.magosso@unibo.it

Abstract. Attention to internal representations is crucial in tasks involving temporary storage and mental manipulation of information (working memory tasks), to suppress concurrent interfering processes, such as sensory-intake processes. Theta activity (4–8 Hz), especially frontal, has been extensively associated to working memory tasks, increasing with task complexity and error monitoring. Alpha rhythm (8–14 Hz) has been hypothesized to enhance processes within the attentional focus via inhibition of task-irrelevant regions (alpha increase) and engagement of task-relevant regions (alpha decrease). Here, we provide further contribution to the investigation of alpha and theta activity by computing electroencephalographic (EEG) alpha and theta power from 10 participants who performed a resting phase (no attention) and a mental arithmetic task (internal attention), under two different visual conditions i.e., pictures and no picture presentation. Pictures stimulated external attention and acted as distractors during the math task but not during rest. Results show that frontal theta power increased during the math task vs rest, more largely in presence of visual distractors. Posterior alpha power increased in the math task vs rest in picture condition, but not in the no picture condition; moreover, it decreased in the picture vs no picture condition during rest but not during the math task. These results support frontal theta as a mechanism implementing cortical control in complex tasks and indicate that EEG alpha power reflects a sophisticate balance between increase in some regions and decrease in other regions, depending on the current goal and external context.

Keywords: Mental arithmetic · Attention · Electroencephalography · Power spectral density · Frontal theta · Posterior alpha

1 Introduction

Electroencephalographic (EEG) oscillations are not merely epiphenomena but have a functional role in cognitive processes and task execution. One cognitive function extensively investigated in relation to EEG oscillations is attention [1]. Attention allows neural enhancement of task-related information while simultaneously filtering out irrelevant distractors. Attention to internal representations is a crucial element of

J. Henriques et al. (Eds.): MEDICON 2019, IFMBE Proceedings 76, pp. 1105–1112, 2020.
https://doi.org/10.1007/978-3-030-31635-8_134

working memory (i.e. transient storage and utilization of information) to promote maintenance of relevant information in memory and their manipulation, and to suppress interfering processes. Two main rhythms have been associated with working memory and attention, theta rhythm (4–8 Hz) and alpha rhythm (8–14 Hz).

EEG theta synchronization (power increase) has been found, mainly frontally, in tasks involving mental effort, working memory and short-term memory [2] during encoding and retention of information, and increases with increased memory load, cognitive and attentional demands [3]. Frontal theta enhancement is a consistent finding across heterogeneous working memory tasks (verbal, visuospatial, mathematical) and may reflect a common mechanism of adaptive control and process monitoring across widespread cortical networks [3, 4].

Alpha oscillations dominate the EEG during rest, especially in the parietal-occipital regions. Alpha activity can respond to a task demand or stimulus with power increase (synchronization) or decrease (desynchronization) [5], and has been hypothesized to play an active inhibitory role that enhances cortical processes within the attentional focus and blocks processing outside this focus, depending on the goal and context [1, 5–7]. Specifically, alpha desynchronization is interpreted to enhance cortical activation via removal of inhibition, since it is observed in regions involved in task execution (task-relevant regions). Indeed, in tasks implicating externally directed attention (sensory intake), alpha power decreases in sensory active regions [1, 5]. Conversely, alpha synchronization is interpreted to implement increase of inhibition, since it occurs in regions associated with task-irrelevant and potentially interfering processes [5–8]. For example, in tasks requiring internally directed attention (sensory rejection), posterior alpha power increases reflecting suppression of sensory-intake processes that may disturb the ongoing task [1, 5].

The present study aims to provide further contribution to the investigation of EEG alpha and theta rhythms in attention and working memory, by exploring a resting phase and a mental arithmetic task phase, both performed in two different visual stimulation conditions, i.e. pictures and no picture presentation. The pictures were irrelevant for the math task and thus potentially interfering with it. We expected disparate effects on theta and alpha power depending on the combination of the phase (rest/math task) and visual condition (picture/no picture). Indeed, the phase modulates the attentional demand, i.e., no attention (rest) or sustained internal attention (math task, requiring number manipulation, transient storage of partial results and continuous updating of the working memory store [9]). The visual condition modulates the external sensory stimulation from poor (no picture) to reach (pictures); hence, it modulates the level of external visual distraction (low/high) during the math task and the corresponding need of sensory rejection.

2 Materials and Methods

2.1 Participants, Experimental Protocol and Data Acquisition

Ten healthy volunteers (3 males, age: 23.8 ± 1.4 years) participated in the study that was approved by the local ethical committee.

The participant seated facing a computer monitor and underwent two sessions, each lasting 10 min and consisting of two consecutive phases: a 5-min resting phase and a 5-min mental arithmetic task. In one session (no picture condition), a gray screen was displayed during the rest and math task. In the other session (picture condition), the same set of 30 emotionally neutral pictures was presented during both the rest and math task; the pictures followed one another every 10 s in a random order. The sessions were separated by a 10-min break and their order was randomized across participants. Within each session, instructions regarding each phase were given just before the beginning of the phase, so the participant remained unaware of the task until its onset. Immediately before the start of the session, the operator told the participant to rest until new instructions were delivered. At the end of the resting phase, a 15-s screen displayed the instructions for the following 5-min phase (task): the participant was asked to perform mental serial subtractions in steps of seventeen starting from a given number, and to communicate the reached final number at the end of the session. The starting number for subtractions was 2500 in the first session; in the second session it was the number reached by the participant at the end of the previous session, to avoid same operations in the two tasks. The emotionally neutral pictures were photographs selected from the catalogue of the International Affective Picture System (IAPS) [10], having normative level of pleasure between 4.5 and 5.5 points on a 9-point scale.

During each session, 32 EEG scalp electrodes (Fp1, Fp2, AF3, AF4, F7, F3, Fz, F4, F8, FC5, FC1, FC2, FC6, T7, C3, Cz, C4, T8, CP5, CP1, CP2, CP6, P7, P3, Pz, P4, P8, PO3, POz, PO4, O1, O2) was recorded (Brainbox® EEG-1166 amplifier, Braintronics) at 128 Hz, against the left earlobe and re-referenced offline to averaged earlobes.

2.2 EEG Data Analysis

For each participant, the two EEG recordings were imported in MATLAB (Math-Works), concatenated along time dimension and high-pass filtered (cut-off frequency 1 Hz) to remove DC offset and slow drift. Then, Independent Component Analysis (ICA) was applied to the concatenated signals using the extended infomax algorithm (EEGLAB toolbox) for artifact reduction [11]. Independent components related to blinking, eye movements, heartbeat, and other clear artifacts were visually identified and removed. The pre-processed EEG signals were subdivided into four portions of 5 min each, corresponding to the following states: no picture rest, no picture math, picture rest, picture math. For each state, power spectral density of each channel was obtained by the Welch's periodogram method, using 5-s Hamming windows at 50% overlap, zeropadded to 10 s (0.1 Hz frequency resolution). Then, the powers in the theta (4–8 Hz) and alpha (8–14 Hz) bands were computed for each channel and the powers of the 32 channels (averaged across participants) were used to create scalp maps of alpha and theta distribution in each state. For statistical analyses, theta power values and alpha power values were topographically aggregated (via arithmetic mean) across frontal channels and parieto-occipital channels respectively (see Results), to obtain frontal theta power and posterior alpha power in each state. Two-way repeated measures analysis of variance (ANOVA) was performed separately for frontal theta power and posterior alpha power with factors *condition* (no picture and picture) and *phase* (rest and math task). Post-hoc Fisher's LSD tests were used for multiple pairwise comparisons. Cut-off significance was set at 0.05.

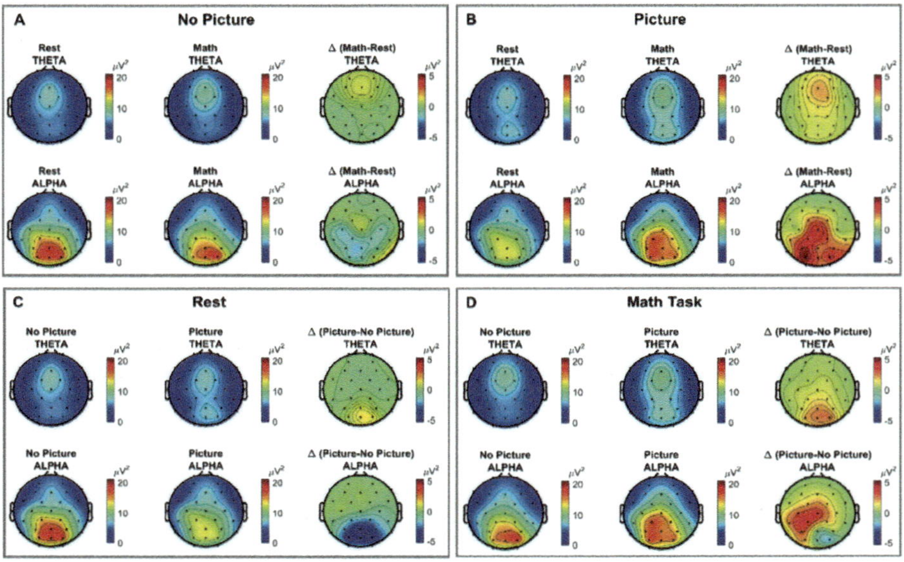

Fig. 1. Scalp maps (averaged across participants) of theta and alpha powers for 2 × 2 contrasting configurations. Within each panel, one factor is kept constant (condition in panels A and B, phase in panels C and D), and the first and second column display the scalp maps for the values of the other factor; the maps in the third column represent the difference (second-first).

3 Results

Figure 1 shows the scalp maps of the theta and alpha power for 2 × 2 contrasting configurations. Theta power exhibited an increase at the frontal channels during the math task vs rest, both in no picture and picture condition (panels A and B). Moreover, picture presentation (vs no picture) induced theta power increase at posterior sites both in rest (panel C) and in the math task (panel D), in the latter case spreading also anteriorly. The math task (vs rest) slightly decreased posterior alpha in the no picture condition (panel A) while markedly increased it in the picture condition (panel B). Moreover, picture presentation (vs no picture) induced a strong posterior alpha decrease in the rest phase (panel C), and a posterior alpha increase (with a leftward bias) in the math task.

Based on scalp maps, channels AF3, AF4, F3, Fz, F4, FC1, FC2 and channels CP1, CP2, P3, Pz, P4, PO3, POz, PO4, O1, O2 were used to compute the aggregated frontal theta power and aggregated posterior alpha power, respectively (Fig. 2). Two way ANOVA on frontal theta yielded a significant interaction effect ($F(1,9) = 5.19$, p = 0.048). Post-hoc comparisons (Table 1) revealed a significantly higher theta power in the math task vs rest in both visual conditions; moreover frontal theta during the math task was significantly higher in the picture vs no picture condition. Two way ANOVA on posterior alpha yielded a close-to-significance interaction effect ($F(1,9) = 4.34$, $p = 0.067$), examined via post-hoc comparisons (Table 1). The latter indicated higher alpha power in the math task vs rest in picture condition, and lower alpha power in picture vs no picture condition in rest (close to significance).

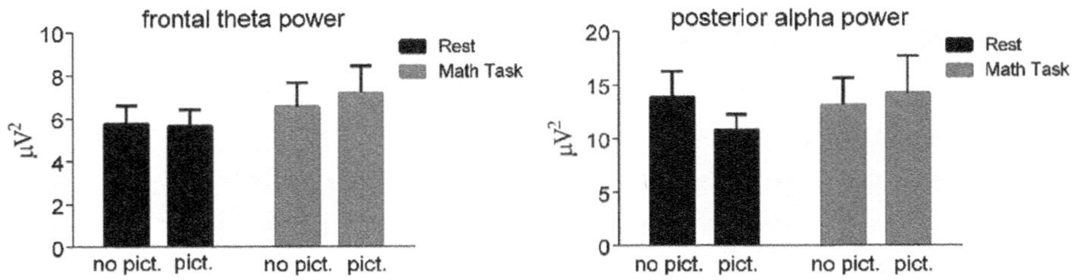

Fig. 2. Frontal theta power and posterior alpha power averaged across participants (mean ± SEM) computed for each of the four states: no picture rest, picture rest, no picture math, picture math.

Table 1. Results of post-hoc comparisons (uncorrected Fisher's LSD tests) performed on frontal theta power and posterior alpha power

Frontal theta power	Posterior alpha power
Rest vs Math task: No picture p = 0.01 Picture p < 0.001	Rest vs Math task: No picture p = 0.63 Picture p = 0.037
No picture vs Picture: Rest p = 0.61 Math task p = 0.025	No picture vs Picture: Rest p = 0.058 Math task p = 0.47

4 Discussion

Investigating attentional effects on EEG oscillations is important to improve understanding of neural rhythms function, and may have impact in clinical practice to comprehend the electrophysiological correlates of attentional disorders (characterized by reduced ability to suppress distracting inputs [7]), and in neuroengineering applications (e.g. neurofeedback and brain-computer interface). Here, we explored how the sensory context (absence/presence of visual stimuli) affects theta and alpha power depending on the subject's concurrent attentional demand (internal attention/no attention) whereby the visual stimulation may or not act as distracting element.

Theta power – Our results support the hypothesis that frontal theta oscillations may implement integrative control and coordination to perform complex tasks distributed across several brain areas [3, 4]. Indeed, EEG frontal theta power was increased during the math task compared to rest in both visual conditions (Fig. 1A and 1B, Fig. 2, Table 1); this finding agrees with previous studies that reported theta increase during mental arithmetic at middle frontal scalp sites [9, 12], likely originating from neural sources in the prefrontal cortex [13]. Moreover, frontal theta synchronization during the math task was larger in picture vs no picture condition (Fig. 1D, Fig. 2, Table 1), the former condition arguably requiring increased cortical control for task execution and maintenance of goal state against the visual distractors. Conversely, no modulation of frontal theta power was induced by picture presentation vs no picture during rest (Fig. 1C, Fig. 2, Table 1), as visual stimulation by pictures (without any oriented goal) did not require widespread integrative control processes.

Alpha power – Alpha oscillatory activity is hypothesized to serve as a mechanism functionally inhibiting cortical areas implicated in unnecessary or conflicting processes with the ongoing task, and disinhibiting task-relevant areas [1, 5–7]; the obtained results are in line with this hypothesis. During our task involving internally directed attention, the simultaneous delivery of task-irrelevant sensory information required active blocking of the corresponding sensory processes to avoid interference. Indeed, in the picture condition, larger posterior alpha power was observed during the math task as opposed to rest (Fig. 1B, Fig. 2, Table 1), an effect interpretable as inhibition of parietal-occipital visual processes interfering with the ongoing numerical computation. This interpretation is also supported by the absence of a similar effect in the no picture condition (Fig. 1A, Fig. 2, Table 1): in this condition, the posterior alpha power did not increase (rather exhibited a decreasing trend) during the task compared to rest, as the absence of visual distractors reduced the need for visual inhibition. Specifically, the posterior alpha power during the math task may reflect an antagonistic balance between recruitment (alpha decrease) of parietal-occipital regions involved in arithmetic problem solving [13–15] and simultaneous inhibition (alpha increase) of sensory parietal-occipital regions. The latter was probably weak in no picture condition because of the absence of external distractors and could be balanced or slightly overcome by the former. Conversely, in picture condition, alpha synchronization associated with inhibition of visual areas was likely stronger to reject distractors, overcoming alpha desynchronization associated with activation of math-related cortical areas. Other studies have evidenced possible antagonistic effects of alpha synchronization/desynchronization in tasks simultaneously involving sensory rejection and internal information processing [12, 16]. Finally, different alpha power changes across conditions emerged in the two phases. During rest, pictures did not need to be 'rejected' as they did not act as distractors, and posterior alpha tended to decrease compared to no picture condition (Fig. 1C, Fig. 2, Table 1) because of visual parietal-occipital stimulation. During the task, the posterior alpha power in the picture condition was not different from the no picture condition (although a trend to increase appeared, Fig. 1D, Fig. 2, Table 1); this emphasizes the tendency of alpha power to inhibit pictures processing when pictures acted as distracting flow of information.

In conclusion, this preliminary work shows that frontal theta increased during the math task, more largely in presence of visual distractors requiring stronger monitoring and control. The effects on alpha power are less straightforward as alpha modulation may reflect balancing between synchronization in some regions and desynchronization in others, depending on the current task and external context. The obtained results motivate future developments: (i) Increasing the sample size to corroborate the significance of the presented results; (ii) Cortical sources and functional connectivity estimation from scalp EEG via resolution of the inverse problem, to identify cortical regions specifically presenting theta synchronization and alpha synchronization/desynchronization and their reciprocal connections.

References

1. Frey, J.N., Ruhnau, F., Weisz, N.: Not so different after all: the same oscillatory processes support different types of attention. Brain Res. **1626**, 183–197 (2015)
2. Klimesch, W.: EEG alpha and theta oscillations reflect cognitive and memory performance: a review and analysis. Brain Res. Rev. **29**(2–3), 169–195 (1999)
3. Sauseng, P., Griesmayr, B., Freunberger, R., Klimsch, W.: Control mechanisms in working memory: A possible function of EEG theta oscillations. Neurosci. Biobehav. Rev. **34**(7), 1015–1022 (2010)
4. Cavanagh, J.F., Frank, M.J.: Frontal theta as a mechanism for cognitive control. Trends Cogn. Sci. **18**(8), 414–421 (2014)
5. Klimesch, W.: Alpha-band oscillations, attention, and controlled access to stored information. Trends Cogn. Sci. **16**, 606–617 (2012)
6. Foxe, J.J., Snyder, A.C.: The role of alpha-band brain oscillations as a sensory suppression mechanism during selective attention. Front. Psychol. 2(Article 154), 1–15 (2011)
7. Jensen, O., Mazaheri, A.: Shaping functional architecture by oscillatory alpha activity: gating by inhibition. Front. Hum. Neurosci. 4(Article 186), 1–8 (2010)
8. Copper, N.R., Croft, R.J., Dominey, S.J., Burgess, A.P., Gruzelier, J.H.: Paradox lost? Exploring the role of alpha oscillations during externally vs. internally directed attention and the implications of idling and inhibition hypotheses. Int. J. Psychophysiol. 47, 65–74 (2003)
9. Sammer, G., Blecker, C., Gebhardt, H., Bischoff, M., Stark, R., Morgen, K., Vaitl, D.: Relationship between regional hemodynamic activity and simultaneously recorded EEG-theta associated with mental arithmetic-induced workload. Hum. Brain Mapp. **28**, 793–803 (2007)
10. Bradley, M.M., Lang, P.J.: The international affective picture system (IAPS) in the study of emotion and attention. In: Coan, J.A., Allen, J.J.B. (eds.) Handbook of Emotion Elicitation and Assessment, pp. 29–46. Oxford University Press, New York (2007)
11. Lee, T.W., Girolami, M., Sejnowski, T.J.: Independent component analysis using an extended infomax algorithm for mixed subgaussian and supergaussian sources. Neural Comput. **11**(2), 417–441 (1999)
12. Williams, C.C., Kappen, M., Hassall, C.D., Wright, B., Krigolson, O.E.: Thinking theta and alpha: mechanisms of intuitive and analytical reasoning. Neuroimage **189**, 574–580 (2019)

13. Kitaura, Y., Nishida, K., Yoshimura, M., et al.: Functional localization and effective connectivity of cortical theta and alpha oscillatory activity during an attention task. Clin. Neurophysiol. Pract. **2**, 193–200 (2017)
14. Magosso, E., Ricci, G., Ursino, M.: Modulation of brain alpha rhythm and heart rate variability by attention-related mechanisms. AIMS Neurosci. **6**(1), 1–24 (2019)
15. De Smedt, B., Grabner, R.H.: Studer, B: Oscillatory EEG correlates of arithmetic strategy use in addition and subtraction. Exp. Brain Res. **195**(4), 635–642 (2009)
16. Sauseng, P., Klimesch, W., Dopplemayr, M., Pecherstorfer, T., Freunberger, R., Hanslmaer, S.: EEG alpha synchronization and functional coupling during top-down processing in a working memory task. Hum. Brain Mapp. **26**, 148–155 (2005)

EEG Motor Execution Decoding via Interpretable Sinc-Convolutional Neural Networks

Davide Borra(✉), Silvia Fantozzi, and Elisa Magosso

Department of Electrical, Electronic and Information Engineering "Guglielmo Marconi" (DEI), University of Bologna, Cesena, FC, Italy
davide.borra2@unibo.it

Abstract. The decoding of brain signals is a fundamental component of a brain-computer interface. Despite the success of deep convolutional neural networks (CNNs) in other fields, only recently these techniques have been applied to electroencephalographic (EEG) signals. One drawback of CNNs is the lack of interpretation of the learned features. In this study we introduce for the first time a sinc-convolutional layer into a CNN for EEG motor execution decoding, allowing a straightforward interpretation of the learned kernels. Furthermore, we apply a gradient-based analysis to assess the most relevant EEG bands for each movement and the most relevant EEG electrodes exploited in these bands. In addition to a slight accuracy improvement from 91.9 to 92.4%, our results suggest that the $high\gamma$ band is the most relevant EEG band, with gradient-based scalp distributions well localized at specific subsets of electrodes.

Keywords: Brain-computer interfaces · Electroencephalogram · Convolutional neural networks · Sinc-convolutional layer

1 Introduction

A Brain-Computer Interface (BCI) enables a direct link between the brain activity and an output command that replaces, restores or enhances the natural central nervous system output [1]. Brain signals can be acquired via electroencephalography (EEG), allowing non-invasive, inexpensive and portable recordings. In the last decades, several BCI applications were proposed [2], including neuroprosthesis control, post-stroke rehabilitation, new ways of communication after motor neuron disease and seizure control.

BCIs generally comprise [3] a data pre-processing step, a features extraction and selection steps, and a classification step. Depending on the predicted class, a feedback is provided to the user completing the closed loop. In a traditional pipeline, these steps are separately performed and exploit a strong a priori knowledge about the characteristics of the signals.

To overcome these limitations, research is moving towards the development of deep convolutional neural networks (CNNs) that operate directly with raw or

J. Henriques et al. (Eds.): MEDICON 2019, IFMBE Proceedings 76, pp. 1113–1122, 2020.
https://doi.org/10.1007/978-3-030-31635-8_135

slightly pre-processed EEG signals to output the predicted class. CNNs automatically extract relevant features and finalize the classification, without keeping these steps separate but optimizing them simultaneously. This aspect is remarkable, since not all the relevant features are known a priori. Some examples are the application of CNNs to EEG signals for motor execution and imagery decoding [4], seizure detection [5], P300 detection [6], driver performance prediction [7] and abnormalities detection [8]. One drawback of deep neural networks regards the lack of interpretation of the learned features, which is particularly important for neuroscientists. To this purpose, visualization techniques can be exploited, such as saliency maps [9]. Recently, Ravanelli and Bengio [10] proposed a temporal filtering layer, called sinc-convolutional layer, for speaker recognition directly from raw waveforms. This layer makes use of a parametrized function to compute the kernel values and forces the CNN to learn more meaningful kernels, such as band-pass filters. This can lead to some advantages compared to a standard convolutional layer, such as a reduction of the number of trainable parameters, a performance improvement and a more straightforward interpretation of the learned parameters.

This study introduces for the first time a sinc-convolutional layer based on the one proposed by Ravanelli and Bengio [10], in a deep CNN for motor execution decoding from EEG data. As baseline architecture we adopted the model proposed by Schirrmeister et al. [4], named DeepConvNet. In addition, a gradient-based analysis was performed to rank the movement-specific EEG bands learned by the sinc-convolutational layer and to identify the most relevant EEG electrodes exploited in these bands.

2 Materials and Methods

This section describes the dataset and pre-processing, the proposed Sinc-DeepConvNet (i.e. the DeepConvNet [4] modified by introducing the sinc-convolutional layer) and the gradient-based analysis. In particular, in order to perform a fair comparison between the Sinc-DeepConvNet and DeepConvNet, the dataset and the hyper-parameters, except for the first temporal convolutional layer, were maintained the same.

The CNNs were trained from scratch using Google Colaboratory and developed in PyTorch.

2.1 Dataset and Pre-processing

The EEG dataset is a 128-channel dataset obtained from 14 healthy subjects (age 27.2 ± 3.6, 6 female, 2 left-handed) by [4] and made freely available[1]. For each subject, roughly 1000 (963.1 ± 150.9 mean $\pm$ std) four-second trials of executed movements were recorded. The four classes were repetitive right and left hand sequential finger tapping, repetitive toes clenching, and rest. This dataset was especially well-suited for extracting information from the *high*γ band [4].

[1] https://web.gin.g-node.org/robintibor/high-gamma-dataset.

We adopted the same pre-processing procedure as in the original DeepConvNet study [4] (Fig. 1). The pre-processing steps are displayed in Fig. 1; the outcome of the pre-processing were cropped trials of shape (1, 523, 44) that represented the CNN input.

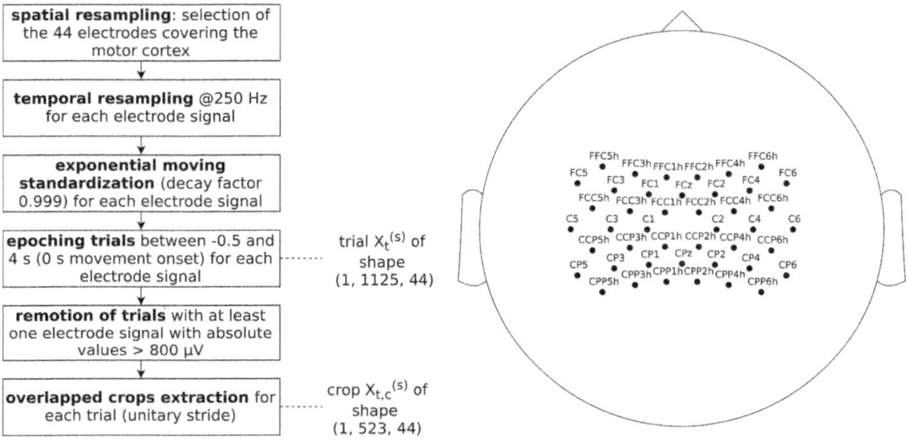

Fig. 1. (Left) Pre-processing steps (s, t, c are the subject, trial and crop indices, respectively). (Right) Position of the 44 EEG channels used.

For each subject, 160 trials (i.e. 40 trials per class) were used as test set and the remaining as training set. From the latter, a validation set (20% of the training set) was selected to perform early stopping.

2.2 Sinc-DeepConvNet

CNN Architecture. The proposed architecture with the details of the main hyper-parameters is described in Fig. 2. The CNN comprehends five convolutional layers. The first convolutional layer performs a temporal filtering of each EEG signal, while the second layer performs a spatial filtering of the first layer output.

After the second, third, fourth and fifth convolutional layers, a batch normalization layer [11], ELU non linearity [12] and max-pooling layer are included in this order. Furthermore, after the first, second and third max-pooling layers a dropout layer [13] is included. The last layer with trainable parameters is a dense layer with four units followed by a log softmax activation function that finalizes the 4-way classification task.

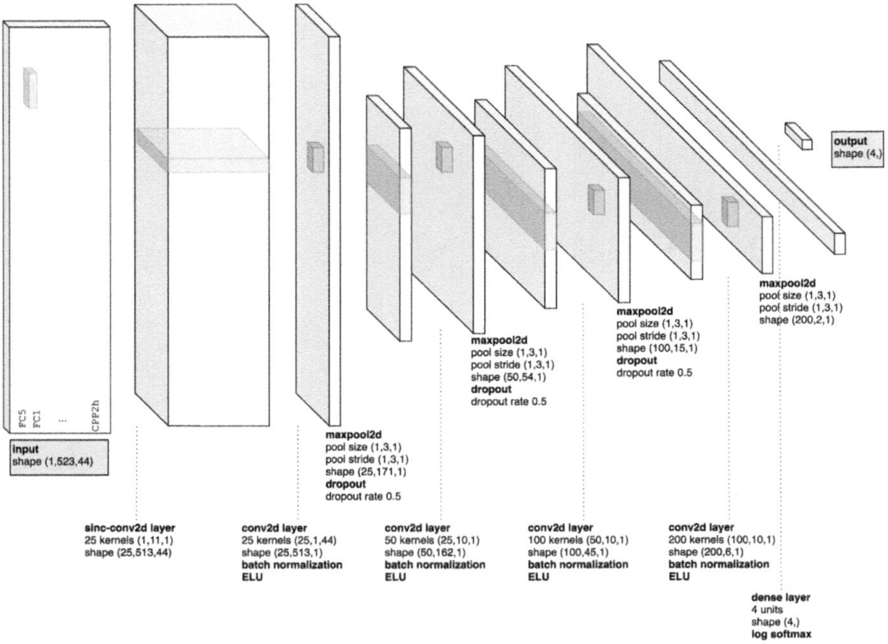

Fig. 2. Sinc-DeepConvNet architecture. This architecture is based on DeepConvNet [4] except the first temporal convolutional layer. The tuple elements refer to the CNN channel, EEG temporal and EEG channel dimensions, respectively.

Sinc-Convolutional Layer. In the original DeepConvNet [4], all the elements of the kernels were learned during the training. The 1-D convolution (Eq. 1) between the i-th EEG signal ($x_i, i \in [0, 43]$) and the j-th kernel ($h^{(j)}, j \in [0, 24]$) of length $N = 11$ is:

$$y_i^{(j)}[n] = x_i[n] * h^{(j)}[n] = \sum_{k=0}^{N-1} x_i[n-k] \cdot h^{(j)}[k], \tag{1}$$

where $y_i^{(j)}[n]$ is the output of the convolution.

Recently, in a CNN for speaker recognition, Ravanelli and Bengio [10] introduced the use of parametrized functions to compute the first layer kernel values, such that each kernel is forced to describe a band-pass filter in the temporal domain.

The proposed sinc-convolutional layer took inspiration from [10] and can be formalized as follows. The kernel values are defined by the following parametrized function $h^{(j)'}\left[n; \theta^{(j)}\right]$, where $\theta^{(j)}$ is a set of a few trainable parameters which characterize the j-th band-pass filter (Eq. 2):

$$y_i^{(j)'}[n] = x_i[n] * h^{(j)'}\left[n; \theta^{(j)}\right]. \tag{2}$$

In the frequency domain, the amplitude $H^{(j)'}$ of these parametrized filters can be expressed as $H^{(j)'}\left[f; f_1^{(j)}, f_2^{(j)}\right] = rect\left(\frac{f}{2f_2^{(j)}}\right) - rect\left(\frac{f}{2f_1^{(j)}}\right)$, where $\theta^{(j)} = \left\{f_1^{(j)}, f_2^{(j)}\right\}$ is the set of the trainable parameters of the j-th kernel which comprise the inferior and superior cutoff frequencies of the band-pass filter, respectively. Hence, the total number of trainable parameters is 2 per kernel, against $N = 11$ per kernel in the traditional formulation. In the temporal domain, the function $h^{(j)'}$ is (Eq. 3):

$$h^{(j)'}\left[n; \theta^{(j)}\right] = h^{(j)'}\left[n; f_1^{(j)}, f_2^{(j)}\right] = 2f_2 sinc\left(2\pi f_2 n\right) - 2f_1 sinc\left(2\pi f_1 n\right). \quad (3)$$

In order to alleviate the effects of the truncation of the function $h^{(j)'}$ on the characteristics of the filters, the function is multiplied by a Hamming window (Eq. 4) [10]:

$$\begin{aligned} h_w^{(j)'}\left[n; \theta^{(j)}\right] &= h^{(j)'}\left[n; \theta^{(j)}\right] \cdot w\left[n\right] \\ w\left[n\right] &= 0.54 - 0.46\cos\left(\frac{2\pi n}{N}\right). \end{aligned} \quad (4)$$

The EEG bands were set as follows: $\delta \in [1, 4)\,Hz$, $\theta \in [4, 8)\,Hz$, $\alpha \in [8, 13)\,Hz$, $\beta \in [13, 30)\,Hz$, $\gamma \in [30, 125)\,Hz$. Then, 5 kernels cutoff frequencies ($f_1^{(j)}, f_2^{(j)}, j \in [0, 24]$) were randomly initialized in the frequency range of each EEG band, resulting in 25 kernel cutoff frequencies randomly initialized inside the EEG bands, maintaining a balanced representation between bands.

While optimizing, the cutoff frequencies were updated in the range $[1, 125)\,Hz$ forcing $f_2^{(j)} > f_1^{(j)}$; furthermore, bandwidths shorter than 2 Hz were set to 2 Hz.

Training. The loss function used was a cross entropy-based function introduced by [4], named tied sample loss function. Adam [14] was used as optimization algorithm with a learning rate of 1e-3, a batch size of 60 entire trials and a maximum number of epochs of 800. The training phase was divided into two runs [15]. During the first run the CNN was trained until the validation loss reached its minimum. The training loss recorded at the first run minimum became the target threshold for the second run. During the second run, the validation set was included in the training set and the training continued until the validation loss reached the threshold loss.

Finally, the adopted performance metric was the accuracy and the values reported in this study always refer to the test accuracy.

2.3 Gradient-Based Analysis

At the end of the Sinc-DeepConvNet training, we interpreted the learned filters by visualizing them and by applying a gradient-based technique [9]. Here, we evaluated within which band and across which electrodes, the temporal filtering was more relevant for the classification.

Considering the t-th test trial of the s-th subject $X_t^{(s)}$, the trial was divided into crops $X_{t,c}^{(s)}$ and these were fed to the CNN. The following steps were performed:

(a) We computed the gradient of the output category y with respect to each neuron of the first layer Z_1 (i.e. band-pass filtered version of the input, second cuboid in Fig. 2) at the current crop $X_{t,c}^{(s)}$, $G_{t,c}^{(s)} = \frac{\partial y}{\partial Z_1}|_{X_{t,c}^{(s)}}$, having shape $(25, 513, 44)$. Then, $|G_{t,c}^{(s)}|$ was computed. These values allow to evaluate how much a perturbation of each first layer neuron affects the last layer neurons (output category).

(b) For each subject and movement, the gradients were averaged across the temporal dimension and across crops belonging to trials of that specific movement (averaged gradient shape $(25, 44)$). Then, the gradients related to the temporal filters belonging to each EEG band were extracted. These bands were δ, θ, α, β, $low\gamma$, $high\gamma$, as the γ-band was further split into $low\gamma \in [30, 60)\,Hz$ and $high\gamma \in [60, 125)\,Hz$. For each band, the filter with a maximum value of the average gradients across electrodes was identified and the 44 gradient values related to this filter were selected (gradients at the level of single electrode).

(c) For each movement and EEG band, the gradients obtained from (b) were averaged across subjects.

3 Results

3.1 Performance Comparison

The two CNNs were trained with the EEG data of each subject and the average accuracies across subjects were $92.4 \pm 9.2\%$ and $91.9 \pm 9.2\%$ for the Sinc-DeepConvNet and DeepConvNet, respectively.

3.2 Temporal Filters Visualization

When the Sinc-DeepConvNet training was completed, the sinc-convolutional layer allowed a visual assessment of the learned temporal filters for each subject-specific classifier. An example of this visualization for a single subject is reported in Fig. 3 (top), while the center frequency distribution across all subjects is showed in Fig. 3 (down). This distribution indicates absence of kernels in θ and α bands and very marginal involvement of the δ band (4 kernels of 3 subjects), while filters belong mainly to the β (17 kernels of 8 subjects), $low\gamma$ (271 kernels of 14 subjects) and $high\gamma$ bands (58 kernels of 14 subjects).

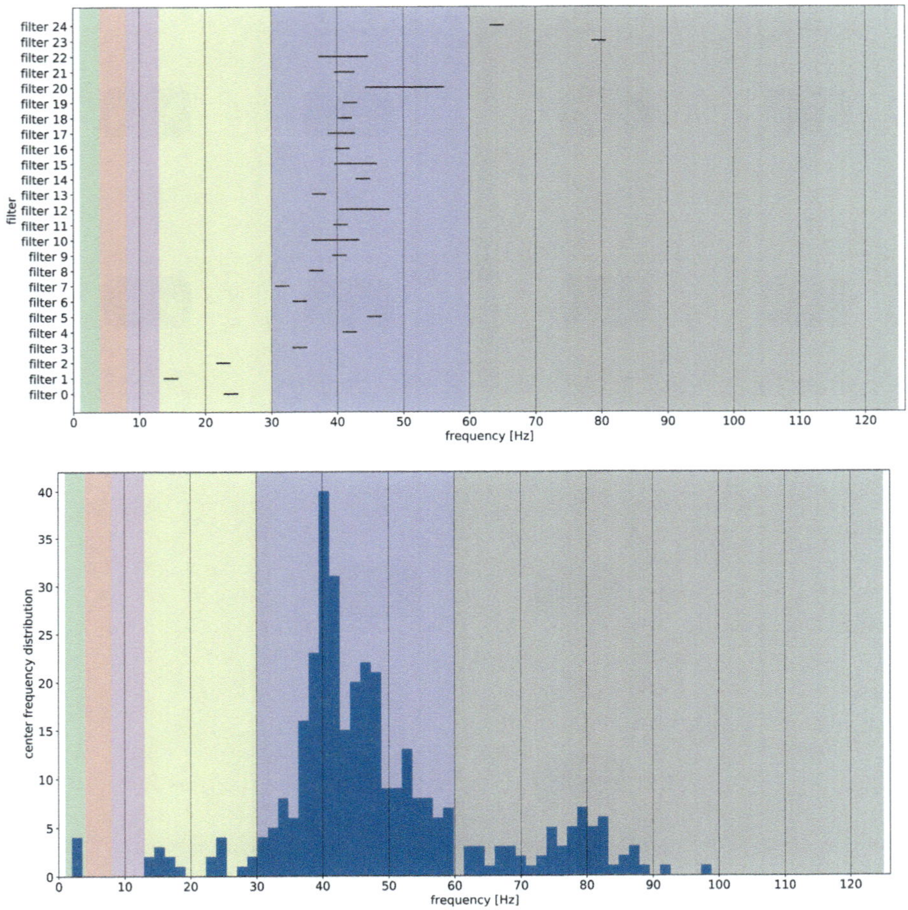

Fig. 3. (Top) Example of the temporal filters learned by the Sinc-DeepConvNet for one subject. The filters are displayed as black lines, with the end points representing $f_1^{(j)}$ and $f_2^{(j)}$. (Bottom) Center frequency distribution of the filters across all subjects. The color-code used is: green-δ, red-θ, magenta-α, yellow-β, blue-$low\gamma$, gray-$high\gamma$.

3.3 Gradient-Based Analysis

Based on previous results, the gradient analysis was limited to the kernels in the β, $low\gamma$ and $high\gamma$ bands. It is interesting to note that the gradient scalp distributions show a large variability from channel to channel (Fig. 4), the gradient peaking at specific electrode clusters.

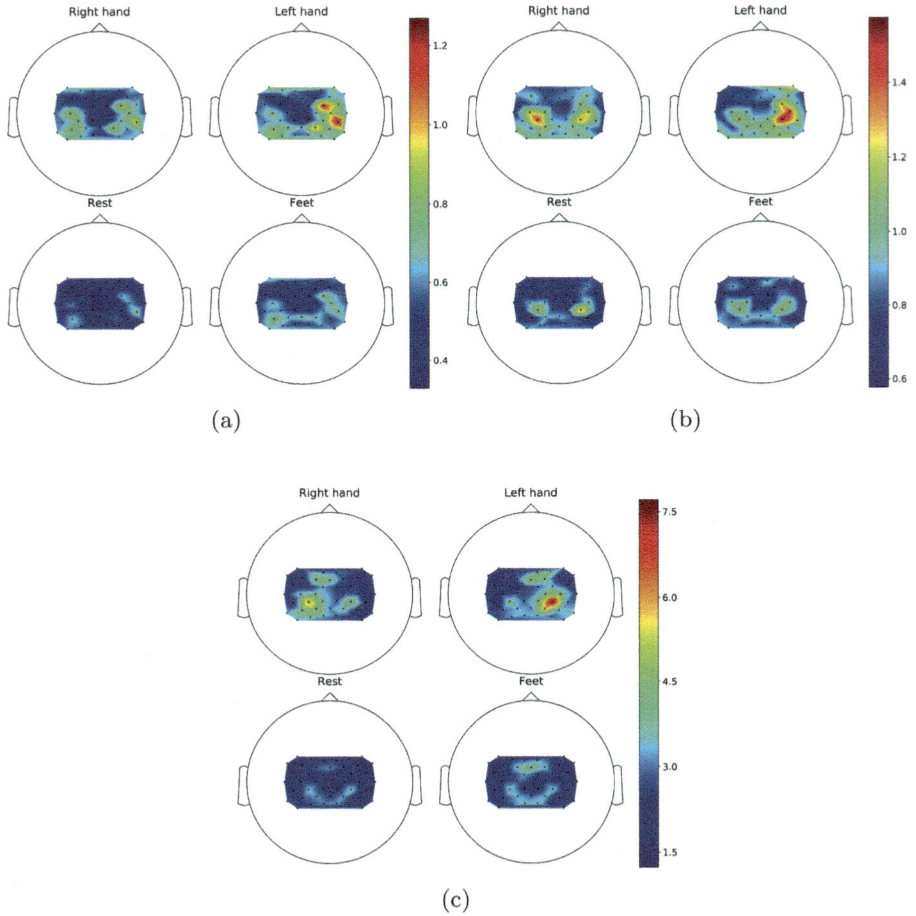

Fig. 4. Average maximum gradient distributions at level of single electrode related to the learned kernels belonging to the β (a), $low\gamma$ (b), $high\gamma$ (c) bands for the decoded classes.

4 Discussion

The introduction of the sinc-convolutional layer in the baseline DeepConvNet architecture leads to a slight accuracy improvement (on average from 91.9 to 92.4%). This indicates that forcing the CNN to learn band-pass filters in the first layer, does not negatively affect the model behaviour. The proposed architectural change involves three main advantages. First, the Sinc-DeepConvNet performs better using the proposed temporal filtering layer having only about 1/5 of the DeepConvNet first layer trainable parameters. Furthermore, the kernel values can be efficiently computed since the function described in the Eq. 3 is an even function and it requires to evaluate only half of the kernel values. Lastly, the Sinc-DeepConvNet allows a direct interpretation of the learned temporal kernels.

Figure 3 shows the cutoff frequencies for a specific subject and the center frequency distribution across all subjects. In order to perform a fair comparison between the original architecture DeepConvNet [4] and the proposed one, the number of kernels was not changed. For each subject, the majority of temporal filters lies in the $low\gamma$ range with similar cutoff frequencies. This result probably indicates that several learned kernels are unnecessary and that the number of kernels could be reduced, resulting in a further reduction of the first layer trainable parameters.

By examining the gradient distributions in Fig. 4 the three EEG bands can be ranked as $high\gamma > low\gamma > \beta$ for each movement. The importance of the temporal filtering in $high\gamma$ and β bands is in line with previous results on the same dataset [4]. Furthermore, for each EEG band the gradient values were similar between the left and right hand movements and were higher than the other two classes (rest and feet). In particular, the gradient values related to the rest condition were the lowest. Furthermore, the distributions were more localized at specific subsets of electrodes. In addition, moving from the EEG band with lower gradients (β) to the band with higher gradients ($high\gamma$), the subset of relevant electrodes became more localized. In particular, regarding the $high\gamma$ band, the distribution appeared strongly contralateral to the moved hand and symmetric for the feet movement, and the temporal filtering looked highly relevant for the channel CCP3h and CCP2h, but also for FC1, FC2 and FCz.

5 Conclusions

The reduction of the CNN first layer capacity introduced by the sinc-temporal convolutional layer is a viable solution for a more interpretable CNN allowing a more direct neuroscientific insights of the decoding task. Indeed, this first layer change allows a more straightforward assessment of the temporal filtering and using a gradient-based analysis the band- and movement-specific importance of this filtering across the electrodes can be directly evaluated. In particular, our results suggest that the $high\gamma$ band is the most relevant, with a scalp distribution of the gradients well lateralized and localized at specific subset of electrodes. Based on these promising results, the sinc-convolutional layer may be perspectively applied to EEG signals to investigate cognitive and/or motor aspects (attention, verbal and mathematical tasks, reaching/grasping) for which the relevant features are less known, taking advantage of its more straightforward interpretability.

References

1. Wolpaw, J., Wolpaw, E.: Brain-Computer Interfaces: Principles and Practice. Oxford University Press, USA (2012)
2. Chaudhary, U., Birbaumer, N., Ramos-Murguialday, A.: Brain-computer interfaces for communication and rehabilitation. Nat. Rev. Neurol. **12**, 513–525 (2016)
3. Nicolas-Alonso, L.F., Gomez-Gil, J.: Brain computer interfaces, a review. Sensors (Basel, Switzerland) **12**(2), 1211–79 (2012)
4. Schirrmeister, R.T., Springenberg, J.T., Fiederer, L.D.J., Glasstetter, M., Eggensperger, K., Tangermann, M., Hutter, F., Burgard, W., Ball, T.: Deep learning with convolutional neural networks for EEG decoding and visualization. Hum. Brain Mapp. **38**(11), 5391–5420 (2017)
5. Acharya, U.R., Oh, S.L., Hagiwara, Y., Tan, J.H., Adeli, H.: Deep convolutional neural network for the automated detection and diagnosis of seizure using eeg signals. Comput. Biol. Med. **100**, 270–278 (2018)
6. Cecotti, H., Graser, A.: Convolutional neural networks for p300 detection with application to brain-computer interfaces. IEEE Trans. Pattern Anal. Mach. Intell. **33**(3), 433–445 (2011)
7. Hajinoroozi, M., Mao, Z., Jung, T.P., Lin, C.T., Huang, Y.: Eeg-based prediction of driver's cognitive performance by deep convolutional neural network. Signal Process. Image Commun. **47**, 549–555 (2016)
8. van Leeuwen, K., Sun, H., Tabaeizadeh, M., Struck, A., van Putten, M., Westover, M.: Detecting abnormal electroencephalograms using deep convolutional networks. Clin. Neurophysiol. **130**(1), 77–84 (2019)
9. Simonyan, K., Vedaldi, A., Zisserman, A.: Deep inside convolutional networks: Visualising image classification models and saliency maps. arXiv preprint (2013)
10. Ravanelli, M., Bengio, Y.: Speaker recognition from raw waveform with sincnet. In: 2018 IEEE Spoken Language Technology Workshop (SLT), pp. 1021–1028, December 2018
11. Ioffe, S., Szegedy, C.: Batch normalization: accelerating deep network training by reducing internal covariate shift. In: Bach, F., Blei, D. (eds.) Proceedings of the 32nd International Conference on Machine Learning, Proceedings of Machine Learning Research, vol. 37, pp. 448–456. PMLR, Lille, France, 07–09 July 2015
12. Clevert, D.A., Unterthiner, T., Hochreiter, S.: Fast and accurate deep network learning by exponential linear units (elus). arXiv preprint (2015)
13. Srivastava, N., Hinton, G., Krizhevsky, A., Sutskever, I., Salakhutdinov, R.: Dropout: a simple way to prevent neural networks from overfitting. J. Mach. Learn. Res. **15**(1), 1929–1958 (2014)
14. Kingma, D.P., Ba, J.: Adam: A method for stochastic optimization. arXiv preprint (2014)
15. Hinton, G.E., Srivastava, N., Krizhevsky, A., Sutskever, I., Salakhutdinov, R.: Improving neural networks by preventing co-adaptation of feature detectors. arXiv preprint (2012)

Central Alpha Bicoherence Is Reduced in Photosensitive Subjects

Stefania Coelli[1], Elisa Visani[2], Giulia Tacchino[1], Ferruccio Panzica[2], Silvana Franceschetti[2], and Anna Maria Bianchi[1(✉)]

[1] Department of Electronics, Information and Bioengineering, Politecnico di Milano, Milan, Italy
{stefania.coelli,annamaria.bianchi}@polimi.it
[2] Neurophysiopathology Department and Epilepsy Centre, Fondazione IRCCS Istituto Neurologico Carlo Besta, Milan, Italy

Abstract. The photic driving response is known to induce epileptiform activity in photosensitive subjects and a recent hypothesis suggests that a reduced inhibitory effect of the alpha oscillations may be associated to this phenomenon. In this work, we studied the linear and non-linear spectral characteristic of the physiological alpha rhythms during a visual stimulation protocol in a group of twelve healthy and eight treated photosensitive subjects. Results confirmed the desynchronization of the individual alpha power during stimulation over the occipital region in both the groups and over the central for the PS subjects. A reduced bicoherence of the central alpha rhythm in PS subjects was found, suggesting the presence of a different behavior in the cortical structures involved in the generation of the alpha oscillations, both at rest and during visual stimulation.

Keywords: Bicoherence · Alpha · Photosensitivity

1 Introduction

Rhythmic visual stimulation can induce a physiological synchronization of the brain waves at a frequency matching and related to the stimulus rate. This physiological phenomenon is called "photic driving" response, maximal over the occipital regions [1, 2], and in predisposed subjects, it may induce seizures [2]. This pathological condition is known as photosensitive epilepsy (PSE). Even though PSE has been recognized for a long time and EEG and fMRI studies have shown the presence of abnormal brain connectivity patterns in rest condition [3] and during photic stimulation [4], the mechanisms generating the PS reflex is still not fully understood.

The role of the alpha oscillations as sensorimotor inhibitory rhythm has been investigated in individuals affected by PSE, suggesting that an impaired inhibitory circuit may be responsible for an uncontrolled synchronization [3, 5]. Over the motor cortex, inhibition is associated to oscillations in the alpha range ($\sim$10 Hz) representing the inactive state of the cortex, which is also characterized by a non-linear mechanism of coupling with its harmonic component ($\sim$20 Hz) reflecting a phenomenon called

J. Henriques et al. (Eds.): MEDICON 2019, IFMBE Proceedings 76, pp. 1123–1128, 2020.
https://doi.org/10.1007/978-3-030-31635-8_136

quadratic phase coupling (QPC), indicating a specific behavior of the inhibited sensorimotor cortex [6]. In this study, using the bicoherence as a quantitative measure of non-linearity and QPC rhythms interaction, we investigated the behavior of the physiological alpha oscillation during a visual stimulation protocol in a group of healthy subjects and PS patients.

2 Materials and Methods

2.1 EEG Recording and Stimulation Protocol

EEG signals from 12 healthy volunteers (6 males, mean age of 31 (21–67) years) with normal or correct to normal vision and no history of photosensitivity or other brain disorders acted as controls. Eight photosensitive (PS) patients (two males; mean age 24 (17–32) years) with paroxysmal response to photic stimulation during their clinical EEG examination were included in this study. All the participants and parents of subjects' younger than 18 years gave informed consent for the participation in the study, previously approved by the Ethical Committee of the C. Besta Foundation in agreement with the Declaration of Helsinki.

All the patients were under treatment during the stimulation protocol.

Stimulation was carried out by means of a full field square wave reversing checkerboard (black/white) on a screen, placed 1.5 m in front of subjects, at three different rates, corresponding to the stimulation frequencies of 7.5 Hz, 15 Hz and 24 Hz. Ten 20-s intervals of stimulation were delivered for each frequency, with a resting period of equal length (black screen). Before each stimulation block, an 80-second-long baseline interval was acquired at rest with eyes open.

EEG was simultaneously recorded by means 19 electrodes placed according to the standard 10/20 International System. The sampling rate was set to 256 Hz with antialiasing filters (1–120 Hz). Signals were acquired with a common reference and re-referenced offline to the mean of the auricular signals.

2.2 EEG Analysis

Signals were band-pass filtered in the 1–45 Hz frequency band applying a FIR, zero-phase filter to avoid line power frequency at 50 Hz. Independent Component Analysis (ICA) algorithm was performed to identify and remove eye-related artefacts. Large movements and muscular artefacts were manually removed. Each stimulation block (block I: 7.5 Hz; block II: 15 Hz; block III: 24 Hz) was divided in the 80-s baseline (B), ten 20-s stimulation (S) and ten 20-s recovery (R) sections. The surface Laplacian transform was applied to reduce the volume conduction effect.

The power spectral density (PSD) of the EEG signal was computed through the Welch's averaged modified periodogram. The non-parametric direct method was employed for the estimation of the bispectral measures [7] dividing the process in K segments on which the bispectrum is computed as $b_i(f_1,f_2) = X(f_1)X(f_2)X*(f_1+f_2)$ where * denotes the complex conjugate and X(f) the DFT of the signal. The expected value over the K segments produces the estimation of the bispectrum of the whole signal as in Eq. 1

$$B(f_1,f_2) = \frac{1}{K}\sum_{i=1}^{K} b_i(f_1,f_2) \tag{1}$$

The quadratic bicoherence, $Bic^2(f_1, f_2)$ computed as in Eq. 2, is the normalized version $B(f_1, f_2)$, it has values limited to the range [0 1], representing the percentage of power at f_1+f_2 due to the coupling between the waves at f_1, and f_2 [8].

$$Bic^2(f_1,f_2) = |B(f_1,f_2)|^2 \Big/ E[|X(f_1)X(f_2)|^2]E[|X(f_1+f_2)|^2] \tag{2}$$

The statistical significance of the bicoherence values was tested through surrogated data-sets (phase randomization) and setting a 99% statistical threshold.

Both spectral and bicoherence analysis was performed averaging 2-s sliding time windows with Hanning tapering and a 1 s of overlap. The Individual Alpha Frequency was determined, for each subject, as the frequency at which the maximum power value within the range 8–13 Hz occurred at rest over the occipital cortex. The relative spectral power was computed as the ratio of the IAF band (i.e., $f_{IAF} \pm 1$ Hz) and the total power in the range 1–45 Hz. In the bi-frequency plane, for the IAF ROI centered in (f_{IAF}, f_{IAF}) the maximum of the significant bicoherence was considered. Measures were averaged within three brain areas: Frontal – F (F3/4, Fz, F7/8), Central – C (C3, Cz, C4) and Parieto-Occipital – PO (P3/4, O1/2, Pz). Before the averaging procedure, bicoherence values underwent to the Fisher's Z transform. A three-way repeated measures analysis of variance (BWW-ANOVA) with 'CONDITIONS' and 'AREAS' as within factors and 'GROUPS' as between factor, was employed for the statistical analysis of all the measures. To test the normality of the data the Shapiro-Wilk test was employed and the Greenhouse-Geisser correction was applied when the assumption of sphericity of the data was violated according to the Mauchly's test. Post-hoc analysis p-values were corrected with the Bonferroni's method.

3 Results

For the relative spectra power the three factors interaction was not significant in any of the stimulation blocks. The overall between factor was also not significant. The interaction AREAS*CONDITIONS was significant in block I and II ($p < 0.0005$), but not in block III, where only the main factor AREAS was significant ($p < 0.0005$; with PO > C > F with $p < 0.001$). In block I and II, a significant difference was found in the central region (I, $p = 0.011$; II, $p = 0.003$), with S < R (I, $p = 0.04$; II, $p = 0.003$), while in the PO region there was a significant difference among conditions (I, II, $p < 0.0005$), with S < B (I, $p = 0.004$; II, $p = 0.001$) and S < R (I, $p = 0.002$; II, $p < 0.0005$). Looking at the trends of the alpha power in the central area reported in Fig. 1, it can be noticed that while controls did not show evident power changes across sections, patients presented a suppression of the alpha power during stimulation.

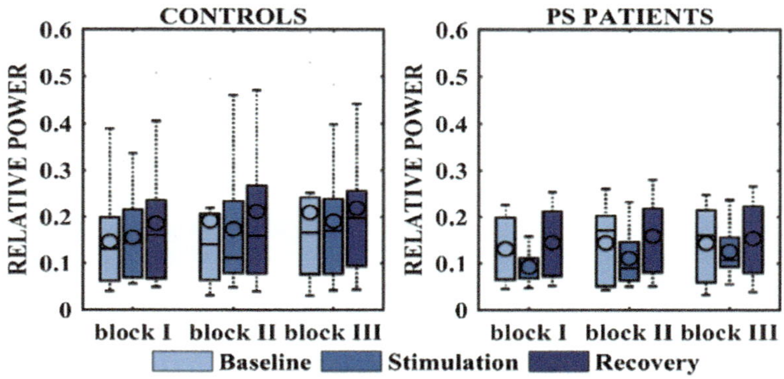

Fig. 1. Box-plots (Median, 25th–75th percentiles) of relative alpha power values averaged over the central regions.

In Fig. 2 the averaged Bic^2 values across subjects in each group presenting a significant bicoherence peak for the IAF is mapped at each channel location. The three factors interaction was not significant in any block, as the GROUPS main effect, while a significant effect AREAS*CONDITIONS was present in blocks I and II (I, $p = 0.001$; II, p 0.002). Precisely, in the PO area significant differences were present among conditions in block I and II (I, $p = 0.004$; II, $p = 0.008$) with S < B (I, $p = 0.026$; II, $p = 0.042$). In block III, the bicoherence in the IAF ROI was characterized by a GROUPS*AREAS interaction ($p = 0.017$) with a significant higher bicoherence in controls than in the PS subjects ($p < 0.0005$) over the central area. Further analysis on the central area alone across all the stimulation blocks showed a significant difference between groups during stimulation ($p = 0.039$).

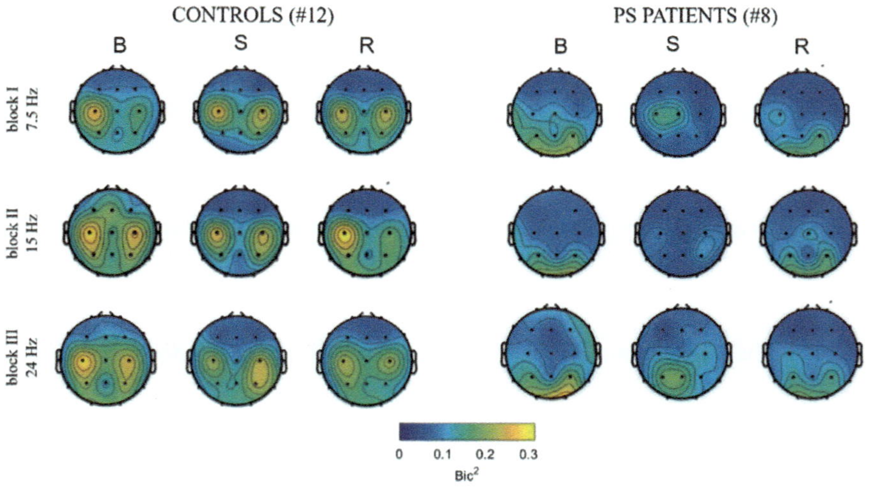

Fig. 2. Topographical representation of the significant bicoherence in the IAF ROI, averaged across controls and patients in each stimulation block

4 Discussion

In this study, we quantitatively confirmed that the visual rhythmic stimulation has a desynchronizing effect of the occipital alpha rhythm in both the analyzed populations. This effect was present also on the central area for the PS subjects. Despite the evident variability across participants, in healthy subjects the significant alpha peak in the bicoherence plane suggested a QPC between the components at $\sim$10 Hz and $\sim$20 Hz (beta range) localized at the occipital and central channels, as previously documented [9], while in PS patients this phenomenon was reduced on the central electrodes both at rest and during stimulation. The alpha rhythm over the central regions is associated to the "idling" rhythm (mu) of the sensorimotor cortex. The QPC relation associated to this rhythm suggests a specific non-linear behavior of the sensorimotor cortex [6]. In PS patients, lower significant bicoherence peaks were detected over the motor area, maybe suggesting a reduced QPC-like behavior with a decreased inhibitory effect. In line with this result, Vaudano found a smaller decrease of the bold response in PS subjects (on treatment) than in PS-negative subjects, suggesting a reduced inhibition of the motor cortex at rest [3]. A reduced inhibition of the visuomotor integration was supported also by studies combining IPS and TMS stimulation showing a higher excitability of the motor cortex in PS subjects [5]. Our results further support, through the quantitative analysis of the non-linear behavior of EEG signals, the hypothesis that photosensitivity may be linked to an abnormal alpha generating network.

Conflict of Interests Declaration. The authors declare that they have no competing interests.

References

1. Herrmann, C.S.: Human EEG responses to 1–100 Hz flicker: resonance phenomena in visual cortex and their potential correlation to cognitive phenomena. Exp. Brain Res. **137**, 346–353 (2001)
2. Visani, E., Varotto, G., Binelli, S., Fratello, L., Franceschetti, S., Avanzini, G., Panzica, F.: Photosensitive epilepsy: spectral and coherence analyses of EEG using 14 Hz intermittent photic stimulation. Clin. Neurophysiol. **121**, 318–324 (2010). https://doi.org/10.1016/j.clinph.2009.12.003
3. Vaudano, A.E., Ruggieri, A., Avanzini, P., Gessaroli, G., Cantalupo, G., Coppola, A., Sisodiya, S.M., Meletti, S.: Photosensitive epilepsy is associated with reduced inhibition of alpha rhythm generating networks. Brain J. Neurol. **140**, 981–997 (2017). https://doi.org/10.1093/brain/awx009
4. Varotto, G., Visani, E., Canafoglia, L., Franceschetti, S., Avanzini, G., Panzica, F.: Enhanced frontocentral EEG connectivity in photosensitive generalized epilepsies: a partial directed coherence study. Epilepsia **53**, 359–367 (2012). https://doi.org/10.1111/j.1528-1167.2011.03352.x
5. Strigaro, G., Prandi, P., Varrasi, C., Monaco, F., Cantello, R.: Defective visual inhibition in photosensitive idiopathic generalized epilepsy. Epilepsia **53**, 695–704 (2012)
6. Pfurtscheller, G., Stancák, A., Edlinger, G.: On the existence of different types of central beta rhythms below 30 Hz. Electroencephalogr. Clin. Neurophysiol. **102**, 316–325 (1997)

7. Nikias, C.L., Mendel, J.M.: Signal processing with higher-order spectra. IEEE Signal Process. Mag. **10**, 10–37 (1993). https://doi.org/10.1109/79.221324
8. Kim, Y.C., Powers, E.J.: Digital bispectral analysis and its applications to nonlinear wave interactions. IEEE Trans. Plasma Sci. **7**, 120–131 (1979). https://doi.org/10.1109/TPS.1979.4317207
9. Bianchi, A.M., Molteni, S.C., Panzica, F., Visani, E., Franceschetti, S., Cerutti, S.: Spectral and bispectral analysis of the EEG rhythms in basal conditions and during photic stimulation, pp 574–577 (2004)

Combined and Singular Effects of Action Observation and Motor Imagery Paradigms on Resting-State Sensorimotor Rhythms

Aleksandar Miladinović[1(✉)], Antonella Barbaro[2], Eddi Valvason[2], Miloš Ajčević[1,3,4], Agostino Accardo[1], Piero Paolo Battaglini[2], and Joanna Jarmolowska[3]

[1] Department of Engineering and Architecture, University of Trieste, Via A. Valerio 10, 34127 Trieste, Italy
aleksandar.miladinovic@phd.units.it
[2] Department of Life Sciences, University of Trieste, Via Fleming 22, 34127 Trieste, Italy
[3] Department of Medicine Surgery and Health Sciences, University of Trieste, Trieste, Italy
[4] NEUROFARBA Department, Neuroscience Section, University of Florence, Florence, Italy

Abstract. In the present study, 30 right-handed participants randomly performed one of three motor neurorehabilitation paradigms: action observation (AO), motor imagery (MI) and combined action observation and motor imagery (AO+MI) of the right arm and hand movement. Resting state electroencephalography (EEG) was acquired for 5 min before and immediately after the motor paradigms session. EEG was recorded from 10 sites over sensorimotor areas, and the average power was calculated for left (FC3, C3, C1, C5, CP3) and right (FC4, C4, C2, C6, CP4) regions in the spectral bands: delta, theta, alpha, mu, low and high beta. Our main finding demonstrates that delta, theta and mu activity decreased significantly on the contralateral regions during MI, while low beta increased significantly. Except for the mu band, the same changes were observed on the ipsilateral side, where delta and theta decreased significantly, while low beta became significantly higher. No relevant effects were observed for AO or combined AO and MI. These findings demonstrate a rapid effect of MI on cortical modulation in sensorimotor areas which is revealed by changes in resting state oscillatory activity and suggest an interesting interplay between MI and AO. The presented findings may be relevant for choosing a proper protocol for clinical motor neurorehabilitation approaches.

Keywords: Motor imagery · Action observation · EEG · Biomedical signal processing

1 Introduction

It is known that action observation (AO) engages almost the same brain regions as action execution [1, 2]. The neurophysiological basis for this hypothesis lies in the presumed human mirror neuron system, where cortical motor regions that are active

J. Henriques et al. (Eds.): MEDICON 2019, IFMBE Proceedings 76, pp. 1129–1137, 2020.
https://doi.org/10.1007/978-3-030-31635-8_137

both when we execute an action and when we observe similar actions being performed by others [1–3]. Recent work has described that "action observation network" involves parietal, premotor, and occipitotemporal brain regions [2, 4]. By sharing motor circuits with action execution, AO may prime the motor system for subsequent motor practice [5, 6]. In that regards, in the last decade, AO has been recommended in the clinical environment as additional practice in neurorehabilitation settings [7, 8].

Motor imagery (MI), on the other hand, is a cognitive process during which the execution of a motor action is internally prepared without any motor output [9, 10]. Although no actual movement is made, functional magnetic resonance imaging (fMRI) and electroencephalography (EEG) studies have shown that the brain regions engaged in action execution are also activated in MI [10]. The cognitive literature proposes that the effective kinesthetic sensations, including movement, effort, heaviness, and position provide information that enables the human system to determine the position of limbs and to identify the origins and the cause of action [11]. Further, it has been proposed that there are functional equivalence and the use of common neural pathways in motor preparation/execution and motor imagery [12, 13]. Thus, because motor preparation/execution and motor imagery involve the same motor representation systems, they likely have the same neuronal substrates [1, 2, 14].

In the neurorehabilitation practice, both AO and MI have shown beneficial effects [15, 16]. Moreover, together with physical exercise, MI does not produce only beneficial effects on athletes [17] and musicians [18], but also improves behavioral outcomes on a clinical population suffering from stroke and other neurological impairments [5, 6, 19].

The cortical activation using either AO or MI alone was studied [20], but until now, the investigations on how to combine MI with AO are quite rare. According to recent studies, combined AO and MI could enhance the activation of motor circuits by producing changes in the EEG [21, 22], suggesting that the combined use of them might be even more useful. On these bases, in order to investigate the possibility of development of new neurorehabilitation protocols, this study examines changes in resting state oscillatory activity in the sensorimotor area after AO, MI and their simultaneous application, in healthy subjects.

2 Materials and Methods

2.1 Subjects and Experimental Protocol

Experiments were performed on 30 right-handed (Edinburgh Handedness Inventory [23] 83.5 ± 19.3) participants (18 females, 12 males; mean ± 1SD age: 21.66 ± 1.18 years), all with normal or corrected-to-normal vision. The subjects' motor capabilities were evaluated by Italian version of Movement Imagery Questionnaire [24, 25]. The research was conducted according to the principles of the Declaration of Helsinki. All participants released their informed consent to participate in the study after all procedures had been fully explained.

Participants were randomly assigned to one of three motor neurorehabilitation paradigms: AO, MI and combined AO+MI, thus yielding three different groups with 10 participants each. The experimental protocol consisted of: (1) pre resting-state recording, (2) one of the mentioned paradigms repeated for 40 times and (3) post resting-state recording (Fig. 1). To avoid subjects fatigue and to increase comfort, after each 10 trials in (2) a long pause was introduced. In each experimental block, participants were seated in a dark room in a front of a computer screen that was located at eye-height in front of the observer's central viewing position.

In the AO paradigm, subjects were asked to watch a video showing a right hand reaching, grasping and moving objects. The video was filmed from the subject perspective, with the aim to create an immersive effect. A male or female hand was displayed in accordance with the subject's gender. Each trial started with the warning "beep" sound followed by two seconds of black screen after which the video-clip started to play automatically. The single video had a duration of 6.5 s and it was presented 40 times. In order to keep subjects attentive, two types of videos with a randomized number of precision or coarse grasp into a single block (of 10 trials) were presented and the subjects were instructed to count the number of appearances of one of them and report it after an experimental block of 10 trials.

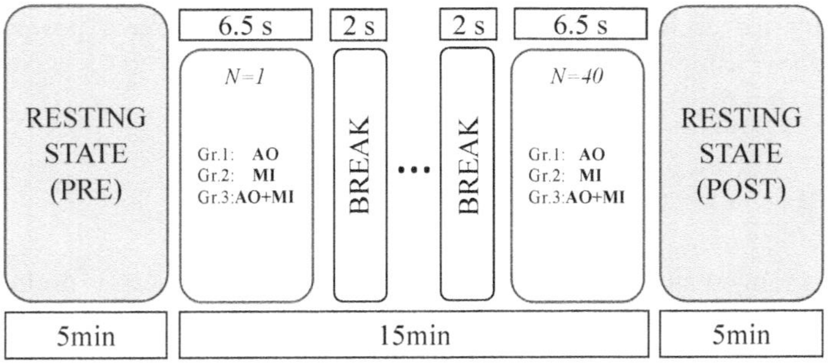

Fig. 1. A schematic diagram of the experimental protocol which begins and ends with the resting-state recordings of 5 min (gray boxes) and administers (for N = 40 repetitions) one of the AO, MI and AO+MI paradigms, to each 10-participant group (denoted by Gr.1-3).

In the MI paradigm, the subjects were trained on how to properly perform motor imagery, simulating their proprioception and adopting the first-person perspective (i.e. imagining the movement of their own hand). After the training, the recording session started with the "beep" warning followed by the still image of a grip (hand movement) as in AO. In this case, subjects were instructed to mentally simulate the action by trying to "experience the same feelings as during the actual execution" facilitating kinesthetic motor imagery approach. As in AO, the subjects had 6.5 s for MI and the process repeats for 40 times.

For the AO+MI condition, subjects observed the same videos presented in AO, but in this case, they were additionally required to perform MI corresponding to the displayed video. The paradigm was performed with the same parameters (duration and repetition) as the previous two.

2.2 EEG Acquisition and Processing

5-minutes resting state EEG for each subject were recorded before and immediately after each session. EEG signals were sampled at 256 Hz by using SAM 32FO amplifier (Micromed, Treviso, Italy) and a prewired headcap with 10 Ag/AgCl electrodes (Spes Medica, Genova, Italy) placed at standard 10-10 locations covering the sensorimotor area (FC3, FC4, C3, C4, C1, C2, C5, C6, CP3 and CP4). The reference electrode was placed in POz, while the ground electrode was placed in AFz. Electrode impedances were kept below 5 kΩ. EOG activity was recorded to identify eye-movement artifacts. EEG off-line analysis was carried out using MATLAB® (The MathWorks Inc., Natick, MA). All channels were digitally filtered with the 0.5–45 Hz 2nd order Butterworth bandpass filter. Artifacts were manually discarded after visual inspection of tracings and 60 s of stationary EEG signal pre and after motion paradigm epochs were selected for spectral analysis. Power spectral density (PSD) was estimated for each channel using Welch's periodogram [26] (averaged on 11 tracts of 10 s each, windowed with a Hamming window, with 50% overlap). Subsequently, for each channel the relative power in each spectral band (delta: 0.5–4 Hz; theta: 4–8 Hz; alpha: 8–13 Hz; mu: 8–10 Hz; $beta_{low}$: 13–18 Hz; $beta_{high}$: 18–30 Hz) was calculated. For each subject and each band average power was calculated for left (FC3, C3, C1, C5 and CP3) and right (FC4, C4, C2, C6 and CP4) sensorimotor area channels. Differences between pre and post activity for each of three motion paradigms and for each power band were determined by using Wilcoxon signed-rank test. P-values < 0.05 were considered statistically significant.

3 Results

Median and range values of relative powers of the considered bands during the resting state before (PRE) and after (POST) each of the three performed paradigms are reported in Table 1. Difference between PRE and POST activity in the case of MI, showed a significant decrease of power in the delta, theta and mu bands on the left contralateral sensorimotor area, while low beta increased significantly. Except for the mu band, similar behavior was observed on right ipsilateral side, where delta and theta also decreased significantly, and low beta became significantly higher. No significant changes were observed for AO in any of the analysed bands. In the case of AO+MI, a significant difference was found only in the theta band, which decreased in both hemispheres.

Table 1. Median (range) values of relative power in delta, theta, beta and sigma bands during the resting state right before and after each of the three paradigms (MI, AO and AO+MI). * statistically different from Pre (p-value < 0.05).

SM area	BAND	Paradigm	PRE	POST	*p*-value
Left contralateral	delta	AO	0.413 (0.275-0.478)	0.437 (0.330-0.565)	0.148
		MI	0.429 (0.226-0.517)	0.377 (0.154-0.545)	0.049*
		AO+MI	0.412 (0.279-0.671)	0.342 (0.263-0.770)	0.82
	theta	AO	0.120 (0.084-0.198)	0.121 (0.092-0.158)	0.461
		MI	0.110 (0.081-0.198)	0.085 (0.059-0.170)	0.004*
		AO+MI	0.118 (0.075-0.171)	0.085 (0.073-0.168)	0.027*
	alpha	AO	0.085 (0.068-0.193)	0.082 (0.062-0.254)	0.844
		MI	0.082 (0.071-0.135)	0.078 (0.054-0.135)	0.557
		AO+MI	0.089 (0.045-0.132)	0.072 (0.038-0.128)	0.203
	mu	AO	0.038 (0.029-0.093)	0.034 (0.020-0.064)	0.109
		MI	0.035 (0.027-0.065)	0.029 (0.021-0.045)	0.048*
		AO+MI	0.037 (0.026-0.058)	0.031 (0.017-0.045)	0.164
	$beta_{low}$	AO	0.098 (0.048-0.394)	0.135 (0.064-0.211)	0.945
		MI	0.127 (0.055-0.342)	0.203 (0.077-0.498)	0.01*
		AO+MI	0.111 (0.060-0.326)	0.142 (0.039-0.349)	0.25
	$beta_{high}$	AO	0.147 (0.116-0.227)	0.132 (0.073-0.176)	0.195
		MI	0.154 (0.062-0.392)	0.137 (0.068-0.408)	0.77
		AO+MI	0.165 (0.041-0.287)	0.118 (0.042-0.304)	0.95
Right ipsilateral	delta	AO	0.383 (0.276-0.571)	0.384 (0.283-0.527)	0.94
		MI	0.444 (0.211-0.526)	0.351 (0.128-0.544)	0.037*
		AO+MI	0.497 (0.263-0.664)	0.375 (0.285-0.688)	0.426
	theta	AO	0.120 (0.084-0.191)	0.113 (0.090-0.170)	0.461
		MI	0.106 (0.072-0.202)	0.086 (0.057-0.174)	0.02*
		AO+MI	0.124 (0.088-0.166)	0.101 (0.060-0.146)	0.008*
	alpha	AO	0.099 (0.069-0.193)	0.095 (0.050-0.234)	0.742
		MI	0.079 (0.061-0.106)	0.076 (0.045-0.136)	0.846
		AO+MI	0.086 (0.042-0.134)	0.073 (0.048-0.132)	0.652
	mu	AO	0.040 (0.031-0.082)	0.040 (0.020-0.057)	0.312
		MI	0.035 (0.024-0.051)	0.032 (0.014-0.041)	0.105
		AO+MI	0.035 (0.018-0.063)	0.032 (0.016-0.050)	0.301
	$beta_{low}$	AO	0.128 (0.042-0.427)	0.172 (0.093-0.280)	0.25
		MI	0.123 (0.057-0.363)	0.163 (0.097-0.533)	0.004*
		AO+MI	0.096 (0.065-0.232)	0.154 (0.062-0.350)	0.25
	$beta_{high}$	AO	0.150 (0.096-0.195)	0.141 (0.101-0.188)	0.945
		MI	0.127 (0.062-0.433)	0.153 (0.060-0.435)	0.922
		AO+MI	0.133 (0.069-0.209)	0.099 (0.059-0.290)	0.652

4 Discussion

The main finding of our study is that motor imagination induces higher activation of motor cortex with respect to action observation or their combination, inducing significant changes of resting-state EEG in mu and low beta bands, also known as sensorimotor (SMR) bands, and in the slower oscillatory bands, in particular delta and theta.

The effect of MI on mu rhythms could be explained by the specific repetitive motor event-related synchronization (ERS) and desynchronization (ERD) occurring during imagination. During ERS/EDR the strongest activity is present on the contralateral hemisphere with respect to the imagined arm [27–29], which can lead to effective changes of the resting-state EEG even after the MI session has been completed.

On the other hand, the significant increase in the lower beta band, in MI in comparison to AO and AO+MI, could be interpreted as a result of repeated motor specific beta rebounds (ERS) occurring immediately after every imagination on both contralateral and ipsilateral hemisphere, and are physiologically related to sustained muscle contraction or voluntary movement [29].

The participants in the present study were explicitly asked to use motor imagery in order to simulate a movement. Therefore, we speculate, that motor imagery can assist in the creation of a vivid simulation of the same feelings as during the actual execution [30, 31], activating in this way, the common neural pathways in motor preparation/execution and motor imagery [12, 13]. Our results suggest that motor imagery induce most efficiently changes of resting-state of SMR.

Regarding the effects of AO on the resting-state, in our case, no significant change was observed. The possible lack of attention during the task has to be disregarded as an explanation for the results, since all subjects had excellent performance on the side-task (counting the repetition of hand movements) that controlled their focus on task. As a matter of fact, the inefficacy of AO to induce significant corticomotor variation in SMR is probably due to a lower activation causing no significant variation of resting-state activity with respect to MI.

Concerning the effects of combined action observation and motor imagery on the resting-state of SMR, in our case, no significant change was founded. Therefore, we must recognize that brain activity during motor imagery with action observation was not simply the sum of these two tasks. It should be noted that in such dual task, the cognitive load requested to the participants was very high and this could have an impact on its efficiency.

Clearly, our results support the idea that MI, AO and AO+MI all cause different changes of SMR. In fact, despite apparent similarity, suggesting a common substrate underlying MI and AO [32], the present results show important differences between them, that could be explained from a different theoretical point of view.

MI is an explicit covert mental state during which participants internally simulate a movement without actually performing it [9]. As during real execution, mental simulation of movement involves anticipations about sensory and motor effects of that action. Precisely, the framework of internal models suggests that, during both actual and imagined actions, the future sensorimotor state is predicted by the given efferent copy of the motor command and the current state of the body [9]. It has been proposed that such kinesthetic feeling related to the limb is typically processed through the parietal region that modulates, in turn, motor cortex facilitation during MI [33].

In contrast, during AO, the visual inputs of the movement performed by the others would not have immediate access to the observer's motor system [9]. In this context, our results are not in contrast with previous findings, which have suggested the existence of a common neural substrate for MI and AO. We believe that both AO and MI activate the motor network but exploit different sensory-motor processes. AO is based on the processing of visual information involved in the movement, indirectly activating the motor cortex [9]. This is a type of bottom-up processing, in which attention is mainly focused on sensory input that is external to the body. On the other hand, MI is related to top-down internal processing. This is the main difference, which in our opinion, can explain why in a short session of only 15 min, MI, but not AO, activated the motor system rapidly, causing a change in the oscillatory activity in the following resting state.

In other words, we suggest that mental simulation of movement, which involves anticipations about sensory and motor effects of that action, induces higher activation of motor cortex with respect to action observation.

We also found that theta power band decreased bilaterally in MI and in AO+MI and that the delta power band decreased contralaterally in MI. The role of the theta band is usually related to memory formation, information processing [34], working memory [35] and sensorimotor integration [36]. It was also reported that the theta band plays a role in tasks of motor imagery [37]. There is some evidence supporting the idea that frontal theta EEG activity correlates negatively with the default mode network in resting state [38]. Another study reports that amplitude increase in low frequency oscillations (e.g. delta, theta) are related to a decrease of BOLD signal. Based on theoretical considerations, it is suggested that higher energy dispersal, and therefore a higher BOLD signal, is related to a relative shift in neuronal activity from lower to higher frequencies [39]. This would result in, for instance, reduced delta and theta and increased beta amplitudes. For this reason, we suggest that default mode network of brain regions could be deactivated during attention-demanding cognitive tasks required by both MI and AO+MI paradigm. Thus, the delta and theta decrease may be considered to reflect the participant's sensory-motor integration and attention load.

In conclusion, our results demonstrate a direct and rapid effect of cortical modulation induced by MI on the EEG resting-state. Moreover, they show important differences between MI and AO. Such information may be used to improve clinical protocols of AO and MI in therapeutic approaches that can include also BCI neurofeedback protocols.

Acknowledgments. A. Miladinović is supported by European Social Fund (ESF) - FVG. M. Ajčević is supported by AIRAlzh Onlus - (ANCC-COOP). Joanna Jarmolowska is supported by the Interreg V-A Italia-Slovenia 2014-2020 program MEMORI-net.

Work partially supported by Master In Clinical Engineering, University of Trieste and Interreg V-A Italia-Slovenia 2014-2020 program MEMORI-net.

Conflict of Interest. The authors have no conflict of interest do declare.

References

1. Di Pellegrino, G., Fadiga, L., Fogassi, L., Gallese, V., Rizzolatti, G.: Understanding motor events: a neurophysiological study. Exp. Brain Res. **91**, 176–180 (1992)
2. Gallese, V., Fadiga, L., Fogassi, L., Rizzolatti, G.: Action recognition in the premotor cortex. Brain **119**(Pt 2), 593–609 (1996)
3. Rizzolatti, G., Craighero, L.: The mirror-neuron system. Annu. Rev. Neurosci. **27**, 169–192 (2004)
4. Lotze, M., Montoya, P., Erb, M., Hülsmann, E., Flor, H., Klose, U., Birbaumer, N., Grodd, W.: Activation of cortical and cerebellar motor areas during executed and imagined hand movements: an fMRI study. J. Cogn. Neurosci. **11**, 491–501 (1999)
5. Pomeroy, V., Aglioti, S.M., Mark, V.W., McFarland, D., Stinear, C., Wolf, S.L., Corbetta, M., Fitzpatrick, S.M.: Neurological principles and rehabilitation of action disorders. Neurorehabil. Neural Repair. **25**, 33S–43S (2011)
6. Ertelt, D., Small, S., Solodkin, A., Dettmers, C., McNamara, A., Binkofski, F., Buccino, G.: Action observation has a positive impact on rehabilitation of motor deficits after stroke. Neuroimage **36**, T164–T173 (2007)
7. Buccino, G.: Action observation treatment: a novel tool in neurorehabilitation. Philos. Trans. R. Soc. B Biol. Sci. **369**, 20130185 (2014)
8. Ste-Marie, D.M., Law, B., Rymal, A.M., Jenny, O., Hall, C., McCullagh, P.: Observation interventions for motor skill learning and performance: an applied model for the use of observation. Int. Rev. Sport Exerc. Psychol. **5**, 145–176 (2012)
9. Jeannerod, M.: Neural simulation of action: a unifying mechanism for motor cognition. Neuroimage **14**, S103–S109 (2001)
10. Sharma, N., Baron, J.-C.: Does motor imagery share neural networks with executed movement: a multivariate fMRI analysis. Front. Hum. Neurosci. **7**, 564–570 (2013)
11. Enoka, R.M.: Neuromechanical Basis of Kinesiology, 2nd edn. Human Kinetics, Champaign (1994)
12. Decety, J.: Neural mechanisms subserving the perception of human actions. Trends Cogn. Sci. **3**, 172–178 (1999)
13. Jeannerod, M.: The 25th Bartlett Lecture: to act or not to act: perspectives on the representation of actions. Q. J. Exp. Psychol. Sect. A **52**, 1–29 (1999)
14. Porro, C.A., Cettolo, V., Francescato, M.P., Baraldi, P.: Ipsilateral involvement of primary motor cortex during motor imagery. Eur. J. Neurosci. **12**, 3059–3063 (2000)
15. Schuster, C., Hilfiker, R., Amft, O., Scheidhauer, A., Andrews, B., Butler, J., Kischka, U., Ettlin, T.: Best practice for motor imagery: a systematic literature review on motor imagery training elements in five different disciplines. BMC Med. **9**, 75 (2011)
16. Sharma, N., Pomeroy, V.M., Baron, J.-C.: Motor imagery: a backdoor to the motor system after stroke? Stroke **37**, 1941–1952 (2006)
17. Holmes, P., Calmels, C.: A neuroscientific review of imagery and observation use in sport. J. Mot. Behav. **40**, 433–445 (2008)
18. Lotze, M., Halsband, U.: Motor imagery. J. Physiol. **99**, 386–395 (2006)
19. Jackson, P.L., Lafleur, M.F., Malouin, F., Richards, C., Doyon, J.: Potential role of mental practice using motor imagery in neurologic rehabilitation. Arch. Phys. Med. Rehabil. **82**, 1133–1141 (2001)
20. Macuga, K.L., Frey, S.H.: Neural representations involved in observed, imagined, and imitated actions are dissociable and hierarchically organized. Neuroimage **59**, 2798–2807 (2012)

21. Vogt, S., DiRienzo, F., Collet, C., Collins, A., Guillot, A.: Multiple roles of motor imagery during action observation. Front. Hum. Neurosci. **7**, 807 (2013)
22. Eaves, D.L., Riach, M., Holmes, P.S., Wright, D.J.: Motor imagery during action observation: a brief review of evidence, theory and future research opportunities. Front. Neurosci. **10**, 514 (2016)
23. Oldfield, R.C.: The assessment and analysis of handedness: the Edinburgh inventory. Neuropsychologia **9**(1), 97–113 (1971)
24. Hall, C.R., Kathleen, A.M.: Measuring movement imagery abilities: a revision of the movement imagery questionnaire. J. Ment. Imag. **21**, 143–154 (2012)
25. Williams, S.E., Cumming, J., Ntoumanis, N., Nordin-Bates, S.M., Ramsey, R., Hall, C.: Further validation and development of the movement imagery questionnaire. J. Sport Exerc. Psychol. **34**, 621–646 (2012)
26. Welch, P.: The use of fast Fourier transform for the estimation of power spectra: a method based on time averaging over short, modified periodograms. IEEE Trans. Audio Electroacoust. **15**(2), 70–73 (1967)
27. Pfurtscheller, G., Brunner, C., Schlogl, A., Lopes da Silva, F.H.: Mu rhythm (de) synchronization and EEG single-trial classification of different motor imagery tasks. NeuroImage **31**, 153–159 (2006)
28. Pfurtscheller, G., Aranibar, A.: Event-related cortical desynchronization detected by power measurements of scalp EEG. Electroencephalogr. Clin. Neurophysiol. **42**, 817–826 (1977)
29. Van Elswijk, G., Maij, F., Schoffelen, J.-M., Overeem, S., Stegeman, D.F., Fries, P.: Corticospinal beta-band synchronization entails rhythmic gain modulation. J. Neurosci. **30**, 4481–4488 (2010)
30. Sirigu, A., Duhamel, J.R., Cohen, L., Pillon, B., Dubois, B., Agid, Y.: The mental representation of hand movements after parietal cortex damage. Science **273**, 1564–1568 (1996)
31. Papaxanthis, C., Schieppati, M., Gentili, R., Pozzo, T.: Imagined and actual arm movements have similar durations when performed under different conditions of direction and mass. Exp. Brain Res. **143**, 447–452 (2002)
32. Hardwick, R.M., Caspers, S., Eickhoff, S.B., Swinnen, S.P.: Neural correlates of motor imagery, action observation, and movement execution: a comparison across quantitative meta-analyses. Neurosci. Biobehav. Rev. **94**, 31–44 (2018)
33. Tian, X., Poeppel, D.: Mental imagery of speech and movement implicates the dynamics of internal forward models. Front. Psychol. **1**, 166 (2010)
34. McCartney, H., Johnson, A.D., Weil, Z.M., Givens, B.: Theta reset produces optimal conditions for long-term potentiation. Hippocampus **14**, 684–687 (2004)
35. Jensen, O., Tesche, C.D.: Frontal theta activity in humans increases with memory load in a working memory task. Eur. J. Neurosci. **15**, 1395–1399 (2002)
36. Cruikshank, L.C., Singhal, A., Hueppelsheuser, M., Caplan, J.B.: Theta oscillations reflect a putative neural mechanism for human sensorimotor integration. J. Neurophysiol. **107**, 65–77 (2012)
37. Kahana, M.J., Seelig, D., Madsen, J.R.: Theta returns. Curr. Opin. Neurobiol. **11**, 739–744 (2001)
38. Scheeringa, R., Bastiaansen, M.C., Petersson, K.M., Oostenveld, R., Norris, D.G., Hagoort, P.: Frontal theta EEG activity correlates negatively with the default mode network in resting state. Int. J. Psychophysiol. **67**(3), 242–251 (2008)
39. Kilner, J.M., Mattout, J., Henson, R., Friston, K.J.: Hemodynamic correlates of EEG: a heuristic. Neuroimage **28**, 280–286 (2005)

Network Analysis on Overnight EEG Spectrum to Assess Relationships Between Paediatric Sleep Apnoea and Cognition

Gonzalo César Gutiérrez-Tobal[1](✉), Javier Gomez-Pilar[1], Leila Kheirandish-Gozal[2], Adrián Martín-Montero[1], Jesús Poza[1], Daniel Álvarez[1,3], Félix del Campo[1,3], David Gozal[2], and Roberto Hornero[1]

[1] Biomedical Engineering Group, Universidad de Valladolid, Valladolid, Spain
gonzalo.gutierrez@gib.tel.uva.es

[2] Department of Child Health, The University of Missouri School of Medicine, Columbia, USA

[3] Pneumology Service, Río Hortega University Hospital, Valladolid, Spain

Abstract. One major risk of paediatric sleep apnoea-hypopnoea syndrome (SAHS) is the development of cognitive impairments among affected children. Electroencephalography (EEG) is ordinarily used as part of polysomnography, the standard diagnostic test for SAHS. However, how SAHS changes overnight EEG, and its relationships with cognitive performance, remains unclear. In this study, we first analyzed the spectral content of EEG recordings from 294 children to explore possible differences caused by SAHS. Then, a correlation network analysis was conducted to evaluate relationships among different EEG spectral bands and the results from the Differential Ability Scales (DAS) battery of cognitive tests. Our analyses identified up to six new SAHS spectral bands of interest in the EEG. They also showed higher absolute correlations among the different spectral bands as SAHS severity increased. Higher SAHS severity degree also indicated higher absolute correlations with DAS tests. Our results suggest that the spectral content of the overnight EEG is useful to characterize both paediatric SAHS and the cognitive performance of the affected children.

Keywords: Paediatric sleep apnoea · EEG · Spectral analysis · Correlation networks

1 Introduction

The sleep apnoea-hypopnoea syndrome (SAHS) is a highly prevalent disease affecting up to 5% of children [1]. Recurrent upper airway collapse and airflow reductions during the night lead to inadequate gas exchange, hypoxia episodes and fragmented sleep [2]. These undesirable effects may cause negative symptoms among which cognitive and behavioral deficits are regarded as very common and deleterious [2].

Several previous studies focused on establishing relationships between paediatric SAHS and the cognitive performance of affected children. These studies provided evidence of higher prevalence of SAHS among poorly performing students [3],

J. Henriques et al. (Eds.): MEDICON 2019, IFMBE Proceedings 76, pp. 1138–1146, 2020.
https://doi.org/10.1007/978-3-030-31635-8_138

associations between SAHS symptoms and lower cognitive functioning [4], as well as relationships between SAHS presence/severity and lower scoring in neurocognitive tests [2, 5]. Additionally, recent studies have focused on specific electroencephalographic (EEG) events, reaching different conclusions about associations between slow waves and spindle patterns with SAHS and cognition [6, 7]. Although EEG is ordinarily used to evaluate sleep architecture, its overnight analysis in healthy and SAHS-affected children has been scarcely investigated. Hence, in this preliminary study, we propose the analysis of the entire overnight EEG.

Spectral analysis has shown its usefulness in different contexts to automatically process biomedical signals, such as EEG [8] and magnetoencephalography (MEG) [9], including overnight recordings associated with SAHS [10–12]. The recurrent nature of the apneic events, which also induce abnormal recurrent episodes in the EEG, such as respiratory arousals [13], also supports the suitability of this frequency-domain methodology. Accordingly, we hypothesized that the spectral information of the overnight EEG provides useful information about paediatric SAHS and its cognitive implications.

In order to evaluate this assumption, we calculated the relative power (RP) from the conventional EEG frequency bands, as well as from new SAHS-related bands of interest (BOI) identified during the current study. RPs were extracted from each of the typical channels recorded in sleep studies. They were subsequently included in correlation networks, representing different SAHS-severity degrees, along with the General Conceptual Ability (GCA) score. This cognitive indicator is derived from the battery of tests included in the Differential Ability Scales (DAS), and is commensurate with the intelligent quotient (IQ) [2]. Our main objective was to show the evolution of the EEG spectral content as paediatric SAHS worsens, as well as its relationship with the general cognitive ability of children.

2 Materials

Two hundred and ninety four community non-referral children (5–9 years) underwent an overnight polysomnography (PSG) that included 8-channel EEG (F3, F4, T3, T4, C3, C4, O1, and O2, from the international 10–20 system). Apnoea-hypopnoea index (AHI) was derived from PSG by clinicians according to the rules of the American Academy of Sleep Medicine (AASM) [13]. AHI was used to establish both the presence and severity of SAHS: no-SAHS, AHI < 1 event/hour (176 subjects); mild-SAHS, $1 \leq$ AHI < 5 e/h (98 subjects); moderate-to-severe-SAHS AHI $\geq$ 5 e/h (20 subjects). Table 1 shows clinical and demographic data of the subjects under study. No statistically significant differences (Kruskal-Wallis p-value > 0.05) were found among groups in body mass index (BMI) or age.

EEG was recorded at sampling rates of 250 Hz and 500 Hz that were subsequently homogenized to 200 Hz, as recommended by the AASM [14]. A four-step preprocessing stage was also conducted prior to the analysis of the EEG: (i) re-referencing to the average of all EEG channels; (ii) stop-band filter to 60 Hz and band-pass filter (Hamming window) from 0.1 to 70 Hz; (iii) automatic artifact rejection on the basis of epoch-adaptive thresholding; and (iv) rejection of first and last parts of the EEG to avoid initial and final awake periods.

Table 1. Demographic and clinical data in each SAHS severity group (mean ± Sd)

	no-SAHS	mild-SAHS	moderate-to-severe-SAHS
# subjects	176	98	20
Age (years)	6.60 ± 0.75	6.56 ± 0.75	6.6 ± 0.82
Sex (M/F)	104/72	55/43	10/10
BMI (kg/m^2)	17.10 ± 2.97	18.46 ± 4.82	19.48 ± 5.27
AHI (e/h)	0.39 ± 0.28	1.81 ± 0.83	13.61 ± 10.92

*BMI: body mass index; AHI: apnoea-hypopnoea index; M/F: male/female

DAS tests were administered in the morning immediately after the PSG night [2]. It was composed of a battery of neurocognitive tests that measured the ability among a wide range of intellectual activities [2]. Scores derived from each test were computed in accordance with the age of each child and merged into a single standard score that estimates the global intellectual development (GCA score) [2].

3 Methods

3.1 Spectral Analysis

The power spectral density (PSD) of all EEG channels from all subjects were estimated following the Blackman-Tukey method. Thus, a rectangular non-overlapping 6000-sample window was used to split each EEG channel. The PSDs of the epochs were averaged to estimate a single PSD for each channel of each subject. Then, each averaged PSD was normalized by the total power of the corresponding channel.

The RPs of the conventional EEG frequency bands ($Delta_1$: 0.1–2 Hz, $Delta_2$: 2–4 Hz; *Theta*: 4–8 Hz, *Alpha*: 8–13 Hz; $Beta_1$: 13–19 Hz, $Beta_2$: 19–30 Hz; *Gamma*: 30–70 Hz) were obtained from all channels of each subject. RP was computed as the sum of the amplitude values of the normalized PSDs within each frequency band. RPs from six SAHS frequency bands of interests found during the current study were also obtained (Fig. 1). They were found by comparing the grand averaged PSDs amplitude values at each frequency of the different SAHS severity groups using *p*-values from analysis of covariance (ANCOVA) [15].

3.2 Correlation Network Analysis

Weighted correlation network is based on pairwise correlations between variables [16]. In this study, a multi frequency network approach was used in which the RP for each frequency band and channel considered was depicted with a single node in the network. The weighted link between each pair of nodes represents the association between such pair of variables (in terms of monotonic relationship). GCA DAS score was also included as a different node (variable) with the ultimate goal of assessing the interaction between cognitive performance and RPs from EEG.

In order to remove possible bias due to the unbalanced number of subjects in each group, the following split was conducted: (i) 8 subgroups of 22 subjects from the no-SAHS group, (ii) 4 subgroups of 20 subjects and an additional subgroup of 18 subjects from the mild-SAHS group, and (iii) a single group of 20 subjects from the severe-SAHS group. All correlation matrices from a same SAHS-severity group were averaged. Thus, a single matrix was obtained per SAHS-severity group. A non-paranormal transformation was performed for this purpose [17]. Correlation networks were constructed by means of the R package qgraph [18]. Particularly, we used the iterative algorithm described by Fruchterman and Reingold [19], which forces embedded network layouts after 500 iterations.

The whole correlation network analysis was conducted in two cases: (i) using the conventional EEG frequency bands and (ii) the BOIs derived from the statistical analysis of the comparison between groups. Therefore, six networks were obtained (2 cases $\times$ 3 SAHS-severity groups).

3.3 Statistical Analysis

ANCOVA was used to evaluate statistically significant differences when establishing the spectral BOIs (p-value < 0.05). These were corrected for multiple comparisons using the false discovery rate (FDR) methodology. They were also adjusted for age and sex. Spearman's partial-correlation coefficient was used in the correlation networks to assess possible relationships among the RPs of the frequency bands of each channel.

4 Results

4.1 Definition of the SAHS Frequency Bands of Interest

Figure 1 shows the normalized grand average of the PSD (PSDn) from all channels split into the three SAHS severity groups. It also displays the significant p-values (<0.05) reached in the comparison of the PSDn amplitude values from these groups at each frequency. Six BOIs in total were found. The spectral content was mainly concentrated below 5 Hz, where more visual differences arose and up to four BOIs were defined (BOI 1: 0.15–0.38 Hz; BOI 2: 1.17–1.50 Hz; BOI 3: 1.87–2.90 Hz; BOI 4: 3.72–4.63 Hz). All of them were within *Delta 1*, *Delta 2*, and *Theta* bands. However, another two BOIs were present above 39 Hz, within the conventional *Gamma* band (BOI 5: 39.21–58.26 Hz; BOI 6: 61.73–70.0 Hz) with the gap between them coinciding with the values obtained after the 60 Hz stop-band filter.

4.2 Correlation Networks

Figures 2(A)–(F) show correlation networks for the three SAHS severity groups, where the nodes are the RPs of each band and each channel taken as variables. There is another node for the DAS GCA score. Different colors represent each frequency band and DAS. Figures 2(A)–(C) show the conventional EEG frequency bands for each of the SAHS groups, whereas Figs. 2(D)–(F) show the corresponding correlation networks for the BOIs. The wider the lines that connect each node, the higher the absolute Spearman's correlation coefficient. Red and green colors address the sign of the coefficient (negative and positive, respectively). Absolute correlation values below 0.05 are considered residual and are not represented.

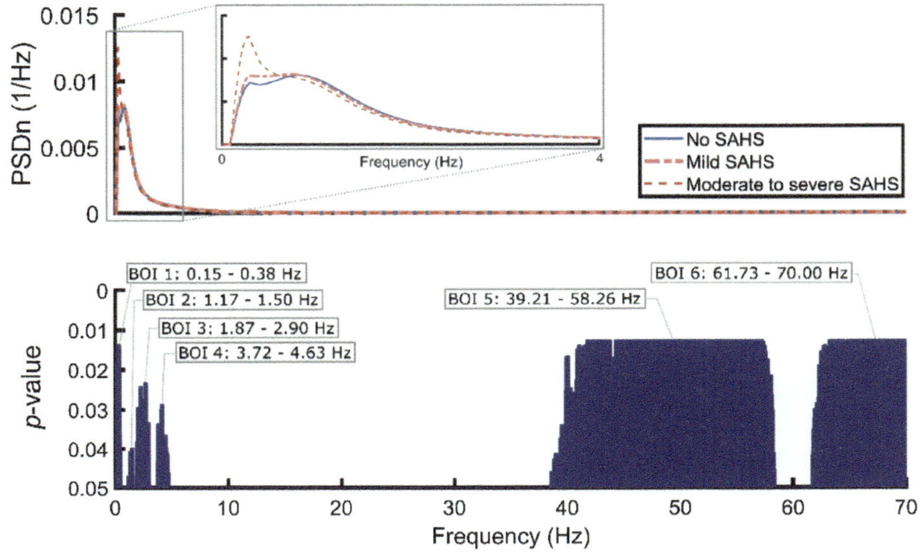

Fig. 1. Normalized grand-average PSD (zoomed in 0–4 Hz) of the three SAHS severity groups and corresponding p-values at each frequency. Bands of interests (BOI) 1 to 6 (p-values < 0.05 Kruskall-Wallis test) are also shown.

Correlation networks of the conventional EEG frequency bands show an ordered network for the controls in Fig. 2(A), where the stronger correlations form clusters of channels representing information from the same frequency band.

Clusters gradually disappear as SAHS severity increases in Figs. 2(B) and (C), suggesting higher degree of relationship among the information of the entire EEG spectrum in SAHS presence. Interestingly, the correlation between GCA DAS score and the EEG spectral information also increases with SAHS severity.

Similar tendencies can be observed in the networks formed with the RPs from the BOIs, with the information among the different bands being increasingly correlated as SAHS worsens. Absolute correlations between the GCA DAS score and the RPs from the EEG BOIs also increase with SAHS severity. In addition, the positively correlated clusters formed with BOI 5 and BOI 6 in Figs. 2(D) and (E) may indicate a continuous BOI only disturbed by the 60 Hz stop-band filter.

5 Discussion and Conclusions

In this preliminary study, we have shown differences in the spectral content of the overnight EEG from three different SAHS severity groups: controls (no-SAHS), mild-SAHS, and moderate-to-severe SAHS. We have also presented new spectral BOIs with the potentiality of being more SAHS-specific than the conventional EEG frequency bands. RPs extracted from both these bands and the six BOIs showed lower clustering behavior in the correlation networks as SAHS worsens, indicating that SAHS may be affecting the EEG spectrum in a wide range. This would be consistent with recent studies that reported a continuous impact on the brain due to abnormal breathing during sleep [20]. In both cases, the DAS GCA score increased the absolute value of correlation with the RPs as SAHS severity increased, showing that the cognitive effects in children induced by SAHS could be reflected in the overnight EEG.

BOI 1 (0.15–0.38 Hz) mainly falls within a frequency range in which arousals have particular influence (0.025–0.3 Hz) [21]. As Fig. 1 shows, PSDn amplitudes of SAHS groups are higher than those from no-SAHS are in that region. In addition, it is well-known that arousals often follow apneic events [14]. These circumstances support the specific relationship of this band with SAHS. Moreover, BOI 2 (1.17–1.50 Hz), BOI 3 (1.87–2.90 Hz) and BOI 4 (3.72–4.63 Hz) are mostly included within *Delta* band, whose relationship with non-rapid eye movement (NREM) deeper stage sleep is commonly accepted [22]. Decreasing PSDn values are shown in mild-SAHS and

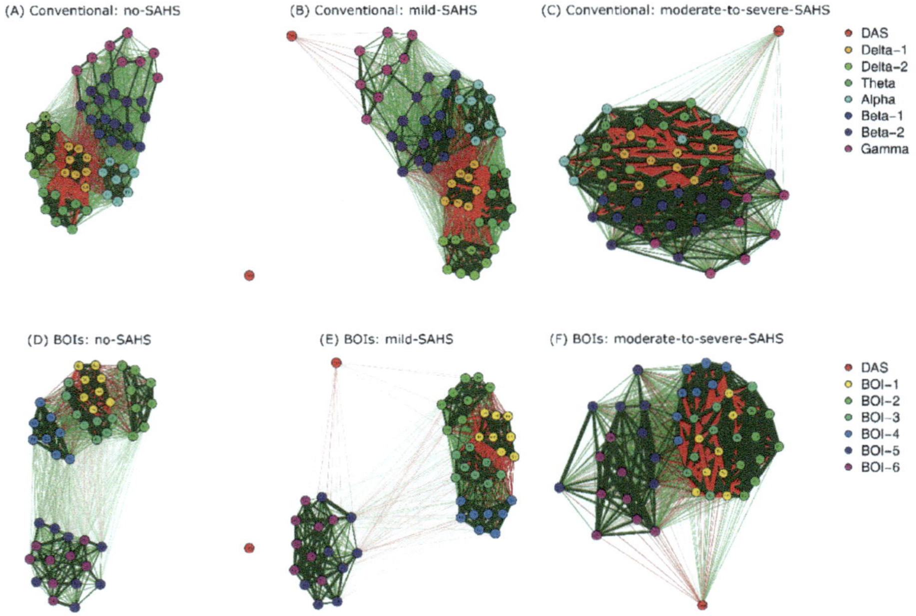

Fig. 2. Correlation networks in the three SAHS severity groups (no-SAHS, mild-SAHS, and moderate-to-severe-SAHS) for the conventional EEG frequency bands (A)–(C) and the new six BOIs found (D)–(F). DAS GCA score is also included in the analysis as a red node.

moderate-to-severe-SAHS groups in these bands, suggesting less deep NREM sleep activity as SAHS worsens. This agrees with the typical sleep fragmentation present in SAHS-affected children [2]. Finally, BOI 5 and BOI 6 are fully included within *Gamma*. This band is associated with active cognitive processing regardless from whether it is generated after external stimuli or internal sources [23]. It has been hypothesized that an association exists between *Gamma* and conscious states [25]. Thus, changes among SAHS groups might be indicating different consciousness degree, which may be due to alterations in the arousal pattern. However, assessment of these hypotheses requires further study.

The cognitive impairment suffered by SAHS-affected children has been well-established [2–5]. However, little research is available regarding possible characteristics in the EEG as reflecting diminished cognitive performance. Weichard et al. [6], reported *Delta* power at the beginning of the night as the key factor underlying specific cognitive and behavioral problems in a cohort of 42 children, a finding which is consistent with the results of the current study (BOIs 2 to 4). On the other hand, Brockman et al. [7], analyzed spindle patterns in both control and SAHS children (33 subjects in total). They found significant correlations between N2-REM stage spindle density from the SAHS group and several neurocognitive tests. These results are also consistent with ours since spindle recurrence and duration are reflected in the band 0.025–0.3 Hz [23], in which BOI 1 is included. These studies, however, focused on single EEG events or frequency bands (*Delta* activity and spindles), whereas our work used a large database to analyze the overnight EEG in an unbiased fashion. This approach allowed us to identify increased absolute correlations of the GCA DAS cognitive score in the entire nocturnal EEG information as SAHS worsens.

Despite the promising results identified in this study, several limitations deserve mention. The proportion of moderate-to-severe subjects should be increased. Although our database is larger than any other used in other state-of-the-art studies, such expansion of the database would be useful for the sake of the generalization of our results. Additionally, subjects from a SAHS clinical referral population would be helpful to validate and enrich our conclusions, as our subjects were selected from the general community. This issue is being currently addressing for future studies. Another future goal is the analysis of our results in accordance with each sleep stage.

To summarize, we have shown that changes in the overnight EEG spectrum arise in the presence of paediatric SAHS. These changes alter both low (BOIs 1 to 4) and high (BOIs 5 and 6) frequencies. The information from the EEG spectrum shows more relationships among all its spectral bands as SAHS worsens. In addition, the spectral information from the EEG has also enhanced relationships with cognitive performance as SAHS severity increases. These results suggest that the spectral information from the overnight EEG, along with its correlation network analysis, can be useful to characterize both paediatric SAHS and identify those children whose cognitive performance is affected by this condition.

Acknowledgements. This work was supported by 'Ministerio de Ciencia, Innovación y Universidades' and 'European Regional Development Fund (FEDER)' under projects DPI2017-84280-R and RTC-2017-6516-1, and by 'European Commission' and 'FEDER' under project 'Análisis y correlación entre el genoma completo y la actividad cerebral para la ayuda en el diagnóstico de la enfermedad de Alzheimer' ('Cooperation Pro- gramme Interreg V-A Spain-Portugal POCTEP 2014–2020'). DG and LKG are supported by US National Institutes of Health grants HL130984 (LKG) and HL140548 (DG).

The authors declare no conflict of interests.

References

1. Marcus, C.L., et al.: Diagnosis and management of childhood obstructive sleep apnea syndrome. Pediatrics **130**(3), e714–e755 (2012)
2. Hunter, S.J., et al.: Effect of sleep-disordered breathing severity on cognitive performance measures in a large community cohort of young school-aged children. Am. J. Respir. Crit. Care Med. **194**(6), 739–747 (2016)
3. Gozal, D.: Sleep-disordered breathing and school performance in children. Pediatrics **102**(3), 616–620 (1998)
4. Gottlieb, D.J., et al.: Sleep-disordered breathing symptoms are associated with poorer cognitive function in 5-year-old children. J. Pediatr. **145**(4), 458–464 (2004)
5. Kheirandish-Gozal, L., et al.: Obstructive sleep apnoea is associated with impaired pictorial memory task acquisition and retention in children. Eur. Respir. J. **36**(1), 164–169 (2010)
6. Weichard, A.J., Walter, L.M., Hollis, S.L., Nixon, G.M., Davey, M.J., et al.: Association between slow-wave activity, cognition and behaviour in children with sleep-disordered breathing. Sleep Med. **25**, 49–55 (2016)
7. Brockmann, P.E., et al.: Sleep spindle activity in children with obstructive sleep apnea as a marker of neurocognitive performance: a pilot study. Eur. J. Paediatr. Neurol. **22**(3), 434–439 (2018)
8. Bachiller, A., et al.: Decreased spectral entropy modulation in patients with schizophrenia during a P300 task. Eur. Arch. Psychiatry Clin. Neurosci. **264**(6), 533–543 (2014)
9. Poza, J., et al.: Evaluation of spectral ratio measures from spontaneous MEG recordings in patients with Alzheimer's disease. Comput. Methods Programs Biomed. **90**(2), 137–147 (2008)
10. Garde, A., et al.: Development of a screening tool for sleep disordered breathing in children using the phone Oximeter™. PLoS ONE **9**(11), e112959 (2014)
11. Hornero, R., et al.: Nocturnal oximetry-based evaluation of habitually snoring children. Am. J. Respir. Crit. Care Med. **196**(12), 1591–1598 (2017)
12. Gutiérrez-Tobal, G.C., et al.: Evaluation of machine-learning approaches to estimate sleep apnea severity from at-home oximetry recordings. IEEE J. Biomed. Health Inform. **23**(2), 882–892 (2019)
13. Berry, R.B., et al.: Rules for scoring respiratory events in sleep: update of the 2007 AASM manual for the scoring of sleep and associated events. J. Clin. Sleep Med. **8**(05), 597–619 (2012)
14. Iber, C., et al.: The AASM Manual for the Scoring of Sleep and Associated Events. Manual, American Academy of Sleep Medicine (2007)
15. Gutiérrez-Tobal, G.C., et al.: Diagnosis of pediatric obstructive sleep apnea: preliminary findings using automatic analysis of airflow and oximetry recordings obtained at patients' home. Biomed. Signal Process. Control **18**, 401–407 (2015)
16. Borsboom, D., et al.: The small world of psychopathology. PLoS ONE **6**(11), e27407 (2011)
17. Liu, H., et al.: The nonparanormal: Semiparametric estimation of high dimensional undirected graphs. J. Mach. Learn. Res. **10**(10), 2295–2328 (2009)

18. Epskamp, S., et al.: qgraph: network visualizations of relationships in psychometric data. J. Stat. Softw. **48**(4), 1–18 (2012)
19. Fruchterman, T.M., Reingold, E.M.: Graph drawing by force-directed placement. Softw. Pract. Experience **21**(11), 1129–1164 (1991)
20. Guilleminault, C., et al.: The nocturnal-polysomnogram and "non-hypoxic sleep-disordered-breathing" in children. Sleep Med. (2018)
21. Vanhatalo, S., et al.: Infraslow oscillations modulate excitability and interictal epileptic activity in the human cortex during sleep. Proc. Natl. Acad. Sci. **101**(14), 5053–5057 (2004)
22. Amzica, F., Steriade, M.: The K-complex: its slow (<1-Hz) rhythmicity and relation to delta waves. Neurology **49**(4), 952–959 (1997)
23. Cantero, J.L., et al.: Gamma EEG dynamics in neocortex and hippocampus during human wakefulness and sleep. Neuroimage **22**(3), 1271–1280 (2004)

Brain Processing During Postural Control – A Study Case

Run Friðriksdóttir[1(✉)], Gunnar H. Karlsson[1], Halldor Á. Svansson[1], Fabio Barollo[1], Kyle J. Edmunds[1], Hannes Petersen[2,3], and Paolo Gargiulo[1,4]

[1] Institute for Biomedical and Neural Engineering, Reykjavik University, Menntavegur 1, Reykjavik, Iceland
run13@ru.is
[2] Department of Anatomy, University of Iceland, Reykjavík, Iceland
[3] Akureyri Hospital, Akureyri, Iceland
[4] Department of Science, Landspítali, Reykjavik, Iceland

Abstract. In humans, postural control is naturally unstable and is considered to be a complex motor skill derived from the interaction of multiple sensorimotor processes to maintain postural orientation. This is a case study based results from a previous research performed on a large cohort [1]. From a technological point of view, the aim is to characterize the cortical responses associated with abrupt balance perturbations (adaptation) and changes in cortical activity during prolonged postural control (habituation).. Also, to compare the cortical activity with the mechanical body sway performance during a balance task.

Proprioceptive vibratory stimulation on calf muscles at 85 Hz was performed to evoke postural perturbation in a closed-eye experimental trial. Thirty seconds of quiet stance (baseline) and 225 s of random vibratory stimulation phases were divided into three epochs of 75 s each, with the impulses synchronized to EEG recording, using a 256-channel EEG cap and sampling rate of 1024 Hz. Channels shown to have statistically significant change in absolute power spectra variation were distinguished over six frequency bands (Δ, θ,α, β, low γ, and high γ). Force and torque actuated by the feet were recorded by a force platform and sampled at 50 Hz. A fluctuation index was computed to determine postural performance during the epochs.

Results show a significant correlation between progress in platform performance and changes in cortical activity. The fluctuation indices decrease (indicating improvement in postural performance), and significant increase was found in absolute power values during adaptation phase in the parietal regions particularly in low gamma and thetaband.

Keywords: Postural response · Balance · Cerebral cortex · EEG · Power spectral density

J. Henriques et al. (Eds.): MEDICON 2019, IFMBE Proceedings 76, pp. 1147–1154, 2020.
https://doi.org/10.1007/978-3-030-31635-8_139

1 Introduction

It has become clear that controlling posture and balance requires a complex interplay within and between the sensory and motor systems and also a strong evidence for a crucial contribution of the cerebral cortex [2]. Certain health conditions can greatly affect postural control. Lesions in the CNS have been found to lead to postural disorders especially lesions affecting the basal ganglia resulting in Parkinson's and Huntington's disease [3, 4]. Disturbed vestibular input can cause vertigo and postural disturbances. Moreover, benign paroxysmal positional vertigo (BPPV) are the most common causes of vertigo and result in spinning sensations upon changes in the position of the head which increases the chance of falls [5]. Motion sickness may also be considered a disorder affecting postural control, and affects many people. Effective postural rehabilitation and improvement of mobility to prevent falls requires a better understanding of the complex mechanisms underlying postural control.

1.1 Electroencephalogram (EEG)

Cortical assessment of postural control requires the EEG recording technique which can measure spontaneous neuronal activity from the scalp of a human. Demonstrating the direct involvement of various brain areas in human postural control is a fairly recent research topic; however, methods for research on cortical assessment vary. Some studies focus on cortical processes associated with continuous balance control using force plates with different levels of surface stability [6, 7], while others investigated changes in evoked activity as time-domain event-related potentials (ERP), time locked to postural perturbation [8].

1.2 Body Mechanic Assessment

Assessment of postural performance requires observing movement of the body in one way or another while maintaining posture. Biomechanical analysis during postural control is a technique that has been used quite persistent assessment the past decades [9].

2 Methods

Measurements were carried out in the Icelandic Center for Neurophysiology laboratory at Reykjavik University. Subjects received a detailed information letter about the measurements and signed a consent form accepting participation. Measurements were performed in accordance with the Helsinki declaration of 1975 and approved by the Icelandic ethical committee. The subject had no history of vertigo, central nervous disease, or injury to the lower extremities. Alcohol beverages had not consumed for at least 24 h before the measurement. The experimental set-up and the protocol was invented for a recent postural control study during data collection and cohort analysis [1].

2.1 Experimental Set-Up

The experimental setup was twofold: (1) Assessment of postural performance by means of posturography; (2) Assessment of EEG recording cortical activity. Spontaneous sway was recorded for 30 s followed by 225 s of vibratory stimulation exposed to the gastronomes muscles of both legs of the subject. Postural performance was recorded using a custom made force platform device, Postcon^{TM}, from Lund Institute of Technology in Sweden, which also generated a sequence of random vibratory stimulation [10]. The EEG signal was acquired using a 256 channel cap. A custom made trigger box was built, connecting the two techniques. When vibratory stimuli to the calves turned on and off, a trigger was generated in both cases to the real-time recorded EEG signal (Fig. 1).

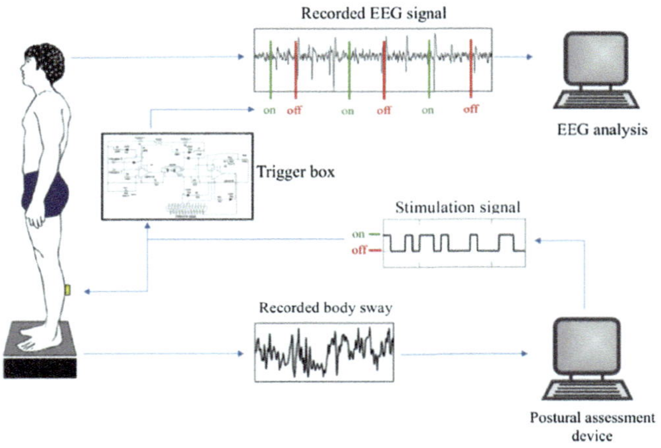

Fig. 1. Schematic view of the experimental set-up

The subject was instructed to maintain upright posture during vibratory balance perturbations evoked by vibrators designed to deliver simultaneous stimulation to both calf muscles. Each vibrator, constructed as cylinders 6 cm in length and 1 cm in diameter, was held in place by elastic straps fastened tightly around the calf muscle. Vibratory stimulation was set to deliver a vibratory stimulation of amplitude 0.1 cm and frequency of 85 Hz. Stimulation was applied according to a pseudorandom binary sequence schedule (PRBS), where each shift had a random duration of 1 s to 6.6 s. EEG signals were recorded from 256 scalp locations. The measurement includes an initial quiet stance phase (30 s) and a stimulation phase (225 s). Measurements were divided into four epochs: QS Q1 Q2 Q3 where QS was designated as the pre-stimulus Quiet Stance (QS). The epochs following QS each contained an interval of75 s. For this study in particular, QS, Q1 and Q3 epochs were isolated to define adaptation and habituation epoch events, respectively.

2.2 Assessment of Postural Control

The assessment of postural performance was recorded through the force platform. Forces and torques actuated by the feet were recorded with six degrees of freedom with an accuracy of 0.5 N, and data sampled at 50 Hz by, a custom-made program, running on a computer equipped with an AD converter [10]. The subject was positioned on the force plate with arms by the side and feet at an angle of about 30° open to the front with and the heels approximately 3 cm apart. Postural sway was evaluated by recording anterior/posterior and lateral stance pressure changes.

2.3 Assessment of EEG Signals

EEG signals were recorded using a 256 channel Ag/AgCl electrode cap based on a standardized 10–20 system montage. An additional electrode was used to measure the electrooculographic signal. The cap was placed on the head of the subject followed by applying conductive electrode gel to reduce impedance to less than 8 kΩ in each channel. The proprioceptive stimulation experiment was then initialized, with EEG data acquired at a sampling frequency of 1024 Hz. Using a custom made trigger system, the vibratory impulses created events in the EEG signal data indicating stimulation *on* and *off*. Measurement trials were 225 s long.

3 Data Analysis

3.1 Platform Performance

The platform data for both open- and closed-eye datasets were segmented into four epochs: QS, Q1, Q2, Q3. As previously stated, the main objective here is to present the correlation of change in mechanical body sway performance with change in cortical activity. Hence, a fluctuation index was computed to determine postural performance during each epoch. Events from the custom made trigger system identified the number of stimuli and length of stimuli in each epoch. The index was calculated by adding together the difference in amplitude between each data point of the anterior/posterior sway (*n* and *m*), including the parameters which characterize each quartile; number of vibratory stimuli ($t_{Q1,Q2,Q3}$) and the time while stimuli was on ($ON_{Q1,Q2,Q3}$), finally normalizing with respect to the subjects' QS.

$$Index = \frac{\sum_{n=1}^{N} |x_n + 1 - x_n| \times \frac{t_{Q_{1,2,3}}}{ON_{Q_{1,2,3}}} \times \frac{1}{75s}}{\sum_{m=1}^{M} |x_{m-1} - x_m| \times \frac{1}{30s}}$$

3.2 EEG Signals

The frequency spectrum analysis aimed to identify changes in spectral power during adaptation and habituation, while quantifying the cortical changes with the results from the mechanical body sway. Adaptation and habituation were determined as follows:

$$Adaptation : Q1-QS$$
$$Habituation : Q3-Q1$$

Using EEGLAB, the raw data was filtered with a bandpass filter from 0.5–0 Hz and an additional stop band filter from 49.5–50.5 Hz. Channels were re-referenced to a common average. Three approaches to artifact rejection were performed on the data. Channels considered to be outliers were manually interpolated (average 4 channels per epoch), automatic continuous artifact rejection including rejecting 1–3 s of continuous data visually, and principle component analysis (PCA) where components were visually inspected, selected and removed manually. Equivalent to the platform data, the data-set was segmented into five epochs: QS (quiet stance), Q1, Q2, and Q3. Subsequently, epochs QS, Q1, and Q3 were further analyzed using custom MATLAB code to present the phase of adaptation and habituation, respectively.

To investigate the cortical changes within the epochs, absolute power (AP) values for six frequency bands, delta (0.5–3.5 Hz), theta (3.5–7.5 Hz), alpha (7.5–12.5 Hz), beta (12.5–30 Hz), low-gamma (30–50 Hz), and high-gamma (50–80 Hz), were calculated using the fast Fourier transformation (FFT, resolution of 0.977 Hz; 10% hanning window). Power spectral density (PSD) values were obtained and the AP values for each channel in all six frequency bands calculated for QS, Q1, and Q3. Normalization of the AP values was performed by subtracting channel values in epochs QS, Q1, and Q3 with channel values in epoch QS. A paired t-test was performed at each channel in corresponding epochs determining if significant power changes were in adaptation phase (between QS and Q1) and/or the habituation phase (between Q3 and Q1), resulting in 256 p-values for each phase. Channels with $p\text{-value} < 0.05$ were considered to have a change in AP over the corresponding frequency. To address the statistical problem of multiple comparisons, false discovery rate (FDR) correction method was employed. Channels that resisted the multiple comparison correction method were listed. Head plots including only channels having $p\text{-values} < 0.05$ were designed, identifying the increase or decrease in AP for the corresponding channel.

4 Results

Figure 2 presents the results of postural performance for the subject during the 225 s of vibratory stimulation. The dots indicate the value of the fluctuation index calculated for the corresponding epoch. The fluctuation index in Q1 significantly increased when vibratory stimulation began. The index increases to 5.8 meaning and increase of 58% in fluctuation compared to the subjects QS. During epoch Q2, the fluctuation index decreases to the level 4.1 and for the final epoch, Q3, the fluctuation index decreases to 3.9 meaning a 39% higher fluctuation index compared to QS. This suggests that the subject has learnt what to expect and started to habituate (Table 1).

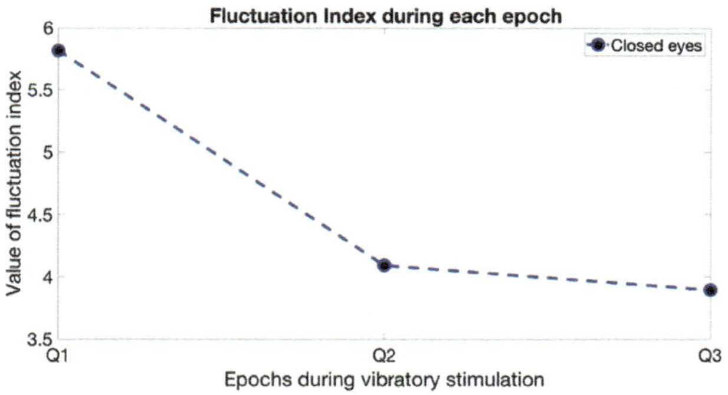

Fig. 2. Fluctuation index for vibratory stimulation epochs

Table 1. Fluctuation index for each epoch

Epoch	Index
QS	1
Q1	5.8
Q2	4.1
Q3	3.9

Figure 3 depict the cortical mapping of grand mean changes in AB solely in channels shown to have a significant difference between epochs. White areas are considered non-significant. As previously mentioned, habituation values are normalized values in epoch Q3 subtracted by values in Q1, and adaptation values are normalized values in epoch Q1 subtracted by normalized values in QS.

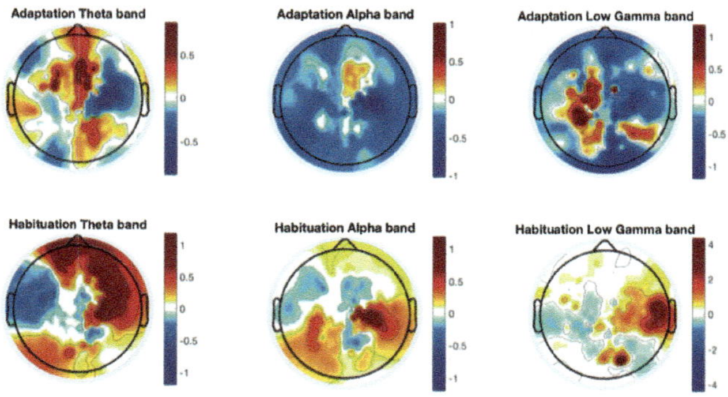

Fig. 3. Regions showing significant changes in absolute power during adaptation (upper row) and habituation (lower row) with eyes closed in theta, alpha and low gamma frequency, $p < 0.05$.

The overall results show a 50% increase of absolute power spectral in adaptation and a 100% increase in habituation in theta band specifically in the frontal and parietal regions of the cortex. A noticeable increase of low gamma activity can be observed during adaptation and habituation in the parietal and temporal regions of the cortex. Channels listed in Table 2 are channels resistant to FDR correction method and considered having a significant change in AP.

Table 2. Channels resistant to FDR correction method

Frequency band and condition	Channels
Adaptation – Theta	Z5Z, R9R, R10R, R3A, L2B, L1B
Habituation – Theta	R10R, R3A, R2A, R1A, R1C, R2D, R2Z, R3Z, L3D
Adaptation – Alpha	R9R, R10R, R4A
Habituation – Alpha	R9R, R10R, R4A, R5A, L8C, L9D
Adaptation – Low gamma	L8C, L7B, L13L, L12L, L10L, L4A, Z8R
Habituation – Low gamma	Z19R, Z18R, R3F, R4E, R10R

5 Conclusions

Research has demonstrated that an increase in theta activity in frontal and parietal regionsis generated during balance control tasks [6]. Also, literature on compensatory backward postural movement demonstrated a burst of gamma activity in the parietal area, preceded in the initiation of postural movement when balance was perturbed [11]. Our results supports both of these theories: theta activity increases in the parietal and frontal area during adaptation and habituation, and an increase of 100% in low gamma activity is well presented during the adaptation phase.

Moreover, the major increase in fluctuation indices between epochs QS and Q1, and further habituation in Q3, demonstrates a direct correlation between the role of the cortex in maintaining upright posture and the kinematic response in body sway [7].

The present study aimed to approach the understanding of postural control from many aspects. It demonstrates the direct involvement of cortical activity in maintaining upright posture during proprioceptive perturbation. Specifically, our results highlight the important role of increased theta and low gamma oscillations in balance control. Understanding the systems underlying postural control can lead to improving approaches to rehabilitating balance disorders.

Acknowledgment. This research was supported by the Institute for Biomedical and Neural Engineering at the University of Reykjavík, the Department of Anatomy at the University of Iceland, and the Icelandic National Hospital (Landspitali Scientific Fund, PI: Paolo Gargiulo) with additional funding support from the Rannis Icelandic Research Fund (Rannsoknasjodur, PI: Paolo Gargiulo).

Conflict of Interest. The authors declare that they have no conflict of interest.

References

1. Friðriksdóttir, R., Karlsson, G.H., Svansson, H.Á., Barollo, F., Edmunds, K.J., Petersen, H., Gargiulo, P.: Cortical- and kinematic response during adaptation and habituation in human upright posture, using HD-EEG, Reykjavik, Iceland (2019)
2. Mierau, A., Pester, B., Hülsdünker, K.S.T., Strüder, H., Witte, H.: Cortical correlates of human balance control. Brain Topogr. **30**(4), 434–446 (2017)
3. Morris, M., Iansek, R., Smithson, F., Huxham, F.: Postural instability in Parkinson's disease: a comparison with and without a concurrent task. Gait Posture **12**(3), 205–216 (2000)
4. Boecker, H., Ceballos-Baumann, A., Bartenstein, P., Weindl, A., Siebner, H.R., Fassbender, T., Munz, F., Schwaiger, M., Conrad, B.: Sensory processing in Parkinson's and Huntington's disease: investigations with 3D H(2)(15)O-PET. Brain J. Neurol. **122**(9), 1651–1665 (1999)
5. Furman, J.M., Cass, S.P.: Benign paroxysmal positional vertigo. New Engl. J. Med. **341**(21), 1590–1596 (1999)
6. Hülsdünker, T., Mierau, A., Neeb, C., Kleinöder, H., Strüder, H.: Cortical processes associated with continuous balance control as revealed by EEG spectral power. Neurosci. Lett. **592**, 1–5 (2015)
7. Fransson, P., Magnusson, M., Johansson, R.: Analysis of adaptation in anteroposterior dynamics of human postural control. Gait Posture **1**(1), 64–74 (1998)
8. Jacobs, J.V., Horak, F.B.: Cortical control of postural responses. J. Neural Transm. **114**(10), 1339–1349 (2007)
9. Buderath, P., Gärtner, K., Frings, M., Christiansen, H., Schoch, B., Konczak, J., Gizewski, E.R., Hebebrand, J., Timmann, D.: Postural and gait performance in children with attention deficit/hyperactivity disorder. Gait Posture **29**(2), 249–254 (2009)
10. Fransson, P.-A.: Analysis of adaptation in human postural control, Head and Neck surgery. PhD Dissertation, Lund University, Sweden, Department of Otorhinolaryngology (2005)
11. Slobounov, S., Hallett, M., Stanhope, S., Shibasaki, H.: Role of cerebral cortex in human postural control: an EEG study. Clin. neurophysiol. Official J. Int. Fed. Clin. Neurophysiol. **116**(2), 315–323 (2005)

Classifying Different Stages of Parkinson's Disease Through Random Forests

Carlo Ricciardi[1,2(✉)], Marianna Amboni[3,4], Chiara De Santis[3], Gianluca Ricciardelli[5], Giovanni Improta[6], Luigi Iuppariello[7,8], Giovanni D'Addio[2], Paolo Barone[3], and Mario Cesarelli[2,7]

[1] Department of Advanced Biomedical Sciences, University Hospital of Naples 'Federico II', Naples, Italy
carloricciardi.93@gmail.com
[2] Istituti Clinici Scientifici Maugeri IRCCS, Telese Terme, Italy
[3] Center for Neurodegenerative Diseases, Department of Medicine and Surgery, University of Salerno, Fisciano, Italy
[4] Istituto di Diagnosi e Cura Hermitage-Capodimonte, Naples, Italy
[5] Azienda Ospedaliera Universitaria OO.RR. San Giovanni di Dio Ruggi d'Aragona - Scuola Medica Salernitana, Salerno, Italy
[6] Department of Public Health, University Hospital of Naples 'Federico II', Naples, Italy
[7] DIETI, University of Naples 'Federico II', Naples, Italy
[8] Department of Neuroscience, Santobono-Pausilipon Children's Hospital, Naples, Italy

Abstract. Parkinson's disease (PD) is a progressive, neurodegenerative and age-related disease whose clinical characteristics include both motor and non-motor symptoms. Gait analysis, a three dimensional, non-invasive and computerized analysis of gait, can analyse walking features and carry out spatial and temporal parameters that can be included in machine learning algorithms. Knime analytics platform is employed to implement Random Forests. The aim of the present research is to distinguish De Novo PD patients (patients in early phase, without treatment) and Stable PD patients (patients in intermediate phase, in stable treatment) using spatial and temporal parameters of gait analysis. The dataset consists of 59 people, 32.2% De Novo and 67.8% Stable patients. Results show high accuracy (84.6%) and capacity to detect De Novo patients (94.9% of sensitivity). Recall and precision got high values, too. Despite needing further investigation, this pilot research should encourage health policy and facilities to introduce machine learning techniques and gait analysis in clinical practice. Moreover, results suggest the existence of gait patterns characterizing each phase of Parkinson's disease.

Keywords: Machine learning · Gait analysis · Parkinson's disease · Random Forests

J. Henriques et al. (Eds.): MEDICON 2019, IFMBE Proceedings 76, pp. 1155–1162, 2020.
https://doi.org/10.1007/978-3-030-31635-8_140

1 Introduction

Parkinson's disease (PD) is the second most prevalent neurodegenerative disease in the world; it consists of a progressive, age-related disorder whose prevalence and incidence will grow due to the increase of life expectancy [1]. Its neuropathological characteristics have been studied for decades and are well known in literature [2, 3]. PD clinical features, including motor and non-motor symptoms, are destined to worsen with the progression of the disease.

Various methods have been employed for the handling and management of chronic diseases [4, 5] and also advanced procedures have been introduced to reduce wastes in healthcare [6, 7] so as to support decision-makers in the assessment of technologies and in the choice of suitable treatments [8, 9].

There are many upgrades in recording quantitative data regarding movements and several researchers have dealt with this thematic in the last years [10, 11]. Particularly, gait analysis is an exam that allows clinicians to analyse patients' gait in a three-dimensional, non-invasive and computerized way, providing them with numerous spatial and temporal parameters regarding gait [12, 13]. In this setting, machine learning can be useful because it exploits all features to find hidden patterns in datasets. This approach was applied for different aims, from the management of patients to the support of clinicians in diagnosis and prognosis [14]. Management of hospitals' resources, improvement in patients' care process, detecting hospital infections are common applications of machine learning in healthcare.

In neurology many attempts to discriminate healthy people from patients with PD were conducted, obtaining good results [15, 16]. Rastegari et al. performed a machine learning analysis using a wide range of algorithms to distinguish healthy elderly, geriatric patients and those affected PD, getting to 100% of accuracy through Random Forests (RF) [17]. Similarly, Mittra et al. applied many algorithms to evaluate the possibility to distinguish PD patients from healthy people [18].

As a result of the literature explored, the aim of this paper is to use RF and gait analysis spatial and temporal parameters to distinguish PD patients in different phases of the disease based on gait features rather than focusing on a comparison between affected and non-affected people.

2 Materials and Methods

2.1 Dataset

Our dataset consisted of 59 patients and included two groups:

- De Novo PD: patients with recent diagnosis of PD never treated with anti-Parkinson drugs.
- Stable PD: patients in early/intermediate stage of PD in stable treatment during the previous four weeks.

The local Ethics Committee approved the study, which was performed in accordance with the Declaration of Helsinki. The average age of patients was 62.0 years for

De Novo PD group and 63.3 years for Stable PD group; the BMI was 26.8 and 27.6 and the percentages of man in the groups were 73.7% and 67.5%, respectively. Characteristics are summarized in Table 1.

Table 1. Dataset characteristics

Characteristics	De Novo PD (n = 19)	Stable PD (n = 40)
Age (years)	62.0 ± 8.6	63.3 ± 8.0
Sex (male/female)	14/5	27/13
BMI	26.8 ± 2.6	27.6 ± 3.8
Disease duration (years)	0	5.5 ± 2.5

2.2 Gait Analysis

Gait analysis is used in different clinical application by researchers [19, 20]; yet, it is not included in clinical practice. Gait of patients was assessed through an optokinetic system (BTS Bioengineering) furnished with 6 infrared cameras, 2 video cameras, a force plate and a data acquisition software (Smart Clinic). Patients' gait was evaluated in 3 conditions (gait tasks):

1. Normal gait (G);
2. Motor dual-Task (M): walking while carrying a tray with two glasses filled of water;
3. Cognitive dual-Task (C): walking while serially subtracting 7s starting from 100.

The straight trajectory was 10 m for each walking and each patient repeated the trial four times allowing us to analyse 40 m per patient.

2.3 Tool, Techniques and Evaluation Metrics

Knime Analytics Platform is a free "business intelligence" or "predictive analytics" tool employed in healthcare, e.g. in radiomics and cardiac studies [21, 22]. It implements the RF, in its version proposed by Breiman et al. [23], and the Smote (Synthetic Minority Over-sampling Technique) technique for imbalanced data, proposed by Chawla et al. [24]. The algorithm is an ensemble technique using both randomization and bagging principles on decision trees: a number of decision trees is chosen and each one is learned on a different set of records and attributes. The training of the decision trees is conducted separately and executed in parallel, the randomness is injected in the training process to reduce the variance of the predictions. To make a prediction on a new instance, the algorithm aggregates the predictions from each decision tree. When making classification, the aggregation is done by majority vote.

Smote, instead, is used when the distribution of classes in a dataset is not balanced. It creates synthetic rows by extrapolating between a real object of a class and one of its nearest neighbours. The 10-fold cross-validation is a procedure used to validate the results: the dataset is divided into 10 groups (folds), 9 are used for training and one for testing repeating the procedure for 10 times. The following metrics were employed to evaluate the algorithms:

- Accuracy: the ratio of correct predictions over the total.
- Error: the ratio of incorrect predictions over the total.
- Precision: a measure of the positive patterns correctly predicted from the total predicted patterns in a positive class.
- Recall: a measure of the fraction of positive patterns that are correctly classified.
- Specificity: "true negative rate" and, generally, the ability to identify people affected by the disease.
- Sensitivity: "true positive rate" and, generally, the ability to identify people not affected by the disease.
- Area Under the Curve Receiving Operating Characteristics (AUCROC): a qualitative indicator for the binary classification ranging from 0 to 1 with 0.5 indicating a classification not better than random guessing.

3 Results

Firstly, since our dataset was imbalanced, smote was applied, making the number of records rise to from 59 to 78. 96 features (16 for each condition of gait and for both left and right side) regarding spatial and temporal parameters of gait were available but, in order to choose an adequate number of features, a matrix of correlation was computed and all the features overcoming a threshold of correlation were filtered: 26 features were included in the study (Table 2).

Table 2. Features with gait task and side (R for right and L for left) included in the algorithm

Gait task	Features	Measure	Gait task	Features	Measure
G	Duration of the cycle (R)	s	M	Stance phase (R)	%
G	Duration of the cycle (L)	s	M	Stance phase (L)	%
G	Dev. Std. Duration of swing (R)	s	M	Double support phase (R)	%
G	Dev. Std. Duration of swing (L)	s	M	Mean velocity	m/s
G	Stance phase (L)	%	M	Variability of step's length (R)	Not applicable
G	Swing phase (L)	%	C	Dev. Std. duration of swing (R)	s
G	Double support phase (R)	%	C	Dev. Std. duration of swing (L)	s
G	Length of the cycle (R)	%height	C	Swing phase (L)	%
G	Length of the cycle (L)	%height	C	Single support phase (L)	%
G	Dev. Std. Length of the step (R)	m	C	Double support phase (R)	%
M	Dev. Std. Duration of swing (R)	s	C	Double support phase (L)	%
M	Duration of swing (L)	s	C	Mean velocity	%height/s

Furthermore, a 10-fold cross-validation was implemented to validate the results. In conclusion, the model was learned and tested through RF and results are reported in Table 3.

Table 3. Scores

Group	Accuracy [%]	Error [%]	Accuracy Max [%]	Precision [%]	Recall [%]	Sensitivity [%]	Specificity [%]	AUCROC
De Novo PD	84.6	15.4	100	78.7	94.9	94.9	74.4	0.962
Stable PD				93.5	74.4	74.4	94.9	

Moreover, a high AUCROC (0.96) proved the goodness of this application, in Fig. 1 there is a representation of the ROC curve.

The accuracy achieved a good percentage (84.6%) with a maximum of 100% more than once during the cross-validation. The sensitivity in De Novo PD group and the specificity in Stable PD group were particularly surprising: PD patients with an early diagnosis were almost perfectly recognized through their gait.

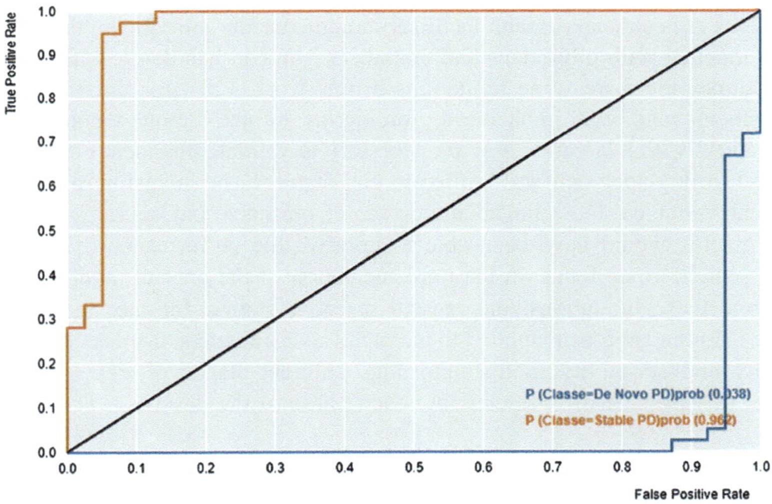

Fig. 1. ROC curve

4 Discussion and Conclusion

In this paper, the authors acquired gait signals for 59 patients affected by PD in different phases of the disease in three conditions and spatial and temporal parameters were extrapolated through an elaboration with Smart Clinic software. These parameters were used in order to classify patients in the correct phase of their pathology (De Novo and Stable) obtaining the metrics reported in Table 3.

High accuracies were obtained with RF when trying to distinguish healthy people and patients with PD [15], Wahid et al. used the same features and algorithm but without focusing on the phases of PD [16]. RF proved to be powerful particularly in this context in the application of Rastegari et al. [17], when PD patients were distinguished by healthy elderly and geriatric patients with 100% of accuracy. Aich et al. reached 98.4% distinguishing PD patients with shuffling gait from older people through an investigation on feature selection algorithms [25]. Nevertheless, none of these studies has focussed in classifying different phases of PD. In particular, the present research shows that gait pattern can distinguish different phases of PD clinical course, even with the presence of the confounding variable represented by the treatment that affects, by improving, the spatial temporal features of gait in PD [26, 27]. The high accuracy, sensitivity and specificity of our model unmask the existence of different gait patterns being peculiar to different disease stages. Despite some challenges that must be faced in the healthcare context regarding collection and quality of data, our results support the notion that quantitative gait assessment in combination with machine learning techniques may represent a reliable, non-invasive tool useful for tracking PD course and characterizing objectively different stages of the disease. Ultimately, the present approach may provide a possible marker of disease progression.

Our results further suggest that machine learning tools may support clinicians in all the contexts of healthcare. Health facilities should consider introducing them in clinical practice together with physicians and engineers who can handle these techniques.

Of course, there are some limitations in this work. First, the dataset had not got many patients and was imbalanced, forcing us to use Smote technique. Further investigations with a larger dataset are necessary to validate this idea. Furthermore, the machine learning approach was limited to gait features; the inclusion in the algorithm of clinical variables, like clinical motor score, presence and severity of non-motor symptoms etc., would have been able to improve the performance of classification: future researches may focus on the implementation of the present model.

Despite these limitations, the present research shows, for the first time, that a machine learning approach, applied to gait analysis parameters, displays high accuracy, specificity and sensitivity in distinguishing different stages of PD, suggesting the existence of gait patterns characterizing each phase of the disease.

Conflict of Interest Declaration. The authors declare they have no conflict of interests.

References

1. Lee, A., Gilbert, R.M.: Epidemiology of Parkinson disease. Neurol. Clin. **34**, 955–965 (2016)
2. Pearce, J.M.: Aetiology and natural history of Parkinson's disease. Br. Med. J. **2**(6153), 1664–1666 (1978)
3. Spillantini, M.G., Crowther, R.A., Jakes, R., et al.: Alpha-Synuclein in filamentous inclusions of lewy bodies from Parkinson's disease and dementia with lewy bodies. Proc. Natl. Acad. Sci. U.S.A. **95**(11), 6469–6473 (1998)

4. Santini, S., Pescapè, A., Valente, A.S., et al.: Using fuzzy logic for improving clinical daily-care of _-thalassemia patients. In: 2017 IEEE International Conference on Fuzzy Systems (FUZZ-IEEE). IEEE, pp. 1–6 (2017)
5. Biondi, M., Crispino, M., Improta, G., Triassi, M., et al.: The condroprotector role in the osteoarthritis of the knee. Giornale Italiano di Ortopedia e Traumatologia **39**, 44–47 (2013)
6. Improta, G., Romano, M., Di Cicco, M.V., et al.: Lean thinking to improve emergency department throughput at AORN Cardarelli hospital. BMC Health Serv. Res. **18**(1), 914 (2018)
7. Improta, G., Balato, G., Ricciardi, C., Russo, M.A., Santalucia, I., Triassi, M., Cesarelli, M.: Lean six sigma in healthcare: fast track surgery for patients undergoing prosthetic hip replacement surgery. TQM J. (2019). https://doi.org/10.1108/TQM-10-2018-0142
8. Improta, G., Russo, M.A., Triassi, M., et al.: Use of the AHP methodology in system dynamics: modelling and simulation for health technology assessments to determine the correct prosthesis choice for hernia diseases. Math. Biosci. **299**, 19–27 (2018)
9. Improta, G., Triassi, M., Guizzi, G., et al.: An innovative contribution to health technology assessment. In: Modern Advances in Intelligent Systems and Tools, pp. 127– 131. Springer (2012)
10. D'Addio, G., Iuppariello, L., Gallo, F.: Comparison between clinical and instrumental assessing using Wii Fit system on balance control. In: 2014 IEEE International Symposium on Medical Measurements and Applications (MeMeA), Lisboa, pp. 1–5 (2014)
11. Iuppariello, L., Romano, M., D'Addio, G., et al.: Comparison of measured and predicted reaching movements with a robotic rehabilitation device. In: Paper presented at the IEEE MeMeA 2014 - IEEE International Symposium on Medical Measurements and Applications, Proceedings (2014). https://doi.org/10.1109/MeMeA.2014.6860056
12. Galna, B., Lord, S., Burn, D.J., Rochester, L.: Progression of gait dysfunction in incident Parkinson's disease: impact of medication and phenotype. Mov. Disord. **30**, 359–367 (2015). https://doi.org/10.1002/mds.26110
13. Rochester, L., Galna, B., Lord, S., et al.: Decrease in Aβ42 predicts dopa-resistant gait progression in early Parkinson disease. Neurology **88**(16), 1501–1511 (2017). https://doi.org/10.1212/WNL.0000000000003840
14. Koh, H.C., Tan, G.: Machine learning applications in healthcare. J. Healthc. Inf. Manag. **19** (2), 65 (2011)
15. Arora, S., Venkataraman, V., Donohue, S., et al.: High accuracy discrimination of Parkinson's disease participants from healthy controls using smartphones. In: 2014 IEEE International Conference on Acoustics, Speech and Signal Processing (ICASSP), pp. 3641–3644. IEEE, May 2014
16. Wahid, F., Begg, R.K., Hass, C.J., et al.: Classification of Parkinson's disease gait using spatial-temporal gait features. IEEE J. Biomed. Health Inf. **19**(6), 1794–1802 (2015)
17. Rastegari, E., Azizian, S., Ali, H.: Machine learning and similarity network approaches to support automatic classification of parkinson's diseases using accelerometer-based gait analysis. In: Proceedings of the 52nd Hawaii International Conference on System Sciences, January 2019
18. Mittra, Y., Rustagi, V.: Classification of subjects with Parkinson's disease using gait data analysis. In 2018 International Conference on Automation and Computational Engineering (ICACE), pp. 84–89. IEEE, October 2018
19. Iuppariello, L., Bifulco, P., Romano, M., et al.: Submovements composition and quality assessment of reaching movements in subjects with Parkinson's disease. In: 2015 IEEE International Symposium on Medical Measurements and Applications (MeMeA) Proceedings, Turin, pp. 329–334 (2015). https://doi.org/10.1109/MeMeA.2015.7145222

20. Iuppariello, L., et al.: The effects of the vibratory stimulation of the neck muscles for the evaluation of stepping performance in Parkinson's disease. In: 2015 IEEE International Symposium on Medical Measurements and Applications (MeMeA) Proceedings, Turin, pp. 606–609 (2015). https://doi.org/10.1109/memea.2015.7145275
21. Mannarino, T., Assante, R., Ricciardi, C., et al.: Head-to-head comparison of diagnostic accuracy of stress-only myocardial perfusion imaging with conventional and cadmium-zinc telluride single-photon emission computed tomography in women with suspected coronary artery disease. J. Nucl. Cardiol. (2019). https://doi.org/10.1007/s12350-019-01789-7
22. Romeo, V., Ricciardi, C., Cuocolo, R., et al.: Machine learning analysis of MRI-derived texture features to predict placenta accreta spectrum in patients with placenta previa. Magnetic resonance imaging (2019). https://doi.org/10.1016/j.mri.2019.05.017. ISSN 0730-725X
23. Breiman, L.: Random forests. Mach. Learn. **45**, 5–32 (2001)
24. Chawla, N., Bowyer, K., Hall, L., Kegelmeyer, W.: SMOTE: synthetic minority over-sampling technique. J. Artif. Intell. Res. (JAIR) **16**, 321–357 (2002). https://doi.org/10.1613/jair.953
25. Aich, S., Pradhan, P.M., Park, J., Kim, H.C.: A machine learning approach to distinguish Parkinson's disease (PD) patient's with shuffling gait from older adults based on gait signals using 3D motion analysis. Int. J. Eng. Technol. **7**(3.29), 153–156 (2018)
26. Smulders, K., Dale, M.L., Carlson-Kuhta, P., et al.: Pharmacological treatment in Parkinson's disease: effects on gait. Parkinsonism Relat. Disord. **31**, 3–13 (2016)
27. Nonnekes, J., Timmer, M.H., de Vries, N.M., et al.: Unmasking levodopa resistance in Parkinson's disease. Mov. Disord. **31**(11), 1602–1609 (2016)

Regular Sessions: Technologies for Preventive Healthcare

Photoplethysmogram Modeling of Extreme Bradycardia and Ventricular Tachycardia

Birutė Paliakaitė[1(✉)], Andrius Petrėnas[1], Andrius Sološenko[1], and Vaidotas Marozas[1,2]

[1] Biomedical Engineering Institute, Kaunas University of Technology, 51423 Kaunas, Lithuania
birute.paliakaite@ktu.lt

[2] Department of Electronics Engineering, Kaunas University of Technology, 51368 Kaunas, Lithuania

Abstract. A model for simulating life-threatening arrhythmias, namely, extreme bradycardia and ventricular tachycardia, in photoplethysmogram (PPG) signals is proposed. The model uses RR interval series as an input, thus publicly available ECG databases can be employed to generate PPG signal database. The model has been originally developed for simulating atrial fibrillation, however, is adjusted to better represent the variation in PPG pulse amplitude, intrinsic to life-threatening arrhythmias. The original and adjusted models are evaluated on 40 recordings containing the episodes of extreme-bradycardia and ventricular tachycardia. By using the adjusted model, the root mean square error between real and modeled PPG signals is reduced, on average, by 20.6% for extreme bradycardia, whereas remains similar for ventricular tachycardia. The simulator is expected to be valuable when developing PPG-based wearable devices for long-term monitoring of patients at high risk of sudden cardiac death due to life-threatening arrhythmias.

Keywords: PPG simulator · Unobtrusive technology · Long-term monitoring · Wearable device · Life-threatening arrhythmias · Sudden cardiac death

1 Introduction

Sudden death due to life-threatening arrhythmias, such as extreme bradycardia or ventricular tachycardia, comprises 40% of known causes of death among end-stage kidney disease patients on hemodialysis [11,15]. Incidence of arrhythmias and mortality are the highest during the long interdialytic period when traditional thrice-weekly hemodialysis is prescribed [9,14,22]. Potentially, patients on hemodialysis would benefit from monitoring for the detection of life-threatening arrhythmias [4]. However, the existing arrhythmia monitoring equipment is uncomfortable for wearing due to disposable electrodes and device-connecting

J. Henriques et al. (Eds.): MEDICON 2019, IFMBE Proceedings 76, pp. 1165–1174, 2020.
https://doi.org/10.1007/978-3-030-31635-8_141

wires, thus making implantable devices the only available technology for long-term monitoring. Nevertheless, implantable devices are not considered as an attractive alternative due to predisposition of hemodialysis patients to post-implantation complications [4].

Advancements in wearable device technology have led to convenient means of acquiring pulsatile photoplethysmography (PPG) signal. This optical signal, usually obtained using a wrist-worn device, has already been used to detect premature ectopic beats [17] and atrial fibrillation [3,7,8,13,19,21]. Differentially from the aforementioned rhythm disturbances, which could be detected relying solely on heart rhythm information, the detection of extreme bradycardia and ventricular tachycardia is more challenging. That is, ventricular tachycardia must be distinguished from non-fatal tachycardias (e.g., atrial tachycardia, AV node tachycardia), whereas, extreme bradycardia must not be confused with the segment of lost signal. An attempt to detect abnormally low or high rhythm has been made in [2], however, the proposed approach is not capable of differentiating among different arrhythmia types.

The development of PPG-based arrhythmia detectors is restrained by the lack of publicly available databases, containing PPG signals with annotated arrhythmia episodes. This problem can be mitigated by developing and testing the detectors on simulated PPG signals that allow to control for various properties, such as signal-to-noise ratio [18]. In the present study, the phenomenological PPG signal simulator, originally developed for simulating sinus rhythm, premature beats and atrial fibrillation [18], is adjusted to model extreme bradycardia and ventricular tachycardia. The main advantage of the proposed model is that publicly available ECG databases (MIT-BIH Arrhythmia, MIT-BIH Malignant Ventricular Arrhythmia, MGH/MF Waveform, Creighton University Ventricular Tachyarrhythmia) [10], containing the annotated episodes of life-threatening arrhythmias, can be employed to generate PPG signal database.

2 Photoplethysmogram Model

An original model $\mathcal{M}_{\mathrm{o}}$ and that adjusted to simulate life-threatening arrhythmias $\mathcal{M}_{\mathrm{a}}$ share the same approach to modeling a single PPG pulse [18]. The modeled PPG pulse $\phi(t;\boldsymbol{\theta})$ is a linear combination of three weighted, time-shifted and time-scaled functions: one log-normal and two Gaussian. Estimation of the model parameter vector $\hat{\boldsymbol{\theta}}$, which minimizes the difference between the PPG pulse template (see Sect. 3) and the modeled PPG pulse $\phi(t;\boldsymbol{\theta})$, is given in [18].

Simulated individual PPG pulses are modified and then connected into a PPG signal based on the durations between the consecutive heart contractions (further referred to RR interval series). The width of systolic and diastolic parts of the PPG pulse is defined by two time scale factors $\beta_{s,k}$ and $\beta_{d,k}$, respectively,

$$\beta_{s,k} = \frac{1}{r_k}, \tag{1}$$

$$\beta_{d,k} = \begin{cases} \dfrac{1}{r_{k+1}}, & \text{if } r_k \text{ is a premature beat} \\ \dfrac{1}{r_k}, & \text{otherwise,} \end{cases} \quad (2)$$

where r_k is the RR interval preceding the k:th PPG pulse.

Since premature beats are especially common among hemodialysis patients and may cause significant hemodynamic alterations, the decision on whether the RR interval is premature or not is made according to [20]. The RR interval r_k is considered premature if at least one of the following rules is satisfied:

$$1.15r_k < r_{k-1} \text{ and } 1.15r_k < r_{k+1} \quad (3)$$

or

$$|r_{k-1} - r_k| < 0.3 \text{ and } r_{k-1} < 0.8 \text{ and } r_k < 0.8 \text{ and } 1.2\times\frac{r_{k-1} + r_k}{2} < r_{k+1} \quad (4)$$

or

$$|r_k - r_{k+1}| < 0.3 \text{ and } r_k < 0.8 \text{ and } r_{k+1} < 0.8 \text{ and } r_{k-1} > 1.2\times\frac{r_k + r_{k+1}}{2}. \quad (5)$$

The original model $\mathcal{M}_o$ assumes that the amplitude of the PPG pulse is linearly proportional to ventricular filling time, which is arguable [5]. Hence, we invoke the following rationale to relate the amplitude of the PPG pulse to the left ventricular stroke volume. Ventricular filling spans almost entire ventricular diastole and can be divided into three phases: early rapid filling, diastasis and atrial systole. Only the first and the third phases, which are independent of the duration of the RR interval, contribute to the stroke volume [5]. On the contrary, diastasis depends on the RR interval, but does not increase stroke volume [5]. As a result, stroke volume is usually constant when the RR interval is longer than 0.75 s [5,12]. Our detailed examination of PPG signals of low heart rate (average $r_k > 1$ s) revealed that such signals still show variation in PPG amplitude. Accordingly, a signal-specific threshold r_η is introduced, which defines the duration of the RR interval corresponding to the maximum stroke volume,

$$r_\eta = \max\left(\frac{\frac{1}{K}\sum_{k=1}^{K} r_k}{1 - \frac{1}{6K}\sum_{k=1}^{K} r_k}, 0.75\right), \quad (6)$$

where K is the total number of RR intervals in the signal. The duration of diastasis becomes negative when $r_k < r_\eta$, thereby leading to an overlap of the phases of early rapid filling and atrial systole, which results in a decreased stroke volume [5].

The change in stroke volume as well as in PPG pulse amplitude is defined as [12]:

$$\Delta\alpha_k = -\int_{1/r_\eta}^{1/r_k} \left(0.136r^2 - 10.044r + 21.18\right) dr. \quad (7)$$

Premature beat results in a reduced stroke volume [6,23]; therefore, the model assumes that the amplitude of the premature pulse changes exponentially [16]. A post-premature beat has a slightly increased pulse amplitude [6,23], which is accounted for by allowing the increase of amplitude even when $r_k \geq r_\eta$. Finally, the amplitude of the normalized PPG pulse is given by

$$\alpha_k = \begin{cases} \left(\frac{r_k}{r_\eta}\right)^{1.32}, & \text{if } r_k \text{ is premature} \\ 1 + \frac{\Delta\alpha_k}{C \times r_\eta}, & \text{if } r_k < r_\eta \text{ or } r_{k-1} \text{ is premature} \\ 1, & \text{otherwise,} \end{cases} \tag{8}$$

where C is the integration constant set to 25 according to [12].

PPG pulse, denoted $s_{a,k}(t)$, is modified by scaling the amplitude of $\phi(t;\hat{\boldsymbol{\theta}})$ with α_k, and the width with either $\beta_{s,k}$ or $\beta_{d,k}$ as defined in [18]:

$$s_{a,k}(t) = \begin{cases} \alpha_k \times \phi\big(\beta_{s,k}(t - t_p);\hat{\boldsymbol{\theta}}\big), & -\infty < t < t_p \\ \alpha_k \times \phi\big(\beta_{d,k}(t - t_p);\hat{\boldsymbol{\theta}}\big), & t_p \leq t < \infty, \end{cases} \tag{9}$$

where t_p is the time instant of the largest positive peak in $\phi(t;\hat{\boldsymbol{\theta}})$. The k:th PPG pulse $s_{a,k}(t)$ is sampled by

$$s_k(n) = s_{a,k}(t = nT), \; n = 0, 1, ... \tag{10}$$

where T is a sampling period. A modeled signal $x(n)$ is obtained by placing PPG pulses at the heart beat occurrence times δ_k, obtained from the RR interval series,

$$x(n) = \sum_{k=1}^{K} s_k(n - \delta_k). \tag{11}$$

3 Database and Performance Evaluation

The original and adjusted models are evaluated on PPG signals from the PhysioNet/CinC Challenge 2015 database training set [10]. The set contains 294 recordings with true life-threatening arrhythmia alarms, namely, asystole, extreme bradycardia, extreme tachycardia, ventricular tachycardia, and ventricular flutter/fibrillation. Recordings with extreme tachycardia were excluded since they contain arrhythmias of supraventricular origin, which are out of scope of our study. From the remaining dataset, recordings that satisfied the following criteria were considered for model evaluation: (1) PPG signal of sufficient quality, (2) at least one ECG lead eligible for R-peak detection, (3) underlying sinus rhythm. No ventricular flutter/fibrillation recordings and a single asystole recording met the aforementioned criteria. In total, 40 recordings, listed in Table 1, are used in the study: 17 recordings with extreme bradycardia (heart rate lower than 40 bpm for 5 consecutive beats) and 23 recordings with ventricular tachycardia

Table 1. Recordings from the PhysioNet/CinC Challenge 2015 database used in the study. The first letter of the file name indicates whether the recording contains episode of extreme bradycardia ("b") or ventricular tachycardia ("v").

b183l	b265l	b516s	b562s	b838s	v206s	v368s	v628s	v696s	v769l
b227l	b299l	b517l	b656s	v132s	v253l	v471l	v630s	v733l	v788s
b228s	b455l	b560s	b764s	v158s	v254s	v573l	v635l	v748s	v806s
b229l	b515l	b561l	b794s	v159l	v318s	v574s	v648s	v758s	v818s

(5 or more ventricular beats with heart rate higher than 100 bpm). The last minute of each recording is used to evaluate the models since this part of the signal contains the episode of life-threatening arrhythmia that caused the alarm.

ECG and PPG signals were sampled at 250 Hz with 12-bit precision. ECGs were used for extracting RR interval series, which served as an input to the simulation model. PPGs were filtered with a bandpass zero-phase finite impulse response filter with cut-off frequencies at 0.5 and 15 Hz. A single PPG pulse template of distinctive morphology, free from noise and artifacts, was extracted from every PPG signal and used for estimating the model parameter vector $\hat{\boldsymbol{\theta}}$.

The adequacy of the models $\mathcal{M}_{\mathrm{o}}$ and $\mathcal{M}_{\mathrm{a}}$ is evaluated by computing the root mean square error between real and modeled PPG signals,

$$E = \sqrt{\frac{1}{N}\sum_{n=1}^{N}\left(u(n) - x(n)\right)^2}, \tag{12}$$

where n is a sample number, N is a total number of samples in the modeled signal, and $u(n)$ is the real signal. Cross-correlation was used to align real and modeled signals. Prior to computing E, both signals were normalized to the unit amplitude of the first PPG pulse. The results are expressed as mean ± s.d.

4 Results

Examples of real and modeled PPG signals with the episodes of extreme bradycardia and ventricular tachycardia are presented in Fig. 1. Compared to the real PPG signal, some pulses of extreme bradycardia, simulated using the original model $\mathcal{M}_{\mathrm{o}}$, is of considerably larger pulse amplitude. Meanwhile, the amplitude is resembled properly by the adjusted model $\mathcal{M}_{\mathrm{a}}$. The episode of ventricular tachycardia is simulated similarly by both models, though neither modeled signal corresponds precisely to the real PPG.

Figure 2 displays E between real and simulated PPG signals. The advantage of the adjusted model is particularly obvious for extreme bradycardia, since E computed within the arrhythmia episode is 20.6% lower compared to $\mathcal{M}_{\mathrm{o}}$. On the other hand, when the error is found for the entire signal with underlying sinus rhythm and the episodes of extreme bradycardia, the difference between

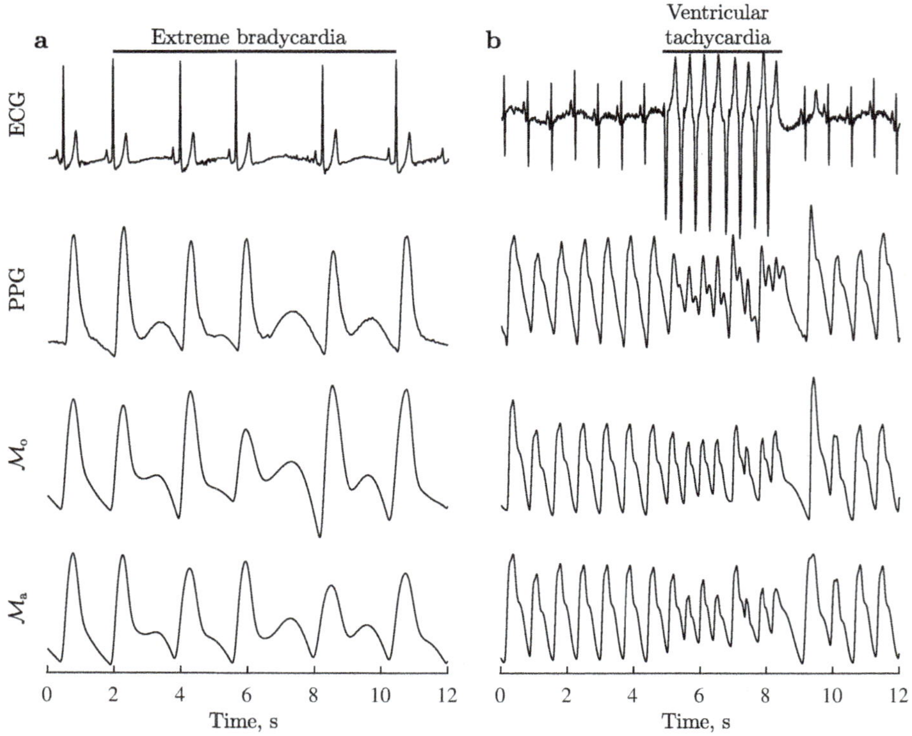

Fig. 1. ECG, PPG and modeled PPG signals with the episode of **a** extreme bradycardia (recording b299l) and **b** ventricular tachycardia (recording v206s).

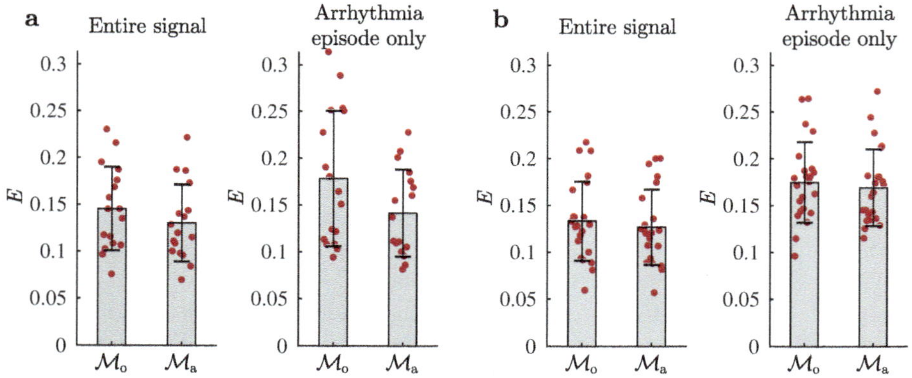

Fig. 2. The root mean square error E between real and simulated signals with the episode of **a** extreme bradycardia and **b** ventricular tachycardia. Error bars show mean ± s.d.

the models is less obvious, suggesting similar performance of the models for other rhythms, i.e., sinus rhythm and premature beats. This can be supported by the observation that $\mathcal{M}_a$ only slightly improves simulation of ventricular tachycardia, showing 4.8% lower error for the entire signal and 3.3% for arrhythmia episode. It should be noted, that the error computed within the arrhythmia episode is larger for both arrhythmia types due to manifestation of noise and artifacts.

5 Discussion

This paper aims to evaluate simulation of PPG signals with life-threatening arrhythmias using the adjusted phenomenological PPG model, originally proposed in [18]. This work is inspired by the widespread use of personal wearable devices, capable of acquiring PPG, which opens the possibility to monitor patients at high risk of life-threatening arrhythmias. Thus far, the development of arrhythmia detectors is hindered by the lack of annotated PPG databases, thereby calling for alternative signal sources.

When developing arrhythmia detectors, it is crucial to distinguish between arrhythmia-induced and artifact-induced changes in PPG signal. This implies that simulated arrhythmia episodes should closely resemble the changes in pulse amplitude and morphology. The original model assumes that the pulse amplitude is inversely proportional to the duration of the preceding RR interval [18], however, this assumption is debatable [5]. Accordingly, the adjusted model takes into account the relationship between the duration of the RR interval and stroke volume to determine the amplitude of the PPG pulse [5,12]. The adjusted model outperforms the original when simulating extreme bradycardia, thanks to signal-dependent protective threshold, which sets the maximum duration of the RR interval that still allows variation in PPG pulse amplitude. The original model was initially developed to simulate PPG signals with atrial fibrillation, which is tachycardia itself [18]. This explains why the error decreased only slightly when ventricular tachycardia was simulated using the adjusted model.

Sudden cardiac death due to arrhythmias accounts for approximately 50% of all cardiovascular deaths [1]. Therefore, the simulation model would be beneficial for developing detectors for monitoring other patient groups, such as patients with acute coronary syndrome or structural heart disease.

6 Conclusions

The present study shows that the adjusted phenomenological PPG model accurately simulates the episodes of extreme bradycardia. The simulator is expected to be valuable when developing PPG-based wearable devices for long-term monitoring of patients at high risk of sudden cardiac death due to life-threatening arrhythmias.

Acknowledgment. This work was supported by EU Structural Funds (agreement No. 01.2.2-LMT-K-718-01-0030).

Conflict of Interest. None declared.

References

1. Al-Khatib, S.M., Stevenson, W.G., Ackerman, M.J., Bryant, W.J., Callans, D.J., Curtis, A.B., Deal, B.J., Dickfeld, T., Field, M.E., Fonarow, G.C., Gillis, A.M., Granger, C.B., Hammill, S.C., Hlatky, M.A., Joglar, J.A., Kay, G.N., Matlock, D.D., Myerburg, R.J., Page, R.L.: 2017 AHA/ACC/HRS guideline for management of patients with ventricular arrhythmias and the prevention of sudden cardiac death. Circulation **138**(13), e272–e391 (2018). https://doi.org/10.1161/CIR.0000000000000549
2. Bonomi, A.G., Eerikäinen, L.M., Schipper, F., Aarts, R.M., de Morree, H.M., Dekker, L.: Detecting episodes of brady-and tachycardia using photoplethysmography at the wrist in free-living conditions. In: 2017 Computing in Cardiology (CinC), pp 1–4 (2017). https://doi.org/10.22489/CinC.2017.271-329
3. Bonomi, A.G., Schipper, F., Eerikäinen, L.M., Margarito, J., van Dinther, R., Muesch, G., de Morree, H.M., Aarts, R.M., Babaeizadeh, S., McManus, D.D., Dekker, L.R.: Atrial fibrillation detection using a novel cardiac ambulatory monitor based on photo-plethysmography at the wrist. J. Am. Heart Assoc. **7**(15), e009351 (2018). https://doi.org/10.1161/JAHA.118.009351
4. Boriani, G., Savelieva, I., Dan, G.A., Deharo, J.C., Ferro, C., Israel, C.W., Lane, D.A., La Manna, G., Morton, J., Mitjans, A.M., Vos, M.A., Turakhia, M.P., Lip, G.Y., Group, E.S.D.: Chronic kidney disease in patients with cardiac rhythm disturbances or implantable electrical devices: clinical significance and implications for decision making-a position paper of the European Heart Rhythm Association endorsed by the Heart Rhythm Society and the Asia Pacific Heart Rhythm Society. EP Europace **17**(8), 1169–1196 (2015). https://doi.org/10.1093/europace/euv202
5. Chung, C.S., Karamanoglu, M., Kovács, S.J.: Duration of diastole and its phases as a function of heart rate during supine bicycle exercise. Am. J. Physiol. Heart Circ. Physiol. **287**(5), H2003–H2008 (2004). https://doi.org/10.1152/ajpheart.00404.2004
6. Cohn, K., Kryda, W.: The influence of ectopic beats and tachyarrhythmias on stroke volume and cardiac output. J. Electrocardiol. **14**(3), 207–218 (1981). https://doi.org/10.1016/S0022-0736(81)80001-5
7. Corino, V.D.A., Laureanti, R., Ferranti, L., Scarpini, G., Lombardi, F., Mainardi, L.T.: Detection of atrial fibrillation episodes using a wristband device. Physiol. Meas. **38**(5), 787–799 (2017). https://doi.org/10.1088/1361-6579/aa5dd7
8. Fallet, S., Lemay, M., Renevey, P., Leupi, C., Pruvot, E., Vesin, J.M.: Can one detect atrial fibrillation using a wrist-type photoplethysmographic device? Med. Biol. Eng. Comput. **57**(2), 477–487 (2019). https://doi.org/10.1007/s11517-018-1886-0
9. Foley, R.N., Gilbertson, D.T., Murray, T., Collins, A.J.: Long interdialytic interval and mortality among patients receiving hemodialysis. N. Engl. J. Med. **365**(12), 1099–1107 (2011). https://doi.org/10.1056/NEJMoa1103313

10. Goldberger, A.L., Amaral, L.A.N., Glass, L., Hausdorff, J.M., Ivanov, P.C., Mark, R.G., Mietus, J.E., Moody, G.B., Peng, C.K., Stanley, H.E.: PhysioBank, PhysioToolkit, and PhysioNet: components of a new research resource for complex physiologic signals. Circulation **101**(23), e215–e220 (2000). https://doi.org/10.1161/01.CIR.101.23.e215
11. Kalra, P.A., Green, D., Poulikakos, D.: Arrhythmia in hemodialysis patients and its relation to sudden death. Kidney Int. **93**(4), 781–783 (2018). https://doi.org/10.1016/j.kint.2017.12.005
12. Melbin, J., Detweiler, D.K., Riffle, R.A., Noordergraaf, A.: Coherence of cardiac output with rate changes. Am. J. Physiol. Heart Circ. Physiol. **243**(4), H499–H504 (1982). https://doi.org/10.1152/ajpheart.1982.243.4.H499
13. Paliakaitė, B., Petrėnas, A., Skibarkienė, J., Mickus, T., Daukantas, S., Kubilius, R., Marozas, V.: Towards long-term monitoring of atrial fibrillation using photoplethysmography. In: Proceedings of the 10th International Joint Conference on Biomedical Engineering Systems and Technologies, pp. 141–146 (2017). https://doi.org/10.5220/0006115601410146
14. Roy-Chaudhury, P., Tumlin, J.A., Koplan, B.A., Costea, A.I., Kher, V., Williamson, D., Pokhariyal, S., Charytan, D.M., Williamson, D., Roy-Chaudhury, P., Tumlin, J., Kher, V., Reddy, V., Prakash, K.C., Charytan, D., Tiwari, S.C., et al.: Primary outcomes of the monitoring in dialysis study indicate that clinically significant arrhythmias are common in hemodialysis patients and related to dialytic cycle. Kidney Int. **93**(4), 941–951 (2018). https://doi.org/10.1016/j.kint.2017.11.019
15. Saran, R., Robinson, B., Abbott, K.C., Agodoa, L.Y., Bragg-Gresham, J., Balkrishnan, R., Bhave, N., Dietrich, X., Ding, Z., Eggers, P.W., Gaipov, A., Gillen, D., Gipson, D., Gu, H., Guro, P., et al.: US Renal Data System 2018 annual data report: epidemiology of kidney disease in the United States. Am. J. Kidney Dis. **73**(3), A7–A8 (2019). https://doi.org/10.1053/j.ajkd.2019.01.001
16. Sarnari, R., Kamal, R.Y., Friedberg, M.K., Silverman, N.H.: Doppler assessment of the ratio of the systolic to diastolic duration in normal children: relation to heart rate, age and body surface area. J. Am. Soc. Echocardiogr. **22**(8), 928–932 (2009). https://doi.org/10.1016/j.echo.2009.05.004
17. Sološenko, A., Petrėnas, A., Marozas, V.: Photoplethysmography-based method for automatic detection of premature ventricular contractions. IEEE Trans. Biomed. Circuits Syst. **9**(5), 662–669 (2015). https://doi.org/10.1109/TBCAS.2015.2477437
18. Sološenko, A., Petrėnas, A., Marozas, V., Sörnmo, L.: Modeling of the photoplethysmogram during atrial fibrillation. Comput. Biol. Med. **81**, 130–138 (2017). https://doi.org/10.1016/j.compbiomed.2016.12.016
19. Sološenko, A., Petrėnas, A., Paliakaitė, B., Sörnmo, L., Marozas, V.: Detection of atrial fibrillation using a wrist-worn device. Physiol. Meas. **40**(2), 025003 (2019). https://doi.org/10.1088/1361-6579/ab029c
20. Tsipouras, M., Fotiadis, D., Sideris, D.: An arrhythmia classification system based on the RR-interval signal. Artif. Intell. Med. **33**(3), 237–250 (2005). https://doi.org/10.1016/j.artmed.2004.03.007
21. Turakhia, M.P., Desai, M., Hedlin, H., Rajmane, A., Talati, N., Ferris, T., Desai, S., Nag, D., Patel, M., Kowey, P., Rumsfeld, J.S., Russo, A.M., Hills, M.T., Granger, C.B., Mahaffey, K.W., Perez, M.V.: Rationale and design of a large-scale, app-based study to identify cardiac arrhythmias using a smartwatch: the apple heart study. Am. Heart J. **207**, 66–75 (2019). https://doi.org/10.1016/j.ahj.2018.09.002

22. Wong, M.C., Kalman, J.M., Pedagogos, E., Toussaint, N., Vohra, J.K., Sparks, P.B., Sanders, P., Kistler, P.M., Halloran, K., Lee, G., Joseph, S.A., Morton, J.B.: Temporal distribution of arrhythmic events in chronic kidney disease: highest incidence in the long interdialytic period. Heart Rhythm **12**(10), 2047–2055 (2015). https://doi.org/10.1016/j.hrthm.2015.06.033
23. Zheng, D., Allen, J., Murray, A.: Determination of aortic valve opening time and left ventricular peak filling rate from the peripheral pulse amplitude in patients with ectopic beats. Physiol. Meas. **29**(12), 1411–1419 (2008). https://doi.org/10.1088/0967-3334/29/12/005

Evaluation in a Real Environment of a Trainable Cough Monitoring App for Smartphones

Carlos Hoyos-Barceló[1,2](✉), José Ramón Garmendia-Leiza[3], María Dolores Aguilar-García[4], Jesús Monge-Álvarez[1], Diego Asay Pérez-Alonso[2], Carlos Alberola-López[2], and Pablo Casaseca-de-la-Higuera[1,2]

[1] School of Computing, Engineering, and Physical Sciences, University of the West of Scotland, Paisley PA1 2BE, UK
Carlos.hoyos@uws.ac.uk
[2] Laboratorio de Procesado de Imagen, ETSI Telecomunicación, Universidad de Valladolid, 47002 Valladolid, Spain
[3] SACYL, Gerencia de Asistencia Sanitaria de Palencia, Complejo Asistencial Universitario, 34005 Palencia, Spain
[4] SACYL, Gerencia de Asistencia Sanitaria de Palencia, CS Villamuriel de Cerrato, 34190 Villamuriel, Palencia, Spain

Abstract. This paper presents SmartCough, an M-health app for Android smartphones that monitors cough trends in patients with respiratory diseases. The app is designed to be battery-efficient, fast, and robust against noise. It relies on efficiently-implemented machine learning algorithms that have been validated in laboratory conditions. Since these conditions are rarely met in a real situation where the user carries the phone inside their pocket or bag, the app features a self-training module that allows easy adaptation to new environments. In this paper, we have evaluated the app with real patients in an outdoor setting to test the performance in real environments that are hostile to cough detection. Our results show that the average sensitivity obtained in laboratory conditions drops significantly (down to 60%) when the baseline configuration is employed. By activating the built-in self-training module, the median sensitivity raises to 85.87% after a small training step, with a bounded false positive rate. The achieved performance is analogous to the one obtained in laboratory conditions, making the app suitable for use in real life scenarios.

Keywords: Cough detection · M-health · Optimization · Android

1 Introduction

The emerging technology field of *mobile-Health* (M-health) advocates for the use of smartphones to deliver ubiquitous telemedicine solutions at reduced production costs [1], taking advantage of the huge investments that have already been done in infrastructures to make this technology globally available. Modern smartphones are wearable devices packed with sensors that can run general-purpose programs, and which most

J. Henriques et al. (Eds.): MEDICON 2019, IFMBE Proceedings 76, pp. 1175–1180, 2020.
https://doi.org/10.1007/978-3-030-31635-8_142

users already carry with themselves for most of the day. This makes them the perfect choice of platform for the development of health monitors for ambulatory tracking of medical conditions even in places with low technology penetration.

We present SmartCough, a cough monitoring app for the Android platform that detects cough in real time and generates trend statistics in the form of short and long-term charts. These charts can be used to reveal symptoms such as a sudden increase in coughing rate, which might be an indication of an impending COPD exacerbation. As most exacerbations are caused by bacterial or viral infection [2], early detection of the symptom would allow the patients to treat the disease before their condition becomes critical, avoiding hospitalisation [3]. The cost of patients with mild to severe cases of COPD were estimated in 2015 to be in order of €96.4 billion per annum in the EU, plus another €283 billion in opportunity costs [4], so a potential reduction in hospitalisation rates and bed-days could turn into huge savings for the national health services. Other uses for the app would be as a support tool to track the response of a patient to medication, and as an aid in the diagnosis of patients with unknown respiratory diseases.

The engineering challenges for a product for real-life deployment are different than those to prove the feasibility of a system in a research lab. For example, the limited CPU power and battery may cause a correct but unoptimised app to be unable to process cough events at the time they occur or will deplete the battery of the device too quickly for practical use. That is why in previous work, we presented an optimised machine learning engine for cough detection that could perform in real-time even on low-end devices [5]. In [6] we proposed additional optimisations to make the app battery-friendly in order to support over three days of continuous operation.

Another problem resides in the fact that scientific research is typically evaluated on unrealistic databases, as the set of samples used for training and testing cannot possibly contain every event that can happen in real life [7], so the systems trained this way rely on optimistic assumptions and may behave poorly outside of that controlled environment (e.g. lab conditions). We previously evaluated the algorithms implemented in SmartCough on a synthetic database [8] and achieved good performance in a variety of noise conditions. After that, we trained the system with a database of real patients recorded in a clinical environment, also with good evaluation results [9]. However, these results may not translate so well on a real-life trial with different hardware, the device carried inside a pocket—which significantly weakens the signal—and unexpected sources of noise such as walking nearby to a construction area. Solving this requires a system able to adapt to data and conditions that it was not trained to handle.

Some studies have shown that adaptively training a system with input from the user significantly improves the performance, with over 30% better recall rates [10, 11]; however the training procedure for most machine learning engines is so computationally expensive [12] that available cough detectors pre-train the system once and leave it as is, forfeiting the use of this technique. We solved this problem by implementing an efficient training mode on our SmartCough app that lets users calibrate the cough detector to better handle their voice. The user only has to press a button and speak to the device, providing input that the app uses to learn in real-time. In this study we measure how the baseline app performs on a real noisy database, and how much our system improves when the accompanying training feature is used.

2 Materials and Methods

Our proposed cough detector is composed of three steps as shown in Fig. 1.

- *Signal processing*: the audio from the microphone is divided in 50 ms data frames with 25 ms overlap, and then their spectrogram is computed. For efficiency reasons, we apply a filter to detect and discard uneventful chunks of data; this filter is based on a minimum energy threshold and dynamically adjusts itself based on the energy of background noise, to quickly adapt to changing environments.
- *Feature Extraction*: for each spectrogram, we efficiently compute a set of features based on image moments that are used to identify cough windows [6].
- *Classification*: we use an efficient k-NN classifier to detect cough events [5]. The choice of k-NN allows the system to learn by adding new samples to our database.

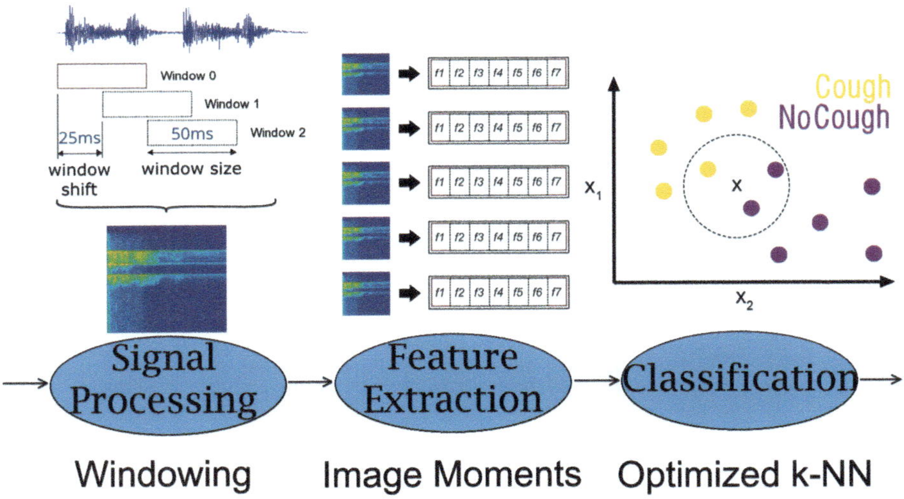

Fig. 1. Diagram of our proposed detector

The testing database is composed of audio recordings of slightly over 24 h from 20 patients with different respiratory conditions. We used a Sony Xperia Z2 android device to collect the data, storing the files in 16-bit WAV format at 44100 Hz. The patients were instructed to carry the device inside their pockets or purses for a full day and spend the day as they normally would, going outside and talking to people, to get samples from real, noisy environments that are hostile to cough detection.

We used the software tool Praat [13] and created a subtitle file to manually label the coughs, which we used to set the ground-truth. If we could not determine with certainty if an audio segment should be marked as a cough or not, we removed it from the database and from further evaluation. We used this subtitle file and Praat to create separate audio clips for each sound event, then passed them to our app to test the cough detector—trained with its default database—and calculated the Sensitivity and the

False Positive Rate (FPR), with the latter measured as False Positives per hour. The default database is composed of patient data from our previous indoor validation [9], augmented with a collection of SFX audio clips from Youtube from common sources of noise.

To test the effect of enabling the self-training module, we configured the app in training mode and then fed it a few seconds of random cough clips from each patient. As this training tends to make the detector overly sensitive to the voice of that patient, we also trained the app with random sound samples from that patient that do not contain any cough. After that, we passed the rest of audio clips to calculate the new values of Sensitivity and the FPR. We trained the app for each patient individually, always reinstalling the app with default settings so that the different training tests were independent from each other.

3 Results and Discussion

Table 1 shows the system performance both in baseline mode, and after enabling the self-training module. The results in baseline mode reveal overall poor detection results and higher variability in sensitivity and FPR. However, both values greatly improve after using the real-time training feature that is built-in with the app. On Fig. 2. we see that the system goes from an unaffordable sensitivity of 60% up to 85% after using the self-training feature, and how it removes outliers. A paired t-test run over the results before and after calibration revealed high significance in sensitivity improvements ($p = 2.7 \cdot 10^{-7}$). The statistical comparison for FPR values was close to significant ($p = 0.073$).

Table 1. Performance of the SmartCough app before and after training.

	Baseline		Trained	
	Sensitivity (%)	FP/h	Sensitivity (%)	FP/h
Average	61.56%	19	80.63%	16
Median	60.00%	15	85.87%	14.5
STD	18.36%	12	15.26%	5

When comparing SmartCough with other automatic cough monitors, Coughsense [3] achieves higher sensitivity (92%) but worse FPR (17 FP/h), whereas the Leicester Cough Monitor reports 71% sensitivity and 13 FP/h [14]. These proposals however, have not been implemented in Smartphones. Considering only smartphone-based cough-detectors, both ADAM [15] and SymDetector [16] reported sensitivities around 83%. ADAM's FPR was lower (11 FP/h) but it was only evaluated indoors, so signals were in principle less noisy than the ones employed in this work. The authors of both systems also reported battery depletion after less than a day of operation, which indicates that their choice of implementation is too resource-intensive for mobile devices.

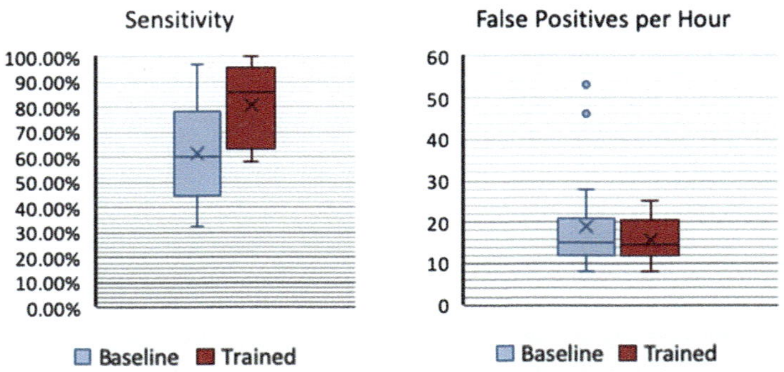

Fig. 2. Sensitivity and FPR boxplots obtained before and after training.

4 Conclusions

After evaluating the SmartCough app over a real-life noisy database we found that even though the baseline performance of the detector is below acceptable levels, we achieve good results both in terms of sensitivity and false positive rate after calibrating the app through its self-training functionality. The achieved performance is comparable to that of other smartphone-based detectors that have only been evaluated over indoor-acquired databases, while surpassing them in terms of efficiency. Also, while it is possible to pre-train the default database with more data to improve the initial results, we have found that user personalisation is a superior approach that enables the system for use in real life scenarios.

Acknowledgement. The authors would like to acknowledge the clinical support provided by the Pneumology Service staff at Complejo Sanitario Asistencial de Palencia (Spain).

Conflict of Interest. We declare no conflicts of interest.

References

1. Adibi, S.: Mobile Health: A Technology Road Map, p. 593. Springer, Cham (2015)
2. Balter, M.S., et al.: Canadian guidelines for the management of acute exacerbations of chronic bronchitis. Can. Respir. **10**(5), 248–258 (2003)
3. Larson, E.C., Lee, T., Liu, S., Rosenfeld, M., Patel, S.N.: Accurate and privacy preserving cough sensing using a low-cost microphone. In: UbiComp 2011, Beijing, pp. 375–384 (2011)
4. European Respiratory Society (ERS). The economic burden of lung disease, in European Lung White Book, Chapter 2, pp. 16–27 (2015). http://www.erswhitebook.org/chapters/the-economic-burden-of-lung-disease/
5. Hoyos-Barceló, C., Monge-Álvarez, J., Zeeshan Shakir, M., Alcaraz-Calero, J., Casaseca-de-la-Higuera, P.: Efficient k-NN implementation for real-time detection of cough events in Smartphones. J. Biomed. Health Inf. (JBHI) **22**(5), 1662–1671 (2018)

6. Hoyos-Barceló, C., Monge-Álvarez, J., Pervez, Z., San-José-Revuelta, L.M., Casaseca-de-la-Higuera, P.: Efficient computation of image moments for robust cough detection using smartphones. Comput. Biol. Med. (CBM) **100**, 176–185 (2018)
7. Nguyen, L.T., Zeng, M., Tague, P., Zhang, J.: I did not smoke 100 cigarettes today!: avoiding false positives in real-world activity recognition. In: UbiComp 2015 Proceedings of the 2015 ACM International Joint Conference on Pervasive and Ubiquitous Computing, Osaka (2015)
8. Monge-Álvarez, J., Hoyos-Barceló, C., Dahal, K., Casaseca-de-la-Higuera, P.: Audio-cough event detection based on moment theory. Appl. Acoust. **135**, 124–135 (2018)
9. Monge-Álvarez, J., Hoyos-Barceló, C., Lesso, P., Casaseca-de-la-Higuera, P.: Robust detection of audio-cough events using local Hu moments. J. Biomed. Health Inf. **23**, 184–196 (2019)
10. Yatani, K., Truon, K.N.: BodyScope: a wearable acoustic sensor for activity recognition. In: UbiComp 2012, Pittsburgh, pp. 341–350 (2012)
11. Könönen, V., Mäntyjärvi, J., Similä, H., Pärkkä, J., Ermes, M.: Automatic feature selection for context recognition in mobile devices. Pervasive Mob. Comput. **6**, 181–197 (2010)
12. Suarez-Tangil, G., Tapiador, J.E., Peris-Lopez, P., Pastrana, S.: Power-aware anomaly detection in smartphones: An analysis of on-platform versus externalized operation. Pervasive Mob. Comput. **18**, 137–151 (2015)
13. Boersma, P., Weenink, D.: Praat: doing phonetics by computer, Phonetic Sciences. University of Amsterdam. http://www.fon.hum.uva.nl/praat/
14. Matos, S., Birring, S.S., Pavord, I.D., Evans, D.H.: Detection of cough signals in continuous audio recordings using hidden Markov models. IEEE Trans. Biomed. Eng. **LIII**(6), 1078–1083 (2006)
15. Sterling, M., Rhee, H., Bocko, M.: Automated cough assessment on a mobile platform. J. Med. Eng. **2014**, 9 pages (2014). Article ID 951621
16. Sun, X., Lu, Z., Hu, W., Cao, G.: SymDetector: detecting sound-related respiratory symptoms using smartphones. In: UbiComp 2015, Osaka, pp. 97–108 (2015)

New Approaches for Personalizing Daily Activity Monitoring in mHealth Applications

Diego Moreno-Blanco[1(✉)], Patricia Sánchez-González[1,2], Francisco J. Gárate[1], Cesar Cáceres[1,3], Javier Solana-Sánchez[4,5,6], José M. Tormos-Muñoz[4,5,6], and Enrique J. Gómez[1,2]

[1] Biomedical Engineering and Telemedicine Centre, ETSI Telecomunicación, Center for Biomedical Technology, Universidad Politécnica de Madrid, Madrid, Spain
dmoreno@gbt.tfo.upm.es

[2] Centro de Investigación Biomédica en Red, Biomateriales y Nanomedicina (CIBER-BBN), Madrid, Spain

[3] Universidad Rey Juan Carlos, Madrid, Spain

[4] Institut Guttmann, Institut Universitari de Neurorehabilitació adscrit a la UAB, Badalona, Spain

[5] Universitat Autònoma de Barcelona, Barcelona, Spain

[6] Fundació Institut d'Investigació en Ciències de la Salut Germans Trias i Pujol, Badalona, Spain

Abstract. Healthy daily activities have a positive influence on many aspects of our lives. Habits have a deep impact on our health, they help to prevent the appearance of chronic and neurodegenerative diseases and will provide a healthy and active aging. This research work is aiming to analyze the need of new approaches on monitoring daily life activities, investigating new technologies and user modelling methods for healthy habits monitoring. mHealth platforms allow to perform a multivariable monitoring for allowing effective and personalized interventions. Data analytics, data mining and gamification methodologies are being applied to investigate user experience models. This user adaptation is commonly focused on personality and mood monitoring. Furthermore, new user models are built based on monitoring data, personality of user and the daily activity patterns extracted from intelligent and personalized monitoring. The final goal is contributing to improve user´s adherence to interventions, quality of life and quality of care in mHealth applications.

Keywords: Healthy habits · Habit management · Monitoring · Machine learning · Daily activity patterns · Emotional status · Data analytics · User model

1 Introduction

Recent researches seem to point in the same direction: healthy habits improve our general health status [1, 2]. However, creating and maintaining healthy habits and routines to achieve a healthy lifestyle are not simple tasks. Many researches and

J. Henriques et al. (Eds.): MEDICON 2019, IFMBE Proceedings 76, pp. 1181–1186, 2020.
https://doi.org/10.1007/978-3-030-31635-8_143

commercial products have tried to achieve this goal [3, 4], but only limited results have succeeded [5, 6]. Furthermore, most of these solutions are only focused on a determinate area, like physical exercise or nutrition, and they do not apply the required multidimensional approaches. Healthy habits have a clear impact against obesity, diabetes, cardiac diseases, neurodegenerative diseases and many other chronic diseases and new approaches are needed for personalized monitoring in mHealth applications supporting these clinical areas [7].

2 New Approaches for Personalised Monitoring

This research work is focused on looking for new technological and methodological solutions to create personalised monitoring services integrated in a mHealth platform. We will integrate innovative multivariable monitoring and interaction technologies, data analysis, gamification, user experience and predictive models to develop this mHealth technological platform.

2.1 Monitoring Technologies Integration: Enriched Monitoring Environment Integrating Devices Outcomes

Nowadays we can find plenty of devices monitoring. Mobile phones and wearables devices are typical examples of them. Smart fitness trackers allow us to develop our applications and access sensors' data and analyze and interpret them [4]. Likewise, although there are many commercial applications and platforms like Google Fit or Apple Health, we are developing our own platform to have total control of the data and guarantee user privacy according to the GDPR standards.

Attention, memory and executive functions can be considered as the main cognitive functions. They play a main role in cognitive activity monitoring, and there are some applications to evaluate them such as the 'Guttmann, NeuroPersonalTrainer®' (GNPT) [8] system developed by the authors' research team.

The mHealth platform will also allow us to monitor sleep habits and routines. Sleep disorders have important implications on daily life and can trigger other serious problems like difficulties to perform professional or social activities [9]. Parameters like duration, efficiency, time to sleep or percentages on REM or deep phases will be analyzed.

Nutrition is one of the hardest monitoring challenges. Although there have been attempts of getting efficient methods with image processing or other techniques, the preferred technique to monitor is questionnaires and reports [10]. The main challenge here is to develop a method simple enough to not frustrate the user, but precise enough to get relevant information from it. Gamification and user experience (UX) techniques will be applied to get the best possible results.

Monitoring these aspects is crucial for providing proper advice oriented to change and promote healthy habits. There are many options available to monitor those behaviors; however there are not many systems focused on monitoring the emotional impact associated to recommendations and advices provided to users. Although we know that motivation plays a key role in the adoption of new habits, or patterns that facilitate the adoption of healthier habits.

Because of that, we will propose to strength the research on personality and mood detection from emotional status. Emotion detection is being highly used nowadays in the health application field. One of the main innovative points in our approach is the monitoring of emotions and affective computing. There are some applications that allow mobile emotions monitoring, like EmotionSense, Affectiva or EmoPy.

2.2 New Non-invasive and Unobtrusive Behaviour Monitoring Methods

The growing amount of accessible information from mHealth applications and devices has opened new possibilities for the creation of models through machine learning. Methodologies like Case Based Reasoning, with a long tradition of health applications, are a good example of combining automatic learning methods with personalization methods. One of the most important challenges from these new decision support systems is considering patients as active actors (empowered). This characteristic is even more relevant to healthy people who are often unaware of the consequences of not taking preventive measures. Thus, to maximize the adoption of healthy life habits we must consider not only the relevant information about an individual but also the perception that he/she has of his/her health, tastes and preferences.

The approach using artificial intelligence techniques applied to the extraction of information and knowledge is an active and expanding field, but it is still a great challenge that is only partially covered by existing tools [11, 12]. Cognitive Computing (CC) platforms are a good example of how combining natural language processing, hypothesis generation and evidence-based learning can help medical professionals in clinical decision making.

When we talk about personality, we talk about the characteristics that define who we are. We can find people more competitive or more cooperative. People more aggressive or more passive, more outgoing or more introvert. These characteristics are usually long lasting. They can change but they usually change in a progressive and not sudden way. We can infer these variables from user's behavior or testing him with different kinds of gamification goals.

Different psychological tests can help us to get personal behavior characteristics. Thus, it is better to get an initial status from these evaluating tests to get an initial starting point. Then, we can infer personality changes from intelligent monitoring results, and we could reinforce our inferences from repeating these tests. For example, an increase of self-confidence could result in a person become more extrovert or more aggressive.

On the other hand, mood status is a volatile variable. We can be so happy in a moment and so sad on the next. It is hard to monitor but it is worth it. If our system must get user's input, is easier that the person collaborates if he/she is in a good state of mind. We can infer user's mood from different techniques detailed below.

- **Image analysis.** We can analyze images of the user's face to infer his state of mind. A neural network processes the image to get facial expression points and then another neural network analyzes the correlation among points to extract the facial expression. If we analyze a consecutive set of frames of a video from the user, we can infer his current mood.

- **Voice analysis.** From audio wave of user's voice, we can also infer his state of mind. Our voice tone changes from sad to neutral or anger. With a neural network, we can analyze audio incomings from users to detect his current state of mind.
- **Speech analysis.** We can also try to guess user's mood from his messages. If it comes in audio form, we can transform it to text and then do a Natural Language Processing (NLP) to look for keywords and sentences construction that denote his current mood. If the message arrives as a text input, we can skip the first step.

Combining all these techniques, we could correlate the different results, and make an efficient measure of in which state of mind the user is. We can obtain all this information from a video message of the user. We are focusing on a two ways approach. On one hand, we are monitoring and detecting emotional status analyzing the text introduced in the application by the user. In each report, we analyze the words and the general tone to extract the mood of the user. On the other hand, we are making another approach via image analysis. User will send records to analyze the image and extract facial expressions in video-messages.

2.3 User Adaptive Monitoring and Intervention

UX in human-machine applications refers mainly to the feelings, emotions and perceptions of the users before or during the use of a system [13]. Personalized monitoring based on technology will have a higher effectivity as well as they are supported with a behavior model [14]. These monitoring processes will include different scopes. Cognitive-social theory refers to subject perception of his capacity to change his behavior. If the subject presents a low self-efficacy, his/her motivation to change will be reduced. Thus, interventions based on intelligent monitoring technologies must provide the user with resources to empower them. Goal setting theory talks about the importance of setting short-term goals to induce behavior changes. Some studies have proved their relevance in nutrition [15], physical exercise [16] or self-management of chronical diseases like diabetes [17]. Self-determination theory helps to comprehend the reasons that make a subject to start an action. The enjoyment while an action is being realized, is a predictor of the subject's will to repeat the same action. This means that, when designing a technology to promote behavior change, the system must consider emotional experience of the user to increase the intervention adherence.

The incorporation of motivational strategies through gamification is being applied more and more in the context of the promotion of healthy behaviors. Up to now, gamification has been used to create an attractive and motivating experience that can increase the commitment. In our proposal, gamification will also be used to point to many results, including intelligent and personalized monitoring, knowledge acquisition, motivation maintenance, adherence increasing and behavior change.

3 Conclusions

This research presents the concept of personalized intelligent monitoring to implement a technological mHealth platform. Current approaches have not fully succeeded in helping to improve quality of life and quality of care through daily activity monitoring. The literature suggests that one of the main reasons is the lack of personalized approaches.

An intelligent and personalized monitoring will be based on different variables adapted to users such as activity and cognitive patterns, nutrition habits or emotional status. This monitoring will allow to create intervention modules which adapt to user needs and goals and will not be generic and impersonal. With gamification and personalization layers we will be able to dispense personalized recommendations and reminders in the correct timings and trying to impact in the prefect way.

We want this platform to be dynamic and can be applied to many different mHealth approaches, from healthy aging and chronic diseases prevention, to health education or rehabilitation fields. Prevention is the key to avoid chronical and neurodegenerative diseases which nowadays are some of the main charges for healthcare systems. The best way of prevention is having a healthy lifestyle, which is formed by healthy habits and routines. An effective way of monitoring focused on habit management would improve global healthcare systems and make them more sustainable saving several funds.

The technological solution that make possible the integration of the majority of monitoring devices or APIs will be successful. This solution will grant the independence of the user preferred devices from the platform but will allow to collect data and information from them. This will make possible to develop a scalable and global solution, valid for everyone.

Acknowledgements. This work has been partially funded by the Spanish Ministry of Science (project DPI2017-86088-C3-1-R, iBC).

Josep M. Tormos-Muñoz was partly supported by INNOBRAIN (COMRDI15-1-0017).

We also want to thank the Barcelona Brain Health Initiative and its research team, which has received funding from "la Caixa" Foundation (grant agreement n° LCF/PR/PR16/11110004), and also from Institut Guttmann and Fundació Abertis. A special thanks is extended to all participants and other partners (Ad-Salutem Institute, Sodexo, I.C.A. Informática y Comunicaciones Avanzadas, Neuroelectrics, Corporació Catalana de Mitjans Audiovisuals, Club Metropolitan, Casa Ametller, and Agència de Qualitat i Avaluació Sanitàries de Catalunya-AQuAS) for their invaluable collaboration.

References

1. Lally, P., Chipperfield, A., Wardle, J.: Healthy habits: efficacy of simple advice on weight control based on a habit-formation model. Int. J. Obesity **32**, 700–707 (2008)
2. Johansson, L., Thelle, D.S., Solvoll, K., Bjørneboe, G.A., Drevon, C.A.: Healthy dietary habits in relation to social determinants and lifestyle factors. British J. Nutr. **81**(3), 211–220 (1990)
3. Habitica. https://habitica.com/static/home. Accessed 08 April 2018

4. Fitbit Homepage. https://www.fitbit.com/home. Accessed 08 April 2018
5. Rothman, A.J., Gollwitzer, P.M., Grant, A.M., Neal, D.T., Sheeran, P., Wood, W.: Hale and hearty policies: how psychological science can create and maintain healthy habits. Perspect. Psychol. Sci. **10**(6), 701–705 (2015)
6. King, D.E., Mainous III, A.G., Geesey, M.E.: Turning back the clock: adopting a healthy lifestyle in middle age. Am. J. Med. **120**(7), 598–603 (2007)
7. Baniasadi, T., Niakan Kalhori, S.R., Ayyoubzadeh, S.M., Zakerabasali, S., Pourmohamadkhan, M.: Study of challenges to utilise mobile-based health care monitoring systems: A descriptive literature review. J. Telemed. Telecare **24**(10), 661–668 (2018)
8. Solana, J., Cáceres, C., García-Molina, A., Opisso, E., Roig, T., Tormos, J., Gómez, E.: Improving brain injury cognitive rehabilitation by personalized telerehabilitation services: Guttmann neuropersonal trainer. IEEE J. Biomed. Health Inform. **19**(1), 124–131 (2015)
9. Sexton, C.E., Storsve, A.B., Walhovd, K.B., Johansen-Berg, H., Fjell, A.M.: Poor sleep quality is associated with increased cortical atrophy in community-dwelling adults. Neurology **83**, 967–973 (2014)
10. Guasch-Ferré, M., Salas-Salvadó, J., Ros, E., Estruch, R., Corella, D., Fitó, M., et al.: The PREDIMED trial, Mediterranean diet and health outcomes: how strong is the evidence? Nutr. Metab. Cardiovasc. Dis. **27**, 624–632 (2017)
11. Etzioni, O., Fader, A., Christensen, J., Soderland, S., Mausam.: Open information extraction: The second generation. In: IJCAI International Joint Conference on Artificial Intelligence, vol. 11, pp. 3–11 (2011)
12. Lok, C.: Literature mining: Speed reading. Nature **463**(7280), 416–418 (2010)
13. Van Der Geest, T., Ramey, J., Rosenbaum, S., Van Velsen, L.: Introduction to the special section: Designing a better user experience for self-service systems. IEEE Trans. Prof. Commun. **56**(2), 92–96 (2013)
14. Fanning, J., Mullen, S.P., Mcauley, E.: Increasing physical activity with mobile devices: A meta-analysis. J. Med. Internet Res. **14**(6), e161 (2012)
15. Shilts, M.K., Horowitz, M., Townsend, M.S.: Goal setting as a strategy for dietary and physical activity behavior change: A review of the literature. Am. J. Health Promot. **19**(2), 81–93 (2004)
16. Cabrita, M., Akker, H. D., Achterkamp, R., Hermens, H., Vollenbroek-hutten, M.: Automated personalized goal-setting in an activity coaching application. In: Proceedings of the 3rd International Conference on Sensor Networks, pp. 389–396 (2014)
17. Patton, S.: Adherence to glycemic monitoring in diabetes. J. Diab. Sci. Technol. **9**(3), 668–675 (2015)

Encouraging Adherence of Chronic Obstructive Pulmonary Disease Patients to Physical Rehabilitation Programs Through Technology

Jorge Calvillo-Arbizu[1,2(✉)], Laura M. Roa-Romero[1,3], and Javier Reina-Tosina[1,3]

[1] Grupo de Ingeniería Biomédica, Universidad de Sevilla, Seville, Spain
jcalvillo@us.es
[2] Departamento de Ingeniería Telemática, Universidad de Sevilla, Seville, Spain
[3] Departamento de Teoría de la Señal y Comunicaciones, Universidad de Sevilla, Seville, Spain

Abstract. Encouraging behaviour changes (e.g., healthy habits) is a recurrent problem addressed nowadays by adopting general interventions in hospital or primary care settings. However, professionals count with no time or resources to develop a long-term follow-up for each patient. The application of information technology (IT) on behaviour change intervention may bring a set of benefits such as personalized intervention, remote supervision and continuous accompaniment, increased efficiency, and optimization of human and material resources.

Chronic obstructive pulmonary disease (COPD) is a worldwide current leading cause of death. Physical activity practice is one of the most effective non-pharmacological interventions for the improvement of COPD patients' health state. However, adherence rate is low because performing home programs requires engaged and proactive patients or a continuous supervision by the professional. Both conditions are difficult to be addressed with current interventions.

This work proposes addressing the high rate of COPD patients that drop out programs implying personalized rehabilitation at home through technology. This solution is approached by means of a controller that supervises a set of variables (collected from smart devices) and compares them with desirable outcomes in order to make a decision about adjusting the treatment, encouraging the patient or making amendments to the intervention. After assessing a first prototype by professionals and patients regarding usability of devices and preliminary interfaces, the current prototype is being piloted in controlled environments to check if it influences COPD patient adherence.

Keywords: Behaviour change · Information technology · COPD · Sensors

J. Henriques et al. (Eds.): MEDICON 2019, IFMBE Proceedings 76, pp. 1187–1194, 2020.
https://doi.org/10.1007/978-3-030-31635-8_144

1 Introduction

Acquiring and maintaining healthy behaviours is a current challenge for healthcare organizations as it may help to prevent cardiovascular diseases and negative effects on functional health and quality of life of citizens [1]. Physical exercise, varied nutrition and proper rest are some of the most effective long-term behaviours. But this kind of habits are difficult to acquire and easy to drop. Thus, encouraging behaviour changes is a recurrent problem addressed nowadays by adopting interventions (e.g., physical exercise, nutrition, healthy habits, etc.) in hospital or primary care settings. The same general interventions are often applied to any individual, so outcomes depend on each user's motivation and adherence. In order to increase interventions efficiency, personalized feedbacks can minimize the risk of dropping out and reinforce behaviour change [2]. Notwithstanding, personal interventions require human resources leading to higher costs, what limits their wide adoption.

The application of IT in behaviour change interventions makes possible personalized actions in home scenarios [3]. Several benefits can be obtained from IT such as a better access to these services (rural patients can be treated in their homes), continuous and remote monitoring allowing health professionals to supervise patient's performance and adherence, closer patient-professional relation, and sustainable costs. Remote interventions for behaviour change based on technologies have been used in different scenarios such as the promotion of physical activity and better nutrition [4–6], rehabilitation programs [7], medication adherence [8], management of long-term conditions [9, 10] and cessation of tobacco use [11, 12].

Chronic obstructive pulmonary disease (COPD) was classified in 1990 as the sixth leading cause of death, and it is assumed it will be the fourth in 2030 [13]. Since it presents secondary systemic manifestations as musculoskeletal disfunction [14], physical activity practice is one of the most effective non-pharmacological interventions for the improvement of physical capacity, symptoms and health state of COPD patients [15]. Nowadays, interventions on ambulatory settings for COPD patients are time-limited and resource-consuming tasks that require professional experience [16]. As an alternative, remotely supervised physical activity programs at home may remove access burdens and reduce costs of such increasing patient population [17]. However, adherence rate of COPD patients on physical activity interventions is low because performing home programs requires engaged and proactive patients or a continuous supervision by the professional. Both conditions are difficult to be addressed with current interventions.

Several studies suggest that remote physical activity interventions supported by technology mean an increase of quantity and quality of physical activity and a reduction of sedentary time when they are compared to control populations [17]. However, they also report that these adherence improvements disappear in the long term if intervention stops. This may be due to the factors that enhance a particular behaviour, which are different from those contributing to its maintenance [18].

This work proposes addressing the high rate of COPD patients that drop out programs implying personalized rehabilitation at home through technology. This solution is approached by means of a controller that supervises a set of variables (collected from smart devices) and compares them with desirable outcomes in order to make a decision about adjusting the treatment, encouraging the patient or making amendments to the intervention. This work presents a first approach to the issue in a controlled environment.

2 Materials and Methods

2.1 Encouraging Behaviour Change Through Technology

Remote personalized intervention for behaviour change can be seen as a control issue in which the measured variables should be within thresholds (i.e., desirable behaviour), in spite of perturbations. According to outcomes, a controller decides at real time if correction actions should be taken to amend deviations. The controlled process is an intervention upon a single individual, that is the target of the personalization.

In the scenario of physical rehabilitation at home for COPD patients, the individual's behaviour (registered by biomedical devices) may serve for extracting information about no visible effects (e.g., improvement of respiratory capacity) and for performing a personalized feedback. Perturbations may influence outcomes and require corrections or feedback to the patient. Perturbations can be external (e.g., malfunction of devices) or internal (e.g., discouragement, low adherence) to the patient (Fig. 1). The professional, acting as the controller, works upon measured data and influences the process for controlling or minimizing perturbations. The manipulated variable will be the stimulus that the professional uses with the patient with the aim of achieving an acceptable level of performance and adherence to the program. The controlled variable will be data collected by a physical activity monitor from the patient's performance (see Sect. 2.2).

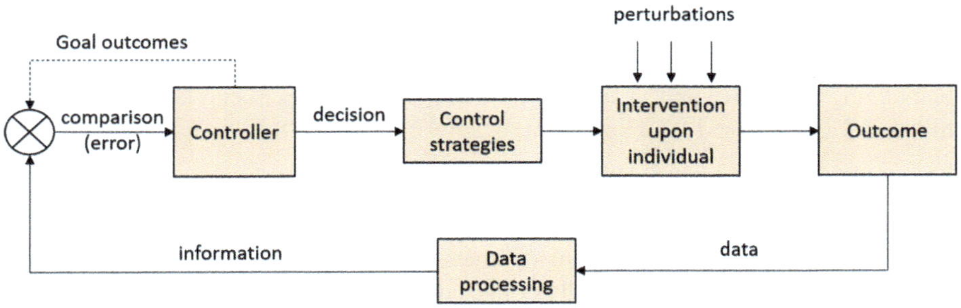

Fig. 1. Control model for remote behaviour change interventions

2.2 Smart Device for Monitoring Physical Activity

Inside the e-EPOC project [19], the authors have developed a smart and non-invasive device for monitoring physical activity, respiratory frequency, and metabolic expenditure [20] (Fig. 2). Movement monitoring is performed by means of several processing modules:

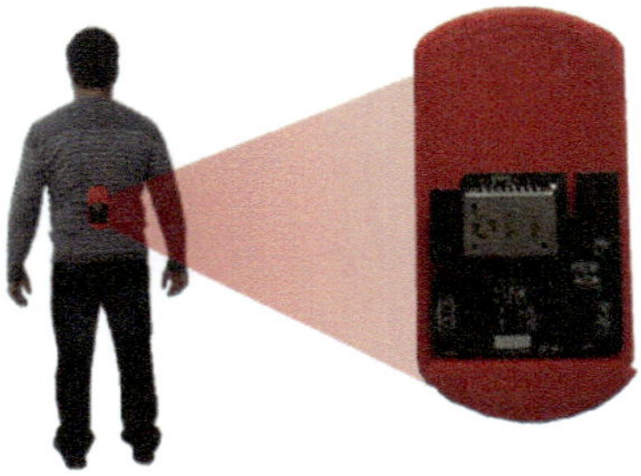

Fig. 2. Smart device for physical activity monitoring

1. Module for classification of activity level: that can be used to get distribution and intensity of in/activity patterns in order to quantify thresholds or improve physical exercise prescription.
2. Module for classifying activities of daily living (ADL): this module is capable of identifying activities such as stair climbing or descending, through the decomposition of acceleration axes.
3. Module for estimating metabolic expenditure: valid for different users with no regard of their anthropometric characteristics.

This device is carried by the patient anytime and registers automatically levels of activity and inactivity, activities performed, and metabolic expenditure. Such data is gathered in real time and sent to a processing element such as a smartphone.

3 Results

Figure 3 shows the control model applied to our scenario of physical rehabilitation for COPD patients at home. The controlled process is the proper execution of rehabilitation programs (i.e., set of physical activities personalized for each patient), focusing on maintaining adherence and increasing physical performance. Each element composing this model is explained below.

3.1 Outcome Variables and Monitoring System

A home monitoring system is the key element for remote supervision of interventions. The more variables and data can be collected, the more suitable the decision to personalize patient rehabilitation. In our approach, we adopted an iterative development process through which variables are included in each iteration. In the current stage, patient monitoring relies on a home and non-invasive monitoring system including the smart device described above. This system collects respiratory frequency, cardiac frequency, and movement of the patient in real time. Data is sent to a smartphone where an app performs a processing to generate information from the data such as exercise performance, respiratory capability and adherence level.

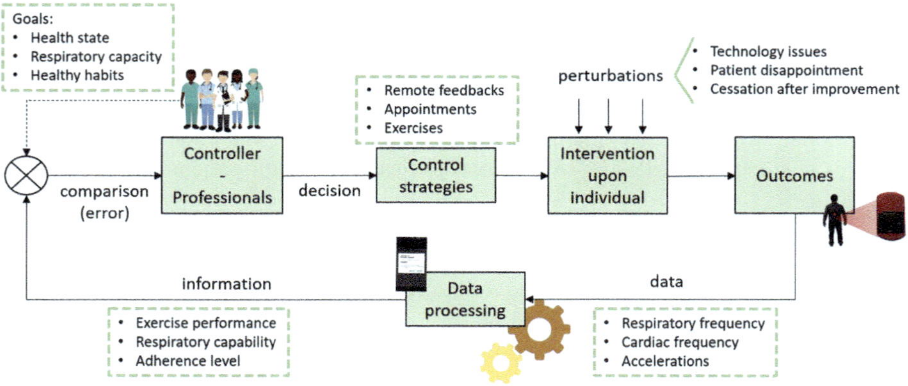

Fig. 3. Home physical rehabilitation for COPD patients supported by technology

When healthy habits are monitored, often data missing is not random but correlated with the patient's activity and her will. External perturbations such as malfunction of technology or improper placement of devices can lead to misreadings. Internal (patient-based) perturbations include disappointment, cessation after improvement of health status, living with unhealthy-habits people that discourage the patient, etc. All these perturbations should be identified and countered in order to enhance patient's performance and adherence. In the data processing step, data may be combined with information from other sources to provide context and generate knowledge useful for the controller.

3.2 Professionals as Controllers

Interventions (i.e., physical exercise, nutrition, healthy habits, etc.) must be prescribed by professionals (pneumologists for COPD patients). Personalization of intervention should be applied according to the severity of patient's condition, adherence level, and patient's context. After starting the personalized intervention, a continuous follow-up of its performance can help to identify deviations or warn risk trends. By comparing intervention goals with real performance, professional can make several decisions such as adjust/change the intervention, reward a proper execution or encourage to resume a dropped intervention. Decisions are implemented as control strategies (e.g., remote communications, new recommendations, rewards, personal appointments for patients in serious risk, etc.).

In the current stage of our work, professionals receive adherence level of their patients in rehabilitation programs (observance of exercise schedule, performance of exercises, and number of repetitions of each exercise) as well as respiratory capability and cardiac frequency. Adherence level is compared with prescribed exercise to conclude if the patient is performing exercises rightly and in the proper sequence. Either if there is a deviation or not, professionals can contact the patient (via telephone or message) to warn about wrong performance, recommend changes or congratulate the patient's work.

Finally, in order to guarantee that patients feel accompanied, bidirectional communication channels must be provided. The channel from professional to patient is, as has been described, for reinforcing good behaviours or amending deviations. The opposite direction (from patient to professional) should be considered but limited to exceptional circumstances (health issues, risk events, questions about interventions or devices use, etc.) in order not to overload professionals with patient demands.

3.3 Ongoing Development

The first iteration of the approached solution was developed in the e-EPOC project [19], and the prototype was assessed by professionals and patients regarding usability of devices and preliminary interfaces. This assessment resulted in new requirements that were addressed on the second development iteration. Currently the prototype is being piloted in controlled environments to check if it influences COPD patient adherence.

The ongoing development includes several extensions to the whole model. Firstly, the inclusion of new variables collected from other devices such as weight, oxygen saturation, temperature, etc. Additional data can complete patient's status and provide a more accurate decision about her treatment. Moreover, it is planned to improve the data processing stage towards the development of a Clinical Decision Support System (CDSS) that integrates data from devices with other information sources related to the patient (e.g., electronic health record -EHR-, environmental data, questionnaires filled by the patient, social record, etc.) in order to generate advanced knowledge about the patient beyond the monitoring data. The CDSS will allow to storage and exploit high volumes of information by means of AI techniques. Outcomes from CDSS will include alerts of risk tendencies, prediction of results from particular interventions (through the comparison with past interventions), automatic generation of reports for professionals and patients, etc. Finally, we presume to design and develop a mobile interface for the patient. This element will allow patients and caregivers to be aware of intervention changes, goals, outcomes, and professionals' recommendations and messages.

4 Conclusion

Maintaining behaviour changes in the long term requires a continuous follow-up by health professionals, what is not feasible nowadays. Technology can ease the deployment of large-scale behaviour change interventions minimizing human and economic costs meanwhile benefits such as continuous accompaniment and increased efficiency are provided to the patient.

In this work, a control system for COPD patients in personalized rehabilitation interventions at home has been approached. Professionals (as controller) supervise a set of variables (e.g., respiratory capability, exercise execution) and compares them with desirable outcomes in order to make a decision about adjusting the treatment, encouraging the patient or making modifications to the intervention. The solution presented is a first stage of the control model for home behaviour change interventions. Future research lines will be focused on including new monitored variables and developing a CDSS that integrates data provided by the monitoring system with other information sources (e.g., EHR, environmental data…) and generates helpful knowledge for professionals and patients.

References

1. Piepoli, M.F., Hoes, A.W., Agewall, S., Albus, C., et al.: 2016 European Guidelines on cardiovascular disease prevention in clinical practice. Eur. Heart J. **37**(29), 2315–2381 (2016)
2. Marteau, T.M., Hollands, G.J., Fletcher, P.C.: Changing human behaviour to prevent disease: the importance of targeting automatic processes. Science **337**, 1492–1495 (2012)
3. Schoeppe, S., Alley, S., Van Lippevelde, W., Bray, N.A., et al.: Efficacy of interventions that use apps to improve diet, physical activity and sedentary behaviour: a systematic review. Int. J. Behav. Nutr. Phys. Act. **13**(1), 127 (2016)
4. Roberts, A.L., Fisher, A., Smith, L., Heinrich, M., Potts, H.W.W.: Digital health behaviour change interventions targeting physical activity and diet in cancer survivors: a systematic review and meta-analysis. J. Cancer Surviv. Res. Pract. **11**, 704–719 (2017)
5. Flores-Mateo, G., Granado-Font, E., Ferré-Grau, C., Montaña-Carreras, X.: Mobile phone apps to promote weight loss and increase physical activity: a systematic review and meta-analysis. J. Med. Internet Res. **17**, e253 (2015)
6. Middelweerd, A., Mollee, J.S., van der Wal, C.N., Brug, J., Te Velde, S.J.: Apps to promote physical activity among adults: a review and content analysis. Int. J. Behav. Nutr. Phys. Act. **11**, 97 (2014)
7. Rawstorn, J.C., Gant, N., Direito, A., Beckmann, C., Maddison, R.: Telehealth exercise-based cardiac rehabilitation: a systematic review and meta-analysis. Heart **102**, 1183–1192 (2016)
8. Mistry, N., Keepanasseril, A., Wilczynski, N.L., Nieuwlaat, R., et al.: Technology-mediated interventions for enhancing medication adherence. J. Am. Med. Inform. Assoc. **22**, e177–e193 (2015)
9. Jackson, B.D., Gray, K., Knowles, S.R., De Cruz, P.: EHealth technologies in inflammatory bowel disease: a systematic review. J. Crohn's Colitis **10**, 1103–1121 (2016)
10. Su, D., Zhou, J., Kelley, M.S., Michaud, T.L., et al.: Does telemedicine improve treatment outcomes for diabetes? A meta-analysis of results from 55 randomized controlled trials. Diabetes Res. Clin. Pract. **116**, 136–148 (2016)
11. Taylor, G.M.J., Dalili, M.N., Semwal, M., Civljak, M., et al.: Internet-based interventions for smoking cessation. Cochrane Database Syst. Rev. **9**, 1–159 (2017)
12. Whittaker, R., McRobbie, H., Bullen, C., Rodgers, A., Gu, Y.: Mobile phone-based interventions for smoking cessation. Cochrane Database Syst. Rev. **4**, 1–54 (2016)
13. Mathers, C., Loncar, D.: Projections of global mortality and burden of disease from 2002 to 2030. PLoS Med. **3**, 2011–2030 (2006)
14. Global Initiative for Chronic Obstructive Lung Disease (GOLD). From the Global Strategy for the Diagnosis, Management and Prevention of COPD. http://www.goldcopd.org. Accessed 11 April 2019
15. ZuWallack, R.: The nonpharmacologic treatment of chronic obstructive pulmonary disease: Advances in our understanding of pulmonary rehabilitation. Proc. Am. Thorac. Soc. **4**, 549–553 (2007)

16. Hekler, E.B., King, A.C., Banerjee, B., Robinson, T., et al.: A Case Study of BSUED: Behavioral Science-informed User Experience Design. http://www.personalinformatics.org/docs/chi2011/hekler.pdf. Accessed 11 April 2019
17. Stockwell, S., Schofield, P., Fisher, A., Firth, J., et al.: Digital behavior change interventions to promote physical activity and/or reduce sedentary behavior in older adults: a systematic review and meta-analysis. Exp. Gerontol. **120**, 68–87 (2019)
18. Rothman, A.J.: Toward a theory-based analysis of behavioral maintenance. Health Psychol. **19**(1), 64–69 (2000)
19. ISCIII PI15/00306 – Sistema de e-Salud para el control supervisado de pacientes con EPOC en programas de mantenimiento domiciliarios (e-EPOC)
20. Naranjo-Hernández, D., Roa, L.M., Reina-Tosina, J., Estudillo-Valderrama, M.A.: SoM: a smart sensor for human activity monitoring and assisted healthy ageing. IEEE Trans. Biomed. Eng. **59**(11), 3177–3184 (2012)

Evaluation of an Environmental Autism Spectrum Disorder Monitoring Device

José María Vicente-Samper[1(✉)], Carolina Blanco-Angulo[2], Ernesto Ávila-Navarro[2], and José María Sabater-Navarro[1]

[1] Department of Systems Engineering and Automation, Miguel Hernández University of Elche, 03202 Elche, Spain
jose.vicentes@umh.es

[2] Department of Materials Science, Optics and Electronic Technology, Miguel Hernández University of Elche, 03202 Elche, Spain

Abstract. Currently, there is a growth in the diagnostic of Autism Spectrum Disorder (ASD) in children. This type of neurocognitive disorder prevents normal development from an early age. In addition, most people suffering from ASD have a Sensorial Processing Disorder (SPD) associated. This makes them perceive objects and environment in a different way and sometimes causes changes in behaviour, causing a crisis at certain times. It is proposed to monitor different physical parameters of the environment and the user to detect the causes that yield these changes in behaviour. For this, the paper shows the development of a first prototype that integrates some of those environmental parameters selected to measure and it is accompanied by a software for the management of the device. Finally, the correct operation of the prototype is validated.

Keywords: Autism Spectrum Disorder (ASD) · Sensory Processing Disorder (SPD) · Monitoring system · Wearable electronic

1 Introduction

According to the World Health Organization (WHO), currently one in 160 children is diagnosed with an Autism Spectrum Disorder (ASD) [1,2]. This estimation represents an average figure. The prevalence of ASD in many low income countries is so far unknown. ASD is present in all races, ethnicities and socioeconomic groups. It is 4 to 5 times more likely among males than females. In addition, between 30% and 50% of people with ASD have an associated intellectual disability (IQ < 70) [3–5].

Sleep, attention, and behaviour are common problems in people with neurocognitive disorders such as ASD. At the present day there is still no drug that treats ASD. However, people with Autism Spectrum Disorder have an intensive use of drugs to mitigate derived problems from the disorder such as sleep problems, hyperactivity, physical problems from motor malpractice or epilepsy. And this medication sometimes contributes to aggravate developmental problems and cause behavioural problems.

J. Henriques et al. (Eds.): MEDICON 2019, IFMBE Proceedings 76, pp. 1195–1204, 2020.
https://doi.org/10.1007/978-3-030-31635-8_145

Sensory integration can be defined as the organization of sensory information for its use [6]. It is a neurological process that allows the individual to make sense of the environment that surrounds him. However, it is estimated that 95% of people with ASD also present a Sensory Processing Disorder (SPD) [7–13]. These problems are related to difficulties in social interaction, learning, and behaviour in a person with ASD. In the Diagnostic and Statistical Manual of Mental Disorders (DSM-V) of the American Psychiatric Association, it recognizes as diagnostic criteria of ASD hypo- or hyper-reactive alterations to sensory stimuli or an unusual interest in sensory aspects of the environment [14].

People with ASD exhibit great heterogeneity in their atypical sensory behaviours. Some individuals do not respond when they perceive particular sounds and others are extremely sensitive to certain sounds considered by the majority as soft sounds. The same applies to some visual, tactile, olfactory and gustatory stimuli. Sensory alterations significantly affect the activities of daily life and contribute to a challenging behaviour and social isolation [15].

The challenging behaviours are mostly the result of a negative stimulus for the person that causes discomfort. Due to a hypersensitivity in some of the senses in people with ASD, any change in the environment can motivate this type of defiant, hysterical, and even aggressive behaviour that must be intervened [16]. When talking about intervention, it is defined as "Any responding method to behaviour that implies some degree of physical force aimed to limiting or restricting movement or mobility" [17], to prevent the individual from injuring himself or others.

It is inevitable that new situations will arise in a person's daily life. But it is important to control most of the circumstances faced by a person with Autism Spectrum Disorder to prevent behavioural crises. Therefore, trying to know the variables of the environment where the person is, as well as some biological variables of the individual, could allow to have a greater control of the possible situations and it could help to detect new causes that produce those behavioural changes in him.

Deciding which are the interesting variables to measure and which one can provide information about the physical state of the individual and the environment is an important task.

The line of work tries to establish which are those physical parameters of the user and the environment that should be measured. Once the variables have been selected, a monitoring device must be developed to obtain the information. Then, the data must be analysed to feedback the system and, thus, it will be able to anticipate possible changes. The paper is structured as follows: in the materials and methods section, the selection of the included variables in the system and the development of a first environmental monitoring prototype, as well as the software to manage it, are presented. The results section shows the validation of the device in a real scenario. Finally, these results are discussed and the next steps in the line of work are presented.

2 Materials and Methods

2.1 Evaluation of Measurable Parameters

Due to the special characteristics of people with ASD and their reluctance to foreign bodies, a modular solution has been chosen for the device. On the one hand, there are the environmental characteristics. The environment has, for example, climatological, luminosity or sonorous conditions. The measurement of these parameters will make it possible to know the situation when a behaviour change occurs. For it, an environmental monitoring device will be developed with all the selected variables from this class. On the other hand, there are the biological variables of the person. It is necessary to control the physical state of the individual to detect a behavioural crisis. Then, it will be possible to relate it to the environment condition change or a problem in the physical state of the person and thus it will be able to anticipate the reason that causes it in order to correct it or to prevent it in the future. The monitoring of these personal variables will be integrated in a customizable wearable device. Thus, attending to the different characteristics of each user, it will be feasible to equip them with more or less parts of the device.

Table 1 groups all the information considered to select the variables that will be included in the monitoring system.

First, each variable is classified depending on whether it is a parameter to be measured from the environment or from the person. Also if this variable is a parameter that will have influence on a change of the user's physical status or whose modification implies that there has been a change in the user's status. Then, the viability of the measurement of each variable is evaluated, as well as the ease of integration of the variable measurement means in the device.

The next column shows a numerical assessment, where 10 means very important, of the importance of each variable in the behaviour of a person with ASD. These values have been assigned in collaboration with professionals from the Infanta Leonor Center for people with ASD in Alicante (Spain).

Finally, it is specified which variables, taking into consideration all the previous information presented, will be incorporated in the monitoring device. Educational and biological variables are considered modifiers that affect the physical and cognitive states of the user and depend on many factors. Therefore, it is not feasible to include them directly in the device. Table 2 presents the main factors that correspond to these variables.

Table 1. Measurable parameters studied for the system

	Personal	Environment	Reason	Consequence	Ease of measurement	Ease of integration	Clinical assignment	Selected variable
Educational variables	✓	✓	✓		No	No	9	No
Biological variables	✓		✓		No	No	8	No
Amount of people in the environment		✓	✓		No	Yes	10	Yes
Relationship with people in the environment		✓	✓		No	No	8	No
Body temperature	✓			✓	Yes	Yes	–	Yes
Ambient temperature		✓	✓		Yes	Yes	8	Yes
Atmospheric pressure		✓	✓		Yes	Yes	0	Yes
Relative humidity		✓	✓		Yes	Yes	0	Yes
Luminosity		✓	✓		Yes	Yes	6	Yes
Heart rate	✓			✓	Yes	Yes	–	Yes
Ambient noise: Sonority		✓	✓		Yes	Yes	10	Yes
Ambient noise: Frequency		✓	✓		Yes	Yes	10	Yes
Electrodermal activity	✓			✓	Yes	No	–	Yes
Textures	✓		✓		No	No	5	No
Human contact	✓		✓		No	No	10	Yes
Smells		✓	✓		No	Yes	8	No
Flavors	✓		✓		No	No	6	No
Physical activity	✓		✓	✓	Yes	Yes	6	Yes
Electromyography activity	✓			✓	Yes	No	–	No
Visual stimulation: Colors		✓	✓		No	Yes	1	Yes
Visual stimulation: Shapes		✓	✓		No	Yes	1	Yes
Visual stimulation: Rate		✓	✓		No	Yes	3	Yes
Body balance	✓		✓		No	No	5	No
Propioception	✓		✓		No	No	8	No

Table 2. Main factors of the educational and biological variables

Biological variables	- Medication - Nutrition - Hormonal fluctuations - Rest
Educational variables	- Schedule - Spontaneous occurrence - Make a choice - Activities

2.2 Device for Environmental Data Acquisition

The development of a first prototype for the environmental measurement device is presented. This prototype was designed with the purpose to evaluate the methodology for measuring in real scenarios with users suffering from ASD.

The device integrates the measurement of four of the selected variables to be monitored by the system: luminosity, atmospheric pressure, relative humidity and ambient temperature. A number of sensors are used for this purpose. An OPT3001 [18] is used to measure the luminosity of the environment. Relative humidity is measured with an HDC1010 [19]. Finally, a BMP280 [20] is used to measure atmospheric pressure and ambient temperature. All of them are connected via I^2C with a CC2650 [21] microcontroller. It incorporates the Bluetooth Low Energy communication protocol (BLE) and a module to manage it in its architecture. In addition, a push button has been integrated into the device. It allows the caregiver to make a manual annotation of timemarkers when a change in the user's behaviour is appreciated. The device is powered by a 260 mAh Li-ion battery. It has a charge management controller MCP7381 [22] with a charging interface through a μUSB connector.

Figure 1 shows an image of the prototype with all the modules protected inside a $90 \times 60 \times 60$ mm casing. The sensors are located at the bottom left. This part of the device must remain uncovered at all times so that the values measured by the sensors are correct. On the right is the push button for adding manual timemarkers. It has a yellow led that turns on to verify that the marker has been added. Finally, at the top is the charging interface and the switch to turn the device power on and off. It has a blue led to indicate that the equipment is powered on and a red led to indicate that it is charging. When the charge is completed the red led switches off.

Fig. 1. Prototype of data acquisition device

2.3 Software for Controlling the Device

The obtained data by the sensors are transmitted through BLE to a smartphone. An Android application has been developed to manage the device and the information. The application acts as a platform between the device and a non-relational MongoDB database in which the measured data by the sensors are stored, as well as the timemarkers added manually with the push button. In addition, the application interface shows the latest sensors values to the user. Figure 2 shows a screenshoot of the interface during applitacion execution when the device is connected.

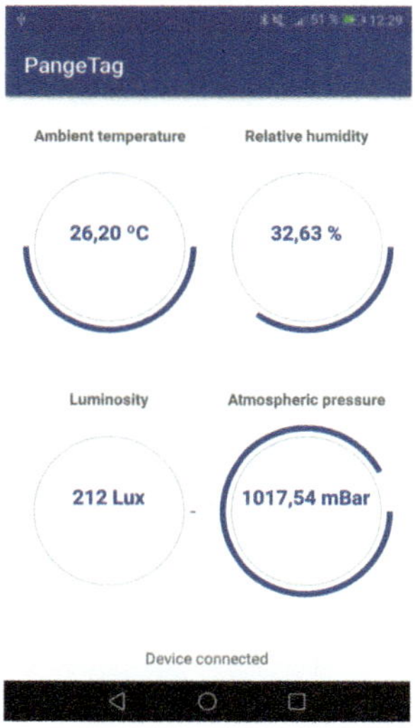

Fig. 2. Android application for controlling the device

The interface warns when the monitoring device connects to the application. From that moment, the periodically received data from the sensors are uploaded to the database through a mobile data connection. In addition, the interface is updated showing the new values.

When the device's battery is low or the device is disconnected from the application a notification is launched to the user on the smartphone.

3 Results

3.1 Validation of the Environmental Device

To evaluate the working method, the developed prototype, the management software and the remote database for data storage, the device was handed over to a family with an 9-year-old child with ASD. The evaluation period was 15 days. During it, the family was asked to place the prototype in the room where the user was and to move it to another room if the user does so. In this way the environment where it is located is measured. Figure 3 shows a time record of 11 h where we can observe that the device is measuring atmospheric pressure, relative humidity, ambiental temperature and luminosity from environment. For instance, the luminosity sensor informs that the room remains dark from 17:30 to 6:30.

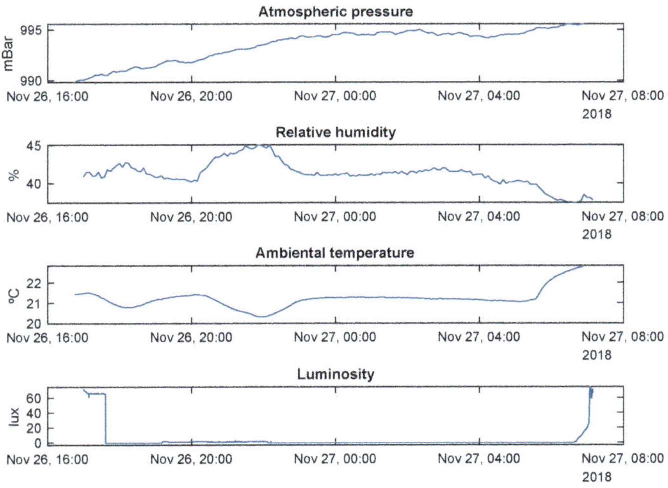

Fig. 3. A recording piece of the sensors measurements

The family was also asked to add a marker through the prototype push button when detecting a behavioural change in the user. This way the event will be registered. Figure 4 shows the detection of a timemarker during environmental monitoring. It can be seen that the marker was added around 15:30. In the absence of biological signals from the user, it is still not possible to determine the cause why the marker was added.

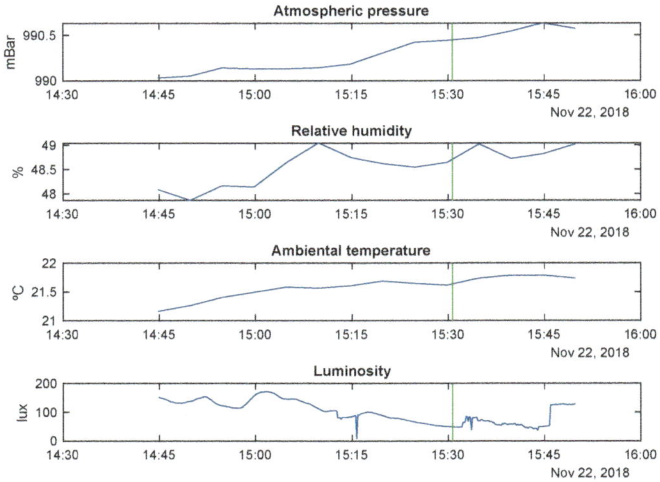

Fig. 4. A timemarker detection during the measurement recording

4 Discussion

It has been possible to verify the correct functioning of the device. Despite being a first prototype that only integrates 4 of the selected environmental parameters, the data collection and storage processes have been verified.

Therefore, the remaining selected environmental variables must be integrated into the device. The personal device for monitoring biological parameters must be also developed. Once the system is fully developed, it will be necessary to obtain much more data in order to design a signal processing algorithm.

Acknowledgement. Authors would like to thank the economical support from Spanish National Research Council through DPI2016-80391-C3-2-R project. And the Infanta Leonor Center and the Autism Parent Association of the Valencian Community for their collaboration and the work they do. José M. Vicente would like to thank the Conselleria d'Educació, Investigació, Cultura i Esport of Generalitat Valenciana for FDEGENT/2018/015.

References

1. National Institute of Mental Health, Autism Spectrum Disorder, NIMH – Autism Spectrum Disorder (2018). https://www.nimh.nih.gov/health/topics/autism-spectrum-disorders-asd/index.shtml. Accessed 1 Feb 2019

2. World Health Organization, Autism spectrum disorders, News center World Health Organization (2017). https://www.who.int/news-room/fact-sheets/detail/autism-spectrum-disorders. Accessed 16 Apr 2018
3. Christensen, D.L., et al.: Prevalence and characteristics of autism spectrum disorder among 4-year-old children in the autism and developmental disabilities monitoring network. J. Dev. Behav. Pediatr. **37**(1), 1–8 (2016)
4. Confederacinñ Espaóola de Autismo, Trastorno del Espectro Autista, Autismo – TEA Confederación Española de Autismo (2016). http://www.fespau.es/autismo-tea.html. Accessed 16 Apr 2018
5. Centers for Disease Control and Prevention, Autism Spectrum Disorder (ASD), Data & Statistics - Autism Spectrum Disorder (ASD) CDC (2018). https://www.cdc.gov/ncbddd/autism/data.html. Accessed 16 Apr 2018
6. Ayres, A.J.: Sensory integration and learning disorders. Western Psychological Services (1972)
7. Pfeiffer, B.A., et al.: Effectiveness of sensory integration interventions in children with autism spectrum disorders: A pilot study. Am. J. Occup. Ther. **65**(1), 76–85 (2011)
8. Baranek, G.T.: Efficacy of sensory and motor interventions for children with autism. J. Autism Dev. Disord. **32**(5), 397–422 (2002)
9. Leekam, S.R., et al.: Describing the sensory abnormalities of children and adults with autism. J. Autism Dev. Disord. **37**(5), 894–910 (2007)
10. Little, L.M., et al.: Activity participation among children with autism spectrum disorder. Am. J. Occup. Ther. **68**(2), 177–185 (2014)
11. Baker, A.E.Z., et al.: The relationship between sensory processing patterns and behavioural responsiveness in autistic disorder: A pilot study. J. Autism Dev. Disord. **38**(5), 867–875 (2008)
12. Baranek, G.T., et al.: Sensory experiences questionnaire: Discriminating sensory features in young children with autism, developmental delays, and typical development. J. Child Psychol. Psychiatry **47**(6), 591–601 (2006)
13. Rogers, S.J., Ozonoff, S.: Annotation: What do we know about sensory dysfunction in autism? A critical review of the empirical evidence. J. Child Psychol. Psychiatry **46**(12), 1255–1268 (2005)
14. American Psychiatric Association (APA). Diagnostic and statistical manual of mental disorders (DSM-5). American Psychiatric Pub (2013)
15. Ortiz Huerta, J.H.: Terapia de integración sensorial en niños con el trastorno de espectro trastorno de espectro autista. TOG (2014). http://www.revistatog.com/num19/pdfs/original5.pdf. Accessed 6 Jan 2019
16. Fundación Uliazpi. Protocolo de actuación ante conductas desafiantes graves y uso de intervenciones físicas (2012)
17. Harris, J., Allen, D., Cornick, M., Jefferson, A., Mills, R.: Physical Interventions: A policy framework. Tizard Learn. Disabil. Rev. **2**(3), 41–42 (1997). https://doi.org/10.1108/1359547199700028
18. Texas Instruments, OPT3001 Ambient Light Sensor (ALS), Texas Instruments (2017). http://www.ti.com/lit/ds/symlink/opt3001.pdf. Accessed 16 Feb 2019
19. Texas Instruments, HDC1010 Low Power, High Accuracy Digital Humidity Sensor with Temperature Sensor, Texas Instruments (2016). http://www.ti.com/lit/ds/symlink/hdc1010.pdf. Accessed 16 Feb 2019
20. Bosch, BMP280 Digital Pressure Sensor, Bosch (2018). https://ae-bst.resource.bosch.com/media/_tech/media/datasheets/BST-BMP280-DS001.pdf. Accessed 17 Feb 2019

21. Texas Instruments, CC2650 SimpleLink Multistandard Wireless MCU, Texas Instruments (2016). http://www.ti.com/lit/ds/symlink/cc2650.pdf. Accessed 16 Feb 2019
22. Microchip Technology Inc., MCP73831/2 Miniature Single-Cell, Fully Integrated Li-Ion, Li-Polymer Charge Management Controllers, Microchip Technology Inc. (2014). http://ww1.microchip.com/downloads/en/DeviceDoc/20001984g.pdf. Accessed 19 Feb 2019

Evaluation and Comparison of Text Classifiers to Develop a Depression Detection Service

Diego Moreno-Blanco[1(✉)], Borja Ochoa-Ferreras[1], Francisco J. Gárate[1], Javier Solana-Sánchez[2,3,4], Patricia Sánchez-González[1,5], and Enrique J. Gómez[1,5]

[1] Biomedical Engineering and Telemedicine Centre, ETSI Telecomunicación, Center for Biomedical Technology, Universidad Politécnica de Madrid, Madrid, Spain
dmoreno@gbt.tfo.upm.es

[2] Institut Guttmann, Institut Universitari de Neurorehabilitació adscrit a la UAB, Badalona, Spain

[3] Universitat Autònoma de Barcelona, Barcelona, Spain

[4] Fundació Institut d'Investigació en Ciències de la Salut Germans Trias i Pujol, Badalona, Spain

[5] Centro de Investigación Biomédica en Red, Biomateriales y Nanomedicina (CIBER-BBN), Madrid, Spain

Abstract. Depression is a mental disorder which can become a serious health problem. This research line is focused on creating a depression detection service from text analysis. Sentiment Analysis and Natural Language Processing methods will be used to develop this service. The service will classify text in positive or negative depending on the emotions inferred from user's input texts. In this study, five classifiers have been evaluated to determine which one fits better for this purpose. Naïve Bayes, Decision Tree, Naïve Bayes based on Bernoulli model and Maximum Entropy are the classifiers analyzed. A specific corpus which includes sentences that fit research needs was designed for testing purposes, whereas Sentiment140's corpus was used with training purposes. An evaluation methodology formed by three tests was designed. Results show a promising starting point, but further analysis will be needed. Future related works will focus on expanding classification to get user's mood and not only a binary classification. It will allow to include this service as input for developing personalized intervention and education mHealth systems.

Keywords: Natural Language Processing · Sentiment Analysis · Machine learning · Depression · Artificial Intelligence

1 Introduction

1.1 Clinical Context: Depression

More than 300 million people of all ages are affected by depression worldwide [1]. Depression is a mental disorder different from usual mood fluctuations and short emotional responses to challenges in everyday life. Depression may become a serious health condition when long-lasting and with moderate or severe intensity.

J. Henriques et al. (Eds.): MEDICON 2019, IFMBE Proceedings 76, pp. 1205–1212, 2020.
https://doi.org/10.1007/978-3-030-31635-8_146

Healthcare providers may offer psychological treatments or pharmacological treatments with antidepressant medication, which are effective for moderate and severe depression [1]. Although these effective treatments, less than a half of affected people in the world receive such therapies [1] because of lack of resources, absence of trained healthcare providers, and social stigma associated with mental disorders. Other obstacle for effective care is an inaccurate assessment of the disease because people who are depressed are not often correctly diagnosed, and others who do not have the disorder are too often misdiagnosed [1].

The aim of this research is to develop a depression detection service based on mHealth. Analyzing text entries from the user, the service is pretended to detect negative sentiments and depression indicators. This service will be based in Natural Language Processing (NLP) techniques to perform a Sentiment Analysis (SA) that determine if the text is positive or negative.

1.2 Sentiment Analysis

Sentiment Analysis, also known as opinion mining, analyzes people's opinions, sentiments, evaluations, attitudes, and emotions towards entities such as products, services, organizations, individuals, issues, events, topics, and their attributes [2]. Subjectivity analysis classify a given text as objective or subjective, separating facts from feelings or beliefs, while polarity estimation tries to determine if a text includes positive or negative connotations. These are the two main tasks traditionally involved in SA [3].

Most of the implemented sentiment analysis approaches employ machine learning algorithms (Naïve Bayes classifier, Support Vector Machines or regression models) in order to create classifiers that assign a polarity class to each text review [4–7].

The most used classification feature is the bag of words (BoW), through which each sentence of a text is represented as a vector, the elements of which are the TF-IDF values of the corresponding words [5, 8]. TF-IDF is defined as the occurrence frequency of a term within a dataset, reflecting the importance of it. It is also very common to use sentiment-based lexicons to detect positive, negative and neutral words within a text, as well as subjective expressions and emotions, whose frequencies of appearance are later used as classification features [9–11].

1.3 Natural Language Processing

Natural Language Processing (NLP) is a subfield of Artificial Intelligence (AI) that focuses on enabling computers to understand and process human languages, bringing computers closer to an understanding of language at a human level [12].

The development of NLP applications is a challenge because computers traditionally require humans to communicate with them in a programming language that is accurate, unambiguous and highly structured, or through a limited number of clearly enunciated text or voice commands. However, human writing and speech are not always precise – they are often ambiguous and the linguistic structure depends on many complex variables, including slang, regional dialects and social context [13].

NLP can be used to interpret free text and make it analyzable. There is a large amount of information stored in free text files, like patients' medical records, for example. Prior to deep learning-based NLP models, this information was inaccessible to computer-assisted analysis, and could not be analyzed in any kind of systematic way. Nowadays, NLP allows analysts to filter massive collections of free text in order to find or extract relevant information in the files [13].

2 Methodology

2.1 Natural Language Toolkit Classifiers

Natural Language Toolkit (NLTK) is a platform for building Python programs to work with human language data. It provides easy-to-use interfaces to over 50 corpora and lexical resources such as WordNet, along with a suite of text processing libraries for classification, tokenization, stemming, tagging, parsing, and semantic reasoning, wrappers for industrial-strength NLP libraries, and an active discussion forum [14]. NLTK provides materials to perform text analysis with different classifiers. Among the classifiers offered in this toolkit, the following ones have been analyzed:

Naïve Bayes. In naive Bayes classifiers, every feature gets a say in determining which label should be assigned to a given input value [14]. To choose a label for an input value, the naive Bayes classifier begins by calculating the prior probability of each label, which is determined by checking frequency of each label in the training set. The contribution from each feature is then combined with this prior probability, to arrive at a likelihood estimate for each label.

Decision Tree. A decision tree is a simple flowchart that selects labels for input values. This flowchart consists of decision nodes, which check feature values, and leaf nodes, which assign labels [14]. To choose the label for an input value, we begin at the flowchart's initial decision node, known as its root node. This node contains a condition that checks one of the input value's features, and selects a branch based on that feature's value. Following the branch that describes our input value, we arrive at a new decision node, with a new condition on the input value's features. We continue following the branch selected by each node's condition, until we arrive at a leaf node which provides a label for the input value.

Naïve Bayes Based on Bernoulli Model. It is a Naïve Bayes classifier, but it uses the Bernoulli model for probabilistic distributions [14]. A Bernoulli distribution is a distribution in which the random variable (X) takes only two possible values. A possible value is 1 (success) with probability p and another value is 0 (failure) with probability (1–p). Here, p denotes the probability of success. Moreover, if there are only two possible outcomes in a random experiment, then it is known as Bernoulli trial.

Support Vector Classifier (SVC). Supervised vector machines (SVM) are machine learning supervised methods that can be used for classification, regression or detection model creation [14]. They are more effective in a high dimensional space. SVCs are able to perform multi-class classification.

Maximum Entropy. The Maximum Entropy classifier uses a model that is very similar to the model employed by the naive Bayes classifier [14]. But rather than using probabilities to set the model's parameters, it uses search techniques to find a set of parameters that will maximize the performance of the classifier. In particular, it looks for the set of parameters that maximizes the total likelihood of the training corpus. Maximum Entropy classifiers choose the model parameters using iterative optimization techniques, which initialize the model's parameters to random values, and then repeatedly refine those parameters to bring them closer to the optimal solution. These iterative optimization techniques guarantee that each refinement of the parameters will bring them closer to the optimal values, but do not necessarily provide a means of determining when those optimal values have been reached. Because the parameters for Maximum Entropy classifiers are selected using iterative optimization techniques, they can take a long time to learn. This is especially true when the size of the training set, the number of features and labels are all large.

2.2 Corpus and Datasets

A corpus is a collection of written or spoken material stored on a computer and used to find out how language is used. Sentiment140 [15] is an API developed by Alec Go, Richa Bhayani, and Lei Huang, from Stanford University. This service allows to discover the sentiment of a brand, product, or topic on Twitter. They provide a corpus for academic purposes which is the one used on this research. This corpus is formed from tweets tagged as positive, negative or neutral.

The main inconvenience of this corpus is that it is based in tweets. Small sentences of 140 characters where it is common to find abbreviatures, emoticons or informal expressions. They usually are so variated and do not follow a specific topic. In this research, users will send text to the service talking about how their day was. There will not be characters limit and it is expected no use of emoticons, informal expressions or shortened words.

Because of this corpus does not fully fit this research needs, other dataset has been created from sentences collected on blogs where people talk about their daily lifes. The selected sentences or texts were specifically chosen because of its content. They must express positive or negative emotions that can represent a confirm or a deny of a depression situation. This dataset created will be used to test if classifiers trained with Sentiment140's corpus fit research's problem needs. This dataset contains 10 text extracts formed by a total of 288 sentences. They were tagged as positive or negative by one person. A text was considered positive if 50% or more of its sentences were positive, and negative otherwise.

2.3 Text Pre-processing Tools

Before classifying texts extracts, it is needed to clean and preprocess these texts. These processes will provide a text easier to classify. The first step of preprocessing texts is to remove words that are usually present in text but do not contribute to the analysis. These words are often prepositions or determinants. These words are declared as 'stop words' and removed from texts.

After that, different preprocessing techniques can be applied. For this research, two preprocessing techniques were selected to be analyzed:

- **Stemming** is the process of reducing inflected (or sometimes derived) words to their word stem, base or root form—generally a written word form [16]. The stem need not be identical to the morphological root of the word; it is usually sufficient that related words map to the same stem, even if this stem is not in itself a valid root.
- **Lemmatisation** is the algorithmic process of determining the lemma of a word based on its intended meaning [17]. Unlike stemming, lemmatisation depends on correctly identifying the intended part of speech and meaning of a word in a sentence, as well as within the larger context surrounding that sentence, such as neighboring sentences or even an entire document.

2.4 Evaluation Methodology

The methodology designed for evaluating the classifiers will consist of three tests.

1. On the first test the classifiers will be trained with different training sets sizes. The original corpus from sentiment140 will be divided in sub-corpus of different sizes to perform the training and testing. The accuracy obtained for each combination of sample size-classifier will be analyzed.
2. On the second test, the classifiers will be trained on the same way, but the test will be done on the specifically created corpus from blogs extracts.
3. On the third test, the difference of the analysis will be the preprocessing technique applied to the text. The same sample will be used for each classifier, but different preprocessing techniques will be applied to the sample. The train and the test will be from the original corpus.

3 Results

The results of the evaluation on test 1 are shown on Table 1. Decision Tree and Maximum Entropy could not be tested with large size samples because of the computational effort they require. Computers available to test these training sessions could not finish these classifiers trainings with sample sizes larger than 4000 registers. On this table it can be observed the accuracy obtained for each classifier for each training and test size.

The results obtained on the second test can be observed on Table 2. As in the previous phase, Maximum Entropy could not be trained and tested with more than 4000 registers, whereas Decision Tree could perform more than in the previous phase. The observed accuracy is significantly lower than the obtained on the first test. The format of the table is the same than in Table 1. The test set was made of 288 sentences as it was previously described.

Results from the third test are presented on Table 3. In this table, it can be observed the accuracy result for each classifier depending on the preprocessing technique used.

Table 1. Accuracy of each classifier with Sentiment140 corpus.

Train/Test sample size	Naïve Bayes	Decision Tree	The Bernoulli model (Naïve Bayes)	Support Vector Classifier (SVC)	Maximum Entropy
500/125	72.00%	47.20%	72.00%	83.20%	60.00%
1000/250	72.40%	64.00%	73.20%	75.60%	72.00%
2000/250	74.80%	67.20%	76.00%	75.20%	49.60%
4000/350	74.29%	–	75.43%	56.86%	48.29%
6000/350	76.86%	–	76.86%	60.57%	–
8000/350	76.29%	–	76.57%	60.29%	–

Table 2. Accuracy of each classifier with blog created dataset.

Train sample size	Naïve Bayes	Decision Tree	The Bernoulli model (Naïve Bayes)	Support Vector Classifier (SVC)	Maximum Entropy
250	53.80%	57.25%	56.05%	52.65%	55.19%
500	53.61%	51.03%	53.61%	47.13%	55.59%
1000	55.66%	56.12%	55.89%	53.45%	54.48%
2000	56.91%	57.47%	55.62%	50.67%	54.37%
4000	57.05%	57.93%	57.34%	51.83%	50.13%
6000	55.24%	50.44%	56.24%	52.28%	–
8000	63.63%	–	62.13%	54.61%	–

Table 3. Accuracy of each classifier with different preprocessing techniques.

Preprocessing technique applied	Naïve Bayes	Decision Tree	The Bernoulli model (Naïve Bayes)	Support Vector Classifier (SVC)	Maximum Entropy
Stemmer	71.20%	46.40%	71.20%	84.00%	59.20%
Lemmatizer	76.80%	45.60%	76.80%	60.00%	57.60%

4 Discussion

From the results obtained on the first phase, it can be observed that Naïve Bayes based classifiers and Maximum Entropy classifier have a similar perform. It is striking the accuracy reported by the SVC classifier. It reports better accuracy with low size samples, it may due to an overtraining caused by the similarity of size of train and test samples. Finally, Decision Tree is the classifier that reports lower accuracy. Maximum Entropy results are also striking because of the variation of its accuracy levels which also are decreasing while the train/test sample sizes increase.

On the second phase, a significant descend of the accuracy can be observed because of the use of the specially prepared dataset as test sample. This dataset is significantly different from the tweets of the corpus, which is the main reason of the decrease of accuracy.

On the third phase the SVC obtains a high accuracy with stemmer, but it could be because the overtraining identified before. Naïve Bayes classifiers obtain a better accuracy with lemmatization processes which is consistent with similar studies results.

Naïve Bayes based classifiers show best and most stable results in all phases. But even these classifiers (Naïve Bayes and Bernoulli Model) report an accuracy lower than expected in phase 2. It is important because phase 2 is the most representative for the research. Phase 2 is where the text used for testing is more similar to real text expected in the service.

From these results, more processing is needed. Main actions to perform are to find a corpus which fits better to the final purpose, make a more complex preprocessing of the text and adjusting classifiers parameters to find a better configuration.

In the final service developed this text emotion recognizer module will be integrated with other modules for emotion detection based on audio and image that will complement the information retrieved and will allow to perform a more robust analysis.

5 Conclusions

This research is focused on depression detection, which is a direct and important application for this technology. Further work is needed to grant an accuracy level enough to trust the service. Different preprocessing text tools will be analyzed, and the use of neural networks will be evaluated.

In future works it is expected to refine the service to detect not only positive or negative emotions. Adding more possible outputs to the service will allow to detect the mood or mind status of the user. Intervention systems and mHealth applications could take advantage of this service and adjust their programs to the user's mood, getting an extra point for personalization their services.

Acknowledgements. This work has been partially funded by the Spanish Ministry of Science (project DPI2017-86088-C3-1-R, iBC).

References

1. World Health Organization. https://www.who.int/en/news-room/fact-sheets/detail/depression. Accessed 21 Apr 2019
2. Liu, B.: Sentiment analysis and opinion mining. Synth. Lect. Hum. Lang. Technol. **5**(1), 1–167 (2012)
3. Carrillo-de-Albornoz, J., Rodríguez Vidal, J., Plaza, L.: Feature engineering for sentiment analysis in e-health forums. PLoS ONE **13**(11), e0207996 (2018)
4. Mullen, T., Collier, N.: Sentiment analysis using support vector machines with diverse information sources. In: Proceedings of the 2004 Conference on Empirical Methods in Natural Language Processing, EMNLP 2004, pp. 412–418 (2004)

5. Pang, B., Lee, L., Vaithyanathan, S.: Thumbs up?: Sentiment classification using machine learning techniques. In: Proceedings of the ACL 2002 Conference on Empirical Methods in Natural Language Processing, EMNLP 2002, vol. 10, pp. 79–86 (2002)
6. Kennedy, A., Inkpen, D.: Sentiment classification of movie reviews using contextual valence shifters. Comput. Intell. **22**(2), 110–125 (2006)
7. Boiy, E., Hens, P., Deschacht, K., Francine Moens, M.: Automatic sentiment analysis in on-line text. In: Proceedings of the 11th International Conference on Electronic Publishing, pp. 349–360 (2007)
8. Salvetti, F., Reichenbach, C., Lewis, S.: Opinion polarity identification of movie reviews. In: Wiebe, J. (ed.) Computing Attitude and Affect in Text: Theory and Applications, pp. 303–316. Springer, Dordrecht (2006)
9. Carrillo-de-Albornoz, J., Plaza, L., Gervás, P.: A hybrid approach to emotional sentence polarity and intensity classification. In: Proceedings of the Fourteenth Conference on Computational Natural Language Learning, CoNLL 2010, pp. 153–161 (2010)
10. Read, J., Carroll, J.: Weakly supervised techniques for domain-independent sentiment classification. In: Proceedings of the 1st International CIKM Workshop on Topic-sentiment Analysis for Mass Opinion, TSA 2009, pp. 45–52 (2009)
11. Turney, P.D.: Thumbs up or thumbs down?: Semantic orientation applied to unsupervised classification of reviews. In: Proceedings of the 40th Annual Meeting on Association for Computational Linguistics, pp. 417–424 (2002)
12. An easy introduction to Natural Language Processing. https://towardsdatascience.com/an-easy-introduction-to-natural-language-processing-b1e2801291c1. Accessed 26 Apr 2019
13. Search Business Analytics. https://searchbusinessanalytics.techtarget.com/definition/natural-language-processing-NLP. Accessed 25 Apr 2019
14. Natural Language Toolkit. https://www.nltk.org. Accessed 26 Apr 2019
15. Sentiment140 - A Twitter Sentiment Analysis Tool. http://www.sentiment140.com. Accessed 26 Apr 2019
16. Lovins, J.B.: Development of a stemming algorithm. Mech. Transl. Comput. Linguist. **11**(1–2), 22–31 (1968)
17. Plisson, J., Lavrac, N., Mladenic, D.: A rule based approach to word lemmatization. In: Proceedings of IS-2004, pp. 83–86 (2004)

Regular Sessions: Biomaterials and Tissue Engineering

Co-encapsulation of Beta Cells and Nanoparticles Containing GLP-1 Greatly Improves Insulin Secretion in Alginate-Based Bioartificial Pancreas

Joana Crisóstomo[1(✉)], Francisca Araújo[2,3], Pedro Granja[2,3,4,5], Cristina Barrias[2,3,5], Bruno Sarmento[2,3], and Raquel Seiça[1]

[1] Laboratory of Physiology, iCBR - Coimbra Institute for Clinical and Biomedical Research, Faculty of Medicine, University of Coimbra, Coimbra, Portugal
joanacrisostomo@hotmail.com

[2] INEB - Instituto de Engenharia Biomédica, Porto, Portugal

[3] i3S - Instituto de Investigação e Inovação em Saúde, Universidade do Porto, Porto, Portugal

[4] FEUP - Faculdade de Engenharia, Universidade do Porto, Porto, Portugal

[5] ICBAS - Instituto de Ciências Biomédicas Abel Salazar, Universidade do Porto, Porto, Portugal

Abstract. The bioartificial pancreas presents a promising methodology to treat diabetes. However, this approach is still associated with a high rate of graft failure caused by poorly functioning cells. The incretin effect of glucagon like peptide-1 (GLP-1) make this peptide an attractive therapeutic agent to improve the success of transplantation of encapsulated beta cells. In present work, we developed a novel model based on the co-encapsulation of beta cells with nanoparticles containing GLP-1 on alginate hydrogels, allowing its delivery and action in the specific target, the beta cells.

INS-1E beta cells were co-encapsulated with poly(lactide-co-glycolide) (PLGA) nanoparticles containing GLP-1 in alginate and were maintained in culture aiming to evaluate the performance of our drug-delivery system and the influence of GLP-1 in cellular performance.

PLGA nanoparticles showed an average size of 169.3 nm, a polydispersity index of 0.05 and an average zeta-potential of −24.3 mV. The average association efficiency of GLP-1 was 65.4% and the in vitro release test showed 71% of GLP-1 delivery after 7 days. The presence of nanoparticles led to increased cellular metabolic activity and 8-fold higher insulin levels. Our results reveal a promising approach to ensure the proper delivery and action of GLP-1 in encapsulated beta cells, ensuring its viability.

Keywords: Bioartificial pancreas · PLGA nanoparticles · GLP-1

J. Henriques et al. (Eds.): MEDICON 2019, IFMBE Proceedings 76, pp. 1215–1222, 2020.
https://doi.org/10.1007/978-3-030-31635-8_147

1 Introduction

Transplantation of encapsulated insulin-producing cells in alginate-based hydrogels has shown potential to become a widely applicable treatment for diabetes [1, 2]. However, maintenance of cellular function and survival over time remains a big challenge. Despite being protected from immune cells and antibodies, encapsulated cells are exposed to low molecular weight entities, such as proinflammatory cytokines or reactive oxygen species, which are able to cross the membrane and interact with cells, leading ultimately to cellular death [1, 3]. Therefore, the lack of proper immune protection is the main difficulty to overcome, and the development of strategies capable of endowing cells the ability to survive in a challenging environment might be the key of success of this cell-based technology.

An attractive therapeutic agent produced after meals in gut L cells is glucagon like peptide-1 (GLP-1). Besides its effects in other organs, the pancreas is the preferential target of this incretin. GLP-1 is able to protect beta cells from apoptosis, to increase their neogenesis and proliferation and to increase glucose-dependent insulin secretion. GLP-1 also inhibits the glucagon secretion from pancreatic alpha cells [4–6].

The usefulness of GLP-1 or mimetics was already demonstrated in the scope of transplantation of insulin-producing cells [7–12]. The transient effect due to rapid degradation of this incretin is one of the challenges that needs to be tackled [8]. Approaches involving the development of polymeric conjugates with GLP-1 have attempted to increase GLP-1 stability but could produce inactive incretin due to the possibility of random chemical reaction in the active centre of peptide [9, 12]. Functionalization of poly(ethylene glycol) diacrylate (PEGDA) with GLP-1 for posterior beta cells and rats islets encapsulation showed to be a methodology able to increase insulin secretion [10].

In this work, we present a novel approach to promote GLP-1 contact with encapsulated cells. For that purpose, poly(lactide-co-glycolide) (PLGA) nanoparticles containing GLP-1 were prepared and co-encapsulated with beta cells in alginate hydrogels. Alginate is the most frequent used material in bioartificial pancreas due to its biocompatibility and facility to produce 3D microenvironments close to native cellular microenvironment [2]. PLGA nanoparticles represent one of the most effective biodegradable polymeric nanoparticles and were approved by Food and Drug Administration (FDA) to use in drug delivery systems due to controlled and sustained release properties, low toxicity, and biocompatibility [13]. With this approach, the releasing of GLP-1 is confined to the local of interest, easily targeting beta cells. As the incretin structure does not suffer any alteration, its efficacy is also guaranteed.

2 Materials and Methods

2.1 Nanoparticles Preparation and Characterization

PLGA (Purasorb PDLG 5004, Purac Biomateriais) was used to produce nanoparticles containing GLP-1 (United Peptides) through a modified solvent emulsification-evaporation method, based on the water-in-oil-in-water (w/o/w) double emulsion technique already described by our group [14].

Nanoparticles' size, polydispersity index and zeta potential were characterized by dynamic light scattering (DLS - Zetasizer Nano ZS, Malvern Instruments). Samples were diluted in MilliQ-water at 25 °C and evaluated with a detection angle of 173°. Nanoparticles' morphology and size were assessed by transmission electron microscopy (TEM - FEI-Tecnai G2 Spirit Biotwin, FEI Company) using an acceleration voltage of 100 kV. Samples were diluted in MilliQ-water, placed on copper coated grids whit Formvar/Carbon and contrasted with 2% (m/v) uranyl acetate before imaging.

2.2 GLP-1 Association Efficiency and in Vitro Release Profile

GLP-1 quantification was assessed using a commercial kit (GLP-1 EIA, Sigma) according to the manufacturer's instructions. To determine GLP-1 association efficiency, the amount of peptide was calculated as the difference between the total amount of GLP-1 used to prepare the systems, and the amount of GLP-1 that remained in the aqueous phase after nanoparticles isolation by centrifugation, according to the follow equation:

$$\text{Association efficiency (\%)} = \frac{(GLP - 1\, inicial\, mass) - (GLP - 1\, mass\, in\, supernatant)}{GLP - 1\, total\, mass} \times 100 \tag{1}$$

For in vitro release test, nanoparticles (corresponding to 50 μg of GLP-1) were added to 25 mL of Tris buffer supplemented with calcium (50 mM Tris, 150 mM NaCl, 7.5 mM $CaCl_2$, pH = 7.6) and remaining under stirring for 7 days, at 37 °C. Aliquots of 800 mL were collected at the same hour of day 1, 4 and 7. Withdrawn volume was replaced with fresh pre-heated buffer, keeping the volume constant and sink conditions. Collected aliquots were centrifuged at 20 000 g for 20 min and the supernatant was used for quantification of GLP-1 released from the nanoparticles over time.

2.3 Cell and Nanoparticles Co-encapsulation in Alginate Hydrogels

INS-1E beta cells from rat insulinoma were maintained in culture at 37 °C under a 5% CO_2 humidified atmosphere, in RPMI 1640 medium supplemented with 10% (v/v) fetal bovine serum, 100 U/mL penicillin + 100 μg/mL streptomycin, 10 mM HEPES, 1 mM sodium pyruvate and 50 μM de β-mercaptoethanol. PLAG nanoparticles Nanoparticles were sterilized by ultraviolet light explosion for 20 min.

For encapsulation sodium alginate with high guluronate content (PRONOVA Ultrapure Sodium Alginate LVG, NovaMatrix, FMC Biopolymers) was prepared in 0.9% NaCl at a final concentration of 2% (m/v). 5×10^6 cells/mL were carefully mixed with alginate solution (Alg 0) or also with PLGA nanoparticles containing approximately 33 nM of GLP-1 (Alg GLP-1) and extruded under a coaxial nitrogen flow using a Var J1 encapsulation unit (Nisco) [15, 16]. Alginate hydrogels were cross-linked in an isotonic solution with 100 mM $CaCl_2$ and 10 mM HEPES for 15 min, washed and cultured under standard conditions. For size and shape evaluation, spheres were observed by microscopy (Stereo microscope SZX10, Olympus).

2.4 Metabolic Activity and Insulin Secretion of Encapsulated Cells

Cell metabolic activity was assessed at days 1, 4 and 7 by resazurin assay. Hydrogels were incubated with 20% (v/v) of resazurin solution (0.1 mg/mL) in RPMI medium for 2 h at 37 °C and supernatant fluorescence was read using a microplate reader (Biotek Synergy MX) with Ex/Em at 530/590 nm.

Glucose-stimulated insulin secretion (GSIS) test was performed at day 4 of culture. Media were discarded and hydrogels were carefully washed with a modified Krebs-Ringer buffer (KRB) (135 mM NaCl, 3.6 mM KCl, 0.5 mM NaH_2PO_4, 5 mM $NaHCO_3$, 10 mM HEPES, 1.5 mM $CaCl_2$, 0.5 mM $MgCl_2$, 0.1% (m/v) BSA, pH = 7.4) and replaced with the same buffer for 1 h. After washes, samples were incubated with KRB supplemented with glucose at low (2 mM) or high concentration (20 mM) for 1 h. All samples were collected, and insulin levels were assessed using a commercial ELISA kit (Ultrasensitive Rat Insulin ELISA, Mercodia), according to the manufacturer's instructions.

2.5 Statistical Analysis

Statistical analyses were performed using Graphpad software. Differences were considered statistically significant for p-values <0.05, tested with two-way ANOVA and post-hoc pairwise comparisons with Sidak correction. All data are presented as mean ± standard deviation (SD).

3 Results

Nanoparticles showed an average size of 169.3 ± 7.25 nm, a polydispersity index of 0.05 ± 0.02 and an average zeta-potential of −24.3 ± 8.82 mV, assessed by DLS (Table 1). TEM imagens showed the spherical shape of nanoparticles (Fig. 1A). The average association efficiency of GLP-1 was 65.4% ± 11.5 (Table 1). The in vitro release test showed a higher release rate between day 1 and 4 and a low rate between day 4 and day 7, reaching 71% ± 0.07 in last day evaluated (Table 1). According to existent literature, PLGA nanoparticles showed to be biocompatible (data not shown).

Alginate hydrogels showed a uniform spherical shape, cells were uniformly distributed inside them and their average diameter (1270 ± 69 μm) did not change with the introduction of nanoparticles (Fig. 1A).

Table 1. Characterization of PLGA nanoparticles containing GLP-1.

Parameter		
Size (nm)		169.3 ± 7.25
Polydispersity index		0.05 ± 0.02
Zeta-potential (mV)		−24.3 ± 8.82
GLP-1 association efficiency (%)		65.4 ± 11.5
GLP-1 release profile (%)	Day 1	17.0 ± 2.82
	Day 4	63.0 ± 1.41
	Day 7	71.0 ± 0.07

Comparing the absence (Alg 0) with the presence of PLGA nanoparticles containing GLP-1 (Alg GLP-1) in INS-1E cells' outcome, it was demonstrated that the presence of nanoparticles greatly improved cellular performance. Following GLP-1 release profile, despite Alg GLP-1 condition showed high levels of metabolic activity at all timepoints, more pronounced increments were found in the first days ($p < 0.001$ at days 1 and 4) and less evident in the last day of experiment ($p < 0.05$ at day 7) (Fig. 1C). Over time profile was different between two conditions, showing a lower metabolic activity increments in Alg GLP-1 ($p < 0.001$ between day 7 and day 1 and $p < 0.05$ between day 7 and day 4) than those observed in Alg 0 ($p < 0.001$ between both days 7 and 4 and day 1 and $p < 0.001$ between day 7 and day 4).

Beta cells' ability to secrete insulin was also positively influenced by presence of nanoparticles (Fig. 1D). Marked increment on insulin levels were found at day 4 in cells co-encapsulated with nanoparticles (Alg GLP-1): comparing 2 mM and 20 mM of glucose-induced insulin secretion, insulin levels increased 5.7-fold ($p < 0.001$), whereas only 1.8-fold in Alg 0 condition. Furthermore, insulin levels secreted after 2 mM and 20 mM of glucose stimulus were 2.5 and 8 times higher, respectively, in Alg GLP-1 condition when compared with Alg 0.

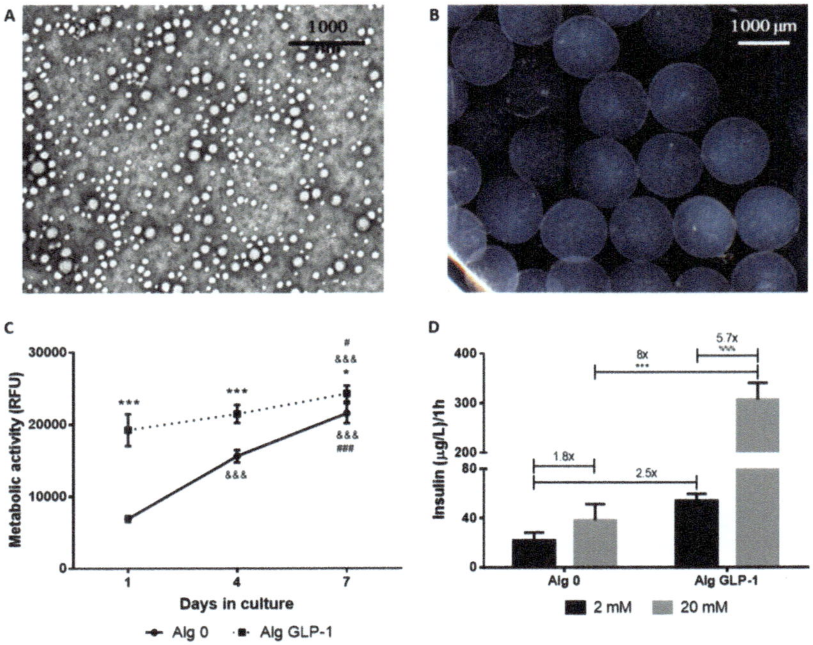

Fig. 1. (A) PLGA nanoparticles' transmission electron microscopy images. (B) Hydrogels' stereoscopic imagens. (C) Cells' metabolic activity in absence (Alg 0) or presence of PLGA nanoparticles containing GLP-1 (Alg GLP-1) at days 1, 4 and 7. (D) Cells' insulin secretion levels in Alg 0 and Alg GLP-1 conditions at day 4, after glucose-stimulated insulin secretion essay (2 mM and 20 mM of glucose stimuli). Data are presented as mean ± SD (n = 3–5). $^{*}p < 0.05$ e $^{***}p < 0.001$ between Alg 0 and Alg GLP-1; $^{\&\&\&}p < 0.001$ between the days 4 and 7 and the day 1, $^{\#}p < 0.05$ $^{\#\#\#}p < 0.001$ between the day 7 and the day 4 and $^{\%\%\%}p < 0.001$ between 2 mM and 20 mM.

4 Discussion and Conclusion

Providing encapsulated beta cells the capacity to survive in challenging conditions and increasing its ability to secrete insulin is betting on functionality and longevity of cellular implants for diabetes treatment. Literature has shown that GLP-1 is a good candidate for that purpose, but the methodologies using this incretin hamper its full action [7–11]. Therefore, we developed a novel approach in which INS-1E beta cells were co-encapsulated with PLGA nanoparticles containing GLP-1 in alginate hydrogels. The noteworthy increments of cellular metabolic activity and insulin secretion certainly confirmed the delivery of GLP-1 from nanoparticles and its proper action in beta cells.

Nanoparticles' physicochemical properties were in accordance with what was already reported by our colleagues [14] and the introduction of this nanosystems in our model did not alter the capacity of alginate to reticulate and produce biocompatible hydrogels with the same characteristics already demonstrated [16]. With this delivery system, it will be possible to solve the problem of GLP-1 short half-live time [6, 8]. The incretin is released inside of alginate hydrogels containing beta cells, accelerating its targeting. Moreover, using these nanoparticles in which GLP-1 is only physically imprisoned, the probability to create ineffective GLP-1 [9, 12] is also overcome. We also postulated that free GLP-1 peptides are more prone to link to their receptors. This hypothesis was corroborated by presented results, showing a significantly increase of beta cells insulin secretion for a lower concentration of GLP-1 than the one reported in the Lin's hydrogels model, where GLP-1 is chemically linked with PEGDA [10]. Actually, in the present study it was not possible to directly assess the behaviour of nanoparticles inside the hydrogels and its release profile. Being encapsulated together with cells in alginate hydrogels, the incretin could be released from nanoparticles and interact with beta cells' receptors. It may also happen, as already reported [14], that due to their nanometric size and spherical geometry, nanoparticles could be free to be internalized and deliver their content inside cells. Either way, the GLP-1 accurately delivery antagonizes whit what happens physiologically, escaping from rapid degradation in liver and by dipeptidyl peptidase-4 (DPP-4) action [6].

Being the most successful 3D model used in beta cells' encapsulation [2, 17, 18], adding new value to alginate hydrogels with the introduction of GLP-1 is a promising approach to improve cell survival and function, and increase the success of this therapeutic strategy for diabetes mellitus. The improvement of glucose dependent insulin secretion was already demonstrated in different models using GLP-1, via stimulation of cAMP-PKA signalling [4, 6, 19]. With this novel model, it was robustly shown the same positive influence of the presence of this incretin in insulin secretion after glucose stimuli, reaching 8-fold higher insulin levels compared with counterpart without GLP-1.

Concluding, these results reveal a promising approach to ensure the proper delivery and action of GLP-1 in encapsulated beta cells. Using this kind of approach, the main issue of encapsulation - the beta cells protection - is not neglected and the therapeutic agent is not altered or physically impeded to act. Moreover, this system is very versatile; it is usable with different attractive therapeutic agents and adjustable regarding the therapeutic agent delivery rate and time, which depends on the nanoparticles' degradation. Therefore, more experiments are required to better characterize and precisely tailor this novel model, which can effectively contribute to bioartificial pancreas success.

Acknowledgements. The present work is a collaboration between IBILI and INEB, supported by grants UID/NEU/04539/2013, UID/BIM/04293/2013 and SFRH/BD/72964/2010.

Conflict of Interest Declaration. The authors report no conflicts of interest.

References

1. Crisóstomo, J., Coelho, J., Seiça, R.: Bioartificial pancreas: in the road to clinical application. In: Coelho, J. (ed.) Drug Delivery Systems: Advanced Technologies Potentially Applicable in Personalised Treatment, pp. 127–151. Springer, Dordrecht (2013)
2. Sun, J., Tan, H.: Alginate-based biomaterials for regenerative medicine applications. Materials **6**(4), 1285–1309 (2013)
3. Wilson, J.T., Chaikof, E.L.: Challenges and emerging technologies in the immunoisolation of cells and tissues. Adv. Drug Deliv. Rev. **60**, 124–145 (2008)
4. Brubaker, P., Drucker, D.: Minireview: Glucagon-like peptides regulate cell proliferation and apoptosis in the pancreas, gut, and central nervous system. Endocrinology **145**, 2653–2659 (2004)
5. Puddu, A., Sanguineti, R., Mach, F., Dallegri, F., Viviani, G.L., Montecucco, F.: Update on the protective molecular pathways improving pancreatic beta-cell dysfunction. Mediators Inflamm. **2013**, 750540 (2013)
6. Holst, J.J.: The Physiology of Glucagon-like Peptide 1. Physiol. Rev. **87**, 1409–1439 (2007)
7. Toyoda, K., Okitsu, T., Yamane, S., Uonaga, T., Liu, X., Harada, N., et al.: GLP-1 receptor signaling protects pancreatic beta cells in intraportal islet transplant by inhibiting apoptosis. Biochem. Biophys. Res. Commun. **367**, 793–798 (2008)
8. Bohman, S., Waern, I., Andersson, A., King, A.: Transient beneficial effects of exendin-4 treatment on the function of microencapsulated mouse pancreatic islets. Cell Transplant. **16**, 15–22 (2007)
9. Kim, S., Wan Kim, S., Bae, Y.H.: Synthesis, bioactivity and specificity of glucagon-like peptide-1 (7-37)/polymer conjugate to isolated rat islets. Biomaterials **26**, 3597–3606 (2005)
10. Lin, C.C., Anseth, K.S.: Glucagon-like peptide-1 functionalized PEG hydrogels promote survival and function of encapsulated pancreatic beta-cells. Biomacromolecules **10**, 2460–2467 (2009)
11. Chepurny, O.G., Holz, G.G.: Over-expression of the glucagon-like peptide-1 receptor on INS-1 cells confers autocrine stimulation of insulin gene promoter activity: a strategy for production of pancreatic beta-cell lines for use in transplantation. Cell Tissue Res. **307**, 191–201 (2002)
12. Kim, S., Bae, Y.: Long-term insulinotropic activity of glucagon-like peptide-1/polymer conjugate on islet microcapsules. Tissue Eng. **10**, 1607–1616 (2004)
13. Sadat, T.M., Nejati-Koshki, K., Akbarzadeh, A., Yamchi, M.R., Milani, M., Zarghami, N., et al.: PLGA-based nanoparticles as cancer drug delivery systems. Asian Pac. J. Cancer Prev. **15**, 517–535 (2014)

14. Araújo, F., Shrestha, N., Shahbazi, M.-A., Fonte, P., Mäkilä, E.M., Salonen, J.J., et al.: The impact of nanoparticles on the mucosal translocation and transport of GLP-1 across the intestinal epithelium. Biomaterials **35**, 9199–9207 (2014)
15. Bidarra, S., Barrias, C., Barbosa, M., Soares, R., Granja, P.: Immobilization of human mesenchymal stem cells within RGD-grafted alginate microspheres and assessment of their angiogenic potential. Biomacromolecules **11**, 1956–1964 (2010)
16. Crisóstomo, J., Pereira, A.M., Bidarra, S.J., Gonçalves, A.C., Granja, P.L., Coelho, J., et al.: ECM-enriched alginate hydrogels for bioartificial pancreas: an ideal niche to improve insulin secretion and diabetic glucose profile. J. Appl. Biomater. Funct. Mater. (in press)
17. de Vos, P., Faas, M.M., Strand, B., Calafiore, R.: Alginate-based microcapsules for immunoisolation of pancreatic islets. Biomaterials **27**, 5603–5617 (2006)
18. Qi, M.: Transplantation of encapsulated pancreatic islets as a treatment for patients with Type 1 Diabetes Mellitus. Adv. Med. **2014**, 429710 (2014)
19. Luo, G., Kong, X., Lu, L., Xu, X., Wang, H., Ma, X.: Glucagon-like peptide 1 potentiates glucotoxicity-diminished insulin secretion via stimulation of cAMP-PKA signaling in INS-1E cells and mouse islets. Int. J. Biochem. Cell Biol. **45**, 483–490 (2013)

Potentialities of LL37 for Wound Healing Applications: Study of Its Activity in Synergy with Biodegradable Composites Made of PVA and CA

Helena P. Felgueiras(✉), Marta A. Teixeira, and M. Teresa P. Amorim

Centre for Textile Science and Technology, University of Minho, Campus de Azurém, 4800-058 Guimarães, Portugal
helena.felgueiras@2c2t.uminho.pt

Abstract. Wound healing is a dynamic and complex process that results from the interaction between cytokines, growth factors, blood components and the extracellular matrix. Conventional dressings made of natural or synthetic materials have only the ability to manage the wound and protect it from repeated trauma. With the advancement of technology, wound dressings have evolved and are now capable of intervening in the healing process by targeting specific features of the wound, aside from protecting the wounded site. In this work, bioactive dressings capable of promoting healing and fighting infection in chronic wounds were explored. Various antimicrobial biomolecules were examined in light of their pathogen fighting skills and immunoregulatory potentialities. Dressing production processes were also investigated. Biodegradable composite dressings made of poly(vinyl alcohol), polycaprolactone, chitosan and cellulose blends were our main focus. The goal was to evaluate the synergistic effect of biomolecules and biodegradable polymeric dressings, considering the local and systemic treatment demands of chronic wounds.

Keywords: Biodegradable polymers · Active biomolecules · Antimicrobial agents · Bioactive dressings · Effective chronic wound healing

1 Introduction

Wound healing is a systemic process that requires the interaction of a complex cascade of cellular events that generate resurfacing, reconstitution and restoration of the tensile strength of the injured skin, establishing a provisional tissue from which a complete regenerated epidermis is built. Skin regeneration in acute wounds evolves by following the phases of healing (coagulation, inflammation, proliferation and remodeling), in a predictable amount of time. To the contrary, chronic wounds can stall in one phase of healing (i.e. inflammation), developing a defective cell matrix colonized by bacteria (biofilms) and characterized by moisture imbalance, that requires much more time to heal [1, 2]. In light of these issues, new generation of wound dressings have been proposed, and engineered aiming a more active intervention in the healing process that surpasses the simple protection/management of conventional, passive dressings.

J. Henriques et al. (Eds.): MEDICON 2019, IFMBE Proceedings 76, pp. 1223–1226, 2020.
https://doi.org/10.1007/978-3-030-31635-8_148

There is a growing interest in the development of biodegradable dressings from composites made of natural and synthetic biopolymers. Natural polymers arise from renewable resources, being highly desirable in wound healing for their biocompatibility and non-toxic nature. On its turn, biodegradable synthetic polymers have shown exceptional mechanical features that when combined with natural resources provide structural integrity [2, 3]. Bioactive dressings are a fairly recent class of wound dressings responsible for promoting wound healing via the incorporation of biomolecules of interest, including antimicrobial peptides (AMPs), proteins, growth factors, ions, plant extracts, etc. [4, 5]. In the present work, we propose to evaluate the impact of bioactive dressings made of combinations of poly(vinyl alcohol) (PVA), polycaprolactone (PCL), chitosan (CS) and cellulose, modified with active biomolecules, for the prospective treatment of chronic wounds.

1.1 Bioactive Dressings and the Incorporated Biomolecules

Bioactive dressings are made of biodegradable, non-toxic biomaterials that play an active role in the normal healing of wounds and new tissue formation. They are based on the principle of creating and maintaining a moist wound environment that prevents the formation of a scab by facilitating exudates absorbency, oxygen exchange and controlling the pH on site. Bioactive dressings usually exhibit a fibrous structure in which are incorporated active compounds for site delivery (Fig. 1). These biomolecules intervene in the wound healing either directly or indirectly as cleansing or debridement agents removing necrotic tissues, as antimicrobials fighting infections, or as growth agents instigating tissue regeneration. Currently, the biomolecules considered most important are the AMPs, proteins and growth factors [4, 5].

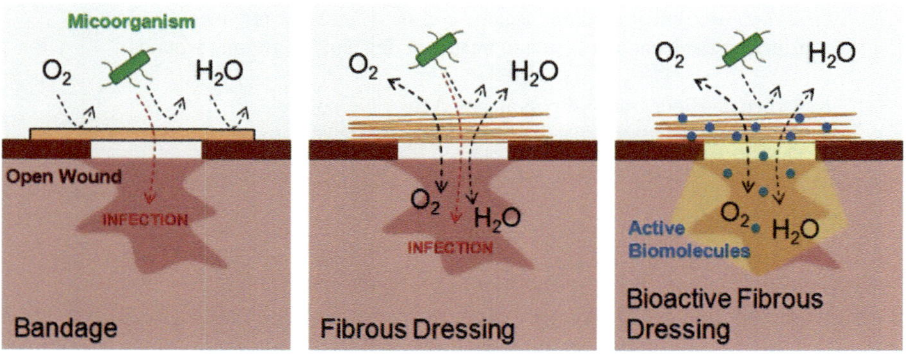

Fig. 1. Bioactive fibrous dressings impregnated with active biomolecules capable of fighting infection and promoting wound healing.

AMPs form an integral part of our innate immune system, being capable of killing pathogens in few seconds and, thus, reducing the likelihood of these microorganisms to develop resistance. They are low molecular weight molecules formed of 5–100 amino acids, with cationic charge. AMPs are usually classified in terms of their target microorganism, being the bacterial AMPs the most commonly employed in chronic wound healing. Because of their effective nature and natural-origin, the AMPs LL37, the only human cathelicidin-derived AMP, the pexiganan and the cecropin are the most researched; these AMPs prevent and/or treat infections and can even aid in the healing process [2, 6]. Regarding proteins, collagen is without match the most important extracellular matrix (ECM) element for wound healing. Collagen exhibits excellent biocompatibility and has been highlighted as uniquely suited for wound healing because of its intervention in all phases of wound healing: promotes wound debridement, stimulates the clotting cascade, acts as a sacrificial substrate for proteolytic enzymes, and instigates tissue remodeling [3]. Finally, the growth factors, which take an active physiological role in chronic wound healing, have been shown to intervene in the stimulation of angiogenesis and cell proliferation, which affect both the production and degradation of the ECM, and in the inflammatory and remodeling phases. A large group of growth factors have been reported to participate in the healing process, including the epidermal growth factor (EGF), the platelet derived growth factor (PDGF), the fibroblast growth factor (FGF) and the transforming growth factor (TGF), to name a few [5].

1.2 Composites Functionalized with Biomolecules for Wound Healing

Bioactive dressings based on natural-origin materials, such as collagen and CS, have been reported to accelerate granulation tissue formation and epithelialization above synthetic dressings. However, their hydrophilicity, which difficults the control of antibiotic release, and their low tensile strength resistance have been pointed as disadvantages [7]. Electrospun PCL and PVA mats are suitable to treat acute and chronic wounds as they mimic the ECM structure. However, their use is limited by their poor antimicrobial properties. Hence, natural and synthetic polymeric combinations are desirable. PCL-collagen composites have been established as excellent triggers for integrin-β1 signaling pathways to regulate the growth of fibroblasts and initiate wound healing [8]. Also, they have both been used in the delivery of EGF factors. CS has been employed in the delivery of EGF as well, for the treatment of full thickness wounds [9]. Coaxial electrospun PCL/PVA-chitosan nanofibrous scaffolds have been engineered for drug and gene delivery [8, 10]. Bioengineered cellulose can be used as scaffolding material for chronic wound dressings, by reducing pain and shortening healing time. Just recently, we have uncovered that porous, flexible mats made of PVA and cellulose blends, processed to resemble actual skin, when functionalized with the AMP LL37 can stimulate healing and decrease microbial action in more than 80%. Preliminary data on the immobilization of this AMP and pexiganan on PVA/cellulose acetate (CA) and PVA/PCL mats has revealed an accelerated clotting time and bactericidal effect against both Gram-positive and Gram-negative bacteria [unpublished work]. Pexiganan-incorporated collagen matrices have also been studied with successful results in infected wounds [8]. The use of functionalized AMPs in wound healing is still quite recent. However, new formulations are being developed everyday and are expected to overcome quickly the limitations of the conventional strategies.

2 Conclusions and Future Work

The role of bioactive dressings is to deliver active biomolecules to the wounded site, to promote healing while fighting infections. At the present, we are studying combinations of PVA/CA, PVA/PCL and PVA/CS impregnated with immunoregulatory peptides and collagen for chronic wound healing applications. Using phase-inversion and electrospinning techniques, flexible, porous films and reticulated networks are being fashioned with properties that mimic the actual skin. In a near future, we intend to examine the synergistic performance of biomolecules by combining different AMPs, with different action modes, in a single formulation. We are confident these new systems aside from being target-effective against pathogens will also contribute effectively to the healing process by accelerating it.

Acknowledgements. Authors acknowledge the Portuguese Foundation for Science and Technology (FCT), FEDER funds by means of Competitive Factors Operational Program (POCI) for funding the projects POCI-01-0145-FEDER-028074 and UID/CTM/00264/2019.

References

1. Martin, P., Nunan, R.: Cellular and molecular mechanisms of repair in acute and chronic wound healing. Br. J. Dermatol. **173**, 370–378 (2015)
2. Felgueiras, H.P., Amorim, M.T.P.: Functionalization of electrospun polymeric wound dressings with antimicrobial peptides. Colloids Surf. B **156**, 133–148 (2017)
3. Felgueiras, H.P., Amorim, M.T.P.: Electrospun polymeric dressings functionalized with antimicrobial peptides and collagen type I for enhanced wound healing. In: 17th World Textile Conference, AUTEX 2017. IOP Conference Series: Materials Science and Engineering, vol. 254, p. 062004 (2017)
4. Georgescu, M., Chifiriu, M.C., Marutescu, L., Gheorghe, I., Lazar, V., Bolocan, A., Bertesteanu, S.: Bioactive wound dressings for the management of chronic wounds. Curr. Org. Chem. **21**(1), 53–63 (2017)
5. Boateng, J.S., Matthews, K.H., Stevens, H.N.E., Eccleston, G.M.: Wound healing dressings and drug delivery systems: a review. J. Pharm. Sci. **97**(8), 2892–2923 (2008)
6. Querido, M.M., Felgueiras, H.P., Rai, A., Costa, F., Monteiro, C., Borges, I., Oliveira, D., Ferreira, L., Martins, M.C.L.: Cecropin-melittin functionalized polyurethane surfaces prevent *Staphylococcus epidermidis* adhesion without inducing platelet adhesion and activation. Adv. Mater. Interfaces **5**, 1801390 (2018)
7. Mogoşanu, G.D., Grumezescu, A.M.: Natural and synthetic polymers for wounds and burns dressing. Int. J. Pharm. **463**(2), 127–136 (2014)
8. Mir, M., Ali, M.N., Barakullah, A., Gulzar, A., Arshad, M., Fatima, S., Asad, M.: Synthetic polymeric biomaterials for wound healing: a review. Prog. Biomater. **7**, 1–21 (2018)
9. Park, J.W., Hwang, S.R., Yoon, I.-S.: Advanced growth factor delivery systems in wound management and skin regeneration. Molecules **22**(8), 1259–1279 (2017)
10. Silva, I.O., Ladchumananandasivam, R., Nascimento, J.H.O., Silva, K.O.S., Oliveira, F.R., Souto, A.P., Felgueiras, H.P., Zille, A.: Multifunctional chitosan/gold nanoparticles synergistic coatings for biomedical textiles. Nanomaterials **9**(8), 1064–1085 (2019). https://doi.org/10.3390/nano9081064

Cellulose Acetate in Wound Dressings Formulations: Potentialities and Electrospinning Capability

Marta A. Teixeira(✉), M. Teresa P. Amorim, and Helena P. Felgueiras

Centre for Textile Science and Technology, University of Minho, Campus de Azurém, 4800-058 Guimarães, Portugal
martaalbertinateixeira@gmail.com

Abstract. Any open wound is a potential site for microorganisms' invasion since their presence around us is inevitable. Skin wound healing relies on a series of complex physiochemical processes that remain a big challenge for healthcare professionals, particularly when the wounds are colonized by bacteria. Wound dressings play a major role in wound healing as they manage the wounded site, controlling the moisture balance and protecting the wound from repeated trauma, and by preventing possible infections from developing into more serious complications. Recently, bioactive dressings loaded with drugs and/or antimicrobial agents, allowing for a continuous and sustainable release of these molecules at the wounded site, have appeared in the market. Antimicrobial resistance is a growing health care problem, requiring more effective solutions than antibiotics. As such, nano- and microfibrous mats produced via electrospinning technique and loaded with natural-origin antimicrobial agents have attracted a lot of attention. Various polymers have been applied to engineer nanofibrous electrospun dressings. However, the environment impact of the synthesis and processing methods of synthetic polymers is undesirable. Therefore, the application of cellulose-derived materials (highly abundant polymer of natural-origin) becomes crucial as a green alternative to produce electrospun wound dressings with superior wettability, breathability and high capacity to promote cell proliferation, at relatively low costs. In this paper, different biomolecules loaded onto cellulose acetate (CA)-based polymeric nanofibers were investigated, and their antimicrobial properties were highlighted as alternatives to conventional antibiotics.

Keywords: Natural-origin polymers · Electrospinning · Biodegradable mats · Antimicrobial agents · Bioactive wound dressings

1 Introduction

Fighting wound infections is a major problem in the field of wound care. Infections delay wound healing and may even cause local exudates formation, systemic problems and, in serious cases, threaten the patient's life [1]. An ideal wound dressing must confer protection from external microorganisms and from chemical and physical aggressions,

J. Henriques et al. (Eds.): MEDICON 2019, IFMBE Proceedings 76, pp. 1227–1230, 2020.
https://doi.org/10.1007/978-3-030-31635-8_149

as well as to promote the healing process by stimulating cell adhesion, proliferation and differentiation [2]. Nowadays, it is imperative to search for efficient alternatives that reduce microbial resistance and offer a platform that allows for the wound to breath, maintain a moisture balance and absorb exudates. Electrospun nanofibrous dressings with incorporated biomolecules meet all these requirements. These dressings mimic the reticulated extracellular matrix, offering a large surface area with oxygen exchange capacity and exudates absorbency, and promoting connective tissue synthesis and angiogenesis for an enhanced epidermal cell migration and growth [3].

Electrospun dressings have been fashioned using a variety of natural and synthetic polymers. Among those, cellulose acetate (CA), which is the acetate ester of cellulose generated by acetylation of the hydroxyl groups, can be electrospun in the form of ultrafine fibers of sub-micrometer scale with desirable properties for wound healing. CA electrospun fibers have shown high capacity to promote cell proliferation due to their hydrophilicity and bioactivity. Also, CA nanofibers have been characterized as possessing good thermal and hydrolytic stability, chemical resistance and being of low cost [4]. In the following sections, we evaluate the impact of CA-based electrospun nanofibers, modified with active biomolecules, for prospective wound healing applications.

2 Antimicrobial Agents Combined with Cellulose Acetate-Based Dressings

Various biomolecules have been functionalized or incorporated within CA-based electrospun dressings for the treatment of skin wounds, including antibiotics, nanoparticles, plant extracts, proteins, antimicrobial peptides (AMPs), etc. For instance, CA/corn starch dressings have been treated with silver sulfadiazine to prevent a chronic inflammatory response and kill Gram-negative bacteria [5]. Thymoquinone (TQ), a natural antibacterial drug, has been used with CA to prevent common clinical infections and to accelerate the rate of wound closure and re-epithelialization [6]. Antibacterial electrospun nanofibers were also prepared by physically blending polyurethane (PU), CA, zein and streptomycin sulfate. This combination revealed desirable properties of hydrophilicity, excellent cell attachment, proliferation and blood clotting ability, together with good antimicrobial features [1]. Even though successful results have been achieved with antibiotics, in the fight against wound infections, their selective and slow action mode against microorganisms has become a disadvantage. In fact, many organisms are now able to resist their action, the currently known antibiotic-resistant pathogens. As such, new biomolecules have been developed and tested on electrospun polymeric dressings, including nanoparticles (NPs, i.e. silver nanoparticles) and natural extracts/polymers (e.g. honey, essential oils and chitosan) [2].

NPs such as silver (AgNPs), titanium dioxide (TiO_2), and zinc oxide (ZnO) are well known effective groups of antimicrobial agents. However, research has been a bit ambiguous regarding NPs advantages and disadvantages. Studies have shown that NPs display no cytotoxicity towards human cells in low concentrations and possess a lower tendency to induce microbial resistance compared with antibiotics. On the other hand, it has been shown that, based on their size alone, NPs can penetrate the circulatory system and even cross the blood-brain barrier in humans [7]. To prevent this from happening, AgNPs nanocomposite particles have been synthesized with dopamine hydrochloride on top of TiO_2 NPs to obtain an improved anchoring to electrospun CA. The composite nanofibers were found desirable for long-term antibacterial applications, demonstrating excellent infection fighting skills [8]. In fact, CA fibers incorporated only with AgNPs have demonstrated significant antimicrobial action against Gram-negative bacteria *Escherichia coli* (*E. coli*) and *Pseudomonas aeruginosa* and Gram-positive *Staphylococcus aureus* (*S. aureus*) [9].

Various plant-derived essential oils (EOs) have been incorporated in CA-based electrospun nanofibers for topical wound care. CA mats containing EOs have demonstrated high cell compatibility and no cytotoxicity. Low amount of EOs immobilized on CA dressings were seen to exert exceptional growth inhibition of *E. coli*. For instance, rosemary and oregano oils successfully immobilized onto CA-based fibers revealed high antimicrobial activity against both *E. coli* and *Candida albicans* [4]. Curcumin (Curc), a biologically active substance with remarkable antibacterial, antifungal, antioxidant, anti-inflammatory, anticoagulant and antitumor activity has also been encapsulated within CA and polyvinylpyrrolidone nanofibers. Curc-containing mats exhibited antibacterial activity against *S. aureus* and, due to an increase of surface hydrophilicity and programmed control release, were determined effective as wound dressings [10].

More recently, a major advance in the manufacture of wound dressings has been uncovered: the incorporation or functionalization of electrospun mats with AMPs [3]. AMPs are low molecular weight cationic molecules that provide a nonspecific defense against a broad spectrum of invaders, including bacteria, virus, and fungi, with low propensity to induce bacterial resistance. They have been used in different applications, including blood-contacting medical devices [11], and more recently have been identified as potent antimicrobial and regenerative agents for wound healing [3]. AMP coatings are characterized as the new, promising step in the manufacture of bioactive dressings for new generation of acute and chronic wound therapies.

3 Conclusion and Future Work

To prevent common clinical infections various antimicrobial biomolecules have been incorporated into nano- and microfibrous electrospun wound dressings. However, the rising of antibiotic-resistant infection agents has increased the need for more effective therapies, namely the AMPs. Little research has yet been presented regarding AMPs' contribution in wound healing when incorporated onto electrospun polymeric mats. In fact, to our best knowledge, dressings that combine CA-based electrospun mats and AMPs have not yet been explored. Currently, we are developing such tests, incorporating immunoregulatory AMPs within the polymeric matrices of composite electrospun scaffolds mats containing CA. Preliminary data revealed the symbiotic nature of these dressings in both accelerating clotting time and fighting bacterial colonization.

It was also demonstrated the thermal and physical stability of the porous mats and their biodegradability in physiological media. The proposed alternatives aim an accelerated healing but as well the development of clean, safer processing approaches to produce wound dressings with minimal environment impact.

Acknowledgements. Authors acknowledge the Portuguese Foundation for Science and Technology (FCT), FEDER funds by means of Competitive Factors Operational Program (POCI) for funding the projects POCI-01-0145-FEDER-028074 and UID/CTM/00264/2019.

References

1. Unnithan, A.R., Gnanasekaran, G., Sathishkumar, Y., Lee, Y.S., Kim, C.S.: Electrospun antibacterial polyurethane-cellulose acetate-zein composite mats for wound dressing. Carbohydr. Polym. **102**, 884–892 (2014)
2. Miguel, S.P., Sequeira, R.S., Moreira, A.F., Cabral, C.S.D.: An overview of electrospun membranes loaded with bioactive molecules for improving the wound healing process. Eur. J. Pharm. Biopharm. **139**, 1–22 (2019)
3. Felgueiras, H.P., Amorim, M.T.P.: Functionalization of electrospun polymeric wound dressings with antimicrobial peptides. Colloids Surf. B Biointerfaces **156**, 133–148 (2017)
4. Liakos, I.L., Holban, A.M., Carzino, R., Lauciello, S., Grumezescu, A.M.: Electrospun fiber pads of cellulose acetate and essential oils with antimicrobial activity. Nanomaterials **7**, 84 (2017)
5. Modolon Zepon, K., Petronilho, F., Soldi, V., Vitor, G., Alberto, L.: Production and characterization of cornstarch/cellulose acetate/silver sulfadiazine extrudate matrices. Mater. Sci. Eng. C **44**, 225–233 (2014)
6. Gomaa, S.F., Madkour, T.M., Moghannem, S., El-Sherbiny, I.M.: New polylactic acid/cellulose acetate-based antimicrobial interactive single dose nanofibrous wound dressing mats. Int. J. Biol. Macromol. **105**, 1148–1160 (2017)
7. Prabhu, S., Poulose, E.K.: Silver nanoparticles: mechanism of antimicrobial action, synthesis, medical applications, and toxicity effects, pp. 1–10. Springer, Heidelberg (2012)
8. Wahab Jatoi, A., Kim, I.S., Ni, Q.: Cellulose acetate nano fi bers embedded with AgNPs anchored TiO 2 nanoparticles for long term excellent antibacterial applications. Carbohydr. Polym. **207**, 640–649 (2019)
9. Xu, F., Weng, B., Materon, L.A., Kuang, A., Trujillo, J.A., Lozano, K.: Fabrication of cellulose fine fiber based membranes embedded with silver nanoparticles via Forcespinning. J. Polym. Eng. **36**(3), 1–10 (2015)
10. Tsekova, P.B., Spasova, M.G., Manolova, N.E., Markova, N.D., Rashkov, I.B.: Electrospun curcumin-loaded cellulose acetate/polyvinylpyrrolidone fibrous materials with complex architecture and antibacterial activity. Mater. Sci. Eng. C **73**, 206–214 (2017)
11. Querido, M.M., Felgueiras, H.P., Rai, A., Costa, F., Monteiro, C., Borges, I., Oliveira, D., Ferreira, L., Martins, M.C.L.: Cecropin – Melittin Functionalized Polyurethane Surfaces Prevent Staphylococcus epidermidis Adhesion without Inducing Platelet Adhesion and Activation. Adv. Mater. Interfaces **5**(24), 1801390 (2018)

Electrospun Collagen Variability Characterized by Tensile Testing

Ján Kužma[1,2(✉)], Lukáš Horný[1], Tomáš Suchý[1,2], Monika Šupová[2], and Zbyněk Sucharda[2]

[1] Faculty of Mechanical Engineering, Czech Technical University in Prague, Technická 4, 160 00 Prague, Czech Republic
kuzma@irsm.cas.cz

[2] Institute of Rock Structure and Mechanics, Academy of Sciences of the Czech Republic, V Holešovičkách 41, 182 09 Prague 8, Czech Republic

Abstract. The objective of this study was to verify repeatability of the preparation process of the collagen nanofibrous layers. The layers were fabricated by means of electrospinning. Constant conditions were held within the entire production process. It means that all samples were fabricated from collagen solution with the same chemical composition. Electrospinning conditions were also adjusted to have the same parameters, i.e. temperature, relative humidity, voltage and distance between electrodes. Finally, cross-linking procedure was also the same for all samples involved in the study. Elasticity of the material was assessed by means of uniaxial tensile test in a fully hydrated state of collagen samples. Internal structure of collagenous layers was studied by means of scanning electron microscopy. Although all technological conditions were held constant, mathematical modelling of the elastic behaviour revealed differences between samples. The results suggest that there is a certain variability of mechanical properties of electrospun hydrated collagen that is difficult to eliminate. By this fact, however, collagen is, to some extent, similar to biological tissue.

Keywords: Collagen · Constitutive model · Electrospinning · Hyperelasticity · Microstructure

1 Introduction

Collagen is one of the most occurring proteins. It makes up to 30% of all proteins in human body [1–6]. Collagen exists in the form of fibres which are formed by fibrils of chained monomers referred to as tropocollagen [5]. Tropocollagen molecule consists of triple-helix built from alpha chains of amino acids. Depending on a specific repeating arrangement of amino acids, at least 29 types of collagen are distinguished [3].

It is generally accepted that collagen provides structural strength and stability to tissues and organs [6, 7]. It has excellent biocompatibility, biodegradability and weak antigenicity [4]. It is considered as an excellent biomaterial and is used in various medical applications [4]. Collagen isolate may lose some properties due to hierarchical

J. Henriques et al. (Eds.): MEDICON 2019, IFMBE Proceedings 76, pp. 1231–1238, 2020.
https://doi.org/10.1007/978-3-030-31635-8_150

changes during the isolation and purification process [2]. Electrospinning is used to imitate collagen's natural fibrillary structure which should restore its physical properties [3].

During the last several years, electrospun collagen has been used in many studies and experiments in research fields covering, for example, regenerative medicine, tissue engineering, and drug delivery systems [8–12]. In applications at which the electrospun material will serve as a load-bearing element, the consistency of the mechanical properties may be crucial. One can name bone implants, where electrospun collagen enhances the rate of osseointegration [13, 14], or vascular prostheses, that may be based on collagenous matrix [15], as examples.

Mechanical properties of collagen materials can be also linked to their drug delivery capability. Since drug releasing mechanism is mostly based on a surface area which relies on the size and diameter of a structural fibre, it is clear that mechanical properties, naturally conditioned by an internal architecture, and drug delivery capability are mutually correlated. From this point of view, studies focused on reproducibility of the electrospinning fabrication of collagenous materials are an ordinary part of the materials science and development of reliable production technologies.

2 Methods

2.1 Material Fabrication

Collagen nanofibrous materials were prepared based on 8 wt% collagen (VUP Medical, Czech Republic) phosphate buffer/ethanol solution to which was added polyethylene oxide as an auxiliary polymer (PEO, Mn 900,000, Sigma-Aldrich, Germany) in an amount of 8 wt% to the weight of collagen.

Electrospun materials were fabricated using a high voltage of 45 kV and the feeding rate was 80 $\mu L \cdot min^{-1}$. The temperature was kept at 26 °C, needle to collector distance was set to 20 cm and the relative humidity was kept at 23–26% (4SPIN, Contipro, Czech Republic). The electrospinning process was enhanced by combination with electroblowing. The flow rate of the preheated air (25 °C) was set at 10 $L \cdot min^{-1}$. All the electrospun materials were collected on a static collector (22 × 29 cm).

The stability of all the collagen layers was enhanced by means of cross-linking with a 95% ethanol solution containing EDC and NHS at a weight ratio of 4:1; the EDC and NHS (Sigma Aldrich, Germany) were used as received. Following a reaction period of 24 h at 37 °C. Finally, all the layers were washed in 0.1 M Na_2HPO_4 (2 × 45 min) and subsequently rinsed with deionised water (30 min). Layers were frozen at −15 °C for 5 h and lyophilised (BenchTop 4KZL, VirTis, U.S.A.).

Five groups of layers were fabricated in the above-mentioned procedure. They were obtained even from a single collagen batch.

2.2 SEM Image Analysis

The inner structure of specimens was characterized by means of scanning electron microscopy (SEM; Quanta 450 Microscope, FEI, U.S.A.) in high vacuum mode. The specimens were cut by a surgical knife prior to SEM analysis. The resulting cross-sections were coated with a thin layer of gold in an ion sputter (Emitech K550X, Quorum Technologies, UK).

The diameter of fibres was measured by means of ImageJ software (Rasband, W.S., ImageJ, US National Institutes of Health, Bethesda, Maryland, USA, http://imagej.nih.gov/ij/, 1997–2015). The manual mode of the ImageJ analyser was used for the measurement of the average diameter of the fibres. At least 10 fibres were assessed at each of six SEM micrographs (mag. 10 000× $n = 3$, and 25 000x $n = 3$) for each layer type.

2.3 Mechanical Testing

Tensile tests were conducted using a Zwick/Roell multipurpose testing machine equipped with a built-in video extensometer. By using contrasting marks on the surface of the samples, the video extensometer automatically determined the reference length and elongation of the samples. Tensile experiments were conducted at a constant clamp velocity of 0.1 $mm{\cdot}s^{-1}$. The loading force was measured by the U9B (±25 N, HBM, Germany) force transducer.

Mechanical properties were evaluated by means of the uniaxial tensile tests of rectangular strips of the layers which were hydrated in physiological solution (the average width and length of the samples were approximately 10 mm and 40 mm, respectively).

During the test procedure, 5 preloading cycles were carried out as preconditioning. The sixth loading sequence was used in the subsequent mathematical modelling. To this end, the engineering strain defined as the ratio of the elongation of the sample to its reference length, and nominal stress defined as the ratio of applied force to the reference cross-section of the sample were determined from experimental records.

2.4 Model

To model fully nonlinear mechanical behaviour of hydrated collagen layers, hyperelastic material description was adopted. Hyperelastic material is characterised by the strain energy density function W. Fung-Demiray exponential model [16], frequently used in soft tissue biomechanics, was employed. Its particular form is expressed in the Eq. (1).

$$W = \frac{\mu}{2\alpha}\left(e^{\alpha(I_1-3)} - 1\right) \tag{1}$$

Here, μ is stress-like material parameter corresponding to infinitesimal shear modulus and α is dimensionless parameter modulating strain-stiffening response. I_1 is the first principal invariant of the right Cauchy-Green strain tensor $\mathbf{C}$ which is obtained from $\mathbf{C} = \mathbf{F}^T\mathbf{F}$. $\mathbf{F}$ is the deformation gradient tensor.

Kinematics of the uniaxial tensile test was assumed to be described by (2). Here, $\boldsymbol{X}$ is a position vector of a material particle in the reference configuration and $\boldsymbol{x}$ is its position vector in the deformed state. In such a case, the deformation gradient $\mathbf{F}$, that is defined as $\mathbf{F} = \partial \boldsymbol{x}/\partial \boldsymbol{X}$, has the form $\mathbf{F} = diag[\lambda_1, \lambda_2, \lambda_3]$ and $I_1 = diag[\lambda_1^2 + \lambda_2^2 + \lambda_3^2]$.

$$x_1 = \lambda_1 X_1 \quad x_2 = \lambda_2 X_2 \quad x_3 = \lambda_3 X_3 \tag{2}$$

In hyperelastic materials, the strain energy density function W serves as a potential function for stress. In case of the nominal (the first Piola-Kirchhoff) stress tensor $\mathbf{P}$, this is expressed in the Eq. (3) which plays a role of the constitutive equation.

$$\mathbf{P} = \frac{\partial W}{\partial \mathbf{F}} - p\mathbf{F}^{-T} \tag{3}$$

In the Eq. (3), the assumption of incompressible behaviour of collagen was adopted. Which means that p is Lagrangean multiplier enforcing incompressibility constraint and has to be determined from a force boundary condition. Combining (1), (2), (3), incompressibility condition $det(\mathbf{F}) = 1$, and isotropy of the material, final expression for the nominal stress sustained by the sample in uniaxial tension applied in the direction 1 is obtained. It is expressed in (4). The relation between λ_1, used in (4), and engineering strain ε_{11}, measured in the experiment, is given by $\lambda_1 = 1 + \varepsilon_{11}$.

$$P_1 = \mu\left(\lambda_1 - \frac{1}{\lambda_1^2}\right)e^{\alpha\left(\lambda_1^2 + \frac{2}{\lambda_1} - 3\right)} \tag{4}$$

Employing least square method, parameters μ and α characterising every tested sample were determined. Least square optimization was carried out in multipurpose computer algebra system Maple (Maplesoft, Canda).

2.5 Statistical Evaluation

Statistical analysis was conducted in Maple (Maplesoft, Canada). The coefficient of determination R^2 was used to evaluate a quality of the regression model (4). As usual, R^2 was considered to be equal to 1 minus ratio of the residual sum of squares to the total sum of squares. The resulting values of R^2 were displayed in a boxplot. Graphical methods were adopted to show whether differences in the mechanical behaviour of groups of samples exist. First, clusters of the material parameters μ and α in 2D phase space were created. Second, a convex hull for each group was added to the figure. Finally, diameters of collagen fibres obtained in SEM were compared by means of Kruskal-Wallis test to estimate if differences in the internal structure of collagenous layers exist. Dunn's test was used in post hoc analysis to show which groups differ. Significance level 0.05 was considered within the entire study.

3 Results and Discussion

A total number of 40 uniaxial tensile tests were conducted. Unfortunately, several samples did not survive the complete testing procedure and were eliminated from the subsequent analysis. Thus, final number of samples in each group was as follows: $n_1 = 7$, $n_2 = 8$, $n_3 = 8$, $n_4 = 5$, and $n_5 = 4$. Index i denotes the group number and $n = \Sigma n_i = 32$.

Figure 1 depicts resulting stress-strain responses and their models described by (4). The boxplot in the lower-right panel of Fig. 1 summarizes coefficients of determination obtained in the least square minimization. Considering that none of R^2 is lesser than 0.97, it can be concluded that (4) is a suitable model for a description of hydrated collagen mechanical response.

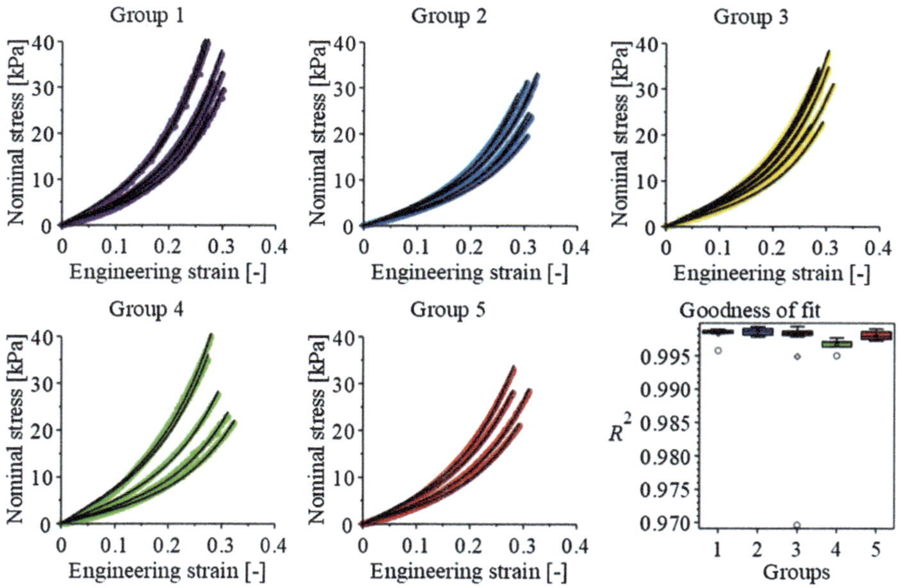

Fig. 1. Nominal stress-engineering strain relationships obtained from the experiment and fitted by the model (4). Bottom-right panel shows the boxplot of resulting coefficients determination.

Table 1. Expected values of the material parameters and their variability. IQR denotes interquartile range and SD denotes sample standard deviation.

Group	α [−]	IQR	μ [kPa]	SD
1	5.27	5.17–5.34	16.97	4.30
2	4.34	4.14–4.50	17.34	4.03
3	4.57	4.44–4.69	16.49	3.66
4	4.18	4.05–4.34	16.42	4.09
5	4.32	4.22–4.46	16.54	4.71

Mean values (arithmetic mean in case of μ and median in case of α) of the material parameters estimated in the nonlinear regression analysis are presented in Table 1.

Figure 2 in its left panel depicts all experimental data. Individual groups are distinguished by the same colours as in Fig. 1. Since individual stress-strain curves mutually overlap, a hypothetical conclusion could be drawn that differences in the mechanical behaviour of the groups would be negligible. However, the right panel in the same figure, where convex hulls of material parameters are depicted, suggests that previous conclusion is false. At least Group 1, depicted with violet colour, seems to be different from others because it has an empty intersection with all other groups. Another conclusion, which can be drawn from the simplified cluster analysis, is that the parameter μ (horizontal axis) seems to fall into approximately the same range in all groups. Thus, in the model (4), it is only α which reflects differences between material groups.

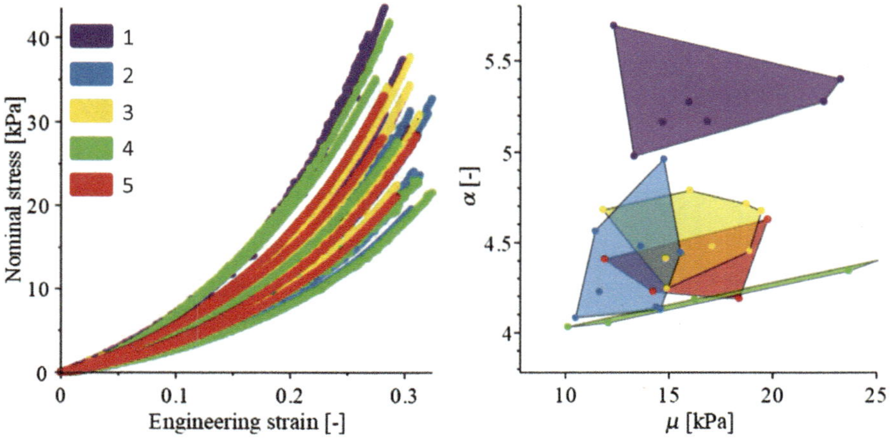

Fig. 2. Nominal stress-engineering strain relationships obtained from the experiment (left panel), two-dimensional graph of material parameters and convex hulls (right panel).

The panels of Fig. 3 show microphotographs obtained via SEM microscopy. Fibrous structure of collagen electrospun is clearly visible. By means of image analysis, 60 measurements of fibre diameters were conducted in every group. Resulting data are compared in the boxplot presented in Fig. 3. The same colour scheme as in Fig. 1 and 2 is used again. The results clearly suggest that Group 1 is different from others (*p-value* < 0.001).

To be able to identify which groups mutually differ, Dunn's post hoc test was employed. Its results showed that Group 1 differs from all other groups (*p-value* $<$ 0.001 in all cases). Also Group 2 differs from 3 and 4 (*p-value* = 0.01 and 0.001, respectively). Remaining pairs of groups did not show significant differences in fibre diameters.

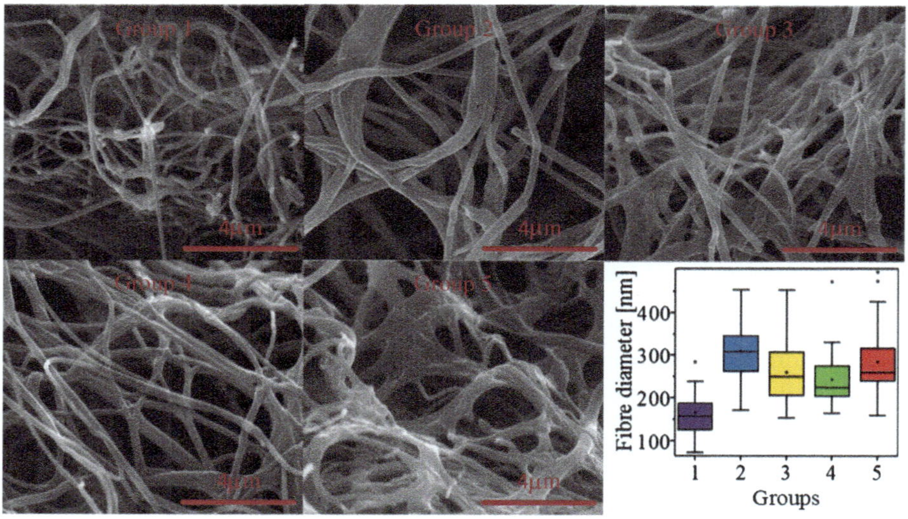

Fig. 3. Representative SEM microphotographs (mag. 25 000x). The boxplot of fibre diameters obtained in the image analysis.

Results suggest that, although constant conditions are kept during electrospinning, and the same collagenous solution is used, the final materials may differ in their mechanical properties and structure. Therefore, some extent of variability in mechanical properties of electrospun collagen should be taken into account in bioengineering applications.

Acknowledgement. This research has been supported via project SGS19/155/OHK2/3T/12.

Conflict of Interest. The authors declare that they have no conflict of interest.

References

1. Shoulders, M.D., Raines, R.T.: Collagen structure and stability. Annu. Rev. Biochem. **78**, 929–958 (2009)
2. Rýglová, Š., Braun, M., Suchý, T.: Collagen and its modifications—crucial aspects with concern to its processing and analysis. Macromol. Mater. Eng. **302**(6), 1600460 (2017)
3. Bazrafshan, Z., Stylios, G.K.: Spinnability of collagen as a biomimetic material: a review. Int. J. Biol. Macromol. **129**, 693–705 (2019)
4. Yannas, I.: Collagen and gelatin in the solid state. J. Macromol. Sci. Rev. Macromol. Chem. **7**(1), 49–106 (1972)
5. Caves, J.M., Kumar, V.A., Wen, J., Cui, W., Martinez, A., Apkarian, R., Coats, J.E., Berland, K., Chaikof, E.L.: Fibrillogenesis in continuously spun synthetic collagen fiber. J. Biomed. Mater. Res. Part B Appl. Biomater. Off. J. Soc. Biomater. Jpn. Soc. Biomater. Aust. Soc. Biomater. Korean Soc. Biomater. **93**(1), 24–38 (2010)
6. Fratzl, P.: Collagen. Springer, US (2008)

7. Gautieri, A., Vesentini, S., Redaelli, A., Buehler, M.J.: Hierarchical structure and nanomechanics of collagen microfibrils from the atomistic scale up. Nano Lett. **11**(2), 757–766 (2011)
8. Song, J.-H., Kim, H.-E., Kim, H.-W.: Electrospun fibrous web of collagen–apatite precipitated nanocomposite for bone regeneration. J. Mater. Sci. Mater. Med. **19**(8), 2925–2932 (2008)
9. Sell, S.A., McClure, M.J., Garg, K., Wolfe, P.S., Bowlin, G.L.: Electrospinning of collagen/biopolymers for regenerative medicine and cardiovascular tissue engineering. Adv. Drug Deliv. Rev. **61**(12), 1007–1019 (2009)
10. Li, M., Mondrinos, M.J., Gandhi, M.R., Ko, F.K., Weiss, A.S., Lelkes, P.I.: Electrospun protein fibers as matrices for tissue engineering. Biomaterials **26**(30), 5999–6008 (2005)
11. Buttafoco, L., Kolkman, N., Engbers-Buijtenhuijs, P., Poot, A.A., Dijkstra, P.J., Vermes, I., Feijen, J.: Electrospinning of collagen and elastin for tissue engineering applications. Biomaterials **27**(5), 724–734 (2006)
12. Cui, W., Zhou, Y., Chang, J.: Electrospun nanofibrous materials for tissue engineering and drug delivery. Sci. Technol. Adv. Mater. **11**(1), 014108 (2010)
13. Suchý, T., Šupová, M., Klapková, E., Horný, L., Rýglová, Š., Žaloudková, M., Braun, M., Sucharda, Z., Ballay, R., Veselý, J., Chlup, H., Denk, F.: The sustainable release of vancomycin and its degradation products from nanostructured collagen/hydroxyapatite composite layers. J. Pharm. Sci. **105**(3), 1288–1294 (2016)
14. Suchý, T., Šupová, M., Klapková, E., Adamková, V., Závora, J., Žaloudková, M., Rýglová, Š., Ballay, R., Denk, F., Pokorný, M.: The release kinetics, antimicrobial activity and cytocompatibility of differently prepared collagen/hydroxyapatite/vancomycin layers: microstructure vs. nanostructure. Eur. J. Pharm. Sci. **100**, 219–229 (2017)
15. Ercolani, E., Del Gaudio, C., Bianco, A.: Vascular tissue engineering of small-diameter blood vessels: reviewing the electrospinning approach. J. Tissue Eng. Regen. Med. **9**(8), 861–888 (2015)
16. Demiray, H.: A note on the elasticity of soft biological tissues. J. Biomech. **5**(3), 309–311 (1972)

Thermal Effect by Applying Laser Heating in Iron Oxide Nanoparticles Dissolved in Distilled Water

Leonardo A. Bermeo Varon[1(✉)], Bruna R. Loiola[2], Luiz A. da Silva Abreu[3], Bernard Lamien[4,6], Nilton Pereira da Silva[5,6], Helcio R. B. Orlande[6,8], and Dilson Silva dos Santos[7,8]

[1] Universidad Santiago de Cali, Engineering, Bioengineering, Street 5 N 62-00, Cali, Colombia
leonardo.bermeo00@usc.edu.co
[2] Department of Mechanical Engineering, Military Institute of Engineering, IME, 80 General Tibúrcio Square, Urca, Rio de Janeiro, RJ 22290-270, Brazil
[3] Department of Mechanical Engineering and Energy, Rio de Janeiro State University, UERJ, Polytechnic Institute, 25 Bonfim Street, Vila Amélia, Nova Friburgo, RJ 28625-570, Brazil
[4] École Polytechnique de Ouagadougou (EPO), 18 BP 234, Ouagadougou, Burkina Faso
[5] Department of Mechanical Engineering, DEMEC, Federal University of Amazonas, UFAM, Av. General Rodrigo Otávio, 1200, Coroado I, 69080-900 Manaus, AM, Brazil
[6] Department of Mechanical Engineering, Federal University of Rio de Janeiro, COPPE, Cid. Universitaria, Cx. Postal: 68503, Rio de Janeiro, RJ 21941-972, Brazil
[7] Department of Metallurgical and Materials Engineering, Federal University of Rio de Janeiro, COPPE, Cid. Universitaria, Cx. Postal: 68503, Rio de Janeiro, RJ 21941-972, Brazil
[8] Department of Nanotechnology Engineering, Federal University of Rio de Janeiro, COPPE, Cid. Universitaria, Cx. Postal: 68503, Rio de Janeiro, RJ 21941-972, Brazil

Abstract. Due to their physical properties and biocompatibility, iron oxide nanoparticles have received particular attention in recent years for the localized hyperthermia therapy, where they are targeted to an organ, tissue or tumor and an external source is used for heating. Several physical, chemical and biological methods have been used to synthetize nanoparticles. A mechanical alloying method was used in this work to manufacture Fe_2O_3 nanoparticles. The size and shape of the nanoparticles were measured by scanning electron microscopy and X-ray diffraction. In this work, experiments were conducted with the nanoparticles dissolved in distilled water and heated by a laser in the near infrared range, with temperature measurements taken by an infrared camera. Numerical simulations were performed with COMSOL Multiphysics and compared to the experimental results. The numerical results agree with the measurements within the experimental uncertainties. The experimental results revealed a larger temperature increase of the sample surface for a larger concentration of nanoparticles.

J. Henriques et al. (Eds.): MEDICON 2019, IFMBE Proceedings 76, pp. 1239–1245, 2020.
https://doi.org/10.1007/978-3-030-31635-8_151

Hence, the Fe_2O_3 nanoparticles manufactured in this work behaved as a thermal agent with potential use for the hyperthermia therapy, including the treatment of cancer.

Keywords: Fe_2O_3 nanoparticles · Hyperthermia · Mechanical alloying

1 Introduction

The use of iron oxide nanoparticles (Fe_2O_3) for medical applications is of great interest, because these materials have biocompatibility characteristics and are non-toxic. In addition, iron oxide nanoparticles can be used as a photothermal agent, in order to generate heat under external electromagnetic excitation (radiofrequency, microwave and laser) for the hyperthermia treatment of cancer [1–8]. Nanoparticles deposited in tumours can improve the localized energy generation resulting from the electromagnetic excitation, thus minimizing thermal damage to the healthy cells surrounding the tumour [9].

There are several methods for manufacturing Fe_2O_3 nanoparticles, e.g., hydrothermal reaction, microwave, sol–gel microemulsion, forced hydrolysis, solid-phase grinding and high-energy ball milling [9]. High-energy ball milling is of great interest, because of its low cost, simple operation and high yield [10]. In fact, different protocols and strategies have been used to produce iron oxide nanoparticles in a planetary ball mill, with excellent results [10–14]. In this work, Fe_2O_3 particles smaller than 5 μm were ground in a planetary mill, with fixed milling time and rotational speed. The obtained powder was then characterized in terms of size reduction and mechanochemical effects by X-ray diffraction (XRD), as well as of particle size and morphology with a field emission gun scanning electron microscopy (FEG-SEM). The capabilities of the manufactured nanoparticles to improve energy absorption and to generate heat were demonstrated in experiments involving water nanofluids, with two different concentrations of nanoparticles. The external heat source consisted of a laser diode at a mean wavelength of 829.1 nm. Numerical simulations were performed in COMSOL MultiPhysics to predict the temperature increase of the nanofluids, as described next.

2 Materials and Methods

The iron oxide nanoparticles were manufactured by loading Fe_2O_3 powder with particles smaller than 5 μm (by Sigma Aldrich) into a planetary ball mill (Fritsch GmbH Pulvirisette 6). The mill contained hardened steel vials with 257.65 cm^3, with six stainless steel balls of 22 mm diameter that were set in rotation at 600 rpm. The ball-to-powder mass ratio was 30:1 [5]. Milling was performed in a dry air atmosphere for 24 h. The morphology and size of the nanoparticles were observed by field emission gun-scanning electron microscopy in a FEG-SEM model FEI Versa 3D. The X-ray diffraction measurements of the produced nanoparticles were performed using a DRX Bruker D8 Discover with Co ($K\alpha$) radiation, in the 2θ range from 20° to 90°, with counting times varying of 2 s for steps of 0.05° [5].

Figure 1 presents SEM micrographs of the Fe_2O_3 nanoparticles, where a variation in size around 50 nm can be observed. Figure 2 presents an X-ray powder diffraction pattern for the initial unmilled sample (bottom) and for sample milled during 24 h (top). This figure shows that the diffraction patterns are consistent with Fe_2O_3, thus indicating that the milling process did not affect the chemical structure of the initial sample.

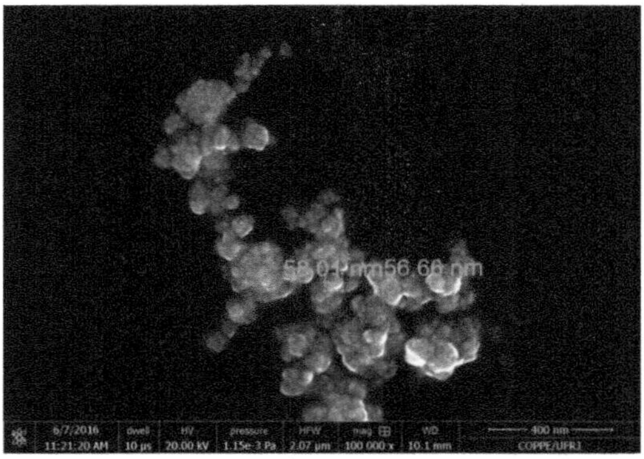

Fig. 1. SEM micrographs of the Fe_2O_3 nanoparticles.

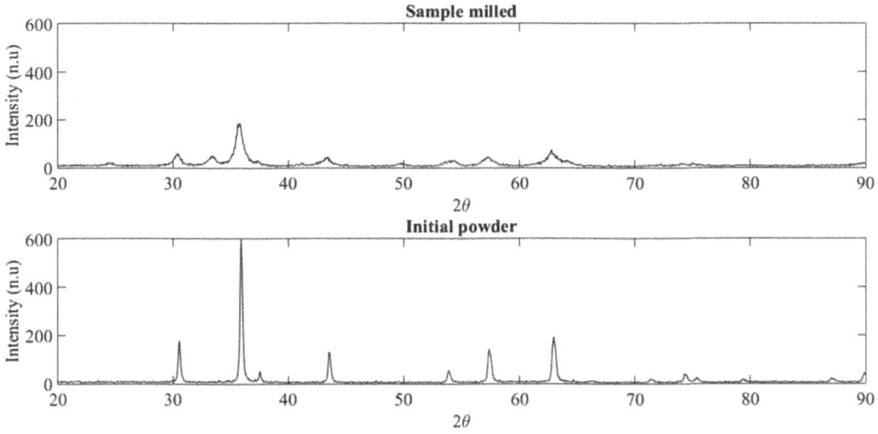

Fig. 2. X-ray powder diffraction pattern [5].

The Fe_2O_3 nanoparticles manufactured in this work were then dispersed in distilled water to produce nanofluids of two different concentrations. An ultrasonic processor (SONICS -VC 505) was used for dispersion of the nanoparticles in distilled water for 30 min, under an input power of 250 W. Two different nanofluids were produced by dissolving 10 mg and 20 mg of nanoparticles in 40 mL of distilled water, with concentrations of 0.025 wt% and 0.050 wt%, respectively.

Samples (each nanofluid and distilled water) with a volume of 10 mL were maintained in Petry plates, with radius 28 mm and height of 13.4 mm, were heated with a laser diode (Newport, model 525B - λ = 829.1 nm) at current of 2200 mA, corresponding to a heating power (P) of 0.327 W in the near infrared range during 100 s. The laser power was previously calibrated and exhibited a Gaussian profile with 11 mm of diameter (THORLABS VRC2 laser card). The temperatures of the heated surfaces of the samples were measured with an infrared camera (FLIR SC 660).

3 Results and Discussions

The temperature measurements were taken by the infrared camera with a frequency of 9 Hz, with associated uncertainty of 0.05 °C. Figure 3 presents the surface temperature measurements for distilled water (a) and for each of the nanofluids (b and c), after 100 s of heating (second row). The initial temperatures for the fluids are also presented at the first row of Fig. 3. Over the fluid surfaces, the mean initial temperatures were of 23.4 ± 0.2 °C, 25.1 ± 0.3 °C and 22.9 ± 0.2 °C, for the distilled water, 0.025 wt% and 0.050 wt% nanofluids, respectively. Figure 3 clearly shows larger temperature increases for the nanofluids than for the distilled water.

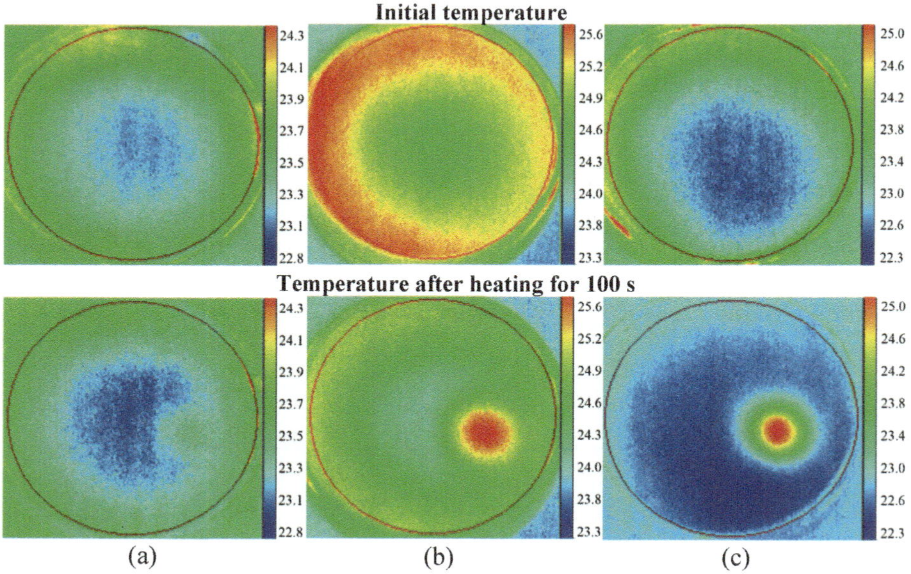

Fig. 3. Surface temperature (°C) measured with the infrared camera for: (a) Distilled water, (b) 0.025 wt% nanofluid, (c) 0.050 wt% nanofluid.

The numerical simulations were performed with the commercial software COMSOL Multiphysics. The mathematical model for the natural convection problem inside the Petry plate assumed Boussinesq's approximation as valid. The simulations were performed in the domain that only contained the liquid (radius $r_0 = 28$ mm and height H = 4.06 mm). The bottom and lateral surfaces were assumed as thermally insulated, while the top surface was heated by the laser beam and exchanged heat with the surroundings by natural convection and linearized radiation ($T_\infty = T_0$ and $h_\infty = 10$ W/(m^2.K)). The shape of the laser beam was fixed in accordance with the calibration procedure described above. The initial condition for the simulations was given by zero fluid flow and uniform temperature inside the domain (23.2 °C, 24.8 °C and 22.7 °C for the distilled water, 0.025 wt% and 0.050 wt% nanofluids, respectively). The interaction of the laser beam and the fluid was accounted for as a source term in the energy conservation equation, while the laser propagation inside the fluid was assumed to follow Beer-Lambert's law, that is,

$$Q(r,z) = (1-R)\mu_a\varphi_0 \exp(-\mu_a z)\exp\left(-\frac{2r^2}{r_l^2}\right) \tag{1}$$

where R is the surface reflectivity for normal incidence, μ_a is the absorption coefficient, φ_0 the laser fluence at the beam center, r_l is the Gaussian beam radius, while r and z are the cylindrical coordinates.

The numerical simulations with COMSOL Multiphysics were performed on a mesh with 6300 elements and a time step of 1 s. Such number of elements and time step were selected based on a grid refinement analysis. The physical properties of water, as well as other parameters used in the numerical simulations, are presented by Table 1. For the numerical simulations of the cases involving the nanofluids, the absorption coefficients were fixed as 29.40 m^{-1} and 88.20 m^{-1}, for the concentrations of 0.025 wt% and 0.050 wt%, respectively, by curve fitting of the temperature measurements at the fluid surfaces. It was assumed that the remaining physical properties were not affected by the addition of the nanoparticles. The numerical temperature distributions at the surfaces of distilled water and 0.025 wt% and 0.050 wt% nanofluids, are presented by Fig. 4a–c, at t = 100 s, respectively. A comparison between Figs. 3 and 4 demonstrate an excellent agreement between the measured and the calculated surface temperatures.

Figures 5a–c present a comparison of the transient variation of the numerical and measured temperatures at r = 10 mm, for distilled water and 0.025 wt% and 0.050 wt% nanofluids, respectively. The maximum temperature variations during this time interval were respectively of 0.28 °C, 0.72 °C and 2.31 °C. The measurement uncertainties at the 99% confidence level are also presented by these figures. We note that the numerical results are within the measurement uncertainties. The local temperature variation increased by approximately 2 °C from the distilled water to the 0.050 wt% nanofluid. Therefore, the nanoparticles manufactured in this work effectively behaved as a thermal agent, thus improving the absorption of the energy imposed by the laser diode in the near infrared range.

Table 1. Data used in the numerical simulations

Parameter	Value	Dimension	Parameter	Value	Dimension
ρ	998.2	kg/m^3	P	0.327	W
c_p	4184.1	J/(kg.K)	R	0.02	–
k	0.6034	W/(m.K)	t_s	100	s
r_l	5.5×10^{-3}	m	μ_{a_water}	1.96	1/m
r_0	28×10^{-3}	m	ν_{water}	1.0×10^{-3}	kg/(m.s)
H	4.06×10^{-3}	m			

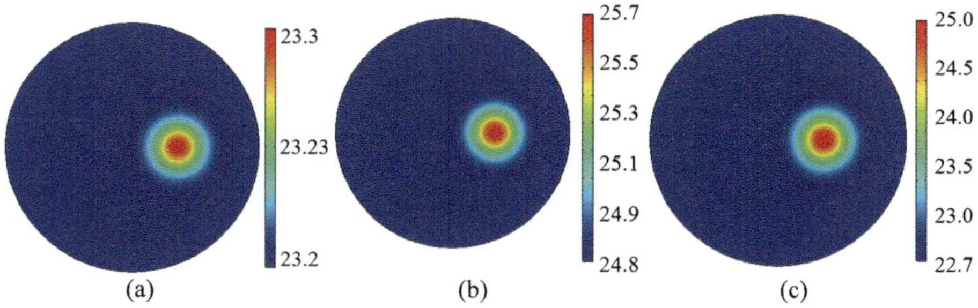

Fig. 4. Surface temperature (°C) measured with the infrared camera for: (a) Distilled water, (b) 0.025 wt% nanofluid, (c) 0.050 wt% nanofluid.

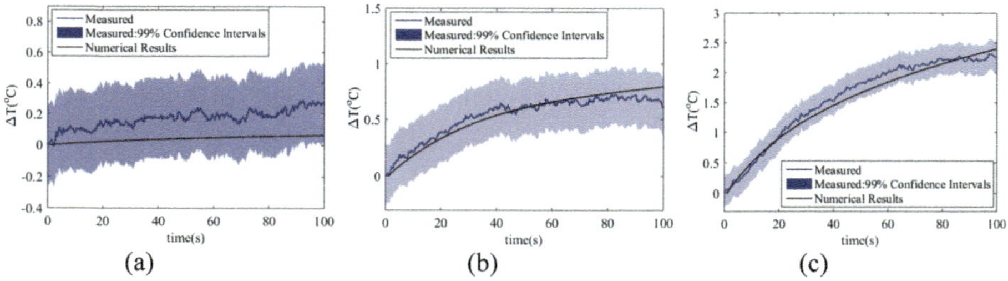

Fig. 5. Transient temperature variation at the center of the laser beam for: (a) Distilled water, (b) 0.025 wt% nanofluid, (c) 0.050 wt% nanofluid.

4 Conclusions

This work demonstrates the improved absorption of the energy imposed by a laser diode in the near infrared range, on nanofluids with two different concentrations of Fe_2O_3 nanoparticles. A sample of the base fluid (distilled water) was also used for comparison with the results obtained with the nanofluids. The nanoparticles used in this work were manufactured by milling of Fe_2O_3 powder during 24 h in a planetary ball mill. The morphology and size of the nanoparticles were characterized by field emission gun-scanning electron microscopy in a FEG-MEV and by X-ray diffraction. The local temperature variation increased by approximately 2 °C from the distilled water to the 0.050 wt% nanofluid. Numerical computations performed with a commercial code are in excellent agreement with the measured temperatures.

Acknowledgments. The authors are thankful for the support provided by Conselho Nacional de Desenvolvimento Científico e Tecnológico; Coordenação de Aperfeiçoamento de Pessoal de Nível Superior - Brasil (CAPES) - Finance Code 001; Fundação Carlos Chagas Filho de Amparo à Pesquisa do Estado do Rio de Janeiro, Brasil; Fundação de Amparo à Pesquisa do Estado do Amazonas – FAPEAM and Dirección General de Investigaciones (DGI) of Universidad Santiago de Cali, Colombia, project No. 819-621118-120.

References

1. Lamien, B., Bermeo Varon, L.A., Orlande, H.R.B., Eliçabe, G.E.: State estimation in bioheat transfer: a comparison of particle filter algorithms. Int. J. Numer. Methods Heat Fluid Flow **27**(3), 615–638 (2017)
2. O'Neal, D.P., Hirsch, L.R., Halas, N.J., Payne, J.D., West, J.L.: Photo-thermal tumor ablation in mice using near infrared-absorbing nanoparticles. Cancer Lett. **209**(2), 171–176 (2004)
3. Bermeo Varon, L.A., Orlande, H.R.B., Elicabe, G.E.: Estimation of state variables in the hyperthermia therapy of cancer with heating imposed by radiofrequency electromagnetic waves. Int. J. Thermal Sci. **98**, 228–236 (2015)
4. Bermeo Varon, L.A., Orlande, H.R.B., Eliçabe, G.E.: Combined parameter and state estimation in the radiofrequency hyperthermia treatment of cancer. Num. Heat Transf. Part A Appl. **70**, 581–594 (2016)
5. Lamien, B., Orlande, H.R.B., Bermeo Varón, L.A., Basto, R.Q.L., Eliçabe, G.E., dos Santos, D.S., Cotta, R.M.: Estimation of the temperature field in laser-induced hyperthermia experiments with a phantom. Int. J. Hyperth. **35**(1), 279–290 (2018)
6. Gas, P.: Multi-frequency analysis for interstitial microwave hyperthermia using multi-slot coaxial antenna. J. Electr. Eng. **66**(1), 26–33 (2015)
7. Gas, P.: Temperature inside tumor as time function in RF hyperthermia. Prz. Elektrotech. **86** (12), 42–45 (2010)
8. Gas, P., Miaskowski, A.: Specifying the ferrofluid parameters important from the viewpoint of magnetic fluid hyperthermia. In: 2015 Selected Problems of Electrical Engineering and Electronics (WZEE), pp. 1–6. IEEE Xplore (2015). 7394040
9. Estelrich, J., Busquets, M.A.: Iron oxide nanoparticles in photothermal therapy. Molecules **23**(1567), 1–26 (2018)
10. Wang, L.-L., Jiang, J.-S.: Preparation of α-Fe_2O_3 nanoparticles by high-energy ball milling. Phys. B Condens. Matter. **390**, 23–27 (2007)
11. Can, M.M., Ozcan, S., Ceylan, A., Firat, T.: Effect of milling time on the synthesis of magnetite nanoparticles by wet milling. Mater. Sci. Eng. B Solid-State Mater. Adv. Technol. **172**, 72–75 (2010)
12. Chen, D., Ni, S., Chen, Z.: Synthesis of Fe_3O_4 nanoparticles by wet milling iron powder in a planetary ball mill. China Part. **5**, 357–358 (2007)
13. Arbain, R., Othman, M., Palaniandy, S.: Preparation of iron oxide nanoparticles by mechanical milling. Miner. Eng. **24**, 1–9 (2011)
14. Kihal, A., Bouzabata, B., Fillion, G., Fruchart, D.: Magnetic and structural properties of nanocrystalline iron oxides. Phys. Procedia **2**, 665–671 (2009)

Special Sessions: Optimization in Medicine and Biology

Exact Linearization Techniques to Analyze the Population Dynamics of the Dengue Fever Vector

Helenice de Oliveira Florentino[1(✉)], Daniela Renata Cantane[1], Célia Aparecida dos Reis[2], Diego Cólon[3], and Suélia Rodrigues Fleury Rosa[4]

[1] Universidade Estadual Paulista UNESP, Botucatu, SP, Brazil
helenice.silva@unesp.br
[2] Universidade Estadual Paulista UNESP, Ilha Solteira, SP, Brazil
[3] Universidade de São Paulo USP, São Paulo, Brazil
[4] Universidade de Brasília UnB, Brasília, DF, Brazil

Abstract. This work presents an analysis of a mathematical model describing the population dynamics of *Aedes aegypti* and the design of a vector control via exact linearization techniques and optimal control. The control strategy optimizes the use of resources for vector control. Theoretical and computational results are presented.

Keywords: Mathematical model · Transmission · Dengue fever · Control

1 Introduction

Dengue fever, chikungunya, and zika are diseases transmitted by the *aedes aegypti* mosquito [7]. There is no fully developed vaccine against these diseases. Therefore, the only way to combat these diseases is via vector population control. However, it is necessary that the control is applied in a correct way, avoiding the generation of high costs and contamination of the environment. On the other hand, for a correct application of the control, it is necessary to understand the population dynamics of the mosquito [1, 7]. In this context, we propose to study the population dynamics of *aedes aegypti* mosquitoes using a mathematical model and to apply vector control techniques for the aquatic population phase of the mosquitoes. This study was developed applying the technique of exact feedback linearization to linearize this model and to design a control law. Using this control law, we propose to act in the population aquatic phase and to reduce the quantity of adult mosquitoes. An optimization procedure is used in order to minimize the cost of the control.

2 Mathematical Model

We consider the life cycle of mosquitoes divided into aquatic (eggs, larvae, and pupae) and terrestrial (male and female adults mosquitoes) phases, where $x_1(t)$ represents the density of the population in aquatic phase, $x_2(t)$ the density of the immature females

J. Henriques et al. (Eds.): MEDICON 2019, IFMBE Proceedings 76, pp. 1249–1255, 2020.
https://doi.org/10.1007/978-3-030-31635-8_152

population, $x_3(t)$ the density of fertilized females population (after mating), and $x_4(t)$ the density of male individuals.

After the egg hatching, the development of mosquitoes into the adult form can take up to 10 days. We denote γ as the per capita rate at which mosquitoes in aquatic phase pass to the adult phase (terrestrial), and a ratio r develop in females and $(1 - r)$ in males. β is the mating rate of immature females with male mosquitoes.

Mortality rates associated with aquatic, immature females, fertilized females and males are denoted respectively by μ_1, μ_2, μ_3, and μ_4. The oviposition rate of a fertilized female is given by $\phi\left(1 - \frac{x_1}{C}\right)$, where ϕ is the rate of intrinsic oviposition and C means the environmental capability.

The variable $u(t)$ acts as a time-varying mortality rate due to application control.

Under these considerations, we present the mathematical model (1), [4].

$$\begin{cases} \dot{x}_1 = -(\gamma + \mu_1)x_1 + \phi x_3 + \left(-\frac{\phi}{C}\right)x_1 x_3 - x_1 u(t) \\ \dot{x}_2 = r\gamma x_1 - (\beta + \mu_2)x_2 \\ \dot{x}_3 = \beta x_2 - \mu_3 x_3 \\ \dot{x}_4 = (1 - r)\gamma x_1 - \mu_4 x_4 \end{cases} \tag{1}$$

3 Exact Linearization Techniques

The procedure of the exact feedback linearization was applied aiming to transform the dynamics of the nonlinear system (1) into a linear dynamics. The advantage of this linearization, when compared with the classic technique is that it can be applied to all the state domain [3, 5, 6]. The equilibrium points and the stability of the dynamic system is analyzed. Using this methodology we determined the control law $u(t)$.

$$u(t) = -\frac{1}{a_1 A_3}\left[\begin{array}{l} (aA_1A_3 + (aA_4 + b\beta)A_3)x_1 + ((aA_4 + b\beta)A_4 - b\beta\mu_3)x_2 \\ + (aA_3\varphi + b\mu_3)x_3 + aA_2A_3x_1x_3 - v \end{array}\right] \tag{2}$$

where $v(x) = -ak_1A_3x_1 - (ak_1A_4 + bk_1\beta + ak_o)x_2 - (bk_o - bk_1\mu_3)x_3$, and k_1, k_o are conveniently chosen such that the polynomial $p(\lambda) = \lambda^2 + k_1\lambda + k_o$ has negatives roots [4].

4 Optimal Control Problem

We propose a methodology to choose the values for the parameters a, b, k_1, and k_0 in order to have: closed-loop stability; optimality related to the use of the minimum possible quantity of aquatic control; and to keep the mosquito population at a low level.

In this way, we propose a bi-objective optimization model (3)–(8) aiming to determine a strategy that uses the minimum possible quantity of aquatic control keeping the mosquito population controlled.

$$Minimize \left[J_1 = \int_{t=0}^{T} u(t)dt, \quad J_2 = \int_{t=0}^{T} (x_1(t) + x_2(t) + x_3(t) + x_4(t))dt \right] \tag{3}$$

Subject to

$$System\,(1), \tag{4}$$

$$Equation\,(2), \tag{5}$$

$$k_o > 0, \tag{6}$$

$$\frac{b}{a} > -\frac{\mu_3}{\beta} \tag{7}$$

$$u(t) \geq 0, x_1(t) \geq 0, \quad x_2(t) \geq 0, \quad x_3(t) \geq 0, \quad x_4(t) \geq 0, \quad t \in [0, T], \tag{8}$$

where $(x_1(t), x_2(t), x_3(t), x_4(t))$ are the state variables of system (1) and $u = u(t, a, b, k_0, k_1)$ is the control defined in (2).

The functional J_1 is related to the control applied in the time interval [0, T] and J_2 is the total density of mosquitoes in this interval. The constraints (8) define the non-negativity of the variables, guaranteeing their biological meaning.

5 Solving the Optimization Model (3)–(8)

We propose a heuristic algorithm VNS-Variable Neighborhood Search [2] to solve numerically the optimization problem (3)–(8), that is detailed next.

Define S_F as the set of feasible solutions to the problem (3)–(8):

$$S_F = \{X = (x_1(t), x_2(t), x_3(t), x_4(t), u(t,s)) \text{ such that } x_1(t) > 0, x_2(t) > 0, x_3(t) > 0, x_4(t) > 0, u(t,s) \geq 0, s \in S_P, t \in [0, T]\},$$

where $S_p = \{s = (a, b, k_0, k_1) \text{ such that } k_0 > 0, (b/a) > (-\mu_3/b), \text{and } k_1 \in R\}$.

A feasible solution $\breve{X} \in S_F$ can be generated by taking some $\breve{s} \in S_P$ and solving the system (1) with $u(t, \breve{s})$ defined as in (2).

To solve the bi-objective optimization problem (3)–(8), it is necessary to determine a pair (s^*, X^*), $s^* \in S_p$, and $X^* \in S_F$, which simultaneously minimizes the functionals J_1 and J_2. We propose the algorithm VNS to search for this solution.

Neighborhood Structure Considered:

Let $D' = (s', X')$, $s' \in S_p$, and $X' \in S_F$, such that $s' = (s'_1 = a', s'_2 = b', s'_3 = k'_0, s'_4 = k'_1)$. A neighborhood $N_k(D')$, $k \in \{1, 2, 3, 4\}$, is defined by fixing value to s'_i *($i \in \{1, 2, 3, 4\}$ and $i \neq k$)*, and by varying the value of s'_k. For example, $k = 1$:

$$N_1(D') = \{(s, X) \text{ such that } s = (s'_1 = \boldsymbol{a}, b', k'_0, k'_1) \in S_p, \boldsymbol{a} \neq a' \text{ and } X \in S_F\}.$$

VNS Algorithm

Let k_{max}=4 and the neighborhood $N_k(D)$, k=1,2,3,4, do:

1. Determine a random initial feasible solution $D_o=(s_o, X_o)$, $s_o \in S_p$ e $X_o \in S_F$;
2. $k \leftarrow 1$, $D' \leftarrow D_o$;
3. While $k \leq k_{max}$
   3.1 Find the ***Best Neighbor*** $D'' \in N_k(D')$;
   3.2 If ($J_1(D'') < J_1(D')$),
      3.2.1 $D' \leftarrow D''$; k $\leftarrow$1;
   3.3 Else
      3.3.1 $k \leftarrow k + 1$;
   3.4 End-If;
4. End-While;
5. Do $D^* = D'$.

Best Neighbor *Algorithm*

Let k, D' and l_{max} do:

1. Generate randomly $D'' \in N_k(D')$, $l \leftarrow 1$
2. While $l \leq l_{max}$
   2.1 Perturb D'',(generate randomly $\underline{D}'' \in N_k(D')$, ($\underline{D}'' \neq D''$))
   2.2 If ($J_1(\underline{D}'') < J_1(D'')$),
      2.2.1 $D'' \leftarrow \underline{D}''$;
   2.3 End-If;
3. l $\leftarrow$l+ 1;
4. End-While;
5. Take back D".

The VNS algorithm is capable of determining an optimized control strategy $u(t)$, in the time interval [0, T] that keeps the population density of mosquitoes at a required low level.

6 Computational Application

Aiming to implement the proposed methodology, some computational experiments were made using MATLAB 7.14.0.739 (R2012a) software and run on a micro-computer Dual Core i5-650 with 4 GB memory and 400 GB hard drive.

Table 1 shows the parameters used in the simulations for a particular case when the average local temperature is 25 °C, [4]. All parameters are measured per day (d^{-1}) except r.

Figure 1 shows the effect of the control on the population dynamic, according to the adopted values of the parameters in Table 1 and $a = b = k_o = k_1 = 1$, $T = 150$ days and using different initial conditions.

In Fig. 1 it is possible to observe that, without optimization, the control actions are capable of stabilizing the system under different initial conditions, but the density is always at an undesirable high level. The control strategy calculated by the VNS approach chooses to apply a higher intensity control during the first days, and then reducing it as the population also decreases (Fig. 2).

Table 1. Parameter

Parameter	Value
C	3.0000
γ	0.0941
R	0.5000
β	0.9270
ϕ	6.3530
μ_1	0.0583
μ_2	0.0202
μ_3	0.0319
μ_4	0.0600

Source: [4]

The control was applied from the period when the critical phase of the proliferation of diseases begins. The determined values for the optimized parameters of $u(t)$ were $a* = 0.0672$, $b* = 6.0197$, $k_0^* = 3.8677$, $k_1^* = -0.8401$. This optimized control strategy was able to reduce the density of all segments of the population (see Fig. 3).

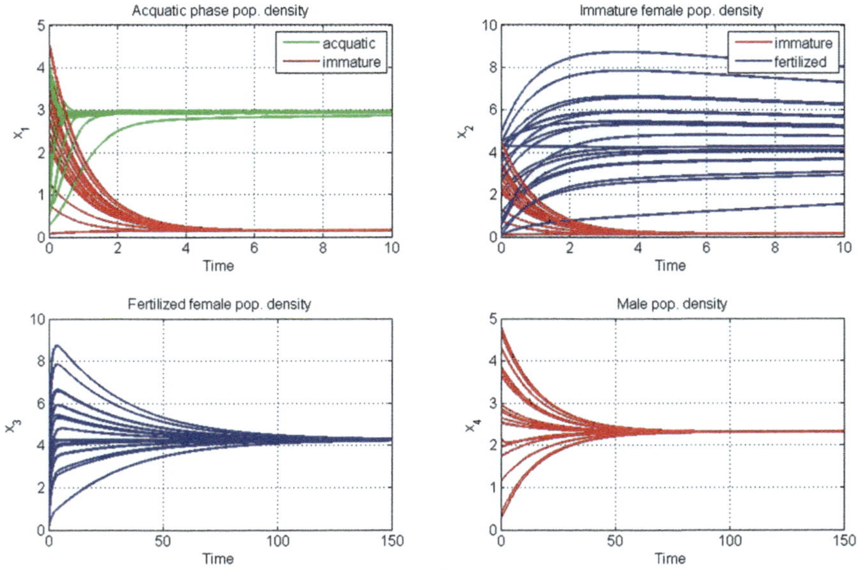

Fig. 1. Simulation for the open-loop Model using different initial conditions [4].

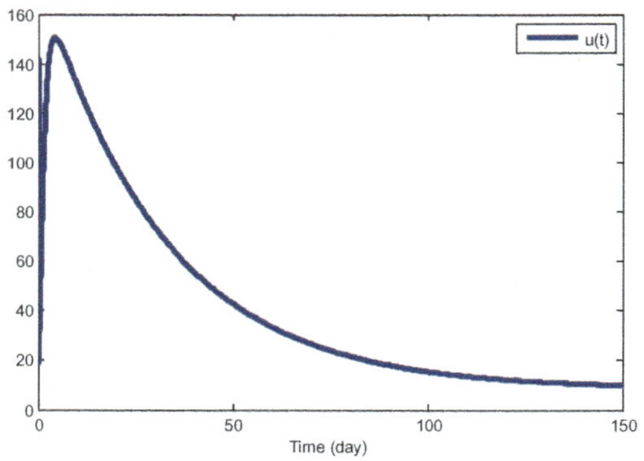

Fig. 2. Optimized control strategy obtained applying the proposed VNS heuristic [4].

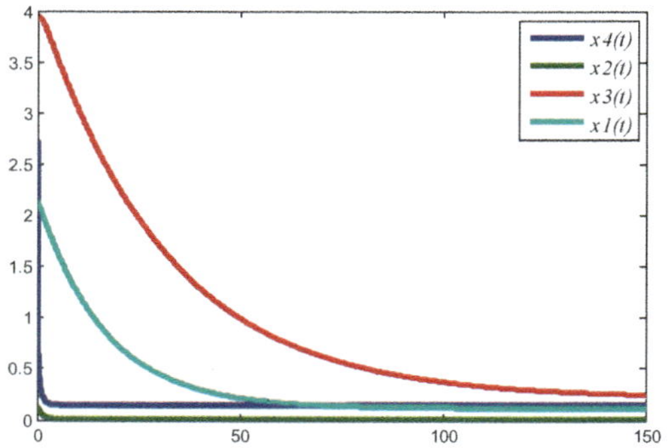

Fig. 3. Population dynamic of the mosquitoes in the phase: $x_1(t)$, $x_2(t)$, $x_3(t)$, $x_4(t)$ under the application of the optimal control [4].

7 Conclusion

We proposed a VNS heuristic aiming to determine optimized control strategies for the aquatic phase of the vector, which favors the reduction of the diseases transmitted by this mosquito, such as dengue fever, chikungunya, zika, and yellow fever, and reducing the number of mosquitoes also reduces the need for chemical control. The computational results show that this methodology has great potential as a tool for the decision makers in the fight against the *aedes aegypti*.

Acknowledgements. We wish to thank FAPESP (Grant No. 2009/15098-0, 2013/07375-0 and 2014/01604-0), FUNDUNESP/PROPG/PROPE/UNESP, CNPq (302454/2016-0), and CAPES for their financial support.

References

1. Florentino, H.O., Bannwart, B.F., Cantane, D.R., Santos, F.L.P.: Multiobjective genetic algorithm applied to dengue control. Math. Biosci. **258**, 77–84 (2014)
2. Hansen, P., Mladenović, N., Perez, J.A.M: Variable neighbourhood search: methods and applications **6**, 319–360 (2008)
3. Isidori, A.: Nonlinear Control Systems. Springer, Roma (1995)
4. Reis, C.A., Florentino, H.O., Cólon, D., Rosa, S.F., Cantane, D.R.: An approach of the exact linearization techniques to analysis of population dynamics of the mosquito *Aedes aegypti*. Math. Biosci. **299**, 51–57 (2018)
5. Silva, G.V.M.: Controlo Não Linear. Escola Superior de Tecnologia Setúbal, Lisboa (2003)
6. Slotine, J., Li, W.: Applied Nonlinear Control. Prentice Hall, New Jersey (1991)
7. WHO Homepage. http://www.who.int/mediacentre/factsheets/fs117/en/. Accessed 23 Feb 2017

Advantage of Beam Angle Optimization in Head-and-Neck IMRT: Patient Specific Analysis

Tiago Ventura[1,2,3](✉), Maria do Carmo Lopes[1,2,3], Humberto Rocha[3,4], Brigida da Costa Ferreira[3,5,6], and Joana Dias[3,4]

[1] Physics Department, University of Aveiro, Campus Universitário de Santiago, 3810-193 Aveiro, Portugal
[2] Medical Physics Department, Instituto Português de Oncologia de Coimbra Francisco Gentil, EPE, Avenida Bissaya Barreto, nº 98, 3000-075 Coimbra, Portugal
tiagoventura@ipocoimbra.min-saude.pt
[3] Institute for Systems Engineering and Computers at Coimbra, Rua Sílvio Lima, 3030-290 Coimbra, Portugal
[4] Economy Faculty, University of Coimbra and Centre for Business and Economics Research, Avenida Dias da Silva 165, 3004-512 Coimbra, Portugal
[5] School Health Polytechnic of Porto, Rua Dr. António Bernardino de Almeida, 400, 4200-072 Porto, Portugal
[6] I3N Physics Department of University of Aveiro, Campus Universitário de Santiago, 3810-193 Aveiro, Portugal

Abstract. Radiation therapy (RT) main purpose is to eliminate, in a controlled way, all tumor cells sparing as much as possible the normal tissues. Intensity-Modulated Radiation Therapy (IMRT) is becoming the standard treatment technique in RT. Beam angle optimization (BAO) has potential to confer more quality to IMRT inverse planning process compared to manual trial and error approaches. In this study, the BAO advantages in head-and-neck patients are highlighted, using a patient specific analysis. Fluence optimization was done with Erasmus-iCycle multicriterial engine and BAO optimization was performed using two different algorithms: a combinatorial iterative algorithm and an algorithm based on a pattern search method. Plan assessment and comparison was performed with the graphical tool SPIDERplan. Among a set of forty studied nasopharynx cancer cases, three patients have been select for the specific analysis presented in this work. BAO presented plan quality improvements when beam angular optimized plans were compared with the equidistant beam angle solution and when plans based on non-coplanar beams geometries were compared with coplanar arrangements. Improvement in plan quality with a reduced number of beams was also achieved, in one case. For all cases, BAO generated plans with higher target coverage and better sparing of the normal tissues.

Keywords: Radiation therapy · Beam angle optimization · Plan assessment · Head-and-neck cancer

J. Henriques et al. (Eds.): MEDICON 2019, IFMBE Proceedings 76, pp. 1256–1263, 2020.
https://doi.org/10.1007/978-3-030-31635-8_153

1 Introduction

Radiation therapy (RT) is one of the most important therapeutic options used in the battle against cancer. It makes use of ionizing radiation to eliminate in a controlled way the tumour cells, sparing as much as possible the adjacent normal tissues. Intensity-Modulated Radiation Therapy (IMRT) is a radiation therapy technique that enables the generation of conformal dose distributions to the target volume by delivering non-uniform intensity fields from multiple directions. IMRT treatment planning uses inverse planning techniques, wherein an objective function containing the desired plan objectives guides the fluence map optimization (FMO) by scoring the goodness of the plan [1]. The FMO will determine the beam intensities for each of the selected angles. Most of the times, an empirical trial-and-error manual tuning of plan parameters (like weights, objectives or beam angles) is done until an acceptable plan is achieved. This planning framework is extremely dependent on the planner skills and experience and on the case complexity [2]. Moreover, it is not possible to guarantee that an optimal plan is found. The IMRT optimization process should be tackled by more reliable methods and algorithms such as multi-criteria optimization, beam angle optimization (BAO) and eventually by machine learning automated techniques.

BAO methods have contributed to the enhancement of the IMRT optimization process. A beam angle selection based on mathematical criteria may lead to important improvements in the quality of the plan dose distribution [3] that can be even more expressive if non-coplanar beam geometries are available in the optimization process [4]. The BAO problem mathematically described as a highly non-convex multi-modal optimization problem with many local minima [5] can be addressed separately or jointly with the FMO problem. For the coupled modality, the FMO solution guides the BAO problem along the optimization [6], where the best beam ensemble can be achieved using heuristic methods in exhaustive combinatorial searches over a discretized space search [6–8] or alternatively using pattern search methods [9] or multistart derivative-free optimization frameworks [4] that continuously explore the search space.

The purpose of this work is to highlight the advantages of BAO for the head-and-neck pathology following a patient specific analysis approach. From a set of forty nasopharyngeal (NPC) studied cases, particular patient cases were selected to show the improvements that can be obtained in plan quality when BAO plans are compared with the equidistant beam angle solution and when coplanar and non-coplanar BAO sets are confronted. To complete this patient specific analysis, a third case where it is possible to get improved plans diminishing the number of beams is also evaluated, to emphasize how the individual patient anatomy may influence the results.

2 Materials and Methods

For this study, three cases were selected among a set of forty NPC clinical cases, previously studied. All cases had a simultaneous integrated boost prescription composed of two dose levels, where the tumour planning target volume (PTV) was prescribed with a dose of 70 Gy and the lymph nodes PTVs with a dose of 59.4 Gy.

Spinal cord, brainstem, chiasm, optical nerves, retinas, lens, parotids, oral cavity, larynx, oesophagus, ear canals, temporal mandibular joints, mandible, brain, pituitary gland, thyroid and lungs were also contoured and defined as organs-at-risk (OAR).

IMRT optimization was performed by Erasmus-iCycle, a multicriterial dose calculation engine guided by a wish-list defined *a priori,* that automatically generates a single Pareto solution for a given set of beams [10]. All plans were initially optimized using the most clinically used beam angle configuration: 7 coplanar equidistant beams (*d7*). BAO was performed for coplanar and non-coplanar beam geometries of 5, 7 and 9 beams with two algorithms. The first (algorithm *i*) is included in Erasmus-iCycle and it is based in a discrete iterative combinatorial approach [10]. The second algorithm, named as algorithm *B*, belongs to the continuous search space class and uses a pattern search method to find the best possible beam angle ensemble [9].

For each selected NPC case, three plans were considered that highlight the BAO advantages. For the first case (patient #1), the *d7* plan was compared with the 9 non-coplanar beams plan of algorithm *B* (*B9nc*). For the second case (patient #2), the 9 coplanar beams plan and the 9 beams non-coplanar plan of algorithm *i* (*i9c* and *i9nc,* respectively) were compared side-by-side. For the third case (patient #3), the 5 non-coplanar beams plan of algorithm *B* (*B5nc*) and the 9 coplanar beams plan of algorithm *i* (*i9c*) were compared.

SPIDERplan, a graphical method based on a scoring approach [11], was used for plan assessment and comparison. Targets and OARs are divided into groups. A score based on pre-defined planning objectives and relative weights defined according to the radiation oncologist clinical preferences is assigned to each structure. Global plan quality is computed as a weighted sum of all structures' scores. Graphically, a customised radar plot is used to represent all dosimetric information, where plan evaluation can be done visualizing all structures, in a Structures Plan Diagram (SPD), or just considering the defined groups, in a Group Plan Diagram (GPD). Each group can also be assessed with more detail in partial group radar plots named Structures Group Diagrams (SGD), where only the structures belonging to the group and a partial group score are displayed.

For this study, SPIDERplan configuration included six groups of structures: the PTV group composed by the PTVs, the Critical group constituted by the spinal cord and the brainstem, the Optics group including the chiasm, the optical nerves, the retinas and the lens, the DigestOral group composed by the parotids, the oral cavity, the oesophagus and the larynx, the Bone group composed of the temporal mandibular joint, the mandible and the ear canals and Other group considering the brain, the pituitary gland, the thyroid and the lungs. The groups were assigned with relative weights of 50%, 30%, 10%, 5%, 3.5% and 1.5%, respectively. The score of each structure is given by the ratio between the clinical tolerance criteria and the planned dose. A score value of one is achieved when the planned dose is equal to the structure tolerance criteria. Higher score values mean that these tolerances were surpassed and the best plans will have scores less than one, tending towards the centre of the radar plot [11].

3 Results

Global plan comparison was done with SPIDERplan GPDs (left diagrams in Figs. 1, 2 and 3), while individual group evaluation was performed with SPIDERplan SGDs (right diagrams in Figs. 1, 2 and 3). The selection of the groups to be depicted in each figure was based on the visual analysis of the GPD and on the corresponding relative weight importance value.

For patient #1, *d7* and *B9nc* plans were compared, Fig. 1. *B9nc* plan achieved the best global plan score with an expressive percent difference of –7% from *d7* score. The largest difference between the tolerance and the planned dose of these plans was obtained for the Optics group. For *d7* plan the score of both lenses was higher than one, meaning that the tolerance dose criteria was not accomplished. Also, for the Critical group, a percent score difference of 9% can be observed between the two plans.

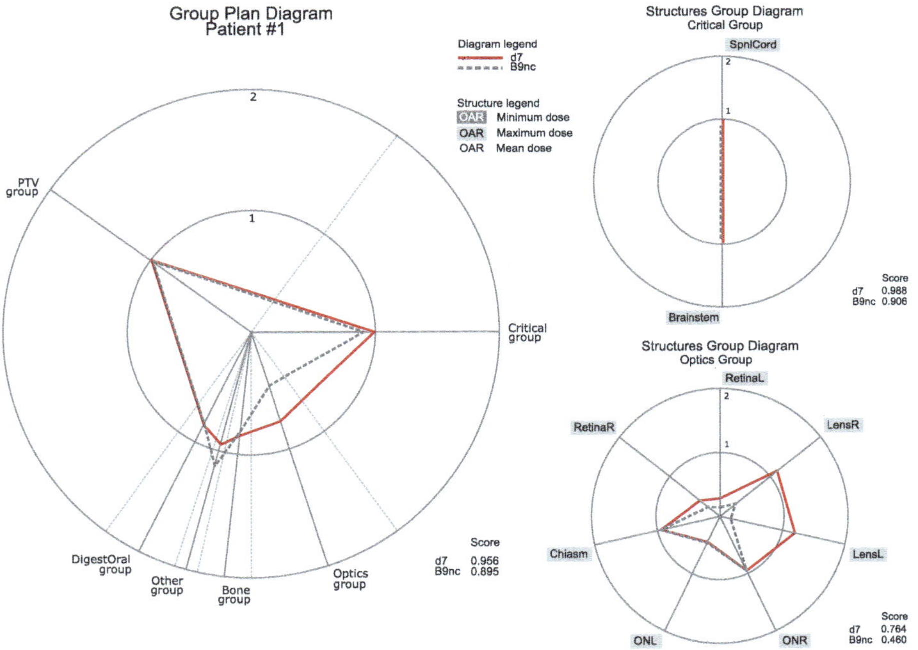

Fig. 1. SPIDERplan group plan diagram of patient #1 and structures group diagrams for Critical and Optics groups for 7 equidistant beam angles (*d7*) and 9 non-coplanar beams plan of algorithm *B* (*B9nc*).

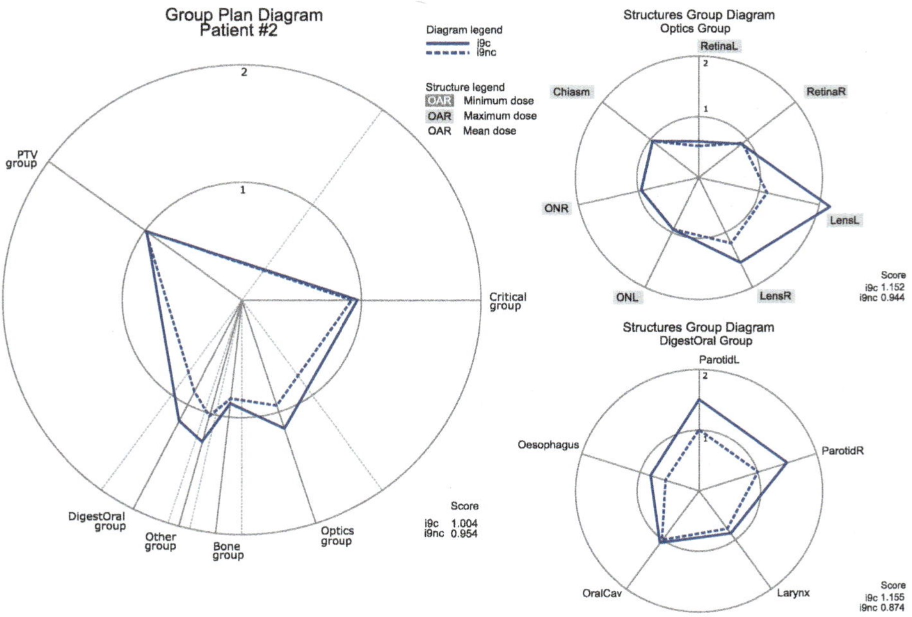

Fig. 2. SPIDERplan group plan diagram of patient #2 and structures group diagrams for Optics and DigestOral groups for algorithm *i* with 9 coplanar beams (*i9c*) and 9 non-coplanar beams (*i9nc*)

For patient #2, the best global plan score was obtained with the non-coplanar plan *i9nc* with a percent difference score of –5% from the coplanar plan i9c, that presented a score of 1.004 (Fig. 2). Also for *i9c,* score values higher than one were achieved by the Optics group, the DigestOral group and the Other group, while for *i9nc* only the Other group score was out of tolerance. The reason for these global results was found in the SGDs. Both plans presented scores very near the tolerance or out of tolerance for the lenses (Optics group SGD), for the parotids and for the oral cavity (DigestOral group SGD). Nevertheless, plan *i9nc* was much better than *i9c* plan, since for the mentioned structures it presented score values near one or just slightly higher than one, while *i9c* got score values well above the defined tolerances.

Figure 3 compares the quality of *B5nc* and *i9c* plans for patient #3. Globally, plan *B5nc* attained the best global plan score with a percent global score difference of –4% compared to *i9c* plan. This result is obtained even if for the DigestOral group *B5nc* plan is worse than plan *i9c,* due to the relatively lower weight of this group. The main reason for the best score of plan *B5nc* is the almost vanishing scores presented for the optics structures (dotted line at the Optics SGD centre). The topographic relation of the PTV and the optics structure, well separated for this patient, explains this result.

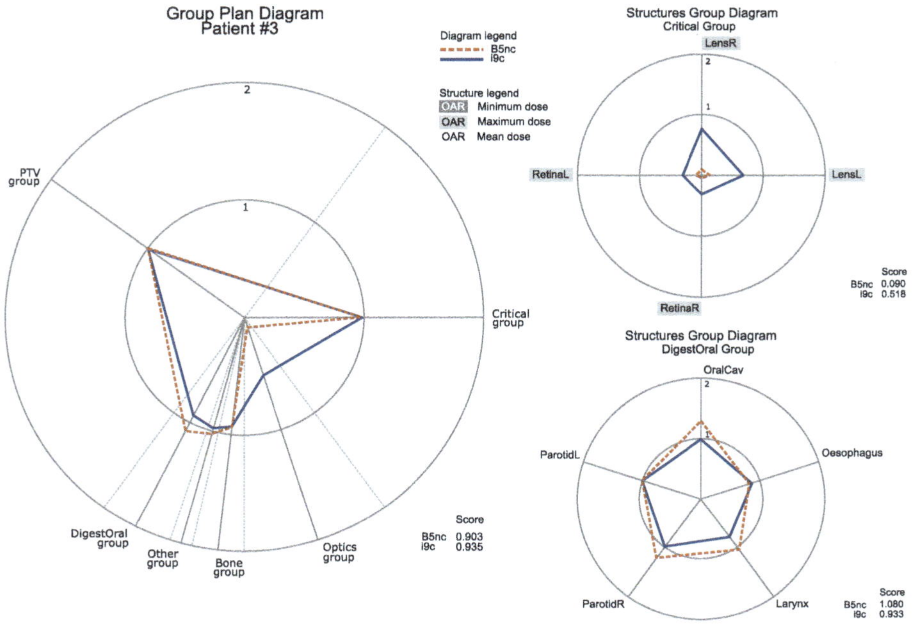

Fig. 3. SPIDERplan group plan diagram of patient #3 and structures group diagrams for Optics and DigestOral groups for algorithm *B* with 5 non-coplanar beams (*B5nc*) and for algorithm *i* with 9 coplanar beams (*i9nc*)

4 Discussion

In this study, patient specific plan assessment was used to highlight the advantages of BAO for head-and-neck cases.

Among a set of forty NPC clinical cases, three patients were chosen. For patient #1, important plan quality improvements were achieved when the plan using optimized directions was compared with the equidistant beam angle solution, the standard beam configuration used in the clinical routine. For this patient, the plan generated with BAO presented non-coplanar beam incidences and a higher number of beams than equidistant beam angle solution. This has improved the quality of dose distribution by further sparing important OARs.

The utilization of non-coplanar beams in RT is a very popular approach among medical physicists and mathematicians. There is an intrinsic perception that non-coplanar beam incidences may enhance the quality of the plan. BAO has a crucial role in the selection process of the best beam arrangement, due to the high number of possible combinations and geometry complexity. Promising results with non-coplanar BAO have been achieved for different pathology sites such as: lung, liver, head-and-neck and intracranial tumours [4, 12–14]. Nevertheless, depending on site complexity, a patient-by-patient assessment may be required to fully evaluate the real impact of the achieved improvements. For patient #2, *i9c* and *i9nc* plans were compared enhancing the advantage of non-coplanarity. The higher score group differences were found for the Optics group (lenses and left retina) and for the DigestOral group (oesophagus, parotids and larynx), where the structures were located very near or even contiguous to the PTVs.

The number of beams needed to achieve a 'good' dose distribution is also an interesting topic due to treatment delivery efficiency reasons. The planner intuition will always predict that plan quality will tend to increase with the number of beams available for optimization. However, the amount of fluence modulation caused by the addition of new beams has a physical threshold, where beyond that level no further plan quality improvements occur. BAO has then an important role to play, achieving better treatment plans with a smaller number of beams. This scenario occurred for patient #3 where plan *B5nc* was better than *i9c*, due to the significant improvement achieved in the optics structures that were completely spared when using fewer non-coplanar beam orientations.

For all patients, PTV group scores were always below one. The high target coverage index may be related with the quality of the FMO and to the definitions of the multicriterial IMRT optimization engine wish-list where a primordial importance was assigned to the PTVs. Also, for the PTV group, small score differences were obtained among the compared plans. For the OARs, the largest score group differences were obtained for the Critical group (patient #1), for the Optics group (patients #1, #2 and #3) and for the DigestOral group (patients #2 and #3), that contained structures of vital importance, as the spinal cord and the brainstem, and with great impact in the patient quality of life, as the lens, the optical nerves, the parotids or the oral cavity. These findings help to support two key aspects: the main contribution of BAO was in further sparing the OARs while maintaining good PTV coverage.

The use of BAO never worsens the treatment plan quality though for some patients, the improvements achieved may be negligible. However, if BAO can be done without the planner's intervention, then it is worth to be considered for all patients, since some of them will greatly benefit from this optimization.

The potential enhancements that an efficient BAO, made in reasonable computing times, may bring to the quality of static IMRT plans, can reopen the discussion about what is the most appropriate inverse planning technique for a given patient treatment.

5 Conclusions

In this study the benefits of BAO for head-and-neck patients were addressed. Patient specific analysis of NPC cases were used to highlight the improvements achieved by plans with BAO when compared with IMRT plans using equidistant beam angle directions. Furthermore, the advantages of non-coplanar over coplanar beam plans and the importance of BAO in the generation of plans with a small number of beams was highlighted. This study also showed that BAO did not compromise the PTV coverage and that the most important contribution was in the sparing of the normal tissues, taking into account the anatomy specificities of each patient.

Acknowledgments. The authors would like to express their gratitude to Sebastiaan Breedveld and Ben Heijmen for making available the multicriterial optimization of IMRT plans system Erasmus-iCycle and their valuable support and coaching along the work.

This work was supported by project grant POCI-01-0145-FEDER-028030 and by the Fundação para a Ciência e a Tecnologia (FCT) under project grant UID/Multi/00308/2019.

No potential conflict of interest nor any financial disclosures must be declared.

References

1. ICRU. International commission on radiation units and measurements: prescribing, recording, and reporting photon-beam intensity-modulated radiation therapy (IMRT). ICRU Report 83. J ICRU, **10**(1), 1–106 (2010). https://doi.org/10.1093/jicru/10.1.report83
2. Thieke, C., Küufer, K.H., Monz, M., et al.: A new concept for interactive radiotherapy planning with multicriteria optimization: first clinical evaluation. Radiother. Oncol. **85**(2), 292–298 (2007). https://doi.org/10.1016/j.radonc.2007.06.020
3. Das, S.K., Marks, L.B.: Selection of coplanar or noncoplanar beams using three-dimensional optimization based on maximum beam separation and minimized nontarget irradiation. Int. J. Radiat. Oncol. Biol. Phys. **38**(3), 643–655 (1997). https://doi.org/10.1016/S0360-3016(97)89489-8
4. Rocha, H., Dias, J., Ventura, T., Ferreira, B., Lopes, M.C.: A derivative-free multistart framework for an automated noncoplanar beam angle optimization in IMRT. Med. Phys. **43**(10), 5514–5526 (2016). https://doi.org/10.1118/1.4962477
5. Ehrgott, M., Holder, A., Reese, J.: Beam selection in radiotherapy design. Linear Algebra Appl. **428**(5), 1272–1312 (2008). https://doi.org/10.1016/j.laa.2007.05.039
6. Craft, D.: Local beam angle optimization with linear programming and gradient search. Phys. Med. Biol. **52**(7), 127–135 (2007). https://doi.org/10.1088/0031-9155/52/7/N02
7. Dias, J., Rocha, H., Ferreira, B.C., et al.: Simulated annealing applied to IMRT beam angle optimization: a computational study. Physica Med. **31**(7), 747–756 (2015). https://doi.org/10.1016/j.ejmp.2015.03.005
8. Bangert, M., Unkelbach, J.: Accelerated iterative beam angle selection in IMRT. Med. Phys. **43**(3), 1073–1082 (2016). https://doi.org/10.1118/1.4940350
9. Rocha, H., Dias, J., Ferreira, B.C., et al.: Beam angle optimization for intensity-modulated radiation therapy using a guided pattern search method. Phys. Med. Biol. **58**(9), 2939–2953 (2013). https://doi.org/10.1088/0031-9155/58/9/2939
10. Breedveld, S., Storchi, P., Voet, P., et al.: iCycle: integrated, multi-criterial beam angle and profile optimization for generation of coplanar and non-coplanar IMRT plans. Med. Phys. **39**(2), 951–963 (2012). https://doi.org/10.1118/1.3676689
11. Ventura, T., Lopes, M.C., Ferreira, B.C., et al.: SPIDERplan: a tool to support decision-making in radiation therapy treatment plan assessment. Reports Pract. Oncol. Radiother. **21**(6), 508–516 (2016). https://doi.org/10.1016/j.rpor.2016.07.002
12. Dong, P., Lee, P., Ruan, D., et al.: 4π non-coplanar liver SBRT: a novel delivery technique. Int. J. Radiat. Oncol. Biol. Phys. **85**(5), 1360–1366 (2013). https://doi.org/10.1016/j.ijrobp.2012.09.028
13. Bangert, M., Ziegenhein, P., Oelfke, U.: Comparison of beam angle selection strategies for intracranial IMRT. Med. Phys. **40**(1), 011716 (2013). https://doi.org/10.1118/1.4771932
14. Lyu, Q., Yu, V.I., Ruan, D.: A novel optimization framework for VMAT with dynamic gantry couch rotation. Phys. Med. Biol. **63**(12), 125013 (2018). https://doi.org/10.1088/1361-6560/aac704

Determining Patient-Specific Dosage Scheme Using Integer Programming

Lars Hellemo[1(✉)] and Vegard Heimly Brun[2]

[1] Operations Research, SINTEF Industry, Oslo, Norway
Lars.Hellemo@sintef.no
[2] Department of Breast and Endocrine Surgery, University Hospital of North Norway, Tromsø, Norway

Abstract. Finding the correct substitution dosage of levothyroxine after total thyroidectomy can be time consuming using current methods. We present a mixed integer quadratic programming model for determining correct dosage given characteristics of a patient. The model is flexible in terms of determining fixed daily dosage regime, repeating schedule (e.g. weekly), and may include a loading dosage phase to achieve full substitution quicker. A similar model is used in an ongoing research project to determine dosages for patients. One challenge of the most sophisticated dosage regime is that it increases the burden for patients in following a dosage regime with frequent changes and where pills need to be divided to achieve more precise dosage.

Keywords: Pharmacokinetics · Mathematical programming · Integer programming · Decision support systems · Dosage regime · Precision medicine

1 Background

Patients who undergo total thyroidectomy need life long substitution treatment with levothyroxine, a manufactured form of thyroxine, a hormone that regulates the metabolism. Predicting correct dosage for substitution treatment with levothyroxine after total thyroidectomy has proven difficult due to large inter-patient variation in dosage needed to achieve euthyroidism Sukumar et al. (2010).

The time to determine correct dosage can be months or even years, and this has motivated work in individually-tailored treatment Del Duca et al. (2015). In determining adequate dosage, the pharmacokinetics considered is dominantly based on steady state considerations Brunton et al. (2011). Dosage adjustments are normally considered only when steady state levels has been reached after 6–8 weeks, meaning that the time to achieve euthyroidism can be very long for some patients.

Several retrospective studies have tried to predict correct dosage for patients based on features known before surgery such as height, weight and sex. A recent

J. Henriques et al. (Eds.): MEDICON 2019, IFMBE Proceedings 76, pp. 1264–1269, 2020.
https://doi.org/10.1007/978-3-030-31635-8_154

example is Zaborek et al. (2019) who tried various machine learning methods and achieved better fit than previously reported with a Poisson regression model.

Even with better statistical models, the variation remains large, and by characterizing the phenotype based on blood samples of uptake and response, we attempt to give patients correct dosage earlier than with current methods. For an overview of the pharmacokinetics of levothyroxine (from a hypothyreosis perspective), see Colucci et al. (2013). For a presentation of different pharmacokinetic models with references, see Bonate (2011).

In an ongoing research project, we seek to characterize patients from measurements of TSH (thyroid stimulating hormone) and FT4 (free thyroxine) levels the first two weeks after surgery. An individual dose-response curve is calculated to create an FT4 target for each patient to reach a certain TSH goal. This allows earlier and possibly more precise dosage adjustment than the standard clinical practice of having a 6–8 week interval between dosage adjustment to make sure steady state levels are reached.

The contribution of this paper is to present a model to determine dosages for an already characterized patient, given the pill strengths available commercially. The model is based on mixed integer quadratic programming (MIQP), and can provide a fixed daily dosage, a repeated schedule, typically weekly, and if appropriate include a loading dosage scheme to achieve the correct level of medication more quickly.

2 MIQP Model Formulation

The following describes the MIQP model to determine dosages. Loosely, the goal is to minimize the deviation from a defined target concentration level Y by changing dosages x_t for days t for a defined period of time $\mathcal{T}_{\mathrm{Dec}}$. We assume the fixed patient-specific parameters reduction factor Δ (reformulated from half time $T_{\frac{1}{2}}$) and absorption factor A to be known. The concentration in period t is given by the estimated total FT4 γ_t divided by distribution volume DV.

$$\min_{\mathrm{x}_t,\gamma_t} \sum_{t\in\mathcal{T}_{\mathrm{Dec}}} \left(\frac{\gamma_t}{\mathrm{DV}} - \mathrm{Y}\right)^2 \tag{1}$$

Alternatively with a (small) penalty C applied to the total amount of medication for preference of lowest possible dosage:

$$\min_{\mathrm{x}_t,\gamma_t} \sum_{t\in\mathcal{T}_{\mathrm{Dec}}} \left(\frac{\gamma_t}{\mathrm{DV}} - \mathrm{Y}\right)^2 + \sum_{t\in\mathcal{T}_{\mathrm{Dec}}} Cx_t \tag{2}$$

Subject to the following constraints:

Total FT4 γ_t in any time period t is given by the level in the preceding period $t-1$, γ_{t-1}, the fixed reduction factor Δ, the intake x_t and the fixed absorption A (converting from μg using the molecular weight of thyroxine, $\mathrm{MW}_{\mathrm{T4}}$):

$$\gamma_t = \Delta\left(\gamma_{t-1} + A\mathrm{x}_t 10^6/\mathrm{MW}_{\mathrm{T4}}\right), \forall t \in \mathcal{T}_{\mathrm{Dec}} \tag{3}$$

In addition, the initial condition must be met:

$$\gamma_1 = \Gamma_0 \tag{4}$$

γ is greater than zero and bounded from above:

$$0 \leq \gamma_t \leq \Gamma^{\text{Max}}, \forall t \in \mathcal{T}_{\text{Dec}} \tag{5}$$

The dosages can be restricted in various ways, depending on the distribution scheme. In the following we present formulations for free dosage, assuming an ideal adjustment of dosages, through various ways of working with pills available in discrete dosage strengths.

2.1 Ideal Dosage Choice

The model can be set to allow free choice of dosages (continous values). This represents an upper bound on performance, but is unrealistic, as patients need to take medication in form of full or half pills, available in discrete dosages.

$$\mathrm{x}_t \geq 0, \forall t \in \mathcal{T}_{\text{Dec}} \tag{6}$$

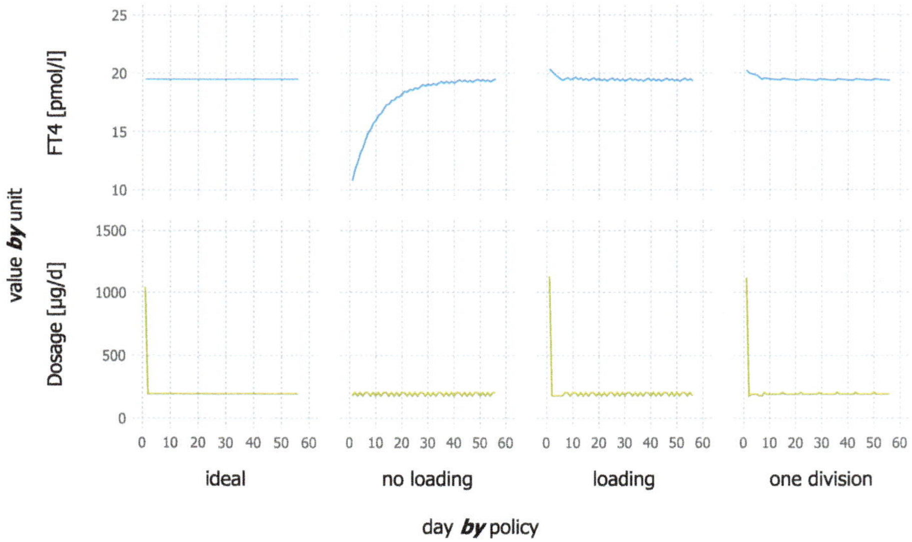

Fig. 1. Illustration of different dosage regimes: Predicted concentration (top) and daily dosages to reach a target FT4 concentration of 19.5 pmol/l with no restrictions on allowed dosages the left, followed by fixed schedule with 2 allowed strengths (no loading) and with loading dosage allowing zero or one division (half pill) to the right.

2.2 Discrete Dosages

In practice, dosages x_t need to be restricted to be chosen from a set of available dosages $\mathcal{X}$. This can be formulated generally with Eq. (7) replacing Eq. (6):

$$\mathrm{x}_t \in \mathcal{X}, \forall t \in \mathcal{T}_{\mathrm{Dec}} \tag{7}$$

However some more level of detail may be useful in selecting allowed dosage regimes, and these are detailed in Sect. 2.3.

2.3 Restricted Dosages

In the following, we describe how a practical dosage scheme can be chosen from a limited number of the strengths commercially available, while still allowing some flexibility. For further adjustments, each pill may be divided once or twice (for half/quarter dosage). Each patient will usually be prescribed a limited number of different pill strengths, and follow a regular dosage regime repeating e.g. every day or every week. This means Eq. (6) is replaced by the following constraints with appropriately chosen parameters.

The following constraints restrict dosages to equal commercially available strengths X_j which may be halved K times ($K = 0$ means no division). To allow the use of loading doses (or zero dose in case of too high initial concentration), the dosage schedule can be selected individually for each day in the first L days, thereafter the schedule for maintenance dosages is repeated every M days, hence binary variables are defined over $t = 1 \ldots L + M$. The formulation below allows up to 3 (2+1) full pills on one day.

$$\mathrm{x}_t = \sum_{j=1}^{J} \sum_{k=1}^{K+2} \xi_{tjk} X_j 2^{2-k}, t = 1 \ldots L + M \tag{8}$$

$$\xi_{tjk} \in \{0, 1\}, t = 1 \ldots L + M, j = 1 \ldots J, k = 1 \ldots K + 2 \tag{9}$$

The indicator variable ρ_j is used to determine which dosages (pill strengths) are in use, and we restrict the total number of distinct strengths in use at the same time to N:

$$\sum_{j=1}^{J} \rho_j \leq N \tag{10}$$

Only allow use of divisions of strengths already selected:

$$\xi_{tjk} \leq \rho_j, t = 1 \ldots L + M, j = 1 \ldots J, k = 1 \ldots K + 2 \tag{11}$$

In addition, the following formulation makes sure the dosage regime is repeated every M days. For the first L days, a nonrepeating schedule is allowed to adjust dosage level quicker:

$$\mathrm{x}_t = \mathrm{x}_{L+(t-1 \bmod M)+1}, t = L + M + 1 \ldots T^{\mathrm{Max}} \tag{12}$$

A non-repeating schedule can be determined by setting $L = T^{\mathrm{Max}}$, setting $M = 0$ and removing constraints defined by Eq. (12).

A repeating schedule also has computational advantages, as ξ_{tjk} and related Eqs. (8) to (11) only need to be defined over the first M time periods ($t = 1 \ldots M$), giving a lower number of binary variables. For a fixed schedule without loading dosages, the objective function Eq. (2) should be defined over a period of time after the concentration has stabilized, e.g. for $t = T^{\mathrm{Max}} - 7 \ldots T^{\mathrm{Max}}$.

3 Discussion

With an optimization model to calculate both the steady state dosage regime and the loading dosage regime, patients can reach the target concentration quicker, see Fig. 1. Providing the scheme with divisions of pills to increase the precision of the dosage beyond the available pill strengths have some practical limitations. Dividing pills precisely and without spilling is not trivial, and makes the pills harder to ingest due to shape, coating and taste. As there is generally a problem of achieving high adherence among patients, this extra burden for patients represents increased risk of inadherence.

As the successful characterization of patients was still unproven at the time of the experimental design, only the simple dosage regime with steady state regime was opted for in the clinical study. The preliminary results from testing the patient-tailored dosages in an ongoing clinical study are promising, and a successful result would open for trying out more advanced dosage schemes. However, due to the extra burden on patients associated with manual splitting and combining of pills with different strengths, this should be combined with some form of innovation in patient-tailored distribution technique. Pre-packaged daily dosages would be more practical for patients, and reduce the risk of inadherence. While implementing a scheme with division of pills may not be worth the extra effort, a loading dosage scheme seems to have interesting potential and should be considered for a follow-up study.

4 Conclusion

We have described a model that can be used to determine patient-specific dosage schemes using a limited number of available strengths. The proposed formulation may be configured to also determine individual loading dosage schemes. Clinical studies are needed to answer if applying the dosage schemes found by solving this model will shorten the time to euthyroidism after total thyroidectomy or after dosage adjustments.

References

Bonate, P.L.: Pharmacokinetic-Pharmacodynamic Modeling and Simulation. Springer, New York (2011)

Brunton, L.L., Chabner, B.A., Knollman, B.: Goodman Gilman's the Pharmacological Basis of Therapeutics, 12th edn. McGraw-Hill, New York (2011)

Colucci, P., Yue, C.S., Ducharme, M., Benvenga, S.: A review of the pharmacokinetics of levothyroxine for the treatment of hypothyroidism. Eur. Endocrinol. **9**(1), 40–47 (2013)

Del Duca, C.S., Santaguida, M.G., Brusca, N., Gatto, I., Cellini, M., Gargano, L., Verga Falzacappa, C., Frattaroli, F.M., Virili, C., Centanni, M.: Individually-tailored thyroxine requirement in the same patients before and after thyroidectomy: a longitudinal study. Eur. J. Endocrinol. **173**(3), 351–357 (2015)

Sukumar, R., Agarwal, A., Gupta, S., Mishra, A., Agarwal, G., Verma, A.K., Mishra, S.K.: Prediction of LT4 replacement dose to achieve euthyroidism in subjects undergoing total thyroidectomy for benign thyroid disorders. World J. Surg. **34**(3), 527–531 (2010)

Zaborek, N.A., Cheng, A., Imbus, J.R., Long, K.L., Pitt, S.C., Sippel, R.S., Schneider, D.F.: The optimal dosing scheme for levothyroxine after thyroidectomy: a comprehensive comparison and evaluation. Surgery **165**(1), 92–98 (2019)

Optimization of Highly Noncoplanar Arc Therapy Trajectories: A Dosimetric Approach

Humberto Rocha[1,2(✉)], Joana Dias[1,2], Tiago Ventura[2,3], Brígida Ferreira[2,4], and Maria do Carmo Lopes[2,3]

[1] CeBER and FEUC, Universidade de Coimbra, 3004-512 Coimbra, Portugal
hrocha@mat.uc.pt, joana@fe.uc.pt
[2] INESC-Coimbra, 3030-290 Coimbra, Portugal
[3] Serviço de Física Médica, IPOC-FG, EPE, 3000-075 Coimbra, Portugal
{tiagoventura,mclopes}@ipocoimbra.min-saude.pt
[4] Escola Superior de Saúde, Politécnico do Porto, 4200-072 Porto, Portugal
bcf@ess.ipp.pt

Abstract. The latest generation of linear accelerators allows the use of noncoplanar trajectories in arc therapy which combine the benefits of noncoplanar intensity-modulated radiation therapy (IMRT) treatment plans, such as improved organ sparing, with the benefits of arc therapy treatment plans, such as short treatment times. In this paper, we propose a two-step approach based on dosimetric criteria for the optimization of noncoplanar arc trajectories. In the first step, an initial set of anchor points (noncoplanar beam directions) is computed using a beam angle optimization (BAO) algorithm. In the second step, anchored in the beam directions already calculated, the noncoplanar arc trajectory is defined by iteratively computing additional anchor points considering the same dosimetric criteria used for the noncoplanar BAO. A nasopharyngeal tumor case already treated at the Portuguese Institute of Oncology of Coimbra (IPOC), is used to illustrate the benefits of the proposed optimization approach.

Keywords: Radiation therapy · Noncoplanar arc therapy · Optimization

1 Introduction

In step-and-shoot IMRT, a linear accelerator mounted on a gantry rotates around the patient, stops at fixed beam directions and delivers non-uniform radiation fields. Noncoplanar beam directions are obtained if the couch is allowed to rotate as well. In arc therapy, irradiation is done continuously while the gantry rotates around the patient with the treatment beam always on. One of the most efficient IMRT arc techniques is volumetric modulated arc therapy (VMAT), particularly in terms of dose delivery time [1–3]. VMAT treatment plans typically use coplanar trajectories, considering a fixed couch angle equal to 0°.

J. Henriques et al. (Eds.): MEDICON 2019, IFMBE Proceedings 76, pp. 1270–1275, 2020.
https://doi.org/10.1007/978-3-030-31635-8_155

The latest generation of linear accelerators allows the use of noncoplanar trajectories in arc therapy which combine the benefits of noncoplanar IMRT treatment plans, such as improved organ sparing, with the benefits of arc therapy treatment plans, such as short treatment times. Several authors have proposed approaches to optimize highly noncoplanar arc trajectories, mainly resorting to geometrical metrics [4,5].

In this paper, we propose a two-step approach based on dosimetric criteria for the optimization of noncoplanar arc trajectories. In the first step, an initial set of anchor points (noncoplanar beam directions) is computed taking advantage of the work previously developed in BAO for step-and-shoot IMRT [6–12]. In the second step, anchored in the beam directions already calculated, the noncoplanar arc trajectory is defined by iteratively computing additional anchor points considering the same dosimetric criteria used for the noncoplanar BAO. A nasopharyngeal tumor case already treated at IPOC is used to illustrate the benefits of the proposed approach. The paper is organized as follows. The strategy proposed for noncoplanar arc trajectory optimization is described in the next Section. In Section three we present the computational results. Conclusions and future work are presented in the last Section.

2 Noncoplanar Arc Trajectory Optimization

In this study, simultaneous gantry and couch rotation is considered while the treatment beam is on, leading to a highly noncoplanar arc trajectory. An optimization approach for the noncoplanar arc trajectory of a VMAT plan, called *ncVMAT*, is proposed and compared with the coplanar arc trajectory of a VMAT plan, called *cVMAT*, and with the typically used equispaced step-and-shoot IMRT plan, called *Equi*. The two-step approach proposed combines two optimization problems, the BAO problem and the arc trajectory optimization, that are quite challenging just by themselves. The dosimetric criteria used to guide these two optimization problems is the optimal value of the fluence map optimization (FMO) problem. Formulation and resolution approaches used to address FMO and BAO problems correspond to the ones detailed in Rocha et al. [11].

A clinical nasopharyngeal tumor case treated at IPOC was used to illustrate and test our approach. Nasopharyngeal tumors are complex cases to plan due to the large number of organs-at-risk (OARs) in the neighborhood of the tumor(s). The spinal cord and the brainstem are two of the main OARs considered. They are serial organs, i.e, they are compromised even if only a small part is damaged. Therefore, maximum-doses are prescribed for spinal cord and brainstem, 45 Gy and 54 Gy respectively. The other OARs considered are the parotids (the larger salivary glands) and the oral cavity (that contains the remaining salivary glands). These are parallel organs whose functioning is not much affected if a small part of the organ is damaged. Thus, mean-doses are prescribed for parotids and oral cavity, 26 Gy and 45 Gy respectively. For safety purposes, the tumor volume is enlarged by adding a margin originating a structure called planning target volume (PTV). Two levels of radiation dose are prescribed: a higher radiation dose of 70 Gy is prescribed to the tumor (called PTV_{70}) and a lower radiation dose of 59.4 Gy is prescribed to the lymph nodes (called $PTV_{59.4}$).

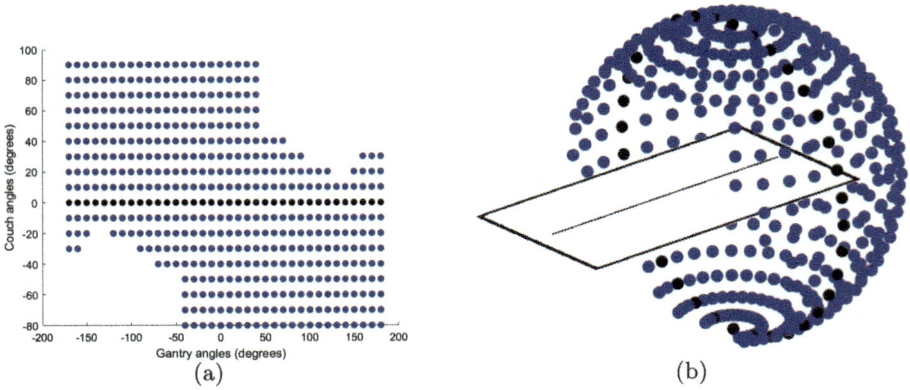

Fig. 1. Candidate beams homogeneously distributed represented in 2D – (a) and the corresponding 3D representation – (b). Black beams correspond to the coplanar candidate beams (couch fixed at 0°) while blue beams correspond to noncoplanar candidate beams.

The second step, arc trajectory optimization, is now described. A grid with equispaced beams, separated by 10° for both the gantry and the couch, is considered in this second step. After exclusion of infeasible couch-gantry angle pairs due to possible collisions of patient and gantry for a nasopharyngeal tumor case, we end up with 472 candidate beams homogeneously distributed as illustrated in Fig. 1.

The initial anchor points corresponding to the 7-beam noncoplanar BAO solution for the nasopharyngeal tumor case at hand are displayed in red in Fig. 2(a). In order to enhance one of the main features of VMAT, short treatment times, the arc trajectory starts at the leftmost anchor point in Fig. 2(a), visit each anchor point from left to right and ends at the rightmost anchor point in Fig. 2(a), with the gantry always rotating towards the next anchor point while the couch might be halted or moving towards the next anchor point. Considering these gantry/couch movement restrictions, the feasible points when calculating a new anchor point are shown in green in Fig. 2(a).

There are different ways of considering the optimal value of the FMO problem to iteratively add novel anchor points, one by one. In this study we consider the most expensive, in terms of computational time, that adds each one of the green points, one at a time, to the existing set of anchor points and then compute the corresponding optimal FMO value considering these beams. The candidate beam that leads to the minimum optimal FMO value when added to the existing anchor beams will be selected as the next anchor point. In Fig. 2(b) the novel

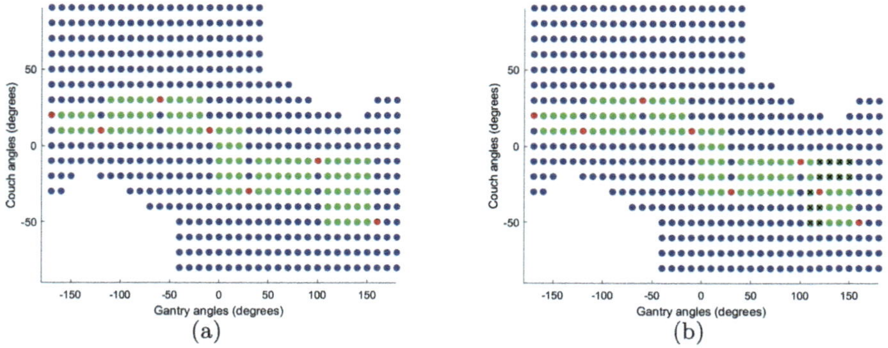

Fig. 2. The 7-beam noncoplanar BAO solution is displayed in red and the feasible points to consider when calculating a new anchor point are displayed in green – (a). A novel anchor point is added and green candidate beams that became infeasible are removed – (b).

Table 1. Target coverage, conformity and homogeneity obtained by treatment plans.

Target	Parameters	*Equi*	*cVMAT*	*ncVMAT*
PTV_{70}	Coverage	0.877	0.882	0.888
	Conformity	0.556	0.523	0.550
	Homogeneity	0.880	0.881	0.881
$PTV_{59.4}$	Coverage	0.907	0.911	0.880
	Conformity	0.784	0.760	0.841
	Homogeneity	0.774	0.773	0.767

anchor point is displayed. This recently added red point leads to the infeasibility of some green beams due to the gantry/couch movements defined. At the end of each iteration, green candidate beams that became infeasible are removed as illustrated in Fig. 2(b). This iterative procedure ends when 20 anchor points are obtained, which is the typical number of anchor points considered in the literature (see, e.g., [13,14]).

3 Computational Results

Computational tests were conducted on a Dell Precision T5600 with Intel Xeon processor 64GB 1600MHz. The noncoplanar arc trajectory obtained is displayed in Fig. 3. In terms of optimal FMO values, *ncVMAT* clearly outperforms the other treatment plans in terms of optimal FMO value, improving 9.8% the value obtained by *Equi* plan while the improvement of *cVMAT* was 5.5%. For a similar tumor volume coverage, displayed in Table 1, the organ sparing, reported in Table 2, also shows the advantage of the *ncVMAT* treatment plan.

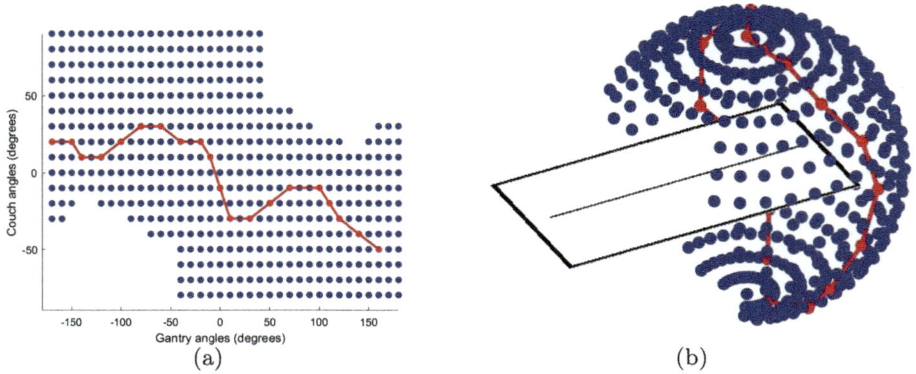

Fig. 3. Trajectory obtained by the noncoplanar arc trajectory optimization approach in 2D – (a) and in 3D – (b)

Table 2. OARs sparing obtained by treatment plans.

OAR	Mean Dose (Gy)			Max Dose (Gy)		
	Equi	*cVMAT*	*ncVMAT*	*Equi*	*cVMAT*	*ncVMAT*
Spinal cord	–	–	–	33.4	29.1	33.1
Brainstem	–	–	–	52.1	49.5	43.4
Right parotid	37.0	34.8	34.5	–	–	–
Left parotid	32.0	29.8	27.6	–	–	–
Oral cavity	26.2	22.3	22.4	–	–	–

4 Conclusions and Future Work

A dosimetric approach for the optimization of highly noncoplanar arc trajectories was described and tested using a complex nasopharyngeal tumor case already treated at IPOC. For the patient tested, the resulting noncoplanar arc plan, *ncVMAT*, clearly outperforms both the coplanar arc plan, *cVMAT*, and the typically used coplanar equispaced step-and-shoot IMRT plan. Although, for the patient at hand, the overall quality of the treatment is undoubtedly greater considering the noncoplanar arc plan, tests with more patients should be conducted. Moreover, other strategies to accelerate both the noncoplanar BAO procedure and the second step that determines the remaining anchor points should be investigated.

Acknowledgments. This work has been supported by project grant POCI-01-0145-FEDER-028030 and by the Fundação para a Ciência e a Tecnologia (FCT) under project grant UID/Multi/00308/2019.

Conflicts of Interest. None.

References

1. Bedford, J.L.: Treatment planning for volumetric modulated arc therapy. Med. Phys. **36**, 5128–5138 (2009)
2. Otto, K.: Volumetric modulated arc therapy: IMRT in a single gantry arc. Med. Phys. **35**, 310–317 (2008)
3. Yu, C.X.: Intensity-modulated arc therapy with dynamic multileaf collimation: an alternative to tomotherapy. Phys. Med. Biol. **40**, 1435–1449 (1995)
4. MacDonald, R.L., Thomas, C.G.: Dynamic trajectory-based couch motion for improvement of radiation therapy trajectories in cranial SRT. Med. Phys. **42**, 2317–2325 (2015)
5. Smyth, G., Bamber, J.C., Evans, P.M., Bedford, J.L.: Trajectory optimisation for dynamic couch rotation during volumetric modulated arc radiotherapy. Phys. Med. Biol. **58**, 8163–8177 (2013)
6. Rocha, H., Dias, J., Ferreira, B.C., Lopes, M.C.: Selection of intensity modulated radiation therapy treatment beam directions using radial basis functions within a pattern search methods framework. J. Glob. Optim. **57**, 1065–1089 (2013)
7. Rocha, H., Dias, J., Ferreira, B.C., Lopes, M.C.: Beam angle optimization for intensity-modulated radiation therapy using a guided pattern search method. Phys. Med. Biol. **58**, 2939–2953 (2013)
8. Rocha, H., Dias, J., Ferreira, B.C., Lopes, M.C.: Pattern search methods framework for beam angle optimization in radiotherapy design. Appl. Math. Comput. **219**, 10853–10865 (2013)
9. Dias, J., Rocha, H., Ferreira, B.C., Lopes, M.C.: A genetic algorithm with neural network fitness function evaluation for IMRT beam angle optimization. Cent. Eur. J. Oper. Res. **22**, 431–455 (2014)
10. Dias, J., Rocha, H., Ferreira, B.C., Lopes, M.C.: Simulated annealing applied to IMRT beam angle optimization: a computational study. Phys. Med. **31**, 747–756 (2015)
11. Rocha, H., Dias, J., Ventura, T., Ferreira, B.C., Lopes, M.C.: A derivative-free multistart framework for an automated noncoplanar beam angle optimization in IMRT. Med. Phys. **43**, 5514–5526 (2016)
12. Rocha, H., Dias, J., Ventura, T., Ferreira, B.C., Lopes, M.C.: Beam angle optimization in IMRT: are we really optimizing what matters? Int. Trans. Oper. Res. **26**, 908–928 (2019)
13. Papp, D., Bortfeld, T., Unkelbach, J.: A modular approach to intensity-modulated arc therapy optimization with noncoplanar trajectories. Phys. Med. Biol. **60**, 5179–5198 (2015)
14. Langhans, M., Unkelbach, J., Bortfeld, T., Craft, D.: Optimizing highly noncoplanar VMAT trajectories: the NoVo method. Phys. Med. Biol. **63**, 025023 (2018)

Dose-Response to Different Radiochemotherapy Regimens in Locally Advanced Pancreatic Cancer

Brígida C. Ferreira[1,2,3(✉)], Joana Dias[2,4], Adriana Gomes[1], Panayiotis Mavroidis[5,6], and Humberto Rocha[2,4]

[1] School of Health, Polytechnic of Porto, Porto, Portugal
bcf@ess.ipp.pt
[2] Institute for Systems Engineering and Computers at Coimbra, Coimbra University, Coimbra, Portugal
[3] I3N, Physics Department of Aveiro University, Aveiro, Portugal
[4] CeBER and FEUC, Coimbra University, Coimbra, Portugal
[5] Department of Radiation Oncology, University of North Carolina at Chapel Hill, Chapel Hill, NC, USA
[6] Division of Medical Radiation Physics, Karolinska Institutet and Stockholm University, Stockholm, Sweden

Abstract. Conformal radiation therapy (RT) delivered concomitantly with chemotherapy including 5-fluorouracil (5-FU) or Gemcitabine (GEM) is a common treatment for patients with unresectable locally advanced pancreatic tumors. In this study, the Poisson model describing tumor response to these two treatment options was derived. Clinical data was retrieved from reports published from 1990 to 2015. Dosimetric and clinical data from 1196 patients treated with RT with concurrent 5-FU or GEM were gathered. RT doses ranging from 3.6–64.8 Gy, delivered in fractions of 1.2–8 Gy, were converted to a 2 Gy fractionation scheme using the Biological Effective Dose concept. The parameters of the Poisson-Linear-Quadratic-Time model were derived using genetic algorithm optimization to minimize the least-square fitting error and a local search was then made using the maximum likelihood method. The goodness of the fit was assessed using the Pearson χ^2-test. For RT+5-FU, D_{50} was 59.8 Gy, γ was 1.3, α/β was 3.2, T_{pot} was 18.6 days and Tk was 25.0 days. For RT+GEM, D_{50} was 54.5 Gy, γ was 1.4, α/β was 4.6, T_{pot} was 34.2 days and Tk was 37.2 days. As expected, RT+GEM showed higher efficacy than RT+5-FU. A RT dose-response effect was obtained showing that treatment strategies allowing a dose-escalation in pancreas tumors should be investigated.

Keywords: Pancreas tumors · Dose-response models · Radiation therapy · Chemotherapy

1 Introduction

Pancreatic tumors are amongst the most challenging cancer types. Patients are generally diagnosed with the disease in advanced stages for which surgery is no longer viable. Different chemotherapy regimens delivered alone or in combination with RT,

J. Henriques et al. (Eds.): MEDICON 2019, IFMBE Proceedings 76, pp. 1276–1282, 2020.
https://doi.org/10.1007/978-3-030-31635-8_156

have largely been used without achieving long-term overall success. Survival rates are thus generally poor and consensus about the best treatment option for each patient has not been established [1].

In concomitant radiochemotherapy, the most common chemotherapy agents are 5-fluorouracil (5-FU) and gemcitabine (GEM). These have frequently been used as single agents but have also been combined with different regimens [2–34]. Better overall survival and progression-free survival was generally obtained with radiochemotherapy based on the GEM agent compared to the 5-FU [35, 36]. Interestingly in a network meta-analysis that compared different GEM regimens, the combination of RT with GEM has shown to be the most effective GEM based treatment compared to the combination of GEM with other chemotherapy drugs [37]. With conformal RT, dose-prescription were mostly confined to values between 50.4–59.4 Gy. A dose escalation appears to be beneficial in terms of local control, progression-free survival and overall survival [27, 33, 35, 38], but limited by the tolerance of surrounding organs at risk. An accurate model describing tumor response to delivered treatments is thus very useful as it may help investigating new treatment approaches for pancreatic cancer.

The aim of this study was to derive the dose-response parameters for the Poisson-Linear-Quadratic-Time model for locally advanced pancreatic tumors for patients that received conformal RT concomitant with 5-FU or GEM using the clinical data reported in the literature.

2 Materials and Methods

Clinical information was retrieved from scientific papers reporting treatment outcome of patients with locally advanced pancreatic cancer (cases with distant metastasis were excluded). Primary treatment was 3D conformal RT concurrent with 5-FU or GEM and chemotherapy alone (simulating zero RT dose). No prior treatments were delivered to these patients. Thirty three papers, which were published from 1990 to 2015, with comprehensive descriptions on patients, delivered treatment (chemotherapy regimen, RT prescription dose, fractionation) and response to therapy (according to WHO or RECIST guidelines [39, 40]) were included in this study [2–34].

The groups RT+5-FU and RT+GEM were composed by a total of 1196 patients treated from about 1988 to 2008. RT doses, ranging from 3.6–64.8 Gy delivered in fractions of 1.2–8 Gy in 2–54 fractions, were converted to a common 2 Gy fractionation scheme using the Biological Effective Dose (BED) concept [41]. The best estimates for the delivered doses were made based on the information described in each report about compliance to treatment. Response Rate (RR), as quantified by the sum of the rate of complete and partial tumor response to therapy, was used as the endpoint of interest for the derivation of the dose-response curve. For pancreatic tumors, this endpoint follows more closely the typical sigmoidal shape of radiobiological models than the endpoint local control (given by the sum of RR and the rate of stable disease) [42].

The probability of tumor response to the delivered treatment, P, was determined assuming an additive effect between chemotherapy and RT using the expression,

$$P = P_{ch} + P_{rt}(1 - P_{ch}) \tag{1}$$

where P_{ch} and P_{rt} are the probability of response to the chemotherapy regimen and the delivered RT, respectively [43]. The probability of response to chemotherapy was calculated as the weighted mean of the RR obtained to chemotherapy delivered as monotherapy [28–31]. The Poisson-Linear-Quadratic-Time model was used to describe tumor response to RT for the groups of patients treated with RT concomitant with GEM (RT+GEM) or RT concomitant with 5-FU (RT+5-FU). The parameters of this model are: D_{50}, the dose that results in a 50% response; γ, the maximum normalized dose-response gradient; α/β, the fractionation sensitivity parameter of the Linear-quadratic model, T_{pot}, the tumor potential doubling time and T_k, the time at which repopulation begins. The parameters: α, β, γ, T_{pot} *and* T_k were derived using genetic algorithm optimization to minimize the least-square fitting error after which a local search was done using the maximum likelihood method. D_{50} and α/β were calculated using the expressions described in Ferreira et al. [41]. The goodness of the fit was evaluated using the Pearson χ^2-test.

3 Results

The local control rates of the patient cohorts that received RT+5-FU varied between 25% and 100%, whereas for those treated with RT+GEM it varied between 33% and 100%. The RR of the group RT+5-FU varied between 5% and 45%, whereas for the RT+GEM group it ranged from 5% to 50%. The average of RR for patients treated with chemotherapy alone with 5-FU was 7.1% and with GEM was 9% (range: 4%–15%). The RR as a function of the delivered dose for a 2 Gy fractionation schedule (EQD2) is shown Fig. 1. The values of the dose-response parameters of the Poisson-Linear-Quadratic-Time model are shown in Table 1.

Table 1. Dose-response parameters for the poisson-linear-quadratic model

Treatment	D_{50}/Gy	γ	α/β/Gy	α/Gy^{-1}	β/Gy^{-2}	T_{pot}/days	T_k/days
RT+5-FU	59.8	1.3	3.2	0.041	0.013	18.6	25.0
RT+GEM	54.5	1.4	4.6	0.053	0.011	34.2	37.2

4 Discussion

Dose-response curves for pancreatic tumors in function of BED values were first derived by Moraru et al. [42] that proposed a biophysical model to describe RR to radiochemotherapy. Durante et al. [43] derived dose-response models for different treatment modalities but have selected as endpoint overall survival. In this study, the Poisson-Linear-Quadratic-Time model describing RR in function of EQD2 was selected instead.

A smaller D_{50} value for concurrent radiochemotherapy with GEM compared with the D_{50} value obtained for RT+5-FU is consistent with the studies showing the higher efficacy of GEM compared to 5-FU [35, 36]. During optimization the parameters α and β were derived, resulting in a α/β smaller than commonly used [42, 44]. A low value of α/β would explain the good local tumor response obtained with hypofractionation schedules used with Stereotactic Body RT [45].

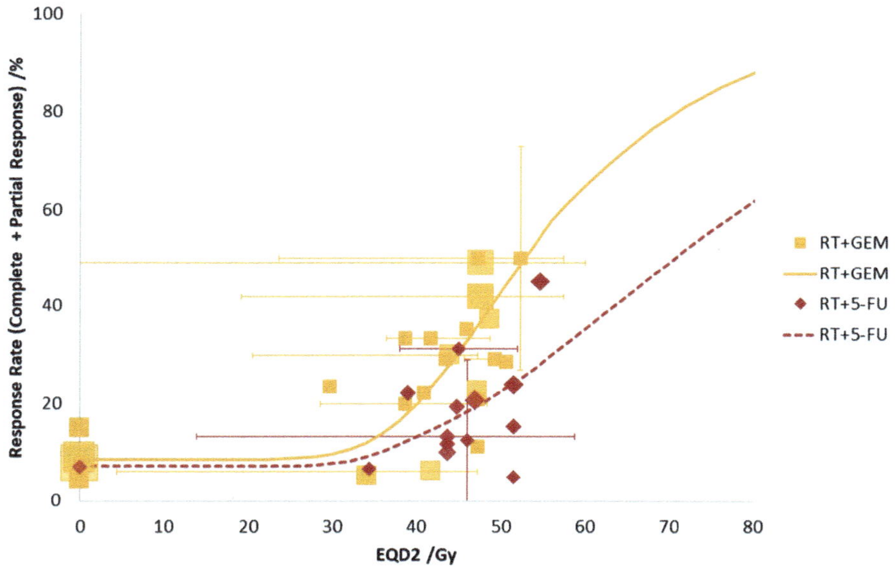

Fig. 1. Dose-response for unresectable locally advanced pancreatic tumors treated with concomitant radiochemotherapy with 5-FU or GEM. Dose-bars show the range of delivered doses and response-bars show its 95% confidence intervals reported by original authors [2–33].

Ideally, for the accurate derivation of dose-response curves, dosimetric data consisting of the delivered dose, corrected for the delivered fractions and the real overall treatment time, should be used for those patients that achieved complete or partial response. Unfortunately, discrimination between the dose delivered to patients that responded or not to the therapy was never made. Furthermore, although in each study dose prescription was uniform, the delivered dose varied due to the lack of compliance to therapy. Most authors reported a compliance to planned RT of approximately 80% but without reporting the delay in RT for each patient [2, 7, 23, 25]. Thus, corrections for the real overall treatment time were not made. Ideally, the 3-dimensional dose-distribution should be used to assess the dose delivered to the target volume. However, with conformal RT patients with pancreatic tumors are generally irradiated with the box technique for which a good dose homogeneity in the planning target volume is obtained. Despite the uncertainties characterizing the input data, certain reasonable approximations had to be made. The result is a good model fitting as indicated by the good correlation between the dose-response curves and their association with the clinical dose-response points (Fig. 1).

5 Conclusion

The parameters for the Poisson model describing tumor response to radiochemotherapy concomitant with the 5-FU and GEM agents were derived. The determined dose-response curves indicate a good correlation between applied treatments and outcome. The generation of large databases integrating the data of all treated patients is largely needed for the development of accurate models describing tumors, and normal tissues, to delivered therapies.

Acknowledgements. This work was partially funded by project grants UID/CTM/50025/2019, POCI-01-0145-FEDER-028030 and by the *Fundação para a Ciência e a Tecnologia* under project grant UID/Multi/00308/2019.

Conflicts of Interest. None.

References

1. Torgeson, A., Tao, R., Garrido-Laguna, I., et al.: Large database utilization in health outcomes research in pancreatic cancer: an update. J. Gastrointest. Oncol. **9**(6), 996–1004 (2018)
2. Fisher, B.J., Perera, F.E., Kocha, W., et al.: Analysis of the clinical benefit of 5-fluorouracil and radiation treatment in locally advanced pancreatic cancer. Int. J. Radiat. Oncol. Biol. Phys. **45**(2), 291–295 (1999)
3. Boz, G., Paoli, A., Innocente, R., et al.: Radiotherapy and continuous infusion 5-fluorouracil in patients with nonresectable pancreatic carcinoma. Int. J. Rad. Oncol. Biol. Phys. **51**(3), 736–740 (2001)
4. Osti, M.F., Costa, A.M., Bianciardi, F., et al.: Concomitant radiotherapy with protracted 5-fluorouracil infusion in locally advanced carcinoma of the pancreas: a phase II study. Tumori **87**(6), 398–401 (2001)
5. Morganti, A.G., Forni, F., Macchia, G., et al.: Chemoradiation of unresectable pancreatic carcinoma: impact of pretreatment hemoglobin level on patterns of failure. Strahlenther. Onkol. **179**(2), 87–92 (2003)
6. Morganti, A., Valentini, V., Macchia, G., et al.: 5-fluorouracil–based chemoradiation in unresectable pancreatic carcinoma: phase I-II dose-escalation study. Int. J. Radiat. Oncol. Biol. Phys. **59**(5), 1454–1460 (2004)
7. Ishii, H., Okada, S., Tokuuye, K., et al.: Protracted 5-fluorouracil infusion with concurrent radiotherapy as a treatment for locally advanced pancreatic carcinoma. Cancer **79**(8), 1516–1520 (1997)
8. Shinchi, H., Takao, S., Noma, H., et al.: Length and quality of survival after external-beam radiotherapy with concurrent continuous 5-fluorouracil infusion for locally unresectable pancreatic cancer. Int. J. Radiat. Oncol. Biol. Phys. **53**(1), 146–150 (2002)
9. Li, C.P., Chao, Y., Chi, K.H., et al.: Concurrent chemoradiotherapy treatment of locally advanced pancreatic cancer: gemcitabine versus 5-fluorouracil, a randomized controlled study. Int. J. Radiat. Oncol. Biol. Phys. **57**(1), 98–104 (2003)
10. Ueno, H., Okusaka, T., Ikeda, M., et al.: Phase I study of hyperfractionated radiation therapy with protracted 5-fluorouracil infusion in patients with locally advanced pancreatic cancer. Oncology **67**(3–4), 215–221 (2004)

11. Wilkowski, R., Boeck, S., Ostermaier, S., et al.: Chemoradiotherapy with concurrent gemcitabine and cisplatin with or without sequential chemotherapy with gemcitabine/cisplatin vs chemoradiotherapy with concurrent 5-fluorouracil in patients with locally advanced pancreatic cancer-a multi-centre randomised phase II study. Br. J. Cancer **101**(11), 1853–1859 (2009)
12. Pipas, J.M., Mitchell, S.E., Barth Jr., R.J., et al.: Phase I study of twice-weekly gemcitabine and concomitant external-beam radiotherapy in patients with adenocarcinoma of the pancreas. Int. J. Radiat. Oncol. Biol. Phys. **50**(5), 1317–1322 (2001)
13. Wolff, R.A., Evans, D.B., Gravel, D.M., et al.: Phase I trial of gemcitabine combined with radiation for the treatment of locally advanced pancreatic adenocarcinoma. Clin. Cancer Res. **7**(8), 2246–2253 (2001)
14. Lange, S.M., van Groeningen, C.J., Meijer, O.W., et al.: Gemcitabine-radiotherapy in patients with locally advanced pancreatic cancer. Eur. J. Cancer **38**(9), 1212–1217 (2002)
15. Okusaka, T., Ito, Y., Ueno, H., et al.: Phase II study of radiotherapy combined with gemcitabine for locally advanced pancreatic cancer. Br. J. Cancer **91**(4), 673–677 (2004)
16. Magnino, A., Gatti, M., Massucco, P., et al.: Phase II trial of primary radiation therapy and concurrent chemotherapy for patients with locally advanced pancreatic cancer. Oncology **68**(4–6), 493–499 (2005)
17. Oya, N., Shibuya, K., Sakamoto, T., et al.: Chemoradiotherapy in patients with pancreatic carcinoma: phase-I study with a fixed radiation dose and escalating doses of weekly gemcitabine. Pancreatology **6**(1–2), 109–116 (2006)
18. Brade, A., Brierley, J., Oza, A., et al.: Concurrent gemcitabine and radiotherapy with and without neoadjuvant gemcitabine for locally advanced unresectable or resected pancreatic cancer: a phase I-II study. Int. J. Radiat. Oncol. Biol. Phys. **67**(4), 1027–1036 (2007)
19. Igarashi, H., Ito, T., Kawabe, K., et al.: Chemoradiotherapy with twice-weekly administration of low-dose gemcitabine for locally advanced pancreatic cancer. World J. Gastroenterol. **14**(34), 5311–5315 (2008)
20. Maemura, K., Shinchi, H., Noma, H., et al.: Comparison of hyper-fractionated accelerated and standard fractionated radiotherapy with concomitant low-dose gemcitabine for unresectable pancreatic cancer. Anticancer Res. **28**(4C), 2369–2372 (2008)
21. Girard, N., Mornex, F., Bossard, N., et al.: Estimating optimal dose of twice-weekly gemcitabine for concurrent chemoradiotherapy in unresectable pancreatic carcinoma: mature results of GEMRT-01 Phase I trial. Int. J. Radiat. Oncol. Biol. Phys. **77**(5), 1426–1432 (2010)
22. Loehrer Sr., P.J., Feng, Y., Cardenes, H., et al.: Gemcitabine alone versus gemcitabine plus radiotherapy in patients with locally advanced pancreatic cancer: an eastern cooperative oncology group trial. J. Clin. Oncol. **29**(31), 4105–4112 (2011)
23. Shibuya, K., Oya, N., Fujii, T., et al.: Phase II study of radiation therapy combined with weekly low-dose gemcitabine for locally advanced, unresectable pancreatic cancer. Am. J. Clin. Oncol. **34**(2), 115–119 (2011)
24. Cardenes, H.R., Moore, A.M., Johnson, C.S., et al.: A phase II study of gemcitabine in combination with radiation therapy in patients with localized, unresectable, pancreatic cancer: a Hoosier Oncology Group study. Am. J. Clin. Oncol. **34**(5), 460–465 (2011)
25. Mattiucci, G.C., Morganti, A.G., Valentini, V., et al.: External beam radiotherapy plus 24-hour continuous infusion of gemcitabine in unresectable pancreatic carcinoma: long-term results of a phase II study. Int. J. Radiat. Oncol. Biol. Phys. **76**(3), 831–838 (2010)
26. Small Jr., W., Berlin, J., Freedman, G.M., et al.: Full-dose gemcitabine with concurrent radiation therapy in patients with nonmetastatic pancreatic cancer: a multicenter phase II trial. J. Clin. Oncol. **26**(6), 942–947 (2008)
27. Huang, P.I., Chao, Y., Li, C.P., et al.: Efficacy and factors affecting outcome of gemcitabine concurrent chemoradiotherapy in patients with locally advanced pancreatic cancer. Int. J. Radiat. Oncol. Biol. Phys. **73**(1), 159–165 (2009)

28. Tada, M., Arizumi, T., Nakai, Y., et al.: Efficacy of gemcitabine for locally advanced pancreatic cancer: comparison with 5-fluorouracil-based chemoradiotherapy. Chemotherapy **54**(4), 302–308 (2008)
29. Yanagimoto, H., Ishii, H., Nakai, Y., et al.: Improved survival with combined gemcitabine and S-1 for locally advanced pancreatic cancer: pooled analysis of three randomized studies. J. Hepatobiliary Pancreat. Sci. **21**(10), 761–766 (2014)
30. Louvet, C., Labianca, R., Hammel, P., et al.: Gemcitabine in combination with oxaliplatin compared with gemcitabine alone in locally advanced or metastatic pancreatic cancer: results of a GERCOR and GISCAD phase III trial. J. Clin. Oncol. **23**(15), 3509–3516 (2005)
31. Evans, T.R.J., Van Cutsem, E., Moore, M.J., et al.: Phase 2 placebo-controlled, double-blind trial of dasatinib added to gemcitabine for patients with locally-advanced pancreatic cancer. Ann. Oncol. **28**(2), 354–361 (2017)
32. Cengiz, M., Zorlu, F., Yalcin, S., et al.: Concurrent gemcitabine and radiotherapy for locally advanced pancreatic cancer. Med. Oncol. **24**(2), 239–243 (2007)
33. Ogawa, K., Ito, Y., Hirokawa, N., et al.: Concurrent radiotherapy and gemcitabine for unresectable pancreatic adenocarcinoma: impact of adjuvant chemotherapy on survival. Int. J. Radiat. Oncol. Biol. Phys. **83**(2), 559–565 (2012)
34. Cullinan, S., Moertel, C.G., Wieand, H.S., et al.: A phase III trial on the therapy of advanced pancreatic carcinoma. Evaluations of the Mallinson regimen and combined 5-fluorouracil, doxorubicin, and cisplatin. Cancer **65**(10), 2207–2212 (1990)
35. Huang, J., Robertson, J.M., Margolis, J., et al.: Long-term results of full-dose gemcitabine with radiation therapy compared to 5-fluorouracil with radiation therapy for locally advanced pancreas cancer. Radiother. Oncol. **99**(2), 114–119 (2011)
36. Zhu, C.P., Shi, J., Chen, Y.X., et al.: Gemcitabine in the chemoradiotherapy for locally advanced pancreatic cancer: a meta-analysis. Radiother. Oncol. **99**(2), 108–113 (2011)
37. Chen, J., Chen, L., Yu, J., et al.: Meta-analysis of current chemotherapy regimens in advanced pancreatic cancer to prolong survival and reduce treatment-associated toxicities. Mol. Med. Rep. **19**(1), 477–489 (2019)
38. Tozzi, A., Comito, T., Alongi, F., et al.: SBRT in unresectable advanced pancreatic cancer: preliminary results of a mono-institutional experience. Radiat. Oncol. **8**, 148 (2013)
39. World Health Organisation. WHO handbook for reporting results of cancer treatment. Offset Pub. 48. World Health Organisation, Geneva (1979)
40. Therasse, P., Arbuck, S.G., Eisenhauer, E.A., et al.: New guidelines to evaluate the response to treatment in solid tumors. European Organization for Research and Treatment of Cancer, National Cancer Institute of the United States, National Cancer Institute of Canada. J. Natl. Cancer Inst. **92**, 205–216 (200)
41. Ferreira, B.C., Lopes, M.C., Mateus, J., et al.: Radiobiological evaluation of forward and inverse IMRT using different fractionations for head and neck tumours. Radiat. Oncol. **5**, 57 (2010)
42. Moraru, I.C., Tai, A., Erickson, B., Li, X.A.: Radiation dose responses for chemoradiation therapy of pancreatic cancer: an analysis of compiled clinical data using biophysical models. Pract. Radiat. Oncol. **4**(1), 13–19 (2014)
43. Durante, M., Tommasino, F., Yamada, S.: Modeling combined chemotherapy and particle therapy for locally advanced pancreatic cancer. Front. Oncol. **5**, 145 (2015)
44. Chapman, J.D.: Can the two mechanisms of tumor cell killing by radiation be exploited for therapeutic gain? J. Radiat. Res. **55**(1), 2–9 (2014)
45. De Bari, B., Porta, L., Mazzola, R., et al.: Hypofractionated radiotherapy in pancreatic cancer: lessons from the past in the era of stereotactic body radiation therapy. Crit. Rev. Oncol. Hematol. **103**, 49–61 (2016)

Comparison of Different Radiotherapy Techniques for Locally Advanced Pancreatic Tumors

Adriana Gomes[1(✉)], Darlene Rodrigues[2], and Brigida C. Ferreira[1,3,4]

[1] School of Health, Polytechnic of Porto, Porto, Portugal
adrianafariagomes.sa@gmail.com, bcf@ess.ipp.pt
[2] Radiotherapy Department, Centro Hospitalar Universitário de São João, Porto, Portugal
[3] Institute for Systems Engineering and Computers at Coimbra, Coimbra University, Coimbra, Portugal
[4] I3N Physics Department, Aveiro University, Aveiro, Portugal

Abstract. Radiotherapy (RT) associated with systemic therapy is the standard treatment for Locally Advanced Pancreatic Cancer (LAPC). The aim of this study was to compare the efficacy of different RT techniques using the clinical data reported in the literature. Clinical data was collected from scientific papers searched in the databases PubMed and ScienceDirect. Thirty-four documents published between 1997 and 2015 were found and met the inclusion criteria: locally advanced adenocarcinoma, unresectable and no metastasis. Values of Complete Response (CR), Partial Response (PR), Stable Disease (SD), Progression Disease, Progression Free Survival (PFS), and Overall Survival (OS) for Three-Dimensional Conformal Radiation Therapy (3DCRT), Intensity Modulated Radiation Therapy (IMRT) and Stereotactic Body Radiotherapy (SBRT) in the treatment of LAPC were collected. For all RT techniques, Response Rate (RR), defined as the sum of CR and PR, was for 3DCRT 25.2% ± 9.5 [range: 5.0%–49.0%], for IMRT 33.5% ± 10.5 [range: 10.6%–55.6%] and for SBRT 52.2% ± 17.7 [range: 13.3%–69.5%]. For all studied techniques, Local Control (LC), defined as the sum of RR and SD, ranged from 47% to 100%; PFS ranged from 4 to 12 months and OS ranged from 6 to 20 months. A significant improvement in overall response rate was obtained with SBRT compared to 3DRCT and IMRT. However, LC, PFS and OS were similar among the three RT techniques.

Keywords: Pancreatic cancer · Radiotherapy techniques · Local control · Survival

1 Introduction

Pancreatic tumors are characterized by their aggressive tumor biology and rapid disease progression. In most cases, these factors lead to diagnoses in advanced stages, since the symptoms in the initial phase are nonspecific. For advanced stages of the pancreatic tumor, surgical resection is no longer possible due to the size of the lesion and the

J. Henriques et al. (Eds.): MEDICON 2019, IFMBE Proceedings 76, pp. 1283–1290, 2020.
https://doi.org/10.1007/978-3-030-31635-8_157

invasion of adjacent structures. Chemotherapy is often used for advanced cases, but this treatment modality does not present satisfactory results [1, 2]. New therapeutic approaches, combining different chemotherapy drugs and/or Radiation Therapy (RT), were used aiming to improve treatment outcome [2].

For unresectable pancreatic cancer, modern RT techniques delivered concomitantly with chemotherapy may be an alternative treatment for advanced disease [3, 4]. In the treatment of pancreatic tumors Tridimensional Conformal Radiation Therapy (3DCRT) was the first RT technique used. With the development of Intensity Modulated RT (IMRT), high dose gradients are achieved in the target volume limits increasing the protection of adjacent risk organs [3, 5]. With Stereotactic Body RT (SBRT), hypofractionation schemes are generally used and therefore the dose in the surrounding normal tissues must be highly redistributed while a high conformation is also produced [6]. Some of the clinical and practical benefits of SBRT potentially being capable of enhancing local control are: the possibility of the escalation of the prescription dose; the lower number of fractions and therefore shorter overall treatment time; lower probability of repopulation of tumor cells during RT and minimal interference with administration of chemotherapy [6].

Given the technological advances in RT planning and delivery, treatment efficacy of 3DCRT, IMRT and SBRT for locally advanced pancreatic adenocarcinoma was compared in this study using the clinical data reported in the literature.

2 Materials and Methods

Clinical data was collected from the literature searching papers in the databases PubMed and ScienceDirect, using the keywords: *Radiotherapy, Pancreas, Survival and Local control.* Thirty-four papers published between 1997 and 2015 met the inclusion criteria: locally advanced adenocarcinoma, unresectable and no metastasis [6, 9–40]. Endpoints used to evaluate the effectiveness of the different RT techniques were: Local Control (LC), defined as the sum of the rate of Complete Response (CR), Partial Response (PR) and Stable Disease (SD); Response Rate (RR), defined as the sum of CR and PR; Progression Disease (PD); Progression-Free Survival (PFS) and Overall survival (OS) - according to RECIST and WHO criteria [7, 8].

This study included 676, 168 and 229 patients irradiated with 3DCRT, IMRT and SBRT, respectively. In concomitant radiochemotherapy, the most commonly used chemotherapy agents were Gemcitabine and 5-fluorouracil but others like Cisplatine, Paclitaxel, Capecitabine and mFOLFOX were also used. Delivered dose in 3DCRT ranged from 30 to 59.4 Gy delivered in 3–33 fractions ranging from 1.8 to 10 Gy. For IMRT, the prescribed dose varied from 45 to 60 Gy delivered in 15–28 fractions of 1.8 to 3 Gy and in SBRT, mostly delivered with the Cyberknife system, the prescribed dose ranged from 22 to 50 Gy delivered in 1–5 fractions of 5 to 22 Gy [6, 9–40].

3 Results

Figure 1 shows the results of each report for the endpoints evaluated. For 3DCRT, the weighted average overall response rate was 25.2% ± 9.5% [range: 5.0%–49.0%], for IMRT it was 33.5% ± 10.5% [range: 10.6%–55.6%] and for SBRT it was 52.2% ± 17.7% [range: 13.3%–69.5%] (Fig. 2). These differences are due to an increase in both complete and partial tumor response and a significant reduction in the rate of stable disease. Average values for complete response for 3DCRT and IMRT were: 1.6% ± 2.5% [range: 0.0%–6.4%] and 2.3% ± 3.8% [range: 0.0%–9.5%], respectively, which increased for SBRT to 8.5% ± 6.9 [range: 0.0%–20.0%]. Average partial response for 3DCRT, IMRT and SBRT was 23.6% ± 9.5, 31.2% ± 11.1 and 43.7% ± 19.7, respectively.

For 3DCRT, IMRT and SBRT, LC was 82.4% ±14.3 [range: 47.0%–100%], 91.6% ± 13.8 [range: 57.1%–100%] and 83.0% ± 10.2 [range: 60.0%–95.0%], respectively.

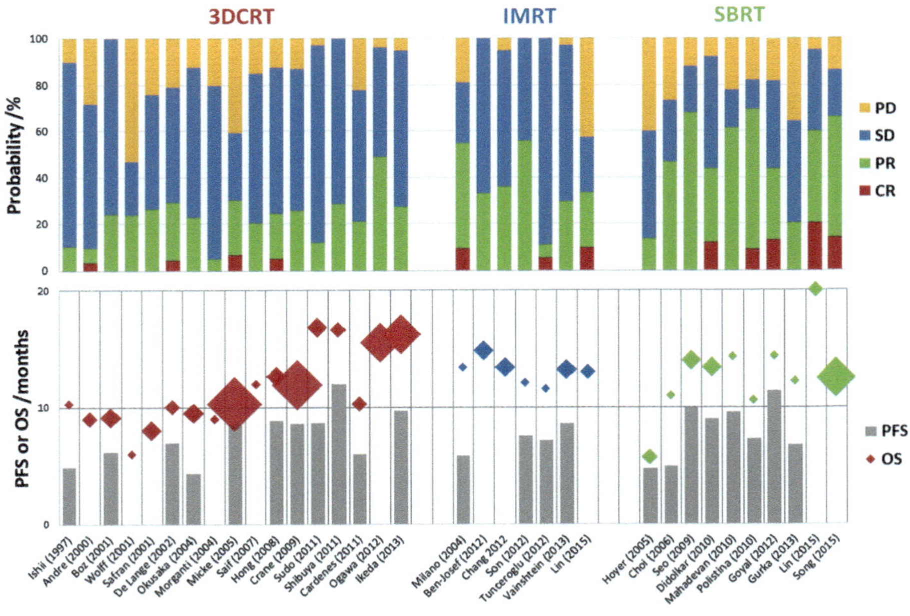

Fig. 1. (Upper panel) Values of PD, SD, CR and PR for 3DCRT, IMRT and SBRT, respectively. (Lower panel), Median OS and PFS time for the three RT techniques evaluated. The dimension of the points is proportional to the number of patients included in each study.

For 3DCRT, median OS ranged from 6.9–16.8 months, for IMRT from 11.6–14.8 months and for SBRT from 5.7–20.0 months (Fig. 1, lower panel). PFS values, reported only in a few studies, were similar among the three techniques varying from 4.4–12.0 months, 5.9–8.6 months and 4.8–11.4 months for 3DCRT, IMRT and SBRT, respectively.

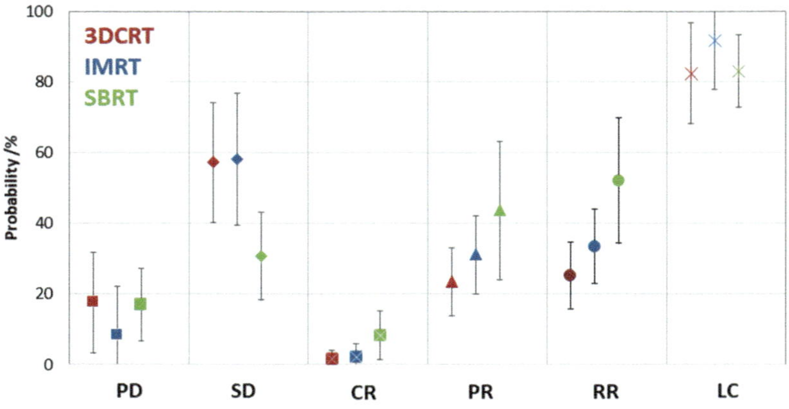

Fig. 2. Weighted average, and standard deviation, for each tumor response endpoint evaluated for 3DCRT, IMRT and SBRT

4 Discussion

Pancreatic cancer is still one of the most challenging oncological cases. The combination of several chemotherapy agents combined or not with RT has shown mild therapeutic advantages [41]. Important advances in RT allowed the clinical implementation of highly sophisticated techniques like IMRT and SBRT that produce a high conformation of the prescription isodose to the target volume and the protection of surrounding normal tissues. The implementation of IMRT increased average overall response rate by 8%, while SBRT increased RR by 27% compared to 3DCRT. This increase in RR was mostly obtained due to an increase in the rate of partial tumor response of about 20% and an increase of 7% in the rate of complete response compared to 3DCRT. However, the rate of disease progression remained similar to those obtained with 3DCRT (Fig. 2). The further improvement in local tumor response obtained with SBRT, compared to IMRT, suggests that hypofractionation schemes might be advantageous in the treatment of pancreatic cancer.

Despite the significant dosimetric benefits obtained with advanced RT techniques, the results presented here indicate that the increase in RR obtained with SBRT did not translate into improvements in LC, PFS and OS. The small differences between the efficacy of the different techniques for these endpoints may be due to the advanced stage of the disease at the time of diagnosis needing more aggressive local and systemic treatments.

The efficacy of a therapy depends not only on tumor cure but also on a low rate of treatment side-effects. Although, toxicity was not evaluated in this study, IMRT reduced RT complications compared to 3DCRT and a further reduction was obtained with SBRT [3, 42]. However, multi-fraction SBRT, using 3 to 5 fractions, was recommended over single fraction schedules [43, 44].

A fair comparison on the efficacy of different therapies should ideally be made within the scope of randomized clinical trials. However, most studies used in this analysis were retrospective and non-randomized, including a small number of patients with significantly different follow-up times. Furthermore, there was a lack of systematization in reporting. Generally, different endpoints, using slightly different definitions, were reported among the several studies generating incomplete research databases (Fig. 1). The homogeneous cohort needed, to eliminate bias resulting also from an unbalanced number of patients, is thus extremely difficult to achieve.

The discrimination between the different chemotherapy regimens and the delivered doses of chemotherapy and RT was not made. Multi-agent chemotherapy is nowadays recommended due to better OS, PFS and RR, compared to single agent, but higher rates of side-effects were reported with polychemotherapy [45, 46]. Furthermore, a distinct radiation dose-response effect was observed with hypofractionation SBRT [47, 48]. These factors may impact significantly in treatment outcome and therefore are being analysed in detail elsewhere [49]. Accurate biological models describing the response of patients with LAPC to the delivered treatments are useful to predict the efficacy of new therapies prior to clinical trials [47]. Additionally, these are also needed for biologically optimized RT [50].

Acknowledgments. This work partially funded by project grant UID/CTM/50025/2019.

Conflicts of interest. None.

References

1. Salgado, M., Arévalo, S., Hernando, O., et al.: Management of unresectable, locally advanced pancreatic adenocarcinoma. Clin. Transl. Oncol. **20**(2), 113–118 (2017)
2. Balaban, E., Mangu, P., Khorana, A., et al.: Locally advanced, unresectable pancreatic cancer: American society of clinical oncology clinical practice guideline. J. Clin. Oncol. **34** (22), 2654–2657 (2016). Asco Special Article
3. Reese, A., Lu, W., William, R.: Utilization of intensity-modulated radiation therapy and image-guided radiation therapy in pancreatic cancer: is it beneficial? Semin. Radiat. Oncol. **24**(2), 132–139 (2014)
4. Wang, F., Kumar, P.: The role of radiotherapy in management of pancreatic cancer. J. Gastrointestial Oncol. **2**(3), 157–162 (2011)
5. Nandi, M., Biswal, S., Bhattacharya, J., et al.: Dosimetric comparison of rapid arc, intensity modulated radiotherapy and 3D conformal radiotherapy plans in periampullary and pancreatic head carcinomas. J. Pancreas (JOP) **17**(5), 466–467 (2016)
6. Chang, D., Schellenberg, D., Shen, J., et al.: Stereotactic radiotherapy for unresectable adenocarcinoma of the pancreas. Am. Cancer Soc. **115**(3), 665–672 (2009)
7. Eisenhauer, E., Therasse, P., Bogaerts, J., et al.: New response evaluation criteria in solid tumours: revised RECIST guideline (version 1.1). Eur. J. Cancer **45**(1), 228–247 (2009)
8. Miler, B., Hoogstraten, B., Winkler, A.: WHO criteria in tumour response. Reporting results of cancer treatment. Cancer **47**(1), 207–214 (1981)

9. Ishii, H., Okada, S., Tokuuye, K., et al.: Protracted 5-fluorouracil infusion with concurrent radiotherapy as a treatment for locally advanced pancreatic carcinoma. Cancer **79**(8), 1516–1520 (1997)
10. André, T., Balosso, J., Louvet, C., et al.: Combined radiotherapy and chemotherapy (cisplatin and 5-fluorouracil) as palliative treatment for localized unresectable or adjuvant treatment for resected pancreatic adenocarcinoma: results of a feasibility study. Int. J. Radiat. Oncol. Biol. Phys. **46**(4), 903–911 (2000)
11. Safran, H., Moore, T., Iannitti, D., et al.: Paclitaxel and concurrent radiation for locally advanced pancreatic cancer. Int. J. Radiat. Oncol. Biol. Phys. **49**(5), 1275–1279 (2001)
12. Boz, G., Paoli, A., Innocente, R., et al.: Radiotherapy and continuous infusion 5-fluorouracil in patients with nonresectable pancreatic carcinoma. Int. J. Radiat. Oncol. Biol. Phys. **51**(3), 736–740 (2001)
13. Morganti, A., Valentini, V., Macchia, G., et al.: 5-fluorouracil–based chemoradiation in unresectable pancreatic carcinoma: phase I-II dose-escalation study. Int. J. Radiat. Oncol. Biol. Phys. **59**(5), 1454–1460 (2004)
14. Okusaka, T., Ito, Y., Ueno, H., et al.: Phase II study of radiotherapy combined with gemcitabine for locally advanced pancreatic cancer. Br. J. Cancer **91**(4), 673–677 (2004)
15. Saif, M., Black, G., Roy, S., et al.: Phase II study of capecitabine with concomitant radiotherapy for patients with locally advanced pancreatic cancer: up-regulation of thymidine phosphorylase. Cancer J. **13**(4), 247–255 (2007)
16. Hong, P., Park, Y., Jeon, J., et al.: Weekly full-dose gemcitabine and single-dose cisplatin with concurrent radiotherapy in patients with locally advanced pancreatic cancer. Br. J. Cancer **98**(5), 881–887 (2008)
17. Crane, C., Winter, K., Regine, W., et al.: Phase II study of bevacizumab with concurrent capecitabine and radiation followed by maintenance gemcitabine and bevacizumab for locally advanced pancreatic cancer: radiation therapy oncology group RTOG 0411. J. Clin. Oncol. **27**(25), 4096–4101 (2009)
18. Shibuya, K., Oya, N., Fujii, T., et al.: Phase II study of radiation therapy combined with weekly low-dose gemcitabine for locally advanced, unresectable pancreatic cancer. Am. J. Clin. Oncol. **34**(2), 115–119 (2011)
19. Cardenes, R., Moore, M., Johnson, S., et al.: A phase II study of gemcitabine in combination with radiation therapy in patients with localized, unresectable, pancreatic cancer: a Hoosier Oncology Group study. Am. J. Clin. Oncol. **34**(5), 460–465 (2011)
20. Sudo, K., Yamaguchi, T., Ishihara, T., et al.: Phase II study of oral S-1 and concurrent radiotherapy in patients with unresectable locally advanced pancreatic cancer. Int. J. Radiat. Oncol. Biol. Phys. **80**(1), 119–125 (2011)
21. Milano, M., Chmura, S., Garofalo, M., et al.: Intensity–modulated radiotherapy in treatment of pancreatic and bile duct malignancies: toxicity and clinical outcome. Int. J. Radiat. Oncol. Biol. Phys. **59**(2), 445–453 (2004)
22. Son, S., Song, J., Choi, B., et al.: The technical feasibility of an image-guided intensity-modulated radiotherapy (IG-IMRT) to perform a hypofractionated schedule in terms of toxicity and local control for patients with locally advanced or recurrent pancreatic cancer. Radiat. Oncol. **203**(7), 1–8 (2012)
23. de Lange, M., van Groeningen, J., Meijer, W., et al.: Gemcitabine-radiotherapy in patients with locally advanced pancreatic cancer. Eur. J. Cancer **38**(9), 1212–1217 (2002)
24. Hoyer, M., Roed, H., Sengelov, L., et al.: Phase–II study on stereotactic radiotherapy of locally advanced pancreatic carcinoma. Green J. Radiotheraphy Oncol. **76**(1), 48–53 (2005)
25. Mahadevan, A., Jain, S., Goldstein, M., Miksad, R., et al.: Stereotactic body radiotherapy and gemcitabine for locally advanced pancreatic cancer. Int. J. Radiat. Oncol. Biol. Phys. **78** (3), 735–742 (2010)

26. Polistina, F., Costantin, G., Casamassima, F., et al.: Unresectable locally advanced pancreatic cancer: a multimodal treatment using neoadjuvant chemoradiotherapy (gemcitabine plus stereotactic radiosurgery) and subsequent surgical exploration. Soc. Surg. Oncol. **17**(8), 2092–2101 (2010)
27. Didolkar, M., Coleman, C., Brenner, M., et al.: Image–guided stereotactic radiosurgery for locally advanced pancreatic adenocarcinoma results of first 85 patients. J. Gastrointestinal Surg. **14**(10), 1547–1559 (2010)
28. Goyal, K., Einstein, D., Ibarra, R., et al.: Stereotactic body radiation therapy for nonresectable tumors of the pancreas. J. Surg. Res. **174**(2), 319–325 (2012)
29. Gurka, M., Colins, S., Slack, R., et al.: Stereotactic body radiation therapy with concurrent full-dose gemcitabine for locally advanced pancreatic cancer: a pilot trial demonstrating safety. Radiat. Oncol. **44**(8), 1–9 (2013)
30. Wolff, A., Evans, B., Gravel, M., et al.: Phase I trial of gemcitabine combined with radiation for the treatment of locally advanced pancreatic adenocarcinoma. Clin. Cancer Res. **7**(8), 2246–2253 (2001)
31. Micke, O., Hesselmann, S., Bruns, F., et al.: Results and follow-up of locally advanced cancer of the exocrine pancreas treated with radiochemotherapy. Anticancer Res. **25**(3), 1523–1530 (2005)
32. Ogawa, K., Ito, Y., Hirokawa, N., et al.: Concurrent radiotherapy and gemcitabine for unresectable pancreatic adenocarcinoma: impact of adjuvant chemotherapy on survival. Int. J. Radiat. Oncol. Biol. Phys. **83**(2), 559–565 (2012)
33. Ikeda, M., Ioka, T., Ito, Y., Tonemoto, N., et al.: A multicenter phase II trial of S-1 with concurrent radiation therapy for locally advanced pancreatic cancer. Int. J. Radiat. Oncol. Biol. Phys. **85**(1), 163–169 (2013)
34. Ben-Josef, E., Schipper, M., Francis, I., et al.: A phase I/II trial of intensity-modulated radiation (IMRT) dose escalation with concurrent fixed- dose rate gemcitabine (FDR-G) in patients with unresectable pancreatic cancer. Int. J. Radiat. Oncol. Biol. Phys. **84**(5), 166–1171 (2012)
35. Tunceroglu, A., Park, J., Balasubramanlan, S., et al.: Dose-painted intensity modulated radiation therapy improves local control for locally advanced pancreas cancer. Int. Sch. Res. Netw. **7**(1), 1–6 (2012)
36. Vainshtein, J., Schipper, M., Zalupski, M., et al.: Prognostic significance of carbobydrate antigen 19-9 in unresectable locally advanced pancreatic cancer treated with dose-escalated intensity modulated radiation therapy and concurrent full-dose gemcitabine: analysis of a prospective phase ½ dose escalation study. Int. J. Radiat. Oncol. Biol. Phys. **86**(1), 96–101 (2013)
37. Lin, J., Jen, Y., Li, M., et al.: Comparing outcomes of stereotactic body radiotherapy with intensity-modulated radiotherapy for patients with locally advanced unresectable pancreatic cancer. Eur. J. Gastroenterol. Hepatol. **27**(3), 259–264 (2015)
38. Choi, C., Kim, M., Cho, C., et al.: Locally advanced, unresectable pancreatic cancer treated by stereotactic radiation therapy. Radiat. Oncol. J. **24**(1), 11–20 (2006)
39. Seo, Y., Kim, M., Yoo, S., et al.: Stereotactic body radiation therapy boost in locally advanced pancreatic cancer. Int. J. Radiat. Oncol. Biol. Phys. **75**(5), 1456–1461 (2009)
40. Song, Y., Yuan, Z., Li, F., et al.: Analysis of clinical efficacy of CyberKnife® treatment for locally advanced pancreatic cancer. OncoTargets Ther. **8**, 1427–1431 (2015)
41. Chen, J., Chen, L., Yu, J., et al.: Meta-analysis of current chemotherapy regimens in advanced pancreatic cancer to prolong survival and reduce treatment-associated toxicities. Mol. Med. Rep. **19**(1), 477–489 (2019)
42. Park, J., Hajj, C., Reyngold, M., et al.: Stereotactic body radiation vs. intensity-modulated radiation for unresectable pancreatic cancer. Acta Oncol. **56**(12), 1746–1753 (2017)

43. Ng, S.P., Koay, E.J.: Current and emerging radiotherapy strategies for pancreatic adenocarcinoma: stereotactic, intensity modulated and particle radiotherapy. Ann. Pancreat. Cancer. **1**(22), 1–10 (2018)
44. Petrelli, F., Comito, T., Ghidini, A., et al.: Stereotactic body radiation therapy for locally advanced pancreatic cancer: a systematic review and pooled analysis of 19 trials. Int. J. Radiat. Oncol. Biol. Phys. **97**(2), 313–322 (2017)
45. Petrelli, F., Coinu, A., Borgonovo, K., Cabiddu, M., Ghilardi, M., Barni, S.: Polychemotherapy or gemcitabine in advanced pancreatic cancer: a meta-analysis. Dig. Liver Dis. **46**(5), 452–459 (2014)
46. Jin, S.F., Fan, Z.K., Pan, L., Jin, L.M.: Gemcitabine-based combination therapy compared with gemcitabine alone for advanced pancreatic cancer: a meta-analysis of nine randomized controlled trials. Hepatobiliary Pancreat. Dis. Int. **16**(3), 236–244 (2017)
47. Moraru, I., Tai, A., Erickson, B., et al.: Radiation dose responses for chemoradiation therapy of pancreatic cancer: an analysis of compiled clinical data using biophysical models. Am. Soc. Radiat. Oncol. **4**(1), 13–18 (2013)
48. Gomes, A.: Impacto da Radioterapia no Tratamento do Adenocarcinoma Localmente Avançado do Pâncreas. Bachelor thesis. School of Health –Polytechnic of Porto (2018)
49. Costa Ferreira, B., Dias, J., Gomes, A., et al.: Dose-response to different radiochemotherapy regimens in locally advanced pancreatic cancer. In: MEDICON2019 (2019)
50. Ferreira, B.C.: Biological optimization of angle of incidence and intensity modulation in breast and cervix cancer radiation therapy. Ph.D. Thesis. Stockholm University, Sweden (2004)

Optimal Location of Novel Robotic Prostrate Cancer Biopsy and Brachytherapy Treatment Devices

Sina Firouzy[1(✉)], Dylan Jones[2], and Ashraf Labib[1]

[1] Faculty of Business and Law, University of Portsmouth, Portsmouth, UK
{sina.firouzy,ashraf.labib}@port.ac.uk
[2] School of Mathematics and Physics,
University of Portsmouth, Portsmouth, UK
dylan.jones@port.ac.uk

Abstract. This paper details research towards the optimal placement of a novel robotic device for the detection and treatment of prostate cancer, via biopsy and brachytherapy respectively. A methodology for analysis of available data in order to determine geographical areas with relatively high prevalence of prostate cancer and low access to treatment is proposed. Areas in the South of the UK with high values of these indices are highlighted. The development of single and multiple criteria optimization models based on the new metric in order to optimally locate a small number of future prototype treatment devices is discussed. Discussions, conclusions and avenues for future research are given.

Keywords: Healthcare technologies · Cancer detection and treatment · Optimization

1 Introduction

This study focuses on finding the optimal locations for a small number of prototype robotic devices in the South of the United Kingdom for the detection and treatment of prostate cancer. This problem is considered as part of a project for the development of the robotic device, funded by the European Union 2-Seas Interreg scheme under the CoBra (Co-operative Brachytherapy) [1] project.

The remainder of this paper is divided into three sections which describe the data analysis (Sect. 2) and model formulation and solution (Sects. 3), and conclusions and suggestions for future work (Sect. 4).

2 Data Analysis

2.1 Identification of the Prostate Cancer Occurrence Hotspot

Two distinct quantities can be used to assess the prevalence of prostate cancer, namely prostate cancer incidence, and prostate cancer mortality rate. Prostate cancer incidence represent the number of patients (per year) who are *diagnosed* with prostate cancer.

J. Henriques et al. (Eds.): MEDICON 2019, IFMBE Proceedings 76, pp. 1291–1295, 2020.
https://doi.org/10.1007/978-3-030-31635-8_158

The diagnosis methods used are not consistent throughout the UK, and the efficacy of different methods are different; this renders incidence rate unsuitable for comparing prostate cancer prevalence in different areas of the UK. Therefore, mortality rate is a better measure for this purpose [2], and is used in this analysis. The only risk factors of prostate cancer currently known are age and ethnicity. Deprivation does not seem to affect the risk of prostate cancer [2, 3]. A quantitative breakdown of the risk is available, for age groups with 5 year increments [3], which has been used in this study. The risk is very low before the age of 50, after which the mortality rate monotonically increases with age.

To exclude the effect of different age profiles in different populations, prostate cancer mortality rates are commonly expressed as Age Standard (AS) rates, which reveal the real differences between populations which can be attributed to factors other than age.

Ethnicity is the other factor relevant to the risk of prostate cancer. Compared to white males, Age-Standard (AS) prostate cancer mortality rate is significantly lower in Asian men, and significantly higher in Black males. For this reason, the ethnic demographics of the population may have to be taken into account in order to obtain a reliable estimate of prostate cancer prevalence in different areas. However, a study of prostate cancer mortality rates in different areas of Britain (analyzed at ward[1] level) revealed a *flat* distribution map; i.e. the AS mortality rate is quite uniform in the UK. This means that the differences in ethnical demographics of different areas within the UK can be ignored in estimating prostate cancer prevalence, and age demographics alone are sufficient for that purpose. Therefore, age has been used as the sole estimating factor in this study, as follows:

1. The geographic area of study (the southern part of England, in the 2 seas zone) was first broken down to the level of postcode district (i.e. PO2).
2. The age demographics of the population in each postcode district were then obtained from the Office of National Statistics [4]. The data contains the number of male residents of an area, within each age range. Each age range spans over five years (e.g. 50–54 years old).
3. The Age-Standard risk rate of prostate cancer for different age ranges [3] was used together with age demographics of each postcode district to produce an estimate. To do this, the number of the male population within each age range was multiplied by the prostate cancer rate corresponding to that age range, as follows:

$$\sum\nolimits_{i=1}^{n} M_i.P_i$$

where M_i and P_i are the male population and the Prostate cancer rate for the i'th age group, respectively, and n is the number of age groups.
This was done for all age ranges, and the sum of all products represents the estimated rate of prostate cancer for the given postcode district.

[1] A ward is a district level geographic division in The UK.

4. The procedure was executed for all postcode district, to produce a map of prostate cancer dispersion in the area studied.

The map of the area studied is shown in Fig. 1, together with the centroids of the postcode districts in the area of study. The centroids are significant because the distance from each postcode district (to a given treatment center) was measured from its centroid. The prostate cancer prevalence map developed in this study is represented in a color intensity graph in Fig. 2.

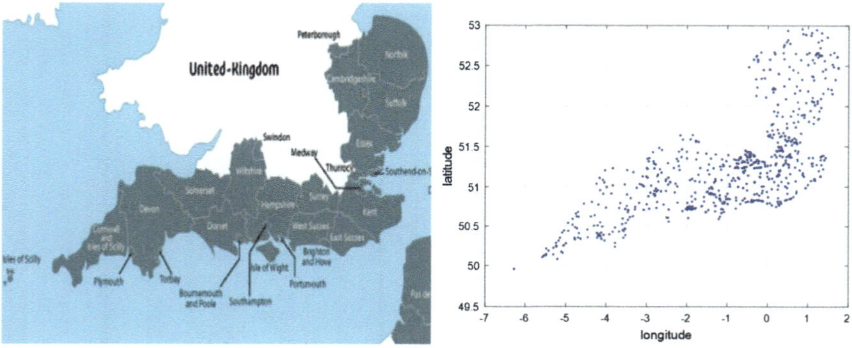

Fig. 1. The Southern parts of England within the 2 Seas area (left), and the centroids of the postcode districts within the 2 seas area in the UK (right)

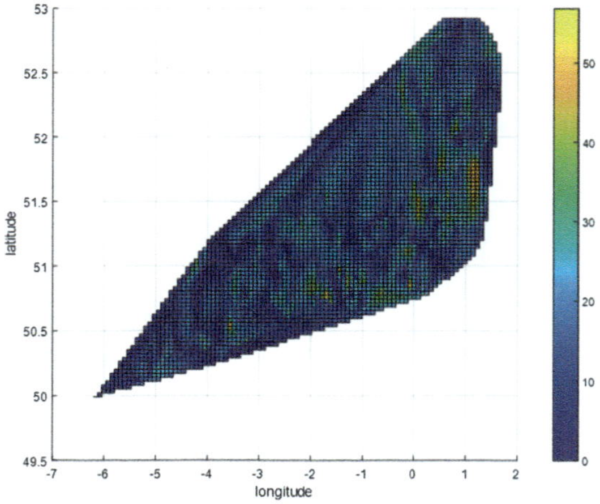

Fig. 2. The color-intensity graph of the estimated prostate cancer rate (patient/year) in each postcode district within the 2 seas zone in the UK

2.2 Identification of the Prostate Cancer Treatment Dry-Spots

The term dry here refers to a relative lack of access to treatment facilities, which has been measured by the distance from the considered area to the nearest prostate cancer treatment facilities. The current version of the analysis is based on the linear (Euclidean) distance rather than road distance, which is one of the limitations of this work. Figure 3 shows the location of the treatment centers for prostate cancer.

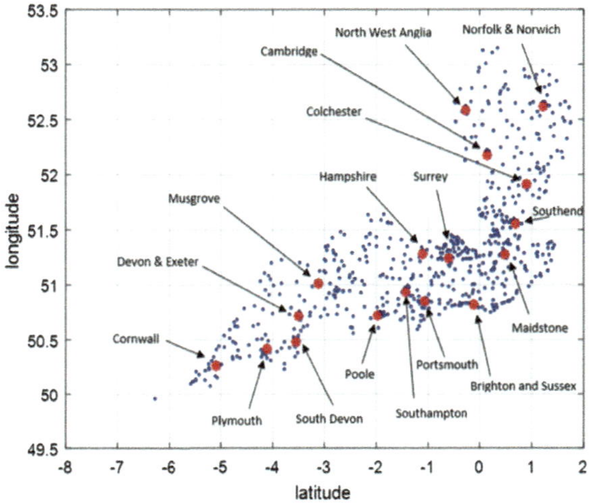

Fig. 3. The location of the treatment centers (red dots), and the centroids of the postcode districts within the 2 seas area in the UK (blue dots)

3 A Strategy to Find the Optimal Location

The method used to solve the problem is as follows:

1. For each of the potential hosting locations (the treatment centers shown in Fig. 3), a total travel time distance was defined, as follows:
   a. For each of the postcode districts, the number of the patients in that district was multiplied by the distance to the given treatment center. The resultant value was calculated and saved for all of the postcode districts.
   b. The sum of the above-mentioned value for all of the postcode districts was saved as the total travel distance for the given treatment center.
2. The total travel distance, explained above, was calculated for all of the treatment center.
3. The treatment center with the smallest value of the total travel distance was identified as the optimal location.

The optimization goal is to minimize the travel distance for all of the patients who may be potential receivers of the treatment provided by the new device. The result shows that the optimal location is the treatment center in Surrey, shown in Fig. 3.

4 Discussion and Conclusion

In this paper, we have developed a methodology towards the development of an optimized facility location modelling that includes geographical areas with relatively high prevalence of prostate cancer against access to treatment centers. We have focused our application on the South of the UK. A detailed mapping has identified age as the dominant factor for the analysis. The development and combination of measures for incidence and ease of access allows the identification of priority postcode areas. These are useful from a policy perspective as they show the areas most in need of enhanced provision, and from a modelling perspective as they allow effective measurement criteria for the development of single and multiple criteria optimization models.

The suggested single and multiple criteria facility location model offers a promising platform for future modelling developments. These relate on a specific level to the European Union project in which the next generation prostate cancer detection and treatment devices are being developed. On a general level, the mapping and proposed optimization models in this paper can be seen as the basis as a wider methodology to identify priority geographical areas for placement of medical resources to detect and treat disease, particularly in the case of a relatively small amount of resources and a large potential treatment population.

Further modelling developments involve the development of multiple criteria facility location models for the allocation of the small number of robotic devices. These could relate to the improvement of access for specific target postcodes with (i) poor access (dryness), (ii) high incidence levels (hotness) or (iii) combination of the both as detailed by the methodology develop in this paper. Specific targets could be developed for these priority postcodes and for the general population with respect to access time, travel distance and associated patient costs. Trade-off with respect to priority versus general access and the cost of locating multiple devices could be explored via a multi-objective weight sensitivity analysis. Given the target-based and geographical nature of this problem, extended network goal programming [10] is a viable modelling methodology.

References

1. The CobRa Project Website. https://cobra-2seas.eu/
2. Hansell, A.L., Ghosh, R.E.: The Environment and Health Atlas for England and Wales. Oxford University Press, Oxford (2014)
3. Cancer Research UK. https://www.cancerresearchuk.org/health-professional/cancer-statistics/statistics-by-cancer-type/prostate-cancer. Accessed Nov 2018
4. Population Age Demographics of UK Postcodes, Table LC1117EW. https://www.nomisweb.co.uk/census/2011/lc1117ew. Accessed Dec 2018

Special Sessions: Electronics and Smart Algorithms for the Effective Lung Monitoring and COPD Management

A Low-Cost USB-Compatible Electronic Stethoscope Unit for Multi-channel Lung Sound Acquisition

Gürkan Yilmaz(✉), Pierre Starkov, Mathilde Crettaz, Josias Wacker, and Olivier Chételat

Centre Suisse d'Electronique et de Microtechnique SA, 2002 Neuchâtel, Switzerland
guerkan.yilmaz@csem.ch

Abstract. The diagnosis of respiratory diseases relies largely on auscultation. To further improve the diagnosis quality, continuous acoustic multi-channel surveillance of the airways is highly desirable. The application of multiple sensors around the thorax implies size and cost constraints for the sensors. In this article we first report on the design and development of a wearable and low-cost electronic stethoscope which is easy to integrate in a garment. We then analyzed the noise and linearity of the recorded sound signal and benchmarked a prototype with a standard stethoscope which is widely used in clinical practice.

Keywords: Electronic stethoscope · Piezoelectric transducer · Wearables · Lung sounds · Auscultation

1 Introduction

Stethoscope auscultation is a prevalent clinical practice in diagnosing lung related diseases, which make up 5 of the 30 most common causes of death [1]. Despite its widespread use in clinic, digitalization and standardization of this method fall behind [2]. However, an increasing incidence rate of respiratory diseases combined with the need and will to increase the efficiency in medical practice have begun to give a thrust to these belated developments, even with additional expectations such as multi-channel recordings and continuous monitoring, especially at remote settings.

In a clinical examination, a physician typically auscultates a patient's thorax on different points. Taking this approach one step further, we envisage to synchronously record several sound signals from the different auscultation points with electronic stethoscopes to enable computerized multi-channel lung sound analysis. The main challenges in the realization of a system containing multiple sensors which record sound signals in parallel are (1) the sensor size, (2) the sensor synchronization, and (3) the costs, eventually inhibiting the use of commercially available electronic stethoscopes in such a system. Additionally, electronic stethoscopes available in the market are optimized for human hearing and most importantly for physicians' listening habits inherited from acoustic stethoscopes. Built-in filters of these devices impose a limitation on the bandwidth of acquired sound signal, which otherwise could be useful for signal processing algorithms.

J. Henriques et al. (Eds.): MEDICON 2019, IFMBE Proceedings 76, pp. 1299–1303, 2020.
https://doi.org/10.1007/978-3-030-31635-8_159

In this work, we propose a low-cost wearable electronic stethoscope unit which can be used as a building block for a wearable multi-channel lung sound acquisition system. The article presents the design of the prototype with an emphasis on electronics development as well as characterization results.

2 Materials and Methods

The core of the presented lung sound acquisition system (Fig. 1) is a piezoelectric transducer which is designed based on the curved-clamped structure presented in [3] to convert the incoming pressure waves to electrical signals. The transducer is realized with a 110 μm-thick PVDF film (TE connectivity, 3-1004346-0) shaped as a square (13×13 mm^2) and curved with 100 mm radius of curvature. This configuration provides 11 mV/Pa sensitivity and a sensor capacitance of 100 pF. The transducer is connected to an impedance converter which drives the input of the analog-to-digital converter (ADC). The data conversion and the USB interface are realized in an audio codec single-chip compliant with USB 2.0 (PCM2912A from Texas Instruments). This off-the-shelf component accelerates the development and provides a plug-and-play solution. The board containing the electronics is equipped with a micro-USB port, thus enabling a standard connection to a computer or a portable device. The open source software Audacity (v2.1.3) is employed to perform the sound recordings. The developed system can perform signal acquisition at sampling rates ranging from 8 ksps to 48 ksps.

The sound acquisition chain begins with the piezoelectric transducer which works as a voltage source whose output amplitude varies as a function of its strain. In the linear region, where the amplitude of the voltage depends linearly on the strain, the sensor is modelled as a voltage source in series with a capacitance (Fig. 1). Consequently any resistive load creates a network with high-pass filter characteristics, while any capacitive load (such as the input capacitance of the impedance converter) behaves like a voltage divider. Hence, it is desired to have a negligible input capacitance to preserve the signal amplitude.

The transducer, modelled as a capacitance, corresponds to a high source impedance at low frequencies where most of the information is located for this application. Thus, the design needs to address the common issues related with high impedance sources: bias current and current noise at the input. In most scenarios, the latter one is overlooked as the contribution of the voltage noise is significantly higher than the one of the current noise, particularly for low and moderate sensor impedance values.

Traditionally JFET input stages have been used to interface with high impedance sensors thanks to their high input impedance and very low (gate) bias currents. Today it continues to be a good option for discrete implementations thanks to their low 1/f noise. Realizing the impedance converter (high impedance of source to a lower impedance) with a single transistor leaves more room for design flexibility on the amplifier stage in terms of selection of components and topologies. Additionally, they enable a slightly easier biasing scheme, which is realized by connecting the gate to the source (or simply to ground for a common-source amplifier), thanks to its negative threshold voltage. In this work, the transducer is interfaced with a common-source JFET amplifier (Fig. 1).

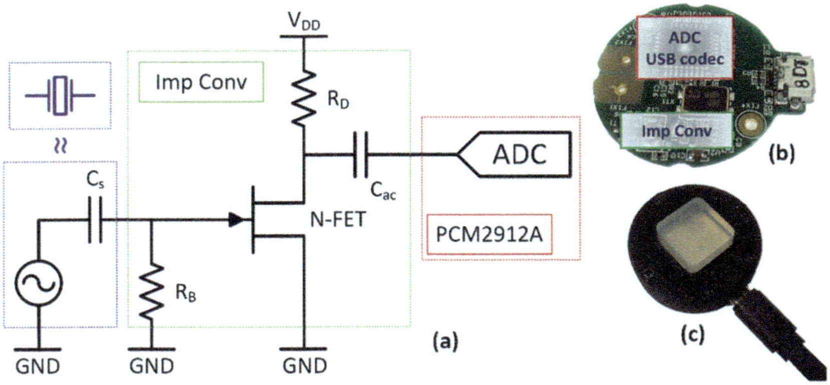

Fig. 1. (a) Simplified schematic of the presented system, (b) fabricated printed circuit board, and (c) assembled prototype (height: 18 mm, diameter: 32 mm) with a USB cable connected to the system

The gain of the impedance converter stage is obtained by multiplying the forward transconductance of the JFET (2SK3666-3 from ON Semiconductor) with the drain resistance. A 1st order high-pass filter (HPF) is created by the sensor capacitance (C_s) and the input resistance of the impedance converter, which is often equal to the gate bias resistor (R_B). Moreover, another HPF is created by the decoupling capacitor (C_{ac}) and the drain resistor (R_D) (Fig. 1). A 3 dB bandwidth of the output signal is limited to 0.454×8 kHz = 3632 Hz by the digital low-pass filter of the ADC [4].

The energy supply for the board is provided via USB which powers the audio codec directly and the impedance converter via a low drop-out (LDO) voltage regulator (3.0 V). The PCM2912A chip incorporates internal LDOs to regulate its analog and digital supply voltages (3.3 V). The overall current demand of the board is dominated by the audio codec and USB interface (85 mA). Given that it is not designed to be an autonomous device at this phase, USB 2.0 is quite sufficient to power the system with its rated current at 500 mA.

3 Results

First, we characterized the noise and linearity performance of the signal acquisition chain (Fig. 1) in the absence of the sensor. A full-scale (FS) input measurement was conducted to quantify the noise floor and the non-linearity of the acquisition chain. The signal generator was connected in series with a 100 pF capacitance, which is equivalent to the sensor capacitance, and the input signal was set to 160 $mV_{p\text{-}p}$ at 1 kHz. Figure 2 shows the spectrum of the full-scale signal. 2nd and 3rd harmonics of the signal are located 48 dB and 56 dB below the fundamental tone, respectively. The noise floor lies around −90 dB.

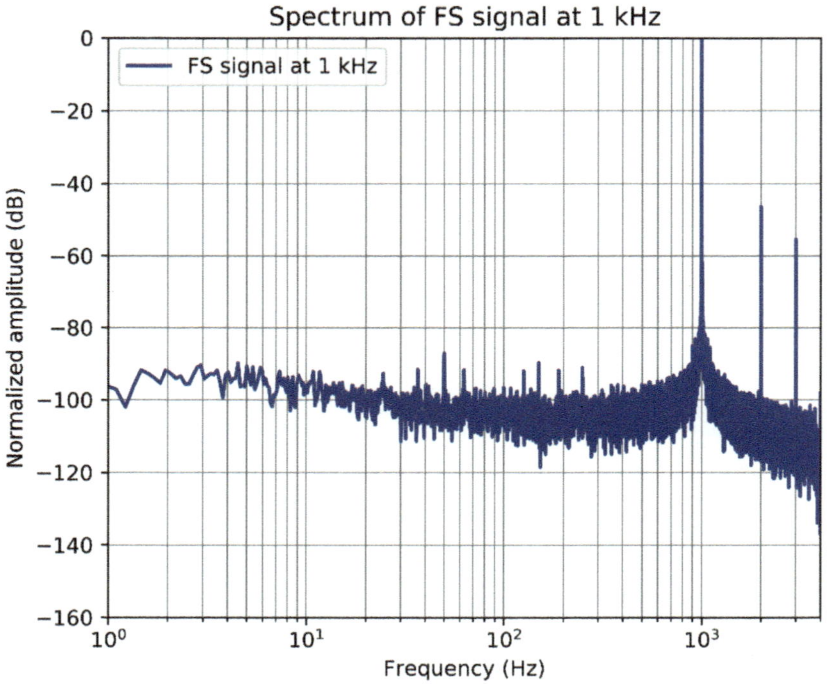

Fig. 2. Spectrum of full-scale signal at 1 kHz

Following the characterization of the system, a comparative study was conducted with a reference electronic stethoscope. The Littmann® 3200 from 3 M is one of the most often used electronic stethoscope in the clinical environment and is capable of transmitting the recorded signals to a computer via Bluetooth. This feature provides a means to compare the signals acquired by our device and the one of Littmann 3200 in the time and frequency domains. Both devices were placed below the right middle lobe next to each other and both were held in place by a chest strap in order to minimize the secondary effects such as hand tremor and contact pressure. A synchronous 10-second lung sound recording has been performed. Figure 3 presents the spectral comparison of the lung sounds acquired by our device and the Littmann 3200 stethoscope in extended mode. As the Littmann 3200 stethoscope is optimized for human hearing, it can be suggested that a high-pass filter around 50 Hz, which creates the deviation between two spectra, is implemented [5]. Peaks observed around 550 Hz and 800 Hz with an anti-resonance dip in between have also been reported in [6] for extended mode. For the rest of the spectrum, a good match is observed. Similar results were obtained from a second run with the positions of the two devices swapped. This shows that the distance between two devices is short enough not to have an effect on the results.

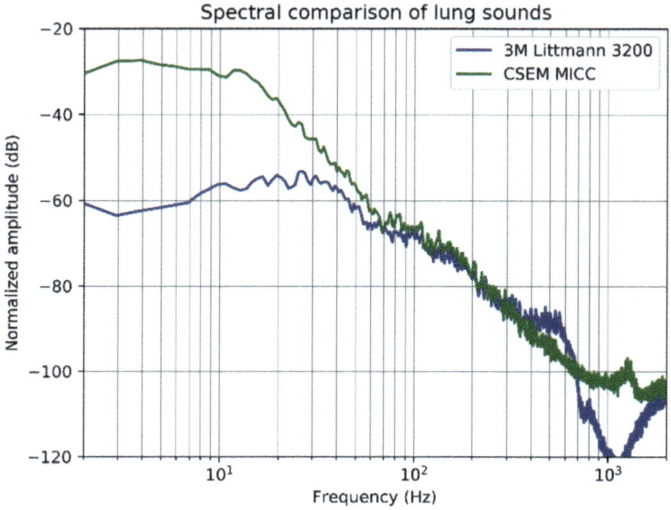

Fig. 3. Spectral comparison of lung sounds acquired by our device (CSEM MICC) and Littmann 3200 stethoscope

4 Discussion and Conclusion

We have designed and developed a low-cost, USB-powered electronic stethoscope prototype as a precursor of a wearable multi-channel lung sound recording system. The system exhibits promising results in terms of signal-to-noise ratio and linearity. Benchmarking results with a clinically-accepted electronic stethoscope showed equivalent signal qualities. It is envisaged to further miniaturize the system by means of integrated circuit technology to facilitate the integration onto garments.

References

1. Global Alliance against Chronic Respiratory Diseases Homepage. https://www.who.int/gard. Accessed 10 May 2019
2. Kraman, S.S.: Comparison of lung sound transducers using a bioacoustic transducer testing system. J. Appl. Physiol. **101**(2), 469–476 (1985)
3. Toda, M.: Contact-type vibration sensors using curved clamped PVDF film. IEEE Sens. J. **6** (5), 1170–1177 (2006)
4. PCM2912A datasheet. http://www.ti.com/lit/ds/symlink/pcm2912a.pdf. Accessed 15 May 2019
5. Drummond, T.E.: US7806226B2 Stethoscope with frictional noise reduction (2010)
6. Oliynik, V.: On potential effectiveness of integration of 3M Littmann 3200 electronic stethoscopes into the third-party diagnostic systems with auscultation signal processing. In: IEEE 35th International Conference on Electronics and Nanotechnology, pp. 417–421. IEEE, Kiev (2015)

Special Sessions: Non-invasive Temperature Assessment Using Ultrasound

Effect of Continuous Application of Heating-Cooling Cycles on Ultrasonic Attenuation of Muscle Tissue

Guillermo Cortela[1](✉), Carlos Negreira[1], and Wagner C. A. Pereira[2]

[1] Ultrasound Laboratory, University of the Republic, 11400 Montevideo, Uruguay
gcortela@fisica.edu.uy
[2] Biomedical Engineering Program/COPPE/Federal University of Rio de Janeiro, Rio de Janeiro 21941914, Brazil

Abstract. In ultrasound therapies with HITU applications, the temperature reached in the tissue depends both on the properties of the tissue to be treated and on the ultrasonic parameters. Ultrasonic attenuation (α) is one such parameters since it plays a fundamental role in the thermal source term of the biothermic equation. The study addresses the behavior of α by subjecting an ex vivo muscle to continuous heating-cooling cycles (15–50 °C). Using the transmission technique of ultrasonic pulse (PU) and a controlled thermal bath, the evolution of the pulses was obtained as a function of temperature. It was observed that, for the first three cycles, the evolution of the attenuation presents a non-reversible behavior. Then, the cycle becomes reversible, but at values very different from the initial value. Disregarding this behavior in the planning of a therapeutic session will result in an over or underestimation of the temperature, and, therefore, compromising the limits of safety and in the effectiveness of the application.

Keywords: Attenuation · Temperature · Ultrasound

1 Introduction

The goal of ultrasonic therapies (UT) is to complement or even replace standard therapies [1]. In addition, it has evolved for new therapies such as the administration of medications through the blood-brain barrier [2]. The safety of UT and the accuracy of their treatment plans as well as their devices are crucial [3] for patient safety. A major limitation of thermal therapies is the lack of detailed temperature information available to guide therapy. Temperatures are routinely measured in an invasive manner, but only limited measurements can be made [4]. The limited number of measurements may result in less information than is necessary to produce satisfactory temperature distributions to adequately assess thermal dosimetry.

Efforts have been made to obtain objective safety criteria during treatment, for example, based on the thermal index [5], the maximum temperature allowed on the surface of the transducer [6]. However, there is a lack of standards to allow predicting

J. Henriques et al. (Eds.): MEDICON 2019, IFMBE Proceedings 76, pp. 1307–1314, 2020.
https://doi.org/10.1007/978-3-030-31635-8_160

the quality of the therapy, or of a single treatment session, before its execution, because there is no reliable way to calculate the doses and estimate the side effects [7]. These facts make the reproducibility of treatment sessions and the comparability of clinical studies virtually impossible [8].

The use of ultrasound to estimate the temperature is not new [9] and its accuracy has been improved successively. Methods of temperature estimation are generally divided into three categories: (1) those that analyze the changes in the thermal expansion of the tissues and the speed of sound [10], (2) those that use the acoustic attenuation coefficient [11] and (3) those that explore the change in backscattered energy in the tissue [12]. To obtain a reliable estimate of the temperature, in each stage of treatment, one must have a correct knowledge of the properties of the tissue under study. In biological tissues the absolute value of the speed of the ultrasound and its dependence on temperature are closely related to the particular composition of the tissue [13]. The stiffness of the tissue changes significantly and irreversibly after thermal ablation [14, 15] and above 60 °C the changes become irreversible due to the denaturation of collagen [16].

Although the dependence of ultrasonic tissue parameters on temperature from *in vitro* analysis of the characteristics of ultrasonic tissues has been widely reported [17], the analysis of these parameters has not been addressed when the tissue under study is subjected to repeated and continuous cycles of temperature change.

This study explores the changes in the ultrasonic attenuation originated in *ex vivo* tissue when subjected to continuous heating-cooling cycles within the temperature range of the ultrasonic therapies.

2 Methodology

2.1 Sample Preparation

Samples of bovine muscle tissue were kept submerged in 0.9% saline at room temperature. They were placed in a vacuum chamber until no air bubbles were released from the tissue, by visual inspection. This method has been successfully tested in the elimination of air inclusions in the liver tissue [18]. The attenuation measures were taken immediately after this procedure to avoid changes in attenuation [19].

The transfer of the tissue sample from the vacuum chamber to the water bath was performed submerged in degassed saline solution to minimize exposure to the air.

For the measurements of attenuation in the muscle tissue sample, the muscle fibers were oriented perpendicular to the ultrasonic beam. The presence of interfaces (fat) in the path of ultrasound is avoided.

2.2 Attenuation Measurement System

The measured attenuation coefficient was based on the transmission and reception method described in the literature [20]. Two non-focused circular immersion transducers (Harisonic I3-0108S, 1 MHz) were aligned along the central axis of the transducers. The separation between them is given by the thickness of the sample (see Fig. 1).

The ultrasonic pulse is generated from a generator-receiver (FG) (Olympus NDT, MA, USA; 5073PR, rise time <2 ns). The pulse width is <0.2 μs, with repetition frequency of 1 kHz providing an adequate temporal resolution. The time base of the oscilloscope is synchronized with the FG trigger (Tektronix, INC, USA, TDS-2024B). The transducers with the sample in the middle were submerged in a degassed saline bath.

A heat exchanger was submerged in the bath to vary the tissue temperature. The exchanger is connected to a thermal bath with adjustable temperature control, the temperature range used is 15–50 °C with a step of 0.5 °C. A Type-J thermocouple was inserted in the tissue away from the ultrasonic path and another in the saline bath. A high-speed multiplexer (NI, National Instruments, NI9213) was used to record temperatures. When the bath and sample temperatures differ by ≤ 0.05 °C, the acquisition is proceeded. Next, control of the thermal bath is increased (decreased) by 1 °C in the case of heating (cooling) at a rate of 0.1 °C per min. The oscilloscope, multiplexer and thermal bath control are all commanded by a program developed in MatLab® platform.

Finally, the data was stored on a computer. The signal that passes through the sample is captured, and based on the decrease in the intensity of the received signal, the behavior of the attenuation as a function of temperature is analyzed. Care was taken to ensure that the attenuation of the ultrasonic pulse was not related to air bubbles, tissue interfaces (fat) or other unwanted inclusions within the ultrasonic path.

The behavior of the attenuation with temperature is evidenced when analyzing the evolution of the amplitude spectrum.

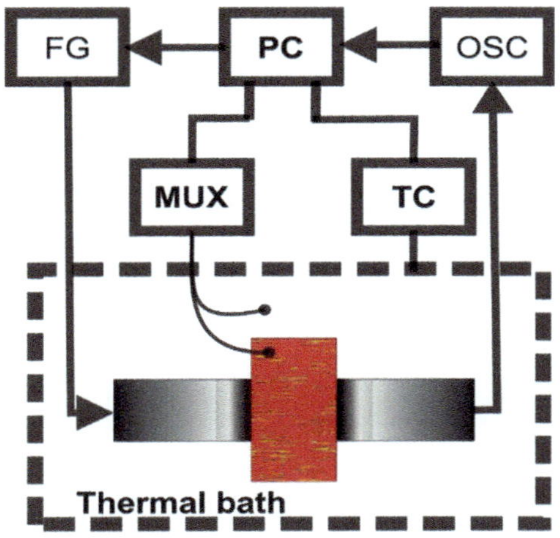

Fig. 1. Schematic diagram of radio frequency detection system. The elements are: Function generator (FG), Oscilloscope (OSC), Thermal controller (TC), Multiplexor and Computer (PC).

3 Results

Figure 2 shows the time evolution of the ultrasonic pulses (US) as a function of the temperature for the heating stage of the second cycle. As the sample of muscle tissue increases (decreases) its temperature the US presents a temporary shift and change in amplitude.

To analyze the evolution of the attenuation, the behavior of the maximum of the amplitude spectrum is followed (Fig. 3).

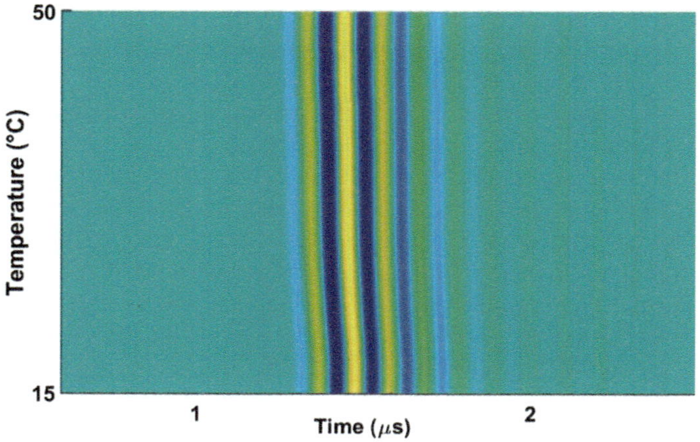

Fig. 2. Temporal evolution of ultrasonic pulses as a function of temperature. The ultrasonic velocity has a linear relationship (yellow and blue corresponds to the maximum and minimum ultrasonic pulse, respectively).

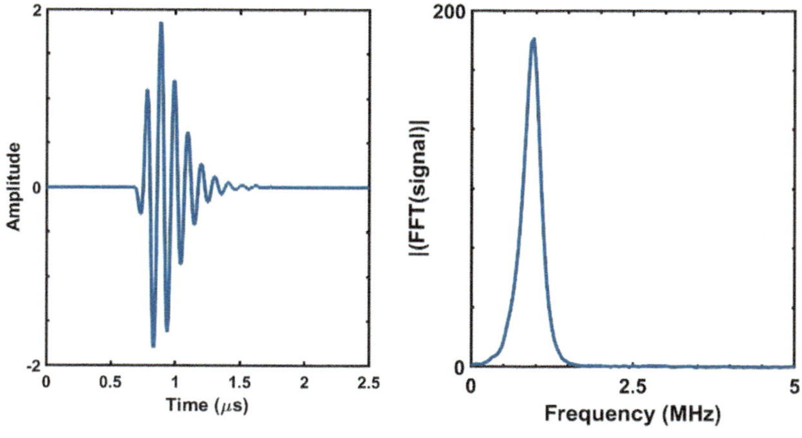

Fig. 3. Ultrasonic pulse at 37 °C (heating phase of the third cycle) and its amplitude spectrum. From the maximum spectral amplitude and its evolution with temperature, the behavior with temperature is analyzed.

Figures 4, 5, 6 and 7 show the evolution of the attenuation for the four heating-cooling cycles. The curve in red corresponds to the heating stage and in the blue the cooling stage.

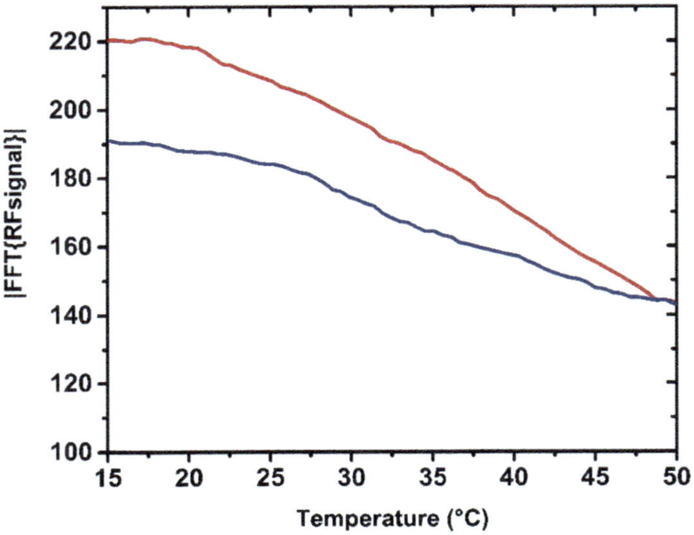

Fig. 4. Maximum spectral peak curves as a function of temperature for Cycle 1.

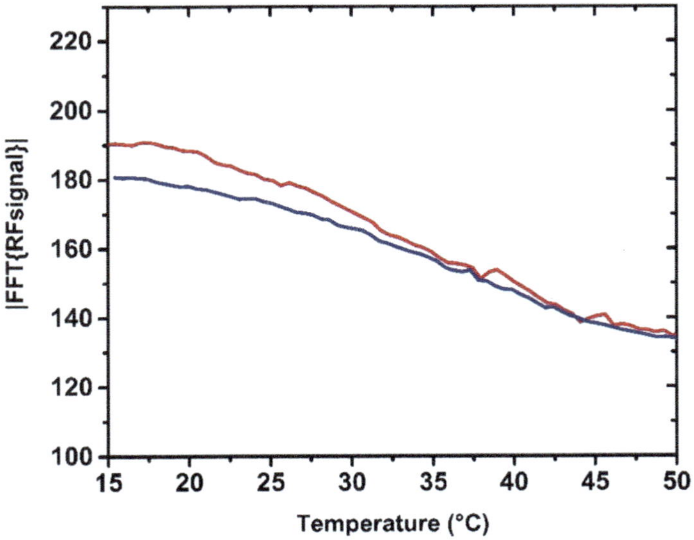

Fig. 5. Maximum spectral peak curves as a function of temperature for Cycle 2.

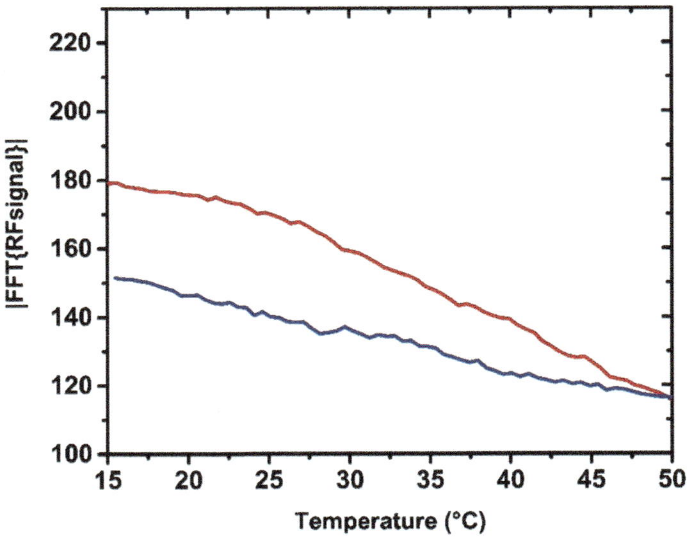

Fig. 6. Maximum spectral peak curves as a function of temperature for Cycle 3.

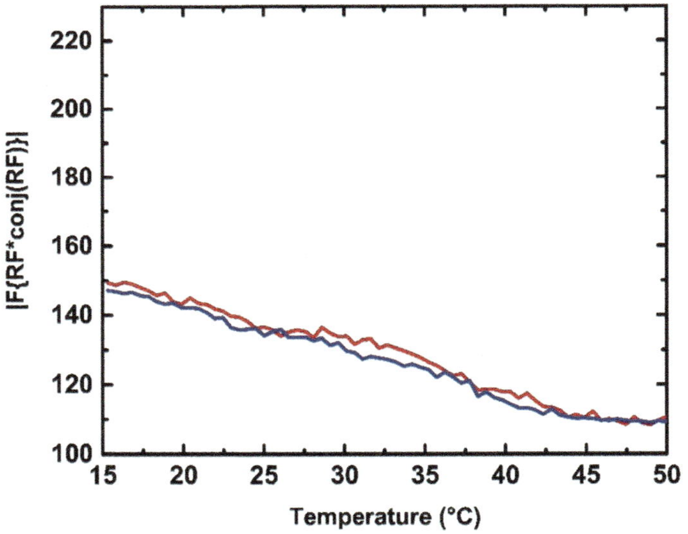

Fig. 7. Maximum spectral peak curves as a function of temperature for Cycle 4.

4 Discussion

It can be seen that the attenuation of the skeletal muscle tissue depends on the temperature in a non-reversible manner for the first three heating-cooling cycles. Such behavior is not present for the fourth cycle. The non-cyclic behavior of the ultrasonic attenuation, for the studied temperature range, is not expected in that temperature range [15, 16]. One possible explanation is that there is a cumulative effect of temperature on the structures and proteins that compose the tissue, despite not achieving high temperatures. This effect is intrinsically linked to the denaturation of proteins, as well as the behavior of elastin associated with the phenomenon called Reverse Transition with Temperature (ITT). It should be noted that the possible denaturation takes place in the range 40–50 °C. The behavior of the attenuation becomes reversible for the fourth cycle can be explained by the total denaturation of proteins that undergo lysis at temperatures lower than 50 °C. A fifth cycle was started, but which was discarded due to a change in the orientation of the muscle fibers and ultrasonic beam during the heating process.

The range of therapeutic applications in humans is 37–45 °C. US attenuation (absorption and scattering) is a fundamental parameter in the determination of temperature [21], therefore, it has a direct effect on the thermal dose, as well as on the planning of the therapy. This fact indicates that the "new" properties of the tissue must be taken into consideration in each treatment session planning with ultrasound. This care should be considered especially in high intensity applications. It can be observed that, once a HITU application has been made, the new value of the attenuation must be estimated not to commit an over (or under) exposure of the patient.

5 Conclusion

The changing of ultrasonic attenuation of the tissue with repeated application of heating-cooling cycles was evidenced despite not reaching temperatures that generate denaturation of the collagen.

Experimentally, there was a cumulative effect of temperature on the protein structures, which caused a non-reversible behavior. Apparently, once the total denaturation of proteins is achieved, the attenuation behavior becomes reversible. Such behavior should be considered in the planning of therapeutic sessions so to avoid over or underestimation of the temperature, and thus, keeping the limits of safety and effectiveness of the application.

More experiences with other types of biological tissues are necessary to confirm and find a relationship between the attenuation-temperature and amount of heating-cooling cycle applied.

References

1. ter Haar, G.: Therapeutic ultrasound. Eur. J. Ultrasound **9**(1), 3–9 (1999)
2. Aryal, M., Arvanitis, C., Alexander, P., McDannold, N.: Ultrasound-mediated blood–brain barrier disruption for targeted drug delivery in the central nervous system. Adv. Drug Deliv. Rev. **72**, 94–109 (2014)
3. Izadifar, Z., Babyn, P., Chapman, D.: Applications and safety of therapeutic ultrasound: current trends and future potential. Clin. Res. Open Access **3**(1) (2017)

4. Arthur, R., Straube, W., Trobaughi, J., Moros, E.: Non-invasive estimation of hyperthermia temperatures with ultrasound. Int. J. Hyperthermia **21**(6), 589–600 (2005)
5. IEC 62359: Ultrasonics–Field characterization–Test methods for the determination of thermal and mechanical indices related to medical diagnostic ultrasonic fields, 2nd edn. International Electrotechnical Commission, Geneva (2010)
6. IEC 60601-2-37: Medical electrical equipment : Particular requirements for the basic safety and essential performance of ultrasonic medical diagnostic and monitoring equipment, 1st edn. International Electrotechnical Commission, Geneva (2015)
7. Shaw, A., Martin, E., Haller, J., ter Haar, G.: Equipment, measurement and dose—a survey for therapeutic ultrasound. J. Ther. Ultrasound **4**(1), 1–7 (2016)
8. ter Haar, G., Shaw, A., Pye, S., Ward, B., Bottomley, F., Nolan, R., Coady, A.: Guidance on reporting ultrasound exposure conditions for bio-effects studies. Ultrasound Med. Biol. **37** (2), 177–183 (2011)
9. Seip, R., VanBaren, P., Cain, P., Ebbini, E.: Noninvasive real-time multipoint temperature control for ultrasound phased array treatments. IEEE Trans. Ultrason. Ferroelectr. Freq. Control **43**(6), 1063–1073 (1996)
10. Seip, R., Ebbini, E.: Noninvasive estimation of tissue temperature response to heating fields using diagnostic ultrasound. IEEE Trans. Biomed. Eng. **42**(8), 828–839 (1995)
11. Bamber, J., Hill, C.: Ultrasonic attenuation and propagation speed in mammalian tissues as a function of temperature. Ultrasound Med. Biol. **5**(2), 149–157 (1979)
12. Teixeira, C., Pastrana-Chalco, M., Simões, R., Pedrosa, A., von Kruger, M., Alvarenga, A. V., Fontes-Pereira, A., Pereira, W.: On the feasibility of ultrasound imaging enrichment by medium temperature changes. Ultrason. Imaging **41**(1), 17–34 (2019)
13. Simon, C., VanBaren, P., Ebbini, E.: Two-dimensional temperature estimation using diagnostic ultrasound. IEEE Trans. Ultrason. Ferroelectr. Freq. Control **45**(4), 1088–1099 (1998)
14. Benech, N., Negreira, C.: Monitoring heat-induced changes in soft tissues with 1D transient elastography. Phys. Med. Biol. **55**(6), 1753–1765 (2010)
15. Sapin, E., Gennisson, J., Pernot, M., Fink, M., Tanter, M.: Temperature dependence of the shear modulus of soft tissues assessed by ultrasound. Phys. Med. Biol. **55**(6), 1701–1718 (2010)
16. Wu, T., Felmlee, J., Greenleaf, J., Riederer, S., Ehman, R.: Assessment of thermal tissue ablation with MR elastography. Magn. Reson. Med. **45**(1), 80–87 (2001)
17. Shore, D., Miles, C.: Attenuation of ultrasound in homogenates of bovine skeletal muscle and other tissues. Ultrasonics **26**, 218–222 (1988)
18. Bamber, J., Nassiri, D.: Effect of gaseous inclusions on the frequency dependence of ultrasonic attenuation in liver. Ultrasound Med. Biol. **11**, 293–298 (1985)
19. Frizzell, L., Cartensen, E., Davis, D.: Ultrasonic absorption in liver tissue. J. Acoust. Soc. Am. **65**, 1309–1312 (1979)
20. Kossoff, G., Kelly-Fry, E., Jellins, J.: Average velocity of ultrasound in the human female breast. J. Acoust. Soc. Am. **53**, 1730–1736 (1973)
21. Cortela, G., Pereira, W., Negreira, C.: Ex vivo determined experimental correction factor for the ultrasonic source term in the bioheat equation. Ultrasonics **82**, 72–78 (2017)

Metrological Approach for Characterizing Ultrasonic Properties of Soft Tissue-Mimicking Material

Raquel Monteiro Souza(✉), Mylena K. Mosqueira de Assis, Rodrigo P. B. Costa-Félix, and André Victor Alvarenga

Institute of Metrology, Quality and Technology, Duque de Caxias, RJ, Brazil
{rmonteiro, avalvarenga}@inmetro.gov.br

Abstract. This paper reports the investigation of the metrological impact of temperature variation on ultrasonic velocity and attenuation coefficient of a standardized tissue-mimicking material (TMM), in temperatures ranging from 20 °C to 45 °C, in steps of 5.0 °C, using a new experimental method. One calibrated thermo-hygrometer was used to measure the temperatures of the water-bath in which the TMM was inserted. The pulse-echo technique was used to measure the TMM's properties. The Guide to the Expression of Uncertainty in Measurement was used to estimate the measurements uncertainties. The experiments were performed under repeatability conditions (n = 4). Based on the results, it was observed that the values of the ultrasonic velocity tend to increase as the temperature increases, whilst the values of the attenuation coefficient tend to decrease. The group velocity varied from 1,533.7 m s^{-1} to 1,575.6 m s^{-1} with expanded uncertainties lower than 6.9 m s^{-1} ($\alpha = 0.05$), for temperature varying from 19.9 °C to 45.8 °C. The attenuation coefficient varied from 2.55 dB cm^{-1} to 1.80 dB cm^{-1} with expanded uncertainties lower than 0.19 dB cm^{-1} ($\alpha = 0.05$) for the same temperature range. The uncertainties found in this study are relevant to provide metrological reliability of the results since it takes into account the influence of the most relevant quantities involved in the measurement process for TMM characterization.

Keywords: Non-invasive thermometry · Tissue-mimicking material · Ultrasonic properties · Metrology

1 Introduction

Knowing the internal temperature of biological tissues by means of non-invasive methods can be very useful and necessary both to assist in therapies that make use of ultrasound to heat the tissue and for the safety evaluation of ultrasound physiotherapy equipment. There is an absence of relevant thermal data to guide thermal therapies [1]. Besides that, there is a concern about the biological effects caused by ultrasound [2]. Among the relevant non-invasive thermometry methods by ultrasound, the following stand out: change in backscattered energy (CBE) [1, 3], the average gray level (AVGL) variation of ultrasound images [4–6], the time shift (TS) [7], and the variation in the attenuation coefficient [1].

J. Henriques et al. (Eds.): MEDICON 2019, IFMBE Proceedings 76, pp. 1315–1328, 2020.
https://doi.org/10.1007/978-3-030-31635-8_161

Many authors have studied these thermometry methods by using biological tissues and phantoms [3–6, 8]. Phantoms or tissue-mimicking materials (TMM) are test object models intended to mimic the thermal and acoustical properties of a specific tissue [9].

According to [1], there are ultrasonic methods, which can be used for estimating temperature if the temperature-dependent ultrasonic properties are properly defined. Ultrasonic velocity and attenuation are highly dependent on temperature and their dependence may vary among different sort of phantoms [10].

Straube and Arthur [11] had predicted CBE behavior under temperature rise for certain medium and scatterer combinations. They used attenuation coefficient and ultrasonic velocity data reported in literature. Later, Arthur et al. [8] validated experimentally the theoretical model proposed in [11].

Similarly, to validate the results obtained by means of AVGL, Alvarenga et al. [6] proposed a method for estimating AVGL values theoretically. For this, the authors carried out the characterization of a TMM, by measuring its ultrasonic group velocity and attenuation coefficient at different temperatures. They estimated the measurement uncertainty of the ultrasonic properties whilst presented the sources of uncertainty about the experimental setup used.

As observed in the literature, few studies involving the characterization of biological tissues or TMM give importance to metrological rigor. To estimate measurements uncertainties is relevant to provide metrological reliability of the results since it takes into account the influence of all the quantities involved in the measurement process.

Based on the foregoing, and aiming to improve and contribute to the state of the art, this paper presents a metrological approach to a new experimental method for characterizing ultrasound propagation velocity and attenuation coefficient of a soft tissue-mimicking material.

2 Materials and Methods

2.1 Manufacturing of the Soft Tissue-Mimicking Material

The preparation of TMM can influence directly on the accurate reporting of its ultrasonic properties [11]. According to the international standard IEC 60601-2-5 [9], a soft tissue model shall have the acoustic and thermal properties similar to human soft tissue. The reference values of the ultrasonic velocity and the attenuation coefficient of the soft TMM are respectively 1,540 m s^{-1} and 0.5 dB cm^{-1} MHz^{-1} [9].

In the research reported in the present paper, it was manufactured a soft TMM based on the standard operating procedure described in [12], which presents a detailed step-by-step procedure to develop an agar phantom, according to the recipe described on [9].

The agar phantom was made from the materials described in [9], as can be seen in Table 1. The second column shows the percentage mass of each pure component. The mass of the component was calculated according to the necessary amount of water to produce the desired batch. Considering 1,123 g of water, the mass of components are show in the last column of Table 1.

Table 1. Mass of components.

Component	Mass %	Mass used [g]
Glycerol	11.21	151.76
Deionized Water	82.95	1,123
Benzalkonium chloride	0.47	6.36
Silicon Carbide	0.53	7.18
Aluminum Oxide (Al_2O_3 (0.3 μm))	0.88	11.91
Aluminum Oxide (Al_2O_3 (3 μm))	0.94	12.73
Agar	3.02	40.89
Sum	100	1,353.83

At the end of the preparation, the mixture was poured into a rectangular mold (300 mm × 150 mm × 28 mm). After the cooling process, the phantom was demolded, cut and stored. Six samples of this phantom were used in the present study, all with the same size (Ø 70 mm × 28 mm) and named as: A; B; C; D; E; and F (Fig. 1).

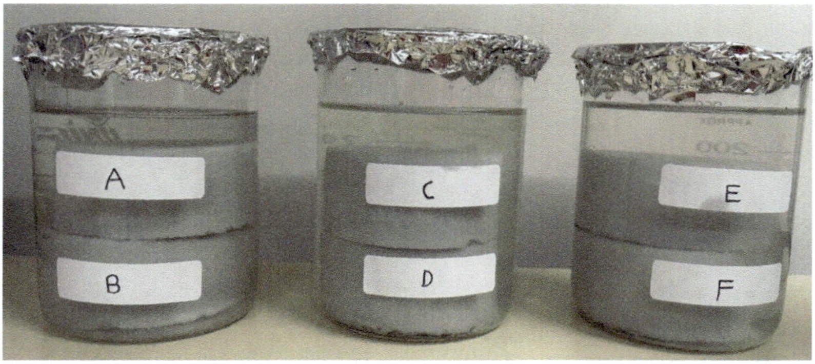

Fig. 1. Six phantoms (A, B, C, D, E, and F) stored in pairs in three beakers.

2.2 Characterization of the Soft Tissue-Mimicking Material

The ultrasonic velocity and the attenuation coefficient of the six phantoms were measured at different temperatures, as disclosed in Table 2. Four replicates were carried out in four days for each temperature. T_0 is the room temperature, whereas a step of approximately 5.0 °C was used for the others temperatures, until reaching T_5, as follows: $T_1 = T_0 + 5.0$ °C; $T_2 = T_1 + 5.0$ °C; $T_3 = T_2 + 5.0$ °C; $T_4 = T_3 + 5.0$ °C; and $T_5 = T_4 + 5.0$ °C.

Measurement System. The measurement system (Fig. 2) consists of a 5 MHz central frequency transducer (model A309S) (Olympus-NDT, USA) that acts as transmitter and receiver and is excited by a waveform generator model 33250A (Agilent, USA)

Table 2. Phantoms arrangement for measurements.

Temperature	1st day	2nd day	3rd day	4th day
T_0	A	E	C	F
T_1	B	F	D	A
T_2	C	A	E	B
T_3	D	B	F	C
T_4	E	C	A	D
T_5	F	D	B	E

with a sine burst of 20 cycles, 20 V peak-to-peak amplitude at 5 MHz. An oscilloscope model DSO-X 3012A (Agilent, USA) digitalizes the acquired signals. The experimental setup includes a temperature controlled Ø 300 mm × 110 mm high water-bath model 557 (Fisatom, Brazil) filled with distilled water, and a stainless steel cylindrical reflector (Ø 63 mm × 10 mm). A positioning system with three linear and two goniometric stages was used to align the transducer [14]. A calibrated thermohygrometer was used to measure the temperatures of water and TMM. In order to preserve their properties during all the process of measurement, inside the water-bath the phantoms were stored in pairs in three beakers containing a water-glycerol-benzalkonium chloride mixture to prevent them from drying out and to avoid air contact [9], as can be seen in Fig. 2. The mixture contains 88.1% (mass) demineralized water, 11.9% (mass) glycerol and 0.5% (mass) solution of benzalkonium chloride.

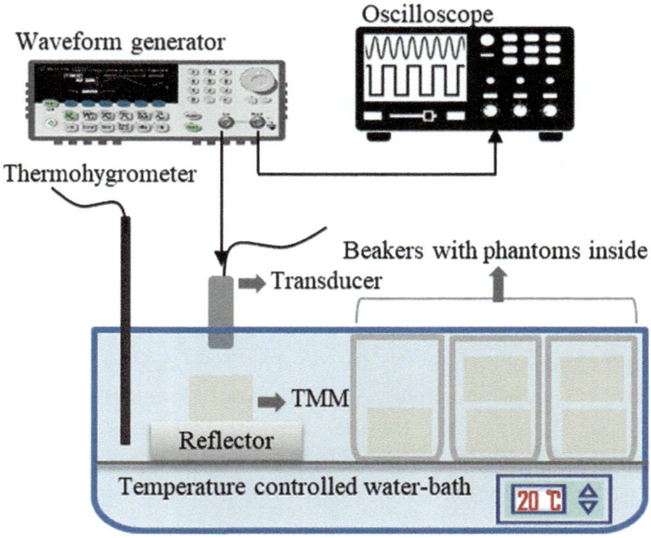

Fig. 2. Experimental setup illustration.

Measurement Procedure. The TMM's properties were estimated by the pulse-echo technique. For the ultrasonic velocity and attenuation coefficient calculation, it was necessary to measure different times of flight and amplitude of the signals using the oscilloscope.

Before starting the measurements at each temperature, all the experimental setup, including the water, the reflector and the phantoms had to reach the thermal equilibrium. Hence, the temperatures of the phantoms and of the water-bath were monitored with the thermohygrometer. To acquire the reference signal, the transducer output beam was normally aligned to the stainless steel reflector surface, maximizing the echo. Then, the time of flight in the water (t_W), without the TMM, was measured (Fig. 3a). After that, the time of flight on the surface of the TMM (t_1) (Fig. 3b) and the time of flight in reflector with the TMM inserted (t_2) were measured (Fig. 3a). In turn, for the attenuation coefficient determination, it was also necessary to measure the time of flight in the water, the time of flight on the surface of the TMM, the amplitude of these signals, (V_{in1}) and (V_{in2}), and the amplitude of signal on the reflector surface with the TMM inserted (V_{in3}) (Fig. 3).

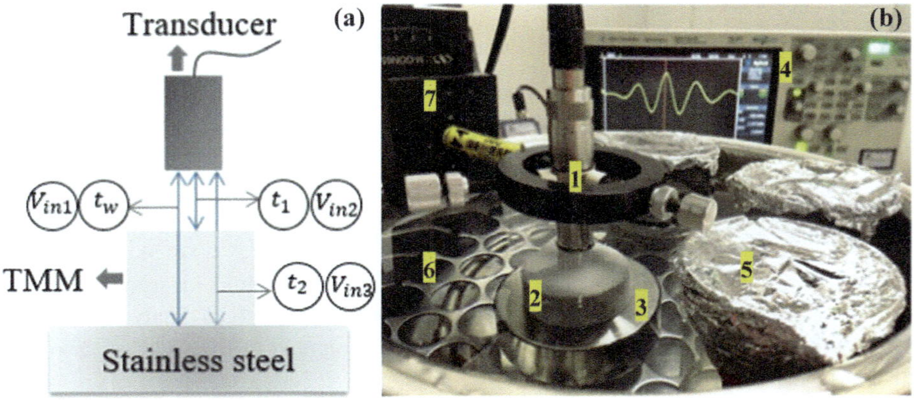

Fig. 3. (a) Pulse-echo technique scheme; (b) Experimental setup: 1: Transducer; 2: phantom; 3: stainless steel target; 4: oscilloscope and the TMM's surface signal; 5: beaker with phantoms inside covered by aluminum to avoid evaporation; 6: temperature controlled water-bath; 7: positioning system for transducer alignment.

The measurements were carried out four times at each temperature, and all this process was repeated four times in four days.

2.3 Measurements Uncertainty

The measurement uncertainties of the ultrasonic group velocity and the attenuation coefficient were calculated based on the Guide to the Expression of Uncertainty in Measurement [13]. The measurements were carried out under repeatability conditions, and the number of repetition was n = 4.

An uncertainty model for velocity and attenuation coefficient in soft TMM has already been proposed by [14], but the experimental setup was different. Thus, this paper presents new uncertainty models of the velocity and attenuation for characterizing TMM and is disclosed below.

Ultrasonic Group Velocity. The mathematical model for group velocity is presented in [14]. The contribution of each source of uncertainty is presented in Ishikawa's diagram in Fig. 4.

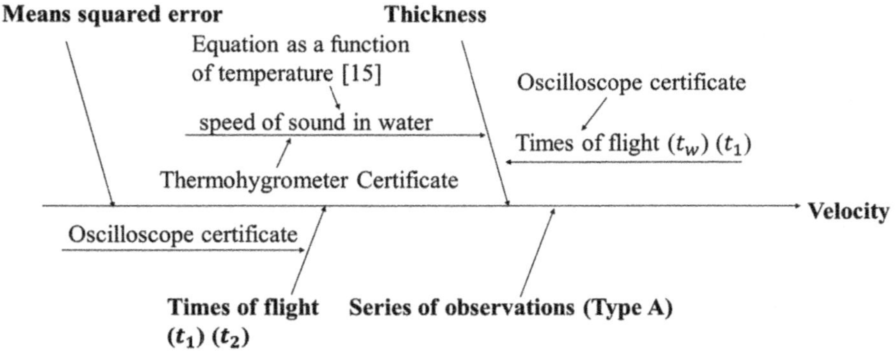

Fig. 4. Ishikawa's diagram: sources of uncertainty for ultrasound group velocity.

As can be observed in the diagram, the following uncertainty sources for group velocity measurements of the TMM were considered:

Type A Standard Uncertainty. The dispersion of speed of sound measurements. The evaluation of standard uncertainty was made from the statistical analysis of group velocity measurements repetition (n = 4) and it is computed as the standard deviation of the mean.

Type B Standard Uncertainty. The ultrasound propagation velocity in water was considered in type B evaluation, which comprises the uncertainty of the equation [15] as a function of the temperature (u = 0.18 m s^{-1}) and the thermohygrometer certificate uncertainty (u = 0.15 °C for temperatures of 20 °C and 25 °C; and u = 0.1 °C for temperatures of 35 °C, 40 °C and 45 °C). Besides that, the uncertainty of the oscilloscope time base (u = 2.4×10^{-8} s) was used to the measurement uncertainties of the time of flight in the water (t_W), the time of flight on the surface of the TMM (t_1) the time of flight in reflector with the TMM over it (t_2). The uncertainty of the measurement of phantom's thickness was also considered and varied from 0.04 mm to 0.11 mm.

Uncertainty Arising from Mean Squared Error. This source was obtained from the regression line fit for all measurements. MSE Uncertainty was calculated by the MSE's root. The linear regression establishes a mathematical relation between the temperature and the ultrasonic parameter.

Attenuation Coefficient. The mathematical model of the attenuation coefficient is described in [14]. The sources of uncertainty are shown in Ishikawa's diagram (Fig. 5).

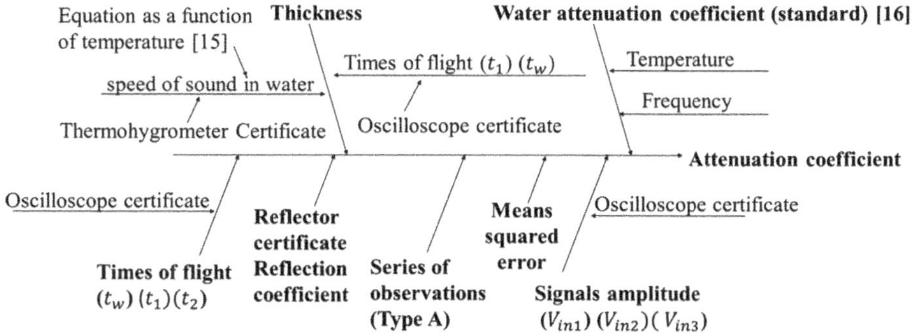

Fig. 5. Ishikawa's diagram: sources of uncertainty of the attenuation coefficient.

As can be seen in the diagram, the following uncertainty sources of the attenuation coefficient measurements of the soft TMM were considered:

Type A Standard Uncertainty. The evaluation of standard uncertainty was made from the statistical analysis of attenuation measurements repetition (n = 4).

Type B Standard Uncertainty. The uncertainty of the attenuation coefficient in the water [16] is 3.5×10^{-3} dB cm^{-1} and was used to estimate the Type B standard uncertainty. The uncertainty of the oscilloscope signal amplitude (5.2×10^{-3} V), and the uncertainty of the reflection coefficient in the stainless steel cylindrical reflector. (8.68×10^{-4}) were also considered in the calculation. In addition, the same ones described for velocity uncertainty were used: the uncertainty of the determination of the ultrasound propagation velocity in water; the uncertainty of the oscilloscope time base and the uncertainty of the measurement of phantom's thickness.

Uncertainty Arising from Mean Squared Error. This source was obtained from the regression line fit for all attenuation coefficient measurements. The MSE Uncertainty was calculated by the MSE's root.

3 Results

The thickness of the phantoms was measured using the pulse-echo technique under repeatability conditions (n = 4) and are disclosed on the Table 3 below.

Tables 4, 5, 6 and 7 show the results arising from the four replicates of the group velocity measurements at each temperature in four different days, the expanded uncertainties (U) together with Type A and Type B standard uncertainties and coverage factor (k).

Table 3. The thickness of phantoms measured under repeatability conditions and its respective expanded uncertainties (p = 0.95), together with Type A and Type B standard uncertainties and coverage factor (k).

Phantom	Thickness [mm]	Uncertainty Type A [mm]	Uncertainty Type B [mm]	Coverage factor (k)	Expanded uncertainty [mm]
A	28.31	0.04	0.01	2.78	0.11
B	28.23	0.03	0.01	2.57	0.08
C	27.98	0.02	0.01	2.36	0.06
D	28.26	0.01	0.01	2.13	0.04
E	27.86	0.02	0.02	2.26	0.06
F	28.06	0.02	0.02	2.26	0.06

Table 4. The results of the first day of the group velocity measurements, its respective expanded uncertainties (p = 0.95), together with MSE uncertainty, Type A and Type B standard uncertainties and coverage factor (k).

Temperature [°C]	Velocity [m s^{-1}]	Type A uncertainty [m s^{-1}]	Type B uncertainty [m s^{-1}]	MSE uncertainty [m s^{-1}]	Coverage factor (k)	Expanded uncertainty [m s^{-1}]
20.3	1,540.45	0.29	0.98	3.21	1.96	6.6
25.8	1,552.28	0.26	1.06		1.96	6.7
29.8	1,556.48	0.53	1.01		1.96	6.7
35.2	1,565.52	0.14	0.99		1.96	6.6
40.3	1,572.14	0.42	0.85		1.96	6.6
45.7	1,574.75	0.26	0.91		1.96	6.6

Table 5. The results of the second day of the group velocity measurements, its respective expanded uncertainties (p = 0.95), together with MSE uncertainty, Type A and Type B standard uncertainties and coverage factor (k).

Temperature [°C]	Velocity [m s^{-1}]	Type A uncertainty [m s^{-1}]	Type B uncertainty [m s^{-1}]	MSE uncertainty [m s^{-1}]	Coverage factor (k)	Expanded uncertainty [m s^{-1}]
19.9	1,533.69	0.30	1.36	3.21	1.96	6.9
25.7	1,545.76	0.23	1.34		1.96	6.8
29.3	1,551.34	0.20	1.04		1.96	6.6
35.3	1,561.96	0.30	1.04		1.96	6.6
40.1	1,567.98	0.20	0.83		1.96	6.5
45.5	1,571.28	0.33	0.85		1.96	6.5

Table 6. The results of the third day of the group velocity measurements, its respective expanded uncertainties (p = 0.95), together with MSE uncertainty, Type A and Type B standard uncertainties and coverage factor (k).

Temperature [°C]	Velocity [m s^{-1}]	Type A uncertainty [m s^{-1}]	Type B uncertainty [m s^{-1}]	MSE uncertainty [m s^{-1}]	Coverage factor (k)	Expanded uncertainty [m s^{-1}]
20.1	1,534.73	0.28	0.99	3.21	1.96	6.6
25.6	1,546.17	0.31	0.96		1.96	6.6
29.8	1,559.82	0.28	0.92		1.96	6.6
35.5	1,567.20	0.30	0.91		1.96	6.6
40.6	1,572.92	0.18	0.81		1.96	6.5
45.8	1,575.56	0.20	0.80		1.96	6.5

Table 7. The results of the fourth day of the group velocity measurements, its respective expanded uncertainties (p = 0.95), together with MSE uncertainty, Type A and Type B standard uncertainties and coverage factor (k).

Temperature [°C]	Velocity [m s^{-1}]	Type A uncertainty [m s^{-1}]	Type B uncertainty [m s^{-1}]	MSE uncertainty [m s^{-1}]	Coverage factor (k)	Expanded uncertainty [m s^{-1}]
20	1,538.49	0.18	1.03	3.21	1.96	6.62
25.7	1,548.97	0.06	1.00		1.96	6.59
30.7	1,557.30	0.12	0.95		1.96	6.57
35.5	1,561.02	0.18	0.95		1.96	6.57
40.8	1,566.97	0.33	0.93		1.96	6.59
45.4	1,574.54	0.24	0.84		1.96	6.53

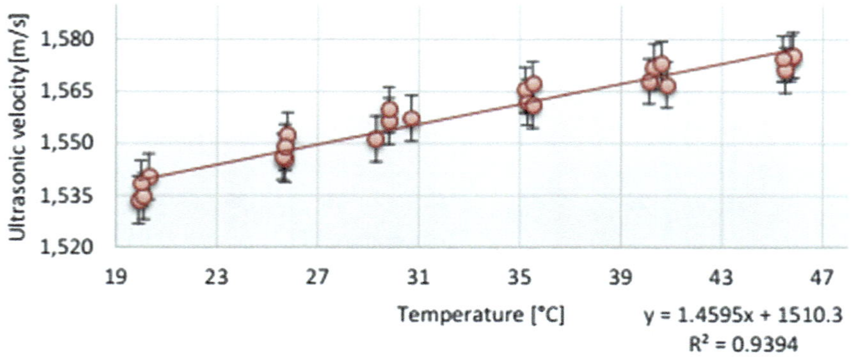

Fig. 6. The results of the four replicates of group velocity measurements at each temperature, the uncertainty ranges and the regression line.

Figure 6 shows the graphic of the group velocity measurements at each temperature in four different days, its respective expanded uncertainties and the regression line.

Tables 8, 9, 10 and 11 show the results arising from the four replicates of the attenuation coefficient measurements at each temperature in four different days, the expanded uncertainties (U) together with Type A and Type B standard uncertainties and coverage factor (k).

Table 8. The results of the first day of the attenuation coefficient measurements, its respective expanded uncertainties (p = 0.95), together with MSE uncertainty, Type A and Type B standard uncertainties and coverage factor (k).

Temperature [°C]	Attenuation coefficient [dB cm^{-1}]	Type A Uncertainty [dB cm^{-1}]	Type B uncertainty [dB cm^{-1}]	MSE uncertainty [dB cm^{-1}]	Coverage factor (k)	Expanded uncertainty [dB cm^{-1}]
20.3	2.54	0.06	0.004	0.06	2.14	0.19
25.8	2.35	0.01	0.004		1.96	0.12
29.8	2.32	0.02	0.004		1.97	0.13
35.2	2.19	0.02	0.004		1.97	0.13
40.3	1.99	0.06	0.003		2.16	0.19
45.7	1.86	0.06	0.003		2.16	0.19

Figure 7 shows the graphic of the attenuation coefficient measurements at each temperature in four different days, its respective expanded uncertainties and the regression line.

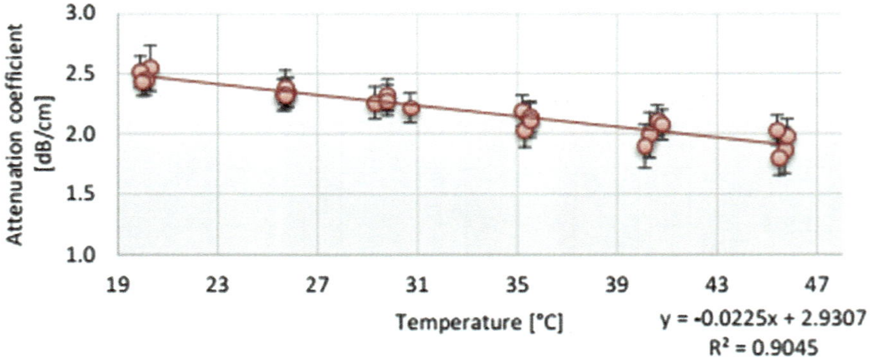

Fig. 7. The results of the four replicates of attenuation coefficient measurements at each temperature, the uncertainty ranges and the regression line.

Table 9. The results of the second day of the attenuation coefficient measurements, its respective expanded uncertainties (p = 0.95), together with MSE uncertainty, Type A and Type B standard uncertainties and coverage factor (k).

Temperature [°C]	Attenuation coefficient [dB cm^{-1}]	Type A uncertainty [dB cm^{-1}]	Type B uncertainty [dB cm^{-1}]	MSE uncertainty [dB cm^{-1}]	Coverage factor (k)	Expanded uncertainty [dB cm^{-1}]
19.9	2.52	0.01	0.004	0.06	1.96	0.13
25.7	2.39	0.03	0.004		1.98	0.14
29.3	2.26	0.03	0.004		1.98	0.14
35.3	2.03	0.04	0.003		2.02	0.15
40.1	1.90	0.06	0.003		2.13	0.18
45.5	1.80	0.04	0.003		2.02	0.15

Table 10. The results of the third day of the attenuation coefficient measurements, its respective expanded uncertainties (p = 0.95), together with MSE uncertainty, Type A and Type B standard uncertainties and coverage factor (k).

Temperature [°C]	Attenuation coefficient [dB cm^{-1}]	Type A uncertainty [dB cm^{-1}]	Type B uncertainty [dB cm^{-1}]	MSE uncertainty [dB cm^{-1}]	Coverage factor (k)	Expanded uncertainty [dB cm^{-1}]
20.1	2.45	0.01	0.004	0.06	1.96	0.13
25.6	2.33	0.02	0.004		1.97	0.13
29.8	2.28	0.01	0.004		1.96	0.13
35.5	2.14	0.02	0.003		1.97	0.13
40.6	2.11	0.02	0.003		1.96	0.13
45.8	1.98	0.03	0.003		1.98	0.14

Table 11. The results of the fourth day of the attenuation coefficient measurements, its respective expanded uncertainties (p = 0.95), together with MSE uncertainty, Type A and Type B standard uncertainties and coverage factor (k).

Temperature [°C]	Attenuation coefficient [dB cm^{-1}]	Type A uncertainty [dB cm^{-1}]	Type B uncertainty [dB cm^{-1}]	MSE uncertainty [dB cm^{-1}]	Coverage factor (k)	Expanded uncertainty [dB cm^{-1}]
20	2.44	0.01	0.004	0.06	1.96	0.12
25.7	2.32	0.01	0.003		1.96	0.12
30.7	2.22	0.01	0.003		1.96	0.12
35.5	2.11	0.03	0.004		2.00	0.14
40.8	2.07	0.02	0.003		1.96	0.13
45.4	2.03	0.01	0.003		1.96	0.13

4 Discussions and Conclusion

This paper presents a new experimental setup for characterizing the ultrasonic properties of a soft TMM at temperature range from 20 °C to 45 °C. It was also made a metrological study, by estimating the uncertainties of measurements of group velocity and attenuation coefficient, taking into account all quantities involved in the measurement process.

Few works care about the metrological rigors of ultrasonic properties measurements at different temperatures [6]. Besides that, some issues require more attention in the experimental setup and may affect directly the results of the ultrasound propagation velocity, for example.

In general, ultrasonic phantom's properties are measured in water. Moreover, to measure the ultrasonic properties, the phantom shall stay underwater for a while repeatedly. On this period, glycerol loss may occur. As the glycerol is responsible for providing the velocity required to the standardized phantom [17], the ultrasonic propagation velocity may be reduced. Therefore, the need to reduce the glycerol loss during the experimental measurements arises.

Herein, an experimental design was proposed to reduce the phantoms exposure to water and temperature variations. Six phantoms were produced from the same batch and stored on beakers contained the preservation solution proposed in the international standard [9] in order to preserve its properties during all measurement process. Each phantom was exposed to water four times during all measurements carried out. If just one phantom was used, the phantom would be submerged in water 24 times.

Based on the results of the present paper, it was noted that the values of the ultrasound propagation velocity tend to increase as the temperature increases, whilst the values of the attenuation coefficient tend to decrease as the temperature increases. The group velocity (Tables 4, 5, 6 and 7) varied from 1,533.7 $m \cdot s^{-1}$ to 1,575.6 $m{\cdot}s^{-1}$ with expanded uncertainties lower than 6.9 $m \cdot s^{-1}$ ($\alpha = 0.05$), for temperature varying from 19.9 °C to 45.8 °C. The attenuation coefficient (Tables 8, 9, 10 and 11) varied from 2.55 $dB \cdot cm^{-1}$ to 1.80 $dB \cdot cm^{-1}$ with expanded uncertainties lower than 0.19 $dB \cdot cm^{-1}$ ($\alpha = 0.05$) for the same temperature range.

Recently, Alvarenga et al. presented a model to predict temperature variations based on AVGL variation and TMM properties. They characterized a TMM at different temperatures, using an experimental setup different from the one presented here (details in [6]). According to the authors [6], there may have been a loss of phantom's glycerol, because the phantom was placed underwater during measurements without protection. However, they used a larger phantom, aimed to reduce that loss, because [18] testified experimentally that the lower the thickness of the phantom, the greater the loss of glycerol. The authors [6] obtained group velocity results from 1,530.9 $m\ s^{-1}$ to 1,555.4 $m\ s^{-1}$ with combined uncertainty of 6 $m \cdot s^{-1}$ ($\alpha = 0.05$), for temperature varying from 26.2 °C to 40.7 °C, whilst the attenuation coefficient varied from 2.3 $dB\ cm^{-1}$ to 1.1 $dB\ cm^{-1}$, for the same range of temperature and the combined uncertainty was estimated as 0.08 $dB\ cm^{-1}$. In comparison, it is perceived that the new experimental setup has reached smaller uncertainties, and it was possible to minimize the glycerol loss.

Considering the means squared error on the uncertainty calculation, it was possible to include the influence of the regression line fit for all properties measurements. Moreover, with the uncertainties obtained on the present paper, as can be seen in Figs. 6 and 7, it is possible to say that all measurements were consistent with each other.

To conclude, this paper aimed to contribute to the state of the art about studies involving non-invasive thermometry methods. With the experimental setup proposed, it was possible to reduce the glycerol loss during the TMM characterizing process, whereas smaller uncertainties were obtained.

Acknowledgements. Thanks to the financial support of the Brazilian National Council for Scientific and Technological Development (CNPq) (Grant: 401685/2016-0).

References

1. Arthur, R.M., Straube, W.L., Trobaugh, J.W., Moros, E.G.: Non-invasive estimation of hyperthermia temperatures with ultrasound. Int. J. Hyperth. **21**(6), 589–600 (2005)
2. Ter Haar, G., Shaw, A., Pye, S., Ward, B., Bottomley, F., Nolan, R., Coady, A.M.: Guidance on reporting ultrasound exposure conditions for bio-effects studies. Ultrasound Med. Biol. **37**(2), 177–183 (2011)
3. Lewis, M.A., Staruch, R.M., Chopra, R.: Thermometry and ablation monitoring with ultrasound. Int. J. Hyperth. **31**(2), 163–181 (2015)
4. Teixeira, C.A., Alvarenga, A.V., Cortela, G., Von krüger, M.A., Pereira, W.C.A.: Feasibility of non-invasive temperature estimation by the assessment of the average gray-level content of B-mode images. Ultrasonics **54**(6), 1692–1702 (2014)
5. Alvarenga, A.V., Teixeira, C.A.D., Von krüger, M.A., Pereira, W.C.A., Costa-Felix, R.P.B.: Uncertainty evaluation from non-invasive estimation of temperature variation using B-mode ultrasonic images from a plastic phantom. Measurement **69**, 189–194 (2015)
6. Alvarenga, A.V., Wilkens, V., Georg, O., Costa-Félix, R.P.B.: Non-invasive estimation of temperature during physiotherapeutic ultrasound application using the average gray-level content of B-mode images: a metrological approach. Ultrasound Med. Biol. **43**(9), 1938–1952 (2017)
7. Maass-Moreno, R., Damianou, C.A.: Noninvasive temperature estimation in tissue via ultrasound echo-shifts: Part II. In vitro study. J. Acoust. Soc. Am. **100**, 2522–2530 (1996)
8. Arthur, R.M., Straube, W.L., Starman, J.D., Moros, E.G.: Noninvasive temperature estimation based on the energy of backscattered ultrasound. Med. Phys. **30**, 1021–1109 (2003)
9. Electrotechnical Commission: IEC 60601-2-5:2015: Medical electrical equipment Part 2-37: Particular requirements for the basic safety and essential performance of ultrasonic physiotherapy equipment, 2.0 ed. Genova (2015)
10. Culjat, M.O., Oldenberg, D., Tewari, P., Singh, R.S.: A review of tissue substitutes for ultrasound imaging. Ultrasound in Med. Biol. **36**(6), 861–873 (2010)
11. Straube, W.L., Arthur, R.M.: Theoretical estimation of the temperature dependence of backscattered ultrasonic power for noninvasive thermometry. Ultrasound Med. Biol. **20**, 915–922 (1994)

12. Souza, R.M., Santos, T.Q., Oliveira, D.P., Souza, R.M., Alvarenga, A.V., Costa-Felix, R.P. B.: Standard operating procedure to prepare agar phantoms. J. Phys. Conf. Ser. **733**, 012044 (2016)
13. BIPM, JCGM:100. Guide to the Expression of Uncertainty in Measurement (GUM). BIPM, Paris (2008)
14. Souza, R.M., Monteiro, R.S., Costa-Felix, R.P.B., Alvarenga, A.V.: Ultrasonic properties of a four years old tissue-mimicking material. J. Phys. Conf. Ser. **975**, 012025 (2018)
15. Lubbers, J., Graaff, R.A.: Simple and accurate formula for the sound velocity in water. Ultrasound Med. Biol. **24**, 1065–1068 (1998)
16. Electrotechnical Commission: IEC 62127-2:2013, ultrasonics – hydrophones – Part 2: calibrations for ultrasonics fields up to 40 MHz, Geneva, Switzerland (2013)
17. Santos, T.Q., Alvarenga, A.V., Oliveira, D.P., Costa-Felix, R.P.B.: Metrological validation of a measurement procedure for the characterization of a biological ultrasound tissue-mimicking material. Ultrasound Med. Biol. **43**, 323–331 (2017)
18. Brewin, M.P., Pike, L.C., Rowland, D.E., Birch, M.J.: The acoustic properties centered on 20 MHz, of an IEC agar-based tissue-mimicking material and its temperature, frequency and age dependence. Ultrasound Med. Biol. **34**, 1292–1306 (2008)

Skin Contribution to Heating by Ultrasonic Field Irradiation: Simulation of a Multilayer Biological Application

Wagner Coelho de Albuquerque Pereira[1(✉)], Thaís Pionório Omena[1,2], and Eduardo Moreno[3]

[1] Biomedical Engineering Program, COPPE/Federal University of Rio de Janeiro, Rio de Janeiro 21941914, Brazil
wagner@peb.ufrj.br
[2] Department of Physiotherapy, Faculty of Medicine, Federal University of Rio de Janeiro, Rio de Janeiro, Brazil
[3] Dpto. Fisica Aplicada, ICIMAF, 15 e/C y D Vedado, La Habana, Cuba

Abstract. Ultrasound diathermy is a very popular therapeutic method to treat musculoskeletal conditions. Nevertheless, there is no validated protocol and the physiotherapists empirically chose the dose to be applied. In this work, a computer simulation of therapeutic ultrasound propagated through a biological medium composed of four plane parallel layers representing skin-fat-muscle-bone was performed. The corresponding thermal field was obtained. Simulation was performed at 1 MHz, with 1.0 and 1.5 W . cm^{-2}, with irradiation time of 120 s. with and without the skin layer (replacing skin by fat layer), to evaluate the role played by this layer. Results indicated that, when skin is present, it is the layer that reaches the highest temperatures, and the muscle layer did not reach temperature at therapeutic levels. When skin is excluded, then the fat/muscle and muscle/bone interfaces reach therapeutic temperatures in a very large region. These findings suggest that the skin cannot be neglected in dose planning. Additional experimental work has to be done to confirm this result.

Keywords: Ultrasound · Temperature · Therapeutic dose · Physiotherapy

1 Introduction

Thermal effects produced by ultrasound (US) were recognized as a potential therapy in the 1930s [1, 2]. Since then this resource has been growing applied in physical therapy scenarios, and recommended for improving pain relief [3, 4], collagen extensibility [5], muscle and tendon elasticity, tendon tension strength, nerve conduction, and acceleration of ligament repair [6–8].

Ultrasound therapeutic dose is a combination of radiation intensity and application time, and it should be able to keep the temperature in the region to be treated within the range of 40–45 °C for a minimum time of 5 min. To reach this temperature, radiation parameters (US frequency and intensity, effective radiation area, duty cycle), as well as medium physical properties (density, specific heat, thermal conductivity, US

J. Henriques et al. (Eds.): MEDICON 2019, IFMBE Proceedings 76, pp. 1329–1336, 2020.
https://doi.org/10.1007/978-3-030-31635-8_162

propagation velocity and attenuation coefficient, tissue temperature) have to be combined together.

Evidences of US therapeutic effects can be found in literature, nevertheless, there is a lack of quantitative scientific information regarding treatment protocols proposed for different patient's clinical conditions. Therefore, in daily practice, physical therapists empirically choose the US dose for the patients based on a general instruction provided by the user's manual of the equipment. Biological safety precautions are observed, in special to avoid overheating of the muscle/bone interface, and the most common ones are continuously move the transducer (at approximately 4 cm . s^{-1}) in a linear (side-to-side) or circular pattern; and apply light pressure to the skin surface [7, 9–15].

Once depicted this scenario of concerns, it is evident there is motivation enough to study US doses, application protocols, and their consequent temperature distributions, and corresponding benefits.

Computer simulation is a valuable tool to study phenomena from all the domains. It is especially useful in the Biomedical field as a first approach to guide the research to following experimental steps. Ultrasound propagation and heating in biological tissues comprise the well-known formulations of wave equation and the biothermic equation, and the thermal and ultrasonic properties of the tissues under consideration.

This work presents a simulation study in Comsol® platform where irradiation at therapeutic level is promoted in a multilayer biological tissue, composed of skin, fat, muscle and bone. The aim of the study is to evaluate the role of the skin in the ultrasonic propagation and heating patterns along the different tissue layers.

2 Methodology

The experiment was performed on Comsol® (COMSOL Multiphysics Inc., MA, USA), which solves the acoustic wave equation and connects it to the by the Biothermic equation. The models are implemented in two dimensions.

Two models were simulated containing 4 layers of tissue (Fig. 1): (1) skin, fat, muscle and bone. On the second case the skin was substituted by fat, with the same thickness so the total thickness is the same for both models.

The biological tissue properties used in the simulation were taken from literature [16–18], and are depicted in Table 1.

Simulations were carried out with a 1-MHz transducer of 30-mm diameter face which was positioned fixed on the surface of the skin (Model 1) and fat (Model 2). Computation resources was a 14″ laptop with an Intel i7 6th generation processor and 8 GB RAM.

The initial temperature was 36 °C, and the irradiation protocol was the same for both cases: continuous wave of intensity 1.0 W . cm^{-2}, irradiation time 120 s, and total acquisition time 200 s. Simulations with intensity of 1.5 W . cm^{-2} were also proceeded in a similar manner.

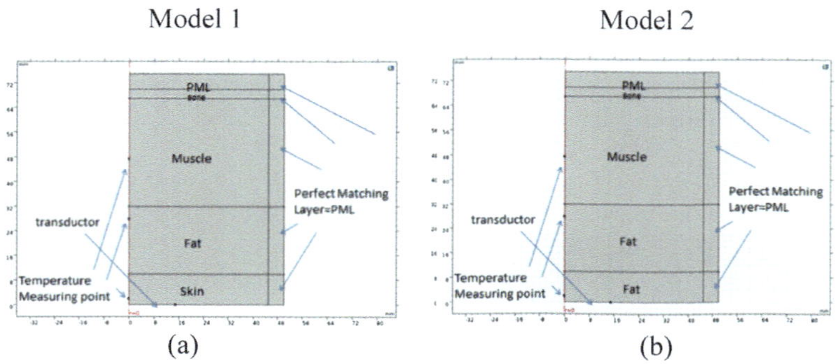

Fig. 1. Models used in the COMSOL Simulation. (a) skin (10,0 mm), Fat (22.00 mm), muscle (35.5 mm), bone (3.0 mm); (b) skin substituted by a 10-mm fat layer. Other layers are the same of Model 1.

Table 1. Properties for the biological tissues used in the simulation

Property	Symbol (unity)	Biological tissue			
		Skin	Fat	Muscle	Bone
Density	ρ (kg . m^{-2})	1200	950	1050	1975
Ultrasound speed	c (m . s^{-1})	1558	1478	1547	3476
Ultrasound attenuation	(Neper . m^{-1})	24	5.58	12.7	80.2
Thermal conductivity	k (W . m^{-1} . K^{-1})	0.23	0.19	0.55	2.3
Heat capacity	C_p (J . m^{-1} . K^{-1})	3590	2670	3640	1250

3 Results

The software generates 2D acoustic intensity and thermal fields, as well as temperature curves for the 200-s observation time.

Results for Model 1 are as follows. In Fig. 2a is the acoustic intensity field, while in Fig. 2b is the temperature curves for skin, fat and muscle. Figures 2c and d present the corresponding temperature field for 60 and 120 s, for intensity 1.0 W . cm^{-2}.

Results for Model 1, now with 1.5 W . cm^{-2} are as follows. In Fig. 3a is the acoustic intensity field, while in Fig. 3b is the temperature curves for skin, fat and muscle. Figures 3c and d present the corresponding temperature field for 60 and 120 s

Results for Model 2 (no more skin layer), with 1.0 W . cm^{-2} are as follows. In Fig. 4a is the acoustic intensity field, while in Fig. 4b is the temperature curves for first layer of fat (that substituted the skin of the previous model), fat and muscle. Figures 3c and d present the corresponding temperature field for 60 and 120 s.

Results for Model 2, now with 1.5 W . cm^{-2} are as follows. In Fig. 5a is the acoustic intensity field, while in Fig. 5b is the temperature curves for first layer of fat (that substituted the skin of the previous model), fat and muscle. Figures 5c and d present the corresponding temperature field for 60 and 120 s.

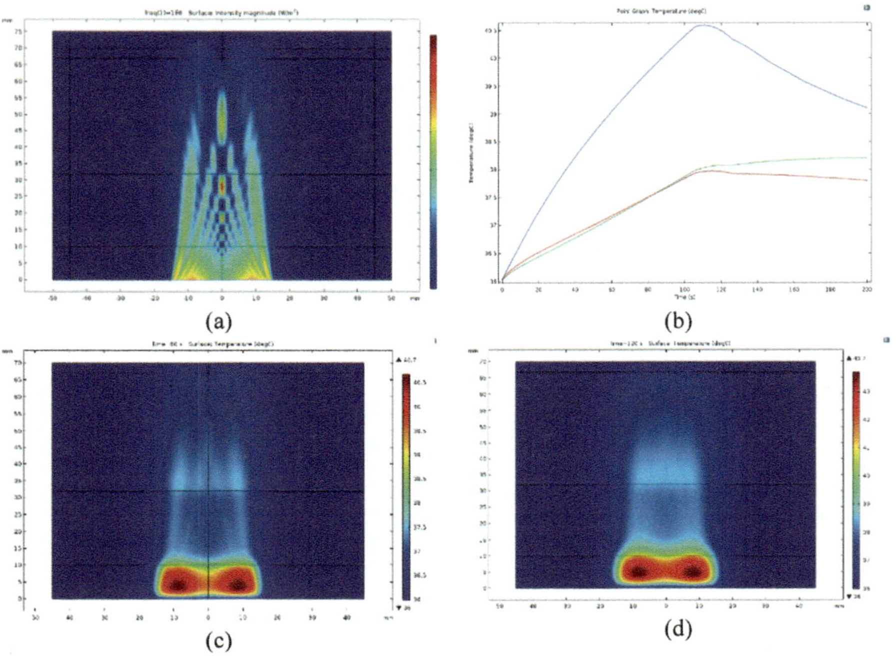

Fig. 2. Model 1, 1 MHz, 1.0 W . cm^{-2}, 120 s irradiation time. (a) Acoustic intensity field; (b) Temperature curves for skin (blue), fat (green), and muscle (red); (c) and (d), temperature field after 60 s and 120 s, respectively with corresponding peak temperature 39.0 °C and 40.7 °C.

4 Discussion

A computer simulation of therapeutic ultrasound propagated through a biological medium composed of four plane parallel layers representing skin-fat-muscle-bone was performed. The corresponding thermal field was obtained.

Regarding Model 1 (Fig. 2), for 1.0 W . cm^{-2}, for 120 s of application time, which is common therapeutic dose, skin has reached the highest temperature, retaining almost all the acoustic energy, and even with the beam focus inside the muscle layer, this tissue did not reach temperature at therapeutic levels (39 °C – 43 °C). So, temperature was higher on the skin than on the other tissues (T max skin: 40.5 °C, T max fat: 38.2 °C, T max muscle: 37.8 °C).

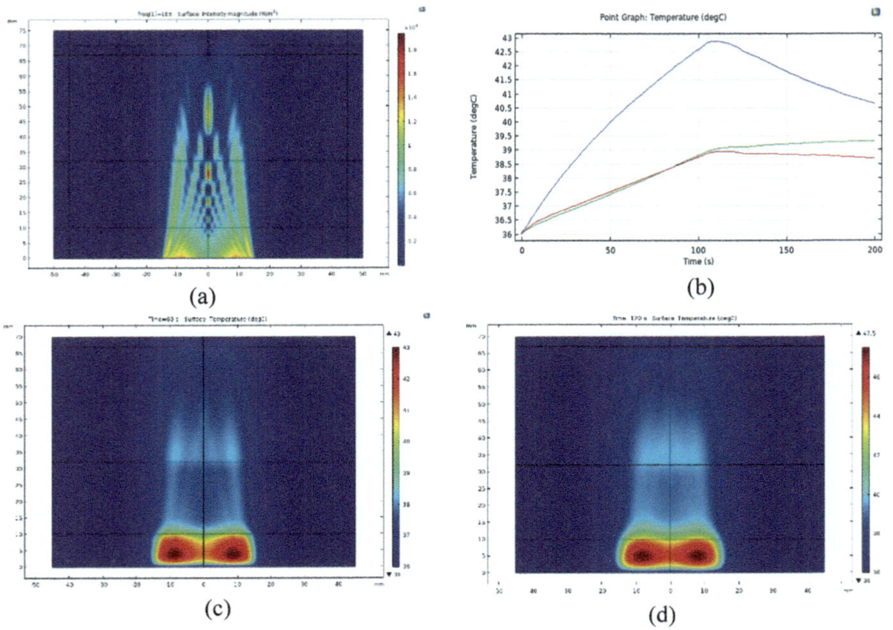

Fig. 3. Model 1, 1 MHz, 1.5 W . cm^{-2}, 120 s irradiation time. (a) Acoustic intensity field; (b) Temperature curves for skin (blue), fat (green), and muscle (red); (c) and (d), temperature field after 60 s and 120 s, respectively, with corresponding peak temperature 43.0 °C and 47.57 °C.

Now, when applying 1.5 W . cm^{-2}, for the same 120 s (Fig. 3), again skin was the tissue that reached the highest temperatures (outside the safety zone), but muscle temperature remained under the therapeutic zone.

Considering Model 2 (skin layer substituted by fat layer of the same thickness), for 1.0 W . cm^{-2}, and 120 s of application time, the beam focus is still inside the muscle layer (Fig. 4). The second fat layer has reached the highest temperature, and muscle has reached 40 °C, which is inside the recommended therapeutic range. Additionally, it is possible to see through the temperature mapping (Figs. 4a and b), that a large region of the muscle, as well as the second layer of fat, has reached the therapeutic temperature. It is also important to point out that the interface between the 2nd fat layer and muscle is included as the region with the highest temperatures. No effect of blood perfusion was simulated in any case. When intensity is changed to 1.5 W . cm^{-2}, the spatial distribution of heat is similar to the one for 1.0 W . cm^{-2} (Fig. 4d), but, this time, peak

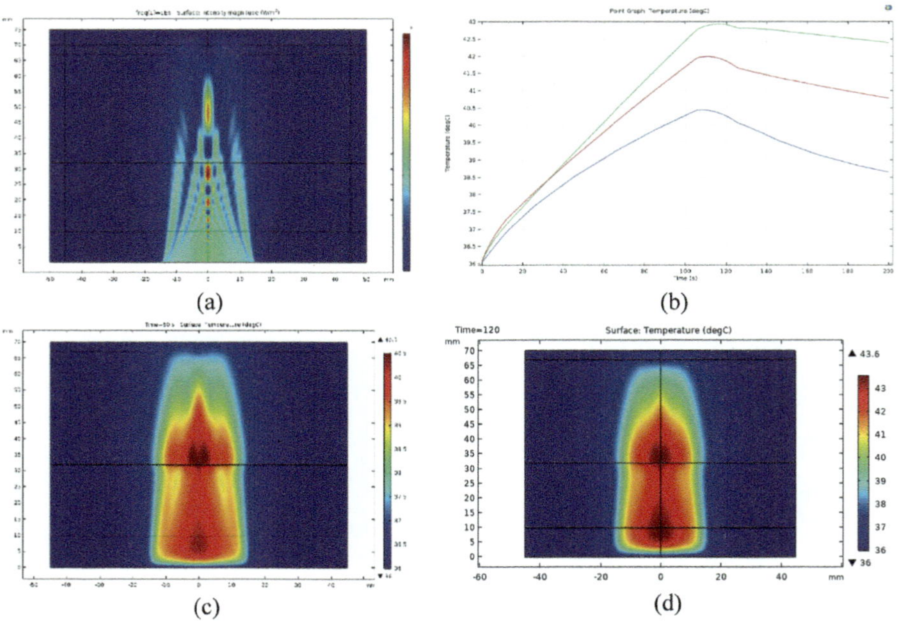

Fig. 4. Model 2, 1 MHz, 1.0 W . cm^{-2}, 120 s irradiation time. (a) Acoustic intensity field; (b) Temperature curves for first fat layer (blue), fat (green), and muscle (red); (c) and (d), temperature field after 60 s and 120 s, respectively, with corresponding peak temperature 40.5 °C and 43.60 °C.

temperature in muscle is 45 °C (upper limit for safe application) and fat layer reaches more then 47 °C, which is above safety levels. At this point, it is important to observe that when temperature reaches upper safety limits, a shorter application time has to be considered.

In literature a similar computational simulation in one dimension was published in 2008 [19], and a similar result was obtained, regarding the importance of the role that the skin layer plays as a means of protection of the whole body.

Omena et al. [20] have published simulated and experimental results with ultrasonic phantoms, in which the main result is that the interface fat/muscle suffers an important heating, thing that is usually neglected in the main didactic books adopted in undergraduate courses on Physical Therapy. This result is also in agreement with ours.

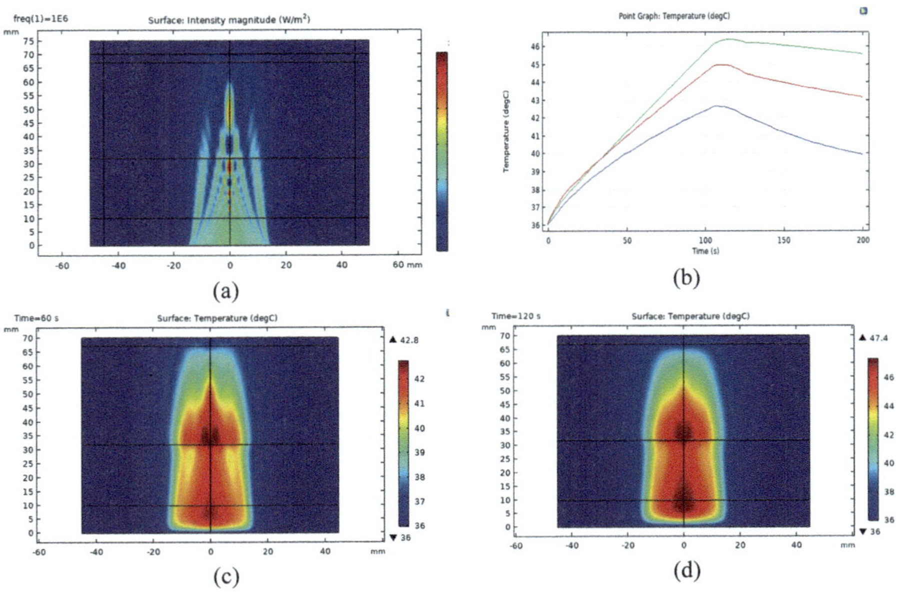

Fig. 5. Model 2, 1 MHz, 1.5 W . cm^{-2}, 120 s irradiation time. (a) Acoustic intensity field; (b) Temperature curves for first fat layer (blue), fat (green), and muscle (red); (c) and (d), temperature field after 60 s and 120 s, respectively, with corresponding peak temperature 42.8 °C and 47.40 °C.

5 Conclusion

Common therapeutic ultrasound doses were simulated, considering biological tissues of plane and parallel layers and with thicknesses compatible with humans. The transducer was kept fixed and no blood perfusion was simulated.

From the results of the simulations performed, we can conclude the skin absorbed most of the ultrasonic energy, minimizing the heating of the other layers (fat and muscle), even when the natural focus of the beam is inside the muscle layer, compromising the intended therapy. It happened for both ultrasonic intensities used.

In the next steps, these results have to be confirmed by experiments, in a more realistic scenario (blood perfusion and moving transducer) so to be used as an important information to be considered in the planning of diathermy with ultrasound.

References

1. Erikson, K.R., Fry, F.J., Jones, J.P.: Ultrasound in Medicine-A review. IEEE Trans. Sonics Ultrason. **21**, 144–170 (1974)
2. Miller, D.L., Smith, N.B., Bailey, M.R., Czarnota, G.J., Hynynen, K., Makin, I.R.S., et al.: Overview of therapeutic ultrasound applications and safety considerations. J. Ultrasound Med. **31**, 623–634 (2012)

3. Maxwell, L.: Therapeutic ultrasound: its effects on the cellular and molecular mechanisms of inflammation and repair. Physiotherapy **78**, 421–426 (1992)
4. Speed, C.A.: Therapeutic ultrasound in soft tissue lesions. Rheumatology **40**, 1331–1336 (2001)
5. Gersten, J.W.: Effect of ultrasound on tendon extensibility. Am. J. Phys. Med. Rehabil. **34**, 362–369 (1955)
6. Itakura, D.A., Magas, V., Neves, E.B., Nohama, P.: Alteração da temperatura nos tecidos biológicos com a aplicação do ultras terapêutico: uma revisão. Fisioter. Em Mov. **25**, 857–868 (2012)
7. Signori, L.U., da Costa, S.T., Neto, A.F.S., Pizzolotto, R.M., Beck, C., Sbruzzi, G., et al.: Haematological effect of pulsed ultrasound in acute muscular inflammation in rats. Physiotherapy **97**, 163–169 (2011)
8. Watson, T.: Ultrasound in contemporary physiotherapy practice. Ultrasonics **48**, 321–329 (2008)
9. Merrick, M.A., Bernard, K.D., Devor, S.T., Williams, J.M.: Identical 3-MHz ultrasound treatments with different devices produce different intramuscular temperatures. J. Orthop. Sports Phys. Ther. **33**, 379–385 (2003)
10. Bélanger, A.: Therapeutic Electrophysical Agents: Evidence Behind Practice. Wolters Kluwer Health/Lippincott Williams & Wilkins, Philadelphia (2009)
11. Cameron, M.H.: Ultrassom. In: Cameron, M.H. (ed.) Agentes Físicos Na Reabil. 3th Ed., pp. 177–206. Elsevier, São Paulo (2009)
12. Grey, K.: Distribution of treatment time in physiotherapeutic application of ultrasound. Physiotherapy **89**, 696–707 (2003)
13. Guirro, E.C.d.O., Guirro, R.: Fisioterapia dermato-funcional: fundamentos-recursos-patologias, Manole (2002)
14. Kitchen, S., Ribeiro, L.B.: Eletroterapia: prática baseada em evidências, Manole (2003)
15. Prentice, W.E.: Modalidades terapêuticas para fisioterapeutas. Artmed (2004)
16. Holmes, K.: Thermal properties. http://users.ece.utexas.edu/~valvano/research/Thermal.pdf. Accessed Oct 2010
17. Goss, S.A., Johnston, R.L., Dunn, F.: Compilation of empirical ultrasonic properties of mammalian Tissue II. J. Acoust. Soc. Am. **68**(1), 93–108 (1980)
18. Lin, W.L., Liauh, C., Chen, Y.Y., Liu, H.C., Shieh, M.J.: Theoretical study of temperature elevation at muscle/bone interface during ultrasound hyperthermia. Med. Phys. **27**(5), 1131–1140 (2000)
19. Maggi, L.E., Omena, T.P., von Krüger, M.A., Pereira, W.C.A.: Didactic software to modeling heating pattern of tissues irradiated by physiotherapeutic ultrasound. Braz. J. Phys. Ther. **12**(3), 204–214 (2008)
20. Omena, T.P., Fontes-Pereira, A., Costa, R.M., Simões, R., von Krüger, M.A., Pereira, W.C. A.: Why we should care about soft-tissue interfaces when applying ultrasonic diathermy: an experimental and computer simulation study. J. Ther. Ultrasound **5**(3), 01–06 (2017)

Sensitivity Study in High Intensity Focused Ultrasound Therapy for Cancer

Laura de Los Ríos Cárdenas[1(✉)], Leonardo A. Bermeo Varón[1], and Wagner Coelho de Albuquerque Pereira[2]

[1] Faculty of Engineering, Bioengineering, Universidad Santiago de Cali, Street 5 N 62-00, Cali, Colombia
lauradelosrioscardenas@gmail.com

[2] Biomedical Engineering Program, COPPE/Federal University of Rio de Janeiro, Rio de Janeiro, RJ 21941-914, Brazil

Abstract. Hyperthermia using High Intensity Focused Ultrasound (HIFU) is an acoustic therapy used in clinical applications to destroy malignant tumors of bone, breast, brain, kidney, pancreas, prostate, rectum and testis. This technique consists in raising the temperature in the tumor or specific zone, to achieve coagulative necrosis and immediate cell death. For having a successful treatment, it is important to monitor and observe what is the tissue behavior, as well as its changes, before, during and after the procedure. In this paper, a sensitivity study is presented to determine the suitable parameters that may be estimated using mathematical modeling, thereby simulating an optimal treatment of cancer by heating induced by High Intensity Focused Ultrasound. The sensitivity analysis indicates the conditions under which temperature profiles are sensitive to changes in thermal conductivity and attenuation coefficient. This information provides the basis for estimation of the parameters in different tissues and for prediction of the thermal responses of these tissues.

Keywords: Cancer · High Intensity Focused Ultrasound · Hyperthermia · Sensitivity analysis

1 Introduction

High Intensity Focused Ultrasound (HIFU) is a technique used in different medical applications such as cancer treatment by ablation or hyperthermia [1, 2]. This technique is based on the use of an acoustic transducer to focus ultrasound energy in a specific zone, as occurs when a magnifying glass is used to focus sunlight. An acoustic lens is used to concentrate multiple intersecting beams of ultrasound on a target. Each individual beam passes through tissue with little effect but at the focal point where the beams converge, energy can have useful thermal effects. Thus, the characterization of temperature field provided by transducers is important for the prediction of these effects [2].

On the other hand, hyperthermia therapy is the result of the absorption of the acoustic energy in the tissue. It consists in raising the temperature of the point to be treated at more than 38 °C (up to 70 °C, depending on the application), modulating its intensity, frequency and the time of exposure, thereby achieving coagulative necrosis

J. Henriques et al. (Eds.): MEDICON 2019, IFMBE Proceedings 76, pp. 1337–1342, 2020.
https://doi.org/10.1007/978-3-030-31635-8_163

and immediate cell death. This absorption can be provided by ultrasound as it propagates through biological tissue [1, 2].

Mathematical simulations are now commonly useful tools for planning and evaluating treatments such as HIFU. A sensitivity study allows to analyze the behavior the system's parameters when temperature profiles are increased by inducing heat from an external source in the tissue. Therefore, appropriate mathematical models are applied to perform the analysis and to identify which parameters are sensitive to thermal changes [3–5].

For this study, acoustic equation coupled with the Bioheat transfer equation is implemented. The solution of the mathematical formulation for this physical problem requires several thermophysical properties and input parameters, like frequency and intensity, as well as the time of exposure that determine the treatment performance considering the difference for each person [6].

In this order, the mathematical formulation was implemented, a sensitivity analysis is performed to determine which parameters can be estimated without having a deep knowledge of the system, monitoring the process before, during and after being applied.

2 Materials and Methods

The acoustic field is simulated by Eq. 1, where r and z are the radial and axial coordinates, ω is the angular frequency, m is the circumferential mode number to setting pressure acoustic equation, q is the inward heat flux of the material, p is the acoustic pressure, ρ_0 is the density and c_0 is the speed of sound propagating through the material [7]:

$$\frac{\partial}{\partial r}\left[-\frac{r}{\rho_0}\left(\frac{\partial p}{\partial r}-qr\right)\right]+r\frac{\partial}{\partial z}\left[-\frac{1}{\rho_0}\left(\frac{\partial p}{\partial z}-qz\right)\right]-\left[\left(\frac{\omega}{c_0}\right)^2-\left(\frac{m}{r}\right)^2\right]\frac{rp}{\rho_0}=rQ \quad (1)$$

The initial acoustic pressure is calculated by Eq. 2, where Z is the material acoustic impedance and I is the acoustic intensity [3]:

$$P=\sqrt{2ZI} \quad (2)$$

The temperature field is calculated by the Bioheat (Eq. 3), where c_p is the specific heat, c_b is the blood specific heat, ρ is the density, ρ_b is the blood density, Q is the external heat source, Q_m is the metabolic heat source, ω_b is the blood perfusion rate, T_b is the blood temperature and k is the thermal conductivity.

$$\rho c_p\frac{\partial T}{\partial t}+\rho c_p\mu_{trans}\cdot\nabla T+\nabla\cdot(-k\nabla T)=\rho_b c_b\omega_b(T_b-T)+Q_m+Q \quad (3)$$

The external heat source is defined by Eq. 4, where the μ_0 is the amplitude of the particle velocity and α is the attenuation coefficient:

$$Q = 2\alpha I = \alpha \rho c \mu_0^2 \tag{4}$$

For all cases examined, a rectangular two-dimensional axisymmetric domain with dimensions $L_x = 85$ mm and $L_y = 12.5$ mm was considered, with the ultrasonic transducer located at the bottom with dimensions $L_x = 12.5$ mm and $L_y = 2$ mm. Table 1 shows the physical properties involved in the models considering the material properties of the domain (muscle) and the HIFU transducer properties (water). The models were defined by using frequency analysis in pressure for the acoustics model and the time-dependent analysis in temperature for the Bioheat transfer model.

Table 1. Defined parameters [8].

Properties	
Nominal Frequency (*f*)	1 MHz
Acoustic intensity (*I*)	0.5 W . cm^{-2}
Initial value of temperature (F)	36.8 °C
Cooling temperature (T_c)	24 °C
Time of exposure	200 s
Physical properties of the domain	
Density (ρ)	1350 kg . m^{-3}
Thermal conductivity (k)	0.275 W . $m^{-1}K^{-1}$
Heat Capacity (c_p)	705 J . $kg^{-1}K^{-1}$
Speed of sound (c)	998 m . s^{-1}
Attenuation coefficient (α)	19.8 m^{-1}
Physical properties of the transducer	
Density (ρ)	1000 kg . m^{-3}
Thermal conductivity (k)	0.6 W . $m^{-1}K^{-1}$
Heat Capacity (c_p)	4178 J . $kg^{-1}K^{-1}$
Speed of sound (c)	1480 m . s^{-1}
Attenuation coefficient (α)	0.025 m^{-1}

A sensitivity study was implemented in Matlab® to identify which parameters of the model show significant variations when applying temperature changes at the system. The goal is to identify which of them are linearly independent or not among them, as well as which one are less sensitive. The method applied is the finite difference approximation defined by Eq. 5, where J is the matrix formed by the results of each sensitivity coefficient vector of each parameter, P_N the parameters that we want to determine and ε is the disturbance that will be 10% [5]:

$$j_{ij} = \frac{T_i(P_1, P_2, \ldots, P_N) - T_i(P_1, P_2, \ldots, P_j - \varepsilon P_j, \ldots, P_N)}{2\varepsilon P_j} \tag{5}$$

3 Results and Discussions

Before examining the sensitive analysis, in the Fig. 1 is shown the temperature distribution in the domain at time t = 200 s for f = 1 MHz and I = 0.5 W . cm^{-2} obtained with Comsol Multiphysics® 5.3, the analysis of this figure indicating an increase of the temperature of 13.9 °C, the rise to values are appropriate for the hyperthermia treatment [9].

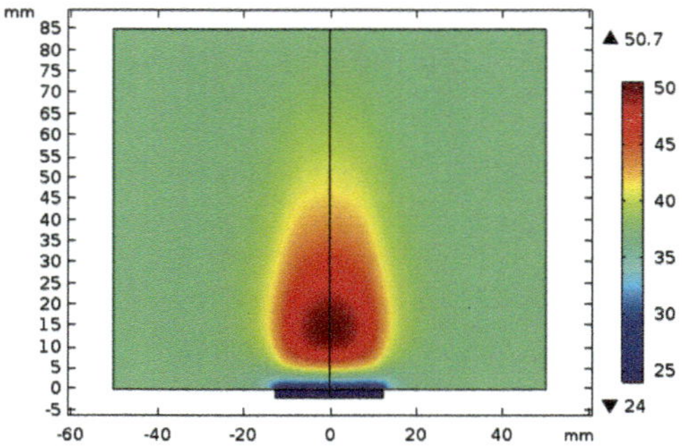

Fig. 1. Temperature distribution in the domain at time t = 200 s.

Figure 2 shows the simulated dynamic profiles of the temperature sensitivity functions in the domain(muscle) with respect to the Attenuation Coefficient *(α)*, Thermal Conductivity (k), Density (ρ), Velocity of the Sound *(c)* and Heat Capacity (C_p), as they exert more influence on the prediction of the model, considering an experimental scenario studied. From these profiles, it is possible to examine their magnitudes and whether they are linearly independent. These sensitivity functions show that the parameters ***c*** and ***α*** for the domain are mutually dependent. In the same way it is observed with the parameters ρ and c_p, depend on each other. Furthermore, the sensitivity function shows that the Thermal Conductivity *(k)* exerts more influence on the temperature profiles, observing that it is linearly independent, which indicates that the parameter can be estimated.

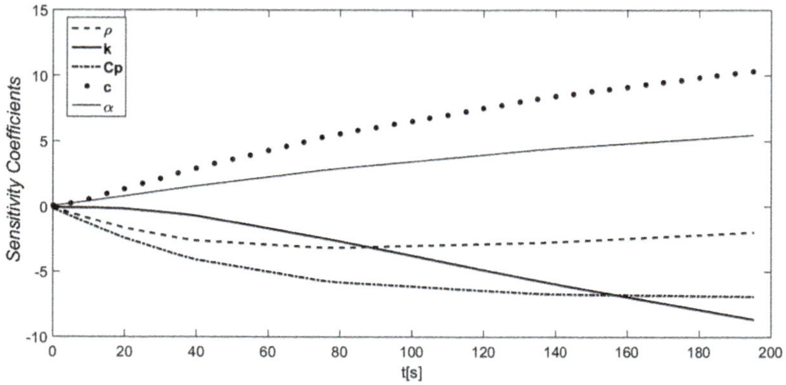

Fig. 2. Simulated dynamic profiles of the temperature sensitivity functions at the domain.

The simulated dynamic profiles of the temperature with respect to the transducer are not shown, it is due to the fact that the sensitivity analysis of these parameters has a minimum or no value, or in the experimental scenario studied, these parameters have little or no influence about the prediction of the model.

4 Conclusion

In this paper a sensitivity study is presented and the suitable parameters to be estimated had been identified. The study can contribute to develop an optimal cancer treatment with simulated and experimental data, inducing heat with High Intensity Focus Ultrasound (HIFU) and considering individual differences. The sensitivity analysis showed that two parameters can be estimated: the thermal conductivity (k) and the attenuation coefficient (α) in the domain. The remaining parameters of the domain are dependent and the parameters of the transducer exert little influence upon temperature profile. This information provides the basis for estimation of the parameters in different tissues and for prediction of the applied of therapy by HIFU.

Acknowledgments. The authors are thankful for the support provided by CNPq, CAPES, FAPERJ (Brazil) and DGI of Universidad Santiago de Cali, Colombia, project No. 819- 621118-120.

Disclosure Statement. No potential conflict of interest was reported by the authors.

References

1. Haar, G.T., Coussios, C.: High intensity focused ultrasound: physical principles and devices. Int. J. Hyperthermia **23**(2), 89–104 (2007)
2. Canney, M.S., Bailey, M.R., Crum, L.A.: Acoustic characterization of high intensity focused ultrasound fields: a combined measurement and modeling approach. J. Acoust. Soc. Am. **124** (4), 2406–2420 (2008)
3. Kałuża, G., Majchrzak, E., Turchan, Ł.: Sensitivity analysis of temperature field in the heated soft tissue with respect to the perturbations of porosity. Appl. Math. Model. **49**, 498–513 (2017). https://doi.org/10.1016/j.apm.2017.05.011
4. Mochnacki, B., Ciesielski, M., Piasecka-Belkhayat, A.: Numerical solution of the bio-heat transfer equation with uncertain parameters using the sensitivity analysis methods. Defect Diffus. Forum **379**, 39–47 (2017). ISSN 1662-9507
5. Pereyra, G.A., Lombera, G., Frontini, G., Urquiza, S.A.: Sensitivity analysis and parameter estimation of heat transfer and material flow models in friction stir welding. Universidad Nacional de Mar de Plata, Mar de Plata (7600), Argentina (2013)
6. Rocha, T., Von Kruger, M., Albuquerque Pereira, W.: Projeto e construção de um transdutor ultrassônico focalizado para terapia. XXIV Brazilian Congress on Biomedical Engineering – CBEB, COPPE/UFRJ, Rio de Janeiro (2014)
7. Datta, A., Rakesh, V.: An Introduction to Modeling of Transport Processes. Applications to Biomedical Systems, 2nd edn. Cambridge University Press, New York (2010)

8. Hasgall, P.A., Di Gennaro, F., Baumgartner, C., Neufeld, E., Lloyd, B., Gosselin, M.C., Payne, D., Klingenböck, A., Kuster, N.: IT'IS database for thermal and electromagnetic parameters of biological tissues, Version 4.0, 15 May 2018. https://doi.org/10.13099/vip21000-04-0.itis.swiss/database
9. Bermeo Varón, L.A., Orlande, H.R., Elicabe, G.: Estimation of state variables in the hyperthermia therapy of cancer with heating imposed by radiofrequency electromagnetic waves. Int. J. Therm. Sci. **98**, 228–236 (2015)

Improving Visual Contrast Between Fat and Muscle Tissues in B-Mode Images Using CBE: A Simulation Study

Mario Pastrana-Chalco[1,2], Wagner C. A. Pereira[1(✉)], and Cesar A. Teixeira[3]

[1] Biomedical Engineering Program, COPPE-Federal University of Rio de Janeiro, Rio de Janeiro, Brazil
wagner.coelho@ufrj.br

[2] Biomedical Engineering Program, Department of Physics, University of Coimbra, Coimbra, Portugal

[3] Centre for Informatics and Systems, Department of Informatics Engineering, University of Coimbra, Coimbra, Portugal

Abstract. Ultrasonic radiation can be used as a non-invasive, ionizing radiation-free, portable and inexpensive imaging technique that allows to acquire images in real time. However, this technique presents some limitations, especially when referring to image quality, for example it is not easy to distinguish some soft tissues. Several methods for improving the visual quality have been developed, which are based, in most of cases, on disturbances of the medium with external stimuli. This work uses the changes in backscattered energy (CBE) to improve the visual difference between the muscle and fat tissue, because some studies show that CBEs calculated from RF signals present different behaviors when scattered by particles with different characteristics. The new image has more visual details when compare with the conventional ultrasound images. Two numerical phantoms were simulated with different kinds of scatterers. Conventional B-mode images were obtained for different temperatures (37 °–40 °C). A new parametric image was proposed by using the angular coefficient of the curve pixel intensity *versus* temperature, obtained for each pixel. The proposed parametric image was able to enhance the visual contrast between the simulated tissues.

Keywords: Ultrasound images · Images processing · Changes in backscattered energy

1 Introduction

Ultrasound images are widely used to help medical diagnostics as they are free of ionizing radiation, easily obtained in real time, the equipment is easy to transport and is not inexpensive when compared with other systems [1]. Gynecology and obstetrics [2], cardiology [3] and ophthalmology [4] are the principal areas on medical diagnostics that use this technology. Nevertheless, the images have some limitations, one of them is the low contrast resolution since distinct tissues may have similar acoustic impedances as in some soft tissues [5]. This problem has led to new ways of acquiring ultrasound

J. Henriques et al. (Eds.): MEDICON 2019, IFMBE Proceedings 76, pp. 1343–1349, 2020.
https://doi.org/10.1007/978-3-030-31635-8_164

images to improve their quality and taking advantage of the system. Some works are based on the disturbance of the medium or the tissue with external stimuli, such as mechanical stimuli in elastography images [6, 7], optical stimuli in photoacoustic images [8], magnetic induction in magneto-acoustic tomography [9] or microbubbles injection for super-resolution ultrasound images [10, 11]. Literature contains some works on changes in backscattered energy (CBE) and they indicate that their behavior with the increase of temperature depends of the characteristic of the tissue scatterers. It was reported that in media with a predominant lipid scattering content CBE increase monotonically with the increase of temperature, while media with predominant aqueous (similar to muscle) scatterers present monotonical decrease of CBE curves [12].

This work explores CBE using computer simulation to verify the possibility to improve visual contrast in ultrasound images. In this case, two kind of scatterers were used to mimic muscle and fat tissue. The study follow the already done in [13], in order to evaluate the limitations of the proposed technique.

2 Methods

For the simulations, the MATLAB open source toolbox for time-domain simulation, k-Wave, was chosen [14].

A grid consisted of a 20 by 20 mm square with a mesh grid of 1024 × 1024 points as presented in Fig. 1 was defined. To prevent from waves to be reflected at boundaries, a perfectly matched layer (PML) is added to work as an absorbing boundary layer: a PML of 40 grid points was added to all boundaries.

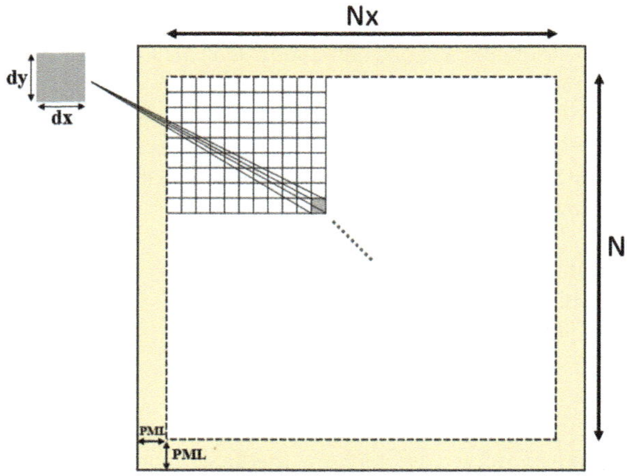

Fig. 1. Spatial grid used to simulate an effective area 20 mm × 20 mm. Nx e Ny area equal to 1024 points, and PML is equal to 20 points in every boundary. Element length is dx = dy = 19.53 μm.

The characteristics of the scatterers using in the simulated phantoms are described on Table 1. The simulations considered phantoms immersed in water because its acoustic properties are similar to soft tissue.

Table 1. Medium constants used in the simulation

	Density (ρ) (kg m^{-3})	Attenuation (α) (dBcm^{-1} MHz^{-1})
Aqueous scatterers (muscle)	1065	0.9
Lipid scatterers (fat)	950	0.6
Water	1000	0.002

Two phantoms were developed with 512 scatterers per mm^2, one of them was composed of a circular inner region with a radius of 2 mm and three concentric circular regions with different thicknesses to prevent constructive and destructive events. Each region has a different scatterer type, as presented in Fig. 2.

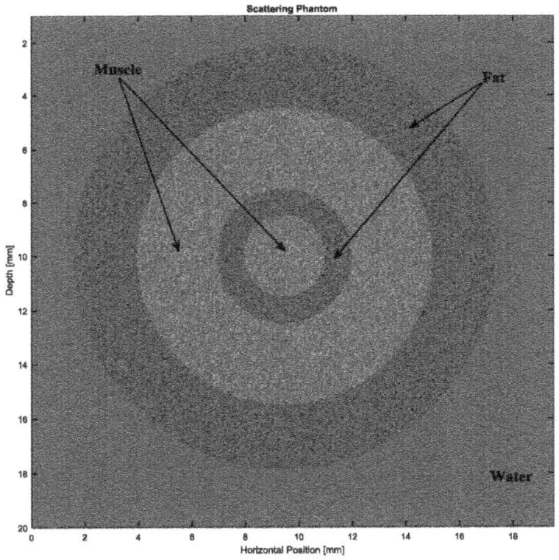

Fig. 2. Phantom with inner circular region and three concentric areas of lipid and aqueous scatterers, surrounded by a water medium

In the second phantom four circular region with one concentric area were considered. These regions are at different depths and the inner radii and the thicknesses of the concentric region of each one are 0.25 mm, 0.5 mm, 0.75 mm and 1 mm, respectively. The inner circular region in all cases has aqueous scatterer and the concentric regions contain fat scatterers, as presented in Fig. 3.

Circular and concentric phantoms are standard for image evaluation as they allow a large range of beam incidence angles, and the depth variation takes into account the influence of medium attenuation as well as the ultrasound beam spatial shape.

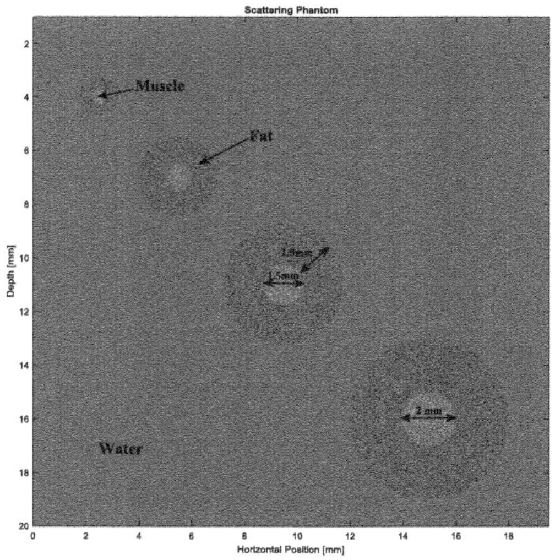

Fig. 3. Simulated four inner circular regions with concentric areas. In all cases, the lipid scatterers are in the outside region. The radius of each region is different, 0.25, 0.5, 0.75 and 1 mm respectively.

When simulating temperature variation, only sound speed was updated along the temperature increasing, because changes in density and attenuation are small for the working temperature range of 37 °C to 41 °C. So, in our simulations, density and attenuation were considered constants, and the sound speed was updated using the polynomial functions: (1), (2) and (3) [12, 15]

$$c_{water}(T) = 1471.4 + 3.9979T - 0.03513T^2 \tag{1}$$

$$c_{aq}(T) = 1471.4 + 3.9979T - 0.03513T^2 \tag{2}$$

$$c_{lip}(T) = 1810.7 - 13.892T + 0.10160T^2 \tag{3}$$

where c_{water}, c_{aq}, and c_{lip} are the temperature-dependent speed-of-sound in water, aqueous scatterers and lipid scatterers, respectively.

The ultrasound frequency used on simulations was 10 MHz with tone bursts of 4 cycles. The B-mode images were formed using the algorithm of the toolbox K-wave.

To evaluate the effect of temperature variation on the images a single-pixel intensity tracking was implemented. A curve temperature versus pixel intensity was built, then a linear model was adjusted to this curve. The angular coefficient of the adjusted model was used to create a parametric image, to possibly discriminate different scatterers types. Figure 4 summarizes this process where each pixel of the final image is the numeric value of an angular coefficient.

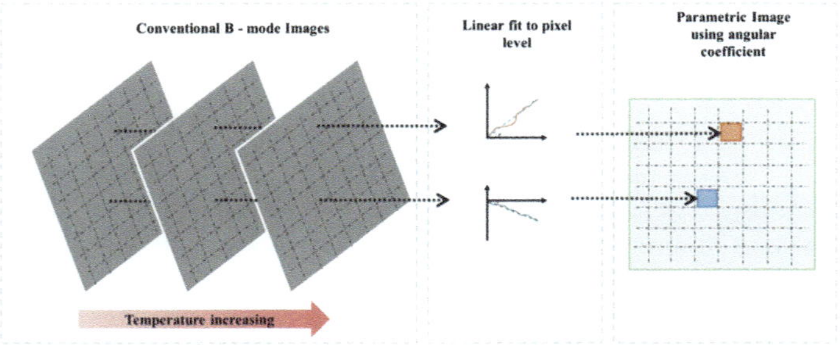

Fig. 4. Parametric image construction. A polynomial fit is applied to the curve of pixels intensities as a function of temperature and the angular coefficient of the model is represented in a matrix with equal dimensions as the original image (adapted from [16]).

3 Results

We simulated conventional B-mode images for a temperature range between 36 °C and 40 °C. This range was discretized by temperature steps of 1 °C. In Fig. 5(a), the simulated B-mode image of the phantom of Fig. 2 is presented. It is difficult, or nearly impossible, to distinguish any region that we see in the phantom. However, when analysing the parametric image presented in Fig. 5(b), it is possible to see all regions, so the structures are perceptible. This result show the usefulness of the proposed parametric imaging. It is possible to recognize the concentric regions in the conventional image of Fig. 5(a) after observing the parametric image of Fig. 5(b).

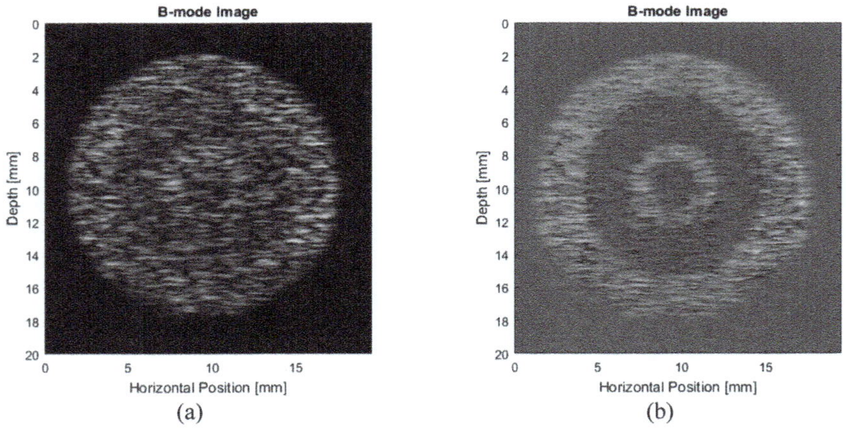

Fig. 5. Simulated conventional B-mode image (a) as compared to the parametric image (b). Both images are from the phantom of Fig. 2.

The simulated B-mode image of the phantom on Fig. 3 is shown in Fig. 6(a). As seen in the previous case, it is hard to distinguish the regions that compose the phantom. However, the parametric image presented in Fig. 5(b) shows distinguishable regions in 3 of the 4 structures. Region with inner radius less than 0.25 mm is not visible on the parametric image.

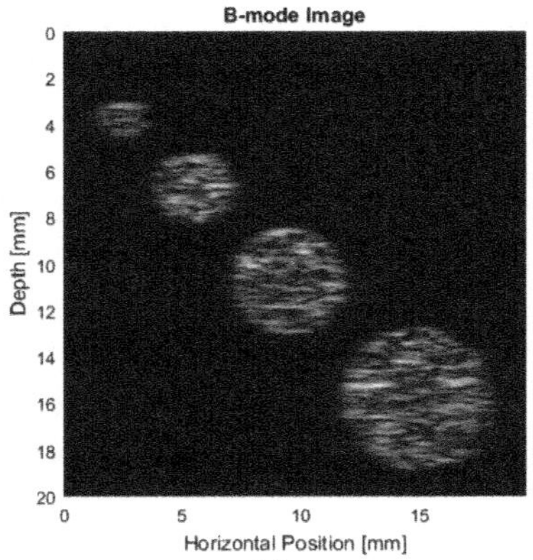

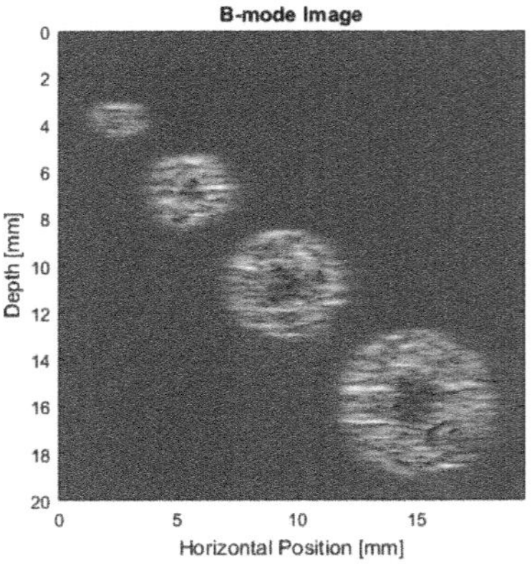

Fig. 6. Simulated conventional B-mode image (a) comparing to the parametric image (b). Both images are from the phantom of Fig. 3.

A quantitative comparison between the conventional images is not possible as they are of different nature. The conventional images are a mapping of the acoustic impedance variations of the medium in a fixed temperature, while the proposed image is a mapping of the angular coefficient related to the pixel behavior within a temperature range.

4 Discussion and Conclusion

It was possible to distinguish lipid and aqueous scattering regions by monitoring pixel intensity changes with temperature. On the second phantom, some structures cannot be visible when these are less than 0.75 mm. It would have happened because the quantity of scatterers was small. It is necessary to study small regions (less than 0.75 mm) with more scatterers inside and also study media with a mixture of the two kinds of scatterers. These new steps will help use valuate the potentialities and limitation of our proposed method. The same patterns will be able to study using real phantoms of both tissues to compare our results.

Acknowledgment. This work had the financial support from the Luso-Brazilian cooperation project CAPES/FCT "Thermo - response" (FCT/CAPES: 10172/13-0) and from CNPq and FAPERJ.

References

1. Deserno, T.M.: Biological and Medical Physics, Biomedical Engineering: Biomedical Image Processing. Springer, Alemanha (2011)
2. Callen, P.W.: Ultrasonography in Obstetrics and Gynecology, 5th edn. Saunders, Philadelphia (2008)
3. Belohlavek, M., Foley, D.A., Gerber, T.C., Kinter, T.M., Greenleaf, J.F., Seward, J.B.: Three-and four-dimensional cardiovascular ultrasound imaging: a new era for echocardiography. Mayo Clin. Proc. **68**(3), 221–240 (1993)
4. Fledelius, H.C.: Ultrasound in ophthalmology. Ultrasound Med. Biol. **23**(3), 365–375 (1997)
5. Azhari, H.: Basics of Biomedical Ultrasound for Engineers. Wiley, Hoboken (2010)
6. Righetti, R., Ophir, J., Ktonas, P.: Axial resolution in elastography. Ultrasound Med. Biol. **28**(1), 101–113 (2002)
7. Manduca, A., Oliphant, T.E., Dresner, M.A., Mahowald, J.L., Kruse, S.A., Amromin, E., et al.: Magnetic resonance elastography: non-invasive mapping of tissue elasticity. Med. Image Anal. **5**(4), 237–254 (2001)
8. Xu, M., Wang, L.V.: Photoacoustic imaging in biomedicine. Rev. Sci. Instrum. **77**(4) (2006)
9. Li, X., Yu, K., He, B.: Magnetoacoustic tomography with magnetic induction (MAT-MI) for imaging electrical conductivity of biological tissue: a tutorial review. Phys. Med. Biol. **61** (18), R249–R270 (2016)
10. Errico, C., Pierre, J., Pezet, S., Desailly, Y., Lenkei, Z., Couture, O., et al.: Ultrafast ultrasound localization microscopy for deep super-resolution vascular imaging. Nature **26**, 499–502 (2015)
11. Christensen-Jeffries, K., Browning, R.J., Tang, M.X., Dunsby, C., Eckersley, R.J.: In vivo acoustic super-resolution and super-resolved velocity mapping using microbubbles. IEEE Trans. Med. Imaging **34**(2), 433–440 (2015)
12. Straube, W.L., Martin, A.R.: Theoretical estimation of the temperature dependence of backscattered ultrasonic power for noninvasive thermometry. Ultrasound Med. Biol. **20**, 915–922 (1994)
13. Teixeira, C.A., Pastrana-Chalco, M., Simões, R.J., Pedrosa, A., von Krüger, M.A., Alvarenga, A.V., Fontes-Pereira, A.J., Pereira, W.C.A.: On the feasibility of ultrasound imaging enrichment by medium - temperature changes. Ultrason. Imaging **41**(1), 17–35 (2019)
14. Treeby, B.E., Cox, B.T.: k-Wave: MATLAB toolbox for the simulation and reconstruction of photoacoustic wave fields. J. Biomed. Opt. **15**(2), 21314 (2010)
15. Arthur, R.M., Straube, W.L., Starman, J.D., Moros, E.G.: Noninvasive temperature estimation based on the energy of backscattered ultrasound. Med. Phys. **30**(6), 1021–1029 (2003)
16. Teixeira, C.A., Fontes-Pereira, A., von Kruger, M.A., Alvarenga, A.V., Pereira, W.C.A.: Expressiveness of temperature-induced changes in backscattered energy in conventional B-mode images. In: Proceedings of Meetings on Acoustics, Montreal, vol. 19, p. 075006 (2013)

Special Sessions: Computational Biology and Medical Applications

CFD Analysis for the Evaluation of Patient-Specific Hemodynamic Parameters in Cerebral Aneurysms

Iolanda Velho[1(✉)], Jorge Tiago[1], Alberto Gambaruto[2], Adélia Sequeira[1], and Ricardo Pereira[3]

[1] Instituto Superior Técnico, and CEMAT/IST, Universidade de Lisboa, Lisbon, Portugal
iolanda.velho@tecnico.ulisboa.pt
[2] University of Bristol, Bristol, UK
[3] CHUC - Centro Hospitalar e Universitário de Coimbra, Coimbra, Portugal

Abstract. Blood flow simulations are now considered a valuable tool for a deeper understanding of the physiopathology of intracranial aneurysms. Many authors built robust computational settings based on accurate computer-assisted registration, segmentation, and 3D geometry reconstruction from medical images of patient-specific cerebral aneurysms, and special techniques to derive appropriate boundary conditions. However, an accurate description of flow mechanics in the near-wall region and its connection with the evolution of the wall disease evolution remains linked to several questions not yet fully understood. Recently, several authors have suggested a lower order approximation of the Lagrangian dynamics in the near-wall region, which allows for a meaningful characterization of both normal and parallel flow direction to the wall. We verify this computational approach in a follow-up case study and try to provide a step further in the understanding of the hemodynamics environment and its possible connection with the aneurysm inception and growth.

Keywords: CFD · Cerebral aneurysms · Near-wall region · WSS critical points

1 Introduction

Due to the asymptomatic behaviour of the majority of the unruptured Intracranial Aneurysms (IA), they are detected incidentally. In most of these cases, the treatments options are proposed, and longitudinal follow-up analyses without any treatment, are not possible.

In the case of no treatment recommendations, the follow-up evaluation should be done. However, this follow-up is limited on time and number of acquisitions which explains why very few studies with longitudinal follow-up are available. Follow-up analysis of stable (without growth) and unstable aneurysms (with morphological changes) provide an interesting comparison between the possible

J. Henriques et al. (Eds.): MEDICON 2019, IFMBE Proceedings 76, pp. 1353–1360, 2020.
https://doi.org/10.1007/978-3-030-31635-8_165

reasons for the morphological aneurysm growth. In [1], these two different IA status were compared, and the results show that unstable IA has more complex flow pattern, a concentrated high WSS and large areas of low WSS, that is a more non-uniform WSS distribution, than stable IA. On the other hand, in [2] the authors found that the region of low WSS is associated with the growth of the aneurysm. Some controversy still remains. The few works with unruptured unstable IA focus on the CFD analysis regarding only flow patterns and WSS magnitude values and its derived parameters. An effort to incorporate more information and mechanical parameters is needed. The near-wall region is where the blood flow interacts with the endothelium cell layer and, therefore, it is extremely important to understand the physiological functions of the vascular system and how the mechanical forces act in the mechanotransduction. In [3], the present authors and others, have applied the methodology presented in this work to an IA cohort, in order to perform a detailed analysis of the near wall region.

In this work, we go further, investigating a follow-up case. The advantage of the follow-up analysis in one individual is to isolate other factors such as the geometrical features, the genetic factors, aneurysms site, among others. It is more robust to compare under the same patient, and then verify hemodynamic factors associated with their anatomic local differences. We propose to analyze an unruptured and untreated IA follow-up patient-specific case, obtained during four-years, with four different acquisitions. With this long-term study, analyzing in detail the hemodynamics through the parameters above described, we can understand the conditions that predispose aneurysms for growth. Up to our knowledge, possible correlations between geometrical changes, along time, and hemodynamical factors, such as near-wall transport, and critical points, have not been investigated yet.

2 Methods

2.1 Study Case

We investigate a 4-year longitudinal follow-up of an IA. The patient is a 53 year-old female with a basilar aneurysm. The first acquisition was done in 2011, the second one in 2012, the third one in 2015 (February) and the last one, in October 2015. The last acquisition allowed for the identification of a thrombus. In Fig. 1 we can see the 3D reconstructions and the morphological changes. In Fig. 2, we can observe the values of the volumes along the follow-up presented in the table, and on the right, the positive (expansion) and negative displacements (contraction) between two consecutive acquisitions. This figure reveals an interesting case study since it is possible to see prominent geometrical variations along all times. This case is a very unstable IA once the volume increased too much along the four years. From the first acquisition until the last one, the volume became around three times bigger (the last volume measure did not take into account the thrombus area).

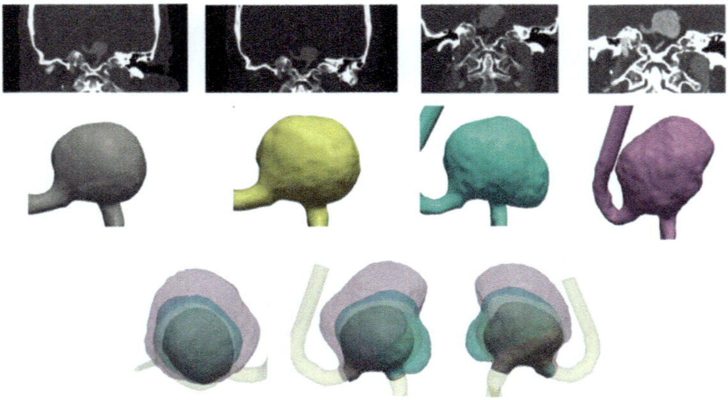

Fig. 1. Top: CT scan for each acquisition and the correspondent 3D reconstruction. Bottom: Overlap of the 4 geometries of the follow-up.

2.2 Geometries

The geometries of IA were reconstructed from medical images obtained *in vivo* from Computed Tomography (CT). The reconstruction of the 3D geometry consists in image segmentation, surface extraction, and surface smoothing and meshing. For this procedure, we used existing open-source software (ITKSnap, MeshLab and Gmsh). Artificial extrusions are created in every inlet and outlets to partially account for that ideal length (in order to be long enough for the fluid to be fully developed). The extruded inlet lengths are 20 times its diameter. The surface is composed by 2D elements, triangles, and the volume by 3D elements, tetrahedra. The number of degrees of freedom is around 3 million.

2.3 Numerical Simulation Setup

We consider the Navier-Stokes equations to model the blood flow, which we assume as an incompressible, homogeneous and Newtonian fluid. To solve the above equations, a Finite Element Method was applied to the space discretization (FEniCS) and the Incremental Pressure Correction Scheme to the time discretization. The Blood Density was 1.056×10^{-3} g/mm^3 and the Blood Dynamic Viscosity 3.45×10^{-3} g/mm.s. Vessel walls were assumed to be rigid, and we imposed the non-slip condition at the wall boundary. Since patient-specific velocity was not available at the inflow boundary, we applied a velocity curve found in the literature. This curve has a great similarity to our case with respect to demographic factors such as sex, age and aneurysm localization. The velocity curve imposed at the inlet was the same (with the same Reynolds number) for all cases of this follow-up. At the outlet of the artery, we applied a homogeneous Neumann condition. Three cardiac cycles were computed using 800 time-steps per cycle (10^{-3} s). The results presented correspond to the third cycle.

Year	Volume (cm^3)
2011	1,68
2012	2,10
2015feb	3,45
2015oct	4,58

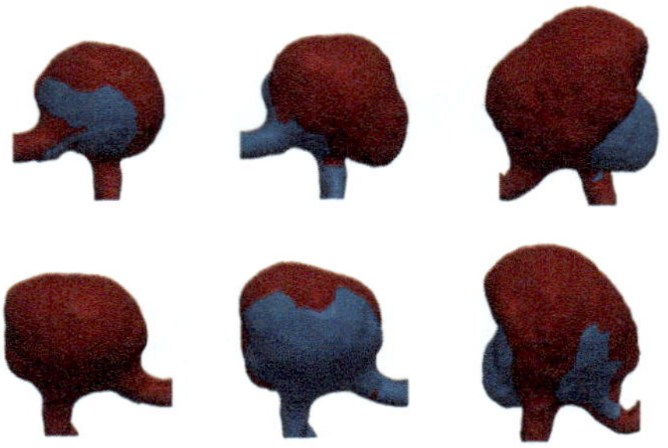

Fig. 2. Left: Table with the volumes along the follow-up. Right: Positive (red) and negative (blue) displacements between every two consecutive acquisitions, in two viewpoints.

2.4 Hemodynamics Parameters

The flow topology within the IA is well accepted to play a fundamental role in their natural history. Besides this analysis, other common indicators derived from the Wall Shear Stress (WSS) are essential to better understand the dynamics between the fluid flow and the wall, as the Relative Residence Time (RRT), used to quantify the near-wall stagnation, the Time Average WSS (TAWSS) and the Oscillatory Shear Index (OSI) which is a dimensionless measure often used to study spatially quantified oscillatory flow behaviour. In [4] one can find the description of these measures. In this work, we analyze the above parameters but also the near-wall region, regarding the near-wall transport perpendicular to the wall and the WSS critical points. The approach here presented is detailed in [3].

The near-wall transport, which provides important information in the arterial hemodynamics from the standpoint of exchange processes and interactions between the fluid flow and the wall, can be derived by Taylor expansion of a fluid element's trajectory in the limit of the wall. In this limit, the tangential component of the transport is given by the WSS (τ) and the normal component is given by the divergence of the WSS (WSSdiv). As a result, the fluid velocity at a small distance normal to the wall δn can be written as:

$$\mathbf{u}_{\pi} = \frac{\tau \delta n}{\mu} + O(\delta n^2) \qquad \text{and} \qquad \mathbf{u}_n = -\frac{\delta n^2}{2\mu} \nabla \cdot \tau + O(\delta n^3) \tag{1}$$

where u_π and u_n represent the tangential and normal components of the velocity vector, respectively, evaluated in a small distance normal to the wall δn, and μ is the dynamic viscosity.

The WSS critical point, also called fixed point, is a point where this vector is null. It can be directly correlated with the fluid flow topology near to the wall, for instance, the flow separation and impingement regions. In this work, we first project the 3-dimensional elements of the boundary mesh into 2-dimensional coordinates. The WSS values computed in each vertex, in each triangle, have two orthogonal directions. The triangles are piecewise-linear and planar and so, the analysis performed is in a 2D domain. In the 2D dimension vector field, there are five possible solutions which involve either real or imaginary eigenvalues. Three of these possible cases originate critical points, which are further reported. We also describe the lines that are aligned to the tangential component of the viscous traction exerted by the flow on the wall, the surface shear lines. They are obtained by integrating the WSS on the surface.

3 Results and Discussion

In this work, we propose to investigate the correlation between the shape changes of the aneurysm, expansion or contraction with the hemodynamic parameters, including the near-wall flow and the WSS critical points.

We performed the analysis in three different instants of time of the cardiac cycle, at the peak of systole, mid-diastole and at the end of diastole. Regarding the flow topology, the velocity magnitude tends to decrease along with the follow-up. The vorticity also decreases. In all cases, the streamlines show a complex and active flow pattern (evident recirculation and vortex) within the IA (laminar and with low velocity values). A prominent jet, with an evident direction, was observed at four instants of time, with higher velocity in the first acquisition. Along the other three years, this jet decreased its intensity. In the last year, the presence of a thrombus changed the flow completely, giving a confirmation that the geometry is one of the most important features to determine the blood flow behaviour inside the aneurysm.

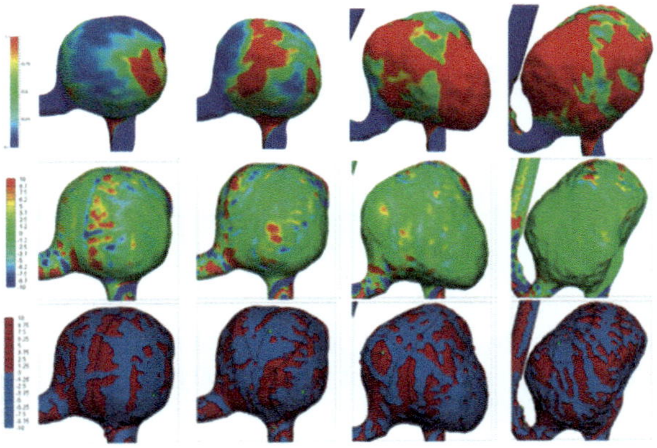

Fig. 3. First row shows the RRT. Second row shows the WSSdiv (peak of systole) and in the third row are presented the critical points (green points) and the surface shear lines that overlap the positive and negative WSSdiv (peak of systole).

We intend to investigate the values which are considerable abnormals and can lead to an abnormal IA wall response [5]. Values of WSS magnitude considered too high (upper than 15 Pa) tend to decrease along with the follow-up. The RRT, contrary, tends to increase the high values and the associated area (see Fig. 3). These results manifest the stagnation process, a probable event related with the growth of IA. Even if is not easy to establish the correspondence between the WSS values and the areas with positive displacement, there was a higher increase in volume at the top of the aneurysm dome. The highest values of WSS magnitude (around 10 Pa) are localized at the top, especially during the systole period. This is not consistent with [2].

In [6] it is shown a negative correlation between RRT and aneurysm inception. Contrary to our results, the RRT is the parameter which highlights, with large areas with high values. The RRT is associated with disturbed blood flow, marked by a low magnitude and high OSI, possibly responsible for atherogenesis and EC perturbed response.

Table 1. Number of the critical points based on its eigenvalues, real numbers and spiralling, during the cardiac cycle, for the follow-up.

Year	Time (s)	# Spiralling Points	# Real Points (node/saddle)
2011	0,182	5	11
	0,390	10	20
	0,790	6	8
2012	0,182	8	14
	0,390	13	24
	0,790	34	57
2015 feb	0,182	5	12
	0,390	9	17
	0,790	23	35
2015 oct	0,182	4	11
	0,390	8	15
	0,790	16	19

The normal convective transport (WSSdiv) takes positive values to indicate flow movement towards the wall and becomes negative if the movement is away from the wall. The WSS critical points, also analyzed, are marked by coloured dots, such that green indicates a spiralling, hence complex conjugate pair solution (spiralling motion), blue indicates a saddle point or node, hence real solution. Red points mark the vertices of small edges identifying the local eigenvectors associated with critical points.

From Fig. 3, it is clear that the near-wall flow field is very homogeneous regarding the range of values, that is between $|\text{WSSdiv}| < 10\,\text{Pa mm}^{-1}$ (green scale). This suggests a slow perpendicular flow. If we compare the near-wall transport in terms of positive/negative values with the WSS magnitude, the near-wall transport is more complex, and undergoes more variations in space.

We computed the area of negative values (flow from the wall) of WSS divergence (compared with positive), but this difference is not too big (55% to 45%, respectively) for all years. The tendency is increasing the positive values during the cardiac cycle (deceleration) and decreasing the negative values (highest in systole peak). However, the percentage of negative values is still dominating, revealing that even in deceleration of the flow, the movement of the fluid is away to the wall, for the mainstream.

Regarding the critical points, there is the presence of a considerable quantity of both, real and spiralling points. During the entire follow-up, the critical points (both natures) increase significantly with the deceleration of the cardiac systole (see Table 1). This can be justified by the fact that the deceleration actually increases eddy motion as a dissipation mechanism. Another interesting result is the consistency of some spiralling WSS critical points, present in the entire cardiac cycle and along the follow-up. These persistent critical points are reported in negative near-wall transport region, that indicates the flow is moving away from the wall. Through the shear lines computed, it is possible to determine that they are a stable spiral (sink) complex pairs, which means that the flow moves from near-wall to the main flow stream, along the vortex.

4 Conclusions

In conclusion, the results suggest the following hypotheses: in stagnation flow, with high RRT values, a complex flow topology behaviour with recirculation with laminar low velocities, promotes an abnormal response from the wall, leading to the wall remodelling and consequently growth of the aneurysm. The near-wall flow shows slow activity relative to transport and reveals a significant quantity of critical points. The analyses of near-wall flow is a new method to describe the flow behaviour in a more detailed way which can provide interesting informations.

Acknowledgments. This project was supported by Fundação para a Ciência e a Tecnologia (Portugal), through the grant SFRH/BD/90339/2012, by the projects UID/Multi/04621/2013 and UID/Multi/04621/2019 of CEMAT/IST and by the Department of Mathematics of the Instituto Superior Técnico, University of Lisbon.

References

1. Sforza, D.M., Kono, K., Tateshima, S., Viñuela, F., Putman, C., Cebral, J.R.: Hemodynamics in growing and stable cerebral aneurysms. J. Neuro. Surg. **8**, 407–412 (2016)
2. Boussel, L., Rayz, V., McCulloch, C., Martin, A., Acevedo-Bolton, G., Lawton, M., Higashida, R., Smith, W.S., Young, W.L., Saloner, D.: Aneurysm growth occurs at region of low wall shear stress: patient-specific correlation of hemodynamics and growth in a longitudinal study. Stroke **39**, 2997–3002 (2008)
3. Ardakani, V., Tu, X., Gambaruto, A., Velho, I., Tiago, J., Sequeira, A., Pereira, R.: Near-wall flow in cerebral aneurysms. Fluids **4**, 89 (2019)

4. Arzani, A., Gambaruto, A., Chen, G., Shadden, S.C.: Wall shear stress exposure time: a Lagrangian measure of near-wall stagnation and concentration in cardiovascular flows. Biomech. Model. Mechanobiol. **16**, 787–803 (2017)
5. Zhou, G., Zhu, Y., Yin, Y., Su, M., Li, M.: Association of wall shear stress with intracranial aneurysm rupture: systematic review and meta-analysis. Sci. Rep. **7**, 5332 (2017)
6. Riccardello Jr., G.J., Shastri, D.N., Changa, A.R., Thomas, K.G., Roman, M., Prestigiacomo, C.J., Gandhi, C.D.: Influence of relative residence time on side-wall aneurysm inception. Neurosurgery **83**, 574–581 (2017)

Wireless Capsule Endoscope Location and a Robotic Validation Experiment

Isabel N. Figueiredo[1(✉)], Luís Pinto[1], Luís Perdigoto[2,3], Marina Oliveira[3], Hélder Araújo[3], and Pedro N. Figueiredo[4,5]

[1] CMUC, Department of Mathematics, FCTUC, University of Coimbra, Coimbra, Portugal
{isabelf,luisp}@mat.uc.pt
[2] ESTG, Polytechnic Institute of Leiria, Leiria, Portugal
luis.perdigoto@ipleiria.pt
[3] Institute for Systems and Robotics, University of Coimbra, Coimbra, Portugal
marinaa.oliveira95@gmail.com, helder@isr.uc.pt
[4] Gastroenterology, CHUC (Centro Hospitalar e Universitário de Coimbra), Coimbra, Portugal
[5] Faculty of Medicine, University of Coimbra, Coimbra, Portugal
pnf11@sapo.pt

Abstract. We present results concerning the validation of a novel approach for wireless capsule endoscope localization, using as ground truth a simulated biological/mechanical environment experiment. The approach relies essentially on image-based methods. It involves a hybrid multiscale affine and elastic image registration procedure which is afterwards appropriately complemented with calibration and visual odometry techniques. The capsule was fixed at the extremity of a robotic arm and moved along a part of an *ex-vivo* mammalian bowel. The first validation results indicate a good correlation between the ground truth velocity and distance traveled by the capsule and the velocity and distance given by the proposed approach.

Keywords: Image processing · Image registration · Calibration · Visual odometry · Endoscopy

1 Background

Wireless Capsule Endoscope (WCE) [1] is a medical device which enables the interior visualization of the gastrointestinal tract for diagnosis purposes. The capsule consists of one or more miniaturized cameras, light sources and a wireless system for the acquisition and transmission of images [2]. In a WCE exam, a patient ingests the capsule, and as it moves in the gastrointestinal tract, propelled by peristalsis (contractions of the bowel muscles that move its contents), images are transmitted to a tape recorder, placed on a belt outside the body. It is particularly important for the examination of the small intestine, since this organ is not easily analyzed by conventional techniques.

J. Henriques et al. (Eds.): MEDICON 2019, IFMBE Proceedings 76, pp. 1361–1365, 2020.
https://doi.org/10.1007/978-3-030-31635-8_166

However, one of the difficulties of this WCE diagnostic technique is the impossibility of knowing the precise location of the capsule when an abnormality is detected. Clearly, location of the lesions is a relevant information for physicians, in order to plan treatment and surgery procedures. Therefore lack of information about the location of the capsule, or alternatively, its distance measured from landmark anatomical points, is a strong limitation of WCE (see for instance [3] for an extended review on WCE location and also [4–6] for other papers on the same topic).

This WCE drawback is even more important for images of the small bowel, which is a long organ (the size is in average 6.5 to 7 m in adults). It does not have any intermediate reference points, except the pylorus, that connects the stomach to the small bowel, and the ileocecal valve, a sphincter muscle valve that separates the small intestine and the large intestine.

2 Methods

Recently, in [7] we have proposed an hybrid multiscale affine and elastic image registration approach towards wireless capsule endoscope location [2]. This methodology [7] enables a fast and global interpretation of a video by converting it into a curve. This curve gives also a qualitative information about the speed of the capsule. In fact, each point in the curve represents the degree of similarity between two consecutive frames in the video, after being registered with the proposed image registration procedure, and by using an appropriate similarity measure. So, in this curve, a very low value indicates that the image content does not change in the corresponding pair of consecutive frames, thus this means that the capsule is almost still or rotates/moves slowly, because it is capturing the same scene. On the other hand, a high value in the curve indicates abrupt scene changes in the two consecutive frames, therefore the image content is changing, and consequently the capsule should have moved, or the background has changed dramatically due to peristalsis.

By combining the above image registration procedure with calibration (for determining capsule specific measurements) and visual odometry approaches (for deriving the change in position over time), see for example [8] for a detailed explanation of these techniques, we evaluate, in an experimental simulated biological/mechanical environment, the ability of this image registration procedure, for estimating the capsule location and motion, in physical units.

3 Preliminary Results

The experimental setup, providing the ground truth information (which is always inaccessible in real-life WCE videos), consisted of a WCE (MiroCam®, with one camera), that was fixed at the extremity of a robotic arm and moved along a part of an *ex-vivo* porcine bowel. The preliminary results show a good agreement between the ground truth path and the estimated path traveled by the capsule, obtained with the proposed image-based methods.

Figure 1 shows three examples of the images of the mammalian bowel, acquired by the capsule during the experiment.

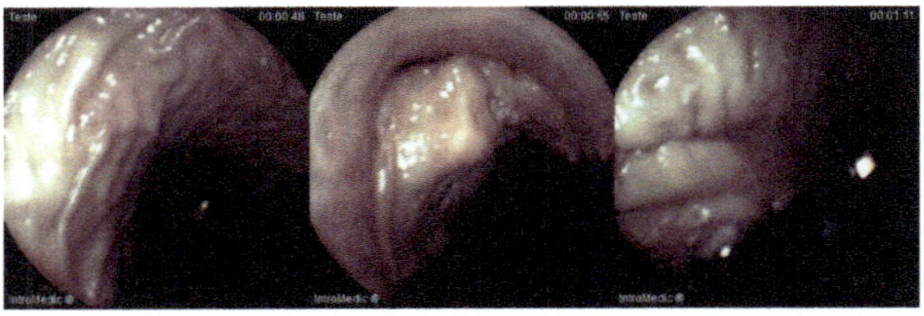

Fig. 1. Examples of WCE images.

Figure 2 shows a comparison between the ground truth capsule velocity and that given by our method [7], for a part of a video generated by the capsule between frames 134 and 257. Figure 3 displays, for the same frames, the comparison concerning the distance traveled by the capsule along the mammalian bowel. In both Figures the horizontal axis represents the 124 frames, numbered from 1 until 124. It is evident the similarity between the ground truth and the results provided by our proposed methodology [7].

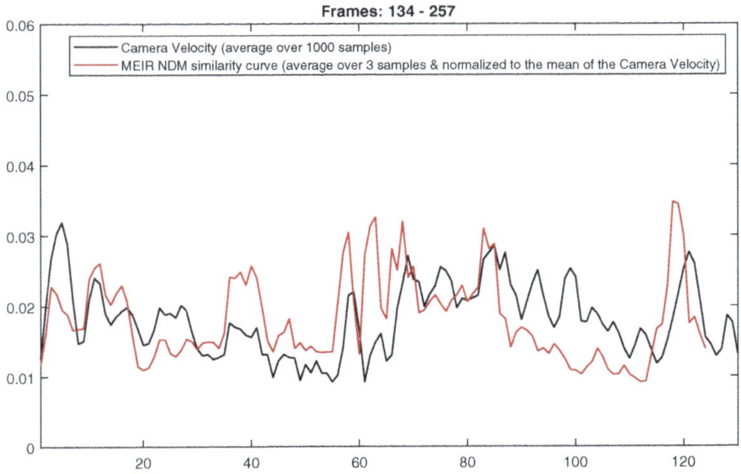

Fig. 2. Comparison between the ground truth WCE velocity (in black) and the WCE velocity given by our method (in red).

Using Structure-from-Motion (SfM) techniques [8] the position of the capsule in three-dimensional (3D) space can be reconstructed from image information, specifically, by using the correspondences between points in consecutive

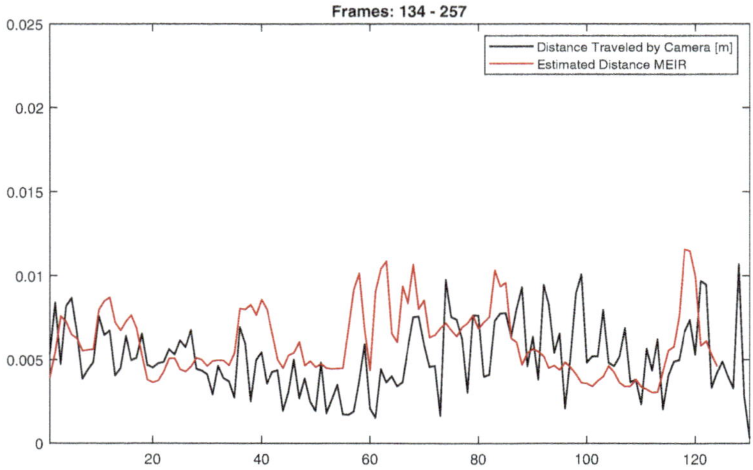

Fig. 3. Comparison between the ground truth distance traveled by the WCE (in black) and the one given by our method (in red).

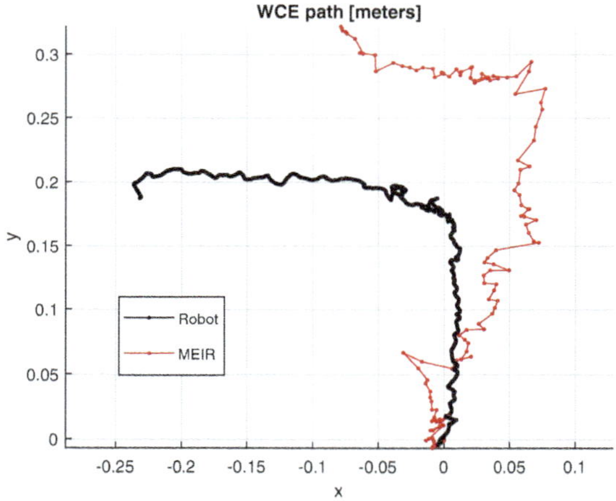

Fig. 4. Comparison between the ground truth trajectory (in black), provided by the robot data, and the recovered capsule trajectory (in red). The figure shows the projection of both trajectories onto the XY plane, which corresponds to the "ground" plane.

images, established from the hybrid image registration of [7]. Figure 4 shows the ground truth WCE path (in black), obtained from the robot data, and the SfM reconstructed path. Both 3D trajectories are projected onto the $Z = 0$ plane, corresponding to the "ground" plane.

It should be noted that the 3D trajectory of the capsule can only be reconstructed up to an unknown scale factor, due to a fundamental limitation of SfM. Despite that, the similarities between ground truth and estimated paths are easily seen, with the overall shape of the real trajectory being successfully reconstructed.

Acknowledgement. This work was supported by the FCT (Fundação para a Ciência e a Tecnologia, Portugal) research project PTDC/EMD-EMD/28960/2017, and also partially by the FCT grant UID/MAT/00324/2019.

References

1. Idan, G., Meron, G., Glukhovsky, A., Swain, P.: Wireless capsule endoscopy. Nature **405**, 417–417 (2000)
2. Moglia, A., Menciassi, A., Dario, P.: Recent patents on wireless capsule endoscopy. Recent Pat. Biomed. Eng. **1**(1), 24–33 (2008)
3. Than, T.D., Alici, G., Zhou, H., Li, W.: A review of localization systems for robotic endoscopic capsules. IEEE Trans. Biomed. Eng. **59**(9), 2387–2399 (2012)
4. Cunha, J., Coimbra, M., Campos, P., Soares, J.: Automated topographic segmentation and transit time estimation in endoscopic capsule exams. IEEE Trans. Med. Imaging **27**(1), 19–27 (2008)
5. Ye, Y., Swar, P., Pahlavan, K., Ghaboosi, K.: Accuracy of RSS-based RF localization in multi-capsule endoscopy. Int. J. Wireless Inf. Networks **19**(3), 229–238 (2012)
6. Spyrou, E., Iakovidis, D.K.: Video-based measurements for wireless capsule endoscope tracking. Meas. Sci. Technol. **25**(1), 015002 (2014)
7. Figueiredo, I.N., Leal, C., Pinto, L., Figueiredo, P.N., Tsai, R.: Hybrid multiscale affine and elastic image registration approach towards wireless capsule endoscope localization. Biomed. Signal Process. Control **39**, 486–502 (2018)
8. Hartley, R., Zisserman, A.: Multiple View Geometry in Computer Vision. Cambridge University Press, Cambridge (2003)

Special Sessions: Smartphone Based, Patient-Centred Technologies

SmartBEAT: A Smartphone-Based Heart Failure Telemonitoring Solution

José Silva-Cardoso[1,2,3(✉)], Emília Moreira[3], Inês Lopes[4], Carla Sousa[2], Sérgio Leite[3,5], Manuel Campelo[2,3], José Maria Sousa[2,3], Manuela Fonseca[2,3], Linda Harnevo[6], Moshè Farin[6], Luís Filipe Azevedo[1,3], and Filipe Sousa[4]

[1] Faculty of Medicine, University of Porto, Alameda Prof. Hernâni Monteiro, 4200-319 Porto, Portugal
silvacardoso30@gmail.com
[2] São João University Hospital Center, Alameda Prof. Hernâni Monteiro, 4200-319 Porto, Portugal
[3] Center for Health Technology and Services Research, R. Dr. Plácido da Costa, 4200-450 Porto, Portugal
[4] Fraunhofer Portugal, R. Alfredo Allen 455, 4200-135 Porto, Portugal
[5] Hospital Senhora da Oliveira, Rua Cutileiros 114, Creixomil, Guimarães, Portugal
[6] LifeonKey Inc., 60 Medinat HaYehudim, Entrance C, 46766 Herzlya, Israel

Abstract. Heart failure (HF) is a chronic syndrome associated with poor prognosis and high healthcare costs, mainly due to frequent hospitalizations. These are preceded by changes in several physiological parameters, creating an opportunity for early detection and treatment of hemodynamic unbalance, and hospitalizations avoidance. Invasive telemonitoring systems have proven efficacy on this detection. Conversely, non-invasive systems need further evidence, mainly due to poor technological adherence.

In this paper we describe SmartBEAT as smartphone-based HF non-invasive home telemonitoring system (NIHTS) and analyze preliminary results of SmartBEAT field test.

SmartBEAT collects, analyses and manages HF patients' data. Biological parameters are collected by external sensors and transmitted to a mobile phone and to a portal, where an algorithm identifies false from true alarms and displays a hierarchical interventional response.

Patients were recruited in a Heart Failure Clinic at a University Hospital and used this system during 3 months, on a daily basis. Adherence to the system was analyzed and pre-/post-test data of usability assessments and perception of benefit were compared.

A total of 10 patients aged ≥ 60 were recruited. Mean age 68 ± 4, 30% women, 8 ± 5 mean education years.

Adherence to monitoring protocol was 97% at 3 months. Patients' evaluation of this NIHTS was very positive in pre- and post-test assessments, reporting it was easy to use, conveyed a higher sensation of safety and was useful for healthcare management.

Keywords: Heart Failure · Non-invasive telemonitoring · Smartphone

J. Henriques et al. (Eds.): MEDICON 2019, IFMBE Proceedings 76, pp. 1369–1376, 2020.
https://doi.org/10.1007/978-3-030-31635-8_167

1 Introduction

Heart Failure (HF) is associated with a very high mortality and morbidity. The increasing number of HF patients is overloading public healthcare systems worldwide, representing 1–2% of healthcare budget in developed countries [1, 2]. This is mainly due to HF hospitalizations, which in turn are associated with high mortality and re-hospitalization rates. Hospitalizations are mainly due to fluid accumulation within the body. This occurs gradually, creating an opportunity for early detection and treatment, thus allowing to avert hospitalizations [3].

Invasive telemonitoring proved this concept as it effectively showed to reduce hospitalizations [4]. Nevertheless, these invasive systems are only feasible in a minority of patients. Non-invasive home telemonitoring system (NIHTS) emerged as a way of extending telemonitoring to a much wider target [4].

Evidence on NIHTS efficacy is still conflicting. Its positive impact on hospitalizations and mortality reduction was documented in two meta-analysis [5, 6]. More recently TIM-HF2 trial also proved its benefit on the reduction of days lost by HF hospitalization or all cause-mortality in a specific HF population [7, 8]. Other studies showed a positive impact of HF NIHTS on patients' empowerment and adherence to self-care behaviors [9]. Nevertheless, more evidence on these systems is needed, for its implementation as a healthcare management tool.

NIHTM systems have improved greatly so far, towards a more automatic data input and interaction with the medical system. The increasing number of mobile phone users over the last years [10], made this device almost universal. This constituted a great opportunity for the development of smartphone-based ICT (Information and Communications Technology/Technologies) solutions for healthcare management in HF and other chronic diseases.

Nowadays, there are several telemonitoring (TM) solutions targeting HF, either on the market or implemented in the scope of research studies [10, 11]. Nonetheless, poor adherence to technology still remains a major obstacle to the adoption of these systems [4, 9].

In this paper we aim to describe SmartBEAT as smartphone-based HF NIHTS and to analyze preliminary results of a field test designed to initially test this system in terms of adherence, usability and perception of benefit.

2 Methods

2.1 SmartBEAT Description

SmartBEAT is a smarphone-based HF NIHT system designed to detect congestion through daily monitoring of HF symptoms, weight, peripheral blood oxygen saturation, blood pressure, heart rate, physical activity, and therapy adherence. This solution comprises a monitoring device and a system to collect, analyze, store, manage and transmit data.

SmartBEAT system integrates an afferent and an efferent loop. The afferent loop consists in a smartphone, equipped with a set of sensors (weight scale, sphygmomanometer, bracelet for heart rate and physical activity monitoring), and a set of small questionnaires designed to screen HF signs and symptoms. Data is daily collected and is exported into the cloud, in order to be analyzed by a Medical Rule and Action Engine. An algorithm was designed and is used to differentiate true from false alarms and help correctly hierarchize medical team feedback to the patient/informal caregiver. The algorithm also takes into account electronic medical record (EMR) data, which enables the system to produce a more personalized analysis and intervention.

In response to the data reflecting the patient's status, the system interprets the risk of fluid accumulation and activates a hierarchical 3-level efferent alert loop comprising: 1 – direct communication with the patient by sms, 2 – nurse contact or, in more severe situations, 3 – doctor contact. The ensuing therapeutic intervention is tailored according to the estimated risk of HF decompensation. This will enable early correction of fluid unbalance and will prevent HF-hospitalizations. Additionally, SmartBEAT comprises specific features specially designed to address technology and therapeutic adherence.

2.2 SmartBEAT Field Test

SmartBEAT first pilots were performed on August 2017, involving 3 HF patients, 3 cardiologists and one nurse. This resulted in the first improvements in the technology and was followed by a field test.

This field test was implemented between June and September 2018, in a HF Clinic at a University Hospital in northern Portugal. Ten patients were randomly selected from the patient list, if they had a HF diagnosis, were aged ≥ 60 years old, acquainted with smartphone technology, or computer/tablet and living in the metropolitan area of the University Hospital.

A standardized methodology was used to assess patients' usability and adherence to technology and monitoring protocol. Adherence to the system was defined as the number of total daily monitoring routines completed vs the total number of expected monitoring routines. We expected that, during the 3-month follow-up, patients performed a total of 92 monitoring routines.

Usability was assessed with an observational grid and a five-Likert questionnaire in which 5 represents the highest and most positive perceptions. Perception of benefit was also assessed with a five-Likert questionnaire with similar design. Patients also participated in depth-interviews. Nevertheless, in this work, we only report data collected with the questionnaires.

SmartBEAT system was first tested by patients in a session at the hospital with the telemonitoring team. Patients tested the system, initially with help, and then autonomously, in order to assure they were able to use it at home. They were then invited to use SmartBEAT at home during 3 months.

Patients daily monitoring routine was performed at wakening. By the end of the first telemonitoring day, patients received a phone-call by the research team to identify any possible difficulties. By protocol, no further contacts initiated by the medical team were established during the 3-month follow-up. On the other hand, patients were encouraged to spontaneously contact the medical team to any difficulty, problem or doubt regarding the telemonitoring system.

At this preliminary stage of SmartBEAT development program, no therapeutic action was implemented based on telemonitoring data. In case of severe parameter deviations or patients' compliance, the latter were prompted to address the HF Clinic.

Participants were evaluated in pre-and post-test assessment sessions. Data regarding adherence, usability and perception of benefit are described and compared.

3 Results

3.1 Patients Characteristics

A total of 10 patients were included (Table 1).

At pre-test assessment, 3 patients were in NHYA class I and 7 in NYHA (New York Heart Association) class II.

Table 1. Patients characteristics.

	Patients (n = 10)
Age (years)	68.4 ± 3.8
Female sex	3 (30%)
Education (years)	8.1 ± 5.0
LVEF	32.3 ± 10.2
NYHA I	3 (30%)
NYHA II	7 (70%)
Caregiver support	10 (100%)

LVEF – Left ventricular ejection fraction; NYHA – New York Heart Association

All participants had a proxy person or a caregiver (usually a relative) who helped them in health management, if needed. In most of the cases, the latter helped with the medication and accompanied patients in medical appointments.

3.2 Adherence to the System, Usability and Perception of Benefit

Adherence to monitoring protocol was 97% at 3 months. Almost all patients used the system every day. They were able to measure correctly their weight, blood pressure, and answer to the Heart Failure symptoms questionnaires. This routine took them on average 5 min in the morning. They were also able to use the bracelet during the day.

During follow-up, a total of 11 calls (mean of 1.1 ± 1.1 per patient) were spontaneously initiated by patients to clarify technological issues.

At post-test, overall patients' evaluation of SmartBEAT (SB) was very positive in both pre- and post-test assessments (Fig. 1). SmartBEAT was evaluated as easy to use and this perception increased along the follow-up. Patients also increased their willingness to use the system at home, even though they detected technical issues that need to be improved.

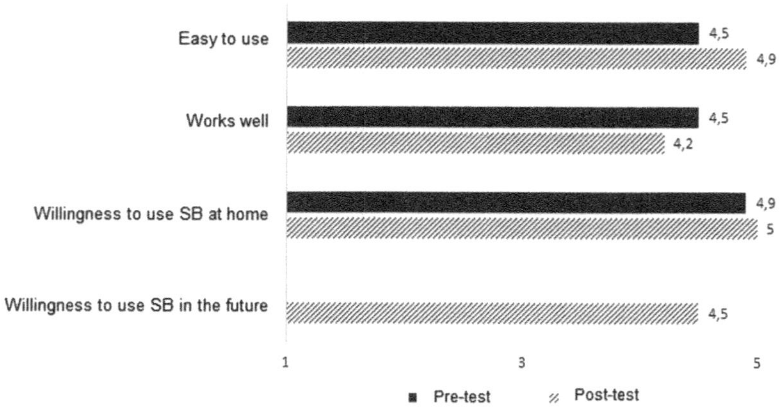

Fig. 1. Patients' perception of SmartBEAT's (SB) usability.

Regarding perception of benefit, patients conveyed the feeling that SmartBEAT (SB) helped them taking a better care of their health and of heart failure symptoms (Fig. 2). They reported that the system helped them feel safer and this perception increased in post-test assessment.

Patients' expectations that SmartBEAT would improve their wellbeing showed a slight decrease from 4.6 to 4.1 points from pre- to post-test.

Daily assessment of hemodynamic parameters did not increase patients' anxiety during follow-up. Moreover, the majority expressed they wished to use SmartBEAT in the future.

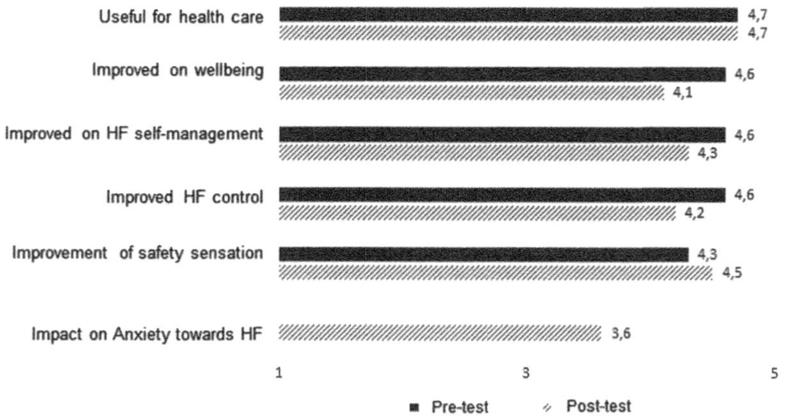

Fig. 2. Patients' perceptions of benefit regarding SmartBEAT (SB).

4 Discussion

This study aimed at describing SmartBEAT protocol and evaluating adherence, usability and perception of benefit regarding this system.

SmartBEAT collects, analyses and manages HF patients' data. Biological parameters are collected by external sensors and transmitted to a mobile phone and to a portal, where an algorithm identifies false from true alarms and displays a hierarchical interventional response. This system is designed to early detect signs and symptoms of HF decompensations, demanding a daily monitoring routine from the patients.

Adherence to SmartBEAT system at three-month follow-up was high, in contrast with current results from large randomized trials, in which maximum adherence is overall 85% [5, 6].

It is noteworthy that despite the natural progressive fatigue of continued daily system use, patient adherence was kept at a high level by the end of follow-up. We hypothesize that initial comprehensive training played a key role. Continued research team availability to solve patients' problems and encouragement was also probably important.

Moreover, patients maintained a global positive perception regarding system's usability during follow-up. Patients identified system's accuracy failures, involving some sensors (i.e. sphygmomanometer). Software instability was identified, signaling the need for further improvement.

Daily monitoring control, far from causing psychological fatigue, conveyed a surprising sensation of being at a higher control of disease and a higher safety sensation. Patients also valued this system as a stimulus for a higher engagement in HF self-management. This corroborates previous results that showed a positive effect of HF NIHTS on improvement of patient empowerment and adherence to self-care behaviors [9].

Patients' expectations that SmartBEAT would improve their wellbeing decreased along the follow-up. Despite valuing the contribution of SmartBEAT in their health management, participants recognized that their wellbeing relied on several other aspects of life. Thus, at the end of the trial they reshaped their expectations regarding SmartBEAT contribution to their wellbeing to a lower level. This may indicate a more realistic evaluation of this system merit, that it is not the sole solution for heart failure per se. They also adjusted their initial perception to a more realistic vision of the disease and treatment.

Nevertheless, they felt safer because they understood the system's capacity for early detection of HF decompensation.

5 Limitations

This paper presents preliminary and partial data regarding SmartBEAT field test. Considering the small sample size, these results may not be representative of the HF population. This limits the generalization of these results to HF population.

The short follow-up duration may have also limited the detection of significant differences between pre- and post- test, as well as the variability in adherence rates. This way further tests with larger samples sizes and longer follow-up periods are needed in order to gather more evidence regarding SmartBEAT adherence, usability and impact.

6 Conclusion

SmartBEAT field test preliminarily results show that this HF NIHTS system is easy to use, accurate, and reliable. Patients conveyed the feeling that it has the potential to be valuable for HF disease management. A few technological problems need nevertheless to be addressed to further improve system's reliability and stability. Overall these preliminary results show promise in future more comprehensive and larger studies.

Funding. This work was supported by

- Smart system for the management of Heart Failure in older adults (SmartBEAT) https://www.smartbeatproject.org/ AAL Joint Programme aal-2014-157;
- ERDF through POCI-01-0145-FEDER-007746, funded by COMPETE2020, National Funds through FCT, within CINTESIS(UID/IC/4255/2013), and Project - NORTE-01-0145-FEDER-000026 – "Symbiotic technology for societal efficiency gains: Deus ex Machina", financed by NORTE2020 under PORTUGAL2020.

Conflict of Interest. None declared.

References

1. Dunlay, S.M., Roger, V.L.: Understanding the epidemic of heart failure: past, present, and future. Curr. Heart Fail. Rep. **11**(4), 404–415 (2014)
2. Heidenreich, P.A., Albert, N.M., Allen, L.A., Bluemke, D.A., Butler, J., Fonarow, G.C., Ikonomidis, J.S., Khavjou, O., Konstam, M.A., Maddox, T.M., Nichol, G., Pham, M., Pina, I.L., Trogdon, J.G.: Forecasting the impact of heart failure in the United States: a policy statement from the American Heart Association. Circ. Heart Fail. **6**(3), 606–619 (2013)
3. Gheorghiade, M., Filippatos, G., De Luca, L., Burnett, J.: Congestion in acute heart failure syndromes: an essential target of evaluation and treatment. Am. J. Med. **119**(12 Suppl 1), S3–S10 (2006)
4. Shochat, M.K., Shotan, A., Blondheim, D.S., Kazatsker, M., Dahan, I., Asif, A., Rozenman, Y., Kleiner, I., Weinstein, J.M., Frimerman, A., Vasilenko, L., Meisel, S.R.: Non-invasive lung IMPEDANCE-guided preemptive treatment in chronic heart failure patients: a randomized controlled trial (IMPEDANCE-HF Trial). J. Cardiac Fail. **22**(9), 713–722 (2016)
5. Klersy, C., De Silvestri, A., Gabutti, G., Regoli, F., Auricchio, A.: A meta-analysis of remote monitoring of heart failure patients. J. Am. Coll. Cardiol. **54**(18), 1683–1694 (2009)
6. Inglis, S.C., Clark, R.A., McAlister, F.A., Ball, J., Lewinter, C., Cullington, D., Stewart, S., Cleland, J.G.: Structured telephone support or telemonitoring programmes for patients with chronic heart failure. Cochrane Database Syst. Rev. **8**, Cd007228 (2010)

7. Koehler, F., Koehler, K., Deckwart, O., Prescher, S., Wegscheider, K., Kirwan, B.A., Winkler, S., Vettorazzi, E., Bruch, L., Oeff, M., Zugck, C., Doerr, G., Naegele, H., Stork, S., Butter, C., Sechtem, U., Angermann, C., Gola, G., Prondzinsky, R., Edelmann, F., Spethmann, S., Schellong, S.M., Schulze, P.C., Bauersachs, J., Wellge, B., Schoebel, C., Tajsic, M., Dreger, H., Anker, S.D., Stangl, K.: Efficacy of telemedical interventional management in patients with heart failure (TIM-HF2): a randomised, controlled, parallel-group, unmasked trial. Lancet **392**(10152), 1047–1057 (2018)
8. Koehler, F., Koehler, K., Deckwart, O., Prescher, S., Wegscheider, K., Winkler, S., Vettorazzi, E., Polze, A., Stangl, K., Hartmann, O., Marx, A., Neuhaus, P., Scherf, M., Kirwan, B.A., Anker, S.D.: Telemedical Interventional Management in Heart Failure II (TIM-HF2), a randomised, controlled trial investigating the impact of telemedicine on unplanned cardiovascular hospitalisations and mortality in heart failure patients: study design and description of the intervention. Eur. J. Heart Fail. **20**(10), 1485–1493 (2018)
9. Ware, P.: Patient adherence to a mobile phone-based heart failure telemonitoring program: a longitudinal mixed-methods study. JMIR mHealth uHealth **7**(2), e13259 (2019)
10. Athilingam, P., Jenkins, B.: Mobile phone apps to support heart failure self-care management: integrative review. JMIR Cardio **2**(1), e10057 (2018)
11. Yun, J., Park, J., Park, H.: Comparative effectiveness of telemonitoring versus usual care for heart failure: a systematic review and meta-analysis. J. Cardiac Fail. **24**(1), 19–28 (2018)

How Secure Is Your Mobile Health?

Ana Ferreira[1(✉)], Rute Almeida[1], and Joana Muchagata[2]

[1] CINTESIS - Centro de Investigação em Tecnologias e Serviços de Saúde, MEDCIDS - Departamento de Medicina da Comunidade, Informação e Decisão em Saúde, Faculty of Medicine of University of Porto, Porto, Portugal
amlaf@med.up.pt

[2] CINTESIS - Centro de Investigação em Tecnologias e Serviços de Saúde, Faculty of Medicine of University of Porto, Porto, Portugal

Abstract. Research suggests that the interactions between a patient and a health professional through an mHealth app (a mobile health application), can improve the efficiency and quality of healthcare. However, the risk of disclosure of personal or health related data can be higher when using such devices, as there is a lack of specific security standards or guidelines for their deployment. Also, there are commonly no controls to comprehensively verify and minimize security and privacy vulnerabilities that may exist in those apps before they are released in the "wild". To make matters worse, the medical record has a significantly higher financial value (on the black market) compared with other personal records (e.g., credit card or bank account details), which obviously increases the motivation for unauthorised accesses and misuse. In order to mitigate these problems, socio-technical security as well as privacy and legal aspects need to be taken into account when developing mHealth apps. In this work, essential recommendations regarding the previously itemized are provided as a means to guide both developers and users alike, into more secure, private and usable mHealth apps.

Keywords: mHealth applications · Security and privacy · Legal requirements · Contextual and visual security · Patient empowerment

1 Introduction and Background

Interactions between a patient and a health professional, through a mobile app, can meaningfully improve the efficiency and quality of healthcare [1, 2]. Mobile health applications (mHealth apps) enhance communication and care coordination among specialists, doctors, nurses, providing access to services independently of time and location. Physicians can access patient records, view test results and prescribe medications [3, 4] while patients can use mHealth apps to have access and update their Electronic Health Records [5, 6], and for self-monitoring their therapeutic and health outcomes, to be empowered in self-managing their condition. Physicians and patients can share information such as diagnoses, prescriptions, lab results and even schedule appointments. The paradigm of mobile anytime/anywhere can also be beneficial in emergency or unanticipated situations [6, 7]. But not all are advantages. Often there is a

J. Henriques et al. (Eds.): MEDICON 2019, IFMBE Proceedings 76, pp. 1377–1384, 2020.
https://doi.org/10.1007/978-3-030-31635-8_168

lack of security standards or guidelines for mHealth developers to follow and no adequate scientific validation. Also, the majority of existing mobile apps provide low or no security protection [1, 8] and may not be properly tested for vulnerabilities before hitting the online stores [9, 10].

It is important to take special care in the collection and treatment of users' personal health data in order to resolve security issues for their protection, as risks include not only data breaches but also patients' physical and mental safety. Current research suggests that mHealth app developers should comply with three basic requirements: (1) accessibility and usability; (2) a certification from domain experts; and (3) privacy and security [10, 11]. This paper aims to give developers and users the main recommendations for the third point, which focus on: (a) legal; (b) technical; and (c) socio-contextual and interactional security requirements, described in the following three sections.

2 Legal Requirements

The majority of mobile apps do not comply with contemporary legislation requirements and do not follow well-known practices and guidelines, thus jeopardizing the privacy of millions of users [12]. For this purpose, recommendations for the legal requirements of the General Data Protection Regulation, the main legislation for data protection in Portugal [13], are summarized in Table 1. This can help both mHealth app users and developers to check whether the processing of personal data is compliant with local regulations, as the fines are heavy and the consequences to the users can be dangerous.

Table 1. Portuguese legislation security and privacy requirements for health personal data.

Title	Articles	Main recommendations for mHealth app developers
Personal data	4	Protect personal data such as name, gender, home address, phone number and photographs, as well as Internet Protocol or email addresses; biometrics; geo-location; mobile/device identifiers; cookies and pseudo-anonymized data
Processing	4	Identify and protect any operation performed on personal data, by automated or manual means, from collection till erasure/destruction (e.g., storage, alteration, retrieval, consultation, use, disclosure, dissemination)
Special categories of personal data	9	**Forbidden to use**: race, ethnicity, political opinions, religious beliefs, genetic and biometric data for identification, health data or sexual orientation **Exceptions:** if the data subject has given explicit consent
Lawfulness	6	Acquire consent to the processing of personal data for one or more specified purposes (a consent for each different purpose)

(continued)

Table 1. (*continued*)

Title	Articles	Main recommendations for mHealth app developers
Consent	7	Provide a clear, informed and unambiguous indication of the data subject's agreement to the processing of personal data. Have in mind that consent can be withdrawn at any time
Right to transparency	12	Take appropriate measures to provide information relating to personal data processing, to the data subject in a concise, transparent, intelligible and easily accessible form, using clear and plain language
Right to be informed	13 14	Describe information to be provided where personal data are collected from the data subject or from another source
Right to be forgotten	17	Provide means (and proof) to erase personal data without undue delay
Protection by design and by default	25	Implement appropriate technical and organisational measures for data-protection principles, and for ensuring that, by default, only personal data necessary for each specific purpose of the processing, are processed
Security of processing	32	Implement appropriate technical/organisational measures to ensure security and to regularly assessing those measures for appropriate compliance
Records	30	Maintain a record of processing activities
Impact assessment	35	Make an impact assessment, prior to the processing, of the envisaged processing operations on the protection of personal data

3 Security Requirements

The security of any computer device is hard, but of mobile devices is even harder. Being a very small and highly portable computer, it carries all the risks and vulnerabilities of a normal computer, and it can propagate those easily to any other equipment to which it connects. Currently, companies still do not focus on mobile computing as part of their information security policy although smartphones are being stolen at an increased rate, they are very easy to lose and many lack secure authentication [14]. Further, the lack of awareness that smartphones are an important criminal target for their data content makes people not having at least the same caution as they have for their PCs or laptops, where they at least install anti-virus or anti-malware software [15]. So, as a standalone device, smartphones carry many security and privacy issues. Mobile applications should not add to this toll.

Although it is not possible to discuss in detail mHealth apps technical security measures, some of the main concerns both users and developers must take into account, drawn from the most recent research literature [16–19], are listed below (Table 2)

Table 2. Main concerns and recommendations for mHealth apps' technical security.

Concern	Recommended action
Mobile devices and data content not integrated into the company's security policy	Include mobile devices and their data into the organization's security policy and provide awareness and training
Users/workers/patients' lack of awareness and caution for their devices	Provide security and privacy awareness and training
Inadequate or non-existent access control features and non-secure management and storage of credentials	Develop, implement and test appropriate access control features for mobile devices including their secure management. Provide awareness and training
Apps do not exercise encryption to prevent unauthorized data usage and theft	Provide secure communication and data encryption on the mobile, including SD cards and other related locations
Man-In-the-Middle attacks [16]	Secure transfer (e.g., encryption) of sensitive information as well as enable continuous updates of their current cryptographic protocol versions (such as TLS)
Client-side injection, when the malicious code leaks from the app into the user's mobile device	User/app supplied data needs validation during input
Unsecure data storage	mHealth apps need to restrict access to patient data via unsecure encoding or file permissions (encryption can help)
Unintended data leakage	Sensitive data needs to be stored in a location where it cannot be accessed by other apps on the same device or by the attacker with physical access to the device (encryption can help)
Storing data in third party servers/services (e.g., Cloud, others)	Make sure third-party services have the required/appropriate security measures before using them
Sensitive data collected by Bluetooth-enabled health devices can be sniffed or injected	Provide Operating System-level protection, which generates secure bonding policies between a device and its official app and enforces these rules when establishing and terminating Bluetooth connections
Sensitive information is put into system logs where it is not commonly secured	Make sure app logs are secured (e.g., encryption, access control)
Sensitive information is stored as unencrypted files on the SD card, publicly accessible by any other app	Make sure SD cards data are secured (e.g., encryption)
Android app components, intended to be private, are set as exported, making them accessible by other apps	Make sure limits are set on what Android app components can/should share

(*continued*)

Table 2. (*continued*)

Concern	Recommended action
Sensitive information can be inferred by an app with side channels, e.g. network package size, sequence, timing, and so on	Pad blank information to network packets to ensure they are fixed-length, or develop off-line strategies for downloading sensitive data. For the motion-sensor-based side channel attack, disable untrusted access to motion sensors whenever a trusted input function (*e. g.* password entry) is being performed

Many solutions can be attained right away from other existing devices while others are specific to mobile devices. More research effort is needed to find additional adequate solutions but till then, re-using what it works, is better than no protection at all.

4 Socio-Contextual Security

Security and privacy of mHealth apps cannot be only focused on the technical and legal/regulatory aspects. Every time users interact with such apps can influence the protection of their personal data. How can a user be searching for information in a mobile device, within a specific place and context, while having the guarantee that the way that information is displayed is the most secure and private? Anyone using an mHealth app on a public space can be easily eavesdropped. Further, if all required and non-required data at a specific moment are travelling via an unsecure channel, attackers listening to public non secure Wi-Fi hotspots can also gather personal data and credentials to later perform more sophisticated attacks. So, beyond warning messages, and taking the new paradigm of anytime/everywhere, security visualization of mHealth apps must adapt to all technical and physical/contextual characteristics relating to a user's specific request.

Previous research has focused on comparing mHealth scenarios such as the one presented in Fig. 1 [18]. This illustrates a scenario where a patient is at a pharmacy looking for data on allergy medication on a prescription. On a normal access, all data is shown.

(Figure 1 on top - BEFORE), but once risk assessment of the request with associated contextual characteristics is performed, security and privacy can be easily enhanced to avoid privacy breaches (Fig. 1 at the bottom - AFTER). In this case, less personal health data is shown by default (only medications and prescriptions, since the patient is at a pharmacy) and a timer is set (to avoid prolonged shoulder-surfing - eavesdropping) on zoom functions (for improved availability and integrity of highly sensitive data), all this for improved confidentiality. Obviously, accessing more information than what is shown is possible.

For this type of scenarios to work, it needs to be complemented with an access control model integrating risk assessment to verify, at each interaction, the context, technology used and previous users' history to identify the most secure, yet available/usable way to show requested resources, for a specific request. Research still needs to advance more on this path but some guiding work in the form of the SoTRAACE model is available here [19]. The authors suggest that these types of issues are tested in mHealth apps for improved security, privacy and trust.

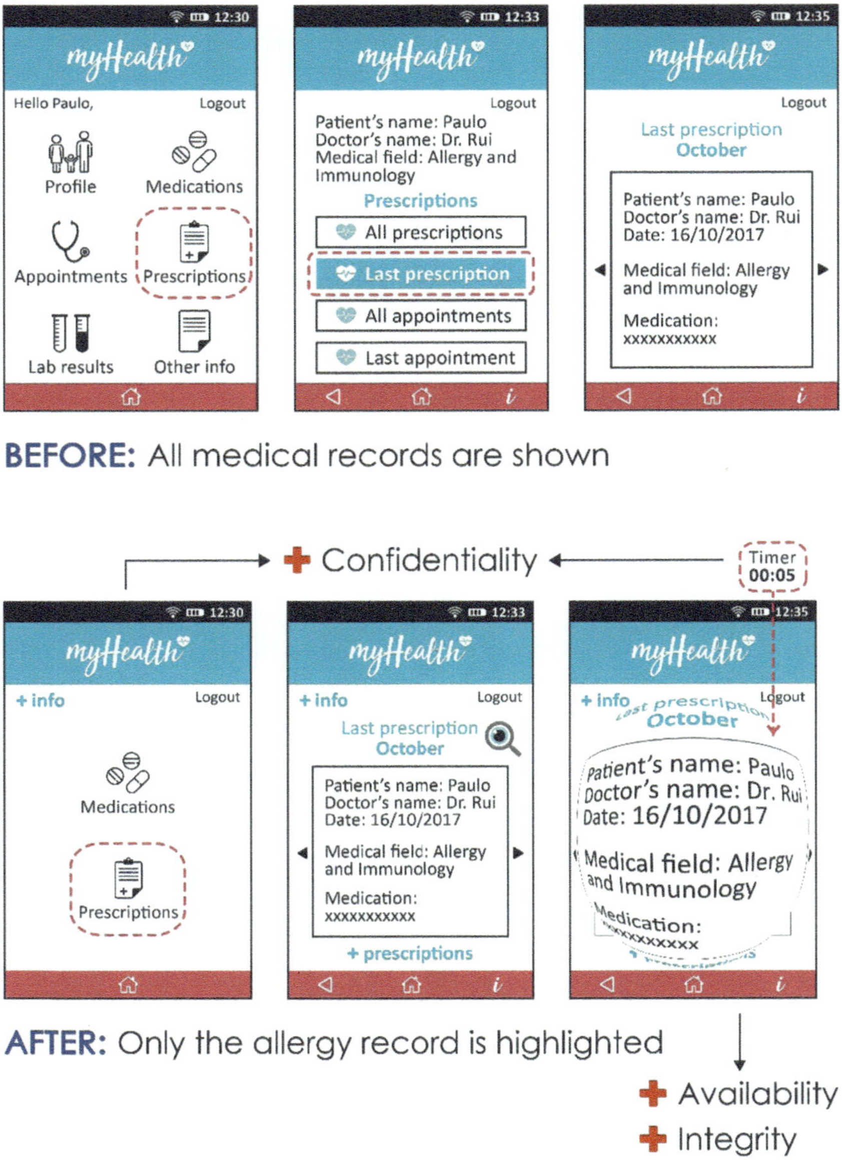

Fig. 1. Improved confidentiality, integrity and availability on a patient's access to a mHealth app. BEFORE any visual adaptation regarding contextual and risk analysis is performed (on top) and, AFTER that is done, only required personal data adapted to the context characteristics, at that specific moment (at the bottom), are available.

5 Discussion and Conclusion

mHealth apps can increase the exposure of many new attack surfaces to the outside world, as they allow a much larger amount of data being collected from the patient, over extended periods of time. Also, a much broader range of health-related data are being collected, as many mHealth apps collect patient activities and lifestyle, physiological data, physical activity, location tracking, eating habits and diet details, social interactions and so on [17].

Based on a study of only 22 mHealth apps, the great variety and sensitive degree of information this small number of apps can comprise is of notice. Sensitive data include personal profiles, app login credentials, long-term health sensor data (e.g., monitoring of diabetes, heart conditions, respiratory diseases), lifestyle data, medical information browsing history, information about health professionals and third-party app data (e.g., Facebook account information) [17]. Depending on the characteristics of mHealth data breaches, disclosure or tampering with these sensitive data may lead to serious consequences, such as profiling, medical identity theft and healthcare decision-making errors.

For all these aspects, security and privacy of mHealth apps need to be taken seriously. In the case of a mobile app, and especially health related, it must comply with existing local/national legislation, which sometimes can be cumbersome when is not easy to interpret. However, in this work the authors suggest, in an objective way, the legal aspects that need to be in conformity. Next, technology related requirements can have similarities to other existing equipment but there are also new aspects to be aware, also here presented. Finally, there is the user experience and interaction with the application, including visualization and user contextual history that can enrich the way security and privacy measures can be adopted, at every moment in time.

The facility, flexibility and high functionality given by mHealth apps contrast with their high difficulty of providing proper security and privacy measures, which can hinder their adoption and use. This work helps to have a summarized integrated view of the most pressing security aspects to take into account when using and developing mHealth apps. If these are timely taken care of, mHealth apps can finally provide all the benefits for which they were initially thought and vouch for a truly useful mobile health for everyone.

Acknowledgments. TagUBig - Taming Your Big Data (IF/00693/2015) from Researcher FCT Program funded by National Funds through FCT - Fundação para a Ciência e a Tecnologia; COMPETE2020, and NORTE-01-0247-FEDER-033275 (AIRDOC, project NORTE-01-0247-FEDER-033275, financed by the North Portugal Regional Operational Programme (NORTE 2020), under the PORTUGAL 2020 Partnership Agreement, and through the European Regional Development Fund (ERDF). AIRDOC is with ITEA 3 16040 PHE - Personal Health Empowerment project consortium).

Rute Almeida is supported by ERDF (European Regional Development Fund) through the operation POCI-01-0145-FEDER-029130 (mINSPIRERS — mHealth to measure and improve adherence to medication in chronic obstructive respiratory diseases - generalisation and evaluation of gamification, peer support and advanced image processing technologies) funded by COMPETE2020 and by National Funds through FCT.

References

1. Mirkovic, J., Bryhni, H., Ruland, C.M.: Secure solution for mobile access to patient's health care record. In: 2011 IEEE 13th International Conference on e-Health Networking, Applications and Services, pp. 296–303 (2011)
2. Ventola, C.L.: Mobile devices and apps for health care professionals: uses and benefits. Pharm. Ther. **39**(5), 356 (2014)
3. Plachkinova, M., Andrés, S., Chatterjee, S.: A taxonomy of mHealth apps - security and privacy concerns. In: 48th Hawaii International Conference on System Sciences, pp. 3187–3196 (2015)
4. Luxton, D.D., McCann, R.A., Bush, N.E., Mishkind, M.C., Reger, G.M.: mHealth for mental health: integrating smartphone technology in behavioral healthcare. Prof. Psychol. Res. Pract. **42**, 505–512 (2011)
5. Ferreira, A., Lenzini, G., Santos-Pereira, C., Augusto, A.B., Correia, M.E.: Envisioning secure and usable access control for patients. In: 2014 IEEE 3rd International Conference on Serious Games and Applications for Health (SeGAH), pp. 1–8 (2014)
6. Pyper, C., Amery, J., Watson, M., Crook, C.: Access to electronic health records in primary care-a survey of patients' views. Med. Sci. Monit. Int. Med. J. Exp. Clin. Res. **10**, SR17–SR22 (2004)
7. Ferreira, A., Chadwick, D., Farinha, P., Correia, R., Zao, G., Chilro, R., Antunes, L.: How to securely break into RBAC: the BTG-RBAC model. In: 2009 Annual Computer Security Applications Conference, pp. 23–31 (2009)
8. Parati, G., Torlasco, C., Omboni, S., Pellegrini, D.: Smartphone applications for hypertension management: a potential game-changer that needs more control (2017)
9. Muchagata, J., Ferreira, A.: Visual schedule: a mobile application for autistic children - preliminary study. In: 21st International Conference on Enterprise Information Systems, pp. 452–459 (2019)
10. Larson, R.S.: A path to better-quality mHealth apps. JMIR Mhealth Uhealth **6**, e10414 (2018)
11. Cook, V.E., Ellis, A.K., Hildebrand, K.J.: Mobile health applications in clinical practice: pearls, pitfalls, and key considerations. Ann. Allergy Asthma Immunol. **117**, 143–149 (2016)
12. Papageorgiou, A., Strigkos, M., Politou, E., Alepis, E., Solanas, A., Patsakis, C.: Security and privacy analysis of mobile health applications: the alarming state of practice. IEEE Access **6**, 9390–9403 (2018)
13. European Union: Regulation (EU) 2016/679 of the European Parliament and of the Council L 119. Official Journal of the European Union (2016)
14. Sampat, B.H., Prabhakar, B.: Privacy risks and security threats in mHealth apps. J. Int. Technol. Inf. Manage. **26**, 126–153 (2017)
15. Maymi, F., Harris, S.: CISSP all-in-one exam guide (2016)
16. Shugalo, I. https://thedoctorweighsin.com/mhealth-apps-security-privacy. Accessed March 2019
17. He, D., Naveed, M., Gunter, C.A., Nahrstedt, K.: Security concerns in Android mHealth apps. In: AMIA … Annual Symposium Proceedings. AMIA Symposium 2014, pp. 645–654 (2014)
18. Muchagata, J., Ferreira, A.: How can visualization affect security? In: ICEIS 2018 - 20th International Conference on Enterprise Information Systems. SCITEPRESS Digital Library, Poster Presentation in Funchal, Madeira – Portugal, vol. 2, pp. 503–510 (2018)
19. Moura, P., Fazendeiro, P., Vieira-Marques, P., Ferreira, A.: SoTRAACE – socio-technical risk-adaptable access control model. In: 2017 International Carnahan Conference on Security Technology (ICCST), Madrid, pp. 1–6 (2017)

Diabetes Management Guidance by a Logical Unit Supported by Data-Mining in a Mobile Application

Diogo Machado[1,4(✉)], Vítor Santos Costa[2,4], Inês Dutra[3,4], and Pedro Brandão[1,4]

[1] Instituto de Telecomunicações, Lisbon, Portugal
[2] INESC-TEC, Porto, Portugal
[3] CINTESIS, Porto, Portugal
[4] Faculdade de Ciências da Universidade do Porto, Porto, Portugal
dmachado@dcc.fc.up.pt

Abstract. Diabetes type I is a chronic disease that requires strict supervision. MyDiabetes is a utility application for diabetic users. This application served as basis to develop a logical unit, composed of logical rules, translated from medical protocols and guidelines, to advise the user. The data in the application is a source of knowledge about the user's health state and diabetes intrinsic characteristics. An existing concern is the weak user adherence and consequential data absence. The implemented solutions were gamification and an interface rework. As later confirmed through a survey, users feel captivated by appealing interfaces, achievements and medals. In a near future, we will resume our work with the S. João's hospital, with a new trial and volunteers. User testing will be used to validate the gamification techniques.

Keywords: Diabetes · mhealth · Expert-systems · Gamification

1 Introduction

Diabetes is a chronic disease that in 2015 affected 415 million people around the world [2]. Despite all the efforts to stale the growth of this disease, the number of people with diabetes is continuously increasing and it is estimated to reach 642 million in 2040 [2]. With these numbers, a series of problems and concerns arise, namely how to avoid new cases and how to help those already affected by diabetes. Diabetes is characterised by high levels of glucose in the blood, caused by the person's pancreas inability to produce enough or any insulin. If glycaemic values are not controlled, diabetes can seriously decrease quality of life. In the worst cases, this disease can lead to blindness, amputation and/or heart problems. There are different types of diabetes where type 2 and the type 1 are the most common. Each diabetes type has a specific treatment associated. Diabetes type 1 requires a strict regime and the continuous verification of glycaemic values. This regime can be adjusted while consulting with the doctor.

J. Henriques et al. (Eds.): MEDICON 2019, IFMBE Proceedings 76, pp. 1385–1393, 2020.
https://doi.org/10.1007/978-3-030-31635-8_169

In these appointments, the medical expert evaluates the evolution of the glycaemic values, among other relevant annotated information and personalises the current regime, to the diabetic's needs.

The solution presented is not optimal as it requires the availability of a medical expert to receive the person, evaluate the data and refine the treatment. Real-time events such as hyperglycaemia or hypoglycaemia, require immediate action which poses a possible problem in the absence of expert knowledge. In the current market, there are two main concepts of diabetes self-management support: mobile applications and monitoring devices, in some instances working together with an insulin pump. Mobile applications aim to help their users to record daily relevant information. This information can be later used in order to review the evolution of glycaemic values or the effectiveness of the current treatment. The approach that uses monitoring devices acts with minimum direct input from the user. To achieve a satisfactory management, companies are developing devices that constantly monitor the user's glycaemic values and work with insulin pumps to react to these values. This system is still far from being fully autonomous. Although, both mobile applications and monitoring devices seem excellent tools for diabetics to have, there is a treatment component missing. Advising and alerting the user is crucial for diabetes management. The user should know and be able to react to the different possible diabetes related occurrences. In our approach we tackle this issue. It is our objective, not only to help our users to better manage diabetes, but also to alert and teach them how to act in different situations.

In previous work we developed a prototype for a rule-based system that addressed self-management advice for diabetes [4]. This prototype was designed as an expert system embedded in the MyDiabetes mobile application. The MyDiabetes application, by itself, allows for the daily record of diabetes related information e.g. glycaemia, carbohydrates, insulin, exercise and HbA1c. The rule-based system, to be added, contains expert knowledge, able to interpret the information recorded in the application. This rule-based system includes a series of medical rules and protocols able of solving certain occurrences, common in people with diabetes. The knowledge included was learned from collaborating with experts, namely, MD Celestino Neves, an endocrinology expert from S. João's Hospital. Each rule has an advice indexed, which is displayed to the user when intervention is required. This advice guides the user, to recover or even avoid undesirable occurrences.

The expert system, is still not a working component of the application, since it requires a port of a Prolog compiler for the Android architecture. This need led to our work on the YAP (Yet Another Prolog) [8], high-performance Prolog compiler, port to Android. While developing this port, we continue to further improve our advice system, by reviewing the existing rules and creating new rules. This way expanding the advising reach of the base system.

Expert knowledge can be used to validate current rules, update older rules and even create new "state of art" rules. We want to give experts the ability to adapt, remove or even, ideally, create new rules, useful to a particular diabetes' cases.

A common issue we also tackled is the user adherence to the application. Normally, people tend to subscribe to new and appealing projects, just to abandon them a few weeks later. In our experience, this happens because people require constant motivation and a feeling of gratification from using the application. We expect, with the already implemented gamification [7] and the rule-based system, to overcome this problem. The gamification component gives the user goals to reach and motivates the user to improve current values. The advising system assists on the flaws the user has and directs the user to a better path to follow.

The data recorded in the mobile device is used to create new and more personalised advice. It is important that the user feels rewarded and motivated to record new information in the application. In a previous work [6] on data-mining, using scarce data gathered from the MyDiabetes application, it was possible to detect basic patterns such as 'at certain days of the week, the user tends to have a meal richer in carbohydrates resulting in higher glycaemic values'. It is our intention to continue our partnership with the S. João's Hospital, recruit volunteers to use our application, collect more data and, through it's analysis, detect other and more complex patterns. We aim to detect not only personal patterns but also common patterns on the user population. We also aim to not only detect new complex patterns, but to also discover their cause. In a previous work we implemented a system capable of finding possible explanations for an occurring crisis, defined by logical rules. Upon the detection of a hyperglycaemia or hypoglycaemia, the system would test a list of possible occurrences that justify the current crisis. If one or more crisis triggers were confirmed, a message containing the system's conclusion would appear, informing the user of what may have caused the current situation. We now want to apply data-mining in order to deepen our explanation for a certain pattern and crisis.

2 The Mydiabetes' Advising Unit

The main concept of our approach is to have an expert system close to the user, providing diabetes' self-management advice throughout the day. This concept was translated into a prototype composed of the MyDiabetes mobile application and an external expert system.

Previous tests with the MyDiabetes application, without the expert system, raised some concerns we addressed. Among these, user motivation to continuously record data. Frequent records are essential for the expert system to intervene appropriately. On these tests, we gathered thirty-one volunteers, from these only eight used the mobile application for more than a week. This fact raises a serious concern, specially to our future data-mining component that requires a big amount data to reach reasonable conclusions.

In order to motivate the user we redesigned the application, as shown in Fig. 1, to have more colour and an overall more appealing design [1]. Along with these changes, we have implemented gamification elements such as a badges, goals, points and level systems. The badge interface, where the user can access all the obtained badges together with the still locked objectives is represented in Fig. 2. We believe that the badges and the tasks, needed to obtain said badges, will give the user a short-term objective to aim for. Short-term objectives are a known method used by doctors with educational purposes and to keep the user motivated [5]. Having these small objectives and rewards (badges) not only serves as daily motivation, but also, in a longer term, can help the user to better manage diabetes. When the user frequently records new information, the current diabetes treatment's effectiveness becomes clearer.

We conducted a survey with the collaboration of the S. João's Hospital to evaluate the impact of the new interface and gamification techniques. We were able to gather fourteen patients with diabetes type I (ages between 21 and 54). We asked the participants to compare the previous mobile application, with the old design and no gamification elements, to the new version. The reaction to the gamification component was positive: 93% of the participants preferred the new interface; only one patient did not feel motivated to use either one of the applications; 78% of the participants showed interest in using the mobile application as their main management control device; 71% of the participants felt compelled to use the badge system; 77% of the participants would like to access an anonymous ranking system. The ranking system is a gamification method still being considered.

The expert system responsible for advising the user was developed as a rule-based system. In this system we integrated a core of rules based on medical protocols and guidelines. These rules evaluate the user's state and, when necessary, trigger an advice. This advice contains important information for the user, encoded in a data structure that can be interpreted by the mobile application. It contains information on how to display the message to the user and extra interactions needed e.g. "schedule a new registry for a near future". The user then receives a message with instructions on how to act. For this we developed the Advice Rule Based System (ARBS). It is a Prolog-based set of rules, that analyses data, based on medical knowledge and guidelines. The main objective of this component is to perceive the current situation based on data from the database and return appropriate advice, defined for that situation. A mock-up example of a possible advice implementation can be seen in Fig. 3, alerting the patient to adjust the insulin dosage, considering the exercise recently done. The implementation of the ARBS is divided in three categories: system, advice query and medical rules.

The System Rules are the structural rules of the ARBS. For a MD the perfect work environment would be one where development and work could be done with clauses that can be read and understood as simple text. Predicates like `atom_concat/3`, used to concatenate two atoms (single data item), are useful while building rule IDs, or rules that filter certain facts, but must be avoided during rule development with medical collaboration in order to avoid unnecessary confusion. Thus, system rules include the rules that support the different sub-components of the ARBS, joining them as a working unit. The system rules have

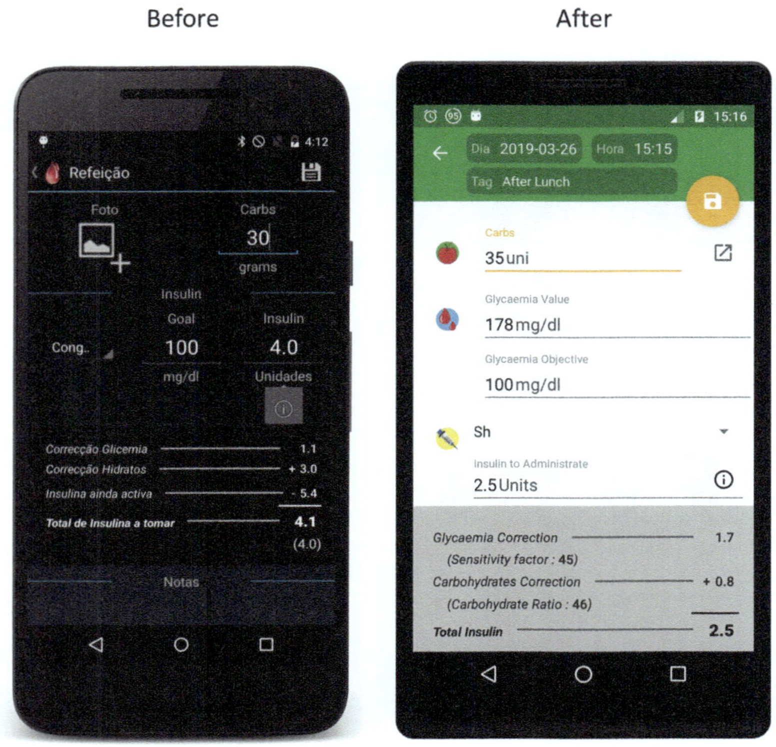

Fig. 1. Depiction of the interface's redesign.

a specific rule called "master rule" constructed to be the main method of the ARBS. This is the method the Android Native Application (ANA) uses to call the ARBS to verify the existence of new advice.

Advice Query Rules are the most important elements in the ARBS, enabling the connection to the database. This arises from the main goal of the application to know the user's current health state so to advise accordingly. The *Advice Query Rules* are Prolog clauses created for accessing the database. They are the link needed to ascertain facts of the user's clinical condition such as the value of the last recorded glycaemia. These rules, used to access the contents of the DataBase (DB), are a direct use of YAP's library Myddas [9] capabilities of importing database tables as terms in the logical program's environment from the sqlite database.

Medical Rules are clauses that represent the possible user's state that may lead to crisis situations. Every one of these rules places the user in a particular situation e.g.: "the patient exercised recently (less than 2 h) and is going to take

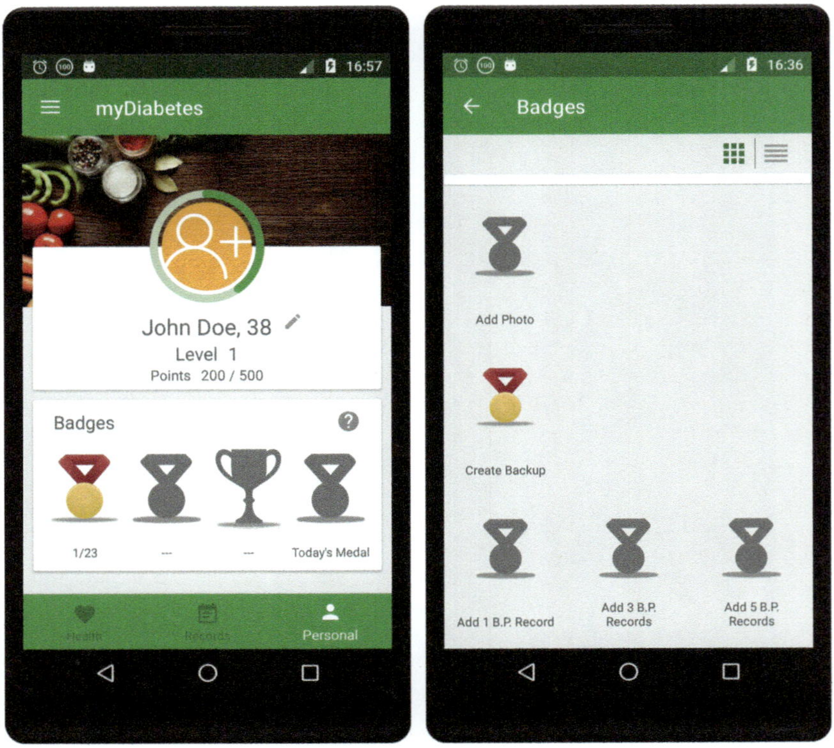

Fig. 2. Personal user page with badge stats (left) and the badge list expansion (right).

insulin". These rules, when occurring, trigger an advice to warn the user for the need to take precautions. Clinical conditions are represented by facts and symptoms e.g.: "Hypoglycaemia is defined by abnormally low blood glucose (blood sugar) levels, normally with values below 70 mg/dl".[1] People with hypoglycaemia may show symptoms such as: shakiness, nervousness, anxiety, sweating, chills and clamminess. Likewise, Medical Rules are composed by conditions and symptoms. In this case every condition is a fact about the user's health state and is represented by an Advice Query Rule (AQR). When a rule's conditions are met, the rule is triggered and the user's situation requires an advice. Every medical rule has a unique **ID**, which, when the rule is triggered, is used to search the corresponding advice's text recommendation, localised in the language file.

The rule's composition makes them understandable for non-experts, since the medical expert can read the different present conditions and recognise the represented situation.

In order to create medical rules, it is necessary to first acquire medical knowledge. The already developed rules resulted, mainly, from meetings with experts.

[1] From American Diabetes Association.

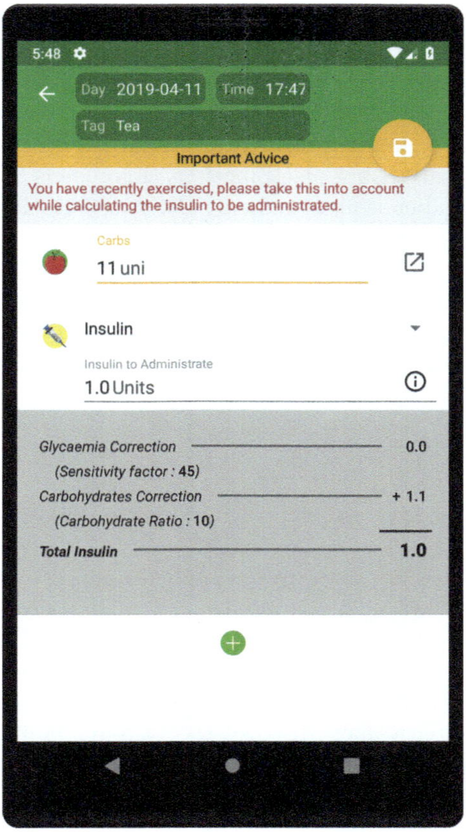

Fig. 3. Mock-up example of a user advice.

As referred in the introduction, we are working with endocrinology experts from S. João's Hospital. Nonetheless, not discarding other sources of knowledge, we also discussed with the Portuguese Diabetics' Protection Association (APDP), in order to gain knowledge and know the opinion of an association with more than 90 years of existence. This introduced new problems and occurrences. Among these, the notion that doctors may sometimes have issues finding a definitive cause to a recurrent crisis. To be more specific, a patient was having recurrent hypoglycaemias. However no justification was found for the crisis. Also the patient, did not remember doing anything out of the ordinary on the referred days. Only after displaying the glycaemia weekly, it became noticeable that the hypoglycaemias were occurring every Wednesday. The patient then remembered that Wednesdays were the days she went to get groceries. This activity was not being translated as exercise and consequently not being compensated. This oversight justified the recorded hypoglycaemias. This case and similar situations can be detected by analysing data. In the context of our project we intend to use data-mining to find patterns in the user's routines that may lead to avoidable crisis situations. The discovered patterns can be easily converted into logical rules that can help and advise the user avoid these situations.

Both data-mining and medical input can not only improve the set of existing rules, but also create a singular set of rules adapted to the user. This set not only serves the user's needs but also functions as a means to analyse the present state of the user's diabetes management. By studying the user's current rule set it is possible to understand the user's main difficulties and advice needs. If the user's current set of rules contains a majority of rules related with exercise, this may indicate a tendency, by the user, to not compensate correctly the effort made. By noticing this, the application may be able to further advise the user or, if the situation requires expert knowledge, ask the user to see an expert on this matter.

3 Conclusion

In this project we aim to advise type 1 diabetics providing guidelines, counsels and information for diabetes' management.

The individual's data recorded in the application is essential to: find patterns and create new, more personalised rules to advise the user; observe the triggered rules: if a rule is regularly being triggered it may represent a fundamental flaw in the user's current treatment regime; analyse the impact of the rule-based system: verify if, with the advice system, there was an improvement in the diabetes' management and related values; discover the system's faults and gaps on aiding the user's self-management. These discovered patterns can not only be used for creating new advice, but also as indicators to medical doctors of current behaviours that lead to mismanagement of the disease.

The data-mining used in a prior work [6], in order to obtain user behavioural patterns, included the use of bayesian networks and association rules. We plan to explore the possibility of including Inductive Logic Programming (ILP) in our project. ILP is described to be the intersection of logic programming and machine learning [3]. The main advantage ILP can give to our project is the ability to read and understand the theories concluded.

Our main concern continues to be the ability to obtain a satisfactory amount of data for the development of our data-mining unit. We expect users to value the application's advice capability and the gamification implemented.

It is important to expand the existing system, studying medical documents and to work with medical doctors.

Until now, the existing concept is still to be tested. We have already scheduled a new trial in cooperation with the S. João's Hospital to gather volunteers and test the newest version of the MyDiabetes application. The feedback from users, will be a valuable tool to evaluate the interactions made with the system and to determine possible interface improvements. We will evaluate the gamification impact on the number of new records, number of objectives fulfilled and number of badges earned. This trial will be also fundamental in order to gather user data, improve our data-mining unit and consequently our rule-based system.

References

1. Gong, J., Tarasewich, P., et al.: Guidelines for handheld mobile device interface design. In: Proceedings of DSI 2004 Annual Meeting, pp. 3751–3756 (2004)
2. International Diabetes Federation: IDF diabetes atlas (2015). www.diabetesatlas.org
3. Klinken, J.B.V.: Inductive Logic Programming, 8(2016), 1–81 (2006). https://doi.org/10.1007/BFb0027303
4. Machado, D.: Apoio ao controlo diabético baseado em registos e protocolos médicos. (2015). https://sigarra.up.pt/fcup/pt/pub_geral.pub_view?pi_pub_base_id=37974
5. Maldonato, A., Bloise, D., Ceci, M., Fraticelli, E., Fallucca, F.: Diabetes mellitus: lessons from patient education. Patient Educ. Couns. **26**(1–3), 57–66 (1995)
6. Paiva, T.: Melhorias no controlo da glicemia através de data mining. dissertação de mestrado (2016). https://sigarra.up.pt/fcup/pt/pub_geral.pub_view?pi_pub_base_id=143124
7. Parra, R.: Aumentar a interaçã com o utilizador – mydiabetes app. dissertação de mestrado (2017). https://mydiabetes.dcc.fc.up.pt/results/teseFinal-RicardoParra-InteraccaoUtilizador.pdf
8. Santos Costa, V., Rocha, R., Damas, L.: The YAP prolog system. TPLP **12**(1–2), 5–34 (2012). https://doi.org/10.1017/S1471068411000512
9. Soares, T., Ferreira, M., Rocha, R.: The MYDDAS programmer's manual. Technical report, Faculdade de Ciências da Universidade do Porto (2005)

Automatic Quality Assessment of a Forced Expiratory Manoeuvre Acquired with the Tablet Microphone

Rute Almeida[1(✉)], Bernardo Pinho[1,2], Cristina Jácome[1], João Fonseca Teixeira[3], Rita Amaral[1,4], Ivânia Gonçalves[1,2,5], Filipa Lopes[2], Ana Catarina Pinheiro[4], Tiago Jacinto[1,4], Cátia Paixão[6], Mariana Pereira[1,2], Alda Marques[6], and João Almeida Fonseca[1,2,5]

[1] CINTESIS - Centro de Investigação em Tecnologias e Serviços de Saúde, MEDCIDS, Departamento de Medicina da Comunidade Informação e Decisão em Saúde, Faculty of Medicine of Univ. of Porto, Porto, Portugal
ruteborgesalmeida@gmail.com, fonseca.ja@gmail.com
[2] MEDIDA, Porto, Portugal
[3] INESC TEC, Campus da FEUP, Porto, Portugal
[4] Porto Health School, Polytechnic Institute of Porto, Porto, Portugal
[5] Allergy Unit, CUF Porto Institute & Hospital, Porto, Portugal
[6] Lab3R, Laboratório de Investigação e Reabilitação Respiratória, Escola Superior de Saúde da Univ. de Aveiro (ESSUA) and Institute of Biomedicine (iBiMED), Univ. de Aveiro, Aveiro, Portugal

Abstract. Evaluation of lung function is central to the management of chronic obstructive respiratory diseases. It is typically evaluated with a spirometer by a specialized health professional, who ensures the correct execution of a forced expiratory manoeuvre (FEM). Audio recording of a FEM using a smart device embedded microphone can be used to self-monitor lung function between clinical visits. The challenge of microphone spirometry is to ensure the validity and reliability of the FEM, in the absence of a health professional. In particular, the absence of a mouthpiece may allow excessive mouth closure, leading to an incorrect manoeuvre. In this work, a strategy to automatically assess the correct execution of the FEM is proposed and validated. Using 498 FEM recordings, both specificity and sensitivity attained were above 90%. This method provides immediate feedback to the user, by grading the manoeuvre in a visual scale, promoting the repetition of the FEM when needed.

Keywords: Lung function monitoring · Smart device · Signal quality

1 Introduction

Chronic obstructive respiratory diseases (CORD), such as asthma and chronic obstructive pulmonary disease (COPD), are high-burden diseases throughout the life cycle, with considerable impact on patients' quality of life [1]. The progressive deterioration of CORD depends primarily on the occurrence of frequent exacerbations, that

J. Henriques et al. (Eds.): MEDICON 2019, IFMBE Proceedings 76, pp. 1394–1398, 2020.
https://doi.org/10.1007/978-3-030-31635-8_170

force patients to have regular medical visits and constant monitoring of their health status. The Forced Expiratory Manoeuvre (FEM), usually done when performing a spirometry, is crucial for the evaluation of lung function in CORD diagnosis, management, and follow-up. A FEM consists of a maximal inspiration to Total Lung Capacity, followed by a forced blast of exhalation, while keeping the mouth open, continued to a complete exhalation, i.e., until reaching Residual Volume. The correct execution of a FEM, in order to comply with acceptability and repeatability established criteria, requires the patient's training by a specialized health professional. The increasing prevalence of CORD led to a rising number of patients with lung function limitations being managed at home and using self-monitoring tools in between clinical visits. Traditional self-monitoring tools require external portable devices (e.g., flow meters and portable spirometer) which are costly, do not guarantee FEM correct execution nor motivate continued engagement.

Smartphones and tablets, aside from being portable, personal, connected and increasingly affordable, can now be found in almost every home, including in the homes of older people. Using embedded microphones of smart devices to collect FEM can allow integrating lung function monitoring tools in everyday patient's life, in any location and without the need for additional devices. Using FEM sound as a surrogate for lung function evaluation has been previously proposed in recent years [2, 3], including in a previous attempt of this group [4]. The major challenges are the sound acquisition and pre-processing components, particularly regarding the validity of the FEM, by ensuring the correct execution of the manoeuvre in the absence of a healthcare professional.

Smart device microphone spirometry presents specific challenges, such as the distance to the microphone, background noise and the absence of a mouthpiece (used in spirometry) which may allow excessive mouth closure, leading to an incorrect manoeuvre. Automatic quality assessment during the pre-processing of this surrogate signal is crucial. Efforts regarding automatic detection of the initial effort (satisfactory start of exhalation) and global effort (complete exhalation) were previously implemented in the framework of the mobile serious game INSPIRERS Arcade (Fig. 1), a medical application developed by the team with useful results [5, 6] (the beta version (https://play.google.com/store/apps/details?id=com.medida.inspirers.arcade&hl=en&pageId=none) is freely available at Google Play). Available information in the problem of automatically classifying blowing sounds is scarce, as noticed by Carbonneau et al. [7], whom have presented a naïve Bayes classifier to identify cold blowing, hot blowing, and no blowing sounds.

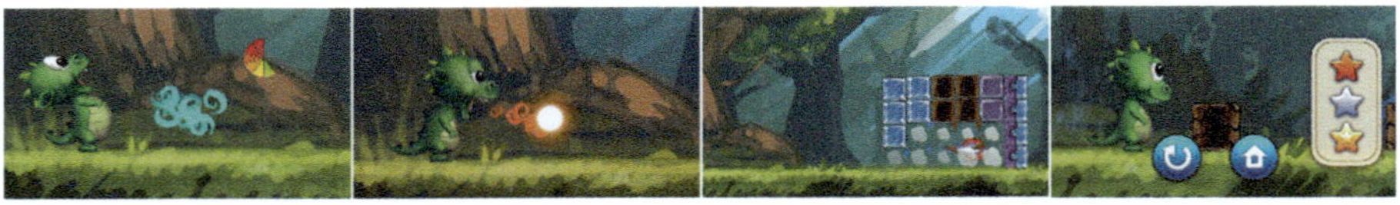

Fig. 1. INSPIRERS Arcade recording of FEM. The player mimics the dragon and is guided to a maximal inspiration and forced blast of exhalation which is recorded. Visual feedback of the quality of the manoeuvre is given by the type of projectile generated and destruction level in the brick hall, reinforced with a star-based reward. Verbal feedback gives advise on how technique can be improved.

In this work, a simple strategy to automatically assess the correct mouth position execution during FEM is proposed and validated.

2 Materials and Methods

Data used was collected using the INSPIRERS Arcade App, and is the same used in [5, 6]. The players interact with the game by blowing (FEM) to the smart device. The FEM is recorded with the microphone and immediately processed for evaluation and provision of visual and verbal feedback (Fig. 1). A total of 498 FEM recordings were collected from 50 healthy children and 4 children with self-reported asthma (5–10 years old) recruited in the community, using the INSPIRERS Arcade App and a LG V700 tablet. Written permissions from legal guardians were collected. Data was pseudo-anonymized by coding each child with an internal ID. Data collection was supervised by a health professional, who verified the correct execution of each manoeuvre regarding mouth position (as yes, not sure or no). A total of 290 recordings were labelled as correct, 198 as incorrect. Last 30% recordings of each label constituted the validation data set, the remaining the development data set.

A sound quality assessment methodology was developed using MATLAB. For each signal in the development set, the peridiogram was estimated using fast Fourier transform algorithm. In Fig. 2, the peridiograms are plotted separately for the signals labelled as incorrect or correct. It can be observed that in case of incorrect mouth position a relatively strong low frequency component appears, when compared to higher frequencies. The ratio between the spectral powers in the bands 500–1000 Hz and <1000 Hz (Low frequency ratio) was calculated. The observed ratios are summarized in the bar plot in Fig. 3. The threshold value of 42 was chosen as maximizing the weighted average: 0.6 Specificity + 0.4 Sensitivity. Weights reflect the higher impact of false positives (incorrect mouth position sounds accepted as correct, reinforcing bad form in the manoeuvre) compared to false negatives (have the child to repeat the manoeuvre).

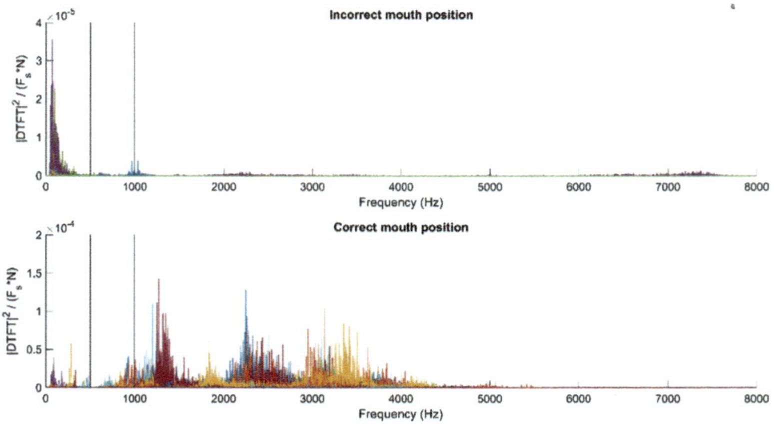

Fig. 2. Peridiograms for the signals in the development set, by reference classification group. Vertical lines stand for thresholds, Fs is the sampling frequency and N the signal length.

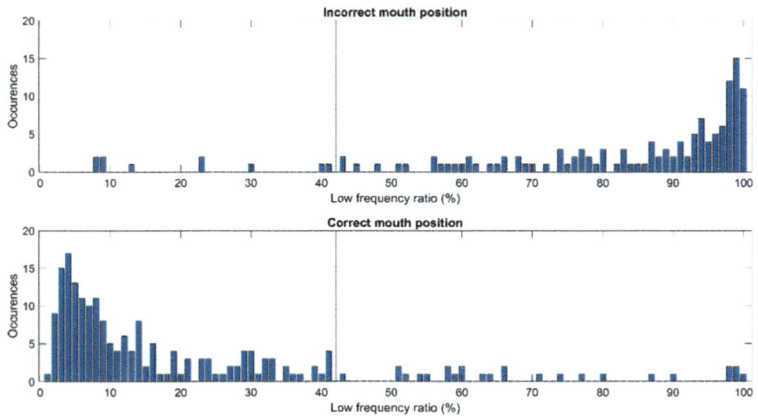

Fig. 3. Low frequency ratio values found for the signals in the development set, by mouth position as classified by the healthcare professional. Vertical lines stand for the thresholds.

3 Results and Discussion

The proposed method achieved good performance over both development and validation datasets, as presented in Table 1.

Table 1. Performance of the proposed method.

Data set	True positive	True negative	False negative	False positive	Specificity	Sensitivity	Accuracy
Development	128	177	10	26	87.2%	92.8%	89.4%
Validation	56	81	4	6	93.1%	93.3%	93.2%

The cold and hot blowing described in [7] have resemblance with our classification of mouth position (incorrect or correct, respectively). Also, the low frequency band is used in one out of the nine features considered in that work. Nevertheless, in our approach a much simpler method, with the Low frequency ratio as the only feature, is used with good results. As the data collection took place in an uncontrolled environment (community) the method is expected to be robust enough to account for ambient noise.

4 Conclusions

The proposed method was valid for assessing automatically the mouth position during the execution of a forced expiratory manoeuvre. The methodology is fully integrable in the INSPIRERS Arcade App, allowing immediate feedback to the user and the correction of the mouth position in subsequent uses. Additional validation in independent data sets is required to evaluate the generalization ability to other devices and age groups.

Funding. This work is supported by NORTE-01-0247-FEDER-033275, AIRDOC, project NORTE-01-0247-FEDER-033275, financed by the North Portugal Regional Operational Programme (NORTE 2020), under the PORTUGAL 2020 Partnership Agreement, and through the European Regional Development Fund (ERDF). AIRDOC is with ITEA 3 16040 PHE - Personal Health Empowerment project consortium.

R Almeida is supported by ERDF (European Regional Development Fund) through the operation POCI-01-0145-FEDER-029130, mINSPIRERS—mHealth to measure and improve adherence to medication in chronic obstructive respiratory diseases - generalisation and evaluation of gamification, peer support and advanced image processing technologies, funded by the COMPETE2020 and by National Funds through FCT (Fundação para a Ciência e a Tecnologia). C. Jácome has a post-doctoral grant (SFRH/BPD/115169/2016) funded by FCT, reimbursed by Fundo Social Europeu and by national funds of MCTES (Ministério da Ciência, Tecnologia e Ensino Superior). C. Paixão and A. Marques are funded by FCT under UID/BIM/04501/2013 and UID/BIM/045, and by Programa Operacional de Competitividade e Internacionalização - COMPETE, through ERDF - FEDER (POCI-01-0145-FEDER-007628).

No additional conflict to be reported.

References

1. Forum of International Respiratory Societies: The Global Impact of Respiratory Disease (2017)
2. Larson, E.C., et al.: SpiroSmart. In: Proceedings of 2012 ACM Conference on Ubiquitous Computing - UbiComp 2012, vol. 4, p. 280 (2012). https://doi.org/10.1145/2370216.2370261
3. Xu, W., Huang, M.-C., Liu, J.J., Ren, F., Shen, X., Liu, X., Sarrafzadeh, M.: mCOPD: mobile phone based lung function diagnosis and exercise system for COPD. In: Proceedings of the 6th International Conference on Pervasive Technologies Related to Assistive Environments (2013). https://doi.org/10.1145/2504335.2504383
4. Teixeira, J., Teixeira, L., Fonseca, J., Jacinto, T.: Lung function classification of smartphone recordings - comparison of signal processing and machine learning combination sets. In: Proceedings of the International Conference on Health Informatics (2015). https://doi.org/10.5220/0005222001230130
5. Pinho, B., et al.: Automatic quality assessment of smart device microphone spirometry. In: Proceedings of the 8th International Joint Conference on Pervasive and Embedded Computing and Communication Systems, pp. 115–122. SCITEPRESS - Science and Technology Publications (2018). https://doi.org/10.5220/0006866701150122
6. Almeida, R., et al.: Quality assessment and feedback of Smart Device Microphone Spirometry executed by children. In: Proceedings of IEEE 6th Portuguese Meeting on Bioengineering (ENBENG), p. 8692452. IEEE (2019). https://doi.org/10.1109/ENBENG.2019.8692452
7. Carbonneau, M.-A., Gagnon, G., Sabourin, R., Dubois, J.: Recognition of blowing sound types for real-time implementation in mobile devices. In: 2013 IEEE 11th International New Circuits and Systems Conference (NEWCAS), pp. 1–4. IEEE (2013). https://doi.org/10.1109/NEWCAS.2013.6573655

Combined Image-Based Approach for Monitoring the Adherence to Inhaled Medications

Pedro Vieira-Marques[1(✉)], João Fonseca Teixeira[2], José Valente[1], Bernardo Pinho[1,3], Rui Guedes[1], Rute Almeida[1], Cristina Jácome[1], Ana Pereira[1,4], Tiago Jacinto[1,3,4], Rita Amaral[1,5], Ivânia Gonçalves[1,3,4], Ana Sá Sousa[1], Mariana Couto[1], Mariana Pereira[1,3], Manuel Magalhães[1,6], Diana Bordalo[1,7], Luís Nogueira Silva[1,8], and J. Almeida Fonseca[1,3,4]

[1] CINTESIS - Centro de Investigação em Tecnologias e Serviços de Saúde, MEDCIDS - Departamento de Medicina da Comunidade, Informação e Decisão em Saúde, Faculdade de Medicina da Universidade do Porto, Porto, Portugal
pmarques@med.up.pt, fonseca.ja@gmail.com
[2] INESC TEC, Porto, Portugal
[3] MEDIDA, Porto, Portugal
[4] Allergy Unit, CUF Porto Institute & Hospital, Porto, Portugal
[5] Porto Health School, Polytechnic Institute of Porto, Porto, Portugal
[6] Centro Hospitalar do Porto - Centro Materno Infantil do Norte, Porto, Portugal
[7] Centro Hospitalar do Médio Ave, Vila Nova de Famalicão, Portugal
[8] Internal Medicine Department, Centro Hospitalar de São João EPE, Porto, Portugal

Abstract. The adherence to inhaled controller medications is of critical importance to achieve good clinical results in patients with chronic respiratory diseases. To objectively verify the adherence, a detection tool was previously developed and integrated in the mobile application InspirerMundi, based on image processing methods. In this work, a new approach for enhanced adherence verification was developed. In a first phase template matching is employed to confirm the inhaler positioning and to locate the dose counter. In a second phase Google ML Kit framework is used for the detection of each numerical dose in the dose counter. The proposed approach was validated through a new detection tool pilot implementation, using a set of images collected by patients using the application in their daily life. Performance of each of the two phases was evaluated for a set of commonly used inhaler devices. Promising results were achieved showing the potential of mobile embedded sensors without the need for external devices.

Keywords: Medication adherence · Smart device · Image

J. Henriques et al. (Eds.): MEDICON 2019, IFMBE Proceedings 76, pp. 1399–1404, 2020.
https://doi.org/10.1007/978-3-030-31635-8_171

1 Introduction

Inhaled controller medications are the cornerstone of asthma treatment [1], with established benefits in decreasing the severity and frequency of symptoms and exacerbations [2, 3]. However, adherence rates are known to be low [4]. With the recognition that nonadherence is a major problem and that patient self-reports of high adherence rates are unreliable, attempts have been made to objectively measure and improve adherence. Some approaches use sound to monitor inhaler usage [5], but this imply attaching additional devices to smart devices. Electronic monitoring devices (EMD) registering adherence rates by logging the time of actuations have been developed and were a step forward gaining some acceptance [6, 7]. But wide dissemination is difficult and costly. Adopting mHealth technologies and embedded mobile device sensors to deliver adherence measurement tools is highly promising, being easily integrated in everyday lives.

Inhalers usually provide visual cues for remaining dosages through numbers or scale. Although not revealing the effectiveness of each inhalation, they provide evidence on the actual use by counting activations. In addition to having different shapes, inhalers also have different presentation approaches to dosage counters, with variation in dimensions and content (some provide sequential numbers, others number steps with pictographic indicators). InspirerMundi application [8], a health serious game, incorporate an image based tool for ubiquitous verification of adherence. The tool was developed for inhalers; uses the smartphone camera; and provides, through image processing techniques, a confirmation of the inhaler presented to the camera, locating the dose counter. To address the challenge of reading the dose counter, we thought of using Machine Learning (ML) approaches, as they have gained momentum in providing methods for data feature extraction and classification. Google made available ML Kit that provides a straightforward away for mobile developers to incorporate ML into apps without the need to fully master the whole process of training models [9]. ML Kit provides out of the box, production ready application programming interface (api) for among others text recognition [10]. Aiming to provide objective verification of adherence, the application detection tool was extended incorporating the ML Kit text mining api for counter number extraction. The detection tool is at a proof-of-concept stage currently under validation of detection approaches and with still additional challenges to solve.

In this work we present the adopted image processing pipeline is presented and preliminary results of validation over 5 types of inhalers.

2 Methods

The adherence monitoring module described in this work is included in the InspirerMundi application [8]. The inserted therapeutic plan and the expected inhalations are used to engage users by requesting them to frame their inhaler in a predefined device outline. The mobile camera is used to capture a video stream, where each frame is processed with the intent of identifying the type of inhaler. Upon success the text detection module is activated filtering the results according to the particular inhaler counter region of interest (ROI). The general pipeline is depicted in Fig. 1.

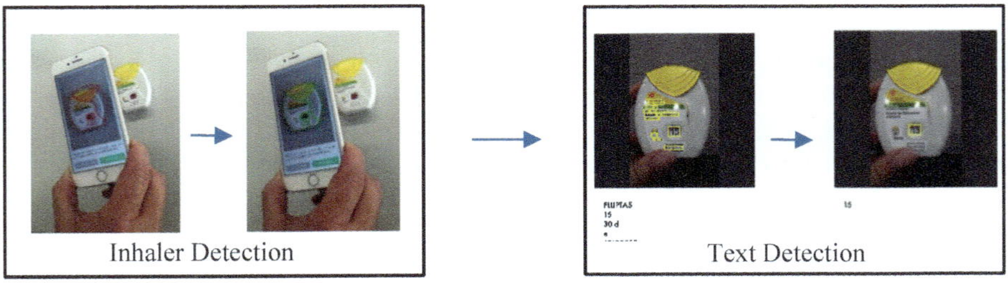

Fig. 1. General pipeline for detection of the number of activations.

The approach used to validate an inhaler is based on template matching (TM). It is employed to confirm that the inhaler is presented to the camera and locate the dose counter. By using one template per inhaler and not performing rotation matches, we restrict acquisition position to a single orientation.

Confirmation of a match can usually be done by single thresholding, although, as the images being acquired may include uncontrolled background, this would not suffice, leading to generating a second threshold. Each template is searched in the image by correlation, gaussian smoothing and selection of the most likely response position (highest value - match). Afterwards, positions are evaluated on distance to expected locations and between each other, in order to fight back spurious detection and guarantee that the inhaler is in the expected position and size. Two templates are used for each inhaler: one with the complete inhaler outline and other with the counter main outline. Both, as well as each frame, are processed by invalidating high saturation pixels, with soft threshold masking over the HSV (Hue, Saturation, Value) space, performing a morphological closing operation and converting to greyscale. Subsequently, object's edges are obtained, becoming binary images [11]. Each edge generally presents jagged lines and unstable contours, thus representative edge maps are manually cleaned, removing spurious pixels and correcting not-connected lines, before being dilated and saved.

Due to inhalers' presentation variability, some regions, are manually removed from the template such as varying brand stickers with serial numbers or the intake counter numbers. Conversely, due to the likeness between some counters and background objects or recurring geometric shapes, some counter templates (some of the smaller ones) include neighbouring parts of the non-counter inhaler edges as well.

Upon successful detection, the frame is passed to the text detection module using the ML Kit api for text mining in images. The module identifies blocks of text and creates a string representation for each along with their location within the image. By using a validated image regarding positioning in the frame we can make use of a predetermined ROI (around the counter, specific for each type of inhaler) to discard all exterior blocks.

3 Images Database

A set of images of the 5 types of inhalers considered in this work was collected by team researchers and real patients using InspirerMundi application [8]. After each scheduled inhalation, the application collected the images corresponding to success of detection if it occurs, and after 50% and 75% of the detection time if no success had occurred before that timings. All collected images were randomly distributed by 16 researchers for visual checking: 925 events were identified, with 101 images with sufficient quality to allow manual recognition of dose counter number (no image for Metered Dose Inhaler-Flutiform was found).

4 Results and Discussion

Regarding the inhaler confirmation using the template match method, it was possible to identify an average of 32% of the inhalers (Table 1). Failures can be related with unclear instructions [12], frequent low quality of the images (blurry/dark), or insufficiency of the template matching method. In future versions of the application an explanatory video description will be included to allow users to better understand what is expected. Alternative computer vision methods, such as HSV channel fusion for image enhancement, adaptive thresholding and morphological filtering will also be considered in the future to cope with quality issues.

Table 1. Performance of the template identification.

Template	Number of events	Number of successes	Success (%)
Ellipta	144	47	33
Metered Dose Inhaler-Flutiform	73	22	30
Novolizer/Genuair	163	70	43
Spiromax	154	39	25
Turbohaler	391	117	30

The counter numbers were detected in 42% of the 101 images (Table 2). To cope with the possibility that other text on the device could interfere, an additional step was included in the method. If no numbers are detected in ROI, a crop is performed around this area and submitted to text analysis again. Preliminary results of the two-step method are presented over a 20 images subset, chosen in order to fulfil strict quality criteria. The first step allowed to retrieve correctly the doses of 13 images (65%). Second step increased the performance to 15 (75%) by cropping the images ROI. Further validation in a larger dataset must be performed. These preliminary tests suggest promising results in some inhalers, although further improvement may result from better image quality acquisition or specific model training.

Table 2. Performance of the doses recognition in the validation set.

Template	Number of images	Success n (%)
Ellipta	24	10 (42%)
Novolizer/Genuair	19	10 (53%)
Spiromax	8	3 (38%)
Turbohaler	50	19 (38%)
Total	101	42 (42%)

Although in several inhalers this approach can extract the text within the ROI, several challenges still need to be addressed given the characteristics of the inhaler dial: multiple numbers present, no numbers present and only symbols, lightening effects and other artefacts that may hinder the correct identification of numbers. To some of these problems the past detections may allow to choose the right one by proximity, in other cases the user might be requested to validate himself the number identified.

5 Conclusions

The proposed image detection method adopted in this pilot implementation revealed to be feasible and promising for acquiring additional evidence regarding inhaler adherence. This approach allows to explore the mobile devices capabilities without the need for external devices or still expensive electronic monitoring devices. Further improvements are still needed to enhance the detection performance. To our knowledge there are not many approaches proposed in the literature that make use of mobile devices to gather effective activation count from inhalers devices, thus making this work relevant to help mitigating the patient unreliable, self-reported adherence.

Funding. This work is supported by ERDF (European Regional Development Fund) throughthe operation POCI-01-0145-FEDER-029130 ("mINSPIRERS—mHealth to measure and improve adherence to medication in chronic obstructive respiratory diseases - generalisation and evaluation of gamification, peer support and advanced image processing technologies") funded by the COMPETE2020 and by National Funds through FCT (Fundação para a Ciência e a Tecnologia). C. Jácome has a post-doctoral grant (SFRH/BPD/115169/2016) funded by FCT, reimbursed by Fundo Social Europeu and by national funds of MCTES (Ministério da Ciência, Tecnologia e Ensino Superior).

References

1. Global Initiative for Asthma: Global strategy for asthma management and prevention (2018)
2. O'Byrne, P.M., Barnes, P.J., Rodriguez-Roisin, R., Runnerstrom, E., Sandstrom, T., Svensson, K., Tattersfield, A.: Low dose inhaled budesonide and formoterol in mild persistent asthma: the OPTIMA randomized trial. Am. J. Respir. Crit. Care Med. **164**, 1392–1397 (2001). https://doi.org/10.1164/ajrccm.164.8.2104102

3. Pauwels, R.A., Pedersen, S., Busse, W.W., Tan, W.C., Chen, Y.-Z., Ohlsson, S.V., Ullman, A., Lamm, C.J., O'Byrne, P.M.: START Investigators Group: early intervention with budesonide in mild persistent asthma: a randomised, double-blind trial. Lancet **361**, 1071–1076 (2003). https://doi.org/10.1016/S0140-6736(03)12891-7
4. Engelkes, M., Janssens, H.M., de Jongste, J.C., Sturkenboom, M.C.J.M., Verhamme, K.M. C.: Medication adherence and the risk of severe asthma exacerbations: a systematic review. Eur. Respir. J. **45**, 396–407 (2015). https://doi.org/10.1183/09031936.00075614
5. D'Arcy, S., MacHale, E., Seheult, J., Holmes, M.S., Hughes, C., Sulaiman, I., Hyland, D., O'Reilly, C., Glynn, S., Al-Zaabi, T., McCourt, J., Taylor, T., Keane, F., Killane, I., Reilly, R.B., Costello, R.W.: A method to assess adherence in inhaler use through analysis of acoustic recordings of inhaler events. PLoS ONE **9**, e98701 (2014). https://doi.org/10.1371/journal.pone.0098701
6. Chan, A.H.Y., Harrison, J., Black, P.N., Mitchell, E.A., Foster, J.M.: Using electronic monitoring devices to measure inhaler adherence: a practical guide for clinicians. J. Allergy Clin. Immunol. Pract. **3**, 335–349.e5 (2015). https://doi.org/10.1016/j.jaip.2015.01.024
7. Pilcher, J., Holliday, M., Ebmeier, S., McKinstry, S., Messaoudi, F., Weatherall, M., Beasley, R.: Validation of a metered dose inhaler electronic monitoring device: implications for asthma clinical trial use. BMJ Open Respir. Res. **3**, e000128 (2016). https://doi.org/10.1136/bmjresp-2016-000128
8. Jácome, C., Almeida, R., Teixeira, J.P., Vieira-Marques, P., Vilaça, R., Fernandes, J.M., Ferreira, A., Fonseca, J.A.: Inspirers: an app to measure and improve adherence to inhaled treatment. In: Proceedings of the International Conference on E-Health, EH 2017 - Part of the Multi Conference on Computer Science and Information Systems 2017 (2017)
9. Curukoglu, N., Ozyildirim, B.M.: Deep learning on mobile systems. In: 2018 Innovations in Intelligent Systems and Applications Conference (ASYU), pp. 1–4. IEEE (2018). https://doi.org/10.13140/RG.2.2.15012.12167
10. Google Developers Blog. https://developers.googleblog.com/2018/05/introducing-ml-kit.htm
11. Canny, J.: A computational approach to edge detection. IEEE Trans. Pattern Anal. Mach. Intell. **PAMI-8**, 679–698 (1986). https://doi.org/10.1109/tpami.1986.4767851
12. Ferreira, A., Almeida, R., Jácome, C., Fernandes, J., Almeida Fonseca, J., Vieira-Marques, P.: How inspiring is your app? A usability take on an app for asthma medication adherence. In: 11th International Conference on e-Health (in press)

Mobile Application to Support Children with Anxiety Disorders

Nuno Fonseca[1](✉), Ana Almeida[1], Maria Moreno[4], Raquel Simões de Almeida[3], Luiz Faria[1], António Marques[2], Paulo Matos[1], Pedro Rocha[1], and Constantino Martins[1]

[1] GECAD Research Group on Intelligent Engineering and Computing for Advanced Innovation and Development, Porto, Portugal
nuf@isep.ipp.pt

[2] School of Allied Health Sciences, Oporto Polytechnic Institute, Porto, Portugal

[3] Faculty of Psychology and Educational Sciences, University of Porto, Porto, Portugal

[4] Departamento de Informática y Automática, Universidad de Salamanca, Salamanca, Spain
mmg@usal.es

Abstract. This paper describes WEBECOOL, a hybrid web/mobile application with the purpose of improving the self-management of anxiety disorders. The development of WEBECOOL was driven by a multidisciplinary collaboration between computing engineers, psychologists and psychiatrics. The first stage of the web/mobile application test was performed to assess the satisfaction with the technological proposal. Also, several components were identified as crucial and were employed in WEBECOOL: stress management, problem solving, medication adherence, symptoms monitoring and social interaction. WEBECOOL also pretends to include caregivers in the process, with the purpose of implementing better help therapy strategies.

Keywords: Mobile application · Anxiety disorders · Assistive technology · Ambient intelligence

1 Introduction

Mental illness, particularly anxiety disorders, comprehends many clinical conditions associated with several changes that include limitations related with social interaction or several task such as sleeping through the night, doing homework, and making friends, thinking capacity and reality understanding, deficits in communication skills and difficulties in developing appropriate emotional and behavioral responses.

Some international guidelines encouraged not only active user's involvement and participation, nuclear values of recovery, as well as the emergence of integrated strategies that contribute not only to symptoms remission, social participation and quality of life promotion [1], but also to establish the need to promote an approach focusing on the children and their needs and choices.

J. Henriques et al. (Eds.): MEDICON 2019, IFMBE Proceedings 76, pp. 1405–1410, 2020.
https://doi.org/10.1007/978-3-030-31635-8_172

Illness self-management is an approach designed to involve children with mental health problems as active agents in their own treatment and recovery, teaching them to monitor their condition and to use appropriate coping strategies. Patients who received self-management education were more likely to improve medication adherence and symptomatology, compared to other patients.

Due to recent technological developments, several researchers designed innovative assistive tools that improve the accessibility and quality of care for people with mental health problems [2, 3].

Mobile applications seem to contribute to help their users to engage in health promoting behaviors outside the clinical context [5, 6] or in other activities such as therapeutic homework, facilitating generalization to the day-to-day life of its users [7].

The use of mobile devices seems to be the logical path for self-management. More than self-management, mobile devices may help alerting patients, therapists and caregivers in more faster way than traditional methods. They also allow to interact in different manners with patients and therapists.

All these aspects may lead to a recommendations system, suggesting activities to minimize some episodes of the children's anxiety.

2 Brief Literature Review

Increasing of health information is generalized, people consult internet in search of information about diseases, symptoms and treatments. The availability of this information tends to shift patient-physicist, allowing patient to question and discuss treatments [9].

However, information overload and wrong or irrelevant information are obstacles for getting conclusions on the personal health status and taking adequate actions [4]. Faced with a large amount of medical information on different channels (e.g., news sites, web forums, etc.) users often get lost or feel uncertain when investigating on their own. In addition, a manifold and heterogeneous medical vocabulary poses another barrier for laymen [4]. Therefore, improved personalized delivery of medical content can support users in finding relevant information [4].

Recommender systems (RS) suggest items of interest to users of information systems or e-business systems and have evolved in recent decades. A typical and well-known example is Amazon's suggest service for products. The idea behind recommender systems can be adapted to cope with the special requirements of the health domain [4, 7].

Health recommendation systems are a specialization of Recommender Services. In a health recommendation system, a recommended item must be a proven scientific information, non-confidential acceptable medical information.

3 WEBECOOL System

The main objective is to build a recommendation system that using specific algorithms and knowledge database may help children with anxiety disorders by reducing the response time to episodes or even prevent those same episodes from happening. The recommendation system will be built on the WEBECOOL mobile app.

WEBECOOL is a system already developed by multidisciplinary team that helps therapists and patients to interact with the purpose of minimize response times to children anxiety episodes it also allows children to self-monitors and give feedback to the therapist. WEBECOOL is accessible as a webpage or android app.

The architecture of WEBECOOL was adapted from weCOPE System [8] (Fig. 1).

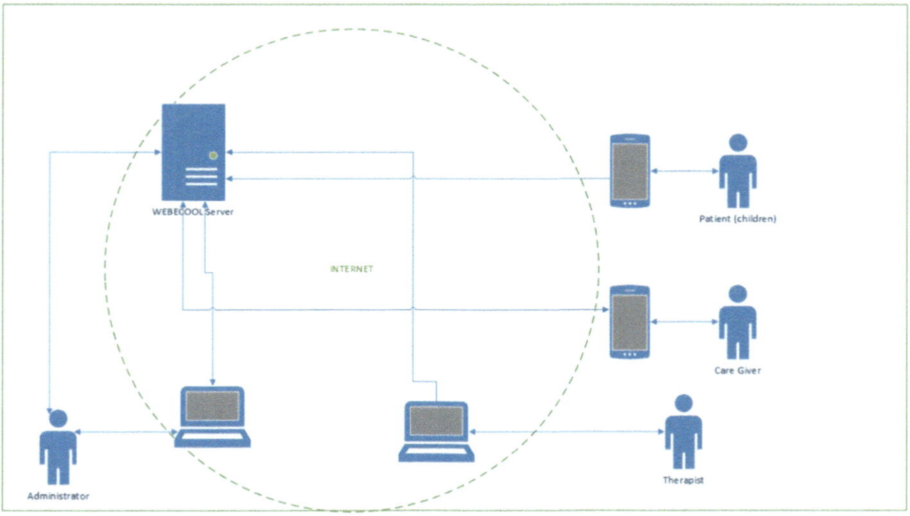

Fig. 1. WEBECOOL system

The WEBECOOL web platform works with a centralized server, meaning that children progress may be registered and consequently monitored through the therapist's interface. The WEBECOOL mobile app supports the main philosophy of the project to provide a platform capable of assisting the rehabilitation process in children with anxiety disorders, which allows the children to follow the therapy course based on computer-aided treatment.

Considering the centralized architecture of the project, the mobile application requires communication with the server in order to acquire, display and record data. The API offers a layer communication with the server and the database. It is, therefore, a service that allows interaction with the various modules/entities that support the data model. RESTful architectures present several advantages, such as the possibility to build a Web service outline resource easily accessible through different platforms (especially useful for developing mobile applications). It creates a layer of abstraction that separates server implementations from client implementations.

3.1 WEBECOOL Web Platform

The web platform has two modules, which are described as: The back-office system for the administrator's interface; The therapist's interface where a professional psychotherapist or psychologist may interact with the patient.

The therapist's interface allows the therapists to oversee their children's inputs in the daily log, objectives and problem-solving modules. This allows them to acquire better knowledge about each children's "battles" and goals, helping them adjust the therapy so as to better increase the children's quality of life.

3.2 WEBECOOL Mobile App

The mobile application implements the features and modules which are considered as crucial for the children's therapy. It leverages on the advantages offered by mobile applications, such as the capability to easily transmit time-sensitive information, the ability to access remote services from several real-world locations and, more important, it may provide immediate therapeutic support. Mental health applications may provide patients with tools that help them overcome their difficulties, increase self-management skills, encourage the definition of goals and improve their overall quality of life.

WEBECOOL Mobile APP is built with React Native. React Native is a framework for building cross-platform mobile applications. Its main feature is the ability to build hybrid applications using languages that are part of the web development standards, such as HTML5 and JavaScript.

Four modules are part of the application and are consequently considered as crucial for the children's therapy.

1. Day-to-day Log: The Day-to-day log is a module in which the children is able to describe and expose a anxiety experience or event that triggered an social problem. This module also works as a record of children's mood over time and allows the review of past entries. For the therapist, it works as a feedback mechanism and may help to determine the best strategies and interventions for symptoms' monitoring. Medication adherence was also included in this module, involving custom notifications for taking medication.
2. Goals and Progress: A children with therapist can define a set of goals or tasks to accomplish. The main objective of this module is to provide a personal motivation element that the children can follow.
3. Anxiety Problem Solving: The anxiety problem solving module describes an introspection and self-management tool. As a way to promote anxiety problem solving, it is necessary for the children to provide tangible solutions to solve the specific problem. After this process, the children must evaluate their own solutions and choose the most appropriate, together with an assessment of their success. Social interaction may be difficult for children with anxiety disorders as they may strive to start a conversation or to appropriately communicate with others. Thus, this

module can be considered as way of interaction and as place where the children may vent or describe a stressful situation in a safe space.

4. Chill Out Tools: This module offers the children not only stress management skills but also a relaxing environment with access to media files to view or listen. This tool incorporates these features in order to provide a calm setting including guided meditation and relaxation instructions. Multimedia content is available dynamically, meaning content can be managed server-side, avoiding the need to change anything client-wise.

4 Conclusions

Anxiety self-management is fundamental for children, allowing them to take control over their lives and fostering genuine collaboration with professionals in their recovery process. New self-management technologies could change health functioning, symptoms, and progress towards recovery. The next step will be to validate our proposal and see the impact of using artificial intelligence algorithms on the WEBCOOL health recommendation system, with the use of those algorithms we expect to have faster responses to children and his anxiety's disorders.

Acknowledgements. This work was supported by National Funds through the FCT – Fundação para a Ciência e a Tecnologia (Portuguese Foundation for Science and Technology) Project UID/EEA/00760/2019.

References

1. Chan, S.R., Torous, J., Hinton, L., et al.: Mobile tele-mental health: increasing applications and a move to hybrid models of care. Healthcare **2**(2), 220–233 (2014). https://doi.org/10.3390/healthcare2020220
2. Álvarez-Jimenez, M., Gleeson, J.F., Bendall, S., et al.: Internet-based interventions for psychosis: a sneak-peek into the future. Psychiatr. Clin. North Am. **35**, 735–747 (2012). https://doi.org/10.1016/j.psc.2012.06.011
3. Ben-Zeev, D., Drake, R.E., Corrigan, P.W., et al.: Using contemporary technologies in the assessment and treatment of serious mental illness. Am. J. Psychiatric Rehabil. **15**, 357–376 (2012). https://doi.org/10.1080/15487768.2012.733295
4. Wiesner, M., Pfeifer, D.: Health recommender systems: concepts, requirements, technical basics and challenges. Int. J. Environ. Res. Public Health **11**, 2580–2607 (2014). https://doi.org/10.3390/ijerph110302580
5. Harrison, V., Proudfoot, J., Wee, P.P., et al.: Mobile mental health: review of the emerging field and proof of concept study. J. Mental Health **20**, 509–524 (2011). https://doi.org/10.3109/09638237.2011.608746
6. Luxton, D.D., McCann, R.A., Bush, N.E., et al.: mHealth for mental health: integrating smartphone technology in behavioral healthcare. Prof. Psychol. Res. Pract. **42**, 505–512 (2011). https://doi.org/10.1037/a0024485

7. Ben-Zeev, D., Drake, R.E., Brian, R.M.: Technologies for people with serious mental illness. In: Marsch, L., Lord, S., Daller, J. (eds.) Behavioral Health Care and Technology: Using Science-Based Innovations to Transform Practice, pp. 70–80. Oxford, New York (2014)
8. de Almeida, R.S., Martins, C., Marques, A., Benevides, D., Costa, A., Queirós, C., Sousa, T., Almeida, A., Fonseca, N., Faria, L.: Development of a hybrid application for psychotic disorders self-management. In: Conference: International Symposium on Ambient Intelligence. Advances in Intelligent Systems and Computing (2017). https://doi.org/10.1007/978-3-319-61118-1_28
9. Proudfoot, J.: The future is in our hands: the role of mobile phones in the prevention and management of mental disorders. Aust. N. Z. J. Psychiatry **47**, 111–113 (2013). https://doi.org/10.1177/0004867412471441

Smartphone Recommendation System to Prevent Potential Injuries in Young Athletes

Paulo Matos[1,3](✉), João Rocha[1], Ramiro Gonçalves[2], Filipe Santos[1,3], Goreti Marreiros[1,3], Daniel Mota[1,3], Nuno Fonseca[1,3], and Constantino Martins[1,3]

[1] Computer Science Department, Engineering Institute – Polytechnic of Porto, Porto, Portugal
{psm,jsr,jpe,mgt,ddm,nuf,acm}@isep.ipp.pt

[2] Trás-os-Montes e Alto Douro University, Vila Real, Portugal
ramiro@utad.pt

[3] GECAD Research Group on Intelligent Engineering and Computing for Advanced Innovation and Development, Porto, Portugal

Abstract. Over the last decades web and mobile technologies are increasingly being used in sports, especially in soccer, but none of them seems to allow to prevent injuries. However, training systems for young athletes do not have, for the most part, learning abilities in order to adapt, evolve and find new training recommendations. It is in this context which the Smart Coach project is presented in this work, and whose main goal is to introduce our mobile training recommendation system allow to young athletes evolve. The training mobile recommendation system is also designed to identify potential injuries risk for each young athlete.

Keywords: Mobile health application · Prevent injuries · Recommender systems

1 Introduction

Information and Communication Technologies (ICTs) namely mobile technologies are increasingly being used in the sport, especially in football, aiming to help the coaching team, enhance athletes training methods, improve the team results or support sports decisions and refereeing. However, training systems, for the most part, don't consider preventing potential injuries in young athletes' considering their characteristics, technique, tactics, physical and mental status, in the training selection and recommendation process.

Additionally, these systems do not use artificial intelligences techniques in order to adapt, evolve and find new training recommendations to allow the young athletes to evolve or to prevent potential injuries. These limitations render the systems results, inadequate, and unfocused on the players needs and specificities. The *SmartCoach* recommendation system intends to fill this gap, expecting to innovate and make impact, using user modelling and recommendation techniques and mobile technology. This adds support to coaches and technical staff,

J. Henriques et al. (Eds.): MEDICON 2019, IFMBE Proceedings 76, pp. 1411–1415, 2020.
https://doi.org/10.1007/978-3-030-31635-8_173

allowing them to analyse better their young athletes' skills and enhance their development and training. In this paper we introduce the *SmartCoach* web mobile system. It allows to represent technical, tactical, physical and/or psychological characteristics of young athletes, and adapt a Dynamic Training Model, defining a training schedule to improve a young sportsman performance, targeting is evolution as a player and to prevent potential injuries.

2 State of the Art

2.1 Recommender System

A Recommender System can be characterized like a collection of different techniques and methods used by different applications or systems to filter and organize their items, in order to select either the best ones or the most suitable ones, according to the user profile [6].

Health Recommender Systems (HRS) are becoming an important tool in health decision making processes in healthcare services [1]. HRS increase usability of technologies and reduce information overload in healthcare decision processes [1].

The mode of operation normally used by recommender systems is to use the user profile as the basis for a series of calculations to infer which are going to be, amongst all the items available, the ones that will better please the user, according to a wide variety of approaches [7].

Recommender systems are essentially based in three types of paradigms (content, collaborative and knowledge-based) and all their possible combinations [3–5,8].

2.2 Context-Awareness Computing

Context-aware computing is a paradigm where applications and services use environmental information to provide relevant information and/or services to the user. The concept of context has been researched over the last decades in software engineering, especially in areas such as Natural Language Processing, and more generally in Human-Computer Interaction. Context makes the interaction with computers easier by adapting the information to the user, discriminating what is relevant and what is not, so that the user in human-machine interactions can focus on high-level tasks, which is very important in scenarios of information overload, especially as we move towards a world of ubiquitous and pervasive computing and the Internet of Things.

Even though the term "context" has been addressed many times in the literature since the 90s, it is still difficult to understand as sometimes there is not a clear boundary between what is context and what it is not. One of the most widely cited definitions for context was proposed by Abowd and Dey (Abowd, 1999), with the following statement: "Context is any information that can be used to characterise the situation of an entity. An entity is a person, place, or object that is considered relevant to the interaction between a user and an application, including the user and applications themselves."

Abowd and Dey (Abowd, 1999) identified four context information types that can be considered more important than others and called them "primary context". They answer basic questions like who (Identity), where (Location), when (Time) and what (Activity or Status). Additionally, they considered "secondary context" any relevant information that can be derived from primary context.

3 *SmartCoach* Web Mobile System

The *SmartCoach* web mobile system (Fig. 1), with learning capacity, was implemented to facilitate the interaction between members of a club technical staff, medical staff and their young athletes, reinforcing the young person counselling, and their potential as an athlete.

The system allows young athlete performance evaluation, considering several attributes (Fig. 1). The data is collected in an intuitive and adaptive manner, using a friendly mobile interface.

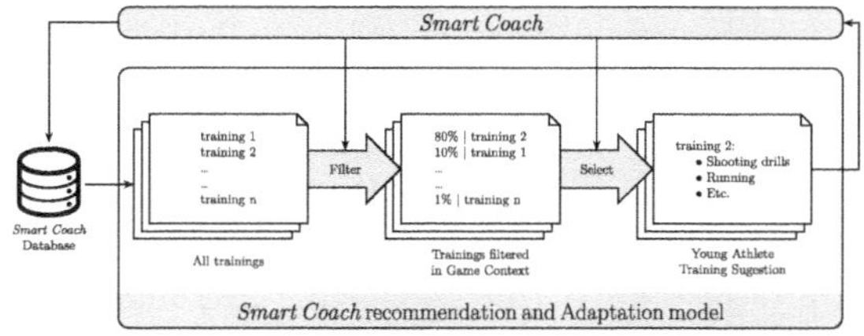

Fig. 1. Smart Coach recommendation and Adaptation model

The *SmartCoach* system, is designed and implemented using an innovative hybrid recommendation algorithm, which allows to relate the representation of the technical, tactical, physical and/or psychological characteristics of the young athlete, with a model of dynamic training, in order to identify and prevent injuries risks, customize training and improve skills that the athlete has some deficit. The application is aimed mainly for young athletes and is now being used in a small club youth academy for their soccer players data collection.

3.1 Smart Coach User Profile Modelling

The athlete profile model is used, to identify the most appropriate training to fulfil young athlete training necessities and to prevent injuries. The *SmartCoach* User Modelling solution involves initially the definition of stereotypes for young athletes. These stereotypes were defined during interviews with several football coaches. Also, we are using the K-Means [2] clustering algorithm application to data obtained during matches and training of several football academies in Portugal. Each cluster was classified with a set of attributes, with diverse weights and mapped according to their relevance in training, and their influence in football tactics and team performance. The clustering outcome is the young football player profile, e.g., goalkeeper, defender, midfielder, striker. Each cluster has a user type, which is classified with several types of attributes/tasks that typically has to perform during a football match. Also, a personal health record is implemented for each young athlete with the goal to supply medical information.

3.2 Context-Awareness Approach

Finding proper patterns and accurately predicting the potential injuries *SmartCoach* can provide better recommendation. Here context has an important role, because the football player behaviour may be influenced by the context (i.e. football boots, pitch types, weather conditions, etc.).

From alerting and warning situations, to special events in the state of the football match, *SmartCoach* can predict several scenarios where the use of context awareness (i.e. football boots, pitch types, weather conditions, etc.) could be relevant to deliver the right information, at the right moment, to the right person with several recommendations to prevent potential injuries.

3.3 *SmartCoach* Injuries Matrix

This tool has components to present data in a matrix format, created from various variable context-awareness groups and the aggregation of information related to frequent injuries in young football players. Also, the injuries matrix construction allows us to produce statistical knowledge presentation. This information helps to support decision-making and the acquisition of data-based knowledge.

3.4 Training Determination Process

The training determination process intends to infer, based on the young athlete state, captured at a certain point of time, the set of training exercises the athlete should execute in the same time window (Fig. 1). This information allows the determination of relevant workouts, appropriate to actions being performed. The action being performed recognition can be an uncertain and imprecise process. The observation of athlete's actions may not have an objective evaluation, and as such, may result in inaccurate conclusions.

In *SmartCoach* web/mobile application, the training exercises, are associated with defined stereotypes in the young athletes' models. The parameters function regarding the young athletes' performance attributes, such as the training and necessary steps to complete it successfully or the context awareness variables like the type of boots, pitch surface/quality and weather conditions. For each stereotype a hierarchical set is defined, that associates all the trainings that are within its competence and the relations between them are represented.

In view of the initial catalogue of all possible training sessions, the adaptation model (Fig. 1) will initially, use the user's stereotype information and the attributes considered relevant by the coaching and medical staffs to perform/apply our filtering algorithm. This will allow, for example, that only trainings related to young athletes' actions and goals are considered. Hence, an ordering is made according to the probability of each activity being performed at a given time from among all possible, using data such as the expected course of training and young athletes' personal performances and injuries history.

4 Conclusions

In this work is considered that the best way to please users is to suggest trainings that can improve their abilities or prevent injuries. The first version of the data collection prototype is being evaluated in two football clubs' academies, which we will designate by club A and club B. The study will run over two months in two teams: one from club A and one from club B. The athletes age is between 15 and 18 years. None of the athletes and staff had any experience using mobile applications to support training. However, they are familiar with mobile apps and personal computers (PCs), namely, social networking and playing videogames.

Acknowledgements. This work was supported by National Funds through the FCT – Fundação para a Ciência e a Tecnologia (Portuguese Foundation for Science and Technology) Project UID/EEA/00760/2019.

References

1. Afolabi, A.O., Toivanen, P., Haataja, K., Mykkänen, J.: Systematic literature review on empirical results and practical implementations of healthcare recommender systems: lessons learned and a novel proposal. Int. J. Healthc. Inf. Syst. Inform. (IJHISI) **10**(4), 1–21 (2015)
2. Anderberg, M.R.: Cluster analysis for applications. Technical report, Office of the Assistant for Study Support Kirtland AFB N MEX (1973)
3. Berka, T., Plößnig, M.: Designing recommender systems for tourism. In: Proceedings of ENTER 2004, pp. 26–28 (2004)
4. Felfernig, A., Gordea, S., Jannach, D., Teppan, E., Zanker, M.: A short survey of recommendation technologies in travel and tourism. OEGAI J. **25**(7), 17–22 (2007)
5. Kabassi, K.: Personalizing recommendations for tourists. Telematics Inform. **27**(1), 51–66 (2010)
6. Porter, J.: Watch and learn: how recommendation systems are redefining the web, vol. 5 (2006). https://articles.uie.com/recommendation_systems/. Accessed 13 Dec 2006
7. Santos, F., Almeida, A., Martins, C., Gonçalves, R., Martins, J.: Using POI functionality and accessibility levels for delivering personalized tourism recommendations. Comput. Environ. Urban Syst. (2017). https://doi.org/10.1016/j.compenvurbsys.2017.08.007. http://www.sciencedirect.com/science/article/pii/S0198971517302880
8. Schafer, J.B., Konstan, J., Riedl, J.: Recommender systems in e-commerce. In: Proceedings of the 1st ACM Conference on Electronic Commerce, pp. 158–166. ACM (1999)

Special Sessions: Computational and Experimental Modelling for Designing Bone Implant Systems

Experimental Assessment of Knee Arthrodesis

Ana M. Amaro[1(✉)], Luis Roseiro[1,2], Maria F. Paulino[1], and Maria A. Neto[1]

[1] CEMMPRE, Department of Mechanical Engineering, University of Coimbra, Coimbra, Portugal
{ana.amaro,maria.paulino,augusta.neto}@dem.uc.pt, lroseiro@isec.pt
[2] Coimbra Polytechnic - ISEC, Coimbra, Portugal

Abstract. This study aims to evaluate experimentally the performance of the eternal fixator AO to promote an efficient knee arthrodesis, in terms of bone alignment. The knee arthrodesis serves as an option of salvage treatment for failed total knee arthroplasty procedures. Knee arthrodesis consists of the artificial formation of a joint between two bones (femur and tibia) through surgery, reducing pain, providing stability of the limb. It is also used to correct malformations or congenital deformations. In this work, the surgical procedure of knee arthrodesis was replicated by removing the knee joint, and having been applied compression loads in the femur and tibia, in order to promote bone fusion. The pressure distribution was evaluated in order to identify the occurrence of the bone fusion. The knee arthrodesis technique used was promoted with external fixation consisting on the union of two or more femoral pins with two or more tibial pins through sidebars. The objective is to verify the best bone alignment, in order to guarantee a good rate of bone fusion. It was possible to observe that the bone alignment presents great influence on the success of the knee arthrodesis, with an increase around 81% in the area of pressure distribution, when compared with the misaligned bones.

Keywords: Arthrodesis · External fixator · Experimental evaluation · Knee joint

1 Introduction

The knee is the largest joint that exists in the human body, and connects the femur to the tibia. It is vulnerable to injuries as it bears an enormous amount of pressure while providing flexible movement. For example, during the simple walk the load on the knees is around 1.5 times the body weight. Due to its function in the human body many knee injuries can occur, include fractures around it, dislocation, and sprains and tears of soft tissues, such as ligaments. Knee arthrodesis as a treatment option in the case of a failed total knee arthroplasty has been used for more than 100 years [1]. Agreeing to Charnley [2] the technique of arthrodesis using external fixation was pioneered by Albert Key in 1932 for the treatment of a knee with tuberculosis. In the year of 1948 this technique was improved by himself. The use of external fixators play a large role in arthrodesis indications, because the installation is straightforward and reproducible,

J. Henriques et al. (Eds.): MEDICON 2019, IFMBE Proceedings 76, pp. 1419–1423, 2020.
https://doi.org/10.1007/978-3-030-31635-8_174

enabling a good adjustment of frontal and sagittal alignments [3]. The knee arthrodesis consists of removal of the knee joint, followed by the application of compression loads on the femur and the tibia, in order to promote bone fusion. Cunningham et al. [4] say that only a compressive load of 200 N is enough to perform a complete fusion between the two bones, femur and tibia, and more than this will result unnecessary bone loss. The knee arthrodesis can be also used in other clinical situations, as, periarticular tumor, posttraumatic arthritis, and chronic sepsis [5]. There are some specific arthrodesis techniques, external fixation, intramedullary rod or compression plates, which may be used. According to Kim et al. [6] the decision about which specific arthrodesis technique to use, should be based on three principal variables such as the infection state, condition of the soft tissue envelope, and preoperative bone loss. Letartre et al. [7] promote a study using the fixator AO and concluded that weight bearing is only allowed after a 3 to 6-month delay and a mean 40 mm shortening is observed. The external fixators show some advantages which include the ability to provide a mechanical stimulus for bone formation as also to adjust alignment of the arthrodesis [8]. However some disadvantages are also observed, such as the pin site infection and pins and clamps failure [9]. Bono and Scott [8] say that the fusion rates in the arthrodesis knee depend on the surgical technique used, and in the case of the external fixators the failure rate is greater than 50%. To increase the success of the fusion rate is necessary promote a correct alignment, where the radiographs obtained before the surgical procedure have an important role.

The aim of this study is developed an experimental device to replicate the knee arthrodesis. In this work it was possible conclude that the bone alignment presents great influence on the success of the knee arthrodesis surgical, reducing the time necessary to a bone fusion.

2 Materials and Methods

A synthetic model tibia-femur from the Sawbones® was used. For the correct cut it was necessary a perfect alignment of the center of the femoral head with the center of the distal portion of the tibia, ensuring the neutrality of the mechanical axis of the lower limb. After this, the synthetic bone model tibia-femur was cut, to replicate the knee arthrodesis, into two phases: the first one corresponds to the cut of the distal epiphysis of the femur, with the saw inclined at an angle of 15°, whereas the second phase consist of the vertical cut of the proximal epiphysis of the tibia. The objective should be to obtain an overall limb alignment with the knee in 5–7° of valgus and $15 \pm 5°$ of flexion [10].

The external fixator used in the present study is AO-type and was initially introduced by Müller in 1969. This is considered a simple fixator, because of the independent connections between each pin and the sidebar. This fixator was used to promote the bone fixation and was clamped to the bones with four pins Steinman in AISI steel 316 L, with 5 mm of diameter and 300 mm of length, sidebars and eight clamps to allow the connection between the pins and the bars in a monoplanar and bilateral position, Fig. 1. As a more homogeneous pressure distribution reflects a more effective bone fusion, the pressure that occurred between the two bones was evaluated with a board of the Arduino type produced by the brand Kitronyx®, where at one end of the board is connected one computer and at the other end a sensory matrix. The technology behind pressure reading consists of the detection of forces by a touch-sensitive matrix sensor. In this case the 16 × 10 matrix has 160 sensory nodes and is designed for low force values, allowing the construction of a pressure mapping system based on the force applied to each cell. The conversion of the signal into pressure values is possible due to the software present in Arduino Snowboard that does the conversion of values.

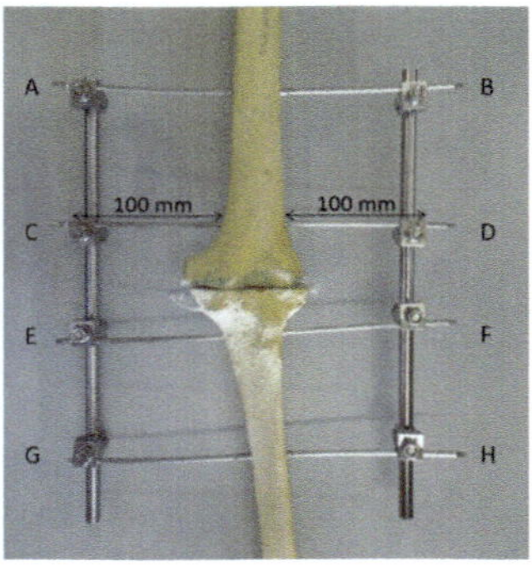

Fig. 1. External fixator AO connected to the bones.

Fig. 2. Experimental assembly.

For a correct positioning of the bones and the fixator, it was developed a mechanical device, where all components have been assembled and properly positioned, Fig. 2. At least five tests in each condition at room temperature were made. A compression load of 100 N was applied in each clamp loaded, using a torque wrench, in order to ensure that the load placed in the clamps was always the same.

3 Results and Discussion

To understand the influence of the bone alignment two different tests were performed. Firstly, it was considered that the bones are misaligned. In the second test, the bones are correctly aligned. The threaded pins in the femur (A, B, C, D) remain attached to the sidebars by clamps (without loading), and the clamps of the pins applied to the tibia are loaded sequentially in the order G, H, E, F (see Fig. 1). In Fig. 3 and Fig. 4 can be observed the pressure distribution during the tests with the bones misaligned, and aligned, respectively. These figures are representative for the others obtain for the same test conditions. According to Somayaji et al. [1] to achieved a good knee fusion it is necessary an extensive preoperative preparation, in which a correct alignment is required. After the complete bone fusion the overall alignment of the limb should be measured using long-leg films. Comparing the Figs. 3 and 4 it is possible to see significant differences in the pressure distribution. The bone alignment provides a greater distribution of pressures, denounced by the occupation of a larger and uniform contact area, and therefore of a greater colored area. The pressure area in the case of the aligned bones, when compared with the misaligned ones, increases around 81%, 72%, 70% and 69% after loaded the pins in the sequence G, H, E, F respectively.

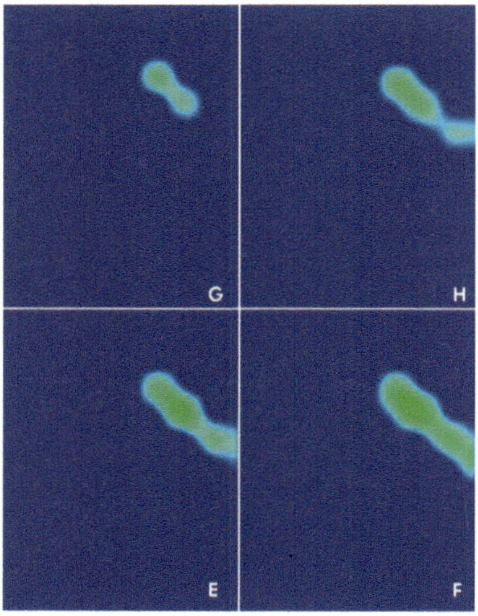

Fig. 3. Pressure distribution in the case of bones misaligned.

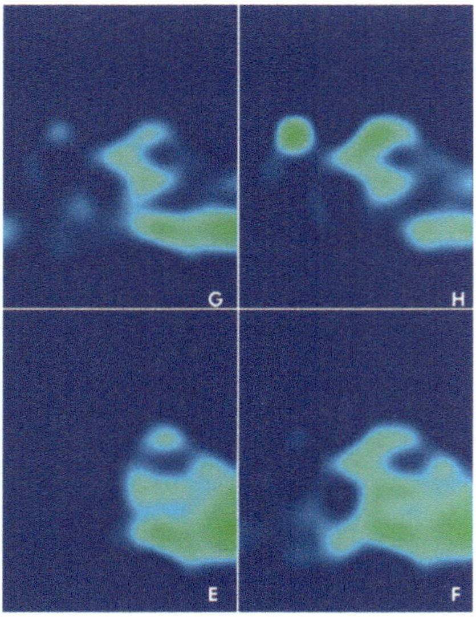

Fig. 4. Pressure distribution in the case of bones aligned

Associated to the fact that a good bone fusion rate occurs not only for a larger size of the contact area, but also for a shade corresponding to more yellowish colors, can be concluded that the bone alignment promotes more effective knee arthrodesis, which corresponds to the surgeons' objectives in order to promote a faster and more efficient recovery of the patient. After the tests, a slight bending of the pins was observed, although with different shapes, due to the demands caused by the various clamps fixing (Fig. 5a). This situation is also observed in- vivo (Fig. 5b). The bone fusion is the key for the knee arthrodesis success. According to Robinson et al. [11] the knee arthrodesis can be considered an achievement if after a radiography a stable fused joint is reached. On the other hand, the knee arthrodesis is considered a failure if patients did not achieve solid bone union.

(a)

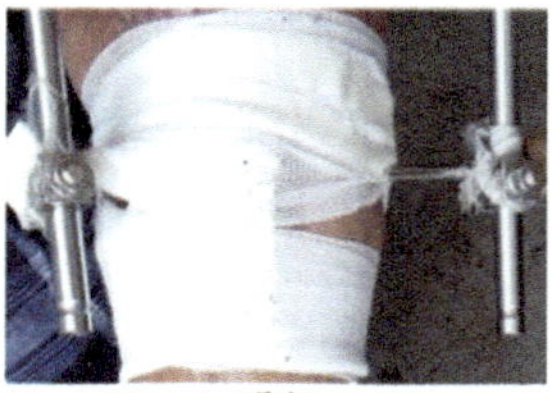

(b)

Fig. 5. Bending in the pins: (a) experimental; (b) in-vivo.

4 Conclusions

The bone alignment provides high and homogeneous distribution of pressures in the sensory matrix, which will cause a higher rate of bone fusion, leading to a greater success of the surgical procedure being studied. The difference observed in the pressure area distribution is around 69% greater in the case of the bones aligned, when compared with the misaligned ones, after the tightening of the 4 pins. After performing the experimental surgical procedure it was observed a certain bending of some of the pins used in the external fixator, but this event is quite natural and verified also in-vivo situations.

References

1. Somayaji, H.S., Tsaggerides, P., Ware, H.E., Dowd, G.S.: Knee arthrodesis–a review. Knee **15**, 247–254 (2008)
2. Charnley, J.C.: Positive pressure in arthrodesis of the knee joint. J. Bone Joint Surg. Br. **30** (B), 478–486 (1948)
3. Riouallon, G., Molina, V., Mansour, C., Court, C., Nordin, J.-Y.: An original knee arthrodesis technique combining external fixator with Steinman pins direct fixation. Orthop. Traumatol. Surg. Res. **95**, 272–277 (2009)
4. Cunningham, J.L., Richardson, J.B., Soriano, R.M., Kenwright, J.: A mechanical assessment of applied compression and healing in knee arthrodesis. Clin. Orthop. Relat. Res. **242**, 256–264 (1989)
5. MacDonald, J.H., Agarwal, S., Lorei, M.P., Johanson, N.A., Freiberg, A.A.: Knee arthrodesis. J. Am. Acad. Orthop. Surg. **14**(3), 154–163 (2006)
6. Kim, K., Snir, N., Schwarzkopf, R.: Modern techniques in knee arthrodesis. Int. J. Orthop. **3** (1), 487–496 (2016)
7. Letartrea, R., Combes, A., Autissier, G., Bonnevialle, N., Gougeon, F.: Knee arthodesis using a modular customized intramedullary nail. Orthop. Traumatol. Surg. Res. **95**, 520–528 (2009)
8. Bono, V.B., Scott, R.D., Author, T.: Revision Total Knee Arthroplasty, 2nd edn. Springer, Cham (2018)
9. Gunes, T., Sen, C., Erdem, M.: Knee arthrodesis using circular external fixator in the treatment of infected knee prosthesis: case report. Knee Surg. Sports Traumatol. Arthrosc. **13**, 329–334 (2005)
10. Conway, J.D., Mont, M.A., Bezwada, H.P.: Arthrodesis of the knee. J. Bone Joint Surg. Am. **86-A**(4), 835 (2004)
11. Robinson, M., Piponov, H.I., Ormseth, A., Helder, C.W., Schwartz, B., Gonzalez, M.: Knee arthrodesis outcomes after infected total knee arthroplasty and failure of two-stage revision with an antibiotic cement spacer. JAAOS Glob. Res. Rev. **2**(1), e077, 1–9 (2018)

Implant-Assisted Removable Partial Dentures in Mandibular Kennedy Class I Patients: The Impact of Implant Positioning

Ana Messias[1,4(✉)], Pedro Nicolau[1,5], Fernando Guerra[1,5], Ana Amaro[2,4], Luís Roseiro[3,4], and Maria Augusta Neto[2,4]

[1] Department of Dentistry, Faculty of Medicine, University of Coimbra, Coimbra, Portugal
ana.messias@uc.pt
[2] Department of Mechanical Engineering, Faculty of Sciences and Technology, University of Coimbra, Coimbra, Portugal
[3] Coimbra Polytechnic – ISEC, Coimbra, Portugal
[4] Centre for Mechanical Engineering Materials and Processes (CEMMPRE), Coimbra, Portugal
[5] Centre for Innovation and Research in Oral Sciences (CIROS), University of Coimbra, Coimbra, Portugal

Abstract. The installation of two implants to support and retain the prosthesis has been suggested as an option to improve the prognosis of removable rehabilitations in Kennedy class I patients. Using the finite element method, the present study performed a biomechanical analysis of two implant-assisted options (IARPD) with the implants in distinct positions – mesially in the edentulous space or in a distal position. The installation of a dental implant in each of the bilateral edentulous regions promotes the vertical and anterior-posterior stabilization of the prosthesis, leading to low displacement. IARPDs with the implant placed adjacent to the abutment teeth retain a pattern of displacements comparable to conventional RPDs, with sinking of the prosthesis. Addition of implants to the denture bearing areas, regardless of the position, results in low strain in the edentulous bone and high strain in the peri-implant bone, indicating that load is distributed to the bone through the implants.

Keywords: Implant-assisted removable partial dentures · Bone strains · Displacements · Finite element analysis

1 Introduction

Mandibular removable partial dentures with bilateral distal extension present poor prognosis due to the limited stability and retention provided by the combined tooth-mucous support. In 1993, Herman Keltjens reported that the use of bilateral implant anchorage would stabilize the RPD in a vertical direction and could provide additional retention, reduce stress on the natural abutment teeth and the number of needed clasps, prevent alveolar bone resorption beneath the denture base and improve comfort for the patient [1]. These are called implant-supported or implant-assisted removable partial

J. Henriques et al. (Eds.): MEDICON 2019, IFMBE Proceedings 76, pp. 1424–1430, 2020.
https://doi.org/10.1007/978-3-030-31635-8_175

dentures whether simply providing support or providing both support and retention with attachments, respectively.

Though other studies point out the viability of the use of implants to improve stability, retention, aesthetics and masticatory efficiency [2–4], no evidence can be found in the literature regarding the optimal implant positioning within the edentulous saddle.

Thus, the objective of the present study is to conduct a finite element analysis to evaluate the different patterns of stress in the prosthetic structure and strain in the supporting bone induced by the variation of the implant position within the edentulous span.

2 Materials and Methods

2.1 Patient Data Acquisition and 3D Geometry Modelling

One female patient with mandibular posterior bilateral edentulism (Kennedy Class I) and rehabilitated with a removable partial denture was selected to undergo a CBCT scan (i-Cat® FLX Cone Beam 3D system, Kavo-Kerr, Germany), with a single scanning time of 26.9 s for a FOV of 16 cm × 10 cm and 0.25 mm voxel size, and to do impression casts, in order to build two different configurations of implant-assisted removable partial dentures (IARPD).

The reconstruction of the mandible geometry was performed by image segmentation methods applied to the raw medical data (Mimics 19.0, Materialise, Leuven, Belgium), with application of specific masks determined by individual pixel Hounsfield units for cortical bone, cancellous bone and teeth. The soft tissue contour was obtained by scanning the plaster models of the patient (inEOS® X5, Sirona Dental Systems, Inc.). All components, bone and teeth masks as well as soft tissue point cloud object, were further processed in a modelling software (Geomagic Studio 2013.0.1, Geomagic Inc.) to create geometric models with non-uniform rational B-spline (NURBS). The 3D geometric model of the removable partial denture was created over the scanned plaster model using the 3Shape Dental System™ software. The implants were designed in a 3D parametric modelling software (Solidworks® 2014, Dassault Systèmes SOLIDWORKS Corp.) after feature extraction of the specific geometrical dimensions from official images of Straumann Standard Plus implants (Ø 4.1 mm) with regular platform and 10 mm long provided by the manufacturer (Institut Straumann AG). Using the extracted features, the implant profile was sketched on a plane and posteriorly converted to a 3D solid. The ball attachment (patrix) and corresponding matrix were designed using the same principles.

The non-parametric components were assembled into a single file and then two IARPD configurations were built considering the implants placed in the mesial position, adjacent to the abutment teeth (IARPD PREMOLARS) or considering the implants placed in a distal position within the edentulous saddles, in the former position of the molar teeth (IARPD MOLARS).

The two models were imported to ADINA System for linear and nonlinear finite element analysis (ADINA AUI version 9.3.1, ADINA R&D Inc., Watertown, USA).

Assignment of the mesh density to the solid bodies of each model was made by promoting equally spaced subdivisions of the bodies using the desired element edge length, with refinement in critical areas. Discretization of the domains was made using the Delaunay free-form meshing algorithm to generate 8-node hexahedral (brick) elements with mixed interpolation formulation. This formulation resulted in meshes with 889884 elements and 472936 nodes for the IARPD PREMOLARS model and 857671 elements and 478075 nodes for the IARPD MOLARS model.

2.2 Material Properties

Linear elastic and homogenous material properties were attributed to bone, teeth, prosthesis components and implants, as detailed in Table 1.

Table 1. Mechanical properties of the isotropic linear elastic materials.

Material	Young modulus (Pa)	Poisson coefficient
Cortical bone	20×10^9 [5–7]	0.30
Cancellous bone	1.37×10^9 [5, 8]	0.30
Teeth (dentin)	13×10^9 [9]	0.366
Chromium-cobalt	211×10^9 [10]	0.30
Acrylic	2.2×10^9 [10]	0.31

The present study considered the oral mucosa and the periodontal ligament as hyperelastic materials with specific nonlinear constitutive models. Several models have been proposed to model the hyperelastic effects. For the case of biological soft tissues, the approximation proposed by Ogden [11, 12] provides the closest matching to experimental data and is herein considered for the modelling of both the oral mucosa and the periodontal ligament.

In the case of the oral mucosa, the strain energy density function was set based on the fitting of a uniaxial stress-strain curve from experimental data reported by Kishi et al. in 1972 *cited in* Chen, Ahmad [13].

The periodontal ligament material properties were obtained from the data of experimental testing of porcine periodontal ligaments subjected uniaxial tensile tests were performed along the fibber direction, as reported by Natali, Carniel [14] and Nishihira, Yamamoto [15].

2.3 Loading and Boundary Conditions

Load was applied as a uniform pressure, equivalent to a purely vertical 120 N force, over the acrylic portion of the IARPDs. A 100 steps constant time function of magnitude 1 was defined so that the loads were applied gradually to avoid dynamic effects and convergence problems.

All three degrees of freedom were fixed in the condyles of the mandible and in the areas of insertion of the mastication muscles, assuming that muscle contraction promotes the stabilization of the mandible during function.

3 Results

3.1 Displacement of the IARPDs

For the IARPD PREMOLARS, it was noticeable a trend for sinking of the acrylic portion of the RPD, with displacements increasing with the increase of the distance to the supports (implant). In the IARPD MOLARS model, on the contrary, the least vertical displacement was registered in the region of the implants (plotted in yellow) and the largest displacements in the apical direction were associated with the anterior region, where the framework contacts the abutment teeth. Mean values for the two models were, respectively, -14 ± 6.4 μm and -4 ± 0.81 μm. The corresponding maximum values were -28.4 μm and -5.5 μm (see Fig. 1).

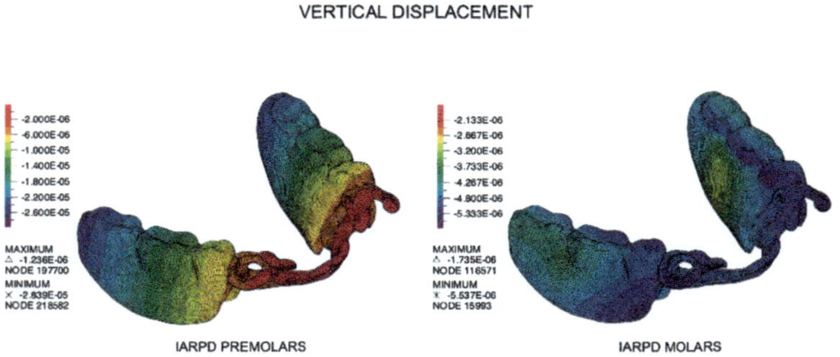

Fig. 1. Band plot of the vertical displacement for the two models analyzed: IARPD PREMOLARS (left), IARPD MOLARS (right). Non-matching color scales.

3.2 Stress in the Framework of the IARPDs

There was no high stress accumulation in the direct retainers or major connector. The highest stress accumulation was in the interior portion of the matrix and surrounding material, as observed in Fig. 2.

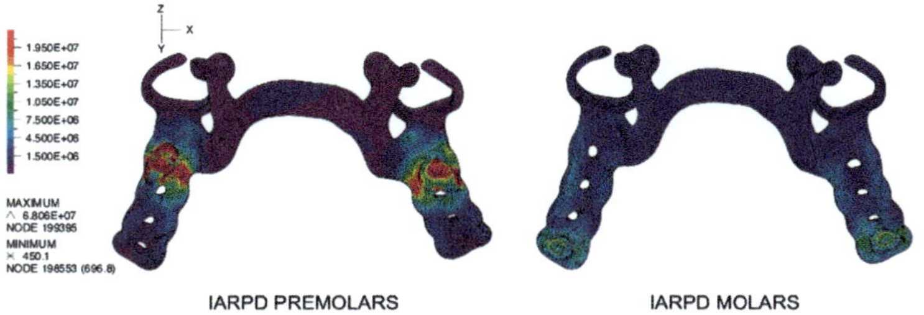

Fig. 2. Effective stress distribution in the framework of the two models analyzed: IARPD PREMOLARS (left), IARPD MOLARS (right). Matching color scales.

3.3 Bone Strain in the Edentulous Areas

The IARPD PREMOLARS model presents a wide area of strains well above 100 με in the edentulous area posterior to the implant, as well as concentrated strains in the cortical bone surrounding the implant. Tensile strains are almost absent from the IARPD MOLARS model but can be observed in the area posterior to the implant (Fig. 3).

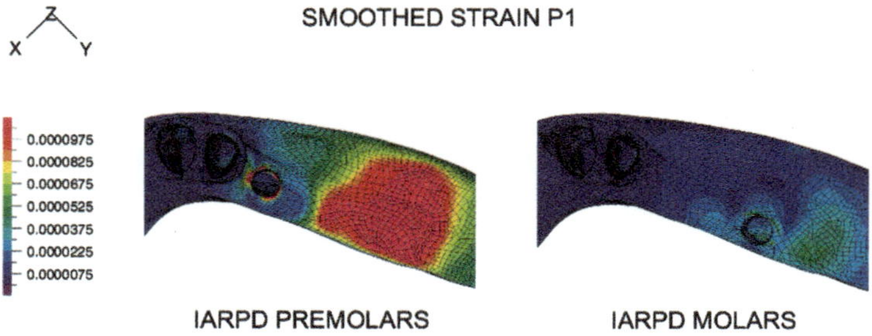

Fig. 3. Distribution of principal tensile strains in the edentulous areas and alveolar bone supporting the abutment teeth. Similar scale for the three models with band plot adjusted for positive values with a maximum of 100 με. Purple elements do not undergo tensile deformation.

A large area of compression is evident in the edentulous region of the IARPD PREMOLARS model, but intensive compression is only noticed around the implant site. Similarly, high compression is noticed around the implant site of the IARPD MOLARS model while the remaining model is neutral (Fig. 4).

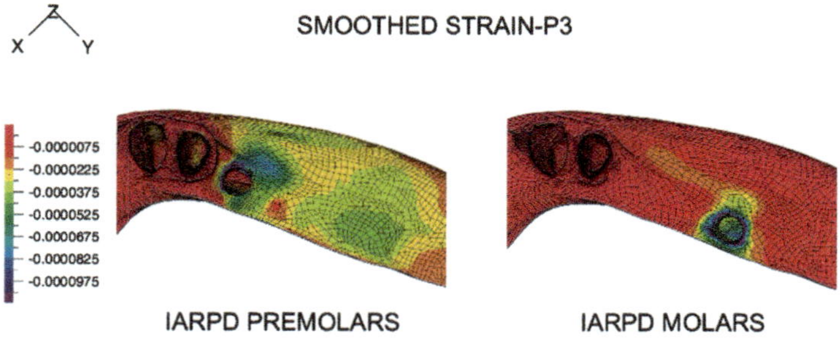

Fig. 4. Distribution of principal compressive strains in the edentulous areas and alveolar bone supporting the abutment teeth. Band plot adjusted for negative values with a maximum of −100 με. Red elements do not undergo compression.

4 Discussion

The present study focused on the evaluation of the stress and deformation patterns associated with distal extension implant-assisted removable partial dentures when submitted to static bilateral and homogenous loading of the artificial acrylic teeth.

The pattern of displacements presented by the prosthesis in the IARPD PREMOLARS model can be interpreted as a Euler-Bernoulli beam problem [16, 17], in which the implant-matrix assembly placed in the mesial portion of the edentulous span acts as a hinge support with the remaining section of the acrylic unsupported similarly to an overhanging beam. The highest displacement of the prosthesis of IARPD MOLARS model were registered in the unsupported region, between the direct retainers of the framework, where the support is non-rigid allowing some free movements, and between the support provided by the implant, where the least displacements are found [10].

According to Frost's mechanostat theory [18], bone presents structural adaptations to mechanical usage (SATMU). Mechanical loading generates strain, which is a small deformation throughout the calcified matrix that induces stretching of osteocytes to the same extent as the surrounding bone tissue. In the IARPD PREMOLARS model load generates strains within the modelling minimum effective strains threshold (50–100 $\mu\varepsilon$) thus normal mechanically-controlled modelling is expected to occur with neutral variation and no marginal bone loss. In the IARPD MOLARS model, alveolar strains are very small and could surprisingly fall into the disuse window which, according to the mechanostat theory, would contribute for bone weakening and removal. However, one must take into account that the present study only considers the deformations resulting from the homogenous loading of the acrylic teeth of the removable partial denture.

5 Conclusion

Implant-assisted removable partial dentures (IARPDs) with the implant placed in the most mesial/anterior position, adjacent to the abutment teeth retain a pattern of displacements comparable to that of conventional RPDs.

The lowest effective stress of the framework is obtained with the implant in the most distal/posterior position.

No absolute conclusions can be taken regarding the possibility of bone remodeling in the edentulous regions.

References

1. Keltjens, H.M., et al.: Distal extension removable partial dentures supported by implants and residual teeth: considerations and case reports. Int. J. Oral Maxillofac. Implants **8**(2), 208–213 (1993)
2. Goncalves, T.M., et al.: Mastication improvement after partial implant-supported prosthesis use. J. Dent. Res. **92**(12 Suppl), 189S–194S (2013)

3. Jensen, C., et al.: Implant-supported mandibular removable partial dentures: functional, clinical and radiographical parameters in relation to implant position. Clin. Implant Dent. Relat. Res. **19**(3), 432–439 (2017)
4. Wismeijer, D., Tawse-Smith, A., Payne, A.G.: Multicentre prospective evaluation of implant-assisted mandibular bilateral distal extension removable partial dentures: patient satisfaction. Clin. Oral Implant Res. **24**(1), 20–27 (2013)
5. Schwartz-Dabney, C.L., Dechow, P.C.: Edentulation alters material properties of cortical bone in the human mandible. J. Dent. Res. **81**(9), 613–617 (2002)
6. Odin, G., et al.: Determination of Young's modulus of mandibular bone using inverse analysis. Med. Eng. Phys. **32**(6), 630–637 (2010)
7. Rapoff, A.J., et al.: Elastic modulus variation in mandibular bone: a microindentation study of Macaca fascicularis. Am. J. Phys. Anthropol. **135**(1), 100–109 (2008)
8. Schwartz-Dabney, C.L., Dechow, P.C.: Variations in cortical material properties throughout the human dentate mandible. Am. J. Phys. Anthropol. **120**(3), 252–277 (2003)
9. Zhang, Y.R., et al.: Review of research on the mechanical properties of the human tooth. Int. J. Oral Sci. **6**(2), 61–69 (2014)
10. Shahmiri, R., et al.: Finite element analysis of an implant-assisted removable partial denture during bilateral loading: occlusal rests position. J. Prosthet. Dent. **112**(5), 1126–1133 (2014)
11. Ogden, R.W.: Nonlinear elasticity with application to material modelling. Lecture Notes-AMAS (2003)
12. Ogden, R.W., Saccomandi, G., Sgura, I.: Fitting hyperelastic models to experimental data. Comput. Mech. **34**(6), 484–502 (2004)
13. Chen, J., et al.: A comparative study on complete and implant retained denture treatments: a biomechanics perspective. J. Biomech. **48**(3), 512–519 (2015)
14. Natali, A.N., et al.: A visco-hyperelastic-damage constitutive model for the analysis of the biomechanical response of the periodontal ligament. J. Biomech. Eng. **130**(3), 031004 (2008)
15. Nishihira, M., et al.: Mechanics of periodontal ligament. In: Natali, A. (ed.) Dental Biomechanics. Taylor & Francis, London (2003)
16. Labuschagne, A., van Rensburg, N.F.J., van der Merwe, A.J.: Comparison of linear beam theories. Math. Comput. Model. **49**(1–2), 20–30 (2009)
17. Neto, M.A., et al.: Engineering Computation of Structures: The Finite Element Method. Springer, Cham (2015)
18. Frost, H.M.: Bone "mass" and the "mechanostat": a proposal. Anat. Rec. **219**(1), 1–9 (1987)

Residual Ridge Resorption in Mandibular Kennedy Class I Denture Wearers: Proposal of a Pressure-Induced Mechanism Based on a Finite Element Analysis

Ana Messias[1,4(✉)], Pedro Nicolau[1,5], Fernando Guerra[1,5], Ana Amaro[2,4], Luís Roseiro[3,4], and Maria Augusta Neto[2,4]

[1] Department of Dentistry, Faculty of Medicine, University of Coimbra, Coimbra, Portugal
ana.messias@uc.pt
[2] Department of Mechanical Engineering, Faculty of Sciences and Technology, University of Coimbra, Coimbra, Portugal
[3] Coimbra Polytechnic - ISEC, Coimbra, Portugal
[4] Centre for Mechanical Engineering Materials and Processes (CEMMPRE), Coimbra, Portugal
[5] Centre for Innovation and Research in Oral Sciences (CIROS), University of Coimbra, Coimbra, Portugal

Abstract. Removable partial dentures (RPDs) have been considered an acceptable cost-effective solution to restore oral function in partially edentulous patients. However, in mandibular Kennedy class I patients, the RPDs present unfavourable prognosis due to the combined tooth-mucous support with no posterior elements of retention and continuous resorption of the residual ridge. This process is inevitable but highly variable from one individual to the other and modified by both uncontrollable factors and functional or prosthetic factors that determine load distribution and might be subject of modification by the professional. Using the finite element method, the present study performed a biomechanical analysis of a mandibular distal extension removable partial denture regarding prosthesis displacements, strains in the supporting bone and pressure in the mucosa. The results suggest that bone resorption in the alveolar regions is a strain-mediated process whereas in the edentulous areas it is a process induced by pressure in the mucosa.

Keywords: Removable partial dentures · Bone strains · Displacements · Hydrostatic pressure · Finite element analysis

1 Introduction

In 1925, Kennedy proposed a method of classification of partially edentulous arches that comprised four major categories, denoted Class I through Class IV, attributed according to a decreasing order of frequency [1]. Kennedy class I arches are characterized by the presence of bilateral edentulous areas located posterior to the remaining natural teeth and are frequently named free end situations.

J. Henriques et al. (Eds.): MEDICON 2019, IFMBE Proceedings 76, pp. 1431–1440, 2020.
https://doi.org/10.1007/978-3-030-31635-8_176

In Kennedy class I situations, fixed implant-supported dental prostheses are the only option capable of *restitutio ad integrum* the function and aesthetics of every tooth missing in the edentulous areas with reported 10-year survival rates of 93% [2, 3]. However, the option is not always feasible due to the high economic cost or local anatomical conditions that would require extensive surgical treatments to which the patient might not be disposed to.

Conventional distal extension removable partial dentures are the least invasive treatment option with an acceptable cost-efficiency relation, and might even be the best practice therapy in long edentulous spans or challenging clinical scenarios that are associated with unfavourable maxillary relations due to significant loss hard and soft tissues that require support of the orofacial structures [4, 5].

Even though patients present good acceptance of removable partial dentures, approximately 10, 25 and 50% of distal extension base RPDs are no longer in service after 1, 5 and 10 years, respectively, due to replacement or discarding [6]. These low survival rates when compared to other types of RPDs have been attributed to the combined tooth-mucous support and absence of posterior elements to provide retention. In distal extension RPDs, the framework is supported and retained anteriorly by the abutment teeth whereas the extension bases are only supported by the mucosa overlying the edentulous ridge. This tooth-mucous support with no posterior retention allows some freedom of movements. During mastication, the extension bases are displaced towards the residual ridge, compressing the alveolar mucosa, which can be perceived as discomfort or even pain. The sensation of discomfort during function tends to increase over the years due to progressive misfit of the extension base, which is dictated by a process of resorption of the residual ridge, highly variable from patient to patient.

Functional and prosthetic factors are intricately related in the process of residual ridge resorption as it is through the extension base of the removable partial denture that the forces transmitted to the residual ridge, thus dictating the mechanical stimulus to the edentulous regions. The functional factors, such as average bite force or frequency of movements generated during mastication, determine the magnitude, duration and direction of occlusal loading, whereas the prosthetic factors control its distribution, which was corroborated *in vitro* [7–9].

Until 2011, the biomechanical analysis of distal extension removable partial dentures available in the literature was sparse and based on model simplifications that compromised the clinical extrapolation of the results. Up to this moment no study has performed a three-dimensional evaluation of conventional removable partial dentures, thus the aim of the present work is to identify the biomechanical mechanisms responsible for the resorption of the residual ridge by the evaluation of strains in the bone of the edentulous regions and pressure in the oral mucosa resulting from loading of the extension base of a mandibular Kennedy class I conventional removable partial denture.

2 Materials and Methods

2.1 Patient Data Acquisition and 3D Geometry Modelling

One female patient with mandibular posterior bilateral edentulism (Kennedy Class I) and rehabilitated with a removable partial denture was selected to undergo a CBCT scan (i-Cat® FLX Cone Beam 3D system, Kavo-Kerr, Germany), with a single scanning time of 26.9 s for a FOV of 16 cm × 10 cm and 0.25 mm voxel size, and to do impression casts, in order to build a conventional removable partial denture (CONVENTIONAL RPD).

The reconstruction of the mandible geometry was performed by image segmentation methods applied to the raw medical data (Mimics 19.0, Materialise, Leuven, Belgium), with application of specific masks determined by individual pixel Hounsfield units for cortical bone, cancellous bone and teeth. The soft tissue contour was obtained by scanning the plaster models of the patient (inEOS® X5, Sirona Dental Systems, Inc.). All components, bone and teeth masks as well as soft tissue point cloud object, were further processed in a modelling software (Geomagic Studio 2013.0.1, Geomagic Inc) to create geometric models with non-uniform rational B-spline (NURBS). The 3D geometric model of the removable partial denture was created over the scanned plaster model using the 3Shape Dental System™ software.

All components were assembled into a single file and exported into a parasolid binary file (.x_b). The model (CONVENTIONAL RPD) was imported to ADINA System for linear and nonlinear finite element analysis (ADINA AUI version 9.3.1, ADINA R&D Inc., Watertown, USA). Assignment of the mesh density to the solid bodies of each model was made by promoting equally spaced subdivisions of the bodies using the desired element edge length (generally 0.5 mm except for the cortical and trabecular bone – 1 mm), with refinement in critical areas (0.25 mm). Discretization of the domains was made using the Delaunay free-form meshing algorithm to generate 8-node hexahedral (brick) elements with mixed interpolation formulation. This formulation resulted in a mesh with 871851 elements and 506375 nodes.

2.2 Material Properties

Linear elastic and homogenous material properties were attributed to bone, teeth and prosthesis components, as detailed in Table 1.

Table 1. Mechanical properties of the isotropic linear elastic materials.

Material	Young modulus (Pa)	Poisson coefficient
Cortical bone	20×10^9 [10–12]	0.30
Cancellous bone	1.37×10^9 [10, 13]	0.30
Teeth (dentin)	13×10^9 [14]	0.366
Chromium-cobalt	211×10^9 [15]	0.30
Acrylic	2.2×10^9 [15]	0.31

The present study considered the oral mucosa and the periodontal ligament as hyperelastic materials with specific nonlinear constitutive models. Several models have been proposed to model the hyperelastic effects. For the case of biological soft tissues, the approximation proposed by Ogden [16, 17] provides the closest matching to experimental data and is herein considered for the modelling of both the oral mucosa and the periodontal ligament.

In the case of the oral mucosa, the strain energy density function was set based on the fitting of a uniaxial stress-strain curve from experimental data reported by Kishi et al. in 1972 *cited in* Chen, Ahmad [18] depicted in Fig. 1. The closest Ogden approximation was obtained using the method of singular value decomposition and a 9th order approximation [16, 17].

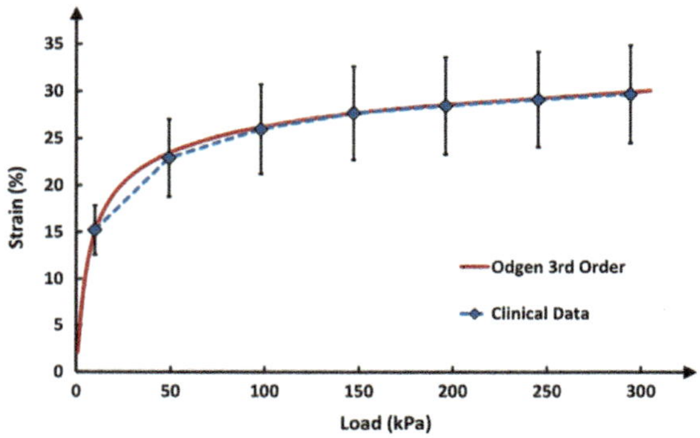

Fig. 1. Stress-strain clinical data of the oral mucosa obtained from the work of Kishi et al. (blue markers) used for the for the approximation of the hyperelastic characteristics of the oral mucosa. Adapted from Chen, Ahmad [19].

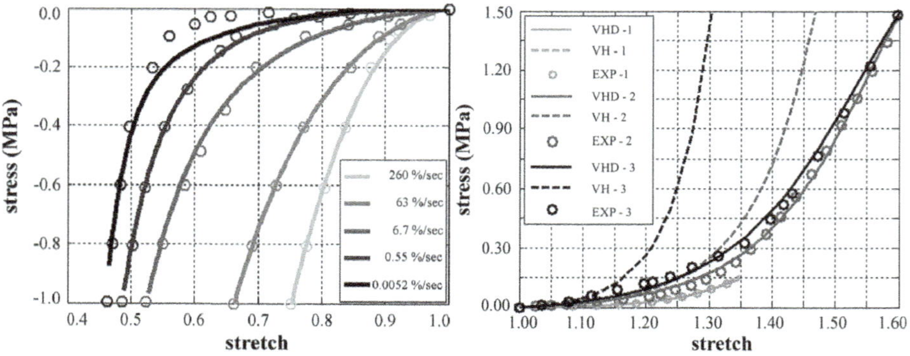

Fig. 2. Stress-strain data of the uniaxial tensile tests of the periodontal ligament of the work of Natali et al., which was used for the for the approximation of the hyperelastic characteristics of the periodontal ligament. On the left, compressive behaviour at different strain rates. On the right results tensile mechanical response of the periodontal ligament at different strain rates.

The periodontal ligament material properties were obtained from the data of experimental testing of porcine periodontal ligaments subjected uniaxial tensile tests were performed along the fibber direction, as reported by Natali, Carniel [20] and Nishihira, Yamamoto [21] and represented in Fig. 2.

The curve fitting procedure for the Ogden function was also obtained by singular value decomposition and 9th order approximation [16, 17], defining 18 material constants.

2.3 Loading and Boundary Conditions

Load was applied as a uniform pressure, equivalent to a purely vertical 120 N force, over the acrylic portion of the prosthesis. The automatic time stepping procedure was used so that the loads were applied gradually to avoid dynamic effects and to improve convergence rate.

All three degrees of freedom were fixed in the condyles of the mandible and in the areas of insertion of the mastication muscles, assuming that muscle contraction promotes the stabilization of the mandible during function.

3 Results

3.1 Displacement of the RPD

The CONVENTIONAL RPD model revealed significant displacement in the sagittal plane, particularly noticed on the bottom of the acrylic where the highest absolute values were identified. All displacement occurred in the posterior-anterior direction (towards the natural teeth) and was the least in the occlusal surfaces of the acrylic teeth and on the occlusal rests that seat on the premolars (Fig. 3), with a mean value of -190 ± 36 μm. Displacement in Z (vertical displacement in the sagittal plane) was the main responsible for the displacement magnitude of the RPD, accompanying the direction of the applied force, i.e. downwards, in the direction of the underlying soft tissues. It was noticeable a trend for sinking of the acrylic portion of the RPD (Fig. 3), with displacements increasing with the increase of the distance to the supports (abutment teeth) and a mean value of -710 ± 84 μm.

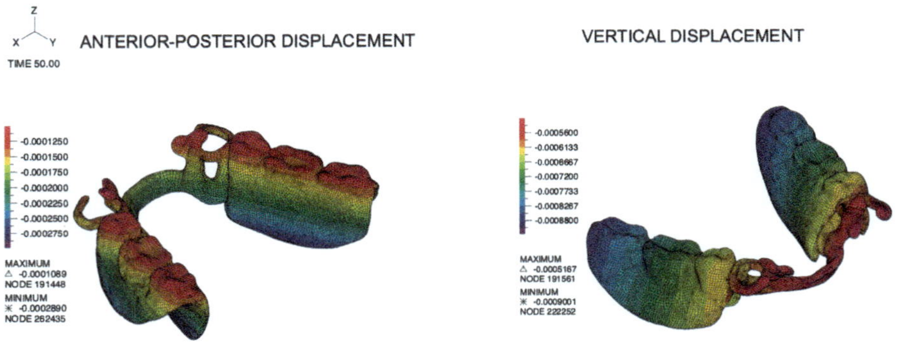

Fig. 3. Band plot of the displacement of the CONVENTIONAL RPD: anterior-posterior (left), vertical (right). Non-matching color scales.

3.2 Strains in Alveolar Bone and Residual Ridge

Principal tensile strains (P1) spread all through the alveolar bone and edentulous areas that underlie the acrylic portion of the RPD (Fig. 4). In the abutment teeth area, the strains exceeded 500 με in the marginal region of the alveolar bone, with values rising to 6734 με.

The CONVENTIONAL RPD model presents a global state of residual compression. Though a large area of compression is evident in the edentulous region, notorious peaks of elements in compression are visible in the alveolar bone that supports the abutments, plotted in purple (Fig. 4).

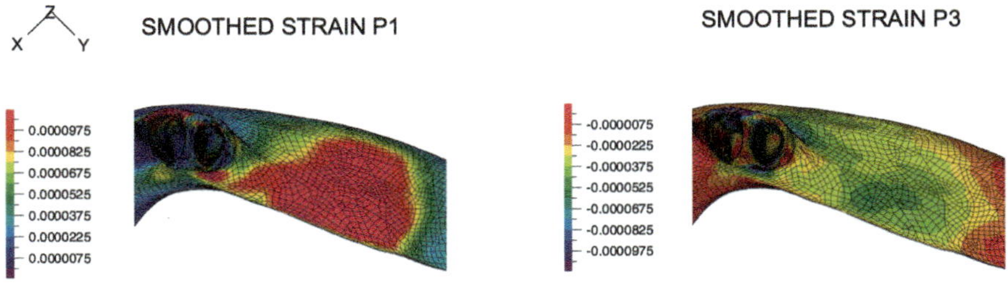

Fig. 4. Distribution of principal strains in the edentulous areas and alveolar bone supporting the abutment teeth. Tensile strains (left) with the band plot adjusted for positive values with a maximum of 100 με. Purple elements do not undergo traction. Compressive strains (right) with the band plot adjusted for negative values with a maximum of −100 με. Red elements do not undergo compression.

3.3 Hydrostatic Pressure in the Mucosa

The compression of the mucosa due to loading by the acrylic base induces pressure in the tissue, represented in Fig. 5. The mean value in the CONVENTIONAL RPD model was 83.64 ± 52.11 kPa, registering a maximum of 779 kPa.

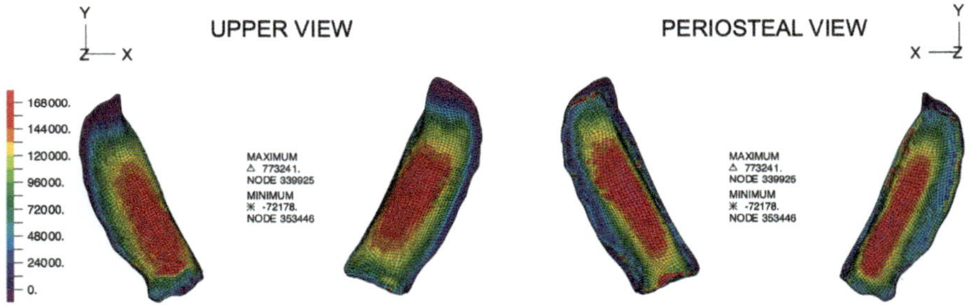

Fig. 5. Pressure in the mucosa. Left side representing denture contact surface of the mucosa (upper view) and the right side presenting the periosteal surface of the mucosa. Plotting scales based on the average systemic blood pressure (120 mmHg or 16 kPa) with maximum of 160 kPa.

4 Discussion

The present study focused on the evaluation of the displacements, deformation and pressure patterns associated with distal extension removable partial dentures when submitted to static bilateral and homogenous loading of the artificial acrylic teeth. The present work is, up to our knowledge, the first non-linear analysis presenting a three-dimensional evaluation of a conventional extension base removable partial denture.

The pattern, consistency and magnitude of the displacements registered to the buccal, mesial and apical directions were not unexpected for a prosthesis with non-rigid retainers. The clasp systems applied to the abutment teeth are usually designed to have a stress-releasing effect, allowing the tissueward rotation of the RPD bases around the axis passing through the occlusal rests, without torqueing the clasped teeth. When the distal extension base is loaded, the occlusal rests of the prosthesis act as hinge support resisting to vertical and horizontal loads but are unable to resist to the moment generated by the vertical load. As a consequence, the prosthesis and tooth ensemble can be interpreted as a Euler-Bernoulli beam problem [22, 23].

As seen in Fig. 4, the marginal alveolar bone undergoes large deformation, both tensile and compressive, when the acrylic portion of the prosthesis is loaded. According to Frost's mechanostat theory [24], bone presents structural adaptations to mechanical usage (SATMU). Mechanical loading generates strain, which is a small deformation throughout the calcified matrix that induces stretching of osteocytes to the same extent as the surrounding bone tissue. Different levels of strain promote different bone responses, categorized as: disuse window, adapted window, mild overload window, and pathologic overload window. The strain windows are separated by narrow threshold intervals of remodelling minimum effective strains (MESr), modelling minimum effective strains (MESm) and pathologic minimum effective strains (MESp). The author claimed that the disuse-mode remodelling would take place below the MESr threshold of 50–100 $\mu\varepsilon$ for which the lack of mechanical stimuli would promote bone removal and weakening. The physiological window would be within a lower bound established by the MESr and an upper bound limited by the MESm, that is up to 1000–1500 $\mu\varepsilon$, bone undergoes controlled modelling. Above the MESm and up to 3000 $\mu\varepsilon$, which corresponds to the MESp, bone would be in the mild overload window and undergo increased modelling. Above that upper threshold bone would enter the pathological overload window and accumulate microdamage. The remodelling BMUs usually repair the microdamage and keep it from accumulating but when strains surpass 3000 $\mu\varepsilon$, the repairing mechanisms might be overwhelmed by the amount of microdamage and can cause small fatigue failures. The largest normally (or safe) allowed peak bone strains lies in the 1500 $\mu\varepsilon$ region.

Though the mean of the equivalent strains of the CONVENTIONAL RPD model is smaller, a large number of elements surpass the 1500 $\mu\varepsilon$ threshold either by compression or tension (Fig. 4), particularly in the marginal region of the alveolar bone of the abutment teeth. Elements undergoing deformation above the MESm threshold (above 1000–1500 $\mu\varepsilon$) and the elements in the vicinity indicate that such bone undergoes increased modelling (mild overload window).

Application of homogenously distributed loads to the acrylic teeth of a tooth-mucous supported removable partial denture and compression of the underlying mucosa are intimately connected to hydrostatic pressure-mediated resorption of the residual ridge [19].

The oral mucosa is a highly vascularized tissue with considerate amounts of interstitial fluid. On the one hand, when the denture base compresses the mucosa, the collagen fibbers are forced to align along the direction of the mechanical stress [25] creating an initial effect of mechanical cushioning that passively contributes for the protection of the underlying bone [26, 27].

On the other hand, the compression of the mucosa also promotes pumping of the interstitial fluid to the neighbouring tissues not in direct contact with the acrylic, which induces a state of isotropic stress within the mucosa known as the interstitial fluid pressure or, mechanically, hydrostatic pressure [19]. The mucosa behaves as a soaked sponge under compression, releasing the liquid content. As an outflux of interstitial fluid is created in the mucosa, an influx is created in the underlying bone, which can be considered a stimulus for the bone-modelling cells.

Additionally, the increase of the interstitial fluid pressure in the mucosa results in a constriction of the vessels and reduction of the blood flow, potentially leading to localized ischaemia [28–30] and compromising the nourishment of the underlying bone. The blood flow reduction favours the accumulation of nitrite oxide and other metabolites that promote a pro-inflammatory reaction in the bone underlying the denture base [31] and that could also contribute for the mucosa-induced bone resorption.

The works of Chen, Ahmad [18] and Ahmad, Chen [32] reported an excellent correlation between the finite element modelling of hydrostatic pressure in the mucosa tissue and clinical measurement of bone remodelling.

As reported, in the CONVENTIONAL RPD model studied the compression of the mucosa generates a mean hydrostatic pressure of 85.6 $\pm$ 52.1 kPa, which is 5 times the average systolic pressure for the healthy adult (120 mmHg or 16 kPa). Though the analysis here presented is static and shows a possible momentary effect, it reveals that some limitation of the blood flow is expected in the tissue. Clinically, the continuous use of the prosthesis delays the possibility of blood flow recovery of the mucosa, leading to local anoxia and accumulation of pro-inflammatory metabolites and low nutrient delivery to the bone cells, eventually causing residual ridge resorption.

5 Conclusion

Loading of conventional Kennedy class I RPDs leads to significant vertical displacement of the extension bases towards the underlying soft tissues, similar to a cantilever beam under flexion.

Hydrostatic pressure in the mucosa underlying the extension bases of a conventional RPD largely surpasses the average systolic blood pressure values, suggesting blood-flow limitation and a pressure-induced mechanism of residual ridge resorption. In the alveolar region, the resorption mechanism seems to be strain-mediated.

References

1. Basker, R.M.: After Kennedy. Int. Dent. J. **52**(2), 99–112 (2002)
2. Pjetursson, B.E., et al.: A systematic review of the survival and complication rates of implant-supported fixed dental prostheses (FDPs) after a mean observation period of at least 5 years. Clin. Oral Implants Res. **23**(Suppl 6), 22–38 (2012)
3. Aglietta, M., et al.: A systematic review of the survival and complication rates of implant supported fixed dental prostheses with cantilever extensions after an observation period of at least 5 years. Clin. Oral Implants Res. **20**(5), 441–451 (2009)
4. Bohnenkamp, D.M.: Removable partial dentures: clinical concepts. Dent. Clin. North Am. **58**(1), 69–89 (2014)
5. Campbell, S.D., et al.: Removable partial dentures: the clinical need for innovation. J. Prosthet. Dent. **118**(3), 273–280 (2017)
6. Vermeulen, A.H., et al.: Ten-year evaluation of removable partial dentures: survival rates based on retreatment, not wearing and replacement. J. Prosthet. Dent. **76**(3), 267–272 (1996)
7. Itoh, H., et al.: Effect of direct retainer and major connector designs on RPD and abutment tooth movement dynamics. J. Oral Rehabil. **35**(11), 810–815 (2008)
8. Jin, X., et al.: Influence of loading positions of mandibular unilateral distal extension removable partial dentures on movements of abutment tooth and denture base. J. Med. Dent. Sci. **51**(3), 155–163 (2004)
9. Mizuuchi, W., et al.: The effects of loading locations and direct retainers on the movements of the abutment tooth and denture base of removable partial dentures. J. Med. Dent. Sci. **49** (1), 11–18 (2002)
10. Schwartz-Dabney, C.L., Dechow, P.C.: Edentulation alters material properties of cortical bone in the human mandible. J. Dent. Res. **81**(9), 613–617 (2002)
11. Odin, G., et al.: Determination of Young's modulus of mandibular bone using inverse analysis. Med. Eng. Phys. **32**(6), 630–637 (2010)
12. Rapoff, A.J., et al.: Elastic modulus variation in mandibular bone: a microindentation study of Macaca fascicularis. Am. J. Phys. Anthropol. **135**(1), 100–109 (2008)
13. Schwartz-Dabney, C.L., Dechow, P.C.: Variations in cortical material properties throughout the human dentate mandible. Am. J. Phys. Anthropol. **120**(3), 252–277 (2003)
14. Zhang, Y.R., et al.: Review of research on the mechanical properties of the human tooth. Int. J. Oral Sci. **6**(2), 61–69 (2014)
15. Shahmiri, R., et al.: Finite element analysis of an implant-assisted removable partial denture during bilateral loading: Occlusal rests position. J. Prosthet. Dent. **112**(5), 1126–1133 (2014)
16. Ogden, R.W.: Nonlinear elasticity with application to material modelling. Lecture Notes-AMAS (2003)
17. Ogden, R.W., Saccomandi, G., Sgura, I.: Fitting hyperelastic models to experimental data. Comput. Mech. **34**(6), 484–502 (2004)
18. Chen, J., et al.: A comparative study on complete and implant retained denture treatments: a biomechanics perspective. J. Biomech. **48**(3), 512–519 (2015)
19. Chen, J., et al.: Biomechanics of oral mucosa. J. R. Soc. Interface **12**(109), 20150325 (2015)
20. Natali, A.N., et al.: A visco-hyperelastic-damage constitutive model for the analysis of the biomechanical response of the periodontal ligament. J. Biomech. Eng. **130**(3), 031004 (2008)
21. Nishihira, M., et al.: Mechanics of periodontal ligament. In: Natali, A. (ed.) Dental Biomechanics. Taylor & Francis, London (2003)
22. Labuschagne, A., van Rensburg, N.F.J., van der Merwe, A.J.: Comparison of linear beam theories. Math. Comput. Modell. **49**(1–2), 20–30 (2009)

23. Neto, M.A., et al.: Engineering computation of Structures: the finite element method. Springer (2015)
24. Frost, H.M.: Bone "mass" and the "mechanostat": a proposal. Anat. Rec. **219**(1), 1–9 (1987)
25. Kydd, W.L., Daly, C.H.: The biologic and mechanical effects of stress on oral mucosa. J. Prosthet. Dent. **47**(3), 317–329 (1982)
26. Sawada, A., et al.: Viscoelasticity of human oral mucosa: implications for masticatory biomechanics. J. Dent. Res. **90**(5), 590–595 (2011)
27. Mori, S., et al.: Effect of continuous pressure on histopathological changes in denture-supporting tissues. J. Oral Rehabil. **24**(1), 37–46 (1997)
28. Akazawa, H., Sakurai, K.: Changes of blood flow in the mucosa underlying a mandibular denture following pressure assumed as a result of light clenching. J. Oral Rehabil. **29**(4), 336–340 (2002)
29. Atasever, N.E., et al.: Effect of wearing complete dentures on human palatal mucosal blood flow measured by 133Xe clearance. Arch. Oral Biol. **36**(9), 627–630 (1991)
30. Kocabalkan, E., Turgut, M.: Variation in blood flow of supporting tissue during use of mandibular complete dentures with hard acrylic resin base and soft relining: a preliminary study. Int. J. Prosthodont. **18**(3), 210–213 (2005)
31. Suenaga, H., et al.: Bone metabolism of residual ridge beneath the denture base of an RPD observed using NaF-PET/CT. J. Prosthodont. Res. **56**(1), 42–46 (2012)
32. Ahmad, R., et al.: Investigation of mucosa-induced residual ridge resorption under implant-retained overdentures and complete dentures in the mandible. Int. J. Oral Maxillofac. Implants **30**(3), 657–666 (2015)

Numerical Evaluation of the Knee Arthrodesis Using a Modified External Fixator

Maria A. Neto[1,2(✉)], Luis M. Roseiro[2,3], Maria F. Paulino[1,2], and Ana M. Amaro[1,2]

[1] Mechanical Engineering Department, Coimbra University, Coimbra, Portugal
{augusta.neto, ana.amaro}@dem.uc.pt, maria.paulino@uc.pt
[2] CEMMPRE, Department of Mechanical Engineering, University of Coimbra, Coimbra, Portugal
[3] Coimbra Polytechnic - ISEC, Coimbra, Portugal
lroseiro@isec.pt

Abstract. This work presents the numerical results of arthrodesis using a modified bilateral-monoplanar external fixator for the treatment of septic sequelae of the knee joint. The knee arthrodesis serves as an option of salvage treatment for failed total knee arthroplasty procedures [1]. The arthrodesis technique used in this work was promoted with external fixation consisting on the union of two or more femoral pins with two or more tibial pins through bended lateral bars. Biplanar fixation has higher sagittal stability and higher fusion rates than the monoplanar fixation, but the level of compression on the fusion area as well as the extension of contact area can have a great impact on the bone union. Hence, the main purpose of this work is to give the surgeons additional information on how to influence positively the knee arthrodesis, namely relatively to the fixation level, the homogeneity of contact and to the compression level, but also about some possible complications. The arthrodesis technique was implemented using the CAD Solidworks® software and the numerical analysis was carried out on ADINA® software. With this methodology it was possible to recreate the mechanical procedure of the knee arthrodesis and the results indicate that the arthrodesis technique is not only influenced by the parameters associated with the kind of externatal fixator used, but also with the type of supported used to perform the external fixation assembly.

Keywords: Knee · Arthrodesis · Finite element method

1 Introduction

Knee arthrodesis is a surgical procedure that removes the knee joint and promotes de bone fusion between femur and tibia members. This chirurgical procedure may be the only alternative to the amputation of human leg in certain indications, especially in infected knee prosthesis. Success in these situations depends on achieving good fixation assembly stability and optimal compression of fusion surfaces as well as on the bone contact surface as large as possible [2]. The compression can be done by an external fixator or other methods of fixation, such as: intramedullary nail, screws, a plate or two

J. Henriques et al. (Eds.): MEDICON 2019, IFMBE Proceedings 76, pp. 1441–1445, 2020.
https://doi.org/10.1007/978-3-030-31635-8_177

plates. The techniques for obtaining arthrodesis are classified by the type of fixation used, the quality and amount of bone being determinant for their selection, as well as the experience of the surgeon. Nevertheless, only external fixation and intramedullary nailing are recommended where arthrodesis is indicated for total knee arthroplasty (TKA) infection [1].

In this work we chose a bilateral external fixation assembled with bended lateral bars in the sagittal plane. The hypothesis was that external fixation using lateral bars bended and boundary conditions applied in the head of femur would improve the contact area required for satisfactory bone fusion.

2 Numerical Model

To simulate the arthrodesis surgery with an external fixator, the CAD assembly of the arthrodesis, femur-tibia-fixator, was performed, respecting the leg orientations suggested for the technique. This development includes the geometrical model # 3034 (Sawbones®) of the femur, the geometrical model # 3034 (Sawbones®) of the tibia and the bended external fixator with bilateral assembly.

The femur and tibia models are composed of cortical and trabecular bone and the division is already assured in CAD geometries. Since this is a qualitative analysis, bones are considered to have isotropic characteristics and the material properties are presented in Table 1. The complete assembly of the arthrodesis (see Fig. 1(a)) follows the orientations mostly used by surgeons.

Table 1. Mechanical properties of CAD components.

Component	Material	Young modulus [GPa]	Poisson's ratio
Femur	Cortical	16	0.28
	Trabecular	0.155	0.3
Tibia	Cortical	15	0.33
	Trabecular	1.1	0.33
External fixator	Steel	193	0.27

In order to perform the numerical simulation, it is necessary to define boundary conditions that must represent the interaction of the real word with your model. Hence, it is important to understand how surgeons perform the assembly of the external fixator. In this work we use the two different boundary conditions: (1) restrain all three translations at the femur head; (2) restrain all the three translation in the point where the lateral bars are connected, see Fig. 1(b).

The assembly of the external fixator is assured by means of loads that are applied in some clamps of the external fixator that are free to translate through the lateral bars while other clamps are rigidly attached to the lateral bars, see Fig. 2. The two clamps of the Steinman pin that is placed in the tibia, just below the line joint, are the first loaded with a 200 N load each one, followed by the loading of the two clamps of the pin that is placed in the region of condyles of femur.

Numerical models were obtained using ADINA® software and the same finite element type. The discretization of the geometric model was obtained using linear, solid three-dimensional tetrahedral elements with 4 nodes. The average size for each element was 2 mm, and was chosen after a convergence study considering the displacement.

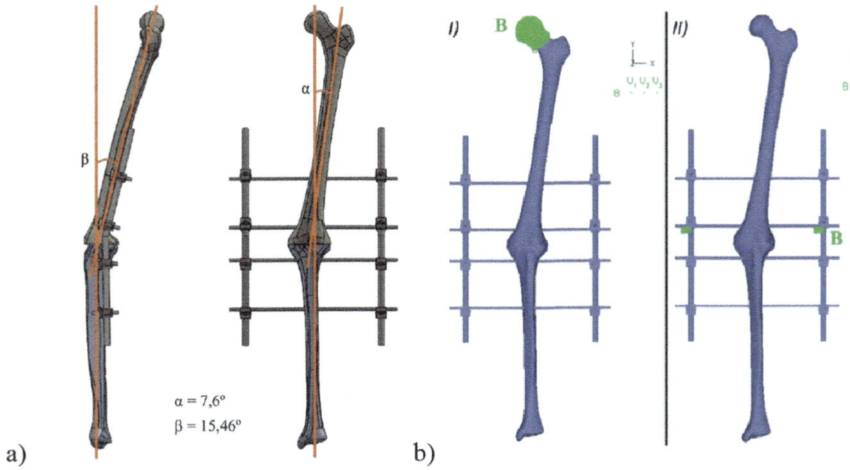

Fig. 1. (a) Geometrical model of the arthrodesis; (b) boundary conditions of the numerical model.

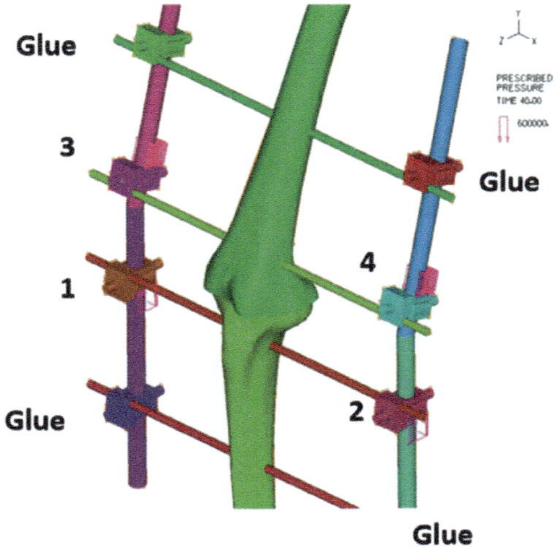

Fig. 2. Sequence of loading.

3 Results

The results of the relative displacement between the two surfaces of fusion, femur-tibia, showed no significate sensibility to the boundary conditions used, with values about 0.3 mm.

The von Mises stresses in the femur bone using the two boundary conditions are presented in Fig. 3. The higher level of stress appears in the region where the pins are loaded, on both support conditions, and the level of stress is quite similar, the higher difference appears in the proximal region. This difference is justified by the support condition, because in the first case the femur head can't move and, therefore, shows higher deformation and stresses. The level of stress in the tibia is also similar on both models. The main difference appears in the contact pressure over the fusion surfaces. This difference can be seen in Fig. 4.

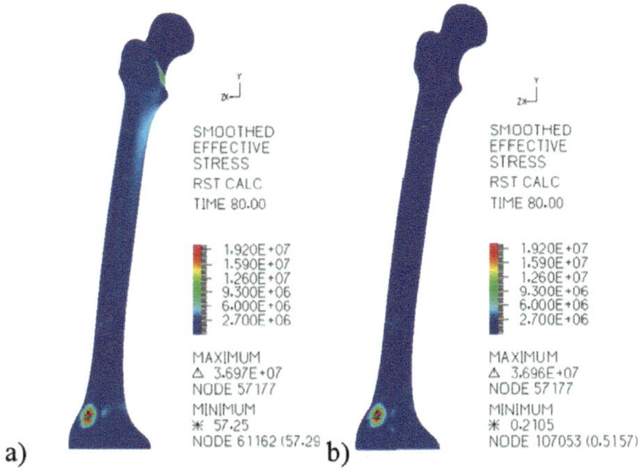

Fig. 3. Von Mises stresses: (a) using boundary conditions in the femur head; (b) using boundary conditions in the lateral bars of the external fixator.

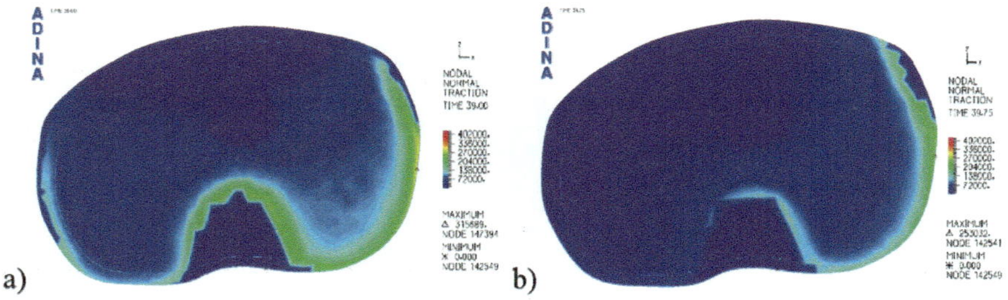

Fig. 4. Contact pressure in the fusion surface: (a) using boundary conditions in the femur head; (b) using boundary conditions in the lateral bars of the external fixator.

4 Conclusions

This results show that the type of boundary conditions can have a significant effect on the distribution of the contact pressure and, therefore, on the level of stress to which the bone is exposed. Moreover, it seems that using the fixation of femur head tends to improve the homogeneity of contact and therefore can contribute to the bone fusion.

References

1. Somayaji, H.S., Tsaggerides, P., Ware, H.E., Dowd, G.S.E.: Knee arthrodesis—a review. Knee **15**, 247–254 (2008). https://doi.org/10.1016/J.KNEE.2008.03.005
2. Riouallon, G., Molina, V., Mansour, C., Court, C., Nordin, J.-Y.: An original knee arthrodesis technique combining external fixator with Steinman pins direct fixation. Orthop. Traumatol. Surg. Res. **95**, 272–277 (2009). https://doi.org/10.1016/J.OTSR.2009.04.006

Finite Element Comparison of Two Implants for the Treatment of Unstable Trochanteric Femur Fractures

Maria A. Neto[1,3(✉)], Luis M. Roseiro[2,3], Maria F. Paulino[1,3], and Ana M. Amaro[1,3]

[1] Mechanical Engineering Department, Coimbra University, Coimbra, Portugal
{augusta.neto, ana.amaro}@dem.uc.pt, maria.paulino@uc.pt
[2] Department, Coimbra Polytechnic – ISEC, Coimbra, Portugal
lroseiro@isec.pt
[3] CEMMPRE, Department of Mechanical Engineering, University of Coimbra, Coimbra, Portugal

Abstract. Fixation of unstable intertrochanteric fractures in osteoporotic bones often fails, due to lag screw cut-out. This study correlates the numerical results of a new Trochanteric Plate of Contention (TPC), which may improve the resistance to cut-out failure of internal fixation in osteoporotic bone, and the well-known Dynamic Hip Screw (DHS) system. Biomechanical properties of the new TPC met the ASTM F384-12 guideline requirements. Generally, it is well accepted that the DHS is the implant of choice in the treatment of stable intertrochanteric femur fractures [1], as well as is considered the implant that any new design should be compared with [2]. The histograms of the bone tissue level loading distribution, represented by the three principal strains in the femur head, were selected as comparative indices.

Keywords: Internal fixation · Femur · Finite element analysis

1 Introduction

Hip fracture is a main concern of the elderly, most of whom suffer from osteoporosis in addition to various medical and mental diseases that result in their being a high-risk population [3]. A large number of internal fixation devices have been developed for the purpose of stabilize intertrochanteric femur fractures. Nevertheless, despite the great number of devices that is available, they can be separated into two main categories: extramedullary (sliding hip screws) fixation devices and intramedullary (nails) fixation devices. Unstable intertrochanteric femur fractures are commonly fixed using intramedullary fixation devices, such as proximal femur nail antirotation (PFNA) and reverse LISS, while extramedullary fixation devices are most suitable to treatment of stable intertrochanteric femur fractures. Kouvidis et al. [2] tested if resistance to cut-out failure can be improved by using a dual lag screw implant in place of a single lag screw implant, concluding that the use of double screw provided significantly greater resistance against

J. Henriques et al. (Eds.): MEDICON 2019, IFMBE Proceedings 76, pp. 1446–1450, 2020.
https://doi.org/10.1007/978-3-030-31635-8_178

varus collapse and neck rotation in comparison to a standard DHS lag screw implant. Nevertheless, whether one or another of these devices is superior over the other remains controversial, particularly as their introduction coincided with a decline in complication rates due to increasing surgeon experience [4].

Moreover, although a great number of comparison studies have been addressed in the literature, no concrete conclusions can be drawn from existing publications regarding unstable fractures, especially of the reverse obliquity variety. The purpose of the present study was to evaluate the biomechanical properties of a new Trochanteric Plate of Contention (TPC). The design principle of this new medical device group concepts of both intramedullary nails and fixed-angle plate devices. Furthermore, because in this new PTC fixation device the lag screw has an additional support area, close to the femoral neck, it is expected superior rotational stability. The Biomechanical properties of the new prototype were evaluated by laboratorial tests, which met the ASTM F384-12 guideline requirements. The numerical results of the TPC finite element model were compared with those obtained for the DHS system.

2 Finite Element Model

The new Trochanteric Plate of Contention (TPC) system is comprised of a blade plate, a lag screw and an anti-rotation pin, see Fig. 1(a). The system is constructed of stainless-steel 316L alloy. The innovative blade plate has a geometry that gets inside of femur in the region of greater trochanter and, after inside, is crossed by the lag screw.

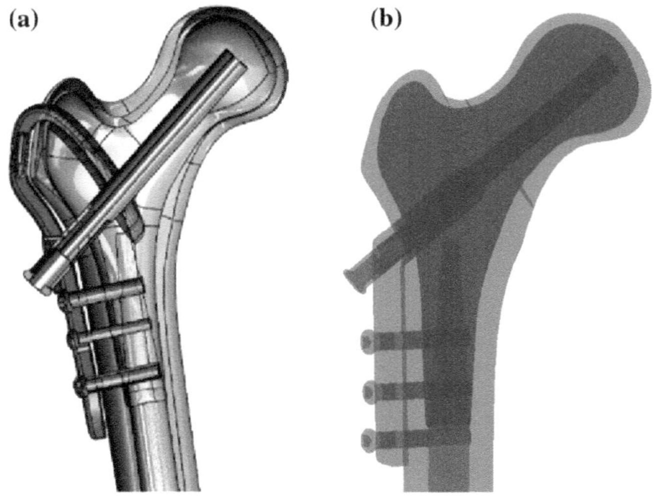

Fig. 1. Geometrical assemblage: (a) TPC system; (b) DHS System.

In Fig. 1(b) is also presented the geometrical assemblage of the DHS system. In order to perform the numerical simulation, it was necessary to define boundary conditions that represent the interaction of the real word with the numerical model. In this work we fix all three translations of all finite element nodes placed at the condyles femur in both models, and the loading was assured by means of two forces, joint-reaction and abductor-muscle forces, applied on the femur. The forces were set to 700 N and 300 N, respectively [5].

The femur is composed of cortical and trabecular bone and the division is already assured in CAD geometries, (model #3908, Sawbones, Vashon, WA, USA). Linear elastic isotropic and transversely isotropic material properties were applied. The principal axes of orthotropy were cylindrical with axial directions parallel to the anatomical axes of diaphysis and neck regions, respectively. The longitudinal and transversal Young's modulus of composite cortical bone were defined according the manufacturer's data, while shear modulus and Poisson's ratios were defined taking into account the relevant literature [6].

3 Results

The histogram of the bone tissue level loading distribution represented by the three principal strains in the cortical bone head, presented in Fig. 2, show the differences between the TPC and the DHS systems. For the TPC system the distribution curve is wider, indicating that more fraction of tissue is under high strains than for the DHS device. Figure 3 shows the histograms of the tissue level loading distribution on the head of a trabecular bone for both systems. For this type of bone the distribution curve that is wider is that associated with the DHS system, thus more material is under higher strains than for the TPC device.

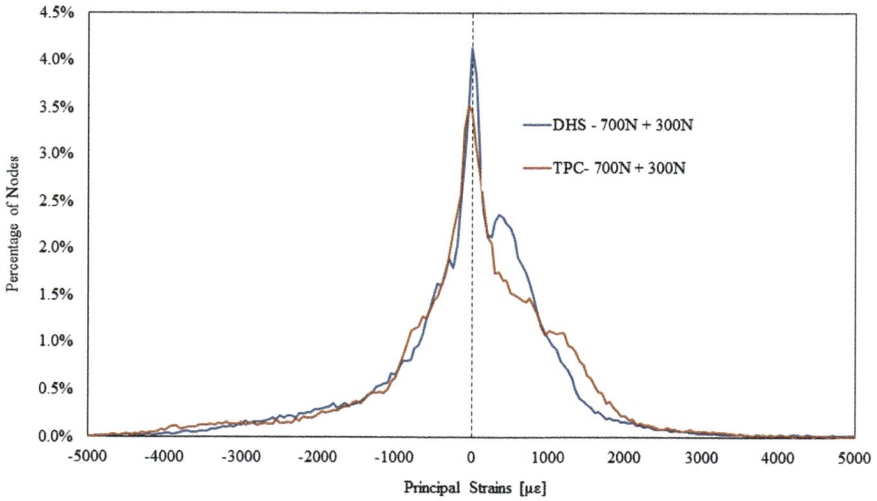

Fig. 2. Histograms of the distributions found for all three principal strains on the head cortical bone.

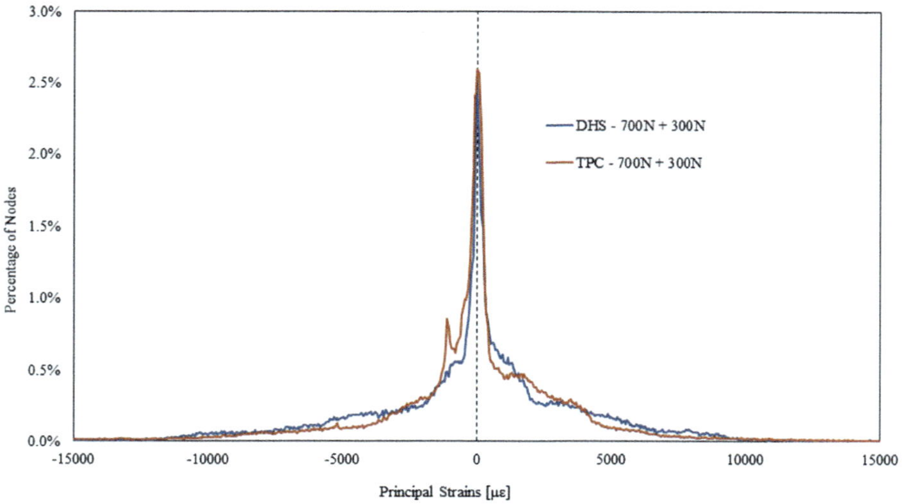

Fig. 3. Histograms of the distributions found for all three principal strains on the head trabecular bone.

These results of the principal strain field distributions predicted inside the bone by the two numerical models were similar, but the TPC device showed higher fractions of the cortical tissue loaded beyond 1000 με than the DHS system. Nevertheless, the fraction of the trabecular tissue loaded beyond the 4000 με is lower in the TPC device than in the DHS system.

4 Conclusions

The results reported on Figs. 2 and 3 related with performance comparisons of the two medical devices are not conclusive. Hence, a further analysis, changing the loading conditions, should be done, in order to evaluate its effect on the bone strain level verified with the two medical devices.

References

1. Kokoroghiannis, C., Aktselis, I., Deligeorgis, A., Fragkomichalos, E., Papadimas, D., Pappadas, I.: Evolving concepts of stability and intramedullary fixation of intertrochanteric fractures—a review. Injury **43**, 686–693 (2012). https://doi.org/10.1016/J.INJURY.2011.05.031
2. Kouvidis, G.K., Sommers, M.B., Giannoudis, P.V., Katonis, P.G., Bottlang, M.: Comparison of migration behavior between single and dual lag screw implants for intertrochanteric fracture fixation. J. Orthop. Surg. Res. **4**, 16 (2009). https://doi.org/10.1186/1749-799x-4-16

3. Steinberg, E.L., Blumberg, N., Dekel, S.: The fixion proximal femur nailing system: biomechanical properties of the nail and a cadaveric study. J. Biomech. **38**, 63–68 (2005). https://doi.org/10.1016/J.JBIOMECH.2004.03.014
4. Bhandari, M., Schemitsch, E., Jönsson, A., Zlowodzki, M., Haidukewych, G.J.: Gamma nails revisited: Gamma nails versus compression hip screws in the management of intertrochanteric fractures of the hip: a meta-analysis. J. Orthop. Trauma **23**, 460–464 (2009). https://doi.org/10.1097/BOT.0b013e318162f67f
5. Chang, C.-W., Chen, Y.-N., Li, C.-T., Peng, Y.-T., Chang, C.-H.: Role of the compression screw in the dynamic hip–screw system: a finite-element study. Med. Eng. Phys. **37**, 1174–1179 (2015). https://doi.org/10.1016/J.MEDENGPHY.2015.10.001
6. Lopes, V.M.M., Neto, M.A., Amaro, A.M., Roseiro, L.M., Paulino, M.F.: FE and experimental study on how the cortex material properties of synthetic femurs affect strain levels. Med. Eng. Phys. **46**, 96–109 (2017). https://doi.org/10.1016/j.medengphy.2017.06.001

Special Sessions: Artificial Organs: Extracorporeal Blood Circulation Medical Devices

Computational Fluid Dynamics and Experimental Analysis of Blood Gas Transport in a Hollow Fiber Module

Michael Harasek(✉), Benjamin Lukitsch, Paul Ecker, Christoph Janeczek, Martin Elenkov, and Margit Gföhler

TU Wien, Getreidemarkt 9, Vienna, Austria
michael.harasek@tuwien.ac.at

Abstract. Membrane oxygenators or artificial lungs have become an important, reliable and lifesaving clinical technique. To limit side effects on blood platelet parameters and to reduce hemolysis the gas exchange in such devices has to be improved while the size of the membrane packing has to be reduced to meet geometric constraints. Computational fluid dynamics (CFD) provides a spatial and temporal resolution of the membrane oxygenation process and enables systematic optimization of artificial lungs. An innovative CFD approach was developed to examine the gas exchange performance of oxygenators. Blood and sweep gas flow in the fiber packing as well as blood gas exchange through the membrane between blood and sweep fluid were fully resolved and simulated. The results were compared to in vitro experiments comprising determination of blood side pressure loss and CO_2 exchange performance of a prototype membrane module. This simulation approach provides a sound basis for the design of future artificial lungs.

Keywords: Artificial lung · Membrane oxygenator · Computational fluid dynamics · Hollow fiber membrane

1 Introduction

Artificial lungs or extracorporeal membrane oxygenators (ECMOs) are medical devices to supplement the respiratory function during cardiopulmonary bypass, or support patients suffering from respiratory failure [1]. In state-of-the-art oxygenators, blood gas exchange is enabled by mass transfer through hollow fiber membranes from and to a secondary phase, usually a gas phase (air or oxygen). In these devices blood flows on the shell side of the packing, while the fiber lumen is swept with O_2. CO_2 and O_2 are exchanged through the membrane following the partial pressure gradient. To provide sufficient gas exchange a membrane area of up to 2.5 m^2 is built into oxygenators. This large extrinsic surface area has serious effects on platelets and clotting factors [2]. Therefore, the focus in optimization of such contactors lies in minimizing membrane area while maximizing the gas exchange rates. Using experimental methods for the necessary development iterations is usually very time consuming, expensive and limited to pointwise data. In contrast, computational fluid dynamics (CFD) can provide

J. Henriques et al. (Eds.): MEDICON 2019, IFMBE Proceedings 76, pp. 1453–1458, 2020.
https://doi.org/10.1007/978-3-030-31635-8_179

a rather fast and flexible platform for the investigation of different designs and operational parameters. CFD also delivers a spatial and temporal resolution of the process.

In the last years, various CFD simulation approaches have been proposed to describe the gas exchange process in oxygenators. An early approach was presented by Zhang et al. [3] modeled the fiber packing as homogenous porous medium using Darcy's law to modify the momentum equation. Gas exchange was then implemented using a uniform source term in the packing. The source term was computed based on Sherwood correlations which in turn have to be determined experimentally for different packing densities and flow arrangements [4]. As the blood flow through the fiber packing is not resolved, the main blood gas transport resistance in oxygenators – blood gas concentration polarization layer attached to the membrane surface [1] – cannot be examined. Due to the high number of computational cells which are necessary to represent the geometry of a fiber packing, simulations that compute the flow and blood gas transport in the packing are computationally expensive. Additionally, the strong concentration gradients of the blood gases in the polarization layer make a high cell refinement aligned with the fiber surface mandatory. Thus, gas transport simulations are limited to microscopic geometries such as single fibers and periodic fiber arrangements. To reduce computational effort and simulation complexity gas transport in the membrane and sweep gas flow are usually neglected. Instead a fixed uniform partial pressure is applied on the blood side fiber surfaces as boundary condition for the material balances. Ambitions to simulate the gas transport in the lumen or the membrane are rare [5]. Here, the membrane gas transport resistance is considered, but the common boundary condition method leads to an overprediction of oxygen exchange rates as membrane resistances are not accounted for.

In this work a alternative CFD approach is presented. In addition to blood side and membrane wall also the fiber lumen side is added to the simulation domain. Sweep gas and blood flow as well as gas transport in blood, membrane and sweep fluid can thus be simulated simultaneously. This allows to evaluate the CO_2 enrichment of sweep flow and to assess different fiber geometries and membrane materials. In order to carry out these simulations the in-house multi-region solver `membraneFoam` developed at TU Wien [6] was utilized. To limit computational costs, the simulated geometry was reduced to a representative packing segment of a prototype hollow fiber membrane module. To determine adequate velocity boundary conditions blood flow simulations of the whole prototype module were conducted in a preliminary study. Pressure drop in dependency from the blood flow rate as well as gas exchange performance of the membrane module were evaluated experimentally during in vitro tests using bovine blood. Results of both simulations are compared with the experimental data which are taken from [7].

2 Materials and Methods

For all CFD simulations, the opensource toolbox OpenFOAM® 4.1 (ESI Group, Paris, France) was used. As a preliminary study, blood flow through the whole prototype module was simulated to gain pressure loss and velocity distribution. Due to the narrow space between fibers and a Reynolds number of about 1600 in the inlet pipe the flow was assumed laminar and no turbulence model was applied. Second order discretization schemes were utilized. A uniform inlet velocity at the beginning of the inlet pipe was set and varied to match inlet flow rates ranging from 800 to 3000 mL/min. At all walls, including membrane surfaces, a no-slip condition was applied. At the end of the outlet pipe a fixed uniform value of 0 Pa was set. For the remaining boundaries zero gradient conditions were imposed. The influence of the shear thinning rheological behavior of blood on the flow field was examined by conducting simulations with two viscosity models, Newtonian and power law. For Newtonian viscosity, the kinematic viscosity of bovine blood was used. For the power law model coefficients for human blood were chosen. Figure 1 shows the algorithmic implementation for the transport of CO_2.

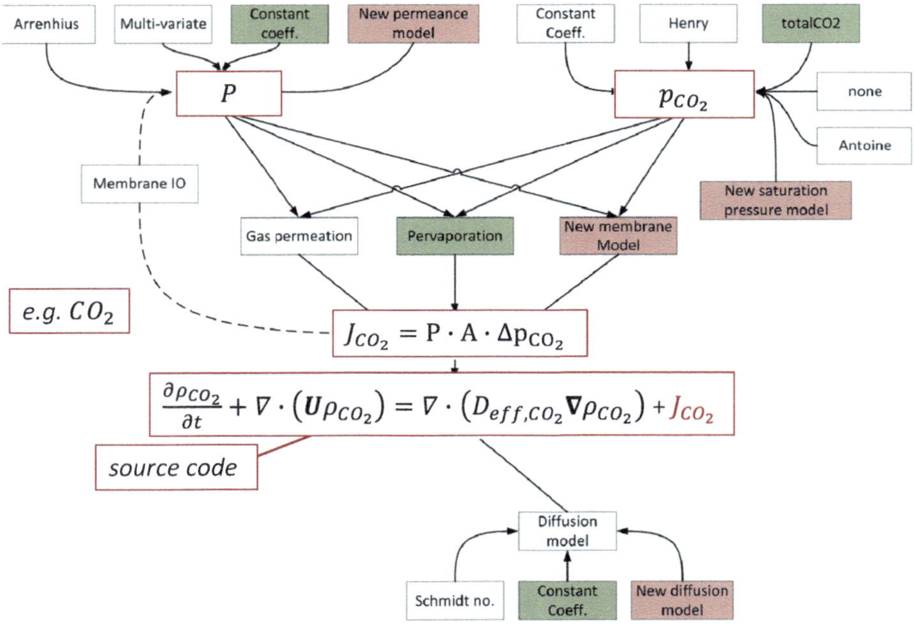

Fig. 1. Transmembrane flux is implemented as volumetric source term in an in-house developed multi region solver, `membraneFoam` [6]

The experimental in-vitro setup is shown in Fig. 2. An ECMO is used to saturate blood with CO_2. The blood is then transported to the prototype module. CO_2 mass transfer is balance through sweep gas flow and sweep gas concentration. Further details as well as experimental results on blood flow and pressure drop can be found in [7].

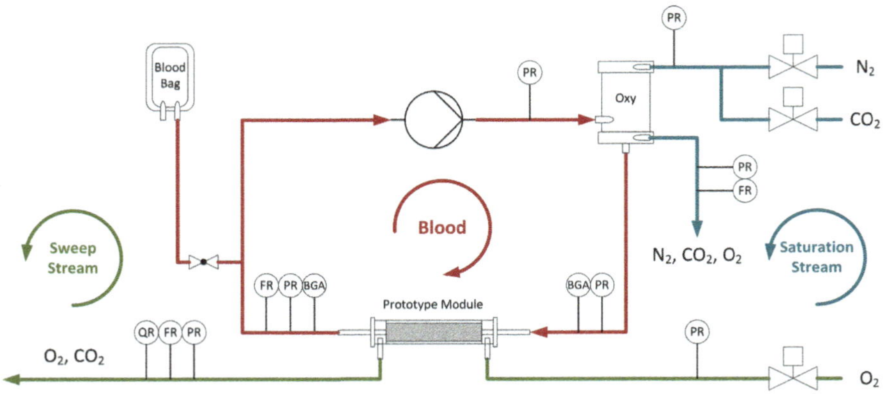

Fig. 2. Scheme of in vitro recirculation loop with prototype module, oxygenator and pump.

3 Results

In a preliminary study, CO_2 mass transfer across the membrane through the blood side boundary layer was investigated. Figure 3 shows a typical concentration profile in the near membrane region. The boundary concentration polarization layer is the dominating mass transfer resistance.

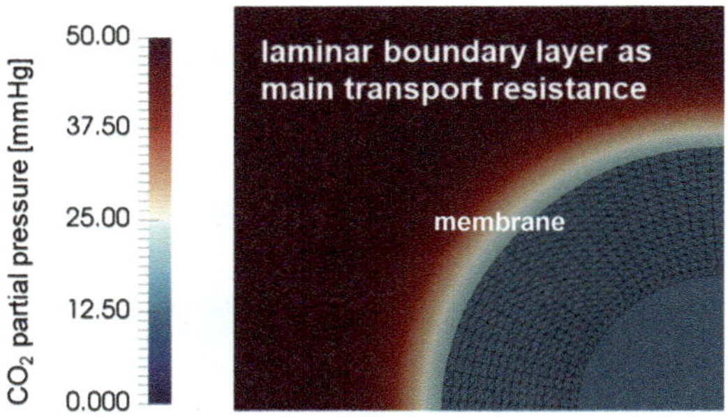

Fig. 3. CO_2 partial pressure across hollow fiber membrane – O_2 is applied on the lumen side.

CO_2 partial pressure in blood decreases from 50 mmHg outside the boundary layer to around 15 mmHg at the membrane surface. CO_2 partial pressure drop over the selective membrane layer totals to roughly 12 mmHg. In a second step, a section of the fiber bundle was simulated (Fig. 4, left). From this, suitable boundary conditions were derived to investigate transverse blood flow and mass transfer along an idealized sequence of fibers. As can be clearly seen, fibers in the flow shadow show more pronounced concentration polarization layers resulting in locally lower CO_2 mass transfer.

In comparison to the simulation, in-vitro results show over-prediction of specific CO_2 flux reflecting the reduction of module geometry to a packing segment with idealized cross flow set-up and lower blood viscosity in the bovine blood used in-vitro.

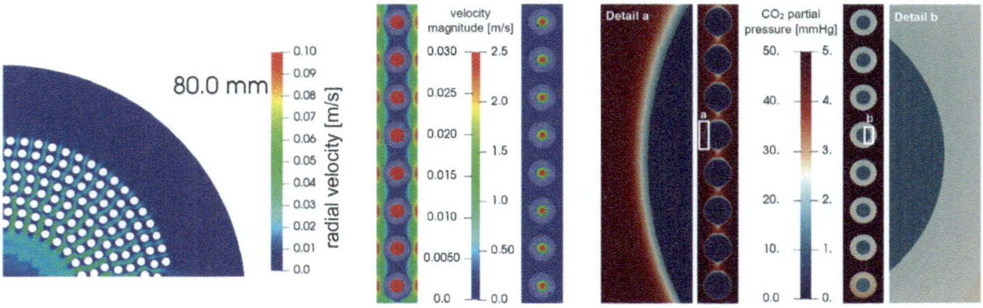

Fig. 4. Velocity profiles through the fiber bundle on the blood side, 80 mm from inlet (left), and flow and concentration profiles in an idealized fiber arrangement to analyze flow shadowing, stagnant zones and concentration polarization layer around the fibers (middle/right)

4 Conclusion

Local velocity and concentration profiles can be calculated with the proposed CFD approach. This gives close insights into local effects of the membrane mass transfer which may help to improve the design of future artificial lungs.

References

1. Federspiel, W.J., Henchir, K.A.: Lung, artificial: basic principles and current applications. In: Encyclopedia of Biomaterials and Biomedical Engineering, 2 edn., vol. 1, pp. 1661–1672. CRC Press (2008). https://doi.org/10.1081/e-ebbe2-120007349
2. Jaffer, I.H., Reding, M.T., Key, N.S., Weitz, J.I.: Hematologic problems in the surgical patient, hematology, pp. 2304–2312. Elsevier (2018). https://doi.org/10.1016/b978-0-323-35762-3.00159-1
3. Zhang, J., Nolan, T.D.C., Zhang, T., Griffith, B.P., Wu, Z.J.: Characterization of membrane blood oxygenation devices using computational fluid dynamics. J. Membr. Sci. **288**(1), 268–279 (2007). https://doi.org/10.1016/j.memsci.2006.11.041
4. Low, K.W.Q., Loon, R.V., Rolland, S.A., Sienz, J.: Formulation of generalized mass transfer correlations for blood oxygenator design. J. Biomech. Eng. **139**(3), 031007 (2017). https://doi.org/10.1115/1.4035535
5. Taskin, M.E., Fraser, K.H., Zhang, T., Griffith, B.P., Wu, Z.J.: Micro-scale modeling of flow and oxygen transfer in hollow fiber membrane bundle. J. Membr. Sci. **362**(1–2), 172–183 (2010). https://doi.org/10.1016/j.memsci.2010.06.034

6. Haddadi, B., Jordan, C., Miltner, M., Harasek, M.: Membrane modeling using CFD: Combined evaluation of mass transfer and geometrical influences in 1D and 3D. J. Membr. Sci. **563** (2018). https://doi.org/10.1016/j.memsci.2018.05.040
7. Harasek, M., Lukitsch, B., Ecker, P., Janeczek, C., Elenkov, M., Keck, T., Haddadi, B., Jordan, C., Neudl, S., Krenn, C., Ullrich, R., Gföhler, M.: Fully resolved computational (CFD) and experimental analysis of pressure drop and blood gas transport in a hollow fiber membrane oxygenator module, accepted for publication in Chemical Engineering Transactions (2019). ISSN 2283-9216

Online Urea Concentration Estimation from Spent Dialysate Using Optical Sensor

Kristjan Pilt[1(✉)], Jürgen Arund[1], Annika Adoberg[2], Liisi Leis[2], Merike Luman[2], and Ivo Fridolin[1]

[1] Tallinn University of Technology, Ehitajate tee 5, 19086 Tallinn, Estonia
kristjan.pilt@taltech.ee
[2] North Estonian Medical Centre, Sütiste tee 19, 13419 Tallinn, Estonia

Abstract. The aim of this study was to estimate urea concentration on-line in the spent dialysate by a novel miniaturized optical sensor during hemodialysis and hemodiafiltration with different settings, and to compare the results with urea concentrations measured in laboratory. Ten end-stage kidney disease patients were enrolled for this study and the dialysate samples were collected, and signals were recorded with optical sensor during 5 midweek dialysis sessions with 5 different modalities for each patient. Using samples and optical signals from calibration set, the linear model was calculated to estimate urea concentration. For linear model the determination coefficient R^2, the systematic error as BIAS and the standard error (SE) were 0.876, 0.00 ± 1.60 mM/L (calibration set) and 0.841, 0.01 ± 1.68 mM/L (validation set), respectively. In summary, the novel miniaturized optical sensor enables on-line estimation of urea concentrations during dialysis sessions and accuracy outperforms the measurement accuracy of UV–absorbance based urea concentration method from a previous study.

Keywords: Hemodialysis · Optical sensor · Urea concentration · Online monitoring · Spent dialysate

1 Introduction

The need to enable dialysis to the end-stage renal disease patients is growing steadily worldwide [1]. Because dialysis is a time consuming and expensive procedure, the dialysis adequacy is followed by measuring specific uremic markers. Urea is considered the most suitable marker for uremic toxins in the range of small molecular weight water-soluble compounds. As urea is abundant, its concentration in the blood is easy to measure, and it is easily dialyzed. Urea Kt/V, and its simplification, urea reduction ratio (URR), are the most commonly exploited parameters of dialysis adequacy [2].

The advantages of the on-line monitoring of the dialysis dose by dialysate-based sensors are avoidance of the blood sampling and the treatment quality can be observed in near real-time, making it possible to react to any deviances immediately. On-line assessment of urea-based dialysis quality parameters Kt/V and URR can be accomplished by monitoring ultraviolet (UV) -absorbance of spent dialysate [3–5] or ionic dialysance [6]. However, current dialysis adequacy monitors are not capable to monitor

J. Henriques et al. (Eds.): MEDICON 2019, IFMBE Proceedings 76, pp. 1459–1464, 2020.
https://doi.org/10.1007/978-3-030-31635-8_180

urea concentration on-line, and Kt/V and URR are monitored indirectly [7]. Furthermore, knowledge about the urea removal in the spent dialysate offers the possibility to assess dietary protein intake related to patients' nutritional status [8, 9].

A novel, specifically designed compact and robust on-line dialysis monitoring device capable for urea concentration measurement is a step closer to automatic, on-line optical assessment of dialysis quality and nutritional parameters in dialysis.

The aim of this study was to estimate urea concentration on-line in the spent dialysate by a novel miniaturized optical sensor during hemodialysis (HD) and hemodiafiltration (HDF) with different settings.

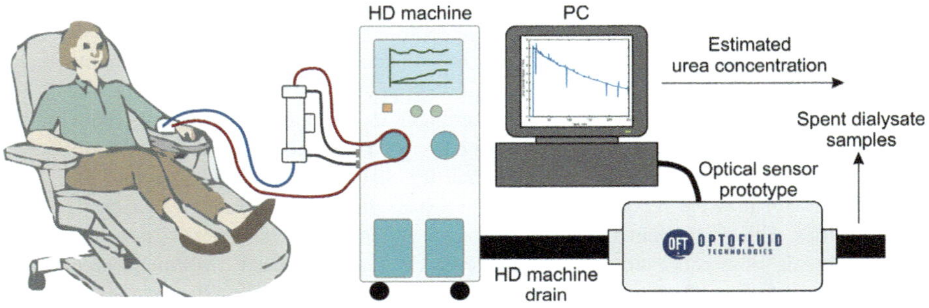

Fig. 1. Overview of the experiment setup.

2 Methods

After approval of the protocol by the Tallinn Medical Research Ethics Committee at the National Institute for Health Development, Estonia, the study was carried out at the Department of Dialysis and Nephrology, North Estonia Medical Centre, Estonia. Ten end-stage kidney disease patients, 6 male and 4 female with average age of 60.2 ± 16.8 years from Tallinn, Estonia on chronic HDF were enrolled into the study with arteriovenous fistula ($N = 9$) or graft ($N = 1$) as vascular access. 5 midweek dialysis sessions with 5 different modalities (Table 1), and with duration of 240 min were included in this study, while the dialysis sessions performed at other days of weeks were kept as prescribed. One dialysis for each patient was performed by 1.8 m^2 dialyzer (FX 800), one by 1.5 m^2 dialyzer (Xevonta Lo15), and three procedures with 2.2 m^2 dialyzer (FX 1000). Blood flow varied from 200 to 390 mL/min and dialysate flow from 300 to 800 mL/min. The average substitution volume over all HDF treatments was >14.81 L, being the highest for the last HDF treatment (>20.74 L).

Spent dialysate from the drain was monitored on-line by using optical sensor prototype (Optofluid Technologies OÜ, Estonia). The spent dialysate was guided through the cuvette of the optical sensor (Fig. 1). The fluorescence and transmitted light intensities were measured at different UV wavelengths. The light absorbance was calculated based on the transmitted light intensities using the pure dialysate as a reference [3]. The optical measurement principle and selected UV wavelength range are based on our earlier studies [10, 11].

For the reference measurements, samples from the spent dialysate drain tube of the dialysis machine were taken at the following moments during the dialysis: 7, 60, 120, and 180 min after the start and at the end of the session (240 min). The concentrations of urea in dialysate samples were determined in the Clinical Chemistry Laboratory (Synlab Eesti OÜ, Estonia) using standardized methods.

Total number of collected samples was 250. However, some samples (50) were left out due to the self-tests of the machine and concentration values being below the laboratory detection limits. The samples were randomly divided into a calibration set (50% of the material, $N = 100$) and a validation set (50% of the material, $N = 100$).

A linear model was calculated using samples from the calibration set. The reference urea concentrations from the laboratory and measured fluorescence light intensities and calculated absorbances from optical sensor at the sampling times 7, 60, 120, 180 and 240 min after the start of the dialysis session were used in model computation. The urea concentration values were estimated for validation set using the calculated model. Linear correlation relationships between the reference and the model estimated urea concentrations for calibration and validation set were analyzed and coefficients of determination were calculated. In addition, the standard error (SE) and systematic error (BIAS) were calculated based on the calculated residuals.

Table 1. Dialysis modalities used in this study.

Dialysis modality	M1: HDF (as prescribed)	M2: HD	M3: HDF	M4: HDF	M5: HDF
Dialyser	Fresenius FX800	Xevonta Lo15	Fresenius FX1000	Fresenius FX1000	Fresenius FX1000
Blood flow mL/min	300	200	300	300	390 (max)
Dialysate flow, mL/min	500	300	800	800	800
Total ultrafiltration (average of 10 patients), mL	2080	2530	2360	2000	2350
V substitution (average of 10 patients), L	14.9	0	14.81	14.85	20.74

3 Results

The estimated urea concentrations from the model using the optical signals during a single dialysis session are shown in Fig. 2. In addition, the reference concentrations are given at the sampling times 7, 60, 120, 180 and 240 min after the start of the dialysis session. The relationship between the model estimated and the laboratory measured urea concentrations for calibration and validation set is shown in Fig. 3a and b, respectively. In Fig. 3c and d are shown the Bland-Altman plots of the residuals with SE and BIAS lines for calibration (BIAS $\pm$ 1.96·SE = 0.00 $\pm$ 1.60 mM/L) and validation (BIAS $\pm$ 1.96·SE = 0.01 $\pm$ 1.68 mM/L) sets.

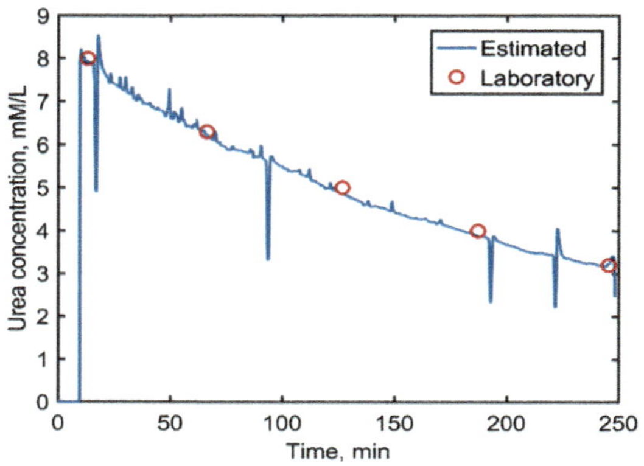

Fig. 2. Online dialysis monitoring session of the subject nr. 5 with urea concentrations. The estimated urea concentration by optical sensor (blue line) and laboratory measured urea concentrations (red circles) at sampling times of 7, 60, 120, 180 and 240 min after the start of the dialysis session.

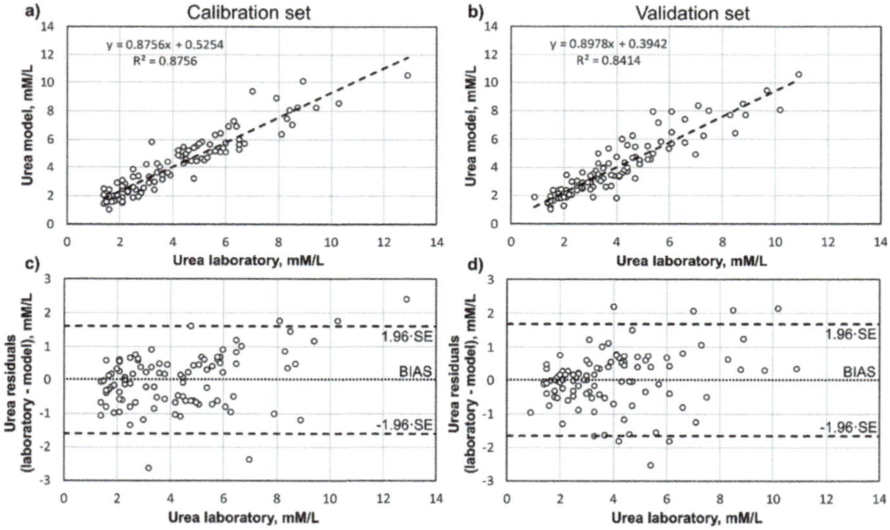

Fig. 3. Relationship between model estimated and laboratory measured urea concentrations for (a) calibration and (b) validation sets. Bland-Altman plot of the residuals for (c) calibration and (d) validation sets.

4 Discussion

The aim of this study was to estimate urea concentration from spent dialysis using optical sensor, and to compare the results with urea concentrations measured in laboratory. The strong linear relationship between optically estimated and lab urea concentrations was found both for the calibration set ($r = 0.94$) and for the validation set ($r = 0.92$). Furthermore, the BIAS and SE are in the same magnitude for both data sets, which shows that the optical method has a rather stable and general performance considering accuracy.

In Fig. 2, the raw signal of urea concentration during HD is shown. The urea concentration decreases during the session with exponential manner. However, there are visible spikes in the signal, which are caused by the hemodialysis machine self-tests when only the pure dialysate is pumped through the dialyzer (filter).

In our previous study, a mathematical model was developed for the estimation of urea concentration in the collected spent dialysate samples through UV-absorbance, which was measured using double-beam spectrophotometer [12]. The results achieved by the optical sensor outperform the urea concentration measurement accuracy compared to the earlier study [12]. In addition, compared to the earlier study, the optical sensor enables online urea concentration monitoring. The linear relationship between the estimated and the measured urea concentrations is stronger for the validation set in this study, compared to the earlier study ($r = 0.92$ vs $r = 0.87$). Furthermore, found BIAS and SE outperform results from the previous study [12] about 1.2 times for validation set. The reasons are that in addition to the absorbance, fluorescence of the spent dialysate is also included in the optical model, and a novel calibration method is applied, taking quenching effect into account.

5 Conclusions

This study demonstrated that the novel miniaturized optical sensor enables on-line estimation of urea concentrations during dialysis sessions. The urea concentration estimation accuracy outperforms the measurement accuracy of UV–absorbance based urea concentration method from a previous study. The next studies should investigate the clinical performance of the novel optical sensor for on-line monitoring of dialysis quality parameters.

Acknowledgment. The research was funded partly by the European Union through the European Regional Development Fund H2020 – SMEINST – 2 – 2017, OLDIAS2 - On-line Dialysis Sensor Phase2 project, the Estonian Ministry of Education and Research under institutional research financing IUT 19-2, and by Estonian Centre of Excellence in IT (EXCITE) funded by European Regional Development Fund.

Conflict of Interest. The authors have no conflict of interest.

References

1. Jager, K.J., Fraser, S.D.S.: The ascending rank of chronic kidney disease in the global burden of disease study. Nephrol. Dial. Transplant. **32**(suppl 2), ii121–ii128 (2017)
2. Daugirdas, J.T., et al.: KDOQI clinical practice guideline for hemodialysis adequacy. Am. J. Kidney Dis. **66**(5), S884–S930 (2015). National kidney foundation

3. Uhlin, F., Fridolin, I., et al.: Dialysis dose (Kt/V) and clearance variation sensitivity using measurement of ultraviolet-absorbance (on-line), blood urea, dialysate urea and ionic dialysance. Nephrol. Dial. Transplant. **21**(8), 2225–2231 (2006)
4. Luman, M., Jerotskaja, J., et al.: Dialysis dose and nutrition assessment by optical on-line dialysis adequacy monitor. Clin. Nephrol. **72**(4), 303–311 (2009)
5. Castellarnau, A., Werner, M., Günthner, R., Jakob, M.: Real-time Kt/V determination by ultraviolet absorbance in spent dialysate: technique validation. Kidney Int. **78**(9), 920–925 (2010)
6. Manzoni, C., Di Filippo, S., Corti, M., Locatelli, F.: Ionic dialysance as a method for the on-line monitoring of delivered dialysis without blood sampling. Nephrol. Dial. Transplant. **11**(10), 2023–2030 (1996)
7. Daugirdas, J.T., Tattersall, J.E.: Automated monitoring of hemodialysis adequacy by dialysis machines: potential benefits to patients and cost savings. Kidney Int. **78**(9), 833–835 (2010)
8. Fouque, D., Vennegoor, M., Wee, P.T., et al.: EBPG guideline on nutrition. Nephrol. Dial. Transplant. **22**(Suppl 2), ii45–ii87 (2007)
9. Garred, L., Canaud, B., Argiles, A., Flavier, J., Mion, C.: Protein catabolic rate determination from a single measurement of dialyzed urea. ASAIO J. **41**(3), M804–M809 (1995)
10. Arund, J., Tanner, R., Uhlin, F., Fridolin, I.: Do only small uremic toxins, chromophores, contribute to the online dialysis dose monitoring by UV absorbance? Toxins **4**(10), 849–861 (2012)
11. Arund, J., Luman, M., Uhlin, F., Tanner, R., Fridolin, I.: Is fluorescence valid to monitor removal of protein bound uremic solutes in dialysis? PLoS ONE **11**, 1–12 (2016)
12. Tomson, R., Fridolin, I., Uhlin, F., Holmar, J., Lauri, K., Luman, M.: Development of the model for the optical monitoring of urea in spent dialysate. In: IEEE Proceedings on Electronics Conference (BEC), 2012 13th Biennial Baltic, pp. 179–182. IEEE, Tallinn (2012)

Synthesis of Composites of Polyurethane Membranes/Polycaprolactone Fibers for Membrane Blood Oxygenators

Tiago Eusébio[1](✉), Mónica Faria[1], Viriato Semião[2], and Maria Norberta de Pinho[2]

[1] Department of Chemical Engineering, Universidade de Lisboa Instituto Superior Tecnico, CeFEMA, Av. Rovisco Pais, 1049-001 Lisbon, Portugal
{tiago.eusebio,monica.faria}@tecnico.ulisboa.pt
[2] Department of Mechanical Engineering, Universidade de Lisboa Instituto Superior Tecnico, LAETA, IDMEC, Av. Rovisco Pais, 1049-001 Lisbon, Portugal

Abstract. Membrane blood oxygenators (MBOs) assure in extracorporeal blood circulation the metabolic functions of failing lungs. The technical and medical progress of the MBOs depends on two major factors: (1) hemocompatibility of the membrane/blood interfaces, and (2) enhancement of the flow management/mass transfer associated to the metabolic functions of the lung. Integral asymmetric bi-soft segment polyurethane (PEUU) membranes with polycaprolactone (PCL) as the second soft segment have been studied and showed enhanced hemocompatibility properties. In the present work, integral asymmetric PEUU membranes, containing 10%wt. of PCL (PEUU10), were synthetized by a modified phase inversion method. The synthesis of the composites, PEUU10/PCL, is carried out by the electrospinning of the PCL fibers over the top surface of the PEUU10 membranes dense layer. The electrospinning was performed in an in-house built set-up by applying an electric potential between the metallic syringe-tip with PCL solution and the target plate with a PEUU10 membrane. The solution was continuously fed to the syringe-tip at a constant flow rate and accelerated by the electric field to be deposited over the target. The PEUU10 membrane was attached to a cylindrical support target that allows to tune the fiber deposition conditions by varying the rotation speed. The PEUU10 membranes and the PEUU10/PCL composites were characterized by Scanning electron microscopy (SEM).

The performance of the PEUU10/PCL composites in terms of boundary layer disruption, mixing promotion and subsequent oxygen mass transfer enhancement is evaluated in a surrogate system of a MBO.

Keywords: Membrane blood oxygenator · Bi-soft segment polyurethane membranes · Polyurethane membranes/polycaprolactone fibers composites · Gas transfer

J. Henriques et al. (Eds.): MEDICON 2019, IFMBE Proceedings 76, pp. 1465–1468, 2020.
https://doi.org/10.1007/978-3-030-31635-8_181

1 Introduction

Membrane blood oxygenators (MBOs) can provide partial or total support to the failing lung assuring its metabolic functions by extracorporeal blood circulation. One of the major factors in the technical and medical progress of the MBOs is the enhancement of the gas mass transfer. Integral asymmetric bi-soft segment polyurethane (PEUU) membranes with polycaprolactone (PCL) as the second soft segment have been studied and showed enhanced hemocompatibility properties [1]. A model of the oxygen mass transfer (Eq. 1) in a system surrogate of a MBO (Fig. 1) that allows the direct quantification of the experimental oxygen fluxes was developed and validated [2]. The model assumes that the O_2 mass transfer results from the convective contribution in the z-direction and the diffusive contribution in the y-direction; the flow is fully developed, unidimensional and characterized by a parabolic velocity profile; due to the very small distance of penetration of O_2 in the y-direction the velocity profile can be approximated by a linear profile; the liquid phase can be approximated to a semi-infinite medium. In the present work, the oxygen fluxes were predicted by Eq. 1 for different oxygen feed pressures and water flow rates. Furthermore, considering the resistances in series of the membrane and of the liquid boundary layer, the partial and the global mass transfer coefficients were predicted and compared with the experimental values. Composites PEUU/PCL were synthesized by the electrospinning of PCL fibers over the surface of the PEUU membranes dense layer. The effect of PCL fibers, that act as boundary layer disruptors and mixer promoters, in the oxygen transfer in a MBO surrogate system is under investigation. The experimental performance of the PEUU/PCL composites was compared with the PEUU membranes and with the model predictions.

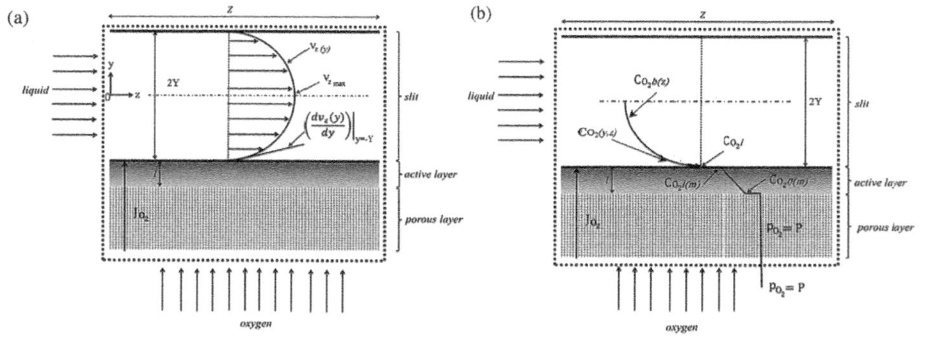

Fig. 1. Scheme of a cross-section of the system surrogate of a MBO; (a) Schematic representation of the velocity profile of the water flow; (b) Schematic representation of the oxygen concentration profile.

$$\left\langle J_{O_2}\big|_{y=0} \right\rangle = \frac{3D_{O_2l}(C_{O_2l} - C_{O_20})}{2\Gamma\left(\frac{4}{3}\right)} \left(\frac{a}{9D_{O_2l}Z}\right)^{1/3} \tag{1}$$

2 Methods

Integral asymmetric PEUU membranes were synthetized by a modified phase inversion method [3]. PEUU/PCL composites were prepared by the electrospinning of the PCL fibers over the surface of the PEUU membranes dense layer. The electrospinning was performed in an in-house built set-up by applying an electric potential between the metallic syringe-tip with PCL solution and the target plate with a PEUU membrane. The PEUU membrane and the PEUU/PCL composites were characterized by Scanning electron microscopy (SEM). Oxygenation experiments were performed in a custom made MBO surrogate system composed of a slit for water circulation as surrogate of blood flow chamber and an oxygen chamber kept at constant pressure separated by the membrane in study. The water flow and the oxygen feed pressure were varied.

3 Results and Discussion

The experimental measurements of oxygen permeation fluxes show that the evolution with time of the experimental oxygen average concentration follows a three-regime pattern: (i) a linear variation of O_2 vs. time for the initial short times range where the driving force to oxygen mass transfer is maximal; (ii) a nonlinear variation in the intermediate time range; (iii) a plateau for larger times where the oxygen saturation in the liquid stream is reached. Equation 1 allows the prediction of the O_2 fluxes in the regime of linear variation of O_2 concentration with time. Table 1 shows the predicted O_2 permeation fluxes, the partial and the global mass transfer coefficients, the experimental O_2 permeation fluxes and the experimental global mass transfer coefficient obtained for the PEUU membrane varying the water flow rates from 2.0 to 3.0 L/min and the oxygen feed pressures of 30 and 60 kPa. Figure 2 shows the SEM micrographs of the top surfaces of the PEUU membrane and of the PEUU/PCL composite.

Table 1. Predictions of the average O_2 permeation fluxes $\langle J_{O_2} \rangle$, of the liquid mass transfer coefficient (Kliq), of the membrane mass transfer coefficient (Kmemb) and of the global mass transfer coefficient (K_G) and the experimental O_2 permeation fluxes J_{O_2} and the global mass transfer coefficient (K_{GExp}) obtained for the PEUU membrane for water flow rate of 0.10 and 0.25 L/min and 20 and 40 kPa of oxygen feed pressure.

Q (L/min)	pO2 (kPa)	Predictions				Experimental	
		$\langle J_{O_2} \rangle (10^{-10}$ molcm^{-2}s$^{-1})$	Kliq (10^{-3} cm/s)	Kmemb (10^{-3} cm/s)	K_G (10^{-3} cm/s)	J_{O_2} (10^{-10} molcm^{-2}s^{-1})	K_{GExp} (10^{-3} cm/s)
2.0	30	13	6.6	7.0	3.4	9	2.3
2.5	30	14	7.1	7.0	3.5	9	2.3
3.0	30	14	7.5	7.0	3.6	10	2.6
2.0	60	26	6.6	7.0	3.4	19	2.4
2.5	60	27	7.1	7.0	3.5	19	2.4
3.0	60	28	7.5	7.0	3.6	20	2.6

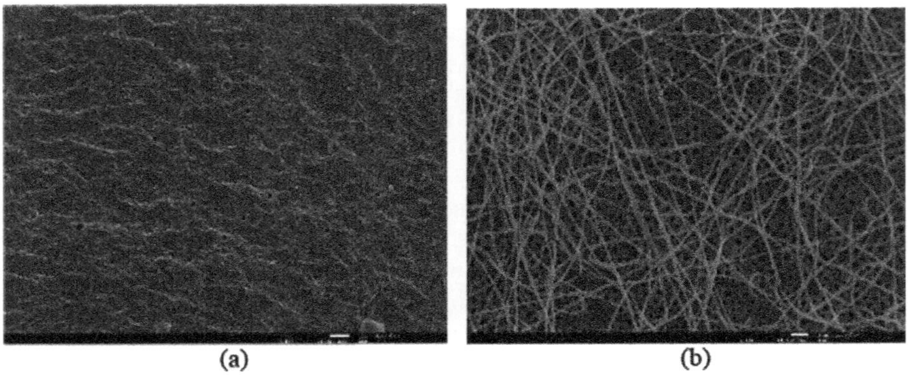

Fig. 2. SEM micrographs of the top surfaces of: (a) PEUU membrane; (b) PEUU/PCL composite.

4 Conclusion

PEUU/PCL composites were synthesized by depositing electrospinning PCL fibers over the membrane. The experimental oxygen permeation fluxes of the PEUU membranes and of the PEUU/PCL composites was attained in a surrogate system of a MBO. The enhancement of the O_2 mass transfer by the PCL fibers is investigated by comparing the experimental fluxes obtained for the PEUU membranes and for the PEUU/PCL composites. Experimental oxygen permeation fluxes and mass transfer coefficients are compared with the ones predicted by the model.

References

1. Faria, M., Brogueira, P., de Pinho, M.N.: Sub-micron tailoring of bi-soft segment asymmetric polyurethane membrane surfaces with enhanced hemocompatibility properties. Colloids Surf. B Biointerfaces **86**(1), 21–27 (2011)
2. Faria, M., Moreira, C., Eusébio, T.M., de Pinho, M.N., Brogueira, P., Semião, V.: Oxygen mass transfer in a gas/membrane/liquid system surrogate of membrane blood oxygenators. AIChE J. **64**(10), 3756–3763 (2018)
3. de Pinho, M.N.: Process of synthesis asymmetric polyurethane based membranes with hemocompatibility characteristics and membranes obtained by said process. US 2010/0111761 A1 (2010)

Hybrid Integral Asymmetric Cellulose Acetate/Silicon Dioxide Ultrafiltration Membranes for Uremic Blood Purification

Mónica Faria[1(✉)], Pedro Brogueira[2], and Maria Norberta de Pinho[1]

[1] Department of Chemical Engineering, Universidade de Lisboa Instituto Superior Tecnico, CeFEMA, Av. Rovisco Pais, 1049-001 Lisbon, Portugal
monica.faria@tecnico.ulisboa.pt

[2] Department of Physics, Universidade de Lisboa Instituto Superior Tecnico, CeFEMA, Av. Rovisco Pais, 1049-001 Lisbon, Portugal

Abstract. Monophasic hybrid cellulose acetate/silica ($CASiO_2$) integrally skinned membranes with silica contents between 5–18 wt.% were synthetized by an innovative method which combines the phase inversion and sol-gel techniques. The morphological and topographical characterization was performed by scanning electron microscopy (SEM) and atomic force microscopy (AFM). Permeation experiments were performed to determine the hydraulic permeability and rejection coefficients to reference solutes pertaining to the metabolic functions of the kidney. SEM confirmed asymmetric membrane cross-section structures and AFM showed that the introduction of silica reduced the submicron surface roughness at least 3 times compared to the pure CA membrane reaching a roughness mean value below 2.5 nm. Permeation studies show that the integration of silica into CA membranes increased hydraulic permeability of the hybrid $CASiO_2$ membranes by a factor of ~ 2 and that all of the hybrid membranes fully permeate urea and totally reject albumin. In terms of hemocompatibility, all of the $CASiO_2$ membranes are non-hemolytic, low thrombogenic and do note promote the highest stages of platelet activation.

Keywords: Monophasic hybrid membrane · Blood purification · Hemocompatibility

1 Introduction

Artificial organs associated to clinically well-established membrane-based treatments assure in extracorporeal blood circulation devices the metabolic functions of a failing organ like the hemodialyzer for substitution of the kidney.

Patients suffering from End Stage Renal Disease must undergo intermittent hemodialysis (HD) treatment, typically for 4 h 3 times week, in order to remove accumulated uremic toxins (UTs) and extra water from the blood their blood. Current HD systems are very effective in removing small (<500 Da), water soluble metabolites and some middle (500 Da–15 kDa) molecular weight toxins through diffusion and convection but fail to remove significant amounts of protein-bound uremic toxins (PBUTs) that accumulate once renal excretory function declines. Growing

J. Henriques et al. (Eds.): MEDICON 2019, IFMBE Proceedings 76, pp. 1469–1473, 2020.
https://doi.org/10.1007/978-3-030-31635-8_182

evidence shows that retention of PBUTs such as indoxyl sulfate, and p-cresyl sulfate, exert toxic effects on many vital organs, including the kidney, the liver, blood vessels, and heart [1]. Convective therapies such as hemodiafiltration, which improve the removal of β2microglobulin (β2M), surrogate markers for middle size UTs, do not show an improvement in all-cause mortality [2], mainly due to the inefficient clearance of PBUTs.

The hemodialysis (HD) membrane is the key component of the artificial kidney as it is responsible for removing accumulated uremic toxins, excess ions and water while simultaneously retaining vital components from the blood of End Stage Renal Disease (ESRD) patients [3]. The technical and medical progress of the artificial kidney depends on two major factors: (1) hemocompatibility of the membrane/blood interfaces, and (2) enhancement of the flow management/mass transfer associated to the metabolic functions of the artificial kidney.

The present study aims the synthesis and characterization of hemocompatible high flux ultrafiltration membranes for the artificial kidney with retention of albumin and permeation of urea.

2 Materials and Methods

2.1 Membrane Synthesis

Monophasic hybrid integrally skinned asymmetric $CASiO_2$ membranes were synthesized [4] by a modified version of the phase inversion technique [5] coupled to sol-gel techniques [6]. Casting solutions containing silica contents 0 wt.%, 5 wt.%, 11 wt.% and 18 wt.% rendered four membranes: $CASiO_2$-0, $CASiO_2$-5, $CASiO_2$-11 and $CASiO_2$-18. The casting solution of the pristine CA membrane, $CASiO_2$-0, is composed of 17 wt.% cellulose acetate (Sigma-Aldrich), 30 wt.% formamide ($\geq$99.5%, Carlo Erba) and 53 wt.% acetone (99.7%, Labchem). The ratio of these constituents is maintained in the casting solutions of the hybrid $CASiO_2$ membranes to which the precursor to silicon dioxide, Tetraethyl orthosilicate (TEOS), water and HNO_3 (60%, JMGS) are added. Table 1 shows the compositions of the casting solutions prepared for the synthesis of the $CASiO_2$ membranes.

Table 1. Casting solutions composition of the $CASiO_2$ membranes.

Compound (wt.%)	$CASiO_2$-0	$CASiO_2$-5	$CASiO_2$-11	$CASiO_2$-18
Cellulose acetate	17.0	16.4	15.8	15.2
Formamide	30.0	29.0	27.9	26.7
Acetone	53.0	51.1	49.2	47.2
TEOS	–	3.0	6.1	9.3
H_2O	–	0.5	1.0	1.6

2.2 Membrane Characterization

The surface morphology and topography of the membranes were characterized by Scanning Electron Microscopy (SEM) and Atomic Force Microscopy (AFM). Permeation experiments were carried out to yield Lp, MWCO and rejection coefficients (f) to a set of reference solutes including urea and Bovine Serum Albumin (BSA). In-vitro hemocompatibility was evaluated under static conditions in terms of hemolysis index, thrombosis degree and platelet interaction (adhesion and activation) according to the ISO 10993-4:2002 [7] standard using anticoagulated pooled sheep blood.

3 Results

3.1 Membrane Synthesis and Cross Section Structure

Three hybrid membranes of cellulose acetate (CA) and silicon dioxide (SiO_2), CA/SiO_2, containing different concentrations of SiO_2 were synthesized by a novel method that combines phase inversion with acid-catalyzed sol-gel process. The existence of a very thin dense active layer and a much thicker porous layer were confirmed by SEM micrographs of the membrane cross sections. Figure 1 displays the SEM image of a monophasic asymmetric cross section of the CA/SiO_2 membrane synthesized with 5 wt.% of silica.

3.2 Membrane Surface Characterization

AFM revealed that the introduction of silica affects the topography of the active layer surface of the $CASiO_2$ membranes by reducing the submicron surface roughness at least 3 times compared to the pure CA membrane reaching values between 1.0–2.5 nm.

3.3 Membrane Permeation

Permeation studies show that the integration of silica into CA membranes by the innovative phase inversion/sol-gel method results in high flux ultrafiltration membranes with enhanced mass transfer properties pertaining to the metabolic functions of the kidney. The high flux property is confirmed by the values of the hydraulic permeability of the hybrid $CASiO_2$ membranes, which is approximately 2 times higher than the value of the pure CA membrane and the essential total permeation of urea and total rejection of albumin is assured by all of the hybrid membranes.

3.4 Hemocompatibility

Hemocompatibility assays shows that all of the membranes are non-hemolytic and that even for an extended static blood/membrane surface contact time of 15 min the thrombosis degree of the $CASiO_2$ membranes is lower than 30% of the value observed for glass (positive control). Platelet adhesion quantified by the global parameters of platelet deposition (PD) and platelet coverage (PC) show that the introduction of silica enhances the hemocompatibility by reducing PD from $\sim$6 platelets/10,000 μm^2 for the

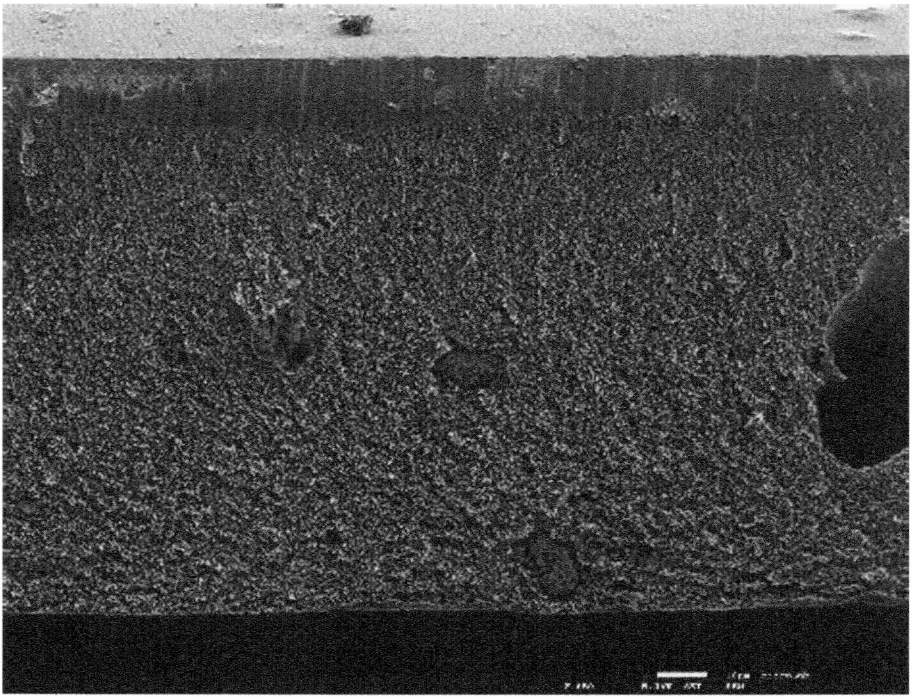

Fig. 1. SEM image of the cross sections (500×) of the hybrid integral asymmetric $CASiO_2$-5 membrane.

pure CA membrane to ~2.5 platelets/10,000 μm^2 for the hybrid $CASiO_2$ membranes and lowering the PC of the hybrid $CASiO_2$ membranes by approximately one half of the PC value found for the CA membrane. Platelet activation studies show that none of the membranes present platelets in the highest activation stage (V-fully spread) and that the number of platelets in the second highest stage (IV-spread) was greatly reduced by the introduction of silica from ~16% for the pure CA membrane to <3% for the $CASiO_2$ hybrid membranes.

References

1. Lekawanvijit, S., Kompa, A.R., Krum, H.: Protein-bound uremic toxins: a long overlooked culprit in cardiorenal syndrome. Am. J. Physiol. Renal Physiol. **311**, F52–F62 (2016). https://doi.org/10.1152/ajprenal.00348.2015
2. Grooteman, M.P.C., van den Dorpel, M.A., Bots, M.L., Penne, E.L., van der Weerd, N.C., Mazairac, A.H.A., den Hoedt, C.H., van der Tweel, I., Lévesque, R., Nubé, M.J., ter Wee, P. M., Blankestijn, P.J.: CONTRAST Investigators, Effect of online hemodiafiltration on all-cause mortality and cardiovascular outcomes. J. Am. Soc. Nephrol. **23**, 1087–1096 (2012). https://doi.org/10.1681/ASN.2011121140

3. Vienken, J.: Membranes in hemodialysis. In: Peinemann, K.-V., Nunes, S.P. (Eds.) Membranes for the Life Sciences, pp. 1–48. Wiley-VCH Verlag GmbH & Co. KGaA, Weinheim (2007). https://doi.org/10.1002/9783527631360.ch1
4. Mendes, G., Faria, M., Carvalho, A., Gonçalves, M.C., de Pinho, M.N.: Structure of water in hybrid cellulose acetate-silica ultrafiltration membranes and permeation properties. Carbohydr. Polym. **189**, 342–351 (2018). https://doi.org/10.1016/j.carbpol.2018.02.030
5. Kunst, B., Sourirajan, S.: An approach to the development of cellulose acetate ultrafiltration membranes. J. Appl. Polym. Sci. **18**, 3423–3434 (1974). https://doi.org/10.1002/app.1974.070181121
6. Brinker, J., Scherer, G.: Sol-gel science: the physics and chemistry of sol-gel processing presents the physical and chemical principles of the sol-gel process. Gulf Professional Publishing, Houston (1990). https://books.google.com/books/about/Sol_gel_Science.html?hl=pt-PT&id=V2vRvsTaCwMC. Accessed 18 September 2017
7. ISO - International Organization for Standardization. ISO10993-4 - Biological evaluation of medical devices, Part 4: Selection of tests for interactions with blood., (n.d.). http://www.iso.org/iso/iso_catalogue/catalogue_tc/catalogue_detail.htm?csnumber=3208

Special Sessions: Diabetes and Cardiovascular Diseases: Ibero-American Trends

Pulse Transition Time Method for Unobtrusive Blood Pressure Estimation

Maria G. Ruano[1,2(✉)], Amir Sadat Fazel[3], Ana Jiménez Martín[3], António Ruano[1,4], and Juan Jesús García Domínguez[3]

[1] University of Algarve, Faro, Portugal
mruano@ualg.pt
[2] CISUC, University of Coimbra, Coimbra, Portugal
[3] Electronics Department, University of Alcalá, Alcalá de Henares, Spain
[4] IDMEC, Instituto Superior Técnico, Universidade de Lisboa, Lisbon, Portugal

Abstract. Pulse Transition Time (PTT) is being increasingly used for estimating blood pressure unobtrusively aiming at continuous assessment of cardiovascular diseases (CVD) and illness monitoring using portable and low-cost equipment. Several methods are available for PTT estimation, each one accepting its own constraints and physical parameters' assumptions, besides the panoply of definitions attributed to PTT calculus. Defining PTT as the time difference between the time instant were peak R occurred in an electrocardiogram (ECG) cardiac cycle and the time instant were the photoplestimography (PPG) signal of correspondent cardiac cycle presents the inflexion point of its maximum slope, a method of PTT estimation based on measured ECG and PPG signals during several cardiac cycles is proposed. This method was tested on ECG and PPG data collected from 7 individuals for periods ranging from 30 min to 8 h. For the sake of PTT algorithm performance evaluation, ECG and PPG signals were collected with a commercial system, together with the PTT signals provided by the same equipment resultant from automatic computation with unknown algorithm. Preliminary results so far obtained encourage the use of the proposed PTT estimation algorithm on future internet of things (IoT) e-health system development aiming at elderly blood pressure estimation and CVD assessment. Accuracy of the proposed algorithm (0.9 correlation with reference signal) may be improved by further studies on raw data filtering parametrizations.

Keywords: Pulse transition time · Photoplestimographer · Electrocardiogram

1 Introduction

Cardiovascular diseases (CVD) are responsible for 31% of the global deaths [1], and according to the same source, hypertension is the CVD aetiology for which only less than 1 in 5 adults is controlled. Hypertension is the most important risk factor to evaluate CVD and stroke [2] constituting a diagnostic feature of other pathologies. Blood pressure (BP) varies between systolic BP and diastolic BP in each heartbeat, being as so an indicator of hypertension (high BP). Continuous BP monitoring enables identification of variations on cardiovascular functioning, being prognostically relevant

J. Henriques et al. (Eds.): MEDICON 2019, IFMBE Proceedings 76, pp. 1477–1484, 2020.
https://doi.org/10.1007/978-3-030-31635-8_183

for heart rate (HR) variability detection, CVD prevention and control [3]. Typically, BP is evaluated invasively using a catheter, requiring clinical intervention and patient's displacement to hospital environment. On the other hand, this procedure cannot be used to monitor BP at a continuous mode, for instance, during 24-h ambulatory examination, neither in less critical situations like elderly health assessment, who's health condition may require deeper BP analysis but, due to the costs involved on hospitalization and clinical essays, the BP analysis is generally pretermit besides the importance of its early detection to reduce mortality and morbidity.

The most common noninvasive method to measure BP is the auscultatory method based on an inflatable cuff. This simple and inexpensive method is however very uncomfortable and unhandy for continuous BP monitoring providing, in general, only intermittent readings. Alternative strategies of assessing BP, based on less intrusive equipment have been proposed. To pinpoint, usage of photoplestimographer (PPG) to compute HR variability and pulse rate (PR) variability, as published in [4, 5] also employed PPG signals and presented a comparison of the obtained PR variability results with those obtained through direct computation from the electrocardiogram (ECG) signal; Ding and co-authors [6, 7] developed methods of BP estimation using PPG and researched the use of pulse transit time (PTT) as a method for continuous cuff less BP estimation. From these and other published researches, together with the evidence of internet of things (IoT) developments for diagnostic and monitoring technologies ([8, 9]) the PPG, together with ECG are the basic equipment employed to non-intrusively and costless CVD prognosis, making use of PTT calculations.

PTT becomes a challenge research issue since there are different approaches to the definition of PTT, dependent on the definition of pulse wave (PW) variability (or PW velocity) and the body location where PPG sensor is allocated. For [10] PTT is the time spent by the PW to travel a distance which can be defined as the time interval between the ECG's R-peak till the occurrence of its correspondent event on the PW collected at the body measuring point (typically the finger) or, it might be considered as the time delay observed when measuring PW at different body sites. Present article assumes the first PTT definition since it enables usage of the same ECG sensor of the IoT system to process other clinical variables and produce other diagnostic events rather than just HR, for instance, sleep disorders [9], another key diagnostic feature to accomplish the major goals of the research project on which this work is included.

In addition, algorithms for calculus of accurate PTT values based on PW and ECG parameters still need being researched since ECG signals are known to be noise corrupted, and PPG signals are sensitive to factors such as: variations in pathlength between the source and detector, optical density of blood, venous volume fluctuations, vasomotor activity, thermoregulation and patient's ventilation [5, 11]. Examples of proposed PPG calculation algorithms include improving the accuracy of diagnosis by eliminating corrupted pulsed waves [12] or by minimizing the influence of respiratory movements (high frequencies) on BP calculations [6]. Another work to be mentioned, developed aiming at a ubiquitous application, calculates the PTT by combining signals from ECG, PPG and ballistocardiogram devices [13], but obliges patient to be standing on the ballistocardiogram scale, which is a handicap for elderly patients.

Present paper presents a new approach to compute PTT based on PPG and ECG signals collected from unobtrusive sensors of an IoT system under development envisaging assessing BP of elderly people. The calculation of BP based on PTT is currently under development and therefore not yet included. To reliably assess the proposed algorithm, PPG and ECG signals were collected by a commercial equipment.

The structure of the paper is as follows: next section describes the proposed methodology, specifying the strategy applied for artifacts and noise removal, and detailing the proposed PTT calculation algorithm; Sect. 3 exposes the origin and the type of data used for testing the proposed method; Sect. 4 presents the results obtained and finally, in Sect. 5 are drawn conclusions and future research steps are suggested.

2 Methods

2.1 Overview

As previously mentioned, we propose to compute PTT based on the analysis of PPG and EEG signals. At this stage of development, PPG and ECG were provided by a commercial device, FDA approved, the SOMNOtouch™ NIBP [14]. PPG was provided by a plethysmography clump located around the index finger, and the ECG was obtained through three dry electrodes located on chest: one under the right clavicle near right shoulder within the rib cage frame, another under the left clavicle near left shoulder within the rib cage frame and the third one located on the left side below pectoral muscles lower edge of left rib cage. These signals are corrupted, so the first step is to pre-process them, and only afterwards the PTT proposed algorithm is applied. Since the SOMNOtouch™ NIBP device automatically computes PTT signals, these signals were also collected to be used as ground truth reference to evaluate the performance of the proposed algorithm.

All algorithms were implemented in Matlab (The MathWorks, Inc).

2.2 ECG and PPG Preprocessing

The ECG and PPG signals were acquired from SomnoTouch with sampling frequencies of 512 and 256 Hz respectively. ECG and PPG data vectors are reduced to the same data length by decimation.

Then, a Butterworth high-pass filter with a cut-off frequency of 0.01 Hz is applied to eliminate the DC component, and a low pass filter of moving average is applied to smooth the signals. After testing different window lengths, a 5-point sliding window was the one providing best results.

As a final stage of preprocessing the signals are manually verified to eliminate those cardiac cycles that are still corrupted, indicating incomplete cardiac cycles or the presence of patient's movement or other external artifact which was not eliminated during filtering procedure. Then, ECG and PPG data containing valuable information on several cardiac cycles is stored in Matlab vectors.

2.3 Strategy for PTT Calculation

We propose the calculus of PTT as the time difference between the time instant were peak R occurred in ECG signal and the time instant were PPG presents the inflexion point of its maximum slope. To help computation of this inflexion point we compute the first derivative of PPG signal and the peaks of this latter signal are the time points to match with the ECG's R-peak. Figure 1 clarifies this strategy.

Following this approach, the next step is to automatically detect the peaks of the ECG signal and those from the first derivative of PPG, as accurately as possible.

Detection of Peaks of ECG and PPG Signals

Both ECG and PPG signals contain several cardiac cycles, and, in this study, we propose a peak detection algorithm based on the temporal difference between peaks (NN) of two signals. Each signal (ECG and PPG) presents several peaks. The time between peaks for each signal needs as so to be computed.

The first three peaks for each signal are estimated from amplitude thresholds, similarly to other authors' strategy. We employed a modified version of Gutiérrez-Rivas's dynamic threshold algorithm [15]. For the remaining peaks of each signal, the peak N is identified from the peak N−1, by searching for a maximum amplitude value in the interval (NN_{n-1}−250 ms) and (NN_{n-1} +250 ms). Therefore, the peak's searching range is dependent on the time location of previous maxima (peak) and ±250 ms. The value of 250 ms is empirical, based on the assumption that it would, in general, correspond to one fourth of a cardiac cycle, assumption also adopted by other authors [16, 17].

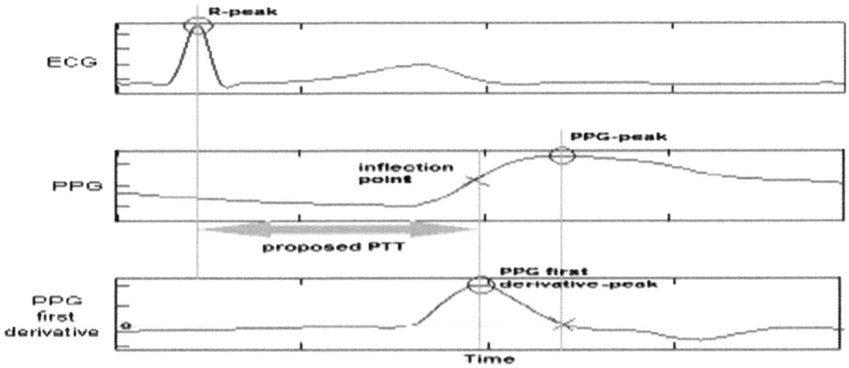

Fig. 1. Proposed strategy to compute PTT within a cardiac cycle: time difference between R-Peak of ECG waveform and the time instant of occurrence of the inflection point on PPG, computationally determined by the time instant of the peak location of the PPG's first derivative.

Synchronizing Peaks of ECG and First Derivative PPG

The PTT is determined as the time difference between the R-peak of the ECG and the moment of arrival of that heartbeat reaction to the periphery (in this case, the finger) marked in the PPG signal. Therefore, we must avoid that the value of PTT is negative or greater than the duration of the pulse it represents, since both cases correspond to wrong evaluations. To ensure synchronization between the ECG and PPG first derivative signal, there are authors who fix the wave reference time to the 50 ms interval between the R peak under study and 80% of the average interval between R peaks [12].

In this work, synchronism is sought from thresholds, comparing the times $t_{ECG\ R\text{-}peak}$ (n) and $t_{PPG\ maxSlope}$ (n). If it does not match, it is checked if $t_{PPG\ maxSlope}$ (n + 1) complies. If $t_{PPG\ maxSlope}$ (n + 1) does not comply either, $t_{ECG\ R\text{-}peak}$ (n) is discarded and the operation is repeated for $t_{ECG\ R\text{-}peak}$ (n + 1) and $t_{PPG\ maxSlope}$ (n + 1).

3 Testing Data

To carry out this study, a set of physiological samples of 7 individuals was acquired. Six of them were healthy young people (22 $\pm$ 2 year old) whose samples were collected when lying in supine position at a bed in our laboratory. After experiencing these preliminary tests, physiological data from an elderly individual (70 years old) with obstructive and central apneas diagnosed was collected at added to the set of study. Data from the 7th individual was recorded during night sleep at his home.

As previously mentioned at this stage the SomnoTouch ™ tool was used and parameters such as ECG, PPG, pulse wave, airflow, blood pressure and PTT values, were gathered for future testing the IoT system under development.

For the work hereby presented, ECG and PPG signals with a minimum duration of 30 min and a maximum duration of 8 h were employed. Both signals were collected at body locations specified in subsection 2.1.

4 Results

After running the algorithms representing the proposed methodology, according to which PTT is estimated by PTT(n)= $t_{PPG\ maxSlope}(n) - t_{ECG\ R\text{-}peak}(n)$ and comparing the obtained PTT estimations with the PTT values provided by SomnoTouch ™ tool we observed that the proposed algorithm was able to estimate PTT very similarly to the PTT signal (gold standard) provided by the commercial instrument.

Aiming at improved results, a smoothing of the estimated PTT was performed, using a moving average filter with 3 sample points windows producing the results observed in Fig. 2.

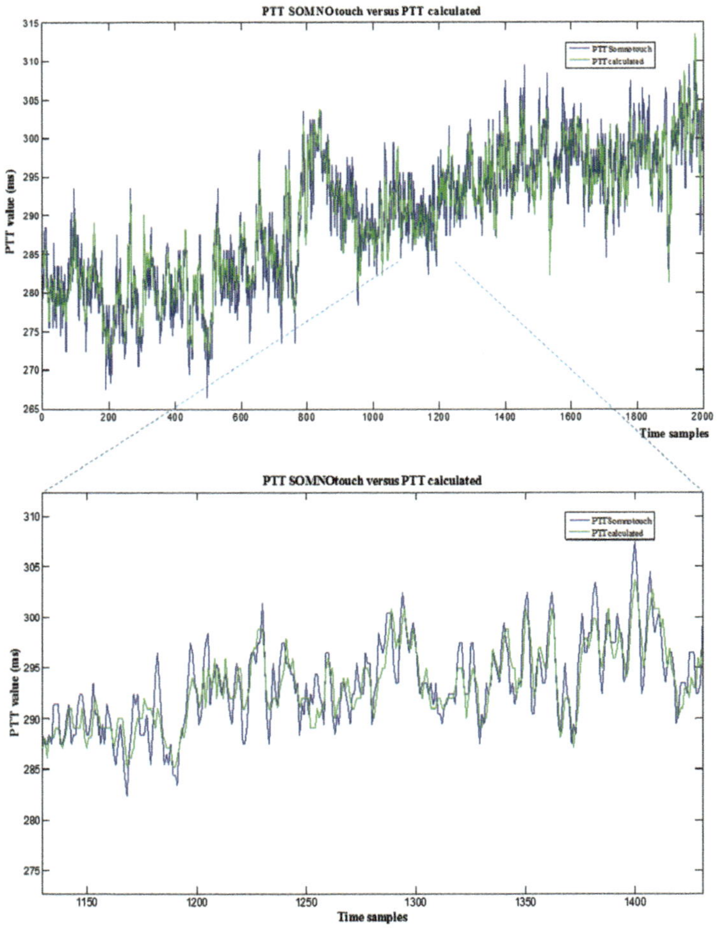

Fig. 2. Calculated PTT (green line) after being filtered by a moving average filter against commercially produced PTT (blue line) on top figure and a zoom of a few time instants below.

5 Concluding Remarks

Present paper exposes a new approach to compute PTT based on ECG and PPG signals. The relevance of accurate estimation of PTT is concerned with precise prognosis of HR, CVD and other illnesses using IoT resources to avoid invasive and intrusive BP assessment aiming at elderly monitoring, major goal of the project framing current research. Commercially available equipment (SomnoTouch ™) was employed at this stage of research and development, to guarantee acquisition of testing data as well as gold standard reference PTT signals which could be used to evaluate the performance of the proposed methodology.

The proposed PTT calculation algorithm presents a correlation factor of 0.90 in comparison with the reference value provided by the FDA approved equipment.

Authors recognize the need for tuning the filtering process of both ECG and PCG signals, since such may avoid having to smooth the estimated PTT curve as mentioned in Sect. 4. Algorithms should be also tested on other data, namely data provided by other equipment and clinically annotated to ensure the performance of the proposed method. In near future the current PPT estimations will be confronted with those provided by data-driven models (work in progress). Nevertheless, assuming current results as preliminary, we consider the proposed method to be a valuable PTT estimator for integration on the BP computation module of the diagnostic system under development.

Acknowledgements. Authors would like to thank Fundação para a Ciência e Tecnologia grants SFRH/BSAB/142998/2018 and SFRH/BSAB/142997/2018, Junta de Comunidades de Castilla La Mancha for supporting this work (FrailCheck project SBPLY/17/180501/000392) and the medical doctors from Guadalajara Hospital that are collaborating in this study.

Conflict of Interest. There is no conflict of interest.

References

1. WHO. https://www.who.int/cardiovascular_diseases/en/. Accessed 12 March 2019
2. Chow, C.K., et al.: Prevalence, awareness, treatment, and control of hypertension in rural and urban communities in high-, middle-, and low-income countries. JAMA **310**, 959–968 (2013)
3. Parati, G., et al.: Assessment and management of blood-pressure variability. Nat. Ver. Cardiol. **10**, 143–155 (2013)
4. Pinheiro, N., et al.: Can PPG be used for HRV analysis? In: Conference of the Proceedings IEEE Engineering in Medicine and Biology Society, pp. 2945–2949, IEEE, Orlando (2016). https://doi.org/10.1109/embc.2016.7591347
5. Schäfer, A., Vagedes, J.: How accurate is pulse rate variability as an estimate of heart rate variability?: a review on studies comparing photoplethysmographic technology with an electrocardiogram. Int. J. Cardiol. **166**(1), 15–29 (2013)
6. Ding, X.-R., et al.: Continuous blood pressure measurement from invasive to unobtrusive. IEEE J. Biomed. Health Inf. **20**, 1455–1465 (2016)
7. Ding, X.-R., et al.: Pulse transit time based continuous cuffless blood pressure estimation: a new extension and a comprehensive evaluation. Sci. Rep. **7**, 11554 (2017). https://doi.org/10.1038/s41598-017-11507-3
8. Ullah, K., et al.: Effective ways to use internet of things in the field of medical and smart health care. In: Proceedings of ICISE. IEEE (2016). https://doi.org/10.1109/intelse.2016.7475151
9. Alberto, M.C. et al.: Sensory system for the sleep disorders detection in the geriatric population. In: Proceedings Experiment@ Conference, Faro, Portugal (2017). https://doi.org/10.1109/expat.2017.7984384
10. Zhou, X., et al.: Validation of new and existing decision rules for estimation of beat-to-beat pulse transit time. BioMed Res. Int. (2015). https://doi.org/10.1155/2015/306934
11. Allen, J.: Photoplethysmography and its application in clinical physiological measurement. Physiol. Meas. **28**(3), R1–R39 (2007)

12. Velzen, M.H., et al.: Increasing accuracy of pulse transit time measurements by automated elimination of distorted photoplethysmography waves. Med. Biol. Eng. Comput. **55**, 1989–2000 (2017). https://doi.org/10.1007/s11517-017-1642-x
13. Carek, A.M., Inan, O.T.: Robust sensing of distal pulse waveforms on a modified weighting scale for ubiquitous pulse transit time measurement. IEEE Trans. Biomed. Circ. Syst. **11**(4), 765–772 (2017)
14. Somnomedicis. https://somnomedics.eu/products/cardiology/24h-bloodpressure-24h-ecg/somnotouch-nibp/. Assessed 12 March 2019
15. Gutiérrez-Rivas, R., et al.: Novel real-time low-complexity QRS complex detector based on adaptive thresholding. IEEE Sens. J. **15**(10), 6036–6043 (2015)
16. Chatterjee, S.: Time domain analysis of heart rate variability – an attempt towards characterization of trends. IJETSR **4**(1), 41–49 (2017)
17. Conde, C.A.: Heart rate variability parameters and the PNNX family as discriminators between athletes and sedentary people. JPES **16**(4), 1316–1325 (2016)

Improved Spectral Method to Obtain Strains of an Ex-Vivo Membrane Tissue and Its Performance Under Elevated SNRs

Ivonne Bazán[1(✉)], Antonio Ramos[2], and Carlos Negreira[3]

[1] Biomedical Eng. Depart., Eng. Sciences Center, Univ. Autonoma Aguascalientes, Aguascalientes, Mexico
ivonne.bazan@edu.uaa.mx

[2] Ultrasonic Systems & Tech. Group. Inst. Phys. & Information Technol, CSIC, Madrid, Spain

[3] Ultrasonics Lab. Physics Inst. Fac. Ciencias, Univer. de la Republica, Montevideo, Uruguay

Abstract. In the recent years, ultrasound has become an inspection tool under investigation to explore laminated biological structures to found mechanical or structural changes related to early stages of diseases. The resolution capability of ultrasound to explore these micro-changes not only depends on the frequency, but also in the processing methods applied to extract information from ultrasonic (US) signals. In this paper an analysis is presented to obtain strain values of the wall of a carotid artery segment (from an ex vivo porcine model) during a cyclic variation of the internal liquid pressure. Two methods were used to estimate the wall strain: the cross-correlation, which was considered as the gold-standard option up to 2018, and the SERHH spectral method which has previously demonstrated to obtain ultra-high resolution (0.1%) results on estimating temperature (2009) and wall thickness (2018) variations into biological tissues. Distinct results for the strain variations during the circulatory pressure cycle were detected with both methods (between 0 and 3.5 $10^{-2} \pm 0.6741$ for the cross correlation and from 0 up to 3.0 $10^{-2} \pm 0.0488$ for our SERHH method). This represents a considerable error of 17% if the cross-correlation options were used Moreover, a precise analysis for evaluating robustness of the SERHH method performance was performed under noised conditions. The original measured signals were added with different noise levels, and then new noise influenced strain values were predicted, for filtered and non-filtered noised signals. Results show that strains obtained for filtered signals almost maintain their original value, presenting small errors for the worst case.

Keywords: Strain · Carotid artery · Ultrasound · Spectral analysis · Noised signals

1 Introduction

In recent years, several tools for cardiovascular diseases diagnosis have been developed based on the characterization of mechanical properties of vessels, as the compliance, distensibility and stiffness [1–10]. Parameters as elastic modulus, or strain-stress

J. Henriques et al. (Eds.): MEDICON 2019, IFMBE Proceedings 76, pp. 1485–1491, 2020.
https://doi.org/10.1007/978-3-030-31635-8_184

relationship are used to determine the vessel state and relate it to a specific disease [2, 5, 7, 9, 10]. Variations in these parameters, detected with an appropriate resolution, could be related to vessel disorders. One common method to evaluate vessel condition is to estimate the wall thickness and its variations during cardiac cycles [8]. In this work, we present the analysis performed to US signals, acquired from an ex-vivo carotid artery (porcine), to obtain the strain of a vessel wall by means of two methods: The classical Cross Correlation Method (CCM) and the Spectral Estimation of Resonant High Harmonics (SERHH). The influence of SNR level on results of strain was evaluated for our SERHH method, which has demonstrated a very high resolutions (of 0.1%) for wall thickness estimation [3].

2 Methodology

2.1 Experimental Setup and Signal Acquisition

Carotid artery segment from pig was used to acquire a set of US signals during 9.8 s. This specimen was immersed in distilled and degassed water at a temperature of 20 °C, and it was connected to a flexible silicone tubing system through which fluid is circulating, as it is shown in Fig. 1. A dynamic regimen of internal pressure (IP) was established inside artery segment, using a device to emulate cardiovascular system [3]. IP varies into a range between 30 and 110 mmHg in an oscillatory cyclic pattern. The US scan was performed, using a LeCoeur Electronique Transceiver to drive a transducer (A326-S SU Panametrics (Olympus)) with a central frequency of 5 MHz. Received signals were digitized with a sampling frequency of 80 MHz and a resolution of 12 bits. A total of 100 US signals were acquired. Pressure through the artery segment was measured with a Koninsberg micro-transducer.

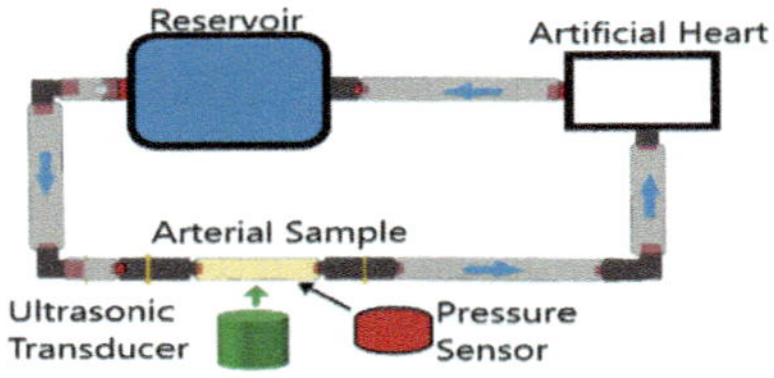

Fig. 1. Scheme of set up used to acquire US signals from arterial sample

2.2 Processing Methodology

In this section, the basic principle of two methods used to process signals and to obtain strain values for the artery wall is exposed.

The strain is defined as the relation of the extension/compression of a body to a reference dimension:

$$\varepsilon = l / £ \tag{1}$$

where l is the extension or compression and £ is the initial (reference) dimension. In order to estimate strain of vessel wall a linear model was considered [5] and the strain was computed as:

$$\varepsilon = [\xi o - \xi(P)] / \xi o \tag{2}$$

where, ξo is the wall thickness at initial pressure and $\xi(P)$ the thickness at pressure P.

The wall thickness values were obtained from acquired US signals, based on the estimation of the period between the two echoes produced by the anterior wall: the echo produced by its outer surface and the one produced by its inner surface (Fig. 2). To obtain the temporal separation between echoes, two methods were applied: the cross correlation and spectral estimation of resonant high harmonics (SERHH). Both, previously (recently) evaluated to estimate Young modulus of arterial walls [3].

Cross-correlation Method. The well-known cross correlation method is based on detecting the time of occurrence of the maximum peak in a correlation sequence, which corresponds to the temporal lag between two echoes [11]. The sequence is computed from the correlation of the two signal segments containing an echo of interest each one. The maximum value indicates when the two echoes are overlapped, and the corresponding temporal lag indicates the period of time between echoes.

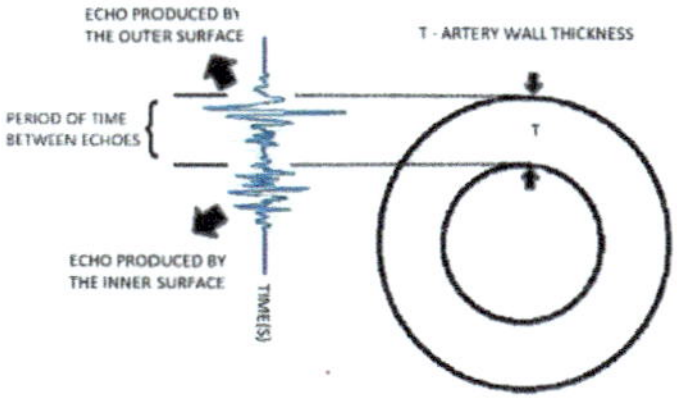

Fig. 2. Scheme representing a transversal section of artery and the relation between artery wall thickness and the US signal from reflection from the outer and inner surfaces of the wall

Once the time between echoes was obtained, the wall thickness is computed as:

$$\xi = c * (\Delta\tau/2) \tag{3}$$

where c is the ultrasound velocity on arterial wall and $\Delta\tau$ is the time between echoes.

Spectral Estimation of Resonant High Harmonics Method. In the SERHH method, the repetition frequency (and its harmonics) related to the time between the pair of wall echoes is used, instead this time, to obtain the wall thickness. This method estimates the frequency components present in the segment of US signal containing both echoes from the anterior wall, based on a high-resolution AR spectral estimation [3, 12].

Harmonics of the frequency related to the temporal separation of echoes are detected and the wall thickness is computed by means of:

$$\xi = n * [c/(\Upsilon_n * 2)] \quad (4)$$

where, Υ_n is the n harmonic of the frequency related to echoes temporal separation. As it is appreciated on Eq. 4, any change on the wall thickness produces an inversely proportional variation in the fundamental frequency (n = 1) and its harmonics.

Once thickness and its variations during one cycle were estimated with cross correlation and SERHH methods, the strain values were computed based on Eq. 2.

Assessment of the Impact of SNR on SERHH Results. To evaluate the influence of the noise level contained in the US signals on the strain estimation by means of the SERHH method, a random white noise was generated and added to the original signals. Noise level was defined so that SNR value varies in a range from 25 to 40 dB in steps of 5 dB. Prior to signal processing, all signals (with and without added noise) were filtered with a bandpass Butterworth filter of order 4 with a central frequency Fc = 8 MHz and bandwidth BW = 10 MHz. Signals with added noise were processed, using the SERHH option, without previous filtering too.

3 Results

Strain values were estimated for the 100 acquired signals using both analyzed methods. In Fig. 2 the estimated strain values for a time interval of 450 ms (that corresponds to one cycle of pressure variation) is shown. Figure 3a depicts the results obtained by means of cross correlation; a maximum strain value of $3.5\ 10^{-2} \pm 0.6741$ just at the end of the cycle is appreciated. Results found with SERHH method are shown in Fig. 3b for the same time interval; in this case a maximum strain value of only $3\ 10^{-2} \pm 0.0488$ on both extremes of the cycle was obtained. On both cases a strain value equal to cero appears near to 1.45 s that matches with the reference value of thickness occurrence.

On Fig. 4a and 4b, strain values obtained from signals with added noise so that SNR = 40 dB for the same interval of 450 ms is illustrated. As it can be appreciated, the strain values of filtered signal are almost identical to the original values shown in Fig. 3b. However, values obtained processing unfiltered signal present a large error compared with the original measured in the vessel; actually, the expected cero value of strain when no thickness variation occurred at 1.45 s was no longer detected, and the maximum strain value detected is $2\ 10^{-2}$.

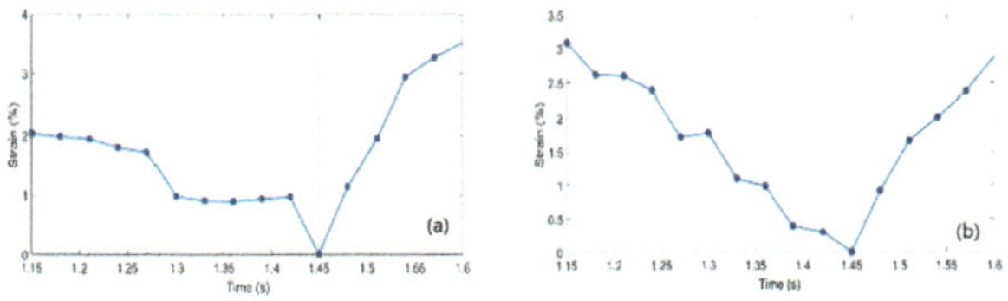

Fig. 3. Strain values of the carotid anterior wall obtained during one cycle of pressure variation with (left) cross correlation and (right) spectral estimation methods.

In the case of noised signals with a SNR = 25 dB (the most severe simulated case), the Fig. 4c and 4d show results obtained for strain values during the same described period of time. As it is observed in Fig. 4c, the strain values obtained from filtered signals still maintain the tendency of the original values (Fig. 2b) although showing some error in the curve minimum that was not present in the previous case (SNR = 40 dB) for filtered data. In Fig. 4d, values corresponding to strains for non-filtered signals are graphed, as in the case of SNR = 40 dB, and there are a great difference between behavior of these values and the filtered ones: the zero-value strain is no longer detected, and the maximum strain value is 2 10^{-2}.

4 Discussion

Strain values of arterial wall estimated by the correlation method during one cycle of pressure variation fall in a range between 0 and 0.035 ± 0.6741. Results obtained with the spectral method confined in a range between 0 and 0.03 ± 0.0488. This represents a notable improvement in accuracy (17%) in relation to the cross-correlation classical method.

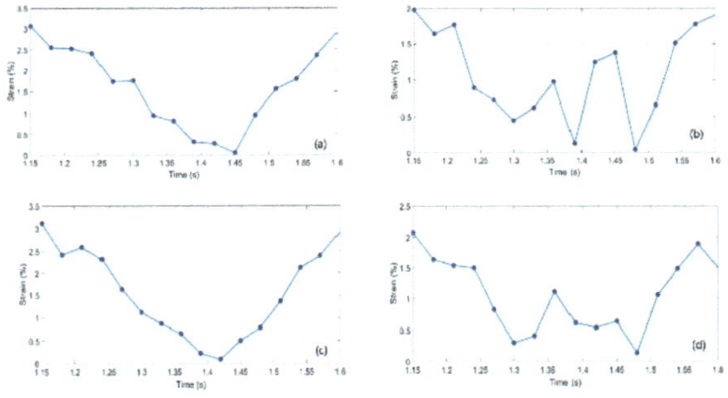

Fig. 4. Strain values of carotid anterior wall. SERHH method applied to signals with SNR = 40 dB, results with filtered (a) and non-filtered signals (b), and with SNR = 25 dB, results obtained with filtered (c) and non-filtered signals (d).

When signals with added noise were processed by our spectral method, without pre-filtering, the strain values present certain errors (a mean error of 0.82%, a maximum error of 1.65% and a correlation coefficient of 0.73 for a SNR = 40 dB) from their expected values (original measured values estimated without added noise) even for a low noise level of SNR = 40 dB (see Table 1). Nevertheless, when signals were filtered before processing, strain values did not present large error from the original ones, including for the relative high-level noise corresponding to a SNR = 25 dB.

Table 1. Maximum and mean errors of strain values regard with original strain curve

SNR	Maximum Error (%)	Mean Error (%)	Maximum Error (%)	Mean Error (%)
Non filtered signals			Filtered signals	
40	1.4879	0.8504	0.2076	0.0710
35	1.4327	0.8066	0.1958	0.0912
30	1.4934	0.7930	0.2267	0.0876
25	1.4785	0.7045	0.6557	0.2020

5 Conclusions

Strain values during a cycle of pressure variation were obtained based on thickness change estimation of a porcine artery wall by means of two methods: (a) with our new method, the SERHH analysis, the strain values vary between 0 and a maximum value of only 3 10^{-2} ± 0.0488, and the zero-value of strain match with the non-thickness variation occurrence; (b) with the classical cross-correlation method, where strain values vary between 0 and a maximum value of 3.5 10^{-2} ± 0.6741, which represents a considerable error of 17%. An analysis to determine the performance of the SERHH spectral method under influence of noise present in the signal was performed adding original signals with different levels of noise and then filtering them. Filtered and non-filtered signals were processed by means of our spectral method and results show that, for the non-filtered signal, light errors (around 1%) in strain were predicted respect to the measured signal case, even for an SNR = 40 dB, nevertheless for the filtered signals no errors occurred in the case of SNR = 40 dB and only a small error was found for the worst case of an SNR = 25 dB.

Acknowledgments. Authors would like to thank at the Spanish National Plan project DPI 2017-90147-R. We also appreciate the additional funding for the development of this work from the Iberoamerican Network CYTED-DITECROD-218RT0545, and of the Universidad Autónoma de Aguascalientes (to fund proyects PII18-4 and PII19-8).

References

1. Brum, J., Bia, D., Benech, N., Balay, G., Armentano, R., Negreira, C.: Set up of a cardiovascular simulator: application to the evaluation of the dynamical behavior of atheroma plaques in human arteries. Physics Procedia **03**, 1095–1101 (2010)
2. Bazan, I., Negreira, C., Ramos, A., Brum, J., Ramirez, A.: A new high-resolution spectral approach to noninvasively evaluate wall deformations in Arteries. Comput. Math. Methods Med. 15 p (2014)
3. Bazan, I., Ramos, A., Balay, G., Negreira, C.: Power spectral estimation of high-harmonics in echoes of wall resonances to improve resolution in non-invasive measurements of wall mechanical properties in rubber tube and ex-vivo artery. Ultrasonics **87**, 133–144 (2018)
4. Fung, Y.C., Liu, S.Q.: Determination of the mechanical properties of the different layers of blood vessels in vivo. In: Proceedings of the National. Academy of Science USA, vol. 92, pp. 2169–2173, March 1995. Physiology

5. Brum, J., Balay, G., Bia, D., Benech, N., Ramos, A., Armentano, R., Negreira, C.: Improvement of Young modulus estimation by ultrasound using static pressure steps. Physics Procedia **3**, 1087–1094 (2010)
6. Fung, Y.C.: Mechanical properties of blood vessels. In: Biomechanics. Springer, New York (1981)
7. Mokhtari, D.M., Nikanjam, N., Saberi, H.: Application of static pressure change in estimation of elastic parameters of rabbit's artery by doppler ultrasound. Iran. J. Radiol. **2** (3,4), 141–146 (2005)
8. Ramos, A., Bazán, I., Negreira, C., et al.: Estimation of PSD shifts for high-resolution metrology of thickness micro-changes with possible applications in vessel walls and biological membrane characterization. Sensors **12**, 15394–15423 (2012)
9. Larsson, M., Verbrugghe, P., Smoljkić, M., Verhoeven, J., Heyde, B., Famaey, N., Herijgers, P., D'hooge, J.: Strain assessment in the carotid artery wall using ultrasound speckle tracking validation in a sheep model. Phys. Med. Biol. **60**, 1107–1123 (2015)
10. Larsson, M., Heyde, B., Kremer, F., Brodin, L., D'hooge, J.: Ultrasound speckle tracking for radial, longitudinal and circumferential strain estimation of the carotid artery – an in vitro validation via sonomicrometry using clinical and high-frequency ultrasound. Ultrasonics **56**, 399–408 (2015)
11. Proakis, J.G., Manolakis, D.G.: Digital Signal Processing, 3rd edn. Prentice Hall, Upper Saddle River (1998)
12. Marple Jr., S.L. (ed.): Digital Spectral Analysis with Applications, Prentice Hall, Upper Saddle River (1987)

Instrumental Proposal to Determine the State of Health of the Patients with Diabetic Foot

Ilse Anahi Torres[1(✉)], Lorenzo Leija[1], Arturo Vera[1], Josefina Gutiérrez[2], and Antonio Ramos[3]

[1] Electric Engineering Department, Bioelectronics Section, CINVESTAV-IPN, Mexico City, Mexico
iatorres@cinvestav.mx
[2] Institute National of Rehabilitation, Luis G. Ibarra Ibarra (INR), Mexico City, Mexico
[3] Group Ultrasonic Systems and Technologies, Institute ITEFI, CSIC, Madrid, Spain

Abstract. This work describes an instrument to make measurements of electrical impedance over the skin of the foot in Diabetes Mellitus patients, in order to provide an early detection of diabetic foot, a very severe complication of that disease. These measurements must be made in the area of the foot most prone to suffer ulceration in diabetic people. This electrical instrument is connected to a concentrator and distributor of multiple diagnostic information, whose purpose is to obtain and store the data from several measuring sub-systems related to a number of related physical parameters. This computerized global system permits the gradual expansion of the distinct foot parameters to be analyzed. Preliminary results indicate that there are significant differences in the measurements of electrical impedance in the very-low frequency frame, when diabetic vs. non-diabetic curves are compared.

Keywords: Diabetic foot · Electrical impedance · Early diagnosis

1 Introduction

Some complex interplay of different factors induced by, mainly, the influence of a sustained hyperglycemia, creates certain changes in the physical properties of the diabetic foot. Currently there are not systems for early detection of diabetic foot. In the 20th century, some passive electrical properties in biological tissues were studied. These tissues are conductive and theirs impedances change with frequency [1]. The electric impedance of tissues depends on the cell type and the tissue density. For making this measurement, a strip of electrodes that cover the entire perimeter of the patient's chest and then the voltage is measured.

This article presents a new non-invasive electronic instrument, which materializes one of the distinct measuring techniques of an integrated multi-tool computational system, specifically developed in order to obtain some physical and physiological foot characteristics, to facilitate a pre-diagnosis helping the specialized doctor to determinate the foot conditions and, in consequence, propose a treatment for avoiding the

J. Henriques et al. (Eds.): MEDICON 2019, IFMBE Proceedings 76, pp. 1492–1498, 2020.
https://doi.org/10.1007/978-3-030-31635-8_185

occurrence of a situation of diabetic foot. The system has a portable design, which includes electronic and computational parts. It is non-invasive, and contains independent modules.

The particular instrumental module, described here, was developed in order to measure electrical impedances over the foot skin of diabetic or pre-diabetic patients, for detecting and evaluating symptoms previous to the developing of a diabetic foot.

2 Instrumental Methodology

In this work, an early diagnosis proposal of diabetic foot, based on the anticipated measurement of changes on the impedance behavior of the foot skin, is proposed. More specifically, the response in frequency of this parameter is analyzed, by using electrical spectroscopy. A specific instrumental testing is applied, with that purpose.

2.1 Electrical Impedance Spectroscopy

In general terms, the electrical impedance represents the opposition that presents a determined material to the current passing through it. The characteristic electrical impedance (Z) between two points, in a biological tissue, like as it happens with any other material, is a complex number (module and phase) defined as the ratio between the measured voltage (V) across these points and the total current flowing (I) through them [2]. This parameter can to infer information about a biological tissue, when this presents a change in dimension, their dielectric properties or its conductivity. The electrical behavior of the biological tissues has a dependence of their dielectric parameters with the frequency of the applied current flowing through them. When this frequency increases, the conductivity of the most tissues increases from a low value when a direct current is considered (which depends on the extracellular volume) up to a constant level in the frequency range from 10 to 100 MHz.

2.2 Three-Electrodes Measuring Method (3EM)

This method involves the application of a current (Io) through two electrodes: one of them injects the currency (Ze_1) and the other is a reference electrode (Ze_2). This current flows through the tissue. Then, a voltage (Vo) is detected across a third electrode (Ze_3) and the electrode (Ze_2), in the Fig. 1. The difference in voltage (Vo) corresponds to the fall caused by the current applied to the impedance of interest (Zx), respect to the second electrode (Ze_2) [3]. This "three electrodes" method (3EM) being responsive to the impedance across the two electrodes ($Ze_{1\text{-}2}$), i.e., the impedance of the skin under these two electrodes, has been used to measure the hydration of the skin, by means of the electrodes characterization (Riu), in the calibration of electrical impedance systems [3].

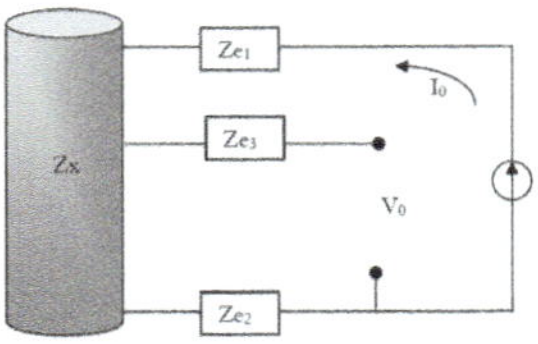

Fig. 1. Configuration of the three electrodes method for impedance measuring.

The overall system for measuring, storing and displaying the electrical impedance values contains a LCD display to show the graphical user interface, four buttons for system control (reset, enter, 1 ↑ and 2 ↓), ON/OFF button, LED indicators, RS232 output, input electrodes, and two rechargeable batteries, Fig. 2a. The device is connected to a concentrator and distributor of information, whose purpose is to obtain and store the data of several sub-systems for measuring a number of physical parameters, which permits to gradually expand the variables to be analyzed (See Fig. 2b).

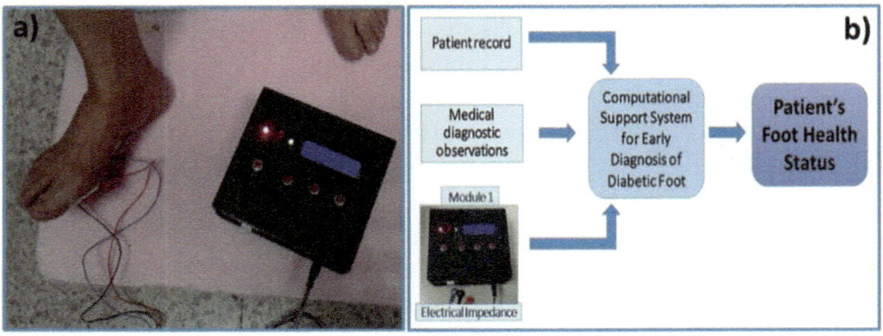

Fig. 2. (a) Measuring technique of complex electrical impedances over the foot skin, (b) Diagram of the hub of measuring information to facilitate an early diagnosis of diabetic foot.

The measurement of electrical impedance is performed in the area with the highest risk of ulceration in people with diabetes, which is, in the first metatarsal [4]. The technique to perform the measurement is the technique of three electrodes, which are distributed in the form of a triangle (see Fig. 3a). The red electrode is in charge of the current injection to the tissue. The black electrode is the reference electrode and the blue electrode registers the generated potential to know the electrical impedance of the tissue.

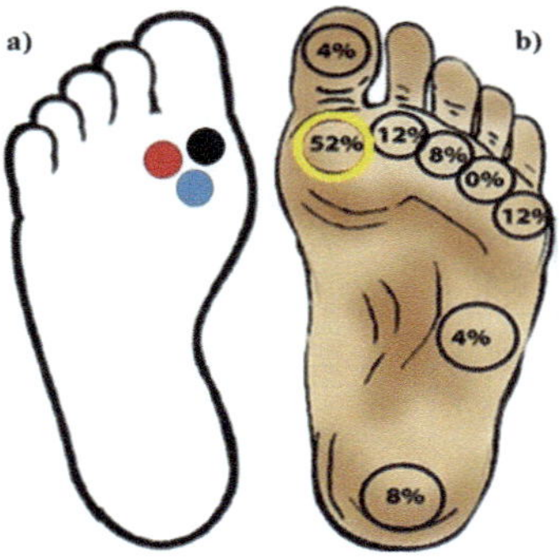

Fig. 3. (a) Electrode placement technique for electrical impedance measurement. (b) Area to be measured with the electric impedance measurement system. This area was chosen (first metatarsal) since it is the area with the highest risk of ulceration in the diabetic foot [4].

2.3 Data Acquisition in Patients

A database was created with diabetic patients from the National Institute of Rehabilitation (INR, in México City) in order to obtain indicators about a possible developing of a future diabetic foot complication. During the system testing, we had the support of a specialist doctor in diabetic foot to give a clinical diagnosis for each patient. It provides support in the realization of the electrical impedance measurements over the sole of the foot with which they worked. To perform the measurement and surveys in each one of the patients, they were first informed about the purpose of this project. They signed a letter of consent, where they are informed about the measurements to be taken. And they facilitated some personal information, in order to keep a file of each patient. All patients were part of this project voluntarily.

3 Obtained Results and Their Discussion

Measurements were made in 25 patients with diabetes and 9 healthy cases in an age range from 25 to 85 years. All patients met the inclusion criteria. The data stored in the monitoring system were transmitted to a computer using the Matlab R2014a software.

3.1 Electrical Impedance System

A comparison of the averages of the measurements, in both feet of all diabetic patients, against the healthy cases, was performed. In this comparison, significant differences were found ($P < 0{,}001$, two-way ANOVA) between two signs that may point to a range of frequencies useful for the differentiation of both groups. This comparison is shown in Fig. 4. The initial frequencies were removed to appreciate the separation between both sets of data.

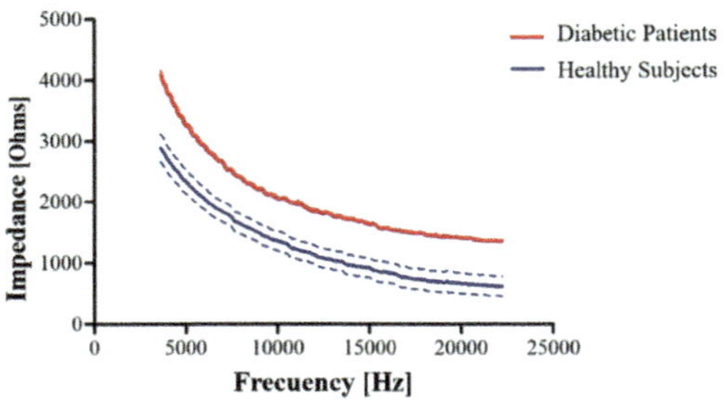

Fig. 4. Comparison of electrical impedance, between diabetic patients and healthy cases. The differences between groups are statistically significant in frequency range shown ($P < 0.001$). Dotted lines (shown only in healthy subjects) indicate the standard error of the mean (SEM).

Comparison of the relationship, between the measured voltages and the injected voltage are shown in Fig. 5, with a statistical significance level of $P < 0.0001$ for most of frequencies (some frequencies of $P < 0.001$, using two-way ANOVA).

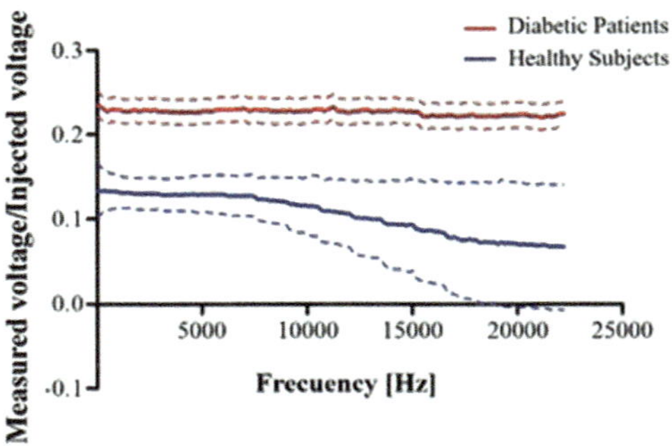

Fig. 5. Comparison of the relationship, between the measured voltages and the injected voltage, in diabetic patients and in healthy cases. The dotted lines indicate the standard deviation (SD). The SEM is not shown as it appears very close to the average values. The results are statistically significant ($P < 0.0001$).

3.2 Some Aspects About the Data Acquisition in Patients

In the data acquisition over diabetic patients, the 78% of them are above normal corporal weight, even though the 72% have a healthy diet and the 68% exercise to take care of their weight and health. In spite of they are patients who mostly care for diabetes mellitus, however they suffer from overweight and obesity. 94% of patients are taking some medication to control diabetes. All patients with diabetes have macules of some kind, in legs and feet, where 64% of these macules are due to vascular problems, this being an indicator of alert to prevent circulatory problems. In addition to the typically acquired data, we can focus the research in order to obtain other parameter, through the circulatory system, helping to achieve an early detection of diabetic foot.

It is necessary to perform a longitudinal study of the measured variables (>10 years at least). It is intended to monitor each patient for several years to know how he is changing the measured parameters and so to determine a trend in the changes being observed that would be linked to the deterioration, or not, of the lower limbs.

4 Conclusions

From the partial results obtained for the electrical impedance measurements, in the feet sole of a number of cases, it can be concluded that these foot impedances in diabetic patients and healthy people, appear as quite different. A more deep analysis of measured values in this electrical parameter will allow us to deduce hypotheses around some change suffered in the structure of the foot tissues once we dispose of more variety of measurement and of more specific clinical analyses.

With the data that we have, nowadays, about the complex electrical impedances in the feet of diabetics and healthy persons, still it cannot be concluded if some kind of structural damage is present in the feet sole of patients with a first-level diabetes, but we can say that they are significant differences among the feet impedances of both patient groups, in the particular frequencies range of 5 to 22 kHz.

The impedance comparisons between left and right foot of the same group (diabetic or healthy), were not significantly different. However, a larger sample is required to make these conclusions more generalizable.

Up to this moment, impedance measurements were made in 25 diabetic patients, of which 60% belong to the female gender, with which, it seems that, in a first approximation, the risk of suffering diabetic foot is greater in this gender.

Of the 100% of patients, 77% are above normal weight, 58% of these are above the optimal levels of HbA1c and 72% have high triglycerides. Maculae in legs and feet are present in all the patients with diabetes; and 64% of the macules recorded are due to vascular problems, this being an indicator of warning to prevent serious circulation problems.

Acknowledgments. Special thanks to MSc. José Hugo Zepeda Peralta and MSc. Rubén Pérez Valladares for their collaboration in this work. Authors would like to thank the National Council for Science and Technology (CONACYT, Mexico) for the support received with project CONACYT-Salud 2013-I-201590, 201256; and to the Bilateral Cooperation Project CSIC-COOPB20166. We appreciate very much the funding, for the development of this work, of the European doctoral project EraNet-EMHE 200022, and of the Iberoamerican Network CYTED-DITECROD-218RT0545.

References

1. Rigaud, B., Morucci, J.P., Chauveau, N.: Bioelectrical impedance techniques in medicine. Part I: Bioimpedance measurement. Second section: impedance spectrometry. Crit. Rev. Biomed. Eng. **24**(4–6), 257–351 (1996)

2. Muñoz, Y.S.: Caracterización de tejidos cardíacos mediante métodos mínimamente invasivos y no invasivos basados en espectroscopía de impedancia eléctrica. Universitat Politècnica de Catalunya (2004)
3. Martinsen, O.G., Grimnes, S., Karlsen, J.: Electrical methods for skin moisture assessment. Skin Pharmacol. **8**(5), 237–245 (1995)
4. Comité de Educación Pública de la American College of Foot and Ankle and Surgeons, "Neuropatía Diabética Periférica", pp. 1–2 (2007)

A CYTED Network: New Non-invasive Ways for an Early Diagnosis of Chronic and Degenerative Diseases: Diabetes and Cardiovascular

Antonio Ramos[1](✉), Lorenzo Leija[2], Carlos Negreira[3], Eduardo Moreno[4], M. G. Ruano[5], Wagner Coelho[6], Ivonne Bazán[7], Fernando Merchan[8], César Yegros[9], and Juan Prohias[10]

[1] ITEFI-CSIC, Madrid, Spain
aramos@ia.cetef.csic.es
[2] CINVESTAV, Mexico City, Mexico
[3] Univ. URU, Montevideo, Uruguay
[4] ICIMAF, Havana, Cuba
[5] Univ. Algarve, Faro, Portugal
[6] Univ. UFRJ-COPPE, Rio de Janeiro, Brazil
[7] Univ. UAAC, Aguascalientes, Mexico
[8] Univ. Techn., Panama City, Panama
[9] Univ. Nac. & Tec., San Lorenzo, Paraguay
[10] Hosp. CardioCentro, Santiago de Cuba, Cuba

Abstract. In this work, an analysis of the principal characteristics, scientific aims and some initial results, of the Iberoamerican R&D network "Ditecrod" (Project of CYTED) is made. This network is based on an international agreement to propitiate an efficient trans-disciplinary cooperation in a scientific area where currently significant research efforts are being made for their high potential impact on future health and quality of life: The non-invasive Early medical diagnosis of Cardiovascular and Diabetic Foot diseases. Both chronic & degenerative pathologies are endemic today in many American countries. The main project objectives are the research, application and diffusion of new non-invasive diagnostic methods, having a low technologic cost and an easy portability, in relation to nowadays disposable approaches. The main results (in the first year of operation) of this multinational network are described.

This project also seeks to promote the cooperation working among 11 working teams integrated in six R&D groups with large experience and previous innovations on the subject, and 2 emerging university groups until now focused mainly on the university teaching; in total 13 teams of 8 countries: 5 of Bio-medical Engineering, 4 of Biophysics, 2 hospitals and 2 companies.

Keywords: R&D network · Early Diagnosis · Noninvasive · Diabetes · Cardiovascular pathology

J. Henriques et al. (Eds.): MEDICON 2019, IFMBE Proceedings 76, pp. 1499–1505, 2020.
https://doi.org/10.1007/978-3-030-31635-8_186

1 Introduction

CYTED is an intergovernmental organization on Science and Technology for promoting the development in Ibero-America. It was created in 1984 through an inter-institutional framework agreement signed by the countries that make up the Ibero-American Community of Nations. It was included in the Cooperation Programs dependent on the Summits of Ibero-American Heads of State and Government in its "V Summit" held in San Carlos de Bariloche, October 16–17, 1995.

Its purpose is to contribute to the harmonious and sustainable development of Ibero-America through cooperation in science, technology and innovation.

– Thematic areas of CYTED:

♣ Energy ♣ Agri-food ♣ Industrial development ♣ Health ♣ Information and communication technologies ♣ Environment and sustainable use of resources ♣ Science and society.

– Member countries of CYTED in Ibero-America. They are twenty two:

2 Ditecrod Cyted Network (2018–2021): New Non-invasive Pathways of Early Diagnosis in Chronic and Degenerative Diseases

This R&D network is based in an international agreement of 8 countries to make a (multi & trans) - disciplinary cooperation about the scientific subject "Low-cost, non-invasive early medical diagnosis of two endemic pathologies: Cardiovascular pathology and Diabetic Foot". There are 11 teams integrated in six national groups which are innovators on these subjects, and 2 emerging groups: 5 of Bio-medical Engineering, 4 of Biophysics, 2 hospitals and 2 biomedical companies.

2.1 Participating Groups

Among the participants, there are several experienced groups, one in Bio-Ultrasonics of ITEFI-CSIC (Spain) and five of Ibero-America with a clear bio-medical orientation, which have joint experience since 1998 in scientific projects of biomedical research

[CYTED, CE-ALFA, CONACYT, CICYT, CSIC, PEDECIBA, CONICYT, ANII, CAPES, CNPq] which are the basis of the network subject: 25 European-American Cooperation Projects, and joint training of 14 doctors and several masters on: Ultrasonic (US) Noinvasive (MicroFlowmetry, Micrometry, and Thermometry); US Vascular Elastography; Tissue characterization; Acoustic temporal inversion; US- tissue interaction; Electronic Collimation of US beams & X-waves; and US therapy in hemostasis & HIFU; all of them with hospital collaborations.

There are an intense connections among the consolidated groups, CSIC, CINVESTAV, LAU-URU, COPPE, ICIMAF and Univ. Faro In addition to the 8 responsible for R&D in the countries involved, other 27 doctors + 3 engineers participate (Total: 38). Two specialized medical centers participate in this project: [Inst. Nac. Rehabilitation of Mexico (Diabetes)]; and [CardioCentro of Havana (Cardiac bypasses)]. And two companies (Pusonics and ICYB S.C) could to use some results (of technological and training type).

2.2 Objectives of This Network

Generals

(a) Generation, integration and dissemination of training tools, by the consolidate-groups, on their research results in Engineering/BioMedical Physics & Applications with high incidence in quality of life (early diagnosis of cancer and arterial diseases).
(b) Adaptation and complementation of previous results in R & D and hospital trials, of the experienced groups, for transfers to other groups, medical centers and biomedical companies of the region, on: New Non-invasive/Low Cost/High Resolution Methods for Early Diagnosis of the Cardiovascular and Diabetic Foot pathologies.
(c) It will also be possible for the emerging groups to merge spin-off companies to achieve their own technology transfer in developing countries at the medium term.

Specifics

(I) Create tools for training in using these new engineering methodologies aimed at preventive diagnoses, valid for groups and companies with different levels of development (Seminars, Courses and Conferences). Encourage the creation of postgraduate courses as seeds for new R & D groups and spin-offs for early non-invasive diagnosis.
(II) Adapt and validate new early markers generated in the groups (spatial, fluxometric, thermal), and define ultrasonic technologies for non-invasive detection and mea- suring. The aim is to enable future early diagnoses, alternative to conventional ones (CT, PET, High-Resolution Ultrasound, MRI, analytical), and which provide the use of preventive therapies in the precursory stages of the 2 pathologies, thus saving costs.
(III) Implement cutting-edge scientific-technological training applicable to R & D groups, hospitals and companies, and their subsequent dissemination in Latin

America (publications), on these methods for early diagnosis of said diseases, using innovative R&D solutions and new ultrasonic non-invasive techniques.

There is previous experience (20 years) in cooperation among the groups; e.g., in Consortium Medical Project SUCODIC - Cyted, and in pre/post-doctoral stays of 7 of the researchers participants in Cinvestav and CSIC. US and Computational techniques will be integrated to: (a) estimate wall thickness and elasticity for early diagnosis of arteriosclerosis; (b) non-invasive quantification of tissue thermal gradients, for early diagnosis of liver cancer; (c) methods to detect diabetic foot in preliminary stages.

3 Achievements Obtained by Network "Ditecrod" in Its First Year

The Network began its work in the first quarter of 2018, and its results reached until December 2018, can be summarized in these points:

(I) Benefits and impact on the target countries (2018):

- Training activities in Biomedical Engineering and Diagnosis, aimed at local entities in developing countries, developed in Panama, designing specific didactic media. End users: Universidad Tecnol. Panama, and some hospital and company on Biomedical Engineering; during the 2nd network Meeting, C. Panama (19-24) - XI - 2018.
- Scientific Seminars, in Mexico, April 2018, by consolidated groups of the Network, in order to integrate and complement their ultrasonic detection techniques towards non-invasive diagnostics. End users: the 8 groups of the Network, and Mexican centers: Cinvestav, UAA and INR hospital. There were 2 seminars in 2018:

• Seminar **A**: "Latest research to achieve an Early Non-invasive Diagnosis of the mechanical state in Arterial Walls".
    A1 - Carlos Negreira (Uruguay): US arterial elastography in the time domain.
    A2 - Ivonne Bazán (Mexico-UAA): Cross-correlation and Spectral Analysis of HR in US medical diagnosis.
    A3 - Antonio Ramos. US micrometric estimate of thickness and hardness with HR

• Seminar **B**: "US measurement of blood micro-flows through Doppler & TTFM. Simulation of blood flow and interaction with Ultrasound".
    B1 - Eduardo Moreno, Cuba. US measurement & flow simulation. Doppler method
    B2 - Antonio Ramos: US TTFM system to measure flow. Narrow and wide bands.

(II) Impact on training. Contribution in forming human resources (2018)

During the 9 working months of the first year (from Constitutive Meeting in April):

- 8 Exhibitions of participating Laboratories of the Network to present the projects in which they are working in their institutions - 1st meeting of the network (México).
- 1 Postgraduate Training Course on selected Themes of the Network, taught at Cinvestav (México), by professors from Mexico and Portugal, during the 1st Meeting.

Aimed at attendees of Emerging Groups, of the Medical Center "National Institute of Rehabilitation" of Mexico, and was open to doctors and local engineers of Cinvestav, and Department of Biomedical Engineering in UAA University.

The Course included 4 specialized sessions on subjects of Postgraduates:

- Ilse A. Torres, A. Ramos, L. Leija, A. Vera (México, Spain). Early detection of diabetic foot complications-
- R. Bayareh, J. Gutiérrez, L. Leija, A. Vera (México). Development of a thermographic image system for the study of the diabetic foot.
- H. Maldonado, A. Vera L. Leija (México). Diagnostic system of the state of health of the diabetic foot.
- Antonio Ruano (Portugal). Signal analysis; biomedical cardiovascular applications.

(III) Dissemination: Conferences and Seminaries in R&D Centers/Hospitals (2018)

- A divulgative Conference of the CYTED network "DITECROD" was given in an IEEE Congress (2018 GMEPE, Porto).
- Three Seminars and one course in Mexico & Panamá. Tools for future Training are being now defined for different levels: Postgraduate studies and Master's in developing countries, and Improvement of Doctoral programs in other partner countries.

(IV) Some research results and scientific publications of the network (2018)

Several important contributions to non-invasive diagnosis [1–7] have been made, by means of network group cooperation, for instance in relation to:

- Impulsive US *Elastography* with Very-high Resolution (0.1%) in artery walls, improving in two magnitude orders the "state of the art".
- Non-invasive US *Micrometry* of thickness in tissues with very HR (0.6 μm).
- US *microflowmetry* by broadband TTFM with HR (20 μL/s) in coronary vessels.
- Non-invasive tissue *Thermometry* of HR (0.08 °C) into the humane body.
- Non-invasive US & I-R Image and Detection of anomalies in the *Diabetic foot*.
- During 2018, 6 collaborating papers were published & indexed in Databases (IEEE-Xplore) and in Impact Journals (Ultrasonics-Elsevier).

- Moreover, as a didactic medium for cutting-edge training, a CD-Rom with ISBN was published, about: *Early Non-invasive Diagnosis of Degenerative and Chronic Diseases using: US Elastography, Micrometry in tissues, Vascular micro-flowmetry, and Thermal Detection.* Many of the 25 works here included are focused on detecting vascular or diabetic problems in previous precursor stages.
- In addition, along 2018, a book was programmed on the network subject, to be written and published in the future by a prestigious publisher, for global diffusion.

4 Responsible Doctors in the 8 Countries and Network Contact

- Spain: Antonio Ramos (ITEFI - CSIC). International Coordinator of the network.
- Mexico: Lorenzo Leija (CINVESTAV), Ivonne Bazán (Univ. UAA)
- Uruguay: Carlos A. Negreira (Univ. URU)
- Cuba: Eduardo Moreno (ICIMAF). Juan Prohías (EcoCentro-HHA)
- Portugal: Ma. Graca Ruano (Univ. do Faro).
- Brazil: Wagner Coelho (Univ. UFRJ, COPPE)
- Panamá: Fernando Merchán (Univ. Tecnológica de Panamá)
- Paraguay: César Yegros (Univ. Nacional & Univ. Politécnica de Asunción)
- Contact information: Antonio Ramos - International Coordinator of DITECROD Head of the R & D Group "Ultrasonic Systems & Technologies". Institute of Physical and Information Technologies. (CSIC), Serrano 144, 28006. Madrid. Spain.

  Tlfn: 34-91-5618806, Ext 920214. e-mail: aramos@csic.es
  https://www.researchgate.net/profile/A_Ramos
  https://www.linkedin.com/in/antonio-ramos-fern%C3%A1ndez-b629b76a/

Acknowledgments. We appreciate very much the funding for the development of this work from: Projects: CONACYT-Salud 2013-I-201590; R&D Spanish National Plan Retos-DPI2017-90147-R; European doctoral project EraNet-EMHE 200022, Bilateral Cooperation Project CSIC-COOPB20166; and the Iberoamerican Network CYTED-DITECROD-218RT0545. And also, to all the different Group and Centers.

References

(Scientific publications of the Network in 2018)

1. Bazán, I., Ramos, A., Balay, G., Negreira, C.: Power spectral estimation of high-harmonics in echoes of wall resonances to improve resolution in non-invasive measurements of wall mechanical properties in rubber tube and ex-vivo artery. Ultrasonics **87**, 133–144 (2018)
2. Torres, I.A., Leija, L., Vera, A., Maldonado, H., Bayareh, R., Gutiérrez, J., Ramos, A.: Computational support system for early diagnosis of diabetic foot. In: 2018 Global Medical Engineering Physics Exchanges (GMEPE), pp. 66–70. IEEE Catalog number 2018 & CFP1818G-CDR. ISBN: 978-1-5386-5473-6

3. Gutiérrez, M.I., Vera, A., Ramos, A., Gutiérrez, J., Leija, L.: Measuring curvature and radiation distribution of spherically focused ultrasound transducers using the electric continuity and the emitted pressure. In: Global Medical Engineering Physics Exchanges (GMEPE), pp. 164–167. IEEE Catalog number 2018 CFP1818G-CDR. ISBN: 978-1-5386-5473-6
4. Ramos, A., Bazán, I., Ruiz, A., Negreira, C.: Measurement in vascular tissues by a new H-R spectral estimation of transient resonant high over-tones in noisy echo-signals. In: 2018 Global Medical Engineering Physics Exchanges (GMEPE), pp. 168–172. IEEE Catalog number 2018 CFP1818G-CDR
5. Torres, I.A., Leija, L., Ramos, A., Vera, A., Gutiérrez, J.: Instrument to measure temperature and electrical impedance in the foot sole to assess the health of the diabetic foot. In: I2MTC 2018 International Instrumentation & Measurement technology Conference, 14–17 May, Houston, Texas, USA, pp. 1201–1206. 978-1-5386-2222-3/18/$31.00 ©2018 IEEE
6. Design of Acoustic Impedance measuring equipment. Presentation of Mst. Ing. Ilse Torres, about advances on her doctoral thesis (co-directed at ITEFI-CSIC and Cinvestav), in a scientific seminar of the Section Biolectronica of Cinvestav
7. Annual Report of "Ditecrod" Network (Ref-218RT0545) to Cyted Evaluating Committee of the Thematic Area 2 - Health: Conference of A. Ramos, in the Plenary Meeting CYTED, Brasilia, 23 September 2018

Computational Strategy for the Generation of the Clinical Histories of Patients with Diabetic Foot

Ilse Anahi Torres[1(✉)], Lorenzo Leija[1], Arturo Vera[1], Josefina Gutiérrez[2], and Antonio Ramos[3]

[1] Electric Engineering Department, Bioelectronics Section, CINVESTAV-IPN, Mexico City, Mexico
ilse_torresa@hotmail.com

[2] Institute National of Rehabilitation Luis Guillermo Ibarra (INR), Mexico City, Mexico

[3] Group Ultrasonic Systems and Technology, Institute of Physical and Information Technology, CSIC, Madrid, Spain
aramos@ia.cetef.csic.es

Abstract. In patients, suffering diabetes mellitus by long time, the pathology of the diabetic foot is a consequence of that serious disease, appearing in many cases. This problem can be prevented through the evaluation and tracking of certain physical characteristics into tissues composing the patient feet, in such a way that, specific therapies could be applied. For instance, the blood flow in some foot zones is correlated with the tissues health status. Moreover, it would be possible to follow the effectiveness of some applicable preventive therapies, through a protocol based on regular clinical measurements. Here we explain the strategy used to generate a computer system, designed to dynamically create and maintain the medical records of patients with diabetic foot. The system can be used through a Web site, specialized in diabetic foot, including clinical and researching institutions having the possibility of sharing a complete information about many patients with diabetes: (i) data logging about patient medical histories, (ii) medical diagnostic observations, (iii) distinct therapeutic prescriptions, (iv) particular notes of the treating medical specialist, (v) and to follow-up the evolution of ulcers or predictive changes in the areas most prone to ulcerations. This powerful computational tool was designed for an easy handling, and an efficient inter-communication of a main server with a number of user computers connected through a medical network. Some results for "diabetic foot" tracking are shown. From its use by medical specialists and their successive observations, adjustments in the initial programming and possible extensions of this clinical tool are expected in the future.

Keywords: Clinical histories generator · Diabetic foot · Early diagnostic & therapy · Blood flow

J. Henriques et al. (Eds.): MEDICON 2019, IFMBE Proceedings 76, pp. 1506–1512, 2020.
https://doi.org/10.1007/978-3-030-31635-8_187

1 Introduction

The diabetic foot is one of the most severe and degenerative complications of the diabetes mellitus. In fact, the diabetic patients are exposed to an amputation risk that can be 25 times higher than in people without diabetes [1]. The diabetic-foot disease includes ulceration, infection and/or gangrene, neuropathy, and peripheral arterial disease, due to interactions of different metabolic factors. But, by making a pre-programmed preventive control, an important part of the nowadays amputations in feet could be avoided. Even when a person suffers a first amputation, the remaining leg, and its life, can be preserved by means of a good care and being follow-up by a multi-disciplinary team specialized in theses aspects.

All those above commented complications can be prevented from a wide disease knowledge, which would allow punctual early actuations in a timely and decisive way during the management and very especially during the pre-pathogenic period.

Among the systems proposed for detecting physiological changes in diabetic foot, a temperature monitoring system was developed in our laboratory for measuring the temperature of the feet areas more prone to suffer ulceration: Hallux finger, first metatarsal, and arch & heel of the foot [2]. And an impedances measurement system, over the area with greater risk for ulceration (first metatarsal) was also developed [3]. And, in order to make efficient data storage of these thermal results united to all the other measurement results, now in developing under our R&D projects, for the early detection of diabetic foot, an information concentrator was developed in our lab, seeking to help for determining the health status of patients with diabetic foot. In addition, this concentrator performs a statistical study to determine the physical characteristics of the patient foot that could be indicators of possible future damages like ulcerations. All these integrated tools are being strategically created for the finality of the prevention of the onset of each diabetic foot case or further typical complications appearing when the initial diabetic indications are already presented. The idea is helping to the specialized medical doctors to make a pre-diagnosis and decide an exclusive medical treatment, particularized to each patient [4].

In this terrain, it was necessary to integrate strategies of prevention and of early diagnosis at the primary level of health care, with participation of an expert in angiology. We are working now on a global virtual platform for the specific characterization of feet tissues, which has an internet connection to other hospital users, so allowing the construction of tools for the integrated creation de patient clinical histories, complemented with an interactive control network (by internet) to improve the current hospital levels in early detection and/or monitoring of the diabetic foot pathologies.

2 Characteristics of the Distributed Computational System

2.1 Architecture of the Developed Information System

The computational system designed for the generation, maintaining and modification of clinical histories, related to patients with diabetic foot pathology, has a basic structure composed by a distributed global computer system, with a central information unit to control and centralize, in real time, the measurements devices in patient feet (temperatures, impedances, etc.), also developed in the Cinvestav and installed in the INR hospital. This structure has also the objective of acquiring and storing all the data, results and individualized medical information about the early diagnosis of each patient, the prescribed therapy, the changing measurements, and the periodical evaluation by the medical specialists. The central unit and the remote instruments are complemented by the associated Web platform, a network system created for bidirectional communications & control and the efficient clinical use in the hospital departments.

In addition to the central unit, a specific network has been designed for concentration of all the measuring information, which is based on the integration of *N* specialized instrumental modules with complementary strategies carefully intended to characterize the tissues state related to the current situation and distinct evolutions in each patient with diabetic foot. Each module provides precise indicators to contribute to the prevention and periodic control of this particular diabetic-mellitus complication. This is the result of an added R&D step to previous lab designs [4], seeking to ensure the availability of these new technologies in hospitals, intended for providing an appropriate more global diagnosis of diabetes mellitus and its complications, including a low-cost and non-invasive multi-parametric tissues evaluation.

The implemented hub of information helps to determine the health status of a diabetic patient's foot; it concentrates the data recorded in each individual module aimed to measure every feet parameter. This central unit makes also a statistical analysis to determine the indicators of possible damages or ulcers before their occurrence. In this way the physicians could to determine a preventive treatment to avoid complications.

The central unit system was implemented with a low-cost reduced plate computer (Raspberry Pi ®) having a touch screen. It has the capability to synchronize with external devices via USB and RS232, for reading, ordering and post-processing all the involved data. With this system, patient's information can be captured by a user interface and a touch screen, concentrating all the information of the physical variables measured with the other clinical history data and medical diagnoses. This information is stored in a database updated continuously with all the new measurements. This hub system is portable (as shown in Fig. 1.) and friendly to use.

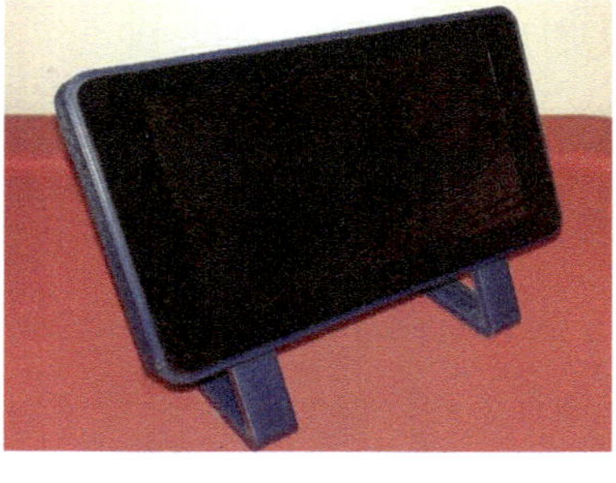

Fig. 1. System for central control, concentration, storing and distribution, of remote information from patients. The cover of this information system was printed with a 3D printer.

2.2 Associated Web Platform for Remote Measuring

Our web platform is a network system for support, clinical use, and centralized instrumentation, by internet, from the central unit; in this sense it can be named "virtual measuring platform". Its design counted with the assistance and guidance of a group of physician, specialists in diabetic foot. So it was determined the contents of the clinical data and factors that are considered nowadays important (of risk) in the revision of the feet of diabetic patients. Some of these risk factors are ulcer locations & classification, deformities, treatment, etc.

The websites are a complex conjunction of computational systems integrated with each other (databases, servers, networks, backup and security components, etc.). They are a collection of related web pages and common to an Internet domain or subdomain in the World Wide Web, and generally accessible by Internet HTTP Protocol.

Our "virtual" platform incorporates patient personal data, evaluation results from clinical diagnoses of patients with diabetic foot, and sampling data from our measuring modular systems for early detection. In addition, a tracking of the patients can be performed including the appropriate treatment prescribed for each one of them. The platform is divided into the sections shown in Fig. 2.

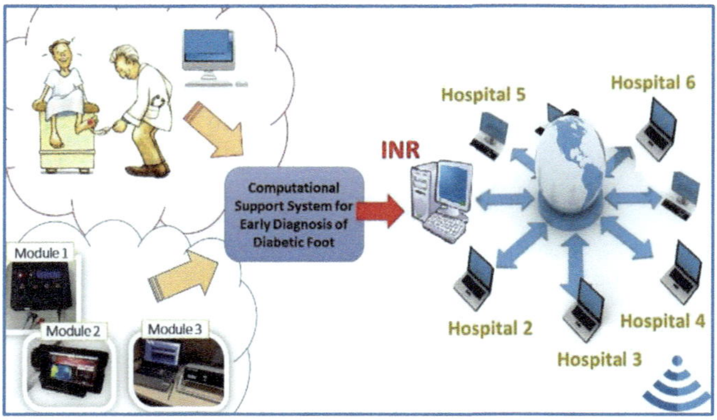

Fig. 2. Diagram of the virtual platform for diabetic foot.

2.3 Options Now Available

(a) *Patient record:* The medical specialist may enter personal data, antecedents and the type of diabetes that the patient suffers, in order to have a complete control of the state of each one.

(b) *Patient foot ulcers register:* text, binary and image data from physician's visual assessment can be entered in relation to the patient's legs and feet.

(c) *Measurements from remote electronic equipment for early detection of diabetic foot:* The aim is to add other data of early detection of diabetic foot. By the moment: external temperature and electrical impedance.

(d) *Database:* This section makes the storage of all the data entered to the platform, and then, they can be used by the physicians.

3 Programs for Managing Clinical Histories and Diagnostic Information

In Fig. 3, we can see the operation of the system for information hub; a serial communication is employed to perform the reception of all the data from the modules of early detection of diabetic foot. The data processing is made by using the graphical interface of Fig. 4. By pressing the "Results" button in each sub-system, those are graphically presented. With this graphical interface, it is possible to add the information to generate the clinical history of the patient and go to the safe information.

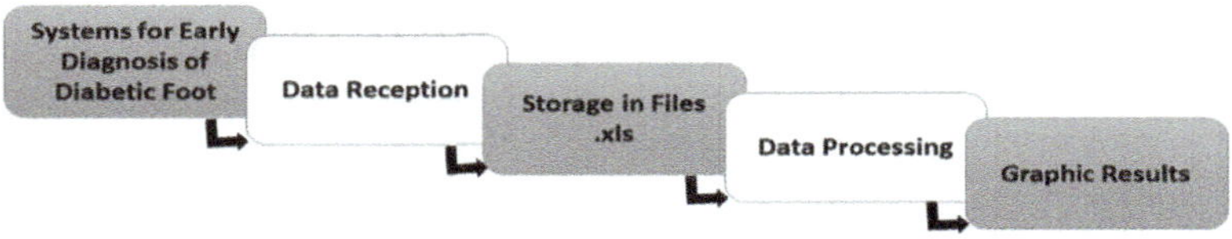

Fig. 3. Block diagram of virtual platform operation for diabetic foot.

In the web platform, the software used for the design and construction of the virtual platform for remote measuring of diabetic foot parameters is shown in Fig. 5. This is a system easy to use and each of the software used to build this platform are free and easy to work. For this first approximation, we considered that these software meet the requirements to perform a favorable job.

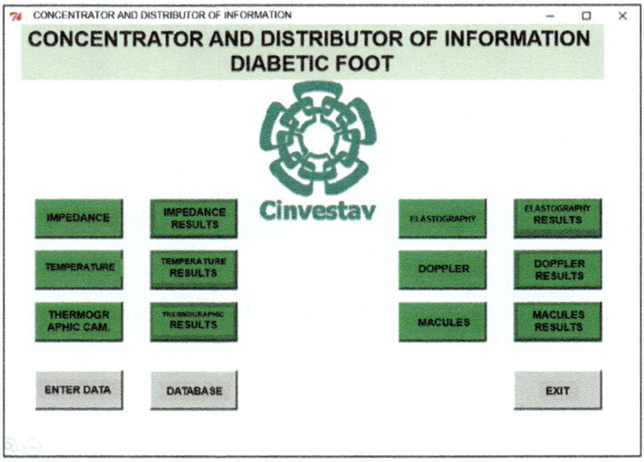

Fig. 4. Main screen of the information hub.

Fig. 5. Block diagram of software used for the construction of the virtual measuring platform.

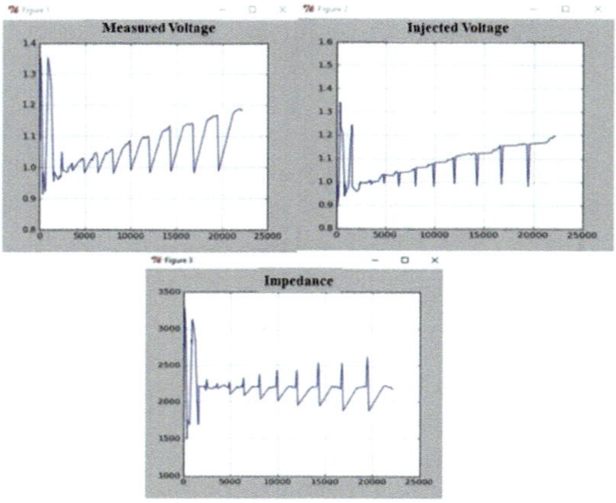

Fig. 6. Electric impedance results graphs. (a) Measured voltage, (b) Injected Voltage and (c) Electrical impedance.

Figure 6 shows some preliminary results of the reception of data of the electrical impedance equipment, where the behavior of the impedance, voltage injected and measured in a measurement in a diabetic patient can be seen.

4 Conclusions

A remote measuring platform has been created for medical users specialized in diabetic foot, including several hospitals and researchers in the area, all of them linked to the same internet network server. It permits the flexible creation and control of patient clinical histories.

The system is controlled by an user administrator, responsible for authorizing new users and keep their track access. Any new medical user of the platform creates his account by self-managing and under the authorization of the user administrator. The platform allows the concentration of measuring results provided by specific external electronic modules, of personal data, and results of diagnostic & evaluation. In fact, an open system permits the successive incorporation of a number of electronic units for early detection of the diabetic foot complications. The management of the information and monitoring of the patients is performed through a web connection, limited to the network of the established hospitals.

The results shown of the processing of impedance data are the first step in the processing of these, it is intended to incorporate adequate processing techniques for each system with which one is working. This computational system has a flexible strategic structure that permits to add and adapt observations of the physicians, in order to provide a joint support to the patients diagnoses to be determined by the specialist physician.

Acknowledgment. We thank to the group of Biomedical engineering and medical specialists in diabetic foot of the INR-LGII for their support and guidance in the definition and elaboration of this virtual platform for diabetic foot. Thanks to MSc. José Hugo Zepeda Peralta and MSc. Rubén Pérez Valladares for their important collaboration. We appreciate the funding of projects: R&D Spanish National Plan DPI2017-90147-R, EraNet-EMHE-CSIC 200022 and the R&D Network CYTED-Ditecrod-218RT0545.

References

1. Reiber, G.E., Boyko, E.J., Smith, D.G.: OMS. Lower-extremity foot ulcers and amputations in diabetes. In: Diabetes in America, 2nd edn. Bethesda, MD: National Institutes of Health, pp. 9–28, no. 95-1468. NIH Publication (2006)
2. López Murillo, F.E.: Proposal of a non-invasive electronic medical aid equipment for the early detection by temperature of foot complications in diabetic patients, Master's thesis, Electric Ing. Program, Cinvestav-IPN, Mexico City, Mexico (2014)
3. Torres, I.A., Leija, L., Ramos, A., Vera, A., Gutiérrez, J.: Instrument to Measure Temperature and Electrical Impedance in the Foot Sole to Assess the Health of the Diabetic Foot. In: I2MTC 2018 International Instrumentation & Measurement Technology Conference, 14–17 May 2018, Houston Texas, USA, pp. 1201–1206. IEEE (2018). 978-1-5386-2222-3/18/ $31.00 ©2018
4. Torres, I.A.: Hub of information and set-up of different instruments aimed at early detection of diabetic foot complications, Master's thesis, Electric Ing. Program, Cinvestav-IPN, Mexico City, Mexico (2016)

Special Sessions: Smart Robotic Assistant for Minimally Invasive Surgery: The SMARTsurg Project Experience

Towards Finger Motion Tracking and Analyses for Cardiac Surgery

Mohammad Fattahi Sani[1], Sajeeva Abeywardena[1], Efi Psomopoulou[1], Raimondo Ascione[2], and Sanja Dogramadzi[1](✉)

[1] Bristol Robotics Laboratory, Coldharbour Lane, Bristol, UK
sanja.dogramadzi@brl.ac.uk

[2] Bristol Heart Institute, University of Bristol, Bristol, UK

Abstract. Robot Assisted Surgery is attracting increasing amount of attention as it offers numerous benefits to patients as well as surgeons. Heart surgery requires a high level of precision and dexterity, in contrast to other surgical specialties. Robot assisted heart surgery is not as widely performed due to numerous reasons including a lack of appropriate and intuitive surgical interfaces to control minimally invasive surgical tools. In this paper, finger motion of the surgeon is analyzed during cardiac surgery tasks on an ex-vivo animal model with the purpose of designing a more intuitive master console. First, a custom finger tracking system is developed using IMU sensors, which is lightweight and comfortable enough to allow free movement of the surgeon's fingers/hands while using instruments. The proposed system tracks finger joint angles and fingertip positions for three involved fingers (thumb, index, middle). Accuracy of the IMU sensors has been evaluated using an optical tracking system (Polaris, NDI). Finger motion of the cardiac surgeon while using a Castroviejo instrument is studied in suturing and knotting scenarios. The results show that PIP and MCP joints have larger Range Of Motion (ROM), and faster rate of change compared to other finger/thumb joints, while thumb has the largest Fingertip WorkSpace (FWS) of all three digits.

Keywords: Robot Assisted Surgery · Finger tracking · Cardiac surgery

1 Introduction

Minimally invasive surgery brought a remarkable improvement over open surgery in terms of hospitalization time and patients' scarring [1]. On the other hand, this method has complicated surgery performance due to a range of problems such as fulcrum effect, inaccurate scaling, and lack of precision, dexterity and intuitive handling [2]. Robot Assisted Surgery (RAS) has addressed some of these problems. Among many available robotic systems, Da Vinci is currently the only widely used in hospitals [3]. In the field of cardiac surgery, however, RAS has not been widely adopted [3] and its application is mainly limited to mitral valve repair [2–4] and in smaller extent, coronary artery revascularization

J. Henriques et al. (Eds.): MEDICON 2019, IFMBE Proceedings 76, pp. 1515–1525, 2020.
https://doi.org/10.1007/978-3-030-31635-8_188

[2,3], closure of simple septal defects, tricuspid valve repair, and cardiac tumor removal [4]. In the literature, cost and steep learning curve of currently available systems are counted as reasons for this [1,3,5,6]. Despite Da Vinci system's 7 DOF master station and a wide range of tools, the control of fine movements required in heart surgeries is still an issue [6–8]. In addition, the lack of haptic feedback [1,5,7,8], and the system bulkiness [1,7] motivate research on more intuitive control of the robotic surgical tools for this surgical area [8]. The most important step towards designing a teleoperation system that deals with the restricted space of heart surgeries and a finer set of motions required is to gain a thorough understanding of the characteristics of finger/ hand motions of cardiac surgeons. The characteristics of finger motion required for complex surgical tasks are necessary for developing better, more customized master-slave robotic tools for cardiac surgery. Therefore, the aim of our work is to track and analyze finger/hand motion during cardiac surgery with the purpose of extracting ROM for each finger and thumb joint as well as FWS during different maneuvers of cardiac surgery.

2 Related Works

Finger motion tracking can be implemented for many purposes such as tele–operating a robotic hand [9], or patient motion analysis in order to study diseases [10,11]. Different methods to implement finger tracking include using: (1) Inertial Measurement Unit (IMU) based sensors [12,13] (2) optical tracking systems [9,11,14–17] (3) exoskeleton based systems including anthropomorphic exoskeletons attached to fingers, or highly redundant exoskeleton attached on top of the hand and wrist [18], (4) magnetic sensing [19], (5) flex sensor based [20], and (6) fusing several methods together [21]. When it comes to surgeon fingers tracking, optical tracking methods are highly vulnerable to occlusion [22]. Exoskeleton based finger-tracking methods, on the other hand, are usually bulky for surgery, and in some models suffer from inaccuracy due to misalignments of the exoskeleton and the finger joints [18]. IMU sensors, are relatively small, lightweight and occlusion free method of acquiring three rotation angles for each link of the hand/finger. Nevertheless, their values might drift after a while, and experience some magnetic interferences.

Hand motion analysis is a useful tool in different applications. Researchers in [21] analyzed three different methods of data glove, force sensors and EMG sensors in order to study human hand motion. The hand motion analysis is typically used to study physical impairments [10,11], electrical stimulation of hand [15], analyze joint loads [16], gestures [23], or to find comfort zone of fingers when interacting with smartphones [17].

Surgical tool tracking has already been studied in the literature using either 2D or 3D image processing [24], or mounting optical tracking markers [25]. Therefore, it is apparent that developing a reliable platform for analyzing finger motions is necessary.

Our Contributions: We propose a tracking system customized for hand tracking of surgeon which is occlusion-free, precise and lightweight to collect data during specific cardiac surgery tasks. We have identified ROM and FWS of the three digits (Middle, Index and thumb) in different stages of the surgery.

3 Hand Kinematic Model

In order to study hand/finger motions, a good understanding of biomechanics of hand is essential. A full hand consists of 27 bones including fingers, thumb, palm, and wrist [22]. Each finger is comprised of three parts of bones called phalanges. Fingers are attached to Metacarpals through Metacarpophalangeal (MCP) joint, which is followed by Proximal interphalangeal joint (PIP) and Distal interphalangeal joint (DIP). Thumb, however, which plays a crucial role in human manipulation capability, consists of two phalanges followed by a Metacarpal bone attached to Carpus [18]. Various approaches have been put forward to model hand/finger kinematics [26,27]. In this study, we are mainly interested to model middle finger, index finger and thumb, therefore we utilize simple model proposed in [27], which can be seen in Fig. 1.

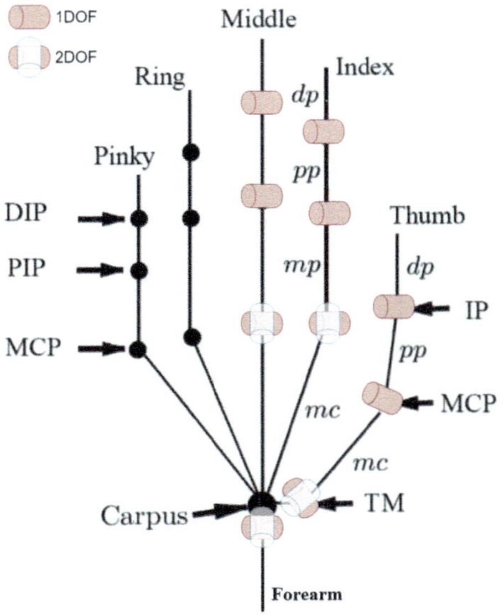

Fig. 1. Finger joint models used in our method

As it can be seen in Fig. 1, DIP and PIP joints in index and middle fingers as well as IP and MCP joints in thumb all have one revolute joint, whereas MCP and CMC joints have two DOF revolute joints. We measure absolute orientation of each phalange by a single IMU. Therefore, the joint angles can be calculated by simply subtracting the orientation of two consecutive IMU sensors. Sensors are aligned with the finger/thumb lengthwise (Y axis) while the sensors Z axis is perpendicular to the fingers/thumb. Let us assume ϕ, θ, ψ for rotation around Z, Y and X axis, respectively. Therefore, for instance, ϕ_{MCP} and θ_{MCP} which show ϕ and θ angles for MCP joints, can be calculated as follows:

$$\phi_{MCP} = \phi_{MC} - \phi_{MP}, \theta_{MCP} = \theta_{MC} - \theta_{MP} \qquad (1)$$

Having all the required joint angles allow calculation of fingertip positions for the two fingers and the thumb using forward kinematics [26]. Denavit–Hartenberg (DH) parameters of the fingers and the thumb and forward kinematics calculations are used according to SynGrasp toolbox [28].

4 Experimental Setup

The tracking system consists of 10 BNO055 IMU sensors (4 on the thumb and 3 on index and middle fingers). IMU sensors are connected to an acquisition board comprised of multiple Microprocessor Units (MCU) running Arduino firmware. Each MCU connects to two IMU sensors on a single I2C bus. Orientations from each sensor are sent to the computer with frequency of 50 Hz through a serial port. Figure 2 shows the experimental implementation of the data acquisition system.

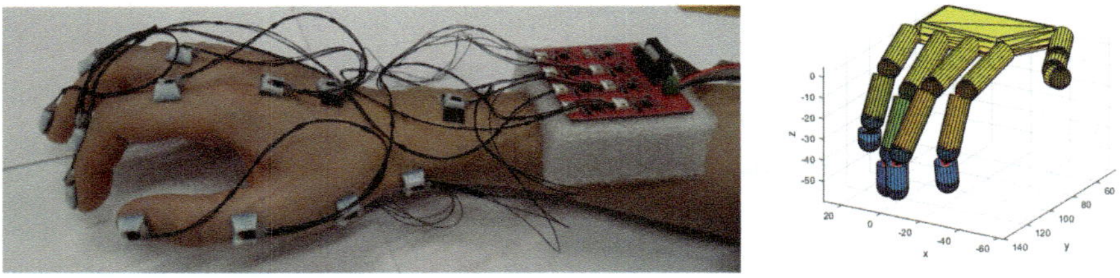

(a) Our custom Hand Tracking system.

(b) Reconstructed model of the hand in computer.

Fig. 2. Experimental setup for hand/finger tracking.

In addition to the sensor's internal calibration process, we remove the offset by measuring the angle values at the beginning of the test.

4.1 Accuracy Assessment of IMU Finger Tracking System

In order to validate accuracy of the hand tracking system, digit orientations and joint angles estimated by the IMUs were compared to a ground truth. As it can be seen in Fig. 3a, an angled circle is used as a ground truth to test the accuracy of IMU sensors for each orientation. Results show that average error is 4.1° with a standard deviation of 2°. Researchers in [12] assessed accuracy of the same sensors in a planar pose and reported average error of $3^\circ \sim 6^\circ$ with standard deviation of 1.7° for different angles. The researchers in [13] reported in Static angle errors of $\leq 2^\circ$ for their developed IMU-based hand tracking system.

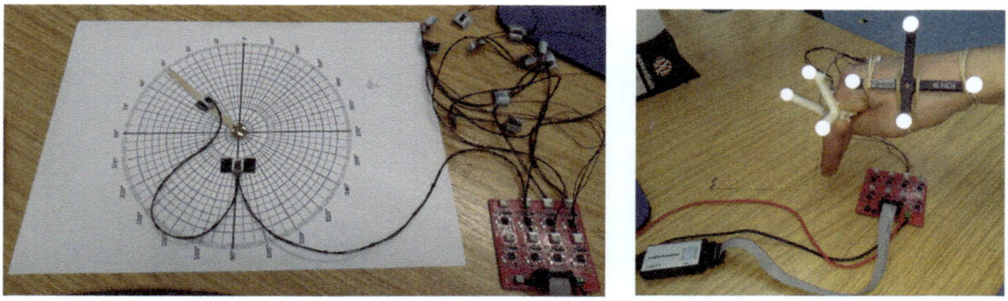

(a) Accuracy assessment setup in planar mode. (b) Accuracy assessment in 3D space.

Fig. 3. Accuracy assessment setup.

In addition, a dynamic angle verification was carried out using Polaris Spectra system (NDI) to verify the accuracy of the tracking system. A custom tool with markers (Fig. 3b), which also houses IMU sensors, is designed and attached to the Index finger, and orientations were measured relative to the marker and the sensor fixed on the palm. Figure 4 shows simple flexion extension movements of the index finger and their corresponding error values. According to experimental tests, the measurement error is less than 8° for rotating around Z, Y and X axis of the sensor.

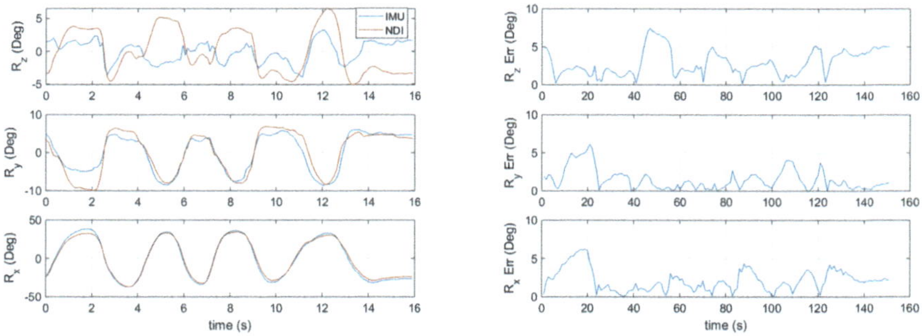

(a) Finger joint angles extracted from IMU sensors and NDI Polaris system. (b) Joint angle error with respect to NDI Polaris system.

Fig. 4. Finger motions: Flexion, Extension. IMU sensors compared with NDI Polaris motion capture system.

4.2 Cardiac Surgery Data Collection

An experienced cardiac surgeon performed typical mitral valve surgery and aorta suturing tasks on an ex-vivo pig's heart specimen with the tracking system fitted to his hands, as shown in Fig. 5. The tasks were performed using 7" Castroviejo Needle Holder Plier Straight and 7" Castroviejo Micro Scissor.

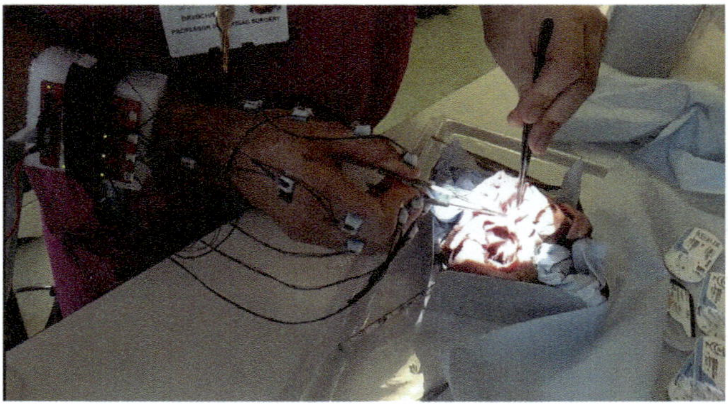

(a) Finger/Hand data collection using our developed system during the Mitral Valve surgery.

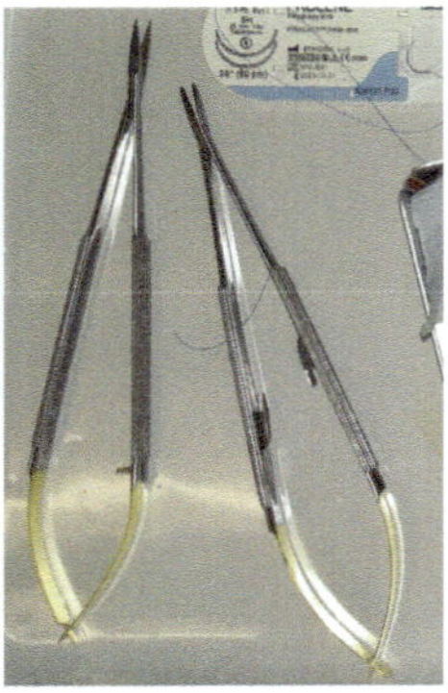

(b) A set of Castroviejo instruments used in the experiment.

Fig. 5. Cardiac surgery data collection setup and instruments.

4.3 Results and Discussion

ROM and FWS are two important features of finger motion. They were extracted in [11] and [15] in order to study human functional abilities and to explore hand grasp patterns, respectively. In our study, ROM for each joint of the surgeon's fingers has been extracted and is shown in Table.1. Figures 6, 7, 8 and 9 demonstrate joint angles for a sample suturing operation.

Table 1. Range of motion (ROM) for surgeon's fingers. (Degrees)

Action Type	Index DIP	Index PIP	Index MCP		Middle DIP	Middle PIP	MiddleMCP		Thumb DIP	Thumb MCP	ThumbCMC		Wrist	
IMU angle	ϕ	ϕ	ϕ	θ	ϕ	ϕ	ϕ	θ	ϕ	ϕ	ϕ	θ	ϕ	θ
Knotting	23	40	30	10	50	85	60	21	35	38	25	15	25	41
Suturing	15	25	19	12	39	60	55	30	22	15	15	20	45	45

* ϕ stands for flexion–extension movement of fingers, whereas θ stands for abduction movements.

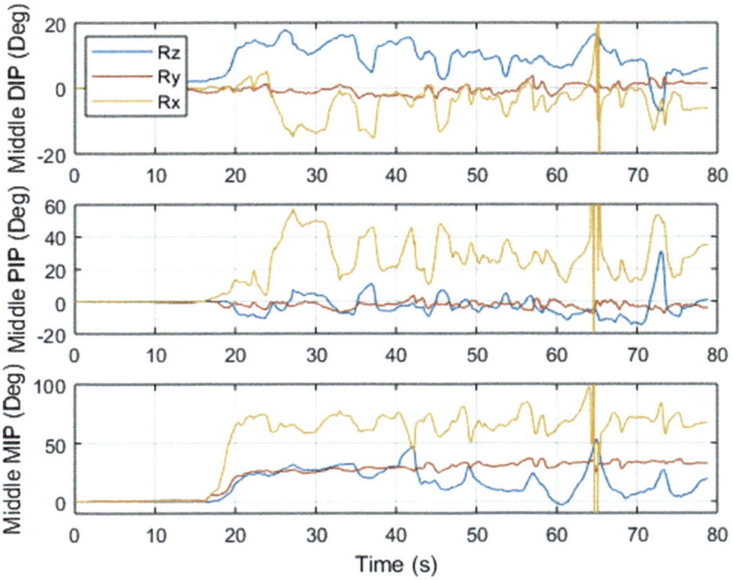

Fig. 6. Middle finger joint angles

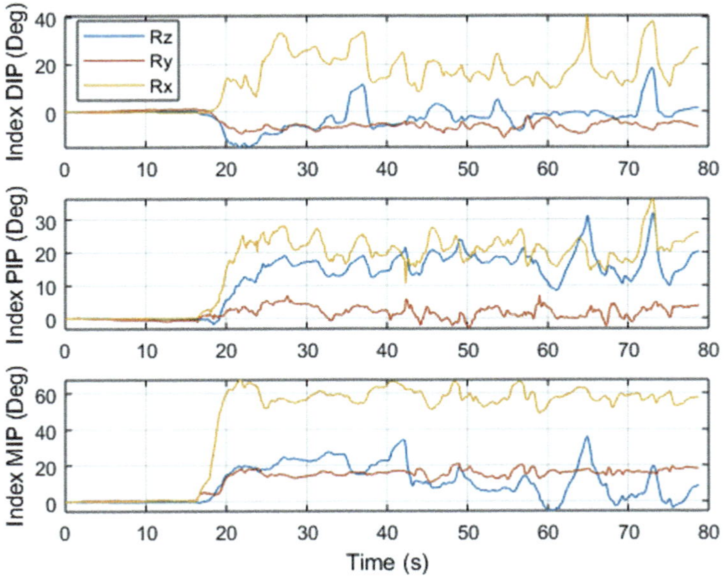

Fig. 7. Index finger joint angles

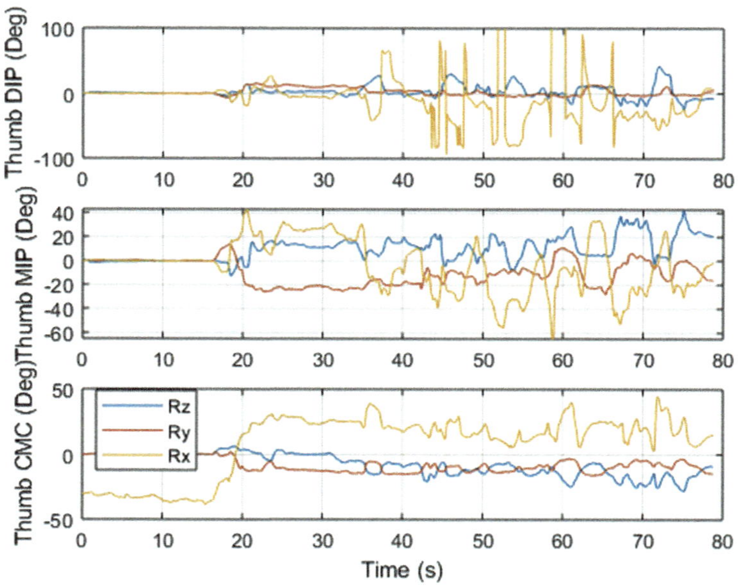

Fig. 8. Thumb joint angles

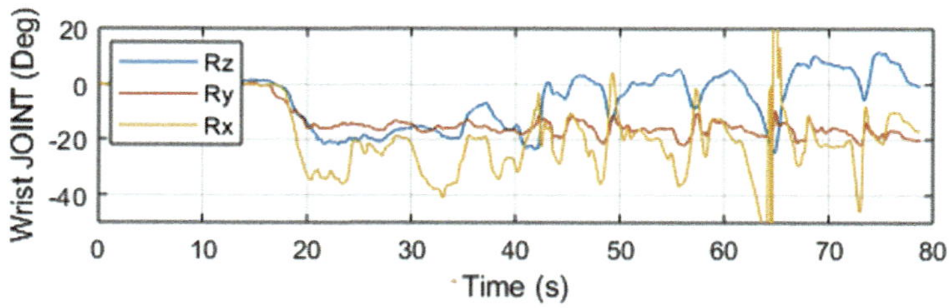

Fig. 9. Wrist joint angles

Rate of change is another valuable characteristic of finger joint motion which demonstrates how fast each joint is moving. Assuming $\phi_{[n]}$ as a discrete–time series of joint angles, the angular velocity (ω) is calculated as follows.

$$\omega_{[n]} = (\phi_{[n]} - \phi_{[n-1]})/Fs \tag{2}$$

Where Fs = 50 is the sampling frequency. Now, the average angular velocity (ω_{avg}) of a series with N samples is calculated as follows:

$$\omega_{avg} = 1/N \sum_{n=1}^{N} \omega_{[n]} \tag{3}$$

Figure 10 shows rate of change for two different surgical scenarios. In addition, FWS during the surgery is shown in Fig. 11.

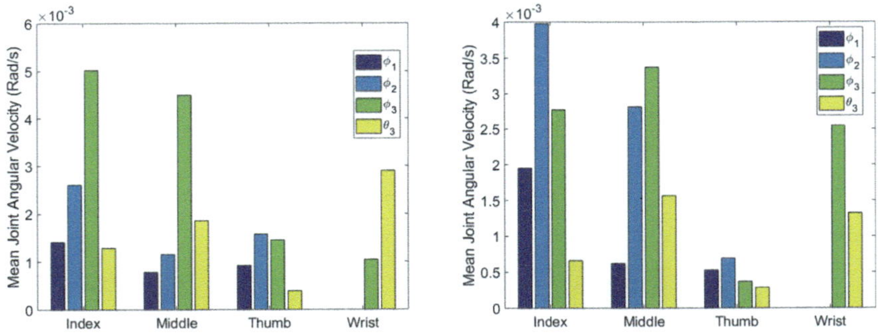

(a) Knotting. (b) suturing.

Fig. 10. Mean rate of change for each joint (joints are numbered from 1 for DIP joint).

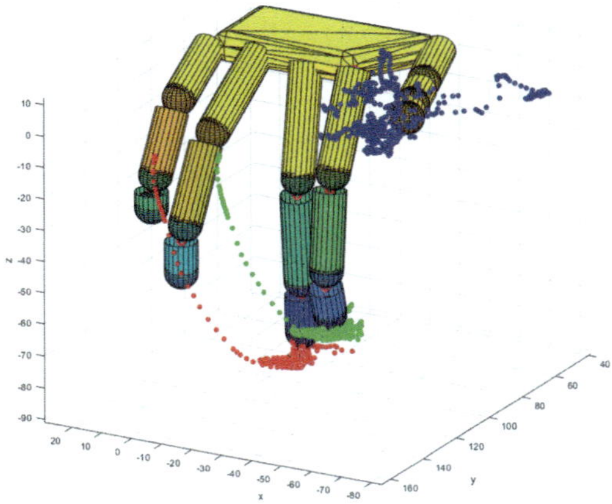

Fig. 11. Fingertip position during the suturing process with Castroviejo tool

A close look into the ROM results show that MCP and PIP joints in middle finger have a relatively higher range of motion compared to others. Furthermore, according to Fig. 10, PIP and MCP joint are the ones that are changing fastest. Finally, Fig. 11 shows that FWS for the thumb has larger space compared to the middle and index fingers.

5 Conclusion

In this paper, motion of the fingers and the thumb during typical cardiac surgery tasks was studied. A custom IMU–based finger tracking system has been developed which is lightweight, comfortable, and precise enough to track the surgeon's finger motions. The captured data during the heart surgery shows that PIP and

MCP joints are moving faster and more than the other tracked joints. In addition, the thumb has larger FWS in this study. The outcome of this research will be further utilized in designing a new intuitive way of controlling tele–operated surgical robot tools for heart surgeries. In future steps, surgical tools will also be monitored and their motion analyzed together with finger/thumb motions.

References

1. Peters, B.S., Armijo, P.R., Krause, C., Choudhury, S.A., Oleynikov, D.: Review of emerging surgical robotic technology. Surg. Endosc. **32**, 1636–1655 (2018)
2. Rodriguez, E., Chitwood, W.R.: Robotics in cardiac surgery. Scand. J. Surg. **98**, 120–124 (2009)
3. Quint, E., Sivakumar, G.: The role of robotic technology in cardiac surgery. Univ. West Ont. Med. J. **87**, 40–42 (2019)
4. Dearani, J.A.: Robotic heart surgery: Hype or hope? J. Thorac. Cardiovasc. Surg. **155**, 77–78 (2018)
5. Pettinari, M., Navarra, E., Noirhomme, P., Gutermann, H.: The state of robotic cardiac surgery in Europe. Ann. Cardiothorac. Surg. **6**, 1–8 (2017)
6. Scholar, M.D., Ashford, H.: Use of Robots on Cardiac Surgery (2015)
7. Simorov, A., Stephen Otte, R., Kopietz, C.M., Oleynikov, D.: Review of surgical robotics user interface: What is the best way to control robotic surgery? Surg. Endosc. **26**, 2117–2125 (2012)
8. Freschi, C., Ferrari, V., Melfi, F., Ferrari, M., Mosca, F., Cuschieri, A.: Technical review of the da Vinci surgical telemanipulator. Int. J. Med. Robot. **6**, 468–472 (2010)
9. Cerulo, I., Ficuciello, F., Lippiello, V., Siciliano, B.: Teleoperation of the SCHUNK S5FH under-actuated anthropomorphic hand using human hand motion tracking. Rob. Auton. Syst. **89**, 75–84 (2017)
10. Medicine, P.: Kinematic Motion Analysis in Upper Extremity Cerebral Palsy (2016)
11. Leitkam, S.T., Reid Bush, T.: Comparison between healthy and reduced hand function using ranges of motion and a weighted fingertip space model. J. Biomech. Eng. **137**, 041003 (2014)
12. Liu, H., Xie, X., Millar, M., Edmonds, M., Gao, F., Zhu, Y., Santos, V.J., Rothrock, B., Zhu, S.-C.: A Glove-based System for Studying Hand-Object Manipulation via Joint Pose and Force Sensing
13. Lin, B.-S., Lee, I.-J., Chiang, P.-Y., Huang, S.-Y., Chih, P.W.: A modular data glove system for finger and hand motion capture based on inertial sensors. J. Med. Biol. Eng. https://doi.org/10.1007/s40846-018-0434-6
14. Cerveri, P., De Momi, E., Lopomo, N., Baud-Bovy, G., Barros, R.M.L., Ferrigno, G.: Finger kinematic modeling and real-time hand motion estimation. https://doi.org/10.1007/s10439-007-9364-0
15. Shin, H., Watkins, Z., Hu, X.: Exploration of hand grasp patterns elicitable through non-invasive proximal nerve stimulation OPEN. https://doi.org/10.1038/s41598-017-16824-1
16. Vignais, N., Cocchiarella, D.M., Kociolek, A.M., Keir, P.J.: Dynamic assessment of finger joint loads using kinetic and kinematic measurements. Digit. Hum. Model. **2013**, 1–5 (2012)

17. Le, H.V., Mayer, S., Bader, P., Henze, N. (2018) Fingers' range and comfortable area for one-handed smartphone interaction beyond the touchscreen, pp. 1–12
18. Sarakoglou, I., Brygo, A., Mazzanti, D., Hernandez, N.G., Caldwell, D.G., Tsagarakis, N.G. (2016) Hexotrac: A highly under-actuated hand Exoskeleton for finger tracking and force feedback. In: IEEE International Conference Intelligent and Robotic Systems, 1033–1040, November 2016
19. Chen, K.-Y., Patel, S., Keller, S.: (2016) Finexus: Tracking precise motions of multiple fingertips using magnetic sensing. CHI **2016**, 1504–1514 (2016)
20. Saggio, G., Pallotti, A., Sbernini, L., Errico, V., Di, P.F.: Feasibility of commercial resistive flex sensors for hand tracking applications. Sens. Transducers **201**, 17–26 (2016)
21. Ju, Z., Liu, H., Member, S.: Human Hand Motion AnalysisWith Multisensory Information. IEEE/ASME Trans Mechatronics **19**, 456–466 (2014)
22. Wheatland, N., Wang, Y., Song, H., Neff, M., Zordan, V.: Jörg S state of the art in hand and finger modeling and animation
23. Tits, M.: Expert gesture analysis through motion capture using statistical modeling and machine learning Mickaël Tits (2018). https://doi.org/10.13140/RG.2.2.36839.50084
24. Zhang, L., Ye, M., Chan, P.-L., Yang, G.-Z.: Real-time surgical tool tracking and pose estimation using a hybrid cylindrical marker. Int. J. CARS **12**, 921–930 (2017)
25. Ye, M., Zhang, L., Giannarou, S., Yang, G.-Z.: Real-time 3D Tracking of Articulated Tools for Robotic Surgery (2016)
26. Van Der Hulst, F.P.J., Schätzle, S., Preusche, C., Schiele, A.: A functional anatomy based kinematic human hand model with simple size adaptation. In: Proceedings of IEEE International Conference Robot Automation, pp. 5123–5129 (2012)
27. Li, K., Chen, I.-M., Yeo, S.H., Lim, C.K.: Development of finger-motion capturing device based on optical linear encoder. J. Rehabil. Res. Dev. **48**, 69 (2011)
28. Malvezzi, M, Salvietti, G, Gioioso, G, Prattichizzo, D.: SynGrasp: A Matlab Toolbox for Underactuated and Compliant Hands Version 2.2 User Guide (2015)

Surgeon Training with Haptic Devices for Computer and Robot Assisted Surgery: An Experimental Study

Salih Ertug Ovur(✉), Marisa Cobanaj, Luca Vantadori, Elena De Momi, and Giancarlo Ferrigno

Department of Electronics Informatics and Bioengineering, Politecnico di Milano, Milan, Italy
salihertug.ovur@mail.polimi.it

Abstract. Development of robot-assisted minimally invasive surgery increasingly demands for efficient training methods. This paper describes an experiment exploring the use of haptic interaction for the purpose of skill training (haptic training). Various experiment tasks that include simple and complex tool paths have been developed for this purpose. 105 acquisition sessions distributed in 7 different tasks from 27 naive subjects and one surgeon performed teleoperated exercises with Omni Phantom. Task's learning curves with and without robotic assistance and effects of damping have been discussed. For the force guidance case, assistance level gradually decreased the applied force as the training progressed. In the controversial scenario of robotic assistance for motor-learning benefits, this study shows such assistance improve the rate of learning for both simple and complex tasks.

Keywords: Active constraints · Haptics · Surgical robotics training · Virtual-Reality · Augmented-Reality

1 Introduction

The advent of robot-assisted minimally invasive surgery (RAMIS), has the downside of a reduced, or, in the majority of the cases, completely absent sense of touch. Surgeons, therefore, have to rely on the visual cue to compensate for this lack and can exploit haptic training in order to improve the learning curve [1]. *Haptics* and computer-based *Virtual Reality* (VR) platforms together might play a key role in RAMIS to enhance the capabilities of the operator to improve the outcome of the surgical procedure in terms of safety, accuracy and cognitive load [2]. Substantial progress in the learning curve, indeed, has been achieved by using repetitive VR tasks before practicing on patients [3]. Moreover, computer-based VR training offers objective performance assessment during and after training able to accelerate the learning process [4, 5].

A brief video of this work is available on-line at: https://drive.google.com/open?id=1Qj92bBNK2-JwA34MBKrPoXK0ZVNBqh6lg.

J. Henriques et al. (Eds.): MEDICON 2019, IFMBE Proceedings 76, pp. 1526–1535, 2020.
https://doi.org/10.1007/978-3-030-31635-8_189

To implement haptic feedback, systems for the intraoperative definition of active constraints have been developed. Latest researches about *Active constraints* focused to limiting the tool to within a certain space (forbidden region constraints) or attempt to guide the tool in moving along the desired path (guidance constraints) [6].

Benefits of the simulated environments in the case of forbidden region active constrain have been largely discussed and demonstrated [7, 8]. Force feedback systems for RAMIS typically measure or estimate the forces applied to the patient by the surgical instrument and generate a repulsive force on the robotic tool when it tries to enter the constrained, preventing common complication are accidental damages to nerves, veins and arteries [2].

Few works instead have found significant benefits of a simulated environment that enhance a training process with augmented forces [9], such as guidance constrain. The performance of attractive force applied to the robotic tool getting away from the constrained have been judged beneficial [10], null [11] and even detrimental [12].

2 Related Works

The guidance hypothesis [13] confirms that the guiding properties of augmented feedback are beneficial for motor learning when knowledge about the outcome is used to correct errors and improve subsequent performance, but detrimental when guiding affects processes important for learning. For example, practice with a high frequency of augmented feedback can affect learning by excessive reliance on external cues [12]. According to Schmidt [13], dependency on the robotic assistance can be addressed by gradually decreasing the guidance constraints intensity as the user learns. It provides the performer with an appropriate amount of haptic guidance throughout training [14]. This assistance-as-needed approach [15] prevents assistance from minimizing voluntary control, which has been shown to reduce the effectiveness of learning.

Moreover, haptic training supports passive learning, where subjects are not actively performing the task. In general, passively learning is not as efficient as actively one, committing errors and elaborating new strategies to accomplish the skill [16]. Therefore, haptic training strategies should involve the subjects' in actively participating and draw his attention to promote learning.

Finally, haptic guidance can considerably alter the dynamics of the task and affect motor learning. The guidance constraints can lead to unpredictable and unwanted motion of the tool. Indeed, the stored potential energy in common elastic links pulls the tool toward the desired path as long as the tool is away from the path. When the user relaxes his grasp on the haptic device, the stored energy can cause a sudden unintended motion of the tool and diminish the surgeon's overall control of the procedure. To overcome these issues, a solution has been proposed by [17]. A viscosity-based guidance active constraint is introduced: it attempts to guide the motion towards the desired path in the least aggressive way and without limiting the freedom of the user. Experimentation confirmed an improvement in accuracy of dynamic path following and a decrease in user's distraction.

This paper presents an experimental study of haptic guidance support in training for teleoperated tasks. The work sums up solutions claimed by the mentioned experimental studies and aims to discuss such use of augmented forces for improving the training learning curve. Thus, it finally assesses the benefits of guidance constraints. For this purpose, the participants to the study have been trained in a VR environment with and without haptic feedback, and with damping force: their performance, during and after training, has been measured in order to enhance eventual improvements of subjects whose training has been assisted by guidance constraint.

3 Materials

3.1 Omni Phantom Robot

The PHANTOM Omni Touch Firewire was used for this research. Omni is six degree-of-freedom haptic manipulator with haptic feedback on the first three axes. Forward&Inverse kinematics, haptic feedback controllers and damping parameters are implemented in C++. 100 Hz is used for the haptic device updating rate. The tools' pose, forces, and task options were sent through common roscore to the laptop for visualization software.

3.2 Visualization Software

Visualization software has been realized in ROS (Robot Operating System) environment with C++ code. The graphics are produced using the Visualization Toolkit (VTK) [18] and OpenGL [19] libraries with a refresh rate of 30 Hz. Visualization software with VTK libraries makes possible to create various simulations for both VR and augmented reality (AR) environments.

3.3 Experimental Setup

Omni Phantom is integrated with control hardware and software that allows for reading the tool tip position and sending commands to the joints of the robot. Experimental setup while a participant surgeon is following tasks is given in Fig. 1.

The software architecture includes two main ROS-nodes: Omni phantom and visualization software. To prevent the computational load of the graphics from affecting the haptic control performance, these two were executed on separate computers: desktop computer running Ubuntu 14.04 LTS for haptic constraint control and laptop, featuring an Intel Core i7-7700HQ CPU and NVidia GTX 1050 GPU with Ubuntu 16.04 LTS, used for visualization software.

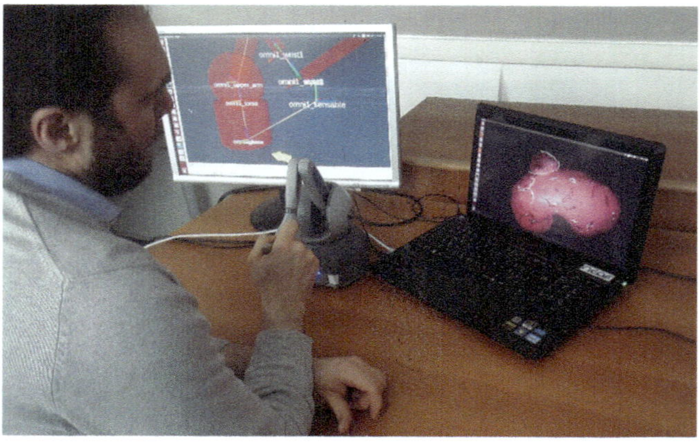

Fig. 1. The experimental setup

4 Methods

Designed tasks, acquisition procedure and finally, performance criteria of our study are introduced in this section.

4.1 Tasks

7 tasks are designed for simulating shapes in AR and VR but also with basic geometrical shapes such as rectangle, circle. Created shapes were followed by using Omni Phantom. All subjects trained on the same order of tasks (see Fig. 2): the contour of a kidney tumor on an AR scenario, a circle, a rectangle, an elliptical and an infinity shape, the contour of the renal artery during nephrectomy procedure on VR scenario and finally the first task again were performed. To keep subjects engaged and motivated intuitive performance color map feedback is provided through contour color shift from green to yellow and red, regarding the distance from the desired path Task 6 in VR has been introduced in order to make the subject dealing with virtual anatomy. Also, this more complicated shape could be employed to verify if guidance constraint on complicated paths can improve without the performance, as suggested in the literature.

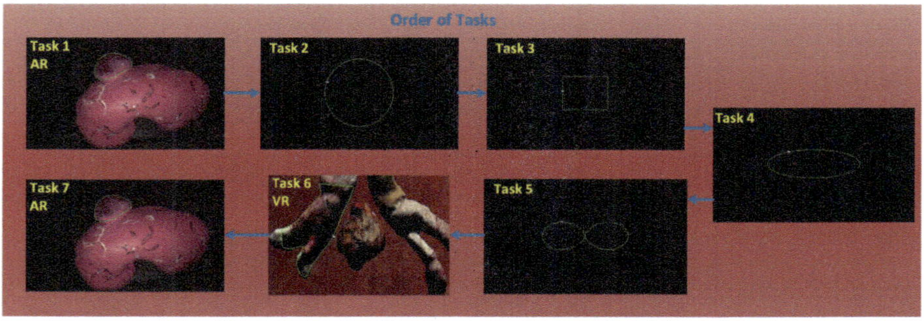

Fig. 2. Experiment structure design

The shape contours of AR and VR tasks have been obtained from well-trained users. Filters have been applied in order to remove the effects of noise. All other shapes have been defined mathematically.

4.2 Acquisition Procedure and Performance Metrics

To enhance the effects of without the haptic guidance training, 27 non-medical participants (22 to 40 years old, 13 males and 14 females, all right-handed) were randomly divided into three groups of 9: a group that performed the tasks with no augmented force, a second one assisted by a damping effect and a third one with augmented force support for guidance towards the path. A medical participant instead accomplished all cases This way three groups of 10 participants each have been defined. Each group proceeded in sessions of 15 repetitions for each of the 7 tasks, lasting an average of 40 min. Guidance force has been decreased so that at the end of the task no augmented force was applied: from a unitary value force, gain (K_o in Fig. 3) was decreased each 3 repetitions by a factor 0.25 so that in the final 3 repetitions augmented force intensity was null. For the performance metrics, distance between cursor point and closest point on the followed path is computed.

4.3 Robotic Assistance Method

For the force assistance training, Fuzzy-PD controller has been used. It performs a control on the joint angle error, computed as the difference between the measured joint angles and desired joint angles:

$$\theta_{error} = \theta_{desired} - \theta_{measured} \tag{1}$$

Desired joint angles are obtained by using inverse kinematics on the desired path, as applied in [20]. θ_{error} and its derivative value $\dot{\theta}_{error}$ are used for computing the output ($\tau_{controller}$), as seen in Fig. 4. The controller Scaling Factors (SFs) K_e and K_d correspond respectively to the spring and damping constants. The input SFs together with saturation limits are used to normalize the inputs θ_{error} and $\dot{\theta}_{error}$ into the Membership Functions (MFs). The output U is converted into the control signal of the Fuzzy-PD as $\tau_{controller} = K_o U$ where K_o is the output of SFs [21]. Tuned parameters for SFs: ($K_e = 0.18$, $K_d = -1$, $K_o = 2.2$).

Fig. 3. General Scheme of the design

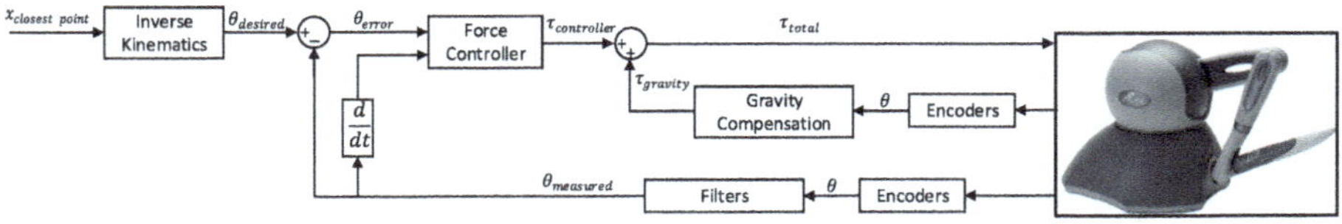

Fig. 4. Schematic joint space control diagram

The Fuzzy-PD is composed of a 5 × 5 rule base as given in Table 1. The MFs are defined with triangular MFs and are defined with Negative Big (NB), Negative Medium (NM), Zero (Z), Positive Medium (PM) and Positive Big (PB) (Fig. 5). The consequent parts of the rules are defined with crisp singletons which are Positive Big ($PB = 1$), Positive Medium ($PM = 0.7$), Zero ($Z = 0$), Negative Medium ($\mathrm{NM} = -0.7$) and Negative Big ($NB = -1$). The resulting control surface is given in Fig. 6.

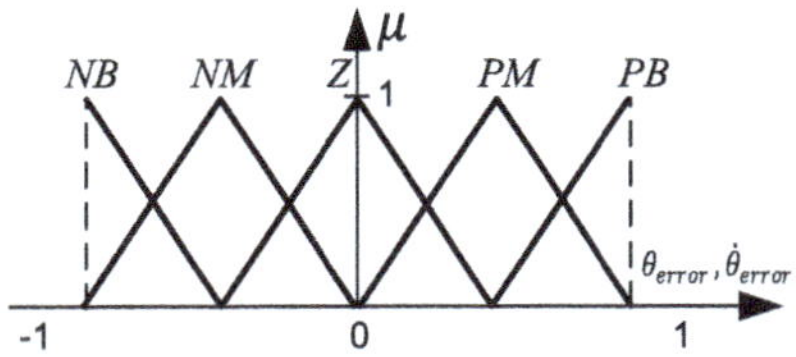

Fig. 5. Illustration of the MFs

Table 1. Rule base of Fuzzy-PD controller

$X_e/\dot{X}$	PB	PM	Z	NM	NB
PB	PB	PB	PM	PM	Z
PM	PB	PM	PM	Z	NM
Z	PM	PM	Z	NM	NM
NM	PM	Z	NM	NM	NB
NB	Z	NM	NM	NB	NB

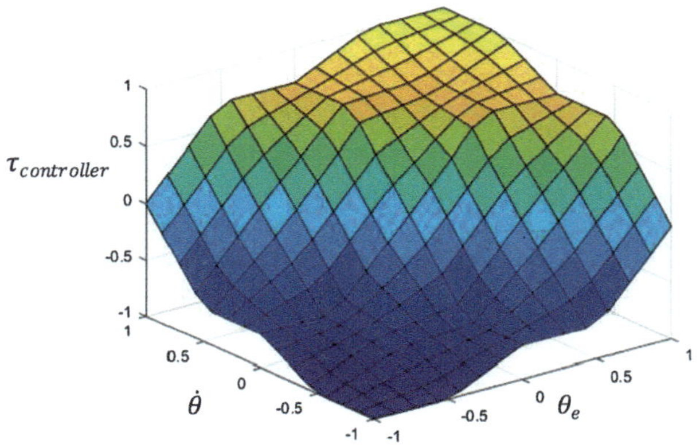

Fig. 6. Control surface of the Fuzzy-PD

Gravity compensation provided for all experiments. Algorithm introduced at [22] is used for the compensation of gravity.

5 Results and Discussion

Results have been obtained for 3 different subject group, each one of 10 subjects (9 naive and the surgeon). Each participant performed for every 105 acquisitions, including 7 tasks and 15 repetitions for each task. During tasks, distance between cursor point from end effector of Omni and closest point on desired path are used for performance criteria. Error computed for each experiment has been computed according to the Root-Mean-Square method. Before proceeding, all obtained data have been normalized between maximum and minimum errors by excluding outliers (Normalized Error). Then, data have been collected for 10 people and their 7 tasks. This gives us distribution of error with respect to repetitions. For each repetition 70 data are considered. In order to comment on relations easier, performance criteria considered instead of an error. Maximum performance means minimum error and minimum performance equals maximum error. Obtained results are given in Fig. 7.

$$\text{Performance Criteria} = 1 - \text{Normalized Error} \tag{2}$$

In order to comment on obtained statistical results, normality tests are applied to learn if data present a normalized distribution. Firstly, one-sample Kolmogorov-Smirnov test has been applied on data for each repetition. As a result of the test, all data have a normal distribution ($p < 0.005$) for each repetition ($N = 70$). T-test and Shapiro-Wilk tests gave similar results. According to normally distributed data we could use mean and other main statistical parameters to comment on performance. However, we had also to check if there was a significant improvement between different subject groups. In order to find relations of different subject groups, t-test is used for studying relations of force enhanced, damping assisted and null subject groups (Table 2). T-test's hypothesis is true ($h = 1$) when two sample's mean values are different.

Table 2. Statistical test results

Compared subject groups	Hypothesis	p-value (average)
Force Assisted - Null	h = 1 for (15/15 repetitions)	0.005
Force Assisted - Damping Assisted	h = 1 for (14/15 repetitions)	0.01
Damping Assisted - Null	h = 0 for (15/15 repetitions)	0.6

Results of these tests between subject groups show that there is a meaningful difference between force assisted group with respect to null and damping assisted groups. However, there is no guaranteed distinction between damping assisted with respect to null subject groups. Moreover, these relations are also verified with Bonferroni Correction. For the analytical comparison, only the last 3 repetitions have been considered, where force-assistance is totally removed. For the comparison, mean and % 95 Confidence Interval (CI) has been used (see Table 3).

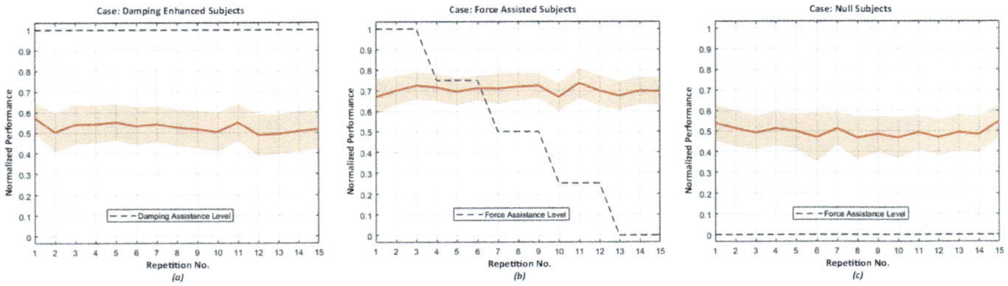

Fig. 7. The performance analysis of (a) damping enhanced, (b) force assisted and (c) null subjects. Dark orange line shows the mean and light-orange area represents the %95 confidence interval for 7 tasks and 10 people for each repetition.

Table 3. μ and CI of all groups performance

Repetition no.	Damping asst.		Force asst.		Null	
	μ	CI (%95)	μ	CI (%95)	μ	CI (%95)
13	0.49	0.59	0.68	0.76	0.49	0.58
		0.39		0.63		0.40
14	0.51	0.61	0.70	0.77	0.48	0.57
		0 41		0.63		0.39
15	0.52	0.61	0.70	0.75	0.55	0.63
		0.42		0.61		0.46

Figure 7 visually shows, and Table 3 statistically proves us there are significant improvement occurred for the force assisted subjects. Although force assistance level reduced repetitions show decrease in performance, subjects could tolerate the changes and they could keep the same level of performance. Moreover, the normalized error has been compared for different tasks performed by the same subject and it was noticed that for all subjects, the highest improvement in performance occurs for the most complex task (Task 6). This confirms what supported by the state-of-art. However, damping assistance did not significantly increase performance which could be caused by, damping was only applied for only the first three joints.

6 Conclusions

This work proposed a haptic guidance method in surgical robotic training. According to the literature findings and our conclusion, applied haptic guidance during training increases the performance. The method was put to test through a comparison with a no force and damping, whose efficacy has been supported in some studies we wanted to validate, assisted training. 30 experiments sessions, in which 27 novice participants and a surgeon followed 7 different tasks with 15 repetitions for each of them, were considered. Results confirmed that after the force removal, the performance of subjects with haptic guidance training kept at a higher rate during all experiments with respect to one of the subjects with no force and damping supported training. Therefore, it is possible to assess that in this study the guidance constraints have a beneficial effect on improving the training learning curve. These results are promising, and further investigation is needed to validate this work with different tasks. The study was performed on a relatively small sample due to the difficulties in recruiting subjects for a quite long experiment and a larger sample might lead to more significant results.

Acknowledgment. We received assistance on the software from Hang Su. We thank our subjects for kindly participating in the training experiment. This work was supported by European Union's Horizon 2020 research and innovation programme (grant agreement No. 732515 "SMARTsurg").

References

1. Wagner, C.R., Stylopoulos, N., Jackson, P.G., Howe, R.D.: The benefit of force feedback in surgery: examination of blunt dissection. Presence Teleoperators Virtual Environ. **16**, 252–262 (2007)
2. Vantadori, L., Mariani, A., Chupin, T., De Momi, E., Ferrigno, G.: Design and evaluation of an intraoperative safety constraints definition and enforcement system for robot-assisted minimally invasive surgery. In: Proceedings of Computer/Robot Assisted Surgery CRAS (2018)
3. Enayati, N., De Momi, E., Ferrigno, G.: Haptics in robot-assisted surgery: challenges and benefits. IEEE Rev. Biomed. Eng. **9**, 49–65 (2016)
4. Sanchez, B.R., Mohr, C.J., Morton, J.M., et al.: Comparison of totally robotic laparoscopic Roux-en-Y gastric bypass and traditional laparoscopic Roux-en-Y gastric bypass. Obes. Surg. **16**, 690–696 (2006)
5. Seymour, N.E., Gallagher, A.G., Roman, S.A., et al.: Virtual reality training improves operating room performance: results of a randomized, double-blinded study. Annu. Surg. **6**, 458–463 (2002)
6. Lau, J.W., Yang, T., Toe, K.K., Huang, W., Chang, S.K.: Can robots accelerate the learning curve for surgical training? An analysis of residents and medical students. Ann. Acad. Med. Singap. **1**, 29–35 (2018)
7. Okamura, A.: Haptic feedback in robot-assisted minimally invasive surgery. Curr. Opinion Urol. **19**, 102–107 (2009)
8. Bowyer, S.A., et al.: Active constraints/virtual fixtures: a survey. Trans. Robot. **30**, 138–157 (2014)
9. Sigrist, R., Rauter, G., Riener, R., Wolf, P.: Augmented visual, auditory, haptic, and multimodal feedback in motor learning: a review. Psychon. Bull. Rev. **20**, 21–53 (2013)
10. Enayati, N., Okamura, A.M., Mariani, A., Pellegrini, E., Coad, M.M., Ferrigno, G., De Momi, E.: Robotic assistance-as-needed for enhanced visuomotor learning in surgical robotics training: an experimental study. In: IEEE International Conference on Robotics and Automation (ICRA) (2018)
11. Jacobs, S., Holzhey, D., Strauss, G., Oliver Burgert, O., Falk, V.: The impact of haptic learning in telemanipulator-assisted. Surg. Laparosc. Endosc. Percutaneous Tech. **17**, 402–406 (2007)
12. Coad, M.M., et al.: Training in divergent and convergent force fields during 6-DOF teleoperation with a robot-assisted surgical system. In: IEEE World Haptics Conference (WHC) (2017)

13. Schmidt, R.A.: Frequent augmented feedback can degrade learning: evidence and interpretations. In: Tutorials in Motor Neuroscience, vol. 62, pp. 59–75. Springer, Dordrecht (1991)
14. Li, Y., Huegel, J.C., Patoglu, V., O'Malley, M.K.: Progressive shared control for training in virtual environments. In: 3rd Joint EuroHaptics Conference and Symposium on Haptic Interfaces for Virtual Environment and Teleoperator Systems. World Haptics, pp. 332–337 (2009)
15. Marteniuk, R.G.: Retention characteristics of motor short-term memory cues. J. Motor Behav. **5**, 249–257 (1973)
16. Crespo, L.M., Reinkensmeyer, D.J.: Haptic guidance can enhance motor learning of a steering task. J. Motor Behav. **40**, 545–557 (2014)
17. Enayati, N., Alves Costa, E.C., Ferrigno, G., De Momi, E.: A dynamic non-energy-storing guidance constraint with motion redirection for robot-assisted surgery. In: IEEE International Conference on Intelligent Robots and Systems (IROS), Daejeon Convention Center, Korea (2016)
18. VTK Homepage. https://vtk.org/. Accessed 26 Apr 2019
19. OpenGL Resources page. https://www.opengl.org/resources/. Accessed 26 Apr 2019
20. Gudino-Lau, J., Chavez-Montejano, F., Alcala, J., Charre-Ibarra, S.: Direct and inverse kinematic model of the OMNI PHANToM. ECORFAN-Bolivia J. **5**, 25–32 (2018)
21. Hong, M., Rozenblit, J.W.: Haptic guidance with fuzzy control in simulation. In: Proceedings of the Symposium on Modeling and Simulation in Medicine, Alexandria, Virginia, pp. 25–30 (2015)
22. Koul, M.H., Kumar, P., Singh, P.K., Manivannan, M., Saha, S.K.: Gravity compensation for PHANToMTM Omni® haptic interface. In: First Joint International Conference on Multibody System Dynamics (2015)

Augmented Reality Toolkit for a Smart Robot-Assisted MIS Platform

Georgios Zampokas, Konstantinos Tsiolis, Georgia Peleka(✉), Angeliki Topalidou-Kyniazopoulou, Ioannis Mariolis, Sotiris Malasiotis, and Dimitrios Tzovaras

Information Technologies Institute,
Centre for Research & Technology - Hellas, Thessaloniki, Greece
{gzampokas,ktsiolis,gepe,angeliki.topalidou,
ymariolis,malasiot,tzovaras}@iti.gr

Abstract. Minimally invasive surgery (MIS) offers important benefits to the patient, however it introduces additional complexity and constraints to the surgical workflow. Thus, the development of guidance systems which use Augmented Reality (AR) in order to provide clinicians with important information, has become a popular research topic in recent years. In this work, a visualization toolkit is presented, in which pre-operative 3D structures are overlaid to the intra-operative surgical view, during Robot-Assisted MIS. These structures are extracted from pre-operative Magnetic Resonance Imaging (MRI) or Computed Tomography (CT) scans and are registered into the endoscopic view through a semi-automatic pipeline, that includes minimal user interaction.

Keywords: Robot assisted minimally invasive surgery · 3D reconstruction · 3D registration

1 Introduction

In most cases MIS is preferred to open surgery, given its multiple benefits for patients. It relies on much smaller wounds and incisions, resulting in decreased blood loss, pain and post operational healing time. However, considering doctors perspective, MIS introduces important limitations and constraints. Compared to conventional open surgery, where doctors manipulate organs and surgical instruments directly, control in MIS is complicated and haptic feedback is limited or even non-existent. Moreover, the most important shortcoming of MIS is its visual feedback. The input comes solely from an endoscopic camera. Hence, the visual field is rather limited and often includes blood, smoke, occlusions and challenging lighting conditions. On top of that, MIS procedures last longer than open surgeries, considerably amplifying the strain and effort of the surgeon.

As a consequence, researchers are focusing on circumventing those issues and providing a better and more complete visual experience for surgeons. In that direction, there has been a considerable effort on enriching the visual content

J. Henriques et al. (Eds.): MEDICON 2019, IFMBE Proceedings 76, pp. 1536–1544, 2020.
https://doi.org/10.1007/978-3-030-31635-8_190

that the surgeon receives, by augmenting it with additional information. This can be achieved using AR to display structures which are not visible directly by the camera, but have been acquired by pre-operative images, such as CT or MRI scans.

This paper presents a general AR visualization framework, that is able to estimate the position of internal structures and superimpose them over the endoscopic feed in real-time. The main contributions of the paper are threefold: (a) a 3D reconstruction algorithm that estimates the depth of the structures inside the surgical scene with minimal error while maintaining real-time speed, (b) a 3D-to-3D registration method, which tracks and registers the pre-operative data to the current view, after having initially applied a manual registration step through an intuitive custom graphical user interface, and (c) an AR visualization toolkit displaying the superimposed 3D information over the real view produced by an endoscope mounted on a robotic arm, including an extension for virtual 3D view from a Virtual Reality (VR) headset.

2 Related Work

The use of AR during surgery has been an important topic for research in recent years. Due to the complexity and variety of surgical operations, most approaches in the literature focus on a specific type or set of operations. However, the general pipeline is somewhat similar, where the most important component in order to produce meaningful and accurate AR visualization, is the estimation of the transformation (registration) between the camera and the pre-operative models.

Most methods rely on some manual intervention, most often in the initialization stage in order to produce an accurate initial registration. Considering the technique and materials required to perform the registration, methods are divided into two main categories. The first contains methods that rely on additional materials besides the endoscope, such as fiducials or markers and the second category consists of methods that perform the task using only the endoscopic camera. Muller et al. [1] uses fiducial markers in order to perform the registration between images acquired from CT scans and the camera. Once the registration is achieved, the structures of interest are superimposed on a tablet semi-transparently, that is positioned with its camera facing the surgical area. Simpfendorfer et al. [2] proposes an AR navigation system for laparoscopic radical prostatectomy (LRP), that displays structures acquired from Transrectal Ultrasonography (TRUS), in real-time during the operation. In order to achieve correct registration, a set of needles with colored heads, called navigation aids, are placed in the surface of interest as markers. Then, the TRUS scan is performed, and the navigation aids are segmented. During the operation, these markers are tracked in every frame and using a 2D-to-3D point correspondences, the models generated from TRUS are superimposed via AR in real time video feed.

The aforementioned methods require additional equipment, which is not practical, since endoscopic operations are already complicated and including

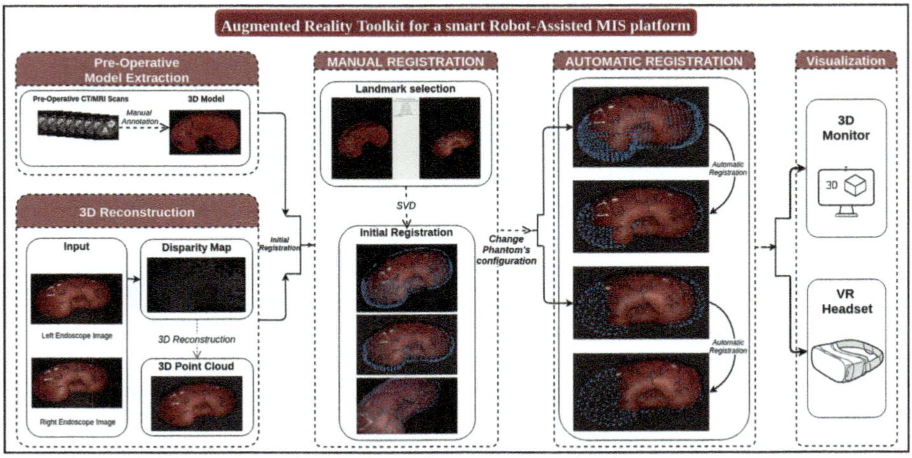

Fig. 1. Augmented Reality Toolkit for a smart Robot-Assisted MIS platform: complete pipeline

extra modules may interfere with the surgical workflow. Thus, there is an increased interest towards methods that rely solely on the standard surgical equipment. Collins et al. [3] introduced an AR guidance system for uterine laparosurgery, to locate myoma. A set of MRI scans are used, and through a semi-automatic framework, are overlaid on intra-operative video feed. Registration is performed using multiple images simultaneously, using a deformable model that enforces constraints from occluding contours. Su et al. [4] presented a markerless system for augmenting live laparoscopic view with pre-operative structures acquired from CT scans. The pipeline includes three manual steps before the intra-operative tracking. First, 3D models of kidney and tumor are segmented from CT scans. Then, they are manually calibrated in order to get the best visual fit with the stereoscopic view. Next, a set of surface points is manually selected to be tracked during the operation and finally, a variant of Iterative Closest Point algorithm is applied for the final registration.

3 System Overview and Methods

Our proposed system includes several modalities, both in terms of hardware and software. A Richard Wolf ENDOCAM Epic 3DHD stereo endoscopic camera is used to acquire pairs of images of the intra-operative scene. Using these pairs we retrieve disparity maps which are then converted into 3D point-clouds. An additional module is used to extract 3D models from manually annotated pre-operative

MRI/CT scans. The aforementioned modules are used from the registration modality in order to create the AR composite view. The last, can be visualized through either a 3D monitor or a Virtual Reality (VR) headset (Figs. 1 and 2).

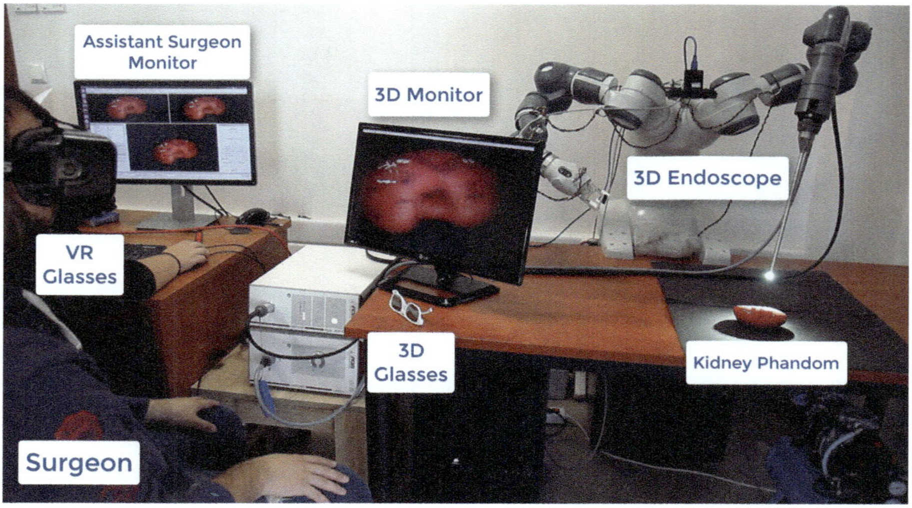

Fig. 2. System overview of the Augmented Reality Toolkit for a smart Robot-Assisted MIS

Finally, a computer with Intel Xeon E5-1650 CPU and NVIDIA TitanX GPU is used for testing all the proposed methods.

3.1 GPU Quasi-Dense 3D Reconstruction

Stereo 3D reconstruction is a well-studied field where many methods have been developed. However, applying these methods to MIS data is very challenging due to significant limitations introduced by the presence of blood, smoke, tissue deformation, reflections as well as occlusions from the surgical instruments and adjacent organs. In addition, lack of real time performance is deemed as a deterrent against applying stereoscopic methods to MIS data. Concerning this, in the current toolkit, we decided to use a Graphics Processing Unit (GPU)-implemented Quasi-Dense 3D reconstruction method which not only reaches real time performance, but also presents similar results compared to state-of-the-art methods [5]. This method uses stereo-laparoscopic images in order to calculate a disparity map, this forms the base for the reconstruction of a 3D point-cloud displaying the operation field as illustrated in the left-bottom part of Fig. 1. Disparity map is estimated in two phases, namely Sparse Matching and Dense Matching, presented below.

Sparse Matching. The algorithm creates a strong base from sparse robust matches across the image pair. Specifically, strong features located in the left image are able to be detected using Good Features to track proposed by Shi-Tomasi [6]. Then, Lucas-Kanade method [7] is used in order to estimate optical flow which subsequently indicates the corresponding points in the right image.

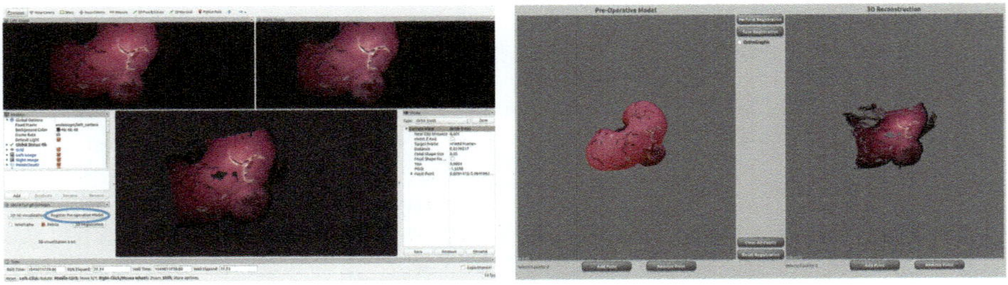

Fig. 3. Left: Evoking Pre-Operative initial registration GUI through Rviz interface, Right: The initial registration GUI. The user can see and manipulate the 3D pre-operative model and the 3D reconstructed area of the surgical scene

Dense Matching. Having established a set of sparse robust matching pairs, it is possible to propagate disparity calculation for the pixels within spatial neighbourhoods of these pairs, which are used as seeds. If a seed match consists of a sparse pixel $p_0(x, y)$ in the left image and the corresponding pixel $p_1(x', y')$ in the right image, then the spatial neighborhoods around them are defined by

$$\begin{aligned} N(p_0) &= \{(x - n_x, y - n_y) : n_x, n_y \in [-N, N]\} \\ N(p_1) &= \{(x' - n_x - d_x, y' - n_y) : n_x, n_y \in [-N, N], d_x \in [-D_g, D_g]\}, \end{aligned} \tag{1}$$

where $(x - n_x, y - n_y)$ denotes the coordinates of each pixel within a spatial window of $(2N + 1) \times (2N + 1)$ pixels centered at seed pixel p_0 and can be matched with a candidate pixel $(x' - n_x - d_x, y' - n_y)$, which is placed within a spatial area of $(2D_g + 1)$ pixels centered at $(x' - n_x, y' - n_y)$ in the right image. Multiple Compute Unified Device Architecture (CUDA) kernels are launched for every seed matching pair in order to calculate Zero Mean Normalized Cross Correlation (ZNCC) between pixels located within the aforementioned neighbourhoods. Furthermore, matching costs are sorted so the algorithm can retrieve the highest matching score for each pixel within a seed's neighborhood. Pairs which are classified as valid are subsequently treated as seeds for the next iteration [5]. This repetitive process terminates when no new valid pairs exist. Having the disparity map, the algorithm is able to calculate depth for every pixel and finally generate a 3D point-cloud as illustrated in the left-bottom part of Fig. 1.

GPU Optimizations. The proposed method includes several computation intensive processes, such as correlation cost calculation for every pixel within seeds' spatial neighbourhoods as well as pixel-wise operations, such as down-sampling and filtering. For this reason, we take advantage of modern GPU capabilities in order to assure real-time performance. Initially, GPU kernels were created in order to calculate and sort multiple correlation scores simultaneously. Shared memory is used as cache memory in this process as it is much faster than global memory. Considering that the algorithm is repetitive, the most important feature we use is multi-streaming parallel programming. This feature enables data transferring and kernel execution overlapping, which is a contributory factor in real-time performance [5]. Currently, the system is able to produce 3D reconstructed point clouds at a rate of 14 Hz.

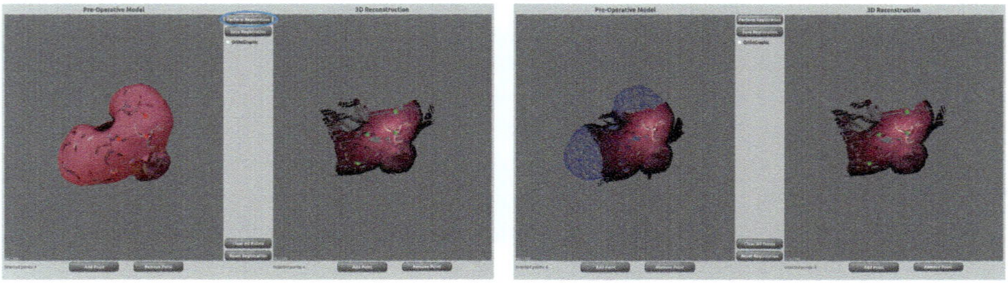

Fig. 4. Left: Corresponding landmark points marked on both the pre-operative 3D model and the 3D reconstruction of the intra-operative scene, Right: The registration result is presented by superimposing the pre-operative model (blue) with the 3D reconstruction of the surgical scene.

3.2 Pre-operative to Intra-operative Registration

A lot of research efforts were devoted to overcoming some of the current limitations applying in MIS, by using AR. The universal demand is the on-the-fly fusion of the pre-operative data and the MIS video, creating a composite view that comprises additional a priory information (such as the location of malignant structures). For this toolkit, several methods have been examined in order to select the most robust and accurate method for registering the pre-operative data to the intra-operative scene [8].

The registration framework we propose consists of two separate phases, the initial manual registration and the automatic registration procedure. The first, acts as initialization step for the automatic phase.

Initial Manual Registration. The key to a good registration result within the automatic registration phase is a good initialization. A dedicated Graphical User Interface (GUI) was created to address the need of this step. The manual registration GUI we created allows the surgeon to manually start the process when a sufficient view of the organ of interest is achieved. When the organ is exposed (even partially), in this early version of the toolkit, the surgeon can evoke the initial registration GUI through a dedicated button in the Rviz interface as shown in the left part of Fig. 3. The module captures the current intra-operative scene and presents its 3D reconstruction side by side with the pre-operative model for comparison (right part of Fig. 3).

The user can manipulate the model and the 3D reconstruction separately by rotating and moving the meshes in the 3D space in a way that common landmarks on both of them can be identified. After that, the user should mark these landmarks by clicking on them (Fig. 4). At least 4 points should be selected before performing the registration step. The corresponding pairs of points are then used to perform the initial registration phase that treats the models as rigid objects. We are using Singular Value Decomposition (SVD) [9] to calculate the transformation matrix.

After this first step, the registration result is presented on the left window in the GUI by superimposing the pre-operative model (in wire-frame mode) with the intra-operative reconstruction. In this way the user can review the registration result and either accept it or discard it and restart the procedure. When an acceptable registration is achieved, it can then be saved. The registered pre-operative model is overlaid onto the live 3D reconstruction of the intra-operative scene in the Rviz interface, since the whole framework is currently integrated into ROS (Fig. 5).

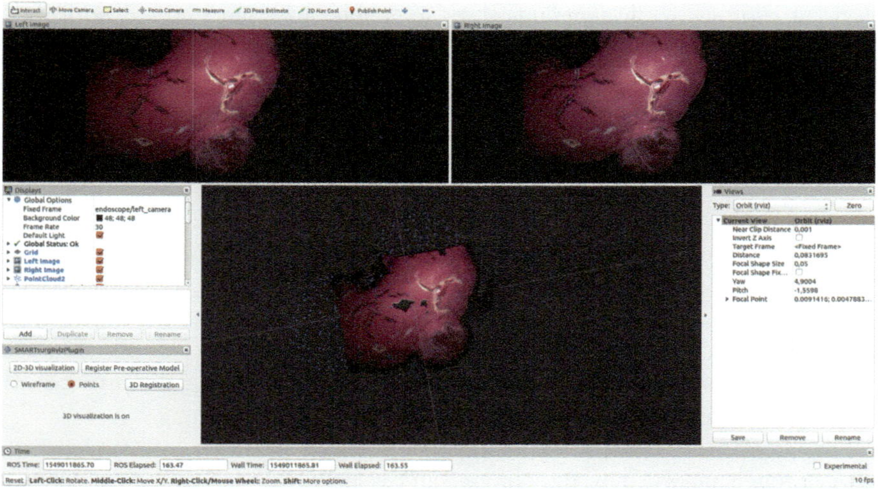

Fig. 5. The pre-operative model is shown (in blue) fully registered with the live 3D reconstruction of the intra-operative scene, as a result of the manual initial registration procedure.

Automatic Registration and Tracking. The second phase is the automatic registration and tracking process. The transformation extracted from the previous phase is used as an initialization step in order to track the movement of the organ and adjust the pre-operative model to that. We investigated several state-of-the-art methods that could be applied in this phase [8] including methods for point-to-point and surface-to-surface registration. We concluded after their evaluation that Coherent Point Drift (CPD) method is more accurate in our case, producing acceptable registration results in the 78.8% of the cases with Mean Square Error of 1.76 mm.

The pre-operative model is continuously superimposed within the intra-opera-tive view and it can track the changes in the position of the organ within the real-time framework at approximately 4.5 Hz. The pre-operative model is registered not only to the 3D reconstruction, but also to the whole system (camera, and master-slave robot arms). This way the registration remains in accordance with the camera movements. The outcome is visible in various mediums. It can be seen inside the Rviz environment overlaid with the 3D reconstruction point cloud. Additionally, it is visible on a passive 3D monitor accessible from the assistant surgeon, and it is also visible in a VR headset (we used Oculus Rift and HTC Vive for testing purposes) for the operating surgeon (Fig. 6).

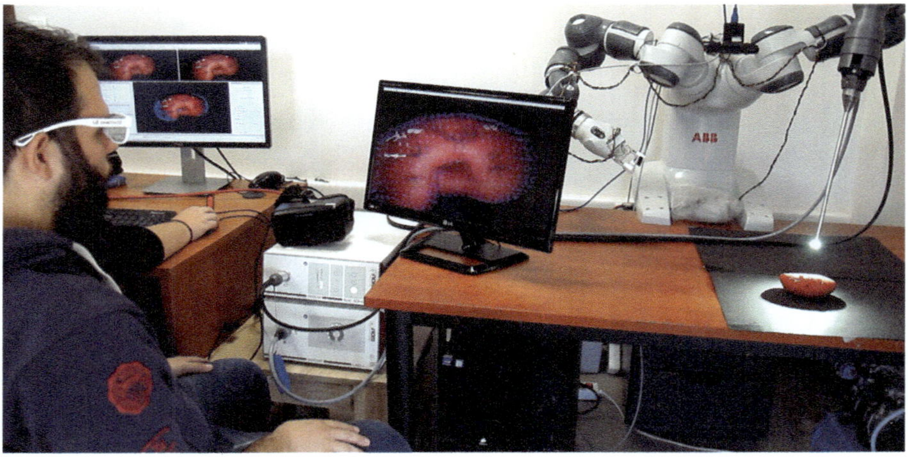

Fig. 6. The composite view visible to the assistant surgeon on the 3D monitor and the VR headset.

4 Conclusion and Future Work

In this paper, an AR visualization toolkit for MIS is presented. It overlays 3D models of structures which are acquired pre-operatively, semi-transparently to the live endoscopic video. The process of registering those models to the current endoscopic frame is semi-automatic, requiring minimal user interaction. An intuitive GUI has been developed, making the presented visualization toolkit suitable for use by non-technical users.

During this research, we examined methods concerning only rigid transformations for the registration. We continue our efforts on the development of our own proposal for a novel surface registration method, that will be appropriate for SMARTsurg's constraints and limitations. We intent to develop a method that will be suited for on-the-fly registration and will perform sufficiently taking into account noisy data and deformations of the organ tissue. Furthermore, registration accuracy is highly dependent upon 3D reconstruction. If wrong depth estimates, or outliers are produced, the success of the registration will be compromised. MIS video sequences mostly contain smooth and untextured areas, which are proven to be very challenging to accurately reconstruct through stereo matching. Hence, a set of methods that are able to accurately estimate depth in such areas are currently being examined.

Acknowledgments. The research leading to these results has received funding from the European Community's Framework Programme Horizon 2020 under grant agreement No. 732515 SMARTsurg.

References

1. Müller, M., Rassweiler, M., Klein, J., Seitel, A., Gondan, M., Baumhauer, M., Teber, D., Rassweiler, J.J., Meinzer, H., Maier-Hein, L.: Mobile augmented reality for computer-assisted percutaneous nephrolithotomy. Int. J. Comput. Assist. Radiol. Surg. **8**(4), 663–675 (2013)
2. Simpfendörfer, T., Baumhauer, M., Müller, M., Gutt, C.N., Meinzer, H.-P., Rassweiler, J.J., Guven, S., Teber, D.: Augmented reality visualization during laparoscopic radical prostatectomy. J. Endourol. **25**(12), 1841–1845 (2011)
3. Collins, T., Pizarro, D., Bartoli, A., Canis, M., Bourdel, N.: Computer-assisted laparoscopic myomectomy by augmenting the uterus with pre-operative MRI data. In: 2014 IEEE International Symposium on Mixed and Augmented Reality (ISMAR), pp. 243–248 (2014)
4. Su, L.M., Vagvolgyi, B.P., Agarwal, R., Reiley, C.E., Taylor, R.H., Hager, G.D.: Augmented reality during robot-assisted laparoscopic partial nephrectomy: toward real-time 3D-CT to stereoscopic video registration. Urology **73**, 896–900 (2009)
5. Zampokas, G., Tsiolis, K., Peleka, G., Mariolis, I., Malasiotis, S., Tzovaras, D.: Real-time 3D reconstruction in minimally invasive surgery with quasi-dense matching. In: 2018 IEEE International Conference on Imaging Systems and Techniques (IST), pp. 1–6. IEEE (2018)
6. Shi, J., Tomasi, L.: Good features to track. In: Proceedings of IEEE Computer Society Conference on Computer Vision and Pattern Recognition, CVPR (1994)
7. Kanade, T., Okutomi, M.: A stereo matching algorithm with an adaptive window: theory and experiment. TPAMI **16**(9), 920–932 (1994)
8. Peleka, G., Zampokas, G., Mariolis, I., Malasiotis, S., Tzovaras, D.: Intra-operative 3D registration of MIS reconstructed surfaces to pre-operative models. In: 2018 IEEE International Conference on Imaging Systems and Techniques (IST), pp. 1–6. IEEE (2018)
9. Golub, G.H., Reinsch, C.: Singular value decomposition and least squares solutions. In: Linear Algebra, pp. 134–151. Springer, Berlin (1971)

Control of a da Vinci EndoWrist Surgical Instrument Using a Novel Master Controller

Sajeeva Abeywardena(✉), Efi Psomopoulou, Mohammad Fattahi Sani, Antonia Tzemanaki, and Sanja Dogramadzi

Bristol Robotics Laboratory, University of the West of England, Bristol, UK
sajeeva.abeywardena@brl.ac.uk

Abstract. A novel master controller for robot-assisted minimally invasive surgery (RAMIS) is introduced and used to control a da Vinci EndoWrist instrument. The geometric model of the master mechanism and its mapping to the geometry of the EndoWrist tool are derived. Experimental results are conducted to open and close the jaws of an EndoWrist tool, and show that the developed mapping algorithm is accurate with a root mean square error of 0.7463 mm.

Keywords: Exoskeleton · Robot-assisted minimally invasive surgery · Geometric models · Mapping

1 Introduction

Robot-assisted minimally invasive surgery (RAMIS) has gained popularity in recent decades through use of the da Vinci master-slave surgical system [1]; offering improved vision, precision and patient recovery time compared to traditional MIS [2]. Typically in RAMIS, the surgeon controls the surgical instrument external from the patient's body via a master robotic controller that is not kinematically similar to the instrument tip, i.e. the da Vinci system [1] and other investigations in literature [3–6]. As such, the control of the surgical instrument is not an intuitive motion consistent with those used in open surgical techniques, i.e. grasping and cutting. Researchers have developed master controllers for uses in medical robotic interventions that are kinematically similar to the slave mechanism for more intuitive control [7,8]. However, these are not compatible for use with the da Vinci EndoWrist instruments [11]. To offer surgeons a more anthropomorphic control interface similar to those they experience in open surgery and be able to control EndoWrist instruments, commercial data gloves or exoskeletons such as CyberGrasp [9] and Manus VR [10] could be used. However, the bulkiness and weight of such devices will hinder the performance of surgeons during RAMIS.

In this paper, a novel anthropomorphic master controller for use with daVinci EndoWrist instruments is introduced. The master controller and its geometric model are discussed in Sect. 2; the mapping of the geometry of the master to a da

J. Henriques et al. (Eds.): MEDICON 2019, IFMBE Proceedings 76, pp. 1545–1550, 2020.
https://doi.org/10.1007/978-3-030-31635-8_191

Vinci EndoWrist surgical instrument is derived in Sect. 3; experimental procedure and results presented in Sect. 4; and conclusion and future work summarised in Sect. 5.

2 Master Controller

The developed master controller is a 3D printed wearable exoskeleton that tracks the motion of the thumb as well as the index and middle finger. Design considerations focussed on minimising the footprint and weight of the exoskeleton. Hall-effect sensors are utilised to measure the angle of the distal interphalangeal (DIP) and proximal interphalangeal joints (PIP) of the fingers; and the interphalangeal joint (IP) of the thumb. Further, nine-degrees-of-freedom (dof) Bosch BNO-055 inertial measurement units (IMUs) are used to track the orientation of the metacarpophalangeal joints (MCP) of the thumb, index and middle fingers; as well as the carpometacarpal (CMC) joint of the thumb and the wrist. A reference IMU is placed on the dorsal side of the hand, as shown in Fig. 1, for which relative orientation is calculated and thus joint angles can be derived. The exoskeleton was affixed to the wearer's hand via flexible 3D printed straps; and the CMC, wrist and reference IMUs stuck onto the hand with double-sided tape.

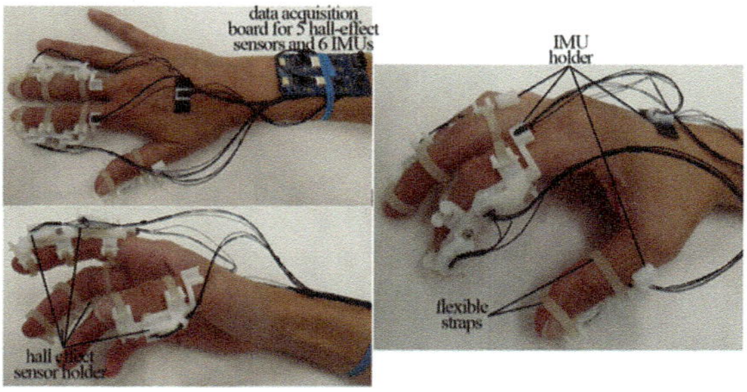

Fig. 1. The novel master wearable exoskeleton

The Direct Geometric Model (DGM) calculates the position of the thumb, index and middle fingers with respect to a global base frame. However, it is important to note that as sensors cannot be directly attached to the joints of the thumb and fingers, the DGM calculates the motion of the wearable control interface which is assumed to provide a reasonable approximation to the motion of the fingers. The modified Denavit-Hartenberg (DH) notation is applied to the model of the links of the exoskeleton from which the position of the digits can be calculated.

3 Mapping Algorithm

The da Vinci EndoWrist surgical instruments consists of four-dof—roll, pitch, yaw and grasp. Roll and pitch are directly controlled by individual motors; whilst the yaw and grasp width of the instrument are dictated by the location of the independently controlled jaws. To map between the exoskeleton and the EndoWrist instrument, the geometric models—direct and inverse (IGM)—of the instrument are required.

3.1 Direct Geometric Model

The DGM of the jaws of the EndoWrist instrument can be derived using the modified DH parameters from which the transformation between the global base frame $\mathcal{F}_0$ and the frame of the individual jaws $\mathcal{F}_{3i}$, ${}^0\mathbf{T}_{3i}$ can be derived. To prescribe a unified motion between the jaws, a frame $\mathcal{F}_C$ was defined at the centroid of the triangle formed by vertices B_1, B_2 and B_3, defined in Fig. 2. The co-ordinates of the vertices in $\mathcal{F}_0$ can be determined from the transformation matrices previously defined. Specifically, the co-ordinates in terms of the joint-space variables $\boldsymbol{\theta} = [\theta_1, \theta_2, \theta_{31}, \theta_{32}]$ are

$$ {}^0\mathbf{b}_{1\theta} = {}^0\mathbf{T}_2{}^2\mathbf{l}_1 \qquad {}^0\mathbf{b}_{2\theta} = {}^0\mathbf{T}_{31}{}^{31}\mathbf{l}_2 \qquad {}^0\mathbf{b}_{3\theta} = {}^0\mathbf{T}_{32}{}^{32}\mathbf{l}_2 \tag{1} $$

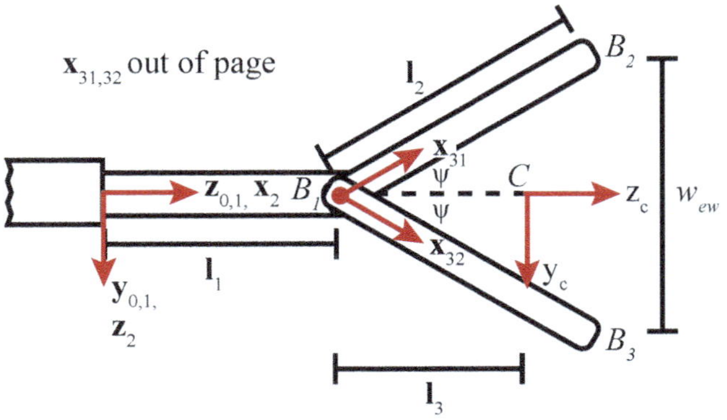

Fig. 2. Frame assignment and geometric parameters of EndoWrist instrument

As such, the origin of this frame—C—is the control point of the EndoWrist instruments. The pose of this point expressed in the base frame is ${}^0\mathbf{T}_C$, i.e.

$$ {}^0\mathbf{T}_C = \begin{bmatrix} {}^0\mathbf{Q}_C & {}^0\mathbf{p}_C \\ \mathbf{0}_{1\times 3} & 1 \end{bmatrix} \qquad {}^0\mathbf{Q}_C = [\mathbf{i}\ \mathbf{j}\ \mathbf{k}] \qquad {}^0\mathbf{p}_C = \frac{{}^0\mathbf{b}_{1\theta} + {}^0\mathbf{b}_{2\theta} + {}^0\mathbf{b}_{3\theta}}{3} \tag{2} $$

$$ \mathbf{k} = \frac{{}^0\mathbf{p}_C - {}^0\mathbf{b}_{1\theta}}{\|{}^0\mathbf{p}_C - {}^0\mathbf{b}_{1\theta}\|} \qquad \mathbf{j} = \frac{({}^0\mathbf{b}_{3\theta} - {}^0\mathbf{b}_{1\theta}) - \mathbf{k}\mathbf{k}^T({}^0\mathbf{b}_{3\theta} - {}^0\mathbf{b}_{1\theta})}{\|({}^0\mathbf{b}_{3\theta} - {}^0\mathbf{b}_{1\theta}) - \mathbf{k}\mathbf{k}^T({}^0\mathbf{b}_{3\theta} - {}^0\mathbf{b}_{1\theta})\|} \qquad \mathbf{i} = \mathbf{j} \times \mathbf{k} $$

3.2 Inverse Geometric Model

The IGM determines the set of joint angles $\boldsymbol{\theta}$ that will attain a specific pose ${}^0\mathbf{T}_C$. As the EndoWrist instruments have four-dof, only four-dof of ${}^0\mathbf{T}_C$ can be directly specified. The chosen task space variables are the orientation ${}^0\mathbf{Q}_C$ and the width, w_{ew}, between the jaws of the EndoWrist instrument. The orientation is specified as the orientation of the wrist of the exoskeleton wearer, whilst the width of the jaws is correlated to the width between the tips of the index and middle finger. This width, w_{exo}, is mapped to w_{ew} as

$$w_{ew} = \frac{max_{ew} - min_{ew}}{max_{exo} - min_{exo}} w_{exo} + \frac{min_{ew} max_{exo} - min_{exo} max_{ew}}{max_{exo} - min_{exo}} \tag{3}$$

where max_j/min_j is the maximum/minimum width of the exoskeleton and EndoWrist instrument.

From Fig. 2, vertices B_1, B_2 and B_3 can be defined in $\mathcal{F}_C$ in terms of the angle ϕ, i.e.

$$^C\mathbf{b}_{1C} = \begin{bmatrix} 0 \\ 0 \\ -\frac{2}{3} l_2 c_\phi \end{bmatrix} \qquad ^C\mathbf{b}_{2C} = \begin{bmatrix} 0 \\ l_2 s_\phi \\ \frac{1}{3} l_2 c_\phi \end{bmatrix} \qquad ^C\mathbf{b}_{3C} = \begin{bmatrix} 0 \\ -l_2 s_\phi \\ \frac{1}{3} l_2 c_\phi \end{bmatrix} \tag{4}$$

where $c_\phi = \cos\phi$, $s_\phi = \sin\phi$ and $\phi = \arcsin\left(\frac{w_{ew}}{2l_2}\right)$.

Equation (4) are converted into $\mathcal{F}_0$ using ${}^0\mathbf{T}_C$, whereby the position components are expressed in terms of the specified orientation ${}^0\mathbf{Q}_C$ and jaw angle ϕ, i.e.

$${}^0\mathbf{p}_C = {}^0\mathbf{Q}_2 {}^2\mathbf{l}_2 + {}^0\mathbf{Q}_C {}^C\mathbf{l}_3 \tag{5}$$

where ${}^0\mathbf{Q}_2$ only takes into account roll and pitch of ${}^0\mathbf{Q}_C$, ${}^2\mathbf{l}_2$ is the vector from the origin of $\mathcal{F}_0$ to vertex B_1 and ${}^C\mathbf{l}_3$ is the vector from the vertex B_1 to the origin of $\mathcal{F}_C$.

As the vertices B_1, B_2 and B_3 are specified in terms of the joint angles and task space variables through Eqs. (1) and (4), the following holds

$${}^0\mathbf{b}_{iC} = {}^0\mathbf{b}_{i\theta} \tag{6}$$

for $i = 1, 2, 3$. As such, there are 9 equations in terms of 4 unknowns—the joint space variables. Thus, solving these equations provides the solution to the IGM.

4 Experimental Setup

The master controller was used to control an 8 mm EndoWrist ProGrasp Forceps, as shown in Fig. 3, which was actuated by four Maxon brushed DC motors. The motors interfaced with a custom built motherboard that consisted of ePos2 control modules. A master-slave Arduino setup was utilised for the master controller. The sensors interfaced with a custom built motherboard that consisted of four Arduino Uno chips that communicated via SPI with a master Arduino Uno board. This board in turn communicated with a PC via serial. The exoskeleton and instrument were interfaced in Linux and communicated via ROS.

Fig. 3. ProGrasp Forceps (left) and example capture of instrument tip (right)

4.1 Results

Investigation into the accuracy of the mapping algorithm was conducted, specifically focussing on grasp width. A camera was used to track the motion of the jaws of the ProGrasp forceps in a two-dimensional plane, with an example capture provided in Fig. 3. The results from an experiment in which the jaws were continuously opened and closed are shown in Fig. 4. This experiment yielded a root mean square error of 0.7463 mm. The EndoWrist instruments are cable driven with base-located motors, i.e. motors are not directly attached to the joints of the instrument-tip. Thus, cable tension and losses such as friction and hysteresis will affect the motion transmitted to the tool tip, i.e. why desired and measured width differ. As evident in Fig. 4, this is most apparent when closing the jaws. Nevertheless, the magnitude of error shows the mapping algorithm to be relatively accurate for the grasping dof.

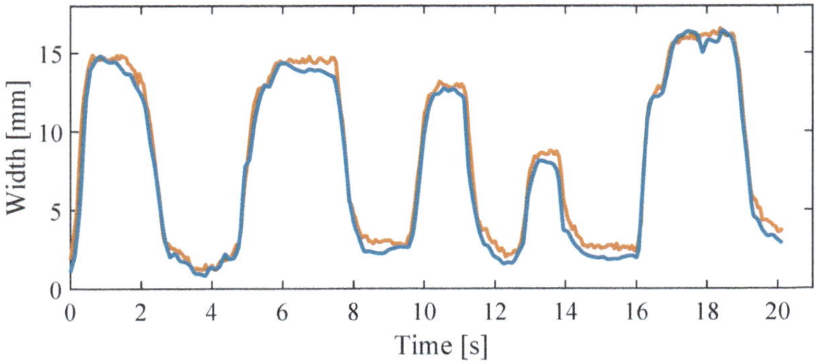

Fig. 4. Experimental results: command width (blue), measured width (orange)

5 Conclusion and Future Work

A novel master controller for RAMIS was introduced which provides surgeons with a more intuitive control interface compared to existing solutions. The developed control interface is akin to the surgeons performing RAMIS directly using their hands. The geometric model of the master mechanism was summarised and its mapping to the motion of the EndoWrist tool derived. Experimental results showed that the mapping algorithm is accurate in the grasping dof, providing promise for the use of the developed master in RAMIS. In the future, the master controller will be integrated with a Haption Virtuose6D desktop device to allow for the depth of the EndoWrist instrument to be controlled. Further, the accuracy and robustness to parameter variation of the mapping algorithm and the efficacy of the developed master controller compared to the standard da Vinci master controller will be investigated. Additionally, the use of the wearable exoskeleton to control a novel anthropomorphic surgical instrument will be explored.

Acknowledgements. This work was supported by the European Union's Horizon 2020 research and innovation programme under grant agreement No. 732515.

References

1. https://www.intuitive.com/
2. Lanfranco, A.R., Castellanos, A., Desai, J., Meyers, W.: Robotic surgery: a current perspective. Ann. Surg. **239**(1), 14–21 (2004)
3. Arata, J., Mitsuishi, M., Warisawa, S., Tanaka, K., Yoshizawa, T., Hashizume, M.: Development of a dexterous minimally-invasive surgical system with augmented force feedback capability. In: 2005 IEEE/RSJ International Conference on Intelligent Robots and Systems, pp. 3207–3212 (2005)
4. Thielmann, S., Seibold, U., Haslinger, R., Passig, G., Bahls, T., Jrg, S., Nickl, M., Nothhelfer, A., Hagn, U., Hirzinger, G.: Mica - a new generation of versatile instruments in robotic surgery. In: 2010 IEEE/RSJ International Conference on Intelligent Robots and Systems, pp. 871–878 (2010)
5. Hannaford, B., Rosen, J., Friedman, D.W., King, H., Roan, P., Cheng, L., Glozman, D., Ma, J., Kosari, S.N., White, L.: Raven-II: an open platform for surgical robotics research. IEEE Trans. Biomed. Eng. **60**(4), 954–959 (2013)
6. Talasaz, A., Trejos, A.L., Patel, R.V.: The role of direct and visual force feedback in suturing using a 7-DOF dual-arm teleoperated system. IEEE Trans. Haptics **10**(2), 276–287 (2017)
7. Ikuta, K., Hasegawa, T., Daifu, S.: Hyper redundant miniature manipulator "hyperfinger" for remote minimally invasive surgery in deep area. In: 2003 IEEE International Conference on Robotics and Automation, vol. 1, pp. 1098–1102 (2003)
8. Zhang, L., Zhou, N., Wang, S.: Direct manipulation of tool-like masters for controlling a master-slave surgical robotic system. Int. J. Med. Robot. Comput. Assist. Surg. **10**(4), 427–437 (2014)
9. http://www.cyberglovesystems.com/cybergrasp
10. https://manus-vr.com
11. https://www.intuitive.com/en-us/products-and-services/da-vinci/instruments

Toward a Neural-Symbolic Framework for Automated Workflow Analysis in Surgery

Hirenkumar Nakawala[1(✉)], Elena De Momi[2], Roberto Bianchi[3], Michele Catellani[3], Ottavio De Cobelli[3], Pierre Jannin[4], Giancarlo Ferrigno[2], and Paolo Fiorini[1]

[1] Department of Computer Science, University of Verona, Via Str. le Grazie 15, 37134 Verona, Italy
hirenkumarchandrakant.nakawala@univr.it
[2] Department of Electronics, Information and Bioengineering (DEIB), Politecnico di Milano, Piazza Leonardo da Vinci 32, 20133 Milan, Italy
[3] Department of Urology, European Institute of Oncology, Via Guiseppe Ripamonti 435, 20141 Milan, Italy
[4] LTSI, INSERM U1099, Universitè de Rennes 1, Rennes 35000, France

Abstract. Learning production rules from continuous data streams, e.g. surgical videos, is a challenging problem. To learn production rules, we present a novel framework consisting of deep learning models and inductive logic programming (ILP) system for learning surgical workflow entities that are needed in subsequent surgical tasks, e.g. "what kind of instruments will be needed in the next step?" As a prototypical scenario, we analyzed the Robot-Assisted Partial Nephrectomy (RAPN) workflow. To verify our framework, first consistent and complete rules were learnt from the video annotations which can classify RAPN surgical workflow and temporal sequence at high-granularity e.g. steps. After we found that RAPN workflow is hierarchical, we used combination of learned predicates, presenting workflow hierarchy, to predict the information on the next step followed by a classification of step sequences with deep learning models. The predicted rules on the RAPN workflow was verified by an expert urologist and conforms with the standard workflow of RAPN.

Keywords: Production rules · Robot-Assisted Partial Nephrectomy · Surgical workflow · Deep learning · Inductive logic programming

1 Introduction

Learning production rules from continuous data streams, e.g. surgical videos, is a challenging problem. In our previous work [1, 2], we used production rules, i.e. "if-then" rules, inside an ontology to define the surgical workflow and to discretize surgical activities through symbol grounding by assigning semantic meaning to the objects in the environment. However, pre-defining a fixed set of production rules limits the scalability of expert systems in dynamic situations because it limits understanding of the streaming data and the real-time reasoning on it. The production rules are generally very discriminative and easily comprehensible by humans. The latter could

J. Henriques et al. (Eds.): MEDICON 2019, IFMBE Proceedings 76, pp. 1551–1558, 2020.
https://doi.org/10.1007/978-3-030-31635-8_192

be helpful to understand the surgical workflow, for example displaying information on the consequent surgical activities which are automatically extracted from the rules. Rules could also be helpful as constraints in automation of the surgery, e.g. verification of step sequences for the safe task execution. A full surgical activity is expected to consist of surgical entities, e.g., *phases*, *steps*, *instruments*, *actions* and *anatomy*. Recognition of a surgical activity needs a large set of annotated data on these entities as well as robust neural network architectures. On the contrary, scene recognition is still outside the capabilities of symbolic systems which are good at recognizing relations between entities in surgical activities. Integrating neural and symbol systems [3] for analyzing surgical workflows may provide complementary benefits interpreting a surgical activity.

Inductive Logic Programming (ILP) is formed at the intersection of machine learning and logic programming. ILP systems learn predicate descriptions, first-order causal theories, in the form of rules from the examples, e.g. real-world instances, and background knowledge. Examples, e.g. surgical step sequences, background knowledge, which are predicates describing relational information on surgical entities, and final descriptions i.e. a hypothesis or a rule all are described as logic programs. Recently, ILP has been extensively used in classification, data mining and information extraction (IE) tasks [4]. In the medical field, ILP has been applied to diagnostics e.g. to detect the breast cancer [5] and to do knowledge discovery from the mammography reports [6]. As far as our knowledge allows, ILP is yet to be applied for the analysis of surgical workflows which can be used to predict information about workflow steps. In [7], ILP system is used with an instance detection system, using deep learning models, for interpretation of the scene. In this manuscript, we present a preliminary analysis of a use of ILP to analyze surgical workflow for learning complete and consistent rules for classification of RAPN workflow as well as its use in a pipeline with an instance detection system which was used to recognize surgical steps and sequences.

As a prototypical scenario, in this work, we automatically analyzed the surgical workflow of Robot-Assisted Partial Nephrectomy (RAPN). RAPN is performed to remove the kidney tumor. RAPN is associated with high complicate rates from 12.3 to 33% with different surgical approaches [8]. Learning production rules to classify relations between the surgical entities, following a recognition of an instance e.g. a sequence of surgical step, would be of great importance to surgical assistance, especially for novice surgeons, by providing information on a workflow which is expected to improve surgical outcomes.

2 Methods

In our previous work [1], we extracted features from 9 videos (996,373 frames) representing 10 RAPN steps and classified it using the deep learning models. As shown in Fig. 1, the deep learning models were consisted of Convolutional Neural Network (CNN) and Long Short-Term Memory (LSTM) cells to classify "Current step" and "Next step" respectively. In this work, we combined predictions of these two networks as predicates representing the current step and corresponding subsequent step, for example "hasNextStep"(mobilization, dissection), and used correctly predicted step sequences, as positive examples, in an ILP system. We then applied ILP to classify a dataset on RAPN workflow entities to find relational information between instances of surgical entities, for example, instances of instruments needed in the next step.

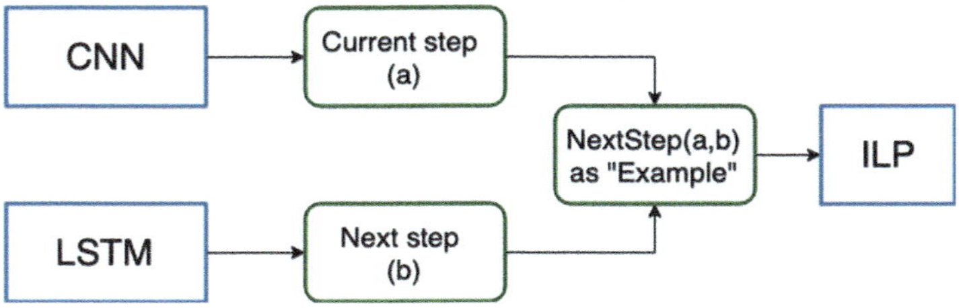

Fig. 1. A prototypical framework consisting of two components, i.e. (1) deep learning models – CNN and LSTM, and (2) ILP. In the framework, outputs of the deep learning models, instances of current and next step, are forwarded as relational predicates "Example" to an ILP system.

Terminology

The first-order logic alphabet is composed of *predicate* symbol (e.g., "step"), *function* symbols (e.g., "hasAction"), *constant* (e.g., "dissection") and *variables* (e.g., *x*). A *term* is any constant, variable, or function applied to a term (e.g., *dissection, x, has Action (x))*. An *atomic formula* is a predicate symbol together with its arguments, each argument being a "term". A *ground atom (or fact)* is an atomic formula with no variables (e.g., *hasNextStep (mobilization, dissection))*. The dataset created from video annotations are *ground atoms* and they also constitute the *background knowledge*. A *literal* is an atomic formula and a clause is a disjunction of literals whose values are assumed to be universally quantified. "Horn clause" is a clause with exactly one positive literal.

Aleph

ILP classifiers attempt to learn a set of rules (definite clauses), given a dataset of positive and negative instances, that will correctly discriminate between the sets. These rules cover most or all of the positive instances, and little or none of the negative instances. The aim of ILP is to find a theory that is complete and consistent. More formally, given a positive example $pos(x_i)$, e.g. a predicate representing surgical step sequence, let $\bot_i$ be the bottom clause, consisting of action, phase and instruments for example *i*. $\bot_i$ is the most specific hypothesis that together with the background knowledge *B*, a procedural knowledge base, entail $x_i : (B \wedge \bot_i \wedge x_i) \vdash pos(x_i)$. In this work, we used "Aleph" (A learning engine for proposing hypotheses) [9], a type of ILP classifier, to learn first-order rules in the form of Horn clauses. Aleph also uses bottom clauses, i.e. the most specific hypothesis, to guide the search. Moreover, it runs also without any negative examples. The predicate in the rule *head* is not in the background knowledge. We followed the four basic steps in the Aleph greedy learning approach:

(1) Select a positive example. Each instance of the relation can be seen as pair of step sequences, e.g. *hasNextStep*(mobilization, dissection).
(2) Build bottom clause. The bottom clause is conjunction of all the relations as pair of action and step, step and instrument, phase and step and between the steps.

For example, if (mobilization *hasNextStep* A) then (*actionperformedinstep* in A are dissect and cut) and (A *stephasinstrument* monopolarCurvedScissors and fenestratedBipolar and sucker and hemolokClipApplicator) and (hilumDissection *phasehasstep* A) and (A *haspreviousstep* mobilization).

(3) Search. This step uses greedy best-first search to find a clause consistent with the data.

(4) Removed covered positive examples. Coverage of each hypothesis is counted and the positive examples from the training set are removed.

As a refinement operator, we only considered examples where we cannot find a consistent generalization as exceptions. In those cases, we added the bottom clause as the consistent rule.

Dataset From Video Annotation and Background Knowledge Construction
To extract the background knowledge, 9 videos at 24 Hz with an approximate length of 82.49 $\pm$ 37.54 min on RAPN were acquired using the da Vinci Xi surgical system and the da Vinci Xi endoscope (Intuitive Surgical Inc., CA, USA) at European Institute of Oncology (Milan, Italy). As mentioned in [1], we used Anvil annotation tool [10] to annotate the videos, consisting the frame-by-frame annotations of "Phase", "Step", "Instrument", "Assistant-Instrument1", "Assistant-Instrument2", "Anatomy", and "Actions". Each track specifies different surgical workflow entities in synchrony without specifying any relations between them. Definitions of classes of annotations and more details can be found in [1]. Video annotations were saved in a comma separated values as a database, where the relational predicates were extracted considering the time-series in each video. To form a background knowledge, we extracted tuples specifying relations between *phase*, *step*, *action*, *instrument* and *anatomy* and temporal relations between *step*. Background knowledge is represented in Prolog language. Each relational predicate is represented in the form: predicate (ModeType, Modetype, ..), where ModeType is one of: (1) + ModeType specifies the input, (2) − ModeType specifies the output, and (3) # Modetype specifies a constant. Examples of Prolog facts derived from video annotations are shown in Table 1.

Experimental Protocols
The goal of the experimental protocols is to assess the complete and consistent rules to solve the problem of classification of surgical workflow and prediction of information in the next step following a classification of steps and sequences from the videos.

1. As shown in Table 1, "Rule" i.e. hypotheses 1, 2 and 3 are evaluated if they are consistent and complete over a set of 5 annotations. In "EP1", all the datasets are combined, and redundant examples are removed. Then, the dataset is divided into 65%, a total of 17 examples, as a training set and 35%, a total of 9 examples, as a testing set to get the baseline results (a single split). Furthermore, in "EP2", we did fivefold LOOCV (Leave-one-out cross-validation) to learn the rules and to check whether they are useful for classifying the workflow.
2. In "EP3", predicted instances of steps predicates and sequences from deep learning models are forwarded to Aleph, with background facts and hypothesis as shown in "Rule-4" in Table 1, to predict workflow entities of the next step. The predictions are qualitatively assessed by an expert urologist "MC".

Table 1. Predicates, "ground atoms", used in background knowledge

Rules	Type of information	Predicates	Description	An exemplary predicate showing examples
1	Step transitions induced by temporal relations of actions in steps	"hasNextStep" (+step, −step) "stepHasAction" (+step, −action) "actionPerformedInPreviousStep" (+action, −step)	Relations between actions and steps for predicting next step	*hasNextStep*(mobilization, dissection) as a positive example *hasNextStep*(dissection, mobilization) as a negative example
2	Relations between step and phase induced by relations between phase, actions and instruments	"phaseHasStep" (+phase, −step) "phaseHasInstrument" (+phase, −instrument) "actionPerformedbyInstrument" (+action, −instrument) "actionPerformedInStep" (+action, −step)	Relations between actions, instrument and phases to classify steps in phases	*phaseHasStep*(mobilization, mobilize) as a positive example *phaseHasStep*(mobilization, suturing) as a negative example
3	Temporal relations	"hasPreviousStepOfPreviousStep" (+step, −step) "hasPreviousStep" (+step, −step) "hasPreviousStep" (+step,−step)	Classifying previous steps	*hasPreviousStepOfPreviousStep* (removal, drainage) as a positive example *hasPreviousStepOfPreviousStep* (dissection, drainage) as a negative example
4	Step transitions induced by the instances of workflow entities.	"hasNextStep" (#step, +step) "actionPerformedInStep" (#action, +step) "stepHasInstrument" (+step, #instrument) "phaseHasStep" (#phase, +step) "hasPreviousStep" (+step, #step)	Classify instances of workflow entities, e.g. actions, instruments, phases and step of a next step	*hasNextStep*(mobilization, dissection) as a positive example

3 Results and Discussion

Aleph's learning is done by restricting the clause length from minimum of 2 literals to maximum of 8 literals. The results are summarized as below:

Table 2. Induced rules, i.e. for hypothesis representing rule-4, from the instances recognized from deep learning models. The rules are represented in horn clauses and show instances of surgical entities that are required in the next step ("A").

Predicted step sequence	Generated rules
hasnextstep (mobilization, dissection)	hasnextstep(mobilization, A):- actionperformedinstep(dissect, A), actionperformedinstep(cut, A), stephasinstrument(A, monopolarCurvedScissors), stephasinstrument(A, fenestratedBipolar), stephasinstrument(A, sucker), stephasinstrument(A, hemolokClipApplicator), phasehasstep (hilumDissection, A), haspreviousstep(A, mobilization)
hasnextstep(dissection, ultrasound)	hasnextstep(dissection, A):- actionperformedinstep(detect, A), stephasinstrument(A, fenestratedGraspingForcep), stephasinstrument(A, monopolarCurvedScissors), stephasinstrument(A, fenestratedBipolar), stephasinstrument(A, laparoscopicUltrasoundProbe), phasehasstep(tumorExposure, A), haspreviousstep(A, identificaion)
hasnextstep(marking, clamping)	hasnextstep(marking, A):- actionperformedinstep(clamp, A), actionperformedinstep(dissect, A), stephasinstrument(A, monopolarCurvedScissors), stephasinstrument(A, fenestratedBipolar), stephasinstrument(A, laparoscopicBulldog), phasehasstep(tumorResection, A), haspreviousstep(A, marking)
hasnextstep(clamping, resection)	hasnextstep(clamping, A):- actionperformedinstep(resect, A), stephasinstrument(A, monopolarCurvedScissors), stephasinstrument(A, fenestratedBipolar), stephasinstrument(A, fenestratedGraspingForcep), stephasinstrument(A, sucker), stephasinstrument(A, metalClipApplicator), phasehasstep (tumorResection, A), haspreviousstep(A, clamping)
hasnextstep(resection, suturing).	hasnextstep(resection, A):- actionperformedinstep(suture, A), stephasinstrument(A, roboticLargeNeedleDriver), stephasinstrument(A, monopolarCurvedScissors), stephasinstrument(A, hemolokClipApplicator), phasehasstep (renorraphy, A), haspreviousstep(A, resection)
hasnextstep(suturing, unclamping).	hasnextstep(suturing, A):- actionperformedinstep(unclamp, A), stephasinstrument(A, monopolarCurvedScissors), stephasinstrument(A, roboticLargeNeedleDriver), stephasinstrument(A, laparoscopicBulldog), phasehasstep (renorraphy, A), haspreviousstep(A, suturing)
hasnextstep (reconstruction, drainage)	hasnextstep(reconstruction, A):- actionperformedinstep(put, A), stephasinstrument(A, fenestratedGraspingForcep), stephasinstrument(A, roboticLargeNeedleDriver), stephasinstrument(A, monopolarCurvedScissors), phasehasstep (closure, A), haspreviousstep(A, reconstruction)

Quantitative Evaluation: Rule 1 and 2 represent hypothesized relations between the workflow entities, while rule-3 verifies the temporal relations at higher-granularity level, i.e. steps. With both experiments, we obtained 100% accuracy, precision and recall predicting hypotheses with rule-2 and rule-3. In "EP-1", with rule-1, we obtained 71% accuracy, 33% sensitivity, and 100% specificity. However, in "EP-2", for the Rule 1, we obtained 100% accuracy, precision, and recall. While doing these experiments, we excluded other 4 video annotations considering some missing data. The results present that the relations between surgical entities can be used to classify the hierarchical RAPN surgical workflow. The latter is also supported by the Discrete-Time Markov Chain (DTMC) constructed for RAPN step transitions from the similar set of video annotations in our previous work [1].

Qualitative Evaluation: In "EP3", since we extracted constants in "Rule-4", i.e. relations between the instances of workflow entities, we inputted prediction of deep learning models as examples to check the hypothesis whether Aleph can predict consistent and complete information of the subsequent step. As shown in Table 2, the generated rules represent surgical resources, e.g. instruments, needed in the next step and information on actions and phases. Generated rules on the workflow were verified by an expert urologist ("MC") and found it to be true for the sequence of steps in RAPN workflow.

4 Conclusions

In this feasibility study, we have shown that implementation of ILP with an instance detection system can be useful to learn explanatory rules on the RAPN surgical workflow. We have tested four different types of hypotheses: (1) relations between surgical workflow entities; (2) temporal relations; (3) relations between instances of workflow entities and (4) relations for verifying the step sequences. The system was be able to learn rules for RAPN workflow representing all these hypotheses. The learned predicates represent workflow decomposition from higher-granularity e.g. phases to lower-granularity e.g. actions. Instruments can be considered as task resources to execute actions in steps. The relational information between surgical entities can be useful to predict information on the next step.

In future, we will find the exact temporal sequence especially at lower granularity, e.g. prediction of sequences of actions in the next predicted step. We will consider rules as constraints while training the deep learning networks [3] for recognizing surgical entities and ontology to learn the strict-order. Moreover, we will also compare it with the performance of other rule learners, such as associative rule learning [11].

Acknowledgement. This research is funded by the European Research Council (ERC) under the European Union's H2020 research and innovation programme (grant agreement No. 742671 "ARS"). This work was supported by European Union's Horizon 2020 research and innovation programme (grant agreement No. 732515 "SMARTsurg").

Conflict of Interest. The authors confirm that there are no known conflicts of interest associated with this publication.

References

1. Nakawala, H., Bianchi, R., Pescatori, L.E., De Cobelli, O., Ferrigno, G., De Momi, E.: "Deep-Onto" network for surgical worflow and context recognition. Int. J. Comput. Assist. Radiol. Surg. **14**(4), 685–696 (2019)
2. Nakawala, H., Ferrigno, G., De Momi, E.: Developement of an intelligent training system for Thoracentesis. Artif. Intell. Med. **84**, 50–63 (2018)
3. Garcez, A., Besold, T.R., de Raedt, L., Földiak, P., Hitzler, P., Icard, T., Kühnberger, K.-U., Lamb, L.C., Miikkulainen, R., Silver, D.L.: Neural-symbolic learning and reasoning: contributions and challenges. In: 2015 AAAI Spring Symposium, Palo Alto, California (2015)
4. Bock, H.-H.: Data Mining Tasks and Methods: Classification: the Goal of Classification. Handbook of Data Mining and Knowledge Discovery. Oxford University Press, New York (2002)
5. Dutra, I., Nassif, H., Page, D., Shavik, J., Strigel, R.M., Wu, Y., Elezaby, M.E., Burnside, E.: Integrating machine learning and physician knowledge to improve the accuracy of breast biopsy. In: AMIA Annual Symposium Proceedings, pp. 349–355 (2011)
6. Burnside, E.S., Davis, J., Costa, V.S., de Castro Dutra, I., Kahn Jr., C.E., Fine, J., Page, D.: Knowledge discovery from structured mammography reports using inductive logic programming. In: AMIA Annual Symposium Proceedings, pp. 96–100 (2005)
7. Evans, R., Grefenstette, E.: Learning explanatory rules from noisy data. J. Artif. Intell. Res. **61**(1), 1–64 (2018)
8. Hu, J.C., Treat, E., Filson, C.P., McLaren, I., Xiong, S., Stepanian, S., Hafez, K.S., Weizer, A.Z., Porter, J.: Technique and outcomes of robot-assisted retroperitoneoscopic partial nephrectomy: a multicenter study. Eur. J. Urol. **66**, 542–549 (2014)
9. Srinivasan, A.: The Aleph Manual (1999). https://www.cs.ox.ac.uk/activities/programinduction/Aleph/aleph.html
10. Kipp, M.: Anvil–a generic annotation tool for multimodal dialogue. In: Proceedings of the 7th European Conference on Speech Communication and Technology (Eurospeech), pp. 1367–1370 (2001)
11. Agrawal, R., Imieliński, T., Swami, A.: Mining association rules between sets of items in large databases. In: Proceedings of the 1993 ACM SIGMOD International Conference on Management of Data - SIGMOD 1993, p. 207 (1993)

Manipulation of a Whole Surgical Tool Within Safe Regions Utilizing Barrier Artificial Potentials

Theodora Kastritsi[2], Iason Sarantopoulos[1,2], Sotiris Stavridis[1,2], Dimitrios Papageorgiou[1,2], and Zoe Doulgeri[1,2(✉)]

[1] Information Technologies Institute (ITI), Center of Research and Technology Hellas (CERTH), 57001 Thessaloniki, Greece
{iasons,sotistav,dimpapag,doulgeri}@iti.gr
[2] Department of Electrical and Computer Engineering, Aristotle University of Thessaloniki, 54124 Thessaloniki, Greece
tkastrit@ece.auth.gr, {iasons,sotistav}@auth.gr, {dimpapag,doulgeri}@eng.auth.gr

Abstract. Active constraint enforcement in robotic-assisted surgery is critical for reducing the intra-operative risk of unintentionally damaging sensitive tissues by the surgical instrument. This work considers surgical instruments which can be circumscribed by a geometric capsule and forbidden regions which can be approximated by point clouds in order to produce a repulsive wrench by the control action to guarantee manipulation within safe regions. This work details the control scheme which is based on barrier artificial potentials when considering the whole tool extending our previous results on the tool point. A proof of the control system's passivity and non constraint violation is provided together with experimental results using a 7-dof KUKA LWR4+ manipulator as a master device in a virtual surgical scene in order to demonstrate the effectiveness of the proposed scheme.

Keywords: Surgical robotics · Physical human-robot interaction · Active constraints

1 Introduction

Barrier artificial potentials are positive scalar functions defined in a subset of a system's state with the sole purpose of enforcing actively i.e. via the control action, the confinement of the system evolution within this region. They take infinite values on the boundaries of the region thus creating a barrier and they are minimized at the system's equilibrium that can be an isolated point or even a whole set of states. Utilizing the gradient of these potentials in a control law is

We certify that there is no actual or potential conflict of interest in relation to this article.

J. Henriques et al. (Eds.): MEDICON 2019, IFMBE Proceedings 76, pp. 1559–1570, 2020.
https://doi.org/10.1007/978-3-030-31635-8_193

equivalent of creating a repulsive force away from the boundaries of the forbidden regions; the closest to the boundary the higher the repulsive force.

These type of artificial potential fields were initially proposed for obstacle avoidance in robot's autonomous operation [1]. In physical human robot interaction, they were also utilized to impose spatial forbidden regions for ensuring the safety of the robot and the environment. A study on the passivity, state boundedness, constraint satisfaction and performance of the system under the exertion of a human force is studied in [2]. The study is restricted to the case of spatial constraints defined analytically in the robot's world frame and for only the manipulator's tool tip position.

Being a promising solution for confining a robot within a restricted area, these type of potentials have been proposed for use in the application area of robotic surgery in order to guard the safety of sensitive regions like vessels surrounding the organ that is being operated from accidental penetrations. They are one way of implementing active constraints (AC), an extensive review of which can be found in [3]. Surgical sites consisting of organs, vessels and surrounding tissue have complex geometries and analytical expressions of sensitive areas are not available. What is available during robotic surgery is an image of the surgical site provided by the endoscopic camera usually in the form of a dense point cloud. For this reason, subsets of these 3D-point clouds are used to describe the constrained regions or even the operating regions as reported in several related previous works [4–6]. The point cloud characterization can be performed by the doctors via an appropriate interface. In our previous work [6], an active constraint controller was proposed based on a barrier artificial potential applied in point cloud regions that are characterized as forbidden areas. The work is restricted for the case of constraining the tool tip so that it does not touch or penetrate the sensitive areas and includes a proof of the passivity and constraint satisfaction. However, the protection of the sensitive area requires the consideration of the whole tool body which can be a scalper with a cutting edge running along one of its edges. Previous work on AC enforcement on the whole tool is reported in [4] but it does not include any theoretical justification. We have extended our method for the whole tool and initial experimental results are reported in [5]. The latter however does not elaborate on the whole tool method neither provides a stability analysis for the case of the whole tool.

In this work, we detail our proposed method for the whole tool and include a stability analysis which proves the passivity and non constraint violation given point cloud representation of the forbidden regions. The method involves tool modelling by a capsule and finding the points of the capsule that are closer than a prespecified distance from the points of the forbidden area. The proposed active constraint controller essentially aggregates the repulsive force signal at all these points producing a total force and torque on the robot's wrist. Such a force can be visualized or even fed back for haptic transparency. Experimental results are presented with a KUKA robot and a virtual slave site.

2 Active Constraint Control

The robot manipulator must be controlled in such a way that, firstly, will actively resist the human operator by exerting a repulsive generalized force when he guides the instrument near the forbidden area not allowing to enter it and, secondly, will maintain the passivity property and state boundedness irrespective of whether its instrument is away or near the constrained space.

The surgical instrument (e.g. Endowrist Permanent Cautery Spatula or Scalpels) is assumed to be rigidly fixed in the robot wrist frame $\{W\}$ whose position $\mathbf{p}_w \in \mathbb{R}^3$ and orientation $\mathbf{R}_w \in SO(3)$ is known in the inertia world frame $\{O\}$. Let the instrument be circumscribed by a capsule which is described by a line segment with length $L \in \mathbb{R}^+$ and a radius $r \in \mathbb{R}^+$ (Fig. 1). These parameters can be easily found from the surgical instrument's specifications. Without loss of generality, assume that one of the line segment's end points coincides with the wrist position $\mathbf{p}_w$ when the instrument is fixed at the robot's wrist. Let the forbidden area be provided by a 3-D point cloud $\mathcal{O}_s$ from the endoscopic camera with a known and homogeneous density of ρ points per cm^3. For the purposes of the AC enforcement we consider spheres $S(\mathbf{p}_i, d_{c_1})$, with radius $d_{c_1} = \frac{\sqrt{3}}{2\sqrt[3]{\rho}}$ centered at each point $\mathbf{p}_i \in \mathcal{O}_s$ [5] so that empty spaces between the points of $\mathcal{O}_s$ are covered. The forbidden region is thus mathematically described by the set $\bigcup_{\mathbf{p}_i \in \mathcal{O}_s} \{\mathbf{p} \in \mathbb{R}^3 : \|\mathbf{p} - \mathbf{p}_i\| \leq d_{c_1}\}$. The control objective for constraint enforcement is thus achieved if the instrument does not touch the surface of the forbidden region i.e the surface of the overlapping spheres. In order to achieve this objective, for each point $\mathbf{p}_i \in \mathcal{O}_s$, we need first to find the nearest point on the capsule's line segment $\mathbf{p}_i^* \in \mathbb{R}^3$ (see Fig. 1) and then impose a repulsive force on it. Calculation of the nearest point on the capsule's line segment, the barrier artificial potential and its related control signal are detailed in the following sections.

2.1 Nearest Point

Given the length L and the unit direction of the surgical tool ${}^w\mathbf{n} \in \mathbb{R}^3$ expressed in the wrist frame $\{W\}$ as denoted by the left superscript, we can express the position of all points belonging to the capsule's line segment with respect to the inertia frame $\{O\}$ by the following parametric equation:

$$\mathbf{p}_s(\sigma, t) = \mathbf{p}_w(t) + \mathbf{R}_w(t)^w\mathbf{n}L\sigma \tag{1}$$

where $\mathbf{n} = \mathbf{R}_w(t)^w\mathbf{n}$ expresses the main unit axis of the capsule w.r.t $\{O\}$ and $\sigma \in [0, 1]$.

In each control cycle and for each point $\mathbf{p}_i \in \mathcal{O}_s$, we find the nearest point $\mathbf{p}_i^* = \mathbf{p}_s(\sigma_i^*, t)$ on the line segment of the capsule, where σ_i^* is found by:

$$\sigma_i^* = \min_{\sigma;\ 0 \leq \sigma \leq 1} ||\mathbf{p}_s - \mathbf{p}_i||^2 \tag{2}$$

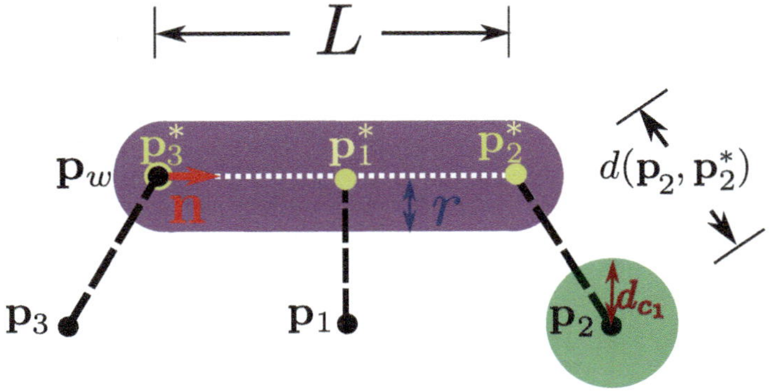

Fig. 1. The nearest point on the capsule's lime segment $\mathbf{p}_i^*$

Notice that given the analytic expression of the line segment allows the analytic calculation of σ_i^* as follows:

$$\sigma_i^* = \begin{cases} \zeta_i, & \text{if } 0 \leq \zeta_i \leq 1 \\ 1, & \text{if } \zeta_i > 1 \\ 0, & \text{if } \zeta_i < 0 \end{cases} \tag{3}$$

where

$$\zeta_i = \frac{1}{L}\mathbf{n}^T(\mathbf{p}_i - \mathbf{p}_w).$$

Figure 1 illustrates the three possible cases of Eq. (3). Notice that $\mathbf{n}^T(\mathbf{p}_i - \mathbf{p}_w)$ is the length of the orthogonal projection of $(\mathbf{p}_i - \mathbf{p}_w)$ on $\mathbf{n}$.

2.2 Controller Design

A barrier artificial potential initially proposed in [5] for the tool tip case is then applied to each pair $(\mathbf{p}_i, \mathbf{p}_i^*)$. It is given by the following equation and illustrated in Fig. 2:

$$V_i(\mathbf{p}_i, \mathbf{p}_i^*) = \begin{cases} \frac{1}{4}\ln\left(\frac{1}{1-\psi_i}\right)^2 & \text{if } d_c < d(\mathbf{p}_i, \mathbf{p}_i^*) \leq d_c + d_0 \\ 0 & \text{if } d(\mathbf{p}_i, \mathbf{p}_i^*) > d_c + d_0 \end{cases} \tag{4}$$

with

$$\psi_i = \frac{(d(\mathbf{p}_i, \mathbf{p}_i^*) - (d_0 + d_c))^2}{d_0^2} \tag{5}$$

where $d(\mathbf{p}_i, \mathbf{p}_i^*) = ||\mathbf{p}_i^* - \mathbf{p}_i|| \in \mathbb{R}$ is the Euclidean distance between $\mathbf{p}_i$ and $\mathbf{p}_i^*$, $d_c = d_{c_1} + r$ moves the point of interest on the capsule's surface and $d_0 \in \mathbb{R}$ is a positive constant which determines the area of influence of the repulsive potential. Notice that as $d(\mathbf{p}_i, \mathbf{p}_i^*)$ tends to d_c i.e as a sphere of the forbidden region approaches the capsule surface, the potential (4) tends to infinity.

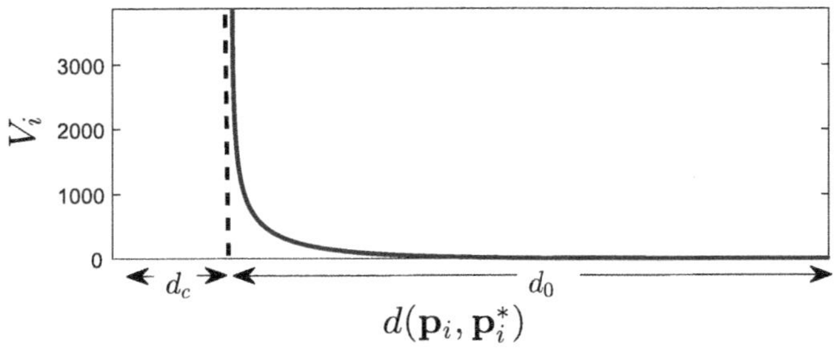

Fig. 2. The barrier artificial potential.

Remark 1. $V_i < \infty$ if and only if $\psi_i < 1$. Hence, considering the first case of (4), $\psi_i < 1$ if and only if $d(\mathbf{p}_i, \mathbf{p}_i^*) > d_c$.

For ensuring the safety of the sensitive forbidden region regarding the whole tool, the following inequality constraints should be satisfied at all times, $d(\mathbf{p}_i, \mathbf{p}_i^*) > d_c$ for each $\mathbf{p}_i$. To impose actively these constraints the negative gradient of the function V_i is used as a repulsive force $\mathbf{f}_i = -\dfrac{\partial V_i}{\partial \mathbf{p}_i^*}$ acting on $\mathbf{p}_i^*$, and is analytically given by:

$$\mathbf{f}_i(\mathbf{p}_i, \mathbf{p}_i^*) = \begin{cases} \frac{d_0 + d_c - d(\mathbf{p}_i, \mathbf{p}_i^*)}{d_0^2(1-\psi_i)} \ln\left(\frac{1}{1-\psi_i}\right) \mathbf{e}_i, & \text{if } d_c < d(\mathbf{p}_i, \mathbf{p}_i^*) \leq d_c + d_0 \\ \mathbf{0}, & \text{if } d(\mathbf{p}_i, \mathbf{p}_i^*) > d_c + d_0 \end{cases} \tag{6}$$

where $\mathbf{e}_i \in \mathbb{R}^3$ is the unit vector with direction pointing from the center of the sphere towards to $\mathbf{p}_i^*$:

$$\mathbf{e}_i = \frac{\mathbf{p}_i^* - \mathbf{p}_i}{d(\mathbf{p}_i, \mathbf{p}_i^*)} \tag{7}$$

Then, we aggregate the effect of all these repulsive force by transferring them to the origin of the wrist frame or any other frame of interest. In fact the action of each $\mathbf{f}_i$ acting on $\mathbf{p}_i^*$ can be produced by the pair of force $\mathbf{f}_i$ and torque $\boldsymbol{\tau}_i = (\mathbf{p}_i^* - \mathbf{p}_w) \times \mathbf{f}_i$ acting on the wrist $\mathbf{p}_w$. The total repulsive wrench can be found by summing the repulsive forces $\mathbf{f}_i$ and the generated torques $\boldsymbol{\tau}_i$ using (1) for $\sigma = \sigma_i^*$:

$$\mathbf{f} = \sum_i \mathbf{f}_i, \; \boldsymbol{\tau} = \sum_i \mathbf{n} L \sigma_i^* \times \mathbf{f}_i. \tag{8}$$

The proposed control signal for enforcing active constraints for the whole tool body is therefore given by:

$$\boldsymbol{u}_c = k[\mathbf{f}^T \; \boldsymbol{\tau}^T]^T \in \mathbb{R}^6 \tag{9}$$

where $k > 0$ is a scalar gain. Using this control signal, it is guaranteed that the robot's instrument does not violate the forbidden region and that the system is passive. The proof is given in the Appendix.

3 Experiment

The experimental set up to validate the proposed method is consisted of a 7-dof KUKA LWR4+ robotic manipulator acting as the master device. On the wrist of the master device a handle is attached which is used to guide the tool in a virtual slave scene. This scene consists of an identical virtual manipulator to be used as the slave manipulator; its joint positions copy the joint positions of the real KUKA arm. In the virtual scene a 3-D point cloud of the vessels (forbidden area) and a 3-D point cloud of the kidney (operative area) are imported and registered with respect to the slave arm (Fig. 3). The repulsive wrench calculated in the virtual scene is applied as a wrench input in the wrist of the master device to simulate the haptic rendering of a real master-slave architecture. A wrist force/torque sensor (ATI Mini40) is utilized at the master device to measure the exerted force for visualization purposes. The application of the repulsive wrench input in the master is implemented utilizing KUKA's FRI library with control frequency $f_s = 500\,\text{Hz}$ via joint impedance control mode. In the slave scene a capsule circumscribing the surgical tool is considered with radius $r = 0.009\,\text{m}$ and length $L = 0.08\,\text{m}$ (see Fig. 3).

Fig. 3. The experimental set up consisting of the master device (real KUKA) and the virtual surgical scene with the slave device and the point cloud of the human organs.

As the purpose of the surgeon is to be able to remove a tumor from the kidney avoiding to touch the surrounded vessels, the repulsive artificial potential field is designed so that the minimum distances between the sphere of each point of the forbidden area and the capsule are less than 2.3 cm, in other words $d_0 = 2.3\,\text{cm}$. Thus the surgeon is able to move freely in the operating area and feels a repulsive force only when accidentally guides the tool closer to the vessels by this distance. The controller gain is set to the value $k = 0.014$. In order to increase the computational performance, the point cloud of the vessels is down sampled and the radius of

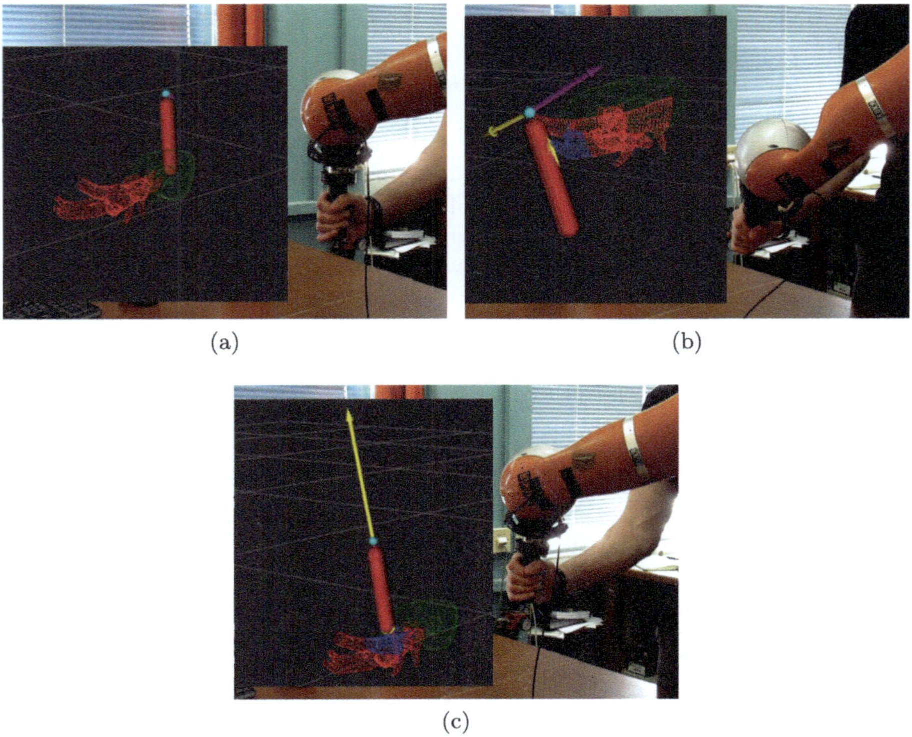

Fig. 4. Experiment emulating a surgical operation on a kidney shown in green and its surrounding vessels shown in red that should not be penetrated. Purple arrows visualize the force exerted by the operator and yellow arrows the repulsive force **f** generated by the controller and felt by the user. The red capsule is circumscribing the surgical instrument and the cyan dot the slave's wrist. Blue points denote the points in $\mathcal{O}_s$ that are activated and yellow points the points on the capsule nearest to the blue points; (a) The user guides the instrument in the area of the kidney where no repulsive force is generated. (b) The user guides the instrument near the vessels starting to feel the resistance of the generated repulsive force. (c) The user guides the instrument closer to the vessels by exerting higher guiding forces inducing even higher repulsive forces.

the spheres to cover the empty spaces is calculated to the value of $d_{c_1} = 2.18\,\text{mm}$. To reduce the computational complexity of finding the $\mathbf{p}_i^*$ by Eq. (3) the search space is reduced via a k-d tree search [7,8] to a sphere around the capsule centered at $\mathbf{p}_s(0.5, t)$ with radius $d_0 + d_c + \frac{L}{2}$. Then Eq. (2) is utilized for finding the nearest points $\mathbf{p}_i^*$ within the points of $\mathcal{O}_s$ that belong to this reduced space.

As proposed in [4] and in order to make the repulsive force norm independent from the density of the point cloud, the following formula is utilized for the calculation of the total repulsive force:

$$\mathbf{f} = \frac{\max(\|\mathbf{f}_i\|)\sum_i \mathbf{f}_i}{\|\sum_i \mathbf{f}_i\|}, \boldsymbol{\tau} = \frac{\max(\|\mathbf{f}_i\|)\sum_i (\mathbf{p}_i^* - \mathbf{p}_w) \times \mathbf{f}_i}{\|\sum_i \mathbf{f}_i\|}. \tag{10}$$

Notice that in this way the norm of the repulsive force depends only on the nearest repulsive pair $\mathbf{p}_i, \mathbf{p}_i^*$; thus the repulsive force that is generated does not depend on the number of the cloud points that are activated and hence on the local geometry of the activated forbidden area.

To validate the proposed controller, the user was asked to guide the tool initially on the kidney and then close to forbidden area (vessels). In the first case no repulsive forces are generated (Fig. 4a) while in the second repulsive forces are produced (Fig. 4b). As the operator tries to come even closer by exerting high guiding forces, even higher repulsive forces on the wrist are produced that cannot be practically overcome thus keeping the vessels safe from penetration (Fig. 4c). The minimum distance from the capsule and the repulsive force and torque exerted in the master's wrist are shown in Fig. 5. It is observed that when the minimum distance decreases the repulsive force increases and vice-versa. Notice that the minimum distance is never less than $d_{c_1} = 2.18$ mm as expected. Regarding the repulsive torque notice that it depends not only on the minimum distance but also on the direction of the tool's axis and the point of action of each repulsive force. Hence, as it can be seen in Fig. 5 from 50–55 s, the torque remains at low levels although the repulsive force increases owing to the decrease of the minimum distance.

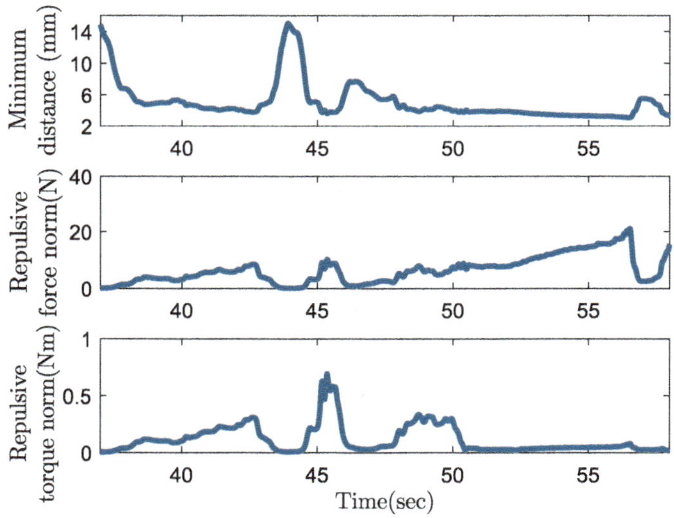

Fig. 5. The minimum distance between the capsule and the points on the forbidden region, the norm of the repulsive force and the respective torque norm during the experiment.

4 Conclusion

In this paper, details are provided on a control scheme that guarantees that the tool body will never violate the forbidden area. The proof of the passivity and constraint non-violation of the control system is given. Experiments are performed with a 7-dof KUKA LWR4+ manipulator used as master device and a virtual scene consisted of an identical virtual manipulator as a slave arm and point clouds of a kidney and its surrounding vessels characterized as forbidden regions. The results demonstrate how the repulsive force and torque produced by the control action and exerted at the wrist of the master manipulator cannot be practically overcome thus keeping the vessels safe from penetration. Our future work aims at extending the methodology for the case of articulated tools.

Acknowledgement. This work is funded by the EU Horizon 2020 research and innovation programme under grant agreement No. 732515, project SMARTsurg. Regarding Theodora Kastritsi: «This research is co-financed by Greece and the European Union (European Social Fund- ESF) through the Operational Programme «Human Resources Development, Education and Lifelong Learning» in the context of the project "Strengthening Human Resources Research Potential via Doctorate Research" (MIS-5000432), implemented by the State Scholarships Foundation (IKY)».

Appendix

System's Stability

Let us consider a 6-DOF revolute joint manipulator with gravity compensation and an additive dissipation term under the kinesthetic guidance of a human generalized force $\mathbf{F}_h \in \mathbb{R}^6$.

The dynamic model of the robot in the Cartesian space can be written in the task space as follows:

$$\mathbf{\Lambda}_x \dot{\mathbf{v}} + (\mathbf{C}_x + \mathbf{D}_d)\,\mathbf{v} - \mathbf{u}_c = \mathbf{F}_h, \tag{11}$$

where

$$\mathbf{\Lambda}_x = [\mathbf{J}(\mathbf{q})\mathbf{\Lambda}^{-1}(\mathbf{q})\mathbf{J}^T(\mathbf{q})]^{-1}, \tag{12}$$

$$\mathbf{C}_x\mathbf{v} = \mathbf{\Lambda}_x \left(\mathbf{J}(\mathbf{q})\mathbf{\Lambda}^{-1}(\mathbf{q})\mathbf{C}(\mathbf{q},\dot{\mathbf{q}}) - \dot{\mathbf{J}}(\mathbf{q})\right)\dot{\mathbf{q}}, \tag{13}$$

with $\mathbf{q}, \dot{\mathbf{q}} \in \mathbb{R}^6$ being the robot joint position and velocity, $\mathbf{x} = [\mathbf{p}_w^T\, \mathbf{Q}_w^T]^T \in \{\mathbb{R}^3 \times \mathbb{S}^3\}$ the pose of the wrist, where $\mathbf{p}_w$ is its position and $\mathbf{Q}_w$ its orientation expressed as a unit quaternion, $\mathbf{v} = [\dot{\mathbf{p}_w}^T\, \boldsymbol{\omega}_w^T]^T \in \mathbb{R}^6$ the generalized velocity of the wrist with $\dot{\mathbf{p}}_w$, $\boldsymbol{\omega}_w$ the translational and angular velocities, $\mathbf{D}_d \in \mathbb{R}^{6\times 6}$ being a positive definite matrix of the desired damping, $\mathbf{\Lambda}(\mathbf{q}) \in \mathbb{R}^{6\times 6}$ being the manipulator's inertia matrix, $\mathbf{C}(\mathbf{q}, \dot{\mathbf{q}}) \in \mathbb{R}^{6\times 6}$ being the Coriolis and centripetal matrix, $\mathbf{J}(\mathbf{q}) \in \mathbb{R}^{6\times 6}$ being the Jacobian of the manipulator which maps the velocity of the joints to the wrist velocity and $\mathbf{u}_c \in \mathbb{R}^6$ the virtual constraint control input. Notice that the unit Quaternion $\mathbf{Q}_w$ can be deduced given the rotation matrix $\mathbf{R}_w$, $\mathbf{\Lambda}_x$ is positive definite and $\dot{\mathbf{\Lambda}}_x - 2\mathbf{C}_x$ is skew symmetric.

To proceed with the stability analysis, we initially write system (11) in state-space. Utilizing the state vector:

$$\mathbf{s} = [\mathbf{v}^T \ \mathbf{x}^T \ \xi]^T \in \mathbb{R}^6 \times (\mathbb{R}^3 \times \mathbb{S}^3) \times \mathbb{R} \tag{14}$$

with $\xi = \sum_{\mathbf{p}_i \in \mathcal{C}} V_i(\mathbf{p}_i^*)$, the system (11) can be written as:

$$\dot{\mathbf{s}} = \mathbf{g}(\mathbf{s}, \mathbf{F}_h), \mathbf{s}_0 = \mathbf{s}(t_0) \in D \tag{15}$$

where

$$D \triangleq \mathbb{R}^6 \times \Omega \times \mathbb{R},$$

$$\Omega = \bigcup_{\mathbf{p}_i \in \mathcal{O}_s} \{\mathbf{x} \in \mathbb{R}^3 \times \mathbb{S}^3 : d(\mathbf{p}_i, \mathbf{p}_i^*) > d_c\},$$

$$\mathbf{g}(\mathbf{s}, \mathbf{F}_h) = \begin{bmatrix} \mathbf{\Lambda}_x^{-1}(-(\mathbf{C}_x + \mathbf{D}_d)\mathbf{v} + \mathbf{F}_h + \mathbf{u}_c) \\ \frac{1}{2}\mathbf{J}_x \mathbf{v} \\ \sum_i \frac{\partial V_i(\mathbf{p}_i^*)}{\partial \mathbf{p}_i^*}^T \dot{\mathbf{p}}_i^* \end{bmatrix}$$

with $\mathbf{J}_x = \text{diag}(2\mathbf{I}_3, \ \mathbf{J}_Q^T)$ where $\mathbf{J}_Q$ is the mapping between the angular velocity of the frame $\{W\}$ and the unit quaternion rates. Notice that Ω is the union of all wrist configurations in which the instrument does not touch the spheres $S(\mathbf{p}_i, d_{c_1})$.

Theorem 1. *For system* (15) *under the exertion of the external human force* $\mathbf{F}_h$ *the forbidden area is not violated and the system is strictly output passive with respect to the wrist velocity* $\mathbf{v}$.

Proof. Let the initial state be such that $\mathbf{u_c} \neq \mathbf{0}$. Utilizing the following candidate Lyapunov-like function:

$$V = \frac{1}{2}\mathbf{v}^T \mathbf{\Lambda}_x \mathbf{v} + k \sum_i V_i(\mathbf{p}_i^*), \tag{16}$$

its time derivative, substituting $\mathbf{\Lambda}_x \dot{\mathbf{v}}$ from (11) and utilizing the skew-symmetry of matrix $\left(\dot{\mathbf{\Lambda}}_x - 2\mathbf{C}_x\right)$, yields:

$$\dot{V} = -\mathbf{v}^T \mathbf{D}_d \mathbf{v} + \mathbf{v}^T \mathbf{F}_h + \mathbf{v}^T \mathbf{u}_c - k \sum_i \dot{\mathbf{p}}_i^{*T} \mathbf{f}_i. \tag{17}$$

The time derivative of the nearest point, using the time derivative of (1) for $\sigma = \sigma*$, given by:

$$\dot{\mathbf{p}}_i^* = \frac{\partial \mathbf{p}_s(\sigma, t)}{\partial t}|_{\ \sigma = \sigma_i^*} + \frac{\partial \mathbf{p}_s(\sigma_i^*, t)}{\partial \sigma_i^*}\dot{\sigma}_i^* = \dot{\mathbf{p}}_w(t) + \hat{\boldsymbol{\omega}}_w(t)\mathbf{n}L\sigma_i^* + \mathbf{n}L\dot{\sigma}_i^*$$

with $\hat{\boldsymbol{\omega}}_w(t) = \dot{\mathbf{R}}_w(t)\mathbf{R}_w(t)^T$ the wrist's angular velocity expressed in the inertia frame and $(\hat{})$ operator denoting the skew symmetric matrix of a vector. Substituting $\dot{\mathbf{p}}_i^*$ in (17) yields:

$$\dot{V} = -\mathbf{v}^T\mathbf{D}_d\mathbf{v} + \mathbf{v}^T\mathbf{F}_h + \mathbf{v}^T\mathbf{u}_c - k\sum_i (\dot{\mathbf{p}}_w(t) + \hat{\omega}_w(t)\mathbf{n}L\sigma_i^* + \mathbf{n}L\dot{\sigma_i^*})^T\mathbf{f}_i \quad . \tag{18}$$

Utilizing (9) in (8) yields:

$$\begin{aligned}\dot{V} =& -\mathbf{v}^T\mathbf{D}_d\mathbf{v} + \mathbf{v}^T\mathbf{F}_h + k\dot{\mathbf{p}}_w^T\sum_i \mathbf{f}_i + Lk\boldsymbol{\omega}_w^T\sum_i \sigma_i^*\mathbf{n}\times\mathbf{f}_\mathrm{i} \\ &-k\sum_i (\dot{\mathbf{p}}_w(t) + \hat{\omega}_w(t)\mathbf{n}L\sigma_i^* + \mathbf{n}L\dot{\sigma_i^*})^T\mathbf{f}_i.\end{aligned} \tag{19}$$

Using the distributive property of the dot product and the property $\mathbf{a}(\mathbf{b}\times\mathbf{c}) = (\mathbf{a}\times\mathbf{b})\mathbf{c}$ the above equation becomes:

$$\dot{V} = -\mathbf{v}^T\mathbf{D}_d\mathbf{v} + \mathbf{v}^T\mathbf{F}_h - kL\sum_i \dot{\sigma_i^*}\mathbf{n}^T\mathbf{f}_\mathrm{i}. \tag{20}$$

For the first case of (3) utilizing (1) for $\sigma = \sigma_i^*$ yields that $\mathbf{n}^T(\mathbf{p}_i^* - \mathbf{p}_i) = 0$ and for the other two cases $\dot{\sigma_i^*} = 0$. Therefore:

$$\mathbf{n}^T(\mathbf{p}_i^* - \mathbf{p}_i)\dot{\sigma}_i^* = 0.$$

As $\mathbf{f}_i$ is in the direction of $\mathbf{p}_i^* - \mathbf{p}_i$ and using the above property in (20) the time derivative of V becomes:

$$\dot{V} = -\mathbf{v}^T\mathbf{D}_d\mathbf{v} + \mathbf{F}_h^T\mathbf{v} \leq \mathbf{F}_h^T\mathbf{v}. \tag{21}$$

Hence, system (15) is strictly output passive (see Definition 6.3 in [9]). Rewriting (21) by completing the squares, yields:

$$\begin{aligned}\dot{V} =& -||\sqrt{\mathbf{D}_d}\mathbf{v} - \frac{1}{2}\sqrt{\mathbf{D}_d}^{-1}\mathbf{F}_h||^2 + \frac{1}{4}\mathbf{F}_h^T\mathbf{D}_d^{-1}\mathbf{F}_h \\ \leq& \frac{1}{4}\mathbf{F}_h^T\mathbf{D}_d^{-1}\mathbf{F}_h.\end{aligned} \tag{22}$$

Notice that $\mathbf{F}_h$ represents the force applied by the human to guide the robot. Thus, $\mathbf{F}_h$ is bounded and therefore $\mathbf{F}_h^T\mathbf{D}_d^{-1}\mathbf{F}_h$ is also a bounded function of time. Additionally the human forces have bounded energy so by integrating equation (22) we get:

$$V \leq V(t_0) + \int_{t_0}^{t} \frac{1}{4}\mathbf{F}_h^T\mathbf{D}_d{}^{-1}\mathbf{F}_h < \infty. \tag{23}$$

Thus, states ξ and $\mathbf{v}$ are bounded under the exertion of human force or otherwise there exist compact sets Ω_ξ, Ω_v such that $\xi \in \Omega_\xi \subset \mathbb{R}$ and $\mathbf{v} \in \Omega_v \subset \mathbb{R}^6$. As a consequence, there exists a positive constant $\bar{\varepsilon}_i$ such that $V_i \leq \bar{\varepsilon}_i$, which for the logarithmic function defined in (4) yields:

$$0 \leq \psi_i \leq 1 - \frac{1}{e^{2\sqrt{\bar{\varepsilon}_i}}} < 1, i = 1, ..., M,$$

and this means that $\|\mathbf{p}_i^* - \mathbf{p}_i\| > d_c$ i.e. the instrument will never touch the sensitive area.

In the case that all $\mathbf{p}_i$ in $\mathcal{O}_s$ are such as $d(\mathbf{p}_i, \mathbf{p}_i^*) \geq d_c + d_0$ i.e. $\mathbf{u}_c = \mathbf{0}$ the Lyapunov function take the form $V = \frac{1}{2}\mathbf{v}^T\mathbf{\Lambda}_x\mathbf{v}$ and the time derivative $\dot{V} = -\mathbf{v}^T\mathbf{D}_d\mathbf{v} + \mathbf{v}^T\mathbf{F}_h \leq \frac{1}{4}\mathbf{F}_h^T\mathbf{D}_d^{-1}\mathbf{F}_h$. Thus, the system is strictly output passive w.r.t $\mathbf{v}$. Notice that when $\mathbf{F}_h = \mathbf{0}$, $\mathbf{v}$ converges exponentially to zero. Hence, using Lemma 1 of the [5], we can conclude that $\mathbf{x}$ converges to a constant.

References

1. Khatib, O.: Real-time obstacle avoidance for manipulators and mobile robots. Int. J. Robot. Res. **5**(1), 90–98 (1986)
2. Kastritsi, T., Papageorgiou, D., Doulgeri, Z.: On the stability of robot kinesthetic guidance in the presence of active constraints. In: 2018 European Control Conference (ECC), pp. 622–627, June 2018
3. Bowyer, S.A., Davies, B.L., Rodriguez y Baena, F.: Active constraints/virtual fixtures: a survey. IEEE Trans. Robot. **30**(1), 138–157 (2014)
4. Leibrandt, K., Marcus, H.J., Kwok, K., Yang, G.: Implicit active constraints for a compliant surgical manipulator. In: 2014 IEEE International Conference on Robotics and Automation (ICRA), Hong Kong, pp. 276–283, May 2014
5. Kastritsi, T., Papageorgiou, D., Sarantopoulos, I., Stavridis, S., Doulgeri, Z., Rovithakis, G.: Guaranteed active constraints enforcement on pointcloud-approximated regions for surgical applications. In: IEEE International Conference on Robotics and Automation (ICRA), Montreal, Canada, pp. 8346–8352, 20–24 May 2019
6. Kastritsi, T., Papageorgiou, D., Sarantopoulos, I., Doulgeri, Z., Rovithakis, G.: Stability of active constraints enforcement in sensitive regions defined by point-clouds for robotic surgical procedures. In: European Control Conference (ECC), Naples, Italy (2019)
7. Bentley, J.L.: Multidimensional binary search trees used for associative searching. Commun. ACM **18**(9), 509–517 (1975)
8. Rusu, R.B., Cousins, S.: 3D is here: Point Cloud Library (PCL). In: IEEE International Conference on Robotics and Automation (ICRA), Shanghai, China, 9–13 May 2011
9. Khalil, H.: Nonlinear Systems, 3rd edn. Prentice Hall, Upper Saddle River (2002)

Evaluation of Force Feedback for Palpation and Application of Active Constraints on a Teleoperated System

Efi Psomopoulou[1(✉)], Raj Persad[2], Anthony Koupparis[2], Sajeeva Abeywardena[1], Mohammad Fattahi Sani[1], Chris Melhuish[1], and Sanja Dogramadzi[1]

[1] Bristol Robotics Laboratory, Coldharbour Lane, Bristol, UK
efi.psomopoulou@brl.ac.uk
[2] Bristol Urological Institute, Southmead Hospital, Bristol, UK

Abstract. A desktop haptic device is used to teleoperate an industrial redundant and compliant robotic arm with a surgical instrument mounted on its end-effector. The master and slave devices are coupled in a bilateral position-position architecture. Force feedback is provided by the master haptic device to the user, from the position of the slave's wrist. A surgical task (palpation) that involves force feedback is presented and tested in a user study with surgeons and non-medical participants. Results show that users easily discern between three different materials during palpation given minimal familiarisation time. Active constraint enforcement is also integrated with the system as a sensitive area around the palpation samples which the slave instrument is prohibited to enter.

Keywords: Force feedback · Teleoperation · Medical robotics · Palpation · Active constraints

1 Introduction

Minimally Invasive Surgery (MIS) has become an established alternative to the conventional open approaches in a significant number of surgical fields and teleoperated surgical systems have been increasingly used in MIS procedures over the past decade providing 3D vision and better ergonomics for the surgeon. However, further developments are needed to create more sensitive robotic systems with new teleoperation interfaces to restore the dexterity and haptic feeling of open surgery, thus improving their efficacy and expanding their applicability to more complex surgical procedures. Teleoperated systems with force feedback have been used in various robotic applications over the years. However, current commercially available systems for MIS robots do not include significant haptic feedback to the operator, mainly because of the associated stability issues [1]. A significant amount of effort has been put into this research area by the

J. Henriques et al. (Eds.): MEDICON 2019, IFMBE Proceedings 76, pp. 1571–1580, 2020.
https://doi.org/10.1007/978-3-030-31635-8_194

scientific community. Some works look to improve the currently used RA-MIS systems like the DaVinci Surgical System by adding force feedback [2]. There is always a trade-off between transparency and stability of a teleoperated system with force feedback [3] and there are stability issues when non negligible time delays are introduced [4]. Other works investigate the combination of visual, tactile and force feedback towards the improvement of telepresence [5,6] showing that depending on the task, multi-modal feedback can improve the accuracy of perception of the remote environment.

One of the main tasks that require force feedback in remote medical examinations or MIS procedures is tissue palpation [7,8]. Our study uses a lightweight haptic device as a master and a redundant industrial compliant robotic arm with a daVinci instrument as a slave in a bilateral teleoperation architecture to evaluate the force feedback during a palpation task and test a safety feature in the form of active constraints enforcement using the methodology proposed in [9].

Section 2 provides the description of the teleoperated system setup; experimental results on the palpation of three different sample materials are given in Sect. 3 and a user study evaluating this surgical task is provided in Sect. 4; Sect. 5 gives preliminary results on the enforcement of active constraints around the area of palpation; the conclusion and future work are presented in Sect. 6.

2 System Description

The system is comprised of kinematically dissimilar master and slave robotic manipulators. The master arm is the Haption Virtuose 6D Desktop, shown in Fig. 1a, which is a haptic device that provides 6 degrees-of-freedom (DOF) of force feedback with continuous force of 3 N and continuous torque of 0.2 Nm. A load cell (6 DOF force/torque sensor) is mounted on the Haption end-effector to measure the force that the user applies on the master and a pedal is used as a clutch during teleoperation. The slave arm is the KUKA LBR iiwa, a serial link manipulator with 7 DOF and a payload of 14 kg. The slave arm is mounted upside down on a gantry and a daVinci Endowrist instrument with monopolar curved scissors is attached to the KUKA flange, Fig. 1b. The instrument is not actuated in this work but its wrist (the point just before the jaws) is used as the tool centre point (TCP) of the slave arm.

Bilateral Teleoperation with Force Feedback

A position-position control architecture is utilised to provide force feedback where both robots are controlled to track each other's TCP [10]; the wrist of the haptic device controls the wrist of the instrument, meaning that the force feedback is provided from the position of the instrument's wrist to the TCP of the master. The coupling gains between the two robots can be interpreted as a spring and damper between their end-effectors. Since the two robots are substantially different in terms of size and dynamics, the master-slave forces are

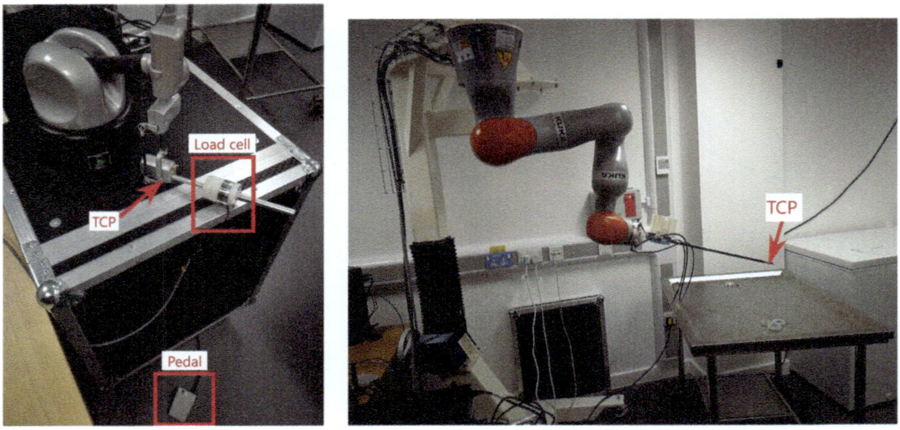

(a) Master haptic device: Haption Virtuose 6D Desktop with a load cell attachment and a pedal clutch.

(b) Slave manipulator: KUKA LBR iiwa mounted on a gantry with a daVinci tool attached on the flange.

Fig. 1. Master-slave teleoperation system

scaled by a factor of 0.4 and their velocities are scaled by 1.5. This means that the movement of the slave is larger than the movement of the master and the force felt at the master is smaller than the force measured at the slave. By construction, this architecture provides to the user the external environment forces as well as the slave's controller forces which include forces associated with the spring-damper and slave inertia.

The teleoperation control loop is implemented on a machine with a real time kernel patch for Linux at a 2 ms rate. Experimental results show that any time delays introduced by the communication network between the controller and the robots do not exceed a control loop of 2.7 ms.

3 Palpation of Soft Tissue-Like Materials

First, a palpation task was considered, where the TCP of the slave was controlled to press gently and repeatedly the surface of tissue-like materials. Three samples from bolus material [11] were placed inside an area of 10.5 × 10.5 cm, as shown in Fig. 2a, and on a horizontal table within the workspace of the slave arm. The bolus material is made of a synthetic gel with specific gravity close to that of water (1.02), approximating tissue equivalence very well [11]. The properties of the samples are given in the table of Fig. 2b. Sample #2 is the softest material of the three while samples #1 and #3 are made of a stiffer material. However, because of the difference in their thickness, sample #1 feels softer than #3 during palpation.

Material	Thickness	Young's Modulus
#1	10 mm	184709 Pa
#2	5 mm	103539 Pa
#3	5 mm	171404 Pa

(a) Bolus materials with properties shown in Table 2b.

(b) Properties of the materials used in the palpation task.

Fig. 2. Materials used for palpation task

Figure 3 presents the experimental results of the task, when palpating consecutively the three samples. Figure 3a shows the z-axis components of the force as measured by the load cell at the master side (F_s), the control input to the slave (F_c) and the external environment force as estimated by the joint torque sensors of the slave (F_e). It is clear that the external forces are exerted in the opposite direction to that of the control forces of the slave, as was expected. Moreover, the force being felt at the master is much smaller since a factor of 0.4 was applied, amplifying the forces at the slave. It is also easy to see that the forces are proportional to the stiffness of the samples, as shown by their Young's modulus; the softest sample (#2) produces the smallest forces, etc. In order to better highlight this, the magnitude of each force was calculated from the recorded generalised vector as shown in Fig. 3b and the mean values for each sample are given in Table 1. Figure 3c shows the z-axis component of the position error between the master and the slave where it is clear that larger forces correspond to larger position errors and to stiffer materials. The mean value of this error is also included in Table 1.

Table 1. Mean values for each sample

	Sample #1	Sample #2	Sample #3
$\|F_c\|$ $[N]$	4.74	1.49	4.88
$\|F_e\|$ $[N]$	4.9	1.5	4.99
$\|F_s\|$ $[N]$	1.56	0.64	1.65
$\|e_z\|$ $[m]$	0.013	0.004	0.015

4 Evaluation of the Usability and Force Feedback Perception of the Teleoperated System

A user study regarding the palpation experiment was conducted in order to evaluate the perception of the provided force feedback as well as the ease of

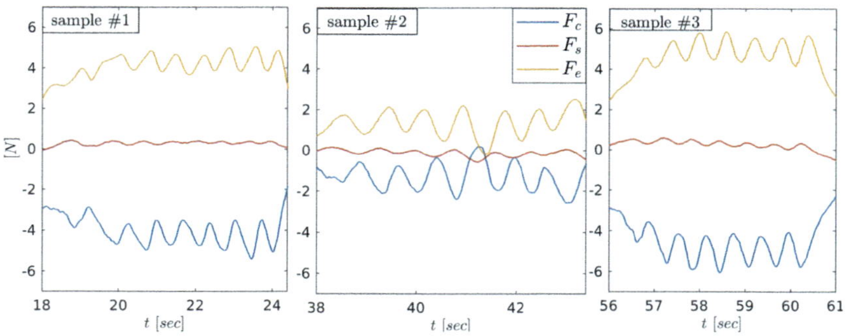

(a) Applied and measured forces on the z axis. F_s is the force as measured by the load cell at the master side, F_c is the control input to the slave and F_e is the external environment force

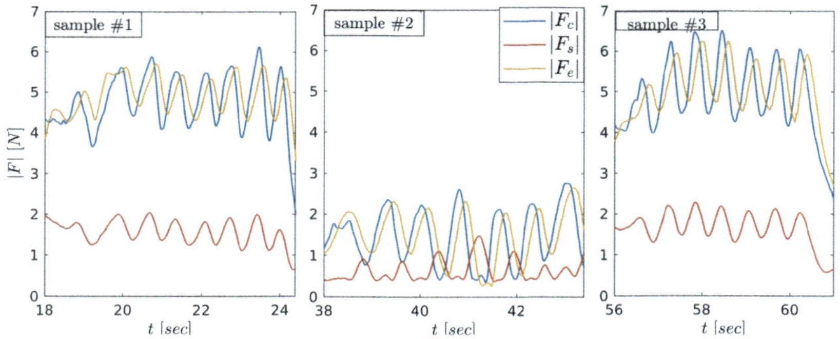

(b) Forces magnitude

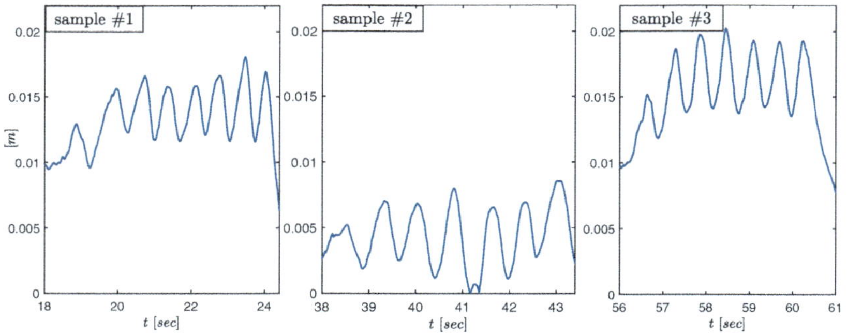

(c) Position error between master and slave on the z axis

Fig. 3. Experimental results while palpating consecutively the three samples.

use of the system. Eleven subjects were asked to perform the palpation task. Two of the users were surgeons with more than 10 years of experience in using the DaVinci surgical system, one of the users had previous experience with the developed system and most users had little to no experience with haptic feedback in virtual or teleoperated environments.

Each subject was asked to palpate all three samples one time for as long as they felt necessary to appraise the softness of each sample. Then, they were asked to repeat the procedure two more times. The choice of the palpation technique was left to the discretion of the subject. The most common technique involved pressing the wrist of the tool on the samples repeatedly while there were a few subjects who dragged the wrist of the tool across the surface of the samples (Fig. 4). All subjects were positioned at a ~1 m distance from the operating table and had direct visual contact with the samples (i.e. not through glasses or a camera). All subjects were given approximately 2 min to familiarise themselves with the operation of the teleoperated system prior to the palpation task.

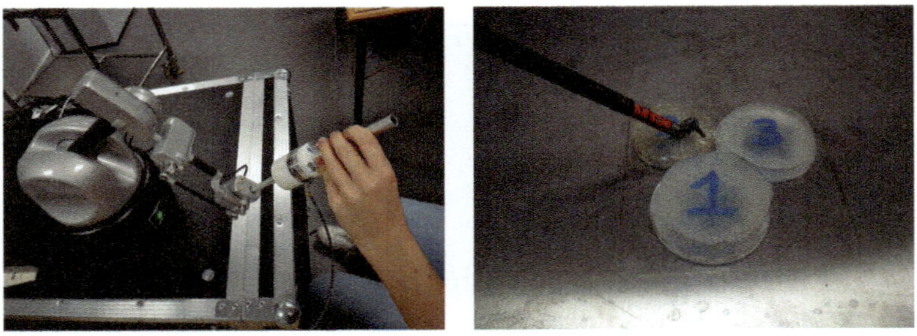

(a) User operating the master device for a palpation task. (b) Slave in contact with soft material.

Fig. 4. User palpating sample #2

After the completion of the task, all subjects were asked to rate the softness of each sample, whether their perception changed after the first trial and the level of confidence in their final perception. They were also asked to rate the ease of use of the teleoperated system and if the provided force feedback felt sufficient.

Perception of Environmental Stiffness

Results showed that 54% of the subjects accurately perceived the order of softness of all three samples. 72% of the subjects accurately appraised sample #2 as the softest but 25% of them had difficulty discerning the difference between samples #1 and #3 due to their similar properties. Moreover, 27% of the subjects rated sample #2 as the hardest. This is because that particular sample is very soft and thin, and when applying a large force on its surface, the TCP can easily reach the bottom of the sample resulting in the user perceiving the stiffness of the table underneath instead. Only 18% of the subjects didn't feel they could easily perceive differences between the samples and were among the group that rated sample #2 as the hardest. The data collected during the trials showed that in these cases, the subjects were applying much higher forces on the samples overall, e.g. a mean value of $F_c = 9N$ and $F_s = 3N$ for the stiffest material while $F_c = 5N$ and $F_s = 2.1N$ for the softest respectively. Further discussions with the subjects after the completion of the task revealed that these cases were caused by different expectations of how a teleoperated system works.

45% of the subjects felt confident in their final perception and 80% of them were indeed accurate. 33% of the subjects that didn't feel confident were however accurate in their perception.

Ease of Use

The subjects were asked to rate the ease of use of the system on a scale of 1 to 10. 72% of them rated it as more than 6 and 38% of them as more than 8. 27% of the subjects rated the system as less than 4. This was mainly because of the point of view of the user with respect to the area of operation, an issue that can be addressed by using a camera to provide a better viewpoint. Moreover, only 9% of the subjects suggested that the force feedback could be stronger while the rest were satisfied with the feeling of the force provided by the master.

5 Preliminary Application of Active Constraints Enforcement

In the previous sections, the slave manipulator had unconstrained access to the samples as they could be approached from any direction above the operating table. However, in RA-MIS procedures, this is not always the case. A core issue is to ensure the safety of the patient by minimizing the risk of unintentional tissue damage. This objective can be fulfilled by determining the forbidden areas of operation in the tissue and by constraining the instrument path within the acceptable regions. A methodology applied to this end is the implementation of active constraints.

Consequently, a virtual wall is created around the samples and active constraints are enforced to produce repulsive forces on the slave when it approaches the wall. The wall emulates a sensitive area that a surgeon using this system would not be allowed to enter for safety reasons. The repulsive forces due to the active constraints are calculated using the methodology proposed in [9]. The virtual wall is introduced as a point cloud of a torus with inner diameter 10.5 cm, outer diameter 24 cm and height 10 cm. The centre of the torus is located on the surface of the operating table meaning that its height above the samples is 5 cm.

Figure 6 shows the real world scene where the TCP of the slave is approaching the samples on the table while contacting the virtual restricted area as seen in

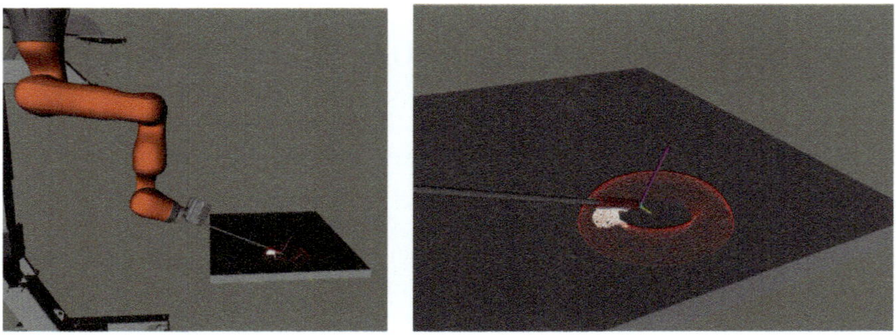

(a) Virtual representation of the configuration of slave while in contact with active constraints.

(b) Focus on the area of the active constraints.

Fig. 5. Representation of the slave manipulator in a virtual environment while in contact with the active constraints.

(a) Actual configuration of slave while in contact with active constraints.

(b) Focus on the area of the active constraints.

Fig. 6. Actual slave manipulator in contact with the active constraints.

Fig. 5b. Figure 5 shows the representation of the slave manipulator in the virtual environment RViz at the same instance where a portion of the tool shaft has approached the virtual wall and repulsive forces are generated. The white area consists of the virtual wall points that are activated due to the proximity of the tool shaft and are generating repulsive forces using the methodology proposed in [9]. These repulsive forces are represented by the purple arrow in the virtual environment and are added to the control input F_c of the teleoperation.

Figure 7a depicts the z-axis components, for simplicity reasons, of the system's forces, i.e. the force as measured by the load cell at the master side (F_s), the external environment force as estimated by the joint torque sensors of the slave (F_e) and the control input to the slave with ($F_{c_{ac}}$) and without the repulsive forces (F_c). The active constraints are enforced at the time instant $t = 636$ s (dashed line in Fig. 7). Figure 7b shows the respective position error between the master and the slave. It is clear that at the same time instant, the position error increases, thus, based also on the results of the previous section, the force fed back to the user is larger once active constraints are activated.

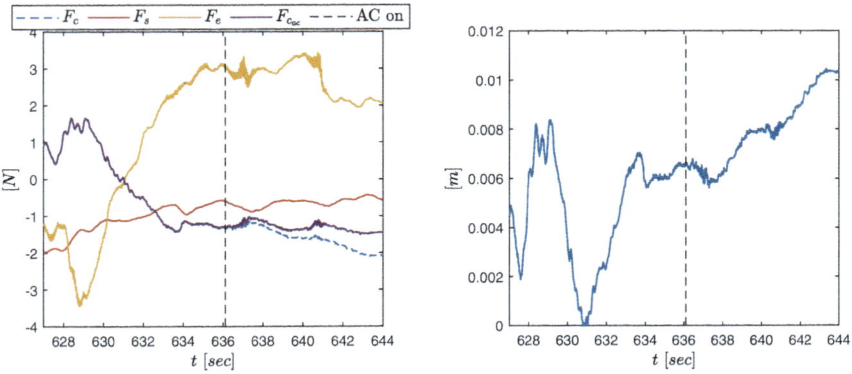

(a) Applied and measured forces on the z axis. The dashed line indicates the time instant when the slave contacts the active constraints area.

(b) Position error between master and slave on the z axis.

Fig. 7. Active constraints enforcement results.

It is important to note that the processing related to the calculation of the repulsive forces increases the time delay of the control loop to an average rate of 6 ms. Preliminary testing showed that this delay as well as the use of a long tool on the slave manipulator narrow the stability limits of the system. In order to achieve stability, the magnitude of the force feedback was reduced. This can be displayed by the smaller magnitude of the control input force F_c with respect to the magnitude of the external force F_e in Fig. 7a compared to the same forces during the palpation experiment (Fig. 3a). Consequently, further investigation is required towards an improved experience of the active constraints.

6 Conclusion

A teleoperated system comprised of a desktop haptic device and an industrial redundant and compliant robotic arm with a surgical tool mounted on its end-effector is used in a user study for a palpation task. Results show 72% success in discerning the difference between different materials when the applied force allows for palpation on the surface of the samples. Active constraints enforcement was also applied on the system to emulate a sensitive no-go area around the palpation samples providing repulsive force feedback when in proximity. Future work will include obtaining user feedback on the active constraints enforcement, evaluation of the system when teleoperating through a trocar and investigation of stability issues related to communication and processing delays.

Acknowledgements. This work was supported by the European Union's Horizon 2020 research and innovation programme under grant agreement No. 732515.

References

1. van der Meijden, O.A.J., Schijven, M.P.: The value of haptic feedback in conventional and robot-assisted minimal invasive surgery and virtual reality training: a current review. Surg. Endosc. **23**(6), 1180–1190 (2009). https://doi.org/10.1007/s00464-008-0298-x
2. Gibo, T.L., Deo, D.R., Okamura, A.M.: Effect of load force feedback on grip force control during teleoperation: a preliminary study. In: IEEE Haptics Symposium (HAPTICS), pp. 379–383 (2014). https://doi.org/10.1109/HAPTICS.2014.6775485
3. Zarrad, W., Poignet, P., Cortesao, R., Company, O.: Stability and transparency analysis of a haptic feedback controller for medical applications. In: 46th IEEE Conference on Decision and Control, pp. 5767–5772 (2007). https://doi.org/10.1109/CDC.2007.4434677
4. Takhmar, A., Polushin, I.G., Talasaz, A., Patel, R.V.: Cooperative teleoperation with projection-based force reflection for MIS. IEEE Trans. Control Syst. Technol. **23**(4), 1411–1426 (2015). https://doi.org/10.1109/TCST.2014.2369344
5. Talasaz, A., Patel, R.V., Naish, M.D.: Haptics-enabled teleoperation for robot-assisted tumor localization. In: IEEE International Conference on Robotics and Automation, pp. 5340–5345 (2010). https://doi.org/10.1109/ROBOT.2010.5509667
6. Mahvash, M., Gwilliam, J., Agarwal, R., Vagvolgyi, B., Su, L., Yuh, D.D., Okamura, A.M.: Force-feedback surgical teleoperator: controller design and palpation experiments. In: Symposium on Haptic Interfaces for Virtual Environment and Teleoperator Systems, pp. 465–471 (2008). https://doi.org/10.1109/HAPTICS.2008.4479994
7. Filippeschi, A., Jacinto Villegas, J.M., Satler, M., Avizzano, C.A.: A novel diagnostician haptic interface for tele-palpation. In: 27th IEEE International Symposium on Robot and Human Interactive Communication (RO-MAN), pp. 328–335 (2018). https://doi.org/10.1109/ROMAN.2018.8525667
8. Torabi, A., Khadem, M., Zareinia, K., Sutherland, G.R., Tavakoli, M.: Application of a redundant haptic interface in enhancing soft-tissue stiffness discrimination. IEEE Robot. Autom. Lett. **4**(2), 1037–1044 (2019). https://doi.org/10.1109/LRA.2019.2893606
9. Kastritsi, T., Papageorgiou, D., Sarantopoulos, I., Stavridis, S., Doulgeri, Z., Rovithakis, G.: Guaranteed active constraints enforcement on point cloud-approximated regions for surgical applications. In: IEEE International Conference on Robotics and Automation, pp. 8346–8352 (2019). https://doi.org/10.1109/ICRA.2019.8793953
10. Niemeyer, G., Preusche, C., Hirzinger, G.: Telerobotics, pp. 741–757. Springer, Heidelberg (2008). https://doi.org/10.1007/978-3-540-30301-5_32
11. Eckert and Ziegler BEBIG, Superflab: Bolus material for external beam radiation therapy. https://www.bebig.com/home/products/radiotherapy_accessories/superflab/

A Knowledge-Based Graphical Interface for Modeling Surgical Workflows in Robot-Assisted Minimally Invasive Surgery

Christos Papadopoulos[1(✉)], Angeliki Topalidou-Kyniazopoulou[1], Ioannis Mariolis[1], Aristotelis Sideridis[2], Emmanouel Papacostas[2], and Dimitrios Tzovaras[1]

[1] Centre of Research & Technology – Hellas, 6th Km Charilaou - Thermi, 57001 Thessaloniki, Greece
pap-x@iti.gr

[2] Thessaloniki Minimally Invasive Surgery Centre, Papakyritsi 8, Kalamaria, Thessaliniki, Greece

Abstract. This paper presents a knowledge-based graphical interface for modelling surgical workflows in robot-assisted minimally invasive surgery (R-A MIS). The interface provides an effortless way for the surgeon to generate a preoperative protocol for the surgery, while the knowledge base adds semantic information and rules to the surgery-specific ontologies. The surgeon can either modify a pre-existing prototype workflow for a known type of R-A MIS surgery or create a completely new surgical workflow for unknown types of R-A MIS. The workflow consists of phases and steps in consecutive order, incorporating anatomies, instruments and actions that the surgeon can modify. Any modification adheres to the rules of the knowledge base where information of incorporated robotic interactions is included. The interface is built as a web application that can be operated through a computer or portable device. The interface had positive reviews by orthopedic surgeons that were requested to create surgical workflows for R-A MIS arthroscopic procedures and evaluate its usability.

Keywords: Surgical workflow modeling · Medical graphical interface · Medical knowledge base

1 Introduction

In recent years, many advances have been made in the field of Information Technologies assistance in the operating room. Different technologies create a highly complex surgical environment, while the introduction of robotic assistance systems establishes a demand for modeling of the surgical workflow and protocols for achieving the highest outcome quality [1]. Even though there were some advances in the field of automatic workflow generation from video or sensor data, most of these approaches required simple procedures and could not provide a high level reference point for guidance and safety during a Robot-Assisted MIS [2–4]. On the other hand, manual creation of a workflow can be a time consuming process or suffer from errors

J. Henriques et al. (Eds.): MEDICON 2019, IFMBE Proceedings 76, pp. 1581–1589, 2020.
https://doi.org/10.1007/978-3-030-31635-8_195

and personal bias [5]. The latest advances in R-A MIS procedures require a robust modeling tool that can aid in training inexperienced surgeons, an inhibitor of R-A MIS success [6], while easing the transition from traditional to modern R-A surgeries.

Although knowledge-based surgery assistance systems are already employed and utilized in the operating rooms and many advancements are being made in their utilization [7], a new approach to preoperative surgical planning is needed, using the latest technologies in human-computer interaction and knowledge-based application development. All these while taking into account the demands of a modern approach to the operating room that can include advanced instruments and robotic tools.

In this paper, we present a user-friendly and intuitive graphical interface for modelling surgical workflows for R-A MIS with the help of a knowledge-based ontology infrastructure. The interface has been developed as a web application for portability and ease of use, while a semantic database of interconnected instruments, anatomies and actions, aiming to aid the surgeon to preoperatively generate the surgical workflow. The process can be performed during the surgical planning, where patient-specific and case-specific information should be taken into account to generate a protocol that can act as a guidance and safety instrument in the modern and technologically rich surgical environment.

The main contribution of this work can be summarized as follows:

- A user-friendly and intuitive graphical interface for preoperative workflow management
- A knowledge-based modeling approach to the surgical workflow
- Inclusion of robotic instruments to the surgical workflow
- Web-based interface that can be accessed through a computer or portable device.

2 Overall Architecture

Our approach is part of a larger system that aims to improve complex minimally invasive surgical operations by developing a novel robotic platform for assisting surgeons in such tasks. While the platform has advanced features that provide a seamless experience to the surgeon, a high-level modeling of the surgery workflow is essential for the guidance and safety of the procedure. The interface is incorporated into the preoperative surgery planning procedure and helps the surgeons to modify the protocols based on patient and operation-specific requirements, while taking into account the advanced tools and robotic arms that are going to be integrated. Furthermore, the interface is interconnected with a knowledge base that allows the surgeon to easily modify the surgery properties without worrying about the low-level details and avoid vital mistakes.

We define the surgical workflow as a process that consists of consecutive phases that describe an important stage of the workflow. Each phase consists of consecutive steps that include smaller states of the workflow, whereas each step is divided in smaller substeps. Each substep is defined by an action, at least one surgical instrument and one human body anatomy, whereas instruments may be assigned to either a robotic arm or an assistant surgeon. Taking into consideration the complexity of generating a specific workflow based on the aforementioned elements, we have developed a knowledge base where each element constitutes an ontology with specific rules, properties and interconnections, so the surgeon can work without concern of following the required protocols. A detailed description of the knowledge base is offered in Sect. 2.2.

The graphical interface is built using a server-client architecture for increased portability and centralized control. Different technologies in both the client and the server should work in unison to create an integrated environment. The client side must be intuitive and lightweight so it can be deployed even on a mobile device, while the server side should be able to handle all client requests and connect to the knowledge base. Aiming to increase the system's portability, we have developed a RESTful API for the knowledge base that can respond to multiple modification requests, while providing the advantages of the ontologies' semantic interconnections and rules. This API can be used by a variety of preoperative tools to present a coherent ontology-based overview of the surgery components to the surgeon (Fig. 1).

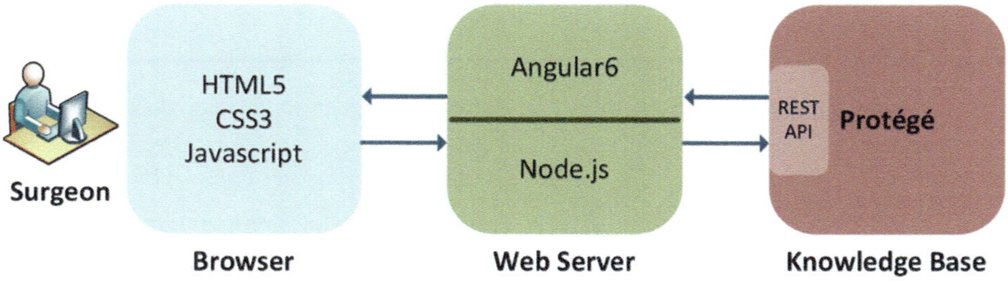

Fig. 1. The overall architecture of the interface

2.1 Graphical Interface

The graphical interface is built using the latest web technologies to provide a seamless experience to the surgeon. The graphical components of the interface are developed in HTML5 and CSS3, with simple and easy to use buttons and icons, while the server runs on Node.JS using the Angular6 framework for advanced DOM manipulation and improved performance.

The visual components of the interface are designed to provide a detailed overview of the surgical workflow while at the same time allowing the interface to be easy and intuitive to use. There is a clear division between the phases, the steps and the substeps of the workflow, whereas the surgeon is only presented with the detailed view of a step at a time (as seen in Fig. 2). Surgical images can be added to each substep for improved visualization or training of a new surgeon.

The interface is connected with the knowledge base so that the rules and semantic interconnections of every ontology can be applied inside the interface. Using *HTTP post* and *get* calls through Angular, various aspects of the protocol ontologies in the knowledge base can be *loaded*, *added*, *deleted* or *modified*, while at the same time multiple checks approve the desired outcome based on the ontology rules. This way,

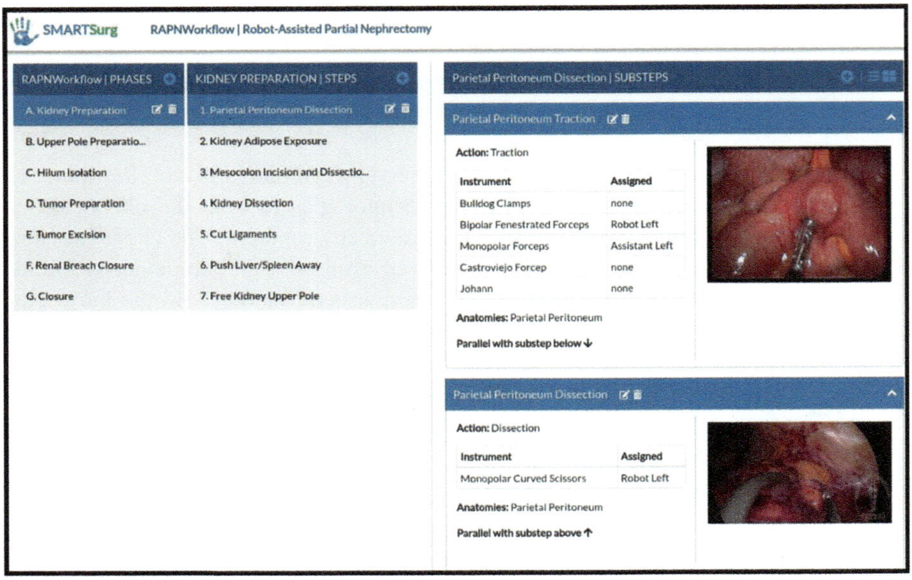

Fig. 2. The main screen of the interface that allows the surgeon to navigate through the R-A MIS workflow

the surgeon can effortlessly construct a coherent protocol preoperatively, avoiding errors or missing information that could later interfere with the progress of the surgery.

2.2 Knowledge Base

The presented Knowledge Base for R-A MIS workflows has been developed in Protégé [8] with OWL (Web Ontology Language) and SWRL (Semantic Web Rule Language) rules and reasoning. OWL provides an easy way to create classes, object properties that connect class instances and data properties. With these properties and class instances SWRL rules can be formed that are used consequently for object property inference by a reasoner. Reasoning in a knowledge base is essential as it provides object properties not explicitly defined and checks if explicitly defined properties are consistent with the knowledge base rules. Thus, a cohesive and consistent knowledge base can be designed that accurately represents an R-A MIS procedure workflow.

Ontologies for seven different R-A MIS procedures have been created with Protégé and are part of the knowledge base. A prototype workflow is available for R-A Partial Nephrectomy, R-A Prostatectomy, R-A Cystectomy, R-A Partial Lateral Meniscectomy, R-A Repair of Lateral Meniscus Tear, R-A Mitral Valve Repair/Replacement and R-A Coronary Artery Bypass Grafting. These protocols can be used as a template or guide and can be adapted to each case. Their purpose is to provide a starting point for the surgeon in the creation and editing of new surgical protocols (Fig. 3).

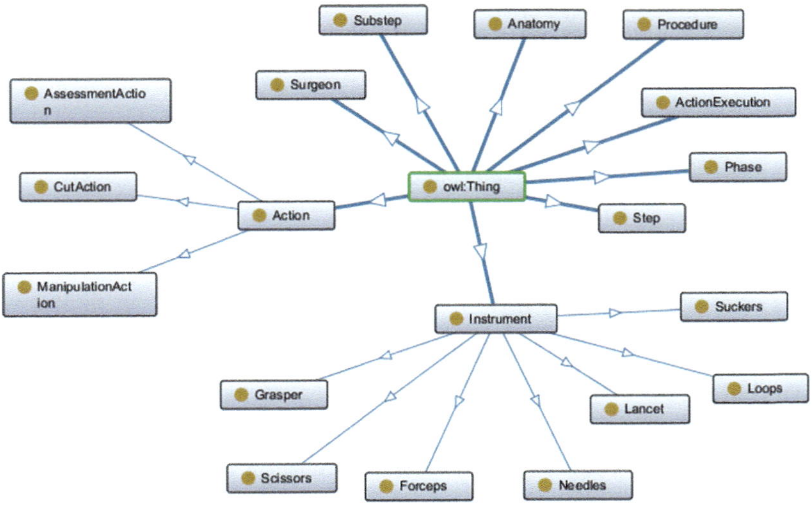

Fig. 3. The Knowledge Base classes

3 Implementation of the Interface

The developed graphical interface's modules enable the surgeon to construct a coherent protocol preoperatively, including robot specific information and advanced semantic-based interconnections. The surgery workflow is presented in a detailed but simple arrangement that provides the surgeon with easy-to-use tools to modify the workflow as intended.

The first step of the process involves the creation of a new protocol or the modification of an existing one. The surgeon is presented with a simple interface where there are options between creating a new workflow or modifying an existing workflow. The system provides prototype protocols for a variety of operations that the surgeon can modify to the needs of the specific case. A new name has to be given to the workflow for every newly created protocol, in order for it to be saved and accessed later (Fig. 4).

After loading the protocol the surgeon is presented with the main interface, where the view is divided into three components-menus. The first component contains the phases of the workflow that the surgeon can select, the second contains the steps that belong to the selected phase and the third contains the substeps that belong to the selected step. After selecting a phase or a step the surgeon can either delete it or modify its name clicking on the respective icons. New phases and steps can be added clicking on the plus icon where the surgeon defines the new entity's name and its position in the workflow sequence. For a new step to be defined it has to include at least one substep, while a new phase has to include at least one step (Fig. 2).

The substeps are the core of the workflow, where each one is defined by its name, an action, an anatomy and the surgical instruments involved in order to complete the action. Each instrument should be assigned to either the left arm or right arm of the robot, or one of the assistant's arms respectively. The surgeon can also add a

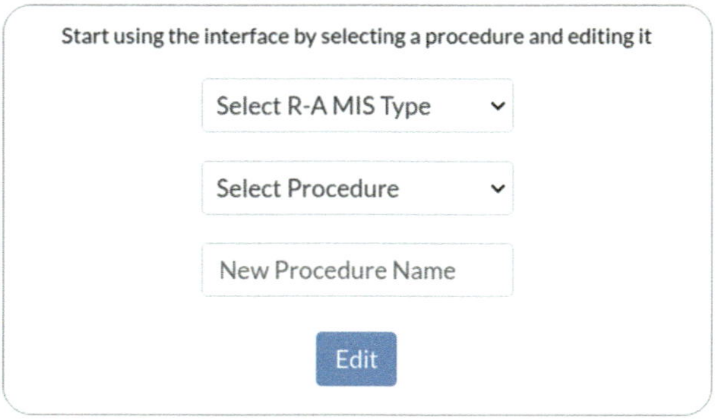

Fig. 4. The workflow loading interface

representative image of the action for future reference or training purposes and extra details that could involve the usage of a specific type of sutures or tubes for example. The substeps are not only sequential but they can be in parallel order, where the surgeon performs an action while the assistant surgeon performs another.

A surgeon can either add new substeps inside a step or modify existing substeps accordingly clicking on the respective icons. The interface provides easy-to-use buttons and input fields that are validated automatically through the knowledge base. Every action has a defined set of instruments that can be employed, while each instrument can be either used only by a robot, an assistant surgeon or both (Fig. 5). These features prevent the surgeon from making any errors and enable the effortless creation of a coherent surgical protocol.

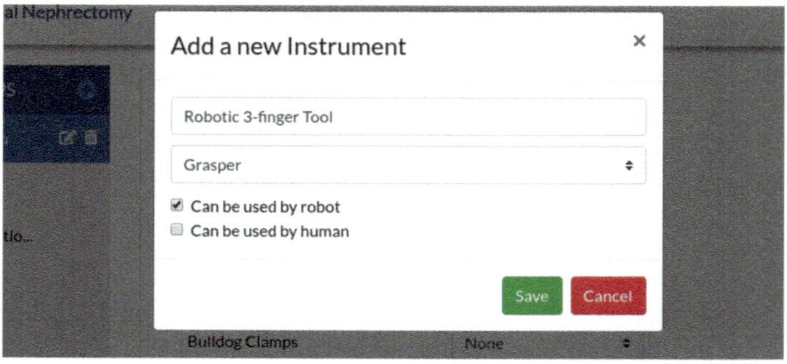

Fig. 5. The dialog for adding a new Instrument to the Knowledge Base

Last but not least, new anatomies and instruments can be added to the knowledge base, so the surgeon can create surgery workflows for other parts of the body or add advanced instruments that can be later employed in new protocols. Additionally, when a new instrument is being defined, the surgeon should take extra care to provide the system with the type of the instrument (forceps, clamp etc.) so it can be assigned to the appropriate set of actions in the knowledge base. It is also required for the surgeon to define whether the instrument can be assigned to a robotic arm, an assistant surgeon or both in order to prevent any errors when the instrument is employed in a future workflow (Fig. 6).

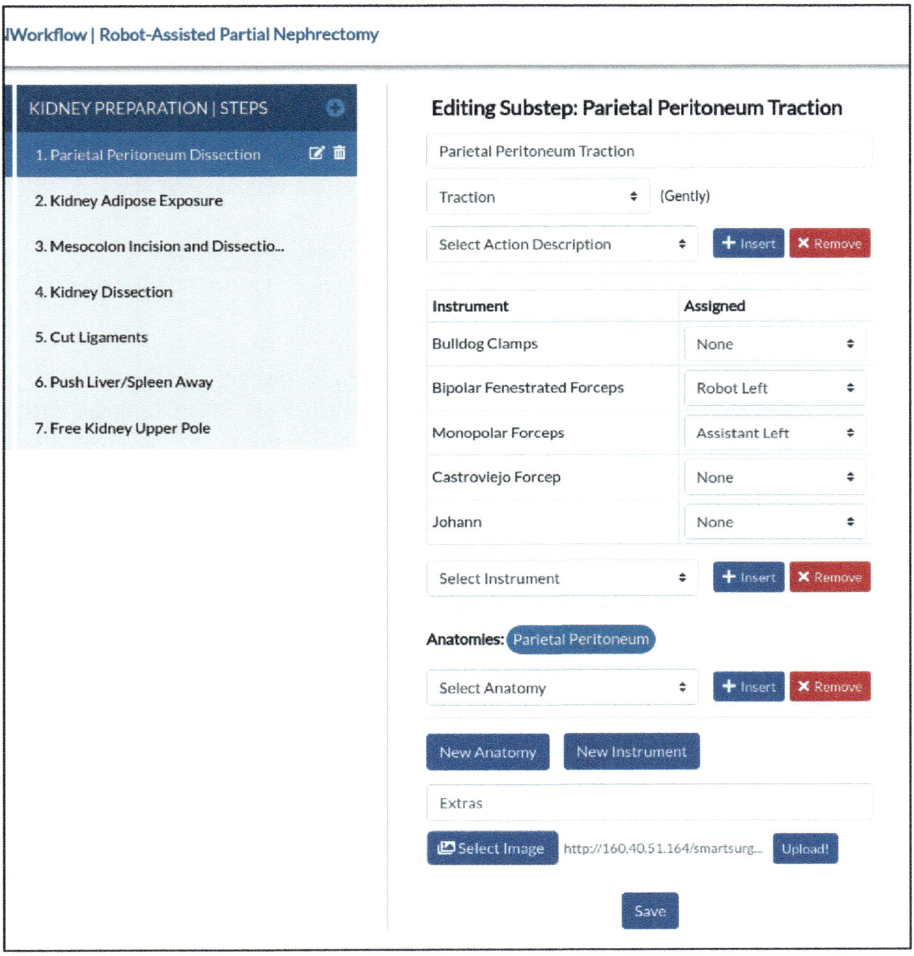

Fig. 6. The form for adding or editing a substep

4 Experiments and Evaluation

In this section, we present some results from the experiments we performed with 5 orthopedic surgeons that were requested to use the interface to create a perioperative surgical workflow for two types of arthroscopic meniscus surgery. The doctors had a brief introduction of the system's functionalities and used a laptop PC to perform the desired modifications. The modifications were made on provided prototype protocols where the surgeons could add new phases or steps while editing their sequence, add new instruments or anatomies to the protocol and incorporate robotic assistance. Furthermore, after completing the task, to evaluate the usability of the interface, the surgeons had to answer on a five-point Likert scale the questions in Table 1.

Table 1. List of questions used in the evaluation of the interface

1	An inexperienced surgeon can easily create or modify a protocol with the GUI
2	I understood what buttons I needed to press to perform each action
3	I found the interface easy to use
4	The interface was pleasant to use
5	The interface would be helpful in the creation of a R-A MIS surgical protocol
6	The information was clearly arranged on the screen
7	I would like to employ such an interface to my preoperative workflow
8	The interface would help in the training of new surgeons in R-A MIS procedures

The answers had the following options: strongly disagree (=1), disagree, no opinion, agree, and strongly agree (=5). The average scores and standard deviations for each question are presented in Table 2. From interviews the doctors evaluated positively the experience and were confident with the interface if they were asked to create a new surgical workflow or modify an existing one without any assistance. The average score for all the questions the doctors had to answer was **4.53** out of **5**.

Table 2. Median, Average and St. Deviation of question scores on a five-point Likert scale

Likert scores\Question	1	2	3	4	5	6	7	8	All
Median	4	4	5	5	5	5	5	5	**5**
Average	4	4.2	4.6	4.6	4.8	4.6	4.6	4.8	**4.53**
St. Deviation	0	0.84	0.55	0.55	0.45	0.55	0.55	0.45	**0.49**

The doctors were also asked to answer the **Questionnaire for User Interface Satisfaction** that was introduced in [9], where the not applicable questions were removed. All the questions were rated on a 0 to 9 scale where 0 represents the lowest score and 9 represents the highest. The average score for all the questions was **7.77** out of **9** with a standard deviation of **0.73**.

5 Conclusion and Future Work

In this paper, we presented the design and implementation of a knowledge-based graphical interface for modeling surgical workflows involving Robot-Assisted MIS procedures. The interface is built as a web application for portability and can be easily integrated to the modern preoperative surgery planning. To this end, many advanced technologies are employed offering a seamless experience to the surgeon that has to create or modify a surgery protocol for advanced guidance and safety during operation, while easing the transition from traditional surgeries to robot-assisted MIS. We evaluated the usability of the system by having 5 orthopedic surgeons use the interface to modify arthroscopic R-A MIS, for which we had positive reviews. In the future, we aim to add new features to the interface like additional modification options for a broader range of surgeries, while planning of incorporating verification of the surgical process into the interface by implementing automated workflow recognition by video and sensor data analysis during the R-A MIS operation.

Acknowledgment. The research leading to these results has received funding from the European Community's Framework Programme Horizon 2020 – under grant agreement No 732515 – SMARTSurg.

References

1. Schumann, S., Bühligen, U., Neumuth, T.: Outcome quality assessment by surgical process compliance measures in laparoscopic surgery. Artif. Intell. Med. **63**(2), 85–90 (2015)
2. Lalys, F., Jannin, P.: Surgical process modelling: a review. Int. J. Comput. Assist. Radiol. Surg. **9**(3), 495–511 (2014)
3. Ahmadi, S.A., Sielhorst, T., Stauder, R., Horn, M., Feussner, H., Navab, N.: Recovery of surgical workflow without explicit models. In: International Conference on Medical Image Computing and Computer-Assisted Intervention, pp. 420–428. Springer, Heidelberg, October 2006
4. Padoy, N., Blum, T., Essa, I., Feussner, H., Berger, M.O., Navab, N.: A boosted segmentation method for surgical workflow analysis. In: International Conference on Medical Image Computing and Computer-Assisted Intervention, pp. 102–109. Springer, Heidelberg, October 2007
5. Blum, T., Padoy, N., Feußner, H., Navab, N.: Workflow mining for visualization and analysis of surgeries. Int. J. Comput. Assist. Radiol. Surg. **3**(5), 379–386 (2008)
6. Sridhar, A.N., Briggs, T.P., Kelly, J.D., Nathan, S.: Training in robotic surgery—an overview. Curr. Urol. Rep. **18**(8), 58 (2017)
7. Schoch, N., Engelhardt, S., Zimmermann, N., Speidel, S., De Simone, R., Wolf, I., Heuveline, V.: Integration of a biomechanical simulation for mitral valve reconstruction into a knowledge-based surgery assistance system. In: Medical Imaging 2015: Image-Guided Procedures, Robotic Interventions, and Modeling, vol. 9415, p. 941502. International Society for Optics and Photonics, March 2015
8. Musen, M.A.: The protégé project: a look back and a look forward. AI Matters **1**(4), 4 (2015)
9. Chin, J.P., Diehl, V.A., Norman, K.L.: Development of an instrument measuring user satisfaction of the human-computer interface. In: Proceedings of the SIGCHI Conference on Human Factors in Computing Systems, pp. 213–218. ACM, May 1988

Augmented and Virtual Reality in Minimally Invasive Surgery, State of the Art and Future Prospects

Michele Catellani[1], Giovanni Cordima[1], Ottavio de Cobelli[1], Efthymios Papasoulis[2], Emmanuel Papacostas[2], Aristotelis Sideridis[2], Georgia Peleka[3(✉)], Georgios Zampokas[3], Konstantinos Tsiolis[3], Angeliki Topalidou-Kyniazopoulou[3], Ioannis Mariolis[3], Sotiris Malasiotis[3], and Dimitrios Tzovaras[3]

[1] IEO European Institute of Oncology, Milan, Italy
[2] Thessaloniki Minimally Invasive Surgery, The M.I.S. Orthopaedic Center, Thessaloniki, Greece
[3] Information Technologies Institute, Centre for Research & Technology, Thessaloniki, Greece
gepe@iti.gr

Abstract. Computer assisted guidance systems which utilize Augmented Reality (AR) and/or Virtual Reality (VR) during Minimally invasive surgery (MIS) have turned into a prominent research subject in the recent years. These technologies provide noteworthy advantages to the patients and the clinicians, by providing significant additional data during MIS. In this way, a priori information known to the surgeon pre-operatively (segmented anatomical structures of interest and lesions) can be superimposed semi-transparently on the directly visible tissues. This process can lead to higher preservation rates of healthy tissue and shorter post surgical recovery periods. The paper at hand is a survey investigating research efforts that have been conducted in the recent years in terms of utilization of AR and VR in laparoscopic and arthroscopic MIS.

Keywords: Minimally invasive surgery · Laparoscopy · Arthroscopy · Augmented Reality · Virtual Reality

1 Introduction

AR and VR within the content of Minimally Invasive Surgery (MIS) is the real time alignment or superimposition of pre-operative or intra-operative imaging onto patients' actual images or video [1]. This allows surgeon to assimilate important visual information obtained with imaging modalities (i.e., ultrasound, computed tomography (CT) or magnetic resonance imaging (MRI) that usually are captured pre-operatively. The images are displayed allowing a seemingly transparent visual anatomy of the internal structures or lesions through the overlying tissues. The most of the research in AR has focused on orthopedic surgery where

J. Henriques et al. (Eds.): MEDICON 2019, IFMBE Proceedings 76, pp. 1590–1595, 2020.
https://doi.org/10.1007/978-3-030-31635-8_196

bony landmarks provide fixed and constant reference points for registration and tracking [2].

2 Augmented and Virtual Reality in Laparoscopic Surgery

The use of AR/VR in laparoscopic surgery is more challenging because of the movement or deformation of the organs due to the heartbeat, ventilation, pneumoinsufflation, and surgical manipulation [3]. At present, the AR environment in urology has been involved in the field of surgical navigation, which was mainly focused on nephrolithotomy and partial nephrectomy (PN). CT scan has been the main modality used for the creation of the roadmap providing the basis for the surgical navigation [4]. Ukimura and Gill reported the application of a combination of ultrasound and CT acquired imaging during percutaneous ablative techniques for renal cancer. The same procedure was tested during laparoscopic partial nephrectomy (PN) and radical prostatectomy (RP) [5]. Rassweiler et al. reported a new iPad-assisted marker-based navigation as a tool for assisting the percutaneous puncture of the kidney during percutaneous nephrolithotomy (PCNL) [6].

A 3D reconstruction model acquired from the CT scan or from 3D laser scan has been used by Su and colleagues and Drewniak et al. respectively, during robotics PN. They concluded that AR was helpful to the surgeon during the procedures in order to save healthy parenchyma without compromising the oncological effectiveness [7,8]. An image-processing application was used by Nakamura and colleagues to create a 3D volume-rendered image of the renal vessels, kidneys, and surrounding organs which was simultaneously fused onto live video [9]. Kleinjan et al. used a hybrid navigation system based on the integration of preoperative 3D SPECT/CT with real-time fluorescence imaging acquired during a lymph node biopsy [10]. Simpfendorfer and colleague presented an initial human in vivo application of an AR navigation system. They superimposed, during RP, virtual prostate models generated from trans-rectal ultrasonography onto the laparoscopic video [11]. Thompson et al. re-proposed the experiment overlaying unprocessed T2-weighted MRI scans over the prostate anatomy during the surgery [12].

Ruppert et al. implemented a hand tracking and gesture recognition system, which simulates mouse movements with the freehand movements. This system enables the surgeon to touchlessly navigate the CT scan data during the surgery [13]. In 2019 Bertolo et al. performed a systematic literature review on the different techniques of AR and their clinical impact for urological interventions. Only one comparative study was found, with nineteen studies reporting single-arm outcomes of the use of AR. They identify very limited evidence supporting the adoption of AR as a way to improve the outcomes of urological interventions, probably due to the characteristics of the studies. The major limitation of AR-assisted surgery is inaccuracy in registration, which is translating into a poor navigation precision [14]. They conclude that a strict collaboration between clinicians and engineers is needed, to focus on the clinical necessities and to avoid the creation of technologies without real-world application or utility.

Nevertheless several studies evaluate the quality feedback from the surgeons exposed to the use of AR. The utility of smart glasses was investigated by García-Cruz et al. They evaluate the use of smart glasses in eighty urologists during their daily surgical and clinical practice obtaining an appreciation score of seven and half on a ten-point scale [15]. Six experienced robotic surgeons were invited to use a software during their nerve-sparing robotic RPs [16]. Enthusiastic results were obtained from the evaluation of each analyzed step, with a median rating of nine out of ten. Several studies measured the impact of the 3D anatomy reconstruction in improving the preoperative counseling before complex renal masses PN [17, 18]. There was a clear preference in performing the robotic PN with the 3D reconstruction than with the CT scan image evaluation. All these results confirm the increasing interest for a new generation of image-guided surgery.

3 Augmented and Virtual Reality in Arthroscopic Surgery

In orthopaedics, the aim of computer-assisted arthroscopy is to provide assistance with orientation in the joint, triangulation of the instruments used, and avoidance of soft tissue injury. The computer-assisted arthroscopy system should be able to reconstruct the preoperative magnetic resonance images into a three-dimensional pre-operative model. Localization of the knee joint tissues, as well as the tools inserted into the joint is performed with a registration module. Navigation and localization within the joint typically require the use of optical and/or mechanical sensors, but mechatronic [19], and electromagnetic localization systems [20] have also been used. A human-machine interface should be able to provide visual and haptic data of what is going on in the joint in real-time. Force-feedback (haptic) is an essential tool, as - in combination with the image - it allows recognition of the structures within the joint, and their pathology for diagnostic purposes, during intervention [21,22].

However, application of VR systems in clinical arthroscopy is still in its infancy. The major application of arthroscopic VR simulators so far is surgical skills development for training purposes. Arthroscopy has a long learning curve, and navigation systems have been used in order to shorten it [23,24]. Surgeons with greater experience tend to do better with the simulators [25,26]. There are studies however, that have found that VR training did not transfer as readily to the benchtop (BT) simulator [27], and question if it can be transferred into OR [24] or whether it would eventually improve patient safety. Additionally, residents practicing on cadaveric specimens improved twice as fast as those utilizing a high-fidelity simulator; nevertheless, based on cost estimation, the simulator may be more cost-effective [28].

Arthroscopic virtual reality simulators have been used in most big joints, such as the hip [29], the knee [30], the shoulder [30], and the wrist [20]. Several simulators for training purposes exist, some of the most commonly used are the Arthro S, Arthro-Sim, Arthro VR, and ARTHRO Mentor [31].

4 Augmented and Virtual Reality for Smart Robot-Assisted MIS

Within the context of Horizon2020 research project SMARTsurg [32], an innovative visual guidance system for MIS has been developed. It is able to overlay structures which are captured pre-operatively to the intra-operative surgical view. First, the 3D models are extracted from pre-operative MRI or CT scans. Then, the frames of a stereo endoscopic camera are used as input to a 3D reconstruction module, which is able to estimate the depth of the surgical scene. The method from [33] is chosen since it operates within real-time framework, while producing accurate depth maps.

For a successful intra-operative registration, a good initialization step is required. Thus, a dedicated graphical user interface has been developed, where the reconstructed point cloud and the pre-operative model are visualized. The user manually selects a few correspondence points between them and an initial rigid registration is performed. However, during the surgical procedure, the registration process must be automatic for every frame. After extensive evaluation of relevant methods [34], Coherent Point Drift achieved high registration accuracy with adequate execution speed and is selected as the online registration method between the 3D reconstructed scene and the pre-operative model. The last, is superimposed into the intra-operative surgical view, while its position and pose are updated with every frame at a rate of 4.5 Hz. The main visualization medium is a passive 3D monitor via AR but the view has also been extended to support visualization to a VR headset.

Acknowledgments. The research leading to these results has received funding from the European Community's Framework Programme Horizon 2020 under grant agreement No 732515 SMARTsurg.

References

1. Tang, S.L., Kwoh, C.K., Teo, M.Y., Sing, N.W., Ling, K.V.: Augmented reality systems for medical applications. IEEE Eng. Med. Biol. Mag. **17**(3), 49–58 (1998)
2. Sim, H.G., Yip, S.K.H., Cheng, C.W.S.: Equipment and technology in surgical robotics. World J. Urol. **24**(2), 128–135 (2006)
3. Westwood, J.D.: Real-time anatomical 3D image extraction for laparoscopic surgery. In: Medicine Meets Virtual Reality 2001: Outer Space, Inner Space, Virtual Space, vol. 81, p. 18 (2001)
4. van Oosterom, M.N., van der Poel, H.G., Navab, N., van de Velde, C.J., van Leeuwen, F.W.: Computer-assisted surgery: virtual-and augmented-reality displays for navigation during urological interventions. Curr. Opin. Urol. **28**(2), 205–213 (2018)
5. Ukimura, O., Gill, I.S.: Imaging-assisted endoscopic surgery: Cleveland clinic experience. J. Endourol. **22**(4), 803–810 (2008)
6. Rassweiler, J.J., Müller, M., Fangerau, M., Klein, J., Goezen, A.S., Pereira, P., Meinzer, H.P., Teber, D.: iPad-assisted percutaneous access to the kidney using marker-based navigation: initial clinical experience. Eur. Urol. **61**(3), 628 (2012)

7. Su, L.M., Vagvolgyi, B.P., Agarwal, R., Reiley, C.E., Taylor, R.H., Hager, G.D.: Augmented reality during robot-assisted laparoscopic partial nephrectomy: toward real-time 3D-CT to stereoscopic video registration. Urology **73**(4), 896–900 (2009)
8. Drewniak, T., Rzepecki, M., Juszczak, K., Kwiatek, W., Bielecki, J., Zieliski, K., Ruta, A., Czekierda, Ł., Moczulskis, Z.: Augmented reality for image guided therapy (ARIGT) of kidney tumor during nephron sparing surgery (NSS): animal model and clinical approach. Folia Med. Crac. **51**(1–4), 77–90 (2011)
9. Nakamura, K., Naya, Y., Zenbutsu, S., Araki, K., Cho, S., Ohta, S., Nihei, N., Suzuki, H., Ichikawa, T., Igarashi, T.: Surgical navigation using three-dimensional computed tomography images fused intraoperatively with live video. J. Endourol. **24**(4), 521–524 (2010)
10. KleinJan, G.H., van den Berg, N.S., van Oosterom, M.N., Wendler, T., Miwa, M., Bex, A., Hendricksen, K., Horenblas, S., van Leeuwen, F.W.: Toward (hybrid) navigation of a fluorescence camera in an open surgery setting. J. Nucl. Med. **57**(10), 1650–1653 (2016)
11. Simpfendörfer, T., Baumhauer, M., Müller, M., Gutt, C.N., Meinzer, H.P., Rassweiler, J.J., Guven, S., Teber, D.: Augmented reality visualization during laparoscopic radical prostatectomy. J. Endourol. **25**(12), 1841–1845 (2011)
12. Thompson, S., Penney, G., Billia, M., Challacombe, B., Hawkes, D., Dasgupta, P.: Design and evaluation of an image-guidance system for robot-assisted radical prostatectomy. BJU Int. **111**(7), 1081–1090 (2013)
13. Ruppert, G.C.S., Reis, L.O., Amorim, P.H.J., de Moraes, T.F., da Silva, J.V.L.: Touchless gesture user interface for interactive image visualization in urological surgery. World J. Urol. **30**(5), 687–691 (2012)
14. Bertolo, R., Hung, A., Porpiglia, F., Bove, P., Schleicher, M., Dasgupta, P.: Systematic review of augmented reality in urological interventions: the evidences of an impact on surgical outcomes are yet to come. World J. Urol., 1–10 (2019)
15. García-Cruz, E., Bretonnet, A., Alcaraz, A.: Testing smart glasses in urology: clinical and surgical potential applications. Actas Urológicas Españolas (English Edition) **42**(3), 207–211 (2018)
16. Porpiglia, F., Bertolo, R., Amparore, D., Checcucci, E., Artibani, W., Dasgupta, P., Montorsi, F., Tewari, A., Fiori, C.: Augmented reality during robot-assisted radical prostatectomy: expert robotic surgeons' on-the-spot insights after live surgery. Minerva Urologica Nefrologica = Ital. J. Urol. Nephrol. **70**(2), 226–229 (2018)
17. Antonelli, A., Veccia, A., Palumbo, C., Peroni, A., Mirabella, G., Cozzoli, A., Martucci, P., Ferrari, F., Simeone, C., Artibani, W.: Holographic reconstructions for preoperative planning before partial nephrectomy: a head-to-head comparison with standard CT scan. Urol. Int. **102**(2), 212–217 (2019)
18. Bertolo, R., Autorino, R., Fiori, C., Amparore, D., Checcucci, E., Mottrie, A., Porter, J., Haber, G.P., Derweesh, I., Porpiglia, F.: Expanding the indications of robotic partial nephrectomy for highly complex renal tumors: urologists' perception of the impact of hyperaccuracy three-dimensional reconstruction. J. Laparoendosc. Adv. Surg. Tech. **29**(2), 233–239 (2019)
19. Dario, P., Carrozza, M.C., Marcacci, M., D'Attanasio, S., Magnami, B., Tonet, O., Megali, G.: A novel mechatronic tool for computer-assisted arthroscopy. IEEE Trans. Inf. Technol. Biomed. **4**(1), 15–29 (2000)
20. Zemirline, A., Agnus, V., Soler, L., Mathoulin, C.L., Liverneaux, P.A., Obdeijn, M.: Augmented reality-based navigation system for wrist arthroscopy: feasibility. J. Wrist Surg. **2**(04), 294–298 (2013)

21. Zivanovic, A., Dibble, E., Davies, B., Moody, L., Waterworth, A.: Engineering requirements for a haptic simulator for knee arthroscopy training. Stud. Health Technol. Inform. **94**, 413–418 (2003)
22. Lyu, S.R., Lin, Y.K., Huang, S.T., Yau, H.T.: Experience-based virtual training system for knee arthroscopic inspection. Biomed. Eng. Online **12**(1), 63 (2013)
23. Banaszek, D., You, D., Chang, J., Pickell, M., Hesse, D., Hopman, W.M., Borschneck, D., Bardana, D.: Virtual reality compared with bench-top simulation in the acquisition of arthroscopic skill: a randomized controlled trial. JBJS **99**(7), e34 (2017)
24. Aïm, F., Lonjon, G., Hannouche, D., Nizard, R.: Effectiveness of virtual reality training in orthopaedic surgery. Arthrosc. J. Arthrosc. Relat. Surg. **32**(1), 224–232 (2016)
25. Ziegler, R., Fischer, G., Müller, W., Göbel, M.: Virtual reality arthroscopy training simulator. Comput. Biol. Med. **25**(2), 193–203 (1995)
26. Pedowitz, R.A., Esch, J., Snyder, S.: Evaluation of a virtual reality simulator for arthroscopy skills development. Arthrosc. J. Arthrosc. Relat. Surg. **18**(6), 1–6 (2002)
27. Middleton, R.M., Alvand, A., Roberts, P.G., Hargrove, C., Kirby, G., Rees, J.L.: Simulation-based training platforms for arthroscopy: a randomized comparison of virtual reality learning to benchtop learning. Arthrosc. J. Arthrosc. Relat. Surg. **33**(5), 996–1003 (2017)
28. Camp, C.L., Krych, A.J., Stuart, M.J., Regnier, T.D., Mills, K.M., Turner, N.S.: Improving resident performance in knee arthroscopy: a prospective value assessment of simulators and cadaveric skills laboratories. JBJS **98**(3), 220–225 (2016)
29. Bauer, D.E., Wieser, K., Aichmair, A., Zingg, P.O., Dora, C., Rahm, S.: Validation of a virtual reality-based hip arthroscopy simulator. Arthrosc. J. Arthrosc. Relat. Surg. **35**(3), 789–795 (2019)
30. Yari, S.S., Jandhyala, C.K., Sharareh, B., Athiviraham, A., Shybut, T.B.: Efficacy of a virtual arthroscopic simulator for orthopaedic surgery residents by year in training. Orthop. J. Sport. Med. **6**(11), 2325967118810176 (2018)
31. Martin, K.D., Akoh, C.C., Amendola, A., Phisitkul, P.: Comparison of three virtual reality arthroscopic simulators as part of an orthopedic residency educational curriculum. Iowa Orthop. J. **36**, 20 (2016)
32. SMart weArable Robotic Teleoperated surgery SMARTsurg Horizon2020 project. http://www.smartsurg-project.eu/
33. Zampokas, G., Tsiolis, K., Peleka, G., Mariolis, I., Malasiotis, S., Tzovaras, D.: Real-time 3D reconstruction in minimally invasive surgery with quasi-dense matching. In: 2018 IEEE International Conference on Imaging Systems and Techniques (IST), pp. 1–6. IEEE (2018)
34. Peleka, G., Zampokas, G., Mariolis, I., Malasiotis, S., Tzovaras, D.: Intra-operative 3D registration of MIS reconstructed surfaces to pre-operative models. In: 2018 IEEE International Conference on Imaging Systems and Techniques (IST), pp. 1–6. IEEE (2018)

Special Sessions: Intelligent Computational Systems in Biomedical Engineering

Modeling and Objectification of Skiagraphy Image Quality Deterioration Caused by X-Ray Secondary Irradiation on Mobile X-Ray Device

Klara Fiedorova[1(✉)], Martin Augustynek[1], Jan Kubicek[1], Marek Penhaker[1], Andrea Vodakova[1], and Karol Korhelik[2]

[1] FEECS, VSB-Technical University of Ostrava, 17. Listopadu 15, K450 Ostrava-Poruba, Czech Republic
{klara.fiedorova, martin.augustynek, jan.kubicek, marek.penhaker, andrea.vodakova.st}@vsb.cz

[2] Faculty of Medicine, Institute of Diagnostic Imaging, University of Ostrava, Syllabova 19, Ostrava-Zábřeh, Czech Republic
karol.korehelik@osu.cz

Abstract. Nowadays, the use of X-ray is an important part of imaging techniques used in medicine. The resulting skiagraphy images are kept in the workplace and can be influenced by secondary radiation, which occurs as a secondary effect when irradiated with X-rays. This secondary radiance has not to effect only for the skiagraphy images but also for the human body. The effect of secondary radiation is irreversible image acquisition, which can lead to poor diagnosis and poor clinical information. Currently, there are no techniques and methods available to determine the extent of secondary damage to the skiagraphy cassette. In the work, we analyse the influence of secondary radiation on the quality of the skiagraphy images using mobile X-ray devices. During image analysis, a multi-region segmentation method is used to determine pixels that are affected by irradiation. By this method, we are able to derive the characteristics of the image, determine the level of irradiation and estimate the effect of secondary radiation.

Keywords: X-ray · Secondary radiation · Computed radiography cassette · Imaging plate · Multilevel Otsu method

1 Introduction

During X-ray examination, part of the body is irradiated with the primary beam passes through the body and it has deviated from the original direction [1]. Sources of radiation load are shown in Fig. 1. X-rays can penetrate any matter, the passage through reduces its intensity [1–3]. Radiation attenuation depends on the radiation quality and the properties of the mass. X-ray radiation is attenuated, scattered and deviated from the primary beam. Each radiated object is the source of secondary radiation. The influence of secondary radiation decreases with the distance from the radiation source. For example, if the patient is at a distance of 1 m from the radiation source during the test, the secondary radiation load is 0.1–0.15% and the radiation dose is around 0.1 μGy/1. For a distance of 2 m is dose about 0.005 μGy/1 [1, 4].

J. Henriques et al. (Eds.): MEDICON 2019, IFMBE Proceedings 76, pp. 1599–1608, 2020.
https://doi.org/10.1007/978-3-030-31635-8_197

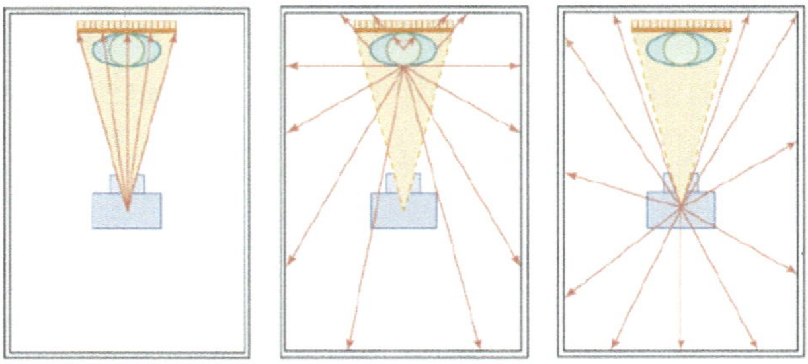

Fig. 1. From the left: primary, scattered, leaking radiation

During the method, the secondary radiation inappropriately affects the patient, healthcare personnel, and the quality of the displayed image. Changes in the image are reflected by the contrast of the image. In radiology, this phenomenon is addressed by subsequent image processing and editing [1, 4–6, 14].

1.1 Processing of Medical Images

Image processing is an important process for an order to obtain other important insights. Preprocessing images is first step for basic image editing [8–10]. Computer algorithms for displaying anatomical structures and their area of interest play a very important role in specific radiological tasks [8]. The most common segmentation methods are Edge Detection, Region-Based Method or Statistical Methods and Hybrid Methods [7–10, 13].

2 Experimental Measurement

Experimental measurements were carried out at The University Hospital in Ostrava at the Institute of Radiodiagnostics, where the possibility of modeling and analysis of the secondary radiation on the scintigraphic cassettes with memory foils.

Due to the harmfulness of X-rays on the body, a phantom with artificial joints and human organs replacing the human body was used for experimental measurements.

For monitor radiation exposure is used dosimeter DMC 3000 dosimeter. To measure the effective dose, measurements need to be carried out over a period of several minutes to months.

The Siemens Mobile Mira digital handheld device was used for the measurement to provide a high-quality. For the patient's bedside skigeography, protection is mainly used to maintain the distance from the radiation source and to keep the primary radiation beam directly to the irradiated object.

In this experiment were used: a CR cassette with a memory foil (skiagraphy cassette) using the principle of indirect digital radiography, a piece of thin sheet metal with a thickness of 1 mm and a size of 55 mm × 45 mm and lead content of 99.97%.

2.1 Implementation of Measurements

The measurements were made to obtain a digital image representing the secondary radiation produced by the patient and determining the influence of this radiation.

The measurement process was carried out under constant conditions and parameters. The phantom and CR cassette on the stand were 75 cm high, with the X-ray being 85 cm away from the phantom. The central ray of the primary beam of X-ray radiation was directed to the phantom chest. The CR cassette was positioned perpendicular to the phantom. For the measurement results to be well interpretable, the cassette was provided with a piece of lead. This creates a contrast in the resulting image. In order to determine the effective dose, a dosimeter is located perpendicular to the CR cassette. The CR cassette together with the dosimeter was placed at a distance of 1 m, 1.5 m, 2 m. The exposure values were: 77 kV and 11 mAs. 10 cassettes of the same type were used to measurements.

2.2 Measurement Results for Mobile X-Ray Machine

Table 1 includes the measured dose values (mSV) from the dosimeter and the value of the surface kerma (ionizing radiation energy) (cGy · cm2n) from the RTG device for the three distances. The planks are not dependent on the distance from the X-rays source, this confirms the general fact that applies to point sources.

Table 1. Measured values.

	Distance (m)	Exposure		Spatial kerma (cGy · cm2n)	Effective dose (mSv)
		Anode voltage (kV)	Exposition (mAs)		
1	1	77	11	63.67	0.0010
2	1.5	77	11	63.73	0.0009
3	2	77	11	63.69	0.0005

3 Evaluation and Determination of the Influence of Secondary Radiation

The primary image generated by radiation on CR cassette was the latent image on the memory film. Image digitization was performed using a red laser.

3.1 Preprocessing the Acquired Image

The digitized image was in DICOM format, it had to be converted to png. [11] After converting the image, image processing was performed in MATLAB [12]. The processing involved extracting an area of interest (RoI), where parts of the lead image were removed, resulting in a size of 1008 × 832 pixels [13, 14]. The example for different distances is in the picture Fig. 2.

Fig. 2. Digital images for 1 m, 1.5 m, 2 m.

3.2 Image Segmentation

The images were processed using the multilevel Otsu method. It is determining the threshold value by scattering from histogram of image. Output is multiregional segmentation model [7–10, 15]. From this image, a histogram was created to derive a dependency of the frequency of occurrence of individual pixels at a given gray level to the color intensity distribution, i.e. 0 to 255. The histogram (Fig. 5) represents two peaks corresponding to two classes of objects in the image, differing in brightness - the lead (lighter) and the background (darker) and in the histogram is determined the threshold. The number of areas on which the histogram will be divided equals the number of required threshold levels. The final value of the scattering limit is calculated from the equation [9, 14]:

$$\sigma^2 = W_{01} * W_1 * (\mu_0 - \mu_1)^2 \tag{1}$$

In the work, 8 segmentation classes were applied, where the threshold levels <5 represented the area of interest with an assigned value of 1 corresponding to the secondary radiation. And threshold levels ≥ 5 equal to objects with an assigned value of 0 corresponding to the leaded object (Fig. 3). The resulting image is a binary image with the preservation of secondary radiation (Fig. 4).

Fig. 3. From the left: original image, multiregional segmentation output.

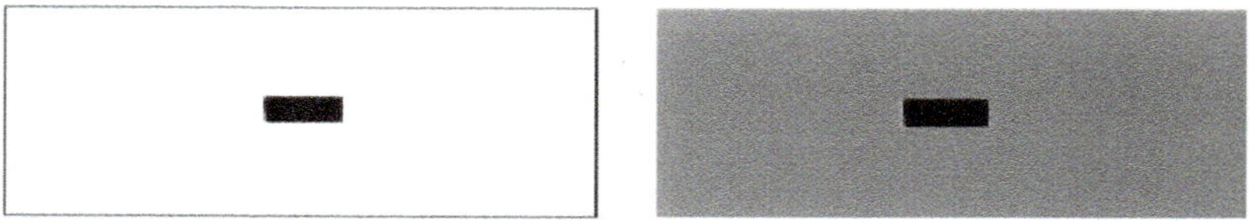

Fig. 4. From the left: binary form, segmented image.

Figure 5 compares the histograms of the original image with the image after segmentation. The histogram of the original image is characterized by the peaks of the lead. These peaks are suppressed. Peaks characterizing the pixels corresponding to secondary radiation are left.

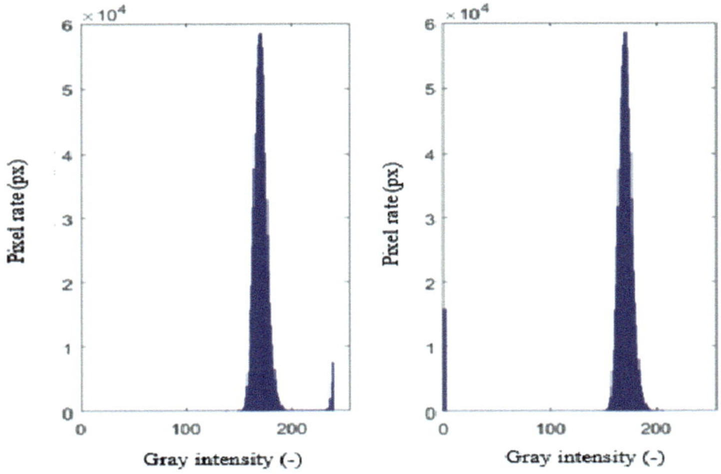

Fig. 5. From the left: histogram of the original image, histogram of the image after segmentation.

3.3 The Band Intensity of Secondary Radiation

The curve determined by the dominant peak was applied to the generated histogram after the segmentation. The curve includes two depressions that characterize different levels of secondary radiation intensity. To find these points, use the peak detector (Fig. 6). Then we calculate the pixels below the histogram that corresponds to the secondary radiation. The following Table 2 shows the sample of elements after extraction.

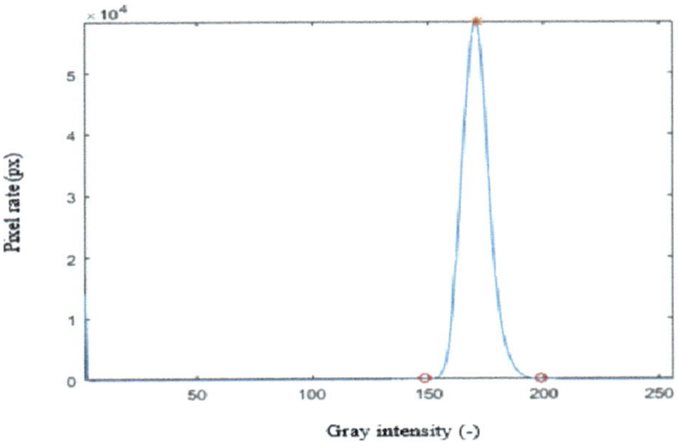

Fig. 6. The curve representing the number of the pixels for single levels of gray.

Table 2. Selected elements after extraction.

	Distance (m)	Number of pixels below the histogram curve (px)	Maximum peak intensity (−)	Lower bandwidth limit (−)	The upper limit of the bandwidth (−)
1	1	823115	82	61	111
2	1.5	820991	135	112	160
3	2	823450	174	153	205

4 Statistical Analysis

The obtained and calculated data were subjected to a discrete statistical analysis. [16, 17] The statistical analysis showed that the data did not come with a normal distribution. Table 3 contains statistical pixel count indicators below the histogram curve.

Table 3. Statistical indicators of the number of pixels below the histogram curve.

	1 m	1.5 m	2 m
Number of measurements	10	10	10
Minimum	820000	820000	820000
Lower quartile	820000	820000	820000
Average value	820000	820000	820000
Median	820000	820000	820000
Upper quartile	820000	820000	820000
Maximum	820000	820000	820000
Variance	380000	570000	760000
Std. deviation	619.591	753.767	872.163

Percentage of pixels belonging to secondary radiation was counted. The median value used for the calculation was the same for all distances. The result confirms the theoretical assumption that the secondary ray pixels occupy the same pixel on the whole image in all 10 images at all three distances.

The following table (Table 4) contains statistical indicators of the peak intensity grayscale.

The median peak intensity of the gray peak is increased along with the increasing distance of the CR cassette from the source. This implies that secondary radiation loses its influence with increasing distances. The following table (Table 5) contains statistical indicators for the lower bandwidth range and the upper bandwidth limits.

Table 4. Statistical indicators of the maximum peak intensity.

	1 m	1.5 m	2 m
Number of measurements	10	10	10
Minimum	82	135	168
Lower quartile	92	135	169
Average value	102.20	150.30	171.80
Median	96.50	137,50	172.50
Upper quartile	103	157	174
Maximum	136	193	177
Variance	329.29	552.23	9.96
Std. deviation	18.15	23.50	3.16

Table 5. Statistical indicators for the lower bandwidth range and the upper bandwidth limit.

	1 m	1.5 m	2 m
Number of measurements	10	10	10
Minimum	111	160	193
Lower quartile	126	175	199
Average value	138.60	185.80	201.90
Median	132.50	181.50	204.00
Upper quartile	139	202	205
Maximum	187	219	207
Variance	513.82	450.18	26.77
Std. deviation	22.67	21.22	5.17

By calculating the median of the lower and upper limit of the intensity range, it was possible to determine the mean bandwidth for each distance (Table 6). For the device being analyzed, the bandwidth is approximately 60 grayscales.

Table 6. Grayscale strips.

Distance (m)	Interval	Number of grayscales
1	<73; 133>	61
1.5	<117; 182>	66
2	<151; 204>	54

5 Effective Dose and Surface Kerma

The intensity of the secondary radiation depends on the distance from the radiation source, decreasing the increasing distance. Tables 7 and 8 contain computed statistical pointers for the effective dose and for the area kermas.

Table 7. Effective dose indicators.

	1 m	1.5 m	2 m
Number of measurements	10	10	10
Minimum	0.0010	0.0008	0.0005
Lower quartile	0.0010	0.0009	0.0005
Average value	0.0010	0.0009	0.0005
Median	0.0010	0.0010	0.0005
Upper quartile	0.0010	0.0010	0.0005
Maximum	0.0010	0.0010	0.0005
Variance	0	0	0
Std. deviation	0	0.0001	0

The effective dose was measured with 100% accuracy, except for 1.5 m with a small standard deviation.

Table 8. Surface kerma indicators.

	1 m	1.5 m	2 m
Number of measurements	10	10	10
Minimum	63.41	63.41	63.69
Lower quartile	63.55	63.71	63.76
Average value	63.579	63.695	63.806
Median	63.585	63.725	63.83
Upper quartile	63.62	63.73	63.85
Maximum	63.67	63.77	63.87
Variance	0.005	00105	0.0034
Std. deviation	0.0711	0.10255	0.0582

Surface kerma corresponds to the multiple of absorbed dose and the irradiated area. From the data, the surface kerma is independent of the distance of the CR cassette from the radiation source.

6 Correlation Analysis

Correlation analysis was applied to the data obtained to determine the existence of a correlation between the distance of the CR cassette from the radiation source and the image radiation by the secondary radiation. Shapiro-Wilk's test was used to verify normal distribution (significance level of 0.05), with a null hypothesis H_0: The data comes from the normal distribution and the alternative hypothesis H_A: $\neg H_0$. Table 9 outlines the results of normality verification.

Table 9. Test to verify the assumption of normality.

Parameter	p-value
Distance	<0.001
Shade intensity of the highest peak	<0.010
Lower limitation of intensity band	<0.006
Upper limitation of intensity band	<0.008

Data analysis has shown that data do not come from normal distribution. For this reason, Spearman's correlation coefficient was used, where the null hypothesis H_0: $\rho = 0$ and the alternative hypothesis H_A: $\rho \neq 0$ was determined. The results of the analysis are shown in Table 10.

Table 10. Spearman correlation coefficients.

Parameter	p-value
Shade intensity of the highest peak	0.8313 $\ll$ 0.001
Lower limitation of intensity band	0.8451 $\ll$ 0.001
Upper limitation of intensity band	0.7932 $\ll$ 0.001

The results show that the correlation is low but statistically significant. Lower correlation was due to greater variability of detected data at individual distances. The statistical analysis confirmed that the insertion of the skiagraphic cassette by unwanted secondary radiation is dependent on the distance of the cassette from the radiation source.

7 Conclusion

If CR cassettes are stored in a room where an X-ray is used, these cassettes are exposed to secondary radiation and can be degraded. Our proposed method, which is based on Otsu segmentation, can separate the area of shadow characterizing secondary radiation. The secondary radiation pattern can be determined by determining the intensity of the incident secondary radiation depending on the distance from the radiation source. To illustrate the effect of secondary radiation on the resulting image, the following features were selected: the number of pixels under histograms, the intensity of the highest frequency, the lower bandwidth limitations and bandwidth limitation. All the information obtained serves for created a predictive model that implies a deterioration of image quality depending on the distance from the radiation source.

Acknowledgment. The work and the contributions were supported by the project SV450994 Biomedical Engineering Systems XV'. This study was also supported by the research project The Czech Science Foundation (GACR) 2017 No. 17-03037S Investment evaluation of medical device development run at the Faculty of Informatics and Management, University of Hradec Kralove, Czech Republic. This study was supported by the research project The Czech Science Foundation (TACR) ETA No. TL01000302 Medical Devices development as an effective investment for public and private entities.

References

1. Choy, H., et al.: Radiation therapy systems that include primary radiation shielding, and modular secondary radiation shields. U.S. Patent Application No 14/688,751 (2016)
2. Pisano, S., et al.: Single and double spin asymmetries for deeply virtual Compton scattering measured with CLAS and a longitudinally polarized proton target. Phys. Rev. D **91**(5), 052014 (2015)
3. Moreh, R., Finkelstein, Y., Vos, M.: Comparison between electron and neutron Compton scattering studies. In: EPJ Web of Conferences, EDP Sciences, p. 02011 (2015)
4. Goldman, A.H., Hoover, K.B.: Source-to-detector distance and beam center do not affect radiographic measurements of acetabular morphology. Skeletal Radiol. **46**(4), 477–481 (2017)
5. Mori, S., Amano, S., Furukawa, T., Shirai, T., Noda, K.: Effect of secondary particles on image quality of dynamic flat panels in carbon ion scanning beam treatment. Br. J. Radiol. **88**(1047) (2015). Art. no. 20140567
6. Yamamoto, S., Koyama, S., Yamauchi-Kawaura, C., Komori, M.: Luminescence imaging of biological subjects during X-ray irradiations lower energy than Cerenkov-light threshold. Opt. Rev. **24**(3), 428–435 (2017)
7. Valeri, G., et al.: Open source software in a practical approach for post processing of radiologic images. Radiol. Med. (Torino) **120**(3), 309–323 (2015)
8. Mahmood, N., et al.: Image segmentation methods and edge detection: an application to knee joint articular cartilage edge detection. J. Theoret. Appl. Inf. Technol. **71**(1), 87–96 (2015)
9. William, W., et al.: A review of applications of image analysis and machine learning techniques in automated diagnosis and classification of cervical cancer from pap-smear images. In: 2018 IST-Africa Week Conference (IST-Africa), pp. 1–11. IEEE (2018)
10. Prabha, D.S., Kumar, J.S.: Performance evaluation of image segmentation using objective methods. Indian J. Sci. Technol. **9**, 1–8 (2016)
11. Oladiran, O., Gichoya, J., Purkayastha, S.: Conversion of JPG image into DICOM image format with one click tagging. In: International Conference on Digital Human Modeling and Applications in Health, Safety, Ergonomics and Risk Management, pp. 61–70. Springer, Cham (2017)
12. Patil, P., Patil, S.V., Chaudhari, D.D.: Research on conversion of DICOM multi-frame medical image in to multimedia format using MATLAB. Int. J. Curr. Eng. Technol. **5**, 156–160 (2015)
13. Pal, S.K., Ghosh, A., Kundu, M.K.: Soft Computing for Image Processing. Physica-Verlag, New York (2000). ISBN 3790812684
14. Kubicek, J., Penhaker, M., Cerny, M., Augustynek, M., Korhelik, K., Vodakova, A., Oczka, D.: Analysis and modeling of secondary X-ray effect on structure of skiagraphy images. In: BIODEVICES 2018 - 11th International Conference on Biomedical Electronics and Devices, Proceedings; Part of 11th International Joint Conference on Biomedical Engineering Systems and Technologies, BIOSTEC 2018, vol. 1, pp. 30–38 (2018)
15. Yuan, X-c, Wu, L-s, Peng, Q.: An improved Otsu method using the weighted object variance for defect detection. Appl. Surf. Sci. **349**, 472–484 (2015)
16. Busk, P.L., Marascuilo, L.A.: Statistical analysis in single-case research. In: Single-Case Research Design and Analysis (Psychology Revivals): New Directions for Psychology and Education, p. 159 (2015)
17. Bain, L.: Statistical Analysis of Reliability and Life-Testing Models: Theory and Methods. Routledge, New York (2017)

Segmentation of Blood Vessels from Fundus Retinal Images by Using Gabor Transformation

Alice Krestanova(✉), Jan Kubicek, and Jana Kosturikova

VSB - Technical University of Ostrava, FEI K450, Ostrava, Czech Republic
{alice.krestanova, jan.kubicek, jana.kosturikova}@vsb.cz

Abstract. In this paper, it was using segmentation based on Gabor transformation for extraction the retinal blood vessels. The algorithm has been applied on images from freely accessible DRIVE and STARE databases. The segmentation algorithm provides the possibility of automatic modeling and consequent extraction of the retinal vascular system. Resultant images are in the form of a binary map. An algorithm for calculating the value of the curvature of the retinal vessels and thus objectifying the tortuosity can be applied to these images.

Keywords: Segmentation · Extraction · Blood vessels · Retina · Fundus camera · Gabor transformation

1 Introduction

Retinopathy of prematurity (ROP) affects predominantly premature infants with a birth weight of less than 1500 g born before 32 weeks of gestation. The disease can lead to blindness. The most important pathological changes occur on the retina at the border between its non-vascular and vascular part. There are scarring changes and growth of pathologically altered vessels. At this interface, further retinal development stops [1]. Tortuosity, or curved arteries, and veins are commonly found in the human body and in animals. While mild tortuosity is a common anomaly without clinical symptoms, severe tortuosity leads to severe symptoms. Tortuosity is investigated in atherosclerosis, hypertension, genetic defects, diabetes mellitus, and in the elderly [2].

Retinal vessels are often affected by retinopathy tortuosity. It further affects arterioles and capillaries in skeletal muscle, myocardium, brain, and tumors. Coronary arteries are altered in patients with hypertension and myocardial infarction [2].

2 Related Works

Hough Transformation can be used for detection of the retinal blood vessel because the method is based on analytical shape detection such as lines, circles, or ellipses. This algorithm has its weaknesses, the success of the method depends on the quality of the input images [3].

J. Henriques et al. (Eds.): MEDICON 2019, IFMBE Proceedings 76, pp. 1609–1619, 2020.
https://doi.org/10.1007/978-3-030-31635-8_198

In different access, it can be implemented a procedure for segmentation of the retinal vascular system using two-dimensional Gabor filters amplified by adaptive histogram equalization (CLAHE). This technique seems to be useful without unnecessarily increasing errors. Provides great precision and sensitivity. However, the properties of images can be improved by changing values in the algorithm, e.g. for vessel width, threshold value, or better eye image [4].

A method based on Gabor filters can be used for the automatic detection and segmentation of coronary arteries in X-ray angiograms In reference 5, they tested various parameters directly on angiograms to select those most suitable for the Gabor filter [5].

The Morlet wavelet transform solution is a method for detection images from the STARE and DRIVE databases. Detection by this method is not 100%, it contains errors in some images due to the presence of noise, optical disc, hemorrhage or other high-contrast pathologies [6].

Thanks to the Gabor filter, papillary lines can be highlighted [7].

3 Design of Proposed Segmentation Algorithm

The segmentation algorithm consists of Gabor's transformation and linear regression, cross validation of both models using the "gold standard" and the maximum response to Gabor's transformation. Database DRIVE uses the results of manual segmentation created by observers under the supervision of experienced ophthalmologists as a "golden standard" and in STARE database, there are resultant binary images after segmentation by Valentina Kouznets as "golden standard". The proposed algorithm is shown in Fig. 1.

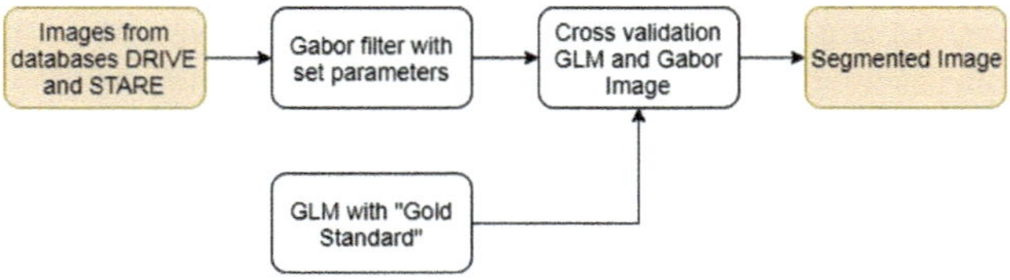

Fig. 1. Design of segmentation algorithm based on Gabor Transformation.

3.1 Gabor Transformation

This is a convolution of the input image with one of the components of the Gabor 2D function. In practice, images in computer graphics are represented as a discrete point grid, so they cannot be filtered by a continuous filter. It is necessary to create the so-called core filter K of dimensions IxJ which can be used to perform discrete convolution of the image by dilatation, translation, and rotation [8, 9].

In the next step, it is used extraction the symptoms from the image. This is the procedure where the input image is transformed by a bank N Gabor filters, resulting in 2 N responses at each point of the image. One response is for the imaginary part, the other for the real one. Subsequently, the magnitude of the total response of the imaginary part combinations is calculated, so the result has only a N response [8].

The image (Fig. 2) demonstrate the response to the core of the Gabor wavelet consisting of real and imaginary parts with given parameters. The Gabor filter can be defined:

$$g(x, y;\ \lambda, \theta, \sigma, \gamma) = exp\left(\frac{-x'^2 + \gamma^2 y'^2}{2\sigma^2}\right) exp\left(i\left(2\pi\frac{x'}{\lambda}\right) + \psi\right) \quad (1)$$

Where λ represents the cosine factor bandwidth, orientation, ψ phase shift, σ is the Gaussian envelope sigma, and γ is the spatial aspect ratio.

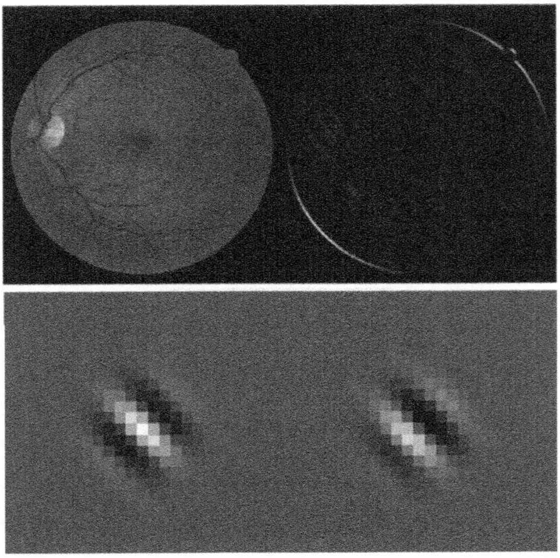

Fig. 2. Testing and Gabor Image (up), the response to the Gabor wavelet with variable kernel parameters real and imaginary part (bottom) in MATLAB from the DRIVE database

3.2 Generalized Linear Model

Generalized Linear Model (GLM) includes various variants including linear regression. This model has long been used for data analysis. We can write it using the matrix expression:

$$Y = X\beta + \varepsilon \quad (2)$$

Where Y is the vector n of the values of the explained variable, X is the matrix of the values of the explanatory variables, β is the vector p of the unknown parameters and ε is the vector n of the random component values [10].

In this work desirable to examine the relationship between two images and their pixels in a binary image. One of the variables is the so-called independent variable x, which affects the second, the so-called dependent variable y. These variables are assumed to be continuous. The first step in the analysis is plotting data into a point graph, the so-called correlation field, and verifying that there is a real dependence between the quantities, the so-called regression. The simplest form of dependence is a simple linear regression assuming a linear dependence between two variables. It can be expressed by:

$$y = \beta_0 + \beta_1 . y_1 + \varepsilon \tag{3}$$

Where β_0 and β_1 are parameters of the regression equation and ε is called residual deviation or error. The theoretical values of the parameters are denoted by the Greek letters (β_0 and β_1), the predicted values are b_0 and b_1. The desired regression line will take the form: [11, 12]

$$y = b_0 + b_1 . y_1 \tag{4}$$

In the following figure (Fig. 3), the output of the GLM model from the "golden standard" is shown, which is then compared with the resultant image after the Gabor wave filtration.

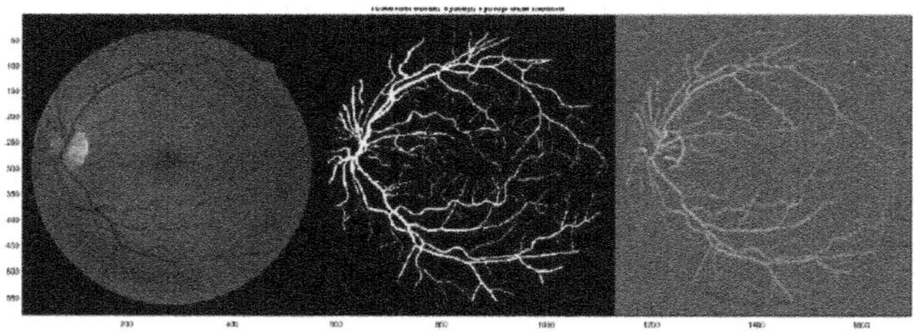

Fig. 3. Testing image (left), "Gold Standard" (in the middle), Output of GLM model (right).

4 Testing of the Implemented Segmentation Algorithm

In this paper were used databases DRIVE and STARE for testing the proposed segmentation algorithm.

4.1 Used Databases

The DRIVE (Digital Retinal Images for Vessel Extraction) database was established as a standardized database to support the research of segmentation of the blood vessels in retinal images. 400 diabetic subjects were observed between 25 and 90 years of age. Forty of them were randomly selected, 33 subjects showed no evidence of retinopathy, 7 yes. All images are compressed into JPEG format. They were captured using Canon CR5 with a non-mydriatic 3CCD camera with a 45° field of view. All images are RGB with 8-bit color depth and a resolution of 768 to 584 pixels [13].

The STARE (Structured Analysis of the Retina) offers 400 images of the retina with a specified diagnosis. The database contains 20 images for retinal vascular segmentation, 10 of which have a pathological finding. The TopCon TRV-50 fundus camera was acquired with a 35° field of view. The color depth of the images is like the previous 8 bits with a resolution of 605 to 700 pixels. They are compressed into PPM format [14, 15].

4.2 Testing on the Real Data

The algorithm based on the Gabor transformation for segmentation of the retinal blood vessels was applied to 40 images, more precisely 20 image data comes from the DRIVE database and the last 20 images from the STARE database. Resultant images after the segmentation algorithm based on a linear regression and resultant images after previous mentioned segmentation, were comparing with "golden standard" of retinal vessels. If there is a match between the vessels in both models, this will be indicated in the predicted model (output). In the DRIVE database, such as "golden standards" are already available, so they are drawn directly from the source. In the STARE database, this is also the case with the fact that the segmentation by Valentina Kouznets has been chosen. During testing were changed parameters of setting Gabor wave.

Table 1. Set parameters of Gabor filter n. 1

Parameters of Gabor wave	
Frequency (Fs)	0.1:0.1:0.3
Size of wavelet (δ)	[2:2:8].*scale Scale = 6
Angle (θ)	$\frac{\pi}{8}:\frac{\pi}{8}:\pi-\frac{\pi}{8}$

The First Set of Parameters for Gabor Transformation

The algorithm with parameters (see Table 1) was applied to these image data (Fig. 4). Due to the fact that Gabor's transformation is a computationally demanding process, MATLAB records the times of this procedure at given parameters. 20 images from the DRIVE database were tested, 10 of the test scores and 10 of the training session images. In addition, all the images from the STARE database for 20 images.

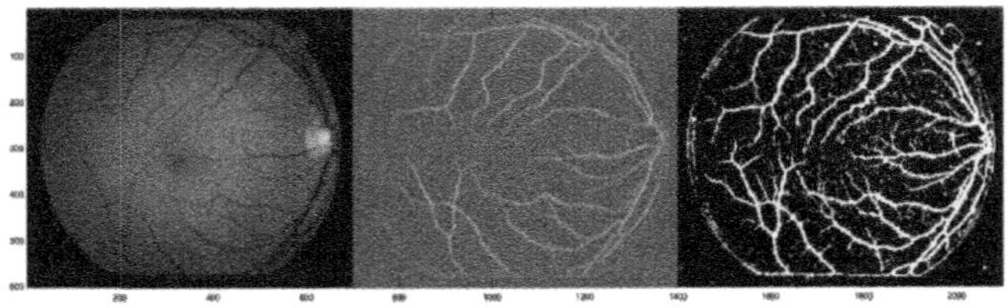

Fig. 4. Example of optimized output with set parameters n. 1 (from left): native image, GLM model and output image.

Table 2. Averages and medians of computational time from database DRIVE and STARE with setting parameters n. 1

	Average	Median
DRIVE	13.5429	13.3143
STARE	15.7368	15.6986

An overview of average and median computational times in the following table (Table 2). Obviously, this computational time is affected by the pixels of the image and therefore is longer at the fundus image than images with low resolution.

Table 3. Set parameters of Gabor filter n. 1

Parameters of Gabor wave	
Frequency (Fs)	0.1:0.1:0.4
Size of wavelet (δ)	[1:1:8].*scale Scale = 6
Angle (θ)	$\frac{\pi}{9}:\frac{\pi}{9}:\pi-\frac{\pi}{9}$

The Second Set of Parameters for Gabor Transformation

Testing also took place when the original parameters were changed (Table 3). Overall, the aim was to increase the sensitivity of vessel detection by reducing the angle of rotation, by mitigating the steps in the wavelength setting and by slightly increasing the frequency. As a result of this setting, the computational difficulty increased (Table 4). The newly set parameters can be seen in the following table, then the newly measured times. This change in parameters has had a positive effect on the elimination of obstruction in the resulting image (Fig. 5).

The Third Set of the Gabor Filter

At the last change of values, only 6 images from each database were tested due to the most time-consuming. All outputs including ROC (receiver operating characteristic) curves (see below) lasted about 25 to 35 min. The final times only Gabor transforms are recorded again in the tables (Table 5). ROC curve allows you to assess the ability of a diagnostic test depending on its sensitivity and specificity, thus eliminating the consequences of wrong decisions. These two components measure the validity of the test.

Table 4. Averages and medians of computational time from database DRIVE and STARE with setting parameters n. 2

	Average	Median
DRIVE	39.7923564	39.702765
STARE	49.590628	49.43119

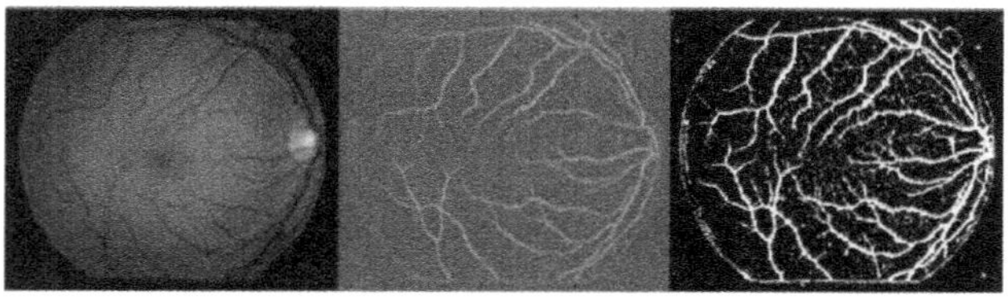

Fig. 5. Example of optimized output with set parameters n. 2 (from left): native image, GLM model and output image.

Table 5. Averages and medians of computational time from database DRIVE and STARE with setting parameters n. 3

	Average	Median
DRIVE	95.302777	93.4412625
STARE	123.01918	121.92658

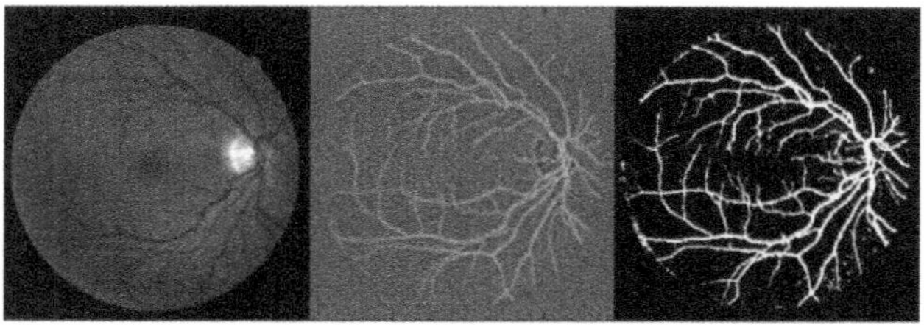

Fig. 6. Example of optimized output with set parameters n. 3 (from left): native image, GLM model and output image.

During the last set of the Gabor filter, it was changed the frequency from 0.05 to 0.35 with step 0.05. The size of wavelet was the same as in the preview settings. The angle was smaller from $\pi/12$ to $\pi - \pi/12$ with step $\pi/12$. Below it is shown images after applied algorithm with setting parameters n. 3 (Fig. 6).

5 Evaluation of Testing

For the overview, all the times that have been mentioned so far are graphically recorded. First for the DRIVE database and then for the STARE database separately (see Figs. 7 and 8).

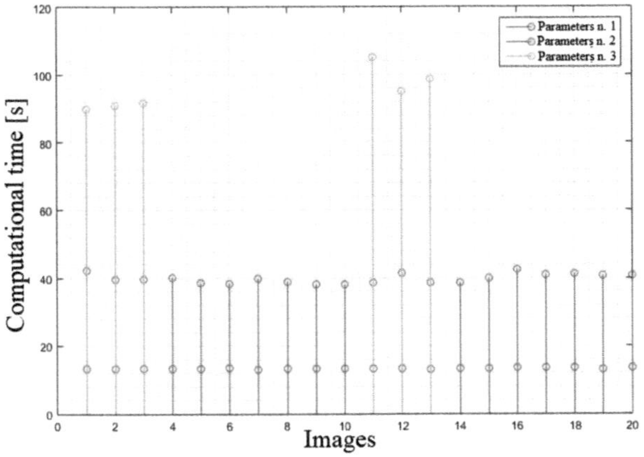

Fig. 7. The computational difficulty of calculation Gabor transformation images from database DRIVE with different parameters

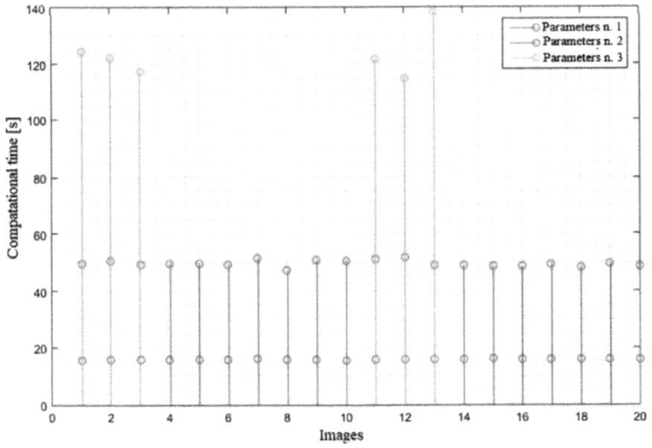

Fig. 8. The computational difficulty of calculation Gabor transformation images from database STARE with different parameters

It has been found in the STARE database that only one "gold standard" is not suitable for detecting blood vessels in different patients. Images were divided into groups according to the location of the optical disc (left, right, and center). Then, for each group, a "gold standard" were randomly selected and according to him, the segmentation of the vascular bed was taking place. As already mentioned, it is not appropriate to use a universal "golden standard". Failure was apparently due to individual vessel deployment on the retina, as well as the heterogeneity of the location of the optical disk or the leading vessels.

In addition, the extent to which results were influenced when applying image preprocessing techniques to images from the STARE database was tested. The results were expected to be largely unaffected, but the result is surprising. The original images had on average a 22.3% better ROC curve rating (more on the ROC curve in the next chapter) than images that were preprocessed, except for G channel extraction and monochromatic image conversion.

The results from the databases were evaluated based on ROC analysis. In the test of classification by Tapea, there are results, in this case, are in evaluation EXCELLENT. The average is 0.9605 and median is 0.96 (see Fig. 9).

These results, according to the above-mentioned classification, range from GOOD to EXCELLENT. The mean value is 0.9155 and the median 0.92. At the first change of parameters, the area under the curve slightly increased. Subsequently, results with images with parameters 2 will be demonstrated.

The average AUC (area under the curve) of the images from DRIVE database with parameter setting 2 is 0.967 and the median equals 0.97. There was an increase in AUC compared to those with parameters 1 and ±0.01, i.e., about 1%. For STARE, the average value is 0.938 and the median equals 0.94 (Fig. 10). There was an increase in AUC of ±0.02, by 2%. The third parameter setting was DRIVE The mean AUC is 0.9733 and the median 0.927a for the STARE database. The mean value equals 0.927 and the median is 0.92.

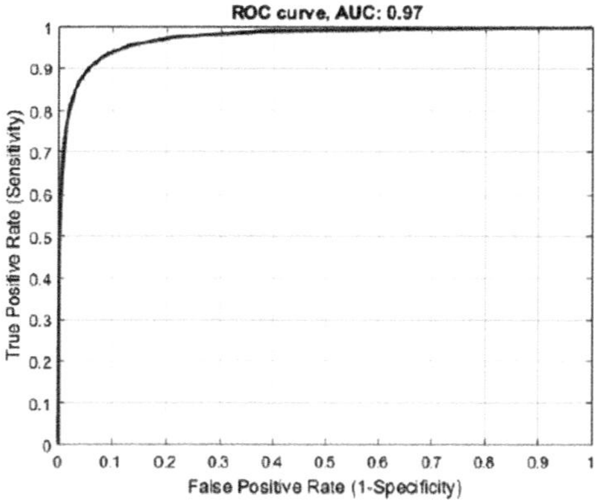

Fig. 9. ROC curve with value AUC 0.97 from DRIVE database with parameters n. 1

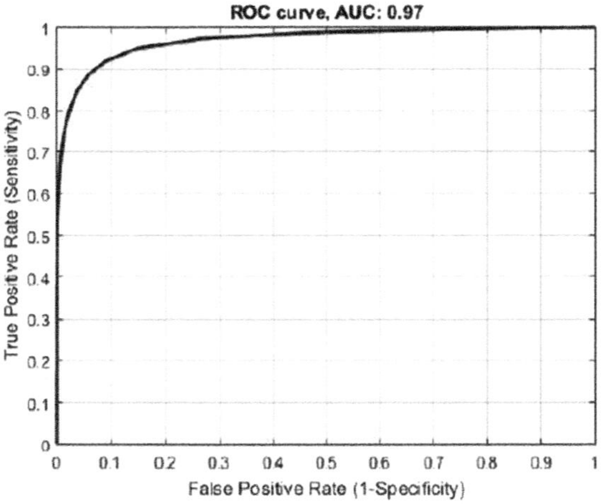

Fig. 10. ROC curve with value AUC 0.97 from STARE database with parameters n. 1

6 Conclusion

Fundus images have sufficient contrast to the retinal background and resolution. During the testing, the Gabor filter parameters were gradually changed for all image data. By reducing the angle of rotation of the Gabor filter and improving the other parameters, better results were obtained for background images. Because of all the data, the computational intensity has increased due to fewer revolutions and thus increased the number of records.

The ROC analysis also showed an increase in most cases. It has been experimentally investigated if "Golden Standard" can be used with images from the STARE database.

In addition, it has been tested to what extent the pre-processing of basic records from the STARE database will be affected, which have not been performed due to a larger number of pixels.

The result was not the expected slight increase in AUC, but the decrease by more than 20%. The reason is data bias due to this preprocessing algorithm. Since tortuosity has not been parameterized so far, the proposed algorithm, based on the calculation of the 1st derivative, is suitable as regards the display of a certain curvature based on the threshold value chosen by the physician.

Acknowledgments. The work and the contributions were supported by the project SV450994 Biomedical Engineering Systems XV'. This study was also supported by the research project The Czech Science Foundation (GACR)2017 No. 17-03037S Investment evaluation of medical device development at the Faculty of Informatics and Management, University of Hradec Kralove, Czech Republic. This study was supported by the research project The Czech Science Foundation (TACR) ETA No. TL01000302 Medical Devices development as an effective investment for public and private entities.

References

1. Rozsíval, P., et al.: Oční lékařství, 1. vyd. Galén, Praha (2006). ISBN 9788072624041
2. Han, H.: Twisted blood vessels: symptoms, etiology and biomechanical mechanisms. J. Vasc. Res. **49**(3), 185–197 (2012). https://doi.org/10.1159/000335123. ISSN 1423-0135
3. Floryček, J.: Experimenty se snímky sítnice oka, Bakalářská práce. Vysoké učení technické v Brně, Brno (2010)
4. Ramlugun, G.S., Nagarajan, V.K., Chakraborty, Ch.: Small retinal vessels extraction towards proliferative diabetic retinopathy screening. Expert Syst. Appl. **39**(1), 1141–1146 (2012). https://doi.org/10.1016/j.eswa.2011.07.115. ISSN 09574174
5. Cruz-Aceves, I., Oloumi, F., Rangayyan, R., Maviña-Cervantes, J.G., Arturo, H.-A.: Automatic segmentation of coronary arteries using Gabor filters and thresholding based on multiobjective optimization. Biomed. Signal Process. Control **25**, 76–85 (2016). https://doi.org/10.1016/j.bspc.2015.11.001. ISSN 17468094
6. Soares, J.V.B., Leandro, J.J.G., Cesar, R.M., Jelinek, H.F., Cree, M.J.: Retinal vessel segmentation using the 2-D Gabor wavelet and supervised classification. IEEE Trans. Med. Imaging **25**(9), 1214–1222 (2006). https://doi.org/10.1109/TMI.2006.879967. ISSN 0278-0062
7. Cejpek, T.: Detekce přerušení papilárních linií otisku prstu a jejich rekonstrukce, Bakalářská práce. Vysoké učení technické v Brně, Brno (2011)
8. Zamazal, Z.: Ohodnocení okolí bodů v obraze, Diplomová práce. VUT v Brně, Brno (2011)
9. Vlach, J.: Metody zpracování obrazu pro časově náročné úlohy, Disertační práce. Technická Univerzita v Liberci, Liberec (2012)
10. Veselý, M.: Aplikace GLM modelu v provozní praxis, Diplomová práce, Brno (2007)
11. Jednoduchá lineární regrese. http://homel.vsb.cz
12. Lineární regrese. http://mdg.vsb
13. Image Sciences Institute. DRIVE: Digital Retinal Images for Vessel Extraction (2017)
14. STructured Analysis of the Retina. What is STARE? (2013)
15. Retinal Image Computing & Understanding. DRIVE: Digital Retinal Images for Vessel Extraction (2010)

Evaluation of System for Simultaneous Measurement of Physiological Parameters: Potential for Determination of Age-Related Cardiovascular Status

Honoka Koga[1(✉)], Jihyoung Lee[1], Peter Rolfe[2,3], Ken-ichi Yamakosh[4,5], Akira Kamiya[6], and Takehiro Yamakoshi[1]

[1] Information Systems and Engineering, Graduate School of Engineering, Fukuoka Institute of Technology, Fukuoka-shi, Ishikawa, Japan
h.koga1122@gmail.com, yamakoshi@fit.ac.jp
[2] Department of Automatic Measurement and Control, Harbin Institute of Technology, Harbin, China
[3] Oxford BioHorizons Ltd., Maidstone, UK
[4] College of Science and Engineering, Kanazawa University, Kanazawa-shi, Ishikawa, Japan
[5] Showa University School of Medicine, Ohta-ku, Tokyo, Japan
[6] Tokyo Institute for Interdisciplinary Science, Kamakura-shi, Kamakura, Japan

Abstract. We have developed a system for simultaneously determining arterial parameters by performing measurements on just one finger. The purpose of the present study was to evaluate the ability of this system to identify a state of ill health. The system can obtain eight physiological parameters: the systolic blood pressure (SBP), mean blood pressure (MBP), diastolic blood pressure (DBP), pulse pressure (PP), pulse rate (PR), normalized pulse volume (NPV) as an index of sympathetic activity, along with volume elastic modulus (E_v) and vascular volume change ratio (η_v). In this study, we attempted to identify poor physical condition and perform a clinical evaluation using these two parameters in particular. Poor physical condition includes the general deterioration in health due to age. We investigated whether it was possible to determine differences in health between old and young people. All participants were women because we wanted to eliminate errors due to gender, and measurements were performed under the same conditions. The results indicated that there were differences in the measured SBP, PP, E_v and η_v values for the older and younger groups, suggesting that the proposed system can be used to determine the health condition of a subject.

Keywords: Arterial parameters · Arteriosclerosis · Poor health

1 Introduction

World Heal Organization proclaimed that there was cardiovascular disease at the top of the death. Prevention such disease was globally significant. We therefore needs to know peripheral arterial condition to prevent cardiovascular disease with extraordinary

J. Henriques et al. (Eds.): MEDICON 2019, IFMBE Proceedings 76, pp. 1620–1623, 2020.
https://doi.org/10.1007/978-3-030-31635-8_199

speed. Our group therefore thought about measuring the cardiovascular parameters with the one finger.

We previously developed a system for evaluating the cardiovascular condition of a subject [1]. It was based on measurements of the systolic blood pressure (SBP), mean blood pressure (MBP), diastolic blood pressure (DBP), pulse pressure (PP), pulse rate (PR), and normalized pulse volume (NPV), which reflects peripheral sympathetic nerve activity [2], in addition to the finger-artery elasticity index (FEI) [3]. A verification experiment on subjects in a sleep-deprived state indicated the potential of this system for health screening [1]. However, since it is difficult to understand intuitively the parameter of FEI as arterial elasticity, we modified the system so as to measure and visually monitor the arterial elasticity (E_v: volume elastic modulus) [4] and vascular volume change ratio (η_v) as a function of transmural pressure. In this preliminary study, we investigate the potential for identifying a decrease in health using SBP, MBP, DBP, PP, PR, NPV, E_v and η_v.

2 Methods

2.1 Measurement Principle

SBP and MBP were determined using the volume-oscillometric method [1], and DBP was derived from these measured values. PP was calculated as the difference between SBP and DBP. PR was obtained from the pulse wave interval during 10 s of measurements. NPV was calculated as the ratio of the AC to DC components of the photoelectric sensor signal [3].

For the theoretical evaluation, E_v and η_v were obtained based on our previous report [2, 3], as schematically shown in Fig. 1. Conclusively, E_v and η_v can be derived from the following equation as:

$$E_v = k \cdot \exp(VEI \cdot P_{tr}), \quad (1)$$

$$\eta_v = k' \cdot \exp(DI \cdot P_{tr}). \quad (2)$$

where VEI; elasticity index. DI; volume change index, k and k′; constant parameters, and P_{tr}; arterial transmural pressure (MBP – applied cuff pressure (Pc)).

2.2 Experimental Method

The study was approved by the ethics commission of Fukuoka Institute of Technology. The participants were 10 healthy young women (20–22 years) and 10 aged women (75–85 years). They agreed to take part in the study voluntarily and signed an informed consent statement. Each participant sat on a chair quietly in a temperature-controlled, soundproofed room, and measurements were performed continuously five times for each after 5 min rest condition. One measurement required about 60 s. The room temperature and humidity were approximately 23.9 ± 2.1 S.D. °C and 36.2 ± 7.8 S.D.%, respectively.

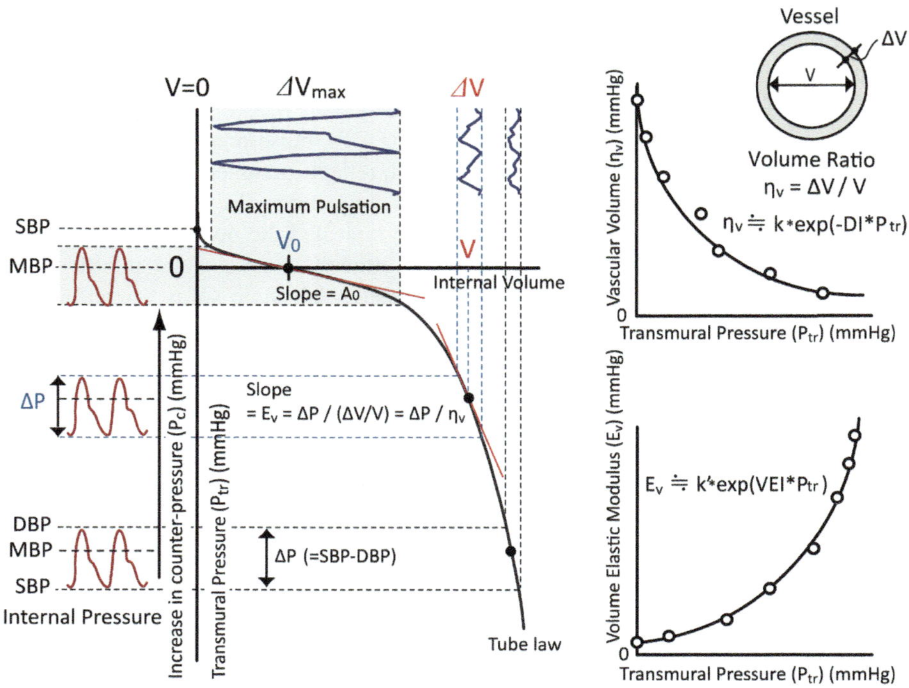

Fig. 1. Method for determining arteriosclerosis and intra-arterial deposits.

3 Results and Discussion

The results showed that the proposed system could detect differences in health between the older and younger participants in view of significant changes in SBP, PP, E_v and η_v values. The other parameters showed no significant differences. Figure 2 shows typical examples of E_v and η_v. According to the Japanese Ministry of Health, Labor and Welfare (JMHLW), old people are more likely to have high blood pressure and arteriosclerosis, and this was reflected in the results of the present study. It was also reported by the JMHLW that the pulse rate tends to decrease due to aging. However, no significant difference in PR was found in this experiment. This is likely to be because all of the aged women participated in the study were accustomed to frequent exercise. Thus, only changes in blood pressure and degree of arteriosclerosis were observed between the two age groups.

Generally speaking, PR decreases with age, as mentioned above. However, in the present study, no change in PR was observed. Therefore, we need to perform additional experiments using aged people, including men, who do not perform regular exercise. We should also consider using a wider age group.

In this study, we indicated the changes in the arterial elasticity by PP, E_v and η_v. To determine the arterial elasticity is very meaningful. Because, the arteriosclerosis is caused from the peripheral vessels [5].

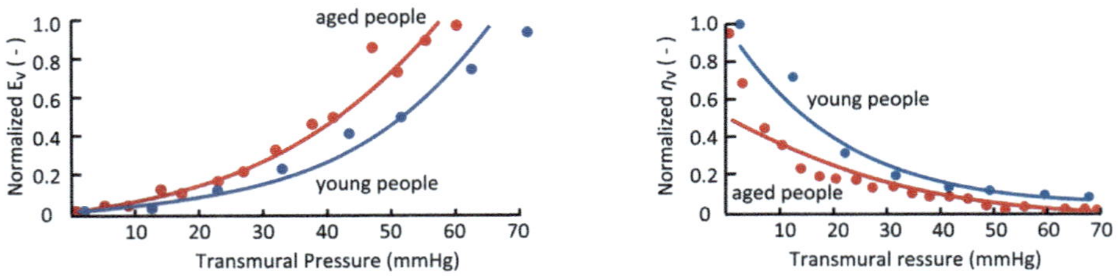

Fig. 2. Typical examples of E_v and η_v normalized by the maximum value in each parameter obtained from an aged (red circles) and a young (blue circles) subject.

Conflicts of Interests. The authors declare that there are no conflicts of interests.

References

1. Yamakoshi, T., Matsumura, K., Rolfe, P., Hanaki, S., Ikarashi, A., Lee, J., Yamakoshi, K.: Potential for health screening using long-term cardiovascular parameters measured by finger volume-oscillometry: pilot comparative evaluation in regular and sleep-deprived activities. IEEE J. Biomed. Health Inform. **18**(1), 28–35 (2014)
2. Sawada, Y., Tanaka, G., Yamakoshi, K.: Normalized Pulse Volume (NPV) derived photoplethysmographically as a more valid measure of the finger vascular tone. Int. J. Psychophysiol. **41**(1), 1–10 (2001)
3. Tanaka, G., Yamakoshi, K., Sawada, Y., Matsumura, K., Maeda, K., Kato, Y., Horiguchi, M., Ohguro, H.: A novel photoplethysmography technique to derive normalized arterial stiffness as a blood pressure independent measure in the finger vascular bed. Physiol. Meas. **32**(11), 1869–1883 (2011)
4. Shimazu, H., Yamakoshi, K., Kamiya, A.: Noninvasive measurement of the volume elastic modulus in finger arteries using photoelectric plethysmography. IEEE Trans. Biomed. Eng. **33**(8), 795–796 (1986)
5. Cohn, J.N., Duprez, D.A., Grandits, G.A.: Arterial elasticity as part of a comprehensive assessment of cardiovascular risk and drug treatment. Hypertension **46**(1), 217–220 (2005)

Special Sessions: INT4DAT - Intelligent Systems and Technologies for Diagnostic, Assistance, and Therapeutics

Expressive Robotic Head for Human-Robot Interaction Studies

Ricardo Pereira[1(✉)], Luís Garrote[1,2], Tiago Barros[1], Carlos Carona[3], Luís C. Bento[1], and Urbano J. Nunes[1,2]

[1] Institute of Systems and Robotics, University of Coimbra, Coimbra, Portugal
{ricardo.pereira,garrote,tiagobarros,conde,urbano}@isr.uc.pt
[2] Department of Electrical and Computer Engineering, University of Coimbra, Coimbra, Portugal
[3] Cerebral Palsy Association of Coimbra, Center for Research in Neuropsychology and Cognitive-Behavioural Intervention, University of Coimbra, Coimbra, Portugal
ccarona@fpce.uc.pt

Abstract. In this paper we present an improvement of the ISR-RobotHead prototype interaction capabilities having in view future Human-Robot Interactions studies. To accomplish that, a new hardware and software architectures and new facial expressions were developed. This new prototype (ISR-RobotHead v2.0) uses LCDs to display six human facial emotional expressions, avoiding the use of interlinked mechanical systems. The prototype also incorporates cameras, microphones, speakers, and LEDs on the robot's cheeks. To validate the new cartoon facial expressions, an exploratory study was performed with children (4–6 years old) and adults (23–24 years old), who were asked to identify the expression displayed by the ISR-RobotHead v2.0; for children, the evaluation of accuracy in the detection of emotional expressions was performed through a corresponding number of cards picturing child's emotional facial expressions. Preliminary results successfully validate the proposed robot's non-verbal communication.

Keywords: Human-robot interaction · LCD-based robotic head · Facial expressions · Speech recognition

1 Introduction

For several years, robots were considered best suitable to work on production lines and heavy duty jobs, but nowadays, robots have been increasingly present in our society due to its promising interaction capabilities that have been developed in recent years (e.g., facial emotional expressions, speech recognition, face recognition) [2,14]. From a pragmatic point of view, robots can be classified into two categories: assistive robots (e.g. industrial robots, medical robots) and social robots [10].

Develop a social robot to engage humans' attention, especially children, remains a great challenge. There are many aspects difficult to combine, such as, robot's design, and the way how to interact with the humans [2]. Within the field of psychological intervention, social robots that are trying to emulate

J. Henriques et al. (Eds.): MEDICON 2019, IFMBE Proceedings 76, pp. 1627–1637, 2020.
https://doi.org/10.1007/978-3-030-31635-8_200

human facial emotion expressions are seen as promising education tools [3,7], therapeutic caregivers [16], and tools in rehabilitation sessions [6]. To reach this level, social robots started by making non-verbal communication (e.g. gestures, human facial expressions) and currently have artificial intelligence modules that allow them to have multiple interactions with humans [18].

This paper presents a Human-Robot Interaction (HRI) evaluation using the LCD-based ISR-RobotHead v2.0 [17] prototype developed at Institute of Systems and Robotics (ISR), University of Coimbra. The current study was aimed at extending previous work [12] targeting the emotional expressiveness of the ISR-RobotHead, by improving hardware and software architectures, encompassing new cartoon facial emotional expressions, integrating speech recognition systems and combining a remote control interface. The results of the research reported here constitute an intermediate step to validate this prototype in view of future applications and studies in the HRI area. To interact with humans, ISR-RobotHead v2.0 uses non-verbal communication (new six cartoon facial emotion expressions) and is able to receive user commands. Since this robot does not use interlinked mechanical systems to perform its expressions, it is easier to improve any expression and add new ones.

2 Related Work

Social robots where designed to operate closer to humans in order to entertain and help them. Among several features, these robots can express and recognize human facial emotional expressions, communicate using dialog or non-verbal communication (e.g. facial emotions, gestures) making them promising tools to be used in the promotion of children's development [6]. Interaction between robots and humans is strongly influenced by the communication capabilities and the physical appearance of the robots. The robots' appearance should be carefully chosen to prevent their fall into the *uncanny valley* [15] (a common unsettling feeling people experience humanoid robots closely resemble humans in many respects but are not quite convincingly realistic). *M. Mori* [15] also concluded that children prefer robots with toy's appearance rather than human's appearance.

2.1 Robotic Heads

Robotic heads have been proposed in many HRI research works whose main goal is trying to capture person's attention. To accomplish that, the machine is able to give some "familiar" design and some interesting features (e.g. facial emotions).

Kismet [4] was the first robotic head developed. It is a mechanical head with fifteen degrees of freedom to move its eyes, eyebrows, ears, lips, and mouth. This robot is also able to express nine emotional expressions. After Kismet, several mechanical heads capable to express facial emotions started to appear. Some examples are EDDIE [19] with the dragon appearance, Flobi [13] with an anthropomorphic appearance and the EMYS [8] inspired by cartoons, and series *Teenage Mutant Ninja Turtles*.

Another technical approach used in robots to convey emotions is LCD-based robotic heads. The idea of these robots is to use LCDs as a way to implement the eyes and even the mouth, while avoiding the use of interlinked mechanical systems with several degrees of freedom, allowing the development of an easier cartoon facial emotional expression. Some examples are QTrobot [6], Cozmo [20], Buddy Robot [5], and ISR-RobotHead [12]. QTrobot and Buddy have a tablet-like screen to display its expressions while the ISR-RobotHead uses one LCD for each eye and one LCD for mouth expression. Cozmo is a small commercial robot-toy developed by *Anki* with a small LCD to display its eyes. Another robotic head with a different look is PIL Head Robot [1] that only has a huge LCD to display a virtual anthropomorphic head. The works developed by Loureiro *et al* [12] and Costa *et al.* [6], where children with disabilities were involved to carry out their experimental results, have shown that it is possible to use these robots to facilitate children's development.

The robotic heads mentioned above, mechanical or LCD-based, are equipped with cameras, microphones, and speakers to receive environmental information, which enable the development of learning modules, such as, speech recognition and emotional facial recognition. These robots aim to interact with people, either children or adults, through the exhibition of facial emotional expressions. Due to the hardware limitations or design of each one, some expressions were well recognized and others were not, but all of these works show that these kinds of robots can interact with people, especially with children in different developmental settings, such as daily living assistance, educational support, and rehabilitation therapy [3–7].

3 Robotic Head System

3.1 Hardware

The new hardware architecture (presented in Fig. 1) is based on a central control module, Raspberry Pi Compute Module 3, which controls all actions of the remaining peripherals [17]. This architecture can be divided in three modules: facial expression system, stereo camera, and sound system.

(1) Raspberry Pi Compute Module: The Raspberry Pi is a small and affordable computer that operates with a Debian-based Linux distribution operating system, and it allows interconnection with external hardware through several General-Purpose Input/Output (GPIOs) available, including UART, SPI, and I^2C communication channels. The main features of Raspberry Pi Compute Module 3 (RPiCM3) make it the main module, such as its 1.2 GHz CPU clock, 1 GB RAM, Micro SD Card slot for data storage, 120 GPIOs available, two Camera Serial Interfaces (CSI), and two Display Serial Interfaces (DSI).

(2) Facial Expressions system: Composed by three LCDs for facial expressions display (two for the eyes and one for mouth expression), one LED on each cheek, and an Arduino. For the eyes' expressions, two uLCD-32PTU LCDs were

used. Each LCD is controlled by a PICASO processor, has a 240×320 pixel resolution with 16-bit color, serial communication protocol, touch screen, and a Micro-SD card socket for data storage. Each uLCD-32PTU LCD is connected individually to the main module (RPiCM3) allowing an independent control of each LCD. For the mouth's expression, an ITDB02-1.8SP was used. It has a ST7735 micro-controller embedded, 128×160 pixels resolution with 16-bit color, and a Serial Peripheral Interface protocol (SPI) to connect to the main module. To give color in some expressions, two NeoPixels digital RGB LEDs were used, one on each side of the robot face, more precisely, on the cheek. To control the color of the NeoPixeis, a PWM signal with 800 KHz of frequency is required, and for this, an *Arduino Duemilanove* micro-controller was used.

(3) Stereo camera: The RPiCM3 supports stereo vision since it has two independent CSI ports. The ISR-RobotHead has two standard Raspberry Pi Camera Module V2. This camera has a Sony IMX219 8-Megapixel sensor and supports 1080p30, 720p60, and VGA90 video modes.

(4) Sound system: The sound system is composed by two modules: input and output sound. For the input, a Matrix Voice kit and a Raspberry Pi 3 Model B were used. The Matrix Voice (MV) is a small and affordable component controlled by a Raspberry Pi or an ESP32 micro-controller. This component has 8 microphones, 16 RGB LEDs, 40 Input GPIOs (compatible with Raspberry Pi), ESP32 micro-controller, a Xilinx Spartan-6 FPGA, and 24 Output GPIOs for FPGA data. The 8 microphones are distributed throughout the board, which allows to capture sound and to determine its orientation. In this stage, a Raspberry Pi 3 Model B was used to control the Matrix Voice hardware. For output sound, the 3.5 mm output jack available on the same Raspberry Pi 3 Model B used to control the Matrix Voice, two speakers, and an Adafruit ST 2012 audio amplifier were used.

3.2 Software

The new software architecture implemented (presented in Fig. 2) in this work was mainly developed to develop and display six human facial emotional expressions. In this software architecture, was also integrated a system based in voice commands and a remote control interface, to choose and control, which expression to display [17]. Each external hardware (RPi3 and remote control device) is connected with the central module through a TCP Socket connection. In order to control all the modules in real time, a six-threaded software was implemented: one for external hardware communication, one for Speech Recognition software, and four threads to perform a synchronized facial emotion display controlling the time of each frame for each LCD. The main software developed in this work can be divided in three modules: facial expressions, speech recognition, and remote control interface.

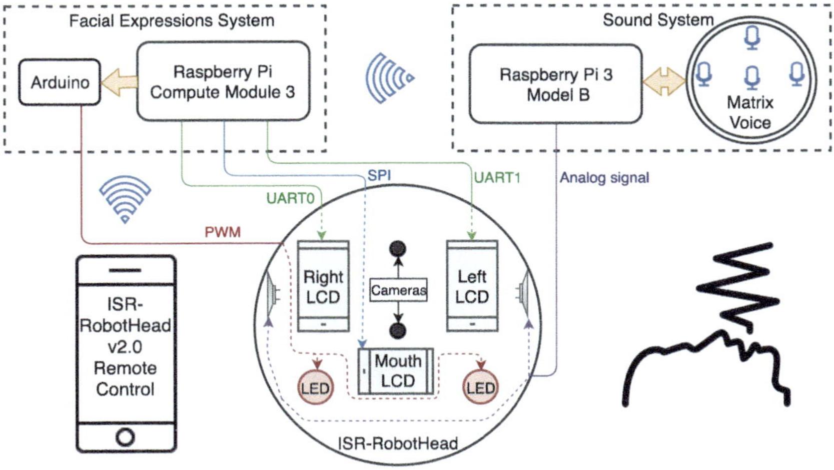

Fig. 1. ISR-RobotHead v2.0 hardware architecture.

(1) Facial Expressions: The new eyes' expressions were drawn individually in each LCD through some computer graphics primitives available in the PICASO processor and stored in a micro-SD card available in each LCD. The mouth expressions were developed in an external computer, converted to a raw 16 bits image through an image quantization method and stored in the RPiCM3 memory. Animations between normal state and all other emotional states were developed to perform a smooth transition between different emotional states. Each animation uses five intermediate frames. In order to capture the human's attention, the ISR-RobotHead v2.0 is able to blink its eyes in the normal state and a reaction is activated when one of the eyes' LCDs is touched.

(2) Speech Recognition: At this stage of HRI study, we chose to integrate and evaluate in our system, five Speech Recognition software systems already developed and trained: Microsoft Bing Speech, Google Speech API, Wit.Ai, IBM, and CMU Sphinx. Most of all data processing of these systems are in their servers, and for this, an online login is required. CMU Sphinx is the only software system with offline mode and it is possible to process the data in Raspberry Pi.

(3) Remote Control Interface: In order to be easier to configure some parameters of the ISR-RobotHead v2.0, an Android remote control interface was developed allowing to choose an emotional expression to be displayed, to control the Cartesian coordinates of each eye, the transition time between two emotional expressions, and also the gaze orientation of the eyes. To communicate with the central module, a TCP connection is required.

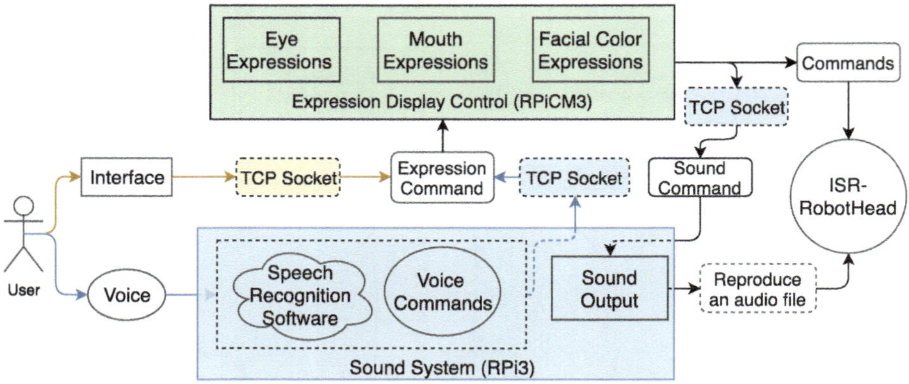

Fig. 2. ISR-RobotHead v2.0 software architecture.

4 Experimental Results

ISR-RobotHead v2.0 is able to display six cartoon human facial emotional expressions (disgust, fear, happiness, anger, sadness, and surprise), presented in Fig. 3. In the next sections, detailed experimental evaluations of the implemented work are described.

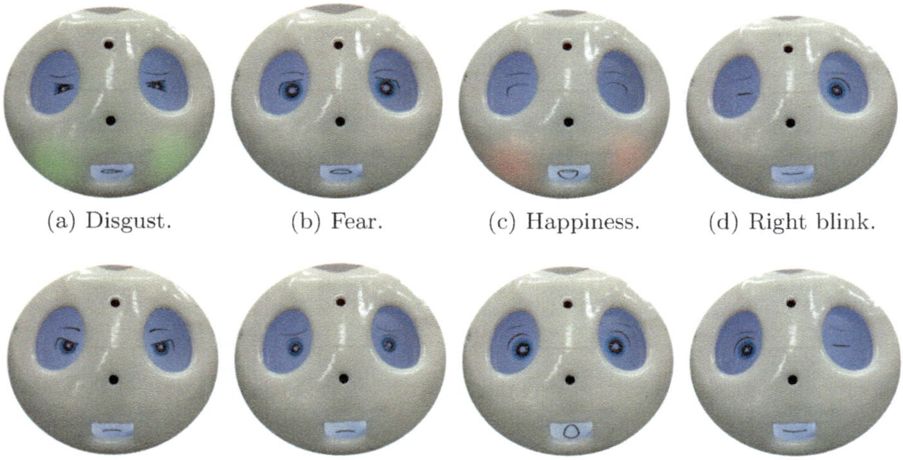

(a) Disgust. (b) Fear. (c) Happiness. (d) Right blink.

(e) Anger. (f) Sadness. (g) Surprised. (h) Left blink.

Fig. 3. ISR-RobotHead v2.0 showing the six facial expressions and LCD-eye touch event facial changes.

4.1 Facial Expressions Experimental Evaluation

In order to validate the facial expressions displayed by ISR-RobotHead v2.0, a survey was performed at the preschool of CASCI ("Centro de Ação Social do Concelho de Ílhavo") with a sample of 24 children (N=24, 10 female and 14 male), aged between 4 and 6 years old, with informed assents and consents being respectively obtained from children and their parents, in agreement with the Portuguese legal framework.

Since the children were very young, and thus with limited verbal ability, a matching test to evaluate the new human facial emotional expressions displayed by the ISR-RobotHead v2.0 was performed. The experiments were conducted individually (i.e., one child at a time). Six cards were available for each child, with each card depicting a child's facial emotional expression, as retrieved from the facial emotional expressions data set provided by LoBue *et al.* [11]. The expressions presented on the cards were the same that the ISR-RobotHead v2.0 was able to express. Next, each child was asked to pick the card that he/she thought to depict the emotion represented by the robotic head. For every child, each expression was consecutively displayed three time by the robotic head.

As shown in Table 1, all participants correctly identified the emotional state of "Surprise" and most participants (n=22) correctly detected the emotions of "Happiness" and "Sadness". However, the worst result of the matching test was obtained with the emotion of "Disgust" (n=16) followed by "Anger" (n=17) and "Fear" (n=19). These results show that the cartoon facial expressions can be improved since it is possible to verify what emotions were confused, especially in the emotional state of "Angry", where all the wrong answers went to the emotional state of "Disgust". Similar preliminary evaluations (also using a matching exercise for the detection of emotional facial expressions) were performed with other robotic heads, namely Kismet [4] and EDDIE [19], where recognition accuracy values of $\approx 73\%$ (with 17 participants) and $\approx 58\%$ (with 24 participants) were respectively observed.

Table 1. Confusion matrix obtained by the matching test performed.

		Emotional state presented					
		Happiness	Sadness	Fear	Disgust	Angry	Surprise
Emotional state answered	Happiness	**22**	0	0	1	0	0
	Sadness	1	**22**	2	4	0	0
	Fear	0	1	**19**	2	0	0
	Disgust	0	1	1	**16**	7	0
	Angry	1	0	1	1	**17**	0
	Surprise	0	0	1	0	0	**24**

4.2 Speech Recognition Software Systems

To ascertain which, among the aforementioned speech recognition software systems, achieved best performance in our setup, six sentences were chosen from the research of Veton Këpuska *et al.* [9]. The sentences were chosen in order to contain a greater lexical and phonetic field and they have been spoken by six different people (one female and five male) in the same conditions (same background noise).

The main metric used to evaluate the accuracy of a Speech Recognition system is through the Word Error Rate (WER): WER = (I+D+S)/N where I is the number of words inserted, D is the words deleted, S is the words substituted and N is the total number of words in the original sentence. Table 3 shows that the Google Speech Recognition system has the lower WER and the lower response time in the English language.

Afterwards, the same experience was performed for the Portuguese language. For this, only two of the five speech recognition systems have been trained for the Portuguese language, Google's and Microsoft's systems. The conditions were the same, six sentences were chosen (see Table 2) and the results are presented in Table 3. This time, Microsoft's speech recognition software has presented a lower WER. It should be noted that for Portuguese recognition, both systems have a WER and a response time lower than the English recognition.

4.3 Speech Recognition Commands

In order to test the expression commands through voice, it was asked to a group of seven adults one female and six male), aged between 23 and 24 years old, to verbally identify the six emotional states represented by the ISR-Robothead v2.0. For each person, 18 expressions were randomly displayed, which would give a mean of 3 times for each emotional state and the person had to verbally say which emotional state was being represented. This experience allows to validate the cartoon facial emotional expressions with adults and evaluate the voice commands recognition rate.

Table 4 shows the number of expressions well recognized along with the number of expressions that have been misinterpreted. Table 5 shows the number of

Table 2. Portuguese sentences used to evaluate the Speech Recognition systems.

Portuguese sentences
Por favor, leve esta toalha de mesa suja para a lavandaria por mim
Quem autorizou uma conta de despesas ilimitada?
O peixe começou a saltar freneticamente na superfície do pequeno lago
A raposa castanha salta rapidamente sobre o cão preguiçoso
O rato roeu a rolha da garrafa do rei da Rússia
O Pedro pregou um prego na porta preta

Table 3. WER and response time obtained for English and Portuguese languages recognition.

	English language		Portuguese language	
	WER (average)	Response time (average)	WER (average)	Response time (average)
Google	0.174	2.261 s	0.167	1.506 s
IBM	0.172	4.125 s	–	–
Sphinx	0.590	13.938 s	–	–
Microsoft	0.233	3.144 s	0.104	2.769 s
Wit.Ai	0.205	3.936 s	–	–

Table 4. Confusion matrix obtained by the evaluation performed with adults.

		Emotional state presented					
		Happiness	Sadness	Fear	Disgust	Angry	Surprise
Emotional state answered	Happiness	**7**	0	0	0	0	0
	Sadness	0	**5**	2	2	0	0
	Fear	0	2	**5**	0	0	0
	Disgust	0	0	0	**4**	0	0
	Angry	0	0	0	1	**7**	0
	Surprise	0	0	0	0	0	**7**

voice commands correctly recognized by each speech recognition software system. In general, the emotional states were well recognized by adults and well recognized by the speech recognition systems. Note that during the test, the expression "Sad" was the most difficult to be verbally recognized because the word was commonly misinterpreted with phonetically similar words, such as "said".

Table 5. Correct verbal acknowledgments through speech recognition software systems.

	Emotions						
	Disgust	Fear	Happiness	Anger	Sadness	Surprise	Sum
Appeared	22	22	18	28	14	22	126
Google	8	5	16	22	4	21	76
IBM	0	7	13	26	1	21	68
Microsoft	11	9	17	27	4	16	84
Sphinx	1	3	2	11	0	14	21
Wit.Ai	6	6	13	22	3	21	71

5 Conclusion

In this paper, we presented an improvement of the ISR-RobotHead's Human-Robot Interaction along with a brief overview of its current hardware and software modules. The facial expressions experimental results were promising since the children were able to recognize most of the expressions represented by the prototype. The integration of speech recognition systems in this work allows to increase the Human-Robot Interactions in order to choose which expression should be displayed. In general, through the WER study and voice expression commands evaluation, most of the speech recognition systems used (Google, IBM, Microsoft, and Wit.Ai) achieved good performances. However, these systems have some constraints, such as the requirement of an internet connection paired with the fact that data processing is not performed in real time, which ends up delaying for a few seconds the ISR-RobotHead v2.0 display of the requested expression.

For future work and considering the promising results presented in this work, we intend to improve the interaction of the robotic head, by means of adding artificial intelligence modules (e.g., facial expressions recognition), and integrating a "pan&tilt" system to facilitate the performance of more complex facial and sound tracking (e.g. moving the head around). This would ultimately improve not only the robot's emotional expressiveness but also its empathetic abilities in emotion recognition, which may then facilitate the development of its ability to perform an autonomous response (e.g., displaying a facial expression according to the situation), thus becoming increasingly proactive in Robot-Human Interaction. Although the cartoon facial emotional expressions were well recognized by children and adults in the current study, the increased complexity in tracking systems would ultimately allow the testing of more robust research designs, including the examination of accuracy in the expression and detection of a greater variety of emotions.

Acknowledgements. The authors would like to thank Professor Rosa Roldão for facilitating the conduction of experimental tests with children at "Centro de Ação Social do Concelho de Ílhavo". This work was partially supported by FCT Project AMS-HMI12 (https://sites.google.com/site/amshmi12/), ISR-UC and University of Coimbra, and by UID/EEA/00048/2013 with FEDER funding through programs QREN and COMPETE. It was also supported by Center for Research in Neuropsychology and Cognitive-Behavior Intervention (UID/PSI/00730/2019) and Cerebral Palsy Association of Coimbra.

References

1. Ahn, H.S., Kim, P.J., Choi, J.H., Mansoor, S.B., Kang, W.S., Yoon, S.M., Na, J.H., Baek, Y.M., Chang, H.J., Song, D.S., Choi, J.Y., Ko, H.S.: Emotional head robot with behavior decision model and face recognition. In: 2007 International Conference on Control, Automation and Systems, pp. 2719–2724, October 2007
2. Anzalone, S.M., Boucenna, S., Ivaldi, S., Chetouani, M.: Evaluating the engagement with social robots. Int. J. Soc. Robot. **7**, 465–478 (2015)

3. Belpaeme, T., Kennedy, J., Ramachandran, A., Scassellati, B., Tanaka, F.: Social robots for education: a review. Sci. Robot. **3**(21), eaat5954 (2018)
4. Breazeal, C.: Emotion and sociable humanoid robots. Int. J. Hum.-Comput. Stud. **59**, 119–155 (2003)
5. Buddy Robot. http://www.bluefrogrobotics.com/en/buddy/. Accessed 20 February 2019
6. Costa, A., Steffgen, G., Lera, F., Nazarikhorram, A., Ziafati, P.: Socially assistive robots for teaching emotional abilities to children with Autism Spectrum Disorder. In: 3rd Workshop on Child-Robot Interaction at HRI (2017)
7. Han, J., Jo, M., Park, S., Kim, S.: The educational use of home robots for children. In: ROMAN 2005. IEEE International Workshop on Robot and Human Interactive Communication, pp. 378–383, August 2005
8. Kędzierski, J., Muszyński, R., Zoll, C., Oleksy, A., Frontkiewicz, M.: EMYS–emotive head of a social robot. Int. J. Soc. Robot. **5**, 237–249 (2013)
9. Këpuska, V., Bohouta, G.: Comparing speech recognition systems (Microsoft API, Google API And CMU Sphinx). Int. J. Eng. Res. Appl. **7**, 20–24 (2017)
10. Libin, A.V., Libin, E.V.: Person-robot interactions from the robopsychologists' point of view: the robotic psychology and robotherapy approach. Proc. IEEE **92**, 1789–1803 (2004)
11. LoBue, V., Thrasher, C.: The child affective facial expression (CAFE) set: validity and reliability from untrained adults. Front. Psychol. **5**, 1532 (2015)
12. Loureiro, R., Lopes, A., Carona, C., Almeida, D., Faria, F., Garrote, L., Premebida, C., Nunes, U.J.: ISR-RobotHead: Robotic head with LCD-based emotional expressiveness. In: 2017 IEEE 5th Portuguese Meeting on Bioengineering (ENBENG) (2017)
13. Lütkebohle, I., Hegel, F., Schulz, S., Hackel, M., Wrede, B., Wachsmuth, S., Sagerer, G.: The bielefeld anthropomorphic robot head Flobi. In: 2010 IEEE International Conference on Robotics and Automation, pp. 3384–3391, May 2010
14. McColl, D., Hong, A., Hatakeyama, N., Nejat, G., Benhabib, B.: A survey of autonomous human affect detection methods for social robots engaged in natural HRI. J. Intell. Robot. Syst. **82**, 101–133 (2016)
15. Mori, M., MacDorman, K.F., Kageki, N.: The uncanny valley [From the Field]. IEEE Robot. Autom. Mag. **19**, 98–100 (2012)
16. Orejana, J.R., MacDonald, B.A., Ahn, H.S., Peri, K., Broadbent, E.: Healthcare Robots in Homes of Rural Older Adults. In: Tapus, A., André, E., Martin, JC., Ferland, F., Ammi, M. (eds) Social Robotics. ICSR 2015. Lecture Notes in Computer Science, vol. 9388, pp. 512–521. Springer, Cham (2015)
17. Pereira, R.: ISR-RobotHead v2.0: New hardware and software modules. Master degree Dissertation, Coimbra, University (2018)
18. Schulz, T., Tørresen, J., Herstad, J.: Animation techniques in human-robot interaction user studies: a systematic literature review. In: Clinical Orthopaedics and Related Research (2018)
19. Sosnowski, S., Bittermann, A., Kuhnlenz, K., Buss, M.: Design and Evaluation of Emotion-Display EDDIE. In: 2006 IEEE/RSJ International Conference on Intelligent Robots and Systems, pp. 3113–3118, October 2006
20. Tan, X.Z., Vázquez, M., Carter, E.J., Morales, C.G., Steinfeld, A.: Inducing bystander interventions during robot abuse with social mechanisms. In: Proceedings of the 2018 ACM/IEEE International Conference on Human-Robot Interaction, pp. 169–177 (2018)

Machine Support to Discrimination of Parkinson's Disease and Essential Tremor

José Ignacio Serrano[1], Julián Benito-León[2], Aleš Holobar[3], and Eduardo Rocon[1](✉)

[1] Centro de Automática y Robótica, UPM-CSIC, 28500 Madrid, Spain
e.rocon@csic.es
[2] Department of Neurology, University Hospital, 12 de Octubre, Madrid, Spain
[3] Faculty of Electrical Engineering and Computer Science, University of Maribor, 2000 Maribor, Slovenia

Abstract. Pathological tremor is a common but also highly complex movement disorder, affecting about 5% of population over the age of 65. Different methodologies have been proposed for its quantification and analysis. Nevertheless, the discrimination between Parkinsonian and Essential tremor remains a looming challenge in clinical practice, greatly impacting both patient treatment and the development of new interventions. In this study, the discriminative powers of cortical thickness features, extracted from magnetic resonance imaging of 21 ET, 15 PD patients and 18 healthy control subjects have been presented and discussed. A total of 129 volumetric features (whole brain, except cerebellum) and 152 cortical thickness features (average plus standard deviation of the thickness of the different cortical areas) have been extracted from each subject. Data mining of these characteristics has been employed to identify the most discriminative structural features for tremor diagnosis. These features were then combined in advanced classification models that yielded accuracy of up to 100% in discrimination of examined ET and PD patients, with the volume and thickness of enthorinal cortex as the most discriminative feature. Noteworthy, when compared to healthy age-matched controls, the identified MRI-based features indicate the structural changes of the brain and cortex in both PD and ET group of patients, potentially illuminating the neurodegenerative nature of both pathologies studied. This is an important finding by itself as the neurodegenerative aspects of ET have not yet been rigorously proven.

Keywords: Tremor · Parkinson disease · Essential tremor · Machine support to discrimination

1 Introduction

Pathological tremors constitute the most extended movement disorder, affecting up to 15% of people with age ranging between 50 and 89 years [1]. Moreover, more than 65% of population suffering from upper limb tremor reports serious difficulties in performing their activities of daily living (ADL), greatly decreasing their independence and quality of life [2]. Pathological tremors –simply referred to as tremors in the remainder of the document– arise due to various conditions, making it very

J. Henriques et al. (Eds.): MEDICON 2019, IFMBE Proceedings 76, pp. 1638–1643, 2020.
https://doi.org/10.1007/978-3-030-31635-8_201

complicated to differentiate them according to their aetiology [3]. Since their exact underlying mechanisms have not been elucidated, none of them is completely understood [4]. This motivates the occurrence of exceedingly common misdiagnosis; for instance, 30% to 50% of patients diagnosed with essential tremor (ET) do not have ET [5]. Moreover, since treatment of tremors is disease specific, accurate diagnosis plays a fundamental role in tremor management. Supporting clinical diagnosis with novel computer approaches based on objective evaluation of multisource neurophysiological and biomechanical information may constitute a remarkable step forward in this direction.

In the recent years, intelligent systems proved to be very useful in decision support, data mining and knowledge discovery across different fields, especially in medicine where the ability of decision support systems to set an early and accurate diagnosis of various diseases is of paramount importance. Data mining and knowledge discovery is of similar, but not so immediate importance in clinical practice. The decision support tools facilitate the testing of the existing hypotheses and contribute to discovery of new knowledge. Thus, in a longer term, they support the advancement of medicine as a scientific discipline. Not only new prediction and diagnostic methods, but also new treatments of different pathologies or disorders have been derived from advanced decision support systems [6].

Many decision support systems have been developed in the field of medical diagnostics [6]. They successfully cope with huge amount of medical records, available from daily clinical practice and use classification models and data mining techniques to extract the most vital diagnostic parameters. The latter are utilized to suggest the most probable diagnosis and to objectively quantify the differential properties of different patient groups. Clearly, black-box classification methods (neural networks, for example) are not appropriate for this kind of task, because the clinical experts need to evaluate and validate the decision making process, induced by computer-aided tools, before putting it into everyday clinical practice. On the other hand, the evaluation of the induced classifiers, as produced by the computerized tools, by a clinical expert can be an important source of new knowledge on how to make a diagnosis based on the available attributes. In order to facilitate the utilization of this source of information, the computer-aided classification process should be easily understandable by humans. In other words, while different kinds of knowledge discovery methods are appropriate for computer-aided support to clinical diagnostics, the ones that are based on symbolic approaches (such as the decision trees, decision rules, etc.) provide a very important advantage – the possibility of explaining the decisions in a way understandable to clinicians. Therefore, the simplicity and knowledge representation have to be watchfully considered when evaluating the induced models for decision support.

The work presented herein explores the structural characteristics of ET and PD based on neuroimage analysis in order to provide the clinicians and neurophysiologists with new evidences that contribute to the differential diagnosis between ET and PD. These findings serve to generate basic knowledge on tremor [7]. Despite the relatively large number of works addressing neural connectivity and metabolic changes in ET, relatively few studies have recently focused on the characterization of its structural features as revealed by analysis of neuroimaging. Most of these studies search for the white matter (WM) and the grey matter (GM) differences in ET, finding volume loss in cerebellum, thalamus and primary motor cortex [8]. However, there is a lack of research in cortical thickness characterization of ET. The research presented in this paper is motivated by the overlapping of ET and cognitive disorders shared with Alzheimer and Parkinson disease (PD). Furthermore, PD has shown cortical thickness differences with healthy subjects [9], even serving as biomarker of disease stage [10].

2 Methodology

2.1 Experimental Protocol

Patients: Structural images were recorded from 18 control subjects (62 ± 10 years old, 8 women), 23 ET patients (66 ± 11 years old, 11 women) and 15 PD patients (64 ± 11 years old, 10 women). Age differences between conditions were not significant ($F(2, 53) = .568$, $p = .570$). Control subjects did not suffer from any movement disorder. Control subjects presented an overall educational level slightly higher than ET and PD patients.

Recordings and Procedure: Both patients and controls were immobilized with a custom-fit blue bag vacuum mold (Medical Intelligence, Inc.) to prevent image artifacts. Magnetic Resonance Image (MRI) data were acquired on each patient and control using a GE Signa 3.0 T scanner (General Electric Medical Systems, Milwaukee, WI) with a standard quadrature birdcage headcoil, using an axial 3D T1-weighted inversion-recovery fast gradient echo sequence (TR = 5.0 ms; TE = 2.2 ms; FlipAngle = 12°; TI = 750 ms; NEX = 1.0). A total of 176 contiguous 1-mm slices were acquired with a 240 × 240 matrix with an in-plane resolution of 1 × 1 mm, resulting in isotropic voxels. Standard sequences of the MRI scans were checked before inclusion of a patient or control. Those with structural abnormalities in the brain, affecting gray or white matter, were excluded prior to the image analysis. MRI studies and imaging processing were performed by a neuroradiologist and a physicist who were blinded to the clinical diagnosis.

Image Processing: MRI images were processed to extract two types of information: volumetric features and cortical thickness features. Cortical reconstruction and volumetric segmentation was performed with the Freesurfer image analysis suite, which is documented and freely available for download online (http://surfer.nmr.mgh.harvard.edu/). Briefly, this processing includes motion correction and averaging of multiple volumetric T1 weighted images (when more than one is available), removal of non-brain tissue using a hybrid watershed/surface deformation procedure [11], automated Talairach transformation, segmentation of the subcortical white matter and deep gray matter volumetric structures (including hippocampus, amygdala, caudate, putamen, ventricles) intensity normalization, tessellation of the gray matter white matter boundary, automated topology correction, and surface deformation following intensity gradients to optimally place the gray/white and gray/cerebrospinal fluid borders at the location where the greatest shift in intensity defines the transition to the other tissue class [12]. Once the cortical models are complete, a number of deformable procedures can be performed in further data processing and analysis, including surface inflation, registration to a spherical atlas which utilizes individual cortical folding patterns to match cortical geometry across subjects, subdivision of the cerebral cortex into units based on gyral and sulcal structure [13], and creation of a variety of surface based data, including maps of curvature and sulcal depth. This method uses both intensity and continuity information from the entire three dimensional MR volume in segmentation and deformation procedures to produce representations of cortical thickness, calculated as the closest distance from the gray/white boundary to the gray/CSF (Cerebrospinal Fluid) boundary at each vertex on the tessellated surface [12]. The maps are created using spatial intensity gradients across tissue classes and are therefore not simply reliant on absolute signal intensity. The maps produced are not restricted to the voxel resolution of the original data and are, thus, capable of detecting submillimeter differences between groups. The cortical thickness features are average values for each region. Additionally, for each cortical region considered, the standard deviation of the cortical thickness was also calculated as a measure of roughness. The above processing steps yielded 129 WM and GM volumetric features of the whole brain (excepting cerebellum) and 152 cortical thickness features (average plus standard deviation of the thickness), having a total of 281 structural features from each subject.

Data Analysis: Given the volumetric and cortical thickness features, and the corresponding diagnostic condition for each subject (Control, Essential and Parkinson), several data mining methods were applied. First, a linear regression method was applied in order to linearly characterize the differences between each pair of subject conditions. The features were selected by following the stepwise method, where features are added to the model taking into account their marginal contribution according to Akaike's Information Criterion Corrected (entropy measurement). After each feature is added to the model, the algorithm examines to see if any of the previously selected variables could be dropped without appreciably increasing the residual sum of squares. This step allows the discovering of the set of features that best distinguishes pairs of subject conditions. Second, a rule learner was applied again to each pair of subject conditions. The rule algorithm searches for IF-THEN rules that cover the maximum number of subject in a condition and do not cover any subjects belonging to the other condition. This is done by gradually adding features to the IF part of the rule until this criterion is reached. This method allows describing and characterizing clusters of subjects inside each condition, as well as detecting special cases and outliers. This method also gathers relations between features with respect to particular conditions. Third, each feature was scored according to six information criteria for each pair of conditions: Chi-square statistic, Rule accuracy, Information Gain, Information Gain Ratio, Uncertainty and G_{ini} index. All these criteria give a measure of how good any single feature is in separating the two target conditions. The criteria measures are normalized between 0 and 1, and finally each feature is assigned the average score of the six criteria. After putting the features in descending order according to the assigned final score, the most discriminant single features for each pair of conditions were revealed. Ultimately, predictive tests were run by following a 10-cross-fold validation. The entire set of subjects was divided in 10 non-overlapping, equally distributed parts. Then 9 out of 10 parts were used for training a rule-based classifier, and the remaining part was used as validation set. The process was repeated 10 times using a different part each time as validation set. The average of the ten classification results was given as the classifier predictive performance.

3 Results

The performance accuracy of the 10-cross-fold validation of the rule-based classifier is independently presented for the diagnosis of the two pathologies. Precision and recall values for each condition are also showed together with the corresponding confusion matrix (Tables 1, 2 and 3).

Table 1. Healthy vs. Essential Tremor. Confusion matrix, precision and recall values and average accuracy of 10-cross-fold validation of a rule-based classifier for the diagnosis of ET.

	Actual Healthy	Actual Essential tremor	Precision
Predicted HEALTHY	18	0	100%
Predicted ESSENTIAL TREMOR	1	23	95.83%
Recall	94.74%	100.00%	97.73 (±3.94)

Table 2. Healthy vs. Parkinson. Confusion matrix, precision and recall values and average accuracy of 10-cross-fold validation of a rule-based classifier for the diagnosis of Parkinson.

	Actual Healthy	Actual Parkinson	Precision
Predicted HEALTHY	19	0	100%
Predicted PARKINSON	0	15	100.00%
Recall	100.00%	100.00%	100.00%

Table 3. Essential Tremor vs. Parkinson. Confusion matrix, precision and recall values and average accuracy of 10-cross-fold validation of a rule-based classifier for the differential diagnosis between Essential Tremor and Parkinson.

	Actual ET	Actual Parkinson	Precision
Predicted ET	21	1	95.45%
Predicted PARKINSON	2	14	87.50%
Recall	91.30%	93.33%	92.11 (±4.67)

4 Discussion

Finally, the predictive classification results shows that the structural features from MRI are informative enough to be used in diagnosis. Parkinson condition was the easiest to predict, followed by Essential Tremor condition and the discrimination between the two conditions. In Essential Tremor, only one misclassification was produced, and it corresponds to the patient that was latter on re-diagnosed as Parkinson by the clinicians. In fact, also the rest of error cases from the differential diagnosis classification are currently being re-evaluated by the clinicians.

In summary, structural features from MRI have been proven to be highly informative for diagnosis and characterization of tremor pathologies. Particularly, cortical thickness related features are really useful to characterize and discriminate the two pathological conditions. These features likely represent the cognitive deficits associated to each pathology. Despite the fact that cognitive deficits seem to overlap between conditions at the behavioral level, they are clearly distinguished at the structural level. Moreover, they might be used as a fine-grained marker of disease stage. Therefore, they represent a rich alternative to the traditional cerebellar structures and a helpful complement for the clinical tests.

References

1. Wenning, G.K., et al.: Prevalence of movement disorders in men and women aged 50-89 years (Bruneck Study cohort): a population-based study. Lancet Neurol. **4**, 815–820 (2005)
2. Rocon, E., Belda-Lois, J.M., Sanchez-Lacuesta, J.J., Pons, J.L.: Pathological tremor management: Modelling, compensatory technology and evaluation. Technol. Disabil. **16**, 3–18 (2004)
3. Helmich, R.C., Toni, I., Deuschl, G., Bloem, B.R.: The pathophysiology of essential tremor and parkinson's tremor. Curr. Neurol. Neurosci. Rep. **13**, 378 (2013)
4. Elble, R.J.: Tremor: clinical features, pathophysiology, and treatment. Neurol. Clin. **27**, 679–695, v–vi (2009)
5. Benito-León, J., Louis, E.D.: Essential tremor: emerging views of a common disorder. Nat. Clin. Pract. Neurol. **2**, 666–678 (2006). quiz 2p following 691
6. Povalej Bržan, P., et al.: New perspectives for computer-aided discrimination of parkinson's disease and essential tremor. Complexity **2017**, 1–17 (2017)
7. Benito-León, J., et al.: Tremor severity in Parkinson's disease and cortical changes of areas controlling movement sequencing: a preliminary study. J. Neurosci. Res. 1–12 (2018). https://doi.org/10.1002/jnr.24248
8. Louis, E.D., Huang, C.C., Dyke, J.P., Long, Z., Dydak, U.: Neuroimaging studies of essential tremor: how well do these studies support/refute the neurodegenerative hypothesis? Tremor Other Hyperkinet. Mov. (N. Y.) **4**, 235 (2014)
9. Jubault, T., et al.: Patterns of cortical thickness and surface area in early Parkinson's disease. Neuroimage **55**, 462–467 (2011)
10. Zarei, M., et al.: Cortical thinning is associated with disease stages and dementia in Parkinson's disease. J. Neurol. Neurosurg. Psychiatry **84**, 875–881 (2013)
11. Ségonne, F., et al.: A hybrid approach to the skull stripping problem in MRI. Neuroimage **22**, 1060–1075 (2004)
12. Fischl, B., Dale, A.M.: Measuring the thickness of the human cerebral cortex from magnetic resonance images. Proc. Natl. Acad. Sci. U.S.A. **97**, 11050–11055 (2000)
13. Desikan, R.S., et al.: An automated labeling system for subdividing the human cerebral cortex on MRI scans into gyral based regions of interest. Neuroimage **31**, 968–980 (2006)

Investigating Whole-Brain MRI Markers in Multiple Sclerosis – Emerging Dimensions in Morphometric Space

Júlia Soares[1,4(✉)], Teresa Sousa[2,3,4], Otília C. d'Almeida[2,3,4,6], Sónia Batista[3,5,6], Lívia Sousa[5,6], Miguel Castelo-Branco[2,3,4,6], and João Valente Duarte[2,3,4]

[1] Faculty of Sciences and Technology, University of Coimbra, Coimbra, Portugal
julii.f.soares@gmail.com
[2] Institute of Nuclear Sciences Applied to Health, University of Coimbra, Coimbra, Portugal
[3] Coimbra Institute for Biomedical Imaging and Translational Research, University of Coimbra, Coimbra, Portugal
[4] CNC.IBILI, Faculty of Medicine, University of Coimbra, Coimbra, Portugal
[5] Department of Neurology, Centro Hospitalar e Universitário de Coimbra, Coimbra, Portugal
[6] Faculty of Medicine, University of Coimbra, Coimbra, Portugal

Abstract. Magnetic resonance imaging (MRI) has become a crucial criterion for diagnosis and prognosis in early Multiple sclerosis (MS). Grey matter (GM) pathology has been described. However, little is known about the cause of cortical atrophy in MS. This study aims at identifying structural neuroimaging biomarkers of interest for the investigation of MS pathophysiology. We used structural MRI-based features - GM and WM volume, cortical thickness and gyrification index - to investigate patterns of deficits in MS. We analyzed images from 59 MS patients and 64 age-matched healthy controls. Imaging data underwent univariate statistical analyses, namely voxel-based morphometry (VBM) and surface-based morphometry (SBM), to investigate regional morphometric differences. Multivariate pattern analysis (MVPA) using a Support Vector Machine (SVM) classifier was also applied to explore pattern recognition in neuroimaging as a tool with potential for development of medical imaging biomarkers. Either VBM and SBM analyses yielded several morphological disease-related changes, which can be used to highlight disease effects in MS. SVM classification yielded an accuracy of 86,51% (sensitivity 74,58%, specificity 98,44%) with GM volume, while with WM data, SVM correctly classified 82,34% of participants (sensitivity 67,80%, specificity 96,88%). As such, MVPA can be a useful predictive biomarker with potential in assisting diagnosis in MS.

Keywords: Multiple Sclerosis · Neuroimaging biomarkers · Multivariate pattern analysis · Brain structure

M. Castelo-Branco and J. V. Duarte—Shared last authors.

J. Henriques et al. (Eds.): MEDICON 2019, IFMBE Proceedings 76, pp. 1644–1652, 2020.
https://doi.org/10.1007/978-3-030-31635-8_202

1 Introduction

Multiple Sclerosis (MS) is a chronic, autoimmune, progressive, inflammatory neurological disease which disturbs the central nervous system (CNS), mainly in young adult, affecting 2.3 million people in the world [1, 2]. MS is characterized by myelin degradation of axons' sheath. Demyelinated regions, known as white matter (WM) lesions, cause motor, sensory and cognitive degradation in as many as 65% of patients [1, 2].

The detection of WM lesions based on magnetic resonance imaging (MRI) is a key feature to proper MS diagnosis according to McDonald's criteria and the MAGNIMS consensus guidelines [2, 3]. However, MS misdiagnosis is still a concern [4]. There are other structural changes such as regional WM and grey matter (GM) atrophy in the brain, associated with cognitive deficit and disease severity. Moreover, its cause, how early it begins, and which regions are primarily affected remains unclear [5, 6]. Thus, it is important to identify biomarkers to provide more accurate diagnostics and cues about the pathophysiology of MS disease.

Structure of the brain can be accessed in a non-invasive way through MRI, which is a good starting point to carry out methodical analysis to identify structural biomarkers [7, 8]. Univariate analysis methods such as Voxel-Based-Morphometry (VBM) [9] and neuroimaging pattern recognition using multivariate pattern analysis (MVPA) techniques [8] such as automatic classification have been used to identify morphological alterations in MS [10–12]. VBM is an automated statistical test performed in each voxel individually, which evaluates whether there is a difference between groups of participants in the tissue volume. This type of analysis is usually applied based on GM distribution, but it can be also applied to WM (lower sensitivity). Since VBM is an hypothesis-driven univariate test, it will only detect a predicted effect (typically, a mean difference between two groups) in individual voxels (massive multiple comparisons). In contrast, multivariate methods are able to identify unpredicted data patterns reflecting group (many times individually subtle) differences, in the absence of a priori hypotheses, spread across different voxels [8, 13].

Additionally, Surface-Based-Morphometry (SBM) [14] can be an useful and innovative approach to identify new MS structural biomarkers, based in surface variations. Like VBM, SBM is a voxel-wise univariate method to test the hypothesis of mean differences between groups in a particular surface measure, such as the cortical thickness and the gyrification index [15]. It is known that there is an increase in gyrification during early stages of brain development; after childhood, there is stabilization and with aging, there is a decrease of cortical thickness and gyrification [15, 16]. Investigation of how these measures behave age-independently, in the context of MS, can provide valuable information for diagnosis improvement and comprehension of disease.

In order to better understand the brain structural patterns involved in MS, we applied VBM in GM and WM volumes, SBM in cortical thickness and gyrification index, and multivariate classification of GM and WM volumes with a Support Vector Machine (SVM). Despite univariate analysis and MVPA methods cannot be directly comparable [13], as classification is performed using volumetric measures, it allows us to compare the regions highlighted by the weight maps of SVM classification with regions identified by VBM analysis.

2 Methods

Participants. The study included 59 MS patients (39F; mean age: 37,00 ± 7,24 years) and 64 controls (42F; mean age: 36,84 ± 9,79 years). Patients were recruited at the Coimbra Hospital and University Centre (CHUC) and met the criteria for MS diagnosis according to McDonald Criteria [2]. All participants gave written informed consent. Local ethics committee approved the study.

MRI Data Acquisition and Processing. Data were collected with a Siemens Magnetom TIM Trio 3 T scanner (Siemens, Munich, Germany) with a phased array 12-channel birdcage head coil. We acquired a 3D anatomical T1-weighted MRI pulse sequence (TR 2530 ms; TE 3.42 ms; TI = 1100 ms; flip angle 7°; 176 slices with no gap, with isotropic voxel size of 1 mm; FOV 256 mm) of all participants. Data were pre-processed using SPM12 software [17] and CAT12 [18] toolbox in MATLAB (version R2019a) [19]. Orientation of structural scans was manually corrected by aligning each image onto the axis of the anterior and posterior commissures. CAT12 was used to perform high-dimensional spatial normalization to a template stereotaxic space, segmentation of each scan in GM and WM tissue images, and modulation to compensate for the effect of spatial normalization. This last step involves scaling, by the amount of expansion or contraction, so that the total amount of GM/WM in the modulated images remains the same, as it would be in the original images [20].

Voxel Based Morphometry (VBM). Modulated and normalized GM and WM outputs were smoothed using an 8 mm full-width-at-half-maximum (FWHM) isotropic Gaussian kernel, required to guarantee normality of the data for subsequent statistical analysis. The statistical model was built as an independent-samples t-test (two groups), with 3 covariates - gender, age and total intracranial volume (TIV) - to correct for different individual brain sizes. We applied a 0.1 absolute masking threshold to the GM/WM data and considered differences with $p < 0.05$ (FWE corrected) as significant. Clusters of contiguous voxels with significant differences were overlaid onto a template T1 brain image for visualization and the corresponding brain region was identified by their coordinates in standard space.

Surface Based Morphometry (SBM). The image processing was performed as for VBM and allowed the extraction of additional surface parameters: cortical thickness and gyrification index. The extracted cortical thickness and gyrification maps were smoothed using 15 mm and 20 mm FWHM kernel, respectively. The statistical model for analysis was built similarly to VBM, with 2 groups and 2 covariates: gender and age (surface measures do not depend on volume measures so TIV is not needed as covariate in this model). Results of surface differences were overlaid as maps in a template reconstruction of the brain and a threshold of $p < 0.05$ (FWE corrected) was applied.

Support Vector Machine (SVM) Classification. We used the Pattern Recognition for Neuroimaging Toolbox (PRONTO) [21] in MATLAB to build a classification model using a SVM: a supervised learning method that requires labelled training data to perform the classification of unseen test data [13]. In our study, the labels represent groups, MS patients and controls, whereas the features are the volume of GM or WM in each voxel. Each brain image is assigned to MS or control group and each voxel is labelled according with the image label. Leave-one-subject-per-group-out cross validation (LOO-CV) was implemented, leaving one image per group in each iteration out for testing. During the training phase, in order to learn the relationship between the features and the labels a mathematical decision function is determined, and a weight is assigned to each feature, representing its importance for discrimination. This decision function will be the criterion to classify the unseen images in the test set, and then obtain performance values (accuracy, sensitivity and specificity). To estimate the significance of the classification, we considered the null hypothesis that there is no information about the labels in the data, i.e., there is no difference throughout the voxels of all images, hence there is no discrimination between groups to be done. To calculate this probability, we applied a permutation test, by repeating 1000 times the same classification procedure with randomized labels each time. After this procedure, a level of statistical significance was obtained and expressed as the p-value. The smaller the p-value, the more significant is the performance of our classifier [13, 22, 23]. This classifier was applied in GM and WM volume data to assess whether MS patients can be accurately differentiated from healthy controls. Model performance was evaluated, and we also extracted weigh maps to explore which brain regions contributed more to the discrimination.

3 Results

3.1 VBM

In both VBM analyses with GM and WM we tested the contrasts controls > MS, which yields several significant clusters ($p < 0.05$, FWE corrected), showing GM and WM atrophy in MS patients (these are illustrated in Figs. 1 and 3, respectively). The inverse contrast, MS > controls, did not reveal any region where patients have significantly higher tissue volume than controls (data not shown).

Grey Matter VBM Results. We found a decrease of GM volume in MS patients relatively to controls in clusters with standard space coordinates corresponding to regions in right thalamus and lentiform nucleus, right cingulate gyrus, left medial frontal gyrus, middle frontal gyrus, right rectal gyrus, left lingual gyrus and in right fusiform gyrus (see Fig. 1).

White Matter VBM Results. We found a decrease of WM volume in MS patients in clusters that include regions in both right and left parahippocampal gyrus and regions in frontal and temporal lobe (see Fig. 3).

3.2 SVM Classification

SVM classification applied to GM volume achieved an overall accuracy of 86.51% (sensitivity 74.58%; specificity 98.44%) and a p-value of 0.001. In what concerns WM, SVM correctly classified 82.34% of cases (sensitivity 67.80%; specificity 96.88%), with a p-value of 0.001. The GM and WM weight maps show voxels with higher positive weights in regions similar to those identified with VBM analysis (see Figs. 2 and 3).

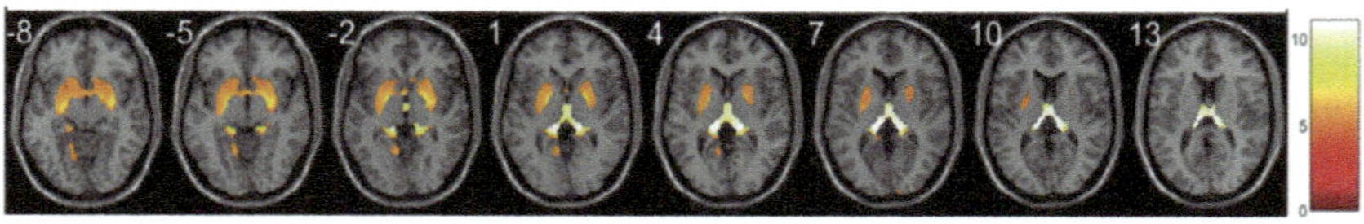

Fig. 1. Results of VBM analysis of GM (controls > MS). Results are presented at a voxel-level p-value < 0.05, FWE corrected. Color bar scale represents t-values.

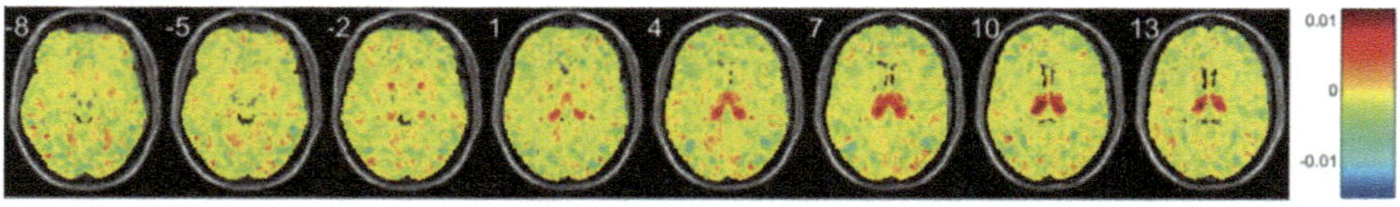

Fig. 2. Whole-brain representation of the discriminative map from SVM classification of GM images. In color bar scale, positively weighted voxels are displayed in orange/red and negatively weighted voxels are displayed in blue/green.

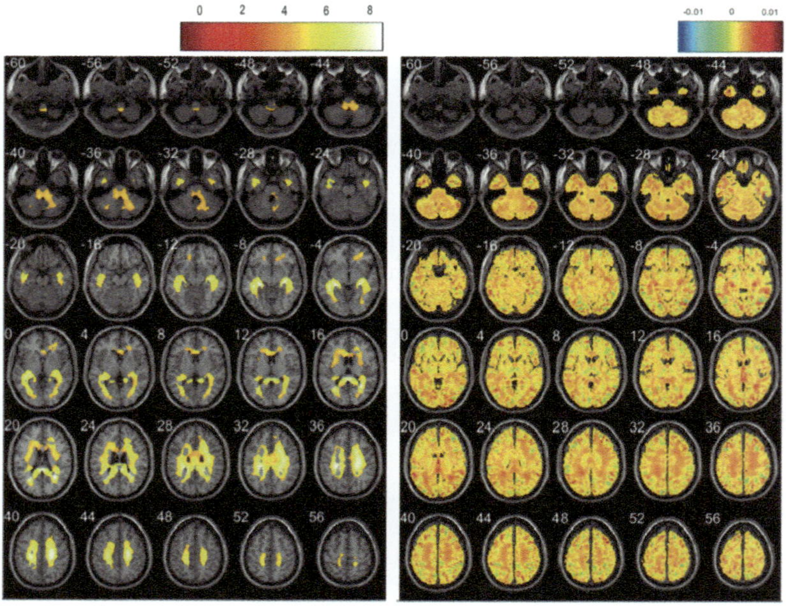

Fig. 3. Results of VBM analysis of WM (controls > MS), on the left panel. Results are presented at a voxel-level p-value < 0.05, FWE corrected. Color bar scale represents t-values. The right panel represents the whole-brain discriminative map from SVM classification of WM images. In color bar scale, positively weighted voxels are displayed in orange/red and while negatively weighted voxels are displayed in blue/green.

3.3 SBM

The same contrasts, as defined in VBM, were tested to study differences of cortical thickness and gyrification index between groups with on SBM analysis.

Cortical Thickness. SBM analysis of cortical thickness yielded several significant clusters ($p < 0.05$, FWE corrected), shown in Fig. 4. We found lower values of cortical thickness in MS patients, as compared to controls, in the following regions: left postcentral gyrus; right and left precentral gyrus; left paracentral lobule; right and left middle frontal gyrus; right superior frontal gyrus; right and left inferior parietal lobule; right precuneus; right and left superior temporal gyrus; right and left middle temporal gyrus; right cuneus; right middle occipital gyrus; left parahippocampal gyrus; right posterior cingulate; insular lobe; cerebellum. The symmetric contrast (MS > controls) did not reveal regions with significant difference (data not shown).

Gyrification Index. Since this is an exploratory study, we show gyrification maps with thresholds at $p < 0.001$ (uncorrected), as with $p < 0.05$ (FWE corrected) there were no significant differences between groups. We found decreased gyrification in MS patients in: precentral gyrus; medial frontal gyrus; postcentral gyrus; superior temporal gyrus; cuneus; left superior occipital gyrus; right cingulate gyrus and in insular cortex (see Fig. 5). We found increased gyrification in MS patients in: inferior frontal gyrus; superior frontal gyrus; rectal gyrus; middle frontal gyrus; subcallosal gyrus; paracentral lobule; postcentral gyrus; precuneus; inferior temporal gyrus; superior temporal gyrus; lingual gyrus; cuneus; parahippocampal gyrus; cingulate gyrus and insular cortex (see Fig. 5).

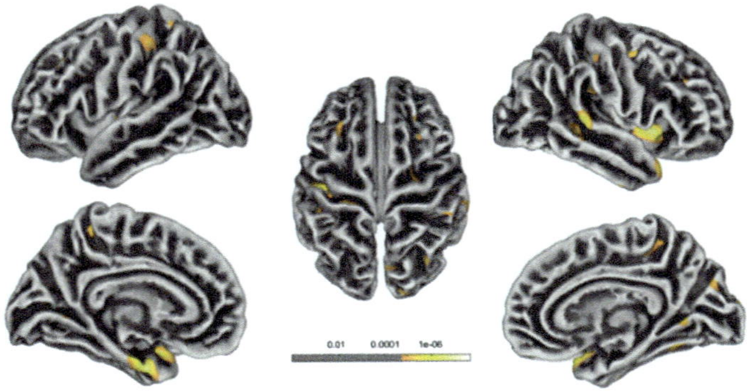

Fig. 4. Results of SBM analysis of cortical thickness (controls > MS). Results are presented at a voxel-level p-value < 0.05, FWE corrected. Color bar scale represents log p-values.

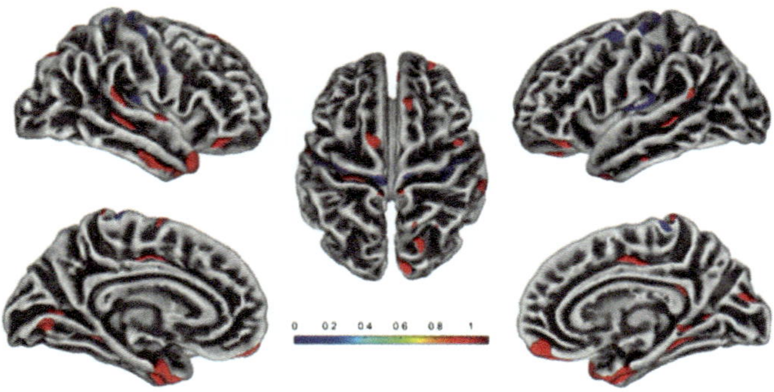

Fig. 5. Results of SBM analysis of gyrification. Results are presented at a voxel-level p-value < 0.001 (uncorrected). Blue and red indicates decreased (controls > MS) and increased gyrification (MS > controls), respectively. Color bar scale represents p-values.

4 Discussion and Conclusions

We performed both univariate and multivariate analyses of structural MRI scans of MS patients compared to control subjects. We aimed to identify different brain structural changes that occur in MS and to demonstrate the potential of these approaches to differentiate effects of this disease. Our VBM results revealed GM and WM atrophy. GM atrophy is in accordance with previous literature [11, 24]. On the other hand, WM atrophy is not commonly studied. As such, our results add to the understanding of the disease by demonstrating that WM is in fact a feature in MS. The more significantly affected region is thalamus, which is highly associated with cognitive deficit, a disease effect [25]. Since there is atrophy in MS patients, it is not expected any increase of volume relatively to controls. The few cases that we found are related to physiognomy differences in brain or hyperintensity of images.

SVM classification results demonstrated reliable detection of MS patients, which reveals this is indeed a useful predictive tool that could be used as biomarker with potential for support MS diagnosis. Weight maps from SVM classification highlighted similar regions to those found with VBM analysis, suggesting that differences in such regions are crucial for the discrimination between patients and control groups.

SBM analysis revealed a substantial decrease of cortical thickness in MS patients, which stresses the need of investigate its relationship with the disease effects. Regarding gyrification, both decreased and increased gyrification indexes were observed in different brain regions of MS patients. Nevertheless, only the increased gyrification cases survived to statistical correction for multiple comparisons. Studies addressing correlations between surface measures and aging have suggested that increased gyrification is linked to brain plasticity mechanisms to facilitate brain connectivity and functional development [16]. On the other hand, it is believed that MS patients have their brain connectivity altered [26]. As such, increased gyrification might be related with this MS effect, reflecting compensatory mechanisms.

In all analyses, clusters with between-group differences in GM/WM/cortical thickness/gyrification index between groups or voxel-wise classification weights are assigned a brain location, which is identified by their anatomical correspondence with coordinates in a standard space. These allow a qualitative description of disease discrimination based on macrostructural measures from MRI. Quantitative analysis of the statistical maps, e.g. number of voxels per cluster, peak localization, is required for further investigation of disease pathophysiology.

This study is, to the best of our knowledge, the first to apply VBM, SBM analysis of less common morphometric measures, and classification analysis of structural MRI scans in the same group of MS patients. Furthermore, these data-driven analyses indicated spatially distributed networks of brain regions with abnormal structure in individuals with MS providing important clues for the pathophysiology of this disorder.

Funding. This research was supported by grants from Biogen, FCT and COMPETE [projects BIGDATIMAGE (CENTRO-01-0145-FEDER, 000016), POCI-01-0145-FEDER-007440, UID/NEU/04539/2013, and CENTRO-01-0145-FEDER-031973]. The sponsors did not participate in any aspect of the design or performance of the study, including the data collection, management, analysis, and interpretation or preparation of the manuscript. The authors declare no competing financial or other conflict of interest.

References

1. Goldenberg, M.M.: Multiple sclerosis review. P&T Peer Rev. J. Formul. Manag. **37**, 175–184 (2012)
2. Thompson, A.J., Banwell, B.L., et al.: Diagnosis of multiple sclerosis: 2017 revisions of the McDonald criteria. Lancet Neurol. **17**, 162–173 (2018)
3. Filippi, M., Rocca, M.A., et al.: MRI criteria for the diagnosis of multiple sclerosis: MAGNIMS consensus guidelines. Lancet Neurol. **15**, 292–303 (2016)
4. Kaisey, M., Solomon, A.J., et al.: Incidence of multiple sclerosis misdiagnosis in referrals to two academic centers. Multiple Sclerosis Relat. Disord. **30**, 51–56 (2019)
5. Ontaneda, D., Fox, R.J.: Imaging as an outcome measure in multiple sclerosis. Neurotherapeutics **14**, 24–34 (2017)
6. Sima, D.M., Loeckx, D., Smeets, D., Jain, S., Parizel, P.M., Hecke, W.V.: Use case I: imaging biomarkers in neurological disease. Focus on multiple sclerosis. In: Imaging Biomarkers, pp. 169–180 (2017)
7. Jain, S., Sima, D.M., Ribbens, A., et al.: Automatic segmentation and volumetry of multiple sclerosis brain lesions from MR images. NeuroImage Clin. **8**, 367–375 (2015)
8. Linn, K.A., Gaonkar, B., Doshi, J., Davatzikos, C., Shinohara, R.T.: Addressing confounding in predictive models with an application to neuroimaging. Int. J. Biostat. **12**, 31–44 (2015)
9. Ashburner, J., Friston, K.J.: Voxel-based morphometry—the methods. NeuroImage **11**, 805–821 (2000)
10. Ceccarelli, A., Rocca, M.A., et al.: A voxel-based morphometry study of grey matter loss in MS patients with different clinical phenotypes. NeuroImage **42**, 315–322 (2008)
11. Bodini, B., Khaleeli, Z., et al.: Exploring the relationship between white matter and gray matter damage in early primary progressive multiple sclerosis: an in vivo study with TBSS and VBM. Hum. Brain Mapp. **30**, 2852–2861 (2009)
12. Bendfeldt, K., Klöppel, S., et al.: Multivariate pattern classification of gray matter pathology in multiple sclerosis. NeuroImage **60**, 400–408 (2012)

13. Duarte, J.V., Ribeiro, M.J., Violante, I.R., Cunha, G., Silva, E., Castelo-Branco, M.: Multivariate pattern analysis reveals subtle brain anomalies relevant to the cognitive phenotype in neurofibromatosis type 1. Hum. Brain Mapp. **35**, 89–106 (2014)
14. Structural Brain Mapping Group. http://www.neuro.uni-jena.de/cat/index.html#SBM
15. Spalthoff, R., Gaser, C., Nenadić, I.: Altered gyrification in schizophrenia and its relation to other morphometric markers. Schizophr. Res. **202**, 195–202 (2018)
16. Hogstrom, L.J., Westlye, L.T., Walhovd, K.B., Fjell, A.M.: The structure of the cerebral cortex across adult life: age-related patterns of surface area, thickness, and gyrification. Cereb. Cortex **23**, 2521–2530 (2013)
17. Wellcome Trust Centre for Neuroimaging, Institute of Neurology, UCL, London, UK. http://www.fil.ion.ucl.ac.uk/spm/software/spm12/
18. Gaser, C.: Structural Brain Mapping Group, Jena University Hospital, Jena, Germany. http://dbm.neuro.uni-jena.de/cat/
19. Mathworks. https://www.mathworks.com/
20. Structural Brain Mapping Group. http://www.neuro.uni-jena.de/vbm/segmentation/modulation/
21. Schrouff, J., Rosa, M.J., et al.: Pattern recognition for neuroimaging toolbox. Neuroinformatics **11**, 319–337 (2013)
22. Kaplan, J., Meyer, K.: Multivariate pattern analysis reveals common neural patterns across individuals during touch observation. Neuroimage **60**, 204–212 (2012)
23. Nichols, T.E., Holmes, A.P.: Nonparametric permutation tests for functional neuroimaging: a primer with examples. Hum. Brain Mapp. **15**, 1–25 (2001)
24. Prinster, A., Quarantelli, M., et al.: Grey matter loss in relapsing–remitting multiple sclerosis: a voxel-based morphometry study. NeuroImage **29**, 859–867 (2006)
25. Fama, R., Sullivan, E.V.: Thalamic structures and associated cognitive functions: relations with age and aging. Neurosci. Biobehav. Rev. **54**, 29–37 (2015)
26. Tahedl, M., Levine, S.M., et al.: Functional connectivity in multiple sclerosis: recent findings and future directions. Front. Neurol. **9**, 828 (2018)

Computational Intelligence Generation of Subject-Specific Knee and Hip Healthy Joint Angles Reference Curves

Pedro Sá Cunha[2](✉), João Ferreira[1,2](✉), A. Paulo Coimbra[2], and Manuel Crisóstomo[2]

[1] Department of Electrical Engineering, Superior Institute of Engineering of Coimbra, 3030-199 Coimbra, Portugal
ferreira@isec.pt

[2] Department of Electrical and Computer Engineering, Institute of Systems and Robotics, University of Coimbra, 3030-290 Coimbra, Portugal
pedro.de.sa.cunha@gmail.com

Abstract. Backpropagation Neural Network (BNN) and Extreme Learning Machine (ELM) can generate subject-specific joint angle reference profiles based on a subject's height, weight, age and walking speed. These reference profiles are useful in various fields such as biomechanics and medicine, for detection of gait pathologies and rehabilitation. A common procedure used to identify abnormal gait is comparing an individual's knee and hip curves against healthy reference curves. These reference curves are usually obtained from a heterogeneous sample of healthy subjects and might lack the specificity required to obtain accurate results. It is why the generation of reference curves according to an individual's height, weight, age and gait speed shall result in a better comparison and diagnosis. The main objective of the present study is to observe which of the two Computational Intelligence (CI) methods, BNN and ELM, present more accurate results when used to generate reference curve profiles based on subject height, weight, age and gait speed for the knee and hip joint angles.

Keywords: BNN · ELM · Human gait · Subject-specific profiles

1 Introduction

Gait is one of the most basic, fundamental human characteristics. The way we walk allows us to have mobility and live life as we know it. Nevertheless, some pathologies can arise at an early age, during or towards the end of our life, as a result of mental impairments, physical injuries and/or neurodegenerative diseases [1–3]. Independently of its cause, gait pathologies can severely affect our quality of life [4, 5].

Gait analysis and assessment is used to detect abnormalities in human gait, propose a treatment and observe its effectiveness [6]. Traditional methods based themselves in the qualitative clinician observation [7], but a plethora of methods based on the

J. Henriques et al. (Eds.): MEDICON 2019, IFMBE Proceedings 76, pp. 1653–1668, 2020.
https://doi.org/10.1007/978-3-030-31635-8_203

quantitative measurements such as spatiotemporal, kinetic and kinematics have been applied in the identification and classification of pathological gait [2, 8–10]. With the fast growth of hardware computational power, the application of CI methods in broader fields, including biomedical engineering, were made possible [11]. CI methods offer the capability of modeling complex systems such as the human body [12] in a fast-inexpensive way, having by now demonstrated to be capable of identifying and classifying gait pathologies [13, 14].

A common procedure used to identify gait pathologies is comparing the kinematic data, obtained using vision acquisition systems and body markers on a subject walking to a healthy-standard joint angle curve [15, 16]. The lower body joint angles profile, namely the hip and knee, are crucial and commonly used in several different marker placements [17, 18] both in visual acquisition systems and wearable technologies.

The generation of the reference healthy joint angle curve is based on very distinct subjects in such way that it can serve as a standard joint angle curve and so it lacks the specificity necessary for an accurate comparison, seen that subject specific features such as weight, height and age influence gait pattern [19–22]. Furthermore, such reference profile does not commonly take in consideration the gait speed, which is known to have great impact on joint angle's profile [23, 24].

CI can generate healthy joint angle curve based on subject specific parameters. Luu et al. [25] used anthropometric data and gait parameters of subjects to feed a Generalized Regression Neural Network (GRNN) which was trained to target Fourier coefficients of joint angle's waveform. Posteriorly, inverse Fourier transform can be applied to obtain the lower limb joint angle waveform.

Alaskar et al. [8] used signals from force sensitive resistors of continuous walk from healthy subjects and with Parkinson's Disease (PD). Multilayer neural network, trained with backpropagation using the Levenberg-Marquardt optimization, was used to detect if subjects had PD. Classification was performed using statistical and frequency features from data, clinical information (age, height, gait speed, weight and gender), and a mixture of both, resulting in the following classification accuracies of 64%,81% and 91% respectively.

Yoo et al. [13] described a method for human recognition through gait using BNN. The 2D human skeleton trajectory is extracted from dataset of indoor and outdoor walking subjects. The mean and variation of the gait angles for single sequence of each subject are extracted as features. After being trained, BNN is used as a classifier, identifying if the individual is recognized, acting as a biometric system.

Rani et al. [9] used ELM in multicategory classification of gait abnormalities in children, obtaining accuracies of 97.98% when using Principal Component Analysis (PCA) and 99.21% accuracy when used T-Test.

Utomo et al. [14] showed that ELM performed better as a generalization classifier model, when compared with BNN in diagnosing breast cancer. ELM resulted in sensitivity, specificity and accuracy results of 94.8%, 97.4%, 96.4% respectively. Meanwhile, BNN showed considerably lower sensitivity (84.3%) and accuracy (92.1%) and slightly higher specificity (98%).

Ma et al. [26] used Swarm Particle Optimization (SPO) to choose input bias, weights and number of neurons of hidden layer and trained the ELM algorithm to identify falls, among other activities, out of low-cost Kinect depth camera. This strategy denominated by the author as variable-length particle swarm optimization was able to achieve up to 86.83% fall detection accuracy.

ELM was chosen due to its simplicity and fast results when compared to commonly used regression algorithms, while BNN served as baseline [26]. Support Vector Machine (SVM) was not included in the comparison, since results showed to be slower in such applications [9, 28, 29].

2 Dataset

A public dataset from the Laboratory of Biomechanics and Motor Control at the Federal University of ABC, Brazil [18] was used. The dataset is composed by raw and processed lower-body kinematic data of 42 healthy volunteers, 24 young adults and 18 older adults obtained through a motion-capture system with 12 cameras (Raptor-4; Motion Analysis Corporation, Santa Rosa, CA, USA) on 10 meters over-ground walk and using dual-belt, instrumented treadmill (FIT; Bertec, Columbus, OH, USA).

Overground trials were made at 3 different speeds: self-selected comfortable speed, slow (30% slower) and fast (30% slower). Each participant's comfortable speed was determined during the 10 meters overground walking familiarization period. The 45 healthy subjects' characteristics (age, height, weight) are displayed in Table 1. The Young/Older groups division was employed by dataset authors and comprises the ranges of 27–37 and 50–84 years old, respectively.

Table 1. Subjects characteristics.

Age group	Gender	Number of persons	Mean age	Mean height (cm)	Mean weight (kg)
Older	F	8	60.1	154.76	61.094
Older	M	10	64.8	167.49	71.58
Young	F	10	25.9	162.91	58.915
Young	M	14	28.8	176.92	75.171

Figure 1 shows the anatomical markers used for obtaining kinematic data.

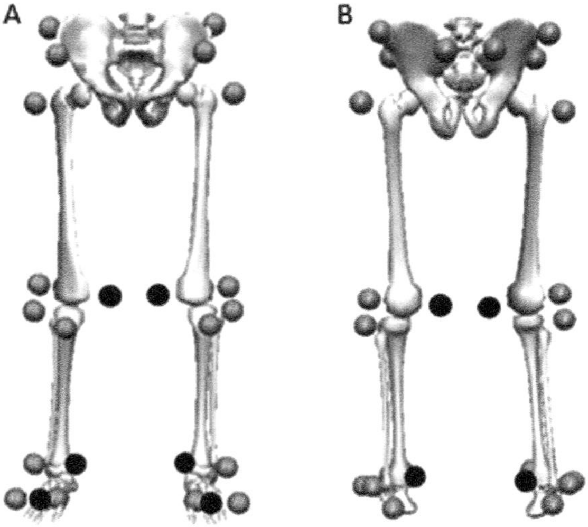

Fig. 1. Markers protocol. Anterior (A) and posterior (B) views. Markers shown black are removed during for the trial recording. Image from [18].

3 CI Methods and Training Procedures

3.1 BNN

An artificial neural network is composed of several layers containing multiple neurons per layer. Every neuron has a weight, bias and chosen activation function. The weight and bias are used to calculate a weighted sum of the input of the neuron which is then applied to the activation function.

The hidden layer neurons have non-linear activation functions, while input and output are usually chosen linear. Initially the weights, as well as the bias of the hidden neurons are attributed randomly. The neural network training consists in feeding inputs to the network and comparing the output results with the expected results. The error of every iteration is used to change the weights for the next iteration in the attempt of improving results. This procedure is referred as backpropagation and consists of a gradient descendent optimization algorithm [30, 31]. In this particular case the regression algorithm used for networks was the Levenberg Marquardt.

Deep Neural Network (DNN) comprises an extended number of hidden layers and neurons. This type of network can learn more complex data relations than NNs however, they require high processing power and memory [32]. Nevertheless, single hidden layer neural networks, with sufficient hidden units, can attain high accuracies and model complex problems [33, 34]. Figure 2 shows the structure of the single hidden layer neural network where θ_0 to θ_{100} represents the angle values from 0% to 100% of the gait cycle. One typical angle curve is represented in Fig. 3.

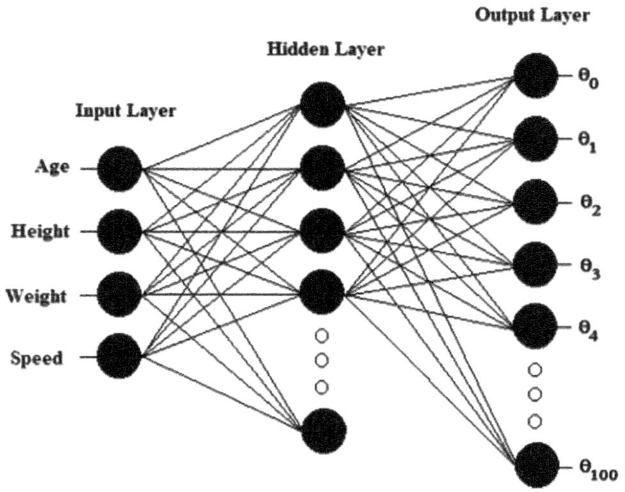

Fig. 2. Single hidden layer neural network structure.

3.2 ELM

ELM algorithm consists, in a simplistic way, of a BNN, where the input layer, weights and bias of hidden neurons are attributed randomly and kept fixed. This method uses the least squares solution and the Moore-Penrose generalized inverse matrix to obtain the weights between the hidden layer and the output (see Eq. 1).

$$\widehat{\omega}_o = H^{\dagger}T \tag{1}$$

$\widehat{\omega}_o$ is the smallest least square solution of the linear system $H\omega_o = T$ and represents the output weights. T represent the targets and $H^{\dagger}$ the Moore-Penrose generalized inverse of the hidden layer output matrix [27].

ELM can be used for classification and regression achieving the smallest training error, being a clear advantage when comparing with backpropagation algorithms that stop at local minimums. Besides its simplicity, applications in several areas have shown that ELM is much faster when compared with gradient based algorithms and produces good generalization [27, 35–37].

3.3 Training Procedures

Kinematic data, specifically the knee and hip angles, of male subjects (young and older) on overground walking for 3 different speeds (slow, comfortable and fast) was used as target to train the BNN and ELM. This data consisted on a target matrix of 42 columns (14 young males × 3 speeds) by 101 rows (marker angles normalized to 101 points) for the younger group and 54 columns (18 males × 3 speeds) by 101 rows for the older group.

The inputs fed to the CI's were age, height, weight and speed of the subject trial, all normalized between 0 and 1. The amplitudes of the marker's angles were normalized for improved results in CI methods and reliable comparison.

The accuracy was measured in terms of Mean Squared Error (MSE) for both algorithms and consists in the average squared error between the BNN output and the target. Since both network output and target are angular curves, MSE will be the mean of the angular distance between BNN results and targets. It's given by:

$$MSE = \frac{1}{N}\sum_{i=1}^{N}\left(\frac{1}{101}\sum_{k=0}^{100}\left(\theta_{t,k} - \theta_{CI,k}\right)^2\right) \tag{2}$$

Where N is the number of test samples and $\theta_{t,k}$ and $\theta_{CI,k}$ are respectively, the target value and the CI output at k% of the gait cycle.

Table 2 shows the data division used for the two methods and resumes the previous given information.

Table 2. Data normalization ranges.

Method	Input range	Target range		Accuracy measurement	Data division		
		Knee	Hip		Train (%)	Test (%)	Validation (%)
BNN	[0,1]	[0,1]	[−1,1]	MSE	70	25	5
ELM	[0,1]	[0,1]	[−1,1]	MSE	75	25	–

Different networks were generated for dominant/non-dominant limb, younger/older group for both knee and hip angles.

The CI's were implemented using MATLAB in a computer with an Intel i7-6700, 2.90 GHz processor and 16,0 GB RAM. For the neural network's creation was used the Deep Learning Toolbox™ and for ELM a self-written script.

The BNN number of hidden units was iterated from 4 to 50 and each network topology was trained and tested 3 times. For ELM the number of hidden neurons were iterated from 4 to 50 and trained/tested 50 times.

In both algorithms the activation function of the hidden layer neurons used, was the sigmoid and the output activation functions were kept linear.

4 Results and Discussion

The most accurate results for the BNN and ELM are shown in Tables 3 and 4 for young and older man dominant limb and young and older man non-dominant limb respectively.

Table 3. BNN's and ELM best accuracies, young and older man dominant limb.

Limb	Method	Age group	Train accuracy MSE	Test accuracy MSE	Validation accuracy MSE
Knee	BNN	Young	**0.0001**	**0.0022**	0.0009
		Older	**0.0010**	**0.0022**	0.0027
	ELM	Young	0.0036	0.0061	-
		Older	0.0045	0.0089	-
Hip	BNN	Young	**0.0015**	**0.0069**	0.0023
		Older	**0.0025**	0.0784	0.0027
	ELM	Young	0.0463	0.0215	-
		Older	0.0330	**0.0349**	-

Table 4. BNN's and ELM accuracies, young and older man non-dominant limb.

Limb	Method	Age group	Train accuracy MSE	Test accuracy MSE	Validation accuracy MSE
Knee	BNN	Young	**0.00004**	**0.0010**	0.0006
		Older	**0.00008**	**0.0011**	0.0013
	ELM	Young	0.0037	0.0078	-
		Older	0.0034	0.0094	-
Hip	BNN	Young	**0.0010**	**0.0021**	0.0029
		Older	**0.0002**	**0.0030**	0.0038
	ELM	Young	0.0394	0.0330	-
		Older	0.0378	0.0544	-

The best test results for the dominant knee and hip networks are shown to be for BNN except for the results for the older group hip, where ELM performed slightly better. For non-dominant hip and knee BNN presented better results for every situation. Nevertheless, in both dominant and non-dominant cases, similar performances were obtained with the exception of the hip results for young subjects.

Table 5. BNN's and ELM topology for best test accuracies.

Limb	Age group	Dominant/Non-Dominant	Nº hidden neurons	
			BNN	ELM
Knee	Young	D	16	5
		ND	17	4
	Older	D	8	4
		ND	20	6
Hip	Young	D	20	4
		ND	11	4
	Older	D	9	4
		ND	15	4

Table 5 presents the number of hidden neurons that resulted in the best accuracies for the several BNN and ELM.

Two subjects outside of the training and validation set were used to test CI's capabilities of generating accurate and smooth subject adapted joint angle curves. Choice of the subjects for every net was based on the overall difference from his characteristics (age, height, weight, gait speed) to the mean characteristics of subjects used in the train. The test subject that showed to have lower differences between his characteristics and the train mean characteristics will be referred to as, the closest to the range test subject. On the contrary, the subject who showed greater difference will be identified as the furthest from range subject. Table 6 contains the MSE of the joint angle curves generated by the CI's to the real test subject joint angle curve for the subject closest to the range. Table 7 contains the MSE for the subject furthest from range.

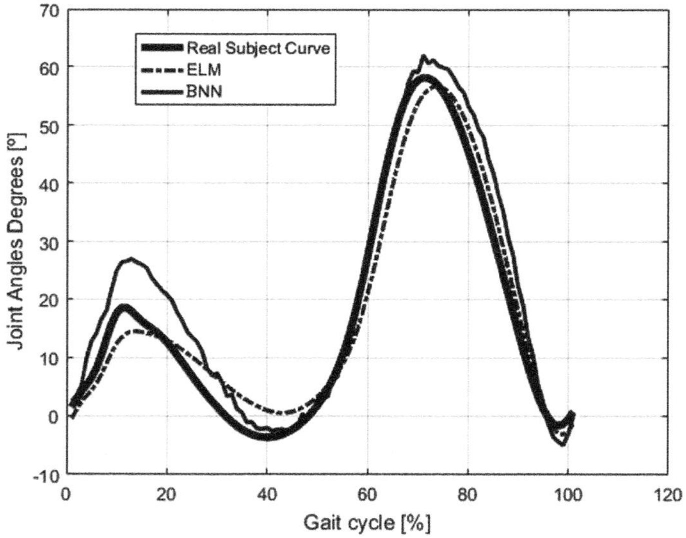

Fig. 3. Dominant joint angle knee curve generation for young subject and real subject joint angle knee curve.

Figures 3 and 4 show the real flexion/extension for dominant knee and hip of the test subject furthest from the range. Both figures also have the respective subject CI generated joint angle curve. Figures 4 and 5 present the same data for test subject furthest from the range.

For the case of the test subject furthest from mean characteristic, all knee results are better or very similar to the ELM. Hip results show better performance of the BNN. For the closest to the range test subject most results are favorable to the BNN. Nevertheless, in both dominant and non-dominant cases, BNN's weren't capable of generating smooth joint angle curves neither for knee nor the hip (see Figs. 3, 4, 5 and 6). ELM presents, in every case, smooth joint angle curves.

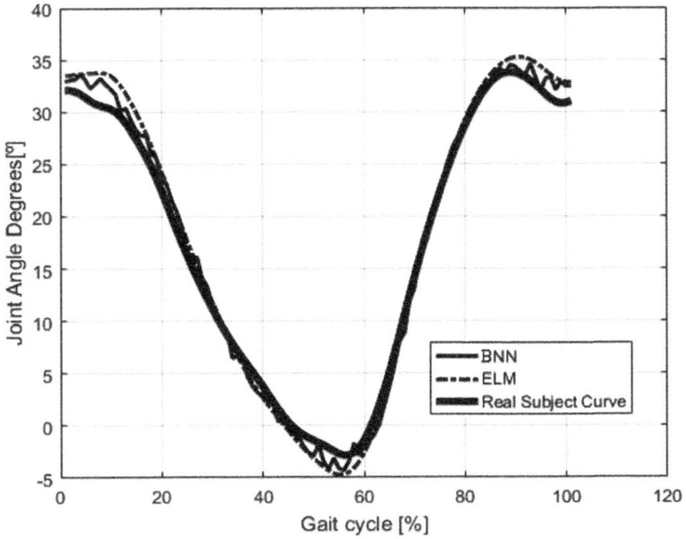

Fig. 4. Dominant joint angle hip curve generation for young subject and real subject joint angle hip curve.

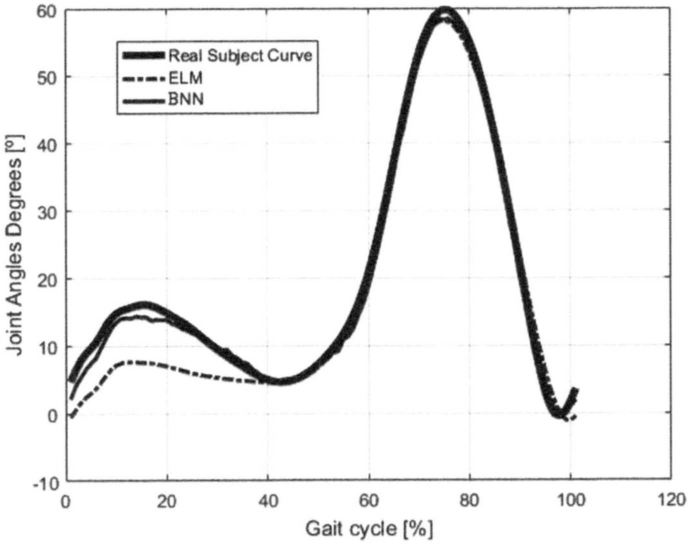

Fig. 5. Dominant joint angle knee curve generation for young subject and real subject joint angle knee curve.

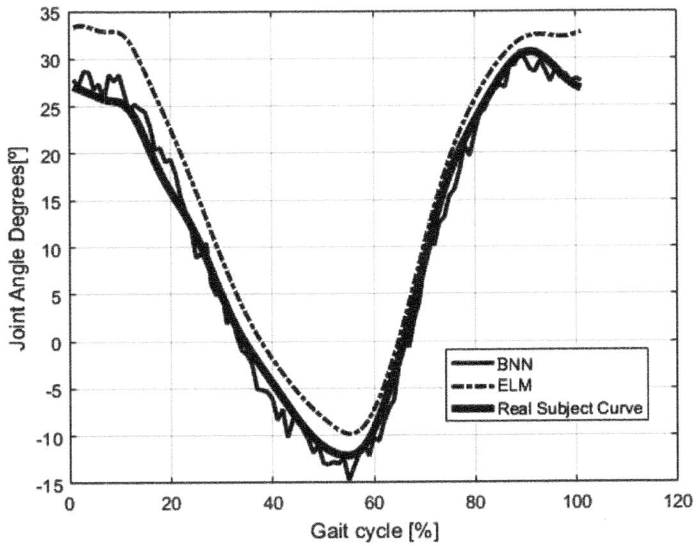

Fig. 6. Dominant joint angle hip curve generation for young subject and real subject joint angle hip curve.

Figures 7, 8, 9 and 10 show the joint angles prediction for test subjects (age 33, height 179 cm, weight 75 kg in the case of the knee and age 25, height 174 cm, weight 83 kg for hip) at the 3 distinct speeds (slow, comfortable, fast) together with literature reference joint angles curve. CIs predictions and the literature joint angle curve were both compared to the real subject joint curves and Table 8 contains the MSE results as well as the trial concrete speeds.

Table 6. Error for test subject furthest from subject's mean characteristics.

Network information			Subject input information				CI's MSE	
Limb	Age Group	D/ND	Age	Height (cm)	Weight (kg)	Gait Speed (m/s)	BNN	ELM
Knee	Young	D	30	171.0	95.4	1.27	0.0053	**0.0028**
		ND	24	184.0	61.1	1.65	**0.0004**	0.0028
	Older	D	55	165.6	79.1	0.89	0.0044	**0.0006**
		ND	84	155.5	66.4	1.00	0.0009	**0.0008**
Hip	Young	D	36	182.5	64.0	1.34	0.027	**0.0021**
		ND	37	155.0	69.6	0.69	**0.001**	0.031
	Older	D	84	155.5	66.4	1.00	**0.0009**	0.043
		ND	84	155.5	66.4	1.00	**0.002**	0.003

Table 7. Error for test subject closest from subject's mean characteristics.

Network information			Subject input information				CI's MSE	
Limb	Age group	D/ND	Age	Height (cm)	Weight (kg)	Gait speed (m/s)	BNN	ELM
Knee	Young	D	33	179.3	75.9	0.96	**0.0003**	0.003
		ND	33	179.3	75.9	0.96	**0.0007**	0.002
	Older	D	68	167.0	70.3	1.30	0.029	**0.005**
		ND	68	167.0	70.3	1.30	**0.0001**	0.0016
Hip	Young	D	31	172.9	77.9	1.48	**0.001**	0.014
		ND	33	179.3	75.9	0.68	**0.005**	0.024
	Older	D	68	167.0	70.3	1.58	**0.001**	0.023
		ND	62	164.5	70.5	0.78	**0.005**	0.094

For the test subject used BNN and ELM presented significant improvements when compared with the literature curve. The literature MSE increased for faster gait speeds. BNN shown lowest error results specifically for the knee, and MSE decreased for faster gait speed. Nevertheless, joint angle curves originated were not smooth. ELM presented better or similar results as the BNN for knee, as seen in [29], and slightly worse results for the hip.

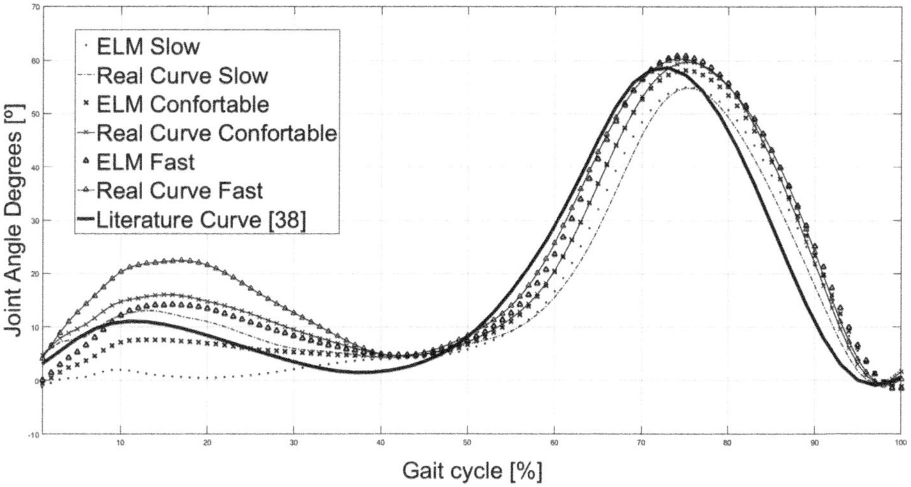

Fig. 7. Dominant joint angle knee curve generation for young subject at several speeds and literature reference curve.

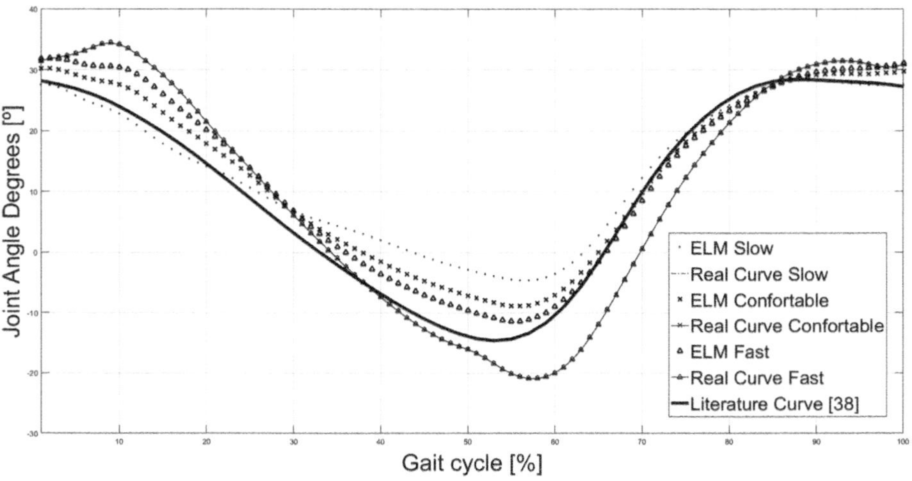

Fig. 8. Dominant joint angle hip curve generation for young subject at several speeds and literature reference curve.

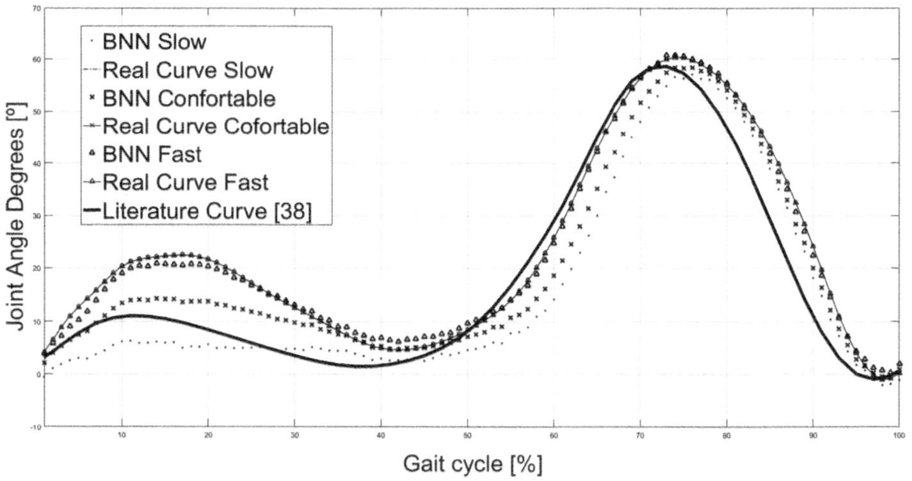

Fig. 9. Dominant joint angle knee curve generation for young subject at several speeds and literature reference curve.

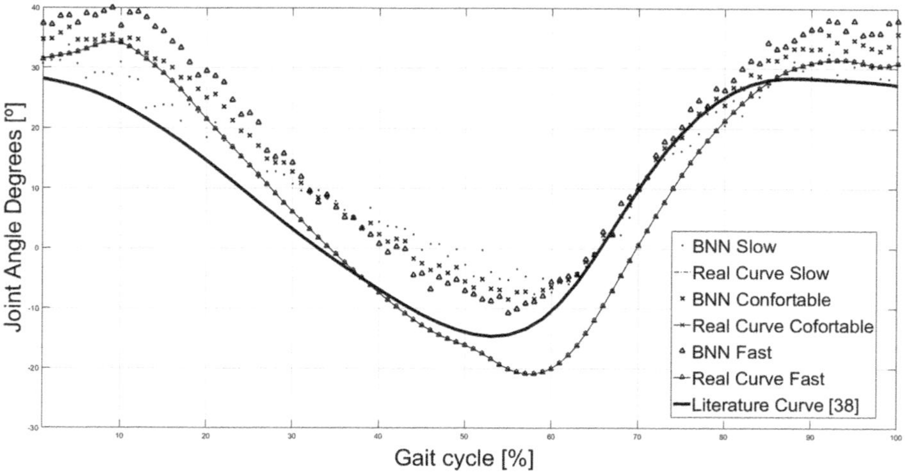

Fig. 10. Dominant joint angle hip curve generation for young subject at several speeds and literature reference curve.

Table 8. CI's and literature MSE at different gait speeds for young subjects.

Limb	Gait speed (m/s)	Literature [38] MSE	CI's MSE	
			BNN	ELM
Knee	0.68	0.017	0.014	**0.005**
	0.96	0.016	0.004	**0.003**
	1.29	0.021	**0.0002**	0.003
Hip	0.92	0.16	0.024	**0.011**
	1.32	0.12	**0.004**	0.022
	1.56	0.11	**0.004**	0.026

5 Conclusions

Both CIs generated reference curves closer to real curves than literature standard curve. The capability of generation of this healthy profile for subject characteristics and gait speed permit the use of this methodology in gait assessment in a broad range of impairments. Such results were reinforced for increasing gait speeds.

For the BNN, the lack of smoothness of the curves for subjects at the edge of the train data set was a recurrent issue, although accuracies were higher than with ELM for most cases. For the knee, ELM presented similar or better accuracy than BNN and showed better generalization capacity. Curves generated with ELM were always smooth, and the results of the furthest from mean subjects were better than with BNN, suggesting that this CI has better generalization and is the indicated CI to generate knee, subject-specific, healthy joint angle curves for both dominant and non-dominant limbs. BNN was better suited for subjects closest from mean of the data set, although generated curves need to be smoothed.

For future work, subject-specific reference joint angle curves and knee-hip graphs will be generated for patients with gait pathologies. It is expected that this methodology will allow an easier gait pathology diagnosis.

Compliance with Ethical Standards

Conflict of Interest. All authors declare that they have no conflict of interest.

References

1. Buckley, C., Alcock, L., McArdle, R., Rehman, R.Z.U., Del Din, S., Mazzà, C., Yarnall, A.J., Rochester, L.: The role of movement analysis in diagnosing and monitoring neurodegenerative conditions: insights from gait and postural control. Brain Sci. **9**, 34 (2019)
2. Zeng, W., Wang, C.: Classification of neurodegenerative diseases using gait dynamics via deterministic learning. Inf. Sci. **317**, 246–258 (2015)
3. Morris, R., Lord, S., Bunce, J., Burn, D., Rochester, L.: Gait and cognition: mapping the global and discrete relationships in ageing and neurodegenerative disease. Neurosci. Biobehav. Rev. **64**, 326–345 (2016)
4. Minakaki, G., Canneva, F., Chevessier, F., Bode, F., Menges, S., Timotius, I.K., Kalinichenko, L.S., Meixner, H., Müller, C.P., Bjoern, B.M., Casadei, N., Ries, O., Schröder, R., Winkler, J., Xiang, W., Hörsten, S., Klucken, J.: Treadmill exercise intervention improves gait and postural control in alpha-synuclein mouse models without inducing cerebral autophagy. Behav. Brain Res. **363**, 199–215 (2019)
5. Santamato, A., Ranieri, M., Cinone, N., Stuppiello, L.A., Valeno, G., De Sanctis, J.L., Fortunato, F., Solfrizzi, V., Greco, A., Seripa, D., Panza, F.: Postural and balance disorders in patients with Parkinson's disease: a prospective open-label feasibility study with two months of action observation treatment. Parkinsons Dis. (2015)
6. Ham, T.R., Farrag, M., Soltisz, A.M., Lakes, E.H., Allen, K.D., Leipzig, N.D.: Automated gait analysis detects improvements after intracellular σ peptide administration in a rat hemisection model of spinal cord injury. Ann. Biomed. Eng. **47**, 744–753 (2019)
7. Alaqtash, M., Sarkodie-Gyan, T., Yu, H., Fuentes, O., Brower, R., Abdelgawad, A.: Automatic classification of pathological gait patterns using ground reaction forces and machine learning algorithms. In: Annual International Conference of the IEEE Engineering in Medicine and Biology Society 2011, pp. 453–457. IEEE (2011)
8. Alaskar, H., Hussain, A.: Prediction of Parkinson disease using gait signals. In: 11th International Conference on Developments in eSystems Engineering (DeSE) 2018, Cambridge, United Kingdom, pp. 23–26 (2018)
9. Rani, M., Arumugam, G.: Children abnormal GAIT classification using extreme learning machine. Glob. J. Comput. Sci. Technol. [S.l.] (2010)
10. Joshi, S., Shenoy, D., Rrashmi, P.L., Venugopal, K.R., Patnaik, L.M.: Classification of Alzheimer's disease and Parkinson's disease by using machine learning and neural network methods. In: Second International Conference on Machine Learning and Computing 2010, Bangalore, pp. 218–222 (2010)
11. Chen, X.W., Lin, X.: Big data deep learning: challenges and perspectives. IEEE Access **2**, 514–525 (2014)

12. Lai, D.T., Begg, R.K., Palaniswami, M.: Computational intelligence in gait research: a perspective on current applications and future challenges. IEEE Trans. Inf. Technol. Biomed. **13**(5), 687–702 (2009)
13. Yoo, J.H., Hwang, D., Moon, K.Y., Nixon, M.S.: Automated human recognition by gait using neural network. In: 2008 First Workshops on Image Processing Theory, Tools and Applications, pp. 1–6. IEEE (2008)
14. Utomo, C.P., Kardiana, A., Yuliwulandari, R.: Breast cancer diagnosis using artificial neural networks with extreme learning techniques. Int. J. Adv. Res. Artif. Intell. **3**(7), 10–14 (2014)
15. Manal, K., Stanhope, S.J.: A novel method for displaying gait and clinical movement analysis data. Gait Posture **20**, 222–226 (2004)
16. Toliopoulos, P., Hagemeister, N., Fuentes, A., Desmeules, F., Vendittoli, P.A.: Normal knee gait kinematics: describing a normal cohort and illustrating differences between genders. Clin. Neurophysiol. **46**, 283 (2016)
17. Horst, F., Lapuschkin, S., Samek, W., Müller, K.-R., Schöllhorn, W.I.: A public dataset of overground walking kinetics and full-body kinematics in healthy individuals. Mendeley Data v1 (2018)
18. Fukuchi, C.A., Fukuchi, R.K., Duarte, M.: A Public Dataset of Overground and Treadmill Walking Kinematics and Kinetics in Healthy Individuals (2018)
19. Summa, S., De Peppo, F., Petrarca, M., Caccamo, R., Carbonetti, R., Castelli, E., Adorisio, D.O.: Gait changes after weight loss on adolescent with severe obesity after sleeve gastrectomy. Surg. Obes. Relat. Dis. **15**, 374–381 (2019)
20. Begg, R., Kamruzzaman, J.: A machine learning approach for automated recognition of movement patterns using basic, kinetic and kinematic gait data. J. Biomech. **38**(3), 401–408 (2005)
21. Horst, F., Mildner, M., Schöllhorn, W.I.: One-year persistence of individual gait patterns identified in a follow-up study–A call for individualized diagnose and therapy. Gait Posture **58**, 476–480 (2017)
22. Fathima, S.S.S., Banu, W.R.: Abnormal walk identification for systems using gait patterns. Biomed. Res. India **27**, 112–117 (2016)
23. Schwartz, M.H., Rozumalski, A., Trost, J.P.: The effect of walking speed on the gait of typically developing children. J. Biomech. **41**(8), 1639–1650 (2008)
24. Kang, H.G., Dingwell, J.B.: Separating the effects of age and walking speed on gait variability. Gait Posture **27**(4), 572–577 (2008)
25. Luu, T.P., Low, K.H., Qu, X., Lim, H.B., Hoon, K.H.: An individual-specific gait pattern prediction model based on generalized regression neural networks. Gait Posture **39**(1), 443–448 (2014)
26. Ma, X., Wang, H., Xue, B., Zhou, M., Ji, B., Li, Y.: Depth-based human fall detection via shape features and improved extreme learning machine. IEEE J. Biomed. Health Inf. **18**(6), 1915–1922 (2014)
27. Huang, G.B., Zhu, Q.Y., Siew, C.K.: Extreme learning machine: a new learning scheme of feedforward neural networks. Neural Netw. **2**, 985–990 (2004)
28. Vieira, A., Ribeiro, B., Sobral, H., Coimbra, A.P., Crisóstomo, M., Ferreira, J.P., Liu, T.: Dynamic human gait VGRF reference profile generation via extreme learning machine. In: International Joint Conference on Neural Networks (IJCNN) 2018, pp. 1–7. IEEE (2018)
29. Ferreira, J.P., Vieira, A., Ferreira, P., Crisostomo, M., Coimbra, A.P.: Human knee joint walking pattern generation using computational intelligence techniques. Neural Comput. Appl. **30**, 1701–1713 (2018)
30. Li, H., Zhang, Z., Liu, Z.: Application of artificial neural networks for catalysis: a review. Catalysts **7**(10), 306 (2017)

31. Mair, C., Kadoda, G., Lefley, M., Phalp, K., Schofield, C., Shepperd, M., Webster, S.: An investigation of machine learning based prediction systems. J. Syst. Softw. **53**(1), 23–29 (2000)
32. Abdelsalam, A.M., Boulet, F., Demers, G., Langlois, J.P., Cheriet, F.: An efficient FPGA-based overlay inference architecture for fully connected DNNs. In: International Conference on ReConFigurable Computing and FPGAs (ReConFig) 2018, pp. 1–6. IEEE (2018)
33. Hinton, G.E., Srivastava, N., Krizhevsky, A., Sutskever, I., Salakhutdinov, R.R.: Improving Neural Networks by Preventing Co-Adaptation of Feature Detectors (2012)
34. Osmanović, A., Halilović, S., Ilah, L.A., Fojnica, A., Gromilić, Z.: Machine learning techniques for classification of breast cancer. In: World Congress on Medical Physics and Biomedical Engineering 2018, pp. 197–200. Springer, Singapore (2019)
35. Matias, T., Souza, F., Araújo, R., Antunes, C.H.: Learning of a single-hidden layer feedforward neural network using an optimized extreme learning machine. Neurocomputing **129**, 428–436 (2014)
36. Tang, J., Deng, C., Huang, G.B.: Extreme learning machine for multilayer perceptron. IEEE Trans. Neural Networks Learn. Syst. **27**(4), 809–821 (2016)
37. Karakuzu, C., Yüzgeç, U.: A simple heuristic approach to improve performance of extreme learning machine. In: International Conference on Advanced Technologies, Computer Engineering and Science (ICATCES 2018), 11–13 May, Safranbolu, Turkey (2018)
38. Darras, N.: Deviation graphs and deviation for clinical gait evaluation. https://sites.google.com/site/gaitanalysisadplot/file-cabinet. Accessed 26 Jan 2019

Assistive Smart Cane (ASCane) for Fall Detection: First Advances

Pedro Mouta(✉), Nuno Ferrete Ribeiro, Cristina P. Santos, and Rui Moreira

Center for MicroEletroMechanical Systems (CMEMS), University of Minho, 4800-058 Guimarães, Portugal
pmfmouta@gmail.com, nuno.ferrete.ribeiro@gmail.com, a71335@alunos.uminho.pt, cristina@dei.uminho.pt

Abstract. The development of fall detection systems with the capability of real-time monitoring is necessary considering that a large amount of people die and suffer severe consequences from falls. Due to their advantages, daily life accessories can be a solution to embed fall-related systems, and canes are no exception. In this paper, it is presented a cane with fall detection abilities. The ASCane is instrumented with an inertial sensor which data will be tested with three different fixed multi-threshold fall detection algorithms, one dynamic multi-threshold and machine learning methods from the literature. They were tested and modified to account the use of a cane. The best performance resulted in a sensitivity and specificity of 96.90% and 98.98%, respectively.

Keywords: Fall detection · Machine learning · IMU · Activities of daily living

1 Introduction

Falls are the second main reason of death by accident worldwide [1]. The estimated medical costs in the United States of America (U.S.A.) concerning fatal and nonfatal falls, in 2015, were approximately $32 billion, and the annual average cost of treating individuals due to injuries from a fall is approximately $20.000 [2,3]. By 2020, expenses linked to injuries from falls to senior citizens are expected to cost $43.8 billion [4].

Researchers have proposed several different solutions regarding fall-related systems. Most of the developed projects focus on fall detection and employ methods supported by vision, wearable and environmental approaches. Commonly, different sensors are attached to the subject's body. However, the wearable system weights on the individual and hinder its flexibility. On the contrary, both optical and environmental device approaches free the subject of sensors, but require a pre-built infrastructure restricting the subject movements [5].

Populations with motor impairment normally use assistive devices that auxiliary their gait. Nowadays, more than 4 million people in the U.S.A. use a cane and its usage will increase with the elderly population growth. The size is another advantage to choose canes as fall-related systems [6–8].

J. Henriques et al. (Eds.): MEDICON 2019, IFMBE Proceedings 76, pp. 1669–1684, 2020.
https://doi.org/10.1007/978-3-030-31635-8_204

Bourke et al. [9], developed a threshold based algorithm using exclusively accelerometry data from the trunk or thighs for the computation of the acceleration Sum Vector Magnitude. If the upper or the lower individual threshold is surpassed, a fall is detected with an accuracy of 100%.

Bourke et al. [10], also identified three gyroscopic features from the trunk, namely the Sum Vector Magnitude (ω_{res}), the resultant angular acceleration (θ_{res}) and the resultant change in trunk-angle (α_{res}). When their thresholds are exceeded, a fall is also detected with an accuracy of 100%.

The algorithm introduced by Kangas et al. [11] is based on the analysis of 5 acceleration parameters from the wrist, head or waist. With the combination of the Sum Vector Magnitude, the Dynamic Sum Vector, SV_{TOT}, the Vertical acceleration, Z_2, the SV_{maxmin} and the final posture, which is detected 2 s after the impact, a sensitivity of 97.5% and a specificity of 100% was obtained for the waist.

Fixed threshold-based algorithms can be insufficient to achieve the main goal of fall-related systems due to inter and intra-variability of subjects, and limited sample size [9–12]. Thus, these methods should be adaptive and account for variability. Otanasap et al. [13] developed a dynamic threshold algorithm by the means of accelerometry data. A Fixed Threshold (FT) is computed based in the data acquired from the subject while performing Activities of Daily Living (ADL), ADL_{acc}. Secondly, the Dynamic Threshold (DT) is formulated by the FT added by a standard deviation calculated with the data gathered in the last second. The algorithm outputs a percentage which discriminates the possibility of a fall, reaching results of 97.4%, 99.5% and 95.3% for accuracy, sensitivity and specificity, respectively.

Xu et al. [5] reviewed and compared fall detection algorithms on the most cited works. It was found an increase of machine learning algorithms in recent years, namely the Support Vector Machine (SVM) and the Decision Tree. The accuracy of the algorithms are relatively high since most accuracies are above 90%, ranging between 79.6% and 100%.

The main objective of this work is to develop a cane able to detect falls. This manuscript focus on applying the described algorithms over the data collected on a cane being used as an assistive device for eleven healthy users. The algorithms are modified to account for the use of a cane. The remainder of this paper is organized as follows. Section 2 provides a complete system overview which includes the hardware installed into the cane, its purpose, the different fall detection algorithms considered and the experimental protocol for data acquisition. In Sect. 3, the results attained with the different fall detection algorithms are presented. Finally, in Sect. 4, a discussion of the achieved results is accomplished and the paper is concluded in Sect. 5.

2 Methods and Materials

2.1 System Overview

The ASCane, Fig. 1, uses an inertial sensor located on the upper part, an 9-axis Motion Processing Unit (MPU-9250), which is widely embedded in systems related to human motion [3,5,12,14,15]. Chen et al. [16] studied acceleration readings in different places of a cane under the same falling process and concluded that the amplitudes of the acquired data in the upper part of the device were higher than the other locations. The higher the variation, the easier is to observe discriminative characteristics of the signal.

As full scale range, it was used ± 8 g for the accelerometer and $\pm 300°/s$ for the gyroscope [15,17]. When the cane is being used, all its sensors are operating at 200 Hz, which is higher then the sampling rate in other cane related works [3, 5,16]. The acquired data was saved in a memory card and the microcontroller STM32F303K8 was used.

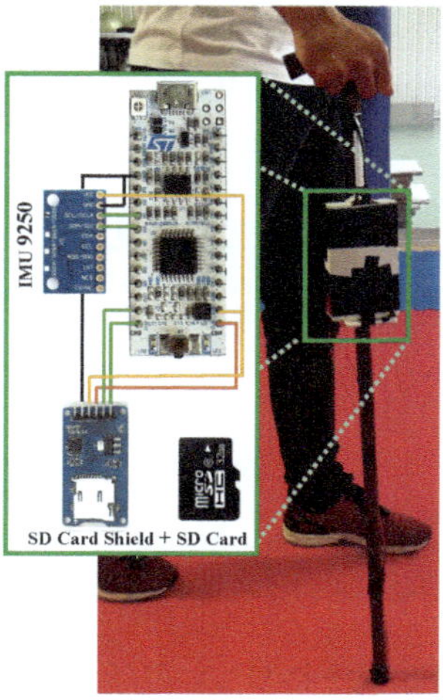

Fig. 1. Set-up of the ASCane and its location in a healthy user.

2.2 Fall Detection Algorithms

The complete fixed threshold algorithms are exposed in Fig. 2 and the original threshold values for each feature, body location and study are summarized in Table 1. The algorithm introduced by Otanasap et al. [13], which will also be tested and adapted to the ASCane, is presented in Fig. 3.

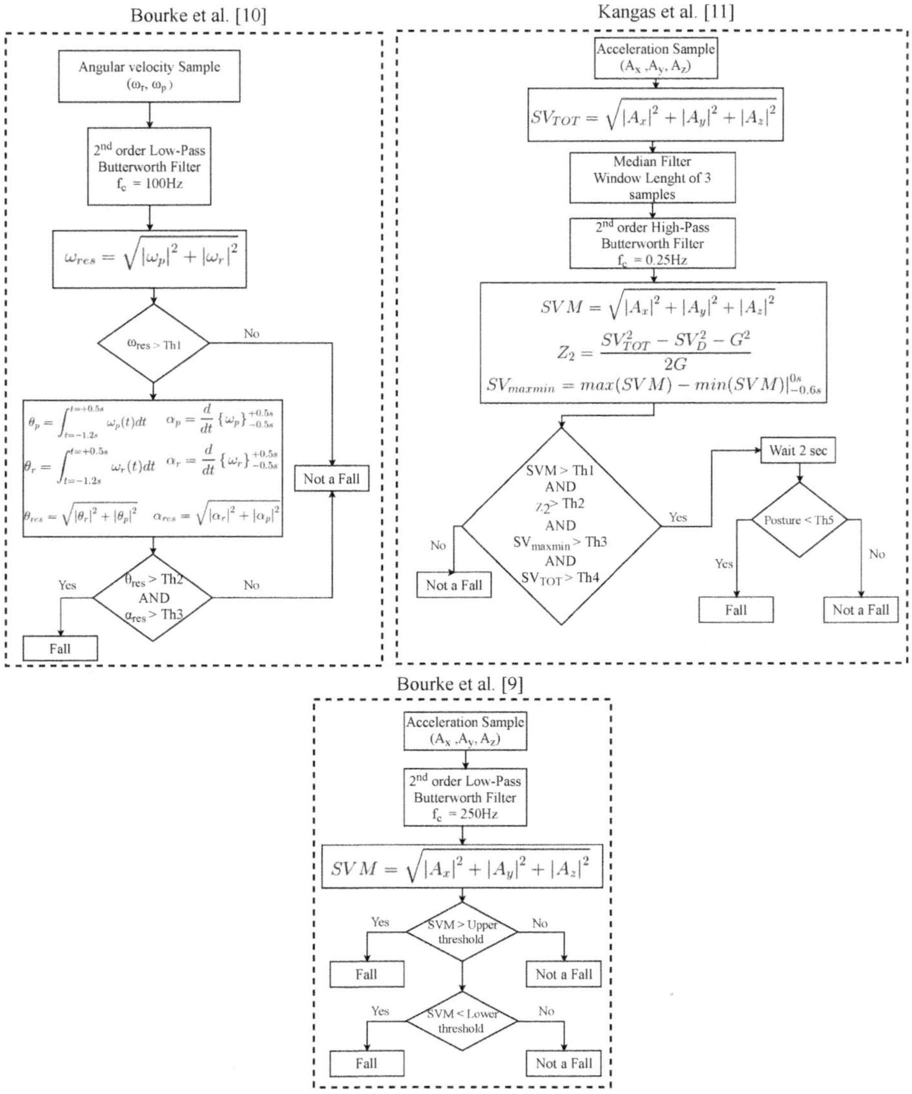

Fig. 2. Three different fixed threshold algorithms implemented into the ASCane.

A search was conducted in order to uncover which machine learning algorithms authors use to classify falls. As result, 4 articles were selected [18–21], and their corresponding features are presented in Table 2. Data were then divided in two different classes: Fall and ADL samples. Afterwards, 70% of each data were used to train the classifier and 30% to test it as indicated in Fig. 4.

Table 1. Threshold values for the different fixed threshold fall detection algorithms [9–11].

Study	Parameter	Location	Value	Type of Threshold*
Bourke et al. [9]	SVM (g)	Trunk	3.52	UFT
			0.41	LFT
		Thigh	2.74	UFT
			0.60	LFT
Bourke et al. [10]	ω_{res} (rads/s)	Trunk	3.1	UFT
	α_{res} (rads/s^2)		0.05	
	θ_{res} (rad)		0.59	
Kangas et al. [11]	SVM (g)	Waist	2.0	
		Head	2.0	
		Wrist	5.2	
	SV_D (g)	Waist	1.7	
		Head	1.2	
		Wrist	5.1	
	Z_2 (g)	Waist	1.5	
		Head	1.8	
		Wrist	3.9	
	SV_{MaxMin} (g)	Waist	2.0	
		Head	1.7	
		Wrist	6.5	

*UFT – Upper Fall Threshold. LFT – Lower Fall Threshold

Different tests were accomplished by varying the kernel type and proportion of samples between classes in the SVM Classifier. However, the best set of parameters was determined by enabling the "OptimizeHyperparameters" option in MATLAB.

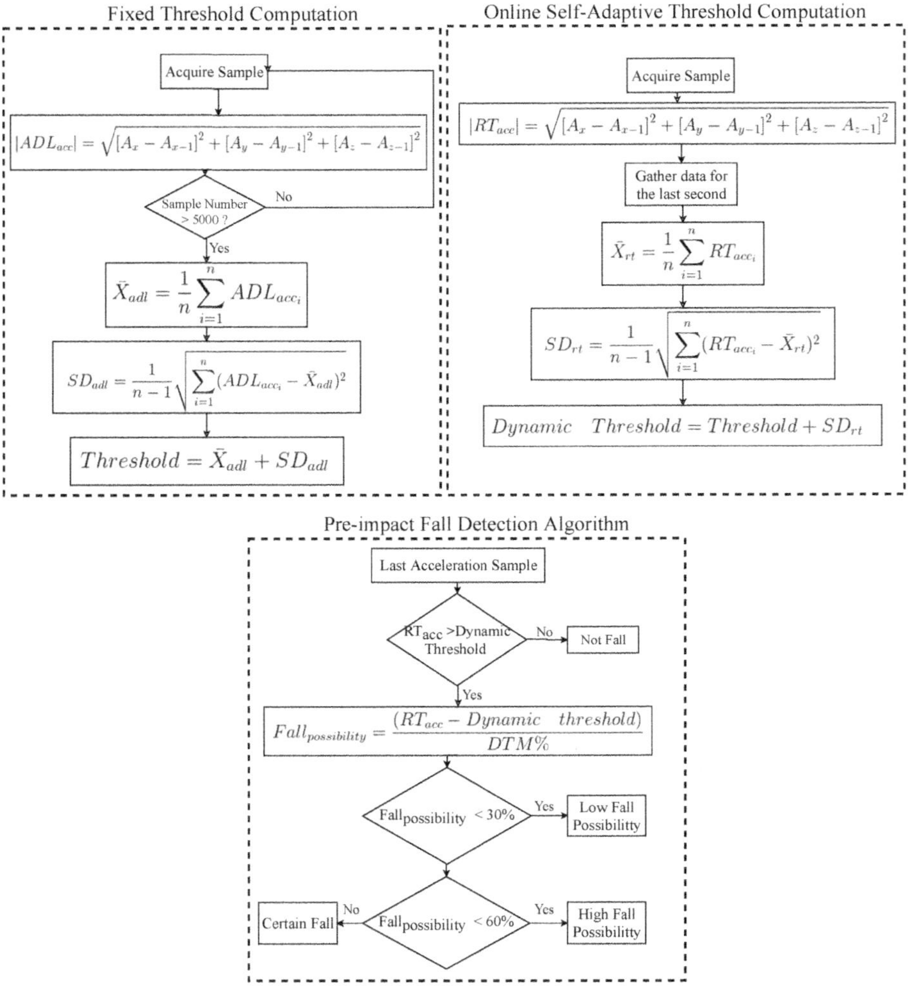

Fig. 3. Self Adaptive Threshold Algorithm presented by Otanasap et al. [13]

Features regarding ADL and falls were labeled using the parameter CV_{Fast} to mark the falling range [19]. The maximum CV_{Fast} of each fall trial was calculated and multiplied by 0.87. The samples higher than $0.87CV_{Fast}$ were considered a fall and labeled as 1.

Table 2. Summary of the features which may correlate with falls-risk in the selected fall detection algorithms [18–21]

Study	Feature name
Shibuya et al. [18]	Range of angular velocity for each individual axis (Rω.x . Rω.y and Rω.z)
	Range of acceleration for each individual axis (RA.x . RA.y and RA.z)
Liu et al. [19]	Sum Vector Magnitude (SVM)
	Fast Changed Vector (CVFast)
	Vertical Acceleration (Z_2)
Chen et al. [20]	Sum Vector Magnitude (SVM)
	Rotation angle (RA)
	Slope (SL)
	The acceleration in the xy − plane (A_{xz})
Putra et al. [21]	Sum Vector Magnitude (SVM)
	Maximum Sum Vector Magnitude (Max_{SVM})
	Minimum Sum Vector Magnitude (Min_{SVM})
	Average Sum Vector Magnitude (Avg_{SVM})
	Root mean square of the acceleration vector magnitude (RMS_{SVM})
	Acceleration exponential moving average (EMA)
	Signal magnitude area (SMA)

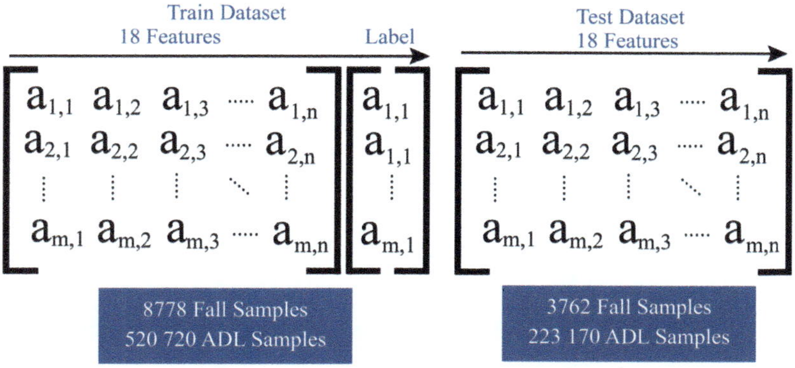

Fig. 4. Data structures achieved for each class (Fall + ADL) after differentiating the percentage for train and test of the classifier.

2.3 Experimental Protocol

A set of activities (Table 3 and Fig. 5) was executed by eleven volunteers which ranged from 22 to 29 years (24.2 ±2.6 years), with a body mass between 52 and 80 kg (70.8 ±8.23 kg) and a height of 1.51 to 1.83 m (1.73 ±0.086 m). All participants provided their written consent. Each activity was performed three times. A total of 132 simulated falls were recorder with 66 combining the subject and cane (Activities 6 and 7) and 66 only with the cane (Activities 4 and 5). Also, 99 ADL were registered (Activities 1, 2 and 3). The algorithms were implemented offline using the Matlab 2017b version.

Table 3. Activities simulated with the ASCane Prototype

Activity no.	Description
1	Walking at Normal Speed and 180° rotation (Subject + Cane)
2	Walk forward and turn right (Subject + Cane)
3	Walk forward and turn left (Subject + Cane)
4	Free Falling (Cane)
5	Thrown out (Cane)
6	Falling Forward (Subject + Cane)
7	Falling Sideways (Subject + Cane)

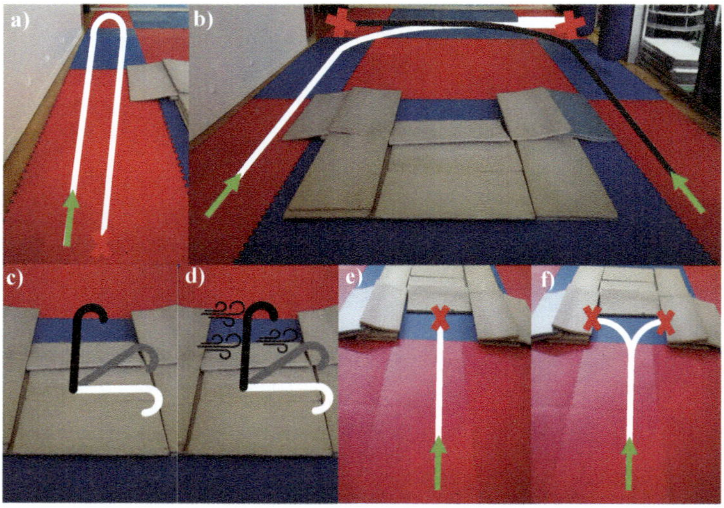

Fig. 5. Activities performed for data acquisition: (a) Activity 1; (b) Activities 2 and 3; (c) Activity 4; (d) Activity 5; (e) Activity 6; (f) Activity 7.

2.4 Evaluation of Classification Performance

Several performance indicators were calculated such as accuracy [18–20], precision [21], sensitivity [18–21], specificity [18–20], cohen kappa [22], and Matthews Correlation Coefficient (MCC) [22] to compare the different fall detection algorithms. The True Negatives (TN) correspond to the number of ADL correctly classified. True Positives (TP) are the falls proper classified. False Positives (FP) are the ADLs not right classified. Finally, False Negatives (FN) are the falls incorrectly classified.

3 Results

3.1 Original Algorithms

The algorithms were tested with the acquired data using the original thresholds. The results of the different performance indicators are summarized in Table 4.

Table 4. Performance Indicators of fall detection algorithms

Type	Study	Details	Accuracy	Precision	Sensitivity	Specificity	MCC	Kappa
Fixed Threshold	Bourke et al. [9]	[1]Trunk	0.5746	0.5708	1	0.0202	0.1074	0.023
		[1]Thigh	0.5658	0.5658	1	0	NaN[6]	0
	Bourke et al. [10]	[1]Trunk	0.8114	0.9388	0.7132	0.9394	0.6534	0.6296
	Kangas et al. [11]	[1]Waist	0.5789	0.5740	0.9922	0.0404	0.1105	0.0367
		[1]Head	0.5658	0.5658	1	0	NaN[6]	0
		[1]Wrist	0.5789	0.9714	0.2636	0.9899	0.3485	0.2282
Dynamic Threshold	Otanasap N. [13]	[2]0.0740	0.5658	0.5658	1	0	NaN[6]	0
Machine Learning	Support Vector Machine	[3]1:60 [4]RBF	0.9913	0.9744	0.4863	0.9998	0.6852	0.6449
		[3]1:1.6 [4]RBF	0.9154	0.9390	0.8347	0.9660	0.8211	0.8178
		[3]1:1.6 [4]Linear	0.9105	0.9329	0.8273	0.9627	0.8106	0.8070
		[3]1:1.6 [5]Optimized	0.9121	0.9358	0.8289	0.9643	0.8141	0.8105

[1]Location; [2]Fixed threshold; [3]ADL:Fall Proportion; [4]Kernel Function; [5]Optimized with MATLAB; [6]Not a Number

The algorithm introduced by Bourke et al. [9] presented similar results for the two sets of thresholds described (Table 4). It detected a fall in 100% of the cases.

However, all or almost all the ADLs performed were also considered a fall with a Specificity of 0 and 2.02% for the thighs and trunk, respectively. With the method presented by Kangas et al. [11], the results are similar to the ones reached by Bourke et al. [9] in the three different sets of thresholds (Table 4). Nevertheless, while with the waist and head thresholds a fall is detected in 99.22% and 100% of the cases, respectively, the thresholds for the wrist detected only 26.36% of falls. Using the algorithm from Bourke et al. [10], it resulted in an overall higher performance compared to the remaining fixed threshold algorithms (Table 4), achieving an accuracy of 81.14%. Similar to Bourke et al. [9] and Kangas et al. [11], with the dynamic algorithm proposed by Otanasap et al. [13], a fall was spotted 100% of the cases, yet, the entirely ADL dataset was also assessed as a fall (Table 4). With the machine learning approach, the best set of parameters achieved an accuracy of 91.54%, sensitivity of 83.47% and specificity of 96.60%. The results for all accomplished tests are revealed in Table 4.

3.2 Modified Algorithms

Both falls and ADLs present a similar acceleration maximum as identified in Table 5 and Fig. 6, which explains why the algorithm by Bourke et al. [9] was not able to detect ADLs. Thus, the algorithm was tested with a single lower threshold. The corresponding results are presented in Table 6. On the contrary, the (ω_{res}) does not exhibit the same behavior as the acceleration.

Table 5. Maximum, minimum, mean and standard Deviation of the acceleration Sum Vector Magnitude and the angular velocity for the intentional falls and ADL trials

Features	Type of activity	Maximum	Minimum	Mean	Standard deviation
SVM (g)	ADL	13.8357	0.1351	1.0557	0.3427
	Fall	13.8980	0.0681	3.8644	3.8296
$\omega_{res(rad/s)}$	ADL	3.5636	0	0.6711	0.5440
	Fall	12.6706	0	2.7512	1.89002

The algorithm present by Otanasap et al. [13] was also not able to detect ADLs. An analysis of the features behaviour throughout the trials was accomplished, Fig. 7, and the algorithm was tested with several different FT which results are indicated in Table 7.

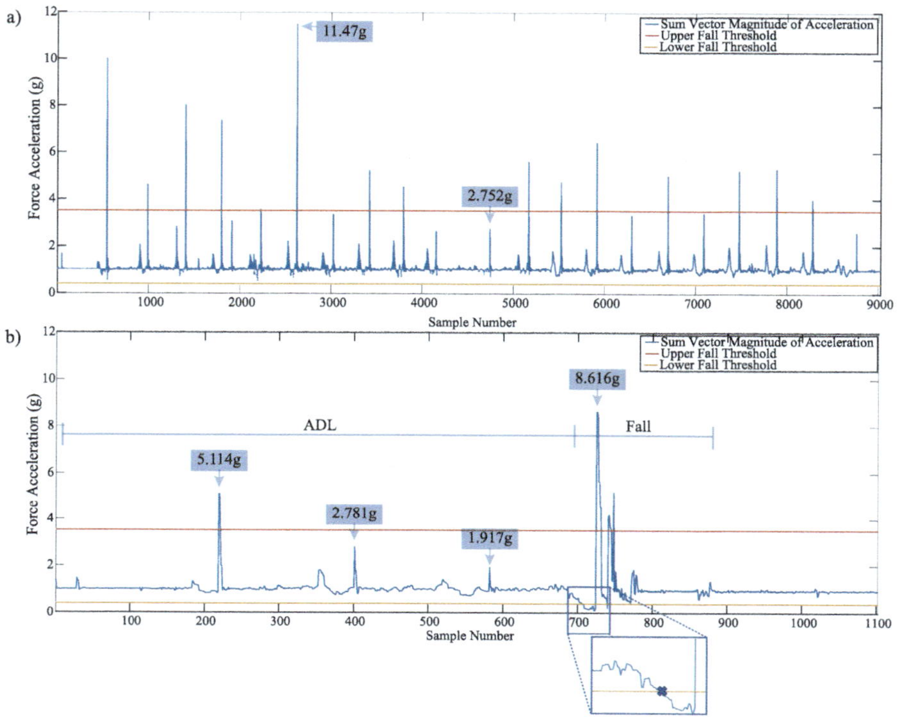

Fig. 6. Sum Vector Magnitude for: (a) One ADL trial; (b) One intentional fall trial with the corresponding fall detection as a result of the lower threshold of 0.41 g.

Table 6. Performance indicators of the fall detection algorithm proposed by Bourke et al. [9] tested only with a single lower threshold

Lower threshold	Accuracy	Precision	Sensitivity	Specificity	MCC	Kappa
0.41	0.9190	0.8815	0.9917	0.8222	0.8406	0.8312
0.2	0.9781	0.9920	0.9690	0.9898	0.9559	0.9555

Table 7. Performance Indicators of the fall detection algorithm proposed by Otanasap et al. [13] tested with different FT

FT	Accuracy	Precision	Sensitivity	Specificity	MCC	Kappa
7	0.8478	0.8444	0.9157	0.7455	0.6796	0.6756
7.2	0.8229	0.8488	0.9125	0.7679	0.6945	0.6914
7.4	0.8636	0.8750	0.8974	0.8148	0.7167	0.7163
7.6	0.8837	0.9155	0.8784	0.8909	0.7648	0.7639
7.8	0.8819	0.9104	0.8714	0.8947	0.7633	0.7624
8	0.8810	0.9206	0.8529	0.9138	0.7643	0.7619

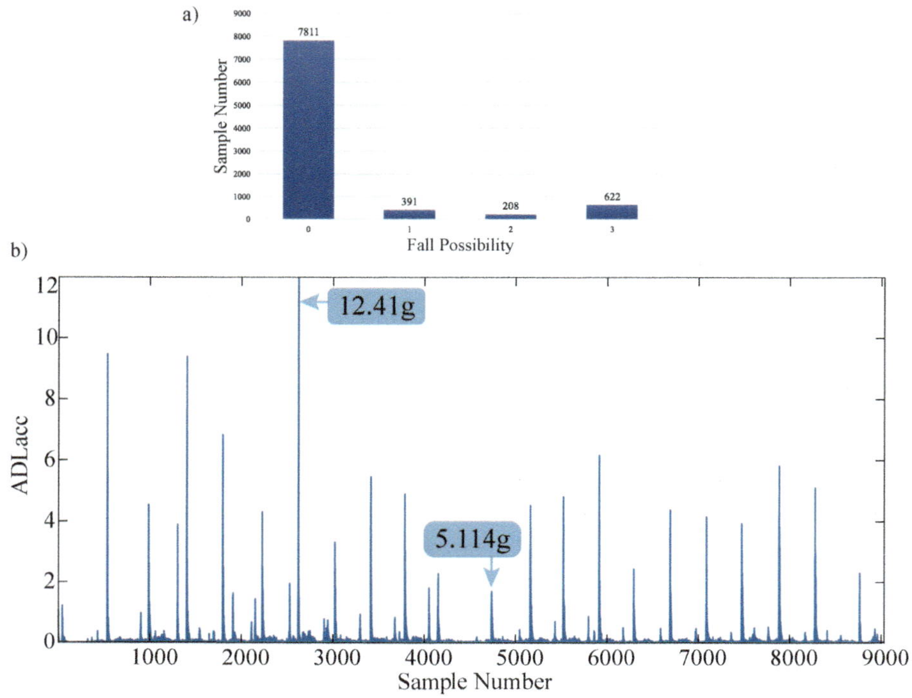

Fig. 7. (a) Fall possibility computed by the algorithm proposed by [13] during an ADL trial (b) ADL_{acc} of the same trial.

4 Discussion

The algorithm introduced by Bourke et al. [9] considered a fall in almost all ADL trials. This indicates that the original thresholds are not appropriate or adapted to canes considering that when the cane hits the ground, there is a substantial increase in the Sum Vector Magnitude, Fig. 6(a), similarly to the trials of falls, Fig. 6(b). Since the upper threshold is frequently surpassed when the cane hits the ground, contrarily to the lower threshold, Fig. 6(a), the algorithm was tested with different lower thresholds. Consequently, the performance was significantly higher, Table 6. Thus, the upper thresholds with Sum Vector Magnitude on canes are not recommended due to the aforementioned problem. This feature is also directly related to the force applied to the cane for each strike with the floor, and it is different for every gait cycle (Fig. 6(a)).

Regarding the study from Kangas et al. [11], none of the set of thresholds are suitable to canes. Both waist and head thresholds detect falls in almost ADL trials and the wrist thresholds only detects a fall in 26% of the cases (Table 3). Considering that the five features used to evaluate the trial are accelerometry based, all of them will be affected when the cane hits the ground. Therefore, using this algorithm with the original thresholds is inefficient. Proposing a new set of threshold requires a more complex analysis of the data.

Due to the fact that peak values of ω_{res} for the recorded ADLs and falls are different (5), the first threshold of 3.1 rad/s (ω_{res}) is hardly ever surpassed as can be seen in Fig. 8 for one trial. Thus, the algorithm described by Bourke et al. [10] presented the best results among the fixed threshold fall detection algorithms. However, when using a single lower acceleration threshold of 0.2 g, the accuracy increased to 97.81%, which is better than the results attained by the aforementioned algorithms.

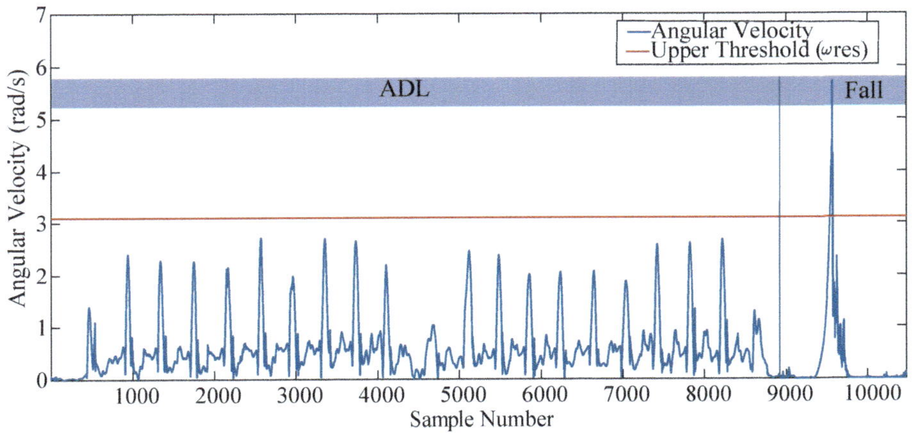

Fig. 8. Angular Velocity of an ADL trial versus a simulated fall trial.

Since the algorithm introduced by [13] is mainly based upon the ADL_{acc}, it is expected a lower performance compared to the results stated in this study because this feature is accelerometry based.

As seen in Fig. 7(b), during an ADL trial, the ADL_{acc} surpasses the fixed threshold numerous times as proven by the fall possibility computed and plotted in Fig. 7(a). Thus, this method is not optimized for cane systems with the original FT. Consequently, the dynamic algorithm was tested with several different FT (Table 7). The best performance was achieved by a new FT of 7.6 g.

Class imbalance is a commonly problem faced in data mining due to imbalanced datasets [21]. In this situation, the amount of samples from ADL is immensely larger than the number of fall samples with a proportion of 60:1. From Table 4, when the classifier was trained with an imbalanced dataset, it achieved an accuracy of 99.13%. However, the classifier is overfitting the data. Afterwards, when the classifier was trained with a proportion of 1:1.6 (Table 4), the sensitivity improved by almost 40% in the three other cases. However, when using the RBF kernel, the best result in this domain was achieved with a specificity and sensitivity of 96.60% and 83.47%, respectively.

Comparing the MCC and Kappa values from the implemented algorithms, the embedment of a single lower threshold of 0.2 g is more desirable (MCC = 95.59%; Kappa = 95.55%). This method surpasses the values of the machine learning implementation which has a range of MCC between 0.68 and 0.82 and a Kappa between 0.69 and 0.82.

5 Conclusions

This paper describes and analyses the results of five fall detection algorithms implemented in the ASCane. When using the original fixed thresholds, falls were not detected effectively because accelerometry based features were affected by the impact of the cane on the ground on each gait cycle. This event has similar acceleration values to a fall and justifies the low accuracy results. The dynamic threshold method was also inefficient in detecting ADLs since it always considered them as a fall. The application of Support Vector Machine achieved great results when compared to the dynamic and fixed threshold algorithms. After class balancing, a sensitivity, specificity and MCC of 83.47%, 96.60% and 82.11% were obtained, respectively. However, the best performance was achieved by the algorithm proposed by Bourke et al. [9] that was modified by the authors. With a single lower threshold of 0.2 g, values of sensitivity, specificity and MCC were 96.90%, 98.98% and 95.59%, respectively. Results obtained from the machine learning classifier were lower when compared to the proposed method by the authors likely because of the sample labeling method used, the CV_{Fast}. This method could be inappropriate for data acquired with a cane, and may need to be improved.

Acknowledgment. This work is supported by the FCT - Fundação para a Ciência e Tecnologia - with the scholarship reference PD/BD/141515/2018, with the reference project UID/EEA/04436/2013, by FEDER funds through the COMPETE 2020 - Programa Operacional Competitividade e Internacionalização (POCI) - with the reference project POCI-01-0145-FEDER-006941.

Conflict of Interest. The authors declare that they have no conflict of interest.

References

1. World Health Organization: Falls - WHO (2018). http://www.who.int/news-room/fact-sheets/detail/falls. Accessed 1 Nov 2018
2. Burns, E.R., Stevens, J.A., Lee, R.: The direct costs of fatal and non-fatal falls among older adults United States. J. Saf. Res. **58**, 99–103 (2016). https://doi.org/10.1016/j.jsr.2016.05.001
3. Lachtar, A., Val, T., Kachouri, A., Lachtar, A., Val, T., Kachouri, A.: 3DCane: a monitoring system for the elderly using a connected walking stick. Int. J. Comput. Sci. Inf. Secur. **14**(8), 1–8 (2017)

4. Cates, B., Sim, T., Heo, H.M., Kim, B., Kim, H., Mun, J.H.: A novel detection model and its optimal features to classify falls from low- and high-acceleration activities of daily life using an insole sensor system. Sensors (Switzerland) **18**(4) (2018). https://doi.org/10.3390/s18041227
5. Xu, T., Zhou, Y., Zhu, J.: New advances and challenges of fall detection systems: a survey. Appl. Sci. **8**(3), 418 (2018). https://doi.org/10.3390/app8030418
6. Shi, G., Chan, C.S., Li, W.J., Leung, K.S., Zou, Y., Jin, Y.: Mobile human airbag system for fall protection using mems sensors and embedded SVM classifier. IEEE Sens. J. **9**(5), 495–503 (2009). https://doi.org/10.1109/JSEN.2008.2012212
7. Bateni, H., Maki, B.E.: Assistive devices for balance and mobility: benefits, demands, and adverse consequences. Arch. Phys. Med. Rehabil. **86**(1), 134–145 (2005). https://doi.org/10.1016/j.apmr.2004.04.023
8. Luz, C., Bush, T., Shen, X., Pruchno, R.: Do canes or walkers make any difference? Nonuse and fall injuries. Gerontologist **57**(2), 211–218 (2017). https://doi.org/10.1093/geront/gnv096
9. Bourke, A.K., O'Brien, J.V., Lyons, G.M.: Evaluation of a threshold-based tri-axial accelerometer fall detection algorithm. Gait Posture **26**(2), 194–199 (2007). https://doi.org/10.1016/j.gaitpost.2006.09.012
10. Bourke, A.K., Lyons, G.M.: A threshold-based fall-detection algorithm using a bi-axial gyroscope sensor. Med. Eng. Phys. **30**(1), 84–90 (2008). https://doi.org/10.1016/j.medengphy.2006.12.001
11. Kangas, M., Konttila, A., Lindgren, P., Winblad, I., Jämsä, T.: Comparison of low-complexity fall detection algorithms for body attached accelerometers. Gait Posture **28**(2), 285–291 (2008). https://doi.org/10.1016/j.gaitpost.2008.01.003
12. Ashfak Habib, M., Mohktar, M.S., Bahyah Kamaruzzaman, S., Seang Lim, K., Maw Pin, T., Ibrahim, F.: Smartphone-based solutions for fall detection and prevention: challenges and open issues. Sensors (Switzerland) **14**(4), 7181–7208 (2014). https://doi.org/10.3390/s140407181
13. Otanasap, N.: Pre-impact fall detection based on wearable device using dynamic threshold model. In: Proceedings of Parallel and Distributed Computing, Applications and Technologies, PDCAT, pp. 362–365 (2017). https://doi.org/10.1109/PDCAT.2016.083
14. Yan, Q., Huang, J., Luo, Z.: Human-robot coordination stability for fall detection and prevention using cane robot. In: 2016 International Symposium on Micro-NanoMechatronics and Human Science, MHS 2016 (1) (2017). https://doi.org/10.1109/MHS.2016.7824171
15. Bouten, C.V., Koekkoek, K.T., Verduin, M., Kodde, R., Janssen, J.D.: A triaxial accelerometer and portable data processing unit for the assessment of daily physical activity. IEEE Trans. Biomed. Eng. **44**(3), 136–147 (1997). https://doi.org/10.1109/10.554760
16. Chen, P.H., Li, Y.H., Chiou, C.W., Lee, C.Y., Lin, J.M.: A smart safety cane for human fall detection. Int. J. Ad Hoc Ubiquitous Comput. **20**(1), 49–65 (2015). https://doi.org/10.1504/IJAHUC.2015.071662
17. Özdemir, A.T.: An analysis on sensor locations of the human body for wearable fall detection devices: principles and practice. Sensors (Switzerland) **16**(8) (2016). https://doi.org/10.3390/s16081161
18. Shibuya, N., Nukala, B.T., Rodriguez, A.I., Tsay, J., Nguyen, T.Q., Zupancic, S., Lie, D.Y.: A real-time fall detection system using a wearable gait analysis sensor and a Support Vector Machine (SVM) classifier. In: 2015 8th International Conference on Mobile Computing and Ubiquitous Networking, ICMU 2015, pp. 66–67 (2015). https://doi.org/10.1109/ICMU.2015.7061032

19. Liu, S.H., Cheng, W.C.: Fall detection with the support vector machine during scripted and continuous unscripted activities. Sensors (Switzerland) **12**(9), 12301–12316 (2012). https://doi.org/10.3390/s120912301
20. Chen, K.H., Yang, J.J., Jaw, F.S.: Accelerometer-based fall detection using feature extraction and support vector machine algorithms. Instrum. Sci. Technol. **44**(4), 333–342 (2016). https://doi.org/10.1080/10739149.2015.1123161
21. Putra, I.P.E.S., Brusey, J., Gaura, E., Vesilo, R.: An event-triggered machine learning approach for accelerometer-based fall detection. Sensors (Switzerland) **18**(1), 1–18 (2018). https://doi.org/10.3390/s18010020
22. Misiunas, A., Meskauskas, T., Samaitien, R.: Accuracy of different machine learning type methodologies for EEG classification by diagnosis, pp. 441–448 (2019). https://doi.org/10.1007/978-3-030-10692-8_50

Multi-view Robust Gesture Recognition for Assistive Interfaces

João Paulo(✉), Pedro Girão, and Paulo Peixoto

Institute of Systems and Robotics - University of Coimbra, Coimbra, Portugal
{jpaulo,pedro.girao,peixoto}@isr.uc.pt

Abstract. In this paper, we propose a gesture recognition approach using a multi-view setup for assistive device applications. As smart assistances become a reality, the need to interact with them in a natural fashion, as we do with other humans, becomes crucial. Gestures are a fundamental modality of human interaction, being natural and intuitive. We propose a gesture recognition approach relying on upper-body joints' motions, so that individuals suffering from motor dysfunctions, that need to use wheelchairs or cannot stand, can as well interact with their smart assistive devices. To achieve this goal, we propose a robust multi-view skeleton fusion through a Kalman filtering technique, followed by an upper-body handcrafted feature extraction process. Gestures are classified using a support vector machine (SVM) classifier. Experiments with our captured dataset revealed a strong generalization from our method, and an increased performance of our multi-view fusion over the individual views.

Keywords: Assistive robotics · Skeleton fusion · Gesture recognition · Machine learning

1 Introduction

Gestures are a fundamental modality of human interaction. In the last decade, we have started to see a common presence, in our daily lives, of technologies like computer interfaces, robotics, and smart surveillance, which have been seeing a rapid advancement. In the assistive robotics domain, in particular, gesture recognition helps individuals interact with their assistive devices without the need for direct physical interaction, a major advantage for disabled people. To interact with such technologies, as we do between humans, gesture recognition comes as a means to interpret human intentions and relaying them to the machine [27].

Gesture recognition has been a long-standing challenge in computer vision. Major modalities used for gesture recognition include RGB videos [8,23,26], optic flow [11,24], and skeleton sequences. When comparing with RGB videos and optic flow, skeleton sequences are computationally efficient. Furthermore, skeleton sequences allow for a better representation of a dataset of actions or gestures, due to being invariant to information pertaining to clothing, illumination, or background context since skeleton data does not incorporate these [14,22].

Taking into account the skeleton representation, in this paper, we propose an upper-body gesture recognition based on a robust multi-view skeleton fusion. The skeletons

J. Henriques et al. (Eds.): MEDICON 2019, IFMBE Proceedings 76, pp. 1685–1695, 2020.
https://doi.org/10.1007/978-3-030-31635-8_205

retrieved from the RGB-D cameras are fused according to each joint's confidence value through a Kalman filter technique. Then, using upper-body feature extraction on a sliding temporal window, dynamic gestures are classified using a support vector machine (SVM) approach.

In this work the methodological contributions are two-fold: (1) The proposal of a robust skeleton fusion representation. (2) Upper-body gesture recognition based on simple spatiotemporal features. This approach when compared to the state-of-the-art of gesture recognition, specifically addressing assistive technologies, contributes by offering several technical advantages. First, it relies only on the user's upper-body, contrary to conventional full-body techniques that introduce a limitation to sitted or lied down individuals. Second, our approach is independent of specific camera's SDK for skeleton generation, like in the case of works dependent on the Microsoft's Kinect SDk. Our approach is only based on RGB-D technology. Finally, since our approach is multi-view-based, it allows for a more robust skeleton through fusion techniques which minimize modeling errors with occlusions and incomplete skeletons, due to view obstructions. This translates into a more accurate gesture recognition.

The proposed approach was tested using our own captured dataset, comprised of 6 dynamic gestures, yielding a satisfactory performance. When compared to a single-view, our approach surpassed the performance of all the single-view classifications, validating our skeleton fusion.

2 Related Work

In this work, we address two research domains that inevitably complement each other: (i) multi-view skeleton fusion, (ii) and gesture recognition. In this section, we provide an understanding of the previous works that address each domain. First of, to better clarify the terms, single-view means a user being captured through a single camera, opposite to a multi-view approach, which captures the user using several cameras in different perspectives.

Considering the multi-view skeleton fusion domain, skeleton tracking algorithms can be classified into single-view-based models [15,21,31] and multi-view ones [4]. Considering the single-view scenario, the estimated 3D body pose usually is associated with problems to determine positions of joints due to self-occlusions. Therefore, approaches utilizing multi-view scenarios began to receive increased attention. Zhang *et al.* [29] fused single depth images into a fused point cloud, achieving pose estimation through a particle filtering approach. Another work, Liu *et al.* [19] proposed a multi-view markerless motion-capture approach which reconstructs the skeletal motion and detailed surface geometries of two individuals closely interacting. These previous works focused on image fusion and later obtaining a pose estimation. In this work, we focus on creating a more robust and accurate real-time body pose estimation, fusing multiple 3D body poses from multiple views.

Few works can be found using this approach. Masse *et al.* [20] presented a framework based on multiple Kinect sensors which output skeletal representations which are combined using a gated Kalman Filter. In this approach, the gated Kalman filter rejects measurements that present a residual lower than the gated threshold. Yeung *et al.* [28]

proposed a fusion using two Kinect sensors that captured the human motion from different views, and then are fused through a method that estimates confidence for each 3D skeleton based on mathematical optimization. Other works deal with skeleton fusion but using naive approaches such as weighted average of the positions by joint [9, 17].

The other research domain, previously mentioned, is gesture recognition. This domain has been vastly explored and is of increased relevance in our rapidly growing technological environment. A good perspective of this domain can be obtained through the recent surveys [1, 2, 16]. In this paper, we focus on skeletal data-based gesture recognition methods. Skeleton features have the advantage of being invariant to the camera location and the subject appearance. Since we focus on hand-crafted features, specifically the upper-body for disabled people applications, we provide representative works in this field. Recently, new low-level features have been created, such as the distance vector between each joint to the torso joint [7], the joint movement normals [10], and the skeleton shape trajectories [3]. In [25], different human body parts have been defined and dynamic motion features of all body parts from short temporal segments have been calculated as low-level features. The human body has also been divided into different limbs in [30] and the normalized relative orientation (NRO) between adjacent joints in each limb has been employed as the low-level feature.

3 Methodology

3.1 System Overview

The pipeline of our proposed system is presented briefly to provide a general overview of how the described modules in the following subsections interact. Figure 1 shows a representation of the pipeline as a snapshot-based view.

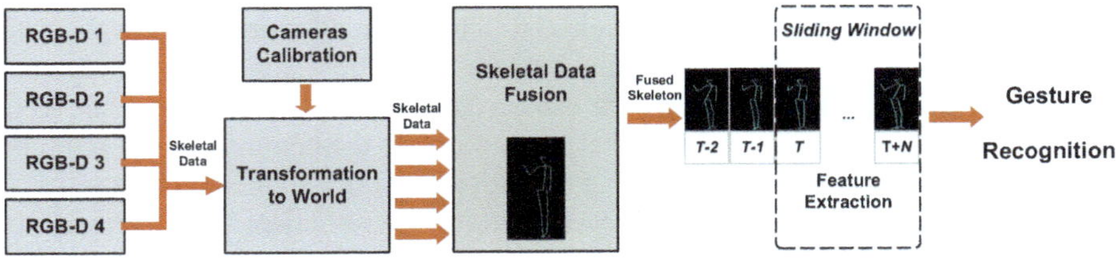

Fig. 1. Proposed method's schematic overview, where the skeletal data obtained from each RGB-D camera is fused and then, within a sliding temporal window, the fused skeleton is fed to the SVM classifier.

As previously mentioned in Sect. 1, the data modality here used is skeletal data. In this work, we obtained the RGB-D data using 4 RGB-D cameras placed on each corner of a squared room. The skeleton of each view was computed using the open library OpenPose [6]. We take advantage of both infrared (IR) and depth images. Since both are mapped on the same space, first, the IR image is processed through the OpenPose library, resulting in 2D skeletal data. This step is followed by a mapping of these 2D skeleton joint points into the depth image domain, obtaining a 3D skeleton of the user in the scene. To fuse the skeletons, first, a one-time camera calibration is required, then, the 4 skeletons transformed to a common reference system are fused using a Kalman filter approach [12]. This fusion takes into account the confidence of each skeleton's joint provided by the OpenPose library. The last module in the pipeline is gesture recognition. Here, frames of fused skeletons are grouped in a sliding temporal window approach and spatiotemporal features are extracted per window. The features are fed to a SVM classifier which discriminates the gestures.

3.2 Skeleton Fusion

This process aims to produce a more robust skeleton with the contribution of the single-view skeletons. The goal is to fuse the single-view skeletons using their joint confidence values. However, first, we need to compute the single-view skeletal data representations. For this step, the system relies on RGB-D cameras. For each camera, we use in combination IR and depth images. The IR image is processed by the open source library OpenPose, which outputs skeleton joint coordinates in the 2D image space, where $J_i \in \mathbb{N}^2$ is the i-th skeleton joint with image coordinates (u_i, v_i), which represent width and height, respectively. The number of joints is 25 resulting from the body/foot keypoint estimation used in the OpenPose library, where the model was trained using the COCO dataset [18]. Next step, is the 3D projection of the joint coordinates. Since the IR image and depth image have the same resolution and can be mapped directly, the joint coordinates directly become $J_i(u,v) \rightarrow J_i(u,v,d)$, where d is the depth value in the depth image for the corresponding pixel with coordinates (u,v). Using the camera's calibration parameters, we obtain the transformation of the joint coordinates into the world domain $J_i(x,y,z) \in \mathbb{R}^3$. However, the OpenPose library does not only provide skeleton joints positions, but as well confidence values per joint, making it possible to have, after the previous description, $J_i(x,y,z,c)$, where c is the confidence value in percentage.

To fuse the skeletons, first, it is required to have them all represented in the same world reference system. Hence, an initial camera calibration is necessary to estimate intrinsic and extrinsic camera parameters, used to calibrate each sensor's internals (comprised of focal length and principal point) and to obtain the common world reference system. This step was implemented using the previous work detailed in [13].

The skeleton fusion itself is achieved using a measurement fusion model and Kalman filter-based approach. We use a measurement fusion model approach to fuse the 4 views for each joint. We consider a linear system with the following discrete-time state-space model for each view (sensor):

$$\mathrm{x}_k = \mathrm{A}_k \mathrm{x}_{k-1} + \mathrm{B}_k \mathrm{u}_k + \mathrm{w}_k \tag{1}$$

$$\mathrm{y}_k = \mathrm{C}_k \mathrm{x}_k + \mathrm{v}_k \tag{2}$$

where k represents the discrete-time index, x the state vector, y the measurement vector, u the input control vector, A the state transition matrix, B the input transition matrix, and C the measurement matrix. We assume that w and v are the process noise vector and measurement noise vector, respectively, both with zero mean and a covariance matrix $W = E\{ww^T\}$ and $V = E\{vv^T\}$, respectively. We also consider, in this paper, an uncorrelated covariance matrix, Q and R, these becoming diagonal matrices represented by:

$$W = \begin{cases} W_{ii} = E\{ww^T\} \\ W_{ij} = 0 \end{cases} \tag{3}$$

$$V = \begin{cases} V_{ii} = E\{vv^T\} \\ V_{ij} = 0 \end{cases} \tag{4}$$

Considering that we have 4 different measurements ($N = 4$) for each joint, we can fuse them, getting the vectors:

$$y = [(y_k^1)^T \ \ldots \ (y_k^N)^T]^T \tag{5}$$

$$v = [(v_k^1)^T \ \ldots \ (v_k^N)^T]^T \tag{6}$$

This leads to the measurement fusion model represented by the following:

$$\overline{V}_k = (\sum_{i=1}^{N} (V_k^i)^{-1})^{-1} \tag{7}$$

$$\overline{C}_k = \overline{V}_k (\sum_{i=1}^{N} (V_k^i)^{-1} C_k) \tag{8}$$

$$\overline{y}_k = \overline{V}_k (\sum_{i=1}^{N} (V_k^i)^{-1} y_k) \tag{9}$$

where $\overline{V_k}$ is the fused measurement covariance, $\overline{C_k}$ is the fused measurement matrix, and $\overline{y_k}$ is the fused measurement vector. the fusion results from the weighted sum of the individual measurement vectors, where the inverse of the measurement covariance matrix $(V_k^i)^{-1}$ serves as weight. Here, we use the inverse of the confidence value of each measurement (view) to compute the covariance matrix, thus, with a small observation confidence (large variance), there is a weak contribution of that observation when determining the fused measurement. Otherwise, the observation might have a significant influence on the fused value.

Having our fused model, we implement a Kalman filter with a constant velocity model [12]. Considering the state-space model described here, the Kalman filter provides an unbiased and optimal estimate of the state-vector by minimizing estimate covariance. Figure 2 shows the result of the fusion of the 4 single-views in a frame.

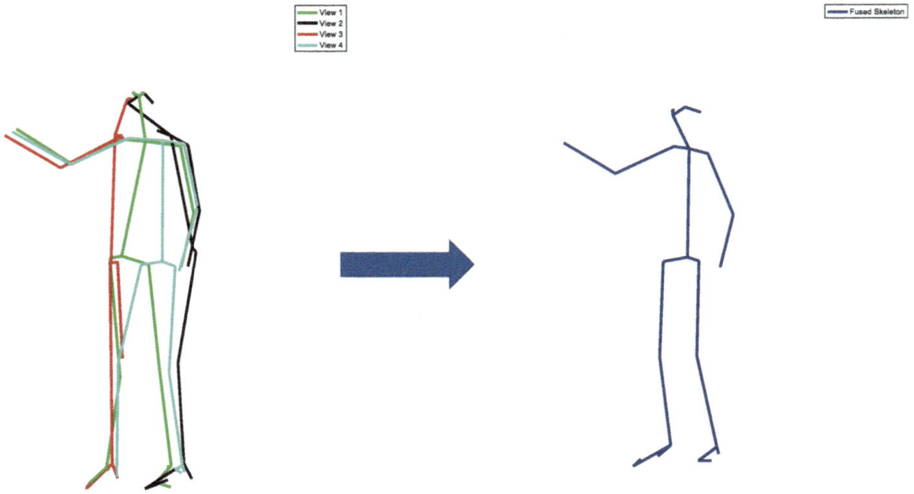

Fig. 2. On the left the 4 skeleton views which are fused resulting in the skeleton seen on the right.

3.3 Gesture Recognition

The aim of this work is to recognize gestures performed by the user to facilitate accurate human-machine interaction. The classification strategy adopted here takes advantage of a supervised SVM classifier. To achieve classification, first, features need to be extracted. Having a robust skeleton representation, we compute the feature extraction from a sequence of frames (sliding window). The sliding window approach allows to capture the dynamic characteristics of gestures, offering a spatiotemporal feature extraction approach. We chose a statistical approach of basic features since we consider these are efficient both, computationally, and completely describe the kind of interaction gestures needed to translate intention to an assistive device. Typically, these gestures are based on repetitive linear motions that do not introduce complex geometrical relations between joints, and present low variability. The features that are considered in this paper are presented in Table 1, which are extracted per joint. Since we are considering disabled individuals, the joints that we consider in this work to be viable for gesture recognition are:

- Left/right wrist;
- Left/right elbow;
- Left/right shoulder.

Each joint is represented by a 3D coordinate. For each frame all joints are transformed to the reference frame where the torso's joint is the origin.

Addressing our classification strategy to discriminate the collected gestures, each specific gesture has its own characteristics with specific variations of the joints' movements. A classifier handles the obtained feature space and labels posterior observations according to a set of trained gesture classes.

Table 1. Spatiotemporal features per joint within a sliding window.

Feature	Definition
Mean	$\mu\,(J_i(x,y,z))$
Standard deviation	$\sigma\,(J_i(x,y,z))$
Maximum value	$max\,(J_i(x,y,z))$
Minimum value	$min\,(J_i(x,y,z))$

The classifier used is a support vector machine (SVM) [5]. The feature set is normalized to improve classification performance. The SVM was trained according to the strategy "one-against-one", using a soft margin parameter set to 1.0. The chosen kernel is the Gaussian kernel. For this work, we consider testing concerning "unseen" users, in comparison with the ones from the training.

4 Results

4.1 Dataset

To adequately evaluate our proposed dynamic gesture recognition approach, we have collected a real-life dataset with the help of 11 volunteers, 8 males and 3 females with an average age of 23 with a standard deviation of 7. In order to obtain a meaningful and randomly distributed dataset, none of these volunteers were involved in the research being conducted nor had been previously involved in similar tasks. Moreover, we only provided them with an intuitive and semantic description of the way gestures should be performed, encouraging the volunteers to naturally express their freedom of motion, allowing for the introduction of variability in the dataset collection process. This way, we feel that the collected gestures were as "natural" as possible. The dataset is composed of 6 dynamic gestures:

- Push right;
- Push left;
- Swipe right;
- Swipe left;
- Clap;
- Wave.

The volunteers were asked to continuously perform each gesture for 1 min. The volunteers directly faced one of the 4 cameras at a distance of 2 m. The direction and distance which the volunteers positioned themselves in the room is here presented, however, it is not an evaluation consideration. Since our proposed approach fuses the 4 camera views, the users positioning would not be a limitation, being the classification performance of equal results. For this experiment, we used a sliding window of 15 frames, with a camera frame rate of 30 fps.

4.2 Experimental Results

For this experiment, we trained our gesture recognition approach using the data from 9 volunteers in a "leave-one-out" cross-validation scheme, and tested with 2 "unseen" ones. The results of the classification are presented in Tables 2 and 3 for each single-view and for our fusion approach, respectively.

Looking at the results, the individual results for each single-view present a poor performance. However, when the skeletons are fused, the results demonstrate that even with the same classification approach, performance is greatly improved. We attribute this behavior to the poor individual skeletons, which are plagued with occlusions and inconsistencies. By fusing these skeletons, we obtained a strong performance from our classification. This validates, first, our fusion approach, and secondly our classification strategy, which when provided with consistent tracking features performs robustly.

Table 2. Confusion matrix in percentage of accurate classifications for each individual view.

	View 1						View 2					
	Push_r	**Push_l**	**Swipe_r**	**Swipe_l**	**Clap**	**Wave**	Push_r	Push_l	Swipe_r	Swipe_l	Clap	Wave
Push_r	65	0	11.67	0	0	23.33	80.42	0	17.08	0	0	2.5
Push_l	0	61.67	0	1.67	36.67	0	0	95	0	5	0	0
Swipe_r	4.58	0	58.33	0	0.42	36.67	16.67	0	38.75	0	0	44.58
Swipe_l	0	30	0	35	35	0	0	15.83	0	62.5	21.67	0
Clap	0	0	0	15	85	0	0	0	0	8.33	91.67	0
Wave	25.42	0	0.42	0	0	74.17	11.67	0	2.92	0	0	85.42
	View 3						View 4					
Push_r	35.42	0	6.67	0	0.42	57.5	0	0	11.25	0	0.42	88.33
Push_l	0	48.33	0	49.12	2.08	0.42	0	51.25	0	32.5	16.25	0
Swipe_r	3.75	0	3.75	0	0	92.5	0.42	0	7.92	0	0	91.67
Swipe_l	0	23.75	0	45.83	28.75	1.67	0	10.83	0	73.75	15.42	0
Clap	0	0	0	0	100	0	0	0	0	14.17	85.83	0
Wave	0	0	0	0	0	100	1.67	0	1.25	0	0	97.08

Table 3. Confusion matrix in percentage of accurate classifications for the proposed fusion approach.

	Push_r	**Push_l**	**Swipe_r**	**Swipe_l**	**Clap**	**Wave**
Push_r	93.75	0	0.0625	0	0	0
Push_l	0	96.67	0	3.33	0	0
Swipe_r	11.25	0	88.75	0	0	0
Swipe_l	0	16.25	0	83.75	0	0
Clap	0	0	0	0	100	0
Wave	0	0	0	0	0	100

5 Conclusions

In this paper, we proposed an approach for dynamic gesture recognition for assistive devices applications, with focus on gait impaired individuals, who need to use wheelchairs or are confined to beds. Assistive devices that rely on interaction with humans, focusing on human intention recognition, need to interpret modalities of human communication such as gestures.

Our proposed approach presented a gesture recognition based on a robust multi-view human skeleton fusion, to recognize user commands. Results comparing our classification approach in single-view against our fused skeleton, reveal that our fusion approach greatly improves classification results, due to a more robust and accurate skeleton. At the same time, with a consistent skeleton, our classification approach shows a satisfactory accuracy.

In the future, we plan to further validate our skeleton fusion using a ground-truth based on a professional motion-capturing system. This approach will allow us to directly assess the fusion's accuracy. We also aim to introduce more complex features that can better discriminate gestures with a less expressive motion range, unlike the ones in this work. We are also pursuing the use of deep learning approaches, that could deal with a skeleton representation without the need for hand-crafted features.

Acknowledgments. This work was supported by the project POCI-01-0247-FEDER-017644 HTPDIR - "Human Tracking and Perception in Dynamic Immersive Rooms" financed by the Portugal2020 program and European Union's structural funds.

References

1. Aggarwal, J.K., Ryoo, M.S.: Human activity analysis: a review. ACM Comput. Surv. (CSUR) **43**(3) (2011). Article no. 16
2. Aggarwal, J.K., Xia, L.: Human activity recognition from 3d data: a review. Pattern Recogn. Lett. **48**, 70–80 (2014)
3. Amor, B.B., Su, J., Srivastava, A.: Action recognition using rate-invariant analysis of skeletal shape trajectories. IEEE Trans. Pattern Anal. Mach. Intell. **38**(1), 1–13 (2016)
4. Baak, A., Müller, M., Bharaj, G., Seidel, H.P., Theobalt, C.: A data-driven approach for real-time full body pose reconstruction from a depth camera. In: Consumer Depth Cameras for Computer Vision, pp. 71–98. Springer (2013)
5. Boser, B.E., Guyon, I.M., Vapnik, V.N.: A training algorithm for optimal margin classifiers. In: Proceedings of the Fifth Annual Workshop on Computational Learning Theory, pp. 144–152. ACM (1992)
6. Cao, Z., Simon, T., Wei, S.E., Sheikh, Y.: Realtime multi-person 2d pose estimation using part affinity fields. In: CVPR (2017)
7. Cippitelli, E., Gasparrini, S., Gambi, E., Spinsante, S.: A human activity recognition system using skeleton data from RGBD sensors. Comput. Intell. Neurosci. **2016**, 21 (2016)
8. Donahue, J., Anne Hendricks, L., Guadarrama, S., Rohrbach, M., Venugopalan, S., Saenko, K., Darrell, T.: Long-term recurrent convolutional networks for visual recognition and description. In: Proceedings of the IEEE Conference on Computer Vision and Pattern Recognition, pp. 2625–2634 (2015)

9. Du, H., Zhao, Y., Han, J., Wang, Z., Song, G.: Data fusion of multiple kinect sensors for a rehabilitation system. In: 2016 38th Annual International Conference of the IEEE Engineering in Medicine and Biology Society (EMBC), pp. 4869–4872. IEEE (2016)
10. Eweiwi, A., Cheema, M.S., Bauckhage, C., Gall, J.: Efficient pose-based action recognition. In: Asian Conference on Computer Vision, pp. 428–443. Springer (2014)
11. Feichtenhofer, C., Pinz, A., Zisserman, A.: Convolutional two-stream network fusion for video action recognition. In: Proceedings of the IEEE Conference on Computer Vision and Pattern Recognition, pp. 1933–1941 (2016)
12. Gan, Q., Harris, C.J.: Comparison of two measurement fusion methods for Kalman-filter-based multisensor data fusion. IEEE Trans. Aerosp. Electron. Syst. **37**(1), 273–279 (2001)
13. Girão, P., Paulo, J., Garrote, L., Peixoto, P.: Real-time multi-view grid map-based spatial representation for mixed reality applications. In: De Paolis, L.T., Bourdot, P. (eds.) Augmented Reality, Virtual Reality, and Computer Graphics, pp. 322–339. Springer International Publishing, Cham (2018)
14. Herath, S., Harandi, M., Porikli, F.: Going deeper into action recognition: a survey. Image Vis. Comput. **60**, 4–21 (2017)
15. Hofmann, M., Gavrila, D.M.: Multi-view 3d human pose estimation in complex environment. Int. J. Comput. Vision **96**(1), 103–124 (2012)
16. Ke, S.R., Thuc, H., Lee, Y.J., Hwang, J.N., Yoo, J.H., Choi, K.H.: A review on video-based human activity recognition. Computers **2**(2), 88–131 (2013)
17. Kitsikidis, A., Dimitropoulos, K., Douka, S., Grammalidis, N.: Dance analysis using multiple kinect sensors. In: 2014 International Conference on Computer Vision Theory and Applications (VISAPP), vol. 2, pp. 789–795. IEEE (2014)
18. Lin, T.Y., Maire, M., Belongie, S., Hays, J., Perona, P., Ramanan, D., Dollár, P., Zitnick, C.L.: Microsoft COCO: common objects in context. In: European Conference on Computer Vision, pp. 740–755. Springer (2014)
19. Liu, Y., Gall, J., Stoll, C., Dai, Q., Seidel, H.P., Theobalt, C.: Markerless motion capture of multiple characters using multiview image segmentation. IEEE Trans. Pattern Anal. Mach. Intell. **35**(11), 2720–2735 (2013)
20. Masse, J.T., Lerasle, F., Devy, M., Monin, A., Lefebvre, O., Mas, S.: Human motion capture using data fusion of multiple skeleton data. In: International Conference on Advanced Concepts for Intelligent Vision Systems, pp. 126–137. Springer (2013)
21. Park, S., Trivedi, M.M.: Understanding human interactions with track and body synergies (TBS) captured from multiple views. Comput. Vis. Image Underst. **111**(1), 2–20 (2008)
22. Poppe, R.: A survey on vision-based human action recognition. Image Vis. Comput. **28**(6), 976–990 (2010)
23. Qiu, Z., Yao, T., Mei, T.: Learning spatio-temporal representation with pseudo-3d residual networks. In: 2017 IEEE International Conference on Computer Vision (ICCV), pp. 5534–5542. IEEE (2017)
24. Simonyan, K., Zisserman, A.: Two-stream convolutional networks for action recognition in videos. In: Advances in Neural Information Processing Systems, pp. 568–576 (2014)
25. Tao, L., Vidal, R.: Moving poselets: a discriminative and interpretable skeletal motion representation for action recognition. In: Proceedings of the IEEE International Conference on Computer Vision Workshops, pp. 61–69 (2015)
26. Tran, D., Bourdev, L., Fergus, R., Torresani, L., Paluri, M.: Learning spatiotemporal features with 3d convolutional networks. In: Proceedings of the IEEE International Conference on Computer Vision, pp. 4489–4497 (2015)
27. Turaga, P., Chellappa, R., Subrahmanian, V.S., Udrea, O.: Machine recognition of human activities: a survey. IEEE Trans. Circuits Syst. Video Technol. **18**(11), 1473 (2008)
28. Yeung, K.Y., Kwok, T.H., Wang, C.C.: Improved skeleton tracking by duplex kinects: a practical approach for real-time applications. J. Comput. Inf. Sci. Eng. **13**(4), 041007 (2013)

29. Zhang, L., Sturm, J., Cremers, D., Lee, D.: Real-time human motion tracking using multiple depth cameras. In: 2012 IEEE/RSJ International Conference on Intelligent Robots and Systems, pp. 2389–2395. IEEE (2012)
30. Zhu, G., Zhang, L., Shen, P., Song, J.: Human action recognition using multi-layer codebooks of key poses and atomic motions. Sig. Process. Image Commun. **42**, 19–30 (2016)
31. Ziegler, J., Nickel, K., Stiefelhagen, R.: Tracking of the articulated upper body on multi-view stereo image sequences. In: 2006 IEEE Computer Society Conference on Computer Vision and Pattern Recognition (CVPR 2006), vol. 1, pp. 774–781. IEEE (2006)

Virtual Interface for an Active Motorized Pedal Exerciser for Human Leg Rehabilitation

João Ferreira[1,2](✉), A. Paulo Coimbra[2], Manuel Crisóstomo[2], and Tao Liu[3]

[1] Department of Electrical Engineering, Superior Institute of Engineering of Coimbra, Coimbra, Portugal
ferreira@isec.pt

[2] ISR - Department of Electrical and Computer Engineering, University of Coimbra, Coimbra, Portugal

[3] State Key Laboratory of Fluid Power and Mechatronic Systems, School of Mechanical Engineering, Zhejiang University, Hangzhou 310027, China

Abstract. The ability to move the lower limbs is crucial for most daily activities so there is a permanent need to develop new methods to improve people's mobility. Common leg rehabilitation devices are pedal exercisers and static bicycles. This paper presents a tool for legs rehabilitation that is based on a motor assisted static bicycle, which is gradually triggered according to the pressure exerted on force sensors located in the pedals. It allows compensation for a leg with mobility limitations permitting the expected cycling movement to be performed easily.

This device has also a sensor to monitor the patient's heart rate. It is used to ensure the treatment's efficiency and the patient's safety. Each leg can be trained individually using different parameters. This is particularly useful for stroke patients. If the patient has a leg with mobility problems, the system compensates that leg impairment.

The physiotherapist can manage the training parameters (speed, force on each pedal, heart rate) on a computer interface and monitor the training session. There were developed two interfaces for the patient to keep him motivated to stay close to the targeted values. The first interface consists of bar graphs that represent the values of feet forces, speed and heart rate, in real time. The second interface is a virtual game, a bicycle on a road, symbolizing the performance of the user as it would be in real life. The farther the patient's performance is to the reference values, the fastest he or she will lose points. The game goal is to have the bicycle upright and the road with no inclination. Through ergonomic questionnaires it was determined that the virtual game was the preferred interface.

Keywords: Lower limb rehabilitation · Active Pedal Exerciser · Assistive technologies · Virtual game

J. Henriques et al. (Eds.): MEDICON 2019, IFMBE Proceedings 76, pp. 1696–1705, 2020.
https://doi.org/10.1007/978-3-030-31635-8_206

1 Introduction

The ability to walk is one thing that we take for granted. We rely on it for every task in our daily lives. The lost or impairment of the ability to walk does not threaten our lives, but it makes them a lot harder.

Maintaining a normal gait requires three crucial components: locomotion, which embodies initiation and maintenance of rhythmic stepping, balance, and capacity to adjust to the environment [1]. If one or more of these components is not operational, the gait will be disturbed.

Gait can be affected by many factors, ranging from accidents, to diseases and age related problems. Some of them will be mentioned bellow.

In elderly populations, gait disorders are very common, and increase drastically with age. Gait disorders can lead to falls, which can have serious consequences [1].

Some of the pathologies that cause asymmetries in locomotion are multiple sclerosis (MS), stroke, unilateral total hip replacement and osteoarthritis (OA) [2]. Individuals with these pathologies pedal with an asymmetrical pattern, rendering asymmetrical training more important than normal cycling.

Robotic devices have been implemented in treatments for both upper and lower limbs rehabilitation. They provide a safe, intensive and task-oriented rehabilitation for patients with mild to severe motor impairments, and are an affordable therapy. Some of their major advantages are: controllable assistance or resistance, good repeatability and objective measurement of the subject's performance [3].

Cycling improves coordination, balance and physical condition. Cycling can be distinguished between active and passive. In active cycling, the person steps on the pedals do perform the motion, whereas in passive cycling, the person follows the movement of the pedals. Both types are efficient for therapy, due to reciprocal movement with alternation between flexion and extension of the joints. As a result, the muscles become stronger [4].

In the beginning of rehabilitation, it is often preferred passive rehabilitation to reduce swelling, alleviate pain and restore range of motion. The following stage is often an active-assistive movement phase, involving the use of external support to assist the muscles in moving the joint in order to reestablish neuromuscular control. The last stages aim at returning an individual to normal activities through resistance exercises focused on regaining muscle strength [5].

Several diseases and accidents can provoke a diminished or loss of the mobility of the limbs, either inferior or superior. Some of these pathologies cause a discrepancy between the mobility of both limbs, making it necessary to train each leg, or arm, individually, for the training to be as effective as possible. Medical rehabilitation's goal is to reduce deficiency and disability, resistance to work, aerobic exercise, practice of balance and coordination, active and passive mobility and training in activities of every-day life.

The goal of this work was to develop a rehabilitation device to address these needs. The device is based on a stationary bike with an electrical motor, which is controlled according to the force exerted on the force sensors existent on the pedals. This device is now called the Active Pedal Exerciser (APE). In addition, it was developed a virtual interface, to guide the patient's training, in order to make it as productive as possible.

This paper is organized as follows. Section 2 presents the Active Pedal Exerciser (APE) for lower limb rehabilitation and its user interfaces. Section 3 presents the methods and results obtained in the testing phase. Conclusions and future work are discussed in the final section.

2 Active Pedal Exerciser

The Active Pedal Exerciser is aimed at the rehabilitation of the lower limbs. It is an adaptation of a stationary bike, where the goal of design is a pedal exerciser instead of a stationary bike [7].

A DC motor, a gear box, a power supply, a data acquisition system, three force sensors in each pedal, an ESCON controller, and a sensor to monitor the heart rate were added to the stationary bike (Fig. 1).

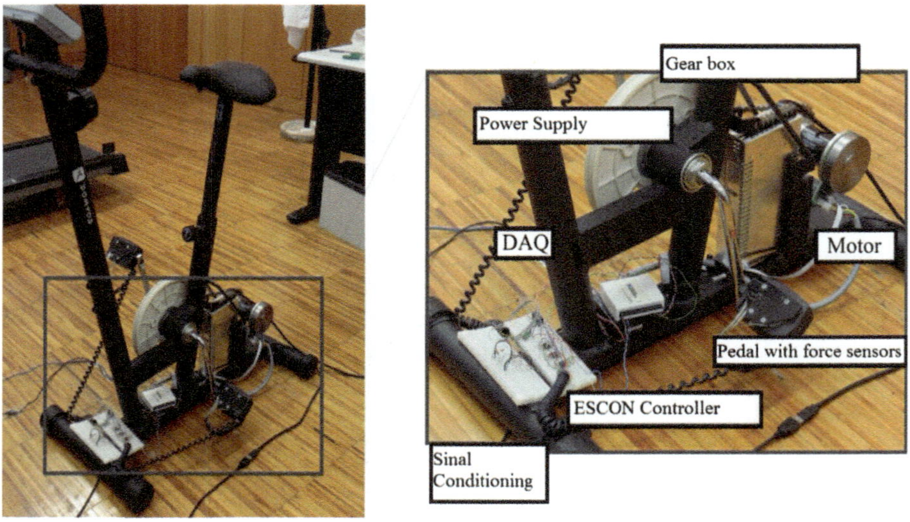

Fig. 1. Active Pedal Exerciser

The APE's motor is gradually activated, depending on the pressure exerted on the force sensors existing on the pedals, allowing to compensate the leg with mobility problems, helping it to perform the expected cycling movement. The data acquired is transmitted to the software in real time.

It has the ability to train each leg in a different way, through the prior introduction of training parameters, on the physiotherapist interface. It can also compensate for a missing or impaired limb by mimicking the performance of the healthy leg.

The training can occur in passive or active mode, depending on the patient's capabilities. The modes are named after what the bicycle does. So, in the passive mode, the motor works as a brake because the person can mobilize the leg to do the intended training. In the active mode, the motor aids the leg to perform the movement. The bicycle does not have to be in either one mode or the other, it can be in passive mode on the healthy leg and in the active mode for the impaired one.

A computer interface allows the physiotherapist in charge to set the required parameters for each patient. The patient also has an interface to keep track of his or her progress and stay motivated.

2.1 Physiotherapist Interface

The physiotherapist interface has three blocks (Fig. 2) The first block (Fig. 2 - blue) is for the user's data. All the fields are editable, except the BMI (Body Mass Index), that is calculated using the height and weight of the person. It uses the standard formulae, in which the BMI is the ratio between the weight in kilograms and the square height, in meters.

The second block (Fig. 2 - green) is the crucial one. The first step of the physiotherapist is to choose the control mode. The possible therapeutics are the control by force and the control by speed. After this selection, some options will be enabled (those that are needed for the control) and other will be disabled (those not pertinent to the control mode.) This editable text catches errors. It only accepts integer numbers in a pre-set range. This range was not yet evaluated by a healthy physician, so it might not be appropriated for therapy. When the control by force is selected, the editable fields are reference force for each leg, the percentage of operability for each leg and the maximum heart rate. When is selected the control by speed, the enabled fields are the reference speed and the maximum heart rate.

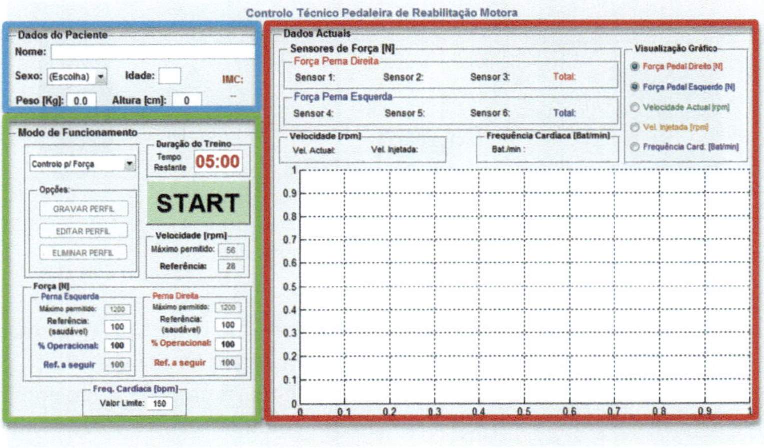

Fig. 2. Physiotherapist interface

The section on the right side (Fig. 2 - red) of the screen is for visualization only. It shows the force in each sensor, for each pedal, and the sum of the total force for each pedal in front, in Newtons (N), as well as the current and desired speeds, in rotations per minute (rpm), and the current heart rate, in beats per minute (bpm). In the section Graph Visualization, the physiotherapist can select which graphs he wants to see in the area below.

2.2 Patient Bar Interface

Patient's interface (Fig. 3) only receives information and shows it on the screen. The information received comes from the physiotherapist interface.

It has a similar field of user data to the physiotherapist interface. All the data is imported from there. The field for training duration is also imported. The timer is initialized by the physiotherapist.

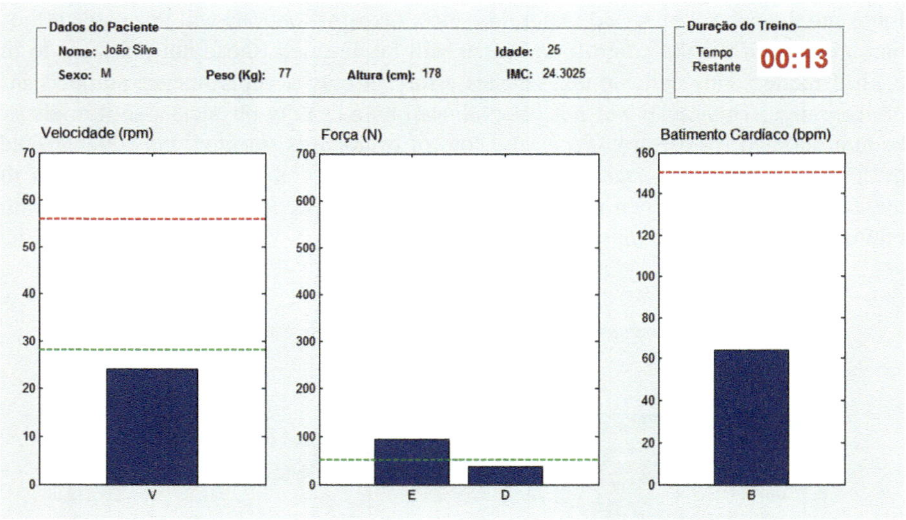

Fig. 3. Patient bar interface

The last field is composed by three bars. The first one is the speed in rotations per minute (rpm) on the pedals. The second is the force exerted by the user. Each pedal has three sensors, so the force shown is the sum of the three. Within the force graph, the left column is the force from the left leg and the right column is of the force from the right leg. The third bar is the heart rate in beats per minute (bpm). When these values are within the therapeutic range, the bars are green. If the patient is above the maximum limit, the bars will go red, as to instruct the patient to slow down. The reference lines, in green and red, represent the reference and the maximum value, respectively.

2.3 Virtual Environment

The virtual environment presented in Fig. 4, or virtual game, is composed by a bicycle on a road in a forest. There were used two programs to develop the game. The model of the bicycle and the trees were found on the internet [6]. After these models were obtained, the complete interface was developed in Matlab 3D World Editor® and Cinema 4D®. The road was designed from scratch in Cinema 4D®.

The main goal of this game was to motivate the user to stay close to the reference values. The farther the patient was from the reference values the faster he or she lose points. The score starts in 5000 points and diminishes from there. The user can also chose the level he wants to play at, from the three available: Easy, intermediate and hard. The difference between levels is the speed in which the points are lost.

The user is on the reference value when the bicycle is in the same position as the Fig. 4. If the user exerts more force in the left pedal, the bicycle tilts left, and if the force on the right pedal is bigger, it tilts right. If the speed is lower than the reference, the road inclines upward, and if it is greater, it inclines down.

Fig. 4. Virtual environment

2.4 System and Interface Usability Evaluation

After the development of the two patient's interfaces, it became necessary to evaluate if one was clearly preferred over the other, or if there was no preference. The choice between the two patient's interfaces was made based on the answers from usability questionnaires. The decision to opt for one of the patient's interfaces was made by measuring the ergonomics of each one, and the system as a whole. It was also wanted to know if there were other improvements the patients could suggest, before starting the process of finalizing the prototype.

3 Methods and Results

During the development of the APE, two separate phases of tests were performed. The first one had the goal of searching for patterns in cycling that could be of use for the development of better control modes or interfaces. The second one was meant to evaluate the two patient's interfaces, to select one and develop only that one in future phases of the project.

3.1 Searching for Useful Patterns

In this first trial, the goal was to test the initial control modes, and also, to look for patterns in cycling that could be useful for further improvement of the control modes. The test subjects had to perform a total of twelve tests of one minute each. This one minute tests were performed at speeds of 600 rpm, 1300 rpm and 2000 rpm, for a resistance of 0, 0.2, 0.4 and 0.7. The resistance was being interpreted as a normalized value, so it is the scale from 0 to 1, and does not represent physical quantity. The users were told the speed that they should cycle at, and they had to try and maintain it while guiding themselves with a graph. The speed mentioned is the motor's speed. It is related to the speed on the pedals by a factor of 53. There were performed tests with 8 users.

3.2 Results of the Usability Questionnaire

Each test subject filled two forms, one System Usability Scale (SUS) questionnaire for each interface. The SUS score was calculated as

$$SUS = 100 - \left[\left(\sum (x_{2n-1} - 1) + \sum (5 - x_{2n})\right) \times 2.5\right] \quad (1)$$

where n is the number of the question, from 1 to 10. So 2n–1 represents the odd questions and 2n the even questions and x is the value, of the answer to the question, between 1 (completely agree) and 5 (completely disagree). The results are presented in Tables 1 and 2.

The questions 4 and 10 are said to estimate the "learnability". These questions are negatively quoted. Since "learnability" is a positive parameter, the percentage was converted to calculate it, but not to interpret the answer. For the virtual interface the "learnability" is 69% = ((100% – 48%) + (100% – 15%))/2 and for the game is 79% = ((100% – 33%) + (100% – 10%))/2. So, the users felt that it was more or less needed technical assistance to use the system, but that it wasn't necessary to learn a lot to be able to use it. The most meaningful difference between the interfaces is that the users found that it was required more assistance for the use of the bar interface, rather than the game.

Table 1. SUS questionnaire answers and results, for the bar interface

Question											
User	1	2	3	4	5	6	7	8	9	10	SUS
1	1	5	1	2	1	4	1	4	1	5	87.5
2	2	5	2	4	2	3	2	5	1	5	82.5
3	2	4	1	4	2	4	3	4	2	5	77.5
4	2	2	3	1	2	3	3	2	3	1	40.0
5	1	4	2	4	1	4	2	3	2	5	80.0
6	3	5	3	2	5	1	1	5	3	4	55.0
7	2	4	2	2	3	4	2	3	4	4	60.0
8	2	4	2	2	2	3	1	4	3	5	70.0
Average	1.9	4.1	2.0	2.6	2.3	3.3	1.9	3.8	2.4	4.3	69.1
%	63	18	60	48	55	35	63	25	53	15	Average SUS score

Table 2. SUS questionnaire answers and results, for the game interface

Question											
User	1	2	3	4	5	6	7	8	9	10	SUS
1	1	5	1	2	1	4	1	4	1	5	87.5
2	1	5	3	2	2	4	3	5	1	5	77.5
3	2	5	2	3	2	4	3	4	2	5	75.0
4	3	3	2	4	2	3	2	4	3	4	65.0
5	1	4	1	4	1	3	2	4	2	4	80.0
6	4	4	1	4	3	1	2	4	2	5	65.0
7	2	4	2	4	2	5	1	3	3	4	75.0
8	3	4	1	4	2	3	2	3	2	4	70.0
Average	2.1	4.3	1.6	3.4	1.9	3.4	2.0	3.9	2.0	4.5	74.4
%	58	15	68	33	63	33	60	23	60	10	Average SUS score

The users felt that they would rather use the interface (63% against 58%), but that the game was easier to use (60% against 68%). The users did not feel the systems were complex (18% and 15%). On the other hand, they somewhat found that there was inconsistency in the systems (35% and 33%).

According to the results, the functions are better integrated in the game (55% for the bar interface, 63% for the game), but still need refinement. The users also found that other people would probably learn to use it quickly (63% and 60%). They did not feel particularly uncomfortable, while using the systems (25% and 23%) and they felt fairly confident while using the APE (53% and 60%).

The averages for the results obtained were: 69 for the patient bar interface and 74 for the virtual game. A system is acceptable with a score higher than 68. This was verified in both cases. In the questionnaire were also included open-ended questions for the users to give their opinion.

In the writing segment of the questionnaires, the users stated that the best parts of the patient bar interface where the visualization of the movement in real time, the variation of each parameter, the ease of use and intuitiveness, and the establishment of goals and limits. On the other hand, it was appointed that there should be a better way to strap the feet to the pedals.

For the virtual game, the best parts were said to be the visualization of the bicycle in the virtual environment, the score acted as a motivation, the quick response to changes and the simplicity and intuitiveness. As negative aspects mentioned were that there should be tips in the interface to indicate if the position is correct, the bicycle could be more prominent on the screen and again, better synchronization.

4 Conclusions

Given the importance of the ability to drive the lower limbs to perform most daily activities for all people and knowing that there is a constant need to develop new ways to help people who do not fully make use of their legs, either by external or physical causes, it was proposed to develop a tool for motion rehabilitation for people with this limitation. This tool is based on a kind of motor assisted bicycle, which will gradually be triggered, depending on the pressure exerted on the force sensors existing on the pedals, allowing to compensate a leg with mobility problems, helping to perform the cycling movement. This pedal exerciser was connected in real time to the Matlab software and to the motor controller. For the professional environment, was also developed a computer interface, using Matlab, which allows the clinician in charge to make an assisted management and monitoring of the rehabilitation.

Two interfaces for the patient were developed, a bar interface and a virtual game. Once there was no need for two different interfaces for the patient, ergonomic tests were performed in order to decide on one of them. The System Usability Scale (SUS) scores were better for the virtual game.

In future work a computational intelligent method will be introduced to determine the level of the leg mobility limitations using the data of the force sensors and automatically adjusting the controller force reference for better rehabilitation experience. Also, some effort will be put in the development of a more attractive virtual game.

Acknowledgments. The Fundação para a Ciência e Tecnologia (FCT) and the COMPETE 2020 program are gratefully acknowledged for funding this work with the grant PTDC/EEI-AUT/5141/2014.

Compliance with Ethical Standards

Conflict of Interest. All authors declare that they have no conflict of interest.

References

1. Snijders, A.H., van de Warrenburg, B.P., Giladi, N., Bloem, B.R.: Neurological gait disorders in elderly people: clinical. Lancet Neurol. **6**, 63–74 (2007)
2. Ranky, R.G., Sivak, M.L., Lewis, J.A., Gade, V.K.: Modular mechatronic system for stationary bicycles interfaced with virtual environment for rehabilitation. J. NeuroEng. Rehabil. **93**, 11 (2014)
3. Belda-Lois, J.-M., Horno, S.M.-d., Bermejo-Bosch, I., Moreno, J.C., et al.: Rehabilitation of gait after stroke: a review towards a top-down approach. J. NeuroEng. Rehabil. **8**, 66 (2011)
4. Blumenstein, T., Zeitlmann, H., Alves-Pinto, A., Turova, V., Lampe, R.: Optimization of electric bicycle for youths with disabilities. Springer Plus **3**, 646 (2014)
5. Constantinos, M., Jason, N., Brian, W., Gil, D., Katherine, J., Philip, P., Jennifer, P., Ryan, S., Roberto, A., Matt, L., Robert, P., Andrew, P., Dan, Y.: Smart portable rehabilitation devices. J. NeuroEng. Rehabil. **18**, 2 (2005)
6. vrml models, 13 Agosto (1997). http://www.ocnus.com/models/models.html. Accessed Jan 2019
7. Garcia, F., Ferreira, J., Ferreira, P., Cruz, S., Crisostomo, M., Coimbra, A.: Active pedal exerciser for leg rehabilitation. In: 2015 IEEE 4th Portuguese Meeting, Porto, Portugal (2015)

Special Sessions: Upper Limb Exoskeletons for a Better Quality of Life: What is Real, What is Useful, and What is Next?

Research Technologies for Assistance During Daily Life Activities

Marta Gandolla, Alberto Antonietti(✉), Valeria Longatelli, Stefano Dalla Gasperina, Emilia Ambrosini, and Alessandra Pedrocchi

NearLab - Neuroengineering and Medical Robotics Laboratory, Department of Electronics, Information and Bioengineering, Politecnico di Milano, Milan, Italy
{marta.gandolla,alberto.antonietti}@polimi.it

Abstract. An overview of upper limbs exoskeletons aimed at supporting daily life activities of persons with different levels of disabilities is presented, with a particular focus on the functional gain the devices possibly deliver to the user. Observed differences in terms of effective functional improvement and self-perceived functional improvement will be discussed, with particular attention to deriving requirements for a successful assistive device.

Keywords: Assistive devices (AD) · Activities of daily living (ADL) · Functional improvement · Externally assessed measurements · Self perceived scales

1 Background

Severe muscular weakness and chronic disability derived from neuromuscular (e.g., muscular dystrophy (MD), spinal muscular atrophy, spinal cord injuries (SCI) or stroke) or neurodegenerative (i.e., multiple sclerosis (MS), amyotrophic lateral sclerosis) diseases lead to the unavoidable loss of the possibility to perform even simple actions, such as walk, eat, up to change limbs posture. The patients suffer the consequences in terms of independence, quality of life, and self-esteem, given their need to continuously rely on assistance from their caregivers. The development of technologies offers to the patients a way to compensate for their muscular weakness through assistive devices (ADs), which empowers the user in the execution of daily life activities, and are designed to maintain or to improve the functional capabilities of individuals with disabilities. ADs for lower limbs, such as wheelchairs and electric wheelchairs, have been successfully developed and diffused to deal with the deambulation issue. On the other side, the support of upper limbs related activities is more versatile and challenging (e.g., feeding, hygiene, grooming, etc.), and with the increased life expectancy, upper extremity function becomes more and more important to support, especially given the underdevelopment of upper-limbs ADs. Non-ambulant neuromuscular patients identified arm function as their highest priority, including repositioning at night, bring hands to mouth, shift while seated, using the

J. Henriques et al. (Eds.): MEDICON 2019, IFMBE Proceedings 76, pp. 1709–1713, 2020.
https://doi.org/10.1007/978-3-030-31635-8_207

wheelchair joystick and the keyboard of a computer, and personal hygiene as priority functions to be regained [9]. The currently existing upper limb assistive devices can be categorized in (i) end-effector devices, and (ii) exoskeletons. As for end-effector devices, robotic manipulators have as human-machine interface (HMI) point user's forearm/hand, and they drive the position of the HMI point. The driving technology can be either passive – working through pre-stored mechanical energy - or active – working with motors, and able to exert greater forces or control movements more precisely. The main disadvantage of robotic manipulator devices is the impossibility of controlling upper limb joints directly: position changes of the end effector result in unexpected movements of shoulder and elbow joints. As for exoskeletons, they are external electromechanical structures worn by the patient, with joints and links corresponding to those of the human body, in order to work in parallel. These devices can have non-powered – using counterweights, springs or elastic bands - or powered controllers, and are usually mounted on a wheelchair. Patients usually prefer exoskeleton as AD, given that these devices help not only to execute the function, but they increase the perception of a self-executed movement. In this work, an extensive literature search has been conducted to identify upper limb ADs, which have been properly tested to assess functional improvement in the execution of daily life activities for neuromuscular patients.

2 Identified Studies

Studies with participants affected by upper limb disability induced by neurological or neuromotor pathologies were included. The investigated pathologies involve MD, MS, SCI, chronic stroke or in general myopathies that leave patients with muscular weakness. No restrictions have been put in terms of sex and age.

We identified 16 studies, involving 213 subjects. The number of participants in each study ranged from 3 to 55. All included studies compared patients' ability to perform functional tasks with and without ADs. Most of the studies (11) included at least one subject affected by MD, involving overall 73 MD subjects [2,4–6,8, 10,11,13,15–17], while 3 studies included 33 stroke subjects [7,12,14], one study included 3 SCI patients [1], and another study included 4 MS patients [3].

3 Assistive Devices Used

All the included studies involved the use of wearable ADs for the upper limb, in particular, seven studies used passive devices [4,6,7,10,13–15], five studies involved active devices [1,3,5,11,12], while other four studies included both active and passive ones [2,8,16,17]. ADs used in thirteen studies are available on the market [2–4,6–8,11–17], while the ones involved in the other three are still laboratory prototypes [1,5,10]. The passive device that has been employed in most of the studies [4,6,7,14,15] is Wrex, a passive, body-powered antigravity exoskeleton that uses linear elastic bands both for balance and to assist movements in three dimensions against the effect of gravity. In addition to Wrex, other passive commercial ADs were Balancer by Focal Meditech [16], Mechanical Top/Help [8]. Another study involved A-gear, a passive arm support still prototype [10]. The active devices, instead, included commercial ADs equipped with at least one actuator, as MyoPro Motion-G by Myomo [12], Armon Products such as the Ayura [11,16], and HapticMaster by Moog FCS Robotics [3]. Two studies used a prototype based on Functional Electrical Stimulation together with an exoskeleton [1] and a completely actuated exoskeleton called BRIDGE/EMPATIA [5] (Fig. 1).

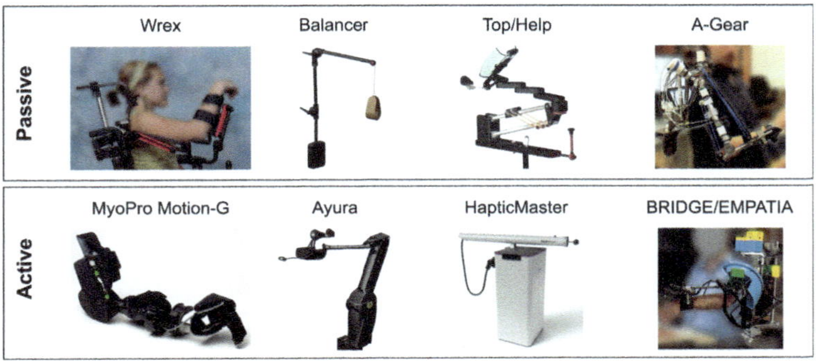

Fig. 1. Different commercial and research upper limb ADs, passive systems in the first row and active ones in the second row

4 Outcome Measures Analyzed

The included studies considered different outcome measures, here we have considered only the primary outcome measures. Three studies used the Performance of the Upper Limb scale (PUL), a new tool recently designed for specifically assessing arm functionality in Duchenne MD patients [2,5,10], three studies used the Range Of Motion (ROM) of the shoulder and elbow joint [3,4,17], two studies used the upper extremity section of Fugl-Meyer scale [12,14], while the other studies used the Five-point Likert scale [6], the Canadian Occupational Performance Measure [15], the Perceived Functional Benefit [16], the Jebsen Hand Function Test [13], the Fraction Of Reach scale [7], the Action Research Arm Test [8], the Individually Prioritized Problem Assessment [11], and the Root Mean Square Error between the actual and the target angle during elbow flexion and extension [1].

5 Assistive Device Effects

All involved ADs improved patients' ability in performing ADL in subjects affected by neuromuscular diseases. We have computed for each study an indicator of the AD efficacy (ΔPerf), which measured the percentage of functional change with the AD, compared to the baseline condition (without the AD). Thanks to these devices, patients increased their actual and perceived capability with the upper limb. In general, the mean improvement was equal to 25.78% (± 20.13%). Studies were characterized by different changes, varying from a gain of 5% to 67.5%. The highest improvements were related to studies that measured the ROM, showing how it really increased thanks to the AD. However, this could implicate an extended working volume, but it was not specifically related to daily life activities. On the contrary, studies characterized by functional outcome measures, such as the PUL, which tested the impact of the AD when performing daily tasks, shown a lower enhancement of the performance indicator (Table 1).

Table 1. Percentage improvement in the performances (ΔPerf) measured with the AD with respect to the performances without the AD (mean ± standard deviation).

ID	ΔPerf	ID	ΔPerf	ID	ΔPerf	ID	ΔPerf
[1]	58.9 ± 54.1%	[2]	6.5 ± 17.0%	[3]	65.3 ± 48.8%	[4]	67.5 ± 33.4%
[5]	15 ± 15.7%	[6]	12.4 ± 10.3%	[7]	22.0 ± 23.8%	[8]	7.7 ± 21.8%
[10]	18 ± 34.4%	[11]	19.5 ± 10.8%	[12]	13.2 ± 10.8%	[13]	10.8 ± 44.2%
[14]	5.0 ± 22.7%	[15]	37.4 ± 19.9%	[16]	29.6 ± 21.7%	[17]	23.8 ± 46.1%

6 Conclusions

The present work attempted to investigate the functional effects of upper limbs exoskeletons ADs for neuromuscular patients, both in terms of effectively and perceived regained functionality. The identified studies indicated an important functional gain while using ADs. However, the quality of the included studies was mainly low, given the restricted number of participants involved in the single studies, limitations of participants selection and blindness of outcome assessors, which lead to potentially high risk of bias. On the other hand, all studies were characterized by a positive functional effect across different levels of disability, going towards strong recommendations to use upper limb ADs for neuromuscular patients. The perceived improvements were wider with respect to externally assessed scales, which may indicate a better technology acceptance from the users. In addition, a patient could be motivated and encouraged to use these devices even if the actual gain is perceived only by himself/herself, for example, because he/she can complete the same actions that he/she would complete without support, but with less effort. Future studies may concentrate on the conduction of rigorous trials with the proper number of participants.

Conflict of Interest Declaration. The Authors declare that there is no conflict of interest regarding the publication of this contribution.

References

1. Ambrosini, E., et al.: A myocontrolled neuroprosthesis integrated with a passive exoskeleton to support upper limb activities. J. Electromyogr. Kinesiol. **24**(2), 307–317 (2014)
2. Antonietti, A., et al.: Clinical benefits and acceptability of two commercial arm exoskeletons for patients with muscular dystrophy. In: Converging Clinical and Engineering Research on Neurorehabilitation III, pp. 31–35 (2019)
3. Bastiaens, H., et al.: Facilitating robot-assisted training in MS patients with arm paresis: a procedure to individually determine gravity compensation. In: ICORR 2011, p. 5975507 (2011)
4. Estilow, T., et al.: Use of the Wilmington robotic exoskeleton to improve upper extremity function in patients with Duchenne muscular dystrophy. Am. J. Occup. Ther. **72**(2), 1–5 (2018)
5. Gandolla, M., et al.: BRIDGE—behavioural reaching interfaces during daily antigravity activities through upper limb exoskeleton: preliminary results. In: ICORR, pp. 1007–1012 (2017)
6. Gunn, M., et al.: User evaluation of a dynamic arm orthosis for people with neuromuscular disorders. IEEE TNSRE **24**(12), 1277–1283 (2016)
7. Iwamuro, B., et al.: Effect of a gravity-compensating orthosis on reaching after stroke: evaluation of the therapy assistant WREX. Arch. Phys. Med. Rehabil. **89**(11), 2121–2128 (2008)
8. Jan Burgers, M.J.: Upper limb training with dynamic arm support in boys with Duchenne muscular dystrophy: a feasibility study. Int. J. Phys. Med. Rehab. **03**(02) (2015)
9. Janssen, M., et al.: Patterns of decline in upper limb function of boys and men with DMD: an international survey. J. Neurol. **261**(7), 1269–1288 (2014)
10. Kooren, P., et al.: Design and pilot validation of A-gear: a novel wearable dynamic arm support. J. NeuroEng. Rehab. **12**, 83 (2015)
11. Lund, K., Brandt, R., Gelderblom, G.J., Herder, J.L.: A user-centered evaluation study of a mobile arm support. In: ICORR, Kyoto, Japan, pp. 582–587, June 2009
12. Peters, H., et al.: Giving them a hand: wearing a myoelectric elbow-wrist-hand orthosis reduces upper extremity impairment in chronic stroke. Arch. Phys. Med. Rehabil. **98**(9), 1821–1827 (2017)
13. Rahman, T., Sample, W., Seliktar, R., Scavina, M.T., Clark, A.L., Moran, K., Alexander, M.A.: Design and testing of a functional arm orthosis in patients with neuromuscular diseases. IEEE Trans. Neural Syst. Rehabil. Eng. **15**(2), 244–251 (2007). A publication of the IEEE Engineering in Medicine and Biology Society
14. Sanchez, R., Reinkensmeyer, D., Shah, P., Liu, J., Rao, S., Smith, R., Cramer, S., Rahman, T., Bobrow, J.: Monitoring functional arm movement for home-based therapy after stroke. EMBC **7**, 4787–4790 (2004)
15. Shank, T.: Outcome measures with COPM of children using a wilmington robotic exoskeleton. Open J. Occup. Ther. **5**(1) (2017)
16. van der Heide, L., de Witte, L.: The perceived functional benefit of dynamic arm supports in daily life. J. Rehab. Res. Dev. **53**(6), 1139–1150 (2016)
17. Van der Heide, L.A., et al.: Is it possible to assess the effects of dynamic arm supports on upper extremity range of motion during activities of daily living in the domestic setting using a portable motion capturing device? - A pilot study. Tech. Disab. **29**(1–2), 91–99 (2017)

Clinical Needs and Possible Perspectives in Rehabilitation Context

Franco Molteni(✉), Roberto Ballarati, and Eleonora Guanziroli

Villa Beretta Rehabilitation Center, Ospedale Valduce Como, Costa Masnaga, Italy
fmolteni@valduce.it

Abstract. One of the major fields of application of upper limb exoskeletons in rehabilitation medicine is to reduce disability enhancing recovery of functions in patients with Upper Motor Neuron Syndrome.

Robotic treatment has received significant attention because it can provide high-intensity and repetitive movement therapy. There is much evidence that these technologies do not just provide assistance to perform functional exercises but are also involved in promoting re-learning processes by stimulating neuroplasticity. Human-machine interactions, control strategies, driving modes, mechanical design and training modes are scientific and technological challenges open in the community to improve the efficacy of robotics in neurorehabilitation to restore ability to manage daily life activities.

Keywords: Upper limb rehabilitation · Robotic rehabilitation · Exoskeleton

1 Introduction

One of the most affected body functions in patients with Upper Motor Neuron Syndrome is the ability to control voluntary movements [1], especially of the upper limb, which determines a negative impact on the autonomy in performing Activities of Daily Living (ADL) [2], in working activities and in social relations with the consequent onset of anxiety, depression and a perception of poor quality of life [3, 4]. Therefore the recovery of functions of the upper limb represents one of the main goals in rehabilitation of patients with Upper Motor Neuron Syndrome.

When approaching the rehabilitation of the upper limb it is necessary to consider that, at this level, movements are multi-joints and complex without any cyclic or stereotypical pattern, as in the case of lower limbs during walking [5]. The upper limb is also a redundant system from a kinematic perspective with multiple degrees of freedom so that the same goal-directed movement can be performed with many different strategies [6]. The shoulder is deputized to the orientation of the limb into the space, the elbow and the wrist allow movements of approaching/turning away objects to/from the subject's body and the hand allows to grasp and manipulate objects. In this complexity of movements robotic rehabilitation can provide a guide for motor recovery influencing neuronal plasticity by providing controlled, repetitive and variable patterns that cannot be implemented with "conventional" treatment. Plasticity is the ability to modify the organization of neural components in response to anatomical or functional

J. Henriques et al. (Eds.): MEDICON 2019, IFMBE Proceedings 76, pp. 1714–1721, 2020.
https://doi.org/10.1007/978-3-030-31635-8_208

changes in the brain. Plasticity phenomena can occur after central nervous system damages and involve the perilesional tissue and the affected hemisphere, the contralateral hemisphere, subcortical and spinal regions [7]. The reorganization of the connections can be influenced by experiences [8, 9] and therefore, a rehabilitation treatment with a high intensity and high number of repetitions, which involves the execution of task-oriented exercises and the combination of top-down and bottom-up approaches as well as an adequate feedback during exercise, can promote plasticity and functional recovery. In this perspective, robotic systems have increasingly attracted the interest of researchers and have been progressively introduced in rehabilitation, even in the early stages, as they can guarantee higher intensity, greater repeatability and longer treatment duration compared to other methods.

2 Robotic Devices

In rehabilitation we can first identify two types of robots, the so-called "rehabilitation devices" and the "assistive devices" [10]. The aim of the former is to allow recovery through specific exercises based on the execution of different tasks, the purpose of the latter is to compensate a lost or impaired function. The choice of using a rehabilitation device or an assistive device is strictly based on the patient's characteristics, in particular the degree of limb functions impairment, residual abilities, treatment goals, evolution of the pathology and also patient's and caregivers' expectations. Electromechanical devices can also be divided on the basis of the assistance they provide during exercises, distinguishing between active, passive, active-assisted and resistive robots. Active devices provide a patient's active assistance to the movement, passive devices support the limbs without providing further assistance, active-assisted devices help the patient to complete a movement after it has been actively started. Resistive devices offer an opposition to the subject's movement execution. The choice of the suitable device is closely related to the characteristics of the patient and to his residual abilities. Robotic devices for upper limb rehabilitation can also be distinguished according to the segment of the body they are applied to. The upper limb, without considering finger joints, has 9 degrees of freedom: five degrees at the level of the shoulder, two at the elbow and two at the wrist. Since there is currently no device capable of working simultaneously on all these joints, we can identify robots for the proximal district (shoulder), for the shoulder-elbow district, for the elbow and/or for the distal district (wrist and fingers) [11].

A further distinction is between unilateral and bilateral treatment (Bilateral Arm Training, BAT) of the upper limb [12]. The latter involves the simultaneous treatment of both upper limbs with the execution of the same movements independently of each other [13]. This treatment can be performed in different modes (in phase or in anti-phase) and may include an active, passive and active-assisted training modality [14]. The beneficial effects of BAT are thought to arise from a coupling effect in which both limbs adopt similar spatio-temporal movement parameters leading to a sort of coordination [15]. Furthermore, most ADL are performed with both upper limbs simultaneously and therefore BAT could result in a more physiological recovery of motor control. Although different studies have evaluated the efficacy of bilateral treatment, a review of 2015 concluded that clinical data in the reviewed literature was insufficient to confirm the effectiveness of BAT when compared to other therapies for the upper limb [16].

2.1 End-Effectors and Exoskeletons

Robotic devices for the rehabilitation of the upper limb can be also divided in end-effector and exoskeletons devices [17]. With end-effectors devices the interaction between patient and machine is only at the end-effector level, which can be represented by a lever or a knob; therefore, these devices apply mechanical forces only to the distal segments of the limbs, with the advantage of being quite simple to set and more adjustable to different patients. However, the main limitation of end-effector systems is to allow a poor control of the proximal section of the limbs with the consequent risk of developing abnormal patterns of movement. Exoskeletons are wearable systems in which the interface between patient and machine is more extensive, even up to the whole limb. Such robotic systems, presenting the same degrees of freedom of the body district they are applied to, allow a direct control over each single joint, minimizing the risk of pathological muscular activations, in particular co-contractions, the development of anomalous postures or the execution of incorrect movements. The main limitation of exoskeletons is that they are more expensive and more complex to adjust to different patients. Robotic exoskeletons also offer software settings that allow clinicians to adjust the level of motor assistance as the patient's condition progresses. Each training session can be customized according to patient's characteristics. An important role in this process is played by the fine-tuning procedure of the robot's control system. This procedure is crucial to guarantee the best power transfer between the subject and the robot and can be set according to the residual functional abilities of the patient. Surface electromyography of key muscles controlling upper limbs movements, performed during the execution of spontaneous movements or during the robotic treatment, represents an effective way to noninvasively define the quantity of intervention by the robotic system in terms of power, speed and amplitude of movement. Robotic exoskeletons can also be adjusted to provide the minimum amount of motor power required, allowing to maximize patients' strength and motor function while controlling movement quality. The patient is thus "guided" in the execution of the task so that the exoskeleton can provide the perception of near-physiological movements and induce positive mechanisms of adaptive plasticity, avoiding maladaptive plasticity phenomena. It has been described how, when a movement is performed under robotic control, in addition to the typical motor areas, brain areas such as the insula, the amygdala and other neural circuits connected to the deeper centers are activated. These centers are important in determining movement memory and motivation to the movement. The interaction between exoskeleton and patient is therefore not only physical, but also cognitive [18]. The execution of movements made possible by the robotic system provides a visual feedback that seems to determine an activation of the mirror-neuron system that can promote the reorganization of cortical motor areas and increase motor learning [19–22]. For this reason robotics should not be considered a simple replacement of physiotherapists to provide only intensive and repetitive stereotypical movement patterns, perhaps with poor patient participation and poor emotional involvement. On the contrary, robotic treatment should be considered a useful rehabilitation tool to generate a more complex and controlled multisensory stimulation of the patient and to modify the plasticity of neural connections through the experience of movement. It is necessary to keep in mind that the sensory components play a crucial role, perhaps as important as the motor components, in the enhancement of a performance. A substantial part of literature is devoted to the description of the motor system plasticity induced by robotic training, while very little attention is placed on how robotics could affect the sensory domain and how the acquisition or improvement of robot-mediated sensory-motor functions could be accompanied by plastic changes in the central nervous system.

2.2 Robotic Devices in Association with Other Technologies

Robotic devices can also be used in association with other technologies in order to increase patient's active participation and to improve the results of the treatment. In this regard, for example, robotic devices can be used in association with virtual reality programs which increase the patient's attention and participation; they can also be used in association with graphical interfaces that provide visual and/or acoustic feedback which is a further stimulus in the execution of the exercise; they can finally be used in combination with functional electrostimulation devices. These systems, classified as "hybrids", have gained attention in recent years as they could compensate impaired motor functions through the combined action of muscle activation, with FES, and mechanical support, with exoskeletons. The increase in proprioceptive feedback guaranteed by the use of FES seems to induce changes in cortical excitability and to stimulate neuroplasticity at cortical level [23, 24]. The hypotheses on the mechanisms underlying these changes induced by the FES are different and are supported by evidence described in literature [25, 26]. The electrical stimulation devices can provide the impulse automatically or use a triggering mode based on the patient's residual muscular activity recorded by electromyography. It has been demonstrated how the use of a residual EMG activity to trigger movement assistance is able to promote motor learning [27]. This activation mode of the electro-stimulator was also associated with the achievement of better functional performances than those obtained with passive electrostimulation [28]. The electrical stimulation triggered by electromyographic activity can help the patient in performing movements and it is also able to offer the patient additional information useful for improving muscle activation and motor control. The effectiveness of FES is also supported by functional imaging studies in which it was described how it determines an increase in cortical excitability and perfusion of ipsilesional sensorimotor cortex [28–30] and a reduction in the activity of the contralateral areas [31]. Recent studies have shown that the application of precisely synchronized peripheral assistance with the patient's residual motor activity induces a long-term potentiation in the corticospinal pathway both in healthy subjects [32, 33] and in subjects with stroke [34]. The main challenge for hybrid robotic systems is the development of an optimal interaction between the FES, the exoskeleton and the patient to promote and enhance the effects of the therapy. The correct association between stimulation and voluntary muscle activation has indeed proved to be an important factor to promote neuromuscular recovery [32, 35]. Today only a few hybrid systems described in literature have addressed the associative paradigm using voluntary activity recorded with electromyography to activate robotic assistance [36–38].

3 Robotic Treatment Limitations

Robotic rehabilitation presents some limitations. First, in several studies it has been reported how robotic treatment could lead the so-called "slacking effect", that is a passive attitude of the patient who lets himself be guided by the technological tool without taking an active part and without performing any effort, leading to a potential reduction in motor function improvements [39, 40]. However, in a recent review [5] the authors underlined how this treatment modality, described as "passive", can determine the activation of sensorimotor cortical areas and stimulate their reorganization with a consequent positive effect on motor recovery. What is considered by many to be a limitation of robotic treatment in stroke patients could therefore be re-evaluated in a different perspective and allow the identification of new rehabilitative perspectives. The "slacking effect" could eventually be minimized by using technologies such as virtual reality and FES in order to increase patient's active participation. Moreover, Another critical issue related to robotic treatment lies in the difficulty to verify its efficacy due to the lack of standardized data measurement and evaluation procedures [41] that would allow a correct clinical and instrumental functional stratification of patients and the possibility of a direct comparison between the results of the different studies. Furthermore, the relatively small number of studies and their methodology, as well as the small number of subjects tested, make it difficult to get to generalized conclusions. The inclusion of a control group in studies to demonstrate the efficacy of a robotic rehabilitation system could provide important information but it is difficult to implement due to the high number of patients required for this type of study.

4 Conclusions

Robotic devices for the rehabilitation of the upper limbs, and in particular exoskeletons, represent an important tool to promote motor recovery in patients affected by UMNS. There is a lot of evidence that these technologies do not just provide assistance to perform functional exercises but are also involved in promoting re-learning processes by stimulating neuroplasticity. These devices are allowing us to modify the conventional treatment in terms of intensity and complexity of therapeutic exercises as well as in terms of "personalization" of the treatment, which is increasingly tailored on each patient according to his residual abilities, the evolution of the pathology and the rehabilitation goals. The variability of the robotic devices for the upper limb rehabilitation available today allows us not only to choose between the most suitable options for each patient, but also to gradually adjust the treatment on the basis of the functional evolution of the patient, particularly for exoskeletons, thanks to the possibility they offer to be finely tuned and to be used in association with technologies such as virtual reality, visual/acoustic feedback and FES. Human-machine interactions, control strategies, training modes and intensity, mechanical design are some scientific and technological challenges still open in the community in order to improve the efficacy of the treatment and to restore the ability to manage activities of daily living. We can provide adequate answers to these questions only by implementing innovative and large scale research programs.

References

1. Langhorne, P., Coupar, F., Pollock, A.: Motor recovery after stroke: a systematic review. Lancet Neurol. **8**(8), 741–754 (2009)
2. Sveen, U., Bautz-Holter, E., Sodring, K.M., Wyller, T.G., Laak, K.: Association between impairments, self-care ability and social activities 1 year after stroke. Disabil. Rehabil. **21**, 372–377 (1999)
3. Morris, J.H., van Wijck, F., Joice, S., Donaghy, M.: Predicting health related quality of life 6 months after stroke: the role of anxiety and upper limb dysfunction. Disabil. Rehabil. **35**, 291–299 (2013)
4. Franceschini, M., La Porta, F., Agosti, M., Massucci, M.: Is health-related-quality of life of stroke patients influenced by neurological impairments at one year after stroke? Eur. J. Phys. Rehabil. Med. **46**(3), 389–399 (2010)
5. Molteni, F., Gasperini, G., Cannaviello, G., Guanziroli, E.: Exoskeleton and end-effector robots for upper and lower limbs rehabilitation: narrative review. PM R **10**(9), S174–S188 (2018)
6. Caimmi, M., Guanziroli, E., Malosio, M., et al.: Normative data for an instrumental assessment of the upper-limb functionality. Biomed. Res. Int. **2015**, 484131 (2015)
7. Alia, C., Spalletti, C., Lai, S., et al.: Neuroplastic changes following brain ischemia and their contribution to stroke recovery: novel approaches in neurorehabilitation. Front. Cell. Neurosci. **11**, 76 (2017)
8. Stroemer, R.P., Kent, T.A., Hulsebosch, C.E.: Acute increase in expression of growth associated protein GAP-43 following cortical ischemia in rat. Neurosci. Lett. **162**, 51–54 (1993)
9. Li, S., Carmichael, S.T.: Growth-associated gene and protein expression in the region of axonal sprouting in the aged brain after stroke. Neurobiol. Dis. **23**, 362–373 (2006)
10. Lum, P., Godfrey, S., Brokaw, E., Holley, R., Nichols, D.: Robotic approaches for rehabilitation of hand function after stroke. Am. J. Phys. Med. Rehabil. **91**, 242–254 (2012)
11. Lo, H., Xie, S.: Exoskeleton robots for upper-limb rehabilitation: state of the art and future prospects. Med. Eng. Phys. **34**, 261–268 (2012)
12. Gandolfi, M., Formaggio, E., Geroin, C., Storti, S.F., Boscolo Galazzo, I., Bortolami, M., Saltuari, L., Picelli, A., Waldner, A., Manganotti, P., Smania, N.: Quantification of upper limb motor recovery and EEG power changes after robot-assisted bilateral arm training in chronic stroke patients: a prospective pilot Study. Neural Plast. **2018**(26), 8105480 (2018)
13. Stewart, K.C., Cauraugh, J.H., Summers, J.J.: Bilateral movement training and stroke rehabilitation: a systematic review and meta-analysis. J. Neurol. Sci. **244**(1–2), 89–95 (2006)
14. Choo, P.L., Gallagher, H.L., Morris, J., Pomeroy, V.M., van Wijck, F.: Correlations between arm motor behavior and brain function following bilateral arm training after stroke: a systematic review. Brain Behav. **5**(12) (2015). Article e00411
15. Swinnen, S.P.: Intermanual coordination: from behavioural principles to neural-network interactions. Nat. Rev. Neurosci. **3**(5), 348–359 (2002)
16. Sheng, B., Zhang, Y., Meng, W., Deng, C., Xie, S.: Bilateral robots for upper-limb stroke rehabilitation: state of the art and future prospects. Med. Eng. Phys. **38**, 587–606 (2016)

17. Mehrholz, J., Pohl, M.: Electromechanical-assisted gait training after stroke: a systematic review comparing end-effector and exoskeleton devices. J. Rehabil. Med. **44**(3), 193–199 (2012)
18. Mehrholz, J., Thomas, S., Werner, C., Kugler, J., Pohl, M., Elsner, B.: Electromechanical-assisted training for walking after stroke. Cochrane Database Syst. Rev. **5**, CD006185 (2017)
19. Rizzolatti, G., Craighero, L.: The mirror-neuron system. Annu. Rev. Neurosci. **27**, 169–192 (2004)
20. Fadiga, L., Fogassi, L., Pavesi, G., Rizzolatti, G.: Motor facilitation during action observation: a magnetic stimulation study. J. Neurophysiol. **73**(6), 2608–2611 (1995)
21. Ertelt, D., Small, S., Solodkin, A., Dettmers, C., McNamara, A., Binkofski, F., et al.: Action observation has a positive impact on rehabilitation of motor deficits after stroke. Neuroimage **36**, 164–173 (2007)
22. Garrison, K.A., Winstein, C.J., Aziz-Zadeh, L.: The mirror neuron system: a neural substrate for methods in stroke rehabilitation. Neurorehabil. Neural Repair **24**(5), 404–412 (2010)
23. Maffiuletti, N.A., Minetto, M.A., Farina, D., Bottinelli, R.: Electrical stimulation for neuromuscular testing and training: state-of-the art and unresolved issues. Eur. J. Appl. Physiol. **111**(10), 2391–2397 (2011)
24. Popović, D.B.: Advances in functional electrical stimulation (FES). J. Electromyogr. Kinesiol. **24**(6), 795–802 (2014)
25. Ridding, M., Brouwer, B., Miles, T., Pitcher, J., Thompson, P.: Changes in muscle responses to stimulation of the motor cortex induced by peripheral nerve stimulation in human subjects. Exp. Brain Res. **131**, 135–143 (2000)
26. Rushton, D.: Functional electrical stimulation and rehabilitation–an hypothesis. Med. Eng. Phys. **25**, 75–78 (2003)
27. Chae, J.: Neuromuscular electrical stimulation for motor relearning in hemiparesis. Phys. Med. Rehabil. Clin. N Am. **14**, 93–109 (2003)
28. Barsi, G.I., Popovic, D.B., Tarkka, I.M., Sinkjær, T., Grey, M.J.: Cortical excitability changes following grasping exercise augmented with electrical stimulation. Exp. Brain Res. **191**(1), 57–66 (2008)
29. Hara, Y., Obayashi, S., Tsujiuchi, K., Muraoka, Y.: The effects of electromyography-controlled functional electrical stimulation on upper extremity function and cortical perfusion in stroke patients. Clin. Neurophysiol. **124**(10), 2008–2015 (2013)
30. Joa, K.L., Han, Y.H., Mun, C.W., Son, B.K., Lee, C.H., Shin, Y.B., et al.: Evaluation of the brain activation induced by functional electrical stimulation and voluntary contraction using functional magnetic resonance imaging. J. Neuroeng. Rehabil. **9**(1), 1 (2012)
31. Kimberley, T.J., Lewis, S.M., Auerbach, E.J., Dorsey, L.L., Lojovich, J.M., Carey, J.R.: Electrical stimulation driving functional improvements and cortical changes in subjects with stroke. Exp. Brain Res. **154**(4), 450–460 (2004)
32. Mrachacz-Kersting, N., Kristensen, S.R., Niazi, I.K., Farina, D.: Precise temporal association between cortical potentials evoked by motor imagination and afference induces cortical plasticity. J. Physiol. **590**(7), 1669–1682 (2012)
33. Xu, R., Jiang, N., Mrachacz-Kersting, N., Lin, C., Asin Prieto, G., Moreno, J.C., et al.: A closed-loop brain-computer interface triggering an active ankle-foot orthosis for inducing cortical neural plasticity. IEEE Trans. Biomed. Eng. **61**(7), 2092–2101 (2014)
34. Mrachacz-Kersting, N., Jiang, N., Stevenson, A.J.T., Niazi, I.K., Kostic, V., Pavlovic, A., et al.: Efficient neuroplasticity induction in chronic stroke patients by an associative brain-computer interface. J. Neurophysiol. **115**(3), 1410–1421 (2016)
35. Ethier, C., Gallego, J.A., Miller, L.E.: Brain-controlled neuromuscular stimulation to drive neural plasticity and functional recovery. Curr. Opin. Neurobiol. **33**, 95–102 (2015)

36. Fujiwara, T., Kasashima, Y., Honaga, K., Muraoka, Y., Tsuji, T., Osu, R., et al.: Motor improvement and corticospinal neuromuscular dynamic stimulation (HANDS), pp. 125–132 (2009)
37. Hu, X.L., Tong, R.K.Y., Ho, N.S.K., Xue, J.J., Rong, W., Li, L.S.W.: Wrist rehabilitation assisted by an electromyography-driven neuromuscular electrical stimulation robot after stroke. Neurorehabil. Neural Repair **29**(8), 767–776 (2015)
38. Barker, R., Brauer, S., Carson, R.: Training of reaching in stroke survivors with severe and chronic upper limb paresis using a novel nonrobotic device: a randomized clinical trial. Stroke **39**, 1800–1807 (2008)
39. Reinkensmeyer, D., Wolbrecht, E., Chan, V., Chou, C., Cramer, S., Bobrow, J.: Comparison of 3D, assist-as-needed robotic arm/hand movement training provided with Pneu-WREX to conventional table top therapy following chronic stroke. Am. J. Phys. Med. Rehabil. **91**, 232–241 (2012)
40. Marchal-Crespo, L., Reinkensmeyer, D.J.: Review of control strategies for robotic movement training after neurologic injury. J. Neuroeng. Rehabil. **6**(1) (2009)
41. Huang, V.S., Krakauer, J.W.: Robotic neurorehabilitation: a computational motor learning perspective. J. Neuroeng. Rehabil. **6**(1), 1–13 (2009)

Upper-Limb Exoskeletons for Stroke Rehabilitation

Emilia Ambrosini[1(✉)], Stefano Dalla Gasperina[1], Marta Gandolla[2], and Alessandra Pedrocchi[1]

[1] NearLab, Department of Electronics, Information and Bioengineering, Politecnico di Milano, Milan, Italy
emilia.ambrosini@polimi.it
[2] NearLab@Lecco, Polo Territoriale di Lecco, Politecnico di Milano, Milan, Italy

Abstract. Upper-limb exoskeletons provide high-intensity, repetitive, task-specific, interactive and individualized training, making effective use of neuroplasticity for functional recovery in neurological patients. Most exoskeletons have robot axes aligned with the anatomical axes of the subject and provide direct control of individual joints. Recently, novel mechanical structures and actuation mechanisms have been proposed, but still result in bulky and heavy exoskeletons, limiting their applicability into clinical practice. Technological efforts are needed to promote light and wearable exoskeletons that implement active-assistive controllers, providing "assisted-as-needed" rehabilitation therapy, towards patient's motivation and self-esteem. An overview of upper-limb exoskeletons, including mechanical design and control algorithms, will be provided. Special focus will be put on the current evidence about the efficacy of wearable robotic technologies on motor recovery and about other therapies that can be combined with exoskeletons to improve their therapeutic effects.

Keywords: Exoskeleton · Upper-limb · Rehabilitation · Stroke · Functional electrical stimulation · Virtual reality · Randomized controlled trial

1 Introduction

About 80% of stroke survivors suffer from upper-limb paresis [15], which strongly affects their capability to perform activities of daily life (ADL), decreasing their independence and quality of life. Neuroplasticity is the basic mechanism underlying improvements in functional outcomes after stroke. Rehabilitative interventions should make effective use of neuroplasticity for functional recovery, proposing high-intensity, repetitive, task-specific, interactive and individualized training [13]. Technology-supported rehabilitation is emerging as a solution to support therapists in providing such training for a long duration, allowing the participants to progress in task difficulty and to achieve the desired movement,

J. Henriques et al. (Eds.): MEDICON 2019, IFMBE Proceedings 76, pp. 1722–1728, 2020.
https://doi.org/10.1007/978-3-030-31635-8_209

so as to increase their motivation [5]. In the last thirty years, several robotic devices for upper-limb rehabilitation have become commercially available [2]. Therapeutic robots can be divided into two categories: end-effector devices (e.g. InMotion Arm Robot, known as MIT-Manus [12]) and exoskeleton-type devices (e.g. Armeo Power, Hocoma [20]). End-effector devices hold the patient's hand or forearm at one point and generate mechanical forces at the interface to move the arm. They are easy to setup but they suffer from limited control of the proximal joints, which could result in abnormal postures. Conversely, exoskeleton-type devices are characterized by robot axes aligned with the anatomical axes of the subject and provide direct control of individual joints, which can minimize abnormal posture or movement and provides a better control of the arm motion [16]. These advantages are paid by a more complex design of mechanical structure and control algorithms. This contribution focuses on arm exoskeletons for stroke rehabilitation dealing with mechanical and control design, current evidence about their efficacy, alone or in combination with other therapies, up to potential areas for future research.

2 Mechanical Design and Actuation

The anatomical structure of the upper limb is very complex and flexible, permitting the human arm to perform a wide range of movements leveraging on 9 Degrees of Freedom (DOFs). A successfully designed upper-limb exoskeleton operates alongside the human arm smoothly supporting its natural kinematics. In literature, arm exoskeletons mainly provide support at the shoulder and at the elbow, only few of them are provided with handles to support the motion of wrist and hand [16]. Regardless the number of supported DOFs, the design of upper-limb exoskeletons suggests that robot axes aligned with anatomical axes of the human arm improve robot's wearability and safety. However, since the instantaneous center of rotation of the shoulder changes during motion and since an incorrect axis alignment leads to undesirable interaction forces and uncomfortable postures, additional actuated or passive DOFs have been added to adjust and self-compensate misalignments [11,20,27].

Regarding the actuation mechanisms, three main technologies can be identified: pneumatic, hydraulic and electric actuators. Pneumatic and hydraulic actuators provide higher power to weight ratio with respect to electric motors, but their efficiency strongly depends on pressure losses due to friction or to fluid leaks. Pneumatic and hydraulic, due to their nature, are inherently safe and compliant, thus suitable for applications involving human-robot interaction. These actuators have been widely tested in exoskeletons in the past years, however due to their power consumption, low efficiency and limited control bandwidth, they did not find great consensus [10]. On the other side, due to their ability to produce large amount of torque, their commercial availability and their intrinsic precision in controlling motion, current solutions mainly implement stiff joints actuated by electric motors. Nowadays, the interest in compliant control for human-robot interaction has led to the development of new approaches that involve the use of electric motors in series with elastic elements, the so-called series elastic actuators (SEAs) [26]. SEAs were initially developed to improve force control accuracy, meeting interest in the rehabilitation field given provided advantages in terms of safety, tolerance to mechanical shocks and indirect torque measurements [11]. Motors can be both co-located with respect to the axis of rotation of the physiological joints, which leads to a simpler control strategy, but worsens the encumbrance and portability of the system; or positioned behind the exoskeleton, leveraging on the cable-driven approaches, which use wires and pulleys to transmit the power to the actuated joints [22]. With the current solutions, however, it is clear that such complex mechanical design has led to powerful, bulky and heavy exoskeletons (Fig. 1). This drawback strongly limits their applicability into clinical practice and still requires technological efforts to promote light and wearable exoskeletons for robot-assisted rehabilitation therapy.

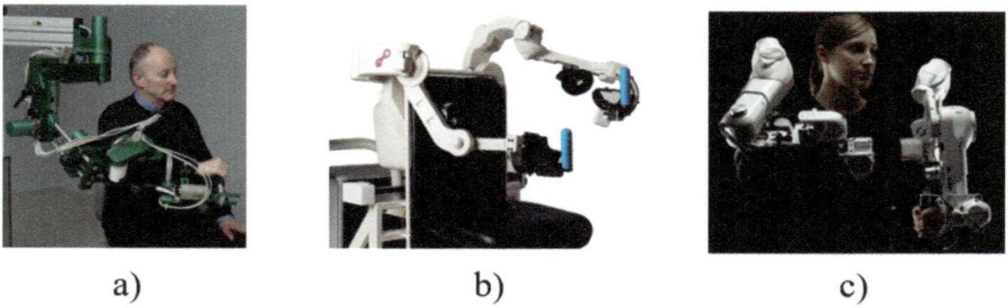

Fig. 1. Upper-limb exoskeletons. (a) ARMin III [20] (b) AleX [22] (c) Harmony [11]

3 Control Algorithms

According to Proietti et al. robot-assisted rehabilitation for upper-limb motor recovery has led to promising results, however most publications do not deal with control aspects [23]. Most exoskeletons are controlled in passive or active-assistive mode [2]. In passive mode, the robot performs the movement regardless of subject's response to therapy. However, repetitive passive motion of the arm has come out with limited effects on neuroplasticity. Conversely, more complex control strategies based on subject's involvement could lead to better results, at least after the first stages of the rehabilitation process when intensive, repetitive and task-oriented training are the key-features for an effective recovery. This is the case of active-assistive mode, where, after the patient has recovered a sufficient amount of motor control, the controller assists the arm motion in a human-like way, exploiting any residual ability of the patient and letting the user drive the movement, improving patient's motivation and self-esteem [23]. Many approaches have been proposed to implement assisted-as-needed strategies, but the most widespread rely on admittance control or impedance control, where the controller shapes the dynamical relation between position and force, as a compromise between tracking the position of the end-effector and controlling the effort imposed by the robotic arm [4]. More advanced controllers take advantage from friction and gravity compensation techniques to counterbalance the weight and the inertia of the exoskeleton during free-space motion. Other controllers involve the use of surface EMG signals to estimate muscle activation and provide EMG-modulated forces, towards patient's active involvement during exercises [16].

4 Effects of Exoskeletons on Upper-Limb Motor Recovery

The effects of robotic devices on rehabilitative outcomes after stroke have been strongly investigated through randomized controlled trials (RCT) and, in the latest years, through systematic reviews and meta-analysis [19,28,29]. Robotic training can be used in addition to usual care to help both therapists and patients in the management of the paralyzed upper-limb and to achieve better rehabilitation outcomes [29]. A recent Cochrane review, including 34 trials (1160 participants), demonstrated that stroke patients who receive electromechanical and robot-assisted arm and hand training after stroke might improve their ADL, arm and hand function and strength [19]. Meta-analyses of 38 trials (1206 participants) showed significant but small improvements in motor control and muscle strength of the paretic arm and a negative effect on muscle tone when robotic training is compared to usual care [28]. Similar results were achieved also in [3], where a comparison between conventional therapy and robot-assisted rehabilitation showed that the latter is more effective in improving upper-limb motor function recovery, especially in chronic stroke patients. It is important to underline that these positive effects on the recovery of arm functions have been mainly achieved using end-effector robotic devices, while there is still a lack of evidence regarding exoskeletons, recently investigated by two systematic reviews [3,28]. They both concluded that so far there is no evidence about the superiority of exoskeleton-assisted training with respect to usual care; however, the effects of exoskeletons have been insufficiently investigated and final conclusion cannot be drawn. Finally, few studies compared different training modalities, but available data indicated that active-assistive mode, stressing the patient's active contribution, led most consistently to improvements in arm function [2].

5 Combination Between Exoskeletons and Other Therapies

Several therapeutic approaches have been combined with upper-limb exoskeletons, making robotic based-interventions more functional and task-oriented. In this contribution, we focus on the use of FES and VR in combination with exoskeletons.

5.1 Functional Electrical Stimulation

FES has been strongly used to enhance functional recovery of the paretic arm in stroke survivors [8]. When FES is combined with the patient's residual voluntary effort [1], cortical plasticity seems to be enhanced, having therefore the potential to improve the therapeutic effects of FES [6]. The combined action of robot-assisted therapy supporting elbow and wrist motion and EMG-triggered FES has shown positive effects on muscle coordination in subacute [24] and chronic stroke survivors [25]. FES has been combined also with passive anti-gravity exoskeletons, e.g. ArmeoSpring (Hocoma) [18] or with custom-built passive exoskeleton [21]. In a recent RCT [9], 68 stroke patients were randomized in an experimental group, training with a passive anti-gravity exoskeleton combined with an arm EMG-triggered neuroprosthesis in addition to conventional therapy, and a control group performing only conventional therapy of equal dose. Preliminary results showed the superiority of the training with the hybrid robotic system, supporting the use of exoskeleton devices combined with FES.

5.2 Virtual Reality

VR is an interactive and individualized treatment modality, which can provide sensorimotor training in enriched environments, so as to maximize patient's engagement [17]. Several systematic reviews have shown the superiority of VR training with respect to usual care in improving motor function and ADL independence, mainly when VR is exploited to increase the time spent in therapy [14]. Several groups have combined VR with multi-joint upper-limb exoskeletons in order to prevent the slacking effect which may characterize robotic rehabilitation when assistance is too supportive [7]. These pilot studies showed promising results in terms of functional improvement and increments of active and passive ranges of motion. In [7], the level of difficulty for the exercise was adjusted by a performance-dependent adaptation algorithm, which may facilitate motor learning by progressively challenging the subject in accordance with the individual capacity. However, more extensive RCTs are needed to demonstrate the superiority of VR combined with upper-limb exoskeletons with respect to VR and robotic rehabilitation alone, as well as to usual care.

6 Perspectives and Challenges

Overall, exoskeleton-type devices have entered the clinical practice of arm stroke rehabilitation. Robotic devices enable several-to-one therapy paradigms, allowing high-dosage and high-intensity treatments, which are recognized as the major positive elements for the recovery process. However, the superiority of exoskeleton-assisted training with respect to usual care is still a matter of challenge. To fill this gap, well-design RCTs should be conducted and should include cost-effectiveness analysis to evaluate the economical sustainability of these technological solutions. A crucial aspect of exoskeleton design is the human-robot interaction, which should let the human take the lead role, guaranteeing at the same time that the robot closely observes the human and decides when to provide corrective actions. Integration of biological signals, such as EMG, is mandatory to achieve this interaction. Despite many significant advances, effective strategies, which can realize minimal assistance paradigms, are still under investigation. Rehabilitation exoskeletons can also be integrated with multiplayer games, which are recently emerged as a promising approach to increase patients' motivation. In order to favor the transfer of motor gains to ADL, it is important to integrate distal and proximal arm training; indeed, so far, exoskeleton devices have been focused mainly on the proximal joints but to make robotic training more functional it is mandatory the integration of robotic hand modules. Finally, the development of home-based rehabilitation systems is the ultimate goal for achieving high intensity training in post-stroke therapy. Home-based exoskeletons would assure a safe, intensive, and controlled training with limited supervision, under a remote control of a human therapist.

Conflict of Interest Declaration. The Authors declare that there is no conflict of interest regarding the publication of this contribution.

References

1. Ambrosini, E., et al.: A myocontrolled neuroprosthesis integrated with a passive exoskeleton to support upper limb activities. J. Electromyogr. Kinesiol. **24**(2), 307–317 (2014). Official Journal of the International Society of Electrophysiological Kinesiology
2. Basteris, A., et al.: Training modalities in robot-mediated upper limb rehabilitation in stroke: a framework for classification based on a systematic review. J. NeuroEng. Rehabil. **11**(1), 111 (2014)
3. Bertani, R., et al.: Effects of robot-assisted upper limb rehabilitation in stroke patients: a systematic review with meta-analysis. Neurol. Sci. **38**(9), 1561–1569 (2017)
4. Calanca, A., et al.: A review of algorithms for compliant control of stiff and fixed-compliance robots. IEEE/ASME Trans. Mechatron. **21**(2), 613–624 (2016)
5. Chang, W.H., Kim, Y.-H.: Robot-assisted therapy in stroke rehabilitation. J. Stroke **15**(3), 174–181 (2013)
6. Gandolla, M., et al.: The neural correlates of long-term carryover following functional electrical stimulation for stroke. Neural Plast. **2016**, 1–13 (2016)
7. Grimm, F., et al.: Closed-loop task difficulty adaptation during virtual reality reach-to-grasp training assisted with an exoskeleton for stroke rehabilitation. Front. Neurosci. **10**, 518 (2016)
8. Howlett, O.A., et al.: Functional electrical stimulation improves activity after stroke: a systematic review with meta-analysis. Arch. Phys. Med. Rehabil. **96**(5), 934–943 (2015)
9. Immick, N., et al.: Hybrid robotic system for arm training after stroke: preliminary results of a randomized controlled trial. In: International Conference on NeuroRehabilitation, pp. 94–97 (2019)

10. Islam, M.R., et al.: A brief review on robotic exoskeletons for upper extremity rehabilitation to find the gap between research prototype and commercial type. Adv. Robot. Autom. **06**(03), 1–12 (2018)
11. Kim, B., Deshpande, A.D.: An upper-body rehabilitation exoskeleton Harmony with an anatomical shoulder mechanism: Design, modeling, control, and performance evaluation. Int. J. Robot. Res. **36**(4), 414–435 (2017)
12. Krebs, H.I., et al.: Robot-aided neurorehabilitation: a robot for wrist rehabilitation. IEEE Trans. Neural Syst. Rehabil. Eng. **15**(3), 327–335 (2007)
13. Langhorne, P., et al.: Motor recovery after stroke: a systematic review. Lancet Neurol. **8**(8), 741–754 (2009)
14. Laver, K.E., et al.: Virtual reality for stroke rehabilitation. In: Laver, K.E. (ed.) Cochrane Database of Systematic Reviews. Wiley, Chichester (2011)
15. Lawrence, E.S., et al.: Estimates of the prevalence of acute stroke impairments and disability in a multiethnic population. Stroke **32**(6), 1279–1284 (2001)
16. Lo, H.S., Xie, S.Q.: Exoskeleton robots for upper-limb rehabilitation: state of the art and future prospects. Med. Eng. Phys. **34**(3), 261–268 (2012)
17. Mazzoleni, S., et al.: Combining upper limb robotic rehabilitation with other therapeutic approaches after stroke: current status, rationale, and challenges. BioMed Res. Int. **2017**, 1–11 (2017)
18. Meadmore, K.L., et al.: Functional electrical stimulation mediated by iterative learning control and 3D robotics reduces motor impairment in chronic stroke. J. Neuroeng. Rehabil. **9**, 32 (2012)
19. Mehrholz, J., et al.: Electromechanical and robot-assisted arm training for improving activities of daily living, arm function, and arm muscle strength after stroke. Cochrane Database Syst. Rev. (11), CD006876 (2015)
20. Nef, T., et al.: ARMin III - arm therapy exoskeleton with an ergonomic shoulder actuation. Appl. Bionics Biomech. **6**(2), 127–142 (2009)
21. Pedrocchi, A., et al.: MUNDUS project: Multimodal Neuroprosthesis for Daily Upper limb Support. J. NeuroEng. Rehabil. **10**(1), 66 (2013)
22. Pirondini, E., et al.: Evaluation of the effects of the arm light exoskeleton on movement execution and muscle activities: a pilot study on healthy subjects. J. NeuroEng. Rehabil. **13**(1), 1–21 (2016)
23. Proietti, T., et al.: Upper-limb robotic exoskeletons for neurorehabilitation: a review on control strategies. IEEE Rev. Biomed. Eng. **9**, 4–14 (2016)
24. Qian, Q., et al.: Early stroke rehabilitation of the upper limb assisted with an electromyography-driven neuromuscular electrical stimulation-robotic arm. Front. Neurol. **8**, 447 (2017)
25. Rong, W., et al.: A Neuromuscular Electrical Stimulation (NMES) and robot hybrid system for multi-joint coordinated upper limb rehabilitation after stroke. J. NeuroEng. Rehabil. **14**(1), 34 (2017)
26. Sensinger, J.W., Weir, R.F.F.: Improvements to series elastic actuators. In: Proceedings of the 2nd IEEE/ASME International Conference on Mechatronic and Embedded Systems and Applications, MESA 2006 (2007)
27. Stienen, A.H.A., et al.: Self-aligning exoskeleton axes through decoupling of joint rotations and translations. IEEE Trans. Rob. **25**(3), 628–633 (2009)
28. Veerbeek, J.M., et al.: Effects of robot-assisted therapy for the upper limb after stroke. Neurorehabil. Neural Repair **31**(2), 107–121 (2017)
29. Zhang, C., et al.: Robotic approaches for the rehabilitation of upper limb recovery after stroke. Int. J. Rehabil. Res. **40**(1), 19–28 (2017)

Industrial Wearable Robots: A HUMANufacturing Approach

Gaia Salvadore, Edoardo Rota, Elena Corsi, and Giuseppe Colombina(✉)

COMAU S.p.A, 10095 Grugliasco, TO, Italy
giuseppe.colombina@comau.com

Abstract. In the last decades, exoskeletons have mostly been developed and studied for applications in the medical field, as rehabilitation or assistive devices for patients with movement disabilities. Recently, given the high performance of emerging wearable technologies, new applications have been proposed including the every-day support of able-bodied subjects such as workers. The execution of repetitive operations or actions that require excessive effort are the main causes of musculoskeletal injuries in people working in production lines or construction sites. The Industry 4.0 program is bringing companies to re-think their processes by considering human factors, ergonomics and sustainability issues. This is leading to a new trend in automation, which place the workers at the center of a modern smart factory, allowing them to take advantages of new interconnected tools. This new tendency fully embodied Comau's vision. The company, in fact, has coined a term for better picturing its vision: HUMANufacturing.

In this framework exoskeletons have the potential to become more and more adopted by industries as tools to provide support to the workers, preventing the rising of musculoskeletal diseases. This paper provides an overview of the main drivers of this nascent technology. Specifically, it aims to define the requirements that led to the development of an industrial exoskeleton, considering both the end-users and the manufacturers perspective, and showing how the HUMANufactuing approach has a role during the development of a new product.

Keywords: Industry 4.0 · Upper limbs exoskeleton · Ergonomics · HUMANufacturing

1 Exoskeletons: From Medicine to Industry

Exoskeletons are wearable external structure that enhance the wearers physical strength. Exoskeletons work in conjunction with the wearers, reproducing his/hers biomechanical movements. Exoskeletons has their origin, in the late 60ies, both in the medical and in the military field. In the first case they were born as assistive technologies for handicapped patients, in the second cases they had the aim to augment the ability of military users [1]. In the past five years exoskeleton technology has been transposed to the industrial market. Since it is a new technology and the application for

J. Henriques et al. (Eds.): MEDICON 2019, IFMBE Proceedings 76, pp. 1729–1733, 2020.
https://doi.org/10.1007/978-3-030-31635-8_210

industry are still under investigation, data regarding exoskeleton market in industry are few and tend to be different.

The reasons why this technology is now being under use for industrial application are plenty. They go from the possibility of minimizing absenteeism from work, to providing factory accessibility to employee of all ages, and/or assist in reducing strain and stress on joints, and finally, and equally important, they started to be adopted with the aim of improving the operators work quality [2]. Several long-term studies are necessary to further investigate the previously mentioned benefits, which are not the target of this work. On the contrary, this paper focus on the main drivers that brought this kind of technology to the industrial field, providing an overview of the requirements necessary to build an industrial exoskeleton.

2 Market Drivers

2.1 Work-Related Musculoskeletal Disease

Musculoskeletal diseases (MSDs) are injuries and disorders that affect the musculoskeletal system; the apparatus that provide support, movement and stability to the human body. Some of these disease, include, for example, problems such tendonitis, ligament sprain, digital neuritis, carpal tunnel, rotator cuff syndrome and others that may concern bones, cartilage, tendinous ligaments and joints.

Despite in the past year, the health and well-being of the operator has always been an employer priority, the rate of work-related disease in general is still high, one every 5 workers in Europe. Of these disease, almost 50% can be classified as MSD, as shown in Fig. 1, meaning that almost half of European workers (75–80 million) suffer from Work Related Muskoskeletal Diseases (WMSDs) [3].

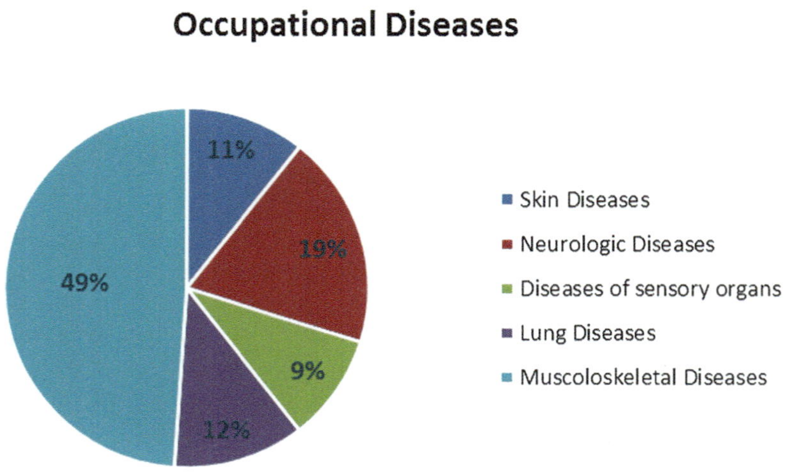

Fig. 1. Distribution of occupational diseases in Europe

In general, work related MSD can be divided depending on the body part they are acting on: lower limbs, backs or upper limbs. According to the UK department of labor, 45% of MSD concern the upper limbs. The most common ULMSD include: rotator cuff syndrome, shoulder bursitis, carpal tunnel syndrome, myofascial 7 pain of upper neck [4]. Those that concern the shoulder are the most critical since they may require a long recuperation period. A risk factor that may cause shoulder disease is any kind of job that required to work with elevated arms, in particular any work performed at or above the acromion level, this is motivated by the fact that this kind of movement imposes high intramuscular pressure and biomechanical demand, loading the soft tissue on the shoulder complex [5].

In US, the direct cost associated to MSD from the companies have been about 50 billion dollars, while the indirect cost can be up to five times the direct cost. It has been documented that shoulder injuries, may results in about 23 lost work days comparing to other kind of injuries that requires from 7 to 9 days off [6].

In UK, an estimated of 3 million working days were lost due to UL-WRMSD, an average of 13 days per case. The total costs to the companies were about 500 million pounds per years. These costs are the sum of work-reorganization, sickness payment, insurance, administration and legal costs.

The economic and human costs of MSDs are unnecessary since Musculoskeletal disorders are "preventable".

2.2 Ageing of the Population

Another key driver is the ageing of the population that need to be active and independent for longer time. According to Bureau of Labor Statics people are not retiring in their 50 anymore. In US the number of workers aged 45 and older increased from 34% to 44%. Those employees may start to experience physical and mental changes with vision, hearing, mental health and stress level. The ageing it may be also related with the lead of sprains and strains and injuries in the different job sector [7].

3 Development Requirements

Most of industrial exoskeleton currently available on the market, have the aim of providing support to the upper limb. When talking about passive exoskeleton, the intention is most of the time to provide a re-distribution of the load, trying to preserve the shoulder joint, being one of the most exposed joint to high biomechanical risk. Below the main design requirements for the development of a passive upper-limb exoskeleton are recapped.

3.1 End-Users Perspective

We will consider as end-users all those operators that work in manufacturer scenarios and that are subject to overhead and repetitive movements. Ideally, they will use the device every day for their entire work-shift. Therefore, they would need a device to be light, or, at least, with almost no perceived additional weight to his current equipment.

As well, they would like the device to be breathable, comfortable, and at the same time it should not to impede any kind of movements he/she usually performs during the working hours. Since the job considered, most of the time, may be an exhausting job for the operators, the final aim of the exoskeletons should the one of bringing relief to the users by alleviating their daily fatigue [8]. If the benefit of using the device overcome the possible negative residual drawbacks (e.g. dressing and undressing time) the operator may be willing to accept a compromise in wearing it.

It is necessary, though, to consider the following two points: first of all, since for most of the end-users, an industrial exoskeleton is something far from their imaginary and everyday life it is fundamental to create awareness and educate them on the actual features and benefits exoskeletons may bring. In order to let them gain confidence with this kind of devices and collect more specific input, it is suggested to involve the end users since the very beginning of the development, and keep the same level of involvement during pilot test with each versions of the device. This is, in fact, reflecting the 'Humanufacturing' approach in the developing phase of a new product made for workers.

Second, it is necessary to consider that, since, at the moment, the technology is not yet mature, the operators are the key figures who will have the final word on the adoption of this new technology.

3.2 Manufacturer Perspective

From the manufacturers perspective, different factors are coming into play. First of all the necessity to guarantee and provide the well-being of the operator. In a sense, exoskeleton, may be seen as a way to make the factory more attractive to work, since they are a tangible symbol of the fact that the employer is taking care of his/her employee. Second the device should respect the plant safety requirements. E.g. if there are aerial line or dynamic line, it is important that the risk of getting stuck with the surrounding environment is minimized. This may be translated in avoiding any kind of cables or tubes protruding from the device itself. Third, it is necessary to satisfy the technical specification of the material of the clothes and uniform usually used in the plant considered. Of course this may change from industry to industry. Food and Medical Industry are the more restricted in terms of uniform and clothes requirement, while industry such as foundry industry may require a fire-resistant device.

Fourth, the manufacturers would also require a minimum time of dressing and undressing. This time, in fact, is considered time of 'non-added value' for the production.

Finally, and equally important, the manufacturer would require a cost-benefit analysis in the long term. This analysis would be depended on the specific industry and company that decide to adopt the technology. Since, among other factors, it is also correlated with the potential number of operators that perform an application where this technology may be used.

3.3 Clinical Perspective

Even if we are talking about industrial devices, exoskeletons, for its nature, have a strong potential clinical impact. During the development phase it is important to ensure that the requirements from a clinical and physiological perspective are met.

The functional target of the device is to reduce the operator fatigue during his/her daily activity. This is possible when there is a reduction of the muscular activation when the operator is wearing the exoskeleton compare to when he/she is not wearing it. The target of a passive upper limb exoskeleton are the muscles of the shoulder chain. Therefore, it need to be verified, through clinical studies, that this hypothesis is valid. Another important point is the avoidance of overloading other muscles of the human body. This, may ensure as well, the avoidance of other kind of discomfort to the users.

Finally it is necessary to verify that the use of exoskeleton do not provide change in the subject posture and/or in the kinematic of the persons.

Both acute studies and long term clinical studies are necessary to investigate these points. At the moment, in the literature, no long-term studies are currently present given the recent years of the adoption of this technology in the industrial field.

4 Discussion and Conclusion

The aim of the work was to first explain the reasons that are leading a technology such an exoskeleton to be exploited in industrial fields, secondly to show the challenges in the definition of the requirements for the development of an exoskeleton for industrial purposes. Answering the new emerging need and exploiting the benefits coming from its background, Comau is the first industrial company adopting the Humanufacturing approach, both for its vision and for its development of new technologies.

As result, using the rationale and the requirements described in this paper, Comau lunched on the market a revolutionary upper limb exoskeleton called "MATE", which stands for muscular aiding tech exoskeleton.

References

1. Dollar, A.M., Herr, H.: Design of a quasi-passive knee exoskeleton to assist running. In: IEEE/RSJ International Conference on Intelligent Robots and Systems (2008)
2. Looze, M.P.d., Bosch, T., Krause, F., Stadler, S., O'Sullivan, L.W.: Exoskeleton for industrial application and their potential effects on physical work load. Ergonomics (2016)
3. European Occupational Diseases Statistics (EODS) 2001–2007
4. Wan Der Windt, D.A.: Occupational risk factors for shoulder pain: a systematic review Occup. Environ. Med. (2000)
5. Coehn, R.B., Williams Jr., G.R.: Impingement syndrome and rotator cuff disease as repetitive motion disorders. Clin. Orthop. Rel. Dis. (1998)
6. Employer reported workplace injuries and illness Bureau of Labor Statistic US department of Labor (2014)
7. Butler, T.: Exoskeleton technology: making worker safest and more productive (2015)
8. Rashedi: Assessing the influence of a passive, upper extremity exoskeletal vest for task requiring arm elevation: Part I 'Expected' effects on discomfort shoulder muscle activity and work task performance. J. Appl. Ergon. (2018)

Upper Limb Exoskeletons for a Better Quality of Life: What Is Currently Available, and What Is Missing in the Market

Marta Baratto[1(✉)], Claudio Ceresi[1], and Valeria Longatelli[2]

[1] E.M.A.C. S.r.L., Genoa, Italy
m.baratto@emac.it
[2] NearLab - Neuroengineering and Medical Robotics Laboratory, Department of Electronics, Information and Bioengineering, Politecnico di Milano, Milan, Italy

Abstract. The market of medical devices for assisting disabled people at home and for rehabilitation in hospital and/or at home is very active and promising. The marketing approach, so far, has been based on experience-based report but a strong need to move to an evidence-based approach is one of the major changes that is expected in the next years. In addition, regulations are pushing towards this direction. How can we get ready for this challenge?

Keywords: Exoskeleton · Upper-limb · Market

1 Introduction

Movement disorders significantly reduce patient's quality of life, as an individual's capacity to move is necessary to perform basic activities of daily living (ADL). Considering that the expectation of life is increasing and that the epidemiology of pathologies affecting the motor control system is critically rising, the impact on the medical care system and on the social network is going to be huge in the next future.

Furthermore, the existing shortage of therapists and caregivers assisting physically disabled individuals is expected to increase and become a serious problem that needs to be addressed.

From early 1990s, researchers have been trying to study solutions to this problem and many robotic devices, for rehabilitation or assistive function, have been developed till to date.

Robotic devices have the potential to give a significant support in a system that risks to collapse, however the availability of these devices in clinical settings is limited leaving plenty of room for improvement.

1.1 Numbers of the Need

Stroke is one of the main causes of acquired disability in adulthood and it is estimated that the number of people with pathologies resulting from stroke is likely to increase,

J. Henriques et al. (Eds.): MEDICON 2019, IFMBE Proceedings 76, pp. 1734–1740, 2020.
https://doi.org/10.1007/978-3-030-31635-8_211

considering the ongoing demographic changes such as the increase of life expectancy and improvements in health conditions [1].

By 2030 stroke is foreseen to be the fourth cause of disability in Western countries [2], approximately 785,000 persons experienced a new or recurrent stroke annually in the United States [3].

In particular, three months after the acute event, a high percentage of patients do not fully recover the motor control functions and such residual alterations concern 93% of the function of the upper limb. Within this series, in about 56% a severe paresis of the upper limb persists, in 37% a partial functional recovery occurs and only 7% see a complete resolution [4].

Focusing on upper limb disability, motor function can be lost also due to traumatic events [5] and full or partial loss of function in the shoulder, elbow or wrist are a common impairment in the elderly people. This overall impairment yields several impacts on health system, domestic life, social life as well as economy. Every year the total cost from lost future productivity is estimated to be $124.5 billion in United States due to stroke [6].

From this quick overview, it is clear that in the future the demand for robotic devices will increase as they represent an effective and operative support for patient rehabilitation. The final goal is to return those individuals as soon as possible into their domestic and social life, capable to perform activities of daily living (ADL).

1.2 Why Robots and Exoskeleton for Rehabilitation

Robotic rehabilitation is a field of research dedicated to the understanding and augmentation of rehabilitation through the application of robotic devices. Rehabilitation using robotics is generally well-tolerated by patients and has been found to be an effective adjunctive therapy for individuals suffering from motor impairments due to neurological disorders [7].

Indeed, such effectiveness relies on intensive task-specific training in which the patient practices physical movements repetitively, using correct form, to boost neural plasticity and functional recovery [8].

Individuals that undergo a stroke need constant medical care and intensive rehabilitation often requiring one-on-one manual interaction with the physical therapist [9]. Robotic devices have the potentiality to provide therapy for a longer time irrespective of skills and fatigue compared to manual therapy [10]. Thus, integration of robotic therapy into current practice could increase the efficiency and effectiveness of therapists by alleviating the labor-intensive aspects of physical rehabilitation [11].

This has a double positive effect on the healthcare system considering that budget restrictions make the duration of rehabilitation program shorter: with the support of robotic devices, the patient recovery time is shorter and therapist's time management is optimized.

Furthermore, the use of robotic devices came along with an important availability of measures and data that allows the clinician to support classical subjective evaluation of the patient status with more precise and objective numbers. This can be used to define the baseline condition of the patient and allows to monitor with very high resolution the functional recovery.

2 What Is Currently Available on the Market

Nowadays the market offers a wide range of possibilities when looking for a robotic device either to support rehabilitation process or for assistive purpose.

The first approach from researchers that resulted then in commercialized products was dedicated to rehabilitation of lower limbs. Lower limbs needed a less complex technical solution compared to the upper limbs due to a narrower range of motion and a lower number of degrees of freedom involved.

Prototypes were developed and products commercialized and a cultural innovation was made in the rehabilitation approach, integrating the manual physical therapy with robot assisted therapy.

Researcher have then translated the validated solutions to the upper limb, but limited to a bidimensional movement and a subset of degrees of freedom involved.

The next step for researchers and the industry is the realization and commercialization of exoskeletons for upper limbs, as technology has finally developed solutions that allow machines to cover the wide range of motion and all the degrees of freedom of the shoulder and the entire upper limb. The first approach on the market was made by ArmeoPower (Hocoma), followed recently by Alex and Aramis.

3 What Is Missing

Despite the use of robotic devices is ongoing since 1990s, commercial products availability is still not so wide and is foreseen to grow significantly in the next years.

The market of certified devices to support robotic lower limbs rehabilitation is quite covered, but for upper limbs is a little behind as most of commercialized devices cover only bidimensional mobilization. What is really still at the first step is the commercialization of exoskeleton for upper limb mobilization.

Price is the main issue. Technology needed to reproduce such a complex joint as the shoulder is still very expensive.

Solved the technical challenge, upper limb exoskeletons will have to face the clinical and validation challenge.

3.1 Technical Challenge

Robotic devices have been incorporated in physical therapy and rehabilitation program for stroke patients from two decades ago [12]. Researchers have developed end effector devices and exoskeleton for rehabilitation of upper limb, but only a few of them are in use for rehabilitation program and commercially available.

However, the rapid advancement in technology has accelerated the development of the upper limb exoskeleton. Researchers have been working on control method and input, actuation, power transmission, sensor technologies etc. to make reliable assistive robotic devices, but there is still significant room for improvement in compactness and weight reduction. The exoskeleton needs high power to weight ratio with high bandwidth, simpler mechanism for transmission with minimum loss, finer coordination of joint torques but preventing the device to become complex and induce a weird posture.

Another important topic is the backdriveable exoskeleton: this kind of control is crucial in the tuning of the rehabilitation session as it allows the device to provide as much as needed torque so that patients can optimize his contribution in motion achievement during therapy [13].

Safety is the top priority issue: an exoskeleton must be stopped in any moment. The controller should be acting like a therapist and adjusting the operation if patients feel pain [14].

The characteristic to peruse to get to a more usable and market oriented device are: reduced weight, excellent power/weight ratio, compensation for gravity, compactness, greater range of motion, safe operation, reliability in all operations, low engineering and construction cost, simple fitting and removal, comfort in wearing, low/no maintenance, portability, modularity, home use [15].

The structure of solutions nowadays available on the market has to become more comfortable to wear, in order to improve acceptability by patients and reduce the setup time for the therapist. The improvement of this single feature of the exoskeleton involves the solution of a wide quantity of complex technical issues.

Unfortunately the optimal availability today of these characteristics came along with significantly high costs.

3.2 Operator Challenge

The number of stroke patients is projected high and one to one rehabilitation program would be impossible due to shortage of therapists. The use of robotic devices can support and cover this need in a specific way, contributing in improving patient quality of recovery and therapist quality of work.

The approach to robotic treatment necessitates a change in the mentality of the clinician and support of the patient in acceptability of the solution.

Patients should not be intimidated by an "invasive" device, nor be afraid to be driven by it, feel safe during the session and experience no pain due to the exoskeleton structure. On the operator side, often physical therapist see robotic devices as a threat to their job, as if they could be replaced by them. They have to be guided in understanding that robots can perform the heavy part of their work, improving the patient involvement and participation thanks to more stimulating interaction (serious gaming) with consequent faster and better recover. The robotic therapy must happen under the expert supervision of the therapist that has to monitor the patient, check the settings and that concentrate on the patient response and consequently improve the therapeutic pathway.

Researchers have developed many robotic devices and exoskeletons for rehabilitation program, which are technologically advanced, but their efficacy and reliability in therapy is still unproved because of their lack of solid clinical evidence. This, with cost constraints, limits the commercialization of also exoskeletons already developed.

3.3 Clinical Validation Challenge

There is evidence in previous literature that robot-assisted rehabilitation in post-stroke individuals have advantages in recover arm motor function in comparison with conventional treatment [11, 16–19]. Performing repetitive movements with the affected limb obtains functional gain thereby an increase of motivation which helps them to use arm further on [20].

There has been a significant effort during the last two decades to improve the design and control strategy of robotic rehabilitation devices, yet less have been done to prove the efficacy of such systems in rehabilitation setting, as the results of clinical evaluation of therapy applying robots are still sparse.

ISO-SPREAD guidelines (Italian Stroke Organization - Stroke Prevention and Educational Awareness Diffusion) foresee the use of robotic devices, better if with assisted-as-need control, combined with conventional treatment in order to improve muscle strength, motor control and promote functional recovery of the limb. Exoskeleton allows intensive and early treatment of patients, even in the presence of severe deficiency in motor control or in ability to deliver strength, thanks to the fine settings allowed by these devices.

Moreover, attention deficits are one of the most frequently observed cognitive deficits after stroke and significantly interfere with patient's participation in motor and cognitive treatment. The use of exoskeleton allows the patient to interact with serious gaming, virtual and real objects: this can help compensating the lack of attention and participation of the patient to the ongoing therapy, keeping the involvement and results to high levels.

Another topic where exoskeleton can provide a significant support is in the assessment and evaluation of the spasticity. The multidisciplinary management of the patient with spasticity includes the definition of spasticity severity both in static and dynamic conditions, the measurement of active and passive movement ranges, all variables that a robotic device can measure and record with high precision.

Even if all the above mentioned advantages of the exoskeleton use are commonly recognized, there are currently no shared rehabilitation protocols that optimize the use of robotic and/or electromechanical devices in rehabilitation.

Further studies should be conducted by aiming at identifying appropriate outcome measures for the assessment of efficacy of rehabilitation techniques with exoskeleton, since the selection of these outcome measures is still a critical element.

As there is not a standard in the exoskeleton technical and biomechanical characteristics and the possibility of range of motion coverage as well as number of degree of freedom actuation can vary significantly between devices, a fundamental starting point is the definition of a grid to combine technical skills and possibility of function to rehabilitate. This can support the standardization of a validation protocol to be shared possibly in multi centric studies, which allow the statistics to be more significant thanks to higher involved numbers.

4 Conclusion

Despite the limited number of commercialized exoskeleton and the limited diffusion of centers equipped with these technologies, the interest in this kind of devices is very high in researcher, either from engineering or clinical field. The proof is that beside the lack of standards, some studies are now starting to be published with the aim of acknowledging the efficacy of robotic devices and exoskeletons.

Some preliminary study on the use of exoskeleton in rehabilitation process shows that it could efficiently induce new connections and facilitate plasticity phenomena potentiation. Knowledge of the potential brain plasticity reservoir after brain damage constitutes a prerequisite for an optimal rehabilitation strategy. Data suggest that the use of an exoskeleton may improve upper limb motor function recovery as predicted by the reshaping of cortical and transcallosal plasticity, according to the baseline cortical excitability [21].

Don Gnocchi Foundation has performed an important multicenter study with the goal of measuring the effectiveness of the use of robotics technology in the rehabilitation of the upper limb in stroke patients. This is an unprecedented search for the number of patients involved, which compared the data collected from patients treated with new technologies (altogether over 250) with data collected from patients treated according to traditional therapies. The results show that rehabilitation with robotic technologies is certainly effective in recovering the upper limb after stroke. Not only that, but for some aspects - such as hand gripping, forearm flexion on the arm and shoulder abduction - rehabilitation with robotic technologies proves to be more effective than conventional rehabilitation, allowing the patient to achieve important motor recovery goals earlier.

The demonstration of the efficacy rationale of new rehabilitative approaches developed in the light of current knowledge on neuroplasticity mechanisms is one of the main topics on which research on stroke rehabilitation is being focused.

On the basis of the expanded knowledge of the neurobiological mechanisms of brain plasticity, approaches are currently favored which confirm the intensity, repetitiveness, significance of an exercise and multisense-stimulation.

Conflict of Interest. The authors declare that they have no conflict of interest.

References

1. Feigin, V.L., et al.: Update on the global burden of ischemic and hemorrhagic stroke in 1990-2013: the GBD 2013 study. Neuroepidemiology **45**(3), 161–176 (2015)
2. Donnan, G.A., et al.: Lancet **371**(9624), 1612–1623 (2008)
3. Benjamin, E.J., et al.: Heart disease and stroke statistics 2017 update. A report from the American Heart Association. Circulation **136**, 230–235 (2017)
4. Mayo, et al.: Disablement following stroke. Disabil. Rehabil. **21**(5–6), 258–268 (1999)
5. Reid, D.C.: Sports Injury Assessment and Rehabilitation. Churchill Livingstone, New York (1992)
6. Mozaffarian, D., et al.: Heart disease and stroke statistics-2015 update: a report from the American Heart Association. Circulation **131**, e29–e322 (2015)
7. Calabrò, R.S., et al.: Robotic gait rehabilitation and substitution devices in neurological disorders: where are we now? Neurol. Sci. **37**, 503–514 (2016)
8. Bertani, R., et al.: Effects of robot-assisted upper limb rehabilitation in stroke patients: a systematic review with meta-analysis. Neurol. Sci. **38**, 1561–1569 (2017)
9. Poli, P., et al.: Robotic technologies and rehabilitation: new tools for stroke patients' therapy. Biomed. Res. Int. **2013**, 153872 (2013)
10. Teasell, R.W., et al.: What's new in stroke rehabilitation. Stroke **35**, 383–385 (2004)

11. Lum, P.S., et al.: Robotassisted movement training compared with conventional therapy techniques for the rehabilitation of upper-limb motor function after stroke. Arch. Phys. Med. Rehabil. **83**, 952–959 (2002)
12. Kommu, S.S.: Rehabilitation Robotics. Itech Education and Publishing, Vienna (2007)
13. Piña-Martínez, E., et al.: A novel exoskeleton for continuous monitoring of the upper-limb during gross motor rehabilitation. In: Converging Clinical and Engineering Research on Neurorehabilitation II, pp. 1199–1203 (2017)
14. Benitez, L.M., et al.: Exoskeleton technology in rehabilitation: towards an EMG-based orthosis system for upper limb neuromotor rehabilitation. J. Robot. **2013**, 13 (2013)
15. Rahman, M., et al.: Development and control of a robotic exoskeleton for shoulder, elbow and forearm movement assistance. Appl. Bion. Biomech. **9**, 275–292 (2012)
16. Bayona, N.A., et al.: The role of task specific training in rehabilitation therapies. Top. Stroke Rehabil. **12**, 58–65 (2005)
17. Prange, G.B., et al.: Systematic review of the effect of robot-aided therapy on recovery of the hemiparetic arm after stroke. J. Rehabil. Res. Dev. **43**, 171–184 (2006)
18. Kwakkel, G., et al.: Effects of robot-assisted therapy on upper limb recovery after stroke: a systematic review. Neurorehab. Neural Repair **22**, 111–121 (2007)
19. Masiero, S., et al.: Robotic assisted rehabilitation of the upper limb after acute stroke. Arch. Phys. Med. Rehabil. **88**, 142–149 (2007)
20. Fazekas, G., et al.: Robot-mediated upper limb physiotherapy for patients with spastic hemiparesis: a preliminary study. J. Rehab. Med. **39**, 580–582 (2007)
21. Calabrò, R.S., et al.: Who may benefit from Armeo Power treatment? A neurophysiological approach to predict neurorehabilitation outcomes. PM R **2**(4) (2016)

Special Sessions: Neurosystems and Connectivity

Optimization of a Motor Imagery Paradigm for Self-modulation of Bilateral Premotor Interhemispheric Functional Connectivity in fMRI Neurofeedback

João Pereira[1,2(✉)], Bruno Direito[1,2], Alexandre Sayal[1,2], Carlos Ferreira[1,2], and Miguel Castelo-Branco[1,2]

[1] Institute for Biomedical Imaging and Life Sciences (CNC.IBILI), Faculty of Medicine, University of Coimbra, Coimbra, Portugal
joaoandrefpereira@gmail.com
[2] CIBIT, Coimbra Institute for Biomedical Imaging and Life Sciences, Institute of Nuclear Sciences Applied to Health (ICNAS), University of Coimbra, Coimbra, Portugal

Abstract. Neuromodulation of a single brain region in an fMRI Neurofeedback experiment is a widely used technique with reported training effects on neuroplasticity and in behavioral changes. Recently, experiments aiming for modulation of functional relationship between segregated brain areas have been tested for their feasibility to understand underlying connectivity-based processing brain mechanisms. We performed a series of pilot studies to optimize an experimental protocol for interhemispheric functional connectivity modulation between bilateral premotor areas in a fMRI Neurofeedback setup, using a motor imagery paradigm. We have found that a better differentiation of correlation coefficient distributions between conditions was achieved with an optimized strategy with a reference task that consisted of the imagination of hand movement with gradual variation in the frequency of the imagined movements. Results also show that distinction between bimanual and unimanual motor imagery is not ideal for correlation-based differentiation using this paradigm. With this proof-of-concept study we conclude the feasibility of this setup as well as hypothesize that in future acquisitions, the reference tasks for an up-regulation condition should be based on this adaptive strategy and down-regulation condition should be optimized to promote the absence of activation in premotor areas.

Keywords: Neurofeedback · fMRI · Functional connectivity

1 Introduction

Functional magnetic resonance imaging (fMRI) is a non-invasive technique that uses the blood oxygenation level dependent (BOLD) signal for neurofeedback (NF) applications, allowing participants to have real-time access to measures of their own brain activity in order to learn to volitionally self-modulate a specific brain region, combination of regions, or connectivity between regions [1]. Since the first real time fMRI-NF experiment [2], studies have reported single region of interest (ROI) self-modulation related to

J. Henriques et al. (Eds.): MEDICON 2019, IFMBE Proceedings 76, pp. 1743–1751, 2020.
https://doi.org/10.1007/978-3-030-31635-8_212

particular cognitive functions [3, 4]. In order to confirm the effect on brain function, neuroplasticity and cognitive improvement, specific behavioral tests were performed to document training effects [5]. Complex cognitive processes (e.g., emotion processing, motor coordination, language, etc.) imply a coordinated activity between several brain regions. In spite of the positive evidence favoring the self-regulation of a specific brain region and induced behavioral changes, underlying connectivity based processing mechanisms can be often overlooked in this context [6]. Recently, neurofeedback experiments focused on the modulation of the functional relation of more than one brain region, i.e., their connectivity in visual parietal networks [7], between inferior frontal gyrus (IFG) and the superior temporal gyrus (STG) [8], between dorsolateral prefrontal cortex (dlPFC) and the ventromedial prefrontal cortex network (vmPFC) [9] and ipsilesional cortical-subcortical connectivity [10].

Motor imagery (MI) is defined as a dynamic state in which the subject mentally simulates a particular action so that he/she can feel himself as if performing the imagined action [11]. Functional imaging studies [12–14] have shown an overlap between the neural networks involved in MI and motor performance (MP), reporting coincident activations in several areas such as the supplementary motor area (SMA), premotor cortex (PMC), parietal areas, the basal ganglia and the cerebellum. Nonetheless there are areas with greater activation during MI as compared with real execution such as the ventral part of the premotor cortex (vPMC), important for movement planning [15]. Interhemispheric functional interactions between motor-related areas in both MI and MP tasks have been previously reported [12, 13] and its relevance in terms of movement coordination highlighted. In this proof-of-concept experiment, we aim to design an experimental protocol that best suits a real-time modulation of interhemispheric functional connectivity between bilateral premotor cortices in an fMRI neurofeedback setup using a MI paradigm. This would be an important step towards the development of functional connectivity-based neurofeedback as a possible therapeutic tool in motor rehabilitation.

2 Materials and Methods

2.1 Participants

Four volunteers were recruited, for repeated assessment with different protocols. All had normal or corrected-to-normal vision and no history of neurological or psychiatric diseases. All volunteers gave informed consent before participating. This study was conducted in accordance to the declaration of Helsinki.

2.2 Real Time fMRI NF Data Acquisition

Each session consisted of an anatomical run and six functional runs - a functional localizer and five imagery runs. The first and the last imagery runs were performed without feedback information. The setup includes a 3T Siemens Magnetom TrioTim scanner with a 12-channel head coil. Participants underwent a T1 high-resolution magnetization-prepared rapid acquisition gradient echo (MPRAGE) sequence for co-registration of functional data (176 slices; Echo Time (TE): 3.42 ms; Repetition time (TR): 2530 ms; voxel size: 1 × 1 × 1 mm; Flip Angle (FA): 7°; Field of view (FOV): 256 × 256 mm). Functional imaging focused on motor-related areas consisted of an echo-planar imaging (EPI) sequence with 26 slices, in-plane resolution: 3 × 3 mm, FOV: 210 × 210 mm, slice thickness: 3.5 mm, FA: 75°, TR = 1500 ms and TE = 30 ms. Real-time data processing was performed using Turbo-BrainVoyager 3.2 (TBV) (Brain Innovation, Maastricht, The Netherlands). The software implements online 3D motion correction and alignment between functional runs. An incremental General Linear Model (GLM) based on a recursive squares algorithm accessed the functional data as soon as available. Stimuli were presented on an LCD screen (70 × 39.5 cm, resolution of 1920 × 1080 pixels, 60 Hz refresh rate) which the participants viewed through a mirror mounted above their eyes at an effective distance of 156 cm.

2.3 Functional Regions of Interest (ROIs) Definition

We used a mapping procedure based on previous tasks [16] that included two activation conditions: motor imagery (MI) and motor performance (MP) as well as baseline periods in a block design experiment. The mapping procedure included a total of 17 30-s blocks - nine baseline blocks, four MI blocks and four MP blocks. Each participant was instructed to execute or imagine performing a specific bilateral finger tapping sequence at a specific frequency (2 Hz). The ROIs (bilateral PMC) were functionally defined according to the contrast MI > Baseline, guided by anatomical landmarks [17].

2.4 Feedback Calculation and Stimulus Presentation

Feedback was calculated using a 8-point sliding window Pearson's correlation coefficient between the mean BOLD values of the online selected ROIs (Eq. 1), updated at every time point (1.5 s). The visual feedback was presented through a thermometer bar divided into 20 blocks, each block representing a 0.1 correlation increment (values range from −1 to +1).

$$p = \frac{\sum_{i=1}^{n}(x_i - \bar{x})(y_i - \bar{y})}{\sqrt{\sum_{i=1}^{n}(x_i - \bar{x})^2.(y_i - \bar{y})^2}} = \frac{cov(X, Y)}{\sqrt{var(X).var(Y)}} \quad (1)$$

2.5 Neurofeedback Reference Strategies

The imagery runs included three conditions: 'Bimanual', 'Unimanual' and 'Baseline'. During the 'Bimanual' condition (presented 4 times per run) the participants were asked to fill up the thermometer bar. On the opposite, during the 'Unimanual' and 'Baseline' blocks (presented 4 and 8 times per run), they were instructed to implement a strategy that would keep the thermometer as close to zero as possible. During this protocol optimization phase, different strategies were used in the reference tasks for both activation conditions. The reference for the 'Baseline' condition, where subjects were asked to remain in a resting state, did not change. Initially, for the 'Bimanual'/ 'Unimanual' conditions, the reference task was a constant bimanual/unimanual imagined movement. With the optimization of the experimental protocol, the reference tasks consisted in bimanual/unimanual motor imagery movement with gradual frequency variation (a first increase until midblock followed by a decrease in frequency, preferably near 0 Hz at the end of the block). An auditory beep at midblock on both conditions was added.

3 Results

3.1 Initial Strategy

Higher correlation values are reported during 'Bimanual' condition in two NF runs (Table 1) with distributions significantly higher than during the 'Baseline' condition (Table 2). On the contrary, during the NF run 2 the median value of the correlation distribution for the 'Baseline' condition was higher than for the activation conditions ('Bimanual' and 'Unimanual'). Less successful differentiation is present between 'Unimanual'/'Baseline' and 'Bimanual'/'Unimanual' conditions, only significant in one run each. Block-average per time point within a condition block of the mean BOLD activation for lPMC and rPMC using as reference strategy a constant motor imagery task is reported in Fig. 1.

Table 1. Median values for the correlation distributions for 'Bimanual', 'Unimanual' and 'Baseline' for each neurofeedback run performed during the first pilot study.

Run\Condition	Median		
	Bimanual	Unimanual	Baseline
NF run 1	0,66	0,41	0,44
NF run 2	0,48	0,47	0,53
NF run 3	0,64	0,42	0,35

Table 2. Wilcoxon test results for differentiation of correlation distributions between conditions in each neurofeedback run for the first pilot study with the initial strategy.

Run\ Condition	Bimanual - Baseline			Unimanual - Baseline			Bimanual - Unimanual		
	p	h	stats	p	h	stats	p	h	stats
NF run 1	5,12E–03	1	2,80E+00	8,59E–01	0	−1,78E–01	5,45E–02	0	1,92E+00
NF run 2	7,80E–01	0	–2,80E–01	9,76E–01	0	–2,95E–02	8,90E–01	0	−1,38E–01
NF run 3	1,47E–06	1	4,81E+00	4,58E–02	1	2,00E+00	1,13E–02	1	2,53E+00

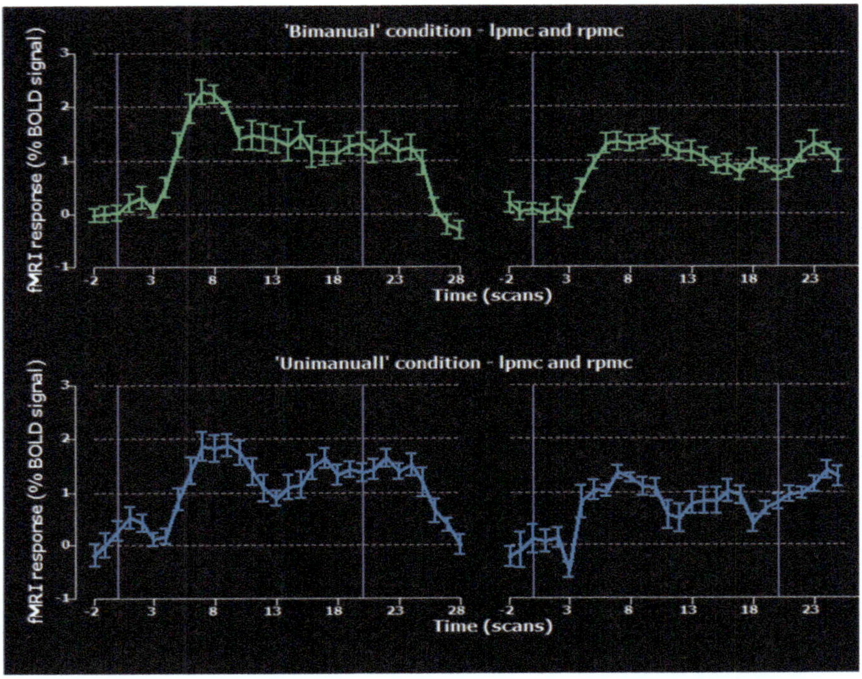

Fig. 1. Block-related average of the mean values of BOLD for each time point (lPMC on the left and rPMC on the right) for the three NF runs in the first pilot for both 'Bimanual' and 'Unimanual' conditions.

3.2 Optimized Strategy

Higher correlation values in activation conditions (Table 3) result in distributions significantly higher than 'Baseline' condition in two runs each. Differentiation between activation conditions is only reported in one run. In one neurofeedback (NF 3), the 'Baseline' correlation values were significantly higher comparing with the 'Unimanual' condition (Table 4). Block-average per time point within a condition block of the mean BOLD activation for lPMC and rPMC using as reference strategy a gradual variation in the frequency of the imagined movements is reported in Fig. 2.

Table 3. Median values for the correlation distributions for 'Bimanual', 'Unimanual' and 'Baseline' for each neurofeedback run performed during the last pilot study.

Run\Condition	Median		
	Bimanual	Unimanual	Baseline
NF run 1	0,65	0,71	0,57
NF run 2	0,59	0,62	0,31
NF run 3	0,44	0,23	0,38

Table 4. Wilcoxon test results for differentiation of correlation distributions between conditions in each neurofeedback run for the last pilot study with the optimized strategy.

Run\Condition	Bimanual - Baseline			Unimanual - Baseline			Bimanual - Unimanual		
	p	h	stats	p	h	stats	p	h	stats
NF run 1	3,27E–04	1	3,59E+00	5,36E–06	1	4,55E+00	4,46E–01	0	–7,63E–01
NF run 2	1,18E–06	1	4,86E+00	5,00E–10	1	6,22E+00	8,09E–02	0	–1,75E+00
NF run 3	1,62E–01	0	1,40E+00	4,82E–02	1*	–1,98E+00	8,90E–03	1	2,62E+00

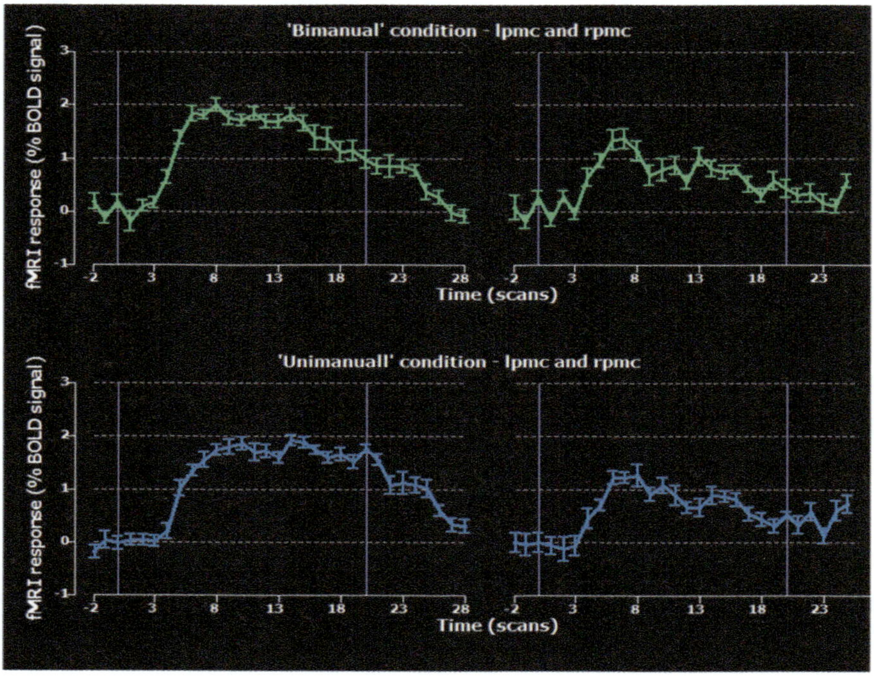

Fig. 2. Block-related average of the mean values of BOLD for each time point (lPMC on the left and rPMC on the right) for the three NF runs in the last pilot study for both 'Bimanual' and 'Unimanual' conditions.

4 Discussion

The main achievement of this work was the design and optimization of a strategy in a real time functional magnetic resonance imaging neurofeedback setup that facilitates self-modulation of functional connectivity.

To do so, several pilot studies were performed, testing different strategies as reference tasks for the imagery runs. Using the initial strategy as a reference task, Fig. 1 shows an approximation to a common boxcar function (constant bilateral PMC activation after an initial peak) for both 'Bimanual' and 'Unimanual' conditions. Considering the functional connectivity metrics, this activation pattern makes correlation values decrease to values near zero midblock, because of the plateau of activity present during this period. Therefore, low feedback values could be fed back to the participant even though it is performing a motor imagery task. Additionally, most of the variance in both ROIs was derived from the protocol (block design), resulting in increases in connectivity values during blocks transitions and in the initial time points of an expected down-regulation condition.

To optimize the experimental protocol a different strategy was needed to differentiate correlation distributions between conditions. Having into consideration the properties of the feedback measurement we hypothesized that consistent variations in the premotor cortex BOLD signal within a condition block would maximize differences between conditions. In that sense, an optimized strategy was used that consisted of the imagination of hand movement with gradual variation in the frequency of the imagined movements, within each 'Bimanual' and 'Unimanual' condition block. This assumption is made since the activation level within the whole motor network increases with increasing finger tapping speed [18]. The rationale of this strategy is to avoid activity 'plateaus' where correlations become non-discernible, and to promote activity variations which favor detection of correlation patterns. This is visible in Fig. 2 where the activation pattern for both 'Bimanual' and 'Unimanual' conditions and both premotor areas resembles a triangular shape time course of the mean BOLD value, increasing the likelihood of identifying functional correlations. This is corroborated with an increase in neurofeedback runs where the activation conditions presented higher correlation values than the 'Baseline'.

The 'Unimanual' condition, with unimanual imagined tasks, thought to have as consequence a decrease in functional connectivity, has correlation values statistically different from 'Baseline' condition in various functional runs. Also, the differentiation of its correlation distribution to the condition 'Bimanual' distribution is not conspicuous. This, alongside the plots of variation of BOLD signal, shows that in unimanual hand motor imagery tasks there is also an activation of the ipsilateral premotor (to the reference hand) which results in higher correlation values. This activation might be caused by the PMC action, reported by [19], of inhibition of the ipsilateral primary motor cortex (M1), even in motor imagery (MI) tasks.

To achieve an optimized protocol and taking into consideration the results from this proof of concept protocol optimization studies, we hypothesized that distinction between 'Bimanual' and 'Unimanual' conditions is not ideal for correlation modulation using this paradigm.

Having this in mind, for future acquisitions functional runs should be composed only by two conditions: an up-regulation and a down-regulation block, in which the subject is told to increase and decrease his own functional connectivity between bilateral premotor cortices. The reference task for the increased connectivity blocks should be similar to the optimized 'Bimanual' condition, with a gradual variation in the frequency of the imagined bimanual movements. The reference task for the decrease connectivity blocks should resemble the previous 'Baseline' condition task, resting state or another mental task that does not activate premotor areas.

References

1. Sulzer, J., Haller, S., Scharnowski, F., Weiskopf, N., Birbaumer, N., Blefari, M.L., Bruehl, A.B., Cohen, L.G., DeCharms, R.C., Gassert, R., et al.: Real-time fMRI neurofeedback: progress and challenges. Neuroimage **76**, 386–399 (2013)
2. Weiskopf, N., Mathiak, K., Bock, S.W., Scharnowski, F., Veit, R., Grodd, W., Goebel, R., Birbaumer, N.: Principles of a brain-computer interface (BCI) based on real-time functional magnetic resonance imaging (fMRI). IEEE Trans. Biomed. Eng. **51**, 966–970 (2004)
3. Caria, A., Sitaram, R., Veit, R., Begliomini, C., Birbaumer, N.: Volitional control of anterior insula activity modulates the response to aversive stimuli. A real-time functional magnetic resonance imaging study. Biol. Psychiatry **68**, 425–432 (2010)
4. Rota, G., Sitaram, R., Veit, R., Erb, M., Weiskopf, N., Dogil, G., Birbaumer, N.: Self-regulation of regional cortical activity using real-time fMRI: the right inferior frontal gyrus and linguistic processing. Hum. Brain Mapp. **30**, 1605–1614 (2009)
5. Weiskopf, N.: Real-time fMRI and its application to neurofeedback. Neuroimage **62**, 682–692 (2012)
6. Friston, K.J.: Functional and effective connectivity: a review. Brain Connect. **1**, 13–36 (2011). https://doi.org/10.1089/brain.2011.0008
7. Koush, Y., Rosa, M.J., Robineau, F., Heinen, K., Rieger, S.W., Weiskopf, N., Vuilleumier, P., Van De Ville, D., Scharnowski, F.: Connectivity-based neurofeedback: dynamic causal modeling for real-time fMRI. Neuroimage **81**, 422–430 (2013)
8. Ruiz, S., Buyukturkoglu, K., Rana, M., Birbaumer, N., Sitaram, R.: Real-time fMRI brain computer interfaces: self-regulation of single brain regions to networks. Biol. Psychol. **95**, 4–20 (2014)
9. Spetter, M.S., Malekshahi, R., Birbaumer, N., Lührs, M., van der Veer, A.H., Scheffler, K., Spuckti, S., Preissl, H., Veit, R., Hallschmid, M.: Volitional regulation of brain responses to food stimuli in overweight and obese subjects: a real-time fMRI feedback study. Appetite **112**, 188–195 (2017)
10. Liew, S.-L., Rana, M., Cornelsen, S., de Barros Filho, M.F., Birbaumer, N., Sitaram, R., Cohen, L.G., Soekadar, S.R.: Improving motor corticothalamic communication after stroke using real-time fMRI connectivity-based neurofeedback. Neurorehabil. Neural Repair **30**, 671–675 (2016)
11. Decety, J.: The neurophysiological basis of motor imagery. Behav. Brain Res. **77**, 45–52 (1996)
12. Liuzzi, G., Hörniss, V., Zimerman, M., Gerloff, C., Hummel, F.C.: Coordination of uncoupled bimanual movements by strictly timed interhemispheric connectivity. J. Neurosci. **31**, 9111–9117 (2011)
13. Gao, Q., Duan, X., Chen, H.: Evaluation of effective connectivity of motor areas during motor imagery and execution using conditional Granger causality. Neuroimage **54**, 1280–1288 (2011)
14. Gerardin, E., Sirigu, A., Lehéricy, S., Poline, J.B., Gaymard, B., Marsault, C., Agid, Y., Le Bihan, D.: Partially overlapping neural networks for real and imagined hand movements. Cereb. Cortex **10**, 1093–1104 (2000)
15. Guillot, A., Di Rienzo, F., Collet, C.: The neurofunctional architecture of motor imagery. In: Advanced Brain Neuroimaging Topics in Health and Disease - Methods and Applications (2014). http://dx.doi.org/10.5772/58270

16. Xie, F., Xu, L., Long, Z., Yao, L., Wu, X.: Functional connectivity alteration after real-time fMRI motor imagery training through self-regulation of activities of the right premotor cortex. BMC Neurosci. **16**, 29 (2015)
17. Schubotz, R.I., Anwander, A., Knösche, T.R., von Cramon, D.Y., Tittgemeyer, M.: Anatomical and functional parcellation of the human lateral premotor cortex. Neuroimage **50**, 396–408 (2010)
18. Rao, S.M., Bandettini, P.A., Binder, J.R., Bobholz, J.A., Hammeke, T.A., Stein, E.A., Hyde, J.S.: Relationship between finger movement rate and functional magnetic resonance signal change in human primary motor cortex. J. Cereb. Blood Flow Metab. **16**, 1250–1254 (1996)
19. Grefkes, C., Eickhoff, S.B., Nowak, D.A., Dafotakis, M., Fink, G.R.: Dynamic intra- and interhemispheric interactions during unilateral and bilateral hand movements assessed with fMRI and DCM. Neuroimage **41**, 1382–1394 (2008)

Special Sessions: Therapeutic Applications of Imaging and Neuro-Stimulation

A Hybrid Brain-Computer Interface Fusing P300 ERP and Electrooculography

João Perdiz[1,2(✉)], Aniana Cruz[1], Urbano J. Nunes[1,2], and Gabriel Pires[1,3]

[1] Institute of Systems and Robotics, University of Coimbra, Coimbra, Portugal
joao.perdiz@isr.uc.pt
[2] Department of Electrical and Computer Engineering, University of Coimbra, Coimbra, Portugal
[3] Engineering Department, Polytechnic Institute of Tomar, Tomar, Portugal

Abstract. An Electrooculography-based method is used to correct misclassification of P300 event related potentials in a Lateral Character Speller (LSC) Brain-Computer Interface (BCI). The LSC speller's circular layout allows us to combine P300 detection with the detection of eye movements to improve symbol detection reliability. We separately classify the vertical and horizontal components of Electrooculography signals from shifts in user gaze during intertrial intervals, determining the quadrant of the character the participant will focus on in the next trial. A P300 EEG-based classification decision can then be corrected using quadrant information, selecting the character with the highest probability on that quadrant. This paper focuses on the implementation of the EOG quadrant detector. Preliminary results show good lateral identification but a lower selection accuracy. Empirically, it was possible to conclude that a relatively high percentage of P300 classification errors were corrected using lateral information alone, significantly increasing LSC character selection accuracy.

Keywords: Electrooculography · Character Speller · Event-related potentials · EEG

1 Introduction

Brain-Computer Interfaces (BCIs) have enabled people with severe motor disabilities to interact with the outside world using only their brain signals [8,12]. However, BCI systems still face many challenges such as low number of control commands, low classification accuracy and low transfer rates. In order to overcome these issues and to broaden the target population that can benefit from BCIs, hybrid brain-computer interfaces (hBCI) technology combining brain signals with other physiological signals have been proposed in different ways [5,9],

J. Henriques et al. (Eds.): MEDICON 2019, IFMBE Proceedings 76, pp. 1755–1766, 2020.
https://doi.org/10.1007/978-3-030-31635-8_213

namely: (a) combination of different physiological signals, such as Electroencephalography (EEG) with Electromyography (EMG) [7], EEG with Electrooculography (EOG) [15] or EEG and functional Near Infrared Spectroscopy (fNIRS) [1]; (b) combination of different neural signals, for example, P300 with motor imagery or with steady state visual evoked potentials (SSVEP) [18]; and (c) combination of different stimulation modalities such as visual and auditory stimuli [3]. Hybrid BCI approaches have been used in many different applications, such as speller system [17], wheelchair control [16], robotic arm control [6], or controlling a personal computer (e.g., select, open, and close a file) [2]; however, it was concluded that most of the hBCIs proposed in more recent years have combined EEG with EOG [5]. For people who still maintain different functionalities, even if residual, hBCIs can be more advantageous than a single-modality BCI, as all their available functionalities can be used to provide a more reliable BCI output. The majority of motor disabled people are capable of minor eye movements, thus the EEG/EOG-based hybrid BCI systems might be useful for these BCI target users.

In this work, we present a new EEG/EOG-based hybrid BCI system that takes advantage of the layout of the P300 paradigm by integrating the direction of the user's gaze, extracted from the EOG signal, in the P300-based BCI speller. The new method combines the EEG detection methods already in use in our speller [12] with an EOG-based algorithm that assists the EEG-based core method in selecting the letter the user has chosen during the last spelling iteration. Quadrant detection will be useful for correcting LSC P300 detection errors, since that from datasets collected in [4], we concluded that about 60% of the P300 errors the detected target are in a different quadrant, and about 30% of the errors occur on the side opposite from the target.

2 Background

The Lateral Single Character speller (LSC) is a P300-based BCI communication speller developed in our research group [12] and that has already been used in different scenarios, for example, combined with detection of error related potentials (ErrPs) [4]. LSC's paradigm is significantly different from those commonly found in other character spellers. Specifically, instead of presenting characters in a 6×6 matrix, it has a circular and symmetrical arrangement as shown in Fig. 1. It encodes twenty-eight symbols comprising all letters of the alphabet and the 'space' and 'del' keys. These symbols flash individually according to an oddball paradigm with a flash duration of 75 ms and that alternate between the left and right sides of the screen. The number of event repetitions (N_{rep}) is usually adjusted individually according to user's performance achieved in the calibration session to attain a classification accuracy around 90%.

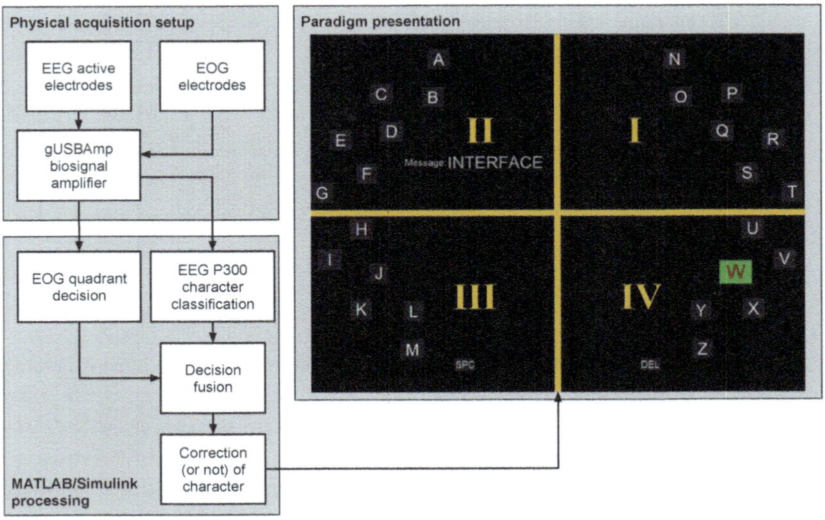

Fig. 1. Physical acquisition setup, classification framework and printscreen of the visual LSC speller [12]. All characters flash for a set number of times, randomly – the highlighted character is flashing green. The message sequence is displayed between the two main groups of characters. Quadrant numbers and divisions shown by yellow markings are not part of the paradigm, but are related to EOG detection.

3 Materials and Methods

3.1 Biosignal Acquisition and Classification Framework

EEG and EOG signals are acquired with a g.USBamp bioamplifier. Signal processing and classification, visual stimuli and visual feedback are implemented in a Matlab/Simulink framework as depicted in Fig. 1. EEG signals were recorded using 12 electrodes (Fz, Cz, C3, C4, CPz, Pz, P3, P4, PO7, PO8, POz and Oz) placed according to the international extended standard system. The right or left earlobe was used as reference and the AFz electrode was the ground. The EEG signals were filtered using a 1–10 Hz bandpass filter and a 50 Hz notch filter and sampled at 256 Hz.

EOG signals are acquired from a pair of bipolar electrodes in vertical and horizontal layouts around the eyes. They are sampled at 256 Hz and filtered with a 50 Hz notch filter. Signal sections relevant for quadrant detection are smoothed using a 5th order Savitzky-Golay filter with a frame size of 51 elements.

3.2 Hybrid EEG-EOG Paradigm

The LSC letters flash randomly according to an oddball paradigm, thus in each round (flashing sequence) there is one target symbol and 27 standard symbols. Each trial consists of a set of rounds. To select a symbol/letter the user focuses his attention on the target symbol and mentally counts them, ignoring all other standard symbols. The experiment consisted of two phases: calibration and online session. Participants performed the calibration session to gather the EEG data to train the classifier. In this session they had to attend the letters of the sentence "INTERFACES-VISUAIS" which flashed 9 times, gathering 162 target epochs and 4374 non-target epochs. During the online session the participants spelled the Portuguese sentence "ESTOU-A-ESCREVER-COM-UMA-INTERFACE-BCI".

The interval between trials (ITI) was set to 4 s. During the ITI, just after a trial ends, participants are shown which character was selected by the classification algorithm. They are then required to center their gaze on the cross marking the paradigm's center. After centering the gaze, they choose which character to target next, and then they divert their gaze to the character they want to select. These changes in gaze are monitored using EOG signals, and used to alter the character selection made by the P300 ERP classifier whenever there is a mismatch between the quadrant detected by the EOG detector and the quadrant of the target letter detected with the P300 classifier. The ITI's event progression pipeline is shown in Fig. 2.

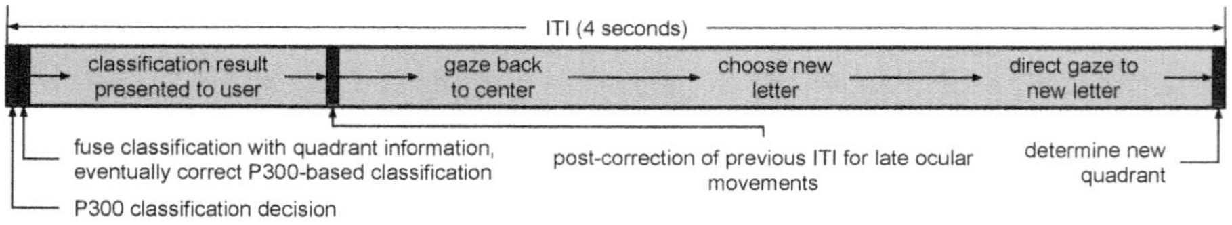

Fig. 2. Sequence of events and processes that occur during an intertrial interval (ITI), from left to right. Shaded sections of the bar show when processing and classification methods are executed, with text indicating the sequence of processes.

Quadrant-Based Decision Correction The combination of horizontal and vertical EOG decisions points us to the quadrant (Q_1, Q_2, Q_3 or Q_4, as shown on Fig. 1) on which gaze is placed before each set of flashing sequences (trial). If this quadrant matches the first choice of the ERP-based classifier in the next trial, no correction is made to the classified character. However, if it conflicts with that outcome, the secondary choice of the ERP classification algorithm is compared and, if matching with the found quadrant, is chosen as the character selected by the user. This method is described more thoroughly in Sect. 3.5.

3.3 EEG Training and Classification

Classification of the P300 ERPs is based on a statistical spatial filter called Fisher criterion beamformer (FCB) [11]. EEG data is transformed into projections, from which the two most discriminative projections are classified by a Naive Bayes classifier, which gives a probability (P_s) for each symbol $j = 1, \cdots, N_s$, where

$N_s = 28$ is the number of symbols. From this classifier's decision we choose the four symbols with the highest probabilities, according to the rule in Eq. 1:

$$S_i \equiv \arg\max P_j, \quad j \in \{1, \cdots, N_s\} \backslash \{\emptyset, \cdots, S_{i-1}\}, \quad i \in \{1, 2, 3, 4\} \qquad (1)$$

3.4 Detecting Quadrant-Related EOG Signal Changes

The EOG detector detects the quadrant at the end of each ITI, by identifying all the saccades and blinks that occur during that interval. The EOG detector follows an approach similar to that of [10]. However, here, more than one EOG event can occur, since the ITI is much longer than the detection window used in [10]. Moreover, here blinks are undesirable events that need to be eliminated while in [10] they are used as control commands. Blinks are defined as signal peaks, usually detected in vertical EOG, sided by positive and negative first derivative peaks, and exceeding a threshold relative to its neighborhood that is much larger than that used in vertical saccade detection. These characteristics can be seen in Fig. 3. Detected blinks must be at least 0.5 s (128 samples) apart. Signal neighborhoods – from both EOG channels – of detected blinks are "flattened" through the use of spline interpolation. All signal points less than 60 samples away from the blink's location are replaced by the interpolation with the rest of each signal's window.

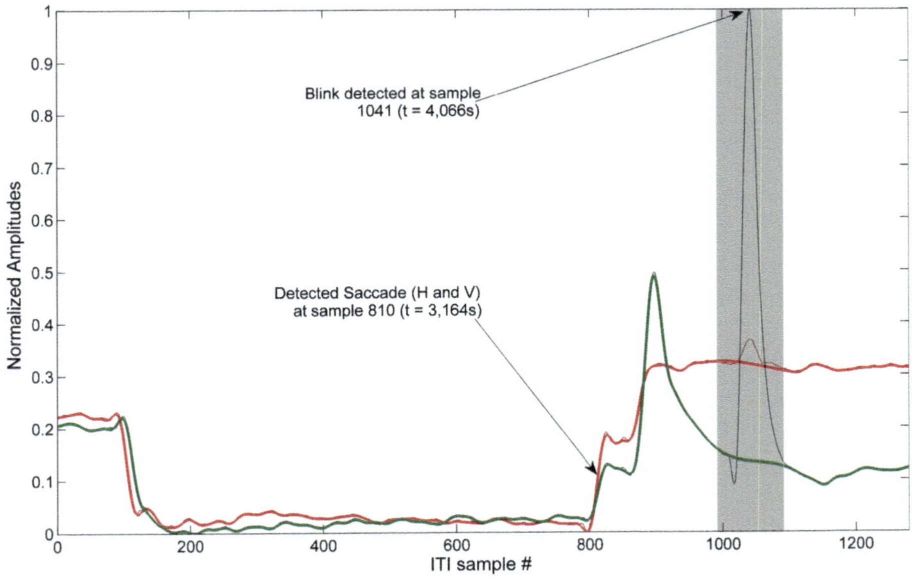

Fig. 3. Typical Horizontal (red) and Vertical (green) EOG signals occurring during one ITI, normalized to the highest amplitude. Thin lines show the signals before smoothing and eliminating blinks. The shaded area shows where each signal was replaced by the spline interpolation of the rest of its window in order to eliminate the blink. Afterwards, horizontal and vertical saccades were found around sample 810.

After blink elimination we search for horizontal saccades. Their clear signature in the horizontal EOG channel is not significantly affected by vertical components of the ocular movement. Horizontal saccades are identified using the signal's first derivative's peaks, and signal averages before and after these peaks are subtracted and compared to a threshold, meaning that a saccade signal must be step-shaped. We also record the vertical signals for the neighborhoods of valid horizontal events, and locate vertical events in those neighborhoods. Contrary to what happens for horizontal signals, both peaks and steps can be detected in vertical signals. The procedure is described in Algorithm 1.

Moreover, a 1 s window of the EOG signal (the first second of the previous trial) can be analyzed after the trial ends to detect late ocular events occurring after the end of the ITI; if found, these can affect EOG-based quadrant selection. This post-correction (see Fig. 2) is executed on the ITI following the trial, using the same methods as the main analysis, and may in the same way change the correction issued using the fusion method.

Algorithm 1. Saccade detection algorithm executed upon the horizontal and vertical EOG signal windows during the intertrial interval. The algorithm is applied after the blink elimination algorithm is applied.

Let X_H and X_V be the horizontal and vertical EOG signal during an ITI, splined from raw signals if blinks were previously found; X' denotes a derivative of X
Define horizontal and vertical EOG amplitude thresholds for saccades, Δ_H and Δ_V
Function 1 Find horizontal saccades and record neighborhoods
Find m_p peaks from $|X'_H|$ in locations $p_i : \forall p_i, p_{i+1} - p_i \geq 128 \wedge \forall p_b, p_i - p_b \geq 64$, with p_b marking blink locations
for $m = 1$ **to** $m = m_p$ **do**
  Calculate neighborhood amplitudes before and after peak $p(m)$, $s_{bef} = \overline{X[p(m) - 60, p(m) - 10]}$ and $s_{aft} = \overline{X[p(m) + 10, p(m) + 60]}$
  if $s_{aft} - s_{bef} \geq \Delta_H$ **then**
    Store saccade location $p(m)$, amplitude $a(m)_h = s_{aft} - s_{bef}$, signal $X[p(m) - 127, p(m) + 128]$
  end if
end for

Function 2 Record vertical signals in the neighborhoods of valid horizontal saccades

for $m = 1$ **to** $m = m_p$ **do**
  Find peak location $p_v(m)$ and first derivative peak location $p_{v'}(m)$ of vertical signal in the neighborhood of horizontal saccade in $p(m)$
  if $|p_v(m) - p(m)| \leq 6$ **then**
    Store location $p_v(m)$, signal $X[p_v(m) - 127, p_v(m) + 128]$
  else if $|p_{v'}(m) - p(m)| \leq 6$ **then**
    Store location $p_{v'}(m)$, signal $X[p_{v'}(m) - 127, p_{v'}(m) + 128]$
  end if
end for

Dynamic Time Warping (DTW) is used to calculate the accumulated distances between vertical EOG signals from saccades and six different pre-set functions: a Heaviside, a Logistic, and a Gaussian function, and their y-axis-inverted variants (Fig. 4), all adapted to 256 samples and normalized to a [0; 1] interval. We model each type of vertical transition as one of these functions, based on the minimization of DTW distance for labelled transitions obtained during calibration.

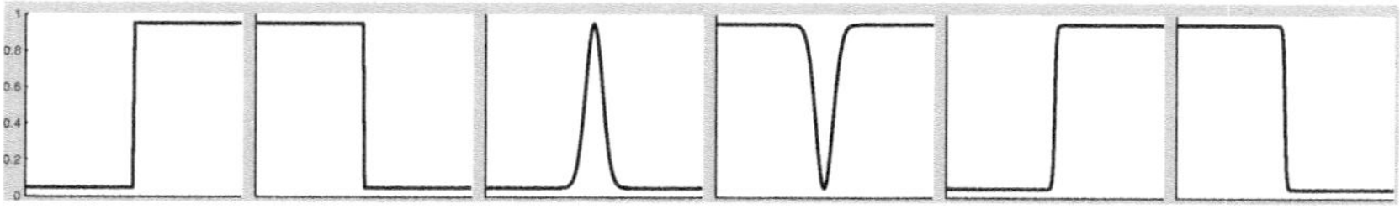

Fig. 4. The six comparison function windows used to train the DTW-based vertical decision method.

After each ITI, a vertical saccade window's DTW distance to each of the model functions is used to classify it as an upward or a downward transition. Horizontal saccades are classified using a linear Bayesian classifier employing

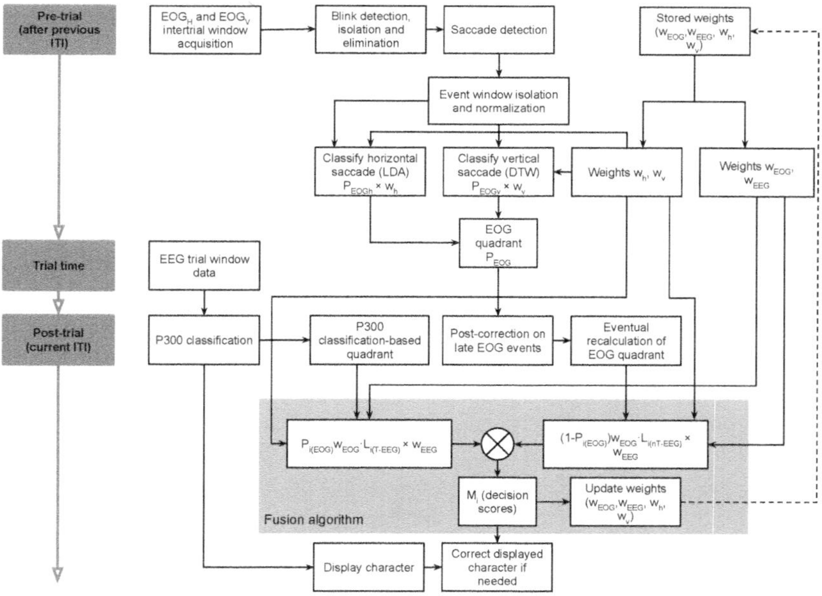

Fig. 5. Full flow diagram of the saccade classification method, the fusion classifier (shaded area), and its influence on character selection on the LSC Speller. Shaded boxes on the left show the stages of the paradigm, and their progression in time. Weights are updated from each fusion branch's success or failure in matching the fusion decision's outcome. Final decision is given by the character that maximizes the vector M of four candidates, chosen through their EEG scores. Dashed lines show weight updates stored for the next ITI.

horizontal EOG amplitudes as its features. To get the final gaze direction and obtain P_{EOG_H} and P_{EOG_V} for each quadrant (based on vertical EOG scores and horizontal classification probabilities), we only consider the last ocular saccades identified in each direction using the procedure in Algorithm 1.

For each quadrant a gaze probability is calculated as follows:

$$P_{i_{EOG}} = (P_{i_{EOG_H}} \times w_h) \cdot (P_{i_{EOG_V}} \times w_v), \quad i \in \{Q_1, Q_2, Q_3, Q_4\} \tag{2}$$

Here $P_{i_{EOG_H}}$ only varies between quadrant pairs (1, 4) and (2, 3), while $P_{i_{EOG_V}}$ varies between pairs (1, 2) and (3, 4). Weight w_h denotes the weight for the horizontal probability classification, while w_v is the weight for the vertical EOG-based decision. Both are initially set to 0.5.

A diagram of the complete procedure as it integrates into the complete information flow and fusion classification method is shown in the non-shaded area in Fig. 5.

3.5 Fusion of EEG and EOG Decisions

The decision of the selected target is based on the weighted fusion of EEG and EOG scores. The latter are obtained through methods described in Sect. 3.2, while the former are derived from the normalized target ($L_{T_{EEG}}$) and non-target ($L_{nT_{EEG}}$) likelihoods calculated for each character.

This method is a variable-weights fusion method adapted from [14], where weights are updated according to each classifier's agreeance with the fusion final decision. Because EEG-based classification results are not constructed around quadrant information, to convert these results into quadrant information we obtain four graded classification decisions, $S_1 \cdots S_4$, based on Eq. 1.

These four EEG-based decisions form the alternatives that may be chosen when analyzing the EOG events. Their target (T) and non-target (nT) scores are considered and multiplied by EOG-based probabilities and dynamic weights. For every alternative S_n, the $P_{i_{EOG}}$ probability is that of quadrant i where S_n is located.

A final decision vector M is then calculated using the weights attributed to the EEG and EOG signals (Eq. 3), where $w_{EEG} + w_{EOG} = 1$. Final fusion decision is determined from Eq. 4. If the quadrant of the fusion decision is different from the quadrant of the EEG-based character decision, the former replaces the latter as the final character decision for that event, with the weights w_{EEG} and w_{EOG} being updated accordingly.

$$M_i = L_{i_{T_{EEG}}} w_{EEG} \times P_{j_{EOG}} w_{EOG} - L_{i_{nT_{EEG}}} w_{EEG} \times (1 - P_{j_{EOG}}) w_{EOG}, \quad i \in \{Q_1, Q_2, Q_3, Q_4\} \tag{3}$$

$$S_{final} = S_{M_{final}} \quad : \quad M_{final} = \arg\max M_i \tag{4}$$

Weights are updated based on agreeance between signal's decision and the fusion decision. For example, to increment score c_h from Eq. 3, the final fusion decision quadrant and the EOG-determined quadrant must belong to the same side of the speller, i.e. have the same horizontal component. Two pairs of j weights, (w_{EEG}, w_{EOG}) and (w_h, w_v), are updated at event i using the rules in Eq. 5:

$$m_{j_i} = \frac{c_{j_i}}{c_{fusion}}, \quad w_{j_i} = \frac{m_{j_i}}{\sum_{j=1}^{2} m_{j_i}} \tag{5}$$

Here c is a weight-specific success counter updated as shown in Eq. 6, with c_{fusion} serving as the fusion decision's event counter.

$$c_{j_i} = \begin{cases} c_{j_{i-1}} + 1 & \text{if } Q_{signal} = Q_{fusion} \\ c_{j_{i-1}} & \text{if } Q_{signal} \neq Q_{fusion} \end{cases}, \quad signal = \{\text{EEG, EOG}\} \tag{6}$$

The full information flow during an intertrial interval and between consecutive intertrial intervals is included in Fig. 5.

4 Results and Discussion

In this Section we show the impact of both EOG-based quadrant or side selection on P300-based classification. Table 1 shows the decision accuracies for EOG-based quadrant or side selection for four participants, and the success rate of the vertical EOG component quadrant decision. Quadrant and side selection-based decisions form the basis for Hybrid 1 and Hybrid 2 modes, respectively, whose results are shown further ahead.

From Table 1 we can see that side selection, based on Horizontal EOG, has the highest accuracy. Accuracy of vertical selection is slightly lower than overall quadrant selection accuracy. In Table 2 we show the impact of both EOG components on fusion decision and final character selection accuracy. Here, "Hybrid 1" refers to full quadrant-based weighing of EOG-based decisions and "Hybrid 2" refers to a bilateral, left-right simplified approach.

Table 1. Accuracy of EOG-based quadrant selection and side selection, and vertical component decision. Quadrant selection takes vertical signals into account, while side selection does not.

	EOG accuracy (%)		
Participant	Quadrant selection	Side selection (L-R)	Up-Down selection
1	86,49	100	84,21
2	89,19	100	86,84
3	86,49	100	84,21
4	56,76	86,49	57,89
Average	**79,73**	**96,62**	**78,29**

Table 2. Register of the accuracies of P-300 based character decisions, on their own and with two different correction modes. Also shown are the percentage of errors corrected (from the EEG classifier) and induced (on the final error count) by each method. The online session consists of 38 events.

	Accuracy (%)			Errors corrected (%)		Errors induced (%)	
Participant	EEG only	Hybrid 1	Hybrid 2	Hybrid 1	Hybrid 2	Hybrid 1	Hybrid 2
1	78,95	86,84	92,11	37,50	62,50	0,00	0,00
2	86,84	84,21	86,84	0,00	0,00	16,67	0,00
3	84,21	92,11	92,11	66,67	66,67	0,00	0,00
4	81,58	78,95	81,58	14,29	28,57	22,22	25,00
Average	**82,90**	**85,53**	**88,16**	**29,62**	**39,43**	**9,72**	**6,25**

Combined with the results shown in Table 1, values in Table 2 confirm that, overall, both of the proposed EOG-based correction modes improve on P300-based character selection accuracy. Also, it can be verified that the vertical channel of EOG has a more important effect in degrading the performance of the EOG-based correction than its horizontal counterpart. On average, the mode without vertical EOG had an additional 2.64% points of accuracy over the quadrant-based method. One of the possible reasons might be the elimination of blinks inadvertently executed during gaze shifts, leading to the exclusion of relevant sections of the signals that cannot then be used to characterize gaze shift. Besides, the participant may be executing the ocular movement towards the next character too late within the ITI, difficulting the acquisition of the relevant signal window on both the horizontal and vertical channels. The impact of this latter issue is larger on vertical signals because of the need for the event to be centered on the window that is inputted to the DTW method. The effects of these issues on vertical accuracy lead us to conclude that the most adequate approach is to only have lateral, left-right-based decisions influencing the P300-based classification.

5 Conclusion

The EOG-based method we have proposed for identifying and correcting errors derived from P300-based classification of EEG signals seems to be valid, with some limitations. We initially envisaged a means to provide quadrant-based information that would be calculated from two channels of EOG signals. However, given the higher error rate of vertical EOG-based decisions, a second method reliant on horizontal EOG was tested and implemented, and found to be better than the quadrant-based process. Lateral identification based on this method has shown very high classification accuracies, which are a prerequisite for identifying errors from P300 given this method's high base accuracy in [4]. Future iterations of the method presented here will require some changes, namely new blink identification procedures able to work in the absence of vertical EOG signals.

Acknowledgments. This work was partially funded by the Portuguese Foundation for Science and Technology (FCT) under the Ph.D. Scholarship SFRH/BD/104985/2014 of João Perdiz and Ph.D. Scholarship SFRH/BD/ 111473/2015 of Aniana Cruz and was developed at the Institute of Systems and Robotics – University of Coimbra. This work was partially supported by Project B-RELIABLE: SAICT/30935/2017, with FEDER/FNR/OE funding through programs CENTRO2020 and FCT, and by UID/EEA/ 00048/2013 with FEDER funding, through programs QREN and COMPETE.

References

1. Ahn, S., Jun, S.C.: Multi-modal integration of EEG-fNIRS for brain-computer interfaces-current limitations and future directions. Front. Hum. Neurosci. **11**, 503 (2017)
2. Bai, L., Yu, T., Li, Y.: A brain computer interface-based explorer. J. Neurosci. Methods **244**, 2–7 (2015)
3. Barbosa, S., Pires, G., Nunes, U.: Toward a reliable gaze-independent hybrid BCI combining visual and natural auditory stimuli. J. Neurosci. Methods **261**, 47–61 (2016)
4. Cruz, A., Pires, G., Nunes, U.J.: Double ErrP detection for automatic error correction in an ERP-based BCI speller. IEEE Trans. Neural Syst. Rehabil. Eng. **26**(1), 26–36 (2018)
5. Hong, K.S., Khan, M.J.: Hybrid brain-computer interface techniques for improved classification accuracy and increased number of commands: a review. Front. Neurorobotics **11**, 35 (2017)
6. Hortal, E., Iáñez, E., Úbeda, A., Perez-Vidal, C., Azorín, J.M.: Combining a brain-machine interface and an electrooculography interface to perform pick and place tasks with a robotic arm. Robot. Auton. Syst. **72**, 181–188 (2015)
7. Leeb, R., Sagha, H., Chavarriaga, R., del R Millán, J.: A hybrid brain-computer interface based on the fusion of electroencephalographic and electromyographic activities. J. Neural Eng. **8**(2), 025011 (2011)
8. Mak, J.N., Wolpaw, J.R.: Clinical applications of brain-computer interfaces: current state and future prospects. IEEE Rev. Biomed. Eng. **2**, 187–199 (2009)
9. Müller-Putz, G., Leeb, R., Tangermann, M., Höhne, J., Kübler, A., Cincotti, F., Mattia, D., Rupp, R., Müller, K.R., Millán, J.d.R.: Towards noninvasive hybrid brain-computer interfaces: framework, practice, clinical application, and beyond. Proc. IEEE **103**(6), 926–943 (2015)
10. Perdiz, J., Garrote, L., Pires, G., Nunes, U.J.: Measuring the impact of reinforcement learning on an electrooculography-only computer game. In: 2018 IEEE 6th International Conference on Serious Games and Applications for Health (SeGAH), pp. 1–8, May 2018
11. Pires, G., Nunes, U., Castelo-Branco, M.: Statistical spatial filtering for a P300-based BCI: tests in able-bodied, and patients with cerebral palsy and amyotrophic lateral sclerosis. J. Neurosci. Methods **195**(2), 270–281 (2011)
12. Pires, G., Nunes, U., Castelo-Branco, M.: Comparison of a row-column speller vs. a novel lateral single-character speller: assessment of BCI for severe motor disabled patients. Clin. Neurophysiol. **123**(6), 1168–1181 (2012)

13. Riccio, A., Leotta, F., Bianchi, L., Aloise, F., Zickler, C., Hoogerwerf, E., Kübler, A., Mattia, D., Cincotti, F.: Workload measurement in a communication application operated through a P300-based brain-computer interface. J. Neural Eng. **8**(2), 025028 (2011)
14. Taher, F.B., Amor, N.B., Jallouli, M.: A multimodal wheelchair control system based on EEG signals and eye tracking fusion. In: 2015 International Symposium on Innovations in Intelligent SysTems and Applications (INISTA), pp. 1–8 IEEE (2015)
15. Usakli, A., Gurkan, S., Aloise, F., Vecchiato, G., Babiloni, F.: A hybrid platform based on EOG and EEG signals to restore communication for patients afflicted with progressive motor neuron diseases. In: 2009 Annual International Conference of the IEEE Engineering in Medicine and Biology Society, pp. 543–546. IEEE (2009)
16. Wang, H., Li, Y., Long, J., Yu, T., Gu, Z.: An asynchronous wheelchair control by hybrid EEG-EOG brain-computer interface. Cogn. Neurodyn. **8**(5), 399–409 (2014)
17. Yin, E., Zeyl, T., Saab, R., Chau, T., Hu, D., Zhou, Z.: A hybrid brain-computer interface based on the fusion of P300 and SSVEP scores. IEEE Trans. Neural Syst. Rehabil. Eng. **23**(4), 693–701 (2015)
18. Zhang, R., Li, Y., Yan, Y., Zhang, H., Wu, S., Yu, T., Gu, Z.: Control of a wheelchair in an indoor environment based on a brain-computer interface and automated navigation. IEEE Trans. Neural Syst. Rehabil. Eng. **24**(1), 128–139 (2016)

Non-invasive Spinal Cord Stimulation: Relevance of Modelling Studies in Clinical Protocol Design

Sofia Rita Fernandes[1,2(✉)], Mariana Pereira[2], Mamede de Carvalho[2], and Pedro Cavaleiro Miranda[1]

[1] Instituto de Biofísica e Engenharia Biomédica, Faculdade de Ciências, Universidade de Lisboa, 1749-016 Lisbon, Portugal
srcfernandes@fc.ul.pt
[2] Instituto de Fisiologia, Instituto de Medicina Molecular, Faculdade de Medicina, Universidade de Lisboa, Avenida Professor Egas Moniz, 1649-028 Lisbon, Portugal

Abstract. Over the past two decades a growing interest has arisen in non-invasive neuromodulation of the spinal cord. The application of transcutaneous electric currents is a possible alternative therapy to many invasive procedures. This neuromodulatory technique may also be applied jointly with other therapies, such as pharmacologic substances, cognitive training and physiotherapy, to enhance the outcomes in the treatment of several neural dysfunctions.

Electric stimulation originates electric fields in neurons that can change the transmembrane potential, facilitating or inhibiting neural responses. Varying stimulation parameters, geometry and placement of the stimulation devices (e.g. electrodes, coils) will result in different electric field patterns and, consequently, in different spinal neuron responses. We will address how computational modelling methods using realistic human models can be useful to optimize stimulation parameters for specific clinical purposes.

Keywords: Spinal cord · tsDCS · Computational modelling

1 The Advent of Non-invasive Neuronal Stimulation

The last decades have witnessed a growing interest in non-invasive stimulation as an alternative or complementary therapy for the treatment of nervous systems diseases and dysfunctions, especially through the development of transcranial direct current stimulation (tDCS). tDCS generates electric fields (EF) by delivering electric currents through electrodes placed on the scalp. The EFs can change neuron transmembrane potential, resulting in excitability changes. Extensive research has shown the tDCS potential for cortical excitability modulation with promising clinical applications [1, 2].

Computational modelling studies on tDCS are extremely useful to understand the biophysical mechanisms of stimulation effects. These vary with different stimulation parameters, such as electrode number, location, design, current intensity and polarity (anodal/cathodal). tDCS protocol characteristics are shown to influence the spatial

J. Henriques et al. (Eds.): MEDICON 2019, IFMBE Proceedings 76, pp. 1767–1773, 2020.
https://doi.org/10.1007/978-3-030-31635-8_214

distribution of the induced EF, and consequently, the effects on the neuronal membrane potential. Accurate predictions of the EF induced in the brain are essential to determine the electrode montages that optimize stimulation for specific cortical targets [3].

Considering the findings on non-invasive brain stimulation, research turned towards transcutaneous spinal direct current stimulation (tsDCS) as a possible stimulation technique to address spinal dysfunctions [4]. tsDCS exploratory studies in humans reported polarity-dependent effects in sensory and motor spinal responses when stimulating cervical and thoracic spinal regions [4]. As in tDCS, computational studies also reveal the influence of stimulation characteristics in the EF spatial distribution during tsDCS application [5–7]: the maximum EF magnitude occurs between electrodes; diverse electrode montages resulted in different current and EF distributions. tsDCS modelling predictions were consistent with changes in physiological responses observed in combined experimental and modelling studies of cervical and lumbar tsDCS [8, 9].

Our aim is to present a brief overview of our modelling results on tsDCS, focusing on the effect of electrode position on EF magnitude distribution. We have modelled cervical, thoracic and lumbar stimulation, with a total of eleven different electrode montages and inferred on the possible clinical applications of this technique according with electrode positioning.

2 Computational Modelling Methods

The methods applied for the computational modelling studies on the EF distribution described here are briefly summarized. A more complete description can be found in Fernandes et al. [7].

Building a Realistic Human Model. A realistic human male model was designed using a database available on the internet. Fourteen tissues were included in the model (Fig. 1): skin, fat, muscle, bone, heart, lungs, viscera (composed by stomach, liver, pancreas, small intestine, large intestine), vertebrae, intervertebral disks, dura mater, CSF, cerebellum, brainstem and spinal cord (SC). Cerebellum and brainstem were homogeneous, without distinction between white matter (WM) and grey matter (GM). Tissues' surface meshes were optimized with the 3-MATIC module from MIMICS software (v16) for successful tetrahedral volume mesh creation. A spinal-GM was artificially designed, considering general knowledge on SC anatomy and relative measurement from the Visible Human Data Set (VHD data) of the National Library of Medicine (NLM) and the Visible Human Project® (www.nlm.nih.gov/research/visible/visible_human.html).

The electrodes were modelled as rectangular prism layers of gel and rubber pad with metallic connectors in each pad represented as a rectangle on the upper surface of the rubber pad, considering the dimensions of the electrodes available in our lab [7]. The electrode placements considered, with one electrodes placed over lumbar, thoracic or cervical spinous processes (s.p.), and the other electrode placed either over the right deltoid (rD), umbilicus (U), right iliac crest (rIC), cervicomental angle (CMA), T8 or T3 s.p., resulted in eleven different montages: T10-rD, T10-U, C7-rD, C7-CMA,

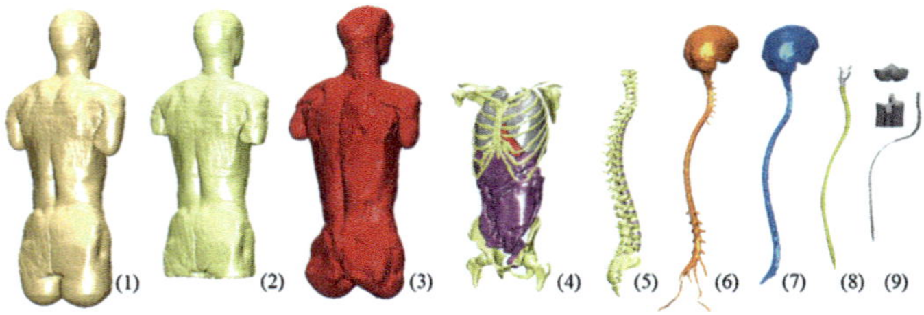

Fig. 1. Tissue surface meshes included in the human model.

C4-CMA, already used in experimental studies [4, 8, 10–12]; T10-rIC, T8-U, T8-rIC, L2-rD, L2-T8, C3-T3, not applied at present.

Calculation of the EF Distribution. Isotropic electric conductivity values of tissues and gel were compiled after carrying out a review of the literature on DC current tissue properties [7]. Spinal-WM and muscles are composed of longitudinal cable-like fibers that conduct electric current preferentially along their length. Thus, anisotropy of muscle and spinal-WM conductivity was modelled by tensors using information about the known spatial orientation of their fibers. The rubber pad conductivity was measured and estimated to be 44 $\pm$ 1 S/m [7].

The EF induced in the SC and all other tissues in model was calculated using the AC/DC module in COMSOL Multiphysics software, which solves the Laplace equation for the electric potential ϕ ($\nabla^2\phi = 0$), using the finite element method, considering the boundary conditions in Miranda et al. [13]. The potential difference between the anode and cathode was adjusted, so that the current injected through the electrodes was 2.5 mA, as usually applied in the experimental studies. The EF was calculated in all nodes of the mesh elements by taking the gradient of the electric potential ($\mathbf{E} = -\nabla\phi$).

In lumbar montages, the first electrode was defined as the anode and the second electrode as the cathode (e.g., in L2-T8 montage, L2 is the anode and T8 the cathode), and the reverse was considered in cervical montages, since in cervical tsDCS studies, a cervical cathode resulted in motor responses facilitation. Reversing polarity will only affect the direction of the EF, not its magnitude [2].

Simulations for each montage resulted in 2.6×10^7 degrees of freedom, with a solution time of about 150 min per simulation on a computer with 2 quad-core Intel® Xeon® processors clocked at 3.2 GHz and 48 GB of RAM. EF analysis was performed in MATLAB. Modelling studies on tDCS reported average EF values greater than 0.15 V/m in the motor cortex when reproducing clinical settings with observed neuromodulatory effects [13]. Thus, we will consider that tsDCS effects may occur wherever the average EF in the spinal-GM exceeds this value.

3 Main Findings and Implications for Clinical Practice

The EF magnitude distribution in the spinal-WM reaches its maximum value in the spinal segments that lie between the two electrodes for all montages (Fig. 2).

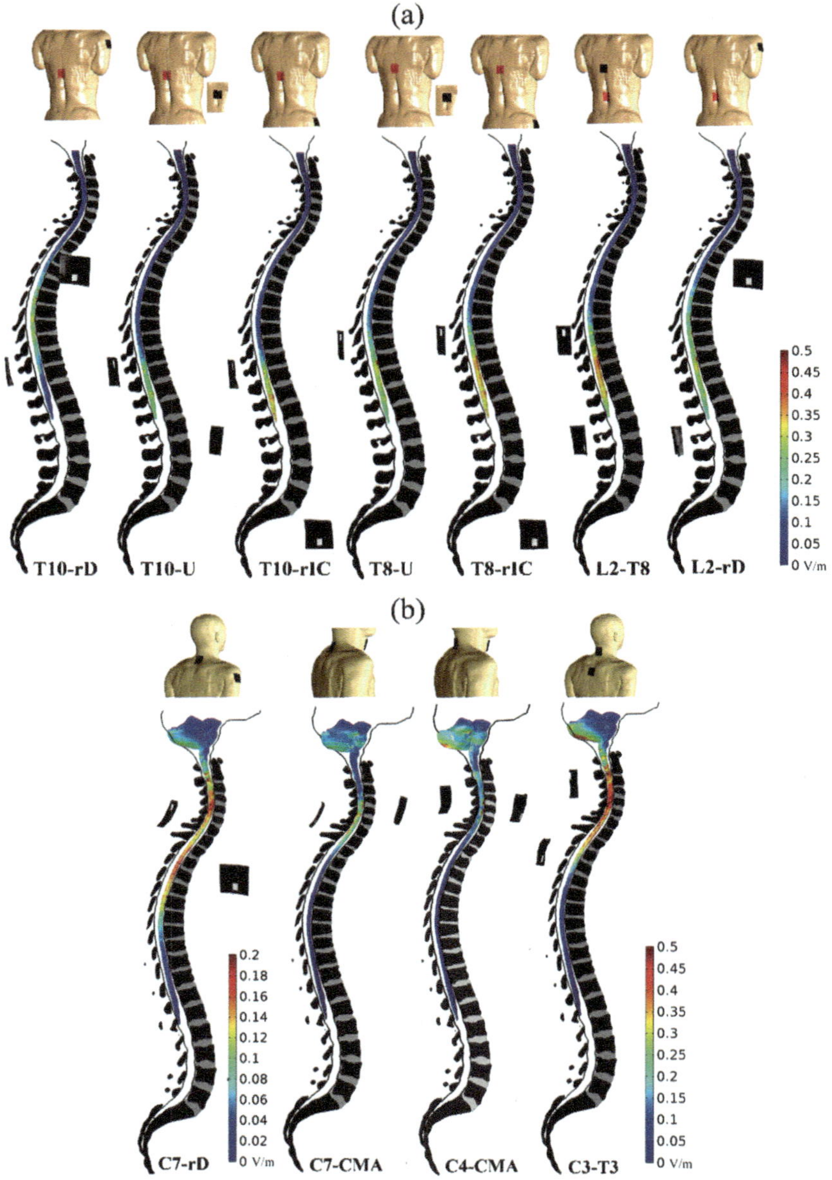

Fig. 2. EF magnitude in the spinal-WM, considering: (a) thoracic and lumbar montages; (b) cervical montages. Magnitude color scale is represented on the right of each row in V/m. C7-rD presents a different color scale to enhance EF spatial distribution features. A sagittal slice of vertebrae, disks and dura is presented in grey and black to illustrate the anatomical relations with spinal-WM. Electrodes are in black, with active connectors in white when visible. Electrode placement in the model is represented above each distribution [7, Fernandes et al., accepted].

T10-rD and T10-U are two montages used in previous experimental studies with observed modulation of sensory and motor responses of the lower limb [4, 10]. EF magnitude predicted for T10-rD and T10-U is above 0.15 V/m in thoracic segments (EF_{max} = 0.67 V/m) and in lumbosacral (LS) segments (EF_{max} = 0.44 V/m), respectively (Fig. 2a). Considering new montages with electrodes at T8 and/or rIC slightly higher values were predicted in the LS segments: T10-rIC, EF_{max} = 0.56 V/m; T8-U, EF_{max} = 0.63 V/m; T8-rIC, EF_{max} = 0.72 V/m. Considering rIC has also the advantage of avoiding an electrode placement over the abdominal region. T8 position increased the field in the lower thoracic segments of the SC. Although the EF maximum values are not as high as in T10-rD, L2-rD addresses both the thoracic and the LS segments (EF_{max} = 0.59 V/m). Placing both electrodes over the SC, at L2-T8, generated an EF distribution similar in magnitude and length to T8-rIC (EF_{max} = 0.76 V/m). L2-T8 and T8-rIC montages may thus be the more adequate for lower limb related spinal circuitry modulation since these produced the highest EF in the LS and lower thoracic segments, and may thus result in stronger effects on motor and sensory responses compared to T10-rD and T10-U (Fig. 2a).

For cervical tsDCS, all montages result in EF magnitude above 0.15 V/m in a varied range of spinal segments, which is consistent with the different observed effects on motor responses of the upper limb reported in experimental studies [8, 11, 12]. C7-rD is the montage with the lower EF values but still above 0.15 V/m in a few lower cervical and upper thoracic spinal segments, whereas CMA montages present local maxima only in the cervical segments under the electrodes (Fig. 2b). C3-T3 originates the highest values (EF_{max} = 0.50 V/m) and in the widest spinal region when compared to the other montages (Fig. 2b). Higher cervical placements (C3-T3, C4-CMA) originate EF above 0.15 V/m in the posterior regions of the cerebellum and brainstem. Since our model considers these tissues as isotropic and homogeneous, the EF distribution predicted lacks accuracy. Even so, these results indicate that neuromodulation of neural circuits in cerebellum and brainstem should be addressed when using higher cervical electrode placement in tsDCS.

Neuromodulatory changes in neuron membrane potential are linearly dependent on the applied EF magnitude during DC stimulation and there is no clear threshold value for stimulation effectiveness. However, assuming a threshold value of 0.15 V/m, all montages investigated in this study may induce neuromodulation effects in the SC. As varying electrode positions result in different target spinal segments due to changes in EF spatial distribution, possible clinical targets can then be pointed for each montage (Table 1). Modelling predictions may thus guide future tsDCS clinical applications, by predicting the appropriate choice of electrode montage according to spinal clinical target, just as observed in previous modelling and experimental studies in tDCS [3].

Table 1. Spinal segments with EF magnitude greater than 0.15 V/m in the spinal-GM for each montage [7, Fernandes et al., accepted].

Montage	Spinal segments	Related body area
C3-T3	C1-T4	Cerebellum, brainstem, neck, diaphragm, shoulder, upper limb, wrist, hand, upper thoracic region
C4-CMA	C2-T1	Cerebellum, brainstem, neck, diaphragm, shoulder, upper limb, wrist, hand
C7-CMA	C4-T1	Neck, shoulder, upper limb, wrist, hand
C7-rD	C6-C7, T2-T5	Upper limb, upper thoracic region
T8-U	T11-Filum	Lower thoracic region and back muscles, pelvic floor, lower limb
T8-rIC	T10-Filum	Lower thoracic region and back muscles, pelvic floor, lower limb
T10-U	L1-Filum	Lower back muscles, pelvic floor, lower limb
T10-rIC	L1-Filum	Lower back muscles, pelvic floor, lower limb
T10-rD	T5-T12	Middle and lower thoracic region
L2-rD	T6-Filum	Middle and lower thoracic region and back muscles, pelvic floor, lower limb
L2-T8	T10-Filum	Lower thoracic region and back muscles, pelvic floor, lower limb

4 Conclusion

This brief review outlines the importance of modelling studies in tsDCS to optimize experimental protocols based on the EF spatial distribution in the spinal cord. This combined modelling-experimental approach may be relevant in the near future for targeting specific clinical spinal targets. This can contribute decisively for a more personalized patient functional recovery using non-invasive spinal stimulation as a coadjutant therapy.

Acknowledgements. The research presented here was supported by UID/BIM/50005/2019, project funded by Fundação para a Ciência e a Tecnologia (FCT)/Ministério da Ciência, Tecnologia e Ensino Superior (MCTES) through Fundos do Orçamento de Estado, and also by the FCT in the scope of the FCT-IBEB Strategic Project UID/BIO/00645/2013. S. R. Fernandes was supported by a FCT grant, reference SFRH/BD/100254/2014.

References

1. Nitsche, M.A., Cohen, L.G., Wassermann, E.M., et al.: Transcranial direct current stimulation: state of the art 2008. Brain Stimul. **1**(3), 206–223 (2008)
2. Ruffini, G., Wendling, F., Merlet, I., et al.: Transcranial current brain stimulation (tCS): models and technologies. IEEE Trans. Neural Syst. Rehabil. Eng. **21**(3), 333–345 (2013)

3. Salvador, R., Wenger, C., Nitsche, M.A., Miranda, P.C.: How electrode montage affects transcranial direct current stimulation of the human motor cortex. In: Conference Proceedings of the IEEE Engineering in Medicine and Biology Society 2015, pp. 6924–6927 (2015)
4. Cogiamanian, F., Ardolino, G., Vergari, M., et al.: Transcutaneous spinal direct current stimulation. Front. Psychiatry **3**, 63 (2012)
5. Parazzini, M., Fiocchi, S., Liorni, I., et al.: Modeling the current density generated by transcutaneous spinal direct current stimulation (tsDCS). Clin. Neurophysiol. **125**(11), 2260–2270 (2014)
6. Kuck, A., Stegeman, D.F., Van Asseldonk, E.H.F.: Modeling trans-spinal direct current stimulation for the modulation of the lumbar spinal motor pathways. J. Neural Eng. **14**(5), 056014 (2017)
7. Fernandes, S.R., Salvador, R., Wenger, C., de Carvalho, M.A., Miranda, P.C.: Transcutaneous spinal direct current stimulation of the lumbar and sacral spinal cord: a modelling study. J. Neural Eng. **15**(3), 036008 (2018)
8. Dongés, S.C., Bai, S., Taylor, J.L.: Concurrent electrical cervicomedullary stimulation and cervical transcutaneous spinal direct current stimulation result in a stimulus interaction. Exp. Physiol. **102**(10), 1309–1320 (2017)
9. Pereira, M., Fernandes, S.R., Miranda, P.C., de Carvalho, M.: Neuromodulation of lower limb motor responses with transcutaneous lumbar spinal cord direct current stimulation. Clin. Neurophysiol. **129**, 1999–2009 (2018)
10. Vergari, M., Cogiamanian, F., di Bisceglie, M., et al.: Additive after-effects of spinal and cortical DC stimulation on human flexion reflex pathways. J. Peripheral Nerv. Syst. **17**, S57–S57 (2012)
11. Bocci, T., Vannini, B., Torzini, A., et al.: Cathodal transcutaneous spinal direct current stimulation (tsDCS) improves motor unit recruitment in healthy subjects. Neurosci. Lett. **578**, 75–79 (2014)
12. Niérat, M., Similowski, T., Lamy, J.: Does trans-spinal direct current stimulation alter phrenic motoneurons and respiratory neuromechanical outputs in humans? A double-blind, sham-controlled, randomized, crossover study. J. Neurosci. **34**(43), 14420–14429 (2014)
13. Miranda, P.C., Mekonnen, A., Salvador, R., Ruffini, G.: The electric field in the cortex during transcranial current stimulation. Neuroimage **70**, 48–58 (2013)

Special Sessions: Value-Based Health Technology Assessment

Integrating HTA Principles into Procurement of Medical Devices: The Italian National HTA Programme for Medical Devices

Giuditta Callea[1(✉)], Carlo Federici[1], Oriana Ciani[1], Fabio Amatucci[1], Ludovica Borsoi[1], Rosanna Tarricone[1,2], and Marcella Marletta[3]

[1] Cergas SDA Bocconi School of Management, Milan, Italy
giuditta.callea@unibocconi.it
[2] Department of Policy Analysis and Public Management, Bocconi University, Milan, Italy
[3] Directorate General of Medical Devices and Pharmaceutical Service, Ministry of Health, Rome, Italy

Abstract. Managing the adoption and diffusion of technological innovation is a key challenge for healthcare systems. In 2015, Italy has started designing and implementing a National Health Technology Assessment (HTA) Programme for Medical Devices (MDs), a network aimed at promoting the use of HTA tools, and safety, effectiveness and cost-effectiveness principles in the decision-making process of medical technologies. Since 2017, all the relevant stakeholders involved in HTA (i.e., patients and citizens' representatives, scientific organizations, industry, healthcare organizations, academic researchers) are supporting a Steering Committee active at the Ministry of Health in the design and implementation of the Programme. Several working groups were established, one of them - based on a review of best practices - was instructed (i) to develop methods and procedures to integrate the results of HTA into decisions of procurement and clinical pathways and (ii) to propose a purchase request form template to be used by health care professionals. The final goal of the HTA Program is to reduce the heterogeneity across practices which translate into uneven access to innovative health technologies. This work provides an overview of the rationale and participatory process undertaken in Italy, and illustrates the working group main findings and recommendations.

Keywords: Health Technology Assessment · Medical Devices · Procurement

1 The Italian National HTA Programme for Medical Devices

1.1 Introduction

In the European Union, there is a diffused awareness that governing the introduction of new technologies represents an irremissible element to improve citizens' health. The Union supported and facilitated cooperation on Health Technology Assessment (HTA) through the Directive 2011/24. Coherently with the EU Directive, a Steering Committee has been established in 2015 at the Italian Ministry of Health (MoH) with the

J. Henriques et al. (Eds.): MEDICON 2019, IFMBE Proceedings 76, pp. 1777–1782, 2020.
https://doi.org/10.1007/978-3-030-31635-8_215

aim to define instruments for governing the diffusion and use of medical devices (MDs) in the Italian National Healthcare System (NHS). Among these instruments, a national network called National Programme of HTA for MDs has been established. The vision shaped by a number of Italian legislative documents is that of a national, centralised evaluation program that guarantees the equity of access to new medical technologies across the country based on safety, effectiveness and cost-effectiveness through the coordination and cooperation between central and regional governments [1].

Given the decentralized architecture of the Italian NHS, in order to ensure the broadest acceptance of the HTA Programme, the Steering Committee promoted the active involvement of all stakeholders into the design and the implementation of the program. Accordingly, several Working Groups (WG) and Subgroups (SG) were established to support the Steering Committee (see Fig. 1).

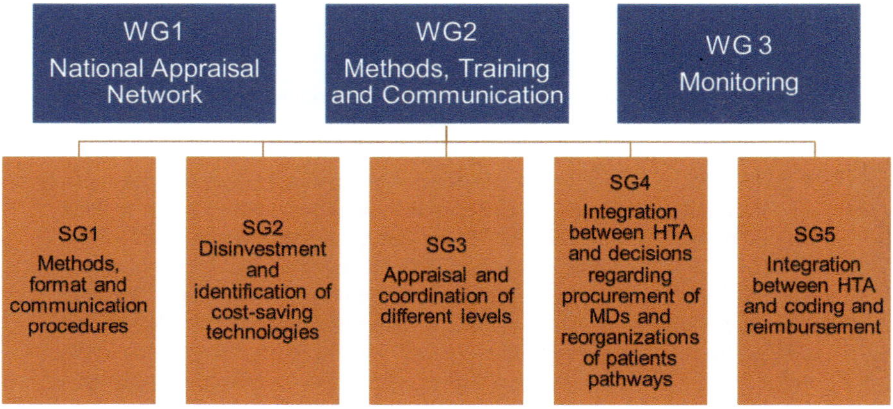

Fig. 1. Working groups supporting the Steering Committee (source adapted from [2]).

The initial phase of the National HTA Programme for MDs (March 2017 – August 2018) was dedicated to the definition of methods and tools. In March 2017 five SGs were established within WG2 Methods, Training and Communication. Each SG is coordinated by a member of the Steering Committee and is composed by a scientific advisor and several co-authors, who were appointed among the members of the Innovation Working Group. The latter is a forum of stakeholders where patients and citizens' representatives, scientific associations, healthcare organizations, industry, academic researchers are represented. The Innovation WG members not actively involved in the SGs act as reviewers. Representatives of the MoH and of the Italian National Agency for Regional Healthcare Services (AGENAS) acted as observers.

1.2 Rationale, Methods and Key Recommendations of the Sub-group 4: Integration Between HTA and Decisions Regarding Procurement of MDs and Reorganization of Patients' Pathways

Rationale. Among the WG2 subgroups, SG4 was instructed to develop methods and procedures to integrate the results of HTA into decisions regarding the procurement of healthcare technologies and their use within clinical pathways. In addition, to guarantee the compliance of future procurement processes with the objectives and recommendations of the National Programme of HTA for MDs SG4 was required to propose a purchase request form template to be used by health care professionals, based on a review of best practices.

Governance. The following table shows SG4 governance (see Table 1).

Table 1. SG4 governance.

Role	Institution
Coordinator	Veneto Region
Scientific Advisor	Italian Federation of Public Local Health Authorities, Public Hospital Trustees and Italian Municipalities (Federsanità)
Co-authors	Italian Association of Clinical Engineers (AIIC) Centre for Research on Health and Social Care Management (Cergas), SDA Bocconi School of Management Italian Federation of Public Health trusts and Hospitals (FIASO) Italian National Institute of Health (ISS) Italian Society of Health Technology Assessment (SIHTA)
Reviewers	Italian Association for Biomedical and Diagnostic Technology (Assobiomedica) Active Citizenship Association (Cittadinanzattiva) Italian Federation of Scientific Associations (FISM) Italian Association of Hospital Pharmacy (SIFO) Lombardy Region

Methods. In order to base their recommendations on best-practices, the SG4 members performed a review of scientific and grey literature addressing the link between HTA and procurement of MDs, and searched for international, national and regional experiences of value-based procurement. Value-based procurement was defined as any practice aimed at setting procurement targets that lead to efficiency improvements, deliver improved patient outcomes and patient pathways, increase savings opportunities and wider operational efficiencies. Moreover, in order to develop a purchase form template, the SG4 members analysed the National HTA Programme submission form and several purchase request forms adopted by Italian hospitals.

Results and Key Recommendations. The literature review did not identify structured HTA programmes that routinely integrate economic evaluations and HTA assessments to respond to procurement needs. Therefore, a *de novo* integrated model was proposed based on the following points:

1. Implementation of an electronic, standard purchase request form to be submitted by the end users of the medical technologies for which a purchase request is made (e.g., physicians, nurses), supported by administrative staff (procurers and supply officers, hospital pharmacists, biomedical engineers). The main information proposed for the form are as follows: (i) applicant's personal information; (ii) type of request (e.g., urgent/not urgent, non-fungible technology); (iii) requested technology and current alternatives; (v) evidences supporting the request (clinical, economic, guidelines); (vi) costs and reimbursement data; (vii) organizational impact of the technology.
2. Design of a three-stage evaluation process (see Fig. 2) consisting of the following phases. First, the end users submit the purchase request form as previously detailed. Second, an *ad hoc* committee checks the state of the art of the requested technology by searching existing HTA recommendations by either the Steering Committee of the National Programme of HTA, individual Italian Regions, the European Network for HTA (EUnetHTA) or national/regional HTA agencies in other countries. Lastly, the response to the purchase request is provided accordingly to the results of this evaluation phase. In case a national or regional recommendation has been formulated within the National HTA Program, the response to the purchase request will be consistent with the Program's recommendation (see Table 2). If the only available recommendations were published by foreign HTA agencies, the HTA reports may be adapted according to the methods approved by the Steering Committee. Lastly, if

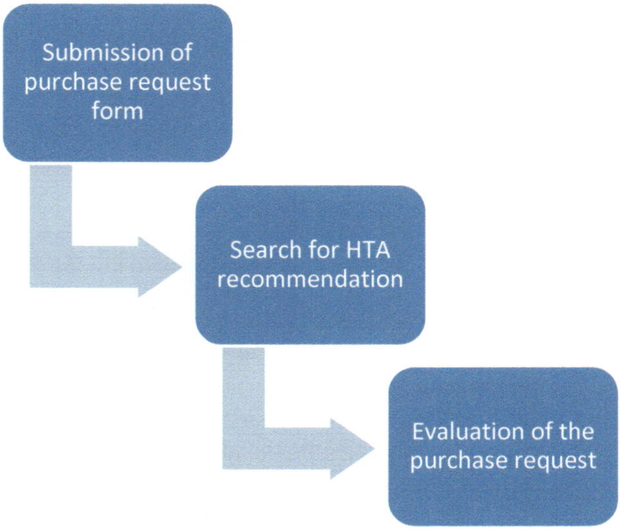

Fig. 2. Main steps of the purchase request evaluation process.

Table 2. Possible decisions on purchase request.

Steering Committee Recommendation	Decision on purchase request
Reject	The purchase request is rejected
Approve	The purchase request is approved
Only in research	The hospital asks to be admitted to the evidence generation plan
Approval with research	The hospital asks to be admitted to a coverage with evidence development plan

no recommendation exists, additional information may be requested to the applicant.

3. Implementation of a regional unit in charge of (i) collecting the purchase requests of new technologies made by all regional hospitals professionals, (ii) selecting the technologies that may be relevant for the Italian NHS according to the prioritization criteria defined by the Steering Committee, and (iii) submit to the Steering Committee the evaluation request form for the selected technologies.
4. Use HTA reports as a data source to inform the tender in terms of lots, target population and quantities to be purchased.

State of the Art of the National HTA Program. The five SGs active in WG2 summarised their findings and recommendations in reports that were submitted to the Steering Committee in August 2018. The Steering Committee analysed, harmonised and approved the five SGs reports. A set of final coherent recommendations was published in June 2019 [3].

1.3 Conclusions

Despite few years of delay with respect to other European countries, Italy is now setting up a structured and modern national system for the HTA of medical technologies that takes into account the decentralized nature of the NHS in the country.

Overall, the National Programme of HTA for MDs promotes an integrated approach to the governance of healthcare technologies along their lifecycle, from first adoption into clinical practice to disposal. Key features of the Programme are the participatory design, the link with the coding and reimbursement phase and, as discussed in detail in this paper, the link with the procurement system.

Through the work of the different SGs the design process of the national program has ensured the pervasiveness of HTA within all decisional levels in the NHS. For example, by specifying the criteria for the disinvestment of obsolete technologies and the identification of cost-saving technologies, the work done by the SG2 may inform healthcare planning decisions at the regional level. Specifically, based on the minimum requirements in terms of procedure volumes and providers experience to generate savings following the adoption of a new technology, regional decision makers will be able to design more efficient service delivery models, which consider both the cost-effectiveness profiles of the technologies and their expected budget impact according to the model chosen.

In our systematic literature review, very few experiences were found that directly establish a systematic link between HTA recommendations and the purchasing processes of technologies at different level within the healthcare systems. Therefore, the work of the SG4 aimed to design an innovative, rational and transparent approach to link evidence-based decisions made by a national or regional HTA body to the procurement process and the design of patient clinical pathways in the Italian NHS. In addition, the SG4 also suggested the criteria to assess the role of new technologies within integrated care pathways across primary, secondary and community care services.

There still exist many unresolved issues that need to be addressed by the program. For example, while the SG3 has defined the overall principles, methods and processes for the appraisal of medical devices at the central or regional level, no clear decision rules have been defined yet to inform which type of recommendations should be issued (e.g., based on an explicit value of cost-effectiveness threshold or other criteria used in other jurisdictions such as cost-effectiveness efficiency frontiers, multiple-criteria decision analysis frameworks or measurement of the added clinical benefit).

Other critical aspects will certainly emerge once the pilot phase will be conducted, including the adherence and participation of all the relevant stakeholders, the adequacy of the designed processes for the assessment, appraisal and adoption of the technologies, and the relevance and adequacy of the IT infrastructure which serves as the backbone for all the designed flows of information within and across the national and regional health authorities.

References

1. Ciani, O., Tarricone, R., Torbica, A.: Programma nazionale HTA per dispositivi medici: quali sfide per l'implementazione?. In: CERGAS (eds.) Rapporto OASI 2016, Egea, Milano (2016)
2. Tarricone, R., Ciani, O., Amatucci, F., Armeni, P., Borsoi, L., Callea, G., Costa, F., Federici, C., Torbica, A., Marletta, M.: Programma nazionale di HTA per dispositivi medici: prove tecniche di implementazione. In: CERGAS (eds.) Rapporto OASI 2017, Egea, Milano (2017)
3. Cabina di Regia Programma Nazionale di HTA dei Dispositivi Medici, Documento finale del Gruppo di lavoro 2 Metodi, Formazione e Comunicazione. http://www.salute.gov.it/portale/documentazione/p6_2_2_1.jsp?lingua=italiano&id=2855 (2019). Accessed 4 September 2019

Multiple Criteria Decision Analysis for Health Technology Assessment of Medical Devices: A Winning Hospital-Based Experience

Martina Andellini(✉), Roxana di Mauro, Francesco Faggiano, Pietro Derrico, and Matteo Ritrovato

Health Technology Assessment Unit, Bambino Gesù Children's Hospital, IRCCS, Rome, Italy
martina.andellini@opbg.net

Abstract. Health technology assessment (HTA) refers to the systematic evaluation of properties, effects, and/or impacts of health technologies. It is a multidisciplinary process to evaluate the social, economic, organizational and ethical issues of a health intervention or health technology, aiming to inform a policy decision making and to support decisions pertaining to the allocation of resources.

In order to empower decision makers to choose more knowingly between the different alternatives, offering a more precise and more structured output as well as contextualized evidence for a specific technology, we developed a standardized methodological approach that integrates Multicriteria Decision Analysis (MCDA) with the EuNetHTA Core Model®.

This new assessment approach, compared to the Core Model®, supplies a timelier as well as contextualized evidence for a specific technology. It makes it possible to obtain data which are more relevant and easier to interpret, and therefore more useful for decision makers to make investment choices with greater awareness. The application of MCDA to HTA has been mainly devised to address decision-making issues at hospital levels as the numerical values of the technologies' performances cannot be always easily predictable and/or measurable being affected by the robustness of the evidences, which supports them. To overcome these issues, the proposed method, combines the static MCDA method with a dynamic Montecalo simulation allowing a structured and more precise output, giving the decision makers the possibility to knowledgeably choose between the different alternatives considered, often in a very short time.

Keywords: Health Technology Assessment · Hospital decision-making · MCDA

1 Background

HTA represents a multidisciplinary process and a method of evidence synthesis and it aims to support complex health technologies decision-making processes. It considers different aspects such as clinical effectiveness, safety, costs, technical, organizational

J. Henriques et al. (Eds.): MEDICON 2019, IFMBE Proceedings 76, pp. 1783–1791, 2020.
https://doi.org/10.1007/978-3-030-31635-8_216

and, when broadly applied, social, ethical, and legal aspects of health technologies. Its application at, so called, meso level and in particular its implementation at hospital level, defined as Hospital-Based Health Technology Assessment (HB-HTA), is essential when considering the adoption or rejection of health technologies in a hospital, as it is aimed at contextualizing both evidence and decisions. However, available scientific evidence cannot always supply useful answers to the questions raised by local decision makers. For this reason, according to the literature, complex decision making processes have been usually carried out through Multi-criteria Decision Analysis (MCDA) method in several fields, even though very few applications to guide resource allocation decisions in health care have been recorded, so far [1]. MCDA represents an effective and reliable mathematical tool that provides support and facilitates the final choice of a healthcare device, identifying the criteria and their weights for priority setting and finally resulting in a rank ordering of interventions under examination. This work aims to show how it is possible to integrate MCDA within the standard theoretical HTA evaluation [2], as an analytical quantitative instrument focused on supporting the decision-making process between alternative products.

Notwithstanding the MCDA final result provides a quantitative output, it represents a mean value which does not take into account the complexity and the high variability of hospital processes. This paper, also aims to illustrate how this innovative approach for health care evaluation could be extended and become more sophisticate through the integration of a probabilistic approach. Taking the key performance indicators' values as inputs a Montecarlo simulation could be run in order to improve the reliability of technologies' performances expected values, including their confidence interval (CI).

2 Methods

The proposed approach is based on the integration of EUnetHTA Core Model [3] with Analytic Hierarchy Process (AHP) [4, 5]. The EunetHTA Core Model is an essential and profitable guide to support decision-making processes, identifying the different types of questions suitable to give appropriate answers about technology considered. It is useful to decompose the decision problem into its constituent parts, analyzing nine evaluation domains: Health problem and current use of technology, description and technical characteristics of technology, safety, clinical effectiveness, costs and economic evaluation, ethical analysis, organizational aspects, patient and social aspects, legal aspects. Each one describes the different aspects of the technology or the consequences or implications of its use, or any other aspect relevant for the assessment. AHP is a multi-criteria decision support technique, which allows quantifying opinions and transforming them into a coherent decision model assessing all factors and their interactions, leading to define a final rank of a number of health interventions.

Using the EUnetHTA Core Model® as a guidance, it is possible to identify which specific "assessment element" is most suitable for describing the technology or the consequences of its use by supplying the information needed to decide about its hypothetical introduction.

A scientific literature review is carried out, in order to acquire a deeper knowledge of the research question and to gather scientific evidence about the alternative health technologies under examination. The aim of this methodological approach is to identify all the components of the decisional hierarchy structure (the evaluation criteria, which describe the assessment problem) and, in agreement with the AHP method, their topological arrangement within it. The most relevant domains for the specific technology under evaluation are selected among those singled out in the EUnetHTA Core Model. They are described in turn by several key performance indicators, which consist in a list of detailed indicators, which describe the technology under assessment. They are technology-specific and they are identified from the synthesis of scientific evidence, from the results of the specific context analysis and from expert opinion. Once a hierarchical structure has been made available, AHP method and tasks can be easily accomplished. Following AHP method, all indicators and domains are judged, through pairwise comparisons, in terms of relative weights and performances values, obtaining the final weight and performance scores of each element of the decision three. It leads to the definition of the final rank of the alternative technologies compared, in terms of their overall preferences [2]. At the end of the process, a sensitivity analysis is usually carried out to test the stability and the robustness of the results, identifying the elements that represent a source of uncertainty and quantifying the impact of this variation on the stability of alternatives' ranking [6]. However, some issues affect MCDA results. Firstly, the numerical values of the technologies' performances cannot be always easily measurable and they are affected, indeed, by the robustness of the evidences, which supports them. Secondly, no historical time series data are explicitly taken into account for the Organizational and Economic aspects assessment, which are instead evaluated, through experts' hypothetical previsions.

To overcome these problems, a Montecarlo simulation (taking into account the MCDA results as input) is introduced as an integral part of the assessment process, in order to improve the reliability of technologies' performances expected values.

The probability distributions are built using two different techniques depending on which data are involved to.

Regarding the costs and economic evaluation and the organizational aspects domains, economic and organizational historical time series data could be used to estimate the related performances' values of the technologies involved in the study with high accuracy. For this reason, we developed an innovative approach, which takes into account the historical time series of Economic and Organizational data and builds a forecast mixed model combining the ARIMA and the Exponential Smoothing algorithms [7–11]. It makes it possible to obtain the short-term expected values and their Confidence Intervals (CI) with high accuracy, basing on real world data. From the forecast values and their CIs, Gaussian distributions per each criterion have been built, with mean values as the expected performance values and standard deviations calculated from the CIs.

The probability distributions of the other domains (Safety, Clinical Effectiveness, Health problem and current use of technology, description and technical characteristics of technology, ethical analysis, patient and social aspects and legal aspects), due to the unavailability of historical time series data, are evaluated in a different way. For these parameters, the majority of information, which define the indicators performances values, comes from the literature review.

Final Key Performance Indicators' values might be supported by strong literature evidences or, in the contrary, could be based only on expert opinion. In this way, applying the GRADE approach [12], it means grading the quality of the evidences, associating to them different weights according to strength of their recommendation.

Per each GRADE level (one to four), a Gaussian distribution is built, with the mean value as the performance percentage value and the standard deviation inversely proportional to the GRADE level. The higher the GRADE level is, the lower the related standard deviation is, indicating the high reliability of sources of information.

Taking into account all the different Gaussian distributions, for each indicators, a Montecarlo simulation, with thousands runs will be carried out. In this way, each performance value will vary in a broader way according to the linked Gaussian distribution, leading to an estimation of the accuracy of the final solution.

3 Results

To show the results of this innovative health technology assessment approach, we proposed a hypostatical case study. Let's assume that decision makers have to choose about the introduction in a hospital of an innovative health technology (Technology B) instead of the ongoing health technology (Technology A).

Let's suppose that after carrying out a scientific literature review and a context analysis, 5 Domains (Safety, Clinical Effectiveness, Technical characteristics of Technology, organizational aspects and cost and economic evaluation) and 12 Performances Indicators have been defined and after conducting the AHP process their global weights have been calculated as shown in Table 1.

To complete the assessment process, the performances values have to be set up. The safety, clinical effectiveness and technology performances' values are defined from the AHP process whereas the organizational and cost preferences' scores resulted from the forecast analysis.

According to AHP method, the safety, clinical effectiveness and technology indicators are judged in terms of performances' values through pairwise comparisons. However, the qualitative judgements are converted into consistent numerical values, through the Saaty scale, ranged from 1 (equal performance) to 9 (extremely higher performance) [4]. In this way, it is possible to obtain the elements' performance values, which will be weighted by the related element's weight.

The forecast analysis provides the expected organizational and cost parameters' values after a time t of the Technology B introduction, as well as those related to the ongoing technology -Technology A-at the same time. These values are translated into absolute performances values through the Saaty scale (Absolute technology A and B performances values). From these data, and taken into account the parameters' weights (evaluated before through AHP process) it is possible to calculate the weighted performances values.

This mixed static process (AHP and forecast) leads to a final rank of the alternatives technologies compared, providing a priority ranking of all alternatives in terms of their overall preferences.

Table 1. Shows the hypothetical domains and the twelve performance indicators, with their hypothetical weights and performances values, the GRADE scores and the related standard deviation.

Domains	Key performance indicators	Weights	Technology B weighted performances	Technology A weighted performances	Grade	Technology 1 absolute performances	Technology 2 absolute performances	Standard deviation
Safety		**33,73%**	**33,72%**	**19,40%**				
	Safety_1	15,64%	15,64%	8,80%	3	64,00	36,00	5
	Safety_2	10,12%	10,12%	7,95%	2	56,00	44,00	14,5
	Safety_3	7,96%	7,96%	2,65%	2	75,00	25,00	14,5
Clinical effectiveness		**23,51%**	**19,40%**	**15,87%**				
	Clinical Effectiveness_1	4,85%	4,85%	3,97%	3	55,00	45,00	5
	Clinical Effectiveness_2	14,55%	14,55%	11,90%	4	55,00	45,00	1,5
Technical characteristics of technology		**16,20%**	**13,86%**	**10,80%**				
	Technical Characteristics of technology_1	9,53%	9,53%	9,53%	4	43,00	57,00	1,5
	Technical Characteristics of technology_2	6,67%	1,27%	1,27%	3	84,00	16,00	5
Organisational aspects		**14,76%**	**14,62%**	**9,34%**				
	Organizational aspects_1	9,00%	9,34%	4,04%		69,00	31,00	23
	Organizational aspects_2	2,97%	2,97%	2,43%		55,00	45,00	6
	Organizational aspects_3	2,87%	2,65%	2,87%		48,00	52,00	12
Costs and economic evaluation		**11,81%**	**10,40%**	**9,61%**				
	Costs and economic evaluation_1	7,20%	7,20%	5,00%		59,00	41,00	9
	Costs and economic evaluation_2	4,61%	3,20%	4,61%		41,00	59,00	11
TOTAL		**100%**	**92%**	**65%**				

In order to carry out a Montecarlo simulation the indicators' probability distributions have been built.

With regards the safety, clinical effectiveness and technology aspects, a grade level, and consequently the relative Gaussian distribution, was associated to each indicator according to the robustness of scientific evidence by which it is supported. The mean values are represented by the Absolute technology A and B performances values and the standard deviation depends on the grade level. More specifically, four probability distributions were built (Fig. 1), according to the normal probability density function (1), as shown in Table 2.

$$f(x) = \frac{1}{\sigma\sqrt{2\pi}} e^{-\frac{(x-\mu)^2}{2\sigma^2}}, \text{ with } x = \mu \quad (1)$$

Furthermore, regarding the organizational and economic aspects, the Gaussian distributions are built starting from the forecast analysis results.

Forecast analysis, apart from the expected values (Absolute technology A and B performances values) which represent the Gaussian distributions' mean values, provides 95% CIs from which it is possible to calculate the standard deviations.

Given the mean values and the standard deviations, all the indicators' Gaussian distributions are described. In this way, the Montecarlo simulation can be run and the results, after 10000 runs, are shown in Fig. 2. Each Montecarlo simulation run, calculates the performances expected values modulated by the related Gaussian distributions. Figure 2 shows the total performances values per each run over the mean distance between the simulated performance scores and the Technologies Weighted Performances calculated through the static methods (AHP and forecast).

To sum up, based on the static approach, as showed in Table 1, the Solution B seems to be the best option comparing with the other one. In detail, the final performance value of the Solution B has led 27% over Solution A (92% vs. 65%). In

Fig. 1. Shows the four Gaussian distributions built per each GRADE level over the Saaty scale, with the standard deviation inversely proportional to the GRADE level.

Table 2. Illustrates how the Gaussian distributions per each GRADE level are built. More specifically, four probability distributions were built, according to the normal probability density function.

GRADE	Max probability	Standard deviation
1	0,2	1,994
2	0,43	0,92
3	0,667	0,598
4	0,9	0,44

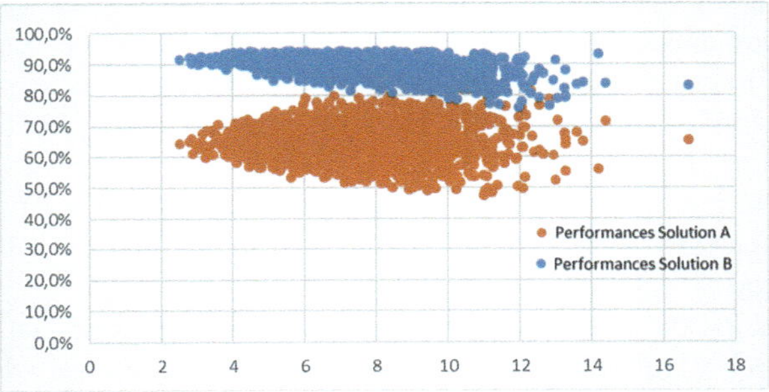

Fig. 2. Shows the total performances values per each Montecarlo simulation's run over the mean distance between the simulated performance scores and the Technologies Weighted Performances calculated through the static methods.

addition, statistical analysis results, depicted in the dispersion graph, show the general stability of the results. Simulated performances values appear perfectly distinguished between the two health technologies compared, highlighting and confirming the robustness and the reliability of the final rank.

More specifically, if the distance between the simulated performance scores and the Technologies Weighted Performances calculated through the static methods (AHP and forecast) is less than 6% points, the two solutions appear perfectly distinguished, whereas as the distance increases, the clouds of markers slightly overlap and only in few cases, they reverse.

4 Discussion and Conclusions

In health care field, very few applications of complex decision-making processes have been carried out through Multi-criteria Decision Analysis (MCDA) method [1]. As MCDA results an effective and reliable tool able to identify the criteria and their weights for priority setting, resulting in a rank ordering of health technologies, we proposed an innovative method, which integrates MCDA and other mathematical approaches within the HTA evaluation [2], as an analytical quantitative instrument focused on supporting the decision-making process between alternative products.

Firstly, conducting an HTA process integrated with AHP, as outlined in Do-HTA method [2], provides a structured outcome able to support the decision-making process, integrating evidences gathered from the literature review with professionals' expertise. It is especially useful for Hospital-Based Health Technology Assessment, where the context analysis and the main hospital stakeholders' judgements play a pivotal role within the evaluation.

At the end of the assessment process, a sensitivity analysis is usually carried out to test the stability and the robustness of the alternatives' ranking, and to identify the elements representing source of uncertainty. It quantifies the minimum possible changes on judgements per each domain needed to reverse the current ranking of alternative technologies [6]. One of its most relevant limits is that sensitivity analysis evaluates the domains' global errors without understanding the robustness of each indicator's performances value. Each indicator's performance score is determined by the synthesis of evidences when available, or by provisional algorithms when historical time series data exist.

The errors on indicators judgements are random and could not be quantified only by sensitivity analysis, whereas the stability of results could be better assessed if a statistical approach is applied.

The idea behind the proposed approach is related to the fact that each indicator's value and its error might be described by probability distributions, where the standard deviation represents the accuracy of the final results.

In this way, designing probability distributions in order to model the accuracy of the final performance scores, it represents an advanced and sophisticated technique [13–17].

We distinguished between data gathered from literature review and those obtained from registries and databases (Organizational and Economic) for which historical time series data are usually available. The former are evaluated and modeled by Gaussian distributions built on the relative GRADE level. The latter, thanks to the availability of their historical time series, are analyzed through the application of forecasting algorithms. Obtaining their forecast values together with the CIs at a time t, it gives the possibility to design the related Probability distributions. The application of time series analysis through forecast models in healthcare sector is proved to be a milestone because it provides short term expected values with significant accuracy.

Comparing with the static approach, the application of the statistical method allows Decision Makers to rationally, conscientiously and objectively evaluate the technologies under assessment, providing a different and more detailed way to carry out an accurate risk analysis, quantifying the error rates around the judgements. Before the acquisition of a technology, it is considered essential to make the best, most rational and reliable decision. Carrying out an evidence based approach, which integrates the literature review, the context analysis and the different points of view and skills of several professionals with an analytical and statistical method, significantly improve the robustness, the reliability and the reproducibility of the solution, recommending decision makers the best solution available within the modern-day technological market.

Conflict of Interest. The authors declare that they have no conflict of interest.

References

1. Triantaphyllou, E., Mann, S.H.: Using the analytic hierarchy process for decision making in engineering applications: some challenges. Int. J. Ind. Eng. Appl. Pract. **2**, 35–44 (1995)
2. Ritrovato, M., et al.: Decision-oriented health technology assessment: one step forward in supporting the decision-making process in hospitals. Value Health **18**(4), 505–511 (2015)
3. HTA core model for medical and surgical interventions. European Network for Health Technology Assessment (EUnetHTA)
4. Saaty, T.L.: The Analytic Hierarchy Process, Planning, Priority Setting, Resource Allocation. McGraw-Hill International Book Co., New York, London (1980)
5. Saaty, T.L.: Decision making with the analytic hierarchy process. Int. J. Serv. Sci. **1**(1), 83–98 (2008)
6. Triantaphyllou, E.: A sensitivity analysis approach for some deterministic multi-criteria decision-making methods. Decis. Sci. **28**(1), 151–194 (1997)
7. Luo, L., et al.: Hospital daily outpatient visits forecasting using a combinatorial model based on ARIMA and SES models. BMC Health Serv. Res. **17**, 1 (2017)
8. Ireneous, N.S., et al.: An overview of health forecasting Environ Health. Prev. Med. **18**, 1–9 (2013)
9. Kasapoglu, O.A.: Selection of the forecasting model in health care. J. Hosp. Med. Manage. **2** (2) (2016)
10. Zhou, L., et al.: Time series model for forecasting the number of new admission inpatients. BMC Med. Inform. Decis. Mak. **18**, 39 (2018)
11. Dritsakis, N., Klazoglou, P.: Time series analysis using ARIMA models: an approach to forecasting health expenditure in USA. Economia Internazionale/International Economics Camera di Commercio Industria Artigianato Agricoltura di Genova **72**(1), 77–106 (2019)
12. Zhang, Y., et al.: Using patient values and preferences to inform the importance of health outcomes in practice guideline development following the GRADE approach. Health Qual. Life Outcomes **15**, 52 (2017)
13. Jalalpour, M., Gel, Y., Levin, S.: Forecasting demand for health services: development of a publicly available toolbox. Oper. Res. Health Care **5**, 1–9 (2015)
14. Astolfi, R., et al.: Informing policy makers about future health spending: a comparative analysis of forecasting methods in OECD countries. Health Policy **107**, 1–10 (2012)
15. Pasupathy, K.S.: Forecasting model for strategic and operations planning of a nonprofit health care organization. In: Lawrence, K.D., Klimberg, R.K. (eds.) Advances in Business and Management Forecasting. Advances in Business and Management Forecasting, vol. 7, pp. 59–69. Emerald Group Publishing Limited, Bingley (2010)
16. Istvan, M.M., et al.: Modeling and forecasting health expectancy: theoretical framework and application. Demography **50**, 673–697 (2013)
17. Zhao, J.: Forecasting health expenditure: methods and applications to international databases (2015)

Biodegradation Behavior of Magnesium Alloy During Exposure to the Conditions of Human Body Environment

Radek Sedlacek[1(✉)], Tomas Suchy[1,2], and Zdenek Padovec[1]

[1] Faculty of Mechanical Engineering, Department of Mechanics, Biomechanics and Mechatronics, Czech Technical University in Prague, Technicka 4, 160 00 Prague, Czech Republic
radek.sedlacek@fs.cvut.cz

[2] Institute of Rock Structure and Mechanics, Department of Composites and Carbon Materials, The Czech Academy of Sciences, V Holesovickach 41, 182 09 Prague, Czech Republic

Abstract. In this study, biodegradation behaviour of WE43 magnesium alloy have been studied and compared during exposure to three different media commonly used to simulate the conditions of human body environment. Magnesium alloys emerged as a new class of bioresorbable implant materials. Their applications reduce certain risks associated with conventional permanent implants. Biodegradation behaviour of the WE43 magnesium alloy was observed under Dulbecco Modified Eagle Medium (DMEM, Sigma Aldrich) supplemented with 5% fetal bovine serum and gentamicin antibiotic as standard. The samples were stored in the medium at 37 °C and in a 5% CO_2 atmosphere. The second type of medium was Hank's Salt Balanced Solution (HBSS, Sigma Aldrich), which simulates the inorganic composition of blood plasma. HBSS was tempered at 37 °C. The last solution was an acidic solution of HCl + NaCl (pH2) with pH ~ 2 (0.01 M HCl and 0.14 M NaCl) tempered at 37 °C. The acidic type of solution was used to simulate a local acidic environment associated with osteoclast activity during bone remodelling. Changes in the mechanical properties of the samples during exposure to simulated body conditions were observed.

Keywords: Bioabsorbable metals · Biodegradation · Magnesium · Implant

1 Introduction

Magnesium alloys emerged as a new class of bioresorbable implant materials [1–3]. Their applications reduce certain risks associated with conventional permanent implants [1]. The need of reliable in vitro models to predict degradation in human body is increasing. In this study, biodegradation behaviour of WE43 magnesium alloy have been studied and compared during exposure to three different media commonly used to simulate the conditions of human body environment. Changes in the mechanical properties and surface area of the samples during exposure to simulated body conditions were observed.

J. Henriques et al. (Eds.): MEDICON 2019, IFMBE Proceedings 76, pp. 1792–1795, 2020.
https://doi.org/10.1007/978-3-030-31635-8_217

2 Materials, Methods and Results

Biodegradation behaviour of the WE43 magnesium alloy was observed under three different simulated body conditions. The first was Dulbecco Modified Eagle Medium (DMEM, Sigma Aldrich) supplemented with 5% fetal bovine serum and gentamicin antibiotic as standard. The samples were stored in the medium at 37 °C and in a 5% CO_2 atmosphere. The second type of medium was Hank's Salt Balanced Solution (HBSS, Sigma Aldrich), which simulates the inorganic composition of blood plasma. HBSS was tempered at 37 °C. Both solutions are physiologically neutral or slightly alkaline. The last solution was an acidic solution of HCl + NaCl (pH2) with pH ~ 2 (0.01 M HCl and 0.14 M NaCl) tempered at 37 °C. The acidic type of solution was used to simulate a local acidic environment associated with osteoclast activity during bone remodelling. Samples (n = 5) in dimensions of 6 mm in diameter and 9 mm in length were exposed to the medium for one and two weeks (see Fig. 1). The solutions were completely replaced every 24 h. The pH of the solutions was checked throughout the experiment.

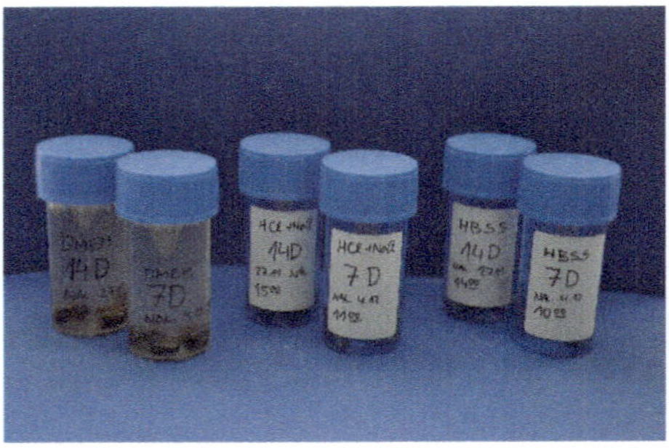

Fig. 1. Cylindrical specimens placed in boxes.

Fig. 2. Testing system MTS 858 Mini Bionix.

The effect of sample exposure to media was observed by mechanical compression tests (see Fig. 2). Furthermore, the sample surface was scanned on a scanning electron microscope (SEM) and optical microscope. The results of mechanical tests are illustrated in Fig. 3.

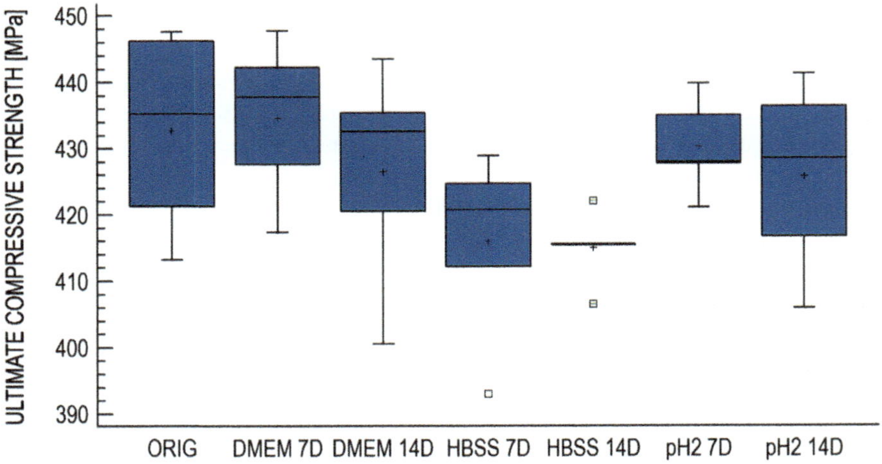

Fig. 3. Compressive strength of magnesium alloys before (ORIG) and seven days (7D) and fourteen days (14D) after exposure to Dulbecco Modified Eagle Medium (DMEM), Hank's Salt Balanced Solution (HBSS) and acidic solution of HCl + NaCl (pH2). There are no statistically significant differences between the samples in the multiple comparison of all groups (Games-Howell test, 0.05).

The results show that the short-term exposure used in all ingested media has no statistically significant effect on the observed mechanical properties ($p = 0.05$). However, the results obtained may also be affected by the sample sizes used. The image analyses showed that the short-term exposure was seen mainly on the surface of the samples, while the reduction of the mechanical properties of the cylindrical samples with a diameter of 6 mm and a height of 9 mm has no effect.

3 Conclusions

The aim of the biodegradation study was to assess the potential changes in the mechanical properties of cylindrical samples made from magnesium alloy WE43. The need of reliable in vitro models to predict degradation in human body is increasing. At present time, there is practically no general consensus on which medium would best and comprehensively simulate events occurring under in vivo conditions. Thus, three types of media commonly used to simulate the body environment were selected for the purpose of the study. Two weeks after the exposure, only local surface degradation was observed while no statistically significant effect on mechanical properties was determined. For next investigation of the short-term degradation-induced loss of mechanical integrity, specimens with smaller dimensions should be used, e.g. specimens with comparable size to that of implants such as cardiovascular stents and orthopaedic wires [1]. The differences among the three kinds of simulated body conditions were not observed within two weeks period.

Acknowledgements. This study was supported by the grant project awarded by the Technology Agency of the Czech Republic No. TACR TH03010418.

References

1. Galvin, E., Jaiswal, S., Lally, C., MacDonald, B., Duffy, B.: In vitro corrosion and biological assessment of bioabsorbable WE43 Mg alloy specimens. J. Manuf. Mater. Process., 1–8 (2017). https://doi.org/10.3390/jmmp1010008
2. Zhou, X.H., Jiang, L., Wu, P.P., Sun, Y., Yu, Y.D., Wei, G.Y., Ge, H.L.: Effect of aggressive ions on degradation of WE43 magnesium alloy in physiological environment. Int. J. Electrochem. Sci. **9**, 304–314 (2014)
3. Jiang, L., Xu, F., Xu, Z., Chen, Y., Zhou, X.H., Wei, G.Y., Ge, H.L.: Biodegradation of AZ31 and WE43 magnesium alloys in simulated body fluid. Int. J. Electrochem. Sci. **10**, 10422–10432 (2015)

Special Sessions: International Collaborative on Medical Devices Assessment

HTAi's Role in the International Collaborative on Medical Device Assessments

Rebecca Trowman[1] and Julie Polisena[2](✉)

[1] Health Technology Assessment International (HTAi), Edmonton, Canada
[2] Health Canada, Ottawa, Canada
julie.polisena@canada.ca

Abstract. HTAi – Health Technology Assessment international, was established in 2003 with the main objective to advance and promote HTA. The Society provides an open platform for global collaboration to leverage the shared and collective experience of its members to improve health outcomes globally. HTAi represents 82 organizational members, and 2,500 individual members from 65 countries. Members include leading HTA professionals and stakeholders, including researchers, policy makers, industry, academia, health service providers, agencies, and patients. The Society's flagship event is an Annual Meeting that provides a platform for the global HTA community to come together to elevate global health decision-making in an open and collaborative environment. HTAi is also home to the Global Policy Forum, which offers an opportunity for senior-level individuals from public and private sector organizations using HTA to support decisions or recommendations about product development and coverage to interact with one another, HTAi Board members, and invited international experts. As well, HTAi has established 10 Interest Groups (IGs) to share international experiences and expertise among HTA users and producers globally. As a non-state actor in official relations with the WHO, HTAi is uniquely positioned to bring insight and expertise to further the work of the ICMDA. The newly established Medical Devices Interest Group and the HTA in Developing Countries IG also offers a platform within HTAi for discussion of the HTA needs, challenges, and solutions pertaining medical device use and HTA in developing countries.

Keywords: Medical Device · Value assessment · Health Technology Assessment (HTA) · World Health Organization (WHO) · Lower-Middle Income Countries (LMICs)

1 Overview of Health Technology Assessment International (HTAi)

HTAi was established in 2003 to advance and promote HTA and is located in Edmonton, Canada. The Society is a Member driven organization that has adopted an open platform for global collaboration to leverage the shared and collective experience of its members to improve health outcomes worldwide. HTAi is governed by a Board of Directors, comprised of leading HTA experts from around the world reflecting the diverse membership of the Society. HTAi represents 82 organizational members, and

J. Henriques et al. (Eds.): MEDICON 2019, IFMBE Proceedings 76, pp. 1799–1801, 2020.
https://doi.org/10.1007/978-3-030-31635-8_218

2,500 individual members from 65 countries. Its membership is encompasses leading HTA professionals and stakeholders, including researchers, policy makers, industry, academia, health service providers, agencies, and patients, who partake in unique international networking and partnership opportunities.

The Society's flagship event is an Annual Meeting that provides a platform for the global HTA community to come together to elevate global health decision-making in an open and collaborative environment. Over 1,000 attendees from around the globe participate in the HTAi Annual Meeting each year. HTAi also offers educational scholarships to people from lower-middle income countries (LMICs), who are studying HTA or a closely related field who receive financial assistance to also present the results of their research at the HTAi Annual Meeting. Furthermore, HTAi provides numerous travel grants for members coming from LMICs to attend the Annual Meeting.

Another key HTAi activity is the Policy Forum meetings. Launched in 2003, the Global Policy Forum gives a unique opportunity for senior-level individuals from public and private sector organizations using HTA to support decisions or recommendations about product development and coverage to interact with one another, members of the HTAi Board, and invited international experts. The Global Policy Forum represents countries with mature HTA systems. This format has been extended to both Latin America and Asia where the regional Policy Forum meetings provide the same opportunity for strategic discussions about the present state of HTA, its development and implications for health systems, industry, patients and other stakeholders in regional settings. The WHO and regional WHO offices participate actively in the Policy Forum meetings and are an active partner in these activities. HTAi is also involved in areas of capacity building in LMICs, as evidenced by the recent workshop in Ghana on priority setting and the attainment of Universal Health Care (UHC).

Members of HTAi are invited to join one or more of the 10 Interest Groups (IGs) to share international experiences and expertise among HTA users and producers worldwide. The IGs serve as professional and scientific forums for all those who undertake and use HTA in health care, business, government, academic institutions, and consultancies. Moreover, the IGs provide a unique environment for HTAi members to collaborate on projects of mutual interest throughout the year. Members share topic updates, participate in the development of new research initiatives and relevant materials, and conduct workshops and training sessions. The IGs are as follows: Disinvestment and Early Awareness, Early Career Network, Ethics, Hospital-based HTA, Information Retrieval, HTA in Developing Countries, Medical Devices, Patient and Citizen Involvement, Public Health and Regulatory Interactions and Conditional Coverage.

2 Development of the International Collaborative on Medical Device Assessments (ICMDA)

HTAi's strategic objective and overall vision and mission is very much aligned with the development of the ICMDA. In particular, as a non-state actor in official relations with the WHO, HTAi is uniquely positioned to bring insight and expertise to further the work of the ICMDA. In addition, the newly established Medical Devices Interest Group and the HTA in Developing Countries IG provides a platform within HTAi for discussion of the HTA needs, challenges, and solutions pertaining medical device use and HTA in developing countries. Along with the rest of the HTAi Society, these IGs also aim to promote the understanding of the HTA concept, facilitate the initiation of HTA activities, and promote linkages and partnerships among individual and institutional HTAi members in developing and developed countries and with international organizations, including the WHO and Pan-American Health Organization, and public and private sector national and regional bodies in developing countries.

Reference

1. HTAi website. www.htai.org. Accessed 07 May 2019

Special Sessions: Ocular Imaging

Towards Improving Human Corneal Care Using Two-Photon Imaging

Ana Batista[1(✉)], Hans Georg Breunig[1,2], Berthold Seitz[3], and Karsten König[1,2]

[1] Department of Biophotonics and Laser Technology, Saarland University, 66123 Saarbruecken, Germany
ana.batista@blt.uni-saarland.de
[2] JenLab GmbH, Johann-Hittorf-Strasse 8, 12489 Berlin, Germany
[3] Department of Ophthalmology, Saarland University Medical Center, Saarland University, 66421 Homburg, Saar, Germany

Abstract. The human cornea is the tissue of the eye that contributes most to its refractive power. A healthy tissue is highly important to maintain visual acuity. In fact, diseases affecting the cornea are one of the major causes of blindness. Therefore, efficient methods to analyze the status of the healthy and diseased cornea are of outmost importance. Current clinical devices are, however, mostly limited to the morphological analysis of the tissue. The simultaneous analysis of tissue morphology, metabolism, and stromal structural organization using two-photon imaging (TPI) could improve corneal examination. In this study, two systems with TPI capabilities equipped with ultra-short near-infrared Ti:sapphire lasers were optimized for corneal imaging and their advantages for tissue examination demonstrated. Additionally, we show that by using TPI, disease diagnosis, follow-up after medical procedures, and corneal evaluation prior to transplantation could be improved. Thus, a future clinical device based on TPI could enhance the current state corneal examination and improve patient diagnosis and care.

Keywords: Cornea · Two-photon imaging · Second-harmonic generation · Fluorescence lifetime imaging · Multiphoton tomography

1 Introduction

The cornea is the outermost transparent tissue of the eye. It is the eye structure that most contributes to its refractive power and therefore has a great role in visual acuity [1]. In fact, the second leading cause of blindness in the world is corneal diseases [2]. A fast and accurate diagnosis is the successful and cost-effective way to prevent this outcome (blindness). However, when such cannot be avoided, corneal transplantation is the only possible way to restore the patient's sight. This is the most commonly performed transplantation, despite the shortage of donor corneal samples (worldwide, only one available for every seventy needed) [3].

In clinical practice, the human cornea is evaluated using methodologies that focus on the morphological evaluation of the tissue, such as slit lamp biomicroscopy and

J. Henriques et al. (Eds.): MEDICON 2019, IFMBE Proceedings 76, pp. 1805–1815, 2020.
https://doi.org/10.1007/978-3-030-31635-8_219

in vivo confocal microscopy. Meanwhile, the analysis of the cellular layer's metabolic activity and of the structural organization of the stroma, which accounts for 90% of the total corneal thickness [1], is disregarded. However, metabolic changes are known to proceed morphological consequences. In addition, changes to the stromal structural organization can be detected prior to changes in tissue transparency. Therefore, the simultaneous analysis of the morphology, metabolism, and structural organization may improve the clinical evaluation of the cornea.

In this study, two-photon imaging (TPI) was used to evaluate the human cornea. The tissue morphology and metabolism were assessed using two-photon excited fluorescence (TPEF) combined with fluorescence lifetime imaging (FLIM) and its structural organization using second-harmonic generation (SHG) imaging. The diagnosis and treatment of keratoconus, a corneal ectatic disorder that causes a progressive thinning and steepening of the cornea leading to the formation of a cone [4, 5], is shown as an example. Using TPI, it was possible to discriminate between non-pathological and keratoconus samples [6]. Moreover, we showed that, using TPI, the assessment of the clinical outcome of accelerated corneal collagen crosslinking (ACXL), a medical procedure used to halt the progression of ectatic corneal diseases such as keratoconus, can be performed much sooner than using current clinical devices [7]. Finally, we showed the potential of TPI for the evaluation of human donor cornea in the eye banks prior to corneal transplantation [8], the treatment chosen when all other treatment options have failed.

Since TPI can provide additional information on the tissue that can be helpful for corneal diagnosis, for patient follow-up after treatment, and for evaluating donor cornea prior to transplantation, a novel medical device based on TPI can improve the current state corneal examination and improve patient diagnosis and care.

2 Methods

2.1 Human Donor Corneas

Human donor corneas, unsuitable for transplantation, were provided by the Lions Cornea Bank Saar-Lor-Lux, Trier/Westpfalz at the Department of Ophthalmology, Saarland University Medical Center, Homburg/Saar, Germany.

Keratoconic corneas were provided by the Department of Ophthalmology, Saarland University, Medical Center, Homburg/Saar, Germany. Samples were collected immediately after penetrating keratoplasty.

All samples were stored at 34 °C under a 5% CO_2 atmosphere in culture Medium II with Dextran T500 (#F9017, Biochrom GmbH, Berlin, Germany) supplemented with 5% new-born calf serum (#S0415, Biochrom GmbH, Berlin, Germany) until image acquisition.

For image acquisition, samples were rinsed in PBS and mounted in individual glass bottom petri dishes or into an artificial anterior chamber. Live cell imaging solution (#A14291DJ, Life Technologies, California, USA) was used to maintain tissue hydration.

This study was conducted according to the principles of the World Medical Association's Declaration of Helsinki for the scientific use of human tissue. Patients have provided written and informed consent for the use of their tissue for research purposes.

2.2 Two-Photon Imaging Acquisition

Two devices with TPI imaging capabilities were used for human corneal imaging: (i) a custom-build laser-scanning microscope with five-dimensionality (5D microscope), and (ii) the commercial clinical tomograph MPT*flex* (JenLab GmbH, Berlin, Germany). The instrumental setups of both devices have already been described in detail elsewhere [8–11].

Briefly, the 5D microscope (Fig. 1A) is equipped with a mode-locked NIR Ti: sapphire laser Integral Pro 400 (FemtoLasers Produktions GmbH, Austria) generating ~10 fs pulses at a frequency of 85 MHz. This laser has a broad emission spectrum (~95 nm bandwidth) with a central wavelength of 800 nm, which enables the simultaneous excitation of multiple fluorophores. Autofluorescence and SHG signals are detected in reflection geometry by a spectral photomultiplier module (PML-16C) in the spectral range between 400 and 600 nm and processed by a time-correlated single-photon counting (TCSPC) SPC-150 module (Becker & Hickl, Berlin, Germany) for FLIM acquisition.

The MPT*flex* (Fig. 1B) is a certified clinical device equipped with an 80-MHz-Ti: sapphire 100 fs laser tunable between 710 and 920 nm. Autofluorescence and SHG signals are simultaneously detected in reflection geometry by two photomultiplier tubes (PMTs). For FLIM acquisition, a TCSPC SPC-830 module (Becker & Hickl GmbH, Berlin, Germany) is employed.

2.3 Image Analysis

Fluorescence lifetime images were analyzed using the commercial software SPCimage (Becker & Hickl GmbH, Berlin, Germany). The TCSPC histograms were fitted, after deconvolution with the instrument response function, with a bi-exponential function of the form $F(t) = a_1 e^{\frac{-t}{\tau_1}} + a_2 e^{\frac{-t}{\tau_2}}$, where t is the time and τ_1 and τ_2 are the fluorescence lifetimes with relative contributions a_1 and a_2, respectively. The mean autofluorescence lifetime weighted by the lifetime's relative contributions is computed as: $\tau_m = a_1\tau_1 + a_2\tau_2$.

The structural organization of the stroma was analyzed using the peak prominence (PP) obtained from fast Fourier transform (FFT) analysis of SHG images [6, 8]. The PP value is a measure for the degree of alignment of collagen fibers. A high value indicates that all fibers have similar orientations (organized). A small PP value results when the orientation distribution histogram is disperse, indicating that fibers have multiple orientation (less organized).

Statistical analysis of the data was performed using the commercial software GraphPad Prism version 6.05 (GraphPad Software Inc., California, USA). Statistically significant differences were computed using the *t-test*. When a Gaussian distribution was not observed and/or for small sample sizes, the non-parametric equivalent, *Mann-Whitney U test*, was used. ρ values lower than 0.05 were considered statistically significant.

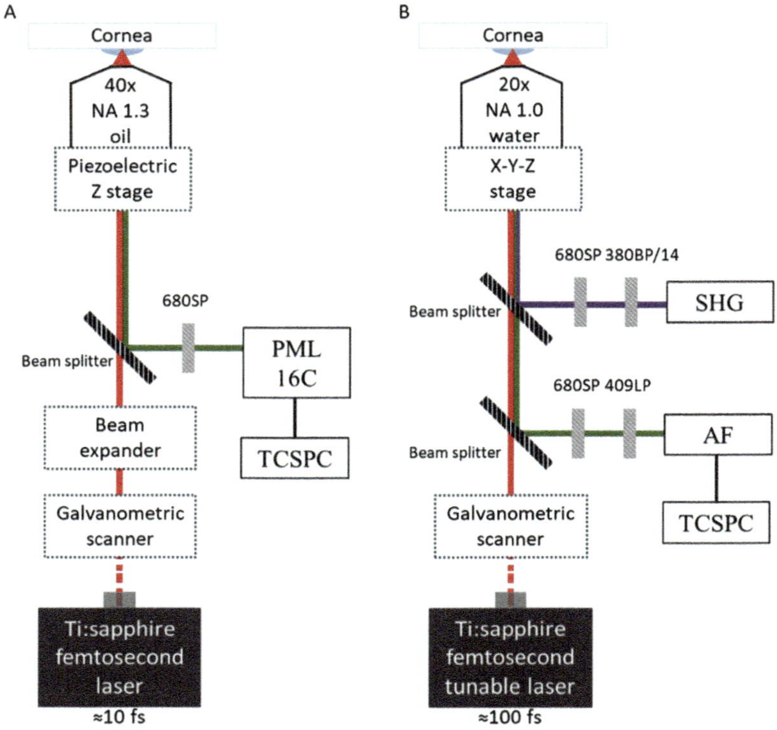

Fig. 1. Simplified schematic representation of the instrumental setups of the custom-built laser-scanning microscope with five-dimensionality (A) and of the MPT*flex* (B; JenLab GmbH, Berlin, Germany).

3 Results

Using TPI, we were able to image and discriminate all corneal layers. Figure 2 shows representative TPI images of the non-pathological corneal epithelium superficial (Fig. 2A), wing (Fig. 2B), and basal (Fig. 2C) cells, corneal stroma (Fig. 2D, E), and endothelial layer (Fig. 2F). In both cell layers, individual cells are visible, which allows their classification based on their morphology. In the corneal stroma, SHG signals, generated by collagen fibers in this layer, are present.

Human corneal samples from patients suffering from keratoconus were imaged within 24 h after surgery (n = 5) and compared with non-pathological samples (n = 4) to demonstrate the diagnostic capabilities of TPI [6].

For all corneal epithelial cell types keratoconus-induced morphological changes were detected (Fig. 3A, B, C) when compared with non-pathological samples (Fig. 2A, B, C). The analysis of the tissue metabolic activity based on the ratio of free (a_1) to protein-bound (a_2) NAD(P)H showed a significant decrease in the cells' metabolism induced by this pathology (Fig. 4A).

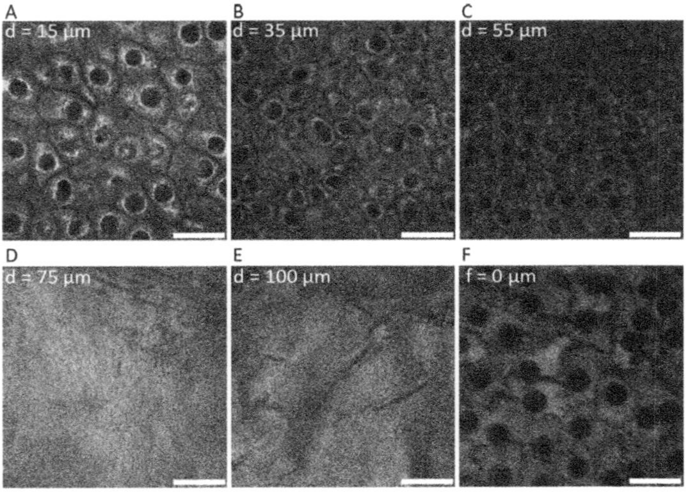

Fig. 2. Autofluorescence intensity of the non-pathological human corneal epithelium superficial (A), wing (B) and basal (C) cells, the corneal endothelial cells (F), and second-harmonic generation images of the corneal stroma (D, E) at multiple depths. The distances *d* and *f* are measured relatively to the epithelial and endothelial layers. Scale bars = 30 µm.

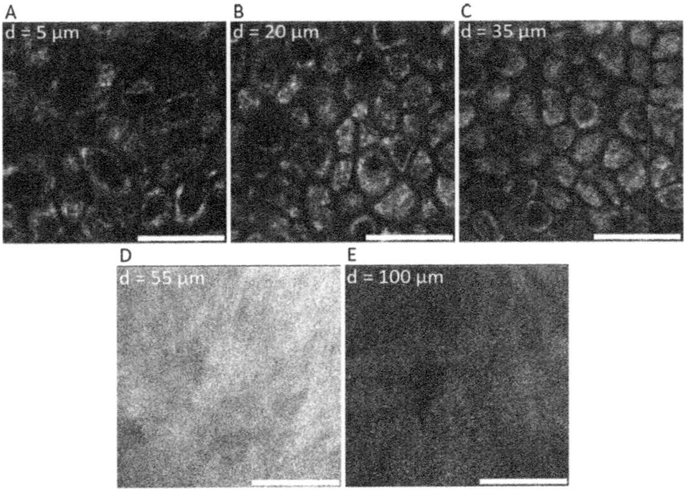

Fig. 3. Autofluorescence intensity of the keratoconic human corneal epithelium superficial (A), wing (B) and basal (C) cells and second-harmonic generation images of the corneal stroma (D, E) at multiple depths. Scale bars = 30 µm.

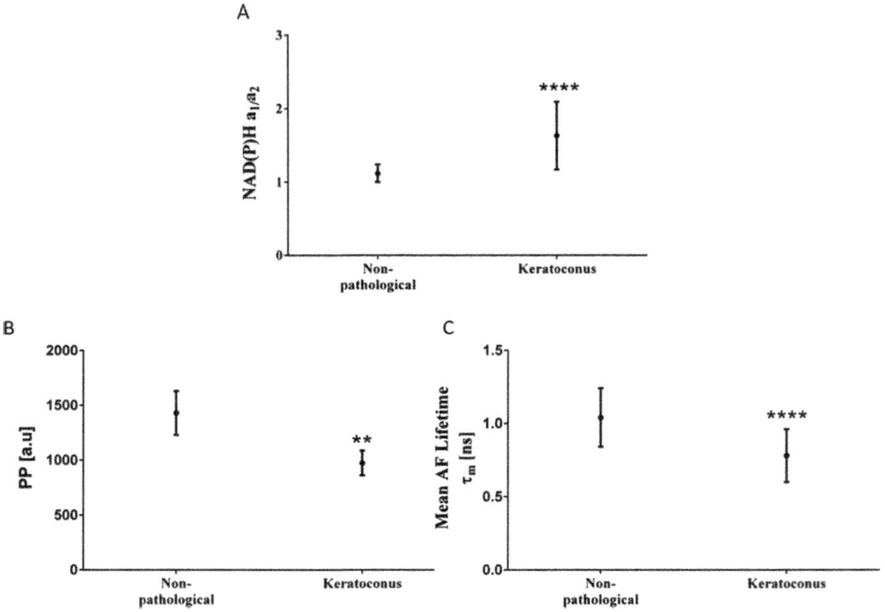

Fig. 4. NAD(P)H free to protein-bound (a_1/a_2) ratios of the non-pathological and keratoconic epithelial cells (A), as well as the peak prominence (B) and mean AF lifetimes of the stroma (C) of the same samples. The non-parametric *Mann-Whitney U test* was used for statistical analysis. ** $\rho < 0.01$; **** $\rho < 0.0001$.

In the corneal stroma, qualitative changes on the collagen organization were observed due to keratoconus (Fig. 3D, E) when compared with non-pathological corneas (Fig. 2D, E). A quantitative analysis of these changes based on PP obtained by FFT showed a significant decrease in the stromal structural organization induced by this pathology (Fig. 4B). Additionally, a significant decrease in the stroma mean autofluorescence lifetime was observed for keratoconus samples (Fig. 4C).

To assess the feasibility of TPI to evaluate the outcome of ACXL, a total of 14 human corneal donor samples, unsuitable for transplantation, were treated and imaged prior to ACXL and 2 h, 1, 3 as well as 6 *days* after treatment. ACXL was performed, after tissue de-epithelialization, by applying 0.1% riboflavin solution (Vibex rapid™ Avedro, Inc., MA, USA) to the samples for 20 min (drops every 2 min) followed by 10 min irradiation with a UVA light source emitting at 365 nm. The total energy dose applied was 7.2 J/cm^2 corresponding to an irradiance of 12 mW/cm^2.

As controls, 9 de-epithelialized human donor corneas were infused without UVA irradiation (RFN), and 10 samples were de-epithelialized without further treatment (Control).

As soon as 2 h after ACXL, significant increases in the corneal stroma autofluorescence intensity and lifetime were observed (Fig. 5). Both parameters remained constant up to at least 6 days after treatment. No significant differences were observed for the RFN and control groups (Fig. 5). Interestingly, a higher increase in the stroma autofluorescence due to treatment was observed in the anterior stroma (Fig. 5A).

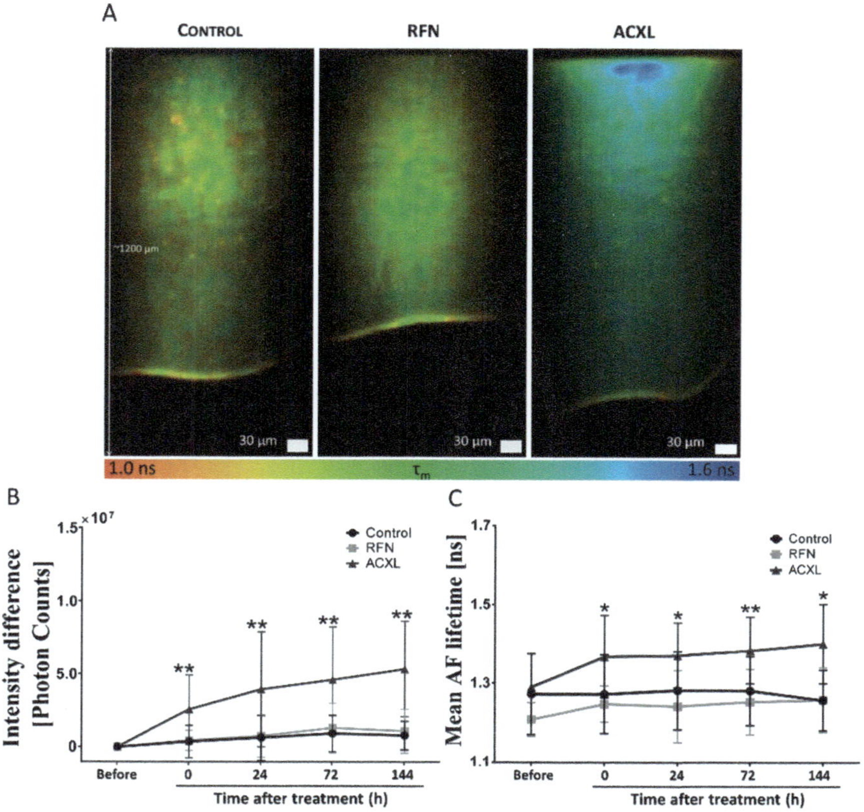

Fig. 5. Cross-sectional autofluorescence lifetime images of the corneal for control, RFN, and ACXL corneas 24 h after treatment (A) and average variation in the mean autofluorescence intensity (B) and lifetime (C). Statistical significance obtained using *t-test*. * $\rho < 0.05$; ** $\rho < 0.01$.

The feasibility of TPI to evaluate human corneal donor samples prior to transplantation was established by comparing samples with different storage times. Based on the time of storage, the samples were classified as short-term (below 1 week storage, STS, n = 4), medium-term (3 to 4 weeks storage, MTS, n = 3), and long-term storage (8 to 13 weeks storage, LTS, n = 3) [8].

Because of the lack of autofluorescence from the cell membranes and nuclei, cells could be clearly discriminated, and we were able to segment individual endothelial cells and provide information on endothelial cell morphology and density. An enlargement of cells accompanied by a decrease in cell density was observed with storage time [8]. Additionally, a significant decrease in the metabolic activity of both epithelial and endothelial cells was observed over time based on the ratio between free (a_1) and protein-bound (a_2) components of NAD(P)H (Fig. 6A, B).

For stromal keratocytes, a significant decrease in the NAD(P)H a_1/a_2 ratio was observed after long periods of storage (Fig. 6C), indicating an increase of these cells' metabolism. The analysis of the stromal structural organization revealed a significant decrease in collagen alignment with storage time (Fig. 6D).

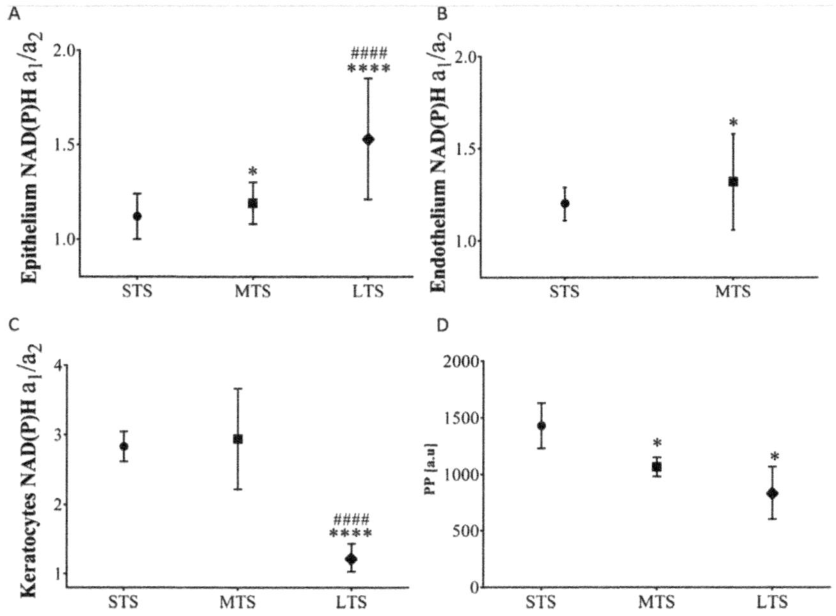

Fig. 6. NAD(P)H free to protein-bound (a_1/a_2) ratios of the corneal epithelium (A), endothelium (B), and stroma keratocytes (C) for short-term (STS), medium-term (MTS) and long-term storage (LTS), as well as the peak prominence (PP) overtime (D). Statistical differences assessed using the non-parametric *Mann-Whitney U test*. * $\rho < 0.05$ when compared with STS samples; **** $\rho < 0.0001$ when compared with STS samples; #### $\rho < 0.0001$ when compared with MTS samples.

4 Discussion

Using TPI, human corneal characterization and discrimination between non-pathological and keratoconic corneas was possible based not only on the cells morphology but also based on their metabolic activity, the stromal structural organization and its mean autofluorescence lifetime [6, 11, 12]. We were able to observed characteristic changes induced by this pathology using the tissue morphology, such as cell enlargement and elongation [4, 5]. The analysis of the cells' metabolism indicates a decrease in the metabolic activity induced by the pathology. In the corneal stroma, a significant decrease in the autofluorescence lifetime was observed, which can be correlated with a decrease in the number of crosslinks between collagen fibers and lamellae [13, 14]. The anticipated decrease in collagen organization was also observed using SHG imaging. The feasibility of this imaging modality to detect keratoconus-induced changes to the stromal structural organization has been previously demonstrated [15–18].

In clinical practice, the progression of keratoconus can be halted using the photodynamic treatment ACXL. In this study, we demonstrated the feasibility of TPI to assess the outcome of ACXL much sooner than current clinical devices.

Efforts to evaluate the outcome of ACXL soon after treatment are not new. Brillouin microscopy, an imaging modality that can provide the Young's modulus of the tissue, has been shown capable of assessing ACXL-induced changes soon after treatment [19–22]. However, this imaging modality requires long image acquisition times (30 to 120 min per image) [19–21]. SHG imaging has also been used to investigate changes induces by ACXL to the collagen organization [23–27].

In this study, we focused on the evaluation of the tissue autofluorescence [7]. We observed that ACXL leads to a significant increase in the stroma autofluorescence intensity and lifetime, noticeable as soon as 2 h after treatment. Similar changes to the stroma autofluorescence were also found for rabbit corneas, two weeks after treatment [28]. This increase in the stroma autofluorescence lifetime is likely related to an increase in the number of crosslinks between collagen [13, 14], and, therefore, indicates a positive treatment outcome.

Higher increases in the autofluorescence intensity and lifetime were observed in anterior than posterior stroma. This indicates that ACXL was, due to the limited penetration depth of the UVA light, more efficient anteriorly.

Keratoconus treatment may ultimately involve tissue transplantation if disease progression cannot be halted. In this study, we also demonstrated the feasibility of TPI to evaluated human donor corneas prior to transplantation [8].

Using TPI, samples can be evaluated not only based on morphological parameters such as the density of endothelial cells but also on the cells' metabolism and tissue structural organization. With storage time the metabolic activity of epithelial and endothelial cells decreased, indicating reduced tissue viability. In contrast, the metabolic activity of keratocytes increased with storage time. These cells are usually in a quiescent state and are activated in case of injury [29]. The observed metabolic increase during storage may also indicate keratocyte activation due to adverse conditions.

Based on SHG, a decrease in stromal structural organization was observed with storage, indicating a decrease in tissue transparency. Although essential for proper corneal function, tissue transparency is currently evaluated prior to transplantation only subjectively [30].

5 Conclusion

TPI provides important additional information on the corneal tissue that can complement current clinical devices and can significantly improve the examination of this tissue in clinical practice. TPI may be used to improve corneal disease diagnosis, to enhance the follow-up patients after treatment, and to provide a better assessment of tissue viability prior to transplantation. Therefore, a novel medical device based on TPI is of great interest and it can lead to improved patient corneal care.

References

1. Remington, L.A.: Clinical Anatomy of the Visual System, 2nd edn. Elsevier, St. Louis (2005)
2. Whitcher, J.P., Srinivasan, M., Upadhyay, M.P.: Corneal blindness: a global perspective. Bull. World Heal. Organ. **79**(3), 214–221 (2001)
3. Gain, P., Jullienne, R., He, Z., Aldossary, M., Acquart, S., Cognasse, F., Thuret, G.: Global survey of corneal transplantation and eye banking. JAMA Ophthalmol. **134**(2), 167–173 (2016)
4. Khaled, M.L., Helwa, I., Drewry, M., Seremwe, M., Estes, A., Liu, Y.: Molecular and histopathological changes associated with Keratoconus. Biomed. Res. Int. **2017**, 1–16 (2017). Article ID 7803029
5. Feder, R.S., Gan, T.J.: Noninflammatory ectatic disorders. In: Krachmer, J.H., Mannis, M.J., Holland, E.J. (eds.) Cornea - Fundamentals, Diagnosis and Management, vol. 1, 3rd edn. Mosby/Elsevier, Maryland Heights, Missouri (2011)
6. Batista, A., Breunig, H.G., König, A., Schindele, A., Hager, T., Seitz, B., König, K.: High-resolution, label-free two-photon imaging of diseased human corneas. J. Biomed. Opt. **23** (03), 1–8 (2018). 036002
7. Batista, A., Breunig, H.G., Hager, T., Seitz, B., König, K.: Follow-up of accelerated-crosslinking non-invasively and label-free using multiphoton tomography. In: Proceedings of SPIE, vol. 10858, p. 108581X–1:7 (2019)
8. Batista, A., Breunig, H.G., König, A., Schindele, A., Hager, T., Seitz, B., Morgado, A.M., König, K.: Assessment of human corneas prior to transplantation using high-resolution two-photon imaging. Investig. Opthalmol. Vis. Sci. **59**(1), 176–184 (2018)
9. Batista, A., Breunig, H.G., Uchugonova, A., Morgado, A.M., König, K.: Two-photon spectral fluorescence lifetime and second-harmonic generation imaging of the porcine cornea with a 12 femtosecond laser microscope. J. Biomed. Opt. **21**(3), 1–11 (2016). 036002
10. Földes-Papp, Z., König, K., Studier, H., Bückle, R., Breunig, H.G., Uchugonova, A., Kostner, G.M.: Trafficking of mature miRNA-122 into the nucleus of live liver cells. Curr. Pharm. Biotechnol. **10**(6), 569–578 (2009)
11. König, K., Batista, A., Hager, T., Seitz, B.: Multiphoton tomography of the human eye. In: Periasamy, A., et al. (eds.) Proceedings of SPIE, vol. 10069, p. 1006906:1-8 (2017)
12. Batista, A., Breunig, H.G., Donitzky, C., König, K.: Two-photon microscopy and fluorescence lifetime imaging of the cornea. In: König, K. (ed.) Multiphoton Microscopy and Fluorescence Lifetime Imaging, 1st edn., pp. 301–320. De Gruyter, Berlin (2018)
13. Manickavasagam, A., Hirvonen, L.M., Melita, L.N., Chong, E.Z., Cook, R.J., Bozec, L., Festy, F.: Multimodal optical characterisation of collagen photodegradation by femtosecond infrared laser ablation. Analyst **139**(23), 6135–6143 (2014)
14. Lutz, V., Sattler, M., Gallinat, S., Wenck, H., Poertner, R., Fischer, F.: Impact of collagen crosslinking on the second harmonic generation signal and the fluorescence lifetime of collagen autofluorescence. Ski. Res. Technol. **18**(2), 168–179 (2012)
15. Tan, H.Y., Sun, Y., Lo, W., Lin, S.J., Hsiao, C.H., Chen, Y.F., Huang, S.C.M., Lin, W.C., Jee, S.H., Yu, H.S., Dong, C.Y.: Multiphoton fluorescence and second harmonic generation imaging of the structural alterations in keratoconus ex vivo. Investig. Ophthalmol. Vis. Sci. **47**(12), 5251–5259 (2006)
16. Morishige, N., Wahlert, A.J., Kenney, M.C., Brown, D.J., Kawamoto, K., Chikama, T., Nishida, T., Jester, J.V.: Second-harmonic imaging microscopy of normal human and keratoconus cornea. Invest. Ophthalmol. Vis. Sci. **48**(3), 1087–1094 (2007)

17. Morishige, N., Shin-Gyou-Uchi, R., Azumi, H., Ohta, H., Morita, Y., Yamada, N., Kimura, K., Takahara, A., Sonoda, K.H.: Quantitative analysis of collagen lamellae in the normal and keratoconic human cornea by second harmonic generation imaging microscopy. Investig. Ophthalmol. Vis. Sci. **55**(12), 8377–8385 (2014)
18. Mercatelli, R., Ratto, F., Rossi, F., Tatini, F., Menabuoni, L., Malandrini, A., Nicoletti, R., Pini, R., Pavone, F.S., Cicchi, R.: Three-dimensional mapping of the orientation of collagen corneal lamellae in healthy and keratoconic human corneas using SHG microscopy. J. Biophotonics **83**(1), 75–83 (2016)
19. Scarcelli, G., Pineda, R., Yun, S.H.: Brillouin optical microscopy for corneal biomechanics. Investig. Ophthalmol. Vis. Sci. **53**(1), 185–190 (2012)
20. Scarcelli, G., Kling, S., Quijano, E., Pineda, R., Marcos, S., Yun, S.H.: Brillouin microscopy of collagen crosslinking: Noncontact depth-dependent analysis of corneal elastic modulus. Investig. Ophthalmol. Vis. Sci. **54**(2), 1418–1425 (2013)
21. Kwok, S.J.J., Kuznetsov, I.A., Kim, M., Choi, M., Scarcelli, G., Yun, S.H.: Selective two-photon collagen crosslinking in situ measured by Brillouin microscopy. Optica **3**(5), 469–472 (2016)
22. Webb, J.N., Su, J.P., Scarcelli, G.: Mechanical outcome of accelerated corneal crosslinking evaluated by Brillouin microscopy. J. Cataract Refract. Surg. **43**(11), 1458–1463 (2017)
23. Bueno, J.M., Gualda, E.J., Giakoumaki, A., Perez-Merino, P., Marcos, S., Artal, P.: Multiphoton microscopy of ex vivo corneas after collagen cross-linking. Invest. Ophthalmol. Vis. Sci. **52**(8), 5325–5331 (2011)
24. Tan, H.Y., Chang, Y.L., Lo, W., Hsueh, C.M., Chen, W.L., Ghazaryan, A.A., Hu, P.S., Young, T.H., Chen, S.J., Dong, C.Y.: Characterizing the morphologic changes in collagen crosslinked-treated corneas by Fourier transform-second harmonic generation imaging. J. Cataract Refract. Surg. **39**(5), 779–788 (2013)
25. McQuaid, R., Li, J., Cummings, A., Mrochen, M., Vohnsen, B.: Second-harmonic reflection imaging of normal and accelerated corneal crosslinking using porcine corneas and the role of intraocular pressure. Cornea **33**(2), 125–130 (2014)
26. Mercatelli, R., Ratto, F., Tatini, F., Rossi, F., Menabuoni, L., Nicoletti, R., Pini, R., Pavone, F.S., Cicchi, R.: Characterization of the lamellar rearrangement induced by cross-linking treatment in keratoconic corneal samples imaged by SHG microscopy. In: Proceedings of SPIE, vol. 10045, p. 100450L:1-6 (2017)
27. Zyablitskaya, M., Takaoka, A., Munteanu, E.L., Nagasaki, T., Trokel, S.L., Paik, D.C.: Evaluation of therapeutic tissue crosslinking (TXL) for myopia using second harmonic generation signal microscopy in rabbit sclera. Investig. Opthalmol. Vis. Sci. **58**(1), 21–29 (2017)
28. Steven, P., Hovakimyan, M., Guthoff, R.F., Hüttmann, G., Stachs, O.: Imaging corneal crosslinking by autofluorescence 2-photon microscopy, second harmonic generation, and fluorescence lifetime measurements. J. Cataract Refract. Surg. **36**(12), 2150–2159 (2010)
29. Nishida, T., Saika, S.: Cornea and sclera: anatomy and physiology. In: Krachmer, J.H., Mannis, M.J., Holland, E.J. (eds.) Cornea - Fundamentals, Diagnosis and Management, vol. 1, 3rd edn. Mosby/Elsevier, Maryland Heights, Missouri (2011)
30. Parekh, M., Ferrari, S., Ruzza, A., Pugliese, M., Ponzin, D., Salvalaio, G.: A portable device for measuring donor corneal transparency in eye banks. Cell Tissue Bank. **15**(1), 7–13 (2014)

Characterization of the Retinal Changes of the 3xTg-AD Mouse Model of Alzheimer's Disease

Hugo Ferreira[1(✉)], João Martins[1,2], Ana Nunes[1,2], Paula I. Moreira[2,3], Miguel Castelo-Branco[1,2,3], António Francisco Ambrósio[2,3,4], Pedro Serranho[1,2,5], and Rui Bernardes[1,2,3]

[1] Coimbra Institute for Biomedical Imaging and Clinical Research (CIBIT), Institute of Nuclear Sciences Applied to Health (ICNAS), University of Coimbra, Coimbra, Portugal
hugo.ferreira3000@gmail.com, rmbernardes@fmed.uc.pt
[2] CNC.IBILI Consortium, University of Coimbra, Coimbra, Portugal
[3] Faculty of Medicine, University of Coimbra, Coimbra, Portugal
[4] Coimbra Institute for Clinical and Biomedical Research (iCBR), Faculty of Medicine, University of Coimbra, Coimbra, Portugal
[5] Mathematics Section, Department of Science and Technology, Universidade Aberta, Lisbon, Portugal

Abstract. In this work, we imaged the retina of wild-type and triple-transgenic mice model of Alzheimer's disease (AD) (3xTg-AD), at the ages of one and two months, by optical coherence tomography. Texture analysis of calculated fundus images, for the six most anterior layers of the retina, present widespread differences between groups, demonstrating that retinal changes in the animal model of AD are notorious across the neuroretina. These results and the consistency of data pave the way for the identification of texture biomarkers of disease onset and progression with potential application to the human retina.

Keywords: Optical coherence tomography · Retina · Biomarkers · Texture · Alzheimer's disease · Mouse model · Diagnosis

1 Introduction

Alzheimer's disease (AD) is one of the most common neurodegenerative disorders, and its prevalence is expected to rise due to a growing ageing population [9]. Moreover, the onset of AD is thought to happen much earlier than the clinical symptoms, so a reliable diagnosis can only be performed after significant brain damage [7].

The use of mouse models offers the opportunity of collecting information through the progression of AD, at regular intervals, for the entire lifespan of the mouse, intending to identify biomarkers for the progression of the disease. Though no mouse model can fully emulate the human form of AD, there are several which replicate its most relevant characteristics [3]. In this work, we use the triple-transgenic animal models of AD (3xTg-AD) whose genetic modifications allow simulating the pathophysiology of the disease [3, 11].

Retina imaging allows the search for image-based biomarkers of AD [8], as a window to the central nervous system accessible by optical means. In this regard, several

J. Henriques et al. (Eds.): MEDICON 2019, IFMBE Proceedings 76, pp. 1816–1821, 2020.
https://doi.org/10.1007/978-3-030-31635-8_220

studies were performed using optical coherence tomography (OCT) focusing on thickness variations of the retina, or some of its layers, as biomarkers for AD [2,6]. A recent study [1] has suggested a process of texture analysis to provide OCT-based biomarkers other than thickness.

In this work, we study the texture biomarkers of retina OCT data from wild-type (WT) and 3xTg-AD mice in the first two months of age. Our objective is to identify the texture biomarkers presenting statistically significant differences between groups at different time-points, illustrating the potential of this technique for early AD diagnosis.

2 Materials and Methods

2.1 Sample and Study Groups

The mouse model of AD used in this work is the 3xTg-AD, while the healthy controls used for comparison are WT (C57BL6/128S background) mice.

This study was approved by the Animal Welfare Committee of the Coimbra Institute for Clinical and Biomedical Research (iCBR), Faculty of Medicine, University of Coimbra. All procedures involving mice were conducted as per the Association for Research in Vision and Ophthalmology statement for animal use, and in agreement with the European Community Directive Guidelines for the care and use of non-human animals for scientific purposes (2010/63/EU), transposed into the Portuguese law in 2013 (DL113/2013).

The OCT data used was gathered (when possible) from both eyes of both 35 WT and 3xTg-AD mice, yielding 68 and 65 eyes respectively for the age of one month and, 60 and 64 at the age of two months.

2.2 Data Acquisition

The data used in this work was obtained by the Micron IV OCT System (Phoenix Research Labs, Pleasanton, CA, USA). All eye volume scans, made of 512 non-compressed B-scans TIF files (of 512 A-scans each), were taken at the same retinal location, horizontally centred with the optic disc and vertically above the optic disc's border.

2.3 OCT Data Segmentation by Retinal Layers

A segmentation method was created to analyze the data from each mouse retinal layer independently. This method anchors on two interfaces that are easily automatically segmented: the inner boundary of the retinal nerve fibre layer (RNFL) and the outer boundary of the outer nuclear layer (ONL). Since the relative thickness of the distinct retinal layers is coherent across the literature, we segmented the remaining interface's using the histological image provided in [2]. Therefore, we consider the space between the two anchor interfaces to be filled in by the RNFL (6%), the ganglion cell layer (GCL) (3%), the inner plexiform layer (IPL) (30%), the inner nuclear layer (INL) (15%), the outer plexiform layer (OPL) (6%) and the ONL (40%).

2.4 Texture Features of OCT Data

A mean value fundus (MVF) image [4] was computed for each of the six layers mentioned above. Each pixel of each MVF image is computed as the average of the A-scan's values between the two retinal interfaces delimiting the layer. In this work, MVF images were split into 24 × 24 adjacent blocks of 20 × 20 pixels each, cropping the image to 480 × 480 pixels from the original 512 × 512 pixels.

Before calculating texture features for each block, based on grey-level co-occurrence matrices (GLCM) [5], the number of grayscale levels was reduced from 65536 (16 bits) to 16 (4 bits). Furthermore, four directions (0°, 45°, 90° and 135°) with symmetry ON (i.e., 180° apart angles are considered the same) and scale (pixel distance) of one pixel, lead to four GLCM per block. For each of these, 20 features were determined as in [10] and [5]. Each feature is considered as the maximum across the four directions to decouple features' values from the orientation, resulting in 20 features per block.

Blocks were then aggregated into quadrants by defining each quadrant feature's value as the average of 12 × 12 blocks. In this way, a total of 480 (20 feature × 4 quadrants × 6 layers) features characterize each eye scan.

2.5 Statistical Analysis

The Kolmogorov-Smirnov normality test was applied to the 480 features. Since the results indicated that the distributions are not normal for most features, we used the non-parametric Mann-Whitney test to determine differences between groups, considering significances of 5%, 1%, and 0.1%.

All data processing was performed using Matlab R2018a (The MathWorks Inc., Natick, MA, USA), and the statistical treatment was made using IBM SPSS 25.

3 Results

The number of features with significant statistical differences between groups, per layer and time-point, are presented in Table 1.

The spread of statistically significant differences across the layers of the retina demonstrates the direct effect of changes associated with Alzheimer's disease on the central nervous system. Although widespread, the INL seems to be particularly affected, with 90% of the features presenting statistically significant differences at the age of two months for 5% significance and with 75% for 0.1% significance. Other texture metrics may show further differences.

3.1 Disease Progression Analysis

The disease progression is well documented in Table 1 by the increase in the number of features presenting statistically significant differences between groups from the age of one to two months.

Furthermore, we note that 178 (37%) features present statistically significant differences at both time-points, as shown in Table 2, demonstrating the consistency of the method and data gathered from the same set of animals.

Table 1. Number of significant features per layer, at the ages of one and two months. For each layer, there are a total of 80 features.

	One month old			Two months old		
	Number of significant features			Number of significant features		
	$p \leq 0.001$	$p \leq 0.01$	$p \leq 0.05$	$p \leq 0.001$	$p \leq 0.01$	$p \leq 0.05$
RNFL	19 (23.8%)	29 (36.3%)	35 (43.8%)	29 (36.3%)	32 (40.0%)	39 (48.8%)
GCL	27 (33.8%)	38 (47.5%)	50 (62.5%)	40 (50.0%)	52 (65.0%)	61 (76.3%)
IPL	25 (31.3%)	38 (47.5%)	47 (58.8%)	46 (57.5%)	59 (73.8%)	65 (82.5%)
INL	48 (60.0%)	61 (76.3%)	69 (86.3%)	59 (73.8%)	67 (83.8%)	71 (88.8%)
OPL	33 (41.3%)	48 (60.0%)	55 (68.8%)	19 (23.8%)	25 (31.3%)	35 (43.8%)
ONL	33 (41.3%)	41 (51.3%)	47 (58.8%)	46 (57.5%)	53 (66.3%)	64 (80.0%)
Total	185 (38.5%)	255 (53.1%)	303 (63.1%)	239 (49.8%)	288 (60.0%)	336 (70.0%)

Table 2. Features with statistically significant differences between wild-type and transgenic mice groups at both time-points. The green-coloured circles (●) represent a p-value ≤ 0.05, the orange-coloured squares (■) show p-values ≤ 0.01 and the red-coloured asterisks (*) represent a p-value ≤ 0.001. The number of symbols in each time-point identifies the number of quadrants where the significance was observed for each of the features shown.

Layer	Feature	Age (months) 1	Age (months) 2
RNFL	Entropy	■●	**
	Dissimilarity	**●	***
	Inertia	**■	***
	Cluster Prominence	*	■
	Sum Entropy	*	●
	Difference Variance	**■	***
	Difference Entropy	**■	***
	IMC1	■■	*●
GCL	Homogeneity	●■	●●
	Uniformity	■*	■■
	Entropy	**	**
	Dissimilarity	***	***
	Inertia	●***	****
	Autocorrelation	■*	**
	Cluster Shade	■*	**
	Maximum Probability	■*	■■
	Sum Of Squares	**	**
	Sum Average	**	**
	Sum Variance	**	**
	Difference Variance	●***	****
	Difference Entropy	●***	****
	IMC1	■***	****
	INN	■	●
	IDN	■	●
IPL	Homogeneity	■*	■*
	Uniformity	■*	●●
	Entropy	●	●
	Dissimilarity	●●**	*■**
	Inertia	●■**	****
	Correlation	■*	■*
	Autocorrelation	**	**
	Cluster Prominence	●■	●*
	Maximum Probability	●	■
	Sum Of Squares	**	**
	Sum Average	**	**
	Sum Variance	**	**
	Sum Entropy	●*	●*
	Difference Variance	●■**	****
	Difference Entropy	●■**	*■**
	IMC1	■**	***
	IMC2	■*	■*
	INN	■*	■*
	IDN	■*	■*
INL	Homogeneity	■***	■***
	Uniformity	**	■■
	Dissimilarity	■***	****
	Inertia	■***	****
	Correlation	■***	■***
	Autocorrelation	■■**	****
	Cluster Prominence	****	■***
	Maximum Probability	■■	●●
	Sum Of Squares	●●**	****
	Sum Average	●●**	****
	Sum Variance	●●**	****
	Sum Entropy	■***	****
	Difference Variance	■***	****
	Difference Entropy	■***	****
	IMC1	****	****
	IMC2	■***	****
	INN	●***	■***
	IDN	●***	■***
OPL	Homogeneity	■	*
	Entropy	*	●
	Dissimilarity	■	■
	Inertia	■	■
	Correlation	*	*
	Autocorrelation	■●	●*
	Cluster Prominence	■*	*●
	Sum Of Squares	■	●
	Sum Average	■	●
	Sum Variance	■	●
	Sum Entropy	***	●*■
	Difference Variance	■	■
	Difference Entropy	■	■
	IMC1	****	****
	IMC2	***	*●●
	INN	■	*
	IDN	■	*
ONL	Homogeneity	**	**
	Uniformity	■*	■*
	Dissimilarity	*■**	■***
	Inertia	*■**	****
	Correlation	**	**
	Autocorrelation	●	*
	Cluster Prominence	**	**
	Maximum Probability	*	●
	Sum Entropy	**	**
	Difference Variance	*■**	****
	Difference Entropy	*■**	■***
	IMC1	■■**	●***
	IMC2	**	**
	INN	**	**
	IDN	**	**

4 Conclusions

In this work, we demonstrated that retinal changes associated with AD are possible to identify based on the texture analysis of computed fundus images. Also, this work demonstrates that at the initial stages of the disease, changes are notorious and widespread across the neuroretina. Furthermore, these results demonstrate that both the proposed methodology and data gathered are consistent, paving the way for the identification of texture biomarkers of disease onset and progression with application to the human retina, based on a non-invasive optical imaging technique.

5 Funding

This study was supported by The Portuguese Foundation for Science and Technology (FCT) through PTDC/EMD-EMD/28039/2017, PEst-UID/NEU/04539/2019 and UID/Multi/04621/2013, and by FEDER-COMPETE through POCI-01-0145-FEDER-028039 and POCI-01-0145-FEDER-007440.

Conflict of Interest. The authors declare no conflicts of interest.

References

1. Bernardes, R., Silva, G., Chiquita, S., Serranho, P., Ambrósio, A.F.: Retinal biomarkers of Alzheimer's disease: insights from transgenic mouse models. In: 14th International Conference on Image Analysis and Recognition (ICIAR) (2017). https://doi.org/10.1007/978-3-319-59876-5_60
2. Dysli, C., Enzmann, V., Sznitman, R., Zinkernagel, M.S.: Quantitative analysis of mouse retinal layers using automated segmentation of spectral domain optical coherence tomography images. Transl. Vis. Sci. Technol. **4**(4), 9 (2016). https://doi.org/10.1167/tvst.4.4.9
3. Elder, G.A., Gama Sosa, M.A., De Gasperi, R.: Transgenic mouse models of alzheimer's disease. Mount Sinai J. Med. J. Transl. Personal. Med. **77**(1), 69–81 (2010). https://doi.org/10.1002/msj.20159
4. Guimarães, P., Rodrigues, P., Lobo, C., Leal, S., Figueira, J., Serranho, P., Bernardes, R.: Ocular fundus reference images from optical coherence tomography. Comput. Med. Imaging Graph. **38**, 381–389 (2014). https://doi.org/10.1016/j.compmedimag.2014.02.003
5. Haralick, R.M., Shanmugam, K., Dinstein, I.: Textural features for image classification. IEEE Trans. Syst. Man Cybern. **SMC–3**(6), 610–621 (1973)
6. Hart, N.J., Koronyo, Y., Black, K.L., Koronyo-Hamaoui, M.: Ocular indicators of alzheimer's: exploring disease in the retina. Acta Neuropathol. **132**(6), 767–787 (2016). https://doi.org/10.1007/s00401-016-1613-6
7. Krantic, S., Torriglia, A.: Retina: source of the earliest biomarkers for alzheimer's disease? J. Alzheimer's Dis. **40**(2), 237–243 (2014). https://doi.org/10.3233/JAD-132105
8. London, A., Benhar, I., Schwartz, M.: The retina as a window to the brain–from eye research to CNS disorders. Nat. Rev. Neurol. **9**(1), 44–53 (2013). https://doi.org/10.1038/nrneurol.2012.227
9. Shah, H., Albanese, E., Duggan, C., Rudan, I., Langa, K.M., Carrillo, M.C., Chan, K.Y., Joanette, Y., Prince, M., Rossor, M., Saxena, S., Snyder, H.M., Sperling, R., Varghese, M., Wang, H., Wortmann, M., Dua, T.: Research priorities to reduce the global burden of dementia by 2025. Lancet Neurol. **15**(12), 1285–1294 (2016). https://doi.org/10.1016/S1474-4422(16)30235-6

10. Soh, L.k., Tsatsoulis, C.: Texture representation of SAR sea ice imagery using multi-displacement co-occurrence matrices. In: International Geoscience and Remote Sensing Symposium (1996). https://doi.org/10.1109/IGARSS.1996.516261
11. Sterniczuk, R., Antle, M.C., LaFerla, F.M., Dyck, R.H.: Characterization of the 3xTg-AD mouse model of Alzheimer's disease: Part 2. Behavioral and cognitive changes. Brain Res. **1348**, 149–155 (2010). https://doi.org/10.1016/j.brainres.2010.06.011

Distinguishing Functional from Non-functional Pituitary Macroadenomas with a Machine Learning Analysis

Ricciardi Carlo[1(✉)], Cuocolo Renato[1], Cesarelli Giuseppe[2,3], Ugga Lorenzo[1], Improta Giovanni[4], Solari Domenico[5], Romeo Valeria[1], Guadagno Elia[1], Cavallo Luigi Maria[5], and Cesarelli Mario[6,7]

[1] Department of Advanced Biomedical Sciences, University of Naples "Federico II", Naples, Italy
carloricciardi93@gmail.com, renato.cuocolo@gmail.com
[2] Department of Chemical, Materials and Production Engineering, University of Naples "Federico II", Naples, Italy
[3] Istituto Italiano di Tecnologia, Naples, Italy
[4] Department of Public Health, University of Naples "Federico II", Naples, Italy
[5] Department of Neurosciences, Reproductive and Odontostomatological Sciences, Division of Neurosurgery, University of Naples "Federico II", Naples, Italy
[6] DIETI, University of Naples "Federico II", Naples, Italy
[7] Istituti Clinici Scientifici Maugeri IRCCS, Telese Terme, BN, Italy

Abstract. Pituitary adenomas are rare intracranial tumors that are often found incidentally in MR images. On the other hand, radiomics is a new field whose aim is converting images in mineable data; particularly, texture analysis is a postprocessing technique extracting quantitative parameters from the heterogeneity of pixel grey level. In this scenario, machine learning can be applied in order to classify these adenomas into functional and non-functional starting from features extracted through texture analysis on MRI images acquired through a protocol including a coronal T2-weighted Turbo Spin Echo sequence. The boosting of J48, a multinomial logistic regression and K nearest neighbour are implemented employing Knime analytics platform. Excluding J48 whose accuracy was 83.0%, multinomial logistic regression and K nearest neighbour achieved accuracies beyond 92.0% and the Area Under the Curve Receiving Characteristic Operator till 98.4%. Diagnosing correctly this delicate disease is crucial in order to achieve the best management as well as the most appropriate cure for patients. The novelty of this paper lies in proving the ability of the combination of radiomics and machine learning to pre-operatively predict tumoral behavior. Prior to this analysis it was believed that only blood tests or histopathological analysis could provide this information.

Keywords: Machine learning · Classification · Texture analysis · Pituitary adenomas · MRI

J. Henriques et al. (Eds.): MEDICON 2019, IFMBE Proceedings 76, pp. 1822–1829, 2020.
https://doi.org/10.1007/978-3-030-31635-8_221

1 Introduction

Pituitary adenomas are rare intracranial tumors, presenting with a prevalence of 1/1500 in the general population. In most cases, they are benign lesions, whose clinical manifestations are related to mass effect signs - depending on tumor size and/or to hormone hypersecretion syndromes. On the other hand, low dimension intrasellar adenomas may be clinically silent and their diagnosis often comes as an incidental finding on MR scans [1–5].

Radiomics, consisting in the conversion of images into mineable data and subsequent analysis for decision support, is an emerging field allowing tumor classification [6]. In particular, texture analysis is a postprocessing technique for quantitative parameter extraction from pixel grey level heterogeneity. It consists of statistical analysis based on both simple intensity value distribution histograms and more complex gray level distribution matrix analyses which also retain information on spatial distribution of voxel intensities [7].

In this setting, machine learning can be applied in order to predict the outcome of patients and help clinicians in decision-making [8–11]. There is a wide range of applications of machine learning in different areas of medicine, from cardiology to radiology [12, 13]. In particular, studies applying machine learning on texture analysis according to the "radiomic process" were described by Kumar et al. [14]: Zacharaki et al. classified brain tumor type and grade using MRI texture and shape through Linear Discriminant Analysis with Fisher's discriminant rule, k-nearest neighbour (KNN), nonlinear Support Vector Machine (SVM) and employing leave one out cross-validation [15]; Juntu et al. differentiated benign from malignant soft-tissue tumours in T1-MRI images testing three classifiers (neural networks, decision tree and SVM) [16]; Romeo et al. characterized adrenal lesions on unenhanced MRI images [17]; finally, Stanzione and colleagues have recently demonstrated the potential of this approach in prostate cancer local staging [18]. Moreover, recent studies investigated the relevance of first and second order histogram features obtained by diffusion-weighted imaging magnetic resonance in differentiating functional from non-functional pituitary macroadenoma through a classic statistical analysis [19].

Therefore, the aim of this study is to apply machine learning algorithms on parameters obtained by texture analysis on MRI images in order to distinguish functional from non-functional pituitary macroadenomas.

2 Materials and Methods

2.1 Subjects

We retrospectively reviewed data of 50 patients, who received so-called standard endoscopic endonasal approach for the removal of a pituitary adenoma, between January 2013 and December 2017, at the Division of Neurosurgery of the University of Naples 'Federico II' in Italy. All of them underwent preoperative MRI at our Institution prior to the surgical procedure. Demographic data, preoperative assessment - i.e. endocrinological and visual status and presenting signs - tumor features, prior treatments, surgical results and complications, were retrieved from our electronic database (Filemaker Pro 11 - File Maker Inc, Santa Clara, CA, USA).

2.2 MRI Acquisition and Texture Analysis

All exams were acquired on a 1.5-Tesla scanner (Gyroscan Intera, Philips, Eindhoven, The Netherlands). The imaging protocol always included a coronal T2-weighted Turbo Spin Echo sequence (TR/TE: 2600/89 ms, FOV: 180 $\times$ 180 mm; matrix: 288 $\times$ 288; thk: 3 mm) used for the following radiomic feature extraction.

First of all, lesions were detected by an expert neuroradiologist who then performed their manual contouring by means of a bidimensional polygonal ROI after selection of the slice where it showed maximum extension (Fig. 1). Further editing with a brush tool was performed, when needed. This process was carried on using a freely available segmentation software (ITKSnap v3.6.0) [20].

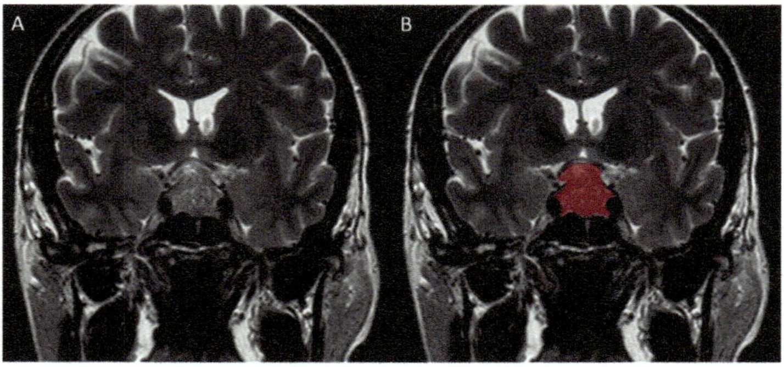

Fig. 1. Coronal T2-weighted MRI exam showing the maximum extension slice of a functioning pituitary macroadenoma (A). Image B depicts the result of the image annotation of the region of interest to be employed for subsequent texture feature extraction

Image pre-processing and feature extraction were performed on an open-source Python radiomics software (Pyradiomics v2.1.2) [21]. The first step consisted of image gray level normalization with a scale of 100). This step was mandatory since T2-weighted images are not quantitative and intensity values are not absolute in contrast to T2 maps. The latter were not available as only routine clinical scans were selected for the analysis, also in order to guarantee reproducibility of the results in the clinical setting. Subsequently, all volumes and corresponding lesion masks were resampled to a 2 $\times$ 2 $\times$ 2 mm voxel resolution. The next pre-processing step consisted of intensity value discretization. For this task, a fixed bin width approach was chosen, obtaining an ideal bin count between 16 and 128, as suggested in previous studies [22].

The use of wavelet decomposition, yielding all possible combinations of High and Low pass filtering in the x, y and z dimensions, and edge enhancement Laplacian of Gaussian (LoG) filters, emphasizing gray level change at different texture coarseness, allowed additional feature extraction from the derived images.

Finally, in relation to texture parameter extraction, together with bidimensional shape and first order statistics, we also obtained higher order class parameters. In detail, these were derived from the symmetrical Gray Level Co-occurrence Matrix (GLCM), Gray Level Size Zone Matrix (GLSZM), Gray Level Run Length Matrix (GLRLM, Neighboring Gray Tone Difference Matrix (NGTDM) and Gray Level Dependence Matrix (GLDM).

2.3 Tool

Knime analytics platform (v. 3.7.1) was chosen to conduct this machine learning study, as it is a well-known open source platform implementing a wide range of machine learning algorithms and integrated with Weka, Python and other software; moreover, it was already employed in literature for other studies [23, 24]. The algorithms used in this paper are briefly presented in the next section.

2.4 Algorithms and Evaluation Metrics

J48 is the Java implementation of a C4.5 decision tree [25], which consists of the evolution of the ID3 algorithm. It is an easy structure made up of leaves, representing classes, and nodes, representing test phases over an attribute. Multinomial Logistic Regression (MLR) with ridge estimator is applied through the "Logistic" node of Weka that follows the implementation of Le Cessie, and van Houwelingen [26, 27]. K Nearest Neighbour (KNN) is an easy instance-based classifier that assigns a label basing its choice on the dominance of a class in the nearest neighbours [28]. For all these algorithms, "smote" (Synthetic Minority Over-sampling Technique) was applied [29]. Smote generates artificial data by extrapolating between a real object of a given class and one of its nearest neighbours (of the same class). Boosting was implemented for J48, it converts weak learners into strong learners that predict with higher accuracy; it selects only the parameters that can improve the predictive ability of algorithms during the training phase, making the complexity in terms of dimension decrease and improving execution time [30]. The evaluation metrics employed in this study are:

- Accuracy: correct classifications over the total;
- Error: misclassifications over the total;
- Recall: the ratio of positives correctly classified;
- Precision: the ratio of positives correctly predicted in the positive class;
- Sensitivity: capacity to detect true positives;
- Specificity: capacity to detect true negatives.

Moreover, Area Under the Curve Receiving Characteristic Operator (AUCROC) was computed for each algorithm and for both bagging and boosting groups.

3 Results

Of the included lesions, 25 were functioning adenomas (5 Adreno Cortico Tropic Hormone, 8 Growth Hormone, 5 Growth Hormone/Prolactin, 6 Prolactin and 1 Thyroid-Stimulating Hormone secreting) and 25 non-functioning. A total of 1128 features was extracted from each patient.

Due to the small number of patients, smote technique was applied to make the number of records rise from 50 to 100. Then, a procedure of feature selection was applied to reduce the number of features extracted by the images: the matrix of correlation was computed among all variables and a threshold of correlation of 0.4 was chosen: all the variables with a correlation higher than the threshold were excluded because they did not add information to the classifiers. It allowed us to reduce the number of features from more than one thousand to 28. As the number of patients was not so high, leave one out was applied for all the implemented algorithms. J48, MLR and KNN were implemented together with the boosting node of Knime. Results are summarized in Table 1 while Table 2 shows the features used to build the models.

Table 1. Scores for each algorithm

	Accuracy [%]	Error [%]	Recall [%]	Precision [%]	Sensitivity [%]	Specificity [%]	AUCROC [%]
Boosting J48	83.0	17.0	82.0	83.7	82.0	84.0	89.8
MLR	97.0	3.0	96.0	98.0	96.0	98.0	98.4
KNN	90.0	10.0	90.0	90.0	90.0	90.0	90.0

MLR obtained the highest accuracy, recall, precision, sensitivity, specificity and AUCROC among the three implemented algorithms. Despite getting the lowest accuracy (83.0%), J48 reached an AUCROC comparable to the KNN's one.

Table 2. Features used to build the predictive models

log-sigma-2-0-mm-3D_gldm_LargeDependenceLowGrayLevelEmphasis	wavelet-LHH_firstorder_Skewness
log-sigma-2-5-mm-3D_glcm_InverseVariance	wavelet-LHH_firstorder_Entropy
log-sigma-2-5-mm-3D_firstorder_Range	wavelet-LLH_glcm_ClusterShade
log-sigma-3-0-mm-3D_gldm_LargeDependenceLowGrayLevelEmphasis	wavelet-LLH_glcm_Contrast
log-sigma-3-5-mm-3D_gldm_LargeDependenceLowGrayLevelEmphasis	wavelet-LLH_firstorder_Kurtosis
log-sigma-3-5-mm-3D_gldm_LargeDependenceHighGrayLevelEmphasis	wavelet-HLH_glcm_Id

(continued)

Table 2. (*continued*)

log-sigma-3-5-mm-3D_firstorder_Maximum	wavelet-HLH_firstorder_Skewness
wavelet-HLL_glcm_ClusterProminence	wavelet-HLH_firstorder_Median
wavelet-HLL_firstorder_Skewness	wavelet-HHH_firstorder_Skewness
wavelet-HLL_firstorder_Maximum	wavelet-LLL_glcm_ClusterShade
wavelet-LHL_glcm_Correlation	wavelet-LLL_firstorder_Kurtosis
wavelet-LHL_firstorder_Skewness	original_glcm_Correlation
wavelet-LHL_firstorder_Mean	original_firstorder_Skewness
wavelet-LHH_gldm_DependenceVariance	original_firstorder_Kurtosis

4 Discussion

First, the MRI acquisition of 50 patients was performed at the department of Advanced Biomedical Sciences of the University Hospital "Federico II" of Naples. Furthermore, a texture analysis was conducted to extract more than one thousand quantitative features from the MRI images. The machine learning analysis was finally performed in order to carry out some evaluation metrics as regards the algorithms.

Mentioning other studies that employed radiomics and machine learning, Romeo et al. [17] characterized adrenal lesions with a diagnostic accuracy of 80%, while Juntu et al. [16] distinguished benign from malignant tumors with an accuracy of 93%; Zacharaki et al. [15] obtained 85% of accuracy classifying type and grade of brain tumours. Although a direct comparison with other studies would not be completely fair (due to the use of different datasets), this study shows greater capacity to correctly make classifications (functional and non-functional pituitary macroadenomas), exploiting features extracted through texture analysis. A comparison may be done with the study of Sanei et al. [19] who distinguished functional from non-functional pituitary macroadenomas with lower scores than those obtained through a machine learning analysis.

The functional status of pituitary lesions has a significant influence on the clinical manifestations of disease: the correct diagnosis and management is crucial for the selection of the correct therapeutic strategy and therefore cure this multifaceted disease. Although a previous study has shown the promise of Apparent Diffusion Coefficient values of pituitary lesions in this assessment [20], Diffusion Weighted Imaging is not routinely performed in the imaging of the sellar region. It is known that this area is potentially more prone to artefacts on echo-planar imaging and this technique is time consuming. For these reasons, an approach that obtains similar results while employing routine MRI sequences has more potential for its application in the current clinical setting.

Of course, this study is affected by some limitations: the dataset was augmented with artificial data in order to improve its size allowing us to perform the analysis. Major dataset could be studied, and machine learning analysis could be performed to reach 100% accuracy in this classification. Nevertheless, machine learning proved to be the best way to distinguish functional from non-functional pituitary macroadenomas using texture analysis on MRI images.

This paper proved that the combination of radiomics and machine learning can be used to predict tumoral behaviour pre-operatively while only blood tests or histopathological analysis were known as providers of this information.

Conflict of Interests. The authors have no conflict of interests.

References

1. Gittleman, H., Ostrom, Q.T., Farah, P.D., et al.: Descriptive epidemiology of pituitary tumors in the United States, 2004–2009. J. Neurosurg. **121**(3), 527–535 (2014)
2. Ostrom, Q.T., Gittleman, H., Farah, P., et al.: CBTRUS statistical report: primary brain and central nervous system tumors diagnosed in the United States in 2006–2010. Neuro Oncol. **15**(Suppl 2), ii1-56 (2013)
3. Thapar, K., Laws, E.R.J.: Pituitary tumors. In: Kaye, A.W., Jr. LER, (eds.) Brain Tumors 2001, pp. 804–854. Churchill Livingstone, London (2001)
4. Wilson, P.J., Omay, S.B., Kacker, A., et al.: Endonasal endoscopic pituitary surgery in the elderly. J. Neurosurg. **128**, 1–8 (2017)
5. Solari, D., Cavallo, L.M., Cappabianca, P.: Surgical approach to pituitary tumors. Handb. Clin. Neurol. **124**, 291–301 (2014)
6. Gillies, R.J., Kinahan, P.E., Hricak, H.: Radiomics: images are more than pictures, they are data. Radiology **278**(2), 563–577 (2015). https://doi.org/10.1148/radiol.2015151169
7. Lubner, M.G., Smith, A.D., Sandrasegaran, K., et al.: CT texture analysis: definitions, applications, biologic correlates, and challenges. Radiographics **37**(5), 1483–1503 (2017). https://doi.org/10.1148/rg.2017170056
8. Malik, M.M., Abdallah, S., Ala'raj, M.: Data mining and predictive analytics applications for the delivery of healthcare services: a systematic literature review. Ann. Oper. Res. **270**(1–2), 287–312 (2018)
9. Santini, S., Pescapè, A., Valente, A., et al.: Using fuzzy logic for improving clinical daily-care of β-thalassemia patients. In: 2017 IEEE International Conference on Fuzzy Systems (FUZZ-IEEE), pp. 1–6. IEEE (2017).. https://doi.org/10.1109/fuzz-ieee.2017.8015545
10. Improta, G., Russo, M.A., Triassi, M., et al.: Use of the AHP methodology in system dynamics: modelling and simulation for health technology assessments to determine the correct prosthesis choice for hernia diseases. Math. Biosci. **299**, 19–27 (2018). https://doi.org/10.1016/j.mbs.2018.03.004
11. Romano, M., D'Addio, G., Clemente, F., et al.: Symbolic dynamic and frequency analysis in foetal monitoring. In: 2014 IEEE International Symposium on Medical Measurements and Applications (MeMeA), pp. 1–5. IEEE (2014). https://doi.org/10.1109/memea.2014.6860122
12. Johnson, K.W., Soto, J.T., Glicksberg, B.S., et al.: Artificial intelligence in cardiology. J. Am. Coll. Cardiol. **71**(23), 2668–2679 (2018)

13. Choy, G., Khalilzadeh, O., Michalski, M., et al.: Cur-rent applications and future impact of machine learning in radiology. Radiology **288**(2), 318–328 (2018)
14. Kumar, V., Gu, Y., Basu, S., et al.: Radiomics: the process and the challenges. Magn. Reson. Imaging **30**(9), 1234–1248 (2012). https://doi.org/10.1016/j.mri.2012.06.010
15. Zacharaki, E.I., Wang, S., Chawla, S., et al.: Classification of brain tumor type and grade using MRI texture and shape in a machine learning scheme. Magn. Reson. Med. Off. J. Int. Soc. Magn. Reson. Med. **62**(6), 1609–1618 (2009)
16. Juntu, J., Sijbers, J., De Backer, S., et al.: Machine learning study of several classifiers trained with texture analysis features to differentiate benign from malignant soft-tissue tumors in T1-MRI images. J. Magn. Reson. Imaging **31**(3), 680–689 (2010)
17. Romeo, V., Maurea, S., Cuocolo, R., et al.: Characterization of adrenal lesions on unenhanced MRI using texture analysis: a machine-learning approach. J. Magn. Reson. Imaging **48**(1), 198–204 (2018)
18. Stanzione, A., Cuocolo, R., Cocozza, S., et al.: Detection of extraprostatic extension of cancer on biparametric MRI combining texture analysis and machine learning: preliminary results. Acad. Radiol. (2019). https://doi.org/10.1016/j.acra.2018.12.025
19. Sanei, M.T., Kimia, F., Mehrnahad, M., et al.: Accuracy of diffusion-weighted imaging-magnetic resonance in differentiating functional from non-functional pituitary macro-adenoma and classification of tumor consistency. Neuroradiol. J. **32**, 74–85 (2018). https://doi.org/10.1177/1971400918809825
20. Yushkevich, P.A., Piven, J., Hazlett, H.C., et al.: User-guided 3D active contour segmentation of anatomical structures: significantly improved efficiency and reliability. Neuroimage **31**(3), 1116–1128 (2006). https://doi.org/10.1016/j.neuroimage.2006.01.015
21. Van Griethuysen, J.J., Fedorov, A., Parmar, C., et al.: Computational radiomics system to decode the radiographic phenotype. Cancer Res. **77**(21), e104–e107 (2017). https://doi.org/10.1158/0008-5472.CAN-17-0339
22. Leijenaar, R.T., Nalbantov, G., Carvalho, S., et al.: The effect of SUV discretization in quantitative FDG-PET radiomics: the need for standardized methodology in tumor texture analysis. Sci. Rep. **5**, 11075 (2015). https://doi.org/10.1038/srep11075
23. Mannarino, T., Assante, R., Ricciardi, C., et al.: Head-to-head comparison of diagnostic accuracy of stress-only myocardial perfusion imaging with conventional and cadmium-zinc telluride single-photon emission computed tomography in women with suspected coronary artery disease. J. Nucl. Cardiol. (2019). https://doi.org/10.1007/s12350-019-01789-7
24. Romeo, V., Ricciardi, C., Cuocolo, R., et al.: Machine learning analysis of MRI-derived texture features to predict placenta accreta spectrum in patients with placenta previa. Magnetic Resonance Imaging (2019). https://doi.org/10.1016/j.mri.2019.05.017. ISSN 0730-725X
25. Quinlan, J.R.: Bagging, boosting, and C4. 5. In: AAAI/IAAI, vol. 1, pp. 725–730, August 1996
26. le Cessie, S., van Houwelingen, J.C.: Ridge estimators in logistic regression. Appl. Stat. **41** (1), 191–201 (1991)
27. Aha, D.W., Kibler, D., Albert, M.K.: Instance-based learning algorithms. Mach. Learn. **6**(1), 37–66 (1992)
28. Keller, J.M., Gray, M.R., Givens, J.A.: A fuzzy k-nearest neighbor algorithm. IEEE Trans. Syst. Man, Cybern. **4**, 580–585 (1985)
29. Chawla, N., Bowyer, K., Hall, L., et al.: SMOTE: synthetic minority over-sampling technique. J. Artif. Intell. Res. (JAIR) **16**, 321–357 (2002). https://doi.org/10.1613/jair.953
30. Freund, Y., Schapire, R.E.: A decision-theoretic generalization of on-line learning and an application to boosting. J. Comput. Syst. Sci. **55**(1), 119–139 (1997)

Sexual Dimorphism of the Adult Human Retina Assessed by Optical Coherence Tomography

Ana Nunes[1,2(✉)], Pedro Serranho[1,2,3], Hugo Quental[1,2], António Francisco Ambrósio[2,4,5], Miguel Castelo-Branco[1,2,4], and Rui Bernardes[1,2,4]

[1] Coimbra Institute for Biomedical Imaging and Clinical Research (CIBIT), Institute of Nuclear Sciences Applied to Health (ICNAS), University of Coimbra, Coimbra, Portugal
anatcnunes@gmail.com
[2] CNC.IBILI Consortium, University of Coimbra, Coimbra, Portugal
[3] Mathematics Section, Department of Science and Technology, Universidade Aberta, Lisbon, Portugal
[4] Faculty of Medicine, University of Coimbra, Coimbra, Portugal
rmbernardes@fmed.uc.pt
[5] Coimbra Institute for Clinical and Biomedical Research (iCBR), Faculty of Medicine, University of Coimbra, Coimbra, Portugal

Abstract. In this work, we imaged the retina of healthy age-matched female and male subjects with optical coherence tomography to assess the existence of localised sex-related differences in the retina. Texture analysis revealed statistically significant differences spread over the different layers of the neuroretina, particularly at the ganglion cell, inner plexiform and inner nuclear layers, and also spread in the macula. These results suggest that sexual dimorphism is expressed in the retina at the microstructural level.

Keywords: Sexual dimorphism · Retina · Optical coherence tomography · Texture analysis

1 Introduction

Sexual dimorphism in the visual system is a well-established phenomenon, recently attributed to gonadal hormone differences in the retina [8]. Sex-related dissimilarities in the retinal structure have been studied via optical coherence tomography (OCT) [4].

In OCT-based studies, retinal layer thickness measurements are traditionally the metric of choice. Across the literature, the reported findings on the sex-related differences in retinal thickness vary: while most studies report differences in thickness between males and females [4], inconclusive results [9] have been reported as well.

In recent years, our group has been exploring new approaches to look at OCT data, based on the texture analysis of fundus images computed from individual

J. Henriques et al. (Eds.): MEDICON 2019, IFMBE Proceedings 76, pp. 1830–1834, 2020.
https://doi.org/10.1007/978-3-030-31635-8_222

retinal layers [1,7]. In this study, we employ the same methodology by computing texture features from OCT data of healthy age-matched female and male subjects, to gain further insight into how sexual dimorphism is expressed across the retinal structure. To best of the authors' knowledge, no prior studies have addressed the existence of sex-based differences using OCT data of the ocular fundus based on texture image analysis.

2 Methods

All processing and statistical analysis were performed using Matlab R2018a (The MathWorks Inc., Natick, MA, USA).

2.1 Participants

OCT data from 50 healthy controls, 25 female and 25 male, with no previous retinal pathologies, were gathered from the authors' institutional database. The data collection protocol used for gathering OCT data was approved by the Ethics Committee of the Faculty of Medicine of the University of Coimbra and executed in accordance to the tenets of the Declaration of Helsinki [11]. Participants were selected aiming to age-match between groups, and so that both groups present an even distribution across adult age (19–74 yrs, mean $\pm$ std: 44.9 ± 15.8 yrs).

2.2 OCT Imaging and Image Processing

Retinal imaging was performed using the Cirrus SD-OCT 5000 (Carl Zeiss, Meditec, Dublin, CA, USA). The 512×128 Macular Cube protocol was executed to scan the macular region of the retina.

The gathered data were processed using the OCT Explorer software (Retinal Image Analysis Lab, Iowa Institute for Biomedical Imaging, Iowa City, IA, USA) to segment the six innermost layers of the retina, namely the retinal nerve fibre layer (RNFL), the ganglion cell layer (GCL), the inner plexiform layer (IPL), the inner nuclear layer (INL), the outer plexiform layer (OPL) and the outer nuclear layer (ONL). All segmentations were manually corrected where necessary.

A mean value fundus (MVF) [5] image was computed for all layers and texture-analysed. Left-eye images were horizontally flipped to match the right ones, and the images were down-sampled to 128×128 pixels, to obtain isotropic sampling in both the horizontal and vertical directions. None of these images presents regular patterns, nor any visually discernible differences between any two eyes.

2.3 Texture Analysis

Texture features were determined for each MVF image, following an approach used previously [1,7] based on the grey-level co-occurrence matrix (GLCM) [6] and the dual-tree complex wavelet transform (DTCWT) [2]. The GLCM was applied to identify localised grey-level variations, and the DTCWT was applied to reveal coarse features across the macula.

Eight-bit images were converted to four-bit ones to reduce the size of the GLCMs. Images were split into an uneven number of blocks (7×7) to be independently analysed, to exclude the fovea, where the layers under study are significantly reduced or do not exist. Four GLCMs were computed for each block, corresponding to four different orientations (0°, 45°, 90° and 135°) with a distance of one pixel and considering 180 degrees apart angles to be the same. The maximum across the four GLCM orientations was used. From each GLCM, 20 features were computed, as listed in [7] and defined in [3,6,10].

Features were aggregated by retinal quadrant, namely the superior-temporal (Q1), the superior-nasal (Q2), the inferior-temporal (Q3) and the inferior-nasal (Q4) quadrants, by computing the average of 3×3 blocks (the middle row and column passing through the fovea were left out of the present analysis). In this way, a total of 80 (20 features $\times$ 4 quadrants) GLCM-based metrics characterise each of the six layers of the retina.

The DTCWT was applied to the entire macula, i.e. to MVF images. The variance and entropy of the magnitude of the DTCWT complex coefficients were computed for all image sub-bands, as in [2]. Exploratory analysis showed that the variance of these coefficients, computed for six sub-bands ($\pm 15^\circ$, $\pm 45^\circ$ and $\pm 75^\circ$) at decomposition level one, leads to the best performance when discriminating between the two subject classes. These six directional variance features were appended to the 80 GLCM features, yielding 86 features per layer.

2.4 Statistical Analysis

All features were tested for normality using the Kolmogorov-Smirnov test. As none of them presents a normal distribution, the Mann-Whitney test was used to evaluate differences between groups with p-values of $p \leq 0.05$, $p \leq 0.01$ and $p \leq 0.001$.

3 Results and Discussion

Table 1 shows the list of features presenting statistically significant differences between groups by retinal quadrant for all layers. None of the global (DTCWT-based) features present statistically significant differences, at any layer.

No features revealed statistically significant differences between the female and male groups in any of the quadrants of the RNFL.

The OPL is the layer where group differences are the most consistently observed across the four quadrants, albeit mostly moderate ($p \leq 0.05$), followed by the INL, where statistically significant differences are found in three quadrants, but with higher significance levels ($p \leq 0.01$ and $p \leq 0.01$) for eight features. In the IPL, the statistically significant differences found are concentrated in Q2 and Q3. The GCL is the layer where a larger number of features present statistically significant differences between groups, concentrated mostly on Q3, but spread over Q1 and Q4 as well. Other texture metrics may show further differences.

Table 1. Features presenting statistically significant differences between the female and male groups for the retinal nerve fibre layer (RNFL); ganglion cell layer (GCL); inner plexiform layer (IPL); inner nuclear layer (INL); outer plexiform layer (OPL), and; outer nuclear layer (ONL). Hyphens (–) represent a p-value > 0.05 (non-significant difference), the green-coloured circles (●) represent a p-value ≤ 0.05, the orange-coloured squares (✱) represent a p-value ≤ 0.01 and the red-coloured asterisks (■) represent a p-value ≤ 0.001. Q1 – Superior-temporal quadrant, Q2 – Superior-nasal quadrant, Q3 – Inferior-temporal quadrant and Q4 – Inferior-nasal quadrant.

Layer	Feature	Q1	Q2	Q3	Q4
RNFL	–	–	–	–	–
GCL	Cluster Prominence	●	–	●	●
	Correlation	●	–	■	–
	Difference Entropy	–	–	■	●
	Difference Variance	–	–	■	–
	Dissimilarity	–	–	■	●
	Entropy	–	–	●	–
	Homogeneity	–	–	✱	●
	IDN	–	–	✱	–
	IMC1	●	–	■	●
	IMC2	●	–	■	●
	Inertia	–	–	■	–
	INN	–	–	✱	●
	Maximum Probability	–	–	–	●
	Sum Average	–	–	–	●
	Sum Entropy	●	–	–	–
	Uniformity	–	–	●	●
IPL	Cluster Prominence	–	–	■	–
	Correlation	●	■	✱	–
	Difference Entropy	–	■	●	–
	Difference Variance	–	●	●	–
	Dissimilarity	–	●	●	–
	Homogeneity	–	●	●	–
	IDN	–	■	■	–
	IMC1	–	■	■	–
	IMC2	–	●	■	–
	Inertia	–	●	●	–
	INN	–	■	■	–
	Sum Entropy	–	–	■	–
INL	Cluster Shade	–	●	–	–
	Correlation	■	■	–	–
	Difference Entropy	■	●	●	–
	Difference Variance	✱	●	●	–
	Dissimilarity	■	●	●	–
	Homogeneity	●	■	■	–
	IDN	■	■	●	–
	IMC1	●	–	–	–
	IMC2	●	●	–	–
	Inertia	✱	●	●	–
	INN	■	■	●	–
	Maximum Probability		–	●	–
OPL	Cluster Shade	●	–	–	–
	Correlation	●	●	–	●
	Difference Entropy	●	●	●	●
	Difference Variance	●	●	●	●
	Dissimilarity	●	●	●	●
	Homogeneity	●	■	●	●
	IDN	●	●	●	●
	IMC1	–	●	–	–
	IMC2	–	●	–	●
	Inertia	●	●	●	●
	INN	●	■	■	●
ONL	Autocorrelation	–	–	●	–
	Cluster Prominence	–	●	–	●
	Cluster Shade	–	–	●	–
	Correlation	–	●	–	–
	IMC1	–	●	–	–
	IMC2	–	●	–	–
	Sum Average	–	–	●	–
	Sum of Squares	–	–	●	–
	Sum Variance	–	–	●	–

Correlation, *IMC1* and *IMC2* are the features that most frequently present statistically significant differences, in five out of six layers. Overall, Q3 is the quadrant where the largest number of features presenting statistically significant differences was found (45 in 60–75% – of the feature/layer pairs).

4 Conclusions

Differences are found to be spread over the different layers of the neuroretina and also spread in the macula, although the level of significance is higher for the GCL, IPL and INL. Since when using the whole macular data, no features presented statistically significant differences between groups, these results strongly suggest that differences exist at the microstructural level.

Funding Information. This study was supported by The Portuguese Foundation for Science and Technology (PEst-UID/NEU/04539/2019), by FEDER-COMPETE (POCI-01-0145-FEDER-007440 and POCI01-0145-FEDER-016428), and by Centro 2020 FEDER-COMPETE (BIGDATIMAGE, CENTRO-01-0145-FEDER-000016).

Conflict of Interest. The authors declare no conflicts of interest.

References

1. Bernardes, R., Silva, G., Chiquita, S., Serranho, P., Ambrósio, A.F.: Retinal biomarkers of Alzheimer's disease: insights from transgenic mouse models. In: 14th International Conference on Image Analysis and Recognition (ICIAR) (2017). https://doi.org/10.1007/978-3-319-59876-5_60
2. Celik, T., Tjahjadi, T.: Multiscale texture classification using dual-tree complex wavelet transform. Pattern Recognit. Lett. **30**, 331–339 (2009). https://doi.org/10.1016/j.patrec.2008.10.006
3. Clausi, D.A.: An analysis of co-occurrence texture statistics as a function of grey level quantization. Can. J. Remote Sensing **28**(1), 45–62 (2002). https://doi.org/10.5589/m02-004
4. Çubuk, M., Kasım, B., Koçluk, Y., Sukgen, E.A.: Effects of age and gender on macular thickness in healthy subjects using spectral optical coherence tomography/scanning laser ophthalmoscopy. Int. Ophthalmol. **38**(1), 127–131 (2018). https://doi.org/10.1007/s10792-016-0432-z
5. Guimarães, P., Rodrigues, P., Lobo, C., Leal, S., Figueira, J., Serranho, P., Bernardes, R.: Ocular fundus reference images from optical coherence tomography. Comput. Med. Imaging Graph. **38**, 381–389 (2014). https://doi.org/10.1016/j.compmedimag.2014.02.003
6. Haralick, R.M., Shanmugam, K., Dinstein, I.: Textural features for image classification. IEEE Trans. Syst. Man Cybern. **SMC–3**(6), 610–621 (1973). https://doi.org/10.1109/TSMC.1973.4309314
7. Nunes, A., Ambrósio, A.F., Castelo-Branco, M., Bernardes, R.: Texture biomarkers of Alzheimer's disease and disease progression in the mouse retina. In: IEEE International Conference on Bioinformatics and Bioengineering, pp. 41–46 (2018). https://doi.org/10.1109/BIBE.2018.00016
8. Nuzzi, R., Scalabrin, S., Becco, A., Panzica, G.: Gonadal hormones and retinal disorders: a review. Front. Endocrinol. **9**(6) (2018). https://doi.org/10.3389/fendo.2018.00066
9. Ooto, S., Hangai, M., Tomidokoro, A., Saito, H., Araie, M., Otani, T., Kishi, S., Matsushita, K., Maeda, N., Shirakashi, M., Abe, H., Ohkubo, S., Sugiyama, K., Iwase, A., Yoshimura, N.: Effects of age, sex, and axial length on the three-dimensional profile of normal macular layer structures. Invest. Ophthalmol. Vis. Sci. **52**(12), 8769–8779 (2011). https://doi.org/10.1167/iovs.11-8388
10. Soh, L.k., Tsatsoulis, C.: Texture representation of SAR sea ice imagery using multi-displacement co-occurrence matrices. In: International Geoscience and Remote Sensing Symposium (1996). https://doi.org/10.1109/IGARSS.1996.516261
11. World Medical Association: Declaration of Helsinki - ethical principles for medical research involving human dubjects. J. Am. Med. Assoc. **310**(20), 2191–2194 (2013). https://doi.org/10.3917/jib.151.0124

Challenges: IFMBE Scientific Challenge Competition

Convolutional Neural Network for a P300 Brain-Computer Interface to Improve Social Attention in Autistic Spectrum Disorder

Davide Borra(✉), Silvia Fantozzi, and Elisa Magosso

Department of Electrical, Electronic and Information Engineering "Guglielmo Marconi" (DEI), University of Bologna, Cesena, FC, Italy
davide.borra2@unibo.it

Abstract. A Brain-Computer Interface (BCI) relies on machine learning algorithms to decode the brain signals. An accurate detection of P300 response in electroencephalography (EEG) data can be used to design P300-based BCIs to improve social attention in Autistic Spectrum Disorder (ASD). Recently, there was a growing interest in the application of Convolutional Neural Networks (CNNs) to decode P300 in an end-to-end fashion. However, the complexity of these models needs to be carefully taken into account. In this study, a lightweight CNN previously validated for P300 detection (EEGNet) was used to decode which object ASD participants were paying attention to in a virtual environment. Two learning strategies were deepened: within-session and cross-session trainings. Cross-session training resulted in a higher target object accuracy scoring 92.27% on average across sessions and subjects, and in a lower decoding variability across sessions.

Keywords: Brain-computer interfaces · P300 detection · Electroencephalogram · Convolutional neural networks

1 Introduction

P300 evoked response in electroencephalography signals (EEG) was first reported by Sutton et al. [1], i.e. a positive voltage deflection at 300 ms after a rare stimulus such as a flashing object or character, with a maximum amplitude in the parietal region. The decoding of P300 signals can be exploited for the design of Brain-Computer Interfaces (BCIs) [2] - P300-based BCIs - to provide an alternative way of communication for people who are unable to communicate (locked-in patiens) [3] or to improve social attention in Autistic Spectrum Disorder (ASD) [4]. The main challenge in the P300 detection is the low signal-to-noise ratio [5]. Carefully designed features extraction and selection steps, and traditional machine learning classifiers, such as Support Vector Machine (SVM) and Linear Discriminant Analysis (LDA) have been used for the P300 detection [6]. The three main limits of a such workflow are (1) the a priori knowledge exploited to design the features, (2) the subject-to-subject variability and (3) the separate execution of the steps.

J. Henriques et al. (Eds.): MEDICON 2019, IFMBE Proceedings 76, pp. 1837–1843, 2020.
https://doi.org/10.1007/978-3-030-31635-8_223

Recently, Convolutional Neural Networks (CNNs) have achieved outstanding performance in computer vision. These models automatically learn the most important features and finalize a classification task in an end-to-end fashion using slightly pre-processed data. Cecotti et al. [7] were the first to develop a simple CNN for P300 detection. This CNN was the first of a family of architectures in which the spatial convolution is performed at first, automatically extracting spatial features from EEG data, and then temporal convolution is performed in the deeper layers [7–9]. Compared to these architectures, better results have been found moving the temporal convolution to the first layer performing a mixed spatio-temporal convolution [5]. A promising alternative to the previous CNN family is EEGNet, a lightweight CNN proposed by Lawhern et al. [10], where temporal convolution was used at first, followed by a spatial convolution and a deeper temporal convolutional layer.

In this study, EEGNet was employed to detect P300 adopting two different training strategies: within-session and cross-session. EEG signals were acquired from participants with ASD [4] and provided to the challengers of the IFMBE 2019 Scientific Challenge Competition of the 15th Mediterranean Conference on Medical and Biological Engineering and Computing.

2 Methods

This section describes the dataset provided for the challenge, pre-processing, single-trial P300 detection and target object detection steps. CNNs were developed in PyTorch [11] and trained from scratch using a workstation equipped with a AMD Threadripper 1900X, NVIDIA TITAN V and 32 GB of RAM.

2.1 Dataset Provided for the Challenge

The dataset was obtained in a feasibility clinical trial in which a P300-based BCI was tested on 15 youngsters with ASD to follow social cues [4] provided to the user in a virtual environment with 8 flashing objects. Each participant's EEG trials were acquired during 7 sessions; each session contained calibration data (20 blocks of recordings) and online data (50 blocks). For each session, the training set consisted of the calibration data, while the test set of the online data. Each block contains a repetition of N runs of target object flashing. For the training set $N = 10$, while for the test set N varies across subjects and sessions. In each run, 8 trials were recorded while the 8 target objects randomly flashed in the scene. Only one of these trials contains the P300 response related to the target object the user was looking at. Therefore, the overall session data contains an important class unbalancing (7:1 for the non-P300 and P300 classes). The objective of the challenge was the design of the best model to decode the objects the participants were paying attention to during the online phase, minimizing the misclassification across the 50 blocks for each session.

EEG signals were recorded at 250 Hz from C3, Cz, C4, CPz, P3, Pz, P4, POz and the reference was placed at the right ear and the ground electrode at AFz. These signals were acquired notch filtered at 50 Hz and pass-band filtered between 2 and 30 Hz. Finally, trials were extracted by epoching signals from −200 to 1200 ms relative to the event stimulus and provided for the competition.

The competition was divided into two phases. Sessions from 1 to 3 were provided for the phase I and the remaining for the phase II. The challenge final ranking was based on the phase II results. Therefore, in this study the phase II results are reported and briefly discussed.

2.2 Pre-processing

At first, data samples were selected from −100 to 1000 ms as done in [12]. Furthermore, in order to reduce the computational cost of the CNN-based P300 detection, signals were resampled at 128 Hz, resulting in 140 time samples for each trial. Then, each electrode signal was normalized to be zero-mean and unit-variance. After these pre-processing steps, the single run trial shape was $(1, 8, 140)$. These spatio-temporal patterns represented the CNN input of the P300 decoding stage.

The CNN training was performed in two different ways: within-session and cross-session. For the first one, each session set was treated separately ($8 * 10 * 20 = 1600$ training trials). For the second one, all the session trials of a specific subject were merged together obtaining an augmented training ($8 * 10 * 20 * 7 = 11200$ training trials) and test sets. To perform early stopping in the P300 detection, a validation set containing 20% of trials was randomly selected from the training set maintaining the class proportion. In particular, for the cross-session training these trials were equally distributed between sessions.

2.3 Single-Trial P300 Detection

The single-trial P300 detection consists of a binary classification between the classes "non-P300" and "P300".

CNN Architecture. To decode the P300 response at single trial level, a previously validated architecture for P300 decoding was used [10]. EEGNet is a compact CNN for EEG-based BCIs, carefully designed to reduce the number of training parameters by including Depthwise and Separable convolutions.

The CNN includes 3 blocks. In the first block, two convolutional layers performing temporal and depthwise spatial convolutions are included, respectively. The second block includes a separable convolutional layer and the last block a fully-connected layer which finalizes the 2-ways classification. Batch Normalization [13], Dropout [14], Exponential Linear Units (ELUs) [15] are used and the main hyper-parameters are shown in the Table 1.

Training. The optimizer used was Adam [16] with the default parameters and a learning rate of 5 * 1e-4, minimizing the categorical cross-entropy loss function. A mini-batch size of 64 was used and the maximum number of epochs set to 1000. To address the non-negligible class unbalance, a single mini-batch of data contained a proportion of 50–50% of the classes randomly selecting the trials within the dataset. Early stopping was performed during training, by interrupting the optimization when the validation loss reached its minimum.

Table 1. Main EEGNet hyper-parameters. K and F are the number and the size of the kernels, respectively. P is the padding size, D the depth multiplier, N the number of units and finally p the dropout rate. *max norm constraint set as in [10]. **p = 0.25 for cross-session training and p = 0.50 for within-session training.

Block	Layer	Hyper-parameters	# parameters	Output	Activation
1	Input			(1, 8, 140)	
	Conv2D	K = 8, F = (1, 65), P = (0, 32)	520	(8, 8, 140)	linear
	BatchNorm2D		16	(8, 8, 140)	
	Depth-Conv2D*	D = 2, K = 16, F = (8, 1), P = (0, 0)	128	(16, 1, 140)	linear
	BatchNorm2D		32	(16, 1, 140)	
	Activation			(16, 1, 140)	ELU
	AvgPool2D	F = (1, 4)		(16, 1, 35)	
	Dropout**	$p = 0.25$ or $p = 0.50$		(16, 1, 35)	
2	Sep-Conv2D	K = 16, F = (1, 17), P = (0, 8)	528	(16, 1, 35)	linear
	BatchNorm2D		32	(16, 1, 35)	
	Activation			(16, 1, 35)	ELU
	AvgPool2D	F = (1, 8)		(16, 1, 4)	
	Dropout**	$p = 0.25$ or $p = 0.50$		(16, 1, 4)	
3	FC	N = 2	130	(2,)	linear
	Activation			(2,)	softmax

2.4 Target Object Detection

The target object detection is the task of interest for the competition. This is based on top of the previous classifiers output by aggregating the trial probabilities. Once the previous CNNs were trained, test trials were fed into the models. The CNNs output are the probabilities $P(y = 0|\mathbf{X})$ and $P(y = 1|\mathbf{X})$, where y is the output category ("0" non-P300 and "1" P300 class). Trials of each run were sorted following the object flashing random order provided in each dataset.

Let $\mathbf{X}_{(i,j)}^{(k)}$ be the i-th trial (corresponding to the i-th object) of the j-th run belonging to the k-th block, where $0 \leq i \leq 7$, $0 \leq j \leq N-1$ and $0 \leq k \leq 49$. Denoting as $P_{(i,j)1}^{(k)}$ the probability that the trial $\mathbf{X}_{(i,j)}^{(k)}$ belongs to the class "1", the target object of a specific block was computed as:

$$P_{(i)1}^{(k)} = \frac{1}{N}\sum_{j=0}^{N-1} P_{(i,j)1}^{(k)} \tag{1}$$

$$o^{(k)} = \operatorname*{argmax}_{i} P_{(i)1}^{(k)}, \tag{2}$$

where $P_{(i)1}^{(k)}$ (Eq. 1) is the object-specific average probability across runs of a specific block and $o^{(k)}$ (Eq. 2) is the decoded target object of a specific block.

To evaluate the CNNs performance on the test set we used the metric recommended by the organizers (i.e. target object accuracy), computed as the ratio between the number of correct decoded blocks and the total decoded blocks, taking into account all sessions and subjects.

3 Results

In this section, the phase II results related to the best predictions are reported for each pipeline. In particular, the within-session and cross-session pipelines scored a target object accuracy of 0.8443 and 0.9227, respectively. It is worth

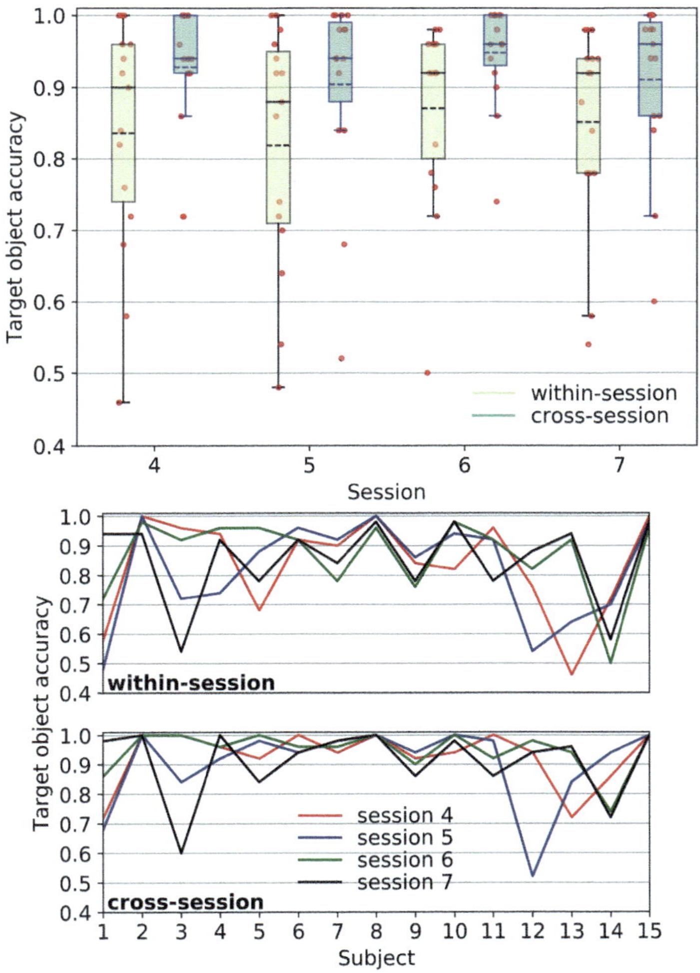

Fig. 1. Target object accuracy: session (top) and subject (down) variability. In the top panel red points represent subject accuracies, dashed lines represent mean values and continuous lines represent median values.

to mention that despite the subject-specific variability across sessions, the cross-session training results in higher performance. For each pipeline, Fig. 1 shows the variability of the predictions across sessions and subjects.

4 Discussion and Conclusions

The results obtained in phase II show an important average accuracy improvement from 0.8443 to 0.9227 (averaging between subjects and sessions) when training subject-specific CNNs merging the session training sets (cross-session training). Furthermore, this training strategy led to an overall reduction of the session accuracy variability and to an increase of the session accuracies (upper panel of Fig. 1). The lower panel of Fig. 1 clearly shows that all the classifiers performed well above chance level (0.1250) and only two sessions scored accuracies $\leq$60% during cross-session learning, e.g. session 7 of subject 3 and session 5 of subject 12. These two worst cases are related to test sets with a low number of runs and this may have affected the decoded object performance.

The results suggest that the CNNs benefit more from cross-session training (subject-specific CNNs) compared to within-session training (session-specific CNNs): despite the cross-session variability, the larger training dataset (resulting from merging session datasets) allows a better subject-specific features learning, providing a more accurate target object detection. Furthermore, the knowledge learned in the cross-session pipeline can be transferred to a new session via transfer learning, reducing the number of training examples needed.

Future developments include the application of deeper neural networks, the comparison between architectures including a spatial convolutional layer as the first layer and the use of transfer learning for new sessions.

Acknowledgments. We gratefully acknowledge the support of NVIDIA Corporation with the donation of the TITAN V used for this research.

References

1. Sutton, S., Tueting, P., Zubin, J., John, E.R.: Information delivery and the sensory evoked potential. Science **155**(3768), 1436–1439 (1967)
2. Wolpaw, J., Wolpaw, E.: Brain-Computer Interfaces: Principles and Practice. Oxford University Press, USA (2012)
3. Sellers, E.W., Donchin, E.: A p300-based brain-computer interface: initial tests by ALS patients. Clin. Neurophysiol. **117**(3), 538–548 (2006)
4. Amaral, C., Mouga, S., Simões, M., Pereira, H.C., Bernardino, I., Quental, H., Playle, R., McNamara, R., Oliveira, G., Castelo-Branco, M.: A feasibility clinical trial to improve social attention in autistic spectrum disorder (ASD) using a brain computer interface. Front. Neurosci. **12**, 477 (2018)
5. Shan, H., Liu, Y., Stefanov, T.: A simple convolutional neural network for accurate P300 detection and character spelling in brain computer interface. In: Proceedings of the 27th International Joint Conference on Artificial Intelligence, pp. 1604–1610. IJCAI 2018, AAAI Press (2018)

6. Fazel-Rezai, R., Allison, B., Guger, C., Sellers, E., Kleih, S., Kübler, A.: P300 brain computer interface: current challenges and emerging trends. Front. Neuroeng. **5**, 14 (2012)
7. Cecotti, H., Graser, A.: Convolutional neural networks for P300 detection with application to brain-computer interfaces. IEEE Trans. Pattern Anal. Mach. Intell. **33**(3), 433–445 (2011)
8. Liu, M., Wu, W., Gu, Z., Yu, Z., Qi, F., Li, Y.: Deep learning based on batch normalization for P300 signal detection. Neurocomputing **275**(C), 288–297 (2018)
9. Manor, R., Geva, A.B.: Convolutional neural network for multi-category rapid serial visual presentation BCI. Front. Comput. Neurosci. **9**, 146 (2015)
10. Lawhern, V.J., Solon, A.J., Waytowich, N.R., Gordon, S.M., Hung, C.P., Lance, B.J.: EEGNet: a compact convolutional neural network for EEG-based brain-computer interfaces. J. Neural Eng. **15**(5), 056013 (2018)
11. Paszke, A., Gross, S., Chintala, S., Chanan, G., Yang, E., DeVito, Z., Lin, Z., Desmaison, A., Antiga, L., Lerer, A.: Automatic differentiation in pytorch. In: NIPS-W (2017)
12. Amaral, C.P., Simões, M.A., Mouga, S., Andrade, J., Castelo-Branco, M.: A novel brain computer interface for classification of social joint attention in autism and comparison of 3 experimental setups: a feasibility study. J. Neurosci. Methods **290**, 105–115 (2017)
13. Ioffe, S., Szegedy, C.: Batch normalization: accelerating deep network training by reducing internal covariate shift. In: Bach, F., Blei, D. (eds.) Proceedings of the 32nd International Conference on Machine Learning. Proceedings of Machine Learning Research, vol. 37, pp. 448–456. PMLR, Lille, France, 07–09 July 2015
14. Srivastava, N., Hinton, G., Krizhevsky, A., Sutskever, I., Salakhutdinov, R.: Dropout: a simple way to prevent neural networks from overfitting. J. Mach. Learn. Res. **15**(1), 1929–1958 (2014)
15. Clevert, D.A., Unterthiner, T., Hochreiter, S.: Fast and accurate deep network learning by exponential linear units (elus). arXiv preprint (2015)
16. Kingma, D.P., Ba, J.: Adam: A method for stochastic optimization. arXiv preprint (2014)

Deep Learning Architecture Based on the Combination of Convolutional and Recurrent Layers for ERP-Based Brain-Computer Interfaces

Eduardo Santamaría-Vázquez(✉), Víctor Martínez-Cagigal, Javier Gomez-Pilar, and Roberto Hornero

University of Valladolid, Valladolid, Spain
eduardo.santamaria@gib.tel.uva.es
http://www.gib.tel.uva.es

Abstract. In this study, we present a novel deep learning architecture for brain-computer interfaces based on event related potentials (ERP). The topology of the neural network combines convolutional and recurrent layers in order to learn high-level spatial and temporal features. Specifically, our model uses a convolutional layer, intended to detect spatial patterns over the scalp in short periods of time, followed by two bidirectional long-short term memory (BLSTM) layers to extract long-term temporal dependencies within the data. To the best of our knowledge, this is the first time that BLSTM layers are explored for ERP classification. This study takes part in the MEDICON 2019 IFMBE scientific challenge. The model has been evaluated using the provided dataset for the competition (15 subjects with autism spectrum disorder, 7 BCI sessions), achieving an average accuracy of 84% in command selection. In the course of our experiments, this approach outperformed traditional methods, such as step-wise linear discriminant analysis (SWLDA), and other deep learning architectures.

Keywords: Brain-computer interface · Event related potential · Deep learning · Recurrent neural network

1 Introduction

Brain-computer interfaces (BCI) allow control over devices directly translating the cerebral activity of the user into commands [1]. Electroencephalography (EEG) is the most used technique to register brain signals in BCI, using non-invasive electrodes placed at different locations over the scalp to capture the electrical activity triggered by neurons. Inherently, EEG is a complex signal that results from the superposition of the activity of millions of neurons. In addition, it has low signal-to-noise ratio, being affected by a number of artefacts, such as muscular movements or eye blinking [1]. For these reasons, decoding thoughts

J. Henriques et al. (Eds.): MEDICON 2019, IFMBE Proceedings 76, pp. 1844–1852, 2020.
https://doi.org/10.1007/978-3-030-31635-8_224

directly from the EEG is currently a chimera, and most of BCIs rely on a control signal to determine user's intentions. Control signals are waveforms triggered under certain conditions that can be detected through signal processing methods. Among them, event related potentials (ERP) elicited by visual stimuli are the most used [1]. ERPs are the natural response of the brain to a stimuli (e.g. a brief flash) [2]. Typically, ERP-based BCIs use the rapid serial visualization presentation (RSVP) paradigm to stimulate the user. RSVP presents several commands that are randomly highlighted, while the user has to stare at one of them. When the target command is highlighted, it provokes an ERP in the user's EEG. Once all commands have been highlighted (one or more times), the system decides which command the user wanted to select by detecting the ERPs in epochs of signal just after each stimulus [2,3].

Recent research has improved accuracy and speed of ERP-based BCIs [4–7]. However, the performance of these systems have levelled off, and traditional methods for feature extraction and classification are unlikely to increase it significantly in the future. In this regard, deep learning architectures have emerged as a promising alternative to the classical machine learning methods. In recent years, deep neural networks have revolutionized fields such as computer vision or speech recognition, due to their ability to learn high-level features automatically from raw data [8]. However, few studies have used these models for ERP classification, most of them using convolutional neural networks (CNN) [8–12]. CNNs can handle 2D data and have demonstrated their ability to detect common patterns in a wide variety of problems, which makes them suitable for EEG analysis [9,10]. Although CNNs have showed higher performance in comparison with classical machine learning approaches, they fail when it comes to learn long-term high-level time distributed features [13]. Recurrent neural networks (RNN) could overcome this limitation. These models have shown great performance in complex time series classification, and are capable to learn long-term temporal patterns. Specifically, long-short term memory (LSTM) are the most used type of RNN [13]. These networks have increased performance in language and speech recognition [14]. Furthermore, they have been successfully adapted to the analysis of physiological signals, such as electromyography (EMG), electrocardiography (ECG), electrooculography (EOG) and EEG [13]. However, to the best of our knowledge, they have not been used yet for command selection in ERP-based BCIs [13].

This study takes part in the MEDICON 2019 IFMBE scientific challenge. The objective is to maximize the command selection accuracy for the provided dataset, which contains data of 15 subjects using an ERP-based BCI. In the following sections we describe the approach that reached the highest accuracy, which combines convolutional and recurrent layers to take advantage of high-level spatial and temporal features. We also provide comparison with other methods that were used in the course of experiments, such as step-wise linear discriminant analysis (SWLDA) and state-of-the-art CNNs.

2 Materials and Methods

2.1 Dataset

This study used the dataset provided by MEDICON 2019 for the IFMBE scientific challenge [15,16]. EEG was registered with a g.Nautilus (g.tec, Guger Technologies) using 8 electrodes (C3, Cz, C4, CPz, P3, Pz, P4, POz). The reference was placed at the right ear and the ground electrode at AFz. Sampling rate was set to 250 Hz. Signals were filtered between 2 and 30 Hz in order to improve the signal-to-noise ratio.

Fifteen subjects with autism spectrum disorder participated in the experiments, which comprised 7 different sessions. RSVP paradigm with 8 targets was used to elicit the ERPs. Each session was divided into calibration and online stages. The calibration stage consisted of 20 blocks of 10 runs per subject and session, making a total of 1600 observations (20 blocks $\times$ 8 targets $\times$ 10 runs). Therefore, the calibration set had a total of 168000 observations (1600 stimuli $\times$ 15 subjects $\times$ 7 sessions). The online stage consisted of 50 blocks per subject and session. The number of runs vary in each session depending on the user's performance during the corresponding calibration. Each observation was the epoch of EEG signal from 0 ms (stimulus onset) to 1000 ms, with dimensions 250 samples $\times$ 8 channels. Baseline normalization (z-score) was applied, using the 200 ms previous to the stimulus.

The dataset was divided into three sets, according to phase 2 the scientific challenge: (i) the training set comprised calibration data of the seven sessions, which had a total of 168000 observations; (ii) the development set was formed by online data from sessions 1, 2 and 3; and (iii) the test set comprised online data from sessions 4, 5, 6 and 7.

2.2 Models

This section describes our approaches for the IFMBE scientific challenge. We used scikit-learn framework for SWLDA models, and Keras framework to design, train and test deep learning approaches (Intel Core i7-7700 @ 3.67 GHz, 32 GB RAM, Windows 10, NVIDIA GTX 1050, Python v3.6, CUDA v9.0, Tensorflow v11.0).

Approach 1. SWLDA. SWLDA has been widely used in ERP-based BCIs, and thus, it is a suitable benchmark for more advanced models [17]. As preprocessing stage, common average reference (CAR) filter was applied. Then, epochs of signal were subsampled to 20 Hz and channels were concatenated to make an observation. Step-wise regression was applied to select the 60 most relevant features for each user (p-in = 0.10; (p-out = 0.15). Then, an LDA was used for the classification stage. More information about this methodology can be found in [4].

Approach 2. CNN. The architecture of this approach is an adaptation of the model proposed by Manor and Geva [9] for 8 EEG channels, instead of 64. CNNs are little prone to overfitting, which makes them suitable for dealing with noisy signals, such as EEG [10]. In addition, since convolutional layers share the kernel through the entire input, they can detect patterns in different spatial and temporal positions, reducing the impact of inter-subject variability. The first layer of this model is a convolutional 2D layer designed to learn spatial patterns across EEG channels. In contrast, the second and third convolutional layers are used to learn features in the temporal axis. The pooling layers were designed to reduce the feature space dimensionality. Dropout regularization was used to avoid overfitting. The architecture of this model is detailed in Table 1.

Approach 3. CNN-LSTM. The first layer of this model is a convolutional 1D layer, which is designed to learn spatial patterns through EEG channels, as well as reducing the feature space dimensionality [12]. The second and third layers of this architecture are LSTM layers, which are the most common type of RNN [13]. In comparison with CNNs, LSTM layers are more suitable to learn high-level long-term temporal features [13]. Dropout regularization was applied to avoid overfitting. The architecture of this approach is detailed in Table 2.

Table 1. Approach 2 architecture (CNN)

N	Type	Description	Input dim.	Output dim.
1	Convolutional 2D	96 filters kernel size = (1,8) activation = ReLu	250 × 8	250 × 1 × 96
2	MaxPooling	Pool size = (2,2) stride = 2	250 × 1 × 96	124 × 1 × 96
3	Convolutional 2D	128 filters kernel size = (6,1) activation = ReLu	124 × 1 × 96	119 × 1 × 128
4	MaxPooling	Pool size =(2,2) stride = 2	59 × 1 × 128	59 × 1 × 128
5	Convolutional 2D	128 filters kernel size = (6,1) activation = ReLu	54 × 1 × 128	54 × 1 × 128
6	Dense	4096 neurons activation = ReLu	6912 × 1	4096 × 1
7	Dense	2048 neurons activation = ReLu	4096 × 1	2048 × 1
8	Dense	1 neuron activation = sigmoid	2048 × 1	1 × 1

Approach 4. CNN-BLSTM. The architecture of this network is very similar to the third approach, but in this case, layers 2 and 3 are bidirectional LSTM (BLSTM). BLSTM layers are, in fact, two LSTM layers, one that processes the training sequence forwards and one that processes it backwards, connected to the same output layer [18]. This architecture allows using past and future information in order to determine if one specific pattern in the EEG is an ERP. BLSTM networks have demonstrated higher performance than LSTM in solving problems such as speech recognition [18]. Dropout regularization was applied to avoid overfitting. The topology of this network is detailed in Table 3.

Table 2. Approach 3 architecture (CNN-LSTM)

N	Type	Description	Input dim	Output dim
1	Convolutional 1D	32 filters kernel size = (4,8) activation = relu stride = 4	250 × 8	62 × 32
2	LSTM	16 neurons activation = tanh	62 × 32	62 × 32
3	LSTM	8 neurons activation = tanh	62 × 32	16 × 1
4	Dense	1 neuron activation = sigmoid	16 × 1	1 × 1

3 Results

Several experiments were conducted to obtain the optimum dropout probability for approaches 2, 3 and 4. This parameter was varied from 0.0 to 0.5. Then, the configuration that reached higher epoch classification in the development set was used. For the CNN model, dropout probability was set to 0.1, 0.3, 0.3, 0.4 and 0.4 for layers 1, 3, 5, 6 and 7, respectively. For the CNN-LSTM and CNN-BLSTM models, dropout probability was set to 0.1, 0.1, 0.1 for layers 1, 2 and 3, respectively.

SWLDA models were trained using the calibration data for each participant. The training procedure for approaches 2, 3 and 4 was: (i) fitting the model with the whole calibration set, mixing sessions and subjects; (ii) refining the model for each subject, using the data from the calibration sessions; (iii) testing for each subject with the refined model. Results are detailed in Table 4. The accuracy was calculated as the average accuracy for command selection in the test set.

Table 3. Approach 4 architecture (CNN-BLSTM)

N	Type	Description	Input dim	Output dim
1	Convolutional 1D	32 filters kernel size = (4,8) activation = relu stride = 4	250 × 8	62 × 32
2	BLSTM	16 neurons activation = tanh	62 × 32	62 × 32
3	BLSTM	8 neurons activation = tanh	62 × 32	16 × 1
4	Dense	1 neuron activation = sigmoid	16 × 1	1 × 1

4 Discussion

In this study, we evaluated 4 models for the MEDICON 2019 IFMBE scientific challenge. Approach 4, which uses a convolutional 1D layer followed by 2 BLSTM layers, achieved the highest accuracy, reaching a final score of 84% in the competition. In addition to this suitable score, to the best of our knowledge, this is the first time that BLSTMs are used for ERP classification. In our experiments, approach 4 improved performance in comparison to SWLDA, which achieved an average accuracy of 75%. Moreover, this model outperformed approach 2, that reached an score of 81%. In fact, by using BLSTM layers we were able to improve our model, reaching greater accuracies for ERP classification and command detection than those achieved with classical LSTMs (82%). The refining procedure also proved to increase the performance of approaches 2, 3 and 4 significantly, adapting the network specifically for each user, while keeping some general features shared across subjects.

This study presents some limitations that should be considered. In this work, only the dropout parameter was optimized for approaches 2, 3 and 4. However, the optimization of the rest of parameters (i.e. number of hidden layers, kernel size, stride and number of neurons) could increase the performance of our model. In addition, deep neural networks have the ability to manage very large datasets [13]. In this regard, a larger training set could allow the model to learn more complex features, improving the accuracy.

Table 4. Target detection accuracy in the test set

Subject	Appr. 1 (%)	Appr. 2 (%)	Appr. 3 (%)	Appr. 4 (%)
S01	62	38	45	56
S02	91	96	97	98
S03	59	69	67	68
S04	89	93	95	96
S05	76	78	82	80
S06	64	90	86	88
S07	73	88	85	87
S08	92	99	98	99
S09	78	69	78	82
S10	78	92	92	93
S11	76	85	85	88
S12	78	81	79	80
S13	68	76	76	81
S14	61	72	65	72
S15	88	97	99	99
Mean	**75**	**81**	**82**	**84**

5 Conclusion

In this study, we analyzed four models for ERP classification, as part of the MEDICON 2019 IFMBE scientific challenge. The approach that reached the highest accuracy combines convolutional and BLSTM layers, achieving a final score of 84% in the competition. Our experiments showed that this topology achieves higher performance than traditional methods, such as SWLDA (75%), and other state-of-the-art deep learning models based on CNNs (81%). These results suggest that the combination of convolutional and recurrent layers is suitable for learning complex spatial and temporal features of the EEG, and could lead to significant improvements in classification accuracy for ERP-based BCIs.

Acknowledgments. This study was partially funded by projects DPI2017-84280-R of 'Ministerio de Ciencia, Innovación y Universidades' and 'European Regional Development Fund' (FEDER), and the project 'Análisis y correlación entre el genoma completo y la actividad cerebral para la ayuda en el diagnóstico de la enfermedad de Alzheimer' (Inter-regional cooperation program VA Spain-Portugal POCTEP 2014-2020) of the European Commission and FEDER. V. Martínez-Cagigal was in receipt of a PIF-UVa grant of the University of Valladolid. E. Santamaría-Vázquez was in a receipt of a PIF grant from the Consejería de Educación de la Junta de Castilla y León and the European Social Fund.

References

1. Wolpaw, J., Wolpaw, E.W.: Brain-computer Interfaces: Principles and Practice. OUP, New York (2012)
2. Wolpaw, J.R., Birbaumer, N., McFarland, D.J., Pfurtscheller, G., Vaughan, T.M.: Brain computer interfaces for communication and control. Clin. Neurophysiol. **4**(113), 767–791 (2002)
3. Farwell, L.A., Donchin, E.: Talking off the top of your head: toward a mental prosthesis utilizing event-related brain potentials. Electroencephalogr. Clin. Neurophysiol. **70**(6), 510–523 (1988)
4. Martínez-Cagigal, V., Santamaría-Vázquez, E., Gomez-Pilar, J., Hornero, R.: Towards an accessible use of smartphone-based social networks through brain-computer interfaces. Expert Syst. Appl. **120**, 155–166 (2019)
5. Aydin, E.A., Bay, O.F., Guler, I.: P300-based asynchronous brain computer interface for environmental control system. IEEE J. Biomed. Health Inform. **22**(3), 653–663 (2018)
6. Schettini, F., Aloise, F., Aricò, P., Salinari, S., Mattia, D., Cincotti, F.: Self-calibration algorithm in an asynchronous P300-based brain-computer interface. J. Neural Eng. **11**(3) (2014)
7. Rezeika, A., Benda, M., Stawicki, P., Gembler, F., Saboor, A., Volosyak, I.: Brain-computer interface spellers: a review. Brain Sci. **8**(4) (2018)
8. Cecotti, H., Gräser, A.: Convolutional neural networks for P300 detection with application to brain-computer interfaces. IEEE Trans. Pattern Anal. Mach. Intell. **33**(3), 433–445 (2011)
9. Manor, R., Geva, A.B.: Convolutional neural network for multi-category rapid serial visual presentation BCI. Front. Comput. Neurosci. **9**, 1–12 (2015)
10. Schirrmeister, R.T., Springenberg, J.T., Fiederer, L.D.J., Glasstetter, M., Eggensperger, K., Tangermann, M., Hutter, F., Burgard, W., Ball, T.: Deep learning with convolutional neural networks for EEG decoding and visualization. Hum. Brain Mapp. **38**(11), 5391–5420 (2017)
11. Liu, M., Wu, W., Gu, Z., Yu, Z., Qi, F.F., Li, Y.: deep learning based on batch normalization for P300 signal detection. Neurocomputing **275**, 288–297 (2018)
12. Lawhern, V.J., Solon, A.J., Waytowich, N.R., Gordon, S.M., Hung, C.P., Lance, B.J.: EEGNet: a compact convolutional neural network for EEG-based braincomputer interfaces. J. Neural Eng. **15**(056013), 1–17 (2018)
13. Faust, O., Hagiwara, Y., Hong, T.J., Lih, O.S., Achaya, U.R.: Deep learning for healthcare applications based on physiological signals: a review. Comput. Methods Programs Biomed. **161**, 1–13 (2018)
14. Graves, A., Mohamed, A.R., Hinton, G.: Speech recognition with deep recurrent neural networks. In: Proceedings of IEEE International Conference on Acoustics, Speech and Signal Processing, ICASSP, vol. 6, pp. 6645–6649 (2013)
15. Amaral, C., Mouga, S., Simões, M., Pereira, H.C., Bernardino, I., Quental, H., Playle, R., McNamara, R., Oliveira, G., Castelo-Branco, M.: A feasibility clinical trial to improve social attention in Autistic Spectrum Disorder (ASD) using a brain computer interface. Front. Neurosci. **12**(477), 1–13 (2018)
16. Amaral, C.P., Simões, M.A., Mouga, S., Andrade, J., Castelo-Branco, M.: A novel brain computer interface for classification of social joint attention in autism and comparison of 3 experimental setups: a feasibility study. J. Neurosci. Methods **290**, 105–115 (2017)

17. Krusienski, D.J., Sellers, E.W., McFarland, D.J., Vaughan, T.M., Wolpaw, J.R.: Toward enhanced P300 speller performance. J. Neurosci. Methods **167**(1), 15–21 (2008)
18. Wöllmer, M., Eyben, F., Keshet, J., Graves, A., Schuller, B., Rigoll, G.: Robust discriminative keyword spotting for emotionally colored spontaneous speech using bidirectional LSTM networks. In: Proceedings of IEEE International Conference on Acoustics, Speech and Signal Processing, ICASSP 2009, pp. 3949–3952 (2009)

Slow Cortical Potential BCI Classification Using Sparse Variational Bayesian Logistic Regression with Automatic Relevance Determination

Aleksandar Miladinović[1(✉)], Miloš Ajčević[1,3,4], Piero Paolo Battaglini[2], Giulia Silveri[1], Gaia Ciacchi[2], Giulietta Morra[2], Joanna Jarmolowska[2], and Agostino Accardo[1]

[1] Department of Engineering and Architecture, University of Trieste, Via A. Valerio 10, 34127 Trieste, Italy
aleksandar.miladinovic@phd.units.it
[2] Department of Life Sciences, University of Trieste, Via Fleming 22, 34127 Trieste, Italy
[3] Department of Medicine Surgery and Health Sciences, University of Trieste, Trieste, Italy
[4] NEUROFARBA Department, Neuroscience Section, University of Florence, Florence, Italy

Abstract. Detecting P300 slow-cortical ERPs poses a considerable challenge in signal processing due to the complex and non-stationary characteristics of a single-trial EEG signal. EEG-based neurofeedback training is a possible strategy to improve the social abilities in Autism-Spectrum Disorder (ASD) subjects. This paper presents a BCI P300 ERPs based protocol optimization used for the enhancement of joint-attention skills in ASD subjects, using a robust logistic regression with Automatic Relevance Determination based on full Variational Bayesian inference (VB-ARD). The performance of the proposed approach was investigated utilizing the IFMBE 2019 Scientific Challenge Competition dataset, which consisted of 15 ASD subjects who underwent a total of 7 BCI sessions spread over 4 months. The results showed that the proposed VB-ARD approach eliminates irrelevant channels and features effectively, producing a robust sparse model with 81.5 ± 12.0% accuracy in relatively short modeling computational time 19.3 ± 1.4 s, and it outperforms the standard regularized logistic regression in terms of accuracy and speed needed to produce the BCI model. This paper demonstrated the effectiveness of the probabilistic approach using Bayesian inference for the production of a robust BCI model. Considering the good classification accuracy over sessions and fast modeling time the proposed method could be a useful tool used for the BCI based protocol for the improvement of joint-attention ability in ASD subjects.

Keywords: BCI · Modeling · P300 · ERP · Variational Bayesian inference · Automatic Relevance Determination · Autism-Spectrum Disorder

J. Henriques et al. (Eds.): MEDICON 2019, IFMBE Proceedings 76, pp. 1853–1860, 2020.
https://doi.org/10.1007/978-3-030-31635-8_225

1 Introduction

Autism Spectrum Disorder (ASD) is a set of pervasive and sustained neurodevelopmental conditions characterized by persistent deficits in social interaction, alongside restricted, repetitive patterns of behavior, interests, or activities [1]. Several studies found that using EEG-based neurofeedback training is a possible way to improve the social skills in ASD subjects [2]. Electroencephalography (EEG) based brain-computer interface (BCI) is a system that acquires brain signals and provides feedback according to the performance of the participant, and it has been used, not only for the attention improvement [3], but also for motor neurorehabilitation [4–8]. In addition, the employment of Virtual Reality (VR) improves the effectiveness of BCI in general, and also the applicability of these skills in real life [9–11].

A novel social attention training paradigm (VR P300-based BCI paradigm) based on P300 ERPs for the enhancement of joint-attention skills was proposed in the papers [12, 13]. A P300 is a huge positive voltage with a peak latency around 300 ms after the appearance of a cognitive attended rare stimulus, and it can be used to monitor the attentional performance of ASD subjects.

In the studies [12, 13] the P300-ERP based paradigm was used to acquire the data, in the VR immersive environment where the subjects had to follow a non-verbal social agent cue performed by an avatar. The avatar moved its head toward a target object signaling to the participants where to focus on a so-called joint attention task [14]. Then the participants received feedback based on the P300 acquired signal allowing them to adjust and improve this social attentional ability. Therefore, the BCI system allows to detect the target object toward which the participants were focusing their attention. Different machine learning methods can be applied to produce a model which adequately provide an output that indicates the subject's mental state. However, without pre-processing, a high number of channels and features can lead to overfitting. To overcome this challenge, a technique that eliminates irrelevant channels and features is preferable.

Consequently, in this paper, we investigated a technique based on variational Bayes automatic relevance determination that is able to prune irrelevant inputs, together with a logistic regression approach with a standard regularization method to create an accurate and robust BCI model based on P300 ERPs.

2 Materials and Methods

2.1 The EEG Dataset

The proposed approach was evaluated utilizing the IFMBE 2019 Scientific Challenge Competition dataset [12, 13], which was acquired from 15 subjects who underwent a total of 7 sessions in the period of 4 months. This dataset represents the complete EEG recordings of a feasibility clinical trial (clinical-trial ID: NCT02445625 — clinical-trials.gov). The recording was performed using g.Nautilus (gTEC, Austria) amplifier coupled with a VR system (Oculus Rift Development Kit 2 headset); the data were acquired from 8 dry electrodes positioned on the standard locations - C3, Cz, C4, CPz,

P3, Pz, P4, POz, with the reference placed on the right ear and the ground electrode placed at AFz with a sampling rate of 250 Hz. The participants were immersed in a virtual environment, a bedroom with a series of objects that were used as targets. The paradigm was subdivided into two parts: (1) calibration phase, where the object target was explicit; (2) online phase, where the participants had to participate in the joint attention task after which they received the feedback according to their performance.

For each session, the training data consisted of 20 blocks (with 10 runs each) while the online phase of 50 blocks (with a variable number of runs each) was used as a test set. In each run 8 objects, flashing in random order, were presented in the virtual scene (1. a wooden plane hanging from the ceiling; 2. a printer on a shelf; 3. a corkboard on the wall; 4. a laptop on a table; 5. a ball on the ground; 6. a radio on top of a dresser; 7. a picture on the wall; 8. books on a shelf). In both phases, the participants had to mentally count how many times the target object flashed, creating an EEG response later used for BCI.

On the basis of the training dataset, it is possible to build a BCI model capable of accurately identifying, which target object was in participants' focus during the online phase.

2.2 Data Processing and BCI Modeling

The data has been analyzed using Matlab (version R2018b, The MathWorks Inc., MA, USA) and BCILAB framework [15]. The dataset had been already pre-filtered from 2 to 30 Hz, therefore no additional frequency filter is applied by our approach at this stage. The window means approach [16], that has been widely applied to event-related slow-changing brain dynamics, such as, the perception of self-induced errors [17], machine-induced errors and/or surprisal [18, 19], prediction of movement intent [20], or (c)overt attention [21], was used. The paradigm consisted of computing 50 ms windows, without overlap, on the segment from 100 ms to 1000 ms from stimulus onset, producing 18 features per channel used in the following machine learning stage. Since the dimensionality of the feature space was high (number of channels * number of windows), the robust classifiers with advance irrelevant feature pruning had to be applied. At the same time, it was important that the BCI model using sparse features can be produced in a reasonable time (i.e. at most within 5–15 min after a training session), so preferable approach would be one without a priori unknown parameters obtainable using time-consuming repetitive cross-validation based optimization procedures. Consequently, in this work we investigated the use of Sparse variational Bayesian logistic regression with Automatic Relevance Determination (VB-ARD) [22], to classify ERPs in ASD subjects and compare it to the logistic regression approach that utilizes standard L1-norm regularization strategy by multiple cross-validations. Briefly, Automatic Relevance Determination (ARD) consists of assigning an individual hyper-prior to each regression coefficient separately and determines the relevance of each of them throughout the optimization of the feature-related weights in order to effectively eliminate irrelevant ones. This approach conversely to [23–25] does not tune hyper-parameters by maximizing the marginal likelihood but applies the full Bayesian treatment and finds the ARD hyper-posteriors by variational Bayesian inference [22]. Therefore, functions associated with these lower weights play no role in the predictions made by the model and are effectively pruned out, resulting in a feature sparse model. Furthermore, the fact that the approach does not compute only a point estimate of the weights, but a full posterior distribution, enables us to use more advanced post-processing of the model and outputs. The Bayesian framework offers optimal sparsity and enables the number of features and channels to be estimated automatically from the training set without requiring time-consuming cross-validation. The VB-ARD model was compared with an L1-norm logistic regression (L1-LOG) model regularized by chronological 5-fold cross-validation techniques with 5 indices safety margin.

2.3 Postprocessing and Classification

The resulting regression binary output, with corresponding class belonging probabilities, provides information for the existence or non-existence of the target ERPs in each run. Finally, the mean probability of all runs in the block had been calculated for each label, and the label with the highest probability has been chosen as the output. In that way, the result averaging has been performed in the post-processing stage.

2.4 Performance Measure and Statistical Analysis

The performance of two approaches was measured in terms of model accuracy on each subject in all 7 sessions and computation time on desktop PC with Intel Core i7 CPU @ 2.67 GHz with 16 GB RAM and Windows 10 OS was measured. A paired t-test approach was used to assess the differences in model accuracy and model computation time between VB-ARD and L1-LOG methods. A value of $p < 0.05$ was regarded as statistically significant.

3 Results

Mean $\pm$ 1SD model accuracy and model computation time for each subject for VB-ARD and L1-LOG approaches are reported in Table 1. In general, both approaches showed high accuracy, however, VB-ARD presented statistically higher accuracy compared to L1-LOG (81.5 $\pm$ 12.0 vs 79.0 $\pm$ 14.5%; $p = 0.0014$) and extremely lower model computation time (19.3 $\pm$ 1.4 vs 643.3 $\pm$ 83.0 s; $p < 0.0001$). The scatter plot for each subject for the accuracy of both approaches was reported in Fig. 1. The topoplot of BCI models calculated with two approaches on the single subject are reported in Fig. 2. It can be observed that VB-ARD approach conversely to the L1-LOG for different time windows produced a sparse model by assigning weights close to zero to less relevant channels and highlighting the relevant ones with higher non-zero values.

Table 1. Mean ± 1SD values of model accuracy and model computation time for each subject for VB-ARD and L1-LOG approaches.

Subject	VB-ARD		L1-LOG	
	Accuracy Mean ± SD (%)	Comp. time Mean ± SD (sec)	Accuracy Mean ± SD (%)	Comp. Time Mean ± SD (sec)
1	71 ± 15	19.44 ± 1.87	58 ± 24	730.62 ± 131.04
2	89 ± 7	19.94 ± 0.86	91 ± 6	633.1 ± 93.26
3	77 ± 17	18.2 ± 1.55	77 ± 16	632.29 ± 82.35
4	89 ± 8	20.02 ± 0.75	87 ± 6	679.38 ± 100.91
5	86 ± 12	19.58 ± 0.51	83 ± 14	661 ± 72.41
6	82 ± 6	18.95 ± 1.58	79 ± 8	584.59 ± 18.4
7	77 ± 7	19.68 ± 0.97	75 ± 4	673.84 ± 41.44
8	87 ± 14	19.39 ± 0.9	85 ± 14	617.24 ± 47.56
9	78 ± 9	17.31 ± 0.7	72 ± 13	572.8 ± 32.27
10	86 ± 8	20.05 ± 1.12	84 ± 8	604.5 ± 61.67
11	85 ± 8	20.46 ± 1.23	85 ± 9	609.31 ± 52.03
12	82 ± 13	18.09 ± 1.4	81 ± 17	600.5 ± 59.64
13	74 ± 17	20.05 ± 0.85	71 ± 19	703.2 ± 92.13
14	73 ± 12	18.24 ± 1.43	73 ± 13	652.99 ± 87.72
15	87 ± 6	19.91 ± 1.61	87 ± 7	694.73 ± 66.3

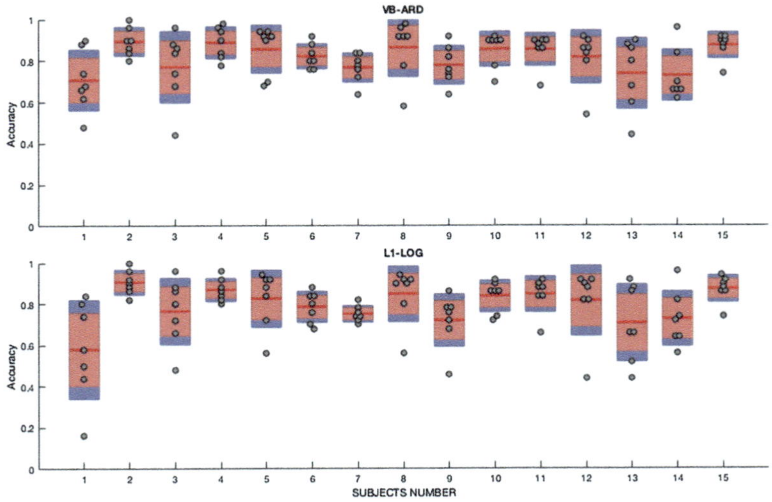

Fig. 1. Plot of accuracy for each subject and each session (points). Points are layed over a 1.96 SEM (95% confidence interval) in red and a 1 SD in blue. Upper panel - VB-ARD model results; Lower panel - L1-LOG model results.

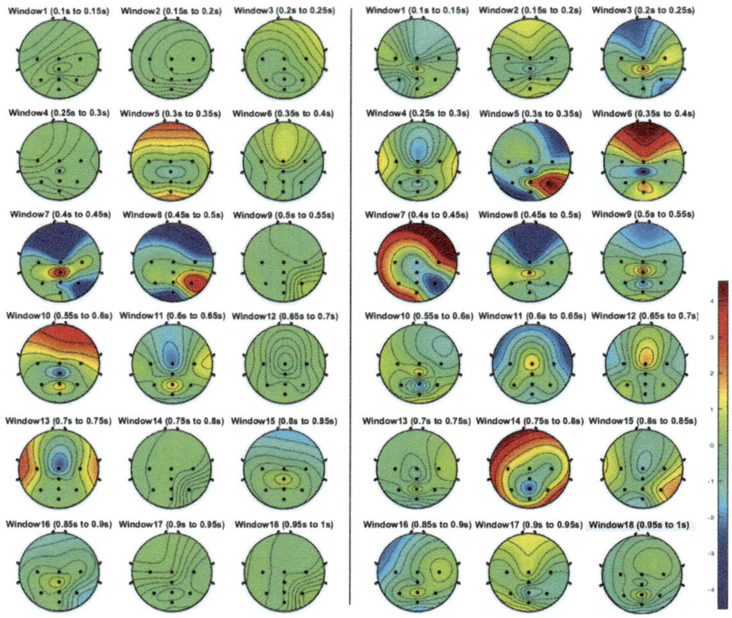

Fig. 2. BCI spatial-temporal feature models produced on one subject. Left panel - Model produced with VB-ARD approach. Right panel - Model produced with L1-LOG method.

4 Discussion

In order to perform a P300 BCI protocol for the improvement of joint-attention skills in Autism-Spectrum Disorder (ASD) subjects, an appropriate approach for the BCI classification is needed. Detecting P300 slow-cortical ERPs poses a considerable challenge to signal to process due to the complex and non-stationary characteristics of a single-trial EEG signal.

The main finding of this study is the high performance of a BCI protocol optimization using a robust logistic regression with automatic relevance determination based on full variational Bayesian inference. Our results showed that the proposed VB-ARD approach eliminates irrelevant channels and features effectively, producing a robust sparse model with 81.5 ± 12.0% accuracy in relatively short modeling computational time 19.3 ± 1.4 s, and it outperforms the standard regularized logistic regression in terms of accuracy and speed needed to produce the BCI model.

Comparison of two BCI models estimated with two different methods revealed that as expected VB-ARD in comparison to L1-LOG approach selects only the relevant time windows and channels that can be also used to study and evaluate the neurophysiological plausibility of produced model. Considering the parsimony of the calculated model we can expect that it represents rather generalized solution resulting in higher robustness, that have to be evaluated in the future studies.

In conclusion, we have presented a BCI computational method based on Bayesian variational inference for P300 ERP detection, which is able to extract and effectively individualize the informative features for binary regression/classification problem of target and non-target objects in the joint attention experiment in ASD subjects. As the approach provides a full posterior distribution it is expected that more advanced post-processing may further improve the performance. Finally, our approach does not average waveforms of every single trial in the pre-processing step but directly extract features with their corresponding class belonging probabilities, so the number of needed runs in the online phase can be adjusted based on so-far estimated discriminatory properties.

Acknowledgment. The authors would like to thank the organizers (IFMBE Scientific Challenge; http://www.medicon2019.org/scientific-challenge/) and the providers of the BCI Competition Dataset [12, 13].

A. Miladinović is supported by European Social Fund (ESF) – FVG

Work partially supported by Master in Clinical Engineering, University of Trieste.

Conflict of Interest Statement. The authors declare that the research was conducted in the absence of any commercial or financial relationships that could be construed as a potential conflict of interest.

References

1. American Psychiatric Association: Diagnostic and Statistical Manual of Mental Disorders (DSM). CoDAS (2013)
2. Friedrich, E.V.C., Suttie, N., Sivanathan, A., Lim, T., Louchart, S., Pineda, J.A.: Brainâ €"computer interface game applications for combined neurofeedback and biofeedback treatment for children on the autism spectrum. Front. Neuroeng. (2014). https://doi.org/10.3389/fneng.2014.00021
3. Rohani, D.A., Sorensen, H.B.D., Puthusserypady, S.: Brain-computer interface using P300 and virtual reality: a gaming approach for treating ADHD. In: 2014 36th Annual International Conference of the IEEE Engineering in Medicine and Biology Society, EMBC 2014 (2014). https://doi.org/10.1109/EMBC.2014.6944403
4. van Dokkum, L.E.H., Ward, T., Laffont, I.: Brain computer interfaces for neurorehabilitation-its current status as a rehabilitation strategy post-stroke. Ann. Phys. Rehabil. Med. (2015). https://doi.org/10.1016/j.rehab.2014.09.016
5. Miladinović, A., Barabaro, A., Eddi, V., Ajčević, M., Accardo, A., Battaglini, P.P., Jarmolowska, J.: Combined and singular effects of action observation and motor imagery paradigms on resting-state sensorimotor rhythms. In: XV Mediterranean Conference on Medical and Biological Engineering and Computing (2019, in press)
6. Ang, K.K., Guan, C.: Brain-computer interface for neurorehabilitation of upper limb after stroke. Proc. IEEE (2015). https://doi.org/10.1109/JPROC.2015.2415800
7. Tang, N., Guan, C., Ang, K.K., Phua, K.S., Chew, E.: Motor imagery-assisted brain-computer interface for gait retraining in neurorehabilitation in chronic stroke. Ann. Phys. Rehabil. Med. (2018). https://doi.org/10.1016/j.rehab.2018.05.431
8. Vourvopoulos, A., Pardo, O.M., Lefebvre, S., Neureither, M., Saldana, D., Jahng, E., Liew, S.-L.: Effects of a brain-computer interface with virtual reality (VR) neurofeedback: a pilot study in chronic stroke patients. Front. Hum. Neurosci. **13** (2019). https://doi.org/10.3389/fnhum.2019.00210

9. Wainer, A.L., Ingersoll, B.R.: The use of innovative computer technology for teaching social communication to individuals with autism spectrum disorders (2011). https://doi.org/10.1016/j.rasd.2010.08.002
10. Bekele, E., Crittendon, J., Zheng, Z., Swanson, A., Weitlauf, A., Warren, Z., Sarkar, N.: Assessing the utility of a virtual environment for enhancing facial affect recognition in adolescents with autism. J. Autism Dev. Disord. (2014). https://doi.org/10.1007/s10803-014-2035-8
11. Georgescu, A.L., Kuzmanovic, B., Roth, D., Bente, G., Vogeley, K.: The use of virtual characters to assess and train non-verbal communication in high-functioning autism. Front. Hum. Neurosci. (2014). https://doi.org/10.3389/fnhum.2014.00807
12. Amaral, C.P., Simões, M.A., Mouga, S., Andrade, J., Castelo-Branco, M.: A novel brain computer interface for classification of social joint attention in autism and comparison of 3 experimental setups: a feasibility study. J. Neurosci. Methods (2017). https://doi.org/10.1016/j.jneumeth.2017.07.029
13. Amaral, C., Mouga, S., Simões, M., Pereira, H.C., Bernardino, I., Quental, H., Playle, R., McNamara, R., Oliveira, G., Castelo-Branco, M.: A feasibility clinical trial to improve social attention in Autistic Spectrum Disorder (ASD) using a brain computer interface. Front. Neurosci. (2018). https://doi.org/10.3389/fnins.2018.00477
14. Bakeman, R., Adamson, L.B.: Coordinating attention to people and objects in mother-infant and peer-infant interaction. Child Dev. **55**, 1278–1289 (1984)
15. Kothe, C.A., Makeig, S.: BCILAB: a platform for brain–computer interface development. J. Neural Eng. **10**, 056014 (2013). https://doi.org/10.1088/1741-2560/10/5/056014
16. Blankertz, B., Lemm, S., Treder, M., Haufe, S., Müller, K.R.: Single-trial analysis and classification of ERP components-a tutorial. Neuroimage (2011). https://doi.org/10.1016/j.neuroimage.2010.06.048
17. Blankertz, B., Schäfer, C., Dornhege, G., Curio, G.: Single trial detection of EEG error potentials: a tool for increasing BCI transmission rates. In: Presented at the Artificial Neural Networks–ICANN 2002 (2002). https://doi.org/10.1007/3-540-46084-5_184
18. Ferrez, P.W., Del, R., Millán, J.: Error-related EEG potentials generated during simulated brain-computer interaction. IEEE Trans. Biomed. Eng. (2008). https://doi.org/10.1109/TBME.2007.908083
19. Zander, T.O., Kothe, C., Welke, S., Roetting, M.: Utilizing secondary input from passive brain-computer interfaces for enhancing human-machine interaction. In: Lecture Notes in Computer Science (including subseries Lecture Notes in Artificial Intelligence and Lecture Notes in Bioinformatics) (2009). https://doi.org/10.1007/978-3-642-02812-0_86
20. Blankertz, B., Curio, G., Müller, K.-R.: Classifying single trial EEG: towards brain computer interfacing. In: Advances in Neural Information Processing Systems Conference (2002). https://doi.org/10.1.1.19.8038
21. Treder, M.S., Blankertz, B.: (C)overt attention and visual speller design in an ERP-based brain-computer interface. Behav. Brain Funct. (2010). https://doi.org/10.1186/1744-9081-6-28
22. Drugowitsch, J.: Variational Bayesian inference for linear and logistic regression. ArXiv e-prints (2013)
23. MacKay, D.J.C., Systems, N.: Bayesian interpolation. Neural Comput. (1992). https://doi.org/10.1162/neco.1992.4.3.415
24. Neal, R.M.: Bayesian learning for neural networks (1996). https://doi.org/10.1007/978-1-4612-0745-0
25. Tipping, M.E.: Sparse Bayesian learning and the relevance vector machine. J. Mach. Learn. Res. (2001). https://doi.org/10.1162/15324430152748236

A Feasible Classification Algorithm for Event-Related Potential (ERP) Based Brain-Computer-Interface (BCI) from IFMBE Scientific Challenge Dataset

Haifeng Zhao[1(✉)], Shiduo Yu[1], Joseph Prinable[1], Alistair McEwan[1], and Petra Karlsson[1,2]

[1] University of Sydney, Camperdown, NSW 2006, Australia
haifeng.zhao@sydney.edu.au
[2] Cerebral Palsy Alliance, Sydney, NSW, Australia

Abstract. Event-Related Potential (ERP) based Brain-Computer Interfaces (BCI) have been widely investigated as an alternative human computer interaction solution. For people with neurological diseases and severe disabilities like amyotrophic lateral sclerosis (ALS) and stroke, BCI may be their only access method for communication. For people with neurodevelopmental disorders, such as autism spectrum disorder (ASD), BCI is also considered to be a potential rehabilitation and education assistance method. Although these are promising developments, further work is required to optimize current classification and filtering methods to improve reliability and enhance the user experience. The aim of this project is to investigate the four most commonly used classification algorithms: Linear Discriminant Analysis (LDA), Support Vector Machine (SVM), Convolutional Neural Network (CNN) and Long-Short Term Memory (LSTM) with a novel and personalised filter design. Results of the classification tasks in Phase I and Phase II, of the IFMBE Scientific Challenge, show a 73% accuracy for phase I and 67% accuracy for Phase II.

Keywords: Event-Related Potential (ERP) · Brain-Computer Interface (BCI) · Machine learning · Linear Discriminant Analysis (LDA) · Single Vector Machine (SVM) · Convolutional Neural Network (CNN) · Long-Short Term Memory (LSTM)

1 Introduction

Brain-computer interfaces (BCI) are an effective interaction method between humans and computers without any inputs from peripheral neural system [1]. BCI is widely used among people with neurological diseases and severe disabilities who may not have other means to communicate. Cerebral palsy (CP) is one

J. Henriques et al. (Eds.): MEDICON 2019, IFMBE Proceedings 76, pp. 1861–1868, 2020.
https://doi.org/10.1007/978-3-030-31635-8_226

of the most common developmental physical disabilities and effects the person's control of movement. Worldwide the incidence of cerebral palsy is 1 in 700 births. There are 17 million people in the world who have CP, of those, one in four of people cannot speak, and one in three have difficulties controlling movements and cannot walk [2]. In addition to the physical disability about 7% of children with cerebral palsy are also diagnosed with autism [3]. For many, BCI holds the promise to provide one of the most efficient access methods for communication [4] and social integration [5].

Furthermore, an emerging trend is to utilize BCI as a game controller. It is believed that, for people with ASD, this BCI application will facilitate a better understanding of the condition itself, while providing a novel gaming experience [5].

There are additional motivations for providing mobility. Recently a continuous pursuit task and associated training paradigm and the spatial resolution of noninvasive neural data through EEG source imaging demonstrated significant improvements in BCI leading to control of a robot arm [6].

Single Vector Machine (SVM) and Linear Discriminant Analysis (LDA) are two of the most popular algorithms used in P300 classification [7]. However, after 10 years evolution of machine learning algorithms, deep learning, and in particular Convolutional Neural Network (CNN) has attracted more attention in respect to BCI classification tasks, becoming a new trend [8].

Traditional electroencephalogram (EEG) systems are considered to be time consuming to set up and maintain performance due to the use of conductive gel [9,10]. Dry EEG electrodes are believed to be one possible solution. However, although studies have illustrated the differences of above algorithms for conventional EEG [7,8], there are only a few papers that report on a comparison of various algorithms for dry electrode EEG. The purpose of this project is to investigate the performance of four most commonly used classification algorithms: LDA, SVM, CNN and Long-Short Term Memory (LSTM) with design of novel and personalised filters, in dry EEG.

2 Methods

Multiple classification methods LDA, SVM, CNN, LSTM were investigated prior to model building. The challenge dataset (Event-related potential (ERP)-EEG) was provided by Amaral et al. [11,12]. The data was collected from 15 participants with ASD [12]. See workflow diagram in Fig. 1.

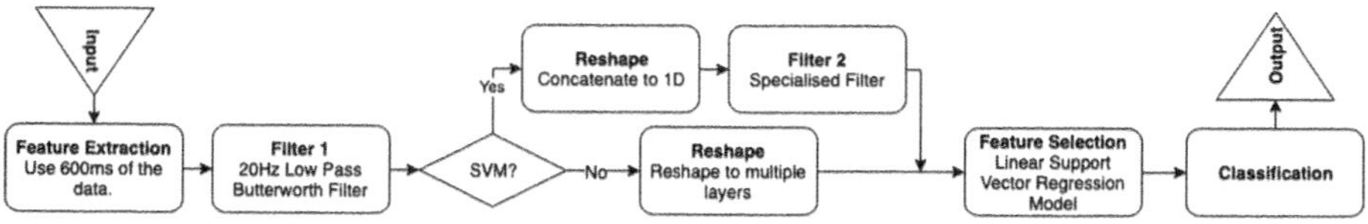

Fig. 1. Classification workflow diagram

2.1 Feature Extraction

Before data processing, ERP feature extraction was carried out. Although the P300 has been the major feature used in previous studies [13], other components of the ERP were investigated to see if they provided valuable features to extract. Optimal duration of feature extraction was determined to be 600 ms after stimulus, consistent with the cumulative time duration of all ERP components [13]. When the data was processed by the SVM, 8 channels were concatenated in the channel number order. While using the CNN, the input data was organized as 8 layers/channels.

2.2 Preprocess Method: Filters

Butterworth Low Pass Filter: The EEG data was treated as a signal wave with various features. The classification algorithms purpose is to use the feature differences between target signals and non-target signals. In order to emphasize the signal features, noises were minimized to achieve a better accuracy. According to Kolev et al.'s study [14], P300 signals which were part of ERP signals are much more recognizable in delta (0.5–4 Hz) and theta (4–7 Hz) rhythms. Although the original signals have been filtered by a 2–30 Hz band-pass filter [11,12], there was still noise located in comparatively higher frequencies. To mitigate this, an additional 20 Hz Butterworth low-pass filter was applied, achieving a rhythm set including delta (0.5–4 Hz), theta (4–7 Hz) and alpha (8–12 Hz) rhythms. See Fig. 2.

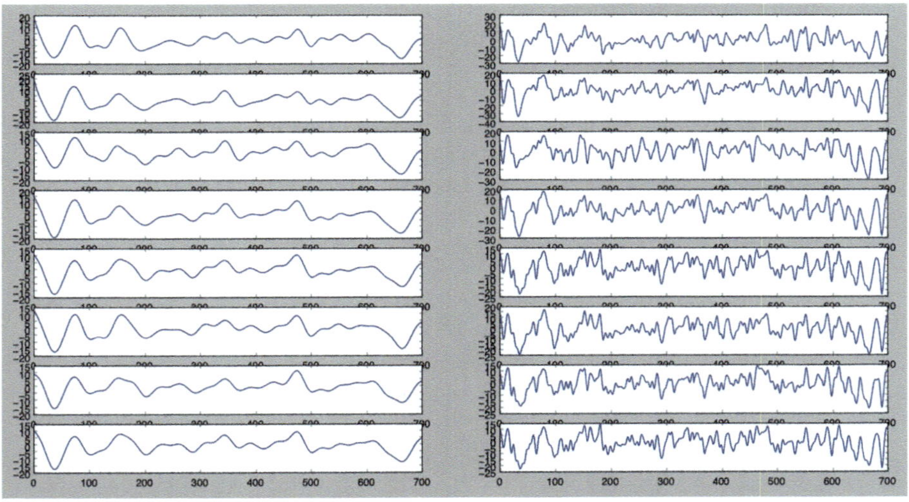

Fig. 2. Epoch 5 and epoch 6, a visualised 8-channel target EEG signals wave. To the right, the original EEG signals wave and to the left, the 20 Hz Butterworth low-pass filtered EEG signals wave.

Noise Feature Based Filter: After observations of the EEG signals, it was found that the target and non-target ERP epochs did not have expected significant differences. Consequently, a general filter would not be sufficient. Therefore, in addition to the low-pass filter, an additional personalised and specialised filter was designed. This filter was designed to address each individual's session's noise features. The non-target ERP signals from training set were averaged and a Fourier transform was applied to address any fluctuating background noise impacting the data flow. This spectrum was deduced from each input data before they entered into the classification models.

$$x_n(c) = \frac{1}{E}\sum_{i=1}^{E} x_i(c)(\text{if i is not target}) \tag{1}$$

$$X_k(c) = \sum_{n=0}^{N-1} x_n(c)e^{-i2\pi kn/N} \tag{2}$$

$$X_k^{Input}(c) = X_k^{Original}(c) - X_k^{Non\text{-}target}(c) \tag{3}$$

$$x_i = \frac{1}{N}\sum_{n=0}^{N-1} X_k^{input} e^{\frac{2\pi j}{N}ni} \tag{4}$$

2.3 Feature Selection

Feature selection is essential to depress the irrelevant time slots and result in a better accuracy. A linear support vector regression (Linear-SVR) classification was used as a selection criteria in this offline classification task, as time consumption was not a priority. The penalty parameter was set to be 0.01.

2.4 Classification

In this project we assessed LDA, SVM, CNN and LSTM. For the LDA, a Bayes' rule based linear decision boundary was used. For the SVM, one of the most frequently used kernel [14], the radial basis function (RBF) kernel was used [15].

$$K(x, x') = exp(-\frac{||x - x'||^2}{2\sigma^2}) \tag{5}$$

A 1D-CNN was trialled in this project. The CNN layers were set to three layers with 128 nodes, 64 nodes and 32 nodes respectively, following a flatten layer, and a binary output dense layer with a softmax activation. As the EEG signals changed rapidly, a small kernel size was used, set to 4. Although the result of a LSTM algorithm is rarely used as a IFMBE competition submission ascribed to its low accuracy, the LSTM was investigated and used in this study after Alhagry et al.'s found in their study [16], that the LSTM outperformed others' in the EEG emotion recognition task. For LSTM, it is set to be 100 nodes, following a 50% dropout layer and a softmax activated binary dense layer. All deep learning methods used in this project were trained with 30 epochs.

2.5 Validation

The test results from the classification were validated in two ways. First, it was compared with the baseline results and then with the black-box validations. Results in Phase I have been compared with baseline results provided by the IFMBE competition Committee. A kernel size 8 was first trialled, however was abandoned. As the EEG signals were changing rapidly, a smaller kernel size of 4 was found to provide improved accuracy.

3 Results

At the outset, the SVM was found to outperform other algorithms, see Table 1. However, with the feature selection and filters applied, LDA improved more significantly than SVM.

Table 1. Classification Methods Comparison

		Plain[a] (%)	Feature Selection(%)	Feature selection and filters (%)
Phase I	LDA	47.73	71.24	73.07
	SVM	52.31	61.60	68.04
	CNN (kernel size = 8)	37.24	Not Applicable	Not Applicable
	CNN (kernel size = 4)	51.24	56.31	50.58
	LSTM	36.89	25.29	26.94
Phase II	LDA	Not Applicable[b]	Not Applicable[b]	67.00
	SVM	45.00	60.00	64.00
	CNN (kernel size = 4)	Not Applicable[b]	54.00	Not Applicable[b]

[a] "Plain" means raw data inputs.
[b] "Not Applicable" - In Phase II, only methods which showed best accuracy and efficiency in Phase I were included.

The feature selection improved the accuracy by 5–10% of other algorithms except LDA. However, in this study, the feature selection contributed more to LDA (23.51%). For SVM, the filter combination of low-pass and noise feature spectrum contributed to around 7% accuracy improvement with the data in Phase I, and 4% in Phase II. In regards to deep learning methods, the overall performance was found to be not applicable. The filters used could not enhance the CNN classifier, however, a smaller kernel size were found to be useful.

The best score we achieved in this competition is 67%. The solution for this record is a combination of a 20 Hz Butterworth low-pass filter, our specially designed personalised filter, Linear-SVR pre-selector and LDA. See Fig. 3.

4 Discussion

The overall classification accuracy is encouraging as all sessions classified by our best solution are above chance level (Table 2). The filters proposed in this

Table 2. Best classification scores

Sessions	Phase I				Phase II					Overall
	01 (%)	02 (%)	03 (%)	Avg. (%)	04 (%)	05 (%)	06 (%)	07 (%)	Avg. (%)	Avg. (%)
SBJ.01	60.00	58.00	46.00	54.67	32.00	44.00	64.00	58.00	49.50	51.71
SBJ.02	88.00	86.00	80.00	84.67	86.00	92.00	86.00	78.00	85.50	85.14
SBJ.03	90.00	56.00	74.00	73.33	82.00	40.00	62.00	20.00	51.00	60.57
SBJ.04	78.00	62.00	74.00	71.33	82.00	68.00	76.00	88.00	78.50	75.43
SBJ.05	88.00	76.00	82.00	82.00	38.00	74.00	86.00	68.00	66.50	73.14
SBJ.06	66.00	70.00	66.00	67.33	62.00	64.00	54.00	86.00	66.50	66.86
SBJ.07	56.00	56.00	66.00	59.33	62.00	64.00	62.00	72.00	65.00	62.57
SBJ.08	80.00	78.00	48.00	68.67	86.00	88.00	88.00	90.00	88.00	79.71
SBJ.09	78.00	60.00	74.00	70.67	68.00	60.00	44.00	52.00	56.00	62.29
SBJ.10	78.00	88.00	84.00	83.33	50.00	82.00	60.00	90.00	70.50	76.00
SBJ.11	86.00	76.00	88.00	83.33	62.00	80.00	80.00	60.00	70.50	76.00
SBJ.12	84.00	68.00	74.00	75.33	70.00	42.00	82.00	82.00	69.00	71.71
SBJ.13	68.00	82.00	84.00	78.00	38.00	48.00	78.00	74.00	59.50	67.43
SBJ.14	70.00	54.00	68.00	64.00	52.00	58.00	28.00	48.00	46.50	54.00
SBJ.15	68.00	86.00	86.00	80.00	80.00	84.00	76.00	78.00	79.50	79.71

paper primarily promoted all linear classification algorithms (SVM and LDA). However, from a single session perspective, not all of the sessions were increased by filters. This could be attributed to the unhandled overlap EEG time slots. The inter-stimulus interval was 200 ms [11,12], but the best noise extraction time duration was found to be 600 ms. The overlapping part between the target and non-target EEG data will mistakenly decrease distinctness of target ERP signals.

SVM outperformed LDA without feature selections and filters. However, LDA increased the accuracy more than SVM with our designed pre-processing methods. This may imply that LDA are more sensitive to noise than SVM. The CNN did not perform as well as LDA and SVM, see Table 1.

Recently, deep learning algorithms have evolved, in particular in the image and sound processing area [17,18]. However, the physic features of brain signals are not as intuitive as in images and sounds. The fundamental mechanisms of brain functions remain unclear, which results in a challenge to design an efficient CNN structure. Thus, EEG signals can only be treated as either an image or sound. In this project a 1D shaped CNN was trialled with three layers with 128 nodes, 64 nodes and 32 nodes respectively.

In this project the feature selection could slightly raise classification accuracy but the filters did not. A conceivable reason is that the CNN could weight every time fragment. Pre-applied selection and filters appeared to be less efficient due to inconsistent time fragments discriminated by the neurons. Linear machine learning methods attributed to an in increased performance compared to the deep learning methods used, the CNN and LSTM.

The successful classification tasks in Phase I and Phase II, dataset, was achieved by a combination of a 20 Hz Butterworth low-pass filter and our specially designed filter, linear support vector regression pre-selector and LDA.

Fig. 3. Classification Accuracy in the dataset for Phase II. Classifications for subject 2 and subject 8 achieved an accuracy over 85%, for subject 2, session 5, the accuracy reached 92%.

It is natural to consider using LSTM for this problem as it is regarded as the premier algorithm for resolving temporal problems. It is unexpected that it has a rather lower accuracy compared to the other trialled methods. On the other hand, the preceding ERP events should be unrelated to the following events as usually the ERP BCI would use an oddball paradigm. This could be a possible explanation of the uselessness of recurrent neural network (RNN) in this task. Even so, we still hold high hopes that a well-designed RNN should be able to accomplish a better result with further refinement and increased training examples.

References

1. Shih, J.J., Krusienski, D.J., Wolpaw, J.R.: Brain-computer interfaces in medicine. Mayo Clin. Proc. **87**(3), 268–279 (2012)
2. Novak, I., Hines, M., Goldsmith, S., Barclay, R.: Clinical prognostic messages from a systematic review on cerebral palsy. Pediatrics **130**(5), e1285–e1312 (2012)
3. Christensen, D., Van Naarden Braun, K., Doernberg, N.S., Maenner, M.J., Arneson, C.L., Durkin, M.S., Benedict, R.E., Kirby, R.S., Wingate, M.S., Fitzgerald, R., Yeargin-Allsopp, M.: Prevalence of cerebral palsy, co-occurring autism spectrum disorders, and motor functioning – autism and developmental disabilities monitoring network, USA, 2008. Dev. Med. Child Neurol. **56**(1), 59–65 (2014)
4. Brumberg, J.S., Pitt, K.M., Mantie-Kozlowski, A., Burnison, J.D.: Brain-computer interfaces for augmentative and alternative communication: a tutorial. Am. J. Speech Lang. Pathol. **27**(1), 1–12 (2018)

5. Friedrich, E.V.C., Suttie, N., Sivanathan, A., Lim, T., Louchart, S., Pineda, J.A.: Brain-computer interface game applications for combined neurofeedback and biofeedback treatment for children on the autism spectrum. Front. Neuroeng. **7**, 21 (2014)
6. Edelman, B.J., Meng, J., Suma, D., Zurn, C., Nagarajan, E., Baxter, B.S., Cline, C.C., He, B.: Noninvasive neuroimaging enhances continuous neural tracking for robotic device control. Sci. Robot. **4**(31), eaaw6844 (2019)
7. Lotte, F., Congedo, M., Lécuyer, A., Lamarche, F., Arnaldi, B.: A review of classification algorithms for EEG-based brain-computer interfaces. J. Neural Eng. **4**(2), R1 (2007)
8. Lotte, F., Bougrain, L., Cichocki, A., Clerc, M., Congedo, M., Rakotomamonjy, A., Yger, F.: A review of classification algorithms for EEG-based brain-computer interfaces: a 10 year update. J. Neural Eng. **15**(3), 031005 (2018)
9. Bleichner, M.G., Debener, S.: Concealed, unobtrusive ear-centered EEG acquisition: cEEGrids for transparent EEG. Front. Hum. Neurosci. **11**, 163 (2017)
10. Lin, C.-T., Liao, L.-D., Liu, Y.-H., Wang, I.-J., Lin, B.-S., Chang, J.-Y.: Novel dry polymer foam electrodes for long-term EEG measurement. IEEE Trans. Biomed. Eng. **58**(5), 1200–1207 (2010)
11. Amaral, C.P., Simões, M.A., Mouga, S., Andrade, J., Castelo-Branco, M.: A novel brain computer interface for classification of social joint attention in autism and comparison of 3 experimental setups: a feasibility study. J. Neurosci. Methods **290**, 105–115 (2017)
12. Amaral, C., Mouga, S., Simões, M., Pereira, H.C., Bernardino, I., Quental, H., Playle, R., McNamara, R., Oliveira, G., Castelo-Branco, M.: A feasibility clinical trial to improve social attention in autistic spectrum disorder (ASD) using a brain computer interface. Front. Neurosci. **12**, 477 (2018)
13. Sur, S., Sinha, V.: Event-related potential: an overview (contemporary issue). Ind. Psychiatry **18**(1), 70–73 (2009)
14. Kolev, V., Demiralp, T., Yordanova, J., Ademoglu, A., Isoglu-Alkaç, U.: Time-frequency analysis reveals multiple functional components during oddball p300. Neuroreport **8**(8), 2061–2065 (1997)
15. Chang, Y.-W., Hsieh, C.-J., Chang, K.-W., Ringgaard, M., Lin, C.-J.: Training and testing low-degree polynomial data mappings via linear SVM. J. Mach. Learn. Res. **11**, 1471–1490 (2010)
16. Fahmy, A.A., Alhagry, S., El-Khoribi, R.A.: Emotion recognition based on EEG using LSTM recurrent neural network. Int. J. Adv. Comput. Sci. Appl. **8** (2017)
17. Guo, Y., Liu, Y., Oerlemans, A., Lao, S., Wu, S., Lew, M.S.: Deep learning for visual understanding: a review. Neurocomputing **187**, 27–48 (2016)
18. Salamon, J., Bello, J.P.: Deep convolutional neural networks and data augmentation for environmental sound classification. IEEE Signal Process. Lett. **24**(3), 279–283 (2017)

Linear vs Nonlinear Classification of Social Joint Attention in Autism Using VR P300-Based Brain Computer Interfaces

Lucia de Arancibia[1(✉)], Patricia Sánchez-González[1,2], Enrique J. Gómez[1,2], M. Elena Hernando[1,2], and Ignacio Oropesa[1]

[1] Biomedical Engineering and Telemedicine Centre, ETSI Telecomunicación, Center for Biomedical Technology, Universidad Politécnica de Madrid, Madrid, Spain
lucia.dearancibia.casillas@alumnos.upm.es, {psanchez, egomez, mariaelena.hernando, ioropesa}@gbt.tfo.upm.es
[2] Centro de Investigación Biomédica en Red, Biomateriales y Nanomedicina, Madrid, Spain

Abstract. VR P300-based BCI has proven to be a suitable method for training social attention skills in youngsters with autism spectrum disorder (ASD). In this study, we present a method that could be used in such an application to identify which object the user is paying attention to in a virtual environment by means of EEG recordings only. Temporal and time-frequency features were explored. Furthermore, the prediction accuracy of linear and nonlinear classification methods was assessed and compared, along with their computational times and complexity, and linear discriminant analysis (LDA) yielded the best overall performance (82%). The successful predictions and low computational times demonstrate the feasibility of the proposed solution for a VR-BCI neurorehabilitation tool.

Keywords: BCI · P300 · Autism Spectrum Disorder · Linear discriminant analysis · Support vector machines · Virtual reality · Neurorehabilitation

1 Introduction

Autism Spectrum Disorder (ASD) is currently one of the most prevalent serious developmental conditions in developed countries. In Europe, average prevalence is of 1 in 89 children aged 7–9 years, and estimated direct annual costs range from 1,534€ to 22,378€ per individual [1]. Moreover, predictions suggest that by 2025 the total costs derived from ASD in the USA will rise to over $450,000 million [2].

ASD symptoms are present from early childhood and limit everyday activities throughout patients' life. A distinctive symptom is impaired social joint attention: failure to share the focus of attention on an object with other individuals. This poses greater difficulties in understanding social cues and thus interacting with others [3].

J. Henriques et al. (Eds.): MEDICON 2019, IFMBE Proceedings 76, pp. 1869–1874, 2020.
https://doi.org/10.1007/978-3-030-31635-8_227

Amaral et al. [4] introduced a promising, novel VR P300-based BCI paradigm for the neurorehabilitation of youngsters with ASD. On the hypothesis that subjects with ASD can learn joint attention skills by internalizing the response to a given social cue, they created a task where attentional mental state is measured through the detection of oddballs, which in turn induces a P300 signal. This training tool has demonstrated to be beneficial and very promising in the field of social cognition training in ASD [5]. In order to develop it, it is necessary to design and implement a fully functioning BCI that can correctly identify the P300 component of the EEG. Different studies have tackled this specific problem for BCI applications.

The first P300-based BCI made use of temporal information of filtered EEG data such as peak values, area under the curve and covariance evaluation with a P300 template to discriminate between target and non-target signals [6]. More recent studies include more complex features based on temporal, time-frequency or spatial domains as input to machine learning classifiers [7–11]. Additionally, in Krusienski et al. [12] one can find a comparison of several classification algorithms for P300 detection. Their results suggest that linear classifiers may be sufficient to correctly identify P300 signals. However, classification accuracies vary widely among subjects and thus it is not clear whether simple classifiers are able to provide good generalization.

In this study, we propose a combination of temporal and frequency features to compare performance of linear and nonlinear classification approaches to the problem and describe our best solutions.

2 Methods

In order to carry out this study, data from clinical trial NCT02445625 was used, as described in [4]. The dataset includes the EEG recordings from 15 participants that underwent 7 sessions each throughout 4 months.

The paradigm proposed by Amaral et al. reduces an 8-group classification problem to a binary discrimination problem, i.e., the discrimination between target events (1 in 8) and non-target events (remaining 7 events in a run). For our predictions, we averaged all EEG events corresponding to the same object in the same block and obtained 8 averaged signals per block (number of signals averaged is equal to the number of runs per block). We then extract features from the averaged signals and calculated the likelihood of them comprising a P300 component. The object whose corresponding signal yields maximum probability is chosen as the predicted target object of the block.

2.1 Feature Extraction

- **Features based on time domain.** The P300 peak typically appears around 200 ms to 300 ms after the stimulus. For each channel in the subset, the 250 samples corresponding to the time slot from stimulus onset (0 ms) until 1000 ms later were extracted and downsampled by average (decimation factor = 10), and the resulting samples from each channel were concatenated obtaining a single vector of 200 samples from each observation. Downsampling allowed us to reduce the number of features and decrease classification time without missing important information.

- **Features based on continuous wavelet transform (CWT).** The CWT was computed with a Mexican Hat wavelet on scales corresponding to the delta (0.5 to 4 Hz) and theta (4 to 8 Hz) bands, following [7]. To further reduce computation time as well as extract the most differential features and eliminate redundancies, the t-CWT technique [8] was applied. This technique applies Student's t-statistics to find points of maximum statistical difference between average P300 and non-P300 in the time–frequency domain. In our case, 128 points were selected for each of the 8 channels and concatenated, adding a total of 128 * 8 = 1024 time-scale features.
- **Feature reduction techniques.** The temporal and frequency features make a total of 1224 features per signal to be used as input to a classifier. Such a large number can slow down the training process and cause overfitting. Therefore, principal component analysis (PCA) reduction was introduced. The first 120 principal components were used, since they are enough to cover 99% of the variance in the data.

2.2 Classification Algorithms and Training Considerations

To solve the EEG classification problem, we explored linear machine learning approaches such as linear discriminant analysis (LDA) and support vector machines with linear (LSVM) kernel, and a more complex support vector machine with radial (RSVM) kernel. For the latter, hyperparameters gamma and C were selected for each patient during training through bayesian optimization with 5-fold cross validation. The search was evaluated for values ranging from 10^{-3} to 10^{3} for both hyperparameters.

The main problem found for classification of the EEG signals was that the data is highly imbalanced: only one in every 8 EEG segments contains a P300. Thus, by classifying all inputs as non-target, an accuracy of 87.5% would be achieved. To avoid this problem, oversampling and boosting methods were explored.

In order to train the classifiers, two approaches were attempted and evaluated. At first, inter-subject classification (train a single classifier using all the training data from all subjects' calibration phases) was approached. However, this alternative ignored subject variability, and the obtained models yielded lower overall accuracy results. To test whether this was due to excessive generalization, we approached the problem as a subject-specific classification problem (train an individual classifier for every subject using only his/her individual calibration data). This approach accounts only for intra-subject variability and fits to the patient's particularities. It demonstrated better overall accuracies and was used for the final predictions.

In order to validate the implemented algorithms and choose the best method for phase II predictions, they were validated on phase I online (test) data. The features and classifiers that yielded best performances (measured in terms of low training times and accurate target object detection) were then used to make predictions for phase II.

3 Results and Discussion

3.1 Phase I Predictions and Validation

Table 1 shows the training, P300 classification and target object prediction performances of the different methods used and tested with phase I online data.

Table 1. Prediction of phase I online data. F_1 score refers to average value for P300 detection (binary classification problem), while online accuracy refers to average target object detection accuracy. In parenthesis maximum and minimum values per session are given. Training and prediction times are an average value per session. In bold: best result (LDA)

Features	Classification	Other methods	Training time	F_1 score	Online accuracy (%)
Temporal, t-CWT features and PCA	**LDA**	–	**0.037 s**	**0.76 (0.38–0.94)**	**80 (38–100)**
	LSVM	–	2.82 s	0.70 (0.41–0.91)	75 (40–94)
	RSVM	–	23.6 s	0.76 (0.5–0.96)	79 (50–94)
	LDA	Oversampling	0.1 s	0.72 (0.28–0.91)	78 (32–94)
	LDA	Boosting	5.2 s	0.73 (0.34–0.94)	79 (34–100)

For subject-specific training, LDA was the most accurate as well as the fastest classification method. Non-linear RSVM did not yield better performance than simpler linear classifiers. However, it did improve the accuracy for subject 13, something that was not achieved through other proposed methods. This suggests that linear models might be too generalist and not capable of particularizing in the most complex cases.

Oversampling and boosting were applied to see whether they improved results of the best classifier (LDA). Both methods slightly improved the accuracy for some sessions but did not improve the overall performance for machine learning classifiers.

3.2 Prediction of Phase II Online Data

The above-mentioned features with PCA and an LDA classifier trained with subject-specific data yielded a final overall phase II online data prediction accuracy of 82%. Although for some sessions accuracy was as low as 50%, we accomplished an accuracy above 80% for 41 sessions (out of 60).

The results for P300 signal classification are above average for some comparable state-of-the-art studies such as [9], which achieves a mean classification accuracy of 48.15% with 5 EEG epochs; [10], that reached accuracies up to 79%; or [11], which reports a P300 detection F1 score of 0.52.

Moreover, we achieve an improvement in target object detection with respect to the study by Amaral et al. [4]. Their highest accuracy was 72% for subjects with ASD compared with ours of 100%, accomplished by averaging only 5 EEG events. These results are also higher than their accuracies for healthy subjects (mean accuracy of 80%).

Our study suggests that subject-specific classification outperforms inter-subject models. This implies that the BCI should be trained for every new patient. However, our methods accomplish classification and object detection in a considerably short time. Around 30 s are needed to train a classifier with enough calibration data. Less than 0.1 s are needed to classify an EEG event. A total of around 0.1 s are needed to detect the target object of a block. This means the user needs to wait less than 1 s since the block is over until he/she receives feedback. Such a result is very advantageous for the pursued neurorehabilitation application for patients with ASD.

In conclusion, even though the proposed methods can be improved, the best overall results demonstrate their feasibility. The desirable next step would be to test it live in a real situation.

References

1. ASDEU. Autism Spectrum Disorders in the European Union. Executive summary (2018)
2. Leigh, J.P., Du, J.: Brief report: forecasting the economic burden of Autism in 2015 and 2025 in the United States. J. Autism Dev. Disord. **45**, 4135–4139 (2015)
3. World Health Organization. Autism spectrum disorders. https://www.who.int/news-room/fact-sheets/detail/autism-spectrum-disorders. Accessed 20 June 2019
4. Amaral, C.P., Simões, M.A., Mouga, S., Andrade, J., Castelo-Branco, M.: A novel brain computer interface for classification of social joint attention in autism and comparison of 3 experimental setups: a feasibility study. J. Neurosci. Methods **290**, 105–115 (2017)
5. Amaral, C., Mouga, S., Simões, M., Pereira, H.C., Bernardino, I., Quental, H., Playle, R., McNamara, R., Oliveira, G., Castelo-Branco, M.: A feasibility clinical trial to improve social attention in autistic spectrum disorder (ASD) using a brain computer interface. Front. Neurosci. **12**, 477 (2018)
6. Farwell, L.A., Donchin, E.: Talking off the top of your head: toward a mental prosthesis utilizing event-related brain potentials. Electroencephalogr. Clin. Neurophysiol. **70**, 510–523 (1988)
7. Demiralp, T., Ademoglu, A., Schürmann, M., Basar-Eroglu, C., Basar, E.: Detection of P300 waves in single trials by the wavelet transform (WT). Brain Lang (1999)
8. Bostanov, V., Kotchoubey, B.: The t-CWT: A new ERP detection and quantification method based on the continuous wavelet transform and Student's t-statistics. Clin. Neurophysiol. **117**(12), 2627–2644 (2006)
9. Agapov, S.N., Bulanov, V.A., Zakharov, A. V., Sergeeva, M.S.: Wavelet algorithm for the identification of P300 ERP component. arXiv:1611.00033 [q-bio.NC] (2016)

10. Manor, R., Geva, A.B.: Convolutional neural network for multi-category rapid serial visual presentation BCI. Front. Comput. Neurosci. **9**, 146 (2015)
11. Liu, M., Wu, W., Gu, Z., Yu, Z., Qi, F., Li, Y.: Deep learning based on batch normalization for P300 signal detection. Neurocomputing **275**, 288–297 (2018)
12. Krusienski, D.J., Sellers, E.W., Cabestaing, F., Bayoudh, S., McFarland, D.J., Vaughan, T. M., Wolpaw, J.R.: A comparison of classification techniques for the P300 speller. J. Neural Eng. **3**(4), 299 (2006)

Linear SVM Algorithm Optimization for an EEG-Based Brain-Computer Interface Used by High Functioning Autism Spectrum Disorder Participants

Mayra Bittencourt-Villalpando(✉) and Natasha M. Maurits

Department of Neurology AB51, University Medical Center Groningen, University of Groningen, 9700RB Groningen, The Netherlands
m.bittencourt.villalpando@umcg.nl

Abstract. Machine-learning algorithms can be used for data classification on EEG-based Brain-computer interfaces (BCIs). Here, we used an algorithm based on linear support vector machine (SVM) to identify the presence of the P300 component in datasets from 15 young adult participants with autism spectrum disorder that were provided for the IFMBE Scientific Challenge 2019. We optimized the parameters and inputs for a linear SVM model throughout the ten attempts of the challenge and compared them in terms of accuracy. The highest score (mean accuracy) of 82% was achieved by a procedure that was customized per session per participant. When using a similar procedure for classification model generation and configuration of parameters for all sessions and participants, the highest score achieved was 77%. The results showed that adding data from targets from different calibration sessions from the same participants to the training dataset resulted in a significant increase in accuracy. In all attempts, the mean accuracy was above 70%, which is considered the minimum classification level for the controllability of a BCI. These are promising results for future use of BCIs as a tool for attention training in ASD participants.

Keywords: Brain-computer interface (BCI) · EEG · P300 · Support vector machine (SVM)

1 Introduction

In autism spectrum disorder (ASD), deficits in the interpretation of intentions of others or other social cues are well recognized. Brain-computer interfaces (BCIs) are systems that are designed to allow the communication between the Central Nervous System (CNS) and an external device and potential use for training of social cognition skills in ASD patients has been suggested and investigated [1].

In EEG-based BCIs electrodes are placed on the scalp to measure the electrical signals from the brain. Amongst brain activity currently used for BCI operation, the

---

Electronic supplementary material The online version of this chapter (https://doi.org/10.1007/978-3-030-31635-8_228) contains supplementary material, which is available to authorized users.

J. Henriques et al. (Eds.): MEDICON 2019, IFMBE Proceedings 76, pp. 1875–1884, 2020.
https://doi.org/10.1007/978-3-030-31635-8_228

P300 event-related potential (ERP) is a typical, or "naïve", brain response to a desired choice. The P300 response is evoked by the occurrence of a rare (less probable) stimulus in a sequence of stimuli and it is most frequently elicited within the framework of the "oddball paradigm". The P300 appears as a large positive deflection in the EEG signal occurring approximately 300 to 800 ms following the stimulus onset. It reflects attention related processes and it is generally observed most strongly over centroparietal brain areas [2, 3].

Studies that use machine-learning algorithms for BCI data classification have achieved impressive results for offline data analysis for healthy participants. However, the evaluation of their applicability to real life scenarios is more difficult. BCI challenges have been organized to address such problems related to BCI research and advance the knowledge in the field [4].

In this study, we used an algorithm based on linear Support Vector Machine (SVM) to discriminate segments where the P300 is present from the ones where it is not present. This algorithm was previously used by us [5] on a 4-choice BCI that was operated by healthy participants. We adapted and optimized it to identify which object a participant is paying attention to in a virtual scene, as part of the IFMBE Scientific Challenge 2019[1].

2 Methods

Determining the presence or absence of a P300 ERP can be considered a binary classification problem with a discriminant function [6]. An SVM was designed to determine the hyper-plane that maximizes the separating margin between the two classes of a binary classification between targets and non-targets (classes defined as 1 and 0, respectively).

2.1 The Dataset

The dataset with the EEG recordings of 15 participants with s = 7 sessions each was provided. Data from sessions 1 to 3 were provided in Phase I and data from sessions 4 to 7 were provided in Phase II of the challenge. This dataset represents the complete EEG recordings of the feasibility clinical trial (clinical-trial ID: NCT02445625 – clinicaltrials.gov) that tested a P300-based BCI to train young adults with ASD to follow social cues [1, 7]. The EEG data provided were acquired from 8 channels (C3, Cz, C4, CPz, P3, Pz, P4 and POz). The reference was placed at the right ear and the ground electrode was placed ad AFz. Sampling rate was set at 250 Hz. Data were acquired notch filtered at 50 Hz and pass band filtered between 2 and 30 Hz.

The experimental design was based on a virtual scene where eight objects could be intensified. The intensification of one object is considered an event. The competition dataset was divided into train and test sets. The train set was made available with the target labels (1 out of the 8 possible events) for each block. Each block consists of a

[1] http://www.medicon2019.org/scientific-challenge/.

number of runs where all of the 8 possible events are intensified once in a pseudo-random order. In the dataset, data from all channels are available from −200 ms to 1200 ms relative to the onset of each event. Within each session, the train set consisted of 20 blocks with 10 runs each (with a total of 1600 events: 200 targets, 1400 non-targets) and the test set consisted of 50 blocks with a variable number of runs (Nrun) each (from 3 to 10, provided with the dataset). The challenge was to predict the labels for each block of the test set. The detailed explanation about the data acquisition process and the datasets is available elsewhere [1, 7].

2.2 IFMBE Challenge Scoring

The global score per team in the competition was defined by the attempt with the highest total accuracy and was computed using the formula:

$$\text{Score}\,(\%) = \#\text{TotalCorrectLabels}/\#\text{TotalLabels} * 100 \tag{1}$$

where #TotalCorrectLabels is the total number of correct labels from all participants in sessions 4 to 7 in one attempt and #TotalLabels is the number of labels of all participants in sessions 4 to 7 in one attempt. In this challenge, #TotalLabels = 3000 (15 participants * 4 sessions * 50 blocks).

For each attempt of label prediction, the accuracy was provided per participant and session using the formula:

$$\text{Accuracy}\,(\%) = \#\text{CorrectLabels}/\#\text{Labels} * 100 \tag{2}$$

where #CorrectLabels is the number of correct labels from one session of a participant in one attempt and #Labels is the number of labels in one session per participant in one attempt. In this challenge, #Labels = 50.

2.3 Training: Model Construction

The input to the SVM classifier is a feature vector formed by concatenating vectors from all eight channels. Each channel vector consists of the potentials at the time points of the ERP measured at that channel. From the training datasets provided, the data from −200 ms to 0 s relative to event onset (baseline) and from 1000 ms to 1200 ms relative to event onset were discarded. Therefore, for each channel in the dataset, datapoints from 0 to 1000 ms following each event onset were extracted, resulting in a vector length of 250 elements. The resulting channel vectors were concatenated for each event and participant, creating a single feature vector for training the classifiers. As a result, the initial training set with all 8 channels resulted in a feature vector length of l = 2000 elements (250 samples * 8 channels) for each participant per event.

During the initial phase and also throughout the challenge, new parameters were defined and adjusted in an attempt to improve the classification performance. All of the parameters used are described below.

Due to the scarcity of target events in one session (1 target for 7 non-targets), two strategies were used in some of the models to increase the number of targets for the classifier:

(1) Add data from target events from other sessions from the same participant: the number of training sessions (from 1 to 7) selected to build a training set was defined as k. Therefore, the total number of targets in one training set is nT = k * 200.
(2) Artificially increasing (x2) the number of targets in one session by pseudorandomly averaging the data from this session with two different pseudorandom matrices. In this case, nT = 2 * k * 200.

The matrix $\boldsymbol{T}_s$ (200 × l) consists of all feature vectors from target events in one session s. The nT feature vectors in $\boldsymbol{T}_s$ were pseudorandomly averaged for signal to noise reduction (using a randomly generated matrix $\boldsymbol{R}$(200 × 200) with r ones and (nT-r) zeros per column to generate a new matrix $avg\boldsymbol{T}_s$ (200 × l), where r is the number of vectors selected for averaging).

The matrix $\boldsymbol{T}$ (nT × l) consists of the horizontal concatenation of the matrices $\boldsymbol{T}_s$ from all of the k sessions selected for the training set.

The matrix $\boldsymbol{NT}_s$ (1400 × l) consists of all feature vectors from non-target events in one session.

The classification feature matrix $\boldsymbol{X}$ (N × l) is the horizontal concatenation of $\boldsymbol{T}$ and $\boldsymbol{NT}_s$, where N = nT + 1400. The respective class labels vector y(N) contains the class (1 for target, 0 for non-target) of the event for each feature vector in $\boldsymbol{X}$.

The SVM training algorithm ("fitcsvm", MATLAB and Statistics and Machine Learning Toolbox Release 2017b, The MathWorks, Inc., USA) returns trained SVM classifiers for binary classification based on a matrix with data from predictive features with their respective labels. $\boldsymbol{X}$ and y were used as input to the SVM training algorithm ("fitcsvm", MATLAB 2017b) using a linear kernel function, automatic kernel scaling, the default Box Constraint and standardized data predictor. The returned classifier was used for class prediction on the testing data.

2.4 Testing: Label Prediction

Similarly to the training data, for the testing datasets the datapoints from 0 to 1000 ms following each event onset were extracted. For each participant, the resulting data segments from all of the 8 channels were concatenated for each event, run and block, creating a testing matrix $\boldsymbol{Y}$(Nobs × l), where Nobs = 8(events) * Nrun * 50(blocks). For the observations of every possible event, the feature vectors from all runs in one block were averaged according to the number of runs per block, resulting in the reduced averaged testing matrix $avg\boldsymbol{Y}$((Nobs/Nrun) × l). This step was done for: (a) signal noise reduction, (b) size reduction of the test matrix and (c) simplification of the process described below to identify which event corresponds to the target.

The classification algorithm ("predict", MATLAB and Statistics and Machine Learning Toolbox Release 2017b, The MathWorks, Inc., USA) returned a matrix of scores that indicated the likelihood that a label came from a particular class (1 for target, 0 for non-target). For each observation in $avg\boldsymbol{Y}$, the predicted target corresponds to the event with the highest score.

2.5 Model Parameters and Optimization Procedure

The model parameters defined for attempt 1 were based on preliminary tests performed in Phase I of the challenge.

The following parameters were adjusted, as indicated in Table 1 below, throughout Phase II of the challenge:

(a) The number of training sessions (*k*) selected for the training set: from 1 to 7.
(b) Artificially increasing the number of targets as described in Sect. 2.3: yes/no
(c) The number of feature vectors *r* used to generate $avg\boldsymbol{T}_s$: 10 or 20.
(d) The solver algorithm used as input to 'fitcsvm': Sequential Minimal Optimization (SMO, default) or L1 soft-margin minimization by quadratic programming (L1QP).
(e) The weight used as input to 'fitcsvm'.
(f) The 'Priors' used as input to 'fitcsvm': prior = [0.875 0.125] or auto (based on the class probability distribution from the training set, which was modified if the targets from other sessions were added as described in Sect. 2.3, item 1).
(g) The cost used as input to 'fitcsvm': [0 1; 1 0] (auto), [0 1; 2 0] or [0 2;1 0].

Table 1. Parameters for training of classification model per attempt

ATTEMPT	k	Training sessions	r	Weight	nT artificially increased?	Solver algorithm	Prior	Cost
1	1	same as current testing session	10	1 for all events	No	SMO	auto	auto
2	7	all training sessions from the participant	10	7 for targets from current session 1 for targets from other sessions 1 for non-targets	No	SMO	auto	auto
3	3	"best performing" sessions (highest accuracy) from the participant	10	1 for all events	No	SMO	auto	auto
4	1	same as current testing session	10	1 for all events	No	L1QP	auto	auto
5	1	same as current testing session	20	1 for all events	No	SMO	auto	auto
6	customized			1 for all events	No	customized	auto	auto

(continued)

Table 1. *(continued)*

ATTEMPT	k	Training sessions	r	Weight	nT artificially increased?	Solver algorithm	Prior	Cost
7	7	all training sessions from the participant	10	7 for targets from current session 1 for targets from other sessions 1 for non-targets	No	SMO	prior	auto
8	7	all training sessions from the participant	10	7 for targets from current session 1 for targets from other sessions 1 for non-targets	Yes	SMO	prior	[0 1; 2 0]
9	7	all training sessions from the participant	10	7 for targets from current session 1 for targets from other sessions 1 for non-targets	Yes	SMO	prior	0 2; 1 0]
10	Selection of best performing model per participant and per session							

From attempts 1 to 5 and 7 to 9, the parameters were set according to Table 1 to generate the classification SVM models. In attempt 5, the parameters were customized per participant and per session based on an evaluation of which adjustments previously resulted in an increase in accuracy. In the last attempt, the labels from the best performing attempt per participant and per session were selected.

3 Results

The highest score of 82% was achieved in attempt 10 by selecting the best performing model that was customized per session per participant. Boxplots of the accuracies per attempt are shown in Fig. 1. The average accuracies of all attempts and of all sessions (except for session 7 from participant 3 in attempt 3) were above chance level (12.5%).

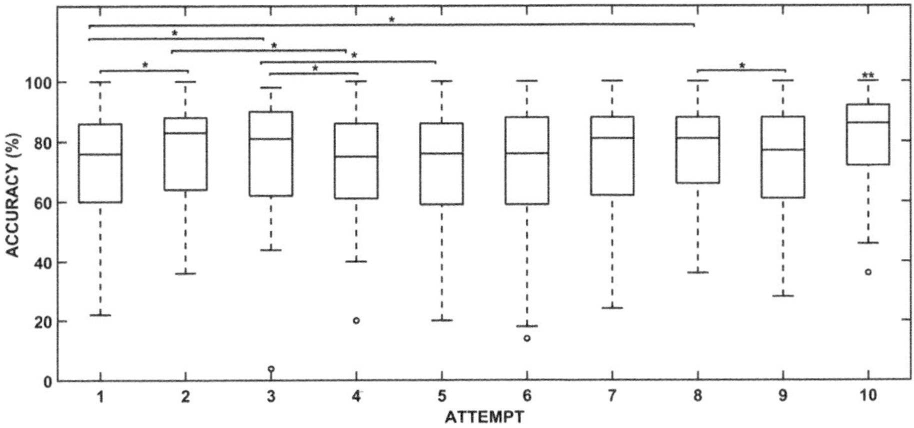

Fig. 1. Boxplots of accuracy (%) per attempt. Asterisks indicate significant differences between attempts (Repeated measures ANOVA, *p < 0.01, **p < 0.001, post-hoc comparisons Bonferroni corrected)

The highest score achieved in the competition by using the same model generation procedure for all sessions for all participants was 77% for the model used in attempt 2.

Boxplots of the accuracies per participant for attempt 10, providing an overview of the variability between and within participants are given in Fig. 2. The highest accuracy in one session of 100% was achieved for participant 4. The accuracies were above chance level (12.5%) for all participants.

Boxplots of the accuracies per session for attempt 10 are given in Fig. 3. According to a repeated measures ANOVA test, there were no significant differences in accuracy across participants between sessions ($\alpha = 0.05$).

The final results from the best attempt and the model selected per session and per participant are provided in the *Supplementary Material.*

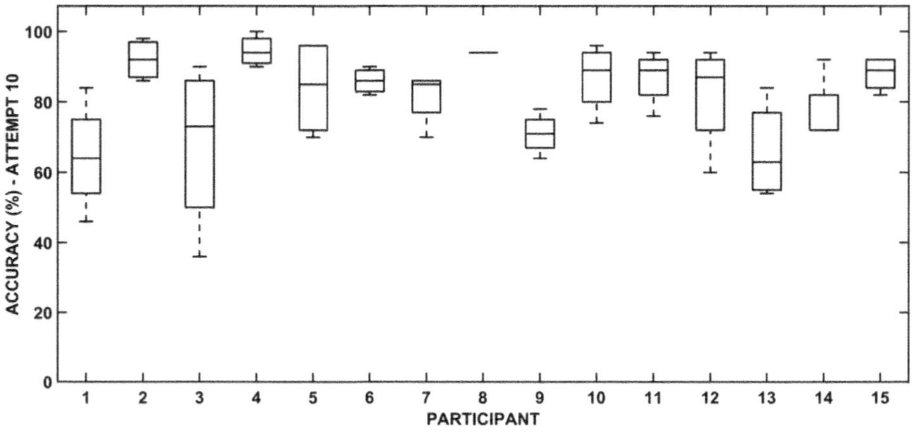

Fig. 2. Boxplots of accuracy (%) per participant in attempt 10.

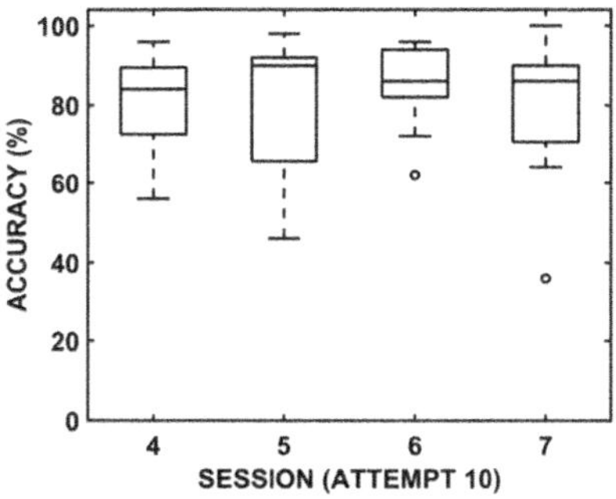

Fig. 3. Boxplots of average accuracy (%) per session across participants for attempt 10.

4 Discussion

The main goal was to maximize the correct prediction of labels for the blocks provided in the IFMBE challenge datasets.

We optimized the parameters and inputs for a linear SVM model throughout the attempts and compared them in terms of accuracy. For the best performing attempt (attempt 10), we provided an overview of the accuracy per participant and compared the accuracy per session.

In all attempts, the average accuracy achieved was above chance level (12.5%) and the score of all ten attempts was above 70%, which is considered the minimum classification level for the controllability of a BCI [8]. Those are promising results for future use of BCIs as a tool for attention training in ASD. However, the accuracy level varied across participants and participants 1 and 13 did not achieve 70% accuracy, in any of the attempts. This might be explained by attention deficits from those specific participants given their clinical condition, although specific information about clinical condition was not made available with the dataset.

From the different optimizations that were attempted, adding data from targets from calibration sessions from the same participants resulted in a significant increase in accuracy, which is evident by the superior performance of attempt 2. Possible reasons could be (a combination of): (a) the scarcity of data from targets in a single session (200 targets vs. 1400 non-targets) to train the algorithm, leading to a classification bias for the non-targets class; (b) reduced performance of specific training sessions due to technical aspects (e.g. changes in the quality of the acquired signal), (c) reduced attention of the participant at specific training sessions. However, it is worth mentioning that the reasons mentioned in items (b) and (c) should not explain the improvement in performance alone, given that no significant difference in accuracy was found between sessions.

There were no significant differences in accuracy when (a) changing the solver algorithm from SMO to L1QP, (b) increasing the number of events for pseudorandom averaging the targets from 10 to 20 or assigning priors (to 0.875 for non-targets and 0.125 to targets) as concluded by comparing attempt 1 to attempt 4, attempt 1 to attempt 5 and attempt 2 to attempt 7, respectively. It could not be concluded from this analysis if assigning a higher cost for misclassification of targets could increase the accuracy.

Finally, the best score was achieved in attempt 10, with a very customized solution consisting of the selection of the best performing model per session and per participant. The accuracy in attempt 10 was significantly higher than the accuracy in all other attempts. This suggests that it is very difficult to find a generic solution that fits all participants and provides stable performance across sessions. Furthermore, the computational cost involved in the different attempts differed: adding data in calibration sessions, for example, increased the accuracy but also the computational cost for training the model. This information can be relevant for the development of real-life applications. Although studies using machine learning algorithms as linear SVM have provided promising results in the literature, the optimization process remains challenging. There are several parameters that can be adjusted and their interpretation is not always straightforward. Positive results seem to depend on the extensive evaluation of the datasets and trial and error processes that are still difficult to automate and further investigation in the field is required.

References

1. Amaral, C., et al.: A feasibility clinical trial to improve social attention in Autistic Spectrum Disorder (ASD) using a brain computer interface. Front. Neurosci. **12**, 477 (2018)
2. Pontifex, M.B., Hillman, C.H., Polich, J.: Age, physical fitness, and attention: P3a and P3b. Psychophysiology, 46(2), 379–387 (2009)
3. Polich, J.: Updating P300: an integrative theory of P3a and P3b. Clin. Neurophysiol. **118**(10), 2128–2148 (2007)

4. Blankertz, B., et al.: The BCI competition III: validating alternative approaches to actual BCI problems. IEEE Trans. Neural Syst. Rehabil. Eng. **14**(2), 153–159 (2006)
5. Bittencourt-Villalpando, M., Maurits, N.M.: Stimuli and feature extraction algorithms for brain-computer interfaces: a systematic comparison. IEEE Trans. Neural Syst. Rehabil. Eng. **26**(9), 1669–1679 (2018)
6. Krusienski, D.J., et al.: A comparison of classification techniques for the P300 Speller. J. Neural Eng. **3**(4), 299 (2006)
7. Amaral, C.P., Simões, M.A., Mouga, S., Andrade, J., Castelo-Branco, M.: A novel brain computer interface for classification of social joint attention in autism and comparison of 3 experimental setups: a feasibility study. J. Neurosci. Methods **290**, 105–115 (2017)
8. Pfurtscheller, G., et al.: The hybrid BCI. Front. Neurosci. **4**, 30 (2010). https://doi.org/10.3389/fnpro.2010.00003

Classification of P300 Component Using a Riemannian Ensemble Approach

Dominik Krzemiński[1], Sebastian Michelmann[2], Matthias Treder[3], and Lorena Santamaria[1(✉)]

[1] CUBRIC, School of Psychology, Cardiff University, Cardiff CF24 4HQ, UK
{KrzeminskiDK,santamariacovarrubiasL}@cardiff.ac.uk
[2] Princeton Neuroscience Institute, Princeton University, Princeton, NJ 08544, USA
[3] School of Computer Science and Informatics, Cardiff University, Cardiff CF24 3AA, UK

Abstract. We present a framework for P300 ERP classification on the 2019 IFMBE competition dataset using a combination of a Riemannian geometry and ensemble learning. Covariance matrices and ERP prototypes are extracted after the EEG is passed through a filter bank and an ensemble of LDA classifiers is trained on subsets of channels, trials, and frequencies. The model selects a final class based on maximum probability of evidence from all ensembles. Our pipeline achieves an average classification accuracy of 81.2% on the test set.

Keywords: Brain-computer interface · ERP · P300 · Classification · Riemannian geometry · Ensemble learning · LDA · Logistic regression

1 Introduction

The P300 is a positive deflection in the event-related potential (ERP) focused on centro-parietal electrode sites, usually observed around 300 ms after a stimuli was presented [10]. It typically follows the occurrence of a rare or unexpected sensory stimulus. Unlike the mismatch negativity, it is mediated by the conscious deployment of attention. For this reason, it has been one of the first control paradigms to be used in brain-computer interfaces [7].

The classical approach to P300 feature extraction is the use of spatio-temporal features [7,11]. Recently, the Riemannian framework - originally used for oscillatory (e.g. motor imagery) data - has been applied to ERP classification [8]. The framework was developed as a method to directly use covariance matrices as features for classification. Symmetric positive definite (SPD) matrices such as non-singular covariance matrices form a Riemannian manifold in the space of matrices. Since the space is non-linear, it has been suggested to first map the covariance matrices into the linear tangent space with respect to a reference SPD matrix, usually the geometric mean of the training data. The approach outperformed state-of-the-art methods such as the Common Spatial Patterns [3].

J. Henriques et al. (Eds.): MEDICON 2019, IFMBE Proceedings 76, pp. 1885–1889, 2020.
https://doi.org/10.1007/978-3-030-31635-8_229

In this paper, we combine the Riemannian framework with the ensemble learning approaches. A separate classifier was constructed for each subject and session according to the following pipeline. First, the EEG is bandpass filtered using different frequency bands in the range 1–20 Hz. Second, ERP templates are extracted and concatenated with the covariance matrix of an individual trial. Third, covariance matrices are mapped into tangent space using a Riemannian approach. Fourth, an ensemble of Linear Discriminant Analysis (LDA) classifiers with shrinkage [5] or Logistic Regression classifiers is constructed using random subsets of features and trials.

2 Methods

The data was provided as part of the 2019 IFMBE scientific challenge competition organized by the 15th Mediterranean Conference on Medical and Biological Engineering and Computing (Medicon 2019). The dataset represents the 8-channels EEG recordings of a feasibility clinical trial to train youngsters with Autism Spectrum Disorder (ASD) to follow the social cues [1,2]. It consists of 7 sessions divided into training and test subsets each. In our analysis we refer to the test data as a set consisting of testing part of sessions 4–7 (as used in the competition evaluation). The task of a participant was to count how many times one out of eight objects was highlighted in a virtual environment [1].

All analyses were performed in MATLAB (Natick, USA). The code to reproduce our results is available on the public code repository[1].

We denote the i-th trial of a multivariate N-channels dataset measured at a set of time instants t as $\mathbf{x}_i(t) = [x_1(t), x_2(t), \ldots, x_n(t)]^T \in \mathbb{R}^N$, where for $t = 1, 2, \ldots, T$ the data can be considered as a wide-sense static data array. Collecting all time points for the i-th trial in a matrix $\mathbf{X}_i \in \mathbb{R}^{N \times T}$ and assuming zero-mean data, this process can be extensively expressed by its spatial covariance matrix as $\mathbf{C}_i = 1/(T-1)\mathbf{X}\mathbf{X}^T$. A scheme of the pipeline to obtain the features for the classifier assemble is presented on Fig. 1.

Several pipelines were constructed with different combinations of filter frequencies and classifiers. However, they all followed the same general structure. First, the data was bandpass filtered and segmented into seven different combinations (red dotted rectangle of Fig. 1) to find the most discriminating features for the classifier. Two different filters were selected (1–20 Hz or 1–8 Hz) and two variations of trial length (whole signal or the first 600 ms after stimuli onset). Additionally, three subsets of electrodes were chosen (all, central or posterior electrodes). The combination of these 3 possibilities (number of channels, window length and filter band) resulted in seven inputs. Lastly, the eighth combination added was the raw signal where no filtering or windowing was performed.

Then, ERP prototypes were created by calculating the ERP for each channel [4]. For robustness, a trimmed mean discarding the 5% largest and 5% smallest values was used. ERP prototypes were then concatenated to the single-trial

[1] https://github.com/dokato/bci-challange

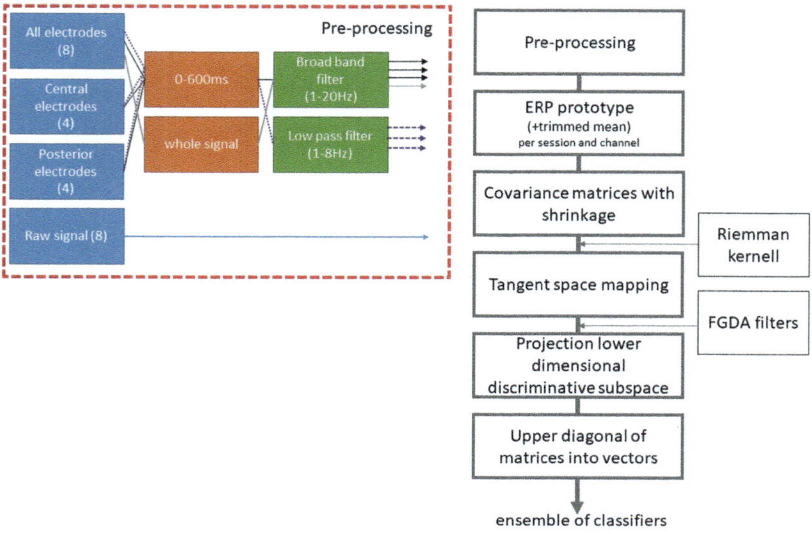

Fig. 1. Depiction of the classification pipeline. The left side (red dotted rectangle) expands on the details of the pre-processing pipeline. The right side (grey rectangles) traces all the steps of the pipeline from pre-processing and feature extraction to classifier training.

data, effectively doubling the number of channels. Regularized covariance matrices were calculated using a shrinkage approach [5]. In shrinkage, the empirical covariance matrix $\mathbf{C}$ is replaced by $\mathbf{C}_{\text{shrink}} = (1-\gamma)\mathbf{C} + \gamma\nu\mathbf{I}$, where $\gamma \in [0,1]$ is a regularization parameter, $\mathbf{I}$ is the identity matrix, and $\nu = \text{trace}(\mathbf{C})$. The Ledoit-Wolf estimate was used to estimate the optimal γ [9].

The resultant covariance matrices were projected into the tangent space of a reference matrix. The reference matrix was calculated as the Riemannian geometric mean of the training covariance matrices [3,6]. Next, Fisher Geodesic Discriminant Analysis (FGDA) was used to project the matrices in tangent space to a lower-dimensional discriminative subspace [3]. The resultant projections were flattened to vectors and used as input features to an ensemble learning algorithm as described in the next section.

3 Results

We tested several combinations of classifiers and ensembles. The outcomes of classification were tested on a subset of the data from sessions 4–7. The details of the strategies are listed below.

- **Ensemble of data features** - 8 different signal features were extracted as follows: (1) All electrodes with broad band frequency spectrum (1–20 Hz); (2) Central electrodes (C3, Cz, C4, CPz) with broad band frequency spectrum;

Table 1. Ensemble models' parameters and their accuracies on the test set.

Classifier	Evidence	Accuracy
LDA	majority vote	73.6%
LogReg	majority vote	69.8%
MDM	majority vote	71.0%
LDA	probability	**81.2%**
LogReg	probability	78.8%
MDM	probability	80.5%

(3) Posterior electrodes (P3, Pz, P4, POz) with broad band frequency spectrum; (4) All electrodes with low-pass filtered signal (1–8 Hz); (5) Central electrodes with low-pass filtered signal; (6) Posterior electrodes with low-pass filtered signal; (7) All electrodes with broad band frequency spectrum. (8) Raw signal. In the features from (1) to (6) a time window of 0–600 ms was extracted from the signal. The final result was the combination of all pieces of evidence from 8 classifiers.

- **Ensemble of classifiers** - ensemble of 400 learners was used taking 40% of data samples and 60% of features each. The optimal number of learners was determined using grid search.
- **Classifier** - type of classifier: linear discriminant analysis (LDA), logistic regression (LogReg), or minimum distance to mean (MDM).
- **Evidence accumulation** - evidence for voting for a final class was either based on majority vote per one trial in block, or based on accumulated probability of voting for a particular class.

The performance of a final multi-class accuracy was computed using a ratio of correctly predicted labels to their total number. The best classification accuracy

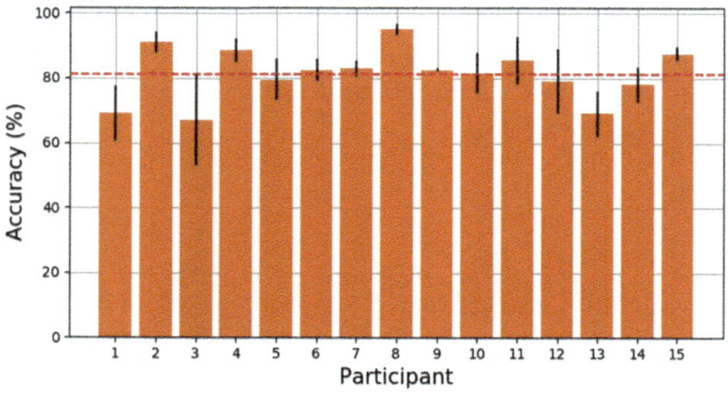

Fig. 2. Performance of the best model per participant of the experiment. Standard error computed over 4 sessions from the test set. The red dashed horizontal line denotes the average accuracy (81.2%).

(81.2%) was achieved by the model utilizing ensemble of signal features and ensemble of LDA classifiers with cumulative evidence for each of the considered classes (Table 1). The performance of the model per subject is depicted on Fig. 2.

4 Discussion

In this study we tested a combination of Riemannian space features extracted from EEG P300 time-series with ensemble approach. The best performance of 81.2% was achieved by the model combining ensemble of data features with ensemble of LDA classifiers. This result corroborates the feasibility of the proposed approach. Future work will focus on exploring a larger range of classifiers in combination with Riemannian kernels to further increase the accuracy and robustness of the ERP classification.

References

1. Amaral, C.P., Simões, M.A., Mouga, S., Andrade, J., Castelo-Branco, M.: A novel brain computer interface for classification of social joint attention in autism and comparison of 3 experimental setups: a feasibility study. J. Neurosci. Methods **290**, 105–115 (2017)
2. Amaral, C.P., Mouga, S., Simões, M., Pereira, H.C., Bernardino, I., Quental, H., Castelo-Branco, M.: A feasibility clinical trial to improve social attention in autistic spectrum disorder (ASD) using a brain computer interface. Front. Neurosci. **12**(July), 1–13 (2018)
3. Barachant, A., Bonnet, S., Congedo, M., Jutten, C.: Riemannian geometry applied to BCI classification. International Conference on Latent Variable Analysis and Signal Separation, pp. 629-636 (2010)
4. Barachant, A., Congedo, M.: A Plug& Play P300 BCI Using Information Geometry. arXiv preprint: https://arxiv.org/pdf/1409.0107.pdf (2014)
5. Blankertz, B., Lemm, S., Treder, M., Haufe, S., Müller, K.R.: Single-trial analysis and classification of ERP components - a tutorial. NeuroImage **56**(2), 814–825 (2011)
6. Congedo, M., Barachant, A., Bhatia, R.: Riemannian geometry for EEG-based brain-computer interfaces; a primer and a review. Brain-Comput. Interfaces **4**(3), 1–20 (2017)
7. Farwell, L., Donchin, E.: Talking off the top of your head. Electroencephalogr. Clin. Neurophysiol. **70**(6), 510–523 (1988)
8. Korczowski, L., Congedo, M., Jutten, C.: Single-trial classification of multi-user P300-based Brain-Computer Interface using Riemannian geometry. In: 2015 37th Annual International Conference of the IEEE Engineering in Medicine and Biology Society (EMBC), pp. 1769–1772. IEEE (2015)
9. Ledoit, O., Wolf, M.: Honey, I shrunk the sample covariance matrix. Economics Working Papers (2003). https://ideas.repec.org/p/upf/upfgen/691.html
10. Linden, D.E.J.: The p300: where in the brain is it produced and what does it tell us? Neuroscientist Rev. J. Bringing Neurobiol. Neurol. Psychiatry **11**(6), 563–576 (2005)
11. Treder, M.S., Purwins, H., Miklody, D., Sturm, I., Blankertz, B.: Decoding auditory attention to instruments in polyphonic music using single-trial EEG classification. J. Neural Eng. **11**(2), 026009 (2014)

Using Time Domain and Pearson's Correlation to Predict Attention Focus in Autistic Spectrum Disorder from EEG P300 Components

V. Sophie Adama[1(✉)], Benjamin Schindler[2], and Thomas Schmid[1,2]

[1] Computer Engineering Group, Universität Leipzig, Augustusplatz 10, 04109 Leipzig, Germany
adama@informatik.uni-leipzig.de

[2] Machine Learning Group, Universität Leipzig, Augustusplatz 10, 04109 Leipzig, Germany

Abstract. Patients with Autistic Spectrum Discorder are known to have deficits in interpreting others' intentions from gaze-direction or other social attention. Here, we use electroencephalography data recorded in virtual reality experiments with patients to predict one out of eight objects that was focused on. Correct labels for these objects were known from parallel eye-tracking measurements. We extracted features from the time domain and from Pearson's correlation and applied both statistical and neuro-inspired supervised machine learning algorithms. Using a multi-layer perceptron, we achieved 65.4% accuracy on the validation data set and 70.0% accuracy on the test data set.

Keywords: Brain-computer interface · Electroencephalogram · P300 · Event-related potentials · Feature extraction · Machine learning · Decision tree · Random forest · Support vector machine · Neural network

1 Introduction

Autism spectrum disorder (ASD) is a set of neurological disorders that affect an individual's social behavior and communication, and includes, e.g., what used to be known as Asperger syndrome. ASD begins in the early childhood and leads, e.g., to a reduced ability for non-verbal coordination of attention of two individuals toward a third object or event [1]. For neuro-rehabilitation of ASD patients, virtual reality and brain-computer interfaces (BCI) based on electroencephalography (EEG) are currently being investigated as novel approaches [2].

Data used in this study was recorded from ASD patients in order to study attentional responses. The subjects were presented eight objects in a virtual reality environment and instructed to focus on one of them. During the experiment, each of these eight objects are flashed in a randomized order to elicit EEG P300 components. The goal of this study is to use this data to train a classifier, which will be employed afterwards to predict one of the eight objects, to which

J. Henriques et al. (Eds.): MEDICON 2019, IFMBE Proceedings 76, pp. 1890–1893, 2020.
https://doi.org/10.1007/978-3-030-31635-8_230

subjects are attending in an online session. Since there are eight possible targets, the problem was approached as a eight-class classification problem.

2 Methods

2.1 Data

The EEG data used in the following was provided through the IFMBE Scientific Challenge of the MEDICON 2019 conference[1]. The experiment setup and recording of the data is described in [1,2]. All EEG measurements were recorded with a sampling rate of 250 Hz from eight channels placed at C3, Cz, C4, CPz, P3, Pz, P4 and POz. No additional preprocessing were performed since the provided data were already notch filtered at 50 Hz and passband filtered between 2 and 30 Hz. No electrode selection was done at this stage.

All measurements were assigned to a specific session and subject. For training classifiers, training data from all subjects in all sessions were used as a single training set consisting of 2100 samples with given labels. Test data from all subjects from Session 1 to Session 3 were combined as well and used as validation set with 2250 samples with given labels. Test data provided by the IFMBE Scientific Challenge consisted of 3000 samples, where no labels were given.

2.2 Feature Extraction

For training classifiers, we preprocessed the raw data in MATLAB R2019a in order to extract the most relevant features for the given task. All extracted features were scaled with respect to the training data.

Time Domain Features. Time domain features were extracted following partly the method described in [3]. For each time window from 200 to 450 ms following the stimulus presentation, the signal from each channel is averaged with respect to each flashed target, and afterwards moving average filtered and decimated with a factor equals to 12. This time window was chosen since the P300 reaches its maximum amplitude between this interval [4]. The decimation process produces 7 values for each of the 8 potential targets, resulting in 56 values per channel. The total size of the feature vector is 448 (56 * 8 channels).

Pearson's Correlation Method. Pearson's correlation coefficient r, also known as Product Moment Correlation Coefficient (PMCC), is used to measure the linear relationship between two variables. Its value ranges between -1 and $+1$. A value equals to 0 means that there is no correlation between the two variables, -1 means total negative correlation and $+1$ means total positive correlation.

The coefficient was computed for the averaged signal from each flashed target and between the time window of interest and the time window preceding stimulus presentation. For two time series $X = x_i, 1 \leq i \leq L$ and $Y = y_i, 1 \leq i \leq L$:

$$r = \frac{L \sum x_i y_i - \sum x_i \sum y_i}{\sqrt{\left(L \sum {x_i}^2 - (\sum x_i)^2\right)\left(L \sum {y_i}^2 - (\sum y_i)^2\right)}} \tag{1}$$

[1] http://www.medicon2019.org/scientific-challenge/.

where L is the length of the signal. One value of the Pearson's correlation is obtained for each of the 8 possible targets. This resulted in 64 features.

2.3 Classification Methods

Several supervised learning approaches were evaluated using Python and the Scikit-learn framework[2]. Time domain and correlation features were used first separately and then combined to train the following classifiers:

- **Decision tree.** An implementation of Classification and Regression Trees (CART, [7]), used with Gini impurity as attribute selection criterion.
- **Random forest.** An implementation of the bootstrapping-based ensemble method by Leo Breiman [8], used with 500 trees.
- **Support vector machine.** A Python binding to the widely recognized LIBSVM implementation by Chang and Lin [9], used with a linear kernel, penalty parameter C=1.0 and shrinking heuristic.
- **Multi-layer perceptron.** A feed-forward neural network with a six-layer architecture (n-1024-512-128-32-8 for n inputs) and a hyperbolic tangent activation function. For training, the stochastic gradient-based optimizer by Kingma et al. [10] was applied with learning rate 0.001 (max. 200 iterations).

The method and feature set with best accuracy on the validation data was then trained with both training and validation data. The resulting classifier was used to predict labels for the test data with unknown labels.

3 Results

Prediction accuracy for the validation data ranged from 13.1 to 65.4% (Table 1). Using both time domain and Pearson's correlation, the multi-layer perceptron yielded maximum accuracy. For comparison, Table 1 includes a dummy classifier that generates predictions randomly from the training set's class distribution.

After re-training the multi-layer perceptron with time domain features and Pearson's correlation from both training and validation data, predictions for the test data set were obtained with an accuracy of 70.0%; using only time domain features, an accuracy of 63.0% was achieved. This final result represents an increase of almost five percent points compared to the accuracy obtained on the validation data set after training exclusively with the training data set.

Table 1. Accuracy obtained for validation data with trained classifiers [%].

Method	Feature sets		
	Time domain	Pearson	Combined
Dummy classifier	11.7	13.0	12.8
Decision tree	24.1	13.1	23.9
Random forest	42.1	13.4	42.1
Support vector machine	58.9	14.1	59.1
Multi-layer perceptron	65.2	12.9	65.4

[2] https://scikit-learn.org/.

4 Conclusions

With this work, we showed that time domain features can be used to predict targets of attention in ASD patients from EEG P300 components. Using a multi-layer perceptron, we achieved an accuracy of 63.0% for predictions on a test data set, which samples were unknown to our trained classifier. Using Pearson's correlation as additional features, training the multi-layer perceptron yielded predictions with 70.0% accuracy. We consider the achieved degree of accuracy an indication that this multi-class problem on EEG data is non-trivial. Considering that training and validation data sets contained only 2,000 and 2,250 samples, however, it can be suspected that prediction accuracy may be increased substantially by using larger numbers of samples in further studies. Future work may also include employing other morphological and frequency-related features or automated feature extraction by a using a convolutional neural network.

Acknowledgements. We would like to thank Prof. Dr. Martin Bogdan for discussing preprocessing steps for the given data set, and Marlo Kriegs for evaluating hyperparameter optimization techniques for support vector machines.

References

1. Amaral, C.P., Simões, M.A., Mouga, S., Andrade, J., Castelo-Branco, M.: A novel Brain Computer Interface for classification of social joint attention in autism and comparison of 3 experimental setups: a feasibility study. J. Neurosci. Methods **290**, 105–115 (2017). https://doi.org/10.1016/j.jneumeth.2017.07.029
2. Amaral, C., Mouga, S., Simões, M., Pereira, H.C., Bernardino, I., Quental, H., Playle, R., McNamara, R., Oliveira, G., Castelo-Branco, M.: A feasibility clinical trial to improve social attention in autistic spectrum disorder (ASD) using a brain computer interface. Front. Neurosci. **12**(July), 1–13 (2018). https://doi.org/10.3389/fnins.2018.00477
3. Krusienski, D.J., Sellers, E.W., Cabestaing, F., Bayoudh, S., McFarland, D.J., Vaughan, T.M., Wolpaw, J.R.: A comparison of classification techniques for the P300 speller. J. Neur. Eng. **3**, 299–305 (2006). https://stacks.iop.org/JNE/3/299
4. Polich, J.: Updating P300: an integrative theory of P3a and P3b. Clin. Neurophysiol. **118**(10), 2128–2148 (2007)
5. Hoffmann, U., Vesina, J.-M., Ebrahimi, T., Diserens, K.: An efficient P300-based brain-computer interface for disabled subjects. J. Neurosci. Methods **167**(1), 115–125 (2008)
6. Abootalebi, V., Hassan, M., AliKhalilzadeh, M.M.: A new approach for EEG feature extraction in P300-based lie detector. Comput. Methods Programs Biomed. **94**(1), 48–57 (2009)
7. Breiman, L., Friedman, J., Olshen, R., Stone, C.: Classification and Regression Trees. Chapman and Hall, Wadsworth (1984)
8. Breiman, L.: Random forests. Mach. Learn. **45**(1), 5–32 (2001)
9. Chang, C., Lin, C.: LIBSVM: a library for support vector machines. ACM Trans. Intell. Syst. Technol. **2**(3), 27:1–27:27 (2011)
10. Kingma, D., Ba, J.: Adam: a method for stochastic optimization. arXiv preprint arXiv:1412.6980 (2014)

Performance Evaluation of Manifold Algorithms on a P300 Paradigm Based Online BCI Dataset

Bipra Chatterjee[1(✉)], Ramaswamy Palaniappan[2], and Cota Navin Gupta[1]

[1] Neural Engineering Lab, Department of Biosciences and Bioengineering, Indian Institute of Technology, Guwahati, India
chatterjee_bipro@hotmail.com

[2] Data Science Research Group, School of Computing, University of Kent, Chatham, UK

Abstract. Healthcare field is highly benefited by incorporating BCI for detection and diagnosis of some health related detriment as well as rehabilitation and restoration of certain disabilities. An EEG dataset acquired from 15 high-functioning ASD patients, while they were undergoing a P300 experiment in a virtual reality platform, was analysed in this paper using three algorithms. Performance of Bayes Linear Discriminant Analysis (BLDA) was predominant over Convolutional Neural Network (CNN) and Random Undersampling (RUS) Boosting. BLDA rendered 73% overall accuracy in predicting target and the best accuracy for each subject using CNN or BLDA yielded an overall accuracy of 76%.

Keywords: Brain-Computer Interfaces · CNN · BLDA · RUSBoosting

1 Introduction

Brain-Computer Interfaces (BCI) is an integration of suitable hardware for acquisition of neural activities (having low signal-to-noise ratio), cognitive paradigms in order to do some tasks and machine learning algorithms for signal processing. It is defined as, a system that measures central nervous system (CNS) activity and converts it into artificial output that replaces, restores, enhances, supplements, or improves natural CNS output and thereby changes the ongoing interactions between the CNS and its external or internal environment [1].

In the current study, a visual evoked potential based BCI incorporating P300 paradigm has been reported. P300 is an event related potential (ERP) component, usually elicited 300 ms after presentation of an infrequent target stimulus while undergoing an oddball experiment. Since the real time realization of P300 speller, researchers have explored its application to numerous other scenarios. We have applied various supervised classification algorithms and performed an offline analysis, to implement the task of target prediction over a few trials,

R. Palaniappan and C. N. Gupta—Both the senior authors have equal contribution to this work.

J. Henriques et al. (Eds.): MEDICON 2019, IFMBE Proceedings 76, pp. 1894–1898, 2020.
https://doi.org/10.1007/978-3-030-31635-8_231

in an online BCI dataset. To achieve a high classification accuracy, one needs appropriate temporal and/or spatial feature extraction techniques followed by machine learning algorithms to separate target and non-target classes.

A linear and two non-linear classifiers were applied here to determine the class of a trial and detect a target among eight objects after a few trials. Bayes Linear Discriminant Analysis (BLDA), an adaptive classifier most frequently and successfully used in P300 BCI, has been implemented. Convolutional Neural Network (CNN), a deep learning algorithm, which shows empirical evidence to extensively capture the temporal and spatial dependencies, motivated us to make an use of it here. Random Forest, an ensemble of weak learners like Decision Trees, renders satisfactory performance to classify linearly inseparable data, though produce biased results when applied to a skewed data. So a different ensemble aggregation approach, called Random Undersampling (RUS) Boosting, was applied to restrain the problem of class imbalance and bolster the influence of minority class (target). At last, an ensemble of all these classifiers was also implemented to check whether it outnumbers the accuracy of individual performances.

2 Methods

A dataset collected from MEDICON 2019 Scientific Challenge, is analysed here. It was acquired from 15 subjects, suffering from high-functioning Autism Spectrum Disorder (ASD), in order to train social cognition skills. The detailed description of the experiment and dataset is given in [2,3]. EEG was recorded from eight electrode positions in the parieto-central region with a sampling rate of 250 Hz. Our objective was to predict the target from each of the 50 blocks in a session given in online phase, where number of runs in a block varied. The framework for target prediction is below.

2.1 Pre-processing

Pre-processing steps extract temporal features of the data, to enhance the differences between two classes of data, target and non-target. It was done in the order mentioned below.

1. **Pre-stimulus mean removal:** Each epoch starts from -0.2 s and ends at 1.2 s relative to the event trigger. Baseline value of the pre-stimulus samples were calculated and post-stimulus samples were taken into account to construct the feature vectors after removing the pre-stimulus mean.
2. **Band-pass filtering:** The data given was already band-pass filtered between 2 to 30 Hz. We used MATLAB function *buttord* in order to determine the suitable order of the filter corresponding to pass-band frequencies 2 and 12 Hz, stop-band frequencies 0.5 and 30 Hz, stop-band attenuation of 40 dB, pass-band ripple of 3 dB; followed by a Butterworth band-pass filter using function *butter*.

3. **Downsampling:** The filtered data was decimated by a factor of 10 in order to remove high frequencies still left as well as reduce the dimension of the feature vector.
4. **Normalisation:** Each of the eight channels data was normalised epoch-wise to the interval [−1,1].

2.2 Classification

After extracting the relevant features, we need to determine the hyper-plane to separate target and non-target class. Epochs obtained after pre-processing were of dimension $8 * 30$, which were concatenated into $240 * 1$ column vectors associated with a label, before training the classifier. We explored three supervised classification algorithms which are discussed below:

Bayes Linear Discriminant Analysis: LDA has been extensively and successfully used in BCI. It aims to maximise the between-class over the within-class variance. The idea of BLDA is an extension of Fisher's LDA. FLDA searches for discriminant vectors that yield a large distance between the projected means (target and non-target) and small variance around the projected means. BLDA was used to check over-fitting to high-dimensional and noisy data like the one presented here. It assumes that, labels and feature vectors are linearly correlated by set of weights with additive white Gaussian noise and perform regression. BLDA exhibited better performance than FLDA, when applied in a P300 BCI data [4]. All the necessary MATLAB codes to implement BLDA were taken from [4] and no manual intervention was needed to determine the hyper-parameters.

RUSBoost: P300 BCI data is usually attributed with high class imbalance. Deviant stimuli, which triggers P300 potentials, occur infrequently than the abundant standard stimuli and this gives rise to a skewed number of trials. Random Forest, an ensemble of Decision Trees, becomes biased because of the inadequate target class. To overcome this problem, we have followed a boosting approach presented in [5]. We used MATLAB Classification App *RUSBoosted Trees* and optimal parameters (Maximum number of splits- 20, Number of weak learners- 450 (Decision Trees), Learning rate- 0.8) are obtained from simulations.

Convolutional Neural Network: Convolution Neural Network (CNN) has traditionally been used in a variety of image processing problems [6]. It is particularly fitting for images due to the two dimensional structure of the input layer. In recent years, multi-dimensional biomedical time series data such as EEG from multi-channels have also been analysed using CNN [7]. CNN, especially deep CNN is very useful for these datasets as they circumvent the stage of feature extraction and in some cases, even pre-processing to reduce noise. It is usually time-consuming and also require a lot of experimentation to decide on the suitable network parameters and even so, the extracted features may not capture all the necessary discriminatory information. Hence, it is more appropriate to use architectures such as CNN to automatically learn the discriminatory features from the dataset [8].

Each pre-processing step was implemented except downsampling and the input size kept as 300×8. After some preliminary simulations using the provided test data labels from Phase 1, the CNN layer sizes were decided as 64, 32 and 16. Each layer had batch normalisation, rectified linear unit and max pooling. As we found overfitting occurred, we used a dropout layer before the max pooling layer. Dropout percentages were decided as 0.3, 0.2 and 0.1. The filter size was set at 3×3, which was also determined from preliminary studies. A fully connected layer of size 128 was used prior to the final classification layer of two nodes.

All the codes for processing were written in MATLAB version 2019a. CNN and BLDA outputs were obtained in the form of a regressed variable. Those continuous scores were summed over trials and the object corresponding to maximum of the summed values was selected as the target of the run. After aggregation of information from the runs, majority voting was used from the respective number of runs for each block and the output object was predicted.

3 Results and Discussion

Classification was done on phase 2 data i.e. four sessions (S4-S7). Individual accuracy for each classifier is computed by averaging all sessions and subjects accuracy, which are 73%, 67% and 63% for BLDA, CNN and RUSBoost respectively. Figure 1 shows the target prediction accuracy averaged over sessions for all subjects, using three classifiers mentioned above. A weighted (weights determined on the basis of previous performances) voting to predict a target from the ensemble of three classifiers rendered an overall accuracy of 73%. The best average accuracy for a subject, obtained from one of the classifiers, enhanced

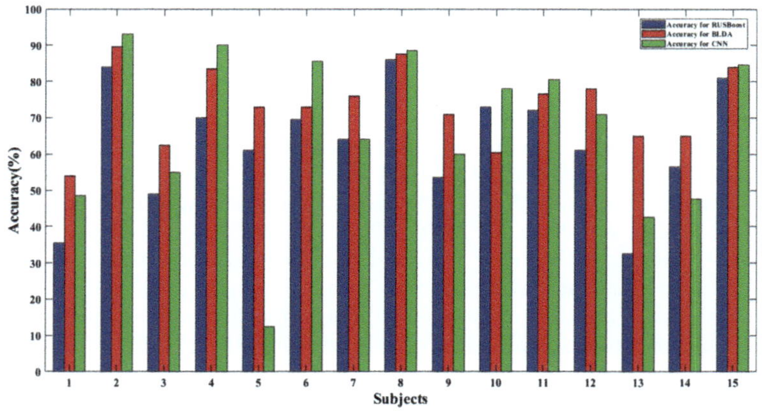

Fig. 1. Target prediction accuracy averaged over four sessions of phase 2 data using three classifiers for all subjects

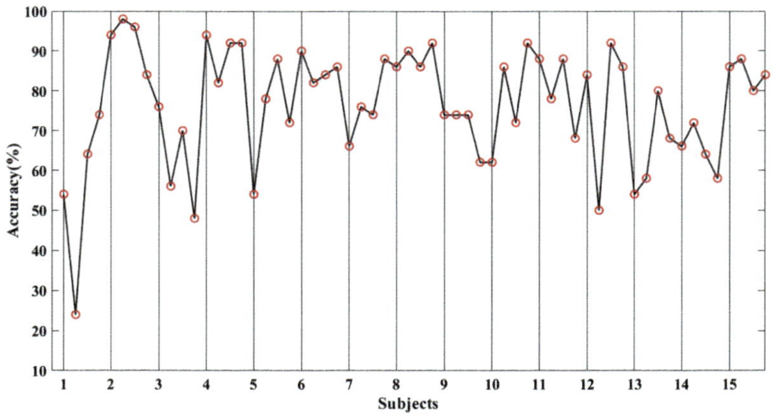

Fig. 2. Session-wise accuracy after selecting the best average performance between CNN and BLDA for each subject (for four sessions)

the overall accuracy to 76%. Figure 2 depicts best accuracy of a subject through sessions, yielded by CNN or BLDA. It is evident from Fig. 1 that, though CNN produced very good accuracy for few subjects, BLDA performed best among all, in terms of accuracy and consistency. It is possible with skewness of the data corrected and deeper structures, CNN may produce improved results. The work has also shown that optimal classifier is subject dependent but it is possible to obtain high accuracy for some of the ASD patients, thereby showing the promise of the BCI approach.

References

1. Wolpaw, J.R., Wolpaw, E.W.: Brain-computer interfaces: something new under the sun. Brain-computer interfaces: Principles and practice, pp. 3–12 (2012)
2. Amaral, C., et al.: A feasibility clinical trial to improve social attention in autistic spectrum disorder (ASD) using a brain computer interface. Front. Neurosci. **12**, 477 (2018)
3. Amaral, C.P., et al.: A novel brain computer interface for classification of social joint attention in autism and comparison of 3 experimental setups: a feasibility study. J. Neurosci. Methods **290**, 105–115 (2017)
4. Hoffmann, U., et al.: An efficient P300-based brain-computer interface for disabled subjects. J. Neurosci. Methods **167**(1), 115–125 (2008)
5. Seiffert, C., et al.: RUSBoost: a hybrid approach to alleviating class imbalance. IEEE Trans. Syst. Man Cybern.-Part A Syst. Hum. **40**(1), 185–197 (2009)
6. Tajbakhsh, N., et al.: Convolutional neural networks for medical image analysis: full training or fine tuning? IEEE Trans. Med. Imaging **35**(5), 1299–1312 (2016)
7. Cecotti, H., Graser, A.: Convolutional neural networks for P300 detection with application to brain-computer interfaces. IEEE Trans. Pattern Anal. Mach. Intell. **33**(3), 433–445 (2010)
8. Lawhern, V.J., et al.: EEGNet: a compact convolutional neural network for EEG-based brain-computer interfaces. J. Neural Eng. **15**(5), 056013 (2018)

Challenges: Fraunhofer Best Portuguese PhD Thesis Competition in Biomedical Engineering

Magnetic Carbon Nanostructures and Study of Their Transport in Microfluidic Devices for Hyperthermia

Raquel O. Rodrigues[1,2(✉)], Rui Lima[3], Helder T. Gomes[4], and Adrián M. T. Silva[1]

[1] Faculdade de Engenharia, Universidade do Porto, Rua Dr. Roberto Frias, 4200-465 Porto, Portugal
adrian@fe.up.pt
[2] Center for MicroElectromechanical Systems (CMEMS-UMinho), University of Minho, Campus de Azurém, 4800-058 Guimarães, Portugal
raquel.rodrigues@dei.uminho.pt
[3] MEtRICs, Mechanical Engineering Department, University of Minho, Campus de Azurém, 4800-058 Guimarães, Portugal
rl@dem.uminho.pt
[4] Centro de Investigação de Montanha (CIMO), cInstituto Politécnico de Bragança, 5300-253 Bragança, Portugal
htgomes@ipb.pt

Abstract. Cancer incidence and mortality are growing worldwide at an alarming pace, emphasizing the urgent need for new strategies to combat this disease. One of the frontiers of cancer research is currently focused on the design of multifunctional magnetic nanoparticles capable to achieve the synergistic cancer theranostics (both diagnosis and therapy). Although the potentiality that these multifunctional nanosystems represents to nanomedicine, cancer treatment and diagnostic, there are still many challenges that must be addressed in a near future before this approach became a reality. The development of efficient multifunctional magnetic nanosystems able to selectively destroy cancer cells in detriment of healthy ones, is one of the main challenges that have damped the spread of this technology into clinical applications. The limited biological and biophysical studies between the biomedical nanosystems and cells/tissues/organs is another challenge that has to be addressed. With these two main challenges in mind, the present Ph.D. work was focused in the development of: (1) *Multifunctional magnetic carbon nanostructures as multifunctional nanosystems for the treatment of cancer,* and (2) *New advanced microfluidic devices capable to give new insights over the developed nanosystems and human cells.*

1 Introduction

In spite of all the progress achieved in the last decade, the translation of multifunctional therapeutic nanosystems for clinical applications, including hyperthermia and drug delivery, has been slow, especially due to the lack of robust preclinical tissue culture

R. O. Rodrigues—Ph.D. scholarship SFRH/BD/97658/2013 granted by FCT.

J. Henriques et al. (Eds.): MEDICON 2019, IFMBE Proceedings 76, pp. 1901–1918, 2020.
https://doi.org/10.1007/978-3-030-31635-8_232

platforms able to mimic the *in vivo* conditions found in the human body and to predict the performance of the developed nanomaterials [1]. Indeed, it should be noticed that nanomaterials designed for theranostic applications have been developed to be applied directly into the bloodstream. So, the understanding of these complex cell-nanoparticle interactions is of utmost importance, especially for the development of multifunctional nanoparticles as drug delivery nanocarriers.

Currently, toxicological screening processes are performed in *in vitro* culture systems, by seeding cells within multi-well dishes. Although their merits, this methodology is not always able to predict the function and effect of drugs *in vivo*, resulting in a large number of failed drug candidates in animal experiences and clinical trials [2]. The main shortcoming of these *in vitro* cellular models, especially for the evaluation of new nanomaterials, is the inability to mimic the complex three-dimensional (3D) *in vivo* microenvironment, wherein the cells and extracellular matrix (ECM) exist in well-organized architectures [2].

Microfluidic devices, which can be defined as the set of technologies handling and processing small fluid volumes (e.g., μL, nL and pL) through microchannel geometries, with dimension of tens to hundreds of micrometers embedded in a chip [3], have several advantages to assess the biophysical properties of cells, when compared to conventional techniques. For instance, this technology allows the use of small sample volumes, integration capability of electronic systems, biocompatibility environment for cell studies, accuracy and fast response. In addition, in the recent decades, a significant advance of microfluidic technologies has allowed the design of microchannel geometries in the sizes of small vascular capillaries, which can mimic the micro-rheological properties of the blood cells in *in vitro* microvascular environment. Furthermore, the advances in computers, optics and digital image processing techniques [4], allow the assessment of haemodynamic phenomena recorded in those *in vitro* microvascular environments. More recently, a new concept of biomicrofluidic devices, called as organ-on-a-chip platforms, has been developed to mimic complex human organ functions at the microscale level. These integrated microfluidic networks, with 3D tissue engineered models, have shown the ability to reduce the discrepancies between the preclinical and clinical trials [2]. Moreover, these 3D microfluidic platforms have several advantages over the *in vivo* animal models, such as lower costs and less time-consuming, end of animal ethical concerns, visualization of the theranostic agents in the target tissues and accuracy to predict human responses.

The present Ph.D. work aimed to join these multidisciplinary approaches, envisaging new insights related to many scientific aspects of the utilization of smart magnetic nanosystems in biomedical applications, namely mechanistic and toxicological phenomena, which are essential to boost the spread of multifunctional nanoparticles as the next generation of therapeutic bio-agents in medicine.

2 Hemocompatibility of Iron Oxide Magnetic Nanoparticles for Theranostic Application: A High Sensitivity Microfluidic Tool

2.1 Development of Superparamagnetic Nanoparticles for Theranostic Applications

The development of superparamagnetic nanoparticles for nanomedicine has attracted much attention in the last decades, especially due to their remarkable physicochemical properties acquired at the nanoscale. In particular, superparamagnetic iron oxide nanoparticles have gained increasing importance in the field due to the possibility to combine diagnosis and therapy of cancer, i.e. as contrast agents for MRI and as nanoheaters in magnetic hyperthermia (MH). MH is considered a promising therapeutic technique for the treatment of cancer since it implements the remarkable nanoscale physicochemical properties of magnetic nanoparticles (MNPs) to generate heat under an alternating magnetic field (AMF). These nanoheaters can be designed to preserve healthy cells while destroying selectively tumoral ones, inherently more sensitive to mild-temperatures changes of 5 to 7 °C above the body temperature [5, 6]. Nevertheless, the poor heating efficiency of most reported MNPs, together with the lack of biocompatibility and hemodynamic studies, have hamper the spread of multifunctional nanoparticles as the next generation of theranostic bio-agents in medicine.

Herein, MNPs where explored as theranostic nanosystems for the simultaneous detection (contrast agent in MRI) and treatment of cancer (MH).

(1) *Synthesis*

Two samples of magnetite were synthesized by alkaline co-precipitation, mixing the iron precursors in the stoichiometric ratio 1:2 of iron (II) and iron (III) salts under vigorous magnetic stirring. Then, the resultant mixture was heated until the desired temperature (30 or 55 °C). At that point, a basic NH_4OH (1 M) solution was added dropwise until pH $\sim$ 9, promoting the co-precipitation of the magnetite nanoparticles. The resulting materials were labelled as MAG30 and MAG55, depending on the heating temperature used (30 or 55 °C, respectively).

Afterwards, hydrophilic MNP ferrofluids were obtained after addition of a solution of alendronic acid (Al, 50 mM, pH 10) into the colloidal suspensions of MAG30 or MAG55, by using a ratio 4:1 (v/v). The resulting MNPs@Al (MAG30@Al and MAG55@Al) were dispersed in water, sonicated for 30 s and stored at room temperature.

(2) *Results and Discussion*

(a) *Size, crystal structure and chemical composition*

The results obtained by transmission electron microscope (TEM) analysis of MAG30 and MAG55 are shown in Fig. 1, revealing that the magnetic core of the synthesized nanoparticles is nearly spherical for both samples. The increase of the synthesis temperature lead to a growth in the average diameter of the magnetic core, from 11.1 (MAG30) to 17.7 nm (MAG55), which is in good agreement with the crystallite sizes obtained by X-ray powder diffraction (XRD) analysis (data not shown).

All the reflection peaks of the XRD patterns of MAG30 and MAG55 were indexed by considering magnetite phase. No reflection peaks were observed as impurities or secondary phases. The cubic cell lattice parameter resulted to be $a \sim 8.37$ Å, in good agreement with the values reported for nanosized magnetite [8].

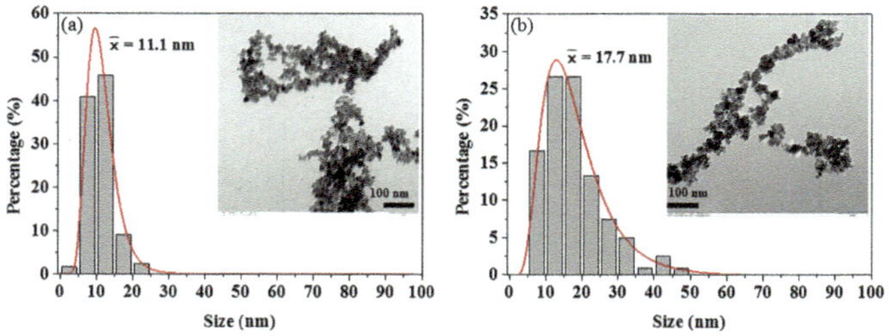

Fig. 1. Histogram and mean diameter of (a) MAG30 and (b) MAG55. Inset shows the corresponding TEM images. Reprinted from [7]. Copyright © 2016, with permission from Springer Nature.

(b) *Magnetic and hyperthermia properties*

The main magnetic parameters of MAG30 and MAG55, namely saturation magnetization, coercivity and remanence, were calculated from the analysis of the magnetization curves as a function of the applied magnetic field obtained for powdered samples (cf. Table 1 and Fig. 2).

Table 1. Magnetic properties of the nanoparticles synthesized by co-precipitation at different synthesis temperatures (30 and 55 °C): Saturation magnetization (M_s), Coercivity (H_c), Saturation remanence (M_r), Specific absorption rate (SAR) and intrinsic loss power (ILP). Reprinted from [7]. Copyright © 2016, with permission from Springer Nature

Sample	M_s (emu g^{-1})	H_c (Oe)	M_r (emu g^{-1})	SAR[a] $\left(W g^{-1} F\right)$	ILP (nHm^2 kg^{-1})
MAG30	56.19 ± 0.02	33.65	3.77	100.18	0.57
MAG55	77.68 ± 0.03	18.33	1.94	565.68	3.23

[a]Specific absorption rate (SAR) calculated for $H = 15.95$ kA m^{-1}, $f = 688$ kHz, C = 0.40 g $_{Fe3O4}$ L^{-1}

Both MAG30 and MAG55 hysteresis loops show a superparamagnetic-like behavior. MAG30 showed a saturation magnetization, M_s, of 56 emu g^{-1}, whereas the increase of the synthesis temperature in MAG55 led to an increase of the particle size that subsequently resulted in a higher M_s. On the other hand, the zero-field-cooling – field-cooling (ZFC–FC) magnetization curves shown in Fig. 2(b) indicate a shift of the blocking temperature (T_B), defined as the temperature above which the MNPs show superparamagnetic behavior, from −98.2 to 26.9 °C for MAG30 and MAG55, respectively. Hence, the confirmation of a superparamagnetic state for both MAG30 and MAG55 at room temperature make these samples suitable for MRI and hyperthermia applications.

(c) *Magnetic resonance relaxivity studies*

The contrast enhancement efficiency of MNPs@Al was evaluated in water through relaxivity measurements by plotting the relaxation rates ($R_1 = 1/T_1$ and $R_2 = 1/T_2$) of water protons in the presence of the nanoparticles, against the Fe content.

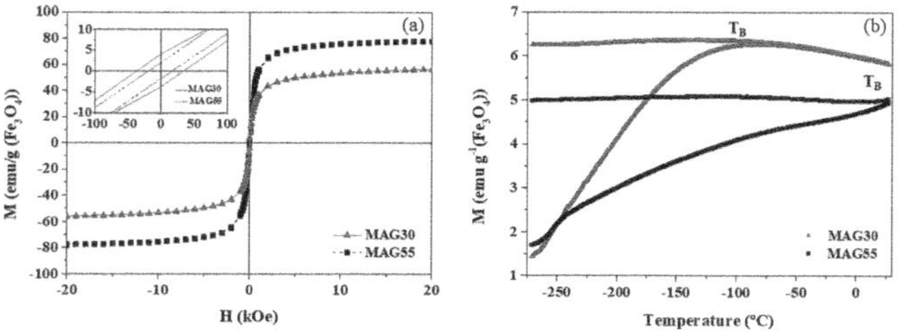

Fig. 2. Magnetic characterization of MAG30 and MAG55. (a) Hysteresis loops at room temperature up to ±20 kOe; the inset is a zoom in the low-field region; (b) ZFC-FC magnetization curves measured at 50 Oe. Reprinted from [7]. Copyright © 2016, with permission from Springer Nature.

Both MAG30@Al and MAG55@Al ferrofluidic samples showed a far superior enhanced r_2 values than currently available intravenous iron oxide nanoparticle contrast agents, i.e. 248.17 mM^{-1} s^{-1} for MAG55@Al and 226.08 mM^{-1} s^{-1} for sample MAG30@Al, indicating that both samples are promising candidates for high-efficiency T_2-weighted MR imaging.

2.2 A New Microfluidic Tool to Assess Hemocompatibility of Theranostic Nanosystems

In this study, and for the first time, a high-sensitivity microfluidic tool capable to assess the deformability of blood cells was used to evaluate the impact of nanoparticles on the human RBCs. This methodology comprises a microfluidic device having a hyperbolic-shaped contraction microchannel and a high-speed video microscopy system. This combined system enables the recording, in real time, of the biomechanical microfluidic phenomena, which can be further used to analyze valuable toxicological data.

In this type of microfluidic devices (Fig. 3), the velocity increases almost linearly and the strain rate is maintained approximately constant, even with the increment of the shear rate [9].

As a result, the technique allows the deformation of the blood cells along the microchannel, causing the RBC stretching by the larger fluid shear stress, without tumbling and rotational motions of the cells, providing a more accurate way to determine the deformation index (DI) of the blood cells, mimicking *in vitro* microvascular environments.

Figure 4 shows a healthy RBC flowing along the axial position, x, of the microchannel with a hyperbolic-shaped contraction followed by a sudden expansion, which was subdivided in four sections to assess the DI evolution of the RBCs along the microchannel.

RBCs are important for scientific and clinical purposes because they can be used as an indicative of many pathological conditions. It has been shown that changes in their normal DI are correlated to blood disorders. Thus, DIs of RBCs in contact with different concentration of MNPs at different incubation time where assessed to infer the hemocompatibility of the developed nanosystems. The DIs of the RBCs (control versus MNPs) were measured flowing along the centreline of the microchannel that was subdivided in four sections (cf. Fig. 4). The results are displayed in Fig. 5.

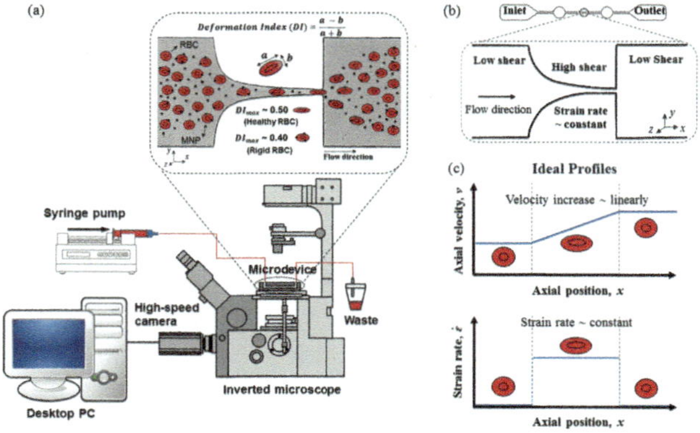

Fig. 3. Microfluidic studies for the hemocompatibility of the RBCs in contact with MNPs, (a) experimental setup; (b) microchannel device geometry with hyperbolic channel; (c) graphical representation of the fluid-induced conditions profiles that occur in the hyperbolic channel. Reprinted from [7]. Copyright © 2016, with permission from Springer Nature.

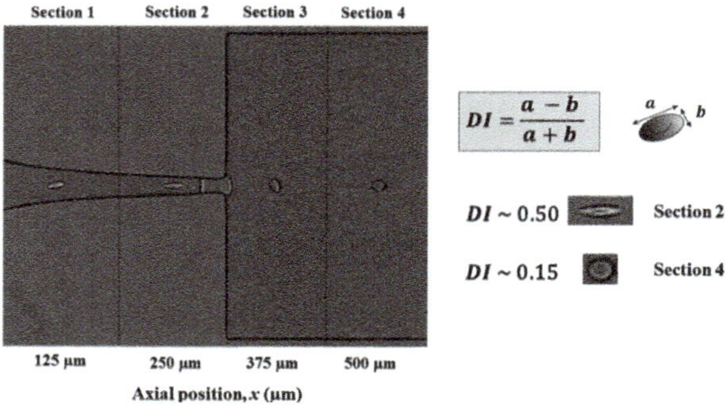

Fig. 4. Representation of a healthy RBC flowing along the axial position of a microchannel with a hyperbolic-shaped contraction followed by a sudden expansion. Reprinted from [7]. Copyright © 2016, with permission from Springer Nature.

Overall, the DI results for the RBCs samples in contact with MNPs show that the incubation time influences directly the deformability of those blood cells. Hence, the DI results allow to conclude that the increase of the incubation time between RBCs and

MNPs, lead to the decrease in DI, or in other words, to the increase of rigidity of those cells. This observation is even more obvious for the sample with the higher content of MNPs and exposure time (34.8 μg $_{Fe3O4}$ mL^{-1}, 60 min), where the DI decreased to 0.34 ± 0.02, which is a significant difference when compared with the control DI (0.45 ± 0.02).

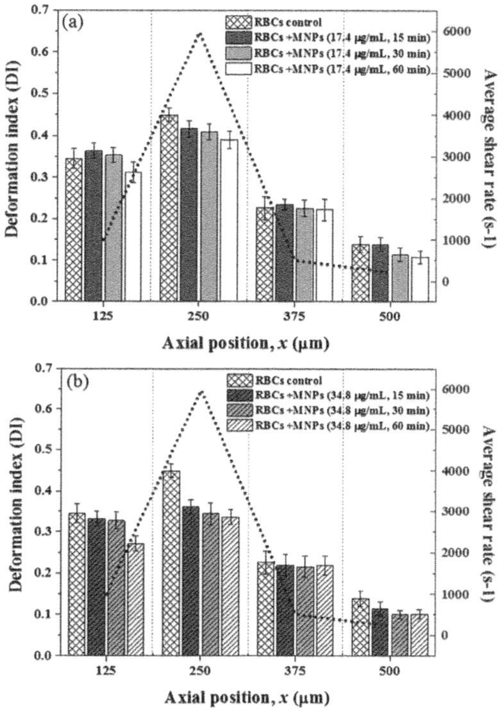

Fig. 5. DI of blood cells flowing along the axial position of the hyperbolic channel during 15, 30 and 60 min: (a) MNPs at a final concentration of 17.4 μg $_{Fe3O4}$ mL^{-1}; (b) MNPs at a final concentration of 34.8 μg $_{Fe3O4}$ mL^{-1}. Error bars show a 95% confidence interval (n = 56 blood cells). Reprinted from [7]. Copyright © 2016, with permission from Springer Nature.

This work was the first experimental test reported in the literature that corroborate the numerical work performed by Curtis et al. [10], which refers to the uptake of the surrounding MNPs by the RBC membranes, as the main reason for the increasing rigidity observed in the RBCs. In this recently numerical study, it was also predicted a dependency on the MNP size, shape and composition with the RBC membrane interaction.

This study was published in the *Journal of Nanoparticle Research* [7].

3 Graphene-Based Magnetic Nanoparticles for the Treatment of Cancer

The frontiers of cancer research are currently focused on the design of multifunctional magnetic nanoparticles capable to achieve the synergistic cancer treatment, by combining the heat effect induced by hyperthermia and their unique drug delivery properties [11]. For this purpose, many magnetic nanoparticle-based drug delivery systems have been developed in the last decade [12]. However, the major bottleneck for their combined clinical achievement has been the low drug loading capacity and poor controlled-drug release of the developed magnetic nanoparticles [13], as well as the low heating efficiency of the composite nanosystems in MH.

Herein, the synthesis and optimization of graphene-based yolk-shell magnetic nanoparticles for nanomedicine are described, as well as its efficiency as multifunctional nanosystems.

3.1 Synthesis

The synthesis procedure of graphene-based yolk-shell magnetic nanoparticles (GYSMNPs) was divided in two main steps: (i) *synthesis of the superparamagnetic core*, and (ii) *formation of the graphene-based yolk-shell architecture*, as shown in Fig. 6.

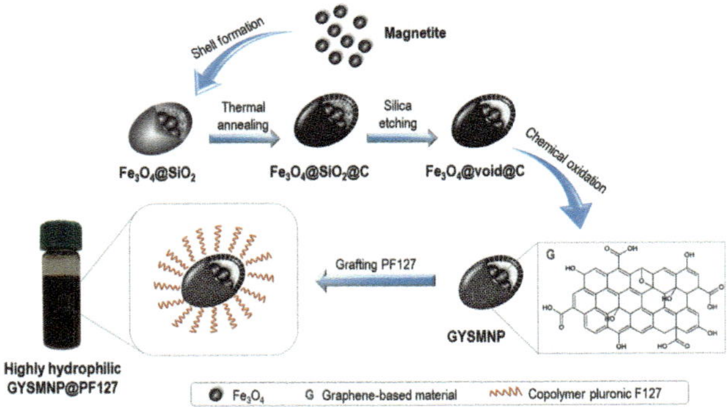

Fig. 6. Schematic overview of the steps involved in the development and grafting of the hydrophilic graphene-based yolk-shell magnetic nanoparticles (GYSMNP@PF127).

In the first step, the superparamagnetic magnetite (Fe_3O_4) core, with mean diameter of 18 nm, was synthesized through the procedure detailed in Sect. 2A, at 55 °C. Afterwards, in the second step, the graphene-based yolk-shell architecture was achieved via one-pot strategy of hydrolysis and polymerization of the precursors resorcinol, formaldehyde and TEOS. Sample was labelled as GYSMNP@PF127 after functionalization with pluronic F127 (PF127).

3.2 Results and Discussion

It is important to note that for nanomedicine, the proposed nanosystem has to show excellent colloidal stabilization in the presence of electrolyte solutions, optimal average hydrodynamic dimension (D_H), low hemotoxicity and good biocompatibility. In this context, the GYSMNP@PF127 nanosystem was further characterized taking into account those properties. D_H of GYSMNP@PF127 was found to be 180 nm, with polydispersion index (PDI) of 0.14 (data not shown). Due to the EPR effect, nanoparticles with size range of 10–200 nm are ascribed to be preferentially accumulated into tumor rather than in healthy tissues [14].

D_H of GYSMNP@PF127 shows an optimal dimension for suitable intravenous administration and EPR effect with prolonged blood circulation.

Figure 7 shows the Fourier transform infrared (FTIR) spectra of the samples collected at different stages of the preparation process.

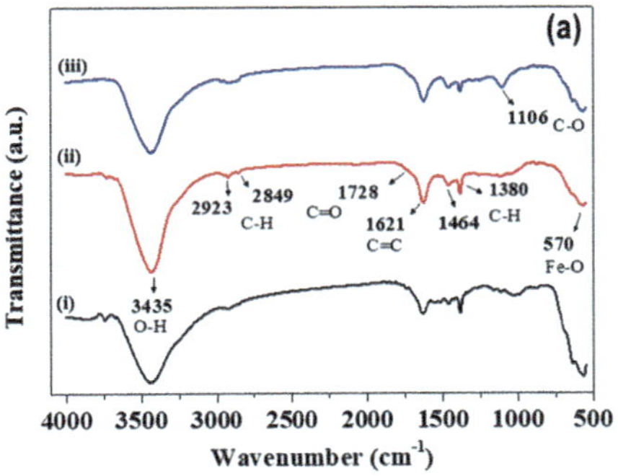

Fig. 7. FTIR spectra of (i) as-synthesized GYSMNP, (ii) GYSMNP after activation with nitric acid, (iii) GYSMNP after grafting with pluronic F-127 (GYSMNP@PF127).

FTIR characterization of GYSMNP@PF127 shows a variety of oxygen-containing functional groups that are favorable for the encapsulation, transport and release of guest molecules, such as chemotherapeutic drugs for drug delivery applications.

(1) *DOX loading studies*

Doxorubicin (DOX) is a cationic chemotherapeutic drug and one of the most commonly used against cancer. However, it presents lethal side effects on healthy cells too. Targeted drug delivery can be an effective method to overcome this issue by delivering the cargo specifically into tumors after being triggered by its abnormal acidic microenvironment. Impressively, the drug loading capacity of GYSMNP@PF127 reached an outstanding 910 $\mu g_{DOX}\ mg^{-1}{}_{GYSMNP@PF127}$, which represents a 91% of drug loading efficiency (1:1, w/w). This ability can be ascribed to the π-π stacking interactions between the aromatic rings of DOX and the carbonaceous structure of graphene-based nanoshells, as well as to some surface groups. Besides the π-π stacking, the high DOX loading on GYSMNP@PF127 is also a consequence of the electrostatic interactions and the mesoporous structure of the graphene-based shell.

(2) *Dual pH- and temperature-dependent drug release*

It is well-known that tumor microenvironment presents a weak acidic pH value, around 6.0–7.2, at the tumor tissue due to the Warburg effect, which dramatically decrease to severe acidic pH values, around 4.5–5.5, in tumor endosome/lysosome environment [15]. The pH differences between the tumor microenvironment and normal tissues, pH around 7.4, motivate the development of pH-dependent/responsive nanocarriers that can be specifically designed as smart drug delivery systems triggered by acidic pH values. Therefore, the cumulative pH-dependent DOX release of GYSMNP@PF127-DOX was studied under different pH conditions (7.4, 6.0 and 4.5) either at 37 °C (physiological temperature) or 45 °C (hyperthermia temperature), cf. Fig. 8.

Figure 8 reveals that GYSMNP@PF127 has a pronounced pH- and temperature-dependency drug delivery behavior. The drug release at these conditions takes place in two stages: (I) a rapid DOX release in the first 8 h, followed by (II) a slower rate release till 48 h.

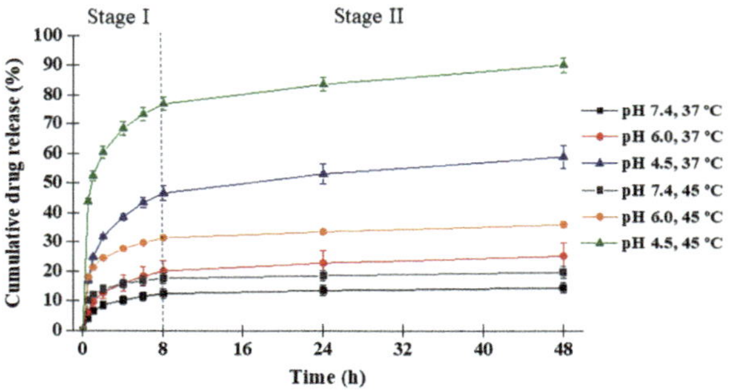

Fig. 8. pH and thermo-responsive release of doxorubicin (DOX) from drug loaded GYSMNP@PF127 under different pH values (7.4, 6.0 and 4.5) and temperatures (37 and 45 °C). Standard deviation of triplicate drug release tests (n = 3).

Once again, the main reasons behind this promising pH/temperature-responsive drug release behavior of the graphene-based material comes from the π-π stacking between the graphene-based nanostructures and the aromatic DOX molecules, which can be easily disrupted under a mild acidic environment [16], but also due to the increased solubility of DOX caused by the protonation process [17]. This will lead to a reduction of the electrostatic attraction between GYSMNP@PF127 and DOX protonated molecules, increasing the progressive drug release.

Overall, graphene-based magnetic nanocarriers shows a great ability to be applied as efficient drug encapsulation and delivery systems, with an especial good pH sensitivity for an intelligent and on-demand drug release on acidic tumor microenvironments.

(3) *Dual pH- and temperature-dependent drug release*

The response of GYSMNP@PF127-DOX as dual pH- and temperature-sensitive drug release nanocarrier was assessed under AMF (f = 340 kHz and H = 21.0 kA m^{-1}) during 30 min, Fig. 9.

Mild hyperthermia temperature (40–43 °C) was reached between 10 and 13 min and maintained until the end of the hyperthermia experiment (30 min) (Fig. 9). The results revealed that after 30 min under AMF, just 6.8% of DOX is released at pH 7.4, while 45.6% at pH 4.5. Therefore, these results suggest that under an AMF, the local heat generated on the GYSMNP@PF127 surface can help the selective diffusion of loaded DOX into the acidic tumor microenvironment through MH. Thus, GYSMNP@PF127 material show high promise to act as dual endogenous pH- and exogenous AMF/temperature-responsive drug nanocarrier for controlled drug delivery as well.

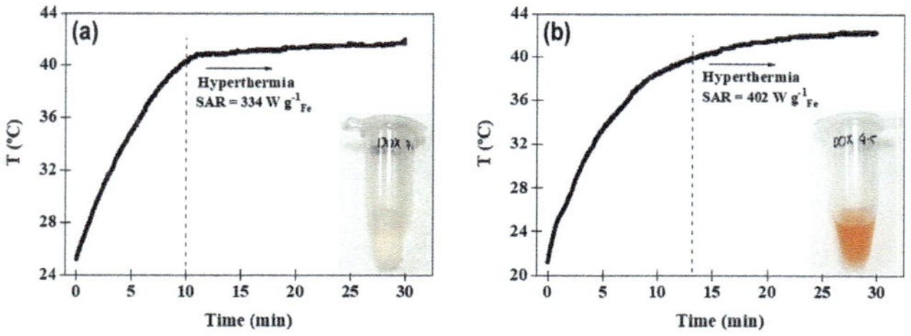

Fig. 9. Magnetic-induced thermal response curve of GYSMNP@PF127-DOX for dual pH- and thermal-responsive drug delivery: (a) dispersed in phosphate buffer at 7.4 to mimic physiological microenvironment, and (b) dispersed in phosphate buffer at 4.5 to mimic tumour microenvironment. Inset shows the drug-containing supernatant resulting from each hyperthermia experiment.

(4) *Cell cytotoxicity assays and cellular uptake of drug nanocarriers*

For a better understanding of the therapeutic effect of the DOX released by the GYSMNP@PF127 material over the tumor cells, HepG2 cells were incubated with the drug nanocarrier for 2, 4 and 6 h (Fig. 10). Cells were observed using confocal microscopy images obtained with Zeiss Laser Confocal Scanning Microscope.

It can be observed (Fig. 10) that the red fluorescence intensity from DOX increases along the incubation time between the nanocarriers and the tumor cells. This observation reveals that the DOX is continuously released from GYSMNP@PF127-DOX nanocarriers. After 6 h, the cell nuclei exhibit a strong DOX red fluorescence intensity, suggesting that a high content of DOX is released into the cell cytoplasm and reaches the nucleus after nanoparticle internationalization by endocytosis. The cellular uptake of drug nanocarriers is not only very important for enhancing the drug delivery efficiency, but it is also beneficial for magnetic hyperthermia therapy, allowing a combinatorial thermo-chemotherapeutic effect.

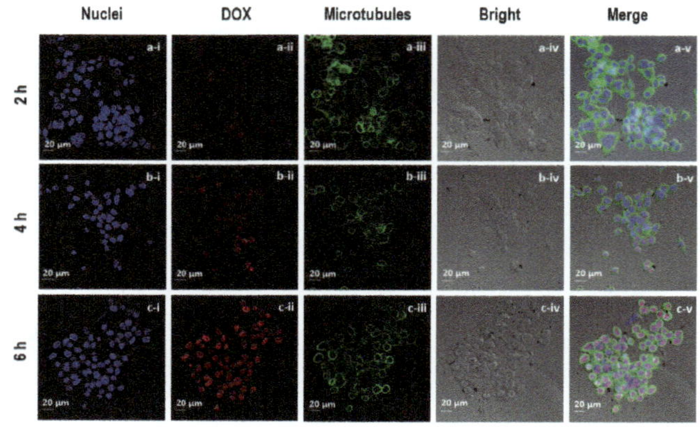

Fig. 10. Confocal microscopy images of HepG2 cells incubated with GYSMNP@PF127-DOX at a concentration of 5 μg mL^{-1} during 2 h (a), 4 h (b) and 6 h (c). (i) Cell nuclei stained with Hoechst and excited at 405 nm; (ii) DOX released in HepG2 cells and excited at 561 nm; (iii) Cell microtubules immunolabeled with the monoclonal anti-tubulin antibody (B512) combined with a secondary goat anti-mouse IgG antibody conjugated with Alexa Fluor 488 excited at 488 nm; (iv) HepG2 cells using bright contrast; (v) Cell merged images.

This combined strategy represents an important step forward in the fight against cancer and highlight this new generation of responsive graphene-based nanocarriers as suitable therapeutic actuators on it.

The compilation of this study was published in the journals *C – Journal of Carbon Research* [18], and *Material Science Engineering C* [19].

4 Dual-Organ-on-a-Chip Platform Developed for Drug Screening Studies in Nanomedicine

Organ-on-a-chip platforms are novel microfluidic tools that mimic complex human organ functions at the microscale level and predict with high accuracy human responses to drugs and theranostic agents. Herein, a novel dual organ-on-a-chip platform integrated with multiplexed electrochemical analysis is described to evaluate with high accuracy the performance of new drug nanocarriers, such as the developed GbMNP, as intelligent and on-demand drug delivery system with pH-dependent controlled release.

This work was developed under the framework of a Fulbright Research Grant 2017, at the Harvard-MIT Division of Health Sciences and Technology (Cambridge, Massachusetts, USA), headed by Prof. Ali Khademhosseini and supervised by Dr. Su Ryon Shin. This bioengineering leading group has a vast experience in developing organ-on-a-chip systems that aims to mimic human response to various chemicals.

4.1 Dual Organ-on-a-Chip

A dual organ-on-a-chip platform, compressing cardiac and breast cancer organoids, with the aim to evaluate the effect of chemotherapy over those organs, in continuous and during several days is shown in Fig. 11. Due to the versatility of the proposed 3D tissue/organ platform, the evaluation of novel drug nanocarriers, such as the GYSMNP@PF127-DOX developed in Sect. 3, is explored in contrast with free chemotherapy.

The dual organ-on-a-chip platform was obtained by embedded cardiac iPSCs (healthy cardiac model) and SkBr3 spheroids (breast cancer model) inside a gelation methacryloyl (GelMA) hydrogel photo-cross linked, generating the human cardiac- and breast cancer-organ-on-a-chip bioreactor (Fig. 12).

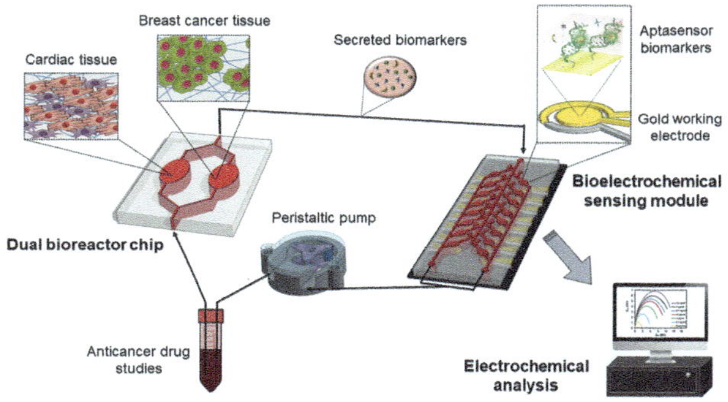

Fig. 11. Schematic representation of the feedback loop of dual organ platform (comprising breast cancer and cardiac organoids) integrated with multiplexed electrochemical analysis for the chemotherapy study effect of doxorubicin using the free or encapsulated drug.

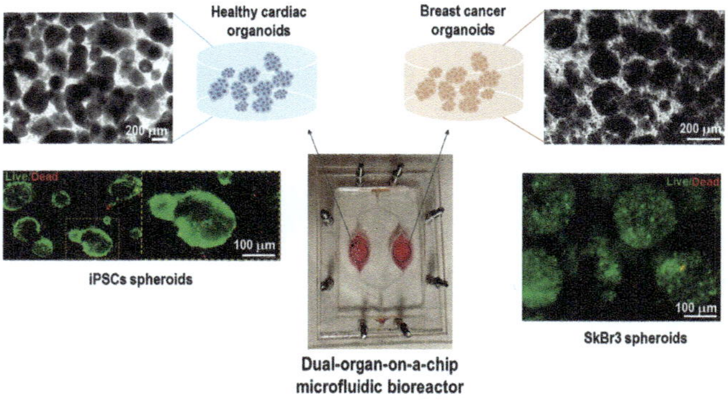

Fig. 12. Schematic representation of the dual organ-on-a-chip microfluidic bioreactor developed to monitor the effect of the chemotherapeutic drug DOX (free or encapsulated in nanocarriers) over the breast cancer and cardiac healthy organoids. Life/dead confocal images from iPSCs and SkBr3 spheroids showing the high viability of these organoids inside the GelMA hydrogel after UV photo-cross linking. Live cells stained green and dead cells stained red.

The development of the biomimetic dual organ-on-a-chip system allows the simultaneous monitoring and investigation of the effect of chemotherapeutic drugs over aggressive breast cancer (such SkBr3 cell line) and healthy cardiac tissues. This non-invasive continuous monitoring of multi-organ systems offers the possibility to obtain new insights over the cellular metabolic activity, as well as to develop new drug models or adjust chemotherapeutic doses in a personalized manner. To date, and to the best of our knowledge, this was the first study developed to determine the effect of anticancer drugs over multi-organs-on-a-chip platforms using biomolecular sensing (being prepared to publication).

4.2 A Label-Free Electrochemical (EC) Sensing Method

The immobilization process of the selected aptamers in the biosensors was performed as described elsewhere [20]. Two cardiac biomarkers (CKMB and Troponin T), and one breast cancer biomarker (HER2) were exploited.

Aptamers, which are short single-stranded oligonucleotides capable of binding molecules with high affinity and specificity, were used in this study to target specific antigens (troponin-T, CKMB and HER2), in a highly reproducible manner (Fig. 13).

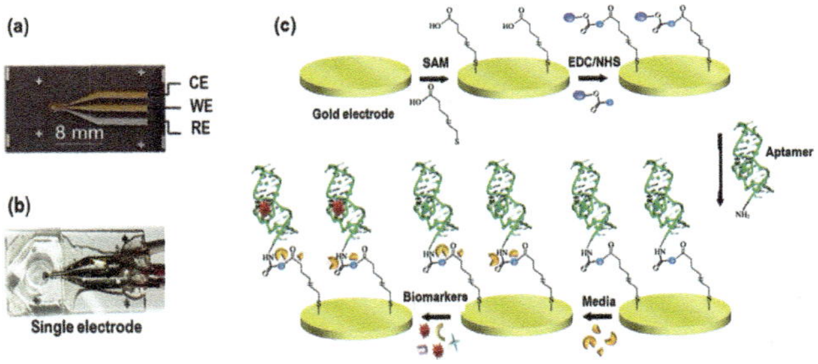

Fig. 13. (a) Image of the microfabricated electrode set containing a reference electrode (RE), a counter electrode (CE) and a working electrode (WE); (b) image of a single electrode as the one used in this study; (c) schematic diagram of the immobilization steps of the aptamers into the microelectrode.

4.3 Results and Discussion

Samples collected from drug-induced experiments along 5 days were measured in individual aptamer-based biosensor for each biomarker by electrochemical impedance spectroscopy (Fig. 14(b–d)). As expected, free DOX shows to induce a dramatic toxicity over the breast cancer tissue, as revealed by the decrease of HER2 biomarkers along the 5 days of chemotherapeutic treatment, but also reveals cardiac injury to the heart organoid with increasing detection of both cardiac diseases' biomarkers (i.e., Troponin-T and CKMB). On the other hand, the encapsulated DOX released by the nanocarrier GYSMNP@PF127-DOX shows a more protective behavior over the healthy cardiac organoids in comparison with the free DOX, and at the same time maintaining the severe toxicity over the breast cancer spheroids. These exciting results attest the previous screening results detailed in Sect. 3, where the developed graphene-based magnetic nanoparticles were tested as smart and on-demand drug nanocarriers triggered by the abnormal acidic pH found in tumor microenvironment.

In addition to the aptamer-based biosensing analysis, the cell viability of the cardiomyocytes and breast cancer organoids after the chemotherapeutic treatment (day 5) were analyzed by live/DOX assay (cf. Fig. 15).

The confocal images subscribe the results obtained from the aptamer-based biosensing analysis, showing that free DOX is delivered in the same proportion on both organoids (healthy cardiac tissues and breast cancer), which cause the damage of cardiac tissues and the release of the cardiac biomarkers (Troponin-T and CKMB). On the other hand, DOX encapsulated in the graphene-based nanocarriers shows high-target delivery over the breast cancer organoids compared to cardiac spheroids, as ascribed by the low amount of cardiac biomarkers detected with the biosensor system. The results confirm the potentiality of the developed GYSMNP@PF127 material as smart and on-demand drug nanocarriers for the treatment of cancer, and the extraordinary ability to use the aptamer-based electrochemical biosensor to detect trace amounts (pg mL^{-1}) of several biomarkers secreted from multi-organ systems, which are undetectable by using conventional methods, such as ELISA (ng mL^{-1}).

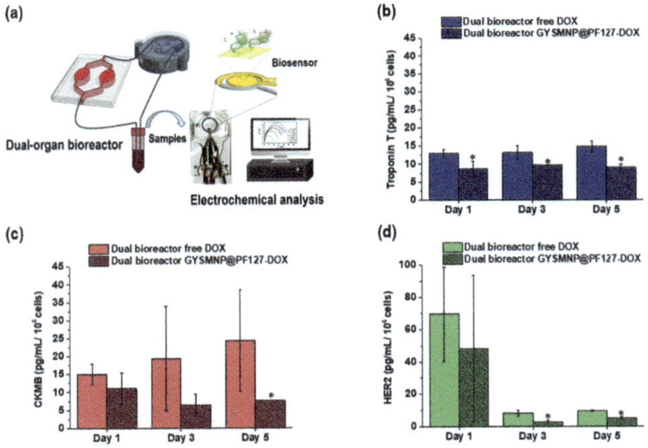

Fig. 14. (a) Schematic representation of the feedback loop system used during the drug induced experiments on dual organ-on-a-chip microfluidic bioreactor along 5 days. Electrochemical sensing of selected biomarkers released from dual organ-on-a-chip bioreactor: (b) cardiac biomarker Troponin-T; (c) cardiac biomarker CKMB; (d) breast cancer biomarker HER2. Error bars show standard deviation of individual single microsensors (n = 3) and the asterisks (*) represent statistical significance in comparison with free DOX assay for $p < 0.05$ determined by Student's t-test.

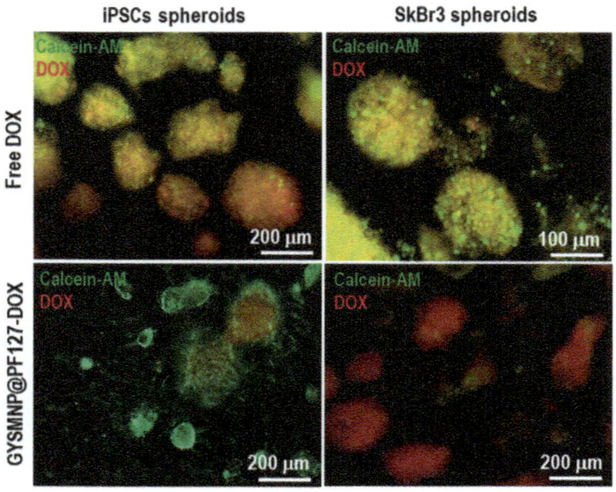

Fig. 15. Cell viability assays of cardiomyocytes (iPSCs) and breast cancer (SkBr3) spheroids obtained after 5 days of chemotherapeutic treatment (free or encapsulated DOX) in dual organ-on-a-chip bioreactors. Live cells stained with green Calcein-AM dye and red cells showing DOX uptake.

5 Conclusion and Future Perspective

The main objectives of this work were the development of carbon-based magnetic nanoparticles suitable for nanomedicine and their microfluidic study using new bio-platforms, aiming new insights over the complex biophysical and toxicological interaction of these nanosystems with human cells (i.e. blood, healthy and tumor cells). The successful achievement of these objectives is involved in complex multidisciplinary approaches, namely chemistry, physics, materials science, biology, bioengineering and biomechanics, which were accomplished by subdividing these objectives in subsequent tasks.

Overall, the developed carbon-based nanosystems, as well as the proposed microfluidic platforms, described in this work, allows to contribute to a better understanding of the biophysical and biological interaction of new biomedical nanosystems designed for nanomedicine.

It was shown that the evolution of smart and multifunctional nanomaterials in nanomedicine could be extremely beneficiated by the parallel evolution of new microfluidic platform model systems to study their safety, bio-distribution and efficacy.

On this demand, organ-on-a-chip platforms can have a significant impact in this field. There are still many challenges to be addressed on the development and optimization of these advanced 3D microfluidic platforms, but the potential and possibilities are far more enthusiastic. Therefore, and as predicted by Professor Langer and Khademhosseini, pioneers in drug delivery and tissue engineering [21]: "with sufficient time and research the promise of nanotechnology based medicine may become a reality".

Acknowledgment. The successful accomplishment of the multidisciplinary tasks considered in this Ph.D. work, was supported by important collaborations that were strengthened at different stages in the frame of this Ph.D. project, namely INL - International Iberian Nanotechnology Laboratory (Braga, Portugal); CeRiCol - Centro Ricerche Colorobbia Consulting (Vinci, Italy); CIMO – Centro de Investigação da Montanha (Bragança, Portugal) and Harvard-MIT Division of Health Sciences and Technology (Cambridge, USA).

R.O. Rodrigues acknowledge the Ph.D. scholarship SFRH/BD/97658/2013 granted by Fundação para a Ciência e a Tecnologia (FCT), as well as a Fulbright Research Grant 2017, granted by Fulbright Portugal.

References

1. Bhise, N.S., Ribas, J., Manoharan, V., Zhang, Y.S., Polini, A., Massa, S., Dokmeci, M.R., Khademhosseini, A.: Organ-on-a-chip platforms for studying drug delivery systems. J. Control. Release **190**, 82–93 (2014). Official Journal of the Controlled Release Society
2. Zhang, Y.S., Zhang, Y.N., Zhang, W.: Cancer-on-a-chip systems at the frontier of nanomedicine. Drug Discov. Today **22**, 1392–1399 (2017)
3. Halldorsson, S., Lucumi, E., Gómez-Sjöberg, R., Fleming, R.M.T.: Advantages and challenges of microfluidic cell culture in polydimethylsiloxane devices. Biosens. Bioelectron. **63**, 218–231 (2015)
4. Lima, R., Wada, S., Tanaka, S., Takeda, M., Ishikawa, T., Tsubota, K., Imai, Y., Yamaguchi, T.: In vitro blood flow in a rectangular PDMS microchannel: experimental observations using a confocal micro-PIV system. Biomed. Microdevice **10**, 153–167 (2008)
5. Chicheł, A., Skowronek, J., Kubaszewska, M., Kanikowski, M.: Hyperthermia – description of a method and a review of clinical applications. Rep. Prac. Oncol. Radiother. **12**, 267–275 (2007)
6. Jordan, A., Scholz, R., Wust, P., Fähling, H., Roland, F.: Magnetic fluid hyperthermia (MFH): cancer treatment with AC magnetic field induced excitation of biocompatible superparamagnetic nanoparticles. J. Magn. Magn. Mater. **201**, 413–419 (1999)
7. Rodrigues, R.O., Bañobre-López, M., Gallo, J., Tavares, P.B., Silva, A.M.T., Lima, R., Gomes, H.T.: Haemocompatibility of iron oxide nanoparticles synthesized for theranostic applications: a high-sensitivity microfluidic tool. J. Nanopart. Res. **18**, 1–17 (2016)
8. Aphesteguy, J.C., Kurlyandskaya, G.V., de Celis, J.P., Safronov, A.P., Schegoleva, N.N.: Magnetite nanoparticles prepared by co-precipitation method in different conditions. Mater. Chem. Phys. **161**, 243–249 (2015)
9. Oliveira, M.S.N., Alves, M.A., Pinho, F.T., McKinley, G.H.: Viscous flow through microfabricated hyperbolic contractions. Exp. Fluids **43**, 437–451 (2007)
10. Curtis, E.M., Bahrami, A.H., Weikl, T.R., Hall, C.K.: Modeling nanoparticle wrapping or translocation in bilayer membranes. Nanoscale **7**, 14505–14514 (2015)
11. Sasikala, A.R.K., Thomas, R.G., Unnithan, A.R., Saravanakumar, B., Jeong, Y.Y., Park, C. H., Kim, C.S.: Multifunctional nanocarpets for cancer theranostics: remotely controlled graphene nanoheaters for thermo-chemosensitisation and magnetic resonance imaging. Sci. Rep. **6**, 20543 (2016)
12. Li, Z., Ye, E., David, Lakshminarayanan, R., Loh, X.J.: Recent advances of using hybrid nanocarriers in remotely controlled therapeutic delivery. Small **12**, 4782–4806 (2016)
13. Chen, D., Dougherty, C.A., Zhu, K., Hong, H.: Theranostic applications of carbon nanomaterials in cancer: Focus on imaging and cargo delivery. J. Control. Release **210**, 230–245 (2015)

14. Hervault, A., Dunn, A.E., Lim, M., Boyer, C., Mott, D., Maenosono, S., Thanh, N.T.K.: Doxorubicin loaded dual pH- and thermo-responsive magnetic nanocarrier for combined magnetic hyperthermia and targeted controlled drug delivery applications. Nanoscale **8**, 12152–12161 (2016)
15. Liu, J., Luo, Z., Zhang, J., Luo, T., Zhou, J., Zhao, X., Cai, K.: Hollow mesoporous silica nanoparticles facilitated drug delivery via cascade pH stimuli in tumor microenvironment for tumor therapy. Biomaterials **83**, 51–65 (2016)
16. Mohapatra, S., Rout, S.R., Das, R.K., Nayak, S., Ghosh, S.K.: Highly hydrophilic luminescent magnetic mesoporous carbon nanospheres for controlled release of anticancer drug and multimodal imaging. Langmuir **32**, 1611–1620 (2016)
17. Li, S., Zheng, J., Chen, D., Wu, Y., Zhang, W., Zheng, F., Cao, J., Ma, H., Liu, Y.: Yolk-shell hybrid nanoparticles with magnetic and pH-sensitive properties for controlled anticancer drug delivery. Nanoscale **5**, 11718–11724 (2013)
18. Rodrigues, R.O., Baldi, G., Doumett, S., Gallo, J., Bañobre-López, M., Dražić, G., Calhelha, R.C., Ferreira, I.C.F.R., Lima, R., Silva, A.M.T., Gomes, H.T.: A tailor-made protocol to synthesize yolk-shell graphene-based magnetic nanoparticles for nanomedicine. C **4**, 55 (2018)
19. Rodrigues, R.O., Baldi, G., Doumett, S., Garcia-Hevia, L., Gallo, J., Bañobre-López, M., Dražić, G., Calhelha, R.C., Ferreira, I.C.F.R., Lima, R., Gomes, H.T., Silva, A.M.T.: Multifunctional graphene-based magnetic nanocarriers for combined hyperthermia and dual stimuli-responsive drug delivery. Mater. Sci. Eng., C **93**, 206–217 (2018)
20. Shin, S.R., Zhang, Y.S., Kim, D.-J., Manbohi, A., Avci, H., Silvestri, A., Aleman, J., Hu, N., Kilic, T., Keung, W., Righi, M., Assawes, P., Alhadrami, H.A., Li, R.A., Dokmeci, M.R., Khademhosseini, A.: Aptamer-based microfluidic electrochemical biosensor for monitoring cell-secreted trace cardiac biomarkers. Anal. Chem. **88**, 10019–10027 (2016)
21. Khademhosseini, A., Langer, R.: Nanobiotechnology: drug delivery and tissue engineering. Chem. Eng. Prog. **102**, 38–42 (2006)

TESSEE – Tool for Early Stem Cells Economic Evaluation

Cátia Bandeiras[1(✉)], Joaquim Manuel Sampaio Cabral[1], Stan Neil Finkelstein[2], and Frederico Castelo Ferreira[1]

[1] Institute of Bioengineering and Biosciences, Department of Bioengineering, Instituto Superior Técnico, Universidade de Lisboa, Lisbon, Portugal
catia.bandeiras@tecnico.ulisboa.pt, {joaquim.cabral, frederico.ferreira}@ist.utl.pt

[2] Institute for Data, Systems and Society, Massachusetts Institute of Technology, Cambridge, MA 02139, USA
snfinkel@mit.edu

Abstract. Stem cell therapies are promising for diverse indications. However, there are manufacturing and reimbursement challenges that must be addressed toward widespread adoption. Relying on bioprocess and/or health economics models, early health technology assessment (eHTA) has provided insights on manufacturing innovations towards reducing the manufacturing cost of goods (CoG), and the long-term value that prospective stem cell therapies must provide to secure reimbursement. These economics models have focused on commercial tools or proprietary code. In order to increase awareness to the usefulness of eHTA in stem cell engineering, this thesis presents a new open-source tool, TESSEE (Tool for Early Stem Cells Economic Evaluation). Three relevant case studies for mesenchymal stem/stromal cell (MSC) and pluripotent stem cell (PSC) based therapies are presented as applications of TESSEE for decision-making in the implementation of stem cell engineering innovations. A study on the choice of a culture media supplement for autologous MSC therapy manufacturing determined that human platelet lysate (hPL) reduces CoG, in comparison to fetal bovine serum (FBS), for 97% of donors. An expansion system focused case study showed that the implementation of a new vertical wheel reactor (VWR) for the microcarrier-based culture of MSC increases the number of cells per batch and reduces costs of goods per dose by up to 48% from typical two-dimensional flasks for expansion. eHTA of devices containing PSC-derived beta cells for treatment of type 1 diabetes patients showed that the transplantation of the devices could be very effective at improving the quality of life. However, a price reduction, accompanied by reduction of manufacturing costs, is required to achieve widespread cost-effectiveness. This thesis highlights the versatility of TESSEE and builds evidence for eHTA for rational implementation of innovations in stem cell engineering, towards more cost-effective stem cell therapies.

Research supported by Fundação para a Ciência e Tecnologia (FCT) through the PhD Fellowship PD/BD/105868/2014.

J. Henriques et al. (Eds.): MEDICON 2019, IFMBE Proceedings 76, pp. 1919–1932, 2020.
https://doi.org/10.1007/978-3-030-31635-8_233

1 Introduction

Stem cell-based therapies are promising for diverse clinical indications, such as hematological, autoimmune, cardiovascular, neurological, and musculoskeletal conditions [1]. The first approved stem cell-based therapy was the transplantation of hematopoietic stem cells, being now a widespread procedure for the treatment of hematological malignancies in diverse countries. Relatively to other types of stem cells, there are over 1,700 active clinical trials using these cells as treatments, and 12 products approved in specific countries [1, 2]. These therapies have very high manufacturing costs and are produced at a much lower scale than small molecule and biological drugs [3]. This is a result of the low number of stem cells found in native tissues. In order for the stem cell-based products to have clinical effectiveness, high numbers of cells (>1 million cells/patient kg) need to be transplanted, which pertains the need for scaling up the manufacturing of stem cells [4].

The scaling-up of stem cell manufacturing is impaired by manufacturing systems with small lot sizes, labor-intensive tasks, long processing times, and costly process components. Currently reported costs of goods per manufactured dose (CoGs) of a stem cell therapy are in the range of $1000/dose [3, 5]. Accordingly, the proposed reimbursement prices are very high, in order to ensure economic viability of the production of these therapies. For stem cell therapies, reported market entry prices range between $2,000–$200,000, depending on the location and clinical indication [1].

The value of new therapies is often assessed by cost-effectiveness analyses using quality-adjusted life-years (QALYs) as a measure of effectiveness. A costlier therapy to a healthcare provider is cost-effective if the additional direct medical costs per QALY, in comparison to the standard of care, are below a specified willingness to pay (WTP) threshold [6]. Due to the high foreseen costs of stem cell therapies, the economic factors make stem cell based therapies vulnerable to the market dynamics.

The early health technology assessment (eHTA) is an emergent field aiming at determining the commercial viability of a new medical technology in the research and development stage. eHTA models include several uncertainty variables to provide informed estimates for the innovation and development of a new technology [7, 8]. The modeling approach is also useful to evaluate the cost-effectiveness of technology innovation to reduce manufacturing costs and assess prospective cost-effectiveness in the clinical setting [9]. In the field of stem cell-based therapies, eHTA studies usually involve bioprocess models and/or disease state models. These modeling studies are actually based in commercial software and/or proprietary code.

2 Evaluating Methodologies for Early Health Technology Assessment in Stem Cell Therapies

While the field of bioprocess and/or cost-effectiveness modeling aiming the economic evaluation of future stem cell therapies is on its infancy, the available studies provide insights on the types of technologies and methodologies and on the key attributes for evaluating cost-effectiveness of stem cell therapies (Table 1).

Table 1. Review of simulation environments used for early health technology assessment of stem cell engineering innovations

Simulation environment	Pros	Cons	References
Microsoft Excel + VBA	Simplicity Minimal programming Iterative algorithms for modeling	Not practical for queuing for equipment Limited time sensitive parameters	[10–12]
C#	Versatile programming language Inclusion of variability and queuing equipment	Licensed language	[3, 13, 14]
SuperPro Designer	Building blocks for bioprocess modeling Process scalability	Limited equipment Expensive license Not compatible with disease state modeling	[15]
TreeAge	Simple decision making software Inclusion of variability in clinical outcomes	Not compatible with bioprocess modeling Expensive license	[8]

The main aim of this thesis work was to perform eHTA of academic laboratory innovations in stem cell engineering, taking into account the biological factors that are assessed from these innovations, and learn what is required to scale these innovations to a clinical setting in an economically sustainable way. A key factor is the biological variability in cell growth using different process components. Since the cells are the product itself, the biological variability is a critical attribute for the time and costs involved in manufacturing cell based products [16]. Therefore, using a flexible programming language, that would allow to code and simulate these uncertainties in a straightforward way, is an attractive paradigm.

Additionally, since the push for this work came from combining a comprehensive bioprocess model for scalability of technological innovation in stem cell engineering with cost-effectiveness analyses, frameworks that do not enable an easy transition between the bioprocess and health economics models were also avoided. Furthermore, in order to increase the dissemination of the framework, an open source methodology is desirable [17]. For this reason, a freely available programming language that would allow flexible programming across a range of different case studies was seen as the best solution. Therefore, a new tool was implemented to better answer these challenges.

3 Presenting TESSEE

TESSEE (Tool for Early Stem cellS Economic Evaluation) was developed as a new freely available and versatile solution for connecting engineering innovation in the development of stem cell-based solutions to the economic sustainability of the development of products based on such innovations. The source code is licensed under a General Public License (GPL). The model is open-source and cross-platform, implemented in the open source programming language Python. The model outputs can be stored as CSV files. The model modules allow for both deterministic and stochastic simulation, either in the bioprocess economics or health economics modules. Stochastic simulation, is applied through the Monte Carlo method. The ranges and shape of parametric distributions are determined from experimental data.

The tool has the following main components: a database, a fixed capital investment (FCI) model, a bioprocess economics model, and a health economics model. The database contains general input parameters, where some assumptions vary according to the case studies. Two FCI options were employed in the case studies presented in this thesis: a cost-per-area model and a Lang factor based model. The utilization of each model is dependent on the most appropriate choice for the specific case study.

The FCI model and the bioprocess specific database components are used by the bioprocess economics model to evaluate the costs of goods (COG) per dose, per batch, or per donor. The bioprocess model optimizes the choice of the number of cell culture systems (either cell culture flasks or bioreactors) in order to minimize equipment occupation and reduce facility footprint. The COG per dose are used to derive the price of a given therapy, assuming that COG is a given percentage of the final reimbursement price (Fig. 1).

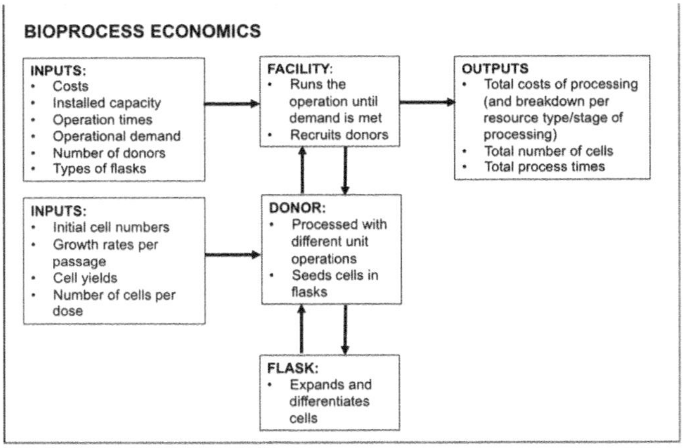

Fig. 1. Basic structure for the bioprocess economics model.

The health economics model calculates, using an appropriate time span, the total direct medical costs, related with managing the disease using a stem cell therapy, vs management with the current standard of care, combined with the clinical effectiveness as quality adjusted life years (QALY). The difference in total costs is divided by the difference in QALY to provide an incremental cost-effectiveness ratio (ICER) of the stem cell therapy. The ICER per patient is then compared with the willingness to pay threshold (WTP) of a particular payer to determine cost-effectiveness for a specific patient (Fig. 2).

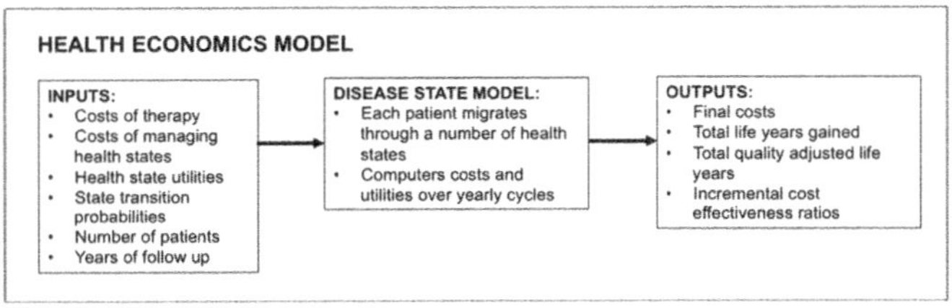

Fig. 2. Basic structure for the health economics model.

The source code, specific model equations, and tutorials, are on GitHub [18]. Anyone can run and modify the code to their needs. In order to run TESSEE, an implementation of Python 3 is necessary, along with a text editor. The code was built on the Anaconda implementation of Python.

4 Case Studies

During this PhD thesis, four representative case studies were developed. Two of these case studies focused only on the evaluation of technologies for stem cell bioprocessing, while the other case studies involved a combined bioprocessing and disease state model for evaluation of prospective stem cell-based therapies. All the case studies were developed in collaboration with academic groups, hospitals sponsoring clinical trials in cell therapies, and companies in the field of stem cell manufacturing. Several experts in Europe and in Boston, USA, were interviewed to refine the modeling assumptions. The three case studies that are already published in the literature are presented in this thesis overview.

4.1 Transition of Mesenchymal Stem/Stromal Cell (MSC) Manufacturing to a Xeno-Free Culture Medium

NOTE: The contents of this case study are partially presented in this paper. The complete version of this case study is published in [19].

Mesenchymal stem/stromal cells (MSC) are attractive for therapeutic purposes due to their immunosuppressive and anti-inflammatory properties. These cells can be found in diverse sources in the human body, including the bone marrow, adipose tissue, umbilical cord blood or matrix, among others, and are isolated by adherence to plastic surfaces. Clinically relevant numbers are on the order of 1 million cells/kg or higher; however, MSCs isolated directly from the native sources are found in low numbers. Typical bone marrow aspirate volumes for clinical trials range from 10–50 ml and can be isolated by density gradient centrifugation or through plating directly the whole bone marrow aspirate, yielding mononuclear cells in the range of 10^6–10^7 cells/ml, from which approximately 1 in 10^4–10^5 are MSCs. Consequently, the cells have to undergo several generations of replications in vitro in order for clinically relevant cell numbers to be achieved. This process is a labor, time and money consuming task due to the need for a facility complying with Good Manufacturing Practices (GMP), highly costly clinical grade reagents, long process times, labor-intensive tasks, impractical scaling up of production and high costs of quality control.

The choice of culture medium and consequent supplementation are among the most critical decisions in MSCs bioprocessing for clinical applications, both from the cost and the product quality points of view. Typically, MSCs are cultured in defined basal media, such as DMEM (Dulbecco's modified Eagle medium) or alpha-MEM (Minimum Essential Media), and supplemented with fetal bovine serum (FBS) for the supply of multiple growth factors and cytokines. As of 2014, approximately 80% of clinical trials were performed with MSCs expanded in media with FBS supplementation. The use of animal origin serum has several disadvantages, such as the lack of chemical composition definition, high batch-to-batch variability, safety issues with the transmission of prion or viral diseases, possible host immune response, supply chain limitations and animal welfare concerns. Additionally, directives by the European Medicines Agency (EMA) and the US Food and Drug Administration (FDA) recommend the replacement of the use of animal origin reagents whenever possible. Under these considerations, the interest in supplementation with human origin products, such as human serum (HS) and human platelet lysate (hPL) has grown. While hPL is viewed as safer than FBS, its current price is higher than animal-based supplementation and concerns remain in terms of batch-to-batch variability. Possible coagulation, and risk of human pathogen transmission are additional concerns.

The aim of this thesis chapter was to use the TESSEE framework to evaluate the costs of producing autologous (cells are collected from the patient, expanded, and then administered back to the same patient) MSC based therapies with a dose of 75 million cells/patient, either with FBS or hPL as the culture media supplement. This process encompassed donor and culture media supplement variability in the isolation yields of bone marrow MSC and their expansion yields per passage from published experimental data. The process and associated costs of goods per dose were simulated for 1000 people using a Monte Carlo approach. The key model inputs can be found elsewhere. The price per volume of hPL was considered to be twice the price of FBS, according to listed commercial prices.

The key results of this case study indicated that, although hPL is a more expensive culture medium supplement, the fact that it promotes a higher isolation yield of MSC and higher numbers of MSC cells per passage, allows cost savings in 97% of the considered donors (Fig. 3A). On average, the costs per dose manufactured with hPL are 84% of the costs with FBS as a supplement (Fig. 3B). There is a very strong correlation between the total number of passages for achieving the final cell dose that were averted by the use of hPL and the total costs per dose (Fig. 3C). It was also shown that, for donors that do not have any significant biological advantage with hPL supplementation, the reduction in costs is not reached. The most common case is the reduction of two passages per donor (Fig. 3D).

These results indicate that use of hPL promotes the isolation of a higher number of cells after their isolation from bone marrow and after multi-passage growth, which

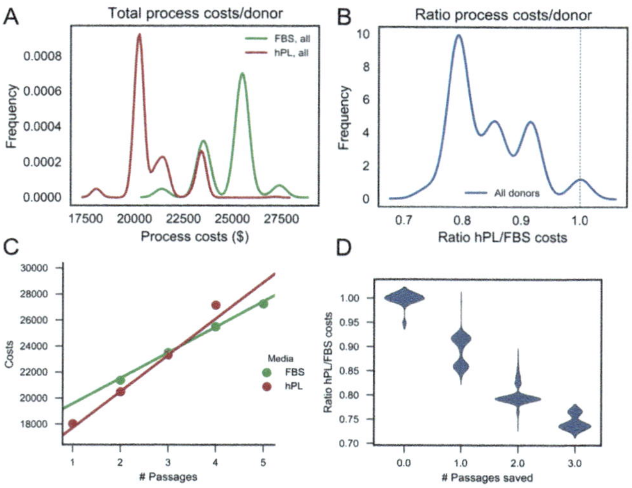

Fig. 3. Human platelet lysate (hPL) is a cost saving alternative in comparison with fetal bovine serum (FBS) for bioprocessing of autologous BM-MSCs therapies. A – Frequency distribution of the processing costs of 1000 donors with FBS (green) or hPL (red) until reaching the final dose of 75 million cells per donor. B – Ratio of hPL vs FBS processing costs per donor. hPL reduces the processing costs of 97% of the donors. C – Both FBS and hPL show a strong relationship between the number of passages for processing per donor and the total manufacturing costs, with a steeper increase for hPL cultures. Points represent the mean of the distributions per passage and per culture media. D – The cost savings per donor when transferring the culture for supplementation with hPL are dependent on the number of passages saved. For donors with no passages saved, there are no cost savings, while the cost savings become more relevant with an increasing number of passages saved for up to 30% of the FBS value

translates in the need of lower passages to achieve the targeted therapeutic dose of cells per patient required. The decrease in the number of passages, in turn, results in cost savings of 16%, on average, per donor, when compared with supplementation with FBS. The impact of additional passages required to achieve the therapeutic dose creates a multimodal cost distribution with average additional costs per passage per donor of $1970 for FBS, and of $2802, on average, for hPL. TESSEE was an effective model for the determination of the full costs of MSCs bioprocessing, assuming a xeno-free supplement and encourages steps to reduce variability in the efficacy of the supplements for autologous based cell culture, since the donor's own cells are the products.

4.2 Scale-Up of MSC Manufacturing to a Vertical Wheel Bioreactor

NOTE: The contents of this case study are partially presented in this paper. The complete version of this case study is published in [20].

In the previous section, a case study for an autologous therapy was evaluated. In the manufacturing of autologous therapies, multiple donors can be processed in parallel, and the required numbers of cells per donor to be expanded are in the range of a few million. Hence, they can be expanded in two dimensional culture systems. For reduction of manufacturing costs, when applicable, allogeneic therapies are sought after, since one donor is expanded to provide therapies for several patients. In this process configuration, annual demands of 10^{11}–10^{14} are envisioned for specific clinical indications. Therefore, scaling up production of MSC in traditional cell culture flasks, such as T-175 and cell stacks, become impractical due to space, labor, logistical and cost challenges. For this reason, suspension cultures are desired in order to approximate the volumes of production of MSC to the volumes of reactors used in small molecule and biologics production. MSC require an adhesion surface for culture, which can be achieved in suspension systems through the use of microcarriers.

The most traditional bioreactor system for expansion of MSC in microcarriers is the stirred-tank reactor (STR). Expansion of MSC in working volumes up to 35L was successfully achieved in this configuration. However, the STR setup promotes shear stress gradients that have negative effects on cell proliferation and potency. A new bioreactor configuration, with a vertically rotating wheel, provides faster and more efficient mixing. This chapter aimed at evaluating the cost-effectiveness of the bioprocess transfer from traditional two-dimensional MSC culture to expansion with the vertical wheel bioreactor for allogeneic therapy purposes. For this goal, a simulation of the maximum expansion potential of adipose tissue (AT) and umbilical cord matrix (UCM) MSC per donor with the two technologies was used. Experimental data comparing the two systems showed that the use of the vertical wheel bioreactor significantly increases the number of expanded cells per unit of area, in comparison with two-dimensional flasks.

A batch of cells is the number of cells terminally expanded from each expansion run of a single donor, and multiple doses, of 70 million cells each, are produced from this batch. The model was run until all the cells isolated from a single donor were expanded. The transfer to a vertical wheel bioreactor was only performed in the fifth (and final) expansion passage. The simulation resulted in the total costs of producing a batch, which was divided by the number of doses to obtain the costs per dose. The initial number of cells per donor for expansion were 5 million AT MSC or 0.5 million UCM MSC. Each two-dimensional flask requires an initial number of 525,000 cells, while each vertical wheel bioreactor requires an initial number of 5 million cells. Other key model inputs can be found elsewhere.

The number of total final product batches is reduced with the transition of the final expansion step to PBS-0.1MAG, from 13 to 9 (AT MSC) and from 8 to 4 batches (UCM MSC). Finally, the number of total doses of 70 million cells produced increased from 26 to 45 (AT MSC) and from 23 to 25 (UCM MSC). The utilization of PBS 0.1 MAG also leads to a reduction in the total process costs of 11% and 30%, for UCM MSC and AT MSC, respectively. Costs per dose decrease from \$17.0 K to \$11.1 K for UCM MSC and from \$21.5 K to \$11.1 K for AT MSC (Table 2).

The economic evaluation, presented in this case study, supports the process transition to vertical wheel bioreactors in a cost-effective manner, as long as safety and clinically meaningful quality attributes are similar to planar technologies. This case study provided advancement in the development of TESSEE due to the implementation of intermediate banking steps and introduction of a choice of microcarrier based culture in bioreactors. As this process was not developed, at this moment, with a specific

Table 2. Total processing costs, number of batches per process, and total doses obtained for each process

Cell source	T-175	VWR
Total process costs		
AT MSC	$559,730	$499,657
UCM MSC	$391,906	$278,081
# of batches		
AT MSC	13	9
UCM MSC	8	4
# of doses		
AT MSC	26	45
UCM MSC	23	25
Total costs per dose		
AT MSC	$21,258	$11,103
UCM MSC	$17,039	$11,123

prospective clinical application, the application of TESSEE here was purely in bioprocess economics.

4.3 Early Health Technology Assessment of a Stem Cell-Based Device Containing Islet Cells for Type 1 Diabetes Management

NOTE: The contents of this case study are partially presented in this paper. The complete version of this case study is published in [21].

Type 1 diabetes mellitus (T1D) is an autoimmune disease destroying beta cells, which impairs both the secretion of insulin by the pancreas and the control of glucose levels in the blood. Disease onset occurs most frequently between 6 and 12 years old. With an estimated population of 1.1 million patients, the economic burden of T1D in the United States is considerable. For instance, total lifetime medical costs of T1D management are $133.7 billion, with a total income loss of $289.2 billion. These patients are insulin dependent and are at increased risk of related complications, such as amputation, blindness, and kidney failure. Most patients are managed on insulin intensive therapy (IIT) for as long as the clinical complications are not prohibitive.

Therapeutic interventions restoring insulin independence could mitigate related complications, and reduce healthcare expenditure. Whole pancreas and islet transplantation have been performed successfully, but are limited due to donor supply issues. This hurdle could be solved by functional insulin-secreting cells obtained from directed pluripotent stem cell (PSC) differentiation. Pre-clinical and early clinical stage research is ongoing, aiming at the implantation of PSC-derived beta cells or pancreatic progenitors, differentiated from either human embryonic stem cells (hESC) or human induced pluripotent stem cells (hiPSC). Long and inefficient processes for expansion of PSCs and differentiation into the pancreatic lineage are cost prohibitive. Therefore, approaches suggesting cues for optimization of manufacturing costs, while retaining clinical effectiveness, are desirable.

In this section, the bioprocess economics model reported in the previous case studies was extended for the expansion and differentiation of PSC into islet cells. The incorporation of a health economics model of T1D supports optimization of manufacturing and clinical effectiveness. This approach aims at providing input for technological innovation on the manufacturer's side, with the goal of producing cost-effective therapies for restoring insulin independence.

The simulated therapy consisted on a device containing approximately 100 million islet cells differentiated from PSC. Each patient is transplanted with 5 devices, for a total final dose per patient of 500 million cells. The first scenario was one patient per batch, 5 patients/year. Then, a second scenario of 50 patients, 250 devices/year was derived in agreement with a Phase I/II clinical trial (NCT02239354). A case where 50 patients/year, 1 patient/batch are produced was simulated, where several batches are processed in parallel, and there was a staggering of the utilization of the purification equipment, aiming at reduction of costs. Further optimization of utilization of the equipment capacity was performed through a case with 50 patients/year, 10 patients/batch. The early assessment comprised a bioprocess economics model and a disease state cost-utility analysis. The bioprocess economics model was used to calculate the devices' cost of goods (CoGs), which were then linked to the clinical cost-effectiveness through a disease state model comparing stem cell-based therapy with IIT.

In the disease state model, five states were defined for the beta cell transplant model: (1) Full insulin independence; (2) partial graft function without complications (insulin dependent but producing other relevant factors for glucose control); (3) graft failure without complications (4) diabetes-related complications (after graft failure) and (5) death. The IIT arm had three states: (1) IIT without complications; (2) diabetes-related complications; and (3) death (Fig. 4). It was considered that patients could suffer complications from 5 main groups: hypoglycemia, cardiovascular, neuropathy, nephropathy, and ophthalmological. Each complication was computed with increased medical costs and a utility decrement per patient. Yearly costs and quality of life scores associated with each state were computed. Moreover, patients could move through states according to event probabilities. Key model inputs are reported elsewhere.

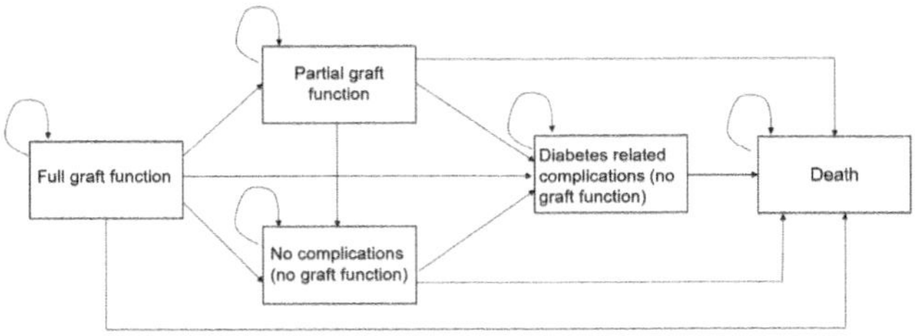

Fig. 4. Disease state model of type 1 diabetes, with or without graft

After the follow-up period, the total direct medical costs were computed, as well as the sum of the utilities per year yielding the total quality-adjusted life years (QALYs). The cost-effectiveness of the new treatment was assessed as an incremental cost-effectiveness ratio (ICER), a ratio of the total difference in costs to the difference in

QALYs between the two therapies. Stem cell-based therapy was considered cost-effective if its ICER was below a given willingness to pay (WTP) threshold in cost/QALY.

The bioprocess economics model estimated that the increase in the annual demand and the number of therapies processed per batch would decrease the costs of goods per patient from $427,231 (5 patients/year, 1 patient/batch) to $358,818 (50 patients/year, 1 patient/batch) and to $163,720 (50 patients/year, 10 patients/batch). These costs were assumed to be 25% of the total reimbursement price, and drove the cost-effectiveness analysis of the therapy, in comparison with IIT. Quality of life was significantly improved at the population level, as 96.4% of the patients had higher QALY with the stem cell-based therapy. The transplant was a cost-saving alternative for only 1.6% of the patients, as the treatment with the stem cell-based therapy resulted in lower direct medical costs over a 20-year follow-up period, in comparison with IIT (Fig. 5A). For these patients, end-stage renal disease was averted with the transplantation, irrespective of the occurrence of other complications.

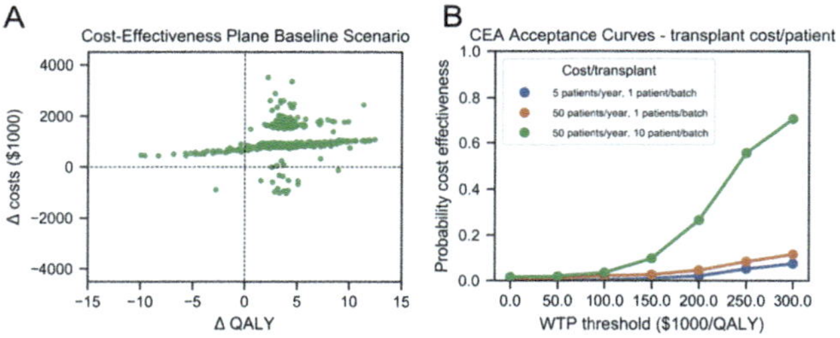

Fig. 5. Cost-effectiveness of beta cell transplants is sensitive to cost-effectiveness acceptance thresholds. A: Cost-effectiveness acceptance plane after a 20 year follow up period from the transplant, assuming a price per transplant of $650,000 (considering that the manufacturing costs per patient of the 50 patients/year, 10 patients/batch scheme represent 25% of the final transplant price per patient). Points are 1000 randomly sampled individual patients. Differences in cost and quality-adjusted life years show that, for most patients, the transplant is more effective in providing a better quality of life, but generally with higher direct medical costs. B: Cost-effectiveness acceptance probability curve relative to the willingness to pay (WTP) thresholds employed by the payer for transplants using devices manufactured under the 5 patients/year, 1 patient/batch ($1.7 M final transplant price), 50 patients/year, 1 patient/batch ($1.45 M final transplant price) and the 50 patients/year, 10 patients/batch ($650,000 final transplant price) strategies. Probabilities are calculated as the ratio of the number of patients with an incremental cost-effectiveness ratio (ICER) for stem cell-based therapy below each WTP threshold by the total number of patients.

The cost-effectiveness for the patients with higher cost and QALY under the transplantation scheme is dependent on the ICER and the willingness to pay (WTP) threshold. At a WTP threshold of $50,000/QALY, the stem cell-based therapy is cost-effective for only 1.9% of the patients. At this threshold, while the cost-effectiveness probability for the 50 patients/year, 10 patients/batch is low, it represents an improvement from the manufacturing strategies with 1 patient/batch. For

$100,000/QALY, the intervention is cost-effective for about 3.4% of the patients. For the manufacture strategies of 1 patient/batch, the cost-effectiveness probability was still very low at high WTP thresholds of $300,000/QALY, since the transplant would be cost-effective for only 10% of the patients, in comparison with IIT. The 50 patient/year, 10 patient/batch strategy indicates, at high WTP thresholds, a vast improvement from the 1 patient/batch strategies, as 70% of patients would benefit from a cost-effective transplant, at the $300,000/QALY threshold (Fig. 5B).

This study suggests that reduction of manufacturing costs, while retaining clinical effectiveness and ensuring prevention of costly complications, is fundamental for cost-effectiveness of new stem cell-based therapies for type 1 diabetes. The finding that the therapy would be cost-saving for patients, for whom end-stage renal disease is avoided, is consistent with the current clinical development of devices, which allows direct vascularization for treatment of patients with high risk for this complication. The development of predictive models of diabetes-related complications would help optimize the allocation of resources of these high value, high-cost therapies under budget limitations.

This case study was critical in the advancement of TESSEE as a fully integrated early health technology assessment framework, applicable to stem cell-based therapies, due to the integration of a compatible disease state model. Furthermore, it also extended the bioprocess modeling capabilities by the inclusion of a differentiation module for therapies derived from pluripotent stem cells. It provides a framework to other combined analysis of early stage prospective therapies, scaling from the laboratory production until the eventual clinical effectiveness, having cost as a critical attribute to consider in the design of these therapies.

5 Conclusions and Future Work

The usefulness and versatility of TESSEE for economic evaluation of innovations in stem cell engineering was demonstrated by different relevant case studies. This tool was well received in presentations in academic groups, conferences, meetings with companies that found it useful for evaluation of their products, and also by medical centers involved in clinical trials in the field of stem cell therapies.

While the pricing and reimbursement schemes for healthcare are different in Europe and in the United States, the methodology was well received in both locations [22], eliciting the potential for applicability of TESSEE for technologies developed for different markets.

This paper presents three case studies performed in the scope of the thesis. The fourth modeling case study, also included in this thesis, is related to the early health technology assessment of mesenchymal stem/stromal cells as a therapy for cystic fibrosis [23, 24]. This case study aims at directly evaluating a technology already in a Phase 1 clinical trial, and will be submitted for publication soon.

This thesis was sound in terms of scientific outputs, with three published scientific manuscripts [19–21], two of these as first author, one accepted review article, and another first author manuscript in preparation. Additionally, this thesis led to an editorial piece in a relevant publication in the field of Regenerative Medicine [25], seven conference presentations, and three invited lectures, both in Portugal and the United States.

Future work involves the dissemination of the tool for education of students and researchers, through the development of a graphical user interface to improve the utilization experience of users with no programming language. Additionally, the creation of a community of different stakeholders, alike public committees for value based pricing, such as the Institute for Clinical and Economic Review (ICER) [26], and the open health economics modeling of therapeutics, such as the Innovation and Value Initiative (IVI) [27], are sought after.

Acknowledgment. The authors acknowledge funding from Institute for Bioengineering and Biosciences (UID/BIO/04565/2013) and Programa Operacional Regional de Lisboa 2020 (Project No. 007317) and also through the project PRECISE (PAC-PRECISE-LISBOA-01-0145-FEDER-016394). C. Bandeiras also acknowledges the support from the MIT Portugal Program in Bioengineering, namely through the research stays in the Institute for Data, System and Society – MIT and the Division of Clinical Informatics, Beth Israel Deaconess Medical Center, Harvard Medical School.

References

1. Shukla, V., Seoane-Vazquez, E., Fawaz, S., Brown, L.M., Rodriguez-Monguio, R.: The landscape of cellular and gene therapy products: cost, approvals, and discontinuations. Hum. Gene Ther. Clin. Dev., April 2019. humc.2018.201
2. Home - ClinicalTrials.gov. https://clinicaltrials.gov/. Accessed 03 Jul 2019
3. Simaria, A.S., et al.: Allogeneic cell therapy bioprocess economics and optimization: single-use cell expansion technologies. Biotechnol. Bioeng. **111**(1), 69–83 (2014)
4. Ikebe, C., Suzuki, K.: Mesenchymal stem cells for regenerative therapy: optimization of cell preparation protocols. Biomed. Res. Int. **2014**, 951512 (2014)
5. Mizukami, A., et al.: Technologies for large-scale umbilical cord-derived MSC expansion: experimental performance and cost of goods analysis. Biochem. Eng. J. **135**, 36–48 (2018)
6. Banta, D.: The development of health technology assessment. Health Policy **63**(2), 121–132 (2003)
7. Ijzerman, M., Wissing, T., de Koning, E.: Early health economic evaluation of bioartificial organs: involving users in the design of the bioartificial pancreas for diabetes. In: Stamatialis, D. (ed.) Biomedical Membranes and (Bio)artificial Organs. World Scientific Publishing Co. Pte Ltd., New Jersey (2017)
8. de Windt, T.S., Sorel, J.C., Vonk, L.A., Kip, M.M.A., Ijzerman, M.J., Saris, D.B.F.: Early health economic modelling of single-stage cartilage repair. Guiding implementation of technologies in regenerative medicine. J. Tissue Eng. Regen. Med. **11**(10), 2950–2959 (2017)
9. Wallner, K., et al.: Stem cells and beta cell replacement therapy: a prospective health technology assessment study. BMC Endocr. Disord. **18**(1), 6 (2018)
10. Jenkins, M., Bilsland, J., Allsopp, T.E., Ho, S.V., Farid, S.S.: Patient-specific hiPSC bioprocessing for drug screening: bioprocess economics and optimisation. Biochem. Eng. J. **108**, 84–97 (2016)
11. Weil, B.D., et al.: An integrated experimental and economic evaluation of cell therapy affinity purification technologies. Regen. Med. **12**(4), 397–417 (2017)

12. Pereira Chilima, T.D., Moncaubeig, F., Farid, S.S.: Impact of allogeneic stem cell manufacturing decisions on cost of goods, process robustness and reimbursement. Biochem. Eng. J. **137**, 132–151 (2018)
13. Hassan, S., Simaria, A.S., Varadaraju, H., Gupta, S., Warren, K., Farid, S.S.: Allogeneic cell therapy bioprocess economics and optimization: downstream processing decisions. Regen. Med. **10**(5), 591–609 (2015)
14. Harrison, R.P., Medcalf, N., Rafiq, Q.A.: Cell therapy-processing economics: small-scale microfactories as a stepping stone toward large-scale macrofactories. Regen. Med. **13**(2), 159–173 (2018)
15. Darkins, C.L., Mandenius, C.-F.: Design of large-scale manufacturing of induced pluripotent stem cell derived cardiomyocytes. Chem. Eng. Res. Des. **92**(6), 1142–1152 (2014)
16. Placzek, M.R., et al.: Stem cell bioprocessing: fundamentals and principles. J. R. Soc. Interface **6**(32), 209–232 (2009)
17. Sampson, C.J., et al.: Transparency in decision modelling: what, why, who and how? Pharmacoeconomics, pp. 1–15, June 2019
18. TESSEE - Tool for Early Stem cellS Economic Evaluation Source Code. https://github.com/catiabandeiras/TESSEE. Accessed 10 Jul 2019
19. Bandeiras, C., Cabral, J.M., Finkelstein, S.N., Ferreira, F.C.: Modeling biological and economic uncertainty on cell therapy manufacturing: the choice of culture media supplementation. Regen. Med. **13**(8), 917–933 (2018)
20. Sousa Pinto, D., et al.: Scalable manufacturing of human mesenchymal stromal cells in the vertical-Wheel^{TM} bioreactor system: an experimental and economic approach. Biotechnol. J., p. 1800716, April 2019
21. Bandeiras, C., Cabral, J.M.S., Gabbay, R.A., Finkelstein, S.N., Ferreira, F.C.: Bringing stem cell-based therapies for type 1 diabetes to the clinic: early insights from bioprocess economics and cost-effectiveness analysis. Biotechnol. J., p. 1800563, May 2019
22. Faulkner, S., et al.: Pricing and reimbursement experiences and insights in the European Union and the United States: Lessons learned to approach adaptive payer pathways. Clin. Pharmacol. Ther. **100**(6), 730–742 (2016)
23. Sutton, M.T., et al.: Antimicrobial properties of mesenchymal stem cells: therapeutic potential for cystic fibrosis infection, and treatment. Stem Cells Int. **2016**, 1–12 (2016)
24. Bandeiras, C., Koc, J.R., Ma, Y., Samberg, M., Finkelstein, S., Ferreira, F.: Cost effectiveness analysis of allogeneic, just-in-time expansion of mesenchymal stem cells with PLUS^{TM} human platelet lysate for a clinical trial. Cytotherapy **20**(5), S60 (2018)
25. Bandeiras, C., Cabral, J.M.S., Finkelstein, S.N., Ferreira, F.C.: Logical process designs for stem cell manufacturing: computational support tools for improved cost-effectiveness. RegMedNet (2017). https://www.regmednet.com/users/3641-regmednet/posts/19843-logical-process-designs-for-stem-cell-manufacturing-computational-support-tools-for-improved-cost-effectiveness
26. ICER - Institute for Clinical and Economic Review. https://icer-review.org/. Accessed 10 Jul 2019
27. Innovation and Value Initiative. https://www.thevalueinitiative.org/. Accessed 10 Jul 2019

Challenges: Fraunhofer Best Portuguese MsC Thesis Competition in Biomedical Engineering

Community Finding with Applications on Phylogenetic Networks

Luís Rita[1(✉)], Alexandre Francisco[1], João Carriço[2], and Vítor Borges[3]

[1] Instituto Superior Técnico, University of Lisbon, Lisbon, Portugal
{luis.rita,aplf}@ist.utl.pt

[2] Institute of Molecular Medicine, Faculty of Medicine, University of Lisbon, Lisbon, Portugal
jcarrico@fm.ul.pt

[3] Department of Infectious Diseases, National Health Institute Dr. Ricardo Jorge, Lisbon, Portugal
vitor.borges@insa.min-saude.pt

Abstract. With the advent of high-throughput sequencing methods, new ways of visualizing and analyzing increasingly amounts of data are needed. Although some software already exist, they do not scale well or require advanced skills to be useful in phylogenetics.

The aim of this thesis was to implement three community finding algorithms – Louvain, Infomap and Layered Label Propagation (LLP); to benchmark them using two synthetic networks – Girvan-Newman (GN) and Lancichinetti-Fortunato-Radicchi (LFR); to test them in real networks, particularly, in one derived from a *Staphylococcus aureus* MLST dataset; to compare visualization frameworks – Cytoscape.js and D3.js, and, finally, to make it all available online (mscthesis.herokuapp.com).

Louvain, Infomap and LLP were implemented in JavaScript. Unless otherwise stated, next conclusions are valid for GN and LFR. In terms of speed, Louvain outperformed all others. Considering accuracy, in networks with well-defined communities, Louvain was the most accurate. For higher mixing, LLP was the best. Contrarily to weakly mixed, it is advantageous to increase the resolution parameter in highly mixed GN. In LFR, higher resolution decreases the accuracy of detection, independently of the mixing parameter. The increase of the average node degree enhanced partitioning accuracy and suggested detection by chance was minimized. It is computationally more intensive to generate GN with higher mixing or average degree, using the algorithm developed in the thesis or the LFR implementation. In *S. aureus* network, Louvain was the fastest and the most accurate in detecting the clusters of seven groups of strains directly evolved from the common ancestor.

1 Introduction

According to *The Economist*, data has become the most important resource in the world [1]. New ways to storage, analyze and visualize increasing amounts of information are becoming a pressing need.

J. Henriques et al. (Eds.): MEDICON 2019, IFMBE Proceedings 76, pp. 1935–1953, 2020.
https://doi.org/10.1007/978-3-030-31635-8_234

Interactions among different elements of real systems can be seen as networks [2]. From their topology or pattern of connections, functional knowledge can be extracted. Although empirical analysis can be useful in small networks, for big data it is indispensable the use of algorithms to uncover their properties. Consequently, mathematical models called graphs are used to simplify their representation. Graph theory provides all the tools available to analyze them. This representation implies all the elements of the network and their relations must be acquired. Generally, the closer to the human population is the analysis, the more difficult it gets to gather data. The two main reasons are the lack of non-invasive acquisition methods or privacy issues [3]. In fact, in health, legislation is becoming progressively stricter in respect to patient's data protection [4]. On one hand, it is fundamental to assure ethical use of medical data, on the other, it may constitute a hurdle for investigating the causes of many pathologies. In the next lines, the discussion will focus in the specific topic of infectious diseases.

As stated by the *World Health Organization*, Antimicrobial resistance (AMR) is in the top ten of major threats demanding attention in 2019 [5]. In the last years, AMR has been potentiated by the use of drugs intended to treat diseases caused by pathogenic organisms. Depending on the organism, it is subdivided in antiviral (viruses), antibiotic (bacteria), antifungal (fungi) or antiparasitic (parasites) resistance. Although resistance is a natural evolutionary process, in which the most competent strains get selected over time, the frequency of use, or even misuse, of antimicrobial drugs is accelerating the growth of resistance rates for several microbial species [6]. This is particularly relevant not only among the human population, but also in farms where cattle are being given preventive doses of antibiotics as growth adjuvants. Moreover, the frequency people are travelling around the world is facilitating the spread of resistant organisms. Due to all of this and the high mortality of infectious diseases, it is fundamental to empower central health authorities with scalable and easy-to-use tools to allow their intervention in short time scales [7].

Next Generation Sequencing methods, developed in the last years, have been providing us massive quantities of microbiological data [9]. In fact, due to the high discriminative power of whole genome sequencing methods, it has been possible to infer more precise evolutionary relationships among strains. Consequently, allelic profiles with higher number of genes keep increasing and, with them, more complete phylogenetic trees or networks depicting evolutionary distances among strains can be assessed. All of this has been useful in the multiple stages of disease prevention: vaccine design, assess the pathogenicity of a sequenced strain, detect outbreaks and in infection control. As well as diagnosis, enhancing the detection of mixed infections (caused by multiple genetically distinct organisms) [10].

Generally, the goal of this thesis was to extract functional knowledge from the topological structure of phylogenetic networks. For instance, to be able to infer properties like resistance to antimicrobials, based on the similarity and characteristics of the other organisms in the network. Contrarily to random networks, real ones present an asymmetrical distribution of edges. From this asymmetry, clusters of nodes can be identified. By definition, communities are regions in the network with higher density of edges. Nonetheless, this definition is not restrictive enough to undoubtedly identify them in graphs. This raises the following question: how can we detect communities if it is not clear what we are looking for? To answer this question, community finding theory has been developed in the past years. In spite there has been a considerable advance in terms of accuracy and speed of detection, some algorithms perform better than others in different networks.

1.1 Objectives

The aim of the thesis was to implement three community finding algorithms in JavaScript – Louvain, Infomap and Layered Label Propagation (LLP); to benchmark them, in terms of accuracy and speed, against two synthetic networks – Girvan-Newman (GN) and Lancichinetti-Fortunato-Radicchi (LFR) networks; to perform additional testing using an Amazon, Zachary's Karate Club and *Staphylococcus aureus* Multilocus Sequence Typing (MLST) Single Locus Variant (SLV) network; to compare different visualization frameworks in terms of their tools and robustness to big data - D3.js (using SVG and Canvas elements) and Cytoscape.js, and to create a web application to allow network visualization (before and after running community finding algorithms) and plot all data obtained from the benchmark tests.

1.2 Structure Outline

In Methodology, the implemented community finding algorithms, implemented benchmark tools, real test data and data visualization libraries are introduced. In Results, community finding algorithms are analyzed in terms of accuracy and speed. The speed of generation of benchmark networks is determined based on the different properties of the networks. Different web visualization frameworks are compared. And a *S. aureus* MLST SLV network is partitioned using different algorithms and input parameters. In Conclusion, a retrospective analysis on whether the goals were attained or not is made, main difficulties, future work to be performed in INSaFLU web application is detailed and a set of prospective improvements to the work already done is suggested.

2 Methodology

2.1 Community Finding

(1) *Louvain Algorithm*

This algorithm, which was implemented in the thesis, is divided in 2 phases: Modularity Optimization and Community Aggregation [11]. Together, they are considered 1 pass. After the first step is completed, the second follows. Both are executed until there are no more changes in the network and maximum modularity is achieved.

Modularity Optimization – the algorithm will randomly order all nodes in the network such that, one by one, it will remove and insert them in a different community C. This will continue until no significant increase of modularity (input parameter) is verified:

$$\Delta M = \left[\frac{\Sigma_{in} + k_{i,in}}{2m} - \left(\frac{\Sigma_{tot} + k_i}{2m}\right)^2\right] - \left[\frac{\Sigma_{in}}{2m} - \left(\frac{\Sigma_{tot}}{2m}\right)^2 - \left(\frac{k_i}{2m}\right)^2\right] \tag{1}$$

Let Σ_{in} be the sum of the weights of the links inside C, Σ_{tot} the sum of the weights of all links to nodes in C, k_i sum of the weights of all links incident in node i, $k_{i,in}$ the sum of the weights of links connecting node i and nodes in the community C and m is the sum of the weights of all edges in the graph.

Community Aggregation – After finishing the first step, all nodes belonging to the same community are merged into a single giant node. Links connecting giant nodes are the sum of the ones previously connecting nodes from the same different communities. This step also generates self-loops which are the sum of all links inside a given community before being collapsed into one node.

(2) *Infomap Algorithm*

Similarly to Louvain, Infomap algorithm was also implemented. It partitions the network using a different quality function – minimum description length [12]:

$$\begin{aligned} L(M) = & \left(\sum_{m \in M} q_m\right) log\left(\sum_{m \in M} q_m\right) - 2\sum_{m \in M} q_m log(q_m) \\ & - \sum_{\alpha \in V} p_\alpha log(p_\alpha) + \sum_{m \in M}\left(q_m + \sum_{\alpha \in m} p_\alpha\right) log\left(q_m + \sum_{\alpha \in m} p_\alpha\right) \end{aligned} \tag{2}$$

Let q_m be the exit probability of a module m and p_α the relative weight w_α that is computed dividing the total weight of the edges connected to α by twice the total weight of all links in the graph.

(3) *Layered Label Propagation Algorithm*

LLP algorithm development was based in Label Propagation (LP) [13]. The latter starts by assigning a different community to each node in the network and then, iteratively, it modifies the respective community based on the dominant one from the nodes in the immediate neighborhood. In each iterate, this step is executed for all nodes, in a previously randomized order.

LLP not only takes into account the nodes in the neighborhood, but also the ones in the remaining network [14]. The iterative process is the same, the difference resides in the value it optimizes. In this case, the assigned community is the one that maximizes:

$$k_i - \gamma(v_i - k_i) \tag{3}$$

Being k_i the number of nodes with label λ_i in the neighborhood of a given node and v_i the total number in the whole graph labeled the same way.

This approach, which was also implemented, buffers the number of nodes belonging to a certain community (in the node's neighborhood), by its usual presence across the whole network.

Table 1. Time complexity of community finding algorithms.

Algorithm	Time complexity	Reference
Louvain	$O(L)$	[11]
Infomap	$O(N \log N)$	[12]
LP	$O(L)$	[13, 15]
LLP	$O(L)$	[14]

2.2 Benchmark

Different community finding algorithms partition the same graph in different ways. Consequently, it is important to understand which algorithms are the most indicated.

An accurate way to gain this insight is by partitioning networks whose community structure is known. GN and LFR benchmark networks are presented in the next subsections, as well as NMI which estimates the similarity between the original and the detected partitions.

Community finding algorithms were not only tested in terms of accuracy, but also speed. All benchmark tests were repeated 10 times. The mean values and corresponding confidence intervals at 95% were included in Results.

(1) *Girvan-Newman Network*

In GN network, 128 nodes are divided in 4 groups, each one with exactly 32 nodes [16]. Then, a mixing parameter, chosen by the user, defines the probability of nodes from different communities being connected. This probability is given by:

$$\mu = \frac{k^{ext}}{k^{int} + k^{ext}} \tag{4}$$

Each node is connected to 16 others (Fig. 1). The algorithm hereby implemented not only allows the mixing parameter to vary, but also the average degree. This led to an increased number of benchmark tests using these networks.

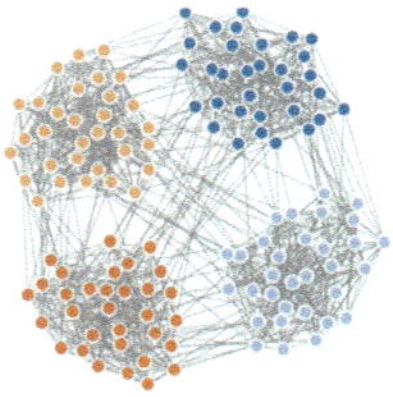

Fig. 1. GN synthetic network. $N = 128$, $\mu = 0.1$ and $k = 16$. Represented using D3.js and SVG.

(2) *Lancichinetti-Fortunato-Radicchi Network*

LFR benchmark networks (Fig. 2) [16, 17] try to better approximate the real ones. This means that not only they consider a power-law distribution of community sizes with coefficient ζ (5), but also a power-law distribution of node degrees with coefficient γ (6):

$$p_{N_c} \sim N_c^{-\zeta} \tag{5}$$

$$p_k \sim k^{-\gamma} \tag{6}$$

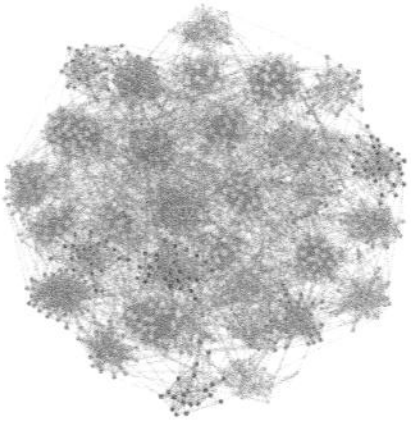

Fig. 2. LFR synthetic network. $N = 1000$, $\mu = 0.1$, $k_{avg} = 15$, $k_{max} = 50$, $k_{max} = 50$, $c_{min} = 20$ and $c_{max} = 50$. Represented using D3.js and SVG.

(3) *Normalized Mutual Information*

The clustering quality of the community finding algorithms was tested using Normalized Mutual Information (NMI) [18]. An algorithm was developed, in JavaScript, in the thesis and tested in several networks. It receives an input of 2 arrays, with the same length, and returns the corresponding NMI:

$$NMI(Y,C) = \frac{2 \times I(Y;C)}{[H(Y) + H(C)]} \tag{7}$$

$$I(Y;C) = H(Y) - H(Y|C) \tag{8}$$

NMI is dependent on the mutual information I, the conditional entropy $H(Y|C)$ and the entropy of the labeled $H(Y)$ and clustered set $H(C)$.

2.3 Real Test Data

The higher the number of tests performed using the implemented algorithms, the higher the certainty they are working properly for the broadest number of networks. This way, disconnected networks (sampled from an Amazon network), small sized ones

(Zachary's Karate Club) and a phylogenetic network (*S. aureus* MLST SLV) were additionally considered. Nevertheless, no benchmark tests were performed using those.

(1) *Amazon Network*

This network was obtained after crawling Amazon website [19]. Each node represents a given product available in the store and a connected pair means they are frequently bought together. The original network contains 334 863 nodes and 925 872 links. Only the first 5 000 links along the respective nodes were used in the tests (Fig. 3).

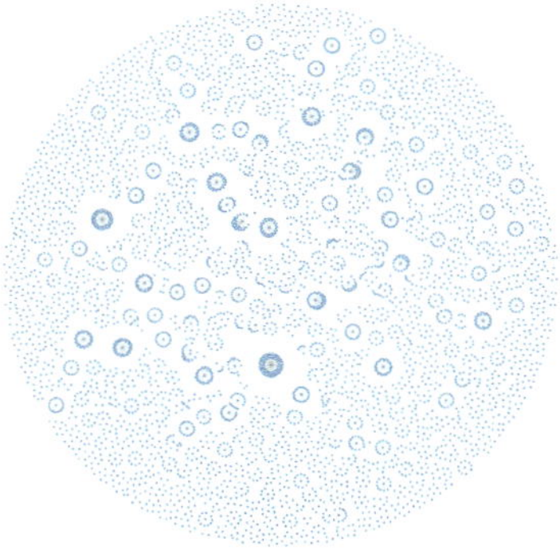

Fig. 3. Amazon product co-purchasing network sampled from a 5 000 links graph. Represented using D3.js and SVG.

(2) *Zachary's Karate Club Network*

This is a social network representing 77 relations between 34 individuals from a university karate club [20].

Node 1 is the instructor and 34 is the president (Fig. 4). Dark and light blue nodes represent the split inside the club that was created due to a conflict between them. Each link connects 2 members when they used to meet regularly outside the club, before splitting.

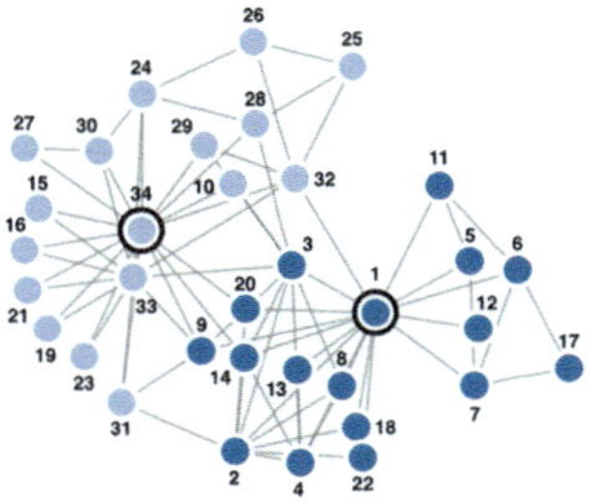

Fig. 4. Zachary's karate club network. Represented using D3.js and SVG.

(3) *Staphylococcus aureus*

To test the application of the community finding algorithms in a phylogenetic network, an MLST database of allelic profiles for *S. aureus* was used (Fig. 5) [21]. The MLST profiles contain information about seven loci (*arcC*, *aroE*, *glpF*, *gmk*, *pta*, *tpi* and *yqiL*) and, at the time of download, it had 5199 MLST profiles (ST). A graph was created by linking each ST to all of its SLVs and the largest component was used for the benchmark of the community finding algorithms. The choice of this test network was based on the fact it can emulate similar patterns to the ones that are created using cgMLST or wgMLST profiles, which have many more loci than MLST.

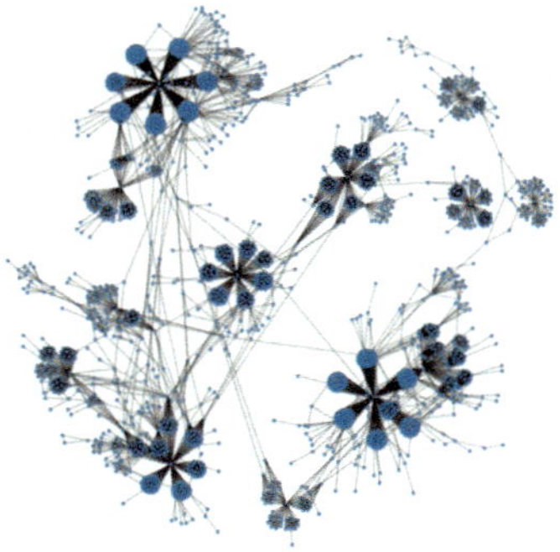

Fig. 5. *S. aureus* MLST SLV network. Represented using D3.js and Canvas.

2.4 Visualization Frameworks

In the visualization application, the user can opt by visualizing every network using D3 or Cytoscape JavaScript libraries. In the case of the first framework, there is the possibility of visualizing data through a Canvas or an SVG element. All networks were plotted using a force directed implementation. This mode requires the developer to set features like the gravity, repulsion or spring constant which will affect how the network is displayed. Depending those parameters, node and link overlapping should be reduced in order to get an enhanced view of the graph (Fig. 6).

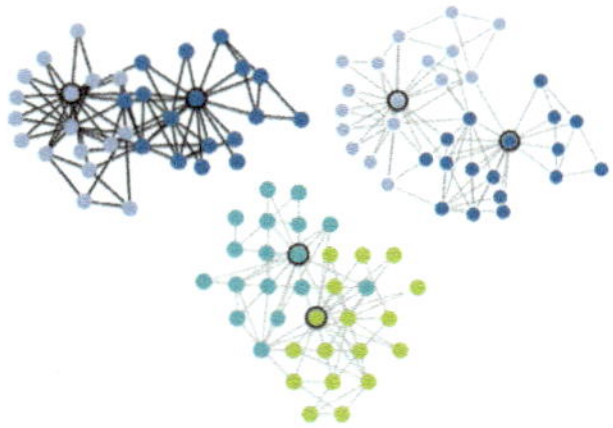

Fig. 6. Zachary's Karate Club network represented using D3.js Canvas (top-left), D3.js SVG (top-right) and Cytoscape.js (bottom).

Regarding benchmark plotting, D3.js was used. Its flexibility in terms of visual representation, performance and the number of practical examples available on the web, determined this choice.

3 Results

3.1 Web Application

In order to facilitate the analysis and visualization of the results after the execution of the previous algorithms, a web application was implemented (mscthesis.herokuapp.com).

Its backend was conceived using Node.js and it runs in a Heroku server which is linked to a GitHub repository (github.com/warcraft12321/Thesis) containing all the implementations and documents related to the thesis. A Digital Object Identifier (DOI) was attributed to this repository using Zenodo (zenodo.org/badge/latestdoi/162063699). An image of the app is available at Docker Hub (cloud.docker.com/u/warcraft12321/repository/docker/warcraft12321/thesis). The following actions are strictly executed in the cloud:

Generation of GN and LFR benchmark networks. The latter are generated using a binary from a C++ implementation [22]. This was possible after importing a package from NPM which simulates the command line in the server;

The real test networks had their data properly processed in the cloud, so that it could be sent and displayed in the frontend;

Once the execution of Louvain, Infomap and LLP algorithms is not performed in the user's browser, this may raise some privacy issues. Although, it was not feasible to run Infomap in the user side and still keep the web application functional in all devices;

The communication between the app and NPM was established in the server, so that it is possible to import and plot the statistics from each package uploaded to NPM in the application;

Finally, all benchmark data was obtained after running the algorithms in the cloud.

The frontend is both static and dynamic. The static counterpart uses HTML, CSS and JavaScript. The dynamic execution is managed by Node.js, which runs in the server-side.

3.2 Community Finding Algorithms

In terms of accuracy, Louvain outperformed all others for weakly mixed networks ($\mu < 0.4$). It is followed by Infomap, LP and LLP ($\gamma = 0.5$ for $\mu < 0.3$ or Infomap, LLP and LP ($\gamma = 0.5$) for $0.3 \leq \mu \leq 0.4$. By increasing μ, LLP, which is able to detect communities based on the panorama of the whole network, consistently becomes the most effective for $\mu < 0.4$. Comparing Louvain and Infomap when $\mu \geq 0.5$, their performance is similar. Based on the CI 95%, it is not possible to state one performs better than the other. Another point to highlight is that Louvain and Infomap are very sensitive, consequently, unreliable detecting communities in networks with $\mu \approx 0.5$. LLP and LP do not present any specific value in which community detection is steeply affected. These conclusions are all valid to GN and LFR networks (Figs. 7 and 8).

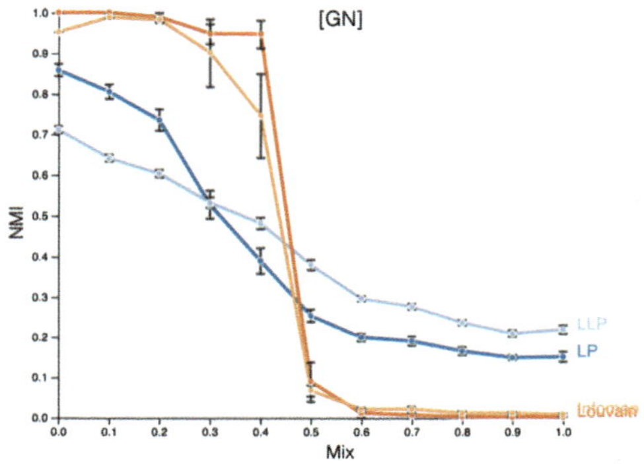

Fig. 7. Clustering quality of each algorithm: Louvain, Infomap, LP and LLP in terms of the mixing parameter in GN networks, with the remaining properties as in Fig. 1.

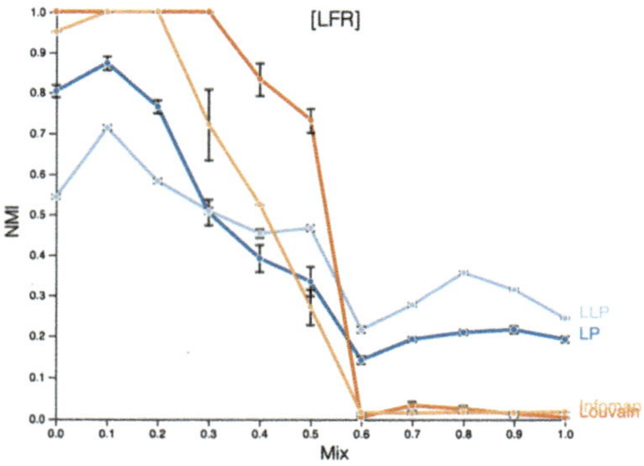

Fig. 8. Clustering quality of each algorithm: Louvain, Infomap, LP and LLP in terms of the mixing parameter in LFR networks, with the remaining properties as in Fig. 2.

Having considered only two gamma values in LLP, a more precise analysis was needed to check whether there could be others achieving better results.

In the case of GN network (Fig. 9), it is possible to detect communities with higher accuracy in networks less mixed. The performance decreases whenever we increase γ for $\mu < 0.5$. When $\mu \geq 0.5$, the NMI between the detected partition and the one in the original network is higher as long as we keep increasing γ. Another point to highlight is the possibility of detecting communities with higher precision in networks with higher μ using appropriate γ than in others with lower μ but with a non-optimal γ. This is valid for networks with $\mu > 0.5$.

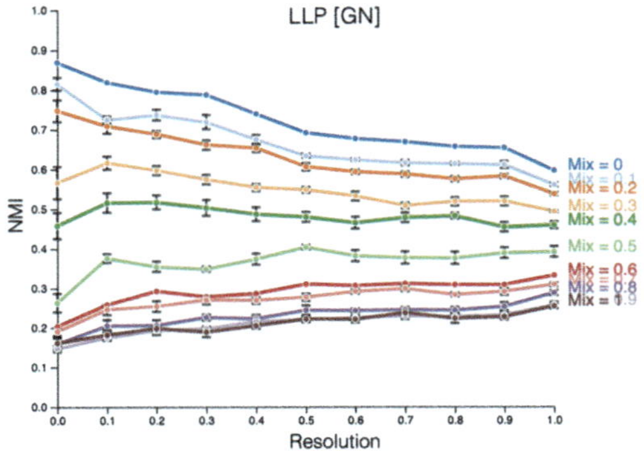

Fig. 9. LLP algorithm accuracy in terms of the resolution parameter. Analysis performed for mixing parameters 0–1 (blue - brown) in GN network, with the remaining properties as in Fig. 1.

In LFR networks, increasing gamma always leads to a decrease in the NMI (Fig. 10).

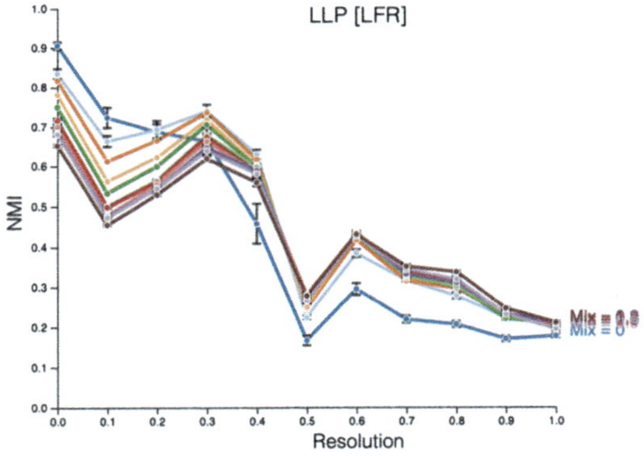

Fig. 10. LLP algorithm accuracy in terms of the resolution parameter. Analysis performed for mixing parameters 0–1 (blue - brown) in LFR network, with the remaining properties as in Fig. 2.

The influence of the average node degree was tested to check whether it affects the capacity of each algorithm to identify communities.

In the case of GN networks, for $\mu \leq 0.5$, an increased average degree enhances the capacity of Louvain, Infomap, LP and LLP to differentiate communities. For $\mu > 0.5$, NMI clustering quality is lower as higher is the average node degree. Assuming it is not expected the algorithms to identify the original communities for $\mu \approx 1$, this suggests that higher average degrees tends to decrease detection by chance. This observation can be applied to GN and LFR networks (Fig. 11).

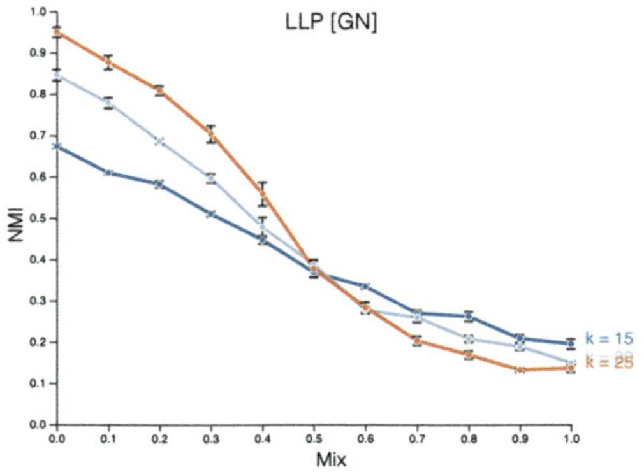

Fig. 11. LLP algorithm accuracy in terms of the mixing parameter. GN network with average node degrees: 15, 20 and 25. The remaining properties are similar to the ones in Fig. 1.

In terms of speed, Louvain algorithm performed significantly better than any other. One important property that contributes to its efficiency is that it only calculates modularity variation at each iteration of Modularity Optimization phase (1). In Infomap, minimum description length is recalculated at the end of each iteration of the optimization step (2). Due to this computationally intensive step, Infomap presented the poorest performance in the benchmark tests. Based on Table 1, it was expected the difference of execution time between Infomap and all others to be logarithmic. Due to inefficient steps in the implementation, this was not achieved. Comparing LLP and LP, the fastest was the one that needed to optimize the simplest equation – LP (Fig. 12). Although, time complexity appears to be the same, as predicted in Table 1. Again, using the same table, it was expected LLP and LP to execute about as fast as Louvain. The small difference observed in Fig. 12 might be explained by some non-optimized step in the implementation of LLP/LP.

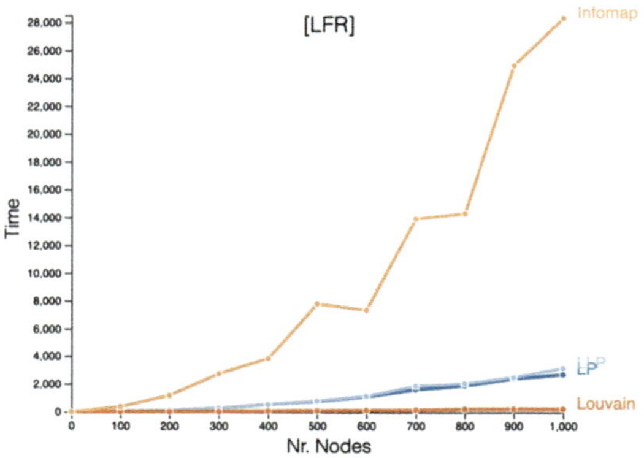

Fig. 12. Time that Louvain, Infomap, LLP and LP took to finalize the analysis in terms of the size of the LFR networks. The remaining properties are similar to the ones in Fig. 2.

3.3 Benchmark Networks Algorithms

Previous GN and LFR networks were generated so the previous tests could be performed. The algorithm to create the former was implemented in the thesis. The time it takes to generate such networks increases exponentially with the average node degree and the mixing parameter (Fig. 14). Becoming unfeasible to execute it for average degrees higher than 20, in an ordinary computer. Its performance was compared to the Fortunato's implementation which is highly scalable and allows the generation of networks with much higher number of links in shorter times (Fig. 13). It was verified an approximately linear increase in the time of generation of networks with progressively higher mixing parameters, contrarily to the one implemented in the thesis. Just like before, generating networks with more edges, requires additional computational power.

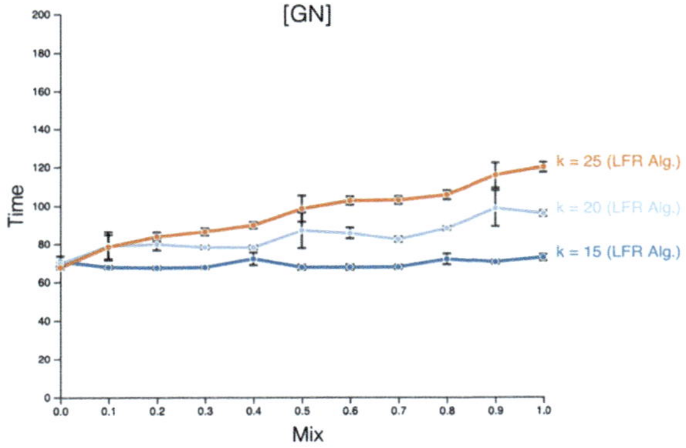

Fig. 13. Time needed to generate GN networks (using LFR implementation) in terms of their mixing parameter. Analysis performed in networks with average node degrees: 15, 20 and 25. The remaining properties are similar to the ones in Fig. 1.

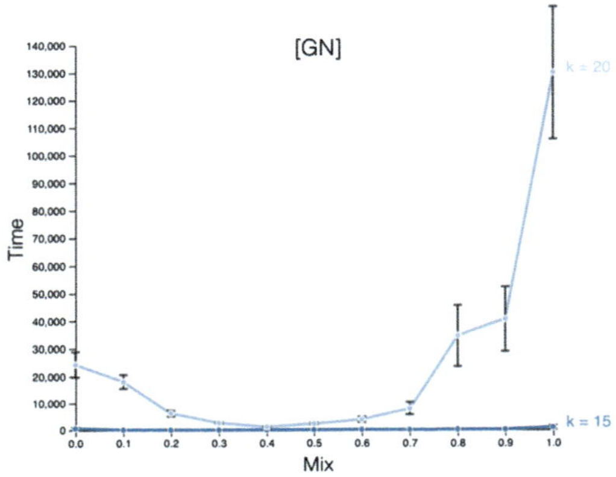

Fig. 14. Time needed to generate GN networks (using thesis implementation) in terms of their mixing parameter. Analysis performed in networks with average node degrees: 15 and 20. The remaining properties are similar to the ones in Fig. 1.

Additional tests were performed in LFR networks. The execution time was tested against the number of nodes and the mixing parameter of the network. In both cases, it was verified an approximately linear relation (Fig. 15 and Fig. 16). The number of nodes was not considered in GN networks, once they have a fixed number of elements per community and of communities. In spite only the mixing parameter and the number of nodes were analyzed, the user can still choose to adjust nodes maximum degree, the exponents of the nodes degree distribution and communities size distribution, the maximum/minimum number of nodes per community, the number of overlapping nodes per community (covers were not considered in the thesis) and the number of memberships for the overlapping nodes [22].

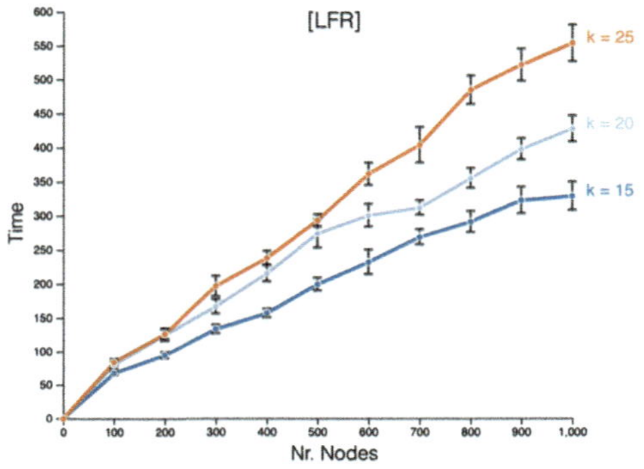

Fig. 15. Time needed to generate LFR networks in terms of their number of nodes. Analysis performed in networks with average node degrees: 15, 20 and 25. The remaining properties are similar to the ones in Fig. 2.

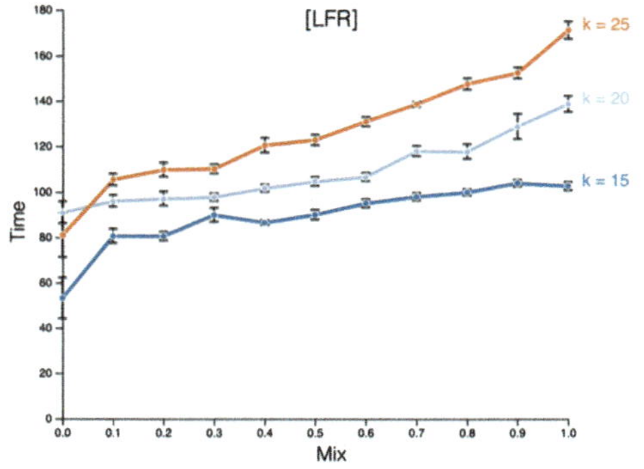

Fig. 16. Time needed to generate LFR networks in terms of their mixing parameter. Analysis performed in networks with average node degrees: 15, 20 and 25. The remaining properties are similar to the ones in Fig. 2.

Every community finding, benchmark network generator, accuracy measure and related algorithms were made available in NPM (Table 2). This way, they can be used in any user-specific application.

Table 2. Algorithms implemented in the thesis were made available in NPM.

Algorithm	Package name	URL
Louvain	Louvain-algorithm	npmjs.com/package/louvain-algorithm
Infomap	Infomap	npmjs.com/package/infomap
LLP	Layered-label-propagation	npmjs.com/package/layered-label-propagation
GN network generator	Girvan-newman-benchmark	npmjs.com/package/girvan-newman-benchmark
NMI	Normalized-mutual-information	npmjs.com/package/normalized-mutual-information
Hamming distance	Hamming-dist	npmjs.com/package/hamming-dist

3.4 Visualization Frameworks

D3.js (SVG) allowed the user to visualize all networks used in the web application, as well as to represent the benchmark plots with the highest resolution possible (limited by density of pixels in the user's device). In terms of performance, it is the second fastest.

D3.js (Canvas) is the fastest among the analyzed frameworks. Nevertheless, when compared to the previous, the nodes and links from the network lose resolution.

Cytoscape.js is advisable when network analysis will be performed along with its representation. There is a considerable number of algorithms, including community finding ones, that can facilitate the study of the graph. The lack of working examples and the slow processing of network data made Cytoscape.js the less efficient. Representation of *S. aureus* SLV network and others with similar or higher number of nodes/edges overflows the web application.

3.5 Staphylococcus Aureus

Louvain, Infomap and LLP algorithms were executed in the biggest component of the SLV graph obtained from the MLST dataset of *S. aureus*.

Louvain algorithm was executed using three modularity variation thresholds: 0.1, 0.02 and 0.01. It is possible to verify that high modularity partitions are rapidly achieved (2 passes needed to obtain final result), as predicted in Introduction. This characteristic makes modularity-based algorithms fast.

Using Infomap algorithm, several minimum description length thresholds were considered: 0.1, 0.02 and 0.01. For all values, every small dense group of nodes descending from the same ancestor was separated in different communities. Thus, it was not further considered.

LLP algorithm considered three values for the resolution parameter: 0, 0.5 and 1. Similarly to Infomap, it partitioned the network in several small and dense communities.

From the benchmark analyses of Louvain, Infomap and LLP against GN network (Figs. 7 and 8), it was predictable Louvain would perform better than the others, once it run in a network with well-defined communities. After adjusting the stopping parameter of Louvain, such that the communities identified were the most pertinent – nodes from densely connected regions descendent from the same ancestor belonging to the same community, a final partition was obtained (Fig. 17). It is important to note that some clusters analogous to the one highlighted with a square were subdivided by the algorithm. One explanation is the presence of high recombination rates between strains of different sub-groups.

By uploading to PHYLOViZ Online the generated metadata file with the nodes labeled accordingly to the inferred communities, one may observe the original phylogenetic tree with the same communities identified.

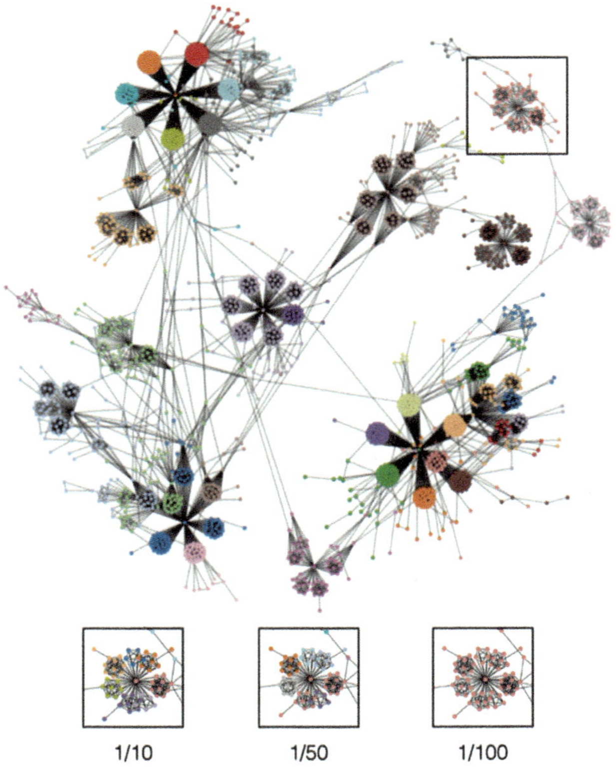

Fig. 17. Partition obtained after running Louvain algorithm (considering different thresholds for the modularity variation) on Clonal Complex 0 (as identified in PHYLOViZ 2) of *S. aureus* MLST SLV network. Represented using D3.js and Canvas.

4 Conclusion

Louvain, Infomap and LLP algorithms were implemented in JavaScript. Unless otherwise stated, next conclusions are valid for GN and LFR networks. In terms of speed, Louvain outperformed all others. In terms of accuracy, in networks with well-defined communities, Louvain was the most accurate. For higher mixing, LLP was the best. Contrarily to weakly mixed, it is advantageous to increase the resolution parameter in highly mixed GN networks. In LFR networks, higher resolution decreases the accuracy of detection, independently of the mixing parameter. The increase of the average node degree enhanced partitioning accuracy, suggesting detection by chance was minimized. It is computationally more intensive to generate GN networks with higher mixing or average degree, using the algorithm developed in the thesis or the LFR implementation. In the *S. aureus* MLST SLV network, Louvain was the fastest and the most accurate in detecting the clusters of 7 groups of strains evolved from the same ancestor. The lower the modularity variation threshold, the better the detection.

Along with the three community finding algorithms, GN benchmark network generator, NMI and hamming distance algorithms were made available in NPM. The web application is hosted in the Heroku cloud platform. An image containing all the required modules to run the app in a local machine can be download from Docker Hub. All the implementations and a complete roadmap of the thesis was made available in GitHub. A Digital Object Identifier (DOI) was attributed to this repository using Zenodo.

One of the main difficulties encountered was running community finding algorithms in the user-side. A feature that would rise less privacy concerns. It was necessary to run the algorithms in the server, so that the application would not crash. It was not possible to reach, in the Infomap implementation, the computational efficiency predicted by its time complexity. The online visualization of networks with more than 10 000 nodes was unfeasible, even using the framework that showed the best performance – D3.js (using Canvas element).

During the next months, some of the tools developed in this thesis will be implemented in INSaFLU. New visualization frameworks for displaying phylogenetic trees, along with the uploaded metadata, will enhance the traceability of the evolutionary path among influenza strains. A geographical map dynamically varying with time will allow the user to precise the moment and location at which each sample was obtained and enhance flu surveillance during the critical season. An interactive histogram displaying each nucleotide belonging to a consensus sequence obtained from a sample of influenza virus with the SNPs highlighted, as well as with their effects in the phenotype of the strain specified. New flu surveillance dynamic reports will allow the visualization of increased quantities of data in a minimalistic way.

As future improvements, the implemented algorithms can be optimized in terms of memory and speed. Network and chart visualization may run faster in a more efficient implementation. The previous updates will make the web application to load quicker and run smoother. To benchmark community finding algorithms with a more representative set of networks (spanning a wider range of topological properties) would make possible to infer, with a higher certainty, if any of the community finding algorithms consistently performs better in phylogenetic networks. The integration of community finding functionalities developed in the thesis in a phylogenetic/surveillance-oriented analysis web application like PHYLOViZ Online or INSaFLU would be beneficial.

References

1. The world's most valuable resource is no longer oil, but data. The Economist, 6 May 2017. economist.com/leaders/2017/05/06/the-worlds-most-valuable-resource-is-no-longer-oil-but-data. Accessed 28 May 2019
2. Barabási, A.-L.: Network science book. networksciencebook.com. Accessed 15 May 2019
3. Patil, H.G.S., Babu, A.N., Ramkumar, P.S.: Non-invasive data acquisition and measurement in bio-medical technology: an overview. In: Maximizing Healthcare Delivery and Management through Technology Integration. IGI Global (2016)
4. Health. European Data Protection Supervisor. edps.europa.eu/data-protection/our-work/subjects/health_en. Accessed 9 June 2019
5. Ten threats to global health in 2019. World Health Organization. who.int/emergencies/ten-threats-to-global-health-in-2019. Accessed 4 June 2019
6. Antimicrobial resistance. World Health Organization, 15 February 2018. who.int/en/news-room/fact-sheets/detail/antimicrobial-resistance. Accessed 29 May 2019
7. Memish, Z.A., Venkatesh, S., Shibl, A.M.: Impact of travel on international spread of antimicrobial resistance. Int. J. Antimicrob. Agents **21**(2), 135–142 (2003)
8. Top 10 Leading Causes of Death Globally. theatlas.com/charts/HkLaDreuW. Accessed 12 May 2019
9. Ribeiro-Gonçalves, B., Francisco, A.P., Vaz, C., Ramirez, M., Carriço, J.A.: PHYLOViZ online: web-based tool for visualization, phylogenetic inference, analysis and sharing of minimum spanning trees. Nucleic Acids Res. **44**(1), 246–251 (2016)
10. Motro, Y., Moran-Gilad, J.: Next-generation sequencing applications in clinical bacteriology. Biomol. Detect. Quantif. **14**, 1–6 (2017)
11. Blondel, V.D., Guillaume, J.-L., Lambiotte, R., Lefebvre, E.: Fast unfolding of communities in large networks. J. Stat. Mech. **P10008**, 12 (2008)
12. Rosvall, M., Axelsson, D., Bergstrom, C.T.: The map equation. Eur. Phys. J. Spec. Top. **178**(1), 13–23 (2009)
13. Raghavan, N., Albert, R., Kumara, S.: Near linear time algorithm to detect community structures in large-scale networks. Phys. Rev. E **76**(3) (2007). 25th Anniversary Milestones
14. Boldi, P., Rosa, M., Santini, M., Vigna, S.: Layered label propagation: a multiresolution coordinate-free ordering for compressing social networks. In: WWW 2011 Proceedings of the 20th International Conference on World Wide Web (2011)
15. Šubelj, L.: Label propagation for clustering. In: Advances in Network Clustering and Blockmodeling. Wiley, New York (2018)
16. Girvan, M., Newman, M.E.J.: Community structure in social and biological networks. Proc. Nat. Acad. Sci. U.S.A. **99**(12), 7821–7826 (2002)
17. Lancichinetti, A., Fortunato, S., Radicchi, F.: Benchmark graphs for testing community detection algorithms. Phys. Rev. E: Stat. Nonlin. Soft Matter Phys. **78**(4) (2008)
18. Lancichinetti, A., Fortunato, S., Kertesz, J.: Detecting the overlapping and hierarchical community structure of complex networks. New J. Phys. **11**, 033015 (2009)
19. Yang, J., Leskovec, J.: Defining and evaluating network communities based on ground-truth. In: Proceedings of 2012 IEEE International Conference on Data Mining (ICDM) (2012)

20. Zachary, W.: An information flow model for conflict and fission in small groups. J. Anthropol. Res. **33**, 452–473 (1976)
21. *Staphylococcus aureus* MLST Databases. PubMLST, 5 June 2019. pubmlst.org/saureus/. Accessed 5 June 2019
22. Lancichinetti, A., Fortunato, S.: Benchmarks for testing community detection algorithms on directed and weighted graphs with overlapping communities. Phys. Rev. E: Stat. Nonlin. Soft Matter Phys. **80**, 016118 (2009)

Functional Electrical Stimulation for Gait Rehabilitation

Ana Correia[1(✉)], Jorge M. Martins[2], and Cristina P. Santos[3]

[1] Industrial Electronic Engineering and Computers, University of Minho, 4800-058 Guimarães, Portugal
a71473@alunos.uminho.pt

[2] Department of Mechanical Engineering, Instituto Superior Técnico, 1049-001 Lisbon, Portugal
jorgemartins@tecnico.ulisboa.pt

[3] Department of Electronic Engineering, University of Minho, 4800-058 Guimarães, Portugal
cristina@dei.uminho.pt

Abstract. Conditions that can lead to a **full or partial motor function loss**, such as stroke or multiple sclerosis, leave people with disabilities that may interfere severely with lower body movements, such as gait. **Drop Foot (DF)** is a gait disorder that results in a reduced ability or total inability to contract the *Tibialis Anterior (TA)* muscle, **causing an inability to raise the foot during gait**. One of the most effective methods to correct DF is **Functional Electrical Stimulation (FES)**. FES is a technique used to reproduce the activation patterns of functional muscles, in order to **create muscular contractions through electrical stimulation of the muscle's nervous tissue.**

FES has first been introduced in 1961. However, the available commercial FES systems still do not take into account the fact that the **gait differs from subject to subject**, depending on their **physical condition**, **muscular fatigue** and **rehabilitation stage**. Therefore, they are **unable to provide a personalized assistance to the user**, delivering constant stimulation pulses that are only based on gait events. Consequently, they promote the **early onset of fatigue** and **generate coarse movements**. This dissertation aims to tackle the aforementioned issues by developing a **FES system for personalized DF correction**, tailored to each individual user's needs through the use of a **Neural Network (NN)**.

A Non-Linear Autoregressive Neural Network with Exogenous inputs (NARX Neural Network) was used to model the dynamics of the electrically stimulated TA muscle, in a **novel approach that uses both the foot angle and the foot velocity**. The model was combined with a **Proportional Derivative controller** to help compensate for any external disturbances. In order to create more natural movements, **reference trajectories** were obtained by recording the foot angle and velocity of **healthy subjects walking at different speeds**.

The system has been validated with a healthy subject walking at 3 different speeds on a treadmill: 1 km/h, 1.5 km/h and 2 km/h. It was **able**

J. Henriques et al. (Eds.): MEDICON 2019, IFMBE Proceedings 76, pp. 1954–1966, 2020.
https://doi.org/10.1007/978-3-030-31635-8_235

to track the desired trajectory for every speed, thus creating a **more natural movement** and **effectively correcting DF gait**.

1 Introduction

Conditions that can lead to full or partial motor loss, as is the case of Cerebral Palsy, Stroke, Multiple Sclerosis, Amyotrophic Lateral Sclerosis, Muscular Dystrophy, Poliomyelitis, Muscular Atrophy and Spinal Cord Injuries, leave people with disabilities that interfere severely with their daily activities [1]. According to the World Health Organization, of the 15 million people that suffer a stroke worldwide every year, 5 million remain permanently disabled. The Portuguese Society of Cerebrovascular Accident reported that 16% of stroke patients develop slight to moderate disabilities while 41% lose their independence completely. Furthermore, a study conducted in the United Kingdom concluded that the yearly rehabilitation cost for each stroke patient averaged at 13 000 €. When it comes to multiple sclerosis, the World Health Organization estimates that it affects 2.5 million people worldwide, while in Portugal, it affects over 8 thousand people. The disabilities that come as a consequence of these conditions can affect upper limb movements, which is the case of grasp, or lower limb movements, such as gait. They affect the patient's quality of life and, in some cases, may even lead to a loss of independence. Hence, there is a growing need for more economic and effective rehabilitation methods. Functional Electrical Stimulation (FES) can replace the central nervous system, generating the necessary potential to induce a muscular contraction. It works by stimulating the nervous tissue of the paralyzed muscle so that it contracts, consequently generating movement [2]. The pulses with which they are stimulated can have different parameter values, such as amplitude, frequency or width. Being able to vary the pulse's parameters in real-time is important, given the non-linear behavior muscles exhibit while being electrically stimulated [3]. It should also be taken into account that different muscles have different characteristics and, therefore, need different stimulation patterns. Drop Foot (DF) is a gait disorder that results in a reduced ability or total inability to contract the Tibialis Anterior (TA) muscle. Subjects affected by DF are more prone to falls, since they cannot lift the foot off the ground completely. In the case of the elderly, this lack of mobility can even be fatal. It is estimated that 75% of the people that survive a stroke are affected by some kind of lower limb weakness and, of those, 20% are affected by DF [4]. If the DF is caused by damaged nerves, FES can have a crucial part in achieving a partial or full recovery, since the electric pulses delivered to the muscle help strengthen the synaptic connections of the spinal cord [5]. Even though FES has been used since 1961, when the first system to correct drop foot was created, most of the recently developed systems still use pre-programmed movements to correct DF [3]. Taking into account the fact that muscular activation patterns are different for every subject, depending on their physical condition, muscular fatigue and rehabilitation stage, it is important to provide a personalized assistance for each user. Therefore, the purpose of this dissertation was the creation of a wearable

FES system for real-time DF correction, so patients can achieve a healthy gait pattern. The system was required to be personalized to each user's needs, be able to provide assistance at different walking velocities and be comfortable for daily use.

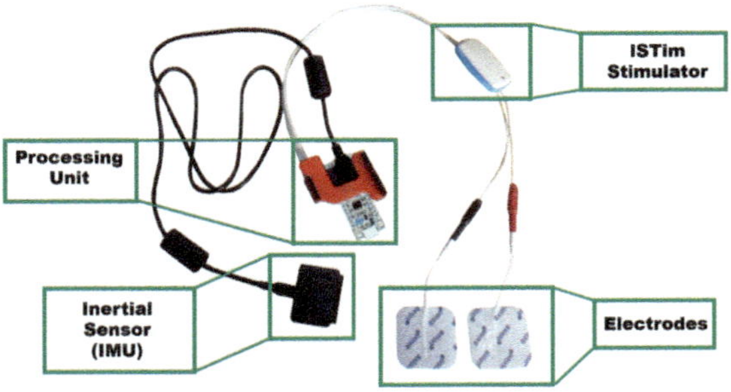

Fig. 1. FES system used for DF correction.

2 Contributions

This project has contributed with the following aspects:

- The use of a NN to create a novel dynamic model for the TA muscle, using both the foot angle and the foot angular velocity, which distinguishes this project from both the commercial and research prototypes available for DF correction.
- A personalized experience for the users, since the model is tailored to the specific dynamics of their electrically stimulated TA muscle, allowing the system to adapt the stimulation parameters to the user's anatomy.
- Won the 1st place in the Fraunhofer Portugal Challenge in the MsC category;
- Yielded an article that has been submitted to the Robotics and Automation Letters

3 Methods

3.1 Designed System

FES systems are meant to be small and practical, since they should be used on a daily basis to help its users perform everyday tasks that cannot be correctly performed otherwise and they must also not interfere with any of the user's movements. FES systems for DF correction are comprised of a stimulator and electrodes, to stimulate the TA muscle, a sensor to acquire the foot's kinematic parameters and a microcontroller unit (MCU) to implement the control strategy.

Based on these requirements the chosen components for this project were the *STM NUCLEO-32F303K8®* MCU used to implement the control strategy; a Modular Stimulation Unit (MSU) from the ISTim Modular Stimulation System developed by IST [6,7], used to deliver the stimulation pulse to the TA muscle; and the *MPU6050®* Inertial Measurement Unit (IMU) used to acquire the foot's kinematic parameters. All the components for the designed DF correction system can be seen in Fig. 1.

3.2 Drop Foot Correction Control Strategy

Following the choice of the system's components, it was necessary to outline the subsequent steps to be taken in order to develop the DF correction control strategy. The proposed trajectory tracking control strategy can be seen in Fig. 2 and it is divided in two main elements: the Neural Network (NN) model and the Proportional Derivative (PD) controller.

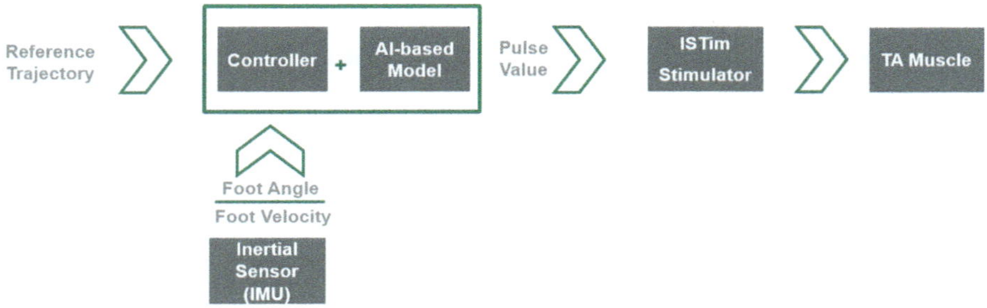

Fig. 2. Implemented trajectory tracking control strategy.

Based on these two main elements, the project was divided into two main stages, each comprised of a set of tasks:

I Modelling of the Electrically Stimulated TA muscle: (i) choosing the most appropriate NN; (ii) defining the excitation signals to capture the full dynamics of the TA muscle; (iii) defining the calibration routine for the data acquisition; (iv) training the NN with different types of data and parameters.

II Trajectory Tracking Control Strategy: (i) implementation of the gait phase detection algorithm; (ii) acquisition of a gait trajectory from healthy subjects walking at different speeds; (iii) implementation of a PD controller; (iv) system validation with the stand-alone model; (v) static validation with model combined with the controller; (vi) system gait validation with subject walking at different speeds.

Modelling of the Electrically Stimulated TA Muscle. NNs have been extensively used in FES research projects to model electrically stimulated muscles, be it for upper limb movements [8–10] or for lower limb movements [11–16], given they can model the non-linear behavior of the electrically stimulated muscles. Although in the past, it was more common to use feedforward neural networks with FES [8,12,13], the most recent projects use recurrent neural networks, in particular Non-Linear Autoregressive Neural Networks with Exogenous inputs (NARX NN) [9–11,14–16]. These models are useful for FES research purposes because of the small number of required parameters and their ability to represent the nonlinear dynamic behavior of the electrically stimulated muscle. Considering that the main goal of this project was to correct DF during real-time gait and since the use of a non-linear model provides a more accurate model for the control of the full range of motion of a joint, a NARX NN was chosen to model the electrically stimulated TA muscle in this project. The use of a NARX NN to create a novel dynamic model for the TA muscle, using both the angle and the angular velocity of the foot, distinguishes this project from both the commercial and research prototypes available for DF correction. The implemented NARX NN can be seen in Fig. 3.

In FES systems, pulse frequency usually remains the same throughout the stimulation, since it is directly linked to the increase of muscular fatigue. Hence, for this project the pulse parameter to be updated could either the pulse amplitude or the pulsewidth, since they are both related to muscle fiber recruitment. Since the pulsewidth range available in the MSU spans through a wider range of values, when compared to the amplitude, that was the parameter chosen to be controlled.

In order to model the dynamics of the electrically stimulated muscle it was necessary to acquire real-time data of the foot angle correspondent to the stimulation pulsewidth, that covered the full range of motion of the foot. This was necessary because DF affects the swing phase of gait. Since the model is tailored to each user's needs, a calibration should be done whenever a new user intends to use the system. The calibration worked by applying different stimulation pulsewidths to the user's TA muscle, so the full dynamics of the muscle can be captured by the model. The proposed calibration method can be done with disabled subjects, since it consists only of being sat down on a chair without having the foot touch the floor, as can be seen in Fig. 4.

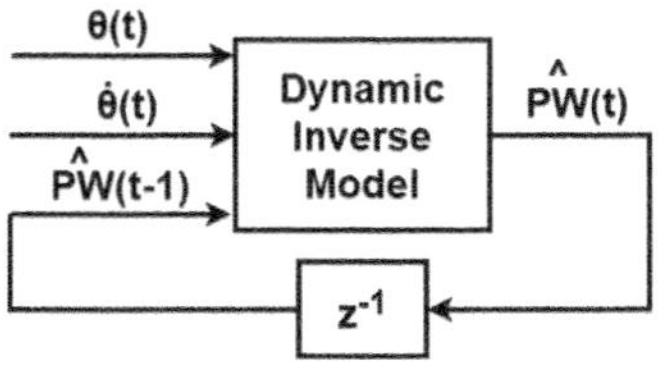

Fig. 3. Inverse dynamic NARXNN diagram.

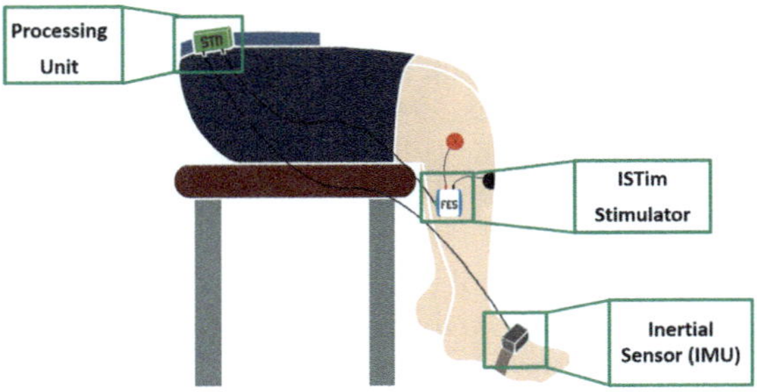

Fig. 4. Proposed calibration method.

With the obtained data, initially a direct model was created as a basis for comparison, having as input the pulsewidth and as output the foot angle. Afterwards, a basic inverse model was created, which had as input the foot angle and as output the pulsewidth. Finally, a dynamic inverse model was created. This model used as input, both the foot angle and the foot angular velocity. The models were created with different parameters, using NARX NNs with 1 or 2 hidden layers and 5, 10 or 20 neurons, to determine which NARX NN configuration generated better predictions.

Trajectory Tracking Control Strategy. The trajectory tracking control strategy is responsible for correcting DF during real-time gait. A literature analysis demonstrated that combining a model with a controller provides the best tracking performance, since the controller is able to compensate for any modeling errors that inevitably occur in real-time and the model avoids time delays inherent to the controller [12,13,13,17–21]. Thus, the chosen dynamic inverse model was combined with a PD controller, in order to help reduce fatigue and compensate for any external disturbances, providing a more robust and effective control strategy.

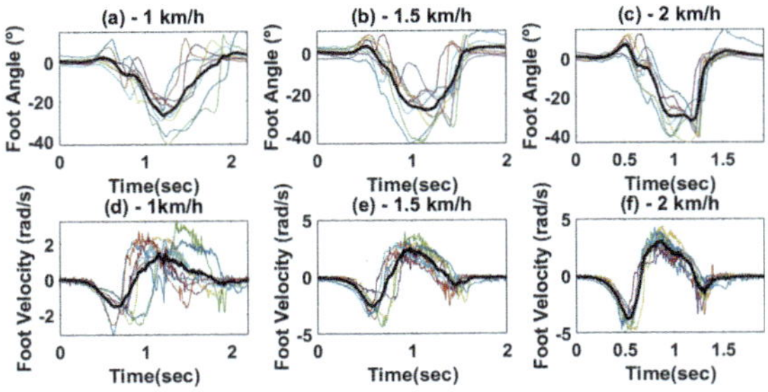

Fig. 5. Acquired gait trajectories: (a) and (d) foot angle and velocity for 1 km/h, respectively; (b) and (e) foot angle and velocity for 1.5 km/h, respectively; (c) and (f) foot angle and velocity for 2 km/h, respectively. The thin lines show each subject's step and the thick line represents the resulting averaged reference trajectory.

Since DF only affects the swing phase of gait, it was necessary to identify when the swing phase ends and when it begins, during real-time gait. For this purpose, the toe-off and heel strike gait phases were identified during real-time gait, using the foot's angular velocity. This was obtained with the IMU that was also being used to acquire the foot kinematics, in order to reduce the number of sensors placed on the subject.

The trajectory tracking control strategy makes the foot follow a desired trajectory, which in this case was the foot angle and foot angular velocity trajectories correspondent to the swing phase, since this is the gait phase affected by DF. In order to generate a biological-like movement for the ankle of DF subjects, the reference trajectories input to the control strategy were created using the foot trajectories of 10 healthy subjects during real-time gait. The subjects walked on a treadmill at different velocities (1 km/h, 1.5 km/h and 2 km/h), and the resultant reference trajectory was input to the control strategy so that the foot of DF subjects mimics healthy gait patterns. The obtained reference trajectories can be seen in Fig. 5.

In order to establish a comparison for the performance of the implemented trajectory tracking control strategy, an open-loop control strategy using only the dynamic inverse NARX NN model was used. The same trajectory was input to the open-loop control strategy and to the trajectory tracking control strategy, to compare the performance of both strategies.

4 Results

4.1 Modelling of the Electrically Stimulated TA Muscle

The results obtained with the dynamic inverse model, which can be seen in Table 2, show the model was able to predict the pulsewidth values with more accuracy, when compared to the results of the basic inverse model, which can be seen in Table 1.

The prediction time responses of the inverse models were also studied, showing that both models had the lowest response time for NARX NNs with 10 neurons, as can be seen in Table 3.

Thus, since there is only 1 ms difference in response time, from the basic inverse model to the dynamic inverse model, and knowing that the dynamic inverse model provided higher average accuracy predictions for all cases, the model chosen for the control strategy was a dynamic model with 2 hidden layers and 10 neurons.

4.2 Drop Foot Correction Control Strategy

Two configurations were used to test the trajectory tracking control strategy: a static configuration, with the subject sitting on a chair, and a gait configuration, with the subject walking on a treadmill. The static configuration was used to determine if the control strategies were able to track the reference trajectories and to compare the performance of the open-loop control strategy with the trajectory tracking control strategy. The trajectory tracking control strategy

Table 1. Results for basic inverse model with 5, 10 or 20 neurons and 1 or 2 hidden layers.

No. Neurons	No. Layers	Trial 1	Trial 2	Trial 3	Average	Transitions
5	1	61.31%	53.02%	57.69%	57.34%	Fast
	2	62.96%	55.82%	62.64%	**60.47%**	
10	1	53.51%	49.95%	−24.31%	26.38%	
	2	57.85%	54.96%	63.34%	**58.72%**	
20	1	58.74%	52.93%	65.44%	59.04%	
	2	62.23%	58.20%	62.33%	**60.92%**	
5	1	71.17%	65.50%	63.01%	66.56%	Medium
	2	72.94%	66.77%	68.45%	**69.39%**	
10	1	78.18%	66.18%	62.89%	69.08%	
	2	78.43%	67.96%	65.25%	**70.55%**	
20	1	77.04%	62.83%	54.48%	64.68%	
	2	80.10%	70.92%	61.52%	**70.85%**	
5	1	46.33%	36.16%	56.77%	46.42%	Slow
	2	67.68%	68.66%	78.99%	**71.78%**	
10	1	45.24%	−57.71%	74.30%	20.61%	
	2	45.14%	72.07%	75.40%	**64.20%**	
20	1	33.81%	52.50%	77.69%	54.67%	
	2	71.08%	70.37%	78.61%	**73.35%**	

[a]Best results for each No. of neurons are colored green.

was also tested in a static configuration with a weight strapped to the foot, to see if the control strategy was able to compensate external disturbances. The gait configuration was used to determine if the trajectory tracking control strategy was able to correct DF during real-time gait.

For the open-loop validation, the dynamic inverse model predicted the correct pulsewidth values for the 1 km/h trajectory accurately, resulting in the foot following the desired trajectory with an accuracy of 68.16%. This showed the model was tailored to the user's characteristics. By pairing the PD controller with the dynamic inverse model, the tracking accuracy increased by 9.71%. The results can be seen in Fig. 6.

For the weighted validation with a 0.4 kg weight strapped to the foot, for the 1.5 km/h trajectory, the control strategy was able to match the performance of the unrestrained validation. The accuracy decreased only by 3.45% when compared to the unrestrained validation, thus being able to compensate the external disturbances. The results can be seen in Fig. 7.

For the gait configuration, for every velocity a pseudo-DF gait pattern was compared to the foot angle generated by the trajectory tracking control strategy in real-time. This showed that, contrary to the DF gait, the foot was able to rise above 0°, generating a foot movement similar to the healthy one. This prevents

Table 2. Results for dynamic inverse model with 5, 10 or 20 neurons and 1 or 2 hidden layers.

No. Neurons	No. Layers	Trial 1	Trial 2	Trial 3	Average	Transitions
5	1	61.74%	51.89%	54.35%	55.9%	Fast
	2	63.75%	53.14%	55.25%	**57.38%**	
10	1	64.03%	49.21%	51.81%	55.02%	
	2	66.01%	66.64%	60.57%	**64.41%**	
20	1	60.34%	52.70%	67.14%	60.06%	
	2	69.49%	64.27%	70.44%	**68.07%**	
5	1	53.05%	70.51%	66.18%	63.25%	Medium
	2	72.42%	65.44%	72.81%	**70.22%**	
10	1	70.52%	63.26%	76.43%	70.07%	
	2	84.38%	67.67%	62.26%	**71.44%**	
20	1	81.31%	68.48%	75.65%	75.15%	
	2	83.90%	73.95%	71.49%	**76.45%**	
5	1	58.83%	64.38%	71.30%	64.84%	Slow
	2	84.78%	68.58%	78.53%	**77.30%**	
10	1	87.15%	73.32%	71.48%	77.32%	
	2	86.89%	73.18%	81.86%	**80.64%**	
20	1	88.06%	73.91%	78.41%	80.13%	
	2	89.05%	72.65%	82.54%	**81.41%**	

[a]Best results for each No. of neurons are colored green.

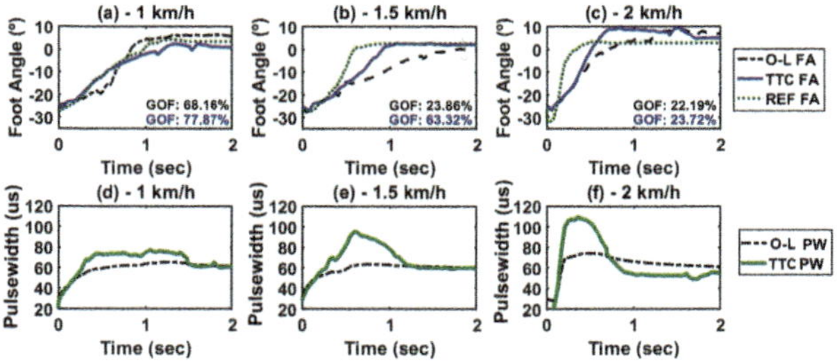

Fig. 6. Performance comparison between open-loop and trajectory tracking control. Reference foot angle trajectory and real foot angle for (a) 1 km/h, (b) 1.5 km/h and (c) 2 km/h. Applied pulsewidth for (d) 1 km/h, (e) 1.5 km/h and (f) 2 km/h. O-L stands for open-loop control, TTC stands for trajectory tracking control, PW stands for pulsewidth and FA stands for foot angle.

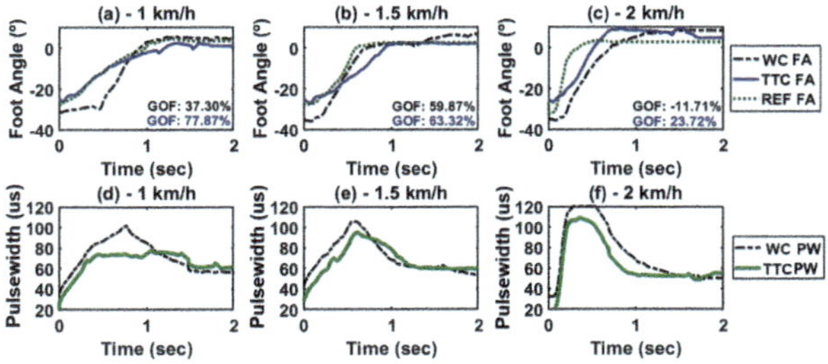

Fig. 7. Performance comparison between unrestrained and weighted trajectory tracking control. Reference foot angle trajectory and real foot angle for (a) 1 km/h, (b) 1.5 km/h and (c) 2 km/h. Applied pulsewidth for (d) 1 km/h, (e) 1.5 km/h and (f) 2 km/h. WC stands for weighted control, TTC stands for trajectory tracking control, PW stands for pulsewidth and FA stands for foot angle.

Table 3. Time response for one prediction for the inverse models with 2 hidden layers.

Inverse Model	No. of Neurons		
	20	10	5
Basic	6.5 ms	**6 ms**	6.5 ms
Dynamic	7.5 ms	**7 ms**	7.5 ms

[a]Best prediction results for each type of transition are colored green.

the subjects from dragging the foot on the ground, thus decreasing the likelihood of falling and tripping when walking. The implemented control strategy was also able to reduce step length.

5 Discussion

In this dissertation a novel DF correction control strategy was developed, with the purpose of increasing the quality of life of subjects with gait impairments. The study of FES methods to help subjects with motor disabilities, used in the commercial and research areas, helped to identify what approach had yet been taken for DF correction.

The favorable results found in the literature and the recent rise in the use of artificial intelligence methods for more user centered FES rehabilitation techniques, made the NARXNN the chosen model in this project. The chosen components for the system that was designed for this project, resulted is a light and small system, that is easily wearable and inserted into daily activities. The use of an IMU for the acquisition of foot kinematics limits the number of sensors placed on the subject, increasing its simplicity, which increases user comfort and facilitates the setup of the system.

The innovative feature of this strategy is the dynamic inverse model, that uses the foot angle as well and the angular velocity to model the electrically stimulated TA muscle, for DF correction. This results in a more robust model, when compared to the basic inverse model, that can be used in the real-time tracking of reference trajectories. Furthermore, the implemented calibration method necessary to create the personalized model for each subject is simple enough to be carried out with DF patients. The chosen model to implement in the trajectory tracking control strategy was a dynamic inverse model with 2 hidden layers and 10 neurons in each layer.

By using a reference trajectory obtained from the gait of healthy subject and by combining the dynamic inverse model with a PD controller, the control strategy increased the accuracy of the reference trajectory tracking, when compared to the dynamic inverse model in open-loop, and allowed the tracking of the reference trajectory during real-time gait. For any of the speeds with which the control strategy was validated on a treadmill, 1 km/h, 1.5 km/h and 2 km/h, the control strategy was able to correct DF, by having the foot at a positive angle every time heel strike occurred.

6 Conclusions

The implemented model is able to capture the dynamics of the electrically stimulated TA muscle of each user, by acquiring user-specific data using a calibration routine. In order to provide a comfortable experience for the user, the system used light and small components, so it would not constraint any user movements. Furthermore, a long-lasting experience is only possible if the muscular fatigue is delayed. Therefore, in this project a trajectory tracking control strategy, composed by a PD controller paired with a dynamic inverse NARX NN model, was used. Thus, only the necessary stimulation pulse is applied at any given moment, depending on the desired reference trajectory. To ensure a natural gait pattern the reference trajectory was obtained from healthy subjects walking at different speeds. This way, the DF gait is corrected not with a constant stimulation pulse, but with a gradual movement that matches the natural one.

For any of the speeds with which the trajectory tracking control strategy was validated on a treadmill, 1 km/h, 1.5 km/h and 2 km/h, the control strategy was able to correct DF. For future improvements, the system should perform an automatic speed selection by calculating the gait velocity in real-time and adapt the reference trajectory accordingly. An wireless IMU should also be used, to allow a complete freedom of movements. This would create a more robust DF correction system, that could be fully integrated in the user's daily activities.

References

1. Melo, P.L., Silva, M.T., Martins, J.M., Newman, D.J.: Technical developments of functional electrical stimulation to correct drop foot: sensing, actuation and control strategies. Clin. Biomech. **30**(2), 101–113 (2015)

2. Kesar, T., Chou, L.W., Binder-Macleod, S.A.: Effects of stimulation frequency versus pulse duration modulation on muscle fatigue. J. Electromyogr. Kinesiol. **18**(4), 662–671 (2008)
3. Hunt, K.J., Munih, M., Donaldson, N.D.N., Barr, F.M.D.: Investigation of the hammerstein hypothesis in the modeling of electrically stimulated muscle. IEEE Trans. Biomed. Eng. **45**(8), 998–1009 (1998)
4. Johnson, C.A., Burridge, J.H., Strike, P.W., Wood, D.E., Swain, I.D.: The effect of combined use of botulinum toxin type A and functional electric stimulation in the treatment of spastic drop foot after stroke: a preliminary investigation. Arch. Phys. Med. Rehabil. **85**(June), 902–909 (2004)
5. Brend, O., Freeman, C., French, M.: Multiple-model adaptive control of functional electrical stimulation. IEEE Trans. Control Syst. Technol. **23**(5), 1901–1913 (2015)
6. Luzio de Melo, P.: A novel functional electrical stimulation system and strategies for motor rehabilitation. Ph.D thesis, Universidade de Lisboa - Instituto Superior Técnico (2014)
7. Luzio de Melo, P., da Silva, M.T., Martins, J., Newman, D.: A microcontroller platform for the rapid prototyping of functional electrical stimulation-based gait neuroprostheses. Artif. Organs **39**(5), E56–E66 (2015)
8. Science, I., Hospital, M.N.: Learning control of hand posture with neural network in FES for hemiplegics, vol. 20, no. 5, pp. 2588–2589 (1998)
9. Imatz-ojanguren, E., Irigoyen, E., Valencia-blanco, D., Keller, T.: Electrical Stimulation in Able-bodied and Hemiplegic Subjects, vol. 0, pp. 1–9 (2016)
10. Popov, N.S., Dozić, D.J., Stanković, M., Krajoski, G.M., Stanišić, D.: Development of a Closed Loop FES System Based on NARX Radial Based Network, pp. 70–74 (2015)
11. Previdi, F.: Identification of black-box nonlinear models for lower limb movement control using functional electrical stimulation. Control Eng. Pract. **10**(1), 91–99 (2002)
12. Chang, G.C., Luh, J.J., Liao, G.D., Lai, J.S., Cheng, C.K., Kuo, B.L., Kuo, T.S.: A neuro-control system for the knee joint position control with quadriceps stimulation. IEEE Trans. Rehabil. Eng. **5**(1), 2–11 (1997)
13. Chen, Y.L., Chen, S.C., Chen, W.L., Hsiao, C.C., Kuo, T.S., Lai, J.S.: Neural network and fuzzy control in FES-assisted locomotion for the hemiplegic. J. Med. Eng. Technol. **28**(1), 32–38 (2004)
14. Azura, N., Bin, S., Kamaruddin, A., Mohamed, N., Mohamed, N.B.: The Quadriceps Muscle of Knee Joint Modelling Using Neural Network Approach: Part 1, pp. 52–57 (2016)
15. Yassin, I.M., Jailani, R., Syahirul, M., Megat, A., Baharom, R., Huzaifah, A.: Comparison Between Cascade Forward and Multi-Layer Perceptron Neural Networks for NARX Functional Electrical Stimulation (FES) -Based Muscle Model, vol. 7, no. 1, pp. 215–221 (2017)
16. Yilei, W., Qing, S., Xulei, Y., Li, L.: Recurrent Neural Network Control of Functional Electrical Stimulation Systems, pp. 400–404 (2006)
17. Ferrarin, M., D'Acquisto, E.: An experimental PID controller for knee movement restoration with closed loop FES system. In: Proceedings of 18th Annual International Conference of the IEEE Engineering in Medicine and Biology Society, pp. 453–454 (1996)

18. Qiu, S., He, F., Tang, J., Xu, J., Zhang, L., Zhao, X., Qi, H., Zhou, P., Cheng, X., Wan, B., Ming, D.: Intelligent algorithm tuning PID method of function electrical stimulation using knee joint angle. In: Conference Proceedings: Annual International Conference of the IEEE Engineering in Medicine and Biology Society, Annual Conference, vol. 2014, pp. 2561–2564 (2014)
19. Basith, A.L., Arifin, A., Arrofiqi, F., Watanabe, T., Nuh, M.: Embedded fuzzy logic controller for functional electrical stimulation system. In: 2016 International Seminar on Intelligent Technology and Its Applications (ISITIA), pp. 89–94 (2016)
20. Quintern, J., Riener, R., Rupprecht, S.: Comparison of simulation and experiments of different closed-loop strategies for functional electrical stimulation: experiments in paraplegics. Artif. Organs **21**(3), 232–235 (1997)
21. Tu, X., Li, J., Li, J., Su, C., Zhang, S., Li, H., Cao, J., He, J.: Model-based hybrid cooperative control of hip-knee exoskeleton and FES induced ankle muscles for gait rehabilitation. Int. J. Pattern Recognit. Artif. Intell. **31**(09), 1759019 (2017)

Deep Aesthetic Assessment of Breast Cancer Surgery Outcomes

Tiago Gonçalves(✉), Wilson Silva, and Jaime Cardoso

Faculty of Engineering, University of Porto, Porto, Portugal
{up201607753,wilson,jsc}@fe.up.pt

Abstract. Breast cancer is a highly mutable and rapidly evolving disease, with a large worldwide incidence. Even though, it is estimated that approximately 90% of the cases are treatable and curable if detected on early staging and given the best treatment. Nowadays, with the existence of breast cancer routine screening habits, better clinical treatment plans and proper management of the disease, it is possible to treat most cancers with conservative approaches, also known as breast cancer conservative treatments (BCCT). With such a treatment methodology, it is possible to focus on the aesthetic results of the surgery and the patient's Quality of Life, which may influence BCCT outcomes. In the past, this assessment would be done through subjective methods, where a panel of experts would be needed to perform the assessment; however, with the development of computer vision techniques, objective methods, such as BAT© and BCCT.core, which perform the assessment based on asymmetry measurements, have been used. On the other hand, they still require information given by the user and none of them has been considered the gold standard for this task. Recently, with the advent of deep learning techniques, algorithms capable of improving the performance of traditional methods on the detection of breast fiducial points (required for asymmetry measurements) have been proposed and showed promising results. There is still, however, a large margin for investigation on how to integrate such algorithms in a complete application, capable of performing an end-to-end classification of the BCCT outcomes. Taking this into account, this thesis shows a comparative study between deep convolutional networks for image segmentation and two different quality-driven keypoint detection architectures for the detection of the breast contour. One that uses a deep learning model that has learned to predict the quality (given by the mean squared error) of an array of keypoints, and, based on this quality, applies the backpropagation algorithm, with gradient descent, to improve them; another which uses a deep learning model which was trained with the quality as a regularization method and that used iterative refinement, in each training step, to improve the quality of the keypoints that were fed into the network. Although none of the methods surpasses the current state of the art, they present promising results for the creation of alternative methodologies to address other regression problems in which the learning of the quality metric may be easier. Following the current trend in the field of web development and with the objective of transferring BCCT.core to an online format, a prototype of a web application for the automatic keypoint detection was

J. Henriques et al. (Eds.): MEDICON 2019, IFMBE Proceedings 76, pp. 1967–1983, 2020.
https://doi.org/10.1007/978-3-030-31635-8_236

developed and is presented in this document. Currently, the user may upload an image and automatically detect and/or manipulate its key-points. This prototype is completely scalable and can be upgraded with new functionalities according to the user's needs.

1 Introduction

1.1 Context

Breast cancer is considered a systemic disease from diagnosis, so, it is extremely important to determine cancer staging (how widespread cancer is at the time of diagnosis), a factor which will influence the treatment decision from that moment [1,2]. Surgery is usually the primary approach in terms of treatment and its main objectives are to remove cancer and to determine the stage. It is possible to specify surgery in the following types:

- **Partial (segmental) mastectomy or lumpectomy**: which is based on the removal of the cancerous tissue with a rim of healthy tissue (free margin). This approach is also known as Breast Cancer Conservative Surgery (BCS) or Breast Cancer Conservative Treatment (BCCT) [1–3].
- **Simple or total mastectomy**: which includes the removal of the entire breast, but not of the lymph nodes under the arm or muscle tissue from beneath the breast [1,2,4].
- **Modified radical mastectomy**: based on the removal of the entire breast without the chest muscle and removal of the first two stages of lymph nodes under the arm [1,2].
- **Radical mastectomy**: which consists of the extensive removal of the entire breast, lymph nodes, and chest wall muscles under the breast [1,2,5].

Patients usually undergo radiation or systemic therapies in order to increase the effectiveness of the treatments. It is important to take into account that the duration of the therapy and the type of radiation that will be used will always depend on the type of performed surgery [6].

1.2 Motivation

Despite being a highly mutable and rapidly evolving disease, it is estimated that most breast cancers (approximately 90% of the cases) are treatable and curable if detected on early staging and given the best treatment [7]. Nowadays, with implemented breast cancer routine screening habits [8,9], it is possible to do proper management of the disease, in case it appears, and to develop a better clinical plan to treat it. This means that most breast cancers can now be treated using conservative approaches, i.e., BCCT, contributing, thus, for local control of the disease, with similar survival rates to those obtained with mastectomy, although, with better cosmetic results [4,7,10]. Moreover, the definition of an objective evaluation standard for the cosmetic outcome of BCCT has become crucial, due to the fact that with the development of new oncoplastic techniques it becomes necessary to have an objective method to compare cosmetic results [7,11]. There is a need for an automatic method that is able to perform this assessment by simply receiving data (i.e., images) as input. To accomplish this, one must develop an algorithm that is able to automatically detect breast fiducial points (i.e., keypoints), which are needed to compute asymmetry measurements and to perform the assessment.

2 Literature Review

2.1 BCCT Assessment

Although BCCT has an identical meaning worldwide, it is not a standardized procedure [12], due to the fact that there are some technical variations related both to surgery and radiotherapy. Generally, the aesthetic result of BCCT is done by an observer who identifies and evaluates colour, shape, geometry, irregularity and roughness of the treated breast in comparison with the untreated breast. This relies on the assumption that better results correspond to more similar breasts; this is considered to be a much more practical approach and a method that fits the evolution and the appearance of new emerging oncoplastic techniques, where, usually, both breasts are submitted to surgery, leading, consequently, to a more demanding comparison [7,13]. Historically, it was Jay Harris, in 1979, the one who had introduced a subjective overall cosmetic score, that would become a standard in conservation breast procedures: the Harvard scale [14]. This scale classifies the overall cosmetic result in four classes:

(1) **Excellent**: if the treated breast is nearly identical to the untreated one.
(2) **Good**: if the treated breast has some differences when compared with the untreated one.
(3) **Fair**: if the treated breast is clearly different from the untreated one, but not seriously distorted.
(4) **Poor**: if the treated breast is seriously distorted.

2.2 Computer-Aided Aesthetic Classification of BCCT Outcomes

Later, software-based methods came up with the main idea of predicting the global aesthetic result since they are based on different individual characteristics which are automatically and objectively extracted from patient photographs. These approaches explore the ability that computational methods have to provide an effective, easy, fast, reliable and reproducible tool to evaluate the consequences of breast cancer patient care [7]. Regarding this topic one must talk about Breast Analysing Tool - BAT© from Fitzal *et al.* [15] and BCCT.core from Cardoso and Cardoso [16]. On behalf of BAT©, Fitzal *et al.* proposed the Breast Symmetry Index (BSI) to evaluate the cosmetic outcome of BCCT. With BAT©, they were able to measure differences between left and right breast sizes from a patient's digital picture. The sum of all differences will output the BSI score, which has the capability of measuring size differences in between breasts. Concerning BCCT.core, it is important to state that it is a computer-aided medical system that detects breast keypoints and performs automatic feature extraction from patient photographs, in order to capture some of the relevant factors for the overall cosmetic result. To do a proper classification, it is necessary to obtain a concise representation of a BCCT image, based on asymmetry, colour differences and scar visibility features. To describe breast asymmetry, several features are taken into consideration [16] and, in order to extract them, it is necessary to mark some specific keypoints in the image, using BCCT.core software: sternal notch, the scale mark, the nipples and left and right breasts contours, adjusted with an active contour based on splines with control points.

2.3 A Deep Learning Approach to Keypoint Detection

Recently, Silva *et al.* have proposed the use of a deep neural network (DNN) for the keypoints detection task. With deep learning, it is possible to follow an integrated learning approach that uses the context information [17]. Actually, regarding keypoint detection with deep learning, there are two relevant works that presented interesting ideas: Cao *al.* [18], which brought the idea of learning *part confidence maps* and *part confidence fields* as a mean to detect keypoints in the end; Belagiannis *et al.* [19] which also proposed an architecture that first learns how to regress *heatmaps* (i.e., each keypoint is modelled with a Gaussian distribution) and, after iterative tuning on the training of heatmap regression, it is able to predict keypoint localizations. Taking these ideas into account, Silva *et al.* have proposed the generation of an intermediate representation (i.e., *heatmaps*), which consists on a fuzzy localization for the keypoints that are intended to be detected. The heatmaps are obtained using the segmentation model, U-Net [20]. For the keypoint regression, the strategy starts with the multiplication of the image with the refined output of the previous module, as a way of improving the fuzzy localization of keypoints, which will then enhance the exact keypoint detection. Therefore, a regression module predicts the keypoints' coordinates.

3 Image Segmentation for Breast Contour Detection

3.1 Background

Typically, in biomedical applications, the employment of a proper segmentation method is of utmost importance, being one of the most relevant steps in the pipeline [21,22]. With the advent of novel artificial intelligence techniques, it is possible to divide image segmentation into traditional approaches, such as **thresholding** [23,24], **edge based** [25,26], **region based** [27–30], **deformable models** [30–32] or **graph cuts** [33,34], or deep learning approaches which make use of the capabilities of DNNs to succeed in these tasks. Regarding the scope of this thesis, it is relevant to mention four architectures:

- **Fully Convolutional Network** (FCN): developed by Long *et al.* and trained end-to-end and/or pixel-to-pixel on semantic segmentation [35]. All of the following architectures are based on this one.
- **U-Net**: proposed by Ronneberger *et al.* and applied in biomedical image segmentation [20].
- **Global Convolution Network** (GCN): proposed by Peng *et al.* and applied in localization and classification tasks in image semantic segmentation [36].
- **DeepLabv3+**: published by Chen *et al.* in 2018, is, in fact, an extension of DeepLabv3 [37], with the addition of an encoder-decoder module to refine segmentation results. This work is based on the use of *atrous convolution*, or dilated convolution, and is the state of the art for image segmentation [38].

3.2 Segmentation for Breast Contour Detection

The main hypothesis behind this approach is that it should be easier to detect breast contours if one is able to properly detect the breast first. This is a challenge of semantic segmentation, where one needs to properly separate both breasts (i.e., *foreground*) from the other image components (i.e., *background*). In this case, the segmentation task would work as means to an end; ideally, within this approach, the pipeline would be composed of two main processes: breast segmentation and breast contour detection, where the second is fully-dependent on the first. In this experience, the main goal was to study the capabilities of state of the art segmentation DNNs on the specific task of segmenting both breasts from the images of the dataset. It is important to take into account that one can only move to the subtask of breast contour detection if good performance is achieved in the segmentation task.

3.3 Implementation and Results

For this study, it was decided to use state of the art architectures, such as: U-Net, GCN and DeepLabv3+. The dataset (221 images) was divided into train (107 images), validation (47 images) and test (67 images) sets. All images were first resized to the dimensions of 512×384. It was used the Dice Coefficient [39] (DC) as performance metric, which is very common in image segmentation tasks; DC values near 1 mean that there is high similarity between both ground-truth and predicted masks. It can be written as

$$DC = \frac{2\sum_i^N p_i g_i}{\sum_i^N p_i^2 + \sum_i^N g_i^2} \tag{1}$$

where N represents the total number of pixels, p_i is the predicted pixel label, g_i is the ground-truth pixel label and $i \in \{0, 1\}$, meaning that this is a binary-class problem. Regarding loss function, the *Dice Coefficient Loss* (DCL) was chosen and monitored during training. It can be written as

$$DCL = 1 - DC \tag{2}$$

where DC is the Dice Coefficient. U-Net was implemented in Keras [40] and GCN was implemented in PyTorch [41]. Both were trained during 300 epochs, with Adadelta [42] as optimization function. During training phase, data augmentation techniques such as rotation, horizontal and vertical shifting, shear mapping, zoom in and zoom out or horizontal and vertical flips were employed. The model with the lower loss value on the validation set was chosen during training and saved. It was then used to perform inference and evaluation on the test set. Results on test set are shown in Table 1; each DC value on the table is actually the average DC value obtained with each model in test set. Due to the lack of computational resources it was not possible to train the TensorFlow implementation of DeepLabv3+.

Table 1. Average Dice Coefficient results on test set for U-Net and GCN models. Best result is highlighted in bold.

	Average Dice Coefficient
U-Net	0.8689
GCN	**0.8937**

3.4 Discussion and Conclusions

Although it was possible to obtain DC results near 1, for this specific task, they do not contribute as much as it was initially expected. Here, the main goal was to generate breast binary masks from a given image in order to detect breast contours, being a two-class segmentation problem, where breasts are foreground and the rest of the image is the background. Taking this into account, it is possible to explain high results on the DC score: masks are mostly composed by background pixels (imbalanced-class problem), so it is easier for the network to predict background instead of foreground due to the fact that it has naturally seen more background pixels, during training. On the other hand, it can be said that, during inference, the trained models are able to predict the most probable breast localization. The main issue is that this is still a fuzzy localization, which means that it will not help in finding breast contours. Also, when images contain small or undefined breasts, it becomes even harder to obtain reliable breast image masks, which would lead to increased difficulty in finding breast contours. For these reasons, post-processing experiences to detect contours after segmentation were not performed. On the other hand, since breast contour detection fully depends on excellent results on image segmentation, this approach, at the moment, is not a viable option.

4 Quality-Driven Keypoint Detection

4.1 State of the Art on Deep Quality-Driven Architectures

Generally, deep learning architectures for image analysis take an image as input and return a mask (image segmentation) or a value (image classification) as output; these deep models are then regularized in a supervised manner on the object of interest. On the other hand, the need for great amounts of training data or the necessity of re-training/fine-tuning such models in order to apply them in different contexts may be problematic [43]. To overcome these issues, two recent works from Fernandes *et al.* and from Rebelo *et al.* are presented and explained.

Deep Image Segmentation by Quality Inference. In their original work, Fernandes *et al.* started by defining the quality metric for the learning phase; they ended up choosing DC, which is widely used in this type of tasks as a similarity metric. Considering that modelling of utility functions is usually inspired in pairwise preferences [44] and/or cardinal/ordinal functions [45], the authors proposed a model that is capable of learning the quality (i.e., DC), given an image and mask pair; in this case, the utility function is precisely this measure of correspondence between a mask and an image. To achieve such a model, the authors proposed a deep architecture, *Gossip Network*, which receives an image and mask pair as input and has two streams that try to model the foreground and background, respectively by the input mask. Moreover, as a strategy to increase/decrease the network's confidence in the recognition of their corresponding regions, these streams communicate (i.e., to gossip) between each other. After training, the *Gossip Network* is used to predict the quality of a given image and mask pair and, based on this prediction, the mask is iteratively refined by backpropagation until it reaches a stop criteria (e.g., quality value, number of iterations).

Quality-Based Regularization for Iterative Deep Image Segmentation. Rebelo *et al.* have presented a new methodology, which uses the notion of quality as a regularization method during the training of a network for direct segmentation refinement [46]. They proposed a network with an encoder-decoder structure which takes an image and a mask as inputs, and returns quality and a refined segmentation as output, following a multi-task learning paradigm. Although this work follows a similar strategy to the one presented by Fernandes *et al.* [43] (e.g., the quality metric is the same), the main focus in this case was the study of a DNN which could receive the predicted mask from a state of the art network and iteratively improve the segmentation, while predicting the quality of the input segmentation, at the same time; this parallel quality prediction can be understood as a regularization method that will favour the learning of features connected to the degree of correction necessary to improve the input segmentations. To start the training, it is required to feed the network with an image and an initial segmentation mask (obtained from any other model); this initial mask will be refined according to an iterative method, where the network will use its own output as a new input for further improvement, leading, thus, to higher predicted quality values. In their original work, Rebelo *et al.* use U-Net [20], trained on the same dataset, to obtain the initial segmentation masks, which are then concatenated to the input image as an additional channel.

4.2 Deep Keypoint Detection by Quality Inference

The main idea behind this first approach is the application of the main concepts from [43] to keypoint detection, instead of image segmentation. To achieve such a model, one must first define a suitable quality metric. Following the work developed in [17], mean squared error (MSE) was chosen as the quality metric. It is written as:

$$MSE = \frac{1}{n}\sum_{i=0}^{n-1}(k_i - \hat{k}_i)^2 \tag{3}$$

where n is the number of samples (i.e., number of coordinates in each keypoints array), i is the index of the value, $\hat{k}$ is the predicted keypoints array and k is the ground-truth keypoints array. The main objective is to have a model that receives an image and an array with initial keypoints candidates as inputs and that outputs a quality value (i.e., MSE between the candidate keypoints array and the ground-truth keypoints array). The image is given as an input to a CNN, a latent representation is obtained and then concatenated with the candidate keypoints array. This concatenation is then fed to a Multilayer Perceptron (MLP) which will output the quality value predicted, related to the image and the candidate keypoints. The CNN Module is composed by VGG16 [47] as backbone without its dense layers, followed by four convolutional layers and three dense layers; the latent representation of the image is obtained from the last dense layer. This latent representation is then concatenated with the candidate keypoints array which is then fed to a MLP; this MLP has three dense layers. The quality value is obtained from the last dense layer. After training the model on the quality inference task, the next step would be the use of the backpropagation algorithm to improve the predicted quality, leading, thus, to coordinate values near ground-truth. In order to do this, one must first to compute the gradients of the predicted quality, $\hat{q}$, with respect to the given candidate keypoints, k', and use gradient descent (important to remind that the quality metric is given by an error metric, so, what is aimed is the minimization of this value) with a learning rate, α, as follows:

$$k' \leftarrow k' - \alpha\frac{\partial \hat{q}}{\partial k'} \tag{4}$$

After several steps, it would be expected that the real MSE between the changed candidate keypoints and the ground-truth keypoints would be lower than the real MSE between the initial candidate keypoints and the ground-truth keypoints.

4.3 Deep Iterative Refinement of Keypoint Detection

This second approach aims to apply the methodology proposed in [46] to the task of keypoint detection, through the iterative refinement of the given candidate keypoints. Similarly to the work developed by Rebelo *et al.*, the main objective here is to use an iterative approach to improve the quality (in this case, minimize its value, since it is a MSE value) of the given candidate keypoints; the quality value acts here as a form of regularization, during training. When compared to the architecture presented in Subsection 4.2, this one has some modifications. The CNN Module keeps the same structure, however, three new MLPs are added to the model in order to fulfill the main objective. The Intermediate MLP receives as input the result of the concatenation between the candidate keypoints and the Image Embedding and generates what is called an Intermediate Embedding, which has lower dimensions than the first one. This latent representation is then given to the Keypoints MLP, which has the task of predicting the keypoints coordinates, and to the Quality MLP, which has to learn how to predict the quality value, related to the image and candidate keypoints pair.

4.4 Implementation and Results

Deep Keypoint Detection by Quality Inference. This model was implemented in Keras [40]. All images were first resized to the dimensions of 512×384; keypoints were also resized accordingly and were normalized by the width of the image, in order to be between 0 and 1. The dataset was divided into train, validation and test sets, as previously explained. To ensure that the model was robust, data augmentation techniques were employed during training, in an online fashion, to both images and ground-truth keypoints: translations, rotations and flips were used. Since the main objective is to teach the model on quality inference, it is important to ensure that a large range of quality values is shown to the network, during training; in order to achieve that, one must be certain that to the initial candidate keypoints were applied the correct diversity of transformations, so that, the correspondent qualities cover a wide range of values. To do this, the data augmentation strategy presented by Fernandes *et al.* in [43] was adapted and used; in this case, the transformations that were applied to the keypoints were translations, rotations and horizontal flips. Once again, all candidate keypoints and their correspondent qualities are generated in an online setting to keep dynamic training. Regarding the training phase, the loss function was defined as follows:

$$\mathcal{L}_{model} = \lambda_1 \mathcal{L}_{imageembedding} + \lambda_2 \mathcal{L}_{quality} \tag{5}$$

where λ_1 and λ_2 represent non-negative weights attributed to each loss. The loss of the image embedding is computed as the MSE between predicted image embedding and the ground-truth keypoints. The intuition behind this supervised approach is to guide the CNN Module to learn a latent representation that is near the ground-truth keypoints in order to facilitate the learning of the quality by the MLP. Regarding quality prediction, the loss function is also computed as the MSE between the predicted quality value and the ground-truth quality for the candidate keypoints. The model was trained during 300 epochs, with Adadelta [42] as optimization function, and the model with lower loss on the validation set was saved. After training, several initial candidate keypoints for test set were generated, following three possible situations: candidate keypoints were the result of the prediction obtained with the model described in [17] in images of the test set; candidate keypoints were the result of random generated values; candidate keypoints were coordinates with value 0. The quality was then inferred from the initial candidate keypoints, using the trained model, and then the backpropagation algorithm proposed was applied. Regarding backpropagation, the presented results use a learning rate $\alpha = 0.001$, selected taking into account the order of magnitude of the errors. The number of backpropagation iterations was 1000 and 5000. Regarding initial candidate keypoints and adapted candidate keypoints, the average MSE, its standard deviation and the maximum and minimum errors were obtained. Tables 2, 3 and 4 show results for the three possible initial candidate keypoints status: candidate keypoints were the result of the prediction obtained with the model described in [17] in images of the test set; candidate keypoints were the result of random generated values; candidate keypoints were coordinates with value 0. **Note:** In the following Tables, GT stands for ground-truth keypoints, IC for initial candidate keypoints, AC for adapted candidate keypoints and STD for standard deviation.

Table 2. Results for the backpropagation applied to the candidate keypoints which were the result of the prediction obtained with the model described in [17] in images of the test set. Best results are highlighted in bold.

Real MSE	Mean	STD	Max	Min	Steps
GT & IC	**0.00116**	**0.00150**	0.00659	**0.00013**	-
GT & AC	0.00118	0.00151	**0.00651**	0.00014	**1000**
GT & AC	0.00139	0.00158	0.00656	0.00017	**5000**

Table 3. Results for the backpropagation applied to the candidate keypoints which were the result of random generated values. Best results are highlighted in bold.

Real MSE	Mean	STD	Max	Min	Steps
GT & IC	0.11286	0.01700	0.15130	0.07088	-
GT & AC	0.11087	0.01639	0.14704	0.07051	**1000**
GT & AC	**0.10519**	**0.01492**	**0.14063**	**0.06954**	**5000**

Table 4. Results for the backpropagation applied to the candidate keypoints which were coordinates with value 0. Best results are highlighted in bold.

Real MSE	Mean	STD	Max	Min	Steps
GT & IC	0.27012	0.03186	0.34395	0.21046	-
GT & AC	0.25837	0.03128	0.33107	0.20054	**1000**
GT & AC	**0.21929**	**0.02865**	**0.28671**	**0.16724**	**5000**

Deep Iterative Refinement of Keypoint Detection. This model was implemented in PyTorch [41]. The data preparation process was the same as previously explained. Regarding training phase, the model was trained during 300 epochs, with Adadelta [42] as optimization function, and the model with lower loss on the validation set was saved. For this approach, a different strategy for the loss function was designed:

$$\mathcal{L}_{model} = \lambda_1 \mathcal{L}_{quality} + \lambda_2 \mathcal{L}_{refinedkeypoints} \tag{6}$$

where λ_1 and λ_2 represent non-negative weights attributed to each loss. The loss of the quality (regularization term) is computed the same way as presented above. The loss of the refined keypoints (main goal) is calculated as the MSE between the predicted refined keypoints and the ground-truth keypoints. Tables 5, 6 and 7 show results for the three possible initial candidate keypoints status: candidate keypoints were the result of the prediction obtained with the model described in [17] in images of the test set; candidate keypoints were the result of random generated values; candidate keypoints were coordinates with value 0.

Table 5. Results for the iterative refinement model applied to the candidate keypoints which were the result of the prediction obtained with the model described in [17] in images of the test set.

Real MSE	Mean	STD	Max	Min
GT & IC	**0.00116**	**0.00150**	**0.00659**	**0.00013**
GT & AC	0.00218	0.00164	0.00750	0.00028

Table 6. Results for the iterative refinement model applied to the candidate keypoints which were the result of random generated values.

Real MSE	Mean	STD	Max	Min
GT & IC	0.11286	0.01700	0.15130	0.07088
GT & AC	**0.00233**	**0.00150**	**0.00633**	**0.00033**

Table 7. Results for the iterative refinement model applied to the candidate keypoints which were coordinates with value 0.

Real MSE	Mean	STD	Max	Min
GT & IC	0.27012	0.03186	0.34395	0.21046
GT & AC	**0.00345**	**0.00294**	**0.01293**	**0.00045**

4.5 Discussion and Conclusions

Both proposed models were functional, however, they were not able to improve the state of the art results, i.e., the ones published by Silva *et al.* in [17]. Regarding the first approach, explained in Subsection 4.1, there are some considerations that have to be taken into account: inference of quality has always an associated error amplitude, which means, that the inference will always have some deviations from the real quality value; this will have implications when performing backpropagation in the candidate keypoints, due to the fact, that, in each step, the inferred quality value will always diminish, but that does not necessarily mean that the real quality value has diminished. Also, although results are presented for different values of backpropagation steps, an optimal value still needs to be found; this is crucial, since, following the main conclusions from Fernandes *et al.* in [43], too much backpropagation steps may lead to a deterioration of the adapted candidate keypoints. On behalf of the second approach, the refined keypoints predictions have better performances, but still need further improvements, regarding the fact that they did not improve the results obtained with the model proposed by Silva *et al.* in [17]. It is important to notice that, even with data augmentation, the dataset used to train the models may not be large enough to address these specific tasks; the applied data augmentation parameters may also not be enough, and, it is possible that the models need an even higher variety of examples to properly learn. It is also important to mention that this was the first work which used a quality-driven approach for a regression task; this means that there is space for improvements. Also, it is still not clear that this approach will lead to success, considering this kind of tasks. For all these reasons, these models still need further research and development in order to be considered viable options to solve this thesis' main problem.

5 A Web Application for Automatic Keypoint Detection

5.1 Motivation

Currently, in order to use BCCT.core [16], users must download, install and set-up the software on their computers; these three steps may be considered a burden for some of the users and they also imply that the software will use their computational resources to properly function. On the other hand, fundamental services such as software maintenance (updates and/or upgrades) or support could be difficult in an offline setting. To overcome these issues, and taking into account that most of the healthcare institutions have an Internet connection [48], it was decided that the best option would be to move BCCT.core into a web-based application. This way, users would only need to deal with a simple website registration in order to use all the functionalities of the software; the rest would be handled in the background, on the server where the application is hosted. Moreover, the field of web development is in a trending evolution, being able to keep up with most of the market's new requirements [49]. Use-cases and the proposed architecture for such an application are further described.

5.2 Use-Cases and Proposed Architecture

The starting point for the establishment of use-cases for this web-based application for automatic keypoint detection was BCCT.core [16]; the main functionalities will not require major improvements, however, the transition from offline to online setup has a big impact on the new practicalities of the application, and this should be taken into account on the redesign of the software. The identified use-cases, from the user side, were:

(1) The user should log in to enter the application: with the application in an online setting, it is crucial that measures that take into account the privacy of the clinical data are implemented, and, as such, it is believed that a login system for user authentication inside the application is of utmost importance.
(2) The user may upload medical photos: to perform the keypoint detection, the application should have an intuitive interface so users can upload the photos that are going to be subjected to evaluation.
(3) The user may consult and change the detection records: it is important that the user has access to the evaluated photos and to the predicted keypoints and he/she should be able to manually correct the model's predictions.
(4) The user may erase all the records from the system: once again, since one is dealing with sensitive clinical data, it is important that the users have the possibility to reset their records.

From the server side, the system should have a database. This database will contain:

(1) User's information: name, e-mail and password.
(2) Uploaded Photo's information: location, predicted keypoints and the associated user.

Regarding the type of relationship, it can be seen that it is a one-to-many relationship, i.e., an user can upload many photos. Another fundamental requirement is a back-end capable of interacting with deep learning models created in Python; the server should have enough space to host such models.

5.3 Implementation and Prototype

The web application prototype was developed with Django, a free and open source high-level Python web framework that promotes fast development, security and scalability. The main reason behind this choice is indeed related with the fact that the deep learning models are written in Python, and since Django is also powered by the same programming language, this application-models interaction would be simpler. Moreover, Django permits the creation and manipulation of databases through Python, simplifying, once again, all the development that requires interaction between the front-end and the back-end. With everything on the back-end being handled by Django, one has to deal with the front-end. The front-end of an application is directly related to its usability and user experience. In terms of development, it was used HTML and CSS, which are the standard code languages for web development. One of the main objectives is to have an interface that allows users to perform their tasks (i.e., keypoint detection and/or aesthetic assessment) as simple as possible; to do that, it was decided to incorporate all the functionalities as buttons, where the user clicks and sees the results, without the need to check what is happening in the background. An administration interface was also developed, since, in reality, such a system requires a professional who can properly manage, update and maintain it, while giving some kind of support to the users.

5.4 Discussion and Conclusions

It was possible to achieve a fully working prototype, capable of fulfilling all the previously stated use-cases and representing, thus, a reliable proof of concept that can act as a starting point for the final deployment of a final version of the web application for automatic keypoint detection. Regarding the fact that the system is in development mode, one of the next steps would be, for example, changing the database framework (currently the system uses SQLite) to a scalable one (for example, PostgreSQL), which is supported by Django; important to remind, that the method of interacting with the database stays almost untouched. Once again, Django handles all the background work.

6 Future Work

Regarding future work recommendations, it is proposed to:

- Use DeepLabV3+ [38] for breast image segmentation;
- Explore recurrent neural networks [50] (RNN) or graph neural networks [51] (GNN) for keypoint prediction;
- Test different metrics for quality evaluation;
- Acquire more data to improve training performance;
- Deploy the Web application for healthcare professionals.

References

1. Ely, S., Vioral, A.N.: Breast cancer overview. Plast. Surg. Nurs. **27**, 128–133 (2007)
2. Street, W.: Breast Cancer Facts & Figures 2017–2018, p. 44 (2017)
3. Grisotti, A.: Immediate Reconstruction After Partial Mastectomy, p. 12 (1994)
4. Fisher, B., Anderson, S., Bryant, J., Margolese, R.G., Deutsch, M., Fisher, E.R., Jeong, J.-H., Wolmark, N.: Twenty-year follow-up of a randomized trial comparing total mastectomy, lumpectomy, and lumpectomy plus irradiation for the treatment of invasive breast cancer. New Engl. J. Med. **347**, 1233–1241 (2002)
5. Fisher, B., Montague, E., Redmond, C., Barton, B., Borland, D., Fisher, E.R., Deutsch, M., Schwarz, G., Margolese, R., Donegan, W., Volk, H., Konvolinka, C., Gardner, B., Cohn, I., Lesnick, G., Cruz, A.B., Lawrence, W., Nealon, T., Butcher, H., Lawton, R., Investigators, O.N.: Comparison of radical mastectomy with alternative treatments for primary breast cancer: a first report of results from a prospective randomized clinical trial. Cancer **39**, 2827–2839 (1977)
6. E. B. C. T. C. G. (EBCTCG): Effect of radiotherapy after breast-conserving surgery on 10-year recurrence and 15-year breast cancer death: meta-analysis of individual patient data for 10 801 women in 17 randomised trials. The Lancet **378**, 1707–1716 (2011)
7. Oliveira, H.P., Cardoso, J.S., Magalhaes, A., Cardoso, M.J.: Methods for the aesthetic evaluation of breast cancer conservation treatment: a technological review. Curr. Med. Imaging Rev. **9**, 32–46 (2013)
8. Grady, K.E., Lemkau, J.P., McVay, J.M., Reisine, S.T.: The importance of physician encouragement in breast cancer screening of older women. Prev. Med. **21**, 766–780 (1992)
9. Smith, R.A., Haynes, S.: Barriers to screening for breast cancer, p. 11 (1992)
10. Veronesi, U., Cascinelli, N., Mariani, L., Greco, M., Saccozzi, R., Luini, A., Aguilar, M., Marubini, E.: Twenty-year follow-up of a randomized study comparing breast-conserving surgery with radical mastectomy for early breast cancer. New Engl. J. Med. **347**, 1227–1232 (2002)
11. Cardoso, M.J., Cardoso, J.S., Vrieling, C., Macmillan, D., Rainsbury, D., Heil, J., Hau, E., Keshtgar, M.: Recommendations for the aesthetic evaluation of breast cancer conservative treatment. Breast Cancer Res. Treat. **135**, 629–637 (2012)
12. Christiaens, M., van der Schueren, E., Vantongelen, K.: More detailed documentation of operative procedures in breast conserving treatment: what good will it do us? Eur. J. Surg. Oncol. (EJSO) **22**, 326–330 (1996)
13. Asgeirsson, K., Rasheed, T., McCulley, S., Macmillan, R.: Oncological and cosmetic outcomes of oncoplastic breast conserving surgery. Eur. J. Surg. Oncol. (EJSO) **31**, 817–823 (2005)
14. Harris, J.R., Levene, M.B., Svensson, G., Hellman, S.: Analysis of cosmetic results following primary radiation therapy for stages I and II carcinoma of the breast. Int. J. Radiat. Oncol. Biol. Phys. **5**, 257–261 (1979)
15. Fitzal, F., Krois, W., Trischler, H., Wutzel, L., Riedl, O., Kühbelböck, U., Wintersteiner, B., Cardoso, M., Dubsky, P., Gnant, M., Jakesz, R., Wild, T.: The use of a breast symmetry index for objective evaluation of breast cosmesis. The Breast **16**, 429–435 (2007)
16. Cardoso, J.S., Cardoso, M.J.: Towards an intelligent medical system for the aesthetic evaluation of breast cancer conservative treatment. Artif. Intell. Med. **40**, 115–126 (2007)

17. Silva, W., Castro, E., Cardoso, M.J., Fitzal, F., Cardoso, J.S.: Deep keypoint detection for the aesthetic evaluation of breast cancer surgery outcomes. In: Proceedings of the IEEE International Symposium on Biomedical Imaging (ISBI 2019) (2019)
18. Cao, Z., Simon, T., Wei, S.E., Sheikh, Y.: Realtime Multi-Person 2D Pose Estimation using Part Affinity Fields, arXiv:1611.08050 [cs], November 2016
19. Belagiannis, V., Zisserman, A.: Recurrent Human Pose Estimation, arXiv:1605.02914 [cs], May 2016
20. Ronneberger, O., Fischer, P., Brox, T.: U-Net: Convolutional Networks for Biomedical Image Segmentation, arXiv:1505.04597 [cs], May 2015
21. Fasihi, M.S., Mikhael, W.B.: Overview of current biomedical image segmentation methods. In: 2016 International Conference on Computational Science and Computational Intelligence (CSCI), Las Vegas, NV, USA, pp. 803–808. IEEE, December 2016
22. Sharma, N., Ray, A., Shukla, K., Sharma, S., Pradhan, S., Srivastva, A., Aggarwal, L.: Automated medical image segmentation techniques. J. Med. Phys. **35**(1), 3 (2010)
23. Saha, P., Udupa, J.: Optimum image thresholding via class uncertainty and region homogeneity. IEEE Trans. Pattern Anal. Mach. Intell. **23**, 689–706 (2001)
24. Otsu, N.: A threshold selection method from gray-level histograms. IEEE Trans. Syst. Man Cybern. **9**, 62–66 (1979)
25. Kekre, D.H.B., Gharge, S.M.: Image Segmentation using Extended Edge Operator for Mammographic Images, vol. 02, no. 04, p. 6 (2010)
26. Zanaty, E.A.: Improved region growing method for magnetic resonance images (MRIs) segmentation. Am. J. Remote Sens. **1**(2), 53 (2013)
27. Shan, J., Cheng, H., Wang, Y.: A completely automatic segmentation method for breast ultrasound images using region growing. In: Proceedings of the 11th Joint Conference on Information Sciences (JCIS). The Harbin Institue of Technology, Shenzhen, China. Atlantis Press (2008)
28. Tamilselvi, P.R., Thangaraj, D.P.: Segmentation of Calculi from Ultrasound Kidney Images by Region Indicator with Contour Segmentation Method, p. 10 (2011)
29. Day, E., Betler, J., Parda, D., Reitz, B., Kirichenko, A., Mohammadi, S., Miften, M.: A region growing method for tumor volume segmentation on PET images for rectal and anal cancer patients: a region growing method for tumor segmentation. Med. Phys. **36**, 4349–4358 (2009)
30. Davis, J.B., Reiner, B., Huser, M., Burger, C., Székely, G., Ciernik, I.F.: Assessment of 18f PET signals for automatic target volume definition in radiotherapy treatment planning. Radiother. Oncol. **80**, 43–50 (2006)
31. Kass, M., Witkin, A., Terzopoulos, D.: Snakes: active contour models. Int. J. Comput. Vis. **1**, 321–331 (1988)
32. Xu, C., Prince, J.: Snakes, shapes, and gradient vector flow. IEEE Trans. Image Process. **7**, 359–369 (1998)
33. Thongnuch, V., Uyyanonvara, B.: Automatic Optic Disk Detection From Low Contrast Retinal Images of ROP Infant Using GVF Snake, p. 13 (2007)
34. Boykov, Y., Funka-Lea, G.: Graph cuts and efficient N-D image segmentation. Int. J. Comput. Vis. **70**, 109–131 (2006)
35. Shelhamer, E., Long, J., Darrell, T.: Fully Convolutional Networks for Semantic Segmentation, arXiv:1605.06211 [cs], May 2016
36. Peng, C., Zhang, X., Yu, G., Luo, G., Sun, J.: Large Kernel Matters – Improve Semantic Segmentation by Global Convolutional Network, arXiv:1703.02719 [cs], March 2017

37. Chen, L.-C., Papandreou, G., Schroff, F., Adam, H.: Rethinking Atrous Convolution for Semantic Image Segmentation, arXiv:1706.05587 [cs], June 2017
38. Chen, L.-C., Zhu, Y., Papandreou, G., Schroff, F., Adam, H.: Encoder-Decoder with Atrous Separable Convolution for Semantic Image Segmentation, arXiv:1802.02611 [cs], February 2018
39. Milletari, F., Navab, N., Ahmadi, S.-A.: V-Net: Fully Convolutional Neural Networks for Volumetric Medical Image Segmentation, arXiv:1606.04797 [cs], June 2016
40. Chollet, F., et al.: Keras (2015)
41. Paszke, A., Gross, S., Chintala, S., Chanan, G., Yang, E., DeVito, Z., Lin, Z., Desmaison, A., Antiga, L., Lerer, A.: Automatic differentiation in PyTorch (2017)
42. Zeiler, M.D.: ADADELTA: An Adaptive Learning Rate Method, arXiv:1212.5701 [cs], December 2012
43. Fernandes, K., Cruz, R., Cardoso, J.S.: Deep image segmentation by quality inference. In: 2018 International Joint Conference on Neural Networks (IJCNN), Rio de Janeiro, pp. 1–8. IEEE, July 2018
44. Fernandes, K., Cardoso, J.S., Palacios, H.: Learning and ensembling lexicographic preference trees with multiple kernels. In: 2016 International Joint Conference on Neural Networks (IJCNN), Vancouver, BC, Canada, pp. 2140–2147. IEEE, July 2016
45. Ellsberg, D.: Classic and current notions of "measurable utility". Econ. J. **64**, 528 (1954)
46. Rebelo, J., Fernandes, K., Cardoso, J.S.: Quality-based Regularization for Iterative Deep Image Segmentation, p. 4 (2019)
47. Simonyan, K., Zisserman, A.: Very Deep Convolutional Networks for Large-Scale Image Recognition, arXiv:1409.1556 [cs], September 2014
48. Séror, A.C.: Internet infrastructures and health care systems: a qualitative comparative analysis on networks and markets in the British national health service and kaiser permanente. J. Med, Internet Res. **4**, e21 (2002)
49. Ricca, F., Tonella, P.: Analysis and testing of web applications. In: Proceedings of the 23rd International Conference on Software Engineering, ICSE 2001, Toronto, Ont., Canada, pp. 25–34. IEEE Computer Society (2001)
50. Wang, J., Yang, Y., Mao, J., Huang, Z., Huang, C., Xu, W.: CNN-RNN: a unified framework for multi-label image classification. In: 2016 IEEE Conference on Computer Vision and Pattern Recognition (CVPR), Las Vegas, NV, USA, pp. 2285–2294. IEEE, June 2016
51. Battaglia, P.W., Hamrick, J.B., Bapst, V., Sanchez-Gonzalez, A., Zambaldi, V., Malinowski, M., Tacchetti, A., Raposo, D., Santoro, A., Faulkner, R., Gulcehre, C., Song, F., Ballard, A., Gilmer, J., Dahl, G., Vaswani, A., Allen, K., Nash, C., Langston, V., Dyer, C., Heess, N., Wierstra, D., Kohli, P., Botvinick, M., Vinyals, O., Li, Y., Pascanu, R.: Relational inductive biases, deep learning, and graph networks, arXiv:1806.01261 [cs, stat], June 2018

Deep Learning for Interictal Epileptiform Discharge Detection from Scalp EEG Recordings

Catarina Lourenço[1,2(✉)], Marleen C. Tjepkema-Cloostermans[2,3], Luís F. Teixeira[1], and Michel J. A. M. van Putten[2]

[1] Faculty of Engineering of the University of Porto, Porto, Portugal
c.dasilvalourenco@utwente.nl, luisft@fe.up.pt
[2] University of Twente, Enschede, The Netherlands
{c.dasilvalourenco,m.j.a.m.vanputten}@utwente.nl
[3] Medisch Spectrum Twente, University of Twente, Enschede, The Netherlands
M.Tjepkema-Cloostermans@mst.nl

Abstract. Interictal Epileptiform Discharge (IED) detection in EEG signals is widely used in the diagnosis of epilepsy. Visual analysis of EEGs by experts remains the gold standard, outperforming current computer algorithms. Deep learning methods can be an automated way to perform this task. We trained a VGG network using 2-s EEG epochs from patients with focal and generalized epilepsy (39 and 40 patients, respectively, 1977 epochs total) and 53 normal controls (110770 epochs). Five-fold cross-validation was performed on the training set. Model performance was assessed on an independent set (734 IEDs from 20 patients with focal and generalized epilepsy and 23040 normal epochs from 14 controls). Network visualization techniques (filter visualization and occlusion) were applied. The VGG yielded an Area Under the ROC Curve (AUC) of 0.96 (95% Confidence Interval (CI) = 0.95 − 0.97). At 99% specificity, the sensitivity was 79% and only one sample was misclassified per two minutes of analyzed EEG. Filter visualization showed that filters from higher level layers display patches of activity indicative of IED detection. Occlusion showed that the model correctly identified IED shapes. We show that deep neural networks can reliably identify IEDs, which may lead to a fundamental shift in clinical EEG analysis.

1 Introduction

Epilepsy is the fourth most prevalent neurological disorder in the world. It is a brain disease that entails a predisposition to generate seizures, encompassing a plethora of syndromes and clinical phenomenology, some similar to other diseases [1–3]. Distinguishing a non-epileptic paroxysmal event from a seizure is clinically difficult, and the rate of misdiagnosis for epilepsy is reported to be up to 30% [4, 5]. This may result in an increased risk of recurrent seizures due to lack of adequate treatment or prescription of potentially harmful medication to patients with other disorders [6, 7].

While ictal EEGs (i.e. EEG signals recorded during seizures) can almost unequivocally distinguish an epileptic seizure from a non-epileptic one, the availability

J. Henriques et al. (Eds.): MEDICON 2019, IFMBE Proceedings 76, pp. 1984–1997, 2020.
https://doi.org/10.1007/978-3-030-31635-8_237

of these signals is low given that seizures occur at unknown times. In many patients, the interictal EEG (i.e. the EEG signal when no seizure is occurring) shows Interictal Epileptiform Discharges (IEDs): transient patterns that indicate an increased likelihood of seizures [1, 8, 9]. Assessment of their presence is done by visual analysis, which has been the gold standard in the clinic for almost over a century [10]. Yet, the learning curve is long, review times are significant, visual assessment is subjective and inter and intra-individual variability ranges from 5 to 25% [11, 12]. Despite these limitations, visual assessment of the EEG still outperforms current computer algorithms in detecting IEDs. These automated approaches are mostly based on 'hand-made' features that aim to mimic the characteristics used by experts in visual analysis [13–17], but they have ultimately failed to substitute the traditional approach to this task.

Deep learning methods learn from experience, creating a hierarchical representation based on raw data. Features are automatically extracted by successive layers, making artificial neural networks unbiased regarding the features used in visual analysis [17–19]. While deep neural networks are not easily interpretable, which has been pointed out as one of the drawbacks of deep learning methods [20], network probing and visualization techniques have been developed to provide some insight into the inner workings of the behavior of these models, making them more understandable and, thus, empirically reliable [21–23].

We explore whether deep neural networks can detect IEDs in EEG signals, automating EEG analysis for this task. In turn, this would render it more objective, eliminating variability and reducing the time and resources spent on this task. We also aim to provide some understanding regarding how the models reach their conclusions based on the filters and the provided inputs.

2 Methods

2.1 EEG Data and Pre-processing

EEG data from 217 patients between 4 and 72 years of age was used. The recordings were made with twenty-one silver/silver chloride cup electrodes placed on the scalp according to the international 10–20 system. All EEGs were obtained as part of routine care, and anonymized before further analysis. The patients were randomly selected from the digital database of the Medisch Spectrum Twente, in the Netherlands. This dataset included interictal EEGs from patients with focal (50 patients) and generalized (49 patients) epilepsy, containing interictal epileptiform discharges. EEGs with non-epileptiform abnormalities ("abnormal", 51 patients) and normal EEGs (67 patients) were also included. This was done based on the diagnosis from the electroencephalographers (e.g. searching for "focal epilepsy" or "normal EEG" in the conclusion of the report). The complete clinical report and the EEG recording itself were reviewed by experts (MvP and MTC). EEGs with epileptiform discharges were annotated by the experts so that IEDs could be easily identified. Epochs in which there was uncertainty regarding the occurrence of an IED were not labeled, ensuring that all the annotations corresponded to the unequivocal presence of an epileptiform discharge.

We filtered the EEG data in the 0.5–35 Hz range to reduce artefacts and down-sampled it to 125 Hz to reduce input size (and consequently computational complexity). Subsequently, the signals were re-referenced to a longitudinal bipolar montage. Each recording was split into 2 s non-overlapping epochs, yielding a 18 × 250 (channels × time) matrix for each epoch. The pre-processing routine was implemented in Matlab R2019a (The MathWorks, Inc., Natick, MA).

The data was randomized and split into a training/validation set containing 80% of the recordings and a test set comprised of the remaining 20%. All epochs from a patient were used either for training or testing. We applied five-fold cross validation on the training/validation set, further partitioning it in each iteration.

We created three different datasets using this method. The positive class was the same across datasets, being comprised of IEDs of focal and generalized epilepsy patients. The negative class of the first set included normal epochs from the EEGs of epilepsy patients, as well as normal EEGs (Set A). The second one was comprised only of normal EEGs (Set B), while the third included EEGs with non-epileptiform abnormalities and normal EEGs (Set C). Table 2 in the Appendix provides more details concerning the created datasets.

2.2 Deep Learning Models

A VGG C convolutional neural network (CNN) [24] was implemented in Python 2.7 using Keras 2, Theano and a CUDA-enabled NVIDIA GPU (GTX-1080), running on CentOS 7. Stochastic optimization was performed using an Adam optimizer [25] with a learning rate of $2 * 10^{-5}$, $\beta1 = 0.91$, $\beta2 = 0.999$, and $\varepsilon = 10^{-8}$. A sparse categorical cross entropy function was employed to estimate the loss. A batch size of 64 and weights of 100:1 were used (100 corresponding to the positive class).

2.3 Performance Evaluation

Receiver Operating Characteristic (ROC) curves were calculated for each of the cross-validation iterations using 101 discretizations. This was then averaged, yielding an average ROC curve for each set. The area under the curve (AUC) was calculated. Confidence Intervals (CIs) at 95% were calculated for the ROCs and AUCs. The sensitivity, specificity, false positive and true positive rate were assessed at a threshold where the values of sensitivity and specificity were as similar as possible (achieved by calculating the minimum of the difference) and at 99% specificity. Confidence Intervals at 95% (CIs) were calculated for these parameters. For each EEG on the test set, the sensitivity and specificity were assessed at a classification threshold of 0.5. These routines were implemented in Matlab R2019a (The MathWorks, Inc., Natick, MA).

2.4 Visualization Techniques

Filter visualization was applied by calculating and normalizing the gradients of the input with regard to the loss for each filter of each convolutional layer. Then, starting with a synthetically generated gray image with random noise with the same dimensions as the EEG epochs, gradient ascent was performed for 10 thousand iterations with a step size of 0.1.

Occlusion consisted in applying a grid to each sample and, in each iteration, setting the contents of a patch to zero, leaving the remaining sample untouched. The probability resulting from the network's prediction was stored in the center of the occluded patch so that it could be compared to the prediction without occlusion. After going through the whole image, the difference between the value in each grid patch center and the original prediction of the network was calculated. The patches with higher differences were plotted with warmer colors. The dimensions of the grid varied between 10 and 50 for the time axis and between 1 and 6 for the channel axis.

3 Results

Figure 1 shows the ROC curves yielded by the VGG model when distinguishing IEDs from normal EEG epochs (Set B). The AUC values were 0.99 (CI = 0.99 − 1.00) on the training set and 0.96 (CI = 0.95 − 0.97) on the test set. Figure 4 in the Appendix shows the analogous results for Sets A and C. When epochs containing non-epileptiform abnormalities were introduced in the dataset (Set C), the AUC value yielded by the VGG on the test set was 0.90 (CI = 0.89 − 0.90).

When IEDs were detected against the normal class (Set B), the intersection between sensitivity and specificity occurred at 98% for the training set and 93% on the test set. The true positive rates per hour were 91.03 (CI = 80.35–100.22) and 47.72 (CI = 45.60 − 49.84), respectively, with corresponding false positive rates of 22.30 (CI = 6.32 − 38.28) and 122.41 (CI = 27.63 − 217.20) (see Table 1). Increasing the specificity threshold to 99%, true positive detections became 88.71 (CI = 78.40 − 99.02) on the training set and 40.27 (CI = 38.43 − 42.12) on the test set. The false positive rates were 16.50 (CI = 4.95 − 28.06) and 32.31 (CI = 15.15 − 49.46) per hour, respectively. Table 1 also presents the equivalent values obtained when non-epileptiform abnormalities were included in the negative class (Set C).

Table 3 shows the results of the VGG using a threshold of 0.5 for the classification of each EEG on the test set of Set B (in which the network aims to distinguish IEDs from normal EEGs). Four out of the 14 patients with a normal EEG had the full EEG classified with 100% specificity (no false positive detections). The average specificity was 98.64%.

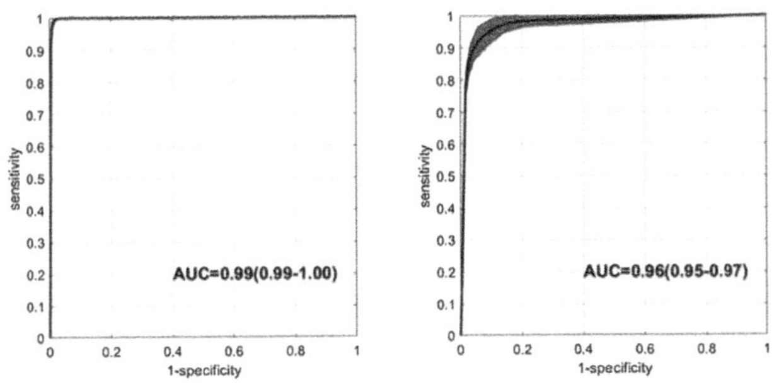

Fig. 1. ROC curves for the VGG model on the training set (left) and test set (right) of Set B. The 95% Confidence Interval (CI) of the ROC curve is shown as a shaded area. The AUC value and the corresponding CI are also presented.

Table 1. True positive and false positive rates per hour on the training and test sets of Set B and Set C. The values were assessed at a threshold where the sensitivity and specificity are equal and at a specificity of 99%. The 95% Confidence Intervals (CIs) of these rates are also presented.

Set	Spec	Sens	TP/hour	FP/hour
	Train			
B	98%[a]	98%	91.03 (80.35–100.22)	22.30 (6.32–38.28)
	99%	97%	88.71 (78.40–99.02)	16.50 (4.95–28.06)
C	85%[a]	85%	41.69 (33.75–49.62)	247.32 (116.88–377.76)
	99%	26%	12.60 (3.68-21.52)	23.14 (6.12–40.17)
	Test			
B	93%[a]	93%	47.72 (45.60–49.84)	122.41 (27.63–217.20)
	99%	79%	40.27 (38.43–42.12)	32.31 (15.15–49.46)
C	79%[a]	79%	49.30 (40.99–57.60)	348.60 (111.15–586.06)
	99%	37%	23.16 (15.10–31.22)	17.09 (1.65–32.53)

[a]sensitivity=specificity

In the focal and generalized classes (total of 20 patients on the test set), 12 patients had the full EEG classified with 100% sensitivity (all IEDs were detected). The average sensitivity and specificity values were 93.28% and 91.11%.

Considering the same analysis for Set C (which additionally included EEGs with non-epileptiform abnormalities), presented in Table 4, 4 out of the 14 patients on the test set had the full EEG classified with 100% specificity, with an average specificity of 98.47% in this class. The EEGs with non-epileptiform abnormalities (14 patients on the test set) were classified with an average specificity of 97.19%, with 100% specificity being reached for one patient. The focal and generalized IEDs (20 patients) were classified with an average sensitivity and specificity of 78.66% and 88.04%, respectively, with the IEDs of one of the patients not being detected (0% sensitivity) and the IEDs of 5 patients being detected with 100% sensitivity.

Figure 2 shows examples of the results obtained with filter visualization, which illustrate the native shape of the filters (before convolution and gradient descent). These figures can be understood as images with the same dimensions as the 2 s, 18-channel EEG epochs used as input, since those dimensions were also applied to the synthetic input used in the algorithm. Filters in lower level layers show more consistent patterns over time, while higher level layers contain filters with activity patches and low response on the remaining epoch. In the two panels of the bottom row of Fig. 2, it is possible to see that some of these patches are vertical (i.e. across channels), suggesting the detection of generalized IEDs. It is also possible to see some smaller patches, around one or two channels (e.g. upper right corner of the last filter shown in Fig. 2), suggesting the detection of focal discharges.

Figure 3 illustrates the results of occlusion with an example of each classification outcome for the set where IEDs are detected against the normal class (Set B). The scale shows the difference between the probability for each epoch and the value obtained with an occluded patch. Larger differences are plotted in warmer colors, showing that removing the patch centered in that area led to a significant change in classification, indicating that that area is important to the networks decision process. Figures 5 and 6

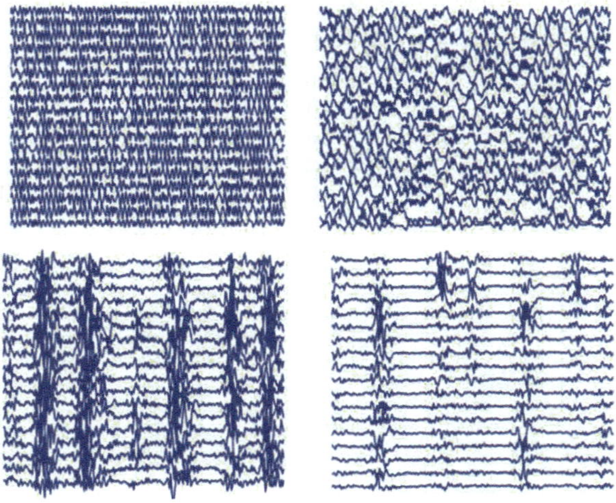

Fig. 2. Examples of results of filter visualization applied to the VGG model. The first row shows filters of lower level layers and the second row shows filters of higher level layers. These have the same dimensions as the 2 s, 18-channel EEG epochs used as input. They show the native shape of each represented filter, before any forward pass of the network.

in the Appendix show analogous examples for the set including the full EEG of epilepsy patients (Set A) and for the set containing non-epileptiform abnormalities (Set C). The detection of IED patterns in true positive cases is visible in true positive panels of Figs. 3, 5 and 6.

Comparing Fig. 6 with Figs. 3 and 5, it is possible to see that the true negative pattern corresponds to an abnormality and not to a normal EEG epoch. The false positive panel of Fig. 5 showcases an IED.

4 Discussion

We show that IEDs can be successfully identified by a VGG model. The best results were obtained when normal EEG epochs constituted the negative class (Set B). On the test set, the intersection of the values of sensitivity and specificity was at 93% and an AUC of 0.96 (CI = 0.95 − 0.97) was obtained.

Our results are up to par with the current literature regarding computer assisted IED detection, surpassing many of the described approaches. Scheuer et al. [26] applied three algorithms based on explicit feature extraction (Persyst 11, 12 and 13), obtaining sensitivities of 18%, 19% and 44%, respectively, worse than our intersection of sensitivity and specificity at 93%. Persyst 13, which yielded the highest sensitivity, detected 1.65 false positive samples per minute, which is lower than our 2.04 false positive rate per minute. However, the sensitivity threshold at which this value was calculated was vastly different. The dataset used by the authors, which consisted of records of 35 patients and 5 controls, with a total duration of 253 h, also differed greatly from ours.

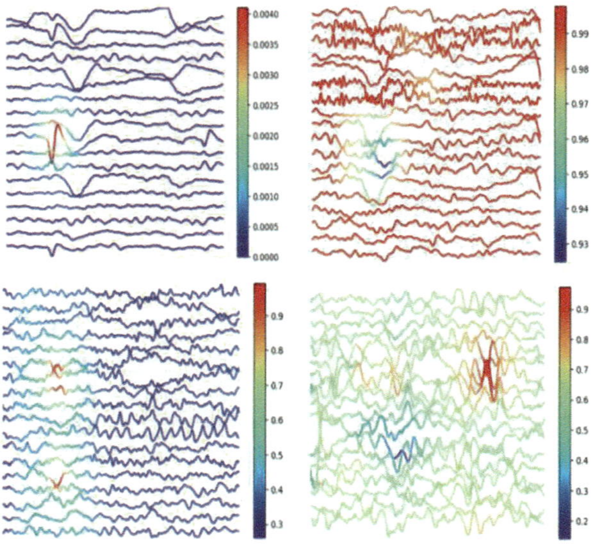

Fig. 3. Examples of the application of occlusion to the VGG model trained with set B (aiming to distinguish IEDs from Normal EEG epochs). First row: true positive (left) and true negative (right). Second row: false positive (left) and false negative (right). The scale shows the difference between the probability assigned to the epoch and what is obtained when a patch is occluded, and warmer colors are assigned to higher differences. Thus, areas plotted in warmer colors are more important for classification.

Using template matching, Lodder et al. [15] achieved 90% mean sensitivity and 2.36 false detections per minute, worse than our 2.04 false detections per minute at a sensitivity of 93%. The dataset used in Lodder et al.'s work included 20–30 min recordings from 23 epilepsy patients. With data from only 3 patients, Nonclercq et al. [16] were able to get a similar sensitivity value with template matching and k-means, again, not surpassing our results. Pang et al. [27] compared three different ANN-based methods ([13, 17, 28, 29]) using the same dataset consisting of 8 channel EEG signals from 7 epilepsy patients and 6 normal controls. This study revealed that Webber et al.'s algorithm [17], based on using mimetic features as input for a simple ANN, led to the best performance. It yielded 86.61% sensitivity and 86.32% selectivity, which is still inferior to the VGG's intersection of sensitivity and specificity at 93%.

Thomas et al. [30] used a CNN as a feature extractor, followed by an SVM as a classifier, which led to a mean AUC of 0.935 across 4 cross-validation folds, lower than the 0.96 (CI = 0.95 − 0.97) obtained on the test set by the VGG. Thomas et al.'s model was trained on 30 min recordings of 63 controls and 93 epilepsy patients. Using a 5 layer CNN as an end-to-end classifier, Johansen et al. [31] obtained an AUC of 0.947, slightly lower than the 0.96 (CI = 0.95 − 0.97) yielded by the VGG. Johansen et al. trained their network on 30 min EEG recordings from 5 epilepsy patients.

Tjepkema et al. [32] used a set of 50 patients and 50 controls to train a 19 layer, 2D CNN. The results were validated on a set of 5 patients and 12 controls, leading to 0.94 AUC for the test set, with 0.6 false detections per minute at 98% specificity.

While these results were close to those of the VGG (AUC of 0.96 and 0.5 false detections at 99% specificity), they are less reliable since the validation set is smaller.

Similarly to the approach described by Lodder et al. [33], by choosing a specificity value, the expert can limit the epochs selected for manual review. Starting with a very high threshold, only the epochs with the highest probability of containing an IED are shown. This allows the expert to diagnose a patient based on a reduced number of samples (limiting the time spent in the analysis). If these are not enough for accurate diagnosis, lowering the threshold will lead to more samples being reviewed.

We also show that the VGG can distinguish IEDs from other types of abnormalities. Abnormal EEGs were classified with an average specificity of 87.76% on the test set of set C, where abnormal EEGs (signals containing non-epileptiform abnormalities) were added to the negative class. Normal EEGs were classified with an average specificity of 98.47%, which represents a difference of less than 1% when compared to the average specificity of classification of normal EEGs on the test set of Set B. It is also important to point out that this decrease in performance was not only caused by the added complexity of the problem, but was also due to the increased class imbalance between the positive (i.e. IEDs) and negative (normal or normal and abnormal EEGs) classes, since the test set went from 8.8 h to 11.4 h and the training set increased from 24.3 h to 40.5 h, with the same number of positive samples.

The VGG model was able to detect IEDs when normal epochs from epileptic EEGs were included in the negative class (Set A), albeit with a lower performance when compared to set B (AUC of 0.91 against 0.96 on the test set of Set B, compare Figs. 1 and 4). While the increase in class imbalance contributed to this decrease in performance, similarly to what happened with the abnormal EEGs, occlusion showed that another cause for this was the mislabeling of some of the normal epochs. Figure 5 in the Appendix shows an example of a 'false positive' detection which corresponded to the occurrence of an IED. This happened since only epochs where there was no doubt about the presence of an IED were labeled. Using the information provided by this visualization technique, it was possible to account for human error and remove the normal epochs, creating Set B and improving the model's performance.

Furthermore, occlusion showed that the VGG was detecting IED shapes correctly (refer to the true positive panels of Figs. 3, 5 and 6). This result is of the utmost importance, since it proves that the network is able to classify the signal based on the correct pattern and not on spurious features, making its results more understandable and reliable.

Filter visualization provided further insight into the model's architecture, showing that filters of lower-level layers can be visually distinguished from those of higher-level layers, which include patches resembling IED detectors for focal and generalized discharges.

5 Conclusion

We show that it is possible to detect Interictal Epileptiform Discharges using a deep artificial neural network. Training a deep learning model with a larger database would contribute to an increase in model performance and robustness.

Occlusion allowed human error to be accounted for when creating the datasets and showed that IED shapes were the most important parts of the input for classification, making the classification process more transparent and contributing to the reliability of the model.

Computer assisted EEG analysis with deep learning can support clinical neurophysiologists in IED detection, reducing workload and potentially changing the way EEG interpretation is currently carried out.

Appendix

Table 2. Description of the training and test sets of each dataset, including its duration, the number of epochs and the number of epochs in the positive class (i.e. IEDs).

Set	Train			Test		
	Duration (h)	Epochs	Positive epochs	Duration (h)	Epochs	Positive epochs
A	62.6	112747	1977	13.2	23774	734
B	24.3	43867	2220	8.8	15886	452
C	40.5	72279	2015	11.4	20470	658

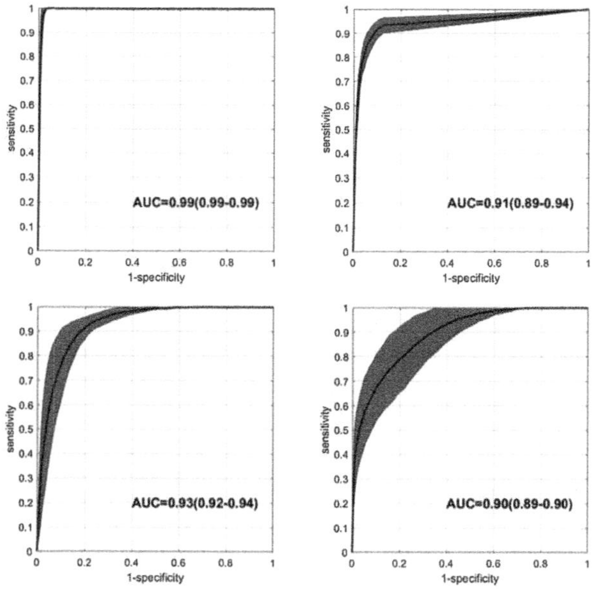

Fig. 4. ROC curves for the VGG model trained with Set A (first row) and Set C (second row). The first column shows the results on the training set and the second column shows the results on the test set. The 95% Confidence Interval (CI) of the ROC curve is shown as a shaded area. The AUC value and the corresponding CI are also presented (right).

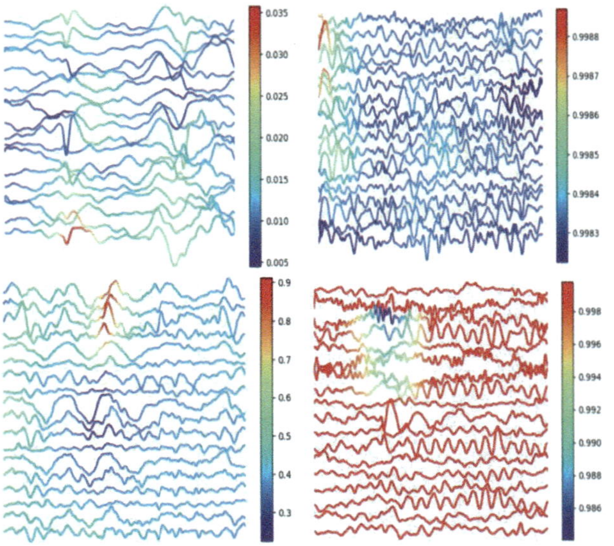

Fig. 5. Examples of the application of occlusion to the VGG model trained with set A (aiming to distinguish IEDs from Normal EEG epochs and normal epochs in epileptic EEGs). First row: true positive (left) and true negative (right). Second row: false positive (left) and false negative (right).

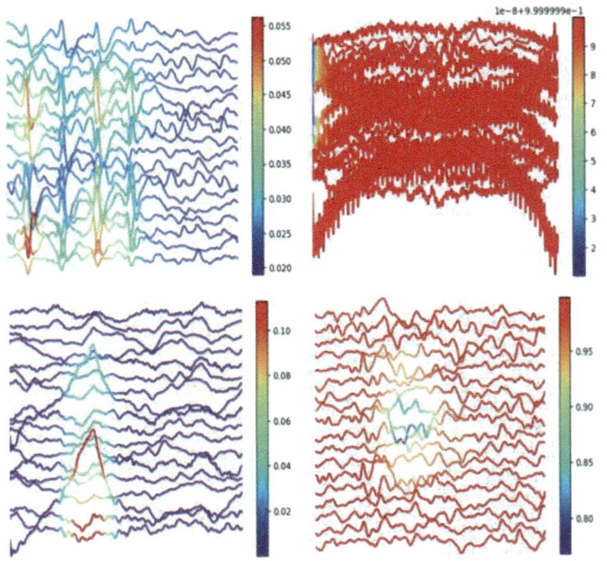

Fig. 6. Examples of the application of occlusion to the VGG model trained with set C (aiming to distinguish IEDs from Normal EEG epochs and epochs from EEGs containing non-epileptiform abnormalities). First row: true positive (left) and true negative (right). Second row: false positive (left) and false negative (right).

Table 3. Number of epochs (Epochs), number of IEDs (IEDs), Sensitivity (Sens) and Specificity (Spec) in each recording on the test set of set B, classified by the VGG with weights 100:1, at a threshold of 0.5.

	Epochs	IEDs	Sens (%)	Spec (%)
Normal	603	0	–	100.00
	538	0	–	98.88
	673	0	–	100.00
	748	0	–	95.45
	643	0	–	99.84
	598	0	–	97.49
	753	0	–	99.73
	2053	0	–	99.22
	2067	0	–	97.53
	2760	0	–	95.11
	1521	0	–	99.67
	598	0	–	97.99
	1273	0	–	100.00
	586	0	–	100.00
Focal	280	117	95.73	47.24
	1154	39	69.23	94.05
	1162	3	100.00	94.05
	204	3	66.67	95.52
	583	15	100.00	98.24
	2903	65	81.54	95.28
	665	3	100.00	99.40
	614	1	100.00	97.72
	663	10	100.00	80.86
	613	10	100.00	90.38
Generalized	618	43	97.67	87.83
	594	8	100.00	95.05
	592	9	100.00	94.34
	590	9	77.78	96.90
	626	34	91.18	95.81
	631	34	91.18	95.81
	195	4	100.00	79.06
	604	21	85.71	94.51
	588	36	100.00	88.77
	638	3	100.00	97.95

Table 4. Number of epochs (Epochs), number of IEDs (IEDs), Sensitivity (Sens) and Specificity (Spec), in each recording on the test set of set C, classified by the VGG with weights 100:1, at a threshold of 0.5.

	Epochs	IEDs	Sens (%)	Spec (%)
Normal	793	0	–	99.50
	533	0	–	98.31
	563	0	–	99.11
	673	0	–	100.00
	603	0	–	98.84
	663	0	–	99.25
	663	0	–	100.00
	593	0	–	100.00
	2053	0	–	96.64
	1895	0	–	98.63
	1723	0	–	97.34
	598	0	–	92.47
	1273	0	–	100.00
	610	0	–	98.52
Abnormal	623	0	–	91.49
	658	0	–	100.00
	533	0	–	97.75
	393	0	–	87.79
	668	0	–	97.01
	429	0	–	94.33
	477	0	–	99.16
	637	0	–	99.37
	512	0	–	96.09
	633	0	–	97.00
	981	0	–	91.03
Focal	577	39	69.23	93.49
	523	95	36.84	94.86
	611	5	80.00	98.18
	883	11	63.64	94.61
	534	67	91.04	86.08
	606	3	100.00	98.01
	639	53	0.00	99.83
	550	43	46.51	91.72
	581	23	86.96	70.61
	612	6	66.67	99.34
Generalized	618	43	100.00	85.22
	594	8	87.50	95.39
	654	20	100.00	71.29
	489	122	88.52	72.48
	195	4	100.00	61.16
	589	11	100.00	93.77
	604	21	76.19	94.34
	1121	34	94.12	89.33
	614	12	91.67	97.67
	510	89	94.38	73.40

References

1. Da Silva, F.L., Blanes, W., Kalitzin, S.N., Parra, J., Suffczynski, P., Velis, D.N.: Epilepsies as dynamical diseases of brain systems: basic models of the transition between normal and epileptic activity. Epilepsia **44**, 72–83 (2003)
2. Fisher, R.S.: ILAE official report: a practical clinical definition of epilepsy. Epilepsia **55**(4), 475–482 (2014)
3. Tatum, W.: Clinical utility of EEG in diagnosing and monitoring epilepsy in adults. Clin. Neurophysiol. **129**(5), 1056–1082 (2018)
4. Beach, R., Reading, R.: The importance of acknowledging clinical uncertainty in the diagnosis of epilepsy and non-epileptic events. Arch. Dis. Child. **90**(12), 1219–1222 (2005)
5. Zaidi, A., Clough, P., Cooper, P., Scheepers, B., Fitzpatrick, A.P.: Misdiagnosis of epilepsy: many seizure-like attacks have a cardiovascular cause. J. Am. Coll. Cardiol. **36**(1), 181–184 (2000)
6. Nowack, W.J.: Epilepsy: a costly misdiagnosis. Clin. Electroencephalogr. **28**(4), 225–228 (1997)
7. Torres, F.: Atlas and classification of electroencephalography. Pediatr. Neurol. **22**(4), 332 (2000)
8. Pillai, J., Sperling, M.R.: Interictal EEG and the diagnosis of epilepsy. Epilepsia **47**, 14–22 (2006)
9. Smith, S.: EEG in the diagnosis, classification, and management of patients with epilepsy. J. Neurol. Neurosurg. Psychiatry **76(**Suppl. 2), ii2–ii7 (2005)
10. Lodder, S.S., Askamp, J., van Putten, M.J.: Computer-assisted interpretation of the EEG background pattern: a clinical evaluation. PLoS ONE **9**(1), e85966 (2014)
11. Lodder, S.S., van Putten, M.J.: Automated EEG analysis: characterizing the posterior dominant rhythm. J. Neurosci. Methods **200**(1), 86–93 (2011)
12. Lodder, S.S., van Putten, M.J.: Quantification of the adult EEG background pattern. Clin. Neurophysiol. **124**(2), 228–237 (2013)
13. Artameeyanant, P., Chiracharit, W., Chamnongthai, K.: Spike and epileptic seizure detection using wavelet packet transform based on approximate entropy and energy with artificial neural network. In: The 5th 2012 Biomedical Engineering International Conference, pp. 1–5. IEEE (2012)
14. Liu, Y.-C., Lin, C.-C., Tsai, J.-J., Sun, Y.-N.: Model-based spike detection of epileptic EEG data. Sensors **13**(9), 12536–12547 (2013)
15. Lodder, S.S., Askamp, J., van Putten, M.J.: Inter-ictal spike detection using a database of smart templates. Clin. Neurophysiol. **124**(12), 2328–2335 (2013)
16. Nonclercq, A., et al.: Cluster-based spike detection algorithm adapts to interpatient and intrapatient variation in spike morphology. J. Neurosci. Methods **210**(2), 259–265 (2012)
17. Webber, W., Litt, B., Wilson, K., Lesser, R.P.: Practical detection of epileptiform discharges (EDs) in the EEG using an artificial neural network: a comparison of raw and parameterized EEG data. Electroencephalogr. Clin. Neurophysiol. **91**(3), 194–204 (1994)
18. LeCun, Y., Bengio, Y., Hinton, G.: Deep learning. Nature **521(**7553), 436 (2015)
19. Schmidhuber, J.: Deep learning in neural networks: an overview. Neural Netw. **61**, 85–117 (2015)
20. Qin, Z., Yu, F., Liu, C., Chen, X.: How convolutional neural network see the world-A survey of convolutional neural network visualization methods. arXiv preprint arXiv:1804. 11191 (2018)

21. Nguyen, A., Yosinski, J., Clune, J.: Multifaceted feature visualization: Uncovering the different types of features learned by each neuron in deep neural networks. arXiv preprint arXiv:1602.03616 (2016)
22. Yosinski, J., Clune, J., Nguyen, A., Fuchs, T., Lipson, H.: Understanding neural networks through deep visualization. arXiv preprint arXiv:1506.06579 (2015)
23. Zeiler, M.D., Fergus, R.: Visualizing and understanding convolutional networks. In: European Conference on Computer Vision, pp. 818–833. Springer (2014)
24. Simonyan, K., Zisserman, A.: Very deep convolutional networks for large-scale image recognition. arXiv preprint arXiv:1409.1556 (2014)
25. Kingma, D.P., Ba, J.: Adam: a method for stochastic optimization. arXiv preprint arXiv: 1412.6980 (2014)
26. Scheuer, M.L., Bagic, A., Wilson, S.B.: Spike detection: inter-reader agreement and a statistical turing test on a large data set. Clin. Neurophysiol. **128**(1), 243–250 (2017)
27. Pang, C.C., Upton, A.R., Shine, G., Kamath, M.V.: A comparison of algorithms for detection of spikes in the electroencephalogram. IEEE Trans. Biomed. Eng. **50**(4), 521–526 (2003)
28. Kalayci, T., Ozdamar, O.: Wavelet preprocessing for automated neural network detection of EEG spikes. IEEE Eng. Med. Biol. Mag. **14**(2), 160–166 (1995)
29. Özdamar, Ö., Kalayci, T.: Detection of spikes with artificial neural networks using raw EEG. Comput. Biomed. Res. **31**(2), 122–142 (1998)
30. Thomas, J., Comoretto, L., Jin, J., Dauwels, J., Cash, S.S., Westover, M.B.: EEG CLassification via convolutional neural network-based interictal epileptiform event detection. In: 2018 40th Annual International Conference of the IEEE Engineering in Medicine and Biology Society (EMBC), pp. 3148–3151. IEEE (2018)
31. Johansen, A.R., Jin, J., Maszczyk, T., Dauwels, J., Cash, S.S., Westover, M.B.: Epileptiform spike detection via convolutional neural networks. In: 2016 IEEE International Conference on Acoustics, Speech and Signal Processing (ICASSP), pp. 754–758. IEEE (2016)
32. Tjepkema-Cloostermans, M.C., de Carvalho, R.C., van Putten, M.J.: Deep learning for detection of focal epileptiform discharges from scalp EEG recordings. Clin. Neurophysiol. **129**(10), 2191–2196 (2018)
33. Lodder, S.S., van Putten, M.J.: A self-adapting system for the automated detection of interictal epileptiform discharges. PLoS ONE **9**(1), e85180 (2014)

Feedback-Error Learning Control for Powered Assistive Devices

Pedro Nuno Fernandes[1(✉)], Joana Figueiredo[1], Juan C. Moreno[2], and Cristina Peixoto Santos[1]

[1] Center for MicroElectroMechanical Systems (CMEMS), University of Minho, Guimarães, Portugal
{a68385,id6003}@alunos.uminho.pt, cristina@dei.uminho.pt

[2] Neural Rehabilitation Group, Cajal Institute, Spanish National Research Council, Madrid, Spain
jc.moreno@csic.es

Abstract. Active orthoses (AOs) are becoming relevant for user-oriented training in gait rehabilitation. This implies efficient responses of AO's low-level controllers with short time modeling for medical applications. This thesis investigates, in an innovative way, the performance of Feedback-Error Learning (FEL) control to time-effectively adapt the AOs' responses to user-oriented trajectories and changes in the dynamics due to the interaction with the user. FEL control comprises a feedback PID controller and a neural network feed-forward controller to promptly learn the inverse dynamics of two AOs. It was carried out experiments with able-bodied subjects walking on a treadmill and considering external disturbances to user-AO interaction. Results showed that the FEL control effectively tracked the user-oriented trajectory with position errors between 5% to 7%, and with a mean delay lower than 25 ms. Compared to a single PID control, the FEL control decreased by 16.5% and 90.7% the position error and delay, respectively. Moreover, the feedforward controller was able to learn the inverse dynamics of the two AOs and adapt to variations in the user-oriented trajectories, such as speed and angular range, while the feedback controller compensated for random disturbances. FEL demonstrated to be an efficient low-level controller for controlling AOs during gait rehabilitation.

Keywords-: Gait rehabilitation · Robotic assistive devices · Human-centered robotics · Feedback error learning control · Neural network

This work has been supported in part by the Fundação para a Ciência e Tecnologia (FCT) with the Reference Scholarship under Grant SFRH/BD/108309/2015, and part by the FEDER Funds through the Programa Operacional Regional do Norte and national funds from FCT with the project SmartOs - Controlo Inteligente de um Sistema Ortótico Ativo e Autónomo - under Grant NORTE-01-0145-FEDER-030386, and by the FEDER Funds through the COMPETE 2020—Programa Operacional Competitividade e Internacionalização (POCI)—with the Reference Project under Grant POCI-01-0145-FEDER-006941.

J. Henriques et al. (Eds.): MEDICON 2019, IFMBE Proceedings 76, pp. 1998–2013, 2020.
https://doi.org/10.1007/978-3-030-31635-8_238

1 Introduction

Gait rehabilitation has been proposed as a more appropriate intervention to restore motor abilities, which are boosted by user-oriented and repetitive gait training [1, 2]. Wearable robotic assistive devices, such as active orthoses (AOs), are becoming a prominent intervention for fostering user-oriented gait training by tuning the AOs' mechanical work in accordance with the real-time evaluation of the user's needs, approaching assist-as-needed (AAN) strategies [2, 3]. Additionally, AOs have embodied assistive strategies, such as the trajectory tracking control based on user-oriented trajectories to improve the movement coordination [4].

Nowadays, the principles of the human motor control system start to influence the design of these active orthotic systems, and it has been proposed bioinspired architectures with emphasis on the human-AOs interaction to deliver an effective response in accordance with the user's needs [2].

Feedback and feedforward controllers play an important role in the low-level layer of these bioinspired architectures. They should guarantee that the AOs' actuation system has accurate and time-effective responses to track the position- or torque-based reference trajectory and the stability of the human-AOs interaction. Feedback controllers, such as the Proportional-Integral-Derivative (PID) and its variants, are the low-level controllers mostly applied in AOs due to their feasibility and mathematical straightforwardness, without an arduous implementation process [5–7]. Nevertheless, adopting user-oriented assistance strategies merely based on PID control might present limitations, such as steady-state errors, measurement noise in the feedback loop, and low robustness to disturbances and human-AO dynamic changes. This limited adaptability may lead the human-AO interface to instability [8]. On the other hand, optimal and robust controls have been used to minimize the influence of disturbances over the control response and to confer adaptability in different rehabilitative scenarios [5]. However, the applicability of these controls for clinical purposes is limited by the considerable complex mathematical understanding and modeling [9].

The development of time-effective low-level controllers for AOs capable of providing low steady-state errors, emphasizing adaptiveness to different walking scenarios without requiring system modeling, becomes imperative. To address these features, Feedback-error learning (FEL) control, proposed by Kawato [10], has been applied in robotic systems [9–11]. FEL, bioinspired on the learning process of the human motor cortex, is a hybrid control combining a feedback controller with a feedforward controller capable of learning the inverse dynamics of the robotic system using the feedback control command as the error signal for the learning (feedback error). The feedforward controller usually includes regression techniques to simplify and shorten the learning of the inverse dynamics for real-time and medical applications not requiring an explicit model of the robotic system neither separating the learning and control phase [9, 11]. FEL allows the feedback controller to control the system while the inverse model is being learned in real-time. The learned inverse model is capable to track the reference trajectory. Consequently, this reduces the feedback contribution to the final control command. The overall performance of FEL beneficiates from the adaptive and anticipatory features of the feedforward control to adapt to changes in AOs' dynamics and enable time-effective fast movements while the feedback controller is able to compensate for disturbances [11].

So far, the application of FEL in real robotic assistive devices has been limited to upper limb assistance by functional electrical stimulation [11] and neuroprosthetics [9] and has not yet been tested in lower limb assistive devices such as AOs. This thesis innovatively investigates the FEL effects in AOs for gait rehabilitation. The main goal of this thesis is to present the real-time implementation of FEL control to time-effectively adapt the AOs' responses to user-oriented trajectories and changes in the dynamics due to the interaction with the user. It was applied an artificial neural network (ANN) as the feedforward controller [10] to learn the inverse dynamics of the AO taking the output of a PID command (feedback controller) as an error signal. The PID control was also used to control the AOs during the real-time learning phase of ANN (performed at 1 kHz), and to handle with disturbances as the learning is completed. The FEL control was tuned and validated in two AOs, a powered knee orthosis (PKO) and a powered ankle-foot orthosis (PAFO), using position-based trajectory tracking assistive strategies for gait rehabilitation of the knee and ankle joints, respectively. These AOs are embodied into a smart, active orthotic system – SmartOs – controlled by a bioinspired architecture. The ANN was designed using normalized inputs to allow variations in the user-oriented trajectory, such as gait speed (0.8, 1.0, and 1.2 km/h) and trajectory angular range.

2 Methods

This thesis presents the FEL as a new low-level controller for the smart, active lower limb orthotic system - SmartOs. It includes a right PKO and a right PAFO synergistically linked to a wearable motion lab (with biomechanical and muscular sensors) for the post-stroke robotic-based gait rehabilitation with a user-oriented, repetitive gait training. SmartOs was designed with a bio-inspired control architecture structured into three levels [2]. Currently, this control architecture includes a set of AAN assistive strategies, namely: adaptive trajectory tracking control, adaptive impedance control, and myoelectric control, using the PID as the low-level control.

Figure 1 illustrates the general overview of the SmartOs modules involved in this thesis. The position-based trajectory tracking control was selected as the control assistive strategy.

2.1 PKO and PAFO Assistive Devices

The used PKO and PAFO (see Fig. 1(a)) are modules of the lower limb H2-exoskeleton (Technaid S.L., Spain) [4]. Each AO has embedded sensors (potentiometer, strain gauges, and hall effect sensors) and an actuation technology comprising an electrical actuator (brushless DC motor EC60-100 W (Maxon)) coupled to a gearbox CSD20-160-2A (Harmonic Drive). The Control Area Network (CAN) was used to establish a deterministic communication among the AOs and the low-level processing unit, the STM32F407VGT microcontroller running freeRTOS OS that manages the low-level controllers at 1 kHz. The reference trajectories are set in the high- and mid-level controllers running at 100 Hz in a central processing unit, the Raspberry Pi 3 [12] with the Ubuntu Mate OS [13]. A serial interface was used, through Universal Asynchronous Receiver-Transmitter (UART), to manage the communication between the central processing unit and the low-level processing unit. The system was powered by a LiFePO4 battery of 24 V and 3 Ah, which enables at least 10 h of autonomy.

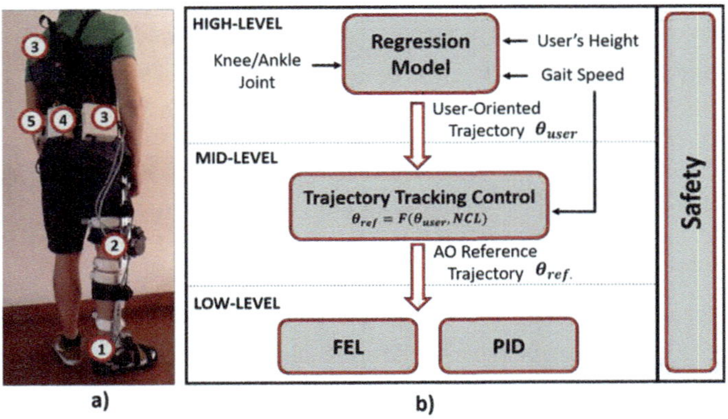

Fig. 1. SmartOs overview: (a) User wearing SmartOs modules: 1 - PAFO, 2 - PKO, 3 - Power supply system, 4 - central processing unit, 5 - low-level processing unit; (b) hierarchical control architecture.

2.2 High- and Mid-Level Controllers

For the position-based trajectory tracking control, the high-level includes a regression model dependent on the user's height and gait speed [14] to generate user-oriented position trajectories (θ_{user}) for the knee and ankle joints. It was set the gait speeds as 0.8 km/h, 1.0 km/h, and 1.2 km/h as they are the target speeds to be used with end-users.

As shown in Fig. 1, the mid-level controller sets the AO's reference trajectory $(\theta_{ref.})$ as the user-oriented angular trajectories (θ_{user}) parameterized in accordance with the gait speed. As such, (1), empirically found, sets the Number of Control Loops (NCL), each lasting 1 ms, that must occur to update the value of the reference trajectory. This speed-parameterization aims to ensure user-AO coordination.

$$NCL = -34.62 * Gait\ Speed + 107.31 \qquad (1)$$

2.3 Low-Level Controller

In the scope of this thesis, the low-level controller covers a position-based control loop to ensure that the AO's angular position matches the user-oriented trajectory. Previously to the FEL control, it has been achieved through PID control. Both controls were implemented in SmartOs as follows.

(1) *PID Control:* It was tuned using the Ziegler-Nichols method with a similar procedure as was reported in [15]. This tuning considered the practical application of PKO and PAFO in a rehabilitation scenario to avoid abrupt movements that can cause discomfort to the user and to avoid oscillations and overshoot in the actuator's response. It was used a $K_p = 90$ and $K_i = K_d = 1.5$ as PID gains. The PID controller implements the control law presented in (2) to compute the feedback command (u_{fb}), where e_k and e_{k-1} correspond to the current and previous error between the reference $(\theta_{ref.})$ and measured (θ_{meas}) angular position by a potentiometer embedded on the AO, as illustrated in Fig. 2.

$$u_{fb} = K_p e_k + K_i \sum\nolimits_{n=1}^{k} e_n \Delta t + K_d \frac{e_k - e_{k-1}}{\Delta t} \tag{2}$$

(2) *FEL Control:* It implies a feedback and a feedforward controller to command the AO in the desired way, as depicted in Fig. 2. The PID feedback controller provides control commands and guarantees stability during the real-time learning of the AOs' inverse dynamics model and to compensate disturbances as the learning

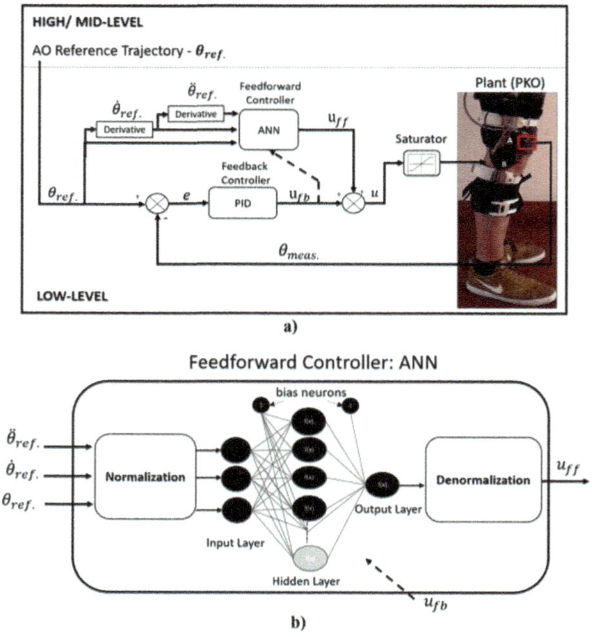

Fig. 2. FEL control: (a) FEL control loop for PKO. $\theta_{ref.}$ is the reference angular position; $\dot{\theta}_{ref.}$ is the reference angular velocity; $\ddot{\theta}_{ref.}$ is the reference angular acceleration $\theta_{meas.}$ is the measured angular position; e is the position error; u_{fb} is the feedback command; u_{ff} is the feedforward command; u is the final control command; A is the potentiometer and B is the actuator, (b) ANN-based feedback controller.

phase is completed. The feedforward controller learns the inverse dynamics model using an ANN as the regression technique, as proposed in [9, 10, 16]. It was involved a supervised ANN to infer the inverse dynamics due to its proper estimation performance, good generalization, and its capability to map non-linearities [17]. Note that two inverse models were learned, one for each AO.

The ANN was trained in real-time following a supervised learning training method, running at 1 kHz. The Multi-Layer Perceptron (MLP) was the adopted structure for the ANN, as in [9, 16]. As illustrated in Fig. 2(b), each ANN has 3 layers as follows: the input layer with 3 neurons for the 3 input trajectories, the reference position (θ_{ref}), the speed $(\dot{\theta}_{ref})$ and acceleration $(\ddot{\theta}_{ref})$; the hidden layer with 4 and 5 neurons for PKO and PAFO, respectively; and, the output layer with 1 neuron, i.e., the feedforward command (u_{ff}). The hyperbolic tangent and the identity were the used activation functions for the hidden and output layers, respectively. It was included bias nodes in the hidden and output layers (see Fig. 2(b)) to increase the flexibility of the ANN to fit the input data.

As the learning method, it was used the backpropagation algorithm that includes a forward and a backward phase. The forward phase predicts feedforward commands (u_{ff}) for the given inputs and based on the current weights. The backward phase updates the weights based on the gradient descent (i.e., minimization) of the current feedback command (u_{fb}, PID command serves as the cost function) with respect to the network's weights change. A stochastic gradient descent (SGD) approach was implemented such that the weights are updated based on inputs presented in random order on each training iteration [18]. To implement SGD in real-time and to meet the temporal requirement of 1 ms (1 kHz), it was applied the Adaptive Moment Estimation (ADAM) optimizer [19] that is an SGD approach for computationally efficient stochastic optimization that only requires first-order gradients with little memory requirements [19]. ADAM used adaptive learning rates per weight connection, providing adaptability to the ANN training and decreasing the training time.

Table 1 presents the empirically found set-up for training the ANN for PKO and PAFO, namely the number of neurons in the hidden layer, the maximum learning rate, and the initial weights in the hidden (w_{hid}) and output (w_{out}) layers. This parameters' setting resulted from a trade-off between the ANN performance and the real-time temporal requirements (1 kHz).

Table 1. ANN's set-up for training phase

AO	Hidden neurons	Initial weights	Learning rate
PKO	4	$-0.058 \leq w_{hid} \leq 0.058$ $-0.05 \leq w_{out} \leq 0.05$	Adaptive (≤ 0.0001)
PAFO	5	$-0.058 \leq w_{hid} \leq 0.058$ $-0.045 \leq w_{out} \leq 0.045$	Adaptive (≤ 0.00001)

For both training and recall phases, the input signals of ANN $\left(\theta_{ref}, \dot{\theta}_{ref}, \ddot{\theta}_{ref}\right)$ were normalized between [−1; 1] in an attempt to provide adaptability and versatility to the ANN. This enables to handle with variations in the reference trajectories, such as gait speed and trajectory angular range, and the application of the tuned ANN in different user-oriented trajectories. Moreover, it reduces the estimation error and accelerates the training phase [20]. As presented in Fig. 2(b), the predicted output of the ANN (u_{ff}) was denormalized to the maximum operating magnitude of the control commands, experimentally set to [−2500; 2500], i.e., to the maximum values of the AOs' pulse-width modulation.

Moreover, in both training and recall phases, the computed feedback (u_{fb}) and feedforward (u_{ff}) commands are summed to get the final control command (u) to be applied to the AOs. The final command was limited to [−2500; 2500] with a saturator (Fig. 2(b)) to protect the operability of AOs. When three gait cycles are performed, the contribution of the feedback control command $(u_{fb}(\%))$ to the final control command (u) is estimated, as given by (3), considering $\overline{u_{ff}^2}$ as the mean squared feedforward contribution.

$$u_{fb}(\%) = 100\% \times \frac{\overline{u_{ff}^2}}{u} \tag{3}$$

The training phase ends when the contribution of the feedback controller is equal or lower to 5% $(u_{fb} \leq 5\%)$ since it was empirically verified that the feedback contribution did not reach much lower than 5%. Then, the recall phase starts and the learned inverse dynamics models of PKO and PAFO are able to predict time-effective control commands to timely track the reference trajectory (θ_{ref}) while the feedforward controller is released from this task to compensate for disturbances.

All software interfaces for FEL real-time implementation were done in C language in the STM32F407VGT microcontroller, using its hardware abstraction layer libraries and following the time-effectiveness architecture of SmartOs.

2.4 Safety Measures

Different safety measures were included throughout the control architecture (as illustrated in Fig. 1(b)) to safeguard the human-AO interaction. The mechanical limits, i.e., the range of motion (ROM) of PKO and PAFO were limited to [3°; 98°] and [−18°; 18°], respectively. These intervals are contained within the ROMs of the user's joints so that the AOs do not compromise the users' joints integrity by applying overextension or over flexion motions. Additionally, a proper and comfortable attachment and alignment of the AOs with the user's lower limbs and joints, respectively, were considered. Lastly, the therapy can only be interrupted or stopped in a smooth and safe manner when the user is touching on the ground (tracked by force resistive sensors embedded on AOs).

3 FEL Control Performance

3.1 Experimental Tuning

The tuning of FEL control focused on the feedforward controller. No tuning was made in the feedback controller since it was used the already tuned PID control. The tuning of the feedforward controller consisted of the real-time training of the ANN by considering the normalized inputs randomly presented to the ANN (SGD approach) and the actual PID commands as the feedback error to be minimized. This procedure was separately performed for the PKO and PAFO with the respective reference trajectories adjusted with NCL for 1 km/h and 0.8 km/h (speeds randomly selected), respectively.

During the real-time training of the feedforward controller, the ANN is trying to learn the AOs' inverse dynamics models. Consequently, the feedforward commands that contribute to the final command (u) could lead the AOs to exceed their mechanical limits and compromising their integrity. Therefore, for the first training phase of the ANN, it was decided to modify the original reference position trajectory. It was added an offset of 15° to the original knee reference trajectory and smoothed the original ankle reference trajectory with an attenuation gain of 40% and 4° of offset. With this procedure, it is was possible to get the inverse dynamics models of PKO and PAFO while operating far from their mechanical limits. As illustrated in Fig. 3, the trajectory modification was possible since the normalized signals of the original and modified trajectory are equal. This finding shows that the ANN would receive equal inputs in both situations. Note that the initial AOs' angular position was set close to the first value of the non-normalized input signals.

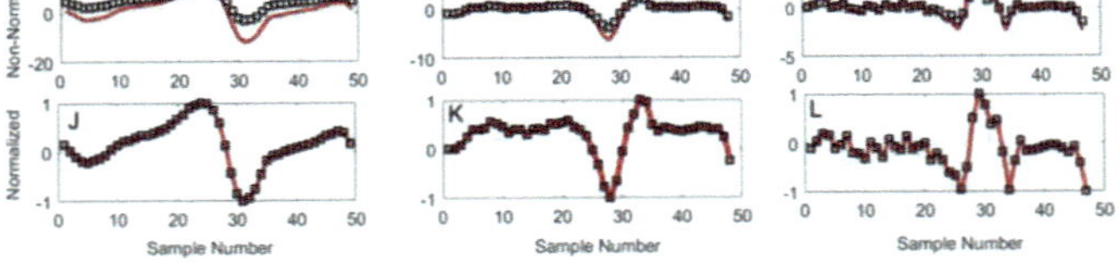

Fig. 3. Original (red) and modified (black) input signals for PKO (A–C non-normalized signals, D–F normalized signals) and PAFO (G–I non-normalized signals, J–L normalized signals).

The pre-trained ANNs were subsequently retrained using the original reference trajectories to get the final configuration for the ANNs. The pre-trained state is an important advantage to clinical application once it may decrease the training time for new user-oriented trajectories, as reported in [11].

3.2 Experimental Validation

FEL control was validated in PKO and PAKO following three experimental procedures. In the first one, it was validated the robustness, repeatability, and time-effectiveness of the FEL control, with focus on the ANN-based feedforward controller. Moreover, it was studied the stability of FEL control commands along the time such that to investigate the possible inclusion of FEL as a low-level controller in SmartOs. For these purposes, a set of experiments were separately performed for the PKO and PAFO without involving the user, considering the original reference trajectories tuned for 0.8 km/h, 1.0 km/h, and 1.2 km/h. Each experiment lasted 5 min.

The second procedure covers the validation of FEL control considering the user-AO interaction to investigate the FEL adaptability to changes in the dynamics due to the interaction with the user. For this purpose, two able-bodied subjects (a male and a female) with 25.5 $\pm$ 0.71 years old, the height of 1.69 $\pm$ 0.1 m, and the weight of 64.50 $\pm$ 14.84 kg were involved. They gave their informed consent to take part in the experiment. They were asked to walk at 0.8, 1.0, and 1.2 km/h in level-ground on a treadmill with the PKO and PAFO. Each participant performed 3 trials with a duration of 5 min.

With the third procedure, it was studied the FEL response to external perturbations to the user-AO interaction. Under the same conditions as the ones taken for the second procedure, the participants were informed and asked to counteract the PKO and PAFO in the terminal stance and in the initial stance phase, respectively, preventing both AOs from increasing their position. The participants performed the disturbances in self-selected gait cycles. These disturbances were selected given their high prevalence during a gait therapy.

3.3 Performance Measures and Metrics

To evaluate the FEL performance while comparing with the single PID feedback controller, it was collected the position trajectory measured by the potentiometer (θ_{meas}), the feedback (u_{fb}) and feedforward commands (u_{ff}), for PKO and PAFO. All signals were sampled at 100 Hz. As performance metrics, it was computed the phase delay (ms), the angular position error (°) and its normalized root mean square error (NRMSE (%)), and the feedback command contribution $(u_{fb}(\%))$.

4 Results and Discussion

4.1 FEL Tuning

It was analyzed the AOs' state during FEL tuning, by considering the evaluation of the measured position trajectory, the control commands, and the angular position error. For instance, Fig. 4 presents the results achieved during the ANN training using the modified trajectory for the PKO at 1 km/h under three different periods of FEL control: Initial Phase, Middle Phase, and Final Phase.

During the Initial Phase (first 11 s), the feedforward controller is starting to tune its ANN. Consequently, feedforward contribution to the final control command is not significant, being the PID control responsible to track the reference trajectory (Fig. 4B). It is possible to see in Fig. 4A that the measured trajectory is delayed 210 ms comparatively to the reference one. Hence, the angular position error varies from -20° to 20°, as demonstrated in Fig. 4C.

In the Middle Phase (between the first 11 s until 40 s), the FEL controller is learning the inverse dynamics model of the PKO. Therefore, the measured trajectory starts to decrease its phase difference to the reference signal to a mean value of 6 ms (considering three gait cycles presented in Fig. 4D). To correct this delay, the feedforward controller produces commands that when summed with PID commands lead the PKO to perform a trajectory with 10° more than the reference trajectory, as illustrated in Fig. 4D. This happens because the ANN has not learned the inverse dynamics model with the best performance yet. Figure 4E highlights an increase and decrease in the feedforward command and the PID command, respectively, relatively to the Initial Phase. As depicted in Fig. 4F, the position error decreased, varying from -20° to 10°.

For the Final Phase (after 40 s), when the ANN already learned the inverse dynamics of the PKO, it was verified that the FEL control has successfully aligned the PKO trajectory with the reference one (Fig. 4G) and corrected the amplitude divergence previous observed. The position error decreased by 75% compared to the Initial Phase. In this phase, the feedback controller contribution was 4.4% (Fig. 4K).

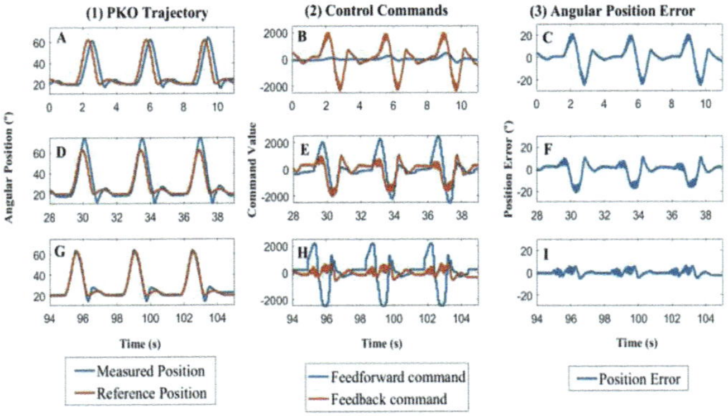

Fig. 4. FEL tuning over three periods: Initial Phase (A–C), Middle Phase (D–F), and Final Phase (G–I). Random example for PKO using the modified trajectory tuned for 1 km/h.

In overall, the findings of FEL tuning indicate that the designed ANN was capable of correctly learning the inverse dynamics for both AOs. When the learning phase finished, the ANN can timely track the reference trajectory, discharging the feedback controller for this task. It was observed that the learning phase lasted 90 s (approximately 25 gait cycles) and 315 s (approximately 70 gait cycles) for PKO and PAFO, respectively. The temporal difference is mainly explained by the different learning rates used in both cases (Table 1). Additionally, it was verified that the run time of FEL control loop (0.25 ms) is lower than the one required by the bioinspired control architecture of SmartOs (1 ms). These temporal findings suggest that the techniques applied for approaching a real-time implementation were effective to avoid long-time periods in the training phase. Furthermore, few iterations were required to tune the pre-trained ANN.

It is important mentioning that the modification made in the reference trajectory for the first learning moment protected the AOs. Figure 4D shows a practical situation, during the Middle Phase, where a 10º-deviation would put the device out of its lower operating limit (3º). The 10º-deviation did not damage the PKO since the added 15º-offset was enough to deviate the PKO from its lower mechanical limit (3º). A similar effect was observed in PAFO for the lower and upper limits.

4.2 FEL Validation

This sub-section presents the validation of FEL control for PKO and PAFO considering their ANNs tuned for the original reference trajectory. It was also compared the FEL performance with the single PID low-level controller of the SmartOs.

(1) *PKO validation:* Table 2 presents the results for FEL and PID controls with and without the user's interaction (first and second procedures, respectively), namely the NRMSE, phase delay, and the contribution of the feedback control (u_{fb}).

For all tested speeds in the first procedure, the FEL control presented a mean NRMSE of 5.13% and a mean phase delay of 13 ms, with the feedback controller only contributing 5.9% for the final control command. Additionally, it was observed that the ANN tuned for 1 km/h well-predicted the feedforward commands for 0.8 and 1.2 km/h speeds by only changing the normalization limits. This finding is supported by the low NRMSE (5.17 ± 0.04% and 4.68 ± 0.03%) and delays (20 ± 0.001 ms) found for 0.8 and 1.2 km/h speeds, respectively. These outcomes state that the FEL control provides robust and time-effective control commands in a repetitive manner along the gait cycle independently of the speed variation.

Regarding the validation with the user-PKO interaction (second procedure), a mean NRMSE of 5.87%, a mean phase delay of 12.5 ms, and a mean contribution of 6.52% of the feedback controller were yielded. Figure 5 also shows the consistency of FEL performance, presenting low position error and delay, and repetitive time-effective feedforward commands even when considering the user-PKO interaction under different gait speeds. Comparing both procedures, i.e., in the presence or not of interaction with the user, it was noted that (i) the phase delay decreased by 0.5 ms; (ii) the mean NRMSE increased by 0.74%; and, (iii) the mean feedback contribution increased 1% to the final control command. The two last findings could result from the load added (i.e., the user) in the PKO, that may difficult the controller to follow the reference trajectory.

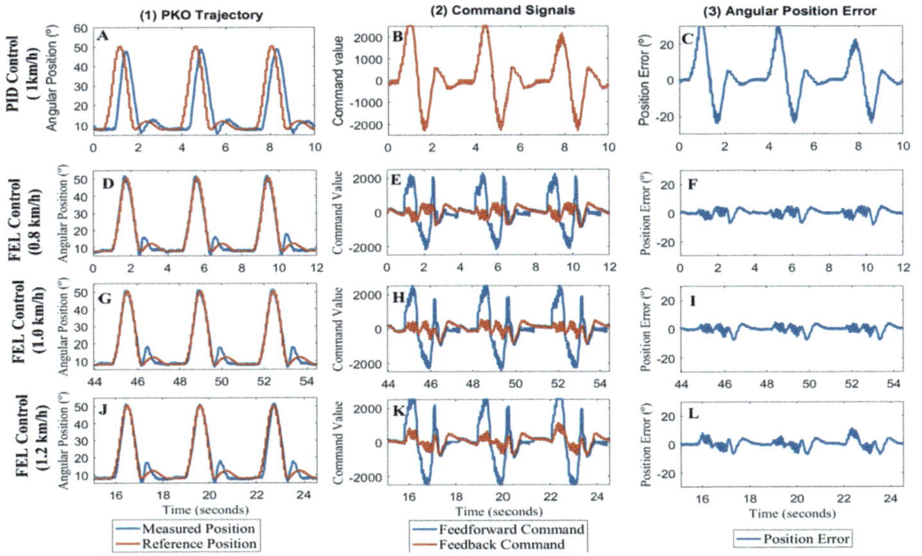

Fig. 5. Results of PID (A–C) and FEL (D–L) controls in the recall phase considering the user-PKO interaction for 0.8, 1, and 1.2 km/h speeds.

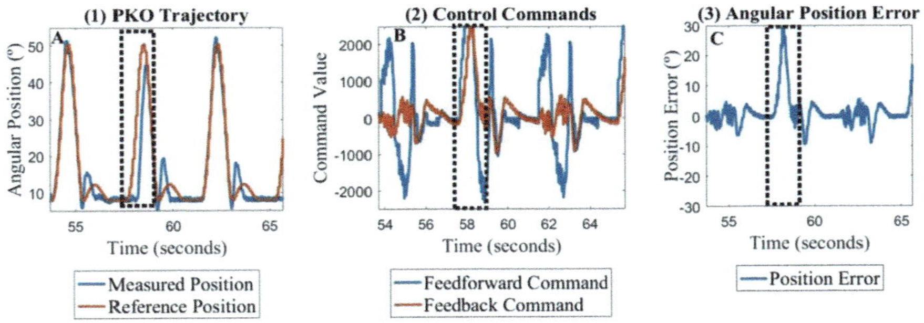

Fig. 6. Results of FEL control to external disturbances to the user-PKO interaction for 1 km/h.

Furthermore, it was also evaluated the FEL performance in the presence of external disturbances to the user-PKO interaction. This analysis was limited to disturbances in the terminal stance phase, like the one marked in Fig. 6 at 58 s. Figure 6B shows that at the instant of the disturbance the feedback command increases once the position error grows (see Fig. 6C) due to the disturbance. The increased contribution of the feedback control aimed to prevent the FEL control from fall into an instability state and from damaging the user-PKO interaction. On the other hand, the feedforward controller command stayed periodic (Fig. 6B), as the reference signals and the learned inverse dynamics model did not change.

(2) *PAFO validation:* The performance of FEL and PID controls with and without the user's interaction (first and second procedures, respectively) was evaluated considering the NRMSE, phase delay, and the contribution of the feedback control, as presented in Table 3.

Table 2. Results (Mean ± std) of FEL and PID controls for first (I) and second (II) procedures with PKO at different speeds

	Speed (km/h)	Control	NRMSE (%)	Delay (ms)	u_{fb} (%)
I	0.8	PID	21.58 ± 0.17	260 ± 0.046	–
		FEL	5.17 ± 0.04	20 ± 0.001	6.9 ± 3.6
	1.0	PID	22.33 ± 0.16	210 ± 0.024	–
		kEL	5.54 ± 0.08	1 ± 0.001	4.5 ± 0.15
	1.2	PID	22.0 ± 0.16	230 ± 0.05	–
		FEL	4.68 ± 0.03	20 ± 0.001	6.3 ± 0.53
II	0.8	PID	21.58 ± 0.17	260 ± 0.046	–
		FEL	5.55 ± 0.04	15 ± 0.007	6.94 ± 0.67
	1.0	PID	22.33 ± 0.16	210 ± 0.024	–
		FEL	5.69 ± 0.04	1 ± 0.004	6.1 ± 0.95
	1.2	PID	22.0 ± 0.16	230 ± 0.05	–
		FEL	6.37 ± 0.013	22.5 ± 0.004	6.52 ± 0.23

Results for the first procedure including all tested speeds demonstrate that FEL achieves a mean phase delay of 23.3 ms, a mean NRMSE of 6.6% with low position errors (0.8 km/h, from -5° to 5°; 1.0 km/h, from -5° to 5°; 1.2 km/h, from -6° to 4°). These outcomes were observed under a predominant contribution of the feedforward controller over the feedback controller (mean contribution of 8.3% for all speeds). Thus, the assistance is almost given entirely by the feedforward controller. FEL control

Table 3. Results (Mean ± std) of FEL and PID controls for first (I) and second (II) procedures with PAFO at different speeds

	Speed (km/h)	Control	NRMSE (%)	Delay (ms)	u_{fb} (%)
I	0.8	PID	22.28 ± 0.14	250 ± 0.28	–
		FEL	5.63 ± 0.04	20 ± 0.03	8.5 ± 0.47
	1.0	PID	24.53 ± 0.18	260 ± 0.27	–
		FEL	6.54 ± 0.04	20 ± 0.03	7.9 ± 0.07
	1.2	PID	26.99 ± 0.17	250 ± 0.28	–
		FEL	7.68 ± 0.05	20 ± 0.03	8.6 ± 0.94
II	0.8	PID	22.28 ± 0.14	250 ± 0.28	–
		FEL	6.95 ± 0.008	30 ± 0.01	6.2 ± 0.01
	1.0	PID	24.28 ± 0.18	260 ± 0.27	–
		FEL	5.99 ± 0.002	20 ± 0.001	6.9 ± 0.3
	1.2	PID	26.99 ± 0.17	250 ± 0.28	–
		FEL	6.58 ± 0.004	25 ± 0.007	4.84 ± 1.9

was able to learn the inverse dynamics of PAFO, providing accurate and time-effective control commands in a repetitively and independently of the speed variation. It was verified that the same set-up of FEL control can successfully be applied for different gait speeds due to the equal temporal performance with slight deviations in NRMSE under speed variation.

Moreover, it was evaluated the FEL performance considering the user-PKO interaction (second procedure). Figure 7 shows the obtained results, randomly selected for one subject, walking on a treadmill for 0.8, 1.0 and 1.2 km/h. In this procedure, it was observed a mean NRMSE of 6.51%, a mean phase delay of 25 ms and a mean contribution of 5.8% of the feedback controller to the final control command. It is possible to verify that FEL performs similarly in the presence or not of the interaction with the user given that: (i) mean NRMSE decreased 0.09%; (ii) the phase delay increased 1.7 ms; and, (iii) the feedback contribution decreased 2.5%. The decrease in the NRMSE and consequent reduction of feedback contribution might be due to the load added to the PAFO that stopped it to exceed the reference signal since it was observed in the first procedure a general tendency to the PAFO position exceeds -5° and 5° the reference trajectory. In overall, the findings show that FEL can be used as an adaptive control to deal with dynamics changes due to interaction with the user. It also yielded time-effective commands since the AO's position gets the reference position at least 40 ms before the reference position is updated (every 65 ms for 1.2 km/h, the more demanding condition).

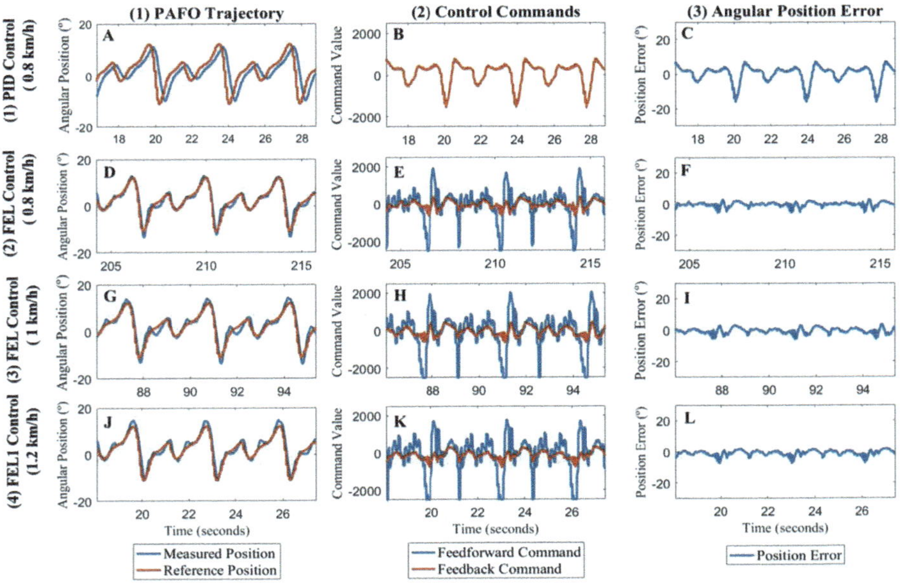

Fig. 7. Results of PID (A–C) and FEL (D–L) controls in the recall phase considering the user-PAFO interaction for 0.8, 1, and 1.2 km/h speeds.

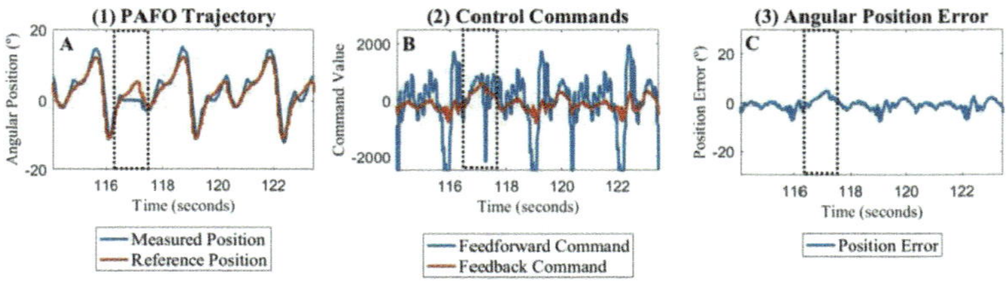

Fig. 8. Results of FEL control to external disturbances to the user-PAFO interaction for 1 km/h.

Furthermore, it was validated the FEL performance in the presence of external disturbances to the ones introduced by the user-PAFO interaction, as depicted in Fig. 8. Such disturbances increased the position error due to the displacement between the reference and measured position. In responding to this disturbance, the FEL control augments the feedback controller contribution (increased PID command) while the feedforward command stayed periodic since the PAFO dynamics did not change. These outcomes demonstrate that the feedback control is charged to compensate for small disturbances while the feedforward control has the higher actuation command to drive the desired movement.

4.3 Comparative Analysis with PID Control

For the assistance provided by the PKO under all tested speeds, the PID control achieved an NRMSE around 22% and a phase delay of 230 ms. Consequently, it was observed that the FEL control decreased, approximately, the NRMSE 16.5% and 93% for the phase delay. In the ankle-foot assistance by PAFO, the PID achieved a mean NRMSE of 24%, a position error varying from -16° to 7°, and a phase delay of 250 ms for all speeds. For PAFO, the FEL control decreased by 17.5% and 90.7% the NRMSE and delay, respectively. In overall, the FEL control can provide more accurate and time-effective assistance, with lower position error and phase delay than single PID feedback control. Therefore, the hybrid arrangement offered by FEL control can control limbs with a small delay and this holds great importance for gait rehabilitation.

5 Conclusion and Future Work

This thesis innovatively implemented a FEL control in assistive devices for gait rehabilitation. The designed ANN was able to learn the inverse dynamics of PKO and PAFO with an acceptable learning time for clinical applications. Furthermore, the learning time can be decreased by using the pre-trained ANN of AOs. Due to its learning capabilities, the feedforward control was able to handle with variations in the user-oriented trajectories. FEL control time-effectively adapts the AOs' responses to the changes in the dynamics due to the interaction with the user. Moreover, the feedback control compensated for random disturbances. FEL control substantially decreased the tracking error for acceptable ranges and the response delay compared to PID control. This thesis shows that FEL control is an adaptive and time-effective low-level controller for AOs, such as the ones embedded in SmartOs, which may yield to an efficient gait rehabilitation.

Future work comprises the validation with more subjects, disturbances, and AAN strategies. Evaluation of the FEL control with patients is also necessary from a practical perspective. It is also relevant to compare the FEL control with a hybrid control.

References

1. Meng, W., Liu, Q., Zhou, Z., Ai, Q., Sheng, B., Xie, S. (Shane).: Recent development of mechanisms and control strategies for robot-assisted lower limb rehabilitation. Mechatronics **31**, 132–145 (2015)
2. Tucker, M.R.: Control strategies for active lower extremity prosthetics and orthotics: a review. J. Neuroeng. Rehabil. **12**, 1 (2015)
3. Bergmann, J., Krewer, C., Muller, F., Koenig, A., Riener, R.: Virtual Reality to control active participation in a subacute stroke patient during robot-assisted gait training. In: IEEE International Conference on Rehabilitation Robotics, vol. 2011, p. 5975407 (2011)
4. Bortole, M.: The H2 robotic exoskeleton for gait rehabilitation after stroke: early findings from a clinical study. J. Neuroeng. Rehabil. **12**(1), 54 (2015)
5. Viteckova, S., Kutilek, P., Jirina, M.: Wearable lower limb robotics: a review. Biocybern. Biomed. Eng. **33**(2), 96–105 (2013)
6. Yan, T., Cempini, M., Oddo, C.M., Vitiello, N.: Review of assistive strategies in powered lower-limb orthoses and exoskeletons. Rob. Auton. Syst. **64**, 120–136 (2015)
7. Anam, K., Al-jumaily, A.A.: Active exoskeleton control systems: state of the art. In: International Symposium on Robotics Intelligent Sensors, pp. 988–994 (2012)
8. Lu, W., Yang, J., Liu, X.: The PID controller based on the artificial neural network and the differential evolution algorithm. J. Comput. **7**(10), 2368–2375 (2012). SPL.ISS
9. Resquín, F., Gonzalez-vargas, J., Ibáñez, J., Brunetti, F., Pons, J.L.: Feedback error learning controller for functional electrical stimulation assistance in a hybrid robotic system for reaching rehabilitation. Eur. J. Transl. Myol. **26**(3), 255–261 (2016)
10. Kawato, M.: Feedback-error-learning neural network for supervised motor learning. Adv. neural Comput. **6**, 365–372 (1990)
11. Kurosawa, K., Futami, R., Watanabe, T., Hoshimiya, N.: Joint angle control by FES using a feedback. IEEE Trans. Neural Syst. Rehabil. Eng. **13**(3), 359–371 (2005)
12. Raspberry Pi 3 Model B
13. Ubuntu MATE for the Raspberry Pi 2 and Raspberry Pi 3
14. Koopman, B., van Asseldonk, E.H.F., Van der Kooij, H.: Speed-dependent reference joint trajectory generation for robotic gait support. J. Biomech. **47**(6), 1447–1458 (2014)
15. Bortole, M.: Robotic Exoskeleton With an Assist-as-Needed Control for Gait Rehabilitation After Stroke (2014)
16. Miyamoto, H., Kawato, M., Setoyama, T., Suzuki, R.: Feedback-error-learning neural network for trajectory control of a robotic manipulator. Neural Netw. **1**, 251–265 (1988)
17. Meireles, A.P.E., Magali, R.G.: A comprehensive review for industrial applicability of ANNs. IEEE Trans. Ind. Electron **50**(3), 585–601 (2003)
18. Ruder, S.: An overview of gradient descent optimization, pp. 1–14 (2016)
19. Kingma, D., Ba, J.L.: Adam: a method for stochastic optimization. In: International Conference on Learning Representations (ICLR), pp. 1–15 (2015)
20. Sola, J., Sevilla, J.: Importance of input data normalization for the application of neural networks to complex industrial problems. IEEE Trans. Nucl. Sci. **44**(3 Part 3), 1464–1468 (1997)

CopyRobot: Interactive Mirroring Robotics Game for ASD Children

Laura Santos[1,2](✉), Alice Geminiani[1], Ivana Olivieri[3], José Santos-Victor[2], and Alessandra Pedrocchi[1]

[1] Department of Electronics, Information and Bioengineering (DEIB) NEARLab - Neuroengineering and Medical Robotics Laboratory, Politecnico di Milano, Piazza Leonardo da Vinci 32, 20133 Milan, Italy
laura.d.santos@tecnico.ulisboa.pt

[2] Institute for Systems and Robotics, Instituto Superior Tecnico, Lisboa, Portugal

[3] CARELab, IRCCS Fondazione Don Carlo Gnocchi, Milan, Italy

Abstract. The family of disorders commonly known as autism is characterized by a deficit in social interaction and restricted repetitive and stereotyped patterns of behaviours, activities and interests. Motor disturbances are not part of the diagnosis of the children with autism but some studies have estimated that between 80 and 90% of children with Autism Spectrum Disorder (ASD) demonstrate some degree of motor impairments. Several therapies have been used for the improvement of motor skills, always leading to behavioural improvements as side-effects, demonstrating the importance of motor interaction and stimulation for the case of autism. Recent studies have shown that motor, imitation and social abilities are all related in people with autism. In this work, a humanoid robot is used to create a therapy that unites all these areas. The system involves a robot (NAO), a Kinect camera and Personal Computer, with the goal of facilitating the interaction between therapist and a child with ASD during a physical therapy session. To improve the imitation abilities of the child, the robot was programmed to mirror both the child and the therapist movements. After testing different tracking methodologies, the Kinect sensor was selected as the best compromise of quality and cost. Two protocols were developed, depending on who plays the role of the main actor. In the first protocol, the robot is the master and leads the interaction. It decides the exercise to execute and gives feedback to both the therapist and the child. In the second protocol, the choice of the exercise sequence is the therapist's responsibility. To promote interaction further during clinical tests, the protocol was changed to include gesture imitation. For the robot master protocol, the space theme was chosen. For the therapist master protocol, the theme of sports, that was already performed by the children in the usual therapy, was adopted. The system was tested in realistic conditions with two different autistic children. The reaction was different in each case but it demonstrated the importance of these imitation games in the treatment of this disease.

J. Henriques et al. (Eds.): MEDICON 2019, IFMBE Proceedings 76, pp. 2014–2027, 2020.
https://doi.org/10.1007/978-3-030-31635-8_239

1 Introduction

Autism Spectrum Disorder (ASD) appears on subjects in very different ways. The incidence has been growing and currently affects 1 in 160 children [1]. It is characterized by a symptomatic triad: difficulties in social interaction, communication deficit, and presence of repetitive behaviours. The cause for autism is still unknown and there is no cure. The treatment, as the Spectrum itself, is multifaceted, ranging from the standard Applied Behaviour Analysis [2], to experiments in the pool and with horses [3]. All treatments try to adapt to the needs of each child and many of them centre on the recovery of the social part. Recently, studies with ASD children proved the relation between imitation, social and motor abilities of these kids [1]. This unleashed an increasing interest in the motor and imitation recovery of ASD children. The main motor problem is related with the praxis of the movement [1], while the imitation problem (also related with this last one) centres on a difficulty of reproducing the same kinetics of another individual, although the same target can be achieved [4].

These capabilities can be developed through the use of robots, in a form of embodied mirroring (Fig. 1). This is justified by two facts. On one hand, research on mirror neurons has revealed the existence of neurons which are activated both during the execution of goal-directed movements, as well as during the observation of those movements performed by others, revealing the importance of mirroring as a form of motor training [5]. On the other hand, robots and physical mirroring are more effective than training with graphical Avatars or other forms of Virtual Reality, [6]. It was verified that the number of people that had a spontaneous imitation reactions was significantly larger when a robot was physically present, than in the situation where the robot was presented through video. Moreover, robots have been extensively used in the social therapy of children with Autism in the form of Social Assistive Robots, where they were mainly used as interactive toys, due to the connection of these children with several kinds of technology [7].

Fig. 1. Embodied mirroring

Mirroring has already been used widely in robotics requiring three implementation steps: capture of the observed human motion; mapping to robot joint angles; control the robot with the calculated angles. Most of the works have used wearable sensors and vision based without markers devices to capture the movement [8]. This was due to the low cost and high portability of both [9]. Inertial Measurement Units are the principal wearable sensors used for robot mimicry. They are constituted by a 3D gyroscope, a 3D accelerometer and a 3D magnetometer, which are responsible to measure the position and velocities of these units in relation to a global reference frame. They have been used in several works for the robotic mirroring of the upper limbs and lower limbs of a human. In terms of marker-less vision based devices, a breakthrough sensor was the Microsoft Kinect, from now on referred to only as Kinect [9]. It is able to reconstruct the 3D joint positions of a human skeleton. The first version was based on a combination of a standard camera and an infrared sensor and it was highly susceptible to noise. The second version solved this problem with a time of flight camera.

Some of the works done with Kinect in the field of embodied mirroring were [10–12]. There are two key control strategies of the robot in these works. The first is based on the end-effector trajectories, obliging the robot end effector to follow the same trajectory as the human which is controlling it. They use inverse kinematics strategies to discover the angles from the position of the end effector [8]. The second is centred on the joint space, trying to convert directly the joint angles of the human in the joint angles of the robot. The spatial vector method is one of the most used. It calculates the links between the several joints and determines the angle between them and a certain reference frame [10].

The majority of these works were mainly used for an easier control of the robot and rarely were presented in a therapeutic environment. They centre on one master to control the robot. Therefore, the main goal of this project is to develop an imitation system that can be used in a therapy for ASD children and that will allow the inclusion of a second master. In this way, child and therapist will be able to control the robot. Consequently, a setup for embodied mirroring with several exercises is implemented and tested to permit a simultaneous training of the motor, imitation and social capabilities of these children, a tool missing in the literature.

2 Methods

2.1 Mirroring System

The robot chosen for this imitation system was NAO from Aldebran Robotics. It is a small humanoid robot that was already used in some social therapies for ASD children due to their simple expressions that resemble a human without being one. It is constituted by 24 degrees of freedom but just 3 of them were used for the physical mirroring of the upper limbs: Elbow Roll, Shoulder Roll and Shoulder Pitch. Therefore, the basic arm movements allowed are the flexion and extension, abduction and adduction, horizontal abduction and adduction of the arms and the flexion and extension of the forearms. NAO has also LEDS and loudspeakers which can be used to transmit visual and audio signals, providing a bigger interaction with people.

As a tracking device the system chosen was the Kinect. Comparing to the wearable sensors, Kinect presents the big advantage of not being intrusive, which is especially important for children with autism. The 3D joint positions are obtained directly from the camera with a frame rate of 30 Hz. In total 25 keypoints are extracted to form a skeleton. A median filter with a window of 5 samples is applied in order to reduce the noise. Then the Kinect reference frame is transformed on the NAO reference frame (Fig. 2) through two rotations, one around the y axis and another around the x axis, summarized by the rotation matrix presented below.

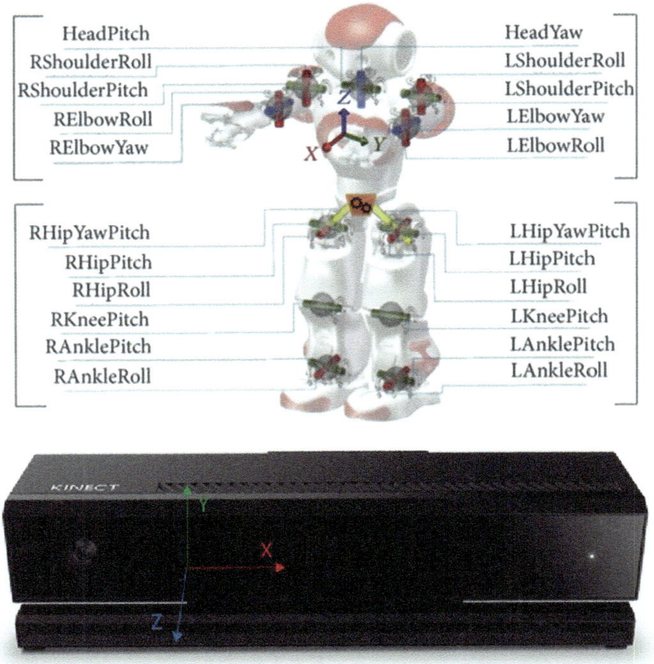

Fig. 2. NAO reference frame and Kinect reference frame

$$R_T = \begin{bmatrix} 0 & -1 & 0 \\ -1 & 0 & 0 \\ 0 & 0 & 1 \end{bmatrix} \tag{1}$$

Then, vectors are created between the 3D joint positions which constitute the limbs that can be seen in Fig. 3. The angles are calculated using Eq. 2 and the vectors represented in Table 1, where **y** and **z** represent the unit vectors of the y and z axes.

$$angle = \arccos\left(\frac{\mathbf{a}\cdot\mathbf{b}}{\|\mathbf{a}\|\,\|\mathbf{b}\|}\right) \quad (2)$$

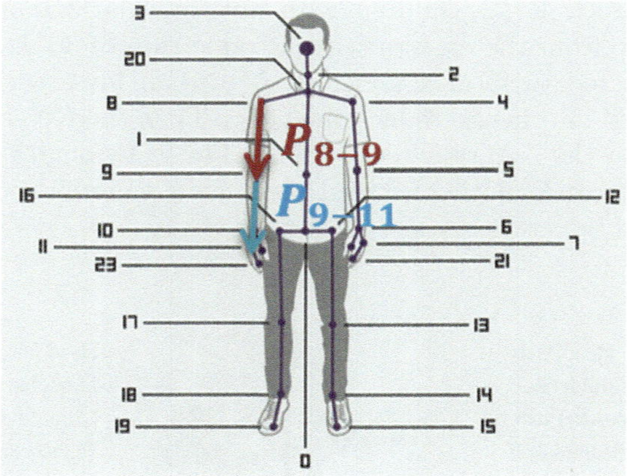

Fig. 3. 3D joints calculated by the Kinect strategy with the vectors representing the right arm ($\mathbf{P_{8-9}}$) and forearm ($\mathbf{P_{9-11}}$).

Table 1. Shoulder and Elbow angles calculated from kinect results.

Angles	a	b
Right shoulder roll	$\mathbf{P_{8-9}}$	$-\mathbf{y}$
Right shoulder pitch	$\mathbf{P_{8-9}}$	$-\mathbf{z}$
Right elbow roll	$\mathbf{P_{8-9}}$	$\mathbf{P_{9-11}}$
Left shoulder roll	$\mathbf{P_{4-5}}$	$\mathbf{y}$
Left shoulder pitch	$\mathbf{P_{4-5}}$	$-\mathbf{z}$
Left elbow roll	$\mathbf{P_{4-5}}$	$\mathbf{P_{5-7}}$

This Kinect-based system was compared with a standard measurement system of 5 inertial measurements units. The complete description of this comparison is presented on the conference paper that was written from this thesis [13]. Basically, the units were put in the chest and in the middle of both arms and forearms of a subject. Five simultaneous acquisitions of three different movements were recorded. The root mean square error between the IMU angles and the Kinect angles was calculated.

The same was done to evaluate the angles of the NAO movements. In this case the sensors were put on the arms, forearms and chest of the robot. The angles given to NAO were calculated through the movement of a subject recorded by Kinect. Then, the true angles done by the robot were measured by the 5 inertial measurement units. The root mean square error was used to compare these angles with the motion angles given to NAO.

2.2 Protocols

As this mirroring system was designed to improve the interaction between the child and the therapist, Kinect should distinguish each subject. Actually, Kinect is able to track both and the identification of each person is based on the position of the left hand. The person with the left hand more at the left of Kinect is considered the therapist. After this identification two protocols can be used: one in which the robot is the master and another where the therapist is the master. On the first protocol, NAO presents a movement. Then the therapist starts doing it. At the same time NAO is mirroring him/her. After, it is the turn of the child of executing the movement and being mirrored by NAO. If the movement is performed correctly, NAO gives a vocal feedback and its LEDs turn out to green. If the child or the therapist do not start the movement the LEDs on NAO eyes are red and it gives a negative feedback. The protocol block diagram is represented on Fig. 4a.

This protocol was adapted to be more interactive to the child, through the collaboration with IRCCS Fondazione Don Carlo Gnocchi. Actually, the movements presented by NAO are part of a story about NAO, whose goal is to train specific gestures. The story is set on the theme of the space, being NAO someone from another planet. In this way the specific gestures are to wave, move out clouds from the sky (reaching and back with two arms), take out stars (reaching and back with one arm) and finally point to the NAO planet. The feedbacks are given when a target position is achieved. This target position is related with the several parts of the movements, and it is constituted by specific target angles. The target angles were decided through the acquisition of the several movements on a group of 28 people. They were optimized in order to all the people have a positive feedback, after the correct execution of the movement.

For the second protocol, in which the therapist is the master, he/she and the child teach NAO the sports from Earth. The chosen sports were basket, bowling, swimming, tennis and skiing. In this case, first the therapist shows the movement during a certain interval, then NAO replicates it, and finally it is the turn of the child. At this moment, NAO also mirrors the child (Fig. 4b). No feedback is given by NAO because the therapist will do it.

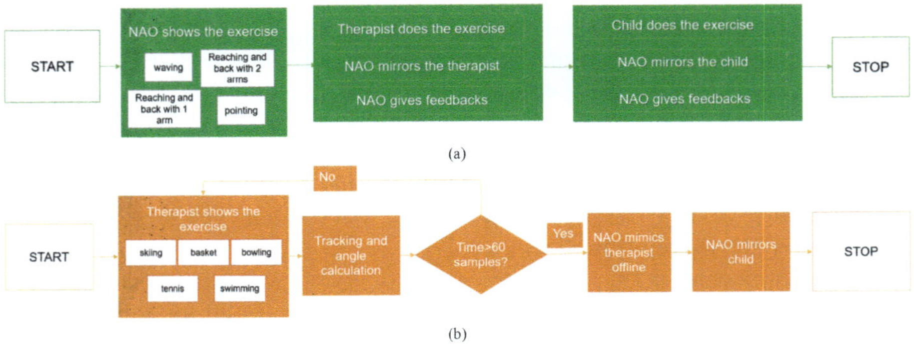

Fig. 4. Implemented protocols - Robot Master (a) and Therapist Master (b)

These protocols, in their clinic/game version were tested in two autistic children and in one healthy child. Some parameters were extracted to allow the future evaluation of the children evolution along the sessions. For the robot master protocol, it was evaluated the starting time of the movement of the children ($\mathbf{t_s}$), the error of the peak angle of the movement in relation to the target position ($\mathbf{\Delta E}$), the mean velocity on the peak angle ($\mathbf{\Delta v}$) and the maximum velocity ($\mathbf{v_i}$) and respective arrival time ($\mathbf{t_i}$). For the second protocol, the chosen parameters were calculated in relation to the whole signal since the gold standard movement is known, the movement of the therapist. In this way, the latency in relation to the beginning of the movement, the root mean square error (RMSE) between the signal of the child and the signal of the therapist and the duration of the movement were calculated. On both protocols, in order to quantify the characteristic random movement of the ASD children, it was counted the number of samples in which the change in the joints angles was larger than 0.1 rad. This parameter was called random samples.

3 Results

From Table 2, it is possible to observe that the root mean square error between the signals obtained with a Kinect and the gold standard given by the IMU sensor is generally low, as well as its variability, for both child and adult acquisitions. This means that the acquisitions done with Kinect are sufficiently similar to the ones measured with IMU. Moreover, Kinect is less invasive, it is easier to assemble the whole system and the cost is smaller in relation to the IMU system. It also allows the tracking of two people, which is impossible with the IMU, since it would require more sensors. In the same table, it is also possible to notice that the mean of the error of the adult is lower than the error of child, which can be justified by a different position of the Kinect camera, or a better performance of the person who was representing the therapist, than the child. Focusing on the angles calculated by Kinect and the angles done by the robot (Fig. 5), a difference lower than 25° can be observed. This means that the movement of the robot is sufficiently similar to the person it is mimicking, even considering that the robot has different range of motions, degrees of freedom and safety controls with respect to the human limbs. Therefore the Kinect system is a good substitute of the IMUs in a mirroring system.

The clinical acquisitions done on two boys with ASD had different outcomes. One of them had a negative reaction after the complexity of the movements requested increased and did not want to continue the session. The other boy, in the beginning was not comfortable with the presence of the robot but after he completely engaged on the therapy. While in the beginning the therapist was helping him to do the exercises, in the end he was teaching the therapist how to do correctly the movements.

The movements of this last child were compared with the movements of a healthy one to understand which parameters could be useful for the evaluation of the performance of the ASD child on future sessions. An example of the

Table 2. RMSE between Inertial Measurement Unit (IMU) and Kinect-based control signals for the adult and the child. Values are reported as mean ± SD (Standard Deviation).

Movements	RMSE (rad)	
	Adult	Child
Lateral abduction/Adduction shoulder	0.23 ± 0.12	0.41 ± 0.19
Abduction/Adduction shoulder	0.35 ± 0.07	0.41 ± 0.04
Flexion extension elbow	0.31 ± 0.10	0.30 ± 0.19

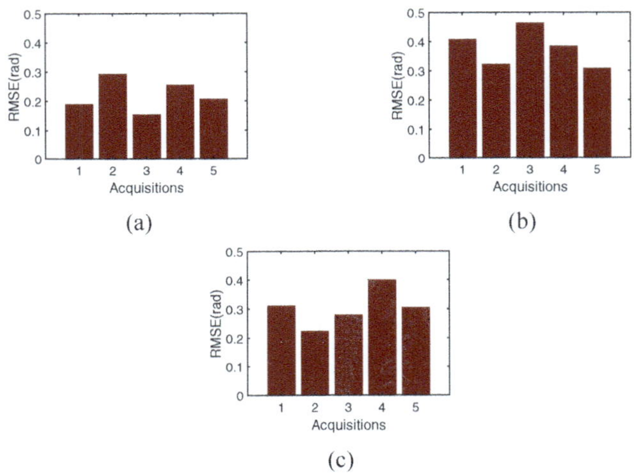

Fig. 5. Root mean square error for the 5 acquisitions done for (a) horizontal abduction and adduction, (b) abduction and adduction of the arms and (c) flexion and extension of the forearms, comparing the angles done by the robot and the angles calculated by the Kinect

Robot wave movement is shown on Fig. 6 and all parameters calculated for the several movements for the ASD child (ASD) and for the healthy child (H) are presented in Table 3. In the figure, the black straight line marks the beginning of the movement. The dashed lines represent the target angles (TA) for the internal (dark green) and external (light green) rotation and the dark green and light green stars the peak angles achieved for the internal and external rotation. From the figure, it is possible to observe the brisk movements done by the ASD child. As a consequence and also for the other movements his velocity parameters are much higher than the parameters of the healthy child. However, this reflects in a higher absolute error. In this example, it is shown for the external rotation of the arm, where one can notice the large distance between the light green star and the light green dashed line. It is also clear that the starting time is longer for the ASD child in comparison to the healthy child, reflecting his longer reaction period. The number of random samples also differs between the healthy and the ASD child being higher for the latter.

Table 3. Parameters calculated for all the movements done in the robot master protocol.

	$\mathbf{t_s}$(s)		$\mathbf{\Delta E}$(rad)		$\mathbf{\Delta v}$(rad/s)		$\mathbf{v_i}$ (rad/s)		$\mathbf{t_i}$(s)		**Random**	
	ASD	H	ASD	H	ASD	H	ASD	H	ASD	H	ASD	H
Wave	13.9	8.9	1.765	0.770	−0.197	−0.039	86	49	19.3	16.3	30	17
Clouds	9.6	8.9	2.467	0.388	−1.468	−0.133	31	6	12.1	28.0	320	49
Stars	17.4	1.1	0.391	0.271	−1.709	−1.314	13	7	26.5	13.7	14	4
Point	6.8	4.2	0.047	0.145	−3.405	−3.181	10	7	6.8	11.6	162	61

Similar conclusions could be taken from the analysis of the parameters of the therapist master protocol (Table 4). The latency tends to be smaller on the healthy child than on the ASD child, showing a larger reaction time for the latter. Observing Fig. 7, an example of the basket movement, in which both the therapist shoulder angle movement and the child movement are presented, it is clear that the shape of the signal for the ASD child differs more from the therapist signal than for the healthy child. Therefore, the root mean square error is larger in comparison with the healthy child, which happens for all the movements tested. In this case, this is due a faster execution of the movement by the ASD child. However, the duration is one of the parameters that varies between ASD child and healthy child for the several protocols. The most differentiating parameter is the number of random samples, which is always larger on the ASD child in comparison to the healthy child.

Table 4. Parameters calculated for all the movements done in the therapist master protocol.

	Latency (s)		Duration (s)		RMSE		Random	
	ASD	H	ASD	H	ASD	H	ASD	H
Basket	1.70	2.09	1.48	2.02	0.93	0.29	91	24
Swimming	1.81	0.33	0.51	2.38	0.51	0.42	101	74
Bowling	2.99	1.91	4.62	0.94	0.92	0.87	120	46
Skiing	0.90	0.40	4.33	1.26	0.99	0.40	195	45
Tennis	8.84	0.07	0.43	1.48	0.17	0.23	144	86

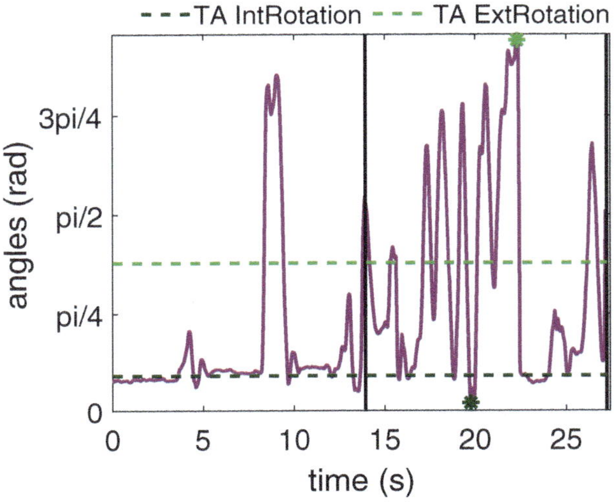

(a)

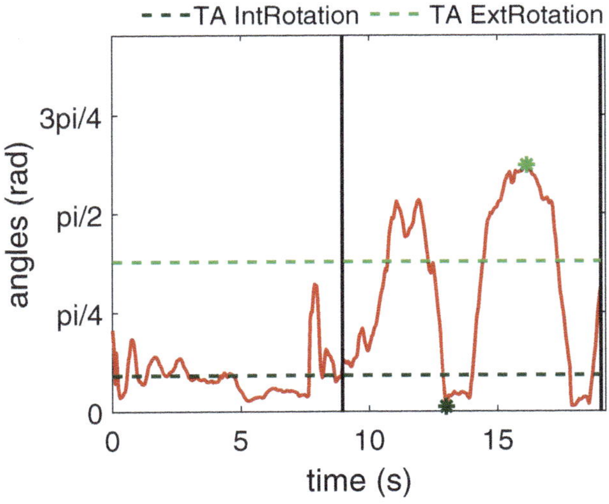

(b)

Fig. 6. Acquisitions of the robot wave movement for the ASD (a) and Healthy (b) children.

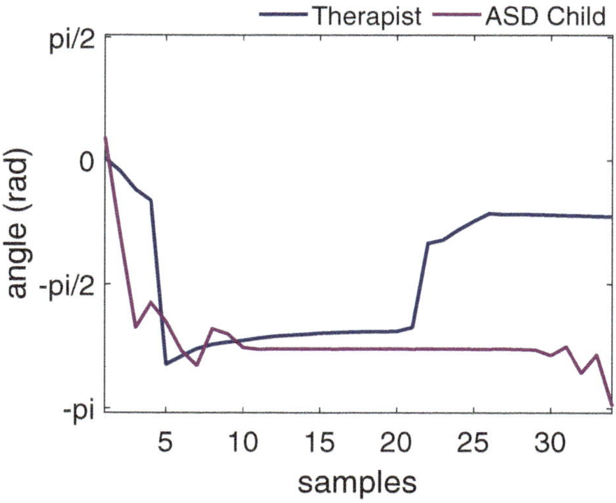

(a)

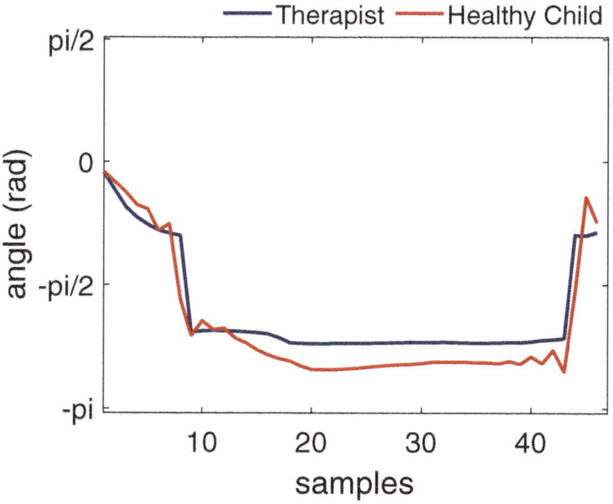

(b)

Fig. 7. Shoulder pitch angle of the basket movement done by the child and therapist where on (a) is the graph of the ASD child and on (b) the one of the healthy child.

4 Discussion and Conclusions

ASD therapies have been mostly focused on social development of these children, and only recently have included the training of motor abilities. The use of the NAO robot is a promising additional resource to stimulate motivation and participation of ASD children in the therapy. The proposed protocol promotes a global intervention to these children: (i) motor training through the execution of motor tasks in a controlled way, (ii) cognitive stimulation, induced through the different modes of mimicking games, and (iii) a social dimension, by proposing a triadic interaction between the child, the therapist and the robot. This three-fold interaction was particularly innovative in relation to existing literature, where dyadic interaction (child, robot) is the most common form of interaction.

The first part of the results has shown that the Kinect is a good substitute to the more precise IMU sensors. However the thorough validation of the Kinect still requires a larger set of movements and subjects, especially children. It has allowed to the successful implementation of an interactive mirroring system with the NAO humanoid robot, both online (Robot Master Protocol) and offline (Therapist Master Protocol). Not all movements worked equally well, namely the Clouds movement on the first protocol and the Skiing and Bowling movements on the second protocol. The main problem on the clouds movement was the inter-subject variability of the movement. The definition of the target angles for this movement was done through the use of a population of 28 adults. Probably, a similar test should be done with children, for defining target angles adapted to children. On the Skiing and Bowling movements, occlusions of the subject arms with the torso created difficulties to the computation of the subject motion angles, which occasionally blocked the robot arms behind its torso. The problem could be avoided by limiting/filtering the reference joint angles to control the NAO robot.

The whole system was tested with two ASD children, who reacted in different ways. One of them positively interacted with the robot, in a very successful and encouraging test. The other child also reacted well in the beginning, and showed interest and empathy towards the robot, but became scared by the difficulty of the movements. These different reactions are related to the different level of autism of the two children, higher in the child with less positive reactions and engagement. This case demonstrated the need of improving the system for different degrees of autism. A possibility could be to adapt the system to mirror just one person, so that the therapist could help the child to execute the movements. Alternatively, an initial step may include mirroring the child only, focusing on the imitation skills of the child. At a later stage, also the sessions with these children should continue in order to evaluate the relevance of the kinematic parameters calculated. We are currently initiating tests of the system with more children, to compare the results to the ones of the standard therapy. This will allow the complete evaluation of the efficacy and effectiveness of this type of intervention. Overall, this work shows also the importance of the clinical inputs to drive the system design (a camera and a robot) towards an attractive and impact-full therapy to autistic children.

Acknowledgment. This work was supported by Portuguese Foundation for Science and Technology (proj. UID/EEA/50009/2019) and CerebNEST, Partnering Project of the Human Brain Project.

This master thesis was awarded the top grade under a double diploma agreement between the Politecnico di Milano (Italy) and the Instituto Superior Técnico (Lisbon, Portugal), under the supervision of Professor Alessandra Pedrocchi, Eng. Alice Geminiani and Professor José Santos-Victor, to whom the author is profoundly grateful. The support of Eng. Paolo Borzatta (European House-Ambrosetti) for providing access to a NAO robot, and Doctor Ivana Olivieri (IRCCS Fondazione Don Carlo Gnocchi) for guidance in the system design towards clinical use, is highly appreciated.

A part of the work was selected to showcased in the exhibition 'Brain, wider than the sky' promoted by the Fundação Calouste Gulbenkian, and the technical work is described in a paper accepted for presentation at the International Conference of Engineering in Medicine and Biology [13].

References

1. Anzulewicz, A., Sobota, K., Delafield-Butt, J.T.: Toward the autism motor signature: gesture patterns during smart tablet gameplay identify children with autism. Sci. Rep. **6**(31107), 1–13 (2016)
2. Colebourn, J.A., Golub-Victor, A.C., Paez, A.: Developing overhand throwing skills for a child with autism with a collaborative approach in school-based therapy. Pediatr. Phys. Ther. **29**(3), 262–269 (2017)
3. Caputo, G., Ippolito, G., Mazzotta, M., Sentenza, L., Muzio, M.R., Salzano, S., Conson, M.: Effectiveness of a multisystem aquatic therapy for children with autism spectrum disorders. J. Autism Dev. Disord. **48**(6), 1945–1956 (2018)
4. Forbes, P.A.G., Pan, X., Hamilton, A.F.d.C.: Reduced mimicry to virtual reality avatars in autism spectrum disorder. J. Autism Dev. Disord. **46**(12), 3788–3797 (2016)
5. Fadiga, L., Fogassi, L., Pavesi, G., Rizzolatti, G.: Motor facilitation during action observation: a magnetic stimulation study. J. Neurophysiol. **73**(6), 2608–2611 (1995)
6. Hofree, G., Ruvolo, P., Bartlett, M.S., Winkielman, P.: Bridging the mechanical and the human mind: Spontaneous mimicry of a physically present android. PLoS ONE **9**(7), e99934 (2014)
7. Scassellati, B., Admoni, H., Matarić, M.: Robots for use in autism research. Annu. Rev. Biomed. Eng. **14**(1), 275–294 (2012)
8. Alibeigi, M., Rabiee, S., Ahmadabadi, M.N.: Inverse kinematics based human mimicking system using skeletal tracking technology. J. Intell. Robot. Syst. **85**, 27–45 (2017)
9. Spasojević, S., Ilić, T.V., Milanović, S., Potkonjak, V., Rodić, A., Santos-Victor, J.: Combined vision and wearable sensors-based system for movement analysis in rehabilitation. Methods Inf. Med. **56**, 95–111 (2017)
10. Yang, N., Duan, F., Wei, Y., Liu, C., Tan, J.T.C., Xu, B., Zhang, J.: A study of the human-robot synchronous control system based on skeletal tracking technology. In: 2013 IEEE International Conference on Robotics and Biomimetics (ROBIO), pp. 2191–2196 (2013)

11. Zannatha, J.M.I., Tamayo, A.J.M., Sánchez, A.D.G., Delgado, J.E.L., Cheu, L.E.R., Arévalo, W.A.S.: Development of a system based on 3D vision, interactive virtual environments, ergonometric signals and a humanoid for stroke rehabilitation. Comput. Methods Programs Biomed. **112**(2), 239–249 (2013)
12. Almetwally, I., Mallem, M.: Real-time tele-operation and tele-walking of humanoid robot NAO using kinect depth camera. In: 2013 10th IEEE International Conference on networking, sensing and control (ICNSC), pp. 463–466, April 2013
13. Geminiani, A., Santos, L., Casellato, C., Farabbi, A., Farella, N., Santos-Victor, J., Olivieri, I., Pedrocchi, A.: Design and validation of two embodied mirroring setups for interactive games with autistic children using the NAO humanoid robot. In: 41st IEEE Engineering in Medicine and Biology Society, Berlin, Germany (2019). Accepted with oral presentation

Neuromechanical and Environment Aware Machine Learning Tool for Human Locomotion Intent Recognition

Simão Carvalho(✉), Joana Figueiredo, and Cristina P. Santos

Center for MicroElectroMechanical, University of Minho, Braga, Portugal
simaopedrocarvalho17@gmail.com,
id6003@alunos.uminho.pt, cristina@dei.uminho.pt

Abstract. Current research suggests the emergent need to recognize and predict locomotion modes (LMs) and LM transitions to allow a natural and smooth response of lower limb active assistive devices such as prostheses and orthosis for daily life locomotion assistance. This Master dissertation proposes an automatic and user-independent recognition and prediction tool based on machine learning methods. Further, it seeks to determine the gait measures that yielded the best performance in recognizing and predicting several human daily performed LMs and respective LM transitions. The machine learning framework was established using a Gaussian support vector machine (SVM) and discriminative features estimated from three wearable sensors, namely, inertial, force and laser sensors. In addition, a neuro-biomechanical model was used to compute joint angles and muscle activations that were fused with the sensor-based features. Results showed that combining biomechanical features from the Xsens with environment-aware features from the laser sensor resulted in the best recognition and prediction of LM (MCC = 0.99 and MCC = 0.95) and LM transitions (MCC = 0.96 and MCC = 0.98). Moreover, the predicted LM transitions were determined with high prediction time since their detection happened one or more steps before the LM transition occurrence. The developed framework has potential to improve the assistance delivered by locomotion assistive devices to achieve a more natural and smooth motion assistance.

1 Introduction

Due to the increasing integration of robotics in the field of medicine over the last few years, a larger number of people with locomotion disabilities are given a chance to live a life with better quality. According to the World Health Organisation, every year 15

This work has been supported in part by the Fundação para a Ciência e Tecnologia (FCT) with the Reference Scholarship under Grant SFRH/BD/108309/2015, and part by the FEDER Funds through the Programa Operacional Regional do Norte and national funds from FCT with the project SmartOs -Controlo Inteligente de um Sistema Ortótico Ativo e Autónomo- under Grant NORTE-01-0145-FEDER-030386, and by the FEDER Funds through the COMPETE 2020—Programa Operacional Competitividade e Internacionalização (POCI)—with the Reference Project under Grant POCI-01-0145-FEDER-006941.

J. Henriques et al. (Eds.): MEDICON 2019, IFMBE Proceedings 76, pp. 2028–2042, 2020.
https://doi.org/10.1007/978-3-030-31635-8_240

million people suffer a stroke (20000 in Portugal) and 80 people in a million worldwide suffer a spinal cord injury (58 per million in Portugal), leaving people with motor disabilities [1, 2].

Advancements towards the use of assistive devices (e.g., active orthosis, exoskeletons and prosthesis) in daily life situations, as well as the ongoing integration of compliant robotic devices in neurorehabilitation generates new challenges. Firstly, the timely adaptation of these devices to diverse locomotion modes (LMs) such as, level-ground walking, climbing stairs and ramps. Secondly, to tune the assistance to the patient's locomotion intention by monitoring the human-robotic interface through wearable sensors [3, 4].

To provide efficient and suitable gait assistance and to decrease fall risk, LM recognition is essential. LM recognition is challenging for gait assistance devices due to the limited sensors and the need for a time effective response [5]. Commercially available devices such as prosthesis and orthosis can switch LMs executing manual commands such as button presses, as used in the ReWalkTM system (Argo Medical Technologies Ltd. Israel) [6] or specific motions such as used in Power KneeTM (Össur, Iceland) [6]. The user's intent motion prediction is fundamental to achieve a natural, user-independent and automatic device response during all locomotion phases. However, the existing manual solutions require the user's training before using the device and a major cognitive effort from the user during locomotion.

Previous studies have proposed promising methods to recognize and predict LMs and LM transitions, through artificial intelligence methods, using several wearable sensors. One approach proposed by Chen et al. [7] exploited signals from two-foot pressure insoles and three inertial measurement units (IMU) as biomechanical features to develop a locomotion intent prediction strategy based in a linear discriminate classifier (LDA). Six kinds of LMs were accurately recognized (99.71% $\pm$ 0.05), however no LM prediction was addressed [7]. The electromyography signal (EMG) is another widely used measure for user's intent recognition during locomotion since it represents muscle activations that precede the limb's motion. Some studies addressed the sensor fusion between EMG and mechanical signals from the user's assistive device such as moment, velocity, angle, gait phase and ground reaction force to be used in machine learning algorithms such as LDA and Support Vector Machines (SVMs) [8, 9]. These studies report an error lower than 2.3% $\pm$ 0.7% for detecting 8 different LM transition.

Despite being widely used, wearable sensors such as IMUs and EMG present some major drawback such as the presence of noise, which introduces drift errors in IMUs' measures [10–12]. Moreover, EMG-based approaches present some practical limitations, namely the EMG electrode shift due to the subject movement or sweat and require an expert and a lengthy installation [13, 14]. The EMG signal is extremely dependent on the electrode's position and orientation. Placing electrodes in the wrong locations will reduce signal strength and quality. Neuro-biomechanical models (combining biomechanics and neurophysiology) have provided accurate kinematic, dynamic estimations, and muscular activation estimations patterns using wearable sensor information such as IMUs and EMGs and resemble a promising method to overcome the mentioned issues [15, 16].

Research efforts have been made to further improve the motion intent recognition and prediction, including environment-aware strategies to infer about the interaction between the user and the world, inspired by the human ability to see the surrounding environment. In [17], an infrared laser sensor was placed in the user's waist in order to acquire terrain change information during the subject's gait. Study [17] proposed a fusion between an LM recognition module and a terrain reconstruction module, using an LDA classifier. The terrain reconstruction module provided an accuracy higher than 98% and detections 500 ms in advance. The anticipatory nature of environment-aware strategies still needs to be further explored to provide robust and time-effective LM transition prediction systems for timely controlling assistive devices.

This dissertation tackles the mentioned challenges. It proposes an automatic and user-independent recognition and prediction tool for classifying daily encountered LMs and LM transitions using machine learning methods. Furthermore, different types of wearable sensors, namely IMU and force sensing resistor sensors (biomechanical gait measure), and an infrared laser sensor (environment-aware gait measure), were explored to find an effective and reliable recognition and prediction tool. Additionally, the Xsens biomechanical model and an OpenSim modified musculoskeletal model are used to generate neuro-biomechanical features such as the lower limbs joint angles and muscle activation patterns, respectively. The tool aims at an efficient classification of the LMs commonly encountered in daily life, with variations in walking speed and terrains (i.e., flat, ascending and descending stairs, climbing up and down ramp). The tool also approaches transitions from/to those terrains using the user's self-selected lower limb to minimize the user's cognitive effort during locomotion. The validation of the proposed tool included 10 able body subjects performing different LMs and the LM transitions at different terrains, directions, and gait speeds. The obtained results may contribute to the optimization of the assistance delivered by the lower limb active assistive devices.

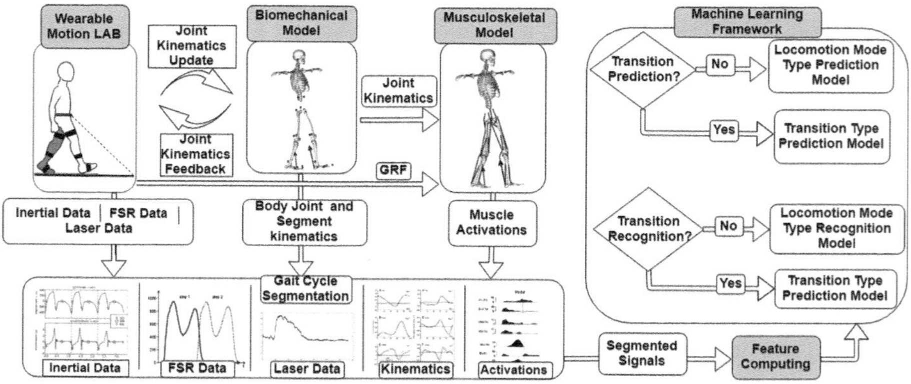

Fig. 1. Schematic overview of the LM and LM recognition and prediction framework

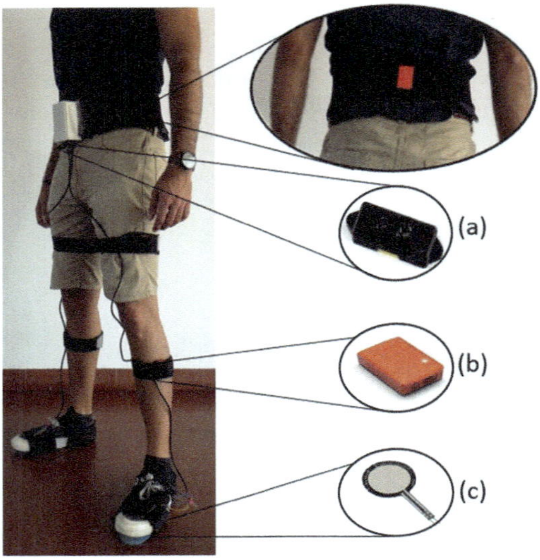

Fig. 2. Wearable Motion LAB sensor placement. (a) Infrared laser sensor. (b) Xsens IMU sensor. (c) FSR ground reaction force sensor

The proposed work approached two main research questions, as follows: (i) what are the best gait measures for LMs and LM transition classification? and (ii) Would the environment-aware measure help to increase LM and LM transition prediction performance? These questions are explored in Sect. 3.

2 Methods

The LM and LM transition recognition and prediction tool proposed in this dissertation is represented in Fig. 1. The proposed methodology is divided into five main modules, namely the wearable motion LAB (WML), biomechanical model, musculoskeletal model, feature computing and the machine learning framework.

2.1 Wearable Motion LAB

The WML includes ergonomic, stand-alone, wearable sensor systems placed on the subject's body, namely the infrared laser sensor, the inertial sensor system and the force sensing resistor (FSR) system. The first sensor refers to the infrared laser sensor (TF Mini LiDAR), placed on the user waist pointing at the ground with and angle of proximally 45° to the user's abdomen vertical plane (Fig. 1(a)). This laser sensor provides a signal that indicates the distance between the subject and the ground or possible obstacles and terrain changes. The second system is composed by 7 wireless IMU (Xsens Technologies BV, Enschede, Netherlands) placed on the outer side of both thighs and shanks, on top of the feet, and one IMU was placed on the lower back (Fig. 2(b)). The IMUs provide acceleration, angular velocity and magnetic field regarding each lower limb segment. The third system consists of 4 FSRs (406 FSR, Interlink Electronics), two per foot, one placed in the and one in the heel. The FSRs provide an electrical signal that is directly related to the pressure applied, measuring the ground reaction force (GRF). All systems were synchronized by wires and the data was collected at 100 Hz.

2.2 Biomechanical Model

The biomechanical model is accountable for the 3D joint angle estimations using IMUs' measures as input. The Xsens well-established motion tracking method, was used to guarantee accurate joint angle estimations, as the lower limb joint angles represent a significative feature of the human gait, and consequently, they are relevant for LM and LMT classification [15]. The Xsens biomechanical model is in constant communication with the wireless IMUs, by sending them the real body segment positions and simultaneously receiving new acquisitions to predict the subsequent motions.

2.3 Musculoskeletal Model

The third module of the tool depicted in Fig. 1 includes a virtual musculoskeletal model able to compute the motion forces and muscle activations involved in locomotion from the given kinematic data (lower limb spatial displacement) and the force of interaction with the external world (from the FSRs). Since during locomotion, the subject only contacts the ground, the GRF (from the FSRs) is the only considered external force applied.

The virtual model consists of a modified version of the gait10dof18musc OpenSim model, which includes the trunk, pelvis, and leg segments with 10 degrees of freedom. The musculoskeletal system incorporates 18 Hill-type muscles (9 in each leg), where 5 muscles are monoarticular (*Soleus, Tibialis Anterior, Gluteus Maximus, Vastus Intermedius, Rectus Femoris)* and 3 muscles are biarticular *(Gastrocnemius, Biceps Femoris, Lateral Hamstring, Iliopsoas).* Literature suggests that the used musculoskeletal system underlies the sufficient characteristics to impose a natural human gait, capable of tolerating small terrain disturbances, being computationally efficient [18].

The motion forces (e.g., net forces and torques) at each joint are computed through inverse dynamics. The algorithm solves the dynamic equation of motion presented in (1), where q, $\dot{q}$ and $\ddot{q}$ refer to the position, velocity, and acceleration of the generalized coordinates; M, C, G are the mass, centrifugal and Coriolis, and gravitational forces matrices, respectively; and T is the vector of unknown motion forces. This algorithm derives the minimal forces responsible for generating the desired motion.

$$M(q)\ddot{q} + C(q, \dot{q}) + G(q) = T \tag{1}$$

With the purpose of computing the muscle activations, a static optimization algorithm was applied to compute muscle activations by minimizing its sum of squared.

2.4 Feature Computing

The feature computing module converts the input sensor and neuro-biomechanical data, to an output feature vector containing meaningful information segmented per gait cycle. The time window's size is adaptive according to the gait speed. The period of each time-window corresponds to the duration of each gait cycle. The gait cycle segmentation was determined by automatically detect gait events using an algorithm based on the angular velocity of each foot [19].

The developed framework performs both LM and LM transition recognition and prediction. For recognition purposes, gait cycles were segmented between consecutive heel-strike events while for prediction purposes, gait cycles were segmented between consecutive toe-off events.

For neuro-biomechanical data, several features were computed such as the mean, standard deviation, range, maximum, minimum and the first and last value. Regarding the laser sensor data, some additional features were computed with the purpose of taking maximum advantage of the signal characteristics, namely, the signal slope, slope module, previous slope, and previous slope module. Table 1 lists the computed features per gait measure.

Table 1. Feature computing representation considering sensor and neuro-biomechanical data

<table>
<tr><th>Sensor/Tool</th><th>Measure</th><th>Body zone</th><th>Features</th></tr>
<tr><td>IMU</td><td>Angular Velocity</td><td>Back, right/left Thigh, right/left Shank and right left foot</td><td rowspan="4">Mean, Standard Deviation, Range of Motion, Maximum, Minimum and the First and Last Value</td></tr>
<tr><td>Xsens Biomechanical Model</td><td>Joint Angles</td><td>Right/Left Hip, Right/Left Knee, Right/Left Ankle</td></tr>
<tr><td>Opensim Musculoskeletal Model</td><td>Muscle Activations</td><td>Soleus, Tibialis Ant., Gluteus Max., Vastus Int., Rectus Fem, Gastrocnemius, Biceps Fem., Lateral Ham., Iliopsoas</td></tr>
<tr><td>FSRs</td><td>Ground Reaction Force</td><td>Right/left Heel and Toe</td></tr>
<tr><td>Laser</td><td>Distance between subject and ground</td><td>Waist</td><td>Mean, Standard Deviation, Range of Motion, Maximum, Minimum and the First and Last Value Slope, Slope Module, Previous Slope and Previous Slope Module</td></tr>
</table>

2.5 Machine Learning Framework

The machine learning framework aims at recognizing and predicting LM and LM transitions to infer about the subject's locomotion intentions in diverse terrain types encountered in daily life. This machine learning framework unfolds in four subsections, namely, Data Pre-processing, Data Labelling, Model Building, and Classification Evaluation.

(1) Data Pre-processing

Pre-processing techniques are applied to the unprocessed feature vector in order to maximize model performance and reduce its training time. The applied pre-processing techniques include data normalization and feature selection. During data normalization, features are processed using the min-max scaling method. This process aims to convert all features to a common range such that features with larger value range do not reduce the influence of features with smaller ranges. The min-max method provides zero-mean and unit-variance characteristics to each feature and limits range between –1 and 1.

Additionally, a feature selection method was used to select the minimum number of features. For this purpose, the mRMR plus forward selection method was used. This method selects the most relevant features from those that have minimally redundancy, i.e. selects features that are maximally dissimilar to each other. This method creates a ranking immune to redundancy, in which a feature is kept if it increases the tool's performance, otherwise it is removed [20].

(2) Data Labelling

For recognition and prediction purposes, labelling was done on features computed considering time-windows segmented between heel strike events and toe off events, respectively. Moreover, prediction was achieved by classifying events one gait cycle before it's occurrence (i.e., on the previous time window as demonstrates by blue time-zones in Fig. 3). An example of the time displacement of recognition (red) and prediction time-windows (blue) used for labelling and feature computing, are represented in Fig. 3.

(3) Model Building

The followed approach assumes that the decision making is empowered by four classification models for recognition and another four for prediction running simultaneously. Figure 1 depicts a simplified scheme of this classification workflow. The classification procedure was constructed as a chain sequence of classification models as demonstrated in Fig. 4.

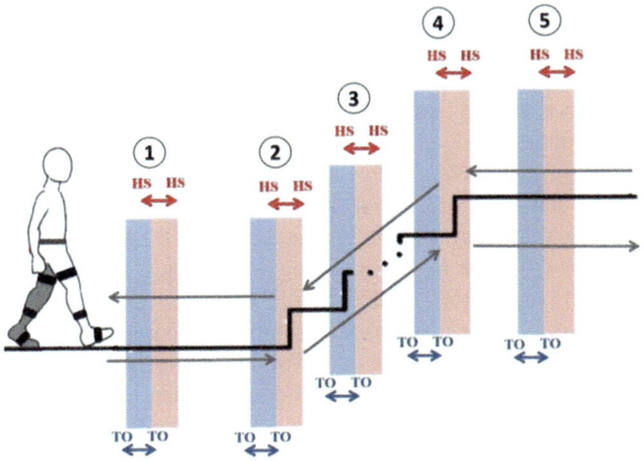

Fig. 3. Considered time-windows for LM (1, 3 and 5) and LM transition (2 and 4) recognition (red) and prediction (blue). Prediction time-windows are limited by toe-off events (TO) and recognition time windows are limited by heel strike events

The workflow starts with the walking direction classification using the Direction Classification Model. This classification model aims at classifying forward, anticlockwise and clockwise directions. The remaining classification sequence only proceeds if forward direction is detected. The following classification model is the binary Steady-State/Transition model and it aims at distinguishing LM steady-states (i.e. during a certain LM) from LM transition states (during an LM change), regardless of what LM or LM transition type it represents. In case this classification model detects a steady-state, it proceeds to the Steady-State Type model to determine its type. The Steady-State Type model classifies five LM, namely, level-walking, stair ascend/descend, and ramp ascends/descend. On the other hand, if the Steady-State Transition detects an LM transition, it proceeds to the Transition Type model to determine the transition type. The Transition Type model classifies eight transitions from level ground to stair/ramp ascend/descend and from to stair/ramp ascend/descend to level ground. To note that this chain sequence of classification is used for both recognition and prediction purposes.

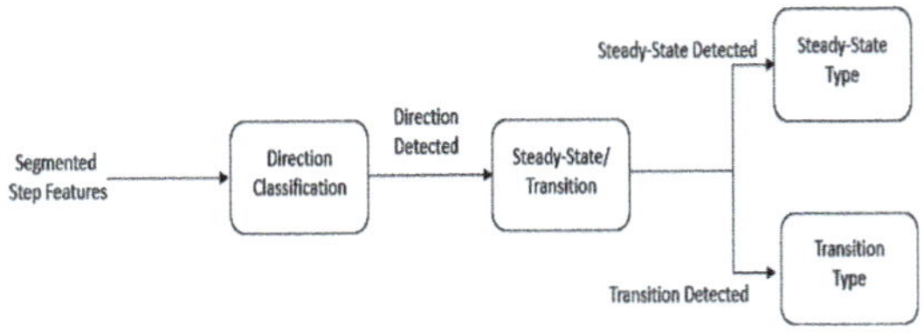

Fig. 4. Classification chain sequence

This scheme allows the easy incorporation of additional LM and/or LM transition in the classification, adding versatility to the proposed framework to act as a benchmark tool. Additionally, it seems to be advantageous regarding the classification strategy proposed by Young *et al.* [21] since it demands the implementation of fewer models, decreasing the computational load and the susceptibility to sequential errors, and allows easier integration of further LMs and/or LM transition.

The SVM classifier with a Gaussian kernel was used for all classification models. A hyperparameter optimization was performed to tune the kernel box constraint (C) and scale parameter (σ). The hyperparameter optimization involved a grid search to determine the combination the yielded the highest performance. This process was conducted by building a model for each combination, performing its evaluation iteratively. The values for each hyperparameter varied according to a range that goes from 2^{-10} to 2^{10}. To note that all classification models are built considering features retrieved from all users which conveys a user-independent characteristic.

(4) Classification Evaluation

The classification evaluation was conducted using cross-validation. The evaluation procedure is used for comparing models using different input parameters such as varying hyperparameters or feature combinations. The evaluation is particularly important to infer the classification performance of unseen data. Each model's performance is evaluated using repeated 10-fold cross-validation, repeated 10 times. To evaluate the classification results, the Matthew's correlation coefficient (MCC) was used due to its good representative properties of unbalanced classes [22], as for the case of this work. The accuracy (ACC) was also computed for comparing the results with the literature's findings [23].

2.6 Experimental Conditions

The experimental trials were performed with 10 healthy subjects (7 males and 3 females with a mean age of 23.9 ± 1.64 years, mean height of 174.9 ± 7.07 cm and mean weight of 69.7 ± 7.14 kg). The participants were asked to ascend and descend a ramp and ascend and descend a staircase 5 times per each terrain elevation (total of 20 terrain elevations per subject). In these trials, subjects started on level-terrain then transited to the ramp or staircase, walk the full ramp/staircase length and back to level-terrain walking. When in level-terrain, the users walked 3–4 m. The staircase had 8 steps with 17 cm of height, 31 cm of depth and 110 cm width. The ramp was 10.3 m with a 10° inclination. Figure 5 depicts the terrain's conditions for level, ramp and stair walking. The subjects could freely perform the LM transitions with any leading leg to enable transition seamlessly and intuitively between LMs. To note that the participants were always instrumented with all sensors that constitute the wearable motion LAB.

As ground truth, an assessor marked the LM transition instants during the experimental trials using a button synchronized with the system.

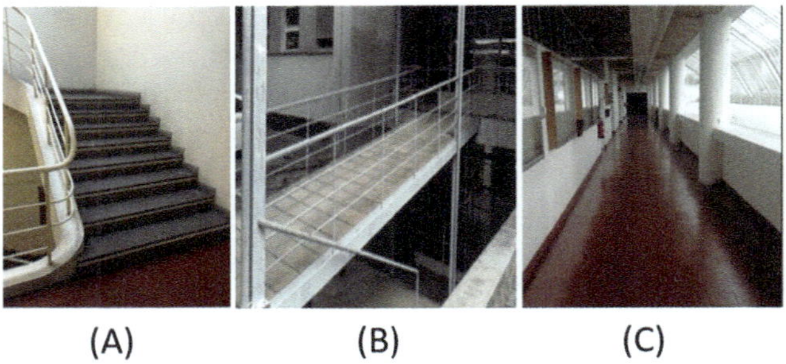

Fig. 5. Pictures of the terrain conditions used in the trials (a) staircase, (b) slope terrain, (c) level ground terrain

3 Results and Discussion

In order to tackle these dissertation goals, three studies were performed combining a specific group of gait measures, namely, (i) biomechanical measures, (ii) neuro-biomechanical measures, and (iii) an environment-aware/biomechanical based measures. All the mentioned studies approached the LM and LM transition recognition and prediction.

The first study aimed to determine if features computed from gait measures (joint angles and angular velocity at sagittal plane) of the Xsens system alone provide good performance. The results obtained from this study are demonstrated in Table 2 for recognition and prediction. For this study, a total of 66 features were computed and subsequently pre-processed.

Table 2. Biomechanical based feature recognition and prediction

	Classification model	Hyperparameters	N° selected features	MCC	ACC
Recognition	Steady-State/Transition	C = 64 σ = 4	34	0.873 ± 0.008	0.957 ± 0.003
	Transition Type		23	0.922 ± 0.041	0.982 ± 0.01
	Steady-State Type		24	0.989 ± 0.004	0.997 ± 0.001
Prediction	Steady-State/Transition		27	0.84 ± 0.006	0.948 ± 0.002
	Transition Type		22	0.83 ± 0.04	0.958 ± 0.01
	Steady-State Type		23	0.948 ± 0.05	0.984 ± 0.02

From the observation of Table 2, it is possible to acknowledge that all classification models show high recognition and prediction performance (ACC > 0.948). However, the MCC values for recognition and prediction of the steady-state/transition classification model are a little bellow when comparing with the remaining (MCC = 0.8729 and MCC = 0.8482 respectively). Furthermore, the recognition performance was higher than the prediction performance for the considered classification models. These results point out that the use of biomechanical features may be insufficient to recognize and predict LM transition and are most suited for recognition purposes. The findings are similar with those reported in the literature, that only used IMU-based features, as seen in the work proposed by Jang *et al.* [5], that reported an ACC of 0.98, 0.95 and 0.99 for SA, SD, and LG, respectively. Nonetheless, the proposed work addressed prediction problematic, in opposition to [5]

The second study aimed to determine the influence of adding neuromuscular features (computed from muscle activation patterns of 18 muscles of Opensim model) to the biomechanical features used in the first study. The results obtained from this study are demonstrated in Table 3 for recognition and prediction. This study used a total of 156 features combining the 66 biomechanical features to the 90 neuromuscular features.

Table 3. Neuro-Biomechanical (neuromuscular plus biomechanical) feature recognition and prediction

	Classification model	Hyperparameters	Nº selected features	MCC	ACC
Recognition	Steady-State/Transition	$C = 64$ $\sigma = 4$	29	0.867 ± 0.01	0.955 ± 0.005
	Transition Type		21	0.961 ± 0.02	0.996 ± 0.003
	Steady-State Type		22	0.949 ± 0.02	0.985 ± 0.006
Prediction	Steady-State/Transition		16	0.847 ± 0.01	0.947 ± 0.004
	Transition Type		24	0.871 ± 0.03	0.985 ± 0.005
	Steady-State Type		24	0.921 ± 0.01	0.975 ± 0.004

For this study, it was hypothesized that the muscle activation patterns would provide a significant contribution for predicting LM transitions since these signals have the purpose of controlling muscle activation that generates the movement of the subject [24]. The general appreciation of the obtained results is that the additional neuromuscular features provided a similar performance when compared to the only use of biomechanical features (first study). Although the prediction performance for the Steady-State/Transition model stayed in line with the first study performance, it is noticeable that in this second study, 11 fewer features were used. This finding indicates that it needed much less information to achieve a similar result. Therefore, despite not improving performance considerably, neuromuscular features present meaningful information regarding the LM and LM transition classification.

The presented results stand in line with the literature. Huang et al. [8] fused biomechanical sensors with EMG features and reported an ACC value of proximally 0.978 in LM recognition whereas the herein developed tool rounded an ACC value of 0.995. Moreover, Huang's work predicted LM transitions with an ACC of 1 for similar terrains, which is higher than the 0.947 ACC value obtained. However, is important to mention that in Huang's work the transition steps were not performed with the leading leg. Furthermore, the work developed by Tkach et al. [9] reported an higher ACC (0.977) relatively to the proposed work (0.947) when using EMG data fused with biomechanical data for transition prediction. Nonetheless, both works used biomechanical prosthesis signals which are characterized for being more "robotic" when comparing to walking patterns of different subjects.

The third study tested the LM and LM transition recognition and prediction performance when using biomechanical (GRF data, IMU sagittal angular velocity, lower limb sagittal joint angles) and environment-aware (infrared laser) features. The results obtained from this study are presented in Table 4 for recognition and prediction. This study used a total of 100 features.

Table 4. Environment-aware/Biomechanical based feature recognition and prediction

	Classification model	Hyperparameters	N° selected features	MCC	ACC
Recognition	Direction	C = 64	33	0.995 ± 0.01	0.998 ± 0.002
	Steady-State/Transition	$\sigma = 4$	57	0.887 ± 0.02	0.947 ± 0.01
	Transition Type		19	0.959 ± 0.005	0.990 ± 0.004
	Steady-State Type		20	0.989 ± 0.01	0.983 ± 0.002
Prediction	Direction		52	0.997 ± 0.02	0.999 ± 0.001
	Steady-State/Transition		51	0.865 ± 0.03	0.945 ± 0.004
	Transition Type		36	0.978 ± 0.02	0.995 ± 0.005
	Steady-State Type		37	0.949 ± 0.005	0.983 ± 0.002

Regarding third study, a direction classification model was included, and it demonstrated excellent performance both for recognition and prediction purposes (ACC = 0.998 and ACC = 0.999, respectively). This finding indicates that the information given to the model was adequate to characterize forward, anticlockwise, and clockwise direction of a walking subject. Novak *et al.* [25] proposed a turn detection system using IMUs and reported results similar to the ones achieved in this work for direction prediction (ACC > 0.97 vs ACC = 0.95, respectively).

Furthermore, the combination of GRF and laser-based features to the biomechanical features increased the performance for all classification models in recognition and prediction. Despite performance improvement using the mentioned additional features, the GRF number of features selected is very low. In this sense, the added laser data features represent higher importance in classification as they were the major reason for the classification performance improvement regarding recognition and prediction. The highest increase was related to the recognition and prediction of the type of transition (MCC = 0.959 and MCC = 0.978, respectively). These results might be related to the abrupt changes in the laser signals characteristics when a terrain change occurs. Features computed considering these laser characteristics may provide meaningful information to classify transition types, as observed in a team study [26]. The fusion between biomechanical sensors with laser data was already explored in literature by Liu *et al.* [17], reported an ACC of 99% in LM recognition, equal to the achieved 99% in ACC. Nonetheless, LM transitions were not considered in [17], as innovatively explored in this work.

In summary, the Direction classification model presented in Environment-Aware/Biomechanical based feature study revealed excellent results (ACC = 0.998 and ACC = 0.999) regarding recognition and prediction respectively. Similar finding can be found in [25] with a success rate of 0.97 to 0.99 in ACC values. The Steady-State/Transition classification models revealed the lowest general performance which might be related to the characteristic of having to distinguish between only two classes (binary classification), and because it was trained in an unbalanced way (fewer LM transition features). The Transition Type and Steady-State Type Classification models have shown good classification performance in a general manner.

4 Conclusion

This dissertation proposes an automatic and user-independent machine learning-based tool to recognize and predict daily performed LMs and LM transitions. The work aimed to determine the combination of gait measures collected by wearable sensors that yielded the best classification performance.

The findings showed that the automatic recognition and prediction tool demonstrated the highest performance for environment-aware/biomechanical based feature combination. Moreover, since GRF-based features provided poor significance, it is possible to conclude that the best feature combination for LM and LM transition classification includes biomechanical features from Xsens together with environment-aware based features from the laser sensor.

Furthermore, the environment-aware features demonstrated to be relevant to classify LM transitions, namely they increased the LM transition prediction performance.

These findings contribute to identifying the meaningful gait measures for building an effective user's motion intent recognition. This will allow a natural and smooth response of lower limb active assistive devices according to the user's intentions and needs.

References

1. Spinal cord injury (2013). https://www.who.int/news-room/fact-sheets/detail/spinal-cord-injury. Accessed 07 Mar 2019
2. World Healthy Organization (WHO): "Global burden of stroke," atlas Hear. Dis. stroke, vol. 15, pp. 50–51 (2004)
3. Tucker, M.R., et al.: Control strategies for active lower extremity prosthetics and orthotics: a review. J. Neuroeng. Rehabil. **12**(1), 1 (2015)
4. Li, Y.D., Hsiao-Wecksler, E.T.: Gait Mode Recognition and Control for a Portable - Powered Ankle - Foot Orthosis (2013)
5. Jang, J., Kim, K., Lee, J., Lim, B., Shim, Y.: Online gait task recognition algorithm for hip exoskeleton. In: International Conference on Intelligent Robots and Systems, 2015 December, pp. 5327–5332 (2015)
6. He, Y., Eguren, D., Luu, T.P., Contreras-Vidal, J.L.: Risk management and regulations for lower limb medical exoskeletons: a review. Med. Devices Evid. Res. **10**, 89–107 (2017)
7. Chen, B., Zheng, E., Wang, Q.: A locomotion intent prediction system based on multi-sensor fusion. Sensors (Switzerland) **14**(7), 12349–12369 (2014)
8. Huang, H., Zhang, F., Hargrove, L.J., Dou, Z., Rogers, D.R., Englehart, K.B.: Continuous locomotion-mode identification for prosthetic legs based on neuromuscular - mechanical fusion. IEEE Trans. Biomed. Eng. **58**(10), 2867–2875 (2011)
9. Tkach, D.C., Hargrove, L.J.: Neuromechanical sensor fusion yields highest accuracies in predicting ambulation mode transitions for trans-tibial amputees. In: Proceedings of Annual International Conference of the IEEE Engineering in Medicine and Biology Society, EMBS, pp. 3074–3077 (2013)
10. Diaz, E.M., De Ponte Muller, F., Jimenez, A.R., Zampella, F.: Evaluation of AHRS algorithms for inertial personal localization in industrial environments. In: Proceedings of the IEEE International Conference on Industrial Technology, June 2015, pp. 3412–3417 (2015)
11. Djurić-Jovičić, M.D., Jovičić, N.S., Popović, D.B.: Kinematics of gait: new method for angle estimation based on accelerometers. Sensors **11**(11), 10571–10585 (2011)
12. Lebel, K., Boissy, P., Hamel, M., Duval, C.: Inertial measures of motion for clinical biomechanics: comparative assessment of accuracy under controlled conditions - Changes in accuracy over time. PLoS ONE **10**(3), 1–12 (2015)
13. David Li, Y., Hsiao-Wecksler, E.T.: Gait mode recognition and control for a portable-powered ankle-foot orthosis. In: IEEE International Conference on Rehabilitation Robotics (2013)
14. Young, A.J., Kuiken, T.A., Hargrove, L.J.: Analysis of using EMG and mechanical sensors to enhance intent recognition in powered lower limb prostheses. J. Neural Eng. **11**, 1–12 (2014)
15. Roetenberg, D., Luinge, H., Slycke, P.: Xsens MVN: full 6DOF human motion tracking using miniature inertial sensors. Hand, January 2009, pp. 1–7 (2009)
16. Pizzolato, C., et al.: CEINMS: a toolbox to investigate the influence of different neural control solutions on the prediction of muscle excitation and joint moments during dynamic motor tasks. J. Biomech. **48**(14), 3929–3936 (2015)
17. Liu, M., Wang, D., Helen Huang, H.: Development of an environment-aware locomotion mode recognition system for powered lower limb prostheses. IEEE Trans. Neural Syst. Rehabil. Eng. **24**(4), 434–443 (2016)
18. Geyer, H., Herr, H.: A Muscle-Reflex Model that Encodes Principles of Legged Mechanics Produces Human Walking Dynamics and Muscle Activities (2010)

19. Félix, P., Figueiredo, J., Santos, C.P., Moreno, J.C.: Adaptive real-time tool for human gait event detection using a wearable gyroscope. In: 20th International Conference on Climbing and Walking Robots and Support Technologies for Mobile Machines, pp. 1–9 (2017)
20. Zhang, Y., Ding, C., Li, T.: Gene selection algorithm by combining reliefF and mRMR. BMC Genomics **9**(Suppl. 2), S27 (2008)
21. Young, A.J., Hargrove, L.J.: A classification method for user-independent intent recognition for transfemoral amputees using powered lower limb prostheses. IEEE Trans. Neural Syst. Rehabil. Eng. **24**(2), 217–225 (2016)
22. Jurman, G., Riccadonna, S., Furlanello, C.: A comparison of MCC and CEN error measures in multi-class prediction. PLoS ONE **7**(8), 1–8 (2012)
23. Figueiredo, J., Santos, C.P., Moreno, J.C.: Automatic recognition of gait patterns in human motor disorders using machine learning: a review. Med. Eng. Phys. **53**, 1–12 (2018)
24. Afzal, T., Iqbal, K., White, G., Wright, A.B.: A method for locomotion mode identification using muscle synergies. IEEE Trans. Neural Syst. Rehabil. Eng. **25**(6), 608–617 (2017)
25. Novak, D., Goršič, M., Podobnik, J., Munih, M.: Toward real-time automated detection of turns during gait using wearable inertial measurement units. Sensors **14**(10), 18800–18822 (2014)
26. Carvalho, S., Figueiredo, J., Santos, C.P.: Environment-aware locomotion mode transition prediction system. In: IEEE International Conference on Autonomous Robot Systems and Competitions (ICARSC), pp. 1–6 (2019)

Classification of Patients with Parkinson's Disease Using Medical Imaging and Artificial Intelligence Algorithms

Helena R. Pereira[1,2](✉) and Hugo A. Ferreira[2]

[1] Departamento de Engenharia Eletrotécnica da Faculdade de Ciências e Tecnologia da Universidade Nova de Lisboa, 1099-085 Lisbon, Portugal
hri.pereira@campus.fct.unl.pt

[2] Instituto de Biofísica e Engenharia Biomédica da Faculdade de Ciências da Universidade de Lisboa, 1749-016 Lisbon, Portugal
hugoferreira@campus.ul.pt

Abstract. The diagnosis of Parkinsonian Syndromes (PS) at early-stages is a challenge. PS usually present similar symptoms, and the diagnosis is mainly clinical, causing often misdiagnosis between PS and other movement disorders. Parkinson's Disease (PD) is the most common PS, affecting a large part of the worldwide population. Medical imaging such as Magnetic Resonance Imaging (MRI) and Single Photon Emission Computed Tomography (SPECT) are currently being used to detect changes in anatomy and in the dopaminergic system, respectively. SPECT imaging allowed to find a group of patients diagnosed as having PD but without the characteristic decreased uptake of a dopamine analogue, so called "Scans Without Evidence of Dopaminergic Deficit" (SWEDD). Nowadays, deep learning algorithms, such as Convolutional Neural Networks (CNN), are becoming a useful tool in the medical field to detect patterns relevant to diseases in images. This study proposed an approach using CNN for the classification of MRI and SPECT images from PD, SWEDD, and Control subjects, to identify regions-of-interest related to PD. The proposed model achieved an accuracy of 97.4% using MRI images encompassing the mesencephalon and 93.3% with SPECT slices encompassing the basal ganglia. The results suggest that CNN was able to discriminate Control vs. PD and PD vs. SWEDD, CNN achieved accuracies up to 65.7%. Regarding PD vs. SWEDD, this classification obtained an accuracy of 73.3% using MRI images encompassing the mesencephalon and 93.3% with SPECT slices embracing the basal ganglia. The results suggest that CNN was able to discrimination Control vs. PD and PD vs. SWEDD, but not Control vs. SWEDD supporting the fact that SWEDD patients do not show evidence of dopamine deficit. In addition, the classification allowed the identification of the images comprising the mesencephalon or the basal ganglia as the most relevant for the classification.

Research supported by Fundação para a Ciência e Tecnologia (FCT) under the projects UID/BIO/00645/2019, POCI-01-0145-FEDER-016428 and NVIDIA GPU grant program.

J. Henriques et al. (Eds.): MEDICON 2019, IFMBE Proceedings 76, pp. 2043–2056, 2020.
https://doi.org/10.1007/978-3-030-31635-8_241

1 Introduction

There is a particular interest in the diagnosis of Parkinsonian Syndromes (PS) using medical imaging [1]. Currently, PS is diagnosed based on the clinical history of the patients, symptoms, positive response to PS medicines, and disease progression [2]. At early stages, however, PS present quite similar symptoms. The most common PS is Parkinson's Disease (PD). According to the World Health Organization [3], PD has an incidence rate of 9.7 to 13.8 per 100 000 people per year, and it is estimated that this value will increase in the following years due to population aging [2]. PD and the other PS are characterized by the degeneration of the dopaminergic neurons in the *substantia nigra* associated with motor symptoms. PD patients may also suffer from non-motor symptoms such as cognitive impairment, depression and sleep disorders [2]. Therefore, physicians are starting to use imaging techniques to find relevant biomarkers related to the disease [1, 2]. For instance, Magnetic Resonance Imaging (MRI) allows the discrimination of PD and other PS that present structural changes [1] and the detection of imaging biomarkers associated with volume changes [1, 4]. For instance, some studies found differences in the *substantia nigra* [5] and small volumes changes in putamen and caudate [6] of PD patients in comparison with healthy subjects. Moreover, the analysis of MRI images through Voxel-Based Morphometry (VBM) showed significant structural neocortical changes in the brain of PD patients with mild cognitive impairment and dementia [7]. Using Tensor-Based Morphometry, Tessa et al. [8] identified atrophy of the cortical and subcortical areas in PD patients at early stages. Besides MRI, Single Photon Emission Computed Tomography (SPECT) that uses the radioligand Ioflupane I 123 (123I-FP-CIT) also known as Dopamine Transporter scan (DaTscan) is also relevant to discriminate PD from other motor diseases non-associated with dopaminergic deficit [9, 10]. In DaTscan, the radioligand binds to the DAT, allowing the analysis of the dopaminergic system. PS patients have abnormal DaTscans because they present a reduced number of DAT in the striatum, unlike healthy subjects. Moreover, DaTscan images are relevant to monitoring disease progression because they allow the observation of degeneration of the dopamine neurons [9]. Interestingly, exams such as SPECT allowed the identification of a group of patients early diagnosed as having PD but that present "Scans Without Evidence of Dopaminergic Deficit (SWEDD)" [10]. The term SWEDD is very controversial in the literature and there are divergent opinions regarding the stage and type of the disease of those individuals. On the one hand, studies support the fact that SWEDD are an early stage of PD [11, 12]. On the other hand, the literature also present follow-up studies where SWEDD patients do not evolve to PD [13, 14] supporting the hypothesis that SWEDD subjects have another motor disorder [15]. Thus, the need to discriminate PD from other motor diseases emerged. Recently, Artificial Intelligence algorithms have been applied to several fields, including medical imaging [16]. Hence, imaging biomarkers and their features have increased the attention of the researchers [1, 17] to the classification of PD, SWEDD, and Control individuals. Concerning PD, several studies were focused on the classification of features extracted from MRI or SPECT images using Support Vector Machine (SVM). For instance, to classify PD and Control subjects, Oliveira et al. [18] computed the Striatal Binding Ratio (SBR) from DaTscan images using a voxel-based approach, while Prashanth et al. [19] computed the SBR of the caudate and putamen. In both approaches, the classification was accurate (above 96% accuracy). Using SVM, Singh et al. [20] used the Self-Organizing MAP (SOM) to extract features from the variations of the intensity in voxels of segmented MRI images of PD, SWEDD and Control. More recently, Amoroso et al. [21] proposed a different approach with SVM to classify structural connectivity measures extracted from MRI images. However, these studies have the disadvantage of being hard to implement in clinical practice because the classification is based on features previously extracted, which is usually very time-consuming. With this regard, several studies were focused on the application of deep learning algorithms to the classification of medical imaging [13, 22, 23]. Esmaeilzadeh et al. [22] demonstrated that the classification of MRI images combined with age and sex using Convolutional Neural Networks (CNN) allowed the discrimination of PD and Control images. Besides, when analyzed the heatmaps of the features extracted, *substantia nigra* and basal ganglia were considered relevant Regions-Of-Interest (ROIs) for the classification [22]. Using also CNN, Choi et al. [13] and Martinez-Murtia et al. [23] claimed to have classified 3D DaTscan images of PD and Control subjects accurately (96.0% and 95.5% accuracy, respectively). Besides, using a pre-trained CNN with PD and Control images, they classified baseline and follow-up 3D DaTscan images of SWEDD patients, and thus finding that the majority of patients maintained the DaTscan normal [13]. However, the throwback of the classification using DaTscan images is the fact that motor symptoms appear when about 50% of degeneration has already occurred [24]. Hence, DaTscan images are only used in non-early stage, while an early diagnosis of PD is aimed to adjust treatment to each case and to reduce financial costs.

In this regard, the aim of this study is to classify structural MRI and SPECT images of PD patients, SWEDD patients and Control subjects using CNN, and to identify which regions are more significant for the classification and diagnosis of the disease.

2 Methodology

2.1 Subjects and Imaging Data

The data used in this study was extracted from the Parkinson's Progression Markers Initiative (PPMI) database (www.ppmi-info.org/data). For up-to-date information on the study, visit www.ppmi-info.org. PPMI is a longitudinal and international study that aims to find progressive biomarkers of PD and similar cohorts including SWEDD to improve therapeutics. This study used 3D T1-Weighted (T1-W) MRI and DaTscan images from *de novo* PD, SWEDD and Control cohorts. *De novo* PD patients are subjects early diagnosed with PD and that never took PD medication [25], while SWEDD patients are individuals that were early diagnosed as having PD but present normal DaTscan. In its turn, the participants inserted into the Control cohort are considered healthy individuals without first relatives with PD. The inclusion criteria comprised the following: subjects must have one MRI and one DaTscan image acquired at baseline; the difference between the acquisition date of the two exams should be up to 6–12 months, since the National Institute for Health and Care Excellence of the United Kingdom [26] recommend that patients should be seen regularly in that interval. Moreover, MRI images were acquired at 1.5T and 3T using different machines but based on similar a protocol established by PPMI [25]. The aim of the classification with images acquired with slight different parameters is to guarantee that the model can classify a large variety of MRI images. DaTscan images were acquired about 4 h after the injection of the radiotracer Ioflupane I123 [27]. The Mann-Whitney U and the Chi-Squared tests ($p = 0.05$) [28] were used to verify if samples used in each classification were age and sex-matched, respectively.

2.2 Image Processing

The main aim of this procedure was to guarantee that all the images were spatially normalized and also to remove noise due to field inhomogeneities. Firstly, the format of the DaTscan and MRI images was converted from DICOM to NifTI using dcm2nii and Statistical Parametric Mapping (SPM) 12, respectively [29]. The pre-processing steps of MRI images were performed using SPM12 [29]. Firstly, Anterior Commissure – Posterior Commissure alignment correction was done. Then, bias regularization was also applied to remove artifacts caused by field inhomogeneities [30]. The registration of the MRI images to the standard Montreal Neurological Institute (MNI) brain template [29] was done using affine registration. Finally, a 2.0 mm Full Width at Half Maximum Gaussian filter was applied to all images to suppress noise and to improve the signal-to-noise ratio [31]. The DaTscan images were also processed in the PPMI database. The raw data acquired was reconstructed using a hybrid ordered subset expectation maximization algorithm in Hermes (Hermes Medical Solutions, Stockholm, Sweden) [32] Then, in PMOD (PMOD Technologies, Zurich, Switzerland), an attenuation correction was applied using Chang's Correction and spatial normalization into MNI template was performed [32]. In this study, an additional step was applied to the DaTscan images. For each subject, the SPECT image was co-registered to the matching MRI image using SPM12. After the pre-processing steps, each image had a size of $79 \times 95 \times 79$ mm^3. Then, each image was split into axial slices to identify slices encompassing regions associated with the disease. Finally, Contrast Limited Adaptive Histogram Equalization (CLAHE) [33] was applied to each slice to improve the contrast of MRI images, considering that noisy regions are not overamplified [33].

2.3 CNN Architecture

The CNN used in this study was adapted from the work of Godinho [34] and based on the network created by Lecun et al. [35]. The network was implemented using Caffe, an open source framework with deep learning algorithms such as CNN [36], compatible with NVIDIA Graphics Processing Units (GPU) to accelerate computation. The input data consist of images with three consecutive slices from MRI or DaTscan along the third dimension, each one with a size of $95 \times 79 \times 3$. Each pixel value of the images was scaled into a range of 0 to 1 to normalize the intensity and accelerate the computation [37].

The CNN used in this work comprised two convolutional layers with 2D learning Kernel filters (size of 5×5) to extract features and to create feature maps [35]. After each convolutional layer a one pooling layer (size of 2×2) to reduce the size of the feature maps using the max pooling function (max pooling selects the maximum value from a given number of pixels) [35]. A Fully Connected (FC)x layer followed the last pooling layer to convert the extracted features into one 1D vector. Then, an activation function, the Rectified Linear Unit (ReLU), was applied to the output of the FC layer, to define which features were more relevant to the classification. The last two layers comprised one FC and one Softmax layer. The FC layer contained the most interesting features, while the Softmax function provided a class of predictive probabilities of the input image, ranging from 0 to 1. The hyperparameters of the network were set using Caffe documentation [36] and based on Godinho's work [34]. The number of iterations used in training was 5000, and at every 500 iterations a validation test was done. The computations were performed in a CentOS Linux 7 and using the GPU NVIDIA GeForce GTX TITAN X.

2.4 Data Sets and Classifier Evaluation

The Monte Carlo cross-validation method was used to prevent biased estimations. In this study, five batches containing train, validation, and test sets were created using this method. The batches comprised 50% of the initial data in the train set, 25% in the validation set, and 25% in the test set. In this study, the data sets had the same number of subjects of each group to guarantee a balanced classification. This study focused on the comparison of Control vs. PD, Control vs. SWEDD, and PD vs. SWEDD.

The accuracy, Sensitivity, Specificity, Positive Predictive Value (PPV) and Negative Predictive Value (NPV) were computed. The performance of the classifier was further evaluated using Cohen's Kappa (kappa) and the Area Under Curve (AUC) metric derived from Receiver Operating Characteristic (ROC) analysis.

3 Results

The demographic characteristics of the subjects are summarized in Table 1. As shown in Table 1, the SWEDD group is the cohort with fewer subjects which influenced the number of subjects used in each data set. The p-value obtained in the Mann-Whitney U test indicates that there were no significant age differences in Control vs. PD, Control vs. SWEDD, and PD vs. SWEDD ($p = 0.272$, $p = 0.228$ and $p = 0.623$, respectively). The Chi-Square test indicates that there were no significant gender differences in those groups: Control vs. PD ($p = 0.847$), Control vs. SWEDD ($p = 0.930$), and PD vs. SWEDD ($p = 0.825$). Table 2 expresses the number of subjects used in both MRI and DaTscan classification. As a balanced data set was used, only 30 subjects were tested in Control vs. SWEDD and PD vs. SWEDD. Figure 1 illustrates the MRI and DaTscan slices that lead to higher accuracy. These slices encompasse regions associated with PD, namely the *mesencephalon* and the basal ganglia. Figure 2 illustrates all the images used in the classification of Control vs. PD, Control vs. SWEDD, PD vs. SWEDD and the accuracies achieved.

Table 1. Subjects demographic information

Groups	Control	SWEDD	PD
Number of subjects	168	58	378
Gender (M/F)	59/109	20/38	136/242
Age (y.o) Avg ± std	60 ± 11	62 ± 10	62 ± 10

M = Male, F = Female, Avg = Average, y.o. = years old, std = standard deviation

Table 2. Number of subjects used in each classification group in train, validation and test sets

Classification group	Number of slices		
	Train	Validation	Test
Control vs. PD	84 Control 84 PD	42 Control 42 PD	42 Control 42 PD
Control vs. SWEDD	29 Control 29 SWEDD	14 Control 14 SWEDD	15 Control 15 SWEDD
PD vs. SWEDD	29 PD 29 SWEDD	14 PD 14 SWEDD	15 PD 15 SWEDD

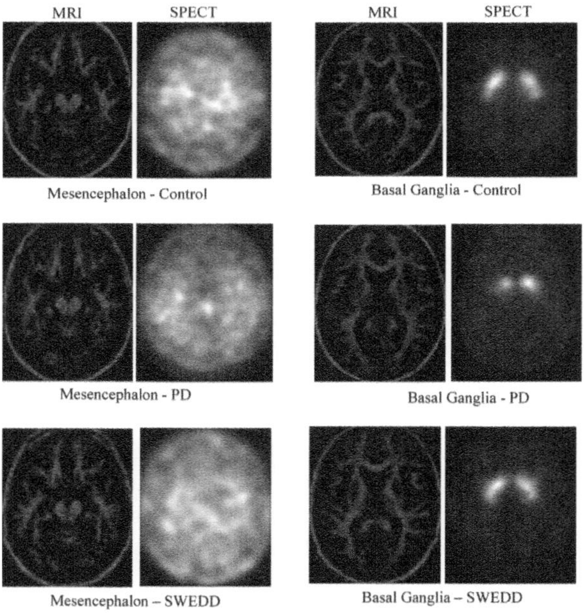

Fig. 1. MRI and DaTscan slices of representative subjects.

As shown in Fig. 2, the accuracy of the classification using DaTscan slices encompassing the basal ganglia is higher than using MRI slices in Control vs. PD, and PD vs. SWEDD. In the classification using DaTscan slices, the accuracy increased until

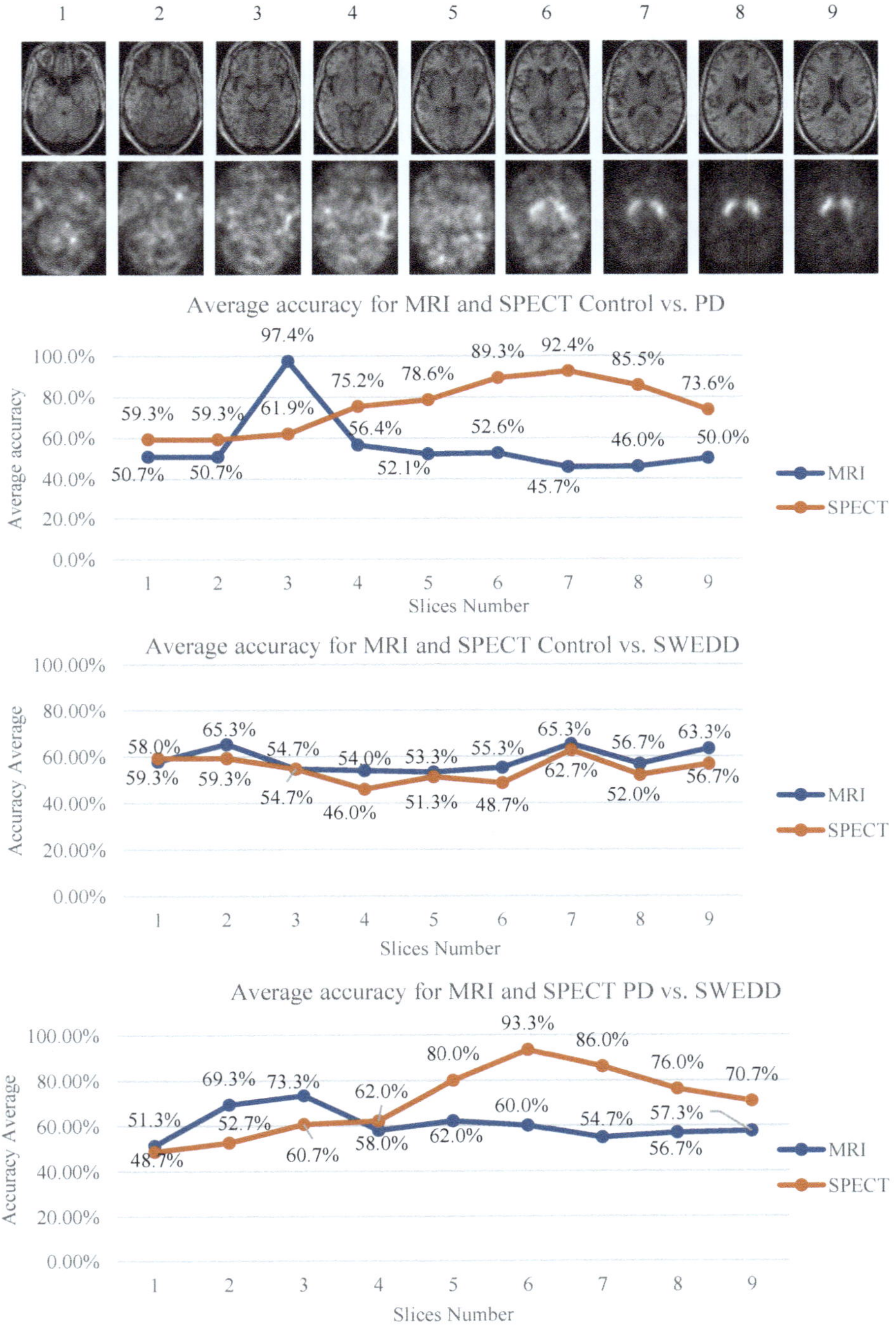

Fig. 2. The average accuracy of the classification using MRI and DaTscan slices

it reaches a maximum and then decreased. This peak was achieved using slices encompassing the basal ganglia. In Control vs. SWEDD, the values of accuracy were similar in the classifications using the two medical imaging modalities.

The results of the classification using slices encompassing the mesencephalon and the basal ganglia are given in Table 3. Slices encompassing the *mesencephalon* achieved the highest accuracy (97.4%) using MRI slices, followed by the DaTscan slices comprising the basal ganglia (92.4%) as shown in Table 3 for Control vs. PD. In the classification of Control vs. SWEDD, the classifier had accuracies up to 65.3%.

Table 3. Classification results

Classification group		Acc (%)	Sen (%)	Spe (%)	PPV (%)	NPV (%)	Kappa Avg ± std. dev.	AUC-ROC CI: 95%
Control vs. PD								
Mesencephalon	MRI	97.4	100.0	94.8	95.1	100.0	0.95 ± 0.03	0.99 [0.96–1.00]
	SPECT	61.9	62.4	61.4	61.8	62.2	0.24 ± 0.11	0.65 [0.54–0.77]
Basal ganglia	MRI	45.7	48.1	43.3	45.9	45.5	−0.09 ± 0.11	0.45 [0.33–0.58]
	SPECT	92.4	92.9	91.9	92.1	92.8	0.85 ± 0.06	0.98 [0.94–1.00]
Control vs. SWEDD								
Mesencephalon	MRI	54.7	58.7	50.7	54.6	54.6	0.09 ± 0.18	0.57 [0.36–0.78]
	SPECT	54.7	49.3	60.0	56.1	53.9	0.09 ± 0.18	0.54 [0.33–0.75]
Basal ganglia	MRI	65.3	65.3	65.3	65.3	65.8	0.31 ± 0.17	0.67 [0.48–0.87]
	SPECT	62.7	68.0	57.3	57.3	63.3	0.25 ± 0.17	0.66 [0.46–0.85]
PD vs. SWEDD								
Mesencephalon	MRI	73.3	65.3	81.3	68.7	70.4	0.47 ± 0.16	0.84 [0.69–0.98]
	SPECT	60.7	58.7	50.7	52.9	52.5	0.09 ± 0.18	0.61 [0.40–0.81]
Basal ganglia	MRI	60.0	53.3	54.8	54.8	55.5	0.09 ± 0.18	0.61 [0.40–0.81]
	SPECT	93.3	85.3	86.5	85.7	86.5	0.72 ± 0.12	0.90 [0.79–1.00]

Acc = Accuracy, Sen = sensitivity, Spe = specificity, PPV = Positive Predictive Values, NPV = Negative Predictive Values, Avg = Average, std. dev = standard deviation, CI: Confidence Interval

Concerning PD vs. SWEDD, the DaTscan slices encompassing the basal ganglia had an accuracy of 93.3%, while MRI slices had an accuracy of 73.3% using the slices encompassing the mesencephalon. The sensitivity and specificity of the slices that showed the highest values are also given in Table 3. In the classification of Control vs. PD, the sensitivity and the specificity were significantly higher using MRI slices encompassing the mesencephalon (100% sensitivity, 94.8% specificity) in comparison with the matching DaTscan slices (62.4% sensitivity, 61.4% specificity). The DaTscan slices encompassing the basal ganglia had higher sensitivity (92.9%) and specificity (91.9%) than the corresponding MRI slices (48.1% sensitivity, 43.3% specificity). These results are similar to those achieved in the discrimination of PD and SWEDD in which DaTscan slices encompassing the basal ganglia achieved higher performance (86.7% sensitivity, 85.3% specificity) than the MRI slices encompassing the same region (56.0% sensitivity, 53.3% specificity). The results of the ROC curve showed a trend of higher AUC values in the classification using MRI slices encompassing the mesencephalon (0.99 [0.96–1.00]) than DaTscan slices encompassing the basal ganglia (0.98 [0.94–1.00]).

The CNN performed the classification of the Control vs. SWEDD with an AUC value of approximately 0.67 [0.48–0.87] in the case with the highest accuracy. Concerning the discrimination of PD vs. SWEDD, the highest AUC values were achieved using MRI slices encompassing the mesencephalon (0.84 [0.69–0.98] and DaTscan slices encompassing the basal ganglia (0.90 [0.79–1.00]). Overall, the classifier achieved higher NPV than PPV, i.e. true negatives were more correctly classified than true positives.

4 Discussion

The proposed approach used in this study was able to discriminate Control vs. PD, and PD vs. SWEDD using MRI and DaTscan slices encompassing regions associated with PD. In addition, the findings yielded in each classification allow the understanding of which slices are more relevant for the comparison of the groups, and the respective imaging biomarkers linked to the disease which allows further analysis of the disease progression. The literature indicates that PD is associated with the neurodegeneration of the *substantia nigra* in the mesencephalon and consequently dopamine deficit in the basal ganglia system [2, 38]. In the classification of Control vs. PD, results suggest that MRI slices encompassing the mesencephalon and the DaTscan slices encompassing the basal ganglia are the most significant because they achieved higher accuracy. Besides, the MRI model was able to discriminate Control vs. PD, showing an almost perfect agreement [39] with the PPMI database by presenting kappa above 0.90. Concerning the comparison of Control vs. SWEDD, the results show that the classifier was not able to discriminate Control vs. SWEDD (AUC value up to 0.67) using either MRI or DaTscan slices supporting the fact that SWEDD patients show similar scans to those of healthy people [9, 40]. The classification of PD vs. SWEDD also suggests that DaTscan slices encompassing basal ganglia are the ones in which the CNN had better performance (AUC = 0.90 [0.79–1.00]) [41]. Although with lower accuracy, the CNN was able to discriminate MRI slices encompassing the mesencephalon (AUC = 0.84 [0.69–0.98]), sustaining the hypothesis that SWEDD patients differ from PD patients [40]. Overall, the sensitivity and the specificity (86.5% specificity and 86.5% sensitivity) were similar in contrast to the classification performed by nuclear imaging experts, as shown in the study of Choi [13], in which the specificity (82.6%) was lower than the sensitivity (96.2%). Besides, the results of the three comparisons are consistent with the study of Minati et al. [42] in which changes in the *substantia nigra* were depicted using MRI, and with the work of Rahmim et al. [43] where DaTscan images allowed the detection of the disease.

The present study achieved favorable results in comparison with other approaches in the literature that also used images from the PPMI database. Besides, it is also important to note that the majority of the studies classified 3D images in contrast to the proposed methodology. For instance, using 3D CNN, Esmaeilzadeh et al. [22] claimed to have achieved 82% accuracy in the classification of 3D MRI images in PD vs. Control, and 100% accuracy when age and sex were added to the model using also batch normalization and one dropout layer. However, such a study used an imbalanced data and data augmentation techniques, while our study balanced data sets were used. Imbalanced data has the disadvantage of favoring the majority group, which may lead to biased estimations [44]. This study differs from the work of Esmaeilzadeh because it achieved higher accuracy (97.4%) using only slices encompassing the mesencephalon, providing more information to aid in the diagnosis. It has been shown that the classification using 3D CNN of 3D DaTscan images from 431 PD patients and 193 Control subjects provided an accuracy of 96% and an AUC value of 0.99 [13]. Using a similar approach, Martinez-Murcia et al. [23] achieved 95.5% with 158 PD patients and 111 Control subjects. Although, in comparison with those studies, this study showed lower accuracy (92.4%), the classification was performed using cross-validations methods, providing a more robust analysis. However, DaTscan images only detect the disease at advances stages in which the neurodegeneration of the dopamine neurons had already occurred [45]. Thus, the classification using this type of images may not be able provide support for an early diagnosis.

Other authors used SVM for the same purpose. Singh et al. [20] claimed to have achieved an accuracy of 97.6% using MRI features from Control subjects and PD patients in similar stages and with the same age, and 87.4% (71.9% Sensitivity, 94.7% specificity) in age unrelated groups. For feature extraction, they used an approach based on the variations of intensity in the segmented grey matter and white matter of PD and Control subjects. However, segmentation is often associated with some minor errors [46] that may influence the classification. Thus, in our study, the MRI and DaTscan slices were directly classified. Amoroso et al. [21] also used features extracted from segmented MRI images to classify PD and Control subjects, obtaining an accuracy of 94% and 97% when clinical data was added. However, unlike Singh et al. [20] they applied Pearson's Correlation to measure the connection between nodes of the brain. The results obtained in our study for the classification using MRI slices encompassing the mesencephalon achieved higher accuracy (97.4%) than those of Amoroso [21] using only features of the Pearson's Correlation, which may be related to the type of features provided to the classifier (94% accuracy). Using also SVM, Oliveira et al. [18] classified striatum binding ratios extracted from DaTscan images of PD patients and Controls. Their approach achieved higher accuracy (97.9%) in comparison with the results yielded in our study using slices encompassing the basal ganglia (92.4%). In that work, the ratio was computed from voxels' values, unlike our study that used only information from pixels. In contrast to the studies that employed SVM, our work only needed one algorithm to classify and extract the most relevant features directly from the images Moreover, our work required no image segmentation, unlike the studies of Amoroso et al. [21] and Singh et al. [20], reducing the computational time. Concerning the classification of Control vs. SWEDD, the literature has only a few studies focused on this discrimination, which difficults the comparison with the state-of-the-art. The study of Singh et al. [20], mentioned above, claimed to have differentiated Control vs. SWEDD using features from 3D MRI images. Using an imbalanced sample (245 Control, 68 SWEDD), they achieved 99.5% accuracy (100.0% sensitivity, 97.2% specificity). The difference between the results may be related to the features used and the number of subjects, which were significantly less in this study. Although the results of the DaTscan slices encompassing the basal ganglia show no evidence of dopamine deficit as shown in this study, 3D MRI may detect significant intensities differences in Control vs SWEDD [20] in contrast to the classification of 2D slices. Using 3D DaTscan images, Choi et al. [13] showed that the majority of the SWEDD patients of the PPMI database remained with the scan normal after a follow-up, supporting the results achieved in this study. To our knowledge, just a few studies performed the classification of PD vs. SWEDD. Singh et al. [20] achieved 98.2% accuracy using an imbalanced sample (518 PD and 68 SWEDD), which could have led to that result, while this work achieved only 73.3% accuracy using MRI slices encompassing the mesencephalon. However, in their study, the Cohen's Kappa was not computed which difficulties the comparison with our study. Martinez-Murcia et al. claimed to obtain 82% in the classification of Control vs. PD vs. SWEDD using DaTscan 3D images, which is lower than the comparison using DaTscan slices encompassing the basal ganglia (93.3%) achieved in our work.

5 Conclusion

The clinical diagnosis of PD is facing a problem in the discrimination of PS and other motor disorders. Although presenting similar symptoms, PS have different pathological characteristics. Previous studies were mainly focused on the classification of 3D images, but in clinical practice, 2D images are relevant to finding regions associated with the disease.

In this study, an approach using CNN to classify MRI and DaTscan images was proposed to differentiate Control vs. PD, Control vs. SWEDD, and PD vs. SWEDD. This model was able to discriminate Control vs. PD and PD vs. SWEDD and to identify the most relevant slices for the classification. The analysis of the results suggests that the MRI slices encompassing the mesencephalon and the DaTscan slices encompassing the basal ganglia allow the discrimination of Control vs. PD and PD vs. SWEDD. Although our approach was not able to discriminate Control vs. SWEDD, suggesting that these two groups have similar imaging patterns. The novelty of this study was the classification of Control vs. SWEDD, and PD vs. SWEDD using 2D slices with 2D CNN. These results suggest that this approach is promising not only for PD but also for other PS and neurodegenerative disease. However, it is worthy to note some limitations. The number of images of the SWEDD patients was quite small in comparison with PD and Control subjects, which may have influenced the classification, especially using MRI slices where the accuracy was lower. Moreover, the classification using 2D slices has the disadvantage of not considering volumes and intensities changes in voxels comprising ROIs. Therefore, future work should include this additional information, and increase the number of images of SWEDD patients to provide more accurate results.

Acknowledgment. The authors would like to thank the financial support from Fundação para a Ciência e Tecnologia (FCT) under the project UID/BIO/00645/2019, Programa Operacional Temático Competitividade e Internacionalização under the project POCI- 01-0145-FEDER-016428, to the NVIDIA GPU Grant Program and to work by PPMI personnel that went into accumulating the data, as well as funding of the study. PPMI – a public-private partnership – is funded by the Michael J. Fox Foundation for Parkinson's Research and industry partners: Abbvie, Allergan, Avid, Biogen, BioLegend, Bristol-Myers Squibb, Celgene, Denali, GE Healthcare, Genentech, GlaxoSmithKline, Lilly, Lundbeck, Merck, Meso Scale Discovery,

Pfizer, Piramal, Prevail Therapeutics, Roche, Sanofi Genzyme, Servier, Takeda, Teva, Ucb, Verily, Voyager Therapeutics and Golub Capital.

References

1. Saeed, U., et al.: Imaging biomarkers in Parkinson's disease and Parkinsonian syndromes: current and emerging concepts. Transl. Neurodegener. **6**, 1–25 (2017)
2. Olanow, C.W., Schapira, A.H.V., Obeso, J.A.: Parkinson's disease and other movement disorders. In: Harrison's Principles of Internal Medicine, 19th edn., pp. 2609–2626. McGraw-Hill Education, New York (2015)
3. World Health Organization. Neurological disorders: public health challenges, Genebra, Switzerland, pp. 140–150 (2006)
4. Salvatore, C., et al.: Machine learning on brain MRI data for differential diagnosis of Parkinson's disease and progressive supranuclear palsy. J. Neurosci. Methods **222**, 230–237 (2014)
5. Kwon, D.-H., et al.: Seven-tesla magnetic resonance images of the substantia nigra in Parkinson disease. Ann. Neurol. **71**(2), 267–277 (2012)
6. Pitcher, T.L., et al.: Reduced striatal volumes in Parkinson's disease: a magnetic resonance imaging study. Transl. Neurodegener. **1**(1), 17 (2012)
7. Beyer, M.K., et al.: A magnetic resonance imaging study of patients with Parkinson's disease with mild cognitive impairment and dementia using voxel-based morphometry. J. Neurol. Neurosurg. Psychiatry **78**(3), 254–259 (2007)
8. Tessa, C., et al.: Progression of brain atrophy in the early stages of Parkinson's disease: A longitudinal tensor-based morphometry study in de novo patients without cognitive impairment. Hum. Brain Mapp. **35**(8), 3932–3944 (2014)
9. Brooks, D.J.: Imaging Approaches to Parkinson Disease. J. Nucl. Med. **51**(4), 596–609 (2010)
10. Eidelberg, D.: Imaging in Parkinson's disease, pp. 3–8. Oxford University Press, New York (2012)
11. Nicastro, N., Garibotto, V., Badoud, S., Burkhard, P.R.: Scan without evidence of dopaminergic deficit: a 10-year retrospective study. Park. Relat. Disord. **31**, 53–58 (2016)
12. Erro, R., Schneider, S.A., Stamelou, M.: What do patients with scans without evidence of dopaminergic deficit (SWEDD) have? New evidence and continuing controversies. J. Neurol. Neurosurg. Psychiatry **87**(3), 319–323 (2016)
13. Choi, H., Ha, S., Im, H.J.: Refining diagnosis of Parkinson's disease with deep learning-based interpretation of dopamine transporter imaging. NeuroImage Clin. **16**(September), 586–594 (2017)
14. Batla, A.: Patients with scans without evidence of dopaminergic deficit: a long-term follow-up study. Mov. Disord. **29**(14), 1820–1825 (2014)
15. Schneider, S.A.: Patients with adult-onset dystonic tremor resembling Parkinsonian tremor have scans without evidence of dopaminergic deficit (SWEDDs). Mov. Disord. **22**(15), 2210–2215 (2007)
16. Hamet, P., Tremblay, J.: Artificial intelligence in medicine. Metabolism **69**, S36–S40 (2017)
17. Lee, J.G., et al.: Deep learning in medical imaging: general overview. Korean J. Radiol. **18**(4), 570–584 (2017)
18. Oliveira, F.P.M., Castelo-Branco, M.: Computer-aided diagnosis of Parkinson's disease based on [123 I]FP-CIT SPECT binding potential images, using the voxels-as-features approach and support vector machines. J. Neural Eng. **12**(2), 026008 (2015)

19. Prashanth, R., Dutta Roy, S., Mandal, P.K., Ghosh, S.: Automatic classification and prediction models for early Parkinson's disease diagnosis from SPECT imaging. Expert Syst. Appl. **41**(7), 3333–3342 (2014)
20. Singh, G., Samavedham, L.: Unsupervised learning based feature extraction for differential diagnosis of neurodegenerative diseases: a case study on early-stage diagnosis of Parkinson disease. J. Neurosci. Methods **256**, 30–40 (2015)
21. Amoroso, N., La Rocca, M., Monaco, A., Bellotti, R., Tangaro, S.: Complex networks reveal early MRI markers of Parkinson's disease. Med. Image Anal. **48**, 12–24 (2018)
22. Esmaeilzadeh, S., Yang, Y., Adeli, E.: End-to-end parkinson disease diagnosis using brain MR-Images by 3D-CNN. CoRR, vol. abs/1806.05233, August 2018
23. Martinez-Murcia, F.J., et al.: A 3D convolutional neural network approach for the diagnosis of Parkinson's disease. Lect. Notes Comput. Sci. (including Subser. Lect. Notes Artif. Intell. Lect. Notes Bioinformatics). LNCS, vol. 10337, pp. 324–333, June 2017
24. Fearnley, M., Lees, A.J.: Ageing and Parkinson's disease: substantia nigra regional selectivity. Brain **114**(Pt 5), 2283–2301 (1991)
25. Marek, K.: The Parkinson Progression Marker Initiative (PPMI). Prog. Neurobiol. **95**(4), 629–635 (2011)
26. National Institute for Health and Care Excellence. Parkinson's disease in adults. (NICE Guideline NG71) (2017)
27. Parkinson's Progression Markers Initiative. MRI - Technical Operations Manual. PPMI (2015)
28. Weave, K.F.: An Introduction to Statistical Analysis in Research. Wiley, Hoboken (2017)
29. Friston, K.J.: Statistical Parametric Mapping: The Analysis of Functional Brain Images. Elsevier/Academic Press, Amsterdam (2006)
30. Jäger, F., Balda, M., Hornegger, J.: Correction of intensity inhomogeneities utilizing histogram-based regularization. In: 4th Russian Conference on Biomedical Engineering 2014, pp. 23–27, March 2008
31. Dubois, E., Mitiche, A.: Digital picture Processing. IEEE Trans. Inf. Theory **30**(4), 694–695 (1984)
32. Marek, K., et al.: The Parkinson's progression markers initiative (PPMI) – establishing a PD biomarker cohort. Ann. Clin. Transl. Neurol. **5**(12), 1460–1477 (2018)
33. Zuiderveld, K.: Contrast limited adaptive histogram equalization. In: Heckbert, P.S. (ed.) Graphics Gems IV, pp. 474–485. Academic Press, Inc., San Diego, CA, USA (1994)
34. Godinho, D.M.: Desenvolvimento de uma aplicação com recurso à unidade de processamento gráfico para classificação de sinais de Perturbações do Espectro do Autismo. M.S. thesis, Dept. Phys, Univ. Nova de Lisboa, Lisboa, Portugal (2016)
35. Lecun, Y., Bottou, L., Bengio, Y., Haffner, P.: Gradient-based learning applied to document recognition. Proc. IEEE, **86**(11), 2278–2324 (1998)
36. Jia, Y., et al.: Caffe: convolutional architecture for fast feature embedding. In: Proceedings of the ACM International Conference on Multimedia - MM 2014, pp. 675–678 (2014)
37. LeCun, Y.A., Bottou, L., Orr, G.B., Müller, K.-R.: Efficient BackProp, pp. 9–48. Springer, Berlin (2012)
38. Williams, D.R., Litvan, I.: Parkinsonian syndromes. Contin. Lifelong Learn. Neurol. **19**(5), 1189–1212 (2013)
39. McHugh, M.L.: Interrater reliability: the kappa statistic. Biochem. Medica **22**(3), 276–282 (2012)
40. Bajaj, N.: SWEDD for the general neurologist. Adv. Clin. Neurosci. Rehabil. **10**(4), 30–31 (2010)
41. Hajian-Tilaki, K.: Receiver operating characteristic (ROC) curve analysis for medical diagnostic test evaluation. Casp. J. Intern. Med. **4**(2), 627–635 (2013)

42. Minati, L., et al.: Imaging degeneration of the substantia nigra in Parkinson disease with inversion-recovery MR imaging. AJNR Am. J. Neuroradiol. **28**(2), 309–313 (2007)
43. Rahmim, A., et al.: Application of texture analysis to DAT SPECT imaging: relationship to clinical assessments. NeuroImage Clin. **12**, e1–e9 (2016)
44. Longadge, R., Dongre, S.: class imbalance problem in data mining review. Eur. J. Intern. Med. **24**(1), e256 (2013)
45. Cheng, H.-C., Ulane, C.M., Burke, R.E.: Clinical progression in Parkinson disease and the neurobiology of axons. Ann. Neurol. **67**(6), 715–725 (2010)
46. Despotović, I., Goossens, B., Philips, W.: MRI segmentation of the human brain: challenges, methods, and applications. Comput. Math. Methods Med. **2015**, 1–23 (2015)

Author Index

J. Henriques et al. (Eds.): MEDICON 2019, IFMBE Proceedings 76, pp. 2057–2067, 2020.
https://doi.org/10.1007/978-3-030-31635-8

F

G

Y

Z

GPSR Compliance

The European Union's (EU) General Product Safety Regulation (GPSR) is a set of rules that requires consumer products to be safe and our obligations to ensure this.

If you have any concerns about our products, you can contact us on ProductSafety@springernature.com

In case Publisher is established outside the EU, the EU authorized representative is:

Springer Nature Customer Service Center GmbH
Europaplatz 3
69115 Heidelberg, Germany

Batch number: 10755090

Printed by Printforce, the Netherlands

IFMBE Proceedings

Volume 76

The IFMBE Proceedings Book Series is an official publication of *the International Federation for Medical and Biological Engineering* (IFMBE). The series gathers the proceedings of various international conferences, which are either organized or endorsed by the Federation. Books published in this series report on cutting-edge findings and provide an informative survey on the most challenging topics and advances in the fields of medicine, biology, clinical engineering, and biophysics.

The series aims at disseminating high quality scientific information, encouraging both basic and applied research, and promoting world-wide collaboration between researchers and practitioners in the field of Medical and Biological Engineering.

Topics include, but are not limited to:

- Diagnostic Imaging, Image Processing, Biomedical Signal Processing
- Modeling and Simulation, Biomechanics
- Biomaterials, Cellular and Tissue Engineering
- Information and Communication in Medicine, Telemedicine and e-Health
- Instrumentation and Clinical Engineering
- Surgery, Minimal Invasive Interventions, Endoscopy and Image Guided Therapy
- Audiology, Ophthalmology, Emergency and Dental Medicine Applications
- Radiology, Radiation Oncology and Biological Effects of Radiation

IFMBE proceedings are indexed by SCOPUS and EI Compendex. They are also submitted for ISI proceedings indexing.

Proposals can be submitted by contacting the Springer responsible editor shown on the series webpage (see "Contacts"), or by getting in touch with the series editor Ratko Magjarevic.

More information about this series at http://www.springer.com/series/7403

Jorge Henriques · Nuno Neves ·
Paulo de Carvalho
Editors

XV Mediterranean Conference on Medical and Biological Engineering and Computing – MEDICON 2019

Proceedings of MEDICON 2019,
September 26–28, 2019, Coimbra, Portugal

Set 1

Springer

Editors
Jorge Henriques
University of Coimbra
Coimbra, Portugal

Nuno Neves
Universidade do Minho
Braga, Portugal

Paulo de Carvalho
University of Coimbra
Coimbra, Portugal

ISSN 1680-0737 ISSN 1433-9277 (electronic)
IFMBE Proceedings
ISBN 978-3-030-31634-1 ISBN 978-3-030-31635-8 (eBook)
https://doi.org/10.1007/978-3-030-31635-8

This Springer imprint is published by the registered company Springer Nature Switzerland AG
The registered company address is: Gewerbestrasse 11, 6330 Cham, Switzerland

Preface

MEDICON 2019 is the XV in the series of regional meetings of the International Federation of Medical and Biological Engineering (IFMBE) in the Mediterranean. MEDICON 2019 will be organized for the first time in Portugal and will be hosted by the UNESCO World Heritage University, the University of Coimbra.

The goal of MEDICON 2019 is to provide updated information on the state of the art on Medical and Biological Engineering and Computing under the main theme "**Patient empowerment**". Patient empowerment has emerged as a new paradigm that positions the patients at the heart of the health system and encourages them to be actively involved in managing their own healthcare needs. Effective patient empowerment requires a holistic approach, combining multiple dimensions of needs and patient contexts. Medical and Biological Engineering and Computing is a discipline at the heart of patient empowerment. Research and development in these areas are impacting the science and technology by advancing fundamental concepts in translational medicine and understanding in human physiology, function and behaviour at multiple levels. This is leading to improved tools and techniques for the detection, prevention, treatment and management of diseases. MEDICON 2019 provides a common platform for the cross fertilization of ideas and to help shape knowledge and scientific achievements by bridging complementary disciplines into an interactive and attractive forum under the special theme of the conference that is "improving health care through holistic patient empowerment".

The programme consists of some approximately 250 invited and submitted papers on new developments around the Conference theme, presented in 8 plenary sessions, 18 parallel scientific sessions and 19 special sessions and also includes a set of competitions and awards.

More specifically, the parallel scientific sessions cover the topics of biomedical signal processing; biomedical imaging and image processing; bio-instrumentation, bio-senso and bio-micro/nanotechnologies; bioinformatics, computational biology and systems biology; biomechanics, robotics and rehabilitation; therapeutic and diagnostic systems, devices and technologies and clinical engineering; information technology in health systems; assistive technologies; technologies for active ageing;

biomedical engineering education and society; clinical engineering and health technology assessment; neuroengineering, neurosystems; technologies for preventive health care; biomedical technologies for developing countries; standardization of open data; biomaterials and tissue engineering.

The special sessions include the topics of optimization in medicine and biology; ontological engineering in biomedical informatics; electronics and smart algorithms for the effective lung monitoring and COPD management; non-invasive temperature assessment using ultrasound; computational biology and medical applications; smartphone-based, patient-centred technologies; computational and experimental modelling for designing bone-implant systems; artificial organs: extracorporeal blood circulation medical devices; diabetes and cardiovascular diseases: Ibero-American trends; smart robotic assistant for minimally invasive surgery: the SMARTsurg project experience; intelligent computational systems in biomedical engineering; INT4DAT - Intelligent systems and technologies for diagnostic, assistance and therapeutics; upper limb exoskeletons for a better quality of life: what is real, what is useful, and what is next?; neurosystems and connectivity; therapeutic applications of imaging and neurostimulation; value-based health technology assessment; international collaborative on medical device assessment; ocular imaging; assessing human error in cognitive/intellectual demanding tasks: case study on software engineering.

The conference programme also includes three competitions and two awards: IFMBE Scientific Challenge competition; Fraunhofer Best Portuguese PhD thesis competition in Biomedical Engineering; Fraunhofer Best Portuguese MSc thesis competition in Biomedical Engineering; Best Student Paper Award; Young Investigator Competition Award.

Furthermore, the conference programme is highlighted by eight plenary sessions: Digitally empowered patients, presented by Aart van Halteren; Cardiovascular modelling and simulations—applications to some clinical studies, presented by Adélia Sequeira; Biomedical signals and images processing: towards innovative paradigms of information integration in the era of precision medicine and big data in health, presented by Sergio Cerutti; Multilingual dictionary of medical physics terms—update and relevance for clinical engineering, presented by Slavik Tabakov; Prevention Engineering: evolving challenges for biomedical and clinical engineering, presented by Luis Kun; Optical coherence tomography: a window into the mechanism of neurodegenerative disorders, presented by Rui Bernardes; Towards a value based healthcare system supported by process mining techniques, presented by Vicente Traver; In silico clinical trials: towards transforming the biomedical industry and the healthcare delivery, presented by Dimitrios Fotiadis.

Particular thanks are expressed to the kind support and effort of a number of external sponsors to which we would like to express our appreciation. Finally, a heartfelt thanks to all of you, the participants for your paper contributions, wishing you every success in your work at the conference. We hope that MEDICON 2019 will offer opportunities for professional growth and establishing new contacts with fellow colleagues. Our intention is to do all we can to make your participation in MEDICON 2019 worthwhile and your stay in Coimbra enjoyable and memorable.

We hope you will appreciate this proceedings volume as much as we are proud of it!

September 2019

Paulo de Carvalho
General Co-chair

Jorge Henriques
Nuno Neves
Program Committee Co-chairs

Organization

Editors

Jorge Henriques University of Coimbra, Portugal
Nuno Neves University of Minho, Portugal
Paulo de Carvalho University of Coimbra, Portugal

Organizers

Thematic Tracks

T1: Biomedical Signal Processing

Anna Bianchi	Politecnico di Milano, Italy
Ana Paula Rocha	University of Porto, Portugal

T2: Biomedical Imaging and Image Processing

Ana Mendonça	University of Porto, Portugal
João Sanches	University of Lisbon, Portugal

T3: Bio-instrumentation, Biosensors and Bio-micro/nanotechnologies

Pedro Vieira	University of Lisbon, Portugal
João Paulo Cunha	University of Porto, Portugal

T4: Bioinformatics, Computational Biology and Systems Biology

Paula Oliveira	University of Coimbra, Portugal
Joel Arrais	University of Coimbra, Portugal

T5: Biomechanics, Robotics and Rehabilitation

Renato Natal Jorge	University of Porto, Portugal
Urbano Nunes	University of Coimbra, Portugal

T6: Therapeutic and Diagnostic Systems, Devices and Technologies and Clinical Engineering

Altamiro Pereira	University of Porto, Portugal
Jens Muehlsteff	Philips Research, Netherlands

T7: Information Technology in Health Systems

Nicos Maglaveras	Northwestern University, USA

T8: Assistive Technologies

Vicente Traver	Universitat Politècnica de València, Spain
Antonio Martinez Millana	Universitat Politècnica de València, Spain

T9: Technologies for Active Ageing

João Malva	University of Coimbra, Portugal
Maria Arredondo	Universidad Politècnica de Madrid, Spain

T10: Biomedical Engineering Education and Society

Shankar Krishnan	Wentworth Institute of Technology, USA
Monique Frize	University of Ottawa, Canada

T11: Clinical Engineering and Health Technology Assessment

Leandro Pecchia	University of Warwick, UK
Ernesto Iadanza	University of Florence, Italy

T12: Neuroengineering, Neurosystems

Miguel Castelo-Branco	University of Coimbra, Portugal
Rita Nunes	University of Lisbon, Portugal

T13: Technologies for Preventive Health Care

Miguel Coimbra	University of Porto, Portugal
Ioanna Chouvarda	Aristotle University of Thessaloniki, Greece

T14: Biomedical Technologies for Developing Countries

Graça Ruano	University of Algarve, Portugal
Martha Zequera	Pontificia Universidad Javeriana, Colombia

T15: Standardization of Open Data

Ratko Magjarevic	University of Zagreb, Croatia

T16: Biomaterials and Tissue Engineering

Birgit Glasmacher	Leibniz Universität Hannover, Germany

Special Sessions

SS01: Optimization in Medicine and Biology

Joana Matos Dias	University of Coimbra, Portugal
Humberto Rocha	University of Coimbra, Portugal

SS02: Electronics and Smart Algorithms for the Effective Lung Monitoring and COPD Management

Nicos Maglaveras	Aristotle University of Thessaloniki, Greece
Andreas Raptopoulos	EXODUS Innovation, Greece
Rui Paiva	University of Coimbra, Portugal

SS03: Non-invasive Temperature Assessment Using Ultrasound

César Teixeira	University of Coimbra, Portugal
André Alvarenga	INMETRO, Rio de Janeiro, Brazil
Wagner Pereira	Federal University of Rio de Janeiro, Brazil

SS04: Computational Biology and Medical Applications

José Ferreira	University of Coimbra, Portugal

SS05: Smartphone-Based, Patient-Centred Technologies

Rute Almeida	CINTESIS, Porto, Portugal
Ana Ferreira	CINTESIS, Porto, Portugal
João Fonseca	University Medical School of Porto, Portugal

SS06: Computational and Experimental Modelling for Designing Bone-Implant Systems

Ana Amaro (Organizer)	University of Coimbra, Portugal

SS07: Artificial Organs: Extracorporeal Blood Circulation Medical Devices

Maria Norberto Pinho	University of Lisbon, Portugal
Mónica Faria	University of Lisbon, Portugal

SS08: Diabetes and Cardiovascular Diseases: Ibero-American Trends

M. Graça Ruano	University of Algarve, Portugal

SS9: Smart Robotic Assistant for Minimally Invasive Surgery: The Smartsurg Project Experience

Sanja Dogramadzi,	University of the West of England, UK
Elana De Momi	Politecnico di Milano, Italy

SS10: Intelligent Computational Systems in Biomedical Engineering

Jan Kubicek	Technical University of Ostrava, Czech Republic
Marek Penhaker	Technical University of Ostrava, Czech Republic
Martin Cerny	Technical University of Ostrava, Czech Republic
Martin Augustynek	Technical University of Ostrava, Czech Republic

SS11: INT4DAT – Intelligent Systems and Technologies for Diagnostic, Assistance and Therapeutics

Joao Ruivo Paulo	University of Coimbra, Portugal
Jerome Mendes	Spanish National Research Council, Spain
Teresa Sousa	University of Coimbra, Coimbra
Eduardo Rocon	University of Coimbra, Portugal

SS12: Upper Limb Exoskeletons for a Better Quality of Life: What Is Real, What is Useful, and What is Next?

Alessandra Pedrocchi	Politecnico di Milano, Italy
Emilia Ambrosini	Politecnico di Milano, Italy
Marta Gandolla	Politecnico di Milano, Italy

SS13: Neurosystems and Connectivity

Miguel Castelo-Branco	University of Coimbra, Portugal
Rita Nunes	University of Lisbon, Portugal

SS14: Therapeutic Applications of Imaging and Neurostimulation

Miguel Castelo-Branco	University of Coimbra, Portugal
Rita Nunes	University of Lisbon, Portugal

SS15: Value-Based Health Technology Assessment

Dan Clark	University of Nottingham, UK
Giuditta Callea	Bocconi University, Italy
Marjan Hummel	Philips Research, Eindhoven, The Netherlands
Martina Andellini	Bambino Gesù Children's Hospital, Italy
Ernesto Iadanza	University of Florence, Italy

SS16: International Collaborative on Medical Devices Assessment

Ernesto Iadanza	University of Florence, Italy
Julie Polisena	Canadian Agency for Drugs and Technologies in Health, Canada
Leandro Pecchia	University of Warwick, UK

SS17: Ocular Imaging

Miguel Morgado	University of Coimbra, Portugal
Rui Bernardes	University of Coimbra, Portugal

SS18: Assessing Human Error in Cognitive/Intellectual Demanding Tasks: Case Study on Software Engineering

Ricardo Couceiro	University of Coimbra, Portugal
Henrique Madeira	University of Coimbra, Portugal

Challenges

IFMBE Scientific Challenge

Marco Simões	University of Coimbra, Portugal
Jorge Henriques	University of Coimbra, Portugal

Fraunhofer Best Portuguese PhD Thesis Competition in Biomedical Engineering

Aurélio Campilho University of Porto, Portugal

Fraunhofer Best Portuguese MsC Thesis Competition in Biomedical Engineering

Aurélio Campilho University of Porto, Portugal

Student Competition

Anna Bianchi Politecnico di Milano, Italy

Young Investigator Competition

Anna Bianchi Politecnico di Milano, Italy

Organizing Committees

General Co-chairs

Mário Secca New University of Lisbon, Portugal
Paulo de Carvalho University of Coimbra, Portugal

Programme Committee

Jorge Henriques University of Coimbra, Portugal
Nuno Neves University of Minho, Portugal

Financial Chair

César Teixeira University of Minho, Portugal

Industrial Track Co-chairs

Miguel Morgado University of Coimbra, Portugal
Hugo Gamboa New University of Lisbon, Portugal

Publicity Co-chairs

Paula Oliveira University of Coimbra, Portugal
Rui Pedro Paiva University of Coimbra, Portugal

Local Arrangement

César Teixeira	University of Coimbra, Portugal
Jorge Henriques	University of Coimbra, Portugal
Márcia Santos	University of Coimbra, Portugal
Miguel Castelo-Branco	University of Coimbra, Portugal
Miguel Morgado	University of Coimbra, Portugal
Paula Oliveira	University of Coimbra, Portugal
Ricardo Couceiro	University of Coimbra, Portugal
Rui Pedro Paiva	University of Coimbra, Portugal

International Advisory Board

Aurélio Campilho	University of Porto, Portugal
Constantinos S. Pattichis	University of Cyprus, Cyprus
Damijan Miklavčič	University of Ljubljana, Slovenia
Dimitrios Fotiadis	University of Ioannina, Greece
James Goh	National University of Singapore, Singapore
José Princípe	University of Florida, USA
Laura Roa	University of Seville, Spain
Lenka Lhotská	Czech Technical University in Prague, Czech Republic
Luis Kun	National Security at CHDS at NDU, USA
Martha Zequera	Pontificia Universidad Javeriana, Colombia
Nitish Thakor	Johns Hopkins University, USA
Pablo Laguna	Universidad de Zaragoza, Spain
Ratko Magjarevic	University of Zagreb, Croatia
Shankar Krishnan	Wentworth Institute of Technology, USA
Timo Jämsä	University of Oulu, Finland
YT Zhang	Chinese University of Hong Kong, Hong Kong

International Programme Committee

Aurélio Campilho	University of Porto, Portugal
Adam Idzkowski	Bialystok University of Technology, Poland
Ákos Jobbágy	Budapest University of Technology and Economics, Hungary
Alan Murray	Newcastle University, UK
Ana Castro	University of Porto, Portugal
Ana Mendonça	University of Porto, Portugal
Ana Paula Rocha	University of Aveiro, Portugal
Ana Paula Rocha	University of Porto, Portugal
Andres Santos	Universidad Politécnica de Madrid, Spain
Andriana Prentza	University of Piraeus, Greece

Marjan Hummel	University of Twente, Netherlands
Martin Cerny	VSB - Technical University of Ostrava, Czech Republic
Martina Andellini	Bambino Gesù Children's Hospital, Italy
Maurizio Schmid	Roma Tre University, Italy
Michal Gála	University of Žilina, Slovakia
Michela Comune	Tel Aviv University, Israel
Miguel Amador	University of Lisbon, Portugal
Miguel Caixinha	University of Beira Interior, Portugal
Miguel Coimbra	University of Porto, Portugal
Monique Frize	Carleton University, Canada
Ofer Barnea	Tel Aviv University, Israel
Olof Lindahl	Umeå University and Luleå University of Technology, Sweden
Panagiotis Bamidis	Aristotle University of Thessaloniki, Greece
Panayiotis Kyriacou	City University London, UK
Paulo Crespo	University of Coimbra, Portugal
Paulo de Carvalho	University of Coimbra, Portugal
Paulo Maia	INESC TEC, Portugal
Paulo Mendes	University of Minho, Portugal
Robert Allen	University of Southampton, UK
Romuald Jolivot	Bangkok University, Thailand
Rui Bastos	University of Minho, Portugal
Rui Bernardes	University of Coimbra, Portugal
Samuel Silva	University of Aveiro, Portugal
Selma Supek	University of Zagreb, Croatia
Simão Paredes	Polytechnic Institute of Coimbra, Portugal
Sofia Rita Fernandes	University of Lisbon, Portugal
Sotiris Pavlopoulos	National Technical University of Athens, Greece
Spyros Kitsiou	University of Illinois at Chicago, USA
Stathis Konstantinidis	University of Nottingham, UK
Stavros Karkanis	University of Thessaly, Greece
Stavroula Mougiakakou	University of Bern, Germany
Stergiani Spyrou	Aristotle University of Thessaloniki, Greece
Styliani Petroudi	University of Cyprus, Cyprus
Stylianos Hatzipanagos	University of West London, UK
Susana Brás	University de Aveiro, Portugal
Susana Catarino	University of Minho, Portugal
Telemachos Stamkopoulos	University Ecclesiastical Academy of Thessaloniki, Greece
Teresa Rocha	Polytechnic Institute of Coimbra, Portugal
Teresa Sousa	University of Coimbra, Portugal
Themis Exarchos	University of Ioannina, Greece

Thomas Penzel	Charite Universitätsmedizin Berlin, Germany
Tomasz Soltysinski	Warsaw University of Technology, Poland
Tomaz Jarm	University of Ljubljana, Slovenia
Vassilis Koutkias	Centre for Research and Technology Hellas, Greece

Contents

Special Sessions: Intelligent Computational Systems in Biomedical Engineering

Special Sessions: INT4DAT - Intelligent Systems and Technologies for Diagnostic, Assistance, and Therapeutics

Special Sessions: Value-Based Health Technology Assessment

Special Sessions: International Collaborative on Medical Devices Assessment

Special Sessions: Ocular Imaging

Challenges: IFMBE Scientific Challenge Competition

Challenges: Fraunhofer Best Portuguese PhD Thesis Competition in Biomedical Engineering

Challenges: Fraunhofer Best Portuguese MsC Thesis Competition in Biomedical Engineering

Regular Sessions: Biomedical Signal Processing

Using Eye Tracking to Analyze Surgeons' Cognitive Workload During an Advanced Laparoscopic Procedure

Juan Francisco Ortega-Morán[1(✉)], J. Blas Pagador[1], Vicente Luis-del-Campo[2], Juan Carlos Gómez-Blanco[1], and Francisco M. Sánchez-Margallo[1]

[1] Jesús Usón Minimally Invasive Surgery Centre, Ctra N-521, km. 41,8, 10071 Cáceres, Spain
jfortega@ccmijesususon.com

[2] Laboratory of Learning and Motor Control, Faculty of Sports Sciences, UNEX, Avd. Universidad s/n, 10003 Cáceres, Spain

Abstract. Introduction: Surgeons' cognitive workload should be controlled during a surgical intervention for a successful and safe performance. Eye tracking technologies could be used for cognitive workload monitoring by analyzing the pupil size and blink frequency of surgeons. This work aims to study the surgeons' cognitive workload watching an advanced laparoscopic video. Methods: 16 surgeons (5 experts, 11 novices) participated in the study watching a colectomy video consisting of eight steps with introductory titles. Surgeons' gaze was recorded with eye tracking glasses while visualizing the video, from which pupil size and blink frequency were analyzed. Results: Pupil diameter of surgeons increased during the visualization of steps and decreased during the titles. Two specific steps of the intervention produced the highest pupil diameter of surgeons. When the pupil diameter increased the blink frequency decreased. Pupil diameter tended to decrease as the video is watched, which is mainly due to expert surgeons, who had a lower pupil diameter than novices. Conclusions: Eye tracking technologies allow monitoring the cognitive workload of surgeons in surgical procedures. Larger pupil size and shorter blink frequency means greater cognitive workload. Such metrics could be used to objectively label the difficult tasks within the surgical procedure. Surgical videos used for training of surgeons should be short, according to micro-learning, since cognitive workload decreases over time while visualizing them. Based on proposed metrics, eye tracking technologies could be used to distinguish the level of experience of surgeons, since cognitive workload is sensitive to the skill level of surgeons.

Keywords: Eye tracking · Cognitive workload · Minimally invasive surgery

1 Introduction

Eye tracking is a powerful tool for objective assessment of visual search strategies in different professional domains (e.g., music, sport, military settings, clinical rehabilitation, and surgery). This technology allows for recording the eye movements and point

J. Henriques et al. (Eds.): MEDICON 2019, IFMBE Proceedings 76, pp. 3–12, 2020.
https://doi.org/10.1007/978-3-030-31635-8_1

of gaze fixation [1]. For example, the gaze has been used in different tasks to address in action observation, action execution, and movement imagery [2]. Eye trackers have been the most common material used to analyze visual behaviors because they provide specific information about the direction of the gaze, the dilation of the pupil, and also the testing of mental processes [3].

Some studies have assessed differences in eye metrics between expert and novices surgeons in open surgery [4] or in basic operations of minimally invasive surgery [5]. In this line, the research has concluded that expert surgeons displayed more fixations on key Areas of Interest (AOI) during basic interventions [6]. The quiet eye (QE) seems to be another discriminatory variable of the skill level in surgery. For example, expert surgeons showed longer QEs during the identification and dissection of the recurrent laryngeal nerve [7]. Recently, we have already performed a study [8] addressing the visual patterns of expert and novice surgeons during the perception of an advanced laparoscopic procedure (i.e., watching a video of a colectomy). Results from this study showed that: (1) experts had more fixations and longer fixation times than novices; (2) expert surgeons focused their fixations on the target of each step of the operation, avoiding looking towards other points; and (3) novice surgeons had a gaze pattern more dispersed.

The pupil dilation and blink rate are also two ocular metrics used to analyze cognitive workload. These variables are related to the cognitive development and plasticity. Pupil dilation is related to the arousal and attention but also with the task difficulty and mental effort. Instead, the blinks could reveal underlying processes to the goal-directed behavior [9]. It is important to take into account the cognitive workload of surgeons because it has numerous implications for safe surgery. Overloaded surgeons may have difficulties in maintaining stable performance in the operating room, or surgeon may lose vigilance to signs of life-threatening conditions during the surgery [10]. Previous studies have indicated that pupillary response is very sensitive to the change of workloads and that pupil size increased with the difficulty in sequential surgical tasks [10], and also that the reduction of blink frequency and shorter blink duration matched the increasing level of mental workload of surgeons [11]. Therefore, the pupil size and blinks measurements can be used for monitoring the mental workload of a surgeon. Studies have been carried out to measure the cognitive workload of experts and novices in advanced surgical tasks, but in a simulator [12], not in live surgical procedures or visualizing them on video.

In this work, we aim to address the cognitive workload of surgeons of different expertise watching a laparoscopic colectomy video, which is an advanced laparoscopic procedure, based on eye tracking metrics related to pupillary response and blink rate.

2 Methods

Sixteen surgeons were recruited to voluntarily participate in the study, of which 5 could be considered as expert surgeons (more than 50 laparoscopic interventions of colectomy attended) and 11 as novices (without experience in laparoscopy). The same experimental procedure was performed with all surgeons to carry out the tests: (1) the participant was informed about the objectives of the study, following a single-blind method to avoid bias in the results; (2) the participant was seated in front of a PC laptop (Fig. 1) inside of a quiet room without noise and with dim light; (3) a 3-point calibration of the eye tracker was carried out to the participant; (4) the colectomy video was shown to the surgeons while their gaze was being recorded with the SMI eye tracking glasses 2w (SensoMotoric Instruments, SMI).

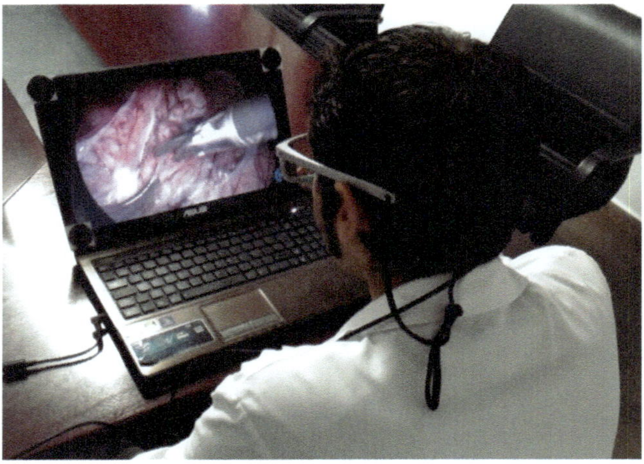

Fig. 1. A surgeon performing the study

The didactic video consisted of a laparoscopic colectomy intervention divided into eight steps, with a brief title and a medical illustration summarizing each step (Fig. 2):

- Step 1 - Exposure of the surgical field, the opening of the mesentery and dissection of the inferior mesenteric artery
- Step 2 - Identification of the left ureter
- Step 3 - Occlusion of the inferior mesenteric artery
- Step 4 - Liberation of the mesorectum
- Step 5 - Superior release of the proximal colic segment
- Step 6 - Distal colic section
- Step 7 - Resection of the piece and anvil arrangement in the proximal segment
- Step 8 - End-to-end circular anastomosis

Subsequently, recordings were analysed with the SMI BeGaze software (SensoMotoric Instruments, SMI) and different metrics related to the surgeons' gaze behaviour were extracted, such as the pupil size and blink frequency of surgeons. Statistical analysis of results was performed with the R software for statistical computing for Windows (https://www.r-project.org/). Analysis by the Shapiro-Wilk test was performed to study the normality distribution of results between expert and novice surgeons. Consequently, the parametric t-Student test was performed to analyse the significant differences between experts and novices in those cases with normal distribution, and the non-parametric Mann-Whitney U (or Wilcoxon) test for non-normal distributions (p-value < 0.05 as statistical significance).

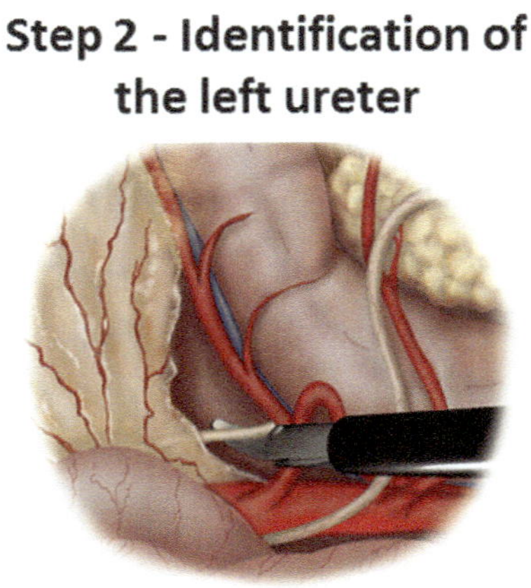

Fig. 2. Title and illustration summarizing step 2

3 Results

3.1 Results from All Participants (Pupil Size and Blinks)

After applying an offset correction of the gaze in those recording required, the results corresponding to the average pupil diameter of participants during the visualization of the video are shown in Fig. 3:

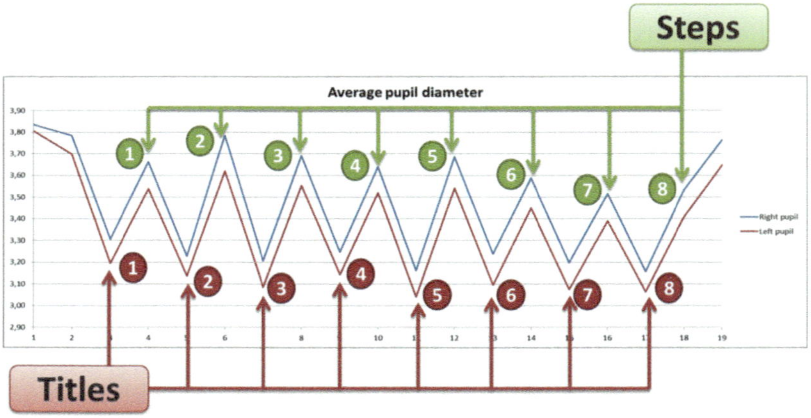

Fig. 3. Average pupil diameter of participants during the visualization of the video

The Fig. 3 shows that the average pupil diameter of surgeons increases during the visualization of steps, and decreases during the visualization of the titles previous of each step.

Specifically, the average pupil diameter during the visualization of steps increases in step 2 (3.78 ± 0.62 mm, 3.62 ± 0.56 mm, right and left pupil, respectively) and step 5 (3.69 ± 0.62 mm, 3.54 ± 0.57 mm, right and left pupil, respectively), and it reaches its minimum value in step 7 (3.52 ± 0.59 mm, 3.39 ± 0.54 mm, right and left pupil, respectively) (Fig. 4). Moreover, the average pupil diameter tends to decrease over time as the video is watched.

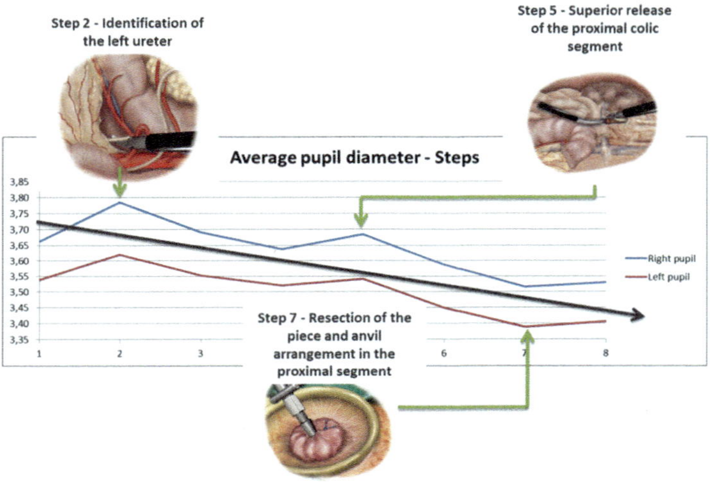

Fig. 4. Average pupil diameter during the visualization of steps

Results also show that when the pupil diameter increases, the blink frequency decreases, and vice versa (Fig. 5).

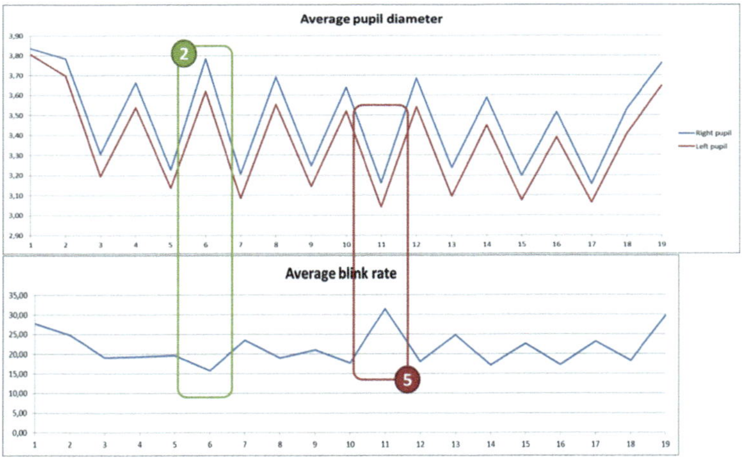

Fig. 5. Average pupil diameter vs average blink rate

We can check it better in extreme cases, when in step 2 the average pupil diameter reaches the maximum value (3.78 ± 0.62 mm, 3.62 ± 0.56 mm, right and left pupil, respectively) and the average blink frequency reaches the minimum value (15.73 ± 18.79 blinks/min). Contrary, in the title of step 5 the average pupil diameter reaches the minimum value (3.16 ± 0.54 mm, 3.04 ± 0.50 mm, right and left pupil, respectively) and the average blink frequency reaches the maximum value (31.44 ± 25.06 blinks/min).

3.2 Comparison Between Expert and Novice Surgeons (Pupil Size)

The analysis comparing skill levels shows that novice surgeons have higher average pupil diameter during the total duration of the video, with only significant differences during visualization of title 2, title 5 and step 6 (Tables 1 and 2).

Table 1. Average pupil diameter (mm) of expert and novice surgeons during steps

	Right pupil			Left pupil		
	Experts	Novices		Experts	Novices	
	Mean ± SD	Mean ± SD	p	Mean ± SD	Mean ± SD	p
Step 1	3.46 ± 0.55	3.76 ± 0.63	0.356	3.29 ± 0.50	3.65 ± 0.52	0.219
Step 2	3.51 ± 0.58	3.91 ± 0.63	0.249	3.34 ± 0.53	3.75 ± 0.55	0.189
Step 3	3.45 ± 0.60	3.80 ± 0.64	0.322	3.27 ± 0.53	3.68 ± 0.56	0.199
Step 4	3.34 ± 0.65	3.78 ± 0.66	0.246	3.18 ± 0.55	3.68 ± 0.57	0.134
Step 5	3.36 ± 0.59	3.84 ± 0.60	0.172	3.18 ± 0.54	3.70 ± 0.53	0.110
Step 6	3.21 ± 0.54	3.76 ± 0.72	0.121	3.01 ± 0.44	3.65 ± 0.63	0.039*
Step 7	3.27 ± 0.45	3.63 ± 0.64	0.227	3.12 ± 0.43	3.51 ± 0.56	0.156
Step 8	3.25 ± 0.53	3.66 ± 0.69	0.230	3.12 ± 0.51	3.54 ± 0.56	0.181

Table 2. Average pupil diameter (mm) of expert and novice surgeons during titles

	Right pupil			Left pupil		
	Experts	Novices		Experts	Novices	
	Mean ± SD	Mean ± SD	p	Mean ± SD	Mean ± SD	p
Title 1	3.20 ± 0.53	3.36 ± 0.42	0.571	3.05 ± 0.43	3.26 ± 0.48	0.407
Title 2	2.97 ± 0.27	3.34 ± 0.57	0.097	2.84 ± 0.30	3.27 ± 0.49	0.049*
Title 3	2.91 ± 0.29	3.34 ± 0.60	0.076	2.79 ± 0.30	3.22 ± 0.53	0.059
Title 4	3.00 ± 0.39	3.36 ± 0.54	0.162	2.85 ± 0.33	3.28 ± 0.49	0.063
Title 5	2.88 ± 0.41	3.29 ± 0.55	0.048*	2.72 ± 0.32	3.18 ± 0.51	0.180
Title 6	2.99 ± 0.35	3.35 ± 0.61	0.166	2.83 ± 0.31	3.22 ± 0.51	0.089
Title 7	2.91 ± 0.40	3.33 ± 0.63	0.138	2.79 ± 0.44	3.21 ± 0.50	0.126
Title 8	2.90 ± 0.33	3.27 ± 0.56	0.118	2.77 ± 0.33	3.20 ± 0.48	0.064

To highlight that during the steps of the procedure, the differences between experts and novice surgeons regarding average pupil diameter is higher (Fig. 6).

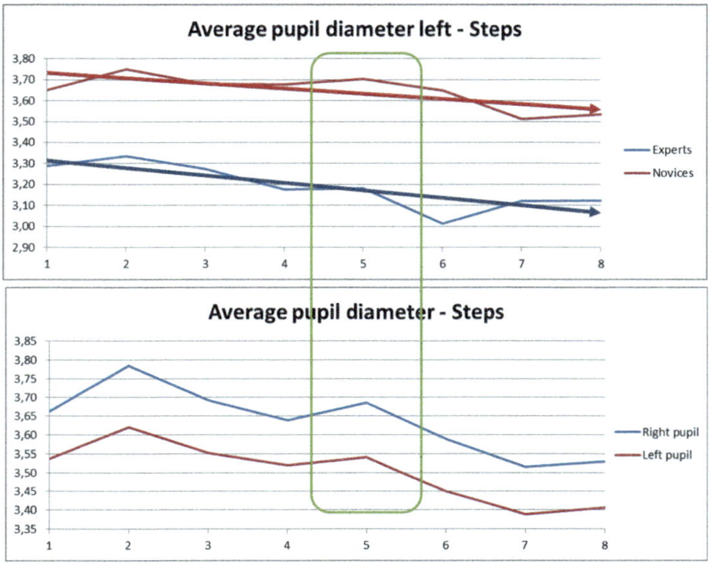

Fig. 6. Average pupil diameter of left eye during the visualization of steps: Experts vs Novices

Comparing these results with the global results taking into account all surgeons (Fig. 6), the increase of pupil diameter of surgeons in step 5 is mainly due to novice surgeons. And moreover, the trend of pupil diameter to decrease during the video is mainly due to expert surgeons.

4 Discussion and Conclusions

The objective of the study was to address different ocular metrics related to the cognitive workload in surgeons with different skill expertise during an advanced laparoscopic procedure. This study shows that eye tracking technologies (through the pupillary response and blink frequency) allow checking the cognitive workload of surgeons in surgical procedures. According to previous studies [9], larger pupil size and shorter blink frequency mean greater cognitive workload.

To highlight that the surgeons' pupil diameter was constrained by the visualization of the information displayed in the video (steps vs titles) or by the type of intervention steps. These changes in pupil diameter were accompanied by a blink rate activity. The study has shown that pupil diameter of surgeons was higher during the visualization of the steps of the intervention than during the reading of the titles. It is expected that surgeons have more cognitive workload visualizing the steps than the titles because they have to process more information and pay more attention. Therefore, our study provides an empirical evidence that the pupil size increases with the cognitive workload, and vice versa.

On the other hand, and although the number of participants must be increased in future studies, it is feasible to think that the information provided by eye tracking metrics could be used to objectively label the difficult tasks or steps within a surgical procedure. In this regard, and concretely in case of the advanced laparoscopic procedure of colectomy, step 2 (identification of the left ureter) and step 5 (superior release of the proximal colic segment) causes the highest pupil size of surgeons, and step 7 (resection of the piece and anvil arrangement in the proximal segment) the lowest pupil diameter. This means that steps 2 and 5 supposes the highest cognitive workload to surgeons, so we can reason that these two steps are the most difficult for surgeons during the intervention of colectomy because they are crucial steps to ensure an expected outcome of the surgical procedure. Therefore, it is plausible to find an increase of the pupil diameter, what it would mean that the surgeons process more information during these temporal moments of the intervention. In this way, this could be useful for concreting what steps of different surgical procedures imply further cognitive efforts in surgeons, in order to train these stressful and high mental situations. Moreover, these results could help the designers and developers of training courses of such surgical intervention to balance the resources (both material and human) used to teach each step of the intervention, giving priority to the most difficult steps using more resources in them, so that they can be somewhat easier for surgeons to learn, making them more attractive and not tedious in their learning.

Results show that the pupil diameter, and consequently the cognitive workload, of surgeons tends to decrease during the visualization of the video. This could be explained because the video lasts a lot, and surgeons lose attention over time. Therefore, as surgical videos are increasingly been used as resources for training in e-learning platforms [13], these should be short videos, according to the new trend in learning, such as micro-learning. The micro-learning is the future of training in the workplace [14], being adequate for participants who cannot devote much time to their training, as is the case of surgeons and other health personnel. Micro-learning is based on the creation of short training content that is within the average attention span of adults [14], and it is often delivered in video format [15] because the video is the most effective medium for communicating information in a short period of time. The optimal length of videos may not last more than six minutes [16], usually between three and five minutes.

As mentioned above, cognitive workload of surgeons decreases during the reading of the titles previous to each step. However, these titles are highly recommended to be included during educational videos to better understand the procedure to which they refer, especially for novice surgeons. And moreover, what is advisable is to incorporate audio into the videos, since the student working memory load can be minimized if visual and verbal inputs are received simultaneously and thus they would be able to 'select relevant information, to build referential connections, and to build integrated mental models for transfer of knowledge' [17].

Finally, results pointed that the cognitive workload, measured through the pupil diameter and blink rates, is sensitive to the skill level of the surgeons. In this line, novices have higher cognitive workload than experts in such advances surgical procedures, so this may further lead to the innovation of designing task-specific training for improving training outcomes in a shorter time. Moreover, such results suggest that eye tracking technologies could be used to distinguish the level of experience of surgeons based on the proposed metrics, such as the pupillary response and blink frequency. However, and as indicated by some previous study [18], machine learning algorithms could differentiate expertise based on workload as indicated by pupil diameter, but this cannot assure that novices and experts are gathering the same visual information for training and assessment. For that, it is recommended to jointly study the cognitive workload and the visual behavior of surgeons, what will be analyzed more in-depth in future works.

Conflict of Interest. The authors declare that they have no conflict of interest.

References

1. Edmondson, M.J., Pucher, P.H., Sriskandarajah, K., Hoare, J., Teare, J., Yang, G.Z., Darzi, A., Sodergren, M.H.: Looking towards objective quality evaluation in colonoscopy: analysis of visual gaze patterns. J. Gastroenterol. Hepatol. **31**(3), 604–609 (2016)
2. Causer, J., McCormick, S.A., Holmes, P.S.: Congruency of gaze metrics in action, imagery and action observation. Front. Hum. Neurosci. **7**, 604 (2013)
3. Liu, H.C., Chuang, H.H.: An examination of cognitive processing of multimedia information based on viewers' eye movements. Interact. Learn. Environ. **19**(5), 503–517 (2011)
4. Tien, T., Pucher, P.H., Sodergren, M.H., Sriskandarajah, K., Yang, G.Z., Darzi, A.: Differences in gaze behaviour of expert and junior surgeons performing open inguinal hernia repair. Surg. Endosc. **29**(2), 405–413 (2015)
5. Wilson, M.R., McGrath, J.S., Vine, S.J., Brewer, J., Defriend, D., Masters, R.S.: Perceptual impairment and psychomotor control in virtual laparoscopic surgery. Surg. Endosc. **25**(7), 2268–2274 (2011)
6. Khan, R.S., Tien, G., Atkins, M.S., Zheng, B., Panton, O.N., Meneghetti, A.T.: Analysis of eye gaze: do novice surgeons look at the same location as expert surgeons during a laparoscopic operation? Surg. Endosc. **26**(12), 3536–3540 (2012)
7. Harvey, A., Vickers, J.N., Snelgrove, R., Scott, M.F., Sheila Morrison, S.: Expert surgeon's quiet eye and slowing down: expertise differences in performance and quiet eye duration during identification and dissection of the recurrent laryngeal nerve. Am. J. Surg. **207**(2), 187–193 (2014)
8. Ortega Morán, J.F., Toribio Silva, A., Pagador, J.B., Sánchez González, P., Gómez, E.J., Sánchez Margallo, F.M.: Eye gaze behaviour watching an advanced laparoscopic procedure: experts versus novices comparison. Int. J. Comput. Assist. Radiol. Surg. (IJCARS) **12**(1), 133–135 (2017)
9. Eckstein, M.K., Guerra-carrillo, B., Singley, A.T.M., Bunge, S.A.: Beyond eye gaze: what else can eyetracking reveal about cognition and cognitive development? Dev. Cogn. Neurosci. **25**, 69–91 (2016)
10. Zheng, B., Jiang, X., Atkins, M.S.: Detection of changes in surgical difficulty: evidence from pupil responses. Surg. Innov. **22**(6), 629–635 (2015)
11. Zheng, B., Jiang, X., Tien, G., Meneghetti, A., Panton, O.N.M., Atkins, M.S.: Workload assessment of surgeons: correlation between NASA TLX and blinks. Surg. Endosc. **26**(10), 2746–2750 (2012)

12. Zheng, B., Cassera, M.A., Martinec, D.V., Spaun, G.O., Swanström, L.L.: Measuring mental workload during the performance of advanced laparoscopic tasks. Surg. Endosc. **24**(1), 45 (2010)
13. Ortega-Morán, J.F., Pagador, J.B., Sánchez-Peralta, L.F., Sánchez-González, P., Noguera, J., Burgos, D., Gómez, E.J., Sánchez-Margallo, F.M.: Validation of the three web quality dimensions of a minimally invasive surgery e-learning platform. Int. J. Med. Inform. **107**, 1–10 (2017)
14. Callisen, L.: Why Micro Learning Is The Future Of Training In The Workplace, eLearning Industry. http://elearningindustry.com/micro-learning-future-of-training-workplace. Accessed 19 Mar 2019
15. Avery, A.: What is the Impact of Blended Learning Including Micro-Learning on Manager Learning and Behavior Change vs. Impact of Classroom Learning? (2016)
16. Optimal Video Length for Training Videos, analysis by Philip Guo, Xcelus. http://www.xcelus.com/optimal-video-length-training-videos/. Accessed 19 Mar 2019
17. Kon, H., Botelho, M.G., Bridges, S., Leung, K.C.M.: The impact of complete denture making instructional videos on self-directed learning of clinical skills. J. Prosthodont. Res. **59**(2), 144–151 (2015)
18. Fichtel, E., Lau, N., Park, J., Parker, S.H., Ponnala, S., Fitzgibbons, S., Safford, S.D.: Eye tracking in surgical education: gaze-based dynamic area of interest can discriminate adverse events and expertise. Surg. Endosc. **33**, 1–8 (2018)

Application of Multivariate Spectral F Test for Somatosensory Evoked Response Detection

Karina Miranda Boson[1,2](✉), Antonio Mauricio Ferreira Leite Miranda de Sá[3], and Danilo Barbosa Melges[1,4]

[1] Electrical Engineering Department, Federal University of Minas Gerais, Belo Horizonte, MG 31270-901, Brazil
karinamboson@gmail.com
[2] Physical Therapy Undergraduate Course, Faculty of Medical Sciences of Minas Gerais, Belo Horizonte, MG 30130-110, Brazil
[3] Biomedical Engineering Program, Federal University of Rio de Janeiro, Rio de Janeiro, RJ 21941-914, Brazil
[4] Graduate Program in Electrical Engineering, Federal University of Minas Gerais, Belo Horizonte, MG 31270-901, Brazil

Abstract. Somatosensory Evoked Potential (SEP) is an important tool for monitoring vascular and spine surgeries, and other clinical applications. However, morphological SEP identification is subjective. Then, statistical techniques, such as Local Spectral F Test (SFT), have been used for response detection. The multivariate extension of SFT employs more than one derivation and has been recently considered advantageous to identify response to visual stimulation. This work aims at evaluating the performance of Multivariate SFT (MSFT) applied to EEG signals from 40 volunteers during stimulation at 5 Hz and different numbers of derivations (*N*), comparing the detection rates (DR). Frequencies of interest f_{o1} = 15 Hz and f_{o2} = 100 Hz were used, as well as $L = 6$ neighbor components at the frequencies from 70 to 95 Hz and a 5%-significance level. The number of derivations varied from $N = 1$ to 6. The detection rates obtained using f_{o1} were higher than those with f_{o2}, which corresponds to false positives, since no response is expected to occur at such frequency. For f_{o1}, half of the volunteers exhibited a monotonic increase of DR as *N* was augmented. For other volunteers, an oscillatory pattern was noticed in DR as new derivations were added, suggesting that raising *N* do not necessarily lead to improvement in MSFT performance. For f_{o2}, raising *N* caused an increase in the false-positives rate above significance level of 5%. This could be explained in part by the correlation among the employed derivations. Finally, MSFT showed promising results at SEP identifying.

Keywords: Somatosensory evoked potential · Multivariate local spectral F test · Objective response detection

J. Henriques et al. (Eds.): MEDICON 2019, IFMBE Proceedings 76, pp. 13–21, 2020.
https://doi.org/10.1007/978-3-030-31635-8_2

1 Introduction

The somatosensory evoked potential (SEP) has been used as an electrophysiological tool to assess functional integrity of pathways from the point of stimulation up to the cortex. The applications include surgical monitoring of vascular (e.g.: carotid endarterectomy) [1] and spine (e.g.: scoliosis fusion) [2]. However, morphological SEP identification is subjective and dependent of the specialist expertise in SEP analysis.

Thus, strategies of response detection based on statistical criteria have been used by means of Objective Response Detection (ORD) techniques [3]. One of such techniques is the Local Spectral F Test (SFT), which has already been applied to auditory and visual evoked potential detection. More recently, a multivariate version of SFT was derived and uses more than one derivation in order to increase the probability of detection in the presence of stimulation response with very low signal-to-noise ratio. In fact, the Multivariate SFT (MSFT) has been considered advantageous to identify response to visual stimulation [4].

Thus, this work aims at evaluating the performance of MSFT applied to EEG signals during somatosensory stimulation, being the detection rates compared for sets with different numbers of EEG derivations employed to estimate MSFT.

2 Methodology

2.1 Local Spectral F Test (SFT)

Zurek [5] described the Local Spectral F Test as the ratio between the power in the frequency component where response is expected (f_o) and the mean power of L neighbor components:

$$\widehat{\emptyset}_y(f_o) = \frac{|Y(f_o)|^2}{\frac{1}{L}\sum_{\substack{i=o-L/2 \\ i\neq 0}}^{o+L/2} |Y(f_i)|^2} \tag{1}$$

where $Y(f_o)$ is the Discrete Fourier Transform (DFT) in frequency f_o and $Y(f_i)$ is the DFT at the ith neighboring frequency to f_o.

The critical value related to the statistical distribution for a chosen significance level α, constitutes the detection threshold. Thus, according to Miranda de Sá *et al.* [6], response detection is determined if $\widehat{\emptyset}_y(f_o)$ is greater than the critical value, which is given as:

$$\widehat{\emptyset}_{ycrit}(f_o) = F_{crit2,2L,\alpha} \tag{2}$$

where $F_{crit2,2L,\alpha}$ is the critical value of Fisher F-distribution with 2 and $2L$ degrees of freedom for a significance level α.

2.2 Multivariate Local Spectral F Test (MSFT)

The extension of the Local SFT for the multivariate case, considers a system with only one input (stimulus signal) and N affected outputs (EEG derivations). For this case, the MSFT is obtained by the ratio between the mean power of N derivations in the interest frequency f_o and the average of the N mean power spectrum estimates of L neighboring frequencies, i.e. [4]:

$$\widehat{\emptyset}_N(fo) = \frac{\sum_{j=1}^{N} |Y_j(fo)|^2}{\sum_{j=1}^{N} \left[\frac{1}{L} \sum_{\substack{i=o-L/2 \\ i \neq 0}}^{o+L/2} |Y_j(f_i)|^2 \right]} \quad (3)$$

where $Y_j(f_o)$ is the DFT of jth EEG signal at the frequency f_o and $Y_j(f_i)$ is the DFT of the ith neighboring frequency to f_o of signal $y_j[n]$.

Thus, according to Rocha *et al.* [4], the detection threshold can be calculated from the critical value of F-distribution with $2N$ and $2NL$ degrees of freedom and significance level α:

$$\widehat{\emptyset}_{Ncrit}(f_o) = F_{crit2N,2NL,\alpha} \quad (4)$$

2.3 EEG Acquisition

EEG signals were collected during somatosensory stimulation from adult volunteers without history of neurological diseases. These signals were recorded according to the 10–20 International System and with all derivations referenced to the earlobe average, using EEG BNT-36 (EMSA, www.emsamed.com.br, Brazil). All volunteers were positioned in supine position with eyes closed, while the stimuli were applied, by means of current pulses with 200 μs duration, to the right tibial nerve at a 5-Hz nominal frequency (real frequency is 4.83 Hz) with motor threshold intensity (lower intensity capable of producing involuntary hallux movements) using the Atlantis Four (EMSA). Further details on the acquisition can be found in Melges *et al.* [3].

Silver electrodes were used to record the EEG and gold electrodes for stimulation, with the ground electrode positioned at the popliteal fossa. The stimulation protocol consisted of the application of 1000 to 1400 stimuli.

The collected EEG signals were analogically bandpass filtered from 0.5 to 100 Hz and digitized with 16-bits resolution and 600 Hz of sampling frequency. The trigger channel, synchronized with stimulation signal, was also recorded.

The research was approved by the Research Ethics Committee of the Clementino Fraga Filho Hospital of Federal University of Rio de Janeiro (UFRJ) (CEP-HUCFF: 114/05), and all patients gave written informed consent to participate.

2.4 Pre-processing

Signals were segmented into 200 ms epochs (inter-stimuli period) synchronized with the stimulation, so that the spectral resolution corresponded to the stimulation frequency.

Knowing that in the first milliseconds of the SEP a broadband stimulus artifact is present, the first and last 7 ms of each epoch were discarded, while the other 186 ms remained. In addition, in order to eliminate windows with high noise influence, epochs in which signal amplitude exceeded 100 μV for more than 10% of the samples were rejected.

2.5 Processing

The Multivariate Local SFT was applied to EEG signals during the somatosensory stimulation previously described, in order to investigate the detection rate for different number of derivations.

The values of $\hat{\phi}_N$ and $\hat{\phi}_{Ncrit}$ were calculated (as described in Sect. 2.2), using f_o = 15 Hz and f_o = 100 Hz as frequencies of interest, L = 6 frequency components from 70 to 95 Hz and significance level α = 5%. The number of derivations used varied from N = 1 (univariate SFT) to N = 6, being gradually added in the following order: P3, C3, P4, C4, Pz and Cz. These derivations were selected because they are regions where the SEP is recorded with higher signal-to-noise ratio, and f_o values were chosen in order to compare detection rates at a frequency in which response to stimulation is expected (15 Hz) and another in which it is not (100 Hz) [7]. Then, for each volunteer, it was verified whether the value of $\hat{\phi}_N$ exceeded $\hat{\phi}_{Ncrit}$, indicating evoked response detection, or not, case in which the null hypothesis of no response would be accepted (absence response to stimulation).

Finally, to verify the performance of the local SFT according to the variation of the frequency of interest, the technique was applied to all volunteers using only derivation P3 (univariate SFT) and the response detection rate was calculated varying the frequency f_o from 5 to 100 Hz.

3 Results

Figure 1a illustrates the use of Multivariate Local SFT applied to C3 and C4 to detect response at frequency f_o = 15 Hz for 884 epochs of volunteer #36. It was observed that practically all $\hat{\phi}_N$ values exceeded the critical value $\hat{\phi}_{Ncrit}$, indicating detection for most epochs and a good sensitivity of the technique. On the other hand, when the multivariate local SFT was applied using f_o = 100 Hz for the same derivations of the same volunteer, few values of $\hat{\phi}_N$ exceeded $\hat{\phi}_{Ncrit}$ among all EEG epochs considered (Fig. 1b).

Figure 2a shows the distribution of detection rates (DR) for all volunteers at frequency f_o = 15 Hz for each set of derivations. As can be seen, the median of DR exceeds 90% from addition of a second derivation, but even when univariate SFT is employed, the median of DR is close to 90%. Some outliers are also observed for all sets of derivations. In addition, there is a reduction of the interquartile range with the increase of number of derivations, and also a subtle increase in the median value. The medians and interquartile ranges (IQR) values are summarized in Table 1a.

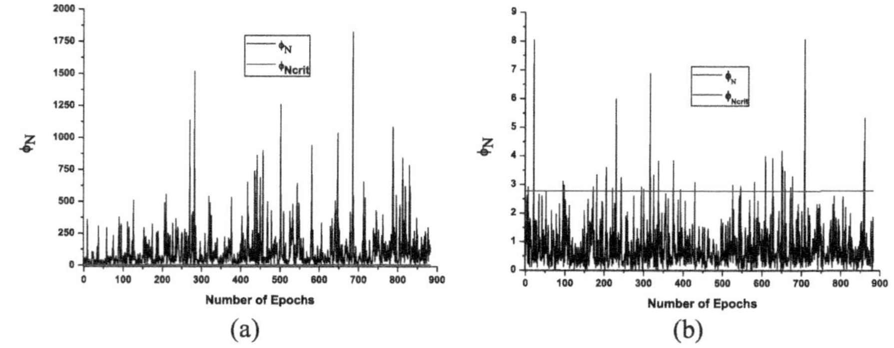

Fig. 1. Critical value $\hat{\phi}_{Ncrit}$ (red horizontal line) and $\hat{\phi}_N$ values obtained in the application of the Multivariate Local SFT for the EEG derivation set C3 and C4 of the volunteer # 36, using (a) f_o = 15 Hz, and (b) f_o = 100 Hz.

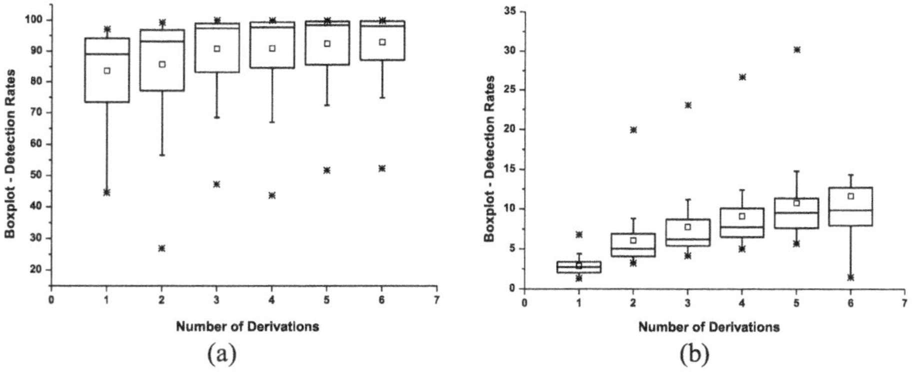

Fig. 2. Distribution of evoked response detection rates as a function of the number of derivations for (a) f_o = 15 Hz, and (b) f_o = 100 Hz.

The distribution of detection rates for frequency f_o = 100 Hz, for which no response detection is expected, is shown in Fig. 2b. As can be observed, in the multivariate case ($N \geq 2$), the median false-positive rate (false detections) exceeds the significance level of the statistical test (5%) involved in the studied technique (MSFT). In addition, for each set of derivations, there are outliers with percentages above 5%. The medians and interquartile ranges (IQR) are shown in Table 1b. As can be noted, there is a gradual increase of false detections for f_o = 100 Hz, the median for N = 6 being around twice the level of significance.

Table 1. Median and IQR of detection rates for each set of derivations at the frequency of (a) f_o = 15 Hz, and (b) f_o = 100 Hz.

	Derivations	D1	D2	D3	D4	D5	D6
(a)	**IQR**	18.23	19.68	15.76	14.76	13.34	12.35
	Median	89.36	94.18	97.76	97.95	98.54	98.48
	Minimum	44.45	26.92	47.24	43.63	51.76	52.35
	Maximum	97.11	99.32	100	100	100	100
(b)	**IQR**	1.41	2.79	3.32	3.57	4.17	4.80
	Median	2.81	5.31	6.72	7.80	9.89	9.96
	Minimum	1.32	3.24	4.18	5.05	5.69	1.48
	Maximum	6.81	19.98	23.09	26.74	30.23	44.35

D1: P3; **D2:** P3 e C3; **D3:** P3, C3 e P4; **D4:** P3, C3, P4 e C4; **D5:** P3, C3, P4, C4, Pz; **D6:** P3, C3, P4, C4, Pz e Cz.

The behavior of detection rates was also assessed for each individual volunteer and results were separated into two different categories according to the evolution of the detection rates for f_o = 15 Hz. Figure 3a shows the result for about half of the volunteers (20), classified in Group 1, which presented monotonically growing detection rates with the increase of N. A trend of stabilization for all curves from $N = 2$ or 3 derivations is also clear.

Figure 3b shows the detection rate curves for the other 20 volunteers (Group 2). As can be seen, for this group, the detection rates oscillated with the increase of N, indicating that the addition of new derivations did not necessarily result in an increase in the performance of the Multivariate Local SFT, although for 12 of such volunteers, DR was greater or equal to 90% for some set of derivations; that is, oscillations occurred around already high detection percentages.

For volunteers #1, #7, #22, #29, #31, #38, #39 and #44 (components of Group 2), results were quite similar to that observed for Group 1, except for a drop in rates with the addition of the sixth EEG derivation. Volunteers #12, #26, #40, and #42, in turn, showed a drop in detection rate for $N = 2$, followed by an upward pattern similar to Group 1 for $N > 2$. In the case of volunteers #8, #11, #15, #20, # 35 and #37, detection rates were increasing, however, with a drop in $N = 4$ succeeded by a new increase up to $N = 6$. The other volunteers included in Group 2 (#30 and #43) had more than one drop in detection rates with the increase in the number of EEG derivations used.

Such behaviors demonstrate that, in some cases, the increase of N from a certain value may worsen the performance of the studied ORD technique, rather than improve it. However, despite the oscillations, some components of Group 2 maintained high detection rates for all considered derivation sets (volunteers #1, #7, #8, #11, #15, #20, #22, #29, #31, #38, #39 and #43).

To verify the performance of the Local SFT with frequency variation, the detection rates were evaluated considering all volunteers for derivation P3 (univariate SFT), varying f_o from 5 to 100 Hz. The result can be observed in the Fig. 4, which shows the percentage detection curves as a function of frequency for each volunteer, as well as the mean curve, highlighted in bold-black.

By observing the traces, it is possible to notice that the technique presented better performance in the detection of the somatosensory response, with good mean detection rate in delta (0–4 Hz), theta (4–8 Hz), alpha (8–13 Hz) and low beta (13–20 Hz) EEG bands, being the frequencies to be chosen for somatosensory response monitoring purposes. For frequencies above 30 Hz, detection percentages were well below 50%, and possibly could not be improved with the multivariate strategy. A high peak in detection rates at the 60 Hz, corresponding to the power line interference frequency, is also observed.

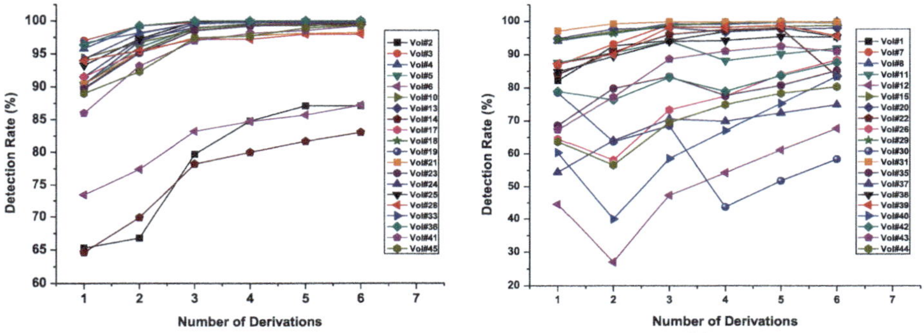

(a) (b)

Fig. 3. Detection rates by number of derivations for volunteers of (a) Group 1 and (b) Group 2.

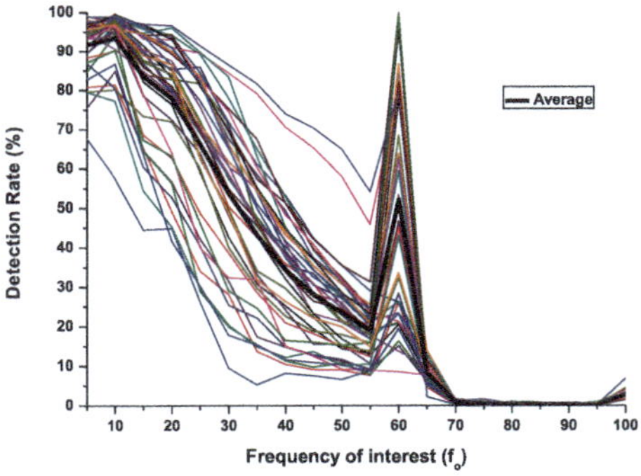

Fig. 4. Detection rates obtained with univariate SFT as a function of the frequency for derivation P3 for all volunteers and their mean (in black).

4 Discussion

Results obtained for Group 1 of volunteers are consistent with the expectation that the increase in the number of EEG derivations used in the Multivariate Local SFT calculus would imply an increase in the detection rate of the somatosensory evoked response. This result agree with those described by [4], who, however, applied this technique to the steady-state visual evoked potential. It was also expected, based on results reported in the literature about other objective response detection techniques (ORD), such as Multiple Coherence [8], that the behavior of detection percentages did not increase linearly as a function of the number of derivations. In fact, a stabilization of such rates was observed in Group 1.

A comparison of the observed behaviors for each group showed that volunteers who did not follow the monotonically increase of detection rates: (i) had already good results (eg: volunteer #31), (ii) always presented low DR values (eg: volunteer #12) or (iii) showed oscillating detection rates (eg: volunteer #30). One can justify the results obtained for Group 2 volunteers based on the theoretical and simulation analysis performed by [8]. Although the technique studied in this work was different, it was observed that an increase in the number of derivations may lead to a decrease in the evoked response detection rates. This can be explained by the fact that a new derivation added in a multivariate ORD approach should present a signal-to-noise ratio higher than a certain value, in order to lead the detection rate to increasing. In cases where the derivation does not meet this requirement, then the maintenance or even reduction of the detection rates may occur, as observed for some volunteers in Group 2.

Concerning the detection rates for frequency f_o = 100 Hz, for which no response to somatosensory stimulation is expected [7], in general, the percentage of false positives exceeded the predefined maximum value of 5% with the increase in the number of EEG derivations. It is believed that this result stems from the fact that, for calculating the degrees of freedom of F-distribution - used to determine the critical value of the Multivariate Local SFT-, the statistical independence of the background activity in the N used channels is assumed. However, such assumption can be considered a simplification since the EEG derivations, particularly those employed, present some type of correlation, because they are at adjacent regions to the primary somatosensory cortex (central and parietal derivations). Thus, the critical value considering more derivations could be altered, resulting in false-positive rates different from those found.

In fact, the evaluation of detection percentages when varying f_o from 5 to 100 Hz, suggests that this technique is effective in detecting somatosensory response at low frequencies (up to 20 Hz), according to the result obtained with univariate Local SFT. Moreover, it is important to note that the detection peaks around the 60 Hz frequency demonstrate that Local SFT is very susceptible to band-concentrated additive noise and it is not recommended in applications involving stimulation frequencies close to the power supply frequency or its harmonics. For this reason, it is common to use fractional stimulation frequencies, such as 4.83 Hz used in this study.

5 Conclusion

This study aimed at evaluating the performance of the Multivariate Local SFT in the detection of response to somatosensory stimulation applied to the right tibial nerve. Therefore, such technique was applied to EEG signals during stimulation of 40 normal adult volunteers, and the response detection rates were evaluated considering the number of derivations ranging from N = 1 to 6.

The results showed that, for part of the volunteers, there was a monotonic increase of the detection rates with the addition of derivations, whereas, for another group, an oscillatory pattern was observed; hence, there was not necessarily gain with the addition of new derivations, although the detection rates were already high for a considerable number of individuals in this second group.

In addition, it was found that the frequency range for which Local SFT showed better performance comprises delta (0–4 Hz), theta (4–8 Hz), alpha (8–13 Hz) and low beta (13–20 Hz) EEG bands. These ranges should be preferred in case of somatosensory response monitoring (e.g.: evaluation during surgery).

Thus, it is concluded that the application of Multivariate Local SFT is promising in the detection of somatosensory evoked response, once high detection rates were reached for the great majority of the volunteers.

Conflict of Interest. The authors declare no conflict of interest.

References

1. Sridharan, N.D., Thirumala, P., Chaer, R., Balzer, J., Long, B., Crammon, D., Makaround, M., Avgerinos, E.: Predictors of cross-clamp-induced intraoperative monitoring changes during carotid endarterectomy using both electroencephalography and somatosensory evoked potentials. J. Vasc. Surg. **67**(1), 191–198 (2018)
2. Thirumala, P.D., Cheng, H.L., Loke, Y.K., Hamilton, D.K., Balzer, J., Crammond, D.J.: Diagnostic accuracy of somatosensory evoked potential monitoring during scoliosis fusion. J. Clin. Neurosci. **30**, 8–14 (2016)
3. Melges, D.B., Miranda de Sá, A.M.F.L., Infantosi, A.F.C.: Tibial nerve somatosensory evoked response detection using uni and multivariate coherence. Biomed. Signal Process. Control **7**(3), 215–220 (2012)
4. Rocha, P.F.F., Felix, L.B., Miranda de Sá, A.M.F.L., Mendes, E.M.A.M.: Multivariate evoked response detection based on the spectral F-test. J. Neurosci. Methods **264**(1), 113–118 (2016)
5. Zurek, P.M.: Detectability of Transient and Sinusoidal Otoacoustic Emissions. Otoacoustic Emissions **13**(5), 307–310 (1992)
6. Miranda de Sá, A.M.F.L., Felix, L.B.: Improving the detection of evoked responses to periodic stimulation by using multiple coherence-application to EEG during photic stimulation. Med. Eng. Phys. **24**, 245–252 (2002)
7. Infantosi, A.F.C., Melges, D.B., Tierra-Criollo, C.J.: Use of magnitude-squared coherence to identify the maximum driving response band of the somatosensory evoked potential. Braz. J. Med. Biol. Res. **39**(12), 1593–1603 (2006)
8. Miranda de Sá, A.M.F.L., Felix, L.B., Infantosi, A.F.C.: A matrix-based algorithm for estimating multiple coherence of periodic signal and its application to the multichannel EEG during sensory stimulation. IEEE Trans. Biomed. Eng. **51**(7), 1140–1146 (2004)

Spatial Cross-Correlation to Determine Atrial Fibrillation Recurrence After Ablation

Raquel Cervigón[1(✉)], Julián Pérez-Villacastín[2], and Javier Moreno[3]

[1] Castilla-La Mancha University, Camino del Pozuelo sn, 16071 Ciudad Real, Spain
raquel.cervigon@uclm.es
[2] Unidad de Arritmias. Hospital Clínico San Carlos, Plaza de Cristo Rey sn., 28040 Madrid, Spain
[3] Unidad de Arritmias. Hospital Ramón y Cajal, Ctra. de Colmenar Viejo km. 9,100, 28034 Madrid, Spain
http://www.uclm.es

Abstract. Atrial fibrillation (AF) is already the most frequent arrhythmia in the clinical practice. Pulmonary vein ablation has emerged as a treatment that is able to make disappear the arrhythmia, nevertheless up to half of ablated AF patients suffer recurrences within the first year. In this study multiple simultaneous intra-atrial bipolar recordings were located at three locations, pulmonary veins and the right and the left atrium before the ablation procedure. Signal organization parameters were determined using cross-correlation analysis along the three main atrial regions. The results showed that before the procedure, patients with recurrence in the arrhythmia showed that left atrial activity was less correlated with other atrial regions compared with patients that maintained sinus rhythm. Moreover, time-lagged correlation coefficient between two time series showed higher delays in patients that experienced recurrence of AF. These results can be helpful for procedures designing to end AF.

Keywords: Atrial fibrillation · Ablation · Cross-correlation

1 Introduction

Atrial fibrillation (AF) is the most frequent arrhythmia associated with a significant risk factor for stroke. AF utilizes significant health resources globally [1], with increased risk of a number of co-morbidities [2]; of particular concern is the five-fold increased risk of stroke [3] and increased mortality risk [4].

AF is characterized by an abnormal excitation of the atria where multiple, rapidly changing and, according to some authors, spatially disorganized activation wavelets sweep across the surface of atria, forming complex and ever-changing patterns of electrical activity.

J. Henriques et al. (Eds.): MEDICON 2019, IFMBE Proceedings 76, pp. 22–28, 2020.
https://doi.org/10.1007/978-3-030-31635-8_3

Although the mechanisms leading to the initiation and self-sustainment of AF have been under extensive investigation, there is no universally accepted theory about its origin, maintenance, and termination. The widely accepted multiple wavelet hypothesis of AF postulates that AF is sustained by numerous coexisting wavefronts of electrical activity that propagate randomly throughout the fibrillating atria [5,6].

Radiofrequency ablation of ectopic beats from pulmonary veins (PVs) is at present, the method more frequently used in patients with paroxysmal AF to restore normal sinus rhythm [7]. It consists of generating electrical barriers in various sites of the atria by altering the tissue properties in the vicinity of the ablating catheter tip.

Despite the development of this technology, mechanisms of AF are multiple and many of them still remain unknown. However, at present we need to have more information on pathophysiology of AF, identification of subjects at high risk of progression is useful for an optimal ablation and to avoid a late procedure when AF becomes permanent.

Moreover, it has been hypothesized that the varying number of wavelets circulating in the atria affects the spatiotemporal organization of AF and defibrillation efficacy, with organization defined as an AF signal with repetitive morphology and high periodicity.

It is just tested that atrial activation sequences within some small region are well-correlated (reflective of their participation in the same wavelet) while activation sequences during AF beyond some distance are no longer well-correlated (reflective of their participation in different wavelets) [8]. We test the hypothesis that atrial activation sequences from different atrial regions, included pulmonary veins has little correlation in patients with recurrences in the arrhythmia compared with patients that maintain normal rhythm after a successful ablation.

In order to estimate atrial organization, we attempt to measure the differential voltage between the two electrodes, bipolar electrograms, collected from a pair of closely spaced electrodes, thus this measure is sensitive to the direction of nearby depolarization and repolarization wavefronts. Some of the extracted features are related to wavefronts similarities between electrograms, in order to find out whether there are differences on the waves morphology between patients with AF recurrence and those that maintenance sinus rhythm. It may also help in designing new approaches to the ablation therapy of AF, by identifying electrophysiological sites candidate to be ablated.

The purpose of this study was, therefore, to explore the atrial electrical activity organization by wavefront correlation measure, through the characterization of the intra-atrial recordings in time-domain with cross-correlation and time lags approaches.

2 Materials

Ten patients with AF were submitted to a catheter ablation procedure and intracardiac recordings were obtained immediately before the intervention. All patients were monitored after the procedure and divided into two groups according to long-term ablation success, defined as absence of arrhythmia 12 months after the ablation. In addition, several parameters are showed in Table 1 such as age, sex, electrical cardioversion (ECV), AF duration longer than six months and patients with paroxysmal AF, using either one (1) or zero (0), expressing "true" or "false", respectively.

Twelve bipolar intracardiac electrograms were digitally recorded at 1 kHz sampling rate (16 bit A/D conversion; Polygraph Prucka Cardio-Lab, General Electric) from the right atrium (RA) and the left atrium (LA) using this catheter. Additionally, a 10-pole catheter (Lasso, 2-5-2 mm electrode spacing) PV sized loop-shaped, was introduced via a transatrial septal long sheath, 5 bipolar intracardiac electrograms from each PV were digitally recorded at 1 kHz sampling rate. Fifty to 60 s Lasso and Orbiter catheters recordings from paroxysmal and persistent AF patients were recorded before the ablation.

Table 1. Patients characteristic.

Patient	Age (years)	Sex	AF > 6months	ECV	Parox.AF	Recurrence
Pat 1	67	Male	0	1	0	0
Pat 2	63	Female	0	1	1	1
Pat 3	32	Male	0	0	1	0
Pat 4	52	Male	0	0	0	0
Pat 5	65	Female	1	1	0	0
Pat 6	24	Male	0	1	1	1
Pat 7	51	Male	1	1	0	1
Pat 8	39	Male	0	1	1	0
Pat 9	57	Female	0	0	1	1
Pat 10	38	Male	1	1	0	1

3 Methods

In order to measure the organization index, the first step was the application of a preprocessing algorithm to the intracardiac electrograms without any ventricular contamination. The recordings were band-pass filtered using a 40 to 250 Hz third-order Butterworth filter. The resulting filtered waveforms were then rectified, and filtered once more using a 20-Hz low-pass third-order Butterworth filter. This algorithm was used to take a complex waveform and transform it to a series of atrial activations while diminishing the effects of changing electrogram morphology and/or amplitude [9,10].

In order to take advantage of stationariness, the series x_i ($i = 1, \ldots N$, with i the temporal index) is reduced to a process with zero mean and unitary variance by means of the normalization

$$x_i = \frac{x(i) - \overline{x}}{\sigma_x} \tag{1}$$

where $\overline{x}$ and σ_x are the mean and standard deviation of the series.

From the normalized series, the correlation function was applied to examine whether it could be identified differences between wave morphologies along the time in different atrial regions. This analysis requires two simultaneous recorded signals x and y from closely spaced bipolar recordings. The series y may be related to past lags of the x activation series or vice versa. The cross-correlation function is helpful for identifying lags of the x that might be useful predictors of y.

The discrete cross-correlation function between discrete time series of data x_i and y_i can be approximated by:

$$R_{xy}(k) = \frac{1}{N}\sum\nolimits_{i=1}^{N} \frac{(x_i - \overline{x})(y_{i+k} - \overline{y})}{\sigma_x \sigma_{y,k}}, \tag{2}$$

where N is the length of the segments to be correlated, k is the time lag between both signals ranging from $-K$ to K (i.e. the range within the correlation function is to be evaluated), $\overline{x}$ and $\overline{y}$ are the mean values of the two time series, and σ_x, $\sigma_{y,k}$ are the standard deviations of the corresponding segments from x_i and y_i, respectively.

In addition, spatial correlation analysis was performed between two electrograms registered from different atrial positions in each patient. The cross-correlation function was calculated over non-overlapping intervals for each electrogram. The maximum absolute peak of each segment was considered the correlation coefficient of two segments and the time lag k, between y_{i+k} and x_i, with k negative, are predictors of y, it is sometimes said that x leads y and with k positive, are predictors of y, it is sometimes said that x lags y.

3.1 Statistical Analysis

As a final step statistical techniques were applied to the extracted parameters. These parameters are expressed in terms of mean and standard deviation values. Independent t-student tests were used for comparison between both groups of results. Results were considered to be statistically significant at $p < 0.05$.

4 Results

The inter-electrode distance, showed a slightly higher correlation in patients that maintain sinus rhythm and at closer distances, with an exponential distribution as was previously demonstrated [8]. The largest spike occurs at time lag when the correlation is maximum, it allows to observe that lag distance between electrodes depends on the wavefronts evolution and on the location. In Fig. 1(a) the distance is lower because this dipoles are located very close, in the same atrial area, and in Fig. 1(b) the distance is higher because the dipoles are located in different atria.

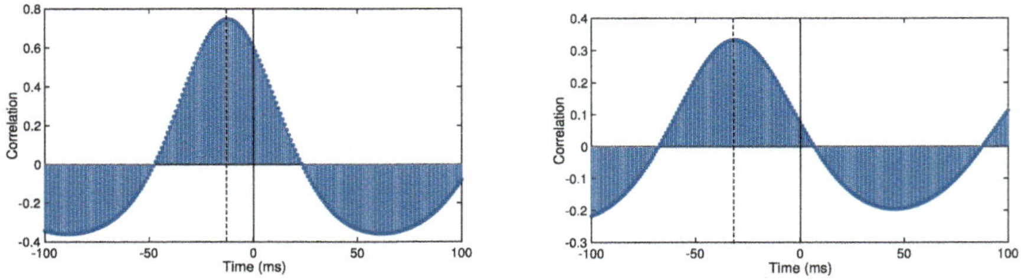

(a) Dip 1-2 (LA) vs. Dip 5-6 (LA) (b) Dip 1-2 (LA) vs. Dip 9-10 (RA)

Fig. 1. Difference of maximum correlation coefficient distance at different inter-dipole separation, dipoles located in the same atria (a) and dipoles located in different atria (b).

The cross-correlation coefficient was calculated from recordings located in the LA to the rest of the areas, other recording from the LA, the RA and in the PVs, in patients that had recurrence in AF and those patients that maintain sinus rhythm. Figure 2 shows the correlation across the patients between one dipole located in the LA and dipoles located in the LA, in the RA and in the PVs (Fig. 2(b)).

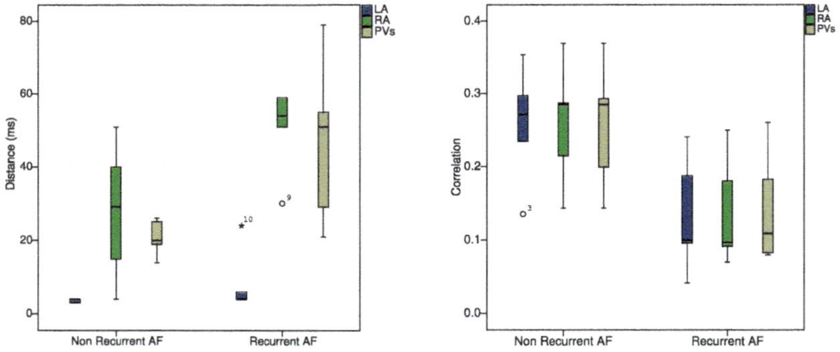

(a) Time-lag (b) Cross-correlation

Fig. 2. The correlation coefficient at different inter-dipole separations across all subjects, with the correlation defined as one at zero distance across AF recurrent and non AF recurrent patients.

From this analysis, we observed that the correlation increased in the group that maintains sinus rhythm compared to the group with recurrence in the arrhythmia (i.e. an increase in the organization degree), 0.26 ± 0.08 in the non-recurrent AF group vs. 0.133 ± 0.07 in the recurrent group ($p = 0.039$) in the LA (Fig. 2(a)). Moreover the time lag between largest spikes showed differences between atrial areas, for instance, in the LA ($p = 0.037$) in the group that maintain sinus rhythm showed lower times (3.6 $\pm$ 0.55 ms) than in patients with recurrence in the arrhythmia (9.80 $\pm$ 7.11 ms) with the same effect in the LA with other regions (Fig. 2).

5 Discussion and Conclusions

Pulmonary vein catheter ablation is a widely used procedure that isolates the pulmonary veins from the rest of the heart and prevents any pulse from these veins from getting into the heart, restoring normal rhythm in patients with AF [11].

However, it has a limited overall success rate and frequently it is necessary to repeat procedure. These limitations can be caused by our current poor understanding of the pathophysiology of the arrhythmia. Nevertheless, previous studies have described numerous clinical variables that were identified as independent predictors of recurrence of AF: non-paroxysmal AF, aging, left ventricular systolic and diastolic dysfunction, LA enlargement, the presence of non-PV triggers, obesity, hypertension [12, 13].

In this paper cross-corrrelation parameter has been determined to quantify the organization of the atrial electrical activity during AF. This parameter measures the spatial organization by analyzing the interaction between different sites of the atria, whereas this measurement reflect the organization of the atrial activity within each single electrode. As shown in the results section, these parameters are able to capture changes in the dynamics of atrial signals along the atria, with higher correlation and smaller distances in the non recurrent AF patients compared with the recurrent AF patients.

As conclusion, correlation coefficient can prognosticate AF recurrences after successful ablation of pulmonary veins. Low correlation and high distance values in AF patients predict a very high-risk for AF recurrence. It suggests that when the atrial electrical activity is more similar along the atria, the reversion to sinus rhythm is less difficult. Similar results were obtained by non-linear morphology measurements [14] and Jensen–Shannon Divergence, with higher dispersion in the group of patients that did not maintain normal rhythm [15]. These findings show that correlation measurements arise as promising indicators of risk for AF recurrence after successful ablation of AF.

References

1. Kim, M.H., Johnston, S.S., Chu, B.-C., Dalal, M.R., Schulman, K.L.: Estimation of total incremental health care costs in patients with atrial fibrillation in the united states. Circ. Cardiovasc. Qual. Outcomes **4**(3), 313–320 (2011)
2. Andersson, T., Magnuson, A., Bryngelsson, I.-L., Frøbert, O., Henriksson, K.M., Edvardsson, N., Poçi, D.: All-cause mortality in 272,186 patients hospitalized with incident atrial fibrillation 1995–2008: a Swedish nationwide long-term case-control study. Eur. Heart J. **34**(14), 1061–1067 (2013)

3. Chugh, S.S., Havmoeller, R., Narayanan, K., Singh, D., Rienstra, M., Benjamin, E.J., Gillum, R.F., Kim, Y.-H., McAnulty Jr., J.H., Zheng, Z.-J., Forouzanfar, M.H., Naghavi, M., Mensah, G.A., Ezzati, M., Murray, C.J.L.: Worldwide epidemiology of atrial fibrillation: a global burden of disease 2010 study. Circulation **129**(8), 837–847 (2014)
4. Wattigney, W.A., Mensah, G.A., Croft, J.B.: Increased atrial fibrillation mortality: United States, 1980–1998. Am. J. Epidemiol. **155**(9), 819–826 (2002)
5. Allessie, M.A., Bonke, F.I.M., Schopman, F.J.G.: Circus movement in rabbit atrial muscle as a mechanism of tachycardia.2. Role of nonuniform recovery of excitability in occurrence of unidirectional block, as studied with multiple microelectrodes. Circ. Res. **39**(2), 168–177 (1976)
6. Moe, G.K.: On multiple wavelet hypothesis of atrial fibrillation. Arch. Int. Pharmacodyn. Ther. **140**, 183–188 (1962)
7. Haissaguerre, M., Shah, D.C., Jais, P.: Electrophysiological breakthroughs from the left atrium to the pulmonary veins. Circulation **102**, 2463–2465 (2000)
8. Botteron, G.W., Smith, J.M.: A technique for measurements of the extent of spatial organization of atrial activation during atrial fibrillation in the intact human heart. IEEE Trans. Biomed. Eng. **42**, 579–586 (1995)
9. Castells, F., Cervigón, R., Millet, J.: On the preprocessing of atrial electrograms in atrial fibrillation: understanding Botteron's approach. Pacing Clin. Electrophysiol. **37**, 133–143 (2013)
10. Thomas, H., Everett, I.V., Kok, L.-C., Richard, H., Vaughn, J., Randall, M., Haines, E.D.: Quantitative assessment of the spatial organization of atrial fibrillation in the intact human heart. Circulation **93**, 513–518 (1996)
11. Calkins, H., Reynolds, M.R., Spector, P., Sondhi, M., Xu, Y., Martin, A., Williams, C.J., Sledge, I.: Treatment of atrial fibrillation with antiarrhythmic drugs or radiofrequency ablation: two systematic literature reviews and meta-analyses. Circ. Arrhythm. Electrophysiol. **2**(4), 349–361 (2009)
12. Sotomi, Y., Kikkawa, T., Inoue, K., Tanaka, K., Toyoshima, Y., Oka, T., Tanaka, N., Nozato, Y., Orihara, Y., Iwakura, K., Sakata, Y., Fujii, K.: Regional difference of optimal contact force to prevent acute pulmonary vein reconnection during radiofrequency catheter ablation for atrial fibrillation. J. Cardiovasc. Electrophysiol. **25**(9), 941–947 (2014)
13. Andrade, J., Khairy, P., Dobrev, D., Nattel, S.: The clinical profile and pathophysiology of atrial fibrillation: relationships among clinical features, epidemiology, and mechanisms. Circ. Res. **114**(9), 1453–68 (2014)
14. Cervigón, R., Moreno, J., Castells, F.: Entropy analysis of atrial activity morphology to study atrial fibrillation recurrences after ablation procedure. In: Francisco F., Ortuño Guzman, M., Rojas, I. (eds.) Proceedings of Bioinformatics and Biomedical Engineering - Third International Conference, IWBBIO 2015, Granada, Spain, 15–17 April 2015. Part I. Lecture Notes in Computer Science, vol. 9043, pp. 146–154. Springer (2015)
15. Cervigón, R., Castells, F., Gómez-Pulido, J., Pérez-Villacastín, J., Moreno, J.: Granger causality and Jensen-Shannon divergence to determine dominant atrial area in atrial fibrillation. Entropy **20**(1), 57 (2018)

Development of a Computer Simulator of the Visual N2 Event-Related Potential Component for the Study of Cognitive Processes

Francesca Marturano, Sabrina Brigadoi, Mattia Doro, Roberto Dell'Acqua, and Giovanni Sparacino(✉)

University of Padova, 35131 Padua, PD, Italy
gianni@dei.unipd.it

Abstract. The importance of visual N2 Event-Related Potential (ERP) component in the study of cognitive processes lies in its interpretation as a measure of the allocation of visual attention to possible targets. Unfortunately, the N2 component has a small amplitude and the domain of validity of methods used for its estimation is difficult to assess if real data are considered. Here, we develop a computer simulator of ERP measurements emulating the variability of the visual N2 component and use the synthetic data to evaluate the performance of four popular literature ERP estimation methods. Results confirmed that a solid simulation framework could allow identifying a reliable method to detect small-amplitude ERP components and quantifying its accuracy.

Keywords: Visual N2 · ERP modeling · EEG signal processing · Gaussian mixture models

1 Introduction

In the study of cognitive processes, the visual posterior N2 Event-Related Potential (ERP) component was shown to be an important index of the allocation of attention in the lateral portion of the visual field [1–3]. Typically, the N2 raises between 200–300 ms after the stimulus onset as a negative peak of few μV and is expected to show a larger negativity over the opposite hemiscalp (contralateral) to the target. The N2 assessment is difficult because of the particularly poor signal-to-noise ratio (SNR) and previous works in the literature have not exhaustively studied the robustness of different estimation techniques on this component.

In the assessment of ERP estimation techniques, answering to questions such as "how many trials do we need to obtain a reliable average response?" requires the use of a controlled synthetic dataset with a known ground truth. A common solution is the use of simple functions, such as sinusoids or Gaussians, to create the gold standard ERP for the data under investigation [4]. A limitation of this approach could be the difficulty to reproduce the real ERP pattern that may be crucial in some investigations. For example, often a Gaussian peak corrupted by random noise is used to represent the P3 positive

J. Henriques et al. (Eds.): MEDICON 2019, IFMBE Proceedings 76, pp. 29–36, 2020.
https://doi.org/10.1007/978-3-030-31635-8_4

deflection [4–6] but less care is usually taken to reproduce the real modulation of this component across participants and conditions. This might bring to a limited and inadequate representation of the data, thus influencing the generalization of the main findings to real data. This issue needs to be carefully addressed depending on the research purposes and the characteristics of the investigated components. In particular, in the study of small amplitude components that are completely covered by noise, reproducing the real parameters variability in simulation might be crucial for the precision of results. Some simulation approaches generate synthetic data using real data, adding a known and as realistic as possible effect on a parameter of the investigated component, for example on the latency. Kiesel et al., adopted a similar procedure to investigate the best performing method for latency estimation [7]. However, a reliable simulation of the real pattern and variability of the ERP traces is the first step that should be fulfilled even before developing any novel technique for ERP quantification.

In the present work, we describe the development of a simulator providing realistic synthetic sweeps with N2 variability derived directly from real data of an experiment eliciting a modulation of the N2 component itself. We then demonstrate a possible example of the application of the simulator by using the simulated sweeps as realistic ground truth signals to evaluate the performance of some commonly used averaging algorithms. In the conclusion, we discuss further applications and extensions of the presented simulator.

2 Dataset for the Simulator Development

2.1 Participants

Fourteen healthy volunteers (mean age 23.2 ± 2.1 years) participated in the study carried out at the Department of Developmental Psychology of the University of Padova. The experiment was a common paradigm employed to elicit the visual posterior N2 component.

2.2 Experimental Procedure

Each participant took part to a visual search experiment eliciting lateralized modulations of the visual posterior N2. Data were collected in a multi-channel setting by 28 electrodes located over the subject's scalp following the 10–20 system of electrodes' positioning; 3 electrodes were placed at the outer canthi and below the left eye (electrooculogram), and 2 electrodes on both the earlobes. During the experiment, 600 visual search arrays were presented to the participant on a screen, 200 per visual hemifield and 200 along the sagittal midline. While fixating a fixation cross placed in the center of the screen, participants were presented 12 little white circles clock-wisely positioned and equally distributed at a distance of 2.5° of visual angle from the fixation cross. The stimulation consisted in a turning-blue circle with a bar (tilted or not) appearing inside of it (Fig. 1). Participants were asked to report the orientation of the bar inside the target circle through a keyboard placed in front of them.

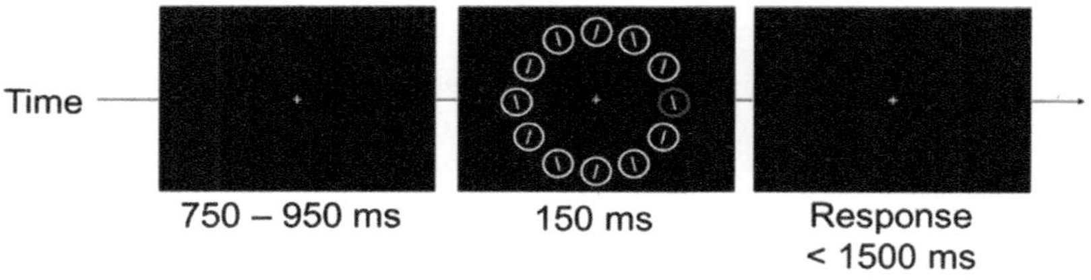

Fig. 1. Experimental paradigm to elicit the N2 component.

2.3 EEG Data Pre-processing

EEGLAB [8] was used for data pre-processing. Data were band-pass filtered (0.1–30 Hz), divided into sweeps (from 0.2 s pre-stimulus to 1 s post-stimulus), re-referenced using the mean of the signals measured on the earlobes and baseline corrected by subtracting the mean amplitude of the pre-stimulus interval. Trials with ocular artifacts, as identified using the electrooculogram, as well as trials from incorrect or no responses and trial associated with the presentation of central stimuli, were discarded from further analysis. Since the N2 component is usually recorded from electrodes located at the posterior sites of the scalp (typically P7/P8 or PO7/PO8), only trials from P7 and P8 (available in our dataset) were selected from the channel pool. Trials were partitioned according to the target side into right-(R) and left-(L) targeted trials. A mean [min, max] of 100 [23, 176] trials per channel-target condition was available among subjects. Finally, data from each participant were standardly averaged for each channel and target side to reduce the noise and obtain the averaged ERP.

3 Simulator Development

3.1 Estimation of N2 Real Variability

To match the morphology and variability of the components of the average ERP obtained from real data, a Gaussian Mixture Model (GMM) was fitted on the averaged data of each subject, separately for each channel and target side to distinguish between contralateral (CL) and ipsilateral (IL) to the target position data. A raw estimation of amplitude and latency of each ERP component was obtained. The GMM model is commonly used to represent ERP components because of their typical morphology of peaks and valleys [4–6]. The optimal number of Gaussians was set to seven using the Akaike criterion [9], five of which were needed to fit the expected voltage deflections of visual ERPs. However, we focused the attention on the components occurring in 200–300 ms from stimulus onset, the typical range of the visual N2, and estimated the distribution of their amplitude and latency across subjects to be used later on to simulate data with a real variability of these components (Fig. 2, lower panels).

3.2 Development of the ERP Template

The ERP model at the basis of the simulator was estimated by fitting the previous GMM model to the grand-averaged (GA) ERP obtained by averaging the individual mean ERP across subjects. A different template was created for each channel and target side, thus obtaining four final templates, here labeled as P7R, P7L, P8R and P8L. These templates were used to create the corresponding simulated single sweeps (Fig. 2, upper panels). Each model was perturbed by adding two to-be-perturbed Gaussians in the 200–300 ms time range to introduce a controlled variability in the amplitude and latency of the components occurring in the same time range (i.e., P2 and N2), differently for each subject and channel, reproducing the same variability range estimated from the real dataset (see Sect. 3.1). Other ERP peaks, occurring outside the time window of interest, were not perturbed.

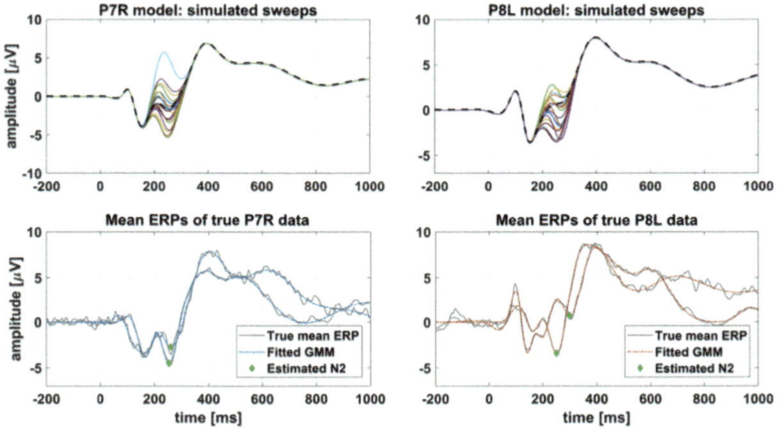

Fig. 2. Upper panels show 20 synthetic noise-free sweeps generated from P7R and P8L templates, respectively. Lower panels show the average ERPs computed from the real data of some representative subjects.

3.3 Model of the Background EEG

The EEG background noise of each simulated sweep was modeled using an autoregressive (AR) model [10, 11], by assuming stationarity within the short time interval of 1.2 s of a sweep and additivity between the EEG noise and the ERP [4]. The EEG noise, x_n, described by an AR model is usually expressed as:

$$x_n = \sum\nolimits_k^P a_k x_{n-k} + \varepsilon_n \tag{1}$$

where the model coefficients a_k, the model order P and the variance of the input white noise ε_n were identified, for each sweep, from the pre-stimulus interval of a random real sweep to match the noise morphology of that actual sweep. Twenty possible model orders were investigated using the built-in MATLAB *ar* function (with argument '*yw*' for the Yule-Walker solving algorithm). The optimal model order, P, was defined as the one giving a mean squared error (MSE) very similar ($\leq \varepsilon$) to the model of order P–1. P was found to vary in the range 3–5 considering the median of the AR model orders across the 100 synthetic sweeps and the subjects of each data group. In addition, the Anderson test was employed to test the model goodness-of-fit by checking the whiteness of the prediction error [12].

The synthetic background EEG was added to the simulated noise-free sweeps obtained as described in Sect. 3.2, thus creating a realistic synthetic dataset with controlled variability in the N2 time range.

4 Assessment of the Simulator

The simulator was first qualitatively evaluated, by submitting a visual test to eight cognitive scientists, expert on EEG data, and asking them to identify the simulated sweep between two randomly placed ERP traces, one real and one synthetic (Fig. 3). The main aim of this first evaluation was to assess the quality of the simulated noise when visually comparing synthetic and real sweeps. Second, the simulated sweeps were statistically compared to the corresponding real ones using the standard deviation (SD) of the baseline (BL) as metric. Ideally, if noise was accurately generated, both synthetic and real sweeps should possess similar statistical properties, at least within the BL interval, where the ERP had not been elicited yet.

Results showed that experts were not able to identify the simulated ERP trace: on average, the final score had a mean of 48.96% and a median [min, max] of 44.44 [36.11, 69.44]%, very close to answer at random. Quantitative results demonstrated a high correlation between the SDs of the BL of synthetic and real sweeps with a Pearson correlation coefficient of 98%.

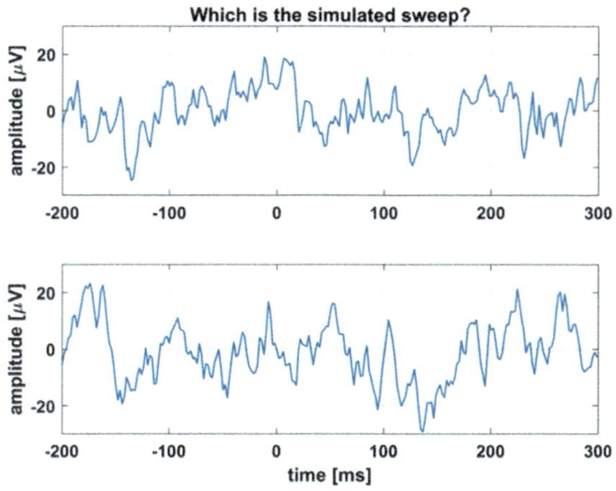

Fig. 3. Exemplary figure of the visual test for the qualitative assessment of the simulator.

5 Example of Simulator Application

5.1 Creation of Synthetic Recordings

Three subsets (here labeled with *S*) differing in the N2 relative amplitude between contralateral (CL, i.e., from P7R and P8L models) and ipsilateral (IL, i.e., P7L and P8R models) to the target side electrodes were created from the generated noisy sweeps. The sweeps of *S*1 had $N2_{CL} < N2_{IL}$ (expected situation), those of *S*2 had $N2_{CL} > N2_{IL}$ and those of *S*3 had $N2_{CL} \approx N2_{IL}$ (both unexpected, but sometimes occurring in reality, behaviors of inverted or similar activity between CL and IL signals). Then, using the three subsets, a synthetic recording of 200 trials (100 per target side) for both channels was created for each participant. In detail, to simulate R-target trials (here P7 is CL, P8 is IL), 70% sweeps were randomly selected from *S*1, 15% from *S*2 and 15% from *S*3. The same selection was performed for L-target trials (where P8 is CL and P7 is IL instead). Synthetic data were standardly pre-processed as the real ones (see Sect. 2.3).

5.2 Assessment of Averaging Methods

The generated synthetic recordings were used to assess the performance of some common averaging methods on the recovery of the mean N2 component. Conventional Average (CA), Sorted Average (SA) and two weighted average methods here labeled with WA and HK (all of them reviewed in [13]) were applied on the simulated sweeps of each channel and for each target side. Methods were compared in terms of mean absolute error on amplitude and latency estimates of the mean N2, which are the parameters commonly estimated from ERPs in studies investigating cognitive processes [4, 14]. The Absolute Error on amplitude (AE_a) and latency (AE_l) across *N* subjects were computed as in Eqs. (2) and (3), where the true values (A_i and L_i) were obtained from the average of the 100 noise-free sweeps of each group and the estimates ($\hat{A}_i$ and $\hat{L}_i$) from the corresponding averaged noisy data.

$$AE_a = \frac{1}{N}\sum\nolimits_{i=1}^{N} |\hat{A}_i - A_i| \tag{2}$$

$$AE_l = \frac{1}{N}\sum\nolimits_{i=1}^{N} |\hat{L}_i - L_i| \tag{3}$$

5.3 Results

Figure 4 shows the error performance of each method on the estimation of the average N2 parameters. Blue dots on the left and right panels represent AE_a and AE_l values, respectively, while bars indicate minimum and maximum error across subjects. CA achieves the best AE_a at P7 while WA at P8R; instead at P8L, methods perform quite similarly but WA achieves the lowest maximum value. WA obtains the best AE_l at P7R and P8R while CA at P7L and SA at P8L. The maximum AE_a is about 1.5 μV (HK and SA at P7L), while the maximum AE_l is about 19 ms (HK at P7L). Error differences among channel-target side groups are due to the different SNR of the noisy sweeps generated for each subject in the group that ranges, on average in the N2 time window, from about 0.13 to 1.1, with a mean value across groups in [0.3–0.5] and a SD in [0.1–0.3].

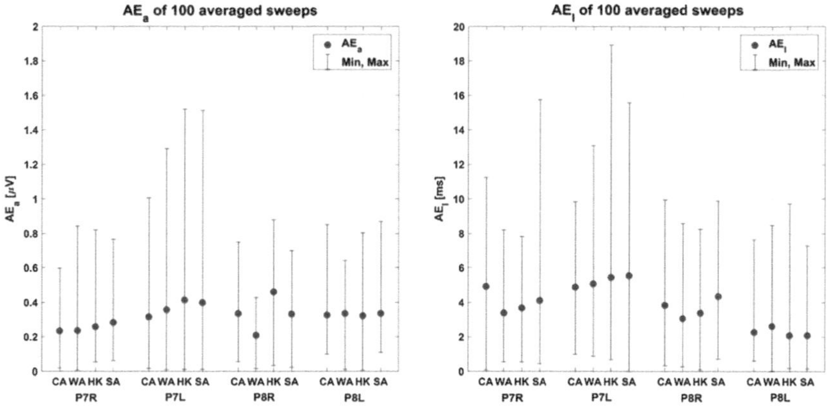

Fig. 4. Error performance of the averaging methods on the estimation of the average N2 parameters for each channel-target side group.

6 Conclusion

In the present work, we developed an ERP simulator reproducing synthetic sweeps with a variability of the N2 component emulating that observed on real data. The ability of the simulator to generate single noisy sweeps as real as possible was evaluated both qualitatively, during a test phase involving eight EEG experts, and quantitatively, by comparing the noise statistical properties between simulated and real sweeps. Results demonstrated that noise had similar morphological and statistical features both in real and synthetic data, thus suggesting the success of the proposed simulator to create realistic data. As a demonstration of the applicability of the simulator, the performance of four popular ERP averaging techniques was assessed in terms of mean absolute error on the estimate of N2 parameters.

Results demonstrated different performances of the averaging techniques depending on the input signal and therefore the importance of being able to simulate realistic data to be used for a priori and controlled validation of proposed methods. Future applications of our simulator could include a quantification of the minimum number of trials required in the average for a reliable N2 recovery and of the relative error of the methods used depending on the SNR. Other challenging applications could include the validation of ad-hoc single-trial estimation techniques by using the generated noise-free sweeps as ground truth signals whose variability of the N2 component emulates that of real traces, thus avoiding overestimated outcomes. Importantly, the proposed simulator could be easily extended to the study of any other ERP component, following the pipeline described in this manuscript.

Conflict of Interest. None

References

1. Luck, S.J., Hillyard, S.A.: Electrophysiological correlates of feature analysis during visual search. Psychophysiology **31**(3), 291–308 (1994)
2. Eimer, M.: The N2pc component as an indicator of attentional selectivity. Electroencephalogr. Clin. Neurophysiol. **99**(3), 225–234 (1996)
3. Hickey, C., Di Lollo, V., McDonald, J.J.: Electrophysiological indices of target and distractor processing in visual search. J. Cogn. Neurosci. **21**(4), 760–775 (2009)
4. D'Avanzo, C., Goljahani, A., Pillonetto, G., De Nicolao, G., Sparacino, G.: A multi-task learning approach for the extraction of single-trial evoked potentials. Comput. Methods Programs Biomed. **110**(2), 125–136 (2013)
5. D'Avanzo, C., Schiff, S., Amodio, P., Sparacino, G.: A Bayesian method to estimate single-trial event-related potentials with application to the study of the P300 variability. J. Neurosci. Methods **198**(1), 114–124 (2011)
6. Ouyang, G., Hildebrandt, A., Sommer, W., Zhou, C.: Exploiting the intra-subject latency variability from single-trial event-related potentials in the P3 time range: a review and comparative evaluation of methods. Neurosci. Biobehav. Rev. **75**, 1–21 (2017)
7. Kiesel, A., Miller, J., Jolicœur, P., Brisson, B.: Measurement of ERP latency differences: a comparison of single-participant and jackknife-based scoring methods. Psychophysiology **45**(2), 250–274 (2008)
8. Delorme, A., Makeig, S.: EEGLAB: an open source toolbox for analysis of single-trial EEG dynamics including independent component analysis. J. Neurosci. Methods **134**(1), 9–21 (2004)
9. Akaike, H.: A new look at the statistical model identification. IEEE Trans. Automat. Contr. **19**(6), 716–723 (1974)
10. Steinberg, H.-W., Gasser, T., Franke, J.: Fitting autoregressive models to EEG time series: an empirical comparison of estimates of the order. IEEE Trans. Acoust. **33**(1), 143–150 (1985)
11. Cerutti, S., Chiarenza, G., Liberati, D., Mascellani, P., Paves, G.: A parametric method of identification. IEEE Trans. Biomed. Eng. **35**(9), 701–711 (1988)
12. Anderson, C.W., Stolz, E.A., Shamsunder, S.: Multivariate autoregressive models for classification of spontaneous electroencephalographic signals during mental tasks. IEEE Trans. Biomed. Eng. **45**(3), 277–286 (1998)
13. Riedel, H., Granzow, M., Kollmeier, B.: Single-sweep-based methods to improve the quality of auditory brain stem responses part II: averaging methods. Zeitschrift für Audiol. **40**(2), 62–85 (2001)
14. Fukami, T., Watanabe, J., Ishikawa, F.: Robust estimation of event-related potentials via particle filter. Comput. Methods Programs Biomed. **125**, 26–36 (2016)

Automatic Segmentation of Ultrasonic Vocalizations in Rodents

Diogo Pessoa[1(✉)], Lorena Petrella[2,3,4,5], Miguel Castelo-Branco[3,4,5,6], and César Teixeira[1]

[1] Center for Informatics and Systems of the University of Coimbra, (CISUC), Coimbra, Portugal
dpessoa@student.dei.uc.pt
[2] Department of Electrical and Computer Engineering (DEEC), University of Coimbra, Coimbra, Portugal
[3] Institute of Nuclear Sciences Applied to Health (ICNAS), University of Coimbra, Coimbra, Portugal
[4] Institute of Biomedical Imaging and Life Science (IBILI), University of Coimbra, Coimbra, Portugal
[5] Coimbra Institute for Biomedical Imaging and Translational Research (CIBIT), University of Coimbra, Coimbra, Portugal
[6] Faculty of Medicine of the University of Coimbra (FMUC), Coimbra, Portugal

Abstract. Ultrasonic vocalizations studies in rodents have increasingly drawn researchers attention as it have been considered a powerful tool to understand the animals behavior and their interactions in different social and environmental contexts.

This paper presents an entropy-based algorithm for accurate and robust segmentation of mouse ultrasonic calls. Instead of using the conventional energy-based features, the spectral entropy is developed to identify the audio segments accurately. The new approach for mice calls detection has been able to detect up to 97% of the vocalizations.

Keywords: Rodents ultrasonic vocalizations · Signal processing · Signal segmentation · Spectral Entropy

1 Introduction

Mouse are typically highly vocal animals which can produce a diverse and vast audio repertoire, mainly ultrasonic (>20 kHz). As seen in the majority of species, vocal interaction between animals can provide a useful insight about how animals interact with each other, and, therefore, studying the vocal calls (syllables) produced by mice, can led to an improved understanding and better characterization of the species. Vocal communication has been documented in different species, namely rodents, and several studies have linked some kinds of ultrasonic vocalizations (USVs) with different environmental and social contexts, which implies that the calls produced by mice can be a measure of social interaction between the animals [10].

J. Henriques et al. (Eds.): MEDICON 2019, IFMBE Proceedings 76, pp. 37–46, 2020.
https://doi.org/10.1007/978-3-030-31635-8_5

It has been long known that both laboratory and wild mice emit ultrasonic vocalizations (USVs), and possess a wide and complex multi-syllabic repertoire [1,2]. Rodents produce both low and high-frequency vocalizations, depending on the context, and typically produce sounds ranging from 200 Hz up to 200 kHz or above. These vocalizations can be acoustically simple, such as distress cries, which are pure tone-like calls that contain few frequency modulations, or can be very complex, such as mouse mating calls, which have rich harmonics and frequency modulations [8]. Furthermore, it has also been shown that specific events, such as isolation distress in pups, social and sexual coordination, nearby predators, anticipation of an aversive occurrence, aggressive encounters, play, and finding food, all give rise to vocalizations [12].

Typically, in studies related to bioacoustics analysis, the first step consists in audio signals segmentation, in order to isolate the syllables. To this end, several algorithms have been proposed, being some of them, the Harma's method [4], time-varying parameters [5,6], the Hilbert follower [9] and the Faster RCNN networks [2]. The Harma's method is a syllable segmentation algorithm based on the iterative amplitude-frequency information. Essentially, this method consists in the computation the spectrogram of an audio segment using Fast Fourier Transform ($S(f, t)$) and then finding f_n and t_n, such that $|S(f_n, t_n)|$ is the maximum value in the spectrogram. From that value, the signal is traced back and forward and when a certain condition is met the instant is stored as the initial or final point of the vocalization, accordingly. After that, the process is repeated iteratively.

In the method used by Holy and Guo [5] and Ivanenko [6], syllables are segmented by computing three time-varying parameters from the audio spectrograms: the mean frequency, the "spectral purity" (fraction of total power concentrated into a single frequency bin), and the "spectral discontinuity". After that, syllables are identified if those three measured parameters meet certain conditions.

The Hilbert follower is essentially an audio activity detector which segments the signals by following the characteristic shape of syllables envelop. This algorithm computes the Hilbert envelope of the signal $x(n)$, $x_h(n) = Hilbert(x(n))$, and then establishes a threshold so that if the envelope is above the threshold it is considered to be audio activity, and, therefore, a vocalization.

Unlike the other three methods described above, the Faster RCNN is based on artificial neural networks, trained with audio segments of the audio signal. Therefore, this method needs to be trained and the input provided for training must be reviewed manually, unlike what happens in the other two algorithms, which are unsupervised methodologies.

The study of ultrasonic vocalizations can be a quiet slow and laborious task, which urges the development of better tools in order to process and analyze the sounds and their vocalizations. In this work, a new approach based on spectral entropy has been developed and presented in this paper to isolate the syllables produced by mice. This algorithm, despite its simplicity, presents good results and a relatively low computational cost, being a very helpful tool in ultrasonic vocalizations studies.

2 Materials and Methods

2.1 Data Acquisition and Dataset

The recording system used to acquire the mouse audio calls consists on a multichannel recording system (*Avisoft-UltraSoundGate 416H, Avisoft Bioacoustics*) whit four channels working with two microphones (*microphone CM16/CMPA, Avisoft Bioacoustics*), one for each test chamber (see Fig. 1). The signals can be digitized at a maximum sampling frequency of 700 kHz. Associated software allows the recording configuration, the real time signal/spectrum visualization and the recording of the signals in the PC memory. Moreover, a playback system, consisting of a speaker (*Ultrasonic Speaker Vifa, Avisoft Bioacoustics*) and a D/A converter(*UltraSoundGate Player 116 (single channel), Avisoft Bioacoustics*), was acquired from the same manufacturer. This playback system is used to reproduce acoustic stimuli. Tests are video-recorded with a webcam positioned at the top of the test cabin, and using a red light to provide a more pleasant condition for mice. The sound-attenuating cabin was projected with 55 cm × 50 cm × 70 cm (H × D × W), to provide enough space for the experimental montage. It was constructed of acrylic plates with a wall thickness of 1.5 cm, to block the external sound. Moreover, the internal side of the cabin was fully covered by sound absorbing foam. The cabin was also designed with a frontal door and holes for ventilation and wiring. An additional acrylic cage was constructed, where mice are placed during tests. It is half separated by an acrylic plate with holds in the inferior part. During tests two animals are placed one in each compartment (with one microphone) for the identification of each mice USV, while the holes allow them to keep communicating. Preliminary tests confirmed the good performance and versatility of the assembled system.

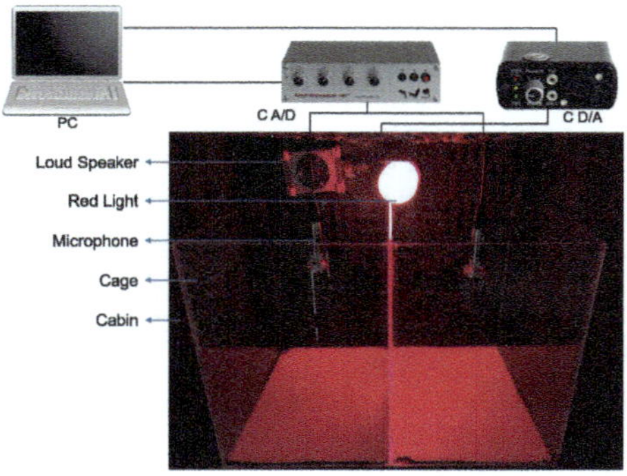

Fig. 1. USV recorder system.

To evaluate and develop the proposed algorithm, 16 audio files (.WAV), sampled at 500 kHz, with 45 s each were considered. The files contain spontaneous vocalizations originated from social interaction between mice. All the signals were acquired by the system described above and were manually segmented, being the initial and final instants of each vocalization recorded for future comparison with the results produced by the automatic segmentation algorithm.

In the manual classification, 11 different types of vocalizations were identified (see Fig. 2), being, all of them, described in the literature [3,10]. The signals used contained a total of 460 vocalizations.

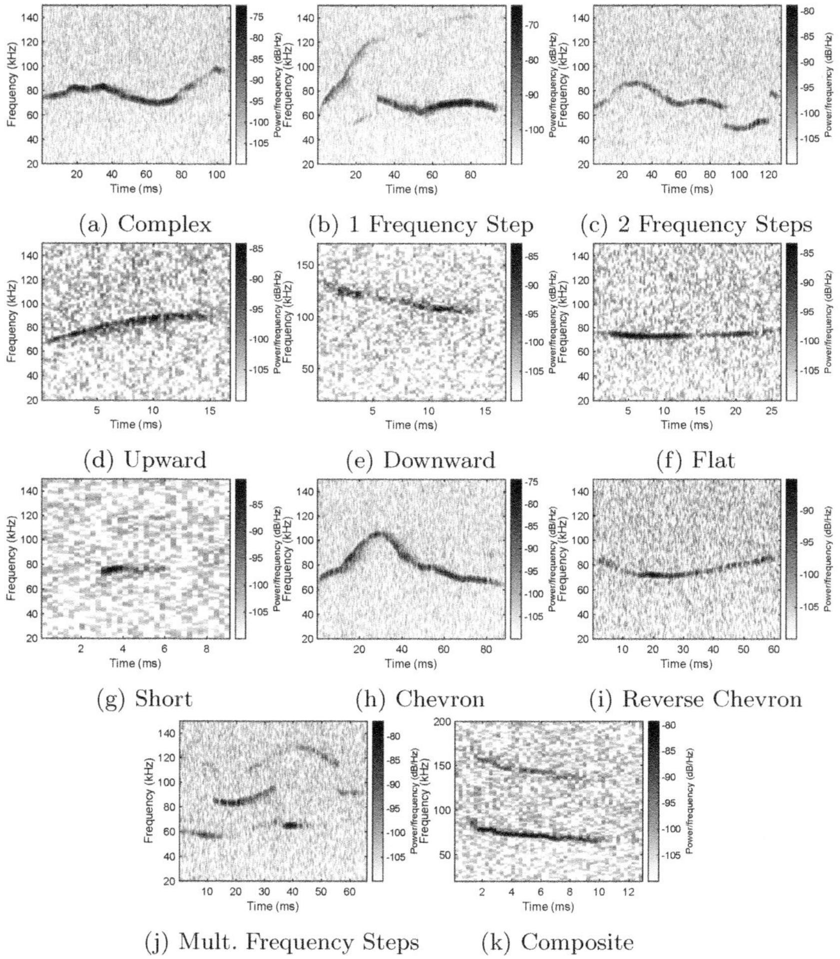

(a) Complex (b) 1 Frequency Step (c) 2 Frequency Steps
(d) Upward (e) Downward (f) Flat
(g) Short (h) Chevron (i) Reverse Chevron
(j) Mult. Frequency Steps (k) Composite

Fig. 2. Time-frequency representation of the 11 types of USVs considered in this study.

2.2 Spectral Entropy Segmentation Algorithm

Spectral entropy is a generalization of information entropy, which was a concept first proposed by Shannon in an attempt to measure information's uncertainty [11]. This concept can be formulated by

$$H = -\sum_{m=1}^{N} P(m) \log_2 P(m) \tag{1}$$

The spectral entropy of a signal is a measure of its spectral power distribution and treats the signal's normalized power distribution in the frequency domain as a probability distribution $P(m)$ [7]. Thus, the Spectral Entropy corresponds to the Shannon Entropy of that distribution and has been proposed to measure the distribution of frequencies of a signal $x(n)$ at a given instant t, defined as:

$$H(t) = -\sum_{m=1}^{N} P(t,m) \log_2 P(t,m) \tag{2}$$

$$P(t,m) = \frac{S(t,m)}{\sum_f S(t,f)} \tag{3}$$

where N is the length of the signal, $S(m) = |X(m)|^2$ and $X(m)$ is the discrete Fourier transform of $x(n)$.

The proposed segmentation algorithm was developed with MATLAB R2018b using an Intel®i7-4700HQ processor and 8 GB of RAM. The segmentation algorithm is essentially composed by two parts: pre-processing and segmentation.

First the signal is filtered with a high-pass Butterworth filter of order 20 with a cut-off frequency of 30 kHz. The high-pass filter aims essentially to reduce the interference of low frequency noise, which can be generated by the mouse walking or touching the box, among other external factors.

After the pre-processing phase, the signal is divided into multiple non overlapping windows of 1 s and for each window the spectral entropy is computed on intervals of 1.9 ms (temporal resolution). Then, the values where spectral entropy vector are below a certain threshold (threshold 1 Fig. 3) are stored as the initial and final points of a vocalization. In addition, the value of the spectral entropy must also be smaller than a second threshold (threshold 2 Fig. 3) at some point between the stored instants, in order reduce the false positive rate of the algorithm. The way these thresholds were estimated is described in Sect. 3. Besides that, vocalizations that followed each other at intervals <15 ms were subsequently merged and considered as a single vocalization [6]. This last step was considered, since in some vocalizations there was a noticeable interruption of the vocalization or loss of power which make the algorithm segment two or more different vocalizations instead of just one.

The output produced by the algorithm is represented in Fig. 3, where the first threshold corresponds to the green line and the second to the red line.

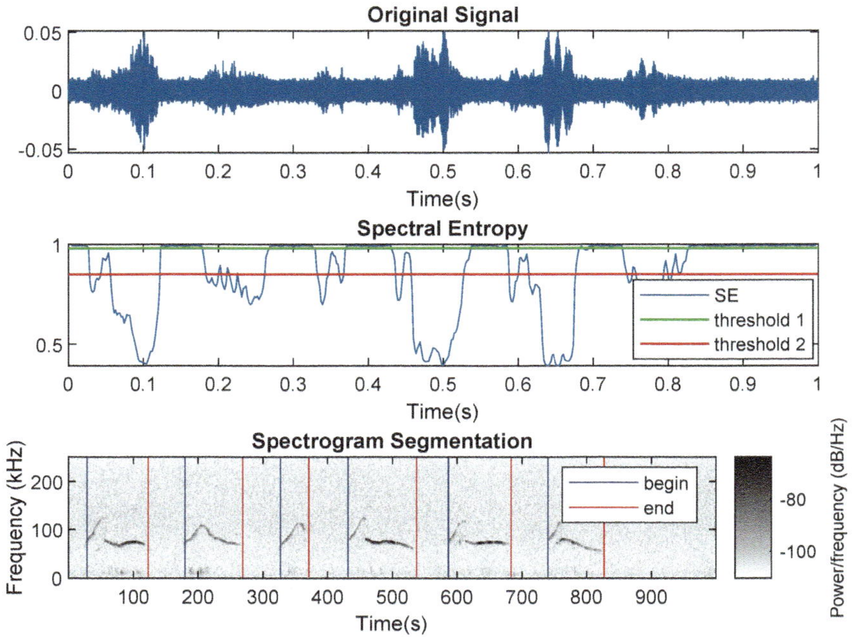

Fig. 3. Example of a signal segmentation.

3 Results

In order to evaluate the algorithm's performance, the results obtained by the automatic segmentation algorithm were compared to the results obtained with the manual annotation, precision and recall were considered as performance values (see Eqs. 4 and 5).

$$Precision = \frac{tp}{tp + fp} \tag{4}$$

$$Recall = \frac{tp}{tp + fn} \tag{5}$$

In Eqs. 4 and 5, the true positives value (tp) represents the number of vocalizations properly segmented, the false positives value (fp) represents the number of instances detected as vocalizations by the automatic algorithm but do not correspond to real vocalizations and the false negatives value (fn) represents the number of real calls that were not detected by the algorithm.

Furthermore, to test the robustness of the algorithm, different levels of Gaussian white noise were added to the signal before being segmented. This way, it was possible to check if there were any significant drops in performance and at which noise levels.

As mentioned in Sect. 2.2, the segmentation algorithm has two different thresholds with different purposes. The first threshold was estimated by observation in order to be as higher as possible without reaching the top baseline of the spectral entropy of the signals (about 0.98) and it was settled to 0.97, whereas the second one was tested with multiple values. The search process for the value of the second threshold is illustrated in Fig. 4, where the values of precision and recall obtained with the segmentation algorithm for the different values of the second threshold are represented.

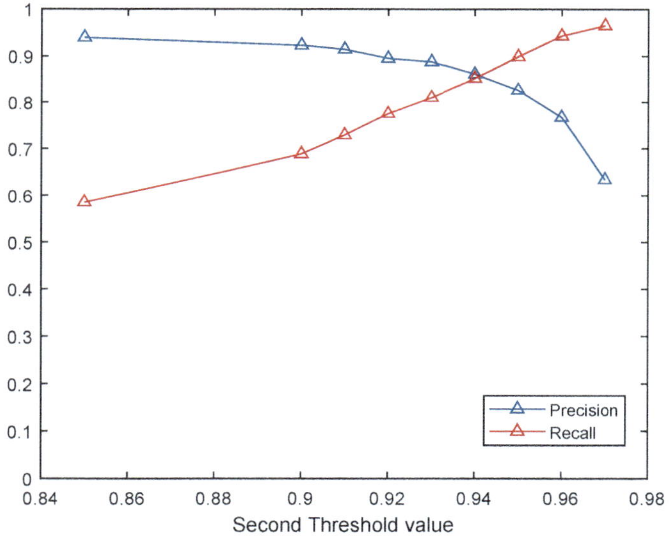

Fig. 4. Precision and recall across the different values tested for the second threshold.

After finding the range of values for the second threshold corresponding to the best performance, {0.94;0.95;0.96;0.97}, the algorithm was tested with Gaussian white noise addition, and several levels of signal to noise ratios (SNRs) were considered. For each value of SNR, the segmentation algorithm was tested 5 times, and then the mean and the standard deviation of the precision and recall values were calculated (see Fig. 5).

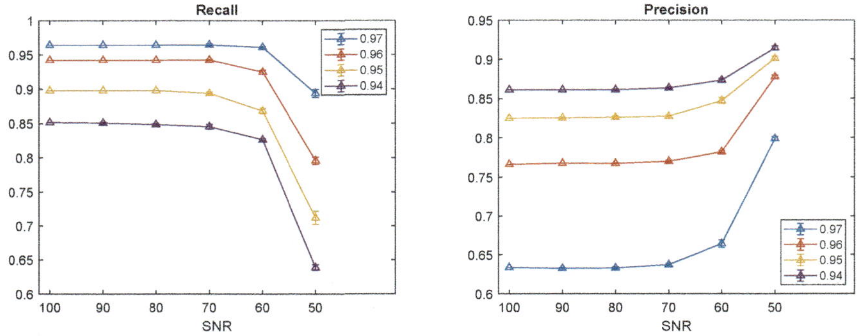

(a) Recall values for different SNRs. (b) Precision values for different SNRs.

Fig. 5. Algorithm's performance with different levels of noise addition.

4 Discussion

Through the analysis of Fig. 4 it is possible to observe that the value chosen for the second threshold establishes a clear trade-off between precision and recall. The closer this value is to the value of the first threshold (0.97), the higher is the number of vocalizations the algorithm is able to recognize, and, therefore, the higher are the true and false positives rates, which corresponds to a higher recall and lower precision values, accordingly. This happens because spectral entropy translates, in some way, the disorder introduced in the system. So, if the second threshold is higher, the algorithm is more permissive since the sounds it recognizes as syllables do not have to be as strong in terms of power. On the other hand, if the value of the second threshold is lower, the precision value is far superior to the recall one, which means that the majority of the sounds recognized are indeed real vocalizations. However, the algorithm is unable to segment a bigger number of calls, especially the ones with lower spectral power, resulting in a lower recall value.

As previously mentioned, the segmentation algorithm was also tested with the addition of Gaussian white noise to the signals. Through the analysis of Fig. 5, it is possible to observe that for values of SNR higher than 60, both the recall and precision values are kept relatively constant for all values of the second threshold used, being the standard deviation between runs very small or even zero. Thus, as the algorithm's performance does not significantly change for low levels of SNR, it is possible to consider that the algorithm is robust to noise.

Using a fully automated segmentation algorithm based on spectral entropy has been proved to be an effective method to isolate mice vocalizations, since the algorithm was able to detected up to 97% of the vocalizations in the signals annotated for this study, using 0.97 as the second threshold value. However, the method proposed in this paper tends to have a higher rate of false positive vocalizations in order to achieve higher recall values, which can be observed in Fig. 4. This phenomenon can be explained by the fact that the signals used in this work contain lots of noisy sounds with similar power when compared to the real vocalizations, which led to the increase of the false positives rate. However, the rate of false positive vocalizations can diminish with the fine tuning of the lowest threshold used in the algorithm, since it was possible to obtain 0.85 and 0.86 of recall and precision values with the threshold set to 0.94, and 0.90 and 0.83 with 0.95, accordingly, which provides a more balanced, yet good, performance.

5 Conclusion

The ultrasonic vocalizations recording and analysis is a valuable tool for the behavioral study of diverse mice models. When studying these bioacoustics events, the detection and isolation of these events is considered one of the crucial steps to undertake.

One major hurdle in USV studies is definitely the detection task, since it can be troubled by different sources of noise, namely produced by the animals themselves. This means that sometimes can be quite difficult to distinguish between an ultrasonic vocalization produced by a mouse and simple noise, even for a trained human.

In this study, we propose a relatively simple and reliable algorithm capable of achieving great performance detecting mice ultrasonic vocalizations, even when artificial noise is added to the signals. With the spectral entropy based algorithm and the tuning of the thresholds used, it is possible to accurately detect most USVs, and, therefore, study the influence of vocal communication in mice. This work will serve as a basis for future development of a classification system capable of segment audio signals and then classify the vocalizations comprised in it, which will allow further studies of animal models in different situations.

Acknowledgments. This work was supported by FCT fellowship SFRH/BPD/112863/2015 to L.I.P and FLAD Life Sciences 2016, BIGDATIMAGE (CENTRO-01-0145-FEDER, 000016) to M.C.B.

References

1. Chabout, J., Sarkar, A., Dunson, D.B., Jarvis, E.D.: Male mice song syntax depends on social contexts and influences female preferences. Front. Behav. Neurosci. **9**, 1–16 (2015)
2. Coffey, K.R., Marx, R.G., Neumaier, J.F.: DeepSqueak: a deep learning-based system for detection and analysis of ultrasonic vocalizations. Neuropsychopharmacology **44**, 859–868 (2019)
3. Grimsley, J.M.S., Monaghan, J.J.M., Wenstrup, J.J.: Development of social vocalizations in mice. PLoS ONE **6**(3), e17460 (2011)
4. Harma, A.: Automatic identification of bird species based on sinusoidal modeling of syllables. In: 2003 IEEE International Conference on Acoustics Speech, Signal Processing 2003, Proceedings, (ICASSP 2003), vol. 5, p. V–545–8. IEEE (2004)
5. Holy, T.E., Guo, Z.: Ultrasonic songs of male mice. PLoS Biol. **3**(12), e386 (2005)

6. Ivanenko, A., Watkins, P., van Gerven, M.A.J., Hammerschmidt, K., Englitz, B.: Classification of mouse ultrasonic vocalizations using deep learning. bioRxiv, p. 358143, January 2018
7. MathWorks: Spectral entropy (2019)
8. Portfors, C.V.: Types and functions of ultrasonic vocalizations in laboratory rats and mice. J. Am. Assoc. Lab. Anim. Sci. **46**(1), 28–34 (2007)
9. Potamitis, I., Ntalampiras, S., Jahn, O., Riede, K.: Automatic bird sound detection in long real-field recordings: applications and tools. Appl. Acoust. **80**, 1–9 (2014)
10. Scattoni, M.L., Gandhy, S.U., Ricceri, L., Crawley, J.N.: Unusual repertoire of vocalizations in the BTBR T+tf/J mouse model of autism. PLoS ONE **3**(8), e3067 (2008)
11. Vajda, S., Shannon, C.E., Weaver, W.: The mathematical theory of communication. Math. Gaz. **34**(310), 312 (1950)
12. Vinet, L., Zhedanov, A.: Rodent Bioacoustics. Springer Handbook of Auditory Research, vol. 67. Springer International Publishing, Cham (2018)

PCG-Decompositor: A New Method for Fetal Phonocardiogram Filtering Based on Wavelet Transform Multi-level Decomposition

Annachiara Strazza, Agnese Sbrollini, Marica Olivastrelli, Agnese Piersanti, Selene Tomassini, Ilaria Marcantoni, Micaela Morettini, Sandro Fioretti, and Laura Burattini(✉)

Department of Information Engineering, Università Politecnica delle Marche, Ancona, Italy
l.burattini@univpm.it

Abstract. Fetal phonocardiography (FPCG) is a non-invasive acoustic recording of fetal heart sounds (fHS). The fHS auscultation plays an important diagnostic role in assessing fetal wellbeing. Typically, FPCG is a non-stationary signal corrupted by the presence of noise. Thus, high-amplitude noise makes detection of FPCG waveforms challenging. Thus, appropriate filtering procedures have to be applied in order to make FPCG clinically usable. In the recent years, Wavelet transformation (WT) filtering has been proposed. In particular, aim of this study is to propose a new method based on WT multi-level decomposition filtering: PCG-Decompositor. To this aim, PCG-Decompositor based on Coiflets mother Wavelet (4th order, 9 levels of decomposition) was applied to 119 real FPCG tracings, all available in Physionet. PCG-Decompositor is a dependent thresholding technique based on FPCG multi-level decomposition analysis. Performances of PCG-Decompositor are computed against soft-thresholding denoising technique (STDT) in terms of Root Mean Square Error (RMSE) and fetal heart rate (fHR). In terms of fHR, PCG-Decompositor and STDT are compared between themselves and also with the so-called annotations, given by the average fHR using a simultaneous cardiotocography analysis. Original signal to noise ratio (SNR) values ranged from 7.1 dB to 24.4 dB; after application of PCG-Decompositor, SNR increased significantly, ranging from 9.7 dB to 26.9 dB ($P < 10^{-7}$). Moreover, PCG-Decompositor showed a lower dispersion than STDT (RMSE: 0.7 dB vs. 1.2 dB), introduced no FPCG signal delay and left fHR unaltered. Thus, PCG-Decompositor could be a suitable and robust technique to denoise FPCG signals.

Keywords: Fetal phonocardiography · Wavelet transform · Multi-level decomposition

1 Introduction

Fetal phonocardiography (FPCG) is a very useful diagnostic technique for the evaluation of fetal health condition. It consists of an electronic recording of the fetal heart sounds (fHS), which are mainly two: the first heart sound (S1) and the second heart

J. Henriques et al. (Eds.): MEDICON 2019, IFMBE Proceedings 76, pp. 47–53, 2020.
https://doi.org/10.1007/978-3-030-31635-8_6

sound (S2) [1], representing the closure of atrioventricular valves and the closure of semilunar valves, respectively. FPCG clinical utility relies on S1 and S2 identification. Still, FPCG signals are heavily contaminated by noise due to fetus movement and maternal heart sounds, breathing, and muscular movements [2]. Additionally, fHS are distorted by the variations in the transmission path due to the presence of amniotic fluid and uterus muscular wall [3–5]. Therefore, automatic detection of S1 and S2 sounds is a challenging task [5] that requires previous noise filtering. Wavelet Transform (WT) is a well-known technique for the analysis of non-stationary signals in time-scale domain [6]. Previous studies showed that WT technique is a promising method for FPCG denoising [5, 7–9], efficacy of which depends on choice of mother wavelet, decomposition level and threshold settings. The Coiflet and Soft-thresholding function are respectively considered the optimal mother Wavelet and rule for FPCG denoising [7, 10], while no proper selection of decomposition level number useful for FPCG reconstruction has been so far identified. Thus, the aim of the present study is to propose PCG-Decompositor a new method for FPCG denoising which uses a 4th-order Coiflet mother Wavelet but performs a multi-level decomposition analysis to select the optimal thresholding rule for FPCG denoising and reconstruction. Performances of PCG-Decompositor is evaluated by comparing it against well-known soft-thresholding denoising technique (STDT) in terms of Root Mean Square Error (RMSE) and of mean fetal heart rate (fHR). STDT performed a scale-based decomposition and a thresholding of FPCG coefficients based on a WT analysis. The noise is represented by WT coefficients at finer scales. Thus, STDT, using Donoho threshold, discarded these noise coefficients [10]. Moreover, PCG-Decompositor and STDT are compared against annotations, given by the average fHR obtained by analyzing simultaneously acquired cardiotocographic (CTG) data [11, 12].

2 Materials and Methods

2.1 Real Data

Clinical data consisted of 119 FPCG recordings belonging to the Shiraz University Fetal Heart Sounds Database (SUFHSDB) freely available on the PhysioNet web site. This database is constituted by recordings made from 109 pregnant women with a digital JABES Electronic stethoscope in wide-band mode. In case of twins the data were collected twice. Average recording duration was 90 s; sampling frequency was usually 16000 Hz (8000 Hz in few cases) with 16-bit quantization [11, 13]. For each FPCG, fHR was indirectly computed by using the simultaneously acquired CTG recordings.

2.2 PCG-Decompositor

PCG-Decompositor is an algorithm for FPCG denoising based on a multi-level decomposition analysis.

In order to remove part of the noise present in FPCG recordings, a 6th order Butterworth band-pass filter (cut-off frequency: 20–120 Hz) is applied (Fig. 1) [14]. Band-pass filtering allows not only to reduce FPCG signal bandwidth before applying

PCG-Decompositor, but also to increase frequency resolution of each WT decomposition level [8]. After this pre-processing step, PCG-Decompositor is applied. PCG-Decompositor performs a FPCG denoising by means of a WT-based procedure. WT is a time-frequency processing method that quantifies temporal changes of the frequency content of non-stationary signals [6]. WT of the input signal x(t) is defined as (Eq. 1):

$$\mathrm{WT}x(a,b) = \int x(t)\Psi a, b(t)dt \quad a \neq 0 \tag{1}$$

where $x(t)$ is the input signal, a the scaling factor, b the translation parameter and $\Psi a, b(t)$ is the transforming function called mother wavelet. The mother wavelet is given by (Eq. 2):

$$\Psi a, b(t) = \frac{1}{\sqrt{a}}\Psi\left(\frac{t-b}{a}\right) \tag{2}$$

In this work, decomposition, denoising and reconstruction were performed by mother wavelet Coiflets of 4th order with 9 levels of decomposition, providing the higher result for signal-to-noise ratio (SNR) for all types of FPCG signal. PCG-Decompositor allows the decomposition of FPCG into approximation (a_n) and detail (d_n) information using a pair of finite impulse response filters, which are low-pass and high-pass, respectively [6]. The process can be described as a tree, which at each step decomposes the low-pass filter into further lower and higher frequency coefficients. The number of steps of this iterative tree procedure is defined by the number of decomposition levels. Thus, a multi-level multiresolution analysis is used to generate an improved reconstruction algorithm. A level-based thresholding criterion is established considering the information contained in each detail. Details, at higher frequencies, are related to noise. In order to obtain a clean FPCG, noise details are set to

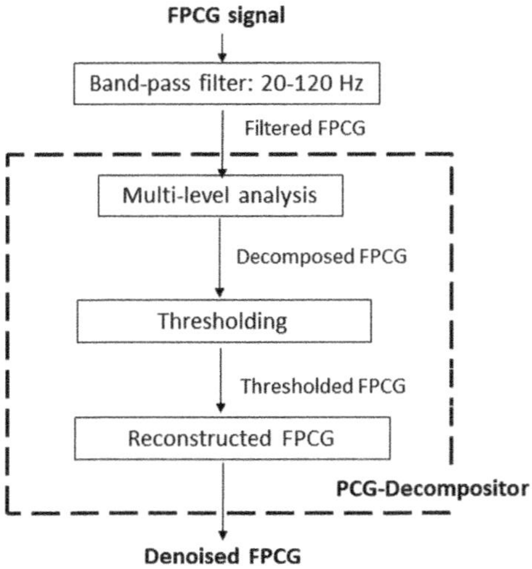

Fig. 1. Block diagram of PCG-Decompositor.

zero. Instead, standard thresholding is applied to details, at lower frequencies, with a higher content of information. Soft thresholding was chosen as standard thresholding [10]. Thus, by means of this thresholding criterion application, PCG-Decompositor discards WT level containing noise and uses the remaining ones to reconstruct the denoised FPCG signal. The block diagram of PCG-Decompositor is depicted in Fig. 1.

2.3 Performance Evaluation and Statistics

Signals denoised by PCG-Decompositor and STDT are evaluated in term of SNR and fHR. Original signal and noise amplitudes were not available to reproduce SNR values reported in Physionet database. Thus, they were recomputed according to the definition reported in [3]. SNR values were calculated before (SNRbefore) and after PCG-Decompositor (SNR_PCG-Dec) and STDT (SNR_STDT).

Then, fHR values are assessed by PCG-Delineator, a novel and efficient algorithm for automatic heart sounds detection in FPCG [14]. PCG-Delineator efficiently identifies, starting from denoised FPCG, S1 sounds. Thus, fHR and the mean fHR are extracted, considering the difference between two consecutive S1 sounds [14]. Values of fHR were calculated after PCG-Decompositor (fHR_PCG-Dec) and STDT (fHR_STDT). In order to evaluate their correctness, the Pearson correlation coefficient (ρ) of SNR was computed, when comparing SNRbefore, SNR_PCG-Dec and SNR_STDT. Performance of PCG-Decompositor is computed in terms of RMSE of SNR when comparing denoised and raw FPCG signals and of mean fHR against well-known STDT. PCG-Decompositor and STDT techniques are compared between themselves and also with the so-called reference annotations, given by the mean fHR using CTG, reported in [11].

Non-normal variable distributions were described in terms of 50th [25th; 75th] percentiles and compared using the Wilcoxon Rank-Sum test for equal medians. Statistical level of significance was set at 0.05 in all cases.

3 Results

An example of raw (grey) and denoised (black) FPCG signal with PCG-Decompositor is shown in Fig. 2. SNRbefore values distribution ranged from (15.2 dB [13.7; 16.7]) and was found to be characterized by a lower value than SNR_STDT (15.6 dB [14.1; 17.8]) and by a significantly lower value than SNR_PCG-Dec (16.9 dB [15.1; 18.9]; $P < 10^{-7}$). Regression lines computed for PCG-Decompositor and STDT methods are reported: SNR_PCG-Decompositor = 0.8 • SNRbefore + 5.2 for STDT, SNR_STDT = 0.9 • SNRbefore + 3.4 for PCG-Decompositor, respectively (Fig. 3). SNRbefore and SNR_PCG-Dec/SNR_STDT were characterized by different RMSE values associated by a high correlation (ρ) and statistically significant (P), as reported in Table 1. Considering fHR of reference annotations [11], fHR of PCG-Decompositor was maintained (140.6 [135.2; 146.3] bpm; Reference annotations: 140.5[135.2; 146.3] bpm; $P > 0.05$; Table 2). Instead, fHR of STDT was altered (146.6[143.7; 150.2] bpm; Reference annotations: 140.5[135.2; 146.3] bpm; $P < 0.05$; Table 2).

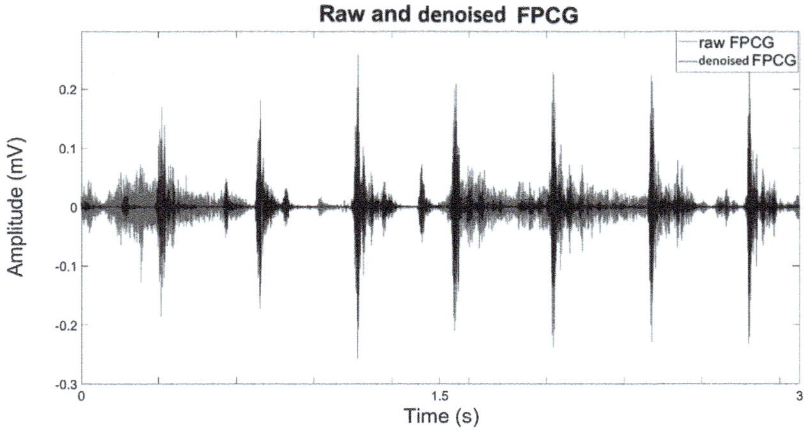

Fig. 2. Raw (grey) and denoised, after PCG-Decompositor application (black), FPCG signal.

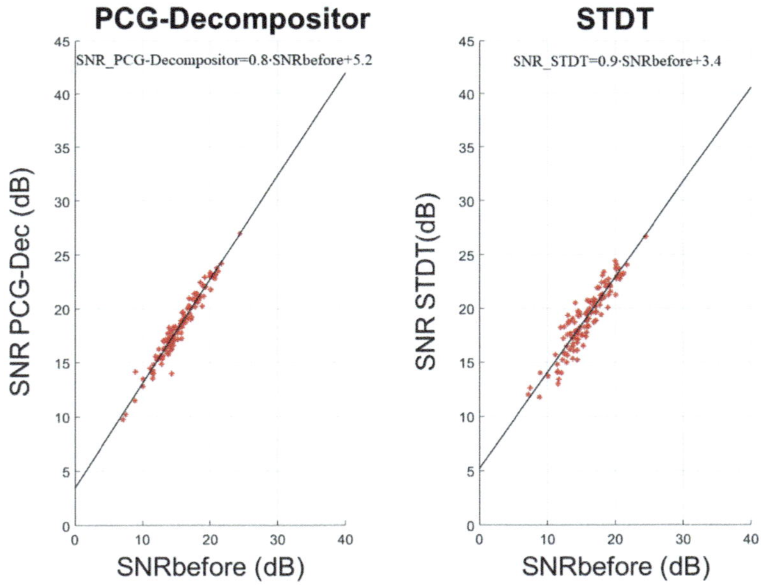

Fig. 3. Regression lines computed for PCG-Decompositor and STDT in terms of SNR. SNR values were calculated before (SNRbefore) and after PCG-Decompositor (SNR_PCG-Dec) and STDT (SNR_STDT).

Table 1. RMSE, correlation coefficient and P-value for STDT and PCG-Decompositor, in terms of SNR.

Method	RMSE (dB)	ρ	P
STDT	1.2	0.924	$<10^{-4}$
PCG-Dec	0.7	0.976	$<10^{-77}$

Table 2. fHR values for Annotations (Ann.), PCG-Decompositor and STDT. *P < 0.05 when performing the Wilcoxon Rank-Sum test for the fHR values obtained with PCG-Dec and STDT comparing with the annotations.

	Ann.	PCG-Dec	STDT
fHR (bpm)	140.5[135.2; 146.3]	140.6[135.2; 146.3]	146.6*[143.7; 150.2]

4 Discussion

The aim of the study was to propose PCG-Decompositor as an efficient method based on a multi-level decomposition analysis, in order to denoise FPCG. For this purpose, 119 real FPCG tracings affected by different kind of noise were analyzed. Coiflet mother wavelet (4th order with 9 levels of decomposition) was chosen for WT analysis. PCG-Decompositor focuses on the choice of a new thresholding rule, based on a proper selection of decomposition level number useful for FPCG denoising and reconstruction. The criterion is based on the application of soft thresholding only to the levels with significant information content (low frequencies). The level associated to noise, high frequencies, were set to zero.

Performance of PCG-Decompositor is computed against STDT in terms of RMSE and fHR. Results showed that SNR_PCG-Dec values increased significantly respect to the SNRbefore; instead SNR_STDT values remained almost unaltered. Computed RMSE values showed a lower dispersion (lower RMSE) of PCG-Decompositor (Table 1). This result confirmed PCG-Decompositor reliability against STDT for FPCG filtering. Moreover, PCG-Decompositor introduced no FPCG signal delay, maintaining unaltered fHR (Table 2).

5 Conclusion

In conclusion, PCG-Decompositor is a novel and efficient method for FPCG filtering, in order to make easier the detection of fHS, clearly contributing to better evaluation of the fetal heart functionality. PCG-Decompositor seems to be a useful tool for clinical purposes.

Acknowledgements. The authors wish to thank Prof. Reza Sameni for sharing the cardiotocographic data, simultaneously recorded to the fetal phonocardiographic data, without which reference values of the fetal heart rate in the experimental study could not be obtained.

References

1. Adithya, P.C., Sankar, R., Moreno, W.A.: Trends in fetal monitoring through phonocardiography: challenges and future directions. Biomed. Signal Process. Control **33**, 289–305 (2017)
2. Chourasia, S., Tiwari, A.K.: A review and comparative analysis of recent advancements in fetal monitoring techniques. Crit. Rev. Biomed. Eng. **36**, 335–373 (2008)
3. Sbrollini, A., Strazza, A., Caragiuli, M., et al.: Fetal phonocardiogram denoising by wavelet transformation: robustness to noise. Comput. Cardiol. **44**, 1–4 (2017). https://doi.org/10.22489/cinc.2017.331-075
4. Messer, S.R., Argzarian, J., Abbott, D.: Optimal wavelet denoising for phonocardiograms. Microelectron. J. **32**, 931–941 (2001)
5. Vaisman, S., Salem, S.Y., Holcberg, G., et al.: Passive fetal monitoring by adaptive wavelet denoising method. Comput. Biol. Med. **42**(2), 171–179 (2012)
6. Mallat, S.G.: A theory for multiresolution signal decomposition: the wavelet representation. IEEE Trans. Pattern Anal. Mach. Intell. **11**(7), 674–693 (1989)
7. Choursia, V.S., Mittra, A.K.: Selection of mother wavelet and denoising algorithm for analysis of foetal phonocardiographic signals. J. Med. Eng. Technol. **33**(6), 442–448 (2009)
8. Koutsiana, E., Hadjileontiadis, L.J., Chouvarda, I.: Fetal heart sounds detection using wavelet transform and fractal dimension. Front. Bioeng. Biotechnol. **5**, 49 (2017)
9. Tomassini, S., et al.: Wavelet denoising of fetal phonocardiography: a comparative analysis. Math. Biosci. Eng. **16**(5), 6034–6046 (2019). https://doi.org/10.3934/mbe.2019302
10. Donoho, D.L.: De-noising by soft-thresholding. IEEE Trans. Inf. Theory **41**(3), 613–627 (1995)
11. Samieinasab, M., Sameni, R.: Fetal phonocardiogram extraction using single channel blind source separation. In: ICEE 2015 - Proceedings of the 23rd Iranian Conference on Electrical Engineering, pp. 78–83 (2015)
12. Sbrollini, A., Agostinelli, A., Burattini, L., et al.: CTG analyzer: a graphical user interface for cardiotocography. In: 39th Annual International Conference of the IEEE Engineering in Medicine and Biology Society, Jeju Island, pp. 2606–2609 (2017)
13. Goldberger, A.L., Amaral, L.A., Glass, L., et al.: PhysioBank, PhysioToolkit, and PhysioNet. Circ. J. **101**(23), 215–229 (2000)
14. Strazza, A., Sbrollini, A., Di Battista, V., et al.: PCG-Delineator: an efficient algorithm for automatic heart sounds detection in fetal phonocardiography. Comput. Cardiol. **45**, 8743707 (2018). https://doi.org/10.22489/cinc.2018.045

Muscular Co-contraction Detection: A Wavelet Coherence Approach

Annachiara Strazza, Federica Verdini, Andrea Tigrini, Stefano Cardarelli, Alessandro Mengarelli, Sandro Fioretti, and Francesco Di Nardo(✉)

Department of Information Engineering, Università Politecnica delle Marche, Via Brecce Bianche, 60131 Ancona, Italy
f.dinardo@univpm.it

Abstract. Muscular co-contraction is defined as the activity of agonist and antagonist muscles around a joint, enhancing joint stability and balance. The quantitative assessment of muscle co-contractions would be meaningful for deepening the comprehension of this physiological mechanism. Thus, the purpose of this work is to quantify muscle co-contraction using a Wavelet transform-based coherence analysis of sEMG signal during straight walking. To this purpose, sEMG from tibialis anterior (TA) and gastrocnemius lateralis (GL) and basographic signals were acquired in five healthy subjects during walking. Basographic signals were analyzed to quantify foot-floor contact. sEMG signals were processed using Wavelet Transform Coherence (WTC) to identify muscular co-contractions. Daubechies (order 4 with 6 levels of decomposition) was chosen as mother wavelet. A denoising algorithm based on a soft thresholding was applied for removing noise from raw signals. Denoised signals were considered to achieve WTC function, a well-known localized statistical assessment of cross-correlation between signals. Thus, in this work, WTC cross-correlation could be considered to assess muscular co-contraction. This methodology applied to TA and GL signals was able to detect GL/TA co-contractions during hell-strike (0–10% of GC) phase and during P-phase (54.2–68.3% of GC), matching with literature. Moreover, WTC approach was able to provide also the frequency band of information content for muscle co-contractions: 32–65 Hz for H-phase co-contraction and 16–32 Hz for P-phase co-contraction. In conclusion, this study proposed WTC analysis as a reliable method to assess muscle co-contraction in time-frequency domain.

Keywords: Muscle co-contraction · Time-frequency analysis · Wavelet coherence · Surface EMG

1 Introduction

Muscle co-contraction, i.e. the concomitant recruitment of agonist and antagonist muscle groups around a joint, represents an important and common strategy for the control of voluntary movement during walking [1]. It is acknowledged that muscle co-contraction has an important role in movement regulation during motor learning activities, enhancing joint stability, in particular ankle and knee joint [2]. Nowadays,

J. Henriques et al. (Eds.): MEDICON 2019, IFMBE Proceedings 76, pp. 54–61, 2020.
https://doi.org/10.1007/978-3-030-31635-8_7

muscular co-contraction represents a field of particular interest in literature as it is considered as a marker for the evaluation of particular pathologies [3]. Thus, quantitative assessment of muscle co-contractions is meaningful for deepening the comprehension of this neurophysiological mechanism and for discriminating with conditions of dysfunction of the central nervous system, such as traumatic brain injury, stroke, cerebral palsy, or Parkinson's disease.

Currently there is no standard method to detect muscle co-contraction. However, surface electromyography (sEMG) has acknowledged as a common experimental tool to assess time duration and magnitude of muscle co-contraction activity. Different methodologies for EMG processing in time domain were developed to this purpose [3, 4], while poor information is available in literature to assess the frequency band where the muscles showed the reported co-contractions. Thus, a reliable approach able to quantify the information included in the concomitant evaluation of time-varying interval of frequency band of the co-contraction could be valuable.

To this aim, the present study is to propose a new method for muscular co-contraction detection based on advanced techniques of EMG-signal processing. Wavelet transform (WT) analysis of sEMG signals seems to be a suitable approach for this purpose, because of its ability to provide the time-frequency representation of a signal without losing resolution in time or frequency [5]. In particular, the Wavelet Transform Coherence (WTC) is a method able to assess the existence and strength of linear coupling between two non-stationary signals [6]. Thus, WTC analysis of sEMG signal is the technique adopted in the present study, in the attempt of developing a novel approach for assessing muscle co-contraction.

2 Materials and Methods

2.1 Real Data

Five healthy young adults (3 females and 2 males) were recruited for the study. Mean ($\pm$SD) characteristics are: age 22.7 ± 0.5 years; height 166 ± 8 cm; weight 53.6 ± 8.1 kg; body mass index (BMI) 18.9 ± 1.5 kg $\cdot$ m^{-2}. Exclusion criteria included pathological or chronic joints pain, neurological pathology, orthopedic surgery within previous year, abnormal gait or BMI ≥ 25. Participants signed informed consent. The present research was undertaken in compliance with the ethical principles of Helsinki Declaration and approved by institutional expert committee.

2.2 Signal Acquisition and Processing

The scheme of the entire experimental processing, from data acquisition to muscle co-contraction detection is reported in Fig. 1.

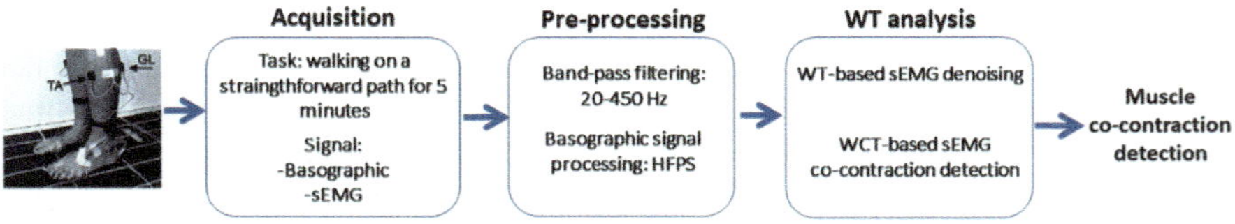

Fig. 1. Scheme of the entire experimental processing, from data acquisition to muscle co-contraction detection.

Participants were bilaterally instrumented with surface EMG (sEMG) probes and three foot-switches. sEMG signals were acquired through the Step32 system (Medical Technology, Italy), with a sampling rate of 2 kHz and a resolution of 12 bit. Fixed geometry sEMG probes (Ag/Ag-Cl disk; electrode diameter 0.4 cm; inter-electrode distance 0.8 cm; gain 1000, input impedance > 1.5 GΩ, CMRR > 126 dB), placed on TA and GL, following the SENIAM recommendations for electrode location-orientation over muscles [7], were used to acquire sEMG signals. Foot-switches were placed beneath the heel and the first and fifth metatarsal heads of the foot; each sensor had a surface of 1.21 cm^2 and an activation force equal to 3 N. After positioning the sensors, subjects were asked to walk barefoot over the floor for 5 min at their natural pace, back and forth over a 12-m long straight track.

Footswitch signals were debounced, converted to four levels, Heel contact (H), Flat foot contact (F), Push-off (P), Swing (S), and processed to segment and classify the different gait cycle (GC). Then, foot-switch signals were processed to identify each gait cycle (GC) [8]. sEMG signals were amplified and band-pass filtered (20–450 Hz). sEMG signals were processed by means of Wavelet Transform (WT). WT is a time-frequency analysis method that quantifies temporal changes of the frequency content of non-stationary signals without losing resolution in time or frequency [9, 10]. WT of the input signal x(t) is defined as the inner product between the analyzed signal *x(t)* and the basis function $\psi_{a,b}(t)$, as in the following expression (Eq. 1):

$$\mathrm{WT}x(a, b) = \int x(t)\psi a, b(t)dt \quad a \neq 0 \tag{1}$$

where *x(t)* is the input signal, *a* the scaling factor, *b* the translation parameter and $\psi a, b(t)$ is the transforming function called mother wavelet. The mother wavelet is given by (Eq. 2) [5, 8]:

$$\psi a, b(t) = \frac{1}{\sqrt{a}}\psi\left(\frac{t-b}{a}\right) \tag{2}$$

In this work, mother wavelet Daubechies of order 4 with 6 levels of decomposition (db4) has been chosen to implement the wavelet transform, for its suitability to detect signal changes and for the shape similarity between its shape and the shape of motor unit action potentials [9]. WT has been applied for removing noise from sEMG [11, 12]. WT denoising is based on decomposition of the signal, modification of detail coefficients with respect to a defined soft threshold [13] and then signal reconstruction with respect to the new coefficients. Noise coefficients are concentrated in low values of scale parameter, corresponding to high frequency ranges while sEMG signal is distributed in high and low frequency ranges (low and large values of scale parameter, respectively). Thus, WT denoising involves adopting a statistical regression of noisy coefficients over time to obtain a nonparametric estimation of the reconstructed signal without noise. The decomposition processing allows removing the level which is seriously corrupted by noise. Coefficients below soft threshold are regarded as noise and are discarded. The signal can then be reconstructed without significant loss of information in the signal content. sEMG signal was reconstructed by revised WT coefficients. The SNR values were computed according to the following definition (Eq. 3):

$$SNR = 10 * \log \frac{\left(\sigma_s^2\right)}{\left(\sigma_n^2\right)} \tag{3}$$

where σ_s and σ_n are the standard deviation of signal and noise, respectively. SNRraw represents the SNR values of sEMG signals before the WT denoising; SNRdenoised is the SNR values of denoised signals after the WT denoising.

The assessment of muscular co-contractions in time-frequency domain has been identified starting from WTC function. WCT analysis provides the measurements of time-varying correlation of two sets of denoised sEMG data as a function of frequency [6]. WTC was computed as following (Eq. 4):

$$R_n^2(s) = \frac{\left|S\left(s^{-1} W_n^{XY}(s)\right)\right|^2}{S\left(s^{-1}\left|W_n^X(s)\right|^2\right) * S\left(s^{-1}\left|W_n^Y(s)\right|^2\right)} \tag{4}$$

where S is a smoothing operator, in time and frequency, W^X and W^Y are the wavelets for each sEMG time series, W^{XY} is the cross wavelet transform. Thus, in this work, WTC time-varying correlation could be considered to assess muscular co-contraction.

3 Results and Discussion

A representative example of raw and denoised sEMG signals from tibialis anterior and gastrocnemius lateralis is presented in Fig. 2, panel A and B respectively. Comparable results were obtained in subjects 2–5.

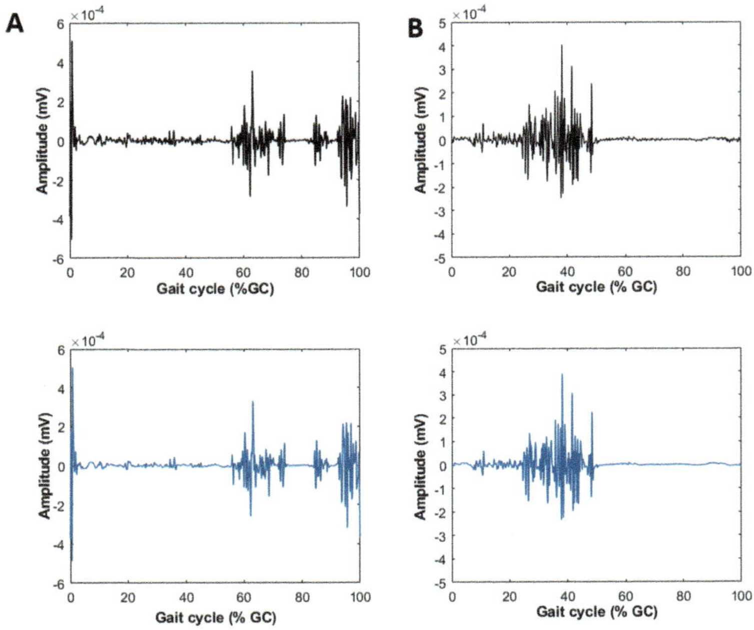

Fig. 2. Example of raw and denoised tibialis anterior (panel A) and gastrocnemius lateralis (panel B) sEMG signal during gait cycle.

SNR values of sEMG signals before the WT filtering (SNRraw) and SNR values of denoised signals after the WT filtering (SNRclean) of TA and GL muscles are reported in Table 1.

Table 1. SNR for tibialis anterior and gastrocnemius lateralis before (SNRraw) and after (SNRclean) WT denoising alghoritm.

Subjects	Tibialis Anterior		Gastrocnemius Lateralis	
	SNRraw	SNRclean	SNRraw	SNRclean
#1	16.4	17.2	17.9	19.1
#2	16.8	17.6	16.7	18.3
#3	14.9	16.2	16.1	17.1
#4	17.7	19.1	18.6	19.2
#5	15.2	17.4	18.9	20.2

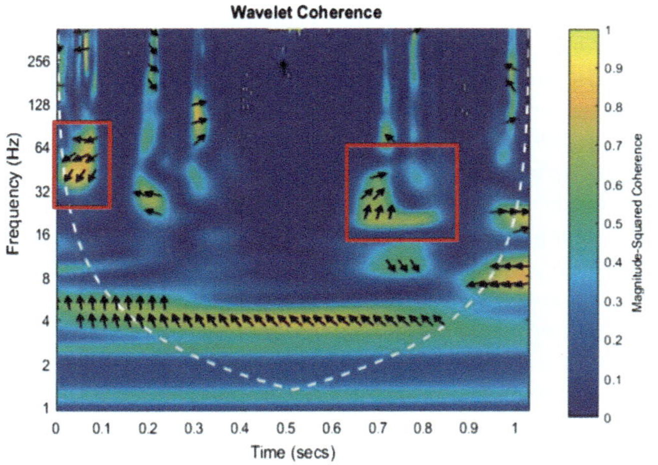

Fig. 3. Wavelet Coherence between TA and GL muscles. Muscle co-contraction is highlighted by a red box.

In a two-dimensional graph, WCT are expressed in function of time (secs) and frequency (Hz). The representation of WCT between sEMG signals from TA, GL is reported in Fig. 3. The white dashed line (Fig. 3) represents the cone of influence, i.e. the region of the wavelet spectrum where the autocorrelation of wavelet power at each scale is maximum.

In time domain, muscular co-contraction is identified as the overlapping periods among activation intervals of the considered muscles in the same strides, such as reported in Fig. 4.

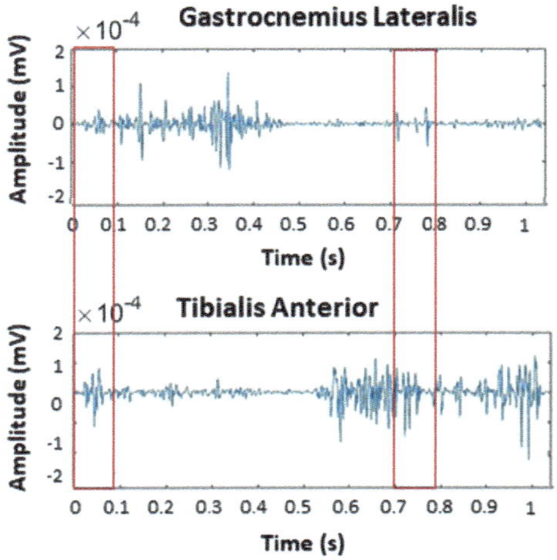

Fig. 4. GL (upper panel) and TA (lower panel) activations. Muscle co-contraction between two sEMG signals is highlighted by a red box

WTC analysis was able to detect GL/TA co-contractions in time and frequency domain. In time domain, the localization of regions with maximum coherence was assessed during H-phase, between the beginning and 0.12 s (0–10% of GC), and during P-phase, between 0.65 and 0.82 s (54.2–68.3% of GC) (Fig. 3). The same result is shown in time domain (Fig. 4). In time domain, muscular co-contraction is identified as the overlapping periods among activation intervals of the considered muscles in the same strides. A muscular activity shorter than 30 ms does not yield an active control on a joint motion during gait [14], thus overlapping periods with a time length lower than 30 ms were discarded. Thus, WTC region of time-varying correlation could be considered as an actual co-contraction, matching with the literature [1, 3]. Moreover, WTC function was able to provide also the frequency band of information content for ankle muscle co-contraction. sEMG signals superimposed in the frequency range of 32–65 Hz for H-phase co-contraction and of 16–32 Hz for P-phase co-contraction (Fig. 3).

4 Conclusion

The present study could be suitable for both supporting WT approach for sEMG denoising and providing a quantitative assessment of muscle concomitant recruitment during walking. In conclusion, this study proposed WT coherence function as a reliable technique for the assessment of muscle co-contraction in time-frequency domain.

References

1. Olney, S.J.: Quantitative evaluation of cocontraction of knee and ankle muscles in normal walking. In: Biomechanics IX: International Congress of Biomechanics, pp. 431–436 (1983)
2. Falconer, K., Winter, D.A.: Quantitative assessment of co-contraction at the ankle joint in walking. Electromyogr. Clin. Neurophysiol. **25**, 135–149 (1985)
3. Rosa, M.C.N., Marques, A., Demain, S., et al.: Methodologies to assess muscle co-contraction during gait in people with neurological impairment - a systematic literature review. J. Electromyogr. Kinesiol. **24**, 179–191 (2014)
4. Di Nardo, F., Mengarelli, A., Maranesi, E., et al.: Assessment of the ankle muscle co-contraction during normal gait: a surface electromyography study. J. Electromyogr. Kinesiol. **25**, 347–354 (2015)
5. Rioul, O., Flandrin, P.: Time-scale energy distributions: a general class extending wavelet transforms. IEEE Trans. Signal Process. **40**, 1746–1757 (1992)
6. Sukiennik, P., Białasiewicz, J.T.: Cross-correlation of bio-signals using continuous wavelet transform and genetic algorithm. J. Neurosci. Methods **247**, 13–22 (2015)
7. Hermens, H.J., Freriks, B., Disselhorst-Klug, C., Rau, G.: Development of recommendations for SEMG sensors and sensor placement procedures. J. Electromyogr. Kinesiol. **10**, 361–374 (2000)
8. Agostini, V., Balestra, G., Knaflitz, M.: Segmentation and classification of gait cycles. IEEE Trans. Neural Syst. Rehabil. Eng. **22**, 946–952 (2014)
9. Farge, M.: Wavelet transforms and their applications to turbulence. Ann. Rev. Fluid Mech. **24**, 395–458 (1992)

10. Strazza, A., Verdini, F., Burattini, L., Fioretti, S., Di Nardo, F.: Time-frequency analysis of surface EMG signals for maximum energy localization during walking. In: IFMBE Proceedings, vol. 65, pp. 494–497 (2017)
11. Reaz, M.B.I., Hussain, M.S., Mohd-Yasin, F.: Techniques of EMG signal analysis: detection, processing, classification and applications (correction). Biol. Proced. Online **8**, 163 (2006)
12. Strazza, A., Verdini, F., Mengarelli, A., Cardarelli, S., Burattini, L., Fioretti, S., Di Nardo, F.: A time-frequency approach for the assessment of dynamic muscle co-contractions. In: IFMBE Proceedings, vol. 68, no. 2, pp. 223–226 (2018)
13. Donoho, D.L.: De-noising by soft-thresholding. Trans. Inf. Theory **41**(3), 613–627 (1995)
14. Bonato, P., D'Alessio, T., Knaflitz, M.: A statistical method for the measurement of muscle activation intervals from surface myoelectric signal during gait. IEEE Tran. Biomed. Eng. **45**(3), 287–299 (1998)

Calculation of Breath-by-Breath Oxygen Uptake in Asthmatic Patients by the "Independent Breath" Algorithm. Comparison with a Classical Approach

Maria Pia Francescato[1(✉)], Miloš Ajčević[2], Valentina Cettolo[1], Mario Canciani[3], and Agostino Accardo[2]

[1] Department of Medicine, University of Udine, Udine, Italy
mariapia.francescato@uniud.it
[2] Department of Engineering and Architecture, University of Trieste, Trieste, Italy
[3] Allergopneumology Clinic and Laboratory, ALPI Association, Udine, Italy

Abstract. Several computation algorithms are available to determine gas exchange on a breath-by-breath basis, each designed on the basis of different theoretical backgrounds, including the newly proposed "Independent breath" algorithm. The new algorithm was tested on the respiratory signals acquired from 11 asthmatic patients and 20 well-matched healthy controls, comparing results also with those provided by a "classical" algorithm commonly applied by other laboratories. Oxygen, carbon dioxide fractions, and ventilatory flow were recorded at the mouth continuously over 26 min in all the volunteers at rest, during unloaded and moderate intensity cycling and subsequent recovery. Average oxygen uptake values calculated for the 4 steady-state conditions (over 3 min), as well as the corresponding standard deviations, were not significantly different between the two groups of subjects (MANOVA, Group effect, p = ns). Almost all the average oxygen uptake values provided by the two algorithms were overlapping, the large majority lying within 5% from the identity line. The corresponding standard deviations obtained for the "Independent breath" algorithm were lower than those obtained for the "classical" algorithm (MANOVA, Algorithm effect, $p < 0.001$), the slope of the regression line between them amounting to 0.672. In conclusion, because of its better precision, with similar accuracy, compared to the "classical" real-time breath-by-breath algorithm, the use of the "Independent breath" algorithm should be recommended, also in asthmatic patients.

Keywords: Respiratory signal processing · Breath-by-breath oxygen uptake · Gas fractions · Exercise · Asthma

1 Introduction

The study of O_2 uptake under various conditions is of interest either for clinicians and for researchers involved in different fields of investigation. Non-invasive measurements are usually made at the mouth, and different computation algorithms are then applied to determine the gas exchange on a breath-by-breath basis.

J. Henriques et al. (Eds.): MEDICON 2019, IFMBE Proceedings 76, pp. 62–74, 2020.
https://doi.org/10.1007/978-3-030-31635-8_8

Fluctuations are observed in breath-by-breath gas exchange data, due to intrinsic noise and, mainly, to the changes in lung gas stores [1]. The behavior of O_2 uptake during transients is of particular interest, since it provides valuable diagnostic and prognostic information. Consequently, the application of a de-noising low pass filter would alter the parameters of interest.

A few algorithms designed and implemented in the past years account for the changes in lung gas stores [2], but an optimal algorithm has not yet been defined. Recently a new breath-by-breath gas exchange calculation algorithm (called "Independent breath") has been proposed [3], which is based on an alternative view of the respiratory cycle [4]. The "Independent breath" algorithm was tested in healthy volunteers under extreme conditions, where changes in lung gas stores definitely occurred [3] and under moderate intensity exercise, where a low signal-to-noise ratio is usually observed. A validation was performed also during step changes in work load, when changes in lung gas stores are expected, thus the gas exchange data are likely more sensitive to the algorithm applied for their calculation [5].

The aim of the present work was to test the "Independent breath" algorithm [3] in respiratory signals acquired from patients suffering of respiratory diseases affecting the resistance to air flow, i.e. asthmatic patients. A comparison with results provided by a "classical" real-time breath-by-breath algorithm, commonly applied by other laboratories, i.e. the algorithm that uses only information obtained during expiration [6, 7], was also performed, as well as a comparison with results obtained on a group of healthy controls.

2 Methods

2.1 Subjects

Eleven asthmatic patients and 20 well matched healthy controls (37.9 $\pm$ 14.6 years old; 1.73 $\pm$ 0.11 m tall; body mass of 71.4 $\pm$ 12.4 kg) were thoroughly informed of the nature, purpose, and possible risks and gave their written consent to participate in the study. The study was approved by the local Ethical Committee of the Friuli-Venezia Giulia (Italy) region and was conducted according to the Declaration of Helsinki.

2.2 Experimental Protocol

Upon arrival in the laboratory, volunteers were equipped with the facemask and then sat quietly on the bicycle ergometer (Excalibur Sport; Lode B.V., the Netherlands).

The trial lasted 26 min and consisted of 4 consecutive steps: (a) 5 min sitting quietly on the ergometer, (b) 5 min zero-load pedaling, (c) 6 min pedaling at 1 $W{\cdot}kg^{-1}$ body mass and (d) 10 min remaining seated quietly on the ergometer. During the two pedaling periods, volunteers were asked to maintain pedaling frequency as close as possible to 60 rpm.

Respiratory flow ($\dot{V}$), O_2 and CO_2 fractions (FO_2 and FCO_2, respectively) at the mouth were continuously recorded during the trial (CPET Metalyzer 3B, Cortex, Germany) and the corresponding traces were saved as text files with a sampling frequency of 50 Hz. The firmware of the metabolic unit temporally synchronized the traces of flow and gas fractions, as well as converted flow to BTPS conditions (i.e. at body temperature, ambient pressure, saturated with water vapor). Gases were sampled through a ~ 2 m long capillary line inserted in the outer frame of the flow meter and analyzed by fast-response sensors embedded in the metabolic cart. Before each test, the analyzers were calibrated with a gas mixture of known composition (FO_2 = 16.0%; FCO_2 = 4.0%; FN_2 as balance) and ambient air; the flow meter was calibrated by means of a 3-L syringe (Cortex, Germany).

2.3 The O_2 Uptake Algorithms

The Classical "Expiration-Only" Algorithm

The volume of O_2 exchanged at the mouth during the j-th breath, i.e. VO_{2j}^{M}, is calculated by subtracting the expired volume of O_2 ($VO_{2E,j}$) from the inspired one ($VO_{2I,j}$):

$$VO_{2j}^{M} = VO_{2I,j} - VO_{2E,j} = \int_{t_{i,j-1}}^{t_{e,j}} \dot{V} \cdot FO_2 \, dt + \int_{t_{e,j}}^{t_{i,j}} \dot{V} \cdot FO_2 \, dt \tag{1}$$

where FO_2 is the instantaneous fraction of O_2 (Fig. 1b), times $t_{i,j}$ and $t_{e,j}$ correspond to the time points where the flow changes direction and the inspiration and expiration start, respectively (Fig. 1a); $\dot{V}$ is the respiratory flow (in STPD conditions), whose sign depends on the phase considered (inspiration or expiration), integrated over the corresponding phases of the *j*-th respiratory cycle.

Measurements are frequently performed using only a single expiratory flow meter [7, 8], and volume of inspiration is estimated from the expiratory volume with the assumption that, apart from O_2 and CO_2, all the other gases, overall labelled N_2, are not exchanged in a single breath ($VN_{2I,j} = VN_{2E,j}$). Hence, under the further assumption that ambient air has a known composition, the inspired volume of O_2 can be estimated as follows:

$$VO_{2I,j} = \frac{FIO_2}{FIN_2} \cdot VN_{2I,j} = \frac{FIO_2}{FIN_2} \cdot VN_{2E,j} \tag{2}$$

where FIO_2 and FIN_2 are set, a priori, to the inspired O_2 and N_2 fractions, assumed constant. Substituting Eq. (2) in Eq. (1), the volume of O_2 exchanged during the *j*-th breath (VO_{2j}^{E}) can be calculated only from information acquired during expiration as follows:

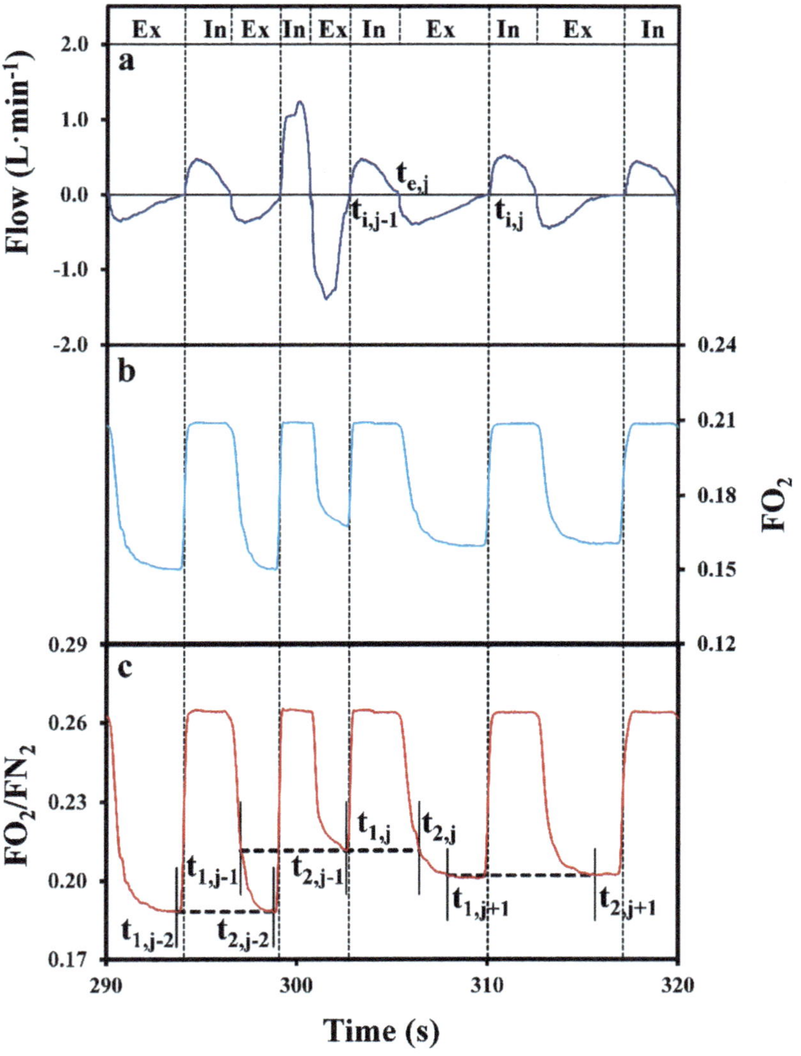

Fig. 1. Raw data for a few consecutive breaths as a function of time. Panels a and b are flow and O_2 fraction traces, as recorded at the mouth. Times $t_{i,j}$ and $t_{e,j}$ correspond to the time points where the flow changes direction, thus where inspiration and expiration start, respectively; according to the classical view of the respiratory cycle, the period between $t_{i,j-1}$ and $t_{i,j}$ corresponds to the duration of the j-th breath. Panel c: trace of the ratio between O_2 and "nitrogen" fractions. This trace is used by the "Independent breath" approach to identify the start ($t_{1,j}$) and end ($t_{2,j}$) time points of the j-th breath; the period between $t_{1,j}$ and $t_{2,j}$ corresponds to the duration of the j-th breath.

$$VO_{2j}^{E} = -\frac{FIO_2}{FIN_2} \cdot \int_{t_{e,j}}^{t_{i,j}} \dot{V} \cdot FN_2 \, dt + \int_{t_{e,j}}^{t_{i,j}} \dot{V} \cdot FO_2 \, dt. \tag{3}$$

where FN_2 is the instantaneous fraction of all the not exchanged gases.

To calculate the O_2 uptake during the *j*-th breath using information obtained only during expiration, i.e. using the classical "Expiration-only" approach ($\dot{V}O_2^{EXP}$), the following equation is used:

$$\dot{V}O_{2j}^{EXP} = \frac{\int_{t_{e,j}}^{t_{i,j}} \dot{V} \cdot FO_2 \, dt - \frac{FIO_2}{FIN_2} \cdot \int_{t_{e,j}}^{t_{i,j}} \dot{V} \cdot FN_2 \, dt}{t_{i,j} - t_{i,j-1}} \tag{4}$$

Notably, the respiratory cycle is delimited by two consecutive starts of inspiration ($t_{i,j\text{-}1}$ and $t_{i,j}$).

The New "Independent Breath" Algorithm

The volume of oxygen taken up at the alveoli during the *j*-th breath, i.e. VO_{2j}^{A}, can be determined from the volume of oxygen exchanged at the mouth minus the volume change of pulmonary oxygen stores (ΔVO_{2j}^{S}):

$$VO_{2j}^{A} = VO_{2j}^{M} - \Delta VO_{2j}^{S} \tag{5}$$

According to Auchincloss et al. [9], the change of volume of pulmonary O_2 stores can be divided into two terms: the first dependent on the variation of the fraction of alveolar oxygen (FO_2^A) from the *j*-1 breath to the *j*-th breath, the second due to the variation of the volume of alveolar gas (ΔV^A) between the two subsequent breaths:

$$\Delta VO_{2j}^{S} = V_{j-1}^{A} \cdot \left(FO_{2j}^{A} - FO_{2j-1}^{A}\right) + FO_{2j}^{A} \cdot \Delta V_j^A \tag{6}$$

where V_{j-1}^{A} is the end-expiratory alveolar volume at the beginning of the current *j*-th breath.

To determine ΔV_j^A we can suppose that the assumptions of Eq. (5) are valid also for all the gases not exchanged, essentially represented by nitrogen [9], and thus that the N_2 volume exchanged at the alveolar level (VN_2^A) during any time interval is null:

$$VN_{2j}^{A} = VN_{2j}^{M} - \left[V_{j-1}^{A} \cdot \left(FN_{2j}^{A} - FN_{2j-1}^{A}\right) + FN_{2j}^{A} \cdot \Delta V_j^A\right] = 0 \tag{7}$$

where FN_2^A is the fraction of alveolar gases overall labelled as N_2. Rearranging Eq. (7):

$$\Delta V_j^A = \frac{1}{FN_{2j}^A}\left[VN_{2j}^M - V_{j-1}^A \cdot \left(FN_{2j}^A - FN_{2j-1}^A\right)\right] \tag{8}$$

By substituting Eq. (8) into Eq. (6) and, in turn, in Eq. (5), the following equation is obtained (after simplifications):

$$VO_{2j}^A = VO_{2j}^M - \frac{FO_{2j}^A}{FN_{2j}^A} \cdot VN_{2j}^M + V_{j-1}^A \left\{\frac{FO_{2j-1}^A \cdot FN_{2j}^A - FO_{2j}^A \cdot FN_{2j-1}^A}{FN_{2j}^A}\right\} \tag{9}$$

In this equation, the only quantity not directly measurable on a breath-by-breath basis is the term V_{j-1}^A [10].

To circumvent the problem that the term V_{j-1}^A in Eq. (9) is not directly measurable on a breath-by-breath basis, it is possible to define the start and end point of the respiratory cycle as the times in two consecutive expirations ($t_{1,j}$ and $t_{2,j}$, respectively) where identical ratios between FO_2^A and FN_2^A are observed [4].

Indeed, two times t_1 and t_2 can be appropriately chosen on two consecutive breaths so as to make the numerator within the curly braces in Eq. (9) becoming zero, i.e.:

$$FO_{2j-1}^A(t_{1,j}) \cdot FN_{2j}^A(t_{2,j}) = FO_{2j}^A(t_{2,j}) \cdot FN_{2j-1}^A(t_{1,j}). \tag{10}$$

Under these conditions, the third part of Eq. (9) disappears, making the knowledge of V_{j-1}^A unnecessary.

The condition described in Eq. (10) can be rewritten as:

$$\frac{FO_{2j-1}^A(t_{1,j})}{FN_{2j-1}^A(t_{1,j})} = \frac{FO_{2j}^A(t_{2,j})}{FN_{2j}^A(t_{2,j})}. \tag{11}$$

This equation shows that the start and end points of each respiratory cycle can be defined in a new way, i.e. as the times in successive breaths where identical FO_2/FN_2 ratios are observed (Fig. 1c). Hence, the calculation of VO_{2j}^A over the interval from $t_{1,j}$ to $t_{2,j}$ is limited to the first two parts of Eq. (9):

$$VO_{2j}^A = VO_{2j}^M - \frac{FO_{2j}^A}{FN_{2j}^A} \cdot VN_{2j}^M = \int_{t_{1,j}}^{t_{2,j}} \dot{V} \cdot FO_2\, dt - \frac{FO_{2j}^A}{FN_{2j}^A} \int_{t_{1,j}}^{t_{2,j}} \dot{V} \cdot FN_2\, dt. \tag{12}$$

The ratio FO_{2j}^{A}/FN_{2j}^{A} is unknown, but if $t_{1,j}$ and $t_{2,j}$ are selected towards the end of the expiration of two consecutive breaths, the fractions of oxygen and of nitrogen at the mouth approximate those in the alveoli [11]:

$$\frac{FO_{2j}^{A}}{FN_{2j}^{A}} \approx \frac{FO_2(t_{1,j})}{FN_2(t_{1,j})}. \tag{13}$$

The alveolar volume of oxygen taken up in a breath can thus be determined by variables all measurable at the mouth:

$$VO_{2j}^{A} = \int_{t_{1,j}}^{t_{2,j}} \dot{V} \cdot FO_2\, dt - \frac{FO_{2j}(t_{1,j})}{FN_{2j}(t_{1,j})} \cdot \int_{t_{1,j}}^{t_{2,j}} \dot{V} \cdot FN_2\, dt. \tag{14}$$

The oxygen uptake during the temporal interval between $t_{1,j}$ and $t_{2,j}$, is then given by:

$$\dot{V}O_{2j}^{IND} = \frac{\int_{t_{1,j}}^{t_{2,j}} \dot{V} \cdot FO_2\, dt - \frac{FO_{2j}(t_{1,j})}{FN_{2j}(t_{1,j})} \cdot \int_{t_{1,j}}^{t_{2,j}} \dot{V} \cdot FN_2\, dt}{t_{2,j} - t_{1,j}}. \tag{15}$$

Indeed, according to Cettolo & Francescato [3], the integration interval between $t_{1,j}$ and $t_{2,j}$ can be identified on the ratio between the FO_2 trace and the FN_2 trace, delimiting each respiratory cycle without taking into account the end time point of the preceding cycle and/or the start time of the subsequent one, hence independently (Fig. 1c).

To delimit each respiratory cycle by its own, the following procedure is applied:

1. the minimum end-expiratory FO_2/FN_2 ratio is searched for the *j*-th breath as well as for the preceding expiration;
2. the higher of the two FO_2/FN_2 values found at point 1 is used as reference ratio;
3. going backwards from the end of the expiration of the *j*-th breath, the first time point where the FO_2/FN_2 ratio corresponds to the reference ratio (as set in point 2) is assumed as the end time point of the respiratory cycle ($t_{2,j}$);
4. going backwards from the end of the preceding expiration, the first time point where the FO_2/FN_2 ratio corresponds to the reference ratio (as set in point 2) is assumed as the start time point of the respiratory cycle ($t_{1,j}$).

With appropriate modifications (i.e. substituting FO_2 with FCO_2), CO_2 exchange ($\dot{V}CO_2^{IND}$ and $\dot{V}CO_2^{EXP}$) can be obtained with the same approaches.

2.4 Respiratory Signal Processing

For all the volunteers, breath-by-breath oxygen uptake was calculated from the original gas fractions and flow traces, as recorded from the metabolic unit, by means of the two algorithms under investigation. Two time series of O_2 uptake values were thus obtained, one for the "Independent breath" algorithm, the other for the "Expiration-only" algorithm. No data were discarded on all the time series before any of the subsequent analyses.

Respiratory signals were processed by software developed in C language under the Unix-like Cygwin environment (Red Hat Cygwin, version 2.5.1); as far as possible, the same C calculation routines were used for both algorithms.

The value of one minus the sum of measured O_2 and CO_2 fractions, labelled FN_2, was assumed to represent the gases not exchanged at alveolar level. To avoid incoherent data, the respiratory cycles were considered valid only if the inspiratory and/or the expiratory volumes were greater than a theoretical dead space, assumed to amount at most to 150 mL [12]; if not, the invalid breath was incorporated with the following one. All the calculated data were converted in STPD conditions.

For both O_2 uptake time series ($\dot{V}O_2^{IND}$ and $\dot{V}O_2^{EXP}$) of each volunteer, average values (over 3 consecutive min), and corresponding Standard Deviations (SD), were calculated over the 4 steady-state time periods composing the trial, during which the O_2 uptake is expected to be constant over each single period.

Statistical analysis was performed using the SPSS package. A repeated measure 2×4 within subjects factors Multivariate Analysis of Variance (MANOVA) was used to detect differences among the 2 investigated algorithms (Algorithm effect), the 4 different conditions (Condition effect), with the diagnosis of asthma (Asthma effect) as between subjects factor. Correlation between variables was assessed by means of the Pearson's correlation coefficient and the least-squares method was applied to calculate the slopes and intercepts (with the asymptotic standard errors) of the corresponding regression lines. Analysis of covariance was used to detect differences among regression lines.

3 Results

The breath-by breath O_2 uptake values obtained using the two algorithms under investigation for one patient are illustrated as a function of time in Fig. 2.

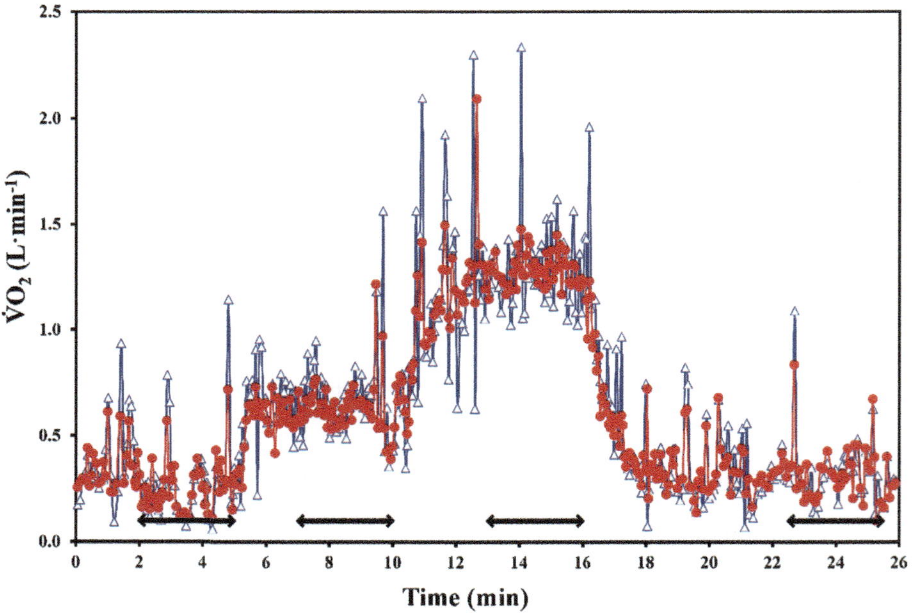

Fig. 2. Breath-by-breath O_2 uptake data obtained for one asthmatic patient applying the "Independent breath" (full dots) or the "Expiration-only" algorithm (open triangles); the same original flow and gas fraction traces were used for both algorithms. Arrows indicate the 3-min averaging periods for the 4 steady-state time periods.

Average $\dot{V}O_2$ values calculated for the 4 steady-state conditions, as well as the corresponding standard deviations, were not significantly different between the two groups of subjects (MANOVA, Group effect, p = ns).

The average $\dot{V}O_2$ values obtained by the "Independent breath" algorithm are plotted against the corresponding values obtained with the "Expiration-only" algorithm in Fig. 3 for both the asthmatic patients and the healthy volunteers and for all the steady-state conditions. It can be observed that almost all the values lie within 5% from the identity line. The relation for the asthmatic patients was not significantly different from that of the healthy volunteers (ANCOVA, p = ns).

The standard deviations obtained for the "Independent breath" algorithm were plotted, for both groups of volunteers and all the steady-state conditions, against the corresponding values obtained for the "Expiration-only" algorithm (Fig. 4). It can be observed that in most of the cases the standard deviations obtained for the O_2 uptakes yielded by the

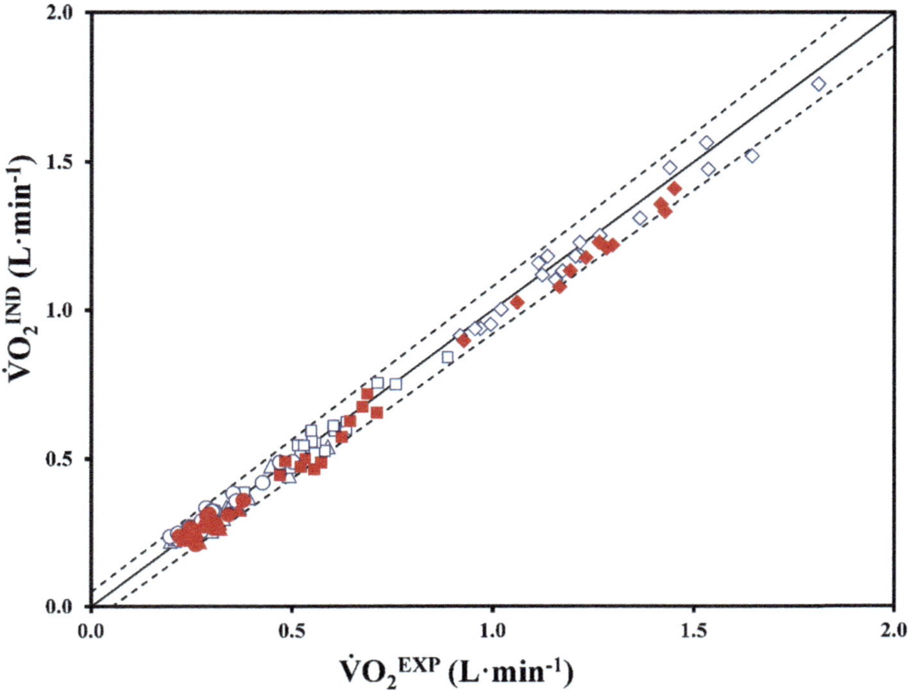

Fig. 3. Average steady state O_2 uptakes of the four investigated conditions as obtained from the $\dot{V}O_2^{IND}$ are plotted against the corresponding values of the $\dot{V}O_2^{EXP}$ time series for both the asthmatic patients (full symbols; n = 44) and the healthy controls (open symbols; n = 80). Dotted lines are 5% change from the identity line. Dots: rest; squares: zero-load pedaling; diamonds: loaded pedaling; triangles: recovery.

"Independent breath" algorithm, were lower as compared to those yielded by the "Expiration-only" algorithm (MANOVA, Algorithm effect, $p < 0.001$). The relation of asthmatic patients was not significantly different from that of healthy volunteers (ANCOVA, p = ns); the overall regression line was: (SD of $\dot{V}O_2^{IND}$) = (0.672 ± 0.033) • (SD of $\dot{V}O_2^{EXP}$) + (0.014 ± 0.004); R = 0.768 (see Fig. 4).

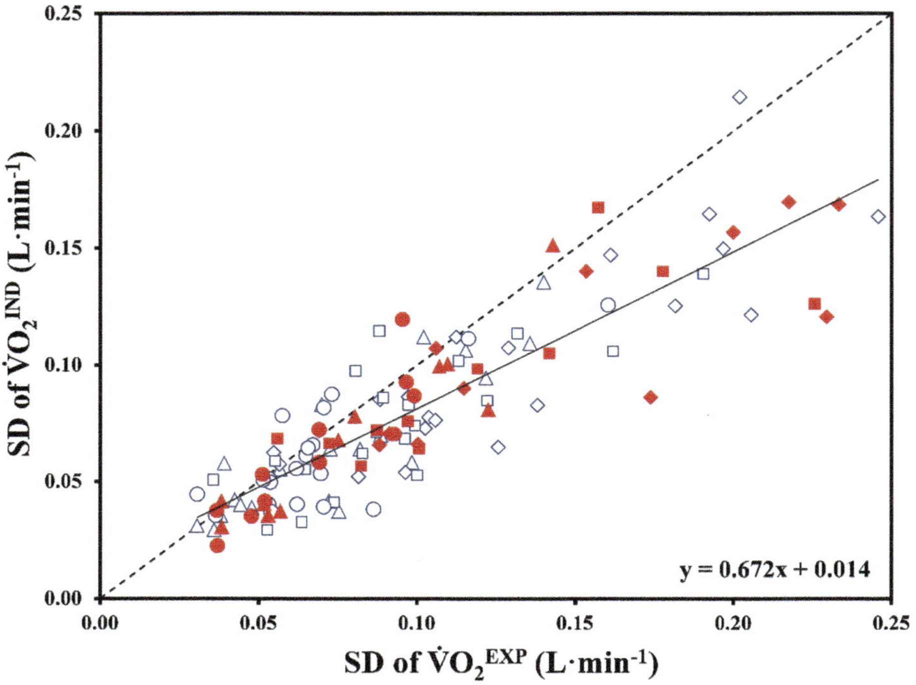

Fig. 4. Standard Deviations (SD) calculated for the O_2 uptakes provided by the "Independent breath" are plotted against the corresponding values obtained for the "Expiration-only" algorithm. Data of all the asthmatic patients (full symbols; n = 44) and of the healthy volunteers (open symbols; n = 80) are plotted together. Dots: rest; squares: zero-load pedaling; diamonds: loaded pedaling; triangles: recovery.

4 Discussion

Main result of the study is that the average values obtained applying the two algorithms under investigation were overlapping for both asthmatic and healthy volunteers and in all the conditions, suggesting that their accuracy is similar. Conversely, the standard deviations were lower for the oxygen uptakes yielded by the "Independent breath" algorithm, supporting the notion of a better precision of this algorithm compared to the "Expiration-only" one. In addition, in patients suffering of a respiratory disease, where the resistance to air flow is likely affected, an increased noise in the breath-by-breath O_2 uptake data can be expected. However, our results show that there was no difference in the standard deviations between the two groups of volunteers for both algorithms. This is the first time that the performance of the two investigated algorithms ("Independent breath" and "Expiration-only", respectively) was studied on respiratory disease patients.

Asthma is a very variable illness, with different levels of severity of the symptoms. A heterogeneous group of patients was deliberately recruited, some of them needing a chronic therapy to control the illness (that was not modified), others having no symptoms during the trials, thus not even requiring drugs on demand. As a consequence, it is likely that all of them had only a minimal bronchoconstriction during the experimental sessions, with no significant effects on the standard deviations of the steady-state breath-by-breath O_2 uptakes. A more precise classification of the patients, based on the actual severity level of bronchoconstriction at the moment of the experimental session, might bring to different results. Further studies are thus warranted to test this hypothesis.

In healthy subjects, the "Independent breath" algorithm provided O_2 uptake data showing less noise also during the transients [5]. The analysis of the transient phases was not investigated in the present work; visual inspection of the individual data, however, suggested us that also in the group of asthmatic volunteers the transients were characterized by reduced noise when the "Independent breath" algorithm was used. Less noisy data might result, also in patients suffering from respiratory disease, in a reduced number of repeated trials required to obtain the same width of the confidence interval when assessing the kinetic parameters of the corresponding O_2 uptake behavior [13].

In conclusion, because of its better precision, with similar accuracy, compared to the "classical" real-time breath-by-breath algorithm commonly applied by other laboratories, the use of the "Independent breath" algorithm should be recommended also in asthmatic patients.

Acknowledgements. We thank Cortex GmbH (Liepzig, Germany) for having provided us with the metabolic unit. Cortex GmbH was not involved in the study design, data collection, analysis or interpretation.

This experimentation was conducted with the unconditioned support of the ALPI Association (Association for the Pediatric Allergies and Pneumopaties), Udine (Italy).

Conflict of Interests. None of the authors has conflicts of interest to declare.

References

1. Capelli, C., Cautero, M., Pogliaghi, S.: Algorithms, modelling and VO_2 kinetics. Eur. J. Appl. Physiol. **111**(3), 331–342 (2011)
2. Golja, P., Cettolo, V., Francescato, M.P.: Calculation algorithms for breath-by-breath alveolar gas exchange: the unknowns! Eur. J. Appl. Physiol. **118**(9), 1869–1876 (2018)
3. Cettolo, V., Francescato, M.P.: Assessing breath-by-breath alveolar gas exchange: is the contiguity in time of breaths mandatory? Eur. J. Appl. Physiol. **118**(6), 1119–1130 (2018)
4. Cettolo, V., Francescato, M.P.: Assessment of breath-by-breath alveolar gas exchange: an alternative view of the respiratory cycle. Eur. J. Appl. Physiol. **115**(9), 1897–1904 (2015)
5. Francescato, M.P., Cettolo, V.: The "independent breath" algorithm: assessment of oxygen uptake during exercise. Eur. J. Appl. Physiol. **119**(2), 495–508 (2019)
6. Roecker, K., Prettin, S., Sorichter, S.: Gas exchange measurements with high temporal resolution: the breath-by-breath approach. Int. J. Sports Med. **26**, S11–S18 (2005)
7. Ward, S.A.: Open-circuit respirometry: real-time, laboratory-based systems. Eur. J. Appl. Physiol. **118**(5), 875–898 (2018)
8. Swanson, G.: Breath-to breath considerations for gas exchange kinetics. In: Cerretelli, P., Whipp, B.J. (eds.) Exercise Bioenergetics and Gas Exchange, p. 211–222. Elsevier/North Holland, Amsterdam (1980)
9. Auchincloss, J.H.J., Gilbert, R., Baule, G.H.: Effect of ventilation on oxygen transfer during early exercise. J. Appl. Physiol. **21**(3), 810–818 (1966)

10. di Prampero, P.E., Lafortuna, C.L.: Breath-by-breath estimate of alveolar gas transfer variability in man at rest and during exercise. J. Physiol. **415**(1), 459–475 (1989)
11. Grønlund, J.: A new method for breath-to-breath determination of oxygen flux across the alveolar membrane. Eur. J. Appl. Physiol. **52**(2), 167–172 (1984)
12. Fowler, W.S.: Lung function studies; the respiratory dead space. Am. J. Physiol. **154**(3), 405–16 (1948)
13. Francescato, M.P., Cettolo, V., Bellio, R.: Assembling more O_2 uptake responses: is it possible to merely stack the repeated transitions? Respir. Physiol. Neurobiol. **200**, 46–49 (2014)

Gait Phase Classification from Surface EMG Signals Using Neural Networks

Christian Morbidoni(✉), Lorenzo Principi, Guido Mascia, Annachiara Strazza, Federica Verdini, Alessandro Cucchiarelli, and Francesco Di Nardo

Department of Information Engineering, Università Politecnica delle Marche, Ancona, Italy
{c.morbidoni,a.cucchiarelli}@univpm.it, lorenzo.principi@mail.polimi.it, mascia.guido@gmail.com, a.strazza@pm.univpm.it, {f.verdini,f.dinardo}@staff.univpm.it

Abstract. Identification and classification of different gait phases is an essential requirement to temporally characterize muscular recruitment during human walking. The present study proposes a Deep-learning methodology for the classification of the two main gait phases (stance and swing), based on the interpretation of surface electromyographic (sEMG) signals alone. Three different Multi Layer Perceptron (MLP) models are tested to this aim. The present approach does not require specific features to be extracted from the signal, differently from previous studies. 12 healthy adult subjects are analyzed during walking over-ground at comfortable speed. sEMG signals from eight leg muscles are selected. Performance of classifiers is tested vs. gold standard, represented by basographic signals measured by means of three foot-switches. A 10-fold evaluation is computed to take into account the possible variability of the results. The direct comparison among the performances of the three different MLP models shows an average high accuracy over the population (around 95%) for all the models, independent from the increasing complexity. Moreover, the accuracy in each single subject does not fall below 92.6% (range of accuracy variability = 92.6–97.2%). This present study suggests that artificial neural networks may be a suitable tool for the automatic classification of gait phases from electromyographic signals, in overall walking tasks.

Keywords: sEMG · Deep learning · Neural networks · Gait phase classification

1 Introduction

Surface electromyography (sEMG) is acknowledged as a non-invasive approach, specifically suitable to monitor muscle activity during dynamic tasks, such as walking [1]. In general, gait events, such as the instant of foot-floor contact and ground clearance, need to be assessed in order to characterize the muscular activity under the temporal and the spatial dimensions.

J. Henriques et al. (Eds.): MEDICON 2019, IFMBE Proceedings 76, pp. 75–82, 2020.
https://doi.org/10.1007/978-3-030-31635-8_9

A fundamental task in this context is to identify and classify the different gait phases. While the entire gait cycle can be divided in 8 distinct sub-phases [1], the two main ones are stance and swing, which identify the functional subdivisions of total limb activity within the gait cycle. For each leg, the stance phase is the time period in which the foot touches the ground while during the swing phase the foot is in the air [1].

Different kind of equipment is typically used to provide synchronization of sEMG signals within gait cycle, such as foot-switch sensors [2], pressure mats [3], stereo-photogrammetric systems [4], and inertial measurements units (accelerometers and gyroscopes) [5]. However, expensiveness of the equipment, limited volume of observation and relative invasive experimental protocol could affect the use of stereo-photogrammetric systems and pressure mats in clinical practice, while wearable sensors have problems of encumbrance and can require particular care for the correct placement and the need of specific calibration procedures, especially for application in pathology.

Furthermore, biological signals (as sEMG) have the advantage, over movement measuring instruments, of being accessible prior to the movement, thus being more suited for specific applications such as robotic exoskeletons, which requires characterization of the motion's intention [6]. For these reasons, novel techniques to detect and classify gait events from sEMG signals alone are desirable. In this paper we investigate the use of (Deep) Neural Networks to classify stance and swing phases from sEMG signals during adult walking at self-selected pace. We train a distinct Neural Network classifier for each subject and we attempt at classifying walking phases in unlearned steps sequences of the same subject. Our experiments resulted in an average classification accuracy over the population of 95.2%, ranging from 92.6% to 97.2%.

Few works in literature uses Machine Learning techniques to addresses this task. [7] applies support vector machines to classify stance and swing phases using novel bilateral muscles features from the sEMG signal, while [8] uses extracts hand crafted features from sEMG signals and provides them as input to a Multi Layer Perceptron (MLP) classifier. Differently from previous works, we investigate a featureless approach, by operating on sEMG signals with only standard pre-processing. Our research direction is to understand if and to what extent a Neural Network can effectively learn hidden features without requiring humans to engineer or select signals features. Furthermore, while previous researches focused on treadmill walking, we evaluate our approach on signals acquired during conditions more similar to daily life walking, i.e. ground walking including straights and curves.

2 Related Works

While several Machine Learning based approaches have been proposed in literature to analyse EMG signals during different tasks, few of them address walking gait phases classification.

Kaczmarczyk et al. [9] applied neural networks for gait classification in post stroke patients. Wang and Zieliska [10] designed an EMG-based method for detecting the variability in gait features depending on footwear, by applying vector quantization classifying network and clustering competitive network.

Few woks attempt to classify stance and swing phases, in a treadmill walking setting, using engineered features extracted from sEMG signals. In [7], the use of novel features, which consider the bilateral activation of 7 muscles for each leg, is proposed. A limitation of this study is the small population (two subjects). A neural network, fed with engineered features was also used in [8,11,12], Multi Layer Perceptrons classifiers are trained to recognize stance and swing phases from time-domain features extracted from sEMG signal during treadmill walking.

Unlike previous works, in this paper we investigate a Deep Learning approach based on learning latent features from the original sEMG signal. In our experiments, no feature engineering and selection is performed and only standard pre-processing is applied. In addition, in the present study we consider a population of 12 subjects and data is captured in a ground walking situation, including direction changes and curves, with the goal of assessing the technique in realistic conditions.

3 Materials and Methods

3.1 Dataset

The dataset included signals recorded from 12 healthy adults (6 females and 6 males), acquired in the Movement Analysis Laboratory of Università Politecnica delle Marche, Ancona, Italy. Mean (±SD) characteristics were: age = 22.9 ± 1.5 years; height = 171 ± 11 cm; mass = 64.1 ± 11.0 kg; body mass index (BMI) = 20.8 ± 2.1 kg/m^2. Subjects with joint pain, neurological pathologies, orthopedic surgery, abnormal gait or a body mass index (BMI) higher than 25 (overweight and obese subjects) were not recruited. The research was undertaken in compliance with the ethical principles of the Helsinki Declaration and was approved by an institutional expert committee. Participants signed informed consent prior to the beginning of the test.

3.2 Signal Acquisition and Pre-processing

sEMG and basographic signals were acquired as reported in [13] and [14], while the approach discussed in [15] was used to process the foot-switch signals as to identify the different gait cycles and phases (stance and swing). As done in related works, we treated the results as a ground truth and measure the performances of our classifier against it.

Electromyographic signals were processed by a band-pass, linear-phase FIR filter (cut-off frequency: 20–450 Hz). Then, sEMG signals were full-wave rectified and the envelope was extracted (2nd-order Butterworth low-pass filter, cut-off frequency: 5 Hz). A min-max normalization of each muscle signal was performed within each individual, thus mapping the values in the [0–1] interval.

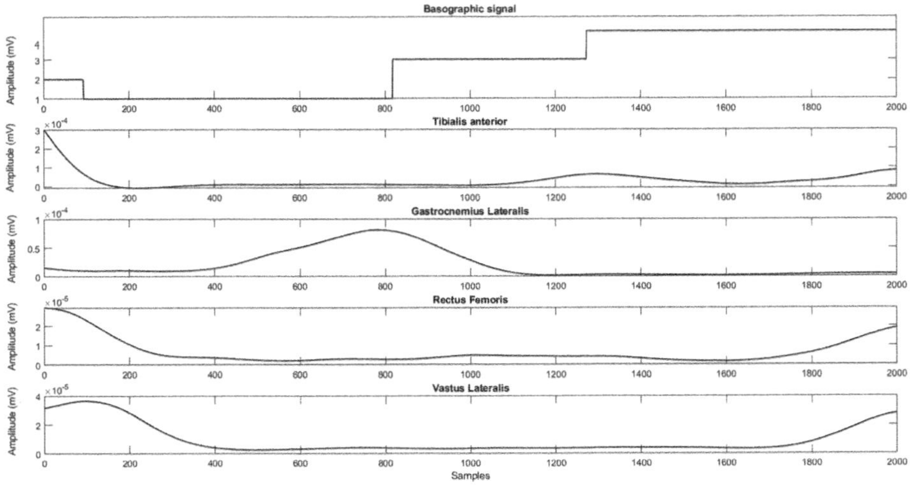

Fig. 1. The basographic signal acquired from foot-switches and the sEMG signals recorded for the 4 muscles of the right leg.

Figure 1 shows an example of the acquired signals, including pre-precessed EMG signals and the basographic signal, corresponding to a complete gait cycle of the right leg of a sample subject.

4 Gait Phase Classification Experiments

4.1 Dataset Preparation

Our classification task is to assign the correct label to 10 ms time segments of the subject walking activity, according to the walking phase it belongs to (label 0 for stance and 1 for swing). For simplicity, in our experiments we consider the phases with respect to the right leg of each subject, but we calculate predictions using signals from both legs. In order to get data into the proper format for feeding the neural networks, we build an input vector for each 10 ms window. Such a vector, depicted in 2, is composed by 20 samples (1 every 0.1 ms). Each sample contains 8 values, corresponding to the activation of the muscles (4 per leg) registered at the sampling instant.

Referring to Fig. 2, $L1_i$, $L2_i$, $L3_i$, and $L4_i$ corresponds to the activation of the following muscles respectively: tibialis anterior, gastrocnemius lateralis, hamstring, vastus lateralis. $R1_i$, $R2_i$, $R3_i$, and $R4_i$ represented the correspondent element for the right leg.

We then assigned a 0 (stance) or 1 (swing) label to each aggregated sEMG segment as follows. If the basographic signal corresponding to all the samples of a sEMG segment are equal to 0 (or 1), we assigned the 0 (or 1) label to the sEMG segment. We exclude those segments that include a phase transition,

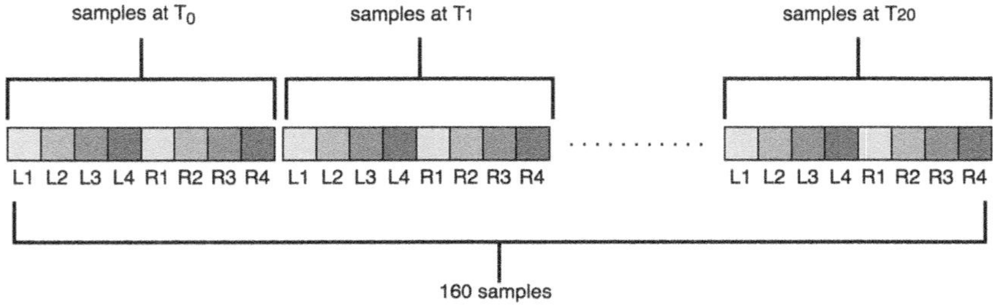

Fig. 2. The structure of EMG vectors fed as input to the ANNs

in this case not all the corresponding basographic samples are 0 or 1, thus we cannot classify it as stance or swing.

In summary, the dataset is composed by around 23 thousand labeled segments for each of the 12 subjects. Each segment corresponds to a 10 ms time slot and contains 20 samples of the sEMG signals corresponding to the 8 muscles (4 per leg).

4.2 Training the Classifiers

After having processed the data as described in Sect. 4.1, we split the data corresponding to each patient into 10 folds. More specifically, we temporally divided the walking activity into 10 equal slots (each one approximately 30 s long). We then perform 10 different training for each subject, each time using a different slot as test set and the remaining 9 as training set. Walking activity might change over different time slots, especially in our experimental conditions (ground walking at self-selected pace), averaging the classification performances on different time slots for each subject, provides a more realistic evaluation of the technique.

As our aim is to provide a first assessment of deep learning approach to gait phase classification, in our experiments we consider three simple Multi Layer Perceptron (MLP) architectures. The first is a shallow network with a single hidden layer composed of 128 units (referred to *mlp(128)*), the second is a two layer network of 256 and 128 units (*mlp(256, 128)*), and the last one is has three layers of 512, 256 and 128 units (*mlp(512, 256, 128)*). We used the ReLU activation function to provide non linearity between hidden layers.

For all the network, we applied the Stochastic Gradient Descent (SGD) optimization algorithm with Binary Cross Entropy (BCE) as a loss function. Finally, we trained the classifiers using a early-stop technique: if accuracy on the validation set (random 10% of the training set) does not increase within 10 learning epochs, we save the best configuration and evaluate it on the test set.

As said, we use the processed signal from foot-switches as a ground truth [16]. It represent the gold standard in gait segmentation since each gait phase can be associated with a specific value of the sensor output [17]. Furthermore, in order to obtain the best accuracy, in this experiments data from three foot-switches (instead of two often used in previous studies) were considered, as suggested in [15].

5 Results and Discussion

The average accuracy obtained with 3 different classifiers is reported in Table 1. In this table, mean accuracy over the 10 folds obtained from each subjects, is averaged over the 12 subjects.

We observe that mlp(128), a very simple network, is able to achieve relatively good classification accuracy (94.471%) and that the performances slightly increase as the complexity of the network grows. The best result is obtained with mlp(512, 256, 128), which reached an accuracy of 95.183%.

In Fig. 3 we show the mean accuracy over the 10 folds for each one of the 12 distinct subjects. From this figure figure one can see that mlp(512,256,128) performed very well for some subjects (e.g., s6 and s8, over 97% of accuracy), but results have a relatively high variability, showing that certain subjects are easier to learn. In the worst case (s12) the mean accuracy (92.58%) is still relatively good.

Table 1. Average classification accuracy (%) obtained with 3 different neural networks.

Model	Accuracy
mlp(128)	94.471 ± 1.504
mlp(256, 128)	95.041 ± 1.542
mlp(512, 256, 128)	**95.183** ± 1.579

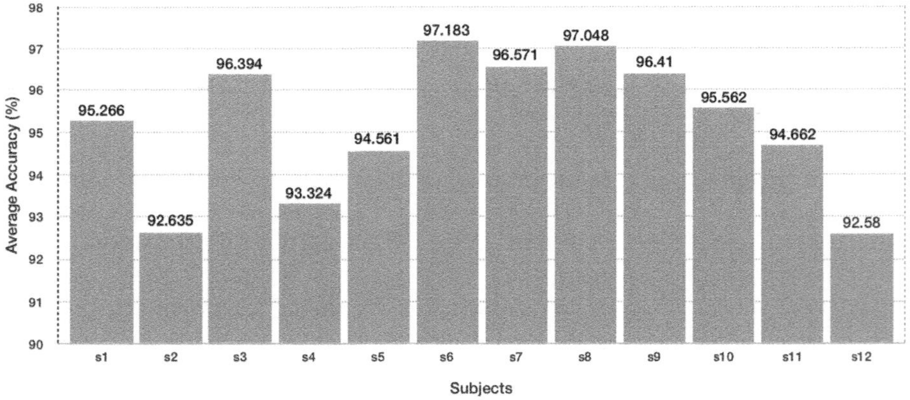

Fig. 3. Variability of average classification accuracy over 10 folds for the 12 subjects

Finally, in Table 2 we distinguish between classification performances obtained for stance and swing phases, reporting precision, recall and F1-score for the two distinct prediction tasks.

In line with what was reported in literature [1], the segments belonging to a stance phase were more frequent (around 60%) that those belonging to a swing phase. This is because in normal walking the stance phase duration is 60% of the gait cycle (while the swing phase duration is the remaining 40%) on average. Despite this, results obtained for swing labeled segments (96.331%) are sensibly better than those obtained for stance labeled segments (93.519).

Table 2. Precision and recall in recognizing (a) swing and (b) stance phases.

Model	Precision	Recall	F1-score
Swing			
mlp(128)	95.635 ± 1.305	94.824 ± 1.304	95.192 ± 1.176
mlp(256, 128)	96.250 ± 1.415	95.199 ± 1.421	95.690 ± 1.226
mlp(512, 256, 128)	**96.331** ± 1.677	**95.402** ± 1.382	**95.824** ± 1.258
Stance			
mlp(128)	92.701 ± 3.036	93.936 ± 2.202	93.245 ± 2.553
mlp(256, 128)	93.238 ± 2.986	94.802 ± 2.320	93.956 ± 2.508
mlp(512, 256, 128)	**93.519** ± 2.830	**94.877** ± 2.674	**94.119** ± 2.535

6 Conclusions and Future Work

The results of present study support the suitability of Artificial Neural Networks in classifying the main gait phases from sEMG signals alone, without needing human intervention in selecting relevant signal features.

In particular, the direct comparison among the performances of three different Multi Layer Perceptron models in our healthy population of 12 healthy subjects showed a high accuracy (around 95%) for all the models, independent from the increasing complexity. A further interesting finding of the study is the higher network performances in the prediction of the swing phase compared to stance phase. Future developments of the study could involve the increase of the experimental population for testing purposes and the attempt of classifying not only swing and stance phases but also the three main sub-phases of stance, i.e. heel strike, flat foot contact, and push off.

Finally, in this study we attempted at automatically classifying signals of subjects whose walking activity was previously recorded (including foot-switches), by using it to properly train a classifier for the given subject. An interesting research directions is that of attempting to classify signals from unlearned subjects, as well as to understand if a classifier trained on more subjects could achieve even better performances.

References

1. Perry, J.: Gait Analysis: Normal and Pathological Function. Slack Inc., USA (1992)
2. Mengarelli, A., Maranesi, E., Burattini, L., Fioretti, S., Di Nardo, F.: Co-contraction activity of ankle muscles during walking: a gender comparison. Biomed. Signal Process. Control **33**, 1–9 (2017). https://doi.org/10.1016/j.bspc.2016.11.010

3. Gurney, J., Kersting, U., Rosenbaum, D.: Between-day reliability of repeated plantar pressure distribution measurements in a normal population. Gait Posture **27**(4), 706–709 (2008). https://doi.org/10.1016/j.gaitpost.2007.07.002
4. Bovi, G., Rabuffetti, M., Mazzoleni, P., Ferrarin, M.: A multiple-task gait analysis approach: kinematic, kinetic and emg reference data for healthy young and adult subjects. Gait Posture **33**(1), 6–13 (2011). https://doi.org/10.1016/j.gaitpost.2010.08.009
5. Caldas, R., Mundt, M., Potthast, W., de Lima Neto, F.B., Markert, B.: A systematic review of gait analysis methods based on inertial sensors and adaptive algorithms. Gait Posture **57**, 204–210 (2017). https://doi.org/10.1016/j.gaitpost.2017.06.019
6. Koller, J.R., Jacobs, D.A., Ferris, D.P., Remy, C.D.: Learning to walk with an adaptive gain proportional myoelectric controller for a robotic ankle exoskeleton. J. NeuroEng. Rehabil. **12**(1), 97 (2015)
7. Ziegier, J., Gattringer, H., Mueller, A.: Classification of gait phases based on bilateral EMG data using support vector machines. In: Proceedings of the IEEE RAS and EMBS International Conference on Biomedical Robotics and Biomechatronics, pp. 978–983, August 2018. https://doi.org/10.1109/BIOROB.2018.8487750
8. Nazmi, N., Abdul Rahman, M., Yamamoto, S.-I., Ahmad, S.: Walking gait event detection based on electromyography signals using artificial neural network. Biomed. Signal Process. Control **47**, 334–343 (2019). https://doi.org/10.1016/j.bspc.2018.08.030
9. Kaczmarczyk, K., Wit, A., Krawczyk, M., Zaborski, J., Piłsudskii, J.: Artificial neural networks (ANN) applied for gait classification and physiotherapy monitoring in post stroke patients. In: Artificial Neural Networks, Chap. 16. IntechOpen, Rijeka (2011). https://doi.org/10.5772/15363
10. Wang, J., Zielińska, T.: Gait features analysis using artificial neural networks - testing the footwear effect. Acta Bioeng. Biomech. **19**(1), 17–32 (2017)
11. Nazmi, N., Yamamoto, S., Rahman, M., Ahmad, S., Adiputra, D., Zamzuri, H., Mazlan, S.: Fuzzy logic for walking patterns based on surface electromyography signals with different membership functions. In: 2016 6th International Workshop on Computer Science and Engineering, WCSE 2016, pp. 636–639 (2016)
12. Nazmi, N., Rahman, M.A.A., Ahmad, S.A.: Generalization of ANN model in classifying stance and swing phases of gait using EMG signals. In: 2018 IEEE-EMBS Conference on Biomedical Engineering and Sciences (IECBES) (2018)
13. Di Nardo, F., Mengarelli, A., Maranesi, E., Burattini, L., Fioretti, S.: Gender differences in the myoelectric activity of lower limb muscles in young healthy subjects during walking. Biomed. Signal Process. Control **19**, 14–22 (2015). https://doi.org/10.1016/j.bspc.2015.03.006
14. Strazza, A., Mengarelli, A., Fioretti, S., Burattini, L., Agostini, V., Knaflitz, M., Di Nardo, F.: Surface-EMG analysis for the quantification of thigh muscle dynamic co-contractions during normal gait. Gait Posture. **51**, 228–233 (2017)
15. Agostini, V., Balestra, G., Knaflitz, M.: Segmentation and classification of gait cycles. IEEE Trans. Neural Syst. Rehabil. Eng. **22**(5), 946–952 (2014). https://doi.org/10.1109/TNSRE.2013.2291907
16. Taborri, J., Palermo, E., Rossi, S., Cappa, P.: Gait partitioning methods: a systematic review. Sens. (Switz.) **16**(1), 66 (2016). https://doi.org/10.3390/s16010066
17. Winiarski, S., Rutkowska-Kucharska, A.: Estimated ground reaction force in normal and pathological gait. Acta Bioeng. Biomech. **11**(1), 53–60 (2009)

Combining Objective Response Detectors Using Genetic Programming

Leonardo Bonato Felix[1,2(✉)], Quenaz Bezerra Soares[1], Antonio Mauricio Ferreira Leite Miranda de Sá[3], and David Martin Simpson[2]

[1] Department of Electrical Engineering, Federal University of Viçosa, Viçosa, Brazil
leobonato@ufv.br
[2] Institute of Sound and Vibration Research, University of Southampton, Southampton, UK
[3] Biomedical Engineering Program/COPPE, Federal University of Rio de Janeiro, Rio de Janeiro, Brazil

Abstract. Many Objective Response Detectors (ORD) have been proposed based on ratios extracted from statistical methods. This work proposes a new approach to automatically generate ORD techniques, based on the combination of the existing ones by genetic programming. In this first study of this kind, the best ORD functions obtained with this approach were about 4% more sensitive than the best original ORD. It is concluded that genetic programming applied to create ORD functions has a potential to find non-obvious functions with better performances than established alternatives.

Keywords: Objective Response Detection · Genetic programming · Evoked responses

1 Introduction

In the detection of evoked responses a number of different methods have been proposed, as Objective Response Detectors (ORDs). Some examples are the Magnitude-Squared Coherence (MSC) [1], the Component Synchrony measure (CSM) [2] and F-tests (Local – LFT [3] and Global – GFT [4]). These methods often use the frequency domain, combining typically real and imaginary parts of responses at the frequency of stimulation (and possibly harmonics) and calculating ratios that increase with the strength of the response, with associated statistical tests of significance. A number of such ratios have been proposed, tested and compared with each other in order to find the 'best', usually defined by showing the highest sensitivity for a given false-positive rate. Best in this sense means that responses can be detected with the lowest possible number of stimuli, leading to short test-times. However, the ratios currently in use have been usually chosen primarily because they are 'reasonable' (i.e. expected to give relatively high values when a response is present) and being statistically tractable to allow for robust statistical test to be applied. What we do not know, is whether these

J. Henriques et al. (Eds.): MEDICON 2019, IFMBE Proceedings 76, pp. 83–92, 2020.
https://doi.org/10.1007/978-3-030-31635-8_10

ratios are the best possible measures, and which combination of spectral parameters might allow for even higher sensitivity.

The current paper thus proposes a data driven approach to optimizing the combination of spectral parameters to maximize sensitivity. This is carried out by using genetic algorithms (GAs) and test data obtained from Monte Carlos simulations. To the best of our knowledge, this is the first time such a data-driven approach is used in optimizing the criterion for detecting responses although GA are used to create optimal designs in a number of other areas, such as aerodynamics [5], thermodynamics [6], optics [7], operations management [8], electronics [9], control [10], image processing [11], controller implementation [12] and time series prediction [13]. We believe this innovative approach also has great potential in many other applications in biomedical engineering (and beyond), where optimal test criteria are unknown. Our approach allows the optimal combination to be estimated directly from the data, avoiding the need for 'manually' searching an ever-increasing library of possible combinations of parameters and instead doing so in a systematic and automated way.

2 Methods

2.1 Objective Response Detectors

Objective Response Detectors are statistical tests that estimate the presence or absence of a response to an input in a system. In our case this is primarily the response in the electroencephalogram (EEG) signal, following auditory stimulation. When the input signal consists of periodic stimuli with a determined frequency, the response will contain a periodic signal with the same frequency, added to noise and other background signals.

The following ORD techniques, based on the output signal in the frequency domain, are used in this work.

$$MSC(f) = \frac{\left|\sum_{i=1}^{M} Y_i(f)\right|^2}{M\sum_{i=1}^{M}|Y_i(f)|^2} \tag{1}$$

$$CSM(f) = \left[\frac{1}{M}\sum_{i=1}^{M}\cos\phi_i(f)\right]^2 + \left[\frac{1}{M}\sum_{i=1}^{M}\sin\phi_i(f)\right]^2 \tag{2}$$

$$GFT(f) = \frac{\sum_{i=1}^{M}|Y_i(f)|^2}{\sum_{i=1}^{M}|Y_i(f)|^2 + \sum_{i=1}^{M}|X_i(f)|^2} \tag{3}$$

$$LFT(f) = \frac{|\bar{Y}(f)|^2}{\sum_{j=-L/2}^{L/2}|\bar{Y}(f_j)|^2} \tag{4}$$

where $Y_i(f)$ is the Fourier transform of the ith window of the signal at which a response is expected, M is the number of windows used, $\phi_i(f)$ is the phase of $Y_i(f)$, $X_i(f)$ is the Fourier transform of the ith window of the signal without stimulation (or more importantly, without a response), L is the number of bins of the Fourier transform used, $\bar{Y}(*)$ is the average of the Fourier transform of M windows and f_j is the jth lateral frequency bin of the tested frequency (i.e. next to stimulus frequency or its harmonics, and thus without a response present). These ORD methods all have established statistical tests associated and will form the basis of the new algorithm to find improved ORD methods. These functions were also used in parts. The CSM was divided into the sum of sine and the sum of cosine, the others were split in numerator and denominator. Furthermore, a modified version of the CSM was estimated using hyperbolic sine and cosine, a normalized LFT and a constant function.

2.2 Evolutionary Algorithm

The main problem is to find new equations for ORDs automatically and our strategy consists of representing a function by means of two binary trees, the main one contains mathematical operations and functions while the second one consists of weights by which each node of the main tree is weighted. Both will be represented by arrays [14, p. 161], where the genetic code of each individual consists of the two concatenated arrays. The Fig. 1 shows an example of a function represented by two trees and the corresponding genetic code as an array.

Binary operations were used, with the operands being the values returned by the two child nodes. The functions used were the ORD techniques and parts of them (e.g. numerator and/or denominator of the defining expression), and such functions had only inputs that come from the data to be tested. Node-combinations that do not generate an output, such as the combination of nodes 6 and 7 in Fig. 1, are ignored. However, these nodes will continue to be part of the genetic code of the individual and may be used in the next generations, since the recombination can place them as children of a node containing mathematical operators.

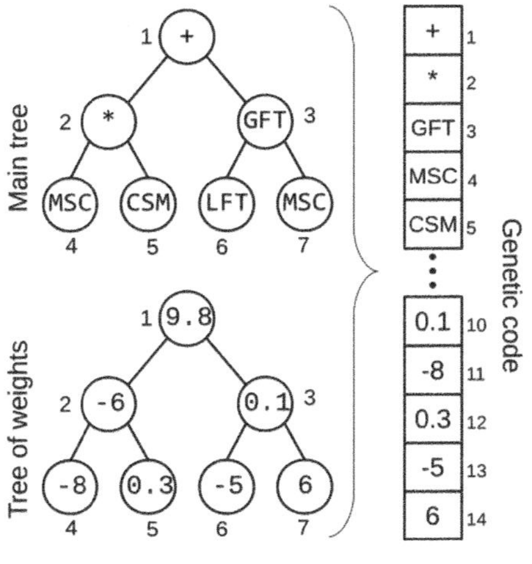

Fig. 1. An example of a tree representation of an expression and its corresponding genetic code array. Here the expression is $9.8(-6(-8MSC * 0.3\,CSM) + 0.1\,GFT)$, containing unused nodes at 6 and 7.

In order to avoid impossible genetic codes, such as mathematical operator nodes in the last nodes, a constraint was used in which only functions can be assigned to the nodes of the last level. An additional constraint imposed on the complexity was that the first k levels should contain only mathematical operators, so the final function will consist of a mathematical combination of at least 2^k functions. The weights have the same constraint for all levels, with all their genes being real numbers between -10 and 10. This value was limited due to computational time limitations. Note that these limits are applied to each individual node. It must be pointed out that expression simplification may lead the weights to be outside the range -10 to 10, as can be seen in the simplified version of the function presented on Fig. 1 that is $141.12 MSC * CSM + 0.98\,GFT$.

The operations and functions used are listed in Table 1, according to the codification used for building the genetic code of the main tree. The functions listed in Eqs. 1 to 4.

Table 1. Main tree gene codes. Negative numbers represent operations, positive number function of the data.

ID	Operator/function
−5	Exponentiation
−4	Division
−3	Multiplication
−2	Subtraction
−1	Addition
0	Unit constant
1	MSC
2	CSM
3	GFT
4	LFT
5	MSC numerator
6	MSC denominator
7	CSM sine
8	CSM cosine
9	GFT numerator
10	GFT denominator
11	LFT numerator
12	LFT denominator
13	Hyperbolic CSM
14	LFT_{mean}

In order to find a new ORD detector from the already existing ones, an adaptation of the genetic programming technique was proposed by means of binary function-generating trees. The evolution was realized through a traditional GA implemented using the Global Optimization Toolbox of Matlab®, using stochastic uniform selection, elitism of 5% and intermediate crossover. As the evolution will be performed through GA, the maximum tree size and hence the maximum complexity of the function will be limited by the size of the genetic code.

2.3 Monte Carlo Simulation

The critical values for each generated detector were determined using Monte Carlo simulations, using 100000 iterations and a significance level of 5% [15]. These simulations consist of independent white Gaussian noise added to a sinusoid in different signal-to-noise ratios (SNRs), where each generated signal has 30 windows of 1 s (256 points in each window, emulating a sampling frequency of 256 Hz), a sinusoidal frequency of 25 Hz and random phase. The signals were normalized so that all had unit variance. Simulated signals were used to calculate the fitness of each individual (in this case, an individual is defined by its genetic code as shown in Fig. 1 and hence the equation generated by the GA).

Only the individuals that were found to give 5% of false positives were used to calculate the Probability of Detection (PD) by applying the individuals to a set of 2000 simulated signals, which had a mean SNR of −15 dB and variance equal to 10. This range is in agreement with the results of SNR in EEG found in [16]. The random SNR was used so that the detectors obtained were not specific for a given SNR. The simulated signals were created with increasing SNR, but the whole time-series is normalized to unit variance, i.e. the total power does not change. The fitness function for the GA was $(1 - PD)$. The set of signals used to calculate the fitness was changed at every 8 generations of the GA, to avoid the detectors becoming too specific for a certain set of simulated signals.

2.4 Finding the Best Detector

Two replicates of each configuration listed in Table 2 were performed. The maximum number of generations was 100.

Table 2. The configuration of level constraints.

Configuration	Tree depth	k levels with only mathematical operators
I	5	1
II	5	2
III	5	3
IV	5	4
V	10	1
VI	10	2
VII	10	4

Since the simulated signals were all reset each 8 generations in order to avoid bias, the fitness calculation may be affected and it is possible that an individual has slightly different fitnesses in two generations. In order to avoid misleading results due to fitness distortion, a histogram of the 1000 best individuals of all generations was drawn, according to the fitness obtained at the time of its evaluation in GA, and the best individual was defined as the one with the highest relative frequency. In the case of a tie, the criterion of the best average fitness was used.

3 Results and Discussion

For a low minimum complexity (i.e. the lowest number of only mathematical operations), as in configuration I and V, fast convergence is observed within the first 100 generations. This convergence is shown in Fig. 2., where it is possible to notice that the average fitness value decreases until generation 40 and then remains practically constant. After every 8 generations there is a disturbance in the fitness values due to the renovation of the simulated signals. This results in a fitness variation even after convergence and also makes it impossible for convergence to occur with fitness value close to the best fitness. When the complexity was increased the convergence took a longer to occur, and may not even occur within the first 100 generations. On average, each execution of 100 generations lasted 65 h, but the execution of configurations with high complexity, as configuration IV and VII, took about 124 h.

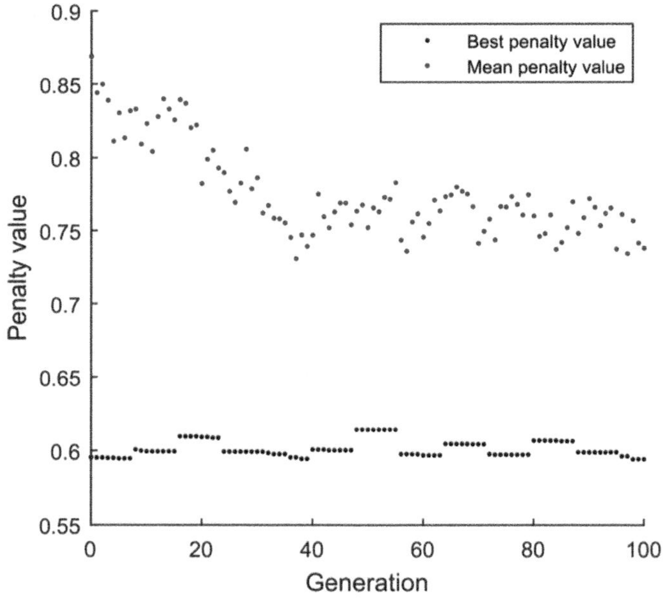

Fig. 2. Fitness throughout the execution of configuration I.

The best detectors were compared by comparing their PD curves with that of the MSC: from all the original techniques, the MSC showed best performance in terms of area under the PD curve, as shown in Fig. 3. In order to carry out this comparison, the ratio between the area under the MSC PD curve and the PD curves of the functions indicated by the genetic algorithm was calculated.

The PD curves of the best results from the GA are shown in Fig. 4, along with the MSC curve, for comparison. The indices a and b refer to each of the two replicates for each configuration. The five different configurations used with the GA algorithm all showed almost identical results. It is possible to note that the probability of detection with the data-driven optimal function functions obtained from the genetic algorithm is slightly greater than the probability of detection of the MSC. The mean PD curve of best result was statistically different from the MSC according to a student T test ($p < 0.5$).

Figure 5 shows the difference between the PD with the best function obtained by the GA (configuration I.a) and the PD from the MSC. Note that, although the simulation was carried out with signals with mean SNR of −15 dB, the greatest increase in the probability of detection was found for an SNR around −8 dB. This is evidence of no biased selection of individuals. Further, there seems to be a ceiling and floor effect in the PD and that at the point of maximum gradient, most improvement can probably be achieved.

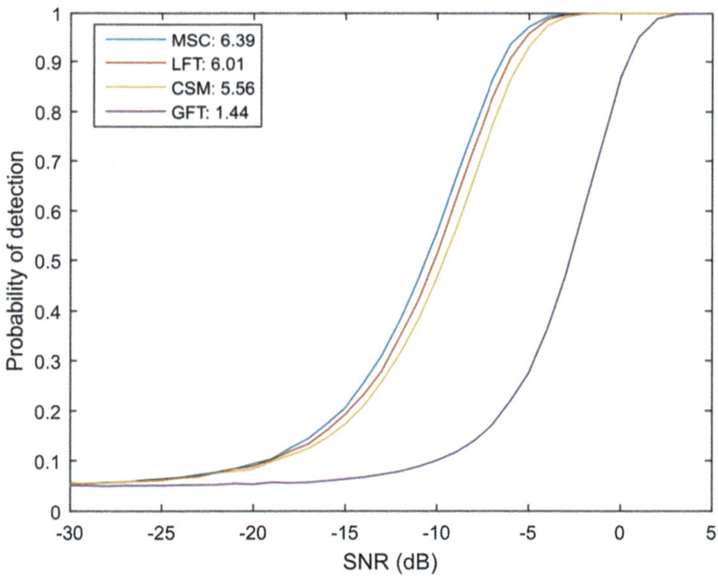

Fig. 3. Probability of detection of the original ORD functions, with their respective area under PD curve.

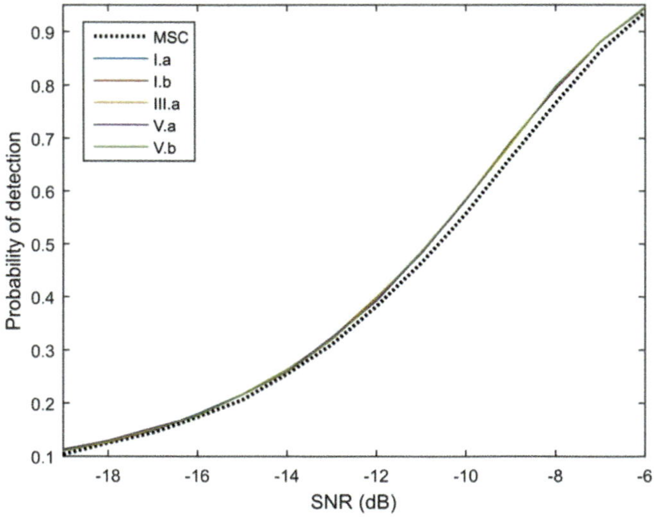

Fig. 4. Probability of detection with the best new detectors. The PD with MSC is also shown as a dotted line in order to allow comparison.

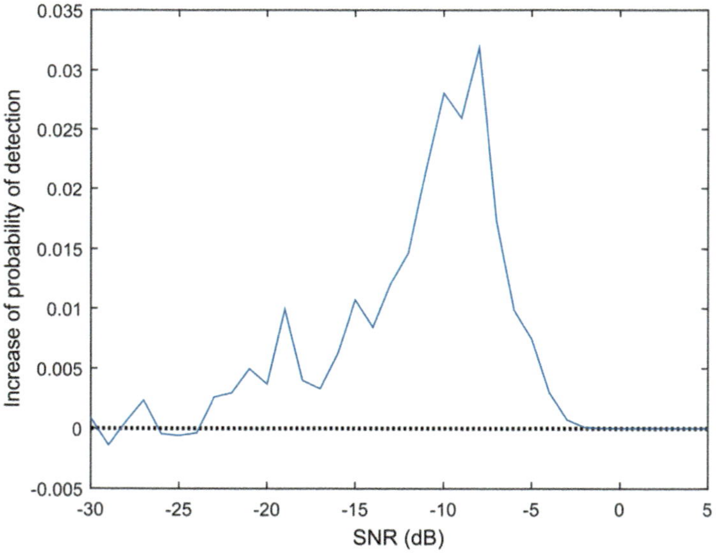

Fig. 5. Increase on detection probability of the best detector compared to MSC.

The best results in terms of the increase in probabilities of detection are found for the configurations with less complexity, that is, I and V, according to Table 3.

Table 3. The relative increase between the areas below the PD curves of each detector compared to MSC.

Configuration	1st execution	2nd execution
I	3,57%	3,35%
II	1,38%	2,26%
III	3,26%	2,87%
IV	2,75%	2,87%
V	3,40%	3,33%
VI	3,00%	2,07%
VII	2,60%	2,46%

3.1 Optimized Detectors

The functions with the best performance are listed below. The subscripts num and den refer to the numerator and the denominator of the original function, respectively. $\mathrm{CSM}_{\mathrm{sin}}$ refers to the sum of sines in the CSM. It is interesting to note that $\mathrm{LFT}_{\mathrm{num}}$ consistently appears in all 5 configurations shown (Table 4).

Table 4. Equations for the best detectors found with the GA.

Execution	Function
I.a	$\boldsymbol{f(y) = -2,75\, LFT + 4,75\, LFT_{num}}$
I.b	$f(y) = 6,48\, LFT_{num} + 3,24\, LFT_{den}$
III.a	$f(y) = -409,08\,\mathrm{CSM} - 660,44\,\mathrm{LFT}_{\mathrm{num}} + 2951,92\,\mathrm{MSC}_{\mathrm{num}} - ((8,64\,\mathrm{CSM}_{\mathrm{sin}} + 193,06\,\mathrm{GFT}) \times (-42,7\,\mathrm{MSC} + 14\,\mathrm{LFT}))$
V.a	$f(y) = 40,32\, LFT_{\mathrm{num}} - 410,13\,\mathrm{LFT} - 111,32\,\mathrm{MSC}$
V.b	$f(y) = 12,18\, LFT_{num} - 6,09\, GFT$

4 Conclusion

The proposed strategy was able to automatically find new ORD detectors. The detection rates of these were higher than those with MSC, and the increase in performance was approximately 4% in the best case. However, the genetic programming technique adopted was not efficient for high complexity configurations, and future studies are needed to make convergence possible in these situations.

The natural continuation of this work involves a number of directions. The first one would be the addition of new functions and/or components of functions calculated from the signal to be tested and the addition of other mathematical operators and/or functions with two operands (e.g. min and max), allowing for a greater range of possible combinations. A second direction would be the use of alternative optimization methods that reduce the constraints imposed by the current choices with the GA algorithm. Finally, the optimal method identified should be tested in a wider range of simulations and test-data, as well as data recorded from human subjects, in order to find evoked responses.

Acknowledgements. This research was supported by the Brazilian agency CNPq, CAPES and FAPEMIG.

Conflict of Interest. The authors declare that they have no conflicts of interest.

References

1. Dobie, R.A., Wilson, M.J.: Analysis of auditory evoked potentials by magnitude-squared coherence. Ear Hear. **10**, 2–13 (1889)
2. Fridman, J., Zappulla, R., Bergelson, M., Greenblatt, E., Malis, L., Morrell, F., Hoeppner, T.: Application of phase spectral analysis for brain stem auditory evoked potential detection in normal subjects and patients with posterior fossa tumors. Audiology **23**, 99–113 (1984)
3. Dobie, R., Wilson, M.J.: Objective response detection in the frequency domain. Electroencephalogr. Clin. Neurophysiol. **88**, 516–524 (1993)
4. Shumway, R.H.: Applied Statistical Time Series Analysis, 1st edn. Prentice-Hall, New Jersey (1988)
5. Ram, K.R., Lal, S.P., Ahmed, M.R.: Design and optimization of airfoils and a 20 kW wind turbine using multi-objective genetic algorithm and HARP Opt code. Renewable Energy **30**, 1e12 (2018)
6. Pak, T.C., Ri, Y.C.: Optimum designing of the vapor compression heat pump using system using genetic algorithm. Appl. Therm. Eng. **147**, 492–500 (2019)
7. Sahin, F.E.: Open-source optimization algorithms for optical design. Optik **178**, 1016–1022 (2019)
8. Lee, C.K.H.: A review of applications of genetic algorithms in operations management. Eng. Appl. Artif. Intell. **76**, 1–12 (2018)
9. Mostafa, N., Horta, N., Ravelo-García, A.G., Morgado-Dias, F.: Analog active filter design using a multi objective genetic algorithm. Int. J. Electron. Commun. **93**, 83–94 (2018)
10. Penchalaiah, D., Kumar, G.N., Gade, M.M., Talole, S.E.: Optimal compensator design using genetic algorithm. IFAC-PapersOnLine **51**, 518–523 (2018)
11. Hernandez-Beltran, J.E., Diaz-Ramirez, V.H., Trujillo, L., Legrand, P.: Design of estimators for restoration of images degraded by haze using genetic programming. Swarm Evol. Comput. **2019**, 49–63 (2019)
12. Verdier, C.F., Mazo, M.: Formal controller synthesis via genetic programming. IFAC-PapersOnLine **50**, 7205–7210 (2017)
13. Mehr, A.D., Nourani, V., Kahya, E., Hrnjica, B., Sattar, A.M.A., Yaseen, Z.M.: Genetic programming in water resources engineering: A state-of-the-art review. J. Hydrol. **566**, 643–667 (2018)
14. Shafer, C.A.: Data structures & algorithm analysis in Java, 3rd edn. Dover Publications, Mineola (2011)
15. Felix, L.B., Rocha, P.F.F., Mendes, E.M.A.M., Miranda de Sá, A.M.F.L.: Multivariate approach for estimating the local spectral F-test and its application to the EEG during photic stimulation. Comput. Methods Programs Biomed. **162**, 87–91 (2018)
16. Goldenholz, D.M., Ahlfors, S.P., Hämäläinen, M.S., Sharon, D., Ishitobi, M., Vaina, L.M., Stufflebeam, S.M.: Mapping the signal to noise ratios of cortical sources in magnetoencephalography and electroencephalography. Hum. Brain Map. **30.4**, 1077–1086 (2008)

Handwriting Kinematic Differences Between Copying and Dictation

Silveri Giulia(✉) and Accardo Agostino

University of Trieste, Via Valerio, 10, 34127 Trieste, Italy
silveri.giulia@phd.units.it

Abstract. Handwriting is a human activity that may be affected by the modality used as input of the information to be written, mainly copying or dictation. Many processes at different levels are involved to produce motor planning and graphomotor automation of handwriting. In order to quantify possible kinematic differences due to the influence of auditory or visual input modalities to these processes, three different tests were proposed to a sample of 101 young students and several kinematics parameters measuring handwriting characteristics were evaluated. The tests required to copy as accurate (CA) and as fast (CF) as possible an Italian sentence and to write the same sentence under dictation (DF). All parameters showed significant differences between each pair of the three tests. The best performance was obtained in the CF test followed by the DF and CA tests; in the latter the greater accuracy required to produce writing yielded lower velocity and automation as well as a longer motor planning time. On the other hand, the dictation response was more similar to that of CF than CA showing a larger planning time, probably due to a different time necessary to correctly identify the words to reproduce. The combination of the two tests could be useful to study the impairment of either visual or auditory input.

Keywords: Handwriting · Kinematic characteristics · Dictation

1 Introduction

Handwriting is a complex motor task requiring visual motor coordination abilities as well as motor planning, perceptual, cognitive and kinesthetic skills. A correct development of handwriting is crucial for autonomy in everyday life and it is a still open challenge the identification of which test and way of administration (copying or dictation) could better identify problems in written production. There are many processes involved in handwriting mainly regulated by three systems: the sensorial (visual, auditory) or cognitive (memory) system, the central processing system (based either on lexical orthographic analysis or on the phonological/non-lexical examination [1]) and the motor system.

On the other hand, advancement of digitizing tablets as well as their increased resolution and reduction of input noise have allowed an objective quantitative kinematic analysis of fine motor movements [2]. The advantages of using kinematic analysis technique are now well recognized, allowing the assessment of the level of graphomotor abilities acquired by a subject. This analysis focuses on the study of

J. Henriques et al. (Eds.): MEDICON 2019, IFMBE Proceedings 76, pp. 93–99, 2020.
https://doi.org/10.1007/978-3-030-31635-8_11

handwriting basic elements as components and strokes, investigating typical kinematic parameters such as duration, length and curvilinear, peak and mean velocities of these elements. These parameters are also frequently used to provide information on the level of graphomotor automation [3, 4].

Dictation and copying are the two main modalities to provide information to be written. However, very poor literature examines changes due to these modalities, mainly related to linguistic, cultural and teaching differences between languages and referred to single word [5]. In particular, Cheng-Lai et al. [6] investigated the relationship between writing to dictation among 45 Chinese children with dyslexia and they found that Chinese character naming is the only parameter associated with dictation and that 60% of children were impaired in performing writing to dictation; same results were confirmed by Tan et al. [7]. In addition, Tucha and Lange [8] analyzed the effect of conscious control on handwriting fluency in healthy students and ADHD children copying a phrase, suggesting that both visual and mental control of handwriting movements affect its automation.

Even if Charvin and Proteau [9] suggested that copying has an advantage because pushes the person to generate the shape based on visual perception, it remains unclear whether either dictation or copying provides better training. Current theories of motor learning do not help to identify which method is the best to learn writing and also to quantify possible writing deficits [10–13]. However, until now no literature exploring possible differences in handwriting kinematic produced by dictation or copying is available.

In order to fill this lack, in this paper, we quantify the kinematic parameters calculated on handwriting of young students subjected to copying and dictation tasks.

2 Methods

A sample of 101 graduated young students, 49 males and 52 females from 18 to 33 years old, without handwriting problems and randomly recruited at University of Trieste were considered. Before starting the test, a written informed consent was obtained from each subject.

In order to study the differences between copying and dictation each student performed a series of three tests. The first two tests concerned the copying in cursive of an Italian sentence as accurate (CA) and as fast (CF) as possible. The student read the test of the sentence, written in script mode, aloud before starting writing. The last test required the student to write in cursive the same sentence under dictation (DF). The dictating voice was recorded so as the speed of dictation was fixed and chosen so that the subjects did not make mistakes.

The sentence, containing all the letters of the Italian alphabet, was: *in pochi giorni il bruco diventò una bellissima farfalla che svolazzava sui prati in cerca di margherite e qualche quadrifoglio* ("in a few days the caterpillar became a beautiful butterfly that fluttered on the grass looking for daisies and four-leaf clover").

Handwriting data was acquired by means of a commercial digitizing tablet (Wacom, In., Vancouver, WA, Model Intuos 2.0), using an ink pen, and a suitable program based on the Software Development Kit (SDK, LCS/Telegraphics). Pen displacement across the tablet was sampled at 200 Hz, both horizontally and vertically, and the spatial resolution was of 5 μm.

Analysis was carried out with a property program written in MATLAB. Several kinematic parameters were calculated from each recording: the number of strokes (#strk), the task duration (DurTot), the mean curvilinear velocity (Vcm), the total length (Ltot), the whole pen lift duration (PLD), the mean length (SL) and duration (SD) of strokes, the mean horizontal (VHpk), vertical (VVpk) and curvilinear (VCpk) peak velocities of strokes.

In order to compare parameter values in the two groups, between each pair of tests, Wilcoxon rank sum test was used with Bonferroni correction. Differences were considered significant for a p-value < 0.05.

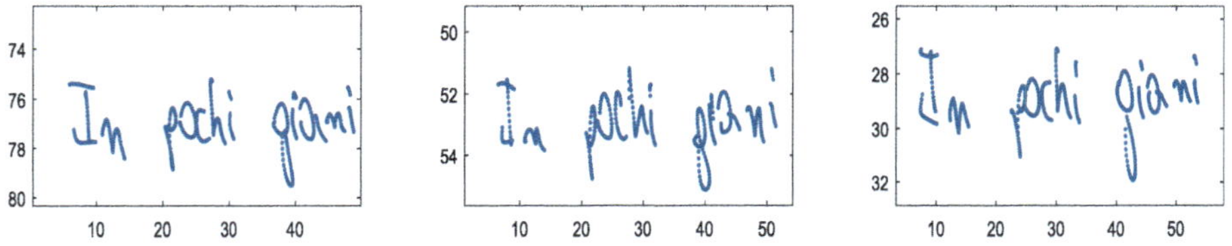

Fig. 1. Examples of a tract of handwriting in the three tests: CA (left), CF (center) and DF (right). Axes are in mm.

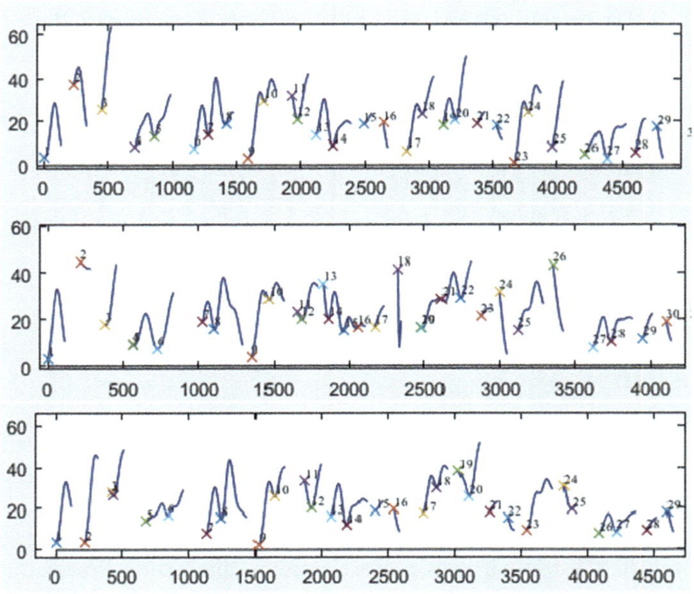

Fig. 2. Examples of curvilinear velocities corresponding to the tracts in Fig. 1: CA (top), CF (middle) and DF (bottom). Horizontal axis is the time in ms and the vertical axis is the velocity in mm/s. The numbers represent the handwriting strokes.

3 Results

Figures 1 and 2 show an example of handwriting production in the three tests together with the corresponding curvilinear velocity profiles. In CA test the letters take up less space than in the other two tests (Fig. 1) while the time to produce them is longer than in the DF ad CF tests although the curvilinear velocity profiles are very similar (Fig. 2).

In Table 1, the values of all parameters for the three tests are reported. All the parameters showed significant p-values of the differences between each pair of the three tests except for the total length (Ltot) that did not present significant changes between CA and DF. In particular, the stroke peak velocities (VHpk, VVpk, and Vcpk) as well as the mean curvilinear velocity (Vcm) and the mean stroke length (SL), i.e. the curvilinear length necessary to write a stroke, showed the highest values in CF condition and intermediate values in DF test. On the contrary, DurTot, #strk, describing the inverse of the fragmentation degree, and PLD, referring to the motor planning time, showed the highest values in CA and the lowest in CF tests. Although SD showed significant differences among tests, the values were very similar and within one sample time (5 ms). Ltot presented a large inter-subject variability with higher values in CF test than in the other two that were not significantly different.

Table 1. Mean values ± 1SD of kinematic parameters calculated in CA, CF and DF tests

	CA	CF	DF
#strk	297 ± 49	250 ± 37	265 ± 43
DurTot (s)	59.41 ± 11.81	41.80 ± 6.32	46.01 ± 9.21
Vcm (mm/s)	26.4 ± 5.6	31.9 ± 6.4	30.3 ± 6.7
Ltot (mm)	934 ± 204	925 ± 221	938 ± 222
PLD (s)	23.61 ± 6.98	12.86 ± 3.96	14.81 ± 5.38
SL (mm)	2.84 ± 0.62	3.32 ± 0.79	3.22 ± 0.80
SD (ms)	119 ± 12	114 ± 10	116 ± 11
VHpk (mm/s)	23.0 ± 5.9	28.1 ± 6.7	26.3 ± 7.0
VVpk (mm/s)	27.2 ± 7.0	30.6 ± 7.7	30.0 ± 7.7
VCpk (mm/s)	32.3 ± 7.1	37.8 ± 8.1	36.4 ± 8.5

4 Discussion

This work represents a first step in the kinematic parameters quantification of handwriting in students under copying and dictation. Since, the written product can be examined as the integration of different functions such as automation, velocity and motor planning time [12], the progressive reduction of the stroke number (#strk) from CA to DF till the CF test suggests a greater motor automation (lower fragmentation) during the tests requiring higher velocity. Moreover, the peak velocity within each stroke as well as the mean velocity through the whole track show a progressive increase from CA to DF until the CF test, confirming the results of Suarez-Coalla and Gonzalez-Martin [13], referred to single word. On the other hand, the total length (Ltot), although significantly different in the three tests, shows a maximum difference among values of only 13 mm in respect of a total distance of over 900 mm (1.4%). Consequently, a higher mean velocity results in a significant decrease (of about 25% between CA and DF and of about 10% between CF and DF) of the total duration (DurTot) necessary to write the sentence. These differences are mainly due to the different requirements of each test asked to the student.

Furthermore, since the stroke duration (SD) remains quite constant in the three tests, the stroke peak velocities increase leads to a longer stroke mean length. Finally, the decrease of mean pen lift duration (PLD) from CA to DF until CF test, suggests a progressive reduction of the motor planning time also confirmed by the reduction of the total duration of the test. Therefore, when greater accuracy is required during writing, lower automation, lower speed and longer motor planning time are necessary, confirming the results obtained by Bonin and Méot on single word [5]. However, the dictation shows intermediate values in respect of CA and CF in all parameters, significantly distinguishing this modality from the other two copying conditions.

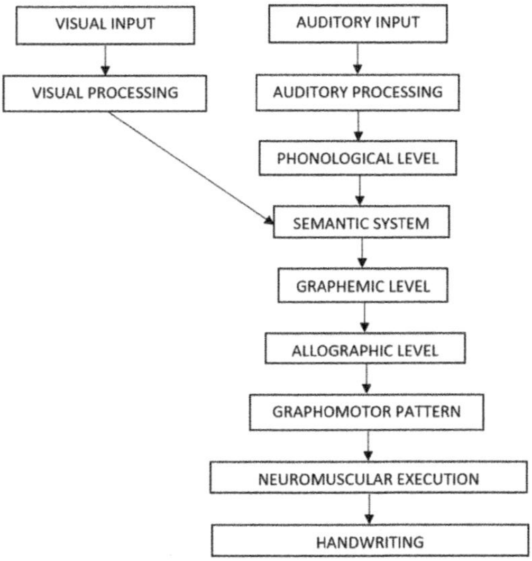

Fig. 3. Model of handwriting production processes starting from auditory or visual input (modified from Bonin [5], Tan [7] and van Galen [14]).

In order to clarify the differences between copying and dictation it is helpful to consider a model of the architecture of written production. Van Galen [14] introduced a hierarchical model to describe the word writing based on linguistic and motor processes. The central process is responsible for the recall and activation of orthographic representations while the peripheral ones are implicated in the motor execution (Fig. 3). In copying, the visual processing is defined as the ability to process two-dimensional visual representations such as shapes, dots and lines [15] so reading is necessary but it is also essential to understand and memorize each word before positioning the pen over the table, as reported by [10]. In dictation, the process starts from the interpretation of auditory sounds as a meaningful word followed by the recovery of the orthographic forms from the mental lexicon [16]. This act involves both a central process concerning retrieving, assembling and selection of an orthographic representation and a peripheral process concerning execution of orthographic codes. The latter is connected to the alphabetic writing system that is based on the idea that orthographic forms are a conventional representation of the word's pronunciation [17]. In a pure phonological alphabet, like the Italian one, there would be a one-to-one correspondence between letters and phonemes. However, in most natural languages the association between letters and sounds is to some extent ambiguous and context dependent [17].

Therefore, the differences found in the speeds between accurate copying and the other two tests could mainly be due to the high level of precision required by this test. This fact produces differences both in the neuromuscular execution of the graphomotor pattern, influencing the velocity and accuracy of the graphemes implementation, and a slower process at all previous levels, due to the greater attention required, producing a longer motor planning time. The difference between DF and CF (in particular measured by PLD) instead suggests a greater involvement at the first levels (slightly slower in the hearing than in the visual input) linked to the recognition of the exact meaning of the word to be reproduced. However, we cannot completely exclude the fact that the speed of dictation, although checked, has also influenced the overall speed.

5 Conclusion

The analysis of handwriting kinematics, allowed by the use of digital tablets and suitable software, give us the opportunity to evaluate the dynamic construction of letters and strokes and to identify the significant differences during copying and dictation. The results highlighted that during accurate copying, when a greater accuracy is required, less automation, lower velocity and a longer motor planning time are present. Moreover, dictation results were similar to those of fast copying but with a larger pen lift time probably due to a different time necessary in the two processes for the correct identification of the word to reproduce. In order to confirm the results of this research, it will be appropriate to perform other studies using different dictation velocities so as to measure the influence of this variable on the handwriting kinematic parameters. The results on a larger number of subjects and tests could be used to investigate possible deficit in the two information inputs.

Acknowledgement. Work partially supported by the Master in Clinical Engineering, University of Trieste.

Conflict of Interest Statement. The authors declare that they have no conflict of interest.

References

1. Ellis, A.W., Young, A.W.: Human Cognitive Neuropsychology: A Textbook. Taylor & Francis Ltd, Milton Park (1996)
2. Rosemblum, S., Chevion, D., Weiss, P.L.: Using data visualization and signal processing to characterize the handwriting process. Pediatr. Rehabil. **9**, 404–417 (2006)
3. Mavrogiorgou, P., Mergl, R., Tigges, P., El Husseini, J., Schroter, A., Juckel, G., Zaudig, M., Hengerl, U.: Kinematical analysis of handwriting movements in patients with obsessive-compulsive disorder. J. Neurol. Neurosurg. Psychiatry **70**, 605–612 (2001)
4. Mergl, R., Tigges, P., Schroter, A., Moller, H.J., Hergel, U.: Digitized analysis of handwriting and handwriting movements in healthy subjects: methods, results and perspectives. J. Neurosci. Methods **90**, 157–169 (1999)
5. Bonin, P., Méot, A., Lagarrigue, A., Roux, S.: Written object naming, spelling to dictation, and immediate copying: different tasks, different pathways? Q. J. Exp. Psychol. **68**(7), 1268–1294 (2015)
6. Cheng-Lai, A., Li-Tsang, C.W.P., Chan, A.H.L., Lo, A.G.W.: Writing to dictation and handwriting performance among Chinese children with dyslexia: relationships with orthographic knowledge and perceptual-motor skills. Res. Dev. Disabil. **34**, 3372–3383 (2013)
7. Tan, Y., Liu, X.: Influence of grapheme and syllable learning on handwriting output of Chinese characters in children with dictation difficulties. Front. Psychol. **9**, 1671 (2018)
8. Tucha, O., Lange, K.W.: The effect of conscious control on handwriting in children with attention deficit hyperactivity disorder. J. Attent. Disord. **1**, 323–332 (2005)
9. Charvin, H., Proteau, L.: Development differences in the processing of afferent information for motor control. Dev. Neuropsychol. **12**, 387–407 (1996)
10. Di Brina, C., Niels, R., Overvelde, A., Levi, G., Hulstijn, W.: Dynamic time warping: a new method in the study of poor handwriting. Hum. Mov. Sci. **27**, 242–255 (2008)
11. Rosemblum, S., Livnech-Zirinski, M.: Handwriting process and product characteristics of children diagnosed with development coordinator disorder. Hum. Mov. Sci. **27**, 200–214 (2008)
12. Accardo, P., Genna, M., Borean, M.: Development, maturation and learning influence on hand-writing kinematics. Hum. Mov. Sci. **32**(5), 999–1009 (2013)
13. Suarez-Coalla, P., Gonzalez-Martin, N., Cuetos, F.: Word writing in Spanish-speaking children: central and peripheral processes. Acta Physiol. **19**, 201–209 (2018)
14. Van Galen, G.P., Stelmach, G.E. (eds.): Handwriting-issue of psychomotor control and cognitive models. Acta Psychol. **82**, 1–3 (1993)
15. Li, H., Shu, H., McBride-Chng, C., Liu, H., Peng, H.: Chinese children's character recognition: visuo-orthograpic, phonological processing and morphological skills. J. Res. Read. **35**, 287–307 (2010)
16. Rapp, B.C., Caramazza, A.: The modality-specific organization of grammatical categories: evidence from impaired spoken and written sentence production. Brain Lang. **286**, 248–286 (1997)
17. Bisani, M., Ney, H.: Joint-sequence models for grapheme-to-phoneme conversion. Speech Commun. **50**(5), 434 (2008)

Bradycardia Assessment in Preterm Infants

Agnese Sbrollini, Martina Mancinelli, Ilaria Marcantoni, Micaela Morettini, and Laura Burattini(✉)

Department of Information Engineering, Università Politecnica delle Marche, Ancona, Italy
l.burattini@univpm.it

Abstract. Prematurity is a severe condition, usually correlated with critical outcomes. One of the major diseases in preterm infants is bradycardia, defined as the heart rate decreasing under 100 bpm for at least two heartbeats in duration. Usually, bradycardia is considered as a manifestation of immature cardiorespiratory control, but no studies investigated its nature in relation to the different clinical features of preterm infants. Thus, aim of this work is to assess the relation between bradycardia features and the main preterm infant clinical features, weight and gestational age. Ten preterm infants were considered, classified according with three criteria: the weight classification, the gestational age classification and the birth size assessment (that combined the two previous classifications). For each preterm infant, bradycardias are automatically identified and characterized in term of bradycardia features: amplitude, duration and area. Moreover, bradycardia events are classified according with their severity. Finally, bradycardia feature distributions of classes that belong to the same classification criterion were compared. Results seems suggesting that bradycardia features differences are more relevant in preterm infants with different weights than in those with different gestational age, contrary to what expected. Anyway, the best results in term of classification were obtained in the birth size assessment; thus, a combined approach that considers both weight and gestational age is preferable. Moreover, a combined evaluation of amplitude and duration for bradycardia characterization can better assess the severity of this arrhythmia and of the preterm infant clinical status.

Keywords: Preterm infants · Bradycardia · Preterm classification

1 Introduction

Preterm birth has an incidence of 10% worldwide [1] and is defined as a birth that occurs before the 37th week of gestational age (GA) [1, 2]. Prematurity is a severe condition, usually correlated with critical outcomes, such as neurosensory disability or cardiovascular diseases [3]. In particular, low birth weight (BW) and low GA are correlated with the most critical preterm infant clinical condition [4].

One of the major diseases in preterm infants is bradycardia. Bradycardia is usually considered as a manifestation of immature cardiorespiratory control [5] due to observed association between bradycardia and apnea. Indeed, according to literature, immaturity of the respiratory system of preterm infants represents an important cause for apnea

J. Henriques et al. (Eds.): MEDICON 2019, IFMBE Proceedings 76, pp. 100–107, 2020.
https://doi.org/10.1007/978-3-030-31635-8_12

that, in turn, provokes hypoxemia and blood desaturation to which the parasympathetic autonomic nervous system reacts by reducing HR, thus triggering bradycardias [6].

Since immaturity of cardiorespiratory system is related to low BW and low GA, it is possible to hypothesize that bradycardia also relates to low BW and low GA. Despite of that, as far as we know, no studies have investigated occurrence and characteristics of bradycardia episodes in relation to BW and GA of preterm infants. Thus, aim of the present work is to assess the relationship between bradycardia and preterm infant BW and GA.

2 Materials and Methods

2.1 Clinical Data

Clinical data consisted of the Preterm Infant Cardio-respiratory Signals (PICS) database [1] by Physionet [7]. PICS database contains electrocardiograms, respiratory signals and beat occurrence of ten preterm infants (from now on identified as I1 to I10), admitted to Neonatal Intensive Care Unit (NICU) of the University of Massachusetts Memorial Healthcare. For each infant, the information about their BW and GA are also available; preterm infants BW ranged from 800 g to 2100 g and GA ranged from 29 to 34 weeks.

2.2 Preterm Infant Classification

Preterm infants can be classified according to three criteria: BW [8], GA [2] and the birth size (BS) [9], where BS can be assessed by using the intrauterine growth curves as functions of BW and GA [9].

In relation to BW, an infant can be classified in low birth weight (LBW), if infant weight is between 1500 g and 2500 g; very low birth weight, (VLBW), if infant weight is between 1000 g and 1500 g); and extremely low birth weight (ELBW), if infant weight is lower than 1000 g [8].

In relation to GA, an infant can be classified in mild preterm (MIP), if birth occurs between the 32nd and the 36th weeks; moderate preterm (MOP), if birth occurs between the 28th and the 31st weeks) and extremely preterm (EXP, if birth occurs before the 28th week) [2].

In relation to BS, an infant can be classified in large-for-GA (LGA), if BS percentile is higher than 90th; appropriate-for-GA (AGA), if BS percentile ranges between 10th and 90th; and small-for-GA (SGA), if BS percentile is smaller than 10th [9].

2.3 Bradycardia Identification and Classification

By considering that the mean heart rate (HR) of a preterm infant range from 120 bpm to 160 bpm [10], a bradycardia occurs when HR decreases under 100 bpm for at least two consecutive beats [1]. In relation to its depth, bradycardias is usually classified as: mild (MI), if HR ranges between 100 bpm and 80 bpm; moderate (MO), if HR ranges between 80 bpm and 60 bpm; and severe (SE), if HR is lower than 60 bpm [11].

In the present study bradycardias were automatically identified and classified. The block diagram for the bradycardia identification and classification procedure is depicted in Fig. 1. Heartbeat occurrence is processed in order to compute the preterm infant heart rate series (PIHRS, bpm) from the time interval between two consecutive heartbeats. PIHRS segments lower than 100 bpm for more than 2 consecutive beats are identified as bradycardia. All bradycardias undergo optimization by merging two consecutive bradycardia events separated by only one PIHRS sample over 100 bpm. Optimized bradycardias are then characterized in terms of mean value (mBR, bpm), duration (dBR, s) and area (ABR, bpm·s), the latter defined as the product between mBR and dBR. Additionally, during bradycardia optimization, the minimum value of PIHRS (mHR) is computed to define bradycardia class (MI, MO or SE) [11].

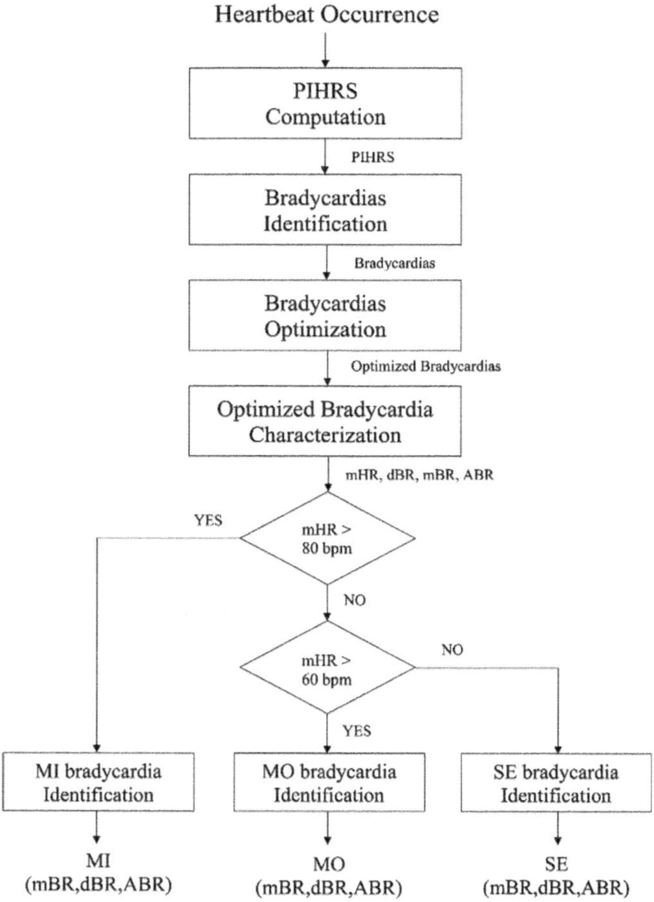

Fig. 1. Block diagram for the bradycardia identification and classification procedure.

2.4 Statistics

For each preterm infant classification criterion (BW, GA and BS), total amount of bradycardias and their distribution over MI, MO and SE classes were computed.

Each group of bradycardias are characterized through the three computed bradycardia features: mBR, dBR and ABR. Normality of features distributions of were evaluated through Lilliefors' test and non-normal distributions were described in terms of $50^{th}[25^{th};75^{th}]$ percentiles. Features belonged to classes of the same criterium were compared using the Wilcoxon rank test for equal medians. Statistical significance was set at 0.05.

3 Results

Preterm infants characterization in terms of BW, GA and BS classification is reported in Table 1. According to BW classification, five infants (I2, I3, I5, I8, I10) were classified as LBW; four (I1, I6, I7, I9) as VLBW and one (I4) as ELBW. According to GA classification, three infants (I5, I8, I10) were classified as MIP, seven (I1, I2, I3, I4, I6, I7, I9) as MOP and none as EXP. Eventually, according to BS classification, no infant was classified as LGA, six (I1, I2, I3, I5, I8; I10) were classified as AGA and four (I4, I6, I7, I9) as SGA.

Bradycardia features distributions are reported in Table 2. Increasing the severity of the preterm infants clinical condition, mBR (from 89 bpm in LWB to 77 bpm in ELWB; from 88 bpm in MIP to 87 bpm in MOP; from 90 bpm in AGA to 81 bpm in SGA) decreases; dBR (from 4 s in LWB to 5 s in ELWB; from 4 s in MIP to 4 s in MOP; from 4 s in AGA to 4 s in SGA) remains essentially constant; and, ABR (from 362 bpm·s in LWB to 316 bpm·s in ELWB; from 364 bpm·s in MIP to 360 bpm·s in MOP; from 361 bpm·s in AGA to 319 bpm·s in SGA) decreases. Moreover, a similar trend can be observed in each bradycardia class.

Table 1. Clinical information and classification of preterm infants.

	GA ($weeks^{days}$)	Weight (kg)	BW classification	GA classification	BS classification
I1	29^3	1.20	VLBW	MOP	AGA
I2	30^5	1.76	LBW	MOP	AGA
I3	30^5	1.71	LBW	MOP	AGA
I4	30^1	0.80	ELBW	MOP	SGA
I5	32^2	1.67	LBW	MIP	AGA
I6	30^1	1.14	VLBW	MOP	SGA
I7	30^1	1.11	VLBW	MOP	SGA
I8	32^3	2.10	LBW	MIP	AGA
I9	30^4	1.23	VLBW	MOP	SGA
I10	34^2	1.90	LBW	MIP	AGA

Table 2. Bradycardia features distributions.

Criteria	Preterm infant classes	Bradycardia classes	#	mBR (bpm)	dBR (s)	ABR (bpm·s)
BW classification	LWB (5 preterm infants)	Total	476	89[82;95]	4[2;9]	362[181;744]
		MI	259	95[91;97]	3[2;5]	241[180;472]
		MO	185	83[79;87]	6[3;13]	542[301;1089]
		SE	32	74[69;79]	18[9;25]	1295[614;1950]
	VLWB (4 preterm infants)	Total	288	88*[79;93]	4[2;8]	328[180;616]
		MI	137	93*[90;95]	2[1;4]	181[120;420]
		MO	80	85*[80;88]	4*[3;6]	361*[242;523]
		SE	71	73[64;77]	14[6;21]	940*[444;1370]
	ELWB (1 preterm infants)	Total	71	77*,§[69;83]	5[2;7]	316[182;508]
		MI	8	90*,§[86;93]	1[1;2]	120*[120;211]
		MO	25	82§[79;85]	2*,§[2;5]	183*,§ [181;378]
		SE	38	70*[63;75]	6*,§[5;9]	440*,§ [313;614]
GA classification	MIP (7 preterm infants)	Total	194	88[80;94]	4[2;8]	364[182;705]
		MI	84	94[91;96]	3[1;5]	271[150;421]
		MO	98	84[80;87]	5[3;9]	483[244;904]
		SE	12	73[65;77]	10[5;19]	889[587;1405]
	MOP (3 preterm infants)	Total	641	87†[80;91]	4[2;9]	360[180;661]
		MI	320	92†[90;95]	3[2;5]	240[120;421]
		MO	192	82†[78;86]	6[3;11]	422[242;729]
		SE	129	70[62;74]	12[9;23]	677[373;1317]
BS classification	AGA (6 preterm infants)	Total	566	90[83;95]	4[2;8]	361[180;720]
		MI	334	94[91;96]	3[2;5]	241[180;481]
		MO	199	83[79;87]	6[3;13]	517[301;1087]
		SE	33	75[69;79]	17[9;25]	1281[605;1923]
	SGA (4 preterm infants)	Total	269	81‡ [73;89]	4[2;8]	319[182;609]
		MI	70	92‡ [89;95]	2‡[1;3]	180‡[120;241]
		MO	91	84[79;86]	4‡[2;6]	304‡[182;469]
		SE	108	72[64;76]	9‡[5;16]	614‡[331;1199]

*P-value < 0.05, comparing bradycardia features of classes VLWB and ELWB in relation to LWB; §P-value < 0.05, comparing bradycardia features of classes ELWB in relation to VLWB; †P-value < 0.05, comparing bradycardia features of classes MOP in relation to MIP; ‡P-value < 0.05, comparing bradycardia features of classes SGA in relation to AGA.

4 Discussion

Aim of this work was to assess the relationship between bradycardia and preterm infant BW and GA. With this aim, bradycardias of ten preterm infants were automatically identified and classified in their level of severity. Due to the particular condition of prematurity, recruitment of many cases is challenging; thus, only ten cases were considered for this pilot study. Future studies will confirm the here presented results on a larger database.

In order to evaluate the bradycardia dependency to different BW and GA, preterm infants were classified according to three standard criteria. According to BW criterion [8], a lower weight is associated with a more severe condition. This evidence is confirmed by our results: bradycardia features (specifically mBR and ABR) decreases passing from LBW to ELBW. According to GA criterion [2], low GA is sign of immaturity, thus of a severe clinical condition. Differently from BW, no differences can be observed in the comparison of GA classes, that present no statistically different bradycardia features distributions. Finally, according to BS criterion [9], a severe preterm infant condition is correlated to low BW and GA. BS classes (AGA and SGA) presents statistically different bradycardia features distributions.

Comparing the different classification criteria, results seem suggesting that bradycardia differences are more relevant in preterm infants with different BW than in those with different GA, contrary to what expected. Anyway, the best results in term of classification were obtained in BS. Thus, a combined approach that considers both BW and GA is preferable for evaluate the preterm infant severity.

Regarding bradycardia features assessment, ABR is defined as a combination of mBR and dBR. This feature is usually computed for fetal decelerations assessment [12], in order to evaluate the fetal well-being. Considering the prematurity conditions and the similarity between fetuses and preterm infants, we hypothesized that it can be a valid feature to be investigated. Considering our results, ABR is statistically different in the same classes in which dBR is statistically different. Thus, this evidence suggests that the difference between the classes is not only guided by mBR, but also by dBR. Nowadays, severity of bradycardia is main classified on the basis of amplitude [11], without defining criteria for bradycardia event durations. On the contrary, this study proves that dBR can be a valid feature for the clinical assessment of bradycardia and, indirectly, for preterm infant clinical condition. Future studies will evaluate the importance of bradycardia classification in relation to their depth and their duration, defining new guidelines for preterm infant bradycardia diagnosis.

Bradycardia assessment remains essential in preterm infants monitoring: it is usually performed extracting the preterm infants heart rate series from cardiac signals. This procedure is reliable, but it provokes discomfort for infants since they are covered by many sensors and electrodes. Innovative solutions to reliably monitoring infants, reducing their discomfort, can be the contactless devices [13]. Future studies will evaluate the reliability of these new technologies in preterm infants bradycardia detection and assessment.

Finally, our automatic algorithm is based on the standard definition of bradycardia, following the standard guidelines [1]. Considering the similarity between fetuses and preterm infants, a baseline approach typically used in cardiotocography [14] can be promising. Thus, future studies will evaluate the applicability of this different approach on preterm infant bradycardia computation.

5 Conclusion

In conclusion, bradycardias are correlated to preterm infant BW and GA; thus a combined evaluation of these to clinical features are suggested. Moreover, a combined evaluation of amplitude and duration for the bradycardia characterization can better assess the severity of these arrhythmias and, consequently, of preterm infant clinical status.

Conflict of Interest. All authors have no financial and personal relationships with other people or organizations that could inappropriately bias the work.

References

1. Gee, A.H., Barbieri, R., Paydarfar, D., Indic, P.: Predicting bradycardia in preterm infants using point process analysis of heart rate. IEEE Trans. Biomed. Eng. **64**(9), 2300–2308 (2017)
2. Moutquin, J.M.: Classification and heterogeneity of preterm birth. BJOG Int. J. Obstet. Gynaecol. **110**(20), 30–33 (2003)
3. Behrman, R.E., Butler, A.S.: Preterm Birth: Causes, Consequences, and Prevention, 1st edn. National Academies Press (2007)
4. Saigal, S., Doyle, L.W.: An overview of mortality and sequelae of preterm birth from infancy to adulthood. Lancet **71**(9608), 261–269 (2008)
5. Southall, D.P., Levitt, G.A., Richards, J.M., Jones, R.A.K., Kong, C., Farndon, P.A., Alexander, J.R., Wilson, A.J.: Undetected episodes of prolonged apnea and severe bradycardia in preterm infants. Pediatrics **72**(4), 541–551 (1983)
6. Veerappan, S., Rosen, H., Craelius, W., Curcie, D., Hiatt, M., Hegyi, T.: Spectral analysis of heart rate variability in premature infants with feeding bradycardia. Pediatr. Res. **47**(5), 659–662 (2000)
7. Poets, C.F., Stebbens, V.A., Samuels, M.P., Southall, D.P.: The relationship between bradycardia, apnea, and hypoxemia in preterm infants. Pediatr. Res. **34**(2), 144–147 (1993)
8. Goldberger, A.L., Amaral, L.A., Glass, L., Hausdorff, J.M., Ivanov, P.C., Mark, R.G., Mietus, J.E., Moody, G.B., Peng, C.K., Stanley, H.E.: PhysioBank, PhysioToolkit, and PhysioNet: components of a new research resource for complex physiologic signals. Circulation **101**(23), e215–e220 (2000)
9. Shum, D., Neulinger, K., O'Callaghan, M., Mohay, H.: Attentional problems in children born very preterm or with extremely low birth weight at 7–9 years. Arch. Clin. Neuropsychol. **23**(1), 103–112 (2008)
10. Olsen, I.E., Groveman, S.A., Lawson, M.L., Clark, R.H., Zemel, B.S.: New intrauterine growth curves based on United States data. Pediatrics **125**(2), e214–e224 (2010)
11. Hansmann, G.: Management of healthy, term newborn infants (vaginal delivery, cesarean section, vacuum extraction, forceps delivery), 1st edn. Cambridge University Press, Boston (2009)
12. Agostinelli, A., Palmieri, F., Biagini, A., Sbrollini, A., Burattini, L., Di Nardo, F., Fioretti, S., Burattini, L.: Relationship between deceleration areas in the second stage of labor and neonatal acidemia. Comput. Cardiol. **43**, 897–900 (2016)

13. Gambi, E., Agostinelli, A., Belli, A., Burattini, L., Cippitelli, E., Fioretti, S., Pierleoni, P., Ricciuti, M., Sbrollini, A., Spinsante, S.: Heart rate detection using Microsoft Kinect: validation and comparison to wearable devices. Sensors **17**(8), 1776 (2017)
14. Agostinelli, A., Braccili, E., Marchegiani, E., Rosati, R., Sbrollini, A., Burattini, L., Morettini, M., Di Nardo, F., Fioretti, S., Burattini, L.: Statistical baseline assessment in cardiotocography. In: Proceedings of the Annual International Conference of the IEEE Engineering in Medicine and Biology Society, EMBS, Institute of Electrical and Electronics Engineers Inc., South Korea, pp. 3166–3169 (2017)

To What Extent Does Heart Rate Alter the Cerebral Hemodynamic Patterns During Atrial Fibrillation?

Stefania Scarsoglio[1(✉)], Luca Ridolfi[2], Andrea Saglietto[3], and Matteo Anselmino[3]

[1] Department of Mechanical and Aerospace Engineering, Politecnico di Torino, Torino, Italy
stefania.scarsoglio@polito.it
[2] Division of Cardiology, "Città della Salute e della Scienza di Torino" Hospital, Department of Medical Sciences, Università di Torino, Torino, Italy
[3] Department of Environmental, Land and Infrastructure Engineering, Politecnico di Torino, Torino, Italy

Abstract. Atrial fibrillation (AF), the most common cardiac arrhythmia leading to irregular and faster heartbeat, has been recently and independently associated to the risk of dementia. A constellation of hemodynamic mechanisms has been proposed to explain the possible link between the two pathologies. However, definitive data still miss, and it is unknown how heart rate (HR) influences the cerebral microcirculation. We propose a computational approach, based on a validated hemodynamic modeling, to compare the cerebral circulation during normal sinus rhythm (NSR) and AF at different HRs. AF is able to trigger a higher variability of the cerebral blood flow variables which grows towards the distal circulation. The alteration of the hemodynamic patterns, inducing the rupture of the signal periodicity and the consequent higher occurrence of extremely high/low values, increases with HR. Awaiting necessary clinical evidences, present findings highlight that a strict rate control strategy could be beneficial in terms of cognitive outcomes in patients with permanent AF.

Keywords: Cerebral hemodynamics · Atrial fibrillation · Cardiovascular modelling · Heart rate

1 Introduction

Atrial fibrillation (AF), causing irregular and faster heartbeat, is becoming a public health problem in developing countries [6]. AF has been recently associated with an increased risk of dementia and cognitive decline, even in the absence of clinical cerebrovascular events [3,4]. Silent cerebral ischemia, microbleeds, and impaired cerebral blood flow have been proposed to explain the link between the two pathologies [2,5]. Among all possible mechanisms, the alteration of cerebral

J. Henriques et al. (Eds.): MEDICON 2019, IFMBE Proceedings 76, pp. 108–116, 2020.
https://doi.org/10.1007/978-3-030-31635-8_13

perfusion during AF is an aspect which remains poorly investigated so far. In fact, current clinical techniques - such as transcranial doppler ultrasonography - do not provide sufficient resolution to fully describe the inner cerebral circle, in terms of flow and pressure signals. In particular, it is to date unknown whether the irregularity of the RR intervals during AF directly influences cerebral hemodynamics, mostly of the deepest cerebral circulation.
Given the scarcity of clinical data, we recently proposed a computational approach to evaluate the cerebral hemodynamics in AF. It emerged that AF - when compared to the same heart rate (HR) in normal sinus rhythm (NSR) - promotes an altered cerebral hemodynamics and a dramatic signal variation, with the onset of transient hypoperfusions and hypertensive events in the cerebral microcirculation [1,10,11].
A signal analysis in terms of cerebral pressure and flow rate time series is here proposed to compare NSR and AF at different HRs, namely from 50 to 130 bpm [7]. In particular, through the percentile and crossing time analyses, we further investigate to what extent the cerebral hemodynamic alterations induced by AF are modulated by means of the heart rate.

2 Methods

2.1 Computational Stochastic Approach

The computational algorithm, as shown in Fig. 1, combines a stochastic extraction of the heart beating, RR [s], with a lumped parameter modeling of the cardiovascular and cerebral circulations. More details, including governing equations, can be found in [7] and therein references.

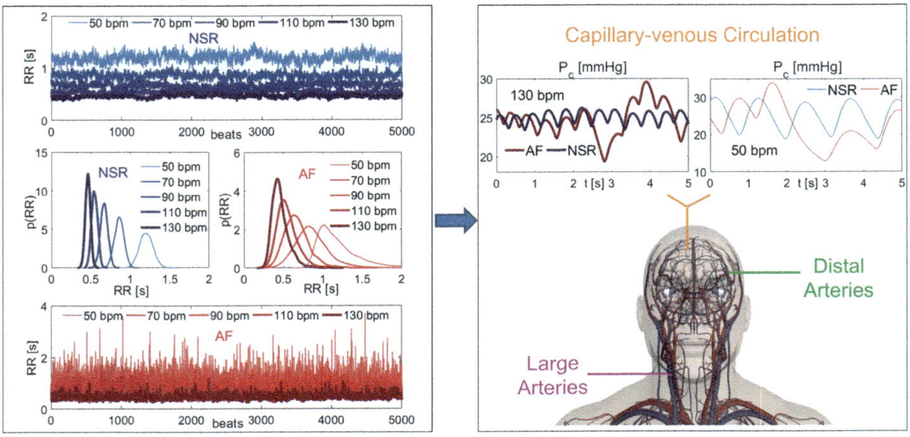

Fig. 1. Scheme of the computational algorithm: (left) RR beats and corresponding probability density functions; (right) Cerebral model with representative capillary pressure signals, P_c.

RR Beating Features in NSR and AF. The artificially extracted RR intervals, based on in vivo beating features and validated over available clinically measured data in NSR and AF [8], span the range of average HR between 50 and 130 bpm. NSR beats are extracted from a Gaussian distribution, which is the typical distribution observed during sinus rhythm. RR extraction is done according to the correlated pink noise temporal structure, the heart rate being a relevant example [8]. Standard deviation values, σ, are determined considering that the coefficients of determination, cv, can be assumed as constant and equal to 0.07 from 50 to 130 bpm. AF distribution can be described by the superposition of two statistically independent times, obtained from a correlated Gaussian and an uncorrelated Exponential distributions. The resulting RR intervals are thus drawn by an exponentially modified Gaussian (EMG) distribution (more details can be found in [8] and therein references). At each HR, the standard deviation, σ, is determined keeping the coefficient of variation, cv, constant and equal to 0.24, while the rate parameter, γ, is taken as a linear function of the mean RR ($\gamma = -9.2\text{RR} + 14.6$) [7,8]. To assure the statistical stationarity of the outcomes, 5000 cardiac cycles are simulated for each configuration. The 5000 beats extracted in NSR and AF conditions and the corresponding RR probability distribution functions are displayed in Fig. 1 (left).

Cardiovascular Model. Once RR intervals are extracted, the cardiovascular model is run to obtain the systemic arterial pressure (P_a). The 0D cardiovascular model, first proposed and validated in AF over more than 30 clinical datasets [8, 9], consists of a network of compliances, resistances and inductances, describing the four contractile cardiac chambers, the systemic and venous circuits. Both atria are imposed as passive to simulate AF conditions, while they can actively contract during NSR. The model includes short-term baroregulation mechanisms [7], accounting for the inotropic effect of both ventricles, as well as the control of the systemic vasculature (peripheral arterial resistances, unstressed volume of the venous system, and venous compliance). The chronotropic effects due to the heart rate regulation, which differently act in NSR and AF, are instead implicitly considered by the RR extraction. The systemic arterial pressure, P_a, is then used as forcing input for the cerebral model.

Cerebral Model. The 0D cerebral model, describing the arterial and venous cerebral circulation, along with the cerebrovascular control mechanisms of autoregulation and CO_2 reactivity [12], has been validated in normal conditions up to the middle cerebral circulation [1], where clinical data are available. A network of compliances and resistances describes the cerebral circulation, which is divided into three main regions: large arteries, distal arterial circulation, and capillary/venous circulation. The left vascular pathway ICA-MCA (i.e., internal carotid artery—middle cerebral artery) is here analyzed as representative of the blood flow and pressure distributions from large arteries to the capillary-venous circulation: left internal carotid artery (systemic arterial pressure, P_a, and left internal carotid flow rate, $Q_{ica,left}$), middle cerebral artery (left middle cerebral artery pressure, $P_{mca,left}$, and flow rate, $Q_{mca,left}$), middle distal district (left middle distal pressure, $P_{dm,left}$, and flow rate, $Q_{dm,left}$), and capillary-venous circulation (cerebral capillary pressure, P_c, and proximal venous flow rate, Q_{pv}). Figure 1 (right) reports examples of capillary pressures, P_c, in NSR and AF at different HRs.

2.2 Data Analysis

Percentile and crossing time analyses are evaluated for 5000 cycles at different HRs. We here extend the definition of extremely high/low values and the temporal persistence of the hemodynamic signal above/below the physiological threshold during AF [1,10] to different HRs, so that each rare event during AF is defined as referred to the corresponding NSR at the same HR. NSR outcomes are thus exploited to define the different reference thresholds of each HR [7]. In order to quantify AF-induced variations, the NSR 5th and 95th percentiles of the cerebral hemodynamic variables are taken as reference thresholds for the

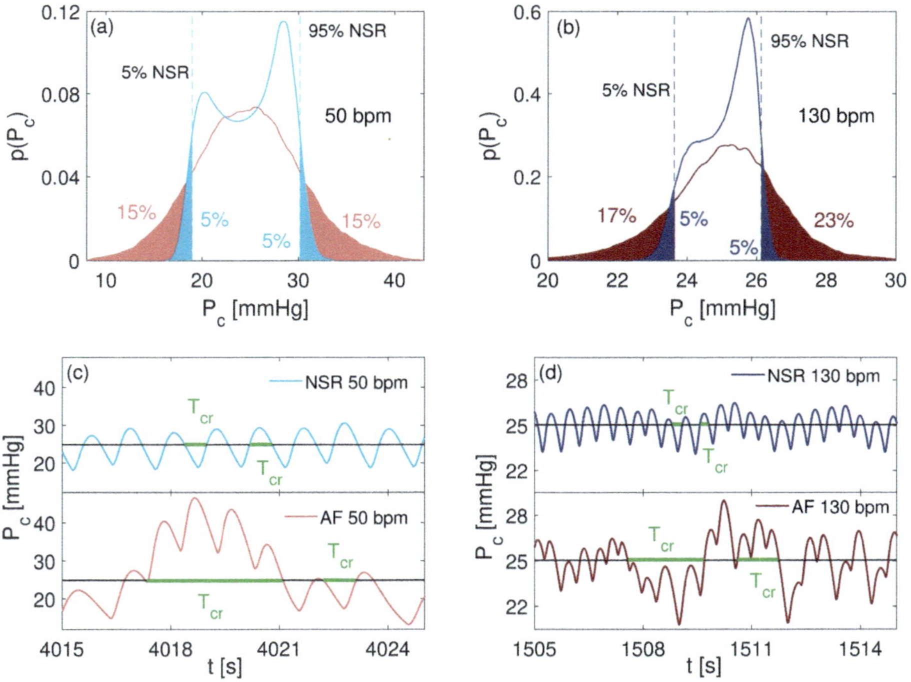

Fig. 2. (a)-(b) Examples of percentile variation for P_c (p(P_c) is the probability density function), at 50 (a) and 130 (b) bpm. NSR thresholds (dashed blue lines) individuate the 5th and 95th percentiles in NSR (blue areas), while they correspond to higher percentiles in AF (red areas). (c)-(d) Examples of T_{cr} evaluation for representative portions of the P_c time series in NSR and AF ((c) 50 bpm, (d) 130 bpm). T_{cr} intervals are indicated in green.

current HR. In AF, we then evaluate to which percentile the NSR reference thresholds correspond, thereby evaluating how AF modifies the probability of assuming extremely high or low values at each fixed HR. Examples of pressure percentile variations between NSR and AF are reported for the capillary region in Fig. 2a and b for HR = 50 and 130 bpm, respectively.
Similarly to the percentile analysis, the crossing time evaluation is carried out throughout the whole temporal series at different HRs. At a fixed HR, the crossing time, T_{cr}, is defined as the temporal interval consecutively spent by the hemodynamic variable above or below the threshold individuated at the same HR by the mean value in NSR. In Fig. 2c and d, illustrative examples of T_{cr} are shown for the capillary pressure, P_c, during NSR and AF, at 50 and 130 bpm, respectively. The T_{cr} intervals show, at each HR, how AF influences the duration of excursions from the reference mean value in NSR (individuated at the same HR).

3 Results and Discussion

The alteration of the cerebral hemodynamic signals due to the effect of HR during AF is here investigated, by means of the percentile and crossing time analyses. All the results involve the hemodynamic variables along the selected left ICA-MCA path. We only account for flow rate variations below the 5th percentile (i.e., hypoperfusions) and pressure variations above 95th percentile (hypertensive episodes), since these configurations are the most meaningful from the hemodynamic point of view. For the crossing time analysis, instead, we do not discern whether the time lapse spent is above or below the reference NSR threshold, since both configurations are equally meaningful to evidence the rupture of the signal periodicity and the variation of its temporal pattern.

3.1 Percentile Analysis

The AF percentiles corresponding to the NSR reference thresholds of each HR along the ICA-MCA pathway are reported in Fig. 3. The percentile alteration during AF suggests a worsening - in terms of higher probability of extreme values - in the microcirculation for increasing HR. In fact, at the large artery level ($Q_{ica,left}$, $Q_{mca,left}$, P_a, $P_{mca,left}$) percentile variations are not substantially influenced by the HR - being all confined below the 10th percentile for flow rates (Fig. 3a) and even above the 95th percentile for pressures (Fig. 3b). This last aspect can be explained considering the slight decrease of mean P_a and $P_{mca,left}$ values during AF with respect to NSR, for all the HRs computed [7]. The scenario definitely changes going towards the distal and capillary circulation. Here, not only percentile variations are more marked (up to the 18th and 75th percentiles for flow rates and pressures, respectively), but they also increase with HR. For all the HRs considered, the distal region is the most prone to low perfusion, since the 5th percentile in NSR corresponds to around the 12th–18th percentiles in AF, for HR = 50 and 130 bpm, respectively. High pressure levels

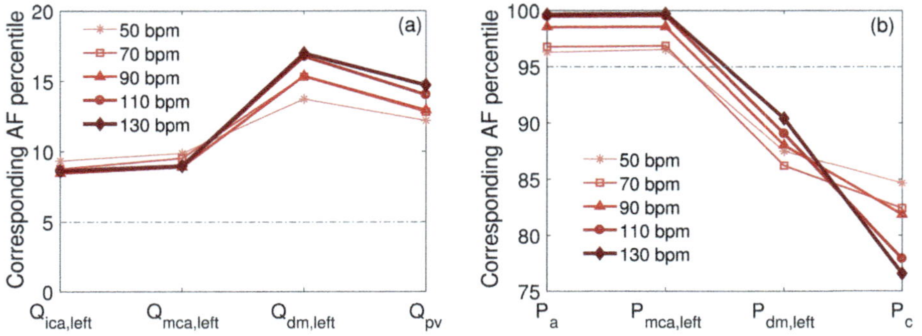

Fig. 3. Percentile analysis corresponding to the NSR threshold of each HR along the ICA-MCA pathway. (a) Flow rate, 5th NSR, (b) Pressure, 95th NSR. Color and thickness (from thin light red to thick dark red) refer to increasing HR (from 50 to 130 bpm), while the abscissa individuates the ICA-MCA regions.

of greater significance are instead more likely to occur in the capillary district, with the 95th NSR percentile corresponding up to the 85th–75th AF percentiles, for HR = 50 and 130 bpm, respectively. Moreover, differently to the large arteries district, mean capillary pressures barely differ between NSR and AF, at all the HRs [7]. Thus, the marked AF percentile variation here observed is all imputable to the greater AF variability, which is not absorbed but even magnified in the microcirculation. What observed so far extends to a wide range of HRs (from 50 to 130 bpm) the higher AF-induced probability of hypoperfusions in the distal region and hypertensive events in the capillary circulation, previously observed only at HR = 70 bpm [1].

3.2 Crossing Time Analysis

The analysis is here presented in terms of the ratio of the mean crossing time in AF, $\mu(T_{cr})_{AF}$, to the mean crossing time in NSR, $\mu(T_{cr})_{NSR}$ (Fig. 4, top panels), and the T_{cr} distributions (Fig. 4, bottom panels). The crossing time analysis shows a behaviour which is qualitatively similar for pressure (left panels) and flow rate (right panels) signals.

For a fixed HR, the ratio $\mu(T_{cr})_{AF}/\mu(T_{cr})_{NSR}$ increases from the proximal to the distal circulation. In particular, in the proximal region the ratio remains unitary, meaning that AF does not modify the signal periodicity and its temporal pattern with respect to NSR. In the distal-capillary compartments, instead, the ratio definitely exceeds one, revealing a higher time lapse spent above/below the reference threshold and thus a pattern alteration. By changing HR, in the proximal region the ratio stays close to one, therefore the HR effect on the signal periodicity and behaviour is negligible. Moving towards the microcirculation, $\mu(T_{cr})_{AF}/\mu(T_{cr})_{NSR}$ ranges from 1.2 (HR = 50 bpm) up to almost 1.8 (HR = 130 bpm). It should be noted that the region mostly involved is the middle distal district ($P_{dm,left}$ and $Q_{dm,left}$), where for HR = 130 bpm crossing times in AF are on average from 50% to 80% longer than in NSR. This result highlights that, in the deepest cerebral circulation, AF signals at higher HR are more prone to lose their periodicity and alter their temporal structure than those at lower HR. To better explore this last aspect, the probability density functions (PDFs) of T_{cr} where the greatest excursions occur (i. e., middle distal district) are analyzed (see Fig. 4c and d). Even if, at the same HR, the main body of the PDF is similarly positioned in NSR and AF, PDFs during AF display much more pronounced right tails and lose the symmetry shown during NSR. This behaviour becomes more evident when HR is increased from 50 to 130 bpm. The right tails are thus responsible for the increase, as HR grows, of the mean crossing time values in AF with respect to NSR (see the legends of Fig. 4c and d).

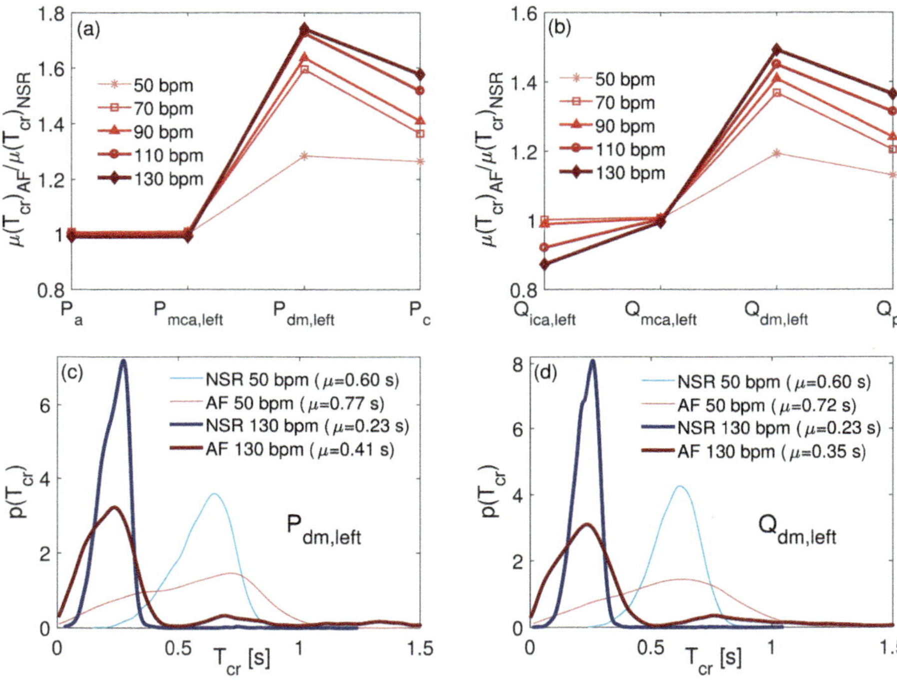

Fig. 4. (a)-(b) Ratio of the mean crossing time in AF, $\mu(T_{cr})_{AF}$, to the mean crossing time in NSR, $\mu(T_{cr})_{NSR}$, at the same HR for (a) pressure and (b) flow rate signals. Color and thickness (from thin light red to thick dark red) refer to increasing HR (from 50 to 130 bpm), while the abscissa individuates the ICA-MCA regions. (c)-(d) Probability density functions (PDFs) of the crossing times, T_{cr}, for (c) $P_{dm,left}$ and (d) $Q_{dm,left}$ during NSR and AF, at 50 and 130 bpm.

4 Conclusions

The present work computationally evaluated how heart rate impacts on the cerebral hemodynamics during AF, through the percentile and crossing time analyses. Critical events - such as the alteration of the cerebral hemodynamic patterns, the loss of the signal periodicity and the related higher occurrence of rare (i.e., extremely high/low) values - emerge in the distal districts and significantly increase in frequency with HR. The trend is basically monotone with HR, thus no optimal HR target to minimize the effects of these transient episodes can be identified. Awaiting further necessary clinical validation and provided that all efforts should be addressed to maintain NSR as long as possible, present findings suggest that a strict rate control strategy could be beneficial in terms of cognitive outcomes in patients with permanent AF.

Acknowledgments. The study was performed thanks to the support of the "Compagnia di San Paolo" within the project "Progetti di Ricerca di Ateneo – 2016: Cerebral hemodynamics during atrial fibrillation (CSTO 160444)" of the University of Turin, Italy.

Conflicts of Interest. The authors declare they have no conflicts of interest.

References

1. Anselmino, M., Scarsoglio, S., Saglietto, A., Gaita, F., Ridolfi, L.: Transient cerebral hypoperfusion and hypertensive events during atrial fibrillation: a plausible mechanism for cognitive impairment. Sci. Rep. **6**, 28635 (2016). https://doi.org/10.1038/srep28635
2. Da Silva, R.M.F.L., Miranda, C.M., Liu, T., Tse, G., Roever, L.: Atrial fibrillation and risk of dementia: epidemiology, mechanisms, and effect of anticoagulation. Front. Neurosci. **13**, 18 (2019). https://doi.org/10.3389/fnins.2019.00018
3. de Bruijn, R.F., Heeringa, J., Wolters, F.J., Franco, O.H., Stricker, B.H., Hofman, A., Koudstaal, P.J., Ikram, M.A.: Association between atrial fibrillation and dementia in the general population. JAMA Neurol. **72**(11), 1288–1294 (2015). https://doi.org/10.1001/jamaneurol.2015.2161
4. Dietzel, J., Haeusler, K.G., Endres, M.: Does atrial fibrillation cause cognitive decline and dementia? Europace **20**(3), 408–419 (2018). https://doi.org/10.1093/europace/eux031
5. Jacobs, V., Cutler, M.J., Day, J.D., Bunch, T.J.: Atrial fibrillation and dementia. Trends Cardiovasc. Med. **25**(1), 44–51 (2015). https://doi.org/10.1016/j.tcm.2014.09.002
6. Morillo, C.A., Banerjee, A., Perel, P., Wood, D., Jouven, X.: Atrial fibrillation: the current epidemic. J. Geriatr. Cardiol. **14**(3), 195–203 (2017). https://doi.org/10.11909/j.issn.1671-5411.2017.03.011
7. Saglietto, A., Scarsoglio, S., Ridolfi, L., Gaita, F., Anselmino, M.: Higher ventricular rate during atrial fibrillation relates to increased cerebral hypoperfusions and hypertensive events. Sci. Rep. **9**, 3779 (2019). https://doi.org/10.1038/s41598-019-40445-5

8. Scarsoglio, S., Guala, A., Camporeale, C., Ridolfi, L.: Impact of atrial fibrillation on the cardiovascular system through a lumped-parameter approach. Med. Biol. Eng. Comput. **52**(11), 905–920 (2014). https://doi.org/10.1007/s11517-014-1192-4
9. Scarsoglio, S., Camporeale, C., Guala, A., Ridolfi, L.: Fluid dynamics of heart valves during atrial fibrillation: a lumped parameter-based approach. Comput. Methods Biomech. Biomed. Eng. **19**(10), 1060–1068 (2016). https://doi.org/10.1080/10255842.2015.1094800
10. Scarsoglio, S., Saglietto, A., Anselmino, M., Gaita, F., Ridolfi, L.: Alteration of cerebrovascular haemodynamic patterns due to atrial fibrillation: an in silico investigation. J. Roy. Soc. Interface **14**(129), 20170180 (2017). https://doi.org/10.1098/rsif.2017.0180
11. Scarsoglio, S., Cazzato, F., Ridolfi, L.: From time-series to complex networks: application to the cerebrovascular flow patterns in atrial fibrillation. Chaos **27**(9), 093107 (2017). https://doi.org/10.1063/1.5003791
12. Ursino, M., Giannessi, M.: A model of cerebrovascular reactivity including the circle of willis and cortical anastomoses. Ann. Biomed. Eng. **38**(3), 955–974 (2010). https://doi.org/10.1007/s10439-010-9923-7

Non-invasive Intrauterine Pressure Estimation Based on Nonlinear Parameters Computed from the Electrohysterogram

Monica Albaladejo-Belmonte[1], Gema Prats-Boluda[1], Yiyao Ye-Lin[1], Carlos Benalcazar-Parra[1], Ángel Lopez[2], Alfredo Perales[2], and Javier Garcia-Casado[1](✉)

[1] Centro de Investigación e Innovación en Bioingeniería, Universitat Politècnica de València, Valencia, Spain
jgarciac@eln.upv.es
[2] Servicio de Obstetricia y Ginecología, Hospital Universitario y Politécnico La Fe de Valencia, Valencia, Spain

Abstract. Monitoring uterine contractions is essential during pregnancy and labor to obtain information on time-to-delivery and maternal and fetal wellbeing Intrauterine pressure (IUP) is considered the "gold standard" to monitor uterine activity, though it requires membrane rupture and is highly invasive. Considering that uterine mechanical activity is a direct consequence of uterine myoelectrical activity, IUP signal can be non-invasively estimated from abdominal electrohysterogram (EHG) recordings. Previous works have reported EHG-based IUP estimates with linear parameters as root-mean-square or Teager energy. Due to non-linear nature of biological processes, the aim of this study was to test the performance of different non-linear EHG parameters to estimate IUP signal. Simultaneous EHG and IUP recordings were conducted in 17 women during labour. Teager energy (TE), Sample entropy (SampEn), Spectral entropy (SpEn), Lempel-Ziv (LZ), and Poincaré parameters: SD1, SD2, SDRR and SD1/SD2 were computed from the EHG. Different window lengths for computation and for a smoothing moving average filter were tested. Mono-variable linear regression models were used to obtain IUP estimates. The best results were obtained with TE and SD1, both computed and filtered with windows of 5 s and 20 s, respectively. In the latter case, the RMSerror was 12.25 ± 4.03 mmHg, which points that non-linear EHG parameters can provide relevant information for non-invasive uterine activity monitoring.

Keywords: Uterine activity · Intrauterine pressure · Electrohysterography · Signal processing · Non-linear analysis

1 Introduction

Pregnant women's uterine activity (UA) is monitored during their pregnancy and labor to provides valuable clinical information on time to delivery, labor progress and on maternal and fetal wellbeing. This information can be obtained by means of different measurement techniques. The first one, surface tocography (TOCO), relies on the use

J. Henriques et al. (Eds.): MEDICON 2019, IFMBE Proceedings 76, pp. 117–124, 2020.
https://doi.org/10.1007/978-3-030-31635-8_14

of a tocodynamometer, which records changes in abdominal contour associated to uterine contractility by using strain gauges incorporated in a set of tight straps disposed over maternal fundus. A second technique consists in the measurement of the intrauterine pressure (IUP) by means of a catheter that registers pressure changes provoked by contractions while introduced into the uterine cavity. Even though it is the gold standard, IUP is highly invasive since it involves the rupture of membranes [1] and can only be used during labor. Furthermore is associated with an increased risk of bacterial infection, maternal fever, and the need for a caesarean delivery. For these reasons, although TOCO involves the loss of some contractions, is uncomfortable because of the placement of straps and its measures often depend on sensor position, it is the most commonly used method in clinical practice since it is non-invasive [1]. Electrohysterography (EHG) has emerged as a promising alternative technique. It is based on the fact that mechanical activity is a direct consequence of electrical activity [2]. Hence, it relies on the use of a set of contact electrodes placed on the abdominal surface to register uterine myoelectrical activity. EHG has been shown to have best capability to pick up uterine contractions than TOCO [1, 3]. However, it has not been used in clinical practice yet since physicians are more familiar with pressure signals. To solve this limitation, IUP morphology and amplitude have been estimated from EHG records by means of linear parameters that value the signal amplitude and frequency such as root means square or Teager energy. Since biological processes are non-linear, the purpose of this study is to explore the potential use of non-linear parameters from EHG signals to work out IUP estimates (IUP-like signals).

2 Materials and Methods

2.1 Signal Acquisition

To perform this study, simultaneous IUP and EHG signals were recorded from seventeen women in the active phase of labor at the *Hospital Universitario y Politécnico de La Fe*, in Valencia, Spain. This study adhered to the Helsinki Declaration and was approved by the local medical ethical board. Women were informed of the purpose of the study, as well as the recording protocol, and provided written informed consent. These signals were acquired from 1 to 4 h. In order to diminish skin-electrode impedance, abdominal skin was exfoliated using gel (Nuprep, Weaver and Company, USA). Four monopolar disposable Ag/AgCl electrodes (3 M red dot 2560) were then located on the patient's abdomen as in [3]: 2 on each side of the median line over the navel at 8 cm inter-electrode distance (M1 and M2), 1 reference electrode on the right hip and 1 ground electrode on the left hip. EHG signal was amplified and filtered between [0.1 Hz, 150 Hz] with an ad hoc biosignal amplifier and digitalized at a sampling frequency of 500 Hz. The IUP signal was recorded simultaneously using a pressure catheter (Intran Plus IUP-450, Utah Medical products Inc, USA) positioned into the uterine cavity and connected to a Corometrics 250cx series monitor (GE HealthCare, General Electric Company, USA). This monitor transmits the data to a PC via serial port at a sampling rate of 4 Hz.

2.2 Signal Processing

Monopolar EHG signals recorded were digitally bandpass filtered between 0.1–4 Hz and then down-sampled at 20 Hz to reduce low-and high-frequency interference and noise and to diminish the data and computational cost. Then a bipolar EHG signal was computed to reduce common mode interference. Motion artifact segments were visually identified by experts and excluded from the study. To perform IUP signal estimates, eight parameters were computed:

- **Teager Energy** (**TE**). Consists in the calculation of the instantaneous energy and it is directly proportional to the square of the amplitude, as well as to the square of the frequency [4].
- **Sample entropy** (**SampEn**) is a complexity measure defined as the negative natural logarithm of the conditional probability that two sequences similar for m points remain similar at the next point, without considering self-matches [5].
- **Spectral Entropy** (**SpEn**). Given a signal $x[n]$ and its Fourier transform, SpEn is the Shanon entropy of the normalized power distribution in the frequency domain of $x[n]$ [6].
- **Lempel-Ziv measure of complexity (LZ).** It evaluates the randomness of finite time series according to the number of different patterns present in it. Signal must be transformed into a finite symbol sequence P, which is usually a binary sequence when applied to biomedical signals [7].
- **SD1, SD2, SD1/SD2, SDRR.** They are extracted from the Poincaré plot, a geometrical representation in which each pair of successive elements of a discrete sequence corresponds to a point. SD1 and SD2 are defined as the standard deviation of the instantaneous and long-term time series variability, respectively, while SDRR is the standard deviation of the time series [8].

Once they were computed, a moving average filtering was applied to EHG parameter sequences in order to smooth them. The computation of each parameter was performed with five values of parameter computing window length (PCWL) - 5, 15, 30, 45, 60 s - and the filtering was carried out with five values of smoothing filter window length (SFWL) - 5, 10, 20, 50, 100 s. For each parameter, the best combination of PCWL and SFWL was determined as the one that provided the highest Pearson correlation coefficient between the IUP record and the parameter sequence.

On the other hand, uterine electrical activity is the origin of its mechanical activity, which is translated into a delay in the IUP signal to respect of the estimated signals (IUP-like signals) obtained from EHG records. In this study, this time lag was calculated as the one that yielded the maximum value of the cross-correlation function between IUP record and TE computed signal. After applying this correction, we performed a global estimation of the IUP record from parameter sequences using a monoparametric approach based in linear regression. Global model was worked out by concatenating the IUP and the corresponding EHG parameter sequence of all patients with leave one out validation. Hence leave-one-out validation was performed: for each patient ‘k’ (the one left for validation) we calculated the coefficients of a linear regression of the concatenated the signals of the rest of patients which were then applied to the EHG parameter sequence of the patient not involved in the model construction to obtain its IUP estimate.

IUP estimations were evaluated according to the root mean square error between the IUP record (IUP) and the IUP-like signal obtained ($\widehat{IUP}$), being N the number of samples of IUP and $\widehat{IUP}$.

$$RMSe = \sqrt{\frac{\sum_{i=1}^{N}\left(IUP_i - \widehat{IUP}_i\right)^2}{N}} \quad (1)$$

3 Results

Teager energy (TE), RMS and unnormalized first moment (UNFM) have commonly been used in the literature to estimate IUP signal from EHG [9]. In this study, we decided to perform the IUP estimation from TE, as well as other nonlinear parameters so as to determine if they could provide additional information which could be useful in the obtention of the IUP-like signal. Hence, we opted to compute some parameters which have proved to be useful in other applications related to the monitoring of uterine activity such as the characterization of the uterine myoelectrical response to labor induction drugs [10] or the prediction of preterm labor [11].

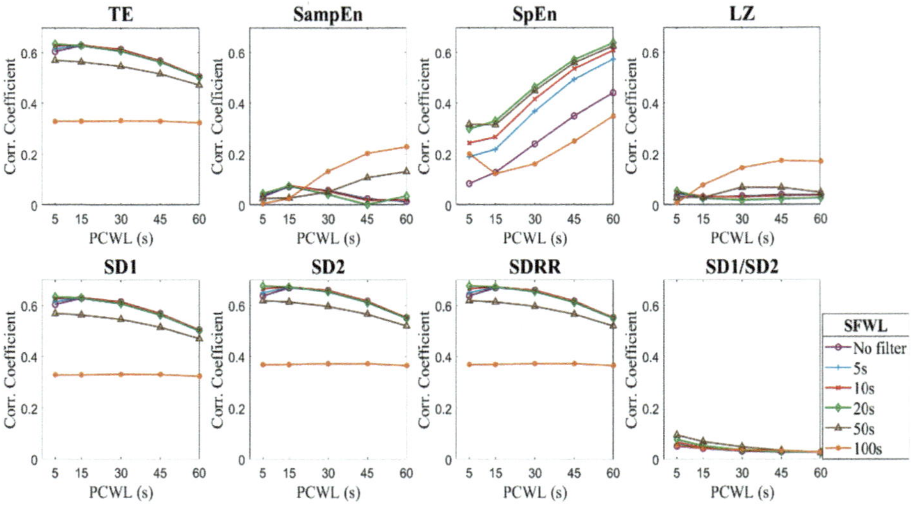

Fig. 1. Pearson correlation coefficient between patient's IUP records and estimated IUP signals for different PCWL and SFWL values.

Depending on the parameter, optimal characteristics of signal preprocessing and windowing may differ. Figure 1 shows Pearson correlation coefficients between IUP-real and IUP-like signals for each pair of PCWL and SFWL values. In each case, optimal combinations have been reported in Table 1 and these windows have been the ones used to compute and filter each EHG parameter sequence for the estimation of the IUP signal. Figure 2 shows that SampEn and SpEn correlation coefficient increases with PCWL value, being maximum for 60 s. This suggests that the use of higher PCWL values could probably bring better estimations of the IUP signal, which can be tested in further studies. In the case of SpEn, this may be due to the fact that the computation of this parameter is based on the power spectral density of the signal, whose resolution in the frequency domain increases with the duration of the signal interval analyzed [12]. In addition, SpEn and SampEn are nonlinear parameters related to the entropy of a system, so their calculation is more suited when it is performed with a larger number of samples [5]. In contrast, TE, SD1, SD2, SDRR and SD1/SD2 provide better estimations when they are computed with a 5 s window.

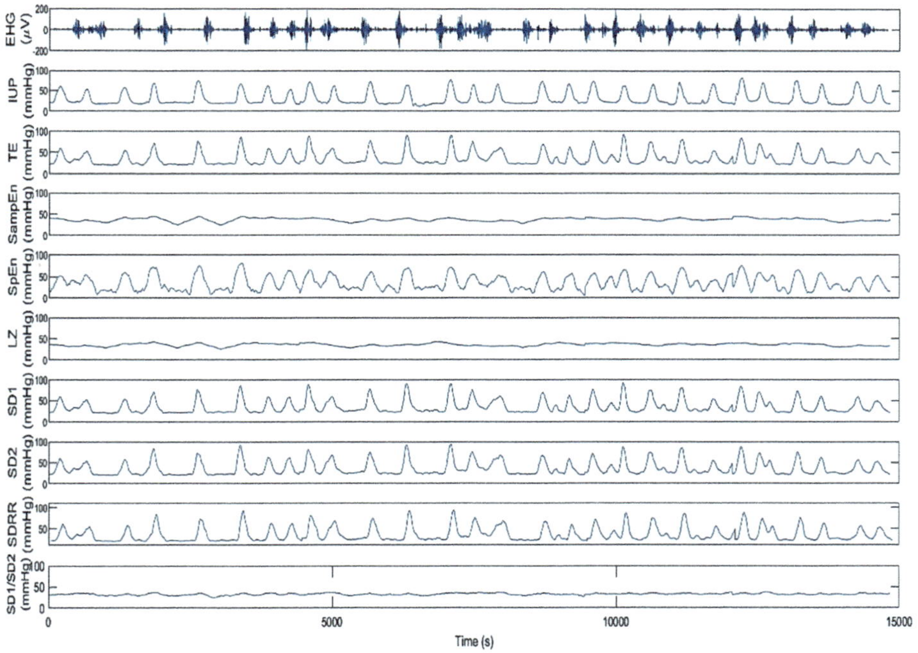

Fig. 2. Simultaneous IUP and bipolar EHG recordings and IUP-like signals obtained from TE, SampEn, SpEn, LZ, SD1, SD2, SDRR and SD1/SD2.

In reference to IUP estimates, Table 1 shows the mean and standard deviation of RMSe values obtained when applying each model to the corresponding excluded patient's parameter sequence (obtained with the optimal PCWL and SFWL values). As in other studies [9], TE provides the best IUP estimation, bringing a RMSe value of 12.25 ± 4.03 mmHg. Similar results are obtained when the model is worked out with SD1, SD2 and SDRR are used to perform the IUP estimation (12.49 ± 3.87 mmHg). In fact, in Fig. 2 it can be observed that the IUP-like signals obtained from these parameters reproduce all contractions that appear in the IUP record. This may be because TE and Poincaré plot parameters consider the amplitude and the frequency of the signal. In the case of SD1, SD2 and SDRR, this is partly implicit in the fact that all of them are metrics of variability. On the other hand, when the model is fitted using SpEn, RMSe values are slightly higher (in this case, 13.95 ± 2.48 mmHg). Figure 2 shows that the IUP-like signal obtained with this parameter reproduces every contraction recorded in the IUP, although the morphology and amplitude of the estimated contractions are not as accurate as in estimations with TE or SD1. As previously mentioned, a longer PCWL could be used to compute SpEn in order to achieve a more accurate estimation of the IUP signal. In reference to SampEn, LZ and SD1/SD2, these parameters did not yield good results, which are observed in Fig. 2 showing that IUP contractions are not reproduced in the IUP-like signals worked out with these parameters. In addition, Table 1 shows that RMSe values obtained for these parameters are too high (almost 18 mmHg or a greater value, depending on the parameter).

Table 1. Optimal PCWL and SFWL for each parameter and RMSe values obtained in the estimation performed with this combination of windows.

Parameter	PCWL (s)	SFWL (s)	RMSe (mmHg)
TE	5	20	**12.25 ± 4.03**
SampEn	60	100	17.96 ± 3.19
SpEn	60	20	13.95 ± 2.48
LZ	45	100	18.09 ± 3.21
SD1	5	20	**12.25 ± 4.03**
SD2	5	20	12.49 ± 3.87
SDRR	5	20	12.49 ± 3.87
SD1/SD2	5	50	18.27 ± 3.04

The obtained results for RMSe with the leave one out validation method point to a high capability of generalization and that non-linear parameters such SpEn, SD2, SD2 and SDRR can be useful in monitoring uterine activity. It is also important to point out that we have only employed recordings from seventeen women, so further studies could attempt to build more robust models with a more comprehensive database. Future studies could also explore the complementary information provided by different parameters by combining them in a multivariable regression.

4 Conclusion

In this study, IUP signal was estimated from EHG signals, working out nonlinear parameters and monoparametric regression linear models. Results revealed that some parameters that are not usually computed from EHG signals, such as those ones extracted from the Poincaré plot, could provide accurate estimations of the IUP signal. In this study, the best IUP-like estimation was obtained with the models performed using TE and SD1, computed and filtered with windows of 5 s and 20 s, respectively. Both estimations yielded a RMSe value of 12.25 ± 4.03 mmH, building bridges to the use of EHG in clinical practice to monitor of uterine contractile activity.

Acknowledgment. Nonetheless, this research has received funding from the Spanish Ministry of Economy and Competitiveness and the European Regional Development Fund (DPI2015-68397-R), the Generalitat Valenciana (GV/2018/104) and UPV-IIS La Fe (UPV_FE-2018-C03). These public entities provided only financial support and did not influence at all in the design, development or publication of the work.

Conflict of Interest. The authors have no conflict of interest in terms of personal financial interests or employment.

References

1. Euliano, T.Y., Nguyen, M.T., Darmanjian, S., McGorray, S.P., Euliano, N., Onkala, A., et al.: Monitoring uterine activity during labor: a comparison of 3 methods. Am. J. Obstet. Gynecol. **208**(66), e1–e6 (2013). https://doi.org/10.1016/j.ajog.2012.10.873
2. Hassan, M.: Analysis of the propagation of uterine electrical activity applied to predict preterm labor. Bioengineering. UTC Compiègne; Reykjavik University (2011). English. fftel-01226162f
3. Benalcazar-Parra, C., Montfort-Orti, R., Ye-Lin, Y., Alberola-Rubio, J., Marin, A.P., Mas-Cabo, J., et al.: Characterization of uterine response to misoprostol based on electrohysterogram. In: Proceedings of 10th International Joint Conference on Biomedical Engineering and System Technology. SCITEPRESS – Science and Technology Publications (2017). https://doi.org/10.5220/0006146700640069
4. Kaiser, J.F.: On a simple algorithm to calculate the "energy" of a signal. In: International Conference on Acoustics, Speech, and Signal Processing, pp. 381–384. IEEE (1990). https://doi.org/10.1109/icassp.1990.115702
5. Richman, J.S., Moorman, J.R.: Physiological time-series analisis using approximate entropy and sample Entropy. Am. J. Physiol. Heart Circ. Physiol. **278**, H2039–H2049 (2000). https://doi.org/10.1152/ajpheart.2000.278.6.h2039
6. Kullback, S.: Information Theory and Statistics. Wiley, New York (1959)
7. Aboy, M., Hornero, R., Abásolo, D., Álvarez, D.: Interpretation of the Lempel-Ziv complexity measure in the context of biomedical signal analysis. IEEE Trans. Biomed. Eng. **53**(11), 2282–2288 (2006). https://doi.org/10.1109/tbme.2006.883696
8. Tayel, M.B., AlSaba, E.I.: Poincaré plot for heart rate variability. Int. J. Biomed. Biol. Eng. **9**(9) (2015). https://waset.org/Publication/10002615

9. Benalcazar-Parra, C., Sempere, C., Marin, A.P.: Improvement of non-invasive intrauterine pressure estimation based on electrohysterogram. In: XXXV Congreso Anual de la Sociedad Española de Ingeniería Biomédica, Bilbao, pp. 225–238 (2017)
10. Benalcazar-Parra, C., Ye-Lin, Y., Garcia-Casado, J., Monfort-Orti, R., et al.: Electrohysterographic characterization of the uterine myoelectrical response to labor induction drugs. Med. Eng. Phys. **56**, 27–35 (2018)
11. Di Marco, L.Y., et al.: Recurring patterns in stationary intervals of abdominal uterine eletromyograms during gestation. Med. Biol. Eng. Comput. **52**, 707–716 (2014)
12. Marple, L.: Resolution of conventional fourier, autoregressive, and special ARMA methods of spectrum analysis. In: IEEE International Conference on Acoustics, Speech, and Signal Processing, ICASSP 1977, pp. 74–77. IEEE (1977). https://doi.org/10.1109/icassp.1977.1170219

Linear and Non-linear Analysis of EEG During Sleep Deprivation in Subjects with and Without Epilepsy

Silvia Marino[1], Giulia Silveri[2], Lilla Bonanno[1], Simona De Salvo[1], Emanuele Cartella[1], Aleksandar Miladinović[2], Miloš Ajčević[2], and Agostino Accardo[2(✉)]

[1] IRCCS Centro Neurolesi "Bonino Pulejo", SS113, C.da Casazza, 98124 Messina, Italy
[2] University of Trieste, Via Valerio, 10, 34127 Trieste, Italy
accardo@units.it

Abstract. EEG has a central role in the diagnosis of epileptiform abnormalities helpful in diagnosing epilepsy. Since irregularities are random and sporadic events, easily activated in the initial phase of sleep but difficult to observe in a standard EEG examination, sleep deprivation is a frequent condition to be used. Thus, in this study the EEG monitoring of 44 subjects, 14 without epilepsy and 30 with epilepsy, afferent to the IRCCS Centro Neurolesi "Bonino Pulejo" of Messina were examined after sleep deprivation the day before performing the registration. EEGs were recorded according to the international setting system using nineteen channels. The normalized power spectral densities in delta (2–4 Hz), theta (4–8 Hz), alpha (8–13 Hz) and beta (13–30 Hz) band were computed and the non-linear parameters such as beta exponent, fractal dimension and zero crossing were considered. The differences between the sleep and awake were significant in almost all the channels in the beta band and in posterior areas for beta exponent, fractal dimension and zero crossing in normal subjects. In epileptic patients they were significant in all the channels in the delta band and for the non-linear parameters, and in several ones in theta and beta bands. Even if in posterior areas all the spectral and the non-linear parameters showed different values between epileptic and healthy subjects, no significant differences were found. The results suggest that analysis of spectral power as well as of complexity, obtained by non-linear parameters, could be used to identify differences between healthy and epileptic patients.

Keywords: EEG · Epilepsy · Sleep deprivation

1 Introduction

The electroencephalography (EEG) is a widely used research tool in the investigation of cognitive process and it plays a fundamental role in the diagnosis of epileptiform abnormalities helpful in diagnosing epilepsy [1]. This disease represents a clinical condition in which an early diagnosis is essential, not only to prevent repeated attacks, but also in light of the harmful effects of convulsions on the brain, especially if prolonged.

J. Henriques et al. (Eds.): MEDICON 2019, IFMBE Proceedings 76, pp. 125–132, 2020.
https://doi.org/10.1007/978-3-030-31635-8_15

However, anomalies are sporadic and random events, difficult to observe in a standard EEG exam, which lasts from 30 to 60 min. Indeed, seizures, even in repeated recordings, can be found only in 10–20% of patients with epilepsy and, conversely, a pathological finding can be observed in 1–2% of subjects with no history of epilepsy.

Moreover, the awake and sleep influence neuronal excitability and the sleep affects the frequency of epileptic discharges [2]. The literature reports that 21% of epilepsy patients had seizures exclusively at night [3, 4]. Since the anomalies tend to be activated in the initial phase of sleep, it would be useful for the patient to fall asleep during registration. Unfortunately, it is unusual that patients fall asleep spontaneously during EEG examination so that, frequently the registration follows a sleep deprivation. In fact, in epilepsy clinics, sleep deprivation is reported to be the most common trigger for seizures regardless of the epilepsy syndrome [5, 6].

In order to examine EEG anomalies due to epilepsy, some clinical methods for calculating the proportion of epileptiform activity during sleep have been proposed [7]. In particular, the spike-wave index (SI) measures the percentage of 1-s epochs containing spikes [8], the interictal epileptiform discharges per hour [9] quantifies the occurrence of interictal epileptiform discharges during sleep deprivation, while the spike index [10, 11] allows the examination of sleep macro- and micro-structure, being significantly higher in sleep deprivation. Other algorithms for automated spike quantification in raw EEG signals are more recently based on wavelets and machine learning [12]. On the other hand, although quantitative EEG analysis may be based on parameters extracted from EEG power spectral density (PSD), in the literature, only few studies investigated these parameters during sleep in healthy subjects. In particular, at the beginning of the sleep, when the need for recovery is greatest, a power increase in the delta and theta bands and a corresponding power decrease in the beta band are present [13, 15, 16].

In addition to PSD linear method, also non-linear methods have been proposed for the examination of the EEG signal [13, 14, 17, 18]. In particular, Ferri et al. [18] studied the non linear dynamic properties of EEG presenting slow-wave forms using a non-linear cross predictor. Moreover, parameters like zero crossing, fractal dimension and beta exponent have shown to be able to distinguish different compartmental states, sleep and resting states, bringing complementary information with respect to the one carried out by linear analysis [14, 17].

However, no studies, exploring possible differences in PSD parameters as well as in non linear ones between patient with and without epilepsy in sleep deprivation, are available in the literature. In order to fill this gap, the purpose of this study is to quantify as linear and non linear parameters extracted from EEG signals recorded during sleep deprivation changes between subjects with and without epilepsy.

2 Materials and Methods

A group of 44 patients who underwent consecutively an EEG monitoring at the IRCCS Centro Neurolesi "Bonino Pulejo" of Messina were retrospectively examined. The sample was composed of 14 subjects without epilepsy (39.0 ± 16 years old) and 30 with epilepsy (39.6 ± 16 years old). In particular, 6 patients were affected by symptomatic focal epilepsy and the remaining 24 were affected by temporal lobe epilepsy. Symptomatic focal epilepsy is determined by a brain injury that usually occurs in a well-defined brain area. Epilepsy diagnosis was based on semiology, clinical history, neuroimaging and EEG findings, according with ILAE recommendations [19]. All the subjects included in this study had given their written consent for the analysis of their clinical data for research purposes.

EEGs were recorded according to the international 10–20 electrode setting system using Ag/AgCl electrodes. Nineteen channels were utilized for the analysis (Fp1, Fp2, F7, F3, Fz, F4, F8, T3, C3, Cz, C4, T4, T5, P3, Pz, P4, T6, O1, O2). All the subjects underwent sleep deprivation the day before performing the registration and no crises have been reported in any subject. At least 10 min of waking state condition and other 10 min in the sleepiness phase were recorded. The EEG signal, coming from each channel, was sampled at 256 Hz and digitally band-pass filtered between 0.4 Hz and 40 Hz. The Power Spectral Density (PSD) was calculated on segments of 5 s each, overlapped of 50% along the whole recording, separately for the two conditions. The power in Delta (2–4 Hz), Theta (4–8 Hz), Alpha (8–13 Hz) and Beta (13–30 Hz) band was then computed and normalized dividing each power for the total power. Beside spectral factors also the following non-linear parameters were considered: Beta exponent, referred to a so-called 1/f-like behavior of EEG a ubiquitous property of complex biological systems [20], Fractal dimension (FD) computed by Higuchi's algorithm [21] and Zero crossing, measuring the changes in dominant frequency [22]. The power-law beta exponent, was calculated as the slope of the regression line in the relationship between the ln(PSD) and ln(frequency). The Higuchi algorithm is based on the estimation of the mean length of the curve, L(k), by using different time intervals, k. If L(k) is proportional to k^{-FD} the EEG time series is fractal with dimension FD, representing the slope of the line fitting the pairs $\{\ln(1/k); \ln(L(k))\}$.

All parameters were averaged among the subjects of the same group, obtaining one value per parameter and channel. For intra-group analysis, Wilcoxon sign rank sum test was used to evaluate the statistical differences between awake and sleep conditions in each group. For inter-group analysis, Mann-Whitney U test, to assess the difference between epileptic and healthy normal group during awake and sleep conditions, was used. A 95% of confidence level was set with a 5% alpha error. Statistical significance was set at $p < 0.05$.

3 Results

Figure 1 shows the normalized spectral power values in the four bands in the 19 EEG channels for healthy and epileptic subjects during awake and sleep. About spectral analysis, the results indicated that all subjects presented higher normalized power in delta band during sleep than during the awake and that the epileptic subjects had lower values than healthy ones. However, the differences between the two conditions were significant in almost all the channels in the beta band in normal subjects (Table 1) and in the delta, theta, and beta bands in epileptic patients (Table 2). The differences between normal and epileptic subjects were not significant in all the cerebral areas mainly because of a large inter-subject variability, especially during sleep.

Table 1. Statistical significance of differences between Awake and Sleep in Normal subjects in the six cerebral areas. BB: Beta Band, BE: Beta Exponent, FD: Fractal Dimension, ZC: Zero Crossing.

	Normal: Awake vs Sleep			
	BB	BE	FD	ZC
Fp1, Fp2	n.s.	n.s.	Fp1$^{\#}$	n.s.
F3,F4,Fz,F7, F8	F3***,F4***,Fz***,F7***,F8***	F7$^{\#}$	Fz$^{\#}$,F7$^{\#}$	F7$^{\#}$
T3, T4, T5, T6	T3**,T4**,T5**,T6**	T3**,T4**,T5**,T6**	T3$^{\#}$,T4$^{\#}$,T6$^{\#}$	T3***,T5***,T6***
C3, Cz, C4	C3**,Cz**,C4**	C3$^{\#}$,Cz$^{\#}$,C4$^{\#}$	C3**,Cz**	C3$^{\#\#}$,Cz$^{\#\#}$,C4$^{\#\#}$
P3, Pz, P4	P3***,Pz***,P4***	P3*,Pz*,P4*	P3***	P3*,Pz*,P4*
O1, O2	O1**,O2**	O1*,O2*	O2**	O1$^{\#\#}$,O2$^{\#\#}$

* p<0.01, ** p<0.02, *** p<0.03, # p<0.04, ## p<0.05, n.s. not significant

Table 2. Statistical significance of differences between Awake and Sleep in Epileptic subjects in the six cerebral areas. DB: Delta Band, TB: Theta Band, BB: Beta Band, BE: Beta Exponent, FD: Fractal Dimension, ZC: Zero Crossing.

	Epileptic: Awake vs Sleep					
	DB	TB	BB	BE	FD	ZC
Fp1, Fp2	Fp1*,Fp2*	Fp1$^{\#}$	Fp1**,Fp2**	Fp1*,Fp2*	Fp1*,Fp2*	Fp1**,Fp2**
F3,F4,Fz	F3*,F4*,Fz*	F3$^{\#\#}$,F4$^{\#\#}$	n.s.	F3*,F4*,Fz*	F3***,F4***,Fz***	F3***,F4***,Fz***
F7, F8	F7*,F8*	F7$^{\#\#}$	F7**,F8**	F7*,F8*	F7***,F8***	F7***,F8***
T3, T4	T3**,T4**	T3***,T4***	T3*,T4*	T3*,T4*	T3*,T4*	T3*,T4*
T5, T6	T5**,T6**	n.s.	T5*,T6*	T5*,T6*	T5*,T6*	T5*,T6*
C3, Cz, C4	C3*,Cz*,C4*	C3$^{\#\#}$,C4$^{\#\#}$	C3*,Cz*,C4*	C3*,Cz*,C4*	C3*,Cz*,C4*	C3*,Cz*,C4*
P3, Pz, P4	P3*,Pz*,P4*	P4$^{*\circ}$	P3*,Pz*,P4*	P3*,Pz*,P4*	P3*,Pz*,P4*	P3*,Pz*,P4*
O1, O2	O1*,O2*	O1$^{\#}$,O2$^{\#}$	O1*,O2*	O1*,O2*	O1*,O2*	O1*,O2*

* p<0.01, ** p<0.02, *** p<0.03, # p<0.04, ## p<0.05, n.s. not significant

Theta band showed very few differences between groups and conditions (Fig. 1), significant only between awake (lower values) and sleep (higher values) states in about half channels of epileptic patients (Table 2). The values of the normalized power in alpha band were very similar (no significant difference) in normal and epileptic subjects during sleep and awake (Fig. 1) and no significant differences between the two groups both during awake and sleep were found. The beta band showed the highest differences between conditions (Fig. 1), significant in almost all the cerebral areas (Tables 1 and 2). The values were higher in healthy subjects than epileptic ones, although without significant differences between the two groups, and during awake than sleep.

Figure 2 shows the non-linear parameters values in the EEG channels, averaged on healthy and epileptic subjects in the two conditions. The beta exponent values were higher in epileptic subjects, during both awake and sleep conditions, than in healthy subjects, with no significant differences between the two groups. Moreover, the values increased from anterior to posterior areas, in all situations. Significant differences between the two conditions were found in normal subjects, from temporal to occipital areas, and in all the cerebral areas in epileptic patients, with higher values during sleep than awake (Tables 1 and 2).

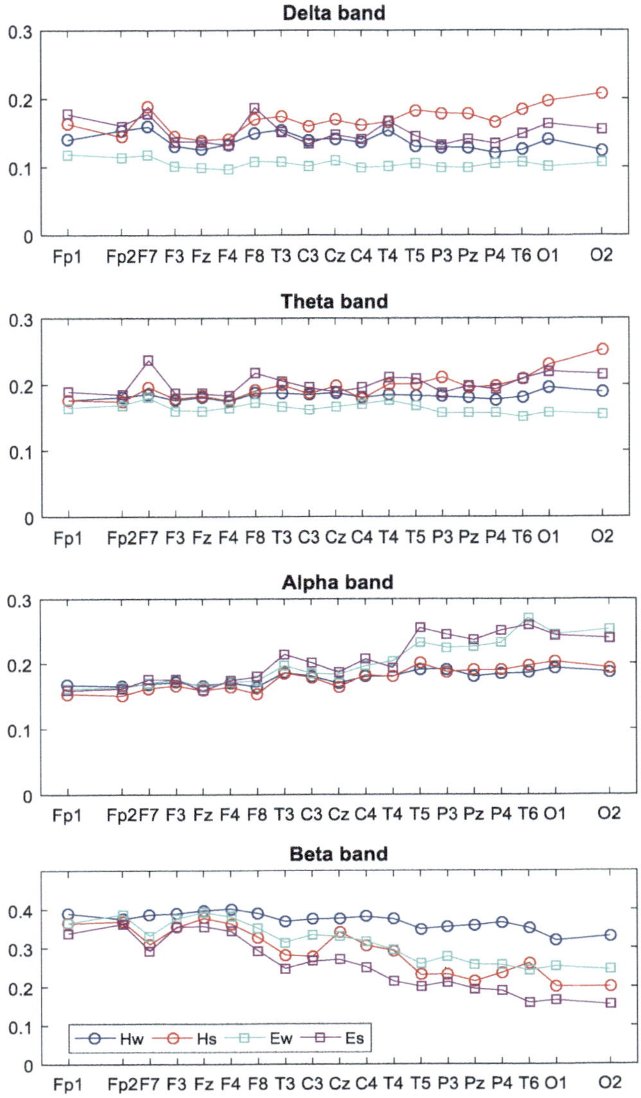

Fig. 1. Normalized spectral power values in the Theta, Delta, Alpha and Beta bands in the EEG channels, averaged on Healthy (H) and Epileptic (E) subjects during awake (w) and sleep (s) conditions.

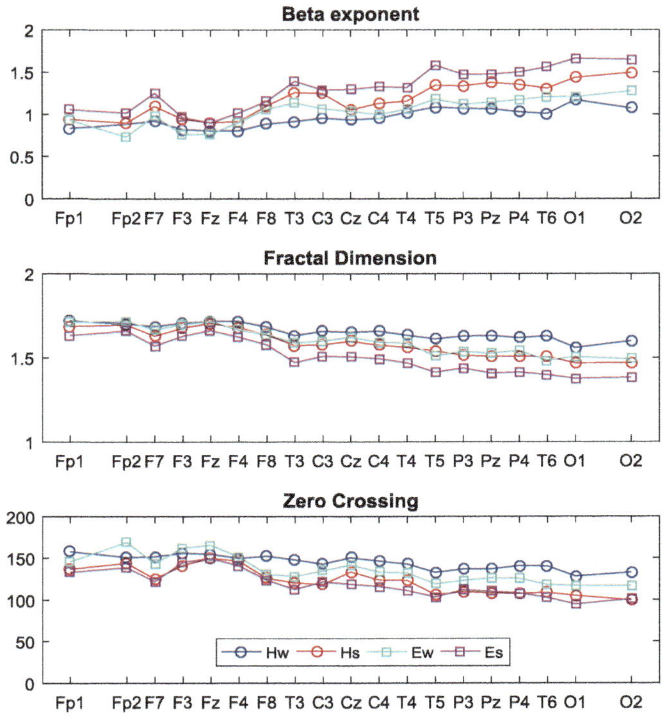

Fig. 2. Non-linear parameters values in the EEG channels, averaged on Healthy (H) and Epileptic (E) subjects during awake (w) and sleep (s) conditions.

The fractal dimension and the zero crossing showed a similar behavior and an inverse one in respect to that of the beta exponent with higher values in healthy subjects than epileptic patients in both conditions and higher values during awake than sleep. These differences between the two states were significant in almost all the cerebral areas for both parameters (Tables 1 and 2) while the differences between the two groups were not significant.

4 Discussion

The features found in this study confirm some previous results present in the literature about changes due to sleep in respect to awake in healthy subjects. In particular, during sleep the relative power in lower frequency bands (delta and theta) significantly increased, confirming the effect of prior wakefulness on the sleep EEG, while in beta band significantly decreases in respect to awake, reflecting sleep arousal level [15, 16]. Furthermore, a similar behavior was found in epileptic patients in which relative powers in delta e theta bands significantly increased during sleep with lower values than in normal subjects. Moreover, in beta band, the values significantly decreased from awake to sleep and from anterior to posterior cerebral areas with higher values present in healthy subjects. However, the differences between normal and epileptic subjects, greater in the posterior cerebral areas, especially in beta band, were not significant probably because of the high inter-subject variability present in both groups and the relatively low number of examined cases. On the other hand, in alpha band no significant differences were found between waking and sleeping nor between the two groups of subjects, with values very similar in all cases except in the parietal-occipital areas in which epileptic subjects showed higher, but not significantly, values than normal ones.

The three non-linear parameters presented a similar behavior between awake and sleep conditions with significant differences in all the channels for epileptic patients and predominantly from temporal to occipital areas for healthy subjects. As for the spectral parameters also for the non-linear ones, differences between the two groups were not significant. In particular, beta exponent showed higher values in epileptic patients than in normal subjects and during sleep, with increasing values from anterior to posterior areas, inversely reflecting the behavior found in particular in the beta band which power decreased from frontal to occipital areas. On the contrary, fractal dimension, measuring the level of complexity of the cerebral system, presented a trend opposed to that of beta exponent, as expected due to the law of self-similarity [21], showing higher values in normal subjects, especially in the anterior areas, and during awake, confirming the results reported in [14] for normal subjects. Zero crossing presented a behavior analogous to that of fractal dimension with slightly decrease from anterior to posterior areas as expected since the relative power in beta band decreases in favor of the power in lower frequencies (in particular alpha e theta bands) which are greater in the posterior areas than in the anterior ones.

In conclusion, this paper compared for the first time the spectral powers in the various cerebral areas in normal subjects with those in epileptic patients, highlighting in the last ones a behavior similar to normal for what concerns the differences between sleep and awake and between the various areas. A lower power in the delta, theta and beta bands and a greater power in the alpha band were found in epileptic patients than in normal subjects. Furthermore, non-linear analysis allowed us to highlight a greater complexity of brain activity in normal subjects than in epileptic patients and in waking than in sleeping, higher in frontal than in occipital areas. Further studies on a larger number of subjects are needed to confirm and provide significance for the differences found between normal and epileptic subjects.

Acknowledgements. Work partially supported by the Master in Clinical Engineering, University of Trieste.

Conflict of Interest. The authors declare that they have no conflict of interest.

References

1. Klimechs, W.: EEG alpha and theta oscillations reflect cognitive and memory performance: a review and analysis. Brain Res. Rev. **29**(2–3), 169–195 (2009)
2. da Silva, A.M., Aarts, J.H.P., Binnie, C.D., Laxminarayan, R., da Silva, F.H.L., Meijer, J.W., et al.: Circadian distribution of interictal epileptiform EEG activity. Electroencephalogr. Clin. Neurophysiol. **58**, 1–13 (1984)

3. Gowers, W.R.S.: Epilepsy, and Other Chronic Convulsive Diseases: Their Causes, Symptoms & Treatment. J&A Churchill, London (1881)
4. Janz, D.: Epilepsy with grand mal on awakening and sleep-waking cycle. Clin. Neurophysiol. **111**(Suppl2), S103–S110 (2000)
5. Foldvary-Schaefer, N., Grigg-Damberger, M.Z.: Sleep and epilepsy: what we know, don't know, and need to know. J. Clin. Neurophysiol. **23**(1), 20–24 (2006)
6. Jain, S.V., Kothare, S.V.: Sleep and epilepsy. Semin. Pediatr. Neurol. **22**(2), 86–92 (2015)
7. Fernandez, I.S., Champman, K., Peters, J.M., Klehm, J., Jacksno, M.C., Berg, A.T., Loddenkemper, T.: Treatment for continuous spikes and waves during sleep (CSWS): survey on treatment choices in North America. Epilepsia **55**(7), 1099–1108 (2014)
8. Weber, A.B., Albert, D.V., Yin, H., Held, T.P., Patel, A.D.: Diagnosis of electrical status epilepticus during slow-wave sleep with 100 seconds of sleep. J. Clin. Neurophysiol. **34**, 65–68 (2017)
9. Giorgi, S.F., Maestri, M., Guida, M., Crnicelli, L., Caciagli, L., Ferri, R., Bonuccelli, U., Bonanni, E.: Cyclic alternating pattern and interictal epileptiform discharges during morning sleep after sleep deprivation in temporal lobe epilepsy. Epilepsy Behav. **73**, 131–136 (2017)
10. Larsson, P.G., Wilson, J., Eeg-Olofsson, O.: A new method for quantification and assessment of epileptiform activity in EEG with special reference to focal nocturnal. Brain Topogr. **22**, 52–59 (2009)
11. Larsson, P.G., Evsiukova, T., Brockmeier, F., Ramm, P., Eeg-Olofsson, O.A.: Do sleep-deprived EEG recordings reflect spike index as found in full-night EEG recordings? Epilepsy Behav. **19**, 348–351 (2010)
12. Chavakula, V., Fernandez, I.S., Peters, J.M., Popli, G., Bosl, W., Rakhade, S., Rotenberg, A., Loddenkemper, T.: Automated quantification of spikes. Epilepsy Behav. **26**, 143–152 (2013)
13. Garrett, D., Peterson, D.A., Anderson, C.W., Thaut, M.H.: Comparison of linear, nonlinear, and feature selection methods for EEG signal classification. IEEE Trans. Neural Syst. Rehabil. Eng. **11**, 141–144 (2003)
14. Acharya, U.R., Faust, O., Kannathal, N., Chua, T., Laxminarayan, S.: Non-linear analysis of EEG signals at various sleep stages. Comput. Methods Programs Biomed. **80**, 37–45 (2005)
15. Borbély, A.A., Baulnann, F., Brandeis, D., Strauch, I., Lehmann, D.: Sleep deprivation: effect on sleep stages and EEG power density in man. Electroencephalogr. Clin. Neurophysiol. **51**, 483–493 (1981)
16. Dijk, D.J., Brunner, D.P., Beersma, D.G.M., Borbély, A.A.: EEG power density and slow wave sleep as a function of prior waking and circadian phase. Sleep **13**, 430–440 (1990)
17. Accardo, A., Affinito, M., Carrozzi, M., Bouquet, F.: Use of the fractal dimension for the analysis of electroencephalographic time series. Biol. Cybern. **77**, 339–350 (1997)
18. Ferri, R., Elia, M., Masumeci, S.A., Stam, C.J.: Non linear EEG analysis in children with epilepsy and electrical status epilepticus during slow-wave sleep (ESES). Clin. Neurophysiol. **112**, 2274–2280 (2001)
19. Engel Jr., J.: International league against epilepsy (ILAE). A proposed diagnostic scheme for people with epileptic seizures and with epilepsy: report of the ILAE task force on classification and terminology. Epilepsia **42**, 796–803 (2003)
20. Bassingthwaighte, J.B., Liebovitch, L.S., West, B.J.: Fractal Physiology. Oxford University Press, New York (1994)
21. Higuchi, T.: Approach to an irregular time series on the basis of the fractal theory. Physica D **31**, 277–283 (1988)
22. Kedem, B.: Spectral analysis and discrimination by zero-crossings. Proc. IEEE **74**, 1477–1493 (1986)

Brain Oscillatory Activity and Neurological Deficit in Hyper-acute Ischemic Stroke: Correlation of EEG Changes with NIHSS

Miloš Ajčević[1,2,3(✉)], Giovanni Furlanis[1], Lara Stragapede[1], Mariana Ridolfi[1], Paola Caruso[1], Marcello Naccarato[1], Agostino Accardo[2], and Paolo Manganotti[1]

[1] Neurology Unit, University Hospital of Trieste, Department of Medicine Surgery and Health Sciences, University of Trieste, Trieste, Italy
majcevic@units.it

[2] Department of Engineering and Architecture, University of Trieste, Trieste, Italy

[3] NEUROFARBA Department, Neuroscience Section, University of Florence, Florence, Italy

Abstract. The accurate identification and prediction of cerebral infarct evolution and clinical outcome are of paramount importance in acute ischemic stroke management. Neuroimaging in acute stroke is mandatory to establish the feasibility of reperfusion therapy, but it is not practical to assess the continuous evolution of brain ischemia. EEG could be an applicable instrument to perform functional monitoring in the hyper-acute phase. EEG activity during ischemic stroke has been widely studied in sub-acute and post-acute phase of ischemic stroke. However, only few studies have focused on the early phase of brain ischemia. The aim of this study conducted at the stroke unit was to investigate stroke-related EEG changes during the earliest phase of ischemic stroke within 4.5 h from symptom onset and to correlate these data with neurological deficit in terms of NIHSS score. We studied 12 patients with ischemic stroke, who underwent EEG recordings within 4.5 h from symptom onset. The EEG signals acquisition was performed bedside without delaying reperfusion treatment, using @64 channels Wi-Fi Be Plus LTM amplifier and 19 channel 10–20 Ag/AgCl electrodes wireless prewired headset. The main finding of this study is a significant positive correlation between stroke-related EEG changes measured by DAR and DTABR parameters and the neurological deficit measured by NIHSS score, during the earliest phase of ischemic stroke. The results of this study highlight the importance of EEG as complementary tool in the assessment of stroke severity and its potential role in acute decision-making and monitoring.

Keywords: EEG · Ischemic stroke · Hyperacute · Neurocoupling · Biomedical signal processing

J. Henriques et al. (Eds.): MEDICON 2019, IFMBE Proceedings 76, pp. 133–141, 2020.
https://doi.org/10.1007/978-3-030-31635-8_16

1 Introduction

Despite improvements in preventive strategies and acute therapy, stroke remains one of the leading causes of disability and death among adults worldwide. Ischemic stroke is a neuroemergency condition that is highly treatable using thrombolysis and thrombectomy in selected patients [1]. The accurate identification and prediction of cerebral infarct evolution and clinical outcome is a key aim for acute stroke therapy and research [2]. Neuroimaging is a cornerstone in ischemic stroke diagnosis and it is crucial for patients' eligibility for reperfusion thrombolysis and thrombectomy treatment [3]. Magnetic resonance imaging (MRI) and CT perfusion (CTP) can quickly and accurately identify the extent of the salvageable tissue and of the ischemic core [4], thus selecting those patients who can most benefit from the treatment [5–7]. DWI-MRI lesion volume and the ischemic territory delineated on PWI-MRI or on CTP have been shown to be correlated with the National Institutes of Health Stroke Scale (NIHSS) [8, 9]. The NIHSS, consisting of 11 items to assess the main neurological functions, is the most adopted tool in the actual medical practice to evaluate stroke severity, assess neurological status [10], and it has proved to be a powerful predictor of clinical outcome [11]. Neuroimaging in acute stroke is mandatory to establish the feasibility of reperfusion therapy, but it is not practical to assess the evolution of brain ischemia in the critical acute post-stroke period.

Electroencephalogram (EEG) in the hyper-acute phase could be an applicable instrument for functional monitoring. Indeed, data can be acquired bedside by portable technologies developed to overcome all inhibiting factors occurring in emergency settings [12, 13]. EEG activity during ischemic stroke has been studied in sub-acute and post-acute phase of ischemic stroke [1, 14–17]. In general, ischemic stroke produces abnormal, slow EEG activity mainly in the delta frequency range and attenuation of faster activity, principally in the alpha frequency range [18]. The bandpower-derived parameters as relative delta power [14], the delta/alpha power ratio (DAR) [14], and (delta+theta)/(alpha+beta) power ratio [15] have proven particularly informative for outcome prognostication and clinical management decisions sub-acute and post-acute phase. However, only few studies have focused on the early phase of brain ischemia and its correlation with neurological deficit on onset.

The aim of this study, conducted in the stroke unit using an EEG wireless system, was to investigate stroke-related EEG changes during the earliest phase of ischemic stroke within 4.5 h from symptom onset, and to correlate these data with neurological deficit in terms of NIHSS score.

2 Materials and Methods

The study was conducted on 12 patients admitted to the Stroke Unit of the University Medical Hospital of Trieste (Italy), between January 2017 and April 2017, with acute focal neurologic symptoms compatible with ischemic stroke. The study included patients who underwent EEG recording at the bedside within 4.5 h from stroke onset. The ischemic stroke diagnosis was verified with a persistent focal neurological deficit after 24 h and/or a positive computed tomography (CT) or magnetic resonance imaging

(MRI) for ischemic lesion 24 h after stroke onset. Conditions such as hematic effusion, previous stroke, history of epileptic seizure, use of medication like benzodiazepines or neuroleptic were exclusion criteria, due to the risk of compromising EEG assessment.

All patients underwent neurological examination including National Institutes of Health Stroke Scale (NIHSS), urgent hematological tests and a multimodal CT imaging protocol, comprising cerebral non-contrast CT, CT angiography and CTP. NIHSS evaluation was carried out by a vascular neurologist trained in performing NIHSS examination. After this assessment, if the conditions were suitable in the emergency setting, EEG was performed with wireless device and a prewired cap. The patients eligible for reperfusion treatment underwent IV thrombolysis and/or IA thrombectomy after the clinical assessment.

This study has been approved by the local ethics committee and has been performed in accordance with the ethical standards laid down in the 1964 Declaration of Helsinki and its later amendments. All participants released their informed consent.

The acquisition of EEG signals was performed bedside within 4.5 h from symptom onset, without delaying reperfusion treatment, using Be Plus LTM amplifier @64 channels Wi-Fi (EBNeuro, Florence, Italy) and a wireless prewired headset (EBNeuro, Florence, Italy) with 19 channel 10–20 Ag/AgCl electrodes positioned at Fz, Cz, Pz, Fp1, Fp2, F3, F4, F7, C3, P3, T3, T5, O1, F4, F8, C4, P4, T4, T6, O2. An electro-conductive gel was used to lower the electric resistance. Electrode impedances were kept below 5 kΩ. The signals were recorded with a 128 Hz sampling frequency and the off-line analysis of 19 unipolar channels was carried out using scripts developed in MATLAB (MathWorks Inc., Natick, MA). All channels were digitally filtered with the 0.4–40 Hz 2nd order Butterworth bandpass filter. Visual artifact rejection was performed and data with muscular, ocular and other types of artifacts were manually discarded, the first 60 s of the artifact free EEG were analyzed. Power spectral density (PSD) was estimated for each channel using Welch's periodogram [19], averaged on 11 tracts of 10 s (n = 1280 samples) each, windowed with a Hanning window, with 50% overlap. Subsequently, the relative power in each spectral band (delta: 0.5–4 Hz; theta: 4–8 Hz; alpha: 8–13 Hz; beta: 13–30 Hz) was calculated for each channel. The relative power for each band was computed as the ratio of summed absolute band-power to total summed power across the 1–30 Hz range. In addition, (delta/alpha ratio) DAR and (delta+theta/alpha+beta ratio) DTABR were computed as the ratios of absolute power for the respective frequency bands of interest. Relative power for each band, DAR and DTABR parameters were averaged over all nineteen scalp electrodes.

Continuous variables were presented as mean±standard deviation or medians (ranges) depending on their distribution (normal or not). The Spearman correlation was used to determine the degree of correlation between the NIHSS scores and EEG spectral band parameters and ratios. The Kruskal-Wallis test with post hoc Tukey test was performed to assess the differences among frequency bands. A level of $P < 0.05$ was regarded as statistically significant.

3 Results

Patient's demographic and clinical data at admission are summarized in Table 1. Twelve patients (6 M/6F, median age 83.5, range 58–89 years) were recruited. Concerning the stroke severity, the median NIHSS at admission was 10 (range 3–21). Direct CT showed median ASPECT score of 10 (range 6–10). Median time between onset of symptoms and hyper-acute EEG recording was 187 min (range 85–255 min).

Relative power of theta, delta, alpha and beta EEG bands, as well as DAR and DTABR calculated ratios are reported in Table 2. The median relative delta, theta, alpha and beta power indices were 0.51 (range 0.43–0.60), 0.19 (range 0.16–0.32), 0.13 (range 0.10–0.30) and 0.14 (range 0.08–0.22), respectively. Relative theta band resulted significantly higher comparted to other bands ($p < 0.018$). The median DAR value was 4.3 (range 1.63–6.00) and median and DTABR values 5.65 (range 1.63–4.01), respectively. Figure 1 reports a diagram between the NIHSS at admission and DAR values. In Fig. 2, NIHSS at admission was plotted against DTABR. NIHSS correlated significantly with DAR (rho = 0.74; $P < 0.006$) and DTBAR (rho = 0.63; $P = 0.028$). In addition, significant negative correlation was found between NIHSS and relative alpha power (rho = −0.69; $P = 0.013$), while NIHSS did not correlate significantly with relative power of theta, delta and beta bands.

Table 1. Patient demographics, lesion vascular territories, ASPECT (Alberta Stroke Program Early CT Score) and NIHSS score. Patient Age (y) L, left; R, right; MCA, middle cerebral artery.

Patient	Age	Gender	Lesion vascular territory	ASPECTS	NIHSS at admission
1	70	M	Right MCA	10	6
2	87	F	Left MCA	10	17
3	79	M	Right MCA	6	8
4	82	F	Left MCA	10	5
5	89	F	Right MCA	9	11
6	89	F	Right MCA	8	18
7	88	M	Left MCA	10	9
8	79	M	Right MCA	10	16
9	58	M	Left MCA	10	3
10	93	F	Left MCA	9	14
11	80	M	Right MCA	10	7
12	85	F	Left MCA	7	21

Table 2. Relative delta, theta, alpha and beta power, DAR and DTABR parameters calculated for each patient.

Patient	Relative delta power	Relative theta power	Relative alpha power	Relative beta power	DAR	DTABR
1	0.55	0.16	0.12	0.17	4.58	2.42
2	0.44	0.32	0.09	0.15	5.02	3.18
3	0.49	0.28	0.10	0.12	4.84	3.42
4	0.47	0.16	0.15	0.22	3.12	1.73
5	0.46	0.23	0.18	0.13	2.61	2.29
6	0.53	0.20	0.09	0.17	6.00	2.82
7	0.60	0.18	0.15	0.07	4.02	3.62
8	0.56	0.15	0.11	0.19	5.26	2.38
9	0.49	0.13	0.30	0.08	1.63	1.63
10	0.43	0.21	0.17	0.20	2.56	1.75
11	0.55	0.17	0.16	0.12	3.40	2.59
12	0.55	0.25	0.10	0.10	5.56	4.01

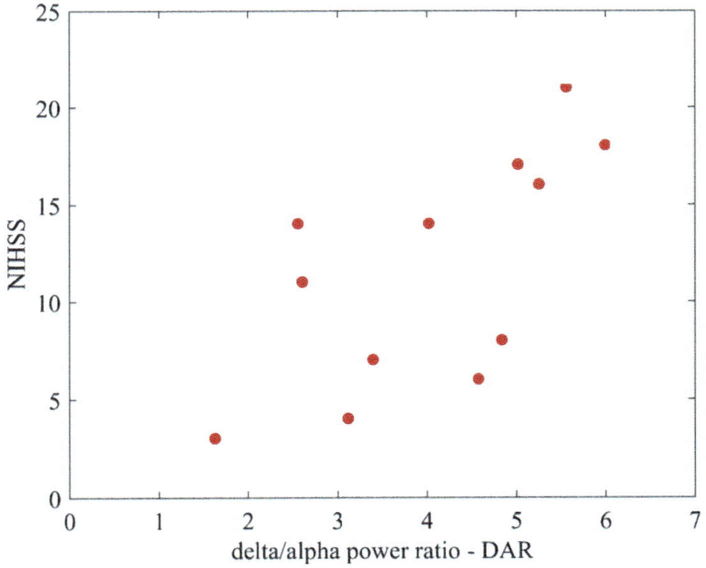

Fig. 1. NIHSS at admission plotted against the delta/alpha power ratio (DAR).

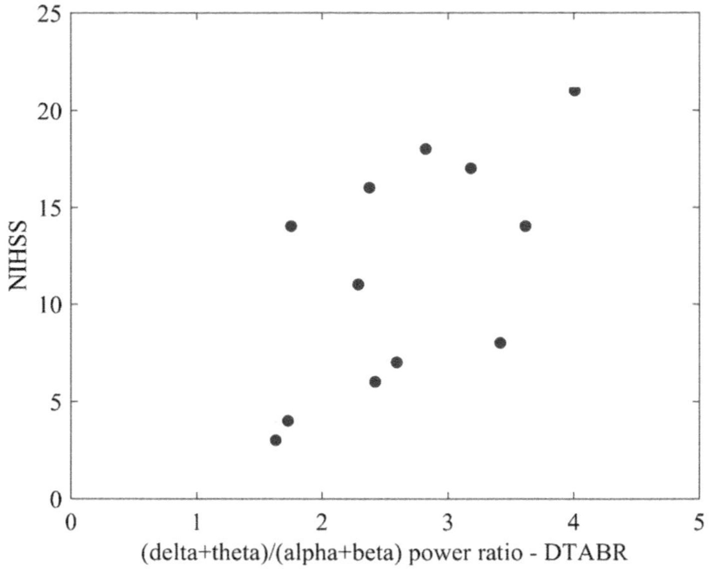

Fig. 2. NIHSS at admission plotted against (delta+theta)/(alpha+beta) power ratio (DTABR).

4 Discussion

In acute ischemic stroke, the progression from ischemia to infarction is a dynamic, rapidly evolving process with irreversible changes occurring within a few minutes to a few hours [15]. Brain oscillatory activity changes occurring in stroke are related to neurophysiological changes in the brain tissue during hypoperfusion and they are manifestation of neurovascular coupling. EEG may reflect changes in cerebral blood flow and metabolism within seconds as these are directly reflected in the neuronal rhythms.

The main finding of this study is a significant positive correlation between stroke-related EEG changes measured by DAR and DTABR parameters, and the neurological deficit measured by NIHSS score, during the earliest phase of ischemic stroke.

The results of this study highlight the importance of EEG as complementary tool in the assessment of stroke severity and its potential role in acute decision-making and monitoring. Neuroimaging has a pivotal role in the diagnosis of ischemic stroke and it is crucial to establish the feasibility of reperfusion therapy [3]. Yet, it cannot be used in functional monitoring to obtain information about the evolution of brain ischemia in the critical acute period. Although the spatial resolution of the EEG is very low compared to advanced neuroimaging techniques, the high temporal resolution of the EEG may enable a fast and sensitive assessment of instantaneous brain functioning [18, 20]. This is the first study which investigates brain oscillatory activity and its relation with NIHSS during the earliest phase of stroke, using EEG wireless recording.

In this study, DAR and DTAR calculated form EEG tracings recorded within 4.5 h correlated significantly with NIHSS at admission. In addition, a negative correlation between relative alpha power and NIHSS at baseline was observed. Our findings are consistent with EEG stroke-related abnormal predominant slow EEG activity mainly expressed in the delta frequency range and reduction of faster activity, principally in the alpha frequency range [18]. Moreover, our data gathered in hyper-acute phase extend previous findings in acute and sub-acute stroke samples with similar EEG changes in terms of DAR [14, 16], DTABR [15] and relative alpha [14], indicating that EEG measures are informative in all phases of ischemic stroke. Recent studies reported that EEG in the subacute setting of ischemic stroke may be of prognostic value for disability after 6 months [15] and that DAR could be the most effective EEG measure for monitoring and prognosticating the evolution of cerebral ischemia in sub-acute phase [14]. In addition, the DAR at 48 h post-stroke onset strongly correlated with NIHSS scores at 30 days post-stroke [14]. In our hyper-acute sample, both DAR and DTBAR correlated with NIHSS at admission and we found lower DAR and DTBAR values compared to those in sub-acute phase reported in a recent studies based on EEG recordings at 48 h [14] and at <72 h post-stroke [15]. The negative correlation between relative alpha power and NIHSS at admission could be explained with the fact that preservation of alpha activity is indicative of neuronal survival in the ischemic regions and good prognosis [20].

Performing an EEG in emergency departments is arduous owing to space, training and complexity of the equipment. In such a setting, portable wireless EEG machines may solve these inhibiting factors [21]. In our study, EEG registrations were carried out at bedside by pre-cabled headset wireless system applied in emergency settings without compromising the treatment protocol. If EEG measures can be acquired during both hyper-acute and sub-acute phase, the EEG spectral parameters might provide information which updates or supplements clinical and neuroimaging data.

In conclusion, the results of this preliminary study showed that stroke-related EEG changes measured by DAR and DTABR parameters correlate with the neurological deficit measured by NIHSS score during the earliest phase of ischemic stroke. These findings support the hypothesis that the use of EEG in acute ischemic stroke may be helpful in clinical practice and could provide valuable information for clinical management and monitoring. The results should be confirmed in a larger clinical study.

Acknowledgements. M. Ajčević is supported by AIRAlzh Onlus - (ANCC-COOP). The authors would like to thank Matteo di Franza for editorial assistance. This study was partially supported by Master in Clinical Engineering, University of Trieste.

Conflict of Interest. The authors have no conflict of interest do declare.

References

1. The National Institute of Neurological Disorders and Stroke rt-PA Stroke Study Group: Tissue plasminogen activator for acute ischemic stroke. N. Engl. J. Med. **333**, 1581–1587 (1995)

2. Finnigan, S.P., Rose, S.E., Walsh, M., Griffin, M., Janke, A.L., McMahon, K.L., Brown, J.: Correlation of quantitative EEG in acute ischemic stroke with 30-day NIHSS score: comparison with diffusion and perfusion MRI. Stroke **35**(4), 899–903 (2004)
3. Lewandowski, C.A., Libman, R.: Acute presentation of stroke. J. Stroke Cerebrovasc. Dis. **8**(3), 117–126 (1999)
4. Peisker, T., Koznar, B., Stetkarova, I., et al.: Acute stroke therapy: a review. Trends Cardiovasc. Med. **27**, 59–66 (2017)
5. Gonzalez, R.G.: Imaging-guided acute ischemic stroke therapy: from "time is brain" to "physiology is brain". AJNR Am. J. Neuroradiol. **27**(4), 728–735 (2006)
6. Nogueira, R.G., Jadhav, A.P., Haussen, D.C., DAWN Trial Investigators, et al.: Thrombectomy 6 to 24 hours after stroke with a mismatch between deficit and infarct. N. Engl. J. Med. 378, 11–21 (2018)
7. Caruso, P., Naccarato, M., Furlanis, G., Ajčević, M., Stragapede, L., Ridolfi, M., Manganotti, P.: Wake-up stroke and CT perfusion: effectiveness and safety of reperfusion therapy. Neurol. Sci. **39**(10), 1705–1712 (2018)
8. Tong, D.C., Yenari, M.A., Albers, G.W., O'Brien, M., Marks, M.P., Mosley, M.E.: Correlation of perfusion- and diffusion-weighted MRI with NIHSS score in acute (6.5 hour) ischemic stroke. Neurology **50**, 864–870 (1998)
9. Furlanis, G., Ajčević, M., Stragapede, L., Lugnan, C., Ridolfi, M., Caruso, P., Manganotti, P.: Ischemic volume and neurological deficit: correlation of computed tomography perfusion with the National Institutes of Health Stroke Scale Score in acute ischemic stroke. J. Stroke Cerebrovasc. Dis. **27**(8), 2200–2207 (2018)
10. Adams Jr., H.P., Davis, P.H., Leira, E.C., et al.: Baseline NIH stroke scale score strongly predicts outcome after stroke: a report of the trial of org 10172 in acute stroke treatment (TOAST). Neurology **53**, 126–131 (1999)
11. Frankel, M.R., Morgenstern, L.B., Kwiatkowski, T., et al.: Predicting prognosis after stroke: a placebo group analysis from the National Institute of Neurological Disorders and Stroke rt-PA Stroke Trial. Neurology **55**, 952–959 (2000)
12. Debner, S., Minow, F., Emkes, R., et al.: How about taking a low-cost, small, and wireless EEG for a walk? Psycopathology **49**(11), 1617–1621 (2012)
13. Hairston, W.D., Whitaker, K.W., Ries, A.J., Vettel, J.M., Bradford, C.J.J., Kerick, S.E., et al.: Usability of four commercially-oriented EEG systems. J. Neural Eng. **11**(4), 046018 (2014)
14. Finnigan, S.P., Walsh, M., Rose, S.E., Chalk, J.B.: Quantitative EEG indices of sub-acute ischaemic stroke correlate with clinical outcomes. Clin. Neurophysiol. **118**(11), 2525–2532 (2007)
15. Sheorajpanday, R.V., Nagels, G., Weeren, A.J., van Putten, M.J., De Deyn, P.P.: Quantitative EEG in ischemic stroke: correlation with functional status after 6 months. Clin. Neurophysiol. **122**(5), 874–883 (2011)
16. Finnigan, S., Wong, A., Read, S.: Defining abnormal slow EEG activity in acute ischaemic stroke: Delta/alpha ratio as an optimal QEEG index. Clin. Neurophysiol. **127**(2), 1452–1459 (2016)
17. Finnigan, S., van Putten, M.J.: EEG in ischaemic stroke: quantitative EEG can uniquely inform (sub-) acute prognoses and clinical management. Clin. Neurophysiol. **124**(1), 10–19 (2013)
18. Jordan, K.G.: Emergency EEG and continuous EEG monitoring in acute ischemic stroke. J. Clin. Neurophysiol. **21**, 341–352 (2004)
19. Welch, P.: The use of fast fourier transform for the estimation of power spectra: a method based on time averaging over short, modified periodograms. IEEE Trans. Audio Electroacoust. **15**(2), 70–73 (1967)

20. Niedermeyer, E.: Cerebrovascular disorders and EEG. In: Niedermeyer, E., Lopes da Silva, F. (eds.) Electroencephalography: Basic Principles Clinical Applications and Related Fields, 3rd edn, pp. 339–362. Williams & Wilkins, Philadelphia (2005)
21. Murri, L., Gori, S., Massetani, R., et al.: Evaluation of acute ischemic stroke using quantitative EEG: a comparison with conventional EEG and CT scan. Neurophysiol. Clin. **28**, 249–257 (1998)

Differences in Circadian Rhythms of Blood Pressure and Heart Rate Among Hypertensive and Normal Blood Pressure Subjects

Silveri Giulia[1(✉)], Pascazio Lorenzo[2], Sabbadini Gastone[2], Guerra Monica[2], and Accardo Agostino[1]

[1] Department of Engineering and Architecture, University of Trieste, 34127 Trieste, Italy
giulia.silveri@phd.units.it

[2] Department of Medical, Surgical and Health Care, CS of Geriatrics, University of Trieste, 34127 Trieste, Italy

Abstract. The purpose of this study was to evaluate the differences in the circadian variations of blood pressure (BP) and heart rate (HR) among three subject groups (hypertensive, normal/high normal and optimal blood pressure). The ambulatory BP and HR of 385 subjects, without clinical evidence of hypertension-related complications, were acquired using a Holter Blood Pressure Monitor and their circadian patterns were analyzed. Systolic, Diastolic and Mean BP showed four different well-defined trends in specific time intervals of the day, similar for the three BP measures and among the three subject groups. Both BP and HR signals presented a decrease between 10:00 and 14:30 and from 19:00 to 2:00 and an increase from 5:00 to 10:00. Between 14:30 and 19:00, BP and HR presented an opposite relationship with decreasing HR and increasing BP, not yet reported in the literature. The behaviors of BP and HR were well approximated in each of the four periods by linear trends in all the three subject groups. On the contrary, in the period between 2:00 and 5:00 both BP and HR showed a quite constant trend. Results support the hypothesis of an independent vagal control during 24-h in respect of HR and BP mean levels, the latter depending on the specific subject group. Moreover, linear approximation in the identified four intervals could be used to quantify the circadian changes for all kind of subjects.

Keywords: Heart rate · Blood pressure · Monitoring and control systems

1 Introduction

Blood pressure (BP) and heart rate (HR) are familiar and routinely measurable clinical parameters of cardiovascular function. High values of BP represent the most important risk factor for cardiovascular disease being consistently associated with the occurrence of stroke, myocardial infarction, heart failure and death [1]. On the contrary, BP-lowering therapies have been associated with significant reduction in cardiovascular morbidity and mortality [2]. In addition, HR values retain a strong predictive significance both in healthy

J. Henriques et al. (Eds.): MEDICON 2019, IFMBE Proceedings 76, pp. 142–149, 2020.
https://doi.org/10.1007/978-3-030-31635-8_17

populations and in patients with various cardiovascular diseases, including those with hypertension [3–5].

Conventional office BP and HR measurements have long been used to evaluate physiological states [4, 5] although they do not provide additional prognostic information about short- and long-term fluctuations during the day [6–9] as estimated by using the 24 h Ambulatory BP Monitoring (ABPM). In fact, the benefits of using ambulatory blood pressure monitoring in addition to office measurement in the diagnosis and in the management of hypertension are well known [10].

Several authors highlighted that the HR during sleep presents lower values than during daytime [11–13] and the BP shows a higher level during the daytime than during the night [14, 15]. Koroboki et al. [16], using ABPM measurements, studied the circadian variation of HR and BPs in 1676 Greek subjects showing that patients with masked hypertension and white coat hypertension presented the same mean pattern as normotensive and hypertensive subjects. The circadian rhythms presented peak values between 10:00 and 14:00, an afternoon decrease with a second peak at about 20:00 and a progressive decrease till 2:00–4:00. Moreover, analyzing only SBP in normotensive Irish adults over 24 h, Madden et al. [17] approximated, by means of piecewise linear splines, the steepest rise and fall occurred just after waking and immediately after falling asleep.

The aim of this paper is the study and the quantification of the differences in the circadian variations of HR and of Systolic, Mean and Diastolic BP among hypertensive, normal/high normal and optimal blood pressure subjects.

2 Methods

2.1 Subjects

The study population consisted of 385 subjects (153 males, 232 females, mean age 65.0 ± 15.6 years) with known or suspected hypertension, referred for BP evaluation to the Ambulatory of Cardiovascular Pathophysiology (Geriatric Department of the University of Trieste, Trieste, Italy) between July 2016 and September 2016. This retrospective study was carried out on subjects selected following these inclusion criteria: (1) no clinical or laboratory evidence of secondary arterial hypertension, (2) absence of clinical evidence of hypertension-related complications, (3) office BP measures with a SBP value between 70 mmHg and 260 mmHg and a DBP value between 40 mmHg and 150 mmHg.

Based on office BP readings, subjects were classified as hypertensive (H) (SBP $\geq$ 140 mmHg and/or DBP $\geq$ 90 mmHg) or non-hypertensive (NH) (SBP < 140 mmHg and DBP < 90 mmHg), according to current Guidelines [15]. NH subjects were further sub-classified as having normal/high normal BP (SBP 120–139 mmHg and DBP 80–89 mmHg) (N_NH) or optimal BP (SBP < 120 mmHg and DBP < 80mmHg) (O_NH). Thus, 215 untreated hypertensive patients were considered together with 112 subjects with normal/high normal BP and 58 subjects with optimal BP.

The measurement procedures were in accordance with the institutional guidelines and all the subjects gave their informed consent.

2.2 Blood Pressure Measurements

The blood pressure was at first measured in office condition, as the average of two consecutive measurements [15]. Trained personnel, using a standard mercury sphygmomanometer with a cuff of appropriate size, evaluated the BP values.

Moreover, an ambulatory measurement along the 24-h was carried out by using a Holter Blood Pressure Monitor (Mobil-O-Graph® NG, IEM gmbh Stolberg, Germany), based on oscillometric technique. The portable monitor was programmed to obtain ambulatory blood pressure and heart rate readings each 15-min interval throughout the daytime (6:00 to 22:00) and each 30-min interval throughout the nighttime (22:00 to 6:00). No patient received additional medication that might affect the circadian blood pressure or heart rate rhythmicity. Participants individually received written instructions on how to use the Holter Blood Pressure Monitor.

The circadian trend of the mean values of SBP, DBP and Mean BP among the subjects was separately examined for H, N_NH and O_NH subject groups analyzing, in each period, the mean and standard deviation and the difference between the maximum and the minimum BP values.

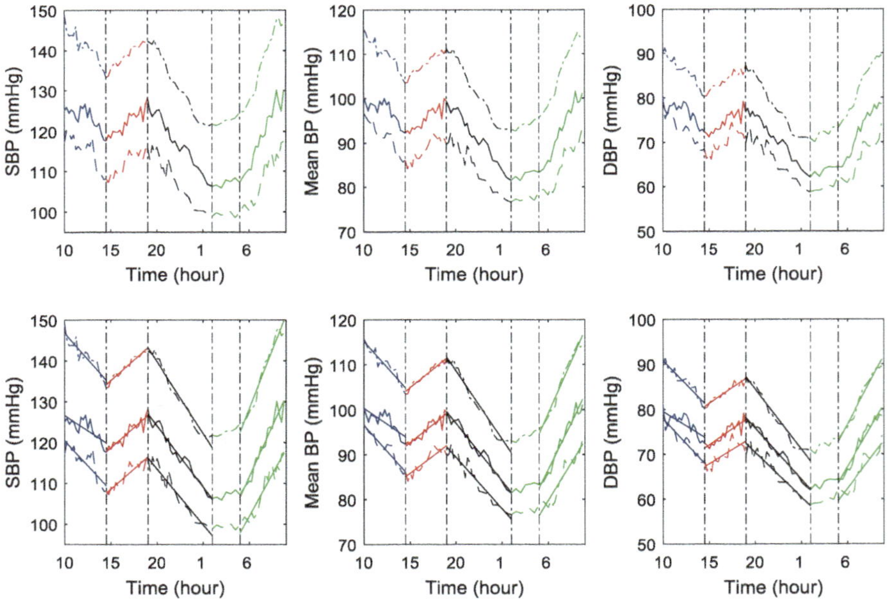

Fig. 1. Top panels: circadian rhythms of SBP, Mean BP and DBP in H (dot-point line), N-NH (solid line) and O-NH (dashed line) subject groups, subdivided in four periods: 10:00–14:30; 14:30–19:00; 19:00–2:00; 5:00–10:00. Bottom panels: Circadian rhythms as in the top panels together with linear regression curves.

The significance of the differences between each pair of the three groups was evaluated by the Wilcoxon rank sum test with the Bonferroni correction. A p-value of

0.05 was used as the level of statistical significance. Since the BP profiles during 24-h showed to be bimodal with two maxima and two minima (Fig. 1), we subdivided the 24-h in four intervals. In each period the quite linear trend was fitted by a regression line and the linear approximation significance was measured by using the R-square statistic and the p-value.

3 Results

The circadian behaviors of SBP, DBP and Mean BP for H, N_NH and O_NH subject groups are shown in Fig. 1. All the nine BP curves decreased between 10:00 and 14:30 then presented a moderate increase between 14:30 and 19:00 followed by a deeper reduction during nighttime between 19:00 and 2:00. An about constant behavior was present between 2:00 and 5:00 followed by a morning quick increase between 5:00 and 10:00. For each BP curve, the mean values (Table 1) were very similar in the four periods, with significant differences ($p < 0.05$) between each pair of groups.

Table 1. Mean values ± 1SD of SBP, Mean BP, DBP and HR and corresponding differences between maximum and minimum values in the four period of time for each subject group.

		Mean values and standard deviations				Delta values			
		10:00–14:30	14:30–19:00	19:00–2:00	5:00–10:00	10:00–14:30	14:30–19:00	19:00–2:00	5:00–10:00
SBP (mmHg)	H	140 ± 14	139 ± 15	133 ± 16	137 ± 17	52 ± 23	48 ± 20	52 ± 20	50 ± 18
	N_NH	123 ± 11	122 ± 12	118 ± 11	119 ± 14	44 ± 18	42 ± 18	47 ± 16	50 ± 26
	O_NH	116 ± 12	112 ± 11	108 ± 11	108 ± 11	43 ± 26	42 ± 24	44 ± 17	46 ± 20
MeanBP (mmHg)	H	109 ± 11	108 ± 12	103 ± 12	106 ± 11	38 ± 17	108 ± 12	42 ± 15	38 ± 13
	N_NH	97 ± 8	95 ± 9	92 ± 8	93 ± 10	31 ± 13	95 ± 9	37 ± 12	37 ± 19
	O_NH	91 ± 10	88 ± 9	85 ± 9	85 ± 8	31 ± 16	88 ± 9	34 ± 10	37 ± 14
DBP (mmHg)	H	85 ± 11	84 ± 11	79 ± 10	82 ± 10	36 ± 17	35 ± 15	40 ± 14	37 ± 14
	N_NH	76 ± 8	74 ± 8	71 ± 7	72 ± 8	31 ± 12	31 ± 14	35 ± 13	34 ± 14
	O_NH	73 ± 9	70 ± 9	66 ± 8	69 ± 8	32 ± 14	31 ± 15	35 ± 10	36 ± 13
HR (bpm)	H	77 ± 12	75 ± 12	70 ± 10	71 ± 11	31 ± 16	28 ± 15	28 ± 13	35 ± 16
	N_NH	77 ± 10	75 ± 11	70 ± 10	70 ± 9	31 ± 13	29 ± 13	29 ± 15	37 ± 15
	O_NH	75 ± 10	74 ± 11	68 ± 10	68 ± 10	31 ± 13	30 ± 14	29 ± 11	39 ± 16

Moreover, the difference between the maximum and the minimum values for each BP curve presented significant difference ($p < 0.05$) between O_NH and H in the first three periods (10:00–14:30, 14:30–19:00 and 19:00–2:00), except for DBP from 10:00 to 14:30. In addition, the differences between N_NH and H over 24 h were always significant ($p < 0.05$) excluding for SBP from 19:00 to 10:00 (Table 1).

The trend of the SBP, DBP and Mean BP values can be well approximated in each of the four periods by a linear relation (Fig. 1) presenting slopes that differ among groups and periods (Table 2). The high values of the R-squared determination coefficient (R2 77%–98% in N_NH, R2 75%–96% in O_NH and R2 75%–99% in H) as well as the p-values < 0.0001 validated the linear approximations. Between 10:00 and

14:30 the slopes of SBP, Mean BP and DBP, were higher in H group than N_NH group. In fact, the BPs decreased more quickly in H than in N_NH. From 14:30 to 19:00, the BP slopes in H group were higher than in the others two groups and BP increased more slowly in O_NH than in N_NH. In the two decreasing and increasing tracts between 19:00 and 10:00, the slopes were higher in H than in O_NH and N_NH being very similar in the latter (Table 2).

Table 2. Slope and Intercept values of the linear regression lines in Fig. 2 for SBP, Mean BP, DBP and HR.

		Slope [mmHg/hour]				Intercept [mmHg]			
		10:00–14:30	14:30–19:00	19:00–2:00	5:00–10:00	10:00–14:30	14:30–19:00	19:00–2:00	5:00–10:00
SBP	H	–2.59	2.01	–3.50	5.74	172.78	104.90	209.96	44.13
	N_NH	–1.49	1.96	–2.99	4.97	141.59	89.23	183.85	–37.81
	O_NH	-2.45	1.97	–2.71	4.11	145.03	78.99	167.82	–21.47
Mean BP	H	–2.33	1.63	–3.03	4.61	138.37	80.37	169.40	–39.28
	N_NH	–1.42	1.62	–2.57	4.16	114.57	68.36	148.45	–38.27
	O_NH	-2.31	1.54	–2.29	3.34	119.67	62.66	135.33	–20.71
DBP	H	–2.09	1.36	2.68	3.84	111.74	60.86	138.20	–38.87
	N_NH	–1.27	1.40	–2.29	3.50	92.27	51.02	121.65	–38.17
	O_NH	-2.21	1.25	–1.99	2.69	100.00	48.97	110.41	–18.55
HR	H	–0.81	–0.51	–1.72	4.64	87.44	82.99	107.63	–75.26
	N_NH	-0.51	–0.36	–2.03	5.14	83.39	80.57	114.48	-91.22
	O_NH	-0.08	–0.38	–1.87	4.56	74.03	80.03	109.79	–76.14

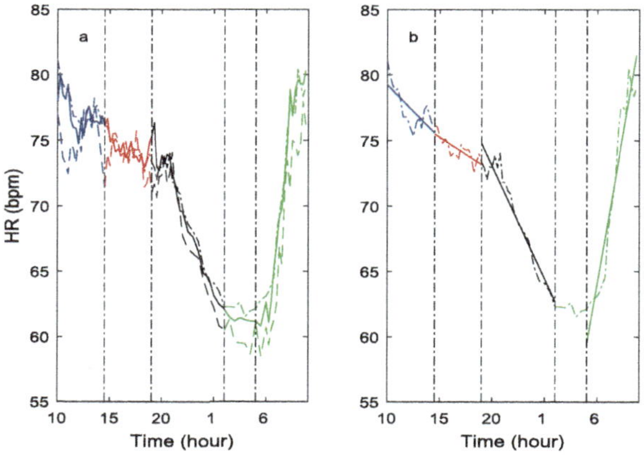

Fig. 2. a: Circadian HR behavior in H (dot point line), N-NH (solid line) and O-NH (dashed line) subjects. b: Circadian HR behavior in H subjects together with the linear regression curves. The 24-h were divided in four periods: 10:00–14:30; 14:30–19:00; 19:00–2:00; 5:00–10:00.

Figure 2a shows the circadian rhythm of HR in the three groups. HR presented a slow decrease from 10:00 to 19:00 followed by a quick decrease from 19:00 to 2:00. From 2:00 to 5:00, HR was about constant while a very quick increase was present during the morning, from 5:00 to 10:00. There were no significant differences in the behavior between N_NH and the other two groups. Since HR curves showed a quite linear behavior in each period, a linear regression best fit of each trend was also calculated in all the three groups (Fig. 2b). Different slopes in the four periods were present for each group (Table 2). In the decreasing period (between 19:00 and 2:00), the slopes for the three groups were similar and lower than from 5:00 to 10:00 in which the slopes presented the maximum values. The corresponding two linear approximations were validated by the high values of R-squared determination coefficients (R2 95% in N_NH, R2 87% in O_NH and R2 93% in H) and the low p-values < 0.0001. In the period between 10:00 and 14:30, the slopes considerably decreased and the linear approximation was not significantly different from zero.

4 Discussion

A large number of external and internal factors, such as, vasoactive hormones, autonomic nervous system tone, cardiovascular system diseases affect the circadian behavior of rhythms of BP and HR [8, 9]. Moreover, the analysis of HR and BP signals along the 24-h demonstrated to better evaluate cardiovascular morbidity and mortality [9, 14, 18] in comparison with office BP measurements [1, 2].

In our study, the circadian rhythms of HR and BP in H, N_NH and in O_NH subjects were analyzed during both daytime and nighttime revealing definite daytime and nighttime profiles. In particular, the circadian profiles showed a remarkable decrease during the night that could be caused by the slowdown rest activities during sleep, and a quick increase during awakening caused by the morning surge (Fig. 1). These findings confirmed previous results [5, 12, 14, 17, 18] in normotensive and hypertensive subjects, concerning lower values of HR and BP during the night than during the day as well as the outcomes of the Kornokobi's study [16]. Moreover, our results corroborated the observations of Litter [13] about larger changes of systolic pressure, during both the day and the night, than of diastolic pressure and higher values of systolic pressure when a greater variability was present.

One of the main findings of our study was that the circadian BP and HR behaviors in hypertensive subjects presented the same profile as in N_NH and O_NH subjects. In a previous study concerning pressure variability examined by using BP standard deviations, Ragot et al. [20] showed similar BP and HR behaviors only in hypertensive subjects. The persistence of the same basic circadian rhythm among the different subject groups supports the evidence that the autonomic nervous system activity is independent from the HR and BP mean levels, as stated by Korokobi [16].

To better analyze the behavior of the BP and HR variability, in our study the 24-h were divided in four intervals, following the two maxima and two minima present in the circadian rhythm (Fig. 1), well approximating the trends by linear relations whose slopes depended on both the interval of the day and the subject group (Fig. 1 and Table 2). The highest values of slopes, present during awakening (5:00–10:00), particularly in the H group (for BPs), correspond to the greatest velocity changes of BPs and HR while the slowest velocities were present for DPB and HR values in the post-prandial period. These slopes could be used to quantify the morning blood pressure surge, associated with acute cardiovascular events [19]. While a corresponding trend between HR and BP signals was present in three of the four periods, in the post-prandial interval, their behaviors were opposite. Hence, we can conclude that, at least in this interval, the mechanisms that regulate the circadian behavior of the two signals are independent.

Acknowledgements. Work partially supported by the Master in Clinical Engineering, University of Trieste.

Conflict of Interest. The authors declare that they have no conflict of interest.

References

1. Lawes, C.M., Vander Hoorn, S., Rodgers, A.: 2008 global burden of blood-pressure-related disease 2001. Lancet **371**, 1513–1518 (2008)
2. Zanchetti, A., Thomopoulos, C., Parati, G.: Randomized controlled trials of blood pressure lowering in hypertension: a critical reappraisal. Circ. Res. **116**, 1058–1073 (2015)
3. Palatini, P., Benetos, A., Grassi, G., Julius, S.: Identification and management of the hypertensive patient with elevated heart rate: statement of a European society of hypertension consensus meeting. J. Hypertens. **24**, 603–610 (2006)
4. Kannel, W.B., Kannel, C., Paffenbarger, R.S.: Heart rate and cardiovascular mortality: the Framingham study. Am. Heart J. **113**, 1489–1494 (1987)
5. Gillman, M.W., Kannel, W.B., Belanger, A.: Influence of heart rate on mortality among persons with hypertension: the Framingham study. Am. Heart J. **125**, 1148–1154 (1993)
6. Pickering, T.G., Shimbo, D., Haas, D.: Ambulatory blood-pressure monitoring. N. Engl. J. Med. **354**, 2368–2374 (2006)
7. Mancia, G.: Short- and long-term blood pressure variability: present and future. Hypertension **60**, 512–517 (2012)
8. Palatini, P., Reboldi, G., Beilin, L.J.: Added predictive value of night-time blood pressure variability for cardiovascular events and mortality: the ambulatory blood pressure-international study. Hypertension **64**, 487–493 (2014)
9. Hansen, T., Thijs, L., Boggia, J.: Prognostic value of ambulatory heart rate revisited in 6928 subjects from 6 populations. Hypertension **52**, 229–235 (2008)
10. Palatini, P., Penzo, M., Racioppa, A.: Clinical relevance of nighttime blood pressure and of daytime blood pressure variability. Arch. Intern. Med. **152**, 1855–1860 (1992)
11. Richard, A.M., Nicholls, M.G., Espiner, E.: Diurnal patterns of blood pressure, heart rate and vasoactive hormones in normal man. Clin. Exp. Hypertens. **A8**, 153–166 (1986)
12. Mancia, G., Ferrari, A., Gregori, L.: Blood pressure and heart rate variabilities in normotensive and hypertensive human beings. Circ. Res. **53**, 96–104 (1983)
13. Littler, W.A., West, M.J., Honour, A.J.: The variability of arterial pressure. Am. Heart J. **95**, 180–186 (1978)
14. Zakopoulos, N., Stamatelopoulos, S.: Effect of hypotensive drugs on the circadian blood pressure pattern in essential hypertension: a comparative study. Cardiovasc. Drugs Therapy **11**, 795–799 (1997)

15. Mancia, G., Fagard, R., Narkiewicz, K.: 2013 ESH/ESC guidelines for the management of arterial hypertension: the task force for the management of arterial hypertension of the European society of hypertension (ESH) and of the European society of cardiology (ESC). Eur. Heart J. **34**(28), 2159–2219 (2013)
16. Koroboki, E., Manios, E., Psaltopoulou, T.: Circadian variation of blood pressure and heart rate in normotensives, white-coat, masked, treated and untreated hypertensives. Hellenic J. Cardiol. **53**, 432–438 (2012)
17. Madden, J.M., Li, X., Kearney, P.M.: Exploring diurnal variation using piecewise linear splines: an example using blood pressure. Emerg. Themes Epidemiol. **14**(1), 1 (2017)
18. Millar-Craig, M.W., Bishop, C.N., Rafery, E.B.: Circadian variation in blood pressure. Lancet **1**, 795–797 (1978)
19. Giles, T.D.: Circadian rhythm of blood pressure and the relation to cardiovascular events. J. Hypertens. Suppl. **24**, S11–S16 (2006)
20. Ragot, S., Herpin, D., Siché, J.P.: Relationship between short-term and long-term blood pressure variabilities in essential hypertensive. J. Hum. Hypertens. **15**, 41–48 (2001)

Ectopic Beat Detection from Wrist Optical Signals for Sinus Rhythm and Atrial Fibrillation Subjects

Serj Haddad[1(✉)], Jarkko Harju[2], Adrian Tarniceriu[1], Tuomas Halkola[3], Jakub Parak[3], Ilkka Korhonen[4], Arvi Yli-Hankala[2,4], and Antti Vehkaoja[3,4]

[1] PulseOn SA, Neuchâtel, Switzerland
serj.haddad@pulseon.com
[2] Department of Anesthesia, Tampere University Hospital, Tampere, Finland
[3] PulseOn Oy, Espoo, Finland
[4] Faculty of Medicine and Health Technology, Tampere University, Tampere, Finland

Abstract. Ectopic beats are abnormal cardiac beats originating from a location different than the sino-atrial node and therefore not being controlled by the autonomous nervous system. Thus, correct heart rate variability analysis inevitably requires accurate ectopic beat detection. Furthermore, an accurate ectopic beat detection is crucial to differentiate irregular cardiac rhythm due to different types of pathological arrhythmias from those caused by isolated ectopic beats. In this paper, we present an algorithm for ectopic beat detection based on wrist plethysmographic (PPG) signals. The proposed algorithm relies on analyzing the inter-beat patterns while considering the heart-rhythm condition; whether sinus rhythm (SR) or atrial fibrillation (AF). We monitor 29 patients recovering from surgery in the post-anesthesia care unit. During the recordings, 15 patients had SR and 14 patients had AF. The proposed ectopic beat detection algorithm achieves a sensitivity of $93.08 \pm 3.83\%$ and a specificity of $97.80 \pm 2.12\%$.

Keywords: Photoplethysmography · Ectopic beat detection · Heart rate variability · Atrial fibrillation

1 Introduction

Periodic impulses from the sino-atrial node cause the contraction and the relaxation of the heart muscle. When these impulses are incurred in unusual/abnormal locations, they result in disturbed and irregular beats, known as arrhythmias. The arrhythmias can consist in isolated irregular beats (generally known as ectopic beats), or pathological conditions such as atrial fibrillation (AF) or atrial flutter (AFl). While the ectopic beats are frequently encountered and do not present a significant health threat, AF or AFl can increase the risk of morbidity or mortality. One way to detect AF is based on inter-beat interval (IBI) variability [1,2]. In such cases, ectopic beats can increase the IBI dispersion, possibly

J. Parak—Affiliation at the time of the research.

J. Henriques et al. (Eds.): MEDICON 2019, IFMBE Proceedings 76, pp. 150–158, 2020.
https://doi.org/10.1007/978-3-030-31635-8_18

causing false alerts. Similarly, ectopic beats could affect other heart rate variability (HRV) analysis applications, such as stress or sleep monitoring. Thus, detecting ectopic beats is of vast importance for algorithms analyzing HRV or extracting various HRV parameters to study cardiac rhythm disturbances or arrhythmias.

Algorithms to detect and correct ectopic beats can be based on analyzing the variation of the instantaneous heart rate [3] or extracting morphological and rhythmical features of the ECG signal [4]. Different methods of ectopic beat detection, such as time and morphological methods, morphological transformation approach, wavelet transform, and their impact on HRV signal parameters are reviewed in [5].

In this paper, we tackle a slightly more challenging problem, which is to detect ectopic beats using photoplethysmography (PPG). PPG presents a simple and efficient alternative method to ECG to estimate heart rate (HR) and HRV [6]. It is a non-invasive technology that utilizes a light source and a photodetector to illuminate the skin and study the amplitude of the transmitted/reflected light to measure the volumetric variations of blood circulation, which, in turn, represents the pumping actions of the heart. This means that we can accurately estimate HR [7,8] and, in the absence of motion, IBI from PPG signals [9,10].

Ectopic beat detection from PPG signals has not been extensively studied, but there is an increasing interest in this domain especially in context of arrhythmia detection techniques. The proposed arrhythmia detection methods investigate the possibility of extracting IBI from PPG signals, usually finger or wrist-based, and taking the decision based on IBI series statistics [1,2].

This paper evaluates an ectopic beat detection algorithm. The algorithm is based on processing IBI extracted from wrist PPG signals. As PPG signals are highly sensitive to motion, varying ambient light, or other interferences, and IBI estimation is not reliable in such situations [11], a method to automatically screen out poor quality data is included as part of the method.

The paper is organized as follows. In Sect. 2, we describe the materials and the methodology used by the study. Section 3 presents the contribution of this work. It describes the proposed algorithm to detect ectopic beats in both sinus rhythm (SR) and atrial fibrillation subjects. The performance of the proposed ectopic beat detection scheme is evaluated in Sect. 4. Finally, Sect. 5 concludes the paper.

2 Study Protocol and Materials

The study involves 29 subjects in the post-anesthesia care unit of Tampere University Hospital, immediately after surgery and prior to completely recovering from the anesthetics effects; 15 subjects (7 females, 67.5 $\pm$ 10.7 years old) belong to the SR group with normal continuous sinus rhythm and 14 subjects (8 females, 74.8 $\pm$ 8.3 years old) belong to the AF group with continuous AF rhythm. The subjects were lying in bed with their hands free to move during the whole duration of the recording. Each recording lasts approximately 1 h and 30 min.

In total, 23.52 h of data is collected from the SR group and 21.96 h of data from the AF group. The IBI values used by the proposed ectopic beat detection algorithm are extracted from the PPG recordings.

The wrist PPG signal and the 3D acceleration vector (Acc) data were recorded using the PulseOn Optical Heart Rate monitor (PulseOn Oy, www.pulseon.fi). The reference ECG signal is simultaneously recorded with GE Carescape^{TM} B850 patient monitor (GE Healthcare Ltd., www.gehealthcare.com). Kubios HRV software (Kubios Ltd., www.kubios.com) is used to obtain the RR intervals (RRI) from the reference ECG signal. Finally, both the ECG signal and the corresponding RR intervals are visually inspected to construct the reference ectopic beats of the 29 recordings performed.

The study is registered at clinicaltrials.gov (NCT03081793). The study protocol, the devices used, and the documentation are all approved by the local ethical review board of Pirkanmaa Hospital District(ETL R17024), the Finnish National Supervisory Authority of Health and Welfare (Valvira), and the hospital technical department. The subjects provide a written consent to participate in this study and have the full right to withdraw from the study at any time. The experimental procedures comply with the principles of the Helsinki Declaration of 1975, as revised in 2000 and 2008.

3 Algorithm Description

In this section, we present the contribution of this paper and explain the proposed ectopic beat detection algorithm with arrhythmia feedback. To highlight the significance of the arrhythmia feedback, we first present ectopic beat detection without feedback (illustrated in Fig. 1). The algorithms used for the sub-blocks ***IBI Extraction***, ***IBI Evaluation***, and ***Arrhythmia Detection***, depicted in Figs. 1 and 3, are presented in [1,2,12].

3.1 Ectopic Beat Detection Without Arrhythmia Feedback

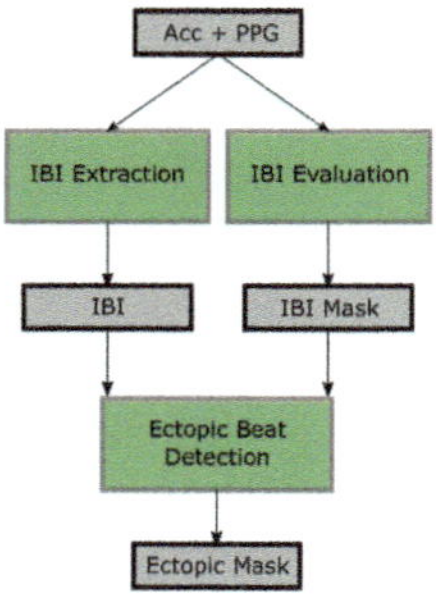

Fig. 1. Ectopic beat detection block diagram

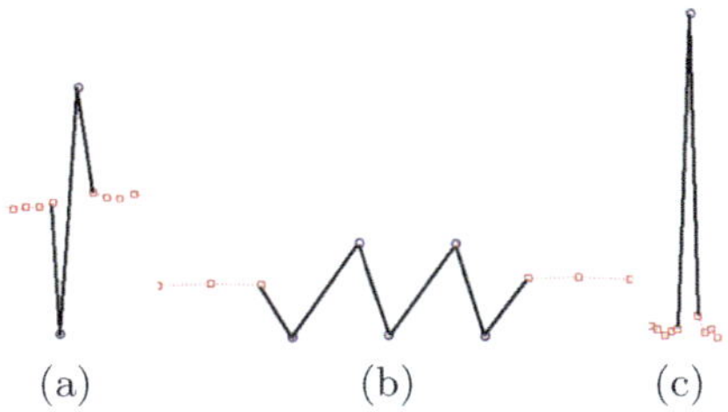

Fig. 2. Examples of inter-beat intervals corresponding to ectopic beats.

The algorithm runs in real-time and its block scheme is presented in Fig. 1. The input signals are PPG and accelerometer data Acc. The main processing sub-blocks are:

- ***IBI Extraction***: estimates the IBI from the raw PPG signals.
- ***IBI Evaluation***: evaluates the reliability of the estimated IBI based on both PPG and Acc signals. It indicates whether the IBI is reliable (mask = 0) or not (mask = 1).
- ***Ectopic Beat Detection***: inspects the variation of the *extracted* IBI values that are marked as *reliable* and compares them with the characteristic patterns of ectopic beats (possible examples of IBI patterns corresponding to ectopic beats are shown in Fig. 2). Based on the similarity between the morphology of the received IBI sequences and the ectopic patterns, it labels the input beats as ectopic or not. In other words, if the IBI beat pattern satisfies certain *morphological conditions*, then the corresponding beats are marked as ectopic. These *conditions* are dynamically modified based on the input IBI values, however, they are independent of the IBI rhythm classification.

The drawback of the described algorithm reveals itself in the compromise it has to make between the *sensitivity* and the *specificity* (as defined in Eq. (1)). In other words, to detect more ectopic beats (high sensitivity), the *morphological conditions* should be more relaxed, which results in increased number of falsely detected ectopic beats (low specificity) especially for AF patients. On the other hand, to reduce the number of falsely detected ectopic beats (high specificity), these *conditions* should be more conservative, which reduces the number of detected ectopic beats (low sensitivity) especially for SR patients. Here we choose to reduce the sensitivity to maintain high specificity for AF patients. Equivalently, we choose conservative *conditioning* to reduce the number of falsely detected ectopic beats.

3.2 Ectopic Beat Detection with Arrhythmia Feedback

To resolve the drawback of the algorithm presented in Sect. 3.1, we propose a feedback loop scheme between the ***Ectopic Beat Detection*** sub-block and the ***Arrhythmia Detection*** sub-block (see Fig. 3):

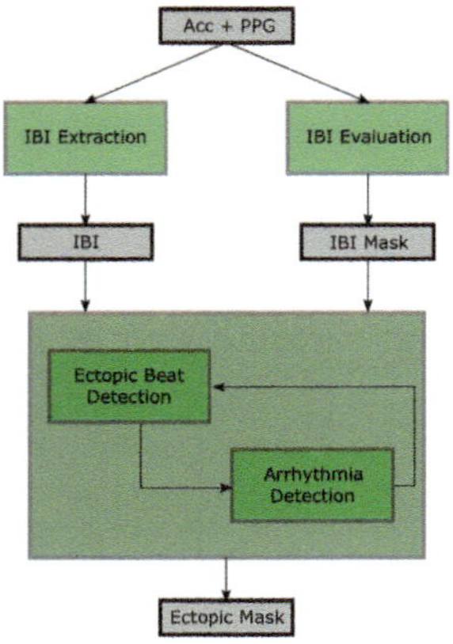

Fig. 3. Feedback loop scheme between ectopic beat detection block and arrhythmia detection block

- ***Arrhythmia Detection***: processes the input IBI values and their corresponding reliability masks to label them as SR or AF using the Markov model approach, similar to the work of Moody and Mark [13].
- ***Ectopic Beat Detection***: operates in a similar fashion as the one without arrhythmia feedback, however, it utilizes the SR/AF labels from the ***Arrhythmia Detection*** sub-block to tune the strictness of the *morphological conditioning* used to label the input beats.

To highlight the significance of the proposed method, consider the scenarios depicted in Figs. 4 and 5. Both figures show IBI signals of similar morphology, however, one IBI signal contains ectopic beats (SR case), while the other contains non-ectopic AF beats (AF case). The proposed algorithm successfully discriminates between the two by strictly analyzing PPG and Acc signals. We show the ECG reference plot corresponding to each signal; the disturbance in the cardiac rhythm shown in Fig. 4 is due to the presence of ventricular ectopic beats, while that shown in Fig. 5 is due to atrial fibrillation.

Other than the improved ectopic beat detection due to the arrhythmia feedback, the proposed scheme also presents an additional advantage in the feedback going from the ***Ectopic Beat Detection*** sub-block to the ***Arrhythmia Detection*** sub-block. Note that the performance of the feedback loop scheme depends on both ***Ectopic Beat Detection*** and on ***Arrhythmia Detection*** sub-blocks: *the ectopic beat detection algorithm is accurate if and only if the arrhythmia detection algorithm is accurate.*

4 Performance

We evaluate the performance of the proposed ectopic beat detection algorithm with arrhythmia feedback using the metrics given in Sect. 4.1. The results are presented and discussed in Sect. 4.2.

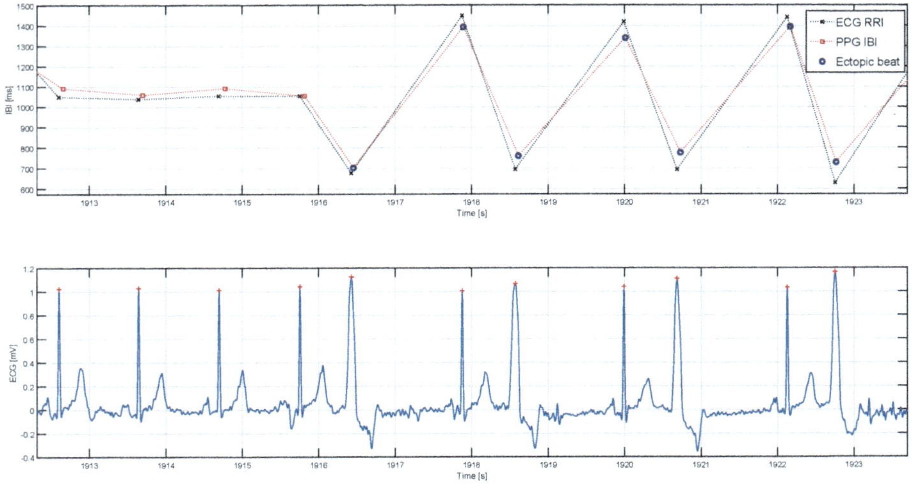

Fig. 4. Example of ectopic beats (marked in blue circles) during sinus rhythm correctly detected by the algorithm with arrhythmia feedback.

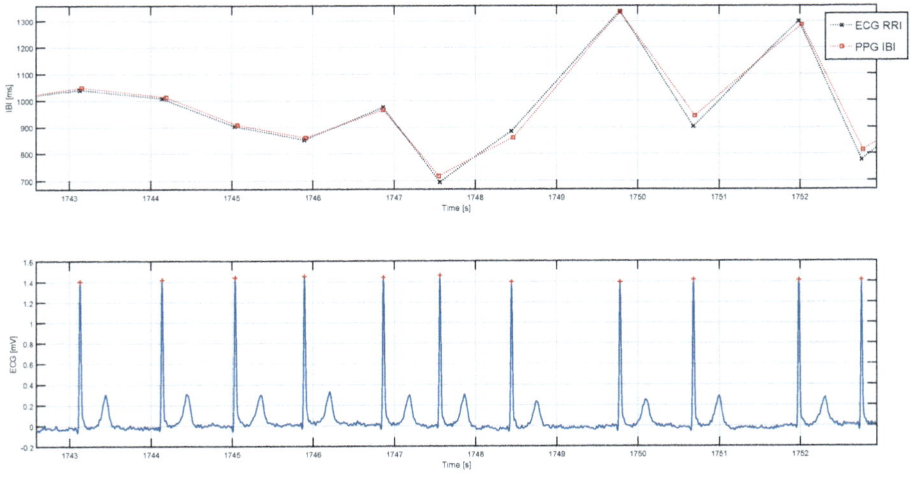

Fig. 5. Example of AF beats correctly not marked as ectopic by the algorithm with arrhythmia feedback.

4.1 Evaluation Metrics

The RR intervals from the ECG signal are labeled as ectopic or not based on close visual inspection of the ECG signal. These labels constitute the reference ectopic beats used to evaluate the performance of the ectopic beat detection algorithms applied to the PPG signal.

The two metrics used to evaluate the performance of these algorithms are the sensitivity and the specificity, defined as,

$$\mathbf{Se} = 100 \times \frac{\mathbf{TP}}{\mathbf{TP} + \mathbf{FN}}$$
$$\mathbf{Sp} = 100 \times \frac{\mathbf{TN}}{\mathbf{TN} + \mathbf{FP}}, \tag{1}$$

where **TP**, **FN**, **TN**, and **FP** denote 'True Positive,' 'False Negative,' 'True Negative,' and 'False Positive,' respectively.

4.2 Results and Discussion

In this section, we discuss the performance of the algorithm with arrhythmia feedback and compare it with that of the algorithm without arrhythmia feedback. For comparison, we also implement the ectopic beat detection algorithm presented by Matteo et al. in [3]. In their work, they propose a method to detect anomalous beat positions based on the variation of the instantaneous heart rate.

Table 1. Ectopic beat detection results

	Method from [3]	Proposed method	
		Without feedback	With feedback
Sensitivity (%)	99.86 ± 0.12	78.83 ± 13.10	93.08 ± 3.83
Specificity (%)	82.08 ± 12.38	99.18 ± 0.70	97.80 ± 2.12

From the results shown in Table 1, one could see that the method proposed in [3] detects ectopic beats with 99.86% sensitivity. However, the shortcoming of this method is that it has low specificity (82.08%), because it only targets SR data and cannot differentiate AF beats from ectopic beats. Consequently, its specificity is low for AF data sets (64.48%) and high for SR data sets (98.20%). Thus, this method accurately detects ectopic beats for SR data sets only. On the other hand, the ectopic beat detection algorithm without arrhythmia feedback detects ectopic beats with 99.18% specificity, because of the conservative *conditioning* discussed in Sect. 3.1. However, it suffers from low sensitivity (78.83%). Indeed, to maintain a high specificity, this algorithm fails to detect ectopic beats efficiently for SR data sets. Finally, our proposed method of ectopic beat detection with arrhythmia feedback guarantees fairly high sensitivity 93.08% and high specificity 97.80%. In other words, it improves the specificity for AF data sets (compared to the method from [3]) without significantly decreasing the sensitivity.

5 Conclusion

This study evaluates the accuracy of an ectopic detection algorithm with arrhythmia feedback for sinus rhythm and atrial fibrillation subjects using wrist PPG and acceleration signals. The sensitivity of the proposed method is $93.08 \pm 3.83\%$ and the specificity $97.80 \pm 2.12\%$.

This work shows that the IBI patterns for both SR and AF subjects can very much look similar, however the causes of this variation/disturbance can be different. To efficiently differentiate between variations caused by ectopic beats and those caused by AF, it is crucial to implement a feedback loop scheme where the ectopic beat detection sub-block and the arrhythmia detection sub-block exchange information.

Finally, given that the proposed algorithm does not require ECG signals and solely relies on wrist optical and acceleration signals, it could be used as a comfortable and unobtrusive method to detect ectopic beats and utilized in HRV or AF detection applications.

References

1. Harju, J., Tarniceriu, A., Parak, J., Vehkaoja, A., Yli-Hankala, A., Korhonen, I.: Monitoring of heart rate and inter-beat intervals with wrist plethysmography in patients with atrial fibrillation. In: Physiological Measurement, IOP Publishing, vol. 39, no. 6, pp. 65007–65019, June (2018)
2. Tarniceriu, A., Harju, J., Yousefi, Z. R., Vehkaoja, A., Parak, J., Yli-Hankala, A., Korhonen, I.: The accuracy of atrial fibrillation detection from wrist photoplethysmography. A study on post-Operative patients. In: 40th Annual International Conference of the IEEE Engineering in Medicine and Biology Society (EMBC), Honolulu, pp. 1–4 (2018)
3. Mateo, J., Laguna P.: Analysis of heart rate variability in the presence of ectopic beats using the heart timing signal. In: IEEE Transactions on Biomedical Engineering, vol. 50, no. 3, pp. 334–343, March 2003
4. Hekler, A., Kikillus, N., Bolz, A.: Detection of ectopic beats in single channel electrocardiograms. In: Vander Sloten, J., Verdonck, P., Nyssen, M., Haueisen, J. (eds.) 4th European Conference of the International Federation for Medical and Biological Engineering (IFMBE) Proceedings, vol. 22. Springer, Heidelberg (2009)
5. Dib, N., Bereksi, R.: Ectopic beats detection and correction methods: a review. In: Biomedical Signal Processing Control, vol. 18, pp. 228–244, April 2015
6. Lemay, M., Bertschi, M., Sola, J., Renevey, P., Parak, J., Korhonen, I.: Application of optical heart rate monitoring. In: Sazonov E., Neuman, M. R. (eds.) Wearable Sensors: Fundamentals, Implementation and Applications, Elsevier, Imprint: Academic Press, pp. 105–129 (2014)
7. Parak, J., Korhonen, I.: Evaluation of wearable consumer heart rate monitors based on photopletysmography. In: 36th Annual International Conference of the IEEE Engineering in Medicine and Biology Society (EMBC), Chicago, pp. 3670–3673 (2014)
8. Delgado-Gonzalo, R., Parak, J., Tarniceriu, A., Renevey, P., Bertschi, M., Korhonen, I.: Evaluation of accuracy and reliability of PulseOn optical heart rate monitoring device. In: 37th Annual International Conference of the IEEE Engineering in Medicine and Biology Society (EMBC), Milan, pp. 430–433 (2015)
9. Renevey, P., Sola, J., Theurillat, P., Bertschi, M., Krauss, J., Andries, D., Sartori, C.: Validation of a wrist monitor for accurate estimation of RR intervals during sleep. In: 35th Annual International Conference of the IEEE Engineering in Medicine and Biology Society (EMBC), Osaka, pp. 5493–5496 (2013)

10. Parak, J., Tarniceriu, A., Renevey, P., Bertschi, M., Delgado-Gonzalo, R., Korhonen, I.: Evaluation of the beat-to-beat detection accuracy of PulseOn wearable optical heart rate monitor. In: 37th Annual International Conference of the IEEE Engineering in Medicine and Biology Society (EMBC), Milan, pp. 8099–8102 (2013)
11. Schäfer, A., Vagedes, J.: How accurate is pulse rate variability as an estimate of heart rate variability? A review on studies comparing photoplethysmographic technology with an electrocardiogram. Int. J. Cardiol. **166**(1), 15–29 (2013)
12. Tarniceriu, A., Harju, J., Vehkaoja, A., Parak, J., Delgado-Gonzalo, R., Renevey, P., Yli-Hankala, A., Korhonen, I.: Detection of beat-to-beat intervals from wrist photoplethysmography in patients with sinus rhythm and atrial fibrillation after surgery. In: IEEE EMBS International Conference on Biomedical & Health Informatics (BHI), Las Vegas, pp. 133–136 (2018)
13. Moody, G. B., Mark, R. G.: A new method for detecting atrial fibrillation using R-R intervals. In: Computers in Cardiology, pp. 227—230 (1983)

Electrocardiographic Alternans: A New Approach

Ilaria Marcantoni, Dalila Calabrese, Giorgia Chiriatti, Roberta Melchionda, Benedetta Pambianco, Giulia Rafaiani, Eleonora Scardecchia, Agnese Sbrollini, Micaela Morettini, and Laura Burattini(✉)

Università Politecnica delle Marche, 60131 Ancona, Italy
l.burattini@univpm.it

Abstract. Alternans is an electrophysiological phenomenon consisting in a beat-to-beat variation of the morphology of an electrocardiographic (ECG) waveform. Literature has particularly studied T-wave alternans (TWA) because it has been widely recognized as a noninvasive and clinically useful index to predict occurrence of malignant ventricular arrhythmias and, eventually, sudden cardiac death. Historically, alternans of other segments of ECG, like P wave (PWA), or QRS complex (QRSA) gained less interest than TWA, but this is an incomplete vision of the action potential (AP). AP is influenced by electrical activity of all myocardial cells, so it is reasonable that all ECG waveforms could be affected by alternans phenomenon. ECG alternans (ECGA) can be intended as the prevalent nature of alternans. This study aimed to use the heart-rate adaptive match filter (AMF) method, previously applied for TWA applications, to detect ECGA. AMF effectiveness was tested on simulated alternating ECG (alternans-amplitude range: 10 μV–200 μV), characterized by single- and multiple-wave alternans (always of the same amplitude and morphology). AMF method proved to be specific, being able to recognize ECGA absence, and particularly sensitive to TWA. In general, in case of singular-wave alternans, AMF correctly identified the type of alternans and correctly determined its amplitude (mean error: 0%). When TWA was combined to PWA or QRSA, only TWA was identified with an overestimation of its amplitude (mean error: 23%). In conclusion, overall AMF proved its effectiveness and specificity in revealing and discriminating ECGA.

Keywords: Electrocardiographic alternans · Adaptive match filter · Electrocardiogram

1 Introduction

Alternans is an electrophysiological phenomenon consisting in a beat-to-beat variation of the morphology – i.e. amplitude, shape and/or polarity - of an electrocardiographic (ECG) waveform (P, QRS, T) occurring at stable heart rate. Amplitude and duration of alternans phenomenon are generally variable, thus alternans is considered as a non-stationary condition. Several experimental and clinical situations have recognized electrical alternans as a marker of cardiac pathology [1–9].

J. Henriques et al. (Eds.): MEDICON 2019, IFMBE Proceedings 76, pp. 159–166, 2020.
https://doi.org/10.1007/978-3-030-31635-8_19

Literature has investigated alternans of a specific ECG wave, i.e. T-wave alternans (TWA). TWA has been recognized as a noninvasive and clinically useful index to predict malignant ventricular arrhythmias occurrence and, eventually, sudden cardiac death [1, 6, 7, 10]. Both microscopic and macroscopic TWA are associated with severe life-threatening arrhythmias, usually due to a reduced left ventricular ejection fraction [2, 3, 10]. On the exact mechanism responsible for TWA, there are many speculative hypotheses: one of them is that TWA occurrence is associated to fluctuations in ionic currents and alternans of cellular action potential (AP) [11]. Besides the specific physiological process underlying TWA evidence, it is linked to an irregularity in the ventricular repolarization.

Historically, alternans of other ECG waves (P wave or QRS complex) elicited less interest than TWA, but this could lead to a partial vision of the AP, since it is strictly linked to electrical activity of all myocardial cells. Thus, both depolarization- and repolarization-activity changes and anomalies of any cell could influence morphology of AP phases. Ventricular arrhythmias are among the most dangerous heart conditions, but this fact should not justify a partial analysis of the cardiac cycle: a complete and exhaustive analysis is necessary. Since AP involves all the heart sites, it is reasonable to conclude that all ECG waveforms are affected by alternans phenomenon.

Regarding P-wave alternans (PWA), in literature nothing but case reports are present; there, PWA has been referred as a rare phenomenon, predictor of atrial fibrillation [4]. In fact, being P wave associated to atrial depolarization, it is plausible to recognize PWA as a warning of atrial electrical instabilities, and eventually, pathologies. Similarly, being QRS complex referred to depolarization of ventricles and repolarization of atria, QRS alternans (QRSA) appeared mostly to arise in case of supraventricular and ventricular tachycardias, as stated in some studies [5, 12]. Few PWA and QRSA data prevent a comparison of results and a complete phenomenon interpretation.

ECG alternans (ECGA) can be understood as the prevalent nature of electrical alternans, i.e. PWA, QRSA or TWA. Thus, finding a method able to estimate them is crucial, singularly. Aim of this study was to adapt the use of the heart-rate adaptive match filter (AMF) method [13–16], firstly designed for TWA detection, for PWA and QRSA detection, validating its effectiveness on simulated ECG. AMF is an effective and robust algorithm introduced for TWA identification and its reliable results in TWA assessment suggested its application to PWA and QRSA study too.

2 Method

2.1 Simulated Data

ECGA was applied to different simulated ECG tracings, affected by uniform rectangular profile modelling alternans equally distributed all along the oscillating wave [17]. Three alternans sets were synthesized, according to different amplitudes (10 µV, 100 µV and 200 µV). Specifically, for each set there are three ECG simulating pure PWA, pure QRSA and pure TWA and four ECG simulating all possible combinations of them. Moreover, an ECG not affected by any alternans was synthetized and considered for ECGA evaluation.

Technically, all these simulated ECG tracings were obtained through the repetition of a single noiseless beat extracted from a real ECG recording. The beat used was a 0.75 s long and sampled at 200 Hz. The beat repetition was done 128 folds. For each ECG wave the onset and offset were estimated. P-wave onset (P_{on}) and offset (P_{off}) were set at 195 ms and 95 ms before the R peak, respectively. Each QRS complex was identified in a window taken 40 ms before (QRS_{on}) and 40 ms after (QRS_{off}) the R peak. T-wave onset (T_{on}) and T-wave offset (T_{off}) were evaluated at 100 ms and 300 ms after the R peak.

2.2 Adaptive Match Filter Method

AMF method [13–16] was designed specifically to detect TWA. Here, it was adapted to PWA and QRSA investigation too, maintaining its theoretical approach.

Ideally, ECGA is characterized by a single frequency equal to half heart rate (HR). Since real ECG signals have physiological HR variability, it was assumed that ECGA frequency (AF) corresponds to a small band around twice mean HR. Therefore, AMF method identifies ECGA through ECG filtration in a narrow passing band around ECGA frequency [13]. The filter is implemented as a 6th-order bidirectional Butterworth bandpass filter with cut-off frequencies of $LPF = AF + dAF$ and $HPF = AF - dAF$, in which $dAF = 0.06$ Hz was empirically found [14]. The squared module of AMF transfer function is expressed by the following equation:

$$|H_{AMF}(\omega)|^2 = |H_{LPF}(\omega)|^2 \cdot |H_{HPF}(\omega)|^2 = \frac{1}{1+\left(\frac{\omega}{\omega_{LPF}}\right)^6} \cdot \frac{\left(\frac{\omega}{\omega_{HPF}}\right)^6}{1+\left(\frac{\omega}{\omega_{HPF}}\right)^6} \quad (1)$$

where $\omega_{LPF} = 2\pi \cdot LPF$ and $\omega_{HPF} = 2\pi \cdot HPF$.

Figure 1 shows the method block diagram. Input signal of AMF is ECG tracing with relative R peaks. After being fed with them, AMF computes the HR value, the corresponding AF and the passing band. The filter effect is to reduce noises, interferences, but also all the ECG components not relating to the frequency band of the analyzed wave alternans. The filter output is a sinusoidal signal called alternans signal (AS) characterized by a constant phase and an amplitude modulation if a possible alternans is detected. A test on the localization of AS zero-derivative points is performed. Indeed, maxima and minima of AS are located over one of the analyzed waves and define the ECGA. Specifically, they define ECGA kind (see Fig. 2): if zero-derivative points are located over the P wave, ECGA is defined as PWA; if they are located over the QRS complex, ECGA is defined as QRSA; and, if they are located over the T wave, ECGA is defined as TWA; otherwise it can't be evaluated as an actual ECGA. In case of presence, ECGA amplitude (ECGAA), particularly discriminated as PWA amplitude, QRSA amplitude or TWA amplitude, is computed basing on local AS amplitude (ASA), otherwise it is considered noise and then ECGAA is set to zero.

Similarly, ECGA absence is defined in case AMF provides a zero-amplitude AS, i.e. a constant output signal. An example to better appreciate the difference between the two kinds of output AS can be seen in Fig. 3. In case of ECGA absence, AMF gives a

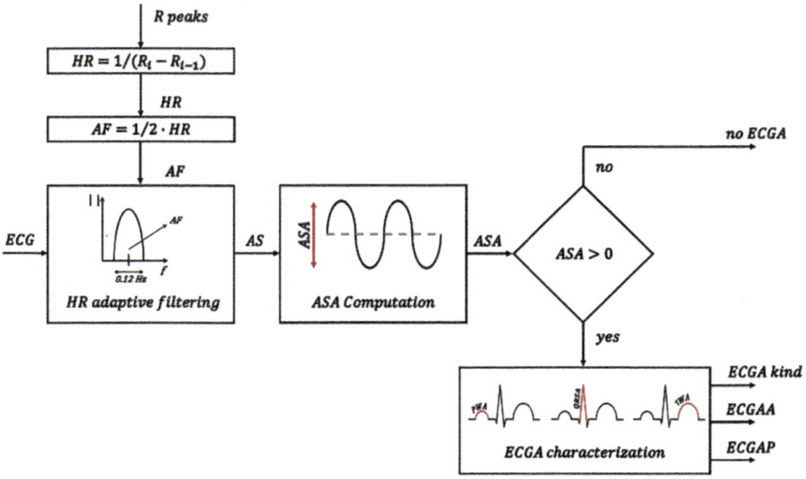

Fig. 1. Block diagram of the heart-rate adaptive match filter (AMF) method adapted to detect any kind of ECG alternans (ECGA). HR: heart rate; Ri: generic R peak; ECG: electrocardiogram; AF: ECGA frequency; AS: alternans signal; ASA: AS amplitude; ECGAA: ECGA amplitude; ECGAP: ECGA position.

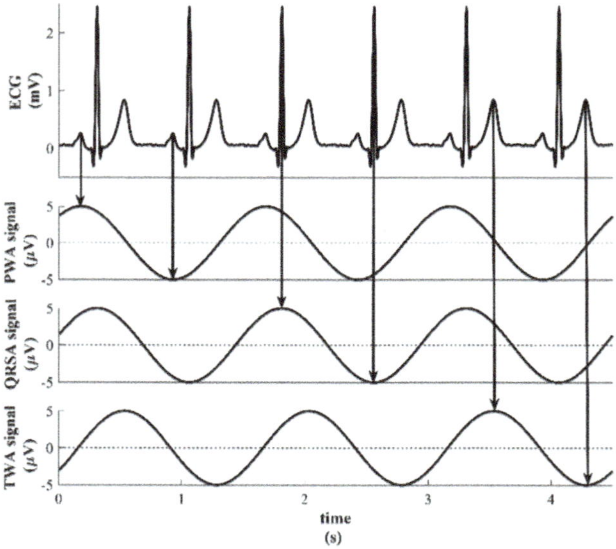

Fig. 2. Alternans signal (AS) in case of P-wave alternans (PWA), QRS alternans (QRSA) and T-wave alternans (TWA) classification.

constant output signal (panel a), while in case of ECGA evidence it gives an amplitude-modulate sinusoid (panel b). Specifically, if simulated alternans is constant for all beats of the input ECG, there is not any modulation.

Finally, in case of presence of ECGA the position of AS maximum inside the analyzed wave is computed with respect to the relative R peak and it is considered as the detected ECGA position (ECGAP): positive values indicate an ECGA following the R peak and vice versa.

3 Results

Results on ECGA quantification by AMF are shown in Table 1: ECGAA and ECGAP were reported for each synthetic ECG signal. Errors computing ECGAA were evaluated as percentage ratio between the estimated ECGAA and the simulated one. In case of ECGA absence, AMF correctly identified and quantified it without errors; ECGAP could not be evaluated. In case of ECGA presence, AMF also correctly identified and discriminated it; ECGAP was estimated. In case of one single alternating wave, results showed percentages of error computing ECGAA that were null in quite all cases: specifically, it is at most 1% in case of 200 μV-amplitude PWA (on average 0%). Moreover, in these cases ECGAP was located around the alternating wave middle part. In case of mixed alternating waves, errors computing ECGAA raised. Its value was always overestimated, with a global mean error of 42%. Specifically, when TWA was combined to PWA or QRSA, only TWA was identified with an overestimation of its amplitude (mean error: 23%); when PWA and QRSA were combined it is PWA to be identified, with an overestimation of its amplitude (mean error: 71%). Moreover, ECGAP was not in the middle of the wave classified as alternating but moved towards one of the wave sides: ECGA was usually defined as TWA and located in the first part of the T wave (this does not happen only in case of complete absence of alternans simulated over the T wave).

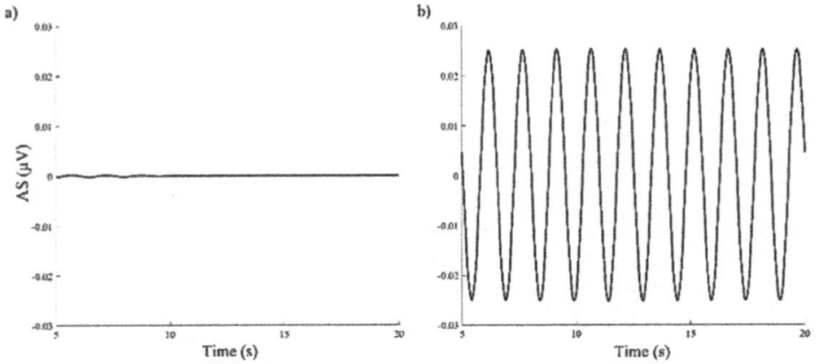

Fig. 3. Example of 15 s output alternans signal (AS) relative to a non-alternating (panel a) and an alternating input electrocardiogram (panel b).

Table 1. Results on simulated data by adaptive match filter (AMF) method adapted to detect any kind of ECG alternans (ECGA).

Alternating amplitude (μV)	Alternating wave	ECG kind					
		PWA		QRSA		TWA	
		ECGAA (μV)	ECGAP (ms)	ECGAA (μV)	ECGAP (ms)	ECGAA (μV)	ECGAP (ms)
0	–	0	–	0	–	0	–
10	T	–	–	–	–	10	+200
	QRS	–	–	10	−1	–	–
	QRS-T	–	–	–	–	13	+149
	P	10	−146	–	–	–	–
	P-T	–	–	–	–	12	+100
	P-QRS	17	−95	–	–	–	–
	P-QRS-T	–	–	–	–	15	+100
100	T	–	–	–	–	100	+200
	QRS	–	–	100	0	–	–
	QRS-T	–	–	–	–	128	+149
	P	100	−145	–	–	–	–
	P-T	–	–	–	–	116	+100
	P-QRS	172	−95	–	–	–	–
	P-QRS-T	–	–	–	–	151	+100
200	T	–	–	–	–	200	+200
	QRS	–	–	200	0	–	–
	QRS-T	–	–	–	–	257	+149
	P	201	−145	–	–	–	–
	P-T	–	–	–	–	232	+100
	P-QRS	343	−95	–	–	–	–
	P-QRS-T	–	–	–	–	301	+100

4 Discussion

This study considered ECGA as the prevalent alternans affecting one of the major ECG waves. In this perspective, AMF method was applied to estimate TWA, PWA or QRSA, after identification of the ECG wave that alternates most.

Results showed AMF method is specific, being perfectly able to recognize and highlight ECGA absence. Moreover, it also proves to be actually very sensitive to recognize TWA in all cases it is present.

The method is based on a very small band filtering, so the output is a sinusoid whose absolute peaks (i.e. local zero-derivative points) correspond to the points located in the center of the distribution of ECGA for each beat, i.e. on its barycenter [18]. In case of singular-wave alternans AMF correctly identified ECGA kind (PWA, QRSA, or TWA) depending on the barycenter position and then quantified quite without errors ECGAA and ECGAP, whatever the wave that is oscillating. For the same reason in case of multiple-wave alternans AMF tends to identify ECGA as TWA if T wave effectively alternates together with other waves. Indeed, ECGA kind definition and quantification are based on the barycenter position and are influenced by the alternating wave length, since simulated alternans always has the same morphology and amplitude. In other words, if T wave alternates, being the longest among the three ECG considered intervals it includes barycenter, so AMF defines ECGA as TWA. In the other considered situations, if TWA is not present, between P wave and QRS complex, P wave was indicated as the longest, thus ECGA is tendentially defined as PWA. ECGAA is always overestimated in case of multiple-wave alternans because the fluctuation of the wave considered as the most oscillating is inevitably influenced by the fluctuation of the other/s and this evidence is confirmed by ECGAP that is no more in the center of the alternating wave but shifted towards the other fluctuating waves. ECGAA overestimation can be considered a negligible issue in this historical moment of ECGA knowledge because reference values discriminating physiologic and dangerous alternans do not exist still. Moreover, the threshold definition could be different for different oscillating waves. Future studies will evaluate a wider dataset to confirm the results of this preliminary study.

ECGA, and especially TWA, is recognize in literature as a meaningful marker for arrhythmic risk in a variety of pathological and critical situations [1, 2, 7–9] The importance of detecting different kinds of alternans and especially the prevalent alternating wave among all the possible ones on the one hand widens the perspective on alternans on the other widens the possibility to investigate different anomalies in the cardiac electrical function. Indeed, a particular kind of ECGA can reveal disfunction in specific part of the heart and in a specific moment of its activity reflecting in the AP and so on the ECG morphology.

In conclusion, AMF proved to be an effective and specific method to both reveal and discriminate ECGA.

Conflict of Interest. All authors have no financial and personal relationships with other people or organizations that could inappropriately bias the work.

References

1. Burattini, L., Man, S., Fioretti, S., Di Nardo, F., Swenne, C.A.: Dependency of exercise-induced T-wave alternans predictive power for the occurrence of ventricular arrhythmias from heart rate. Ann. Noninvasive Electrocardiol. **20**(4), 345–354 (2015)
2. Adam, D.R., Smith, J.M., Akselrod, S., Nyberg, S., Powell, A.O., Cohen, R.J.: Fluctuations in T-wave morphology and susceptibility to ventricular fibrillation. J. Electrocardiol. **17**(3), 209–218 (1984)
3. Bigger, J.T., Bloomfield, D.M.: Microvolt T-wave alternans: an effective approach to risk stratification in ischemic cardiomyopathy?: Commentary. Nat. Clin. Pract. Cardiovasc. Med. **4**(6), 300–301 (2007)
4. Siniorakis, E., Arvanitakis, S., Tzevelekos, P., Giannakopoulos, N., Limberi, S.: P-wave alternans predicting imminent atrial flutter. Cardiol. J. **24**(6), 706–707 (2017)
5. Maury, P., Metzger, J.: Alternans in QRS amplitude during ventricular tachycardia. Pacing Clin. Electrophysiol. **25**(2), 142–150 (2002)

6. Burattini, L., Man, S., Swenne, C.A.: The power of exercise-induced T-wave alternans to predict ventricular arrhythmias in patients with implanted cardiac defibrillator. J. Healthc. Eng. **4**(2), 167–184 (2013)
7. Burattini, L., Bini, S., Burattini, R.: Repolarization alternans heterogeneity in healthy subjects and acute myocardial infarction patients. Med. Eng. Phys. **34**(3), 305–312 (2012)
8. Marcantoni, I., Cerquetti, V., Cotechini, V., Lattanzi, M., Sbrollini, A., Morettini, M., Burattini, L.: T-wave alternans in partial epileptic patients. In: 45th Computing in Cardiology Conference, Maastricht, pp. 1–4 (2018). https://doi.org/10.22489/cinc.2018.043
9. Man, S., De Winter, P.V., Maan, A.C., Thijssen, J., Borleffs, C.J.W., Van Meerwijk, W.P.M., Bootsma, M., Van Erven, L., Van Der Wall, E.E., Schalij, M.J., Burattini, L., Burattini, R., Swenne, C.A.: Predictive power of T-wave alternans and of ventricular gradient hysteresis for the occurrence of ventricular arrhythmias in primary prevention cardioverter-defibrillator patients. J. Electrocardiol. **44**(4), 453–459 (2011)
10. Armoundas, A.A., Tomaselli, G.F., Esperer, H.D.: Pathophysiological basis and clinical application of T-wave alternans. J. Am. Coll. Cardiol. **40**(2), 207–217 (2002)
11. Hopenfeld, B.: Mechanism for action potential alternans: the interplay between L-type calcium current and transient outward current. Heart Rhythm **3**(3), 345–352 (2006)
12. Brembilla-Perrot, B., Lucron, H., Schwalm, F., Haouzi, A.: Mechanism of QRS electrical alternans. Heart **77**(2), 180–182 (1997)
13. Burattini, L., Bini, S., Burattini, R.: Automatic microvolt T-wave alternans identification in relation to ECG interferences surviving preprocessing. Med. Eng. Phys. **33**(1), 17–30 (2011)
14. Burattini, L., Zareba, W., Burattini, R.: Adaptive match filter based method for time vs. amplitude characterization of microvolt ECG T-wave alternans. Ann. Biomed. Eng. **36**(9), 1558–1564 (2008)
15. Burattini, L., Zareba, W., Burattini, R.: Automatic detection of microvolt T-wave alternans in Holter recordings: effect of baseline wandering. Biomed. Signal Process. Control **1**(2), 162–168 (2006)
16. Burattini, L., Bini, S., Burattini, R.: Comparative analysis of methods for automatic detection and quantification of microvolt T-wave alternans. Med. Eng. Phys. **31**(10), 1290–1298 (2009)
17. Morettini, M., Marchesini, L., Pettinari, L.A., Tigrini, A., Marcantoni, I., Sbrollini, A., Burattini, L.: TWA simulator: a graphical user interface for t-wave alternans. In: 45th Computing in Cardiology Conference, Maastricht, pp. 1–4 (2018). https://doi.org/10.22489/cinc.2018.044
18. Bini, S., Burattini, L.: Quantitative characterization of repolarization alternans in terms of amplitude and location: what information from different methods? Biomed. Signal Process. Control **8**(6), 675–681 (2013)

Co-activation of Knee Muscles in Female vs. Male Adults

Francesco Di Nardo(✉), Annachiara Strazza, Andrea Tigrini, Guido Mascia, Stefano Cardarelli, Alessandro Mengarelli, Federica Verdini, and Sandro Fioretti

Department of Information Engineering, Università Politecnica delle Marche, Via Brecce Bianche, 60131 Ancona, Italy
f.dinardo@staff.univpm.it

Abstract. During walking, knee joint mechanics is primarily regulated by thigh-muscle group, i.e. hamstrings and quadriceps femoris. Research purpose was to assess gender-related differences in concomitant recruitment of antagonist knee-joint muscles during ground walking. To this aim, Statistical gait analysis was performed on surface-electromyographic (sEMG) signals from vastus lateralis (VL) and medial hamstrings (MH) in 15 female (F-group) and 15 male (M-group) age-matched able-bodied young adults. sEMG signals from numerous strides (average value ± SD of 452 ± 102 strides for F-group and 440 ± 106 strides for M-group) were analyzed for each subject. Results showed that the same three VL/MH co-activations were found in the gait cycle, irrespective of gender: during early stance (ES), push-off (PO), and swing (SW) phase. No significant gender-related differences ($p > 0.05$) were observed in co-activity duration. Differently, an increase of occurrence frequency was observed in F-group for VL/MH co-activation during PO phase, with respect to M-group (21.9 ± 13.6% vs. 11.3 ± 8.6% of strides, $p = 2.5 \times 10^{-3}$). This increased occurrence of co-activations suggests a more complex muscular recruitment for knee-joint stabilization in females, in particular in PO phase when the control of balance is more awkward because of the final phase of single support. In conclusion, the present study indicates gender as a not negligible factor in evaluating knee-muscle co-activation during walking.

Keywords: Muscle co-activation · Statistical gait analysis · Surface EMG · Human walking

1 Introduction

Knee-joint stability is well-preserved by bones, cartilage, ligaments, tendons and muscles. Each one of these structures is susceptible to disease and injury, both during every day and sport activities [1]. During gait, knee joint mechanics is primarily regulated by thigh-muscle group, i.e. hamstrings and quadriceps femoris (QF). QF and hamstrings are considered antagonist muscle groups for knee joint during walking [2]. Muscular co-activation represents an important strategy of the neuro-muscular system for improving joint stiffness, postural stability and movement accuracy and also to prevent joint injury in dynamic condition [3–5]. In particular, co-activation activity of

J. Henriques et al. (Eds.): MEDICON 2019, IFMBE Proceedings 76, pp. 167–173, 2020.
https://doi.org/10.1007/978-3-030-31635-8_20

thigh muscles in able-bodied individuals during walking is a well-known feature of gait mechanics, acting on knee-joint stiffness and thus on the whole walking stability [6–8]. A recent study of the present group of researchers identified three different kind of co-activations among QF and hamstring muscles in young healthy subjects during walking: throughout weight acceptance, in mid-stance phase and during swing phase [3]. Co-activations during weight acceptance and during swing have been identified very frequently (in around 90% of the total strides), while co-activation during mid-stance has been reported as more infrequent (25% of the total strides). The quantitative estimation of gender effects on co-activation of thigh muscles during gait could be relevant for deepening the comprehension of knee-joint control and stabilization.

A wide literature focused on gender-related differences in temporal gait parameters [9, 10], kinematics [11–13] and EMG activity [13–15] during walking. However, poor information has been provided on influence of gender on muscle co-activation. A recent study reported an overall higher occurrence of ankle-muscle co-activation activity in females [16]. To our knowledge, no information on the effect of gender on muscle co-activation at knee level is available in literature. However, it could be useful for discriminating physiological from pathological behavior typical of the female (or male) population. For example, thigh-muscle co-contractions seems to play a role in the higher female rates of knee sprain and should be likely considering also in evaluating the female increased risk of anterior cruciate ligament injuries [1].

Thus, the objective of the present study was the assessment of possible gender-related differences in the concomitant recruitment (i.e. co-activation) of QF and hamstring muscles in young able-bodied subjects, during walking at self-selected speed and cadence. To this aim, medial hamstrings (MH) were considered as typical of hamstring group. Vastus lateralis (VL) was examined as the representative muscles for QF group.

2 Materials and Methods

2.1 Real Data

Gait data from thirty healthy volunteer subjects were analyzed in this study. Two age-matched groups were considered: M-group and F-group. Characteristics of participants (mean ± SD) were reported in Table 1. Exclusion criteria: necessity of aided walking, neuromuscular diseases, disorders in balance and previous surgical intervention at ankle and/or foot. The present research has been undertaken following the ethical principles of the Helsinki Declaration and was approved by local ethical committee.

Table 1. Anthropometric characteristics and temporal gait parameters of male (M-group) and female (F-group) subjects. Values are mean ± SD. NS = $p > 0.05$. Body mass index (BMI) is defined as the ratio of body weight to the square of height.

Group	Age (years)	Height (cm)	Weight (Kg)	BMI (kg•m^{-2})
M-group (n = 15)	24.2 ± 2.3	180 ± 6	72.4 ± 8.5	22.1 ± 1.7
F-group (n = 15)	23.5 ± 1.3	164 ± 3	51.8 ± 3.2	19.1 ± 0.8
Statistical significance	NS	$p < 0.01$	$p < 0.01$	$p < 0.01$

2.2 Signal Acquisition and Processing

Participants were bilaterally instrumented with surface EMG (sEMG) probes and three foot-switches. Foot-switches were placed beneath the heel and the first and fifth metatarsal heads of the foot; each sensor had a surface of 1.21 cm^2 and an activation force equal to 3 N. Fixed and variable geometry sEMG probes were placed on MH and VL, respectively, following the SENIAM recommendations for electrode location-orientation over muscles [17]. Characteristics of the probes were: Ag/Ag-Cl disk; electrode diameter 0.4 cm; inter-electrode distance 0.8 cm; gain 1000, input impedance > 1.5 GΩ, CMRR > 126 dB for fixed geometry, and Ag/Ag-Cl disks, minimum interelectrode distance: 12 mm, gain: 1000, high-pass filter: 10 Hz, input impedance > 1.5 GΩ, CMRR > 126 dB, input referred noise $\leq$ 200 nV_{rms} for variable geometry. After positioning the sensors, subjects were asked to walk barefoot over the floor for 5 min at their natural pace, back and forth over a 12-m long straight track. Signals were acquired with a sampling rate of 2 kHz and a resolution of 12 bit, through Step32 system (Medical Technology, Italy),

Footswitch signals were debounced, converted to four levels, Heel contact (H), Flat foot contact (F), Push-off (P), Swing (S), and processed to segment and classify the different gait cycle (GC) [18]. sEMG signals were band-pass filtered (FIR filter, 100 taps, 20–450 Hz) for removing motion artifact and high frequency noise and treated by a double-threshold statistical detector, in order to identify myoelectric activation intervals. Briefly, this detector is based on the observation of m successive samples once the signal exceeded the first threshold §; if at least r_0 (second threshold) out of successive m samples are still above §, activation is acknowledged. Detailed algorithm description is reported by Bonato et al. [19]. The analysis of sEMG pattern was achieved by means of the so-called Statistical gait analysis [20]. The approach consisted in characterizing normal gait on the basis of averaging spatial-temporal and sEMG parameters over hundreds of strides. This methodology relies on the cycle-dependency of muscle-activation during walking. Thus, sEMG parameters are averaged only over those cycles including the same number of activations, i.e. over each single activation modality. The activation modality defines the number of times a muscle activates during a single gait cycle: the n-activation modality consists of n active intervals for the considered muscle during a single gait cycle. Onset and offset instants were computed to identify each activation interval during a single gait cycle. Then, muscle activations were grouped according to the number of active intervals, i.e. relative to their activation modality. Onset/offset time instants of each activation modality were averaged over the total strides and then over the whole population. Averaged onset/offset percentage time instants were normalized for gait cycle, in order to limit the effect of the differences detected in the anthropometric parameters (Table 1). This kind of analysis allowed quantifying muscles co-activations also in terms of the occurrence frequency. The occurrence frequency [20] was computed as the number of strides (in percentage) when the muscle is recruited with each specific activation modality.

2.3 Statistics

All data are reported as average and standard deviation (SD). Shapiro-Wilk test was used to assess the normality of data distributions. Student's t-test T was used to compare the two gender groups for normally distributed samples, whereas Kruskal-Wallis test was used in case of not normally distributed samples. Statistical significance was set at 5%.

3 Results and Discussion

Numerous strides were studied for each subject: a mean value (± standard deviation, SD) of 440 ± 106 strides (M-group) and 452 ± 102 strides (F-group) was analyzed. Data were presented as average value over strides from right and left lower limbs. Mean (± SD) duration of H, F, P and S phases was computed. For M-group, H-phase lasted 6.4 ± 3.6% of GC, F-phase 26.6 ± 6.5%, P-phase 24.2 ± 5.1%, and S-phase 42.8 ± 2.9%. For F-group H-phase lasted 6.3 ± 2.3%, F-phase 26.1 ± 6.3%, P-phase 24.0 ± 4.1%, and S-phase 43.5 ± 3.0% of GC. No differences ($p > 0.05$) were observed in gait-phase duration between groups.

SGA of the myoelectric signal indicated that several different activation modalities were adopted for of VL and MH. For each muscle, every subject adopted the same three main activation modalities, covering about 90% of strides, irrespective of gender. Results on concomitant recruitment of thigh muscles showed that the same VL/MH co-activations were found in both groups: during the early stance (ES), push-off (PO) and swing phase (SW). To clarify the meaning of muscles co-activation, example of surface EMG signals of a single subject from VL/MH co-activations were reported in Fig. 1.

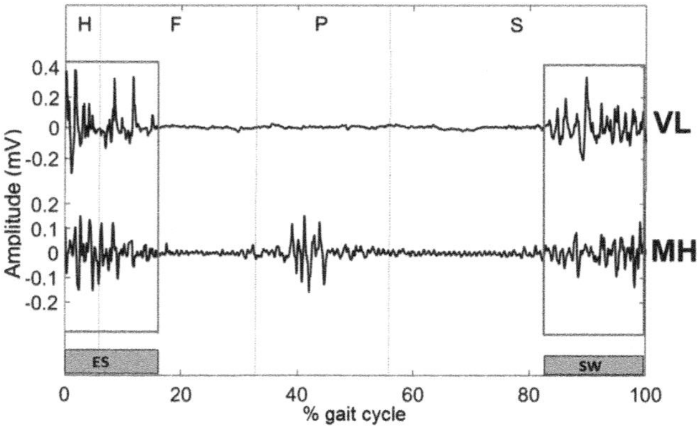

Fig. 1. Examples in a representative subject of ES (Early Stance) and SW (Swing) co-activations between VL and MH. Superimpositions between sEMG signals are highlighted by a box. H, F, P and S phases are delimited by vertical dashed lines.

Table 2. Onset/offset for the three co-activation intervals for the two populations. Values are expressed for female and male population as mean ± standard deviation (SD) in percentage of gait cycle.

VL/M co-activation	Early stance		Push-off		Swing	
	ON	OFF	ON	OFF	ON	OFF
M-group (n = 15)	0.51 ± 0.3	14.0 ± 3.2	37.1 ± 9.7	43.8 ± 10.6	81.1 ± 0.1	99.3 ± 0.6
F-group (n = 15)	0.90 ± 0.2	15.3 ± 3.4	34.9 ± 2.7	46.0 ± 2.4	82.0 ± 0.2	99.1 ± 0.1

Co-activations in ES and SW phase were highlighted with a box. Small differences were found between groups in onset/offset instants of co-activation for VL/MH muscles; details were reported in Table 2. However, no significant differences ($p > 0.05$) were observed in co-activity duration. This meant that the picture of sEMG temporal parameters of co-activations is similar between groups, indicating the absence of relevant gender-related issues. These findings suggested a similar functional purpose of thigh muscle co-activations for males and females. Occurrence frequency of every co-activation was reported in Fig. 2, differentiated between males and females. The analysis of occurrence frequency showed that F-group adopted significantly higher occurrence-frequency values in VL/MH co-activation during PO phase (21.9 ± 13.6% vs. 11.3 ± 8.6%, $p = 0.0025$, Fig. 2). No further gender differences were detected in occurrence frequency of the other co-activations.

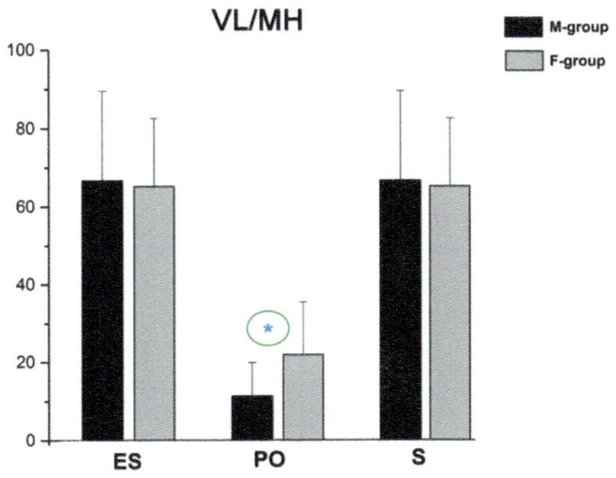

Fig. 2. Occurrence frequency of VL/MH co-activations in the two populations. Data are expressed as percentage of total strides. Black bars represent the frequency of co-activation occurrence in the M-group and grey bars represent the frequency of co-activation occurrence in the F-group. * indicates that differences between F and M groups are statistically significant ($p < 0.05$).

The increased occurrence of VL/MH co-activations in PO phase (Fig. 2) seemed to suggest a more complex recruitment of knee-muscle in females, in line with [14, 16]. In particular, this phenomenon occurred during PO phase of gait, when the control of balance is more awkward because of the final phase of single support. It seemed to reflect a female needing for a higher level of knee-joint stabilization, as reported for ankle joint [16].

4 Conclusion

The present study indicates gender as a not negligible factor in the assessment of muscular co-activation at knee and suggests the necessity of a gender-based approach in developing co-activation reference frameworks, such as the present one. A reference frame for knee muscles co-activations in women could be a useful tool for discriminating physiological from pathological behavior typical of the female population.

References

1. Da Fonseca, S.T., Silva, P.L.P., Ocarino, J.M., et al.: Analyses of dynamic co-contraction level in individuals with anterior cruciate ligament injury. J. Electromyogr. Kinesiol. **14**, 239–247 (2004)
2. Kellis, E.: Quantification of quadriceps and hamstring antagonist activity. Sport Med. **25**, 37–62 (1998)
3. Strazza, A., Mengarelli, A., Fioretti, S., et al.: Surface-EMG analysis for the quantification of thigh muscle dynamic co-contractions during normal gait. Gait Posture **51**, 228–233 (2017)
4. Olney, S.J.: Quantitative evaluation of cocontraction of knee and ankle muscles in normal walking. In: Biomechanics IX: International Congress of Biomechanics, pp. 431–436 (1983)
5. Rosa, M.C.N., Marques, A., Demain, S., et al.: Methodologies to assess muscle co-contraction during gait in people with neurological impairment - a systematic literature review. J. Electromyogr. Kinesiol. **24**, 179–191 (2014)
6. Schmitz, A., Silder, A., Heiderscheit, B., et al.: Differences in lower-extremity muscular activation during walking between healthy older and young adults. J. Electromyogr. Kinesiol. **19**, 1085–1091 (2009)
7. Hortobágyi, T., Solnik, S., Gruber, A., et al.: Interaction between age and gait velocity in the amplitude and timing of antagonist muscle coactivation. Gait Posture **29**, 558–564 (2009)
8. Frost, G., Dowling, J., Dyson, K., Bar-Or, O.: Cocontraction in three age groups of children during treadmill locomotion. J. Electromyogr. Kinesiol. **7**, 179–186 (1997)
9. Bohannon, R.W.: Comfortable and maximum walking speed of adults aged 20–79 years: reference values and determinants. Age Ageing **26**, 15–19 (1997)
10. Oberg, T., Karsznia, A., Oberg, K.: Basic gait parameters: reference data for normal subjects, 10–79 years of age. J. Rehabil. Res. Dev. **30**, 210–223 (1993)
11. Kerrigan, D.C., Todd, M.K., Della Croce, U.: Gender differences in joint biomechanics during walking: normative study in young adults. Am. J. Phys. Med. Rehabil. **77**, 2–7 (1998)
12. Hurd, W.J., Chmielewski, T.L., Axe, M.J., et al.: Differences in normal and perturbed walking kinematics between male and female athletes. Clin. Biomech. **19**, 465–472 (2004)
13. Chumanov, E.S., Wall-Scheffler, C., Heiderscheit, B.C.: Gender differences in walking and running on level and inclined surfaces. Clin. Biomech. **23**, 1260–1268 (2008)

14. Di Nardo, F., Mengarelli, A., Maranesi, E., et al.: Gender differences in the myoelectric activity of lower limb muscles in young healthy subjects during walking. Biomed. Signal Process. Control **19**, 14–22 (2015)
15. Chiu, M.C., Wang, M.J.: The effect of gait speed and gender on perceived exertion, muscle activity, joint motion of lower extremity, ground reaction force and heart rate during normal walking. Gait Posture **25**, 385–392 (2007)
16. Mengarelli, A., Maranesi, E., Burattini, L., et al.: Co-contraction activity of ankle muscles during walking: a gender comparison. Biomed. Signal Process. Control **33**, 1–9 (2017)
17. Hermens, H.J., Freriks, B., Disselhorst-Klugm, C., Rau, G.: Development of recommendations for SEMG sensors and sensor placement procedures. J. Electromyogr. Kinesiol. **10**, 361–374 (2000)
18. Agostini, V., Balestra, G., Knaflitz, M.: Segmentation and classification of gait cycles. IEEE Trans. Neural Syst. Rehabil. Eng. **22**, 946–952 (2014)
19. Bonato, P., D'Alessio, T., Knaflitz, M.: A statistical method for the measurement of muscle activation intervals from surface myoelectric signal during gait. IEEE Trans. Biomed. Eng. **45**, 287–299 (1998)
20. Di Nardo, F., Mengarelli, A., Strazza, A., et al.: A new parameter for quantifying the variability of surface electromyographic signals during gait: the occurrence frequency. J. Electromyogr. Kinesiol. **36**, 25–33 (2017)

Automatic Segmentation of Bipolar EHGs' Contractions Using Wavelet Transform

Amer Zaylaa[1,4](✉), Ahmad Diab[2], Ziad Fawal[3], Mohamad Khalil[4], and Catherine Marque[1]

[1] CNRS UMR 7338, BMBI, Sorbonne Universités, Université de Technologie de Compiègne (UTC), Compiègne, France
zaylaa.amer@gmail.com, Catherine.Marque@utc.fr
[2] Faculty of Public Health, Doctoral School for Sciences and Technology, Lebanese University (UL), Tripoli, Lebanon
ahmaddiab@ul.edu.lb
[3] Faculty of Sciences, Lebanese University (UL), Tripoli, Lebanon
fawal@ul.edu.lb
[4] Faculty of Engineering, Doctoral School for Sciences and Technology, Lebanese University (UL), Tripoli, Lebanon
mohamad.khalil@ul.edu.lb

Abstract. Until recently, many segmentation research trials on uterine EMG have been recorded for the sake of not only automatic detection of contractions but also curtailment of other events presents in the electrohysterogram (EHG). In this study, we use an online segmentation method, which has proven its efficiency and known by Dynamic Cumulative Sum (DCS). The method is first applied on real bipolar EHGs signals then on their details obtained by wavelet transform. The detected rupture instants are driven through an enhanced technique of faulty rupture instants elimination, dynamic selection of multichannel bipolar EHG signals and its details after wavelet transform and event tracking algorithm. Therefore, enhanced method sensitivity and "other events" rate of bipolar EHGs with and without wavelet are computed using Margin validation test in order to classify among events as contractions or not referring to contractions identified by expert. Indeed, enhanced technique of event tracking, proposed in this study, seems to be more efficient comparing to previous techniques. Further studies should be achieved for the sake of fully identifying the uterine contractions from other events and then decreasing the "other events" rate.

Keywords: Multichannel uterine EMG signal · Dynamic cumulative sum · Automatic segmentation · Wavelet transform

1 Introduction

Over the last two decades, the world made substantial progress in reducing mortality among children and young adolescents. Still, in 2017 alone, an estimated 2.5 million deaths occurring in the first month of life, 1.6 million at age 1–11 months [1]. In fact, almost 99% of these deaths are related to complications such as Respiratory Distress

J. Henriques et al. (Eds.): MEDICON 2019, IFMBE Proceedings 76, pp. 174–183, 2020.
https://doi.org/10.1007/978-3-030-31635-8_21

Syndrome (RDS) and intraventricular hemorrhage which are almost a distinctive characteristic of preterm birth [1, 2].

For the recent past, pregnancy and labor uterine contractions were usually monitored by using tocodynamometer, which could monitor the deformation made by the uterus on the abdomen. However, this system is not a reliable technique since the obtained measurements are not fully precise and depend to a large extent on the subjective criteria of the operator [3]. To address this issue, Electrohysterogram (EHG) is widely used to study the electrical activity of the uterus and progressed from dual channel to multichannel study for the sake of obtaining high spatial resolution either in mapping [4] or in features extraction of EHG contractions recorded from such a complex organ [5].

For a proper recording of uterine EMG the electrodes should be positioned along or as near as possible to the median vertical axis, in order to get the better signal-noise ratio (SNR) [5] and to ease the EHGs segmentation.

Many algorithms and methods were addressed to track ruptures changes in non-stationary signals such as Neyman–Pearson algorithm [6], Innovation Whiteness [7], Brandt algorithm [8], Hinkley Divergence Test [9], Cumulative Sum method (CuSum) [10], Dynamic Cumulative Sum method (DCS) [11, 12], nonlinear correlation coefficient (H2) [13].

Implementing a method that could segment contractions automatically from uterine EMG help improve online pregnant woman uterus monitoring which could assist in early preterm birth detection.

In our current study, we have applied the Dynamic Cumulative Sum (DCS) method on bipolar EHG signals opposite to our previous study where monopolar EHGs have been processed [14]. Indeed, bipolar derivations improve signal quality by reduction of the common mode noise [15]. After that, we have applied the DCS on details of bipolar EHG obtained by wavelet in order to take from the signals only the component that includes ruptures and that help us in the segmentation.

The novelty of this study is related to the addition of an elimination stage of faulty rupture instants and the use of a robust event tracking technique, after temporal projection of all instants detected from different channels on the bipolar channel which presents the highest signal to noise ratio. Finally, a comparison of obtained sensitivity and "other events" rate is computed by using Margin validation test, which is detailed in [14]. Indeed, a higher sensitivity and lower "other events" rate have been obtained by applying DCS with implemented techniques on bipolar EHGs without wavelet decomposition.

2 Materials and Methods

2.1 Data

By using a standardized protocol [16], with uterine EMG measurements' approval by the relevant ethical committee (VSN 02-0006-V2) in Iceland and by the regional ethical committee (ID-RCB 2011-A00500-41) of Amiens Hospital in France, EHG signals have been first collected from 36 records of pregnant women by placing an array of 16 electrodes arranged in a 4 × 4 matrix positioned on the woman's abdomen and two reference electrodes on each hip (Fig. 1a), digitized with 200 Hz as sampling frequency then processed to get bipolar EHG signals, in order to get a correct SNR: vertical bipolar signals (BP1..., BP12) were computed as shown in Fig. 1b. For the sake of removing frequencies related to respiration or fetal and maternal movements, a Butterworth filter with cutoff frequencies at 0.3 and 5 Hz [15] is employed prior to apply our segmentation method.

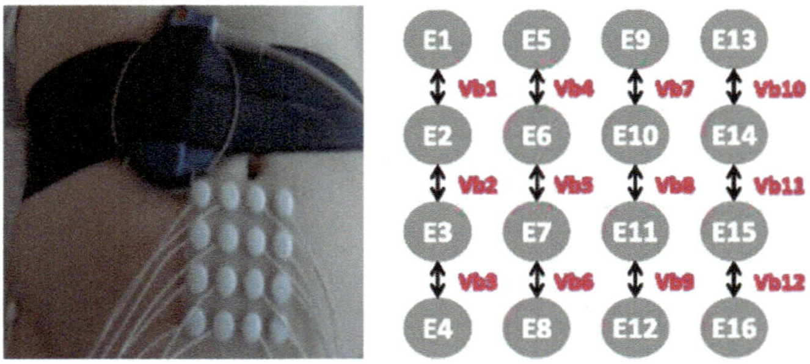

(a) (b)

Fig. 1. (a) 4 × 4 matrix electrodes position with tocodynamometer and reference electrodes (b) Electrodes configuration (Bipolar in red color and monopolar in black color) on the women's abdominal wall.

2.2 Dynamic Cumulative Sum Method

A change in a signal may affect variance, frequency distribution or both at the same time. The dynamic cumulative sum method is a constructed method for detecting changes in signals in the case where the segmentation parameters are unknown and when someone wants to follow local changes in the signals.

This detection approach, called the Dynamic Cumulative Sum (DCS), can be considered as a repeated sequence around the point of change k. It is based on the local cumulative sum of the likelihood ratios between two local hypotheses estimated around the current instant j. These two dynamic hypotheses H_a^j and H_b^j (a: after, b: before) are estimated using two windows of length N, before and after the instant j respectively as follows:

$$\mathrm{H_b^j}: x_i;\ i = \{j - N + 1, \ldots, j\} \text{ follows a density probability distribution } f_{\theta_b}(x_i).$$
$$\mathrm{H_a^j}: x_i;\ i = \{j + 1, \ldots, j + N\} \text{ follows a density probability distribution } f_{\theta_a}(x_i).$$

The parameters of the hypothesis H_b^j and $\hat{\theta}_b^j$ are estimated from N points before the instant j, and from N points after the instant j for the hypothesis H_a^j and $\hat{\theta}_b^j$.

At time j, we define DCS as the sum of the logarithms of the likelihood ratios from the beginning of the new segment of the signal to the instant j:

$$DCS\left(H_a^j, H_b^j\right) = \sum_{i=1}^{j} \ln \frac{f_{\hat{\theta}_a^i(x_i)}}{f_{\hat{\theta}_b^i(x_i)}} = \sum_{i=1}^{j} \hat{s}_i \tag{1}$$

where $\hat{s}_i$ is the logarithm of the likelihood ratio to a local character. The parameters of the two hypotheses are re-estimated at each step in the two windows of N points around the current point j.

It has been demonstrated in [11] that the DCS function reaches its maximum at the time of change k.

The detection function used to estimate the instant of change is expressed by:

$$g_j = \max_{1 \leq i \leq j} \left[DCS(H_a^j, H_b^j) - DCS(H_a^j, H_b^j)\right] \tag{2}$$

The stop time is: $t_s = \text{INF}\ \{j : g_j \geq h\}$; where h is being a fixed threshold.

By applying the DCS method, a series of detected ruptures are obtained. All detected ruptures from all 12 bipolar EHG signals are subject to removing the faulty ruptures by applying a Fisher test algorithm between each 2 consecutives segments in each bipolar channel as shown in Fig. 2. Indeed, a test decision (H) is then computed. The test decision is the null hypothesis if the data in segment 1 and segment 2 come from normal distributions with the same variance. 'H' becomes 0 when the test could not reject the null hypothesis with a 5% significance level (Fig. 2a). Therefore, the detected instant separating 2 consecutive segments could be eliminated (Fig. 2b). The alternative hypothesis is that they come from normal distributions with different variances. 'H' becomes then 1 as shown in Fig. 2a when the test rejects the null hypothesis with a 5% significance level. Hence, the detected instants separating the 2 consecutive segments could not be eliminated (Fig. 2b).

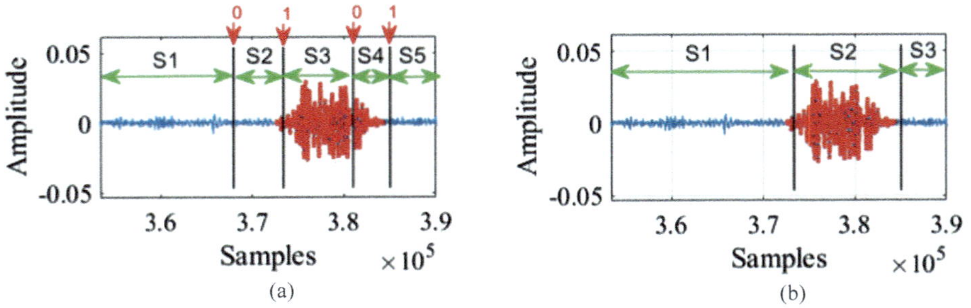

Fig. 2. Computation of decision test H by using fisher test between two consecutive segments separated by detected instant of change (a) detected ruptures before fisher (b) remaining ruptures after fisher.

2.3 Event Tracking Technique

After removing faulty detected instants by using Fisher test algorithm, we obtain new segments for each bipolar signals (Fig. 2b). To begin our event tracking technique, we use the signal to noise ratio detection algorithm (SNRDA) that consists of three steps, starting by computing the signal to noise ratio (SNR) of each new segment. Secondly, we plot the obtained SNR of each channel and then find the SNR peaks (Fig. 3b). Therefore, we obtain segments with higher SNR. Third, we compare the SNR values of each selected segment (higher SNR) to its neighbors (segment after and before). If the neighbor segment SNR value exceeds the half SNR value of the SNR peak, a concatenation of the neighbor segment with the segment of SNR peak is then applied (Fig. 3c). After applying the above steps on each bipolar channel, we project on the time axis of the channel with the highest SNR the detected instants from all bipolar channels. Next, we apply SNRDA on the projected instants in order to track events in the highest SNR channel. Finally, the tracked events undergo Margin Validation test developed previously in order to compare them with contractions identified by expert [14]. Hence sensitivity and other events rate are computed.

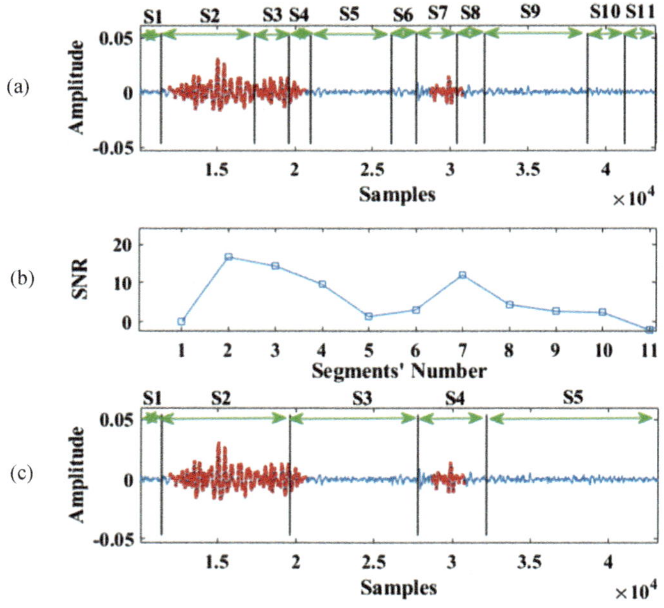

Fig. 3. (a) Obtained segments 'Sx' after applying fisher test algorithm (b) SNR computation of obtained segments (c) Obtained segments 'Sx' after applying signal to noise ratio detection algorithm.

2.4 Dynamic Selection of Wavelet Transform Details

Wavelet and Scaling Level Choice

Frequency distribution is considered an essential element for distinguishing and identifying events in the EHG signal since frequency content of contractions changes from the one of baseline and artifacts. Indeed, the wavelet decomposition permits to extract the useful information from the frequency content.

In this case any change in frequency will result in a change on one or more levels of details.

The continuous wavelet transform of a signal x(t) takes the form:

$$T_x^{\varphi}(a,b) = \int x(t)\varphi_{ab}(t)dt \tag{3}$$

Each signal can be decomposed into details and approximations, and the shape of the scale function is defined by:

$$\varnothing_{\mathrm{mn}}(\mathrm{t}) = 2^{\mathrm{m}/2}\varnothing(2^{\mathrm{m}}\mathrm{t} - \mathrm{n}) \tag{4}$$

where m indicates the scales, n indicates the translation in time.

The choice of the analyzing wavelet and scaling level is a critical problem in wavelet theory. To address this issue, we have referred to [17], where symlet 5 and 5 levels were selected as the nest wavelet decomposition for EHG. Hence the DCS method was applied on the 5 details signals obtained from bipolar EHGs. However, a dynamic selection of details of interest for our analysis is required.

Kullback Leibler Distance

In our study, a dynamic selection of details signals has been applied by using the Kullback Leibler distance which is based on the likelihood principle [18].

$$\widehat{\mathrm{K}_{12}} = \frac{1}{2}\left[\log\left(\frac{\widehat{\alpha_2}}{\widehat{\alpha_1}}\right)^2 + \left(\frac{\widehat{\alpha_2}}{\widehat{\alpha_1}}\right)^2 - 1\right] \tag{5}$$

$$\text{with } \hat{\alpha} = \sqrt{\left(\frac{2}{\mathrm{L}}\sum\nolimits_{\mathrm{i}=1}^{\mathrm{i}=\mathrm{L}} \mathrm{x}_{\mathrm{i}}^2\right)} \tag{6}$$

$$KLD = \widehat{\mathrm{K}_{12}} + \widehat{\mathrm{K}_{21}} \tag{7}$$

The theoretical distribution estimated by histogram of KLD follows an exponential distribution, which is a special case of the gamma distribution, depending only on one parameter λ. Its probability density function is defined as:

$$f(x) = \lambda e^{-\lambda x} \tag{8}$$

Experimental and theoretical distribution of KLD are represented in Fig. 4a. We compare both distributions using Kolmogorov-Smirnov (K-S) test [19]. The K-S statistic D_{max} is the maximum difference between the theoretical and the experimental cumulative distribution. Therefore, we define two hypotheses as:

HK = 0 if $\widehat{K}$ follows an exponential distribution as shown in Fig. 4.

HK = 1 if $\widehat{K}$ does not follow the exponential distribution as shown in Fig. 5.

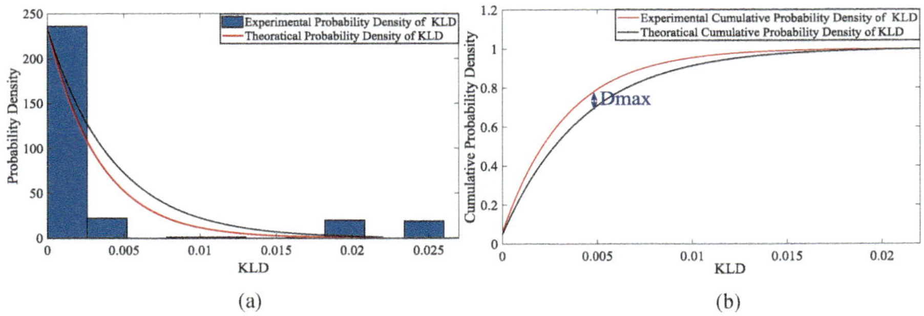

Fig. 4. (a) Histogram of normalized kullback leibler distance (b) Cumulative sum of normalized kullback leibler distance of detail with HK = 0.

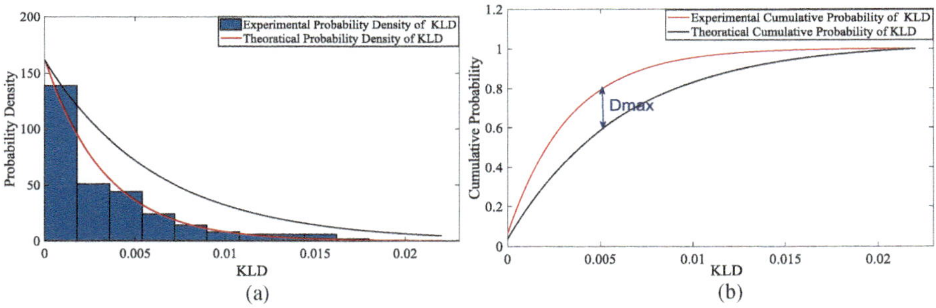

Fig. 5. (a) Histogram of normalized kullback leibler distance (b) Cumulative sum of normalized kullback leibler distance of detail with HK = 1.

Once we obtain the decision test HK for each bipolar channel detail, the sum of all HK of same detail for all bipolar channels and each recording has been computed. If the sum did not exceed 6, hence this detail was rejected and kept otherwise.

3 Results

By applying DCS method followed by the novel implemented technique of event tracking using SNRDA, we notice an enhancement of event detection, indicated by gray rectangles while contractions identified by expert are represented by red color (Fig. 6). The implemented method could detect 15 contractions over 15 plus one other event. By averaging the sensitivity and other events rate of all 36 records, we have obtained a 100% as sensitivity average, 39.77% as other events rate average and 55.05 as precision average on bipolar EHGs. On the other hand, we have obtained 97.27% as sensitivity average, 43.73% as other events rate average and 39.61% as precision average on details EHG signals by applying wavelet decomposition. These results correspond to 204 events identified as totally detected contractions, 185 partially detected and 395 other events when applying our method and techniques on bipolar EHGs, and 94 events identified as totally detected contractions, 280 partially detected and 759 other events by applying them on details signals after wavelet decomposition of these bipolar EHGs.

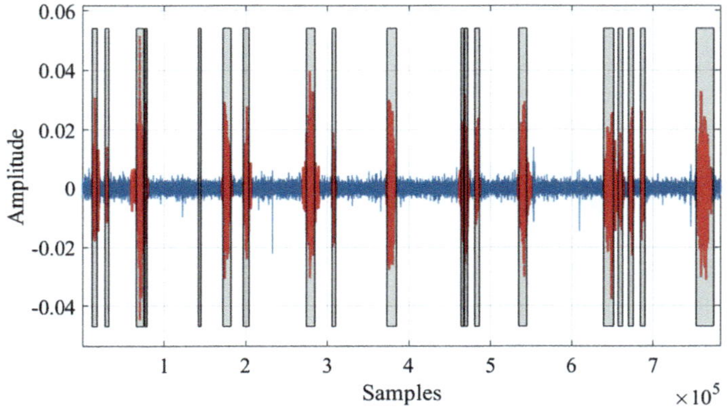

Fig. 6. Contraction and other event detection using DCS followed by SNRDA.

4 Discussion and Conclusion

In the current study, we have presented a comparison of sensitivity and "other events" rate obtained by applying DCS method directly on bipolar EHGs or on their details after wavelet decomposition. In addition, a novel implemented event tracking technique has been described and had proven its efficiency: increase to 100% in sensitivity and decrease to 39.77% in "other events" rate compared to our previous study in [14].

Indeed, in this work, the DCS method was followed by a first step removing faulty ruptures using Fisher test and SNRDA algorithm with dynamic selection of bipolar EHG signal when projecting obtained ruptures. In addition, we have applied a dynamic selection of details signals after wavelet decomposition of bipolar EHGs by using Kullback Leibler distance technique.

Applying our method directly on bipolar EHG seems to be more efficient than applying it on details of same bipolar EHG regarding the decrease in sensitivity average, totally and partially identified contractions and the increase in other events number after wavelet decomposition.

5 Perspectives

We are looking forward to apply the above explained techniques on bipolar EHGs with a multidimensional approach and on a larger database of bipolar EHGs.

Acknowledgement. This research is supported by Lebanese University and Al Koura Hospital.

References

1. Hug, L., Sharrow, D., Zhong, K., You, D.: Levels & Trends in Child Mortality. United Nations Inter-Agency Group for Child Mortality Estimation (UN IGME). UNICEF, New York (2017)
2. Oza, L.S., Hogan, D., Chu, Y., Perin, J., Zhu, J.: Global, regional, and national causes of under mortality in 2000-15: an updated systematic analysis with implications for the Sustainable Development Goals. Lancet **388**(10063), 3027–3035 (2016)
3. Miles, A.M., Monga, M., Richeson, K.S.: Correlation of external and internal monitoring of uterine activity in a cohort of term patients. Am. J. Perinatol. **18**(3), 137–140 (2001)
4. Diab, A., Hassan, M., Laforet, J., Alexandersson, A., Karlsson, B., Marque, C.: Analysis of connectivity strength and direction of simulated and 20 real multielectrode EHG signals. Med. Biol. Eng. Comput. (2014)
5. Alamedine, D., Diab, A., Muszynski, C., Karlsson, B., Khalil, M., Marque, C.: Selection algorithm for parameters to characterize uterine EHG signals for the detection of preterm labor. Signal Image Video Process. 1–10, June 2014
6. Eadie, W.T.: Statistical Methods in Experimental Physics. North-Holland, p. 224 (1971)
7. Basseville, M., Benveniste, M.: Sequential segmentation of non-stationary digital signals using spectral analysis. Inf. Sci. **29**, 57–73 (1983)
8. Brandt, A.V.: Detecting and estimating parameters jumps using ladder algorithms and likelihood ratio test. In: Proceedings of ICASSP, Boston, MA, pp. 1017–1020 (1983)
9. Basseville, M., Benveniste, M.: Sequential detection of abrupt changes in spectral characteristics of digital signals. IEEE Trans. Inf. Theory **29**(9), 709–724 (1983)
10. Basseville, M., Nikiforov, I.: Detection of Abrupt Changes: Theory and Application. Prentice-Hall, Englewood Cliffs (1993)
11. Khalil, M., Duchêne, J.: Une approche pour la détection fondée sur une somme cumulée dynamique associée à une décomposition multi échelle. Application à l'EMG utérin. Dix-septième Colloque Gretsi sur le traitement du signal et des images, Vannes, France (1999)
12. Zaylaa, A., Diab, A., Khalil, M., Marque, C.: Multichannel EHG segmentation for automatically identifying contractions and motion artifacts. In: 2017 International Conference on Advances in Biomedical Engineering (ICABME) IEEE EMBS, Beirut, Lebanon, 19–21 October 2017
13. Muszynski, C., Happillon, T., Azudin, K., Tylcz, J.B., Istrate, D., Marque, C.: Automated electrohysterographic detection of uterine contractions for monitoring of pregnancy: feasibility and prospects. BMC Pregnancy Childbirth **18**, 136 (2018)
14. Zaylaa, A., Diab, A., Khalil, M., Marque, C.: Automatic segmentation of contractions and other events in monopolar EHGs-monodimensional study. ACIT 2018 IEEE 978-1-7281-0385-3/18 (2018)
15. Sammali, F., Kuijsters, N.P.M., Schoot, B.C., Mischi, M., Rabotti, C.: Feasibility of transabdominal electrohysterography for analysis of uterine activity in nonpregnant women. Sage J. **25**(7), 1124–1133 (2018)

16. Karlsson, B., Terrien, J., Gudmundsson, V., Steingrimsdottir, T., Marque, C.: Abdominal EHG on a 4 by 4 grid: mapping and presenting the propagation of uterine contractions. In: 11th Mediterranean Conference on Medical and Biological Engineering and Computing, Ljubljana, Slovenia, pp. 139–143 (2007)
17. Diab, M.O., Marque, C., Khalil, M.: Une approche de classification des contractions utérines basée sur la théorie des ondelettes et la statistique. Leba. Sci. J. **7**(1) (2006)
18. Chendeb, M., Khalil, M., Duchêne, J.: The use of wavelet packets for event detection. In: Proceedings of EUSIPCO, Antalya, Turkey (2005)
19. Press, W.H., Flannery, B.P., Teukolsky, S.A., Vetterling, W.T.: Numerical Recipes in C. Cambridge University Press, New York (1988)

Methods for Removing of Line Noise Artifact from EEG Records with Minimization of Neural Information Loss

Jan Strobl[1,2](✉), Marek Piorecky[1,2], Vlastimil Koudelka[2], Tomas Nagy[1], and Vladimir Krajca[1,2]

[1] Department of Biomedical Technology, Faculty of Biomedical Engineering, Czech Technical University in Prague, Nam. Sitna 3105, 272 01 Kladno, Czech Republic
jan.strobl@fbmi.cvut.cz

[2] National Institute of Mental Health, Topolova 748, 250 67 Klecany, Czech Republic
http://en.fbmi.cvut.cz

Abstract. Line noise is artifact affecting a gamma band of the EEG. Conventional filters are used for removing the line noise artifact, but these filters, also remove physiological activity and could induce spurious oscillations. We tested two alternative methods for removing line noise artifact without removing a physiological activity, specifically Adaptive noise cancellation (ANC) method with Linear regression and modified Independent Component Analysis (ICA). We proposed a new protocol for statistical evaluation of line noise removal effectiveness. The line noise artifact was physically simulated and the protocol for statistical evaluation was verified. The ANC method shows good results according to statistical evaluation. The lowest residuum of simulated line noise artifact corresponds to 0.0005 $\mu V^2/Hz$. Nevertheless, the ANC is very sensitive to the quality of the reference signal, which heavily depends on the reference electrode selection.

Keywords: Line noise artifact · EEG · ICA · ANC · Linear regression · Statistical evaluation

1 Introduction

EEG (Electroencephalography) is an important modality for functional examination of a brain. EEG signal is divided into bands with specific frequencies and typical amplitudes of the signal. Recently, the study of gamma band has become important [4]. Typically, gamma band reaches frequencies between 30 and 80 Hz [4]. EEG signal is often affected by signals that do not come from the brain (so-called artifacts). Line noise artifact occupies frequency within Gamma band specifically at 50 Hz (60 Hz in the USA) and its harmonics [4,6,10].

A large number of studies removed line noise artifact by conventional IIR or FIR filter (for example studies [1,6]). However, EEG signal has nonstationary

J. Henriques et al. (Eds.): MEDICON 2019, IFMBE Proceedings 76, pp. 184–192, 2020.
https://doi.org/10.1007/978-3-030-31635-8_22

nature and using these filters could induce spurious oscillations (so-called ringing) and distort phase of the gamma oscillations [8]. In addition, conventional filters removing all signals within the determined bands (including neural information). In this paper we investigate methods which remove line noise and preserve neural information. In [11] the ICA was used for removing line noise artifact. Another option is using Adaptive noise cancelation (ANC) method [8,10]. Here, we focus on evaluation and comparison of these two commonly used methods.

Statistical evaluation is problematic for the removal of the artifacts because of neural information in the background of artifact isn't known [8]. To the best of our knowledge, a large number of studies don't compute specificity (indicates whether the neural information has been removed), but using only visual or subjective evaluation of the expert (for example [9,11]). We created a validation dataset to compare the line noise removing methods from this reason.

The aim of this study is a comparison of the methods removing line noise artifact with minimizing of the neural information lose using a new statistical evaluation approach.

2 Methods

2.1 Datasets

At first, we developed a validation dataset containing three subjects to identify the best performing method. Then we further evaluated the selected method on the resting state EEG dataset containing 73 subjects.

Validation Dataset. Original measuring of the validation dataset was executed for statistical evaluation. We simulated line noise artifact with frequency 40 Hz. This frequency was chosen to distinguish between simulated line noise and real line noise with frequency 50 Hz. Simulated line noise was generated by sound amplifier connected to a one meter diameter coil lying on the ground below the tested subject. The complete EEG measuring system was placed always on the same place inside the coil (including operational amplifier). The coil produced inside itself an electromagnetic field with characteristics similar to the line noise at frequency 40 Hz.

All measurements were 8 min long resting states with open eyes fixed on the fixed cross. Simulated line noise was applied in 2 min, then followed by 2 min without simulated line noise and this protocol was then repeated one time in the same measuring. In this way, we obtained a baseline of the physiological EEG signal. EGI HD EEG system with 256 electrodes and one ECG channel were used for measuring the validation dataset.

EEG Resting State Dataset. 36 EEG channels with a deployment according to standard system 10–10, ECG, EOG and EMG channels were measured for verifying the results of the best line noise removing method. All measurements were 8 min long resting state with 4 min with open eyes and 4 with closed eyes. Subjects were healthy volunteers.

2.2 Preprocessing

All EEG signals were equally preprocessed. EEG signals were downsampled from 1000 Hz to 500 Hz. An two pass FIR bandpass filter with a cut off frequency 0.5–80 Hz was used. Signals from validation datasets were split into two parts: without simulated line noise (C - control) and with simulated line noise (N - noise). All data processing was performed by a MATLAB R2015a and a FiledTrip toolbox.

2.3 Adaptive Noise Cancellation Method

Adaptive noise cancellation (ANC) methods measure a reference signal which represents line noise. We used channels from mastoids (validation dataset) or from ear lobes (resting state EEG dataset) as reference signal. In one EEG records from the validation dataset was signal from mastoids of poor quality. We used electrode number 230 (placed on cheekbone) like the ANC reference from this reason (see Fig. 4).

FIR filter was applied to obtain only frequency of line noise. We used a Multiple linear regression model in this study for adaptation of the filtering signal (more information in [3]). The filtered reference signal and its shifted version (shift 90°) are entered into the linear regression similarly like in study [8]. Then the adapted reference signal is subtracted from each EEG channel [8, 10].

2.4 ICA for Removing Line Noise

Independent component analysis (ICA) creates statistical independent components (sources of the signal) [2]. An artifact can be suppressed by removing components representing this artifact. We used the Infomax ICA method in this study (for more information see [2, 5]).

ICA creates components which aren't sorted, so it is necessary every time identified components containing artifact [5]. We designed a method for automatic identification of the ICA components containing line noise artifact. We have identified components which have the highest power spectrum in a line noise frequency to compare to others frequencies of the power spectrum. So we are removing components whose power in the frequency of the artifact is the highest with compare to the power in frequencies probably coming from the subject's brain. Parameter upper decile is used for detection of a number of removing components (because dataset hasn't normal distribution).

2.5 Data Evaluation

We compared the power spectrum after removing the line noise by any methods with a power spectrum of a control part "C" of a signal (part of the signal with simulated line noise off). Power spectrum is computed at artifact frequency (40 or 50 Hz). The mean of the obtained difference for all channels in one validation record was computed. If this computed mean is negative, the method removes artifact and part of the physiological signal. If this computed mean is positive, the method does not remove complete line noise artifact.

The winning method was applied to line noise removal in resting state EEG records. Finally, reliability of the method was tested and visually reviewed in all 73 subjects (see Fig. 5).

3 Results

3.1 Verification of Our Validation Protocol

In this section results verifying the correctness of our new protocol are presented. In Fig. 1 can be seen comparing the normalized spectrum of the control part of the measuring signal and a normalized power spectrum of part of the measuring signal with the simulated line noise artifact. This figure depicts whether a control part of the signal can be used for statistical evaluation.

Figure 2 shows the topographic deployment of the simulated and original line noise artifact for all measuring subjects. This figure compares our simulated artifact with the original line noise artifact.

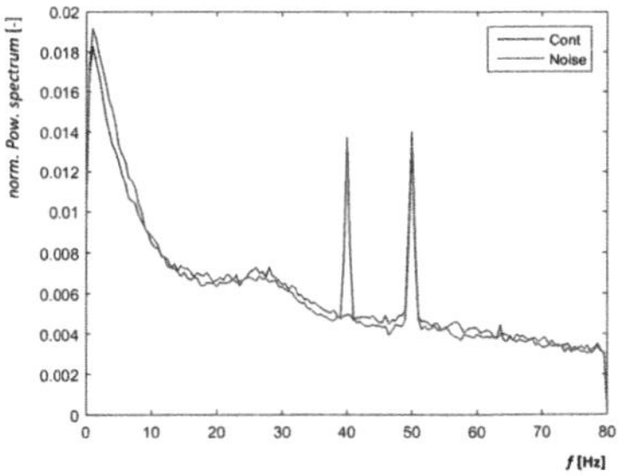

Fig. 1. Example of the normalized power spectrum of control part (black line) without simulated line noise artifact and for the noisy part (blue line) in subject 1. The normalized power spectrum is mean from all channels of the record.

3.2 Validation Dataset

Averaged difference across channels was computed between control part and noisy part (after artifact removal) of the signal at 40 Hz (see Table 1). In Fig. 3 the normalized power spectrum of noisy part before and after artifact removal is depicted.

The ANC method with Linear regression had the best results. Linear regression had a problem with a signal of subject 2. This problem was caused by bad signal of the reference electrode. In Fig. 4 is seen that this reference doesn't distinguish between turning on and off a simulated artifact. After using another electrode the results significantly improved (see Table 1 and Fig. 3).

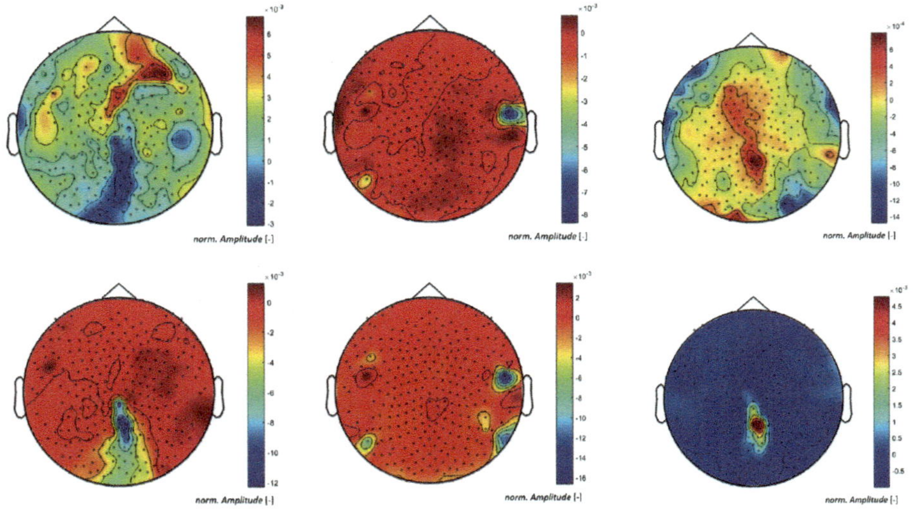

Fig. 2. Topographic maps of the simulated line noise artifact with frequency 40 Hz (up) original line noise artifact with frequency 50 Hz (bottom) for subject 1 (left), subject 2 (middle) and subject 3 (right) in the validation dataset.

Table 1. The mean and a standard deviation (Std.) of the difference between power spectrum control part and the noisy part after using tested methods (ANC with using a Linear regression and modified ICA method) for all records.

Record	Mean ANC [$\mu V^2/Hz$]	Std. ANC [$\mu V^2/Hz$]	Mean ICA [$\mu V^2/Hz$]	Std. ICA [$\mu V^2/Hz$]
Subject 1	0.0005	0.0057	0.0011	0.0056
Subject 2	0.0013	0.0058	0.0199	0.0454
Subject 3	0.0005	0.0008	0.0121	0.0229

3.3 EEG Resting State Dataset

Removing of the line noise artifact of Linear regression method was then tested on original line noise in a resting state EEG records. Figure 5 depicts averaged result of this analysis across 73 records.

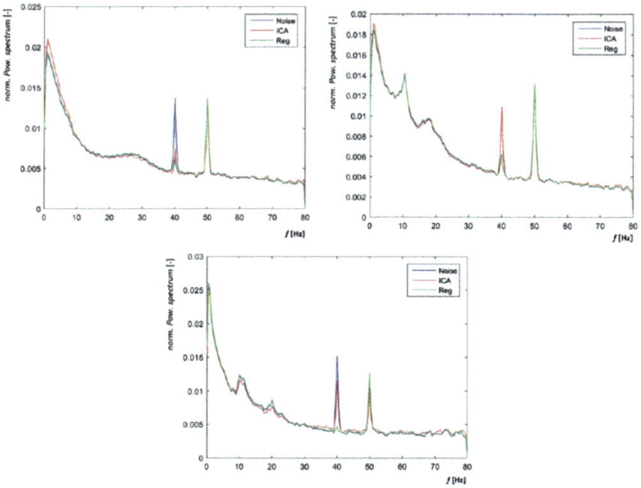

Fig. 3. The normalized power spectrum of noisy part before using artifact removing methods (blue line), after using ANC method (green line) and modified ICA (red line) in subject 1 (top left), subject 2 (top right) and subject 3 (bottom). The normalized power spectrum is mean from all channels of the record.

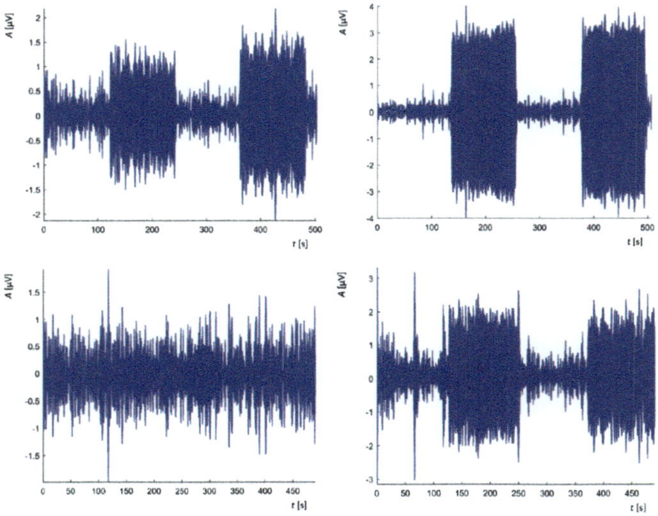

Fig. 4. A graph of the reference signal using for removing a line noise artifact by an ANC method. The reference signal from averaged mastoid electrodes was used in Subject 1 (top left), Subject 2 (down on the left) and Subject 3 (top right). The reference signal from electrode 230 was then used in Subject 2 (down on the right) for getting better results.

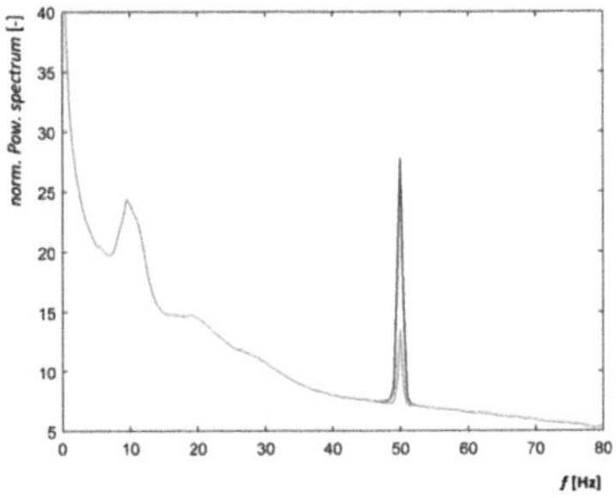

Fig. 5. The power spectrum of resting state EEG before (blue line) and after using ANC method (green line). The power spectrum is obtained like a mean of all channels in all records.

4 Discussion

In this study, we compare methods removing line noise artifact without removing the physiological activity of the brain. ANC with Linear regression and modified ICA were under the scope of this study. We used a new measuring protocol for statistical evaluation of the testing methods.

Physical simulation of the line noise artifact was provided in this protocol (see Sect. 2.1). Figure 1 compares the normalized power spectrum of control (without simulated line noise artifact) part of the signal and noisy (with the simulated line noise artifact) part of the signal in the subject 1.

Topographic deployment of the simulated and original line noise artifact is displayed in Fig. 2. According to these maps, we conclude that line noise artifacts are different in every measuring no matter that simulated artifact has always been equally generated and propagated in space. The electrical impedance of the electrode and deployment of the conductors has probably the main influence for the propagation of the line noise artifact in EEG signal. We can see in Fig. 2 that the power of the simulated and original line noise artifact is different, but they have a similar locations of the local maxima and minima in space.

We computed a difference between the power spectrum of the control part and the power spectrum after using testing methods on the noisy part of the signal. The ANC with Linear regression had significantly better results than the modified ICA method (see Table 1 and Fig. 1). Difference in the power spectrum was near the zero for the ANC method, only in subject 2 was $0.0013\,\mu V^2/Hz$ (18.6% compared to the original signal), modified ICA method had in the same subject the power spectrum $0.0199\,\mu V^2/Hz$ (66.3%). The best result had modified ICA in subject 1, specifically $0.0011\,\mu V^2/Hz$ (15.2%). ANC method had in same subject the power spectrum $0.0005\,\mu V^2/Hz$ (9.1%).

We further focused on the ANC with Linear regression due to the results. ANC method was very sensitive to the good reference signal. In the record of a subject 2, the reference signal didn't react on the turn on and off the simulated line noise artifact (see Fig. 4) and results of the method were insufficient. We changed the reference electrode and the results have improved significantly.

We tested the ANC method with Linear regression on the 73 EEG records with original line noise artifact. Results from Fig. 5 shows that the ANC method removed a large part of the artifact, but left behind a residue. The reason may be a bad reference signal in some records. In future work, we want to find out how to easily identify a bad reference signal.

In future work, we will test the ANC method with Linear regression also on other types of artifacts with typically narrow-band frequencies. For example, Nierhaus et al. alerted in their study [7] on the gradient artifact residue in gamma band after using artifact correction method during EEG/fMRI recording. We want to use the ANC method on higher harmonic components incurred by this artifact from simultaneous recording.

5 Conclusion

We suggested a new measuring protocol for statistical evaluation of removing the line noise artifact. We created simulated line noise artifact with a known characteristic and simulated artifact start time. The results show that the power spectrum of the physiological activity in control part of the signal can be used as a baseline for statistical analysis.

The statistical evaluation shows good result for an ANC method with Linear regression in removing the simulated line noise artifact compared with the modified ICA method. Good quality of the reference signal is essential for the ANC and a proper reference signal should be carefully chosen.

Acknowledgements. This work was supported by the Grant Agency of the Czech Technical University in Prague, registr. numb. SGS19/136/OHK4/2T/17, by the Grant Agency of the Czech Technical University in Prague, registr. numb. SGS18/159/OHK4/2T/17, by the Grant Agency of Czech Republic, register number 17-20480S, by the Grant Agency of Czech Republic, register number 19-15728S and project LO1611 with a financial support from the MEYS under the NPU I program.

Statements by Authors. The authors declare that there is no conflict of interest regarding the publication of this article. The study protocol and patient informed consent have been approved by National Institute of Mental Health ethical committee. The procedures followed were in compliance with the World Medical Association Declaration of Helsinki on Ethical Principles for Medical Research Involving Human Subjects.

References

1. Aspiras, T.H., Asari, V.K.: Analysis of spatio-temporal relationship of multiple energy spectra of EEG data for emotion recognition. In: Computer Networks and Intelligent Computing, pp. 572–581. Springer, Heidelberg (2011). https://doi.org/10.1007/978-3-642-22786-8_73

2. Delorme, A., Makeig, S.: EEGLAB: an open source toolbox for analysis of single-trial EEG dynamics including independent component analysis. J. Neurosci. Methods **134**(1), 9–21 (2004). https://doi.org/10.1016/j.jneumeth.2003.10.009
3. Chatterjee, S., Hadi, A.S.: Influential observations, high leverage points, and outliers in linear regression. Stat. Sci. **1**(3), 379–393 (1986). https://doi.org/10.1214/ss/1177013622
4. Hermann, C.S., Fründ, I., Lenz, D.: Human gamma-band activity: a review on cognitive and behavioral correlates and network models. Neurosci. Biobehav. Rev. **34**(7), 981–992 (2010). https://doi.org/10.1016/j.neubiorev.2009.09.001
5. Laglois, D., Chartier, S., Gosselin, D.: An introduction to independent component analysis: InfoMax and FastICA algorithms. Tutorials Quant. Methods Psychol. **6**(1), 31–38 (2010). https://doi.org/10.20982/tqmp.06.1.p031
6. Lee, S.-H., Kim, S., Shim, M.-S., Kim, D.-W., Im, C.-H.: Dysfunctional patterns of gamma-band activity in response to human faces compared to non-facial stimuli in patients with schizophrenia. Psychiatry Invest. **13**(3), 349–359 (2016). https://doi.org/10.4306/pi.2016.13.3.349
7. Nierhaus, T., Gundlach, C., Goltz, D., Thiel, S.D., Pleger, B., Villringer, A.: Internal ventilation system of MR scanners induces specific EEG artifact during simultaneous EEG-fMRI. NeuroImage **74**, 70–76 (2013). https://doi.org/10.1016/j.neuroimage.2013.02.016
8. Nottage, J.F.: Uncovering gamma in visual tasks. Brain Topogr. **23**(1), 58–71 (2010). https://doi.org/10.1007/s10548-009-0129-y
9. Shen, K., Yu, K., Bandla, A., Sun, Y., Thakor, N., Li, X.: Multiple time-lag canonical correlation analysis for removing muscular artifacts in EEG. In: 35th Annual International Conference of the IEEE Engineering in Medicine and Biology Society (EMBC), pp. 6792–6795 (2013). https://doi.org/10.1109/EMBC.2013.6611116
10. Wang, Z., Roe, A.W.: Trial-to-trial noise cancellation of cortical field potentials in awake macaques by autoregression model with exogenous input (ARX). J. Neurosci. Methods **194**(2), 266–273 (2011). https://doi.org/10.1016/j.jneumeth.2010.10.029
11. Zhou, W., Zhou, J., Zhao, J., Zhao, H., Ju, L.: Removing eye movement and power line artifacts from the EEG based on ICA. In: IEEE Engineering in Medicine and Biology 27th Annual Conference, vol. 6, pp. 6017–6020 (2005). https://doi.org/10.1109/IEMBS.2005.1615863

Pilot Study for Estimating Physical Fatigue Based on Heart Rate Variability and Reaction Time

Ardo Allik(✉), Kristjan Pilt, Moonika Viigimäe, and Ivo Fridolin

Tallinn University of Technology, Ehitajate tee 5, 19086 Tallinn, Estonia
ardo.allik@taltech.ee

Abstract. The aim of this study was to evaluate how heart rate variability (HRV) and reaction time differ when measured in the morning and in the evening of one physically exhausting day in order to explore if these parameters would be sufficient for estimating physical fatigue accumulated during one day. Five different experiment days with fixed schedules were conducted which consisted of measurement set in the morning, physical exercise during the day and measurement set in the evening. The total average reaction time was lower in the morning (228 ± 18 ms) compared to the values measured in the evening (257 ± 22 ms). Both assessed HRV parameters SDNN and RMSSD showed a tendency to decrease during the day (total averages were respectively 61 ± 6 ms and 40 ± 6 ms in the morning vs 37 ± 4 ms and 24 ± 3 ms in the evening). This decrease was more prominent in the heart rate recovery phase compared to the resting heart rate. The results of this study give promising results for new methods for estimating daily physical fatigue and could be a basis for multiple future studies.

Keywords: Heart rate variability · Heart rate recovery · Reaction time · Fatigue

1 Introduction

Fatigue has been used as a term to describe an altered physiological state that results in decreased mental or physical performance which could be caused by sleep loss, circadian changes or high workload [1, 2]. Furthermore, fatigue has been depicted as a state of being tired and weak which reduces arbitrary power output and work capacity, which may cause hazardous situations and also lead to various health problems [3].

Previous studies have found that physical fatigue increases reaction time [1, 3, 4]. Additionally, the extent of how much reaction time declined with physical fatigue has been shown to be dependent of the exercise type [3]. Since it is difficult to measure physical fatigue directly, reaction time could be used as a good reference value for estimating daily accumulated physical fatigue.

Physical fatigue has also been estimated by analysing heart activity. The features of heart activity such as heart rate and heart rate variability (HRV) have been found to be

J. Henriques et al. (Eds.): MEDICON 2019, IFMBE Proceedings 76, pp. 193–200, 2020.
https://doi.org/10.1007/978-3-030-31635-8_23

good indicators of physical fatigue and the assessment of cardiac parasympathetic and sympathetic regulation [1, 5]. HRV is assessed by examining the beat-to-beat variations in normal R-R intervals and has been previously used to estimate both vocational [2, 6, 7] and physical training [8] induced fatigue. Additionally, multiple studies have focused on analysing physical fatigue through heart rate [5, 9] and HRV [5, 10–12] during heart rate recovery phase after exercise. This has been done to estimate physical fatigue accumulated over longer work or training periods. To the best of the authors' knowledge, no study has previously explored if analysing HRV values could be sufficient for estimating accumulated physical fatigue during one day.

The aim of this study was to evaluate how HRV parameters (SDNN and RMSSD) and reaction time differ when measured in the morning and in the evening of one exhausting and physically demanding day.

2 Methods

2.1 Study Overview

As a pilot study, tests were conducted on one healthy male subject. The anthropometric parameters of the subject were: age 28 years, weight 69 kg and height 171 cm. As the aim of the study was to estimate how HRV parameters and reaction time change during one exhausting day, several independent experiment days were conducted where the subject followed a fixed schedule shown in Table 1. The main parts of one experiment day were heart rate and reaction time measurements in the morning (Measurement Set 1 or MS 1) and in the evening (Measurement Set 2 or MS 2) and an exercise session in between them.

The exercise session between MS 1 and MS 2 was done in order to induce fatigue and consisted of both aerobic and anaerobic exercises [3]. For aerobic exercise the subject cycled on a veloergometer constantly at 60 RPM, 60 W for exactly one hour. For anaerobic exercise the subject conducted sets of various gym exercises during one hour (pull-ups, dumbbell rows, dumbbell shrugs, bicep curls). The subject was previously familiar with the chosen gym exercises. There was at least a 2-h rest after gym exercises and before MS2.

Table 1. Schedule of the experiment day

Activity	Approximate time of the day
Waking up	9 AM
Measurement Set 1 (MS 1)	11 AM
Veloergometer cycling	1 PM
Gym training	3 PM
Measurement Set 2 (MS 2)	7 PM

2.2 Reaction Time Measurement

Exactly the same measurement set was done in the morning (MS 1) and in the evening (MS 2) of the experiment days. This measurement set consisted of two sets of reaction time tests and a veloergometer test between them.

Reaction time was measured using a PC program, where the subject had to release the space button when visually cued. The subject was seated comfortably in front of the monitor with their wrist resting on the table. When the subject pressed the button, a small white circle appeared on the middle of the screen. After a period of time varying between three to five seconds [3], the circle increased in size and the subject had to release the button and reaction time was measured. Reaction times were measured during a set of 10 repetitions, before which the subject did one warm-up test.

2.3 ECG Measurement

In-between the reaction time tests the subject completed a veloergometer test during which ECG signals were measured using a Shimmer3 ECG Unit (Shimmer Research, Dublin 11, Ireland). This test was set up to capture heart activity during resting and during heart rate recovery phases. The test schedule is shown in Table 2. Before the test the subject sat on the veloergometer saddle and waited calmly for at least three minutes. To measure the resting phase the test started with the subject sitting on the veloergometer saddle for five additional minutes. Afterwards the subject cycled at 60 RPM on three different power levels (60 W, 90 W, 120 W) for three minutes which were all followed by a five-minute recovery phase [10, 11].

Table 2. Schedule of the ECG measurement test

Activity	Approximate time of the day
Resting	5 min
Cycling @ 60 W	3 min
Recovery	5 min
Cycling @ 90 W	3 min
Recovery	5 min
Cycling @ 120 W	3 min
Recovery	5 min

2.4 HRV Parameter Calculation

HRV parameters were calculated based on the measured ECG signals in order to compare the heart activity between morning and evening measurements sets. In this study assessed parameters were time-domain measures SDNN (the standard deviation of all NN intervals) and RMSSD (the square root of the root mean square of the sum of all differences between successive NN intervals), which are two commonly used HRV parameters for analysing heart activity [13]. In this study these parameters were calculated based on a two-minute-long signal section during heart rate recovery (HRR) 'slow phase' [5], which was chosen between 2.5 min and 4.5 min after cycling. These sections are named HR0 (before cycling), HR@60 W (after cycling at 60 W), HR@90 W (after cycling at 90 W) and HR@120 W (after cycling at 120 W). Figure 1 shows the HR values of one testing day (calculated based on ECG signals) and the sections based on which HRV parameters were calculated.

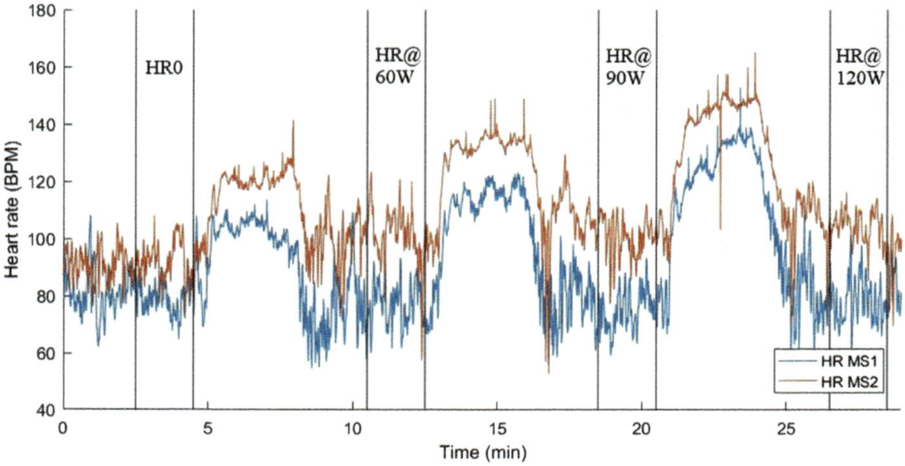

Fig. 1. Heart rate values for different measurement sets (MS 1, MS 2) and sections (HR0, HR@60 W, HR@90 W and HR@120 W) based on which HRV parameters were calculated during veloergometer test of one experiment day.

3 Results

Following figures show the changes in SDNN (Fig. 2) and RMSSD (Fig. 3) values compared to the changes in averaged reaction time values for measurement set conducted in the morning (MS 1) and measurement set conducted in the evening (MS 2) for all five experiment days and their average for different phases of the veloergometer test (HR0, HR@60 W, HR@90 W and HR@120 W) depicted in Fig. 1. The mean values (averaged over all five experiment days) are separately shown in Table 3.

Table 3. The HRV values for different veloergometer test phases and their average (mean ± SD, ms) and reaction time values (mean ± SD, ms) of MS 1 and MS 2 averaged over all experiment days.

Measurement set	HR0	HR@60 W	HR@90 W	HR@120 W	Average	Reaction time
MS 1, SDNN	53 ± 12	66 ± 20	63 ± 11	61 ± 24	61 ± 6	228 ± 18
MS 1, RMSSD	33 ± 7	43 ± 15	38 ± 12	33 ± 15	37 ± 4	
MS 2, SDNN	45 ± 2	44 ± 17	37 ± 12	34 ± 21	40 ± 6	257 ± 22
MS 2, RMSSD	27 ± 4	26 ± 10	24 ± 11	19 ± 12	24 ± 3	

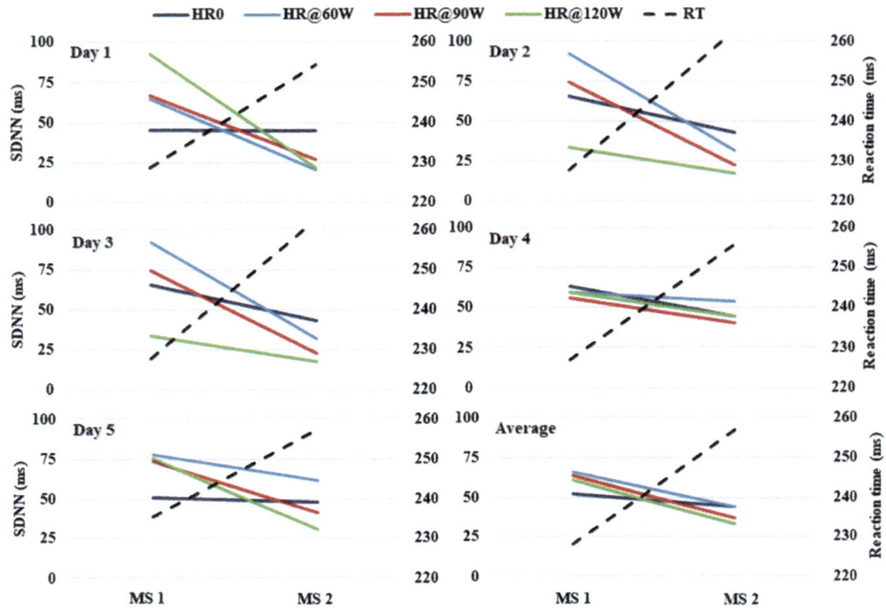

Fig. 2. Average SDNN (for different veloergometer test phases) and reaction time values of MS 1 and MS 2 for each experiment day and their average.

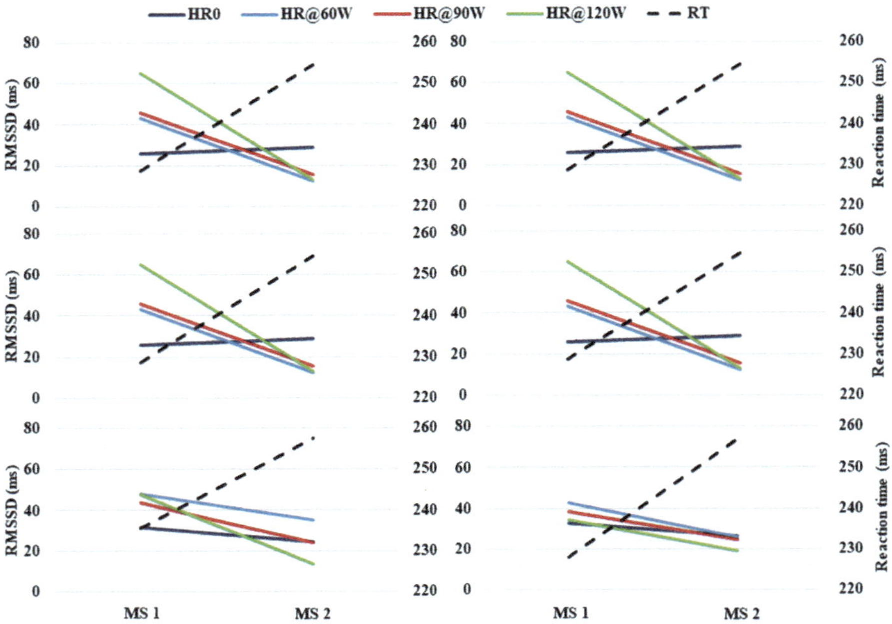

Fig. 3. Average RMSSD (for different veloergometer test phases) and reaction time values of MS 1 and MS 2 for each experiment day and their average.

4 Discussion

In this study it was analysed how reaction time and HRV parameters SDNN and RMSSD change during one exhausting experiment day by measuring them in the morning (MS 1) and in the evening (MS 2) in order to understand if and how these parameters could be used to estimate daily physical fatigue. The main findings of this study were: (1) There was an increase in reaction time between MS 1 and MS 2 during every experiment day (total average 228 ± 18 ms (MS 1) vs 257 ± 22 ms (MS 2)); (2) Both HRV parameters SDNN and RMSSD gave similar results and showed a tendency to decrease (total average 61 ± 6 ms (MS 1) vs 40 ± 6 ms (MS 2) for SDNN and 37 ± 4 ms (MS 1) vs 24 ± 3 (MS 2) for RMSSD); (3) The decrease in SDNN and RMSSD values between MS 1 and MS 2 was larger during heart rate recovery phase after cycling compared to the values measured during rest before cycling.

The reaction time measurement results were very similar for each experiment day. The total average increase in reaction time was about 30 ms between MS 1 and MS 2. Previous studies have similarly shown how physical fatigue causes slower reaction time [3, 4]. This shows that reaction time could be a good indicator and a reference for estimating daily physical fatigue.

To the best of the authors' knowledge no one has previously explored if HRV could be sufficient for estimating daily accumulated physical fatigue. The analysed HRV parameters SDNN and RMSSD showed similar tendencies in this study. Overall it could be noted that both SDNN and RMSSD decreased when comparing values between MS 1 and MS 2, although the values varied between different experiment days and on one experiment day (Day 3, Figs. 2 and 3) the values were higher in the evening compared to the morning. When looking at the HRV values averaged over different experiment days then there is a clear trend of decrease in both SDNN and RMSSD values when comparing MS 1 and MS 2. Similar decrease in HRV has been noted in previous studies where fatigue was accumulated over longer exhausting work [2, 6] or training periods [8]. Additionally, based on the results of this study the decrease in HRV values is more prominent during the HRR phases after cycling compared to the resting heart rate values before cycling. Although HRV is usually measured over longer time periods, then the methods explored in a previous study, where researchers analysed how to measure HRV post-exercise for the assessment of cardiovascular health [5], could also be useful for assessing physical fatigue.

As a pilot study there were some limitations which should be addressed. First, it is important to validate the results with a larger study group. Second, results should be compared to days without physical exercise in order to see how the values would differ. The results of this study could be basis for multiple future studies. The methods could be used to evaluate how the measured values change daily for the people constantly working in physically demanding occupations and whether they would be sufficient to estimate work induced physical fatigue. Additionally, HRV parameter values can be measured constantly in real-time based on PPG or ECG signals. Based on that it would be possible to create an automatic physical fatigue estimation system, where reaction time could be used as a reference data.

5 Conclusion

In this study it was evaluated how HRV parameters (SDNN and RMSSD) and reaction time differ when measured in the morning and in the evening of one exhausting experiment day. While the reaction time increased when comparing morning measurements to the evening measurements, the HRV parameters showed a trend of decreasing which was more prominent in the HRR phase compared to the resting phase. Based on the results of this study both reaction time and HRV values could be used to estimate accumulated physical fatigue during one day.

Compliance with Ethical Standards. The research was funded partly by the Estonian Ministry of Education and Research under institutional research financing IUTs 19-1 and 19-2 and by Estonian Centre of Excellence in IT (EXCITE) funded by European Regional Development Fund and supported by Study IT programme of HITSA.

The authors declare that they have no conflict of interest. All procedures performed in studies involving human participants were in accordance with the ethical standards of the institutional and/or national research committee and with the 1964 Helsinki declaration and its later amendments or comparable ethical standards. This article does not contain any studies with animals performed by any of the authors.

References

1. Mohanavelu, K., Lamshe, R., Poonguzhali, S., Adalarasu, K., Jagannath, M.: Assessment of human fatigue during physical performance using physiological signals: a review. Biomed. Pharmacol. J. **10**(4), 1887–1896 (2017)
2. Shortz, A.E., Franke, M., Kilic, E.S., Peres, S.C., Mehta, R.K.: Evaluation of offshore shiftwork using heart rate variability. In: Proceedings of the Human Factors and Ergonomics Society Annual Meeting, pp. 1036–1039. SAGE, Austin, TX, USA (2017)
3. Sabzi, A.H.: The effect of different fatigue protocols on choice reaction time. Middle-East J. Sci. Res. **12**(8), 1092–1096 (2012)
4. SantAna, J., Franchini, E., Silva, V., Diefenthaeler, F.: Effect of fatigue on reaction time, response time, performance time, and kick impact in taekwondo roundhouse kick. Sports Biomech. **16**(2), 201–209 (2017)
5. Pecanha, T., Bartels, R., Brito, L.C., Paula-Ribeiro, M., Oliveira, R.S., Goldberger, J.J.: Methods of assessment of the post-exercise cardiac autonomic recovery: a methodological review. Int. J. Cardiol. **227**, 795–802 (2017)
6. Patel, M., Lal, S.K.L., Kavanagh, D., Rossiter, P.: Applying neural network analysis on heart rate variability data to assess driver fatigue. Expert Syst. Appl. **38**, 7235–7242 (2011)
7. Pichot, V., Bourin, E., Roche, F., Garet, M., Gaspoz, J.-M., Duverney, D., Andoniadis, A., Lacour, J.-R., Barthelemy, J.-C.: Quantification of cumulated physical fatigue at the workplace. Pflügers Archiv – Eur. J. Physiol. **445**, 267–272 (2002)
8. Pichot, V., Roche, F., Gazpoz, J.-M., Enjolras, F., Antoniadis, A., Minini, P., Costes, F., Busso, T., Lacour, J.-R., Barthelemy, J.C.: Relation between heart rate variability and training load in middle-distance runners. Med. Sci. Sports Exerc. **32**(10), 1729–1736 (2000)
9. Bartels-Ferreira, R., de Sousa, E.D., Trevizani, G.A., Silva, L.P., Nakamura, F.Y., Forjaz, C. L.M., Lima, J.R.P., Pecanha, T.: Can a first-order exponential decay model fit heart rate recovery after resistance exercise? Clin. Physiol. Funct. Imaging **35**, 98–103 (2015)

10. Ng, J., Sundaram, S., Kadish, A.H., Goldberger, J.J.: Autonomic effects on the spectral analysis of heart rate variability after exercise. Am. J. Physiol. Heart Circ. Physiol. **297**, 1421–1428 (2009)
11. Goldberger, J.J., Le, F.K., Lahiri, M., Kannankeril, P.J., Ng, J., Kadish, A.H.: Assessment of parasympathetic reactivation after exercise. Am. J. Physiol. Heart Circ. Physiol. **290**, 2446–2452 (2006)
12. Buchheit, M., Papelier, Y., Laursen, P.B., Ahmaidi, S.: Noninvasive assessment of cardiac parasympathetic function: postexercise heart rate recovery or heart rate variability? Am. J. Physiol. Heart Circ. Physiol. **293**, 8–10 (2007)
13. Achten, J., Jeukendrup, A.E.: Heart rate monitoring, applications and limitations. Sports Med. **33**(7), 517–538 (2003)

Characterization of Eye Gaze and Pupil Diameter Measurements from Remote and Mobile Eye-Tracking Devices

Riccardo Lolatto(✉), Giulia Rocco, Riccardo Mustoni, Chiara Maninetti, Riccardo Pastura, Andrea Pigazzini, and Riccardo Barbieri

Electronic, Informatics and Bioengineering Department, Politecnico di Milano, Via Ponzio 34/5, 20133 Milan, Italy
riccardo.lolatto@polimi.it

Abstract. Eye-tracking technology allows to capture real-time visual behavior information and to provide insights about cognitive processes and autonomic function, by measuring gaze position and pupillary response to delivered stimuli. Over the recent years, the development of easy-to-use devices led to a large increase in the use of eye-tracking in a broad spectrum of applications, e.g. clinical diagnostics and psychological research. Given the lack of extensive material to characterize the performance of different eye-trackers, especially latest generation devices, the present study aimed at comparing a screen-mounted eye-tracker (remote) and a pair of wearable eye-tracking glasses (mobile). Seventeen healthy subjects were asked to look at a moving target on a screen for 90 s, while point of regard (POR) and pupil diameter (PD) were recorded by the two devices with a sampling rate of 30 Hz. First, data were preprocessed to remove artifacts, then correlation coefficients (for both signals) and magnitude-squared coherence (for PD) were calculated to assess signals agreement in time and frequency domain. POR measurements from remote and mobile devices resulted highly comparable ($\rho > 0.75$). PD showed lower correlation and major dispersion ($\rho > 0.50$), besides a higher number of invalid samples from the mobile device with respect to the remote one. Results provided evidence that the two instruments do share the same content at the level of information generally used to characterize subjects behavioral and physiological reactions. Future analysis of additional features and devices with higher sampling frequencies will be planned to further support their clinical use.

Keywords: Eye tracking · Remote eye-tracker · Mobile eye-tracker · Eye gaze · Pupil diameter

1 Introduction

Inspection of the world through vision is crucial to gather information necessary to individuals for visual perception. Moreover, eyes and their movement are known to play a key role in the expression of a person's desires, needs, cognitive processes, emotional states, and interpersonal relations [1]. Hence, visual behavior recording and

J. Henriques et al. (Eds.): MEDICON 2019, IFMBE Proceedings 76, pp. 201–208, 2020.
https://doi.org/10.1007/978-3-030-31635-8_24

measurement takes on particular importance. Eye-tracking is a recently devised technology which allows to capture visual behavior information in real time and obtaining gaze position within stimuli [2]. Specifically, it can provide moment-to-moment measures of the focus of attention underlying cognitive processes, i.e. eye gaze measures, and evaluation of the autonomic function, i.e. pupillary response [3].

Traditionally, eye-tracking technologies required invasive and uncomfortable techniques through supplemental components, such as scleral search coils [4] or electro-oculography [5]. Due to the physical constraints on the subjects, these techniques were confined to specific experimental conditions, thus limited in use. The need to overcome these limitations together with the increase in computational power and graphic boards of standard PCs led to the development of video-based measurements by means of infrared cameras, which are increasingly less intrusive and more accurate [6]. Such devices can be remote (table- or screen-mounted) or mobile, i.e. following the subject's natural head movement (head- or eyeglass-mounted). The evolution to easy-to-use eye tracking technologies is providing devices suitable for a wide range of applications and fields, e.g. psychological research [7], biomarker identification for clinical diagnostics [8], medical decision-making [9].

Given the broad spectrum of applications and further implications, it is extremely important to assess the accuracy with which the eye movement and/or pupil size are measured. Even though some previous studies gave a hint about the possible occurrence of measurement differences among eye-tracking devices [10, 11], there is no extensive material to characterize and compare performance of different eye-trackers, especially for the latest generation devices, such as wearable eyeglasses. For this reason, the present study aimed at comparing two of the most recent eye tracking devices, specifically a screen-mounted eye-tracker and a pair of wearable eye-tracking glasses (i.e. a remote and a mobile device respectively). The two eye-trackers were tested simultaneously during an experimental protocol in which subjects were asked to look at a moving target on a computer screen. The recorded signals consisted of point of regard (POR) x and y coordinates and pupil diameter size (PD), distinctively for the left and right eye. Data were analyzed in both time and frequency domain for a comprehensive assessment in order to evaluate the devices performance and the eventual degree of agreement between them, as described in the following sections.

2 Methods

2.1 Materials

Two different eye-tracking systems were compared in this experiment: the SMI REDn eye-tracker and the SMI Eye Tracking Glasses 2 Wireless, hereafter referred to as remote and mobile eye-trackers respectively. The remote eye-tracker is a USB-powered ultra-light portable camera, which is usually screen-mounted, with a sampling frequency of 30 or 60 Hz. It offers an automatic compensation for head movements and can be comfortably used also by subjects wearing eyeglasses. Data are visualized on the PC with a dedicated software (iViewRED). The mobile eye-tracker is a non-invasive device designed to be used just like a normal pair of glasses, able to record data with a sampling frequency up to 60 Hz. Thanks to automatic parallax compensation, it is claimed to provide reliable real-time data. In this experiment, it was connected to an Android smartphone and interfaced with the PC with a specific software, iViewETG Recording Device.

2.2 Experimental Protocol

Seventeen healthy volunteers aged between 20 and 35 years (10 males and 7 females) were recruited for the present study. All measurements were performed at PHEEL (Physiology Emotion Experience Laboratory), a multi departmental laboratory of Politecnico di Milano. The subjects showed no medical conditions such as epilepsy or colour blindness. If present, visual impairments, e.g. myopia, were corrected via contact lenses to facilitate coupling with both employed instruments. All the recruited volunteers were welcomed and briefly instructed about the experiment. Also, they were asked to sign a written consent module before taking part to the trial. Subjects were comfortably seated in front of a PC monitor, which was used for stimuli delivery.

The experimental protocol design was created using the SMI Experiment Center, a proprietary software which supports the entire trial workflow from experimental design to data analysis. This tool allows to tailor stimulus presentation on the screen according to desired specifications. First of all, both devices were calibrated: the remote eye-tracker benefited from an automatic calibration session by using 9 points, followed by subsequent validation with 4 points; the mobile eye-tracker, instead, was manually tuned with the smartphone by means of a 3 points calibration, by using an image projected on the PC screen. After the previous steps of calibration and preparation of the devices, the volunteer watched a neutral image for 1 min. Neutral image is inserted in the protocol for physiological reasons as it allows the subjects to relax and rest his eyes. Then, the task for each participant consisted in following the path of a mobile target on the screen by changing the direction of the gaze and keeping the head as much still as possible. The target initially appears and disappears in different sections of the screen, while, in a second moment, the target moves with constant speed along the sides and the diagonal of the monitor to induce smooth pursuits. The task lasts 90 s.

Preliminary data acquisition showed extremely non-uniform data sampling for a nominal sampling rate of 60 Hz. As a result, we chose to acquire data at 30 Hz, since non-uniformity was less evident. Moreover, such sampling rate appears adequate for the sake of devices general characterization.

2.3 Data Processing

Both POR and PD signals were processed in Matlab environment. As stated before, the actual temporal difference between two consecutive samples was not constant over time. Thus, signals were resampled using the Piecewise Cubic Hermite Interpolating Polynomial (PCHIP) method. This technique takes advantages of several cubic interpolators to obtain curves with different slopes on each side of the nodes, so that interpolated data complies better to existing ones with respect to other curves computed through different polynomial interpolators [12].

A qualitative signals analysis showed, as expected, a great quantity of blinks and high frequency oscillations. Hence, the need for artifact removal.

First, a simple thresholding process based on physiological and physical constraints was applied:

- Value of each sample (both POR and PD) must be strictly greater than 0;
- PD must not be physiologically lower than 1 mm or higher than 8 mm.

This phase was followed by an adaptive thresholding step where, for each signal, a vector containing the algebraic difference between each sample and the previous one was computed. The mean value of such array was identified as the adaptive threshold. Then, samples having a distance from their following one higher than the threshold were categorised as blinks, removed and subsequently replaced with a cubic interpolation. The whole adaptive thresholding process was reiterated to ensure total artefact removal. Moreover, a tenth order median filter was applied to further reduce high-frequency, low-amplitude oscillations to POR signals only.

While remote eye-tracker automatically starts to record data at the beginning of the experiment, acquisition has to be manually launched for the mobile eye-tracker. Consequently, signals were synchronized in post-processing through the cross-correlation technique.

The correlation coefficients ρ between remote and mobile eye-trackers data (POR in the x and y axes, pupil diameter for left and right eye) were determined in order to evaluate the degree of correlation between signals in the time domain. Furthermore, an analysis in the frequency domain was conducted for the pupillary signals. PD signals were low-pass filtered with a FIR filter at 1 Hz and high-pass filtered with an IIR at 0.025 Hz, to enhance the band of interest of the autonomic nervous system action in analogy with heart rate variability measurements [13]. Afterwards, the Welch periodogram was calculated. Then, the magnitude-squared coherence was computed with the following formula:

$$|K_{12}(\omega)|^2 = \frac{|S_{xy}(\omega)|^2}{S_x(\omega)S_y(\omega)} \tag{1}$$

where $S_{xy}(\omega)$ represents the Cross-Spectrum of the two signals, while $S_x(\omega)$ and $S_y(\omega)$ are the Auto-Spectra of the same signals.

3 Results

Figure 1 shows the correlation coefficients for each participant, whereas all the obtained p-value are lower than 0.01. The correlation coefficients are extremely high for POR with slightly high values for the left eye. Pupil diameter shows high correlation as well, but with a major dispersion: this is probably due to some subjects' recordings, which results more affected by noise.

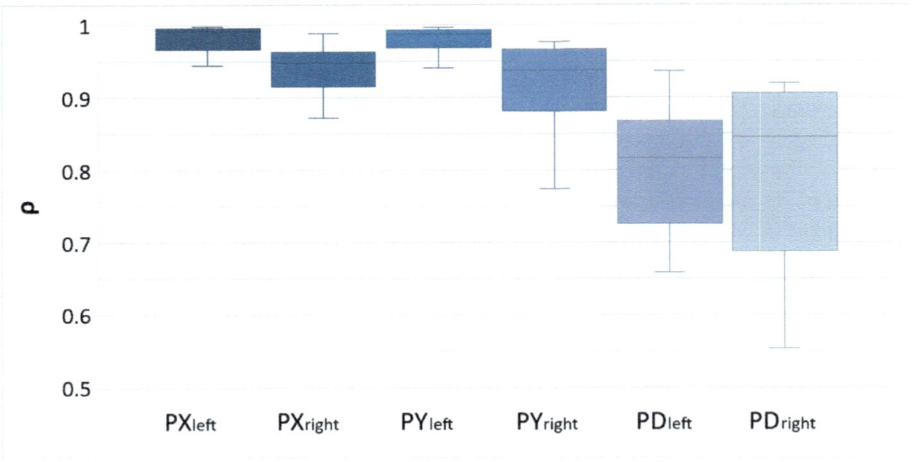

Fig. 1. Correlation coefficients. PXleft (Point of Regard, Left eye, X axis), PXright (Point of Regard, Right eye, X axis), PYleft (Point of Regard, Left eye, Y axis), PYright (Point of Regard, Right eye, Y axis), PDleft (Pupil Diameter, Left eye), PDright (Pupil Diameter, Right eye).

Figure 2 shows the magnitude-squared coherence for each frequency, whereas Fig. 3 displays the mean values of the magnitude-squared coherence for the range 0–1 Hz in accordance with the standard measurements of the influence to the heart of the autonomic nervous system through heart rate variability (generally used in both psychophysiological and clinical settings [13]). From the results it is possible to notice that coherence values are high (on average > 0.5) along most of the frequency range of interest, indicating a good relation between the signals provided by the mobile device and the remote one.

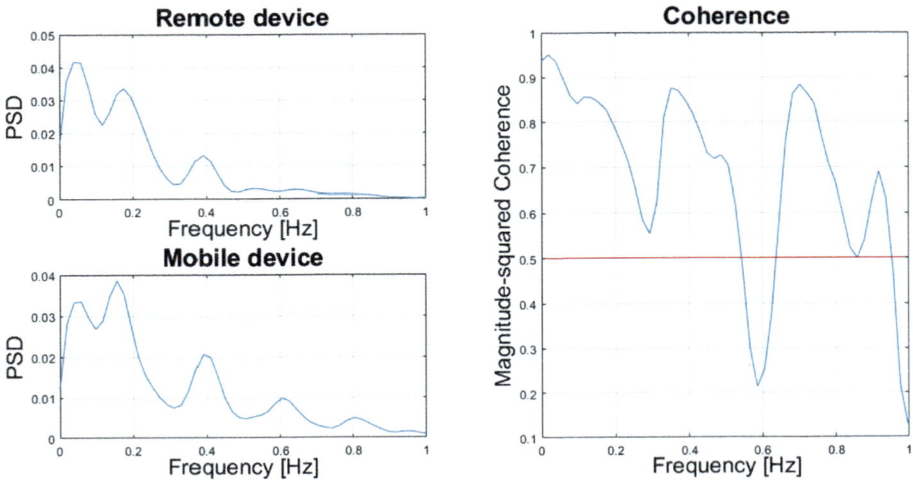

Fig. 2. Example of periodograms estimate via welch and magnitude-squared coherence (PD_{right})

Since the results presented so far were to verify the degree of association between the two eye-trackers signals after application of pre-processing algorithms, we also performed an analysis on the reliability of the raw pupil diameter signal returned by the two devices. Using physiological thresholds mentioned in Sect. 2.3, the samples deemed to be invalid were identified and we calculated the percentage with respect to the total number of samples. Figure 4 collects these percentage values related to pupil diameters. In 28 cases out of 34 (2 eyes for 17 subjects) the first device has a lower percentage of invalid data.

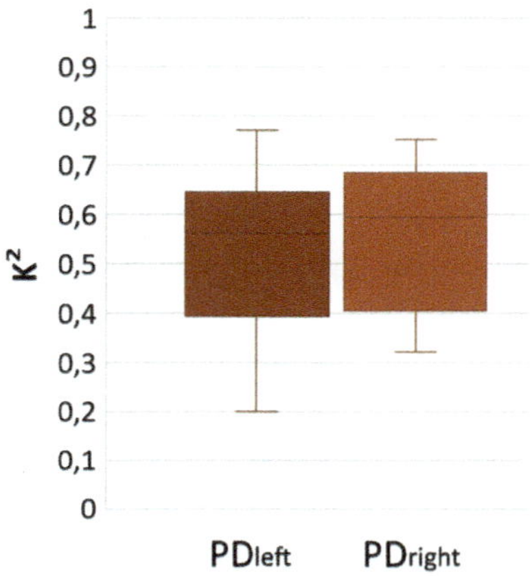

Fig. 3. Mean values of magnitude-squared coherence in the range 0–1 Hz for PD for the left and the right eye between the remote and the wearable eye-trackers

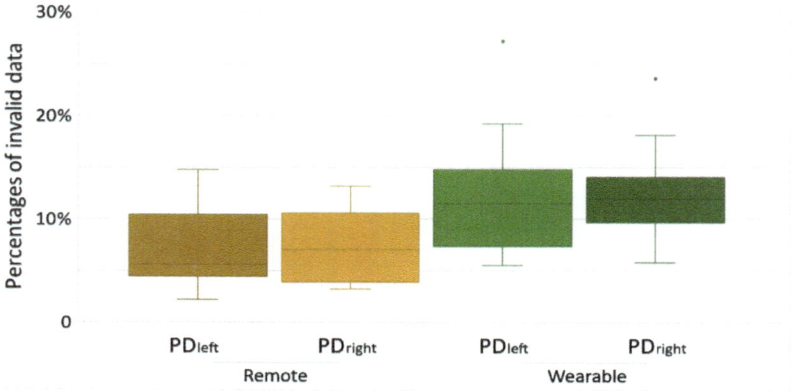

Fig. 4. Percentages of invalid data in the pupil diameter signal recorded by the remote and the wearable eye-trackers.

4 Discussions and Conclusions

The fields in which new eye-tracking technologies are used are increasingly diverse. Depending on the type of experiment or environment, researchers are required to decide whether to use mobile or remote devices. Hence, the purpose of our work was to verify if the information provided by these two macro-categories of instrumentation was comparable, and to what extent, in order to allow experts in the field to eventually use them interchangeably, or at least to choose the most suitable one.

In this study, two types of signals extracted from the instruments were analysed, i.e. point of regards and pupil diameters. The general comprehensive assessment showed that after a pre-processing phase for artifact removal, signals from remote and mobile devices were highly comparable in both time and frequency domain. Nevertheless, the higher correlation coefficients for the left eye POR with respect to the right one ($\rho_{left} > 0.9$, $\rho_{right} > 0.75$, see Fig. 1) showed a slight influence of the subject's head position on the measurements. Instead, by looking at the raw pupil signal, it was possible to notice that, for most of the participants, the mobile device has a higher number of invalid samples than the remote one. It is reasonable to believe that this resulted also in a larger dispersion of the correlation coefficients and coherence (Figs. 1 and 3). Of note, data acquisition revealed that sampling frequency is less uniform also for the wearable eye-tracker.

In conclusion, in a common visual behavior recording the remote eye-tracker provided more stable and reliable data acquisition with respect to the mobile counterpart, given the considerably lower number of invalid data. However, after a simple but functioning pre-processing stage, such as the one presented in this work, we can infer that both signals share the same content at the level of information generally used to characterize the subjects behavioral and physiological reactions. Besides signal accuracy, physical comfort of the observers should also be considered, since wearing eyeglasses might be uncomfortable for someone. Another characteristic to bear in mind is the sampling frequency: these devices are adequate for preliminary data or pilot studies, but such low sampling frequencies are not adequate for all research applications [14]. As a result, a higher sampling frequency capable to capture characteristic pupillary oscillations might be considered for future assessment. The eye-trackers examined here represent only a sample of the available types and models. Thus, it is extremely important for researchers to continue to analyze the capabilities and accuracy of eye-tracking systems in order to determine the validity of the measures that each system provides, together with costs, strengths and weaknesses [15]. Future research should be addressed to examine different devices possibly with higher sampling frequencies by investigating other features (e.g. detection of saccades and fixations) in various experimental setups.

Acknowledgement. The authors acknowledge PHEEL (http://pheel.polimi.it/) for the valuable support in both experimental and methodological development.

Conflict of Interest. The authors declare that there is no conflict of interest regarding the publication of this article.

References

1. Underwood, G.: Cognitive Processes in Eye Guidance. Oxford University Press, Oxford (2005)
2. Hansen, D., Ji, Q.: In the eye of the beholder: a survey of models for eyes and gaze. IEEE Trans. Pattern Anal. Mach. Intell. **32**(3), 478–500 (2010)
3. Eckstein, M.K., Guerra-Carrillo, B., Singley, A.T.M., Bunge, S.A.: Beyond eye gaze: what else can eyetracking reveal about cognition and cognitive development? Dev. Cogn. Neurosci. **25**, 69–91 (2017)
4. Robinson, D.: A method of measuring eye movement using a scleral search coil in a magnetic field. IEEE Trans. Biomed. Electron. **10**(4), 137–145 (1963)
5. Marmor, M.F., Zrenner, E.: Standard for clinical electro-oculography. Doc. Ophthalmol. **85**(2), 115–124 (1993)
6. Holmqvist, K., Nyström, M., Andersson, R., Dewhurst, R., Jarodzka, H., Van de Weijer, J.: Eye Tracking: A Comprehensive Guide to Methods and Measures. Oxford University Press, Oxford (2011)
7. Mele, M.L., Federici, S.: Gaze and eye-tracking solutions for psychological research. Cogn. Process. **13**(1), 261–265 (2012)
8. Pierce, K., Marinero, S., Hazin, R., McKenna, B., Barnes, C.C., Malige, A.: Eye tracking reveals abnormal visual preference for geometric images as an early biomarker of an autism spectrum disorder subtype associated with increased symptom severity. Biol. Psychiat. **79**(8), 657–666 (2016)
9. Al-Moteri, M.O., Symmons, M., Plummer, V., Cooper, S.: Eye tracking to investigate cue processing in medical decision-making: a scoping review. Comput. Hum. Behav. **66**, 52–66 (2017)
10. Nevalainen, S., Sajaniemi, J.: Comparison of three eye tracking devices in psychology of programming research. PPIG **4**, 151–158 (2004)
11. Mello-Thoms, C.: Head-mounted versus remote eye tracking of radiologists searching for breast cancer: a comparison. Acad. Radiol. **13**(2), 203–209 (2006)
12. Carlson, R.E., Fritsch, F.N.: Monotone piecewise bicubic interpolation. SIAM J. Numer. Anal. **22**(2), 386–400 (1985)
13. Camm, A., Malik, M., Bigger, J., Breithardt, G., Cerutti, S., Cohen, R., et al.: Heart rate variability: standards of measurement, physiological interpretation and clinical use. Task force of the european society of cardiology and the north american society of pacing and electrophysiology. Circulation **93**, 1043–1065 (1996)
14. Andersson, R., Nyström, M., Holmqvist, K.: Sampling frequency and eye-tracking measures: how speed affects durations, latencies, and more. J. Eye Mov. Res. **3**(3), 6 (2010)
15. Funke, G., Greenlee, E., Carter, M., Dukes, A., Brown, R., Menke, L.: Which eye tracker is right for your research? Performance evaluation of several cost variant eye trackers. In: Proceedings of the Human Factors and Ergonomics Society Annual Meeting, vol. 60, no. 1, pp. 1240–1244. SAGE Publications, Los Angeles (2016)

Efficacy of Time- and Frequency-Domain Heart Rate Variability Features in Stress Detection and Their Relation with Coping Strategies

Pierluigi Reali[1(✉)], Agostino Brugnera[2], Angelo Compare[2], and Anna Maria Bianchi[1]

[1] Department of Electronics Information and Bioengineering, Politecnico di Milano, Via Golgi 39, 20133 Milan, Italy
pierluigi.reali@polimi.it
[2] Department of Human and Social Sciences, University of Bergamo, Piazzale S. Agostino 2, 24129 Bergamo, Italy

Abstract. Heart Rate Variability (HRV) is a reliable biomarker of the Autonomic Nervous System (ANS) activity and it is widely used to characterize the stress response induced by different laboratory stress tasks. Even if acute stress responses have been previously investigated, researchers are still wondering which HRV indices are more effective for stress assessment. In the present study, we extracted several time- and frequency-domain HRV parameters and investigated which ones are better able to discriminate between a stressful and a non-stressful condition. Moreover, we explored the possibility of a linear correlation between such parameters and certain coping strategies, during three laboratory stress tasks (Montreal Imaging Stress Task, Stroop Color-Word, Speech). The effect size computed for each considered HRV parameter indicate that the average RR interval and the normalized power in low frequency (LF) band were the most effective parameters for the detection of mental stress. As regards the second hypothesis, normalized power in high frequency (HF) band during Speech was found significantly and negatively correlated with specific subscales of the administered questionnaires (CERQ-S and COPE-NVI), suggesting a possible association between the higher use of social support (SS) and other blame (OB) coping strategies and stronger autonomic responses during the Speech task.

Keywords: Heart Rate Variability · Stress · Coping strategies

1 Introduction

Acute stress, or the physiological reaction to internal or external events perceived as stressful (i.e., stressors), has been largely investigated in laboratory settings using psychosocial stress tasks. These tasks lead to a so-called "stress response", characterized by the hyper- or hypo-activation of specific brain areas, increased secretion of cortisol, and by a strong para-sympathetic withdrawal [1–4].

J. Henriques et al. (Eds.): MEDICON 2019, IFMBE Proceedings 76, pp. 209–216, 2020.
https://doi.org/10.1007/978-3-030-31635-8_25

A reliable and well-investigated biomarker of Autonomic Nervous System (ANS) activity during laboratory stress tasks is Heart Rate Variability (HRV). In front of stressors the ANS modulates the heart's rhythm together with other physiological systems, slowing or accelerating the heart rate (HR) through the activation of the parasympathetic or sympathetic nerves, respectively [5]. Through the noninvasive assessment of the variations in the beat-to-beat intervals, HRV provides a picture of the dynamic balance between sympathetic and parasympathetic branches of ANS [5]. Several measures of Heart Rate Variability have been developed. For example, time domain analyses provide information on the variability of HR, applying simple statistical methods on the RR series (e.g., standard deviation). On the contrary, frequency domain analyses quantify the autonomic dynamics through spectral methods, typically by extracting two main components, namely i) the low frequencies (LF; 0.04–0.15 Hz), which reflect the activity of the baroreflex in response to vasomotor tone, and ii) the high frequencies (HF; 0.15–0.40 Hz), considered a well-established marker of vagal activity [5].

Even if acute stress responses to laboratory stress tasks have been previously investigated (see for example [3, 4]), researchers are still wondering which HRV indices are more effective in discriminating between a stressful and a resting condition. Such information could help manufacturers to develop wearable sensors which could focus on fewer and more reliable HRV indices and provide better estimates of individual stress responses during daily lives.

In addition, researchers have also examined the impact of psychological factors on cardiovascular activity, finding that patients with psychiatric illnesses are more likely to show altered HRV [2]. However, to date the effect of coping strategies on HRV, especially during acute stress tasks, is still under-investigated. Coping strategies are a set of cognitive and behavioral efforts to deal with stressors: for example, an individual could distract himself from the stressful event (avoidant coping), could try to regulate his\her emotional responses (emotion-oriented coping), or seek for social support (socially-oriented coping) [6, 7].

Therefore, as part of a larger study [1], we sought to investigate HRV responses to a set of properly-defined stressful and non-stressful conditions in 36 healthy adults who completed a set of 3 different stress tasks. Specifically, we tested (i) which HRV indices are better able to discriminate between a stressful and non-stressful condition, and (ii) if specific coping strategies are associated with altered ANS responses at rest and during specific laboratory stress tasks.

2 Methods

2.1 Experimental Protocol

60 healthy volunteers were recruited for the study. Among them, a randomly selected sample of 36 participants (age: 25.4 ± 3.85 years, 19 females) was considered for the subsequent analyses. The study was conducted in accordance with the Declaration of Helsinki.

Before the beginning of the experiment, participants were asked to sign a written consent. A four electrodes thoracic band was applied on participants' skin to acquire a single-lead electrocardiography (ECG) through a wearable device (Pulse Sensor, MR&D and STMicroelectronics, Italy); the ECG was collected at 256 Hz. After sensor positioning, participants were asked to relax for at least 15 min; the last 5 min of this period were considered for the assessment of the baseline state. The experimental protocol consists of three stress tasks:

- Montreal Imaging Stress Task (MIST) [8];
- Stroop Color-Word task (hereafter denoted "Stroop") [9];
- Speech task [10].

Each task lasted 5 min and was presented to participants in random order to avoid sequence effect. The last-presented task was followed by a 5-min recovery phase.

During MIST participants were asked to solve a series of time-limited arithmetic operations (sums, multiplications and subtractions) that were displayed on a PC screen. The time limit was automatically adapted on the basis of the correctness of the last three responses given by the subjects.

The Stroop task consists in presenting participants with color names written in various ink colors and asking them to name the presented ink color [9]. During the experiment, eight color names were displayed simultaneously on the PC screen; some of them have ink color matching the written color name, while others not. Participants should say aloud all the eight ink colors presented each time on the screen.

During the Speech task participants were asked to defend themselves from a false accusation of shoplifting. To increase task effectiveness, they were informed that a videotape of their speech would have been recorded, in order to be later scored based on subjects' performance.

ECG signal was continuously acquired during the entire procedure.

2.2 Signal Processing

The Pan-Tompkins algorithm [11] was applied on the acquired ECG signals to detect R peaks and the heart rate variability signal (HRV) was obtained by calculating the time distance between the occurrence of consecutive R peaks.

To improve signal quality, HRV signals were processed with an automatic algorithm to detect and replace possible outliers (i.e. too long or too short RR intervals due to non-sinus beats). The SDChg approach proposed in [12] was used for this purpose.

Then time and frequency-domain parameters of interest were extracted from the HRV signals. Specifically, mean, standard deviation and power spectral density (PSD) were computed, for each phase of the experimental protocol (baseline, MIST, Speech, Stroop and recovery). Linear trend was removed from individuals' HRV signals to reduce spectral contribution of the lowest frequencies and highlight faster oscillations [13]. Protocol phases were divided in 2-minutes time windows (sufficient for reliable PSD estimate in LF and HF bands [5]), with 50% overlap, and PSD was estimated separately for each one of them by, first, identifying the optimal autoregressive

(AR) model (according to Akaike's Information Criterion) [13] and, second, computing the power associated to model poles through spectral decomposition method [14]. Power in LF and HF bands was then computed as the sum of the individual contributions of the poles that fall in the frequency range of each band. These two power components were calculated in normalized units too (i.e. band power divided by total power).

2.3 Questionnaires

Two psychological questionnaires of interest were administered one day before the experiment: the Cognitive Emotion Regulation Questionnaire – Short Version (CERQ-S) [6] and the Coping Orientation to the Problems Experienced-25 (COPE-NVI) [7].

The CERQ-S is an 18-item self-report measure of cognitive emotion regulation strategies that people use in response to threatening or stressful life events. It is composed by 9 subscales (Self-blame, Other-blame, Rumination, Catastrophizing, Putting into Perspective, Positive Refocusing, Positive Reappraisal, Acceptance and Planning). Items are rated in a 5-point Likert-type scale (1 to 5), with higher total scores indicating greater use of a specific cognitive coping strategy. In the current study, Cronbach's Alphas were moderate to good (range 0.68–0.88).

The COPE-NVI is a 25-item self-report measure of coping strategies. It is composed by 5 subscales (avoidance strategies, social support, religious coping, positive reinterpretation and growth, focus on problems). Items are rated on a 6-point Likert-type scale (1 to 6), with higher total scores indicating greater use of a specific coping strategy. In the current study, Cronbach's Alphas were moderate to good (range 0.63–0.96).

2.4 Data Analysis

In order to test our first hypothesis, regarding the selection of the most effective HRV features for stress detection, we define two macro-conditions: a "rest" condition, composed by baseline and recovery phases; a "stress" condition, composed by MIST, Speech and Stroop phases.

Each HRV feature was averaged among the phases of the correspondent macro-condition to obtain, for each participant, one value related to "rest" and one related to "stress" condition. Subsequently, a Wilcoxon signed-rank test was performed for each feature to investigate whether a statistically significant difference may be found between "rest" and "stress" condition. In addition, starting from the *z-value* [15], we computed and interpreted Cohen's r, a non-parametric effect size, in accordance with guidelines (≥ 0.1 small, ≥ 0.3 medium, ≥ 0.5 large; [15]).

The significant HRV parameters (i.e. features found statistically different between "rest" and "stress" conditions) were selected for a correlation analysis with the questionnaires scores. Specifically, a Pearson's correlation coefficient was calculated between each HRV feature, computed for the different protocol phases, and the participants' scores reported for every item of the CERQ-S and COPE-NVI self-assessment scales.

3 Results

3.1 Discrimination Between "Rest" and "Stress" Conditions

The results of the statistical analysis carried out on the considered HRV features, grouped in stressful ("stress") and non-stressful ("rest") conditions, are reported in Fig. 1 and Table 1. Mean RR and power in HF band were found significantly ($p < 0.05$, Wilcoxon signed-rank) higher during "rest" than during "stress"; on the contrary, the power in LF band and the power ratio LF/HF resulted significantly higher during "stress". However, only the effect sizes associated to mean RR, normalized power in LF band and power in HF band (in both absolute and normalized units) can be considered large, as they are greater than 0.5.

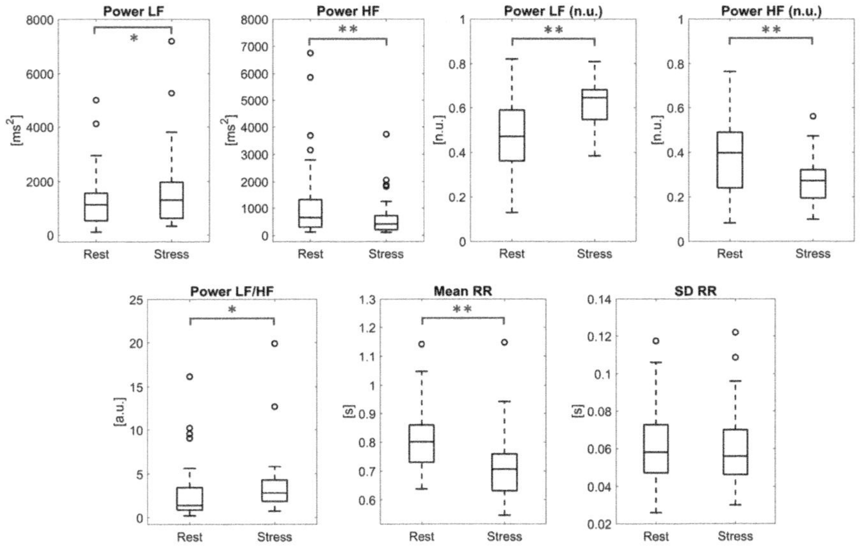

Fig. 1. Boxplots of the considered HRV features. The maximum whisker length is set to 1.5 times the interquartile range. Statistically significant differences: * ($p < 0.05$) or ** ($p < 0.01$).

Table 1. Median and interquartile range of the considered HRV features, sorted in descending order of effect size. Statistically significant differences with "rest": * ($p < 0.05$) or ** ($p < 0.01$).

Feature	Rest	Stress	Effect size (r)
Mean RR [s]	0.801 (0.129)	0.706 (0.127) **	0.862
Power LF [n.u.]	0.471 (0.227)	0.646 (0.137) **	0.757
Power HF [n.u.]	0.398 (0.248)	0.274 (0.128) **	0.592
Power HF [ms^2]	644.7 (1030.7)	407.5 (510.3) **	0.539
Power LF/HF [a.u.]	1.392 (2.592)	2.824 (2.448) *	0.390
Power LF [ms^2]	1114 (1040)	1293 (1369) *	0.340
SD RR [s]	0.058 (0.026)	0.056 (0.024)	0.031

3.2 Correlation Between Psychological Variables and HRV Parameters

Among all subscales of both COPE-NVI and CERQ-S questionnaires, only social support (SS) and other blame (OB) were significantly and negatively correlated with HF normalized power during the Speech task (Table 2). In particular, higher COPE-NVI SS scores led to decreased HF normalized power during Speech. Similarly, higher CERQ-S OB scores were significantly and negatively associated with the same HRV measure. All other correlations were not significant.

Finally, as a follow-up analysis, we explored the possibility of a correlation between significant HRV parameters and the percentage of correct answers provided during MIST (Table 3). This variable was found negatively correlated with normalized power in HF band and positively correlated with LF power and LF/HF power ratio during Stroop, suggesting that participants who were more focused during MIST (i.e., gave more correct responses), experienced higher levels of cognitive efforts and stronger stress responses during another conceptually-similar task, namely the Stroop test.

Table 2. Pearson's correlation coefficients calculated between significant HRV features during Speech task (columns) and questionnaires scores (rows). Statistically significant correlations are marked with * ($p < 0.05$).

	Pow. LF Speech	Pow. HF Speech	Pow. LF (n.u.) Speech	Pow. HF (n.u.) Speech	Pow. LF/HF Speech	Mean RR Speech
COPE-SS	0.355	0.002	0.139	−0.354 *	0.358	−0.110
CERQ-OB	0.096	−0.104	0.122	−0.342 *	0.205	−0.204

Table 3. Pearson's correlation coefficients calculated between significant HRV features during the Stroop task (columns) and the percentage of correct answers during MIST (row). Statistically significant correlations are marked with * ($p < 0.05$) or ** ($p < 0.01$).

	Pow. LF Stroop	Pow. HF Stroop	Pow. LF (n.u.) Stroop	Pow. HF (n.u.) Stroop	Pow. LF/HF Stroop	Mean RR Stroop
MIST correct answers (%)	−0.023	−0.171	0.350 *	−0.533 **	0.434 **	−0.074

4 Discussion and Conclusions

In this work, we extracted several HRV time and frequency-domain features during the execution of three stressful laboratory tasks. These tasks elicit different kinds of psychosocial stress responses (e.g., MIST induce a more cognitive stress response, typical of working environments), allowing for a more comprehensive evaluation of the stress assessment capability of the considered HRV parameters. Such parameters were, first, compared between stressful (MIST, Speech, Stroop) and non-stressful (baseline, recovery) phases of the protocol and, second, analyzed in relation to certain psychological variables of interest.

As regards the first hypothesis, our results indicate that several HRV measures (Table 1) were significantly able to discriminate between stressful and non-stressful conditions. The sign of their variations between "rest" and "stress" conditions are consistent with previous findings and suggest increased sympathetic activation during the stressful condition [4]. Interestingly, among the significant parameters, mean RR and normalized power in LF band showed the highest effect size in discriminating between stressful and non-stressful conditions. This result may be particularly interesting for wearable devices developers, since it supports the idea that even a very simple HRV-derived measure, such as the average RR interval, can provide useful information to evaluate the individual stress state. In fact, there are several wearable devices available on the market (e.g. photoplethysmography wristbands, ECG shirts) that can provide good-quality and low-cost measures of HRV [16] and this may represent one of the advantages of preferring HRV to other biological signals (e.g. electroencephalography [17]) to analyze the physiological stress response.

As regards hypothesis two, we found that only two specific socially-oriented coping strategies were related to a parasympathetic withdrawal during the Speech task. Indeed, the SS subscale of COPE-NVI questionnaire measures the social-support coping strategy, or the tendency to seek social support to obtain advice, while the OB subscale of CERQ-S questionnaire measures the other-blame coping strategy, or the tendency to put the blame for what one has experienced on others. Given that the Speech task has a strong psychosocial component (i.e., requiring participants to defend themselves from a false accusation in front of a video-camera), we suppose that those who are used to seek for positive or negative social support in daily life, during this specific task experience stronger autonomic responses, since they have neither chance to ask for help (in preparing their defense) nor reasons to blame other people for their possible poor performance. These findings are of particular interest, because the Speech task adopted in the current study is similar to one of the most commonly-used stress paradigms, namely the Trier Social Stress Test (TSST): thus, our findings suggest that part of the inter-individual differences resulting from TSST and other similar stress tasks might be attributed to different levels of socially-oriented coping strategies.

Future developments of the present study should include the introduction of non-linear HRV parameters in the analyses. Moreover, our study assessed the ability of each HRV measure, per se, in discriminating between stressful and non-stressful conditions; further analyses are advised to understand if the combination of two or more HRV features could bring to a more accurate stress detection.

Acknowledgements. P. R. thanks LINK project (H2020 Grant Agreement No. 692023) for the given support.

Conflicts of Interest. The authors declare no conflicts of interest.

References

1. Brugnera, A., Zarbo, C., Tarvainen, M.P., Marchettini, P., Adorni, R., Compare, A.: Heart rate variability during acute psychosocial stress: a randomized cross-over trial of verbal and non-verbal laboratory stressors. Int. J. Psychophysiol. **127**, 17–25 (2018)
2. Brugnera, A., Zarbo, C., Adorni, R., Tasca, G.A., Rabboni, M., Bondi, E., et al.: Cortical and cardiovascular responses to acute stressors and their relations with psychological distress. Int. J. Psychophysiol. **114**, 38–46 (2017)
3. Kogler, L., Müller, V.I., Chang, A., Eickhoff, S.B., Fox, P.T., Gur, R.C., Derntl, B.: Psychosocial versus physiological stress—meta-analyses on deactivations and activations of the neural correlates of stress reactions. Neuroimage **119**, 235–251 (2015)
4. Castaldo, R., Melillo, P., Bracale, U., Caserta, M., Triassi, M., Pecchia, L.: Acute mental stress assessment via short term HRV analysis in healthy adults: a systematic review with meta-analysis. Biomed. Signal Process. Control **18**, 370–377 (2015)
5. Malik, M., Bigger, J.T., Camm, A.J., Kleiger, R.E., Malliani, A., Moss, A.J., Schwartz, P.J.: Heart rate variability: standards of measurement, physiological interpretation, and clinical use. Eur. Heart J. **17**(3), 354–381 (1996)
6. Garnefski, N., Kraaij, V.: Cognitive emotion regulation questionnaire–development of a short 18-item version (CERQ-short). Pers. Individ. Differ. **41**(6), 1045–1053 (2006)
7. Foà, C., Tonarelli, A., Caricati, L., Fruggeri, L.: COPE-NVI-25: validazione italiana della versione ridotta della Coping Orientation to the Problems Experienced (COPE-NVI). Psicol. della salute **2**, 123–140 (2015)
8. Dedovic, K., Renwick, R., Mahani, N.K., Engert, V., Lupien, S.J., Pruessner, J.C.: The montreal imaging stress task: using functional imaging to investigate the effects of perceiving and processing psychosocial stress in the human brain. J. Psychiatry Neurosci. **30**(5), 319–325 (2005)
9. Stroop, J.R.: Studies of interference in serial verbal reactions. J. Exp. Psychol. **18**(6), 643–662 (1935)
10. Cacioppo, J.T., Uchino, B.N., Berntson, G.G.: Individual differences in the autonomic origins of heart rate reactivity: the psychometrics of respiratory sinus arrhythmia and preejection period. Psychophysiology **31**(4), 412–419 (1994)
11. Pan, J., Tompkins, W.J.: A real-time QRS detection algorithm. IEEE Trans. Biomed. Eng. **32**(3), 230–236 (1985)
12. Kemper, K.J., Hamilton, C., Atkinson, M.: Heart rate variability: impact of differences in outlier identification and management strategies on common measures in three clinical populations. Pediatr. Res. **62**(3), 337–342 (2007)
13. Kamath, M.V., Watanabe, M., Upton, A. (eds.): Heart Rate Variability (HRV) Signal Analysis: Clinical Applications. CRC Press, Boca Raton (2012)
14. Baselli, G., Porta, A., Rimoldi, O., Pagani, M., Cerutti, S.: Spectral decomposition in multi channel recordings based on multivariate parametric identification. IEEE Trans. Biomed. Eng. **44**(11), 1092–1101 (1997)
15. Fritz, C.O., Morris, P.E., Richler, J.J.: Effect size estimates: current use, calculations, and interpretation. J. Exp. Psychol. Gen. **141**(1), 2–18 (2012)
16. Teng, X.F., Zhang, Y.T., Poon, C.C.Y., Bonato, P.: Wearable medical systems for p-health. IEEE Rev. Biomed. Eng. **1**, 62–74 (2008)
17. Lewis, R.S., Weekes, N.Y., Wang, T.H.: The effect of a naturalistic stressor on frontal EEG asymmetry, stress, and health. Biol. Psychol. **75**(3), 239–247 (2007)

Influence of Physical Models of Electrodes on Rat's Head Forward Modelling

David Kuratko[1(✉)], Jaroslav Lacik[1], Zbynek Raida[1], Daniel K. Wójcik[1,3], and Vlastimil Koudelka[2]

[1] Brno University of Technology, Technická 12, 616 00 Brno, Czech Republic
xkurat01@stud.feec.vutbr.cz
[2] National Institute of Mental Health, Topolová 748, 250 67 Klecany, Czech Republic
[3] Nencki Institute of Experimental Biology of Polish Academy of Sciences, Warsaw, Poland

Abstract. Reliable inverse imaging of source currents in rat's brain requires accurate models of fields and interfaces. Accuracy of field models can be influenced by a proper representation of electrodes used for sensing potentials. In the paper, we study the effect of realistic modelling of electrodes on the distribution of electric field intensity in the head. Fields are evaluated for three commonly used configurations of electrodes using a three-dimensional finite element model of rat's head comprising the brain, the cerebrospinal fluid, and the skull. Simulations show a lower amplitude of electric field intensity if realistic models of electrodes are considered. Moreover, electric field intensity near electrode models is shown to be slightly influenced. Therefore, realistic modelling of electrodes can improve the accuracy of an inverse imaging, and consequently make the localization of current sources in rat's brain more accurate.

Keywords: Forward model · Rat's brain · Electrode models · Electric source imaging

1 Introduction

Brain waves can be measured on the surface of rat's brain in form of electrical potentials [1]. The inverse localization of brain wave sources requires a forward model and an inverse solver [2]. The forward model is needed to calculate the potentials at the electrodes from known source parameters. Solution of the forward problem is then organized into the so-called lead-field matrix, which is essential for solving the inverse problem. The inverse solver is asked to localize wave sources in the brain from the lead-field matrix based on forward modelling.

The brain can be understood as a volumetric source of currents produced by populations of neurons. Various frequencies of electromagnetic waves can be observed due to the complex behavior of the brain. The results of many studies indicate that the dominating frequencies of neural oscillation increase as one moves from the deeper parts of the brain towards the evolutionary younger areas closer to the surface of the

J. Henriques et al. (Eds.): MEDICON 2019, IFMBE Proceedings 76, pp. 217–225, 2020.
https://doi.org/10.1007/978-3-030-31635-8_26

brain [3]. However, a reliable mapping of frequencies is still not available due to the complexity and inter-individual variability of the brain.

In human EEG studies, one is typically interested in brain waves propagating at frequencies from f_{min} = 0.1 Hz to f_{max} = 100 Hz. Since the power of the electroencephalograph (EEG) signal decreases by the factor of $1/f$, we focus on modelling the brain activity at f = 10 Hz where a peak of the spectral power of rat's EEG is assumed.

EEG configurations can be monopolar or bipolar. The monopolar configuration measures the potential at the measuring electrode with respect to the reference electrode. The reference electrode is placed outside of the area of interest (above the olfactory bulb in this study). The bipolar configuration processes the difference between two nearby electrodes [4].

In this paper, we consider the monopolar configuration of the electrodes, the full-wave formulation of brain waves, and a verified brain model. A more detailed comparison of different formulations (static, quasi-static, and full-wave) and various brain models differing in the mesh density, the type of elements, and the geometrical accuracy can be found in [5].

The study is focused on the effect of realistic modelling of electrodes on simulated electric fields. Models of electrodes were considered in three configurations – aligned to the brain surface, nested 1 mm under the surface, and implanted deeply in the brain. Since similar studies are unavailable, experimental verifications on rats have been prepared.

2 Electromagnetic Models and Simulation Setup

2.1 3D Models

The brain model was based on the magnetic resonance imaging (MRI) data, the skull model was developed from computer tomography images (CT), and the cerebrospinal fluid was created manually by filling the gap between the brain and the skull. When creating the model by a segmentation procedure, several kinds of mesh errors can occur due to volume approximation between slices. After removing duplicated surfaces and edges of elements, correcting unevenness, protrusions and sharp edges, and modifying the orientation of mesh elements, a new mesh was generated to preserve the anatomy of models. Finally, the brain model was resized to the length l = 27 mm, the width w = 15 mm and the height h = 10 mm which are the approximate dimensions of the brain of an adult Wistar rat.

For simulation purposes, the brain model was embedded into the cerebrospinal fluid and the skull. The brain model and the skull model come from different Wistar rats. Therefore, the brain was scaled in proportion for the best fitting of the cavity volume in the skull. The free space between the brain and the skull was filled by the cerebrospinal fluid. After that, the brain model, the cerebrospinal fluid and the skull were converted to voxels. Thanks to the conversion, the brain, the fluid and the skull can be aligned (potential gaps are eliminated), and simulations are accelerated.

Figure 1 shows the composite model of rat's head consisting from voxels with the edge length $l_e = 0.7$ mm. The bottom picture on the right proves that all the models are properly aligned thanks to the voxel representation.

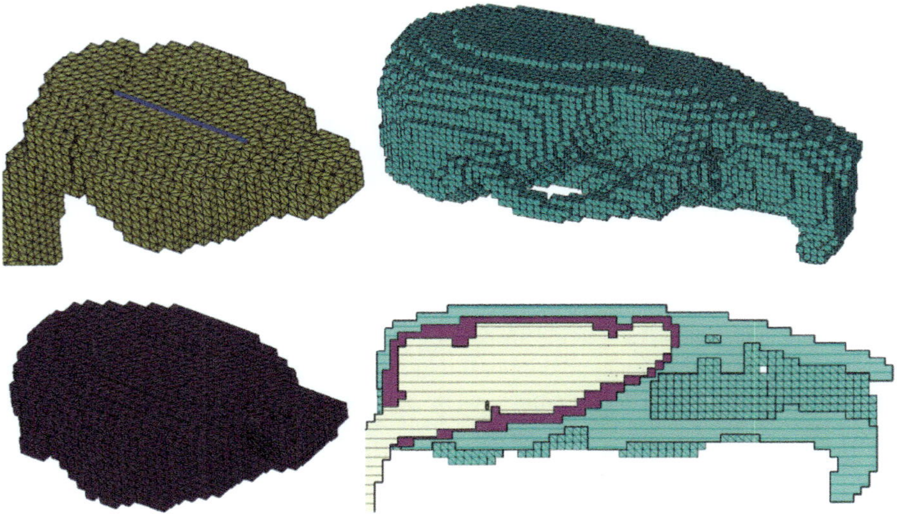

Fig. 1. Voxel models: the brain (top left), the cerebrospinal fluid (bottom left), the skull (top right), vertical cut crossing the center of the composite model (bottom right).

2.2 Simulation Setup

At present, most models of brain waves are based on the quasi-static formulation (Poisson's equation). A recent paper on the full-wave formulation of brain waves [4] showed significant differences between the quasi-static approximation and the full-wave one of simple electro-magnetic structures. Furthermore, the quasi-static formulation neglects displacement currents and time derivative of the magnetic flux density in Faraday's law compared with the full-wave formulation. Therefore, the full-wave formulation in frequency domain is used for simulations.

In order to evaluate the spatial distribution of electric field intensity, the voxel model was imported to the CST EM Studio [6]. Consequently, the full-wave model was evaluated by frequency domain solver computing the following equations:

$$\mathbf{B} = \nabla \times \mathbf{A} \tag{1}$$

$$\nabla \times \left(\nabla \times \frac{\mathbf{A}}{\mu}\right) + j\omega\sigma\mathbf{A} - \omega^2\varepsilon\mathbf{A} = \mathbf{J}_s \tag{2}$$

Here, $\mathbf{A}$ is magnetic vector potential, μ denotes permeability, ω represents angular frequency, ε corresponds to permittivity, σ is electric conductivity, $\mathbf{J}_s$ denotes primary or impressed current and ∇ is the differential operator.

Electrical parameters of tissues (the brain, the cerebrospinal fluid and the skull) were approximated by corresponding parameters of human tissues. The relative permittivity was $\varepsilon_r = 4.07 \cdot 10^7$ [–] for the brain [7], $\varepsilon_r = 1.09 \cdot 10^2$ for the cerebrospinal fluid [7], and $\varepsilon_r = 5.51 \cdot 10^4$ [–] for the skull [7]. The electric conductivity was $\sigma = 0.33$ S/m [2] for the brain, $\sigma = 1.79$ S/m [2] for the fluid, and $\sigma = 0.0174$ S/m [2] for the skull.

The EEG can be measured by surface electrodes when a large group of neurons is simultaneously active. The group of active neurons can be modelled as a current dipole [2]. The dipole produces electrical field propagating to the surface of rat's brain where the potential can be measured by an electrode. In this paper, the dipole of the length $l_d = 0.7$ mm and the width $w_d = 0.1$ mm was used. The dipole length corresponds to the length of the element edge. The dipole in performed simulations was determined to excite the postsynaptic potential of the value –60 mV. The dipole was located in the bottom part of the brain, and the position was fixed for all the simulations.

The simulated measuring electrodes correspond to the system which is used for real EEG recording with rats at the Institute of Mental Health (Klecany, the Czech Republic). The system consists of twelve electrodes implanted into the surface of the cortex in homologous frontal, parietal and temporal regions of the right and left hemispheres [8]. The coordinates of active electrodes were calculated from the bregma, and their approximate positions can be observed in Fig. 2, left. The accurate coordinates of electrodes can be found in [8]. An extra ground electrode for noise cancellation was implanted to the occipital region (the bottom part of the brain close to the cerebellum).

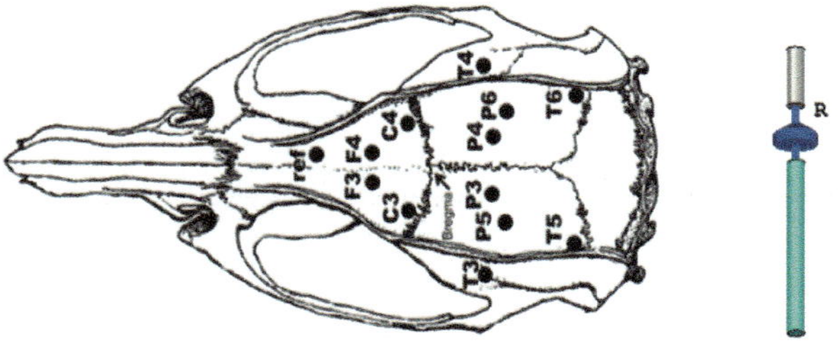

Fig. 2. Positions of electrodes on the skull (left), the model of the electrode in simulations (right) [8].

The model of simulated electrodes (1 reference and 12 measuring ones) is shown in Fig. 2, right from the skull. The model consists of a conductive arm (the light blue cylinder), a lumped resistor, and a ground. The length of the conductive arm equals to $l_e = 3$ mm, and material parameters correspond to the *Stablohm 675 Wire* with conductivity $\sigma = 891$ kS/m. The radius of the arm equals to $r_e = 0.15$ mm, and the lumped resistor is $R = 1$ MΩ. Finally, the ground is represented by perfect electric conductor (PEC).

3 Results

3.1 Electrodes Aligned to the Brain Surface

First, an influence of electrodes touching the brain surface was studied. We compared the electric field intensity simulated at a point (an ideal electrode) against the measurement with the modeled electrode while keeping the remaining parameters fixed.

Simulations at a point and modeled electrodes produce a single value of electric field intensity in a single measuring point. This measuring point corresponds to the center of the bottom surface of the realistic electrode. The measured intensity corresponds to the difference between the measuring electrode and the reference one. An overview of electric field intensities E [μV/m] for simulations with point electrodes and realistic ones is given in Table 1. Comparison of results shows that the simulation including realistic models of electrodes provides a lower level of electrical field.

Table 1. Comparison of electric field intensity E [μV/m] for simulations with point and realistic models of electrodes on rat's brain.

Electrode	F3	F4	C3	C4	P3	P4
Point electrode	0.4756	0.5885	0.5733	0.7342	1.2405	2.0748
Realistic electrode	0.0062	0.1150	0.2754	0.0499	0.8047	1.1396
Electrode	P5	P6	T3	T4	T5	T6
Point electrode	1.2865	1.0917	5.3995	1.8118	3.0674	1.1933
Realistic electrode	0.9016	0.2506	1.7625	1.4703	0.5500	0.5474

In the right part of Fig. 3, electric field intensity depending on the measuring trajectory p_1 is shown. The measuring trajectory p_1 is depicted by the blue line on the top of the brain surface in the left part of Fig. 3. The trajectory p_1 is measured from the left to the right. Obviously, the realistic models of electrodes slightly affect electric field intensity waveform along the trajectory p_1 when compared with the simulation with point electrodes. This influence is the most evident at $p_1 = 2.7$ mm where the measuring trajectory is in the shortest distance from the realistic electrode (P4). Furthermore, the

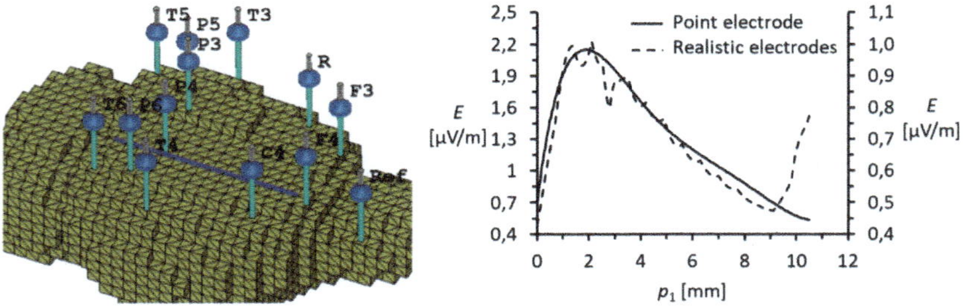

Fig. 3. Left: the measuring trajectory p1 represented by the blue line. Right: electric field intensity along p1 for point electrode (solid line, left vertical axis) and realistic electrode (dashed line, right vertical axis).

simulation with point electrodes (the left vertical axis) provides results with a higher electric field intensity along the whole measuring trajectory p_1.

Figure 4 shows electric field distribution on the surface of rat's brain close to the realistic electrode. We can observe a slight influence of the electrode on the field distribution in the interface between the electrode and the brain.

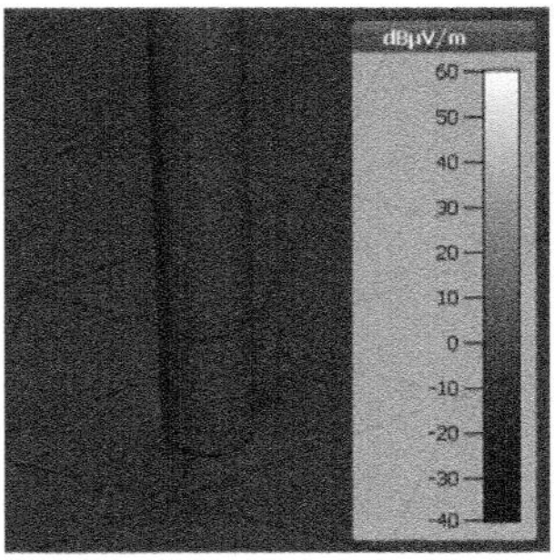

Fig. 4. Electric field distribution close to the electrode.

3.2 Electrodes Nested 1 mm Below the Brain Surface

Now, we study an effect of realistic electrodes nested 1 mm below the brain surface. Electric field intensity is measured in the bottom part of electrodes as in Sect. 3.1. Table 2 shows simulation results obtained for point electrodes and realistic ones. Comparisons indicate that the simulation assuming ideal electrodes produces a higher level of electric field intensity for most electrodes. Nevertheless, the difference between these two simulations is lower compared with the simulations on rat's brain (see Sect. 3.1). Therefore, the increase of the nesting depth decreases the influence of realistic electrodes.

Table 2. Comparison of electric field intensity E [µV/m] for simulations with point and realistic models of electrodes nested 1 mm below the brain surface.

Electrode	F3	F4	C3	C4	P3	P4
Point electrode	0.1896	0.2143	0.6569	0.7951	2.2754	3.3298
Realistic electrode	0.0683	0.2294	0.4023	0.6056	2.8983	3.2134
Electrode	P5	P6	T3	T4	T5	T6
Point electrode	2.3210	2.4323	1.3424	1.7828	2.5170	2.6045
Realistic electrode	1.4445	0.4760	1.0287	1.3428	0.8827	1.4867

Electric field distribution depending on the measuring trajectory p_1 (see the blue line in the left part of Fig. 3) is depicted in the left part of Fig. 5 left for both the cases. Comparison of both the dependencies reveals a higher level of electric field intensity in simulations with point electrodes. However, the difference between both the cases is lower compared with measuring on the brain surface (see Sect. 3.1). As previously, the fluctuations at $p_1 = 2.7$ mm are caused by the shortest distance of measuring trajectory p_1 from the realistic electrode.

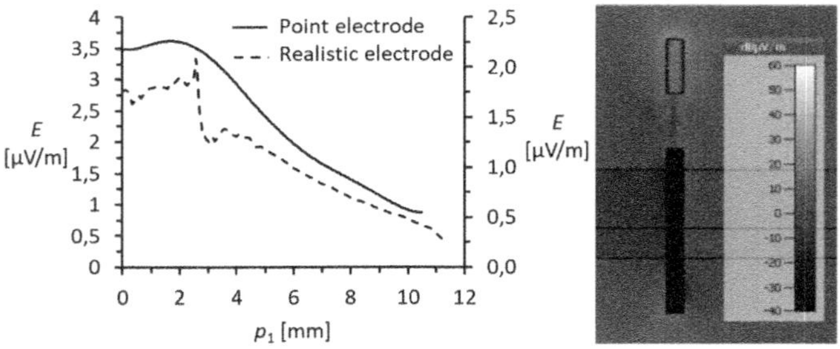

Fig. 5. Left: electric field intensity along p1 for point electrodes (solid line, left vertical axis) and realistic ones (dashed line, right vertical axis) nested 1 mm below the brain surface. Right: electric field distribution in the vertical cut crossing the center of the electrode.

In the right part of Fig. 5, the electric field distribution in the vertical cut crossing the center of the electrode P4 is shown. Obviously, the electrode affects the nearby electric field distribution in all the layers (the brain, the cerebrospinal fluid, and the skull).

3.3 Simulation with Stimulating Electrode

In this simulation, the excitation dipole was replaced by an additional electrode simulating a deep brain stimulation. The electrode is placed 1 mm anterior from the bregma and 1 mm below the brain surface. The excitation amplitude was the same as in the case of the excitation dipole. As previously, the simulations with point or realistic models of electrodes were performed.

Electric field intensity along the measuring trajectory p_1 (see the blue line in the left part of Fig. 3) is depicted In Fig. 6. The results show that the simulation without the model of the excitation electrode considerably attenuates the excited field caused by a different impedance condition of the exciter.

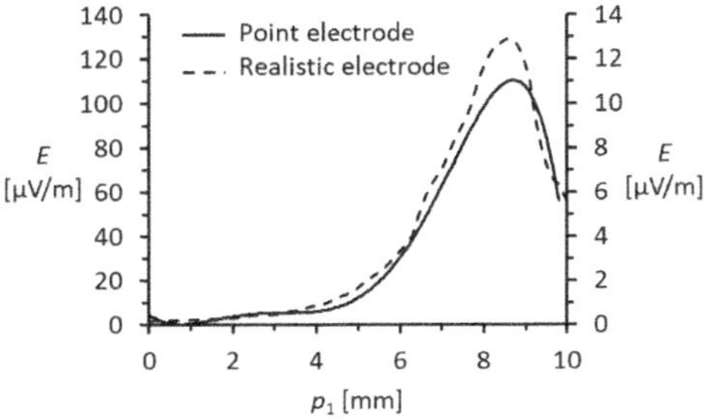

Fig. 6. Electric field intensity along p1 for point electrodes (solid line, right vertical axis) and realistic ones (dashed line, left vertical axis).

Electric field distribution with the stimulating electrode is shown in Fig. 7. A low level of the intensity on cerebellum is caused by the ground electrode. In addition, the different level of electric field intensity on electrodes is due to the inhomogeneous volume conductivity (the cerebrospinal fluid, and the skull).

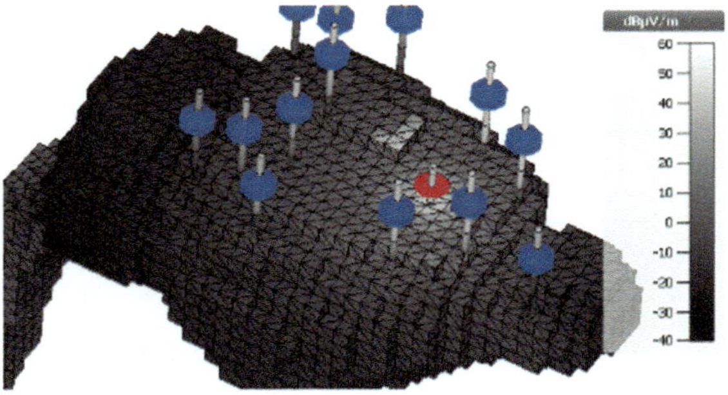

Fig. 7. Distribution of electric field intensity on the brain surface with stimulating electrode.

4 Conclusion

In the paper, the influence of physical models of electrodes on rat's brain forward modelling was studied for three cases.

First, the influence of models of cortical electrodes on the brain surface was studied. The results show that the simulation comprising realistic electrodes provides a lower level of electric field intensity for all measuring electrodes. Moreover, a slight influence on field distribution close to electrodes is observed.

Second, the influence of electrodes nested 1 mm below the brain surface was studied. In these simulations, higher values of intensity were reached for most electrodes. However, the difference between these simulations is lower compared to simulations on rat's brain. Hence, the increase of the nesting depth decreases the influence of electrodes.

Finally, the excitation dipole was replaced by an additional electrode and simulations with point and realistic electrodes were performed. The results indicate considerably attenuated field for simulation with point electrodes. The attenuation is caused by a different impedance condition of the exciter. For this reason, the inclusion of the physical model of the stimulating electrode can provide more accurate results.

Since the inverse estimation is very sensitive to errors in forward model, we consider realistic models of electrodes as an essential part of the whole imaging pipeline.

In near future, the realistic system of electrodes will be created and evaluated by simulations. The realistic model should indicate whether the model comprising the electrode, the resistor and ground is sufficient for forward modelling, or whether it is necessary to consider the whole system.

5 Acknowledgement

The presented work was supported by the Czech Science Foundation under the grant no. 18-16218S. Simulations were performed in the SIX Research Centre thanks to the support of the National Sustainability Program, the grant no. LO1401.

References

1. He, B.: Neural Engineering, 1st edn. Kluwer Academic/Plenum, New York (2005). ISBN: 978-0-306-48609-8
2. Hallez, H., Vanrumste, B., Grech, R., et al.: Review of solving the forward problem in EEG source analysis. J. NeuroEng. Rehabil. **4**(1) (2007). https://doi.org/10.1186/1743-0003-4-46
3. Nunez, P.L., Srinivasan, R.: Electric Fields of the Brain: the Neurophysics of EEG, 2nd ed. Oxford University Press, New York (2006). ISBN 9780195050387
4. Jain, S., Mittra, R., Wiart, J.: Full-wave modeling of brain waves as electromagnetic waves. Progr. Electromagn. Res. **151**, 95–107 (2015). https://doi.org/10.2528/PIER15011404
5. Kuratko, D., Raida, Z., Cupal, M., et al.: Electromagnetic modelling of rat's brain: comparison of models and solvers. In: Proceedings of International Workshop on Computing, Electromagnetics, and Machine Intelligence (CEMi 2018). Stellenbosch (South Africa) (2018)
6. CST Studio Suite (2018). https://www.cst.com/. Accessed 12 Nov 2018
7. Dielectric Properties of Body Tissues. http://niremf.ifac.cnr.it/tissprop. Accessed 25 Mar 2019
8. Páleníček, T., Fujáková, M., Brunovský, M., et al.: Electroencephalographic spectral and coherence analysis of ketamine in rats: correlation with behavioural effects and pharmacokinetics. Neuropsychobiology **63**(4) (2011). https://doi.org/10.1159/000321803

Improvement of Sleep Spindle Detection by Aggregation Techniques

Elizaveta Saifutdinova[1,2(✉)], Daniela Dudysova[1], Vaclav Gerla[2], and Lenka Lhotska[2]

[1] National Institute of Mental Health, Klecany, Czech Republic
saifueli@fel.cvut.cz
[2] Czech Technical University in Prague, Prague, Czech Republic

Abstract. The study focuses on automatic sleep spindle detection. Plenty of methods have been proposed in previous decades. However, there is still space for improvement. In this study, we investigate aggregation methods such as voting to achieve better results. We employ an unweighted model and two weighted voting models in which assigned weights represent reliability of automatic detectors. First weighted model utilizes supervised approach based on logistic regression. The second one applies unsupervised generative Bayesian model often used in crowd-sourcing. Using the expectation maximization algorithm, we uncover hidden true labels and weighs of detectors. We test methods on the real world datasets. The aggregation method overcome single detectors on 10% on average in terms of $F1$. Moreover, a probabilistic explanation of weights could be used in applications for visual analysis.

Keywords: Human sleep EEG · Sleep spindle · Aggregation · Unsupervised methods · Supervised methods

1 Introduction

Sleep spindle is a pattern in sleep electroencephalography (EEG) characterized as a 0.5–2 s long burst in 11–16 Hz frequency range. In sleep medicine, such patterns hallmark human sleep stage 2. Moreover, this sleep EEG pattern is connected to cognitive abilities [16] like learning [5] and memory consolidation [2], aging [6] and mental state [17]. Manual detection of sleep spindle is a challenging task even for experts. Analysis of hours of EEG data makes it tedious in the first place. Moreover, despite the strict definition, sleep spindles are hard to distinguish in sleep EEG. Agreement of different clinical experts may differ for the same data [13]. However, agreement of trained experts favorably differs from agreement of non-experts [3,10,18]. All these reasons lead to the development of automatic methods of sleep spindle detection. The most general approach for the detectors is to apply a detection function sensitive to sleep spindle and then define them by applying a threshold such as the detector calculating root mean square amplitude of the filtered signal (RMS) [12]. The other detector

J. Henriques et al. (Eds.): MEDICON 2019, IFMBE Proceedings 76, pp. 226–234, 2020.
https://doi.org/10.1007/978-3-030-31635-8_27

(RSP) proposed in [3] uses a relative sigma power for sleep spindle detection. S-transform is utilized by the detector proposed in [7] (SIGMA). It defines sigma index computed for obtained frequencies. The last detector transforms the signal with the Teager energy operator (TEO) [1]. Detection of sleep spindles is performed by application of a threshold. There were used implementations of the methods by [13] and [18] and threshold values proposed in [13]. Modern research is mostly focused on improving of existing methods using advanced approaches [9] including optimization [10,14] and combination of methods [4].

The main motivation of this study is to employ previously developed methods and aggregate their outcomes to improve sleep spindle detection. We choose voting as the aggregation model. We compare unweighted and two weighted models in this study. Assuming that weights represent reliability of detectors, we employ supervised and unsupervised methods to assign weights to single detectors. Logistic regression is utilized to derive weights based on supervised approach. Unsupervised method comes from the field of crowd sourcing where there are suggested many raters with different accuracy. Such approach is known for evaluation of human [15] and automatic [8] raters. It applies a generative Bayesian model to uncover a hidden true labels of the data and obtains accuracy of each rater regarding uncovered true labels. Instead of using human raters as it was in [18], we use previously proposed sleep spindle automatic detectors to provide labeling of the data. All three methods were tested on the real world sleep EEG.

2 Materials and Methods

2.1 Data

Three datasets were used in this study. First two datasets are taken from the open-source Dreams database [3]. It has 8 sleep EEG recordings on channel CZ-A1 or C3-A1 with 30 min length. Data of six subjects were recorded at 200 Hz and the two others at 50 Hz and 100 Hz respectively. Sleep spindles were labeled by two experts. The first expert evaluated all 8 recordings and the second one first six. For simplicity, we refer to these data as two different databases Dreams V1 and Dreams V2. We select only sleep stage 2 part of the analysis due to consistency in the statistical properties of spindle parameters in the data. Also, we excluded recording with sampling frequency 50 Hz because of too restricted frequency band-pass. Details on Dreams V1 and Dreams V2 are presented in Table 1. The third dataset NIMH was obtained from the National Institute of Mental Health, Czech Republic. The whole-night EEG as a part of the polysomnography (PSG) was recorded from subjects suffering from insomnia. Sleep stages scoring and artifact rejection was performed by a trained clinician. Sleep spindle labels were scored on C3-A2 EEG channel in sleep stage 2 by an expert. Details on NIMH dataset are presented in Table 1.

Table 1. Datasets description

Dataset	Recordings	fs (Hz)	Time (m)	Sleep spindles
Dreams V1	7	200/100	17.81 ± 3.08	42 ± 21
Dreams V2	5	200/100	16.53 ± 2.63	55 ± 23
NIMH	5	250	139.48 ± 29.39	182 ± 60

2.2 Sleep Spindle Detection

The task of pattern detection assumes that for a signal of N instances there are unknown real labels which can be represented as a binary vector $L = [l_1, \ldots, l_N]$. L contains real pattern labels. Each value l_i is 1 or 0 defining if the instance belongs to a searched pattern. In biomedical signal processing the vector L is often considered as unknown due to complexity of the field and the fact that obtained signals not always could be decoded. However, an expert scoring could be used as an approximation of L.

Unweighted Aggregation. Aggregation approach assumes that X is a binary matrix of observations obtained from one recording of size $N \times M$ where N is the length of binary scoring and M is the number of detectors. Output of aggregation method is a binary detection vector D of length N. For a voting model decision d_i is based on the value $v_i = \sum_{j=1}^{M} w_j X_{i,j}$ where w_j is an impact of each j_{th} detector such as

$$d_i = \begin{cases} 0 & \text{if } v_i < T \\ 1 & \text{if } v_i \geq T \end{cases} \tag{1}$$

where T is a predefined threshold. The unweighted technique assigned equal weights w_i as $1/N$ to all detectors. That is the simplest approach to resolve an unknown weights problem. The unweighted model does not reflect performance differences in the methods which may not be optimal.

Unsupervised Weighted Aggregation. The goal of weighted model is to obtain a final decision which relies on importance of each detector. In this section we describe a fully unsupervised model. As in the previous studies [8], here we employ a Bayesian model to attempt to uncover the performance properties of the detectors in conditions of unknown real labels. We assume independence of each detector given binary vector of real labels L. Assuming that each j_{th} detector performs with true positive rate (TPR) $\alpha_j \in [0, 1]$ and true negative rate (TNR) $\beta_j \in [0, 1]$. Denote a row of X as x_i, it represents an outcome of M detectors for instance i. Then, x_i is dependent on a real label l_i and parameters of the detectors $\alpha = \{\alpha_1, \ldots, \alpha_M\}$ and $\beta = \{\beta_1, \ldots, \beta_M\}$. Also, we assume that values l_i come from Bernoulli distribution with unobserved parameter k. Following the [8], we regularize parameters k by a symmetric Beta distribution $Beta(\theta + 1, \theta + 1)$ for $\theta \geq 0$. All α_j and β_j are regularized by the same parameters φ and ψ using asymmetric Beta priors $Beta(1, \varphi + 1)$ and $Beta(1, \psi + 1)$.

Learning of the model is performed by a standard log-likelihood maximization strategy using expectation maximization method. Here, X is the set of observed values, L is a latent variable and k, α and β are the parameters. Knowing the parameters we can get [8]

$$p(L = 1|X; \alpha, \beta, k) = [1 + \frac{1-k}{k} \prod_j a_j^{X_j} b_j^{1-X_j}]^{-1}, \tag{2}$$

where $a_j = \frac{\beta_j}{\alpha_j}$, $b_j = \frac{1-\beta_j}{1-\alpha_j}$ and X_j is the j_{th} column of X. Maximizing of $\ell(\alpha, \beta, k) = log\ p(X; \alpha, \beta, k)$ provides regularization of parameters [8] such as

$$k = \frac{\theta + \sum_i l_i}{2\theta + N}, \ \alpha_j = \frac{\sum_i l_i X_{i,j}}{\varphi + \sum_i l_i}, \ \beta_j = \frac{\sum_i (1 - l_i) X_{i,j}}{\psi + \sum_i (1 - l_i)}.$$

In the E-step, we update L by Eq. 2 given parameters k, α_j and β_j because L is a vector of Bernoulli variables [8]. The L is initialized as unweighted aggregation by majority voting. Steps are repeated until it converges. We assume convergence when $|k^{\tau} - k^{\tau+1}| + ||\alpha^{\tau} - \alpha^{\tau+1}|| + ||\beta^{\tau} - \beta^{\tau+1}|| < \epsilon$ where τ denotes iteration. We chose ϵ equal 10^{-12}.

Supervised Weighted Aggregation. The idea of supervised aggregation is based on learning a model on a set of provided examples. Due to high variability in sleep spindle density across all subjects, we suggest that a model should be trained on the data from the same patient. Therefore, we divide data of one subject in two successive parts P_1 and P_2 by some point on a time-line. We use P_1 for model training and P_2 is used for model validation. Logistic regression with L_2 regularization was chosen as a model due to its simplicity and a simple interpretation.

3 Experimental Results

First, we applied state-of-the-art sleep spindle detectors RMS, RSP, SIGMA, and TEO to datasets Dreams V1, Dreams V2, and NIMH. The outcome of the detection is represented as a binary matrix. To perform training of weighted voting models, we downsampled obtained binary detection using 0.5-s non-overlapped windows; labels were assigned by the majority of samples. All models were trained for separate subjects individually and do not depend on data of other subjects. The unsupervised model uses detection labels for the whole timeline, whereas supervised approach uses only the first part of continuous data. It contains 50% of sleep spindles for Dreams V1 and V2 datasets and 20% for NIMH dataset. Such values were selected based on signal length and consultancy with experts. Such division into training and testing data keeps the method relevant for practice in terms of data amount, which should be analyzed by an expert. Moreover, we excluded data with less than 25 sleep spindles from testing of the supervised weighed model due to the small number of training examples. Parameters for the unsupervised method were chosen empirically as ψ, φ equal 2 and θ is 1. Such parameters lead to faster convergence. Convergence was achieved not in more than 310 iterations for all datasets. It ended up with equally high (over 0.85 in average for each dataset) TPR and various TNR, which is reflected in the behavior of single detectors regarding false positive results.

Table 2. Performance results for Dreams V1 dataset

Method	F1	Rec	Prec
RMS	0.36 ± 0.15	0.67 ± 0.16	0.27 ± 0.10
RSP	0.43 ± 0.14	0.44 ± 0.18	0.45 ± 0.15
SIGMA	0.44 ± 0.10	0.47 ± 0.13	0.46 ± 0.15
TEO	0.34 ± 0.12	0.32 ± 0.17	0.53 ± 0.13
Unweighted	0.50 ± 0.09	0.61 ± 0.16	0.45 ± 0.14
Weighted unsupervised	0.50 ± 0.08	0.62 ± 0.15	0.46 ± 0.16
Weighted supervised	0.44 ± 0.10	0.52 ± 0.14	0.41 ± 0.17

Table 3. Performance results for Dreams V2 dataset

Method	F1	Rec	Prec
RMS	0.54 ± 0.10	0.75 ± 0.17	0.45 ± 0.13
RSP	0.58 ± 0.08	0.55 ± 0.16	0.67 ± 0.13
SIGMA	0.55 ± 0.11	0.49 ± 0.12	0.65 ± 0.17
TEO	0.41 ± 0.19	0.30 ± 0.18	0.83 ± 0.12
Unweighted	0.65 ± 0.04	0.70 ± 0.14	0.64 ± 0.11
Weighted unsupervised	0.67 ± 0.04	0.69 ± 0.14	0.69 ± 0.14
Weighted supervised	0.68 ± 0.06	0.71 ± 0.13	0.67 ± 0.08

Validation of results is performed by comparison to a labeling provided by the expert. We denote true positives TP as a number of sleep spindle events labeled by an expert which were labeled as a sleep spindle by a detector at least for a 0.3 s, false positives FP as a number of detected events and did not counted in TP, and false negatives FN as a number of sleep spindle events labeled by an expert and did not counted in TP. Presented classes are highly imbalanced. Moreover, we are interested in efficiency of sleep spindle detection. Therefore we use recall $Rec = \frac{TP}{TP+FN}$ and precision $Prec = \frac{TP}{TP+FP}$ and their harmonic mean $F1 = \frac{2TP}{2TP+FP+FN}$. Results for Dreams V1, Dreams V2 and NIMH datasets are presented in Tables 2, 3 and 4 respectively.

Table 4. Performance results for NIMH dataset

Method	F1	Rec	Prec
RMS	0.31 ± 0.11	0.94 ± 0.07	0.2 ± 0.09
RSP	0.50 ± 0.06	0.47 ± 0.13	0.59 ± 0.13
SIGMA	0.54 ± 0.05	0.56 ± 0.11	0.55 ± 0.10
TEO	0.45 ± 0.14	0.80 ± 0.05	0.32 ± 0.13
Unweighted	0.56 ± 0.06	0.62 ± 0.14	0.54 ± 0.13
Weighted unsupervised	0.57 ± 0.05	0.62 ± 0.10	0.54 ± 0.09
Weighted supervised	0.56 ± 0.03	0.64 ± 0.21	0.54 ± 0.12

4 Discussion and Conclusion

The paper describes the aggregation of sleep spindle detector outcomes using weighted and unweighted voting. Methods of automatic sleep spindle detection were applied to three datasets. Obtained results for Dreams datasets are similar to those published before [3,13]. Using only sleep stage 2 could and different evaluation procedure could slightingly change the outcome. Results from single detectors changes for different datasets. For example, for NIMH dataset RMS shows the worst performance and for Dreams V2 it is TEO. Many studies use one single method for discovering sleep spindle [11,12] due to the simplicity of the approach. However, detectors union can discover $68.7 \pm 30.8\%$ of the sleep spindles for the Dreams V1, $80.2 \pm 5.2\%$ for Dreams V2 and $97.4 \pm 2.4\%$ for NIMH dataset. Using aggregation of detectors is a logical step to do. However, obtained results are quite similar in terms of $F1$ due to chosen optimal threshold. Using more detectors and using different threshold strategy will sufficiently change results. Results of aggregation methods sharply differ for all events if they are limited by the performance of separate detectors. A significant number of FP is a typical problem of all sleep spindle detectors and using $F1$ helps to balance both $Prec$ and Rec to obtain detection relevant for practice. Changing the threshold may lead to improved results for visual inspection.

Voting provides a simple interface for method combination in comparison to some recent studies [10]. Strait forward unweighted approach improves the result. However, weighted methods have substantial benefits over the unweighted method. Weights could be interpreted as the reliability of detectors. That gives more insight into the data and detection. The main advantage of the unsupervised method is that they do not need any predefined information. It tries to uncover the true label which is valuable for any pattern EEG detection task. However, having detectors with low performance, it is really hard to achieve very good results. The supervised method tends to outperform unsupervised methods. Performance of the methods is directly connected to expert labeling. While unsupervised method tries to detect the real true labels, the supervised method adapts to the expert decision which may increase $F1$. However, lack of training examples negatively affects results like for Dreams datasets.

Overall, weighted methods are very powerful and may be used for a system for visual investigation. The unsupervised method provides a useful probabilistic explanation of the data which could be interpreted in colors like it is shown in Fig. 1. Weights might be corrected using the supervised method according to an expert decision. Such application design is simple and relevant for practical applications. Computational time is the greatest concern. Processing includes several detection methods and further method aggregation. It takes 7–20 s for unsupervised and 7–12 s for the supervised method for data from NIMH dataset. Testing is performed using python 2.7 and a Windows 10 Pro computer with an Intel Core i7 2.40 GHz CPU and 16 GB of RAM. Future work will be concentrated on the improvement of the presented approach and development of an application for visual investigation. Furthermore, we will study a problem of optimal threshold.

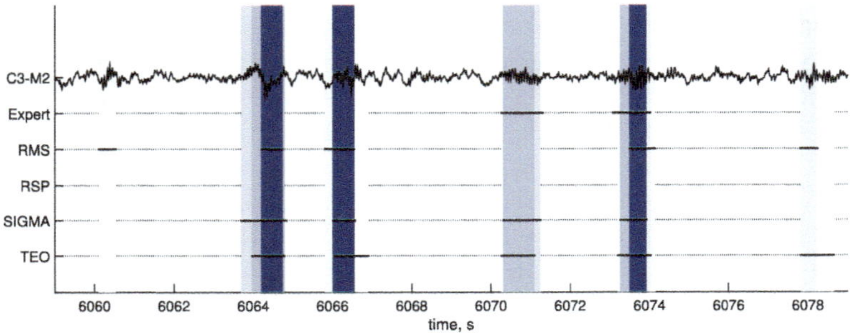

Fig. 1. Example of sleep spindles in C3-M2 channel (C3-M2) labeled by an expert (Expert), detections by sleep spindle detectors (RMS, RPS, SIGMA, TEO) and reliability combination of detection by unsupervised method expressed intensity of blue color (more intense color represent more reliable result)

Acknowledgment. Research was supported by the project Temporal context in analysis of long-term non-stationary multidimensional signal, Register Number 17-20480S of the Grant Agency of the Czech Republic. This work was also supported by the Charles University research program PROGRES Q35, by the project No. LO1611 with financial support from the MEYS under the NPU I program and by the Ministry of Health of the Czech Republic, grant No. NV18-07-00272.

References

1. Ahmed, B., Redissi, A., Tafreshi, R.: An automatic sleep spindle detector based on wavelets and the teager energy operator. In: 2009 Annual International Conference of the IEEE Engineering in Medicine and Biology Society, pp. 2596–2599, September 2009
2. Born, J., Rasch, B., Gais, S.: Sleep to remember. Neuroscientist **12**(5), 410–424 (2006). https://doi.org/10.1177/1073858406292647. pMID: 16957003

3. Devuyst, S., Dutoit, T., Stenuit, P., Kerkhofs, M.: Automatic sleep spindles detection – overview and development of a standard proposal assessment method. In: 2011 Annual International Conference of the IEEE Engineering in Medicine and Biology Society, pp. 1713–1716, August 2011
4. Duman, F., Erdamar, A., Erogul, O., Telatar, Z., Yetkin, S.: Efficient sleep spindle detection algorithm with decision tree. Expert Syst. Appl. **36**(6), 9980–9985 (2009). http://www.sciencedirect.com/science/article/pii/S0957417409001055
5. Gais, S., Mölle, M., Helms, K., Born, J.: Learning-dependent increases in sleep spindle density. J. Neurosci. **22**(15), 6830–6834 (2002). http://www.jneurosci.org/content/22/15/6830
6. Gennaro, L.D., Ferrara, M.: Sleep spindles: an overview. Sleep Med. Rev. **7**(5), 423–440 (2003). http://www.sciencedirect.com/science/article/pii/S1087079202902522
7. Huupponen, E., Gomez-Herrero, G., Saastamoinen, A., Värri, A., Hasan, J., Himanen, S.L.: Development and comparison of four sleep spindle detection methods. Artif. Intell. Med. **40**(3), 157–170 (2007). http://www.sciencedirect.com/science/article/pii/S0933365707000516
8. Kantchelian, A., Tschantz, M.C., Afroz, S., Miller, B., Shankar, V., Bachwani, R., Joseph, A.D., Tygar, J.D.: Better malware ground truth: techniques for weighting anti-virus vendor labels. In: Proceedings of the 8th ACM Workshop on Artificial Intelligence and Security, AISec 2015, pp. 45–56. ACM, New York (2015). http://doi.acm.org/10.1145/2808769.2808780
9. Lacourse, K., Delfrate, J., Beaudry, J., Peppard, P., Warby, S.C.: A sleep spindle detection algorithm that emulates human expert spindle scoring. J. Neurosc. Methods **316**, 3–11 (2019). http://www.sciencedirect.com/science/article/pii/S0165027018302504, methods and models in sleep research: A Tribute to Vincenzo Crunelli
10. Liu, M.Y., Huang, A., Huang, N.E.: Evaluating and improving automatic sleep spindle detection by using multi-objective evolutionary algorithms. Front. Hum. Neurosci. **11**, 261 (2017). https://www.frontiersin.org/article/10.3389/fnhum.2017.00261
11. Martin, N., Lafortune, M., Godbout, J., Barakat, M., Robillard, R., Poirier, G., Bastien, C., Carrier, J.: Topography of age-related changes in sleep spindles. Neurobiol. Aging **34**(2), 468–476 (2013). http://www.sciencedirect.com/science/article/pii/S019745801200320X
12. Mölle, M., Marshall, L., Gais, S., Born, J.: Grouping of spindle activity during slow oscillations in human non-rapid eye movement sleep. J. Neurosci. **22**(24), 10941–10947 (2002). http://www.jneurosci.org/content/22/24/10941
13. O'Reilly, C., Nielsen, T.: Automatic sleep spindle detection: benchmarking with fine temporal resolution using open science tools. Front. Hum. Neurosci. **9**, 353 (2015). https://www.frontiersin.org/article/10.3389/fnhum.2015.00353
14. Parekh, A., Selesnick, I.W., Osorio, R.S., Varga, A.W., Rapoport, D.M., Ayappa, I.: Multichannel sleep spindle detection using sparse low-rank optimization. J. Neurosci. Methods **288**, 1–16 (2017). http://www.sciencedirect.com/science/article/pii/S0165027017301681
15. Raykar, V.C., Yu, S., Zhao, L.H., Jerebko, A., Florin, C., Valadez, G.H., Bogoni, L., Moy, L.: Supervised learning from multiple experts: Whom to trust when everyone lies a bit. In: Proceedings of the 26th Annual International Conference on Machine Learning, ICML 2009, pp. 889–896. ACM, New York (2009). http://doi.acm.org/10.1145/1553374.1553488

16. Reynolds, C., Short, M., Gradisar, M.: Sleep spindles and cognitive performance across adolescence: a meta-analytic review. J. Adolesc. **66**, 55–70 (2018). http://www.sciencedirect.com/science/article/pii/S0140197118300599
17. Wamsley, E.J., Tucker, M.A., Shinn, A.K., Ono, K.E., McKinley, S.K., Ely, A.V., Goff, D.C., Stickgold, R., Manoach, D.S.: Reduced sleep spindles and spindle coherence in schizophrenia: mechanisms of impaired memory consolidation? Bio. Psychiatr. **71**(2), 154–161 (2012). http://www.sciencedirect.com/science/article/pii/S0006322311008146, functional Consequences of Altered Cortical Development in Schizophrenia
18. Warby, S.C., Wendt, S.L., Welinder, P., Munk, E.G., Carrillo, O., Sorensen, H.B.D., Jennum, P., Peppard, P.E., Perona, P., Mignot, E.: Sleep spindle detection: crowdsourcing and evaluating performance of experts, non-experts, and automated methods. In: Nature Methods (2014)

Preprocessing Pipeline for fNIRS Data in Children

Caterina Piazza[1(✉)], Andrea Bacchetta[1], Alessandro Crippa[1], Maddalena Mauri[1], Silvia Grazioli[1], Gianluigi Reni[1], Maria Nobile[1], and Anna Maria Bianchi[2]

[1] Scientific Institute, IRCCS E. Medea, Bosisio Parini, LC, Italy
caterina.piazza@lanostrafamiglia.it
[2] Department of Electronics Information and Bioengineering, Politecnico di Milano, Milan, Italy

Abstract. Functional near-infrared spectroscopy (fNIRS) is a non-invasive neuroimaging technique, largely used in paediatric research. However, there is not a standardized and widely accepted protocol for fNIRS data processing with potential effects on the reliability and replicability of the obtained results. The present study is within this framework aiming at the identification of an adequate pre-processing pipeline to be used for the analysis of children fNIRS datasets. The performance of five different motion correction techniques, based on the principal component analysis and on the wavelet filtering, was evaluated by analyzing fNIRS data recorded in 22 typically developing children (mean age 11.4 years). The results showed that the wavelet analysis combined with a moving average filter achieved the best performance, suggesting that this technique might become a gold-standard approach for motion artifacts correction in fNIRS children's datasets.

Keywords: Functional Near-Infrared Spectroscopy · Children · Motion correction · Principal component analysis · Wavelet filtering

1 Introduction

Functional near-infrared spectroscopy (fNIRS) is an optical neuroimaging technique that measures the changes in tissue hemodynamic related to brain activities [1]. Thanks to its non-invasiveness, portability and ease of use, fNIRS is particularly suitable for studies that involve infants, children and hard-to-test clinical populations (e.g. [2–6]). Although the use of this technique has largely increased over the last decades, no standardized and widely accepted signal processing pipelines for fNIRS data are available and suboptimal data processing might affect the obtained results [7]. Only recently, some studies tried to address this issue evaluating different preprocessing methods using both simulated and real data [8–11]. Since, motion artifacts still remain one of the major problems for the interpretation of fNIRS data, these studies are mainly devoted to assess different motion correction techniques.

One of the most common ways to handle with motion artifacts is the rejection of the trials where they are detected [12]. However, this solution is not suitable when the number of motion artifacts is significantly higher than the number of trials, as it often

J. Henriques et al. (Eds.): MEDICON 2019, IFMBE Proceedings 76, pp. 235–244, 2020.
https://doi.org/10.1007/978-3-030-31635-8_28

happens in pediatric studies. Thus, other methods have been proposed. Some of them are based on complementary measures of motion artifacts that allow their direct removal (e.g. [13, 14]). Whereas, others methods consist in data processing techniques that correct the motion artifacts by using computational approaches (e.g. principal component analysis (PCA) [15], spline interpolation [1] and wavelet analysis [16]). These solutions are the most commonly used, since they can be applied to every experimental design and they do not require the use of additional instrumentations. However, fNIRS signal analysis remains largely data- and subjects-dependent and the lack of standardized methods is particularly problematic in pediatric research.

The present study aims at identifying an efficient pre-processing pipeline for the analysis of children fNIRS datasets. Particular attention is given to the evaluation of motion artifact correction methods.

Among these methods, PCA has been traditionally used in the literature as a signal denoising technique (e.g. [17, 18]) and approaches based on the wavelet filtering seem to be particularly promising [8–10]. Given this preliminary evidence, five different motion artifact correction techniques based on the PCA and on the wavelet analysis were used and compared.

2 Methods

2.1 NIRS Data Acquisition

Twenty-two typically developing children (21 males, mean age 11.4 $\pm$ 1.9) were recruited in the present study. All children were Caucasian, had normal or corrected-to-normal vision, no history of medical treatment and were not taking any medication. Written informed consent to participate to the study was obtained from all parents prior to the inclusion in the protocol; the study was approved by the Ethics Committee of the Scientific Institute IRCCS E. Medea (Bosisio Parini, LC, Italy).

The participants performed a computerized variant of the visuospatial n-back working memory task (n-back VSWM), modified from Cui and colleagues [19].

The paradigm consisted of three blocks: 0-Back (0B), 1-Back (1B) and 2-Back (2B), presented in the following order: rest – 0B – 1B – 2B – rest – 1B – 0B – 2B – rest – 2B – 1B – 0B – rest. The first rest period lasted 60 s, while the other three rests lasted 30 s each. During the rest, children passively viewed an image on a black screen. Each experimental epoch included 32 stimuli, lasting 0.5 s each, with a 1.5 s interstimulus interval. The stimulus was a clown's face displayed in different locations in a 3 $\times$ 3 matrix.

NIRS data were acquired using a continuous-wave NIRS device (NIRx DYNOT compact 9–32, NIRx Medizintechnik, Berlin, DE). 32 channels were used as result of the coupling of 8 emitters and 24 detectors. The channels were placed on the bilateral frontotemporal areas centered at F3 and F4, according to the International 10–20 system. The source-detector distance was 2.7 cm (Fig. 1). Near-infrared spectroscopy recording was performed at two wave lengths (760 and 830 nm).

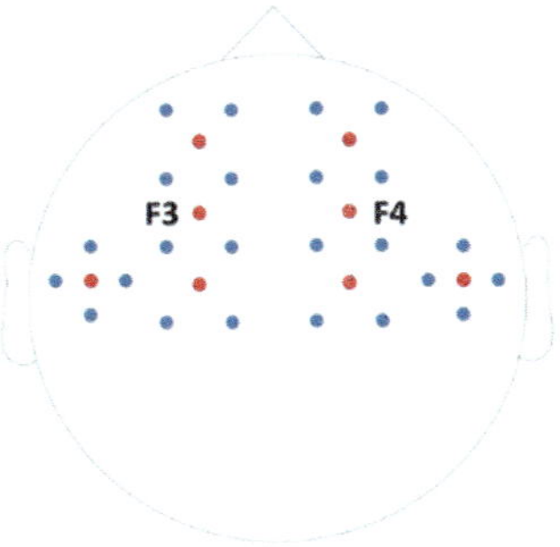

Fig. 1. Sources (red dots) and detectors (blue dots) locations on a simplified 10–20 system.

2.2 Data Pre-processing and Artifact Correction Techniques

NIRS data pre-processing was performed using the Homer2 v2.3 software [20]. Figure 2 depicts the processing pipeline.

First, the channels whose optical intensity signal (d) was too weak, too strong or had a high standard deviation (SD) were discarded (i.e., with mean(d) < 0.03 or > 3; SD(d)/mean(d) > 15). On average 2.6 channels (SD = 2.6, range = 0–9) per subject were rejected.

Then, the raw data were converted into changes in optical density data.

After the conversion, five different motion correction techniques were separately applied to the data. The algorithms used were based on the PCA and on the wavelet filtering.

Specifically, the PCA [15] was performed changing the percentage of variance removed. Two different values of threshold on the variance were used: 90% (PCA90) and 80% (PCA80), as suggested by Hu et al. [10], and Brigadoi et al. [9], respectively.

The third tested technique was the wavelet-based motion artifact correction approach proposed by Molavi and Dumont [16]. This technique is based on the hypothesis that the measured signal is a linear combination of the physiological hemodynamic signal and the artifacts, and that the detailed wavelet coefficients, obtained after the application of the discrete wavelet transform, have a Gaussian probability distribution. Because the hemodynamic response is slower and smoother than the motion artifacts, the distribution of the wavelet coefficients that represents the physiological components of the signal should be centered around zero with a low variance, whereas those corresponding to motion artifact should be the outliers of the probability distribution. The applied algorithm detected the outliers by means of a probability threshold and set them to zero before the wavelet reconstruction. In the present work, a threshold of 0.1 was used [8, 16].

The fourth technique applied consisted in the wavelet denoising described above preceded by a Moving Average filter (MA+Wavelet) performed over 5 s data windows [10].

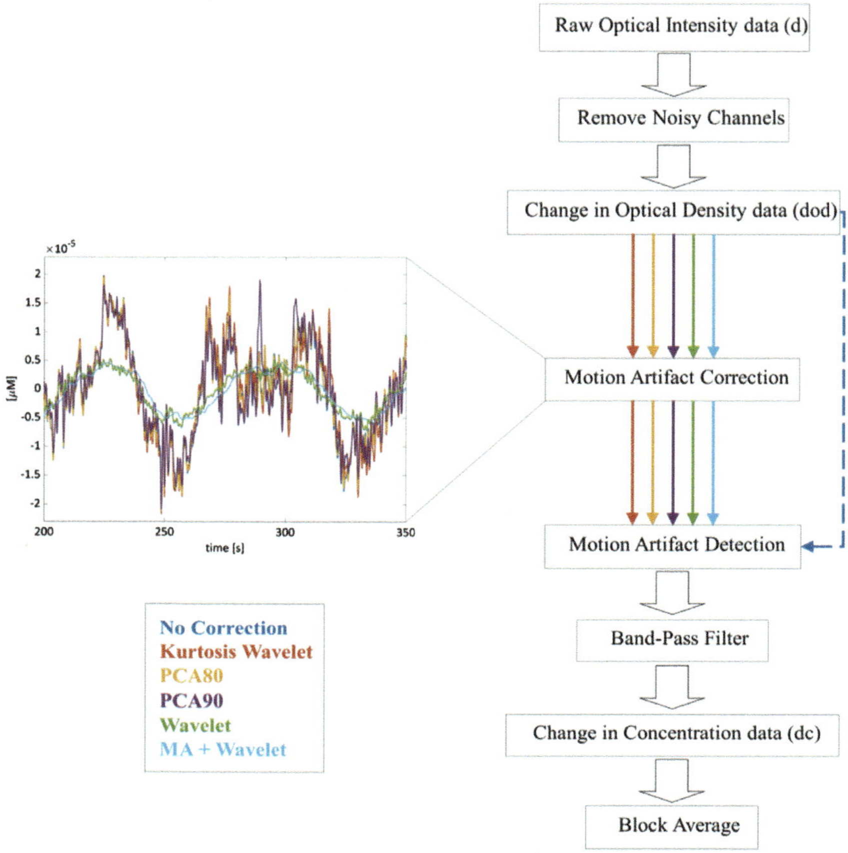

Fig. 2. Right panel: pre-processing pipeline of analysis. The dotted arrow refers to the processing stream without applying any technique for artifact correction (No Correction). Left panel: an example of the application of the different motion correction techniques used vs. uncorrected data is shown. The graph refers to dc data for one representative channel of one subject.

The last algorithm tested was the kurtosis-based wavelet filtering (Kurtosis Wavelet), which is an iterative motion correction method based on the evaluation of the fourth standardized moment (kurtosis) of the discrete wavelet transform coefficients distribution for the chosen decomposition level [21]. For this study, the kurtosis was set to 3.3 as suggested by Chiarelli and colleagues [21].

Finally, we performed an additional processing stream without applying any technique for artifact correction.

For each of the processing stream, the motion detection algorithm implemented in Homer2 was applied. The algorithm identifies the data-points that exceed a threshold in change of amplitude (AMPtresh = 0.6) or in change of standard deviation (STDEV-thresh = 50) within a predefined time interval (1 s).

Data were then filtered with a third order Butterworth 1 Hz low-pass filter, followed by a fifth order Butterworth 0.1 Hz high-pass filter, to further enhance the signal-to-noise-ratio.

Finally, the optical density data were converted into concentration changes (|HbO| and |HbR|) through the Modified Beer-Lambert Law. For the data conversion, the differential path length factor (DPF) was appropriately calculated accordingly to the age of the tested subjects [22]. Specifically, the DPF was set to 5.7 and 5.1 for the 760 nm and the 830 nm wavelength, respectively.

To obtain the task-related hemodynamic responses, data were divided in blocks accordingly to the task conditions (0 Back, 1 Back, and 2 Back) and averaged.

2.3 Metrics of Comparison

Five metrics of comparison were selected in order to quantitatively evaluate the performance of the different motion artifacts correction methods.

The first metric consisted is the percentage of data-points identified through the motion artifact detection algorithm implemented in Homer2 that was ran after the different motion correction techniques, in respect to the artifactual data-points identified in the uncorrected signals.

The second metric was the area under the curve (AUC) computed on the mean hemodynamic response function (HRF) for the first 1.5 s after stimulus onset ($AUC_{0-1.5}$). In particular, it can be assumed that the lower the AUC is, the better is the performance of the artifact correction method applied [9]. The 1.5 value was selected taking into account the time span in which the hemodynamic curve should become significant in children data [10].

The third metric was the ratio (AUC ratio) between the AUC of the mean HRF between 1.5 and 3 s after the stimuli presentation ($AUC_{1.5-3}$) and the $AUC_{0-1.5}$ reported above. This assumes that children hemodynamic response reaches its maximum between 1.5 and 3 s after the stimulus onset [10].

The fourth metric was the mean of the standard deviation (SD) of the hemodynamic data for each subject (within-subject SD). The assumption made is that the variability on the signal is mainly caused by motion artifacts [9].

The last metric used was the mean of the hemodynamic data across all subjects for a given channel (between-subjects SD). This index considers the variability between subjects [9].

For the last two metrics, it can be assumed that lower the standard deviation is, the better the motion artifacts correction technique is performed.

The above-mentioned metrics were computed for both HbO and HbR data, except for the percentage of corrected data-points (first metric), that was calculated on the optical density data.

2.4 Statistical Analysis

The statistical analysis was performed with the IBM SPSS, Statistics 21, software. For each metric, data were averaged across channels and conditions, thus obtaining a unique value for each subject. For between-subjects SD data were averaged across conditions and subjects, thus obtaining a unique value for each channel. The statistical analysis was done separately for HbO and HbR.

To check the distribution of each metric of comparison the Kolmogorov-Smirnov test was used. Since most of the variables showed a non-normal distribution ($p < .050$), the Friedman test was applied for each metric. Significant interaction effects were investigated using Wilcoxon pairwise comparisons.

3 Results

The median of the percentage of the motion artifact removed per method in respect to the number of artifacts identified in the uncorrected data is showed in Table 1. MA+Wavelet, Wavelet and PCA90 showed the best performances, while Kurtosis Wavelet removed less than 1% of the motion artifacts. The Friedman test identified a main effect of technique ($p < .001$) and post hoc comparisons showed statistical differences between Kurtosis Wavelet vs. MA+Wavelet, Wavelet and PCA90 ($p < .001$), PCA80 vs. MA+Wavelet ($p = .001$) and PCA80 vs. Wavelet ($p = .003$).

Table 1. Median, 25th and 75th percentiles of the percentage of motion artifacts removed by each technique in respect to the number of artifacts identified in the uncorrected data

	Median [%]	25th Perc. [%]	75th Perc. [%]
MA+Wavelet	100	100	100
Kurtosis Wavelet	0.75	0.34	1.23
Wavelet	100	100	100
PCA80	82.79	62.90	99.34
PCA90	100	96.27	100

A summary of the other metric results for HbO and HbR, is reported in Figs. 3 and 4. A main effect of technique was found for all the metrics (all Friedman test $p < .002$). The pattern of the results for HbO and HbR are similar.

Considering $AUC_{0-1.5}$, MA+Wavelet and Wavelet exhibited the lower values with low variability as well, compared to the other techniques and to uncorrected data (Fig. 3), as indicated by the Wilcoxon test. Specifically, MA+Wavelet showed $AUC_{0-1.5}$ lower values than the ones obtained with all the other techniques but Wavelet; Wavelet showed $AUC_{0-1.5}$ lower values than the ones obtained with Kurtosis Wavelet and PCA80.

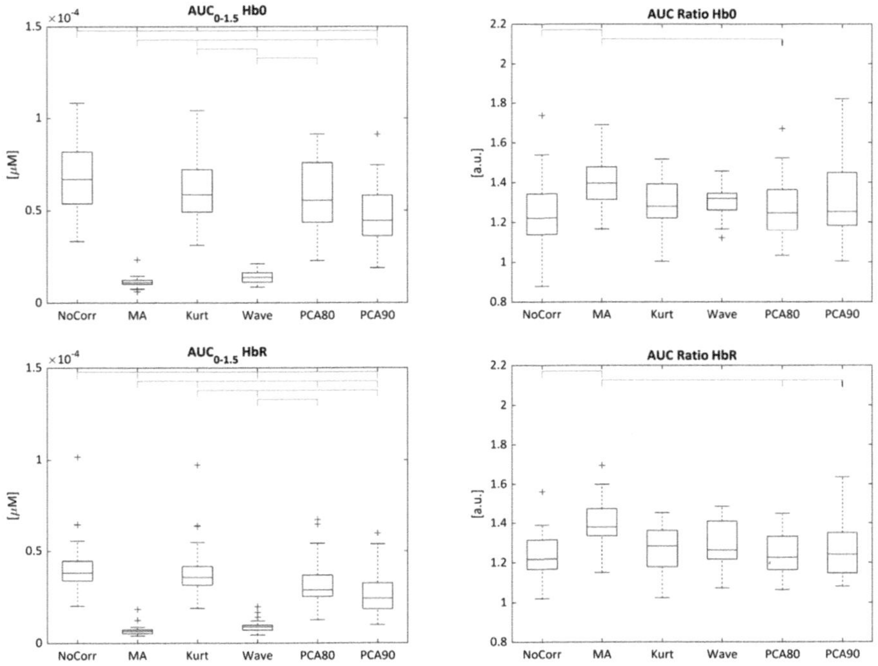

Fig. 3. Box plots of AUC ratio and $AUC_{0-1.5}$ computed for all techniques (from left to right: No correction, MA+Wavelet, Kurtosis Wavelet, Wavelet, PCA80 and PCA90) and for both HbO and HbR. The line in the box plot indicates the median, while the two extremities of the box plot represent the first and third quartile. Crosses indicate outliers. The lines linking the different techniques represent the significant statistical differences calculated with the Wilcoxon test (grey $p < .010$, black $p < .030$).

For AUC ratio, MA+Wavelet produced the highest values corresponding to a better performance (Fig. 3). The statistical analysis shows that only the MA+Wavelet method differed from the other techniques (No Correction, PCA80 and PCA90, the latter only for HbR).

The within-subject SD resulted to be lower in the data corrected with MA+Wavelet and Wavelet (Fig. 4). All techniques but the Kurtosis Wavelet significantly reduced the within-subject SD with respect to uncorrected data. Moreover, data corrected with MA +Wavelet and Wavelet had a significantly lower within-subject SD than data corrected with all other methods.

For the between-subject SD, the same pattern of statistically significant results was observed (Fig. 4).

Overall, the results showed that Wavelet and MA+Wavelet techniques outperformed the other motion artifact correction techniques with respect to the present dataset. However, the MA+Wavelet technique had better results for the AUC ratio metric compared to the Wavelet one.

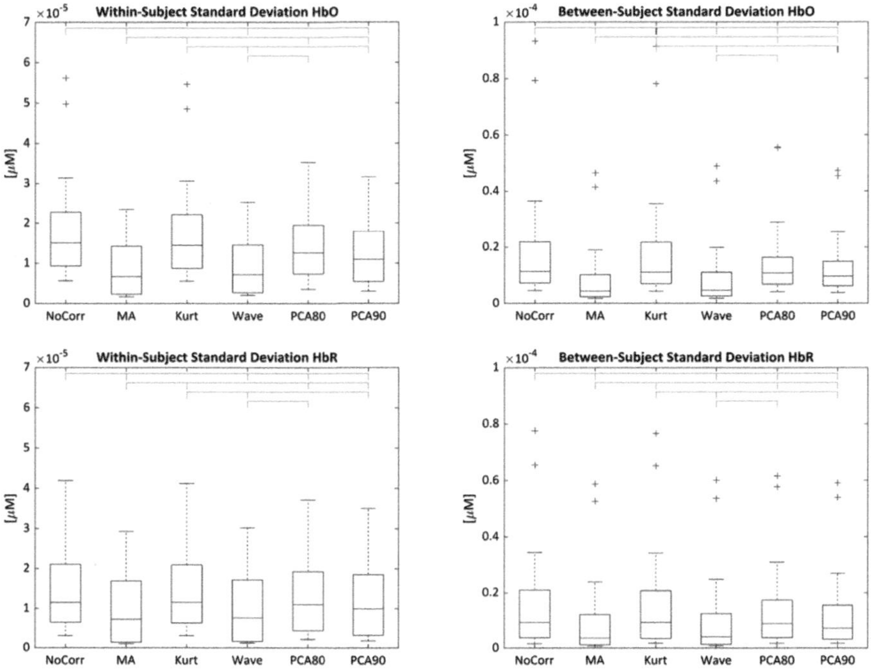

Fig. 4. Box plots of within-subject and between-subject SD computed for all techniques (from left to right: No correction, MA+Wavelet, Kurtosis Wavelet, Wavelet, PCA80 and PCA90) and for both HbO and HbR. The line in the box plot indicates the median, while the two extremities of the box plot represent the first and third quartile. Crosses indicate outliers. The lines linking the different techniques represent the significant statistical differences calculated with the Wilcoxon test (grey $p < .010$, black $p < .030$).

4 Discussion

The goal of this study was to identify the most efficient and reliable pipeline of analysis for the preprocessing of children fNIRS data. The performance of five different motion correction methods was evaluated by analyzing fNIRS data recorded in a group of children. Specifically, two PCA-based techniques and three wavelet-based methods were used. The methods' comparison was performed using five different metrics in order to qualitatively and quantitatively compare the performances of the motion correction techniques.

Previous studies reported that PCA effectively reduce motion artifact [8], although its performance heavily depends on the specific dataset [9]. In this study, we tested the PCA with two different variance thresholds. The PCA80 is expected to be more conservative and should identify only the most pronounced artifacts, whereas the PCA90 should remove more artifacts, with the risk of discharging also part of the brain hemodynamic response. As expected, our results showed a better performance of the PCA90 compared to the PCA80. In particular, the PCA80 obtained poor results in most of the metrics, over performing only the uncorrected data in the within-subject SD of the HbR response and in the between-subject SD of both HbO and HbR responses. Conversely, the PCA90 showed good results in all the metrics, except for the AUC ratio. However, both the PCAs failed to efficiently remove the motion artifacts in a subset of participants. A possible reason for this failure might be that the PCA is a multi-channel approach. This assumes that the artifacts have to be present in multiple channels to be discharged, and this was not always the case for our dataset.

The Kurtosis Wavelet technique yielded minimal improvement in comparison to the non-corrected data and resulted to be the worst correction method among those tested. This is in contrast with the results obtained by Hocke and colleagues [11]. They showed that the Kurtosis Wavelet performed efficiently, although overcompensating the motion artifacts. The discrepancy between the two studies could be due either to the different metrics used for the method comparison or to the differences in the datasets analyzed (children vs. adults; n-back VSWM task vs. finger-tapping task).

As previously reported in the literature [8, 9], our results acknowledge the wavelet technique as the most promising method for artifact correction. Indeed, it significantly reduced the motion artifacts and obtained very good results in all the metrics of comparison, especially in the $AUC_{0-1.5}$ and in the two SD-based metrics.

The MA+Wavelet technique achieved a better performance than the other techniques considered. It efficiently removed all the identified motion artifacts for every subject and achieved good results in all the metrics of comparison. The present results support and expand the previous literature [10], indicating the reliability of this method. The main disadvantage of this technique is that the motion correction process must be applied twice and thus may cause an over-reduction of the magnitude of the signal.

Future investigations of this work should extend the application of the present approach to independent data-sets and to the study of different type of motion artifacts. Nevertheless, if replicated in larger datasets, the present findings could suggest a promising, standardized approach as the gold-standard method for motion artifacts correction in fNIRS children's datasets. This will allow to obtain a more reliable reconstruction of the hemodynamic response, thus contributing to a better understanding/interpretation of children's cognitive states. Moreover, replicability and result comparisons between different studies will be made easier and more consistent.

References

1. Scholkmann, F., et al.: A review on continuous wave functional near-infrared spectroscopy and imaging instrumentation and methodology. Neuroimage **85**, 6–27 (2014)
2. Gallagher, A., et al.: Near-infrared spectroscopy as an alternative to the Wada test for language mapping in children, adults and special populations. Epileptic Disord. **9**(3), 241–255 (2007)
3. Crippa, A., et al.: The utility of a computerized algorithm based on a multi-domain profile of measures for the diagnosis of attention deficit/hyperactivity disorder. Front. Psychiatry **8**, 189 (2017)

4. Mauri, M., et al.: Light up ADHD: I. cortical hemodynamic responses measured by functional near infrared spectroscopy (fNIRS): special section on translational and neur. J. Affect. Disord. **234**, 358–364 (2018)
5. Grazioli, S., et al.: Light up ADHD: II. neuropharmacological effects measured by near infrared spectroscopy: is there a biomarker? J. Affect. Disord. **244**, 100 (2019)
6. McDonald, N.M., et al.: Infant brain responses to social sounds: a longitudinal functional near-infrared spectroscopy study. Dev. Cogn. Neurosci. **36**, 100638 (2019)
7. Pfeifer, M.D., Scholkmann, F., Labruyère, R.: Signal processing in functional near-infrared spectroscopy (fNIRS): methodological differences lead to different statistical results. Front. Hum. Neurosci. **11**, 641 (2018)
8. Cooper, R., et al.: A systematic comparison of motion artifact correction techniques for functional near-infrared spectroscopy. Front. Neurosci. **6**, 147 (2012)
9. Brigadoi, S., et al.: Motion artifacts in functional near-infrared spectroscopy: a comparison of motion correction techniques applied to real cognitive data. Neuroimage **85**, 181–191 (2014)
10. Hu, X., et al.: Comparison of motion correction techniques applied to functional near-infrared spectroscopy data from children. J. Biomed. Opt. **20**(12), 126003 (2015)
11. Hocke, L., et al.: Automated processing of fNIRS data—a visual guide to the pitfalls and consequences. Algorithms **11**(5), 67 (2018)
12. Selb, J.J., et al.: Effect of motion artifacts and their correction on near-infrared spectroscopy oscillation data: a study in healthy subjects and stroke patients. J. Biomed. Opt. **20**(5), 056011 (2015)
13. Robertson, F.C., Douglas, T.S., Meintjes, E.M.: Motion artifact removal for functional near infrared spectroscopy: a comparison of methods. IEEE Trans. Biomed. Eng. **57**(6), 1377–1387 (2010)
14. Virtanen, J., Noponen, T.E., Meriläinen, P.: Comparison of principal and independent component analysis in removing extracerebral interference from near-infrared spectroscopy signals. J. Biomed. Opt. **14**(5), 054032 (2009)
15. Zhang, Y., et al.: Eigenvector-based spatial filtering for reduction of physiological interference in diffuse optical imaging. J. Biomed. Opt. **10**(1), 011014 (2005)
16. Molavi, B., Dumont, G.A.: Wavelet-based motion artifact removal for functional near-infrared spectroscopy. Physiol. Meas. **33**(2), 259 (2012)
17. Wilcox, T., et al.: Using near-infrared spectroscopy to assess neural activation during object processing in infants. J. Biomed. Opt. **10**(1), 011010 (2005)
18. Zhang, X., et al.: Signal processing of functional NIRS data acquired during overt speaking. Neurophotonics **4**(4), 041409 (2017)
19. Cui, X., et al.: A quantitative comparison of NIRS and fMRI across multiple cognitive tasks. Neuroimage **54**(4), 2808–2821 (2011)
20. Homer2. https://homer-fnirs.org/. Accessed Mar 2019
21. Chiarelli, A.M., et al.: A kurtosis-based wavelet algorithm for motion artifact correction of fNIRS data. Neuroimage **112**, 128–137 (2015)
22. Duncan, A., et al.: Measurement of cranial optical path length as a function of age using phase resolved near infrared spectroscopy. Pediatr. Res. **39**(5), 889–894 (1996)

Wavelet Analysis-Based Reconstruction for sEMG Signal Denoising

Annachiara Strazza, Federica Verdini, Alessandro Mengarelli, Stefano Cardarelli, Andrea Tigrini, Sandro Fioretti, and Francesco Di Nardo(✉)

Department of Information Engineering, Università Politecnica delle Marche, Ancona, Italy
f.dinardo@univpm.it

Abstract. Surface electromyography (sEMG) recordings provide a safe, easy, and non-invasive method, allowing objective quantification of the electric activity of muscles. Analysis of sEMG plays an important diagnostic role in assessing muscle disorders. Typically, sEMG is a non-stationary signal contaminated by various noises or artifacts that originate at the skin-electrode interface, in the electronics, and in external sources. Thus, appropriate filtering procedures have to be applied to make sEMG clinically usable, in order to extract the main sEMG features. In the recent literatures, among the best performing denoising methods, Wavelet transformation (WT) denoising has been proposed. In particular, aim of this study is to propose a new denoising method based on WT multi-level decomposition analysis. To this aim, Daubechies mother wavelet (4th order, 9 levels of decomposition) was applied to 5 real sEMG tracings. Tibialis anterior (TA) and gastrocnemius lateralis (GL) signals are considered. This method focusses on the choice of a new thresholding rule for sEMG reconstruction and denoising. Performances of this method are computed against soft-thresholding denoising technique (ST) in terms of Root Mean Square Error (RMSE). After application of WT multi-level denoising technique, signal-to-noise ratio (SNR) increased significantly (TA: 14.5 $\pm$ 6.9 vs. 19.5 $\pm$ 7.1; GL: 14.0 $\pm$ 5.4 vs. 18.7 $\pm$ 6.3). Moreover, WT multi-level denoising technique showed a lower dispersion than ST (RMSE for TA: 0.8 vs. 1.2; RMSE for GL: 0.9 vs. 1.1.), introduced no sEMG signal delay. Thus, this method is a novel and efficient tool for sEMG denoising, that could be used to make easier the detection of sEMG activation onset-offset.

Keywords: Surface Electromyography · Wavelet Transform · Multi-level decomposition

1 Introduction

Surface electromyography (sEMG) is a non-invasive procedure involving the detection, recording and interpretation of the electric activity of muscles at rest and during activity. Analysis of sEMG signals has important uses in various applications including neurological diagnosis, neuromuscular and psychomotor research, sports medicine, prosthetics, and rehabilitation [1–3]. Nevertheless, sEMG signals are heavily contaminated

J. Henriques et al. (Eds.): MEDICON 2019, IFMBE Proceedings 76, pp. 245–252, 2020.
https://doi.org/10.1007/978-3-030-31635-8_29

by undesired effects mostly derived from electrodes, cables, data amplification and acquisition devices, power lines, and even the body itself. These effects are called artifacts, interference, and noise in general, and they can affect the resolution and the shape of biomedical recordings [4, 5]. Conventional filtering techniques such as low-pass, high-pass, and band-pass can be used for reducing line interference, thermal noise, etc. However, conventional techniques cannot remove random noise and they can smooth EMG signal [4, 6]. Recently, novel methods such as wavelet transform have been successfully applied to remove noise from sEMG signals [7, 8]. Generally, the critical point of WT filtering is the choice of mother wavelet, decomposition level and threshold. Soft-thresholding (ST) function is usually a traditional rule used for sEMG denoising [9], while no a proper selection of decomposition level number useful for sEMG reconstruction has been so far identified.

Thus, the aim of the present study is to propose a new method to sEMG denoising based on a WT multi-level reconstruction, focusing on the choice of a proper and new thresholding rule for sEMG denoising and reconstruction.

2 Materials and Methods

2.1 Real Data

Five healthy young adults were recruited for the study. Mean (±SD) characteristics are age 22.7 ± 0.5 years; height 166 ± 8 cm; weight 53.6 ± 8.1 kg; body mass index (BMI) 18.9 ± 1.5 kg•m^{-2}. Exclusion criteria included pathological or chronic joints pain, neurological pathology, orthopedic surgery within previous year, abnormal gait or BMI $\geq$ 25. Each subject signed written informed consent prior the beginning of the tests. The present research was undertaken in compliance with the ethical principles of Helsinki Declaration and approved by institutional expert committee.

2.2 WT Multi-level Decomposition Analysis for sEMG Denoising

sEMG denoising is performed by a multi-level wavelet decomposition analysis. The block diagram is depicted in Fig. 1.

Firstly, baseline removal was performed. In sEMG recordings, to remove artifacts due to subject movement and to the type of electrode used, a 6th order Butterworth band-pass filter (cut-off frequency: 20–450 Hz) has been applied. Band-pass filtering allowed to reduce sEMG signal bandwidth, increasing frequency resolution of each WT level decomposition [5]. This pre-filtering step is followed by sEMG denoising using a WT-based procedure, a flexible approach to signal decomposition. WT is a time-scale method that performs a series decomposition of a non-stationary signals in a set of functions $\Psi a, b(t)$ (mother wavelet), that are different both in the scale factor (a) and in the time shift (b). WT of the input signal sEMG(t) is defined as (Eq. 1):

$$\mathrm{WT}x(a, b) = \int sEMG(t)\Psi a, b(t)dt \quad a \neq 0 \tag{1}$$

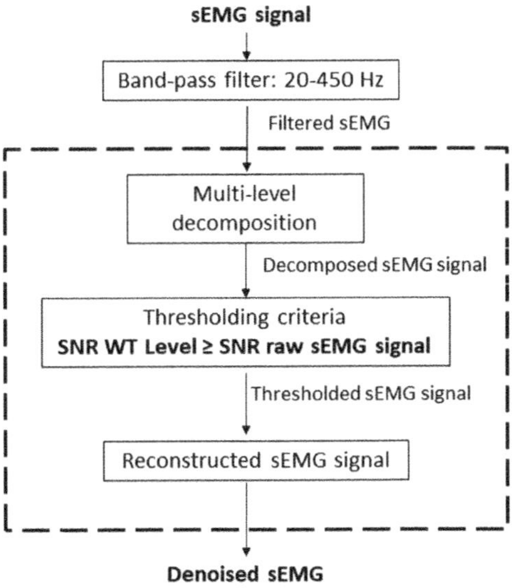

Fig. 1. Block diagram of sEMG denoising based on multi-level decomposition analysis.

The mother wavelet is given by (Eq. 2):

$$\Psi a, b(t) = \frac{1}{\sqrt{a}} \Psi\left(\frac{t-b}{a}\right) \tag{2}$$

where $\Psi a, b(t)$ must satisfy mathematical criteria like finite energy and no zero-frequency component to be admissible [10]. The dilation or scale is inversely proportional to the spectral components. Low scales or high frequencies provide more local information while high scales or low frequencies provide relatively more global information about the signal. This WT multi-resolution property makes this technique appropriate for gait analysis. In this work, decomposition, denoising and reconstruction were performed by mother wavelet Daubechies of 4th order with 9 levels of decomposition [11, 12].

Wavelet Multi-level Decomposition: The decomposition of sEMG is computed using a pair of finite impulse response filters, which are low-pass and high-pass, describing a tree iterative procedure. The outputs give the detail coefficients (from the high-pass filter, y_{high} [n]) and the approximation coefficients (from the low-pass, y_{low} [n]) [10]. The filter outputs are then down-sampled as given by Eqs. 3 and 4, respectively:

$$y_{low} = \sum\nolimits_{k=-\infty}^{\infty} x[k] * g[2n-k] \tag{3}$$

$$y_{high} = \sum\nolimits_{k=-\infty}^{\infty} x[k] * h[2n-k] \tag{4}$$

The number of steps of this iterative tree procedure is defined by the number of decomposition levels. Thus, a multi-level multiresolution analysis was used to generate an improved reconstruction algorithm (Fig. 2).

Thresholding Criteria: A level-based thresholding criterion was established considering the information contained in each detail. Details, at higher frequencies, are related to noise. In order to obtain a clean sEMG, this method considered SNR associated to each WT level; the decomposition level was chosen if SNR associated to each WT level is higher than SNR associated to raw sEMG signal (Fig. 1).

Reconstructed sEMG Signal: Applying this thresholding criterion application, the algorithm discarded WT level containing noise and used the sum of the remaining ones to reconstruct the denoised sEMG signal (Fig. 1).

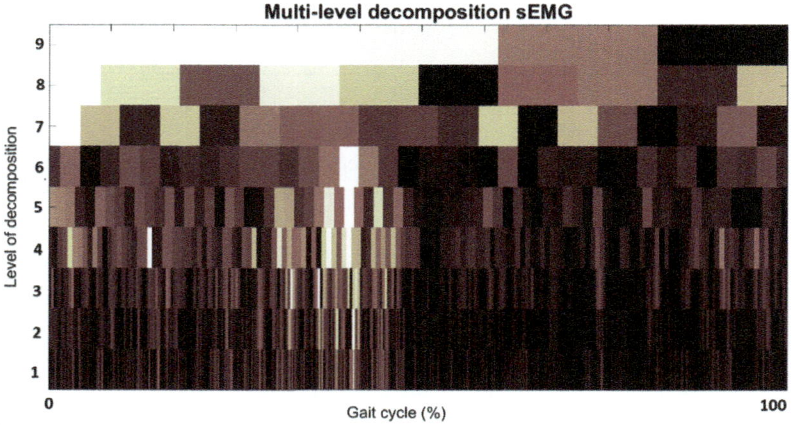

Fig. 2. Multi-level decomposition of sEMG signal. In x-axis, gait cycle expressed in %. In y-axis, number of decomposition level (9).

2.3 Statistics and Performances

Signal-to-noise ratio (SNR) values for sEMG signal were computed according to the following definition:

$$SNR = 10 * \log \frac{(\sigma_s^2)}{(\sigma_n^2)} \tag{5}$$

where σ_s and σ_n are the standard deviation of signal and noise, respectively [11]. SNRraw represents the SNR values of sEMG signals before the two WT denoising procedures; SNR_TA_Dec and SNR_GL_Dec are SNR values of denoised TA and GL signals after the WT multi-level denoising technique, respectively. SNR_TA_ST and

SNR_GL_ST are the SNR values of denoised TA and GL signals after ST, respectively. In order to evaluate their correctness, the Pearson correlation coefficient (ρ) was computed. Performances of this method are computed in terms RMSE against well-known ST, which reconstructs the signal using a fixed default threshold for all decomposition levels [9]. Non-normal variable distributions were compared using the Wilcoxon Rank-Sum test. Statistical significance was set at 0.05 in all cases.

3 Results

An example of raw (grey) and denoised (black) TA signal with WT multi-level denoising technique (panel A) e with soft thresholding method (ST) is shown in Fig. 3. In Fig. 4, raw (grey) and denoised (black) GL signal with WT multi-level denoising technique (panel A) e with soft thresholding method (ST) is reported.

After application of WT multi-level denoising technique, mean SNR values of TA and GL signals increased significantly (Table 1). In Table 1, mean SNR values of TA and GL signals increased but not significantly (Table 1). SNR_TA_Dec, SNR_GL_Dec, SNR_TA_ST and SNR_GL_ST were characterized by different RMSE values associated by a high correlation (ρ) and statistically significant (P), as reported in Table 2.

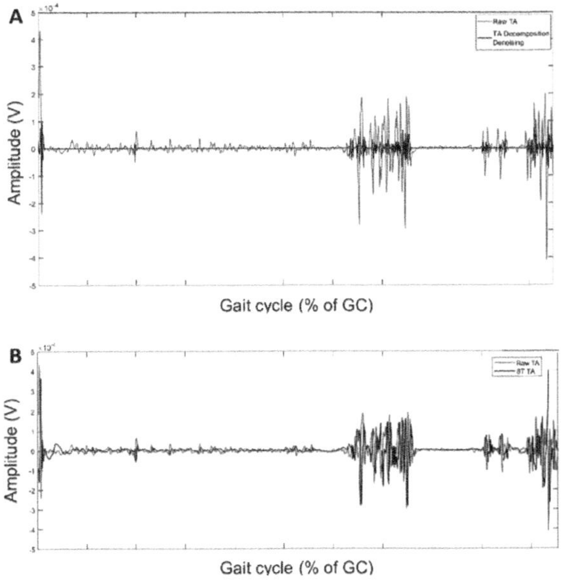

Fig. 3. Raw (grey) and denoised (black) TA signal with multi-level decomposition analysis (panel A) e with soft thresholding method (ST).

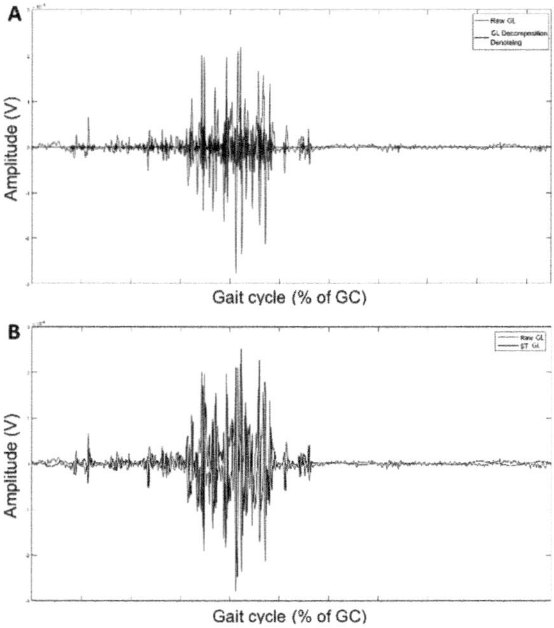

Fig. 4. Raw (grey) and denoised (black) GL signal with multi-level decomposition analysis (panel A) e with soft thresholding method (ST).

Table 1. Mean ± SD of SNR after multi-level decomposition analysis (SNR_ Dec) and after ST (SNR_ ST).

Method	TA	GL
SNRRaw	14.5 ± 6.9	14.0 ± 5.4
SNR_ Dec	19.5 ± 7.1[a]	18.7 ± 6.3[a]
SNR_ ST	15.6 ± 5.8	15.1 ± 4.8

[a]Statistical significance was set at $p < 0.05$

Table 2. RMSE, correlation coefficient of TA/GL for multi-level decomposition technique and for ST.

Method	RMSE	ρ
SNR_TA_Dec	0.8	0,92[a]
SNR_GL_Dec	0.9	0,97[a]
SNR_TA_ST	1.2	0.89[a]
SNR_GL _ST	1.1	0.83[a]

[a]Statistical significance was set at $p < 0.05$

4 Discussion

The aim of the study was to propose WT multi-level denoising technique as an efficient method to denoise sEMG signals. For this purpose, 5 real sEMG tracings were analyzed. Daubechies mother wavelet (4th with 9 levels of decomposition) was chosen for WT analysis. This method focusses on the choice of a new thresholding rule for sEMG reconstruction and denoising. A level-based thresholding criterion was established considering the information contained in each detail. Details, at higher frequencies, are related to noise. In order to obtain a clean sEMG, the new thresholding criterion considered SNR associated to each WT level; the decomposition level was chosen if SNR associated to each WT level is higher than SNR associated to raw sEMG signal.

Performances of this method are computed against ST in terms of RMSE. Results showed that for both TA and GL, SNR_TA_Dec and SNR_GL_Dec values increased significantly respect to the SNRraw; instead SNR_TA_ST and SNR_GL_ST values remained almost unaltered (Table 1). Computed RMSE values showed a lower dispersion of WT multi-level denoising technique (Table 2). Moreover, this method introduced no sEMG signal delay, maintaining qualitatively unaltered onset and offset time activation.

5 Conclusion

In conclusion, this method is a novel and efficient tool for sEMG denoising. A suitable denoising of sEMG could be useful to improve the detection of sEMG activation onset-offset, main features of electromyographic signals.

References

1. Raez, M.B.I., Hussain, M.S., Mohd-Yasin, F.: Techniques of EMG signal analysis: detection, processing, classification and applications. Biol. Proced. Online **8**, 11–35 (2006)
2. Chowdhury, R.H., Reaz, M.B., Ali, M.A., et al.: Surface electromyography signal processing and classification techniques. Sens. (Basel) **13**(9), 12431–12466 (2013)
3. Istenic, R., Kaplanis, P., Pattichis, C.S., et al.: Analysis of neuromuscular disorders using statistical and entropy metrics on surface EMG. WSEAS Trans. Signal Process. **4**(2), 28–35 (2008)
4. Amrutha, N., Arul, V.H.: A review on noises in EMG signal and its removal. Int. J. Sci. Res. Publ. **7**(5), 23–27 (2017)
5. De Luca, C.J., Gilmore, L.D., Kuznetsovb, M., Royb, S.H.: Filtering the surface EMG signal: movement artifact and baseline noise contamination. J. Biomech. **43**(8), 1573–1578 (2010)
6. Doğan, B., Senyurek, V.Y., Varol, H.S., Baspinar, U.: A comparative study of denoising sEMG signals. Turk. J. Electr. Eng. Comput. Sci. **23**(4), 931–944 (2013)
7. Sobahi, N.M.: Denoising of EMG signals based on wavelet transform. Asian Trans. Eng. **1**(5), 2221–4267 (2011)
8. Bhoi, A.K., Tamang, J.S., Mishra, P.: Wavelet packet based denoising of EMG Signal. Int. J. Eng. Res. Dev. **4**(2), 78–83 (2012)

9. Donoho, D.L.: De-noising by soft-thresholding. IEEE Trans. Inf. Theory **41**(3), 613–627 (1995)
10. Mallat, S.G.: A theory for multiresolution signal decomposition: the wavelet representation. IEEE Trans. Pattern Anal. Mach. Intell. **11**(7), 674–693 (1989)
11. Strazza, A., Verdini, F., Mengarelli, A., Cardarelli, S., Burattini, L., Fioretti, S., Di Nardo, F.: A time-frequency approach for the assessment of dynamic muscle co-contractions. In: IFMBE Proceedings, vol. 68, no. 2, pp. 223–226 (2018). https://doi.org/10.1007/978-981-10-9038-7_41
12. Strazza, A., Verdini, F., Burattini, L., Fioretti, S., Di Nardo, F.: Time frequency analysis of surface EMG signals for maximum energy localization during walking. In: IFMBE Proceedings, vol. 65, pp. 494–497 (2017). https://doi.org/10.1007/978-981-10-5122-7_124

An Information-Theoretical Method for Emotion Classification

Susana Brás[1(✉)], João M. Carvalho[1], Filipa Barros[3,4], Claúdia Figueiredo[2], Sandra C. Soares[3,4,5], and Armando J. Pinho[1]

[1] IEETA, DETI, University of Aveiro, Aveiro, Portugal
susana.bras@ua.pt
[2] GOVCOPP, University of Coimbra, Coimbra, Portugal
[3] CINTESIS-UA, DEP, University of Aveiro, Aveiro, Portugal
[4] William James Center for Research, DEP, University of Aveiro, Aveiro, Portugal
[5] Department of Clinical Neurosciences, Division of Psychology, Karolinska Institute, Stockholm, Sweden

Abstract. Identifying the emotion that someone is feeling will allow to improve the experience of the person interaction with environments, devices, and contents. Our body responds to events around us, by emotional responses, reflected in cognitive, behavioral and physiological dimensions. In the present work, we target the electrocardiogram (ECG) response as a mean to express emotions. Its processing is performed using information-theoretical measures, allowing true exploratory data mining. Participants recruited for the experiment watched three video sets in three different days, with a different emotion being induced in each day: fear, happiness, and neutral condition. The method is divided in: (1) conversion of the real-valued ECG record into a symbolic time-series; (2) relative compression of the symbolic representation of the ECG, using the symbolic ECG records stored in the database as a reference; (3) identification of the ECG record class, using a 1-NN (nearest neighbor) classifier. An accuracy of 90% was obtained. A posteriori analysis of the false negative results indicated that there was a relation between the relative dissimilarity measure and the self-reported emotions.

Keywords: Emotion · Affective computing · Classification · Data compression · Kolmogorov complexity

1 Introduction

Emotions interact strongly with decision-making, perception and learning [1]. Thus, studying emotional perception and response allow to understand people's preferences and strategies to adapt and respond across contexts. Emotions regulate bodily changes through activation of the central and peripheral nervous system, being this translated as behavioral and/or physiological responses [2]. Hence, the study of physiological responses allows to establish an empirical model describing a quantitative evaluation of emotion.

J. Henriques et al. (Eds.): MEDICON 2019, IFMBE Proceedings 76, pp. 253–261, 2020.
https://doi.org/10.1007/978-3-030-31635-8_30

Emotion identification allows a better interaction with computers, systems and environments, enabling an improved user experience [3]. Ludic activities and health care devices, for instance, are obvious candidates to benefit from this.

The electrocardiogram (ECG) is a well known signal in biomedical applications. As all biomedical signals, it is influenced by inter and intra variability. The emotion felt by each person is different, so the physiological response is also different between persons [4,5].Nevertheless, the intra variability prints a key reaction to external events and allows emotion identification [6].

One strategy in experimental psychology studies for the induction of emotions is the presentation of movies [7]. Usually, participants are equipped with physiological monitors, which allow to record data, able to be used a posteriori and correlate with self-report measures (e.g., questionnaires) that evaluate the impact of the movies on the participants [7]. One interest topic is the association between these questionnaires response and the physiological alterations evaluated during emotion induction.

In order to explore all data potential, we are exploring parameter free data mining algorithms for emotion physiological evaluation [8]. Nevertheless, when we talk about emotions, we know that, in most of the cases, we are not feeling one isolated emotion. So, in this paper, we present an alternative approach that not only classifies the new record in a class (as presented in our previous work [8]), but also gives a probability of inclusion in the other evaluated classes, based on physiological data processing.

2 Materials and Methods

If an emotion is experienced, then the physiological response of our body will be similar to other physiological records collected in the same conditions, i.e., in the presence of the same emotion. So, in this work, we present a measure of dissimilarity. Basically, we compare a new ECG record with a database representing a group of different emotions, and the new record will be classified as belonging to the class with lower dissimilarity.

The dissimilarity measure is based on compression methods, which are related to the notion of Kolmogorov complexity. A parameter free data mining technique is applied, to overcome most of the problems that affect the traditional ECG based methods (preprocessing error, biases on ECG segments identification, etc.) [9,10]. A true exploratory data mining is allowed by these methods, since there is no assumption about the premises. They do not need to extract features from the data in order to describe it. Instead, they evaluate the global data sequence and compare them, in order to find similarities and dependencies. The Kolmogorov complexity of x, $K(x)$, is defined as the minimum size of a program that produces x and stops [11], where x is a string (e.g., of binary digits). A major drawback of the Kolmogorov complexity (or algorithmic entropy) is that it is not computable. To overcome this limitation, it is usually approximated by some computable measure, such as Lempel-Ziv, linguistic or compression-based complexity measures. These approximations provide upper bounds on the true complexity, allowing, e.g., developing similarity/dissimilarity measures (see [8,12] for recent applications of this concept).

2.1 The ECG Database

Fifty-three participants (19 males and 34 females) aged between 18 and 28 years (21.17 $\pm$ 2.65), volunteered to participate in the study. Only participants with normal or corrected-to-normal visual acuity were included. Regarding education levels, thirty-four participants have completed the secondary education, fifteen had a bachelor and four a master's degree. Forty-five participants were right handed, six were left handed and two reported being ambidextrous. Also, participants were free of any medication or disease that could influence the cardiac functioning (e.g., tricyclic antidepressants or cardiac arrhythmia, respectively) and did not report psychiatric or neurological diagnostics. All participants completed an online sociodemographic questionnaire with relevant information for the study, as well as a Portuguese adaptation of STICSA-Trait, in order to measure their trait anxiety.

The study was approved by the Ethics and Deontology Council of the University of Aveiro, Portugal, and was conducted according to the guidelines of the Declaration of Helsinki and the American Psychological Association. All participants received the necessary information to participate, including their right to withdraw from the study at any time, and provided their informed consent.

Procedure. The study started with an initial interview to collect informed consent and explain the procedure. The emotion-inducing stimuli consisted in three sets of 8–12 movie excerpts, with approximately 30 min of duration, in total. As these sets aimed to induce specifically the emotion of fear, happiness and neutral condition, excerpts of terror, comedy movies and documentaries were selected for each of the inductions, respectively. Also, three documentary excerpts ("baseline" videos), of approximately 5 min, were selected and presented before each set of emotional videos. The movie excerpts of each set, as well as the neutral "baseline" videos, were selected after conducting pilot studies in our lab, which aimed to characterize each set of videos regarding their ability to induce the different emotions mentioned above. The analysis of the participants' subjective responses showed that the selected sets were indeed effective, in our sample, at inducing happiness, fear and neutral emotions.

The three sessions of emotional induction, performed by each participant (i.e., in a within-subject's design) were spaced for at least one week, to avoid emotional contagion, and the order of presentation of each emotional condition was counterbalanced.

In each session, the experimenter asked the participant to seat comfortably on a chair in front of the computer. The participant was informed about the procedure and instructed to place his/her chin on a chin rest. Also, the participant was instructed to be as attentive as possible throughout the task. Four Visual Analogue Scales (VAS) of 100 points were used to measure participants' emotional state (happy, fear, anxiety and stress) before and after the emotional induction. The experience started by the baseline movie presentation, followed by the video set. In the end of the video set, new Video VAS were presented to evaluate the presented video set as a whole and, after this evaluation, the participant had to answer to the Emotional State VAS.

ECG Recording. The cardiac signal of each participant was recorded using BIOPAC MP160 data acquisition system and AcqKnowledge 5 software (BIOPAC Systems, Inc.), with a sampling rate of 1000 Hz. Also, Ag/AgCl disposable vinil electrodes (EL503; BIOPAC Systems, Inc.) and the corresponding conductive gel were used to acquire the signal, following the Lead II configuration [13]. Before the task, the experimenter cleaned the participants' right wrist and the interior part of both the ankles with ethyl alcohol (70%), before placing the electrodes. After the placement, the participant waited at least 10 min before starting the experiment to allow the stabilization of the psychophysiological signal.

2.2 The Normalized Relative Compression Measure

One of the most interesting and important problems that can be addressed using the Kolmogorov complexity theory is the definition of a similarity measure between binary objects. In order to approximate the Kolmogorov complexity, a compression algorithm needs to be able to accumulate knowledge of the data, while the compression is performed. It has to find dependencies, to collect statistics, i.e., it has to create an internal model of the data.

The specific measure that we used is based on the concept of *relative compression* of two objects, denoted by $C(x\|y)$, i.e., the compression of object x is performed using *only* information available from another object, y. We define the normalized relative compression (NRC) of x given y as

$$\mathrm{NRC}(x\|y) = \frac{C(x\|y)}{|x|}, \tag{1}$$

where $|x|$ is the size of the object. Basically, this measure gives information about the fraction of data in x that cannot be described by y. The $C(x\|y)$ uses a combination of finite-context models of several orders (k) to build an internal model of y, which is kept fixed afterwards. Then, x is encoded exclusively using the model built from y. In order to estimate such models, three parameters had to be considered: k, γ, and α (for details on the encoding method, please see [8, 12,14]), considering previous works that already validate the best parameters combination [8], in this work the parameters were γ of 0.2 and a combination of (α, k) pairs taken from set {(1/1000, 5), (1/1000, 6), (1/1000, 7), (1/1000, 8), (1/1000, 9)}.

Using the NRC measure, it is expected that when comparing two ECG segments, lower values would be obtained for the same emotion, rather than when comparing different emotions.

When compression algorithms are used, it is generally preferable the use of symbolic records. Since ECG is a real-valued numerical signal, it was converted into a symbolic time series, by quantization methods.

In particular, it was used the SAX representation [15,16], which depends on two optimization parameters: (1) The alphabet size; (2) The dimension w of the series segments. The parameters combination is achieved by the ones that produces good results. Based on previous works [8], the alphabet size was defined as 20, and the w as 3. The present method is described as follows:

- The real-valued time series is normalized (by subtraction of the mean value and division by the standard deviation);
- The normalized time series is divided into N segments of w data points;
- The mean value of each N_i segment matches a symbol in the new data representation.

2.3 Statistical Analysis

Each ECG record was divided in 4 sub-segments. Then a combination of two sub-segments (15 min duration) was used as reference/training data, while the other two sub-segments (15 min duration) were used as target/test data. Then, we adopted a leave-one-out strategy for the evaluation of the method. A 1-NN classifier was applied, considering as distance measure the NRC value. The idea was to assess the identification performance, which was evaluated by means of sensitivity, specificity and accuracy.

3 Contribution

The classification method was divided into six steps:

1. divide each record in four segments;
2. the reference was a combination of two segments;
3. all possible combinations of reference were used to train the models;
4. the target segments were not in the reference;
5. the NRC was calculated for all combinations of reference/target;
6. the emotion was classified based on the lower NRC value obtained against fear, happy and neutral references.

Besides emotion classification, it was also our goal to present a degree of inclusion of a new record in the database classes:

1. considering each participant data, the standard deviation was calculated for the NRC values obtained in all reference combinations for each emotion;
2. in order to define a comparable measure between classes, the measure may vary between 0 and 1. Therefore, a measure scale change is performed by the normalization of the previously calculated standard deviation. This was achieved by dividing each emotion standard deviation by the sum of participant's emotions standard deviations.

4 Results and Discussion

The use of measures able to classify emotions allow the study and interpretation of physiological responses, contributing for a deeper theoretical understanding of emotions. Emotions are differently experienced by different people. Therefore, a method to classify an emotion independently of the person, in theory, will present higher error. Therefore, in this approach, we intended to present a similarity measure over emotions evaluating segments of ECG from the same person. Table 1 resumes the classification results. A global accuracy of 93%, a sensitivity of 90% and a specificity of 94% were achieved. When the emotions were evaluated separately, the best classification results were associated with the happy emotion. Notwithstanding this result, fear presented a higher counting of false negatives, and the neutral condition presented higher count of false positives.

Table 1. Classification results measured by accuracy, sensitivity and specificity for each evaluated emotion: fear (F), happy (H), and neutral condition (N) and considering the global performance (Total).

	Total	F	H	N
Accuracy	0.93 ± 0.07	0.91 ± 0.09	0.95 ± 0.08	0.92 ± 0.09
Sensitivity	0.90 ± 0.11	0.92 ± 0.14	0.92 ± 0.14	0.87 ± 0.16
Specificity	0.94 ± 0.06	0.90 ± 0.12	0.96 ± 0.07	0.94 ± 0.10

The degree of inclusion allowed not only to classify a record in a class, but also to establish a probability of the record to also be present in another class; this allows the overlap between classes (Table 2). As can be evaluated from Table 2, the average degree in the corresponding emotion was around 76%. With our approach, we can observe that:

1. participant 2 fear record was 86.6% in fear, 7.2% in happy and 6.2% in neutral condition.
2. participant 35 fear record was 33.9% in fear, 54.9% in happy and 11.3% in neutral condition.
3. participant 1 happy record was 5.8% in fear, 82.2% in happy and 11.9% in neutral condition.
4. participant 1 neutral record was 2.3% in fear, 6.3% in happy and 91.4% in neutral condition.

One hypothesis for the obtained errors is the real experienced emotion, i.e., the video was evaluated for example as fear, however, someone may not feel real fear. In order to validate that this might have been the case, we inspecting the subjective responses to VAS regarding emotional state. It was observed that the records catalogued as:

Table 2. Average and standard deviation of the degree of inclusion of records from fear, happy and neutral condition (horizontal), in the corresponding emotion classification (vertical).

	Fear	Happy	Neutral
Fear	0.76 ± 0.17	0.11 ± 0.12	0.13 ± 0.10
Happy	0.09 ± 0.07	0.77 ± 0.18	0.15 ± 0.14
Neutral	0.09 ± 0.08	0.14 ± 0.14	0.77 ± 0.18

1. fear classified in the happy condition presented an increase in correlation between the degree of inclusion in the emotion with happy VAS from the baseline ($R = -0.4$) to the post-video ($R = -0.7$);
2. fear classified in the happy condition presented a similar correlation between the degree of inclusion in the emotion with anxiety VAS from the baseline ($R = 0.7$) to the post-video ($R = 0.8$);
3. fear classified in the happy condition presented a similar correlation between the degree of inclusion in the emotion with fear VAS from the baseline ($R = 0.7$) to the post-video ($R = 0.7$)

These results may be justified by the fact that some participants may usually watch terror movies, and enjoy it, which may indicate that the emotion intended to be elicited was not achieved. As such, it comes as no surprise that the physiological response to the stimuli was not consistent with other participants.

Considering the other emotions, there was not any significant result in the subjective evaluation.

5 Conclusion

Physiological response to emotions is characterized by alterations on the cardiovascular system. The changes on ECG are printed on its waveform. In this work, we propose a method able to find patterns on the ECG waveform to distinguish emotions. This method is based on the principle of relative similarity from a new record to the class, by means of compression algorithms.

The presented method allows emotion classification with similar performance as found in previous work [8]. However, in this work, we also presented a method that allows class overlap, i.e., the idea is to give a probability of inclusion of a new record in all classes. The class where it belongs will have a higher probability of inclusion, in limit 100%, and the class less represented will, in the limit, have a 0% probability of inclusion. When we are measuring emotions, it is not a deterministic question. In the same episode we may experience different emotions simultaneously and with different intensities. So, in this work, we present a new preliminary approach in the way to proceed to a multi-class classification of emotions.

The results found are promising. However, they lack from a systematic validation with other similar algorithms. Also, the design of new protocols with exhaustive self-reported measures for assessing the emotions participants are experiencing is planned. In the present study only 3 conditions were evaluated, which may be limiting, so, a study is planned including all six basic emotions.

The presented method allows a straightforward interpretation of emotion classification based on physiological signals, and its association with subjective questionnaires about emotions.

Acknowledgments. This work is funded by national funds, European Regional Development Fund, FSE through COMPETE2020, through FCT, in the scope of the framework contract foreseen in the numbers 4, 5 and 6 of the article 23, of the Decree-Law 57/2016, of August 29, changed by Law 57/2017, of July 19; and in the scope of the projects UID/CEC/00127/2013 (IEETA/UA), CMUP-ERI/FIA/0031/2013 (VR2Market). J. M. Carvalho acknowledges the Doctoral Grant from FCT, ref. SFRH/BD/136815/2018.

Conflict of Interest. None declared.

References

1. Dolan, R.J.: Emotion, cognition, and behavior. Science **298**(5596), 1191–1194 (2002)
2. Mauss, I.B., Robinson, M.D.: Measures of emotion: a review. Cogn. Emot. **23**(2), 209–237 (2009)
3. Calvo, R., D'Mello, S.: Affect detection: an interdisciplinary review of models, methods, and their applications. IEEE Trans. Affect. Comput. **1**(1), 18–37 (2010)
4. Agrafioti, F., Hatzinakos, D., Anderson, A.K.: ECG pattern analysis for emotion detection. IEEE Trans. Affect. Comput. **3**(1), 102–115 (2012)
5. Barrett, L.F., Mesquita, B., Ochsner, K.N., Gross, J.J.: The experience of emotion. Annu. Rev. Psychol. **58**, 373–403 (2007)
6. Janssen, J.H., Tacken, P., de Vries, J.J.G., van den Broek, E.L., Westerink, J.H.D.M., Haselager, P., IJsselsteijn, W.A.: Machines outperform laypersons in recognizing emotions elicited by autobiographical recollection. Hum. Comput. Interact. **28**(6), 479–517 (2013)
7. Ferreira, J., Brás, S., Silva, C.F., Soares, S.C.: An automatic classifier of emotions built from entropy of noise. Psychophysiology **54**(4), 620–627 (2017)
8. Brás, S., Ferreira, J., Soares, S., Pinho, A.: Biometric and emotion identification: an ECG compression based method. Frontiers Psychol. **9** (2018)
9. Agrafioti, F., Hatzinakos, D.: ECG biometric analysis in cardiac irregularity conditions. SIViP **3**(4), 329–343 (2009)
10. Odinaka, I., Lai, P.-H., Kaplan, A.D., O'Sullivan, J.A., Sirevaag, E.J., Rohrbaugh, J.W.: ECG biometric recognition: a comparative analysis. IEEE Trans. Inf. Forensics Secur. **7**(6), 1812–1824 (2012)
11. Li, M., Vitányi, P.M.B.: An Introduction to Kolmogorov Complexity and Its Applications. Springer Science & Business Media, New York (2009)
12. Pinho, A.J., Ferreira, P.J.S.G., Neves, A.J.R., Bastos, C.A.C.: On the representability of complete genomes by multiple competing finite-context (Markov) models. PLoS ONE **6**(6), e21588 (2011)

13. Berntson, G.G., Quigley, K.S., Lozano, D.: Cardiovascular psychophysiology. In: Handbook of Psychophysiology, 3rd edn., pp. 182–210. Cambridge University Press (2007)
14. Pinho, A.J., Ferreira, P.J.S.G.: Image similarity using the normalized compression distance based on finite context models. In: 2011 18th IEEE International Conference on Image Processing (ICIP), pp. 1993–1996. IEEE (2011)
15. Lin, J., Keogh, E., Lonardi, S., Chiu, B.: A symbolic representation of time series, with implications for streaming algorithms. In: Proceedings of the 8th ACM SIGMOD Workshop on Research Issues in Data Mining and Knowledge Discovery, pp. 2–11. ACM (2003)
16. Lin, J., Keogh, E., Lonardi, S., Patel, P.: Finding motifs in time series. In: Proceedings of the 2nd Workshop on Temporal Data Mining, pp. 53–68 (2002)

Potential Biomechanical Overload on Skeletal Muscle Structures in Students During Walk with Backpack

Giovanni D'Addio[1(✉)], Leandro Donisi[1,2], Luca Mercogliano[3], Giuseppe Cesarelli[4], Paolo Bifulco[1,5], and Mario Cesarelli[1,5]

[1] Istituti Clinici Scientifici Maugeri IRCCS, Pavia, Italy
gianni.daddio@icsmaugeri.it
[2] Department of Advanced Biomedical Sciences, University Hospital of Naples 'Federico II', Naples, Italy
[3] NIT - Naples Institute of Technologies Inc., Naples, Italy
[4] Center for Advanced Biomaterials for Healthcare (IIT@CRIB), Italian Institute of Technology, Naples, Italy
[5] DIETI, University of Naples 'Federico II', Naples, Italy

Abstract. Although a very large number of students in the world use uncomfortable and heavy backpacks, their negative musculoskeletal effects on gait and posture are still not well investigated. Aim of the paper has been the study of differences affecting the kinematic gait parameters during free walk and walk with backpack to evaluate their potential influence on skeletal-muscle disorders. Gait recordings in both conditions on 50 healthy volunteers participating students have been performed by a G-WALK inertial system calculating the main kinematic parameters namely Propulsion index and Initial Double Support, Stance and Swing Phases. ANOVA results between both walking conditions showed that all gait cycle studied values are significantly negatively affected by walking with backpack supposing a potential biomechanical overload on skeletal muscle structures in students exposed to these prolonged conditions.

Keywords: Gait analysis · Inertial device · School backpack

1 Introduction

Idiopathic scoliosis, hyperkyphosis and rachialgie are considered the most common pathologies of the spinal column in children and adolescents. Although at the root of these pathologies there are mainly genetic and physiological components, other significant behavioral factors could play an important role in their evolution as biomechanical loads on the shoulders due to school backpack (BP) consisting of a daily estimated 9.3 kg load carried for an estimated transport time of about 7 min [1–3]. Aim of the paper has been the study of differences affecting the kinematic gait parameters during free walk and walk with backpack in order to evaluate their potential influence on skeletal-muscle disorders.

J. Henriques et al. (Eds.): MEDICON 2019, IFMBE Proceedings 76, pp. 262–266, 2020.
https://doi.org/10.1007/978-3-030-31635-8_31

2 Materials and Methods

2.1 Inertial Device for Gait Analysis

Although the gold standard in gait analysis is still represented by stereophotogrammetric technology, inertial devices are gaining a very increasing use in gait analysis studies, due to their easy experimental setup and data processing procedures, to the lower cost of the instrumentation and to the limited time required for the examinations [4–11]. For this reason, G-WALK System (BTS Bioengineering) has been used in the present study consisting by a commercial inertial device, widely used in rehabilitation settings and clinical gait analysis (Fig. 1), able to compute automatically all kinematic parameters related to gait cycle.

G-WALK System is composed of a single wearable device (G-Sensor) equipped with a 3 axis 16 bit/asse accelerometer, a 3 axis 16 bit/asse gyro, a 3 axis 13 bit/asse magnetometer in wireless Bluetooth communication with a PC running the G-Studio Software able to store the main kinematic parameters (www.btsbioengineering.com).

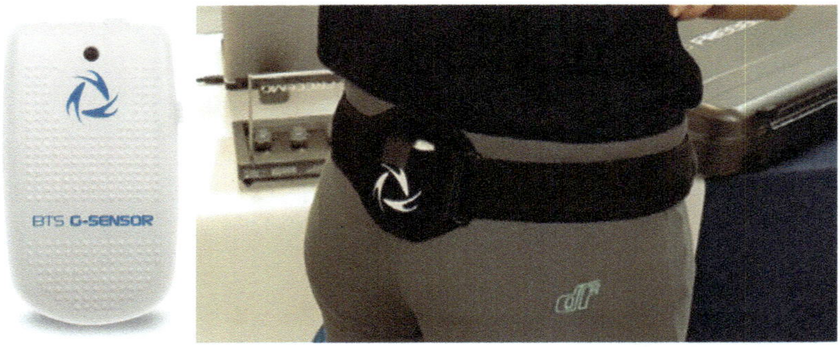

Fig. 1. Inertial sensor of the G-Walk System on the left and its placement on the right

2.2 Population and Study Protocol

The study has been carried out on a population of 50 healthy volunteer participants (students of Scientific High School "Elio Vittorini", Naples, Italy), aging from 14 to 15 years old (60% males). All participants underwent a gait analysis session instrumented by the inertial device G-Sensor placed in the pocket of a belt in a lumbosacral position between S1 and S2 vertebrae (Fig. 1) and wearing sport wears and trainers shoes.

The gait analysis protocol consisted in two different sessions, a first free walking session and a second one walking with a 9.3 kg backpack in accordance with Negrini's study [3]. Each session consisted in four consecutive trials, each of which consisting in a 15 m walk along a straight path, spaced by a pause of at least 1 min. The main kinematic gait parameters calculated by the device and considered in the study has been Propulsion (dimensionless), Initial Double Support Phase (% cycle), Stance Phase (% cycle) and Swing Phase (% cycle) considering for each subject the mean values between the four trials.

2.3 Statistical Analysis

Statistical analysis of results has been first performed testing the Gaussian distribution of all kinematic parameters considered in the study by D'Agostino-Pearson omnibus normality test. Afterwards, differences in mean values of all parameters between the two different study conditions have been studied by appropriate (paired t test or Wilcoxon matched-pairs signed rank test) two tailed ANOVA test. All tests have been performed with a 95% confidence values by GraphPad Prism Software version 7.0 for Windows (www.graphpad.com).

3 Results

ANOVA results for each kinematic parameter studied in the two different conditions are shown in Table 1 and Fig. 2.

Table 1. Anova test between free walk and walk with backpack for each parameter

Parameters	Free walk	Walk with backpack
Propulsion	9.94 ± 2.11	9.21 ± 1.59***
Initial double support phase (%)	9.14 ± 1.21	11.51 ± 1.39****
Stance phase (%)	59.12 ± 1.21	61.46 ± 1.38****
Swing phase (%)	40.88 ± 1.21	38.54 ± 1.38****

* $p < 0.5$, ** $p < 0.01$, *** $p < 0.001$, **** $p < 0.0001$ at ANOVA test

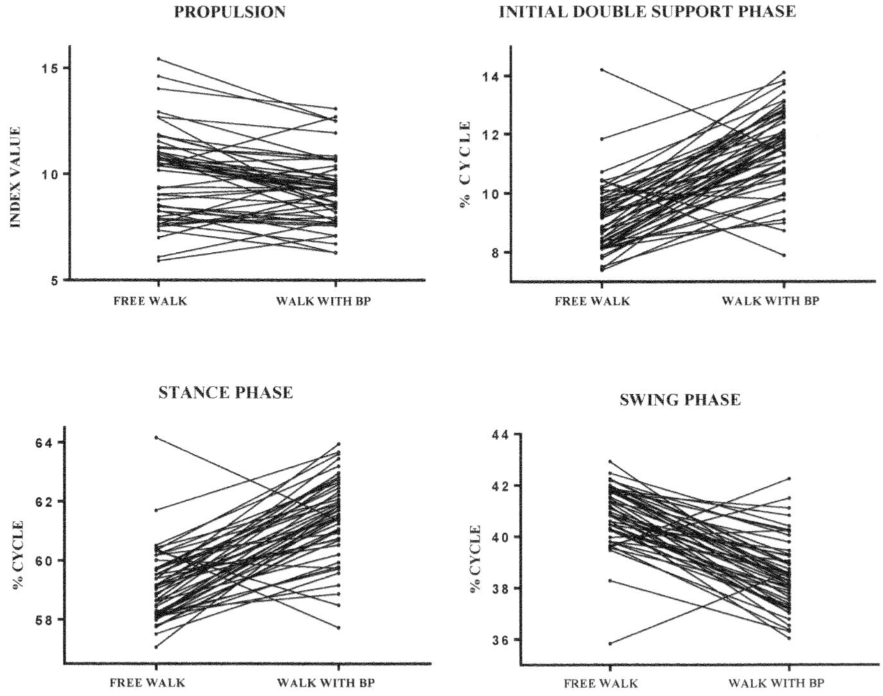

Fig. 2. Plot of all parameters between free walk and walk with backpack

4 Discussion

Results described in the tables and plots allow to highlight the following findings.

The Propulsion Index values related to walking with backpack is significantly lower than free walking one. Propulsion Index represents the slope of the straight line following the rising edge of the antero-posterior acceleration pattern during the single support. Its higher values during free walk indicate a greater capacity during advancement in the Single Support Phase. This result may be associated with a greater effort not able to compensate the backpack weight and then a potential biomechanical overload on skeletal muscle structures. Initial Double Support Phase moves from a normal value of about 10% during free walk to an increase value of about 2.5% in the case of walk with backpack. We could therefore assume that when the load increases, wearing a 9.3 kg backpack, the duration of the Initial Double Support Phase increases, with the subject more anchored to the ground in order to compensate instability generated during the single support phase, when the weight is supported only by a single limb. Similarly, both the Stance and the Swing Phase values move from their normal values, respectively 60% and 40% during free walk, toward a respectively increasing/decreasing value of approximately 2%, with variations resulting significantly away from the normal range of gait cycle's values.

However, it can be observed that all gait cycle studied values are significantly negatively affected by walking with backpack supposing a potential biomechanical overload on skeletal muscle structures in students exposed to these prolonged conditions.

Conflict of Interest Declaration. The authors declare they have no conflict of interests.

References

1. Negrini, S., Carabalona, R.: Backpacks on! Schoolchildren's perceptions of load, associations with back pain and factors determining the load. Spine **27**(2), 187–195 (2002)
2. Negrini, S., Carabalona, R., Pinochi, G., Malengo, R., Sibilla, P.: Backpack and back pain in school children: is there a direct relationship? J. Bone Joint Surg. Br. **80**(3S), 247 (1998)
3. Negrini, S., Carabalona, R., Sibilla, P.: Backpack as a daily load for schoolchildren. The Lancet **354**(9194), 1974 (1999)
4. Iuppariello, L., D'addio, G., Romano, M., Bifulco, P., Pappone, N., Lanzillo, B., Cesarelli, M.: Efficacy of the regent suit-based rehabilitation on gait EMG patterns in hemiparetic subjects: a pilot study. Eur. J. Phys. Rehabil. Med. **54**(5), 705–716 (2018)
5. Iuppariello, L., Bifulco, P., Romano, M., D'Addio, G., Cesarelli, M.: A hybrid decomposition method to infer the sub-movements composition of planar reaching movements. Inform. Med. Unlocked **9**, 210–218 (2017)
6. D'Addio, G., Iuppariello, L., Bifulco, P., Lanzillo, B., Pappone, N., Cesarelli, M.: Validity and reliability of textile system sensoria for posturographic measurements. G. Ital. Med. Lav. Ergonomia **39**(4), 278–284 (2017)
7. D'Addio, G., Iuppariello, L., Pagano, G., Biancardi, A., Lanzillo, B., Pappone, N., Cesarelli, M.: New posturographic assessment by means of novel e-textile and wireless socks device. In: 2016 IEEE International Symposium on Medical Measurements and Applications (MeMeA), pp. 1–5. IEEE (2016)

8. Iuppariello, L., D'Addio, G., Romano, M., Bifulco, P., Lanzillo, B., Pappone, N., Cesarelli, M.: Analysis of reaching movements of upper arm in robot assisted exercises. kinematic assessment of robot assisted upper arm reaching single-joint movements. G. Ital. Med. Lav. Ergonomia **38**(2), 116–127 (2016)
9. Iuppariello, L., Bifulco, P., Romano, M., Cesarelli, M., D'Addio, G.: Submovements composition and quality assessment of reaching movements in subjects with Parkinson's disease. In: 2015 IEEE International Symposium on Medical Measurements and Applications (MeMeA) Proceedings, pp. 329–334. IEEE (2015)
10. D'Addio, G., Iuppariello, L., Romano, M., Lullo, F., Pappone, N., Cesarelli, M.: Kinematic indexes' reproducibility of horizontal reaching movements. In: XIII Mediterranean Conference on Medical and Biological Engineering and Computing, pp. 81–84. Springer, Cham (2014)
11. D'addio, G., Cesarelli, M., Romano, M., Faiella, G., Lullo, F., Pappone, N.: Kinematic and EMG patterns evaluation of upper arm reaching movements. In: 2012 4th IEEE RAS & EMBS International Conference on Biomedical Robotics and Biomechatronics (BioRob), pp. 1383–1387. IEEE, June 2012

Accurate Calculation of Heart Period and Pulse Wave Transit Time

Péter Nagy(✉) and Ákos Jobbágy

Department of Measurement and Information Systems,
Budapest University of Technology and Economics,
Magyar Tudósok krt. 2, 1117 Budapest, Hungary
{nagy, jobbagy}@mit.bme.hu

Abstract. The first step in processing of physiological signals is usually the improvement of the signal-to-noise ratio. However, the applicable filtering methods not only suppress noise but also distort the signal. As a result, parameters derived from filtered physiological signals are also distorted. The paper evaluates filtering methods applied to electrocardiographic (ECG) and photoplethysmographic (PPG) signals in the process of calculating heart period (HP) and pulse wave transit time (PWTT). Band-pass filtering applied in the course of QRS detection causes a shift in R-peak location depending on the QRS waveform. Following QRS detection accurate R-peak localization is suggested using re-filtering of the original signal. The HP was calculated as the time interval between two adjacent R-peaks and also as the time interval between two adjacent local minima in PPG (PPGmin). Both real and simulated data were used for the evaluation. In the latter case the fiducial points of the signals (ECG R-peak and PPGmin locations) were exactly known providing reference positions for the evaluation of filter distortion. Inappropriate filtering shifts the fiducial points. This causes a significant error while calculating parameters by subtraction, like RR interval of heart cycles (t_{RR}, difference of two consecutive R-peak locations) or PWTT (difference of PPGmin and the associated R-peak location). Even more accurate R-peak and PPGmin localization is needed when variation in t_{RR} and PWTT is analyzed. These variations characterize the actual stress level of a person before or during blood pressure measurement more accurately than average heart rate displayed by commercially available blood pressure monitors. Based on accurate R-peak and PPGmin localization it has been confirmed that for the same heart cycle the time interval is different between two R-peaks and two PPGmins.

Keywords: R-peak localization · PPG minimum localization · Heart period · HRV · PWTT

1 Introduction

The signal-to-noise ratio of recorded physiological signals is often very unfavorable. In the paper the following designation will be used:

$$recorded\ signal = signal + noise \quad (1)$$

J. Henriques et al. (Eds.): MEDICON 2019, IFMBE Proceedings 76, pp. 267–275, 2020.
https://doi.org/10.1007/978-3-030-31635-8_32

In the recorded electrocardiograhic (ECG) and photoplethysmographic (PPG) signals signal and noise overlap in the frequency domain. As a result, methods that suppress noise in the frequency domain also distort the signal to be evaluated. The effect of filtering on the signal [1] cannot be easily assessed using recorded physiological signals as only the sum of signal and noise is available. During ECG signal processing the low-pass filtering distorts the shape of the QRS complex that can cause a shift in R-peak location. Simulated ECG and PPG signals are recommended to analyze filter distortion. In this case the exact location of fiducial points are known so the effect of filtering can be quantitatively characterized.

The accurate R-peak and PPGmin designation assures accurate t_{RR} and pulse wave transit time (PWTT) values. Based on these parameters the actual stress level of a person can be assessed. This helps characterize the blood pressure of a person as stress level can have a large impact on the measurement results [2].

Heart rate variability (HRV) is a commonly used measure to determine momentary stress level. It characterizes the variation of the heart period (HP) [3]. PWTT is the time while the pressure wave generated by the heart propagates from the aortic valve to a peripheral part of the body. PWTT is influenced by heart rate, blood pressure and arterial stiffness therefore it is a potential indicator of the complex cardiovascular response to physical and psychological stress. The change in arterial stiffness influencing PWTT is not strictly connected to the change in t_{RR}. Figure 1 shows the relaxation period of a healthy senior male following short physical exercise (20 stairs down and up). Restoration to the resting value takes about 1 min for t_{RR} but 5 min for PWTT.

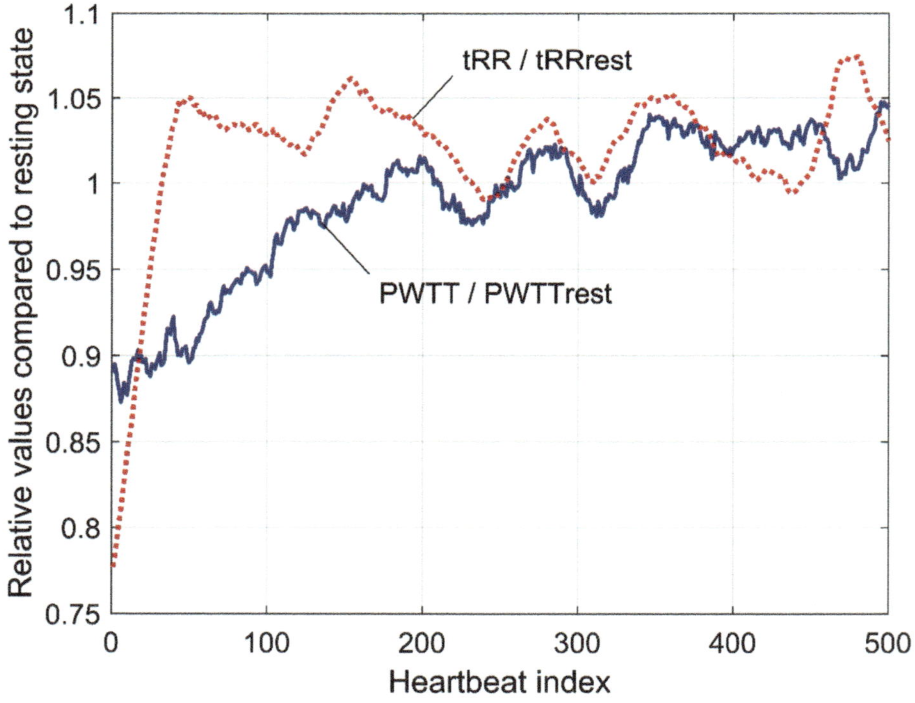

Fig. 1. The relaxation period of a healthy senior male following light physical exercise.

The aim of the paper is to point out the time shift in R-peak and PPGmin location caused by filtering to increase the signal-to-noise ratio.

2 Materials and Methods

The signal processing algorithms were tested using simulated data as well as data recorded using the Home Health Monitoring Device, HHMD [4]. The device has a cuff-based blood pressure monitor. It samples the cuff pressure, ECG in Einthoven I lead and PPG at the index fingertip. Sampling frequency is 1 ksample/s, the A/D resolution is 12 bit.

Data recorded previously with HHMD were evaluated. More than 2.000 records were taken from eight patients (all underwent open chest cardiovascular surgery) and one hundred healthy subjects [5]. Twenty records were selected for the present study.

Simulated data were generated using MATLAB (The Mathworks). Noisy ECG signal was generated based on [6]. First the noise-free ECG and PPG signals were generated and different noises were added later. One period of the noise-free ECG signal was created by averaging normalized periods of an ECG with high signal-to-noise ratio. The single period of the ECG was copied (appended to itself) repetitively to form 100 periods. Beat-to-beat time interval variation was realized by extending each period to a randomly selected length by inserting short constant sections. Maximal increase of the beat-to-beat time interval was 50 ms. Five types of noise artefacts were added to the noise-free signal: power line interference; motion artefact; muscle noise (electromyographic interference); baseline drift and respiration induced amplitude modulation. The resultant signal-to-noise ratio was −7 dB.

Simulated noisy PPG signal was generated similarly to the simulated noisy ECG signal. Power line interference, baseline drift, respiration induced amplitude modulation were added to the noise-free signal.

All frequency domain filtering was realized by the *filtfilt* function of MATLAB. It performs zero-phase filtering and squares the magnitude response.

2.1 Comparing the Effect of Different Filtering on the Location of R-Peak

QRS detection algorithms use a filter stage prior to the actual detection in order to attenuate other signal components and artifacts [7, 8]. For the accurate calculation of PWTT and t_{RR}, not only the QRS complexes but also the R-peaks must be precisely detected. Therefore, we investigated the effect of filtering on the signal in the time domain. For this purpose, we used the simulated noisy ECG signal where the noise-free signal, locations of the R-peaks and the superimposed noise parameters are exactly known. The Pan-Tompkins algorithm [9] was used to detect QRS complexes in the simulated ECG signal. After detection of QRS complexes, the original unfiltered signal was re-filtered (independently of the filtering used in the Pan-Tompkins algorithm) with two notch filters at 50 Hz and 100 Hz (4th order Butterworth), and a low-pass filter at 120 Hz (3rd order Butterworth). Following this pre-filtering the output of three 3rd order Butterworth low-pass filters with corner frequencies 20, 30 and 50 Hz were compared. The effect of median filtering was also investigated. We also assessed the performance of using only a single low-pass filter at 25 Hz without notch filters. R-peaks were detected in the re-filtered signal as local maxima within the ±20 ms neighboring intervals of the time moments provided by the Pan-Tompkins algorithm as QRS locations.

2.2 Comparing the Effect of Different Filtering on the Position of PPGmin

Filtering of the PPG signal is necessary for PPGmin localization and then PWTT calculation. Baseline wandering and motion artifacts have to be removed for the accurate detection of local minima in the PPG signal. Elgendi et al. [10] examined the spectrum of the PPG signal and suggested a 2nd order Butterworth band-pass filter with corner frequencies at 0.5 Hz and 10 Hz for PPG-based HRV analysis. However, the distortion caused by filtering shifts the locations of local minima in the PPG signal. In order to investigate this effect, we used a simulated noisy PPG signal. The simulated signal was filtered with different corner frequencies (3rd order Butterworth band-pass filters were used) and the locations of detected local minima were compared to the reference values.

3 Results

3.1 Evaluation of Signal Processing Methods Using Simulated Data

ECG signals with positive and biphasic QRS complexes and PPG signals were generated as described in Sect. 2. The mean absolute error between the known and the calculated positions and its standard deviation for different filtering are given in Table 1 (R-peak localization) and Table 2 (PPGmin localization). In Table 1 the number of QRS complexes are given in brackets where the absolute error is greater than 10 ms.

Table 1. Effect of different filtering on R-peak detection. Simulated signal and noise, positive and biphasic QRS complexes.

Filter type	Mean abs. error + standard deviation (ms)	
	Positive QRS complex	Biphasic QRS complex
Two notch filters, center frequencies: 50 and 100 Hz	1.55 ± 2.13	2.04 ± 2.77 (8)
Low-pass filter, corner frequency: 25 Hz	0.18 ± 0.43	2.54 ± 0.98 (2)
Median filter	1.15 ± 1.29	1.94 ± 2.13 (1)
No filter	4.50 ± 5.23 (2)	4.13 ± 4.90 (8)
Prefiltering (see text) and 20 Hz low-pass filter	0.16 ± 0.57	4.94 ± 0.68 (4)
Prefiltering (see text) and 30 Hz low-pass filter	0.09 ± 0.33	1.14 ± 0.85 (2)
Prefiltering (see text) and 50 Hz low-pass filter	0.08 ± 0.28	0.28 ± 0.86 (2)
Prefiltering (see text) and 60 Hz low-pass filter	0.10 ± 0.32	0.29 ± 0.87 (2)
Prefiltering (see text) and 70 Hz low-pass filter	0.10 ± 0.35	0.29 ± 0.82 (2)

These QRS complexes were not included in the calculation of the mean absolute error. These large errors result from the coincidence of muscle noise or motion artefact with the R-peak.

Table 2. Effect of different filtering (high (fc_H) and low (fc_L) corner frequencies) on PPGmin detection. Simulated signal and noise.

fc_L	Mean abs. error + standard deviation (ms)					
	fc_H-10 Hz	fc_H-13 Hz	fc_H-14 Hz	fc_H-15 Hz	fc_H-16 Hz	fc_H-17 Hz
0.2 Hz	5.01 ± 0.27	1.01 ± 0.22	0.05 ± 0.22	0.03 ± 0.22	0.89 ± 0.35	1.0 ± 0.14
0.5 Hz	4.68 ± 0.53	0.70 ± 0.49	0.34 ± 0.56	0.49 ± 0.56	0.72 ± 0.55	1.09 ± 0.64

3.2 Evaluation of Recorded Signals

Ten ECG recordings were selected, five with positive and five with biphasic QRS complexes. Altogether 714 (positive) + 756 (biphasic) QRS complexes were analyzed. Low-pass filtering has a different effect on positive and biphasic QRS complexes regarding the shift in R-peak location. The R-peak of positive QRS complexes remain unchanged even after 20 Hz low-pass filtering, see Fig. 2. The R-peak of a biphasic QRS complex is shifted as a result of low-pass filtering. The shift depends on the corner frequency (f_c) of the low-pass filter (LPF). In Fig. 3 the f_c = 30 Hz LPF shifts R-peak by +4 ms, the f_c = 20 Hz LPF by –1 ms, compared to the location determined after an f_c = 50 Hz LPF.

The differences between R-peak positions calculated after applying f_c = 50 Hz, 30 Hz and 20 Hz LPFs are given in Table 3.

Table 3. Effect of low-pass filtering (LPF) with different corner frequencies on R-peak detection. Recorded signals, positive and biphasic QRS complexes.

Recording	Mean abs. difference + standard deviation (ms), positive QRS	
	Difference between 20 Hz and 50 Hz LPF	Difference between 30 Hz and 50 Hz LPF
Record1p	0.60 ± 0.54	0.39 ± 0.52
Record2p	0.64 ± 1.63	0.45 ± 0.69
Record3p	1.06 ± 0.51	0.49 ± 0.50
Record4p	0.56 ± 1.84	0.40 ± 0.76
Record5p	0.43 ± 0.90	0.27 ± 0.50
Recording	Mean abs. difference + standard deviation (ms), biphasic QRS	
	Difference between 20 Hz and 50 Hz LPF	Difference between 30 Hz and 50 Hz LPF
Record1b	2.40 ± 2.97	2.13 ± 2.37
Record2b	2.52 ± 2.21	1.58 ± 1.96
Record3b	3.28 ± 3.14	1.12 ± 1.70
Record4b	4.06 ± 4.41	1.45 ± 1.87
Record5b	5.55 ± 1.94	1.92 ± 1.54

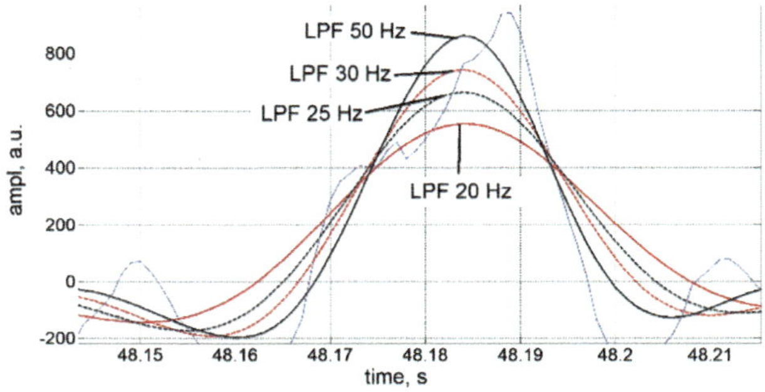

Fig. 2. Effect of low-pass filtering with different corner frequencies on a positive QRS complex.

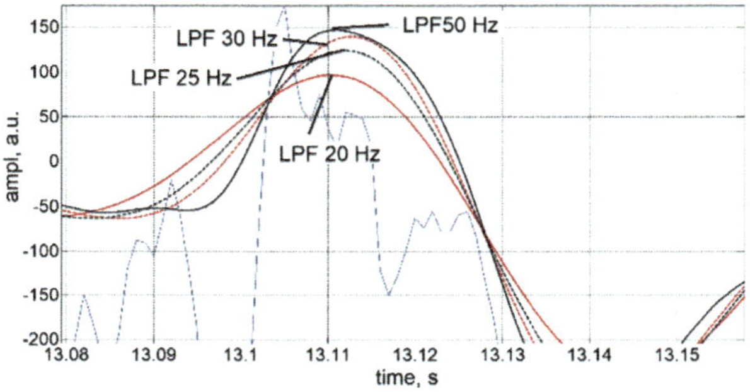

Fig. 3. Effect of low-pass filtering with different corner frequencies on a biphasic QRS complex.

Ten PPG recordings were selected to analyze the effect of different band-pass filtering on PPGmin. Altogether 1597 HPs were evaluated. The differences are given in Table 4.

Table 4. Differences in PPGmin localizations using different filters.

Subjects	Differences between PPGmin locations using two different filters, mean difference + standard deviation (ms)		
	0.2–10 Hz and 0.2–15 Hz	0.2–20 Hz and 0.2–15 Hz	0.5–15 Hz and 0.2–15 Hz
Young subjects (age 22–28)	−5.59 ± 5.37	0.56 ± 3.08	0.25 ± 0.73
	−7.72 ± 1.17	2.39 ± 1.57	0.20 ± 0.57
	−8.34 ± 3.06	2.04 ± 3.44	0.23 ± 0.93
	−8.38 ± 4.35	2.78 ± 2.87	0.49 ± 1.76
	−4.19 ± 2.98	0.59 ± 3.60	0.18 ± 0.66
Senior subjects (age 60–68)	−5.54 ± 1.07	0.99 ± 1.51	0.15 ± 0.59
	−1.43 ± 3.99	−0.45 ± 3.97	−0.31 ± 0.83
	−7.64 ± 6.58	2.18 ± 3.65	0.02 ± 0.93
	−6.09 ± 1.63	0.65 ± 1.83	0.26 ± 0.48
	−6.18 ± 3.44	0.11 ± 4.70	0.56 ± 0.72

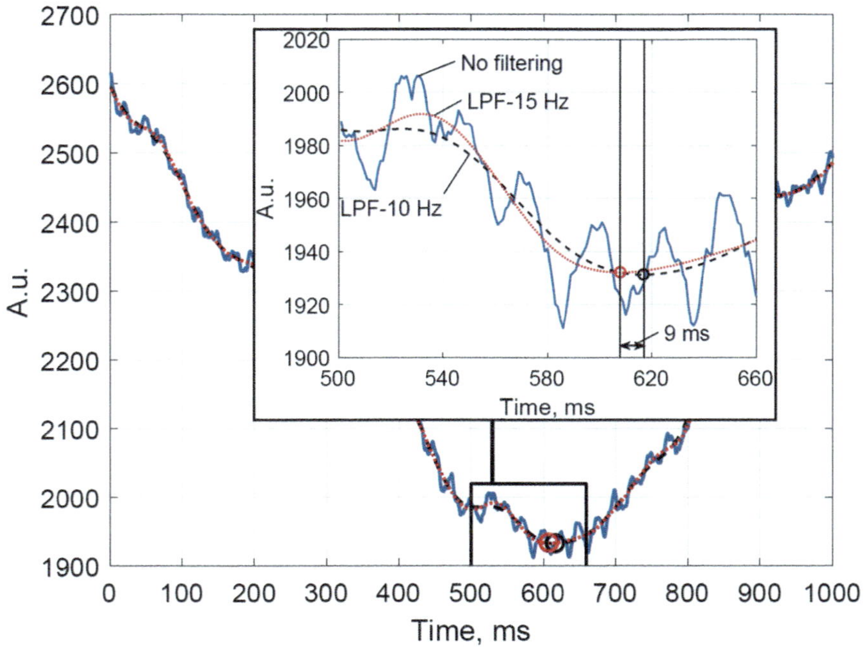

Fig. 4. Effect of low-pass filtering on PPGmin location. The difference between local minima in the two filtered signals is 9 ms

Figure 4 shows the effect of filtering on PPGmin location. Notice that the fluctuation of the signal is not caused by power-line noise.

PWTT is affected by both ECG filtering and PPG filtering. To demonstrate this effect, PWTT was calculated for four different filtering cases. In all cases, ECG was pre-filtered as described in Sect. 2.1, and then low-pass filtering was applied, with corner frequencies of 30 Hz or 50 Hz. For PPG, the two band-pass filters had frequency band 0.2–15 Hz or 0.2–10 Hz. Table 5 shows the results.

Table 5. Differences (in ms) between PWTT values using different filters. Recordings with positive (upper values) and biphasic (lower values) QRS complexes.

Band-pass filtering the PPG signal	Low-pass filtering the pre-filtered ECG signal	
	Corner frequency = 50 Hz	Corner frequency = 30 Hz
0.2–15 Hz	Reference values	0.36 ± 0.52 1.59 ± 1.98
0.2–10 Hz	3.10 ± 4.55 6.21 ± 1.93	3.07 ± 4.72 6.29 ± 2.77

The HPs were determined based on the differences between adjacent R-peaks and also based on the differences between adjacent PPGmin points. The record with positive QRS complexes was taken from a 68 years old male at rest (157 HPs). The record with biphasic QRS complexes was taken from a patient with cardiovascular disease, one year after open chest cardiac surgery (65 years old male at rest, 162 HPs). The results are given in Table 6.

Table 6. HP calculated from ECG and PPG.

Mean + standard deviation	Calculated HP (ms)	
	Record with positive QRS complexes	Record with biphasic QRS complexes
HP calculated from ECG	806.3 ± 17.0	727.2 ± 19.0
Differences of adjacent HPs calculated from ECG	12.8 ± 7.6	10.7 ± 7.4
HP calculated from PPG	807.1 ± 28.8	724.1 ± 50.4
Differences of adjacent HPs calculated from PPG	26.4 ± 32.7	45.8 ± 42.0

4 Discussion

The frequency domain filtering of ECG and PPG signals shifts the R-peak and PPGmin locations. Based on analyzing both simulated and recorded data the filtering described below was found to cause minimum shift in fiducial points. (It must be emphasized that this filtering is not optimal when the aim is to preserve ECG and PPG signal waveform.) We suggest realizing R-peak localization in two steps. First the QRS complex should be designated having applied the usual band-pass filtering (5–12 Hz was appropriate for our records). In the second phase the original recorded signal is re-filtered and the maximum value is searched for in the ±20 ms interval around the point determined in the first step. For both positive and biphasic QRS complexes, filtering with two notch filters at 50 Hz and 100 Hz (4th order Butterworth), and a low-pass filter at 50 Hz (3rd order Butterworth) yielded the best results. Filtering causes significantly greater error for biphasic than for positive QRS complexes.

Band-pass filtering the simulated noisy PPG signal with corner frequencies at 0.2 and 15 Hz produced the smallest time shift in PPGmin locations. Decreasing the corner frequency from 15 Hz to 10 Hz causes a significant time shift in PPGmin both for simulated and for recorded data. Changing the corner frequency from 0.2 to 0.5 Hz does not make significant difference. No unidirectional relation was found between age related differences in the signal shape and the time shift of the local minima caused by filtering.

It is clear that the estimation of the mean value of HPs in a record based on R-peak locations gives the same value as estimation based on PPGmin locations. (Apart from a negligible edge effect). On the contrary, differences of consecutive HPs are significantly different calculated from ECG and PPG signal. The reason is that the PWTT can change beat-by-beat. The small differences require accurate determination of R-peak and PPGmin locations. The comparison of different filtering methods reveals that subtle changes in HP and in PWTT may be masked by inappropriate filtering of ECG and PPG.

Our analysis shows that filtering can cause different time shift of the R-peak at different QRS complexes. Further research work is needed to adjust the R-peak and PPGmin detection algorithm to each heart period.

Acknowledgement. The research has been supported by the European Union, co-financed by the European Social Fund (EFOP-3.6.2-16-2017-00013).

Conflict of Interest. The authors declare that they have no conflict of interest.

References

1. Sörnmo, L., Laguna, P.: Electrocardiogram (ECG) Signal Processing. Wiley Encyclopedia of Biomedical Engineering. Wiley, Hoboken (2006)
2. Floran, J.S.: Blood pressure variability: a novel and important risk factor. Can. J. Cardiol. **29** (5), 557–563 (2013)
3. Bernston, G.G., Bigger Jr., J.T., Eckberg, D.L., Grossman, P., Kaufmann, P.G., et al.: Committee report: heart rate variability: origins, methods and interpretive caveats. Psychophysiol. J. **34**(6), 623–648 (1997)
4. Jobbágy, Á., Csordás, P., Mersich, A.: Blood pressure measurement at home. In: Proceedings of World Congress on Medical Physics and Biomedical Engineering 2006, pp. 3453–3456. Springer, Heidelberg (2007)
5. Csordás, P., Mersich, A., Jobbágy, Á.: Advanced indirect method for measuring blood pressure. Period. Polytech. Electr. Eng. **53**(3–4), 115–121 (2011)
6. Friesen, G.M., Jannett, T.C., Jadallah, M.A., Yates, S.L., Quint, S.R., Nagle, H.T.: A comparison of the noise sensitivity of nine QRS detection algorithms. IEEE Trans. Biomed. Eng. **37**(1), 85–98 (1990)
7. Thakor, N.V., Webster, J.G., Tompkins, W.J.: Estimation of QRS complex power spectra for design of a QRS filter. IEEE Trans. Biomed. Eng. **11**, 702–706 (1984)
8. Kohler, B.-U., Hennig, C., Orglmeister, R.: The principles of software QRS detection. IEEE Eng. Med. Biol. Mag. **21**(1), 42–57 (2002)
9. Pan, J., Tompkins, W.J.: A real-time QRS detection algorithm. IEEE Trans. Biomed. Eng. **32**(3), 230–236 (1985)
10. Elgendi, M., Jonkman, M., DeBoer, F.: Heart rate variability and the acceleration plethysmogram signals measured at rest. In: International Joint Conference on Biomedical Engineering Systems and Technologies, pp. 266–277. Springer, Heidelberg (2010)

Long-Term Stability of EEG Spectral Asymmetry Index – Preliminary Study

Tuuli Uudeberg[1(✉)], Laura Päeske[1], Toomas Põld[2], Jaanus Lass[1], Hiie Hinrikus[1], and Maie Bachmann[1]

[1] Tallinn University of Technology, 19086 Tallinn, Estonia
{tuuude, maie.bachmann}@taltech.ee
[2] Qvalitas Medical Centre, 11312 Tallinn, Estonia

Abstract. The purpose of this preliminary study was to assess the long-term stability of the electroencephalography (EEG) based spectral asymmetry index (SASI) previously proposed as an objective measure for evaluation of depression. Due to high variability between individuals, a long-term study is required to examine how SASI changes over longer period of time for an individual subject. One healthy adult was surveyed over the period of 15 months. The eyes closed resting state EEG was analyzed over 15 sessions, on average once a month. The SASI was calculated from EEG power spectrum density as a relative difference in powers of higher and lower band of the EEG spectrum maximum. The long-term stability of SASI for an individual was compared to inter-individual variability of the measure within the group of healthy subjects in our previous study. The results demonstrated that the individual long-term variability of SASI is less than inter-individual variability within a group, supporting the possibility of application of SASI for evaluation of depression symptoms for an individual. In future, the investigations should be extended on larger number of subjects to study the individual properties of SASI.

Keywords: Electroencephalography · Spectral Asymmetry Index · Stability

1 Introduction

Spectral asymmetry index (SASI), a method based on electroencephalographic (EEG) single channel signal analyses, has been successfully applied for differentiating group of depressive subjects from the group of healthy subjects [1, 2].

A simple cost-effective method for early evaluation of depression symptoms is highly demanded keeping in mind increasing trend of depression in modern society. According to World Health Organization report, unipolar depressive disorder is a leading cause of burden of disease in high- and middle-income countries today and it is presumed to take the first place in the world in 2030 [3].

The diagnosis of depression is currently based mainly on evaluation of the intensity of subjective symptoms using clinical interview and psychiatric questionnaires. On the other hand, any declinations in the brain neuronal activity and mental state are expected to be reflected also in the brain bioelectrical activity and EEG signal even before the subjective symptoms appear. Subjective symptoms become evident in developed stage

J. Henriques et al. (Eds.): MEDICON 2019, IFMBE Proceedings 76, pp. 276–281, 2020.
https://doi.org/10.1007/978-3-030-31635-8_33

of depression and hardly useful for prevention of disorder. Therefore, detection of declinations in brain physiology before the subjective symptoms appear is crucial for early detection of disorders, enabling treatment that is more effective and improving quality of mental health.

Quantitative EEG characteristics have been under consideration for investigation of depression in many studies. Depression has been detected by employing changes in EEG bands powers, coherence, interhemispheric asymmetry or resting sate neurophysiologic connectivity [4–7]. It has been shown that discriminant analysis of quantitative EEG from 21 electrodes classified correctly 91.3% of the patients and controls whereas mainly delta and beta inter-hemispheric coherence, beta intra-hemispheric coherence and alpha intra-hemispheric power asymmetry contributed to the classification [4]. Still, coherence appeared ineffective in the case of smaller number of channels [1], while SASI, based on the evaluation of the balance of powers in two frequency bands selected higher and lower than the spectrum maximum, provided classification accuracy of 85% in single EEG channel [2] and in combination with other single channel measures provided discrimination of depression at the level of multi-channel EEG analysis [8].

To date, studies have found promising results in discriminating between depressive and healthy subjects on a group level. However, due to a very high variability between individuals, the scores of possible measures vary in large scale and distinguishing the mental status for a single individual becomes highly complicated and even hardly possible.

The clinical interview and psychiatric questionnaires give a singular evaluation for a person. However, there are remarkable discordances between the scores by different tests and questionnaires [9]. More that, whereas the questionnaires are based on subjective symptoms, discordances appear in diagnoses of MDD, especially in mild stage, evaluated by different medical specialists.

The applicability of EEG based objective measures of evaluation of depression for an individual has been become a crucial point in early detection of depression,

SASI has been shown to successfully differentiate groups of healthy and depressive subjects. In group comparisons, the individual SASI values are merged, and the state of each subject is not unveiled. The individual temporal variability of SASI has never been studied and the parameters stability is unknown.

The aim of the preliminary study is to evaluate the long-term stability of SASI for an individual evaluating the feasibility to better exploit the marker as an assistant for clinical decisions for revealing depression for an individual.

2 Methods

2.1 The Subject

The subject for the case study was a healthy female aged 26–27 without any history of mental disorders. The mental health was screened by the Emotional State Questionnaire, a self-report questionnaire for depression and anxiety [10]. During the whole experimental period (15 months) no clinically important symptomatology was present.

Participation in this study was voluntary. The experiments were conducted with understanding and written consent of the participant. The study was conducted in accordance with the Declaration of Helsinki and was formally approved by the Tallinn Medical Research Ethics Committee.

2.2 Procedure

The experimental session was conducted always on the same weekday and on the same time, also the subject was asked to maintain same routine 24 h prior to the recording. The whole study consisted of 15 sessions which took place usually in every four weeks (up to six weeks), taking into account that the recording would be performed approximately in the same phase of the menstrual cycle in order to avoid changes in the EEG caused by fluctuations in sex hormone levels [11, 12]. During each session, 10-minute eyes closed resting state EEG was recorded. The recordings were carried out in a dimly lit and sound proof room. During the recording procedure, the subject was lying in relaxed position.

Neuroscan Synamps2 acquisition system was used for the EEG recordings. EEG was recorded using 30 electrodes which were positioned according to the extended international 10–20 electrode placement system. The sampling frequency was 1,000 Hz. Linked mastoids were used as a reference and electrode impedances were kept below 10 kΩ. In addition, two sets of EOG electrodes (vertical and horizontal) were also used.

Before each recording the mental health was screened by Emotional State Questionnaire.

2.3 EEG Analysis

All calculations were performed using MATLAB (The Math-works, Inc.). Signals were digitally filtered (3–47 Hz) using zero-phase Butterworth filter and re-referenced according to the reference electrode standardization technique (REST) [13]. Signals were down-sampled to 200 Hz.

After visual inspection, artifact free segments with the total length of 5 min were kept for further processing.

SASI characterizes EEG spectral asymmetry and describes relative difference between the higher and the lower EEG frequency band power.

In order to calculate the SASI values, power spectral density was estimated for channel Pz for all recorded sessions. The power spectral density estimation was performed by means of Welch's averaged periodogram method. The signal was divided into overlapping epochs (50%), and extracted through a Hanning window.

Next, the SASI was calculated, excluding the EEG alpha frequency band, as a relative difference between the higher and the lower EEG frequency band power. For the lower EEG frequency band 4–7 Hz were selected – close to traditional theta bandwidth, while for the higher EEG frequency band 14–38 Hz were chosen – close to traditional beta bandwidth.

Calculation of SASI comprises the following steps:

Selection of boundary frequencies of the higher (beta band 14 to 38 Hz) and lower (theta band 4 to 7 Hz) EEG frequency bands. Total powers in the theta and beta frequency bands were calculated as

$$P_{\theta} = \sum\nolimits_{f_i=4}^{f_i=7} s, \quad (1)$$

$$P_{\beta} = \sum\nolimits_{f_i=14}^{f_i=38} s, \quad (2)$$

where s is the power of the recorded EEG signal in a unit band at frequency fi. Next, SASI is calculated as

$$SASI = \frac{P_{\beta} - P_{\theta}}{P_{\beta} + P_{\theta}}. \quad (3)$$

The average value and standard deviation of SASI over all recording sessions during 15 months were calculated.

The inter-individual variability of SASI was evaluated from the data of our previous study [2] - the average value and standard deviation over a group of 17 healthy subjects were calculated.

3 Results and Discussion

The calculated SASI values for each experimental session are presented on Fig. 1.

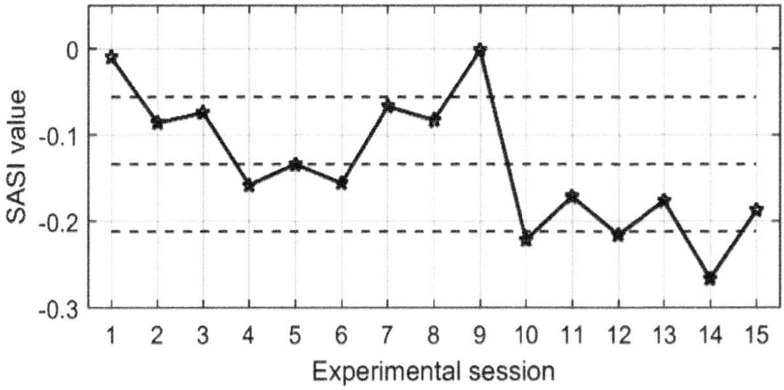

Fig. 1. Subject's SASI values calculated for all experimental sessions, each about 4 weeks apart. Dashed lines represent mean SASI value±standard deviation

The Fig. 1 shows that during the period of 15 months, the SASI values lie between −0.267 and −0.002. This is in accordance with the previous results indicating negative SASI values in parietal channels for healthy subjects [2].

The standard deviation of SASI over all measurements during 15 months was 0.078. The calculated mean±standard deviation of the SASI values for the group of 17 healthy subjects was -0.1253 ± 0.206 [2].

From the values above, the individual variability is more than 2.5 times lower compared to inter-individual variability. This result presents a strong support to the idea, that SASI can be used for evaluation of depression for individuals.

Although the current study did not take into account the subjective stress level, the subject unofficially reported more difficult times during the middle sessions (particularly sessions 8 and 9) and a relaxed period after that. As the SASI values are also higher during the middle sessions it can be speculated that SASI could be useful for revealing stress. While short-term stress, has positive effect by improving the performance by keeping us alert and ready to respond, chronic stress may lead to depression.

Current study involved just one healthy subject. Additional study on larger group is needed to get a better insight of SASI's individual stability in healthy condition. In addition, the future study could share some light whether SASI is also able to detect stress level, giving the possibility to reveal chronic stress, which, while overlooked, might lead to unwanted consequences like depression. In case feasible considering ethical requirements, the SASI long-term behavior in depression would give additional valuable information about the disorder, resulting also in better early detection of the disorder.

4 Conclusion

The preliminary result, that long-term temporal variability of SASI is much lower than inter-individual variability, support the possibility of application of SASI for evaluation of depression symptoms for an individual. Further investigations on larger number of subjects are required.

In future, a method based on analysis of the EEG and integrated in a simple user-friendly device could be used for regular evaluation of the state of brain in family medicine centres and during occupational health examinations to distinguish brain disorder already at the very early stage.

Acknowledgment. This study was financially supported by the Estonian Ministry of Education and Research under institutional research financing IUT 19-2 and by the Estonian Centre of Excellence in IT (EXCITE) funded by the European Regional Development Fund.

Conflict of Interest. The authors declare that they have no conflict of interest.

References

1. Hinrikus, H., Suhhova, A., Bachmann, M., et al.: Electroencephalographic spectral asymmetry index for detection of depression. Med. Biol. Eng. Comput **47**(12), 1291–1299 (2009). https://doi.org/10.1007/s11517-009-0554-9
2. Bachmann, M., Lass, J., Suhhova, A., et al.: Spectral asymmetry and Higuchi's fractal dimension measures of depression electroencephalogram. Comput. Math. Methods Med **2013**, 251638 (2013). https://doi.org/10.1155/2013/251638
3. The global burden of disease: 2004 Update. WHO (2014)
4. Knott, V., Mahoney, C., Kennedy, S., et al.: EEG power, frequency, asymmetry and coherence in male depression. Psychiatry Res. **106**(2), 123–140 (2001). https://doi.org/10.1016/S0925-4927(00)00080-9
5. Leuchter, A.F., Cook, I.A., Hunter, A.M., et al.: Resting-state quantitative electroencephalography reveals increased neurophysiologic connectivity in depression. PLoS One **7**(2), e32508 (2012). https://doi.org/10.1371/journal.pone.0032508
6. Allen, J.J., Urry, H.L., Hitt, S.K., et al.: The stability of resting frontal electroencephalographic asymmetry in depression. Psychophysiology **41**(2), 269–280 (2004). https://doi.org/10.1111/j.1469-8986.2003.00149.x
7. Fingelkurts, A.A., Fingelkurts, A.A., Rytsälä, H., et al.: Composition of brain oscillations in ongoing EEG during major depression disorder. Neurosci. Res. **56**(2), 133–144 (2006). https://doi.org/10.1016/j.neures.2006.06.006
8. Bachmann, M., Päeske, L., Kalev, K., et al.: Methods for classifying depression in single channel EEG using linear and nonlinear signal analysis. Comput. Methods Programs Biomed. **155**, 11–17 (2018). https://doi.org/10.1016/j.cmpb.2017.11.023
9. Cusin, C., Yang, H., Yeung, A., et al.: Rating scales for depression. In: Lee, B., Blais, M.A. (eds.) Handbook of Clinical Rating Scales and Assessment in Psychiatry and Mental Health, pp. 7–35. Humana Press, New York (2010)
10. Aluoja, A., Shlik, J., Vasar, V., et al.: Development and psychometric properties of the emotional state questionnaire, a self-report questionnaire for depression and anxiety. Nord. J. Psychiatry **53**, 443–449 (1999). https://doi.org/10.1080/080394899427692
11. Barth, C., Villringer, A., Sacher, J.: Sex hormones affect neurotransmitters and shape the adult female brain during hormonal transition periods. Front. Neurosci. 9(37) (2015). https://doi.org/10.3389/fnins.2015.00037
12. Gawali, B.W., Rokade, P.B., Janvale, G.B., et al.: Ovarian hormones and the brain signals. Ann. Neurosci. **16**(2), 72–74 (2009)
13. Yao, D.: A method to standardize a reference of scalp EEG recordings to a point at infinity. Physiol. Meas. **22**, 693–711 (2001). https://doi.org/10.1088/0967-3334/22/4/305

Regular Sessions: Biomedical Imaging and Image Processing

Flow Convergence Area Estimation on In Vitro Color Flow Doppler Images Using Deep Learning

Grigorios-Aris Cheimariotis[1(✉)], Kostas Haris[1], Jeesoo Lee[2], Brent E. White[2], Aggelos K. Katsaggelos[3], James D. Thomas[2], and Nikolaos Maglaveras[1,2,3,4]

[1] Lab of Computing, Medical Informatics and Biomedical Imaging Technologies, Aristotle University, Thessaloniki, Greece
ncheimar@gmail.com

[2] Department of Radiology, Feinberg School of Medicine, Northwestern University, Chicago, IL, USA

[3] Department of Electrical and Computer Engineering, Northwestern University, Evanston, Chicago, USA

[4] Department of Industrial Engineering and Applied Science, Northwestern University, Evanston, Chicago, USA

Abstract. We present an automatic method to estimate flow rate through the orifice in in-vitro 2D color-flow Doppler echocardiographic images. Flow rate properties are important for the assessment of pathologies like mitral regurgitation. We expect this method to be transferable to in-vivo patient data. The method consists of two main parts: (a) detecting a bounding box which encloses aliasing contours and its surroundings (namely a region representative of flow convergence area), (b) application of Convolutional Neural Networks for regression to estimate the flow convergence area. Best result achieved is the 5% mean error for validation data which is from other experiments that were used for training. Given the small number of training data, this method shows promising results.

Keywords: Deep learning · Color flow doppler · Flow rate · Mitral regurgitation

1 Introduction

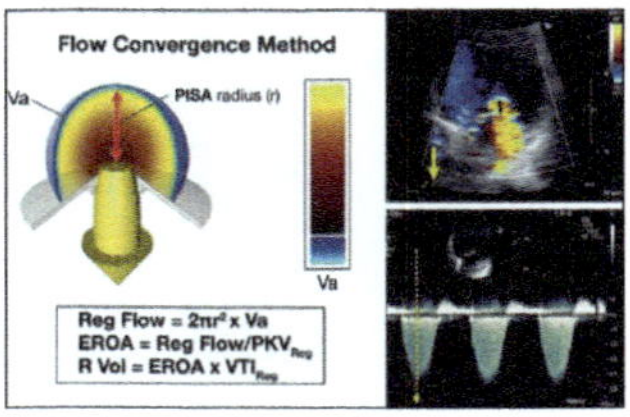

Fig. 1. Schematic illustrating the PFC method from 2017 ASE Valve Regurgitation guidelines. Va, aliasing (contour) velocity; EROA, effective regurgitant orifice area; PKVreg, maximal retrograde velocity; VTI, velocity time interval.

Valvular regurgitation is a large and growing problem to world health with 2% of the American population suffering from chronic moderate or severe mitral regurgitation (MR). Accurate quantification of MR is critical to appropriate medical, surgical, or percutaneous therapy, as documented in recent guidelines from the American College of Cardiology [1] and the American Society of Echocardiography [2]. One of the best

J. Henriques et al. (Eds.): MEDICON 2019, IFMBE Proceedings 76, pp. 285–291, 2020.
https://doi.org/10.1007/978-3-030-31635-8_34

ways to quantify regurgitation is via analysis of the proximal flow convergence (PFC) zone, the area where flow accelerates as it rushes into the regurgitant orifice (Fig. 1), forming hemispheric isovelocity contours from which regurgitant flow rate (Q) and orifice area ROA can be calculated via conservation of mass, termed the proximal isovelocity surface area (PISA) method. For an isovelocity contour with velocity v_a, typically identified by color Doppler aliasing, at a radius r from the orifice, $Q = 2\pi r^2 v_a$, and $ROA = Q/v_0$, where v_0 is the peak regurgitant velocity through the valve (via continuous wave Doppler). Previous work by Dr. Thomas has demonstrated that PISA is subject to a number of technical and operator errors, including nonholosystolic regurgitation [3], contour flattening near the orifice [4], constraint by adjacent geometry [5], elliptical orifices [6], and situations with multiple orifices, such as following percutaneous mitral valve clipping. In a situation of such complexity, it is hypothesized that deep learning [7] from well-annotated datasets could provide guidance to make the PFC method more reliable and robust, even among less than expert cardiologists.

2 Methods

2.1 Materials

A gravity feed in vitro model was designed to allow echo imaging of precisely monitored flow rate (Fig. 2). Four 3D printed models of idealized regurgitant orifices were used to image the PFC. For three cases, three different aliasing velocities were used for imaging while for one case only one aliasing velocity was used. This resulted in 10 series of color flow Doppler images. Each series had around 165-time frames. Flow rate values are known for each time frame since controlled experiments occurred. These values were used as ground truth.

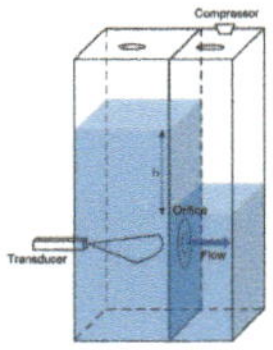

Fig. 2. In vitro model used to generate datasets for initial machine learning, dimensions 8 × 14 × 57 cm.

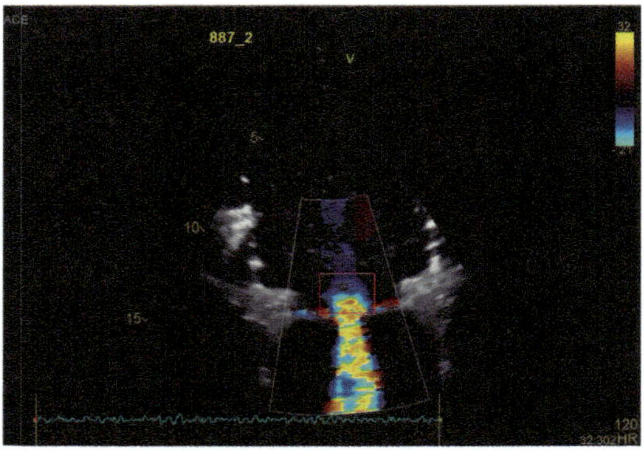

Fig. 3. In-vitro color flow Doppler image. Inside the red rectangle is the automatically detected region of interest. This cropped region is input to the CNNs.

2.2 Flow Convergence Area Detection

Firstly, we detected the part of the image where the transducer is acquiring color flow data. That part is enclosed in a frame that has a characteristic triangular shape (except for the top and bottom) in most echocardiograph machines. We detected this frame points by applying a threshold which keeps the bright white pixels of the image. Then we applied morphological operations mainly to exclude irrelevant bright white pixels and to connect the points of the frame. From this frame we created a mask that excludes irrelevant color parts from the image.

Secondly, the 2D cross-section of the flow convergence area detection was accomplished with the following steps: (a) we extracted the interpolated mean image from each series of images, (b) we kept the green component of the RGB image because its brighter part lies on the flow convergence area in the mean image (c) a threshold was applied to produce a binary image where white is only the flow convergence area part (d) morphological dilation was applied to that part in the upward direction and to the sides to include variability of the flow convergence area (e) the bounding box of the dilated part of this image defined the image input to the convolutional neural networks which were used to estimate the flow convergence area size. In Fig. 3, the result of this procedure is presented inside the red rectangle which is the input to the CNNs.

2.3 Application of CNN for Flow Rate Calculation

We applied two different CNN architectures with a regression output layer to estimate the flow convergence area. The ground truth flow rate was divided by aliasing velocity in each case. The estimated values by the CNNs correspond to the flow convergence area. The reason for using flow convergence area instead of flow rate is that we used as input images with different aliasing velocities.

The one network was a modified version of Alex Net [8] which was fine-tuned to learn to estimate flow convergence area size. Fine-tuning is performed by retraining an already trained AlexNet with a learning rate which is 20 times larger for the last 3 fully-connected layers than for the initial 5 convolutional layers. There was hyperparameter tuning to define CNN properties (e.g., the learning rate was chosen equal to 0.0001). The RMSE did not decrease after 100 epochs, at which point we terminated the training.

The other CNN was originally designed for digit recognition (DRCNN) and was chosen because it generalizes better on certain occasions for processing both grayscale and binary images. We used this CNN to process binary or grayscale images that are outcome of the red component of the original RGB images so that this method can generalize to real patient data where the color scale is usually complementary. In this case, we would apply the in vitro trained network to images which are the outcome of the blue component of the real patient data. Finally, the efficiency of this network was tested with preprocessed images by filtering or morphological operations as a method of denoising the color flow Doppler images.

2.4 Groupings of Training and Validation Data

There were four ways of grouping training and validation data. Given the fact that we had only 4 color flow Doppler series, initially we picked randomly validation data from all frames available that were excluded from the training data. To find out if the proposed architecture can generalize to new data, we picked one of the four color flow Doppler series as validation set and the rest as a training set.

Different experimental results indicated that flow convergence area can be better estimated if the aliasing velocity is low. Therefore, another grouping was formed by randomly picking validation data and training data only from the times series that have the same low aliasing velocity. Also, we picked one of the four color flow Doppler series with low aliasing velocity as validation data and the rest of the data with the same aliasing velocity were the training set.

3 Results

In order to test validation accuracy, we used the root-mean-square error (RMSE) metric. In Table 1, validation RMSE is presented for the two CNN architectures and for the different groupings of data described above. Validation data in the case noted "leave-one out" were from a color flow Doppler series with aliasing velocity equal to 7 cm/s. It can be pointed out that despite the small variability of the training set, there is good estimation in every case. The lowest percentage-wise mean error achieved was equal to 0.46% (cm^2), which shows that this method is very promising if we add more cases to the training set. In addition, we observed that if we train the networks with data with different aliasing velocities, the accuracy is enhanced.

Although DRCNN has generally worse performance than AlexNet, it performs better in the more important cases when applied to new data (leave-one-out)., DRCNN achieves binary leave-one-out RMSE = 0.522 cm^2 while AlexNet achieves leave-one-out RMSE = 0.58 cm^2. If we apply denoising techniques, the performance of DRCNN is even more improved (RMSE = 0.5 cm^2). Better performance with binary images indicates that the most important feature is the shape and size of the area inside the aliasing contour which has red as its major component (blue in other cases). DRCNN which takes binary input has also the advantage of being applicable directly to real-patient data which have complementary colors.

Table 1. Table of average error in flow rate estimation and RMSE in flow convergence area estimation

CNN and grouping of sets	Training set size	Validation set size	RMSE (Area)	Average error percent (Flow Rate)
AlexNet random	1515	174	0.25 cm^2	4.8%
AlexNet low aliasing velocity random	505	174	0.48 cm^2	4.63%
AlexNet leave-one out	1515	174	**0.58** cm^2	**6.95%**
AlexNet low aliasing velocity leave-one out	505	174	0.75 cm^2	5.26%
DRCNN grayscale random	1515	174	0.37 cm^2	7.16%
DRCNN grayscale leave-one out	1515	174	1.26 cm^2	5.6%
DRCNN binary random	1515	174	0.715 cm^2	13.42%
DRCNN binary leave-one out	1515	174	**0.522** cm^2	**4.96%**
DRCNN binary preprocessed leave-one out	1515	174	**0.50** cm^2	**6.76%**

In Fig. 4. we can observe flow convergence area estimation against ground truth for one color flow Doppler time series which was excluded from the training set.

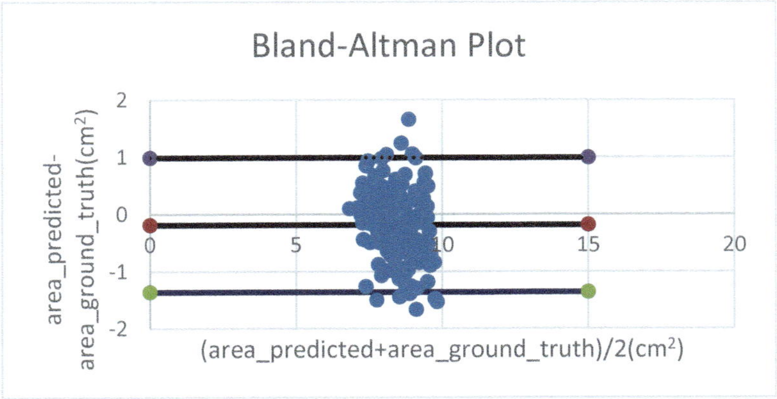

Fig. 4. Bland-Altman plot for flow convergence area estimation that corresponds to DRCNN binary leave-one-out

4 Discussion

We presented a fully automated method which can quantify flow rate through the orifice. The goal is to apply this method to real-patient data. Given the abundance of in-vitro data and the sparsity of in-vivo with ground truth values for flow-rate, we plan to use CNNs trained on in-vitro data to estimate flow rate in in-vivo images, augmented by quantitative labeling by expert cardiologists.

There are several improvements planned to be applied to this method besides increasing the volume of training data. First, the use of deep learning for flow convergence area detection. Secondly, the appearance of color flow Doppler is sensitive to the angle between the transducer and the flow. Therefore, this angle should be considered, and the input images should be aligned. In various occasions in-vitro data are more noisy than real-patient data. We plan to use only those in-vitro frames which are more comparable to real-patient data. We also expect that the CNNs will classify the images depending on the orifice shape. This will require input from two or more image planes and could eventually be extended to 3D Doppler data. Finally, research will be carried out on designing more efficient deep learning architectures.

5 Conclusion

We presented a fully automated system to assess mitral regurgitation through flow convergence area estimation. Results are promising, i.e., 5% error on new data with a relatively small training set. We expect to be able to demonstrate the high potential of applying this method to real-patient data.

Funding Sources. This research is funded by the Greek State Scholarships Foundation and European Social Fund.

References

1. Nishimura, R.A., Otto, C.M., Bonow, R.O., et al.: 2014 AHA/ACC guideline for the management of patients with valvular heart disease: a report of the American college of cardiology/American heart association task force on practice guidelines. J. Am. Coll. Cardiol. **63**, e57–185 (2014)
2. Zoghbi, W.A., Adams, D., Bonow, R.O., et al.: Recommendations for noninvasive evaluation of native valvular regurgitation: a report from the american society of echocardiography developed in collaboration with the society for cardiovascular magnetic resonance. J. Am. Soc. Echocardiogr. **30**, 303–371 (2017)
3. Naji, P., Asfahan, F., Barr, T., et al.: Impact of duration of mitral regurgitation on outcomes in asymptomatic patients with myxomatous mitral valve undergoing exercise stress echocardiography. J. Am. Heart Assoc. **4**, e001348 (2015)
4. Rodriguez, L., Anconina, J., Flachskampf, F.A., Weyman, A.E., Levine, R.A., Thomas, J.D.: Impact of finite orifice size on proximal flow convergence. implications for Doppler quantification of valvular regurgitation. Circ. Res. **70**, 923–930 (1992)

5. Pu, M., Vandervoort, P.M., Greenberg, N.L., Powell, K.A., Griffin, B.P., Thomas, J.D.: Impact of wall constraint on velocity distribution in proximal flow convergence zone. implications for color Doppler quantification of mitral regurgitation. J. Am. Coll. Cardiol. **27**, 706–713 (1996)
6. Mitter, S.S., Wagner, G.J., Barker, A.J., Markl, M., Thomas, J.D.: Impact of assuming a circular orifice on flow error through an elliptical regurgitant orifices: A computational fluid dynamics (CFD) analysis. (submitted for publication)
7. LeCun, Y., Bengio, Y., Hinton, G.: Deep learning. Nature **521**, 436–444 (2015)
8. Krizhevsky, A., Sutskever, I., Hinton, G.: ImageNet classification with deep convolutional neural networks. Adv. Neural Inf

Automated Design of Efficient Supports in FDM 3D Printing of Anatomical Phantoms

Maria Agnese Pirozzi[1,2(✉)], Emilio Andreozzi[2,3], Mario Magliulo[1], Paolo Gargiulo[4,5], Mario Cesarelli[2,3], and Bruno Alfano[1]

[1] Institute of Biostructures and Bioimaging, Italian National Research Council, Naples, Italy
mariaagnese.pirozzi@ibb.cnr.it
[2] Department of Electrical Engineering and Information Technologies, University of Naples Federico II, Naples, Italy
[3] Istituti Clinici Scientifici Maugeri S.p.A. – Società Benefit, Pavia, Italy
[4] Institute for Biomedical and Neural Engineering, Reykjavík University, Reykjavík, Iceland
[5] Department of Science, Landspítali University Hospital, Reykjavík, Iceland

Abstract. Recent improvements in image segmentation techniques enabled the (semi)automatic extraction of biostructures surfaces from 3D medical imaging data. The diffusion of 3D printing technologies promoted their introduction in the medical field, giving rise to several applications, such as the development of 3D-printed anatomical imaging phantoms. These devices provide controlled experimental environments for the improvement of medical imaging techniques, as they mimic the morphological and physiological features of different body parts. However, to obtain a 3D printable model from medical imaging data, different post-processing steps are needed, which require a considerable effort. Supports generation is often a critical task, as it requires to find the minimum amount of support structures necessary to hold a part in place during the printing process. This is particularly difficult for complex anthropomorphic models, for which a high printing level of detail, along with a reasonable number of internal supports, is usually needed. In this paper, an automatic method for the design of efficient support structures is proposed, which is suitable for 3D printing of complex anatomical phantoms, even with non-professional FDM 3D printers. A custom design software was developed, which places paraboloid-shaped shells to support all and only the critical points of the 3D model. This provided different advantages over support generation by means of common slicing software, allowing a reduction of material waste and printing times, along with an easier and faster dissolution of soluble supports for the clean-up of phantoms empty volumes.

Keywords: 3D printing · FDM · Supports design · Anatomical phantoms · Medical imaging

J. Henriques et al. (Eds.): MEDICON 2019, IFMBE Proceedings 76, pp. 292–300, 2020.
https://doi.org/10.1007/978-3-030-31635-8_35

1 Introduction

3D printing is a growing technology that is changing the manufacturing industry. In the last decades, the continuous research of novel materials and methods for 3D printing promoted the rise of new applications. The expiry of earlier patents allowed the development of several entry-level and semi-professional FDM (*Fused Deposition Modelling*) 3D printers, whose availability at lower costs promoted their introduction also in non-industrial settings (i.e. research, laboratories, schools and homes) [1].

The improvements in image segmentation techniques provided more advanced tools for the (semi)automatic extraction of biostructures surfaces from 3D medical imaging data [2, 3]. This promoted the application of 3D printing technologies in the medical field, ranging from surgical planning [4–7], to medical education and training [8, 9], customised prosthesis design [10–13] and medical imaging research [14–16]. Anthropomorphic phantoms of complex organs play a key role in research, as they provide controlled experimental environments for the improvement of medical imaging techniques. In fact, phantoms are designed to accurately reproduce morphological details of organs surfaces, providing at the same time empty volumes to be filled with solutions that mimic their physiological properties. The specific solution depends on the imaging modality of interest, e.g. iodine solutions for Computed Tomography (CT), paramagnetic solutions for Magnetic Resonance Imaging (MRI), radioactive solutions for Nuclear Medicine (NM) studies. In this field, 3D printing offers several advantages over traditional manufacturing technologies, such as the possibility to create objects with complex geometries and internal structures. This allows designing and prototyping highly detailed and complex anthropomorphic phantoms [14–16] without the need of classical manufacturing systems and expertise, making them an inexpensive alternative to commercially available products [2]. Indeed, advanced anatomical phantoms can be extracted from medical data by using voxel-based digital approaches, aimed at representing models as 3D matrices. Then, it is possible to obtain a 3D printable model by using equation-based approaches, which describe the anatomical model with a set of mathematical functions [17]. However, the process to create a 3D anatomical model is not straightforward. In fact, to obtain a 3D-printable model from medical imaging data, a post-processing phase, which requires a relevant effort [5], is always needed. The printing time is also a significant factor as it depends on the level of detail and the number of supports of the model [1]. They are usually very high for anatomical phantoms since they are characterised by complex internal structures which require a detailed printing with a considerable amount of supports. In non-professional 3D printing systems, all these high-demanding requirements are generally left to the designer and may discourage the development of this kind of phantoms. The post-processing phase involves the design of a triangle mesh from imaging data to provide an STL (*STereoLithography interface format*) description suitable for 3D printing. Even if DICOM-to-STL conversion supported by many modern radiology workstations [6] help reducing the time needed for the post-processing, the complexity of some anatomical districts still places a tough challenge to obtain an errorless STL description without designer's intervention [21]. Another critical task which may require manual effort is the internal supports generation. Its main issue is to find the minimum amount of support required to hold a part (or pieces of a part) in place during the printing process [1]. This is necessary to provide efficient support to all the critical points, e.g. in complex and/or convoluted anatomic surfaces.

This work proposes an automatic method for the generation of efficient support structures useful for 3D printing of complex anthropomorphic phantoms, even with non-professional FDM 3D printers.

2 Materials and Methods

The goal was to materialise a physical phantom of the human brain composed by two cerebral compartments (grey matter and white matter), in order to analyse quantitative aspects of imaging systems under conditions very close to their real operating environment. The physical phantom was obtained from the digital brain MRI phantom, *Phantomag* [18] which provides an accurate description of brain structures. It is composed of seventeen segmented tissue compartments of the healthy brain, and one additional compartment representing multiple sclerosis lesions, to simulate the relaxation rate distributions of real tissues in MRI studies. A pre-processing phase was carried out to correct small imperfections of *Phantomag*. Moreover, since voxel size of the digital phantom is 1 mm $\times$ 1 mm $\times$ 1 mm, the images were linearly interpolated to a voxel size of 0.5 mm $\times$ 0.5 mm $\times$ 0.5 mm, in order to design a printable vertical wall with smaller thickness (between a minimum of 0.5 mm and a maximum of 1 mm).

2.1 Extraction of Voxelised Surfaces and Identification of Critical Points

The whole procedure for the automated design of ad-hoc support structures was implemented in MATLAB® (R2018b) using only basic routines (no additional toolboxes are required). Single voxelised surfaces at the interface of compartments of interest were extracted from images of digital brain phantom by selecting voxels (from the outside towards the inside) sharing vertices, edges and faces. Then, they were analysed to identify all those points that need to be supported during 3D printing. These critical points were selected by setting two searching rules:

1. All surface voxels being at a lower level than all neighbouring voxels per side or vertex are to be supported;
2. All voxels not having contiguous voxels at the lowest level, or previously individuated critical points, within a 2.5 mm (5 voxels) neighbourhood along the horizontal or vertical directions must be supported.

Then, the points from which the supports will start have been placed just below the critical points (see Fig. 1a and b).

2.2 Generation of Voxel-Based Supports Structures

In order to minimize the amount of support, it was necessary to create paraboloid-shaped shells, to support all and only the previously identified critical points. Then, paraboloid-like 3D voxels volumes were added to the images. They were first positioned to pin their vertices on critical points, and then truncated in correspondence of voxelised surfaces of the phantom, in order to avoid overlappings with those surfaces. Furthermore, to avoid also intersections between different supports, they were labelled with different labels on the images during their construction. This procedure was designed to extract support volumes which could then be printed empty with a single external shell. The labelling, in fact, allowed distinguishing different support volumes that support each other on the images (see Fig. 1c). The images containing the voxelised surface of the phantom and the labelled voxel-based support structures were then saved in a raster format for subsequent processing.

By defining the coordinates of a generic point to be supported as $p_i = (x_i, y_i, z_i)$, the described paraboloid functions had the following form:

$$z = [a(x - x_i)]^4 + [a(y - y_i)]^4 + z_i \quad (1)$$

with a being a modulating parameter for the width of each paraboloid-like support. For our needs a has been chosen equal to 0.5.

2.3 Extraction of the Phantom and Supports STLs

The 3D surface model and 3D volume of supports were derived from the images using *3D Slicer*, an open source software platform for medical image informatics, image processing, and three-dimensional visualisation [19]. In particular, the *Model Maker* module was used to create 3D surface models from label maps (i.e. segmented image data), employing a high-resolution 3D surface construction algorithm called *Marching Cubes* [20]. The smoothing parameter was set to 0 in order to avoid smoothing operations, which would lead to inaccurate positioning of supports, undermining the correspondence with critical points. The modelled 3D surfaces and supports' volumes were finally saved in different STL files to enable dual-extrusion 3D printing (Fig. 2a).

2.4 Slicing Settings and G-Code Generation

We used *Raise3D N2 Plus* FDM printer to materialise our phantom. The obtained STL files were converted into G-Code format using the *Ideamaker* slicing software of *Raise3D* printers. The anatomical model and the designed supports were printed with two different materials (a soluble material was used for the support). Distinct slicing settings were provided for the model and the supports. The walls of the phantom were printed using at least two juxtaposed shells, each one as large as the nozzle diameter (0.4 mm) and providing a maximum filling at a moderate speed (40 mm/s). The supports were printed at a higher speed (70 mm/s) as empty volumes, using a single shell (Fig. 2b).

3 Results

Our software confirmed the printing complexity of the model, by identifying 10060 critical points for the whole brain phantom. An example is shown in Fig. 1a and b, where the arrows indicate two critical areas on a slice of 3D model and the respective critical points on the next lower slice. Figure 1c shows an explanatory coronal view of labelled paraboloid-like supports structure.

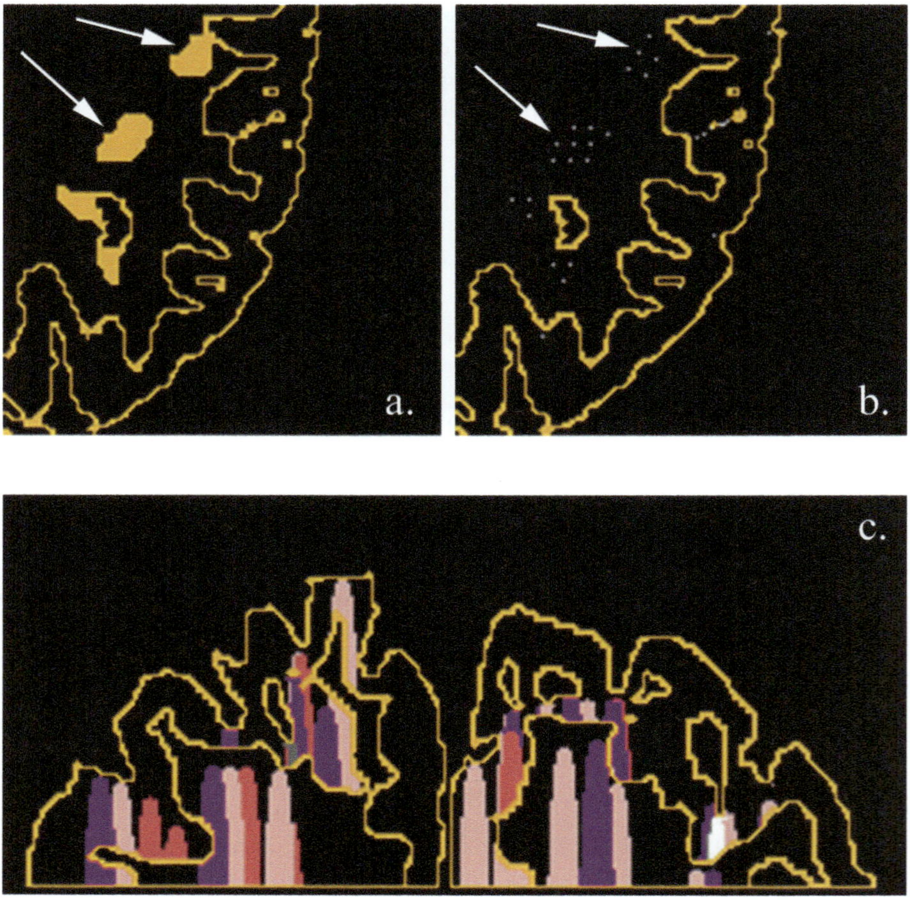

Fig. 1. Horizontal view of critical points and support structures: a. transversal critical areas (arrows); b. their corresponding critical points (arrows) in the next lower slice; c. example of a coronal view of labelled paraboloid-like generated supports.

The STLs of the model and supports were found to be error-free and ready to print. Figure 2a and b shows a detail of the STL and the G-Code. It can be observed how the ad-hoc designed paraboloid-like structures, being pinned only on the critical points, efficiently supported the phantom's complex internal structures, resulting also in a reduced number of supports compared to those automatically generated by common slicing software. In addition, the density of the support material was reduced by lowering the extrusion percentage and by printing the support structures as empty volumes (with a single outer shell).

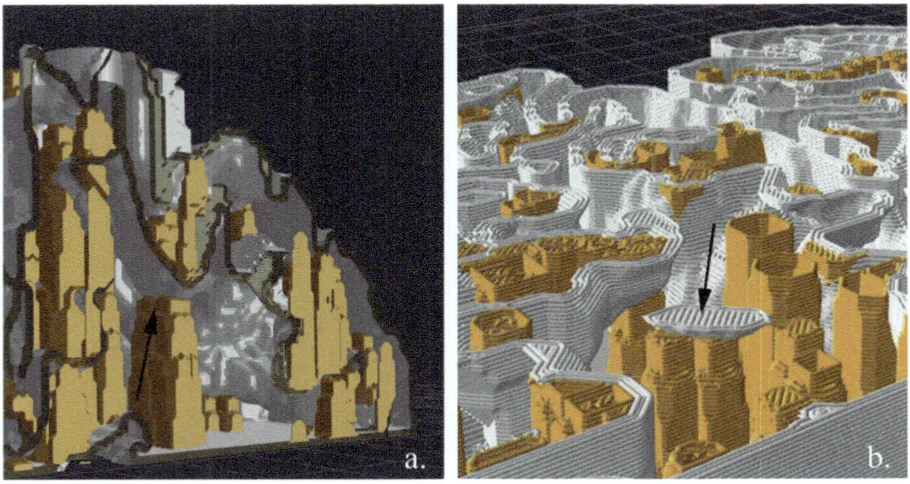

Fig. 2. Details of the internal structures of the phantom: a. STLs, the arrow indicates a critical internal structure; b. G-Code, the same structure is indicated by an arrow.

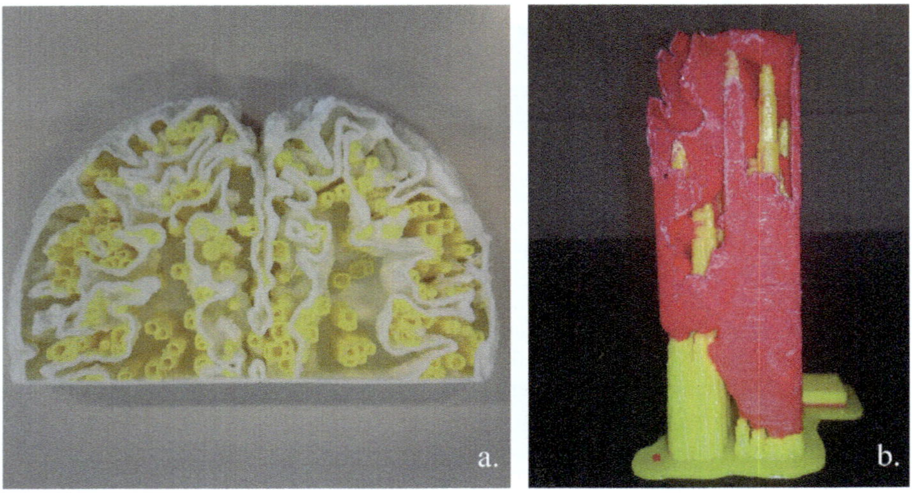

Fig. 3. 3D printing tests: a. double-extruder printing of a quarter of our brain phantom printed with ad-hoc supports; b. a detail of a test part of our phantom with the designed ad-hoc supports (conveniently opened to show internal supports).

Figure 3 shows two partial prints of the model using two different materials for the phantom and the supports. Generally, the dual extrusion greatly extends printing times. In fact, the double-extruder printing with supports generated by the slicing software usually takes almost twice the time of the same single-extruder printing. Instead, using the ad-hoc designed supports, printing times were significantly reduced compared to those required for double-extruder printing of the same piece with grid supports automatically generated by the *Ideamaker* software.

4 Conclusion

This work proposed an innovative automatic method to design efficient supports for 3D printing of complex anatomical phantoms. Common slicing software could not always generate efficient supports, which result often expensive for wasted material and for long printing times. The ad-hoc designed supports efficiently support all and only the critical points, requiring a smaller amount of support material along with a significant reduction in printing times compared to traditional support structures. Moreover, the dissolution of our ad-hoc soluble supports, resulted faster than that of traditional ones, since all supports surfaces were immediately reached by the liquid solvent. This is useful to clean up from internal supports 3D-printed phantoms. However, a limitation of the proposed method is that supports are defined in the design phase of the phantom, so a modification of the model necessarily requires a redesign of the supports. This could be overcome in the future by integrating the procedure into software that extracts STLs from medical images.

Acknowledgment. Funding by the CNR Strategic Project "The Aging: Technological and Molecular Innovations Aiming to Improve the Health of Older Citizens" (http://www.progettoinvecchiamento.it) and by the Italian Ministry for Education, University and Research (Project MOLIM ONCOBRAIN LAB) is gratefully acknowledged.

Conflict of Interest. The authors declare that they have no conflict of interest.

References

1. Oropallo, W., Piegl, L.A.: Ten challenges in 3D printing. Engineering with Computers **32**, 135 (2016). https://doi.org/10.1007/s00366-015-0407-0
2. Bücking, T.M., Hill, E.R., Robertson, J.L., Maneas, E., Plumb, A.A., Nikitichev, D.I.: From medical imaging data to 3D printed anatomical models. PLoS ONE **12**(5), e0178540 (2017). https://doi.org/10.1371/journal.pone.0178540
3. Rengier, F., Mehndiratta, A., von Tengg-Kobligk, H., et al.: 3D printing based on imaging data: review of medical applications. Int. J. CARS **5**, 335 (2010). https://doi.org/10.1007/s11548-010-0476-x
4. Tam, M.D.B.S., Laycock, S.D., Brown, J.R.I., Jakeways, M.: 3D printing of an aortic aneurysm to facilitate decision making and device selection for endovascular aneurysm repair in complex neck anatomy. J. Endovasc. Ther. **20**(6), 863–867 (2013). https://doi.org/10.1583/13-4450MR.1

5. Gargiulo, P., Árnadóttir, Í., Gíslason, M., Edmunds, K., Ólafsson, I.: New directions in 3D medical modeling: 3D-printing anatomy and functions in neurosurgical planning. J. Healthc. Eng. **2017**, 8 (2017). https://doi.org/10.1155/2017/1439643. Article ID 1439643
6. Schmauss, D., Haeberle, S., Hagl, C., Sodian, R.: Three-dimensional printing in cardiac surgery and interventional cardiology: a single-centre experience. Eur. J. Cardio-Thorac. Surg. **47**(6), 1044–1052 (2015). https://doi.org/10.1093/ejcts/ezu310
7. Vukicevic, M., Mosadegh, B., Min, J.K., Little, S.H.: Cardiac 3D printing and its future directions. JACC Cardiovasc. Imaging **10**(2), 171–184 (2017). https://doi.org/10.1016/j.jcmg.2016.12.001
8. Noecker, A.M., Chen, J.F., Zhou, Q., White, R.D., Kopcak, M.W., Arruda, M.J., Duncan, B. W.: Development of patient-specific three-dimensional pediatric cardiac models. ASAIO J. **52**(3), 349–353 (2006). https://doi.org/10.1097/01.mat.0000217962.98619.ab
9. Waran, V., Narayanan, V., Karuppiah, R., Thambynayagam, H.C., Muthusamy, K.A., Rahman, Z.A.A., Kirollos, R.W.: Neurosurgical endoscopic training via a realistic 3-dimensional model with pathology. Simul. Healthc.: J. Soc. Simul. Healthc. **10**(1), 43–48 (2015). https://doi.org/10.1097/sih.0000000000000060
10. Singare, S., Dichen, L., Bingheng, L., Zhenyu, G., Yaxiong, L.: Customized design and manufacturing of chin implant based on rapid prototyping. Rapid Prototyping J. **11**(2), 113–118 (2005). https://doi.org/10.1108/13552540510589485
11. Cohen, A., Laviv, A., Berman, P., Nashef, R., Abu-Tair, J.: Mandibular reconstruction using stereolithographic 3-dimensional printing modeling technology. Oral Surg. Oral Med. Oral Pathol. Oral Radiol. Endod. **108**(5), 661–666 (2009). https://doi.org/10.1016/j.tripleo.2009.05.023
12. Parthasarathy, J.: 3D modeling, custom implants and its future perspectives in craniofacial surgery. Ann. Maxillofac. Surg. **4**(1), 9 (2014). https://doi.org/10.4103/2231-0746.133065
13. Zopf, D.A., Hollister, S.J., Nelson, M.E., Ohye, R.G., Green, G.E.: Bioresorbable airway splint created with a three-dimensional printer. N. Engl. J. Med. **368**(21), 2043–2045 (2013). https://doi.org/10.1056/nejmc1206319
14. Leary, M., Kron, T., Keller, C., Franich, R., Lonski, P., Subic, A., Brandt, M.: Additive manufacture of custom radiation dosimetry phantoms: An automated method compatible with commercial polymer 3D printers. Mater. Des. **86**, 487–499 (2015). https://doi.org/10.1016/j.matdes.2015.07.052
15. Filippou, V., Tsoumpas, C.: Recent advances on the development of phantoms using 3D printing for imaging with CT, MRI, PET, SPECT, and ultrasound. Med. Phys. **45**, e740–e760 (2018). https://doi.org/10.1002/mp.13058
16. Leng, S., Chen, B., Vrieze, T., Kuhlmann, J., Yu, L., Alexander, A., Jane Matsumoto, J., Morris, J., McCollough, C.H.: Construction of realistic phantoms from patient images and a commercial three-dimensional printer. J. Med. Imaging **3**(3), 033501 (2016). https://doi.org/10.1117/1.jmi.3.3.033501
17. Zaidi, H., Tsui, B.M.W.: Review of computational anthropomorphic anatomical and physiological models. Proc. IEEE **97**(12), 1938–1953 (2009). https://doi.org/10.1109/jproc.2009.2032852
18. Alfano, B., Comerci, M., Larobina, M., Prinster, A., Hornak, J.P., Selvan, S.E., Amato, U., Quarantelli, M., Tedeschi, G., Brunetti, A., Salvatore, M.: An MRI digital brain phantom for validation of segmentation methods. Med. Image Anal. **15**(3), 329–339 (2011). https://doi.org/10.1016/j.media.2011.01.004
19. Fedorov, A., Beichel, R., Kalpathy-Cramer, J., Finet, J., Fillion-Robin, J.-C., Pujol, S., Bauer, C., Jennings, D., Fennessy, F., Sonka, M., Buatti, J., Aylward, S.R., Miller, J.V., Pieper, S., Kikinis, R.: 3D slicer as an image computing platform for the quantitative imaging network. Magn. Reson. Imaging **30**(9), 1323–1341 (2012). PMID: 22770690

20. Lorensen, W.E., Cline, H.E.: Marching cubes: a high resolution 3D surface construction algorithm. ACM SIGGRAPH Comput. Graphics **21**(4), 163–169 (1987). https://doi.org/10.1145/37402.37422
21. Friedman, T., Michalski, M., Goodman, T.R., Brown, J.E.: 3D printing from diagnostic images: a radiologist's primer with an emphasis on musculoskeletal imaging putting the 3D printing of pathology into the hands of every physician. Skeletal Radiol. **45**(3), 307–321 (2015). https://doi.org/10.1007/s00256-015-2282-6

Diffusion Weighted Magnetic Resonance Imaging Texture Biomarkers for Breast Cancer Diagnosis

Marialena I. Tsarouchi, Georgios F. Vlachopoulos, Anna N. Karahaliou, and Lena I. Costaridou(✉)

Department of Medical Physics, School of Medicine, University of Patras, 26500 Patras, Greece
costarid@upatras.gr

Abstract. Quantification of breast lesion heterogeneity by means of MRI texture contributes in differentiating benign from malignant breast lesions. This study investigates the diagnostic performance of 1st and 2nd order Texture Analysis descriptors on Apparent Diffusion Coefficient (ADC) lesion maps. 78 histologically verified breast lesions (40 benign, 38 malignant) of 67 patients undergoing DW-MRI at 3.0 T, were analyzed. ADC maps were generated for a slice representative of lesion largest diameter. A two-step segmentation approach was applied on high b-value diffusion image, based on Fuzzy C-Means (FCM) clustering and edge-based contouring, for defining the lesion region contour. Lesion contour was transferred to ADC map and subjected to texture analysis by means of twelve first-order and eleven second-order texture features. Logistic Regression Classifier was employed to assess the diagnostic ability of individual features and feature combinations. Diagnostic performance was evaluated by means of the area under Receiver Operating Characteristic curve (Az). The highest classification performance ($Az = 0.965 \pm 0.024$) was achieved by the combined feature subset 25th Percentile (1storder) and Entropy (2ndorder), suggesting the diagnostic significance of accurately quantifying lesion heterogeneity by texture-based feature combinations on ADC maps. Combined 1st and 2nd order texture biomarkers provide accurate spatial information of lesion ADC heterogeneity and holds potential in differentiating benign from malignant breast lesion status.

Keywords: Texture Analysis · MR-diffusion · Breast cancer diagnosis · Imaging biomarkers

1 Introduction

Quantification of tumor imaging heterogeneity, has the potential to identify (radiomic descriptors) of disease diagnosis, prognosis and response to therapy, non-invasively and cost-effectively [1]. Breast MRI provides functional imaging of breast tumors derived from imaging sequences such as Dynamic Contrast Enhanced (DCE) and Diffusion Weighted Imaging (DWI), mapping tissue vascular permeability/angiogenesis and cellularity/microstructure, respectively. DWI-MRI, quantifies the mobility of water molecules with Apparent Diffusion Coefficient (ADC) maps. Quantitative DWI-MRI

J. Henriques et al. (Eds.): MEDICON 2019, IFMBE Proceedings 76, pp. 301–305, 2020.
https://doi.org/10.1007/978-3-030-31635-8_36

analysis has mainly exploited 1st order Texture Analysis, highlighting features such as mean ADC value, achieving the highest classification accuracy (Az 0.90) between benign and malignant status [1–3]. However, biophysical mechanisms for ADC heterogeneity among different lesions has been further exploited with multi-parametric MRI, combining DCE [4] and DWI image features [5, 6]. Nevertheless, no study has been reported considering 1st and 2nd order texture feature combinations to quantify ADC map heterogeneity for predicting breast cancer clinical diagnosis.

The current study investigates the predictive power of individual 1st and 2nd order texture descriptors and their combinations on ADC lesion maps, in differentiating benign from malignant breast lesions. In order to enhance the proposed pipeline repeatability, image registration and semi-automated segmentation steps were considered.

2 Materials and Methods

2.1 Patient Cohort and MRI Acquisition

This retrospective study was approved by the Institutional Ethics Committee. 78 histologically verified breast lesions were analyzed, originating from 67 female patients undergoing DW-MRI at 3.0 T. Thirty-two female patients with 40 benign findings and thirty-five female patients with 38 malignant breast tumors were included. Patients underwent breast MRI in a 3.0 T MR scanner (Signa HDx, GE Healthcare), with a dedicated breast coil. Imaging included axial T2-weighted fast spin echo sequence, axial short TI inversion recovery sequence, axial diffusion-weighted echo-planar imaging (DWI) sequence (DW-EPI, TR/TE, 6000/63.7 ms; slice thickness, 4 mm; spacing, 0 mm; matrix, 96 × 108; FOV, 360 × 360 mm). Sensitizing diffusion gradients were applied in three orthogonal directions with b values of 0 and 900 s/mm^2. A three-dimensional fat-suppressed T1-weighted dynamic sequence was acquired once before and five times after the intravenous injection of 0.1 mmol/kg of gadopentate dimeglumine followed by a 20-mL flush of saline solution.

2.2 Image Analysis

In order to remove motion artifacts, a registration algorithm was adopted to map high-b value (b = 900 s/mm^2) DW images to corresponding low b value DW images. The applied three-level multi-resolution registration scheme combines two levels of rigid and a final resolution level of b-spline transform, using a Gaussian smoothing pyramid. The applied optimizer is a stochastic gradient descent, while the similarity metric is Mutual Information [7]. The Elastix 4.5 [8] software for intensity-based medical image registration based on the open source software Insight Toolkit (ITK) version 4.0 [9] was used for implementation of the registration scheme.

A semi-automated Fuzzy C-Means (FCM) clustering algorithm [10], was adopted for lesion delineation, having demonstrated improved performance in breast MRI lesion segmentation. Initially, an experienced radiologist defined a loose rectangular region of interest (ROI), containing the whole lesion, in slices representative of lesion's largest diameter, on high b-value diffusion images. FCM was applied the ROI, to build the likelihood membership map (cluster number, 2; weighting exponent, 2; stop criteria, 0.0005; max iteration, 100). Subsequently, an edge-based segmentation model was applied for binarization of the membership map followed by morphological operations to derive the final lesion contour, subsequently transferred to Apparent Diffusion Coefficient (ADC) map (Fig. 1), generated according to the mono-exponential model [2, 3]. Texture Analysis (TA) of ADC lesion maps included two feature sets, composed of twelve 1st order and eleven 2nd order texture features. ADC maps and first order texture analysis were implemented with in-house code in MATLAB R2017b, while second order TA was achieved by MaZda package v4.6 [11]. ADC lesion maps were re-binned to 6 bits/pixel and grey-level normalized ($\mu \pm 3\sigma$) for second order TA. Eleven features from Grey Level Co-occurrence Matrices-GLCM were calculated, considering a distance of one pixel and four directions (0°, 45°, 90° and 135°), averaged in order to obtain rotationally invariant features.

The discriminating ability of extracted features was investigated employing univariate and multivariate Logistic Regression Classifiers, using the publicly available data mining and machine-learning software Weka [12]. The area under Receiver Operating Characteristic (ROC) curve (Az index) was used for evaluating classification performance, employing a 10-fold cross validation training/testing methodology. In case of multivariate logistic regression, the bias of collinearity between features was evaluated using variance inflation factor (VIF) and the corresponding threshold was set at 3.

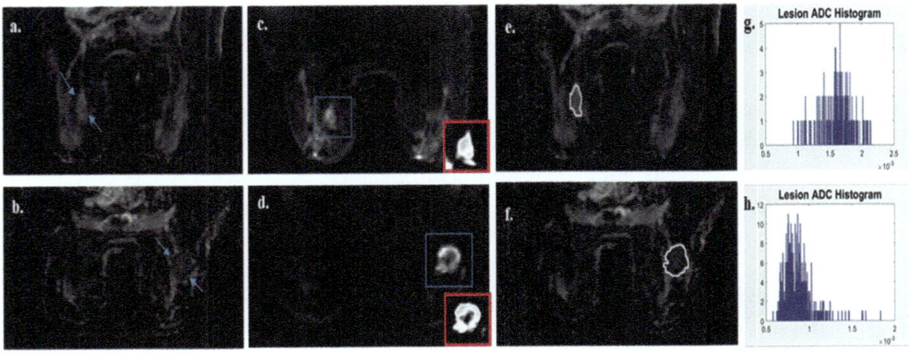

Fig. 1. Benign (upper row) and malignant (lower row) breast lesion; ADC map with blue arrows indicating the lesion site (a, b); lesion ROI (blue outline) containing the lesion on high b-value diffusion weighted image (c, d) and corresponding FCM membership map (red outline); Lesion contour (white outline) on ADC map (e, f) provided by edge-based segmentation of FCM map; Lesion ADC histogram (g, h).

3 Results

Table 1 summarizes classification performance of individual texture features and of feature combinations. Among 1st order texture features, 25th Percentile achieved the highest classification performance (Az = 0.943 ± 0.029), while among individual 2nd order texture features, Entropy achieved the highest classification performance (Az = 0.738 ± 0.057). The highest classification performance (Az = 0.965 ± 0.024) was achieved by the selected fused feature subset [25th Percentile (1st order) and Entropy (2nd order)] suggesting the diagnostic significance of accurately quantifying lesion heterogeneity by second-order texture coupled to ADC histogram analysis.

Table 1. Discriminating ability of individual and selected ADC texture feature subset

1st order ADC features	Az ± SE	2nd order ADC features	Az ± SE
Mean	0.924 ± 0.037	Angular Second Moment	0.737 ± 0.057
Standard Deviation	0.559 ± 0.065	Contrast	0.647 ± 0.062
Skewness	0.748 ± 0.056	Correlation	0.550 ± 0.066
Kurtosis	0.551 ± 0.065	Sum Of Squares	0.612 ± 0.064
Entropy	0.821 ± 0.064	Inverse Difference Moment	0.524 ± 0.067
Minimum	0.888 ± 0.037	Sum Average	0.709 ± 0.059
Maximum	0.831 ± 0.052	Sum Variance	0.605 ± 0.066
25th Percentile	0.943 ± 0.029	Sum Entropy	0.691 ± 0.061
50th Percentile	0.926 ± 0.053	Entropy	0.738 ± 0.057
75th Percentile	0.920 ± 0.038	Difference Variance	0.556 ± 0.066
Range	0.623 ± 0.063	Difference Entropy	0.568 ± 0.065
25th Percentile (1st order) + Entropy (2nd order)			0.965 ± 0.024

4 Discussion and Conclusion

Previously reported studies have demonstrated the ability of histogram ADC features in differentiating malignant from benign breast lesions [1–4]. As histogram ADC features provide a coarse quantification of lesion heterogeneity, 2nd order TA has emerged to extract high-throughput localized description of lesion spatial ADC heterogeneity (tumor phenotype). Results of the current study demonstrated that incorporating 1st and 2nd order ADC texture features into a multivariable diagnostic model in breast DW-MRI improves diagnostic accuracy. These findings agree with previously multiparametric MRI reported studies demonstrating that quantitative information captured by 1st and 2nd order TA is highly correlated to prediction of clinical diagnosis [5, 6]. Limitations of the current study account for small patient cohort and empirically selected parameters of semi-automated segmentation, limiting standardization and validation procedures. Future steps will be focused on validating image registration and segmentation steps on an augmented dataset an on exploiting additional feature selection and classification schemes. The approach of combining 1st and 2nd order ADC texture features is promising, especially in improving diagnostic accuracy, while dealing with cases of non-typical breast cancer morphologies. Towards this direction, a combined DCE and ADC texture analysis approach is currently under investigation.

Acknowledgements. Support by Operational Program "Human Resources Development, Education and Lifelong Learning" and is co-financed by the European Union (ESF) and Greek national funds (MIS:5005772). Special thanks to the Department of Radiology, University Hospital of Larissa, University of Thessaly, Greece, for contributing in this work.

Conflict of Interest Declaration. Authors declare that there is no conflict of interest for the publication of this work.

References

1. Chitalia, R.D., Kontos, D.: Role of texture analysis in breast MRI as a cancer biomarker: a review. J. Magn. Reson. Imaging **49**, 927–938 (2019)
2. Partridge, S.C., Nissan, N., Rahbar, H., Kitsch, A.E., Sigmund, E.: Diffusion weighted breast MRI: clinical applications and emerging techniques. J. Magn. Reson. Imaging **45**, 337–355 (2017)
3. Liu, H.-L., Zong, M., Wei, H., Lou, J.-J., Wang, S.-Q., et al.: Preoperative predicting malignancy in breast mass-like lesions: value of adding histogram analysis of apparent diffusion coefficient maps to dynamic contrast-enhanced magnetic resonance imaging for improving confidence level. Br. J. Radiol. **90**, 2–8 (2017). 20170394
4. Karahaliou, A., Vassiou, K., Arikidis, N.S., et al.: Assessing heterogeneity of lesion enhancement kinetics in dynamic contrast-enhanced MRI for breast cancer diagnosis. Br. J. Radiol. **83**, 296–309 (2010)
5. Parekh, V., Jacobs, M.A.: Intergraded radiomic framework for breast cancer and tumor biology using advanced machine learning and multiparametric MRI. Nat. Breast Cancer **3**, 43 (2017)
6. Jiang, X., Xie, F., Liu, L., Peng, Y., Cai, H., Li, L.: Discrimination of malignant and benign breast masses using automatic segmentation and features extracted from dynamic contrast-enhanced and diffusion-weighted MRI. Oncol. Lett. **16**(2), 1521–1528 (2018)
7. Vlachopoulos, G., Korfiatis, P., Skiadopoulos, S., Kazantzi, A., Kalogeropoulou, C., Pratikakis, I., Costaridou, L.: Selecting registration schemes in case of interstitial lung disease follow-up in CT. Med. Phys. **42**, 4511–4525 (2015)
8. Klein, S., Staring, M., Murphy, K., Viergever, M., Pluim, J.: Elastix: a toolbox for intensity-based medical image registration. IEEE Trans. Med. Imaging **29**(1), 196–205 (2010)
9. Ibanez, L., Schroeder, W., Ng, L., Cates, J.: The ITK Software Guide, 2nd edn. Kitware, Clifton Park (2005). ISBN 1-930934-15-7
10. Chen, W., Giger, M.L., Bick, U.: A fuzzy c-means (FCM)-based approach for computerized segmentation of breast lesions in dynamic contrast-enhanced MR images. Acad. Radiol. **13** (1), 63–72 (2006)
11. Szczpinski, P.M., Strzelecki, M., Materka, A., Klepaczko, A.: MaZda- a software package for image texture analysis. Comput. Methods Prog. Biomed. **94**, 66–76 (2009)
12. Witten, I., Frank, E., Hall, M.: Data Mining: Practical Machine Learning Tools and Techniques, 3rd ed. Elsevier, Amsterdam

Modeling Functional Processes of Brain Tissue: An fMRI Study on Patients with Un-Medicated Late-Onset Restless Leg Syndrome

Amalia K. Ntemou[1], Evanthia E. Tripoliti[1], Persefoni N. Margariti[2], Maria I. Argyropoulou[2], and Dimitrios I. Fotiadis[1,3](✉)

[1] Unit of Medical Technology and Intelligent Information Systems, University of Ioannina, 45110 Ioannina, Greece
ntemou.am@gmail.com, etripoliti@gmail.com, fotiadis@cc.uoi.gr
[2] Medical School of Ioannina, University of Ioannina, 45110 Ioannina, Greece
persamarga@yahoo.gr, margyrop@cc.uoi.gr
[3] Department of Biomedical Research, Institute of Molecular Biology and Biotechnology, FORTH, 45110 Ioannina, Greece

Abstract. In the current study we focus on the modeling of functional processes of brain tissue using functional Magnetic Resonance Imaging (fMRI) data. A brain connectivity analysis of Restless Legs Syndrome (RLS) is presented. Thirteen un-medicated patients with late-onset RLS and six healthy subjects are studied using structural and functional brain images. We compare functional connectivity analysis methods, according to their dependency on models or data, as well as to model effective connectivity. An Independent Component Analysis (ICA) method is implemented and all spontaneously activated areas in resting-state condition in both patients and healthy subjects are recorded and be compared with previous studies. Functional connectivity correlation matrices of both RLS and control subjects are extracted and these functional connectivity measures were compared using a seed-based analysis method. We model the brain tissue, based on the influence that one region exerts over another, using a spectral Dynamic Causal Model (DCM) analysis, which has not yet been implemented for RLS data. Finally, a Bayesian Model Selection is chosen in order to compare the winning model that effectively describes the data. The benefits of each methodology are presented.

Keywords: Brain connectivity analysis · functional Magnetic Resonance Imaging (fMRI) · Restless Legs Syndrome (RLS)

1 Introduction

Over 40% of people worldwide, have experienced abnormal movement behavior during sleep [1]. Restless Legs Syndrome (RLS) is a common sleep disorder which characterized by the irresistible urge to move legs or arms during evening or night hours (in resting state

J. Henriques et al. (Eds.): MEDICON 2019, IFMBE Proceedings 76, pp. 306–311, 2020.
https://doi.org/10.1007/978-3-030-31635-8_37

condition). There is a plethora of knowledge regarding the pathophysiology of RLS while on the other hand little is known about its neuroanatomic basis [2, 3].

MRI studies have indicated low iron levels [4, 5] in specific brain areas, PET and SPECT studies focused on the disturbances in dopamine receptors [6]. fMRI is one of the most used techniques used for understanding the activity of the human brain [7]. It measures the changes in the blood oxygenation levels (BOLD signal) and gives an indirect measure of the neuronal activity in the brain [8]. fMRI can detect brain regions which are responsible for a variety of human brain functions and study how they react according to the stimulus they receive. In addition, clinical applications involve the imaging of activated brain areas which have been injured, the monitoring of rehabilitation or reorganization after an injury and provide biomarkers or patterns for diagnosis of many pathological conditions [9, 10].

Previous resting-state fMRI studies have noticed maximum brain activation for RLS subjects in the cerebellum, the midbrain, the pons, the thalamus, the pars opercularis, the ventral anterior cingulum, the striatum, the inferior and superior parietal lobules, and the dorsolateral prefrontal cortex [11]. Brain connectivity studies have shown reduced default mode network connectivity in the left posterior cingulate cortex, the right orbito-frontal gyrus, the left precuneus, and the right subcallosal gyrus of the RLS subjects. They have also resulted in increased default mode network connectivity in sensory-motor-associated circuits including the right superior parietal lobule, the right supplementary motor area, and the left thalamus [12].

In the current work, we study and compare the different brain connectivity analysis methods related to RLS using fMRI data.

2 Materials and Methods

2.1 Dataset

Thirteen late-onset RLS patients and six control healthy subjects participate in our study. The fMRI data was acquired from MRI (1.5 T, INTERA, Philips Medical System, Best, The Netherlands) in the University Hospital of Ioannina in Greece. The imaging protocol in both patients and control subjects included a T1-weighted high-resolution (0.86 mm $\times$ 0.86 mm $\times$ 1.00 mm) 3D spoiled gradient-echo sequence (TR/TE, 25/4.6 ms), which was used for structural imaging and a single-shot multi-section gradient EPI, which was used for BOLD functional images (TR/TE, 3000/50 ms; flip angle, 40°; matrix 64 $\times$ 64; section thickness, 5 mm; gap, 0 mm). Each session consisted of 160 scans and lasted 480 s. The head of the subject was restrained with the use of cushions to minimize motion artifacts and the patients were advised to keep their eyes closed during the examination to minimize potential visual stimuli. Patients were awake during the examination, and they were instructed to remain still and avoid voluntary movements.

2.2 Preprocessing

fMRI timeseries contain many type of artifacts. Preprocessing steps are performed to enhance the Signal to Noise Ratio (SNR) and reduce the confounding effects of the artifacts. The functional images are realigned spatially to minimize residual head movement and temporally to ensure that the data from any given section were sampled at the same time. The images are then co-registered with the high-resolution anatomic image, normalized with the Montreal Neurological Institute (MNI) template, and smoothed using a Gaussian kernel 8 mm × 8 mm × 8 mm. The preprocessing was performed using Statistical Parametric Mapping (SPM12)[1] [13].

2.3 fMRI Analysis

The fMRI analysis aims to extract functional connectivity from data was accomplished following two different procedures, one based on data and another based on a model [14].

Functional Connectivity Using a Data-Driven Method
The Independent Component Analysis (ICA) is employed. ICA is a widely used method particularly for studies which utilize resting-state fMRI data [15]. The implementation of this method utilized the GroupICA/IVA[2] of the fMRI toolbox (GIFT) version 3.0b. For each separate subject and component, the coordinates of the maximum voxel are recorded according to the z-score and afterwards a spatial sorting of components is achieved using maximum voxel criteria. To estimate the spatial dependency of the automatic neuronal activity among different independent components, a functional connectivity network analysis is implemented using one-sample t-test to produce the functional connectivity correlation matrices for the patients and the control groups.

Functional Connectivity Using Seed-Based Analysis
Following a different approach compared to the previous one, we implemented the CONN[3] functional connectivity toolbox v17 in order to extract functional connectivity measures using seed-to-voxel correlation mapping. This approach calculates the temporal correlation between brain activity of a given region and all other regions using a General Linear Model approach [16]. We compared the functional connectivity of the middle frontal gyrus in the frontal lobe, the occipital lobe and the thalamus in both left and right cerebrum, as well as in the precuneus cortex. The selection of seeds in seed-to-voxel analyses is based on previous knowledge as it has been proven that the regions involving these seeds are responsible for the RLS.

Effective Connectivity Using Spectral Dynamic Causal Modeling
In order to investigate the endogenous effective connectivity, we implemented spectral Dynamic Causal Modeling (DCM) in both RLS and control groups. The aim of this method is to model the BOLD signal when there are no exogenous inputs (driving or modulatory), a fact that renders it suitable for its implementation in resting-state fMRI

---

[1] http://www.fil.ion.ucl.ac.uk/spm/software/spm12/.

[2] http://mialab.mrn.org/software/gift.

[3] https://www.nitrc.org/projects/conn.

data. The spectral DCM analysis is performed using the DCM12 routine implemented in SPM12. The precuneus in right parietal lobe (RPP), the mammillary body of right thalamus (RTM), the right inferior frontal gyrus (RFI), the left middle frontal gyrus (RFM), the parahippocampal gyrus in right limbic lobe (RLP), the culmen in right anterior lobe (RAC), the right occipital lobe (ROM) and the right superior temporal gyrus (RTS) are the eight regions of interest (ROIs) which are selected according to their increased level of involvement in RLS. ROIs are defined as spheres with a radius of 8 mm centered at the peak coordinate of the component map produced by the ICA approach.

Time-series from the ROIs are created as the residuals of a general linear model (GLM). The regressors that are included in this model are the six rigid body realignment parameters to model the movement correlated effects, one contrast regressor to model the baseline that is used to constrain the extraction of BOLD time-series within the brain, cosine basis functions to model possible signal drift and aliased respiratory and cardiac signals, and a high-pass filter of 1/128 Hz to remove possible ultraslow fluctuations [17].

3 Results

3.1 Functional Connectivity Using ICA

The maximum correlation in the RLS subjects involves the middle occipital and lingual gyrus in occipital lobe of the right cerebrum as well as the Culmen in anterior lobe and Uvula in the posterior lobe of the right cerebellum. Additional important functional connectivity is found between the precuneus in parietal lobe and superior frontal gyrus in frontal lobe of left cerebrum, middle frontal gyrus in frontal lobe of the right cerebrum. The maximum correlation in controls is found between the inferior frontal gyrus in frontal lobe of the right cerebrum and the parahippocampal gyrus in limbic lobe of the left cerebrum, as well as in the sub-gyral in temporal lobe of right cerebrum, and in the middle frontal gyrus in frontal lobe of left cerebrum.

3.2 Functional Connectivity Using Seed-Based Analysis

Figure 1 represents the activated areas and their size that resulted from a seed-to-voxel analysis considering as seed the middle frontal gyrus of the left cerebrum.

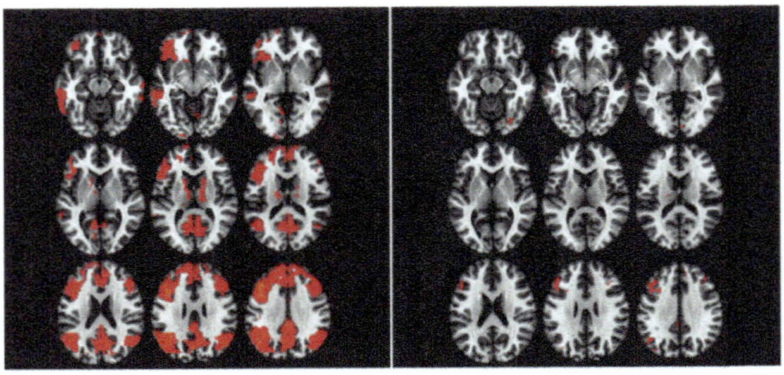

Fig. 1. The activation areas of RLS patients (left) and controls (right) considering the left middle frontal gyrus as seed.

3.3 Effective Connectivity Using Spectral Dynamic Causal Modeling

After extracting the resting-state fMRI time-series from all eight ROIs, a fully-connected model which has bi-directional connections between any pair of ROIs, is specified for each subject. Table 1 includes the effective connectivity parameters of the first RLS subject, where the matrix elements represent the effective influence between regions.

Table 1. Effective connectivity parameters of the first RLS patient.

	RPP	RTM	RFI	RFM	RLP	RAC	ROM	RTM
RPP	–09890	0.7336	–0.0422	0.1822	–0.4870	0.3665	0.0458	–0.0454
RTM	–0.0399	–0.8527	–0.0752	0.0118	–0.1920	–0.0336	–0.2775	–0.0080
RFI	–0.0390	0.1917	–0.8813	0.0729	–0.0024	–0.0801	0.4542	0.00460
RFM	–0.0142	–0.1228	0.0608	0.2077	0.2743	–0.1802	0.0448	–0.0026
RLP	–0.0046	0.9833	0.0327	0.0559	–0.9029	0.0299	–1.3031	–0.0977
RAC	–0.1492	0.7109	0.1707	0.0115	0.3804	–0.7155	–1.1320	–0.0695
ROM	–0.0763	–0.1307	–0.0384	–0.0277	0.2974	–0.0438	–0.5457	0.0364
RTS	–0.0262	0.1929	0.1842	0.0186	–0.1562	0.0321	0.2958	0.1465

Finally, a Bayesian Model Selection is chosen and the fixed effects inference method (FFX) followed to compare the winning model that effectively describes the data, resulting in a large number of models [18].

4 Discussion

Functional connectivity analysis for RLS led to similar findings of previous studies which focused on the correlation of precuneus in the parietal lobe, inferior frontal gyrus in the right frontal lobe, the parahipocampal gyrus in the left limbic lobe or the Culmen in the right anterior lobe. This high correlation among these components does not occur in control subjects' group. Compared with ICA analysis, seed-based analysis presented similar findings and all findings are in line with the bibliography that concerns the investigation of RLS. The use of spectral DCM, which has not yet been implemented for RLS data, was crucial in determining the influence that one region has to the others.

5 Conclusions

In this work, data described the RLS from the perspective of brain connectivity using fMRI, searching the most appropriate method for analysis. The model-based method provides results that aim to be more specific than data-driven methods. The possibility of implementing spectral Dynamic Causal Modeling on data provides useful information going beyond the restricted and often used methods. The selection of the most appropriate method depends on the researchers' incentives and the aim of each study. Since RLS has not been fully investigated yet, there will always be a need for further brain connectivity studies.

References

1. Martinez, D., Lenz, M.D.C.S.: Sleep-related movement disorders. Sleep Sci. **1**(1), 6–14 (2008)
2. Ku, J., Cho, Y.W., Lee, Y.S., Moon, H.J., Chang, H., Earley, C.J., Allen, R.P.: Functional connectivity alternation of the thalamus in restless legs syndrome patients during the asymptomatic period: a resting-state connectivity study using functional magnetic resonance imaging. Sleep Med. **15**(3), 289–294 (2014)
3. Chang, Y., Chang, H.W., Song, H., Ku, J., Earley, C.J., Allen, R.P., Cho, Y.W.: Gray matter alteration in patients with restless legs syndrome: a voxel-based morphometry study. Clin. Imaging **39**(1), 20–25 (2015)
4. Haba-Rubio, J., Staner, L., Petiau, C., Erb, G., Schunck, T., Macher, J.P.: Restless legs syndrome and low brain iron levels in patients with haemochromatosis. J. Neurol. Neurosurg. Psychiatry **76**(7), 1009–1010 (2005)
5. Li, X., Allen, R.P., Earley, C.J., Liu, H., Cruz, T.E., Edden, R.A., van Zijl, P.C.: Brain iron deficiency in idiopathic restless legs syndrome measured by quantitative magnetic susceptibility at 7 tesla. Sleep Med. **22**, 75–82 (2016)
6. Lin, C.C., Fan, Y.M., Lin, G.Y., Yang, F.C., Cheng, C.A., Lu, K.C., Lee, J.T., et al.: 99mTc-TRODAT-1 SPECT as a potential neuroimaging biomarker in patients with restless legs syndrome. Clin. Nucl. Med. **41**(1), e14–e17 (2016)
7. Huettel, S.A., Song, A.W., McCarthy, G.: Functional Magnetic Resonance Imaging, vol. 1. Sinauer Associates, Sunderland (2004)
8. Ogawa, S., Lee, T.M., Kay, A.R., Tank, D.W.: Brain magnetic resonance imaging with contrast dependent on blood oxygenation. Proc. Nat. Acad. Sci. **87**(24), 9868–9872 (1990)
9. Smith, S.M., Matthews, P.M., Jezzard, P. (eds.): Functional MRI: An Introduction to Methods. Oxford University Press (2001)
10. Tripoliti, E.E., Fotiadis, D.I.: Recent developments in computer methods for fMRI data processing. In: Biomedical Engineering. IntechOpen (2009)
11. Margariti, P., Astrakas, L., Tsouli, S., Hadjigeorgiou, G., Konitsiotis, S., Argyropoulou, M.: Investigation of unmedicated early onset restless legs syndrome by voxel-based morphometry, T2 relaxometry, and functional MR imaging during the night-time hours. Am. J. Neuroradiol. **33**(4), 667–672 (2011)
12. Ku, J., Lee, Y.S., Chang, H., Earley, C.J., Allen, R.P., Cho, Y.W.: Default mode network disturbances in restless legs syndrome/Willis–Ekbom disease. Sleep Med. **23**, 6–11 (2016)
13. Chen, J.E., Glover, G.H.: Functional magnetic resonance imaging methods. Neuropsychol. Rev. **25**(3), 289–313 (2015)
14. Maneshi, M.: Resting-state Functional Connectivity: Methods and Application in Epilepsy (Doctoral dissertation, McGill University Libraries) (2015)
15. Li, K., Guo, L., Nie, J., Li, G., Liu, T.: Review of methods for functional brain connectivity detection using fMRI. Comput. Med. Imaging Graph. **33**(2), 131–139 (2009)
16. Schurz, M., Wimmer, H., Richlan, F., Ludersdorfer, P., Klackl, J., Kronbichler, M.: Resting-state and task-based functional brain connectivity in developmental dyslexia. Cereb. Cortex **25**(10), 3502–3514 (2014)
17. Van Den Heuvel, M.P., Pol, H.E.H.: Exploring the brain network: a review on resting-state fMRI functional connectivity. Eur. Neuropsychopharmacol. **20**(8), 519–534 (2010)
18. Rosa, M.J., Bestmann, S., Harrison, L., Penny, W.: Bayesian model selection maps for group studies. Neuroimage **49**(1), 217–224 (2010)

Shift-Compensated Volumetric Interpolation of Tomographic Sequences for Accurate 3D Reconstruction

Chiara Santarelli[1(✉)], Francesca Uccheddu[2], Fabrizio Argenti[1], Luciano Alparone[1], Monica Carfagni[2], and Lapo Governi[2]

[1] Department of Information Engineering, University of Florence, Florence, Italy
{chiara.santarelli, fabrizio.argenti, luciano.alparone}@unifi.it
[2] Department of Industrial Engineering, University of Florence, Florence, Italy
{francesca.uccheddu, monica.carfagni, lapo.governi}@unifi.it

Abstract. In patients affected by craniosynostosis, i.e. a congenital cranial defect, diagnostic evaluation for a prompt surgical treatment is performed using low-dose three-dimensional computer tomography (CT), characterized by a poor spatial resolution (in terms of slice thickness). The limited number of CT images reduces the accuracy of the 3D reconstruction of the skull and leads to a coarser segmentation and modelling. In this paper, Motion Compensated Frame Interpolation (MCFI) techniques are applied for an effective axial interpolation of tomographic images sequences, with the main objective of obtaining a refined 3D reconstruction. The performance of the proposed method was assessed by using high-resolution CT sequences. After downsampling along the axial direction, the missing slices were recovered by using the proposed algorithm, to obtain an estimate of the original sequence. The experimental results show that the 3D models obtained from the downsampled/interpolated sequence are very close to those obtained from the original one thus providing a high-quality 3D skull reconstruction.

Keywords: CT slices · Interpolation · 3D modeling · Motion compensation

1 Introduction

Most cases of congenital cranial deformities fit within the spectrum of craniosynostosis caused by the premature closure of one or more sutures of the cranial bones. It is an uncommon condition occurring in one per 2500 live births [1] and, in some cases, it may lead to severe chronic intracranial hypertension for which a prompt surgical correction is needed. Three-dimensional computed tomography (CT) has high diagnosis accuracy and is considered as the gold standard [2] for the severity assessment of craniosynostosis, even though such an acquisition may need sedation and it is a source of harmful radiation. Young patients are indeed particularly at high risk for delayed radiation-

J. Henriques et al. (Eds.): MEDICON 2019, IFMBE Proceedings 76, pp. 312–321, 2020.
https://doi.org/10.1007/978-3-030-31635-8_38

induced malignancies. For these reasons, the gold standard for the skull radiology of neonatal children is the low-dose acquisition of tomographic imaging, which is achieved by using a less ionizing radiation than a conventional CT scan and by increasing the CT slices thickness to compensate the resulting lower signal-to-noise ratio [3, 4].

CT imaging has been recently proposed not only for the purpose of diagnosis, but also for obtaining 3D physical reproductions by means of 3D printers, which help neurosurgeons in performing a better preoperative strategy [5]. However, a low resolution CT acquisition induces the so-called *Lego effect*, due to the relevant z-jump from one image to the next [6] and affects the quality of the 3D skull reproduction. To overcome such a limitation, an interpolation between the available slices can increment the z-axis resolution, thus limiting such degradation.

In this work, finding inspiration from video processing frame interpolation techniques, the problem of z-axial interpolation of tomographic CT sequences is addressed.

The proposed algorithm is based on the estimation of the displacement vector field (DVF), analogous to a motion vector field in video processing, between two slices by means of an Optical Flow (OF) technique. The DVF models the geometrical evolution of anatomical structures in the axial direction and is used to estimate intermediate slices between the available ones. The proposed method works on a patch-basis: the dense DVF is adapted to small blocks; the displacement information is retrieved and applied both in the forward and backward direction to improve robustness. The experimental setup uses high-resolution CT sequences as the ground truth; after downsampling in the axial direction to simulate low resolution acquisitions, the interpolated slices were recovered to obtain an estimate of the original sequence. The performance of the proposed method is assessed by using both 2D and 3D metrics to evaluate the distortion introduced by the reconstruction process. The experimental results show that the synthetic interpolated images as well as the 3D models obtained from the downsampled/interpolated sequences are very close to those obtained from the original one.

2 Background

Image interpolation covers a wide range of techniques which can be classified as *pixel-based* or *object-based* [7, 8]. The proposed interpolation method focuses on pixel-based techniques that do not require prior segmentation. Motion Compensated Frame Interpolation (MCFI) is a technique used in the field of video encoding and processing to make animation more fluid by increasing the frame rate and reducing artifacts, such as motion blur and jerkiness [6–11].

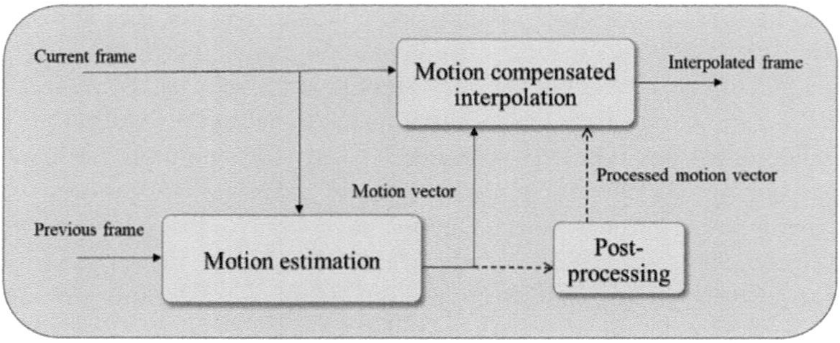

Fig. 1. Motion Compensated Frame Interpolation (MCFI) scheme.

As shown in Fig. 1, MCFI consists of two main steps: motion estimation (ME) and motion compensated interpolation (MCI). The goal of ME methods is estimating the Motion Vector Filed (MVF) between two or more consecutive frames. Most commonly used techniques of ME are Optical Flow [13], Block Matching [14, 15] and Phase Correlation (frequency-domain) [16, 17] These methods calculate the MVFs directly, i.e. by processing and comparing image pixels of subsequent frames. Once the MVFs relative to two consecutive frames have been estimated, and possibly post- processed, the intermediate frame can be constructed by means of a motion compensated interpolation operation [18]. The intermediate frame is generated by combining the current and past frames according to the estimated MVs in the forward and backward direction. In this way, the two estimates complement each other in order to obtain a new interpolated frame of high quality.

In the case of CT sequences, even if the imaged anatomical parts are not subject to movement in adjacent slices, a spatial transformation to relate them in the different sections may be foreseen. In this way, the methods mentioned above can be used to estimate intermediate CT slices starting from two adjacent ones. In the literature, registration methods for 3D images interpolation have been proposed [19–23], as well as applications to 4D medical images where the human anatomy changes shape during a certain period of time [24, 25]. In this paper, an algorithm based on Optical flow and working on a patch basis is proposed to define a spatial transformation (a sort of morphing) that relates the anatomical parts in the different slices. The transformation is then used to create new intermediate slices with the aim of increasing the axial resolution and obtain a more accurate 3D reconstruction.

3 Displacement Compensated Interpolation (DCI)

In this work, techniques inspired by MCFI are applied to axial interpolation of tomographic sequences. Figure 2 shows a schematic view of the proposed method whose steps will be detailed in the following.

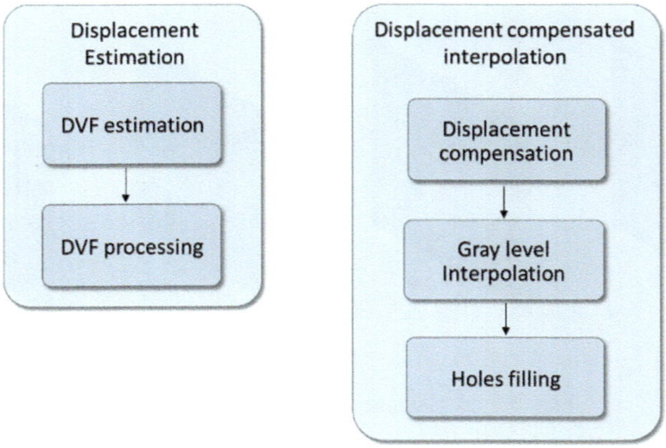

Fig. 2. Block diagram of the proposed method.

3.1 Displacement Estimation

The displacement stage extracts the displacement vector field (DVF) between the pixels belonging to two adjacent sections. Due to the analogy with MCFI, the DVF is assimilated to a motion vector field (MVF) and the problem is solved by using algorithms taken from ME approaches; more specifically, an Optical Flow (OF) technique has been employed. OF methods aim at estimating a dense field of motion vectors (one for each pixel of the images) that, applied to the pixels of one image (reference image), yields a prediction of the second image (target image). The Horn & Schunck algorithm [13], for example, solves this task by minimizing the energy of the prediction error with some regularity constraints. After OF estimation, regularization of DVFs is performed in order to reduce noise and eliminate outliers. The reference image is divided into small patches; a vector median filter (VMF) is applied to the displacement vectors related to a patch, resulting in a single displacement vector (DV) associated to the patch. In order to improve the results, a weighted VMF (WVMF) was also tested. Given a set of displacement vectors $V = \left(\overrightarrow{v_{1,}}\overrightarrow{v_{2,}}\ldots, \overrightarrow{v_{N,}}\right)$ and weights $W = \left(\overrightarrow{w_{1,}}\overrightarrow{w_{2,}}\ldots, \overrightarrow{w_{N,}}\right)$, belonging to a given N-size window, the weighted vector median of these values is the vector such that

$$\vec{v}_{WVM} = \arg\ \min_j\left(\sum\nolimits_{i=1}^{N} w_i \left\|\vec{v}_j - \vec{v}_i\right\|_2\right) \tag{1}$$

The weights calculation is based on the energy of the difference between the displaced reference patch and the target one [26, 27]. The resulting $\vec{v}_{WVM}$ is then scaled by a factor that depends on the position of the slice to be created by interpolation. The estimate of the DVF is performed forward and backward (i.e. swapping the reference images and the target images), so that two different scale factors are considered (see Fig. 3).

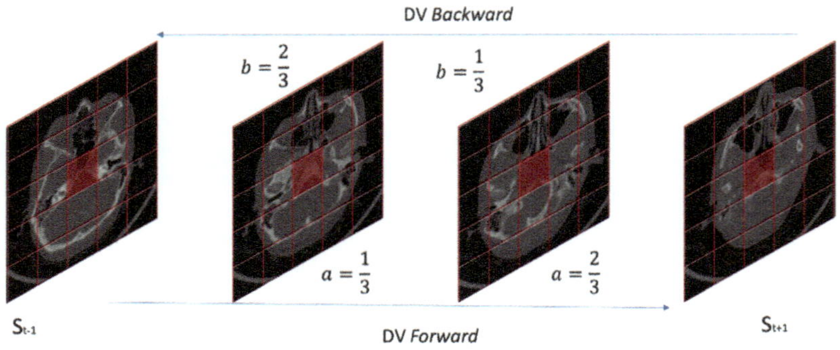

Fig. 3. Displacement compensation process.

3.2 Displacement Compensated Interpolation

This stage of the algorithm consists of three steps. In the first one, the reference patches are displaced by the corresponding scaled DVs according to

$$S_t(x, y) = Q_f S_{t-1}\left(x + av_x, y + av_y\right) + Q_b S_{t+1}\left(x - bv_x, y - bv_y\right) \tag{2}$$

where S_{t-1} and S_{t+1} are two consecutive slices; S_t is the slice to be interpolated; v = (v_x, v_y) is the DV estimated between S_{t-1} and S_{t+1}; Q_f and Q_b are weights to balance the backward and forward estimates. Since the DVs elements are real numbers, the displaced pixels new coordinates as real, and interpolation is needed to achieve the values on the image grid corresponding to integer positions, as shown in Fig. 4.

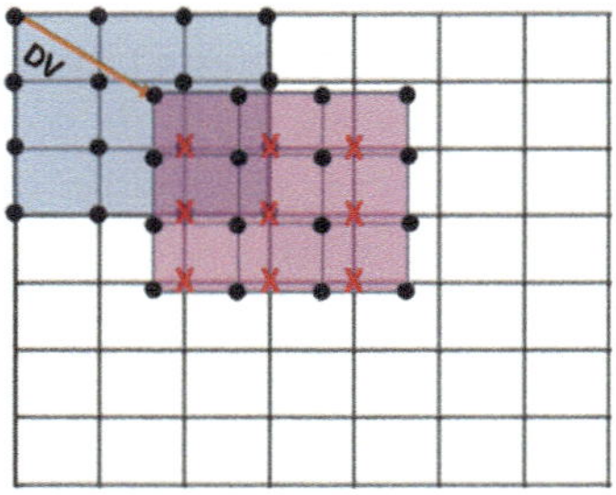

Fig. 4. Interpolation scheme for a single patch. Blue and pink rectangles are original and displaced patch respectively. Red crosses represent interpolated values.

The interpolation of the patches was performed by using both a bilinear and a bicubic function. Since strongly overlapped patches in the reference image are used, overlapping displaced patches are produced; the final pixel value is obtained by averaging all the values corresponding to the same pixel position. Eventually, a filling "holes" procedure is applied; it consists in searching for pixels that were not covered by any patch and assigning them an intensity value obtained as the average of the neighboring pixels.

4 Experimental Results

4.1 Synthetic Low-Resolution CT Sequences

To assess the performance of the proposed method, a high-resolution CT sequence imaging a human skull, was downsampled in the axial direction to simulate a low resolution one and, by using the proposed algorithm, the missing slices were recovered to obtain an estimate of the original sequence. Two downsampling factor (m = 2 and m = 3) were used. The original sequence is composed of 12-bit images. The proposed algorithm was developed in the MATLAB environment. The tests were performed by varying both the type of median filtering and the method for spatial interpolation.

The performance of the proposed method was assessed in different ways. Reconstructed images were compared with the original ones and the normalized. Normalized root-mean-square error (NRMSE), defined as

$$NRMSE = \sqrt{\frac{\sum_{m,n}(x_{true}[m,n] - x_{int}[m,n])^2}{\sum_{m,n}(x_{true}[m,n])^2}} \times 100 \tag{3}$$

was calculated, where x_{true} is the original image and x_{int} is the interpolated one. Table 1 shows the average NRMSE values obtained for different choices of the system configuration parameters (spatial interpolation, DVF filtering type).

Table 1. The average RMSE of each reconstructed sequence.

Spatial interpolation	DVF filtering	NRMSE (m = 2)	NRMSE (m = 3)
Bilinear	VMF	3.182	5.482
Bilinear	WVMF	3.181	5.483
Bicubic	VMF	3.072	5.413
Bicubic	WVMF	3.023	5.324

Although our method is not an object-based technique, the evaluation is also performed by using the 3D model of the cranial anatomy. The 3D model was by using MIMICS, a CE-certified medical software. This software achieved the 3D model by segmenting (and possibly further processing) the anatomical parts of interest. The metrics that were used are the mean value and standard deviation of the distance between the reference mesh and the reconstructed one, obtained by means of Geomagic Design X software. The metrics allow the 3D models obtained from the original high-resolution CT sequences and the reconstructed one to be compared. Tables 2 and 3 show the values resulting from this comparison for $m = 2$ and $m = 3$ respectively. In these tables, the comparison of the 3D models obtained from the downsampled CT vs. the original one is also reported.

Table 2. Average distance and standard deviation values by comparing the 3D model obtained from the original CT sequence with those reconstructed using the proposed method. In this case the reconstruction was made by setting $m = 2$.

Spatial interpolation	Filter type	Average distance (mm)	Standard deviation (mm)
Bilinear	VMF	0.097	0.215
Bilinear	WVMF	0.095	0.218
Bicubic	VMF	0.112	0.231
Bicubic	WVMF	0.096	0.226
3D model from downsampled sequence		0.337	0.502

Table 3. Average distance and standard deviation values obtained by comparing the 3D model obtained from the original CT sequence with those reconstructed using the proposed method. In this case the reconstruction was made by setting $m = 3$.

Spatial interpolation	Filter type	Average distance (mm)	Standard deviation (mm)
Bilinear	VMF	0.172	0.344
Bilinear	WVMF	0.163	0.343
Bicubic	VMF	0.166	0.346
Bicubic	WVMF	0.156	0.347
3D model from downsampled sequence		0.903	0.929

From the experimental results in Tables 1, 2 and 3 some observations about the configuration of the proposed method can be made. From Table 1, we observe that image interpolation benefits from the use of bicubic patch interpolation and weighted VMF with respect to bilinear and standard VMF, respectively. The same behaviour can be inferred from Tables 2 and 3, related to the 3D metrics. It can also be seen that in all the tests that have been performed the proposed approach strongly improved the 3D reconstruction quality with respect to not using interpolation (i.e. reconstruction based on downsampled sequences). Eventually, we can also observe that, as expected, the reconstruction quality degrades when downsampling factor increases.

4.2 Real Low-Resolution CT Sequences

In the case of real CT sequences with low axial resolution, as in the case of neonatal patients, it is possible to apply the proposed method to obtain an accurate 3D reconstruction with limited artefacts (*Lego effect*).

The reconstruction of the neonatal models has been carried out by setting the subsampling factor m according to the number of slices we would like to achieve and of those contained in the original sequence. The reconstruction method that gave the best results (bicubic-WVMF) during the test phase was chosen. In Fig. 5, the 3D models of four neonatal patients are reported: reconstructions using both the original (low-resolution) and the interpolated CT sequences are shown. As ca be seen, the 3D models produced with the proposed method are much more regular then those obtained from the original data.

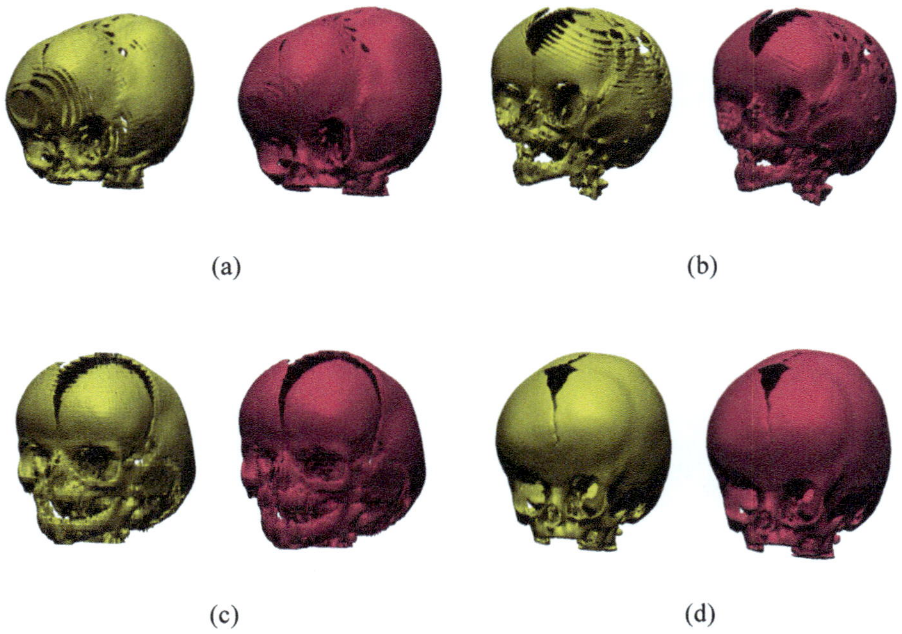

Fig. 5. Reconstruction of 3D skull models of neonatal patients. The models on the left (yellow) are the originals while on the right (pink) are those reconstructed with the proposed method.

5 Conclusions

A fundamental step in the treatment of neonatal patients affected by craniosynostosis is the reconstruction of accurate 3D models from computer tomography (CT) scans; for example, they are used by neurosurgeons to simulate the osteotomies before entering the operating room. In order to reduce harmful radiation, however, low-dose CT are often used, thus reducing the quality of the achievable 3D cranial model. The problem can be solved by increasing the axial resolution by means of interpolation. In this paper, a method for interpolating low-resolution CT sequences has been proposed. The method takes into consideration the works in the field of motion compensated frame interpolation (MCFI), proposed for video processing and is based, basically, on the optical flow estimation among adjacent CT slices. Different choices of the overall system configuration (vector field post-processing, interpolating functions) have been compared in the experimental tests by using synthetically-obtained low-resolution CT scans. These results guided the experiments on real low-resolution CT sequences, showing that accurate 3D reconstructions can be actually achieved.

References

1. Nagaraja, S., Anslow, P., Winter, B.: Craniosynostosis. Clin. Radiol. **68**(3), 284–292 (2013)
2. Vannier, M.W., Hildebolt, C.F., Marsh, J.L., Pilgram, T.K., McAlister, W.H., Shackelford, G.D., Offutt, C.J., Knapp, R.H.: Craniosynostosis: diagnostic value of three-dimensional CT reconstruction. Radiology **173**(3), 669–673 (1989)
3. Kutanzi, K.R., Lumen, A., Koturbash, I., Miousse, I.R.: Pediatric exposures to ionizing radiation: carcinogenic considerations. Int. J. Environ. Res. Public Health **13** (2016)
4. Bahreyni Toossi, M.T., Malekzadeh, M.: Radiation dose to newborns in neonatal intensive care units. Iran. J. Radiol. **9**(3), 145–149 (2012)
5. Volpe, Y., Furferi, R., Governi, L., Uccheddu, F., Carfagni, M., Mussa, F., Scagnet, M., Genitori, L.: Surgery of complex craniofacial defects: a single-step AM-based methodology. Comput. Methods Programs Biomed. **165**, 225–233 (2018)
6. Binaghi, S., Gudinchet, F., Rilliet, B.: Three-dimensional spiral CT of craniofacial malformations in children. Pediatr. Radiol. **30**(12), 856–860 (2000)
7. Raya, S.P., Udupa, J.K.: Shape-based interpolation of multidimensional objects. IEEE Trans. Med. Imaging **9**(1), 32–42 (1990)
8. Joliot, M., Mazoyer, B.M.: Three-dimensional segmentation and interpolation of magnetic resonance brain images. IEEE Trans. Med. Imaging **12**(2), 269–277 (1993)
9. Krishnamurthy, R., Woods, J.W., Moulin, P.: Frame interpolation and bidirectional prediction of video using compactly-encoded optical flow fields and label fields. IEEE Trans. Circuits Syst. Video Technol. **9**(5), 713–726 (1997)
10. Choi, B., Member, S., Han, J., Member, S.: Motion-compensated frame interpolation using bilateral motion estimation and adaptive overlapped block motion compensation. IEE Circuits Syst. Video Technol. **17**(4), 407–416 (2007)
11. Dikbas, S., Altunbasak, Y.: Novel true-motion estimation algorithm and its application to motion-compensated temporal frame interpolation. IEE Trans. Image Process. **22**(8), 2931–2945 (2013)
12. Kim, D., Lim, H., Park, H.: Iterative true motion estimation for motion-compensated frame interpolation. IEEE Trans. Circuits Syst. Video Technol. **23**(3), 445–454 (2013)
13. Horn, B.K.P., Schunck, B.G.: Determining optical flow. Artif. Intell. **17**(1–3), 185–203 (1981)
14. Ertuk, S.: A new perspective to block motion estimation for video compression: high-frequency component matching. IEEE Signal Process. Lett. **14**(2), 113–116 (2007)
15. Puri, A., Hang, H.-M., Schilling, D.: An efficient block-matching algorithm for motion-compensated coding. In: Proceedings of ICASSP, pp. 1063–1066 (1987)
16. Young, R.W., Kingsbury, N.G.: Frequency-domain motion estimation using a complex lapped transform. IEEE Trans. Image Process. **2**(1), 2–17 (1993)
17. Pingault, M., Pellerin, D.: Motion estimation of transparent objects in the frequency domain. Signal Process. **84**(4), 709–719 (2004)
18. Huang, A.-M., Nguyen, T.Q.: A multistage motion vector processing method for motion-compensated frame interpolation. IEEE Trans. Image Process. **17**(5), 694–708 (2008)

19. Goshtasby, A., Turner, D.A., Ackerman, L.V.: Matching of tomographic slices for interpolation. IEEE Trans. Med. Imaging **11**(4), 507–516 (1992)
20. Penney, G.P., Schnabel, J.A., Rueckert, D., Viergever, M.A., Niessen, W.J.: Registration-based interpolation. IEEE Trans. Med. Imaging **23**(7), 922–926 (2004)
21. Grevera, G.J., Udupa, J.K.: An objective comparison of 3-D image interpolation methods. IEEE Trans. Med. Imaging **17**(4), 642–652 (1998)
22. Leng, J., Xu, G., Zhang, Y.: Medical image interpolation based on multi-resolution registration. Comput. Math Appl. **66**(1), 1–18 (2013)
23. Horváth, A., Pezold, S., Weigel, M., Parmar, K., Cattin, P.: High order slice interpolation for medical images. In: Tsaftaris, S., Gooya, A., Frangi, A., Prince, J. (eds.) Simulation and Synthesis in Medical Imaging, SASHIMI 2017, Lecture Notes in Computer Science, vol. 10557. Springer, Cham (2017)
24. Hinkle, J., Szegedi, M., Wang, B., Salter, B., Joshi, S.: 4D CT image reconstruction with diffeomorphic motion model. Med. Image Anal. **16**(6), 1307–1316 (2012)
25. Ehrhardt, J., Werner, R., Säring, D., Frenzel, T., Lu, W., Low, D., Handels, H.: An optical flow based method for improved reconstruction of 4D CT data sets acquired during free breathing. Med. Phys. **34**(2), 21–711 (2007)
26. Alparone, L., Barni, M., Bartolini, F., Cappellini, V.: Adaptively weighted vector-median filters for motion-fields smoothing. In: Proceedings of ICASSP, pp. 2267–2270 (1996)
27. Alparone, L., Barni, M., Bartolini, F., Caldelli, R.: Regularization of optic flow estimates by means of weighted vector median filtering. IEEE Trans. Image Process. **8**(10), 1462–1467 (1999)

Calculating Texture Features from Mammograms and Evaluating Their Performance in Classifying Clusters of Microcalcifications

Marcelo A. Duarte[1], Wagner C. A. Pereira[2], and André Victor Alvarenga[3](✉)

[1] Electronic Engineering Department, UniCarioca University Centre, Rio de Janeiro, Brazil
[2] Biomedical Engineering Program/COPPE, Federal University of Rio de Janeiro, Rio de Janeiro, Brazil
[3] Laboratory of Ultrasound, National Institute of Metrology, Quality and Technology, Rio de Janeiro, Brazil
avalvarenga@inmetro.gov.br

Abstract. In this work, 2432 texture features were calculated from microcalcification clusters presented on 190 images from the Digital Database for Screening Mammography. Mutual information technique was used to rank texture features. Then, an incremental procedure adds top ranked features to the Fisher discriminant analysis to determine the best set of texture features in classifying benign or malignant microcalcification clusters. The result was achieved using 13 texture features ($AUC_{.632+} = 0.945 \pm 0.019$). However, to assure a consistent statistical analysis, at least 30 sample images for each feature added was assumed. The best performance was achieved by a set with 5 texture features ($AUC_{.632+} = 0.884 \pm 0.025$), which is comparable to the ones presented in literature.

Keywords: Microcalcifications · Texture features · Feature selection · Breast cancer · Mammograms

1 Introduction

According to the Brazilian National Institute of Cancer (INCA), breast cancer incidence is increasing in developed and in developing countries. It accounts for 28% new cancer cases that are reported each year [1]. Late diagnosis is the main cause of women death rates related to breast cancer worldwide. Nevertheless, breast cancer is also one of the leading causes of female deaths in Brazil. INCA estimates the occurrence of 59.700 new cases of this disease in 2018 [1]. Clinical examinations and mammography are the best way for detecting the early signs of breast cancer. Particularly the mammography is the screening technique to identify non-palpable breast lesions [2–4].

The most efficient strategies to reduce high death rates related to this kind of cancer are early detection followed by the tumor remotion in the initial stages of the disease.

J. Henriques et al. (Eds.): MEDICON 2019, IFMBE Proceedings 76, pp. 322–332, 2020.
https://doi.org/10.1007/978-3-030-31635-8_39

Although the mammography is considered as the best exam for detecting breast cancer, different factors may impact the diagnostic accuracy, as physician knowledge and experience, equipment quality, and breast densities. The higher is the density, the more difficult is the analysis and diagnosis. Hence, the mammography quality is highly dependent on the breast tissue itself [3].

The existence of microcalcifications in routine mammographic examinations is considered as a relevant sign of the malignant lesions presence, as they occur in most breast cancer cases. Microcalcifications are small granular deposits of calcium and are expressed in mammograms as small bright dots. Microcalcification detection is highly difficult, requiring radiologists to examine the mammogram carefully, as they may be obscured, especially in dense tissues [4]. Even though 60%–80% of microcalcifications are detected via histological examination, only 30%–50% of them are detected via mammographic exams in breast carcinomas [4]. Besides the density of the surrounding tissue, the difficulties in detecting microcalcifications in mammograms are also related to their variation in shape (from granular to rod shapes), brightness, orientation, and diameter [5].

Computer-aided detection (CADe) and computer-aided diagnosis (CADx) systems have been developed [6–8], aiming to minimize false-positive and false-negative rates in breast cancer diagnosis. CADe systems assist radiologist in automatic detecting lesions (including microcalcifications), whereas CADx systems are employed for lesions classification. Usually, CADx systems provide a second opinion, increasing the radiologist's final diagnosis accuracy [3, 4, 9, 10]. Basically, CADx systems involve three steps [11, 12]: (i) segmentation, (ii) features extraction and selection from the segmented microcalcifications and their clusters, and (iii) classification. These systems are commonly based on features extracted from microcalcifications, as roughness, compactness, orientation, average area, and the number of microcalcifications in a cluster [13, 14].

Different texture features combined to morphological features have been studied to quantify microcalcifications clusters characteristics [15], as well as texture features alone [16]. Chan and Sahiner [15] have stated that texture features presented better performance than morphological ones in distinguishing malignant and benign microcalcifications. However, its best result was obtained combining morphological and texture features. Karahaliou et al. [16, 17] have published that texture analysis of the tissue surrounding microcalcifications shows promising performance in diagnosing breast cancer. Concerning breast ultrasound images, Flores et al. [18] studied 1465 texture and 26 morphological features for improving breast lesions classification performance. Their best classification performance, evaluated through the area under ROC curve (AUC) was 0.942, achieved by using a set of five morphological features. Their best texture set was obtained with four features, which was AUC = 0.897.

In this work, similar techniques described by Flores et al. [18] were used to calculate 1216 texture features from microcalcifications clusters presented on 190 images from the Digital Database for Screening Mammography (DDSM) [19]. Hence, the mutual information technique was applied to rank texture features. Then, an incremental procedure adds top ranked features to the Fisher discriminant analysis to determine the best set of texture features in classifying benign or malignant clusters.

2 Materials and Methods

2.1 Texture Features

Due to space limitations, it is not feasible to depict every feature in detail; however, Table 1 illustrates the categorized descriptors, the number of features calculated from each one, and the references for detailed explanation of their computation.

2.2 Microcalcification Dataset and Segmentation

The database was selected from the DDSM, and presented 190 digital mammograms (300 dpi, 8 bits), from 78 patients, being 50 images with malignant clusters and 140 images with microcalcifications benign ones. The segmentation procedure, applied to segment microcalcifications in the digital mammograms, is based on morphological operations as described in [29, 30]. The segmented microcalcifications were used to define the clusters for each image (Fig. 1a). The clusters were defined using two approaches. Firstly, all the microcalcifications most external to the cluster were defined as vertices of the convex polygon that delimits the cluster (Fig. 1b). On a second approach, the four microcalcifications most external to the cluster were used as borderline to construct a rectangle, which delimits the cluster and contains all microcalcifications (Fig. 1c). The texture features presented in Table 1 were calculated twice (totaling 2432 texture features), once for the region defined by the convex polygon (Fig. 1b) and another for the one defined by the rectangle (Fig. 1c), both over the original digital mammogram.

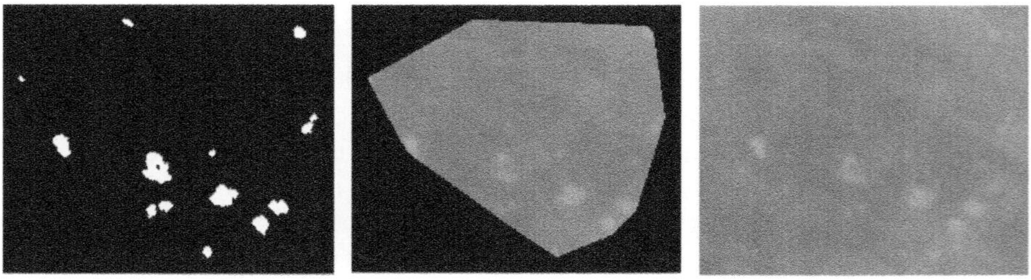

(a) (b) (c)

Fig. 1. (a) Binary image containing a cluster of segmented microcalcifications. (b) ROI defined by the convex polygon in which the vertices are the most external segmented microcalcifications. (c) Rectangular ROI in which the vertices are defined by the most external segmented microcalcifications.

Table 1. Texture features computed in this study.

Technique	Features	Description	Total of features
Histogram-based [20]	Average gray-level, Average histogram, Energy, Modified energy, Entropy, Modified entropy, Standard deviation, Modified standard deviation, Skew, Modified skew	–	10
Complexity curve [21]	Maximum value of transitions, Average value of transitions, Weighted gray-level mean, Weighted gray-level standard deviation, Entropy	–	5
Gray-level cooccurence matrix (GLCM) [22]	Autocorrelation, Contrast, Correlation, Cluster Prominence, Cluster Shade, Dissimilarity, Energy, Entropy, Homogeneity, Maximum probability, Variance, Sum average, Sum variance, Sum entropy, Difference variance, Difference entropy, Information measure of correlation 1, Information measure of correlation 2, Inverse difference normalized, Inverse difference moment normalized	For each image four matrices were determined considering four different distances ($d = 1, \ldots, 4$). Moreover, considering each value of d, the average of four matrices, each one obtained for a different direction ($\theta = 0°$, $45°$, $90°$, $135°$) was determined. This procedure produced a total of 80 features	80 (20 features × 4 average GLCM)
Laws' texture energy [23]	Mean, standard deviation, entropy, kurtosis and skewness	Five statistical texture features are evaluated from Laws' texture energy images resulting from 14 Laws' masks of length 5	70 (5 features × 14 Laws'masks)
Autocovariance [24]	Autocovariance coefficients	–	24
Gray-level average-based texture features [25]	Lesion boundary class, average gray intensity, echo pattern feature	–	3
Block-based coefficients [26]	Block difference of inverse probabilities (16 features), Block variation of local correlation (16 features)	–	32
Binary Gradient Contours and Local Binary Pattern [27]	Single-loop (BGC1), double-loop (BGC2) and triple-loop (BGC3), and local binary pattern (LBP)	–	991
Autocorrelation [28]	Autocorrelation coefficient	–	1
Total of features			**1216**

2.3 Feature Selection and Classification Performance

The feature selection procedure aims reducing the number of texture features while searches for acceptable classification performance. The features exclude should be the ones that do not contribute to distinguishing data of different classes and can be removed without affecting classification accuracy, as well as the features that are redundant. The main idea is to decide which features will be used and how to combine them. Herein, a three-step procedure was implemented for selecting and evaluating the performance of texture features. Initially, the minimal-redundancy-maximal-relevance criterion (MI-mrMR) based on mutual information technique is employed to rank the texture feature set [31, 32]. In the next step, the ranked features are sequentially adds to the classification step based on an incremental procedure. The .632+ bootstrap method [33] is used to assess the classification performance. Herein, 500 bootstrap samples for each feature subset and the local Fisher discriminant analysis (LFDA), with linear kernel, were employed as classifier [34]. The performance index was estimated by calculating the average of the area under ROC curve ($AUC_{.632+}$) considering the 500 bootstraps. In the final step, the feature subset presenting the best classification performance is defined as the "reference subset". A search procedure seeks a subset with fewer features that presents no statistical difference compared to the reference subset, considering a 95% confidence interval. The one-way analysis of variance (ANOVA1) is applied to test whether the mean values between texture features sets were different. Complementarily, a post-hoc analysis was performed using the Scheffé's method to determine if there was statistical significant difference between particular pairs of groups [35]. Details about the procedure steps can be found in [18].

3 Results

The result indicates that the maximum $AUC_{.632+}$ value (0.950 ± 0.021) is achieved considering 18 features, which are presented in Table 2. The result of the post-hoc analysis using the Scheffé's method is presented in Fig. 2. The best result considering the set of 18 texture features (hexagram in Fig. 2) is not statistically different from the set with 17 texture features ($AUC_{.632+} = 0.943 \pm 0.020$; diamond in Fig. 2). Moreover, the set of 16 features ($AUC_{.632+} = 0.941 \pm 0.025$; circle in Fig. 2) is neither statistically different the one with 17 features. The second best performance presenting the less number of features is reached with 13 texture features (square in Fig. 2). One can note that the set with 13 texture features is not statistically different from the other ones with 14 and 15 texture features (squares group in Fig. 2). The third best result is achieved with 11 texture features (right arrow in Fig. 2). The groups with 5 to 9 texture features (left arrows in Fig. 2) are not statistically different among them. The set composed by four features (cross in Fig. 2) achieved a performance of $AUC_{.632+} = 0.874 \pm 0.024$. Finally, the sets with 1, 2 and 3 texture features (pentagram in Fig. 2) are not statistically different among them. Table 3 presents the $AUC_{.632+}$ values, and respective standard deviations, for the sets presenting the best performance and less number of texture features.

Assessing Table 2, one can observe that just two features selected by the proposed technique were estimated considering a rectangle delimiting the cluster (LBP_52 and LBP_245). All other features were calculated considering the region of the cluster defined by its respective convex polygon (16 texture features, see Table 2). Hence, the performance of texture features calculated using just the convex polygon was assessed. The best result was $AUC_{.632+} = 0.950 \pm 0.018$, which is similar to the one achieved using the 18 texture features, previously presented. Moreover, the result achieved using 13 texture features ($AUC_{.632+} = 0.945 \pm 0.019$; just convex polygon) is not statistically different from the one obtained with the 16 texture features.

Assessing the results achieved using 18, 16 or 13 texture features one can observe the similarity among the $AUC_{.632+}$ values. This seems to be an indication that the number of texture features is large considering the number of studied cases. Hence, to assure a consistent statistical analysis, the heuristic rule of a minimum of 30 sample images for each feature added was assumed. Therefore, considering the 190 mammograms used in this work, the maximum number of texture features for an adequate statistical analysis would be 6 (190/30). Based on this assumption, and considering the features calculated to the rectangular and polygonal regions, the set with 5 texture features was the one with the best performance ($AUC_{.632+} = 0.884 \pm 0.025$), which is not statistically different from the one with 6 features.

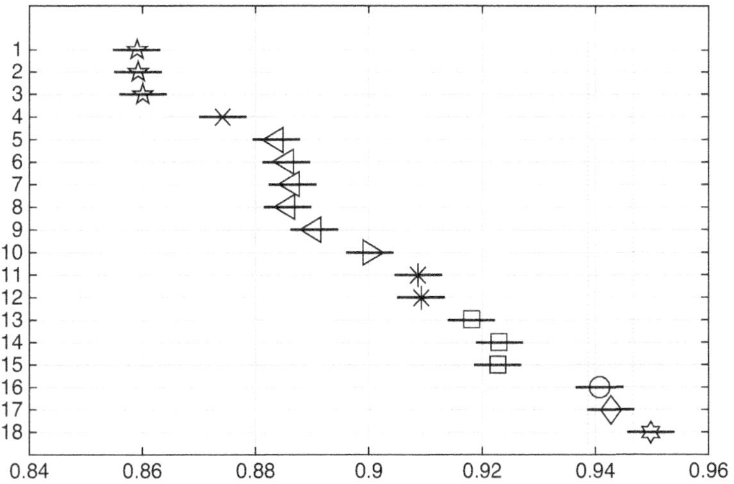

Fig. 2. Result of the post-hoc analysis using the Scheffé's method. The hexagram indicates the best result achieved using the set with 18 texture features. The result achieve using 17 texture features (diamond) is not statistically different from the best result, and the set of 16 features (circle) is neither statistically different the one with 17 features. The other sets that are statistically different from the others sets are presented using different labels (square, asterisk, right arrow, left arrow, cross, and pentagram). The name of the different features can be seen in Table 2.

Table 2. The best eighteen texture features. The rank of features determined by the minimal-redundancy-maximal-relevance criterion is presented in parenthesis, as well as the region (convex polygon or rectangle) considered for texture feature calculation.

Technique	Features	Description (rank/region)	Total of features
Histogram-based [20]	Average gray-level	hAG (14/convex polygon)	1
Gray-level co-occurrence matrix (GLCM) [22]	Information measure of correlation 2	GLCM_inf2h_D3 (2/convex polygon); d = 3	1
Laws' texture energy [23]	Skewness	R5R5_var (5/convex polygon)	1
Gray-level average-based texture features [25]	Pattern feature	EP_ag (4/convex polygon)	1
	Block difference of inverse probabilities	BDIP_sC2 (11/convex polygon)	1
Block-based coefficients [26]	Block variation of local correlation	BVLC_mC6 (12/convex polygon) BVLC_sC7 (8/convex polygon)	2
Binary Gradient Contours and Local Binary Pattern [27]	Single-loop	BGC1_7 (17/convex polygon), BGC1_36 (1/convex polygon), BGC1_141 (3/convex polygon), BGC1_162 (9/convex polygon), BGC1_166 (10/convex polygon)	5
	Triple-loop	BGC3_106 (6/convex polygon)	1
	Local binary pattern	LBP_21 (18/convex polygon) LBP_40 (15/convex polygon) LBP_52 (13/rectangle) LBP_187 (16/convex polygon) LBP_245 (7/rectangle)	5
Total			**18**

Table 3. Values of $AUC_{.632+}$ ($\pm$standard deviation) for the sets presenting the best performance and less number of texture features.

Number of texture features	$AUC_{.632+}$ values
18	0.950 ± 0.021
17	0.943 ± 0.020
16	0.941 ± 0.025
13	0.918 ± 0.022
11	0.909 ± 0.025
10	0.900 ± 0.024
5	0.884 ± 0.025
4	0.874 ± 0.024
1	0.859 ± 0.022

4 Discussions and Conclusion

The results suggested that the feature selection procedure proposed by [18] is useful in determining the smallest subset of features for improving the classification performance in terms of AUC index. The 2432 texture features were reduced to a set of 13 texture features with a performance of $AUC_{.632+} = 0.945 \pm 0.019$. Moreover, to avoid over-fitting and to keep features capacity of generalization, a limit of one texture feature to each 30 image was established. Hence, the best result achieved was $AUC_{.632+} = 0.884 \pm 0.025$, with a set of 5 texture features. Among the five most significant texture features, four feature extraction techniques were relevant (see rank in Table 2): Binary Gradient Contours (single loop), Gray-level co-occurrence matrix (Information measure of correlation 2), Gray-level average-based texture features (Pattern feature), and Laws' texture energy (skewness).

Despite a direct comparison with other texture-based classification studies is not fair due to different classification algorithms, number of features considered, and data sets (MC clusters subtlety and breast density types) used, the results achieved in this work are comparable to the ones presented in literature. Chan and Sahiner [15] stated an AUC value of 0.84 using only texture features calculated based on the gray-level co-occurrence Matrix. By combining morphological and texture features, the result was increased to 0.89, which is similar to the one achieved herein using only 5 texture features (Table 3). Karahaliou et al. [16] assessed the texture of the tissue surrounding microcalcifications using Laws' Texture Energy technique. They stated an accuracy of 89% achieved using the k-nearest neighbor classifier, and employing the leave-one-out training-testing methodology. In a subsequent study, Karahaliou et al. [17] improved the previous performance combining gray-level first-order statistics, gray-level co-occurrence matrices and wavelet coefficients texture to the Laws' Texture Energy, using a probabilistic neural network classifier. The authors stated AUC value of 0.989; however the total number of combined texture features used to achieve that result it is not clear. Dheeba and Tamil Selvi [36] presented a method to detect microcalcification clusters in digital mammograms. Texture features were calculated based on the Law's texture energy measures and applied to Support Vector Machines classifier. The method was tested in 322 digitized mammographic images from the MIAS database and a recognition score of 86.1% was achieved. Tiedeu et al. [37] used the gray-level co-occurrence matrix technique and calculated a set of 13 texture features of microcalcification clusters in mammograms. Nine features were selected as the best to classify the clusters into benign and malignant types, and were used as input to a backpropagation neural network (sixty images were used to train, while 200 were used to test). The result leads to an accuracy of 89% and an AUC of 0.968.

Considering the revised literature, the Laws' Texture Energy measures, gray-level first-order statistics and co-occurrence matrices, as well as wavelet coefficients have been used to calculate texture features from clusters of microcalcifications. However, as far as the authors know, it is the first time that Binary Gradient Contours (BGC) and Local Binary Pattern (LBP) are studied with that objective. Herein, the texture features calculated based on BGC and LBP seem to be relevant in quantifying texture of clusters of microcalcifications (see Table 2). Moreover, considering the region of interest studied, the result indicates that the regions defined by the convex polygon technique work better than the ones defined as a rectangular area.

The number of works in literature using only texture features to classify microcalcification clusters in mammograms is still small compared to the ones using morphometric features. However, the results presented herein, and others stated in literature, indicate that texture features seem to be relevant in classifying microcalcification clusters, and that more studies should be carried out to investigate the comprehensiveness of those features performance. Moreover, this study assess a set of texture features based on Binary Gradient Contours and Local Binary Pattern that were used to classify microcalcification clusters in mammograms for the first time. Among the five best texture features studied here, two of them were calculated from the Binary Gradient Contours technique. Complementarily, texture features calculated to the ROI defined by the convex polygon presented best results than the ones calculated to the rectangular ROI. As far as the authors know, this is the first time that this approach was tested. Finally, comparing different results presented in literature, one can conclude that to limit the number of features, to meet statistical significance and realistic performance of the features, should be a concern to avoid overfitting and to keep the texture features generalization ability.

Acknowledgements. Thanks to the financial support of the Brazilian National Council for Scientific and Technological Development (CNPq) (Grants: 434.858/2016-1, 309717/2014-0, and 308.627/2013-0), and CAPES/PROEX.

References

1. Brasil, Ministério da Saúde, Instituto Nacional de Câncer (INCA), Tipos de Câncer, Câncer de Mama. http://www2.inca.gov.br/wps/wcm/connect/tiposdecancer/site/home/mama
2. Brasil, Ministério da Saúde, Instituto Nacional de Câncer (INCA), Câncer de Mama, Detecção Precoce. http://www2.inca.gov.br/wps/wcm/connect/tiposdecancer/site/home/mama/deteccao_precoce+
3. Jalalian, A., Mashoror, S.B.T., Mahmudb, H.R., Saripan, M.I.B., Ramli, A.R.B., Karasfi, B.: Computer-aided detection/diagnosis of breast cancer in mammography and ultrasound: a review. Clin. Imaging **37**(3), 420–426 (2013)
4. Calas, M.J.G., Gutfilen, B., Pereira, W.C.A.: CAD and mammography: why use this tool? Braz. J. Biomed. Eng. **45**(1), 46–52 (2012)
5. Wei, L., Yanga, Y., Nishikawa, R.M.: Microcalcification classification assisted by content-based image retrieval for breast cancer diagnosis. Pattern Recogn. **42**(6), 1126–1132 (2009)
6. Halkiots, S., Botsis, T., Rangoussi, M.: Automatic detection of clustered microcalcifications in digital mammograms using mathematical morphology and neural networks. Sig. Process. **87**(7), 1559–1568 (2007)
7. Nishikawa, R.M.: Current status and future directions of computer-aided diagnosis in mammography. Comput. Med. Imaging Graph. **31**(4–5), 224–235 (2007)
8. Elter, M., Horsch, A.: CADx of mammographic masses and clustered microcalcifications: a review. Med. Phys. **36**(6), 2052–2068 (2009)
9. Chen, C.H., Lee, G.G.: On digital mammogram segmentation and microcalcification detection using multiresolution wavelet analysis. Graph. Models Image Process. **59**(5), 349–364 (1997)

10. Cheng, H.D., Cai, X., Chen, X., Hu, L., Lou, X.: Computer-aided detection and classification of microcalcifications in mammograms: a survey. Pattern Recogn. **36**(12), 2967–2991 (2003)
11. Arikidis, N.S., Karajaliou, A., Skiadopoulos, S., Korfiatis, P., Likaki, E., Panayiotakis, G., et al.: Size-adapted microcalcification segmentation in mammography utilizing scale-space signatures. Comput. Med. Imaging Graph. **34**(6), 487–493 (2010)
12. Paquerault, S., Yarusso, L.M., Papaioannou, J., Jiang, Y.: Radial gradient-based segmentation of mammographic microcalcifications: observer evaluation and effect on CAD performance. Med. Phys. **31**(9), 2648–2657 (2004)
13. De Santo, M., Molinara, M., Tortorella, F., Vento, M.: Automated classification of clustered microcalcifications by a multiple expert system. Pattern Recogn. **36**(7), 1467–1477 (2003)
14. Veldkamp, W.J., Karssemeijer, N., Otten, J.D.M., Hendriks, J.H.C.L.: Automated classification of clusters microcalcifications into malignant and benign types. Med. Phys. **27**(11), 2600–2608 (2000)
15. Chan, H.P., Sahiner, B.: Computerized analysis of mammographic microcalcifications in morphological and texture feature spaces. Med. Phys. **25**(10), 2007–2019 (1998)
16. Karahaliou, A.N., Boniatis, I.S., Skiadopoulos, S.G., Sakellaropoulos, F.N., Likaki, E., Panayiotakis, G.S., Costaridou, L.I.: A texture analysis approach for characterizing microcalcifications on mammograms. In: Proceeding of IEEE International Special Topic Conference on Information Technology in Biomedicine (ITAB 2006), pp. 251–257, October 2006
17. Karahaliou, A.N., Boniatis, I.S., Skiadopoulos, S.G., Sakellaropoulos, F.N., Arikidis, N.S., Likaki, E.A., Panayiotakis, G.S., Costaridou, L.I.: Breast cancer diagnosis: analyzing texture of tissue surrounding microcalcifications. IEEE Trans. Inf Technol. Biomed. **12**(6), 731–738 (2008)
18. Flores, W.G., Pereira, W.C.A., Infantosi, A.F.C.: Improving classification performance of breast lesions on ultrasonography. Pattern Recogn. **48**(4), 1125–1136 (2015)
19. University of South Florida, DOD Breast Cancer Research Program, US Army Research and Material Command, Digital Database for Screening Mammography (DDSM). http://marathon.csee.usf.edu/Mammography/Database.html. Accessed 03 Jan 2014
20. Zhang, P., Verma, B., Kumar, K.: Neural vs. statistical classifier in conjunction with genetic algorithm based feature selection. Pattern Recogn. Lett. **26**(7), 909–919 (2005)
21. Alvarenga, A.V., Pereira, W.C.A., Infantosi, A.F.C., Azevedo, C.M.: Complexity curve and grey level co-occurrence matrix in the texture evaluation of breast tumor on ultrasound images. Med. Phys. **34**(2), 379–387 (2007)
22. Gomez, W., Pereira, W., Infantosi, A.F.C.: Analysis of co-occurrence texture statistics as a function of gray-level quantization for classifying breast ultrasound. IEEE Trans. Med. Imaging **31**(10), 1889–1899 (2012)
23. Kriti, J.V., Dey, N., Kumar, V.: PCA-PNN and PCA-SVM based CAD systems for breast density classification. In: Applications of Intelligent Optimization in Biology and Medicine: Current Trends and Open Problems, pp. 159–180 (2016)
24. Chang, R.-F., Wu, W.-J., Moon, W.K., Chen, D.-R.: Improvement in breast tumor discrimination by support vector machines and speckle-emphasis texture analysis. Ultrasound Med. Biol. **29**(5), 679–686 (2003)
25. Shen, W.-C., Chang, R.-F., Moon, W.K., Chou, Y.-H., Huang, C.-S.: Breast ultrasound computer-aided diagnosis using bi-rads features. Acad. Radiol. **14**(8), 928–939 (2007)
26. Huang, Y.-L., Wang, K.-L., Chen, D.-R.: Diagnosis of breast tumors with ultrasonic texture analysis using support vector machines. Neural Comput. Appl. **15**(2), 164–169 (2006)
27. Fernandez, A., Alvarez, M.X., Bianconi, F.: Image classification with binary gradient contours. Opt. Lasers Eng. **49**, 1177–1184 (2011)

28. Horsch, K., Giger, M.L., Venta, L.A., Vyborny, C.J.: Computerized diagnosis of breast lesions on ultrasound. Med. Phys. **29**(2), 157–164 (2002)
29. Duarte, M.A., Alvarenga, A.V., Azevedo, C.M., Calas, M.J.G., Infantosi, A.F.C., Pereira, W.C.A.: Segmenting mammographic microcalcifications using a semi-automatic procedure based on Otsu's method morphological filters. Braz. J. Biomed. Eng. **29**(4), 377–388 (2013)
30. Duarte, M.A., Alvarenga, A.V., Azevedo, C.M., Calas, M.J.G., Infantosi, A.F.C., Pereira, W.C.A.: Evaluating geodesic active contours in microcalcifications segmentation on mammograms. Comput. Methods Prog. Biomed. **122**(3), 304–315 (2015)
31. Peng, H., Long, F., Ding, C.: Feature selection based on mutual information criteria of max-dependency, max-relevance, and min-redundancy. IEEE Trans. Pattern Anal. Mach. Intell. **27**(8), 1226–1238 (2005)
32. Pereira, W.C., Alvarenga, A.V., Infantosi, A.F.C., Macrini, L., Pedreira, C.E.: A non-linear morphometric feature selection approach for breast tumor contour from ultrasonic images. Comput. Biol. Med. **11**(12), 912–918 (2010)
33. Sahiner, B., Chan, H.P., Hadjiiski, L.: Classifier performance prediction for computer-aided diagnosis using a limited dataset. Med. Phys. **35**(4), 1559–1570 (2008)
34. Sugiyama, M.: Dimensionality reduction of multimodal labeled data by local Fisher discriminant analysis. J. Mach. Learn. Res. **8**, 1027–1061 (2007)
35. Stoline, M.R.: The status of multiple comparisons: simultaneous estimation of all pairwise comparisons in one-way ANOVA designs. Am. Stat. **35**(3), 134–141 (1981)
36. Dheeba, J., Tamil Selvi, S.: Classification of malignant and benign microcalcification using SVM classifier. In: Proceedings of the 2011 International Conference on Emerging Trends in Electrical and Computer Technology (ICETECT 2011), India, pp. 686–690, March 2011
37. Tiedeu, A., Daul, C., Kentsop, A., Graebling, P., Wolf, D.: Texture-based analysis of clustered microcalcifications detected on mammograms. Digital Sig. Process. **22**(1), 124–132 (2012)

LNDetector: A Flexible Gaze Characterisation Collaborative Platform for Pulmonary Nodule Screening

João Pedrosa[1,2(✉)], Guilherme Aresta[1,2], João Rebelo[3,4], Eduardo Negrão[3,4], Isabel Ramos[3,4], António Cunha[1,5], and Aurélio Campilho[1,2]

[1] Institute for Systems and Computer Engineering, Technology and Science (INESC TEC), Porto, Portugal
joao.m.pedrosa@inesctec.pt
[2] Faculty of Engineering (FEUP), University of Porto, Porto, Portugal
[3] Department of Radiology, S. João Hospital Center, Porto, Portugal
[4] Faculty of Medicine (FMUP), University of Porto, Porto, Portugal
[5] University of Trás-os-Montes e Alto Douro (UTAD), Vila Real, Portugal

Abstract. Lung cancer is the deadliest type of cancer worldwide and late detection is one of the major factors for the low survival rate of patients. Low dose computed tomography has been suggested as a potential early screening tool but manual screening is costly, time-consuming and prone to interobserver variability. This has fueled the development of automatic methods for the detection, segmentation and characterisation of pulmonary nodules but its application to the clinical routine is challenging. In this study, a platform for the development, deployment and testing of pulmonary nodule computer-aided strategies is presented: LNDetector. LNDetector integrates image exploration and nodule annotation tools as well as advanced nodule detection, segmentation and classification methods and gaze characterisation. Different processing modules can easily be implemented or replaced to test their efficiency in clinical environments and the use of gaze analysis allows for the development of collaborative strategies. The potential use of this platform is shown through a combination of visual search, gaze characterisation and automatic nodule detection tools for an efficient and collaborative computer-aided strategy for pulmonary nodule screening.

Keywords: Lung cancer · Low dose computed tomography · Pulmonary nodules · Computer-aided diagnosis

1 Introduction

Lung cancer is the deadliest type of cancer worldwide [13]. In spite of the decreasing trends in incidence and mortality in recent decades, lung cancer is projected to remain the deadliest cancer in the future. Progress in increasing lung cancer survival has also been slow in contrast to other cancer types, mainly due to late

J. Henriques et al. (Eds.): MEDICON 2019, IFMBE Proceedings 76, pp. 333–343, 2020.
https://doi.org/10.1007/978-3-030-31635-8_40

diagnosis. While more than half of the cases are diagnosed at a distant stage - when 5-year survival reaches 5% - early diagnosis allows a more favourable perspective, with a 56% 5-year survival [13]. Low-dose computed tomography (LDCT) has been suggested as an early screening tool and a 20% reduction in mortality has been demonstrated for lung cancer risk groups [14]. Nevertheless, translation to the clinic has been challenging due to equipment and personnel costs as well as the complexity of the task. Lung nodules present a large range of characteristics and their identification is prone to high variability. Computer-aided diagnosis (CAD) systems can thus facilitate the adoption of screening programs by reducing the burden on the clinician.

Extensive research has been conducted on the development of CAD systems for lung cancer screening focusing on different tasks - nodule detection, segmentation and estimation of malignancy likelihood [4,5]. Recently, deep learning methods have shown especially promising results for nodule detection and segmentation [2,15,16] and most of the best-performing methods on the LUNA16 nodule detection challenge incorporate this technique [12].

In spite of the promising results, adoption of CAD systems for screening is not straightforward. The fact that CAD systems are not designed as part of the clinical routine makes them difficult to adopt, representing an additional task and increasing the burden on the clinician. Secondly, the black-box behaviour of most systems hinders their adoption given that no explanation for detection is provided.

As such, the design of clinician-oriented CAD systems is particularly interesting and eyetracking technology especially promising. Recording the radiologist's gaze during image exploration allows to study the perceptual/cognitive process, explain why errors occur and understand differences between readers [8]. Furthermore, collaborative strategies can be designed, combining the strengths of radiologists with those of CAD systems.

As such, the motivation for this work was the development of a platform for the study and development of clinician-oriented CAD systems for lung cancer screening in LDCT. This platform, LNDetector, should allow image exploration and manual annotation as well as automatic or semi-automatic methods for nodule screening. Furthermore, LNDetector should integrate eyetracking technology to enable gaze characterisation and collaborative CAD strategies.

2 LNDetector Software Platform

The layout of LNDetector is shown in Fig. 1. The axial, coronal and sagittal planes are shown and a target indicates the slices being displayed. The image can be explored freely and maximum intensity projection, windowing options and fine tuning of the display are also available, as these are essential for nodule identification [6].

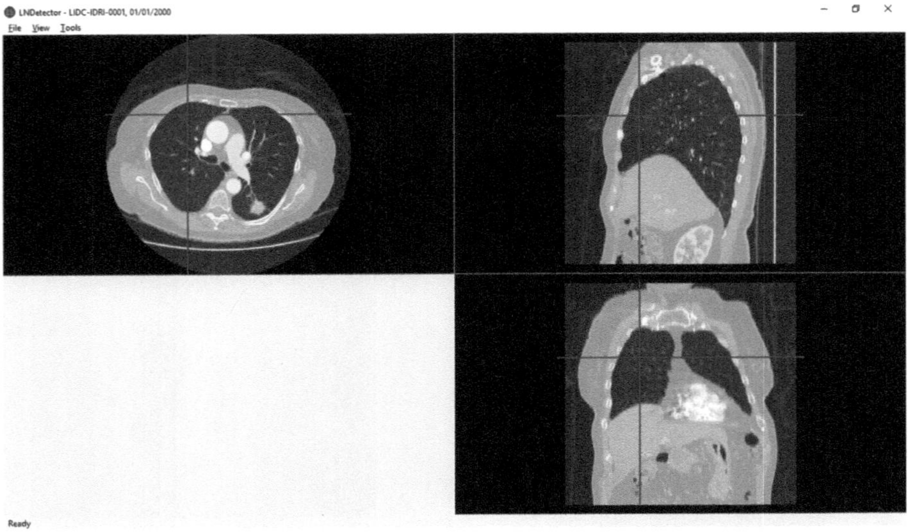

Fig. 1. Layout of LNDetector showing LIDC-IDRI-0001.

To ease the use of this software by the research community, LNDetector follows the annotation system of LIDC-IDRI, a vastly used public dataset for the development of lung cancer CAD systems [3]. Figure 2 shows the annotations for LIDC-IDRI-0001, namely the location and characteristics of `Nodule 001`. Multiple annotations can be compared as shown in Fig. 3, providing a useful tool for exploration of previous annotations. New annotations can also be made and saved according to the LIDC-IDRI format.

Automatic screening tools for nodule detection, segmentation and classification are implemented so that these can be run in a single pipeline or independently in a semiautomatic pipeline to allow comparison of different strategies.

To enable gaze characterisation, LNDetector is compatible with Tobii eye-tracking devices [1]. The screen coordinates of the user's gaze are recorded, as well as the display settings of LNDetector. This data is streamed at 90 Hz and can be used to convert the screen gaze coordinates to image coordinates, allowing to compute where in the image the user was looking at. By concatenating the gaze points throughout a session, a 3D gaze map can be built [9]. Taking into account a visual angle of 5° [10], an attention map can be reconstructed. An example of the gaze coordinates recorded during an annotation task on LIDC-IDRI-1001 is shown in Figs. 4 and 5. Figure 4 highlights the attention devoted to a nodule. Figure 5 on the other hand shows a nodule identified on LIDC-IDRI but missed by the radiologist. The attention map shows that the nodule was missed because this region was not thoroughly analysed.

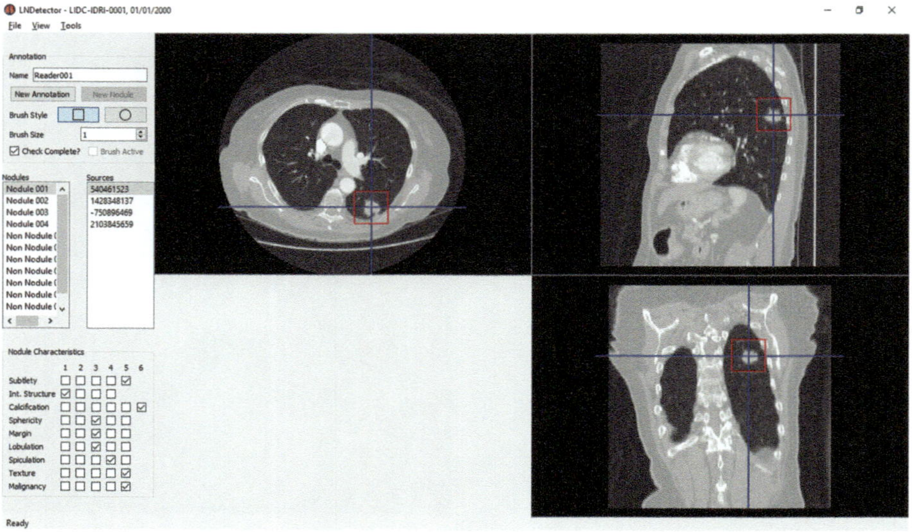

Fig. 2. `Nodule 001` on LIDC-IDRI-0001.

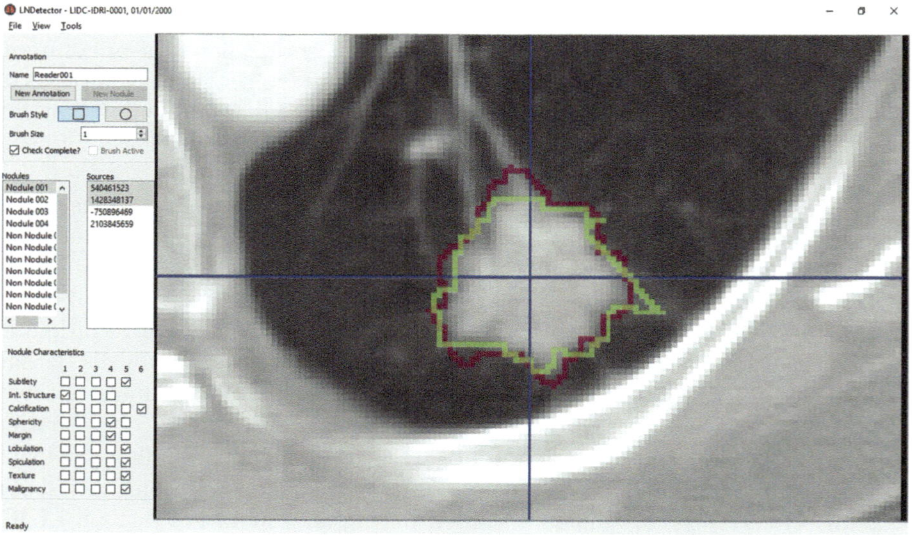

Fig. 3. Segmentation of `Nodule 001` by readers `540461523` and `1428348137` on LIDC-IDRI-0001.

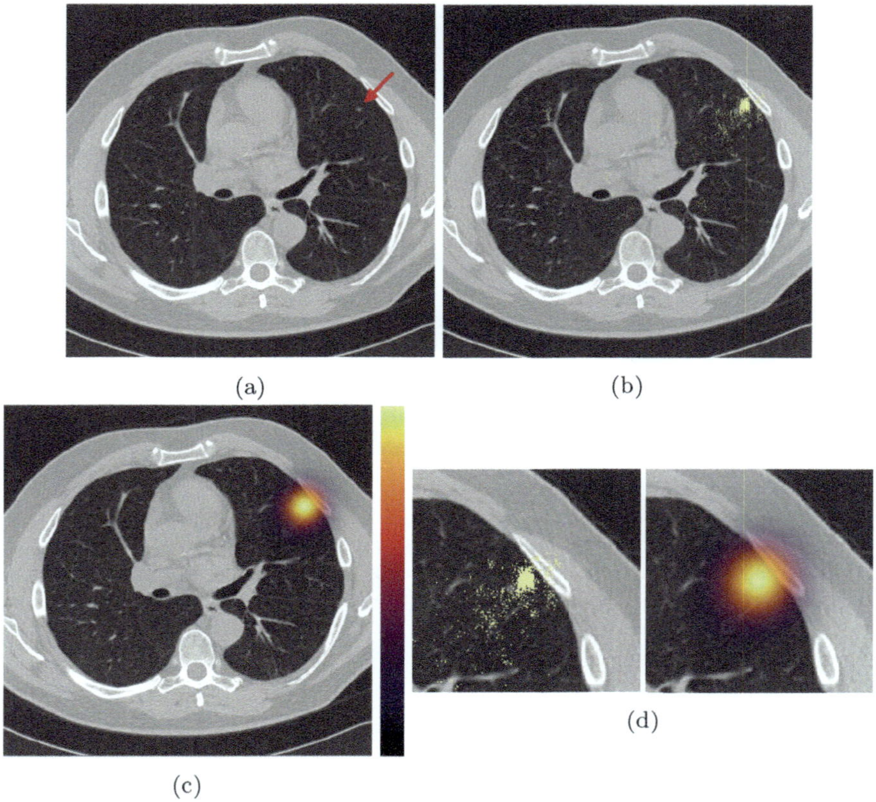

Fig. 4. Gaze characterisation for axial slice of LIDC-IDRI-1001. (a) Red arrow indicates pulmonary nodule; (b) raw gaze data (yellow dots); (c) resulting attention map (black for little/no attention and yellow for high attention); (d) closeup of region around nodule.

Since gaze characterisation can be performed on LNDetector, collaborative strategies can be implemented. Using the attention map after an annotation session one can determine what regions were not analysed and recommend the user to check them. Alternatively, the attention map can be analysed so that an automatic detection algorithm would propose nodule candidates from the areas that were not screened. Such an approach is particularly attractive as it would integrate the CAD system in the screening process, decreasing the time spent analysing the CAD results [7].

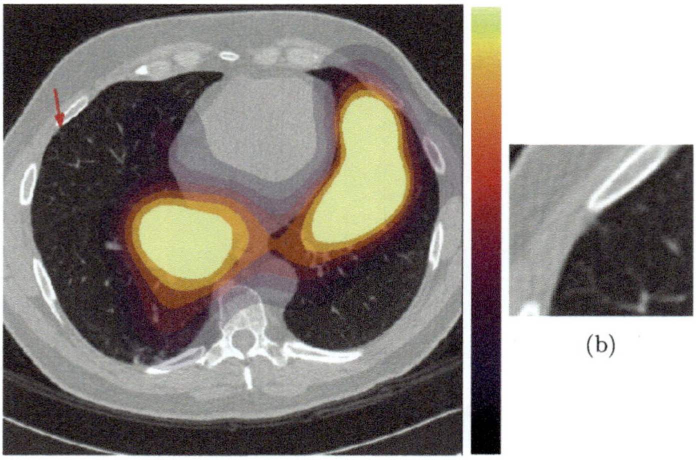

(a)

(b)

Fig. 5. Gaze characterisation results for axial slice of LIDC-IDRI-1001 showing the attention map and a pulmonary nodule (red arrow); (b) closeup of nodule missed. Note that the attention map was saturated to display low attention regions (deep purple).

3 Experiments

To demonstrate the potential uses of LNDetector, an experiment combining manual and automatic analysis, gaze characterisation and a collaborative scenario was designed.

One LDCT acquired at Hospital São João, Porto, Portugal was selected according to the criteria of LIDC-IDRI [3] and signed consent was obtained prior to acquisition. The scan was acquired with a Siemens SOMATOM go. Up scanner with a voxel size of $0.57 \times 0.57 \times 1.0$ mm.

Two radiologists with similar expertise were asked to analyse the LDCT image and annotate pulmonary nodules with greatest in-plane dimension smaller than 30 mm as well as non-nodules, defined as any lesion not consistent with a nodule with greatest in-plane dimension greater than 3 mm [3]. For each finding, each radiologist performed a delineation of the lesion. A description of the nodules according to LIDC-IDRI characteristics (subtlety, internal structure, calcification, sphericity, margin, lobulation, spiculation, texture and malignancy) was performed [3]. During the session, radiologists' gaze was recorded using a Tobii Eyetracker 4C [1].

In addition to manual annotation, automatic nodule detection based on the YOLOv3 architecture [11] was performed. In brief, a pre-trained model was tuned to detect nodules by minimizing a loss function taking into account the width, height and centroid of the prediction in relation to the ground-truth. To account for 3D information, the algorithm uses the axial slice containing the nodule's centroid and two equidistant slices [2]. Predictions are performed by axial slice and candidates merged if their bounding boxes overlap.

Finally, to account for a collaborative scenario, the CAD results were presented to each radiologist and findings that had not been labeled during manual annotation were classified as nodule or non-nodule by each radiologist.

4 Results

The findings of both radiologists, as well as the CAD and collaborative CAD results, are shown in Table 1. Radiologist 1 annotated four nodules and radiologist 2 annotated two nodules and one non-nodule. The CAD returned eight findings, including the four nodules and non-nodule annotated by the radiologists. Non reported findings by each radiologist were reviewed and both radiologists agreed independently that findings 1, 3, 5, 6, and 7 were nodules whereas the rest were false positives.

For each radiologist, gaze characterisation was performed. Gaze data corresponding to lesion delineation and characterisation was discarded as it does not relate to the visual search patterns. Radiologist 1 spent 2 min 7 s in the analysis whereas radiologist 2 spent 4 min 53 s. Lung segmentation was obtained through thresholding and lung coverage was estimated by dividing the observed lung voxels (within a 5° visual angle) by the total lung voxels. Lung coverage for radiologists 1 and 2 was, respectively, 21.4% and 50.5%. The time dedicated per nodule (Table 1) was computed as the time spent in any slice crossing or adjacent to the nodule within a 5° visual angle of the nodule centroid.

Figure 6 shows the radiologists' visual patterns throughout the analysis of the 3D scan. The two lungs are distinguishable as they accumulate the majority of points. Both radiologists use the same strategy, starting with one lung and following a vertical search strategy before moving to the second lung, as shown by the different colors of the two lungs.

5 Discussion

The two radiologists spent different amounts of time and reported a different number of nodules, in spite of using the same visual search strategy. Though radiologist 2 covered more lung area and spent more time, radiologist 2 reported fewer nodules than radiologist 1 missing findings 1 and 8. The limited amount of time dedicated by radiologist 2 to these findings (0 and 0.3 s respectively) indicates that these regions were likely not thoroughly analysed. This is supported by the fact that these findings were labeled as nodules by radiologist 2 in the collaborative scenario.

Regarding the automatic detection, all four reported nodules and four additional findings were given. Upon analysis by both radiologists, findings 2, 4 and 7 were considered to be false positives, whereas finding 3 was a nodule that had been missed by both radiologists in spite of the time devoted to this region by radiologist 2 (4.4 s).

Table 1. Findings after manual analysis (Radiologist 1 and 2), automatic detection (CAD) and collaborative scenario (CAD+Radiologist 1 and 2). *Not reported* indicates that this finding was not annotated as nodule or non-nodule. Time spent on each nodule is shown between parentheses. Images show the axial slice crossing the nodule centroid (50 × 50 mm).

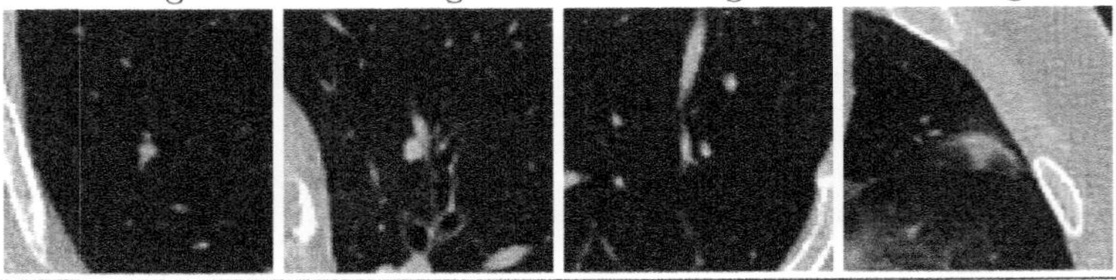

	Finding 1	Finding 2	Finding 3	Finding 4
Radiologist 1	Nodule (12.6s)	Not reported (0.2s)	Not reported (0s)	Not reported (0.4s)
Radiologist 2	Not reported (0s)	Not reported (0.3s)	Not reported (4.4s)	Non-nodule (2.6s)
CAD	Nodule	Nodule	Nodule	Nodule
CAD + Radiologist 1	Nodule	Non-nodule	Nodule	Non-nodule
CAD + Radiologist 2	Nodule	Non-nodule	Nodule	Non-nodule

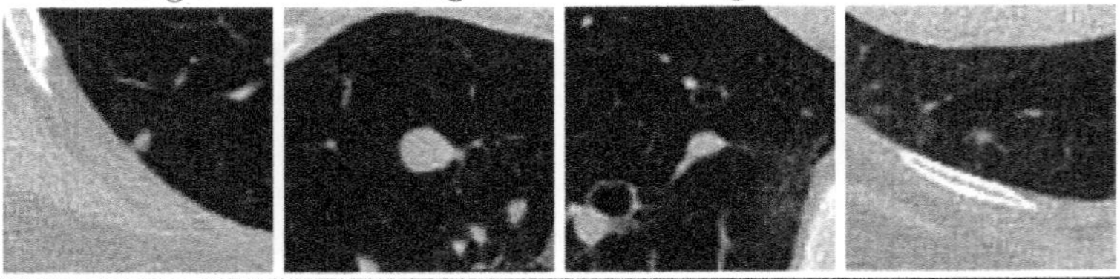

	Finding 5	Finding 6	Finding 7	Finding 8
Radiologist 1	Nodule (2.9s)	Nodule (4.8s)	Not reported (0s)	Nodule (5.0s)
Radiologist 2	Nodule (7.0s)	Nodule (23.2s)	Not reported (1.7s)	Not reported (0.3s)
CAD	Nodule	Nodule	Nodule	Nodule
CAD + Radiologist 1	Nodule	Nodule	Non-nodule	Nodule
CAD + Radiologist 2	Nodule	Nodule	Non-nodule	Nodule

In a traditional CAD scenario, where nodule detection is a parallel process to visual search, radiologists review all CAD findings, whether they have analysed those regions or not. However, a collaborative strategy can be implemented in LNDetector, such that gaze characterisation and nodule detection is performed

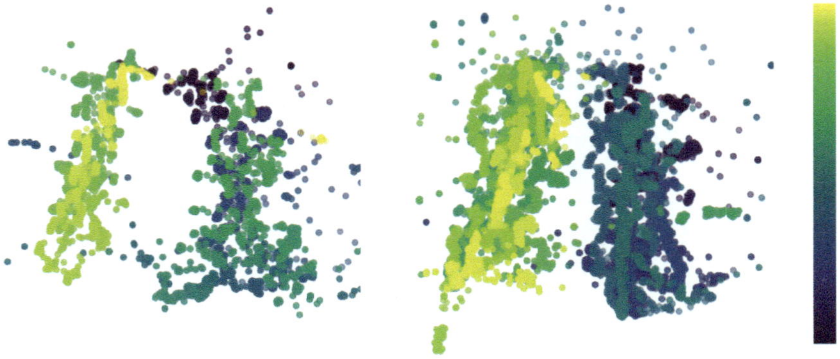

(a) (b)

Fig. 6. Visual search patterns throughout analysis of the 3D image for radiologists 1 (a) and 2 (b). Each dot represents a fixation plotted on the 3D image coordinates. The color of each dot represents time from the beginning (dark blue) to the end (yellow) of the analysis.

after visual search and only findings that were not previously analysed/labeled are shown. Such an approach is less time-consuming than traditional CAD and takes advantage of the qualities of radiologists and CAD systems. CAD systems excel at reviewing images systematically and finding suspicious regions but are prone to high false positive rates. Radiologists, on the other hand, are excellent at distinguishing nodules from other lesions but lack the consistency of CAD systems. It is thus expected that a collaborative strategy presents less false positives when compared to CAD and less missed nodules when compared to the radiologist. Indeed, considering the collaborative scenario as ground truth, radiologists 1 and 2 missed, respectively, one and three nodules, with no false positives, whereas the CAD system did not miss any nodule but returned four false positives.

While it has been shown that a collaborative CAD system would be a promising approach, extensive validation in a large database is needed to confirm these results. A comparison of this strategy to consensus annotations, such as from LIDC-IDRI, would further strengthen the conclusions.

6 Conclusions

In conclusion, LNDetector aims to be a reference platform in pulmonary nodule screening for the study of visual search strategies and integration of efficient CAD systems. This study has shown that valuable information can be retrieved from gaze characterisation and that a combination of visual search and automatic detection can be an efficient approach for the detection of pulmonary nodules.

Acknowledgments. This work was financed by the European Regional Development Fund (ERDF) through the Operational Programme for Competitiveness - COMPETE 2020 Programme and by National Funds through the Portuguese Funding agency, FCT - Fundação para a Ciência e Tecnologia within project: PTDC/EEI-SII/6599 /2014 (POCI-01-0145-FEDER-016673).

Conflict of Interest. The authors declare that there is no conflict of interest.

References

1. Tobii Gaming. https://gaming.tobii.com/
2. Aresta, G., Araújo, T., Jacobs, C., van Ginneken, B., Cunha, A., Ramos, I., Campilho, A.: Towards an automatic lung cancer screening system in low dose computed tomography. In: Image Analysis for Moving Organ, Breast, and Thoracic Images, pp. 310–318. Springer (2018)
3. Armato, S.G., McLennan, G., Bidaut, L., McNitt-Gray, M.F., Meyer, C.R., Reeves, A.P., Zhao, B., Aberle, D.R., Henschke, C.I., Hoffman, E.A., et al.: The lung image database consortium (LIDC) and image database resource initiative (IDRI): a completed reference database of lung nodules on CT scans. Med. Phys. **38**(2), 915–931 (2011)
4. Gonçalves, L., Novo, J., Campilho, A.: Hessian based approaches for 3D lung nodule segmentation. Expert Syst. Appl. **61**, 1–15 (2016)
5. Goncalves, L., Novo, J., Cunha, A., Campilho, A.: Learning lung nodule malignancy likelihood from radiologist annotations or diagnosis data. J. Med. Biolog. Eng. **38**(3), 424–442 (2018)
6. Gruden, J.F., Ouanounou, S., Tigges, S., Norris, S.D., Klausner, T.S.: Incremental benefit of maximum-intensity-projection images on observer detection of small pulmonary nodules revealed by multidetector CT. Am. J. Roentgenol. **179**(1), 149–157 (2002)
7. Khosravan, N., Celik, H., Turkbey, B., Jones, E.C., Wood, B., Bagci, U.: A collaborative computer aided diagnosis (C-CAD) system with eye-tracking, sparse attentional model, and deep learning. Med. Image Anal. **51**, 101–115 (2019)
8. Kundel, H.L.: Reader error, object recognition, and visual search. In: Medical Imaging 2004: Image Perception, Observer Performance, and Technology Assessment, vol. 5372, pp. 1–12. International Society for Optics and Photonics (2004)
9. Machado, M., Aresta, G., Leitão, P., Carvalho, A.S., Rodrigues, M., Ramos, I., Cunha, A., Campilho, A.: Radiologists' gaze characterization during lung nodule search in thoracic CT. In: 2018 International Conference on Graphics and Interaction (ICGI), pp. 1–7. IEEE (2018)
10. Millodot, M.: Dictionary of Optometry and Visual Science E-Book. Elsevier Health Sciences, London (2014)
11. Redmon, J., Farhadi, A.: YOLOV3: an incremental improvement. arXiv preprint arXiv:1804.02767 (2018)
12. Setio, A.A.A., Traverso, A., De Bel, T., Berens, M.S., van den Bogaard, C., Cerello, P., Chen, H., Dou, Q., Fantacci, M.E., Geurts, B., et al.: Validation, comparison, and combination of algorithms for automatic detection of pulmonary nodules in computed tomography images: the LUNA16 challenge. Med. Image Anal. **42**, 1–13 (2017)

13. Siegel, R.L., Miller, K.D., Jemal, A.: Cancer statistics. CA. Cancer J. Clin. **69**(1), 7–34 (2019)
14. The National Lung Screening Trial Research Team: Reduced lung-cancer mortality with low-dose computed tomographic screening. N. Engl. J. Med. **365**(5), 395–409 (2011)
15. Wang, S., Zhou, M., Liu, Z., Liu, Z., Gu, D., Zang, Y., Dong, D., Gevaert, O., Tian, J.: Central focused convolutional neural networks: developing a data-driven model for lung nodule segmentation. Med. Image Anal. **40**, 172–183 (2017)
16. Wu, B., Zhou, Z., Wang, J., Wang, Y.: Joint learning for pulmonary nodule segmentation, attributes and malignancy prediction. In: 2018 IEEE 15th International Symposium on Biomedical Imaging (ISBI 2018), pp. 1109–1113. IEEE (2018)

Physical Breast Phantom Dedicated for Mammography Studies

Firgan Feradov[1(✉)], Stoyko Marinov[2], and Kristina Bliznakova[1]

[1] Laboratory of Computer Simulations in Medicine, Technical University – Varna, 9010 Varna, Bulgaria
csmlab@tu-varna.bg

[2] Department of Imaging and Pathology, KU Leuven, Herestraat 49, 3000 Leuven, Belgium

Abstract. This paper presents a simple methodology for creation of anthropomorphic physical breast phantoms dedicated for x-ray breast imaging. The use of physical phantoms in diagnostic radiology is a well-established approach for patient dose estimation, quality control of developed diagnostic systems and development of new breast imaging techniques. An overview of the design process used for the creation of the presented phantom and evaluation approach is given. A combination of materials (Clear resin, Gray resin and PLA) and methods (fused deposition modelling and stereolithography) are used for the manufacturing of the breast phantom. The phantom was subjected to an evaluation aiming at its suitability for studies with digital mammography technique. In particular, the created phantom was evaluated by using two sets of phantom images, taken using 22 kVp and 28 kVp, which are compared with real mammograms. The comparison of the images is based on extracted statistical parameters – namely skewness, kurtosis, fractals, PSA, GLCMContrast and GLCMEnergy.

Keywords: Physical breast phantom · Suitable materials · Evaluation

1 Introduction

Nowadays the use of breast physical phantoms is related to the applications in diagnostic radiology related to optimization of patient dose, quality control of developed diagnostic systems (Breast Tomosynthesis, breast CT, contrast-enhanced mammography units), as well as for proof of concept and development of new breast imaging techniques [1, 2]. Until recently, the exploited phantoms in the routine clinical work were mainly represented by a homogeneous mixture, incorporating standard test objects dedicated for system performance evaluation.

Recent advanced research in the field of 3D printing materials and technology for printing phantoms dedicated to x-ray breast imaging showed the attempts to produce realistic anthropomorphic breast phantoms from patient images [3] and computational breast models [4]. These are also used in education approach for teaching purposes at advanced classes. An example is the advanced educational course in Medical Physics at European level EUTEMPE-RX, where at several modules the participants are trained to perform virtual studies with anthropomorphic phantoms [5–8].

J. Henriques et al. (Eds.): MEDICON 2019, IFMBE Proceedings 76, pp. 344–352, 2020.
https://doi.org/10.1007/978-3-030-31635-8_41

One major problem with the 3D printing approach is related to the limitations in the current 3D technology, which, however, has made enormous progress during the last two years in terms of technology and materials focused on printing 3D anthropomorphic phantoms. For example, phantoms may now be printed by using inexpensive parchment paper and a radiopaque ink [9].

We have recently studied the absorption and phase-contrast characteristics of several materials commonly used with fuse deposition modelling and stereolithography 3D printers [10, 11]. These studies showed that most of the resins, Nylon, Hybrid, PET-G have absorption properties in the mammography range similar to these of the glandular tissue, while ABS shows absorption characteristics close to these of the adipose tissue. Based on these findings, we further developed a physical phantom, based on these materials. This paper reports on the process, related to the manufacturing of such phantoms and its evaluation for mammography purposes. The evaluated phantoms will be used for assessment of new breast imaging techniques such as phase contrast mammography.

2 Materials and Methods

2.1 Physical Breast Phantom

The main elements in manufacturing a physical breast phantom are depicted in Fig. 1a, while Fig. 1b shows the complete phantom, filled with water. Specifically, the external shape is a semi-cylinder container with sizes: radius of the semi-cylinder base 59 mm, and 49 mm height.

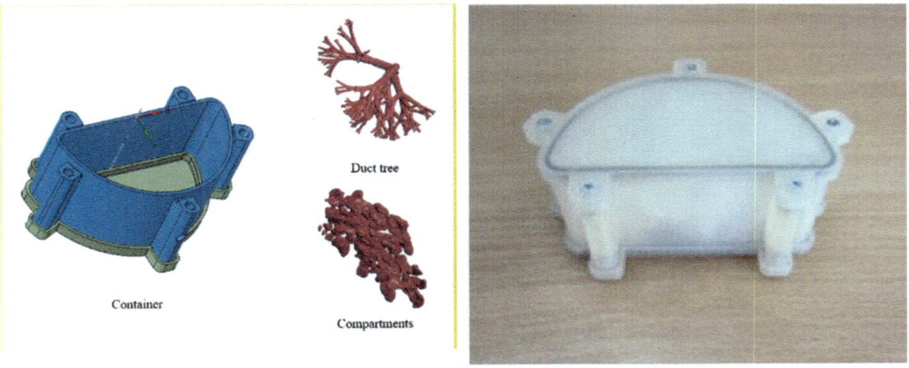

(a) (b)

Fig. 1. Physical breast phantom: (a) main parts, (b) the manufactured physical breast phantom

Two 3D printing technologies are used: fused deposition modelling (FDM) and stereolithography (SLA). The materials used for the production of the phantom are summarized in Table 1. Their properties are taken from Ivanov et al. [10].

Table 1. Materials used for manufacturing of the breast phantom.

Phantom element	Elemental composition						Technology
	Material	H	C	N	O	Density g/cm^3	
Glandular tree	Clear	0.085	0.648	0.059	0.208	1.180	SLA
Adipose compartments	Gray	0.084	0.638	0.052	0.226	1.175	SLA
Skin	PLA	0.058	0.541	0.018	0.383	1.250	FDM

The semicylinder container was modelled with the help of DesignSpark Mechanical 2.0 (https://www.rs-online.com/designspark/mechanical-software), while the glandular tree is modelled with the BreastSimulator [12, 13]. The adipose compartments were based on segmented adipose models from tomography images of the breast [14, 15]. As there are no requirements for breast phantoms to contain pathology, the inclusion of such was not considered at this stage of the development process.

2.2 Experimental Setup

X-ray images of the physical phantom shown in Fig. 1 were acquired with a Siemens digital mammography unit featuring a detector with a pixel size of 0.085 mm × 0.085 mm. Two projection sets, each one comprising of ten images, were taken at tube voltages set to 22 kVp and 28 kVp, respectively. Two such mammography images are shown in Fig. 2.

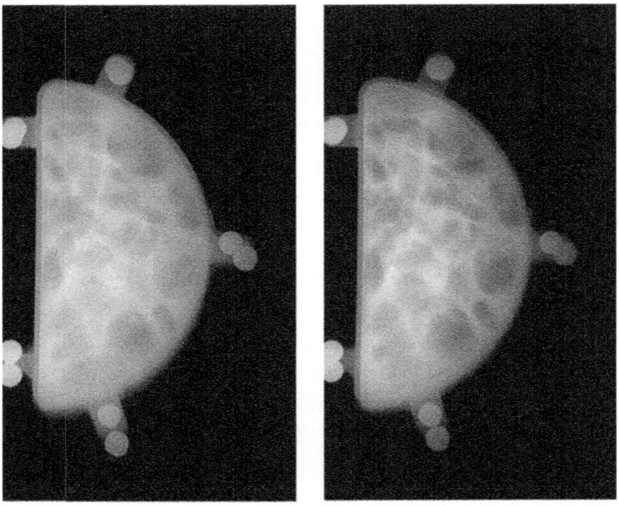

(a) (b)

Fig. 2. Mammography images of the phantom shown in Fig. 1b acquired with (a) 22 kVp (b) 28 kVp.

The source to detector distance was 660 mm, while the incident exposure depended from the phantom thickness.

2.3 Evaluation

The evaluation has been carried out with an in-house developed tool dedicated for feature extraction from x-ray images [16]. Figure 3 shows the main window of the tool. The user interface is implemented in MATLAB 2013a environment using the GUIDE module [17]. In short, the tool is divided into five regions related to different activities and functionalities, such as (a) various possibilities to load DICOM and other types of images, (b) possibilities to adjust contrast and brightness by using sliders, to zoom in/out the image and examine the details of specific pixel, (c) functionalities related to manually or automatically defining the regions of interest (ROI) in the selected image, (d) use of tools for automatic creation of 'mask' image of the area of the breast, where the ROIs are populated, (e) calculation of features based on the selected ROIs.

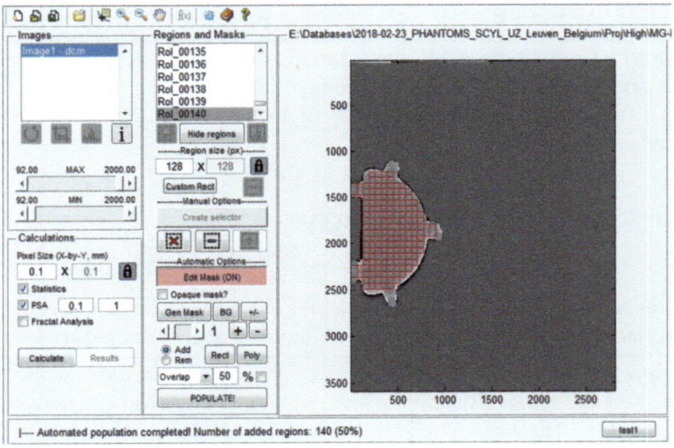

Fig. 3. A screenshot of the main window of the tool, used in the evaluation.

Texture of a breast in mammogram images plays significant role in tissue classifications. This tool is capable of evaluating more than 20 statistical and other feature types. In this study, we used six features of interest, which description is shortly summarized in Table 2.

The evaluation of the phantom has been performed by comparison of features extracted from x-ray mammography images compared to those extracted from 20 mammography patient images. The described features are performed over randomly selected ROIs in the analyzed images and further averaged and expressed by a mean value and standard deviation.

Table 2. Feature description.

Kurtosis. It expresses the relative tailedness, relatively to the Gaussian distribution or flatness of the distribution of the image pixel values:

$$\text{Kurt}[X] = \frac{\mu_4}{\sigma^4} = \frac{E[(X-\mu)^4]}{(E[(X-\mu)^2])^2} \quad [1]$$

where E is the expectation operator, μ_4 is the fourth moment about the mean (μ) and σ is the standard deviation.

Kurtosis is used for evaluation of the local structural texture. Higher values of kurtosis (positive) signify absence of local extreme values and vice versa – which is connected to the homogeneity of the texture and the breast density.

Skewness. It describes the asymmetry of the probability distribution of the image pixel values about their mean. The skewness value can be positive or negative, or undefined.

$$\gamma_1 = E\left[\left(\frac{X-\mu}{\sigma}\right)^3\right] = \frac{\mu_3}{\sigma^3} = \frac{E[(X-\mu)^3]}{(E[(X-\mu)^2])^{\frac{3}{2}}} = \frac{\kappa_3}{\kappa_2^{3/2}} \quad [2]$$

where E is the expectation operator, μ_3 is the third central moment.

This descriptor is used for evaluation of the density of the breast: negative skewness is a sign of fibroglandular tissue, while positive skewness is a characteristic of the fatty tissue.

Fractal Analysis. Computes fractal characteristics of the images, which can be used to describe the smoothness/roughness of the image intensity. The area A(ε) of each image pixel and its difference (treated us surface) with the neighboring pixels right of and below it is computed according to:

$$A(\varepsilon) = \sum_{x,y} \varepsilon^2 + \sum_{x,y} \varepsilon(|i_\varepsilon(x,y) - i_\varepsilon(x,y+1)| + |i_\varepsilon(x,y) - i_\varepsilon(x+1,y)|) \quad [3]$$

The fractal dimension D of an image is then calculated as:

$$D = 2 - \frac{\Delta\log[A(\varepsilon)]}{\Delta\log[\varepsilon]} \quad [4]$$

This descriptor is used for evaluation of the degree of complexity of the breast's textural patterns – low fractal dimensional value implies a simple homogeneous texture/structure.

Power law spectral analysis (PSA). The power spectrum describes the signal and noise into frequency components of which the low frequency range can be characterized by following a power law. The exponent of this power law is a measure of the texture or anatomical clutter.

$$F(k,l) = \frac{1}{MN}\sum_{a=0}^{M-1}\sum_{b=0}^{N-1} f(a,b)\exp\left[-2\pi i\left(\frac{ka}{M}+\frac{lb}{N}\right)\right] \quad [5]$$

where k (0, 1..., M−1) and l (0, 1,..., N−1) are the relative spatial frequencies in the two directions and M×N is the image size in pixels. The power spectrum profile is given by:

$$|SO(f)|^2 \approx \frac{c}{f^{2\beta}} \quad [6]$$

where f is a radial frequency variable, and c is a constant. The logarithm to base 2 of the total power spectrum is further plotted versus the octave number. The trend is almost linear and the data are fitted to a line obtained using linear regression analysis. The slope of the line m is related to the exponential parameter β as: m = β − 2.

This descriptor is used for evaluation of the breast's density.

(continued)

Table 2. (*continued*)

GLCM (Gray-level co-ocurrence matrix) Energy. Returns the sum of squared elements in the GLCM. The *Energy* is also known as uniformity, uniformity of energy, and angular second moment and is calculated as: $$F_{GLCM\ Energy} = \sum_{i=1}^{N_O}\sum_{j=1}^{N_O} p_{ij}^2 \quad [7]$$ where p_{ij} is the probability value for the ij-th gray tone from the calculated GLCM matrix. **This descriptor is used for evaluation of the structural patterns throughout the whole texture.**
GLCM Contrast. Returns a measure of the intensity contrast between a pixel and its neighbor over the whole image. *Contrast* is known as variance and inertia and is calculated as: $$F_{GLCM\ Contrast} = \sum_{i=1}^{N_O}\sum_{j=1}^{N_O}(i-j)^2 p_{ij} \quad [8]$$ **This descriptor is used for evaluation of the local variations throughout the texture.**

3 Results and Discussion

For the calculations of the features, we used ROIs of size 128 × 128 pixels and 50% overlap of the regions. All ROIs are placed "inside" the projected breast, excluding skin, muscle and background. Table 3 summarizes the results for the calculated features for each set of ten images, acquired at 28 kVp and 22 kVp, respectively in comparison to the features extracted from twenty patient mammography images, free of lesions.

Table 3. Comparison of features extracted from phantom and patient images.

Descriptor	Physical phantom 28 kVp	Physical phantom 22 kVp	Mammography images
Kurtosis	3.62 ± 2.88	3.58 ± 0.37	3.05 ± 0.07
Skewness	0.48 ± 0.32	0.41 ± 0.06	0.008 ± 0.10
Fractal	3.08 ± 0.04	3.42 ± 0.01	3.07 ± 0.12
PSA	3.08 ± 0.05	2.78 ± 0.08	2.78 ± 0.17
GLCM Contrast	0.79 ± 0.07	1.10 ± 0.04	0.72 ± 0.21
GLCM Energy	0.11 ± 0.03	0.09 ± 0.01	0.10 ± 0.01

The comparison of features shows a good coincidence between features from phantom and patient images. Better agreement is observed for the phantom images obtained with 28 kVp, which is expected since this energy is traditionally used with mammography units. In addition, comparison of distributions of selected features is shown in Fig. 4a–c.

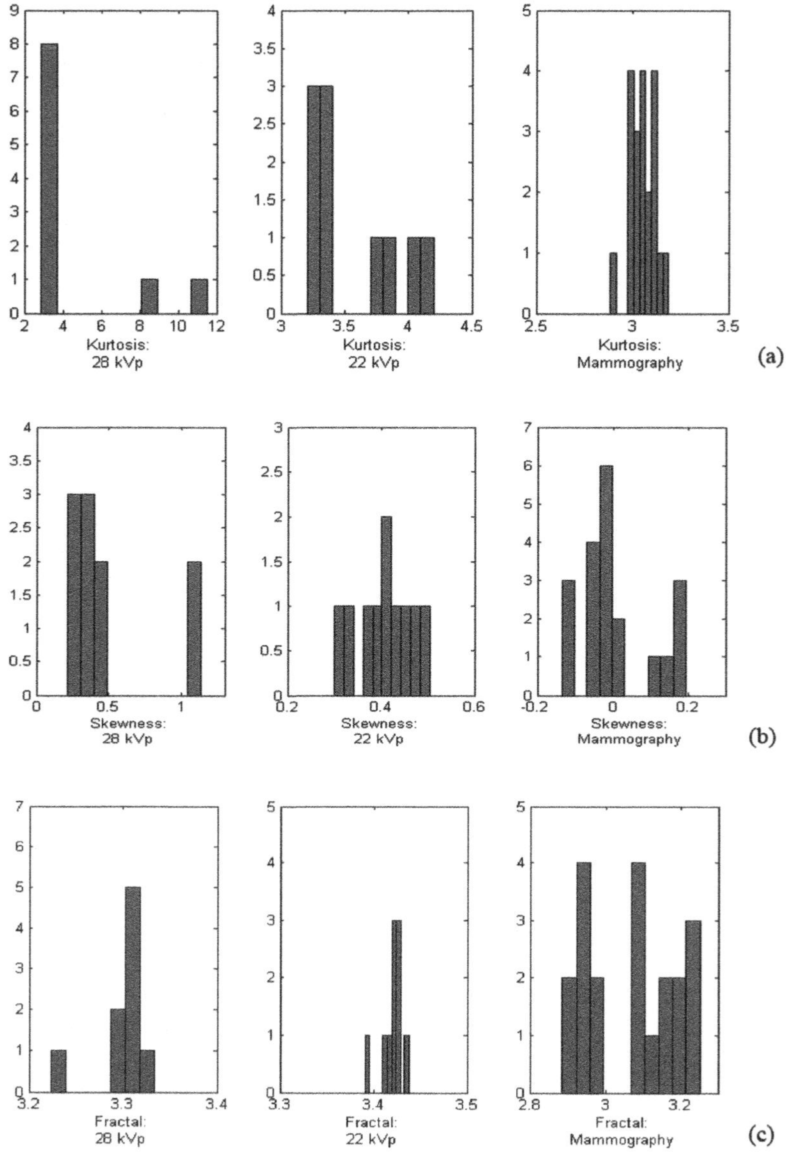

Fig. 4. Distributions of the computed (a) kurtosis, (b), skewness and (c) fractal dimension distributions for the studied images.

4 Conclusions

The conducted comparative evaluation the created physical phantom and the patient mammograms shows similarity between features extracted from their mammography images. From the studied features extracted from mammography images obtained from the two phantom sets of images (taken at 22 kVp and 28 kVp), the use of 28 kVp with the proposed phantom - glandular tree, printed from Gray resin; Adipose compartments, printed from Clear resin and container, printed from PLA - showed close results to these from patient mammography images.

Acknowledgements. This research is supported by the Bulgarian National Science Fund under grant agreement DN17/2.

References

1. Tomal, A., Costa, P.R.: Chapter 56: Phantoms for image quality and dose assessment. In: Russo, P. (ed.) Handbook of X-ray Imaging: Physics and Technology, pp. 1135–1158. CRC Press, Boca Raton (2018)
2. Bliznakova, K.: Chapter 57: Software phantoms for x-ray radiography and tomography. In: Russo, P. (ed.) Handbook of X-ray Imaging: Physics and Technology, pp. 1159–1184. CRC Press, Boca Raton 2018
3. Kiarashi, N., Nolte, A.C., Sturgeon, G.M., Segars, W.P., Ghate, S.V., Nolte, L.W., Samei, E., Lo, J.Y.: Development of realistic physical breast phantoms matched to virtual breast phantoms based on human subject data. Med. Phys. **42**, 4116–4126 (2015)
4. Carton, A.K., Bakic, P., Ullberg, C., Derand, H., Maidment, A.D.: Development of a physical 3D anthropomorphic breast phantom. Med. Phys. **38**, 891–896 (2011)
5. Bosmans, H., Bliznakova, K., Padovani, R., et al.: Eutempe-RX: a new EC supported course for medical physics experts in radiology. Phys. Medica **30**(Supplement 1), e11 (2014)
6. Bosmans, H., Bliznakova, K., Padovani, R., et al.: EUTEMPE-RX, an EC supported FP7 project for the training and education of medical physics experts in radiology. Radiat. Prot. Dosimetry. **165**, 518–522 (2015)
7. Bliznakova, K., Buliev, I., Bosmans, H., et al.: How can biomedical engineers benefit from the new expert level course of the EUTEMPE-RX project. In: 6th European Conference of the International Federation for Medical and Biological Engineering, vol. 45, pp. 765–768 (2015)
8. Bliznakova, K.: Development of breast software phantom dedicated for research and educational purposes. Radiat. Appl. Phys. Chem. Biol. Med. Sci. Eng. Environ. Sci. **2**, 14–19 (2016)
9. Ikejimba, L.C., Graff, C.G., Rosenthal, S., et al.: A novel physical anthropomorphic breast phantom for 2D and 3D x-ray imaging. Med. Phys. **44**, 407–416 (2017)
10. Ivanov, D., Bliznakova, K., Buliev, I., et al.: Suitability of low density materials for 3D printing of physical breast phantoms. Phys. Med. Biol. **63**, 175020 (2018)
11. Esposito, G., Mettivier, G., Bliznakova, K., et al.: Investigation of the refractive index decrement of 3D printing materials for manufacturing breast phantoms for phase contrast imaging. Phys. Med. Biol. **64** (2019)

12. Bliznakova, K., Suryanarayanan, S., Karellas, A., Pallikarakis, N.: Evaluation of an improved algorithm for producing realistic 3D breast software phantoms: application for mammography. Med. Phys. **37**, 5604–5617 (2010)
13. Bliznakova, K., Sechopoulos, I., Buliev, I., Pallikarakis, N.: BreastSimulator: a software platform for breast x-ray imaging research. J. Biomed. Graph. Comput. **2**, 1–14 (2012)
14. Imran, A.A.Z., Bakic, P.R., Pokrajac, D.D.: Spatial distribution of adipose compartments size, shape and orientation in a CT breast image of a mastectomy specimen. In: 2015 IEEE Signal Processing in Medicine and Biology Symposium - Proceedings (2015)
15. Imran, A.A., Pokrajac, D.D., Maidment, A.D.A., Bakic, P.R.: Estimation of adipose compartment volumes in CT images of a mastectomy specimen. In: Medical Imaging 2016: Physics of Medical Imaging, vol. 9783 (2016)
16. Marinov, S., Buliev, I., Cockmartin, L., et al.: Development of a software tool for evaluation of x-ray images: a case study in breast imaging. In: World Congress on Medical Physics and Biomedical Engineering, IUPESMPrague, p. 218, 3–8 June 2018
17. MathWorks: Matlab-MathWorks (2016)

Segmentation of Pulmonary Nodules in CT Images Using the Sliding Band Filter

Joana Rocha[1,2](✉), António Cunha[2,3], and Ana Maria Mendonça[1,2]

[1] Faculdade de Engenharia, Universidade do Porto, FEUP Campus, Dr. Roberto Frias, 4200-465 Porto, Portugal
joana866rocha@gmail.com

[2] INESC TEC – INESC Technology and Science, FEUP Campus, Dr. Roberto Frias, 4200-465 Porto, Portugal

[3] Universidade de Trás-os-Montes e Alto Douro, Quinta de Prados, 5001-801 Vila Real, Portugal

Abstract. This paper proposes a conventional approach for pulmonary nodule segmentation, that uses the Sliding Band Filter to estimate the center of the nodule, and consequently the filter's support points, matching the initial border coordinates. This preliminary segmentation is then refined to try to include mainly the nodular area, and no other regions (e.g. vessels and pleural wall). The algorithm was tested on 2653 nodules from the LIDC database and achieved a Dice score of 0.663, yielding similar results to the ground truth reference, and thus being a promising tool to promote early lung cancer screening and improve nodule characterization.

Keywords: Lung · Nodule · Segmentation · Sliding band filter

1 Introduction

Pulmonary nodules can be associated with several diseases, but a recurrent diagnosis is lung cancer, which is the main cause of cancer death in men and the second cause in women worldwide [1]. For this reason, providing an early detection and diagnosis to the patient is crucial, considering that any delay in cancer detection might result in lack of treatment efficacy. The advances of technology and imaging techniques such as computed tomography (CT) have improved nodule identification and monitoring. In a CAD system, segmentation is the process of differentiating the nodule from other structures. However, this task is quite complex considering the heterogeneity of the size, texture, position, and shape of the nodules, and the fact that their intensity can vary within the borders. Data imbalance also poses a challenge, as in a CT scan less than 5% of the voxels belong to these lesions.

In biomedical image analysis, early methods (generally described as conventional) consisted of following a sequence of image processing steps (e.g. edge/line detectors, region growing) and mathematical models [2]. Afterwards, the idea of extracting features and feeding them to a statistical classifier made supervised

J. Henriques et al. (Eds.): MEDICON 2019, IFMBE Proceedings 76, pp. 353–357, 2020.
https://doi.org/10.1007/978-3-030-31635-8_42

techniques become a trend. More recently, the trend is to use Deep Learning to develop models that are able to interpret what features better represent the data, but these require a large amount of annotated data, and have large computational cost. Among other conventional techniques, lesion detection and segmentation often imply the use of filters; e.g. the Sliding Band Filter can be used to develop an automated method for optic disc segmentation [4] and cell segmentation [5]. Such filter also proved to perform better than other local convergence index filters in pulmonary nodule detection [3]. This work aims to go further and precisely segment pulmonary nodules by implementing a conventional approach based on the Sliding Band Filter (SBF).

2 Local Convergence Filters and the Sliding Band Filter

Local Convergence Filters (LCFs) estimate the convergence degree, C, of the gradient vectors within a support region R, toward a central pixel of interest $P(x, y)$, assuming that the studied object has a convex shape and limited size range. LCFs aim to maximize the convergence index at each image point, which is calculated minding the orientation angle $\theta_i(k, l)$ of the gradient vector at point (k, l) with respect to the line with direction i that connects (k, l) to P. The overall convergence is obtained by averaging the individual convergences at all M points in R, as written in Eq. 1, taken from [4].

$$C(x, y) = \frac{1}{M} \sum_{(k,l) \in R} \cos \theta_i(k, l) \tag{1}$$

LCFs perform better than other filters because they are not influenced by gradient magnitude, nor by the contrast with the surrounding structures. Being a member of the LCFs, the SBF also outputs a measure which estimates the degree of convergence of the gradient vectors. However, the position of the support region, which is a band of fixed width, is adapted according to the direction and the gradient orientation. The SBF studies the convergence along that band, ignoring the gradient's behaviour at the center of the object, which is considered irrelevant for shape estimation. Such feature makes this filter more versatile when it comes to detecting different shapes, even when they are not perfectly round, because the support region can be molded.

The SBF searches each one of the N radial directions leading out of P for the position of the band of fixed width d that maximizes the convergence degree. The search is done within a radial length that varies from a minimum ($Rmin$) to a maximum ($Rmax$) values, and so the filter's response is given by Eq. 2, where $\theta_{i,m}$ represents the angle of the gradient vector at the point m pixels away from P in direction i [4]. The coordinates of the band's support points $(X(\theta_i), Y(\theta_i))$ are obtained using Eq. 3, assuming that the center of the object is (x_c, y_c) [6].

$$SBF(x, y) = \frac{1}{N} \sum_{i=0}^{N-1} \max_{Rmin \leq r \leq Rmax} \left[\frac{1}{d} \sum_{m=r-\frac{d}{2}}^{r+\frac{d}{2}} cos\theta_{i,m} \right] \tag{2}$$

$$\begin{cases} X(\theta_i) = x_c + \underset{Rmin \leq r \leq Rmax}{argmax} \left[\frac{1}{r} \sum_{m=r-\frac{d}{2}}^{r+\frac{d}{2}} cos\theta_{i,m} \right] \times \cos(\theta_i) \\ Y(\theta_i) = y_c + \underset{Rmin \leq r \leq Rmax}{argmax} \left[\frac{1}{r} \sum_{m=r-\frac{d}{2}}^{r+\frac{d}{2}} cos\theta_{i,m} \right] \times \sin(\theta_i) \end{cases} \quad (3)$$

3 Methodology

The following algorithm was implemented on already detected nodules. For each nodule, the 3D volume around its center was split into three anatomical planes (sagittal, axial, and coronal), resulting in three 80×80 pixel images per nodule. For clarity and brevity reasons, the method will be explained for a single plane, in Fig. 1a. The SBF is first applied to get a better estimation of the nodule's center coordinates. Considering most nodules have an overall uniform intensity, the nodules' images were processed by truncating any intensities much higher and lower than the nodule's. To do so, the nodule's average intensity was determined by calculating the mean of a matrix centered in the image. These steps result in a truncated mask, where there is already a very primitive segmentation (Fig. 1b) involving a low computational cost, which now needs substantial refinement.

The SBF takes the original nodule image, as well as the truncated nodule mask, and calculates its response in each pixel around the center of the image. The estimated nodule's center corresponds to the pixel which maximizes the response of the filter. With those coordinates, the SBF then evaluates the corresponding set of support points, returning the N border coordinates marked in Fig. 1c with yellow. To ensure the SBF is as precise as possible, a condition was added to force the cosine of the gradient vector's orientation angle to be null when the pixel which is being evaluated in a certain direction is black in the truncated mask. Ideally, this keeps the SBF from including in the segmentation non-nodular regions within the $Rmin$ and $Rmax$ limits. An outlier attenuation/removal step was implemented, minding the distance between consecutive border coordinates, and afterwards a binary mask with the initial SBF segmentation is created. To further refine the segmentation and specifically select the nodular area, only the intersection of the SBF segmentation mask and the truncated nodule mask is considered, thus eliminating unwanted regions. Any cavities within the intersected binary masks are filled, and the result is displayed in Fig. 1e. By labeling all the different regions present in the intersected masks, which are identified by their connected components, it is possible to eliminate any region that has no connection to the nodule. This can be done by eliminating from the mask all regions that do not encompass the center of the image, as the nodule is always centered. After this step, the final segmentation mask is achieved, as exemplified in Fig. 1d.

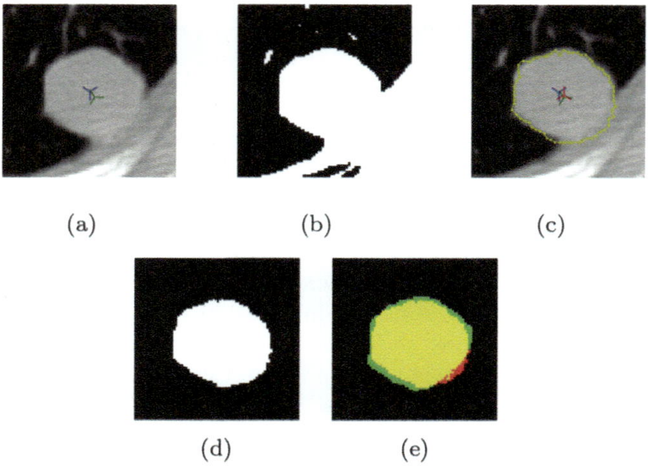

Fig. 1. Exemplification of the methodology steps, where the blue mark is the center of the image, the green mark is the ground truth center of the nodule, and the red mark is the estimated center of the nodule.

4 Results

The method was evaluated on 2653 nodules, using as ground truth the segmentation masks from the LIDC database, which is publicly available and consists of lung cancer screening thoracic CT scans from 1010 patients. These results were obtained with the SBF parameter values $N = 64$, $d = 7$, $Rmin = 1$, and $Rmax = 25$, which were established empirically to maximize the algorithm's performance. The method achieved a Dice score of 0.663, while having Precision and Recall values of 0.710 and 0.732, respectively. Figure 1e compares the achieved segmentation with the ground truth, in which the green pixels belong exclusively to the ground truth mask, the red pixels belong exclusively to the achieved result using the proposed method, and the yellow pixels belong to both - meaning that the yellow pixels mark the correct predictions made by the algorithm. The proposed conventional approach exhibits a highly satisfactory performance when dealing with well-circumscribed solid nodules. The nodules that have a pleural tail generally have the thin structure ignored by the algorithm, which does not include it in the segmentation mask, while the specialists consider the tail as part of the nodule. In spite of vascularized nodules entailing an inherent difficulty when it comes to distinguishing the nodule from the attached vessels, the SBF based approach is frequently able to separate them and create a mask which does not include the non-nodular structures. The algorithm's performance is also satisfying when dealing with nodules whose intensities vary within their border (e.g. calcific and cavitary nodules), as it is able to ignore the cavities/calcific regions during the segmentation process. The main flaws of the algorithm appear when dealing with juxtapleural nodules, since it often does not know where the nodule ends and the pleura begins, thus only being able to estimate the boundary to some extent. Overall, the less satisfying results are mainly due to the unexpected shape of the nodule, or because the nodule does not have a clear margin (e.g. non-solid nodules/ground glass opacities).

5 Conclusion

The segmentation of pulmonary nodules contributes to their characterization, which makes it a key to assess the patient's health state. This way, a segmentation step implemented within a CAD system can help the physician establish a more accurate diagnosis. However, the automation of such task is hampered by the diversity of nodule shape, size, position, lighting, texture, etc. The proposed conventional approach deals with these challenges by implementing the Sliding Band Filter to find the coordinates of the borders, and achieves a Jaccard score of 0.990 when tested with the LIDC database. This is a robust method for different nodule types, particularly well-circumscribed, solid and/or vascularized nodules. Its performance is impaired by juxtapleural lesions, and irregular shaped/non-solid nodules, and so future work includes the refinement of this method to deal with these particular challenges. Finally, it is possible to conclude this is a promising segmentation method, thus contributing to the clinically relevant characterization of pulmonary nodules.

Conflict of Interest. The authors declare no conflict of interest. This work is financed by National Funds through the Portuguese funding agency, FCT - Fundação para a Ciência e a Tecnologia within project: UID/EEA/50014/2019.

References

1. Torre, L.A., Siegel, R.L., Jemal, A.: Lung cancer statistics. In: Lung Cancer and Personalized Medicine, pp. 1–19. Springer, Cham (2016). https://doi.org/10.1007/978-3-319-24223-1_1
2. Litjens, G., Kooi, T., Bejnordi, B.E., Setio, A.A.A., Ciompi, F., Ghafoorian, M., Sánchez, C.I.: A survey on deep learning in medical image analysis. Med. Image Anal. **42**, 60–88 (2017). https://doi.org/10.1016/j.media.2017.07.005
3. Pereira, C.S., Mendonça, A.M., Campilho, A.: Evaluation of contrast enhancement filters for lung nodule detection. In: International Conference Image Analysis and Recognition, pp. 878–888. Springer, Heidelberg, August 2007. https://doi.org/10.1007/978-3-540-74260-9_78
4. Dashtbozorg, B., Mendonça, A.M., Campilho, A.: Optic disc segmentation using the sliding band filter. Comput. Biol. Med. **56**, 1–12 (2015). https://doi.org/10.1016/j.compbiomed.2014.10.009
5. Quelhas, P., Marcuzzo, M., Mendonça, A.M., Campilho, A.: Cell nuclei and cytoplasm joint segmentation using the sliding band filter. IEEE Trans. Med. Imaging **29**(8), 1463–1473 (2010). https://doi.org/10.1109/TMI.2010.2048253
6. Shakibapour, E., Cunha, A., Aresta, G., Mendonça, A.M., Campilho, A.: An unsupervised metaheuristic search approach for segmentation and volume measurement of pulmonary nodules in lung CT scans. Expert Syst. Appl. **119**, 415–428 (2019). https://doi.org/10.1016/j.eswa.2018.11.010

Method for Finding the Limits of Blood Vessel Landmarks in Eye Fundus Images Based on Distances in Graphs: Preliminary Results

Martynas Patašius[1,2(✉)], Jūratė Šimkienė[3], Daivaras Sokas[1], and Andrius Pranskūnas[3]

[1] Kaunas University of Technology, Institute of Biomedical Engineering, K. Baršausko g. 59, Kaunas, Lithuania
Martynas.Patasius@ktu.lt
[2] Department of Applied Informatics, Kaunas University of Technology, Studentų g. 50, Kaunas, Lithuania
[3] Department of Intensive Care, Lithuanian University of Health Sciences, A.Mickeviciaus g. 9, Kaunas, Lithuania

Abstract. The paper proposes a method to process blood vessel landmarks in eye fundus images. The proposed method is based on finding a shortest cycle in a weighted graph, corresponding to a set of possible tracked blood vessel slices and paths between them. In turn, the paths are found by finding a shortest path in a weighted graph, taking gradients into account. The method has been tried out with DRIVE and IOSTAR databases.

Keywords: Eye fundus · Blood vessels · Graph theory

1 Introduction

Retina is one of places where small blood vessels can be observed non-invasively.

Landmarks of blood vessels include blood vessel endpoints, bifurcations and intersections. Their positions can be used for image registration, their geometrical properties are thought have some medical importance [9]. Landmarks are also of importance for blood vessel tracking. Blood vessel landmarks have been described by single points [1,15,17], single points classified by type [5,6], single points with associated blood vessel directions [3].

A common approach is to find the landmarks in the skeleton of the blood vessel mask [2,4,5,11,15].

Alternatively, the landmarks are found while tracking blood vessels, by considering the profile in the half-circle in the tracking direction - thus only finding the branches in the same direction [7].

Somewhat similar approaches are also being used to detect blood vessels in 3D [8,10].

J. Henriques et al. (Eds.): MEDICON 2019, IFMBE Proceedings 76, pp. 358–366, 2020.
https://doi.org/10.1007/978-3-030-31635-8_43

Here we take a different approach, trying to find the area of the landmark with blood vessel slices that could be used to connect to tracked fragments of blood vessels. Ideally, the method would also work for non-landmark points on blood vessels. That would make it easier to use blood vessel tracking methods where segmentation is found to be useful (filling the regions around landmarks in blood vessel mask).

2 Methodology

2.1 Image Preprocessing

Image is preprocessed in several steps. First, the fundus field of view edge is sharpened: the mask consisting of pixels that have intensity of R, G and B channels higher than 10 is morphologically eroded using disk with radius of 3 px as structural element. Then, the image is filtered using median filter (size 3 × 3 px) to remove JPEG artefacts. Finally, blood vessels are emphasised by subtracting the morphologically closed image (using disk with radius of 15 px as structural element) from the image [16].

2.2 Blood Vessel Tracking

Blood vessel tracking is based on the method described in [12,13].

Tracking consists of steps. At each step a single slice is found. The slice connects two points (A and B), its middle point is marked by O. If necessary, the previous slice (end points A_0, B_0, middle point O_0) can be generated from the point (that becomes the middle point), initial direction $\overrightarrow{SDir}$ and expected width $ExpWidth$.

The goal function to be maximised consists of several components:

$$F(A, B) = g \cdot s \cdot d \cdot e \cdot w \cdot t. \tag{1}$$

Furthermore, if any of components is negative, the function is to be negative as well.

Component g further consists of two components:

$$g = g_A \cdot g_B. \tag{2}$$

Each of those takes into account the gradient:

$$g_A = \overrightarrow{OA} \cdot \nabla I(A), \tag{3}$$

$$g_B = \overrightarrow{OB} \cdot \nabla I(B). \tag{4}$$

Component s ensures the consistent step size:

$$s = \frac{1}{0.1 + (ExpStep - |\overrightarrow{O_0 O}|)^2}. \tag{5}$$

Component d ensures consistent direction (step being made in direction indicated by SDir):

$$d = \overrightarrow{SDir} \cdot \overrightarrow{O_0O}. \tag{6}$$

Component e ensures that the slice will be perpendicular to the step direction:

$$e = (\overrightarrow{SDir} \cdot \overrightarrow{ABDir})^4. \tag{7}$$

Component w encourages keeping the width consistent:

$$w = \frac{1}{sqrt0.1 + (ExpWidth - |\overrightarrow{AB}|)^2}. \tag{8}$$

Component t also consists of two components:

$$t = (t_1 \cdot t_2)^2. \tag{9}$$

They ensure that the sides of slice will move in the same direction:

$$t_1 = \overrightarrow{SDir} \cdot \overrightarrow{A_0A}. \tag{10}$$

$$t_2 = \overrightarrow{SDir} \cdot \overrightarrow{B_0B}. \tag{11}$$

The optimisation starts from the slice moved by a single step in given direction.

2.3 Double A*

Method Outline. The method receives an image and a single point approximately marking the center of the special point.

In order to find the limits of the special point several steps are to be performed. First the initial blood vessel slices are to be found. Then the intersections between blood vessel slices are to be removed. Third, edges between the ends of remaining slices are to be found. Finally, a cycle around the special point has to be found.

Finding Initial Blood Vessel Slices. The initial blood vessel slices can be found using simple blood vessel tracking. However, multiple slices are necessary. Also, the method can recover from false positives, but not from missed slices. Thus tracking is to be performed with parameters emphasising high sensitivity, while specificity is not very important.

Tracking is to be performed to various sides. it has been found that 16 directions are sufficient.

The starting point is to be moved a little for this tracking (backwards and to the sides), trying out different positions. However, after moving the point, the goodness of slice is divided by 10. Also, different starting blood vessel widths have to be tried out, as blood vessels of different sizes participate in intersections and especially bifurcations.

Postprocessing Concerning Intersections. The method does not deal well with slices that intersect each other. Therefore, all slices are to be checked for intersections.

Single Edge Tracking. For single edge tracking we want to find the series of points $A_1(x_1, y_1), A_2(x_2, y_2), ..., A_n(x_n, y_n)$ such that several conditions would hold.

First, the gradients in the neighbouring points $A_k(x_k, y_k), A_{k+1}(x_{k+1}, y_{k+1})$ should be similar.

Second, the gradients in the neighbouring points $A_k(x_k, y_k), A_{k+1}(x_{k+1}, y_{k+1})$ should be perpendicular to the vector between those points.

Third, the gradients in the neighbouring points $A_k(x_k, y_k), A_{k+1}(x_{k+1}, y_{k+1})$ should point to the right (or left, if the special point is to be marked in the contrary direction).

Fourth, the neighbouring points $A_k(x_k, y_k), A_{k+1}(x_{k+1}, y_{k+1})$ should be close to each other.

Finding such a series of points can be reduced to finding a shortest path in a constructed weighted graph. Vertices of such graph correspond to pixels. Edges of such graph correspond to possible jumps between pixels. Weights of the edges are to be constructed taking into account the vectors between pixels and the gradients at those points. Several scalar products are to be calculated.

First scalar product takes takes into account the similarity of gradients at the points:

$$SP_M(A, B) = nablaI(A) \cdot nablaI(B). \tag{12}$$

A pair of scalar products takes into account the fact that the gradients are to point to the right of the vector between the points:

$$SP_{R1}(A, B) = nablaI(A) \cdot \begin{bmatrix} 0 & -1 \\ 1 & 0 \end{bmatrix} \overrightarrow{AB}. \tag{13}$$

$$SP_{R2}(A, B) = nablaI(B) \cdot \begin{bmatrix} 0 & -1 \\ 1 & 0 \end{bmatrix} \overrightarrow{AB}. \tag{14}$$

The weight of an edge between vertices corresponding to pixels A and B would be:

$$w(A, B) = |\overrightarrow{AB}| + \frac{ScProdCoef}{(SP_M(A, B) \cdot SP_{R1}(A, B) \cdot SP_R(A, B))^{0.5}}. \tag{15}$$

It is sufficient to consider nearby pixels.

Finding the Minimal Cycle. A new, directed, weighted graph is constructed. Its vertices correspond to the ends of the slices. Some of its edges correspond to the slices. Other edges correspond to the traced paths between endpoints of slices.

Weights of edges corresponding to slices are made of linear part (for making use of A* possible) and part depending on the estimate of goodness of the slice ($DirDG$):

$$w(A, B) = |\overrightarrow{AB}| \cdot DCoef + \frac{NDCoef}{DirDGAdd + DirDG(A, B)}. \tag{16}$$

Weights of edges corresponding to paths between endpoints of slices are calculated from distance between points with coefficient:

$$w(A, B) = PathDist(A, B) \cdot PathDistCoef. \tag{17}$$

The edges corresponding to the traced paths between endpoints of slices that intersect the slices are rejected.

Then a minimal cycle in the resulting graph is found. Each cycle necessarily includes at least some slice. Thus each slice is tried out: the edge that corresponds to it is removed and a shortest path between its ends in resulting graph is found. The length of the cycle is equal to the sum of the length of this path and the weight of the edge. Out of all cycles found in such way the shortest is chosen.

2.4 Evaluating the Method

Sources of Images. The source of images was the DRIVE dataset [14]. The places of landmarks were given in IOSTAR database [1].

Construction of Semi-synthetic Image. In order to investigate the suitability of the method in idealised conditions a semi-synthetic image is constructed (Fig. 1).

The semi-synthetic image is constructed from blood vessel mask and field of view mask. Pixels outside field of view are given black colour (0, 0, 0), pixels outside blood vessels are given colour (250, 200, 150), pixels in blood vessels are given colour (200, 50, 100). Afterwards the two-dimensional Gaussian filter is applied (with sigma 1). Filtering has two main goals: it makes optimisation easier (gradients are non-zero further from the blood vessel), and it makes small blood vessels more vessel-like.

Parameters for Evaluation. Several parameters have been chosen for evaluation of the method: part of pixels outside of blood vessel mask, part of pixels outside of morphologically dilated blood vessel mask (structural element - disk with radius of a single pixel), part of marked starting points that end up outside of detected area, part of landmarks with the right number of slices (3 for bifurcations, 4 for crossings, 1 for endpoints), part of landmarks with the number of slices one more or one less, number of cases when tracking has failed.

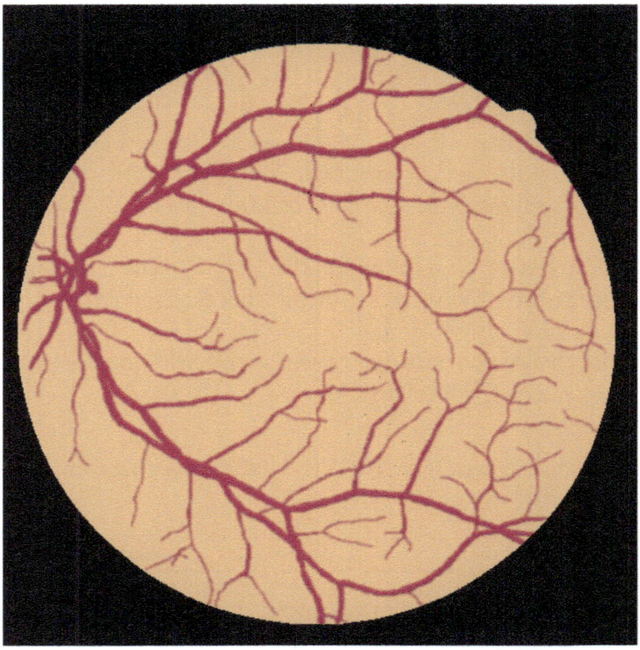

Fig. 1. Example of semi-synthetic image, based on DRIVE database, first training image

3 Results

For the experiments the initial tracking was done in 16 directions, with step 2, 3, 4 and 5 px and expected width of 1, 2, 4 and 6 px. In double A* dot product was given coefficient 100 and weight of non-slice edges was given coefficient of 0.1.

Table 1 shows the evaluations using semi-synthetic images based on DRIVE database (testing sample) with IOSTAR.

Table 1. Results using semi-synthetic images based on DRIVE database (testing sample) with IOSTAR (mean and standard deviation)

Parameter	All	Bifurcations	Crossings	Endpoints
Outside mask	0.2181 (0.0282)	0.1549 (0.0317)	0.1394 (0.0292)	0.3982 (0.0512)
Outside dilated mask	0.0285 (0.0099)	0.0133 (0.0085)	0.0119 (0.0085)	0.0703 (0.0193)
Not including start	0.2806 (0.0849)	0.1286 (0.0502)	0.0917 (0.0475)	0.4741 (0.1245)
Right number of slices	-	0.4993 (0.0548)	0.2883 (0.0761)	0.3603 (0.0807)
One less slice	-	0.3782 (0.0653)	0.3440 (0.0808)	-
One more slice	-	0.0634 (0.0252)	0.0204 (0.0274)	0.4674 (0.1002)
Failed processing	0.0527 (0.0443)	0.0051 (0.0096)	0.0036 (0.0115)	0.1056 (0.0789)

It can be seen that traces do not go far outside of blood vessel mask. The common error with tracking bifurcations and crossings is failing to recognise one branch. The common error in tracking endpoints is finding a second slice, and thus tracking a part of continuous blood vessel.

One cause of such errors can be traced to imperfect correspondence of databases: on average, in testing sample only 68% of points marked in IOSTAR database fall on blood vessel mask (87% of bifurcations, 90% of crossings, 43% of endpoints).

One cause of errors with endpoints is the fact that the endpoints in IOSTAR tend to be marked inside blood vessel, while this method expects such points to be marked just outside of it.

Table 2 shows the evaluations using semi-synthetic images based on DRIVE database (testing sample) with IOSTAR.

Table 2. Results using images from DRIVE database (testing sample) with IOSTAR (mean and standard deviation)

Parameter	All	Bifurcations	Crossings	Endpoints
Outside mask	0.2876 (0.0567)	0.1826 (0.0502)	0.1751 (0.0440)	0.5714 (0.0693)
Outside dilated mask	0.1490 (0.0462)	0.0616 (0.0269)	0.0511 (0.0303)	0.3862 (0.0859)
Not including start	0.5423 (0.0482)	0.3805 (0.0513)	0.3209 (0.0937)	0.7701 (0.0560)
Right number of slices	-	0.2835 (0.0464)	0.1504 (0.0727)	0.6967 (0.0499)
One less slice	-	0.3942 (0.0628)	0.2912 (0.0706)	-
One more slice	-	0.0722 (0.0334)	0.0096 (0.0136)	0.2046 (0.0435)
Failed processing	<0.0001 (0.0014)	<0.0001 (0.0021)	0 (0)	0.0012 (0.0030)

It can be seen that in a normal image the common error tracking endpoints is no longer tracking the blood vessel segment, but tracking the endpoint before it reaches the starting point.

Figure 2 shows examples of well and badly recognised landmarks.

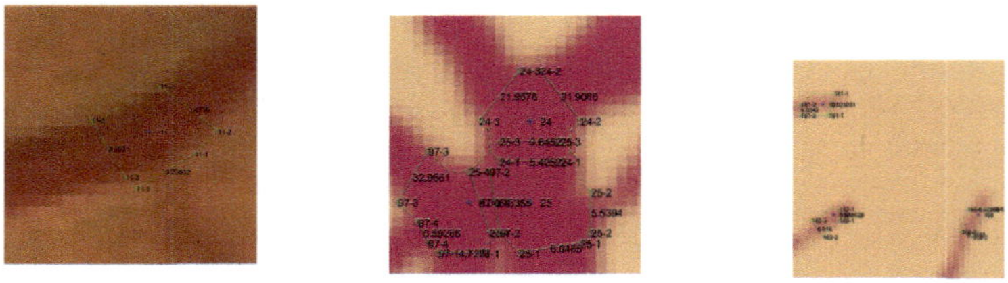

(a) (b) (c)

Fig. 2. Example of (a) well recognised bifurcation, (b) well recognised intersecting bifurcations, (c) badly recognised endpoints with second slices

4 Discussion and Conclusions

A method to process landmarks in blood vessel images has been proposed.

Since the proposed method has a different goal than the other methods, the direct comparison is difficult. Still, there are some similarities to be noted: for example, the proposed method, just like many others, performs search in various directions.

In some cases the proposed method needs precisely chosen starting points. Therefore, one future direction consists in finding ways to improve position of starting points.

Since the method underestimates the number of slices more often than overestimating it, another future direction is evaluation of possibility of merging the processed landmarks or taking into account a trace of blood vessel running into already traced landmark.

It should also be possible to investigate possibilities of finding the parameters for tracking adaptively.

It should be possible to combine the proposed method with a specialised method for finding locations of blood vessel landmarks.

It should be possible to adapt the method for use with other blood vessel tracking methods, provided that they evaluate the slices quantitatively.

Acknowledgements. This research was funded by a grant (No. S-MIP-17-16) from the Research Council of Lithuania.

References

1. Abbasi-Sureshjani, S., Smit-Ockeloen, I., Bekkers, E., Dashtbozorg, B., ter Haar Romeny, B.: Automatic detection of vascular bifurcations and crossings in retinal images using orientation scores. In: 2016 IEEE 13th International Symposium on Biomedical Imaging (ISBI). IEEE (2016). https://doi.org/10.1109/isbi.2016.7493241
2. Aguiar, M., Castano, F., Trujillo, M.: Segmentation and detection of vascular bifurcations and crossings in retinal images. In: Communications in Computer and Information Science, pp. 432–443. Springer International Publishing (2018). https://doi.org/10.1007/978-3-319-98998-3/_33
3. Al-Diri, B., Hunter, A.: Automated measurements of retinal bifurcations. In: O. Dössel, W.C. Schlegel (eds.) World Congress on Medical Physics and Biomedical Engineering, 7–12 September 2009, Munich, Germany, IFMBE Proceedings, vol. 25/XI, pp. 205–208. Springer (2009)
4. Bhuiyan, A., Nath, B., Chua, J., Ramamohanarao, K.: Automatic detection of vascular bifurcations and crossovers from color retinal fundus images. In: Third International IEEE Conference on Signal-Image Technologies and Internet-Based System 2007. SITIS 2007, pp. 711–718 (2007). https://doi.org/10.1109/SITIS.2007.86
5. Bhuiyan, A., Nath, B., Ramamohanarao, K.: Detection and classification of bifurcation and branch points on retinal vascular network. In: 2012 International Conference on Digital Image Computing Techniques and Applications (DICTA). IEEE (2012). https://doi.org/10.1109/dicta.2012.6411742

6. Ghanaei, Z., Pourreza, H., Banaee, T.: Automatic graph-based method for classification of retinal vascular bifurcations and crossovers. In: 2016 6th International Conference on Computer and Knowledge Engineering (ICCKE). IEEE (2016). https://doi.org/10.1109/iccke.2016.7802145
7. Kalaie, S., Gooya, A.: Vascular tree tracking and bifurcation points detection in retinal images using a hierarchical probabilistic model. Comput. Methods Programs Biomed. **151**, 139–149 (2017). https://doi.org/10.1016/j.cmpb.2017.08.018
8. Kang, J., Heo, S., Hyung, W.J., Lim, J.S., Lee, S.: 3D active vessel tracking using an elliptical prior. IEEE Trans. Image Process. **27**(12), 5933–5946 (2018). https://doi.org/10.1109/tip.2018.2862346
9. Luo, T., Gast, T.J., Vermeer, T.J., Burns, S.A.: Retinal vascular branching in healthy and diabetic subjects. Invest. Ophthalmol. Vis. Sci. **58**(5), 2685 (2017). https://doi.org/10.1167/iovs.17-21653
10. Mohan, V., Sundaramoorthi, G., Tannenbaum, A.: Tubular surface segmentation for extracting anatomical structures from medical imagery. IEEE Trans. Med. Imaging **29**(12), 1945–1958 (2010). https://doi.org/10.1109/tmi.2010.2050896
11. Morales, S., Naranjo, V., Angulo, J., Legaz-Aparicio, A., Verdú-Monedero, R.: Retinal network characterization through fundus image processing: significant point identification on vessel centerline. Signal Process. Image Commun. **59**, 50–64 (2017). https://doi.org/10.1016/j.image.2017.03.013
12. Patašius, M.: Akies dugno vaizdu automatinė analizė. Ph.D. thesis, Kauno technologijos universitetas (2010)
13. Patašius, M., Marozas, V., Jegelevičius, D., Lukoševičius, A.: Recursive algorithm for blood vessel detection in eye fundus images: preliminary results. In: Dössel, O., Schlegel, W.C. (eds.) World Congress on Medical Physics and Biomedical Engineering, 7–12 September 2009, Munich, Germany, IFMBE Proceedings, vol. 25/XI, pp. 212–215. Springer, Munich (2009)
14. Staal, J., Abràmoff, M.D., Niemeijer, M., Viergever, M.A., van Ginneken, B.: Ridge based vessel segmentation in color images of the retina. IEEE Trans. Med. Imaging **23**(4), 501–509 (2004). https://doi.org/10.1109/TMI.2004.825627
15. Sutanty, E., Rahayu, D.A., Rodiah, Susetianingtias, D.T., Madenda, S.: Retinal blood vessel segmentation and bifurcation detection using combined filters. In: 2017 3rd International Conference on Science in Information Technology (ICSITech). IEEE (2017). https://doi.org/10.1109/icsitech.2017.8257176
16. Treigys, P.: Grafinių oftalmologinių ir termovizinių duomenų analizės metodų kūrimas ir taikymas. Ph.D. thesis, Matematikos ir informatikos institutas (2010)
17. Tsai, C.L., Stewart, C.V., Tanenbaum, H.L., Roysam, B.: Model-based method for improving the accuracy and repeatability of estimating vascular bifurcations and crossovers from retinal fundus images. IEEE Trans. Inf Technol. Biomed. **8**(2), 122–130 (2004). https://doi.org/10.1109/TITB.2004.826733

Anthropomorphic Physical Breast Phantom Based on Patient Breast CT Data: Preliminary Results

Sivo Daskalov[1(✉)], Nikiforos Okkalidis[2], John M. Boone[3], Stoyko Marinov[4], Zhivko Bliznakov[4], Giovanni Mettivier[5], Hilde Bosmans[4], Paolo Russo[5], and Kristina Bliznakova[1]

[1] Laboratory of Computer Simulations in Medicine, Technical University of Varna, Varna, Bulgaria
csmlab@tu-varna.bg
[2] Centre for Biomedical Cybernetics, University of Malta, Msida MSD 2080, Malta
[3] Department of Radiology, University of California, Davis Medical Center, Sacramento, CA, USA
[4] Department of Imaging and Pathology, KU Leuven, Herestraat 49, 3000 Leuven, Belgium
[5] Dipartimento di Fisica "Ettore Pancini", Università di Napoli Federico II, Via Cintia, 80126 Naples, Italy

Abstract. We propose a novel approach for the fabrication of realistic in shape, size as well as in x-ray absorption properties 3D physical breast models. The approach eliminates the need for segmentation of breast tissues by directly mimicking the radiodensity of each voxel in a computed tomography (CT) image. This is done through a recently published study that enables the use of variable filament extrusion rate when creating 3D phantoms with fused deposition modeling printers. We use the CT image of a real patient's breast to produce a model through the suggested approach. We then validate the fabricated physical breast phantom by obtaining a CT scan of it and comparing the latter to the original CT image of the source patient's breast.

Keywords: 3D breast phantom fabrication · Fused deposition modeling · Variable filament extrusion rate

1 Introduction

Computer and physical anthropomorphic breast models have become a commonly utilized tool used in research for new breast imaging techniques [1], such as breast tomosynthesis, breast CT, and phase-contrast breast imaging. In literature related to physical breast models, much effort has been devoted to the development of new 3D printing materials like photopolymers doped with different concentration of TiO_2, calcium, iodine and zinc. Other research has been directed towards the evaluation of available 3D printing materials such as ABS, different epoxy resin, nylon. A major challenge and requirement in the field of x-ray breast imaging is the use of materials

J. Henriques et al. (Eds.): MEDICON 2019, IFMBE Proceedings 76, pp. 367–374, 2020.
https://doi.org/10.1007/978-3-030-31635-8_44

exhibiting similar x-ray absorption coefficients to those of breast tissue, especially in the typical energy range used in mammography.

As a result, many works have been focused on measurements of the attenuation coefficients of various 3D printing materials used in the manufacturing of physical breast phantoms. In the x-ray mammography range (between 22 kV and 32 kV), Carton et al. [2] found that acrylic-based photopolymers are especially suitable to mimic the properties of the glandular breast tissue. Furthermore, Sikaria et al. [3] and Zhao et al. [4] doped the photopolymer materials with concentrations of the nanoscale TiO_2 and with calcium, iodine and zinc, with the aim to increase the the linear attenuation coefficients of the materials. From the tested ones, they found that zinc-doping can potentially increase attenuation to match and accurately mimic breast density.

Recently, we have studied the properties of available 3D printing materials with both stereolithography (SLA) and fuse deposition modeling (FDM) 3D printers [5]. This study concluded that most of the resins, Nylon, Hybrid, PET-G exhibit absorption properties close to glandular tissue, while ABS shows absorption characteristics similar to those of adipose tissue.

In a recent study, Okkalidis [6] showed that anthropomorphic physical phantoms can be fabricated without tissue segmentation from patient DICOM images through the use of an FDM printer by varying the filament extrusion rate. Instead of using a specific infill density and pattern, the fabrication of the phantom is achieved by covering the whole area of interest and extruding estimated amounts of filament. This method is used to accurately replicate the internal structure of a patient's anatomy.

Based on our study and the development of new approaches in printing anthropomorphic phantoms by Okkalidis [6] we explored the possible application of this new methodology to anthropomorphic breast model fabrication. The aim of this paper is to present the results of utilizing this novel approach in the preparation and manufacturing of realistic in shape and size 3D physical breast models.

2 Materials and Methods

2.1 Overview

The main steps of the proposed approach for the creation of physical anthropomorphic breast phantoms are presented in Fig. 1. The main source of data is the computed tomography (CT) scans of real patient breasts [7, 8]. The novel modified FDM 3D printing method proposed in [6] enables us to print voxels of varying radiodensity. This allows us to more realistically mimic the different types of tissue (glandular, adipose and skin) without requiring segmentation to be performed. The produced breast phantom is then scanned using CT and tomosynthesis. In order to validate the quality of the phantom, we use its CT scan to compare it to the CT scan of the original patient's breast and evaluate the similarities and differences.

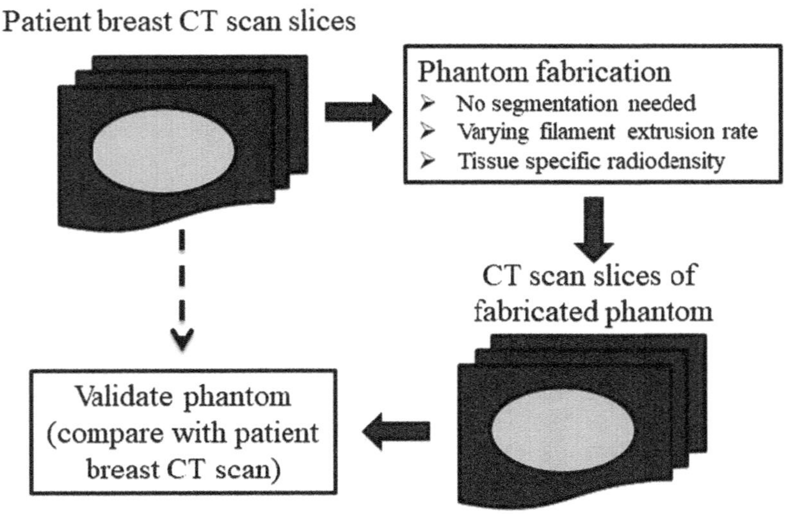

Fig. 1. Overview of the breast phantom creation and validation process.

2.2 Patient Breast Scanning

The patient's breast used as a source for the fabricated phantom is scanned using a Albion CT unit with tube voltage of 80 kV. The unit's resolution of 672 × 656 pixels supports slice thickness and pixel size of 0.273 mm.

2.3 Phantom Creation

Using a Fused Deposition Modeling (FDM) printer, patient-specific phantoms are normally fabricated by employing segmented 3D organ models. The phantoms are printed with a constant, estimated infill density which produces the required average Hounsfield Unit value for each organ. However, a recent method [6] demonstrated that it is possible to avoid long processing delays, such as manual contouring and 3D modeling, increase the achieved Hounsfield Units' range, and produce phantoms close to a specific patient's anatomy. This method involves the identification of the area of interest with a thresholding technique, the implementation of an edge detection algorithm on the resulting image and the pixel-by-pixel 3D printing. This is done by reading the Hounsfield Unit (HU) value for each pixel in the DICOM image, and then controlling the filament extrusion rate to produce the required amount of extruded filament in the respective position in the phantom. This allowed us to fabricate the different types of tissues in the phantom more realistically in terms of radiodensity. The FDM printer used to produce the phantom was model T-Rex 2 (Formbot, China) with printing dimensions 400 × 400 × 470 mm. The filament, polylactic acid (PLA), was chosen due to its suitable printing characteristics and acceptable HU range simulation, as shown in [9]. The filament was obtained from PrimaFilaments, Sweden.

2.4 Phantom Scanning

Computed tomography images of the fabricated physical phantom were acquired with a Siemens unit featuring a detector with a resolution of 512 × 512 and pixel size of 0.85 × 0.85 mm. The tube voltage was 70 kVp. Two CT volumes were reconstructed with 1 mm and 3 mm, respectively. Additionally, the fabricated phantom was scanned with a GE Healthcare tomosynthesis unit. For this scan, both the anode and filter were molybdenum, and the source energy was 30 kVp.

3 Results and Discussion

3.1 Fabricated Breast Phantom

An image of the fabricated physical breast phantom is shown on Fig. 2. The thickness of the phantom was 3 cm and the printing process took approximately 24 h. It is a slow process since the extruder has to cover the whole area for each layer.

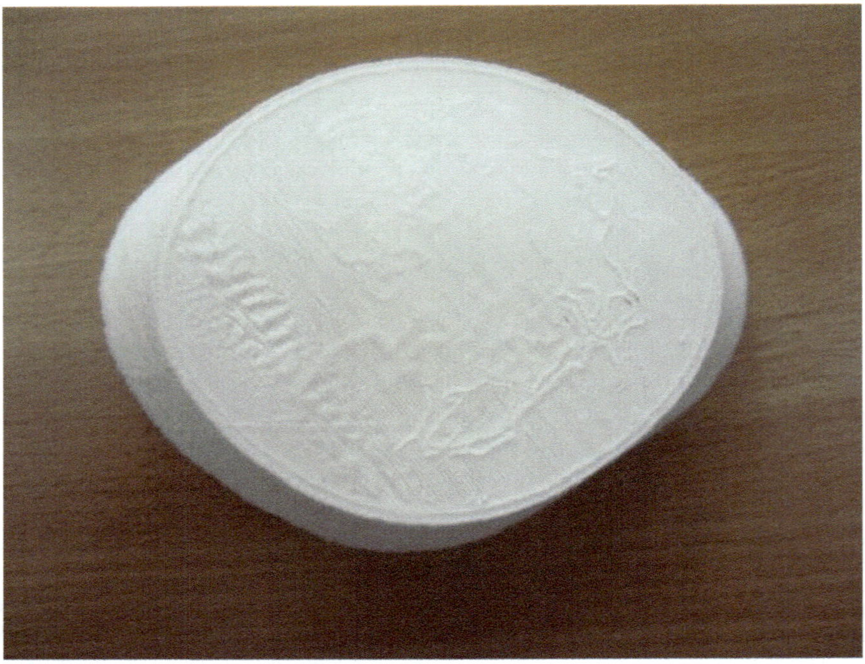

Fig. 2. Image of the fabricated physical phantom.

3.2 Phantom Evaluation

The printed breast phantom was initially scanned at a tomosynthesis unit (Siemens Mammomat Inspiration). A slice from the reconstructed tomosynthesis volume of the fabricated phantom is shown in Fig. 3. As seen from this reconstructed slice, the air in the printed phantom (used to decrease the density of the voxels) results in phase-contrast effects.

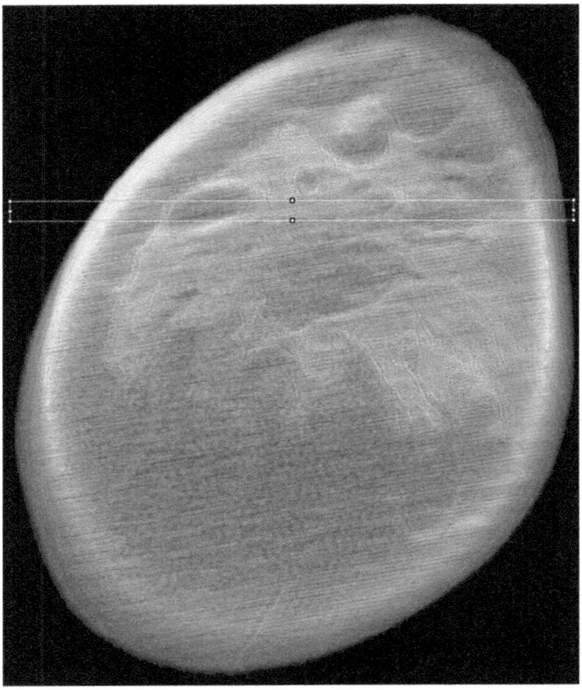

Fig. 3. Example tomosynthesis slice of the fabricated phantom.

Side by side comparison of example CT scan slices is shown in Fig. 4. The leftmost slice is of the source patient breast, while the following two are of the 1 mm and 3 mm CT scans of the fabricated phantom. The similarity between the original and the phantom's CT scan is clearly visible, especially the phantom scan with 3 mm slices (C).

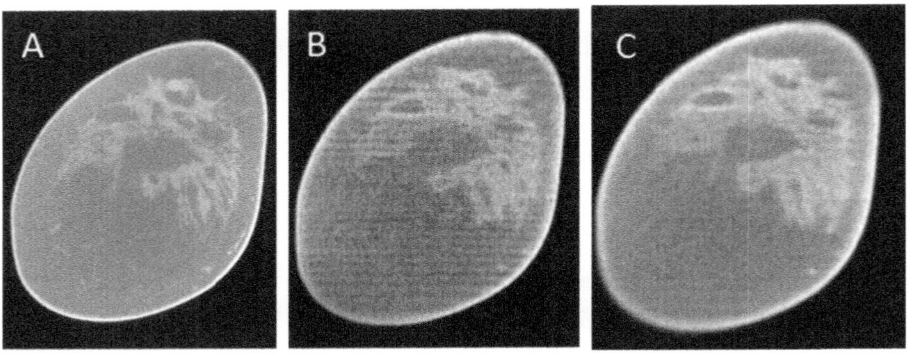

Fig. 4. Example CT slices. Left to right: patient breast (A), phantom 1 mm (B) and 3 mm (C).

We use the profile plotting functionality of *ImageJ* to visually verify the properties of the phantom. The function produces two-dimensional graphs of pixel intensities (column averaged) along a set of lines from the image. Figure 5 shows the profiles of two subsets of 10 lines each chosen from the slices previously defined in Fig. 4.

The column-wise similarities are visibly significant, which validates the phantom fabrication procedure. In the context of computed tomography, the phantom (rows B and C) accurately models the various types of tissues present in the breast and may be considered to be a realistic representation of the source patient breast (row A).

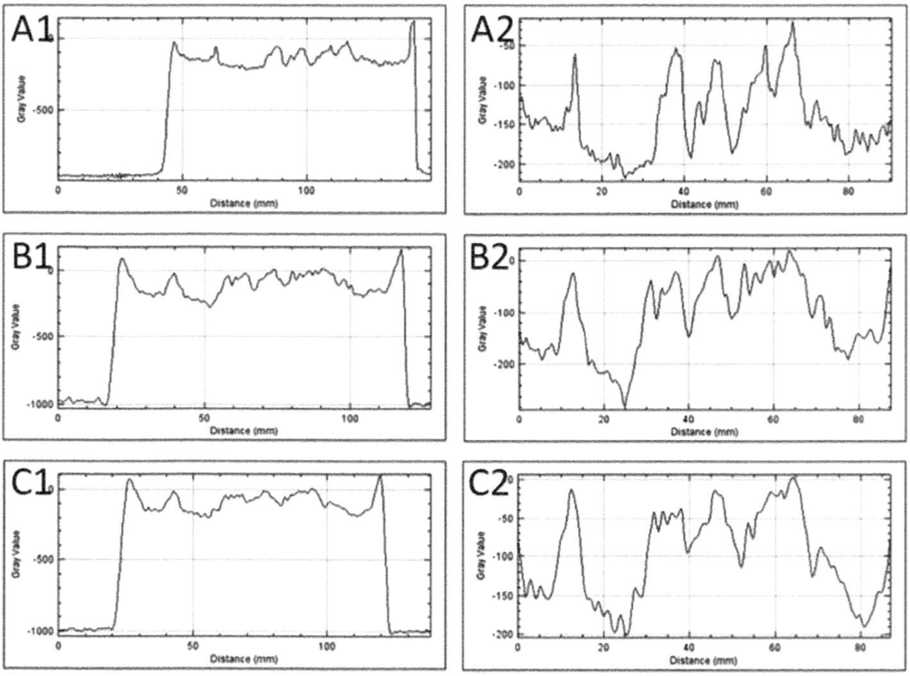

Fig. 5. ImageJ plot profiles of CT scan lines. The two columns result from the selection of two different subsets of 10 lines each. The rows correspond to the CT slices defined in Fig. 4, namely A is the source patient breast, B and C are 1 mm and 3 mm slices of the phantom respectively.

In order to objectively verify the properties of the fabricated phantom numerically, we carried out the following evaluation and comparison procedure. The first step of the process was cropping the images along the respective object's (breast or phantom) bounding box. This was done to ensure uniform positioning in order to obtain more accurate results from the following steps. Then we extracted a vector of 64 features from each of the images through the use of KAZE [10], a 2D feature detection and description algorithm in nonlinear scale spaces. Finally, we compared the feature vector from each image with all other feature vectors, using cosine similarity as a metric for the comparison. The final results from this statistical similarity evaluation are presented on the heatmap shown in Fig. 6.

The values of interest are the KAZE similarities between the real patient breast CT image (BCTNC16) and the CT images of the fabricated phantom, which are CT1MM and CT3MM respectively. Cosine similarity values of 0.58 for the 1 mm slice CT and 0.65 for the 3 mm slice CT are significant because no adjustment has been made to account for possible rotation or image resolution differences when scanning the fabricated phantom.

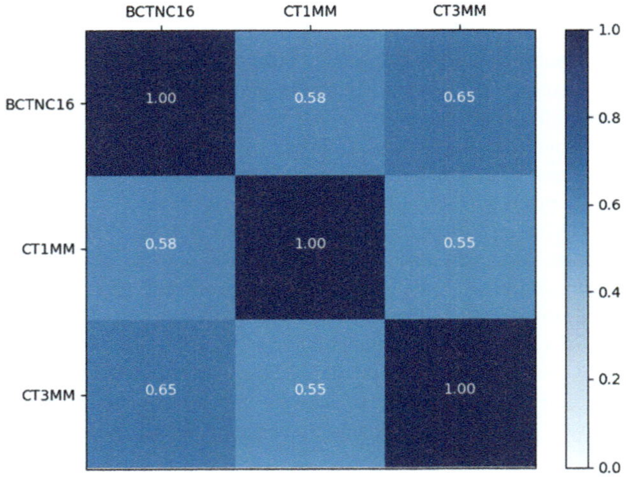

Fig. 6. Cosine similarity between the feature vectors of size 64 extracted from each image through the KAZE algorithm. Labels correspond to the CT slices defined in Fig. 4, namely BCTNC16 is the patient breast, CT1MM and CT3MM are the 1 mm and 3 mm phantom slices.

4 Conclusion

This paper presents the fabrication approach and initial results from tomosynthesis and CT imaging of an ad-hoc anthropomorphic breast phantom obtained with an in-house developed 3D printing software. Comparison of the CT images from the original patient and manufactured breast phantom showed similar visual appearance of the main breast tissues: adipose and gland, which is also well supported by the visual comparison of the *ImageJ* profiles taken across selected slices. Our numerical verification and evaluation procedure, comparing the KAZE feature vectors of the images, achieved promising results as well. Results of this study will be further exploited in the development of a dedicated phantom for breast CT studies and specifically in setting a future experimental setup for an accurate breast CT dosimetry study.

Acknowledgements. This research is supported by the Bulgarian National Science Fund under grant agreement DN17/2.

References

1. Glick, S.J., Ikejimba, L.C.: Advances in digital and physical anthropomorphic breast phantoms for x-ray imaging. Med. Phys. **45**, e870–e885 (2018)
2. Carton, A.K., Bakic, P., Ullberg, C., Maidment, A.D.A.: Development of a 3D high-resolution physical anthropomorphic breast phantom. In: Medical Imaging 2010: Physics of Medical Imaging, San Diego, CA (2010)
3. Sikaria, D., Musinksy, S., Sturgeon, G.M., Solomon, J., Diao, A., Gehm, M.E., Samei, E., Glick, S.J., Lo, J.Y.: Second generation anthropomorphic physical phantom for mammography and DBT: incorporating voxelized 3D printing and inkjet printing of iodinated lesion inserts. In: Medical Imaging 2016: Physics of Medical Imaging, vol. 9783 (2016)
4. Zhao, C., Solomon, J., Sturgeon, G.M., Gehm, M.E., Catenacci, M., Wiley, B.J., Samei, E., Lo, J.Y.: Third generation anthropomorphic physical phantom for mammography and DBT: Incorporating voxelized 3D printing and uniform chest wall QC region. In: Medical Imaging 2017: Physics of Medical Imaging, vol. 10132 (2017)
5. Ivanov, D., Bliznakova, K., Buliev, I., Popov, P., Mettivier, G., Russo, P., Di Lillo, F., Sarno, A., Vignero, J., Bosmans, H., Bravin, A., Bliznakov, Z.: Suitability of low density materials for 3D printing of physical breast phantoms. Phys. Med. Biol. **63**, 175020 (2018)
6. Okkalidis, N.: A novel 3D printing method for accurate anatomy replication in patient-specific phantoms. Med. Phys. **45**, 4600–4606 (2018)
7. Boone, J.: Dedicated breast CT imaging of the breast. Med. Phys. **35**, 1950–1958 (2008)
8. Lindfors, K.K., Boone, J.M., Nelson, T.R., Yang, K., Kwan, A.L.C., Miller, D.F.: Dedicated breast CT: initial clinical experience. Radiology **246**, 725–733 (2008)
9. Okkalidis, N., Chatzigeorgiou, C., Okkalides, D.: Assessment of 11 available materials with custom three-dimensional-printing patterns for the simulation of muscle, fat, and lung hounsfield units in patient-specific phantoms. J. Eng. Sci. Med. Diagn. Therapy **1**(1), 011003 (2018)
10. Alcantarilla, P.F., Bartoli, A., Davison, A.J.: KAZE features. In: Fitzgibbon, A., Lazebnik, S., Perona, P., Sato, Y., Schmid, C. (eds.) Computer Vision – ECCV 2012, Lecture Notes in Computer Science, vol. 7577. Springer, Heidelberg (2012)

Microcalcification Cluster SDNR in Synthesized and 2D Mammography

Andreas Petropoulos, Spyros Skiadopoulos, Anna Karahaliou, Georgios Vlachopoulos, Gerasimos Messaris, and Lena Costaridou(✉)

Department of Medical Physics, School of Medicine, University of Patras, Patras, Greece
costarid@upatras.gr

Abstract. This study aims at quantitative assessment of image quality of microcalcification clusters in synthesized-2-Dimensional (s2D) mammography (i.e. generated from digital breast tomosynthesis-DBT) as compared to 2-Dimensional (2D) mammography, in terms of the signal-difference-to-noise ratio (SDNR). A mammographic phantom (Leeds, TORMAM) simulating microcalcification (MC) clusters located in varying parenchyma density patterns (fatty, glandular and dense) was imaged in 2D and DBT, considering two distinct scattering conditions (phantom thickness 30 mm and 50 mm). An accurate MC cluster segmentation method was used to delineate individual MC cluster particles and define overall cluster SDNR in s2D and 2D images. Pairwise comparisons of each cluster SDNR between 2D and s2D demonstrated no statistically significant difference across the three parenchyma density patterns ($p > 0.05$, Mann Whitney U test), for both scattering conditions. s2D demonstrated constant performance in terms of cluster SDNR with increasing surrounding parenchyma density, while in case of 2D, cluster SDNR is decreasing with substantial parenchyma density increase.

Keywords: Breast tomosynthesis · Synthesized images · Image quality

1 Introduction

Digital breast tomosynthesis (DBT) provides reconstructed tomographic slices of the breast, thus considered to improve breast cancer detection and diagnosis by overcoming the overlapping dense parenchyma patterns in lesion detectability. DBT improves the diagnostic accuracy of 2D mammography in case of low contrast lesions (masses), while detection and characterization of microcalcification (MC) clusters accounts for a challenging task. Currently, DBT imaging seems to not be able to completely replace 2D mammography and is used in clinical routine as an adjunct to 2D imaging, at the expense of increased patient dose.

In order to reduce patient dose, the synthesized 2D (s2D) image has been introduced, by using the DBT image dataset, potentially aiming to replace the 2D acquisition [1]. Recent observer performance studies of clinical data have attempted to compare 2D, DBT, and s2D imaging with respect to lesion detection or visibility [1].

J. Henriques et al. (Eds.): MEDICON 2019, IFMBE Proceedings 76, pp. 375–380, 2020.
https://doi.org/10.1007/978-3-030-31635-8_45

However, only phantom studies can compensate the lack of lesion ground truth, which is an important disadvantage of clinical inter-comparison studies [2].

Previously reported phantom studies have compared image quality between 2D and s2D in terms of observer detectability/visibility ratings [2, 3], while only a few have exploited quantitative image quality indices [4–6]. However, the quantitative image quality evaluation of realistic MC particles in non-uniform parenchyma background is an open research issue.

The aim of the current study is the quantitative assessment of image quality of MC clusters in 2D and s2D images, in terms of the signal-difference-to-noise ratio (SDNR) index, strongly associated to human detectability [9], exploiting a phantom that provides realistic MC cluster mammographic appearance. The effect of increasing parenchyma density on MC cluster SDNR is investigated in 2D and s2D mammography.

2 Materials and Methods

2.1 Mammographic Phantom Image Acquisition

The TORMAM test object (15 mm thick, Leeds Test Objects Ltd, North Yorkshire, United Kingdom) and additional PMMA slabs were used to achieve a total phantom thickness of 30 mm and 50 mm, simulating two distinct scattering conditions. The one half of the TORMAM test object simulates the appearance of breast tissue and contains 6 MC clusters, each one composed of up to 30 particles in realistic geometrical spatial distribution, in 6 distinct phantom locations, corresponding to different parenchymal breast patterns. A 2D and DBT system (Hologic, Selenia Dimensions, Bedford, USA) was used to image the phantom, with Auto-filter setting and ComboHD mode, which enables generation of s2D image (software C-view™ by Hologic), by means of a Maximum Intensity Projection-based method of the reconstructed tomosynthesis images. The current analysis considered DICOM format images (10-bit pixel grey level depth, spatial resolution: 70 μm for 2D, approximately 110 μm for s2D).

2.2 Particle Segmentation and Microcalcification Cluster SDNR Estimation

Two expert radiologists in consensus selected three MC clusters from the heterogeneous part of the TORMAM phantom, surrounded by parenchymal pattern of increasing density: fatty, glandular and dense, consisting of 15, 20 and 30 individual particles, respectively (Fig. 1a, b, c). The two radiologists in consensus performed MC particle segmentation of each MC cluster, defined as the cluster ground truth. A recently proposed home-developed segmentation method [7] was utilized, capable of capturing the varying shape and size of individual particles of a cluster. To validate the radiologists' visual sorting of the three selected MC clusters with respect to increasing parenchymal density (i.e. fatty, glandular, dense), a histogram analysis approach of the parenchyma pattern surrounding a MC cluster was adopted [8], in terms of first-order statistics feature extraction (mean, skewness, kurtosis and entropy), previously exploited for automatic parenchyma density assessment. Figure 1(d, e, f) illustrates the histogram of the parenchyma pattern surrounding each MC cluster depicted in Fig. 1(a, b, c). Quantitative image quality evaluation of each MC cluster was achieved by means of SDNR. The SDNR index of a single MC particle is estimated by [6]:

$$\mathrm{SDNR} = \frac{\mathrm{PV_T} - \mathrm{PV_B}}{\sigma_\mathrm{B}} \tag{1}$$

where $\mathrm{PV_T}$ is the mean pixel grey level value of a segmented MC particle (target region), $\mathrm{PV_B}$ is the mean pixel grey level value of one-pixel width background region surrounding the target and σ_B is the standard deviation of the pixel grey level values within the background region. The width of the background region was suggested by the close proximity of the single particles. MC cluster SDNR was provided by the median value of individual particle SDNR values within a cluster, implemented in Matlab environment (Matlab, R2015a, Mathworks, Natick).

Statistical analysis considered calculation of descriptive statistics (IBM SPSS Statistics, version 24, Chicago, IL, USA) for SDNR values of each MC cluster (surrounded by fatty, glandular and dense parenchyma) in case of 2D and s2D, for two scattering conditions (phantom thickness of 30 mm and 50 mm). Appropriate non-parametric statistical tests were used to compare MC cluster SDNR in s2D vs. 2D, and to study the effect of breast parenchyma density (fatty, glandular, dense) on MC cluster SDNR in 2D and s2D, for the two scattering conditions.

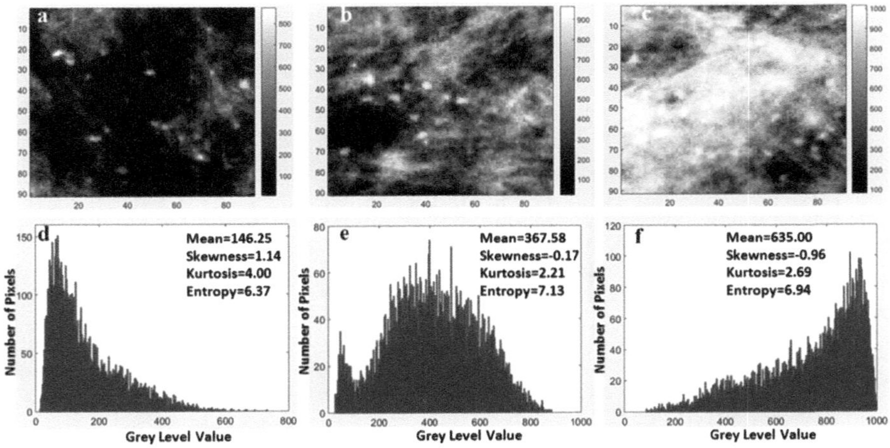

Fig. 1. Microcalcification clusters of the TORMAM phantom surrounded by fatty (a), glandular (b) and dense (c) parenchyma and corresponding surrounding parenchyma histograms (d–f).

3 Results

Table 1 provides SDNR values of the three MC clusters surrounded by increasing parenchyma density of the phantom for two distinct scattering conditions, in 2D and s2D. A corresponding schematic representation of MC cluster SDNR is depicted in Fig. 2. Comparison in terms of MC cluster SDNR between 2D and s2D mammography did not reveal any statistically significant difference ($p > 0.05$, Mann Whitney U test for unpaired data) neither for 30 mm, nor for 50 mm phantom thickness. Furthermore, no statistically significant difference ($p > 0.05$, Mann Whitney U) was found in SDNR of MC clusters surrounded by increasing density patterns for the s2D mode. However, in the 2D mode, MC cluster SDNR was statistically significantly ($p < 0.05$, Mann Whitney U test) decreased as surrounding parenchyma density substantially increased (i.e. comparing fatty vs. glandular and fatty vs. dense parenchyma).

Table 1. Microcalcification cluster SDNR, median and [lower, upper] quartile values for fatty, glandular and dense surrounding parenchyma patterns in 2D and s2D.

Mode	Fatty	Glandular	Dense
Phantom thickness: 30 mm			
2D	4.11 [2.64, 5.23]	2.49 [2.01, 3.20]	2.18 [1.77, 2.95]
s2D	3.60 [2.01, 5.57]	2.43 [1.59, 4.31]	2.63 [1.24, 3.76]
Phantom thickness: 50 mm			
2D	3.79 [2.55, 4.40]	2.90 [1.91, 3.38]	2.02 [1.56, 2.81]
s2D	3.24 [1.39, 4.26]	3.17 [1.12, 4.25]	1.79 [0.95, 3.95]

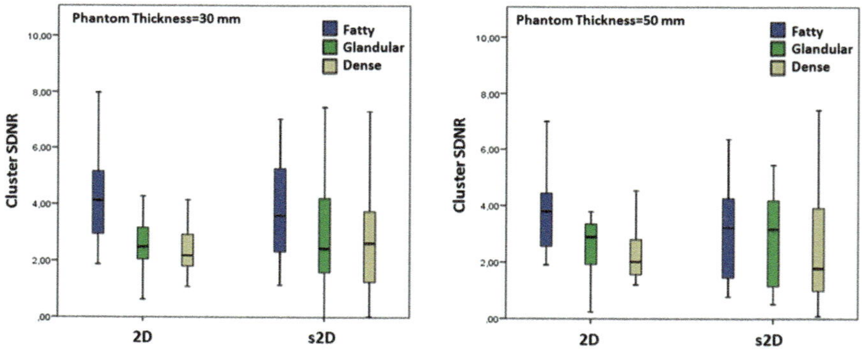

Fig. 2. Boxplots illustrating median, lower and upper quartile values as well as minimum and maximum values of MC cluster SDNR for fatty, glandular and dense surrounding parenchyma patterns in 2D and s2D.

4 Discussion and Conclusions

A phantom-based quantitative image quality evaluation study has demonstrated that s2D provides similar MC cluster SDNR across three increasing density parenchyma patterns, while 2D demonstrated a trend of decreasing MC cluster SDNR performance. Finally, both s2D and 2D seem not to be affected by the two different scatter conditions tested. While a direct comparison to previously reported studies is not feasible due to heterogeneous datasets and image quality indices, results of the current study are in agreement to Baldelli et al. [6] revealing no significant difference in MC cluster image quality in s2D as compared to 2D, however, the effect of breast density on the performance of s2D and 2D has to be further investigated. Despite its pilot character, the current study has exploited an accurate MC particle segmentation algorithm for generating ground truth of realistic MC clusters (with respect to particle shape, number and geometrical distribution of particles and surrounding parenchyma), while previously reported quantitative phantom-based studies have analyzed round objects embedded either in homogeneous [4, 5] or in heterogeneous background [6].

The current study presents a methodology which adopts a quantitative metric (i.e. SDNR), strongly correlated to human observer detection performance [9], leading to results of increased clinical significance. Current results highlight the potential of DBT as a stand-alone imaging modality, allowing generation of s2D for reducing patient dose in mammographic imaging.

Conflict of Interest Declaration. Authors declare that there is no conflict of interest for the publication of this work.

References

1. Wahab, R.A., Lee, S.J., Zhang, B., Sobel, L., Mahoney, M.C.: A comparison of full-field digital mammograms versus 2D synthesized mammograms for detection of microcalcifications on screening. Eur. J. Radiol. **107**, 14–19 (2018)
2. Peters, S., Hellmich, M., Stork, A., Kemper, J., Grinstein, O., Püsken, M., et al.: Comparison of the detection rate of simulated microcalcifications in full-field digital mammography, digital breast tomosynthesis, and synthetically reconstructed 2-dimensional images performed with 2 different digital X-ray mammography systems. Invest. Radiol. **52**(4), 206–215 (2017)
3. Ikejimba, L.C., Glick, S.J., Choudhury, K.R., Samei, E., Lo, J.Y.: Assessing task performance in FFDM, DBT, and synthetic mammography using uniform and anthropomorphic physical phantoms. Med. Phys. **43**(10), 5593–5602 (2016)
4. Nelson, J.S., Wells, J.R., Baker, J.A., Samei, E.: How does c-view image quality compare with conventional 2D FFDM? Med. Phys. **43**(5), 2538–2547 (2016)
5. Fitton, I., Noël, A.: Image quality comparison between digital breast tomosynthesis associated to synthesized image versus 2D FFDM. In: Conference: ECR 2017 EuroSafe Imaging, pp. 1–12 (2017)
6. Baldelli, P., Bertolini, M., Contillo, A., Della Gala, G., Golinelli, P., Pagan, L., et al.: A comparative study of physical image quality in digital and synthetic mammography from commercially available mammography systems. Phys. Med. Biol. **63**(16), 1–19 (2018)

7. Arikidis, N., Vassiou, K., Kazantzi, A., Skiadopoulos, S., Karahaliou, A., Costaridou, L.: A two-stage method for microcalcification cluster segmentation in mammography by deformable models. Med. Phys. **42**(10), 5848–5861 (2015)
8. Karahaliou, A., Boniatis, I., Skiadopoulos, S., Sakellaropoulos, F., Arikidis, N., Likaki, E., Panayiotakis, G., Costaridou, L.: Breast cancer diagnosis: analyzing texture of tissue surrounding microcalcifications. IEEE Trans. Inf. Technol. Biomed. **12**(6), 731–738 (2008)
9. Cockmartin, L., Nicholas, W., Marshall, N., Bosmans, H.: Comparison of SNDR, NPWE model and human observer results for spherical densities and microcalcifications in real patient backgrounds for 2D digital mammography and breast tomosynthesis. In: Fujita, H., Hara, T., Muramatsu, C. (eds.) IWDM 2014, LNCS, vol. 8539, pp. 134–141 (2014)

Enhancing CT 3D Images by Independent Component Analysis of Projection Images

Markus Hannula, Jari A. K. Hyttinen, and Jarno M. A. Tanskanen(✉)

BioMediTech, Faculty of Medicine and Health Technology, Tampere University, Arvo Ylpön katu 34, 33250 Tampere, Finland
{markus.hannula, jari.hyttinen}@tut.fi, tanskanen@ieee.org

Abstract. Computed tomography (CT) is an imaging modality producing 3D images from sets of 2D X-ray images taken around the object. The images are noisy by nature, and segmentation of the 3D images is tedious. Also, detection of low contrast objects may be difficult, if not impossible. Here, we propose an independent component analysis (ICA) based method to process sets of 2D projection images prior to 3D reconstruction to remove noise, and to enhance objects for detection and segmentation. In this paper, a proof-of-concept is provided: the proposed method was able to separate noise and image components, as well as to make visible objects that were not observable in 3D images without processing. We demonstrate our method in object separation with 2D slice image processing simulations, and by enhancing a 3D image of a polymer sample taken with Xradia MicroXCT-400. The method is applicable in any CT tomography for which a number of project image sets with different contrasts can be taken, e.g., in multispectral fashion.

Keywords: Computed tomography · CT · μCT · Micro-CT · Independent component analysis · Image processing · 3D imaging

1 Introduction

In this paper, we propose to enhance 3D computed tomography [1] (CT) images by independent component analysis (ICA) [2–4] based image processing. CT is an imaging modality in which 3D images are produced by a computational reconstruction algorithm from a set of normal 2D X-ray images taken at several angles around the object. Here, in the context of tomography, we call 2D X-ray images projection images. The number of projection images can vary from a few tens to thousands. In general, the more there are projection images, the better the resulting 3D image can be. Normal CT imaging considers usually objects of the size of the order of a human or organ, whereas μCT imaging considers imaging at resolutions of the order of micrometers. All common forms of CT are widely utilized in the image-based analysis of most any radiopaque objects, for example, bio- and industrial materials and their defects, biomaterial compositions, soils, ceramics, and biological tissues and cell constructs.

J. Henriques et al. (Eds.): MEDICON 2019, IFMBE Proceedings 76, pp. 381–389, 2020.
https://doi.org/10.1007/978-3-030-31635-8_46

For CT 3D image reconstruction, good quality projection images are essential; they should be as noiseless and artefact free as possible. As a signal processing rule of thumb, noise and artefact reduction should be done as early as possible in the processing chain to provide the subsequent processing steps as good quality data as possible. Also, processing of 3D images would be computationally much more demanding than processing 2D images. Thus, we propose to enhance 3D μCT images by ICA-based preprocessing the projection images prior to 3D reconstruction.

In CT and X-ray contexts, ICA has previously been used, e.g., for X-ray coronary digital subtraction angiography [5] and removing X-ray scatter [6]. In this paper, we propose a general method to enhance any projection tomography images for which a number of projection image sets with different contrasts is available. We demonstrate the method by MATLAB (MathWorks, Natick, MA, USA) simulations, and by processing multispectral μCT images of polymer samples. This serves as a proof-of-concept of the proposed method.

2 Materials and Methods

2.1 Independent Component Analysis

The heart of ICA is to find independent source signals, i.e., the independent components (ICs) making up the measured signals or recordings. The ICA mixing model [2] is

$$\mathbf{Y}_i = \mathbf{A}_i\mathbf{X}_i \tag{1}$$

where the recordings in the rows of $\mathbf{Y}_i$ are linear combinations of the ICs in the rows of $\mathbf{X}_i$ with $\mathbf{A}_i$ being the mixing matrix and i (here, $i = 1, \ldots, 1600$) is the index of the tomographic projection angle. Numerical iterative algorithms exist to find a separation matrix $\mathbf{B}_i = \mathbf{A}_i^{-1}$ so that the rows of $\mathbf{X}_i$ are mutually statistically independent (down to an error tolerance) based on $\mathbf{Y}_i$. The ICs are found by $\mathbf{X}_i = \mathbf{B}_i\mathbf{Y}_i$. In the sequel, we omit the index i when not referring to our particular application in which ICA is run a number of times.

For ICA, three assumptions must be satisfied [2]: (1) ICs exist, (2) measurements are linear combinations of the ICs (1), and (3) the ICs are non-Gaussian. Observing (1), three ambiguities concerning the ICs are evident: (1) the signs, (2) the energies, and (3) the order of appearance of the ICs in $\mathbf{X}$ are arbitrary. To see the ambiguities explicitly written out, and examples of time signal processing using ICA, see [9].

The intrinsic properties of ICA include that at most as many ICs can be found as there are measurements, and if there are more measurements than ICs, only at most the true number of ICs can be found. ICA can be powerful in recovering minor source signals and eliminating independent noise components. For full discussion, including the description of the ICA algorithm we use here, FastICA [7], see [2]. In this proof-of-concept paper, the choice of ICA algorithm is not important.

The main usages of ICA are (1) to find ICs carrying relevant source signals to utilize them directly, (2) to reconstruct the measurements using only the ICs carrying information of interest, or (3) to reconstruct the measurements omitting the ICs carrying noise or other undesired signal components. Here, we are concerned with the cases 2 and 3. Zeroing the rows of $\mathbf{X}$ corresponding to the uninteresting or noise ICs to get $\mathbf{X}'$, their contributions are eliminated when the measurements are reconstructed as

$$\mathbf{Y}' = \mathbf{A}\mathbf{X}' \tag{2}$$

The measurements and ICs can be any sets of values, such as time signals or images, satisfying the assumptions and (1). Relevant to this paper, note that (1) and nothing above is concerned with the order of samples in the measurements; the corresponding samples must only occur in the same locations in all the measurement vectors. I.e., image pixels can be arranged in the rows of $\mathbf{X}$ (1) in any order, as long as all the images are converted to vectors with the same pixel order.

Principal component analysis (PCA) [2, 8] uses the same mixing model (1), but PCA has been devised from a different point of view: PCA finds uncorrelated components, which in general cannot be associated with source signals. PCA can be powerful, e.g., in noise and dimension reduction tasks [2, 8].

2.2 ICA Usage

All the projection images taken at the same angle with the different CT parameters were collected in to one $\mathbf{Y}_i$ (1), one image per row, and FastICA was executed. This was done for each projection angle separately, i.e., FastICA was run 1600 times with different $\mathbf{Y}_i$s. Thereafter, the independent component images (ICIs) of interest were found in all $\mathbf{X}_i$, and the projection images were reconstructed using (2) with the unwanted ICIs zeroed in the $\mathbf{X}_i$'s. For noise removal, the unwanted ICIs are the ones consisting of (mostly) noise, and for image component separation, the unwanted ICIs can be all ICIs but the ones carrying information from the desired object. 3D reconstruction was then performed normally with the corresponding ICA-reconstructed and enhanced projection images $\mathbf{Y}_i'$. The entire process is illustrated in Fig. 1.

To comply with the ICA mixing model (1), the corresponding projection image pixels from the different imagings must carry X-ray attenuation due to the same material volume. Thus, the phantom remained fixed in the device between the imagings.

For ICA, there should be regions in the sample, whose image contributions change differently from those of the other regions with the changing imaging parameters. This could result from (1) absorption spectrum differences between the materials, (2) accumulation of contrast agent, (3) addition/removal of material, and/or (4) noise differences between the imagings.

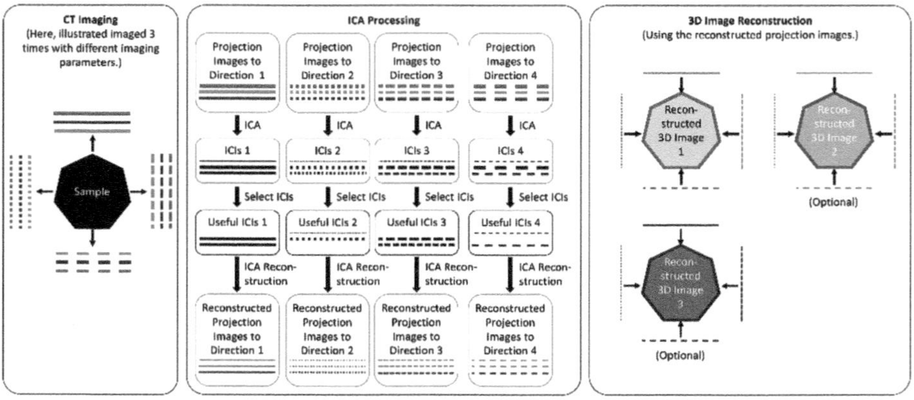

Fig. 1. The steps of the proposed method: CT imaging is performed a number of times with different imaging parameters. Next, ICA processing is applied to the projection images taken to any one direction. Then, ICIs carrying desired information are selected, or ICIs carrying undesirable components zeroed. In this step, it is necessary to identify the ICIs corresponding to each other between the different imaging directions, since the order of appearance of ICIs from ICA is arbitrary; here is illustrated a case in which one ICI carrying similar undesirable information is zeroed for every projection direction. Finally, projection images are ICA-reconstructed from the selected ICIs (here, two for every projection direction). 3D image is reconstructed using the ICA reconstructed projection images; this can be done using all or only select sets of the ICA reconstructed projection images, as desired.

2.3 2D Simulations

To simulate CT image processing in a 2D case (Fig. 2), a human head model slice image was obtained from the MATLAB function `phantom` using a simplified head model. To simulate imagings with different CT settings, object intensities within the slice images were modified ad hoc for each simulated imaging (Fig. 2 Column B). The function `radon` was used to generate 1D projections at 1600 directions equispaced around the slice. Exemplary projections are shown in Fig. 2 Column A. Three tests were conducted: with noiseless images (Fig. 2 Column B), images with slight but different noise added to each mage (noisy originals not shown), and images with the same strong Gaussian uniformly distributed noise added to every image (Fig. 2 Column E); this last case is unnatural but illustrates the workings of ICA.

ICA was calculated over the sets of three projections, i.e., $\mathbf{Y}_i$ (1) had three rows (e.g., the projections in Fig. 2 Column A, one projection per row). All projections $\mathbf{Y}_i'$ were reconstructed using (2) with only one of IC at a time, i.e., $\mathbf{X}_i'$ (2) had only one non-zero row for each ICA reconstruction. This was done separately for each of the 1600 projection directions (c.f., Fig. 1). Finally, a slice image was reconstructed from each set of 1600 reconstructed projections. This was done separately with all three sets of ICA reconstructed projections, resulting in three slice images (c.f., Fig. 2 columns C, D, and F).

2.4 μCT Imaging

We used Xradia MicroXCT-400 (Carl Zeiss Microscopy GmbH, Jena, Germany) with a 4X objective resulting in the pixel size of 44.64 μm. For each 3D image, we acquired 1600 projection images at equispaced angles around the object. The images were cropped to 271 × 401 to ease ICA processing for demonstration purposes. Four different 3D imageries were acquired: at 80 kV and 10 W, 80 kV and 6 W, 40 kV and 10 W, and 40 kV and 6 W. The change in power affected the resulting noise variance, and the change in voltage the photon energy resulting in different absorption characteristics.

2.5 The Phantom

The phantom (c.f., Fig. 3) consisted of a 1 ml syringe filled with water and four polylactic acid (PLA) and two acrylonitrile butadiene styrene (ABS) filaments. Both filaments are common 3D printer materials. Some air bubbles persisted in the tube.

2.6 ICA of the μCT Projection Imagery

The real μCT 3D images were processed analogously to the 2D simulations. Each $\mathbf{Y}_i$ consisted of four projection images (c.f., Fig. 3 Column A), $i = 1, \ldots, 1600$. Pixel values of a 2D projection image entered in the rows of $\mathbf{Y}_i$ from left to right and from top to bottom. ICA was calculated for each $\mathbf{Y}_i$, and $\mathbf{Y}_i'$ was reconstructed with all but one ICIs zeroed in each $\mathbf{X}_i'$. 3D reconstruction was performed separately for all three sets of the 1600 ICA reconstructed projection images. Exemplary 2D slices perpendicular to the imaging direction were extracted for illustration purposes (c.f., Fig. 3 Columns C and D).

3 Results

3.1 Results of ICA Processing of the Simulated Slice Image

The simulation results are shown in Fig. 2. The noiseless original simulated head slices are shown in Fig. 2 Column B, and their 1D projections (the rows of $\mathbf{Y}_i$ (1)) in Column A. The images reconstructed from ICA processed projections $\mathbf{Y}_i'$, with slight noise added to the original images, are shown in Fig. 2 Column C. The corresponding results for the noiseless originals are shown in Fig. 2 Column D. The extremely noisy original simulated slices are shown in Fig. 2 Column E, and the corresponding ICA processed reconstruction results in Fig. 2 Column F.

From Fig. 2 Columns C, D, and F, it is clear that ICA was able to enhance the round region that could not be seen in the original images. In the noise free case (Fig. 2 Column D), this region was quite well separated in an ICs of its own, whereas in the slightly noisy case (Fig. 2 Column C), the corresponding ICs contained contributions also from object edge regions. In the very noisy case (Fig. 2 Columns E and F), the noise quite completely overshadowed the two oval objects, which are probably still embedded in the last reconstructed slice image in Fig. 2 Column F. Still, the round object was nicely separated.

According to our experimentation, the process is noise sensitive. Only low noise levels could be tolerated to recover the round object (Fig. 2 Column C, middle image) when the noise differed between the original images, as it would in reality. In case of the same noise affecting every original image (Fig. 2 Column E), the noise was well captured in one set of ICs (Fig. 2 Column F, bottom image) and could thus be removed, whereas another set of ICs clearly carried the round object (Fig. 2 Column F, uppermost image). The oval objects, however, were not observable; this was probably due to a higher number of true ICs than available measurements.

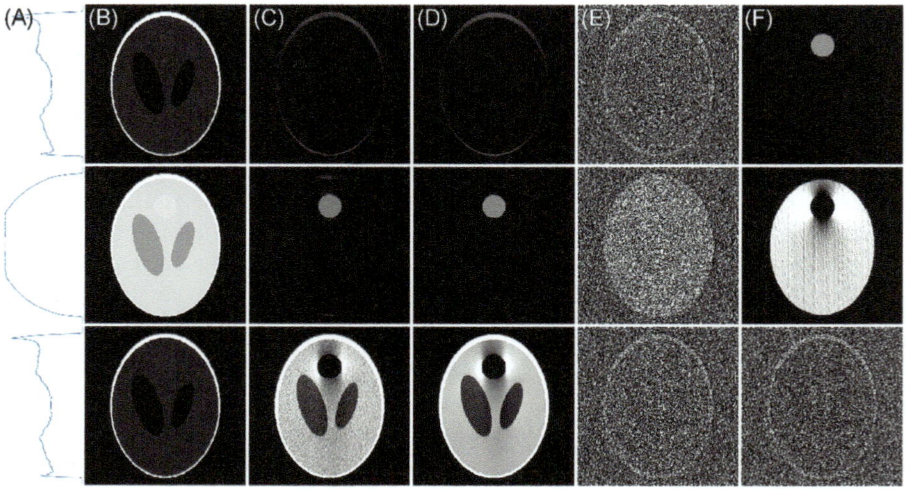

Fig. 2. The simulations and ICA processing results. Column A: Exemplary 1D projections of the original phantom images in Column B. Column C: Slice images reconstructed from ICA processed slightly noisy projections of the slice images in Column B. Column D: Slice images reconstructed from ICA processed noiseless projections of the slice images in Column B. Column E: Highly noisy original phantom images (the same noise in each). Column F: Slice images reconstructed from ICA processed projections of the slice images in Column E.

3.2 Results of ICA Processing of the 3D Phantom Projection Imagery

μCT 3D image processing results are shown in Fig. 3. An exemplary projection image for each of the four imagings with different device settings is shown in Fig. 3 Column A. The corresponding ICIs are shown in Fig. 3 Column B, where noise along with some object features were nicely extracted into the two bottom ICIs, whereas the actual object image information was captured mostly in the two uppermost ICIs.

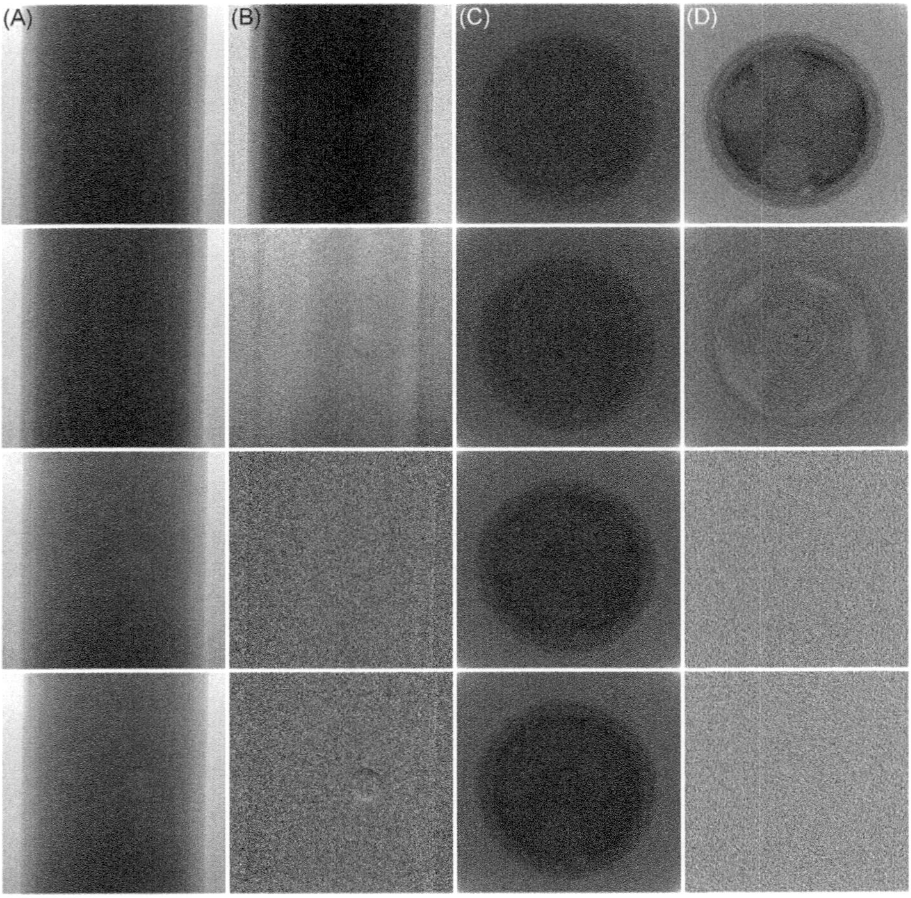

Fig. 3. μCT images and ICA processing results. Column A: Exemplary projection images taken using the four sets of μCT imaging parameters. Column B: Reconstructed ICIs. Column C: Slices of the 3D reconstructions from the raw imaging data. Column D: Slices of the 3D reconstructions from the reconstructed ICIs. The images in Columns C and D have been brightened equally for readability.

An exemplary horizontal slice of the 3D reconstructions of the phantom for the different device settings is shown in Fig. 3 Column C with the corresponding slice from 3D reconstructions made from the ICA reconstructed ICIs in Column D. Although the slice images of the original 3D reconstructions (Fig. 3 Column C) are of quite low contrast and the filaments are not always observable, the contents of the phantom are visually detectable in the slice images produced with the proposed method (Fig. 3 Column D).

After our ICA-based processing (Fig. 3 Column D), two different phantom materials can be distinguished: in the uppermost image, four round objects are clearly observable, whereas in the second image at least five round objects are observable (the object in the middle is obscured by the imaging artefact). From slice images reconstructed without ICA processing (Fig. 3 Column C), the existence of two different materials cannot be observed.

The slice of the two last 3D constructions in Fig. 3 Column D seem to carry mostly noise; these noise components were eliminated from the other reconstructed 3D images.

4 Conclusions

We conclude that ICA processing of multispectral/multienergy X-ray tomography projection images is a potential tool for noise alleviation, object separation, and to distinguish materials in 3D CT reconstructions. However, to enable separation of more image objects and to improve noise alleviation, more image sets with different CT parameters should be available than used here. For object separation, low-noise CT imagery would be preferable.

The proposed method is completely general, and we expect it to be beneficial in several applications, including in enhancing complex biomedical and industrial imagery.

Acknowledgments. The work of J. M. A. Tanskanen has been supported by Jane and Aatos Erkko Foundation, Finland, under the project Biological Neuronal Communications and Computing with ICT. The work of M. Hannula has been supported by the Human Spare Parts Project funded by Finnish Funding Agency for Technology and Innovation (TEKES).

Conflict of Interest Declaration. No conflict of interest.

References

1. Cierniak, R.: X-ray Computed Tomography in Biomedical Engineering. Springer, London (2011). https://doi.org/10.1007/978-0-85729-027-4
2. Hyvärinen, A., Karhunen, J., Oja, E.: Independent Component Analysis. Wiley, Hoboken (2001). https://doi.org/10.1002/0471221317
3. Hyvärinen, A.: Fast and robust fixed-point algorithms for independent component analysis. IEEE Trans. Neural Netw. **10**(3), 626–634 (1999). https://doi.org/10.1109/72.761722
4. Hyvärinen, A., Oja, E.: Independent component analysis: algorithms and applications. Neural Netw. **13**(4–5), 411–430 (2000). https://doi.org/10.1016/S0893-6080(00)00026-5
5. Tang, S., Wang, Y., Chen, Y.-W.: Application of ICA to X-ray coronary digital subtraction angiography. Neurocomputing **79**, 168–172 (2012). https://doi.org/10.1016/j.neucom.2011.10.012
6. Chen, Y.W., Han, X., Oikawa, S., Fujita, A.: Independent component analysis for removing X-ray scatter in X-ray images. In: Proceedings of the 2007 IEEE Instrumentation and Measurement Technology Conference, pp. 1–4. IEEE, Piscataway (2007). https://doi.org/10.1109/IMTC.2007.379278
7. FastICA package for Matlab. http://www.cis.hut.fi/projects/ica/fastica/. Accessed 18 Apr 2019

8. Jolliffe, I.T.: Principal Component Analysis. Springer, Heidelberg (2002). https://doi.org/10.1007/b98835
9. Tanskanen, J.M.A., Viik, J.J.: Independent component analysis in ECG signal processing. In: Millis, R.M. (ed.) Advances in Electrocardiograms, pp. 349–372. InTech, Rijeka (2012). https://doi.org/10.5772/22719

Potentials of OCT in Monitoring Ocular Hemodynamics of Patients with Primary Open Angle Glaucoma

E. N. Iomdina[1], D. D. Khoziev[1], A. A. Kiseleva[2(✉)], P. V. Luzhnov[2], O. A. Kiseleva[1], and D. M. Shamaev[2]

[1] Moscow Helmholtz Research Institute of Eye Diseases, Sadovaya-Chernogryazskaya 14/19, 105062 Moscow, Russia
iomdina@mail.ru
[2] Bauman Moscow State Technical University, 2-nd Baumanskaya 5, 105005 Moscow, Russia
anna14121994@yandex.ru

Abstract. Approaches to using optical coherence tomography (OCT) for ocular hemodynamics monitoring in patients with primary open angle glaucoma (POAG) are presented. The existing OCT systems used in ophthalmology are described. The different OCT systems are compared in accuracy and resolution characteristics, as these are important for clinical studies. The results of ocular hemodynamics monitoring of POAG patients using Spectralis OCT2 with an OCT Angiography module are given.

Keywords: Optical coherence tomography · Optical coherence tomography angiography · Amplitude-decorrelation angiography algorithm · Glaucoma · Ocular blood flow

1 Introduction

In the structure of vision disability, glaucoma is one of the leading conditions [1]. Some of the issues related to the pathogenesis of this disease are still open. No mechanisms that lead to the loss of visual function due to glaucoma have been reliably established. Modern concepts of pathogenesis do not go beyond the multifactorial theory [2]. The cause of glaucoma development is mechanical lesion of the scleral *lamina cribrosa* (LC), which leads to the compression of the axons of the optic nerve and vascular dysfunction of a general and local nature. Deterioration of visual functions was proven to be due to persistent decrease in intraocular pressure (IOP) [3]. Research in ocular hemodynamics could be helpful in in-depth investigation of disease stages, since IOP level is largely dependent on ocular blood flow [4].

Modern methods of eye blood flow studies include color Doppler imaging [5], scanning laser Doppler flowmetry, optical coherence tomography (OCT). Color Doppler imaging is a method that allows a targeted identification of the vessels and trace the occurring changes. The disadvantages of this technique are that we are unable to

J. Henriques et al. (Eds.): MEDICON 2019, IFMBE Proceedings 76, pp. 390–396, 2020.
https://doi.org/10.1007/978-3-030-31635-8_47

measure the diameter of the vessel and obtain an estimate of the volumetric flow rate [6]. Laser Doppler flowmetry provides information about microcirculation oscillation, which is associated with the functioning of the mechanisms of microvascular tone regulation. Again, the method has a significant drawback as it has restricted locality and requires anesthesia during the study [7]. Currently, optical coherence tomography is widely used as it enables a simultaneous qualitative and quantitative assessment of eye tissue. Information content of OCT in glaucoma is the focus of present-day research [8, 9]. While some parameters are more sensitive in the early stages of glaucoma, others become sensitive in later stages. The attainable accuracy of measurements largely depends on the processing algorithms that underlie the tomograph software. The accuracy of diagnostic data depends on which OCT system is used. Accordingly, the correlation of the results obtained by different research groups will depend on the OCT system used, too.

The paper is aimed at comparative analysis of OCT systems, which involves their main technical characteristics and processing algorithms, accuracy parameters and informativeness for monitoring ocular blood supply in primary open-angle glaucoma (POAG).

2 Materials and Methods

OCT is a non-contact method of ophthalmologic examination, which produces images of eye tissues with high spatial resolution (3 to 15 μm). The operating principle of OCT is based on light interferometry. When two or several light waves are superimposed, a spatial redistribution of light radiation energy occurs - the interference of light. The OCT system is based on the Michelson interferometer. A beam of low coherence comes from a superluminescent diode. The first part of this beam is the basis for imaging. It passes through the structures of the eye and is reflected from different depths of the tissue sections. The second part of the beam goes to the movable support holder (mirror). Interference occurs when equal length beam is reflected from the tissue and the reference beam, from the mirror. The obtained interference parameters are recorded and measured by a photodetector. The time delay may vary depending on the tissue, since each tissue has its own reflection parameters. The reference mirror moves when taking measurements at each predetermined point. At each point, for one measurement, one A-scan is obtained. After that, single A-scans are summed up and produce a resulting two-dimensional image. This principle is the basis of all modern OCT systems [13].

In spectral OCT, the beam is decomposed into spectral components and then recorded by the camera. The processing of the obtained interference spectrum and the further formation of the linear A-scan occurs due to the mathematical Fourier transformation. As a result, a linear A-scan is obtained by simultaneously measuring the reflected rays from each individual point.

With OCT systems, it became possible to study in detail the anatomy of the eye structures with the identification of changes in their structure that are beyond the scope of other research methods. However, for the diagnosis of glaucoma it is necessary to have detailed information about the structure of the eye tissue as well as the data on the blood supply. The spatial resolution of OCT systems allows localization of vessels whose size is several microns. The method of obtaining information about the flow rate in OCT is called Doppler optical coherence tomography (DOCT) [14]. This method is available on conventional OCT systems and it is a functional addition to OCT systems. The main difficulty in implementing DOCT algorithms concerns the exact definition of the Doppler angle [15]. An effective way to increase the accuracy of DOCT systems is averaging, which is performed either on the neighboring A-scans or on reconstructed B-scans. In the second case, several scans of one object part are used, which increases the diagnostic time. Averaging a large number of scans increases the scan time, so clinical trials restrict the number of averages to 5–10. In this way, the contrast of the obtained image and the resolution improve from 15 to 5 μm [16].

The accuracy of the DOCT method is reduced when assessing the blood circulation of the retina and choroid, since the blood flow in these tissues has a perpendicular direction to the probing radiation. To overcome this limitation, the functional addition of OCT systems, OCT–angiography (OCT-A) is used. OCT-A makes equal sensitivity to record the transverse and axial blood flow throughout the scan depth. OCT-A combines the capabilities of a high-resolution optical coherent tomograph with non-invasive angiography technique. The OCT-A applies an algorithm which is based on amplitude estimation–split–spectrum amplitude decorrelation algorithm (SSADA). SSADA assesses blood flow in the lumen by measuring the fluctuations of the reflected signal amplitude between successive transverse OCT slices [17]. This algorithm provides the same sensitivity in the axial and transverse directions [18].

A significant advantage of the OCT-A method over other methods is the ability to visualize the complex of vessels, measure their area, determine the shape, branching, and presence of anastomoses and loops. As a result, it is possible to quantify the blood flow index, the area of the perfusion zones, the capillary density index [19]. As part of this work, a comparison has been made of the main technical parameters of modern OCT systems. These are the RTVue-100 (Optivue, Inc., Fremont, CA), Avanti (Optivue, Inc., Fremont, CA) and Spectralis OCT2 (Heidelberg Engineering, Germany). Table 1 presents a comparison of the main technical characteristics of the systems.

The characteristics presented in Table 1 are normalized in accordance with the requirements of ISO 16971: 2015 "Ophthalmic instruments – Optical coherence tomograph for the posterior segment of the human eye". However, the presented characteristics carry information only about the hardware of the system and do not take into account the peculiarities of using different processing algorithms. As a result, the task of this work has been to assess the stage of the disease in a group of patients with glaucoma using OCT and to fix main technical characteristics of our calculations.

In this paper, we assessed the potentials of hemodynamics monitoring in patient with POAG using a Spectralis OCT2 with an Angiography module (Heidelberg Engineering, Germany). Spectralis OCT2 has additional diagnostic modules of Enhanced depth imaging (EDI), used to eliminate the restriction of the choroid and LC visualization with light absorption by pigment epithelium of the retina and its scattering by choroid structures for EDI [20].

Table 1. The comparison performance OCT systems.

System	Longitudinal resolution	Transversal resolution	Scan depth	Scan area	Image acquisition speed
RTVue – 100	Images – no less than 2.9 μm/ optical – no less than 5.0 μm	No less than 8 μm	No less than 2.5 μm	13 mm × 9 mm	26000 A - scans/sec
Avanti	Images – no less than 3.0 μm/ optical – no less than 5.0 μm	No less than 3 μm	3.0 μm	13 mm × 9 mm	70000 A - scans/sec
Spectralis OCT2	3.9 μm per pixel	5.7 μm per pixel	Depends on operating mode	Depends on operating mode	85000 A - scans/sec

The high scanning rate (85,000 kHz), high lateral and axial resolution enables for Spectralis OCT2 accurate layer–by–layer segmentation of the retina, fixing the reflective ability within tissues. Spectralis OCT2 includes an active system of eye movements control in real time, which helps avoid movement artifacts and enables a high resolution of the images. The positioning principle of images in two mutually perpendicular planes for researches is presented on the Fig. 1.

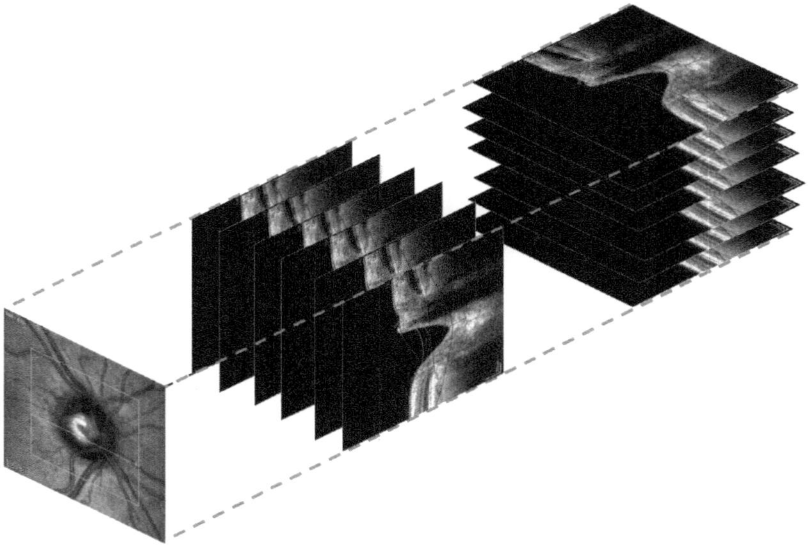

Fig. 1. OCT scans orientation in two mutually perpendicular planes for research

The density of microcapillary meshwork of the superficial (SVL) and deep (DVL) of vascular plexuses was measured in EDI mode using AngioTool software and was defined as the area occupied by vessels and expressed in percentage of the total area of the study domain in a specific cross section.

For definition of LC parameters the algorithm presented on the Fig. 2 was used.

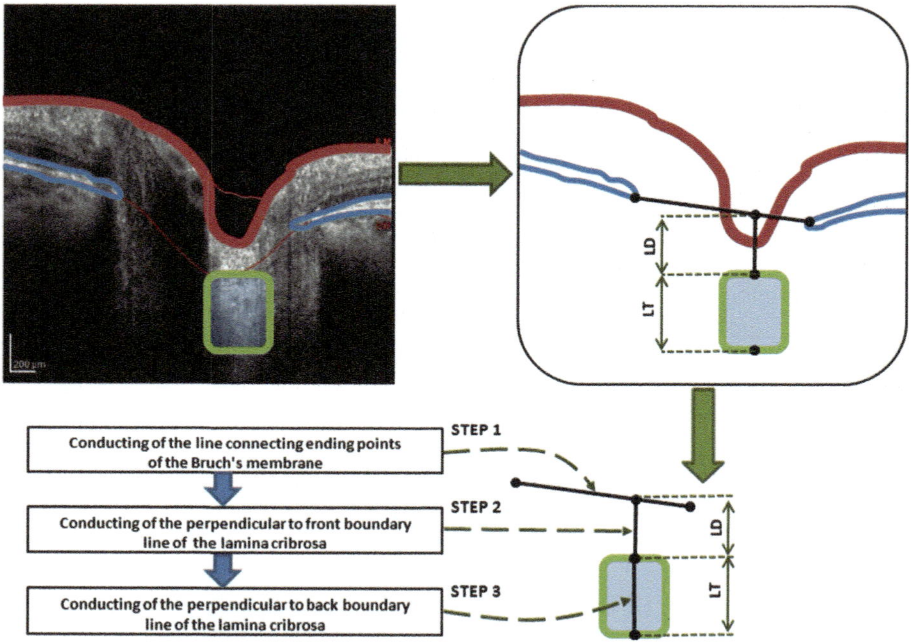

Fig. 2. The algorithm for definition of LC parameters

On the chosen OCT scan were defined geometrically LC depth (LD) and LC thickness (LT). On each algorithm step (steps 1–3) there is an addition measurement error about two pixels (±3.9 μm). The total error of scaling will make thus no less than six pixels or no less than ±11.7 μm.

3 Compliance with Ethical Requirements

This study was performed in accordance with the Declaration of Helsinki and was approved by the Local Committee of Biomedical Ethics of the Moscow Helmholtz Research Institute of Eye Diseases.

4 Results

The study was conducted on 111 eyes of 63 people, including 41 eyes without ophthalmic pathology – control group (mean age 66.9 ± 18.2 years), 22 eyes of patients with stage I of POAG (mean age 65.9 ± 11.3 years), 31 eyes with stage II POAG (mean age 69.2 ± 17.3 years) and 17 eyes with stage III POAG (mean age 69.7 ± 7.3 years).

The results were statistically processed by Statistics 20 which involved the computation of the average value of the parameters and mean deviation (M ± σ), when comparing groups, the Mann-Whitney U-test was used. The differences were considered significant for $p < 0.05$. Table 2 presents SVL, DVL and LC parameters in patients with different POAG stages.

Table 2. Hemodynamic parameters in the control group and in different POAG stages (M ± σ).

Parameters	Control group, n = 41	Stage I of POAG, n = 22	Stage II of POAG, n = 31	Stage III of POAG, n = 17
SVL, %	51.5 ± 4.3	50.1 ± 8.4	47.6 ± 5.7	43.0 ± 6.9
DVL, %	45.9 ± 9.3	43.1 ± 8.2	41.4 ± 4.7	39.3 ± 8.7
LD, μm	386 ± 54	424 ± 46	452 ± 76	501 ± 59
LT, μm	223 ± 18	210 ± 15	203 ± 16	187 ± 25

It has been established that POAG patients display a significant decrease in the density of vascular plexuses of the retina.

5 Conclusions

The research results show a decrease in the density of the vascular plexuses of the retina with POAG stage increase. However, to compare the results obtained with the results of studies obtained using other OCT systems, we need to evaluate the accuracy characteristics for each system. To bring the results of research to a suitable form for comparison, additional analysis is necessary. For example, using phantoms with calibrated geometric characteristics.

Conflict of Interest. The authors declare that they have no conflict of interest. The paper was supported by a grant from RFBR (No. 18-08-01192).

References

1. Morizane, Y., Morimoto, N., Fujiwara, A., et al.: Incidence and causes of visual impairment in Japan: the first nation-wide complete enumeration survey of newly certified visually impaired individuals. Jpn. J. Ophthalmol. **63**(1), 26–33 (2019)

2. Quigley, H.A.: Glaucoma: macrocosm to microcosm, the Frieden wald lecture. Invest. Ophthalmol. Vis. Sci. **46**, 2662–2670 (2005)
3. Fechtner, R.D., Weinreb, R.N.: Mechanisms of optic nerve damage in primary open angle glaucoma. Surv. Ophthalmol. **39**(1), 23–42 (1994)
4. Schmetterer, L.: Ocular Blood Flow, pp. 147–159. Springer, New York (2012)
5. Lieb, W.E.: Color Doppler ultrasonography of the eye and orbit. CurrentOpin. Ophthalmol. **4**, 68–75 (1993)
6. Flammer, J., Orgul, S., Costa, V.P., et al.: The impact of ocular blood flow in glaucoma. Prog. Retin. Eye Res. **21**, 359–393 (2002)
7. Kiseleva, T.N., Kotelin, V.I., Losanova, O.A., Lugovkina, K.V.: Noninvasive methods assessment blood flow in anterior segment and clinical application perspective. Oftalmologiya **15**(4), 283–290 (2017). https://doi.org/10.18008/1816-5095-2017-4-283-290
8. Straubhaar, M., Orgul, S., Gugleta, K., et al.: Choroidal laser Doppler flowmetry in healthy subjects. Arch. Ophthalmol. **118**(2), 211–215 (2000)
9. Kurysheva, N.I., Parshunina, O.A., Shatalova, E.O., et al.: Value of structural and hemodynamic parameters for the early detection of primary open-angle glaucoma. Curr. Eye Res. **42**(3), 411–417 (2017)
10. Weinreb, R., Harris, A.: Ocular Blood Flow in Glaucoma: The 6th Consensus Report of the World Glaucoma Association, pp. 1–159. Kugler Publications, Amsterdam (2009)
11. Liu, L., Jia, Y., Takusagawa, H.L., et al.: Optical coherence tomography angiography of the peripapillary retina in glaucoma. JAMA Ophthalmol. **133**(9), 1045 (2015)
12. Kurysheva, N.I., Maslova, E.V.: Optical coherence tomography angiography in glaucoma diagnosis. Vestn. oftalmol. **132**(5), 98–102 (2016)
13. Popescu, D.P., Choo-Smith, L.-P., Flueraru, C., Mao, Y., Chang, S., Disano, J., Sowa, M.G.: Optical coherence tomography: fundamental principles, instrumental designs and biomedical applications. Biophys. Rev. **3**(3), 155–169 (2011)
14. Grulkpwski, I., et al.: Scanning protocols dedicated to smart velocity ranging in Spectral OCT. Opt. Express **17**, 23736–23754 (2009)
15. Hendargo, H.C., McNabb, R.P., Dhalla, A.H., Shepherd, N., Izatt, J.A.: Doppler velocity detection limitations in spectrometer – based versus swept – source optical coherence tomography. Biomed. Opt. Express **2**, 2175–2188 (2011)
16. Wang, R.K., et al.: Three dimensional optical angiography. Opt. Express **15**, 4083–4097 (2007)
17. Motaghiannezam, R., Fraser, S.: Logarithmic intensity and speacle – based motion contrast methods for human retinal vasculature visualization using swept source optical coherence tomography. Biomed. Opt. Express **3**, 503–521 (2012)
18. Jia, Y., et al.: Split – spectrum amplitude – decorrelation angiography with optical coherence tomography. Opt. Express **20**, 4710–4725 (2012)
19. Tokayer, J., Jia, Y., Dhalla, A.H., Huang, D.: Blood flow velocity quantification using split – spectrum amplitude decorrelation angiography with optical coherence tomography. Biomed. Opt. Express **4**, 1909–1924 (2013)
20. Margolis, R., Spaide, R.F.: A pilot study of enhanced depth imaging optical coherence tomography of the choroid in normal eyes. Am. J. Ophthalmol. **147**(5), 811–815 (2009)

Automatic Segmentation of Bone and Muscle Structures in CT Volumes Using Convex Relaxation and Fine-Tuning

José-Antonio Pérez-Carrasco(✉), Carmen Serrano, and Begoña Acha

Departamento de Teoría de la Señal y Comunicaciones, Universidad de Sevilla, Camino de los Descubrimientos s/n, 41092 Seville, Spain
{jperez2, cserrano, bacha}@us.es

Abstract. Segmentation of muscle and bone structures in CT volumes is a complicated task for physicians and surgeons due to the similarity of Hounsfield values with those within surrounding organs. The lack of efficient and automatic tools for this task makes that most physicians and surgeons prefer to segment bone and/or muscle tissue manually rather than working with specific-focused-tissue segmentation tools. In this work, an automatic tool for the joint segmentation of muscle and bone structures is described. The tool implements the segmentation in two stages. In the first stage an energy minimization approach via convex relaxation is performed. In the second step, a more accurate muscle segmentation is achieved using probability maps. A public database of CT volumes have been used to test the algorithm and different metrics, including Dice, Jaccard, sensitivity and specificity, have been calculated to evaluate the algorithm performance. The results provided outperform recent methods using the same public database.

Keywords: Segmentation · Muscle · Bone · Convex relaxation

1 Introduction

The segmentation of bone and muscle structures in CT volumes is of special interest for physicians, radiologists and surgeons in order to provide the diagnosis of some diseases, to analyze body composition and in surgical planning. Automatic segmentation of bone and muscles is difficult due to several reasons:

1. Processing CT volumes (3D) is a very time-consuming task due to the big size of the data,
2. The similarity of Hounsfield values with those belonging to other organs and structures. Note for instance that cancellous bone can have Hounsfield values close to those present within muscle tissue.
3. Muscle structures can be at any place along the body and they do not have an established pattern. Moreover, there are several types of muscle tissues (skeletal, cardiac and smooth) and they can have a big variability in shape making even more difficult their accurate segmentation. As a consequence, there is a lack of automatic or semi-automatic segmentation tools and thus the need of having to segment them manually.

J. Henriques et al. (Eds.): MEDICON 2019, IFMBE Proceedings 76, pp. 397–404, 2020.
https://doi.org/10.1007/978-3-030-31635-8_48

Thus, there is a growing demand of automatic or semiautomatic software tools to accurately segment these structures in a low (or at least acceptable) computational time. There are many segmentation techniques. They can be roughly classified as: intensity-based techniques, edge-based and/or region-based algorithms, and mathematical model algorithms [1]. The techniques based on intensity are not accurate due to the lack of uniformity in bone and muscle structures. Many studies use thresholding of Hounsfield values without considering the fact that the non-uniformity of these structures and some diseases can consequently modify their values [2]. For instance, in Fig. 1 a CT slice is shown where different Hounsfield values and densities can be observed. Note that the rib and the scapula selected have different densities in their boundaries (compact bone) while the inner bone (cancellous bone) has similar Hounsfield values to those in muscles (pectoralis for instance). Edge-based algorithms aim to find the contour of the structures of interest. However, and as it happens with intensity-based techniques, the presence of diseases can make it difficult the location of these boundaries. Note that, for instance, people suffering of osteoporosis or osteoarthritis present non-uniform cortical bone thickness distribution [3] and, thus, edge-based techniques can fail at locating bone boundaries in those cases. In Region-based techniques, the volume is divided into regions according to some homogeneity criterion. Mathematical model algorithms [4–7] implement the segmentation of structures by minimizing some energy measures, which usually combine region and/or boundary information through different functionals [8]. The main problems found in these techniques are that they may require non-affordable computational times or they can drive to local optima when gradient descent methods are used to solve this optimization problem [6].

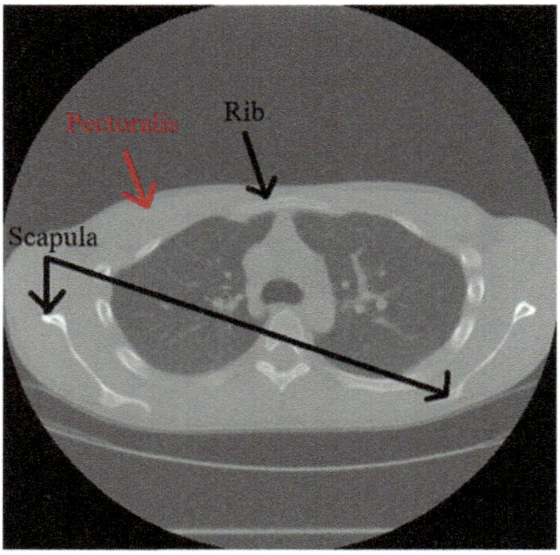

Fig. 1. CT slice where different Hounsfield values and densities can be observed. Note that the rib and the scapula selected have different densities in their boundaries (compact bone) while the inner bone (cancellous bone) has similar Hounsfield values to those in muscles (pectoralis for instance).

Another issue regarding bone and muscle segmentation is that many studies focus their algorithms in some particular bone of muscle structure or to a reduced number of them [8–12]. Consequently there is a clear lack of generalization which is very inconvenient for most surgeons or radiologists as they would need different tools to implement the segmentation of different structures.

In order to help surgeons and radiologists to locate and segment bone and muscle structures in a more general way, a two-stages algorithm is proposed which performs the segmentation independently of the location or type of the structure. The algorithm consists in two stages. The first stage follows the method proposed by Pérez-Carrasco et al. [13] for segmenting bone and muscle based on energy minimization via convex relaxation. In this technique, local histograms at each voxel are computed, and a continuous convex relaxation technique [14, 15] to provide a presegmentation of muscle and bone structures is carried out. This technique [13] has been chosen due to its better performance compared to other state-of-the-art techniques such as active contours and graph-cuts [15]. In the second stage of the algorithm, a fine-tuning stage based on probability maps is performed to remove structures close to muscle tissue in terms of Hounsfield values, such as kidney and liver.

The results obtained in the proposed algorithm outperform the results from [13] in terms of accuracy and sensitivity among other metrics.

2 Materials and Methods

The *Usevillabonemuscle* database [16], which is publicly available at http://grupo.us.es/grupobip/research/research-topics/segmentation-of-abdominal-organs-and-tumors/, has been used to train and to test the algorithm described throughout this paper. The database is composed by 27 CT volumes, with 10 slices each, belonging to 20 different patients aged from 16 to 93 and corresponding to different sections of the body. In the database, manual segmentations by two experts are also provided to be used as ground-truth. Matlab R2017b has been used in all the experiments.

In the database 10 CT volumes has been chosen as training dataset and 17 CT volumes as test dataset.

The algorithm consists of two different stages: presegmentation and fine-tuned muscle segmentation.

2.1 Presegmentation

In the first step, a presegmentation is performed following the steps described in [13]:

1. First, a skin delimitation and removing step is performed, where a neighborhood-connected region growing algorithm is followed by morphological operations to remove skin and other non-body elements.

2. Secondly, thresholding and scaling of the CT volumes are computed. At this stage, the CT volumes are first filtered with a Gaussian filter and two thresholded volumes are obtained by thresholding the volumes using two different Hounsfield windows with different means and ranges. The first Hounsfield window is to roughly select bone structures and the second one is to keep Hounsfield values corresponding to muscle structures. Both windows are wide enough to include all bone and muscle structures. In this way, only Hounsfield values within these two Hounsfield windows are kept and the rest are set to '0'. Finally, the two thresholded volumes are scaled by dividing them by the maximum threshold in each case, so the maximum value is '1'. Thus, two thresholded and scaled volumes denoted as *TSIb* and *TSIm* are obtained.
3. A continuous multi-label max-flow segmentation algorithm is performed. In this substage an energy function which combines statistical information (distance to reference model histograms volumes HDI_b and HDI_m) and gray level information (local Hounsfield values at each voxel) is minimized and three possible labels are considered for each voxel: muscle, bone and other structures. In this substage different weighting parameters *a, b* and *c* and three cost functions (3 labels: bone, muscle and other structures) were computed as:

$$CIT_1 = a\,HDI_m + (1-a)\left(\left|TSI_m - \overline{TSI}_m\right|\right) \tag{1}$$

$$CIT_2 = b\,HDI_b + (1-b)\left(\left|TSI_b - \overline{TSI}_b\right|\right) \tag{2}$$

$$CIT_3 = 1 - \{cCIT_1 + (1-c)(CIT_2)\} \tag{3}$$

Where $\overline{TSI}_m$ and $\overline{TSI}_b$ are the mean values of TSI_m and TSI_b respectively. The energy function to be minimized is computed as:

$$\min_{u_i(x)\in[0,1]} \sum_{i=1}^{3} \int_\Omega (u_{i-1}(x) - u_i(x))CIT_i(x)dx + \int_\Omega C(x)|\nabla u_i(x)|dx \tag{4}$$

$$s.t.\ 1 = u_0(x) \geq u_1(x) \geq u_2(x) \geq u_3(x) = 0$$

In Eq. (4), $u_i(x) \in [0, 1]$ represents the label functions, where i denotes the label and x represents the voxel position. $u_i(x)$ can take real numbers between 0 and 1. $C(x)$ is a penalty function to favor that borders between regions are located in voxels with strong edges.

The results obtained following the steps described above made that some organs (mainly liver and kidney) where labelled as muscle tissue due to the similarity both in Hounsfield values and local histograms models, being this problem important in abdominal CTs. Increasing the number of labels using the former technique with new histogram models was computationally expensive (note that the algorithm requires the computation of histogram distance volumes for each label to be considered) and did not provide better results. Thus, a second stage to improve the segmentation of muscle tissue was performed.

2.2 Fine-Tuned Muscle Segmentation

At this stage, a fine-tuned segmentation stage based on probability maps was performed to improve the segmentation results obtained in the previous step. The groundtruth segmentation of muscle tissue in the training dataset was employed to compute two additional histograms models corresponding to two main structures having Hounsfield values similar to muscle tissue: liver and kidney. Thus, sections corresponding to muscle, liver and kidney within these groundtruth volumes were manually selected and, as done in the former subsection 2.1, thresholded and scaled versions of them were computed (denoted as TSI_{liver} and TSI_{kidney}). Subsequently, histograms of TSI_m, TSI_{liver} and TSI_{kidney} were computed, averaged and normalized to convert them into probability distributions. Subsequently, three different probability distribution functions were built: $P_1(y)$ corresponding to the probability distribution function for voxels corresponding to muscle tissue, $P_2(y)$ corresponding to liver, and $P_3(y)$ corresponding to kidney. In these expressions y denotes the bin number.

Subsequently, for each voxel x belonging to a CT within the test dataset with a gray value belonging to a bin y, a label is assigned according to the following max operation:

$$label_x = max_i\{P_i(y = g(x))\}\ i = 1, 2, 3 \tag{5}$$

Where $g(x)$ represents the gray value at voxel x and y the correspondent bin in the probability distributions. In this stage, only voxels classified as muscle in the presegmentation stage 2.1 were considered.

Finally, average filtering and several morphological operations were performed to remove noise and small unconnected regions.

In Fig. 2 segmentation obtained for three slices belonging to three different CT volumes are shown. Column a) corresponds to the segmentation provided with the method in [13]. Column b) corresponds to the segmentation obtained with the algorithm described in this paper. The groundtruth segmentation is represented by red for bone structures and blue contours for muscle tissue. Note how non-muscular structures are removed with the proposed algorithm whereas other muscle structures such as intestines remain in spite of not having being considered within the groundtruth.

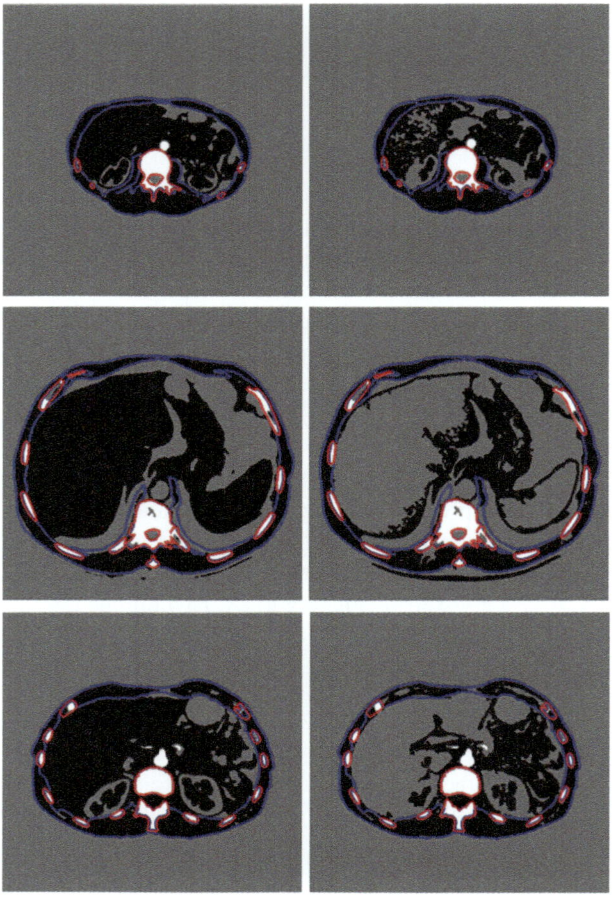

Fig. 2. Segmentation obtained for three slices belonging to three different CT volumes are shown. Column (a) corresponds to the segmentation provided with the method in [13]. Column (b) corresponds to the segmentation obtained with the algorithm described in this paper. Bones are shown in white colour. Muscle tissue is represented in black. Red and blue contours correspond to bone and muscle structures respectively in the groundtruth database.

3 Results

To assess the performance of the algorithm described throughout this paper, the dataset *Usevillabonemuscle* database [16] has been used. The training dataset was used to compute the reference histogram models and the test dataset was used to compute the performance results. In the work by Pérez-Carrasco et al. [13], comparison with several state-of-the-art techniques including level sets (both region-based and edge-based [5]) were performed. In this approach, we have compared the algorithm with two approaches. The first one is the method by Pérez-Carrasco et al. [13] and the second approach corresponds to the algorithm described in this paper but applying fine-tuned segmentation to the segmentation of all tissues instead of only voxels classified as muscle in the presegmentation stage 2.1. Performance metrics including Dice, Jaccard, PPV, specificity and sensitivity have been computed to compare the three techniques. In Table 1 the results provided by each technique are shown. In this table, the technique described in this paper is denoted as *Method 1* and the technique implemented in [13] is denoted as *Method 2*. As it can be seen the method described throughout this paper outperforms the method by [13] in 5 out of the 6 performance metrics.

Table 1. Performance metrics evaluated for the proposed and benchmark algorithms.

	Method 1 (Proposed)		Method 2 [13]	
Parameter	Bone	Bone	Bone	Muscle
Dice	0.88 ± 0.14	**0.82 ± 0.15**	0.88 ± 0.14	0.78 ± 0.12
Jaccard	0.80 ± 0.19	**0.71 ± 0.19**	0.80 ± 0.19	0.65 ± 0.16
PPV	0.90 ± 0.15	**0.79 ± 0.21**	0.90 ± 0.15	0.69 ± 0.18
Sensitivity	0.94 ± 0.15	0.88 ± 0.07	0.94 ± 0.15	**0.94 ± 0.04**
Specificity	1.00 ± 0.00	**0.96 ± 0.04**	1.00 ± 0.00	0.95 ± 0.04
Accuracy	0.99 ± 0.01	**0.96 ± 0.01**	0.99 ± 0.01	0.95 ± 0.03
Comput. Time (per slice)	52.79 ± 29.23		**51.47 ± 30.45**	

4 Conclusions

An algorithm based on statistical information (histograms) that minimizes a convex-relaxed energy function [13] has been used to perform the automatic segmentation of bone and muscle structures in generic CT volumes. In a second stage, a fine-tuned segmentation step has been performed by using probability maps in order to discard other non-muscle structures that are close in terms of Hounsfield values. Average filtering and morphological operation were finally carried out to remove noise and unconnected regions. The tool has been validated using a public databased composed by 27 CT volumes belonging to 20 different patients and different metrics have been computed to measure the performance of the algorithm.

The algorithm outperformed the method described in [13] and thus other state-of-the-art techniques including thresholding and level sets which were compared by Pérez-Carrasco et al. in [13]. As the algorithm is based on the technique described in [13] it suffers of the same problem: a high computational time to create the Histogram Distance Images.

Some of the problems encountered during the segmentation and that justify poor results in some cases are that the groundtruth segmentation is only focused on skeletal muscle, whereas some organs (bladder, stomach, etc.) are also composed partly or entirely of muscular tissue and where not considered in the groundtruth.

Acknowledgment. This research has been supported by project DPI2016-81103-R.

Conflict of Interest. The authors declare that they have no conflict of interest.

References

1. Feeman, T.G.: The Mathematics of Medical Imaging: A Beginner's Guide. Springer Undergraduate Texts in Mathematics and Technology. Springer (2010). https://doi.org/10.1007/978-3-319-22665-1
2. Zou, D., Li, W., Deng, C. et al.: The use of CT Hounsfield unit values to identify the undiagnosed spinal osteoporosis in patients with lumbar degenerative diseases. Eur. Spine. J (2018). https://doi.org/10.1007/s00586-018-5776-9
3. LimFat, D., Kennedy, J., Galvin, R., et al.: The Hounsfield value for cortical bone geometry in the proximal humerus–an in vitro study. Skeletal Radiol. **41**(5), 557–568 (2012). https://doi.org/10.1007/s00256-011-1255-7
4. Haralick, R.A., Shapiro, L.: Computer and Robot Vision, vol. 1. Addison-Wesley Publishing Company, Boston (1992)
5. Li, C., Xu, C., Gui, C., Fox, M.D.: Distance regularized level set evolution and its application to image segmentation. IEEE Trans. Image Process. **19**(12), 3243–3254 (2010)
6. Andersson, T., Läthén, G., Lenz, R., Borga, M.: A fast optimization method for level set segmentation. In: Scandinavian Conference on Image Analysis. LNCS, vol. 5575, pp. 400–409 (2009)
7. Boykov, Y., Funka-Lea, G.: Graph cuts and efficient N-D image segmentation. Int. J. Comput. Vision **70**(2), 109–131 (2006)
8. Klemencic, A., Kovacic, S., Pernus, F.: Automated segmentation of muscle fiber images using active contour models. Cytometry **32**(4), 317–326 (1998)
9. Iglovikov, V., Rakhlin, A., Kalinin, A., Shvets, A.: Paediatric bone age assessment using deep convolutional neural networks. In: Proceedings of the International Workshop on Deep Learning in Medical Image Analysis (DLMIA 2018). LNCS, vol. 11045, pp. 300–308 (2018)
10. Lee, H., Troschel, F.M., Tajmir, S., et al.: Pixel-level deep segmentation: artificial intelligence quantifies muscle on computed tomography for body morphometric analysis. J. Digit. Imaging **30**(4), 487–498 (2017). https://doi.org/10.1007/s10278-017-9988-z
11. Jeffrey, W., Thomas, M., et al.: Anatomically anchored template-based level set segmentation: application to quadriceps muscles in mr images from the osteoarthritis initiative. J. Digit. Imaging **24**(1), 28–43 (2011)
12. Ganesan, K., Acharya, U.R., Chua, K.C., Min, L.C., Abraham, K.T.: Pectoral muscle segmentation: a review. Comput. Methods Programs Biomed. **110**(1), 48–57 (2013)
13. Pérez-Carrasco, J.A., Acha, B., et al.: Joint segmentation of bones and muscles using an intensity and histogram-based energy minimization approach. Comput. Methods Programs Biomed. **156**, 85–95 (2018). https://doi.org/10.1016/j.cmpb.2017.12.027
14. Yuan, J., Bae, E., Tai, X.C.: A study on continuous max-flow and min-cut approaches. In: Proceedings of the IEEE Computer Society Conference on Computer Vision and Pattern Recognition, pp. 2217–2224 (2010)
15. Punithakumar, K., Yuan, J.: A convex max-flow approach to distribution-based figure-ground separation. SIAM J. Imaging Sci. **5**(4), 1333–1354 (2012)
16. Usevillabonemuscle database. http://grupo.us.es/grupobip/research/research-topics/segmentation-of-abdominal-organs-and-tumors/. Accessed 26 Mar 2019

A Comparison of Denoising Algorithms for Effective Edge Detection in X-Ray Fluoroscopy

Emilio Andreozzi[1(✉)], Maria Agnese Pirozzi[1,3], Antonio Sarno[4], Daniele Esposito[1], Mario Cesarelli[1,2], and Paolo Bifulco[1,2]

[1] Department of Electrical Engineering and Information Technologies, University of Naples Federico II, Naples, Italy
emilio.andreozzi@unina.it
[2] Istituti Clinici Scientifici Maugeri S.P.a. – Società Benefit, Pavia, Italy
[3] Institute of Biostructures and Bioimaging, Italian National Research Council, Naples, Italy
[4] National Institute for Nuclear Physics, Naples, Italy

Abstract. X-ray fluoroscopy provides various diagnosis and is widely used in interventional radiology. However, the low-dose involved in fluoroscopy generates an intense Poisson-distributed quantum noise. Object recognition and tracking help in many fluoroscopic applications. Edge-detection is essential, but common derivative operators require noise suppression to provide reliable results. Moreover, homoscedasticity of noise is generally assumed, but is not the case of fluoroscopic images. However, the Anscombe transform can stabilize the quantum noise variance. This study presents a comparison of two denoising algorithms to evaluate their performance in edge-detection for real fluoroscopic sequences. VBM4D is one of best video-processing method for Additive White Gaussian Noise (AWGN), while Noise Variance Conditioned Average (NVCA) is a recent, real-time, algorithm specifically tailored for fluoroscopy. Some real fluoroscopic sequences screening the motion of lumbar spine were processed. Noise parameters were estimated using image sequences of a static scene: the relationship between the luminance and the noise variance was obtained. Generalised Anscombe transform and its inverse were applied to use the VBM4D algorithm. Edge-detection was performed by means of the Sobel operator. The Anscombe transform resulted able to stabilise the noise variance and consequently allow the use of algorithms designed for AWGN. The results show that both approaches provide effective identification of object contours (i.e. vertebral bodies). Despite of its simplicity the NVCA algorithm shows better performances than VBM4D on delineation of boundaries of examined spine fluoroscopic scenes. Furthermore, the NVCA algorithm can be realized in hardware and can offer real-time fluoroscopic processing.

Keywords: X-ray fluoroscopy · Poisson noise · Video denoising · Edge detection

J. Henriques et al. (Eds.): MEDICON 2019, IFMBE Proceedings 76, pp. 405–413, 2020.
https://doi.org/10.1007/978-3-030-31635-8_49

1 Introduction

Low-dose X-ray imaging is a widespread tool in diagnostic [1] and interventional radiology [2]. Fluoroscopy is commonly used to obtain moving images of patients' internal structures. It is an invaluable tool for image-guided surgery [2–4], because is able to track different objects, such as surgical instruments, implantable devices, catheters. Diagnostic applications of fluoroscopy include angiography to image blood vessels using contrast agent, digestive tract examination, orthopedy to evaluate joint motion, vertebral kinematics, etc. [1, 5]. For example, in intervertebral kinematics analysis, fluoroscopy supports quantitative measurement of relative movements between vertebral bodies [5]. This is particularly helpful for the detection of abnormal intervertebral motion in cervical spine pathologies [6] and also for the assessment of spinal segments functionality, whose alterations give information about lumbar spine diseases [5].

To allow patient prolonged exposition, fluoroscopic devices use very low-dose X-ray and image intensifier. However, the lower is the dose the lower is the SNR in fluoroscopic images, degraded by the so-called "quantum noise", a Poisson-distributed, signal-dependent noise emerging from the limited availability of X-ray photons per pixel [5]. This noise is unavoidable and independent from detector performances since it represents the stochastic nature of the photon counting process in the detectors, and so it is intrinsic in the image formation process [7]. Although other noise sources generally contribute to image deterioration, such as thermal noise and A/D conversion noise, in low-dose X-ray images the most dominant contribution to the decrease of signal-to-noise ratio (SNR) is by far the quantum noise [8, 9, 11].

Over the years, several image-processing techniques have been developed to improve image SNR, to enhance image vision and to track objects. Edge-detection plays a pivotal role in most of these techniques, allowing to identify internal structures and to track their movements [12].

Abrupt intensity changes in X-ray images characterise object boundaries. These sharp discontinuities usually correspond to different absorption of projected object. Examples are the contrast between bone and soft tissues, surgical instruments and patient's organs, contrast enhanced structures and surrounding tissues. Common edge-detectors (directional or isotropic) filter out useless data preserving the important structural properties in the image and extracting the edges [13, 14]. A first approach for edge-detection is to use discrete approximations of first-order partial derivatives of the digital image by means of convolution with very small edge-enhancing kernels (e.g. Prewitt, Sobel, Robert, Kirsch) [15, 16]. Alternatively, second-order derivative-based operators can be used, which are approximated by small convolution masks acting as digital Laplacians. These operators are very sensitive to noise because they act as poor digital high-pass filters that amplify high-frequency noise in the resulting images. Therefore, most known Marr-Hildreth [17] and Canny [18] second-order edge-detectors and others [19–22] combine the edge-enhancing kernels with smoothing operators to filter out high-frequency noise, which introduces spurious edges.

In addition, the smoothing operators generally assume homoscedasticity of the noise (i.e. homogeneity of the variances associated to each grey-level), thus making the hypothesis of signal-independent, Additive White Gaussian Noise (AWGN). This is generally not met in fluoroscopic images, since quantum noise is heteroscedastic, as it is signal-dependent because Poisson-distributed.

The aim of this study is the comparison of the performances of two alternative denoising strategies for effective edge-detection in fluoroscopic images.

2 Materials and Methods

2.1 Noise Parameters Estimation

Noise variance estimation is required by almost all denoising algorithms. Since Poisson noise variance depends on the expected value (i.e. the grey level in the X-ray images), it is necessary to identify their functional relationship for the actual noisy image, in order to estimate noise variance associated with every pixel to be filtered. X-ray low-dose noise can be modelled as a Poisson-Gaussian mixture expressed in Eq. 1:

$$G = a \cdot P(\lambda) + N(0, \sigma) \tag{1}$$

where G is the generic pixel grey-level, $P(\lambda)$ is the Poisson distribution with mean and variance λ (average X-ray photons per pixel), $N(0, \sigma)$ is a zero-mean Gaussian distribution and a is the scaling factor, corresponding to the detector gain.

According to Eq. 1, the mean-variance relationship is linear (heteroscedasticity). Its slope and intercept correspond, respectively, to the detector gain and the variance of the Gaussian component.

$$\sigma_G^2 = a \cdot \mu_G + \sigma^2 \tag{2}$$

Noise parameters estimation can be performed in real fluoroscopic sequences from multiple frames of a static scene, because variances of pixels grey levels can be completely ascribed to noise. Therefore, noise parameters can be estimated by computing mean and variance of luminance values assumed by every pixel along the temporal dimension, and then applying a linear regression to evaluate slope and intercept of the mean-variance relationship.

2.2 Noise-Variance Conditioned Average

The NVCA algorithm [7] performs a conditioned spatio-temporal moving average by excluding pixels which do not belong to the local Poisson noise statistics in order to avoid blurring effects and to preserve edges. Pixels in the filter mask are excluded from average if their absolute difference in luminance with the central pixel is greater than a selected threshold, chosen as a multiple of the noise standard deviation corresponding to the grey level of the central pixel. This approach outperformed several state-of-the-art denoising algorithms [23], while maintaining an acceptable computational effort, which allowed hardware real-time implementations [25, 26].

2.3 Variance Stabilisation and Application of AWGN Denoising Algorithm

In literature some techniques have been proposed, which are based on stabilisation of Poisson's noise variance by means of Anscombe transform, in order to obtain an approximation of an AWGN distribution. This allows to use all those denoising methods designed with the homoscedastic AWGN hypothesis, such as Wiener filtering, Non-local Means [27] and VBM4D [28], among others. To obtain the estimated noise-free images, the Anscombe inverse transformation must be applied. However, this introduces a systematic error (bias) for very low intensity values, so it is better to use the asymptotically unbiased inverse or the exact inverse unbiased approximation [29]. In X-ray images, the slope of the mean-variance relationship is modified by detector gain, so the Anscombe transform stabilises the variance to the actual value of the slope. In this case, the generalised Anscombe transform can be used, which assumes for the noise the Poisson-Gaussian mixture model expressed in Eq. 1, requiring the estimation of the noise parameters.

2.4 Processing of Fluoroscopy Sequences

Performance comparison between NVCA and VBM4D algorithms was carried out by processing real fluoroscopic sequences of the lumbar spine.

In order to maximise the efficiency of the noise suppression, noise parameters were estimated on a static scene extracted from the fluoroscopic sequences.

The real sequences were processed with NVCA algorithm (mask size $11 \times 11 \times 7$, noise SD multiplier 3) by providing the estimated noise parameters.

Generalised Anscombe transform was applied to the fluoroscopic sequences considering the previous noise estimation. Then the VBM4D algorithm was applied to the transformed sequences, assuming a noise SD equal to 1. Finally, the generalised Anscombe inverse transform was applied.

Edge detection was performed by applying Sobel operator to the denoised sequences.

3 Results

As example, Fig. 1 shows the mean-variance relationships of one fluoroscopic sequence, before and after the Anscombe Transformation. The black straight line represents the linear regression of the experimental data, which are plotted in grey.

Figure 2 reports the comparison between noisy and filtered sequences on the same single frame. In Fig. 2a a selected Region of Interest (ROI – including L2 vertebral body) is depicted, which is used in Fig. 3 to compare the results of edge detection applied to the fluoroscopic sequences.

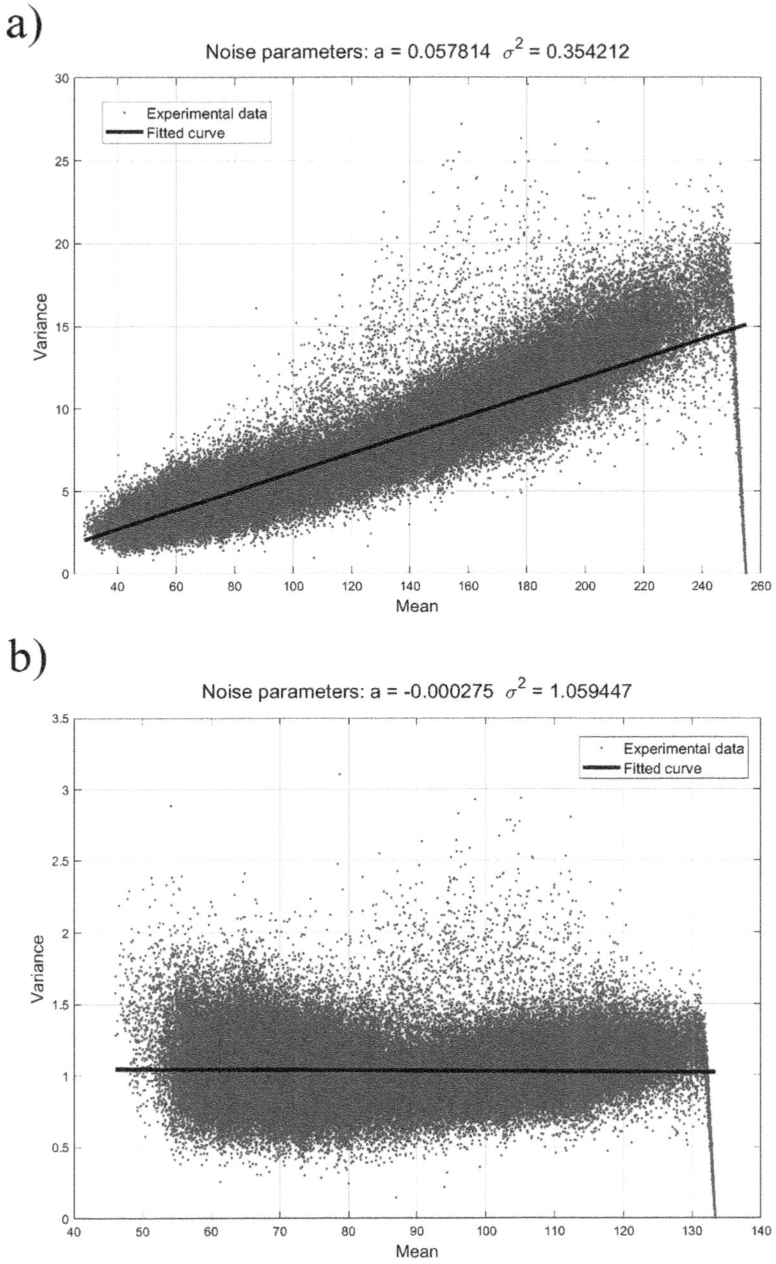

Fig. 1. Mean-variance relationships and noise parameters for: (a) original noisy image; (b) noisy image after generalised Anscombe transformation.

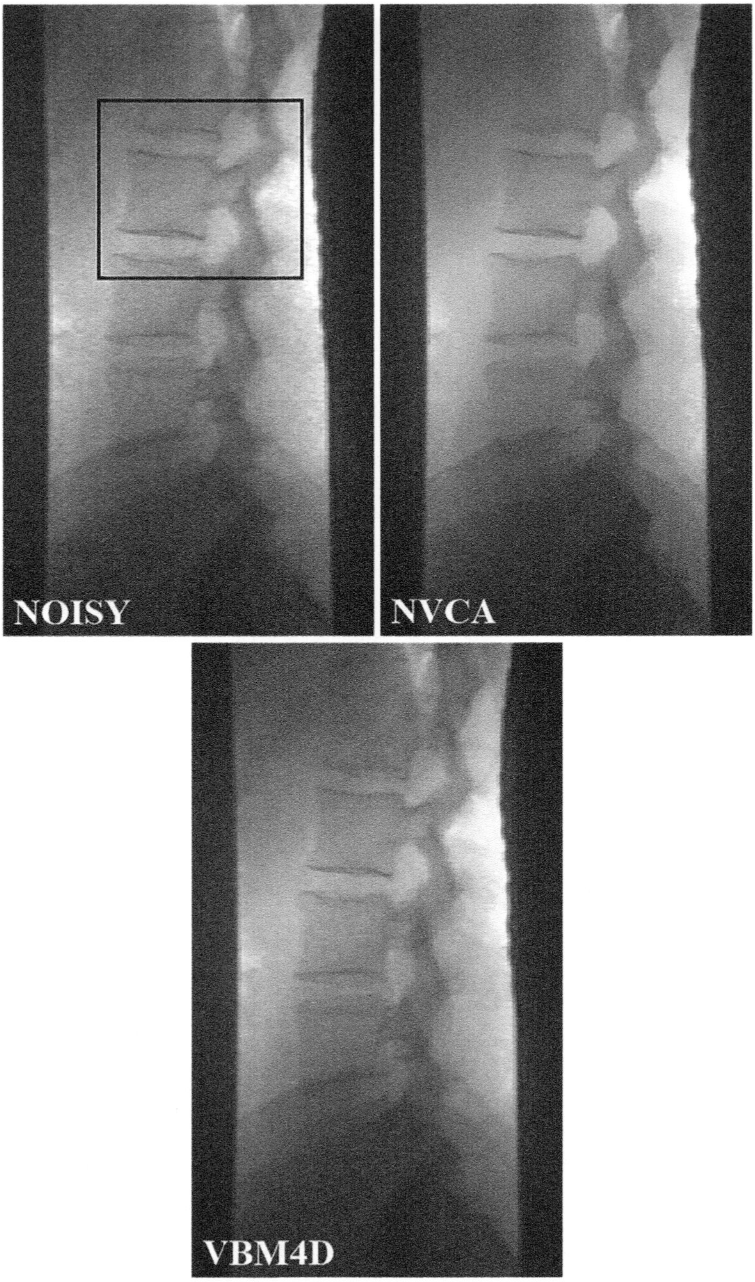

Fig. 2. Comparison of a single frame from: (a) noisy sequence with selected ROI; (b–c) sequences filtered with NVCA and VBM4D.

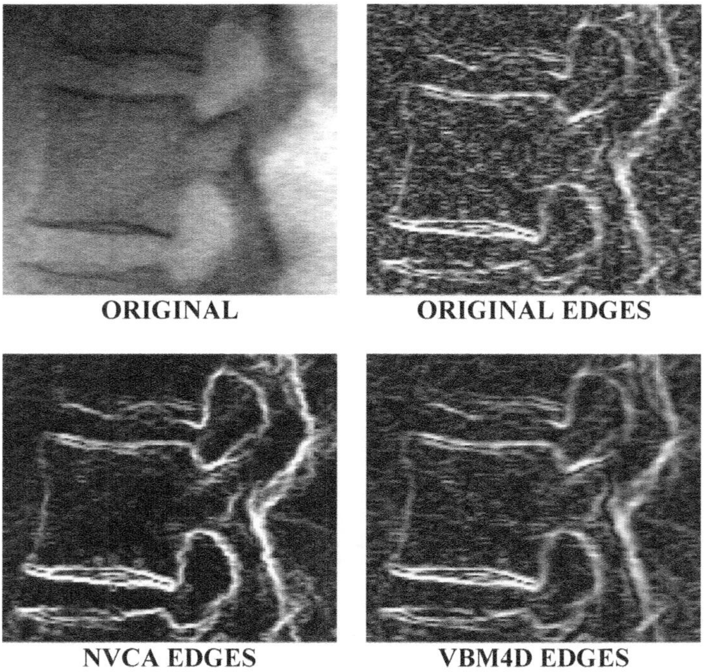

Fig. 3. (a) ROI depicted in Fig. 2a; (b–d) edges extracted in the ROI from the noisy and the filtered sequences.

4 Conclusions

This study compares the performance of VBM4D (nowadays, one of best video-processing methods) and NVCA algorithms in fluoroscopy noise suppression to allow effective edge-detection. Because of their intense noise, fluoroscopic sequences must be pre-processed before applying edge-detection. The results show that both approaches provide effective identification of object contours (i.e. vertebral bodies). Despite of its simplicity, which makes it suitable for real-time hardware implementations, the NVCA algorithm performed better than VBM4D in denoising the examined spine fluoroscopic scenes, allowing a more accurate delineation of vertebral bodies boundaries.

Conflict of Interest. The authors declare that they have no conflict of interest.

References

1. Bifulco, P., Cesarelli, M., Cerciello, T., Romano, M.: A continuous description of intervertebral motion by means of spline interpolation of kinematic data extracted by video fluoroscopy. J. Biomech. **45**(4), 634–641 (2012). https://doi.org/10.1016/j.jbiomech.2011.12.022

2. Weese, J., Penney, G.P., Desmedt, P., Buzug, T.M., Hill, D.L.G., Hawkes, D.J.: Voxel-based 2-D/3-D registration of fluoroscopy images and CT scans for image-guided surgery. IEEE Trans. Inf. Technol. Biomed. **1**(4), 284–293 (1997). https://doi.org/10.1109/4233.681173
3. Yamazaki, T., et al.: Improvement of depth position in 2-D/3-D registration of knee implants using single-plane fluoroscopy. IEEE Trans. Med. Imaging **23**(5), 602–612 (2004)
4. Wang, J., Zhu, L., Xing, L.: Noise reduction in low-dose X-ray fluoroscopy for image-guided radiation therapy. Int. J. Radiat. Oncol. Biol. Phys. **74**(2), 637–643 (2009)
5. Cerciello, T., Romano, M., Bifulco, P., Cesarelli, M., Allen, R.: Advanced template matching method for estimation of intervertebral kinematics of lumbar spine. Med. Eng. Phys. **33**(10), 1293–1302 (2011). https://doi.org/10.1016/j.medengphy.2011.06.009
6. Bifulco, P., Cesarelli, M., Romano, M., Fratini, A., Sansone, M.: Measurement of intervertebral cervical motion by means of dynamic x-ray image processing and data interpolation. Int. J. Biomed. Imaging **2013**, 152920 (2013). https://doi.org/10.1155/2013/152920. Published online 31 Oct 2013
7. Cesarelli, M., Bifulco, P., Cerciello, T., Romano, M., Paura, L.: X-ray fluoroscopy noise modeling for filter design. Int. J. Comput. Assist. Radiol. Surg. 8(2) (2012). https://doi.org/10.1007/s11548-012-0772-8
8. Ma, L., Moisan, L., Yu, J., Zeng, T.: A dictionary learning approach for poisson image deblurring. IEEE Trans. Med. Imaging **32**(7), 1277–1289 (2013). https://doi.org/10.1109/TMI.2013.2255883
9. Lefkimmiatis, S., Maragos, P., Papandreou, G.: Bayesian inference on multiscale models for Poisson intensity estimation: applications to photon-limited image denoising. IEEE Trans. Image Process. **18**(8), 1724–1741 (2009). https://doi.org/10.1109/TIP.2009.2022008
10. Tapiovaara, M.J.: SNR and noise measurements for medical imaging: II. application to fluoroscopic X-ray equipment. Phys. Med. Biol. **38**(2), 1761–1788 (1993)
11. Aufrichtig, R., Wilson, D.L.: X-ray fluoroscopy spatio-temporal filtering with object detection. IEEE Trans. Med. Imaging **14**(4), 733–746 (1995). https://doi.org/10.1109/42.476114
12. Goswami, B., Misra, S.K.: Analysis of various edge detection methods for x-ray images. In: 2016 International Conference on Electrical, Electronics, and Optimization Techniques (ICEEOT), pp. 2694–2699 (2016)
13. Juneja, M., Singh Sandhu, P.: Performance evaluation of edge detection techniques for images in spatial domain. Int. J. Comput. Theory Eng. **1**(5), 1793–8201 (2009)
14. Bhardwaj, S., Mittal, A.: A survey on various edge detector techniques. Procedia Technol. **4**, 220–226 (2012)
15. Pratt, W.K.: Digital Image Processing, 4th edn. Wiley (2007)
16. Bovik, A.: The Essential Guide to Image Processing. Academic Press (2009)
17. Marr, D., Hildreth, E.: Theory of edge detection. Proc. R. Soc. Lond. B Biol. Sci. **207**, 187–217 (1980)
18. Canny, J.F.: A computational approach to edge detection. IEEE Trans. PAMI **8**(6), 679–698 (1986)
19. Haralick, R.M., Lee, J.: Context dependent edge detector. In: Proceedings of CVPR 1988, pp. 223–228, Ann Arbor, Michigan (1988)
20. Deriche, R.: Fast algorithms for low-level vision. IEEE Trans. PAMI **12**(1), 78–87 (1990)
21. Shen, J., Castan, S.: Edge detection by sign correspondence for zero-crossings. Actes du Premier Colloque Image' COrn, Bordeaux, pp. 279–284 (19–21 Nov 1990)
22. Shen, J., Castan, S.: An optimal linear operator for step edge detection. CVGIP Graph. Models. Image Process. **54**(2), 112–133 (1992)

23. Cerciello, T., Bifulco, P., Cesarelli, M., Fratini, A.: A comparison of denoising methods for X-ray fluoroscopic images. Biomed. Signal Process. Control **7**, 550–559 (2012). https://doi.org/10.1016/j.bspc.2012.06.004
24. Sarno, A., Andreozzi, E., De Caro, D., Di Meo, G., Strollo, A.G.M., Cesarelli, M., Bifulco, P.: Real-time algorithm for Poissonian noise reduction in low-dose fluoroscopy: performance evaluation. BioMed Eng OnLine (Article in Press). https://doi.org/10.1186/s12938-019-0713-7
25. Genovese, M., Bifulco, P., De Caro, D., Napoli, E., Petra, N., Romano, M., Cesarelli, M., Strollo, A.G.M.: Hardware implementation of a spatio-temporal average filter for real-time denoising of fluoroscopic images. J. VLSI **49**, 114–124 (2015). https://doi.org/10.1016/j.vlsi.2014.10.004
26. Castellano, G., De Caro, D., Esposito, D., Bifulco, P., Napoli, E., Petra, N., Andreozzi, E., Cesarelli, M., Strollo, A.G.M.: An FPGA-oriented algorithm for real-time filtering of poisson noise in video streams, with application to x-ray fluoroscopy. Circuits Syst. Signal Process. (2019). https://doi.org/10.1007/s00034-018-01020-x
27. Bindilatti, A.A., Mascarenhas, N.D.A.: A non local poisson denoising algorithm based on stochastic distances. IEEE Signal Process. Lett. **20**(11), 1010–1013 (2013). https://doi.org/10.1109/LSP.2013.2277111
28. Maggioni, M., Boracchi, G., Foi, A., Egiazarian, K.: Video denoising, deblocking and enhancement through separable 4-D nonlocal spatiotemporal transforms. IEEE Trans. Image Process. **21**(9), 3952–3966 (2012)
29. Makitalo, M., Foi, A.: Optimal inversion of the anscombe transformation in low-count poisson image denoising. IEEE Trans. Image Process. **20**, 99–109 (2011)

Stereophotogrammetric Basic Framework for Postural Assessment

Alice Fontes(✉) and Mauricio Cagy

Biomedical Engineering Program - Coppe, Federal University of Rio de Janeiro, Rio de Janeiro, Brazil
fontes.alice@poli.ufrj.br, mcagy@peb.ufrj.br

Abstract. Deviations of the vertebral column have become increasingly common. The follow-up of a spine deviation is performed every 2 or 3 months by routine visits and X-ray examinations. X-ray examination produces ionising radiation, which, over time, can cause cancer and is accumulative: this risk increases with the number of exposures. This work describes the conception and initial development of a simple and portable system based on stereoscopy capable of performing a topographic analysis of the patient's back, as well as 3D reconstruction, in order to subsequently allow postural assessment. The stereoscopic system was built and calibrated, then the images underwent a process of rectification and point matching to create a depth map, so that preliminary results point to adequate 3D reconstruction of the back of a plastic phantom. The advantage of this method is the minimization of parallax errors, low cost, easy implementation and portability.

Keywords: Stereo imaging · Postural assessment · Three-dimensional reconstruction

1 Introduction

According to the World Health Organization (WHO), 80% of people will have back pain throughout their lives, possibly as result of postural deviations. During the treatment of a deviation in the spine, it is necessary to observe and follow up. On average, during follow-up, the re-evaluation of the patient's chart is done through routine consultations and requesting new imaging tests every two or three months, verifying the need to change or adapt the treatment [1].

X-ray exams produces ionising radiation, which is a form of radiation with the potential of damaging living tissue, altering the DNA cells and possibly causing cancer. This risk increases with the number of exposures added over the life of the individual, since the radiation is cumulative. Thus, there is a medical concern to submit patients to as few X-ray exams as possible, always balancing it's cost and benefit [2]. Experts give a safe margin, considering tests done with well-calibrated equipment and trained professionals: up to five X-ray examinations per year [2].

As non-ionising alternatives to radiographic examination, many posture assessment approaches have been used. A previous study [3] reviewed some alternatives based on regular photogrammetry, with photos taken of the patient from the side and from the front

J. Henriques et al. (Eds.): MEDICON 2019, IFMBE Proceedings 76, pp. 414–420, 2020.
https://doi.org/10.1007/978-3-030-31635-8_50

in orthostatic position. This review also shows the difficulty to measure some angles and deviations based only on the two-dimensional photos. Furthermore, the review shows that none of the studies reported intra-rater reliability of some angles. Another work [4] addressed studies using different methods for evaluation of body posture used in nowadays researches. Ten evaluation methods were found using different postural assessment methods such as photography, digital photogrammetry, biophotogrammetry and posturogram, all mainly based on two-dimensional photography.

A special case of photogrammetry, the stereophotogrammetry, allows the three-dimensional location of reference points based on two images captured simultaneously from different viewpoints [5, 6]. This technique used to obtain information from three-dimensional space through the analysis of two images obtained at different points is called stereoscopy.

There are commercial systems that uses 3D analysis. As an example, one can cite the "TEMPLO 3D Posture Analysis". Through three cameras, a marked poster and with the prior marking of anatomical points in the patient with small reflective spheres, the system performs the postural analysis [7]. Some disadvantages of this system are the high cost and the need for prior marking of points on the patient's skin.

Previously to the present work, another system for postural analysis was developed by our group. The system was also stereoscopic, but based on previously marking anatomical points in the patient with small reflexive spheres, which serves as points of interest to perform the postural analysis [8]. Only the 3D positions of these marks are estimated in the reconstruction. Although simple and low-cost, such a system requires the prior definition of the anatomical points where the marks will be affixed, like the commercial system cited above. Besides, such marks must be built using reflexive tapes, not easily available in commerce.

In this context, this work aimed at describing the conception and initial development of an even simpler stereoscopic system (aiming at low cost and ease of implementation) able to produce depth maps and a 3D visualization of the patient's back in order to configure a basic framework to yield 3D posture assessment. This system was based on a well-established computer vision library (OpenCV), and does not require the use of reflexive marks. The 3D reconstruction is performed for the entire surface of the patient's back, so that the specialist can later mark the points of interest (using a mouse, for example) in the reconstructed image to calculate the angles of postural deviations.

2 Materials and Methods

The stereoscopic system was built using two low-cost webcam-type cameras connected to a notebook. For camera calibration, one used a sheet with a printed chessboard pattern, attached to a clipboard. For the code development, the programming language Python version 2.7 was used together with the computer vision library OpenCV version 3.4. A clothing mannequin was used as a phantom for tests.

The steps involved in this system are: (*i*) simple and stereo calibration of the two cameras, (*ii*) capture still images of the patient's back using the previous calibrated cameras, (*iii*) rectify and remap the images and, (*iv*) based on these remapped images, generate a depth map and a visualization of the patient's back in 3D.

Based on the correspondence of the pixels related to the points of interest in the photos of the patient's back taken by the two cameras, their respective positions in the three-dimensional space can be further estimated, allowing a diagnosis based on these measurements.

2.1 Simple Calibration

As the camera transforms the 3D world into a 2D image, some notions are lost. Thus, analogous to stereoscopic vision, stereophotogrammetric systems require two cameras (or more), which must be previously calibrated. The calibration of a camera is defined as the process performed to calculate the value of intrinsic and extrinsic parameters and lens distortion coefficients. It is done by observing an object that has a well-known geometry in the 3D space. The goal is to determine a set of camera parameters that reflect the transformation of world coordinates (in meters) to the image system (pixels). The calibration can be performed in several ways [9].

The calibration method chosen in this work is performed using the chessboard pattern method. In this method, a chessboard pattern is moved within the work volume, while acquiring multiple images. After this, using the OpenCV library, algorithms based on the Zhang method [6] are used to detect points of interest in the images and to extract the intrinsic, extrinsic and distortion parameters of the cameras. A sheet with the chessboard pattern can be easily printed anywhere, following the requisites of low cost and portability.

2.2 Stereo Imaging

After calibrating each camera, it is possible to relate points of the physical world with pixels of the image. To reproduce the stereo image, initially, two cameras capture the image of the object of interest at the same time. Then, the features present in an image are matched with the corresponding features in the other image, and the differences are analysed to generate depth information. The disparity effects (triangulation) are used as a means of calculating the distance of objects [9].

The stereo imaging involves the following steps:

Undistortion. Mathematically remove radial and tangential lens distortion. The outputs of this step are undistorted images.

Stereo Calibration. Performed after simple calibration (used as initial "guess"), the stereo calibration is the process of computing the geometric relation between the two cameras in space [9]. This means finding the rotation matrix and the translation vector between the two cameras.

Rectification. In the real world, cameras are almost never exactly parallel and aligned. Instead, image projections and distortion maps that rectify the left and right images can be mathematically found. The goal is mathematically align the two cameras in a single view plane, so that the pixel lines between the cameras are exactly aligned. The purpose of stereo rectification is to redesign the imaging planes of the two cameras so that they reside on the same plane with perfectly aligned image lines.

Stereo Correspondence. This is the process to find the same features in the left and right camera views. The output of this step is a disparity map, where disparities are the differences in the image plane of the same resource viewed in the left and right cameras [9]. OpenCV implements a fast and effective stereo block matching algorithm, called "StereoBM". This algorithm works by using small "sum absolute difference" (SAD) windows to find points between the left and right rectified stereo images. This algorithm finds only strongly corresponding points (high texture) between the two images. In a highly textured scene, as can occur outdoors, each pixel can have computed depth. In a scene with little texture, very few points can record depth.

Reprojection. Knowing the geometric arrangement of the cameras, the map of disparity can be transformed into distances by triangulation [9]. The output of this step is a depth map.

3 Results

Preliminary results were obtained by testing multiple phantom images acquired in different scenarios with multiple light incisions; it became clear that acquiring a good image is essential for a good depth map and good 3D reconstruction. A depth map obtained with the phantom's images and a 3D reconstruction of the phantom are shown in Figs. 1 and 2 respectively.

The phantom used for testing is made of a white and shiny plastic, a very reflexive material. Different points of light incidence were tested in order to try minimizing this reflection.

As practical considerations, the stereoscopic system must be constructed in order to ensure the best parallelism of the cameras as possible. This physical alignment makes mathematical transformations more treatable. Also, when capturing the images that will be used for 3D reconstruction and depth map, the object to be reconstructed cannot be very close to the background. Besides, a good contrast between the object and the background must be guaranteed, for example: light object and dark background. The background should be as clean as possible, without much information that might generate interference. It should be checked whether the image to be captured is well lit.

When capturing the images for calibration, the chessboard was positioned differently in 20 photos, trying to cover the whole area reached by the cameras and to avoid capturing any blurred photos. In addition, should be ensured that the chess pattern is being fully captured by both cameras, and all chess pattern photos should be different from each other. Then, the images of the phantom were captured varying the colour and distance of the background wall.

The selection of parameters for the construction of the depth map configured a critical step. For each scene type, the OpenCV function's parameters should be modified. When the background is dark and the object is lighter, the best parameters, which optimise the depth map, are different from the parameters for a scene with light background and dark object. Different types of lighting also interfere in this choice. Table 1 lists the selected parameters for the function "StereoBM" considering dark background.

Table 1. "Stereo BM" parameter values.

Parameter	Value
minDisparity	0
numDisparities	160
blockSize	1
disp12MaxDiff	2
uniquenessRatio	1
preFilterCap	20

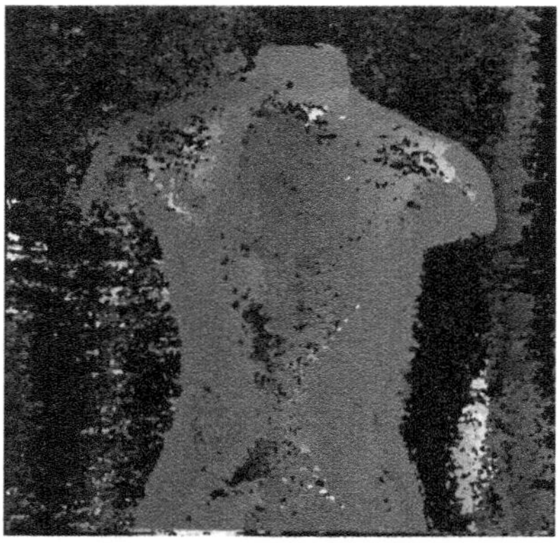

Fig. 1. Depth map obtained with the phantom's images.

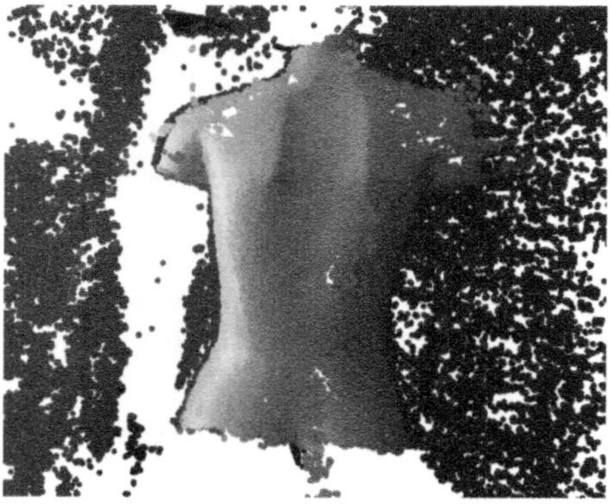

Fig. 2. 3D reconstruction of the phantom.

3.1 Error Analysis

Since future measurement of postural angles will rely on the accuracy in proportionality of measurements, a preliminary evaluation of the accuracy of the method was conducted based on comparing the ratio of two measures (one vertical: neck to coccyx, and one horizontal: shoulder to shoulder) in the real phantom (ratio of 1.39) and in the reconstructed 3D image (1.41), consisting of an error of circa 2%.

4 Discussion and Conclusion

The stereoscopic system has been successfully built according to the basic conception, taking into account the requisites that need to be met for the success and low cost of the system. The stereoscopic system calibration was implemented with a simple, portable and inexpensive technique that demands only a sheet with a printed chessboard pattern. The algorithm capable of generating a depth map and 3D reconstruction was successfully implemented, and preliminary tests were performed to identify the best parameters in order to optimise them.

The calibrated system was applied in a postural phantom to produce the depth map and 3D reconstruction of the phantom. In this way, the 3D reconstruction and the depth map were obtained, making it possible to recognize the overall topography of the patient's back. A preliminary error assessment (considering only frontal plane) has shown a measurements ratio error of circa 2%, which is acceptable for postural angle measurements if compared to inter and intra-rater variability commonly found in simple photogrammetric methods [3]. Nevertheless, due to variations in light reflectance and background contrast, 3D reconstruction presents gaps along the estimated surface, which is potentially deleterious for anatomical analysis. Notwithstanding, one should point out that this work is still under development, so that further stronger results must still be obtained.

Thus, a simple and portable system based on stereoscopy was proposed, capable of performing a topographic analysis of the patient's back, as well as 3D reconstruction, in order to perform a postural assessment. Aiming at simplicity and low cost, this system was designed so that it can be easily implemented in orthopaedic and physiotherapeutic clinics.

Considering robustness for this system, methods to produce adequate light contrast still must be investigated. Later on, an user interface allowing to mark points in anatomical protuberances over the back of the patient shall be developed to yield significant three-dimensionally analysis. Thus, it will be possible to calculate postural angles of interest, helping the specialist in the diagnosis of postural deviations such as scoliosis, kyphosis and lordosis, as well as other postural parameters such as asymmetries and inclinations.

Conflict of Interest. The authors declare that they have no conflict of interest.

References

1. Aleixo, C., Neves, N.: Escoliose degenerativa. Rev. Port. De Ortop. E Traumatologia **21**(3), 271–284 (2013)
2. Lima, R., Afonso, J., Pimentel, L.: Raios-x: fascinação, medo e ciência. Quim. Nova **32**(1), 263–270 (2009)
3. Singla, D., Veqar, Z., Hussain, M.: Photogrammetric assessment of upper body posture using postural angles: a literature review. J. Chiropractic Med. **16**(2), 131–138 (2017)
4. Petermann, X., Meereis, E.: Postural body: a systematic review about assessment methods. MTP&Rehab. J. **14**, 273–282 (2016)
5. Borguese, N.A., Cerveri, P., Rogiroli, P.: A fast method for calibrating vídeo-based motion analysers using only a rigid bar. Med. Biol. Eng. Comput. **39**, 76–81 (2001)
6. Zhang, Z.: A flexible new technique for camera calibration. IEEE Trans. Pattern Anal. Mach. Intell. **22**(11), 1330–1334 (2000)
7. Contemplas. https://www.contemplas.com/posture_analysis_3d.aspx. Accessed 24 Apr 2019
8. Cagy, M., Fontes, A.: Análise Postural Ortostática Usando Estereofotogrametria. In: Nona Semana de Integração Acadêmica, vol. 11, pp. 1–4. Federal University of Rio de Janeiro, Rio de Janeiro (2018)
9. Bradski, G., Kaehler, A.: Learning OpenCV: Computer Vision with the OpenCV Library, 1st edn. O'Reilly, New York (2008)

Dermoscopic Image Segmentation: A Comparison of Methodologies

Paulina Vélez Núñez(✉), Carmen Serrano, Begoña Acha, and José Antonio Pérez-Carrasco

Escuela Técnica Superior de Ingeniería, Departamento de Teoría de la Señal y Comunicaciones, Universidad de Sevilla, 41092 Seville, Spain
pauvelnun@alum.us.es

Abstract. An accurate segmentation of pigmented lesions may improve classification results of Computer Aided Diagnosis (CAD) tools. Thus, finding a reliable segmentation methodology becomes crucial. During the past few years, many segmentation methodologies of dermoscopic images have been proposed. In this paper, a comparison between three methodologies is presented: semantic segmentation with SegNet, histogram-based segmentation via convex optimization and segmentation based on a Fully Convolutional Network (FCN). As a result of evaluating the segmentation results for 600 dermoscopic images from the Test set of ISIC-2017 database, the semantic segmentation provides a 90.12% of accuracy, followed by segmentation using histograms and Fully Convolutional Network, with 86,47% and 81,70% of accuracy, respectively.

Keywords: Dermoscopic images · Semantic segmentation · Convex optimization · Convolutional Neural Network

1 Introduction

The analysis of skin lesions has been a key area of study for decades because the skin cancer is one of the most aggressive and causes many deaths per year. According to the Skin Cancer Foundation, new melanoma cases will increase by 7.7% in 2019. The rate of survival from melanoma is directly related to early detection. Therefore, the accuracy of the CAD methods that help physicians to recognize melanoma from other benign lesions is crucial.

The lesion classification process is bounded to a good lesion segmentation [1]. Concerning the lesion segmentation, it becomes a challenging task due to the variety of the dermoscopic and non-dermocopic images to analyze. The presence of hair, bubbles, different luminance, blurry boundaries, blood vessels, scars or dealing with different skin color, turn the segmentation process into a very delicate and complex process.

Over the years, many techniques to successfully overcome the segmentation challenges have been applied, such as Otsu threshold [2], active contour model,

J. Henriques et al. (Eds.): MEDICON 2019, IFMBE Proceedings 76, pp. 421–426, 2020.
https://doi.org/10.1007/978-3-030-31635-8_51

deep learning [3,10], and the histogram statistical method [6]. In this paper we compare the performance of three methodologies that were tested over ISIC-2017 database.

Section 2 describes the three compared methodologies for dermoscopic image segmentation. Section 3 presents the results after evaluating each methodology. Finally, Sect. 4 is dedicated to expose the conclusions and further research.

2 Segmentation of Dermoscopic Images

2.1 Semantic Segmentation with SegNet

Semantic Segmentation allows to identify an object in an image by classifying each pixel into a labeled class, which is called pixel-wise labeling [4]. In order to achieve the semantic segmentation of dermoscopic images, a Convolutional Neural Network (CNN) architecture named SegNet was applied.

As it is described by Badrinarayanan et al. [4,5], SegNet admits as input a map of features or an image, and the accuracy of its prediction increases as more layers are added to the model. SegNet has a convolutional Encoder-Decoder and omits fully connected layers reducing the parameters of the encoder. The convolutions are performed with a filter bank. The last layer of the Decoder works as a soft-max classifier, which allow to obtain as output the predicted segmentation label for each pixel, where each label is associated to an existing class.

Other important characteristic of a SegNet is that all layers has full connectivity and each layer consists of 64 features, which remains constant.

In order to achieve the Semantic Segmentation of pigmented lesions, a SegNet architecture was created. Two classes were defined, namely healthy skin and lesion. The image size was set at $224 \times 224 \times 3$ (on RGB space). The encoder downsamples the image while the encoder upsamples it, both using the encoder depth D, that was set to 6.

The PixelClassificationLayer was removed from the original graph and replaced by a new Softmax layer, which classify pixel in two classes or labels. The final network has 98 layers, named as: one input layer, 54 encoders, 42 decoders and one softmax layer.

In order to train the SegNet, the stochastic gradient descent with momentum (SGDM) optimizer was applied and the parameters were set as follow: Momentum of 0.99, initial learning rate as 0.001, and weight decay (L2Regularization) as 0.005.

2.2 Histogram-Based Segmentation via Convex Optimization

To achieve the segmentation of dermoscopic images, the methodology proposed by Papadakis and Rabin [6] was applied. Their methodology consists in posing the problem of segmentation as a problem of minimization of a convex energy function. In this energy function, the distance between the histogram of each region within the image and histogram models are evaluated.

Two histogram models were defined for each dermoscopic image, one for the foreground (skin lesion) and one for the background (skin). The optimization and other parameters needed to implement this segmentation methodology remained the same as Papadakis et al. [6,7] .

2.3 Segmentation Based on a Fully Convolutional Network

Fully Convolutional Netwoks (FCN) possess convolutional layers that provide a wide information about global and local features of an image. Hence the deepest convolutional layers contain information of the global, abstract and conceptual features, whereas the lower convolutional layers can give information about the local structure, which is relevant for the segmentation process [8]. Likewise, convolutional layers can be used to obtain the image features [9].

In this paper, a set of images from the lower convolutional layers of VGG-16 network was extracted. Images from the first 4 convolutional layers were normalized and filtered by applying a Gaussian filter with sigma = 2, before being added. Finally, a threshold using Otsu method was applied for each image and morphological operations were applied to obtain a reliable mask as a previous step for the segmentation process.

3 Results

In order to compare the segmentation methodologies described above, the database from ISIC-2017 was considered. This database provides each image with its corresponding ground truth mask, allowing to obtain an accuracy reference to evaluate the results of each methodology.

To achieve the semantic segmentation, the SegNet was trained with 2000 images with their 2000 corresponding ground truth mask. To test each methodology, 600 images were used from the test set of images from ISIC-2017.

The parameters to choose the best methodology were: DICE [11], Jaccard (JACC) [11], Sensitivity (Se), Specificity (Sp) and Accuracy (Acc).

The three methodologies were implemented in Matlab, on a system with Intel Core I7-6700 HQ processor, 12 GB of RAM, and NVIDIA GEFORCE GTX card. While the Semantic Segmentation with SegNet shows the highest computational cost, the Segmentation based on a Fully Convolutional Network consumes 72.6% of that cost. Finally, Histogram-based segmentation via convex optimization consumes 12.33% of the total computational cost.

As it is shown, at Fig. 1, the three methodologies give a very approximated segmented image. To visualize the behavior of each methodology for each dermoscopic image at Fig. 1, Table 1 presents the parameters for each one.

Finally, after applying the three methodologies to the test set of the database ISIC-2017, composed by 600 dermoscopic images, the results shown at the Table 2 highlight the performance of SegNet as the best segmentation methodology for dermoscopic images.

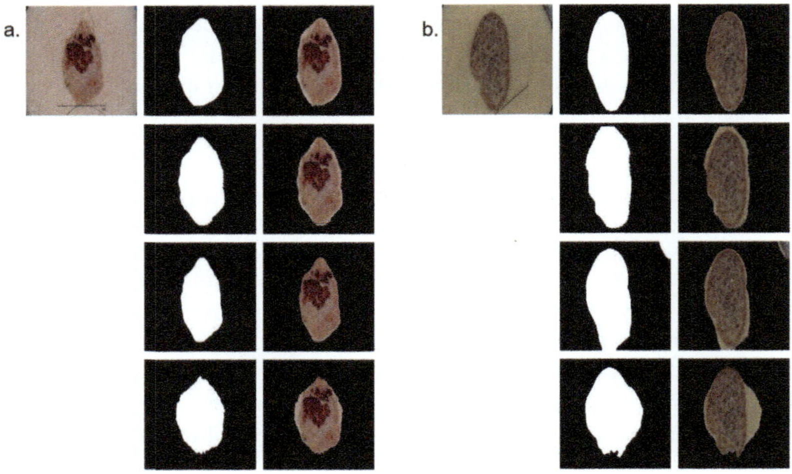

Fig. 1. Segmented Images. For images a, and b: first row shows the original image with its ground truth mask and its corresponding segmentation; second row shows the results of applying Semantic segmentation with SegNet; third row shows the result of Histogram-based segmentation via convex optimization; fourth row shows the result of segmentation based on a Fully Convolutional Network.

Table 1. Segmentation results for Images in Fig. 1.

Image	Methodology	DICE	JACC	Se	Sp	ACC
a.	Semantic segmentation	0.9647	0.9317	0.9599	0.9900	0.9824
	Segmentation by histograms	0.9116	0.8376	0.8382	0.9997	0.9594
	Convolutional images	0.8938	0.8081	0.8383	0.9875	0.9503
b.	Semantic segmentation	0.9178	0.8482	0.9918	0.9409	0.9541
	Segmentation by histograms	0.9181	0.8486	0.9979	0.9386	0.9539
	Convolutional images	0.8534	0.7442	0.9094	0.9223	0.9191

Table 2. Segmentation results.

Methodology	DICE	JACC	Se	Sp	ACC
Semantic segmentation	0.7530	0.6350	0.8218	0.9583	0.9012
Segmentation by histograms	0.5937	0.4927	0.6020	0.9131	0.8647
Convolutional images	0.5853	0.4700	0.8627	0.8690	0.8170

4 Conclusions and Further Research

Until now, CNN were the best option to perform image processing. Depending on the approaches and goals sought with dermocopic images, the architecture of the Neural Network can be modified in order to process the image for lesion classification, feature extraction or lesion segmentation.

Although there are different architectures to perform semantic segmentation, the SegNet architecture not only has shown the best behavior but also provides an easier implementation and fewer resources to be consumed.

SegNet was chosen over UNet because even when both networks share the same functionality, UNet transfers the entire feature map to its corresponding decoders overloading the memory.

As further research, the segnet will be trained by using a deepest encoder and decoder, which will significantly increase the network behavior.

References

1. Jafari, M.H., Karimi, N., Nasr-Esfahani, E., Samavi, S., Soroushmehr, S.M.R., Ward, K., Najarian, K.: 2016 23rd International Conference on Pattern Recognition (ICPR). IEEE Trans. Pattern Anal. Mach. Intell., 337–342 (2016). https://doi.org/10.1109/ICPR.2016.7899656
2. Fan, H., Xie, F., Li, Y., Jiang, Z., Liu, J.: Automatic segmentation of dermoscopy images using saliency combined with Otsu threshold. Comput. Biol. Med. **85**, 75–85 (2017). https://doi.org/10.1016/j.compbiomed.2017.03.025
3. Yuan, Y., Chao, M., Lo, Y.: Automatic skin lesion segmentation using deep fully convolutional networks with jaccard distance. IEEE Trans. Med. Imaging **36**(9), 1876–1886 (2017). https://doi.org/10.1109/TMI.2017.2695227
4. Badrinarayanan, V., Kendall, A., Cipolla, R.: SegNet: a deep convolutional encoder-decoder architecture for image segmentation. IEEE Trans. Pattern Anal. Mach. Intell. **39**(12), 2481–2495 (2017). https://doi.org/10.1109/TPAMI.2016.2644615
5. Badrinarayanan, V., Handa, A., Cipolla, R.: SegNet: a deep convolutional encoder-decoder architecture for robust semantic pixel-wise labelling (2015). arXiv:1505.07293 [cs.CV]
6. Papadakis, N., Rabin, J.: Convex histogram-based joint image segmentation with regularized optimal transport cost. J. Math. Imaging Vis. **59–2**, 161–186 (2017). https://doi.org/10.1007/s10851-017-0725-5
7. Lu, r., Aujol, J.F., Papadakis, N.: A convex formulation for global histogram based binary segmentation. In: Heyden, A., Kahl, F., Olsson, C., Oskarsson, M., Tai, X.C. (eds.) Energy Minimization Methods in Computer Vision and Pattern Recognition, EMMCVPR 2013. LNCS, vol. 8081. Springer, Heidelberg (2013). https://doi.org/10.1007/978-3-642-40395-8_25
8. Yu, L., Chen, H., Dou, Q., Qin, J., Heng, P.: Automated melanoma recognition in dermoscopy images via very deep residual networks. IEEE Trans. Med. Imaging **36–4**, 994–1004 (2017). https://doi.org/10.1109/TMI.2016.2642839

9. Kwasigroch, A., Mikołajczyk, A., Grochowski, M.: Deep convolutional neural networks as a decision support tool in medical problems – malignant melanoma case study. In: Mitkowski, W., Kacprzyk, J., Oprzkedkiewicz, K., Skruch, P. (eds.) Trends in Advanced Intelligent Control, Optimization and Automation, pp. 848–856. Springer, Cham (2017). https://doi.org/10.1007/978-3-319-60699-6_81
10. Huang, L., Zhao, Y., Yang, T.: Skin lesion segmentation using object scale-oriented fully convolutional neural networks. Signal Image Video Process. **13**(3), 431–438 (2019). https://doi.org/10.1007/s11760-018-01410-3
11. Mun, J., Jang, W.-D, Sung, D. J., Kim, C.-S.: Comparison of objective functions in CNN-based prostate magnetic resonance image segmentation. In: International Conference on Image Processing, ICIP, 2017-September, pp 3859–3863 (2018). https://doi.org/10.1109/ICIP.2017.8297005

Quantitative Analysis of Brain ^{18}F-fluordesoxyglucose and Early-Phase ^{18}F-florbetapir Positron Emission Tomography

Alexander P. Seiffert[1(✉)], Adolfo Gómez-Grande[2], Patricia Sánchez-González[1,3(✉)], Walid Dghoughi[1], Alberto Villarejo-Galende[4], Héctor Bueno[5,6], and Enrique J. Gómez[1,3]

[1] Biomedical Engineering and Telemedicine Centre, ETSI Telecomunicación, Center for Biomedical Technology, Universidad Politécnica de Madrid, Madrid, Spain
{aseiffert,psanchez}@gbt.tfo.upm.es
[2] Servicio de Medicina Nuclear, Hospital Universitario 12 de Octubre, Madrid, Spain
[3] Centro de Investigación Biomédica en Red, Bioingeniería, Biomateriales y Nanomedicina (CIBER-BBN), Madrid, Spain
[4] Servicio de Neurología, Hospital Universitario 12 de Octubre, Madrid, Spain
[5] Instituto de Investigación i+12, Hospital Universitario 12 de Octubre, Madrid, Spain
[6] Centro Nacional de Investigaciones Cardiovasculares (CNIC), Madrid, Spain

Abstract. Positron emission tomography (PET) with amyloid binding tracers and ^{18}F-fluordesoxyglucose (FDG) is used in the diagnosis of Alzheimer's disease and other dementias. The dual-phase amyloid PET protocol acquires an early perfusion image immediately after the radiotracer injection that resembles FDG images. We studied the correlation of early phase ^{18}F-florbetapir (eFBP) and FDG PET images by the quantitation of regional radiotracer uptake. Subjects with a neurodegenerative disease and eFBP and FDG PET scans were retrospectively selected. The PET images were normalized to the Montreal Neurological Institute template using Statistical Parametric Mapping version 12. Regional standardized uptake value ratios (SUVR) were extracted with the cerebellum serving as the reference region. Pearson's correlation tests were performed to evaluate the regional correlation between eFBP and FDG PET images based on the quantitative measurements. Overall high correlation coefficients of up to $r = 0.95$ were observed in cortical regions that are in line with other studies or exceed previous results.

Keywords: Alzheimer's disease · Frontotemporal dementia · Amyloid PET · Florbetapir PET · FDG PET · Dual-phase scan · Quantitative analysis

J. Henriques et al. (Eds.): MEDICON 2019, IFMBE Proceedings 76, pp. 427–436, 2020.
https://doi.org/10.1007/978-3-030-31635-8_52

1 Introduction

Positron emission tomography (PET) imaging data is a significant tool in the clinical diagnosis and evaluation of neurodegenerative diseases, especially of Alzheimer`s disease (AD) [1, 2]. The use of specific radiotracers allows for the in-vivo evaluation of the two main types of biomarkers that are defined for the diagnosis of AD, namely the deposition of brain amyloid-beta (Aβ) and decreased cerebral metabolism as a way of describing neuronal degeneration [3]. Brain Aβ plaques can be visualized by radiotracers like ^{11}C-Pittsburgh compound B [4], ^{18}F-florbetapir (FBP) [5] or ^{18}F-florbetaben (FBB) [6] whereas ^{18}F-fluordesoxyglucose (FDG) is used to measure the cerebral metabolism in the diagnosis of neurodegenerative diseases [1].

The so-called dual-phase amyloid PET protocol acquires an early PET image immediately after the injection of the radiotracer and the late "standard" PET image in one session [7]. It has been shown that the early PET data obtained with a radiotracer binding to Aβ result in perfusion images that resemble FDG PET images of cerebral metabolism and thus, could be a surrogate for neuronal degeneration in AD [8–12]. Thus, the dual-phase amyloid PET acquisition allows for the simultaneous inspection of both Aβ deposition and cerebral perfusion with only one PET scanning [7].

We investigated the usefulness of early-phase amyloid PET acquisitions as a surrogate of FDG PET images by performing a correlation analysis in a study group composed of subjects that are suspected of having a neurodegenerative disease. Semi-quantitative standardized uptake values (SUV) were computed for different cortical regions and normalized to a reference region. Regional Pearson's correlation coefficients were calculated based on the respective radiotracer uptake values and the results were compared to previous studies.

2 Materials and Methods

2.1 Subjects

A total of 19 subjects [age 64.7 ± 8.3 years; gender 11/8 (m/f)] were retrospectively selected for image analysis. All subjects were attended at the Hospital Universitario 12 de Octubre, Madrid, Spain. All subjects had to have undergone at least one FDG PET and one dual-phase FBP PET scan offering early-phase FBP (eFPB) PET images of the subjects. All included cases were suspected of either having mild cognitive impairment (MCI), Alzheimer's disease (AD) or frontotemporal dementia (FTD). The subjects were not stratified and composed one study group.

2.2 PET Imaging

eFBP PET Image Acquisition. eFBP PET images were acquired on a Siemens Biograph 6 True Point. Immediately after intravenous injection of 366 ± 69 MBq of AV−45/Amyvid, 10 static PET images were recorded for 10 min, resulting in one scan for every 60 s. The images were reconstructed with the point-spread function

(PSF) algorithm (3 iterations, 21 subsets). A low-dose CT scan was acquired prior to the eFBP PET scan and used for attenuation correction. All reconstructed eFBP PET scans had a matrix size of $168 \times 168 \times 33$ and voxel size of $4.0728 \times 4.0728 \times 5$ mm^3.

FDG PET Image Acquisition. FDG PET images were acquired on a Siemens Biograph 6 True Point after injection of a dose of 258 ± 85 MBq. The FDG PET data was recorded 40–70 min after the radiopharmaceutical injection. The images were reconstructed with PSF (6 iterations, 16 subsets) and attenuation-corrected using a low-dose CT scan that was acquired prior to the FDG image. All reconstructed FDG PET scans had a matrix size of $336 \times 336 \times 55$ and voxel size of $1.0182 \times 1.0182 \times 3$ mm^3.

2.3 Image Processing and Analysis

The general image processing and quantification pipeline is shown in Fig. 1.

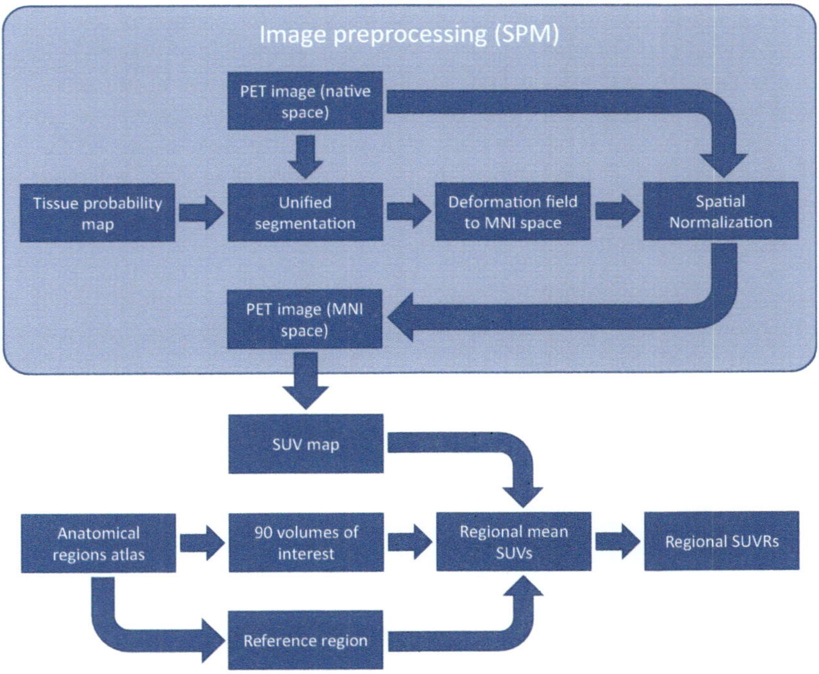

Fig. 1. Image processing and quantification procedure.

All imaging data was processed using Statistical Parametric Mapping (SPM) version 12 (Welcome Department of Cognitive Neurology, Institute of Neurology, London, UK) [13] implemented in MATLAB R2017a (MathWorks Inc., Natick, MA). The individual PET scans were spatially normalized to the image space defined by the Montreal Neurological Institute (MNI) template [14]. The normalization is based on the unified segmentation algorithm implemented in SPM12. A nonlinear deformation field is obtained based on the tissue probability atlas containing 6 tissue probability maps (gray matter, white matter, cerebrospinal fluid, bone, skin and air) provided by SPM12. The individual PET scans were then warped according to the deformation field obtaining spatially normalized PET images in the MNI space with a matrix size of 91 × 109 × 91 and a voxel size of 2 × 2 × 2 mm^3. The resulting normalized images were visually compared to the tissue probability atlas to verify that the images are correctly normalized. All subsequent processing and analyses were performed directly in MATLAB R2017a.

For the quantitative analysis of the PET scans, we based our work on the semi-quantitative Standardized Uptake Value (SUV). SUV maps were computed for each individual eFBP PET and FDG PET scan by calculating the SUVs based on body weight (SUVbw) of each voxel. Following the equation used in [15], we converted the activity concentration of each voxel (Bq/ml) to uptake values (g/ml).

The Automated Anatomical Labeling (AAL) atlas [16] was used to create binary masks of 90 cortical and subcortical volumes of interest (VOI). Moreover, a composite 'Cerebellar Grey Matter' region was constructed based on the AAL atlas. Following [17], we excluded the vermis in the cerebellar VOI. The vermis is a white matter cerebellar region that is located between the two hemispheres of the cerebellum. The 'Cerebellar Grey Matter' VOI was defined as the reference region in our study [18]. The VOI binary masks were then applied to the SUV maps to segment the previously selected regions. Mean SUVs for each VOI were calculated. Regional SUV ratios (SUVR) for each PET scan were then obtained by normalizing to the reference region.

2.4 Statistical Analysis

The correlation between eFBP PET and FDG PET scans was evaluated using a two-tailed Pearson's correlation test. Pearson's correlation coefficients (r) were calculated for each pair (eFBP PET and FDG PET) of regional SUVRs, as well as correlation plots for each anatomical brain region. All statistical analyses were performed using MATLAB R2017a.

3 Results and Discussion

We quantified the radiotracer uptake in 90 cortical and subcortical volumes of interest defined by the AAL atlas in eFBP and FDG PET images. The correlation between both images in a group of 19 subjects was quantitatively assessed by calculating Pearson's correlation coefficients based on the regional SUVRs. We obtained a global correlation of $r = 0.72 \pm 0.18$, calculated as the mean ± standard deviation of 90 regional correlation coefficients.

We selected representative regions from the frontal, parietal and temporal lobe, as well as the posterior cingulate and the precuneus to be shown. The corresponding correlation charts are shown in Figs. 2, 3, 4, 5 and 6. SUVRs and correlation coefficients are shown in Table 1.

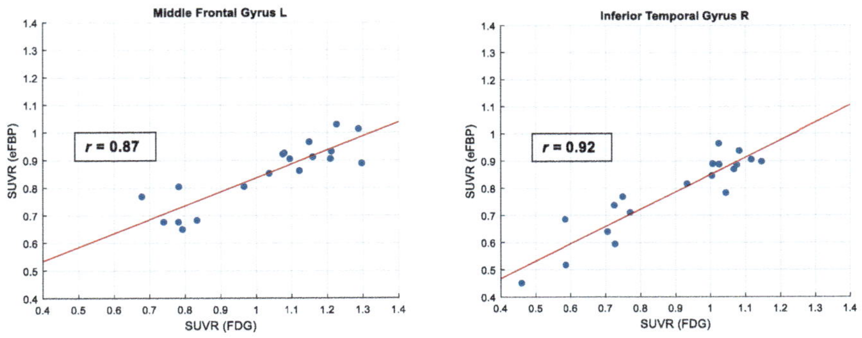

Fig. 2. Correlation charts of the middle frontal gyrus. R: right; L: left.

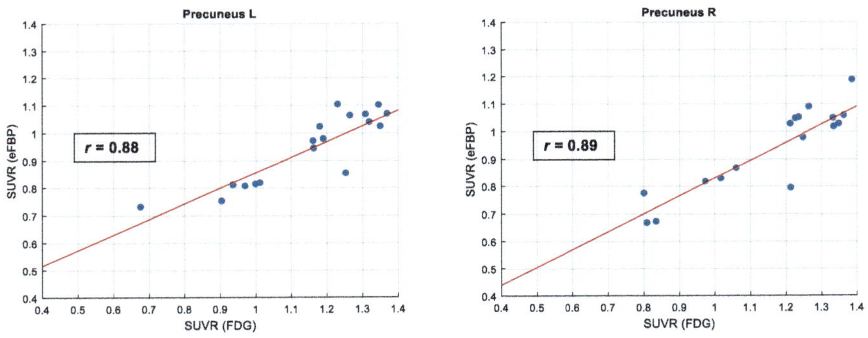

Fig. 3. Correlation charts of the precuneus. R: right; L: left.

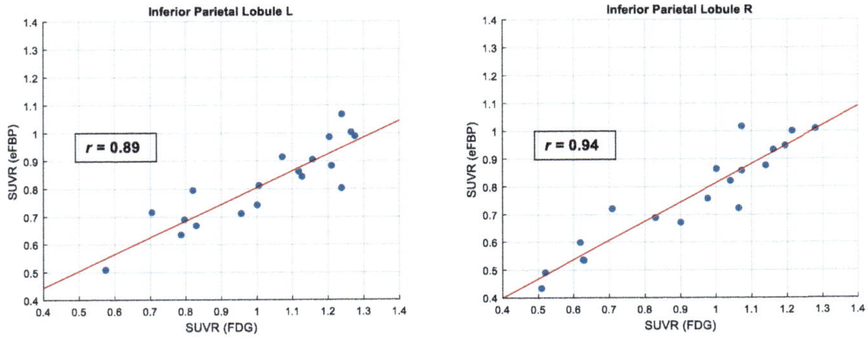

Fig. 4. Correlation charts of the inferior parietal lobule. R: right; L: left.

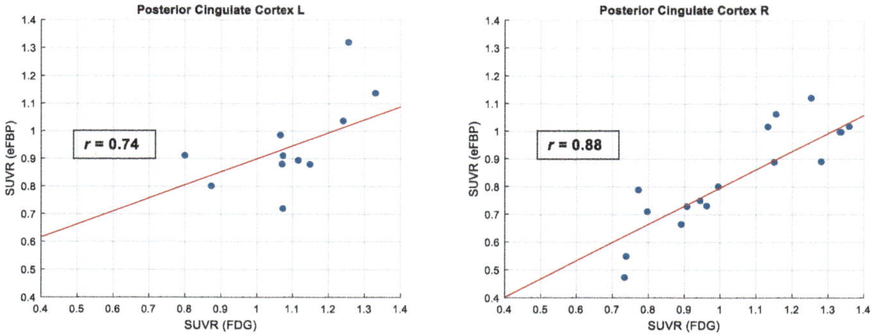

Fig. 5. Correlation charts of the posterior cingulate cortex. R: right; L: left.

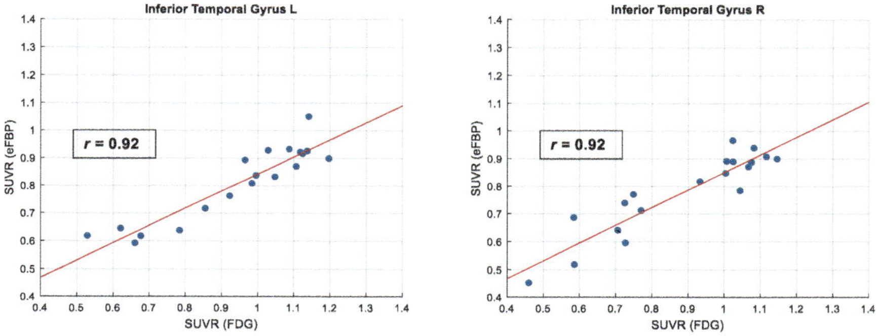

Fig. 6. Correlation charts of the inferior temporal gyrus. R: right; L: left.

Table 1. Regional SUVRs and correlation coefficients. R: right; L: left.

Region	eFBP PET	*r*	FDG PET
	SUVR ± SD		SUVR ± SD
Middle Frontal Gyrus L	0.85 ± 0.12	0.87	1.03 ± 0.20
Middle Frontal Gyrus R	0.87 ± 0.13	0.86	1.06 ± 0.19
Precuneus L	0.95 ± 0.13	0.88	1.16 ± 0.19
Precuneus R	0.95 ± 0.15	0.89	1.19 ± 0.21
Inferior Parietal Lobule L	0.82 ± 0.14	0.89	1.02 ± 0.21
Inferior Parietal Lobule R	0.76 ± 0.18	0.94	0.93 ± 0.25
Posterior Cingulate Cortex L	1.04 ± 0.17	0.74	1.29 ± 0.27
Posterior Cingulate Cortex R	0.85 ± 0.19	0.88	1.10 ± 0.25
Inferior Temporal Gyrus L	0.81 ± 0.14	0.92	0.95 ± 0.20
Inferior Temporal Gyrus R	0.78 ± 0.15	0.92	0.89 ± 0.21

Cortical brain regions showed high correlation coefficients in our study. High VOI-based correlation values of cortical regions between early-phase amyloid and FDG PET images have also been reported in previous studies for ^{11}C-Pittsburgh compound B [10], FBP [8, 19] and FBB [10–12] PET images. Two main differences can be named in relation to previous studies. First, all studies were based on a summed image computed from a dynamic early-phase amyloid PET acquisition whereas we analyzed a static FBP PET acquisition. Second, we calculated correlation coefficients for 90 regions defined in the AAL atlas. Generally, individual regions are grouped to create larger cortical VOIs representing the different lobes. The correlation between early-phase amyloid PET acquisitions and FDG PET images is comparable to those presented by other groups. Moreover, in specific cases our results exceed those in previous studies.

Correlation coefficients in the frontal lobe were generally higher than $r = 0.75$. However, a series of exceptions were observed. Regions with lower correlation coefficients included the left orbital part of the inferior frontal gyrus ($r = 0.60$), the medial orbitofrontal cortex (right: $r = 0.42$; left: $r = 0.65$), the left orbital part of the middle frontal gyrus ($r = 0.63$), the orbital part of the superior frontal gyrus (right: $r = 0.65$; left: $r = 0.67$) and the gyrus rectus (right: $r = 0.44$; left: $r = 0.51$). High correlation coefficients were obtained in the rest of the frontal regions (data not shown). Comparable correlation values in the frontal lobe have also been reported in [10] and [11] when comparing early-phase FBB to FDG PET images. Using cerebellar normalization of SUVs resulted in correlation values of up to $r = 0.609$, and $r = 0.69$, respectively. In the case of early-phase FBP images, the correlation increases slightly as has been shown in [8] (between $r = 0.76$ and $r = 0.9$) and [19] ($r = 0.79$). As has been mentioned above, we included cases of suspected FTD given the hypometabolism patterns this neurodegenerative disease shows in FDG images [1]. The mean SUVR value of the study group in the frontal regions (example in Table 1) also indicates the presence of cases with frontal hypoperfusion or hypometabolism. This suggests that the dual-phase amyloid imaging protocol might be useful in the diagnosis and differentiation of cases with FTD. The usefulness of early-phase FBP in comparison to FDG PET images in the diagnosis of FTD is being investigated and good correlation between both images has been reported [19]. However, further studies need to be conducted to support our findings.

The precuneus, parietal and temporal regions showed overall very high correlation values of up to $r = 0.95$ for the middle temporal gyrus, with the posterior cingulate cortex presenting slightly lower correlation values (see Table 1). Hypometabolic regions in the temporo-parietal cortex are indicative of AD and FDG PET images are therefore used in the diagnostic process [3]. Mean SUVRs were relatively low (0.76 ± 0.18 in the inferior parietal gyrus) of those regions in our study group representing the inclusion of suspected AD cases. Previous studies have reported high visual and quantitative correlations, especially in the parietal and temporal lobes, when comparing the early-phase amyloid PET acquisitions with FDG PET images for the diagnosis of patients with AD and their differentiation from healthy control cases. Concretely, correlation values of over $r = 0.80$ were reported in [8] while analyzing early-phase FBP images for the parietal and temporal lobes, respectively. In studies based on FBB as the amyloid binding radiotracer, correlation coefficients between $r = 0.637$ and $r = 0.88$ were reported for the mentioned cortical regions [10, 11].

The lowest correlation of the 90 analyzed regions was observed in the left putamen (r = 0.14). However, AD and FTD are characterized by cortical metabolism changes in FDG PET images and thus, subcortical regions were not of further interest in our analysis.

Lastly, Segovia et al. [12] reported the highest correlation coefficients in the caudate when investigating early-phase FBB PET images based on a regional analysis with the AAL atlas. In our study, we obtained values of r = 0.89 and r = 0.91 for the right and left caudate, respectively. The regions with the next largest correlation coefficients are also in line with our data, albeit presenting lower values. Moreover, they also reported a preprocessing methodology based on SPM12 similar to the one we employed in our study.

Regarding the quantitative measurements of regional radiotracer uptake, the mean SUVRs of the study group indicate reduced perfusion/metabolism, especially in the parietal and temporal regions. However, these values were expected given the selected subjects for this study. Our data also showed that SUVRs were higher in FDG than eFBP PET images in all regions, which is in line with previous studies [7, 11].

Limitations of this study were the small sample size that could explain partial low correlation coefficients and the absence of a healthy control group for a comparative analysis. Moreover, late-phase "standard" amyloid PET data should be included in the analysis to discriminate amyloid positive from negative subjects and to study the correlation in these groups separately. Another limitation is the analysis of only one early-phase FBP image. All eFBP images should be analyzed to select the time frame resulting in the highest correlation with the FDG PET image. Lastly, the direct segmentation of the SUV map with VOI binary masks could include voxels of cerebrospinal fluid and thus, affecting mean SUVs. Therefore, the influence of the cerebrospinal fluid in the quantification of PET images needs to be investigated.

4 Conclusion

We presented preliminary results of static eFBP and FDG PET image quantification and correlation analyses in a study group of 19 subjects that were diagnosed with a neurodegenerative disease. A preprocessing methodology based on SPM12 normalizing PET images to MNI space for further processing. Regional SUVRs were extracted of 90 cortical and subcortical regions based on the AAL atlas and Pearson's correlation coefficients were computed. The lower SUVRs of eFBP and FDG PET images in several regions indicate reduced perfusion/metabolism in the subjects of our study group. High regional correlation coefficients indicate that both images are comparable. Our results are in line with those presented by other groups and the correlation is larger in specific cases.

Acknowledgments. This work was supported partly by an FPU Fellowship (Beca de Formación de Profesorado Universitario) from the Spanish Ministry of Education, Culture and Sport (MECD) granted to Alexander P. Seiffert (FPU16/06487).

Conflicts of Interest. None.

References

1. Tripathi, M., Tripathi, M., Damle, N., Kushwaha, S., Jaimini, A., D'Souza, M.M., Sharma, R., Saw, S., Mondal, A.: Differential diagnosis of neurodegenerative dementias using metabolic phenotypes on F-18 FDG PET/CT. Neuroradiol. J. **27**(1), 13–21 (2014)
2. Barthel, H., Seibyl, J., Sabri, O.: The role of positron emission tomography imaging in understanding Alzheimer′s disease. Expert Rev. Neurother. **15**(4), 395–406 (2015)
3. McKhann, G.M., Knopman, D.S., Chertkow, H., Hyman, B.T., Jack Jr., C.R., Kawas, C.H., Klunk, W.E., Koroshetz, W.J., Manly, J.J., Mayeux, R., Mohs, R.C., Morris, J.C., Rossor, M.N., Scheltens, P., Carrillo, M.C., Thies, B., Weintraub, S., Phelps, C.H.: The diagnosis of dementia due to Alzheimer's disease: recommendations from the National Institute on Aging-Alzheimer's Association workgroups on diagnostic guidelines for Alzheimer's disease. Alzheimer's Dement. J. Alzheimer's Assoc. **7**(3), 263–269 (2011)
4. Ikonomovic, M.D., Klunk, W.E., Abrahamson, E.E., Mathis, C.A., Price, J.C., Tsopelas, N.D., Lopresti, B.J., Ziolko, S., Bi, W., Paljug, W.R., Debnath, M.L., Hope, C.E., Isanski, B.A., Hamilton, R.L., DeKosky, S.T.: Post-mortem correlates of in vivo PiB-PET amyloid imaging in a typical case of Alyheimer's disease. Brain **131**(6), 1630–1645 (2008)
5. Choi, S.R., Golding, G., Zhuang, Z., Zhang, W., Lim, N., Hefti, F., Benedum, T.E., Kilbourn, M.R., Skovronksy, D., Kung, H.F.: Preclinical properties of 18F-AV-45: a PET agent for Abeta plaques in the brain. J. Nucl. Med. **50**(11), 1887–1894 (2009)
6. Barthel, H., Sabri, O.: Florbetaben to trace amyloid-beta in the Alzheimer brain by means of PET. J. Alzheimer's Dis. **26**(s3), 117–121 (2011)
7. Lin, K.-J., Hsiao, I.-T., Hsu, J.-L., Huang, C.-C., Huang, K.-L., Hsieh, C.-J., Wey, S.-P., Yen, T.-C.: Imaging characteristics of dual-phase 18F-florbetapir (AV-45/Amyvid) PET for the concomitant detection of perfusion deficits and beta-amyloid deposition in Alzheimer's disease and mild cognitive impairment. Eur. J. Nucl. Med. Mol. Imaging **43**(7), 1304–1314 (2016)
8. Hsiao, I.-T., Huang, C.-C., Hsieh, C.-J., Hsu, W.-C., Wey, S.-P., Yen, T.-C., Kung, M.-P., Lin, K.-J.: Correlation of early-phase 18F-florpetapir (AV-45/Amyvid) PET images to FDG images: preliminary studies. Eur. J. Nucl. Med. Mol. Imaging **39**, 613–620 (2012)
9. Rostomian, A.H., Madison, C., Rabinovici, G.D., Jagust, W.J.: Early 11C-PIB frames and 18F-FDG PET measures are comparable: a study validated in a cohort of AD and FTLD patients. J. Nucl. Med. **52**(2), 173–179 (2011)
10. Tiepolt, S., Hesse, S., Patt, M., Luthardt, J., Schroeter, M.L., Hoffmann, K.-T., Weise, D., Gertz, H.-J., Sabri, O., Barthel, H.: Early [18F]florbetaben and [11C]PiB PET images are a surrogate biomarker of neuronal injury in Alzheimer's disease. Eur. J. Nucl. Med. Mol. Imaging **43**, 1700–1709 (2016)
11. Daerr, S., Brendel, M., Zach, C., Mille, E., Schilling, D., Zacherl, M.J., Bürger, K., Danek, A., Pogarell, O., Schildan, A., Patt, M., Barthel, H., Sabri, O., Bartenstein, P., Rominger, A.: Evaluation of early-phase [^{18}F]-florbetaben PET acquisition in clinical routine cases. NeuroImage Clin. **14**, 77–86 (2017)
12. Segovia, F., Gómez-Río, M., Sánchez-Vañó, R., Górriz, J.M., Ramírez, J., Triviño-Ibáñez, E., Carner-Pardo, C., Martínez-Lozano, M.D., Sopena-Pardo, P.: Usefulness of dual-point amyloid PET scans in appropriate use criteria: a multicenter study. J. Alzheimer's Dis. **65**(3), 765–779 (2018)
13. Friston, K., Ashburner, J., Kiebel, S., Nichols, T., Penny, W.: Statistical Parametric Mapping: The Analysis of Functional Brain Images. Academic Press, Amsterdam (2006)

14. Mazziotta, J.C., Toga, A.W., Evans, A., Fox, P., Lancaster, J.: A probabilistic atlas of the human brain: theory and rationale for its development. The International Consortium for Brain Mapping (ICBM). NeuroImage **2**(2A), 89–101 (1995)
15. Kinahan, P.E., Fletcher, J.: Positron emission tomography-computed tomography standardized uptake values in clinical practice and assessing response to therapy. In: Seminars in Ultrasound, CT and MRI, vol. 31, no. 6, pp. 496–505 (2010)
16. Tzourio-Mazoyer, N., Landeau, B., Papathanassiou, D., Crivello, F., Etard, O., Delcroix, N., Mazoyer, B., Joliot, M.: Automated anatomical labeling of activations in SPM using a macroscopic anatomical parcellation of the MNI MRI single-subject brain. NeuroImage **15** (1), 273–289 (2002)
17. Saint-Aubert, L., Nemmi, F., Péran, P., Barbeau, E.J., Payoux, P., Chollet, F., Pariente, J.: Comparison between PET template-based method and MRI-based method for cortical quantification of florbetapir (AV-45) uptake in vivo. Eur. J. Nucl. Med. Mol. Imaging **41**(5), 836–843 (2014)
18. Camus, V., Payoux, P., Barré, L., Desgranges, B., Voisin, T., Tauber, C., La Joie, R., Tafani, M., Hommet, C., Chételat, G., Mondon, K., de La Sayette, V., Cottier, J.P., Beaufils, E., Ribeiro, M.J., Gissot, V., Vierron, E., Vercouillie, J., Vellas, B., Eustache, F., Guilloteau, D.: Using PET with 18F-AV-45 (florbetapir) to quantify brain amyloid load in a clinical environment. Eur. J. Nucl. Med. Mol. Imaging **39**(4), 621–631 (2012)
19. Ashgar, M., Hinz, R., Herholz, K., Carter, S.F.: Dual-phase [18F]florbetapir in frontotemporal dementia. Eur. J. Nucl. Med. Mol. Imaging **46**(2), 304–311 (2019)

Regular Sessions: Bioinstrumentation, Biosenso and Bio-micro/nano Technologies

Cardiac Pacemaker Exposed to Electroporation Pulses – An *Ex Vivo* Study

Tomaz Jarm[1(✉)], Tadej Krmac[1], Damijan Miklavcic[1], and Ratko Magjarevic[2]

[1] Faculty of Electrical Engineering, University of Ljubljana, Trzaska 25, 1000 Ljubljana, Slovenia
tomaz.jarm@fe.uni-lj.si

[2] Faculty of Electrical Engineering and Computing, University of Zagreb, Unska 3, 10000 Zagreb, Croatia

Abstract. Electrochemotherapy (ECT) consists of administration of a chemotherapeutic drug followed by local application of high-voltage electroporation (EP) pulses and is a well-proven method for treatment of various types of tumors. However, as a precaution ECT of deep-seated internal tumors is contraindicated in patients with implanted cardiac pacemakers. To address this limitation a functioning pacemaker programmed in asynchronous D00 mode with the ventricular and atrial leads connected was exposed to EP pulses in a conductive medium mimicking the conditions encountered in clinical setting. EP pulses were generated by the Cliniporator Vitae device and delivered via electrodes for clinical ECT. Application of EP pulses was synchronized with different phases of the pacing cycle, including the potentially most vulnerable phase, the pacing pulse itself. The effects on pacemaker operation were investigated by recording the voltages between the cathode and the anode of the bipolar ventricular lead. EP pulses induced large artefacts far exceeding the amplitude of the pacing pulse itself. This was followed by transiently disrupted shape of the remaining portion of the pacing pulse. However, all effects were strictly limited to this single pacing pulse. No effects at all were observed in subsequent pacing pulses. The pacemaker and its function remained completely immune to electroporation pulses delivered outside the pacing pulse itself.

Keywords: Electroporation · Electrochemotherapy · Cardiac pacemaker · Safety

1 Introduction

Electrochemotherapy (ECT) is a method that can be used successfully for treatment of various types of tumors growing in very different anatomical locations on and inside human body, including in organs and tissues in the thoracic cavity located close to the heart, for example in the liver [1, 2]. ECT involves either local or systemic injection of chemotherapeutic drug (bleomycin or cisplatinum) followed by application of a series (e.g. 8) short-duration (e.g. 100 μs) and high-voltage electric pulses locally to the tumor (at about 1000 V/cm voltage-over-interelectrode-distance ratio) [3, 4]. These pulses induce a temporary increase in the cell membrane permeability for the chemotherapeutic

J. Henriques et al. (Eds.): MEDICON 2019, IFMBE Proceedings 76, pp. 439–446, 2020.
https://doi.org/10.1007/978-3-030-31635-8_53

molecules, a phenomenon called reversible electroporation (EP) [5]. The uptake of the drug by the affected cells is thus increased and its antitumor effectiveness greatly potentiated. This is the main mechanism of antitumor effectiveness of ECT. However, the effect of EP and ECT on tumor blood flow and oxygenation through the blood flow modifying and vascular disrupting mechanisms and the effect on the host immune system are two additional significant pathways that contribute to the overall antitumor effectiveness of ECT [6, 7]. The presented results are relevant to all treatments that involve the use of various types of EP pulses in patients with implanted cardiac pacemakers.

Synchronization of the delivery of EP pulses with the absolute refractory period of the ventricles ensures that ECT can be performed safely even in close proximity of the heart [8] The stimulation of the heart by the EP pulses during the heart's so-called vulnerable period, which may theoretically lead to life-threatening consequences, is thus prevented. An important limitation for the use of ECT in some patients currently still remains however. In patients with implanted cardiac pacemakers application of EP pulses is currently contraindicated as evidenced by the statement to this effect from the manufacturer of the Cliniporator devices approved for clinical generation of EP pulses (IGEA S.p.A., Carpi, MO, Italy). As a precaution measure the company advises that in patients with a pacemakers the ECT treatment should not be given in areas in close proximity to the pacemaker [9].

Here we present the preliminary findings of a study addressing this limitation. Our concern were the effects of the EP pulses on the pacemaker operation in conditions mimicking those when the treatment zone *in vivo* was in immediate vicinity of the heart and hence the implanted pacemaker electrodes. We exposed a functioning pacemaker and its electrodes to high voltage EP pulses in conductive medium and observed the voltage signals recorded at the pacemaker electrodes during the most vulnerable phase for the pacemaker, during generation of the pacing pulse.

2 Materials and Methods

2.1 Experimental Setup

Figure 1 shows the experimental setup. A rectangular glass container was filled with 16 L of physiological saline (depth 14 cm, 0.9% NaCl, conductivity 15.7 mS/cm). The pacemaker (Adapta ADDR01 model, Medtronic, Minneapolis, USA) programmed in asynchronous D00 mode with connected atrial and ventricular bipolar leads was submerged to the bottom in one corner of the container approximately 10 cm from the distal (active) end of the ventricular lead. The distal end of the bipolar ventricular lead (type CapSure Z Novus 5054) was fixed in horizontal orientation at a vertical distance of 5 cm from the bottom of the container and at least 9 cm from the nearest wall and/or the water surface. The cathode (negative electrode) was located at the tip with the ring-shaped anode (positive) at a distance of 1.7 cm from the tip of the lead (Fig. 1). The distal end of the atrial bipolar lead (type CapSure Z Novus 5554) was kept outside the water and used for synchronization.

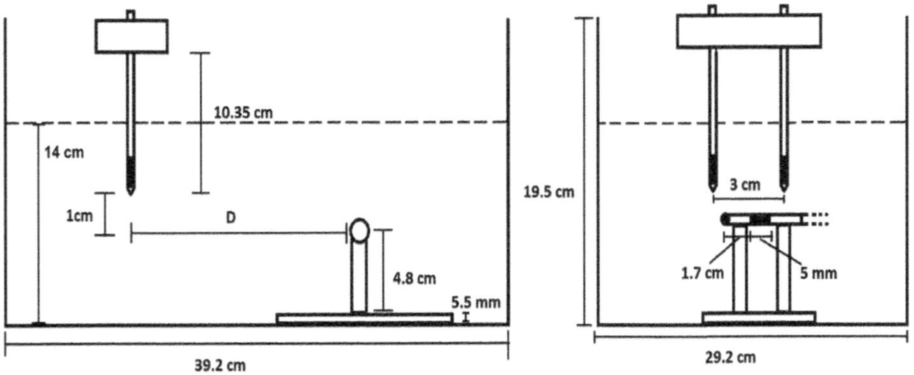

Fig. 1. Relative positions from two perspectives of EP electrodes (suspended vertically from above) and pacemaker electrodes (fixed in horizontal orientation) inside a glass container filled with ca. 16 L of saline (0.9% NaCl). Distance D was adjustable.

Electroporation pulses were generated with the Cliniporator Vitae generator (IGEA, Carpi, MO, Italy) and delivered via two needle electrodes standardly used for clinical ECT (type VG-1230M20; conductive tip length 3 cm, diameter 1.2 mm). These electrodes were inserted in parallel at 3 cm inter-electrode distance into the conductive medium. The vertical distance to the ventricular electrodes was 1 cm (Fig. 1). Horizontal distance between the EP electrodes and the ventricular electrodes (distance D in Fig. 1) was varied between 10 and 1 cm. Rectangular pulses (amplitude 1000 V, duration 100 μs) were generated either individually or in trains of four pulses (repetition rate 5 kHz) at two polarities with respect to the polarity of ventricular pacing pulses. The resulting voltages appearing between the electrodes of the ventricular lead were detected via isolated wires connected inside the water to both ventricular electrodes and led outside the water. Signals were recorded with HDO6104A oscilloscope using the HVD3605 differential high-voltage probes (Teledyne LeCroy, Chestnut Ridge, NY, USA). Another voltage probe of the same type was used to monitor the output voltage of the Cliniporator.

2.2 Synchronization of EP Pulses with Ventricular Pacing Pulses

The purpose of this particular part of our study was to observe the effects of high-voltage EP pules on pacemaker function during the short period when it was in the pacing mode (during generation of the pacing pulse). The assumption was that this was pacemaker's most vulnerable state within the pacing cycle when it was acting as the generator with (relatively) low internal impedance that could potentially lead to harmful current flowing into the device. For this purpose we had to synchronize delivery of EP pulses with the pacing pulse.

We programmed the pacemaker into asynchronous D00 mode in which both atrial and ventricular pacing pulses are generated with no regard to spontaneous electrical activity of the heart. The pacing rate was set to 70 ppm. Ventricular lead voltage was set to 5 V and the pulse width to 1.5 ms (the maximum value available). Ventricular pacing was set to occur 60 ms after the atrial pacing. The atrial pacing pulses (amplitude 3 V, duration 0.52 ms) served as an input to Agilent 33220A arbitrary waveform generator (Agilent, Santa Clara, CA, USA) which was programmed to generate a TTL-compatible rectangular trigger pulses with an adjustable delay. These trigger pulses were then used as an input to the Cliniporator Vitae device. In clinical use a QRS detector would be used to trigger this generator. For safety reasons Cliniporator delivers EP pulses at a fixed and non-adjustable delay of about 50 ms after a validated QRS trigger. The combination of this fixed delay with the adjustable delay of the Agilent function generator allowed us to generate EP pulses at a chosen moment with respect to the ventricular pacing pulse (before pacing, during the stimulation phase of the pacing pulse, during the charge-balancing phase of the pacing pulse).

3 Results

Regardless of the polarity of the EP pulses with respect to the pacing pulse each EP pulse resulted in a distance-dependent voltage artifact of expected polarity and reproducible amplitude at the pacemaker electrodes. The duration of this artefact equaled that of the EP pulse and was the result of the electric field gradient between the pacemaker electrodes during application of the EP pulse. The amplitude of this artifact appeared nearly independent of the phase of the pacing pulse in which the EP pulse was delivered (before, during or after the stimulation phase of the pacing pulse). The amplitude of this artifact at decreasing distances D (see Fig. 1 for reference) was approximately 2.5 V (at 10 cm), 5 V (at 5 cm), 12 V (at 3 cm) and 32 V (at 1 cm).

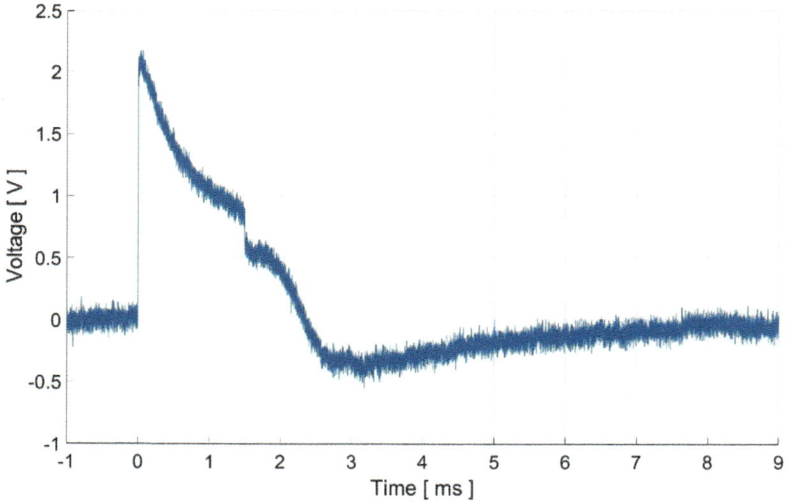

Fig. 2. Unperturbed ventricular pacing pulse

In Fig. 2 an unperturbed ventricular pacing pulse is shown as measured between the cathode and the anode. Note that the pulse is shown with the reversed polarity, the real polarity of the pacing pulse (the cathode vs the anode) is otherwise negative. Both the stimulating (duration 1.5 ms) and the charge-balancing phases of the pulse can be seen. Both phases show exponential decay as a result of the discharge of the main tank capacitor and the DC-blocking capacitance respectively [10].

As an example, Figs. 3, 4 and 5 demonstrate the effect of a train of four EP pulses delivered at 5 kHz repetition frequency for various distances D between the pacemaker and the EP electrodes and for 3 distinct phases of the pacing pulse. When the EP pulse was delivered before the pacing pulse (Fig. 3) there was very little effect on the shape of the subsequent pacing pulse. However, some change in the apparent rate of discharge of capacitances during the stimulating and the charge-balancing phases of the pacing pulse for distances D = 5, 3 and 1 cm was seen. At the distance D = 10 cm the pacing pulse was indistinguishable from that in Fig. 2. When the EP pulse was delivered during the stimulating phase (Fig. 4) the shape of the pacing pulse was changed significantly and very similarly at all distances D. The amplitude was decreased and the dynamics of the charge-balancing phase was also altered, most notably at the smallest distance D = 1 cm. Lastly, when the EP pulses were delivered during the charge-balancing phase, this phase was only mildly affected. For a single EP pulse the effects were of the same characteristics as with four pulses, only less pronounced (not shown).

The most important observation for all conditions in which the effects were investigated was that there was absolutely no effect on the characteristics of pacing pulses that followed the one where the EP pulses were delivered. The pacemaker appeared to be completely immune to EP pulses delivered at any time but during the pacing pulse itself.

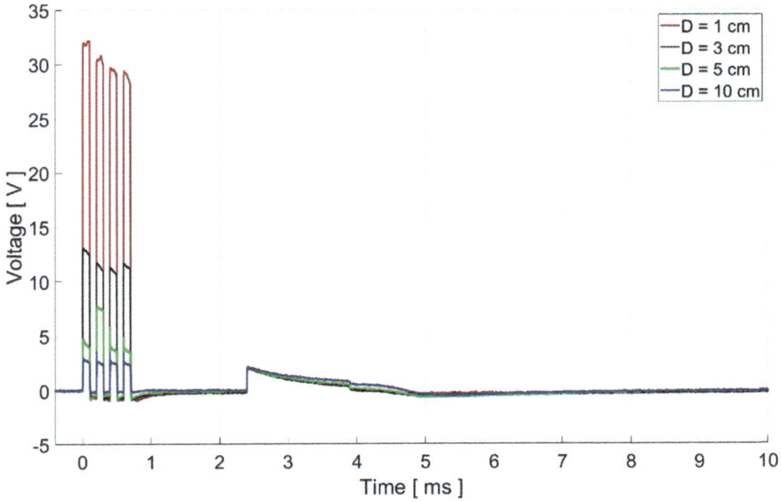

Fig. 3. A train of four EP pulses delivered at 5 kHz before the pacing pulse.

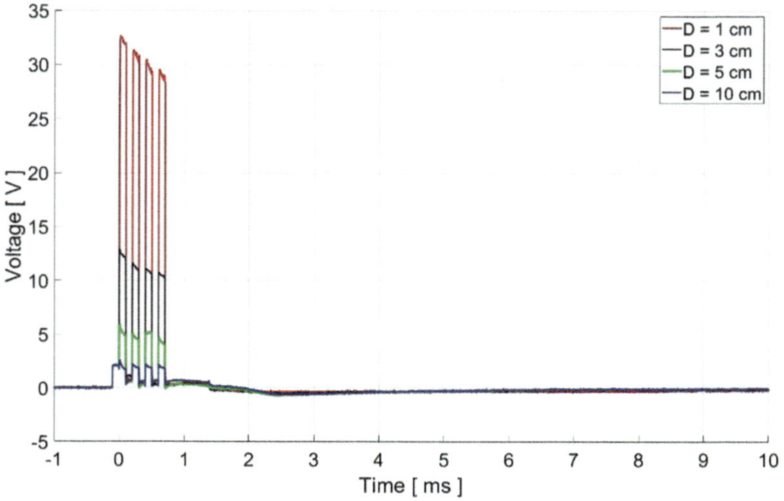

Fig. 4. A train of four EP pulses delivered at 5 kHz during the stimulating phase of the pacing pulse.

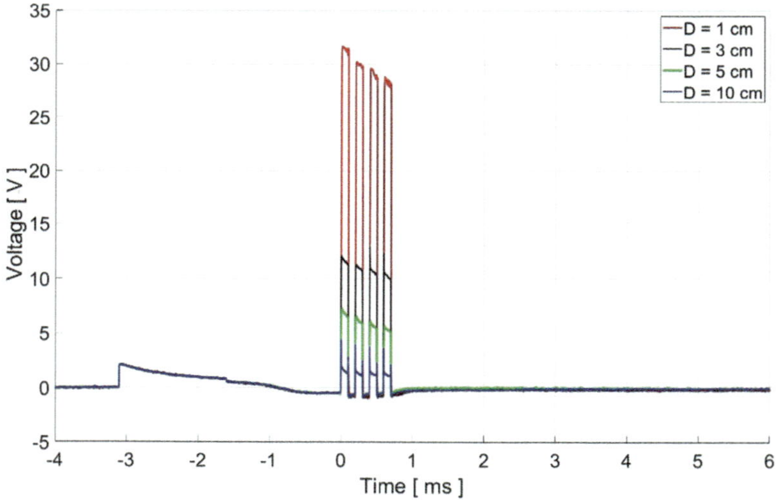

Fig. 5. A train of four EP pulses delivered at 5 kHz during the charge-balancing phase of the pacing pulse.

4 Discussion

The preliminary analysis clearly demonstrated that while the application of EP pulses induced transient changes in the shape of the pacemaker ventricular pacing pulse, most notably when the EP pulses were delivered during the stimulating phase, these effects were confined to the particular pacing pulse only. Subsequent pacing pulses were not affected in any way. This was true for either single EP pulses or for trains of four EP pulses at all tested distances and regardless of the polarity of EP pulse with respect to the orientation of the ventricular lead. In clinical situation the proximity of the electroporation electrodes to the pacemaker electrodes would most probably never be as extremely close as in this preliminary investigation (D = 3 or 1 cm) unless the EP pulses were delivered to the heart muscle itself. In our trial the voltage of EP pulses was limited to 1 kV due to high conductivity of the medium (physiological saline) which resulted in output current intensity of over 40 A at the output of the Cliniporator Vitae (the upper limit for this device is 50 A) [11]. However, higher voltages up to 3 kV are routinely used in clinical practice for ECT of solid tumors where lower conductivities are encountered (conductivity of the liver tissue for example is typically 5× lower than conductivity of the physiological saline).

It is important to note that in realistic clinical application of ECT for treatment of deep-seated tumors in abdominal or thoracic cavity (in relative proximity of the heart and hence the imbedded pacemaker electrodes) the delivery of EP pulses is recommended to be always synchronized with the absolute refractory period of the ventricles. This period occurs after the depolarization of both the atria and the ventricles. Therefore if correctly delivered the EP pulses could never impact on the atrial or ventricular pacing pulse itself. In addition, all pacemakers must withstand the application of defibrillation treatment. The energy of a train of four EP pulses is typically less than one tenth of the energy delivered during external defibrillation.

5 Conclusions

In severe experimental conditions in a highly conductive medium we found no evidence of any adverse effects of EP pulses on functioning of the pacemaker apart from the transient changes in the shape of the particular pacing pulse when the EP pulse (or a train of 4 EP pulses) was delivered during the stimulating phase of the pacing pulse itself. Further investigation in a medium with more tissue-realistic conductivity conditions permitting the use of EP pulses with voltages of up to 3 kV is thus warranted and is currently underway.

Acknowledgement. The research was supported by the Slovenian Research Agency (ARRS) through the Research Programme – Electoporation-based Technologies and Treatments (P2-0249; 2015–2020), the Infrastructural Centre – Cellular Electrical Engineering (part of the Network of research Infrastructural Centers at University of Ljubljana; 2015–2020) and by the Croatian-Slovenian Cooperation in Science and Technology – Safety Aspects of Electrochemotherapy for Patients with Implanted Devices: Numerical Model and Bench Testing (BI-HR/16-17-039; 2016–2017).

Conflict of Interest. The authors declare that they have no conflict of interest. D. Miklavcic was a consultant receiving consultancy fees and grants from IGEA S.p.A.

References

1. Miklavčič, D., Serša, G., Brecelj, E., Gehl, J., Soden, D., Bianchi, G., Ruggieri, P., Rossi, C.R., Campana, L.G., Jarm, T.: Electrochemotherapy: technological advancements for efficient electroporation-based treatment of internal tumors. Med. Biol. Eng. Comput. **50**, 1213–1225 (2012)
2. Campana, L.G., Edhemovic, I., Soden, D., Perrone, A.M., Scarpa, M., Campanacci, L., Čemažar, M., Valpione, S., Miklavčič, D., Mocellin, S., Sieni, E., Serša, G.: Electrochemotherapy – emerging applications technical advances, new indications, combined approaches, and multi-institutional collaboration. Eur. J. Surg. Oncol. **45**, 92–102 (2019)
3. Marty, M., Serša, G., Garbay, J.R., Gehl, J., Collins, C.G., Snoj, M., Billard, V., Geertsen, P.F., Larkin, J.O., Miklavčič, D., Pavlović, I., Paulin-Košir, S.M., Čemažar, M., Morsli, N., Soden, D.M., Rudolf, Z., Robert, C., O'Sullivan, G.C., Mir, L.M.: Electrochemotherapy – an easy, highly effective and safe treatment of cutaneous and subcutaneous metastases: results of ESOPE (European standard operating procedures of electrochemotherapy) study. Eur. J. Cancer Suppl. **4**, 3–13 (2006)
4. Gehl, J., Sersa, G., Matthiessen, L.W., Muir, T., Soden, D., Occhini, A., Quaglino, P., Curatolo, P., Campana, L.G., Kunte, C., Clover, A.J.P., Bertino, G., Farricha, V., Odili, J., Dahlstrom, K., Benazzo, M., Mir, L.M.: Updated standard operating procedures for electrochemotherapy of cutaneous tumours and skin metastases. Acta Oncol. **57**(7), 874–882 (2018)
5. Kotnik, T., Rems, L., Tarek, M., Miklavčič, D.: Membrane electroporation and electropermeabilization: mechanisms and models. Annu. Rev. Biophys. **48**, 63–91 (2019). https://doi.org/10.1146/annurev-biophys-052118-115451
6. Jarm, T., Čemažar, M., Miklavčič, D., Serša, G.: Antivascular effects of electrochemotherapy: implications in treatment of bleeding metastases. Expert Rev. Anticancer Ther. **10**, 729–746 (2010)
7. Miklavčič, D., Mali, B., Kos, B., Heller, R., Serša, G.: Electrochemotherapy: from the drawing board into medical practice. Biomed. Eng. Online **13**, 29 (2014)
8. Mali, B., Gorjup, V., Edhemović, I., Brecelj, E., Čemažar, M., Serša, G., Stražišar, B., Miklavčič, D., Jarm, T.: Electrochemotherapy of colorectal liver metastases – an observational study of its effects on the electrocardiogram. Biomed. Eng. Online **14**(Suppl. 3), S5 (2015)
9. IGEA Homepage: Can Electrochemotherapy be used if the patient has a pacemaker?. https://www.igea.it/en/oncology/information-clinicians/frequently-asked-questions. Accessed 05 May 2019
10. Prutchi, D., Norris, M.: Design and Development of Medical Electronic Instrumentation. A Practical Perspective of the Design, Construction, and Test of Medical Devices. Wiley, Hoboken (2005)
11. Bertacchini, C.: Cliniporator: medical electroporation of tumors. In: Miklavcic, D. (ed.) Handbook of Electroporation. Springer, Cham (2017)

Smart Vest for Respiratory and Physical Activity Monitoring in COPD Patients

David Naranjo-Hernández[1(✉)], Javier Reina-Tosina[1], Laura M. Roa[1], Gerardo Barbarov-Rostán[1], Alejandro Talaminos-Barroso[1], Pilar Cejudo-Ramos[2,3], Eduardo Márquez-Martín[2,3], and Francisco Ortega-Ruiz[2,3]

[1] Biomedical Engineering Group, University of Seville, 41092 Seville, Spain
dnaranjo@us.es

[2] Medical-Surgical Unit of Respiratory Diseases, University Hospital Virgen del Rocío, 41013 Seville, Spain

[3] Biomedical Research Center in Network (CIBER) of Respiratory Diseases (CIBERES), 28029 Madrid, Spain

Abstract. Rehabilitation treatment of patients with chronic obstructive pulmonary disease (COPD) based on muscle training provides clear benefits in health and quality of life. Its positive effects are usually diminished once the training is over, and the effectiveness of home maintenance programs is limited due to the lack of motivation and adherence of patients, as well as by the reduction of clinical supervision. In this work, the preliminary design and first results of a smart vest for the motivation and evaluation of patients with COPD in home maintenance programs are presented. The physical activity (metabolic expenditure and number of repetitions of the exercises) and the respiratory pattern of the patient are the parameters monitored through the built-in sensors in the smart vest. The processing of the information is distributed between the smart vest and a smartphone, which also acts as user interface and link for the management and remote data storage. Bidirectional wireless communications between the smart vest and the smartphone are based on a waveform preprocessing protocol over the standard Bluetooth 4.1, whose purpose is to reduce energy consumption. A preliminary study with healthy volunteers in a controlled environment has shown the feasibility of the proposed solution (100% accuracy in the number of exercise repetitions, close correspondence between the monitoring tests of metabolic expenditure and breathing and the pre-established patterns).

Keywords: Chronic Obstructive Pulmonary Disease (COPD) · Respiratory rehabilitation · Respiratory monitoring · Physical activity · Biomedical sensor device

1 Introduction

Muscle training of respiratory rehabilitation programs improves the exercise capacity, prognosis and quality of life of patients with chronic obstructive pulmonary disease (COPD), while reducing the symptoms [11, 15]. Home maintenance programs aim to

J. Henriques et al. (Eds.): MEDICON 2019, IFMBE Proceedings 76, pp. 447–454, 2020.
https://doi.org/10.1007/978-3-030-31635-8_54

maintain the benefits achieved after the muscle training [11]. However, their effectiveness is limited due to the lack of adherence, motivation and confidence of the patients in the maintenance program, and also because of the reduction of clinical support and supervision [16]. These facts highlight the need to address solutions capable of overcoming these limitations. e-Health systems together with smart devices and systems for the real-time monitoring of relevant clinical parameters are considered a promising approach for the prevention and treatment of respiratory diseases [10], to motivate patients to adopt self-treatment skills [16], thus improving their quality of life and therefore minimizing recurrent exacerbations and hospital re-admission rates [3].

Some studies have shown the positive effects of monitoring physical activity in patients with COPD [2], either by evaluating the number of steps and time spent performing activities [4], analyzing the metabolic expenditure [14] or the number of repetitions of the exercises [6]. The monitoring of the respiratory pattern is usually used to evaluate the health state of the respiratory patient, analyze their evolution or anticipate dangerous events [13]. As alternatives to spirometry and capnography, other less-intrusive technologies have been proposed for respiratory monitoring: ultrasound [7], piezoresistive sensors [5], inductive [9] or capacitive [17, 18] methods. However, and despite these advances, the integration of these technologies in medical practice remains limited. Many of the proposed developments have the disadvantage of requiring a special location of the sensors, often uncomfortable for the user, do not have sufficient sensitivity or require a special processing of the signal that significantly affects the battery life, present failure in the application of user-centric design methodologies, i.e., improper usability and accessibility concerns.

The purpose of this work is to describe a first approach to a smart vest that integrates physical and respiratory activity sensing capabilities designed according to a user-centric design approach for a monitoring application of patients with COPD in home maintenance programs.

2 Materials and Methods

The development of the smart vest has followed a modular user-centered design methodology based on standards to facilitate the integration of new technologies and functionalities. For the non-intrusive monitoring of the respiratory pattern, capacitive sensor technology has been used due to its high sensitivity [17, 18] following the scheme described in [13]. For the monitoring of physical activity, accelerometry has been used, since it allows to evaluate in a simple way the changes of orientation and movements [12]. The processing algorithms and their parameters were developed and adjusted in the Matlab programming environment from the data obtained in preliminary experiments. Finally, C and Java languages were used for their embedded programming in the smart vest and the personal monitoring device, respectively.

A first validation of the smart vest in a controlled environment was carried out through a quasi-experimental study on healthy volunteers emulating the conditions in which the device will finally be used. To evaluate the respiratory pattern, the validation methodology required volunteers to perform an experiment in which the respiratory rate was increased progressively from 6 to 30 breaths per minute (bpm) following a protocol in which the begin and end instants of the inspiration and expiration periods were controlled externally with acoustic warnings. To evaluate physical activity, the volunteers performed some of the typical exercises of respiratory rehabilitation programs [6, 8]: walking, stationary bicycle, costal expansion, lateral arm flexion, front arm raise, lateral torso bend.

3 Results

3.1 Monitoring System Architecture

The monitoring is performed in a distributed way through the following devices arranged in a master-slave architecture:

- **Smart vest**: It is an elastic garment that incorporates a hardware for the sensing and the preprocessing of signals related to the physical activity and the respiratory pattern of the patient (see Fig. 1).

Fig. 1. Smart vest prototype.

- **Personal monitoring device** (master): Data acquired by the smart vest are wirelessly transmitted to a smartphone that performs a second processing to evaluate the number of repetitions of the prescribed exercises, the metabolic expenditure associated with the maintenance activities and the respiratory parameters. This device also serves as user interface and communications gateway for data storage and remote management of the information.

3.2 Smart Vest Hardware

The smart vest hardware follows the following modular design scheme:

- **Respiratory sensing module**: capacitive sensor technology is implemented according to the scheme presented in [13] based on the Colpitts oscillator, with two textile electrodes located on the front and back of the chest.
- **Physical activity sensing module**: A LIS3LV02DQ triaxial accelerometer from STMicroelectronics is used to monitor the user's movement.
- **Processing module**: The PIC32MX440F256H microprocessor from Microchip is employed to perform the preprocessing of the signals provided by the respiratory and physical activity sensing modules.
- **Communications module**: A RN4020 transceiver from Microchip is configured to perform the bidirectional communications of the smart vest based on the Bluetooth 4.1 standard.

3.3 Data Preprocessing

The data from the sensing modules are processed with a latency of 50 ms (sampling period). The data from the respiratory sensor module correspond to the frequency of the Colpitts oscillator, which depends on the value of the electric capacitance of the chest. The data provided by the physical activity sensor module are the triaxial accelerations.

As the energy consumption in communications is an important source of energy loss in body sensor networks, the smart vest performs a preprocessing of the signals to reduce the amount of data transmitted. Given the characteristics of the respiratory pattern, an algorithm for extracting waveform parameters based on the following operations is used:

1. A low pass filtering (cutoff frequency of 1 Hz) of the signal corresponding to the oscillation frequency that generates a smoothed signal of the respiratory component and eliminates high frequency noise (S_1 signal).
2. A second low pass filter (cutoff frequency of 0.2 Hz) of the oscillation frequency that generates a signal related to the evolution of its DC component, not related to respiratory activity (S_2 signal).
3. Identification in S_1 of a 'V' waveform in which the following characteristic points are defined (see Fig. 2):

   - **P4**: Cut point of S_1 with S_2 on its falling edge.
   - **P5**: Cut point of S_1 with S_2 on its rising edge.
   - **P1**: Maximum of S_1 in the time interval before P4.
   - **P2**: Minimum of S_1 in the time interval between P4 and P5.
   - **P3**: Maximum of S_1 in the interval after P5.

The identification of the characteristic points is repeated sequentially so that point P3 corresponds to P1 in the following wave pattern.

The rehabilitation exercises follow a periodic pattern similar to that found in the respiratory signal (see Fig. 2), so that to summarize the information related to the movements, the same algorithm for extracting waveform parameters is also used, but in

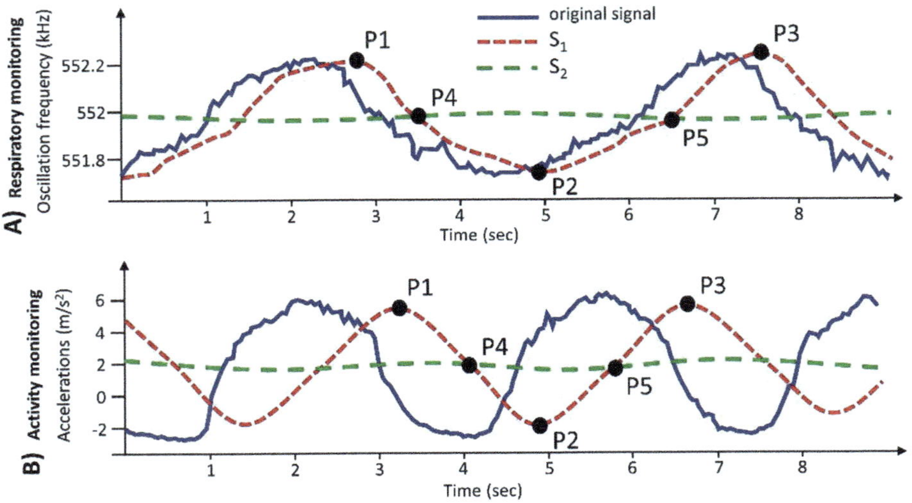

Fig. 2. Example of signals and characteristic points obtained with the respiratory processing.

this case the cutoff frequencies are 0.3 Hz and 0.05 Hz for the smoothing filtering and continuous filtering, respectively. To simplify the processing and decrease the amount of data to be transmitted, the parameter extraction algorithm is only executed on the axis most sensitive to the changes produced by the movements, depending on the type of exercise performed: acceleration x-axis for bicycle and lateral torso bend exercises, y-axis for walking and z-axis for the rest of exercises.

For the estimation of metabolic expenditure, an adaptation of the algorithm presented in [12] is used, executed in a distributed way in the smart vest and the personal monitoring device. The smart vest performs a preprocessing to calculate the E_A parameter related to the accelerations. This parameter is estimated with a latency of 1 s according to a weighted average on the squared acceleration components over a window of 20 samples [12].

3.4 Wireless Communications

Bidirectional wireless communications are based on the Bluetooth 4.1 standard. Data sending from the device is done through the use of a reading-type characteristic in the master device (personal monitoring device) that allows data transfer in blocks of 20 bytes, the maximum allowed by the transceiver to optimize the transfer rate. For the reception of commands another writing-type characteristic is used in the master device, that enables the transfer of 1 byte. Using these characteristics, the information is structured according to the following data frames:

- **Waveform preprocessing frame**, used to send the waveform parameters described in the previous section (P1 to P5) and the values of C1, C2, C3 and C4 counters, corresponding to the number of samples from P1 to P4, P4 to P2, P2 to P5 and P5 to P3, respectively.

- **Activity energy frame**, generated with a latency of 1 s when its estimation is activated, employed to send the value of E_A parameter.
- **Command frame**, used to activate/deactivate the monitoring modes of the smart vest (breathing, estimation of E_A parameter and number or repetitions of the exercises) and configure the exercise to be performed.

3.5 Processing in the Personal Monitoring Device

The respiratory parameters are estimated by the personal monitoring device each respiratory cycle from the data of the waveform preprocessing frame related to the breathing according to the following expressions:

$$T_{\mathrm{E}} = (C1 + C2) \cdot 0.05 \tag{1}$$

$$T_r = (C3 + C4) \cdot 0.05 \tag{2}$$

$$F_R = 60/(T_E + T_I) \tag{3}$$

where T_E is the instantaneous expiration time in seconds, T_I is the instantaneous inspiration time in seconds and F_R is the instantaneous respiratory rate in bpm. The average values of these parameters are also calculated with a latency of one minute.

The metabolic expenditure (*ME*) in kJ is calculated every second from the E_A parameter using the following equation adapted from [12], in which the weight of the user is expressed in kg:

$$ME = 0.003625 \cdot E_A^{0.237} \cdot weight \tag{4}$$

The number of repetitions of the exercises is counted from the recognition of the number of waveform preprocessing frames related to movements with a difference between P1 and P2 greater than 0.05 m/s^2.

3.6 Validation Results in a Controlled Environment

A preliminary evaluation of the smart vest was carried out in a controlled environment on four male volunteers of apparent healthy condition whose anthropometric characteristics are summarized in Table 1.

Table 1. Anthropometric characteristics of the volunteers.

	Mean value	Standard deviation
Age (years)	40	6.1
Height (cm)	176.5	4.6
Weight (kg)	75.7	4.8

The results obtained in the respiratory monitoring tests were faithful to the pre-established respiratory pattern for these experiments. The mean metabolic equivalent calculated from the metabolic expenditure [12] for the rest activity was 1.08 and for the walking activity was 3.53, values very close to those expected for these activities (1 and 3.5 respectively [1]). The number of repetitions of the exercises was detected with an accuracy of 100% (number of steps and number of pedals for walking and cycling exercises, respectively). All volunteers expressed the comfort and non-intrusiveness of the smart vest.

4 Conclusions

The present work has shown a first approach to a smart vest for the non-intrusive surveillance of patients with COPD at home and the assessment of the correct compliance of maintenance exercise programs through the monitoring the number of repetitions of prescribed exercises, the associated metabolic expenditure to the activities performed and the user's respiratory parameters during rest periods between exercises. The information is wirelessly sent to a smartphone that serves as a user interface to motivate the patient to comply with the prescribed rehabilitation program. The results obtained show the technical and functional feasibility of the proposed system. The main advantages over other equivalent systems are non-intrusiveness, the use of standard communications and the combined monitoring of several parameters related to physical activity and respiration.

Acknowledgment. This work was supported in part by the "Fondo de Investigaciones Sanitarias", Instituto de Salud Carlos III, under Grant PI15/00306, in part by the "Fundación Progreso y Salud", Gobierno de Andalucía, under Grant PIN-0394-2017, in part by the "Fundación Mutua Madrileña", under Grant "VÍA-RENAL" and in part by the "Eurostars Programme", European Commission, under Grant "FRAIL".

Conflict of Interests. The authors declare that there is no conflict of interests.

References

1. Ainsworth, B., Haskell, W., Whitt, M., et al.: Compendium of physical activities: an update of activity codes and MET intensities. Med. Sci. Sports Exerc. **32**(9 SUPPL.), S498–S504 (2000)
2. Boeselt, T., Spielmanns, M., Nell, C., et al.: Validity and usability of physical activity monitoring in patients with Chronic Obstructive Pulmonary Disease (COPD). PLoS One **11**(6), e0157229 (2016)
3. Bourbeau, J., Casan, P., Tognella, S., Haidl, P., Texereau, J., Kessler, R.: An international randomized study of a home-based self-management program for severe COPD: The comet. Int. J. COPD **11**(1), 1447–1451 (2016)
4. Demeyer, H., Burtin, C., et al.: Standardizing the analysis of physical activity in patients with COPD following a pulmonary rehabilitation program. Chest **146**(2), 318–327 (2014)

5. Guo, L., Berglin, L., Wiklund, U., Mattila, H.: Design of a garment-based sensing system for breathing monitoring. Text. Res. J. **83**(5), 499–509 (2013)
6. Hansen, H., Bieler, T., Beyer, N., et al.: COPD online-rehabilitation versus conventional COPD rehabilitation - rationale and design for a multicenter randomized controlled trial study protocol. BMC Pulm. Med. **17**(1), 1–14 (2017)
7. Heldt, G., Ward Iii, R.: Evaluation of ultrasound-based sensor to monitor respiratory and nonrespiratory movement and timing in infants. IEEE Trans. Biomed. Eng. **63**(3), 619–629 (2016)
8. Janaudis-Ferreira, T., Hill, K., Goldstein, R., et al.: Resistance arm training in patients with COPD: a randomized controlled trial. Chest **139**(1), 151–158 (2011)
9. Kranjec, J., Begus, S., Drnovsek, J., Gersak, G.: Novel methods for noncontact heart rate measurement: a feasibility study. IEEE Trans. Instrum. Meas. **63**(4), 838–847 (2014)
10. Lahham, A., Mcdonald, C., Mahal, A., et al.: Participation in physical activity during center and home-based pulmonary rehabilitation for people with COPD. J. Cardpulm. Rehabil. Prev. **39**(2), E1–E4 (2019)
11. Liu, X., Li, P., Xiao, L., et al.: Effects of home-based prescribed pulmonary exercise by patients with chronic obstructive pulmonary disease. Trials **20**(1), 41 (2019)
12. Naranjo-Hernández, D., Roa, L., Reina-Tosina, J., Estudillo-Valderrama, M.: SoM: a smart sensor for human activity monitoring and assisted healthy ageing. IEEE Trans. Biomed. Eng. **59**(12 Part 2), 3177–3184 (2012)
13. Naranjo-Hernández, D., Talaminos-Barroso, A., Reina-Tosina, J., Roa, L., Barbarov-Rostán, G., Cejudo-Ramos, P., Márquez-Martín, E., Ortega-Ruiz, F.: Smart vest for respiratory rate monitoring of COPD patients based on non-contact capacitive sensing. Sensors **18**(7), 2144 (2018)
14. van Remoortel, H., Raste, Y., Louvaris, Z., et al.: Validity of six activity monitors in chronic obstructive pulmonary disease: a comparison with indirect calorimetry. PLoS One 7(6) (2012)
15. Riley, C., Sciurba, F.: Diagnosis and outpatient management of chronic obstructive pulmonary disease: a review. J. Am. Med. Assoc. **321**(8), 745–746 (2019)
16. Spencer, L., McKeough, Z.: Maintaining the benefits following pulmonary rehabilitation: achievable or not? Respirology (2019)
17. Takano, M., Yamagishi, S., Ohmuta, T., Fukuoka, Y., Ueno, A.: Non-contact simultaneous measurements of electrocardiogram and respiratory movements using capacitive sheet electrodes. Adv. Biomed. Eng. **6**, 28–36 (2017)
18. White, N., Ash, J., Wei, Y., Akerman, H.: A planar respiration sensor based on a capaciflector structure. IEEE Sens. Lett. **1**(4), 1–4 (2017)

A Prototype of Intelligent Portable Oxygen Concentrator for Patients with COPD Under Oxygen Therapy

Alejandro Lara-Doña[1,2], Daniel Sanchez-Morillo[1,2(✉)], María Pérez-Morales[3], Miguel Ángel Fernandez-Granero[1,2], and Antonio Leon-Jimenez[3]

[1] Biomedical Research and Innovation Institute of Cádiz (INiBICA), Cadiz, Spain

[2] Biomedical Engineering and Telemedicine Research Group, Department of Automation Engineering, Electronics and Computer Architecture and Networks, University of Cadiz, Cadiz, Spain
alejandro.laradona@alum.uca.es, {daniel.morillo,ma.fernandez}@uca.es

[3] Pulmonology, Allergy and Thoracic Surgery Unit, Puerta del Mar University Hospital, Cadiz, Spain
antonio.leon.sspa@juntadeandalucia.es

Abstract. Traditional long-term oxygen therapy devices used in daily-life activities need to be manually adjusted depending on the level of intensity of the activity done. In this study, a system to automatically control a commercial POC devices is proposed with the aim of improving the used and efficacy of the therapy in patients with chronic obstructive pulmonary disease.

A patient unit and a control unit were designed and implemented to control a Inogen G2 oxygen concentrator. The patient unit included an inertial measurement unit to classify in real-time the physical activity performed by the user. The control unit receive via Bluetooth (BLE) the degree of activity and adjusted the oxygen concentrator automatically following the setting provided by clinician.

Data from a group of six volunteers were used to train and validated the system using different machine learning algorithms. A logistic regression model (LR) achieved the best accuracy (93.69%) in classifying the physical activity and consequently adjusting correctly the O_2 dose.

The developed system may improve oxygenation through automatic adaptation to lifestyle, could contribute to improve the patient's therapeutic compliance and may promote physical activity.

Keywords: COPD · Human activity recognition · Portable oxygen concentrator · Long term oxygen therapy

J. Henriques et al. (Eds.): MEDICON 2019, IFMBE Proceedings 76, pp. 455–461, 2020.
https://doi.org/10.1007/978-3-030-31635-8_55

1 Introduction

Long-term oxygen therapy (LTOT) is a medical therapy used to deliver supplemental oxygen (O_2) to patients in order to maintain an adequate blood oxygen level [1].

LTOT is often prescribed for patients with hypoxemic chronic lung disease such as Chronic Pulmonary Obstructive Disease (COPD). Currently, the most common way to deliver O_2 to these patients is to use static sources (such as stationary O_2 concentrators or O_2 cylinders). These sources do not allow the patient to wander under therapy outside the home. For this purpose, lightweight and portable O_2 concentrators (POC) have been developed in order to allow the patient to partially pick up normal daily activity in the home environment [2]. POCs are widely used due to its long autonomy and less maintenance compared to other devices. These devices can also be used in a daily-basis outside home, only restricted by the battery, which can be charged anywhere. In patients that use a POC, the O_2 requirements at each instant vary according to the type, intensity and duration of physical activity performed. Accordingly, the O_2 dosage have to be prescribed by the specialist according to the patient status and the intensity of physical activity that he/she usually performs [3].

POCs can be classified into two groups according to the way they supply the O_2 to the user: continuous flow or pulsed dose [1]. The first ones supply a continuous flow of O_2 to the patient according to the level adjusted in the control panel. On the other hand, the pulsed dose devices supply a preset volume of O_2 in phase with the patient's breathing cycle. Regardless the mode to supply O_2, the user must adjust the dose manually via a keyboard or knob present in the device. In traditional LTOT, this manual adjustment of O_2 dosage is a time-consuming task, requires patient attention and training, and is often subject to errors, which may induce states of hypoxia or hyperoxia, potentially detrimental to the patient's health. Given that advanced average age of patients with COPD under LTOT [4], the probability of inadequate use of the device is significant.

Recent advances in new physiological closed-loop oxygen concentrators are being discarded for real clinical applications, despite of their potential to improve the oxygen therapy. Usability issues, control algorithm robustness, sensors accuracy and the need of fail-safe mechanisms are factors that are hampering application to clinical settings [5].

In this study, we developed the concepts presented in [6] and present a system that allows a POC to be transformed into an intelligent closed-loop device (iPOC), capable of adjust O_2 flow automatically in real time, in accordance with the different physical activities carried out by the patient. In order to control the POC, a sensor and a control unit were designed and implemented. In addition, a classifier was developed using machine learning methods to classify three levels of physical activity, namely: sedentary (sitting or standing), low intensity (walking) and medium intensity (walking upstairs or downstairs). Different machine learning models for activity classification were trained and evaluated using real data recorded from six subjects.

2 Methods

2.1 Intelligent Portable Oxygen Concentrator (IPOC)

Figure 1 details the proposed iPOC system. Feedback control was implemented with a single input-single output (SISO) table look-up controller. The feedback control law T was tabulated by a table look-up function provided by the clinicians and personalized for the specific needs of the patients according to conventional titration procedures [7].

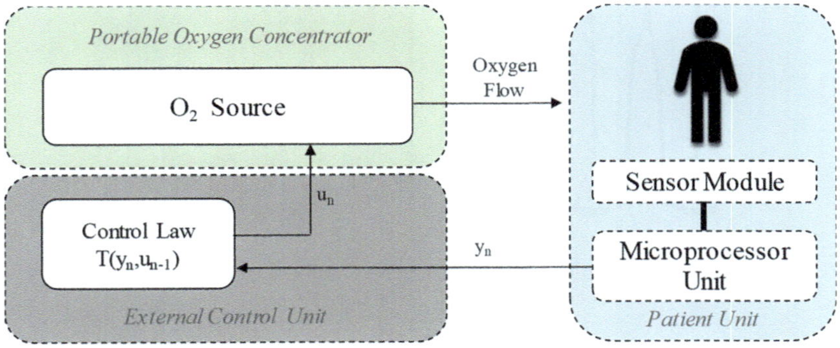

Fig. 1. Control scheme for the proposed intelligent portable oxygen concentrator (iPOC).

Two devices were ad-hoc developed during the research (Fig. 2). First, an external control unit was designed to be connected to a standard POC. The control unit interfaced with the control panel of an Inogen One G2 oxygen concentrator. Second, a patient unit was implemented to: (a) automatically classify the physical activity performed by the user in real-time, and (b) transmit the result (y_n) of the classification process to the control unit via Bluetooth Low Energy (BLE).

When the external control unit received a new activity level from the patient unit, it adjusted the O_2 level according to the established settings. External jumpers were provided for an easier personalization of the unit for different oxygenation patient profiles.

Several safety measures were taken to avoid a misuse of the device. To guarantee that the correct O_2 level was configured according to the activity detected, the control unit reset the O_2 level to the minimum at each beginning of the automated mode. In addition, the control unit was disabled after power-on the POC. Accordingly, the user should start the automated control with a push-button. The automated control could be disabled using the same button or if the user manipulated the POC keyboard.

The activity detector unit was composed of a microcontroller, a 9-axis inertial measurement unit (IMU), and a barometer. Only the accelerometer, gyroscope and barometer data were recorded, discarding the magnetometer signals. The accelerometer and gyroscope signals were acquired at a sample rate of 60 Hz. The sampling rate for the barometer was 0.75 Hz. A state machine was implemented in the control unit in order to save energy and to minimize the errors caused by spurious activity

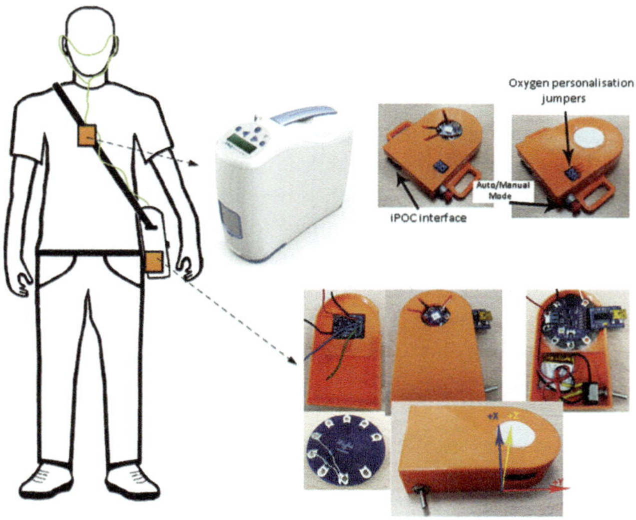

Fig. 2. Detail of the implemented control and patient units.

classification provided by the machine learning classifier. The patient unit only sent a new activity level to the control unit when it was detected for two consecutive cycles and this level was different of the level already adjusted in the POC. The device also limited abrupt changes in the concentrator oxygen supply following recommendations described in [6]. The patient unit was placed in the chest of the user, just at the end of the sternum. This place has been selected based on previous studies since it generates less motion artifacts due to movements were compared to systems placed on the wrist, ankle or the belt [8]. In relation to orientation of the IMU, the positive y-axis pointed to the patient's head, and the z-axis to the front.

2.2 Data, Signal Processing and Features Extraction

In order to build the dataset to train and validate the classification algorithms, signals from six subjects were acquired while sitting, standing, walking, walking upstairs and walking downstairs.

Physical activity level is significantly low in targeted patients. The target group are patients with COPD stage III and IV according to the Global Initiative for Chronic Obstructive Lung Disease [9]. Therefore, intense activity was not considered. Static postures such as sitting, standing still and lying in the bed were considered for the inactive state. Walking was considered as a light intensity activity and walking downstairs and walking upstairs was defined as moderate physical activity. These common activities were selected for automatic recognition because they imply a need to adjust the O_2 dose delivered to the patient.

In this study, only the signals from the accelerometer and gyroscope were used for training. Accuracy provided by the barometer was not enough to be considered. The signals from the accelerometer and gyroscope were high-pass filtered with an infinite impulse response (IIR) filter with a cut-off frequency of 0.15 Hz. Once filtered, features were extracted. Feature selection was limited by the computing power of the micro-controller used in the patient unit for real-time processing. For this rationale, features were calculated from the time-domain using a 3-second window of data collected from the IMU. The features selected were: standard deviation, the average increment, median absolute deviation, largest value, smallest value, average sum of the squares, difference between largest and smallest value, and signal magnitude area.

2.3 Machine Learning Algorithms

Features were used as input to train various supervised machine-learning algorithms such as C4.5 decision tree (DT), linear discriminant analysis (LDA), probabilistic neuronal network (PNN), and logistic regression (LR).

Due to the multiclass classification of the activity into three categories, a simple decision tree was firstly trained to discriminate between inactive (sedentary) and active (light-moderate) states. A second classifier was designed to classify the intensity level (light or moderate). To evaluate the models performance, accuracy, specificity, sensitivity and the active recognition error rate (ARER) proposed in [6] were used. MATLAB software was used for signal processing. DTREG software was used to train and validated the different models. To decrease the overfitting of the models, leave-one-subject-out cross-validation was used.

3 Results

The training dataset comprises 45.3 min of data. Inactivity was classified with a single node decision tree that used only the median absolute deviation for the accelerometer y-axis. 99.94% accuracy was obtained in this stage.

Table 1 report the comparison of accuracy and ARER obtained for each evaluated classifier to discriminate between light and moderate states. The PNN classifier showed the higher (98.79%) accuracy in the training set. However, after applying the safety state machine implemented in the activity detector, LR presented the lowest ARER (6.31%) and the higher (93.69%) accuracy in the validation set.

Table 1. Model performance comparison after safety algorithm.

Model	Training accuracy	Sensitivity	Specificity	Validation accuracy	ARER (%)
DT	71,96%	60,05%	81,30%	42,34%	57,66%
LDA	87,86%	87,44%	88,19%	52,25%	47,75%
PNN	98,79%	98,24%	99,21%	43,24%	56,76%
LR	87,86%	84,45%	96,26%	93,69%	6,31%

4 Discussion and Conclusion

Automated POCs could offer a great benefit to patients. In comparison to closed-loop controlled POC described in scientific literature, this novel iPOC can manage automatically the O_2 supplied to the patient according to the activity intensity that he/she is currently doing in less than 6 s from the start of the activity. This time is lower than the time the user spends updating the configuration manually and will prevent errors caused by the forgetting or misuse of the user to update the O_2 level.

Results obtained in this research lead to recognition algorithms with a 93.69% accuracy using a LR classifier with an ARER lower than the 7%, ensuring a low percentage of the time wrongly classified. At this first approach, basic microcontroller units (MCU) were used to implement the models, limiting the complexity of the features extracted from the signal and restricting the machine-learning model to use. Advanced MCU integrated DSP and multi-core capabilities, what could lead to designs that outperform the one here proposed.

The developed system may improve oxygenation through automatic adaptation to lifestyle, could contribute to improve the patient's therapeutic compliance and might promotes physical activity. Future works include evaluating the iPOC system with patients with COPD in a controlled environment.

Financial Support and Sponsorship. This work has been supported and funded by the EU FEDER ITI Grant for Cadiz Province PI-0006-2017 and by Consejería de Salud y Familias - Junta de Andalucía through Grant PIN-0053-2017.

Conflicts of Interest. DSM, MAFG and ALJ are inventors of the utility model ES U201831680 with the title 'Automatic flow-metering device for oxygen therapy equipment'. ALD and MPM have no conflict of interest.

References

1. Dunne, P.J.: Long-term oxygen therapy (LTOT) revisited: In defense of non-delivery LTOT technology. Rev. Port. Pneumol. **18** (2012)
2. Meena, M., Dixit, R., Kewlani, J.P., Kumar, S., Harish, S., Sharma, D., Singh, M.: Home-based long-term oxygen therapy and oxygen conservation devices: an updated review. Natl. J. Physiol. Pharm. Pharmacol. **5**, 267–274 (2015)
3. Bye, P.T.P., Esau, S.A., Levy, R.D., Shiner, R.J., Macklem, P.T., Martin, J.G., Pardy, R.L.: Ventilatory muscle function during exercise in air and oxygen in patients with chronic air-flow limitation. Am. Rev. Respir. Dis. **132**, 236–240 (1985)
4. British Lung Foundation: Chronic obstructive pulmonary disease (COPD) statistics. https://statistics.blf.org.uk/copd. Accessed 10 June 2019
5. Sanchez-Morillo, D., Olaby, O., Fernandez-Granero, M.A., Leon-Jimenez, A.: Physiological closed-loop control in intelligent oxygen therapy: a review. Comput. Methods Programs Biomed. **146**, 1339–1351 (2017)
6. Sanchez-Morillo, D., Olaby, O., Fernandez-Granero, M.A., Leon-Jimenez, A.: Automatic recognition of daily physical activities for an intelligent-portable oxygen concentrator (iPOC). Lecture Notes Computer Science LNCS, vol. 10305, pp. 212–221 (2017)

7. Egton Medical Information Systems Limited: Use of Oxygen Therapy in COPD, pp. 4–7 (2015)
8. Yang, C.C., Hsu, Y.L.: A review of accelerometry-based wearable motion detectors for physical activity monitoring. Sensors **10**, 7772–7788 (2010)
9. Ainsworth, B.E., Haskell, W.I.L., Whitt, M.C.: e: Compendium of physical activities: an update of activity codes and MET intensities. Med. Sci. Sports Exerc. **32**, S498–S504 (2000)

Optical Metrology of Novel Optically Stimulated Semiconductor Gas Sensor

Yuri Dekhtyar, Maksims Komars and Maksims Sneiders(✉)

Institute of Biomedical Engineering and Nanotechnologies, Riga Technical University, Riga, Latvia
maksims.sn@gmail.com

Abstract. Modern semiconductor sensors which are used for gas detection are based on an MOS (metal–oxide–semiconductor) structure, but the usage of such sensors is limited due to its working principles. Over the last decade, MOS sensors have been combined with a UV-radiation source, but the sensing capabilities of a low gas concentration for those sensors are limited. In previous studies, the possibility to use optically stimulated semiconductors to sense acetone gas at lower concentrations was demonstrated [1]. Another study demonstrated the possibility to sense and determine various solvent gases, where acetone vapour showed the highest signal increase [2].

In this paper, the influence of optical stimulation parameters (irradiance and wavelength) on the output current change and saturation time for an optically stimulated semiconductor gas sensor was studied. It was shown that the maximal deviation of the output current change is achieved at lower tested irradiance $\Phi = 1.6\ \mathrm{mW{\cdot}cm^{-2}}$, and the saturation time decreases with the increase of irradiance. The use of a 367.5 nm LED demonstrated a higher output current change. The saturation time decreased with the increase of the wavelength.

Keywords: Optical stimulation parameters · Acetone gas sensing · Room temperature

1 Introduction

The modern method for measuring volatile organic compounds (VOC) is based on the use of MOS sensors. This method contains a number of drawbacks which are associated with the need to operate the sensor at high temperatures [3, 4].

To eliminate this drawback, MOS sensors have been combined with UV-radiation source, but the sensitivity is insufficient to detect low VOC concentrations. The photon energy of optically stimulated MOS sensors for gas detection should be sufficient to generate free charge carriers in semiconductors, allowing for an excited over the energy gap electron to interact with the target molecule [5, 6].

The intensity (photon flow) affects the performance and sensitivity of the sensor due to the density change of the generated carriers. Increasing the light intensity can improve the sensitivity of the sensor. High optical intensity directly affects the adsorption/desorption rate of the gas molecules and the reaction rate on the semiconductor surface [5, 7], which can lead to the uncontrolled deterioration of sensor sensing properties.

J. Henriques et al. (Eds.): MEDICON 2019, IFMBE Proceedings 76, pp. 462–468, 2020.
https://doi.org/10.1007/978-3-030-31635-8_56

The ability of an optically stimulated semiconductor gas sensor based on the use of p-n junction to detect lower acetone concentrations was shown earlier [8]. To study the effect of optical stimulation parameters on the acetone gas sensing capabilities, the following objectives were defined: a study of an optical stimulation irradiance using a HgXe lamp, and a study of an optimal optical stimulation wavelength using an LED.

2 Materials and Methods

2.1 Test Setup

To achieve the defined objectives, a test set-up was designed and assembled (see Fig. 1). The test set-up consists of three parts: a test chamber to eliminate the influence of external conditions, a source of optical stimulation and an assembly for an acetone gas inlet of a given concentration.

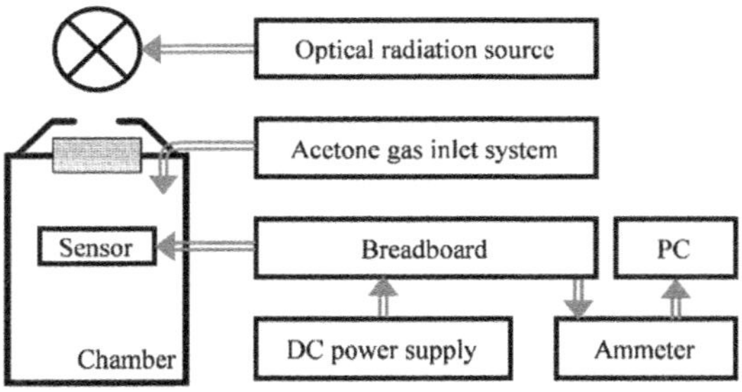

Fig. 1. Experimental test set-up for measurements.

The use of an acetone inlet system allows you to obtain a specific concentration of acetone gas. The method is based on the gas mixture transit method between two volumes until the desired concentration is reached [9, 10].

An optically stimulated semiconductor gas sensor based on a p-n junction was placed in the test chamber. The probability for a free charge carrier to have enough energy to transit over the barrier must follow the Fermi-Dirac distribution function and will be related with the number and energy of the excited electrons [11].

2.2 Influence of the Optical Stimulation Irradiance

An HgXe spotlight source was used for free charge carrier excitation. The source provides a homogeneous flux density on the sensor surface at different irradiance in a range of 0 to 39 mW·cm^{-2}.

2.3 Influence of the Optical Stimulation Wavelength

To study the effect of an optical stimulation wavelength, 10 W LED matrices of different wavelengths from 367.5 nm to 505 nm, normalised to an equal photon flux, were used.

The power of the optical radiation of LED, depending on the current supplied to each individual LED, were converted to the number of photons falling per square unit per second:

$$n = W \cdot \lambda / (h \cdot c) \tag{1}$$

where W – the power of the optical radiation, W; λ – LED wavelength, m; c – speed of light, m/s; h – Planck constant.

Therefore, to conduct tests at different wavelengths for all the LEDs tested, an equal number of photons ($n = 1{,}24 \cdot 10^{16}\ cm^{-2} \cdot s^{-1}$) was taken.

2.4 Sensor Response Measurement

During each measurement in zero-level emission and in the presence of acetone gas, the sensor output signal was recorded. Finally, two values of sensor response on acetone were obtained (see Fig. 2): sensor current change and the saturation time.

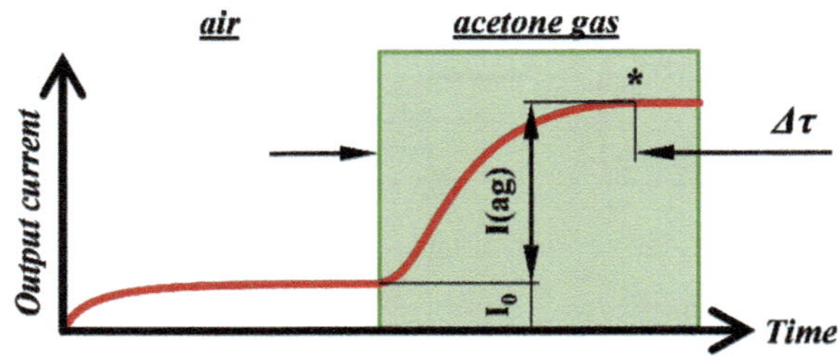

Fig. 2. Sensor signal form for current change and peak current rise time determination, where I_0 – sensor output current value prior to acetone gas inlet, *I(ag)* – sensor output current value, $\Delta\tau$ – the saturation time, * – moment of saturation with acetone gas.

The change in the sensor output current is equal to the ratio of the output current at the time of saturation to the value of the output current in the reference environment.

$$I\% = I(ag)/I_0 \cdot 100\% \tag{2}$$

where I_0 – sensor output current value prior to acetone gas inlet; *I(ag)* – sensor output current value.

In addition, the saturation time ($\Delta\tau$) defined as the time interval between the time of acetone inlet into the test chamber and the time when the output current reaches 95% of the final value.

$$\Delta\tau = \tau(95\%) - \tau_0 \quad (3)$$

where τ_0 – time of acetone inlet, sec; $\tau(95\%)$ – time, when the output current reaches 95% of the final value.

3 Results and Discussion

3.1 Influence of the Optical Stimulation Irradiance

The sensor output current change dependence on optical stimulation irradiance in a range from 0 to 39 mW·cm^{-2} for two different acetone gas concentrations is shown below.

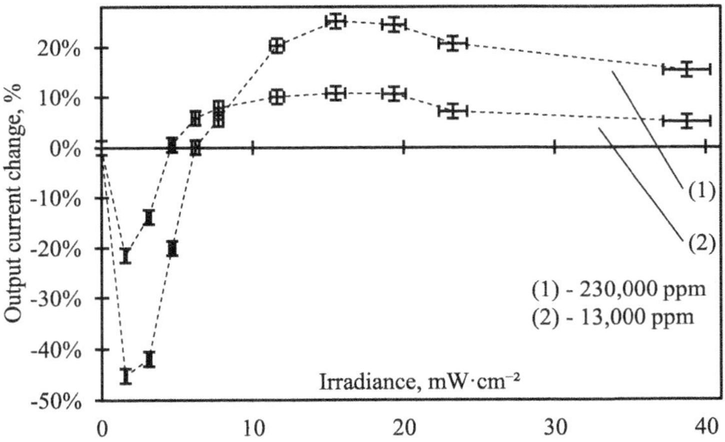

Fig. 3. Influence of optical stimulation irradiance on the output current change.

The results of the effect of optical stimulation irradiance (see Fig. 3) showed an initial decrease of the sensor output current. This was observed in a range from 0 to 6 mW·cm^{-2}. An increase in the output current was observed in a range from 6 to 15.5 mW·cm^{-2} for all the concentrations of acetone gas. For optical irradiance upwards of 15.5 mW·cm^{-2}, a tendency to decrease the output current was observed. The shape of the trendline may represent a different influence of the optical stimulation irradiance on the adsorption/desorption rate of the gas molecules.

The sensor output current saturation time dependence on optical stimulation irradiance in a range from 0 to 39 mW·cm^{-2} for two different acetone gas concentrations is shown below.

The saturation time (see Fig. 4) decreases with the increasing irradiance of optical stimulation (follows logarithmic approximation) and is different for different concentrations of acetone.

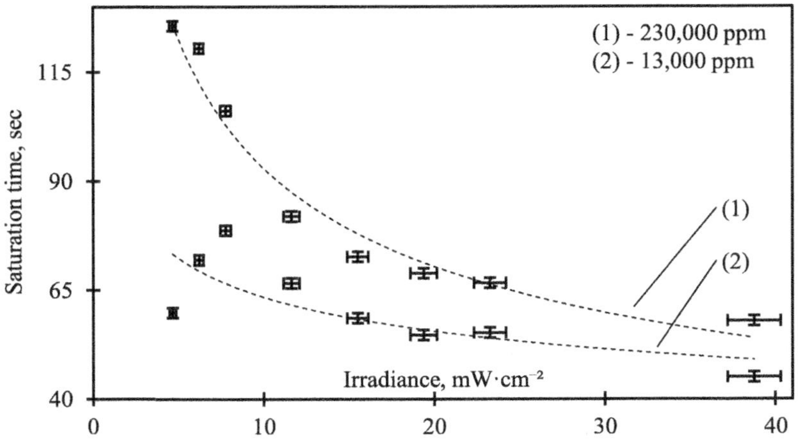

Fig. 4. Influence of optical stimulation irradiance on the saturation time.

3.2 Influence of the Optical Stimulation Wavelength

The sensor output current change dependence on a different emission wavelength of the LED in a range from 367.5 nm to 505 nm in the presence of saturated acetone (230,000 ppm) gas is shown below.

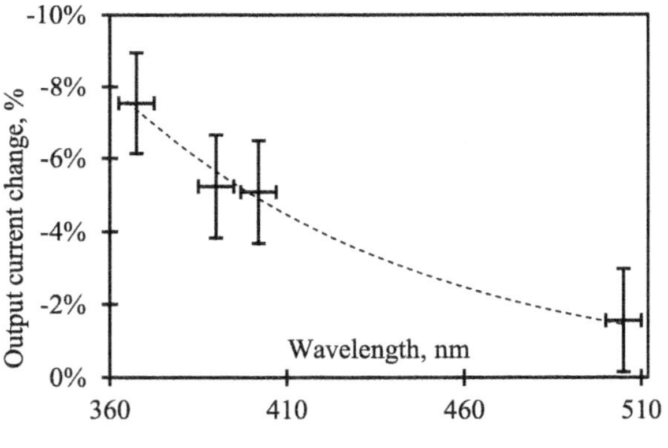

Fig. 5. Influence of an LED optical stimulation wavelength on the output current change.

The results of the influence of an optical stimulation wavelength (see Fig. 5) demonstrate a decrease in the modulus of the output current change at longer wavelengths and can be approximated by an exponential curve. The maximal value was observed at 367.5 nm and a minimal value at 505 nm.

A decrease of the sensor output current (negative output current change) was observed in the presence of acetone gas. This corresponds to the results shown previously (see Fig. 3), as the selected photon flux for the used LED is in the range of lower irradiance (from 0 to 6 mW·cm^{-2}).

The achieved trend follows the exponential Fermi-Dirac distribution function, where the generated by higher photon energy (shorter wavelength) free charge carriers have higher energy, so the probability to overpass the barrier is changing. A decrease of the sensor output current may represent the increase of the barrier height as the result of acetone adsorption, which is to be studied in the future.

The sensor output current saturation time dependence on a different emission wavelength of an LED in a range from 367.5 nm to 505 nm in the presence of saturated acetone (230,000 ppm) gas is shown below.

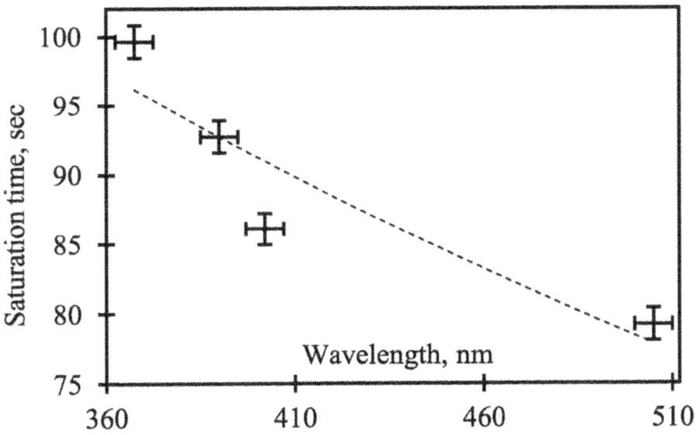

Fig. 6. Influence of an LED optical stimulation wavelength on the saturation time.

The duration of the saturation time (see Fig. 6) decreases with the increasing emission wavelength and can be approximated by a falling regression curve (follows logarithmic approximation). The maximal value was observed at 367.5 nm and the minimal value was 505 nm.

4 Conclusions

The optical stimulation of an optically stimulated semiconductor gas sensor with different irradiance demonstrates two extrema in the output current change values at 1.5 mW·cm^{-2} and 15.5 mW·cm^{-2} with a higher output current change at lower irradiance.

The output current saturation time decreases as the irradiance rises above 15.5 mW·cm^{-2}.

The highest value of output current change achieved using an LED at 367.5 nm. The lower output current change results from using LED radiation compared with ones achieved by using an HgXe lamp may indicate that a shorten (<367.5 nm) wavelengths are required to achieve the better sensing capabilities of an optically stimulated semiconductor gas sensor.

The output current saturation time decreases with the increase of the LED wavelength.

By using an optically stimulated semiconductor gas sensor, it is possible to estimate the presence of acetone by measuring the sensor response time, but measurement uncertainty does not allow you to estimate the gas concentration.

Conflict of Interest. The authors declare that they have no conflict of interest.

References

1. Dekhtyar, Y., Komars, M., Sneiders, M., Selutina, M.: Towards optically induced semiconductor gas sensor: sensing of acetone. In: Jobbágy, Á. (ed.) First European Biomedical Engineering Conference for Young Investigators. IFMBE Proceedings, vol. 50. Springer, Singapore (2015)
2. Dekhtyar, Y., Selutina, M., Sneiders, M., Zunda, U.: Towards optically induced semiconductor human exhalation gas sensor. In: Kyriacou, E., Christofides, S., Pattichis, C. (eds.) XIV Mediterranean Conference on Medical and Biological Engineering and Computing. IFMBE Proceedings, vol. 57. Springer, Cham (2016)
3. Korotcenkov, G., Cho, B.K.: The role of grain size on the thermal instability of nanostructured metal oxides used in gas sensor applications and approaches for grain-size stabilization. Prog. Cryst. Growth Charact. Mater. **58**, 167–208 (2012)
4. Karmakar, M., Mondal, B., Pal, M., Mukherjee, K.: Acetone and ethanol sensing of barium hexaferrite particles: a case study considering the possibilities of non-conventional hexaferrite sensor. Sens. Actuat. B: Chem. **190**, 627–633 (2014)
5. Geng, Q., He, Z., Chen, X., Dai, W., Wang, X.: Gas sensing property of ZnO under visible light irradiation at room temperature. Sens. Actuat. B: Chem. **188**, 293–297 (2013)
6. Ahn, H., Wang, Y., Jee, S.H., Park, M., Yoon, Y.S., Kim, D.: Enhanced UV activation of electrochemically doped Ni in ZnO nanorods for room temperature acetone sensing. Chem. Phys. Lett. **511**, 331–335 (2011)
7. Kamble, V.B., Umarji, A.M.: Defect induced optical bandgap narrowing in undoped SnO_2 nanocrystals. In: AIP Advances, vol. 3 (2013)
8. Dekhtyar, Y., Komars, M., Morozovs, F., Sneiders, M.: Optically induced semiconductor gas sensor: acetone detection range using continuous and cyclic optical irradiation types. In: Eskola, H., Väisänen, O., Viik, J., Hyttinen, J. (eds.) EMBEC & NBC 2017. EMBEC 2017, NBC 2017. IFMBE Proceedings, vol. 65. Springer, Singapore (2018)
9. Chou, J.: Hazardous Gas Monitors: A Practical Guide to Selection, Operation and Applications. McGraw-Hill, New York (2000)
10. Wright, R.S.: EPA Traceability Protocol for Assay and Certification of Gaseous Calibration Standards. EPA, USA (2012)
11. Neamen, D.A.: Semiconductor Physics and Devices: Basic Principles, 4th edn. McGraw-Hill, New York (2011)

Regular Sessions: Bioinformatics, Computational Biology and Systems Biology

Dose–Response Curve: Temporal Dynamics of Respiratory Mechanics in Mice

Otavio Henrique F. Ledesma[1](✉), Renato L. Vitorasso[1], Maria Aparecida de Oliveira[2], and Henrique Takachi Moriya[1]

[1] Laboratory of Biomedical Engineering, Escola Politécnica of the University of São Paulo, São Paulo, Brazil
otavio.ledesma@gmail.br

[2] Department of Pharmacology, Instituto de Ciências Biomédicas, USP, São Paulo, SP, Brazil

Abstract. A very usual practice in literature associated to respiratory mechanics is the challenge of the system, in order to evidence potential differences. Therefore, dose-response curves allow the assessment of respiratory mechanics. Despite the fact that the study of the dose-response curve using peak responses is important and desirable, an evaluation of the temporal dynamics of each dose in a dose-response curve could bring new perspectives about respiratory mechanics. Therefore, the present study aims to understand the kinetics of Newtonian resistance (R_n) of the respiratory system due to transient bronchoconstriction induced by administration of methacholine (MCh) in different concentration. Nine-week female Balb/c mice were studied, nine of the control group and ten of the lung inflammation group (OVA). The doses of MCh were 30, 100 and 300 μg/kg. The constants modelling the single exponential function were compared among of groups and doses. As a result, the increasing values of fitting constants are in direct relation to increasing values of MCh infusion. Additionally, the decay of the curve did follow the increase of the peak value for OVA group. However, it was possible to realize statistically that the temporal dynamics response of bronchoconstriction differs from control to OVA groups only at MCh dose of 100 μg/kg.

Keywords: Assessment of respiratory mechanics · Constant phase model · Respiratory impedance · Dose-response curve

1 Introduction

The assessment of respiratory mechanics is commonly performed through a bronchial challenge induced pharmacologically. The drug used as challenge in this study was methacholine (MCh) and when this drug is administrated in the respiratory system [3, 4], leads to contraction of the smooth muscle of bronchi and bronchioles, i.e. bronchoconstriction.

The literature related to this subject generally presents one value per dose, or concentration, and assesses the whole curve based on these peak values. There are numerous characteristics that may be studied under such approach, such as the

J. Henriques et al. (Eds.): MEDICON 2019, IFMBE Proceedings 76, pp. 471–479, 2020.
https://doi.org/10.1007/978-3-030-31635-8_57

modeling of the effect; the maximum effect; the slope of the curve; the concentration related to the half of the effect [1, 2].

The assessment of the decay or the exponential behavior of the whole [5] and the exponential adjust to a pressure-volume curve [6] are topics well described in the literature. A segmented evaluation of response's temporal dynamics of each dose in a dose-response curve could bring new perspectives about respiratory mechanics. One way to assess the lung mechanical function is through mathematical and computational models.

Therefore, this work aims to adjust an exponential curve in the response pattern of each dose in a dose-response curve and present possible findings on a physiological scenario.

2 Methods

2.1 Animal Preparation

All experiments involving laboratory animals were evaluated and approved by the 'Ethics Committee for Animal Use' (n° 15 on pgs. 16 of Book 3, 03/27 2014) from the Institute of Biomedical Sciences - University of São Paulo, Brazil.

Nine-week female Balb/c mice were studied, 9 of the control group (23.11 $\pm$ 1.27 g) and 11 of the lung inflammation group (OVA), (21.45 $\pm$ 2.16 g), (bronchoalveolar lavage: 2.42 $\pm$ 0.87 vs 15.47 $\pm$ 2.28 $\times$ 10^5 cel/mm^3, control and OVA, respectively). However, data from one of the OVA group was excluded due to the negative value of resistance at the higher dose. In the same way, data set with poor coefficient of determination (COD), COD < 0.9 were also excluded from this study.

The animals were sedated with an intraperitoneal (i.p.) injection of Ketamine (120 mg/kg) and Xylazine (12 mg/kg). When the animal no longer presented sensitivity to pain, tracheostomy and cannulation were performed with a metal cannula 18G (BD Company, Brazil).

2.2 Ovalbumin Sensitization Protocol

In order to induce allergic inflammation by antigen on mice lungs, nine-week-old female Balb/c mice are sensitized on day 0 by i.p. injection of 30 μg ovalbumin (OVA) adsorbed with 1.6 mg aluminium hydroxide (alum) in saline. Second sensitization on day 7 by i.p. injection of 50 μg ovalbumin adsorbed with 1.6 mg aluminium hydroxide (alum) in saline.

Mice are challenged on days 14 and 21 by intranasal (i.n.) instillations of 10 μg and 20 μg respectively of ovalbumin in saline. Third challenge by intratracheal procedure of 20 μg of ovalbumin take place at day 22.

Control animals receive i.n. instillations Phosphate Buffered Saline (PBS), a saline solution, instead of OVA. Administrations were carried out under anesthesia.

2.3 Analysis of Respiratory Mechanics

After the tracheostomy, the animal was connected to a small animal ventilator (flexi-Vent, SCIREQ, Canada) and ventilated in quasi-sinusoidal waveform with a tidal volume of 10 mL/kg, PEEP (Positive End Expiratory Pressure) of 3 cmH_2O and respiratory frequency of 150 breaths per minute. Then, the right jugular vein was dissected by inserting a needle attached to a flexible PVC tube (Critchley Electrical Products PTY, Australia).

The respiratory muscles were blocked with pancuronium bromide (1 mg/kg i.p.). After 7 min, two alveolar recruitment maneuvers were performed. Each recruitment lasted 6 s and consisted of a pressure increase ramp from PEEP up to the value of 30 cmH_2O, and sustained until the end of the maneuver.

After the recruitments, PBS was injected through the right jugular vein. Then, 15 perturbations of 3 s each were performed, in apnea, with 10 s interval between each assessment. The perturbation was composed of a sum of 13 sinusoids (Hz): 1, 1.5, 2.5, 3.5, 5.5, 6.5, 8.5, 9.5, 11.5, 14.5, 15.5, 18.5 and 20.5. After the perturbation, the animals were ventilated for two minutes until the next injection.

The same perturbations were performed for all doses of Methacholine (MCh): 0.03, 0.10, 0.30 mg/kg. Automation routines of the ventilator control program were used to perform the perturbations.

2.4 Mathematical Model

The impedance (Z) data were modeled according to the Constant Phase Model (CPM), first proposed by Hantos Z, Suki B and Daróczy B. in 1987 [7]. The Constant Phase Model is adjusted to the respiratory system input impedance calculated by means of multiple linear regression and in accordance to literature [8, 9].

The respiratory system input impedance by Constant Phase Model is fitted by four-parameters: R_n (airway Newtonian resistance), I (airway inertance), H (elastance, related to energy stored in the tissue) and G (viscance, related to tissue dissipation of energy).

$$Z(f) = R_n + j(2\pi f)I + \frac{G - jH}{(2\pi f)^{\alpha}} \tag{1}$$

Where the parameter α characterize the hysteretic properties of the respiratory tissues:

$$\alpha = \frac{2}{\pi}\tan^{-1}\left(\frac{H}{G}\right) \tag{2}$$

Therefore, the impedance of a normal lung can be characterized by an airway component (R_n and I) and a component due to the tissues (G and H) [8].

2.5 Data Analysis

A single exponential function has been used to represent the decay behavior of Newtonian resistance (R_n). The peak response values of R_n at different doses of MCh were identified and a nonlinear model was adjusted to fit R_n in a decay behavior for each dose. An exponential function of the form:

$$R_n \sim \mathrm{c} + \mathrm{b}e^{-at} \tag{3}$$

where R_n is the airway Newtonian resistance, t is time, a, b and c are constants modeling de function.

The best-fit exponential function is obtained by an iterative least-square method. A nonlinear regression model implemented in R (R Core Team 2015) package *minpack.lm* (Timur V. Elzhov, Katharine M. Mullen, Andrej-Nikolai Spiess, Ben Bolker 2016) has been used. The *minpack.lm* package provides an R interface for solving nonlinear least-squares problems by a modification of the Levenberg-Marquardt algorithm, with support for lower and upper parameter bounds.

2.6 Statistical Analysis

Results are presented as the mean ± standard error of mean (S.E.M.). Statistical significance was assessed by parametric methods using two-way analysis of variance (ANOVA) to determine differences between groups and doses, followed by a Tukey Honest Significant Differences for performing multiple pairwise-comparison between the means of groups. When appropriate, one-way ANOVA was used. Results were considered significant at $p \leq 0.05$ or 5%.

The graphs were prepared with R package ggplot2 (Hadley Wickham, Winston Chang, Lionel Henry, Thomas Lin Pedersen, Kohske Takahashi, Claus Wilke, Kara Woo 2018).

3 Results

The results in Fig. 1, typify the values of R_n obtained from mice studied, in this particular case an animal from Control group. The response to MCh bolus injection begins with sharp peaks in R_n followed by decreasing values of R_n due to relaxation of airway smooth muscle. The plot shows the reactions of R_n to increasing MCh-infusion rates up to 300 µg/kg. The results measurements were in concordance with expectations from literature sources [3, 4, 8].

That means, MCh was administered at time zero and short after, R_n achieves its maximum or peak value, followed by decaying values. The timing of the peak in the response was similar in all cases, from 0 to 20 s.

The data set in Fig. 1 is from Control group of mice and as can be seen, as the MCh dose increases, the bronchoconstriction increases, leading to higher values of R_n. Bronchoconstriction on inflamed (OVA) group of mice is more acute, leading to even higher values of R_n.

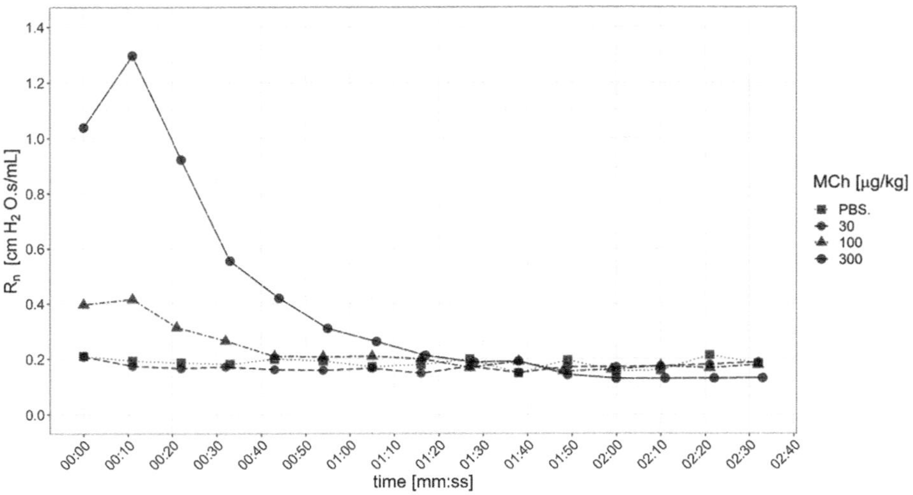

Fig. 1. Typical temporal dynamics of Rn during acute bronchoconstriction in Control group of mice. MCh was administered intravenous at time zero with increasing MCh doses up to 300 μg/kg. The animals were ventilated for two minutes until the next injection with increasing dosage.

The decaying behavior of R_n due to relaxation of airway smooth muscle that follows the sharp peak in R_n has been modelled by an exponential function given by Eq. (3). In Fig. 2, it can be seen the R_n values from the same data set, i.e. same animal and experiment of Fig. 1, and the fitted curve according to the description above.

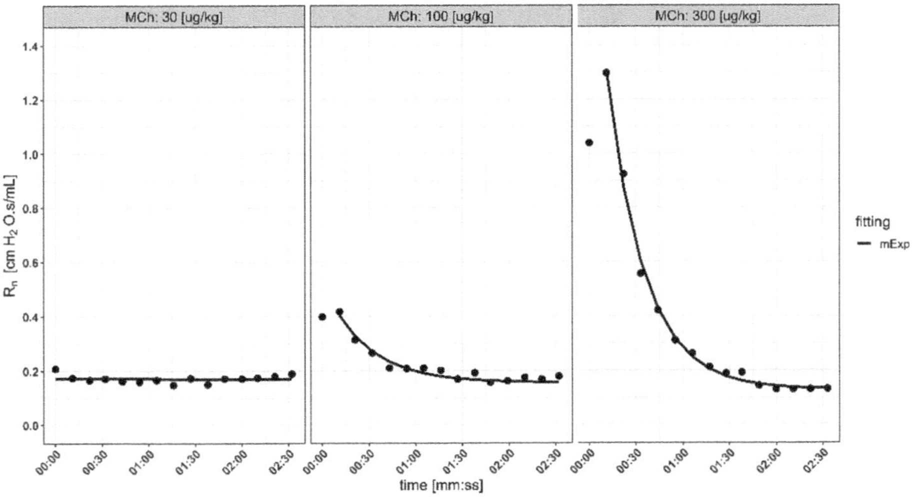

Fig. 2. Typical temporal dynamics of acute bronchoconstriction in mice with nonlinear fitting model, a decay exponential (same data set as Fig. 1).

As can be seen on Fig. 2, as the MCh doses increases the decayment of the curve becomes more similar of an exponential function. Therefore, is possible to notice that a single exponential function can fit the R_n values obtained from CPM.

From Table 1 is possible to observe that significance probability value (p-value) associated with dose factor had showed lower values than the significance level adopted of 5%. Thus, the probability of parameters mean's values been equal across doses is low enough for all used doses. Due to that on Fig. 3 measured data had been analyzed per dose value. Hence, one-way analysis of variance (one-factor ANOVA) had been used regarding group factor and presented on it.

Table 1. Significance probability values from the two-way analysis of variance calculated for the constants modelling the fitted exponential function.

Factors	p-value temporal constant: a $[s^{-1}]$	p-value peak constant: b [$cmH_2O.s.mL^{-1}$]	p-value basal constant: c [$cmH_2O.s.mL^{-1}$]
Dose	0.00013	8.14e-07	0.02070
Group	0.06994	0.00011	0.34830

Fig. 3 presents the summary for the constant values modeling the R_n decaying behavior as an exponential function. Figure 3 exhibits the values of a, b and c constants for both groups (Control and OVA) in all doses presents in this study.

Increasing values of constant b modeling the peak or maximum values of R_n are in accordance with observed higher peak values of R_n as the MCh increases and for inflamed animals. In addition, inverse temporal constant a, also increase values as MCh dose values increases. Similar observation is valid for inflamed mice, presenting higher values of inverse temporal constant a. Regarding basal constant c, is possible to observe a more constant value independent of MCh infusion rate.

4 Discussion

In this study, the bronchoconstriction was induced pharmacologically, leading to a bronchial narrowing. This narrowing is caused by the contraction of the smooth muscles present in the respiratory tract. Narrow airways mean increased airway resistance. Hence, increased values of R_n and I. From its two components, R_n and I, the airway component with higher significance is the R_n due to low values presented by airway inertance, I. Thus, the scope of this study was limited to the Newtonian resistance (R_n).

Dose-response curves are commonly studied using only peak responses [10], however an evaluation that considers others aspects of a particularly subject is highly desirable and to foster the comprehension of respiratory mechanics this work evaluated the dose-to-dose decay of the Newtonian resistance of the respiratory system, R_n.

The increase or decrease of R_n is related with the contraction of airway smooth muscles and smooth muscles relaxation [11] and the reactivity can be assess fitting an exponential curve to the whole dose-response curve [12]. The expected is a more

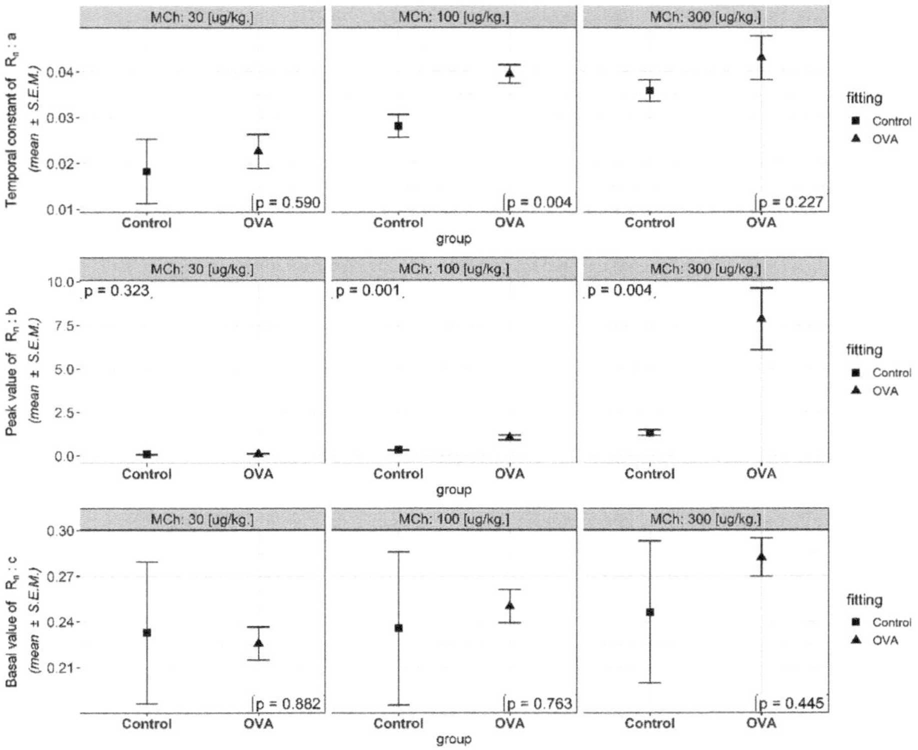

Fig. 3. The parameter *a* is related to time decay of R_n after the application of MCh. It was found difference ($p < 0.05$) when comparing groups (Control vs OVA) at the dose of 100 µg/kg. For parameter *b*, the only dosage of MCh to not have significance difference is 30 µg/kg. The parameter *c* to describe the basal values of R_n presents similar values across MCh doses.

reactive animal in the OVA model, which is translated as an animal with a higher temporal constant of the exponential curve. Additionally, it was expected an increase of these values along the doses in both groups.

It is possible to notice from Fig. 3, that peak constant *b* presents higher values for OVA group when compared with those from Control group, this was expected, since this parameter represents, at the $t = 0$, the R_n peak or maximum value. Besides that, peak constant *b* increases values as bronchoconstriction also increases due to higher doses of MCh.

From Fig. 3 it can be observed that the decay of each dose of both groups was progressively higher through all the concentrations. Visually, the decay of OVA group was always higher than the control group. The decay constant *a*, of inflamed animals in every scenario increase. However, regarding temporal constant *a*, there was statistically difference only at the dosage of 100 µg/kg of MCh. That means, the significance probability value associated with group factor had fail for other doses rather than 100 µg/kg of MCh.

Analysis of variance makes use of F-statistic, being F-statistic a ratio of two variances. Where F-statistic denominator is the within-group variance. As can be seen on Fig. 3 the S.E.M. for temporal constant *a* for MCh dose equal to 300 µg/kg is high. As the error increases, it becomes more likely that the observed differences between group mean at MCh 300 µg/kg are masked by the error not reflecting the actual differences at the population level.

As far as basal constant *c* is possible to observe that as the dosage of MCh increases, the basal or minimum value of R_n presents a slightly increase as well. Higher time of observation and measurements would be required to allow the MCh elimination in order to achieved basal values consistent with PBS measurements. This is particularly true at MCh equal to 300 µg/kg. It can be expected that in a presence of a longer period of experiment and measurement an even sharper decayment would be observed in this scenario. Leading to a higher value of inverse temporal constant *a* for OVA animals. Thus, resulting in a statically distinct value when submitted to an analysis of variance.

In spite the fact, a single exponential function fits and can describe the decaying behavior of R_n due to relaxation of airway smooth muscles after the response to a MCh challenge. The MCh dose of 30 µg/kg is a clear exception of that. From Figs. 1 and 2, there is negligible exponential decayment for this particular dose. Newtonian resistance (R_n) barely shows some increase in value for such MCh dosage. Hence, it was expected that the exponential curve would not fit well this particular dosage of MCh.

From the achieved results, in this study, it is possible to realize that temporal dynamics of bronchoconstriction differs from Control to allergically inflamed (OVA) groups of BALB/c mice only at MCh dose equal to 100 µg/kg.

Walker, Kraft and Fisher [12] advocates that more robust assessment of lung mechanics in mice, both in methodology and scope should yield better clinical translation. Given that human lung mechanics measurements are important to the assessment and prediction of the course of asthma. The optimization of these measurements in mice would be beneficial. Therefore, the current work had aimed to expand the assessment of airway Newtonian resistance beyond of the usual analysis performed employing only the average or peak values of R_n. In order to do so, the current work also perform the analysis including the temporal dynamic behavior of R_n, that takes place at the relaxation phase and is not so common in associated literature.

Conflict of Interest. The authors have no conflict of interest to declare.

References

1. Hardman, J., Limbird, L., Gilman, A.: Goodman & Gilman's The Pharmacological Basis of Therapeutics, 10th edn. McGraw-Hill, New York (2000)
2. Katzung, B.: Basic and clinical pharmacology. Pharmacol. Br. **9**, 911–973 (2012)
3. Peták, F., Hantos, Z., Adamicza, A., Asztalos, T., Sly, P.D.: Methacholine-induced bronchoconstriction in rats: effects of intravenous vs. aerosol delivery. J. Appl. Physiol. **82**(5), 1479–1487 (1997)

4. Jonasson, S., Hedenstierna, G., Hedenström, H., Hjoberg, J.: Comparisons of effects of intravenous and inhaled methacholine on airway physiology in a murine asthma model. Respir. Physiol. Neurobiol. **165**, 229–236 (2009)
5. Peleg, M., Normand, M.D., Damrau, E.: Mathematical interpretation of dose-response curves. Bull. Math. Biol. **59**(4), 747–761 (1997)
6. Salazar, E., Knowles, J.H.: An analysis of pressure-volume characteristics of the lungs. J. Appl. Physiol. **19**, 97–104 (1964)
7. Hantos, Z., Suki, B., Daróczy, B.: Constant-phase modelling of pulmonary impedance. Bull. Eur. Physiopathol. Respir. **12**, 326s (1987)
8. Bates, J.H.T.: Lung Mechanics: An Inverse Modeling Approach, 220p. Cambridge University Press, Cambridge (2009)
9. Moriya, H.T., Moraes, J.C.T.B., Bates, J.H.T.: Nonlinear and frequency-dependent mechanical behavior of the mouse respiratory system. Ann. Biomed. Eng. **31**(3), 318–326 (2003)
10. Arantes-Costa, F.M., Zoriki, S., Santos, M.H.C., Kobata, C.H.P., Vieira, J.E., Martins, M.A.: Effects of ventilation, humidity and temperature on airway responsiveness to methacholine in rats. Eur. Respir. J. **19**(6), 1008–1014 (2002)
11. An, S.S., Bai, T.R., Bates, J.H.T., Black, J.L., Brown, R.H., Brusasco, V., et al.: Airway smooth muscle dynamics: a common pathway of airway obstruction in asthma. Eur. Respir. J. **29**(5), 834–860 (2007)
12. Walker, J.K.L., Kraft, M., Fisher, J.T.: Assessment of murine lung mechanics outcome measures: alignment with those made in asthmatics. Front. Physiol. **3**, 491 (2012)

Influence of Astrocytic Gap Junction Coupling on *in Silico* Neuronal Network Activity

Barbara Genocchi(✉), Kerstin Lenk, and Jari Hyttinen

Faculty of Medicine and Health Technology, BioMediTech, Tampere University, Tampere, Finland
barbara.genocchi@tuni.fi

Abstract. Astrocytes cover a plethora of roles supporting neurons in their maturation and regulating the concentrations of several ions and neurotransmitters. Moreover, astrocytes dysfunctions are, nowadays, suspected to have important implication in several brain diseases, as for example in epilepsy and Alzheimer's disease. The astrocytes themselves are forming a network mediated by gap junctions. A loss of gap junctions between astrocytes has been connected with epilepsy. The aim of this study is to computationally test the influence of astrocytes connectivity in regulating activity in the neuronal network. To conduct the study, it has been used an *in silico* neuron-astrocyte model developed in our group. The model simulates the processes governing the communications between an astrocyte and a pre- and a postsynaptic neuron in the tripartite synapse, as well as between astrocytes through gap junction coupling. The modeled network comprises 250 neurons and 107 astrocytes. Three different astrocytic connectivity levels have been studied – representing 0, 2 and 4 gap junctions on average per astrocyte. Additionally, three different noise levels have been applied to the presynaptic terminal to simulate low, high and hyperactivity. Since the activation of astrocytes is driven by the activity of the neuronal network, the results showed that in case of low activity astrocytes were not activated and did not regulate neuronal activity. In case of high neuronal activity and hyperactivity, astrocytes showed an increased capability of downregulating neuronal activity when increasing the astrocytic connectivity. These results are in accordance with several *in vivo* experiments from different laboratories.

Keywords: Astrocytes · Epilepsy · Simulation · Gap junctions · Connectivity

1 Introduction

Since early 1990s and with the introduction of the concept of the tripartite synapse [1–3], several roles have been acknowledged to astrocytes. Nowadays, astrocytes are known to promote synapse formation, maturation and elimination, and to help in long-distance communication and information integration. They also provide maintenance of ionic and neurotransmitters homeostasis, in order to preserve spatial and temporal encoding of synaptic transmission and to avoid excitotoxicity [4]. Several reviews have focused on astrocyte calcium signaling [5, 6], which can be triggered by neuronal activity and stimulates gliotransmitters release from astrocytes to neurons. Therefore, calcium signaling can in turn modulate the neuronal activity [7]. During the last years,

J. Henriques et al. (Eds.): MEDICON 2019, IFMBE Proceedings 76, pp. 480–487, 2020.
https://doi.org/10.1007/978-3-030-31635-8_58

it has become clear that particular abnormal functionalities in glial cells and astrocytes can be correlated with central nervous system (CNS) diseases [8, 9]. The main astrocytic dysfunctions thought to be related with epilepsy are impaired potassium buffering, abnormalities in astrocytic gap junctions (GJs) and abnormal uptake of glutamate. In particular from *in vivo* experiments, researchers reported spontaneous epileptiform activity in the CNS due to a loss of Cx30 and Cx43. Those two connexins are fundamentals specifically for the formation of astrocytic GJs [10].

During the last decade, the increasing quest for understanding the roles of astrocytes in health and brain diseases together with the large amount of data from *in vitro* and *in vivo* research has stimulated the use of computational tools and models at several scales: from single astrocytes to astrocytes networks towards neuron-astrocytes *in silico* models [11].

The aim of this study is to computationally investigate the role of astrocytic network connectivity in the regulation of neuronal activity. In order to evaluate the influence of astrocytic GJs connectivity, our custom neuron-astrocyte network model has been used with different amounts of GJs between astrocytes [11, 12]. To the best of our knowledge, the role of GJs have not been studied before in biophisically plausible computational models comprising larger networks as well as external stimuli as input. Amiri et al. [13] and Naeem et al. [14] presented network models with neuron and astrocytes as well as GJs between neighboring astrocytes, but lacking a proper network topology; in fact, in both cases the astrocytes were distributed in lines and the cells contacted only the two neighboring astrocytes.

2 Methods

The astrocyte–neuronal network model used in this study is based on a tripartite synapse model [12]. It is a modified version of Tsodyks-Markram presynaptic model [15], used as described by De Pittà et al. [16] and combined with astrocytic effects as described in the latter paper. To simulate the neuronal activity, we used a spiking neuronal network, called INEX model [17], which comprises excitatory and inhibitory neurons connected by synapses and includes a noise term. The neuronal activity is created with Poisson processes. The above mentioned Tsodyks-Markram model simulates the glutamate release from the presynaptic terminal. It assumes that at the time in which a spike arrives only part of the neurotransmitter is ready to be released, and only a small amount of these releasable resources are de facto released.

The astrocyte IP_3 and calcium dynamics are also included in the model as described in De Pittà et al. [16], where intracellular calcium release from the stores follows the IP_3 release. Astrocytes, when active, release glutamate and signal back to the synapses that are connected to them. Calcium release stimulates a wave propagation between neighboring astrocytes through GJs. In our present model, the propagation is implemented according to the model introduced by Lallouette et al. [18].

The model simulates 180 s and the time step used is 5 ms. These two parameters are a satisfactory compromise to consider both, the fast dynamics of neurons and the slower astrocytic response, keeping the simulation time on the computer cluster as reasonable as possible. We simulated each of the below described conditions three times.

The neuronal network, comprising 200 excitatory neurons and 50 inhibitory neurons, was randomly created once and the connections between neurons were created following a Gaussian distribution based on the distance between two neurons. The simulations produced networks with roughly 10,000 excitatory synapses.

Astrocytes cover non-overlapping domains and the average radius of a rodent protoplasmic astrocyte is ~30 μm, considering the soma and the primary processes [19]. Hence, in our model if the distance between astrocytes is shorter than this value the connection is not created. The neuronal network was then combined with three different astrocytic networks, all of them counting 107 astrocytes. On average each astrocyte in the network is connected to 90 nearby excitatory synapses. By varying the parameters of the Gaussian distribution of the astrocyte connections, the chosen average number of GJs for each astrocyte was respectively 0, 2 and 4 (Fig. 1A, B, and C).

To elicit the activation of astrocytes, three different noise levels have been applied to the presynaptic terminals in the neurons. These simulated the three conditions of low, high and hyperactivity. The last one could be considered as an epileptic form. The high noise was three times the noise value used for low activity and the hyperactivity was 8 times the low noise value.

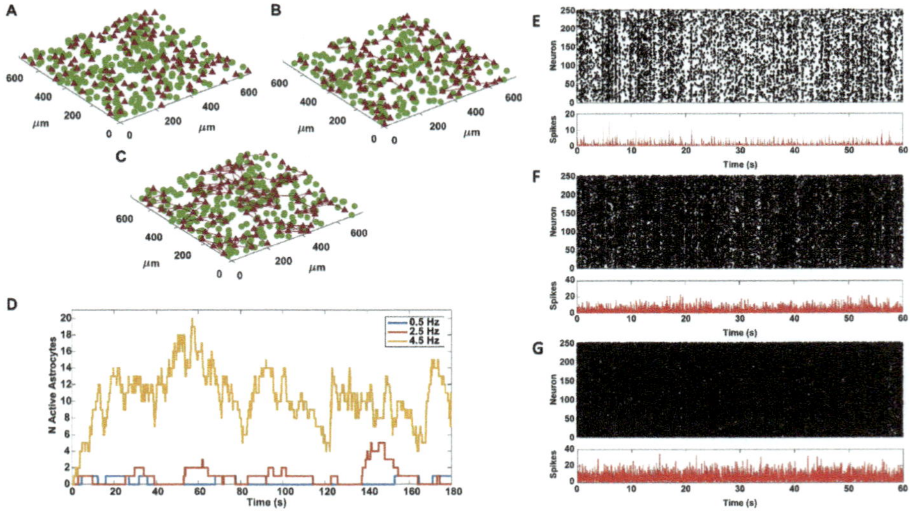

Fig. 1. 2D graphical representation of the three different networks configurations used in the simulations: (A) No gap junction coupling, (B) two GJs, and (C) four GJs per astrocyte on average. Green dots represent excitatory and inhibitory neurons and red triangles represent astrocytes. The red lines represent connections including GJs between astrocytic processes. (D) Comparison between astrocytic activation in the case of low (blue), high (red) neuronal activity and hyperactivity (orange) for the network with no GJs between astrocytes. The plot shows the number of active cells at each time step (5 ms). On the right are shown examples of a raster plot from a simulation with low activity (E), high activity (F) and hyperactivity (G). In the upper part of each subplot, black dots represent spikes occurring in the neurons for the first 60 s of the simulation. In the lower part of each subplot is shown the number of spikes binned to 5 ms.

To conduct neuronal burst analysis, a modified version of an algorithm by Kapucu et al. [20] was used. This tool calculates intrinsic bursting using the cumulative moving average and the skewness of the interspike interval histogram.

3 Results

3.1 Low Noise Condition

To simulate low neuronal activity, a weak noise input was applied to the neurons, resulting in an average presynaptic frequency of 0.5 Hz. The consequent number of active astrocytes is shown in Fig. 1D (in blue).

Since the astrocytic activation is correlated with the presynaptic activity, the average number of active astrocytes was very low. A raster plot and the number of spikes binned per 5 ms of the first 60 s of the simulation for postsynaptic neurons in case of no gap junction coupling between astrocytes is shown as an example in Fig. 1E.

The spike rate, calculated in spikes/min, varied in similar intervals of values for all three configurations (Table 1). In fact, the notches, such as the triangular cuts on the boxes sides, in the box plots in Fig. 2A overlap. This means that the three medians represented by red lines in the plot, cannot be considered statistically different.

In these low activity simulations, bursting is mainly not present in neurons. In Fig. 2A, the medians of the burst rate are all around very low values (Table 1). However, a few neurons presented some bursting, shown by the red pluses. Pluses show data points beyond the whiskers. This probably occurred in neurons with more synaptic connections, enhancing their own activity. Due to this high variability in the bursts rate, the standard deviation appears to be higher than the value itself.

Table 1. Spike rate (spikes/min) and burst rate (burst/min) with 0, 2 and 4 GJs per astrocyte for the simulated low activity.

N of GJs	Avg. spike rate ± std	Avg. burst rate ± std
0	32.05 ± 9.65	0.07 ± 0.16
2	35.54 ± 10.88	0.06 ± 0.16
4	34.75 ± 10.92	0.09 ± 0.20

3.2 High Noise Condition

To simulate diffused high activity in the neuronal network, a noise input was applied to the presynaptic terminal. The noise input was three times larger than the one used for low activity and resulted in a presynaptic activity with a frequency of 2.5 Hz (Fig. 1F). This led to a stronger activation of the astrocytes as can be seen from the red curve in Fig. 1D.

The results show that when increasing the number of GJs, the postsynaptic spike and burst rate were decreased (Table 2, Fig. 2B). The spike rate decrease is clearer, when comparing the no gap junction case to 2 and 4 GJs on average. Between 2 and 4 GJs on average, it is still possible to notice a small increase in the downregulation of the activity.

In addition, the burst rate decreased with increasing number of GJs, but also in this case the high variability of bursting amongst all the 250 neurons resulted in a high standard deviation. Since the notches of the boxes for spike and burst rate of the no gap junction condition and respectively the 2 and 4 GJs do not overlap, we can conclude that the two distributions are statistically different with 95% confidence from the first one.

Table 2. Spike rate (spikes/min) and burst rate (burst/min) with 0, 2 and 4 GJs per astrocyte for the simulated high activity.

N of GJs	Avg. spike rate ± std	Avg. burst rate ± std
0	140.52 ± 42.81	2.46 ± 2.03
2	126.81 ± 37.54	1.74 ± 1.54
4	120.23 ± 34.62	1.59 ± 1.35

3.3 Hyperactivity Condition

To simulate diffused hyperactivity in the neuronal network, a noise input 8 times larger than the one used for low activity was applied to the presynaptic terminal. This resulted in a presynaptic activity with a frequency of 4.5 Hz (Fig. 1G). The stronger activation of the astrocytes can be seen in the orange curve in Fig. 1D. The active astrocytes overexcite the synapses connected to them by releasing glutamate in the extrasynaptic space. In turn, the inhibitory neurons are not anymore capable to downregulate the system.

In order to evaluate, if the spike and burst rate was decreasing when increasing the GJs coupling, it was conducted a paired-sample Student's t-test. For both spike rate (SR) and burst rate (BR) (Table 3, Fig. 2C), the 2 and 4 GJs distributions were significantly decreased in respect to the no gap junction distribution (SR: $p = 0.012$ and $p = 0.002$, respectively. BR: $p = 0.003$ and $p = 4.67e{-}04$, respectively).

Table 3. Spike rate (spikes/min) and burst rate (burst/min) with 0, 2 and 4 GJs per astrocyte for the simulated hyperactivity.

N of GJs	Avg. spike rate ± std	Avg. burst rate ± std
0	280.49 ± 78.21	13.53 ± 8.37
2	272.71 ± 69.65	12.51 ± 7.17
4	271.79 ± 83.97	12.57 ± 8.50

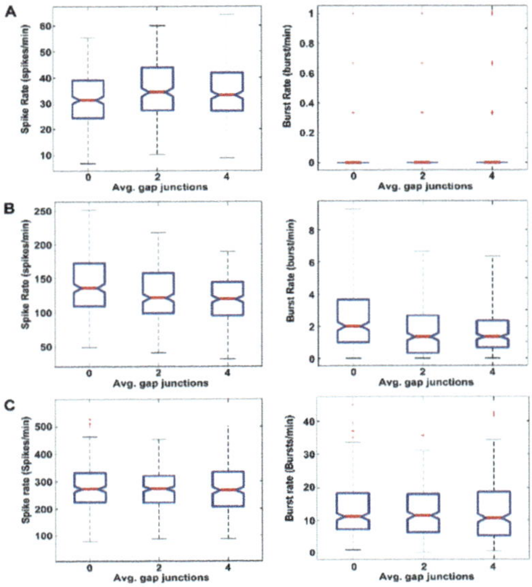

Fig. 2. Box plots of spike rate (spikes/min; left) and burst rate (bursts/min; right) for low activity (A), high activity (B) and hyperactivity (C). The red lines represent the medians of the data distributions. The notches indicate the 95% confidence interval of the median.

4 Conclusions

During the last decade, researchers have investigated whether dysfunctions in astrocytes can cause or partially cause CNS disorders. Several *in vivo* experiments seem to confirm that hypothesis [8–10].

This study aim was to computationally evaluate the role of astrocytic connectivity on the capability to downregulate high activity in an astrocyte-neuron network. Astrocytic connectivity was expressed in terms of average number of GJs per astrocyte. Three different astrocyte network connectivity levels were investigated: 0, 2 and 4 GJs on average per astrocyte in the network.

The activation of astrocytes is triggered by the neuronal network activity. Neuronal activity, in turn, is modulated by the astrocytes controlling the amount of available resources at the synapses through local calcium activity and subsequent glutamate release.

In the low noise condition, as expected, it was not possible to observe any regulatory functions related to the number of GJs (Table 1, Fig. 2A). The input noise resulted in a low number of active astrocytes in the network. In high and hyperactivity, instead, the number of active astrocytes was higher, producing increased modulatory response thus decreasing the spike and burst rates (Fig. 2B and C).

These results demonstrate that the capability of the astrocytic network in downregulating the hyperactivity depends on the number of GJs and thus on the astrocyte network connectivity. These results are in accordance to the experiments in which the deletion of Cx43 and Cx30, resulting in a cell-specific loss of GJs, stimulated the generation of spontaneous epileptiform activity [10].

Our simulation results highlight the possible role of astrocytes in the neuronal regulation that could also affect the susceptibility of the neuronal system to epileptic events. Astrocytes may thus also offer a new drug target for treating neural diseases.

Acknowledgment. This research has received funding from the European Union's Horizon 2020 research and innovation programme under the Marie Skłodowska-Curie grant agreement No. 713645. K. L. is funded by the Academy of Finland (decision no. 314647).

Conflict of Interest. The authors declare that they have no conflict of interest.

References

1. Nedergaard, M.: Direct signaling from astrocytes to neurons in cultures of mammalian brain cells. Science **263**, 1768–1771 (1994)
2. Parpura, V., Basarsky, T.A., Liu, F., et al.: Glutamate-mediated astrocyte–neuron signaling. Nature **369**, 744–747 (1994)
3. Araque, A., Parpura, V., Sanzgiri, R.P., et al.: Tripartite synapses: glia, the unacknowledged partner. Trends Neurosci. **22**, 208–215 (1999)
4. Vasile, F., Dossi, E., Rouach, N.: Human astrocytes: structure and functions in the healthy brain. Brain struct. funct. **222**, 2017–2029 (2017)
5. Bazargani, N., Attwell, D.: Astrocyte calcium signaling: the third wave. Nat. Neurosci. **19**, 182–189 (2016)
6. Perea, G., Navarrete, M., Araque, A.: Tripartite synapses: astrocytes process and control synaptic information. Trends Neurosci. **32**, 421–431 (2009)
7. Savtchouk, I., Volterra, A.: Gliotransmission: beyond black-and-white. J. Neurosci. **38**, 14–25 (2018)
8. Seifert, G., Schilling, K., Steinhäuser, C.: Astrocyte dysfunction in neurological disorders: a molecular perspective. Nat. Rev. Neurosci. **7**, 194–206 (2006)
9. Dossi, E., Vasile, F., Rouach, N.: Human astrocytes in the diseased brain. Brain Res. Bull. **136**, 140–147 (2018)
10. Seifert, G., Carmignoto, G., Steinhäuser, C.: Astrocyte dysfunction in epilepsy. Brain Res. Rev. **63**, 212–221 (2010)
11. Oschmann, F., Berry, H., Obermayer, K., et al.: From in silico astrocyte cell models to neuron-astrocyte network models: a review. Brain Res. Bull. **136**, 76–84 (2018)
12. Lenk, K., Raisanen, E., Hyttinen, J.A.K.: Understanding the role of astrocytic GABA in simulated neural networks. In: Conference Proceedings of the IEEE Engineering in Medicine and Biology Society, pp. 6121–6124 (2016)
13. Amiri, M., Hosseinmardi, N., Bahrami, F., et al.: Astrocyte- neuron interaction as a mechanism responsible for generation of neural synchrony: a study based on modeling and experiments. J. Comput. Neurosci. **34**, 489 (2013)
14. Naeem, M., McDaid, L.J., Harkin, J., et al.: On the role of astroglial syncytia in self-repairing spiking neural networks. IEEE Trans. Neural Netw. Learn. Syst. **26**(10), 2370–2380 (2015)
15. Tsodyks, M.V., Markram, H.: The neural code between neocortical pyramidal neurons depends on neurotransmitter release probability. PNAS **94**, 719–723 (1997)

16. De Pittà, M., Volman, V., Berry, H., et al.: A tale of two stories: Astrocyte regulation of synaptic depression and facilitation. PLoS Comput. Biol. **7**, e1002293 (2011)
17. Lenk, K.: A simple phenomenological neuronal model with inhibitory and excitatory synapses. In: Conference Proceedings of NOLISP 2011. LNAI, vol. 7015 (2011)
18. Lallouette, J., De Pittà, M., Ben-Jacob, E., et al.: Sparse short-distance connections enhance calcium wave propagation in a 3D model of astrocyte networks. Front. Comput. Neurosci. **8**, 1–18 (2014)
19. Oberheim, N.A., Takano, T., Hanet, X., et al.: Uniquely hominid features of adult human astrocytes. J. Neurosci. **29**(10), 3276–3287 (2009)
20. Kapucu, F., Tanskanen, J., Mikkonen, J.E., et al.: Burst analysis tool for developing neuronal networks exhibiting highly varying action potential dynamics. Front. Comput. Neurosci. **6**, 1–14 (2012)

Heart Closed-Loop Model for the Assessment of Cardiac Pacing

Niccoló Biasi[1(✉)] and Alessandro Tognetti[2,3]

[1] Dipartimento di Medicina Clinica e Sperimentale, Universitá di Pisa, Pisa, Italy
biasi.niccolo96@gmail.com
[2] Dipartimento di Ingegneria dell'Informazione, Universitá di Pisa, Pisa, Italy
[3] Centro E. Piaggio, Universitá di Pisa, Pisa, Italy

Abstract. We developed a closed-loop model of cardiac stimulation using a finite element model of the whole-heart embedded in the torso that is proposed as an useful tool for pacemaker design and testing. The electrical activity of the cardiac tissue is reproduced with a bidomain model incorporated with the FitzHugh-Nagumo equations. The finite element model is developed in Comsol Multiphysics, both in two and in three dimensions and then exported in Simulink environment where the pacemaker algorithm is implemented. To validate the model, we chose a demand inhibited pacemaker, which stimulates the myocardium only if the intrinsic activity of the heart is not revealed, but every type of pacemaker can be simulated. The model generates a controlled spontaneous activation in the sinoatrial node and it is also able to reproduce realistic electrocardiographic signals and the effects that the stimulation has on them.

Keywords: Heart modeling · Closed loop pacemaker simulation · Cardiac bidomain · FE models

1 Introduction

Mathematical modelling of electrical activity of the heart is a growing field, with models that are increasingly complex. Nevertheless most of them include only atrium or ventricles, are often focused on specific pathological conditions and are not suitable for closed loop system aimed at simulating and assessing pacemakers [1]. A pacemaker is an implantable medical device that uses electric current to stimulate heart tissue when patients suffer from atrial or ventricular conditions. Using mathematical models to simulate pacemaker is a need arising from two reasons: (i) in the design phase, to evaluate different stimulation patterns, different positions of the stimulation electrode and different morphology of the electrode; (ii) in operative conditions, to prove that the hardware and firmware of the pacemaker are correctly working. For hardware/firmware assessment, it would be important that the model works in real time, but unfortunately the goal is still far away due to the computational complexity of existing models. To

J. Henriques et al. (Eds.): MEDICON 2019, IFMBE Proceedings 76, pp. 488–499, 2020.
https://doi.org/10.1007/978-3-030-31635-8_59

date, pacemakers' testing is performed in open loop by evaluating the pacemaker response to pre-recorded or simulated heart signals. In this way the response of the heart to the stimulation delivered is not taken into account and consequently some possible malfunctions cannot be revealed [2].

The aim of our work was to use a finite-element (FE) model of the heart to develop a closed-loop system that emulates the pacemaker's functioning. We developed a bidomain model of the heart based on the previous works by Sovilj et al. [3,4] and Dokos et al. [5]. The model was developed in Comsol Multiphysics both in two and three dimensions. Our model is based on Fitz-Hugh Nagumo equations (FHN) [6]. We chose FHN equations for their low computational cost (three dependent variables). In addition, the FHN equations can be simply modified to adapt to different regions of the heart. We built the closed-loop pacemaker simulation in Simulink environment by including a block that emulates the pacemaker and implementing our FE heart model in a dedicated S-funciton. We chose a demand inhibited atrial pacemaker as a possible example, but all types of pacemaker could be simulated. Results show that our model is able to reproduce bradycardiac conditions and realistic electrocardiographic signals. In this way, it is possible to assess the behaviour of the pacemaker in this specific condition and the consequent heart response to the stimulation pulses.

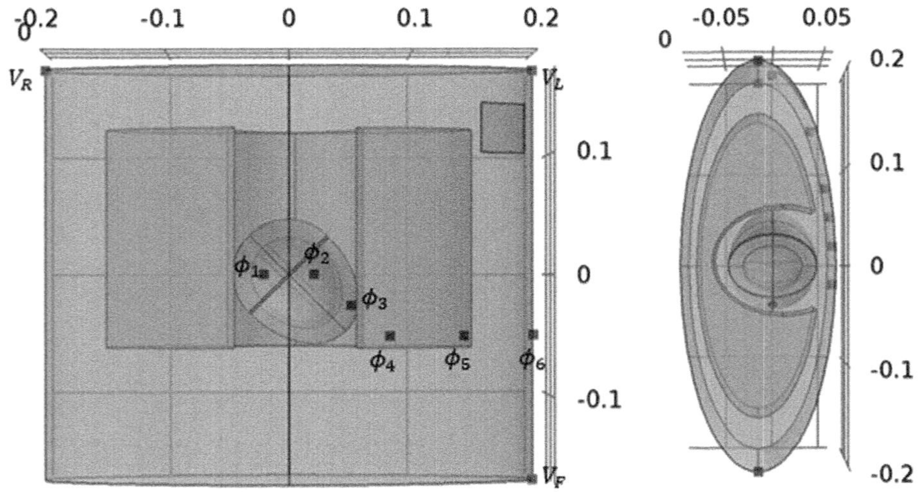

Fig. 1. Model geometry observed in the frontal plane (left) and in the transverse plane (right). The labelled points represent the probes used to simulate ECG, while the rectangle in the left upper side is the dispersive electrode. The measurement unit is meter

2 Methods

2.1 Open-Loop Heart Model

We developed an open-loop heart model composed by different regions defining the torso, lungs, the myocardium and the cardiac chambers (Fig. 1). In particular, the myocardium is divided in sub-regions with different characteristics as described in Fig. 2. We also modeled an atrial stimulation electrode, a dispersive electrode and the standard ECG leads. The model was implemented in FE through Comsol Multiphysics both in two and three dimensions. We used a quite coarse mesh as a compromise between model accuracy and computational speed. In particular, the two-dimensional finite element model is composed by 2153 tetrahedral elements with 10524 degrees of freedom. The three-dimensional finite element model is composed by 6595 tetrahedral elements with 19192 degrees of freedom.

Bidomain Model of the Heart. In a bidomain model the heart tissue is represented by the coupling of two domains: the intracellular space and the extracellular space; they both take up all the heart volume. Considering the two domains continuous and isotropic, the following equations can be obtained by Ohm's law and conservation current law:

$$\frac{\partial \phi_i}{\partial t} - \frac{\partial \phi_o}{\partial t} - \nabla \cdot (\sigma_i \nabla \phi_i) = -I_{ion} \tag{1}$$

$$\frac{\partial \phi_o}{\partial t} - \frac{\partial \phi_i}{\partial t} - \nabla \cdot (\sigma_o \nabla \phi_o) = I_{ion} \tag{2}$$

where ϕ_i and ϕ_o are respectively the intracellular and extracellular potential, while σ_i and σ_o are the intracellular and extracellular conductivities. I_{ion} is the membrane ionic current per unit volume. The membrane capacitance per unit volume was supposed to be unitary. The membrane potential V_M can be defined as:

$$V_M = \phi_i - \phi_o \tag{3}$$

In the sinoatrial node, the ionic current per unit volume is given by the FHN equations:

$$I_{ion} = kc_1(V_M - B)\left(a - \frac{V_M - B}{A}\right)\left(1 - \frac{V_M - B}{A}\right) + kc_2 u \tag{4}$$

$$\frac{\partial u}{\partial t} = ke\left(\frac{V_M - B}{A} - du - b\right) \tag{5}$$

where k, c_1, c_2, A, B, a, b are the model parameters, whose values are shown in Table 1 and are taken by [3] with slight modifications. u is the excitation variable governing the activation in the sinoatrial node. The parameter e regulates the duration of the action potentials and, in the sinoatrial node, can be opportunely tuned to control the heart rate (i.e. as described in Sect. 2.2 we will use e in the closed-loop model to simulate the pathological condition that induces the pacemaker intervention).

In the other regions Eq. 4 is modified according to the Rogers-McCulloch formulation [7]:

$$I_{ion} = kc_1(V_M - B)\left(a - \frac{V_M - B}{A}\right)\left(1 - \frac{V_M - B}{A}\right) + kc_2 u\,(V_M - B) \quad (6)$$

With respect to the original model we introduced a new parameter, ξ, which multiplies each term in the Eqs. (1) and (2), so that the local solutions remain the same, but at the same time, that the current delivered by the heart tissue is multiplied by ξ. In other words this parameter regulates the density of myocytes inside a specific region of the myocardium. In fact a greater number of myocytes implies a greater capacitance per unit volume, a greater conductivity and a greater ionic current per unit volume. In this way the amplitude of the single waves of the simulated electrocardiogram (ECG) can be simply controlled changing ξ in the region that is associated to that wave. For instance, to increase P-wave in amplitude of a certain factor, is sufficient increase ξ of the same factor in the atria. This parameter is set to 1 in all the regions in the two-dimensional model. ξ values for the three-dimensional model are shown in Table 1. In the three-dimensional model, for each region, we fixed the ξ values to obtain realistic amplitude of the ECG signal. Note that in the sinoatrial node and in the atrioventricular node, the ξ value is higher as it affects the stimulation of the surrounding tissues. The initial condition for ϕ_i is set to -60 mV in the sinoatrial node and to -85 mV in the other heart regions. For ϕ_o and u the initial condition is set to 0, as for all the time derivatives.

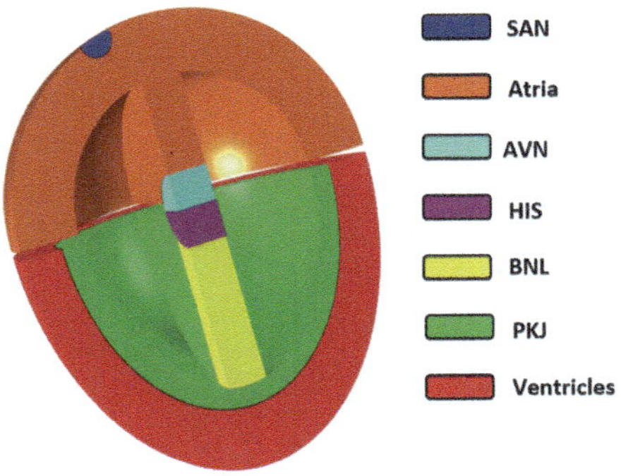

Fig. 2. Cross-sectional view of heart subdomains. SAN: sinoatrial node, AVN: atrioventricular node, HIS: His bundle, BNL: bundle branches, PKJ: Purkinje fibers.

Table 1. Model parameters

Parameters	SAN	ATRIA	AVN	HIS	BNL	PKJ	VENTR
a	-1	0.13	0.13	0.13	0.13	0.13	0.13
b	$-0.29e^{-3}$	0	0	0	0	0	0
c_1	1.9	2.6	2.6	2.6	2.6	2.6	2.6
c_2	$1e^{-3}\,(V)$	1	1	1	1	1	1
d	0	1	1	1	1	1	1
e	0.035 for $\frac{\partial V_M}{\partial t} >= 0$ 0.3 for $\frac{\partial V_M}{\partial t} < 0$	0.02	0.01	0.0045	0.0028	0.0043	0.005
$A\ (V)$	0.035	0.140	0.140	0.140	0.140	0.140	0.140
$B\ (V)$	-0.030	-0.085	-0.085	-0.085	-0.085	-0.085	-0.085
$k\ (s^{-1})$	1000	1000	1000	1000	1000	1000	1000
$\sigma_o\ \left(\frac{mS}{m}\right)$	0.5	4	0.5	10	15	35	4
$\sigma_i\ \left(\frac{mS}{m}\right)$	0.5	4	0.5	10	15	35	4
ξ	2	1.5	2.5	1.5	1.5	1.5	1.5

Parameter values in the different regions of the model. SAN: sinoatrial node, AVN: atrioventricular node, HIS: His bundle, BNL: bundle branches, PKJ: Purkinje fibers, VENTR: Ventricles

Model of the Torso and Boundary Conditions. The torso, the lungs and the cardiac chambers are considered as passive conductors, so the governing equation is:

$$\nabla \cdot (-\sigma \nabla V) = 0 \tag{7}$$

where σ is the tissue conductivity and V is the voltage in the passive volume conductors. The values of conductivity in the different tissue are shown in Table 2 and are taken from [3]. The reference point for the voltage is set at the center of the model, inside the left atrium and the initial condition for V and its temporal derivative is set to 0. All the external edges of the torso are considered electrically insulated so along them:

$$n \cdot J = n \cdot (-\sigma \nabla V) = 0 \tag{8}$$

where n and J indicates the unit outward normal vector and the current density respectively. Instead in all the edges connecting the heart and the passive

Table 2. Conductivity values

Region	Conductivity $\frac{S}{m}$
Thorax	0.2
Lungs	0.04
Blood	0.7

conductors we assumed the continuity between the extracellular space and the external one, while the intracellular space is considered electrically insulated from the passive conductors. In formulas:

$$V = \phi_o \tag{9}$$

$$n \cdot J_i = n \cdot (-\sigma_i \nabla \phi_i) = 0 \tag{10}$$

$$n \cdot J_o = n \cdot (-\sigma_o \nabla \phi_o) = n \cdot J \tag{11}$$

where J_i and J_o represent the current density in the intracellular and in the extracellular space respectively.

ECG Leads. Nine probes are placed on the thorax surface (Fig. 1). Three of them, placed at the verteces of the torso, are used to evaluate the Einthoven standard leads and the Goldberger augmented leads:

$$V_I = V_L - V_R \tag{12}$$

$$V_{II} = V_F - V_R \tag{13}$$

$$V_{III} = V_F - V_L \tag{14}$$

$$aV_R = V_R - \frac{V_L + V_F}{2} \tag{15}$$

$$aV_L = V_L - \frac{V_R + V_F}{2} \tag{16}$$

$$aV_F = V_F - \frac{V_R + V_L}{2} \tag{17}$$

The other six probes are used to compute the precordial leads as the difference between the probe potential and the Wilson central terminal, defined as:

$$V_{CT} = \frac{V_R + V_L + V_F}{3} \tag{18}$$

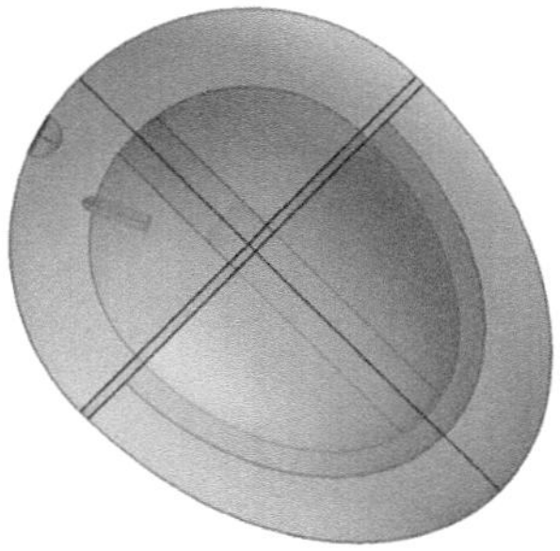

Fig. 3. Stimulation electrode highlighted in blue inside the myocardium

Pacemaker. A stimulation electrode was introduced into the heart model at the level of the sinoatrial node Fig. 3. To reproduce a more realistic situation its active surface is only in part in contact with the myocardium, while the other part is in contact only with the blood inside the cardiac chamber (Fig. 4). The partial contact with the cardiac chamber is a typical problem to consider when the stimulating wave is designed [8]. Since we decided to simulate monopolar stimulation, a dispersive electrode is needed. We placed the dispersive electrode in the upper left side of the thorax. The boundary equations along the active surface of the stimulation electrode separating it by the myocardium is given by:

$$n \cdot J_o = -\frac{I_p}{S_p} \tag{19}$$

$$n \cdot J_i = 0 \tag{20}$$

while for the edge in contact with the blood it is simply:

$$n \cdot J = -\frac{I_p}{S_p} \tag{21}$$

where I_p is the current delivered by the electrode (assumed to be cathodal, hence the minus sign before it) and S_p is its active surface. All the other edges of the pacemaker capsule are electrically insulated. For the dispersive electrode the Eq. 21 change sign and S_p is replaced by the surface of the dispersive electrode S_r:

$$n \cdot J = \frac{I_p}{S_r} \tag{22}$$

2.2 Closed-Loop Simulation

We exported the FE heart model described in Sect. 2.1 in a Simulink S-function (*HeartModelSfun* in Fig. 5) by exploiting the functionalities of Comsol LiveLink for Matlab. HeartModelSfun runs a single step of 1 ms and employs the solution obtained as initial condition for the successive step. The inputs of

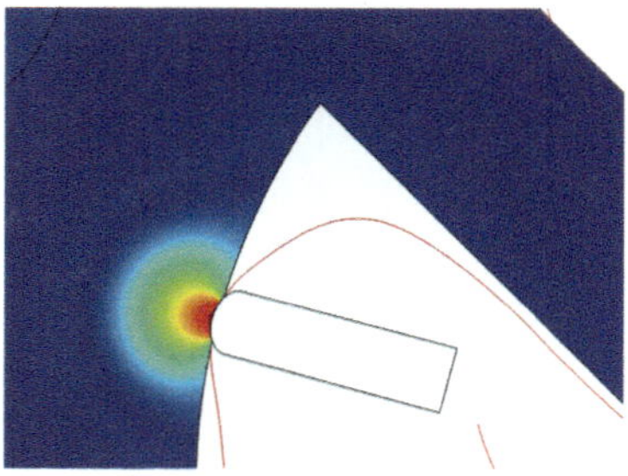

Fig. 4. 2D view of the endocardial electrode during stimulation. The red lines represent the current delivered in the blood

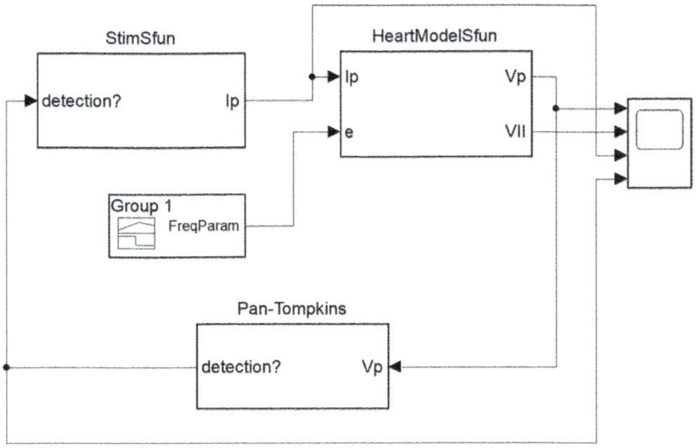

Fig. 5. Closed-loop model in Simulink

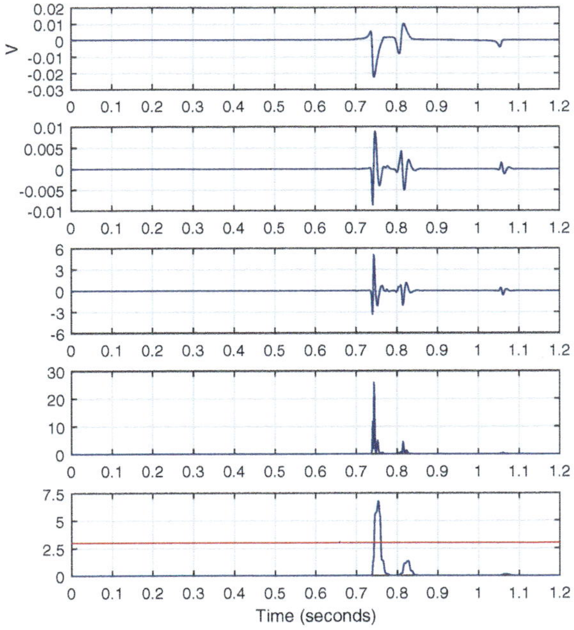

Fig. 6. Step of the process to reveal P-wave in the atrial signal First row: Raw signal; Second row: high-pass filtered signal; Third row: derivative signal; Fourth row: squared signal; Fifth row: Output of the moving average filter and threshold (red line)

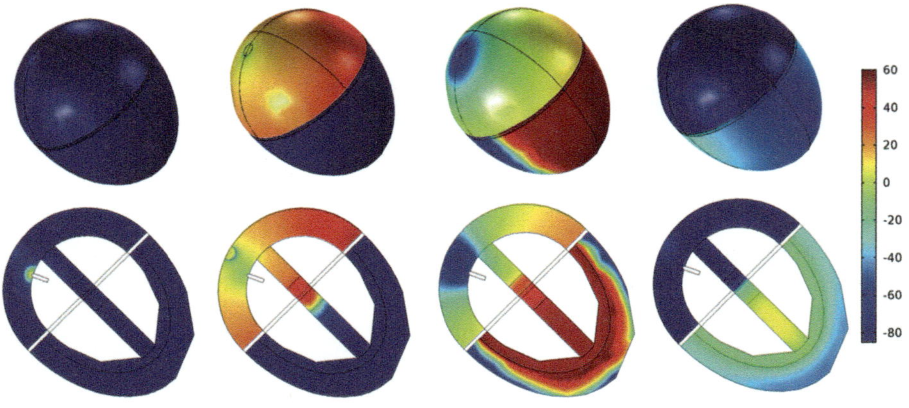

Fig. 7. Membrane potential maps in the principal steps of the cardiac cycle. First column: Stimulation in the atrial region, Second column: Atrial depolarization (final phase), Third column: Ventricular depolarization and simultaneously atrial repolarization, Fourth column: Ventricular repolarization

HeartModelSfun are the current delivered by the pacemaker (i.e, I_p) and the parameter that regulates the heart rate (e defined in Eq. 5). The block *Freq-Param* in Fig. 5 controls e inducing bradycardia and the consequent intervention of the pacemaker. The output of HeartModelSfun is the signal "recorded" by the pacemaker V_p that is computed as the difference between the average extracellular potential on the active surface and the potential on the dispersive electrode.

To extract the correct atrial activation we used a Pan-Tompkins-like algorithm [9] applied to V_p. The results of the single steps are shown in Fig. 6. V_p is high-pass filtered with a 3 dB cutoff frequency equal to 45 Hz. In real cases, also a low-pass filter would be needed but since our simulated signal is noise-free, this is not necessary. Then the square of derivative of the signal is computed. Subsequently a moving average filter is applied to obtain a unique peak in the signal. The width of the window is determined empirically and for our purposes 16 samples are enough (corresponding to 16 ms). Finally the signal is compared to a constant threshold of 3, determined empirically.

The stimulating wave is generated by another S-function (*StimSfun* in Fig. 5). StimSfun generates a predefined current pulse (0.25 mA, 2 ms) if a correct atrial activation is not detected after a time period ΔT (in the simulation we fixed $\Delta T = 1$ s).

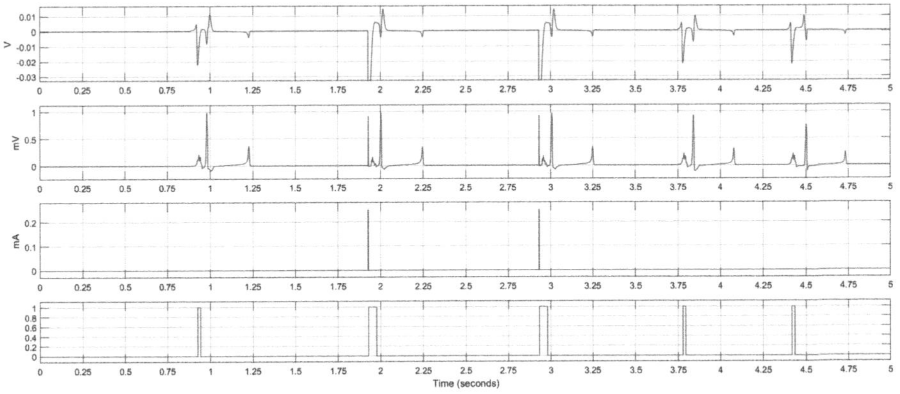

Fig. 8. Results of the simulation. First row: atrial signal (V), second row: II Einthoven lead (mV), third row: stimulation current (mA), fourth row: output of the comparator.

3 Results

3.1 Open-Loop Heart Model

Figure 7 shows the membrane potential maps in the principal steps of a cardiac cycle activated by the pacemaker. The spontaneous activation in the sinoatrial node spreads out through the atria until it reaches the atrioventricular node that introduces a delay in the propagation. Afterwards the atrioventricular node stimulates the His bundle and subsequently the Purkinje fibers and the ventricles. The atrial repolarization is simultaneous to the end of the ventricular depolarization. The ventricular repolarization starts correctly from the external edge of the ventricles and move towards the inner part.

3.2 Closed-Loop Heart Model

Figure 8 shows the results of 5 s of simulation in the closed-loop configuration. On an Intel Core i7-3770K the 2D FE model takes about 2 s to solve 1 ms of simulation. In the same condition, the 3D model takes 10 s to solve 1 ms of simulation. The computational speed is strongly decreased with respect to the open-loop model running directly in Comsol, probably due to the delay in communication between Comsol and Simulink. Indeed, the open loop model is about ten times faster. The model is able to reproduce realistic ECG signal as shown in the example Fig. 8 (second Einthoven lead). Note that the pacemaker pulse produces a spike on the signal before the P-wave.

4 Conclusions

We have developed a open-loop FE model of the heart both in 2D and 3D. The model includes a stimulation electrode and the standard ECG leads. The FE heart model has been integrated in a Matlab/Simulink application for closed loop heart simulations aimed at the assessment of cardiac pacemakers. The model can be used to study the modifications introduced by different positions of the stimulation electrode and by different stimulation patterns. It may be easily adapted to study other type of pacemaker or cardiac defibrillation. In addition, the developed model could be useful for educational and research purposes regarding the study of cardiac bioelectrical phenomena. We used the FitzHugh-Nagumo equations to reproduce the electrical activity of the single cell because they are computationally efficient. However, more complex models could be used to obtain better results. For instance, the Luo-Rudy model [10] is able to reproduce more realistic cardiac AP [11], but it employs 9 dependent variables. A further possible improvement could be the use of anatomically realistic and patient-specific geometry with increased space resolution, which could be obtained from computed tomography (CT) images as shown in [2]. The main limitation of our study is the computational cost in the closed loop configuration (i.e. 10 s for 1 ms in the 3D model, 2 s for 1 ms in the 2D model). With the current performance, real-time hardware in the loop pacemaker assessment is still not possible, however our study enables closed loop emulation of pacemakers or other cardiac stimulation devices.

Conflict of Interest. The authors declared that they have no conflict of interest.

References

1. Trayanova, N.A.: Whole-heart modeling: applications to cardiac electrophysiology and electromechanics. Circ. Res. **108**(1), 113 (2011)
2. Jiang, Z., Radhakrishnan, S., Sampath, V., Sarode, S., Mangharam, R.: Heart-on-a-Chip: a closed-loop testing platform for implantable pacemakers, July 2014
3. Sovilj, S., Magjarević, R., Al Abed, A., Lovell, N.H., Dokos, S.: Simplified 2D bidomain model of whole heart electrical activity and ECG generation. Measur. Sci. Rev. **14**(3), 136 (2014)
4. Sovilj, S., Magjarević, R., Lovell, N.H., Dokos, S.: A simplified 3D model of whole heart electrical activity and 12-lead ECG generation. Comput. Math. Methods Med. **2013**, 134208 (2013)
5. Dokos, S., Cloherty, S.L., Lovell, N.H.: In: 29th Annual International Conference of the IEEE Engineering in Medicine and Biology Society, 2007, EMBS 2007, pp. 908–911. IEEE (2007)
6. FitzHugh, R.: Impulses and physiological states in theoretical models of nerve membrane. Biophys. J. **1**(6), 445 (1961)
7. Rogers, J.M., McCulloch, A.D.: A collocation–Galerkin finite element model of cardiac action potential propagation. IEEE Trans. Biomed. Eng. **41**(8), 743 (1994)
8. Malmivuo, J., Plonsey, R., et al.: Bioelectromagnetism: Principles and Applications of Bioelectric and Biomagnetic Fields. Oxford University Press, USA (1995)

9. Pan, J., Tompkins, W.J.: A real-time QRS detection algorithm. IEEE Trans. Biomed. Eng. **32**(3), 230 (1985)
10. Luo, C.H., Rudy, Y.: A model of the ventricular cardiac action potential. Depolarization, repolarization, and their interaction. Circ. Res. **68**(6), 1501 (1991)
11. Lines, G.T., Grottum, P., Tveito, A.: Modeling the electrical activity of the heart: a bidomain model of the ventricles embedded in a torso. Comput. Vis. Sci. **5**(4), 195 (2003)

Model-Based Assessment of Sex Differences in Glucose Effectiveness and Its Components

Micaela Morettini[1], Ludovica Ilari[1], Christian Göbl[2], Alexandra Kautzky-Willer[2], Andrea Tura[3], Giovanni Pacini[3], and Laura Burattini[1(✉)]

[1] Università Politecnica delle Marche, 60131 Ancona, Italy
l.burattini@univpm.it
[2] Medical University of Vienna, 1090 Vienna, Austria
[3] Metabolic Unit, CNR Institute of Neuroscience, 35127 Padua, Italy

Abstract. Sex differences may assume a key role in condition of impaired glucose metabolism and progression to type 2 diabetes, affecting insulin-dependent processes. However, the presence of sex differences in non-insulin-dependent processes (i.e. glucose effectiveness) has been scarcely investigated. The aim of this study was to detect the presence of sex differences in glucose effectiveness (S_G), as assessed by minimal model analysis, in subjects with different degrees of glucose metabolism impairment. Two groups of subjects ranging from normal (NGR, n = 57, males/females: 31/26) to abnormal glucose regulation (AGR, n = 115, males/females 42/73) underwent a 3-h frequently sampled intravenous glucose tolerance test. Minimal model analysis provided S_G and its components at zero (GEZI) and at basal (BIE) insulin. Values for S_G were 2.52 ± 0.98 10^{-2} min^{-1} and 2.81 ± 1.07 10^{-2} min^{-1} for males and females in the NGR group, and 2.08 ± 1.21 10^{-2} min^{-1} and 2.09 ± 0.98 10^{-2} min^{-1} for males and females in the AGR group. No statistically significant difference was found between males and females in both NGR ($p = 0.29$) and AGR ($p = 0.94$) groups. Sex differences were not detected for GEZI, which provided the major contribution to S_G, either in NGR or AGR group. In conclusion, glucose effectiveness and its components seem to be not affected by sex differences in all glucose tolerance conditions.

Keywords: Mathematical model · Non-insulin-dependent processes · Basal insulin effect

1 Introduction

At present, both basic research and clinical studies have elucidated many interesting sex differences in risk and pathophysiology of Type 2 Diabetes Mellitus (T2DM). As recently reviewed [1], such differences in T2DM risk and pathophysiology can be ascribed to biological factors, like sex hormones, genetic predisposition, and neurohumoral pathways, as well as environmental and behavioral factors.

Moreover, sex differences may assume a key role in condition of impaired glucose metabolism and progression to T2DM, affecting insulin-dependent processes which

J. Henriques et al. (Eds.): MEDICON 2019, IFMBE Proceedings 76, pp. 500–507, 2020.
https://doi.org/10.1007/978-3-030-31635-8_60

accounts for insulin sensitivity (i.e. the ability of insulin to promote glucose uptake) and insulin release; in fact, despite comparable deterioration of insulin sensitivity in women and men, women show a higher early phase insulin release deterioration with age, with respect to men [2].

Differently from insulin-dependent processes, the presence of sex differences in non-insulin-dependent processes has been less investigated although such processes account for about 60–80%, according to the studied species, of whole-body glucose uptake in the fasting state [3] and have been shown to remain unchanged at the actual starting phase of impaired glucose metabolism (i.e. reduced insulin sensitivity but normal tolerance) [4].

Non-insulin-dependent processes are indicated as "glucose effectiveness" (i.e. the ability of glucose *per se* to promote its own uptake and to suppress its own production, even at basal insulin concentration) and represent the target for new antidiabetic pharmacological intervention based on sodium glucose cotransporter 2 (SGLT2) inhibitors [5]. Due to its important therapeutic role, assessment of potential sex effects in glucose effectiveness could be very important in the perspective of personalized diabetes care.

Minimal model analysis of intravenous glucose tolerance test data remains the gold-standard for quantification of glucose effectiveness [6], although new reliable methods for simple assessment have been proposed [7, 8]. Thus, the aim of this cross-sectional study was to detect the presence of sex differences in glucose effectiveness, as assessed by minimal model analysis, in subjects with different degrees of glucose metabolism impairment.

2 Materials and Methods

2.1 Subjects

This study represents a secondary analysis of existing data [9–18]. Data were collected in agreement with the Declaration of Helsinki and upon approval of the respective local ethics committees. Participants included a population of 172 white Caucasian subjects with different clinical characteristics, determining different glucose tolerance regulation: (1) subjects having normal glucose regulation (NGR, n = 57) and (2) subjects with altered glucose regulation (AGR, n = 115), including subjects affected by conditions causing an impairment in glucose metabolism (e.g. thalassemia, obesity, cirrhosis, hypertension, hyperparathyroidism) and subjects affected by T2DM.

To minimize the effects of sex hormones changes, the AGR group was further stratified according to reproductive age in older (AGR-o) or younger (AGR-y) than 50 years as proxy of menopausal age for women or andropausal age for men. All participants underwent a regular 3-h frequently sampled intravenous glucose tolerance (FSIGT) test with measurement of plasma glucose ($\text{mmol}\cdot\text{l}^{-1}$) and insulin ($\text{pmol}\cdot\text{l}^{-1}$) concentrations.

2.2 Calculations

The minimal model of glucose kinetics [19] was applied to fit FSIGT plasma glucose-concentration data, using insulin-concentration data as model input. This procedure provided quantitative information on glucose effectiveness (S_G, min^{-1}), as well as insulin sensitivity (S_I, $min^{-1}/(\mu U \cdot ml^{-1})$).

According to Kahn et al. [20] S_G can be factored out into its two components, namely Basal Insulin Effect (BIE, min^{-1}) depending directly upon basal insulin concentration I_b ($\mu U \cdot ml^{-1}$) and Glucose Effectiveness at Zero Insulin (GEZI, min^{-1}) due to hyperglycemia *per se* on both insulin-sensitive and non-insulin-sensitive tissues:

$$BIE = S_I \cdot I_b \quad (1)$$

$$GEZI = S_G - BIE \quad (2)$$

The percentage contribution of BIE and GEZI to S_G was also computed.

2.3 Statistical Analysis

The Kolmogorov-Smirnov test was used to evaluate the hypothesis that each variable had a normal distribution with unspecified mean and variance. Normally distributed variables were presented as mean ± standard deviation (SD); skewed distributed variables were presented as median [25th percentile; 75th percentile]. Correlation analysis was used to evaluate the linear relationships between variables. Differences in mean values of variables between males and females inside each group were tested by using the unpaired Student's t-test. The two-sided significance level was set at 5% ($p < 0.05$). For skewed distributed variables, tests were applied to the log-transformed values.

3 Results

Participants' characteristics are reported in Table 1.

Table 1. Characteristics of participants.

	NGR		p-value	AGR		p-value
	M	F		M	F	
n°	31	26		42	73	
Age	33 ± 13	32 ± 14	0.76	49 ± 15	46 ± 19	0.34
BMI	23.8 ± 2.9	22.9 ± 4.8	0.42	28.5 ± 5.9	28.9 ± 6.5	0.74
G_b	4.65 ± 0.62	4.37 ± 0.54	0.07	4.39 ± 0.99	4.65 ± 0.80	0.15
I_b	46 [36;62]	43 [36;63]	0.56	72 [56;114]	61 [42;114]	0.22
S_I	4.5 [3.4;7.0]	7.7 [4.7;9.9]	0.02	2.8 [1.7;5.3]	2.7 [1.9;5.5]	0.70

NGR: Normal Glucose Regulation; AGR: Altered Glucose Regulation; M: male; F: female; Age (years); BMI: Body Mass Index ($kg \cdot m^{-2}$); G_b: basal glycemia ($mmol \cdot l^{-1}$); I_b: basal insulinemia ($pmol \cdot l^{-1}$); S_I: insulin sensitivity (10^{-4} $min^{-1}/(\mu U \cdot ml^{-1})$).

3.1 Glucose Effectiveness and Its Components in NGR and AGR Groups

In the NGR group, values for S_G were 2.52 ± 0.98 10^{-2} min^{-1} and 2.81 ± 1.07 10^{-2} min^{-1} for male and female group, respectively (Fig. 1, panel A). No statistically significant difference was detected by using the unpaired Student's t-test in S_G values between the two groups ($p = 0.29$). In the NGR group, no significant linear correlation was found between S_G and age ($p = 0.90$), S_G and body mass index (BMI; $p = 0.99$) and between S_G and S_I ($p = 0.16$). No significant linear correlation was found also when considering males and females in the NGR group, separately.

In the AGR group, values for S_G were 2.08 ± 1.21 10^{-2} min^{-1} and 2.09 ± 0.98 10^{-2} min^{-1} for male and female group, respectively (Fig. 1, panel B). No statistically significant difference was detected by using the unpaired Student's t-test in S_G values between the two groups ($p = 0.94$). In the AGR, statistically significant inverse linear correlation was found between S_G and age ($r = -0.24$, $p = 0.01$), and between S_G and BMI ($r = -0.32$, $p < 0.001$). When considering males separately in the AGR group, statistically significant inverse linear correlation was found between S_G and age ($r = -0.29$, $p = 0.01$), and between S_G and BMI ($r = -0.40$, $p < 0.001$); when considering female group separately, significance in both correlations was lost. However, even when controlling for age and BMI, no statistically significant difference between males and females was found in S_G ($p = 0.88$ and $p = 0.84$, respectively). No significant linear correlation was found between S_G and S_I in AGR ($p = 0.65$) and in males and females' groups.

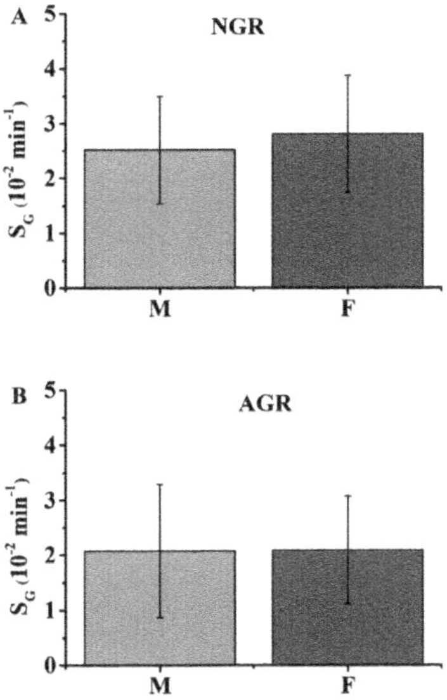

Fig. 1. Glucose effectiveness (SG) values for NGR (Normal Glucose Regulation) group (panel A) and AGR (Altered Glucose Regulation) group (panel B) in male (M) and female (F) subjects.

Glucose effectiveness components BIE and GEZI and their respective percentage contributions for NGR and AGR groups are reported in Table 2.

Table 2. Glucose effectiveness components and their percentage contribution.

	NGR		p-value	AGR		p-value
	M	F		M	F	
BIE ($10^{-2}\cdot min^{-1}$)	0.44 ± 0.22	0.63 ± 0.37	0.03	0.42 ± 0.23	0.39 ± 0.26	0.46
GEZI ($10^{-2}\cdot min^{-1}$)	2.08 ± 0.98	2.18 ± 1.07	0.71	1.65 ± 1.18	1.71 ± 1.01	0.81
BIE (%)	20.2 ± 12.3	24.7 ± 13.7	0.21	27.3 ± 22.9	22.3 ± 16.8	0.22
GEZI (%)	79.8 ± 12.3	75.3 ± 13.7	0.21	72.7 ± 22.9	77.7 ± 16.8	0.22

NGR: Normal Glucose Regulation; AGR: Altered Glucose Regulation; M: male; F: female; BIE: Basal Insulin Effect; GEZI: Glucose Effectiveness at Zero Insulin.

3.2 Glucose Effectiveness and Its Components After Stratification by Age

In AGR-y subgroup, no statistically significant difference was detected between males and females in age (p = 0.38), BMI (p = 0.57) and S_I (p = 0.46). Moreover, values for S_G were 2.62 ± 1.38 and 2.29 ± 1.01 10^{-2} min^{-1} for males and females, respectively (Fig. 2, panel A). No statistically significant difference was detected by using the unpaired Student's t-test in S_G values between males and females (p = 0.36).

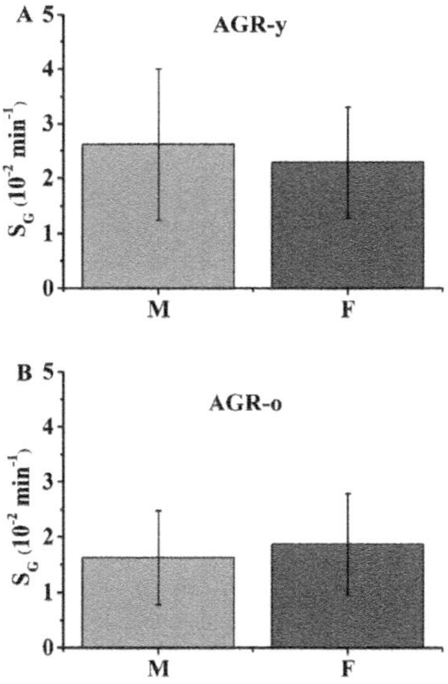

Fig. 2. Glucose effectiveness (SG) values for AGR-y (Altered Glucose Regulation with age ≤ 50 years, n = 58) group (panel A) and AGR-o (Altered Glucose Regulation with age > 50 years, n = 57) group (panel B) in male (M) and female (F) subjects.

Values for BIE were 0.46 ± 0.25 and 0.38 ± 0.25 10^{-2} min^{-1} for males and females, respectively and no statistically significant difference was detected between them (p = 0.27); values for GEZI were 2.16 ± 1.31 10^{-2} min^{-1} and 1.91 ± 1.02 10^{-2} min^{-1} for males and females, respectively and no statistically significant difference was detected (p = 0.46).

In AGR-o subgroup, no statistically significant difference was detected between males and females in age (p = 0.52), BMI (p = 0.27) and S_I (p = 0.92). Moreover, values for S_G were 1.63 ± 0.85 and 1.87 ± 0.92 10^{-2} min^{-1} for males and females, respectively (Fig. 2, panel B). No statistically significant difference was detected by using the unpaired Student's t-test in S_G values between males and females (p = 0.32). Values for BIE were 0.40 ± 0.22 and 0.40 ± 0.27 10^{-2} min^{-1} for males and females, respectively and no statistically significant difference was detected between them (p = 0.99); values for GEZI were 1.24 ± 0.90 10^{-2} min^{-1} and 1.48 ± 0.97 10^{-2} min^{-1} for males and females, respectively and no statistically significant difference was detected (p = 0.34).

4 Discussion

This study analyzed, in a population ranging different glucose tolerance status, the presence of sex differences in glucose effectiveness, which accounts for insulin-independent processes regulating glucose tolerance. Results of this study showed that sex differences do not play a role in determining glucose effectiveness, neither in condition of normal glucose regulation nor when glucose regulation is altered (Fig. 1). Moreover, in condition of impaired glucose regulation ranging from mild impairment to overt T2DM, sex-hormone asset is not supposed to influence this result since no difference between males and females was detected in and beyond reproductive age (Fig. 2).

Glucose effectiveness provides assessment of whole-body effect of hyperglycemia to enhance glucose disposal and to suppress endogenous glucose production at basal insulin. As proposed by Kahn et al. [20], the intrinsic dependency of such phenomena on basal insulin can be quantified by BIE, which describes the compensation for the reduction of insulin sensitivity induced by an increased basal circulating insulin. Although a statistically significant difference in BIE between males and females has been found in normal tolerance conditions, significance is very weak (p = 0.03). Thus, BIE is not supposed to be influenced to a large extent by sex. Moreover, no sex differences were found when considering processes taking place in a total absence of insulin as quantified by GEZI.

BMI has been shown to influence glucose effectiveness and its components [21]. In this study, BMI has not been considered as a possible confounding factor, since in each comparison between male and female groups, no statistically significant difference in BMI has been found.

Glucose effectiveness is important in the light of novel therapeutic approaches for T2DM [5]. In fact, due to its recognized importance in determining glucose tolerance, quantification of glucose effectiveness in metabolic tests is strongly encouraged. Reliable methodologies have been recently proposed to make assessment easier, resorting (as previously done before with insulin sensitivity [22]) to metabolic tests shorter than those used before (3 h or sometimes even more [23, 24]).

In conclusion, the present study provided evidence that glucose effectiveness and its components are not affected by sex differences in all glucose tolerance conditions, ranging from normal to overt T2DM.

Acknowledgements. We would like to thank Angelo Avogaro (University of Padova, Italy), Paolo Cavallo–Perin (University of Turin, Italy), Anton Luger and Michaela Riedl (Medical University of Vienna, Austria), and Giovanni Musso (Gradenigo Hospital Turin, Italy), for granting the permission to use their data collected in previous studies.

Conflict of Interest. The authors declare that they have no conflict of interest.

References

1. Kautzky-Willer, A., Harreiter, J., Pacini, G.: Sex and gender differences in risk, pathophysiology and complications of type 2 diabetes mellitus. Endocr. Rev. **37**(3), 278–316 (2016)
2. Tura, A., Pacini, G., Moro, E., Vrbíková, J., Bendlová, B., Kautzky-Willer, A.: Sex- and age-related differences of metabolic parameters in impaired glucose metabolism and type 2 diabetes compared to normal glucose tolerance. Diabetes Res. Clin. Pract. **146**, 67–75 (2018)
3. Baron, A.D., Brechtel, G., Wallace, P., Edelman, S.V.: Rates and tissue sites of non-insulin- and insulin-mediated glucose uptake in humans. Am. J. Physiol. **255**(6 Pt 1), E769–E774 (1988)
4. Morettini, M., Di Nardo, F., Fioretti, S., Pacini, G., Tura, A., Burattini, L.: No changes in glucose effectiveness in condition of reduced insulin action but preserved glucose tolerance as assessed by minimal model analysis. In: IFMBE Proceedings, vol. 65, pp. 1057–1060 (2017)
5. Seufert, J.: SGLT2 inhibitors—an insulin-independent therapeutic approach for treatment of type 2 diabetes: focus on canagliflozin. Diab. Metab. Syndr. Obes. Targets Ther. **8**, 543–554 (2015)
6. Ader, M., Pacini, G., Yang, Y.J., Bergman, R.N.: Importance of glucose per se to intravenous glucose tolerance: comparison of the minimal-model prediction with direct measurements. Diabetes **34**(11), 1092–1103 (1985)
7. Morettini, M., Di Nardo, F., Burattini, L., Fioretti, S., Göbl, C., Kautzky-Willer, A., Pacini, G., Tura, A.: Assessment of glucose effectiveness from short IVGTT in individuals with different degrees of glucose tolerance. Acta Diabetol. **55**(10), 1011–1018 (2018)
8. Morettini, M., Castriota, C., Göbl, C., Kautzky-Willer, A., Pacini, G., Burattini, L., Tura, A.: Glucose effectiveness from short insulin-modified IVGTT and its application to the study of women with previous gestational diabetes mellitus. Diabetes Metab. J. (in press)
9. Cavallo-Perin, P., Pacini, G., Cerutti, F., Bessone, A., Condo, C., Sacchetti, L., Piga, A., Pagano, G.: Insulin resistance and hyperinsulinemia in homozygous β-thalassemia. Metabolism **44**(3), 281–286 (1995)
10. Cerutti, F., Sacchetti, C., Bessone, A., Rabbone, I., Cavallo-Perin, P., Pacini, G.: Insulin secretion and hepatic insulin clearance as determinants of hyperinsulinaemia in normotolerant grossly obese adolescents. Acta Paediatr. **87**(10), 1045–1050 (1998)

11. Kautzky-Willer, A., Pacini, G., Ludvik, B., Schernthaner, G., Prager, R.: β-cell hypersecretion and not reduced hepatic insulin extraction is the main cause of hyperinsulinemia in obese nondiabetic subjects. Metabolism **41**(12), 1304–1312 (1992)
12. Kautzky-Willer, A., Pacini, G., Niederle, B., Schernthaner, G., Prager, R.: Insulin secretion, insulin sensitivity and hepatic insulin extraction in primary hyperparathyroidism before and after surgery. Clin. Endocrinol. **37**(2), 147–155 (1992)
13. Kautzky-Willer, A., Thomaseth, K., Clodi, M., Ludvik, B., Waldhäusl, W., Prager, R., Pacini, G.: β-cell activity and hepatic insulin extraction following dexamethasone administration in healthy subjects. Metabolism **45**(4), 486–491 (1996)
14. Pacini, G., Tonolo, G., Sambataro, M., Maioli, M., Ciccarese, M., Brocco, E., Avogaro, A., Nosadini, R.: Insulin sensitivity and glucose effectiveness: minimal model analysis of regular and insulin–modified IVGTT. Am. J. Physiol. **274**(4 Pt.1), E592–E599 (1998)
15. Pagano, G., Pacini, G., Musso, G., Gambino, R., Mecca, F., Depetris, N., Cassader, M., David, E., Cavallo-Perin, P., Rizzetto, M.: Nonalcoholic steatohepatitis, insulin resistance, and metabolic syndrome: further evidence for an etiologic association. Hepatology **35**(2), 367–372 (2002)
16. Piccardo, M.G., Pacini, G., Nardi, E., Rosa, M.S., De Vito, R.: Beta-cell response and insulin hepatic extraction in noncirrhotic alcoholic patients soon after withdrawal. Metabolism **43**(3), 367–371 (1994)
17. Piccardo, M.G., Turchetti, G.F., Rosa, M.S., Pacini, G.: Elevated basal insulin secretion and normal dynamic insulin sensitivity in borderline hypertension. Nutr. Metab. Cardiovasc. Dis. **9**(5), 244–249 (1999)
18. Riedl, M., Ludvik, B., Pacini, G., Clodi, M., Kotzmann, H., Wagner, O., Kautzy-Willer, A., Prager, R., Luger, A.: The increased insulin sensitivity in growth hormone deficient adults is reduced by growth hormone replacement therapy. Eur. J. Clin. Invest. **30**(9), 771–778 (2000)
19. Pacini, G., Bergman, R.N.: MINMOD: a computer program to calculate insulin sensitivity and pancreatic responsivity from the frequently sampled intravenous glucose tolerance test. Comput. Methods Programs Biomed. **23**(2), 113–122 (1986)
20. Kahn, S.E., Klaff, L.J., Schwartz, M.W., Beard, J.C., Bergman, R.N., Taborsky, G.J., Porte, D.: Treatment with a somatostatin analog decreases pancreatic B-cell and whole body sensitivity to glucose. J. Clin. Endocrinol. Metab. **71**(4), 994–1002 (1990)
21. Morettini, M., Di Nardo, F., Ingrillini, L., Fioretti, S., Göbl, C., Kautzky-Willer, A., Tura, A., Pacini, G., Burattini, L.: Glucose effectiveness and its components in relation to body mass index. Eur. J. Clin. Invest. **49**(6), e13099 (2019)
22. Tura, A., Sbrignadello, S., Succurro, E., Groop, L., Sesti, G., Pacini, G.: An empirical index of insulin sensitivity from short IVGTT: validation against the minimal model and glucose clamp indices in patients with different clinical characteristics. Diabetologia **53**(1), 144–152 (2010)
23. Burattini, R., Morettini, M., Di Nardo, F., Boemi, M.: Dynamics of insulin action in hypertension: assessment from minimal model interpretation of intravenous glucose tolerance test data. Med. Biol. Eng. Comput. **49**(7), 831–841 (2011)
24. Di Nardo, F., Mengoni, M., Morettini, M.: MATLAB-implemented estimation procedure for model-based assessment of hepatic insulin degradation from standard intravenous glucose tolerance test data. Comput. Methods Programs Biomed. **110**(2), 215–225 (2013)

Insulin Clearance in Women with a History of Gestational Diabetes Assessed by Mathematical Model Analyses of Intravenous Glucose Tolerance Test

Micaela Morettini[1], Christian Göbl[2], Alexandra Kautzky-Willer[2], Giovanni Pacini[3], Andrea Tura[3], and Laura Burattini[1](✉)

[1] Università Politecnica delle Marche, 60131 Ancona, Italy
l.burattini@univpm.it
[2] Medical University of Vienna, 1090 Vienna, Austria
[3] Metabolic Unit, CNR Institute of Neuroscience, 35127 Padua, Italy

Abstract. Circulating concentrations of insulin are determined by a balance between the secretion rate of insulin from pancreatic beta-cells and insulin degradation ("clearance"). However, limited attention has been devoted to the study of insulin clearance in women with former gestational diabetes mellitus (GDM), which are known to be at increased type 2 diabetes risk. The aim of this study was to provide a detailed analysis of insulin clearance in women with former GDM. A population of 156 white Caucasian women, was analyzed early postpartum (4–6 months after delivery) and classified in two groups: women with previous GDM (pGDM, $n = 115$) and women that remain healthy during pregnancy (CNT, $n = 41$). All women underwent a 3-hour Insulin-Modified Intravenous Glucose Tolerance Test (IM-IVGTT). Insulin clearance temporal patterns were derived by mathematical modelling of IM-IVGTT data; average insulin clearance values were also considered during the whole test, and in the first - (0–10 min) and second phase (10–180 min). Insulin clearance temporal patterns were found to be different between CNT and pGDM group ($p < 0.0001$). Average insulin clearance was found different over the second phase of the test ($p = 0.04$), being equal to 0.54 [0.41] and 0.59 [0.41] $l \cdot min^{-1}$ in CNT and pGDM group, respectively. In conclusion, some abnormalities in former GDM women, compared to a group of healthy women were detected. This may be of relevance for more accurate estimation of type 2 diabetes risk.

Keywords: Mathematical model · Insulin extraction · Type 2 diabetes risk

1 Introduction

Insulin plays a fundamental role in regulating many aspects of metabolism, and the decline in its secretion contributes to diabetes pathogenesis and progression. Due to this reason, extensive work has been done to estimate insulin secretion, and pancreatic beta-cell function, this meaning insulin secretion adjusted for the prevailing glycemic levels.

J. Henriques et al. (Eds.): MEDICON 2019, IFMBE Proceedings 76, pp. 508–517, 2020.
https://doi.org/10.1007/978-3-030-31635-8_61

However, there has been insufficient emphasis on the fact that circulating concentrations of insulin are determined by a balance between the secretion rate from pancreatic beta-cells and insulin degradation ("clearance"), which occurs mainly in the liver, but also in kidneys, and other tissues, including skeletal muscles [1]. Thus, changes in plasma insulin levels can be due to alterations not only in insulin secretion/beta-cell function, but also in insulin degradation. Indeed, it has become increasingly evident that insulin clearance is a highly regulated process, also at genetic level, and hence partially genetically coded independently from insulin secretion itself [2–4].

The increasing interest for insulin clearance is due to the fact that a recent study reported altered insulin clearance in subjects with type 2 diabetes (T2D) [5]. This may be of importance from a clinical point of view in those diabetic subjects that undergo insulin therapy, since for a prescribed insulin dose, the plasma circulating levels of insulin, and hence the possibility to develop its action, may be decreased in subjects with higher clearance.

On the other hand, it is well known that there are some conditions that predispose to possible development of T2D. One of these conditions is having a history of gestational diabetes mellitus (GDM) [6, 7]. However, very limited attention has been devoted to the study of insulin clearance in women with former GDM, and at any rate limited to the aspect of hepatic clearance (also indicated as hepatic extraction) [8], which nonetheless was not analyzed in deep. Therefore, the aim of this study is to provide for the first time a detailed analysis of insulin clearance, derived by appropriate mathematical modelling of metabolic data, in women with former GDM, and to compare it to that of a group of healthy women acting as control group.

2 Materials and Methods

2.1 Subjects

This study represents a secondary analysis of existing data [7]. Data were collected in agreement with the Declaration of Helsinki and upon approval of the respective local ethics committees. Participants included a population of 156 white Caucasian women, analyzed early postpartum (4–6 months after delivery), classified in two groups: women with previous GDM (pGDM, n = 115) and women that remained healthy during pregnancy (CNT, n = 41).

All participants underwent a regular 3-hour, insulin-modified intravenous glucose tolerance test (IM-IVGTT) with an intravenous insulin infusion (0.03 U/kg) at 20 min after glucose infusion (0.3 g/kg). Blood samples were collected during 3 h for the measurement of plasma glucose ($mmol \cdot l^{-1}$), insulin ($pmol \cdot l^{-1}$) and C-peptide ($pmol \cdot l^{-1}$) concentrations. For every subject, mean glucose (G_{mean}), insulin (I_{mean}) and C-peptide (CP_{mean}) concentrations during the test were evaluated as the mean areas under the curve during the whole test.

2.2 Assessment of Insulin Action and Beta-Cell Function

A complete characterization of metabolic status was achieved by the assessment of indexes of insulin action and beta-cell function. Insulin action was assessed by using the Homeostasis Model Assessment-Insulin Resistance (HOMA-IR index) [9] according to the following formula:

$$HOMA - IR = \frac{G_b * I_b}{22.5} \quad (1)$$

were G_b and I_b are expressed as (mmol·l^{-1}) and (μU·ml^{-1}), respectively. Beta-cell function was assessed by computing the Acute Insulin Response (AIR) and the Acute C-Peptide Response (ACPR) as the mean of suprabasal insulin and C-peptide curve, respectively in the time interval 3-8 min during the IM-IVGTT test.

2.3 Assessment of Insulin Clearance Temporal Patterns

Procedure to assess insulin clearance temporal patterns ($Cl_{INS}(t)$) during IM-IVGTT is summarized in Fig. 1. $Cl_{INS}(t)$ has been obtained as the ratio, every minute from 0 to 180 min, between the prehepatic insulin secretion rate ISR(t) and the plasma insulin concentration I(t). ISR(t) is equal to the C-peptide secretion rate (CPSR(t)), being C-peptide co-secreted with insulin in equimolar amount but not extracted by the liver to a significant extent. Thus, ISR(t) has been computed according to Van Cauter et al. [10] by deconvolution from plasma C-peptide concentration using individualized C-peptide kinetic parameters computed on the basis of subject 'anthropometric characteristics. I(t) was obtained by linear interpolation of plasma insulin concentrations during IM-IVGTT.

2.4 Assessment of Average Total, First- and Second-Phase Insulin Clearance

For each subject, average total ($Cl_{INS-TOT}$), first- (Cl_{INS-FP}) and second-phase (Cl_{INS-SP}) insulin clearance were computed as the ratio of the area under the ISR(t) and the I(t) curves, computed using the trapezoidal rule in the intervals 0–180 min, 0–10 min, 10–180 min, respectively.

2.5 Statistical Analysis

The Kolmogorov-Smirnov test was used to evaluate the hypothesis that each variable had a normal distribution with unspecified mean and variance. Normally distributed variables were presented as mean ± standard error; skewed distributed variables were presented as median [interquartile range]. Differences in mean values of variables between CNT and pGDM groups were tested by using the unpaired Student's t-test.

Repeated measures analysis of variance (ANOVA) was used to compare $Cl_{INS}(t)$ between CNT and pGDM groups. For skewed distributed variables, tests were applied to the log-transformed values. The two-sided significance level was set at 5% ($p < 0.05$).

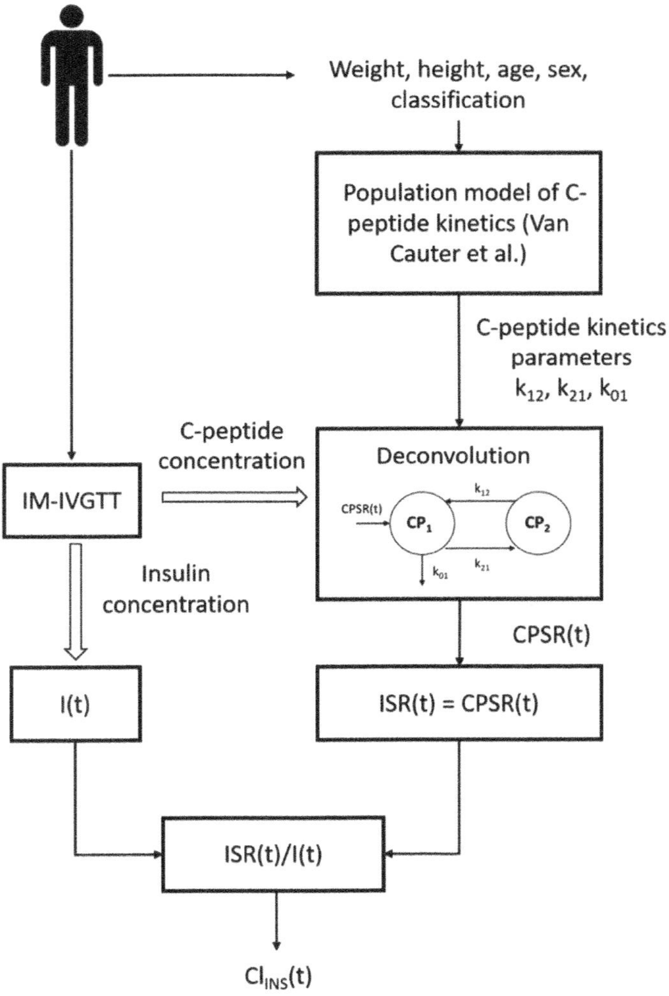

Fig. 1. Procedure to assess insulin clearance temporal patterns (ClINS(t)) from the Insulin-Modified Intravenous Glucose Tolerance Test (IM-IVGTT) data. ISR(t) and CPSR(t) are the Insulin Secretion Rate and the C-peptide Secretion Rate, respectively. I(t) is the insulin concentration curve.

3 Results

Participants' main characteristics and complete characterization of the metabolic status in terms of insulin action, beta-cell function and average insulin clearance are reported in Table 1.

Table 1. Main characteristics of participants and metabolic status.

n°	CNT	pGDM	p-value
	41	115	
Age (years)	31.6 ± 0.8	33.4 ± 0.4	0.06
BMI ($kg \cdot m^{-2}$)	24.5 ± 0.8	27.7 ± 0.5	<0.001
G_b ($mmol \cdot l^{-1}$)	4.56 [0.39]	4.94 [0.78]	<0.001
I_b ($pmol \cdot l^{-1}$)	46 [28]	47 [30]	0.38
CP_b ($pmol \cdot l^{-1}$)	150 [50]	166 [93]	0.06
G_{mean} ($mmol \cdot l^{-1}$)	4.63 [0.68]	5.39 [1.01]	<0.001
I_{mean} ($pmol \cdot l^{-1}$)	197 [113]	218 [116]	0.17
CP_{mean} ($pmol \cdot l^{-1}$)	178 [91]	218 [156]	0.02
HOMA-IR	9.8 [6.2]	10.9 [7.9]	0.12
AIR ($pmol \cdot l^{-1}$)	276 [279]	192 [157]	<0.01
ACPR ($pmol \cdot l^{-1}$)	318 [188]	237 [164]	<0.001
$Cl_{INS\text{-}TOT}$ ($l \cdot min^{-1}$)	0.73 [0.42]	0.75 [0.44]	0.25
$Cl_{INS\text{-}FP}$ ($l \cdot min^{-1}$)	2.97 [1.07]	3.32 [1.36]	0.18
$Cl_{INS\text{-}SP}$ ($l \cdot min^{-1}$)	0.54 [0.41]	0.59 [0.41]	0.04

CNT: women who remain healthy during pregnancy; pGDM: women who developed gestational diabetes mellitus; BMI: Body Mass Index; G_b, I_b, CP_b: basal glycemia, insulinemia and C-peptide concentration; G_{mean}, I_{mean}, CP_{mean}: mean glucose, insulin and C-peptide concentration during the test; HOMA-IR: Homeostasis Model Assessment-Insulin Resistance; AIR: Acute Insulin Response; ACPR: Acute C-peptide Response; $Cl_{INS\text{-}TOT}$, $Cl_{INS\text{-}FP}$, $Cl_{INS\text{-}SP}$: average total, first-phase and second-phase insulin clearance.

Results of the best-fit of plasma C-peptide concentration data and the related ISR(t) in all participants are shown in Fig. 2. Insulin clearance temporal patterns in CNT and pGDM group are reported in Fig. 3. Repeated measures ANOVA results detected a significant group-time interaction ($p < 0.0001$).

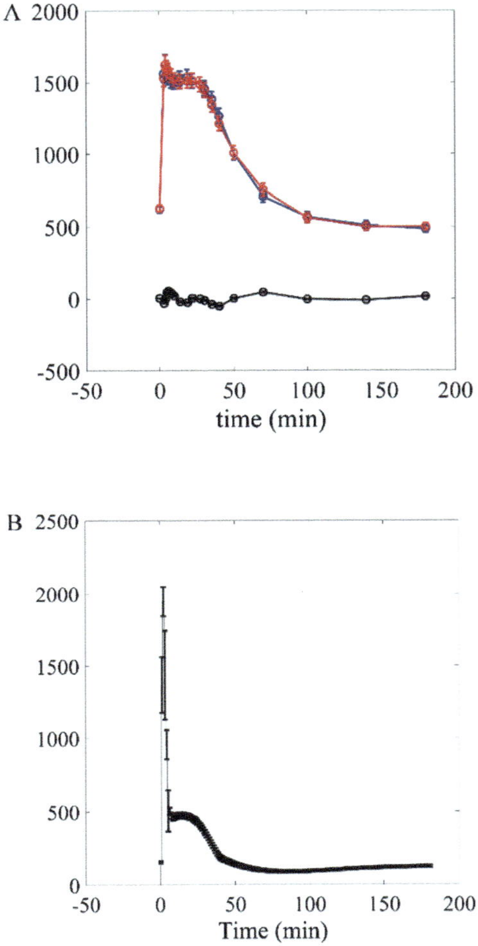

Fig. 2. Results of the best-fit of plasma C-peptide concentration data in pmol·l-1 (panel A; blue: C-peptide experimental data; red: C-peptide model prediction; black: weighted residuals) and prehepatic insulin secretion rate ISR(t) on all subjects (pmol·min-1, panel B). Data are reported as mean ± standard error.

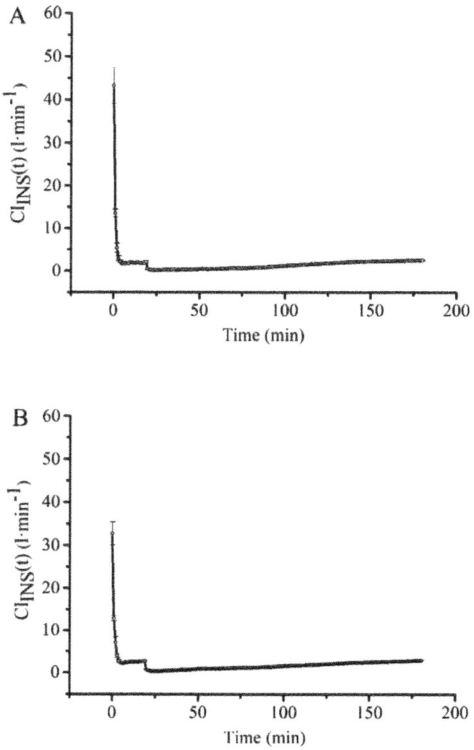

Fig. 3. Insulin clearance mean (± standard error) temporal patterns in the group of women that remained healthy during pregnancy (CNT, panel A) and in the group of women with previous gestational diabetes mellitus (pGDM, panel B).

4 Discussion

In this study, we assessed insulin clearance in women with a history of GDM, by appropriate mathematical modeling applied to data derived from a common test of glucose tolerance (IM-IVGTT). Our main finding is that insulin clearance temporal patterns (biphasic as shown in Fig. 3, due to the presence of insulin infusion during IM-IVGTT) were found different in women with former GDM compared to healthy controls, with average values (at least when calculated over the second phase of the test), significantly (although slightly) increased.

This is of remarkable relevance from a clinical point of view. First, it is possible that in this population insulin clearance may play a role in future T2D development. Indeed, several metabolic markers of future T2D risk are already known in former GDM, such as impairment in insulin sensitivity, beta-cell function and glucose effectiveness [6, 7, 11–13], but the identification of possible further risk markers may be useful to better identify those women that are at the highest risk, thus focusing on them most aggressive prevention strategy.

Furthermore, since it is known that, currently, despite the possible prevention strategies, a not negligible percentage of former GDM women will develop T2D [6, 7], the assessment of insulin clearance may become of relevance even for the pharmacological treatment to adjust the dose for the insulin therapy, as the same insulin dose will translate into different levels of plasma insulin, depending on the individual insulin clearance. In addition, it should be considered the possibility of a modification of the insulin clearance from some specific antidiabetic agents, that is, those acting on the incretin axis (DPP-4 inhibitors [14] and GLP-1 receptor agonists [15]), and this becomes particularly relevant when such type of agent is added to insulin therapy. Indeed, some recent studies have shown in mice that when the incretin receptors are impaired insulin clearance can be altered, this suggesting a connection between incretins, incretin-based pharmacological agents, and insulin clearance [16, 17].

Comparison of our results with those of previous study is difficult, since as previously reported, very few studies analyzed insulin clearance in this population. In the study [8], no significant difference was found in hepatic insulin clearance in women with former GDM compared to a control healthy group. The different findings of our study compared to study [8] can be justified by the lower number of women analyzed in that study. In addition, hepatic insulin clearance is the most relevant portion of the insulin clearance, but other tissues provide a contribution to insulin degradation [1]. In another study, the fasting hepatic insulin clearance was found decreased, rather than increased, in former GDM [18]. However, this finding was based on fasting data only and it was obtained with a basic index (simple ratio between plasma C-peptide and insulin concentrations) and not on mathematical modeling and suffers from the problem of the different kinetics between C-peptide and insulin.

This study has some limitations. First, we assessed the whole-body insulin clearance, without segregating between the hepatic insulin clearance, and that of the other tissues. However, accurate assessment of clearance at different tissues would require complex experiments and estimation procedure [19], hardly feasible in a relatively large dataset. Another limitation is that we have not investigated the possible variation of insulin clearance during a follow-up period, and hence for the moment we cannot provide final conclusions about the possible role of insulin clearance abnormalities as risk factor for T2D. However, this will be addressed in future studies. Finally, BMI could have a role on insulin clearance determination, as already known for other insulin-dependent components of glucose tolerance and recently seen also for the insulin-independent one [20].

In conclusion, for the first time we have analyzed in detail the insulin clearance in women with a history of GDM, and we have identified some abnormalities compared to a group of healthy women. This may be of relevance for more accurate estimation of T2D risk, and for appropriate pharmacological titration in those women that in fact develop T2D.

Conflict of Interest. The authors declare that they have no conflict of interest.

References

1. Polidori, D.C., Bergman, R.N., Chung, S.T., Sumner, A.E.: Hepatic and extrahepatic insulin clearance are differentially regulated: results from a novel model-based analysis of intravenous glucose tolerance data. Diabetes **65**(6), 1556–1564 (2016)
2. Mittelman, S.D., Van Citters, G.W., Kim, S.P., Davis, D.A., Dea, M.K., Hamilton-Wessler, M., Bergman, R.N.: Longitudinal compensation for fat-induced insulin resistance includes reduced insulin clearance and enhanced beta-cell response. Diabetes **49**(12), 2116–2125 (2000)
3. Ader, M., Stefanovski, D., Kim, S.P., Richey, J.M., Ionut, V., Catalano, K.J., Hucking, K., Ellmerer, M., Van Citters, G., Hsu, I.R., Chiu, J.D., Woolcott, O.O., Harrison, L.N., Zheng, D., Lottati, M., Kolka, C.M., Mooradian, V., Dittmann, J., Yae, S., Liu, H., Castro, A.V., Kabir, M., Bergman, R.N.: Hepatic insulin clearance is the primary determinant of insulin sensitivity in the normal dog. Obesity **22**(5), 1238–1245 (2014)
4. Goodarzi, M.O., Langefeld, C.D., Xiang, A.H., Chen, Y.D., Guo, X., Hanley, A.J., Raffel, L. J., Kandeel, F., Nadler, J.L., Buchanan, T.A., Norris, J.M., Fingerlin, T.E., Lorenzo, C., Rewers, M.J., Haffner, S.M., Bowden, D.W., Rich, S.S., Bergman, R.N., Rotter, J.I., Watanabe, R.M., Wagenknecht, L.E.: Insulin sensitivity and insulin clearance are heritable and have strong genetic correlation in Mexican Americans. Obesity **22**(4), 1157–1164 (2014)
5. Ohashi, K., Fujii, M., Uda, S., Kubota, H., Komada, H., Sakaguchi, K., Ogawa, W., Kuroda, S.: Increase in hepatic and decrease in peripheral insulin clearance characterize abnormal temporal patterns of serum insulin in diabetic subjects. NPJ Syst. Biol. Appl. **4**, 14 (2018)
6. Göbl, C.S., Bozkurt, L., Prikoszovich, T., Winzer, C., Pacini, G., Kautzky-Willer, A.: Early possible risk factors for overt diabetes after gestational diabetes mellitus. Obstet. Gynecol. **118**(1), 71–78 (2011)
7. Tura, A., Grassi, A., Winhofer, Y., Guolo, A., Pacini, G., Mari, A., Kautzky-Willer, A.: Progression to type 2 diabetes in women with former gestational diabetes: time trajectories of metabolic parameters. PLoS One **7**(11), e50419 (2012)
8. Kautzky-Willer, A., Prager, R., Waldhausl, W., Pacini, G., Thomaseth, K., Wagner, O.F., Ulm, M., Streli, C., Ludvik, B.: Pronounced insulin resistance and inadequate β-cell secretion characterize lean gestational diabetes during and after pregnancy. Diabetes Care **20** (11), 1717–1723 (1997)
9. Matthews, D.R., Hosker, J.P., Rudenski, A.S., Naylor, B.A., Treacher, D.F., Turner, R.C.: Homeostasis model assessment: insulin resistance and beta-cell function from fasting plasma glucose and insulin concentrations in man. Diabetologia **28**(7), 412–419 (1985)
10. Van Cauter, E., Mestrez, F., Sturis, J., Polonsky, K.S.: Estimation of insulin secretion rates from C-peptide levels: comparison of individual and standard kinetic parameters for C-peptide clearance. Diabetes **41**(3), 368–377 (1992)
11. Bozkurt, L., Göbl, C.S., Tura, A., Chmelik, M., Prikoszovich, T., Kosi, L., Wagner, O., Roden, M., Pacini, G., Gastaldelli, A., Kautzky-Willer, A.: Fatty liver index predicts further metabolic deteriorations in women with previous gestational diabetes. PLoS One **7**(2), e32710 (2012)
12. Tura, A., Mari, A., Prikoszovich, T., Pacini, G., Kautzky-Willer, A.: Value of the intravenous and oral glucose tolerance tests for detecting subtle impairments in insulin sensitivity and beta-cell function in former gestational diabetes. Clin. Endocrinol. **69**(2), 237–243 (2008)

13. Morettini, M., Castriota, C., Göbl, C., Kautzky-Willer, A., Pacini, G., Burattini, L., Tura, A.: Glucose effectiveness from short insulin-modified IVGTT and its application to the study of women with previous gestational diabetes mellitus. Diabetes Metabol. J. in press
14. Ling, J., Ge, L., Zhang, D.H., Wang, Y.F., Xie, Z.L., Tian, J.H., Xiao, X.H., Yang, K.H.: DPP-4 inhibitors for the treatment of type 2 diabetes: a methodology overview of systematic reviews. Acta Diabetol. **56**(1), 7–27 (2019)
15. Goldenberg, R.M., Berard, L.: Adding prandial GLP-1 receptor agonists to basal insulin: a promising option for type 2 diabetes therapy. Curr. Med. Res. Opin. **34**(1), 1–10 (2018)
16. Tura, A., Pacini, G., Yamada, Y., Seino, Y., Ahrén, B.: Glucagon and insulin secretion, insulin clearance, and fasting glucose in GIP receptor and GLP-1 receptor knockout mice. Am. J. Physiol.-Regul. Integr. Comp. Physiol. **316**(1), R27–R37 (2019)
17. Tura, A., Bizzotto, R., Yamada, Y., Seino, Y., Pacini, G., Ahrén, B.: Increased insulin clearance in mice with double deletion of glucagon-like peptide-1 and glucose-dependent insulinotropic polypeptide receptors. Am. J. Physiol.-Regul. Integr. Comp. Physiol. **314**(5), R639–R646 (2018)
18. Koivunen, R.M., Juutinen, J., Vauhkonen, I., Morin-Papunen, L.C., Ruokonen, A., Tapanainen, J.S.: Metabolic and steroidogenic alterations related to increased frequency of polycystic ovaries in women with a history of gestational diabetes. J. Clin. Endocrinol. Metab. **86**(6), 2591–2599 (2001)
19. Di Nardo, F., Mengoni, M., Morettini, M.: MATLAB-implemented estimation procedure for model-based assessment of hepatic insulin degradation from standard intravenous glucose tolerance test data. Comput. Methods Programs Biomed. **110**(2), 215–225 (2013)
20. Morettini, M., Di Nardo, F., Ingrillini, L., Fioretti, S., Göbl, C., Kautzky-Willer, A., Tura, A., Pacini, G., Burattini, L.: Glucose effectiveness and its components in relation to body mass index. Eur. J. Clin. Invest. **49**(6), e13099 (2019)

Computational Models for Predicting Resilience Levels of Women with Breast Cancer

Konstantina Kourou[1], Haridimos Kondylakis[2], Lefteris Koumakis[2], Georgios C. Manikis[2], Kostas Marias[2], Manolis Tsiknakis[2,3], Panagiotis G. Simos[4], Evangelos Karademas[5], and Dimitrios I. Fotiadis[1,6(✉)]

[1] Institute of Molecular Biology and Biotechnology (IMBB), Foundation for Research and Technology Hellas (FORTH), 45110 Ioannina, Greece
fotiadis@cc.uoi.gr
[2] Computational BioMedicine Laboratory, Institute of Computer Science (ICS), Foundation for Research and Technology Hellas (FORTH), 70013 Heraklion, Greece
[3] Biomedical Informatics and eHealth Laboratory, Technological Educational Institute of Crete, 70013 Heraklion, Greece
[4] Department of Psychiatry and Behavioral Science, Medical School, University of Crete, 71003 Heraklion, Greece
[5] Department of Psychology, University of Crete, 74100 Rethymno, Greece
[6] Unit of Medical Technology and Intelligent Information Systems, Department of Materials Science and Engineering, University of Ioannina, 45110 Ioannina, Greece

Abstract. In the current study, a model-based system for predicting resilience in silico, as part of personalizing precision medicine, to better understand the needs for improved therapeutic protocols of each patient is proposed. The computational environment, which is currently under implementation within the BOUNCE EU project ("Predicting Effective Adaptation to Breast Cancer to Help Women to BOUNCE Back"), will help clinicians and care-givers to predict the patient's resilience trajectory throughout cancer continuum. The overall proposed system architecture contributes to clinical outcomes and patient well-being by taking into consideration biological, social, environmental, occupational and lifestyle factors for resilience prediction in women with breast cancer. Supervised, semisupervised and unsupervised learning algorithms are adopted with the inherent ability to represent the time-varying behaviour of the underlying system which allows for a better representation of spatiotemporal input-output dependencies. The in silico resilience prediction approach accommodates numerous diverse factors contributing to multi-scale models of cancer, in order to better specify clinically useful aspects of recovery, treatment and intervention.

Keywords: Breast cancer · Resilience prediction · Psychological and behavioural data · Machine learning models · Model aggregation

J. Henriques et al. (Eds.): MEDICON 2019, IFMBE Proceedings 76, pp. 518–525, 2020.
https://doi.org/10.1007/978-3-030-31635-8_62

1 Introduction

Breast cancer is the most common cancer in women world-wide accounting for 28% of the total cancer cases in the WHO European Region. According to the WHO, despite of all advances in medicine and preventive strategies, the incidence of breast cancer in the developing countries is still rising [1]. In spite of this devastating fact, the mortality of breast cancer in the developing world seems to be decreasing, with five-year relative survival rate for female invasive breast cancer patients rising from 75% in the mid-1970 s to 90% today and in cases of localized breast cancer even reaching 98.5%. In Europe, well-organized screening programmes with good compliance have achieved a reduced breast cancer mortality of at least 20% in women aged over 50, while in 2013 alone more than half a million women were operated in the 21 EU Member States (partial or total mastectomy) in order to prevent or treat breast cancer [2]. The net effect of all these facts is a dramatic rise in both the number of breast cancer sufferers and survivors in Europe. This growth in patient volumes poses new unprecedented challenges to the healthcare ecosystem, as coping with breast cancer becomes a major socioeconomic challenge calling for novel strategies for understanding, predicting and increasing the resilience of women with breast cancer to all the stressful challenges and experiences and ensuring better and faster recovery through personalized interventions.

In the current work, we present a technological infrastructure that is currently under implementation within the BOUNCE H2020 EU project [3]. The vision of BOUNCE is to develop a resilience trajectory predictor taking into consideration biological, social, environmental, lifestyle, occupational and psychosocial status in order to predict levels of resilience of women with breast cancer throughout the cancer continuum and to propose resilience-building interventions to those who need it the most and eventually increase illness adaptation toward optimal clinical, well-being and functionality outcomes. There have been several attempts to model disease progression using advanced computational tools and to develop unified clinical models of modifiable factors associated with optimal disease outcomes in Breast Cancer [4, 5]. In view of the extremely high complexity of the phenomenon that is modelled (encompassing, clinical, physiological, molecular, lifestyle and psychophysiological components), the majority of studies to-date have intentionally intentionally restricted their scope to a limited set of predictors and outcomes with a special focus on survival rate and risk of metastasis.

This work presents the reference architecture in terms of the psychosocial and behavioral trajectory models. The computational modeling and simulation of illness progression within BOUNCE examine the transaction between the resilience trajectory, and the psychosocial/behavioural and medical trajectories of illness. These models are further aggregated at the decision level in order to develop a robust resilience trajectory predictor. In addition, a supporting tool is included in the proposed system for visualization and data exploration purposes. We also present auxiliary components of the infrastructure required for these models to effectively run and we provide an overview of the overall architecture.

The remaining text is structured as follows. In Sect. 2 we briefly outline the concept of resilience as adopted in BOUNCE followed by an overview of the system architecture in Sect. 3. In Sect. 4 we focus on the models that are currently implemented and on the methodology for their aggregation. Finally, Sect. 5 concludes this paper and presents the next steps of the present work.

2 Coping with Resilience

The process of successful adaptation to breast cancer and to the various accompanying stressors can be conceptually defined as the person's resilience, marking the will to "fight for life" and bounce back. When faced with such potentially life-threatening events each person engages coping strategies that can vary widely on their capacity to provide adaptive solutions and ensure optimal recovery with respect to the disease itself as well as to the patient's overall quality of life. It is also worth-stressing that coping strategies frequently vary depending on the stage and timing of illness (e.g. coping strategies when the person is diagnosed with breast cancer vs. after having received treatment). During the past decades considerable progress has been made in the field of Health Psychology towards identifying factors that determine adoption of particular types of coping behaviors highlighting socio-demographic and cognitive-emotional (e.g., the persons own diverse perceptions of their illness), in constant interplay with ever-changing life circumstances [6]. Moreover, it is recognized that intervention and/or prevention efforts should take into account both modifiable attributes as well as (constant) individual characteristics (such as sociodemographic, patient history, and cancer type) that may affect the person's dynamics.

3 System Architecture

A complex architecture is currently implemented towards developing a resilience trajectory predictor that takes into consideration various types of data. The high-level view of this architecture is shown in Fig. 1. It is envisioned as a framework, which integrates several building blocks organized in three tiers:

1. **the applications tier** contains a decision support system based on the developed models, the models repository and the required components for generating, managing, executing and fusing the available models.
2. **the semantic tier** based on hybrid architecture contains data extraction, transformation and serving the applications tier. In this layer a data lake stores the already available data from existing retrospective studies on cancer and is also used to store the various prospective data that are collected. To effectively integrate all these data an ontology is currently under development that eventually provides homogeneous access to the data through security enabled APIs.
3. **the security tier** comprises a privacy framework addressing user privacy and data security which enables the appropriate persons/role to have access to specific data, regulating efficiently and effectively data access.

Below we focus on the computational models that are available through the proposed reference architecture of BOUNCE aiming to achieve resilience prediction through heterogeneous clinical and biological data.

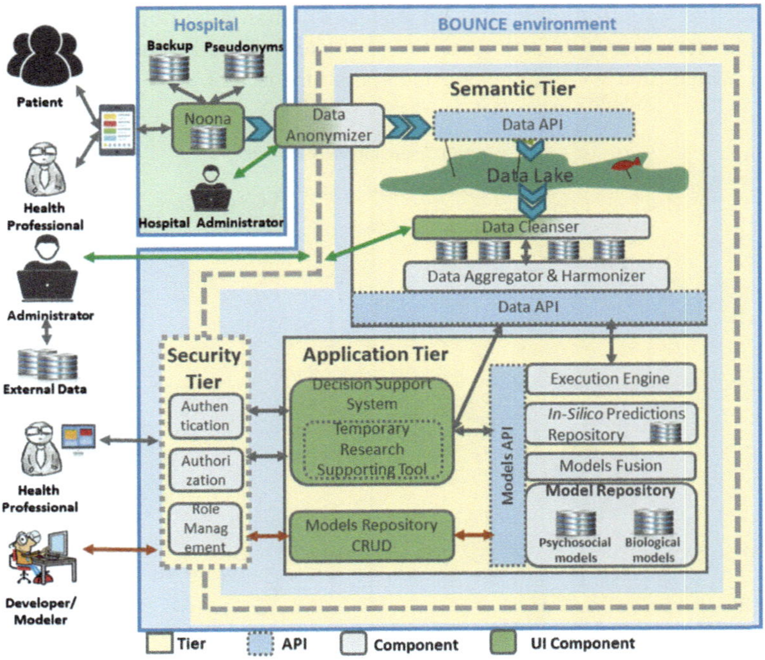

Fig. 1. BOUNCE high-level architecture.

4 Models and Model Aggregation

The main scope of BOUNCE is to develop a comprehensive analysis framework for predicting individual resilience trajectories of measurements taken at 3–6 month intervals during the critical 18-month period following cancer diagnosis. Multimodal, multiscale data comprising specific biological, social, environmental, lifestyle, sociodemographic, and psychosocial variables at successive time points are retrieved from the BOUNCE semantic layer. The main modules of the proposed resilience prediction workflow that are implemented to support the major objective of the general framework are: (i) the Research Supporting Tool, (ii) the Psychological and Behavioural trajectory models and (iii) the Predictor Model aggregation module.

4.1 Research Supporting Tool

The Temporary Research Supporting Tool is the short-term internal tool to facilitate data exploration and visualization of data and scales. The tool should be able to retrieve and visualize anonymized patient data, individual scores on each patient-reported physical symptom and psychosocial variable as well as combinations of values on different biomedical and clinical variables (obtained at the current and/or possible previous assessments). The visualization capabilities of the temporary research tool are based on a generic analytics framework for open health data [7]. As shown in Fig. 2, the interactive visualization tool of the data at different time points provides an overview of the data at once while it gives the option to the healthcare professionals to filter the data in real time with the responsive charts of the tool.

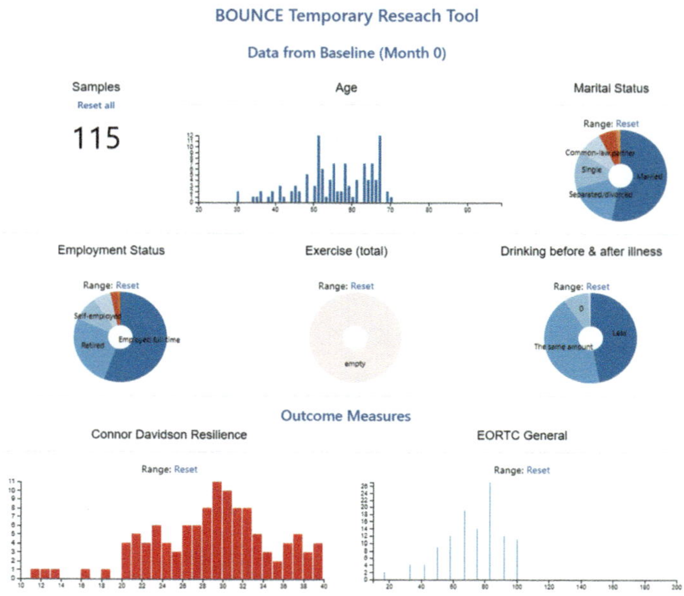

Fig. 2. The temporary research supporting tool of BOUNCE.

4.2 Psychosocial and Behavioural Trajectory Models

Within the BOUNCE computational framework, supervised, unsupervised and semi-supervised models are developed based on state-of-the-art techniques for constructing the Resilience Trajectory Predictor. Using statistical and machine learning methodologies to extract knowledge from clinical, molecular and psychological data that are further used to define patient trajectories, we aim to provide a tool that serves as decision support system for predicting the resilience evolution in women with breast cancer throughout the cancer treatment continuum.

Clustering has emerged as a major technique in the healthcare providing segmented health behavior profiles from the overall examined population that share a certain variance. Patient-specific characteristics can potentially be identified from these profiles, assisting clinicians in receiving significant information for individuals [8]. Within BOUNCE, several clustering techniques are implemented through the 18-month period of the multicenter clinical pilot study (e.g. sparse clustering [9]) aiming to identify psychosocial, life-style, and trait characteristics (i.e. trait resilience) that differentiate the profile of each cluster from the profiles of all the other clusters together both cross-sectionally and longitudinally.

A statistical analysis framework is designed and utilized to address psychosocial and behavioral changes based on repeated measures over time using latent growth curve modelling (LGCM) [10]. LGCM analysis represents a significant advancement in various research areas including supportive care in oncologic patients after treatment [11]. Systematic growth patterns/trajectories are generated over the 18-month study follow-up period and grouped using observed constant variables (i.e. clinical sites, age groups across sites, etc. in order to account for systematic patterns of individual differences in growth over time (i.e. multi-group LGCM based on observed variables). Individual trajectories are generated featuring estimates of latent intercept and slopes of the growth curves. Alternatively, unobserved subpopulations can be estimated from the data to better account for individual differences in adaptation to cancer. Overall model fit is assessed using absolute fit statistics (e.g. root mean square error) and relative model assessment metrics (e.g. Akaike Information Criterion). Alternative techniques for grouping/clustering trajectories will be also tested based on trajectory shape descriptors [12].

Machine learning algorithms utilizing boosted multivariate trees [13] and long short term memory networks (LSTMs) are implemented. Boosted multivariate trees coupled with feature selection techniques may be useful in identifying important features and their interactions over time (i.e., measurement time points) through a cross-validated pipeline to assess the generalization performance of the model. The latent growth models are validated against the end-point outcomes (i.e., clinical status in terms of disease progression, overall quality of life, and functionality). A LSTM network is developed to handle the sequential nature of the provided data and incorporate potential dependencies from measurements at preceding time points as recorded throughout the cancer continuum. Techniques such as dropout [14] are applied to the network to address potential overfitting.

4.3 In-Silico Resilience Trajectory—Predictor Model Aggregation

Once multiple models are generated, a crucial final step is to apply an effective fusion strategy, in order to enhance overall model accuracy. BOUNCE model aggregation takes place at the decision level. In particular, since the output of the models concerns two different types of trajectories, weighted average schemes (e.g. Bayesian combination, Bayesian Model Averaging (BMA) [15], as well as supervised learning methods are applied.

The combination of multiple predictive classifiers to improve accuracy while taking into account data from multiple modalities is increasingly becoming the method of choice in the healthcare research field [16]. In the context of BOUNCE resilience estimation and assessment, different predictive models are developed based on the heterogeneous data sources (i.e. (a) clinical as provided by oncologists, (b) molecular, (c) latent psychosocial data (d) disease- and treatment-related symptoms, (e) lifestyle data). The generated outputs of each predictor are further combined via a weighted average procedure. Linear and Bayesian combinations of the models' decisions will permit the efficient computation of the resilience score based on different data sources and parameters among the two main predictors.

In developing the predictive breast cancer resilience models and integrating them at the decision level, specific methods are considered for averaging the predictions of a set of models. In order to avoid any uncertainty about the "correct" model specification that is introduced by the designed and developed in silico resilience predictors, BMA is implemented. BMA allows the improved out of sample predictive performance.

Bayesian statistics presents all uncertainties in terms of probability theory, and make all inferences by applying the basic rules of probability theory. The posterior distribution of our entity of interest (i.e. aggregation of model's predictions) calculated via BMA represents a weighted average of the posterior distributions of each model. A single model is responsible for generating the whole dataset, and the probability distribution over the total number of models simply reflects our uncertainty as to which that model is. Model combination according to Bayes theorem is implemented in the R package for BMA [17]. In addition, ensemble methods which combine the predictions of several base estimators are implemented with a given learning algorithm to improve generalizability/ robustness over a single estimator. More specifically, VotingClassifier [18] combines conceptually different machine learning classifiers and uses a majority vote or the average predicted probabilities (soft vote) to predict the class labels. Such a classifier can be useful for aggregating several equally well performing models in order to balance out their individual weaknesses.

5 Conclusions

In the present work, the development of a model-based reference system for predicting resilience levels of women with breast cancer, is proposed. The overall architecture of the system compromises a framework with consistency among the different levels and components. The major component of the system, i.e. the resilience trajectory predictor is presented related to the machine learning algorithms considered. The semantic layer ensures the integration of heterogeneous data sources and their effective usage by a plethora of models. The proposed computational scheme, consisting of machine learning predictive models and a research supporting tool enhances the clinical research related to resilience prediction and the patient's capacity to bounce back during the highly stressful treatment and recovery period following disease diagnosis.

Acknowledgment. This work has received funding from the European Union's Horizon 2020 research and innovation programme under grant agreement No. 777167.

Conflict of Interest. The authors declare that they have no conflict of interest.

References

1. WHO, Cancer Key Statistics. https://www.who.int/cancer/resources/keyfacts/en/. Accessed 21 Jan 2019
2. ECIS - European Cancer Information System. https://ecis.jrc.ec.europa.eu/. Accessed 21 Jan 2019
3. Predicting Effective Adaptation to Breast Cancer to Help Women to BOUNCE Back. https://www.bounce-project.eu/. Accessed 21 Jan 2019

4. PREDICT: an online tool that helps patients and clinicians see how different treatments for early invasive breast cancer might improve survival rates after surgery. https://www.predict.nhs.uk/. Accessed 21 Jan 2019
5. Breast Cancer Conditional Outcome Calculator. http://www.lifemath.net/cancer/. Accessed 21 Jan 2019
6. Deshields, T.L., Heiland, M.F., Kracen, A.C., Dua, P.: Resilience in adults with cancer: development of a conceptual model. Psycho-Oncol. **25**, 11–18 (2016)
7. Koumakis, L., Kondylakis, H., Katehakis, D., Iatraki, G., Argyropaidas, P., Hatzimina, M., et al.: A content-aware analytics framework for open health data. In: Precision Medicine Powered by pHealth and Connected Health. Springer, Heidelberg, pp. 59–64 (2018)
8. Lowe, K., Escoffery, C., Mertens, A.C., Berg, C.J.: Distinct health behavior and psychosocial profiles of young adult survivors of childhood cancers: a mixed methods study. J. Cancer Surviv. **10**, 619–632 (2016)
9. Kondo, Y., Salibian-Barrera, M., Zamar, R.: RSKC: an R package for a robust and sparse k-means clustering algorithm. J. Stat. Softw. **72**, 1–26 (2016)
10. Curran, P.J., Obeidat, K., Losardo, D.: Twelve frequently asked questions about growth curve modeling. J. Cogn. Dev. **11**, 121–136 (2010)
11. Brédart, A., Merdy, O., Sigal-Zafrani, B., Fiszer, C., Dolbeault, S., Hardouin, J.: Identifying trajectory clusters in breast cancer survivors' supportive care needs, psychosocial difficulties, and resources from the completion of primary treatment to 8 months later. Support. Care Cancer **24**, 357–366 (2016)
12. Genolini, C., Ecochard, R., Benghezal, M., Driss, T., Andrieu, S., Subtil, F.: kmlShape: an efficient method to cluster longitudinal data (Time-Series) according to their shapes. PLoS ONE **11**, e0150738 (2016)
13. Pande, A., Li, L., Rajeswaran, J., Ehrlinger, J., Kogalur, U.B., Blackstone, E.H., et al.: Boosted multivariate trees for longitudinal data. Mach. Learn. **106**, 277–305 (2017)
14. Srivastava, N., Hinton, G., Krizhevsky, A., Sutskever, I., Salakhutdinov, R.: Dropout: a simple way to prevent neural networks from overfitting. J. Mach. Learn. Res. **15**, 1929–1958 (2014)
15. Hoeting, J.A., Madigan, D., Raftery, A.E., Volinsky, C.T.: Bayesian model averaging: a tutorial. Stat. Sci. **14**, 382–401 (1999)
16. Baltrušaitis, T., Ahuja, C., Morency, L.-P.: Multimodal machine learning: a survey and taxonomy. IEEE Trans. Pattern Anal. Mach. Intell. **41**(2), 423–443 (2018)
17. Raftery, A.E., Painter, I.S., Volinsky, C.T.: BMA: an R package for Bayesian model averaging. R News **5**, 2–8 (2005)
18. Ruta, D., Gabrys, B.: Classifier selection for majority voting. Inf. Fusion **6**, 63–81 (2005)

A Systems Biology Approach to Decipher Genetic Variants in a Canine Model of Sudden Cardiac Death

Martina Vescio[1(✉)], Lia Crotti[2,3,4], Peter Schwartz[2], and Linda Pattini[1]

[1] Department of Electronics Information and Bioengineering, Politecnico di Milano, 20133 Milan, Italy
martina.vescio@polimi.it
[2] Istituto Auxologico Italiano, IRCCS, Center for Cardiac Arrhythmias of Genetic Origin and Laboratory of Cardiovascular Genetics, Via Pier Lombardo, 22, 20135 Milan, Italy
[3] Department of Medicine and Surgery, University of Milano-Bicocca, Via Cadore, 48, 20900 Monza, Italy
[4] Istituto Auxologico Italiano, IRCCS, Department of Cardiovascular, Neural and Metabolic Sciences, San Luca Hospital, Piazzale Brescia 20, 20149 Milan, Italy

Abstract. Sudden cardiac death (SCD) represents a major public health challenge, accounting for approximately 25% of all cardiac deaths. It refers to an unexpected death from ventricular arrhythmia, occurring in individuals with preexisting cardiovascular disorders as well as in individuals not previously diagnosed with heart disease. The identification of genetic variants that increase susceptibility to SCD is fundamental to improve risk stratification and understanding of molecular pathophysiology. In this study, to investigate the molecular mechanisms underlying SCD, a canine model, recapitulating what may happen to patients with a prior myocardial infarction, was used to accomplish a genome-wide association study comparing dogs resulting susceptible or resistant to ventricular fibrillation during submaximal exercise. The identified variants were explored by means of a systems biology approach, which maps human orthologues of mutated genes into a network encompassing co-expression and physical interactions. The paths connecting mutated genes highlighted a subnetwork enriched for genes involved in regulation of cardiac function.

Keywords: Sudden cardiac death · Canine genome-wide association study · Gene networks · Systems biology

1 Introduction

Sudden cardiac death (SCD) is mainly due to ventricular fibrillation (VF), that impairs cardiac pumping and leads to organ failure. SCD is a major cause of death in Western countries and it can occur to patients of any age. Half of the deaths are associated to a

P. Schwartz and L. Pattini—contributed equally as senior authors.

J. Henriques et al. (Eds.): MEDICON 2019, IFMBE Proceedings 76, pp. 526–530, 2020.
https://doi.org/10.1007/978-3-030-31635-8_63

broad range of pre-existent heart conditions. After the age of 45 years, SCD is mainly caused by coronary heart disease, while in younger subjects the major cause are rare inherited cardiac disorders, such as cardiomyopathies and primary electric disorders [1]. However, half of the deaths associated to this disease regards subjects with no previously diagnosed cardiovascular problems. Therefore, the identification of genetic factors related to its occurrence could have a significant impact in SCD management and risk stratification. The aim of this work is to study the molecular determinants of the pathology using a network approach to better understand the potential interaction among different genetic alterations and their impact on the cellular processes. In this study a canine model of the pathology is used. Dogs have been widely used in genetics in the last years as many canine and corresponding human diseases share the same genetic causes. GWASs on canine models have already been successfully used to characterize disease-susceptibility loci for obsessive-compulsive disorder (OCD) and osteosarcoma [2].

2 Materials and Methods

2.1 Animal Model of Ventricular Fibrillation

To better understand the mechanisms underlying sudden cardiac death, a canine model reproducing rather closely what may happen to patients with a prior myocardial infarction was developed [3]. The animals were treated to exhibit known clinical factors associated to an increased risk of undergoing ventricular fibrillation (VF), such as previous myocardial ischemic injury, acute myocardial ischemia and physiologically increased sympathetic activity. An anterior wall myocardial infarction was produced by surgical occlusion of the left anterior descending coronary artery. Susceptibility to VF was assessed 3-4 weeks after surgery during a submaximal exercise. The animals ran on a treadmill until 70% of maximum heart rate was reached. During the last minute of exercise, an acute ischemia was induced by the transient occlusion of the left circumflex coronary artery. The occlusion was maintained for 2 min, the animals undergoing VF during this time were considered susceptible, while the others were considered resistant [3, 4].

2.2 Genetic Variants Identification

Genetic variants significantly associated to disease susceptibility in dogs were identified comparing susceptible to resistant animals in a genome-wide association study (GWAS). The GWAS was carried out applying a linear mixed model approach implemented in GAPIT Version 2 [5]. Linear mixed models are frequently used in GWASs performed on model organisms as they reduce the false positive rates due to population structure and genetic relatedness. In particular, pairwise genetic relatedness is incorporated directly into the model to correct for population structure and to account for the expected higher correlation between genetically similar organisms.

Genetic variants identified on canine models were then mapped on human genes to inspect their role in human interactome. First, we selected dog genes affected by mutations resulting from GWAS, then, we retrieved their human homologs. The obtained list of mutated genes was explored through a systems biology approach.

2.3 Network Analysis

Gene expression data from left ventricle tissue were downloaded from the Genotype-Tissue Expression (GTEx) project website [6]. The dataset contained RNA-Seq data from 303 healthy donors. Read counts were normalized using TMM method implemented in R package edgeR. To reconstruct the transcriptional relationships involving altered genes identified in GWAS, the ARACNe algorithm was applied to gene expression data and was used to calculate mutual information (MI) between mutated genes and each other gene in the dataset. This measure can quantify the mutual dependence between the expression levels of two transcripts. Co-expression networks can identify groups of functionally related genes according to the guilt-by-association principle that states that genes with similar expression pattern are likely involved in the same molecular function. ARACNe estimates MI using a Gaussian Kernel estimator and removes indirect dependencies applying data processing inequality (DPI) [7]. The co-expression network was inferred using $P < 10–55$ as a significance threshold for MI. The transcriptional network was further expanded with protein-protein interactions data retrieved from IntAct database. To focus only on the potential subnetwork including the nodes connecting the genetic variants with each other, the shortest paths were extracted from the integrated network for each pair of mutated genes using PathLinker tool from Cytoscape [8]. Enrichment analysis was carried out on the subnetwork using annotated gene sets from the Molecular Signatures Database (MSigDB) [9]. A hypergeometric test was used to verify the enrichment of each gene set and the final p-values were corrected for multiple testing using the Benjamini-Hochberg procedure.

3 Results

A total of 149 dogs were selected for the genome-wide association study after quality control. The association test was conducted on 92 susceptible dogs and 57 resistant dogs and resulted in a few variants at a suggestive significance level ($p < 5 \times 10^{-5}$). The identified genetic variants were mapped to 5 human genes that were examined in the context of the left ventricle interactome, a network composed of tissue-specific gene interactions. This type of approach is useful to understand the role of the mutated genes in the development of the pathology. The construction of the transcriptional network and its integration with information about molecular interactions resulted in a graph composed by 4597 nodes and 7875 edges. The complete network is shown in Fig. 1. The shortest paths connecting mutated genes were extracted from the complete network to isolate the nodes responsible for the interactions among the genes of interest. The functional enrichment analysis performed on the obtained subnetwork identified various significant terms regarding the regulation of cardiac conduction and contraction, together with the heart development and the regulation of the heart rate. The most relevant terms and their significance are showed in Table 1.

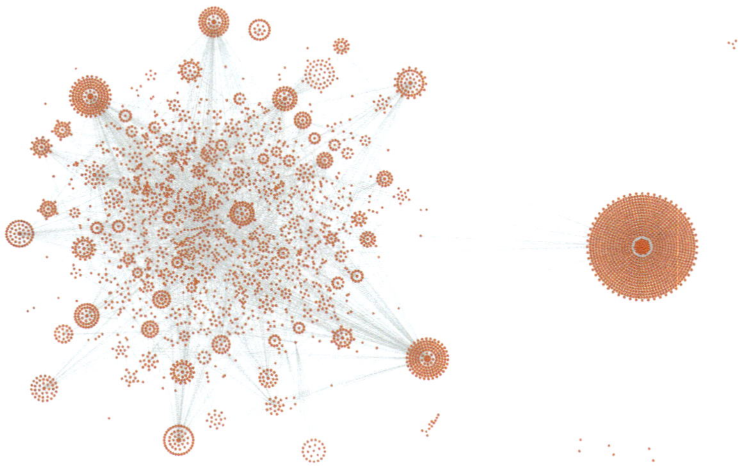

Fig. 1. Gene network obtained using mutated genes as seeds. Nodes represent genes and edges represent statistical dependence between expression levels or interactions retrieved from IntAct database. Nodes dimensions are proportional to their degree.

Table 1. Functional annotations enrichment analysis – most relevant terms related to heart development and function.

Term	FDR
Regulation of cardiac muscle contraction by calcium ion signaling	5.90E-07
Regulation of cardiac conduction	1.39E-06
Regulation of heart rate	4.76E-06
Regulation of heart contraction	3.03E-05

4 Discussion and Conclusions

The aim of this work was to study the molecular mechanisms responsible for sudden cardiac death susceptibility. First, genetic variants distinctive of susceptible subjects were identified in an animal model closely mimicking clinical reality; then, their role in left ventricle interactome was investigated. The canine model described in this work has already proven useful in determining important information about susceptible and resistant subjects, for example the presence of an abnormal autonomic cardiac regulation in susceptible animals [3]. Moreover, this animal model has either confirmed or predicted all clinical studies and trials on antiarrhythmic drugs in post-myocardial infarction patients [10]. Therefore, genomic data derived from this model are highly valuable and they could be crucial for the understanding of the molecular determinants of the disease. If considered individually, the genetic variants resulting from GWAS show a low significance and a limited impact on the pathology. Although they do not have a strong effect individually, they could contribute to the deregulation of the same pathway, inducing disease susceptibility. The analysis of the genetic variants in the context of the tissue specific interactome revealed that altered genes interact closely with genes related to the regulation of cardiac conduction and contraction, heart development and regulation of heart rate. Future works could further enrich the network with the integration of additional omics data.

Acknowledgments. We thank LINK project (H2020 Grant Agreement No. 692023) for the given support. Italian Ministry of Health, Grant GR-2009-1472102 (PI Lia Crotti) "Sudden cardiac death during myocardial ischemia: unmasking genetic factors contributing to ventricular fibrillation".

Conflict of Interest. The authors declare no conflict of interest.

References

1. Bezzina, C.R., Lahrouchi, N., Priori, S.G.: Genetics of sudden cardiac death. Circ. Res. **116**(12), 1919–1936 (2015)
2. Machiela, M.J., Chanock, S.J.: GWAS is going to the dogs. Genome Biol. **15**(3), 1–3 (2014)
3. Schwartz, P.J., Billman, G.E., Stone, H.L.: Autonomic mechanisms in ventricular fibrillation induced by myocardial ischaemia during exercise in dogs with healed myocardial infarction: an experimental preparation for sudden cardiac death. Circulation **69**, 790–800 (1984)
4. Billman, G.E.: A comprehensive review and analysis of 25 years of data from an in vivo canine model of sudden cardiac death: implications for future anti-arrhythmic drug development. Pharmacol. Ther. **111**(3), 808–835 (2006)
5. Tang, Y., et al.: GAPIT version 2: an enhanced integrated tool for genomic association and prediction. Plant Genome **9**(2) (2016)
6. Carithers, L.J., Moore, H.M.: The genotype-tissue expression (GTEx) project. Nat. Genet. **45**(6), 580–585 (2013)
7. Margolin, A.A., et al.: ARACNE: an algorithm for the reconstruction of gene regulatory networks in a mammalian cellular context. BMC Bioinform. (2006)
8. Gil, D.P., Law, J.N., Murali, T.M.: The PathLinker app: connect the dots in protein interaction networks. F1000Research, vol. 6, p. 58 (2017)
9. Subramanian, A., et al.: Gene set enrichment analysis: a knowledge-based approach for interpreting genome-wide expression profiles. Proc. Natl. Acad. Sci. **102**(43), 15545–15550 (2005)
10. Schwartz, P.J.: Do animal models have clinical value? Am. J. Cardiol. **81**(6), 14D–20D (1998)

Computational Fluid Dynamics Study of Inlet Velocity on Extrusion-Based Bioprinting

Juan Carlos Gómez-Blanco[1](✉), Enrique Mancha-Sánchez[1], Juan Francisco Ortega-Morán[1], Antonio Díaz-Parralejo[2], Francisco Miguel Sánchez-Margallo[1], and José Blas Pagador-Carrasco[1]

[1] Jesús Usón Minimally Invasive Surgery Centre, Cáceres, Spain
jcgomez@ccmijesususon.com
[2] University of Extremadura, Badajoz, Spain
https://www.ccmijesususon.com
https://www.unex.es/

Abstract. Extrusion bioprinting is one of the most studied additive manufacturing technologies thanks to the huge possibilities in the creation of engineered tissues that it provides. Despite all research performed about bioprinting there is not enough information on how inlet velocities affect the pressure distribution. In this way, we have carried out CFD simulations to check how different inlet velocities can affect the pressure distribution in the tip of a nozzle.

Keywords: Bioprinting · Computational simulation · Nozzle · Velocity

1 Introduction

The nowadays popularised additive manufacturing technique, known as 3D printing, is a technology that works under a simple principle: the creation of an object layer by layer. The consecutive displacement of a certain material in each layer is the construction method of any additive manufacturing method. This fabrication method allows intricate geometry which means the possibility of fabricating objects that conventional manufacturing methods cannot allow.

The application of 3D printing in biology or health is called bioprinting [1–3]. The difference between this application and conventional 3D printing is the used material, cell-laden or cell-compatible materials. Any kind of material can be used in bioprinting as long as they are at least biocompatible (allow cells to live and proliferate) and biodegradable [4–6]. The most used can be classified in: (1) polymers as Polycaprolactone (PCL) or methylcellulose and (2) hydrogels as Alginate or Hyaluronic Acid (HA). According to the clinical/medical application different stem cells are used, such as Bone Marrow Stem Cells (BMSCs) in bone scaffolds or Mesenchymal Stem Cells (MSCs) in cartilage regeneration. Due to

J. Henriques et al. (Eds.): MEDICON 2019, IFMBE Proceedings 76, pp. 531–540, 2020.
https://doi.org/10.1007/978-3-030-31635-8_64

cells close range of parameters values (sensitivity) they need a proper environment and nutrients to live and proliferate [7]. For this reason, a bioprinter must control, at least, temperature, pressure, CO2 and pH [8].

From all possible deposition methods (Extrusion, InkJet, Laser-based), the most used in bioprinting is extrusion, specifically Pneumatic extrusion [9]. There are two main reasons because extrusion-based techniques are used in bioprinting, its cost-efficiency rate and the possibility of use customized nozzles [10]. So, these kinds of bioprinters use extrusion heads that are usually formed by a syringe and a needle or a conical tip. It is well described in the bibliography that pressure is one of the most important factors that can provoke a low cell survival [11]. In this sense, pressure values in the stream flow can be modified by optimization in geometry and extrusion flow rate.

Computational simulations are widely extended in the calculation of flows through nozzles [12,13]. However, bioprinting related simulations have still room for improvement with only a few research studies [14,15]. For this reason, Computational Fluids Dynamics (CFD) applied in bioprinting technology can provide useful data of how the flow pressure is modified (usually a reduction to enhance cellular viability) by an adequation of geometry and/or extrusion flow rate. According to continuum equation, the flow rate is directly proportional to inlet velocity, so adequation of inlet velocity leads of decreasing the pressure cells must endure during bioprinting extrusion procedures.

Hence, the main objective of this work is to analyse the pressure in the nozzle tip through computational fluid simulation for different inlet velocities.

2 Methods

2.1 Model

The model was designed and simulated in COMSOL Multiphysics through a 2D axisymmetric model and the Two-Phase Flow (TPF) level set interface. The model consists of two different geometries. One is related to the nozzle, modelled by a trapezium of 25 mm of height, 3 mm of the inlet face and 0.2 mm of the outlet face. The rest of the geometry is composed by a rectangle and a trapezium and is related to the air where the nozzle ejects the material.

2.2 Governing Equations

Level set (LS) method is a transport equation which is added to incompressible Navier-Stokes equations to track the interface of two immiscible fluids, in this case bioprinting materials and air, these equations are:

$$\rho\frac{\partial u}{\partial t} + \rho(u \cdot \nabla)u = \nabla \cdot [-pI + \mu(\nabla u + \nabla u^T)] + F + \rho g \tag{1}$$

$$\nabla \cdot u = 0 \tag{2}$$

where, ρ is the density, u is the speed of the fluid, p denotes the pressure, I is the identity matrix, F is all other external forces and g is the gravity force.

In the level set equation, the bioprinting material is expressed by $\phi = 0$, the air is expressed by $\phi = 1$ and the contour lines of level set is expressed by $\phi = 0.5$. The level set equation can be seen as the volume percentage of water in the gas-liquid two-phase flow [16]. Therefore, the migration equation of the gas-liquid interface can be written as follows:

$$\frac{\partial \phi}{\partial t} + u \cdot \nabla \phi = \gamma \nabla \left(\epsilon_{ls} \nabla \phi + \phi(1 - \phi) \frac{\nabla \phi}{\nabla \phi} \right) \tag{3}$$

where ϕ is the contour line of the interface of the gas-liquid two-phase flow, γ is the reinitialization parameter to solve the equation, ϵ is the interface thickness controlling parameter. The physical property parameters need to be processed smoothly to avoid the instability of the numerical calculation. These parameters are the density and viscosity of the fluid near the interface. The smoothing process of the level set equation involves the level set function to express the change in the density and viscosity of fluid during the process of the gas-liquid two-phase flow and are defined as:

$$\rho = \rho_1 + (\rho_2 - \rho_1)\phi \tag{4}$$

and the dynamic viscosity is given by:

$$\mu = \mu_1 + (\mu_2 - \mu_1)\phi \tag{5}$$

where ρ_1 and ρ_2 are the constant densities of bioprinting material and air, respectively, and μ_1 and μ_2 are the dynamic viscosities of bioprinting material and air, respectively. Here, bioprinting material corresponds to the domain where $\phi < 0.5$, and air corresponds to the domain where $\phi > 0.5$.

2.3 Domains

In this model, two different domains were considered. The first domain is related to the nozzle and the second domain is the outside of the nozzle, the air where the biological material is ejected. To make possible a flow an inlet was placed in the top part of the nozzle. Three different inlet velocities were tested. 1 mm/s, 5 mm/s and 10 mm/s were the values selected as inlet velocities. Also, an outlet condition was set at the top part of the air domain to allow bioprinting materials to fill this domain and air to leave. Viscosity and density of the material is 0.83 $Pa \cdot s$ and 1455.08 kg/m^3, respectively. An initial interface boundary condition was placed at the end of the nozzle to define a boundary where the fluid finishes and the air starts. All geometry and boundary conditions can be seen in the Fig. 1.

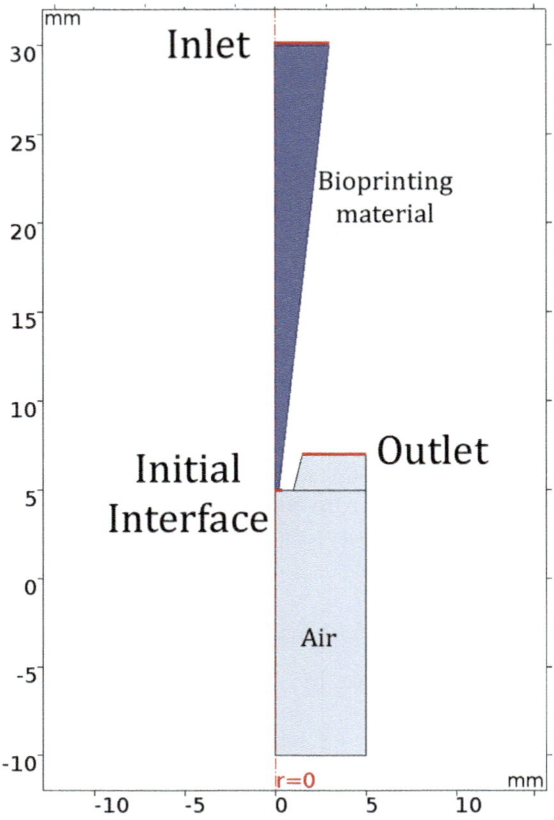

Fig. 1. Representation of the geometry used in the simulations

2.4 Mesh

A 2D triangular mesh was created for the simulations. For better simulations an adaptative mesh refinement was set as can be seen in Fig. 2. This creates multiple meshes for segments of the time-dependent simulation. Initial triangular mesh has 2892 triangular elements for three models, while the remeshing raise the number of elements around the interface between both materials to 4940, 5037 and 5023 elements for 1, 5 and 10 mm/s simulations respectively.

2.5 Simulations

Three 1.5 s simulations (5 ms step) were done, one for every inlet velocity studied (1, 5, 10 mm/s). Each simulation was divided in two steps: Phase Initialization and Time-Dependent. The Phase Initialization step solves for the distance to the initial interface >, Dwi. The Time-Dependent step then uses the initial condition for the level set function according to the following expression:

$$\phi_0 = \frac{1}{1 + e^{D_{wi}/\epsilon}} \tag{6}$$

in domains initially filled with bioprinting material and:

$$\phi_0 = \frac{1}{1 + e^{-D_{wi}/\epsilon}} \tag{7}$$

in domains initially filled with air.

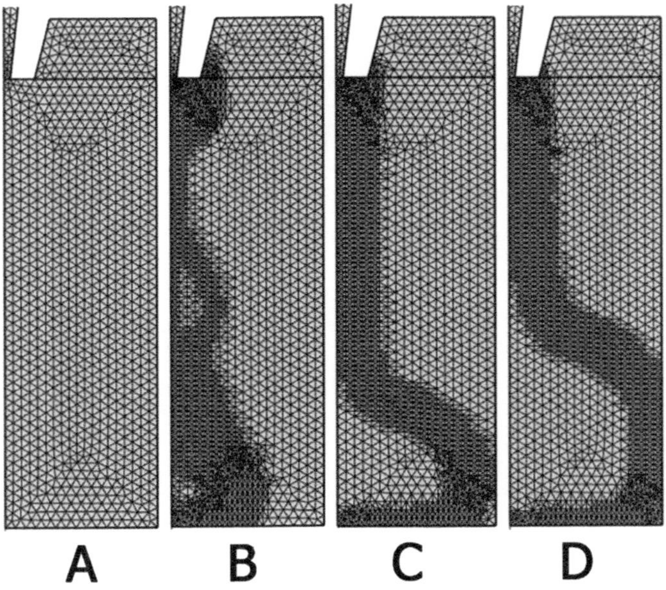

Fig. 2. Original mesh (A), 1 mm/s (B), 5 mm/s (C) and 10 mm/s (D) mesh refinement at 1.5 s of s simulation.

3 Results

3.1 Fluid Volumetric Fraction

The amount of each material is represented by the fluid volumetric fraction. Values 1 means that the 100% of the material is composed by the biomaterial it is extruded. Therefore 0 values mean that all the material is air. The interface between the material and air is defined in the figure with a fluid fraction volume of 0.5.

First inlet velocity set (1 mm/s), Fig. 3, shows the flow formed at 0.695 s of the simulation, just the moment before the drop detached from the jet. In the total time of simulation (1.5 s) two drops are formed and felt down.

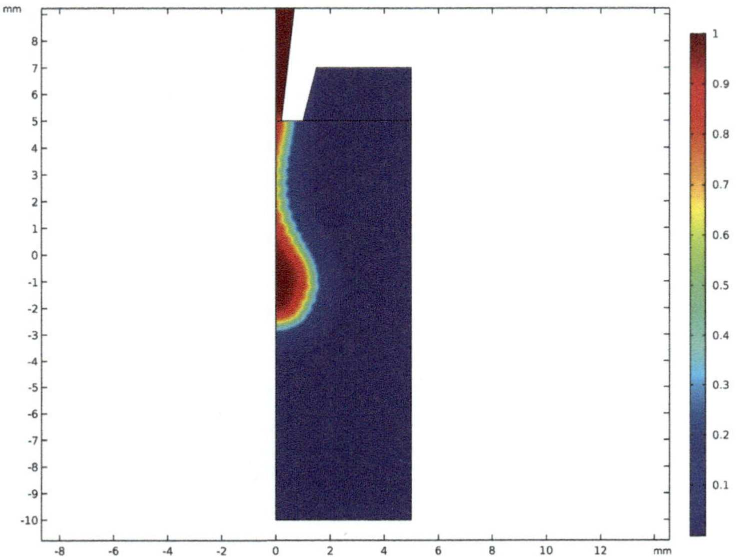

Fig. 3. Fluid Volumetric Fraction of 1 mm/s inlet velocity at 0.695 s

With the second velocity inlet value (5 mm/s), similar results were obtained (Fig. 4). A drop was formed and seems to be detached at 0.215 s. In comparison with 1 mm/s drop is not almost detached form the jet due to the higher inlet velocity.

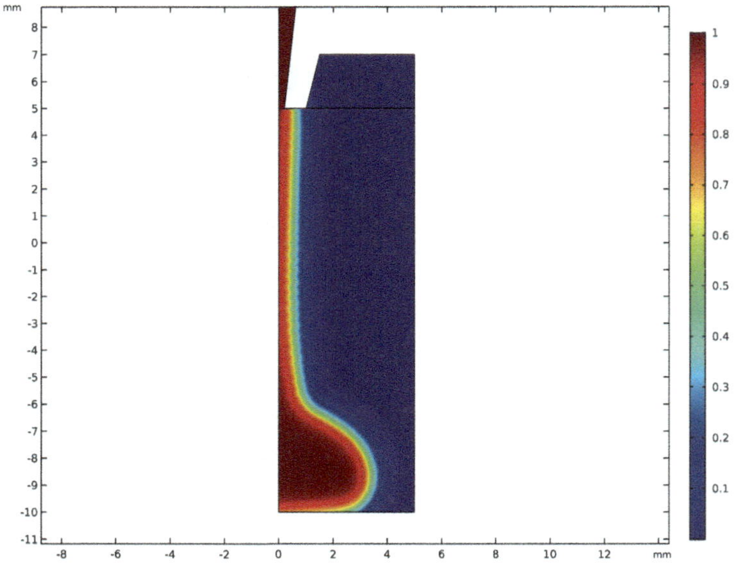

Fig. 4. Fluid Volumetric Fraction of 5 mm/s inlet velocity at 0.215 s

Values higher than 5 mm/s produces a continuous flow of the material extruded as it can be seen in the next inlet velocity value tested (10 mm/s), Fig. 5 In this simulation, a continuous flow is produced during the whole time, without drop detaching from the jet as a result of the increased inlet velocity.

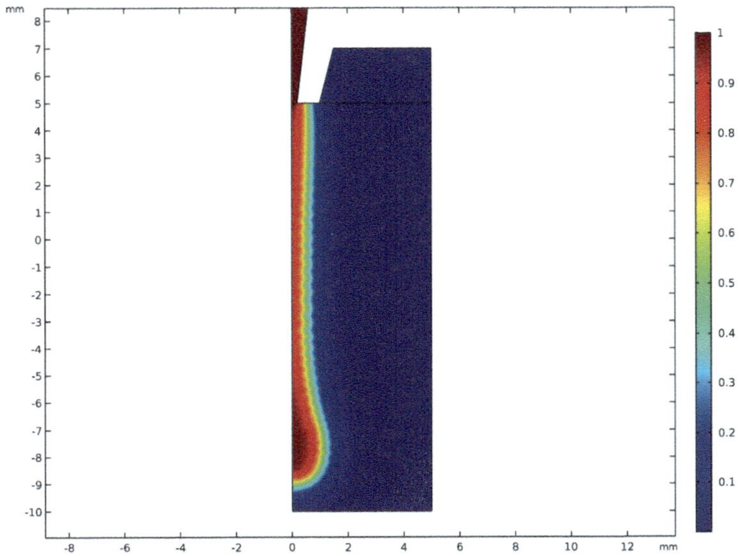

Fig. 5. Fluid Volumetric Fraction of 10 mm/s inlet velocity at 0.1*s*

3.2 Pressures

Pressure plots obtained in the simulations can be seen in Figs. 6, 7 and 8 for the three simulations carried out for 1, 5 and 10 mm/s, respectively. In Fig. 6 pressure increases at the beginning of the drop formation. When the drop is big enough the pressure starts to descend (due to gravity) and, at last, when the drop falls there is a low-pressure peak.

Figure 7 shows the creation of the initial drop (raise and decrease of pressure), from 0 s to 0.15 s, then the low pressure is maintained due to the instability of the jet. It is seen that in 0.6 s the jet stabilised reaching the maximum pressure value. Plot corresponding to 10 mm/s inlet velocity (Fig. 8) has similar behaviour to 5 mm/s inlet velocity plot. The difference between them is the total time of jet instability, reaching the stability at 0.15 s, showing a difference of 0.45 s.

Maximum pressure values are 105.70, 123.16 and 143.28 kPa for 1, 5 and 10 mm/s inlet velocity, respectively. The pressure varies for about 38 kPa comparing the max and min inlet velocity. It can be said that the higher the inlet velocity is the more pressure is needed to extrude it. However, more studies are needed for determining the exact velocity in which the minimum pressure is needed to create and maintain a continuous jet.

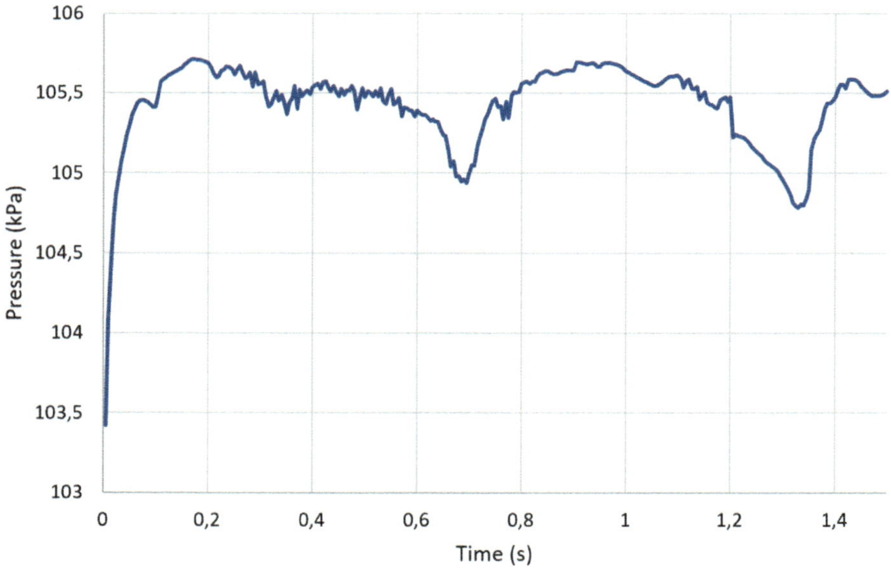

Fig. 6. Pressure plot in the outlet of the nozzle with 1 mm/s inlet velocity

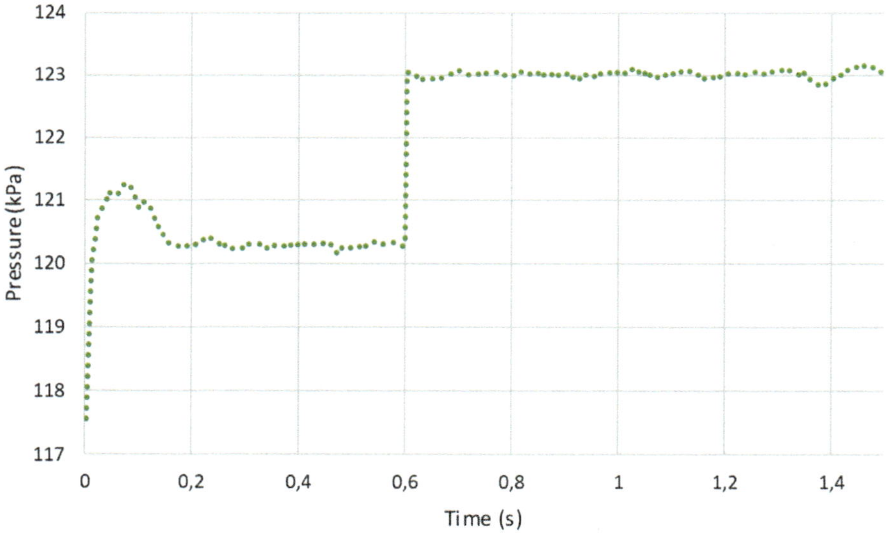

Fig. 7. Pressure plot in the outlet of the nozzle with 5 mm/s inlet velocity

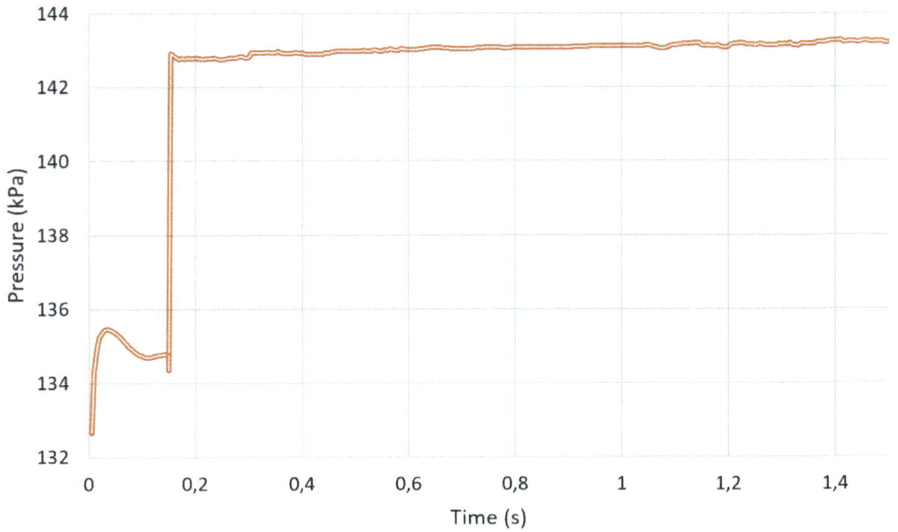

Fig. 8. Pressure plot in the outlet of the nozzle with 10 mm/s inlet velocity

4 Conclusions

In this work, three different flow computational simulations of different inlet velocity through a nozzle have been done. We have checked that the increase of inlet velocity provokes a more stable jet with higher pressures at the nozzle tip. However only three velocities have been simulated, it is necessary further studies with a wide range of velocities to obtain the minimum pressure needed to create and maintain a continuous jet.

Hence, future simulations with modifications in nozzle tip geometry are needed to check if geometry has an important role in pressure distribution. Also, pressure inlet distribution should be studied in order to characterise all bioprinter parameters.

Acknowledgment. This work was supported by Consejeráa de Economía e Infraestructuras, Junta de Extremadura. Project number IB16200 "Optimización y mejora de técnicas de bioimpresión para regeneración de cartílago y prótesis vasculares" and predoctoral grant number PD16067.

Declaration of Interest. The authors declare that they have no conflict of interest.

References

1. Kang, H.-W., Yoo, J.J., Atala, A.: Bioprinted scaffolds for cartilage tissue engineering. Cartil. Tissue Eng. Methods Protoc. **1340**, 981–995 (2015)

2. Hesuani, Y.D., et al.: Design and Implementation of Novel Multifunctional 3D Bioprinter. 3d Print. Addit. Manuf. **3**(1), 65–68 (2016)
3. Cubo, N., Garcia, M., del Cañizo, J.F., Velasco, D., Jorcano, J.L.: 3D bioprinting of functional human skin: production and in vivo analysis. Biofabrication **9**(1), 15006 (2016)
4. Ahmed, E.M.: Hydrogel: preparation, characterization, and applications: a review. J. Adv. Res. **6**(2), 105–121 (2015)
5. He, Y., Yang, F., Zhao, H., Gao, Q., Xia, B., Fu, J.: Research on the printability of hydrogels in 3D bioprinting. Sci. Rep. **6**, 29977 (2016)
6. Panwar, A., Tan, L.P.: Current status of bioinks for micro-extrusion-based 3D bioprinting. Molecules **21**(6), E685 (2016)
7. Alberts, B., et al.: Mollecular biology of the cell. In: Anderson, M., Granum, S. (eds.) Garland Science, Taylor & Francis Group, LLC (2008)
8. Zhao, Y., Li, Y., Mao, S., Sun, W., Yao, R.: The influence of printing parameters on cell survival rate and printability in microextrusion-based 3D cell printing technology. Biofabrication **7**(4), 1–11 (2015)
9. Pati, F., Jang, J., Lee, J.W., Cho, D.W.: Extrusion bioprinting (2015)
10. Ozbolat, I.T., Moncal, K.K., Gudapati, H.: Evaluation of bioprinter technologies. Addit. Manuf. **13**, 179–200 (2017)
11. Ning, L., Chen, X.: A brief review of extrusion-based tissue scaffold bio-printing. Biotechnol. J. **12**(8), 1–16 (2017)
12. Yuan, W., Schnerr, G.H.: Numerical simulation of two-phase flow in injection nozzles: interaction of cavitation and external jet formation. J. Fluids Eng. **125**(6), 963–969 (2004)
13. Zekovic, S., Dwivedi, R., Kovacevic, R.: Numerical simulation and experimental investigation of gas-powder flow from radially symmetrical nozzles in laser-based direct metal deposition. Int. J. Mach. Tools Manuf. **47**(1), 112–123 (2007)
14. Reid, J.A., Mollica, P.A., Johnson, G.D., Ogle, R.C., Bruno, R.D., Sachs, P.C.: Accessible bioprinting: adaptation of a low-cost 3D-printer for precise cell placement and stem cell differentiation. Biofabrication **8**(2), 25017 (2016)
15. Martanto, W., Baisch, S.M., Costner, E.A., Prausnitz, M.R., Smith, M.K.: Fluid dynamics in conically tapered microneedles. AIChE J. **51**(6), 1599–1607 (2005)
16. COMSOL Multiphysics: Theory for the Two-Phase Flow Interfaces - CFD Module User's Guide. Manual, p. 620 (2014)

A System to Assist in the Training of Medical Students in Respiratory Diseases

Alejandro Talaminos-Barroso[1(✉)], Javier Reina-Tosina[1], Laura M. Roa[1], David Naranjo-Hernández[1], Gerardo Barbarov-Rostán[1], Pilar Cejudo-Ramos[2,3], Eduardo Márquez-Martín[2,3], and Francisco Ortega-Ruiz[2,3]

[1] Biomedical Engineering Group, University of Seville, 41092 Seville, Spain
talaminos@gmail.com
[2] Medical-Surgical Unit of Respiratory Diseases, University Hospital Virgen del Rocío, 41013 Seville, Spain
[3] Biomedical Research Center in Network (CIBER) of Respiratory Diseases (CIBERES), 28029 Madrid, Spain

Abstract. This work presents a model to train medical students in the analysis of different pathologies of the respiratory system. The model allows the simulation of dynamic behaviors of some variables associated with the pulmonary function under diverse physiopathological conditions. Variables such as age, gender, anthropometric factors and lung volumes obtained by plethysmography, can be used to personalize the model considering a particular patient. The model has been validated by comparing the results obtained by simulation with works by other authors. The two most common patterns of lung abnormality in clinical practice are presented as an example of applicability of the model, analyzing the relationships between inspiratory-expiratory flows and volumes, flow-volume loop and FEV1/FVC relationship.

Keywords: Computational model · Respiratory system · Education · Pulmonary disease · Emphysema · Simulation

1 Introduction

Lung diseases are within the leading causes of morbidity and represent the third cause of death worldwide [1], with associated health costs exceeding €56 billion per year in the EU alone [2]. Although mortality has decreased slightly in the last 20 years [3], asthma, chronic obstructive pulmonary disease (COPD) and other respiratory disorders will remain a public health problem in the future due to the high rate of smokers and air pollution in large cities. Therefore, from a medical education point of view, it seems appropriate to have a tool to simulate different types of pathologies and their associated pulmonary alterations in terms of static volumes and flow/volume relations.

J. Henriques et al. (Eds.): MEDICON 2019, IFMBE Proceedings 76, pp. 541–550, 2020.
https://doi.org/10.1007/978-3-030-31635-8_65

Computers have proven to be a valuable resource in the study and generation of knowledge in multiple fields of medicine through statistical studies, computational modeling, image analysis, training of medical staff [4] and medicine learning [5], among others. Given the complexity of the respiratory system, the development of easy-to-use computational tools is justified to promote a better understanding of the associated diseases. Although there is a wide bibliography about clinical physiopathology [6], much of the available data are focused on Caucasians, and sometimes it is difficult to analyze, in a simple and visual way, how anthropometric factors influence lung function under different types of pulmonary diseases.

Since the middle of the last century, researchers and clinicians have used computers to generate knowledge, both physiological and clinical, in the context of the respiratory system [7]. From a computational point of view, one of the most studied processes in respiratory physiology is ventilation. In this sense, some models [8, 9] have tried to analyze the dynamic behaviors of pulmonary volumes and flows under different physiological or pathological conditions. Most of these models are based on variables measured in different clinical trials, which help specialists to identify the type of lung disorder [10].

This work presents a first prototype of a model that contributes to a better understanding of the alterations that occur in respiratory variables from respiratory diseases in a simple, fast and low-cost way. The model considers anthropometric factors and the pulmonary disease to be analyzed, allowing to visualize the alterations of variables such as volumes, flows and flow/volume ratios, of great clinical importance. The results obtained have been validated with specialized publications and the clinical experience of the pulmonologists participating in the work.

2 Materials and Methods

The evaluation of lung function in the clinical practice is a fundamental tool for the study of pathologies associated with the respiratory system [11]. Tests such as spirometry, pulmonary diffusion or body plethysmography allow to obtain values associated with physiological variables of a specific patient, which are important for diagnosis and treatment.

Lung function measures have great variability due to the fact that pulmonary volumes are closely linked to personal factors. In this sense, a recent study has analysed the classical factors considered in clinical practice and their relationship with anthropometry [12]. Since most of the existing literature focuses on Caucasians, Table 1 shows the normal values for different lung volumes and capacities in healthy Caucasian men with average age and normal body mass index (BMI). Figure 1 shows the relationship between different lung capacities and volumes.

Other important parameters are the forced expiratory volume in the first second of expiration (FEV1) and the forced vital capacity (FVC), using the FEV1/FVC ratio as a fundamental coefficient for establishing the diagnosis of COPD. Calculation of the flows and volumes has been performed based on Barnea model [13].

Table 1. Normal values of lung volumes and capacities for a healthy male.

Acronym	Description	Normal value (l)
ERV	Expiratory reserve volume	1.2
FRC	Functional residual capacity	2.3
IC	Inspiratory capacity	3.5
IRV	Inspiratory reserve volume	3
RV	Residual volume	1.1
TLC	Total lung capacity	5.8
VC	Vital capacity	4.7
VT	Tidal volume	0.5

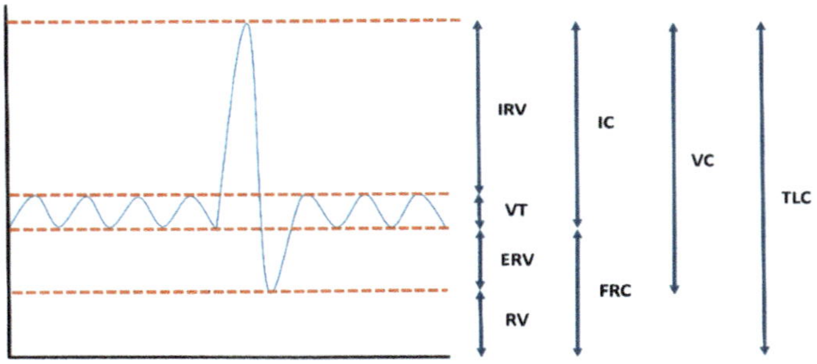

Fig. 1. Volumes and pulmonary capacities.

At the methodological level, the model has been built using functional modules in order to enable the integration of new knowledge in a simple way. Modularity is a desirable feature in software development, abstracting different functionalities and providing a development framework that eases data sharing between models. The model was implemented using the MATLAB tool for system modeling and simulation.

3 Results

3.1 Model Design

The model design is described in the flow diagram presented in Fig. 2.

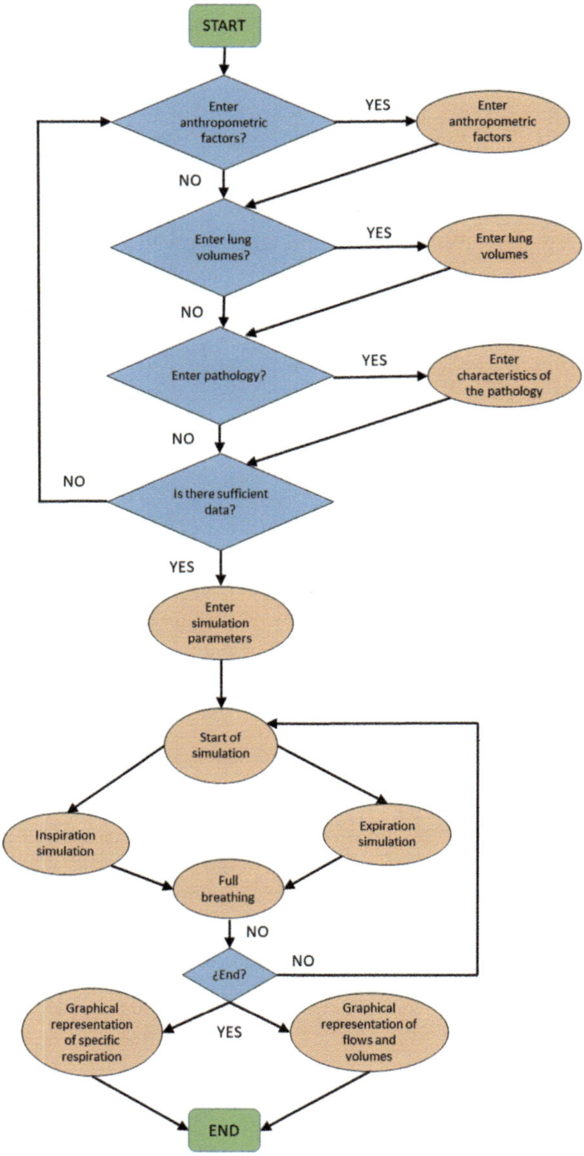

Fig. 2. Flowchart of the proposed model.

The model receives three possible types of input data: anthropometric factors, static lung volumes and the pathology to be reproduced. The introduction of static lung volume values, obtained by plethysmography, allows a customization of the model to a particular patient. If the values of the lung volumes are not introduced, then the anthropometric factors will be considered for the calculation of these volumes. If no pathology is established, a healthy individual will be considered, showing the difference between the graphical representations of the flow-volume loop corresponding to a healthy individual and the data measured in clinical practice. Then, the model checks that all the data entered are correct before continuing with its execution and asks for the simulation options: simulation time or number of respirations to simulate, forced respirations to perform and other options related to the graphical representation.

The variables associated with the static lung volumes are calculated from anthropometric factors, as shown in Table 2.

Table 2. List of variables obtained from anthropometric factors.

Acronym	Equation	Ref.
IBW (kg)	IBW (men) = 50 + 0.91 * (height (cm) − 152) IBW (women) = 45.5 + 0.91 * (height (cm) − 152)	[14]
VT (l)	VT = 6 * IBW (kg)	[15]
RV (l)	RV (men) = 3.380 * height (m) + 0.020 * age (years) − 0.014 * weight (kg) − 3.927 RV (women) = 2.548 * height (m) + 0.017 * age (years) − 3.387	[16]
FRC (l)	FRC (men) = 7.502 * height (m) + 0.009 * age (years) − 0.033 * weight (kg) − 7.608 FRC (women) = 5.867 * height (m) + 0.009 * age (years) − 0.022 * weight (kg) − 5.972	[17]
TLC (l)	TLC (men) = 7.956 * height (m) − 6.948 TLC (women) = 7.107 * height (m) − 6.435	[16]

For the rest of the lung volumes that are not directly calculated (ERV, IRV, IC, VC) the algebraic relationships are established in the Eqs. (1), (2), (3) and (4).

$$\mathrm{ERV} = \mathrm{FRC} - \mathrm{RV} \tag{1}$$

$$\mathrm{IRV} = \mathrm{TLC} - \mathrm{FRC} - \mathrm{VT} \tag{2}$$

$$\mathrm{IC} = \mathrm{TLC} - \mathrm{FRC} \tag{3}$$

$$\mathrm{VC} = \mathrm{TLC} - \mathrm{RV} \tag{4}$$

Due to the importance of the inspiration and expiration phases in respiratory diseases, these two phases have been taken separately in the quantification of the model, considering inspiratory and expiratory flows and volumes (calculated in parallel). These variables are presented in Table 3, together with their acronyms and units.

Table 3. Variables considered in the model associated with inspiratory and expiratory flows and volumes.

Acronym	Description	Units
Q_i	Inspiratory flow	l/s
Q_e	Expiratory flow	l/s
V_i	Inspiratory volume	l
V_e	Expiratory volume	l

The equations for the calculation of the inspiratory flow ($Q_i(t)$) and inspiratory volume ($V_i(t)$) are presented in (5) and (6)

$$Q_i(t) = P(t) \cdot TLC \cdot \left(1 - e^{-\frac{t}{RC}}\right)(FVC - V_i(t));\ 0 \leq t \leq t_i \tag{5}$$

$$\mathrm{V}_i(t) = \int_0^{t_i} Q_i(\xi)d\xi \tag{6}$$

Thus, t_i is the time of inspiration, R is the resistance to the airways, TLC is the total lung capacity and P is the pressure developed in the lungs and calculated from an activation function ($E(t)$) according to (7).

$$P(t) = \mathrm{E(t)(FVC - V_i(t))} \tag{7}$$

The expression proposed by Barnea [13] was used for the calculation of $E(t)$.

For the calculation of expiratory flow ($Q_e(t)$) and expiratory volume $V_e(t)$ Eqs. (8) and (9) have been used

$$Q_e(t) = \frac{E_{max}}{R(V_e)}\left(1 - e^{-\frac{t}{RC}}\right)(FVC - V_e(t));\ 0 \leq t \leq t_e \tag{8}$$

$$\mathrm{V}_e(t) = \int_0^{t_e} Q_e(\xi)d\xi \tag{9}$$

Again, the expression proposed by Barnea [13] was used for the calculation of the $\mathrm{R}(V_e)$ function.

The following results obtained by simulation can be represented graphically: on the one hand, the results of the volumes and flows breath by breath; on the other hand, a specific respiration (specified in the options established before starting the simulation) with its calculated FEV1/FVC ratio.

3.2 Model Validation

In order to verify the validity of the model developed, different types of simulations have been carried out comparing the results obtained with works by other authors [18], under normal or pathological conditions. Figure 3 shows the evolution of total lung volume (V) considering a healthy man, Caucasian, of average weight and height, in resting conditions. The relationships between flows and volumes are presented under normal conditions and under forced inspirations and expirations to cover the whole range of possible physiological values. Although the first respiration is forced, the inspiration phase begins at a resting value for the lung volume, which is the sum of RV and ERV. In contrast, in the fourth respiration the lung volume at the onset of inspiration is equal to RV, as well as at the end of expiration. Therefore, all normal physiological values of the lung volume are traversed in this respiration, from RV to the sum of RV, VT and IRV. The sum of these last three volumes is the TLC. These results are coincident with books on respiratory physiology.

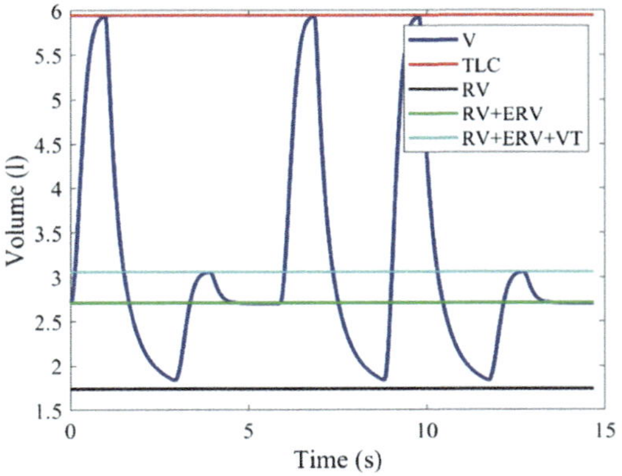

Fig. 3. Simulation of the total lung volume (V) for two normal respirations and three forced respirations considering a healthy man at rest.

3.3 Model Application

Once the model has been validated, several experiments have been carried out to demonstrate its usability and applicability for training medical students, as well as its possible clinical interest. This work presents the results obtained by simulation of the most significant patterns of lung abnormality.

Firstly, the alteration of the FEV1/FVC ratio is presented for obstructive diseases such as COPD or bronchial asthma. FEV1 is reduced much more than FVC, giving a low FEV/FVC. Figure 4 compares the flow/volume loop for a healthy person (blue) and for the same person with obstructive pulmonary disease (red). At the level of lung function, FVC and FEV1 are reduced, although the decrease is less significant for FVC. In particular, Fig. 4 shows a FVC of 3.1 L and a FEV1 of approximately 2.1 L, giving a FEV1/FVC ratio of 68%, lower than in the case of a healthy individual (80%). This value of FEV1/FVC corresponds to an obstructive pulmonary disorder of moderate severity [19].

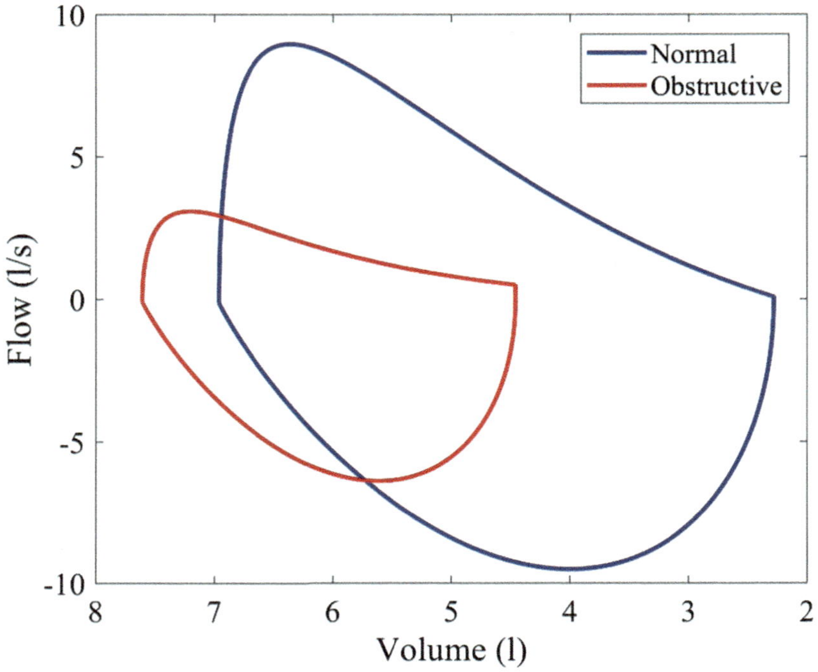

Fig. 4. Simulation of the volume/flow loop considering an individual in healthy conditions (blue) and in the case of an obstructive pulmonary disease (red).

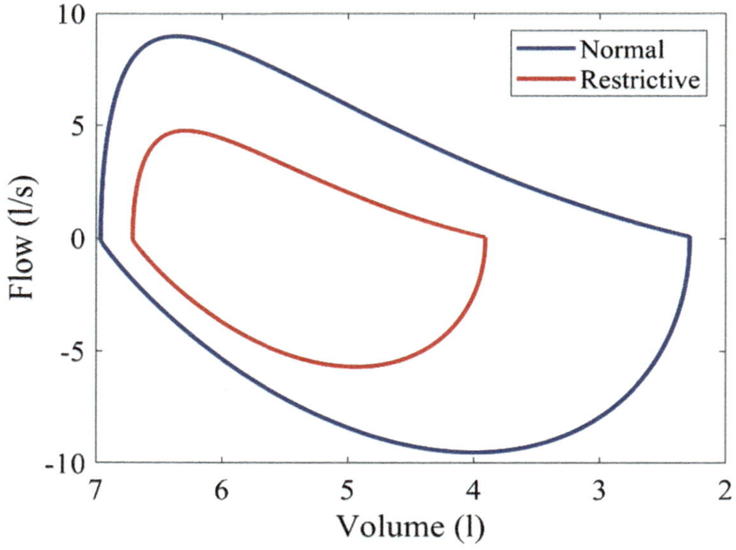

Fig. 5. Simulation of the volume/flow loop considering an individual in healthy conditions (blue) and in the case of a restrictive pulmonary disease (red).

Secondly, the results of simulating a restrictive pattern of lung disease are presented, which are characterized by a decrease in lung volume due to alterations in the thoracic cage or respiratory musculature. Pulmonary fibrosis and sarcoidosis are examples of such diseases. In restrictive pattern, both FEV1 and FVC are reduced, but in a similar proportion, with respect to obstructive pulmonary diseases. Therefore, the FEV1/FVC ratio remains unchanged with respect to the normal case, and can even be increased. Figure 5 compares the flow/volume loop for a healthy person (blue) and for the same person with restrictive lung disease (red). In this case, a FVC of 2.75 L and a FEV1 of 2.36 are shown, which provides an FEV1/FVC ratio of 85%, corresponding to a restrictive pulmonary disease.

4 Conclusions

A computational model has been developed for the training of medical students in respiratory diseases. The applicability and usability of the model have shown its capacity to simulate obstructive and restrictive pulmonary diseases and its possible application to the clinical practice, allowing the calculation of flows, volumes and flow/volume loops in a graphical representation.

A graphical user interface is currently being developed, both for data entry and for graphic representation, following the standards of usability and accessibility required for this type of human-machine interface.

Acknowledgments. This work was supported in part by the "Fondo de Investigaciones Sanitarias" (Instituto de Salud Carlos III, Spain) under Grants PI15/00306 and DTS15/00195, in part by the "Fundación Progreso y Salud" (Government of Andalucía, Spain) under Grant PI-0010-2013, PI-0041-2014 and PIN-0394-2017, in part by "Fundación Mutua Madrileña" under grant VÍA-RENAL.

References

1. Vaughan, A., Stevanovic, S., Jafari, M., Rahman, M., Bowman, R.V., Fong, K.M., Ristovski, Z., Yang, I.A.: The effect of diesel emission exposure on primary human bronchial epithelial cells from a COPD cohort: N-acetylcysteine as a potential protective intervention. Environ. Res. **170**, 194–202 (2019)
2. Burrowes, K.S., Doel, T., Brightling, C.: Computational modeling of the obstructive lung diseases asthma and COPD. J. Transl. Med. **12**, S5 (2014)
3. Burney, P.G.J., Patel, J., Newson, R., Minelli, C., Naghavi, M.: Global and regional trends in COPD mortality, 1990–2010. Eur. Respir. J. **45**, 1239–1247 (2015)
4. Zieliński, K., Stecka, A., Gólczewski, T.: VirRespir—an application for virtual pneumonological experimentation and clinical training. In: Lhotska, L., Sukupova, L., Lacković, I., Ibbott, G.S. (eds.) World Congress on Medical Physics and Biomedical Engineering, pp. 697–701. Springer, Singapore (2019)

5. Naranjo, C.A., Velasco, P.M., Quevedo, W.X., Naranjo, G.R., Rivas-Lalaleo, D., Silva, F.M., Andaluz, V.H.: Virtual reality system for assistance in treating respiratory disorders. In: De Paolis, L.T., Bourdot, P. (eds.) Augmented Reality, Virtual Reality, and Computer Graphics, pp. 118–135. Springer, Heidelberg (2018)
6. Grippi, M.A., Elias, J.A., Fishman, J.A., Pack, A.I., Senior, R.M., Kotloff, R.: Fishman's Pulmonary Diseases and Disorders, 2-Volume Set, 5th edn. McGraw-Hill Education/ Medical, New York (2015)
7. Richesson, R.L., Andrews, J.E., Hollis, K.F.: Introduction to clinical research informatics. In: Richesson, R.L., Andrews, J.E. (eds.) Clinical Research Informatics, pp. 3–15. Springer, Cham (2019)
8. Guler, H., Kilic, U.: The development of a novel knowledge-based weaning algorithm using pulmonary parameters: a simulation study. Med. Biol. Eng. Comput. **56**, 373–384 (2018)
9. Ashworth, L., Norisue, Y., Koster, M., Anderson, J., Takada, J., Ebisu, H.: Clinical management of pressure control ventilation: an algorithmic method of patient ventilatory management to address "forgotten but important variables". J. Crit. Care **43**, 169–182 (2018)
10. Windisch, W., Criée, C.P.: COPD - importance of lung function testing for diagnosis and treatment. Dtsch. Med. Wochenschr. **143**, 593–596 (2018)
11. Huprikar, N.A., Holley, A.B., Skabelund, A.J., Hayes, J.A., Hiles, P.D., Aden, J.K., Morris, M.J., Hersh, A.M.: A comparison of global lung initiative 2012 with third national health and nutrition examination survey spirometry reference values: implications in defining obstruction. Ann. Am. Thorac. Soc. **16**, 225–230 (2019)
12. Talaminos Barroso, A., Márquez Martín, E., Roa Romero, L.M., Ortega Ruiz, F.: Factores que afectan a la función pulmonar: una revisión bibliográfica. Arch. Bronconeumol. **54**, 327–332 (2018)
13. Barnea, O., Abboud, S., Guber, A., Bruderman, I.: New model-based indices for maximum expiratory flow-volume curve in patients with chronic obstructive pulmonary disease. Comput. Biol. Med. **26**, 123–131 (1996)
14. Bukkems, L.H., Roger, C., Hodiamont, C.J., Lefrant, J.-Y., Juffermans, N.P., Roberts, J.A., van Hest, R.M.: Predictive performance of a gentamicin population pharmacokinetic model in two western populations of critically ill patients. Int. J. Antimicrob. Agents **52**, 218–225 (2018)
15. O'Brien, D.E., Kam, J.K., Slater, R.J., Tobin, A.E.: A novel biometric approach to estimating tidal volume. Crit. Care Resusc. **21**, 25–31 (2019)
16. Botsis, T., Mantas, J., Halkiotis, S.: Mathematical modelling and curve fitting for the study of respiratory system parameters. In: Proceedings of the 5th European Systems Science Congress. Crete (2002)
17. Warliah, L., Rohman, A.S., Rusmin, P.H.: Model development of air volume and breathing frequency in human respiratory system simulation. Proc. Soc. Behav. Sci. **67**, 260–268 (2012)
18. West, J.B.: Respiratory Physiology: The Essentials. Lippincott Williams & Wilkins, Philadelphia (2012)
19. Gottlieb, V., Lyngsø, A.M., Nybo, B., Frølich, A., Backer, V.: Pulmonary rehabilitation for moderate COPD (GOLD 2) – does it have an effect? COPD: J. Chronic Obstruct. Pulm. Dis. **8**, 380–386 (2011)

Effects of Arterial and Tracheal Pressures During a Respiratory Mechanics Protocol in Spontaneously Hypertensive Rats

Amanda N. Barros[1(✉)], Vitor A. Takeuchi[1], Felipe Fava de Lima[1], Raissa R. S. Amorim[1], Otavio Henrique F. Ledesma[1], Maria Aparecida de Oliveira[2], Henrique T. Moriya[1], and Renato Vitorasso[1]

[1] Escola Politecnica, University of São Paulo, São Paulo, SP 05508-010, Brazil
manda.n.barros@gmail.com
[2] Biomedical Sciences Institute, University of São Paulo, São Paulo, SP 05508-900, Brazil

Abstract. This work aims to evaluate the response of tracheal and arterial pressures during a Methacholine (MCh) dose-response curve. This drug is a bronchoconstrictor and, additionally, causes vasodilatation in cardiovascular system leading to reduction of arterial pressure. Spontaneously hypertensive rats (SHR) were used and each animal was anesthetized, tracheostomized and connected to a small animal ventilator (flexiVent version 5.2, SCIREQ, Canada). Whereas the animal was ventilated, a device developed by us was connected to the right carotid in order to assess blood pressure. This device is composed of a pressure sensor, a low-pass filter, an amplifier and an analog-to-digital converter. The two main variables studied in this work were the arterial and tracheal pressure, the latter assessed by the ventilator. Firstly, we depicted the temporal dynamics of both pressures. The blood pressure decreased while the tracheal pressure increased with the increment of the MCh doses. In a second analysis, we compared the pressure during (intra) and after (inter) perturbations used to mathematical modeling. There were no differences between intra vs. inter, only among doses for the first three measures of each dose ($p < 0.0001$). Thus, despite the shape of the arterial pressure being different between intra *vs* inter perturbation, the mean value did not alter. Finally, we indicate that the arterial pressure returns to basal after 2 and 4 min after MCh injection, since that there was no statistical difference among doses or in 2 *vs* 4 min comparison ($p > 0.05$), which may guide the recovery time given between doses in a dose-response curve.

Keywords: Methacholine · Blood pressure · Tracheal pressure

J. Henriques et al. (Eds.): MEDICON 2019, IFMBE Proceedings 76, pp. 551–558, 2020.
https://doi.org/10.1007/978-3-030-31635-8_66

1 Introduction

The dose-response curve is frequently used in studies about biomedical engineering [1,2], especially those focusing on mechanics. This curve is used frequently because it shows the response of the studied system to different concentrations of the drug used in the work [3].

One of the systems that can have its mechanics studied is the respiratory system. This system can be evaluated with a mechanical ventilator, an instrument that besides controlling the ventilation, allows system quantities measurements, such as pressure, flow and airway resistance [4].

One prevalent drug used in dose-response curves during respiratory mechanics assessment is the Methacholine (MCh). The MCh has a bronchoconstrictive effect on the respiratory system, *i.e.* constructions of airway smooth muscle. Hence, the MCh is used to stimulate bronchial reactivity causing an increase in pressure and in the airway resistance of the respiratory system [5].

In the cardiovascular system, the administration of MCh leads to vasodilation since the MCh is a parasympathomimetic drug [3]. The resulting vasodilatation is responsible for a temporary reduction of the arterial pressure.

The arterial pressure is noteworthy once it consists in a vital signal. Just as the MCh affects the cardiovascular system, the arterial pressure is also affected, having lower values. In most cases, after the drug application.

Studies of respiratory and cardiovascular systems can be done through animal experimentation [6]. To perform the study in this article, rats of the lineage *Spontaneously Hypertensive Rat* (SHR) were used. Besides that, to measure the values of blood pressure in this study we use a carotid catheter with a transducer and an amplifier.

Given that arterial and tracheal pressure are two vital signals and taking into account the importance of studying the cardiac and respiratory systems under diverse circumstances, this study has been carried out aiming to compare arterial and tracheal pressure under different doses of MCh.

2 Materials and Methods

2.1 Animal Experimentation

All experiments involving laboratory animals were evaluated and approved by the *"Ethics Committee for Animal Use"* (1936211117) from the Institute of Biomedical Sciences - University of São Paulo.

6 male SHR Rats were studied in this protocol (323.8 ± 41.1 g). The animals were anesthetized with an intravenous (*i.v.*) injection of Ketamine (110 mg/kg) and Xylazine (10 mg/kg).

The trachea was cannulated with a 2.5 cm length metallic cannula 14G (Höppner, Brazil) and the animal was connected to the ventilator (flexiVent version 5.2, SCIREQ, Canada) and ventilated with a tidal volume of 10 mL/kg, Positive End Expiratory Pressure (PEEP) of 3 cmH_2O and respiratory frequency of 120 breaths per minute. During the ventilation the jugular vein was dissected with a flexible PVC tube (Critchley Electrical Products PTY, Australia). Also right carotid was cannulated with a polyethylene catheter in order to assess the arterial pressure.

The respiratory muscles were blocked with pancuronium bromide (1 mg/kg *i.v.*). After that, two alveolar recruitment maneuvers (6 s – up to 30 cmH_2O) were performed and the mechanics protocol began.

2.2 Arterial Pressure Measurement

The device used in the experiments to measure the arterial blood pressure is composed by a data acquisition module (HX711, Avia Semiconductor, China) connected to an invasive blood pressure sensor (New NPC 100 Sensor, Amphenol, USA). The module has a gain of 128, a 24-bit resolution and a cutting frequency of 80 Hz. Due to some software processes required by the device's components, this frequency was reduced to 50 Hz.

The instrument calibration is made via software. Pressures of 0 mmHg and 200 mmHg were used as reference, which were set into the device through a mercury column.

2.3 Respiratory Mechanics Assessment

PBS (Phosphate Buffered Saline) was injected and then 15 perturbations of 3 s each were performed with 5 s interval between assessment. This perturbation, used to apply the forced oscillation technique [7], was a preexisting one in the ventilator's software and was composed of a sum of 13 sinusoids. After the measurements, the animals were ventilated for two minutes until the next injection.

The same perturbations were performed for all doses of Methacholine (MCh): 10, 30 and 100 μg/kg through the right jugular vein in bolus. Automation routines of the ventilator control program were used to perform the perturbations and the scheduled time interval.

2.4 Statistical Analysis

The data in the graph were expressed in mean and standard error. In order to compare intra *vs* inter perturbation blood pressures among doses we did a two-way analysis of variance (ANOVA) with Bonferroni post-hoc. The same test was performed in the comparison of arterial pressure at 2 and 4 min after the injection of MCh.

3 Results

The Fig. 1 shows the values obtained of tracheal and arterial pressures according to the phase of the experiment, that are PBS, 10 μg/kg, 30 μg/kg and 100 μg/kg.

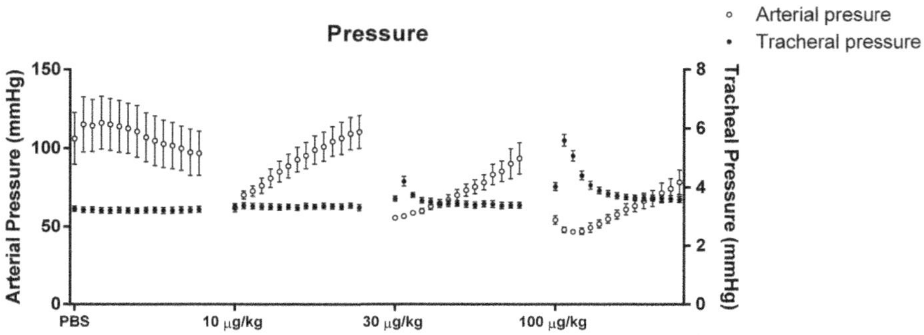

Fig. 1. Values of arterial pressure and tracheal pressures, measures in mmHg, according to the phase of the experiment.

The Fig. 2 depicts the comparison of the arterial pressure obtained at 2 and 4 min after the injection of MCh at the different phases of the experiment.

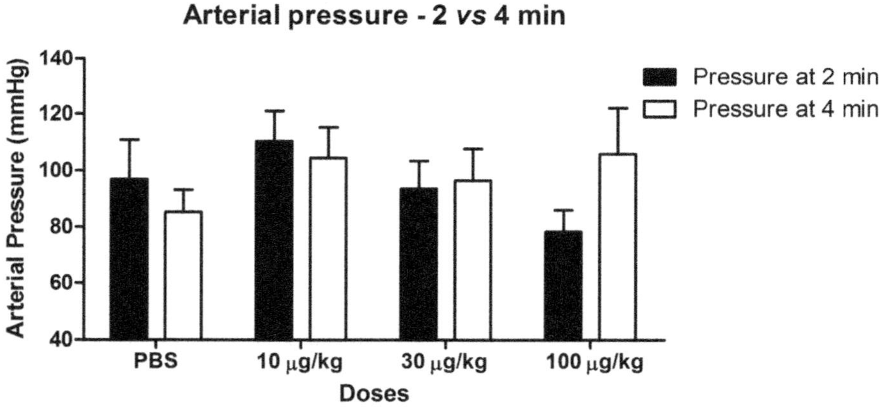

Fig. 2. Values of arterial pressure according to the phase of the experiment. There was no statistical difference ($p > 0.05$) in the arterial pressure in 2 *vs* 4 min comparison.

The Fig. 3 shows the tracheal pressure obtained intra perturbation (during forced oscillation technique) and inter perturbation (after perturbation) at different doses of the experiment.

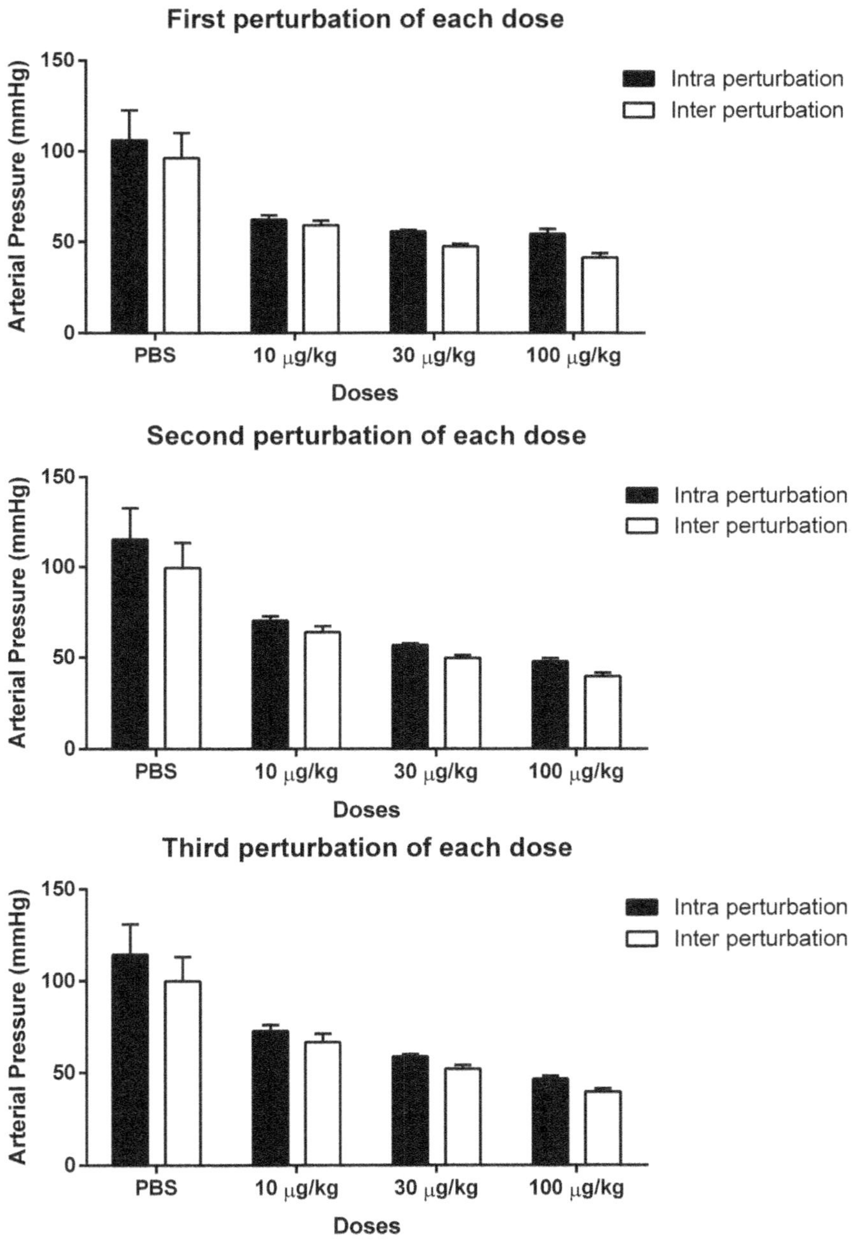

Fig. 3. Values of arterial pressure and tracheal pressure according to the phase of the experiment. There was no statistical difference in intra *vs* inter perturbation comparison, only for doses ($p < 0.0001$) for all graphs. The was difference between PBS and all doses for all graphs ($p < 0.0001$, for all). Additionally, there was difference between the dose of $10\,\mu g/kg$ and $100\,\mu g/kg$ for the last 2 graphs ($p < 0.05$, for all).

4 Discussion

The Spontaneous Hypertensive Rat (SHR) strain is widely applied in works related to arterial blood pressure screening [6]. This lineage was used in this study since the present work is part of an ongoing study related to hypertension and currently we are presenting our findings regarding arterial blood pressure during respiratory mechanics and its association with tracheal pressure.

Dose-response curves is widely used both in Pharmacology and respiratory mechanic studies [8,9]. Basically, dose-response and concentration-effect curves are the assessment of the effects of a drug through different concentrations of pharmacological agent.

One important way to challenge the respiratory system during a dose-response curve is employing Methacholine (MCh). The MCh is a very common cholinergic drug [3], which leads to contraction of the smooth muscle of respiratory system [10].

Once the MCh is a cholinergic drug, it mimics the parasympathetic system through muscarinic receptor, thus also causes vasodilatation and, as consequence, it reduces the arterial pressure. Therefore, in respiratory system the MCh causes the contractions of smooth muscle (bronchoconstriction) and in the cardiovascular system leads to relaxation of smooth muscle (vasodilatation) and diminishes the arterial pressure [3].

This study is not the first work to use or quote the tracheal and arterial pressures. For example, there is a study that uses both blood and tracheal pressures to conduct the experiment [11]. However, up to the author's knowledge this is the first work to describe the relationship between arterial and tracheal pressures during a respiratory mechanics evaluation; and to describe the temporal dynamics of arterial pressure. The latter is relevant regarding the vital signal and drug degradation during the experiment.

In Fig. 1 it is possible to observe the relationship of both pressures during the whole experiment. During the PBS, the arterial pressure presented a drop over time and this should be expected since the animals were blocked and anesthetized [11] before experiment. At the same time interval the tracheal pressure remain constant since there was no challenge to the respiratory system.

In the same figure, we presented the remaining doses of MCh in which we observe a sharp drop of blood pressure with a partial recovery. The sharp drop in blood pressure is related to the short amount of time needed to drug peak action, raging from 5 to 20 s [3,12,13]. The partial recovery can be explained by the fact that the MCh is degraded even if it is less susceptible to Cholinesterase [3] than acetylcholine chloride.

As for the tracheal pressure, the behavior is the opposite and this was expected. MCh constricts the airways and the tracheal pressure is directly associated to the airway resistance [14]. However, it is possible to interpret that the action of MCh on the blood pressure precedes the tracheal pressure response (Fig. 1).

So far, our findings were in tone with our hypothesis. However, we assessed and used the arterial pressure during the respiratory mechanics perturbation (forced oscillation technique) and this method was chosen due the time synchronization with the mean tracheal pressure, which was also measured during perturbation. This perturbation is the sum of sinusoidal signals in which the constant phase model [4] is applied.

Nevertheless, we investigated how, or if, the blood pressure is different during (intra) and immediately after (inter) perturbations, 2 s of measurement in each case. To do so, we have compared (Fig. 3) the arterial pressure intra and the inter pressure among doses and we did this procedures for the first, second and third perturbation of each dose, which is the time interval when the peak response happens.

No statistical differences were found between intra and inter pressures, just among doses. New works should enlighten this matter, however with the data presented so far it is possible to realize that the tendency or the mean value of pressure does no change during perturbation even if the shape of the curve during perturbation does (data not shown). Hence, it does not matter if the arterial pressure is selected during or immediately after the perturbation, in both cases the blood pressure will behave at the same way during a MCh dose-response curve.

Finally, we investigated if the arterial pressure returns to basal level after 2 and 4 min after MCh injection. In Fig. 2 we presented that there was no statistical difference among doses or in 2 *vs* 4 min comparison, even if visually for 2 min there was a tendency of non-recovery. Also visually, the PBS at 4 min was lower than at 2 min, the explanation may rely on the anesthetic supplementation (30% of initial anesthetic dose) between PBS and the first dose do MCh. Hence, statistically and in the scenario presented so far, there was a recovery of arterial pressure after 2 and 4 min of MCh injection.

5 Conclusion

Therefore, we presented in this study the relationship between tracheal and arterial pressure during a respiratory mechanics protocol. Additionally, we assessed the drop in arterial pressure at the first measures of a respiratory mechanics evaluation and we indicated that there was no difference between assessing the blood pressure during perturbation or in the time interval after perturbation. Finally, we showed the recovery of arterial pressure at 2 and 4 min after the MCh injection, which may guide the time of recovery given between doses in a dose-response curve during respiratory mechanics evaluation.

References

1. Arantes-Costa, F.M., Zoriki, S., Santos, M.H.C., Kobata, C.H.P., Vieira, J.E., Martins, M.A.: Effects of ventilation, humidity and temperature on airway responsiveness to methacholine in rats. Eur. Respir. J. **19**(6), 1008–1014 (2002). http://www.ncbi.nlm.nih.gov/pubmed/12108849

2. Hulbert, W.C., McLean, T., Wiggs, B., Paré, P.D., Hogg, J.C.: Histamine dose-response curves in guinea pigs. J. Appl. Physiol. **58**(2), 625–634 (1985). http://www.ncbi.nlm.nih.gov/pubmed/3980365
3. Katzung, B.G., Trevor, A.J., Knuidering-Hal, M.: Katzung & Trevor's Baics & Clinical Pharmacology. Lange Medical Publications, New York (2018)
4. Bates, J.: Lung Mechanics: An Inverse Modeling Approach. Cambridge University Press, Cambridge (2009). https://books.google.com.br/books?id=CNgQ0B7Fo0
5. Peták, F., Kannel, W., et al.: Methacholine-induced bronchoconstriction in rats: effects of intravenous vs aerosol delivery. J. Appl. Physiol. **82**(5), 1479–1487 (1997). https://doi.org/10.1152/jappl.1997.82.5.1479
6. Doris, P.: Genetics of hypertension: an assessment of progress in the spontaneously hypertensive rat. Physiol. Genomics **49**(11), 601–617 (2017)
7. Hantos, Z., Suki, B., Daróczy, B.: Constant-phase modelling of pulmonary impedance. Bull. Eur. Physiopathol. Respir. **12**, 326 (1987)
8. Hardman, J., Limbird, L., Gilman, A.: Goodman & Gilman's The Pharmacological Basis of Therapeutics, 10th edn. McGraw-Hill, New York (2000)
9. Collins, R.A., Gualano, R.C., Zosky, G.R., Atkins, C.L., Turner, D.J., Colasurdo, G.N., et al.: Hyperresponsiveness to inhaled but not intravenous methacholine during acute respiratory syncytial virus infection in mice. Respir. Res. **6**, 142 (2005). https://www.ncbi.nlm.nih.gov/pmc/articles/PMC1325022/
10. Kenakin, T.P.: The classification of drugs and drug receptors in isolated tissues. Pharmacol. Rev. **36**(3), 165–222 (1984)
11. Suki, B., et al.: Airway and tissue constrictions are greater in closed than in open-chest conditions. Respir. Physiol. **108**(2), 129–141 (1997)
12. Bates, J.H.T., Stevenson, C.A., Aliyeva, M., Lundblad, L.K.A.: Airway responsiveness depends on the diffusion rate of methacholine across the airway wall. J. Appl. Physiol. **112**(10), 1670–1677 (2012)
13. Cojocaru, A., Irvin, C.G., Haverkamp, H.C., Bates, J.H.T.: Computational assessment of airway wall stiffness in vivo in allergically inflamed mouse models of asthma. J. Appl. Physiol. **104**(6), 1601–10 (2008)
14. Bates, J., et al.: Temporal dynamics of acute isovolume bronchoconstriction in the rat. J. Appl. Physiol. **82**, 55–62 (1997)

Modeling of Carbohydrates Oxidation Rate During Exercise in Type 1 Highly-Trained Diabetic Patients

Maria Pia Francescato[1], Miloš Ajčević[2(✉)], Alex Buoite Stella[1], and Agostino Accardo[2]

[1] Department of Medicine, University of Udine, Udine, Italy
[2] Department of Engineering and Architecture, University of Trieste, Trieste, Italy
majcevic@units.it

Abstract. Management of Type 1 Diabetes (T1D) in the context of exercise or sports competition still represents a great challenge for athletes living with this disease, due to the wide excursions in blood glucose level with increased risk of life threatening hypoglycemia. Recently, an algorithm called ECRES has been developed to estimate patient-exercise tailored carbohydrates (CHO) supplement required to maintain safe blood glucose levels during physical activity. This method estimates a CHO supplement based on the patient's habitual therapy, the specific patient's insulin sensitivity and the overall amount of CHO oxidized during the specific exercise. The last is based on the glucose pulse relation, i.e. the relation between heart rate (HR) and CHO oxidation rate, already studied in sedentary and moderately-trained subjects, but not in well-trained athletes. This study aimed to model the glucose pulse relation during exercise in type 1 highly trained diabetic patients and in healthy subjects. HR, oxygen consumption and carbon dioxide production were acquired breath-by-breath in seven T1D and seven well-matched healthy highly-trained subjects at four different exercise intensity levels, as well as at rest. Results showed a linear CHOox-HR relation (CHOox = 0.76 · %HRmax - 19.6; n = 70, R^2 = 0.78) with no significant difference between the T1D and healthy athletes (p-value = 0.11). In conclusion, results of this study can be implemented in an updated version of the ECRES algorithm allowing an easy estimate of CHO supplement also in highly trained subjects. This useful support system can enhance the self-management of glycaemia during the training sessions of athletic patients throughout mHealth technologies.

Keywords: Physiological modeling · Glucose pulse · Type 1 diabetes · Glycaemia · Exercise

1 Introduction

Type 1 diabetes (T1D) is a chronic disease characterized by the body's inability, due to autoimmune disorder, to produce enough insulin and adequately regulate blood glucose concentration. Therefore, T1D subjects require chronic injections of exogenous insulin

J. Henriques et al. (Eds.): MEDICON 2019, IFMBE Proceedings 76, pp. 559–566, 2020.
https://doi.org/10.1007/978-3-030-31635-8_67

for survival. Recent progresses in insulin therapy and advancements in blood glucose monitoring improved life expectancy [1] and participation at the highest levels of agonistic athletics [2]. Regular exercise, together with diet and insulin therapy, is an important aspect of T1D management and contributes to enhance insulin sensitivity [3], may help to achieve and maintain better metabolic control [4], as well as reduce the risk of cardiovascular disease [5] and improve psychological well-being [6]. However, lack of insulin regulation during exercise leads to frequent metabolic imbalances [7] and effective management of the T1D subject still represents a great challenge due to the wide excursions in blood glucose concentration [8, 9] with increased risk of hypoglycemia [10]. This challenge is even more enhanced in the case of athletes and individualized management strategies should be foreseen [11, 12].

To overcome this problem in T1D subjects, an algorithm called ECRES was proposed to estimate patient-exercise tailored glucose supplement required to maintain safe blood glucose levels during physical activity [13]. This method is based on the patient's habitual therapy and diet and the specific patient's insulin sensitivity is taken into account through the individual dietary carbohydrates-to-insulin ratio [11]. Actual exercise intensity and duration are used for the estimate and patient's physical fitness is considered when the overall amount of glucose oxidized is estimated. This approach also takes into account the timing of the activity relative to the last meal and insulin bolus, allowing the estimation of the carbohydrates supplement for any time of day the exercise is performed. The method was successfully tested during short [13] and prolonged exercise [14, 15], estimating minor amounts of CHO supplement with respect to a classical approach, while maintaining glycaemia within safe levels [16].

In particular, the ECRES algorithm calculates the amount of carbohydrates supplement required to avoid glycemic imbalances (reqCHO) as a fraction (Ft) of the overall amount of carbohydrates burned during the exercise (CHOox) and further corrects it by subtracting or adding the excess or lack of glucose (Gb) contained in the extracellular compartment.

$$\mathrm{reqCHO} = (\mathrm{CHOox} \cdot \mathrm{Ft}) \pm \mathrm{Gb} \tag{1}$$

The fraction estimation is based on an insulin dependent model that accounts for insulin level and sensitivity throughout the effort [11], while actual exercise duration and intensity, based on heart rate (HR), together with patient's physical fitness level, are considered to estimate the overall amount of glucose oxidized. The estimation of the latter is based on the notion that below the anaerobic threshold, both oxygen consumption and heart rate during exercise increase linearly as a function of exercise intensity [17]. These relations were studied in sedentary and moderately-trained subjects during short [18, 19] and prolonged [20] duration exercise. In well-trained athletes, who would highly benefit of individualized support systems for prevention of exercise related hypoglycemia, the glucose pulse, i.e. the relation between HR and CHO oxidation rate, has not yet been investigated.

The aim of the present study was to model the glucose pulse relation during exercise in type 1 highly trained diabetic patients and in healthy subjects.

2 Material and Methods

Seven T1D highly-trained subjects and seven well-matched healthy athletes (10 M/4F; 34.0 ± 12.1 years old; 1.8 ± 0.1 m tall; body mass of 63.6 ± 11.2 kg; BMI 20.5 ± 1.9 kg/m^2) were recruited on a voluntary basis at the Department of Medicine of the University of Udine after being informed on the nature, purpose, and possible risks of the study. Inclusion criteria for both T1D and healthy subjects was highly trained condition as assessed by estimation of $\dot{V}O_2max$ based on a physical activity questionnaire and subject's anthropometric characteristics, and evaluated according to parameters reported in literature [21]. Additional inclusion criteria for T1D subjects were no clinical evidence of chronic complications, no medication other than insulin, and self-management skills to maintain a good metabolic balance. The included subjects were asked to maintain their usual therapy in terms of diet and insulin dose, as well as to control their glycaemia levels according to standard self-management procedures. This study has been approved by the local ethics committee and has been performed in accordance with the ethical standards laid down in the 1964 Declaration of Helsinki and its later amendments. All participants released their informed written consent.

All subjects underwent the experimental protocol performing the exercise on a bicycle ergometer (Ergomed 839E; Monark, Vansbro, Sweden) 3 h after the meal. The protocol consisted of four 6-min exercise sessions with increasing intensity corresponding to 0.5, 0.8, 1.0, and 1.3 W/kg body mass and 0.6, 0.9, 1.2, and 1.5 W/kg body mass in women and men, respectively.

After the initial 6-min resting state, the subject started pedaling keeping throughout the exercise period the pedaling frequency as close as possible to 60 rpm with the aid of the inbuilt led bar.

In order to avoid the risk of hypoglycemia T1D subjects received appropriate amounts of CHO supplement before and/or during the trial. Blood glucose level was measured before, during and after exercise using a handheld glucometer (Accu-Chek Aviva, Roche Diagnostics). No carbohydrates supplement was given to healthy subjects.

Heart rate (HR) was measured by a commercial heart rate monitor (Polar, Kempele, Finland). Oxygen consumption ($\dot{V}O_2$) and carbon dioxide production ($\dot{V}CO_2$) were recorded breath-by-breath throughout the whole protocol using a metabolic cart (Quark B2, Cosmed, Italy). For each exercise intensity and resting state average values were calculated over the last 2.5 min.

Respiratory quotient, supposed to be identical to the respiratory exchange ratio (RER), was calculated as the ratio between $\dot{V}O_2$ and $\dot{V}CO_2$ [22]. Assuming that net protein utilization did not contribute to any significant extent to energy production during exercise [17], whole-body carbohydrate oxidation rates (CHOox) were calculated applying the following equation [23]:

$$\text{CHOox} = 4.55 \cdot \dot{V}CO_2 - 3.21 \cdot \dot{V}O_2 \quad (2)$$

where $\dot{V}O_2$ and $\dot{V}CO_2$ are expressed in L/min, and CHOox in g/min.

The relation between CHOox per unit body weight ($mg \cdot min^{-1} \cdot kg^{-1}$) and HR, expressed as percentage of maximal theoretical heart rate (%HRmax), was investigated. %HRmax was calculated by following equation:

$$\%HRmax = \frac{HR}{220 - Age} \cdot 100\% \tag{3}$$

To assess whether the CHOox-%HRmax association significantly differed between T1D patients and healthy subjects, analysis of covariance (ANCOVA) was used. A level of $P < 0.05$ was regarded as statistically significant.

3 Results

Mean (±1SD) values of HR and $\dot{V}O_2$ measured at rest and during the four intensity levels of exercise are reported in Table 1. For both variables, increasing values were detected according to exercise intensities, without any significant difference between the two groups.

Table 1. Mean (± 1SD) values of HR and $\dot{V}O_2$ measured at rest and during the four intensity levels, i.e., 0.5, 0.8, 1.0, and 1.3 W/kg body mass and 0.6, 0.9, 1.2, and 1.5 W/kg body mass in women and men, respectively.

	Healthy athletes		T1D athletes	
	HR [bpm]	$\dot{V}O_2$ [mL/min/kg]	HR [bpm]	$\dot{V}O_2$ [mL/min/kg]
Rest	66.9 ± 4.3	5.1 ± 0.6	53.9 ± 6.6	4.4 ± 0.6
Level 1	92.2 ± 14.9	17.3 ± 2.3	73.5 ± 10.2	15.5 ± 1.3
Level 2	100.5 ± 16.3	19.5 ± 2.4	81.2 ± 11.8	17.9 ± 1.0
Level 3	111.4 ± 17.9	23.7 ± 3.7	90.4 ± 13.3	21.2 ± 2.5
Level 4	123.5 ± 18.7	27.9 ± 4.2	99.0 ± 17.2	24.0 ± 1.6

The carbohydrates oxidation rate of the well-trained diabetic patients as well as those of the healthy subjects is plotted in Fig. 1 as a function of the percentage of HRmax. A linear dependency can be observed for both groups of volunteers, without clear distinction between them. Indeed, ANCOVA found no significant difference in CHOox adjusted for %HRmax between the T1D patients and healthy subjects (p-value = 0.11). The overall glucose pulse in highly-trained subjects was modeled by linear regression and the following equation was obtained ($n = 70$, $R^2 = 0.78$):

$$CHOox = 0.76 \cdot \%HRmax - 19.6 \tag{4}$$

where CHOox is carbohydrates oxidation rate per kg of body mass ($mg \cdot min^{-1} \cdot kg^{-1}$).

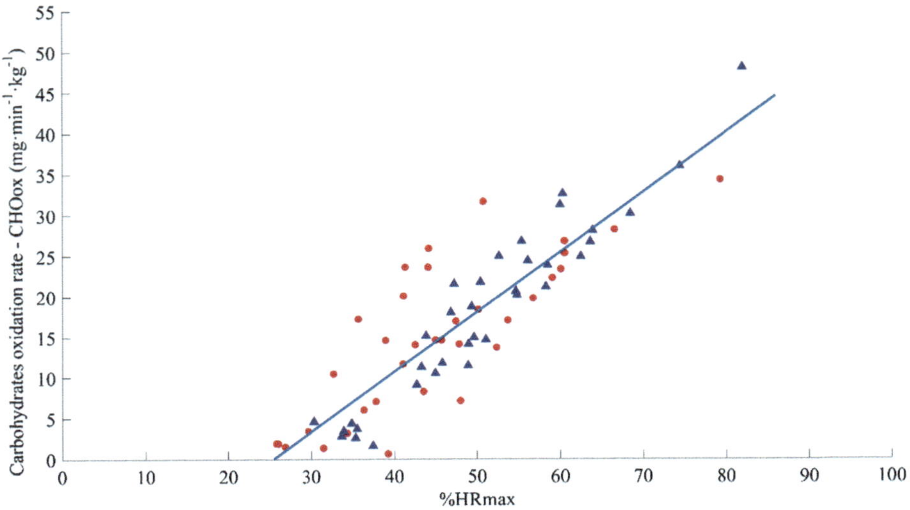

Fig. 1. Carbohydrates oxidation rates per kg of body mass ($mg \cdot min^{-1} \cdot kg^{-1}$) are plotted as a function of the corresponding percentage of maximal heart rate (%HRmax) for T1D athletes (red circles) and healthy athletes (blue triangles).

4 Discussion

There is an increasing need for technology solutions to increase self-empowerment and self-management of chronic patients [24, 25]. In particular, in insulin-dependent patients, personalized strategies should be developed, and implemented using mHealth solutions, to minimize the risk of exercise-induced hypoglycemia [26, 27]. This goal has already been achieved for sedentary and active patients, throughout an algorithm that estimates the amount of required carbohydrates on a patient and situation specific basis [11, 13–16]. Notably, besides the life threatening risk of hypoglycemia, also an excessive amount of administered carbohydrates, might be potentially dangerous leading to unwanted excessively high glycaemia. Interestingly, the majority of T1D athletes still use a trial and error approach to manage their blood glucose around exercise, relying mainly on themselves and endocrinologist for support, while they would prefer to rely more on mobile apps [12]. Nevertheless, an appropriate solution for the training management of T1D athletes is lacking [8].

The main finding of this study is a glucose pulse model based on the estimated CHOox - %HRmax relationship, specifically designed for highly trained individuals, as competitive athletes are, which can be easily introduced in the aforementioned algorithm. Results confirm that the overall physiological response to the investigated exercise intensities was similar between T1D patients and healthy controls, and that heart rate can be a useful physiological parameter to be used to estimate the amount of glucose oxidized during exercise also in highly-trained T1D subjects.

Endurance training has been demonstrated to result in a decreased reliance on carbohydrates during exercise at the same absolute power [28]. The slope of the observed relationship is higher than those observed for sedentary and active patients [18], which might lead to the erroneous conclusion that highly trained patients oxidize greater amounts of glucose during exercise. Nevertheless, highly trained subjects are characterized by lower heart rates at rest and for the same absolute exercise intensity, which is reflected in lower percentages of maximal heart rate [29], in turn leading to lower total glucose oxidation rates.

Results of this study did not revealed any significant difference in heart rates at each exercise intensity level between healthy and T1D athletes. Similar conclusions were reached by Tagougui et al. investigating the maximal oxygen uptake in well controlled patients, i.e. those showing a Hba1c < 7% [30]. Conversely, the same authors found impaired physical fitness level in patients with uncontrolled diabetes. In addition, a recent study showed differences in maximum heart rate between healthy and Type 2 diabetes athletes [29].

In conclusion, the results of this study can be implemented in an updated version of the ECRES algorithm allowing an easy estimate of CHO supplement also in highly trained subjects. This useful support system can enhance the self-management of glycaemia during the training sessions of athletic patients throughout mHealth technologies.

Acknowledgements. The authors wish to thank patients and volunteers for collaboration. Work partially supported by Master in Clinical Engineering, University of Trieste.

Conflict of interest. The authors have no conflict of interest do declare.

References

1. Miller, R.G., Secrest, A.M., Sharma, R.K., Songer, T.J., Orchard, T.J.: Improvements in the life expectancy of type 1 diabetes: the Pittsburgh epidemiology of diabetes complications study cohort. Diabetes **61**(11), 2987–2992 (2012)
2. Neithercott, T.: Rising stars three young athletes with diabetes at the top of their sports. Diabetes Forecast **64**(7), 42–45 (2011)
3. Stallknecht, B., Larsen, J.J., Mikines, K.J., Simonsen, L., Bulow, J., Galbo, H.: Effect of training on insulin sensitivity of glucose uptake and lipolysis in human adipose tissue. Am. J. Physiol. **279**, 376–385 (2000)
4. Diabetes mellitus and exercise: American diabetes association. Diabetes Care **20**, 1908–1912 (1997)
5. Lehmann, R., Kaplan, V., Bingisser, R., Bloch, K., Spinas, G.: Impact of physical activity on cardiovascular risk factors in IDDM. Diabetes Care **20**, 1603–1611 (1997)
6. Steppel, J.H., Horton, E.S.: Exercise in the management of type 1 diabetes mellitus. Rev. Endocr. Metab. Disord. **4**, 355–360 (2003)
7. Hasan, S., Shaw, S.M., Gelling, L.H., Kerr, C.J., Meads, C.A.: Exercise modes and their association with hypoglycemia episodes in adults with type 1 diabetes mellitus: a systematic review. BMJ Open Diab. Res. Care **6**(1), e000578 (2018)
8. Horton, W.B., Subauste, J.S.: Care of the athlete with type 1 diabetes mellitus: a clinical review. Int. J. Endocrinol. Metab. **14**(2), e36091 (2016)

9. Brazeau, A.S., Rabasa-Lhoret, R., Strychar, I., Mircescu, H.: Barriers to physical activity among patients with type 1 diabetes. Diabetes Care **31**, 2108–2109 (2008)
10. Riddell, M.C., Gallen, I.W., Smart, C.E., Taplin, C.E., Adolfsson, P., Lumb, A.N., et al.: Exercise management in type 1 diabetes: a consensus statement. Lancet Diabetes Endocrinol. **5**, 377–390 (2017)
11. Francescato, M.P., Ajčević, M., Accardo, A.: Carbohydrate requirement for exercise in type 1 diabetes: effects of insulin concentration. J. Diabetes Sci. Technol. 1932296819826962 (2019). https://doi.org/10.1177/1932296819826962
12. Dizon, S., Malcolm, J., Rowan, M., Keely, E.J.: Patient perspectives on managing type 1 diabetes during high-performance exercise: what resources do they want? Diabetes Spectrum **32**(1), 36–45 (2019)
13. Francescato, M.P., Geat, M., Accardo, A., Blokar, M., Cattin, L., Noacco, C.: Exercise and glycemic imbalances: a situation-specific estimate of glucose supplement. Med. Sci. Sports Exerc. **43**(1), 2–11 (2011)
14. Francescato, M.P., Stel, G., Stenner, E., Geat, M.: Prolonged exercise in type 1 diabetes: performance of a customizable algorithm to estimate the carbohydrate supplements to minimize glycemic imbalances. PLoS One **10**(4), e0125220 (2015)
15. Buoite Stella, A., Assaloni, R., Tonutti, L., Manca, E., Tortul, C., Candido, R., Francescato, M.P.: Strategies used by patients with type 1 diabetes to avoid hypoglycaemia in a 24 × 1-h marathon. comparison with the amounts of carbohydrates estimated by a customisable algorithm. Can. J. Diab. **41**, 184–189 (2016)
16. Ajčević, M., Francescato, M.P., Geat, M., Accardo, A.: Comparison of ECRES algorithm with classical method in management of diabetes type 1 exercise-related imbalances. In: World Congress on Medical Physics and Biomedical Engineering 2018, pp. 803–806. Springer, Singapore (2019)
17. Astrand, P.O., Rodahl, K., Dahl, H.A., Stromme, S.B.: Textbook of Work Physiology: Physiological Bases of Exercise. McGraw Hill Series in Health Education, Physical Education and Recreation. 4th edn. Human Kinetics, Windsor (2003)
18. Francescato, M.P., Cattin, L., Geat, M., Tosoratti, E., Lazzer, S., Noacco, C., di Prampero, P. E.: Glucose pulse: a simple method to estimate the amount of glucose oxidized during exercise in type 1 diabetic patients. Diabetes Care **28**(8), 2028–2030 (2005)
19. Francescato, M.P., Zanier, M., Gaggioli, F.: Prediction of glucose oxidation rate during exercise. Int. J. Sports Med. **29**(09), 706–712 (2008)
20. Geat, M., Stel, G., Poser, S., Driussi, C., Stenner, E., Francescato, M.P.: Whole-body glucose oxidation rate during prolonged exercise in type 1 diabetic patients under usual life conditions. Metabolism **62**(6), 836–844 (2013)
21. Gibson, A.L., Dale, W., Vivian, H.: Advanced Fitness Assessment and Exercise Prescription, 8th edn. Human Kinetics, Windsor (2018)
22. McArdle, W.D., Katch, F.I., Katch, V.L.: Essentials of Exercise Physiology, 2nd edn. Lippincott Williams & Wilkins, Philadelphia (2000)
23. Frayn, K.N.: Calculation of substrate oxidation rates in vivo from gaseous exchange. J. Appl. Physiol. **55**, 628–634 (1983)
24. Lasorsa, I., D'Antrassi, P., Ajčević, M., Stellato, K., Di Lenarda, A., Marceglia, S., Accardo, A.: Personalized support for chronic conditions. Appl. Clin. Inform. **7**(03), 633–645 (2016)
25. Lasorsa, I., Ajčević, M., D'Antrassi, P., Carlini, G., Accardo, A., Marceglia, S.: Promoting health for chronic conditions: a novel approach that integrates clinical and personal decision support. In: 2016 39th International Convention on Information and Communication Technology, Electronics and Microelectronics (MIPRO) (2016), Opatija, pp. 644–648. IEEE (2016). https://doi.org/10.1109/mipro.2016.7522220

26. Vuattolo, O., Francescato, M.P., Della, V.M., Accardo, A.: A smartphone application for preventing exercise-induced glycemic imbalances in type 1 diabetic patients. In: Studies in Health Technology and Informatics, vol. 180, pp. 1035–1039 (2012)
27. Ajčević, M., De Dea, F., Barbieri, G., Accardo, A.: A mobile app for the self-management of type 1 diabetes as tool for preventing of exercise-associated glycemic imbalances. In: World Congress on Medical Physics and Biomedical Engineering 2018, pp. 475–479. Springer, Singapore (2019)
28. Coggan, A.R., Kohrt, W.M., Spina, R.J., Bier, D.M., Holloszy, J.O.: Endurance training decreases plasma glucose turnover and oxidation during moderate-intensity exercise in men. J. Appl. Physiol. **68**, 990–996 (1990)
29. Ebrahim, N., Özbay, S., Ghazaei, Y.: Evaluating the relation between maximum heart rate and resting heart rate in three groups of people (non-athletes, athletes, and people with type ii diabetes). Int. J. Med. Res. Health Sci. **5**(11), 79–82 (2016)
30. Tagougui, S., Leclair, E., Fontaine, P., Matran, R., Marais, G., Aucouturier, J., et al.: Muscle oxygen supply impairment during exercise in poorly controlled type 1 diabetes. Med. Sci. Sports Exerc. **47**(2), 231 (2015)

Regular Sessions: Biomechanics, Robotics and Rehabilitation

Pressurization of Axially Prestretched Tube: Consequences for Arterial Mechanics

Zdeněk Petřivý(✉) and Lukáš Horný

Faculty of Mechanical Engineering, Czech Technical University in Prague, Technická 4, 160 00 Prague, Czech Republic
zdenekpetrivy@gmail.com, lukas.horny@fs.cvut.cz

Abstract. It is well known that in their in situ position, arteries are axially prestretched. They retract upon excision as a consequence. Studies of the prestretch have shown that it depends on location in arterial tree. Axial prestretch increases with increasing distance from the heart. It has also been shown that the prestretch changes with age. Aging damages internal architecture of arteries and consequently longitudinal pretension is gradually loosen. Results obtained in laboratory experiments document that there is specific value of the prestretch under which axial deformation of the pressurized artery is negligible. Under such conditions, mechanical work performed by the pressure on axial displacements is negligible which is advantageous from biomechanical point of view. In the present study, by adopting the Gent model of the strain energy density function to characterize mechanical properties of the artery and assuming that this artery satisfies assumptions of the membrane theory, it will be shown that the value of prestretch, which minimizes axial displacement, also maximizes internal volume attained in the pressurization of the tube.

Keywords: Artery · Hyperelasticity · Inversion point · Prestretch · Pressure

1 Introduction

Arteries in situ are significantly prestretched in an axial direction [1–4]. It is manifested as a retraction that is observed in autopsy or surgery when arteries are excised. The axial prestretch is expressed as the ratio of *in situ* to *ex situ* length of a cylindrical segment of an artery. Axial prestretch in arteries strongly depends on age [3–5]. It gradually decreases until totally relaxed state and even pre-compressed arteries can be observed [4–6]. The prestretch also depends on anatomical location [1, 2, 4–6]. It increases with increasing distance from the heart and is well correlated with the cross-sectional area of the vascular wall [2].

Ex vivo mechanical tests with enzymatically treated arteries [7, 8] and genetically modified animal models [9, 10] suggest crucial role of elastin for bearing the prestretch. Fundamental role of elastic membranes in the axial prestretching is in accordance with distally increasing magnitude of the prestretch [2]. Caudally decreasing number of elastic lamellae, which sustain approximately constant axial force in the whole aorta [2], lead to increased force per lamella which results in increasing prestretch.

J. Henriques et al. (Eds.): MEDICON 2019, IFMBE Proceedings 76, pp. 569–576, 2020.
https://doi.org/10.1007/978-3-030-31635-8_68

The axial prestretch of arteries has important physiological function. It was found that *in vivo* value of the prestretch enables arteries to sustain minimal changes in axial deformation and axial force during pressure cycle [11–13]. Originally, it was assumed that nearly zero axial deformation during pressure pulse transition is a consequence of significant tethering (attachment to surrounding tissue) [11]. Experimental studies, however, showed that longitudinal immobility of arteries is related to a biological tuning that couples constitutive properties, internal structure and physiological range of arterial loading.

To be more specific, it was found that a typical in vitro inflation behavior of an artery held at constant length is such that there is a value of the axial prestretch above which force–pressure relationship creates increasing curve and under prestretch smaller than this value the force–pressure relationship is decreasing [12–14]. Contrary, pressurization experiments with constant axial load have shown that for small values of the load the arteries elongate during pressurization, whereas with large values of the axial load they shorten [6, 15, 16]. It is generally accepted that in vivo value of the prestretch (and axial force) is exactly the value under which an artery neither shortens nor elongates in pressure cycle. By some authors, this value is referred to as the inversion stretch or the stretch at inversion point [6, 15, 16].

Previous studies of the axial prestretch have revealed that it is mechanically favorable for human arteries to operate in the prestretched state [6, 11–16]. It minimizes mechanical work necessary to transmit pressure pulse because no axial displacements of the wall during pressure cycle means no work spent on these displacements. Since real arteries are rather viscoelastic than elastic, it also means, that there will be no energy dissipated in axial movement of the artery wall. In the present paper we would like to show another interesting property which can be observed when nonlinear elastic tube is inflated at constant axial load. It will be shown that at axial prestretching corresponding to the inversion point, when no axial movement is exhibited during inflation, also maximum internal volume of the pressurized tube is attained. Since large arteries serve as a conduit in blood transport, it suggests that they operate in mechanically optimal regime which allows maximum of the blood to be transferred in one pressure cycle.

2 Methods

Our study will not deal with any specific artery, instead we will focus our attention only to general properties known from arterial elasticity. By means of analytical model, it will be shown that maximal internal volume of the pressurized tube corresponds to loading at inversion axial stretch. To this end, the artery considered in our model is assumed to be made from Gent hyperelastic material [17, 18]. The Gent hyperelastic model is well-known in elastomer mechanics and soft tissue elasticity and is frequently used in biomechanical studies describing physiological principles in arterial mechanics. It was used for example by Horný et al. [19] in their study of aging impact on the pretension sustained by the human abdominal aorta, by Horgan and Saccomandi [20] and Sang et al. [21] to study the inflation-extension behavior of an artery and in [20] to study longitudinal oscillations of arterial strip. It is isotropic nonlinear material model exhibiting large strain stiffening which is a property common to all soft tissues.

Constitutive Model. The specific form of the strain energy density function W corresponding to the Gent model is expressed in (1).

$$W = -\frac{\mu J_m}{2} ln\left(1 - \frac{I_1 - 3}{J_m}\right) \tag{1}$$

Here μ is stress-like material parameter which at infinitesimal strains corresponds to the shear modulus and J_m is dimensionless parameter modulating nonlinear behavior of the material. I_1 is the first principal invariant of the right Cauchy-Green strain tensor $\mathbf{C}$. $\mathbf{C} = \mathbf{F}^{\mathrm{T}}\mathbf{F}$ where $\mathbf{F}$ denotes tensor of the deformation gradient. The material is assumed to be incompressible and its constitutive equation is written in the form of (2).

$$\boldsymbol{\sigma} = \frac{\partial W}{\partial \mathbf{F}}\mathbf{F}^T - p\mathbf{I} \tag{2}$$

In (2), $\boldsymbol{\sigma}$ denotes Cauchy stress tensor, $\mathbf{I}$ denotes second order unit tensor and p is Lagrangean multiplier enforced by incompressibility constraint which has to be determined from force boundary condition.

Kinematics of Inflation–Extension Response. The studied artery is considered to be long thin-walled tube with closed ends which, in the reference configuration, has a middle radius R, thickness H, and length L. Assume that during pressurization, the motion of a material particle located originally at (R, Θ, Z), which is sufficiently distant from the ends, is described by the equations summarized in (3).

$$r = \lambda_{\theta\Theta}R,\ h = \lambda_{rR}H,\ z = \lambda_{zZ}Z,\ \theta = \Theta \tag{3}$$

Here r and h respectively denote deformed middle radius and thickness. The Eqs. (3) express the fact that the tube uniformly inflates and extends and that it does not twist. Stretches λ_{kK} ($k = r, \theta, z$; $K = R, \Theta, Z$) are the components of the deformation gradient $\mathbf{F}$. Consequently, $\mathbf{F} = diag[\lambda_{rR}, \lambda_{\theta\Theta}, \lambda_{zZ}]$. Incompressibility constraint then reads $det(\mathbf{F}) = \lambda_{rR}\lambda_{\theta\Theta}\lambda_{zZ} = 1$.

Equilibrium of Thin-Walled Tube. The equilibrium equations of a thin-walled closed tube loaded by an internal pressure P can be written in the form (4). Here σ_{rr}, $\sigma_{\theta\theta}$, and σ_{zz} respectively, denote the radial, circumferential, and axial component of the Cauchy stress tensor. The radial component was considered to be zero due to the thin-wall assumption.

$$\sigma_{rr} = 0 \qquad \sigma_{\theta\theta} = \frac{rP}{h} \qquad \sigma_{zz} = \frac{rP}{2h} + \frac{F_{red}}{2\pi rh} \tag{4}$$

F_{red} in (4c) denotes additional axial force which ensures longitudinal prestretch of the tube. This loading remains constant during the pressurization of the tube. However, the axial deformation may change according to the mechanical state which results from a combination of equilibrium and constitutive equations and loading by internal pressure.

Geometry and Material Parameters. Constitutive parameters characterizing artery material were adopted from literature. In [19] Horný et al. conducted uniaxial tensile tests with cylindrical segments of the human abdominal aorta and approximated mechanical response by the model (1). For the purposes of our calculations, parameters corresponding to 48 years old female individual were chosen. Specific values used in the study are in Table 1.

Table 1. Material parameters and dimensions.

Age	Gender	J_m	μ	R	H
48	Female	0.22	23	5.81	1.62
Years		–	kPa	mm	mm

Simulation of Inflation-Extension Behavior. Quasistatic inflation-extension behavior of the artery was simulated numerically in Maple 2015. The system of equations describing response to internal pressure and axial loading is obtained by substitution of (1) to (2) and subsequent substitution (2) into (4). Finally deformed radius and thickness in (4) are substituted from (3) and $\lambda_{rR} = (\lambda_{\theta\Theta}\lambda_{zZ})^{-1}$ is used to eliminate the radial stretch. One equation, (4a), is used to determine p coming from the incompressibility constraint. The result is the system of two nonlinear equations which contain material parameters, reference dimensions and P, F_{red}, $\lambda_{\theta\Theta}$, and λ_{zZ}. The final form of the governing equations is expressed in (5) and (6).

$$\left[\lambda_{\theta\Theta}\frac{\partial W}{\partial\lambda_{\theta\Theta}}\right]_{\lambda_{rR}=\lambda_{\theta\Theta}^{-1}\lambda_{zZ}^{-1}} - \left[\lambda_{rR}\frac{\partial W}{\partial\lambda_{rR}}\right]_{\lambda_{rR}=\lambda_{\theta\Theta}^{-1}\lambda_{zZ}^{-1}} = P\lambda_{\theta\Theta}^{2}\lambda_{zZ}\frac{R}{H} \tag{5}$$

$$\left[\lambda_{zZ}\frac{\partial W}{\partial\lambda_{zZ}}\right]_{\lambda_{rR}=\lambda_{\theta\Theta}^{-1}\lambda_{zZ}^{-1}} - \left[\lambda_{rR}\frac{\partial W}{\partial\lambda_{rR}}\right]_{\lambda_{rR}=\lambda_{\theta\Theta}^{-1}\lambda_{zZ}^{-1}} = \frac{P}{2}\lambda_{\theta\Theta}^{2}\lambda_{zZ}\frac{R}{H} + \frac{F_{red}}{2\pi RH}\lambda_{zZ} \tag{6}$$

One may select two variables as independent and remaining quantities are computed from equations modeling the problem. Our solution is obtained in two steps. First, axial prestretch of the tube is induced by assigning $\lambda_{zZ} = \lambda_{zZ}^{ini}$ at $P = 0$. In this step, the system (5–6) is used to compute F_{red} necessary to preelongate the tube. Simultaneously the initial circumferential stretch $\lambda_{\theta\Theta}^{ini}$ is obtained. In the second step, (5–6) is used to compute $\lambda_{\theta\Theta}$ and λ_{zZ} at a given P and F_{red}. The sequence $0.5(i-1)$ kPa for $i = 1, .., 41$ was used to simulate the pressurization of the tube. The axial prestretch λ_{zZ}^{ini} considered in our study was $1 + 0.02(i-1)$ for $i = 1, .., 14$ that includes typical values of the prestretch in the human abdominal aorta expected at 48 years old individual [3, 4, 19].

3 Results and Discussion

The results obtained in our simulation are summarized in Figs. 1, 2 and 3. The inflation-extension response expressed by means of P–$\lambda_{\theta\Theta}$ and P–λ_{zZ} relationships at constant axial load corresponding to $\lambda_{zZ}{}^{ini}$ = 1, 1.02, 1.04, .., 1.26 is in Fig. 1. Colors of the curves uniquely correspond to selected values of $\lambda_{zZ}{}^{ini}$ and specific value of the axial prestretch is evident from the right panel of the figure (P–λ_{zZ}). The figure clearly shows that there is a certain value of the prestretch under which the tube changes its response from inflation–extension to inflation–shortening. This is exactly inversion axial stretch and this mechanical response is highlighted by dotted curve and denoted by $\lambda_{zZ}{}^{inv}$. Numerical computation showed that, in specific case, it is approximately equal to 1.063.

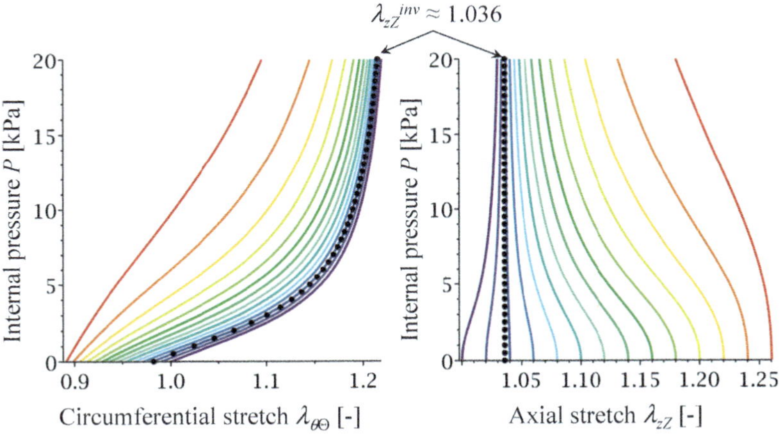

Fig. 1. Inflation-extension response obtained from model. $\lambda_{zZ}{}^{ini}$ = 1, 1.02, 1.04, .., 1.26 was used in calculation. Inversion axial prestretch, under which tube does not move axially, is highlighted by dotted line.

Figure 2 documents a relationship between inflating pressure and relative internal volume v, $v = \pi r^2 l/(\pi R^2 L) = \lambda_{\theta\Theta}{}^2 \lambda_{zZ}$. *It is clear that maximal* volume of the tube is attained when the tube is inflated at inversion axial stretch (dotted curve). If axes of the figure were changed (pressure as vertical axis and volume as horizontal axis), it would be evident that at the same time the inflation at inversion stretch also leads to minimal work performed during pressurization of the tube because it corresponds to minimal area under P–v curve. To be exact, we should talk about work per unit volume instead of work because relative volume instead of volume as such is variable on the axis.

Finally, Fig. 3 shows traces of the kinematics of tube inflation in a phase space of the deformation. Left panel is for λ_{zZ}–$\lambda_{\theta\Theta}$ dependence and right panel shows dependence of relative volume v on λ_{zZ}. Applied axial prestretch is indicated by color of the curve. Both λ_{zZ}–$\lambda_{\theta\Theta}$ and v–λ_{zZ} also contains contours of constant pressure at P = 0, 1,

2.5, 5, 10, and 20 kPa. They are depicted with thin dotted curves. One observes that λ_{zZ}^{inv} intersects contours of the constant pressure at their maximal volume which again documents that the inversion prestretch leads to optimal inflation conditions.

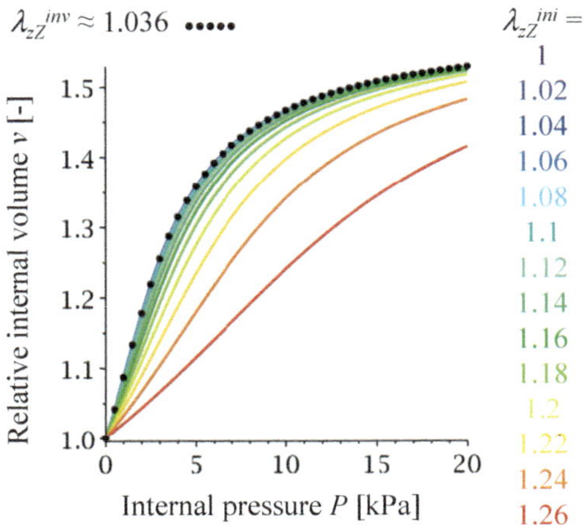

Fig. 2. Relative volume–pressure relationship at various λ_{zZ}^{ini}. Dotted curve corresponds to inversion prestretch under which maximal volume is attained. Additionally to maximal volume, the figure clearly shows that minimal work is also spent when tube is inflated under inversion prestretch λ_{zZ}^{inv}.

Although previous studies focused on the axial prestretch of arteries have pointed out that the prestretch is mechanically favorable, it was not, to the best of our knowledge, explicitly stated that the inversion stretch leads to maximal volume of the inflated tube. It suggests that tuning of the blood transport is highly sophisticated and optimized from the biomechanical point of view.

Our present result is however only one piece of a jigsaw puzzle of mechanical principles acting in our cardiovascular system. One should not overestimate it because there are other questions which have to be considered and were not addressed in our study. Particularly, the present result was obtained for one specific selection of material parameters. In order to reach physiologial relevance it has to be generalized and proved for other, more accurate models of arterial wall which will consider anistropic behavior of the wall. We also have to note that the thin-wall assumption should be accepted with caution because it is known that arteries exhibit residual strains which cannot be modeled under this assumption. The results are also limited by the fact that our study was based on quasistatic formulation but true blood pressure propagation is a wave problem. However, our present contribution is only a preliminary study and in future we plan to refine it in a more accurate computational model.

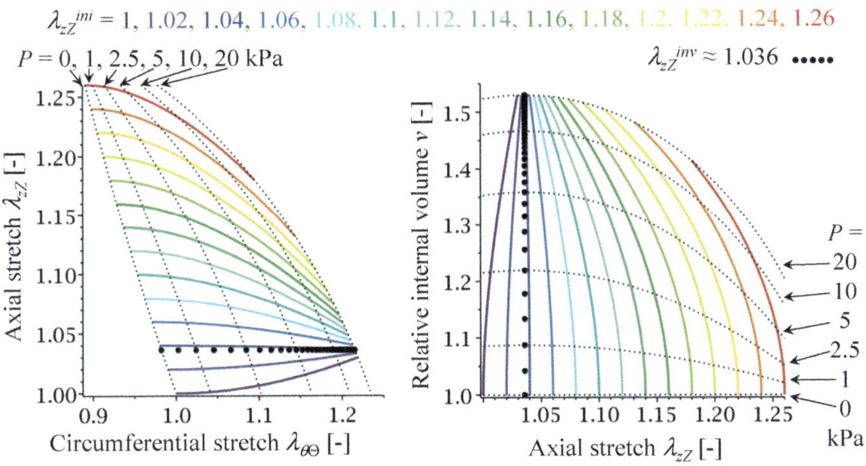

Fig. 3. Kinematics of inflation–extension response depicted in phase space. The left panel shows that $\partial\lambda_{zZ}/\partial\lambda_{\Theta} = 0$ at inversion stretch and right panel documents that contours of constant pressure have maximum v at λ_{zZ}^{inv}.

Finally, it is concluded that results obtained from the simplified model suggest that axial prestrech in human arteries ensures not only the constancy of axial force and axial deformation during the pressure cycle but also the maximum volume attained in the inflation of an artery.

Acknowledgement. This research has been supported by the Czech Science Foundation via project 18-26041S entitled “Effect of axial prestretch on mechanical response of nonlinearly elastic and viscoelastic tubes”.

Conflict of Interest. Authors declare that they have no conflict of interest.

References

1. Learoyd, B.M., Taylor, M.G.: Alterations with age in the viscoelastic properties of human arterial walls. Circ. Res. **18**(3), 278–292 (1966)
2. Han, H.-C., Fung, Y.C.: Longitudinal strain of canine and porcine aortas. J. Biomech. **28**(5), 637–641 (1995)
3. Horny, L., Adamek, T., Kulvajtova, M.: Analysis of axial prestretch in the abdominal aorta with reference to post mortem interval and degree of atherosclerosis. J. Mech. Behav. Biomed. Mater. **33**(1), 93–98 (2014)
4. Horný, L., Adámek, T., Kulvajtová, M.: A comparison of age-related changes in axial prestretch in human carotid arteries and in human abdominal aorta. Biomech. Model. Mechanobiol. **16**(1), 375–383 (2017)
5. Kamenskiy, A., Seas, A., Bowen, G., Deegan, P., Desyatova, A., Bohlim, N., Poulson, W., Mactaggart, J.: In situ longitudinal pre-stretch in the human femoropopliteal artery. Acta Biomater. **32**(1), 231–237 (2016)

6. Schulze-Bauer, C.A.J., Mörth, C., Holzapfel, G.A.: Passive biaxial mechanical response of aged human iliac arteries. J. Biomech. Eng. **125**(3), 395–406 (2003)
7. Dobrin, P.B., Schwarcz, T.H., Mrkvicka, R.: Longitudinal retractive force in pressurized dog and human arteries. J. Surg. Res. **48**(2), 116–120 (1990)
8. Lee, A.Y., Han, B., Lamm, S.D., Fierro, C.A., Han, H.-C.: Effects of elastin degradation and surrounding matrix support on artery stability. Am. J. Physiol. Heart Circulatory Physiol. **302**(4), H873–H884 (2012)
9. Carta, L., Wagenseil, J.E., Knutsen, R.H., Mariko, B., Faury, G., Davis, E.C., Starcher, B., et al.: Discrete contributions of elastic fiber components to arterial development and mechanical compliance. Arterioscler. Thromb. Vasc. Biol. **29**(12), 2083–2089 (2009)
10. Wagenseil, J.E., Mecham, R.P.: Elastin in large artery stiffness and hypertension. J. Cardiovasc. Transl. Res. **5**(3), 264–273 (2012)
11. Van Loon, P., Klip, W., Bradley, E.L.: Length-force and volume-pressure relationships of arteries. Biorheology **14**(4), 181–201 (1977)
12. Weizsäcker, H.W., Lambert, H., Pascale, K.: Analysis of the passive mechanical properties of rat carotid arteries. J. Biomech. **16**(9), 703–715 (1983)
13. Dobrin, P.B.: Biaxial anisotropy of dog carotid artery: estimation of circumferential elastic modulus. J. Biomech. **19**(5), 351–358 (1986)
14. Takamizawa, K., Hayashi, K.: Strain energy density function and uniform strain hypothesis for arterial mechanics. J. Biomech. **20**(1), 7–17 (1987)
15. Ogden, R.W., Schulze-Bauer, C.A.J.: Phenomenological and structural aspects of the mechanical response of arteries. Am. Soc. Mech. Eng. Appl. Mech. Div. **242**, 125–140 (2000)
16. Sommer, G., Regitnig, P., Költringer, L., Holzapfel, G.A.: Biaxial mechanical properties of intact and layer-dissected human carotid arteries at physiological and supraphysiological loadings. Am. J. Physiol. Heart Circulatory Physiol. **298**(3), H898–H912 (2010)
17. Gent, A.N.: A new constitutive relation for rubber. Rubber Chem. Technol. **69**(1), 59–61 (1996)
18. Horgan, C.O., Saccomandi, G.: A molecular-statistical basis for the Gent constitutive model of rubber elasticity. J. Elast. **68**(1–3), 167–176 (2002)
19. Horny, L., Adamek, T., Zitny, R.: Age-related changes in longitudinal prestress in human abdominal aorta. Arch. Appl. Mech. **83**(6), 875–888 (2013)
20. Horgan, C.O., Saccomandi, G.: A description of arterial wall mechanics using limiting chain extensibility constitutive models. Biomech. Model. Mechanobiol. **1**(4), 251–266 (2003)
21. Sang, J., Sun, L., Xing, S.: Analysis of uniaxial tension and circumferential inflation on the mechanical property of arterial wall. Mechanika **21**(2), 112–117 (2015)

A Closed-Loop Multiscale Model of the Cardiovascular System: Application to Heart Pacing and Open-Loop Response

Caterina Gallo(✉), Luca Ridolfi, and Stefania Scarsoglio

Politecnico di Torino, 10129 Torino, Italy
caterina.gallo@polito.it

Abstract. A 1D description of the arterial tree is coupled to a lumped parameter model of the remaining circulatory system, resulting in a closed-loop multiscale model of the cardiovascular apparatus. The regulation of the arterial pressure is also implemented through a short-term baroreceptor model. The proposed framework reproduces well the physiological cardiovascular behaviour of an healthy young man and the modelled baroreflex mechanism is effective in adjusting the hemodynamic responses to both heart pacing and open-loop aortic-carotid sinus control.

Keywords: Computational hemodynamics · Closed-loop multiscale cardiovascular model · Heart pacing · Open-loop control

1 Introduction

Numerous *in-silico* models of the human circulation have been proposed for a variety of clinical applications, helping to explore the genesis and features of cardiac and vascular pathologies, sustain the design of medical devices, foresee the effects of therapeutic actions, and trainee medicine students [1].

Depending on the study to carry out, cardiovascular models of different dimensions, from 0D to 3D, have been developed. In particular, the multiscale modelling approach (i.e., distinct dimensions are used depending on the circulation district to simulate) has demonstrated its effectiveness to describe the whole circulation [2]. In fact, it offers a suitable level of detail for each cardiovascular region at a reasonable computational cost.

In this work, we present a closed-loop multiscale model of the cardiovascular system, which puts together a 1D representation of the arterial network, a 0D description of the rest of circulation (peripheral arterial vessels, venous, cardiac and pulmonary circulations) and a baroreflex model.

We exploit and combine some modelling solutions introduced by other authors [3–6], obtaining a model which is able to reproduce some complex key characteristics of the cardiovascular system: the propagation and reflection phenomena of pressure and flow waves along the arterial tree, the non-ideal behaviour of cardiac valves, the well-organized structure of micro-circulation groups and venous return, and the short-term regulation mechanism. In addition to the previous models [3–6], we explicitly introduce the contribution of the unstressed volumes of each 0D compartment and adequately scale the time-dependent cardiac parameters with the heartbeat duration (RR). The novelty of

J. Henriques et al. (Eds.): MEDICON 2019, IFMBE Proceedings 76, pp. 577–585, 2020.
https://doi.org/10.1007/978-3-030-31635-8_69

this work consists in the integration and completion of different validated models of the main portions of human circulation, aiming at getting a comprehensive and versatile framework to explore the hemodynamic response in various configurations, from pathological states (such as valvular dysfunctions and cardiac arrhythmias) to different posture and acceleration conditions.

2 Multiscale Mathematical Model

A schematic representation of the 0D–1D multiscale model is provided in Fig. 1. In the following subsections the sub-models are detailed and equations listed.

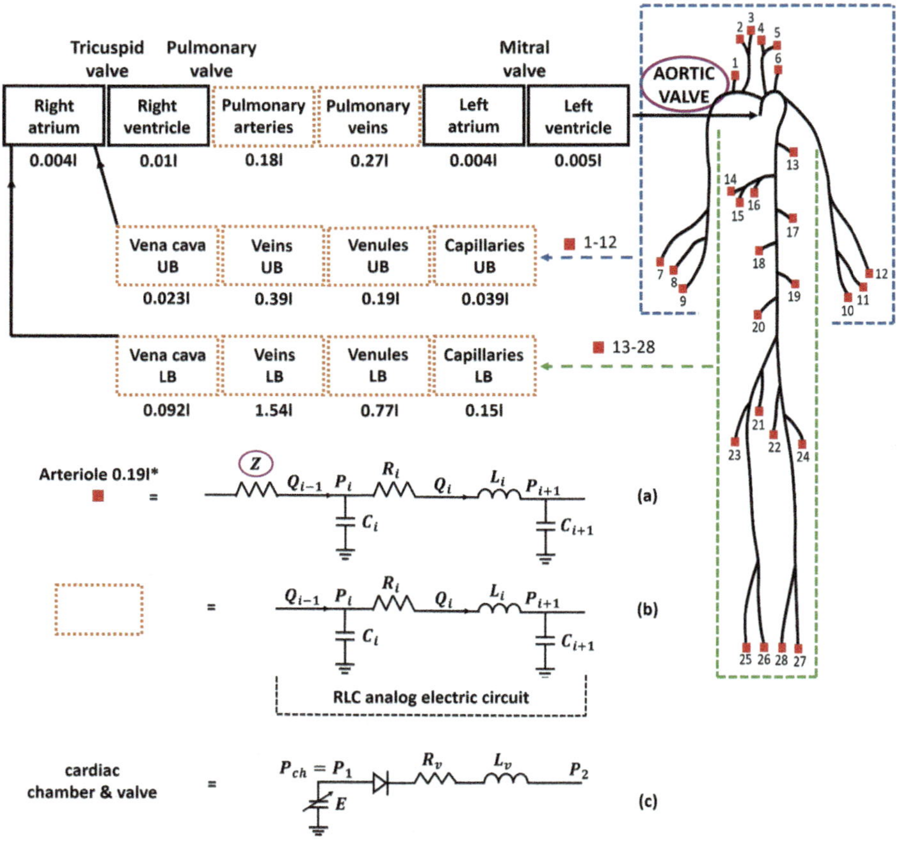

Fig. 1. Schematic representation of the multiscale model, with a sketch of the arterial tree on the right and the structure of all the 0D compartments on the left. The electric circuits corresponding to the lumped elements are given at the bottom (a–c) and the 1D–0D interfaces are highlighted with a purple circle. Values below each 0D compartment are the related unstressed volumes. Regarding the arterioles, the total unstressed volume is indicated (see the asterisk *). This latter is divided among the 28 arteriole groups according to the physiological regional blood volumes [7].

Arterial Tree. Hemodynamics in large-medium arteries is described through the model first proposed by Guala et al. [8], and then used to study the effects of atrial fibrillation on arterial circulation [9]. Mass and momentum balance equations are integrated over the arterial section, leading to the 1D system

$$\begin{cases} \frac{\partial A}{\partial t} + \frac{\partial Q}{\partial x} = 0, \\ \frac{\partial Q}{\partial t} + \frac{\partial}{\partial x}\left(\beta \frac{Q^2}{A}\right) = -\frac{A}{\rho}\frac{\partial P}{\partial x} + N_4, \end{cases} \tag{1}$$

where x and t are the spatial and time coordinates, A is the vascular section, Q is the flow rate, P is the pressure, ρ is the blood density, β is the Coriolis coefficient and N_4 is the viscous term. To close the above system, a constitutive relation for $P = P(A)$, accounting for the anisotropic non-linear viscoelastic behaviour of the arterial walls, is introduced. Details on the parameter settings, which refer to an healthy young man, can be found in [8,9].

Peripheral, Venous and Pulmonary Circulations. Arterioles, capillaries, venous return (venules, veins and venae cavae) and pulmonary circulation (arteries and veins) are simulated adopting different 0D models of similar form (Fig. 1b) [3], provided that arterioles are preceded by the characteristic impedance of the terminating 1D artery (Fig. 1a) [4]. The RLC analog electric circuit of each 0D compartment corresponds to the following system

$$\begin{cases} \frac{\mathrm{d}V_i}{\mathrm{d}t} = Q_{i-1} - Q_i, \\ \frac{\mathrm{d}Q_i}{\mathrm{d}t} = \frac{P_i - R_i Q_i - P_{i+1}}{L_i}, \\ P_i = \frac{V_i - V_{i_0}}{C_i}. \end{cases} \tag{2}$$

For the generic i-th compartment, V_i and V_{i_0} are the total and unstressed volumes, Q_{i-1} and Q_i are the flow rates entering and exiting the compartment, while P_i and P_{i+1} are the pressures at the inlet and outlet of the compartment, respectively. R_i (resistance), L_i (inertance), and C_i (compliance) characterize each compartment and are set according to [4] (for arterioles) and [3,6] (for the other lumped elements). V_{i_0} values are reported in Fig. 1 and, together with the overall stressed volume, give a total blood volume equal to 5.5 l.

Heart Chambers. Pressure, P_{ch}, and stressed volume, $(V_{ch} - V_{ch_0})$, of heart chambers are linked through the equation [3,6]

$$P_{ch} = E(V_{ch} - V_{ch_0}) + S\dot{V}_{ch}, \tag{3}$$

where $S = 0.0005 P_{ch}$ is the viscoelasticity coefficient of the cardiac wall, $\dot{V}_{ch}$ is the time derivative of V_{ch}, and $E = E_A e + E_B$ is the elastance function. In this latter expression, E_A and E_B (the amplitude and baseline values of elastance), as well as e (the time-varying elastance function) are chosen as in [3,6]. The time parameters of e are calculated as functions of RR. In particular, durations of contraction and relaxation periods are: $T_{acp} = 0.17RR$ and $T_{arp} = T_{acp}$ [3,4,6] for atria, and $T_{vcp} = 0.3\sqrt{RR}$ [10] and $T_{vrp} = 0.5T_{vcp}$ [10] for ventricles. Contraction and relaxation start at: $t_{ac} = 0.8RR$ and $t_{ar} = t_{ac} + T_{acp}$ for atria, and $t_{vc} = 0RR$ and $t_{vr} = t_{vc} + T_{vcp}$ for ventricles.

Heart Valves. Each cardiac valve is simulated by a pressure-flow model [6,11]. The non-ideal behaviour of the valve is considered and the valve opening angle, θ_v, is calculated based on the rotational inertia (I_{ao}) and forces acting on the valve leaflets - pressure difference across the valve, F_{pr}, dynamic effect of the blood acting on the valve leaflets, F_{bm}, frictional effects of neighbouring tissue resistance, F_{fr}, and vortices downstream of the valve, F_{vo} - namely

$$\begin{cases} L_v \dfrac{dQ_v}{dt} + R_v Q_v + B_v |Q_v| Q_v = \dfrac{(1-\cos(\theta_v))^4}{(1-\cos(\theta_{max}))^4} (P_1 - P_2), \\ I_{ao} \dfrac{d^2\theta_v}{dt^2} = F_{pr} + F_{bm} - F_{fr} - F_{vo}. \end{cases} \tag{4}$$

Q_v is the flow rate through the valve, $P_1 - P_2$ is the pressure difference across the valve, and L_v, R_v and B_v stand for the inertance, viscous and turbulent flow separation effects, respectively. All parameters are set as in [6].

Baroreflex Model. We adopt the baroreflex model introduced by Ottesen et al. [12]. Sympathetic, n_s, and parasympathetic activity, n_p, are evaluated as

$$n_s(\bar{p}_{acs}) = \frac{1}{1 + \left(\frac{\bar{p}_{acs}}{\bar{p}_{acs,ref}}\right)^{\nu}}, \qquad n_p(\bar{p}_{acs}) = \frac{1}{1 + \left(\frac{\bar{p}_{acs}}{\bar{p}_{acs,ref}}\right)^{-\nu}}, \tag{5}$$

where ν (here set equal to 7) gives the slope of the n_s and n_s responses, while $\bar{p}_{acs}$ is the aortic-carotid sinus pressure averaged over the cardiac cycle. $\bar{p}_{acs,ref}$ is the baseline value of the average aortic-carotid sinus pressure which, determined at a reference heart rate (HR) of 75 bpm, is equal to 92 mmHg. The efferent responses are described as

$$\frac{dx_j}{dt} = \frac{1}{\tau_j} \left(-x_j + \alpha_j n_s(\bar{p}_{acs}) - \beta_j n_p(\bar{p}_{acs}) + \gamma_j\right), \tag{6}$$

where x_j is the generic efferent organ, τ_j is the time delay in the efferent response, α_j and β_j define the weights of n_s and n_p, and γ_j equals x_j when $\bar{p}_{acs} = \bar{p}_{acs,ref}$. Efferent organs are HR, maximum values of right- and left-ventricular elastance ($E_{A_{r,l}}$), arteriole and capillary resistances ($R_{a,c}$), the compliances, and unstressed volumes of venules and veins ($C_{ven,v}$, $V_{0_{ven,v}}$). Baroreflex parameters are indicated in Table 1.

Table 1. Parameters of the baroreflex model for each efferent organ. α, β, γ values and saturation levels (Min. and Max.) of efferent organs are non-dimensional.

Efferent organ	τ [s]	α	β	γ	Min.	Max.
HR	3	0.75	0.75	1	0.25	1.75
$E_{A_{rl}}$	3	0.40	0	0.80	0.8	1.2
$R_{a,c}$	15	0.80	0	0.60	0.6	1.4
$C_{ven,v}$	30	−0.20	0	1.10	0.9	1.1
$V_{0_{ven,v}}$	30	−0.42	0	1.21	0.79	1.21

Numerical Resolution. The complete model is solved through a Runge-Kutta Discontinuous-Galerkin scheme. Space is discretized according to a Discontinuous-Galerkin approach and dependent variables are advanced in time adopting a two-steps Runge-Kutta explicit scheme. Pressure and flow rate values at the 1D–0D interfaces and at each arterial junction are defined by combining physical and compatibility numerical conditions [8]. Physical conditions are given by: the system (4) at aortic valve, the constitutive equation of the characteristic impedance Z at peripheral arterial vessels, and the conservation of mass and total pressure at arterial bifurcations.

3 Results and Conclusions

Physiological Behaviour. The modelled cardiovascular system well reproduces the physiological hemodynamics of a healthy young man at 75 bpm. Figure 2 shows pressure (a) and flow rate (b) signals at specific arterial sites, evidencing that the propagation phenomena of arterial waves are efficiently reproduced. In fact, moving away from the heart, systolic pressures increase, diastolic pressures decrease, pressure signals delay and steepen, and flow rate signals reduce in amplitude [13]. Figure 2c-d display the pressure and flow rate signals for the lower body venous, V_{LB}, and vena cava, VC_{LB}, compartments. As expected [4,14], it emerges that pressure is almost constant, while flow rate is still pulsatile along the venous return. Flow rates through cardiac valves (see Fig. 2e) clearly show the typical trasvalvular flow, with the physiological reversal flow as valves close [11].

Baroreflex Efficacy. To demonstrate the reliability of our model, we report two tests about the baroreflex system: (i) the response to heart pacing and (ii) the so-called open-loop case. In heart pacing, a fixed value of HR is externally forced and the response of the remaining effector organs is determined by the baroreflex mechanism [12]. Instead, in the open-loop case, the ratio $\bar{p}_{acs}/\bar{p}_{acs,ref}$ is externally applied. We simulated eight HRs (35, 55, 75, 95, 115, 135, 155 and 175 bpm) in heart pacing and seven ratios $\bar{p}_{acs}/\bar{p}_{acs,ref}$ (from 0.625 to 1.375) in the open-loop analysis.

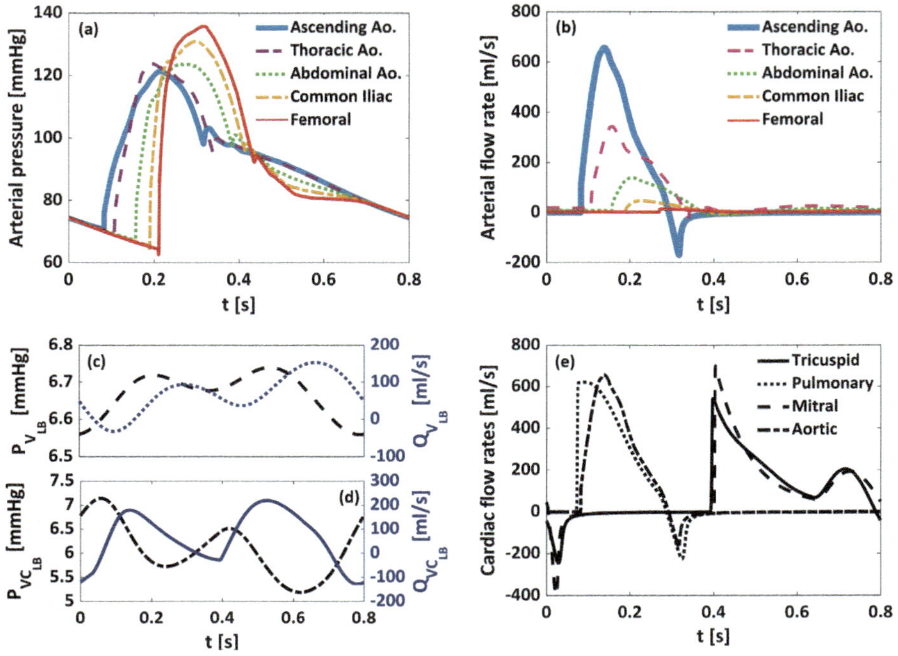

Fig. 2. (a) Arterial pressure and (b) flow rate signals along the aorta. (c)–(d) Pressure (P) and flow rate (Q) signals: (c) lower body venous compartment (V_{LB}), and (d) lower body vena cava compartment (VC_{LB}). (e) Flow rate signals through cardiac valves. All the figures refer to a healthy young man with 75 bpm.

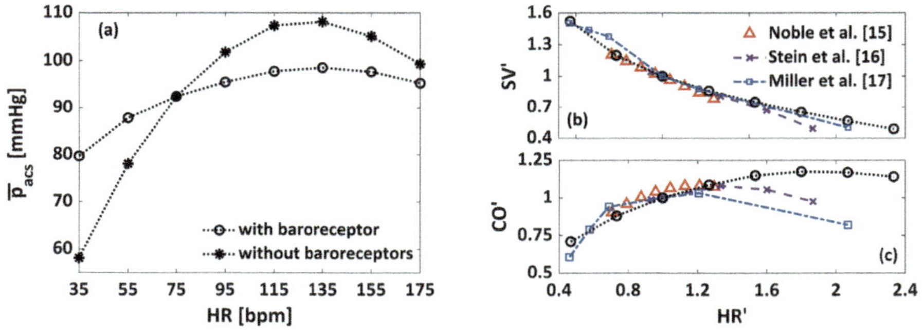

Fig. 3. Heart pacing case. (a) Behaviours of the average aortic-sinus carotid pressure (with and without baroreceptors) with HR. Dependence of the non-dimensional stroke volume (b) and cardiac output (c) on HR. Variables marked with a superscript are non-dimensional. Normalization is referred to the baseline values at 75 bpm for computed results and to the indicated or assumed baseline values for experimental results (provided in dimensional form). Data by Stein et al. [16] refer to human subjects, while data by Noble et al. [15] and Miller et al. [17] are measured on dogs.

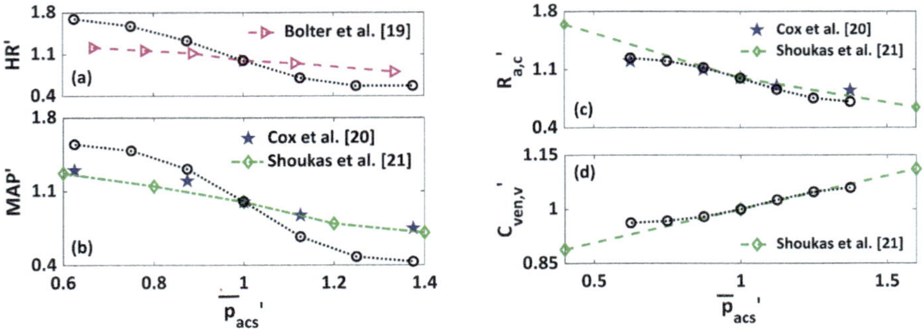

Fig. 4. Open-loop case. Non-dimensional open-loop responses for the (a) heart rate, (b) mean arterial pressure, (c) arterial and capillary resistances, and (d) compliances of venule and venous compartments. Variables marked with a superscript are non-dimensional. Normalization is referred to the baseline values at 75 bpm for computed results and to the indicated or assumed baseline values for experimental results (provided in dimensional form, except in Cox et al. [20]). All experimental data refer to dogs.

Figure 3 shows the results in the case of heart pacing simulation. Panel 3a reports the function $\bar{p}_{acs}$(HR) with and without baroreceptors, demonstrating that arterial pressure variations are limited by baroreflex as cardiac frequency is altered from the reference value (75 bpm). Figure 3b and c show the behaviours of the stroke volume, SV', and cardiac output, CO', as functions of HR' (superscript indicates non-dimensional quantities), together with the experimental data by [15–17]. Results provided by the model are normalized with respect to baseline values at 75 bpm. Instead, dimensional experimental results are normalized with respect to the indicated baseline values (if available) or reasonably assumed baseline values. It is typically reported in literature [18] that SV reduces as HR rises, while CO increases at low HRs, reaches a maximum value for higher HRs (between 90 and 180 bpm) and finally reduces for further HR increments. These trends are reproduced in Fig. 3b and c, where an overall agreement between computed and experimental data is observed. However, the maximum CO predicted by the model (between 135 and 155 bpm) seems to be located at non-dimensional frequencies higher than the ones associated to the experimental measures. This could depend on the the way in which HR is modified in the experimental measures (ventricular pacing, atrial pacing, pharmacological agents, exercise, etc..) as well as the experimental conditions (e.g., conscious or anaesthetized subjects) [16]. Moreover, since experimental results are given in dimensional form, there are some difficulties in comparing data from different sources.

Figure 4 refers to the open-loop analysis. The behaviours of HR, mean arterial pressure (MAP), arterial and capillary resistances ($R_{a,c}$), and compliances of venules and veins ($C_{ven,v}$), as functions of $\bar{p}_{acs}$, are provided in non-dimensional form for computed and experimental results [19–21]. Normalization of both computed and experimental results is done as in the case of heart pacing. One can appreciate a general correspondence between numerical and experimental results, although same discrepancies emerge. In fact, experimental results are obtained considering only the contribution of carotid sinus pressure in the baroreflex action. By contrast, pressure activating the baroreflex in our model is a combination of both aortic arch and carotid sinus pressures.

Based on the previous results, the proposed model satisfactorily reproduces the most important features of the arterial and venous hemodynamics, cardiac dynamics and short-term regulation mechanisms. Thus, the choice of the submodels representing each portion of the circulation, their integration into a unique framework, the parameter setting and the numerical resolution of the complete mathematical formulation, lead to an efficient and accurate *in-silico* model of the cardiovascular system.

Next efforts will consist in considering the effects of orthostatic variations, adding the contribution of the gravity, transmural pressures, and venous valves to guarantee unidirectional blood flow from lower to upper body. These additional features will allow one to investigate space-related medical issues, such as the fallouts of long-term microgravity exposure on the human hemodynamics.

Declaration of Interest. The authors declare that they have no conflict of interest.

References

1. Morris, P.D., Narracott, A., von Tengg-Kobligk, H., Silva Soto, D.A., Hsiao, S., Lungu, A., Evans, P., Bressloff, N.W., Lawford, P.V., Hose, D.R., Gunn, J.P.: Computational fluid dynamics modelling in cardiovascular medicine. Heart **102**, 18–28 (2016). https://doi.org/10.1136/heartjnl-2015-308044
2. Shi, Y., Lawford, P., Hose, R.: Review of zero-D and 1-D models of blood flow in the cardiovascular system. Biomed. Eng. Online **10**, 33 (2011). https://doi.org/10.1186/1475-925X-10-33
3. Liang, F., Takagi, S., Himeno, R., Liu, H.: Multi-scale modeling of the human cardiovascular system with applications to aortic valvular and arterial stenoses. Med. Biol. Eng. Comput. **47**, 743–755 (2009). https://doi.org/10.1007/s11517-009-0449-9
4. Liang, F.Y., Takagi, S., Himeno, R., Liu, H.: Biomechanical characterization of ventricular-arterial coupling during aging: a multi-scale model study. J. Biomech. **42**, 692–704 (2009). https://doi.org/10.1016/j.jbiomech.2009.01.010
5. Blanco, P.J., Trenhago, P.R., Fernandes, L.G., Feijóo, R.A.: On the integration of the baroreflex control mechanism in a heterogeneous model of the cardiovascular system. Int. J. Numer. Meth. Biomed. Engng. **28**, 412–433 (2012). https://doi.org/10.1002/cnm.1474
6. Blanco, P.J., Feijóo, R.A.: A dimensionally-heterogeneous closed-loop model for the cardiovascular system and its applications. Med. Eng. Phys. **35**, 652–667 (2013). https://doi.org/10.1016/j.medengphy.2012.07.011
7. Wayson, M.B., Leggett, R.W., Jokisch, D.W., Lee, C., Schwarz, B.C., Godwin, W.J., Bolch, W.E.: Suggested reference values for regional blood volumes in children and adolescents. Phys. Med. Biol. **63**, 155022 (2018). https://doi.org/10.1088/1361-6560/aad313

8. Guala, A., Camporeale, C., Tosello, F., Canuto, C., Ridolfi, L.: Modelling and subject-specific validation of the heart-arterial tree system. Ann. Biomed. Eng. **43**, 222–237 (2015). https://doi.org/10.1007/s10439-014-1163-9
9. Scarsoglio, S., Gallo, C., Ridolfi, L.: Effects of atrial fibrillation on the arterial fluid dynamics: a modelling perspective. Meccanica **53**, 3251–3267 (2018). https://doi.org/10.1007/s11012-018-0867-6
10. Heldt, T., Shim, E.B., Kamm, R.D., Mark, R.G.: Computational modeling of cardiovascular response to orthostatic stress. J. Appl. Physiol. **92**, 1239–1254 (2002). https://doi.org/10.1152/japplphysiol.00241.2001
11. Korakianitis, T., Shi, Y.: Numerical simulation of cardiovascular dynamics with healthy and diseased heart valves. J. Biomech. **39**, 1964–1982 (2006). https://doi.org/10.1016/j.jbiomech.2005.06.016
12. Ottesen, J., Olufsen, M., Larsen, J.: Applied Mathematical Models in Human Physiology. Society for Industrial and Applied Mathematics, Philadelphia (2004)
13. Caro, C.G., Pedley, T.Y., Schroter, R.C., Seed, W.A.: The Mechanics of the Circulation, 2nd edn. Cambridge University Press, New York (2012)
14. Guyton, A., Hall, J.: Textbook of Medical Physiology, 11th edn. Elsevier Saunders, Philadelphia (2006)
15. Noble, M.I.M., Trenchord, D., Guz, A.: Effect of changing heart rate on cardiovascular function in the conscious dog. Circ. Res. **19**, 206–213 (1966). https://doi.org/10.1161/01.RES.19.1.206
16. Stein, E., Damato, A.N.: The relation of heart rate to cardiovascular dynamics. Circulation **33**, 925–932 (1966). https://doi.org/10.1161/01.cir.33.6.925
17. Miller, D.E., Gleason, W.L., Whalen, R.E., Morris, J.J., Mclntosh, H.D.: Effect of ventricular rate on the cardiac output in the dog with chronic heart block. Circ. Res. **10**, 658–663 (1962). https://doi.org/10.1161/01.RES.10.4.658
18. Berglund, E., Borst, H.G., Duff, F., Schreiner, G.L.: Effect of heart rate on cardiac work, myocardial oxygen consumption and coronary blood flow in the dog. Acta Physiol. **42**, 185–198 (1958). https://doi.org/10.1111/j.1748-1716.1958.tb01551.x
19. Bolter, C.P., Ledsome, J.R.: Effect of cervical sympathetic nerve stimulation on canine carotid sinus reflex. Am. J. Physiol. **230**, 1026–1030 (1976). https://doi.org/10.1152/ajplegacy.1976.230.4.1026
20. Cox, R.H., Bagshaw, R.J.: Baroreceptor reflex control of arterial hemodynamics in the dog. Circ. Res. **37**, 772–786 (1975). https://doi.org/10.1161/01.RES.37.6.772
21. Shoukas, A.A., Connoly-Brunner, M.: Epinephrine and the carotid sinus baroreceptor reflex. Circ. Res. **47**, 249–257 (1980). https://doi.org/10.1161/01.RES.47.2.249

Experimental Study to Improve "Federica" Prosthetic Hand and Its Control System

Daniele Esposito[1,2](✉), Chiara Cosenza[3], Gaetano Dario Gargiulo[4], Emilio Andreozzi[1,2], Vincenzo Niola[3], Antonio Fratini[5], Giovanni D'Addio[2], and Paolo Bifulco[1,2]

[1] Department of Electrical Engineering and Information Technologies, University "Federico II" of Naples, Via Claudio 21, 80125 Naples, Italy
{daniele.esposito, emilio.andreozzi, paolo.bifulco}@unina.it

[2] Istituti Clinici Scientifici Maugeri s.p.a.—Società Benefit, Via S. Maugeri 4, 27100 Pavia, Italy
gianni.daddio@icsmaugeri.it

[3] Department of Industrial Engineering, University "Federico II" of Naples, Via Claudio 21, 80125 Naples, Italy
{chiara.cosenza, vincenzo.niola}@unina.it

[4] The MARCS Institute, Western Sydney University, Penrith, NSW 2751, Australia
g.gargiulo@uws.edu.au

[5] School of Life and Health Sciences, Aston University, Birmingham B4 7ET, UK
a.fratini@aston.ac.uk

Abstract. Modern 3D printing technologies and wide availability of microcontroller boards allow to make active prosthetic devices in a simple way. This is the case of "Federica", a very low-cost, under-actuated, active hand prosthesis. The five fingers of the prosthesis are moved by a single motor through inelastic tendons. The control system of the prosthesis is proportional to muscle contraction: firstly, EMG was used, then mechanical sensors that measure muscle volumetric variation were successfully utilized. This prosthesis proved to be particularly energy efficient and fast; it provided a general grasp function by adapting the exerted forces, thus allowing to easily catch even deformable objects. This study presents further analyses and design improvements of this prosthesis. In particular, a new, extremely simple but effective conditioning system of a force sensor resistor was presented and tested. In addition, the actual three-dimensional kinematics of a single finger was captured by means of high frame rate cameras and then analyzed. The new sensor conditioning system was characterized. It proved to be as effective as the EMG envelope to proportionally control the hand prosthesis motion, and it allowed an easier connection to common microcontroller boards. Kinematic analysis allowed to accurately reconstruct the actual phalanges motion over time.

Keywords: Underactuated prosthetic hand · Muscle contraction sensors · Kinematic analysis

J. Henriques et al. (Eds.): MEDICON 2019, IFMBE Proceedings 76, pp. 586–593, 2020.
https://doi.org/10.1007/978-3-030-31635-8_70

1 Introduction

In recent years there have been many developments and improvements in both the design of the hand prostheses and in their control systems, but despite this, continuous re-search is needed to make the use of the prosthesis even easier and more comfortable. Generally, surface electromyography (EMG) signal is used for controlling electrically powered prosthesis [1–3]. However, EMG has some drawbacks, such as: the need of electrodes, biopotential amplifiers and conductive gel for stabilizing the skin/electrode electrical interface; moreover, raw EMG signal needs to be pre-processed to extract its envelope, which estimate the muscle contraction strength. At last, recordings are very sensitive to external electromagnetic interferences and other noise sources (e.g. motion artifacts, crosstalk with other biopotentials) [4–6]. Acquisition problems remain also by using dry electrodes [7–9].

Hence, previous studies [10, 11] proposed prosthesis control signals related to the muscle mechanical variations as alternative to the EMG. Indeed, EMG reflects the electrical events of muscle contraction and it triggers muscular mechanical activity [6]. Recently a conductive, stretchable, carbon impregnated rubber cord sensor, [10] wrapped around the forearm, successfully sensed the volumetric changes of contracting muscles. This sensor was able to proportionally control a 3D printed, under-actuated prosthetic hand named "Federica" [12–14]. However, using this type of sensor, only the volumetric increments distributed over the whole circumference of the arm can be detected, and it is not possible to sense localized and precise variations on certain muscle groups.

A further recent study [11], proposed to use a Force Sensitive Resistor (FSR) [15], placed on skin by means of a mechanical coupler, for sensing the contraction of a specific muscle. Simultaneous recordings of EMG envelope and FSR signals from the same muscle resulted very similar. Preliminary validation tests [11] showed the ability of the FSR sensor to proportionally control the hand prosthesis "Federica", obtaining EMG comparable performances. An op-amp transimpedance amplifier [11] was used as the conditioning circuit of the FSR, ensuring a constant voltage across the sensor. It was demonstrated that commonly used voltage divider [16] is not convenient for FSR conditioning, because it produces relevant drift on the measurements [17].

In this study, the use of FSR sensor with a re-designed and simplified control system for hand prosthesis was pro-posed and tested. Moreover, the kinematics of the prosthesis fingers was analyzed in more details.

2 Materials and Methods

2.1 Modular Control System for "Federica" Prosthetic Hand

The control system was designed as modular and consists of: an Arduino microcontroller, a metal gear servomotor, a 7.4 V battery pack and an FSR sensor (see Fig. 1a).

The Interlink FSR 402 sensor includes two membranes on which electrodes and conductive polymer are printed respectively. The sensor shows a decrease in resistance as the applied force increases [15]. The FSR was placed on a forearm muscle (flexor

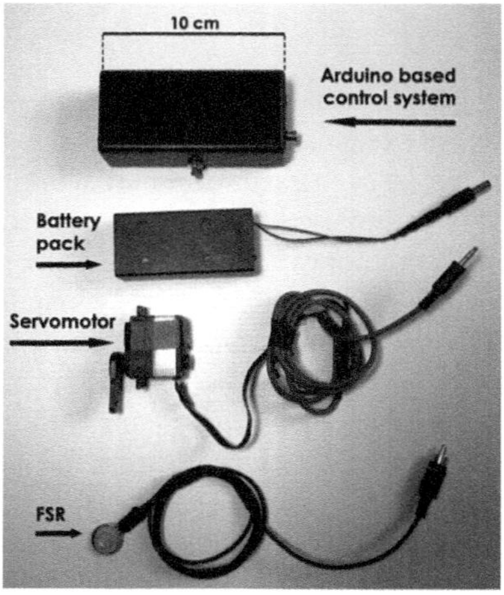

(a)

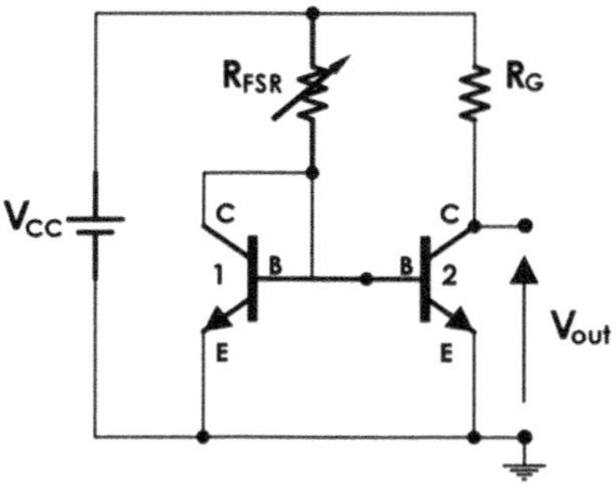

(b)

Fig. 1. (a) Modular prosthetic hand control system; (b) FSR conditioning circuit

carpi ulnaris) by means of a rigid dome [11], in order to improve the force transfer efficiency. The FSR conditioning circuit was re-designed using a simple current mirror (see Fig. 1b), realized by means of a pair of common npn BJT (2N2222), positioned very close to each other.

The circuit replicates the FSR current in the gain resistor R_G (Fig. 1b), providing an output voltage V_{OUT}. The 5 V (V_{CC}) was directly supplied by the Arduino board, and the gain resistor R_G was set to 600 Ω. The output voltage (V_{OUT}) was sampled by the microcontroller to proportionally control the prosthetic hand.

Static calibration of the FSR sensor was carried out to evaluate the relationship between the muscle force and the voltage output V_{OUT} (Fig. 1b). The sensor was placed on a precision electronic scale then different weights were applied on sensor active area perpendicularly to the dome and the corresponding output voltages were recorded.

In order to measure the voltage across the FSR and its variation for different applied loads, the collector-emitter voltage (V_{CE}) of the BJT 1 (Fig. 1b) was acquired. Actually, the voltage across the FSR is given by $V_{FSR} = V_{CC} - V_{CE}$. The output signals were acquired at 1 kHz sampling frequency with 14-bit precision by means of an acquisition board (National Instruments NI USB-6009). The latter analysis was motivated by previous studies recommending the use of constant voltage across the FSR [16, 17].

As reported in a previous study [11], EMG and FSR signals were collected simultaneously from the flexor carpi ulnaris, to compare the EMG envelope and the FSR signals.

2.2 3D Reconstruction of Phalanxes Movements

An experimental method to measure the actual finger kinematic with a quantitative approach was developed. It was based on vision technique properly customized to obtain the finger kinematic over time [14]. A geometrical reconstruction algorithm has been employed to build a three-dimensional, skeletal model of the finger and to compute the phalanxes rotations over time.

The experimental setup comprised an underactuated mechanical finger (Fig. 2a) and a vision acquisition instrument based on the Digital Image Correlation (DIC) DANTEC Dynamics Q-450 system [18], operating at 450 fps (Fig. 2b). The DIC system acquired the three-dimensional shape of the finger phalanxes during a flexion movement. A random pattern of markers was applied on each phalanx, then an image correlation technique estimated their motion in space from the camera stereoscopic acquisitions. Then, for each phalanx, clouds of surface points were available at about every 2 ms and fitted to cylindric shape. Finally, the spatial coordinates of the three cylinders associated with the relative phalanxes are available for subsequent analyzes.

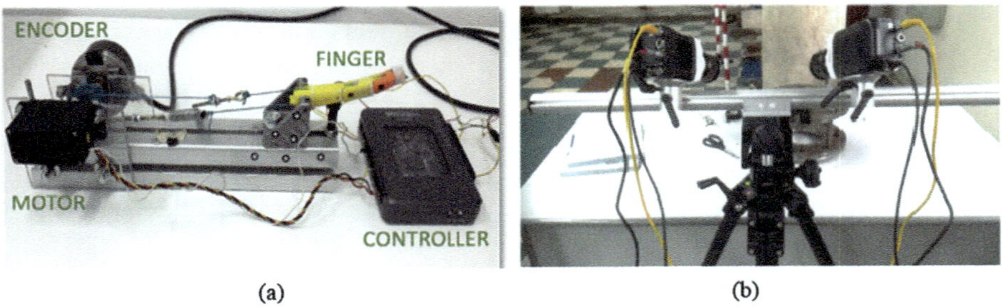

Fig. 2. Experimental test rig: (a) underactuated mechanical finger; (b) vision acquisition instrument

3 Results

Results of the static FSR calibration are showed in Fig. 3. Experimental measurements are represented as circles while linear and exponential regressions are respectively represented as blue and red continuous lines. Considering the weights applied to the sensor as x (kg) and the output voltage from the conditioning circuit as y (V), the equation of the linear regression resulted

$$y = -2.16x + 3.70 \tag{1}$$

with a coefficient of determination R^2 equal to 0.804, whereas the exponential regression resulted

$$y = 0.198 + 5.84\,e^{-2.183x} \tag{2}$$

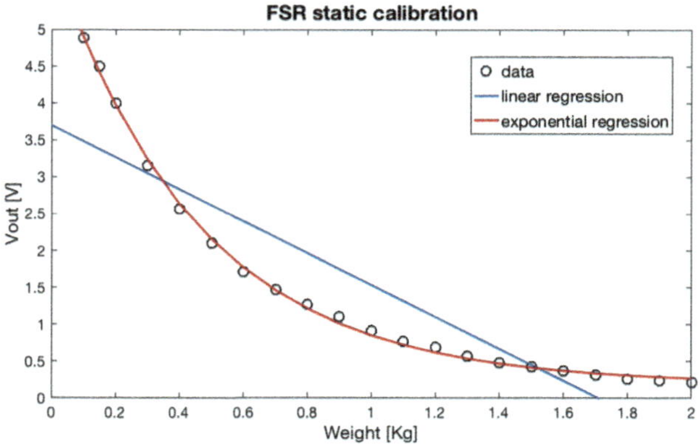

Fig. 3. FSR static calibration: scatter plot of the experimental data (circles), linear regression (blue line) and exponential regression (red line)

with R^2 equal to 0.99. Furthermore, V_{FSR} variations were measured considering different loads between 100 g and 1500 g, mimicking real conditions. The voltage across the FSR changed from 4.43 V to 4.34 V. Hence, the maximum percentage variation was about 2%.

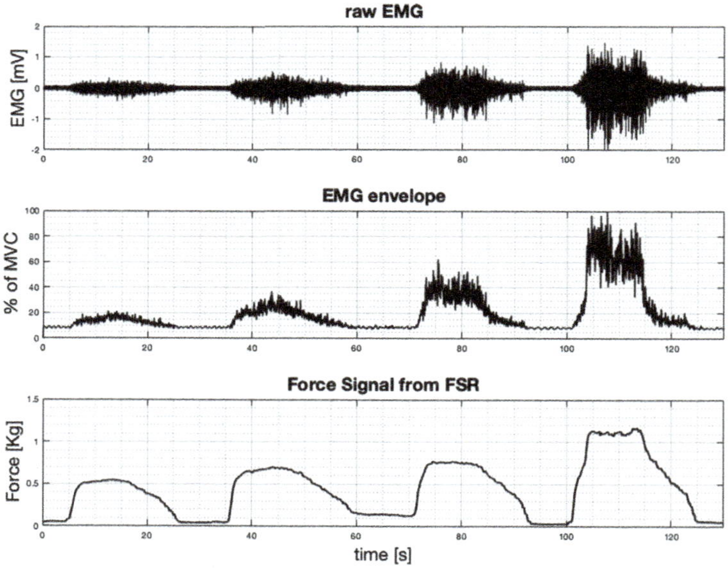

Fig. 4. Simultaneous recordings from flexor carpi ulnaris when performing four grasp actions at increasing strength: (a) Raw EMG signal; (b) EMG envelope; (c) FSR force signal (raw output).

Figure 4 shows an example of simultaneous recordings of EMG and FSR force signal from flexor carpi ulnaris, when performing four grasp actions at increasing strength.

To quantitatively measure the similarity between the EMG envelope and the FSR signal, it was computed the Pearson's correlation coefficient "r" that scored 0.85 (p-value < 0.0001 (two-tailed test)).

As regards the kinematic analysis on the prosthesis finger, the three cylindrical shapes resulting from the fitting of the phalanxes data clouds, at a given time, are shown in the Fig. 5a. Proximal phalanx is depicted in light blue, while the middle and the distal phalanxes are colored in yellow.

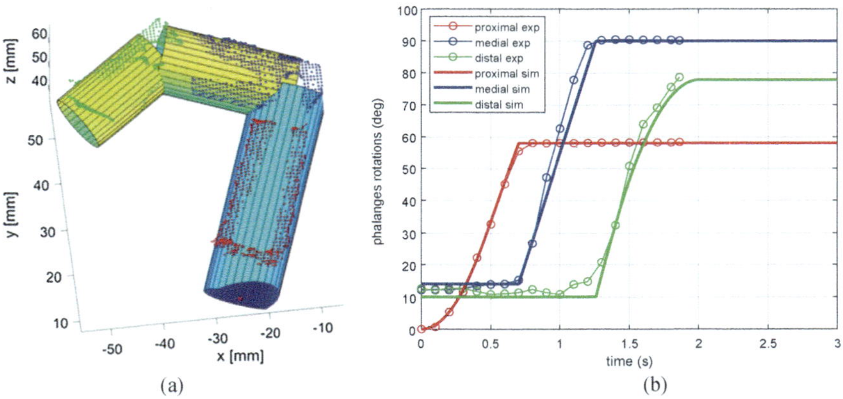

Fig. 5. (a) Finger reconstruction; (b) Phalanxes rotations over time: comparison between experimental (circles) and simulation data (continuous line).

Starting from the cylinder axes, the relative angles be-tween the phalanges over time were computed. Therefore, the actual mechanical behavior of the whole finger was completely known.

These experimental data were compared with the simulated. Figure 5b shows the simulated (continuous- thick line) and the experimental rotations (continuous-thin line and circles) of the three phalanges during a finger closure sequence of 2 s. It resulted that the model enough well fits experimental data. These experimental data allowed to improve the multi-body finger model [19] by taking into account actual friction.

4 Conclusions and Discussion

This study presented further advances in the design of the "Federica" prosthetic hand. The FSR conditioning circuit was much simplified and can be more easily and swiftly connected to any common microcontroller board. The kinematic behavior of the prosthetic fingers was analyzed in detail. The technique developed to measure the kinematic behavior of the fingers, would be of help to improve the "Federica" prototype and to allow more comprehensive mechanical simulations.

Conflict of Interest The authors declare that they have no conflict of interest.

Statement of Informed Consent All subjects gave their informed consent for inclusion be-fore they participated in the study.

References

1. Polisiero, M., Bifulco, P., Liccardo, A., Cesarelli, M., Romano, M., Gargiulo, G.D., McEwan, A.L., D'Apuzzo, M.: Design and assessment of a low-cost, electromyographically controlled, prosthetic hand. Med. Devices: Evid. Res. **6**, 97–104 (2013). https://doi.org/10.2147/MDER.S39604
2. Ulloa, G.D.F., Sreenivasan, N., Bifulco, P., Cesarelli, M., Gargiulo, G., Gunawardana, U.: Cost effective electro Resistive band based myo activated prosthetic upper limb for amputees in the developing world. In: Life Sciences Conference (LSC) 2017, pp. 250–253. IEEE (2017)
3. Neethu Sreenivasan, N., Ulloa Gutierrez, D.F., Bifulco, P., Cesarelli, M., Gunawardana, U., Gargiulo, G.D.: Towards ultra low-cost myoactivated prostheses. BioMed. Res. Int. **2018**, 14 (2018). https://doi.org/10.1155/2018/9634184. Article ID 9634184
4. Nazmi, N., Abdul Rahman, M.A., Yamamoto, S., Ahmad, S.A., Zamzuri, H., Mazlan, S.A.: A review of classification techniques of EMG signals during isotonic and isometric contractions. Sensors **16**, 1304 (2016)
5. Amrutha, N., Arul, V.H.: A review on noises in EMG signal and its removal. Int. J. Sci. Res. Publ. **7**, 23–27 (2017)
6. Basmajian, J.V., De Luca, C.J.: Muscles Alive: Their Functions Revealed by Electromyography. Williams & Wilkins, Baltimore (1985). ISBN 068300414X
7. Gargiulo, G., Bifulco, P., McEwan, A., Nasehi, T.J., Calvo, R.A., Romano, M., Ruffo, M., Shephard, R., Cesarelli, M., Jin, C., Mohamed, A., van Schaik, A.: A dry electrode biopotential recordings. Conf. Proc. IEEE Eng. Med. Biol. Soc. **2010**, 6493–6496 (2010). https://doi.org/10.1109/iembs.2010.5627359
8. Bifulco, P., Cesarelli, M., Fratini, M., Ruffo, M., Pasquariello, G., Gargiulo, G.: A wearable device for recording of biopotentials and body movements. In: IEEE MeMeA, pp. 469–472, Bari (2011). https://doi.org/10.1109/memea.2011.5966735
9. Gargiulo, G., Bifulco, P., Cesarelli, M., Jin, C., McEwan, A., van Schaik, A.: Wearable dry sensors with bluetooth connection for use in remote patient monitoring systems. A. Stud. Health Technol. Inform. **161**, 57–65 (2010)
10. Bifulco, P., Esposito, D., Gargiulo, G.D., Savino, S., Niola, V., Iuppariello, L., Cesarelli, M.: A stretchable, conductive rubber sensor to detect muscle contraction for prosthetic hand control. In: Proceedings of the IEEE E-Health and Bioengineering Conference (EHB), 22–24 Junes Sinaia, Romania (2017). https://doi.org/10.1109/EHB.2017.7995389
11. Esposito, D., Andreozzi, E., Fratini, A., Gargiulo, G.D., Savino, S., Niola, V., Bifulco, P.: A piezoresistive sensor to measure muscle contraction and mechanomyography. Sensors **18**(8), 2553 (2018). https://doi.org/10.3390/s18082553

12. Rossi, C., Savino, S., Niola, V., Troncone, S.: An underactuated mechanical hand: a first prototype. In: IEEE RAAD 2014 - Conference Proceedings, 23rd International Conference on Robotics in Alpe-Adria-Danube Region, Slovakia (2014). https://doi.org/10.1109/raad.2014.7002225
13. Rossi, C., Savino, S., Niola, V., Carbone, G., Gasparetto, A., Quaglia, G.: An underactuated mechanical hand prosthesys by IFToMM Italy. In: 14th International Federation for the Promotion of Mechanism and Machine Science World Congress, Taipei, Taiwan (2015). https://doi.org/10.6567/iftomm.14th.wc.ps13.005
14. Cosenza, C., Niola, V., Savino, S.: Underactuated finger behavior correlation between vision system based experimental tests and multibody simulations. Mech. Mach. Sci. **66**, 49–56 (2019). https://doi.org/10.1007/978-3-030-00365-4_7
15. Interlink Electronics FSR400 Series Datasheet. https://www.interlinkelectronics.com/request-data-sheets. Accessed 07 Mar 2019
16. Paredes-Madrid, L., Matute, A., Bareño, J.O., Parra Vargas, C.A., Gutierrez Velásquez, E.I.: Underlying physics of conductive polymer composites and force sensing resistors (FSRs). a study on creep response and dynamic loading. Materials **10**, 1334 (2017)
17. Paredes-Madrid, L., Palacio, C., Matute, A., Parra Vargas, C.: Underlying physics of conductive polymer composites and force sensing resistors (FSRs) under static loading conditions. Sensors **17**, 2108 (2017)
18. Dantec Dynamics. https://www.dantecdynamics.com/q-450-high-speed-dic-vibration-analysis-and-transient-events. Accessed 07 Mar 2019
19. Cosenza, C., Niola, V., Savino, S.: A mechanical hand for prosthetic applications: multibody model and contact simulation. Proc. Inst. Mech. Eng. Part H J. Eng. Med. **232**, 819–825 (2018). https://doi.org/10.1177/0954411918787548

Study on the Activation Speed and the Energy Consumption of "Federica" Prosthetic Hand

Daniele Esposito[1,2(✉)], Sergio Savino[3], Chiara Cosenza[3], Gaetano Dario Gargiulo[4], Antonio Fratini[5], Giuseppe Cesarelli[6,7], and Paolo Bifulco[1,2]

[1] Department of Electrical Engineering and Information Technologies, University "Federico II" of Naples, Via Claudio, 21, 80125 Naples, Italy
{daniele.esposito,paolo.bifulco}@unina.it
[2] Istituti Clinici Scientific Maugeri S.p.a.—Soc. Benefit, Via S. Maugeri, 4, 27100 Pavia, Italy
[3] Department of Industrial Engineering, University "Federico II" of Naples, Via Claudio, 21, 80125 Naples, Italy
{sergio.savino,chiara.cosenza}@unina.it
[4] The MARCS Institute, Western Sydney University, Penrith, NSW 2751, Australia
g.gargiulo@uws.edu.au
[5] School of Life and Health Sciences, Aston University, Birmingham B4 7ET, UK
a.fratini@aston.ac.uk
[6] Department of Chemical, Materials and Production Engineering, University "Federico II" of Naples, Piazzale Tecchio, 80, 80125 Naples, Italy
giuseppe.cesarelli@unina.it
[7] Istituto Italiano di Tecnologia - Center for Advanced Biomaterials for Healthcare, Largo Barsanti e Matteucci 53, 80125 Naples, Italy

Abstract. Important features of a hand prosthesis are certainly the comfort in wearing it, the ease of use, the activation speed, the low energy consumption and no less important the anthropomorphic aspect. This study focused on the activation speed and the energy consumption of an under-actuated, low-cost, active hand prosthesis named "Federica". The prosthesis is moved by a single servomotor able to rotate 180 degrees. Video acquisitions of complete rotations of the servomotor, when it works freely or fixed to the mechanical components of the prosthesis, were used to compare the different kinematic behaviors of the servomotor. A current sensor was used to measure the absorbed current, i.e. the energy absorption, by the servomotor under different uses of the prosthesis (at rest, grasping objects, raising water bottles, etc.). The comparison between the kinematic behaviors of the servomotor alone or connected to the prosthesis, showed the mechanical efficiency of the prosthesis with very low latencies and small variations in velocity and acceleration profiles. The prosthesis took about half a second from the muscle sensor trigger to the complete closure of the hand, showing a significant speed. Finally, tests on current absorption of the servomotor in various conditions resembling prosthesis daily usage, revealed the capacity to guarantee an autonomy of at least one day when powered by 7.4 V, 3000 mAh battery pack.

J. Henriques et al. (Eds.): MEDICON 2019, IFMBE Proceedings 76, pp. 594–603, 2020.
https://doi.org/10.1007/978-3-030-31635-8_71

Keywords: Muscle controlled hand prosthesis · Grasping speed · Prosthesis energy consumption

1 Introduction

Many features can be considered to assess the performances of active prosthetic hand device [1]. Certainly, appearance with five digits and the color matched to the user's skin, help patient's acceptance of the prosthesis. The device should also be low-weight, easy to be controlled (e.g. by EMG sensors [2–4]) make efficient use of energy (e.g. operate for at least a day using without changing the battery).

Perception of device performance is also based on the user's feeling of not having to wait for the reactivity of the prosthesis with high latencies, and on realistic movements at appropriate speeds. Conventional prostheses that are aesthetically attractive but have too artificial and not anthropomorphic motions, are not desirable by users.

Consequently, users are more likely to accept a hand prosthesis, if it is as anthropomorphic as possible both in shape and in action; it guarantees comfort; it is simple to use with natural movements; it is characterized by proper activation speed; it is low power consumption. Usually, a hand prosthesis that does not meet the requirements considered essential by users, will be rejected soon or later.

This work is focused on the analysis of the activation speed and the energy consumption, of the "Federica" hand, an under-actuated, myo-controlled, prosthetic hand, subject of various other studies previously published [5–9].

Previous studies [10, 11] carried out comparisons of the most well-known hand prostheses (iLimb, Bebionic, Michelangelo, etc.) available on market, reporting specific performances. Hence, it was possible to compare these prosthetic devices in terms of activation speed with the performances of the "Federica" hand. However, no accurate data were found about energy absorption of the above-mentioned commercial prostheses (hindering comparative studies). Concise data declared by the manufacturers of prosthetic devices [12, 13] were not usable for a correct comparison, because they were either too generic (for example, the current absorptions not related to certain actions performed with the prosthesis, but as average consumptions), or in relative units of measurement (for example, the speed of rotation in closing and opening of the prosthesis reported as linear speed (m/s), depending on the reference radius, and not at angular velocity (rad/s)).

2 Materials and Methods

2.1 Kinematic Video Tests on the Prosthesis Motor

Full-HD video acquisitions at 100 fps of a complete rotation of the servomotor arm (Fig. 1a, b) were filmed with a Sony alpha 7III digital camera. Subsequently, the rotation of the prosthesis pulley (Fig. 1c, d) fixed to the same servomotor, was filmed with the settings mentioned above, from the open hand position to the completely closed. This was useful to compare the different kinematic behaviors of the servomotor, when it works freely or linked to the mechanical components of the prosthesis.

The control signals were sent to the servomotor by means of an Arduino board and a LED was used as a light indicator of the servomotor start.

A tool developed in MATLAB allowed the tracking of a red marker applied both on the servomotor arm and on the prosthesis pulley, frame by frame. In this way, it was possible to compute the angular displacements (rad) over time (s) in both the aforementioned cases.

The angular displacements over time were low-pass filtered by means of a Butterworth 2-rd order filter, with 10 Hz cut-off frequency. The angular velocities (rad/s) were obtained as temporal derivatives of the angular displacements, while, the angular accelerations (rad/s^2) as temporal derivatives of the angular velocities.

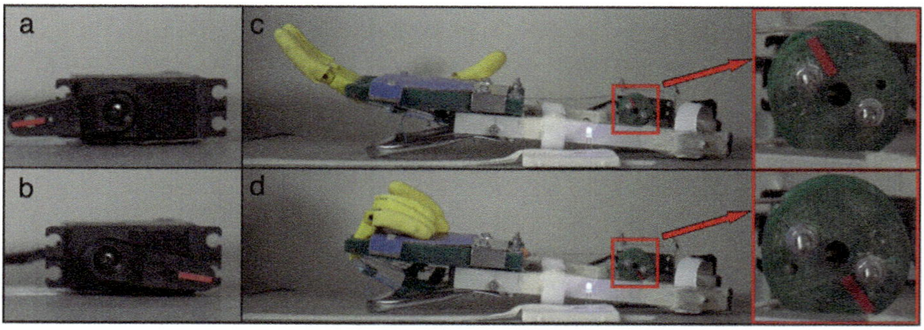

Fig. 1. Setup for video acquisitions: (a) servomotor in start position; (b) servomotor in final position; (c) prosthesis pulley in start position; (d) prosthesis pulley in final position.

2.2 Energy Consumption of the Prosthesis Motor

The underactuated prosthetic hand "Federica" with 15 degrees of freedom [5–9], was moved by a single servomotor (Hitec HSR-5990TG) [14], powered by a 7.4 V battery pack (2x Li-Ion 3000mAh 3.7 V). An Arduino microcontroller board sent control signals to the servomotor by means of digital pulses of variable duration (Pulse Width Modulation), corresponding to the servomotor angular positions of 0 and 180 degrees [15].

The rotation of the servomotor is transmitted through its rotor to the coupled pulley of the prosthesis (Fig. 1c, d), which in turn moves inextensible tendons, allowing the closure and opening of the prosthetic hand.

It was used an Hall-effect current sensor (ACS712) [16] in order to measure the absorbed current by the servomotor, under different conditions of use of the prosthesis (at rest, grasping objects, raising water bottles, etc.). In fact, the greater the load to which the prosthetic hand is subjected, i.e. the servomotor, and the greater the absorbed current by the latter. This sensor detects currents in the ±5A range and has a sensitivity of 185 mV/A.

The servomotor, when holding a position or moving, absorbs current in a non-continuous way (according to its structure and to the servo controller). Thus, the absorbed current continuously shows sudden variations depending on the instantaneous need of energy required by the servo mechanism to hold or to change its angle. Therefore, it was necessary to filter the Hall sensor output signal with a low-pass filter (hardware RC circuit with a 5 Hz cutoff frequency) for the purpose of measuring an averaged value of the absorbed current. The signal was then acquired at 10 kHz sampling frequency with 14-bit precision by means of an acquisition board (National Instruments NI USB-6009). Moreover, it was successively low-pass filtered by means of a Butterworth 3-rd order filter, with 3 Hz cut-off frequency.

The energy consumptions related to the various conditions of use of the prosthetic device, were obtained by integrating the different current absorptions multiplied by the supply voltage for a specific time interval:

$$E = \int_{t}^{t+T} Vcc\, I(t)dt \tag{1}$$

(with E (J) representing the energy consumption, Vcc (V) the constant supply voltage, I (t) (A) the instantaneous current flowing in the servomotor, t (s) the time instant related to the end of the transient phase and the reaching of a stable condition, T (s) integration time: 5 s).

3 Results

Figure 2 shows the results of the servomotor kinematic analysis, when not connected to the prosthesis (Fig. 1a, b). The total angular displacement over time of the rotor (Fig. 2. a) resulted to be about 3.5 radians, and was well represented by a sigmoid regression (S-shape) such as:

$$y = \frac{a}{b + e^{-b(x+c)}} \tag{2}$$

(a = 0.0389, b = 0.01038, c = 126.4 with 95% confidence bound; x = time (ms), y = angle (rad)) with a coefficient of determination R^2 equal to 0.993.

The angular velocity (Fig. 2b) was initially zero, had a rapid rise and remained almost constant at about 8 rad/s, finally, at the end of the rotation, it returned to zero.

The angular acceleration (Fig. 2c) was initially zero, had a rapid rise when the velocity incremented, remained almost null when the velocity was constant, finally, at the end of the rotation, it had a negative peak when the velocity decremented and then returned to zero.

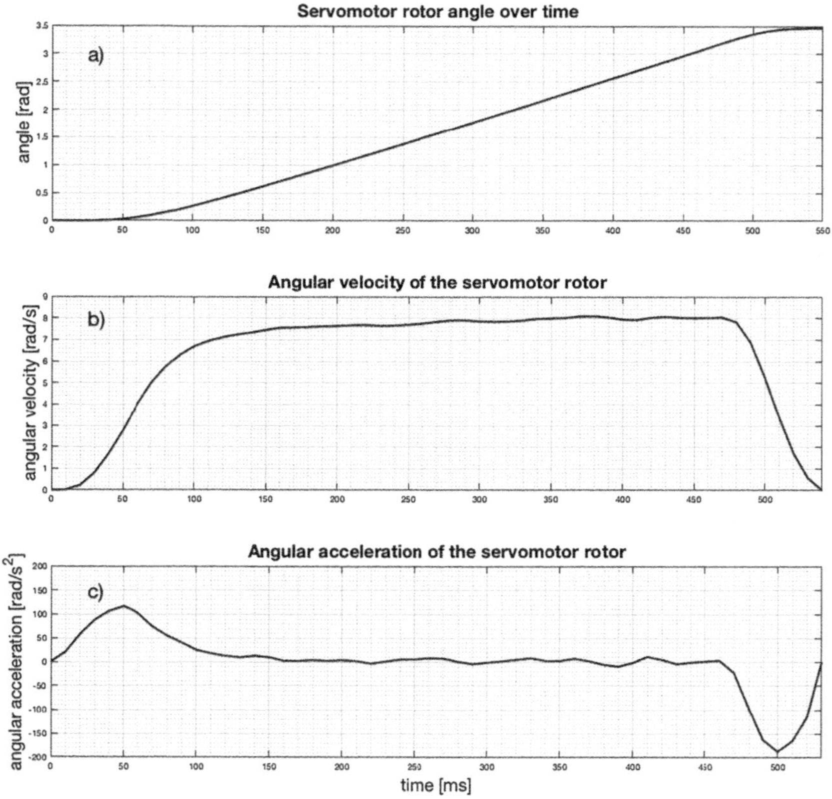

Fig. 2. Kinematic analysis of the servomotor: (a) angular displacement over time of the servomotor rotor; (b) angular velocity of the servomotor rotor; (c) angular acceleration of the servomotor rotor.

Figure 3 shows the results of the kinematic analysis of the prosthesis pulley connected to the servomotor (Fig. 1c, d). The total angular displacement over time of the pulley (Fig. 3a) resulted to be about 3 radians, and was well represented by a sigmoid regression (S-shape) such as:

$$y = \frac{a}{b + e^{-b(x+c)}} \tag{3}$$

(a = 0.03576, b = 0.01133, c = 94.92 with 95% confidence bound; x = time (ms), y = angle (rad)) with a coefficient of determination R^2 equal to 0.994.

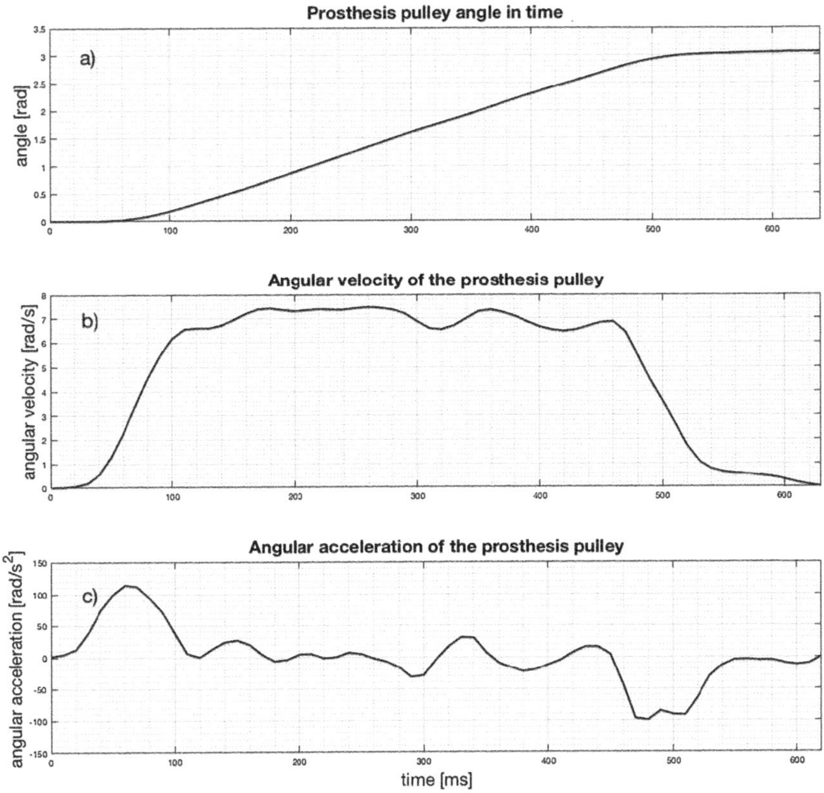

Fig. 3. Kinematic analysis of the prosthesis pulley connected to the servomotor: (a) angle position over time of the prosthesis pulley; (b) angular velocity of the prosthesis pulley; (c) angular acceleration of the prosthesis pulley.

The trend of the angular velocity of the prosthesis pulley (Fig. 3.b) was quite similar to that of the servomotor at the previous point (Fig. 2b), with the differences of being slightly lower (about 7 rad/s) and with greater variations in the central part of the rotation.

Concerning the angular acceleration (Fig. 3c), it resulted to be similar to that of the servomotor (Fig. 2c), especially with regard to the initial and final phases of the rotation, while in the central phase there are much more evident accelerations and decelerations in correspondence to the related variations of velocity.

Figure 4 shows the current absorption trends associated with the various actions, displayed in Fig. 5, performed with the prosthesis.

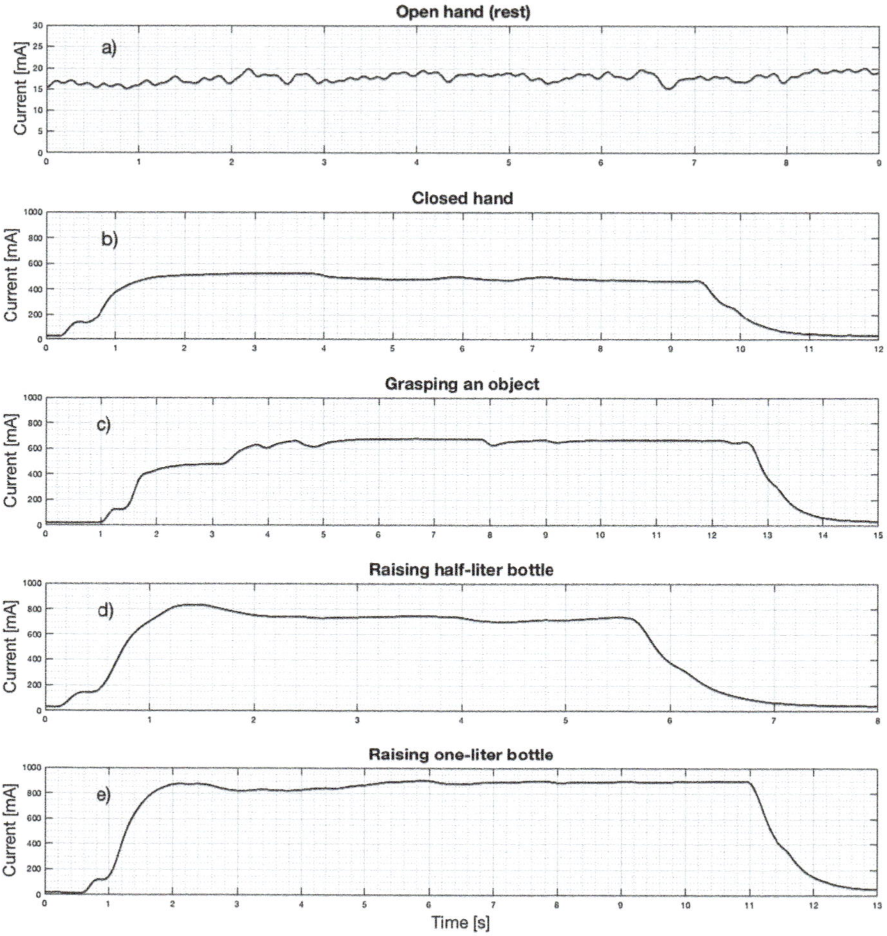

Fig. 4. Trends of the current absorbed by the servomotor in the different actions performed with the prosthesis: (a) open hand (rest condition); (b) closed hand; (c) grasping an object; (d) raising half-liter bottle; (e) raising one-liter bottle.

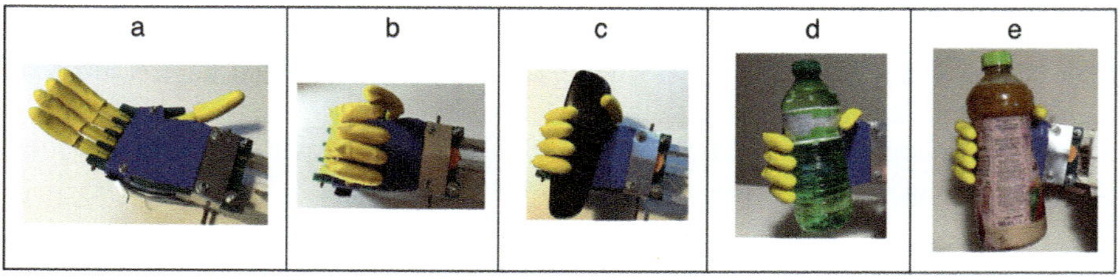

Fig. 5. Images captured during the various tests performed: (a) open hand (rest condition); (b) closed hand; (c) grasping an object; (d) raising half-liter bottle; (e) raising one-liter bottle.

For the time interval T (5 s), corresponding to stable conditions, the average values of current absorptions and the associated energy consumptions were computed for each performed action. The results of these values are shown in Table 1, while in Fig. 6 the related histograms.

The minimum energy absorption occurred in rest condition (less than 20 mA), while in the other tested actions, the maximum absorption achieved (about 860 mA) was obtained by raising a load of 1 kg.

Using batteries of 7.4 V with a capacity of 3000 mAh, the autonomy at rest would be about a week, while constantly raising a load of 1 kg of about 3.5 h. Consequently, considering various daily activities performed by the patient, an autonomy for at least an entire day is predicted.

Table 1. Average values of current absorptions and energy consumptions (in 5 s) for each performed action

Prosthesis activity	Average current absorption [mA]	Energy consumption in 5 s [J]
Open hand (rest)	17.6	0.650
Closed hand	498	18.4
Grasping an object	665	24.6
Raising half-liter bottle	730	27.0
Raising one-liter bottle	859	31.8

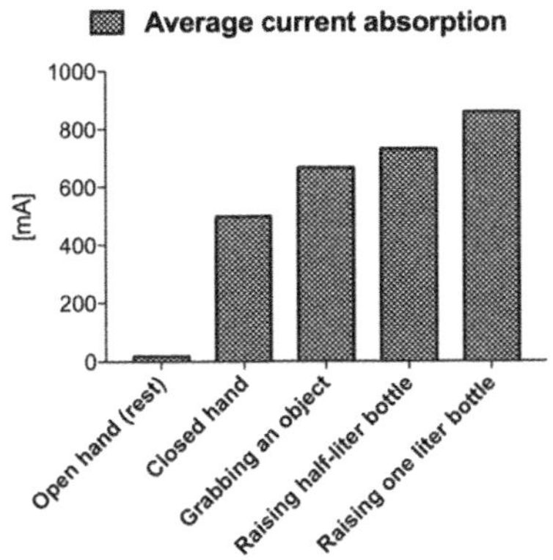

Fig. 6. Histogram showing the average values of current absorptions presented in Table 1.

4 Conclusions and Discussion

This study showed the different kinematic behaviors of the servomotor, actuator of the "Federica" hand prosthesis, when working freely or linked to the mechanical system of the prosthesis. The mechanical efficiency of the prosthesis in the transmission of movement is proved in terms of imperceptible latencies and not noticeable variations in the velocity and acceleration profiles of the motor. The high grasping speed offered by this prosthetic device (about half a second, from the muscle sensor trigger, for a complete closure of the hand) is remarkable, even comparing these performances with the most recent and innovative prostheses on the market (0.35 s for Michelangelo, 0.80 s for iLimb Quantum, 1 s for BiBionic V3) [11].

Regarding the energy absorption of the prosthesis motor in the various conditions of use considered, the estimated autonomy will guarantee an efficient use for at least an entire day.

Conflict of Interest. The authors declare that they have no conflict of interest.

Statement of Informed Consent. All subjects gave their informed consent for inclusion before they participated in the study.

References

1. Muzumdar, A.: Powered upper limb prostheses: control, implementation and clinical application, pp. 176–177. Springer Science & Business Media, Ottawa (2004)
2. Polisiero, M., Bifulco, P., Liccardo, A., Cesarelli, M., Romano, M., Gargiulo, G.D., McEwan, A.L., D'Apuzzo, M.: Design and assessment of a low-cost, electromyographically controlled, prosthetic hand. Med. Dev. Evid. Res. **6**, 97–104 (2013). https://doi.org/10.2147/MDER.S39604
3. Ulloa, G.D.F., Sreenivasan, N., Bifulco, P., Cesarelli, M., Gargiulo, G., Gunawardana, U.: Cost effective electro resistive band based myo activated prosthetic upper limb for amputees in the developing world. In: Life Sciences Conference (LSC) 2017, pp. 250–253. IEEE (2017)
4. Neethu Sreenivasan, N., Ulloa Gutierrez, D.F., Bifulco, P., Cesarelli, M., Gunawardana U., Gargiulo, G.D.: Towards ultra low-cost myoactivated prostheses. BioMed Research International **2018**, 14 (2018). Article ID 9634184. https://doi.org/10.1155/2018/9634184
5. Rossi, C., Savino, S., Niola, V., Troncone, S.: An underactuated mechanical hand: a first prototype. In: IEEE RAAD 2014, Conference Proceedings, 23rd International Conference on Robotics in Alpe-Adria-Danube Region, Slovakia, 7002225 (2014). https://doi.org/10.1109/raad.2014.7002225
6. Rossi, C., Savino, S., Niola, V., Carbone, G., Gasparetto, A., Quaglia, G.: An underactuated mechanical hand prosthesys by IFToMM Italy. In: 14th International Federation for the Promotion of Mechanism and Machine Science World Congress, Taipei, Taiwan (2015). https://doi.org/10.6567/iftomm.14th.wc.ps13.005
7. Bifulco, P., Esposito, D., Gargiulo, G.D., Savino, S., Niola, V., Iuppariello, L., Cesarelli, M.: A stretchable, conductive rubber sensor to detect muscle contraction for prosthetic hand control. In: Proceedings of the IEEE E-Health and Bioengineering Conference (EHB), 22–24 Junes Sinaia, Romania (2017). https://doi.org/10.1109/EHB.2017.7995389
8. Cosenza, C., Niola, V., Savino, S.: A mechanical hand for prosthetic applications: multibody model and contact simulation. Proc. Inst. Mech. Eng. Part. H J. Eng. Med. **232**, 819–825 (2018). https://doi.org/10.1177/0954411918787548
9. Esposito, D., Andreozzi, E., Fratini, A., Gargiulo, G.D., Savino, S., Niola, V., Bifulco, P.: A piezoresistive sensor to measure muscle contraction and mechanomyography. Sensors **18**(8), 2553 (2018). https://doi.org/10.3390/s18082553

10. Belter, J.T., Segil, J.L., Dollar, A.M., Weir, R.F.: Mechanical design and performance specifications of anthropomorphic prosthetic hands: a review. J. Rehabil. Res. Dev. **50**(5), 599–618 (2013). https://doi.org/10.1682/JRRD.2011.10.0188
11. Vujaklija, I., Farina, D., Aszmann, O.: New developments in prosthetic arm systems. Orthop. Res. Rev. **8**(1), 31–39 (2016). https://doi.org/10.2147/ORR.S71468
12. SteeperGroup website. https://www.steepergroup.com/SteeperGroup/media/SteeperGroup Media/Prosthetics/Upper%20Limb%20Prosthetics/Steeper-Upper-Limb-Catalogue.pdf. Accessed 18 Mar 2019
13. Ottobock website. https://www.ottobock.co.uk/media/local-media/brochures/prosthetics/catalogues/upper-limb-catalogue-646k6-masteren_2014.pdf. Accessed 18 Mar 2019
14. Robotstorehk website. http://www.robotstorehk.com/motors/doc/HSR-5990TG.pdf. Accessed 16 Mar 2019
15. Xie, M., Xiong, Y., Xiong, C., Hu, Z.: Intelligent Robotics and Applications: Second International Conference, ICIRA 2009, Proceedings, Singapore, pp. 537–538 (2009). https://doi.org/10.1007/978-3-642-10817-4
16. Allegro micro website. https://www.allegromicro.com/~/media/files/datasheets/acs712-datasheet.ashx. Accessed 16 Mar 2019

New Method to Analyze the Load Propagation on the Plantar Foot Surface During a Walk/Run Using the Smart Sock System

Alexander Okss[1(✉)], Alexei Katashev[2], Peteris Eizentals[2], Sandra Rozenstoka[3], and Dace Suna[3]

[1] Institute of Design Technologies, Riga Technical University, Riga, Latvia
Aleksandrs.Okss@rtu.lv
[2] Institute of Biomedical Engineering and Nanotechnologies, Riga Technical University, Riga, Latvia
[3] Sport Laboratory, Riga, Latvia

Abstract. A new version of DAid® Pressure Sock System (DPSS) for gate monitoring and new method of data processing and analysis are presented. Proposed DPSS has 6 sensing elements, electronic device with sampling rate up to 200 Hz and is able to monitor walking/running gates in outdoor conditions. According to a new method the process of plantar loading during the gate is considered similar to seismic wave propagation and sensors assumed as the reference points. As a result, harvested data after being processed, are represented in the form of loading wave plots (WPs). Obtained WPs show load propagation across the sensors during the stance/gate. Primary tests of developed system and method were made for different types of walk/run and correspondence of conclusions from WPs analysis and diagnoses made by sport medicine doctor during walk/run tests confirms system efficacy and validity of results.

Keywords: Gate monitoring · Smart socks · Data analysis

1 Introduction

Gait is one of the most frequently used forms of human movement during daily activity. Foot pathologies, injuries or overloading can impact foot function during the gait cycle. Accurate reliable knowledge of gait characteristics, monitoring and evaluating them over time, will enable early diagnosis of diseases and their complications. Nowadays gate monitoring and plantar pressure sensing technology has become an indispensable tool for gait analysis in both clinical and research settings. Many different pressure sensing systems are currently available in the market. These systems are based on variety of pressure sensors applied and can be primarily categorized into platform systems and in-shoe systems [1].

Platform systems are comprised of a large number of sensors arranged in a matrix formation, and generally offer a high spatial resolution for the measurement of plantar pressure [2]. However, this kind of systems can be used only in a laboratory/hospital, and only for barefoot measurements.

J. Henriques et al. (Eds.): MEDICON 2019, IFMBE Proceedings 76, pp. 604–609, 2020.
https://doi.org/10.1007/978-3-030-31635-8_72

Much more flexible and applicable for out-of-laboratory and even for outdoor applications are in-shoe systems. Such systems provide possibilities to record the plantar pressure distributions within a shoe and mainly include insoles and socks with imbedded or integrated pressure sensors [3, 4].

Pressure sensitive insoles (PSI) represent wide variety of devices from commercially evaluable, such as F-Scan®, Medilogic® and Pedar® systems, and experimental custom-made devices [5] etc. with different sensors types and quantity. Despite the variety of PSI, all of them have one common drawback–insoles form an additional elastic layer inside the shoe which thickness can be up to several millimeters. Such layer can distort the real data of plantar loading of the foot. In addition, these systems are comparatively expensive and have low reliability to bending load (cannot be used in footwear with supinator).

Abovementioned drawbacks can be overcome by using textile-based systems - such as socks with integrated sensors (Smart socks). This approach was used to develop sock-based systems in medical [6] and sport [7] applications. Unfortunately, most of already developed versions of Smart socks are handmade or utilize complicated manufacturing technology.

The research team of Riga Technical University on the basis on newly developed technology have proposed DAid® Pressure Sock System (DPSS) [4]. The main advantage of DPSS is the possibility to produce textile part of the system (socks with knitted in pressure sensors and electroconductive pathways) using ordinary knitting sock machines. Previous version of DPSS had five knitted in pressure sensors and custom-made electronic devices, that were attached to the socks. The devices provided data harvesting, acquisition and transfer via Bluetooth® to remote computer. DPSS sampling frequency was up to 30 Hz per channel. System tests showed reliable results of gate analysis in case of walking speed up to 4 km/h, which coincides with conclusions [5].

Present paper describes further developments of DPSS and algorithm of gate analysis, shows that new version of the system provides possibility to monitor walking/running gates without limitations of motion type/pace, as well as demonstrates new possibilities to analyze gate patterns.

2 Materials and Methods

2.1 DPSS Structure and Properties

The new version of DPSS consists of the same main elements as previously designed: textile part (socks with knitted in pressure piezoresistive sensors and conductive pathways (Smart socks)), electronic part (data acquisition and transferring systems) and software for data analysis and visualization. In contrast to previous version, due to optimization of Smart socks design and knitting technology number of sensors on the sole part of the socks was increased up to 6 sensors and conductive pathways were redesigned to provide a placement of all sock- electronic device connectors only on the one (lateral) part of socks' cuff. To ensure ability to monitor gate events during fast walking or running, a new version of electronic device was developed as well.

New device is able to collect data with sampling frequency up to 200 Hz for up to 8 h, which is sufficient for rehabilitation or sport trainings.

2.2 Data Processing

Direct analysis of sock sensors signals is a complex and time-consuming procedure and needs highly qualified and experienced observer. To provide effective analyzing tool for signals harvested by DPSS a new method of data processing and visualization is proposed. According to this method the process of foot loading is considered similar to seismic wave propagation. Sensors play role of reference points where loading change versus time is monitored and recorded. To use the method, recorded data must be preprocessed. At first data harvested from each sensor are inverted and then normalized using local normalization with a sliding window according to equation:

$$u_i^{'} = \left(\frac{u_i^{-1} - \min_{a \le i \le b} u_i^{-1}}{\max_{a \le i \le b} u_i^{-1} - \min_{a \le i \le b} u_i^{-1}} \right), \tag{1}$$

where u_i is the corresponding data value and *{a;b}* is a window around *i*-th measurement to which to normalize the signal. Then sample entropy [8] of each sensors' normalized data ($SampEn_i$) is calculated to estimate sensors activity during the gate.

Sample entropy shows signal complexity and has the higher value the more complex signal is analyzed. In our case the less is sensor loading along the gate (for example, heel sensors in case of toe running), the higher is "noisy part" and correspondently complicity of a signal, the higher is $SampEn_i$ value. Evidently, that $SampEn_i$ values are specific for each type of a gate. Thus, values of SampEn_i, calculated for specific gate, can be used as a measure of sensors activity during this gate. Basing on this consideration we proposed to use inversed values of $SampEn_i$ as weight coefficients of sensors activity estimation and apply them to get weighted normalized values of sensors signals:

$$u_i^{''} = u_i^{'}/SampEn_i \tag{2}$$

Further data processing includes application of thresholding procedure to values $u_i^{''}$ calculated by Eq. (2) to exclude "false" steps due to noise or misstep and finalizes by averaging of temporal length of remaining "correct" steps. According to proposed method, processed data are presented as 2D plot which simultaneously performs all sensors loading along the time during one whole gate (or one stance) period. Such plots will be referred further as "wave plots" (WPs). WPs can represent as separate steps, as well as the pattern, obtained by averaging data from step series. Examples of WPs for averaged stance patterns obtained for several different walking and running types are shown in Fig. 1. (For detailed description of Fig. 1, look up Sect. 2.4.)

2.3 Experimental Setup

To verify applicability of a proposed method a set of experiments was made. A volunteer (female, 27 y.o, BM 61 kg) walked/ran on the treadmill with different pacing and gait types while gate data had been gathered using DPSS. Simultaneously volunteer's movement was being observed by a sport medicine doctor (SMD). SMD made notes concerning features of every walking/running type. Analyzed gate types, their description and gate features, diagnosed by SMD, are summarized in Table 1.

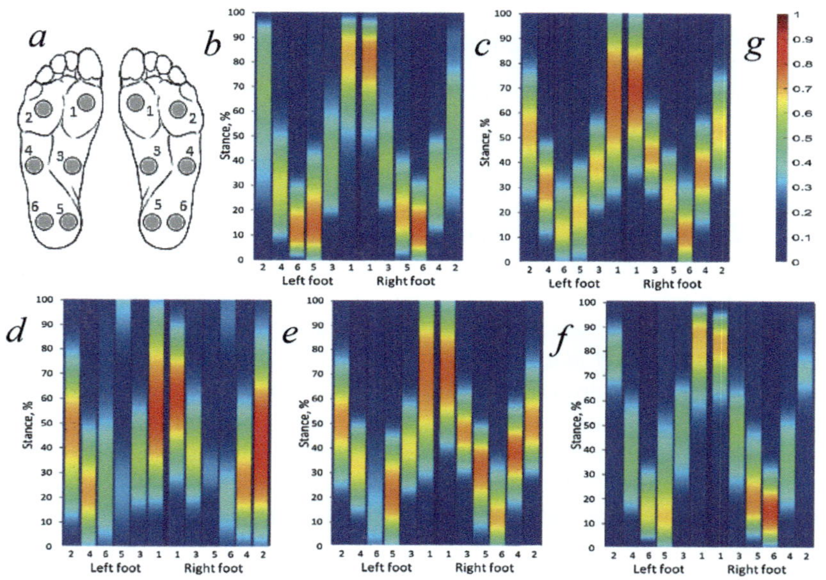

Fig. 1. Wave plots: *a*-sensors placement and numbers, *b*-warm-up walk (case 1), *c*-heel running (case 2), *d*-toe running (case 3), *e*-midfoot running (case 4), *f*-cooldown walk (case 5), *g*-color scale of relative load. Numbers on the horizontal axis correspond to sensors numbers.

Table 1. Gate types and observed features

Case No.	Gate type	Gate characteristic	Gate features, diagnosed by sport medicine doctor (SMD)
1	Walk	Warming up, 5 m/s	Weight bearing during terminal stance on the lateral side of transverse arch.
2	Run	Heel running, 9 m/s	Running step begins by heel striking; flat foot running; left ankle in valgus position/pronation of the foot; the right foot in outer rotation/lateralization during preswing phase
3	Run	Toe running, 11 m/s	Running on tiptoes; right side toe lateralization during toe off phase; foot in pronation; knees in dynamic valgus position
4	Run	Midfoot running, 9 m/s	Flat foot running
5	Walk	Cooling down, 5 m/s	Weight bearing tends to be on right side for balance.

2.4 Results and Discussion

Data, harvested by using DPSS, were processed according to algorithm described above and corresponding WPs were obtained (see Fig. 1). As can be seen from the figures, WP of each foot consists of subWPs constructed for each separate sensor. These subWPs are placed on the total WP according to sensors placement on the Smart sock sole (see Fig. 1). All subWPs are marked by corresponding sensor number below the plot. Vertical direction corresponds to temporal axis and represents stance/gate temporal duration. Colors on the plot reflect relative value of sensor load u_i'' according to provided color scale (Fig. 1g).

Thus, WPs show temporal propagation of feet plantar load during stance/gate and can be useful tool for gate analysis. Let us apply this tool to analyze one of recorded gates, for example, case 3 (see Table 1).

Case 3 (Fig. 1d) WPs of toe running show essential loading of forefoot sensors 1 and 2 and very weak loading of heel sensors, which coincides with SMD diagnosis of "*tiptoe running*". WPs also show asymmetry of gate-higher level of loading of right foot, especially laterally, which also confirms by SMD diagnosis: "*right side toe lateralization during toe off phase*".

WPs of other cases also coincide SMD observations.

In summary, a new version of Smart Sock Pressure System is developed. It provides possibility to monitor different types of walk/run gates in out of lab conditions. High sampling frequency allows to study in details short time events of gate.

A new method of wave plots for gate data visualization and analysis is proposed. The method is verified by comparison of results of WPs analysis with medical conclusions. It was shown that method provides reliable results and can be used as an effective tool for gate analysis.

Acknowledgements. This research is co-financed by the ESF within the project «Synthesis of textile surface coating modified in nano-level and energetically independent measurement system integration in smart clothing with functions of medical monitoring», Project implementation agreement No. 1.1.1.1./16/A/020.

Conflicts of Interest. Hereby authors declare, on the best of their knowledge, absence of any conflicts of interest related to the topic of the present paper.

References

1. Abdul Razak, A.H., Zayegh, A., Begg, R.K., Wahab, Y.: Foot plantar pressure measurement system. Rev. Sens. **12**, 9884–9912 (2012)
2. Orlin, M.N., McPoil, T.G.: Plantar pressure assessment. Phys. Ther. **80**, 399–409 (2000)
3. Bus, S.A., Haspels, R., Busch-Westbroek, T.E.: Evaluation and optimization of therapeutic footwear for neuropathic diabetic foot patients using in-shoe plantar pressure analysis. Diabetes Care **34**, 1595–1600 (2011)

4. Okss, A., Kataševs, A., Zadinans, M., Rancans, M., Litvak, J. Development of smart sock system for gate analysis and foot pressure control. In: IFMBE Proceedings: 14th Mediterranean Conference on Medical and Biological Engineering and Computing (MEDICON 2016), vol. 57, pp. 466–469. Springer, Cham (2016)
5. Lopez-Meyer, P., Fulk, G.D., Sazonov, E.S.: Automatic detection of temporal gait parameters in poststroke individuals. IEEE Trans. Inf. Technol. Biomed. **15**, 594–601 (2011)
6. Perrier, A., Vuillerme, N., Luboz, V., et al.: Smart diabetic socks: embedded device for diabetic foot prevention. Innov. Res. Biomed. Eng. **32**(2), 72–76 (2014)
7. Sensoria Fitness. http://www.sensoriafitness.com/smartsocks/. Accessed 21 Mar 2019
8. Ahmadi, S., Sphere, N., Wu, C., Szturm, T.: Sample entropy of human gait center of pressure displacement: a systematic methodological analysis. Entropy **20**(579), 1–17 (2018)

Intergame Analysis of Upper Limb Biomechanics of Stroke Patients in Real and Virtual Environment

Herta Costa[1], Aline Fernandes[1], Débora Oliveira[1], Jamilson Brasileiro[1], Tatiana Ribeiro[1], Edgar Vieira[2], and Tania Campos[1(✉)]

[1] Universidade Federal do Rio Grande do Norte, Natal, Brazil
campostf@gmail.com
[2] Florida International University, Miami, USA

Abstract. The aim of this study was to realize an intergame analysis of upper limb biomechanics of stroke patients in real and virtual environment. Methods: The sample consisted of 11 hemiparetic patients, mean age of 51 ± 7 years. Participants made 15 attempts in two dart games (real and virtual). Elbow kinematics was video recorded during the dart throwing phase. Analysis was conducted using Kinovea software, paired Student's t-test and Classification Regression Trees. Results: Patients exhibited a higher elbow extension angle ($p = 0.008$) and greater velocity in the real game ($p = 0.005$). In the virtual game patients had longer throwing time ($p = 0.021$) and better performance (fewer absolute errors) ($p < 0.0001$). The decision tree showed that there was a balance between the frequency of patients who played the virtual and real game and displayed elbow extension angles above 157°. Similar frequencies between velocity = 29 cm/s and >87 cm/s for the virtual and real games were found. In regard to dart throwing time, there was greater frequency of patients with time =1.37 s for the real game and >1.37 s for the virtual game. Conclusion: The patients can evolve satisfactorily in terms of angulation, velocity and time during virtual game training. Thus, we propose that the virtual dart game may be a useful tool in the neurorehabilitation of patients with chronic stroke, in line with therapeutic objectives and the patient's clinical condition.

Keywords: Stroke · Upper extremity · Virtual reality · Movement · Biomechanics analysis · Neurorehabilitation

1 Introduction

Stroke is a vascular disease with a high incidence rate and an important cause of neurological impairment [1]. Patients exhibit motor control disorders and as a result, do not produce coordinated and efficient movements in their activities of daily living (ADL) [2]. The recovery of functionality is seen as a significant problem after stroke and continues to be a challenge for neurological rehabilitation. Kinematic measures allow assessment of coordination and motor control using angulation, speed, time and distance of movement trajectories during execution of motor skills [3]. Kinovea

J. Henriques et al. (Eds.): MEDICON 2019, IFMBE Proceedings 76, pp. 610–617, 2020.
https://doi.org/10.1007/978-3-030-31635-8_73

software is a bidimensional video analysis system that can be used by health professionals in kinematic analysis, given its proven inter and intraexaminer reliability [4].

Specialists in rehabilitation have begun to show interest in virtual reality (VR) technologies as a tool to analyze motor control and treat motor deficiencies secondary to stroke [5]. Virtual reality is a multisensory interactive simulation of tasks that can be used in different immersion levels, in 3D (cave systems or head mounted display) and 2D (computer or television screen) [6]. Studies that compare the effects of game training in virtual and real environments on the kinematic patterns of the upper limb are scarce, primarily using decision tree analysis. This analysis helps identify groups, discover relationships between them and predict future events intuitively by applying algorithms, in order to guide the choice of game types in motor rehabilitation programs [7].

As such, the aim of this study was to do an intergame analysis of upper limb biomechanics of stroke patients in real and virtual environment, in order to contribute for neurorehabilitation.

2 Methods

2.1 Participants

The sample consisted of 11 patients with stroke. The sample size was calculated with an effect size of 1.92, alpha of 0.05 and power of 0.99. Eligibility criteria were individuals with clinical diagnosis of stroke, unilateral brain lesion, hemiparesis, time since stroke of more than 6 months, age between 40 and 65 years, Mini-mental State Exam (MMSE) score above 23 points, and the ability to remain in the orthostatic position and perform elbow extension. The exclusion criteria were other neurological diseases, uncorrected hearing or visual impairment, symptoms of malaise, excessive fatigue, pain, hypertensive peak, difficulty understanding game instructions and requesting to leave the study.

2.2 Procedures

This study was approved by institutional Ethics Committee in Research. All participants signed an informed consent form. Patients were evaluated for cognitive status by Mini-mental State Exam (MMSE), neurological impairment level by National Institute Health Stroke (NIHSS), motor function by upper extremity Fugl-Meyer Assessment Scale. After clinical assessments, patients underwent dart game training in a virtual and real environment. The game order was established randomly via a simple draw in order to avoid the influence of one of the games.

The individuals were instructed to bend their elbow as much as possible and place their hand towards the center of the board. When throwing the dart, they were asked to extend their elbow to the maximum. Fifteen throws were made in each game, with a two-minute interval between games. Before the start of the game, participants chose their avatar on the console. During the games, the camera was placed on a tripod perpendicular to the plane of movement on a flat surface, at a height of one meter and four meters away from the individuals.

Patients wore a sleeveless shirt onto which yellow reflective paper markers were placed on the greater tubercle of the humerus, lateral epycondyle of the humerus and the radial styloid process. The virtual game consisted of the Xbox 360 Kinect Sports Season 2 dart game. This device is connected to a 52-in LCD TV placed 2 m from the participants. For the game in a real environment, a 45 cm wide and 4 cm thick wooden professional dart board, covered in suede was used, along with a set of steel-tipped darts. The board was mounted on a wall at the same height as the TV used in the virtual game.

Kinematic analysis was carried out during the game using a digital camcorder. The individuals remained standing and the elbow extension movement was filmed in the sagittal plane. The film was then converted to AVI (Audio Video Interleave) file and exported to Kinovea software for analysis of kinematic variables (elbow extension angle, mean dart velocity and throwing time).

To determine the elbow extension angle, the arm segment was defined as the line formed between the greater tubercle of the humerus and the lateral epicondyle of the humerus; the forearm segment was defined as the line formed between the lateral epicondyle and the radial styloid process. The junction of the segments determined the articular angle, as follows: the elbow extension angle obtained from the association between the "arm" and "forearm" segments.

During video editing in Kinovea software, the fixed points placed on the upper limb markers were located and the system calibrated, supplying the program with the arm length of each individual. The elbow extension angle was shown in degrees, mean elbow extension velocity in centimeters per second and game execution time in seconds.

Performance on the dart game was analyzed by calculating the absolute error (AE) of each attempt, that is, AE = Σ total score/15 (no. of attempts). The score was attributed according to dart location, ranging from 0 to 6. A score of 0 was given when individuals hit the center of the board and between 1 and 6 when they did not.

2.3 Statistical Analysis

Data analysis was carried out using SPSS 22.0 (Statistical Package for Social Science) assigning a 5% significance level for all statistical tests. The Shapiro-Wilk test was applied to check whether data were normally distributed. Next, the paired Student's t-test was used to compare kinematic variables (angle, velocity and time) and performance (absolute error) between virtual and real games. We used Classification Regression Trees (CART), in order to assess how the type of game could be selected for training patients with stroke. The cutoff points found may serve to distinguish the evolution of kinematic measures for each type of game during training and their effects on performance. CART analysis was pruned to the most parsimonious model within one standard prediction error of the tree with the smallest prediction error.

3 Results

The sample consisted of 8 men and 3 women, mean age of 51.45 ± 6.74 years. Sociodemographic, anthropometric and clinical data are illustrated in Table 1.

A significant intergame difference was observed in the maximum elbow extension, the real game exhibiting the higher angle ($p = 0.008$) (Fig. 1A) and greater velocity (p= 0.005) (Fig. 1B). Elbow extension time showed a significant between-game difference, with virtual game participants taking longer to throw the darts (p= 0.021) (Fig. 1C), but committing fewer errors ($p < 0.0001$) (Fig. 1D).

The decision tree constructed for the elbow extension angle contained three nodes with different cutoff points. There was greater frequency of patients with angles $\leq 145°$ and less frequency of those with angles between 145° and 157°. However, there was a balance between the frequency of patients who played the virtual and real game and displayed elbow extension angles above 157°.

During construction of the decision tree for elbow extension velocity, 5 nodes were found with similar frequencies between velocity ≤ 28.8 cm/s and above 86.5 cm/s for the virtual and real games.

In regard to dart throwing time, the decision tree identified 2 nodes, with greater frequency of patients with time ≤ 1.37 s for the real game and above 1.37 s for the virtual game.

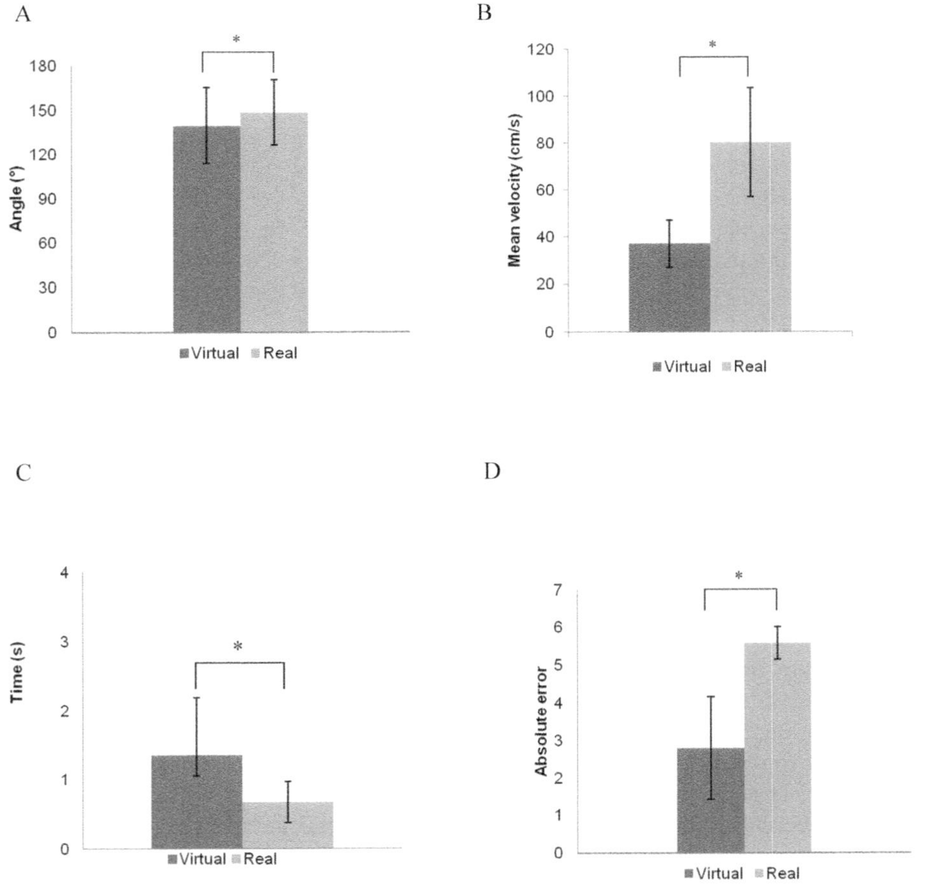

Fig. 1. Significant intergame difference in the elbow extension angle (A), velocity (B), time (C) and performance (D) between real and virtual environment.

4 Discussion

The present study compared elbow kinematics in a virtual versus real environment and constructed decision trees, enabling an important step in understanding how the virtual and physical game can be used in the motor rehabilitation of individuals with post-stroke hemiparesis.

In the real game there was a larger elbow extension angle during dart throwing than in the virtual game. This may have occurred because the virtual dart game provides visual feedback on the dart's final trajectory after being thrown, by displaying a target on the screen. Thus, information on the target and its spatial location may have facilitated the task for patients, causing them to produce lower elbow extension angles in the virtual environment. On the other hand, the lack of this information in the real game may have obliged patients to increase their elbow extension angle to better throw the dart. The coupling of visual feedback (i.e., sight position on the target) with kinesthetic and proprioception feedback (due to arm position) could be useful to re-learn to perform some movements on the initial stages of the neurorehabilitation [8].

Analysis conducted in the present study showed a higher mean elbow extension velocity during dart throwing in the real game. A study with healthy individuals reported that kinematics and muscle activation of the hand is influenced by the final objective of the action and the use of concrete objects [9]. In this respect, the presence of a concrete object in the real game may have contributed to the selection of a kinematic pattern and influenced the velocity of the movement, which did not occur when patients played the virtual game. On the other hand, the lower game velocity in a virtual environment may also represent smoother movement through a strategy of interarticular coordination in an attempt at facilitating the pattern of coordinated movement [10].

In relation to time to complete the games, patients took longer in the virtual game. Studies have shown that the use of virtual games in the training of these patients improves the motor control of the affected arm and its use in daily activities. Patients may have used the strategy to decrease velocity, thereby increasing dart throwing time to adjust the movement and more accurately control their degrees of freedom [11]. Thus, the slower time observed in the virtual environment may be a useful alternative to achieve greater control of movement in individuals with severe motor impairment [10].

Analysis of performance showed that virtual game players had fewer absolute errors than in the real game. The virtual environment may provide greater patients interaction with the activity, helping improve their performance on the task [12]. This results from multisensory stimuli and visual feedback during the game, producing better linkage between the spatial mechanisms of perception-action [13]. Additionally, successful outcomes on the performance of a virtual game task can increase participants'self-confidence and motivation to remain with the treatments promoting a reorganization of neural networks induced by practice [14–16].

The decision tree results suggest that virtual game patients may have increased their elbow extension angle and decreased throwing time during training. In this respect, patients may have made successive corrective movements to hit the target in the virtual game. There is evidence in the literature that training in a virtual environment produces more changes in individuals with moderate and severe motor impairment. Moreover, they manage to develop a motor recovery pattern with less movement compensation compared to training in a real environment [14]. Angles between 145° and 157° and throwing time of around 1.37 s can serve as reference values and eligibility criteria for virtual games in patients with upper limb impairment to be beneficial, and may indicate the degree of clinical evolution in patients.

The decision tree for velocity showed the best stratification index in homogeneous subgroups of patients, for both the virtual and real game. Velocity is a modulation parameter of agonist-antagonist muscle activity, which results in individuals' performing better on a task involving the paretic upper limb [15], suggesting the importance of including this measure in the clinical assessment of patients with stroke.

This study had limitations with regard to the inability to verify correlate affected brain areas, since neuroimaging exams were conducted in different hospitals, not allowing standardization of medical reports. Despite these limitations, the findings convincingly identify the kinematic variables that best demonstrate the importance of each game.

Table 1. Anthropometric and clinical characterization of the sample.

Variables	Mean (SD)
Weight (kg)	72 (15)
Height (cm)	165 (0.14)
Length (cm)	
Arm	29 (3)
Forearm	26 (2)
MMSE (score)	26 (2)
NIHSS (score)	3 (1)
FUGL-MEYER (score)	51 (17)

MMSE: Mini-mental State Exam; NIHSS: National Institute Health Stroke; FUGL-MEYER: Fugl-Meyer Assessment Scale.

5 Conclusion

The present study showed that angle and velocity were better in the real game, while task duration and performance were better in the virtual game. The construction of decision trees showed evidence that the patients can evolve satisfactorily in terms of angulation, velocity and time during virtual game training.

Thus, we propose that the virtual dart game may be a useful tool in the neurorehabilitation of patients with chronic stroke, in line with therapeutic objectives and the patient's clinical condition. However, further studies are needed to understand how the kinematic variables studied should be handled in long-term intervention protocols, in order to optimize upper limb function in individuals post-stroke.

Acknowledgements. This study was supported in part by the Coordenação de Aperfeiçoamento de Pessoal de Nível Superior - Brasil (CAPES) - Finance Code 001 and by the National Council for Scientific and Technological Development (CNPq) [grant numbers 477291/2013-9].

Conflict of Interest. The authors declared no potential conflicts of interest with respect to the research, authorship, and/or publication of this article.

References

1. GBD 2016 Neurology Collaborators. Global, regional, and national burden of neurological disorders, 1990–2016: a systematic analysis for the Global Burden of Disease Study 2016. The Lancet Neurology pii: S1474-4422(18)30499-X (2019). https://doi.org/10.1016/s1474-4422(18)30499-x. [Epub ahead of print]
2. Lang, C.E., Bland, M.D., Bailey, R.R., Schaefer, S.Y., Birkenmeier, R.L.: Assessment of upper extremity impairment, and activity following stroke: foundations for clinical decision making. J. Hand Therapy **26**(2), 104–115 (2013)
3. Bigoni, M., Baudo, S., Cimolin, V., Cau, N., Galli, M., Pianta, L., Mauro, A.: Does kinematics add meaningful information to clinical assessment in post-stroke upper limb rehabilitation? a case report. J. Phys. Therapy Sci. **28**(8), 2408–2413 (2016)
4. El-Raheem, R.M.A., Kamel, R.M., Ali, M.F.: Reliability of using Kinovea program in measuring dominant wrist joint range of motion. Trends Appl. Sci. Res. **10**(4), 224–230 (2015)
5. Levin, M.F., Weiss, P.L., Keshner, E.A.: Emergence of virtual reality as a tool for upper limb rehabilitation: incorporation of motor control and motor learning principles. Phys. Therapy **95**(3), 415–425 (2015)
6. Lledó, L.D., Díez, J.A., Bertomeu-Motos, A., Ezquerro, S., Badesa, F.J., Sabater-Navarro, J. M., García-Aracil, N.: A comparative analysis of 2D and 3D tasks for virtual reality therapies based on robotic-assisted neurorehabilitation for post-stroke patients. Front. Aging Neurosci. **8**, 205 (2016)
7. Lemon, S.C., Roy, J., Clark, M.A., Friedmann, P.D., Rakowski, W.: Classification and regression tree analysis in public health: methodological review and comparison with logistic regression. Ann. Behav. Med. **26**(3), 172–181 (2003)
8. Cohen, R.G., Sternad, D.: State space analysis of timing: exploiting task redundancy to reduce sensitivity to timing. J. Neurophysiol. **107**(2), 618–627 (2012)
9. Naish, K.R., Reader, A.T., Houston-Price, C., Bremner, A.J., Holmes, N.P.: To eat or not to eat? kinematics and muscle activity of reach-to-grasp movements are influenced by the action goal, but observers do not detect these differences. Exp. Brain Res. **225**(2), 261–275 (2013)
10. Michaelsen, S.M., Jacobs, S., Roby-Brami, A., Levin, M.F.: Compensation for distal impairments of grasping in adults with hemiparesis. Exp. Brain Res. **157**(2), 162–173 (2004)
11. Timmermans, A.A., Seelen, H.A., Willmann, R.D., Kingma, H.: Technology-assisted training of arm-hand skills in stroke: concepts on reacquisition of motor control and therapist guidelines for rehabilitation technology design. J. Neuroeng. Rehabil. **6**, 1 (2009)
12. Krakauer, J.W.: Motor learning: its relevance to stroke recovery and neurorehabilitation. Curr. Opin. Neurol. **19**, 84–90 (2006)
13. Saposnik, G., Levin, M.: Virtual reality in stroke rehabilitation: a meta-analysis and implications for clinicians. Stroke **42**(5), 1380–1386 (2011)

14. Liebermann, D.G., Berman, S., Weiss, P.L., Levin, M.F.: Kinematics of reaching movements in a 2-D virtual environment in adults with and without stroke. IEEE Trans. Neural Syst. Rehabil. Eng. **20**(6), 778–787 (2012)
15. Hardwick, R.M., Rajan, V.A., Bastian, A.J., Krakauer, J.W., Celnik, P.A.: Motor learning in stroke: trained patients are not equal to untrained patients with less impairment. Neurorehabil. Neural Repair **31**(2), 178–189 (2017)
16. El Amki, M., Baumgartner, P., Bracko, O., Luft, A.R., Wegener, S.: Task-specific motor rehabilitation therapy after stroke improves performance in a different motor task: translational evidence. Transl. Stroke Res. **8**(4), 347–350 (2017)

The Effect of Perturbation Time on Selected Spatio-Temporal Parameters of Gait

Andrej Olenšek(✉), Matjaž Zadravec, and Zlatko Matjačić

University Rehabilitation Institute, Republic of Slovenia, Linhartova 51, 1000 Ljubljana, Slovenia
andrej.olensek@mail.ir-rs.si

Abstract. The mechanisms that underlie suitable response in fall threatening situation are not well understood yet. Most of the knowledge has been obtained from slip and trip mimicking experiments and from subjecting individuals to external pushes. While most studies focused on balance responses where perturbations were delivered at foot strike in this study we examined how delaying perturbation onset to later stages of stance phase affect step length, step width and step time. The experiments were conducted on five healthy male subjects that were subjected to series of perturbation impulses of different amplitude, direction with respect to stance leg and onset delay. Results show that when perturbed in lateral direction with respect to stance leg healthy subjects increased step length and step time and decreased step width. However, the changes were less pronounced if the perturbation onset was delayed farther in stance phase. When perturbed in medial direction with respect to stance leg they decreased step length and step time and increased step width. However, the changes were less pronounced if the perturbation onset was delayed farther in stance phase. Results suggest that in healthy population perturbation onset plays important role in spatio-temporal responses.

Keywords: Reactive responses · Balance · Perturbation

1 Introduction

Mechanisms that underlie appropriate response in fall threatening situations during walking are not well understood yet. Most of the knowledge on reactive responses to perturbations during walking has been obtained from slip [1, 2] and trip mimicking experiments [3] as well as experiments in which individuals are subjected to external pushes to mimic collision with other persons in a crowd [4, 5]. These studies show that the prevalent strategy used upon perturbation is associated with early muscle activation timing as well as higher peak muscle activation magnitude to counteract altered dynamic conditions and to enable adequate swing foot placement and compensatory stepping [1–5]. Several studies have also shown that responses may extend over several compensatory steps after perturbation and that these compensatory steps may vary in terms of step length, step width and step time to augment restoration of dynamic stability [5, 6]. These studies only considered situations where perturbation onset was

J. Henriques et al. (Eds.): MEDICON 2019, IFMBE Proceedings 76, pp. 618–621, 2020.
https://doi.org/10.1007/978-3-030-31635-8_74

fixed to the time of foot strike and did not contemplate potentially alternative balance mechanisms if perturbation onset had been delayed forward with respect to foot strike.

In present study we wanted to extend the research to examine spatio-temporal parameters of gait when perturbation onset is delayed with respect to foot strike. It is our assumption that step length, step width and step time within compensatory steps following perturbation display distinctive patterns if short force impulses in medio-lateral direction at waist level occur at foot strike, mid-stance and terminal stance.

2 Methods

We have recently developed Balance Assessment Robot (BAR-TM) for assessing postural responses or for balance training while walking on instrumented treadmill (Fig. 1). Operation characteristics of BAR-TM are determined through admittance-based control scheme, which enables rendering of desired mechanical impedance at the pelvis level. Mechanical impedance can be set low to implement transparent mode or can be programmed to provide desired level of assistive/perturbation forces to the pelvis of a walking subjects [6]. Five healthy male adults (age: 31 ± 5 years, height: 180 ± 3.9 cm, mass: 78.7 ± 6.5 kg) with no known neuromuscular or orthopaedic problems were included in this study. The experimental protocol was for each subject split into three experimental blocks distinguished by treadmill speed i.e. 0.4 m/s, 0.6 m/s and 0.8 m/s respectively. Each experimental block started with a period of unperturbed walking followed by perturbed walking. Each perturbation type was completely defined by four parameters: i. medial/lateral direction of perturbation, ii. left/right leg in stance phase at the time of perturbation, iii. delay of perturbation onset with respect to foot strike (0% - foot strike, 30% - mid stance and 60% - terminal stance of gait cycle) and iv. perturbation amplitude (5%, 10% and 15% of body weight) and was repeated seven times.

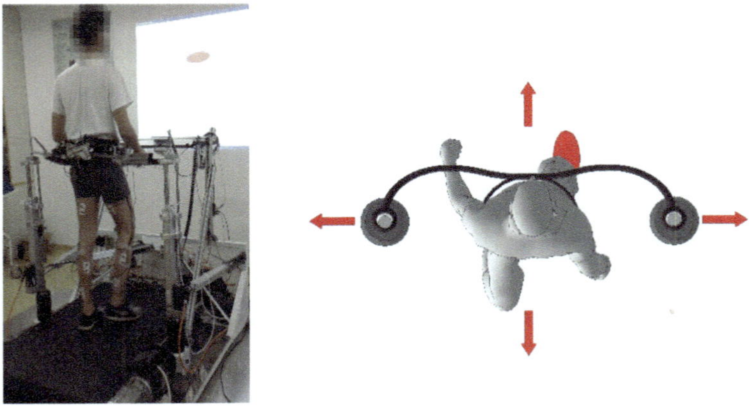

Fig. 1. BAR-TM – a balance assessment robot (left) and perturbation directions with respect to human body (right).

In this study spatio-temporal parameters of responses were investigated in terms of step length, step width and step time in two steps prior to perturbation and four steps following perturbation. Left/right step length and left/right step width were defined as the anterio-posterior and medio-lateral distance between medial malleolis at the time of left/right foot strike respectively. Similarly the step time was defined as the time elapsed between consecutive left/right and right/left foot strikes. The positions of left and right malleolis was recorded by means of Optitrack motion tracking camera (NaturalPoint Inc.) and passive reflective markers that were placed above left and right medial malleois. Spatio-temporal responses of gait were first arranged in sequences comprised of two steps prior to perturbation onset and subsequent four response steps, averaged within each subject and finally also across group of subjects.

3 Results

Figure 2 shows step lengths, step widths and step times when perturbations were delivered during right stance either in right or left direction and at three time instances. Results show that in the two steps before perturbations selected spatio-temporal did not change considerably regardless of perturbation onset or perturbation direction. However when perturbed to the right we notice that in the step following perturbation subjects increased the step length the most if perturbation occurred at the time of foot strike, somewhat less if it occurred in mid stance and did not change it if perturbation was delivered at terminal stance. Step width was on average decreased in all three conditions, however it was decreased the most if perturbation was delivered at mid

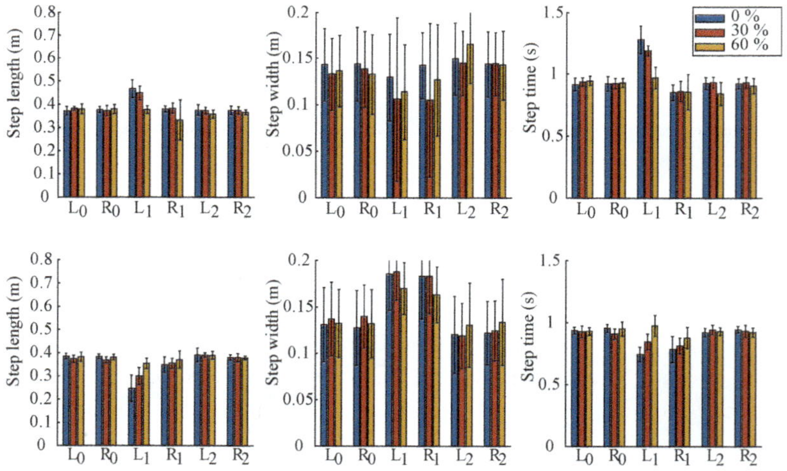

Fig. 2. Step length, step width and step time in two steps prior perturbation (L_0, R_0) and four steps (L_1, R_1, L_2, R_2) after perturbation at foot strike, mid stance and terminal stance (0%, 30% and 60% of gait cycle). Top panel – perturbation in right direction during right stance, Bottom panel – perturbation in left direction during right stance. Perturbation amplitude – 10% of BW. Speed of walking – 0.4 m/s.

stance and the effect of perturbation was evident in two steps following perturbation. The effect of perturbation onset on step time is similar to that of step length – step time increases the most if perturbation is delivered at foot strike and gradually decreases if perturbation onset is delayed farther in stance phase.

If during right stance subjects were perturbed to the left the patterns reversed. Immediately after perturbation step length decreased the most if perturbation was delivered at foot strike, whereas the difference with respect to step length prior to perturbation gradually decreased if perturbation onset was delayed to mid stance or terminal stance. Furthermore results also show that step width increased considerably in two steps following perturbation. Finally, step time decreased the most if perturbation was delivered at foot strike, and gradually decreased if perturbation onset was delayed to mid stance or terminal stance.

4 Conclusion

This study showed that perturbation onset plays important role in how healthy subjects adapt step length, step width and step time after being pushed in medio-lateral direction. Responses to perturbation with various onset delays with respect to foot strike warrant specific adjustments in step length, step width and step time that depend also on the direction of perturbation with respect to stance leg at the time of perturbation. While these adjustments pose reasonably attainable targets in healthy population they are often not attainable in stroke population due to impaired motor control. Delayed perturbation onset therefore may pose additional risk factor for falls in stroke population. Further studies are planned to investigate this hypothesis in more details.

References

1. O'Connell, C., Chambers, A., Mahboobin, A., Cham, R.: Effects of slip severity on muscle activation of the trailing leg during an unexpected slip. J. Electromyogr. Kinesiol. **28**, 61–66 (2016)
2. Inkol, K.A., Huntley, A.H., Vallis, L.A.: Repeated exposure to forward support-surface perturbation during overground walking alters upper-body kinematics and step parameters. J. Mot. Behav. **51**(3), 318–330 (2018)
3. Sharafi, B., Hoffmann, G., Tan, A.Q., Dhaher, Y.Y.: Evidence of impaired neuromuscular responses in the support leg to a destabilizing swing phase perturbation in hemiparetic gait. Exp. Brain Res. **234**, 3497–3508 (2016)
4. Hof, A.L., Vermerris, S.M., Gjaltema, W.A.: Balance responses to lateral perturbations in human treadmill walking. J. Exp. Biol. **213**, 2655–2664 (2010)
5. Matjačić, Z., Zadravec, M., Olenšek, A.: An effective balancing response to lateral perturbations at pelvis level during slow walking requires control in all three planes of motion. J. Biomech. **60**, 79–90 (2017)
6. Olenšek, A., Zadravec, M., Matjačić, Z.: A novel robot for imposing perturbations during overground walking: mechanism, control and normative stepping responses. J. Neuroeng. Rehabil. **13**(55), 1–19 (2016)

Design of a Hybrid Portable System for Measuring the Position of the Spine, Pelvis and Center of Gravity of the Body

Jan Hejda[1(✉)], Petr Volf[1], Monika Bačíková[3], Noa Bar[1], Cestmír Oberman[3], Kristýna Rusnáková[3], Marcela Braunová[3], and Patrik Kutílek[2]

[1] Department of Biomedical Technology, Faculty of Biomedical Engineering, Czech Technical University, Nám. Sítná 3105, 27201 Kladno, Czechia
jan.hejda@fbmi.cvut.cz
[2] Department of Natural Sciences, Faculty of Biomedical Engineering, Czech Technical University, Nám. Sítná 3105, 27201 Kladno, Czechia
[3] CASRI, Podbabská 3, 16000 Praha 6, Czechia

Abstract. Posture and postural stability are important means of assessing the physical condition of a human. The system proposed in this paper allows to measure spine parameters (mutual position of individual vertebrae), mutual position of the pelvis and chest in the lateral plane, center of pressure (COP) under both feet and center of gravity (COG). It uses reflective markers as well as green markers to mark selected anatomical points that are captured by Microsoft Kinect 2 located behind the subject and Microsoft Kinect located to the subject's side. COP and COG are determined using two Nintendo Wii Balance Board, stabilometric platforms which are placed next to each other. A MATLAB GUI application was created to retrieve data from the sensors, determine positions of the individual anatomical points and angles between them and the position of COP and COG. The experimental verification of the system has shown that the measured values, together with the determined statistical variables such as 95% confidence ellipse, can be used to assess the posture and postural stability.

Keywords: Posture · Postural stability · Kinect

1 Introduction

Objective evaluation of posture and postural stability is an important mean for monitoring the physical condition of human. Optical motion capture systems (OMC) are considered to be the gold standard in determining joint angles, as well as angular velocity and moments at selected body parts [1]. This makes them useful for clinical practice, especially in the field of motion analysis and related subfields of gait analysis [2], rehabilitation (e.g. after stroke [3]) and analysis of balance e.g. within postural analysis [4]. OMC systems generally have a frame rate between 100 and 2000 Hz [5, 6].

Most systems currently used today to authenticate the assessment of the postural stability come with considerable financial and spatial requirements. Specific examples

J. Henriques et al. (Eds.): MEDICON 2019, IFMBE Proceedings 76, pp. 622–633, 2020.
https://doi.org/10.1007/978-3-030-31635-8_75

are Vicon [7–9], Qualysis [10] and OptiTrack [11]. OMC are frequently combined with pressure [12] or force [13] platforms. The alternative is to use special pressure measuring shoe inserts which can be used in the conjunction with force platforms [14].

In this paper, an objective method using data collected from Microsoft Kinect (MKV1), Microsoft Kinect 2 (MKV2) and two Nintendo Wii Balance Board (NWBB) stabilometric platforms is stated.

MKV1 has the potential to be used in clinical applications, especially in the form of a low-cost screening method for persons at risk of falling [4]. MKV2 has higher accuracy in depth variable images compared to MKV1 and, above all, its accuracy is constant in measuring range. MKV1 accuracy decreases exponentially with the distance from the object [15]. On the other hand, MKV1 has better accuracy in case of flat objects [15]. Both MKV1 and MKV2 frame rate is 30 Hz [16, 17].

The use of the pair of NWBB force platforms enables measurements of the center of pressure (COP) and center of gravity (COG) [18]. This set-up also allows variability in positioning of the platforms relative to each other in the horizontal and vertical planes, thus it enables to simulate different situations, such as running, walking up the stairs, etc. The NWBB has the potential to become a useful tool in clinical assessments of equilibrium. However, it cannot fully replace the laboratory force platforms since it is able to perform measurements only along the vertical axis and not along the horizontal axis, affecting the evaluation of COP [19]. The measurement uncertainty between the platforms is 9.1 N and 4.1 mm COP. The measurement uncertainty of one platform in case of repeated measurement is 4.5 N and 1.5 mm COP with sampling frequency of approximately 100 Hz [20]. A minimum sampling frequency required for the determination of COP is 40 Hz [21].

The proposed system was designed with regard to ease of transportation and the ability to measure static posturography, especially of spine parameters and the relative position of the pelvis and chest in the sagittal plane. Those parameters also play a role in the transition to erect posture in childhood and are the basis of so-called Neuromuscular Stabilization. These findings are also used to enhance peak performance in sports [22].

2 Methods

MKV1 and MKV2 sensors were used together with a pair of NWBB for measurement purposes. The total system price without computer technology is about 350 USD (2x NWBB, MKV1, MKV2, Kinect V2 PC adapter). All measurements and data processing were performed in MATLAB software 2018a version.

2.1 Camera System

To ensure the correct mutual positioning of MKV1 and MKV2 cameras, the custom made lightweight aluminum profile construction was used (Fig. 1). This design allows the vertical position of both cameras to be set with an adjustable height option for the purposes of measuring subjects of different height. For transportation, it is possible to fold the construction and store it together with the rest of the equipment (2x NWBB, laptop, cables) in a custom made transportation case.

Microsoft Kinect V1. As the camera aimed for right lateral projection, MKV1 was used in RGB mode. MKV1 uses structured-light approach [23] with a 320 × 240 px depth sensor allowing measurement between 1.8–3.5 m and 640 × 480 px RGB camera with Field of View (FOV) of 57° in horizontal and 43° in vertical direction [24]. The standard USB 2.0 interface was used to connect to the PC. Kinect for Windows SDK 1.8 was installed in the Windows 10 environment and the "Image Acquisition Toolbox Support Package for Kinect For Windows Sensor" was installed in the MATLAB 2018a environment.

To evaluate the mutual position of the pelvis and chest in the lateral plane, anatomical points (RIAS, RIPS, RCAJ, TV7, SXS) [25] were used. Green markers of 4 cm diameter were attached to the RIAS, RIPS, TV7, SXS anatomical positions and a green 2.5 cm marker was attached to the RCAJ anatomical position with regard to visibility and possible marker coverage.

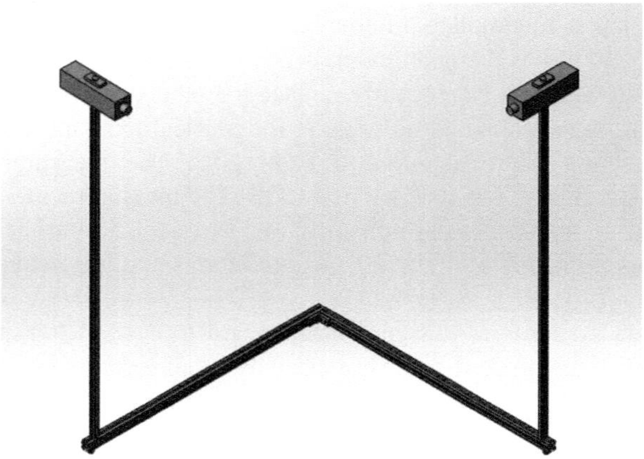

Fig. 1. The custom made lightweight construction to ensure the correct camera positioning

The RGB image returned by the "getsnapshot" function (Fig. 2A) was converted into the Lab color space using the "rgb2lab" [26] function. The empirically found threshold was subsequently applied to lab space's A channel (Fig. 2B). To reduce noise and unwanted image artifacts, the morphological opening and closing with structural element "disk" with 1 px radius provided by the "strel" function was applied on the resulting image using "imopen" and "imclose" functions. Five contiguous regions with the largest area determined by the "regionprops" function with the "area" parameter were labeled as expected markers. To determine regions' centers, the "regionprops" function with the "centroid" parameter was used. Processed image with labeled markers and determined angles between them is depicted on Fig. 3.

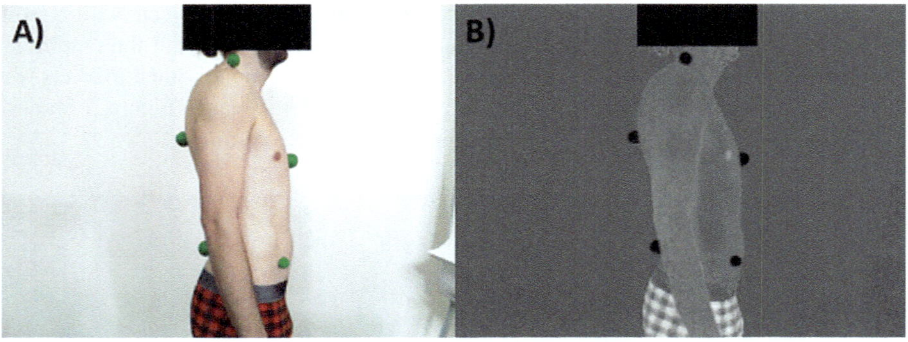

Fig. 2. MKV1 (A – original image, B – Lab channel A)

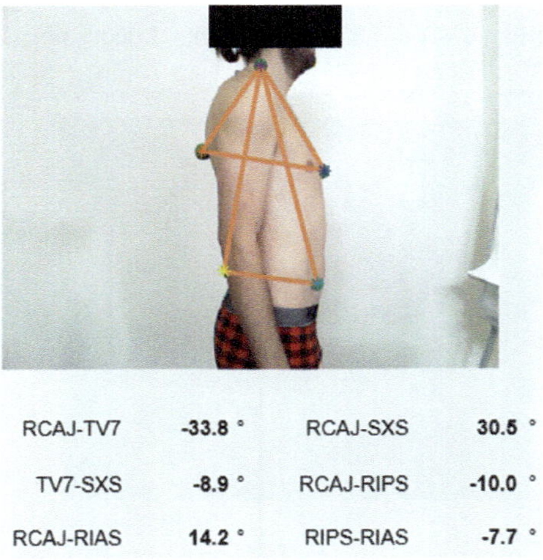

Fig. 3. Identification of markers from the side camera and determined angles between the anatomical points of interest

Microsoft Kinect V2. MKV2 in depth mode was used as a camera for dorsal projection. MKV2 uses Time-of-flight (TOF) approach [23] with a 512 × 424 px depth sensor able to measure between 1.3–3.5 m and a 1920 × 1080 px RGB camera with horizontal FOV of 70° and vertical FOV of 60° [24]. A Kinect V2 PC adapter is required for a connection to PC via standard USB 3.0 interface. Kinect for Windows SDK 2.0 was installed for the necessary communication in the Windows 10 environment.

Anatomical points (CV7, TV2, TV7, LV1, LV3, LV5) [25] were used for evaluation of spine parameters (mutual position of individual vertebrae). Standard reflective markers with a diameter of 1.5 cm appearing in the image as regions with undefined depth were attached to these anatomical points. The pixels with undefined depth were selected (Fig. 4B) from the depth map returned by "getsnapshot" function (Fig. 4A).

2D convolution with convolution kernel for detection of circular objects with a radius of 3 px was applied using "conv2" function. The result after applying the empirically determined threshold is depicted in Fig. 4C. The centers of the individual markers were identified by the same procedure as in case of the lateral plane. Figure 5 shows a processed image with markers and determined angles between them.

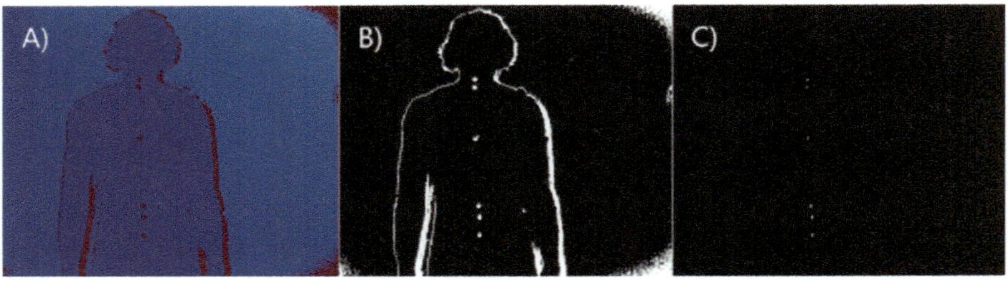

Fig. 4. MKV2 (A – original depth map, B – undefined depth pixels, C – reflective markers)

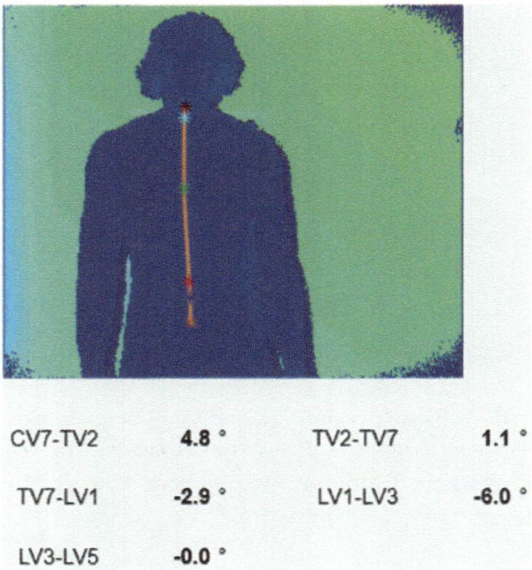

Fig. 5. Identification of markers from the rear camera and determined angles between the anatomical points of interest

2.2 Nintendo Wii Balance Board

The system uses a pair of NWBB platforms to measure the COP under both feet and COG of the body. For connectivity, the "WiiLab" library, originally designed for Nintendo Wiimote drivers [27], was used. The connection is made via Bluetooth. The library "WiiLab" retrieves the output of the COP of the platform or the load of the individual force sensors located in the corners of the platform. However, the values of individual sensors (BBVals) are 4 times higher than the correct value. This is probably due to a library calculation error and multiplication by the number of sensors.

For this reason, the collected BBvals values were divided by 4, and then used for the evaluation of the COP of platforms using the torque method [18] (Eqs. 1–4) and COG for platforms which were placed next to each other (Eqs. 5 and 6). Values 433 and 238 used in the equations indicate the distance between platform sensors in mm. The value 77 is the distance between the sensors of adjacent platforms in mm. Index l is used for the left platform, r is for the right platform.

$$COP_{l,x} = -\left(\frac{77}{2} + \frac{238}{2}\right) + \frac{238}{2}\left(\frac{\sum_{i=1}^{2} \mathrm{BBVals_l}(i) - \sum_{i=3}^{4} \mathrm{BBVals_l}(i)}{\sum \mathrm{BBVals}_l}\right) \quad (1)$$

$$\mathrm{COP_{l,y}} = -\frac{433}{2}\left(\frac{(\mathrm{BBVals_l}(1) + \mathrm{BBVals_l}(3)) - (\mathrm{BBVals_l}(2) + \mathrm{BBVals_l}(4))}{\sum \mathrm{BBVals_l}}\right) \quad (2)$$

$$COP_{r,x} = \left(\frac{77}{2} + \frac{238}{2}\right) + \frac{238}{2}\left(\frac{\sum_{i=1}^{2} \mathrm{BBVals_r}(i) - \sum_{i=3}^{4} \mathrm{BBVals_r}(i)}{\sum \mathrm{BBVals}_r}\right) \quad (3)$$

$$\mathrm{COP_{r,y}} = \frac{433}{2}\left(\frac{(\mathrm{BBVals_r}(1) + \mathrm{BBVals_r}(3)) - (\mathrm{BBVals_r}(2) + \mathrm{BBVals_r}(4))}{\sum \mathrm{BBVals_r}}\right) \quad (4)$$

$$COG_x = \frac{\left(\frac{77}{2} + 238\right)\left(\sum_{i=3}^{4} \mathrm{BBVals_r}(i) - \sum_{i=3}^{4} \mathrm{BBVals_l}(i)\right) + \frac{77}{2}\left(\sum_{i=1}^{2} \mathrm{BBVals_r}(i) - \sum_{i=1}^{2} \mathrm{BBVals_l}(i)\right)}{\sum \mathrm{BBVals_r} + \sum \mathrm{BBVals_l}} \quad (5)$$

$$\mathrm{COG_y} = \mathrm{COP_{r,y}} - COP_{l,y} \quad (6)$$

2.3 The Complete System

The complete hybrid portable system for measuring the position of the spine, pelvis, and COG of the body is shown in Fig. 6. The frame rate and sampling rate depends on the computer used, but it is about 10 Hz in case all mentioned sensors are used.

2.4 Software for Measuring and Processing Data

For data recording purposes, the MATLAB GUI application (Fig. 7) has been designed and used for setting up measurement and storing measured data into "xlsx" format. For statistical processing, scripts for evaluation of statistical parameters such as mean, standard deviation, median, quartile, range, etc., along with the determination of 95% confidence ellipse [28] were implemented.

2.5 Experimental Verification of the System

For the purpose of the experimental verification of the system, a measurement was conducted on a male subject at the age of 30 years without previous balance and spine problems. The subject was placed with the left and right soles on the respective stabilometric platform (Fig. 6) and was instructed to watch a fixation point in front of him while standing upright straight for 60 s, moving as little as possible.

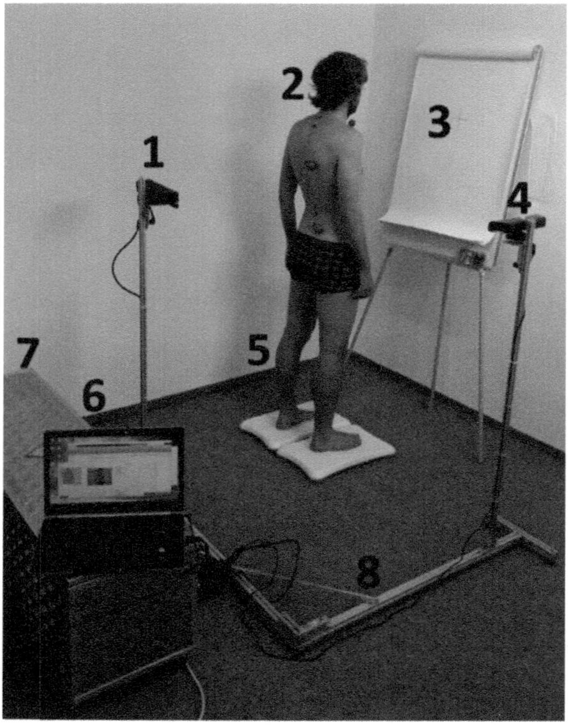

Fig. 6. The hybrid portable system for measuring the position of the spine, pelvis and center of gravity of the body (1 – MKV2, 2 – measured subject, 3 – point to fix the subject's view, 4 – MKV1, 5 – NWBB, 6 – laptop, 7 – transport box, 8 – camera mounting construction)

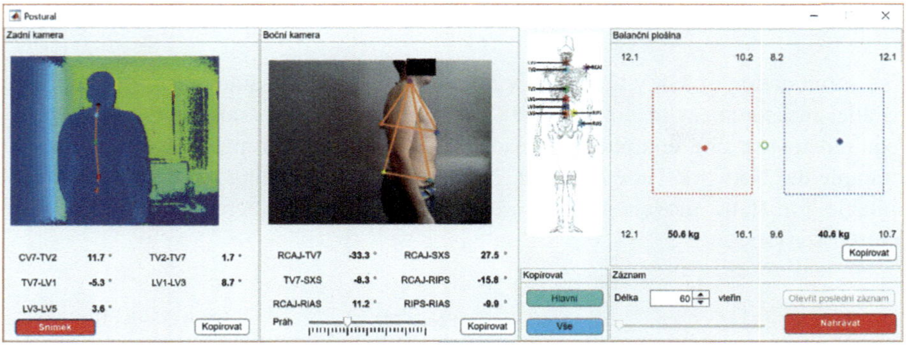

Fig. 7. Data measuring application

3 Results

The following parameters were selected for evaluation:

- 95% confidence ellipse of the COP of the both platforms,
- 95% confidence ellipse of the COG,
- anterior-posterior pelvis tilt in the sagittal plane obtained from MKV1,
- the angle between anatomical points TV7 and LV1 obtained from MKV2.

The illustration of the COP of both platforms and the COG of the entire body with 95% confidence ellipse is depicted in the Fig. 8. Figure 9 is an illustration of a posterior-anterior slope in the sagittal plane. A negative value indicates a downward slope. The tilt of the spine between the points TV7 and LV1 is shown in the Fig. 10. The negative value means the tilt of the sinistrolateral direction in the frontal plane.

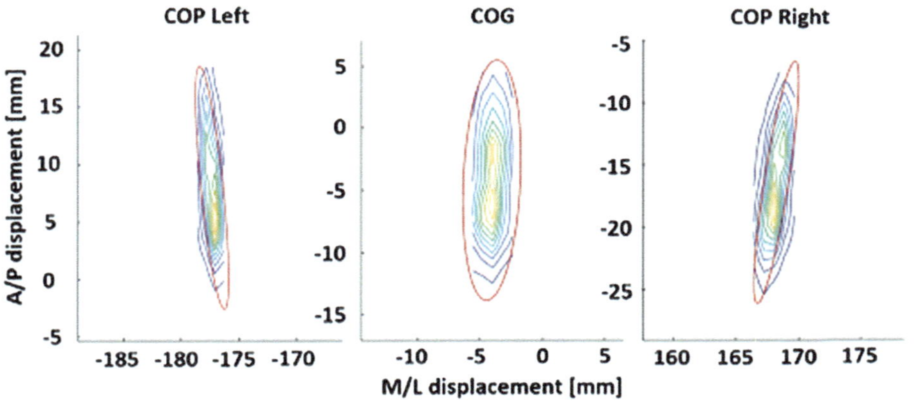

Fig. 8. Evaluation of the COP and the COG (COP_l = 32.12 mm^2, COP_r = 29.86 mm^2, COG = 69.52 mm^2)

4 Discussion

The combination of MKV1 and MKV2 cameras was chosen due to the impossibility of parallel measurement of 2 sets of MKV2 on one PC. For parallel measurement, it is required to use one computer for each MKV2 and synchronize the data, using for example the Network Time Protocol (NTP) [29]. The recording frequency of 10 Hz is suitable for static movements only. Although it can be increased by deactivating unused sensors.

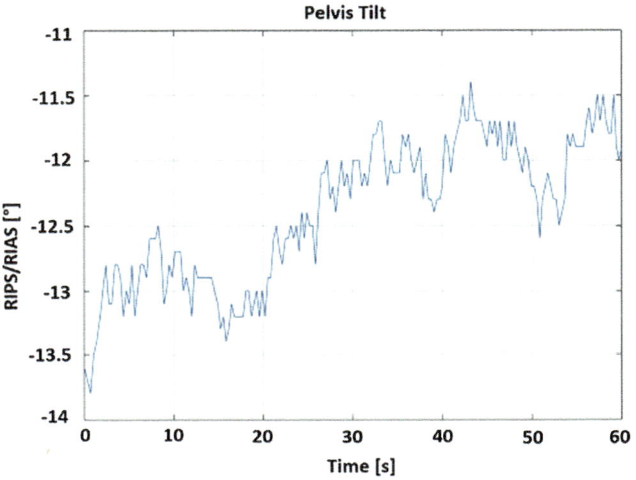

Fig. 9. Evaluation of pelvic tilt (Mean = −12.39°, SD = 0.55)

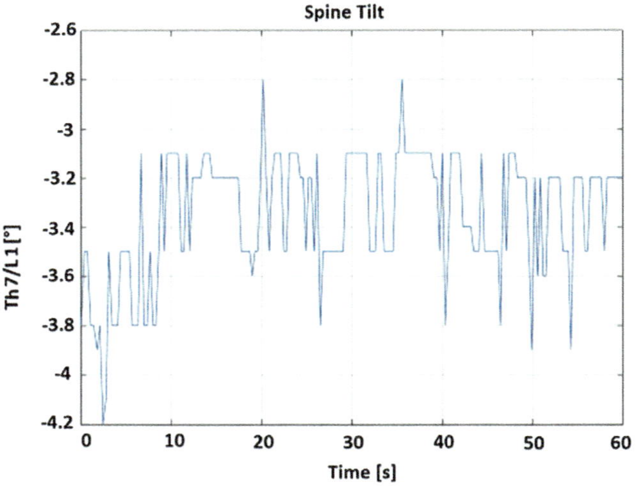

Fig. 10. Evaluation of spinal tilt (Mean = −3.35°, SD = 0.25)

The entire system can be disassembled for easy transport and placed in a transport box along with platforms and a laptop.

The COP and COG results from the platforms show a higher range in the anterior-posterior direction, which is consistent with the assumption of side-by-side position of the legs.

The results of the posterior-anterior direction in the sagittal plane show a downward slope. It can be seen from the graph that the signal quantization value is sufficient. The observed pelvic anteversion is a risk factor for vertebrogenic disorders, especially in the lumbar region of the spine. The pelvic anterior position is due to muscle imbalances in the lower back and pelvis. These imbalances are between muscle pairs such as the weakened mm. glutaei maximi and truncated hip flexors, weakened abdominal muscle, and shortened torso erector, weakened by mm. glutei medii and truncated m. tensor fasciae latae and m. quadratus lumborum.

The results of the tilt of the spine in the frontal plane show that the quantization level is in the level of measurement noise. This is due to the limited depth map resolution of MKV2.

5 Conclusion

The article proposes the hybrid portable system for measuring the position of the spine, pelvis and center of gravity of the body. The experimental verification of the system has shown that the system can be used as a tool in assessing posture and postural stability.

Future research will focus primarily on the possibility of increasing the recording frequency and the possibility of processing dynamic movements, such as a squat. Furthermore, the dependence of stability in the case of optional placement of the platforms in the horizontal and vertical planes, as well as the correlation between muscle tests of affected structures involved in the position of the spine and pelvis, will be evaluated.

Acknowledgement. This work was supported by the Grant Agency of the Czech Technical University in Prague, grant No. SGS18/201/OHK4/3T/17.

Conflict of Interests. None declared.

References

1. Kim, S., Nussbaum, M.A.: Performance evaluation of a wearable inertial motion capture system for capturing physical exposures during manual material handling tasks. Ergonomics **56**(2), 314–326 (2013)
2. Pfister, A., West, A.M., Bronner, S., Noah, J.A.: Comparative abilities of Microsoft Kinect and Vicon 3D motion capture for gait analysis. J. Med. Eng. Technol. **38**(5), 274–280 (2014)
3. Rueda, F.M., Montero, F.R., de Heredia Torres, M.P., Diego, I.A., Sánchez, A.M., Page, J. M.: Movement analysis of upper extremity hemiparesis in patients with cerebrovascular disease: a pilot study. Neurología (English Ed.) **27**(6), 343–347 (2012)

4. Clark, R.A., Pua, Y.H., Fortin, K., Ritchie, C., Webster, K.E., Denehy, L., Bryant, A.L.: Validity of the Microsoft Kinect for assessment of postural control. Gait Posture **36**(3), 372–377 (2012)
5. Motion Capture Systems, VICON. https://www.vicon.com/. Accessed 21 Mar 2019
6. OptiTrack – Motion Capture Systems. https://optitrack.com/. Accessed 20 Mar 2019
7. McMurrough, C.D., Metsis, V., Kosmopoulos, D., Maglogiannis, I., Makedon, F.: A dataset for point of gaze detection using head poses and eye images. J. Multimodal User Interfaces **7**(3), 207–215 (2013)
8. Bonnechere, B., Beyer, B., Rooze, M., Sint, J.S.: What is the safest sprint starting position for american football players? J. Sports Sci. Med. **13**(2), 423–429 (2014)
9. Kramers-de Quervain, I.A., Müller, R., Stacoff, A., Grob, D., Stüssi, E.: Gait analysis in patients with idiopathic scoliosis. Eur. Spine J. **13**(5), 449–456 (2004)
10. Saari, T., Tranberg, R., Zügner, R., Uvehammer, J., Kärrholm, J.: Total knee replacement influences both knee and hip joint kinematics during stair climbing. Int. Orthop. **28**(2), 82–86 (2004)
11. Huntley, A.H., Zettel, J.L., Vallis, L.A.: Effect of aging on dynamic postural stability and variability during a multi-directional lean and reach object transportation task. Arch. Gerontol. Geriatr. **66**, 154–160 (2016)
12. Keijsers, N.L.W., Stolwijk, N.M., Renzenbrink, G.J., Duysens, J.: Prediction of walking speed using single stance force or pressure measurements in healthy subjects. Gait Posture **43**, 93–95 (2016)
13. Terry, K., Gade, V.K., Allen, J., Forrest, G.F., Barrance, P., Edwards, W.T.: Cross-correlations of center of mass and center of pressure displacements reveal multiple balance strategies in response to sinusoidal platform perturbations. J. Biomech. **44**(11), 2066–2076 (2011)
14. Claverie, L., Ille, A., Moretto, P.: Discrete sensors distribution for accurate plantar pressure analyses. Med. Eng. Phys. **38**(12), 1489–1494 (2016)
15. Wasenmüller, O., Stricker, D.: Comparison of kinect v1 and v2 depth images in terms of accuracy and precision. In: Asian Conference on Computer Vision, pp. 34–45. Springer, Cham, November 2016
16. van Diest, M., Stegenga, J., Wörtche, H.J., Postema, K., Verkerke, G.J., Lamoth, C.J.: Suitability of Kinect for measuring whole body movement patterns during exergaming. J. Biomech. **47**(12), 2925–2932 (2014)
17. Corti, A., Giancola, S., Mainetti, G., Sala, R.: A metrological characterization of the Kinect V2 time-of-flight camera. Robot. Auton. Syst. **75**, 584–594 (2016)
18. Kutílek, P., Vítečková, S., Svoboda, Z., Hejda, J.: Application of portable force platforms equipped with a device for measuring position and orientation. Acta Polytechnica **53**(4), 365–370 (2013)
19. Clark, R.A., Bryant, A.L., Pua, Y., McCrory, P., Bennell, K., Hunt, M.: Validity and reliability of the Nintendo Wii Balance Board for assessment of standing balance. Gait Posture **31**(3), 307–310 (2010)
20. Bartlett, H.L., Ting, L.H., Bingham, J.T.: Accuracy of force and center of pressure measures of the Wii Balance Board. Gait Posture **39**(1), 224–228 (2014)
21. Scoppa, F., Capra, R., Gallamini, M., Shiffer, R.: Clinical stabilometry standardization: basic definitions–acquisition interval–sampling frequency. Gait Posture **37**(2), 290–292 (2013)
22. Frank, C., Kobesova, A., Kolar, P.: Dynamic neuromuscular stabilization & sports rehabilitation. Int. J. Sports Phys. Ther. **8**(1), 62 (2013)
23. Sarbolandi, H., Lefloch, D., Kolb, A.: Kinect range sensing: structured-light versus Time-of-Flight Kinect. Comput. Vis. Image Underst. **139**, 1–20 (2015)

24. Amon, C., Fuhrmann, F., Graf, F.: Evaluation of the spatial resolution accuracy of the face tracking system for kinect for windows v1 and v2. In: Proceedings of the 6th Congress of the Alps Adria Acoustics Association, pp. 16–17, October 2014
25. Marker Set Guidelines – Visual3D Wiki Documentation. https://www.c-motion.com/v3dwiki/index.php/Marker_Set_Guidelines. Accessed 24 Jun 2019
26. Tkalcic, M., Tasic, J.F.: Colour spaces: perceptual, historical and applicational background. In: The IEEE Region 8 EUROCON 2003. Computer as a Tool, IEEE, pp. 304–308 (2003)
27. Brindza, J., Szweda, J., Liao, Q., Jiang, Y., Striegel, A.: WiiLab: bringing together the Nintendo Wiimote and MATLAB. In: 2009 39th IEEE Frontiers in Education Conference, IEEE, pp. 1–6 (2009)
28. Chiari, L., Rocchi, L., Cappello, A.: Stabilometric parameters are affected by anthropometry and foot placement. Clin. Biomech. **17**(9–10), 666–677 (2002)
29. Yang, L., Yang, B., Dong, H., El Saddik, A.: 3-D markerless tracking of human gait by geometric trilateration of multiple Kinects. IEEE Syst. J. **12**(2), 1393–1403 (2018)

The Evaluation of the Joint Quasi-Stiffness During the Robot-Assisted Gait Training: A Pilot Study

Luigi Iuppariello[1(✉)], Maurizio Nespoli[1], Fernanda Iammarone[1], Marianna Bertella[1], Ilaria Riccio[1], Marianna Cardillo[1], Angela Natalizio[2], Fabrizio Clemente[3], and Mario Cesarelli[2]

[1] Rehabilitation Unit, AORN Santobono Pausilipon Children's Hospital, Naples, Italy
luigi.iuppariello@gmail.com
[2] Department of Electric Engineering and Information Technologies (DIETI), University of Naples Federico II, Naples, Italy
[3] Institute of Crystallography, Italian National Research Council (IC-CNR), Rome, Italy

Abstract. Walking may be defined as the forward displacement of the body requiring coordination between alternate successions of the swing phase and the stance phase. In literature it is known that, looking at the relationship between external moments and relative angles at lower limb joints during human walking is a new way of analyzing the biomechanical behavior of a joint (stiffness concept). Numerous studies show that the concept of quasi-stiffness applies particularly well to major loading phases of the lower extremity joints during stance phase. This study represents the first attempt to quantify the joint quasi-stiffness during the swing phase of gait over rehabilitation sessions performed by a child with cerebral palsy with the Lokomat system. The obtained results, albeit preliminarily, demonstrate the sensitivity of this new method in evaluating the degree of resistance offered by the patient over the rehabilitation sessions.

Keywords: Quasi-stiffness · Gait rehabilitation · Lokomat

1 Introduction

Walking may be defined as the forward displacement of the body requiring coordination between alternate successions of the swing phase and the stance phase. Consequently, walking can be summarized as the aptitude to produce and control propulsive forces through these alternating successions of double-support and single-support phases in order to move the body forward [1]. Spring-like properties of muscles and joints are believed to play an important role in maintaining human vertical posture, in storing and recoiling elastic energy over a stretch-shortening muscle cycle, particularly during locomotion, and in control of muscular activity [2]. Because joint torque of the musculoskeletal system for movement is generated by the imbalance between tensions of agonist and antagonist muscles which have inherently spring-like properties, stiffness during movement is an important factor in the study of the biological

J. Henriques et al. (Eds.): MEDICON 2019, IFMBE Proceedings 76, pp. 634–639, 2020.
https://doi.org/10.1007/978-3-030-31635-8_76

motion control mechanism [3]. With the term "stiffness" it is possible to indicate the measure of the mechanical resistance of the joints during passive movements. It is represented as the slope of the linear regression line at the joint angle-moment curve. Numerous studies show that the concept of quasi-stiffness applies particularly during stance phase, also children with Cerebral Palsy during walking [4–6]. However, the possibility of evaluating the quasi-stiffness of a joint during active movements can permit the clinicians to better understand, with a quantitative measure, the joint function during an active task, such as driven-gait rehabilitative sessions with Lokomat. This is the first attempt to quantify joint quasi-stiffness during the swing phase (i.e. active movements) of gait over rehabilitative sessions performed with Lokomat in a subject of alteration of the gait patterns following neurological diseases.

2 Methods and Materials

2.1 The Robotic Device for Gait Training - LOKOMAT

The Lokomat (Hocoma AG, Volketswil, Switzerland) is an exoskeletal robotic system used for pathway rehabilitation in patients with movement disorders resulting from different neurological and /or orthopedic disorders. It includes two Driver Gait Orthosis (DGO) used in combination with a body weight support system and a treadmill. The Lokomat system provides standardized assessments to perform a convenient assessment of the main problems of neurological patients, such as stiffness, strength and range of motion. The Lokomat can records biomechanical and technical signals in each session to monitor patient progress. Some of these signals are: joint torques (positive values correspond to the patient flexing the hip or knee and negative values correspond to the patient extending the hip or knee), gait index (it is a numerical indicator to identify the position within the step cycle. The mechanical resistance to displacement, so called stiffness, of the hip and knee joint during flexion and extension movements is assessed with the L-STIFF task while the legs are passively moved in a specific pattern and without touching the ground.

2.2 Evaluation of Quasi-Stiffness

The evaluation of quasi-stiffness at the knee joint during walking with LOKOMAT has been offline performed with a Matlab routine developed by authors, starting from the raw data of gait after a rehabilitation session with Lokomat system, recorded in a text file format. The recorded gait index, knee joint angle (in sagittal plane) and knee torque parameters were considered to assess the quasi-stiffness during swing phase. The gait index is a numerical indicator used to assess the hell strike and toe-off during walking in order to segment the different knee joint angle and knee moment over a single recording. Then, the mean of all knee joint angle and knee moment over a single recording and the event of gait cycles are performed. Finally, a time normalization of all gait cycles it is performed, the original time scale is converted in "percent of cycle" ranging from 0 to 100% of the gait cycle.

Starting from the knee joint angle and knee moment above defined, the quasi-stiffness at knee joint was computed. In this preliminary study only the extension phase (mid and terminal swing) of the knee joint was considered, since the clear linear behavior from the visual examination of the knee angle and knee torque graphs. The quasi-stiffness was computed as the linear regression of the points in the angle-torque plane for knee joint by the Ordinary Least Squares (OLS) method that is the most common estimation method for linear models. The goal of OLS is to closely "fit" a function with the data. It does so with the following equation, obtained by minimizing the sum of squared errors from the data:

$$\hat{\theta} = \left(X^T X\right)^{-1} X^T y$$

where $\hat{\theta}$ is a 2×1 vector containing the estimated parameters (slope and intercept of the regression line), X is the independent variable (angle), y is the dependent variable (torque). The estimated slope represents the joints quasi-stiffness value.

3 Results and Discussion

To test the efficiency of the above described model we considered a patient (age 8 years old, female) with bilateral spastic cerebral palsy, underwent to 6 rehabilitative sessions in a month using the Lokomat system (Hocoma AG, Volketswil, Switzerland) in the Rehabilitation Unit of the paediatric hospital Santobono-Pausilipon in Naples, Italy. The quasi-stiffness for both left and right knee joint movements was computed for each rehabilitative session. Figure 1 reports an example of the averages of the right knee angle joint and the right knee moment over a single recording, with the identification of the eight phases of a gait cycle (vertical lines).

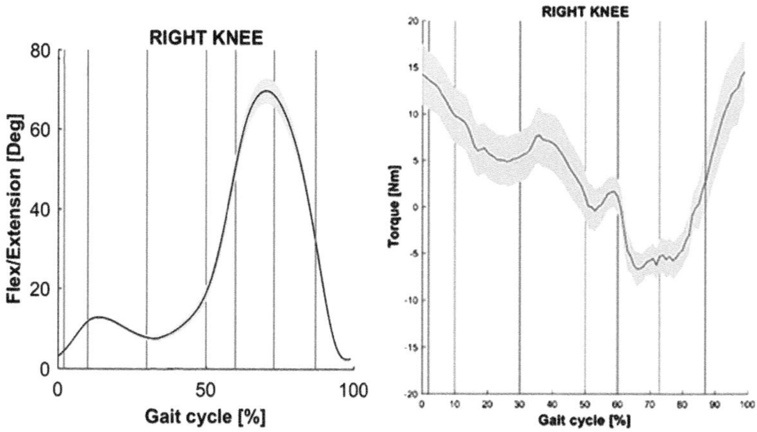

Fig. 1. Averages of the left knee angle (on the left) and knee moment (on the right) joint over a single recording, with the identification of the eight phases of a gait cycle (vertical lines).

As explained above, the quasi-stiffness (Fig. 2) was calculated as the slope of the linear regression line to the angle-torque curve for the extension movements of knee joint during the swing phase.

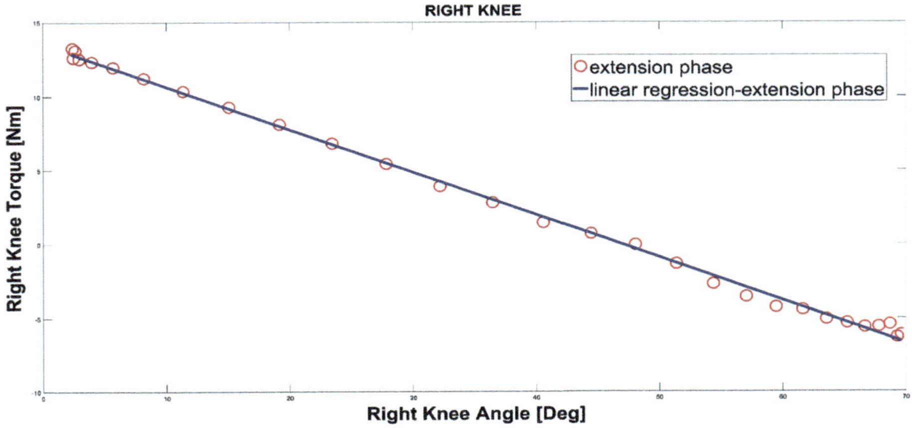

Fig. 2. Example of quasi-stiffness computed as the linear regression of the points in the angle-torque plane, for right knee joint and extension movement, by the Ordinary Least Squares (OLS) method.

The variation between the sessions of the joint quasi-stiffness was reported in Figs. 3 and 4.

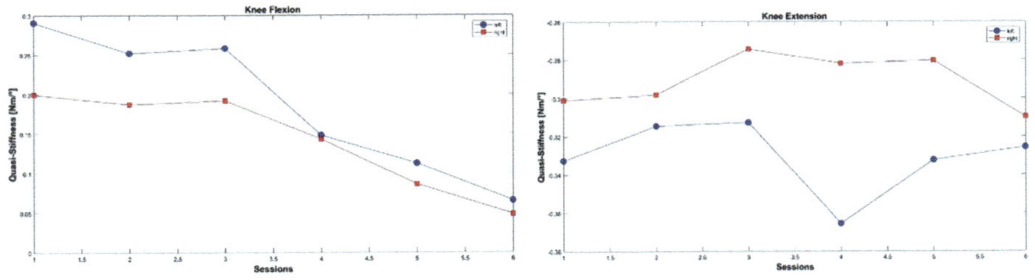

Fig. 3. The quasi-stiffness for the knee joint during the pre-swing phase (flexion movement) on the right and during the terminal swing phase (extension movement) on the left. The change of this value during rehabilitation sessions is shown

For the flexion movement of the left knee, this value ranges from 0.2909 Nm/° at the first session to a value of 0.0654 Nm/° at the last session; for the flexion movement of the right knee it goes from a value of 0.1999 Nm/° at the first session to a value of 0.0484 Nm/° at the last session. Since the rectus femoris is the muscle mainly involved in this phase (flexion in pre-swing) in order to oppose an excessive passive flexion that would prevent to prepare the limb for the swing, from the obtained results it can be deduced that the rectus femoris offers, in the last session, less resistance (i.e. stiffness) to the movement imposed by the orthoses. Finally, due to the extension movement of the right and left knees there are just a few reduction of the quasi-stiffness from the first to last session.

4 Discussion

Walking may be defined as the forward displacement of the body requiring coordination between alternate successions of the swing phase and the stance phase. In literature it is known that, looking at the relationship between external moments and relative angles at lower limb joints during human walking is a new way of analyzing the biomechanical behavior of a joint. The joint stiffness represents one of the clinical aspects of the several robotics systems for rehabilitation which have been developed in last years to help patients with functional disability recover or compensate for loss of limb motor functions [7–12]. One of such system is represented by the Lokomat, a robotic exoskeleton which is able to quantitative evaluate the joint resistance (stiffness) to passive movements, but not to active movements.

This is the first attempt to quantify joint stiffness during active movements (quasi-stiffness) during the swing phase of gait over rehabilitative sessions (active movements) performed with Lokomat. As known, contrary to the stance leg, the swing leg during normal walking approximately undergoes a ballistic movement [13] that does not demand considerable muscular effort, hence the concept of quasi-stiffness is not so substantial. The possibility of evaluating the quasi-stiffness of a joint during active movements can permit the clinicians to better understand, with a quantitative measure, the joint function during an active task (swing phase of driven-gait with Lokomat). The study carried out has allowed, albeit preliminarily, to demonstrate the sensitivity of a new method of evaluation of rehabilitation outcomes with particular reference to the evaluation of joint quasi-stiffness. In fact, the reduction of the values of quasi-stiffness can be interpreted as a less resistance (less stiffness) which the patient offers, to the movement imposed by the orthoses.

Future study will test the efficiency of the described method the described model for the evaluation of quasi-stiffness on a larger sample of patients. Moreover, other phases of swing and other range of motions of joint will be considered.

References

1. Perry, J., Burnfield, J.: Gait Analysis: Normal and Pathological Function. SLACK Incorporated, New Jersey (1992)
2. Latash, M.L., Zatsiorsky, V.M.: Joint stiffness: myth or reality? Hum. Mov. Sci. **12**, 653–692 (1993)
3. Gomi, H., Kawato, M.: Human arm stiffness and equilibrium-point trajectory during multi-joint movement. Biol. Cybern. **76**(3), 163–171 (1997)
4. Davis, R.B., DeLuca, P.A.: Gait characterization via dynamic joint stiffness. Gait Posture **4**(3), 224–231 (1996)

5. Rouse, E., Gregg, R., Hargrove, L., Sensinger, J.: The difference between stiffness and quasi-stiffness in the context of biomechanical modeling. IEEE Trans. Biomed. Eng. **60**, 562–568 (2013)
6. Wang, T.-M., Huang, H.-P., Li, J.-D., Hong, S.-W., Lo, W.-C., Lu, T.-W.: Leg and joint stiffness in children with spastic diplegic cerebral palsy during level walking. PLoS One **10**(12), e0143967 (2015). https://doi.org/10.1371/journal.pone.0143967
7. Iuppariello, L., Bifulco, P., D'Addio, G., Lanzillo, B., Lullo, F., Gallo, F., Romano, M., Cesarelli, M.: The effects of the vibratory stimulation of the neck muscles for the evaluation of stepping performance in Parkinson's Disease. In: 2015 IEEE International Symposium on Medical Measurements and Applications, MeMeA 2015, art. no. 7145275, pp. 606–609 (2015)
8. Cesarelli, M., Iuppariello, L., Romano, M., Bifulco, P., D'Addio, G.: Bioengineering activities in proprioceptive and robotic rehabilitation at Salvatore Maugeri Foundation. In: 2015 AEIT International Annual Conference, AEIT 2015, art. no. 7415277 (2015)
9. Iuppariello, L., Romano, M., D'Addio, G., Bifulco, P., Pappone, N., Cesarelli, M.: Comparison of measured and predicted reaching movements with a robotic rehabilitation device. In: Proceedings of IEEE MeMeA 2014 - IEEE International Symposium on Medical Measurements and Applications, art. no. 6860056 (2014)
10. D'Addio, G., Lullo, F., Pappone, N., Romano, M., Iuppariello, L., Cesarelli, M., Bifulco, P.: Relationships of kinematics indexes with amplitude and velocity of upper arm reaching movement. In: Proceedings of MeMeA 2013 – IEEE International Symposium on Medical Measurements and Applications, art. no. 6549719, pp. 120–123 (2013)
11. Iuppariello, L., Cesarelli, M., Faiella, G., Esposito, S., Nespoli, M., Foggia, L., Clemente, F.: Design of technology-based rehabilitation pathways: The experience of Santobono-Pausilipon Hospital. Acta IMEKO **7**(4), 55–61 (2018)
12. Cesarelli, M., D'Angelo, G., Romano, M., Bifulco, P., Clemente, F.: A new system to deliver vibration, measure motion and EMG reaction for forearm muscles. In: 20th IMEKO TC4 Symposium on Measurements of Electrical Quantities: Research on Electrical and Electronic Measurement for the Economic Upturn, Together with 18th TC4 International Workshop on ADC and DCA Modeling and Testing, IWADC, 15–17 September 2014, Benevento, Italy, pp. 606–609
13. Shamaei, K., Sawicki, G.S., Dollar, A.M.: Estimation of quasi-stiffness of the human knee in the stance phase of walking. PLoS One **8**(3), e59993 (2013). https://doi.org/10.1371/journal.pone.0059993

Design of Device for Measuring the Load of Cross-Country Ski Poles

Jan Hejda[1(✉)], Petr Volf[1], Jakub Mejstřík[1], Ján Hýbl[2], Aleš Tvrzník[3], David Gerych[3], Tomáš Michálek[3], Čestmír Oberman[3], Emil Bolek[3], and Patrik Kutílek[2]

[1] Department of Biomedical Technology, Faculty of Biomedical Engineering, Czech Technical University, Nám. Sítná 3105, 27201 Kladno, Czechia
jan.hejda@fbmi.cvut.cz

[2] Department of Natural Sciences, Faculty of Biomedical Engineering, Czech Technical University, Nám. Sítná 3105, 27201 Kladno, Czechia

[3] CASRI, Podbabská 3, 16000 Praha 6, Czechia

Abstract. Monitoring of physical activity allows the optimization of physical exercise and sports performance. Currently, there are a number of wearable devices designed to measure the acceleration of body segments. However, there are only a small number of experimental systems for evaluation of load and movement of cross-country skiing and Nordic Walking poles. The system proposed in the article allows to measure the force which user is applying on the pole, the pole linear and angular acceleration and the tilt of the pole. The system uses a pair of strain gauge bridges and inertial measurement unit mounted on each pole. The measured data are subsequently processed by the capturing and recording unit, transmitted via Wi-Fi interface and stored on an integrated SD card. Developed software allows analysis of the determined quantities including poling detection. The verification measurement has shown the usability of the system for assessing biomechanical parameters, particularly with regard to the evaluation of training methods and running styles for cross-country skiing and Nordic running. The limitation of the proposed system is the disruption of the structural integrity of the pole due to the bonding of the strain gauge bridges and the associated greater susceptibility to damage in the event of extreme load, e.g. in top sports.

Keywords: Ski poles · Accelerometers · Strain gauges

1 Introduction

Technological development of equipment materials and wearable electronics for monitoring physical condition using heart rate variability and other parameters in sport has an effect on improving sporting results in amateur and professional athletes [1, 2]. It also participates in the optimization of training methods using feedback for athletes and specialists including biomechanics, physiologists, sports psychologists and strength/fitness specialists because biomechanical, physiological and psychological factors contribute to the resulting performance [3, 4]. Biomechanical analysis of kinematics and dynamics enables the optimization and implementation of new techniques [4].

J. Henriques et al. (Eds.): MEDICON 2019, IFMBE Proceedings 76, pp. 640–649, 2020.
https://doi.org/10.1007/978-3-030-31635-8_77

For cross-country skiing, forces are generated by upper and lower limbs using skis and poles. The contribution of the upper half of the body to total force is proven to change with a change of running technique and ranges from 15 to 30% [5]. For measuring the reaction force acting on the pad during the sports performance, the force-measuring platforms can be used. However, these platforms are usually able to measure only a few steps in a limited space and thus systems such as F-Scan and Pedar using special inserts are often used for evaluation of pressure distribution under the foot [6]. In specialized cases, such as the ski pole load and the force acting on the skis during reflection, it is not possible to use these commercial measurement systems. For this reason, the measurement of the rebound force of both cross-country skiing and skiing poles is a challenge in terms of performance improvement due to motion analysis and feedback provided to the athlete and coach [5–7]. Wearable ski binding system using resistance strain gauges for poling force measurement in classic style and skating with a total weight of approximately 1.5 kg (490 g skis, 1050 g recording device) showed changes in classic style during one-hour load. This was probably due to the ski binding design [7]. A 20-m-long measuring platform consisting of meter segments with a sampling frequency of 1000 Hz was used to verify the poling force of skis and poles in vertical and anterior-posterior directions during the classical style [5, 7].

The measured parameters of the poles are: Cycle time, Poling time, Relative poling time, Recovery time, Cycle frequency, Poling frequency, Peak poling force, Impulse of poling force, Average poling force and Average cycle poling force [8].

A single-axis force-measuring sensor with a sampling frequency of 1000 Hz and a 16-bit resolution at a range of ± 1000 N (due to the expected impacts of the pole) placed below the grips of the aluminum poles was used in standard Nordic walking poles [8]. Verification using the Kistler force-measuring platform showed an average absolute error of 1.1 N at a maximum force of 150 N in the case of poling cycles excluding the impacts on the pad. In the case of including the impacts on the pad, the average absolute error was 3 N. Calibration was performed using weights (5–40 kg, 5 kg step) and subsequent regression lines with respect to the voltage to force conversion.

A similar sensor was used for carbon racing poles with length from 115 to 170 cm in 2.5 cm increments [9].

Mechatronic Nordic walking poles were created to determine kinematic and dynamic gait parameters within rehabilitation and the possibility of developing new physiological methods using biofeedback [10]. They use resistance strain gauge bridges bonded with cyanopan adhesive to the top of pole.

To determine the physical load during Nordic walking, the system integrating measurement of linear acceleration, heart rate, pole load and GPS data with sampling frequency of 50 Hz was used [11]. This system allows the measurement of the pole angle during pushing phase with a sufficiently small error only from linear acceleration, but the authors suggest a more sophisticated version using 9 DOF (Degrees of Freedom) sensor.

2 Methods

The proposed system was tested on Leki composite carbon sticks.

2.1 The Device Design

To measure kinematic parameters, the TDK MPU-9250 Inertial Measurement Units (IMU) that meet the requirements of low power consumption, low cost and high performance along with run-time calibration firmware were used. For purpose of measurement, 3-axis gyroscope (range ±250, ±500, ±1000, ±2000 °/s), 3-axis accelerometer (range ±2 g, ±4 g, ±8 g, ±16 g) and integrated temperature sensor with a ±1% deviation in the operating temperature range were used [12].

Foil strain gauges bonded directly on the body of the composite carbon pole were used to measure the dynamics of movement (Fig. 1). The carbon surface was mechanically and chemically cleaned of the protective lacquer layer. The strain gauges were bonded to measure bends in perpendicular directions. The measured signals are amplified by the AD620 amplifiers, shifted and adjusted by the LM741 operational amplifier pairs. This signal processing is performed near the measurement bridges to reduce the effect of noise. The amplified analog signal is digitized by 12-bit AD converters of the used ATxmega processor. The digitized signal, together with the data from the IMU, is stored on the SD card in the form of pairs of text records for both poles (3 × 12 bit acceleration, 3 × 12 bit rotation, 10 bit temperature, 2 × 12 bit bend).

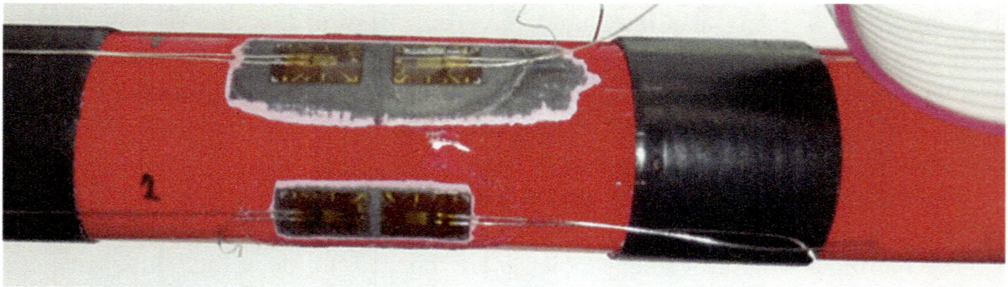

Fig. 1. Mounting of strain gauges.

The unit for synchronous data capturing and recording is housed in a durable case along with batteries for all day measurements (Fig. 2). The case is equipped with LED diods to indicate data recording and errors such as low battery voltage and it can be placed in the sports bag during measurement. The unit is also able to transmit real-time data via Wi-Fi interface.

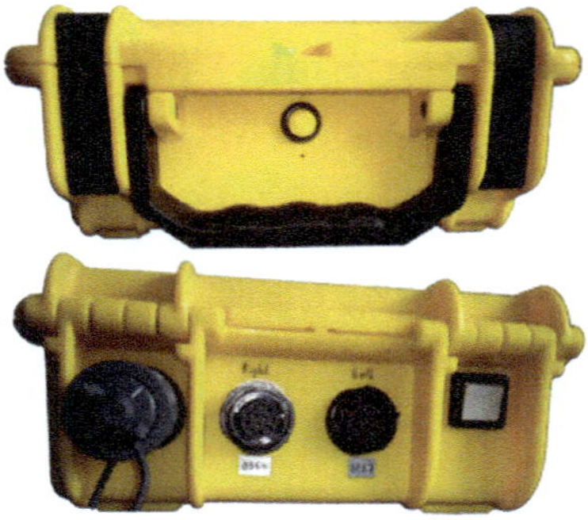

Fig. 2. Case with synchronous capturing and recording unit.

2.2 Calibration

The Vernier FP-BTA force-measuring platform was selected for calibration of the proposed system. The platform with the dimensions 28 × 32 × 5 cm was set on the sampling frequency of 100 Hz with a measurement range of −200 to 850 N, where the positive value represents a compression force with step 0.3 N. For calibration purposes, a load ranging from 0 to 180 N was performed (Fig. 3). The summary voltage (noted *Pole*) was based on the formula (1):

$$Pole = \sqrt{X^2 + Y^2} \tag{1}$$

Where X and Y are the voltages in the strain gauges in the X and Y axis respectively. The force was applied unevenly using the power of the hands. The MS Excel 2016 application was used for determination of linear regression. The equation of the regression line used for calibration and the corresponding coefficient R^2 is depicted in the Fig. 4. Application of the calibration is shown in Fig. 5.

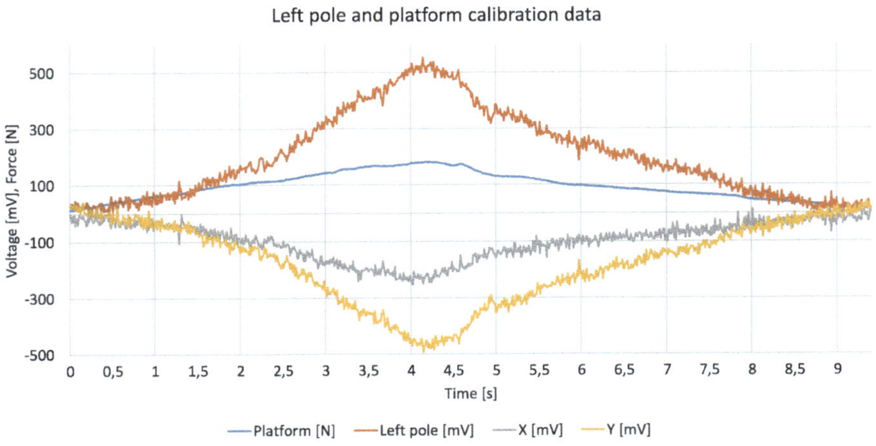

Fig. 3. Process of calibration.

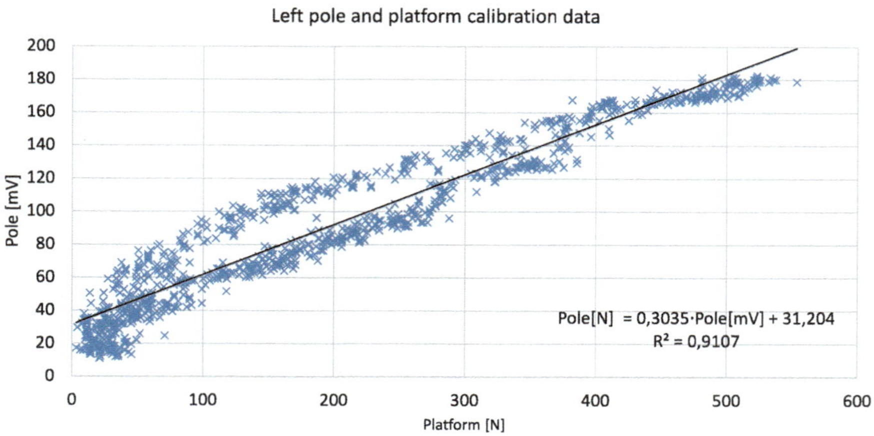

Fig. 4. Calibration – linear regression.

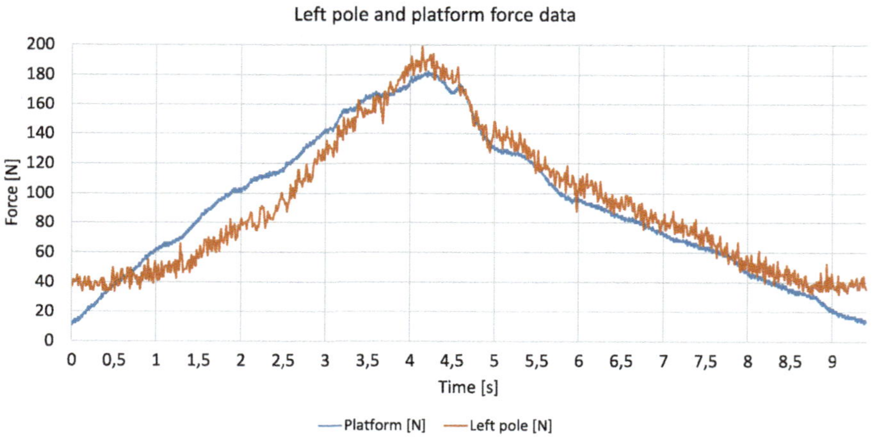

Fig. 5. Calibrated data.

2.3 Data Processing

A graphical user interface application for measured data visualization, analysis and export has been designed and implemented in MATLAB R2018a environment (Fig. 6).

The poles themselves are synchronized at the hardware level and therefore their signal is suitable for measuring the mutual poling time variability. For this purpose, it is possible to evaluate the data using the Poincaré graph and to evaluate SD1 and SD2 parameters that are commonly used to evaluate heart rate variability [13]. For the purposes of statistical processing of data are further used basic statistical parameters

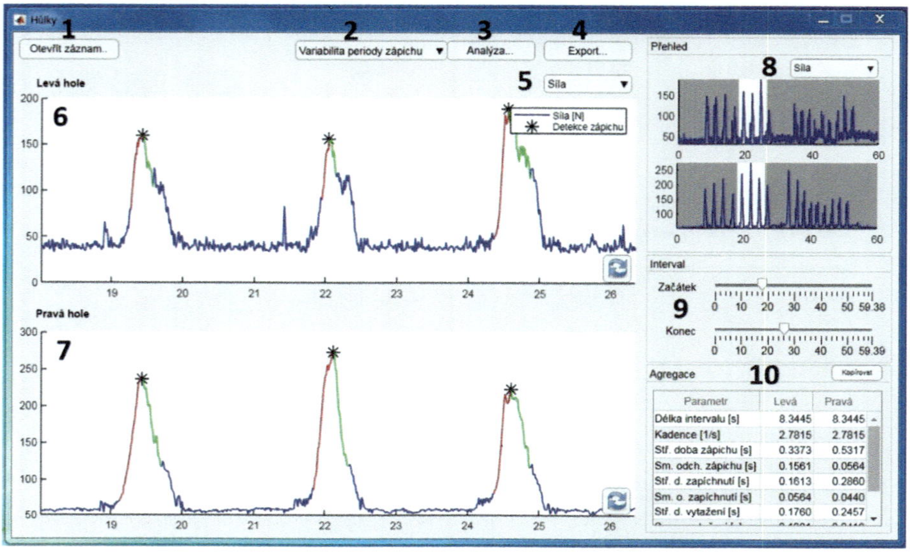

Fig. 6. GUI – (1) open record (2) analysis selection (poling period variability, poling force variability, poling orientation variability) (3) run Poincaré analysis (4) data export (5) value display (poling force, angular acceleration, linear acceleration, tilt) (6) left pole values (7) right pole values (8) cutout selection (9) selection of start and end of cutout (10) calculated parameters from cutout.

such as mean or standard deviation. For poling evaluation, the MATLAB function findpeaks with the option to set the minimum poling force and the minimum distance between two successive peaks was used. The peaks in the graph are noted by symbol *. To determine the poling angle, the Kalman filter using the linear and angular acceleration was applied [14].

2.4 Verification Measurement

For the purposes of the experimental verification of the device, a proband (39 years) with expertise in running biomechanics was selected. The poles were measured for the style "double pole" and the "diagonal stride" for 60 s on a hard pad.

3 Results

The course of the resultant forces in the double pole style with poling detection is shown in Fig. 7. The course of pole tilt in the double pole style is depicted in Fig. 8.

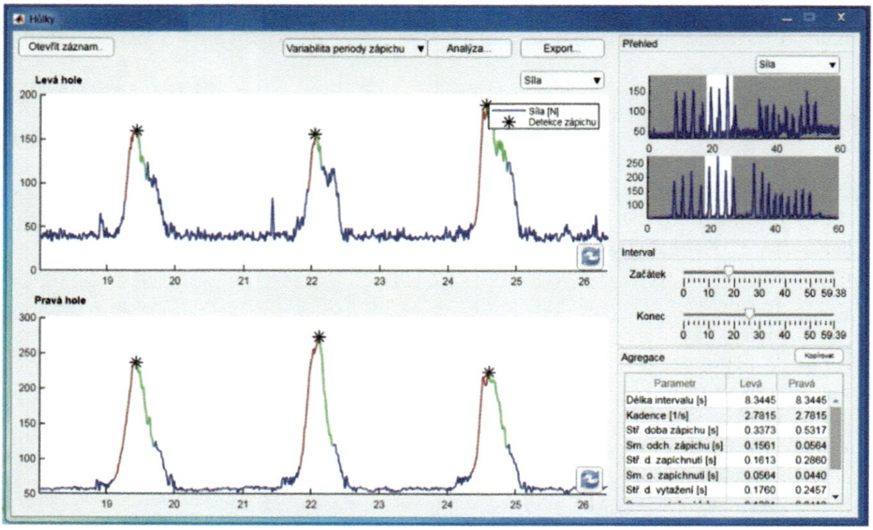

Fig. 7. Evaluation of poling force [N] in time [s] in double pole style. Upper graph – left pole, lower graph – right pole.

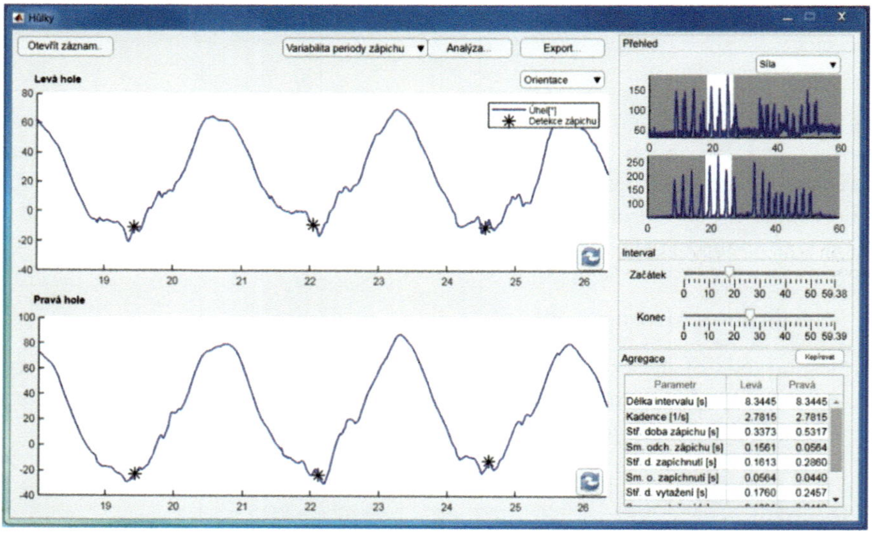

Fig. 8. Evaluation of poling tilt [°] in time [s] in double pole style. Upper graph – left pole, lower graph – right pole.

The course of the resultant forces in the diagonal stride style with poling detection is shown in Fig. 9. The course of pole tilt in the diagonal stride style is depicted in Fig. 10.

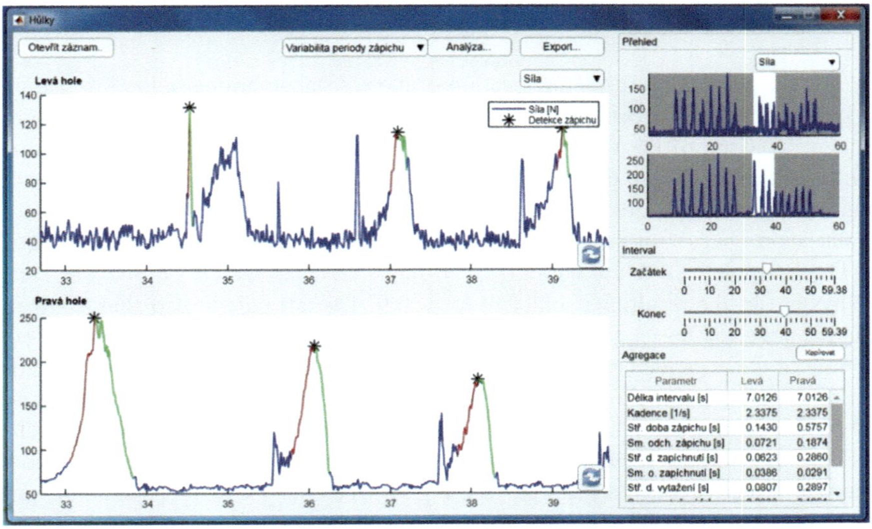

Fig. 9. Evaluation of poling force [N] in time [s] in diagonal stride style. Upper graph – left pole, lower graph – right pole.

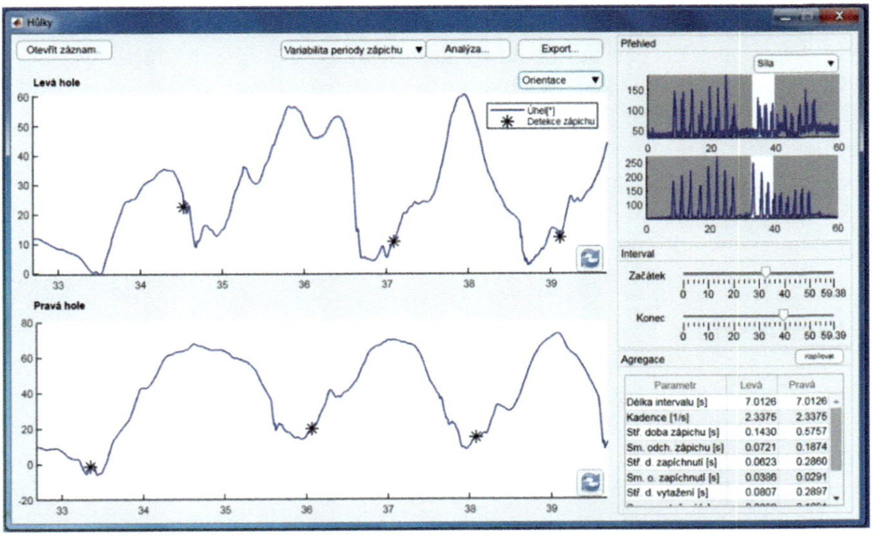

Fig. 10. Evaluation of poling tilt [°] in time [s] in diagonal stride style. Upper graph – left pole, lower graph – right pole.

4 Discussion

The proposed system utilizes the location of the resistance bridge in the lower third of the pole to measure its bend. This fact leads to possible complications due to the more frequent breaking of the pole during top-level sport and the associated application of high forces. However, in the case of Nordic walking and recreational use, this setup is applicable. The measurement system limits the user's movement to the same extent as a regular sports bag, so it can be used to assess existing and practice new techniques.

The left pole calibration curve (Fig. 4) has a certain degree of hysteresis. Despite this hysteresis, the resulting coefficient $R^2 > 0.9$. The left pole is also characterized by a higher level of noise, see Fig. 7. This may be due to the different quality of the bonding of the resistive bridge to the pole surface. It can be seen from the evaluation of the poling force during double pole style as shown in Fig. 7 that a higher force of about 250 N was applied to the right stick, compared to about 170 N at the left stick. This is probably due to the fact that the proband was right-handed. The same phenomenon occurred in the diagonal stride style, see Fig. 9. The data shows a considerable poling synchronism in case of double pole style compared to diagonal stride. In diagonal stride style, peaks appear before the poling itself. These can be caused by clicking the pole on the pad while moving the arm in the direction of running before the force poling itself. Another source is the detection of the mechanical impact of the cable connecting the pole with the case to the pole itself, despite the effort to eliminate the movement of cable. In the case of the first poling of the left pole (Fig. 9), this peak caused incorrect poling detection. This can be solved by using HW or SW filtering methods.

The tilt course of the pole relative to the earth coordinate system, along with the poling detection obtained from determined poling force, shows the typical course of the pole movement (Fig. 8). The same poling position on the left and right pole indicates the correctness of the measurement, which is apparently loaded with an offset that has not yet been determined. In the case of a pole tilt in diagonal stride style, the course of movement of the left and right pole can differ (Fig. 10).

5 Conclusion

The poles' load measuring system has proven its usefulness in assessing biomechanical parameters, particularly with regard to the evaluation of training methods and running styles for cross-country skiing and Nordic running. It can also be used in the evaluation and feedback of crutches with regard to proper rehabilitation. Future research will focus primarily on comparative measurements with a system built-in grips, miniaturizing the system, and measuring other run styles.

Acknowledgement. This work was supported by the Grant Agency of the Czech Technical University in Prague, grant No. SGS18/201/OHK4/3T/17.

Conflict of Interests. None declared.

References

1. Cardinale, M., Varley, M.C.: Wearable training-monitoring technology: applications, challenges, and opportunities. Int. J. Sports Physiol. Perform. **12**(Supplement 2), 55–62 (2017)
2. Morgan, S.J., Mora, J.A.M.: Effect of heart rate variability biofeedback on sport performance, a systematic review. Appl. Psychophysiol. Biofeedback **42**(3), 235–245 (2017)
3. Baca, A., Kornfeind, P.: Rapid feedback systems for elite sports training. IEEE Pervasive Comput. **5**(4), 70–76 (2006)
4. Bideau, B., Kulpa, R., Vignais, N., Brault, S., Multon, F., Craig, C.: Using virtual reality to analyze sports performance. IEEE Comput. Graph. Appl. **30**(2), 14–21 (2010)
5. Vähäsöyrinki, P., Komi, P.V., Seppälä, S., Ishikawa, M., Kolehmainen, V., Salmi, J.A., Linnamo, V.: Effect of skiing speed on ski and pole forces in cross-country skiing. Med. Sci. Sports Exer. **40**(6), 1111–1116 (2008)
6. Salpavaara, T., Verho, J., Lekkala, J., Halttunen, J.: Wireless insole sensor system for plantar force measurements during sport events. In: Proceedings of IMEKO XIX World Congress on Fundamental and Applied Metrology, Lisbon, Portugal, pp. 2118–2123, September 2009
7. Ohtonen, O., Lindinger, S., Lemmettylä, T., Seppälä, S., Linnamo, V.: Validation of portable 2D force binding systems for cross-country skiing. Sports Eng. **16**(4), 281–296 (2013)
8. Krejčí, J., Jakubec, A., Pšurný, M., Janura, M.: Development and validation of system for measuring poling forces during nordic walking. Acta Gymnica **43**(3), 49–56 (2013)
9. Bortolan, L., Pellegrini, B., Schena, F.: Development and validation of a system for poling force measurement in cross-country skiing and Nordic walking. In: ISBS-Conference Proceedings Archive (2009)
10. Szrek, J., Muraszkowski, A., Bałchanowski, J., Wudarczyk, S., Jasiński, R., Niebudek, T., Woźniewski, M.: Force measurement module for mechatronic nordic walking poles. In: International Conference on Computer Aided Engineering, pp. 790–794. Springer, Cham (2018)
11. Mocera, F., Aquilino, G., Somà, A.: Nordic walking performance analysis with an integrated monitoring system. Sensors **18**(5), 1505 (2018)
12. MPU-9250 Product Specification. InvenSense Inc. p. 42, 20 June 2016
13. Volf, P., Kutílek, P., Hejda, J., Smrčka, P., Křivánek, V., Doskočil, R., Farlík, J., Casar, J., Kliková, M.: Heart rate variability during long-term air traffic control in cadets. In: 2018 41st International Conference on Telecommunications and Signal Processing (TSP), IEEE, Athens, pp. 1–5, July 2018
14. Madgwick, S.O., Harrison, A.J., Vaidyanathan, R.: Estimation of IMU and MARG orientation using a gradient descent algorithm. In: 2011 IEEE International Conference on Rehabilitation Robotics, IEEE, Zurich, pp. 1–7, June 2011

Regular Sessions: Therapeutic and Diagnostic Systems, Devices and Technologies and Clinical Engineering

A Risk Stratification Model for Early Cognitive Impairment After Diagnosis of Parkinson's Disease

Kostas M. Tsiouris[1,3], Spiros Konitsiotis[2], Dimitrios D. Koutsouris[1], and Dimitrios I. Fotiadis[3,4(✉)]

[1] School of Electrical and Computer Engineering, National Technical University of Athens, 15773 Athens, Greece
[2] Medical School, University of Ioannina, 45110 Ioannina, Greece
[3] Unit of Medical Technology and Intelligent Information Systems, University of Ioannina, 45110 Ioannina, Greece
[4] Department of Biomedical Research, IMBB-FORTH, 45110 Ioannina, Greece
fotiadis@cc.uoi.gr

Abstract. Cognitive decline is very common in patients with Parkinson's disease (PD) as longitudinal studies have shown that given enough time almost all patients will eventually end up with a diagnosis of Mild Cognitive Impairment (MCI) or dementia. Certain patients however are found to be more susceptible to early cognitive impairment soon after the diagnosis of PD. In this study, baseline evaluation outcomes from newly diagnosed patients from the Parkinson's Progression Markers Initiative are evaluated to identify risk factors for early cognitive impairment using machine learning techniques. Applying an extensive search in the available feature space, consisting of more than 400 baseline features, to isolate the most informative predictors, the proposed methodology can discriminate patients having a diagnosis of MCI or dementia within the first 5 years of PD from those with normal cognition with an accuracy of 80.38%. Older age along with non-motor symptoms including cognitive and memory dysfunction, sleep problems, daytime sleepiness, smell dysfunction, mood impairment and anxiety at baseline are strong determinants of early MCI and dementia.

Keywords: Parkinson's disease · Cognitive impairment · Risk stratification

1 Introduction

Parkinson's disease (PD) is primarily considered a movement disorder due to its motor complications, but it is the non-motor symptomatology which further complicates patient management. Cognitive decline in particular is very troublesome as the diagnosis of Mild Cognitive Impairment (MCI) is considered a milestone in PD and a risk factor for the development of dementia [1, 2]. In fact, longitudinal studies have shown that cognitive decline is almost a certainty in PD. In the Sydney longitudinal study of PD patients, MCI was diagnosed in 84% of the patients in the 15 year follow-up mark while 48% of the remaining patients at 20 years had a diagnosis of dementia [3].

J. Henriques et al. (Eds.): MEDICON 2019, IFMBE Proceedings 76, pp. 653–660, 2020.
https://doi.org/10.1007/978-3-030-31635-8_78

Similar findings were also reported in a longitudinal study in Norway, where 52% of the patients were already diagnosed with dementia at four years, with this figure increasing to 78% at 8 years [4]. In a recent study tracking cognitive dysfunction in patients with PD, the analysis showed that from those having normal cognition at baseline, 9.9% had shifted to MCI after the first year, 23.2% at the third year and 28.9% after 5 years of follow-up [5].

In the search for risk factors for early cognitive decline, advanced age, REM sleep problems and male as gender have been shown to be among the most well established early prognostic factors for MCI and dementia [6, 7]. Hyposmia is another commonly reported factor in newly diagnosed PD patients, as olfactory dysfunction has been associated with increased risk for both MCI and dementia in recent longitudinal studies [8, 9]. Using data from the PPMI study, authors in [10] found that the prognostic accuracy can be further increased when combining the above factors with signs of depression, motor symptoms and fluid biomarkers. Regarding motor symptomatology, the results in [11] showed that postural and gait dysfunction rather than tremor-dominant motor phenotypes are more indicative of latter cognitive decline. Semantic fluency and visuospatial functioning were also suggested as useful features, while anxiety has been also correlated with MCI in another longitudinal study [12].

In this study, a comprehensive evaluation of a wide variety of baseline features is performed to isolate potential prognostic factors of early cognitive decline using open observational data from 423 newly diagnosed PD patients. To the best of our knowledge this is the first time that machine learning methods are used to analyze prognostic baseline features, as all previous studies in the literature relied on statistical analysis. The proposed methodology is able to discriminate patients having a diagnosis of MCI or dementia within the first five years of the disease with an average classification accuracy of 80.38% by selecting the most informative from about 400 baseline features covering every aspect of PD symptomatology.

2 Materials and Methods

2.1 PPMI Dataset

The Parkinson's Progression Markers Initiative (PPMI) is an ongoing prospective observational study collecting and providing open data from longitudinal evaluations. Subjects were enrolled into five cohorts including newly diagnosed PD patients, healthy controls, prodromal subjects, carriers of genetic mutations associated with PD and patients without evidence of dopaminergic deficit. The 423 subjects from the newly diagnosed PD patients (i.e. De Novo cohort) are selected in this study to extract early prognostic factors of cognitive impairment in the first five years after PD diagnosis. From the 423 enrolled patients 65.5% are males while the mean age of the entire population at baseline evaluation was 61.6 years (34–85 years). Following the PPMI protocol, cognitive categorization is performed annually to identify PD patients with

Dementia (PDD), PD with Mild Cognitive Impairment (PD-MCI) or Normal Cognition (PD-NC). The clinician also provides an index of his/hers cognitive diagnosis at 0–9%, 10–49%, 50–89% and 90–100% level of confidence.

2.2 Data Preprocessing

The patients are split into two classes: (a) normal cognition, with no diagnosis of cognitive decline in the first five years of follow-up, and (b) cognitive decline, including patients with a diagnosis of either PD-MCI or PDD in the follow-up period. To minimize the impact of uncertain diagnoses, only the cases where a 90–100% level of confidence is noted by the clinician in the diagnosis of PD-MCI or PDD are considered in the cognitive decline class. In addition, patients having a diagnosis of PD-MCI or PDD at baseline evaluation are excluded. Thus, the initial population of the De Novo cohort is reduced to 333 patients, with 63 patients (18.92%) having a highly confident diagnosis of either PD-MCI or PDD and are assigned to the cognitive decline class, with the remaining 270 patients (81.08%) assigned to the normal cognition class.

2.3 Feature Extraction

The PPMI dataset offers an extensive list of baseline evaluations that were performed to capture multiple aspects of PD. Every available outcome is used in this study to provide a more comprehensive evaluation of early cognitive decline determinants. In total, 601 baseline features are considered from:

- **General patient information:** Demographics, Family history, Socio-economics.
- **Neurological examination:** Cranial Nerves, General neurological examination.
- **Physical examination:** General physical examination, Vital signs.
- **Motor symptoms evaluation:** MDS-UPDRS parts I-IV, Modified Schwab and England ADL, PASE Household Activity.
- **Non-motor symptoms evaluation:** Benton Judgment of Line Orientation, Epworth Sleepiness Scale (ESS), Geriatric Depression Scale (GDS), Hopkins Verbal Learning Test (HVLT), Letter-Number Sequencing (LNS), Montreal Cognitive Assessment (MoCA), Questionnaire for Impulsive-Compulsive disorders (QUIP), REM Sleep Disorder Questionnaire (RBDSQ), Scale for Outcomes in Parkinson's disease for Autonomic Symptoms (SCOPA-AUT), Lexical and Semantic Fluency, State-Trait Anxiety Inventory (STAI), Symbol Digit Modalities (SDM), University of Pennsylvania Smell ID Test (UPSIT).
- **MRI/DatScan-SPECT Imaging:** DatScan-SPECT and MRI results.
- **Laboratory results:** Blood Chemistry and Hematology, Fluid biomarker analysis, Bio-specimen analysis.
- **Genetic analysis:** SNCA multiplication and Genetic Risk Score, NeuroX Genotyping Selected SNPs.

2.4 Feature Selection

Considering the high ratio of features to instances in the PPMI dataset, it is essential to reduce their number to the most informative prior to classification. The first step is to remove baseline features containing many missing values. A cut-off threshold is used to remove features missing more than 20% of their total number of values, reducing the baseline feature vector to 428 features. Then, features containing a single type of values are also removed as they have no discrimination potential (e.g. all patients have bradykinesia at baseline). An extensive feature evaluation scheme is then performed to estimate feature importance by evaluating multiple subsets of features.

Starting from an initial 2-feature node ($f_1 \neq f_2$), their classification performance is evaluated using the training samples. If the two starting features have a sensitivity and specificity above 60%, they are considered as viable features and are expanded by adding more features to improve classification performance, otherwise they are discarded and a different couple of starting features is evaluated. Features are added in each starting node one by one according to their classification accuracy gain through a recursive analysis. In more detail, to add a third feature in a starting node ($f_1 \neq f_2 \neq f_3$) every feature from the feature vector (i.e. excluding f_1, f_2) is evaluated as f_3 and the 3-feature subset that provided the highest classification accuracy is selected. This process is repeated again adding subsequent baseline features (f_4, f_5, etc.), until the classification performance cannot be further improved. The subset that provided the best fit is stored and a new starting node is evaluated until every possible combination of starting nodes is tested. A decision tree model (see Sect. 2.5) is used to evaluate the model's fit to the training dataset during feature selection. Once all feature subsets have been evaluated, feature importance is estimated by measuring the prevalence of each feature in the stored subsets and the 10 more common features are selected for validation with the testing dataset. The effect of selecting different numbers of features in classification performance is evaluated in the results.

2.5 Classification

Decision trees combined with the AdaBoost algorithm are employed for the main classification algorithm using the respective open libraries from the Weka API [13] implemented in Python through the *python-weka-wrapper3* package. Decision trees are among the most common techniques in clinical machine learning applications, as they can be easily interpreted by non-machine learning experts as they can be depicted schematically or transformed into decision rules. The Weka API utilizes the C4.5 algorithm to create a decision tree that is extended to its maximum depth and then pruned to prevent overfitting. The AdaBoost algorithm is an ensemble method that enhances a weak model's inability to adopt in challenging training datasets, by focusing on parts of the sample space that are harder to be correctly classified [14]. The AdaBoost algorithm builds separate decision tree models on different subsets of the training dataset, with the outcome of each one of these models being combined to provide the final weighted output for any given instance of the testing dataset. The number of iterations is set to 100.

3 Results

A 5-fold cross-validation scheme is used to validate the classification performance of the selected baseline features and assess their accuracy in early cognitive decline patient stratification. To overcome the high imbalance between the two classes, as the normal cognition class has more than 4:1 the attributes of the cognitive decline class, the normal cognition class is randomly split into four smaller datasets consisting of 67 or 68 patients (i.e. $67 \times 2 + 68 \times 2 = 270$). The classification performance with each subset of the normal cognition class is presented in Table 1, including a comparison with two popular feature selection methods from the Weka API, namely correlation-based feature subset selection filtering and the wrapper algorithm [13]. The proposed methodology achieved on average a superior classification performance reaching a mean accuracy of 80.38%.

Table 1. Classification results for the prognosis of cognitive decline at baseline evaluation. Subset #1–4: Normal cognition class split, Acc = Accuracy, Sens = Sensitivity, Spec = Specificity.

Normal cognition class split	Proposed feature selection+Decision Tree+AdaBoost			Correlation-based feature selection +Decision Tree +AdaBoost			Wrapper+Decision Tree+AdaBoost		
	Acc. (%)	Sens. (%)	Spec. (%)	Acc. (%)	Sens. (%)	Spec. (%)	Acc. (%)	Sens. (%)	Spec. (%)
Subset #1	78.46	76.19	80.60	66.92	68.30	65.70	65.38	63.56	67.21
Subset #2	81.54	79.36	83.58	71.54	65.10	77.60	60.77	54.00	67.23
Subset #3	79.23	76.19	82.09	72.31	65.10	79.10	65.38	61.95	68.70
Subset #4	82.31	80.95	83.58	70.00	68.30	71.60	58.46	69.83	47.81
Mean:	**80.38**	**78.17**	**82.46**	**70.19**	**66.70**	**73.50**	**62.50**	**62.33**	**62.74**

The most informative baseline features that were selected as prognostic factors of early cognitive decline are shown in Table 2. Features are presented according to the number of normal cognition class subsets they are selected in, starting from features found in all four subsets and followed by those found in three and two subsets, respectively. Features selected in a single subset are discarded as less informative.

The impact of the number of selected features in the classification accuracy is presented in Fig. 1. The AdaBoost algorithm provides an increase in classification accuracy with adding more features until the 10-feature mark, with fluctuating accuracy results and an overall decrease in performance afterwards.

Table 2. Baseline features selected as early cognitive decline determinants. Features are presented according to their prevalence among the normal cognition class subsets.

Feature prevalence	Baseline features
4 subsets	Age (higher), LNS: Sum questions 1–7 (lower), LNS: Derived scale score (lower), SDMT: Total score (lower)
3 subsets	ESS: Sitting quietly after lunch, HVLT: Delayed recall, HVLT: Recognition false positives related, RBDSQ: Total score (higher), SCOPA-AUT: Strain hard to pass stools, UPSIT: Total score (lower), MDS-UPDRS: Cognitive impairment
2 subsets	MDS-UPDRS: Sleep problems, LNS: Trial 3a, Semantic Fluency: Total number of fruits, STAI: Not feeling pleasant, Brain Region: Caudate Right/Left (lower), Benton Judgement of Line Orientation: 15 items (low), MoCA: Visuoconstructional skills-Cube, GDS: Summary score (higher), SCOPA-AUT: Total score (higher), STAI: Feel indecisive

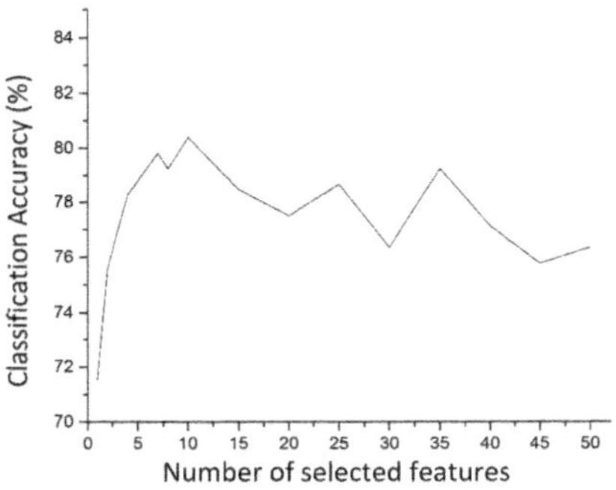

Fig. 1. Effect of the number of selected features on classification accuracy.

4 Discussion

In this study a methodology based on machine learning is proposed to identify prognostic factors for early cognitive decline in newly diagnosed PD patients through multi-parametric evaluations over a comprehensive set of baseline features. Based on the results shown in Table 1, patients having a diagnosis of either PD-MCI or PDD within the first five years can be discriminated from those with normal cognition with 80.38% accuracy. Performing an extensive analysis within the feature space provided a better fit in the training dataset which led to higher classification performance, as the proposed methodology improved total classification accuracy by 10.19% compared to correlation-based feature selection (i.e. 80.38% vs 70.19%) and by 17.88% compared to the wrapper algorithm (i.e. 80.38% vs 62.5%), respectively. It should be noted, however, that the proposed methodology is more computationally intensive since a more comprehensive evaluation of various feature subsets is performed.

Considering the heterogeneity in PD symptomatology, the wide range of selected features presented in Table 2 suggests that there are different factors that should be considered. Higher age at PD diagnosis is among the most commonly found baseline factor for early cognitive decline. Signs of cognitive dysfunction at baseline evaluation are also found to be strong determinants as patients from the cognitive decline class scored lower at Letter-Number Sequencing test assessing working memory, and Symbol Digit Modalities test, which assesses concentration and speed of cognitive processing. These are also complemented by features from other cognitive and memory tests including the Hopkins Verbal Learning test, Lexical and Semantic Fluency test and Montreal Cognitive Assessment scale. These findings suggest that signs of cognitive and memory dysfunction at the early stages of PD that are not currently sufficient for PD-MCI diagnosis, are very likely to evolve leading to a confident diagnosis within the first few years.

Another major category of prognostic factors includes sleep problems and daytime sleepiness as items from Epworth Sleepiness scale, REM Sleep Disorder Questionnaire and MDS-UPDRS sleep evaluation were selected. Mood disturbances including signs of depression and anxiety were also noted at the baseline evaluation of patients with early cognitive decline, as multiple items from the Geriatric Depression scale and the State-Trait Anxiety Inventory were selected. Finally, patients from the cognitive decline class expressed olfactory dysfunction scoring lower in the University of Pennsylvania smell test and reported gastrointestinal problems at baseline evaluation. Motor symptoms were not found as significant factors, nor did the laboratory results from blood analysis and bio-specimens. No association with gender was identified either.

5 Conclusions

The evaluation of multiple baseline features suggests that there are various non-motor symptoms declaring a higher risk for early cognitive decline in PD. Prognostic factors include advanced age at diagnosis, weak memory and speed of cognitive processing, smell dysfunction, sleep problems and mood disturbances.

References

1. Pedersen, K.F., Larsen, J.P., Tysnes, O., Alves, G.: Prognosis of mild cognitive impairment in early parkinson disease: the Norwegian ParkWest study. JAMA Neurol. **70**(5), 580–586 (2013)
2. Hobson, P., Meara, J.: Mild cognitive impairment in Parkinson's disease and its progression onto dementia: a 16-year outcome evaluation of the Denbighshire cohort. Int. J. Geriatr. Psychiatry **30**, 1048–1055 (2015)
3. Hely, M.A., Reid, W.G., Adena, M.A., Halliday, G.M., Morris, J.G.: The Sydney multicenter study of Parkinson's disease: the inevitability of dementia at 20 years. Mov. Disord. **23**, 837–844 (2008)

4. Aarsland, D., Andersen, K., Larsen, J.P., Lolk, A.: Prevalence and characteristics of dementia in parkinson disease: an 8-year prospective study. Arch. Neurol. **60**(3), 387–392 (2003)
5. Pedersen, K.F., Larsen, J.P., Tysnes, O.B., Alves, G.: Natural course of mild cognitive impairment in Parkinson disease: a 5-year population-based study. Neurology **88**(8), 767–774 (2017)
6. Anang, J., Nomura, T., Romenets, S.R., Nakashima, K., et al.: Dementia predictors in Parkinson disease: a validation study. J. Parkinson's Dis. **7**(1), 159–162 (2017)
7. Postuma, R.B., Bertrand, J.A., Montplaisir, J., Desjardins, C., et al.: Rapid eye movement sleep behavior disorder and risk of dementia in Parkinson's disease: a prospective study. Mov. Disord. **27**(6), 720–726 (2012)
8. Baba, T., Kikuchi, A., Hirayama, K., Nishio, Y., et al.: Severe olfactory dysfunction is a prodromal symptom of dementia associated with Parkinson's disease: a 3 year longitudinal study. Brain **135**(1), 161–169 (2012)
9. Fullard, M.E., Tran, B., Xie, S.X., Toledo, J.B., et al.: Olfactory impairment predicts cognitive decline in early Parkinson's disease. Parkinsonism Relat. Disord. **25**, 45–51 (2016)
10. Schrag, A., Siddiqui, U.F., Anastasiou, Z., Weintraub, D., Schott, J.M.: Clinical variables and biomarkers in prediction of cognitive impairment in patients with newly diagnosed Parkinson's disease: a cohort study. Lancet Neurol. **16**(1), 66–75 (2017)
11. Williams-Gray, C.H., Foltynie, T., Brayne, C.E.G., Robbins, T.W., Barker, R.A.: Evolution of cognitive dysfunction in an incident Parkinson's disease cohort. Brain **130**(7), 1787–1798 (2007)
12. Hu, M.T., Szewczyk-Królikowski, K., Tomlinson, P., Nithi, K., et al.: Predictors of cognitive impairment in an early stage Parkinson's disease cohort. Mov. Disord. **29**(3), 351–359 (2014)
13. Hall, M., Frank, E., Holmes, G., Pfahringer, B., et al.: The WEKA data mining software: an update. ACM SIGKDD Explor. Newsl. **11**(1), 10–18 (2009)
14. Freund, Y., Schapire, R.E.: Experiments with a new boosting algorithm. In: ICML, vol. 96, pp. 148–156 (1996)

Upper Limp Movement Analysis of Patients with Neuromuscular Disorders Using Data from a Novel Rehabilitation Gaming Platform

Achilleas Chytas[1], Dimitris Fotopoulos[1], Vassilis Kilintzis[1], Theodoros Loizidis[2], and Ioanna Chouvarda[1](✉)

[1] Lab of Computing, Medical Informatics and Biomedical Imaging Technologies, Aristotle University of Thessaloniki, Thessaloniki, Greece
{achillec,difoto,ioannach}@auth.gr, billyk@med.auth.gr

[2] Theodoros Loizidis Apokatastasi LTM, Thessaloniki, Greece
Loizidis@yahoo.com

Abstract. This paper describes the methodology for analyzing data from a novel platform that utilizes gamification techniques targeting patients with neuromuscular disorders that result in upper-limb movement limitations. The patient is asked to perform a number of sessions as prescribed by the physician for rehabilitation purposes, while the data are used to assess their progress. The hand movement is analyzed, and features are extracted regarding the movement patterns velocity, jitter, and trajectory. A set of sessions derived from healthy individuals have being recorded and analyzed so as to establish a baseline for the metrics. A statistical analysis of the differences between the healthy subjects and the patients is performed, helping us to focus on features of interest to the disease. The results will help determine how the patients' motor skills are improving as the therapy progress, and accordingly adjust the number and type of sessions prescribed in a personalized manner.

Keywords: Serious game · Rehabilitation · Signal-analysis

1 Introduction

Motor control is considered one of the most difficult set of functionalities of the human body. It is a process or a set of sub-processes, during which humans move and coordinate their muscles and limbs so they can perform a motor skill. Every human from the moment of his birth trains in motor control (Motor Learning) through the integration of sensory-motor information. In Motor Learning, movements are consolidated in the Central Nervous System (CNS), at first through observation and then by repeatedly performing them. In some cases, certain pathologies affect the CNS, resulting in the loss of cognitive functions of the brain. This may impact several motor functions and cause partial or complete loss of them. Rehabilitation programs aim to detect any motor deficits of each patient and help them regain control of their movements through motor learning. The standard method of doing that, is the repetition of a movement's correct form, so that it can be permanently stored in the CNS.

J. Henriques et al. (Eds.): MEDICON 2019, IFMBE Proceedings 76, pp. 661–668, 2020.
https://doi.org/10.1007/978-3-030-31635-8_79

The repetitive training of isolated movements is a fundamental principle behind improving the outcome of motor rehabilitation.

Over the recent years, there has been an increasing interest on how technology can assist physical rehabilitation and conventional treatment methods for patients with physical impairments [1, 2]. With recent advancements in technology, a new domain, that combines exergames, gamification mechanisms and traditional rehabilitation methodologies, has emerged [3]. Therapeutic solutions, that combine novel software and hardware components, are implemented to facilitate the rehabilitation for patients with motor deficits. They can facilitate the process of Motor Learning, by offering the ability of executing a series of repetitive and functional movements efficiently [4], in a less monotonous way.

A rehabilitation platform that can be used as a tool for the medical treatment of patients with upper extremity motion difficulties is presented on this paper. The proposed platform was not created with a specific pathological condition in mind and its objective is physical impairments of the upper limbs, regardless of the cause.

The platform will be used to determine concepts and methods that will increment the efficacy of the Motor Learning process. Specifically, the form and the frequency of repetitions that a movement should be performed will be examined, with the purpose of optimizing therapeutic programs.

The main focus of the paper is the analysis of the data gathered during rehabilitation sessions. Certain features are extracted and compared across a sample of patients and healthy subjects, and evaluated as indicators of improvement.

2 Description of Milord

The therapy program conducted within the implementation idea presented above, aims at the rehabilitation of the arms and focuses on the following exercise movements: flexion/extension of the shoulder, horizontal adduction/abduction of the shoulder, and supination/pronation of the forearm. The repetition of these specific movements as many times as possible is the primary goal. In its current state, the platform has one game scenario that the user can perform these rehabilitation exercises in (term "user" refers to a candidate patient, a medical professional or an individual that is testing the game). The scenario follows the flying simulation paradigm. Inside the virtual world of the game, the user controls a red polygonish airplane. The interaction between the user and this game object (avatar) is achieved by the camera tracker Leap Motion Controller [https://www.leapmotion.com], which uses computer vision technology to recognize hands in the scene and calculates a set of measurements that describe them. The user places his hand above the sensor, and can move it along the horizontal and vertical plane, as also rotate it along the depth axis. Inside the virtual world of the game, the airplane mimics the hand's movement. This amplifies the sense of immersion of the user in the game world by reinforcing the game metaphor - the impression that the hand is the airplane. The purpose of the game is for the user to guide the avatar through orthogonal objects (gates) that head towards it (Fig. 1).

The goal is the highest possible number of repetitions, so the condition for the end of a game session is either a time limit or a limit on the number of the gates that the user

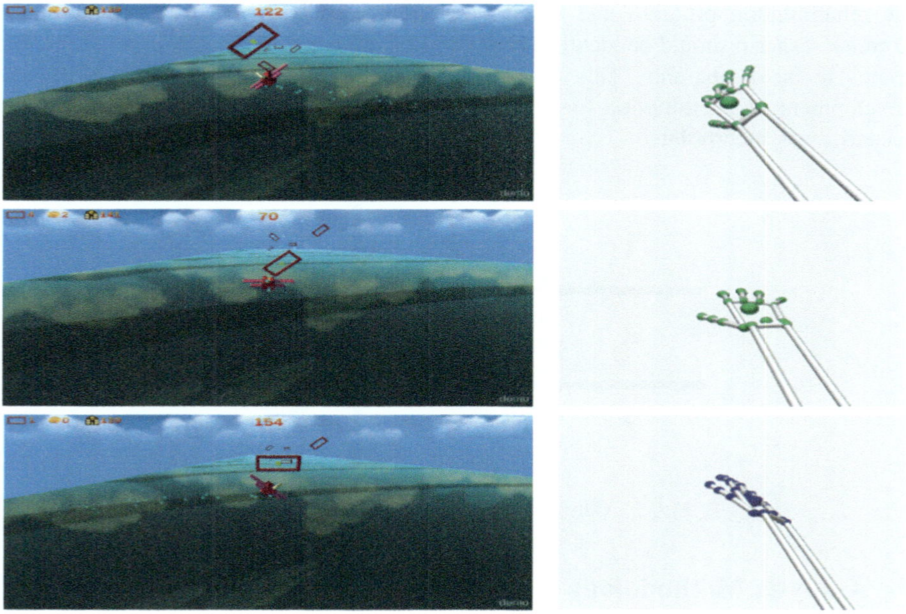

Fig. 1. Movement of the hand and its effect on the virtual world of the game

will successfully guide the airplane through (successful gates). Because of the closely defined set of movement exercises, it was a requirement based design decision that the airplane avatar of the game -and by extension the hand in the real world - cannot move with six degrees of freedom (DOF) [5]. Thus the airplane's movement is restricted to the X and Y plane. The illusion that the airplane advances forward is given by making every game object in the scene move towards the player's avatar. The movement across the horizontal and vertical plane corresponds to these pairs of movement: flexion/extension and adduction/abduction. Pronation and supination of the arm is mapped to the rotation of the airplane on Z axis.

The gate objects that the user leads the aircraft in, appears in a predefined 3×3 grid - meaning there are 9 possible places where gates can appear in the screen (Fig. 2). They were arranged in this manner to reinforce the consistency of the data that are recorded during therapy sessions and respectively to make the analysis process more reliable. This will assist in reaching conclusions about the patterns of the hand's movement and improve the rehabilitation regimen, without adding other influencing factors. The order of display of the gates is random, and is handled by a pseudo-random algorithm, with a modification introduced in it, that ensures each gate will emerge at least once in a game session, given that the overall number of gates will be $N \geq 9$. The randomness is an important factor in this rehabilitation scenario, because it averts the patient from memorizing the order in which the gates appear on the screen, and from completing a game session by memory. Therefore, the results we obtain from the analysis are unbiased in this respect. This is also true for the process of assessment of the patient's functional status, which usually happens initially when the patient joins the rehabilitation program and when she finally completes it. In such a case, the exercise scenario should be identical both for the initial and the final assessment, for the results to be comparable. This assessment case was taken into account during the development and, the option for a physician to be able to choose a previous played scenario was incorporated.

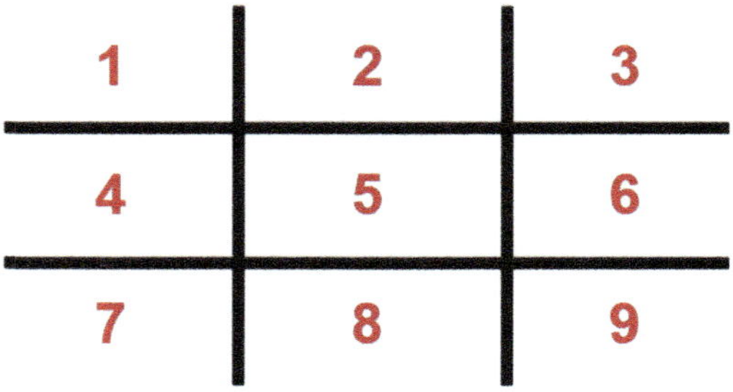

Fig. 2. Gates 3 × 3 grid where the gates appear

3 Analysis Methodology

The main focus of the analysis on the current stage is to determine the variables that are going to be examined and explore the differences between healthy subjects and patients.

The frame rate is capped at 60 fps but minor differences in the time that the frames appear may occur. Since the data collection rate is based on the game frame rate, in order to facilitate an analysis that support exploration in the frequency domain the time-series of the coordinates are interpolated at a steady rate of 16.667 ms.

During the gameplay there are two states, the user is either moving the hand or user tries to keep it steady. For the analysis we divide the gameplay sessions into parts. Each part corresponds to the period of each time window (W_i), between two consecutive gates. Since the gates (G_i) are moving towards the avatar in controlled pace these time windows all have the same duration, with the exception of the first gate which appears a few seconds after the start in order to provide the user ample time to get accustomed to the game.

On each part (W_i) we can distinct 3 different sub-periods.

1. Response (DT1: t0-t1): it refers to the period of time after the user has reached the Gi gate and when they become aware of the upcoming gate Gi + 1 and they begin to move towards it. Steady period.
2. Movement (DT2: t1-t2): it refers to the period of time where the user is moving from Gi towards the upcoming gate G_{i+1}. Movement period.
3. Stabilize (DT3: t2-t3): it refers to the period of time after the user has arrived to the X, Y coordinates that correspond to the G_{i+1} gate and is waiting to reach it (plane pass through the gate). Steady period.

The time points t_{i0}, t_{i1}, t_{i2}, t_{i3} are defined as follow.

- t0: the time point when the avatar collided with the previous gate or had the same Z in case the user missed it.
- t1: the time point after the user has passed through the previous gate and started moving towards the upcoming gate
- t2: the time point when the user considers that has reached the upcoming gate and greatly reduces the movement
- t3: the time point when the avatar collided with the upcoming gate or had the same Z in case the user missed it.

Regarding the t0 and t3 time points, $t_{i-1,3}$ is equal to t_{i0}. For the first target t_1 is defined as the start of the session $t_{10} = 0$. These time points are registered in the raw data that are produced during the gaming session. Time points t1 and t2 on the other hand are calculated based on the movement pattern analysis.

Figure 3 depicts the movement from gate G_{i-1} to $G_{i,}$

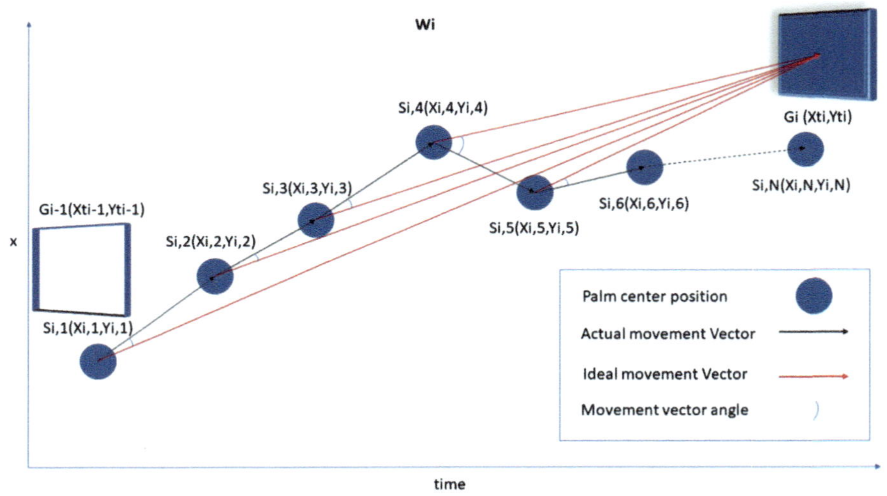

Fig. 3. Movement from gate G_{i-1} to G_i. The $S_{i,j}$ represents the hand position on each j frame for every i window (W_i).

The t1 time point is determined as followed:

1. Determine the periods when the hand is moving
   a. Calculate the instantaneous velocity $V_{i,j}$ (the velocity on j frame of window W_i).
   b. Calculate the rolling average (length = 15) of the instantaneous velocity.
   c. Calculate the distribution of the instantaneous velocities of window W_i.
   d. The hand is considered to be moving the rolling average is above the 75th percentile of the instantaneous velocity
2. Determine when the hand is moving towards the upcoming target G_i
   a. Calculate the Euclidean vector $S_{i,j}$ of every two consecutively points (S_i1 to S_i2, S_i2 to S_i3 etc.)

b. Compare each vector $S_{i,j}$ to the vector starting from $S_{i1}(X_{i1}, Y_{i1})$ to $G_{i1+1}(X_{i1+1}, Y_{i1+1})$ and extract the angle difference, $<S_i 1G_{i1+1}$
c. Calculate the circular angle on a rolling window.
d. T1 is defined as the first data point where the circular angle difference was below a certain threshold (10°) or the first time it was found to be below the 95th percentile

The t2 time is described as the first time after the t1 time point when the movement has been halted or greatly reduced, while also being on X, Y coordinates close the Gi, in case the user successfully hit the target.

Currently the features examined involve the description of the movement in the time domain (velocity, acceleration, trajectory, jitter, etc.)

- Durations of DT1/DT2/DT3: the durations in milliseconds
- Mean/SD Velocity DT1/DT2/DT3 X/Y/both: Mean/Standard deviation of the hand velocity on parts DT1/DT2/DT3 on axis X/Y/ both of them.
- Mean Velocity DT2 start: Mean velocity during the first 0.25 s of the DT2 part.
- Mean/SD Acceleration DT1/DT2/DT3 X/Y/both: Mean/Standard deviation of the hand acceleration on parts DT1/DT2/DT3 on axis X/Y/ both axes.
- Distances total traveled per time window and total, ratio of minimum movement required to actual
- Total Distance/Total Distance DT1/DT2/DT3: actual distance travelled during the whole W_i, on parts DT1/DT2/DT3 respectively.
- Minimum distance (final): minimum distance from the center of the gate during the whole W_i movement/ during the final 0.5 s of the W_i movement
- Movement ratio: actual distance travelled/minimum distance required

The distance and subsequently the velocity and acceleration are measured in the in-game units. Since all four subjects performed their sessions with the calibration settings of the patient, they correspond to the same actual movement.

4 Results

The data consisted of 3 healthy subjects and one subject suffering from a neuromuscular issue that also was under physical rehabilitation at the time those sessions were performed. Each of the subjects had recorded their session within a day, 43 sessions for the patient and 8 for each of the rest.

Since the gates are always visible to the users, in some cases the subject has been found to have started moving towards the next gate a few milliseconds prior to reaching the current gate. This can lead to DT1 segments having 0 duration. For the purpose of this analysis those segments were not taken into account.

The gates were grouped based on the type of movement, vertical, horizontal, diagonal and the direction (e.g. top to bottom etc.) Fig. 2. The features were tested for normality using the Shapiro–Wilk test. In cases where the variables were normally distributed the analysis of variation (AOV) was utilized, while in the other cases the Mann-Whitney U test was preferred. The features that were found be statistically significant ($p < 0.05$) were selected and had their p-values corrected using the Bonferroni correction. The features with corrected p-values above 0.05 were removed. Afterwards, the remaining features where checked for correlation among them. Out of those that were found to have a high degree of correlation (0.8), the best performing features were selected.

Results of the movement analysis from the right side towards the left side, which involved 67 control gate movements and 86 patient game movements, can be found on Table 1.

Table 1. Movement from right towards left

Features	Control		Patient		p-value
	Mean	Sd	Mean	Sd	
DT2	359.20	278.65	702.8	434.5	0.0000005
Mean Velocity DT2 Start	68.85	37.84	47.19	16.66	0.0000107
Mean Velocity DT2X	70.94	31.38	51.55	16.85	0.0000002
Sd Velocity DT2 Y	8.75	13.49	14.25	13.31	0.0000002
Sd Accelaration DT2 X	29.31	21.41	47.74	67.77	0.0000023
Sd Velocity DT3	34.19	16.87	21.60	10.02	0.0000022
Total Distance	108.01	43.84	80.16	27.96	0.0199482
Total Distance T3	60.50	18.81	39.07	10.65	0.0000011
Min Distance Final	4.75	6.68	1.75	1.43	0.0000004
Ratio	4.21	3.68	2.26	0.65	0.0000111

Results of the movement analysis from the bottom towards the top, which involved 23 control gate movements and 148 patient game movements, can be found on Table 2. It is interesting to note that the majority of important features co-incide in the two types of movements.

Table 2. Movement from bottom towards top

Features	Control		Patient		p-value
	Mean	Sd	Mean	Sd	
Mean Velocity DT2 Start	63.72	34.39	25.99	12.94	0.0000005
Mean Velocity DT2 Y	37.65	27.43	23.44	11.34	0.0000107
Sd Acceleration DT2 Y	23.98	60.63	21.00	22.68	0.0000002
Sd Velocity DT3	39.67	24.30	20.31	7.78	0.0000002
Total Distance	111.2	57.79	66.90	20.32	0.0000023
Total Distance T1	24.50	19.90	11.87	11.85	0.0000022
Total Distance T3	63.63	21.73	36.51	14.77	0.0199482
Min Distance Center Final	7.41	5.20	4.43	2.19	0.0000011
Ratio	2.80	1.54	1.61	0.42	0.0000004

5 Conclusions

The meanVelDT2Start, meanVelDT2, sdAccDT2, sdVelDT3, totalDistance, totalDistanceT3 and ratio features are found in both types of movement to have higher values in normal subjects than the patient (with the exception of sdAccDT2X). This is expectable since the patient's neuromuscular disorders lead to limited mobility and as such the patient's hand is moving at a slower pace. Greater value of the patient's sdAccDT2X suggests trouble maintaining a steady horizontal acceleration.

This analysis provides a preliminary proof of concept on how to quantify the movement in a meaningful way and express differences between normal and pathological movement. The next step is to increase the available data (control and patients). More data will enable the refinement of the algorithms while establishing a baseline for the healthy population can lead to a quantifiable evaluation of the patients' progress and response to rehabilitation.

The long-term goal is to provide a gaming as a health service (GaaHS) platform where the user can play games based on physician provided settings, data are collected safely and automatically, analyzed and accessed by the physician who monitors their progress and can readjust the rehabilitation treatment accordingly.

References

1. Meijer, H.A., Graafland, M., Goslings, J.C., Schijven, M.P.: Systematic review on the effects of serious games and wearable technology used in rehabilitation of patients with traumatic bone and soft tissue injuries. Arch. Phys. Med. Rehabil. **99**(9), 1890–1899 (2018). https://doi.org/10.1016/j.apmr.2017.10.018. ISSN 0003-9993
2. Ang, K., Guan, C.: Brain-computer interface in stroke rehabilitation. J. Comput. Sci. Eng. **7**, 139–146 (2013). https://doi.org/10.5626/JCSE.2013.7.2.139
3. Smeddinck, J., Herrlich, M., Malaka, R.: Exergames for physiotherapy and rehabilitation: a medium-term situated study of motivational aspects and impact on functional reach, pp. 4143–4146 (2015). https://doi.org/10.1145/2702123.2702598
4. Veerbeek, J.M., Langbroek-Amersfoort, A.C., van Wegen, E.E.H., Meskers, C.G.M., Kwakkel, G.: Effects of robot-assisted therapy for the upper limb after stroke: a systematic review and meta-analysis. Neurorehabil. Neural Repair **31**(2), 107–121 (2017). https://doi.org/10.1177/1545968316666957
5. Craig, J.J.: Introduction to Robotics: Mechanics and Control. Addison-Wesley, New York (1986)

3D Acquisition of the Ear Anatomy: A Low-Cost Set up Suitable for the Clinical Practice

Rocco Furferi[1], Elisa Mussi[1], Michaela Servi[1](✉), Francesca Uccheddu[1], Yary Volpe[1], and Flavio Facchini[2]

[1] Department of Industrial Engineering, University of Florence, Florence, Italy
{rocco.furferi,elisa.mussi,michaela.servi, francesca.uccheddu,yary.volpe}@unifi.it

[2] Department of Pediatric Surgery, Meyer Children's Hospital, Florence, Italy
flavio.facchini@meyer.it

Abstract. The most common clinical treatment for ear deformities or non-congenital abnormalities is the reconstruction of the missing geometry using autologous costal cartilage. The surgical procedure consists in cutting, sculpting and suturing harvested costal cartilage from the patient to recreate an ear shape which is symmetric to the contralateral ear. During chirurgical operation, surgeons needs an accurate 3D template as reference to reproduce the ear. For this purpose, reverse engineering and additive manufacturing techniques can be employed. Specifically, this works aims to develop a reliable, low-cost and user-friendly system, to acquire the healthy ear geometry in clinical environment avoiding head patient's exposition to radiation (MRI, CT scan). An ideal acquisition setup and device have been selected to achieve accurate results. To this end, a casted model of an ear was created as reference, and the best setup was evaluated by comparing the obtained 3D reconstructions with it. Once the setup has been determined, the anatomies of five volunteers were acquired, to test the methodology on human subjects.

Keywords: Ear · 3D model · Photogrammetry · Microtia · RealSense D415

1 Introduction

Ear reconstruction with autologous tissue is one of the most challenging surgeries that reconstructive surgeon may face due to the complex three-dimensional shape of ears. Mainly, the patients who undergo surgery are:

- subjects with burned ears which represent 30% of the patients with facial burn [1];
- maxillofacial oncologic patients to whom the resection of tumor has led the loss of the auricle;
- subjects with microtia, a congenital malformation of the external ear that involves 1 in 5000 newborns worldwide [2].

Three different clinical approaches are available to restore the lacked anatomical element: (1) silicone ear prostheses, (2) surgical reconstruction with synthetic material implant and (3) surgical reconstruction with autologous costal cartilage. Nowadays,

J. Henriques et al. (Eds.): MEDICON 2019, IFMBE Proceedings 76, pp. 669–678, 2020.
https://doi.org/10.1007/978-3-030-31635-8_80

the reconstruction with autologous cartilage represents the standard clinical treatment despite it requires practice, long experience and a preoperative planning [3].

In detail, such a surgical treatment starts from the harvesting of a portion of costal cartilages from the patient; the portion is then cut and carved to obtain a structure that resembles the three-dimensional shape of an ear. This is subsequently inserted into a skin pocket, directly obtained in the auricular region where the external ear is missing. The challenge is to reconstruct an ear framework that is similar in shape and size to the contralateral healthy ear; this allows to obtain a symmetric and pleasant aesthetic outcome. Intraoperatively, the use of an accurate 3D template as a reference can improve aesthetic outcomes and reduce surgical time [4], with a consequential reduction in healthcare costs and greater patient satisfaction. Reverse Engineering (RE) techniques have the potential to be the best solution for acquiring the 3D anatomy of the healthy ear [5] with an accuracy able to capture the ear complex and detailed geometry. Moreover, RE techniques are not invasive in terms of radiation.

Accordingly, the present work aims at proposing a fast and low-cost system to acquire the anatomical geometry of the ear avoiding the use of diagnostic imaging which presents disadvantages such as radiation exposure, high cost and large size of machines.

In particular, on the basis of the interesting results obtained by Ross et al. [6], the present study focuses on the comparison between two common solutions: low-cost depth camera and photogrammetry.

The aim is to obtain an accurate model of the ear and, at the same time, to facilitate the acquisition procedure, making it accessible to non-technical personnel and usable in an outpatient clinical environment. The paper is organized as follow: the two approaches are devised and validated in Sect. 2. The most performing system is selected and tested on five volunteers in Sect. 3. Finally, conclusions are drafted in Sect. 4.

2 Materials and Methods

As mentioned in the introduction, the main aim of this work is to propose a method for obtaining an accurate replica of the ear geometry. For this purpose, we selected two low-cost alternatives, namely depth cameras and photogrammetry, which proved to be a valid solution [6, 7]. The photogrammetry approach was tested on two devices in order to identify the most accurate and optimized solution: an Apple iPhone X [8] and a Canon EOS 1200D [9]. For the depth-camera tests, the low-cost the latest model of the Intel RealSense series, the D415 [10], was selected. The evaluation of the two devices' performances was carried out following the scheme in Fig. 1, comparing the 3D models with a reference obtained with a professional 3D laser scanner. The diagram is described in detail below.

2.1 Ear Casting

Firstly, a faithful replica of an ear was manufactured by applying cast-based techniques. In particular, a bicomponent and biocompatible Room Temperature Vulcanization

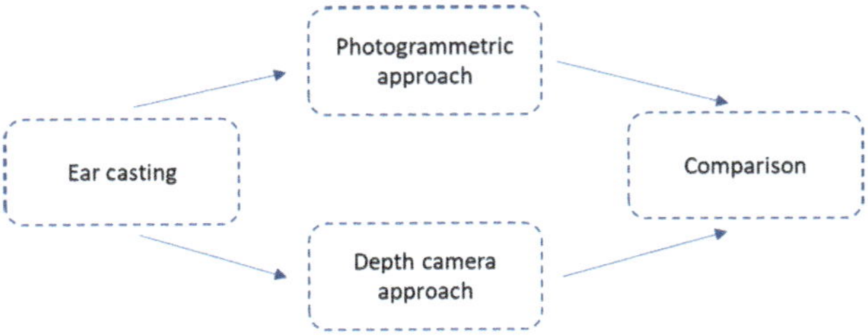

Fig. 1. 3D modelling benchmark process.

(RTV) silicone rubber was poured into a frame around the ear to be captured to create a negative replica used as a mold. Subsequently, the plaster mixed with water was poured into the mold and after an adequate drying time it was demolded to obtain a high-fidelity replica of the ear (Fig. 2).

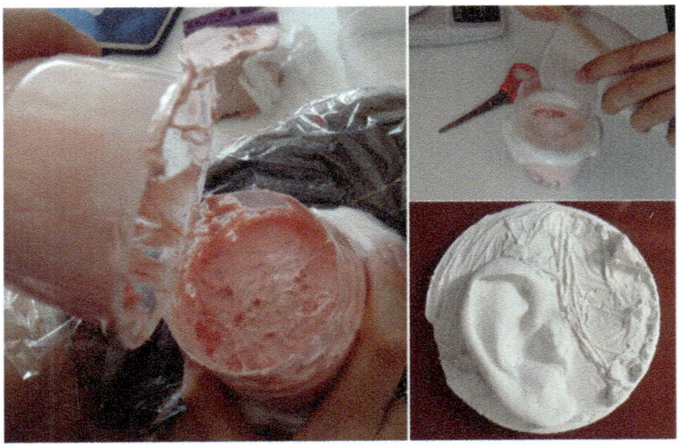

Fig. 2. Fabrication of the casted ear.

During the casting phase, however, the ear's pose may be deformed due to the weight of the casting material. Therefore, not to introduce errors in the validation phase (see Fig. 1), the reconstructions using photogrammetry techniques were carried out directly on the plaster cast.

2.2 Photogrammetric Approach

To obtain an accurate reconstruction of the geometry of interest, the acquisition parameters to maximize the results of the photogrammetric approach were initially defined. Specifically, the following parameters were investigated: camera-subject distance, number of frames and internal device settings.

Referring to the first parameter, distance camera-subject, it is worth noting that a closer focused camera-subject distance generally guarantees a more detailed image of the ear. The iPhone X has a fixed equivalent focal length of 28 mm, that allows a focused camera-subject distance of around 200 mm. The same focused camera-subject distance works with the Canon EOS 1200D used lens thus allowing to keep the same setup for both the systems.

Then, the number of images adequate to obtain a model with sufficient accuracy and usable in common clinical practice was defined: starting from the results obtained by Ross et al. in [6], the objective was to reduce the frame number, to streamline the acquisition process. Specifically, as shown in Fig. 3a, considering a circumference arc of 200° with the subject placed in the center, images were acquired each 10°, to obtain the necessary overlap between consecutive images. The same arc was covered at two different heights for a total of 42 images.

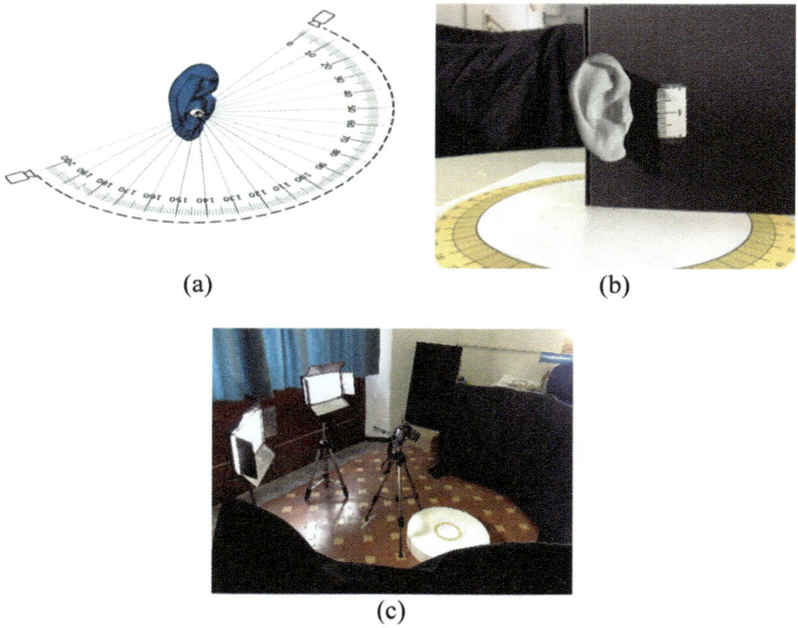

Fig. 3. (a) Acquisition setup scheme; (b) a detail of the devised setup and (c) the complete acquisition setup.

As shown in Fig. 3b, in the acquisition setup a ruler was inserted next to the ear replica to obtain the actual size of the model.

A crucial condition to obtain an accurate and complete reconstruction using photogrammetry techniques is not to have shadow zones, for this reason the setup includes a lighting system consisting of 3 adjustable bi-color LED lights, with color temperature 3200–5600 K and CRI (Color Rendering Index) 96.

In summary, the system consists of a monochrome background (to facilitate the segmentation of the object from the scene), a set of lights, a tripod and a camera.

Table 1 shows the acquisition parameters used for the two devices.

Table 1. Acquisition parameters used in the acquisition phase for the two devices.

Acquisition device	ISO	Focal length	f-stop
iPhone X	50	28 mm	f/1.8
Canon EOS 1200D	200	32 mm	f/5

With such acquired sets of photographs, 3D models were generated using a photogrammetry software. As widely recognized, a photogrammetric acquisition system is based on the estimation of the relative cameras positions and orientation from the given images, and on a 3D triangulation based on adjacent images pixel correspondences. In our setup Agisoft Metashape Professional software [11] was used, with specific reconstruction parameters: the initial cameras alignment was set at *average* accuracy, the dense cloud was set at a *high* reconstruction quality, the depth filtering was set as *mild*, the mesh construction parameters set at a *high* quality and the interpolation mode was enabled. These parameter choices provide a good compromise between processing time and reconstruction results.

2.3 Depth Camera Approach

Among the plethora of commercial depth cameras, the Intel RealSense D415 camera has proven to be among the most suitable device, providing a good compromise between cost, ease of use, compactness and precision [12]. The device includes both a color camera (RGB) and a depth system consisting of two infrared cameras and an infrared projector. Depth data is measured by active stereo triangulation [13]. In addition, the configuration of the device (both for acquisition and post-processing) is easy to perform thanks to the multi-platform open-source library, Intel RealSense SDK 2.0 [14].

After acquiring the plaster replica of the ear from the necessary viewpoints, the three-dimensional model was created aligning the point clouds using the software Geomagic Design X [15].

2.4 Comparison

The reference model for the validation step was created scanning the ear replica with a high precision laser scanner. Specifically, the Romer Absolute Arm 7520 SI/SE [16] was used; such a device has an accuracy of ± 0.063 mm, more than adequate for the purpose of this work.

Photogrammetric Approach
In order to provide an exhaustive examination of this approach, the two devices reconstruction performances were also evaluated using a smaller number of images, so as to find the best and lightest configuration.

For this purpose, the 42 images were halved, moving over the circumference arc by 20° instead of 10°, resulting in a total of 21 images. Moreover, by further reducing the number of images the deviation from the reference model exceeded one millimeter and the model presented many missing areas. Therefore, these results are not here discussed.

Figure 4 shows, as an example, the deviation obtained by comparing camera-based photogrammetric model with the reference model; the deviation values and reconstruction times are presented in Table 2.

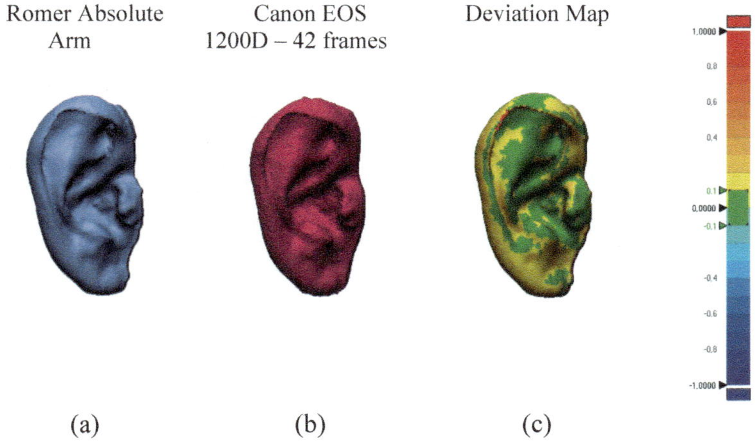

Fig. 4. Ear model acquired (a) with Romer Absolute Arm 7520 SI/SE and (b) ear model acquired with camera. (c) Deviation map between the two models.

Depth Camera Approach

Figure 5 shows the deviation map computed comparing the reference model with the D415 reconstruction. On average, the error obtained is 0.312 mm with a standard deviation of 0.372 mm. The reconstruction has been performed manually by experienced 3D-modelling staff, within ~45 min.

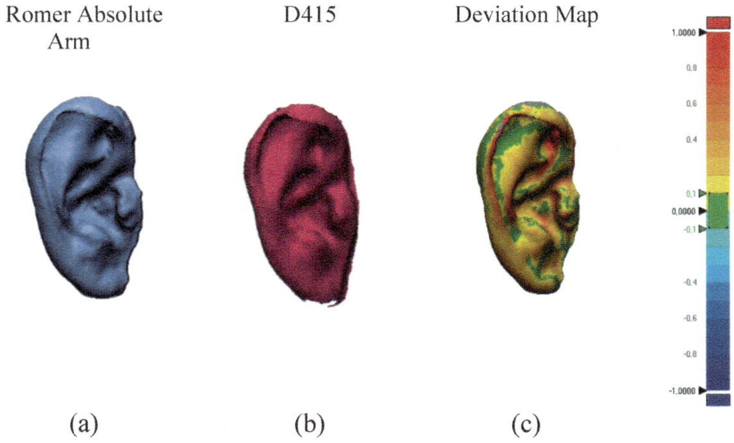

Fig. 5. Ear model acquired with (a) Romer Absolute Arm 7520 SI/SE and (b) RealSense D415. (c) Deviation map between the two models.

Table 2. Mean errors and standard deviation obtained between the models acquired with iPhone X and Canon EOS 1200 D and RealSense D415 with respect to the reference model.

Device	Deviation error [mm]	Reconstruction time [min]
iPhone X - 42 frames	mean: 0.494, std: 0.601	~21
iPhone X - 21 frames	mean: 0.660, std: 0.851	~7
Canon EOS 1200D - 42 frames	mean: 0.152, std: 0.179	~74
Canon EOS 1200D - 21 frames	mean: 0.290, std: 0.414	~12
RealSense D415	mean: 0.333, std: 0.372	~45

Device Selection

In Table 2 are presented the results of the comparison between the identified approaches with the reference model.

From Table 2, considering the trade-off between processing time and reconstruction error, in this work the Canon EOS 1200D result to be the best device to reconstruct the anatomy of the ear.

The photogrammetric reconstruction is in fact less onerous in terms of working time and does not introduce loss of data. On the other hand, although the results obtained with the depth camera are more accurate than those obtained with iPhone X, this type of approach requires a time-consuming phase of manual alignment, which makes it unsuitable for common clinical practice.

The results of the acquisitions on human subjects are presented in the next section.

3 Results

Five healthy volunteers (average age: ~30 years old) were recruited for the study. The preliminary phase of the ear acquisition consisted in making the volunteers wear a swimming cap to hold the hair not to cover the ear; in this phase reference ruler is included in the images to scale the model obtaining its actual dimensions.

The subject was then placed on the graduated seat (Fig. 3(c)) so as to ensure the position of the ear as corresponding with the center of the arc of circumference.

After positioning the subject, the images were acquired according to the setup defined above: the camera was located 200 mm from the ear and a total of 42 images were taken as indicated in Table 1.

Figure 6 shows the reconstructions obtained on two patients as an example. Both 42-frames and 21-frames reconstruction are shown.

Table 3 reports the mean error, standard deviation and computing time of the photogrammetric reconstructions on five healthy volunteers, compared with the 3D models obtained with the abovementioned depth camera. This choice is motivated by the impossibility to acquire on human subjects with the high precision laser scanner: in order to have a reference geometry (although weak) it was decided to use the depth camera, which, as shown in Table 2, produces an acceptable reconstruction error.

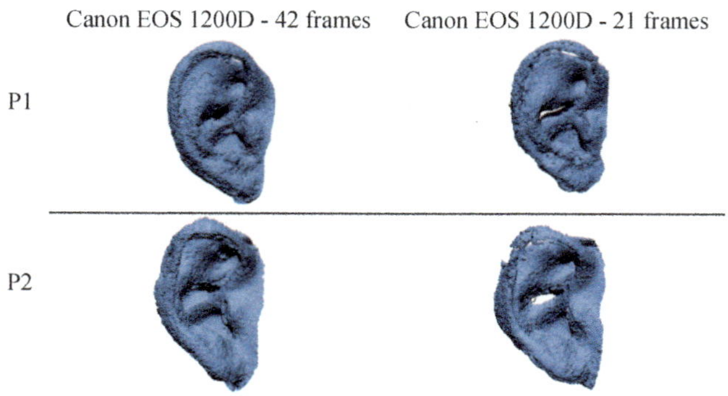

Fig. 6. 3D models obtained with Canon EOS 1200D employing photogrammetric technique.

Table 3. Mean errors, standard deviation and computational time of the models acquired with Canon EOS 1200D on human subjects, compared with the D415 reconstruction.

	42 frames reconstruction		21 frames reconstruction	
	Error [mm]	Time [min]	Error [mm]	Time [min]
P1	mean: 0.517, std: 0.492	~63	mean: 0.583, std: 0.543	~22
P2	mean: 0.482, std: 0.468	~48	mean: 0.785, std: 0.819	~14
P3	mean: 0.386, std: 0.571	~57	mean: 0.675, std: 0.663	~17
P4	mean: 0.492, std: 0.356	~42	mean: 0.731, std: 0.895	~12
P5	mean: 0.315, std: 0.321	~52	mean: 0.564, std: 0.584	~19

The reported mean errors are less than 1 mm that can be considered more than acceptable to be used as a reference during the surgical ear repairing; in fact, the errors introduced by the surgeon during the manual reconstruction of the ear is considerably higher.

Furthermore, it can be noted that to halve the number of images leads to loss of data and a higher average error (either way less than one millimeter) in favor of the processing time.

4 Conclusion

The purpose of this work was to identify the best hardware and software configuration to obtain an accurate replica of the geometry of the ear. The rationale is to treat ear malformations by reconstructing the missing geometry with autologous costal cartilage, avoiding the patient exposition to radiation, as is the case of diagnostic imaging. The aim was therefore to obtain an accurate reproduction of the ear and make the acquisition procedure simple, easily accessible to non-technical personnel and usable in the clinical environment.

The first step was to define a benchmark to allow the identification of the most performing approach for the examined application. An Apple iPhone X and a Canon EOS 1200D were tested, as well as a low-cost depth camera.

The results of the photogrammetric approach with the camera proved to be the most accurate and effective solution.

The final setup, easily reproducible in a clinical outpatient environment, is composed of a monochrome background, a set of lights, a tripod and a camera.

The objective is to allow the surgical reconstruction of an aesthetically pleasing yet symmetrical ear by providing the surgeon with a 3D reference model as a visual reference for reconstruction. In this perspective, the models obtained using the proposed setup are qualitatively and quantitatively acceptable, as the resulting error is neglectable with respect to the potential error introduced by the surgeon during the manual reconstruction of the ear.

References

1. Carrillo-Córdova, J.R., Jiménez Murat, Y., Apellaniz-Campo, A., Bracho-Olvera, H., Carrillo Esper, R.: Reconstruction of the ear in the burns patient. Cirugía y Cir. 454–458 (2017). (English Ed. 85)
2. Luquetti, D.V., Leoncini, E., Mastroiacovo, P.: Microtia-anotia: a global review of prevalence rates. Birth Defects Res. Part A Clin. Mol. Teratol. **91**, 813–822 (2011)
3. Mussi, E., Furferi, R., Volpe, Y., Facchini, F., McGreevy, K.S., Uccheddu, F.: Ear reconstruction simulation: from handcrafting to 3D printing. Bioengineering **6**, 14 (2019)
4. Flores, R.L., Liss, H., Raffaelli, S., Humayun, A., Khouri, K.S., Coelho, P.G., Witek, L.: The technique for 3D printing patient-specific models for auricular reconstruction. J. Cranio-Maxillofacial Surg. **45**, 937–943 (2017)
5. Furferi, R., Governi, L., Uccheddu, F., Volpe, Y.: A RGB-D based instant body-scanning solution for compact box installation. In: Advances on Mechanics, Design Engineering and Manufacturing, pp. 819–828 (2017)
6. Ross, M.T., Cruz, R., Brooks-Richards, T.L., Hafner, L.M., Powell, S.K., Woodruff, M.A.: Comparison of three-dimensional surface scanning techniques for capturing the external ear. Virtual Phys. Prototyp. **13**, 255–265 (2018)
7. Salazar-Gamarra, R., Seelaus, R., da Silva, J.V.L., da Silva, A.M., Dib, L.L.: Monoscopic photogrammetry to obtain 3D models by a mobile device: a method for making facial prostheses. J. Otolaryngol. Head Neck Surg. **45**, 33 (2016)
8. iPhone X - Technical Specifications. https://support.apple.com/kb/SP770?viewlocale=en_US&locale=en_US. Accessed 11 Mar 2019
9. Canon EOS 1200D -Specification - EOS Digital SLR and Compact System Cameras - Canon UK. https://www.canon.co.uk/for_home/product_finder/cameras/digital_slr/eos_1200d/specification.html. Accessed 11 Mar 2019
10. Intel® RealSenseTM Technology. https://www.intel.com/content/www/us/en/architecture-and-technology/realsense-overview.html. Accessed 11 Mar 2019
11. Agisoft Metashape. https://www.agisoft.com/. Accessed 11 Mar 2019
12. Carfagni, M., Furferi, R., Governi, L., Santarelli, C., Servi, M., Uccheddu, F., Volpe, Y.: Metrological and critical characterization of the intel D415 stereo depth camera. Sensors **19**, 489 (2019)

13. Zanuttigh, P., Marin, G., Dal Mutto, C., Dominio, F., Minto, L., Cortelazzo, G.M.: Operating principles of structured light depth cameras. In: Time-of-Flight and Structured Light Depth Cameras, pp. 43–79. Springer International Publishing, Cham (2016)
14. Download Intel RealSense SDK 2.0 - Intel RealSense Depth & Tracking Cameras. https://realsense.intel.com/sdk-2/. Accessed 11 Mar 2019
15. Geomagic Design X Scan-to-CAD solid model software. https://www.3dsystems.com/software/geomagic-design-x. Accessed 11 Mar 2019
16. ROMER Absolute Arm with integrated scanner | Hexagon Manufacturing Intelligence. https://www.hexagonmi.com/en-US/products/portable-measuring-arms/romer-absolute-arm-with-integrated-scanner. Accessed 11 Mar 2019

Machine Learning Classification of Females Susceptibility to Visceral Fat Associated Diseases

Mahmoud Aldraimli[1(✉)], Daniele Soria[3], James Parkinson[2], Brandon Whitcher[2], E. Louise Thomas[2], Jimmy D. Bell[2], Thierry J. Chaussalet[1], and Miriam V. Dwek[2]

[1] School of Computer Science and Engineering, University of Westminster, London W1W 6UW, UK
m.aldraimli@my.westminster.ac.uk
[2] School of Life Sciences, University of Westminster, London W1W 6UW, UK
[3] School of Computing (Medway), University of Kent, Chatham Maritime ME4 4AG, UK
d.soria@kent.ac.uk

Abstract. The problem of classifying subjects into risk categories is a common challenge in medical research. Machine Learning (ML) methods are widely used in the areas of risk prediction and classification. The primary objective of these algorithms is to predict dichotomous responses (e.g. healthy/at risk) based on several features. Similarly to statistical inference models, also ML models are subject to the common problem of class imbalance. Therefore, they are affected by the majority class increasing the false negative rate.

In this paper, we built and evaluated eighteen ML models classifying approximately 4300 female participants from the UK Biobank into three categorical risk statuses based on responses for the discretised visceral adipose tissue values from magnetic resonance imaging. We also examined the effect of sampling techniques on classification modelling when dealing with class imbalance.

Results showed that the use of sampling techniques had a significant impact. They not only drove an improvement in predicting patients risk status, but also facilitated an increase in the information contained within each variable. Based on domain experts criteria, the three best models for classification were finally identified.

These encouraging results will guide further developments of classification models for predicting visceral adipose tissue without the need for a costly scan.

Keywords: Supervised learning · Imbalanced data · UK Biobank · Random under-sampling · Synthetic minority over-sampling technique · Visceral adipose tissue

1 Introduction

Real-world data are often imbalanced and lack uniformly distribution across classes. Classification of imbalanced datasets is one of the challenges across several industrial and research domains [1]. There are multiple approaches to tackle class imbalance [2],

J. Henriques et al. (Eds.): MEDICON 2019, IFMBE Proceedings 76, pp. 679–693, 2020.
https://doi.org/10.1007/978-3-030-31635-8_81

simplest approaches include, for example, Data Enrichment. Others more sophisticated methods include various sampling techniques [3], cost-sensitive learning [4, 5] and feature selection; or more complex strategies including meta-learning [6], combining classifiers [7], and algorithmic modifications [8].

When sampling methods are applied, questions over their suitability are often raised [9]. For example: is the new re-sampled dataset representative of the population in relation to the response variable? Is it acceptable to artificially generate synthetic data of class subjects when training Machine Learning (ML) classification models? It has been argued that by using sampling methods the original class ratio is lost during the training process and that this affects the accuracy metrics [10]. Similarly, training ML models with synthetic data may also compromise accuracy measures by deceiving the process of cross-validations sampling [11].

In this paper we compared the classification performance of six ML algorithms (Naïve Bayes, Logistic Regression, Artificial Neural Network, Decision Tree, Logistic Model Tree, and Random Forest) when using RUS [8] and SMOTE [12] sampling techniques on highly imbalanced data to predict visceral adipose tissue (VAT) disease risk in a multi-class classification problem, and to suggest the most suitable models to meet the domain experts' success criteria. The data imbalance characteristic causing the transition in classifier training performance was monitored visually by Adaptive Projection Analysis (APA) [13] and numerically via Information Gain Attribute Evaluation (IG) [14, 15].

The paper is structured as follows: in Sect. 2, the domain problem and all the methods and approaches used in this study are presented. Then in Sect. 3, the experiments' results are introduced, while Sect. 4 is reserved for discussion and conclusions.

2 Materials and Methods

2.1 The Domain Problem

Obesity affects an increasing number of adults in the UK [16], with obesity-associated changes in adipose tissue (AT) predisposing to metabolic dysregulation [17]. Distribution of AT, in particular the accumulation of VAT and liver fat, are a key factor in determining susceptibility to disease [18, 19]. Excess VAT and liver fat play a significant role in the pathogenesis of type-2 diabetes, dyslipidemia, hypertension and cardiovascular disease [20].

Current strategies for the treatment of obesity and its associated co-morbidities have focused on lifestyle improvements [21, 22], aiming to reduce VAT and liver fat, via exercise, associated with improved insulin sensitivity, decreased blood pressure and lower circulating lipid levels [17, 23, 24]. Large scale analysis of the compartmental distribution of AT is often limited due to the expense and time required to employ requisite imaging techniques. The UK Biobank provides a comprehensive means of assessing the relationship between body composition and lifestyle in a large population-based cohort of adults aged 40–70 years, recruited between 2007 and 2010 [25]. The primary goal of this study was to identify the best model able to predict VAT levels in a cohort of female individuals from the UK Biobank. The UK Biobank had approval from the North West Multi-Centre Research Ethics Committee (MREC) and written consent was obtained from all participants prior to their involvement. The data was acquired through the UK Biobank Access Application number 23889. The study was a cross-sectional assessment of 4327 female individuals from the UK Biobank multimodal imaging cohort [25]; aged 40–73 years and scanned chronologically between August 2014 and September 2016. The analysis of male subjects VAT is outside the scope of this paper.

2.2 Methodology

Multi-class classification ML models were applied with the aim of predicting susceptibility to disease (risk) based on the discretised amount of VAT. A subset of 2292 subjects was randomly selected from the original 4327 females and used to train six ML algorithms using 10-fold cross-validation in three different scenarios. The models were tested on the remaining 2035 cases. Figure 1 shows the methodology: multiple imbalanced datasets with the same predictor variables were modified with sampling techniques, and used for modelling using the six ML algorithms. The accuracies of the models were compared after the training phase. IG was monitored for all predictor variables at every stage.

Information Gain Evaluation Algorithm (IG). Information and entropy levels within independent variables were monitored using an Information Gain Attribute Evaluator Algorithm [15]. This algorithm evaluates the worth of each attribute by measuring information gained with respect to the class in combination with a ranker algorithm which ranks the attributes by their individual influence on the class [14, 15, 26].

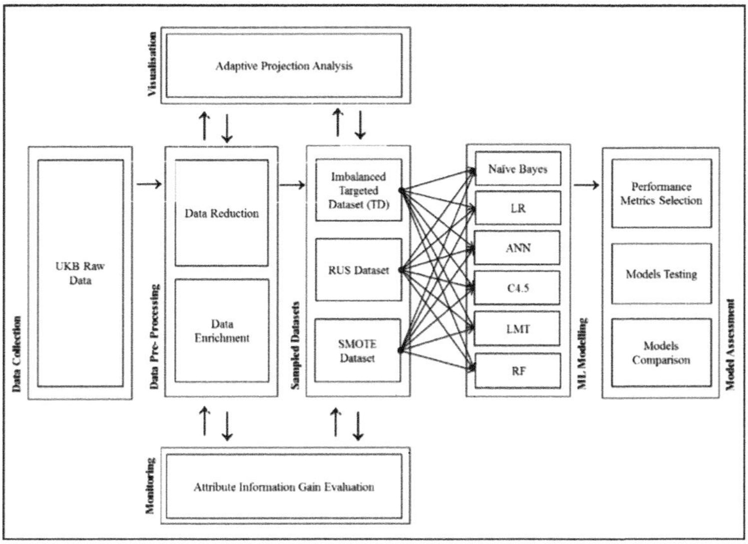

Fig. 1. Methodology adopted in this work, showing the different steps followed.

Adaptive Projection Analysis (APA). APA uses a linear projection to display high dimensional data into 3-dimensions by allowing the user to drag points in an interactive scatter plot to find new views [13]. These views indicate the classes which can be separated, the attribute combinations which are most associated with each class, the outliers, the sources of error in the classification algorithms, and the existence of clusters in the data [27].

2.3 Data Preparation

Random Undersampling (RUS). This approach consisted of selecting a subset of the majority class to balance the data [8]. In this approach some of the majority class records were removed at random. However, it was recognised that deleting records could lead to loss of important information or patterns which may have been relevant to the learning process [28]. Denoting the majority class L and the minority class S, r was defined as the ratio between the size of the minority and majority classes [3]. We performed random under-sampling of L to achieve a value of $r = 0.5$.

$$r = \frac{|S|}{|L|} = 0.5$$

Synthetic Minority Oversampling Technique (SMOTE). SMOTE is an over sampling technique developed by Chawla [12]. It aims to enhance the minority class by creating artificial examples in the minority class. For each data point x in S (the minority class), one of its k-nearest neighbours (k = 5) was identified. The k neighbours were randomly selected, artificial observations were generated and spread in the area between x and nearest neighbours. These synthetic points were added to the dataset in class S. The artificial generation of the data points differed from the multiplication method [16] to avoid the problem of overfitting.

2.4 ML Classification Algorithms

Naïve Bayes (NB). A probabilistic machine learning classifier used for classification tasks. The foundation of the classifier is the Bayes Theorem [29]. It also assumes that predictor variables are independent and that all predictor variables have an equal effect on the response outcome. Despite the simplified assumptions of Naïve Bayes classifiers, they have been reported to be effective in complex real-world situations [30].

Logistic Regression (LR). LR is a deterministic technique which produces a probability-based model that takes into account the likelihood of an event occurring (the value of the class variable) depending on the values of the predictors (categorical or numerical) [31, 32].

Artificial Neural Network (ANN). ANNs are used to fit observed data, especially high dimensional datasets characterised by noise and missingness (pollution). Neural networks comprise elementary autonomous computational units, known as neurons. Neurons are inter-connected via weighted connections and are organised in layers (an input layer, hidden layers and an output layer). In this study, we used a Multilayer Perceptron ANN with a sigmoid activation function [17].

Decision Tree (C4.5). The C4.5 algorithm is used in Data Mining as a Decision Tree Classifier which generates a decision, based on a sample of data. In this method, a new data point is predicted (classified) via a series of tests to determine its class. The tests hierarchically assemble a tree of decisions, hence 'decision tree' [15, 33, 34].

Logistic Model Tree (LMT). LMT is a model with a tree structure but with LR functions at the leaves level. The LMT structure comprises a set of non-terminal nodes and a set of leaves (terminal nodes). LMT has been designed to adapt to small data subsets where a simple linear model offers best bias-variance trade-off [31].

Random Forest (RF). RF is a generalisation of standard decision trees proposed by Breiman based on bagging (Bootstrap Aggregation) from a single training set or random not pruned decision trees [18]. Bootstrap Aggregation is used to combine the predictions of the individual trees [19].

All the six methods used for this study were implemented in Weka [35] (with default parameters), with the C4.5 using the J48 implementation.

2.5 Model Evaluation

In agreement with domain experts, we chose several measures to evaluate the performance of each model. These measures included accuracy (later reported as 'CCI%') true positive rate (also known as sensitivity or recall, 'TPR'), specificity, false positive rate ('FPR'), precision ('Prcn'), area under the receiver operator curve ('ROC'), and F-measure ('F-m') [36–38]. The latter is a harmonic mean of precision and recall. Practically, a high F-measure value indicates that both recall and precision are high, meaning fewer subjects misdiagnosed with a disease or risk of disease. The F-measure is essential to assess the model performance when classifying very imbalanced data [37].

2.6 Experimental Design

The analysis was performed to predict VAT related disease susceptibility based on discretised MRI response labels: Healthy, Moderate and Risk defined according to VAT volume. If VAT volume was ≤ 2 L then the subject was deemed 'Healthy' (H). If VAT volume was >2 L but ≤ 5 L, then the class was 'Moderate' (M). If VAT volume was >5 L, then the patient was at 'Risk' (R) [39]. The training datasets contained ten data variables reported in Table 1, with the VAT in litres being the class determination response variable.

Targeted Dataset (TD). The TD was the first dataset we considered for modelling. The TD contained 2292 female records, from the UK Biobank cohort. Table 1 shows the summary statistics of all TD's variables. The TD was highly imbalanced: class H had 1002 subjects, the M class had 1128 subjects, and the minority R class contained only 162 subjects.

Table 1. Summary statistics of TD variables

Numeric selected dataset variables	Median	Mean	(Min, Max)
Response variable			
Visceral adipose tissue volume (VAT in litres)	2.2	2.5	(0.1, 9.7)
Predictors variables			
Waist circumference (WC in cm)	80.0	81.6	(55.0, 126.0)
Pre-imaging weight (W in Kg)	66.0	68.3	(42.0, 128.0)
BMI (in kg/m^2)	24.8	25.7	(15.5, 48.0)
Hip circumference (HC in cm)	100.0	100.9	(77.0, 147.0)
Standing height (H in m)	163.0	163.0	(141.0, 194.0)
Systolic blood pressure (SBP in mmHG)	133.0	134.5	(87.0, 225.0)
Diastolic blood pressure (DBP in mmHG)	77.0	77.8	(45.0, 120.0)
Physical Activity Index (PAI)	0.5	0.6	(−12.0, 15.5)
Age at recruitment (AGE in years)	55.0	54.6	(40.0, 70.0)

Random Under-Sampled (RUS) Dataset. This dataset was a reduced subset of TD. A subset of each majority class was randomly removed to balance the data. As a result of applying RUS to the TD, each of the H, M and R classes ended up with 162 subjects.

Synthetic Minority Over-Sampled (SMOTE) Dataset. This dataset was obtained as a result of applying SMOTE to the numeric data variables of TD. By doing so, the three VAT classes became more closely balanced. The H class had 1002 subjects, the M class had 1128 subjects and the R class contained 1296 subjects. The effect of SMOTE can be observed via APA visualisation in Fig. 2.

We used the IG Evaluator Algorithm to measure the information levels for independent variables in relation to the class variable. The measurement and ranking of IG in each independent variable in TD, RUS and SMOTE training sets will be presented in Sect. 3.

The Test Dataset. The ML models were tested on the remaining 2035 individuals from the original 4327 UK Biobank cohort. The same ten variables as per the training datasets were available. Table 2 shows their summary statistics. Similarly to TD, the Test Dataset was also highly imbalanced: class H had 823 subjects, the M class had 1039, and class R contained only 173 subjects.

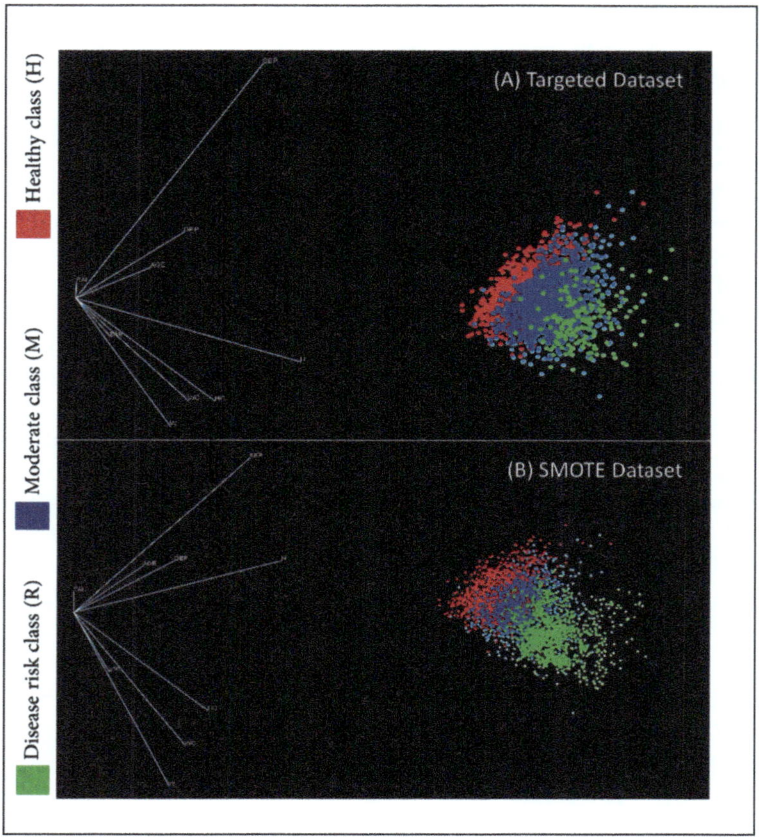

Fig. 2. APA visualisation of SMOTE dataset variables

Table 2. Summary statistics of test set variables

Numeric test dataset variables	Median	Mean	(Min, Max)
Response variable			
Visceral adipose tissue volume (VAT in litres)	2.4	2.7	(0.2, 10.0)
Predictors variables			
Waist circumference (WC in cm)	80.0	81.6	(55.0, 142.0)
Pre-imaging weight (W in Kg)	67.0	68.7	(39.0, 136.0)
BMI (in kg/m^2)	25.2	25.9	(14.4, 54.5)
Hip circumference (HC in cm)	100.0	101.3	(73.0, 156.0)
Standing height (H in m)	163.0	162.7	(145.0, 195.0)
Systolic blood pressure (SBP in mmHG)	129.0	130.4	(87.0, 196.0)
Diastolic blood pressure (DBP in mmHG)	76.0	76.6	(45.0, 115.0)
Physical Activity Index (PAI)	0.0	0.1	(−12.5, 18.0)
Age at recruitment (AGE in years)	55.0	54.6	(40.0, 70.0)

3 Results

3.1 Models Training Results

From Table 3, the model training accuracies (CCI%) of all methods were computed and they showed that resampling methods resulted in an improvement in CCI% as compared to the original TD. Figure 3 shows that LR, ANN, C4.5 and RF models training performances using the RUS dataset worsened compared to the same algorithms trained on TD. The ROC for each of the trained models ranged between 0.783 (for RF on SMOTE) and 0.96 (for C4.5 on TD). These values indicate that the trained models did not sacrifice a lot of precision of the system to get a good recall on the observed data points. The RF model achieved the highest TPR (0.850) when trained on the SMOTE dataset, whilst the C4.5 model achieved the lowest TPR (0.714) when trained on the RUS dataset.

By observing the confusion matrices for all models after training on all the TD and RUS datasets, and bearing in mind that all comprise the same risk group participants, it is clear that the number of incorrectly classified instances for the R class significantly decreased for the models trained on the RUS dataset compared to those trained on the original TD. However, when evaluating the minority class accuracy performance in Fig. 4, it is notable that all trained models benefitted from the sampling methods, exhibiting significant TPR improvement for the R group in each model.

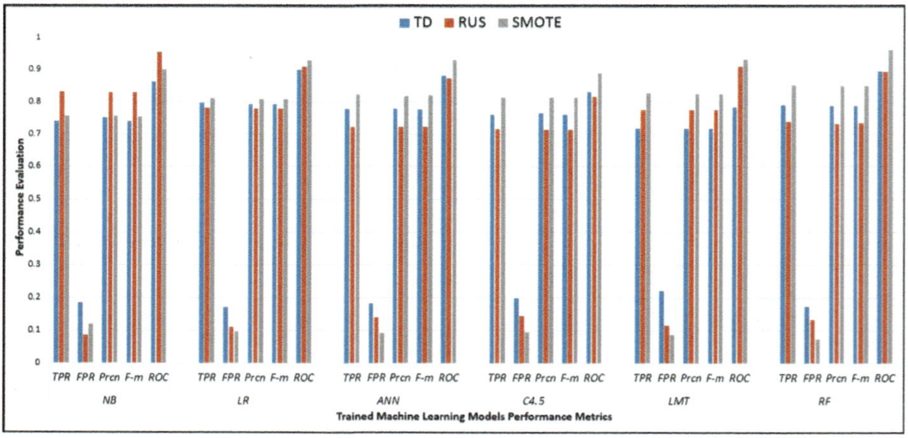

Fig. 3. Comparison of performance metrics across trained models.

3.2 Models Test Results

All models were tested on the same test dataset (N = 2035). When comparing the CCI% for all the models, the CCI% decreased with a maximum degradation of 6.2% when testing the C4.5 model trained on the RUS dataset compared to the same model built on the original TD; this excluded the LMT models which achieved an overall accuracy improvement on test dataset of 6.83% when comparing the SMOTE model to the TD one.

From Fig. 5, it can be observed that in test, RF models achieved the best TPR of 0.770 when trained on TD dataset. LMT model achieved the least TPR of 0.681 when trained on TD dataset. The ROC area across all tested models ranged between 0.786 (for C4.5 on SMOTE) and 0.889 (for LR on TD). These values indicate that also the tested models do not sacrifice a lot of precision to get a good recall on the observed data points.

When observing R, the class of interest, TPR performance results in Fig. 4 show that significant improvements were made in classifying the risk group with the highest level of 0.798 achieved by RF on RUS. RF also achieved the highest TPR improvement in test with a difference of 0.463 between RF on RUS and RF on TD, while NB ranked last, with just 0.121 in minority class TPR improvement between NB on SMOTE and TD. The confusion matrices in Table 3 confirm the above results. The RF model trained on SMOTE correctly classified the highest number of instances (138 of the original 173) in the R group. The model which performed the worst in TPR performance for the minority class R was C4.5 trained on TD, which only correctly classified 43 instances.

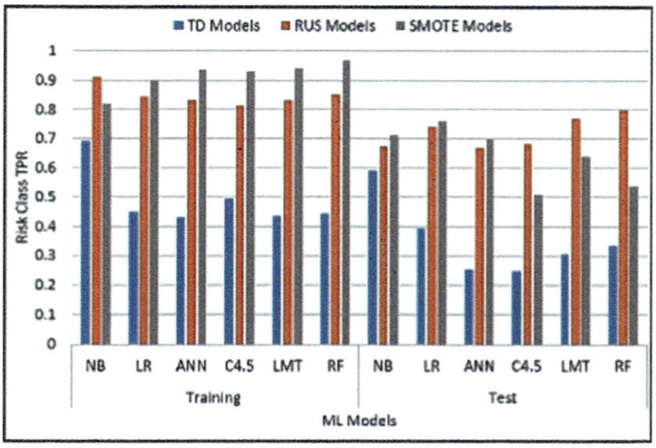

Fig. 4. Risk class TPR performance for trained and tested models per dataset

Table 3. Models confusion matrixes comparison

	TD								RUS Dataset								SMOTE Dataset							
	Training				Test				Training				Test				Training				Test			
	H	M	R		H	M	R		H	M	R		H	M	R		H	M	R		H	M	R	
NB	855	147	0	H	720	103	0	H	143	19	0	H	742	80	1	H	856	146	0	H	721	102	0	H
	287	728	113	M	283	662	94	M	29	113	20	M	342	551	146	M	289	668	171	M	283	600	156	M
	0	50	112	R	2	69	102	R	0	14	148	R	2	54	117	R	0	231	1065	R	2	48	123	R
LR	833	169	0	H	698	125	0	H	136	26	0	H	707	115	1	H	834	167	1	H	699	123	1	H
	180	915	33	M	188	828	23	M	30	106	26	M	211	681	147	M	184	769	175	M	189	704	146	M
	0	89	73	R	1	104	68	R	0	25	137	R	1	44	128	R	0	127	1169	R	1	40	132	R
ANN	809	193	0	H	664	159	0	H	123	38	1	H	671	149	3	H	813	187	2	H	685	133	5	H
	177	907	44	M	152	875	12	M	32	93	37	M	234	678	127	M	162	783	183	M	186	701	152	M
	0	92	70	R	1	128	44	R	1	26	135	R	2	55	116	R	0	83	1213	R	2	50	121	R
C4.5	741	261	0	H	606	216	1	H	125	35	2	H	723	93	7	H	759	239	4	H	680	142	1	H
	151	922	55	M	140	879	20	M	44	90	28	M	299	561	179	M	170	819	139	M	224	710	105	M
	0	82	80	R	1	129	43	R	1	26	132	R	3	52	118	R	2	90	1204	R	2	83	88	R
LMT	765	236	1	H	637	185	1	H	134	28	0	H	727	95	1	H	818	184	0	H	676	145	2	H
	235	808	85	M	284	695	60	M	29	107	26	M	249	634	156	M	180	792	156	M	180	737	122	M
	0	91	71	R	3	117	53	R	0	27	135	R	1	39	133	R	1	75	1220	R	3	59	111	R
RF	823	179	0	H	679	144	0	H	130	32	0	H	649	127	2	H	811	191	0	H	681	142	0	H
	175	916	37	M	183	829	27	M	34	90	38	M	204	649	186	M	169	848	111	M	168	785	86	M
	0	90	72	R	1	114	58	R	0	24	138	R	3	32	138	R	0	43	1253	R	2	78	93	R

3.3 Attribute Information Results

When considering the monitored IG for each variable across all datasets (Fig. 6), it is clear that the information gain increased in each attribute for RUS and SMOTE datasets compared to the TD. By comparing the IG ranking of variables in each dataset, it is apparent that WC achieved the highest IG value in all the three datasets. The dominance in WC ranking was also accompanied by an increase of its values (from TD to RUS and SMOTE) that correlates directly with the increase in R class TPR performance in all trained models except for NB where RUS model overtook SMOTE by a small TPR positive margin of 0.092. From Fig. 6, SMOTE seems to boost the information within each variable. This, in turn, increases the R class separability from other classes in the training datasets which in turn increases the R class TPR (see Fig. 4). Figure 7 displays the APA multi-dimensional visualisation which shows the improved R class separability per dataset.

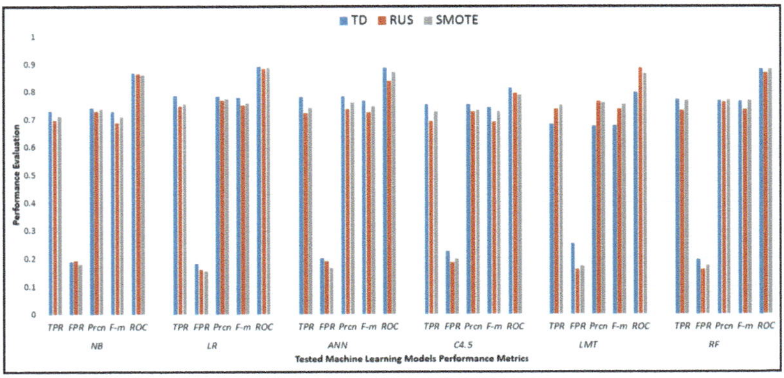

Fig. 5. Comparison of performance metrics across tested models

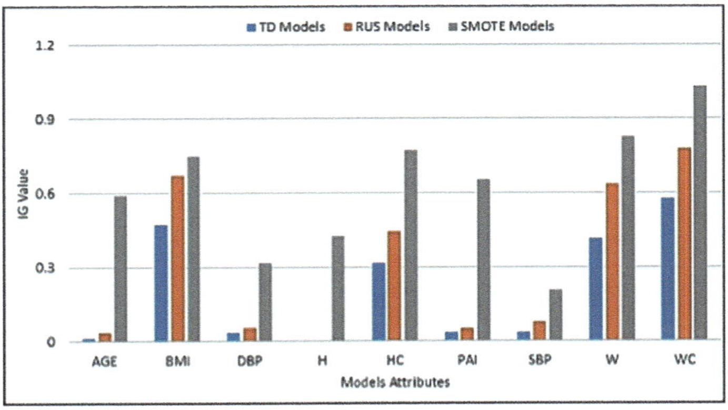

Fig. 6. IG monitoring per variable in each dataset

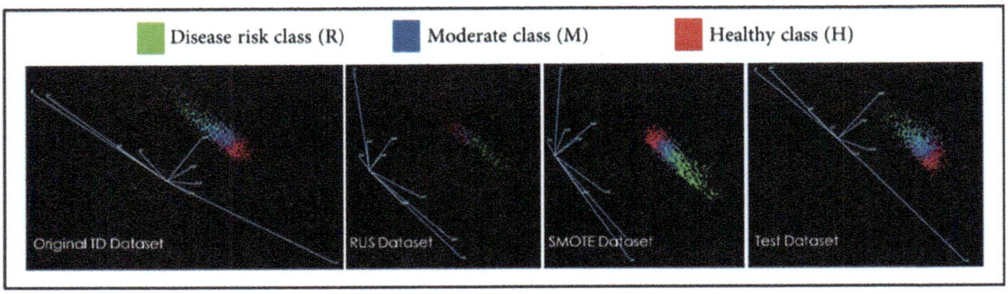

Fig. 7. APA multi-dimensional visualisation

3.4 Domain Experts' Results

During the development of prediction algorithms for use in disease risk-prediction it is important to recognise that the misclassification of subjects, for example false-positive misclassifications, could result in costly and unnecessary follow-up examinations; whereas false-negative misclassifications would result in individuals not receiving interventions to reduce excess VAT. In this particular application, apart from potential cost, there would be few adverse effects associated with healthy/moderate risk subjects being misclassified, as such subjects would be encouraged to undertake interventions to improve their lifestyle. Therefore, in common with other scenarios, the best models to adopt would be those which minimise the number of subjects misclassified as at 'risk' in order that they might initiate interventions at an appropriate time. Confusion matrices play an essential role in defining the best-suited model for use in future trials. When analysing the confusion matrices (Table 3), three models were identified as satisfying the domain experts' criteria. These models are reported in Table 4. They did not occupy the highest ranks when their performance metrics were compared to the others.

Table 4. Domain compliant prediction models. N(LMT RUS Trained) = 486; N(LR SMOTE Trained) = 3426; N(RF RUS Trained = 486). N(all Tested) = 2035. For the F-m metric, m = 1.

Least-Cost Prediction Models

	Model	LMT						LR						RF					
		RUS						SMOTE						RUS					
	Metrics	CCI%	TPR	FPR	Prcn	F-m	ROC	CCI%	TPR	FPR	Prcn	F-m	ROC	CCI%	TPR	FPR	Prcn	F-m	ROC
Performance	Trained	77.37	0.774	0.113	0.774	0.774	0.907	80.91	0.809	0.096	0.807	0.808	0.929	73.66	0.737	0.132	0.731	0.733	0.892
	Tested	73.41	0.734	0.159	0.761	0.734	0.883	75.43	0.754	0.154	0.773	0.757	0.884	72.78	0.728	0.159	0.758	0.731	0.864

		LMT H	LMT M	LMT R		LR H	LR M	LR R		RF H	RF M	RF R	
Confusion Matrix	Trained	134	28	0	H	834	167	1	H	130	32	0	H
		29	107	26	M	184	769	175	M	34	90	38	M
		0	27	135	R	0	127	1169	R	0	24	138	R
	Tested	727	95	1	H	699	123	1	H	649	127	2	H
		249	634	156	M	189	704	146	M	204	649	186	M
		1	39	133	R	1	40	132	R	3	32	138	R

4 Discussion

Imbalanced classes have a significant impact on the performance of standard machine learning algorithms. Classification performance in the training phase is severely impacted by class separability. Training the standard ML algorithms with highly imbalanced overlapping classes without any adjustment to the training set results in an accuracy bias towards the majority class. In this study, two methods (RUS and SMOTE) have been applied to adjust the class imbalance in the classification training phase at the dataset level. It remains unclear as to whether other remedies for imbalanced data classifications, such as Cost-Sensitive and Ensembles Learning (which are implemented at algorithmic level) could result in better performances [4, 6, 40]. The advantages of sampling techniques evaluated here, however, include simplicity and transportability. Nevertheless, they are limited by the amount of IG manipulation as a result of their application resulting in biased prediction towards the minority class. The excessive use of such techniques could result in overfitting of the models.

In this study, traditional ML algorithms were sensitive to higher information gains and tended to produce superb performance results in training, but when testing the models, the overall model accuracy often dropped below the training phase performance. The UK Biobank dataset used in this study showed that applying the correct level of sampling without disrupting the original data distribution, together with the desired choice of performance metrics and slight manipulation of IG levels produced a prediction solution which could be developed further with algorithmic modifications [8]. Among all eighteen models presented in this study, only three satisfied the domain experts' success criteria for this specific domain problem (LMT and RF built with RUS sampled dataset, and LR built with SMOTE sampled dataset).

This domain problem is the first to use the discretised MRI VAT variable ranges to describe the health status of participants and to label instances. It would be impractical to compare the results of this study to any other research from the same domain. Nevertheless, this work will be followed by further analyses where additional methods to improve the outcomes will be investigated. Starting from the best-performing methods in this work (LMT and RF), their combination into ensemble learners will first be considered. Future work will also take into account the predictions from this current paper and compare them to the actual incidence of diseases in the same cohort where data is available.

References

1. Yang, Q., Wu, X.: 10 challenging problems in data mining research. Int. J. Inf. Technol. Decis. Making **5**(4), 597–604 (2006)
2. Gu, J., Zhou, Y., Zuo, X.: Making class bias useful: a strategy of learning from imbalanced data. In: Yin, H., Tino, P., Corchado, E., Byrne, W., Yao, X. (eds.) IDEAL 2007. LNCS, vol. 4881, pp. 287–295. Springer, Heidelberg (2007)
3. More, A.: Survey of resampling techniques for improving classification performance in unbalanced datasets. arXiv:1608.06048 [stat.AP] (2016)

4. Weiss, G.M., McCarthy, K., Zabar, B.: Cost-sensitive learning vs. sampling : which is best for handling unbalanced classes with unequal error costs?. In: Proceedings of the 2007 International Conference on Data Mining, Las Vegas, USA, pp. 35–41 (2007)
5. Bekkar, M., Taklit, A.A.: Imbalanced data learning approaches review. Int. J. Data Min. Knowl. Manag. Process (IJDKP) **3**(4), 15–33 (2013)
6. Ensemble Learning to Improve Machine Learning Results. https://blog.statsbot.co/ensemble-learning-d1dcd548e936. Accessed 19 Feb 2019
7. Dzeroski, S., Zenko, B.: Is combining classifiers better than selecting the best one? In: Proceedings of the Nineteenth International Conference on Machine Learning, San Francisco. Morgan Kaufmann (2002)
8. Choi, J.M.: A Selective Sampling Method for Imbalanced Data Learning on Support Vector Machines. Graduate Thesis and Dissertation, Iowa State University (2010)
9. Unbalanced Data Is a Problem? No, Balanced Data Is Worse. https://matloff.wordpress.com/2015/09/29/unbalanced-data-is-a-problem-no-balanced-data-is-worse/. Accessed 24 Feb 2019
10. When should I balance classes in a training data set? https://stats.stackexchange.com/questions/227088/when-should-i-balance-classes-in-a-training-data-set. Accessed 22 Nov 2018
11. Bharat Rao, R., Fung, G., Rosales, R.: On the dangers of cross-validation. An experimental evaluation. In: Proceedings of the 2008 SIAM International Conference on Data Mining, pp. 588–596 (2008)
12. Chawla, N.V., Bowyer, K.W., Hall, L.O., Kegelmeyer, W.P.: SMOTE: synthetic minority over-sampling technique. J. Artif. Intell. Res. **16**, 321–357 (2002)
13. Faith, J., Mintram, R., Angelova, M.: Gene expression targeted projection pursuit for visualizing gene expression data classifications. Bioinformatics **22**(21), 2667–2673 (2006)
14. Information gain which test is more informative? https://homes.cs.washington.edu/~shapiro/EE596/notes/InfoGain.pdf. Accessed 29 Mar 2019
15. Quinlan, J.R.: Induction of decision trees. Mach. Learn. **1**(1), 81–106 (1986)
16. Wang, Y.C., McPherson, K., Marsh, T., Gortmaker, S.L., Brown, M.: Health and economic burden of the projected obesity trends in the USA and the UK. Lancet **378**(9793), 815–825 (2011)
17. Sam, S., Mazzone, T.: Adipose tissue changes in obesity and the impact on metabolic function. Transl. Res. **164**(4), 284–292 (2014)
18. Dattilo, A.M., Kris-Etherton, P.M.: Effects of weight reduction on blood lipids and lipoproteins: a meta-analysis. Am. J. Clin. Nutr. **56**(2), 320–328 (1992)
19. Fox, C.S., et al.: Abdominal visceral and subcutaneous adipose tissue compartments. Circulation **116**(1), 39–48 (2007)
20. Després, J.-P., Lemieux, I., Bergeron, J., Pibarot, P., Mathieu, P., Larose, E., Rodés-Cabau, J., Bertrand, O.F., Poirier, P.: Abdominal obesity and the metabolic syndrome: contribution to global cardiometabolic risk. Arterioscler. Thromb. Vasc. Biol. **28**(6), 1039–1049 (2008)
21. Chin, S.-H., Kahathuduwa, C.N., Binks, M.: Physical activity and obesity: what we know and what we need to know*. Obes. Rev. **17**(12), 1226–1244 (2016)
22. Golabi, P., Bush, H., Younossi, Z.M.: Treatment strategies for nonalcoholic fatty liver disease and nonalcoholic steatohepatitis. Clin. Liver Dis. **21**(4), 739–753 (2017)
23. Uusitupa, M., Lindi, V., Louheranta, A., Salopuro, T., Lindström, J., Tuomilehto, J.: Long-term improvement in insulin sensitivity by changing lifestyles of people with impaired glucose tolerance. Diabetes **52**(10), 2532–2538 (2003)
24. Brouwers, B., Hesselink, M.K.C., Schrauwen, P., Schrauwen-Hinderling, V.B.: Effects of exercise training on intrahepatic lipid content in humans. Diabetologia **59**(10), 2068–2079 (2016)

25. Sudlow, C., Gallacher, J., Allen, N., Beral, V., Burton, P., Danesh, J., et al.: UK Biobank: an open access resource for identifying the causes of a wide range of complex diseases of middle and old age. PLoS Med. **12**(3), e1001779 (2015)
26. Information gain, mutual information and related measures - Cross Validated. https://stats.stackexchange.com/questions/13389/information-gain-mutual-information-and-related-measures. Accessed 22 Oct 2018
27. Haddow, C., Perry, J., Durrant, M., Faith, J.: Predicting functional residues of protein sequence alignments as a feature selection task. Int. J. Data Min. Bioinform. **5**(6), 691–705 (2011)
28. Drummond, C., Holte, R.C.: C4.5, class imbalance, and cost sensitivity: why under-sampling beats over-sampling. In: Proceedings of the International Conference on Machine Learning, Workshop Learning from Imbalanced Data Sets II (2003)
29. Manning, C., Raghavan, P., Schutze, H.: Introduction to information retrieval. Nat. Lang. Eng. **16**(1), 100–103 (2010)
30. Zhang H.: The optimality of Naive Bayes. In: American Association for Artificial Intelligence (2004)
31. Landwehr, N., Hall, M., Frank, E.: Logistic model trees. Mach. Learn. **59**(1–2), 161–205 (2005)
32. Ayer, T., Chhatwal, F., Alagoz, O., Kahn, C.E., Woods, R.W., Burnside, E.S.: Comparison of logistic regression and artificial neural network models in breast cancer risk estimation. Radio Graphics **30**(1), 13–22 (2010)
33. Quinlan, J.R.: Improved use of continuous attributes in C4.5. J. Artif. Intell. Res. **4**, 77–90 (1996)
34. Witten, I.H., Frank, E.: Data Mining, Practical Machine Learning Tools and Techniques, 2nd edn. Elsevier Inc., Amsterdam (2005)
35. Witten, I.H., Frank, E.: Data Mining: Practical Machine Learning Tools and Techniques with Java Implementations. Morgan Kaufmann, San Francisco (2000)
36. Jonsdottir, T., Hvannberg, E.T., Sigurdsson, H., Sigurdsson, S.: The feasibility of constructing a predictive outcome model for breast cancer using the tools of data mining. Expert Syst. Appl. **34**(1), 108–118 (2008)
37. Maheshwari, S., Agrawal, J., Sharma, S.: A new approach for classification of highly imbalanced data sets using evolutionary algorithms. Int. J. Sci. Eng. Res. **2**(7), 1–5 (2011)
38. Computing Precision and Recall for Multi-class Classification Problems. http://text-analytics101.rxnlp.com/2014/10/computing-precision-and-recall-for.html. Accessed 02 Aug 2018
39. Parkinson, J.R. et al.: Visceral adipose tissue, thigh adiposity and liver fat fraction: a cross sectional analysis of the UK Biobank. UK Biobank (2019)
40. Bagging and Random Forest Ensemble Algorithms for Machine Learning. https://machinelearningmastery.com/bagging-and-random-forest-ensemble-algorithms-for-machine-learning/. Accessed 22 Oct 2018

A Study on Relationship Between Walking Speed and Acceleration of Center of Mass Estimated with Inertial Sensors

Takashi Watanabe(✉) and Yuho Takeda

Graduate School of Biomedical Engineering, Tohoku University, Sendai 980-8579, Japan
t.watanabe@tohoku.ac.jp

Abstract. Since the center of mass (CoM) is related to ground reaction force (GRF), the CoM position is expected to be used in evaluation of walking ability. The purpose of this study was to examine the possibilities of using the CoM acceleration calculated from the CoM trajectory estimated by inertial sensor signals to obtain similar information as the GRF. In this paper, 16 m straight walking measurements were performed under different walking speeds with healthy subjects. CoM acceleration, which was calculated from CoM positions estimated by signals measured with inertial sensors, were analyzed for the steady state gait. Waveforms of the CoM acceleration were similar among subjects. Four peak values (negative peak near the toe off (TO) and positive peak near the initial contact (IC) of the opposite lower limb in the anteroposterior component, and the positive peak before the TO of the opposite lower limb and the negative peak of the mid-stance phase) showed significant correlations with walking speed, which were similar to the relationships between GRF and walking speed in previous studies. These results suggest that anteroposterior and vertical components of CoM acceleration estimated from inertial sensor signals have a possibility of estimating similar information as the GRF during walking. The relationship between positive peak value of the vertical component near the IC of the opposite side and walking speed was seemed to be different from a previous study, although it showed significant correlation with walking speed. Simultaneous measurements with inertial sensors and GRF are expected in the next study.

Keywords: Center of Mass · Ground Reaction Force · Inertial sensor · Walking

1 Introduction

The center of mass (CoM) position is expected to be used in evaluation of walking ability, because gait movement involves three-dimensional movement of the CoM during gait. Walking balance [1] and asymmetry of movement and weight support [2] were evaluated by using CoM position, and CoM position during walking was analyzed in the relation with energy consumption during walking [3]. In addition, calculation of ground reaction force (GRF) by the double differentiation of the CoM trajectory was examined [4]. Peak values of the vertical component of the GRF was

J. Henriques et al. (Eds.): MEDICON 2019, IFMBE Proceedings 76, pp. 694–702, 2020.
https://doi.org/10.1007/978-3-030-31635-8_82

shown to have a correlation with walking speed [5, 6], and rehabilitation intervention for hemiplegic patients showed increase of the propulsive and braking forces (anteroposterior component of GRF) of paralyzed lower limbs [7, 8].

In our previous study, a method of estimating CoM position from signals measured with 7 wireless inertial sensors was proposed [9]. The CoM position was estimated by inclination angles of body segments and feet positions that were estimated by the methods developed in our research group [10, 11]. The CoM estimation method was shown to be useful in estimation of vertical and anteroposterior components of CoM position, in which root mean squared errors were 0.83 ± 0.25 cm and correlation coefficient were 3.8 ± 2.1 cm, and 0.82 ± 0.07 and 0.9999 ± 0.0001 for vertical and anteroposterior components, respectively, in straight forward walking of healthy subjects [9].

The purpose of this study was to examine the possibilities of using the CoM acceleration calculated from the CoM trajectory estimated by inertial sensor signals to obtain similar information as the GRF. In this paper, the CoM positions were estimated from signals measured with inertial sensors during 16 m straight forward walking under different walking speeds of healthy subjects and the CoM acceleration was calculated by the double differentiation of the estimated CoM positions. Relationships between peak values of the CoM acceleration and waking speed were discussed considering the relationships between GRF and walking speed.

2 Estimation Method of Center of Mass Position and Its Acceleration

The CoM position was estimated by reconstructing posture of the body by using posture of each segment of the body and foot positions estimated from signals measured with inertial sensors as shown in Fig. 1. The posture of each segment of the body was represented by a unit vector (segment vector), which was defined in the direction of longitudinal axis of the body segment. Segment vectors were estimated by using rotation matrix calculated by quaternion representing the attitude of each segment in three-dimensional space. Although the quaternion can be calculated from angular velocity measured with an inertial sensor, the calculated quaternion was corrected by applying the Kalman filter using the acceleration vector, because the error was accumulated due to the drift of gyroscope signals.

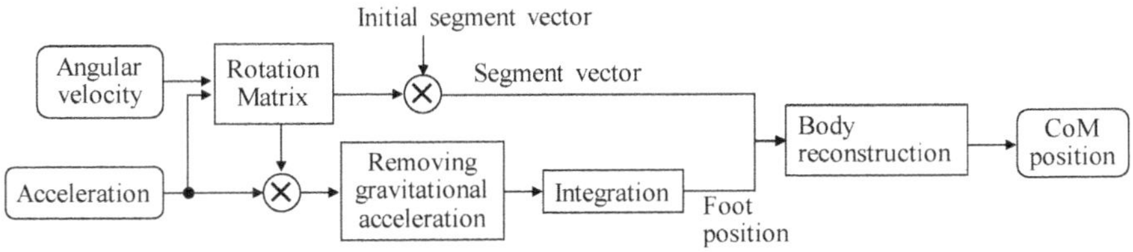

Fig. 1. Outline of estimation method of the center of mass (CoM) position.

Foot position was calculated by double integration of the movement acceleration of the foot segment after removing gravitational acceleration. The movement velocity of the foot was calculated assuming to be 0 m/s at the beginning of the integral. The calculated velocity of the foot was corrected so as to be 0 m/s at the end of the integral by using linear approximation of integral error. The vertical position of CoM was also corrected by using linear approximation of integral error, since the vertical component of the foot position could be regarded as 0 cm for each stationary state in the measurement.

Then, posture of the body at each time was reconstructed using estimated segment vectors from the foot segment to the trunk segment based on the estimated foot position. Position of center of mass of each segment was calculated from the position and attitude of the reconstructed body segment vectors, in which length, weight and center of mass position of each segment were approximately calculated from height and weight of subject based on previous studies [12, 13]. Finally, the CoM position of whole body is calculated by Eq. (1).

$$CoM(t) = \frac{\sum m_i r_i(t)}{\sum m_i} \tag{1}$$

Here, m_i and r_i show mass and center of mass position of each segment, respectively.

Acceleration of the CoM was calculated by the double differentiation of the estimated CoM positions. A zero-phase low-pass filter with a cutoff frequency of 10 Hz was applied to the CoM position in order to remove high frequency noise before the differentiation.

The CoM acceleration can be related approximately to ground reaction force by the following [4]:

$$a_{ap} = \frac{F_{ap}}{m} \tag{2}$$

$$a_v = \frac{F_v}{m} - g \tag{3}$$

Here, a_{ap}, F_{ap}, a_v, and F_v show CoM acceleration and ground reaction force in the anteroposterior and vertical directions, respectively. m and g are weight and gravitational acceleration.

3 Experimental Tests

3.1 Method

Gait movements of 4 healthy male subjects (21–23 years old) were measured with wireless inertial sensors (MPU-9150, Bluetooth class 1) under different three walking speeds. The inertial sensors were attached on both feet, both shanks, both thighs, trunk using stretch bands as shown in Fig. 2. Sampling frequency of the inertial sensors was set to 100 Hz.

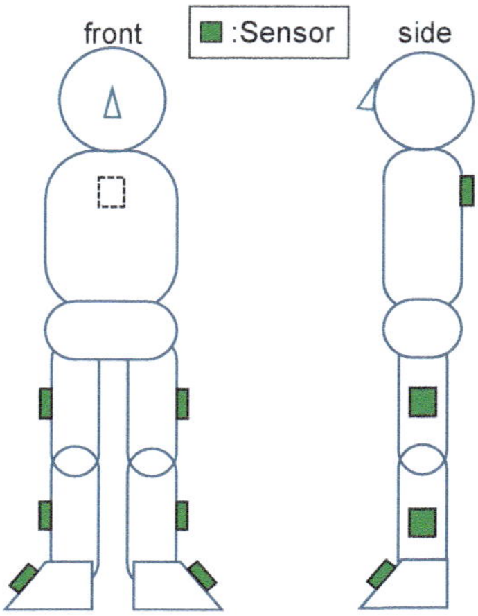

Fig. 2. Attachment positions of inertial sensors.

The subjects walked in a straight line of 16 m on the flat floor. The subjects carried out 10 trials of 3 types walking of moderate speed, slow speed and fast speed, in which walking speeds were determined by subjects themselves. The walking speed was calculated by measuring the required time of 10 m walking removing 3 m from the start and 3 m before the end of walking with a stopwatch. Estimated CoM accelerations were analyzed by excluding two strides just after the start and two strides before the end of walking.

3.2 Results

An example of the CoM acceleration waveform is shown in Fig. 3. Since GRF is produced only at the time of ground contact, the CoM acceleration from the heel contact (IC) to the toe off (TO) of the right foot was shown in the figure, in which time was normalized as 100% for the stance phase. In addition, all trials were averaged for each subject excluding trials in which the integration term for estimating foot position was not determined appropriately. The plot also shows the timing of the TO and the IC of the opposite lower limb (left side) to show the single support phase. The waveforms of anteroposterior and vertical components of the CoM acceleration were similar among 4 subjects.

Figure 4 shows the relationship between the peak value of anteroposterior component of CoM acceleration and the walking speed. Each plot shows average values and standard deviations in a single measurement trial. The analyzed peak values of CoM acceleration were negative peak near the TO of the opposite lower limb and positive peak near the IC of the opposite side because those peak values of GRF were shown to decrease and increase with increase of walking speed respectively [6, 8]. Significant negative correlation ($\rho = -0.834$, $p < 0.01$) and significant positive correlation ($\rho = 0.814$, $p < 0.01$) were shown between walking speed and the average peak values.

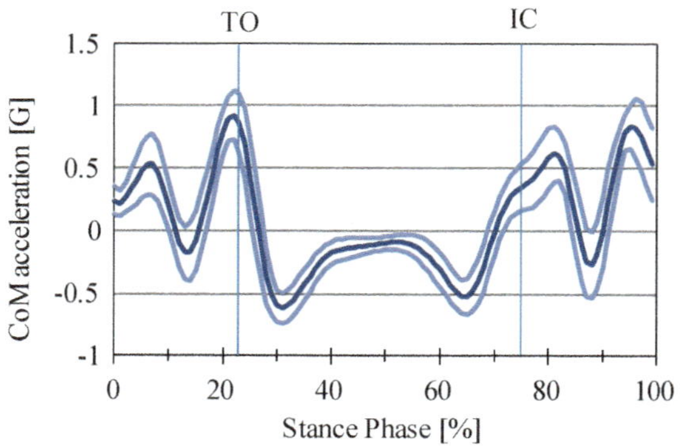

(a) anteroposterior

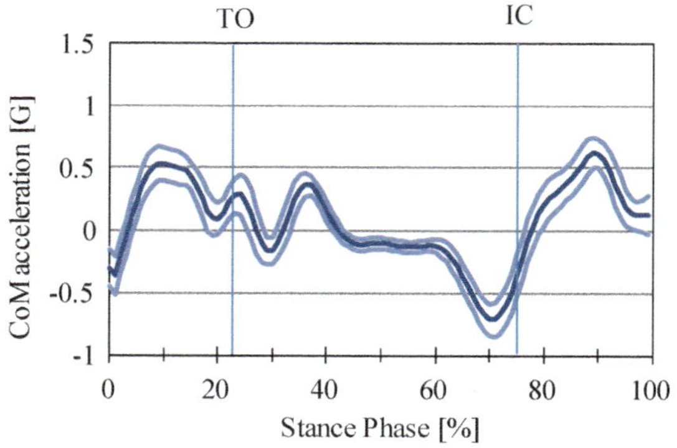

(b) vertical

Fig. 3. Examples of waveform of CoM acceleration (Sub 1, moderate speed). Average waveforms and their standard deviations are shown in the plots. TO and IC timings of the opposite side are also shown.

Figure 5 shows the relationship between the peak value of vertical component of CoM acceleration and the walking speed. The peak values were the positive peak before the TO of the opposite lower limb and the positive peak near the IC of the opposite side, because the positive peak value of the GRF before the TO were shown to increase with increase of walking speed [6]. In addition, the negative peak value of the mid-stance phase were shown. From Fig. 5(a) and (c), the positive peak value before the TO of the opposite lower limb showed a significant positive correlation with walking speed ($\rho = 0.812$, $p < 0.01$), and the negative peak value of the mid-stance phase showed a significant negative correlation ($\rho =$ 0.827, $p < 0.01$). These results were similar to the previous study [5]. In addition, the relationship of the positive peak near the IC of the opposite side showed a significant positive correlation ($\rho = 0.885$, $p < 0.01$) as seen in Fig. 5(b).

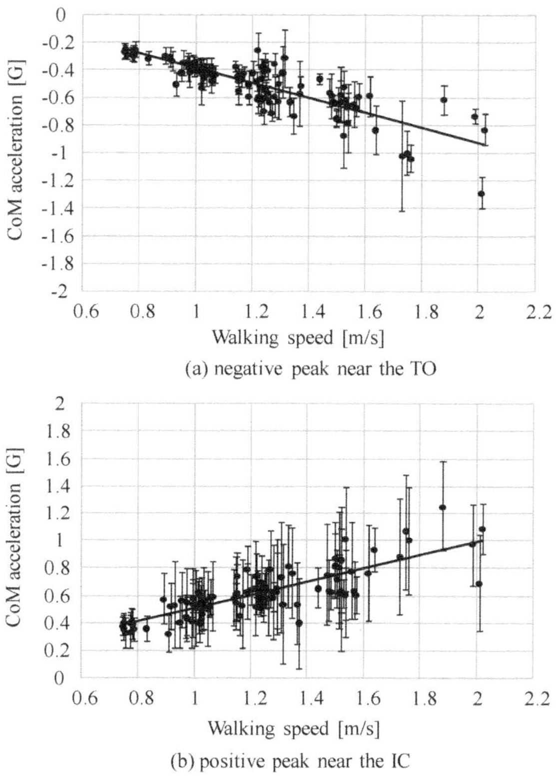

Fig. 4. Relationships between peak values of anteroposterior component of CoM acceleration and walking speed.

4 Discussion

The waveforms of anteroposterior and vertical components of the CoM acceleration, which were calculated by the double differentiation of the CoM positions estimated from signals measured with inertial sensors, were similar among four subjects. All five peak values of the CoM acceleration in the stance phase showed significant correlation with walking speed and relationships of the four peak values of them with walking speed were shown to be similar to the relationships between GRF and walking speed as in the previous studies [5, 6, 8]. These results suggest that the method of CoM estimation with inertial sensors could estimate anteroposterior and vertical components of CoM position during walking of healthy subjects appropriately. Since the similar results as those in the previous studies were obtained, anteroposterior and vertical components of CoM acceleration estimated from signals measured with inertial sensors are considered to have a possibility of estimating similar information as the GRF during walking.

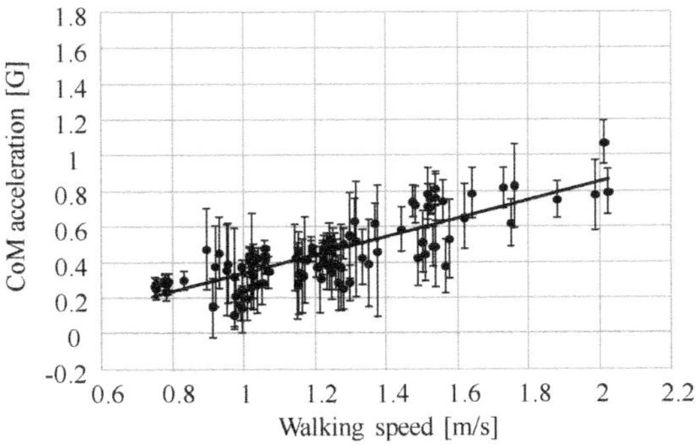

(a) positive peak before the TO

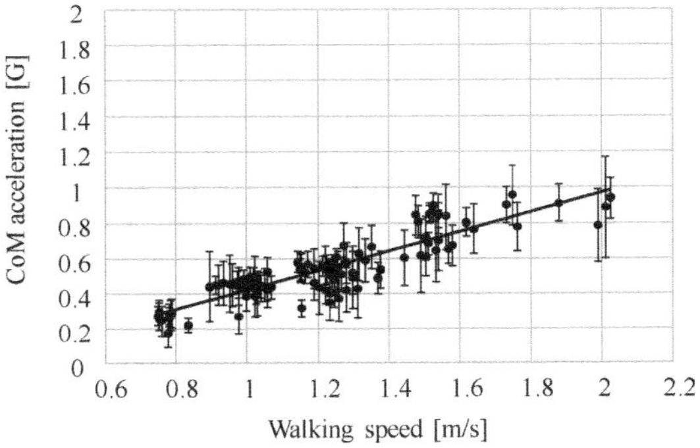

(b) positive peak near the IC

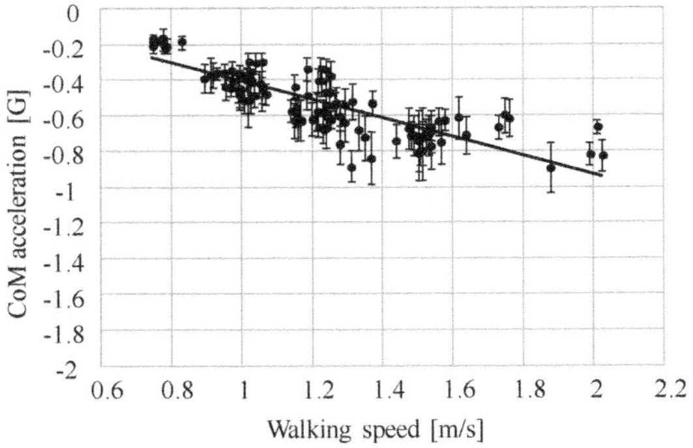

(c) negative peak in the mid-stance phase

Fig. 5. Relationships between peak values of vertical component of CoM acceleration and walking speed.

The relationship in this study showed a significant correlation with the walking speed. However, variations of peak values in a single measurement trial or between trials under the same condition were considered to be large a little bit, especially in the anteroposterior component. Since movements during 16 m walking were measured in order to analyze steady state walking, GRF could not be measured in this paper. The difference of the relationship between CoM acceleration and GRF and variation of peak values have to be examined by conducting simultaneous measurement with inertial sensors and GRF. The number of subjects also has to be increased including subjects in various age ranges, although the number of measurement trials was increased because the number of subjects was small in this paper.

5 Conclusion

In this paper, 16 m straight walking measurements were performed with healthy subjects, and CoM acceleration calculated from CoM positions estimated from measured signals with inertial sensors were analyzed for the steady state gait. Waveforms of CoM acceleration were similar among subjects and relationships between four peak values and walking speed showed similar relationships as those between peak values of GRF and walking speed in previous studies. These results suggest that anteroposterior and vertical components of CoM acceleration estimated from signals measured with inertial sensors have a possibility of estimating similar information as the GRF during walking. The other relationship between peak value and walking speed was seemed to be different from a previous study, although it showed significant correlation with walking speed. Simultaneous measurement with inertial sensors and GRF are expected in the next study.

References

1. Meulen, F.B., Weenk, D., Asseldonk, E.H.F., Schepers, H.M., Veltink, P.H., Buurke, J.H.: Analysis of balance during functional walking in stroke survivors. PLoS ONE, **11**(11) (2016)
2. Kama, D., Kamphuis, J.F., Weerdesteyn, V., Geurts, A.C.H.: The effect of weight-bearing asymmetry on dynamic postural stability in healthy young individuals. Gait Posture **45**, 56–61 (2016)
3. Gordon, K.E., Ferris, D.P., Kuo, A.D.: Metabolic and mechanical energy costs of reducing vertical center of mass movement during gait. Arch. Phys. Med. Rehabil. **90**(1), 136–144 (2009)
4. Winiarski, S., Rutkowska-Kucharska, A.: Estimated ground reaction force in normal and pathological gait. Acta Bioeng. Biomech. **11**(1), 53–60 (2009)
5. Cook, T.M., Farrell, K.P., Carey, I.A., Gibbs, J.M., Wiger, G.E.: Effects of restricted knee flexion and walking speed on the vertical ground reaction force during gait. J. Orthop. Sports Phys. Ther. **25**(4), 236–244 (1997)
6. Nilsson, J., Thorstensson, A.: Ground reaction forces at different speeds of human walking and running. Acta. Physiol. Scand. **136**(2), 217–227 (1989)
7. Yeung, L.-F., Ockenfeld, C., Pang, M.-K., Wai, H.-W., Soo, O.-Y., Li, S.-W., Tong, K.-Y.: Randomized controlled trial of robot assisted gait training with dorsiflexion assistance on chronic stroke patients wearing ankle-foot-orthosis. J. Neuroeng. Rehabil. **15**, 51–62 (2018)
8. Hsiao, H., Awad, L.N., Palmer, J.A., Higginson, J.S., Binder-Macleod, S.A.: Contribution of paretic and non-paretic limb peak propulsive forces to changes in walking speed in individuals poststroke. Neurorehabil. Neural Repair **30**(8), 743–752 (2016)

9. Takeda, Y., Watanabe, T.: A study on estimation of center of mass trajectory using inertial sensors for movement evaluation. IEICE Technical Reports, vol. 118, no. 366, pp. 1–4, MBE2018-50 (2018). (in Japanese)
10. Watanabe, T., Teruyama, Y., Ohashi, K.: Comparison of angle measurements between integral-based and quaternion-based methods using inertial sensors for gait evaluation. Commun. Comput. Inf. Sci. **511**, 274–288 (2015)
11. Watanabe, T., Miyazawa, T., Shibasaki, J.: A study on IMU-based stride length estimation for motor disabled subjects: a comparison under different calculation methods of rotation matrix. In: International Conference on Biomedical and Health Informatics (BHI). https://doi.org/10.1109/bhi.2018.8333372 (2018)
12. Ae, M., Tang, H.P., Yokoi, T.: Estimation of inertia properties of the body segments in Japanese athletes. Biomechanisms **11**, 23–33 (1992). (in Japanese)
13. Okada, H., Ae, M., Fujii, N., Morioka, Y.: Body segment inertia properties of Japanese elderly. Biomechanisms **13**, 125–139 (1996). (in Japanese)

Comparative Assessment Between 3D and Conventional 2D Imaging Systems in Laparoscopic Practice

Juan A. Sánchez-Margallo(✉), Silvia Enciso Sanz, and Francisco M. Sánchez-Margallo

Jesús Usón Minimally Invasive Surgery Centre, Cáceres, Spain
jasanchez@ccmijesususon.com

Abstract. The objective of this study was to evaluate the effect of 3D imaging on laparoscopic suturing performance using a box trainer and during laparoscopic partial nephrectomy. Besides, the safety and usefulness of this technology in clinical practice were also subjectively evaluated. For the study in the training environment participated 30 surgeons and 13 surgeons in the study with the experimental animal model. Participants were organized in a random way into two groups, starting the trials with the 2D imaging system or with the 3D system. Three-dimensional imaging seemed to facilitate the performance of intracorporeal sutures in the box trainer setting. However, these improvements were not as clear during the partial nephrectomy. Surgeons stated that the 3D vision system helps improve depth perception during surgery and facilitates needle handling during intracorporeal suturing. However, they also indicated that its use can lead to the onset of visual discomfort or fatigue. Most of them would use again this vision system in surgical practice. Further studies are needed to more clearly determine the benefits of 3D over standard 2D imaging systems in laparoscopic practice.

Keywords: Laparoscopy · 3D imaging · Suturing · Partial nephrectomy

1 Introduction

Limitations of 2D imaging systems such as depth perception and spatial orientation still remain a challenge in laparoscopic surgery even for the experienced surgeons. Surgeons learn to compensate these limitations by means of visual cues such as shadows and the relative size of organs or textures. Recent technological advances have led to the introduction of new three-dimensional visualization systems in surgery in response to the constant search for new solutions to improve the safety during surgery. This technology has been widely used in robotic surgery, as it is included as a base in almost all robotic platforms for minimally invasive surgery. Several studies have successfully demonstrated the application of 3D vision systems to laparoscopic practice [1]. However, other reports are less enthusiastic and cite that the use of these systems is not superior to the traditional 2D system in terms of efficiency and performance [2, 3]. Therefore, additional research has to be done in order to analyze the effects, safety and utility of 3D vision in laparoscopic practice.

J. Henriques et al. (Eds.): MEDICON 2019, IFMBE Proceedings 76, pp. 703–710, 2020.
https://doi.org/10.1007/978-3-030-31635-8_83

The objective of this study was to evaluate the effect of 3D imaging on laparoscopic suturing performance using a box trainer and during a surgical procedure in an experimental animal model. Besides, the safety and usefulness of this technology in clinical practice were also subjectively evaluated.

2 Materials and Methods

2.1 Subjects

This work consisted of two studies, a laparoscopic simulator study and an experimental animal model study. For the study in the training environment participated 30 surgeons: 16 novice surgeons (less than 10 laparoscopic procedures performed) and 14 experienced surgeons (more than 50 laparoscopic procedures performed). In the study with the experimental animal model participated 13 surgeons: 6 novice surgeons, 2 intermediate experienced surgeons (between 10 and 50 laparoscopic procedures performed) and 5 experienced surgeons. All participants assured that they had no visual disorders concerning stereoscopic vision.

2.2 Tasks and Assessment

In both studies, participants were organized in a random way into two groups: Group A, who started the task with the 2D imaging system (Karl Storz HD; Karl Storz GmbH & Co. KG, Germany) and Group B, who started with the 3D imaging system (Karl Storz).

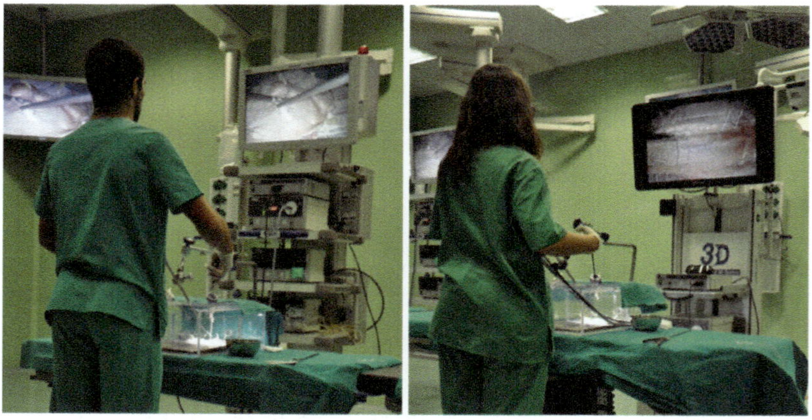

Fig. 1. Experimental setting using the 2D (left) and 3D imaging (right) systems.

For the evaluation in the training environment, participants carried out three laparoscopic sutures with each imaging system on an organic tissue (porcine stomach) using a box trainer (Fig. 1). In every case, they proceeded with identical ports and optics configuration. Participants were asked to carry out one double knot followed by two simple knots in opposite directions. Suturing performance was assessed through a validated suturing specific-checklist [4] by two expert surgeons in a blinded fashion. This checklist scores technical features of the suturing performance such as needle position and driving, pulling through of the suture, and the technique and quality of the knots. Execution time was recorded for each trial.

For the evaluation with the experimental model, participants performed a partial nephrectomy in the left kidney of a porcine model. To accomplish the partial nephrectomy, an artificial tumor model was previously created in the upper renal pole of the kidney. A mixture of saline and alginate was percutaneously injected to reproduce the tumor. This study was reviewed and approved by the competent local Animal Welfare and Ethics Committee. Accommodation of the animals and their handling were done in accordance with the European directive (2010/63/EU) regarding the use of animals for scientific purposes, Spanish laws (RD 53/2013) for animal use and care, ARRIVE guidelines, and according to the Guide for the Care and Use of Laboratory Animals. Surgery time was recorded for each step of the laparoscopic partial nephrectomy: Dissection of the renal artery, Rumel tourniquet, preparation of the renal pole, exeresis, closure of the collecting system, and suture of the renal parenchyma. At the end of the study, participants were asked to complete a subjective questionnaire on a 5-point rating scale with regard to the usability, the perception of visual discomfort, and the quality of image of the 3D imaging system.

2.3 Statistical Analysis

A descriptive analysis for each variable was performed. The Wilcoxon signed-rank test was used to compare the surgeons' execution time and surgical performance when using the 2D and 3D imaging systems. All statistical analyses were carried out using R version 3.4.1 (R Foundation for Statistical Computing, Vienna, Austria). For all tests, $p < 0.05$ was considered statistically significant.

3 Results

3.1 Evaluation in the Training Environment

Results showed statistically significant differences between novice and experienced surgeons for the execution time using both imaging systems (Fig. 2). However, only the expert group performed faster the suturing exercise using the 3D imaging system compared to the conventional one.

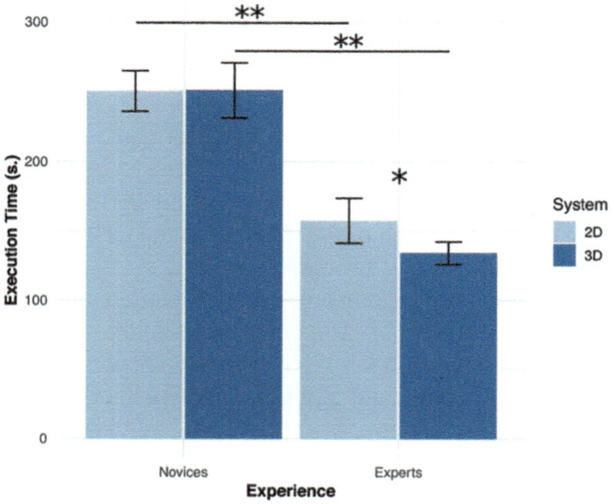

Fig. 2. Comparison of the execution time for the laparoscopic suture in a box trainer between the groups of novice and expert surgeons and between the use of the 3D and 2D vision systems. * $p < 0.05$, ** $p < 0.001$.

Regarding the checklist score, the performance of the suturing task was significantly improved using the 3D vision system compared to the 2D system by both groups of surgeons (Fig. 3). The use of the 3D vision system significantly improved the performance of the pulling maneuver of the suture after driving the needle through the organic tissue.

3.2 Evaluation in the Experimental Animal Model

All surgical procedures were successfully performed, and no complications were registered. Regarding the surgery time, it decreased when using the 3D vision system, but with no significant differences with regard to the 2D system. Although there was also a slight reduction in the operating times of the different surgical steps of the partial nephrectomy performed with 3D vision, except for the suture of the renal parenchyma, there were no statistically significant differences with the results obtained using 2D vision (Table 1).

Surgeons stated that the 3D vision system helps improve depth perception during surgery and facilitates needle handling during intracorporeal suturing (Fig. 4). They also believe that the 3D vision system is safe for its use during hospital activity. However, they also indicated that it can lead to the presence of blurred areas in the image, as well as the onset of fatigue or visual discomfort. If they had to repeat the surgical procedure again, 84.62% of participants stated that they would use the 3D vision system.

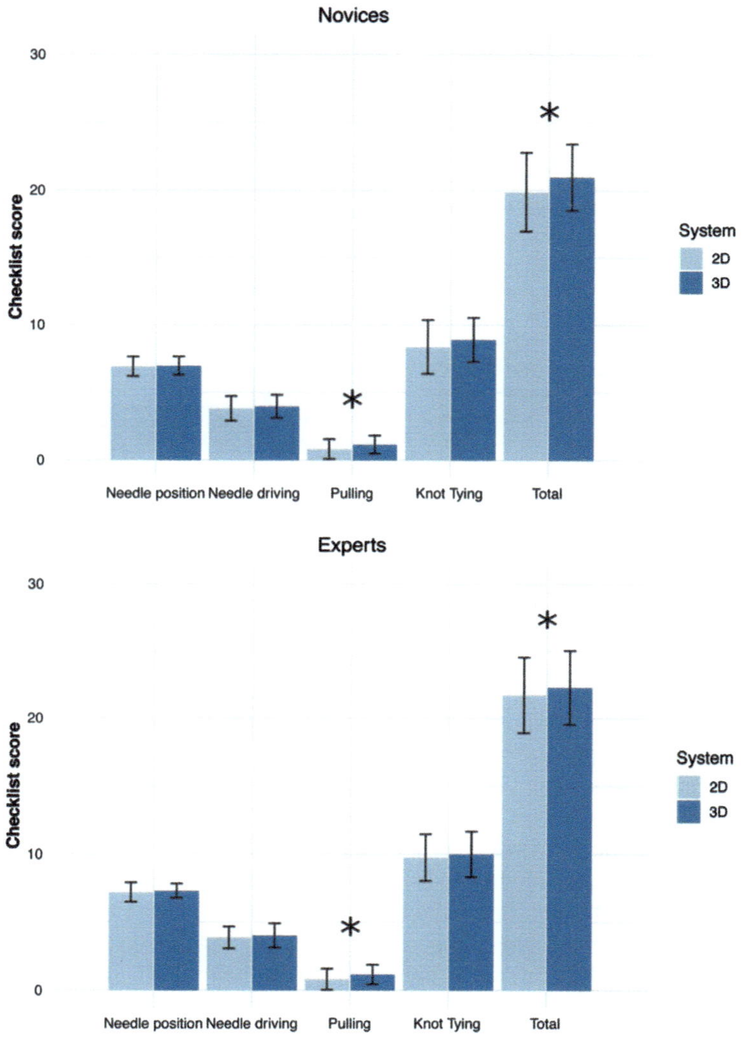

Fig. 3. Suturing checklist score for the groups of novice and expert surgeons using the 2D and 3D imaging systems. * $p < 0.05$.

Table 1. Execution time (in minutes) of each surgical step of the partial nephrectomy performed with the 2D and 3D imaging systems.

Surgical step	2D imaging	3D imaging
Dissection of the renal artery	17.16 ± 10.32	16.27 ± 7.96
Rumel tourniquet	3.77 ± 4.19	3.21 ± 1.60
Preparation of the renal pole	4.24 ± 1.94	3.29 ± 1.34
Exeresis	2.93 ± 0.82	2.65 ± 1.23
Closure of the collecting system	7.93 ± 7.74	6.87 ± 3.55
Suture of the renal parenchyma	13.78 ± 5.20	15.12 ± 5.91
TOTAL	50.18 ± 20.16	46.72 ± 14.84
Ischemia time	28.35 ± 8.71	28.43 ± 9.48

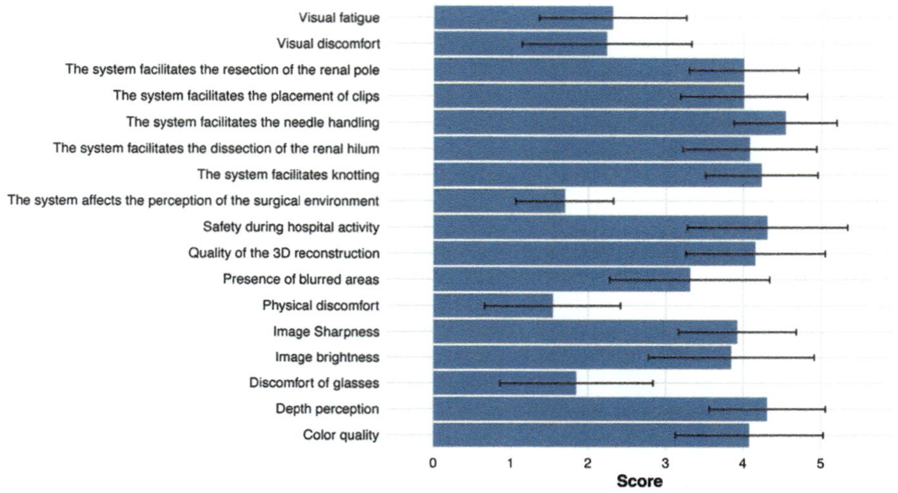

Fig. 4. Results of the subjective questionnaire (5-point rating scale) regarding the safety, image quality and the usefulness of the 3D imaging system.

4 Discussion

Over the past two decades, advances in 3D imaging technology have enabled its rapid application in different laparoscopic disciplines in order to solve the limitations with regard to depth perception. Several clinical and laboratory studies have compared 3D vision with 2D laparoscopic vision showing in some cases contradictory results [2, 5]. Some study groups reported superior performance, while others showed identical results for surgical performance.

The results obtained for the study in the training environment showed that only expert surgeons significantly improved the execution time during laparoscopic suturing by using the 3D vision system compared to the traditional 2D system. These findings have also been found in the study presented by Wilhelm et al. [6]. Buia et al. also obtained a significant improvement in execution time with the 3D vision system only in the group of expert surgeons during the performance of eye-hand coordination exercises [7]. Besides, results showed that the use of 3D imaging in a physical simulator is associated with a significantly better performance than 2D during suturing tasks for both novice and expert surgeons. This has also been proved by other authors, mainly in the case of junior trainees [1]. Some studies have also demonstrated that 3D imaging resulted in a higher degree of accuracy during suturing tasks in a box trainer [5]. However, there are other studies that obtained similar performance during suturing exercises using both imaging systems [2, 3]. In most cases, this may be influenced by the simplicity of tasks performed.

In the study with the experimental model, results showed similar surgery time for both types of vision systems during the partial nephrectomy. None of the surgical steps have seen its execution time significantly decreased with the use of 3D vision technology. Reducing the ischemia time in partial nephrectomy is a critical factor in ensuring the success of the procedure. Cicione et al. also found no difference in terms of operative time between 2D or 3D laparoscopic partial nephrectomy in a porcine model, but they obtained better surgical performance, assessed by a global rating scale, when using the 3D vision system [8]. A randomized controlled trial also obtained no significant differences in surgery time and dissecting time, but the suturing and ischemia times were shorter with 3D vision [9]. This may be a consequence of the better depth perception and spatial orientation provided by 3D laparoscopy [5]. Results from a recent systematic review and meta-analysis by Dirie et al. reported no statistically significant differences in operative time, dissecting time, and suturing time between 2D and 3D laparoscopic systems in partial nephrectomy procedures [10]. However, significant differences were found in warm ischemia time, which was shorter using the 3D system. It appears that the use of 3D vision does not improve operating time in partial nephrectomy, although some authors have shown a reduction in ischemia time. Therefore, additional studies are needed to strengthen the results concerning the benefits of the 3D vision technology in this surgical procedure.

The participants of this study emphasized that apart from the benefits that 3D vision brings to surgical practice, especially during intracorporeal suturing, this technology could occasionally cause visual fatigue or discomfort. This fact has already been reported by other studies [11–13], including the appearance of dry eyes or headache. Therefore, these limitations should be taken into account in future designs of 3D visualization devices and techniques for surgery. Surgeons' ergonomics is a critical factor to bear in mind in order to ensure adequate surgical performance during surgery [14].

Taking into consideration the results obtained in this study and from other authors [9], it seems that the use of 3D vision facilitates the performance of laparoscopic suturing. To reinforce the validity of these results, it would be convenient to analyze the use of this vision technology in more complex surgical procedures involving a high load of intracorporeal sutures such as urethrovesical anastomosis.

There are some limitations in our study, of which we are aware. Our study was limited by the small number of subjects, especially in the experimental model study. Therefore, larger studies should be conducted to further evaluate the clinical value of the use of 3D imaging in laparoscopic surgery. On the other hand, it would be advisable to include an assessment of the quality of the surgical performance, for example through the use of a rating scores such as the R.E.N.A.L nephrometry score (RNS) or the Global Operative Assessment of Laparoscopic Skills (GOALS) rating scale.

4.1 Conclusions

Three-dimensional imaging seems to facilitate the surgical performance of intracorporeal sutures in a box trainer setting. However, these improvements are not as clear in surgical procedures such as partial nephrectomy, in which surgery time is a critical factor. Further studies are needed to more clearly determine the benefits of 3D over standard 2D imaging systems in laparoscopic practice.

Acknowledgements. This work has been partially funded by the Junta de Extremadura, Spain, and FEDER funds (GR18199).

Conflict of Interests. Drs. Juan A. Sánchez-Margallo, Silvia Enciso Sanz and Francisco M. Sánchez-Margallo have no conflicts of interest or financial ties to disclose.

References

1. Arezzo, A., Vettoretto, N., Francis, N.K., Bonino, M.A., Curtis, N.J., Amparore, D., et al.: The use of 3D laparoscopic imaging systems in surgery: EAES consensus development conference 2018. Surg. Endosc **33**, 3251–3274 (2018)
2. Cicione, A., Autorino, R., Breda, A., De Sio, M., Damiano, R., Fusco, F., et al.: Three-dimensional vs standard laparoscopy: comparative assessment using a validated program for laparoscopic urologic skills. Urology **82**, 1444–1450 (2013)
3. Honeck, P., Wendt-Nordahl, G., Rassweiler, J., Knoll, T.: Three-dimensional laparoscopic imaging improves surgical performance on standardized ex-vivo laparoscopic tasks. J. Endourol. **26**, 1085–1088 (2012)
4. Kroeze, S.G., Mayer, E.K., Chopra, S., Aggarwal, R., Darzi, A., Patel, A.: Assessment of laparoscopic suturing skills of urology residents: a pan-European study. Eur. Urol. **56**(5), 865–872 (2009)
5. Lusch, A., Bucur, P.L., Menhadji, A.D., Okhunov, Z., Liss, M.A., Perez-Lanzac, A., et al.: Evaluation of the impact of three-dimensional vision on laparoscopic performance. J. Endourol. **28**, 261–266 (2014)
6. Wilhelm, D., Reiser, S., Kohn, N., Witte, M., Leiner, U., Mühlbach, L., et al.: Comparative evaluation of HD 2D/3D laparoscopic monitors and benchmarking to a theoretically ideal 3D pseudodisplay: even well-experienced laparoscopists perform better with 3D. Surg. Endosc. **28**, 2387–2397 (2014)
7. Buia, A., Stockhausen, F., Filmann, N., Hanisch, E.: 3D vs. 2D imaging in laparoscopic surgery—an advantage? Results of standardised black box training in laparoscopic surgery. Langenbeck's Arch. Surg. **402**, 167–171 (2017)
8. Cicione, A., Autorino, R., Laguna, M.P., De Sio, M., Micali, S., Turna, B., et al.: Three-dimensional technology facilitates surgical performance of novice laparoscopy surgeons: a quantitative assessment on a porcine kidney model. Urology **85**, 1252–1256 (2015)
9. Ruan, Y., Wang, X., Wang, K., Zhao, Y., Xia, S., Xu, D.: Clinical evaluation and technical features of three-dimensional laparoscopic partial nephrectomy with selective segmental artery clamping. World J. Urol. **34**, 679–685 (2016)
10. Dirie, N.I., Wang, Q., Wang, S.: Two-dimensional versus three-dimensional laparoscopic systems in urology: a systematic review and meta-analysis. J. Endourol. **32**, 781–790 (2018)
11. Sakata, S., Grove, P.M., Hill, A., Watson, M.O., Stevenson, A.R.: The viewpoint-specific failure of modern 3D displays in laparoscopic surgery. Langenbecks Arch. Surg. **401**(7), 1007–1018 (2016)
12. Gomez-Gomez, E., Carrasco-Valiente, J., Valero-Rosa, J., Campos-Hernandez, J.P., Anglada-Curado, F.J., Carazo-Carazo, J.L., et al.: Impact of 3D vision on mental workload and laparoscopic performance in inexperienced subjects. Actas Urol. Esp. **39**(4), 229–235 (2015)
13. Zhou, J., Xu, H.J., Liang, C.Z., Zhang, L., Hao, Z.Y., Feng, L.X.: A comparative study of distinct ocular symptoms after performing laparoscopic surgical tasks using a three-dimensional surgical imaging system and a conventional two-dimensional surgical imaging system. J. Endourol. **29**(7), 816–820 (2015)
14. Sánchez-Margallo, F.M., Sánchez-Margallo, J.A.: Ergonomics in laparoscopic surgery. In: Malik, A.M. (ed.) Laparoscopic Surgery, pp. 105–123. InTech, London (2017)

Modeling of Transpalpebral Tonometry System for Parameters Optimization of the Measuring Sensor

P. V. Luzhnov[1(✉)], E. N. Iomdina[2], K. V. Ivanishchev[3], D. M. Shamaev[1], and A. A. Kiseleva[1]

[1] Bauman Moscow State Technical University, 2-nd Baumanskaya 5, 105005 Moscow, Russia
peterl@hotmail.ru
[2] Moscow Helmholtz Research Institute of Eye Diseases, Sadovaya-Chernogryazskaya 14/19, 105062 Moscow, Russia
[3] Research and Development Center of Yelatma Instrument Making Enterprise, Trubezhnaya 16, 390000 Ryazan, Russia

Abstract. This paper presents the method of intraocular pressure (IOP) determination by transpalpebral tonometry (TPT), especially for the diagnosis and the control of glaucoma. TPT can be preferred method for determining IOP in terms of minimizing the contact of the measuring device with the surface of the eye. The biomechanical basis of transpalpebral tonometry is briefly discussed. The two possible schemes of TPT technique which includes a stem with springs as measuring sensor are presented. Modeling of TPT system for parameters optimization of the measuring sensor in case the dynamic measurement method is described. The researches conducted with various combinations of stem weight and diameters are presented for IOP in the range from 10 to 30 mmHg.

Keywords: Intraocular pressure · Transpalpebral tonometry · Glaucoma · Eye · Mechanical oscillations · Modeling

1 Introduction

The determination of intraocular pressure (IOP) is critical for the diagnosis and the control of glaucoma. However, until now, there is still a search for IOP measuring device, which could make it possible to accurately determine the true individual level of IOP with the least discomfort for the patient. The last circumstance is the main, as the patient response to the measurement process (reflex reactions, blinking, etc.) can lead to a deviation of the obtained values from the actual IOP. This fact increases measurement error [1, 2].

Until recently, Goldmann applanation tonometry has been considered as the standard for determining IOP. It is also possible to allocate a dynamic contour tonometry and tonometry using an iCare device, which minimize the contact of the measuring device with the patient's cornea [1, 2]. Numerous randomized comparative studies of tonometers show that if there is a high quality match (up to 95%) of IOP tonometric values (differences in mean values were no more than 2 mmHg), their individual

J. Henriques et al. (Eds.): MEDICON 2019, IFMBE Proceedings 76, pp. 711–717, 2020.
https://doi.org/10.1007/978-3-030-31635-8_84

deviations from the obtained values with other tonometers can be quite significant. Thus, according to [3], individual differences between the dynamic contour tonometry data and the Goldmann applanation tonometry data were in the range from −5.7 to +2.5 mmHg, and for the pneumotonometry data were in the range from −4.1 to +4.7 mmHg. Differences in results are particularly significant in the presence of a thin patient's cornea, as well as in advanced glaucoma stages [4, 5].

Most IOP measurement techniques are based on measuring IOP directly on the cornea of the eye [6–8]. However, the direct instrument contact with the eyeball is not always desirable. Contact of the cornea with the working surface of tonometers leads to corneal microerosion, to the need to use anesthetics, special paint, disinfectants. A prerequisite for tonometry (even contactless) is the lack of lesion of the patient's cornea.

The preferred method for determining IOP can be transpalpebral tonometry (TPT) in terms of minimizing the contact of the measuring device with the surface of the eye. TPT generally excludes contact with the eye surface, and has a contact only with the eyelid. Measuring effects can be made through the eyelid and sclera. This arrangement of the tonometer sensor eliminates the effect of the cornea on the IOP results, that was changed as a result of a refractive keratoplasty or an injury. A measurement can be carried out in contact lenses, which is convenient for a large number of patients using this optical correction tool.

Transpalpebral effect (through the eyelid) extends substantially TPT possible applications, both in the clinic and ambulatory. Such devices allow measuring IOP fairly simply and quickly, which makes them suitable for screening diagnostics. They have a small number of restrictions, can be used in a person's daily activities and in tasks for the home medicine. They can be used together with impedance methods, such as [9, 10]. The transpalpebral tonometers work according to two possible schemes [11]:

- by the principle of "rebound", i.e. elastic interaction of a freely falling stem with an eyeball through the eyelid;
- according to the principle of determining the frequency of mechanical oscillations of the eyeball as an elastic system loaded with a certain weight (the stem's weight).

Static tonometers can not identify the elastic properties of the sclera with sufficient accuracy when measured through the eyelid. Therefore, for the TPT, it is preferable to use a calculation based on the dynamic measurement method. In this paper, we have considered the simulation of a TPT system and an algorithm for determining the TPT sensor parameters performed according to the second scheme.

2 Materials and Methods

The measuring is possible through the eyelid on the sclera for TPT oscillations of the eyeball as an elastic system, loaded with a certain weight. It is perceived by the patient as the soft vibration. The measurement, repeated several times, consists of two phases:

- active TPT-stem displacement phase;
- measuring of free TPT-stem oscillations phase.

In this case, the amplitude of free oscillations is not significant - only the frequency is important, which depends on the weight of the stem and the stiffness of the eye. The biomechanical characteristic of the eye coats stiffness is associated with IOP. The main difference between the TPT and other methods of measuring IOP is the damping effect of the eyelid. This effect is reduced by pre-compression with stem. The amount of eyelid compression affects the weight of the stem and the area of its base. The same parameters affect the measured frequency of the stem oscillations and predetermine the accuracy of TPT results. In this regard, it is necessary to model and calculate the values of the TPT sensor parameters.

During the measurement the IOP is determined by stiffness of an eye coat using dynamic vibration method based on measuring the period of free eye coats oscillations. The last are in contact through the eyelid with the working part – stem (measuring sensor) of the tonometer. The measurement of the own oscillations period is carried out for the oscillatory system "eye–eyelid–stem".

The TPT system consists of a fixed part (inductance coil) and a movable part that can move along the axis of the inductance. The movable part consists of a stem attached to it with the help of two permanent magnets in the form of rings, suspended from two membrane springs on the ends of the coil. The weight of the entire sensitive element provides for overcoming the outer soft layer of the eyelid and minimizing its influence on the measurement result. The weight of the stem with the magnet is part of the oscillation system "eye–eyelid–stem" (see the Fig. 1).

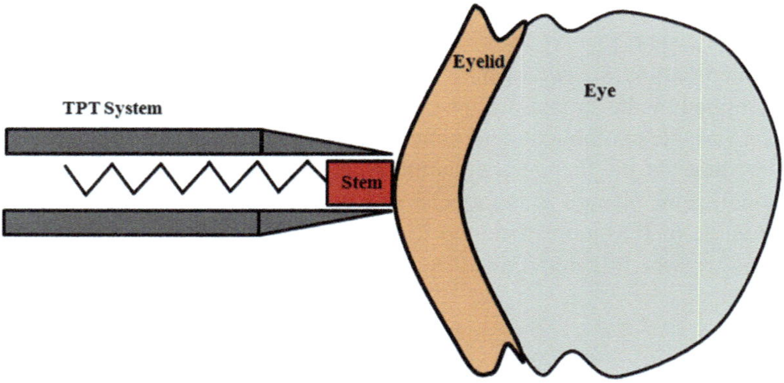

Fig. 1. The oscillation system "eye–eyelid–stem"

The IOP measurement process is as follows.

(1) The stem unit is mounted vertically on the eyelid. At the same time, the stem, having freedom of movement inside the device, drops under the force of gravity. It pushes the "eyelid–eye" system as long as the force of gravity is balanced by the force determined by the stiffness of the system. Since the sensor is mounted on springs, the springs will receive the initial deformation of displacement at the time of mechanical equilibrium establishment. The whole system is formed by a sensor and an "eyelid–eye" system and is in a static state of mechanical equilibrium. It is named the "sensor–eye" system.

(2) An electric pulse is applied to the inductance coil. Under the action of the emerging electromagnetic field, the stem receives an additional displacement Δh. The "sensor–eye" system enters a new static state of mechanical equilibrium.
(3) After the electric pulse the "sensor–eye" system tends to return to its original state in which it was before the supply of external influence in the form of an electric pulse through the inductance coil. The presence of the elastic properties of the "sensor–eye" system causes free damped mechanical oscillations, the frequency f of which is determined by the oscillating weight m (this is the weight of the stem with magnets) and the equivalent stiffness *Keqv* of the "sensor-eye" system.

The equivalent stiffness *Keqv* consists of parallel stiffness *Kv* of the springs and stiffness *Kel-e* of the "eyelid-eye" system:

$$Keqv = Kv + Kel\text{-}e$$

In turn, the stiffness of the "eyelid–eye" system is formed by the serial connection of the stiffness *Kel* of the eyelid and the stiffness *Ke* of the eye and is defined as:

$$Kel\text{-}e = Kel \cdot Ke/(Kel + Ke)$$

The movement of the stem with magnets inside the inductance coil causes an electric signal of alternating current in the form of a damped sinusoid to appear at its terminals. The frequency of the sinusoidal signal is equal to the frequency of free mechanical oscillations, and the amplitude is determined by the additional displacement Δh of the stem relative to the initial equilibrium position.

Equivalent stiffness *Keqv* of the "sensor–eye" system includes a constant component determined by the springs stiffness of the vibrator *Kv* and the eyelid stiffness of the *Kel*, and a variable component determined by the stiffness *Ke* (see the Fig. 2). The last parameter depends on the intraocular pressure. Thus, the output frequency change proportional to the change in intraocular pressure.

Each value of IOP corresponds to its own frequency of free "sensor–eye" system oscillations (*f*), which is determined by the equivalent stiffness *Keqv* (see the Fig. 3):

$$1/f = 1/(2\pi) \cdot (Keqv/m)^{1/2}$$

Equivalent stiffness *Keqv* and stiffness springs *Kv* determine the stiffness of the "eyelid–eye" system *Kel-e*, corresponding to certain IOP values.

To determine the dependence of the *Keqv* parameter and intraocular pressure, it is necessary to carry out numerical modeling, experimental study of the obtained parameters on the calibration phantoms, studies in groups of patients with different levels of IOP.

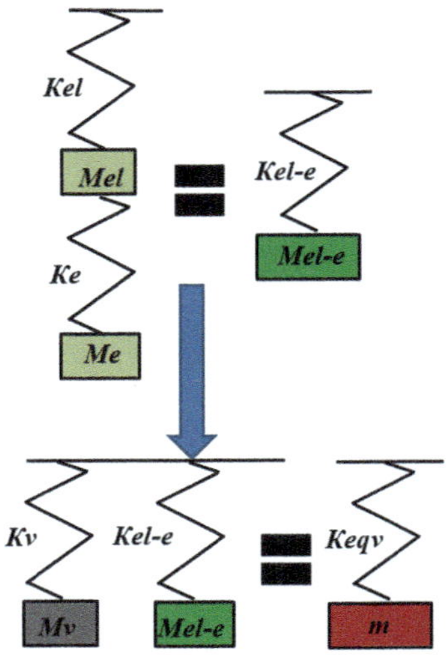

Fig. 2. The modeling of oscillation systems (m) with the weights of each element

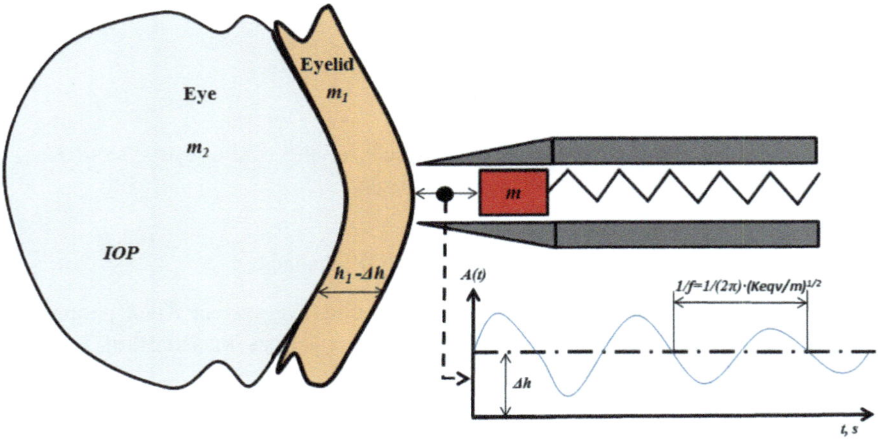

Fig. 3. The determination of the measuring sensor parameters

3 Compliance with Ethical Requirements

This study was performed in accordance with the Declaration of Helsinki and was approved by the Local Committee of Biomedical Ethics of the Moscow Helmholtz Research Institute of Eye Diseases.

4 Results

The researches conducted with various combinations of stem weight and diameter showed good repeatability of results for IOP in the range from 10 to 30 mmHg (see Table 1). The various combinations of stem weight and diameter changed the resulting pressure values not only quantitatively but also qualitatively.

Table 1. The values of measured IOP at different levels of reference pressure depending on the stem parameters, M ± m.

Reference IOP (mmHg)	Measured IOP with d = 1.5 mm and P = 16 g (mmHg)	Measured IOP with d = 3.0 mm and P = 16 g (mmHg)
10	10.7 ± 0.6	10.2 ± 0.2
20	21.0 ± 0.3	20.0 ± 0.6
30	30.6 ± 0.7	31.4 ± 0.9

As shown by the data presented in Table 1, with the stem diameters of 1.5 mm and 3.0 mm and the stem weight of 16 g, there is a good agreement between the reference and measured IOP values.

5 Conclusions

From the obtained results of the stiffness values of the eyelid–eye system *Kel-e* for different IOP, characteristic values can be chosen:

- at the boundaries of IOP taken as normal,
- at the boundaries of the range of possible IOP changes.

For these characteristic values of stiffness eyelid–eye system *Kel-e* parameters of IOP measuring instruments can be calculated, as well as measures of pressure to calibrate the TPT devices.

The developed simulation of a TPT system and the results of determining the optimal TPT sensor parameters allow concluding that TPT can be preferred method for determining IOP in terms of minimizing the contact of the measuring device with the surface of the eye.

IOP measurement with the selected parameters gives a good match with the manometric pressure level in the eye. Therefore, a transpalpebral tonometer with these sensor characteristics can be recommended for a comparative clinical study of IOP in the future. In addition, it is promising for its use in screening surveys, as well as for the purpose of dynamic IOP levels control. It will allow to estimate a IOP value with sufficient accuracy by the transpalpebral method for patients with glaucoma in the future.

Conflict of Interest. The authors declare that they have no conflict of interest.

References

1. Whitacre, M.M., Stein, R.: Sources of error with use of Goldmann-type tonometers. Surv. Ophthalmol. **38**(1), 1–30 (1993)
2. Özcura, F., Yildirim, N., Sahin, A., Çolak, E.: Comparison of Goldmann applanation tonometry, rebound tonometry and dynamic contour tonometry in normal and glaucomatous eye. Int. J. Ophthalmol. **8**(2), 299–304 (2015)
3. Kouchaki, B., Hashemi, H., Yekta, A., Khabazkhoob, M.: Comparison of current tonometry techniques in measurement of intraocular pressure. J. Curr. Ophthalmol. **29**(2), 92–97 (2016). https://doi.org/10.1016/j.joco.2016.08.010
4. Wachtl, J., Töteberg-Harms, M., Frimmel, S., Roos, M., Kniestedt, C.: Correlation between dynamic contour tonometry, uncorrected and corrected Goldmann applanation tonometry, and stage of glaucoma. JAMA Ophthalmol. **135**(6), 601–608 (2017). https://doi.org/10.1001/jamaophthalmol.2017.1012
5. Andreanos, K., Koutsandrea, C., Papaconstantinou, D., et al.: Comparison of Goldmann applanation tonometry and Pascal dynamic contour tonometry in relation to central corneal thickness and corneal curvature. Clin. Ophthalmol. **10**, 2477–2484 (2016). https://doi.org/10.2147/OPTH.S115203
6. Kontiola, A.I.: A new induction-based impact method for measuring intraocular pressure. Acta Ophthalmol. Scand. **78**(2), 142–145 (2000)
7. Kim, K.N., Jeoung, J.W., Park, K.H., Yang, M.K., Kim, D.M.: Comparison of the new rebound tonometer with Goldmann applanation tonometer in a clinical setting. Acta Ophthalmol. **91**(5), e392–e396 (2013). https://doi.org/10.1111/aos.12109
8. Achiron, A., Blumenfeld, O., Avizemer, H., et al.: Intraocular pressure measurement after DSAEK by iCare, Goldmann applanation and dynamic contour tonometry: a comparative study. J. Fr. Ophtalmol. **39**(10), 822–828 (2016). https://doi.org/10.1016/j.jfo.2016.09.009
9. Shamaev, D.M., Luzhnov, P.V., Iomdina, E.N.: Mathematical modeling of ocular pulse blood filling in rheoophthalmography. In: IFMBE Proceedings, vol. 68, no. 1, pp. 495–498 (2018). https://doi.org/10.1007/978-981-10-9035-6_91
10. Luzhnov, P.V., Shamaev, D.M., Kiseleva, A.A., et al.: Analyzing rheoophthalmic signals in glaucoma by nonlinear dynamics methods. In: IFMBE Proceedings, vol. 68, no. 2, pp. 827–831 (2018).https://doi.org/10.1007/978-981-10-9038-7_152
11. Dykin, V.I., Ivanishchev, K.V., Kornev, N.P., Mikheev, A.A., Solomakha, V.N.: Intraocular pressure dynamic type tonometer. Meditsinskaya tekhnika **4**(280), 33–36 (2013)

A Feasibility Test of Evaluation of Gait Movement by Using Center of Mass Estimation with Inertial Sensors

Yuho Takeda and Takashi Watanabe(✉)

Graduate School of Biomedical Engineering, Tohoku University, Sendai 980-8579, Japan
t.watanabe@tohoku.ac.jp

Abstract. It is necessary to control movement of center of mass (CoM) to realize stable walking. Since CoM has been shown to reflect asymmetry of movement, energy cost and ground reaction force (GRF), it is expected that CoM position makes possible to evaluate walking ability. This study aimed at examining the possibility of using CoM in gait evaluation, in which CoM position was estimated from signals measured only with inertial sensors. In this paper, 16 m straight walking of healthy subjects was measured with 7 wireless inertial sensors, and estimated CoM positions for the steady state gait were analyzed. Amplitude of change of vertical position of CoM during gait increased as walking speed increased in all subjects. Waveform of vertical position of CoM for each gait cycle was found to be divided into 2 patterns. The difference in the waveform of vertical position of CoM was considered to involve difference in gait movement to move CoM upward, in which knee joint angle or inclination angle of the shank segment was mainly changed for the term from the double stance phase to the single stance phase. This difference was considered to show the difference in gait strategy. The results of this study suggested that vertical component of CoM position became useful to evaluate gait movements.

Keywords: Center of mass · Inertial sensor · Walking · Knee joint · Shank

1 Introduction

Gait movement involves three-dimensional movement of the center of mass (CoM). Therefore, it is expected to be possible to evaluate walking ability by analyzing CoM position during gait. For example, margin of stability (MoS) based on base of support (BoS) and CoM position in the horizontal plane, and velocity of CoM [1, 2], evaluation of asymmetry of movement and weight support of the paretic leg from the relative positional relationship between CoM and the foot [3], relationship between amplitude of vertical position of CoM and energy consumption during walking [4–6] were studied for evaluation of motor function.

Although three-dimensional motion analysis system is commonly used to estimate CoM position, such system is generally limited to using at large hospital or research institute. On the other hand, a method of using inertial measurement units (IMUs)

J. Henriques et al. (Eds.): MEDICON 2019, IFMBE Proceedings 76, pp. 718–726, 2020.
https://doi.org/10.1007/978-3-030-31635-8_85

equipped with an inertial sensor such as a gyroscope and an acceleration sensor has been studied for measurement of gait movements [7–11]. Since IMU is small and lightweight and has no restriction on measurement environment, it is expected to be introduced into daily life at home in addition to rehabilitation at most of hospitals. However, evaluation of motor function using CoM estimated from signals measured with inertial sensors has not been validated.

This study aimed to make clear a possibility of evaluation of gait movement by using CoM estimated from signals measured with IMUs. Although estimation of CoM position using the inertial sensors was performed in another research group [12], basically the method estimated CoM position approximately with the inertial sensor attached to the sacrum and gait evaluation was not examined. Therefore, in our previous study, CoM position was estimated from signals measured with 7 wireless inertial sensors [13], in which inclination angles of body segments and feet positions were estimated based on the methods developed in our research group [10, 11]. The CoM estimation method was shown to be useful in estimation of vertical and anteroposterior components of CoM position [13]. In this paper, vertical component of CoM position was focused based on previous studies [4–6]. The CoM position was estimated under different walking speeds of healthy subjects and analyzed in relation with lower limb movement during gait in order to examine the possibility of application of CoM position to gait evaluation.

2 Estimation Method of Center of Mass Position

Figure 1 shows the outline of estimation method of CoM position using inertial sensors. First, a unit vector representing the posture of each segment of the body (segment vector) and positions of the feet were estimated. Segment vector was defined in the direction of longitudinal axis of the body segment. Then, posture of the body at each time was reconstructed using estimated segment vectors from the foot segment to the trunk segment based on the estimated foot position. Position of center of mass of each segment was calculated from the position and attitude of the reconstructed body segment vectors, in which length, weight and center of mass position of each segment were approximately calculated from height and weight of subject based on previous studies [14, 15]. Finally, CoM position of the whole body is calculated by Eq. (1).

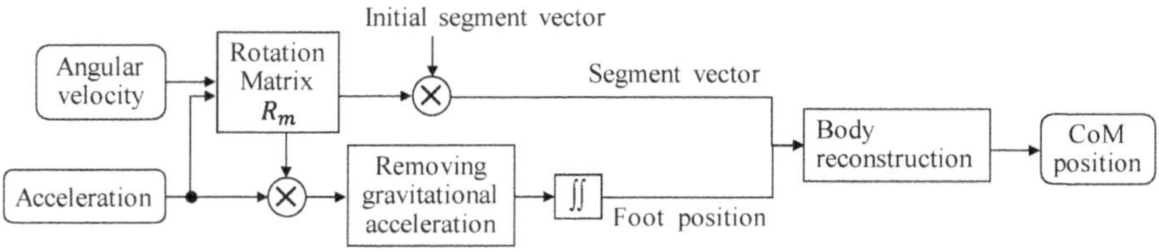

Fig. 1. Outline of estimation method of the center of mass (CoM) position

$$CoM(t) = \frac{\sum m_i r_i(t)}{\sum m_i} \tag{1}$$

Here, m_i and r_i show mass and center of mass position of each segment.

2.1 Calculation Method of Segment Vector

In order to estimate the segment vector, quaternion representing the attitude of each segment in three-dimensional space was used. The quaternion $\boldsymbol{q}_g$ can be calculated from the angular velocity measured with the inertial sensor by using Eq. (2).

$$\boldsymbol{q}_{gk+1} = \frac{1}{2}\begin{bmatrix} 2 & -\Delta t\omega_{xk} & -\Delta t\omega_{yk} & -\Delta t\omega_{zk} \\ \Delta t\omega_{xk} & 2 & \Delta t\omega_{zk} & -\Delta t\omega_{yk} \\ \Delta t\omega_{yk} & -\Delta t\omega_{zk} & 2 & \Delta t\omega_{xk} \\ \Delta t\omega_{zk} & \Delta t\omega_{yk} & -\Delta t\omega_{xk} & 2 \end{bmatrix}\boldsymbol{q}_{gk} \tag{2}$$

Here, ω_x, ω_y and ω_z are three-axis components of measured angular velocities, k is the time, and Δt is the sampling period. The quaternion calculated from the angular velocity was corrected by applying the Kalman filter using the acceleration vector, because the error was accumulated due to the drift of gyroscope signals. The rotation matrix $\boldsymbol{R}_k$ at time k that calculates the current segment vector from the initial segment vector is obtained from the corrected quaternion as shown by Eq. (3).

$$\boldsymbol{R}_k = \begin{bmatrix} 2q_0^2 - 1 + 2q_1^2 & 2q_1q_2 - 2q_0q_3 & 2q_1q_3 + 2q_0q_2 \\ 2q_1q_2 + 2q_0q_3 & 2q_0^2 - 1 + 2q_2^2 & 2q_3q_2 - 2q_0q_1 \\ 2q_1q_3 - 2q_0q_2 & 2q_3q_2 + 2q_0q_1 & 2q_0^2 - 1 + 2q_3^2 \end{bmatrix} \tag{3}$$

Here, quaternion $\boldsymbol{q}$ consists of a real part and 3 imaginary parts, which are shown by the following equation:

$$\boldsymbol{q} = q_0 + q_1\boldsymbol{i} + q_2\boldsymbol{j} + q_3\boldsymbol{k} \tag{4}$$

The posture of segment is expressed by the segment vector. Assuming that the segment vector of the initial posture is $\boldsymbol{s}_0$, the segment vector at time k, $\boldsymbol{s}_k$, is calculated by using the matrix $\boldsymbol{R}_k$ as shown by Eq. (5).

$$\boldsymbol{s}_k = \boldsymbol{R}_k\boldsymbol{s}_0 \tag{5}$$

2.2 Estimation Method of Foot Position

Foot position was calculated by two-step integration of the movement acceleration of the foot segment. Acceleration signal $\boldsymbol{a}$ measured with the inertial sensor attached on the foot was transformed to its value on the reference coordinate system by using the rotation matrix $\boldsymbol{R}_m$ that was calculated from the quaternion representing the foot posture. Then, movement acceleration $\boldsymbol{a}_m$ was calculated by removing the gravitational acceleration $\boldsymbol{g}$. That is,

$$\boldsymbol{a}_m(t) = \boldsymbol{R}_m(t)\,\boldsymbol{a}(t) - \boldsymbol{g} \tag{6}$$

Then, the integral of movement acceleration of the foot was calculated during foot movement excluding the stationary state of the foot. First, the movement state of the foot was detected tentatively by setting a threshold value for the total sum of absolute values of three axes components of acceleration signals (ACC). Then, the movement state was determined by re-searching the minimum value of ACC value in a certain section before the beginning and after the end of the tentative movement state, which were used as the start and the end of the integral term. The movement velocity of the foot was calculated assuming to be 0 m/s at the beginning of the integral. That is,

$$\boldsymbol{v}_m(t) = \int_0^t \boldsymbol{a}_m(\tau)d\tau \tag{7}$$

The calculated velocity of the foot was corrected so as to be 0 m/s at the end of the integral by using linear approximation of integral error. Finally, position of the foot $\boldsymbol{L}$ was calculated by the integral of the corrected movement velocity $\boldsymbol{v}_r$.

$$\boldsymbol{L}(t) = \int_0^t \boldsymbol{v}_r(\tau)d\tau \tag{8}$$

In the case of walking on the flat floor, since the vertical component of the foot position can be regarded as 0 *cm* for each stationary state, the vertical position was corrected by using linear approximation of integral error.

3 Experimental Tests for Evaluation of Gait Movement by Using COM

3.1 Experimental Method

Gait movements of 4 healthy male subjects (21–23 years old) were measured with wireless inertial sensors (MPU-9150, Bluetooth class 1). Figure 2 shows sensor attachment positions. Inertial sensors were attached on both feet, both shanks, both thighs, trunk using elastic bands. The measurement frequency was set to 100 Hz.

The subjects walked in a straight line of 16 m on the indoor flat floor. The subjects carried out 10 trials of 3 types walking of moderate speed, slow speed and fast speed, in which walking speeds were determined by subjects themselves. The walking speed was calculated by measuring the required time of 10 m walking removing 3 m from the start and 3 m before the end of walking with a stopwatch. Movement of steady state walking was analyzed by excluding two strides just after the start and two strides before the end of walking.

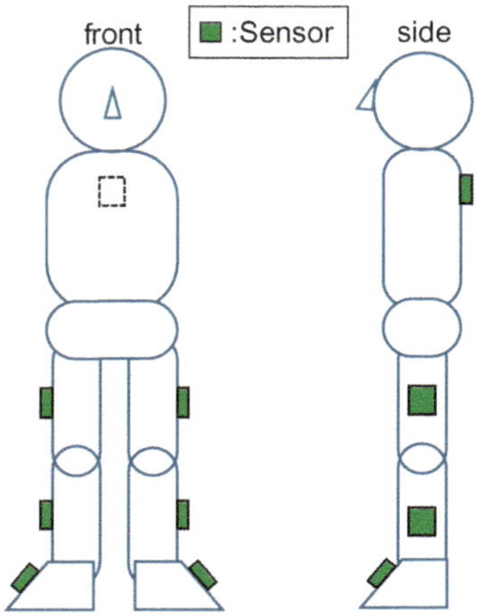

Fig. 2. Attachment positions of inertial sensors

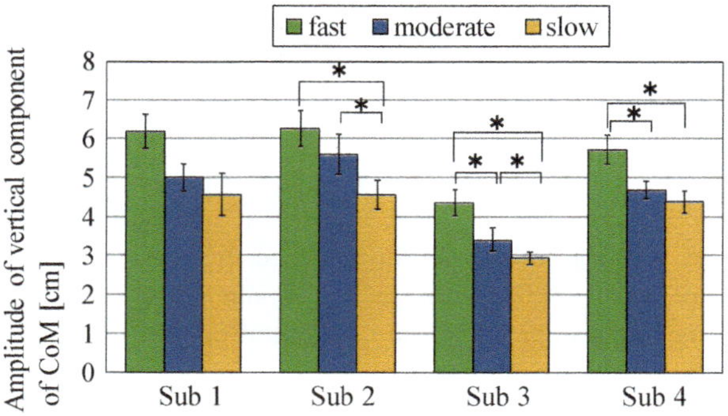

Fig. 3. Amplitude of vertical component of the center of mass (CoM) for 3 different walking speeds (*: $p < 0.05$)

3.2 Results

Figure 3 shows amplitude of change of the vertical component of CoM position calculated from the maximum and the minimum values for each gait cycle. The amplitude of the vertical component increased as walking speed increased in all subjects. Especially, Sub 2, 3 and 4 showed significant difference of the amplitude of vertical component of CoM between walking speeds (paired t-test with Bonferroni correction).

Figure 4 shows examples of average waveforms of the vertical component of the CoM position that was divided into strides by the timing of the initial contact (IC) of the right or the left foot. Here, since there were two patterns of the waveform, waveforms were averaged for each subject and speed after dividing measurement trials visually based on the two patterns. The two patterns were seen in Sub 2 (Pattern A)

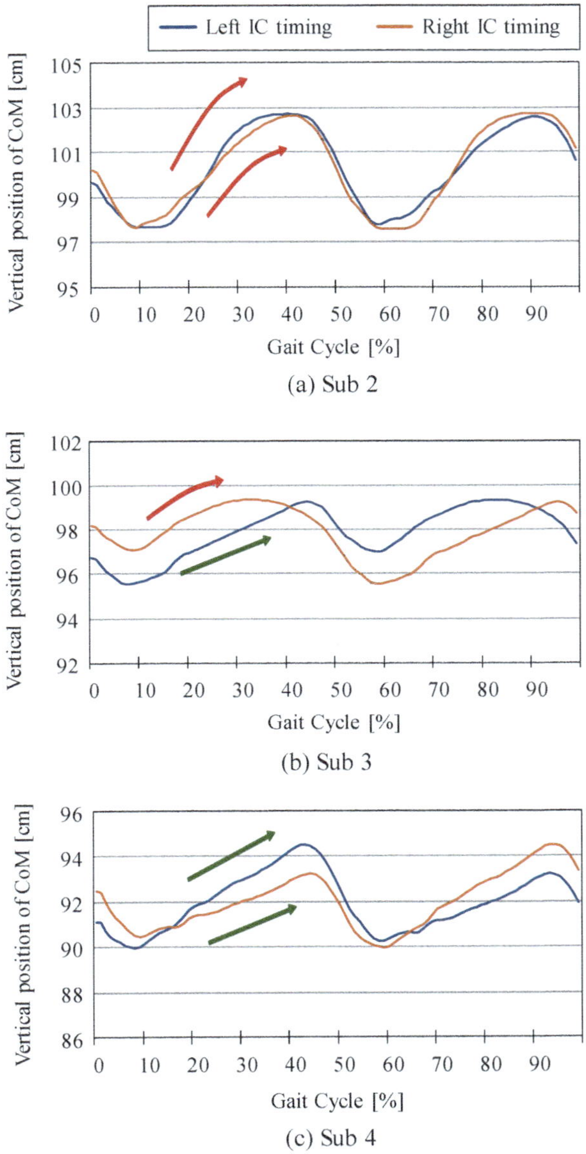

Fig. 4. Examples of average waveforms of vertical component of the CoM positon (moderate speed).

and Sub 4 (Pattern B) in Fig. 4. Sub3 showed Pattern A and B for the right and the left IC timings, respectively. The difference of the waveform between the right and the left IC timings shows difference in movement between the right and the left sides during gait. Sub 2 and 4 showed similar movement waveforms between the right and the left sides, while Sub 3 showed different waveforms between the left and the right sides.

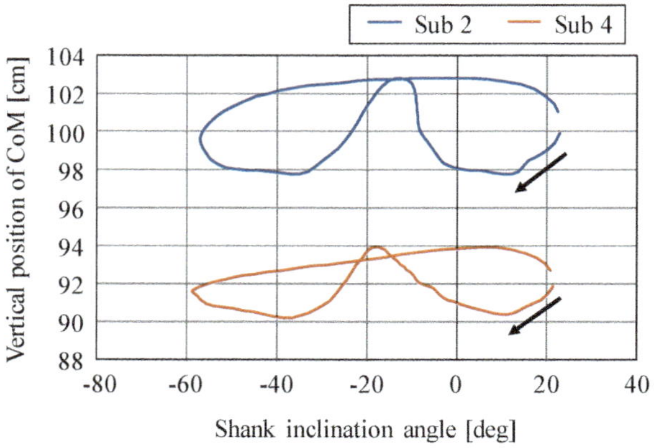

Fig. 5. Example of relationship between vertical position of CoM and shank inclination angle

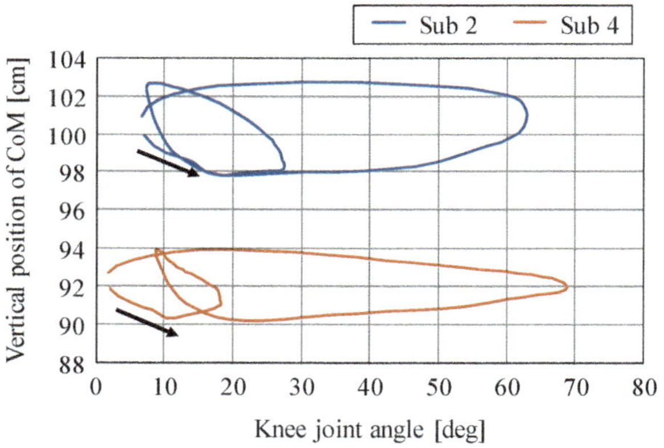

Fig. 6. Example of relationship between vertical position of CoM and knee joint angle

Figure 5 shows the relationships between the shank inclination angle and the vertical component of CoM position for Pattern A (Sub 2) and Pattern B (Sub 4). The relationships showed difference in the increase of the vertical component as the shank inclination angle changed from 0° (the vertical position) to −20° (forward inclination after the IC causing ankle dorsiflexion). That is, the slope of the increase of Sub 2 was larger than that of Sub 4.

Figure 6 shows the relationships between the knee joint angle and the vertical component of CoM position for Patterns A and B. Although there is an elliptical shape inside that is started from about 10°, the minor axis of the elliptical shape of Pattern B (Sub 4) was shorter than that of Pattern A (Sub 2).

4 Discussion

The amplitude of change of the vertical component of CoM position increased with increasing walking speed in all the subjects. It is considered that as the walking speed increased, absorption of the impact at the IC was caused by a decrease in the vertical position of CoM, which corresponded to an increase in double knee action.

Waveforms of vertical position of CoM during gait were divided into 2 patterns. The relationship between the shank inclination angle and the vertical component of CoM showed difference in movement of the shank to move the CoM upward between the 2 patterns. It suggested that Pattern A moved CoM upward without almost changing the shank position and Pattern B did it changing the inclination angle of the shank segment. The relationship between the knee joint angle and the vertical component of CoM also showed difference in lower limb movement to move CoM between Pattern A and B. It suggests that in the case of Pattern A, knee joint movement in the single stance phase is large. That is, the knee joint is greatly changed from flexion to extension when moving the CoM upward in the single stance phase. Therefore, it is considered that CoM positon is moved upward by changing the knee joint in the case of Pattern A, while the shank moved forward in the case of Pattern B. This suggests that waveform of vertical positon of CoM is related to movements of the lower limb during gait, and the waveform of the position and relationships of the positon with the shank inclination angle and knee joint angle have a possibility of showing gait strategy.

5 Conclusion

In this paper, 16 m straight walking measurements were performed with healthy subjects, and estimated CoM positions for the steady state gait were analyzed to examine the possibility of its application to gait evaluation. Amplitude of change of vertical component of CoM position during gait increased as walking speed increased in all subjects. Waveform of vertical position of CoM for each gait cycle was found to be divided into two patterns. The difference in the waveform was considered to involve difference in gait strategy, in which either knee joint angle or inclination angle of the shank segment was mainly changed for the term from the double stance phase to the single stance phase in order to move CoM upward. The results of this paper suggested that vertical component of CoM position estimated with inertial sensors became useful to evaluate gait movements. Quantitative evaluation of differences in gait movement will be studies in the next step.

References

1. Meulen, F.B., Weenk, D., Asseldonk, E.H.F., Schepers, H.M., Veltink, P.H., Buurke, J.H.: Analysis of balance during functional walking in stroke survivors. PLoS One **11**(11), e0166789 (2016)

2. Kama, D., Kamphuis, J.F., Weerdesteyn, V., Geurts, A.C.H.: The effect of weight-bearing asymmetry on dynamic postural stability in healthy young individuals. Gait Posture **45**, 56–61 (2016)
3. Balasubramanian, C.K., Neptune, R.R., Kautz, S.A.: Foot placement in a body reference frame during walking and its relationship to hemiparetic walking performance. Clin. Biomech. **25**(5), 483–490 (2010)
4. Shin, S.-S., An, D.-H., Yoo, W.-G.: Effects of balance control through trunk movement during square and semicircular turns on gait velocity, center of mass acceleration, and energy expenditure in older adults. PM & R **8**(10), 953–961 (2016)
5. Wurdeman, S.R., Raffal, P.C., Stergiou, N.: Reduced vertical displacement of the center of mass is not accompanied by reduced oxygen uptake during walking. Sci. Rep. **7**, 17182 (2017)
6. Gordon, K.E., Ferris, D.P., Kuo, A.D.: Metabolic and mechanical energy costs of reducing vertical center of mass movement during gait. Arch. Phys. Med. Rehabil. **90**(1), 136–144 (2009)
7. Lau, H., Tong, K.: The reliability of using accelerometer and gyroscope for gait event identification on persons with dropped foot. Gait Posture **27**, 248–257 (2008)
8. Sabatini, A.M., Martelloni, C.: Assessment of walking features from foot inertial sensing. IEEE Trans. Biomed. Eng. **52**, 486–494 (2005)
9. Jasiewicz, J.M., Allum, J.H.: Gait event detection using linear accelerometers or angular velocity transducers in able-bodied and spinal-cord injured individuals. Gait Posture **24**, 502–509 (2006)
10. Watanabe, T., Teruyama, Y., Ohashi, K.: Comparison of angle measurements between integral-based and quaternion-based methods using inertial sensors for gait evaluation. Commun. Comput. Inf. Sci. **511**, 274–288 (2015)
11. Watanabe, T., Miyazawa, T., Shibasaki, J.: A study on IMU-based stride length estimation for motor disabled subjects: a comparison under different calculation methods of rotation matrix. In: 2018 IEEE EMBS International Conference on Biomedical & Health Informatics (BHI) (2018)
12. Floor-Westerdijk, M.J., Schepers, H.M., Veltink, P.H., van Asseldonk, E.H.F., Buurke, J.H.: Use of inertial sensors for ambulatory assessment of center-of-mass displacements during walking. IEEE Trans. Biomed. Eng. **59**, 2080–2084 (2012)
13. Takeda, Y., Watanabe, T.: A study on estimation of center of mass trajectory using inertial sensors for movement evaluation. IEICE Technical rep. 118(366), 1–4, MBE2018-50 (2018). (in Japanese)
14. Ae, M., Tang, H.P., Yokoi, T.: Estimation of inertia properties of the body segments in Japanese athletes. Biomechanisms **11**, 23–33 (1992). (in Japanese)
15. Okada, H., Ae, M., Fujii, N., Morioka, Y.: Body segment inertia properties of Japanese elderly. Biomechanisms **13**, 125–139 (1996). (in Japanese)

Controlled Thoracic Motions of an Anthropomorphic Phantom for Myocardial Perfusion Imaging

Sotiris Panagi[1(✉)], Antonis Antoniou[1,2], Isabelle Chrysanthou-Baustert[3], Demetris Kaolis[3], Ourania Demetriadou[4], Costas Kyriacou[5], and Yiannis Parpottas[1,2(✉)]

[1] Department of Mechanical Engineering, Frederick University, Nicosia, Cyprus
sotirispanagi@gmail.com, yparpottas@frederick.ac.cy
[2] Frederick Research Center, Nicosia, Cyprus
[3] Department of Medical Physics, Nicosia General Hospital, Nicosia, Cyprus
[4] Department of Nuclear Medicine, Limassol General Hospital, Limassol, Cyprus
[5] Department of Electrical and Computer Engineering and Informatics, Frederick University, Nicosia, Cyprus

Abstract. The aim of this study was to develop controlled thoracic motions on a cardiac phantom and a breathing phantom, which both are inserted within an anthropomorphic thorax phantom. The cardiac phantom consists of a left ventricle and the breathing phantom of inflatable lungs. The diastole and systole of the left ventricle follows the Wigger diagram while the inflatable lungs follow a sinusoidal exponential pattern. An ECG can gate both the beating left ventricle and the SPECT imaging modality. In addition, the ECG beating left ventricle follows the inferior portion of the inflatable lungs in the cranio-caudal direction at the level of the diaphragm. The lungs and the left ventricle are able to simulate normal and deep breathing conditions. A Programmable Logic Controller precisely controls these motions in time intervals of 10 ms. All motions can be operated simultaneously or independently to each other. A number of motion limitations of other physical phantoms have been overcome. This phantom assembly with variable cardiac defects can be utilized to optimize myocardial perfusion imaging.

Keywords: Breathing phantom · ECG beating cardiac phantom · Cranio-caudal cardiac motion · Myocardial perfusion imaging

1 Introduction

Myocardial perfusion imaging (MPI) has been proven to be able to assess the functional significance of coronary artery stenosis [1]. The motions of the lungs and the heart during respiration may cause artifacts and false positive findings reducing the diagnostic accuracy on MPI [2, 3].

Patient data have been used to study the influence of these motions on MPI. However, physical phantoms with variable cardiac defects and motions, which can

J. Henriques et al. (Eds.): MEDICON 2019, IFMBE Proceedings 76, pp. 727–734, 2020.
https://doi.org/10.1007/978-3-030-31635-8_86

operate simultaneously or independently to each other, are useful to separately study the effect of each motion on physicians' MPI reports.

Previous studies on MPI [4–6] utilized physical phantoms with one motion instead of the three motions of this phantom assembly: (a) ECG beating left ventricle, (b) inflatable lungs, and (c) cranio-caudal motion of the left ventricle during respiration. Two of these studies [4, 5] simulated the cranio-caudal motion of the left ventricle during respiration by moving a torso phantom with static compartments rather than these compartments to perform motions within the torso phantom. In addition, the ECG beating left ventricle of this phantom assembly follows the inferior portion of the inflatable lungs in the cranio-caudal direction at the level of the diaphragm.

This paper presents the precisely controlled motions of the phantom assembly and the procedures followed to validate the motions.

2 Thoracic Motions of the Anthropomorphic Phantom

2.1 ECG Beating Cardiac Phantom

The cardiac phantom is based on an inner-outer membrane system. The inner membrane represents the left ventricle and the closed cavity between the inner and outer membrane represents the myocardial wall of the LV [7–9]. Diastole and systole can be achieved by pumping water into and out the inner membrane using a stepper motor and a piston. Since the myocardial wall cavity is filled with water, which is incompressible, the pressure is transferred from the inner to the outer membrane (see Fig. 1). In addition, cardiac defects of various sizes can be positioned, and radiopharmaceuticals can be injected within the LV myocardial wall.

A touch-screen Programmable Logic Controller (PLC) and the VisiLogic ladder programming language are used to control the stepper motion. The volume of the LV with respect to time $V_{LV}(t)$ is described in cardiovascular physiology by the Wigger diagram. A digital plotter was utilized to fit the Wigger diagram with a six-degree polynomial:

$$V_{LV}(t) = \sum_{i=0}^{i=6} A_i \left(\frac{t}{T}\right)^i \times (V_d - V_s) + V_s \tag{1}$$

where t is the counting time in intervals of 10 ms, Vs is the systolic volume (LV cavity can be filled with 30–50 mL of water), V_d is the diastolic volume in mL and it depends on the value of the ejection fraction $EF = [(V_d - V_s)/V_d] \times 100$, T is the time period for a heartbeat, $A_0 = 1$, $A_1 = -3.23$, $A_2 = 45.35$, $A_3 = 288.14$, $A_4 = -598.93$, $A_5 = 5361.19$ and $A_6 = -176.34$. The volume of the water which is displaced by the piston is equal to:

$$V_{LV} = \pi R_p^2 R_m (1 - \cos\theta) \tag{2}$$

where R_p is the radius of the cylindrical piston, R_m is the distance from the axis of the stepper motor to the piston, as shown in Fig. 1, and θ is the angle of the rotating disk which is attached to the stepper motor. A position sensor is also attached to the stepper motor to define $\theta = 0°$ while a rotary sensor is connected to the axis of the stepper motor to provide feedback on the motor steps. From Eqs. 1 and 2, the angular velocity ω of the stepper motor is:

$$\omega = \frac{d\theta}{dt} = \mathrm{B}\frac{dV_{LV}(t)}{dt} \tag{3}$$

where $\mathrm{B} = \dfrac{(V_d - V_s)}{\pi R_p^2 R_m \sqrt{1 - \left[1 - \frac{(V_d - V_s)}{\pi R_p^2 R_m}\right]^2}}$.

Selecting on PLC, a heart rate value from 30–120 beats per minute and an EF value from 30–70%, the motor steps per second, $f = \omega/2\pi$, are calculated by the PLC using Eq. 3.

When the PLC mode *'ECG'* is in use, the heart rate is selected on a simulator which propagates its signal to an ECG device (see Fig. 1). The high-voltage output signal of an ECG device is then simultaneously propagated to the PLC, to initiate Eq. 1, and to the gating circuit of a SPECT modality. Linear membrane sensors were attached to the cardiac phantom to calculate the time delay between the initialization of Eq. 1 on the PLC and the actual initialization of the cardiac cycle on the phantom.

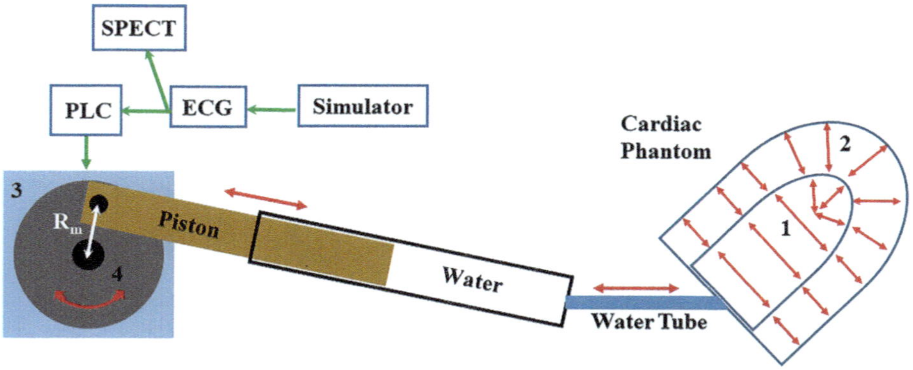

Fig. 1. Block diagram of the mechanical system for the ECG beating cardiac phantom: (1) left ventricle cavity, (2) myocardial wall cavity of the left ventricle, (3) stepper motor, (4) rotating disk of the stepper motor.

2.2 Breathing Phantom

The breathing phantom consists of human-sized inflatable lungs [9]. At end-expiration, the air volumes in the left and right lung are 510 and 565 mL, respectively. Each lung is covered with a thin thermoplastic structure, which forms the lung shape at end-inspiration, to allow inflation in all directions in a physiological correct manner.

The respiratory pattern follows a sinusoidal exponential pattern [10]:

$$V_{\text{LUNGS}}(t) = V_{TIDAL}\ e^{\sin(at)} + V_{EXPIR} \tag{4}$$

where $V_{LUNGS}(t)$ is the air volume in the lungs with respect to time in mL, t is the counting time in intervals of 10 ms, $V_{EXPIR} = 1075$ mL is the air volume in the lungs at end-expiration, $V_{TIDAL} = V_{INSPIR} - V_{EXPIR}$ is the tidal volume, V_{INSPIR} is the air volume in the lungs at end-inspiration, $\alpha = 2\pi/T$ and T is the time period for a breathing cycle.

Figure 2 shows the block diagram of the mechanical system for inhalation and exhalation. It consists of electro-valves, which are controlled by the PLC, a flow pump and a vacuum pump. Two proportional valves regulate the pressure of the air with respect to time, $P_{VALVE}(t)$, entering the lungs for inhalation and escaping from the lungs for exhalation, respectively. A 3/2 valve is placed before the lungs to allow air entering the lungs for inhalation and escaping from the lungs for exhalation, respectively. Also, two 2/2 valves are placed after the pumps to allow air in and out of the system during inhalation and exhalation, respectively.

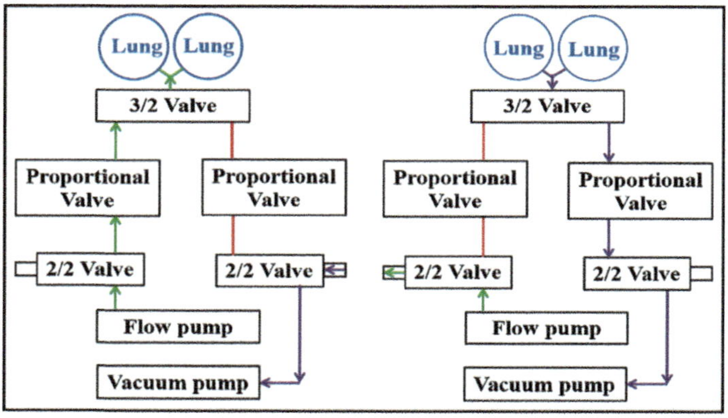

Fig. 2. Block diagram of the mechanical system for inhalation (left) and exhalation (right). The colors of the tubes denote air flow (green), air suction (purple) or none of the previous (red).

To calculate and regulate $P_{VALVE}(t)$ at the exit of each proportional valve according to the $V_{LUNGS}(t)$ of the respiratory pattern, the following equations are used:

$$Q_{LUNGS}(t) = dV_{LUNGS}(t)/dt \tag{5}$$

and

$$Q_{LUNGS}(t) = \frac{\pi R^4}{16\eta L} \times \frac{\left|P^2_{VALVE}(t) - P_o^2\right|}{P_o} \tag{6}$$

where Q_{LUNGS} (t) is the flow of air in and out of the lungs with respect to time, η is the viscosity of air, L is the length of the tube from a proportional valve to the 3/2 valve, R is the radius of the tube, and P_o is the pressure at the entrance of the 3/2 valve when the lungs are at end-expiration. Flowmeters and pressure sensors were utilized to provide feedback on the functionality of the mechanical respiratory system. Selecting on PLC, a tidal volume up to 550 mL and a breath rate up to 14 breaths per minute, P_{VALVE} (t) is calculated by the PLC using Eqs. 4–6.

2.3 Cranio-Caudal Motion of the Cardiac Phantom

The cardiac phantom moves in synchronization with the inferior portion of the inflatable lungs in the cranio-caudal direction at the level of the diaphragm. Figure 3 shows the block diagram of the mechanical system for this motion while Fig. 4a shows the phantom assembly, together with this system, in the SPECT modality of the Limassol General Hospital.

A solid-water rod enters into the thorax, which is filled with water, and holds in place the cardiac phantom (θ = 26°, ϕ = 38° from spline to chest), as shown in Figs. 3 and 4b. At the other end, outside of the thorax, the solid-water rod is connected to a screwed cylinder, which oscillates due to the linear motor of the mechanical system. This oscillatory motion, which is controlled by the PLC, is transferred to the cardiac phantom via the rod. A linear sensor is attached to the mechanical system to provide feedback on the position of the LV (cardiac phantom) due to this motion, while another two sensors define the endpoints of the motion.

This motion is described by the following equation:

$$\mathrm{X_{LV}}(t) = X_{\max} \sin(at + 90°) \tag{7}$$

where X_{LV} (t) is the position with respect to time in mm, X_{max} is the maximum amplitude of the oscillation, which can be set to 15 or 27 mm for normal or deep breathing conditions, respectively [11], t is the counting time in intervals of 10 ms, $\alpha = 2\pi/T$ and T is the time period for the motion. The counting time and the time period are synchronized with the respiratory pattern. The corresponding tidal volumes of the lungs for normal and deep breathing are set to 450 and 550 mL, respectively.

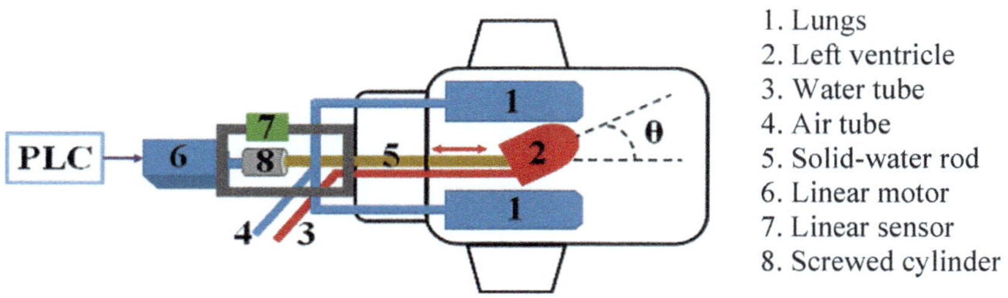

Fig. 3. Block diagram of the mechanical system for the cranio-caudal cardiac motion during respiration.

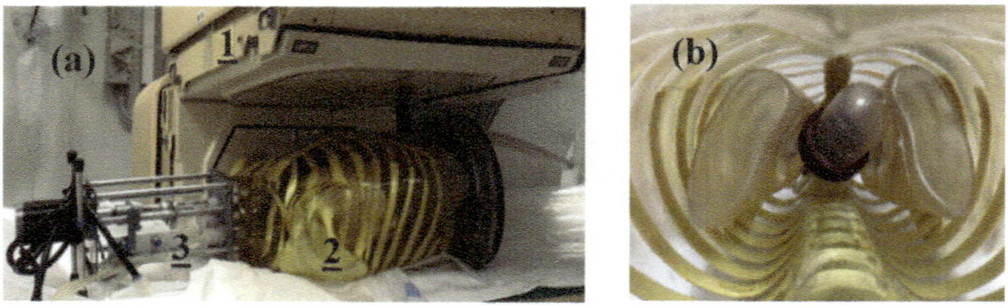

Fig. 4. (a) The phantom assembly in the Philips Forte/Vantage-Pro-Gd gamma camera (SPECT/^{153}Gd) at the Limassol General Hospital: (1) SPECT/^{153}Gd, (2) anthropomorphic phantom, and (3) mechanical system for the cranio-caudal LV motion, (b) The LV and lungs within the thorax.

3 Validation Procedures

The motions of the phantom assembly were validated using medical imaging software.

The end-diastolic and end systolic volumes as well as the EF values were measured by using the Quantitative Gated SPECT (QGS) software. Figure 5a presents an example of these measurements: the volume and flow during the cardiac cycle of this cardiac phantom. A 5% deviation was calculated for EF when compared to the set values on the PLC. Note that the end-systolic volume and the EF value can be selected for this cardiac phantom, as mentioned in a previous section. Figure 5b shows an example of the systole and diastole of the cardiac phantom, as presented by the Myovation software.

The tidal volumes of the lungs were measured using the OsiriX software [12]. A 2% deviation was calculated for the tidal volumes when compared to the set values on the PLC. Figure 6a shows the lungs of the phantom assembly at end-expiration and end-inspiration with a tidal volume of 550 mL.

CT slides, acquired by the Toshiba's Aquilion RXL 16-detector row CT system (0.5 mm slice thickness) at Limassol General Hospital, were utilized to validate the cranio-caudal motion. Figure 6b shows coronal CT slices at end-expiration and end-inspiration with 550 mL tidal volume of the lungs and 27 mm displacement of the LV in the caudal direction.

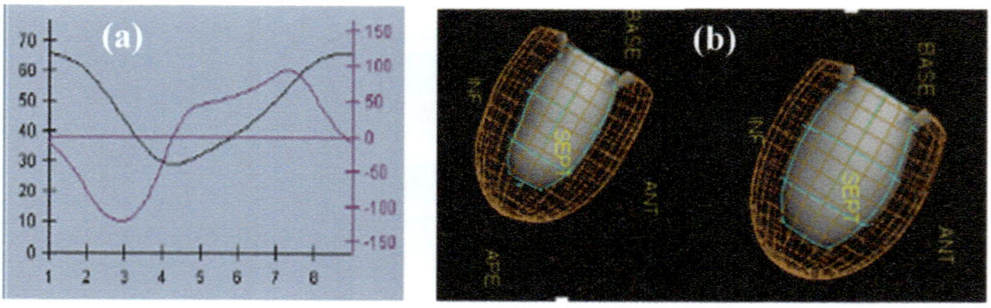

Fig. 5. Examples of (a) the volume (black color) in mL and flow (light purple color) in mL/sec during the cardiac cycle of the cardiac phantom, as measured by using the QGS software, (b) the systole and diastole of the cardiac phantom, as shown by the Myovation software.

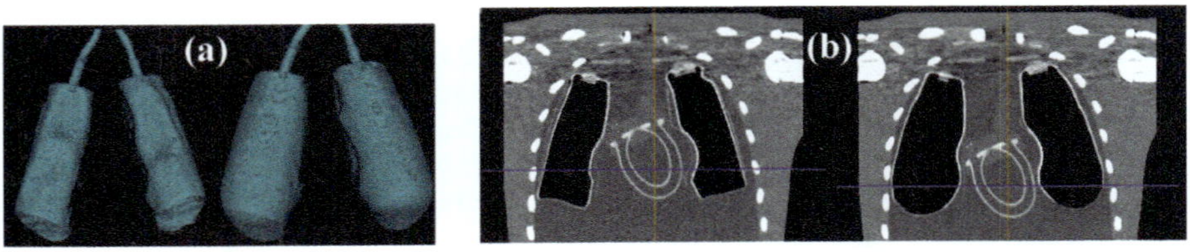

Fig. 6. (a) The lungs of the phantom assembly at end-expiration and end-inspiration with a tidal volume of 550 mL, as obtained by the OsiriX software, (b) Coronal CT slides at end-expiration and end-inspiration with 550 mL tidal volume of the lungs and 27 mm displacement of the LV in the caudal direction.

4 Conclusions

This assembly consists of a cardiac phantom and a breathing phantom, which both are inserted within a tissue-equivalent anthropomorphic thorax phantom. It simulates the breathing of the lungs, the ECG beating of the left ventricle, and the motion of the left ventricle in synchronization with the inferior portion of the inflatable lungs in the cranio-caudal direction at the level of the diaphragm. Its precise motions can be controlled in time intervals of 10 ms and they can be operated simultaneously or independently to each other. A number of motion limitations of other physical phantoms have been overcome.

Cardiac defects of various sizes can be inserted, and radiopharmaceuticals can be injected within the myocardial wall of the left ventricle. Physicians can report on defects' position and severity without having prior knowledge of the phantom parameters. Their reports can be evaluated to assess the diagnostic value of MPI under the influence of attenuation, motions, and artifacts.

This dynamic phantom assembly can be used to optimize MPI without exposing patients to radiation.

Acknowledgment. We gratefully acknowledge the staff of the Department of Nuclear Medicine of the Limassol General Hospital (N. Fasouliotis, A. Panayi, K. Hadjitofi, and D. Demetriou) for the acquisitions. We also thank D. Christodoulou for advising on VisiLogic and A. Kallis for some initial calculations. Part of this work was co-financed by the European Regional Development Fund and the Republic of Cyprus through the Research Promotion Foundation (ΥΓΕΙΑ/ΔΥΓΕΙΑ/0311ΒΙΕ/27).

Conflict of Interest. The authors declare that they have no conflict of interest.

References

1. Baghdasarian, S.B., Heller, G.V.: The role of myocardial perfusion imaging in the diagnosis of patients with coronary artery disease: developments over the past year. Curr. Opin. Cardiol. **20**, 369–374 (2005)

2. Wheat, J.M., Currie, G.M.: Impact of patient motion on myocardial perfusion SPECT diagnostic integrity: part 2. J. Nucl. Med. Technol. **32**, 158–163 (2004)
3. Le Meunier, L., Maass-Moreno, R., Carrasquillo, J.A., Dieckmann, W., Bacharach, S.L.: PET/CT imaging: effect of respiratory motion on apparent myocardial uptake. J. Nucl. Cardiol. **13**, 821–830 (2006)
4. Pitman, A.G., Kalff, V., Van Every, B., Risa, B., Barnden, L.R., Kelly, M.J.: Effect of mechanically simulated diaphragmatic respiratory motion on myocardial SPECT processed with and without attenuation correction. J. Nucl. Med. **43**, 1259–1267 (2002)
5. Könik, A., Kikut, J., Lew, R., Johnson, K., King, M.: Comparison of methods of acquiring attenuation maps for cardiac SPECT in the presence of respiratory motion. J. Nucl. Cardiol **20**, 1093–1107 (2013)
6. Chrysanthou-Baustert, I., Parpottas, Y., Demetriadou, O., Christofides, S., Yiannakkaras, C., Kaolis, D., et al.: Diagnostic sensitivity of SPECT myocardial perfusion imaging using a pumping cardiac phantom with inserted variable defects. J. Nucl. Cardiol. **20**, 609–615 (2013)
7. Chrysanthou-Baustert, I., Polycarpou, I., Demetriadou, O., Livieratos, L., Lontos, A., Antoniou, A., et al.: Characterization of attenuation and respiratory motion artifacts and their influence on SPECT MP Image evaluation using a dynamic phantom assembly with variable cardiac defects. J. Nucl. Cardiol. **24**(2), 698–707 (2017)
8. Polycarpou, I., Chrysanthou-Baustert, I., Demetriadou, O., Parpottas, Y., Panagidis, C., Marsden, K.P., et al.: Impact of respiratory motion correction on SPECT myocardial perfusion imaging using a mechanically moving phantom assembly with variable cardiac defects. J. Nucl. Cardiol. **24**(4), 1216–1225 (2017)
9. Lontos, A., Antoniou, A., Chrysanthou-Baustert, I., Christofides, S., Demetriadou, O., Panagidis, C., et al.: Construction of inflatable lungs to simulate respiratory motion in myocardial perfusion imaging. In: XIV Mediterranean Conference on Medical and Biological Engineering and Computing, pp. 1331–1335. Springer, Cyprus (2016)
10. Cook, L.B.: Rebreathing in the Mapleson A, C, and D breathing systems with sinusoidal and exponential wavefronts. Anaeshesia **52**, 1182–1194 (1997)
11. McLeish, K., Hill, D.L., Atkinson, D., Blackall, J.M., Razavi, R.: A study of the motion and deformation of the heart due to respiration. IEEE Trans. Med. Imaging **21**(9), 1142–1150 (2002)
12. Rosset, A., Spadola, L., Ratib, O.: OsiriX: an Open-Source software for navigating in multidimensional DICOM images. J. Digit. Imaging **17**(3), 205–216 (2004)

3D Printing-Based Pediatric Trainer for Ultrasound-Guided Peripheral Venous Access

Rocco Furferi[1], Lorenzo Guariento[1], Kathleen S. McGreevy[3], Elisa Mussi[1(✉)], Niccolò Parri[2], Francesca Uccheddu[1], and Yary Volpe[1]

[1] Department of Industrial Engineering, University of Florence, via di Santa Marta, 3, 50139 Florence, Italy
{rocco.furferi, lorenzo.guariento, elisa.mussi, francesca.uccheddu, yary.volpe}@unifi.it

[2] Department of Emergency Medicine and Trauma Center, Meyer Children's Hospital, Viale Gaetano Pieraccini, 24, 50139 Florence, Italy
niccolo.parri@meyer.it

[3] Office of International Relations and the Promotion of Innovation, Meyer Children's Hospital, Viale Gaetano Pieraccini, 24, 50139 Florence, Italy
kathleen.mcgreevy@meyer.it

Abstract. Peripheral venous access is an extremely common procedure, crucial in delivering drugs and collecting blood samples. It is associated to high failure rates, especially when pediatric subjects are involved, due to reduced limb size and low cooperation. Ultrasound can sensibly increase success rates and reduce the time required to perform the procedure, though a specific training is necessary to acquire adequate hand-eye coordination and simultaneously handle needle and probe. Commercially available simulators lack of realistic devices that reproduce anatomy and kinematics of pediatric patients. In this work, an echogenic simulator integrating direct 3D printing and silicone casting is proposed. More specifically, it replicates a five years old upper limb's anatomy comprising an articulated skeleton, muscle tissues, skin and an integrated blood circuit. The devised simulator shows its effectiveness in terms of acoustic properties, articular kinematics reproduction and haptic feedback. Furthermore, the simulator can be easily customized according to specific training needs thanks to a highly flexible manufacturing process.

Keywords: 3D printing · Echogenic soft phantom · Peripheral venous access trainer · Silicone casting

1 Introduction

Peripheral venous access is a common clinical procedure used to administer drugs or draw blood samples. Generally, the identification of the target vein is performed by sight and/or palpation guided by anatomical landmarks [1]. This approach often requires multiple attempts; the success rate for peripheral intravenous cannulation at

J. Henriques et al. (Eds.): MEDICON 2019, IFMBE Proceedings 76, pp. 735–745, 2020.
https://doi.org/10.1007/978-3-030-31635-8_87

first attempt ranges between 76% and 91% on adults, while on pediatric population is between 53% and 75% [2]. Indeed, in pediatric cases, the reduced limb size, the poorly developed venous apparatus and the uncooperative patient make the procedure more difficult [3].

The use of ultrasounds to execute the peripheral venous access reduces the failure rate and the average time needed to carry out the procedure increasing patient safety and satisfaction [4]. However, ultrasound-guided intravenous (IV) access requires specific clinicians' skills: hand-eye coordination, simultaneous handling of probe and needle, interpretation of ultrasound images. In fact, ultrasonography is a diagnostic technique characterized by several advantages, including speed of execution, absence of ionizing radiation and low cost but it is highly operator-dependent, requiring expertise and experience to correctly interpret diagnostic images.

Within this context, medical simulators provide a patients-free environment that allows resident physicians to gain clinical experience even before interacting with actual patients, providing a safe environment for a repetitive and standardized practice, enhancing both manual skills and decision-making processes [5, 6].

Several simulators have been developed to successfully perform the ultrasound-guided peripheral venous access. Some of the most widely adopted devices are reviewed in Tables 1 and 2. They can be classified in two main categories: task trainers, designed to teach a specific skill or technique (Table 1), and realistic procedural simulators reproducing the anatomy (i.e. the arm) involved in the procedure (Table 2).

As evinced from Tables 1 and 2, there is a lack of commercial simulators that jointly reproduce articular kinematics, upper limb's geometry, echogenicity and blood flow. This is particularly true in pediatric subjects, even though such features are crucial in making the simulation session engaging and enriching for trainees.

Accordingly, the main aim of the present work is to develop and validate an anatomically realistic and cost-effective simulator for the ultrasound-guided peripheral venous access, reproducing the key aspects of the procedure and overcoming the drawbacks of the devices currently on the market.

For this purpose, a cost-effective system based on CAD/CAM techniques, 3D printing and silicone rubbers casting [14] has been developed. The devised simulator faithfully replicates the geometry and articular kinematics of the arm of a 50th percentile 5-year-old child. Moreover, the simulator is equipped with an artery and a vein in which synthetic blood flows with an internal pressure that provides visual feedback when the puncture succeeds. Eventually, in the volar side of the forearm a patch has been made with material characterized by echogenicity similar to real tissues to perform ultrasound-guided peripheral venous access.

2 Materials and Methods

An articulated skeletal model, a soft anechoic muscular component, a soft echogenic puncture pad, a cephalic vein and brachial artery with a bloodstream and a soft echogenic epithelial glove, composes the devised simulator.

Table 1. Task trainers on the market.

Task trainers				
Model & manufacturer	Picture	Pros	Cons	Price
Branched 4 Vessel Ultrasound Training Block Model CAE Healthcare [7]	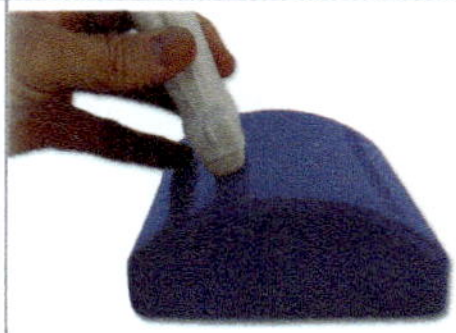	-Ultrasound compatible -Self-sealing material -Pediatric version available -4 different vessels -Refillable vessels	-Unrealistic appearance -Not flowing fluid (vessels are filled with fluid)	550€
SONOtrain 3B Scientific [8]	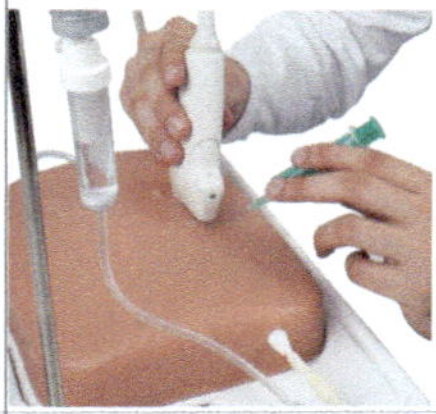	-Ultrasound compatible -Self-sealing material -Adjustable fluid flow	-Unrealistic appearance -Pediatric version not available	360€
HM2.0 Re-Moldable 2 vessel Ultrasound Training Model Humimic Medical [9]	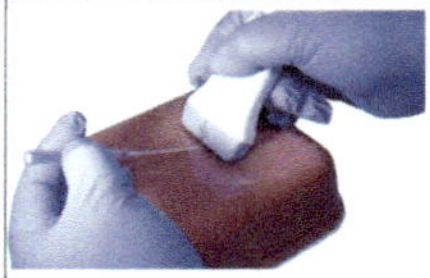	-Ultrasound compatible -Self-sealing material -Refillable vessels	-Unrealistic appearance -Pediatric version not available -Not flowing fluid (vessels are filled with fluid)	250€

Two main manufacturing approaches have been followed: direct 3D printing of hard parts and silicone casting in 3D printed molds for soft components.

The first approach was adopted to manufacture the skeletal model (hard tissue), while the second one was employed to realize the soft components (soft tissues) with materials otherwise unobtainable trough traditional AM techniques.

Figure 1 shows the design and manufacturing workflow of the trainer.

2.1 Skeletal Model

The skeletal model is composed by arm, forearm and hand that have been geometrically simplified. Natural joints have been approximated to mechanical hinges, based on anatomical observations confirmed by scientific medical literature [15–17]. The articulated skeletal model reproduces the upper limb's natural kinematics: flexion-extension of the elbow, pronation and supination of the forearm, pronation-supination and adduction-abduction of the hand. Figure 2 shows the skeletal CAD model.

Table 2. Realistic procedural simulators on the market.

Realistic procedural simulators				
Model & manufacturer	Picture	Pros	Cons	Price
Ultrasound-Guided PICC Training Simulator Kyoto Kagaku [10]	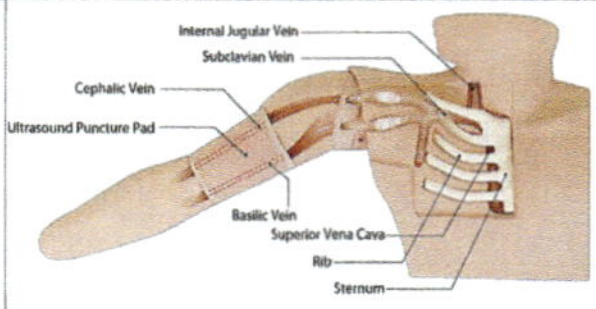	-High fidelity vasculature -Ultrasound compatible puncture Pad -Ribs and right clavicle included	-High price -Small puncture pad -Not articulated arm and hand -Pediatric version not available	1727€
Life/form Advanced Venipuncture and Injection Arm Nasco [11]	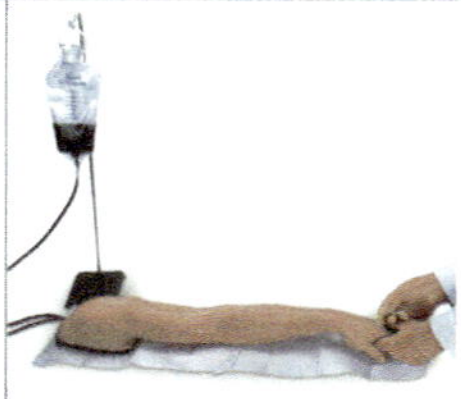	-High fidelity vasculature -Easily replaceable parts	-High price -Ultrasound incompatible -Not articulated -Pediatric version not available	13240€
PICC with I.V. and Arterial Line Vascular Access Ultrasound Trainer CAE Healthcare [12]		-Main blood vessels included -High fidelity anatomy	-Not articulated arm -Manually controlled blood flow -High price -Pediatric version not available	2200€
5-Year-Old Patient Injection Training Arm Gaumard [13]		-Main blood vessels included -High fidelity vasculature -Pediatric	-Ultrasound incompatible -Not articulated -Manually controlled blood flow	530€

The radio-ulnar joint has been assimilated to a spherical hinge, while the humero-ulnar and the two humero-radial joints have been simplified as cylindrical joints [15–17]. The model was printed in PLA with MakerBot Replicator+ (MakerBot, Brooklyn, NY) with a material cost of 2,00€.

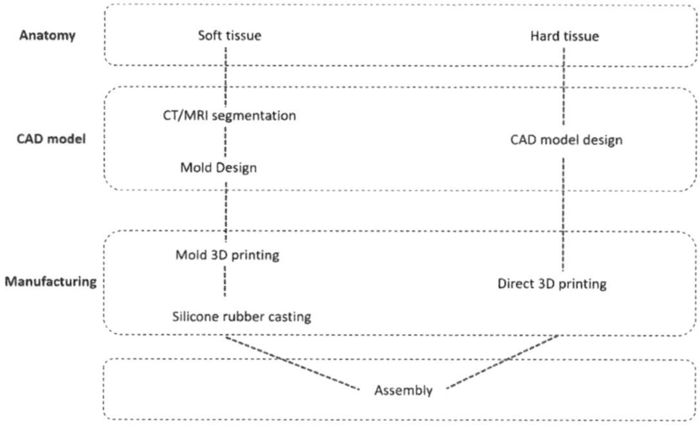

Fig. 1. Design and production workflow

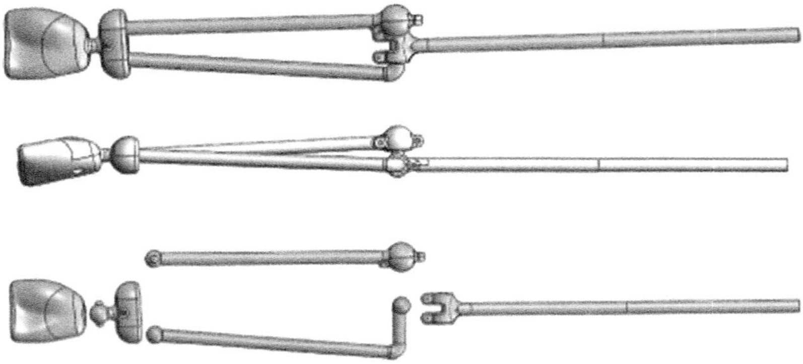

Fig. 2. Skeletal CAD model

2.2 Soft Tissues

Soft tissues of the simulator consist in skin and muscles. An approximately 50th percentile five years-old patient of the Meyer Children's Hospital (Florence) was selected to collect a magnetic resonance (Achieva dStream, Philips, NL) of the right upper limb. The resulting DICOM files were processed with Mimics® (Materialise, Belgium) [18] to reconstruct the upper limb's geometry through a segmentation process with threshold methods [19].

The STL file was imported in Geomagic® Design X (3D Systems, SC, USA) and the measure of each element of the subject's arm, carried out on the 3D model, are compared with the reference 50th percentile values [20] (see Table 3). The small differences between the actual anatomy of the patient and the reference can be considered acceptable for the proposed system.

Table 3. Anatomical reference values of a five years old subject' s upper limb.

	5-years old patient values [mm]	50th Percentile reference values [mm]	Difference [mm]
Arm to shoulder length	21.9	21.9	0
Forearm length	28.4	28.2	0.2
Hand length (wrist-middle finger's tip	11.9	11.8	0.1
Hand length	5.5	5.5	0
Middle finger's length	5.0	4.9	0.1

The CAD model of the muscle and the skin mold have been designed with Geomagic® Design X (see Fig. 3), as well as a core (blue element in Fig. 3) to be introduced inside the mold to create a cavity to successively house the articulated skeletal model and puncture pad.

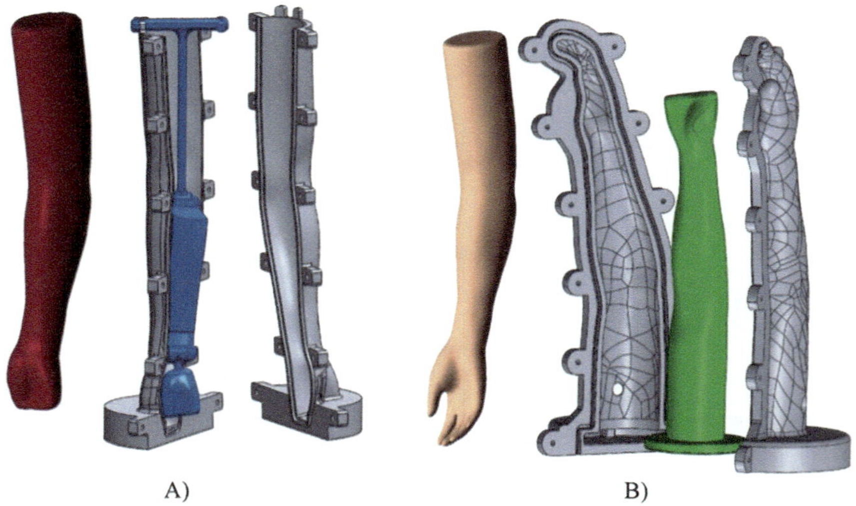

Fig. 3. (A) Muscle mold design; (B) Skin mold design

The molds have been additive manufactured by using the EOS P390 3D printer (EOS, Krailling, Germany).

Once the molds are designed, the main challenge is to identify the materials to reproduce the soft tissues' properties in terms of mechanical and acoustic properties. Reference values for acoustic properties of skin and muscle are reported in Table 4 [21].

A category of materials typically used to resemble the actual human properties are biopolymers or chemically synthetized polymers [21]. Biopolymers show both mechanical and acoustic properties similar to those of soft tissues but a short durability,

Table 4. Reference values of acoustic properties for human tissues. n is an empirical coefficient.

Tissue	Density [kg/m^3]	Speed c [m/s]	Attenuation [dB/(cmMHz^n)]	n
Skin	1100	1631	0,22	1,15
Muscle	1041	1595	1,47	1,15

due to microbial invasion and dehydration [22]. Silicone rubbers are more durable but stiffer than biopolymers, and they need to be mixed with mineral oils and other agents to match the soft tissues' properties [20].

For this work, the Ecoflex 00-50 rubber (Smooth-On, PA, USA) was selected. It consists of a platinum-catalyzed two-component silicones that cure at room temperature by mixing the two reagents in a 1:1 ratio by weigh. To match soft tissue's acoustic properties the pure silicone was mixed with other agents, according to the formula developed by Pacioni et al. [23].

2.3 Puncture Pad

Located in the forearm, the puncture pad is the anatomical region where venipuncture is most frequently performed. Therefore, the pad has been designed to be easily replaced after some puncture simulations are carried out. The pad is manufactured by casting an echogenic mixture in a 3D printed mold shown in Fig. 4. The mold is designed with a housing for the artificial blood vessels. Pad replacement is very easy as it is sufficient to link the silicone tubes to the circuit integrated inside the muscle component. This choice allows reducing the overall number of components for the simulator that are subject to wear, thus reducing also the maintenance costs.

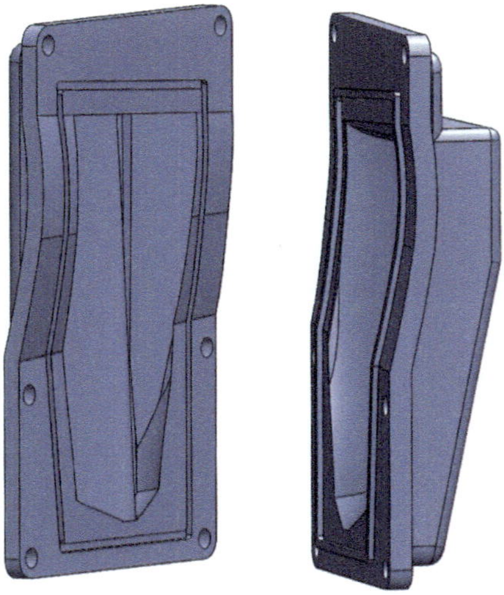

Fig. 4. Puncture pad mold

2.4 Bloodstream

The artificial bloodstream is composed by an integrated circuit powered by an external pump. The blood circuit comprises two small silicone tubes which replicate the basilic vein and the brachial artery with internal lumen of 3 mm and 2 mm respectively. Tubes are placed in the volar area of the arm and are linked through a tube reducer. The artery's free extremity is directly connected to an aquarium sump pump while the vein discharges the fluid into the tank where the pump is placed. The selected fluid to simulate the blood is red stained water.

2.5 Silicone Casting

Muscle, skin and puncture pad have been produced pouring silicone rubber in the corresponding molds, in which the artificial bloodstream is integrated, at room temperature, after an appropriate degassing process to remove air bubbles enclosed when mixing the compound, which can produce artifacts on ultrasound images.

2.6 Simulator Assembly

The assembly process consists of the following steps:

1. insert the articulated skeletal model into the muscle component;
2. connect the puncture pad to the integrated blood circuit;
3. place the skin glove on the muscle for a good aesthetic outcome.

Figure 5 shows the fully assembled trainer and its components.

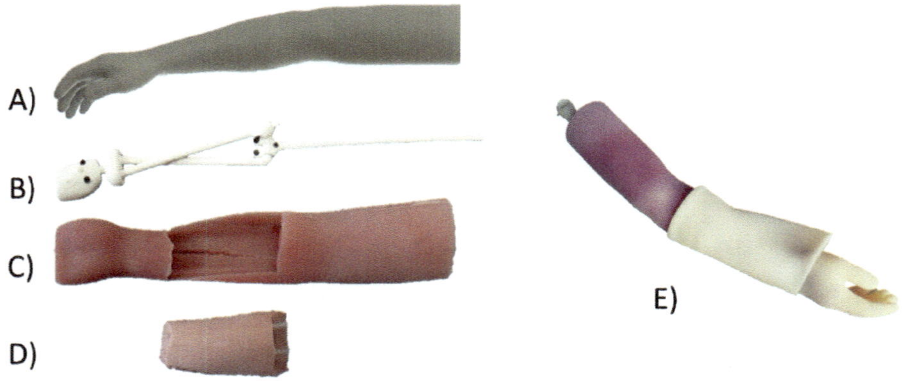

Fig. 5. (A) Skin glove, (B) skeletal model, (C) muscle, (D) puncture pad, (E) assembled trainer.

3 Results

The simulator was tested by the clinicians of the Meyer Children's Hospital, which evaluated the quality of ultrasound images. In order to evaluate the performance of the simulator in terms of acoustic response, ultrasound images of simulator were acquired and compared with the ones of the actual vessels. As shown in Fig. 6 the ultrasound images of vessels and soft tissues of the simulator, performed with standard ultrasound probes and machine settings, are qualitatively comparable to those achievable from human tissues.

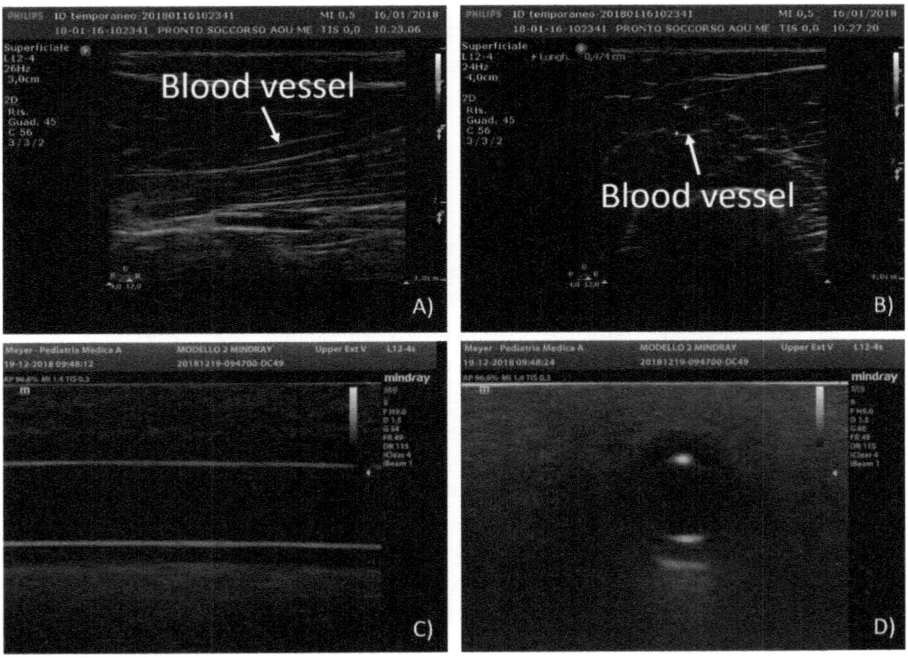

Fig. 6. Ultrasound images comparison: (A) transverse section of human vessels; (B) cross section of human vessels; (C) transverse section of artificial vessel; (D) cross section of artificial vessel

In order to allow ultrasound transmission across the skin-muscle interface a common ultrasound transmission gel has been applied. This allows to avoid artifacts due to the presence of air. Figure 7 underlines that such a solution guarantees a homogenous signal transmission with no distortion and artifacts.

A small group of clinicians from Meyer Children's Hospital has performed ten ultrasound-guided venipunctures. The simulator has been appreciated for a high degree of realism, in terms of articular kinematics reproduction and especially for the haptic feedback during the needle insertion both in soft tissues and blood vessels, features that make the simulation immersive and productive.

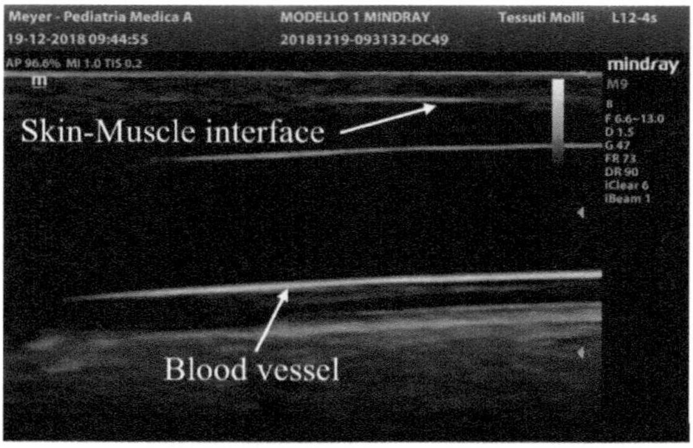

Fig. 7. Ultrasound image of longitudinal section of blood vessel with skin layer and ultrasound gel

4 Conclusion

A novel 3D-printing-based pediatric simulator for ultrasound guided peripheral access training has been presented. The proposed device reproduces both geometric and kinematic features of a five years old subject's upper limb; it is made of soft echogenic silicone, thus enabling clinicians to practice ultrasound guided venous access, improving the hand-eye coordination in a safe and repeatable environment. The simulator is provided with an integrated and tunable artificial blood flow controlled by an external pump that grants a visual feedback when the puncture succeeds. Echogenic material enables for realistic ultrasound imaging of the blood vessels, replicating the real-life procedure. The simulator is cost-effective as the production cost is approximately 50€; the only component subject to wear is the puncture pad, which can be replaced after every training session with the cost of about 4€. Finally, the device can be customized according to trainers' needs as the production process is extremely flexible. Future works will be addressed to the customization of the proposed trainer to be connected to commercially available mannequins, as to provide more immersive training scenarios. Further studies will assess whether the introduction of such a simulator in clinicians' training process will improve clinical skills and consequently healthcare quality. In other words, future studies will be addressed to quantitively assess how ultrasound-guided peripheral access' rate can be increased, as well as how the execution time can be lowered. A standardized evaluation of phantom's ultrasound images will be performed to assess the similarity to human tissues.

References

1. Mitchell, S.E., Clark, R.A.: Complications of central venous catheterization. Am. J. Roentgenol. **133**(3), 467–476 (1979)

2. Costantino, T.G., Parikh, A.K., Satz, W.A., Fojtik, J.P.: Ultrasonography-guided peripheral intravenous access versus traditional approaches in patients with difficult intravenous access. Ann. Emerg. Med. **46**(5), 456–461 (2005)
3. Jones, S.E., Nesper, T.P., Alcouloumre, E.: Prehospital intravenous line placement: a prospective study. Ann. Emerg. Med. **18**(3), 244–246 (1989)
4. Almeida, C.E.: Vascular access: the impact of ultrasonography. Einstein (São Paulo) **14**(4), 561–566 (2016)
5. Alloni, R.: Simulation: an innovative teaching methodology in the education of health-care providers. Methodol. Educ. Clin. Innovation, **23**(2) (2015)
6. Al-Elq, A.H.: Simulation-based medical teaching and learning. J. Fam. Community Med. **17**(1), 35 (2010)
7. Vessel Ultrasound Training Model for venous access. http://www.bluephantom.com/product/Branched-4-Vessel-Ultrasound-Training-Block-Model.aspx?cid=525. Accessed 13 Mar 2019
8. SONOtrain. https://www.3bscientific.it/sonotrain-ultrasound-vein-model-1019637-p120-3b-scientific,p_1397_27465.html. Accessed 13 Mar 2019
9. Hm2.0 Re-Moldable 2 Vessel Ultrasound Training Model. https://humimic.com/product/hm1-0-vessel-ultrasound-training-model/. Accessed 13 Mar 2019
10. Ultrasound-Guided PICC Training Simulator. https://www.kyotokagaku.com/products/detail01/mw18.html. Accessed 13 Mar 2019
11. Life/form Advanced Venipuncture and Injection Arm – Light. https://www.enasco.com/p/Life-form%C2%AE-Advanced-Venipuncture-and-Injection-Arm—Light%2BLF01121U. Accessed 13 Mar 2019
12. Venous and Arterial Vascular Access Ultrasound Training Arm Model. https://www.bluephantom.com/product/Gen-II-PICC-with-IV-AND-Arterial-Line-Vascular-Access-Ultrasound-Trainer.aspx?cid=407. Accessed 13 Mar 2019
13. 5-Year-Old Patient Injection Training Arm. https://www.gaumard.com/s405. Accessed 21 Mar 2019
14. O'Brien, E.K., Wayne, D.B., Barsness, K.A., McGaghie, W.C., Barsuk, J.H.: Use of 3D printing for medical education models in transplantation medicine: a critical review. Curr. Transplant. Rep. **3**(1), 109–119 (2016)
15. Whiteley, G.P.: An Articulated Skeletal Analogy of the Human Upper-Limb (2000)
16. Kapandji, I.A.: The physiology of the joints, volume I, upper limb. Am. J. Phys. Med. Rehabil. **50**(2), 96 (1971)
17. Levangie, P.K., Norkin, C.C.: Joint structure and function: a comprehensive analysis. FA Davis Company
18. Materialise Mimics, https://www.materialise.com/en/medical/software/mimics. Accessed 13 Mar 2019
19. Al-Amri, S.S., Kalyankar, N.V.: Image segmentation by using threshold techniques. arXiv preprint arXiv:1005.4020. Accessed 21 May 2010
20. Snyder, R.G.: Physical characteristics of children as related to death and injury for consumer product safety design. Final report (1975)
21. Wang, Y., Tai, B.L., Yu, H., Shih, A.J.: Silicone-based tissue-mimicking phantom for needle insertion simulation. J. Med. Devices **8**(2), 021001 (2014)
22. Culjat, M.O., Goldenberg, D., Tewari, P., Singh, R.S.: A review of tissue substitutes for ultrasound imaging. Ultrasound Med. Biol. **36**(6), 861–873 (2010)
23. Pacioni, A., Carbone, M., Freschi, C., Viglialoro, R., Ferrari, V., Ferrari, M.: Patient-specific ultrasound liver phantom: materials and fabrication method. Int. J. Comput. Assist. Radiol. Surg. **10**(7), 1065–1075 (2015)

Pectus Excavatum: A New Approach for Monitoring Cup-Suction Treatment

Francesco Buonamici[1], Antonio Marzola[1], Michaela Servi[1(✉)], Francesca Uccheddu[1], Yary Volpe[1], Marco Ghionzoli[2], and Antonio Messineo[2]

[1] Department of Industrial Engineering, University of Florence, Florence, Italy
{francesco.buonamici,antonio.marzola,michaela.servi, francesca.uccheddu,yary.volpe}@unifi.it

[2] Department of Pediatric Surgery, Meyer Children's Hospital, Florence, Italy
{marco.ghionzoli,antonio.messineo}@meyer.it

Abstract. The introduction of the vacuum bell (VB) for the conservative treatment of Pectus Excavatum (PE) has led to a new non-invasive alternative to thoracic surgery. The VB works by elevating the chest as long as a negative differential pressure is internally assured. In recent years studies have been conducted to validate this type of treatment and to outline its correct use; results show a short-term PE improvement when the device is worn for a minimum of 30 min (twice a day) up to a maximum of several hours a day for 12–15 months. Although the worldwide diffusion of VB devices increases year after year, its ability to lift the chest during treatment with respect to the applied pressure has begun to be evaluated only recently. In this paper, a new instrument for measuring chest elevation during treatment is presented and validated. The proposed system consists of two measurement devices: a commercial instrument for the detection of the negative pressure inside the VB, and a specifically developed optical system for the detection of chest movement. The effectiveness of the proposed system, tested on five paediatric patients, paves the way to the objective definition of an optimised patient specific VB scheme of use.

Keywords: Image processing · Optical measurement · Pectus Excavatum · Vacuum bell

1 Introduction

The conservative treatment through the application of a cup suction device, named vacuum bell (VB), is the only alternative nowadays to surgery for the correction of the Pectus Excavatum (PE). Characterized by a depression of the sternum, PE is the most frequent thoracic deformity with an incidence of 1 in 300–1000 births with a male/female ratio of about 4:1 [1–3]. The vacuum bell was initially introduced as a glass bell at the beginning of the last century [4] and was later replaced by a bowl-shaped device made of silicone, easier to apply and use, introduced by Ing. Klobe [5]. The VB body is made of a silicone ring and a transparent polycarbonate window (for inspection). Three different sizes (i.e. 16, 19 and 26 cm in diameter) as well as a model

J. Henriques et al. (Eds.): MEDICON 2019, IFMBE Proceedings 76, pp. 746–754, 2020.
https://doi.org/10.1007/978-3-030-31635-8_88

fitted for young women exist on the market. Inspired by Klobe's device, a number of alternatives have been proposed even tailored on the specific anatomy of a patient [6]. These devices achieve the elevation of the sternum by generating a negative pressure within them; the short-term effects of this type of treatment have already been proved as promising [7, 8] especially when a precise medical guideline is followed by the patient.

Klobe, in the vacuum bell manual user, recommends patients to operate the device twice a day for 30 min for the first 4–6 weeks; later, the period of daily application of the VB can be varied according to the patient's outcomes at the discretion of the physician. Haecker et al. in [7, 8] carried out the measurement of the PE depth before and after the use of the VB by positioning the patient in a standardized supine position and by providing the depth using a designed scaled rod. Consequently, no real-time measurement during the treatment was encompassed. To overcome this limitation, Sesia et al. [9] proposed an electronic device for measuring the elevation of the sternum in relation to the depression within the VB (during the treatment). The device, called *Distance and differential pressure measuring device* (DPMD), allows to measure the negative differential pressure inside the VB and the relative elevation of the sternum during the application of the VB. The DPMD is placed on the upper part of the VB and does not alter its use. The device was designed to be compatible with the aforementioned different models of VB proposed by E. Klobe.

An alternative method for monitoring the chest elevation during the VB treatment is here presented. Unlike the work proposed by Sesia et al., chest uplift is here measured with an external optical device placed above the patient. Such a device can be used for any commercially available or self-produced vacuum bell under the only condition that the chest is visible through the transparent window. Moreover, the proposed system consists of a low-cost solution comprising a commercial RGB-D camera and an appositely devised user-friendly interface that can be exploited on any standard personal computer.

2 Materials and Methods

As mentioned above, the proposed method allows determining the elevation of the chest with respect to the pressure applied inside the vacuum bell.

In detail, the system encompasses two parallel modules providing a measurement of both pressure and elevation. To establish the pressure inside the suction device, a commercially available instrument was used, the PCE-P05 [10], which is linked to the pump tube by means of a junction (Fig. 1). PCE-P05 is a portable digital pressure manometer for measuring the differential pressure in pneumatic systems, compressors, pumps, valves etc. Such a device visualizes the min, max and average values directly on a display and is supplied with real-time data recording capability with an adjustable interval from 1 s to 24 h.

While the pressure values are recorded, the chest elevation is also registered. To calculate such an elevation, a dedicated software, that exploits a low-cost commercial depth sensor, has been developed. 3D laser scanners are not suitable for this application since it is needed a post processing step to add colors on the acquired 3D data [11]; on the contrary, the new generation consumer grade depth sensors are particularly suited to deal simultaneously with both 3D model and colors (needed to track the chest deepest point). In particular the Intel RealSense D415 [12] has been selected to acquire the depth elevation. As demonstrated in [13], the average error in depth evaluation at close distances (i.e. <500 mm), is equal to 2.11 mm that can be considered sufficient for this analysis. The device comprises both a color camera (RGB) and a depth camera system consisting of two infrared cameras and an infrared projector. The depth data are retrieved via active-stereo triangulation [14]. Moreover, the Intel RealSense D400 series is supported by the open-source multi-platform library, Intel RealSense SDK 2.0 [15], which allows the configuration of the camera with different internal settings (both for acquisition and post-processing), and in general facilitates the interface with the device.

To determine the sternum elevation, both RGB and depth streams are used. The point of maximum depression is automatically detected in the color image thanks to an appositely devised image-processing algorithm. At first a self-adhesive circular green-colored marker (Fig. 2a) is manually applied to the patient chest at the point of maximum depression; then, the algorithm detects and tracks such marker using color separation: a green channel threshold is applied to the original image, thus obtaining a binary image in which the circle is identified through a Hough Transform which is a feature extraction algorithm widely used in digital image analysis [16]. It identifies objects with specified shapes by letting all edge pixels in the image "vote" and extract candidates with the highest votes. The classical Hough transform is associated with the identification of straight lines, but with some modification, the algorithm can be used to detect objects of arbitrary shapes (usually circles, ellipses) in an image.

This detecting process allows to identify the pixels of the marker center within the RGB image.

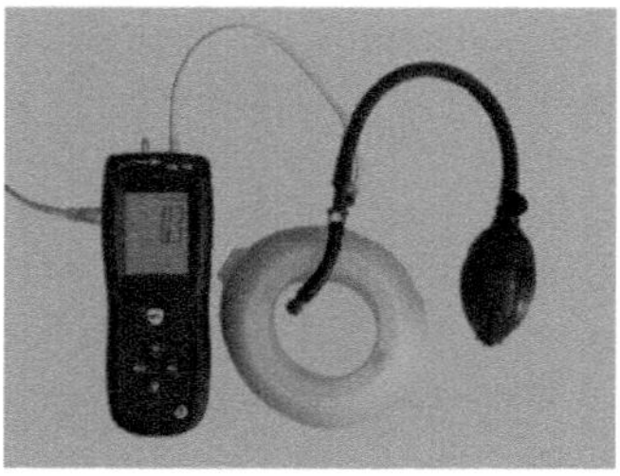

Fig. 1. Example of junction connecting PCE-P05 to vacuum bell.

The corresponding depth value (i.e. the 3D coordinates of the marker center) is then retrieved from the depth stream (Fig. 2b). Since the RGB camera and the depth camera are intrinsically aligned in the D415 device, the pixel coordinates on the RGB image are directly related to the x, y and z values in the depth image.

As said, the aim is the estimation of the chest deepest point elevation during the VB treatment. Such an elevation value is assessed with respect to vertical direction which is orthogonal to the plane the patient is laid on. Therefore, it is necessary to assure the perpendicularity of the optical axis with respect to the mattress that will accommodate the patient. This is possible using the depth camera software [15] which provides the angle between the optical axis and a plane framed by the camera itself.

To perform the elevation measurement the camera is placed at an approximatively 300 mm distance from the patient's chest. This distance was chosen to obtain the necessary accuracy and to allow the algorithm to work on a small portion of the scene to make the marker search more robust. In this work, an appropriate support has been created to hold the camera at a fixed distance from the patient, mounted on the structure previously realized in [17] for the support of the chest severity measurement device.

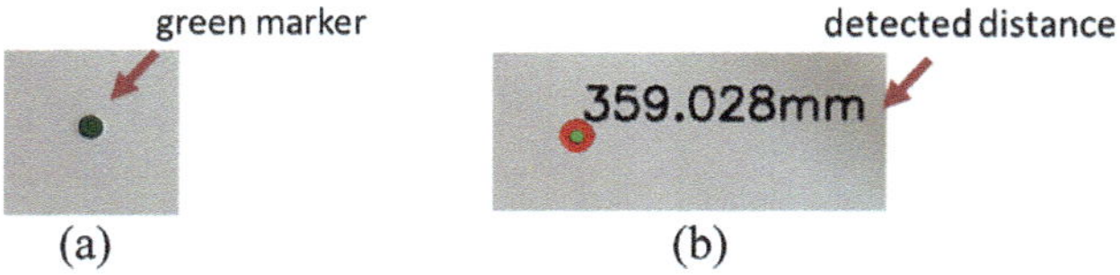

Fig. 2. Green colored marker (a), and output of the implemented algorithm (b).

The so-computed camera-marker distances are saved in a text file, in which the start time of the recording is also noted, to be subsequently aligned with the measured pressure values.

The system was, in the first instance, tested both on static and moving objects, as indicated below, then an initial analysis on 5 patients was carried out.

2.1 Preliminary Accuracy Tests

The first test conducted to evaluate the accuracy of the chest elevation measurement system examines a marker applied to a static calibrated object. After positioning the camera perpendicular to the plane containing the marker, the latter is progressively brought closer to the camera using calibrated blocks of 15 mm height. This allows to define a ground truth and then evaluate the difference between the real displacement of the marker and the one evaluated by the system. The test was repeated by placing the vacuum bell over the marker in order to test if the transparent window could cause alterations in the measurement (Fig. 3).

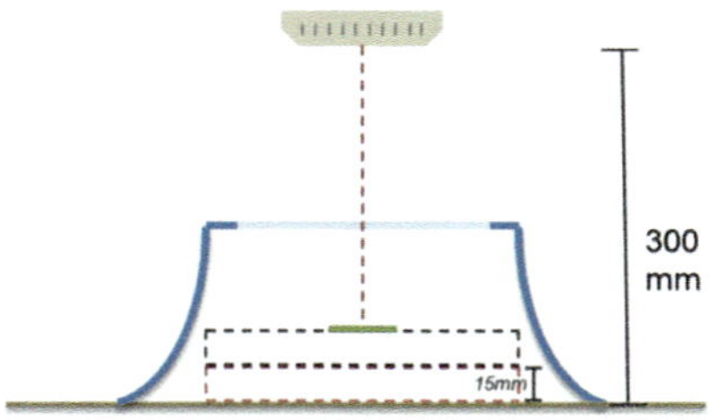

Fig. 3. Acquisition setup for the first test: the marker is installed on a plane that is progressively brought closer to the camera by means of calibrated blocks with height 1.5 mm.

The graphs in Fig. 4 show the trend of the absolute error with respect to the distance of the static object from the camera, defined as the absolute value of the difference between the real excursion and the one evaluated by the system, for the two cases, with (Fig. 4a) and without vacuum bell (Fig. 4b).

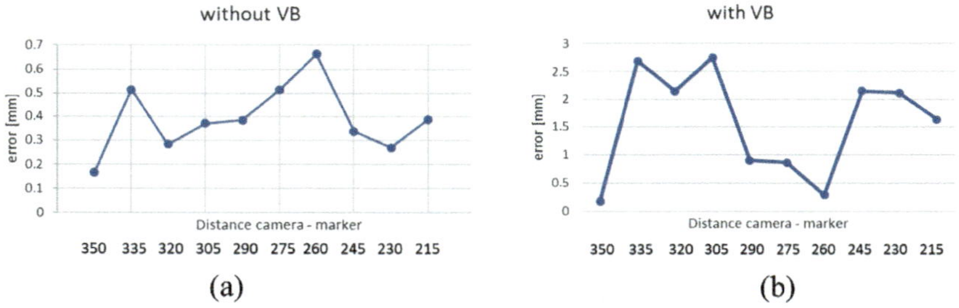

Fig. 4. Results of static test computed as difference between real and measured marker excursion in absence of VB device (a) and with VB between the camera and the marker.

The average error evaluated without the VB device positioned between the marker and the camera is 0.38 mm with a standard deviation 0.14; the average error when the VB device is positioned between the marker and the camera is 1.56 mm with a standard deviation equal to 0.95. Both graphs show that the error does not depend on the camera-marker distance but oscillates around an average value. In addition, although the device's interference produces on average a higher deviation, a mean error of ~1 mm is considered more than acceptable for this application.

The second test studies the behavior of the measurement system in relation to an object in slow and continuous motion. To carry out this second experiment, the target is attached to an object positioned on a sliding linear guide (Fig. 5). The object is progressively moved closer to the camera, to simulate the lifted chest, from a distance of 300 mm up to a distance of 200 mm. The entire shift takes place in one minute. Analogously to the first test, such a test was repeated by placing the vacuum bell between the camera and the marker.

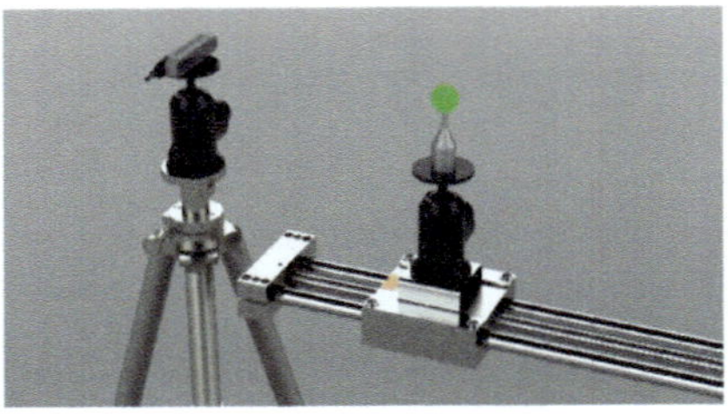

Fig. 5. Acquisition setup for the second test: the marker is installed on a sliding linear guide and the camera is fixed frontally.

The graphs in Fig. 6 show the observed trend in blue and the linear continuous movement in orange. It is visible that the continuous movement of the marker is correctly recorded by the system, with few oscillations in case of the presence of the vacuum bell.

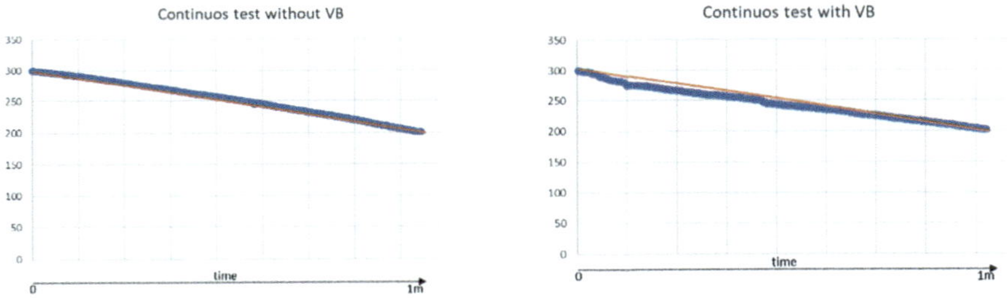

(a) (b)

Fig. 6. Results of dynamic test with and without VB between the marker and the camera (in blue). In orange the real movement of the marker on the linear guide.

3 Results

In general, conservative treatment with vacuum bell is practicable at almost all ages [8]. In this work, however, five patients in pediatric age (patients' ages vary between 10 and 16 years old) have been selected, addressed by the specialized clinic for the deformations of the thoracic wall of the Meyer Children's Hospital. Children, in fact, are the most suitable subjects to receive this type of conservative treatment, since they have more easily deformable bones and cartilages.

The tests were carried out by placing the patient on a semi-rigid mattress and connecting the vacuum bell to the pressure monitoring device. The distance measuring system was then started, and the patient was asked to maintain the position for about 2 min. Figure 7 shows the placement and detection of the green marker during the experimental phase.

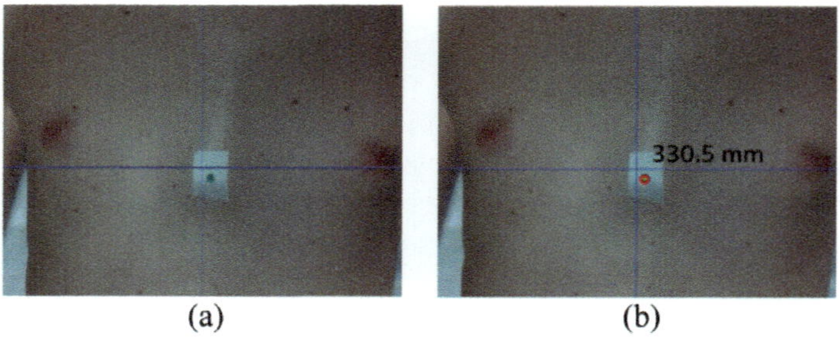

Fig. 7. Example of placement (a) and automatic detection (b) of the green marker on patients during experimental phase.

The depression value inside the vacuum bell was set to a reference value defined by the clinician for each patient. It is important to notice that such a value is different for each patient and depends on age, cartilage strength, etc. During the recording time, when the pressure inside the device reached an excessively high value (e.g. −100 mbar), due to air leakage, it was brought back to the reference pressure. The pressure and distance values were then aligned accordingly to the recording start-time. Figure 8 shows the results obtained on two of the 5 patients which undergo the experimentation.

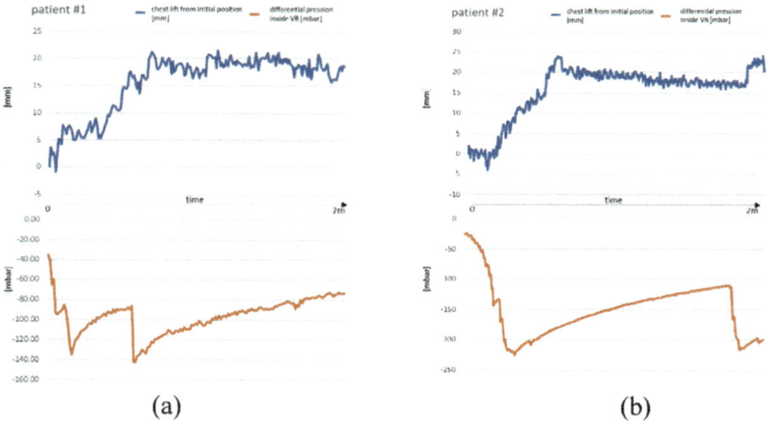

Fig. 8. Results of the use of the proposed method on two patients. In blue the elevation of the chest, in orange the increasing negative pressure inside the VB.

In order to obtain a clear and comprehensible visualization of data, the first recorded depth value is considered as zero and the following values are reported in relation to it. In this way the graph shows how the chest is raised from a starting position.

The pressure values recorded for this first experimental test are consistent with the trial carried out in [9], for patients aged between 11 and 15 years.

4 Conclusions

In this paper, a method for monitoring conservative treatment for patients affected by Pectus Excavatum was studied. The chest lift measurement system is intended to be used during the application of the cup-suction device (the vacuum bell) and is based on two parallel measurements: the negative differential pressure, applied inside the VB, and the chest lift. Differently from other devices proposed in literature, the devised system can be used with any commercially available or self-produced vacuum bell model, as long as the chest is visible through the transparent inspection window.

Preliminary tests were conducted to validate the measurement system, confirming its accuracy. The average error obtained with the static test is on 1.56 mm when the vacuum bell device is placed between the marker and the camera, with a standard deviation equal to 0.95, and, the dynamic test reported the capacity of the system to reproduce and follow the movement of the marker.

As a first trial, the system was then tested on 5 pediatric patients, indicated by the clinic for the deformations of the thoracic wall of the Meyer Children's Hospital. The elevations of the chest in relation to the negative pressure applied inside the device, measured by the proposed system, are consistent with those in the literature. Future developments will involve the use of the proposed system to validate the conservative treatment for both commercially available or customized vacuum bell, and to objectively define an optimised patient specific VB scheme of use.

References

1. Cobben, J.M., Oostra, R.-J., van Dijk, F.S.: Pectus excavatum and carinatum. Eur. J. Med. Genet. **57**, 414–417 (2014)
2. Sarwar, Z.U., DeFlorio, R., O'Connor, S.C.: Pectus excavatum: current imaging techniques and opportunities for dose reduction. In: Seminars in Ultrasound, CT MRI, vol. 35, pp. 374–381 (2014)
3. Tocchioni, F., Ghionzoli, M., Messineo, A., Romagnoli, P.: Pectus excavatum and heritable disorders of the connective tissue. Pediatr. Rep. **5**, e15 (2013)
4. Lange, F.: Thoraxdeformitäten: chirurgie und orthopädie im kindesalter. In: Pfaundler, M., Schlossmann, A. (eds.) Handb. der Kinderheilkd, vol. 157 (1910)
5. Schier, F., Bahr, M., Klobe, E.: The vacuum chest wall lifter: an innovative, nonsurgical addition to the management of pectus excavatum. J. Pediatr. Surg. **40**, 496–500 (2005)
6. Carfagni, M., Facchini, F., Furferi, R., Ghionzoli, M., Governi, L., Messineo, A., Servi, M., Uccheddu, F., Volpe, Y.: A semi-automatic computer-aided method for personalized vacuum bell design. Comput. Aided Des. Appl. **15**, 247–255 (2018)
7. Haecker, F.-M., Mayr, J.: The vacuum bell for treatment of pectus excavatum: an alternative to surgical correction? Eur. J. Cardio-Thorac. Surg. **29**, 557–561 (2006)
8. Haecker, F.-M., Sesia, S.: Vacuum bell therapy. Ann. Cardiothorac. Surg. **5**, 440–449 (2016)
9. Sesia, S.B., Hradetzky, D., Haecker, F.-M.: Monitoring the effectiveness of the vacuum bell during pectus excavatum treatment: technical innovation. J. Pediatr. Surg. **53**, 411–417 (2017)

10. Differential Pressure Manometer PCE-P 05 | PCE Instruments. https://www.pce-instruments.com/english/measuring-instruments/test-meters/manometer-pce-instruments-differential-pressure-manometer-pce-p05-det_61709.htm
11. Uccheddu, F., Del Mastio, A., Remondino, F., Pelagotti, A., Cappellini, V.: Texture mapping of flat-like 3D models. In: 2009 16th International Conference on Digital Signal Processing, pp. 1–6. IEEE (2009)
12. Intel® RealSense^{TM} Technology. https://www.intel.com/content/www/us/en/architecture-and-technology/realsense-overview.html. Accessed 11 Mar 2019
13. Carfagni, M., Furferi, R., Governi, L., Santarelli, C., Servi, M., Uccheddu, F., Volpe, Y.: Metrological and critical characterization of the Intel D415 stereo depth camera. Sensors **19**, 489 (2019)
14. Zanuttigh, P., Marin, G., Dal Mutto, C., Dominio, F., Minto, L., Cortelazzo, G.M.: Operating principles of structured light depth cameras. In: Time-of-Flight and Structured Light Depth Cameras, pp. 43–79. Springer, Cham (2016)
15. Intel RealSense SDK 2.0 - Intel RealSense Depth & Tracking Cameras. https://realsense.intel.com/sdk-2/. Accessed 11 Mar 2019
16. Yuen, H.K., Princen, J., Illingworth, J., Kittler, J.: A comparative study of hough transform methods for circle finding
17. Uccheddu, F., Ghionzoli, M., Volpe, Y., Servi, M., Furferi, R., Governi, L., Facchini, F., Lo Piccolo, R., McGreevy, K.S., Martin, A., Carfagni, M., Messineo, A.: A novel objective approach to the external measurement of pectus excavatum severity by means of an optical device. Ann. Thorac. Surg. **106**, 221–227 (2018)

ARTE Project: EEG Analysis During Robotic Rehabilitation

Alessandra Calcagno[1(✉)], Stefania Coelli[1], Giulia Tacchino[1], Marta Baratto[2], Franco Molteni[3], Eleonora Guanziroli[3], Cosimo Puttilli[4], and Anna Maria Bianchi[1]

[1] Politecnico di Milano, Milan, Italy
alessandra.calcagno@mail.polimi.it
[2] EMAC srl, Genoa, Italy
[3] Ospedale Valduce-Villa Beretta, Costa Masnaga (LC), Italy
[4] AB Medica s.p.a, Milan, Italy

Abstract. The ARTE project has the objective of developing and testing an integrated rehabilitative system composed of a robotic end-effector, a wristband and a wireless EEG system. The main purpose of this integrated system is to monitor patients during the execution of a rehabilitation session to evaluate their performance and the effectiveness of their rehabilitation path. The system integrates the contribution of the *Amadeo* robotic end-effector (Tyromotion), the 8-channel EEG system *Helmate* (ab medica s.p.a) and the *E4* wristband (Empatica). Starting from the EEG signal acquired during the rehabilitation session, indexes about activation of the motor cortex (ERD), level of engagement and brain asymmetry (BSI-*Brain Symmetry Index*) are computed. Moreover, the wristband extracts the patient's heart rate (HR), inter-beat interval (IBI) and galvanic skin response (GSR) during the execution of the rehabilitation task: further information used to assess the subject's condition and performance.

The integrated device has been tested on a small group of healthy users. The obtained results were positive and demonstrated the feasibility of clinical trials.

Keywords: Assistive devices · Neurorehabilitation · Electroencephalography · Rehabilitation robotics

1 Introduction

Brain lesions due to cerebral ictus or brain injuries are one of the major causes of disabilities worldwide. Even after a rehabilitation intervention in the early phases, affected people generally preserve a contralesional impairment at the upper and lower limbs level [1]. In particular, the probability of completely regaining the use of the affected hand is very low: only the 12% of stroke survivors totally restore the contralesional hand function, with a negative influence on their quality of life, and limitation in the autonomous execution of activities in daily life [2].

These evidences suggest that the traditional rehabilitation procedures may not be sufficient to obtain satisfying results in terms of upper limbs functional recovery. In

J. Henriques et al. (Eds.): MEDICON 2019, IFMBE Proceedings 76, pp. 755–760, 2020.
https://doi.org/10.1007/978-3-030-31635-8_89

particular, the development of new hand functional rehabilitation approaches is of great interest.

Literature suggests that the key elements for a more effective neurorehabilitation path are intensive, repetitive, and "task oriented" training and patient's active participation/involvement [3]. The use of a robotic arm can fulfill the requirement of intensive and repetitive exercise by facilitating the execution of precise and repeatable movements. Furthermore, it gives the possibility of providing higher therapy doses [4].

Patient's involvement during the rehabilitation exercise seems to induce the activation of the correct brain areas, promoting brain plasticity and motor recovery [5]. In this regard, electroencephalography (EEG) can allow to investigate non-invasively brain activity in different motor conditions. Specifically, it can provide evidences about the patient's engagement, the timing, and the extent of motor cortex activation during the execution of different motor tasks.

The objective of the ARTE project is to develop and test an integrated rehabilitation platform composed of a robotic hand, an 8-channel EEG helmet and a wristband for the acquisition of vital parameters, such as HR, IBI and GSR. This system has the main purpose of monitoring the patient's state during the rehabilitation session in order to provide quantitative indexes about the goodness of the therapy and to support therapists in the evaluation of the rehabilitation path and in its adaptation to the patient's needs.

Within this project, Politecnico di Milano, Ospedale Valduce-Villa Beretta, ab medica s.p.a, EMAC srl and MCS srl collaborated for the development of the integrated rehabilitation system.

2 System Architecture

The system is composed of the Amadeo rehabilitation robotic arm, the Helmate EEG acquisition system and the E4 wristband. The software "ARTE session manager" integrates all the modules and archives the acquired data: (i) the EEG signal deriving from the Helmate head set, (ii) fingers position and force signals deriving from Amadeo, (iii) vital parameters (i.e., HR, IBI and GSR) from the wristband.

A scheme of the overall system architecture is showed in Fig. 1.

Helmate EEG Device. EEG signals are acquired through the wireless *Helmate* head set (ab medica) from 8 dry electrodes placed according to the international 10–20 system (Fp1, Fp2, Fz, C4/C6, Cz, C3/C5, O1, O2), with a sampling rate set either to 512 or to 1024 Hz. Data are continuously acquired and transmitted via Bluetooth to the PC.

Amadeo Rehabilitation Device. Amadeo is a powered robotic finger-hand device for the neuro-rehabilitation of the hand, developed and produced by Tyromotion. It is composed of an arm-hand rest and five finger supports (one per finger) connected to five sliders. The Amadeo device can work in three different rehabilitation modalities: (i) passive (i.e., the robot does not contribute to the movement), (ii) assistive (i.e., collaboration user-robot), (iii) active (i.e., hand completely moved by the robot). It provides the position signals of the supports and the force applied during the movement. The communication between PC and robotic hand is implemented through a LAN connection.

Empatica Wristband. The E4 wristband (Empatica; www.empatica.com) is a wearable device that offers physiological data acquisition. It is equipped with a temperature sensor, a 3-axis accelerometer, a PPG sensor and a GSR sensor. For this specific application, only parameters of HR, GSR and IBI are used. The E4 wristband transfers this information to the PC thanks to a Bluetooth communication at the end of the rehabilitation session.

Arte Session Manager. Developed by Camelot Biomedical System s.r.l. The software integrates and saves all data streamed by the different involved devices. Then, at the end of the rehabilitation session, it computes all the indexes of interest thanks to an integrated signals elaboration module developed by Politecnico di Milano.

A user interface that guides the operator through the different rehabilitation sessions is also provided.

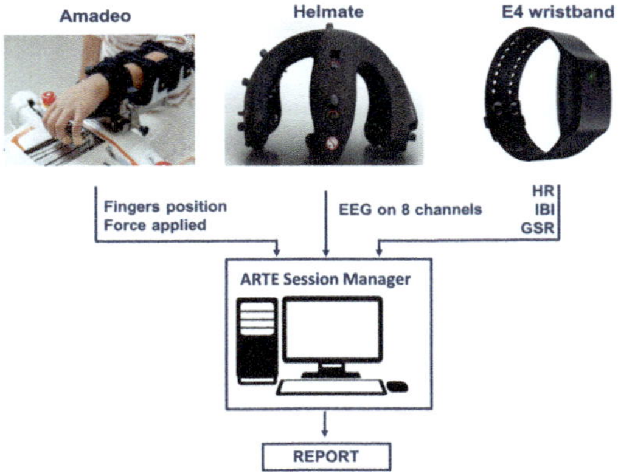

Fig. 1. Integrated system overview

3 Parameters Extraction

Engagement Index. The implemented algorithm for the computation of the EEG-based engagement index was based on the method proposed by Freeman [6]. Since the EEG alpha rhythm (8–13 Hz) can be associated to a relaxed state of the brain (idling state) and the beta rhythm (13–30 Hz) is instead associated to an increased cognitive activity, the engagement index is computed as the power ratio between the beta and the alpha rhythms. Specifically, the proposed engagement index is computed comparing the beta-alpha power ratio measured during the execution of the rehabilitation task to the one measured during a baseline condition (i.e., open eyes EEG at rest) on the same frontal electrodes (Fp1, Fp2 and Fpz) as in Eq. 1.

$$EI(t) = \frac{\beta_{activity}(t)}{\alpha_{activity}(t)} - \frac{mean(\beta_{baseline}(t))}{mean(\alpha_{baseline}(t))} \quad (1)$$

where $\beta_{activity}$ and $\alpha_{activity}$ are the contributes of the three channels summed up. Positive values of the EI will be related to an active participation of the patient, zero and negative values will be related to low level of participation of the patient to the execution of the task.

ERD Index. Different studies demonstrated that immediately before and during the activation of the motor cortex (due to the execution/ imagination of a certain movement) an *Event Related Desynchronization* (ERD) can be observed, i.e., reduction of the power of the mu and beta rhythms [7].

The computed index takes into account the ERD phenomenon, measuring the maximum level of desynchronization of the EEG signal within the considered bands with respect to the baseline. Specifically, the baseline is identified as the first 0.5-s-long segment before movement onset.

The algorithm computes the time course of the ERD following the method proposed by Tacchino, et al [8]. Then, the ERD index is identified as the minimum value assumed by the ERD time course signal.

BSI Index. The BSI index gives a quantitative measure of the level of asymmetry of the two hemispheres.

The algorithm for the computation of this index is based on the method proposed by Van Putten [9] for two bipolar montages:

1. M1: Fp1-C3/C5, C3/C5-O1, Fp1-O1 for the right hemisphere and Fp2-C4/C6, C4/C6-O2, Fp2-O2 for the left hemisphere
2. M2: Fp1-C3, C3-Cz, Fp1-Cz for the right hemisphere and Fp2-C4, C4-Cz, Fp2-Cz for the left hemisphere

The computed index assumes values between 0 and 1: BSI equal to 0 corresponds to perfect symmetry, while a BSI equal to 1 corresponds to a total asymmetry. In general, a healthy subject presents a BSI very close to 0, while in pathological cases the BSI moves towards 1.

4 Pilot Testing Protocol

An open eyes EEG signal at rest was recorded for each subject as baseline signal for the engagement index calculation. Then, six different scenarios were tested (i) ten passive grasping movements (hand passively moved by Amadeo), (ii) ten imagined grasping movements, (iii) ten imagined and passively executed grasping movements, (iv) ten assisted grasping movements, (v) ten completely active grasping movements, (vi) apple picking exercise with no assistance. The integrated system was tested on four healthy volunteers (two males and two females). All subjects gave informed written consent.

5 Results

All pilot volunteers succeeded in the use of the ARTE integrated system. All the modules communicated correctly with the control PC and all the indexes and parameters of interest were computed. At the end of each test, a resuming report was generated for each participant.

The EI assumed for all subjects increasing values passing from the baseline condition to the imagination/execution (passive, active or participatory) of a movement. As an example, Fig. 2(A) and (B) show for S02 and S03 the mean value assumed by EI during three different scenarios. As can be observed, the EI generally increases passing from the execution of passive movements to the execution of tasks that require a higher participation (imagination/ execution of a movement).

Table 1 shows for subject S02 the maximum value of desynchronization saved during passive and assisted exercise. Reasonably, the ERD index assumes negative values, indicating an activation of the motor cortex. It can also be noted that this index generally assumes more negative values during the assisted exercise, when the activation of the motor cortex should be more pronounced.

Finally, the BSI assumed values close to zero for all participants.

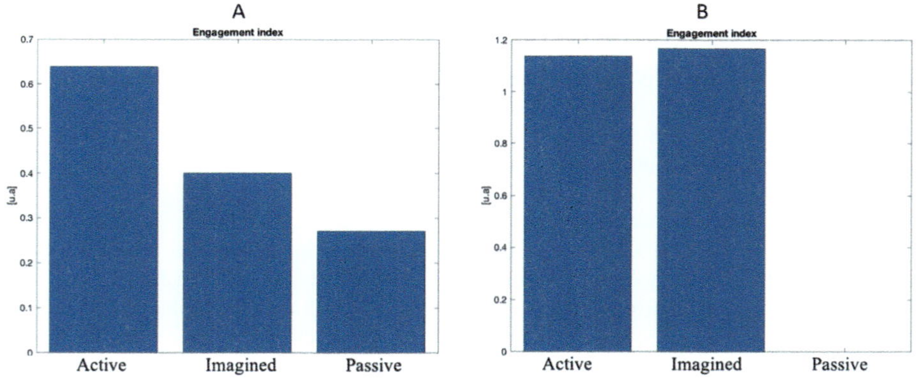

Fig. 2. (A) Mean engagement index of S02 for three tasks; (B) Mean engagement index of S03 for three tasks

Table 1. Values assumed by the ERD/ERS index during the passive and assisted grasping exercise divided in function of channel, frequency and movement type for subject S02

	Passive	Assisted
C3, mu, aperture [%]	−55	−80
C3, mu, closure [%]	−41	−61
C3, beta, aperture [%]	−47	−44
C3, beta, closure [%]	−34	−53
C4, mu, aperture [%]	−52	−38
C4, mu, closure [%]	−27	−69
C4, beta, aperture [%]	−49	−48
C4, beta, closure [%]	−31	−39

6 Discussion and Conclusion

The objective of the ARTE project was to develop and test an integrated system able to monitor patients during the rehabilitation session and to provide quantitative measures about the goodness of the rehabilitation path. This has been done through the use and integration of a robotic rehabilitation end-effector, an EEG helmet and a sensorized wristband. The integrated device was tested on a small group of healthy volunteers and all of them were able to interact with the system. Furthermore, the system was able to extract all the needed parameters, which seemed to assume meaningful values in relation to the tasks executed. These results showed the feasibility of clinical trials, which are already expected within the year.

Acknowledgments. Project "ARTE – Analisi EEG del trattamento robotico" (187230) is supported by Regione Lombardia, Smart Fashion and Design – POR-FESR 2014-2020.

Potential for Conflict of Interest. Potential for conflict of interest: M. Baratto works for EMAC s.r.l.; C. Puttilli works for ab medica s.p.a.

References

1. Daly, J.J., Ruff, R.L.: Construction of efficacious gait and upper limb functional interventions based on brain plasticity evidence and model-based measures for stroke patients. Sci. World J. **7**, 2031–2045 (2007)
2. Lum, P.S., Godfrey, S.B., Brokaw, E.B., Holley, R.J., Nichols, D.: Robotic approaches for rehabilitation of hand function after stroke. Am. J. Phys. Med. Rehabil. **91**, S242–S254 (2012)
3. Luft, A.R., McCombe-Waller, S., Whitall, J., Forrester, L.W., Macko, R., Sorkin, J.D., Schulz, J.B., Hanley, A.P.G.D.F.: Repetitive bilateral arm training and motor cortex activation in chronic stroke: a randomized controlled trial. JAMA **292**, 1853–1861 (2004)
4. Turner, D.L., Ramos-Murguialday, A., Birbaumer, N., Hoffmann, U., Luft, A.: Neurophysiology of robot-mediated training and therapy: a perspective for future use in clinical populations. Front. Neurol. **4**, 184 (2013)
5. Blank, A.A., French, J.A., Pehlivan, A.U., O'Malley, M.K.: Current trends in robot-assisted upper-limb stroke rehabilitation: promoting patient engagement in therapy. Curr. Phys. Med. Rehabil. Rep. **2**, 184–195 (2014)
6. Freeman, F.G., Mikulka, P.J., Prinzel, L.J., Scerbo, M.: Evaluation of an adaptive automation system using three EEG indices with a visual tracking task. Biol. Psychol. **50**, 61–76 (1999)
7. Pfurtscheller, G., da Silva, F.H.L.: Event-related EEG/MEG synchronization and desynchronization: basic principles. Clin. Neurophysiol. **110**, 1842–1857 (1999)
8. Tacchino, G., Gandolla, M., Coelli, S., Barbieri, R., Pedrocchi, A., Bianchi, A.M.: EEG analysis during active and assisted repetitive movements: evidence for differences in neural engagement. IEEE Trans. Neural Syst. Rehabil. Eng. **25**, 761–771 (2017)
9. van Putten, M.J.: Extended BSI for continuous EEG monitoring in carotid endarterectomy. Clin. Neurophysiol. **117**, 2661–2666 (2006)

Bioimpedance, Total Body Water and Phase Angle of Preschool Czech Children: Preliminary Study

Jan Hlubik[1(✉)], Lenka Vyslouzilová[1], Lenka Lhotská[1], Olga Stepankova[1], and Jan Kriz[2]

[1] The Czech Institute of Informatics, Robotics and Cybernetics, Czech Technical University in Prague, Prague, Czech Republic
hlubikjan@post.cz
[2] Faculty of Science, University of Hradec Kralove, Hradec Kralove, Czech Republic

Abstract. This study is a presentation of our ongoing process of obtaining body composition and electrical parameters from sufficient cohort of normal healthy children in Czech Republic that are currently not known. At the moment, 37 children have been fully measured and their parameters evaluated. The measured group consisted of 17 boys and 20 girls and their age was 5 to 6 years. Results are showing that hydration status and electrical parameters of children are approximately the same in the time of leave and in the time of arrival kindergarten. However, changes that are reported through all measurements indicate some specific behavior of bioelectrical values during children day cycle. Bioimpedance seems to act as a possible marker, that can describe in future body composition in another perspective than classical body composition parameters. Further work is needed to obtain sufficient cohort of normal healthy children and to be able to estimate new electrical parameters that may be used as reliable predictors of a health status of the child without using empirical equations or BMI.

Keywords: Bioimpedance · Body composition · Total body water · Phase angle · Children

1 Introduction

For several decades, clinical assessment of children's nutritional and health status has relied mainly on anthropometrical measurements. First sets of children`s data were collected by Quetelet. These data consisted of weight and height and were used for the concept of the "normal distribution" to describe the pattern of human growth [1].

Weight and height are strongly associated in children; various efforts were made to produce charts for weight that took height into account. By the 1970s, opinion was converging on the use of body mass index (BMI) (calculated as weight in kilograms divided by the square of height in meters) as the optimal approach in adults. In children, however, BMI has a characteristic curvilinear shape with age, a scenario for which growth charts provide the ideal solution [2].

J. Henriques et al. (Eds.): MEDICON 2019, IFMBE Proceedings 76, pp. 761–767, 2020.
https://doi.org/10.1007/978-3-030-31635-8_90

Growth charts for BMI formula were firstly introduced for French children in 1982 [3] and for children born in UK in 1995 using the LMS method [4]. These growth charts described the nutritional status of children over time and hence this new standardized approach to an early monitoring of clinical childhood obesity data were possible. Nevertheless, as the approach was replicated in other countries and individual national charts were produced, there was no international consensus on the BMI that is equivalent to obesity at any given age [2]. Other studies were made to address this variations e.g. Cole et al. [5] made analysis in 6 countries and identified age-specific BMI cutoffs that were statistically equivalent to the adult BMI cutoffs of 25 and 30 kg/m^2. International boundaries were made for pediatric overweight and afterwards also boundaries for pediatric underweight followed. Definition of these margins allowed monitoring of nutritional status in children and adolescents worldwide.

1.1 Problems and Limitations of BMI Usage

However, BMI is highly correlated with many different components of weight, such as lean mass (used here synonymously with fat-free mass), skeletal muscle mass, fat mass, and bone mass; yet, it cannot differentiate between them [2].

Nevertheless, a strong correlation of BMI and percentage of body fat (BF) or weight is mainly because of the fact, that low-BMI children to have low fat.% and high-BMI children have high fat.%. In the middle of the range however, children of a given BMI value can have very different %fat. This is shown clearly by disentangling BMI into its fat and lean components; however, before examining this issue it is first helpful to discuss the limitations of using %fat itself as an index of adiposity [2].

This shows clear limitation of BMI to differentiate mainly in groups that would be considered explicit; obese and underweight children. Some studies also used the z-BMI standard deviation (SD) scores, a measure relative to weight and adjusted for child age and sex [6]. Nevertheless, the problem of z-BMI is that this value does not distinguish between fat mass (FM) and fat-free mass (FFM), it is a poor predictor of body fat [7].

Ellis et al. [8] found in their work that there is a large range in BF% for each BMI category and concluded that BMI is not an adequate measure of obesity in children. Forsum et al. [9] found that BMI does not adequately classify children with a high BF % measured by air displacement plethysmography (ADP). Therefore, easy and accurate methods for assessing body composition early in life are needed to identify children with high BF% [10].

1.2 Possibilities of Body Composition Measurements in Small Children

There are many options available for measuring body composition. However, each method has its advantages and disadvantages.

Dual energy X-ray absorptiometry (DXA) is measuring lean body mass, and fat mass from which FM and FFM can be calculated, ADP and isotope dilution is using two component model, where FM and FFM are calculated from measured body volume and total body water (TBW) respectively. All these methods are considered to be quite accurate and stands often as a golden standard. However, they can be very expensive, time consuming, and require subjects to come to a measurement at a specific location and uses ionizing rays.

Bioelectrical impedance analysis (BIA) also uses a 2C (two component) model and can be used to estimate FFM through TBW using the conductive properties of the body's tissues [11]. It is easy-to-use, inexpensive, and very portable. This allows usage in studies where a large number of participants are measured or special conditions have to be made. There are numerous different types of BIA devices.

2 Materials and Methods

Participants of this study were recruited from a population of preschool children located in Central Bohemia district Prague. The children measurement has begun on March 2018 and is still in progress to collect as much of data as possible. All their parents signed written consent to sign for participation in our study.

The weight and height of the children were measured using a digital scale and a wall stadiometer when the children were wearing light clothing and no shoes on. Weight and height were rounded to the nearest 0.5 kg or cm, respectively. Waist circumference (WC) was measured at the midpoint between the last rib and the iliac crest using a tape measure. All these anthropometrical data were taken at 8 a.m. after arrival of a child into kindergarten. Body mass index (BMI) was calculated as the body weight in kilograms divided by the square of height in meters.

All body composition parameters and bioelectrical variables were collected simultaneously during children stay in kindergarten, between 8:00 a.m. and 3:00 p.m. Overall four measurement sessions for every child were done to estimate if a stay in a kindergarten does have influence on children body composition parameters.

First measurement was at 8:00, the second after one-hour break, the third started in two hours later just before lunch and the final measurement was done seven hours after the initial measurement before leaving the kindergarten.

Body composition measurements were taken in a single five-minute individual testing session in supine position by device Bodystat Quadscan 4000 (Bodystat Ltd, British Isles) that was calibrated before each single measuring day. Mainly BF%, total body water (TBW), extra and intra cellular water (ECW, ICW), reactance (5, 50, 100 and 200 kHz), resistance (50 kHz) and phase angle were recorded.

Every measurement was repeated three times to overcome possibility of electrode, sweat and movement biases. All BIA measurements were completed by a trained investigator according to the device manufacturers' instructions. Data were extracted by BIAS software directly from the device.

3 Results

At the moment, 37 children have been fully measured and are evaluable their parameters. The measured group consists of 17 boys and 20 girls and age was 5 to 6 years.

Variable	Mean ± SD	Range
Age	5,40 ± 0,49	5–6
Height	1,17 ± 0,06	1,07–1,32
Weight	22,02 ± 4,62	14–34
BMI (kg/m^2)	16,03 ± 2,19	12,4–22,6
Fat (%)	21,41 ± 6,85	12–19
TBW (%)	60,92 ± 5,26	63–69
Phase angle	4,99 ± 0,49	3,7–6,9

All the parameters were monitored during the whole day to estimate if there are any changes. First two measurements were done in shorter period to test if any influences of kindergarten can be reported within such a short interval.

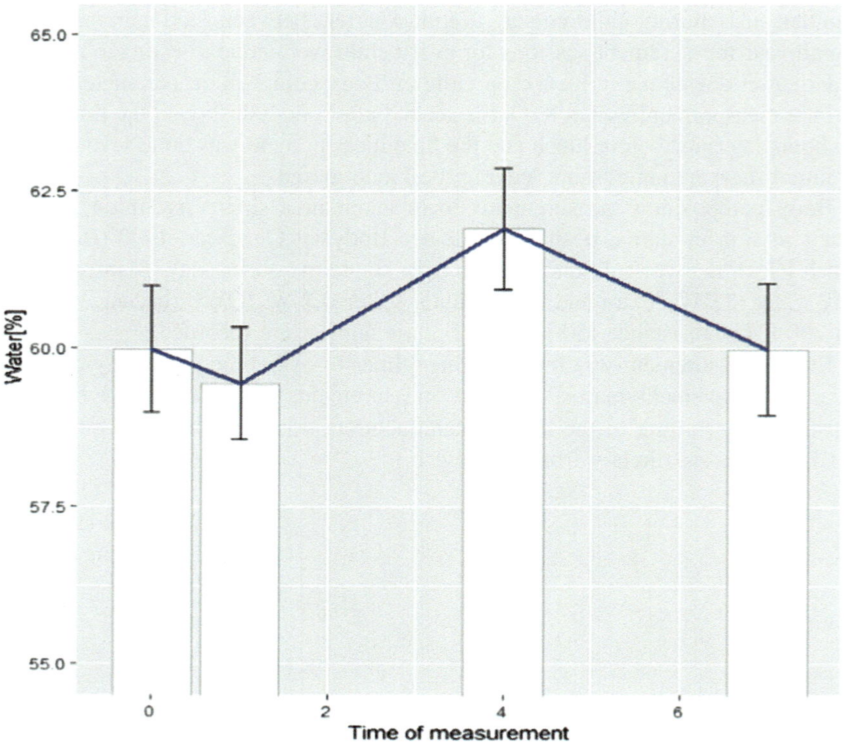

Fig. 1. Changes in TBW% during whole stay (hours) in kindergarten

Figure 1 shows average differences in TWB% during whole day stay. It is visible that after arrival of a child into kindergarten hydration is decreasing. This is probably due to the fact that children are overhydrated from home. First drinks are served to the children after approx. two hours, just after second measurement was taken. Then observed increase in the 3rd measurement is due to the fact, that drinks were served in sufficient amounts. Third measurement is taken just before lunch so we can observe that hydration was highest due to previous intake of drinks and homeostasis that distributed fluids through the body. In the last measurement is seen that hydration is getting back to the normal. This is due to the fact that there is long time gap between third and fourth measurement and also children's rest regime is set after lunch (children are mainly sleeping or doing light exercise or relaxing).

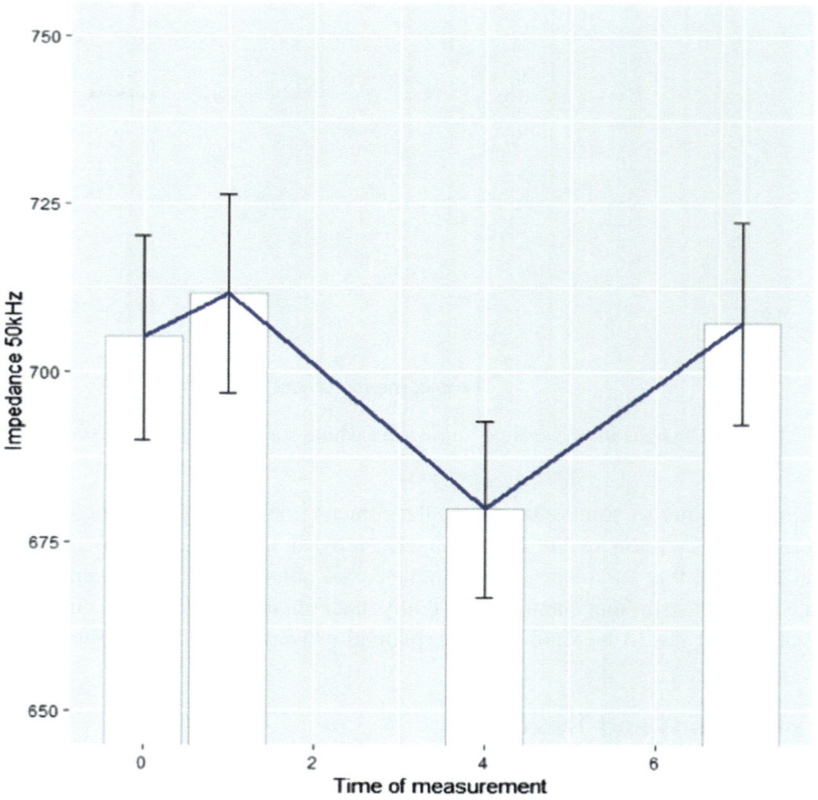

Fig. 2. Changes in Impedance 50 kHz during whole stay (hours) in kindergarten

Figure 2 shows changes in impedance on frequency 50 kHz that are inversely proportional to TBW%. Impedance is changing in dependence on children activity and fluid consumption. These results indicate that impedance measured just before children leaving home is returning back to the value taken just after arrival in the kindergarten.

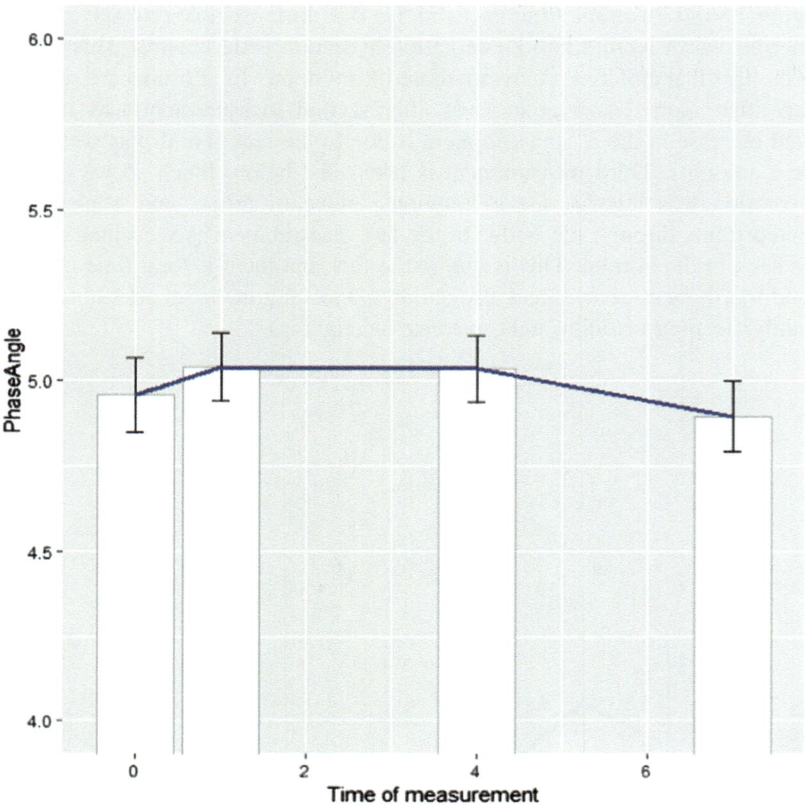

Fig. 3. Changes in phase angle during the whole stay (hours) in kindergarten

Figure 3 points to some changes in the directly measured phase angle during all measurements. Increase in the first hour after arrival into kindergarten may suggest normalization of this parameter. This parameter has constant value between the second and the third measurement and fells slightly later in the day. This is an interesting observation that has to be studied and explained proved in our continuing research.

4 Conclusions and Discussion

Our ongoing research is showing very promising results, that have to be verified using more data. Overall results are showing that hydration status and electrical parameters of children are approximately the same in the time of leave and in the time of arrival. However, changes that are reported indicate some specific behavior of bioelectrical values during children day cycle.

Bioimpedance is showing results as a possible marker, that can describe in future body composition in another perspective than classical body composition parameters that are based on empirical equations for calculation of specific values [2].

Bioimpedace and specially phase angle can be used in future as a predictor of wellbeing of a child. Also this method may be used as one parameter criterion instead of that using three component models [11].

Our work is now in progress and our data collection continues to obtain sufficient cohort of normal healthy children and to be able to estimate new electrical parameters that may be used as reliable predictors of a health status of the child without using empirical equations or BMI.

Acknowledgment. Research was supported by institutional resources of CVUT Prague.

References

1. Bogin, B.: Patterns of Human Growth. Cambridge University Press, Cambridge (1988)
2. Wells, J.C.K.: Toward body composition reference data for infants, children, and adolescents. Adv. Nutr. **5**(3), 320S–329S (2014)
3. Rolland-Cachera, M.F., Sempé, M., Guilloud-Bataille, M., Patois, E., Péquignot-Guggenbuhl, F., Fautrad, V.: Adiposity indices in children. Am. J. Clin. Nutr. **36**, 178–184 (1982)
4. Cole, T.J., Freeman, J.V., Preece, M.A.: Body mass index reference curves for the UK, 1990. Arch. Dis. Child. **73**, 25–29 (1995)
5. Cole, T.J., Bellizzi, M.C., Flegal, K.M., Dietz, W.H.: Establishing a standard definition for child overweight and obesity worldwide: international survey. BMJ **320**, 1240–1243 (2000)
6. Must, A., Anderson, S.E.: Body mass index in children and adolescents: considerations for population-based applications. Int. J. Obes. (Lond.) **30**, 590–594 (2006)
7. Wan, C.S., Ward, L.C., Halim, J., Gow, M.L., Ho, M., Briody, J.N., Leung, K., Cowell, C. T., Garnett, S.P.: Bioelectrical impedance analysis to estimate body composition, and change in adiposity, in overweight and obese adolescents: comparison with dual-energy X-ray absorptiometry. BMC Pediatr. **14** (2014)
8. Ellis, K.J., Abrams, S.A., Wong, W.W.: Monitoring childhood obesity: assessment of the weight/height index. Am. J. Epidemiol. **150**, 939–946 (1999)
9. Forsum, E., Carlsson, E.F., Henriksson, H., Henriksson, P., Lof, M.: Total body fat content versus BMI in 4-year-old healthy Swedish children. J. Obes. **2013** (2013)
10. Delisle Nyström, C., Henriksson, P., Alexandrou, C., Löf, M.: The Tanita SC-240 to assess body composition in pre-school children: an evaluation against the three component model. Nutrients **8**(6), 371 (2016). https://doi.org/10.3390/nu8060371
11. Wells, J.C., Fewtrell, M.S.: Measuring body composition. Arch. Dis. Child. **91**, 612–617 (2006)

TOF-Watch NMB Monitoring Misleading Display Output During Moderate Neuromuscular Blockade

Mafalda Couto[1,2(✉)], Catarina S. Nunes[2,3,4], Pedro Amorim[3], and Joaquim Mendes[1,2]

[1] Faculdade de Engenharia da Universidade do Porto, Porto, Portugal
mafaldacouto88@gmail.com
[2] INEGI-LAETA, Instituto de Ciência e Inovação em Engenharia Mecânica e Engenharia Industrial, Porto, Portugal
[3] Centro de Investigação Clinica em Anestesiologia, Centro Hospitalar e Universitário do Porto, Porto, Portugal
[4] DCeT, Delegation of Porto, Universidade Aberta, Porto, Portugal

Abstract. In general anesthesia, Neuromuscular Blockade (NMB) agents are administered to ensure intubation quality and complete immobilization. Monitoring NMB, based on the evoked response evaluation after a peripheral nerve stimulation, is essential to provide insight on medication dosing and suitable approaches for NMB reversal. Professionals often rely on the data displayed on the monitor screen after TOF stimuli, assuming measurements present the expected fade and a TOF-ratio > 90% is enough to rule out residual paralysis. This can be inaccurate and mislead the clinician to extubate before adequate NMB curarization. Data from 31 patients that underwent general anesthesia were retrospectively analyzed. All patients received a standard dose of rocuronium (0.6 mg/kg) for intubation, during maintenance additional rocuronium boluses were administered if solicited. NMB monitoring was done continuously applying TOF stimuli at the adductor pollicis with TOF-Watch SX®. Two types of monitoring display errors were studied: (1) valid TOF value without fade effect (invalid T1 > T2 > T3 > T4); (2) in the 30 min before extubation a valid TOF-ratio value > 90% (without error type 1) with T1 < 70%. Results show a mean (SD) of type 1 error on 42.0 (17.5)% of valid TOF measurements, type 2 error in 63 (45)%. Only 9 patients (29%) presented a TOF with no errors before extubation. These results may increase the distrust in the NMB monitoring devices, however its use should not be rejected when NMB agents are administered. Careful evaluation of the NMB is recommended and additional efforts should be placed in the accuracy of monitors data display to avoid errors.

Keywords: TOF-Watch · Train-of-four · NMB monitoring

J. Henriques et al. (Eds.): MEDICON 2019, IFMBE Proceedings 76, pp. 768–775, 2020.
https://doi.org/10.1007/978-3-030-31635-8_91

1 Introduction

During general anesthesia neuromuscular blocking agents are administered to allow tracheal intubation and provide muscle relaxation for adequate surgical conditions and immobilization. Monitoring of neuromuscular blockade (NMB) allows to objectively assess muscle relaxation. Usually, to titrate the administration of NMB drugs and adequate recovery of the neuromuscular transmission, the evaluation of the evoked response to a peripheral nerve stimulation is employed. [1]. Train-of-four (TOF) stimulation, which consists of 2 Hz four square pulses, is the standard method used, namely when non-deep NMB is applied, which is the most frequent scenario in clinical practice [2]. This comprises a reference TOF value obtained before NMB drug administration (the responses are individually detected with the same intensity), which followed by the muscle relaxant dose for tracheal intubation, quickly reduces and abolishes all 4 responses. During this phase, the responses to TOF stimuli present fade, which is characterized by the progressive magnitude weakening of the subsequent twitches (T1 > T2 > T3 > T4), so that each response is smaller than the anterior [3]. When the dose effect is overcome the TOF responses recover one by one until all 4 responses are reestablished. Also, the proportion that results from dividing the amplitude of the fourth by the amplitude of the first response, is defined as the TOF-ratio (T4/T1) [4]. The number of TOF responses and TOF-ratio (when four responses are present) are what NMB monitors normally display and what anesthesiologists consider for managing the NMB administration and reversal. At reversal, a TOF-ratio above 90% is required for safe extubation and avoidance of residual blockade [5, 6].

Accelerometry, is the most widely used method in NMB monitoring. Normally, it assesses movement of the thumb produced by the response of the adductor pollicis muscle to ulnar nerve stimulation [5–7]. TOF-Watch, which uses accelerometry, is the only NMB monitor approved for clinical trials and one of the most frequently used during surgical procedures with muscle relaxants [8]. Our observations when using this monitor suggested that measurement errors could be present, a possibility that has been addressed in two studies [2, 9]. Also, it has been reported that an error in TOF interpretation has led to inappropriate conduct that resulted in prolonged ventilation/sedation and delay of curarization (FDA Report Number 3002807818-2012-00001). This led us to conduct a study to evaluate the occurrence of errors when using the TOF-watch monitor in a well-controlled clinical scenario of non-deep NMB for spinal surgeries. In this study two types of errors were defined and their incidence assessed.

2 Materials and Methods

Data from 31 adult patients with ASA score of I to III, undergoing general anesthesia with propofol and remifentanil via target controlled infusion, were retrospectively analyzed. These were patients included in the standard group of a clinical trial (NCT02484651), that randomized patients to deep or non-deep. IRB (institutional review board) approval and informed consent were obtained from the patients enrolled in the study. After loss of consciousness, all patients received a standard dose of rocuronium (0.6 mg/kg), aiming for standard NMB, meaning that during maintenance of anesthesia additional bolus of rocuronium were administered only if required by the surgeon to improve surgical conditions. At the end of surgery, if TOF-ratio was below 90%, NMB was reversed with neostigmine (0.04 mg/kg) given with atropine (0.02 mg/kg).

NMB was monitored at the adductor pollicis using TOF-Watch® SX (Organon, Dublin, Ireland). After degreasing the skin, two electrodes were placed over the ulnar nerve near at the wrist and responses were measured by an accelerometer sensor attached to the tip of the thumb. Before rocuronium, calibration was carried out according to manufacture specifications [2]. After the initial dose of rocuronium tracheal intubation ensued. The assessment of TOF response was performed every 15 s throughout the case and acquired through TOF-link ® software.

The main aim of this investigation study was to characterize and evaluate the incidence of two types of monitoring display errors. Error type 1 was defined by a TOF response in which the fade effect on the four twitches is absent, meaning that T1 > T2 > T3 > T4 was not present, this was named "invalid" response. Error type 2 was studied only for "valid" TOF responses during the last 30 min before extubation for each case. Type 2 error was defined as the occurrence of a TOF-ratio > 90% with a low T1 value. It consists of a valid TOF in which the result is a TOF-ratio value over 90% displayed in the device present T1 inferior to 70%. This was established arbitrarily by the authors as a suitable threshold to assume the recovery of NMB is high enough, considering that adequate reversal might consist not only of a TOF ratio above 90%, but also of a significate T1 value, therefore indicating a superior recovery of the neuromuscular function and increase the safety to perform the tracheal extubation. Therefore, we hypothesized that adequate reversal would consist of a TOF-ratio above 90%, with a T1 above 70% of the initial reference, with the presence of fade between T1 and T4. For each patient, the number of TOF stimulations and number of TOF measurements where 4 responses were present, were assessed.

After the error analysis, an assessment of the cases was conducted in order to evaluate the amount of patients in which TOF monitor provided a correct measurement that fulfill the above-mentioned criteria for a good extubation. In other words, the analysis comprised the assessment of cases in which TOF measured a TOF-ratio over 90% and T1 > 70%, in the 30 min before tracheal extubation. Data is reported as mean (SD).

3 Results

Data analysis showed a duration of the procedures of 129.3 (43.8) min, allowing to perform 361.2 (182.7) TOF stimulations, in which 276.2 (173.9) were positive TOF responses (TOF > 0), valid for error analysis. Thus, error analysis showed 42.0 (17.5)% of the TOF measurements had type 1 errors and 63 (45)% of type 2 errors. In the 30 min before extubation, it was possible to detect TOF-ratio > 90% in all cases and every patient was extubated according to GCP guidelines with no reported complications. Nevertheless, the raw data of these measurements of TOF-ratio > 90% were analyzed, in which 39% of the patients showed a persistent type (1) error, 32% had a type (2) error and only 9 patients presented no errors and, therefore it was possible to verify the achievement of adequate extubation results. In Figs. 1, 2 and 3, examples of errors detection results are illustrated.

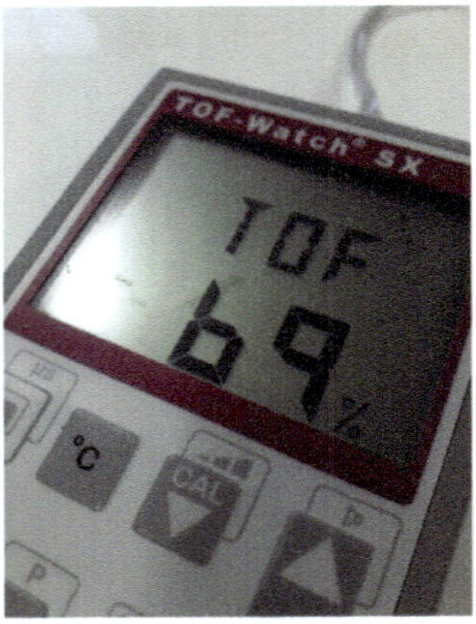

Mode	Tw1	Tw2	Tw3	Tw4	TOF
	%	%	%	%	%
TOF	46	39	25	32	69

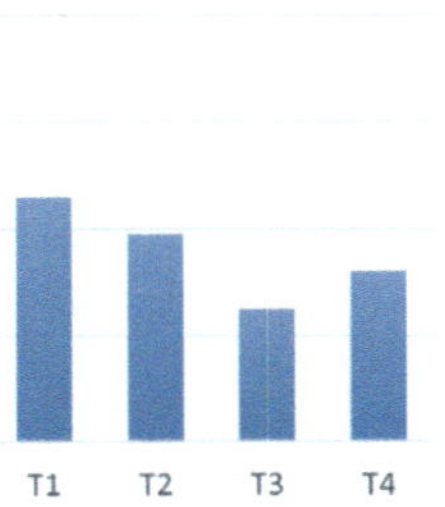

Fig. 1. Illustration of error type 1 example. TOF-Watch SX® displays TOF-ratio of 69% while values of T1 to T4, recorded via TOF-Link, show that fade was not present, since T4 was higher than T3.

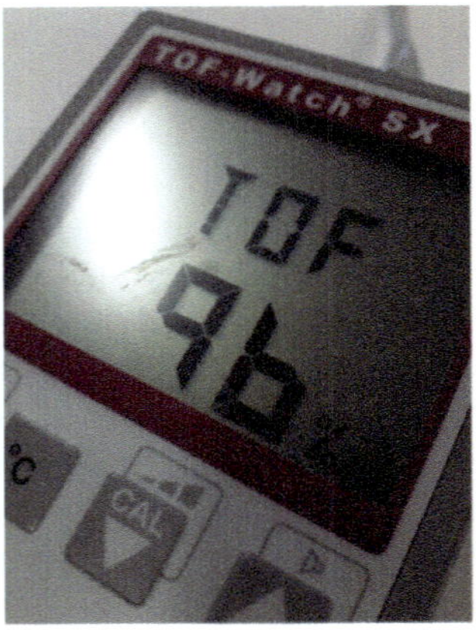

Mode	Tw1	Tw2	Tw3	Tw4	TOF
	%	%	%	%	%
TOF	54	53	52	51	96

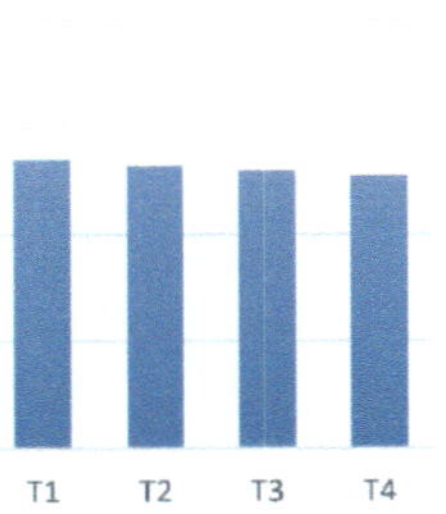

Fig. 2. Illustration of error type 2 example. TOF-Watch SX® display TOF-ratio of 96% while the real value of T1, recorded by the TOF-Link, was below 70% despite the presence of a valid TOF response with fade effect.

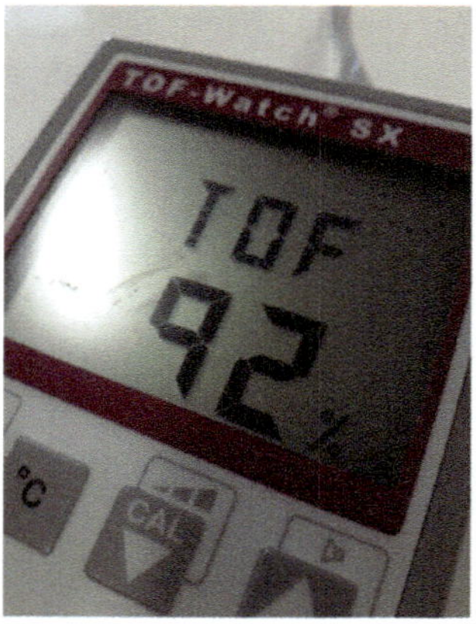

Mode	Tw1	Tw2	Tw3	Tw4	TOF
	%	%	%	%	%
TOF	80	79	76	73	92

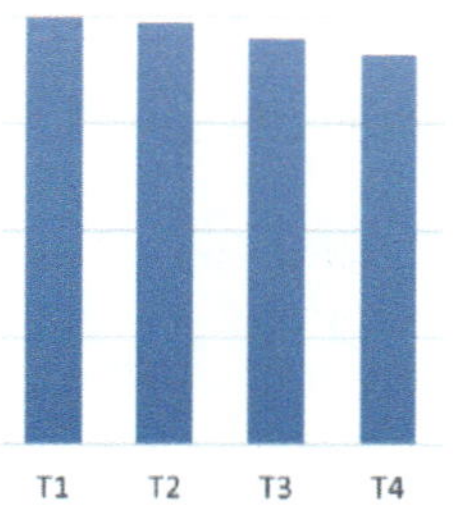

Fig. 3. Illustration of TOF-Watch SX® displaying 92% TOF-ratio in which raw values, recorded by the TOF-Link, prove the presence of fade effect and T1 over 70%.

4 Discussion

Often clinicians rely on the monitors' measurements outcome without ensuring its validity. The information displayed in the screen of the monitoring device, which is limited to the TOF count or TOF-ratio outputs, can often be inaccurate or incomplete and, thereby misguide the anesthesiologist during the maintenance and recovery of the NMB. Professional are able to assume TOF measurements present fade and a TOF-ratio > 90% is enough to rule out residual paralysis. However, this is not always the case. First, non-faded TOF responses are displayed as valid monitoring result without specifying this outcome, and secondly, the fourth (T4) and first (T1) responses can have similar however low amplitudes, indicating TOF-ratio > 90%. This can induce the clinician to admit an adequate recovery and ability to extubate, before achieving higher magnitudes (closer to pre-surgical calibration reference), which may lead to inappropriate curarization of the NMB and residual paralysis complications.

Frequently studies are conducted to investigate sources of errors, being it human or machine. Using TOF-link software is possible to obtain the raw data and verify the data being provided to the monitor and match it with the data displayed. Furthermore, the displayed data can then be matched with the raw data and verified. The aim was to see if the display can instigate clinicians to draw inaccurate conclusions. Error analysis showed type 1 error on 42.0 (17.5)% of TOF measurements, showing that the output of the monitor display can frequently misguide the clinician along the procedure. This may lead to unsuitable approaches on medication dosing. It should be pointed out that, although the several guidelines addressed by international anesthesiology associations for the need for perioperative neuromuscular monitoring [10–12], one of the main reasons why many anesthesiologists do not routinely use an objective method to monitor neuromuscular function is the absence of an ideal and reliable monitoring device [4, 6, 13]. This error analysis results provide some support for these claims.

From the TOF measurements performed in the last 30 min before extubation, that did not include error type 1, it was possible to assess error type 2. These showed an incidence of 63 (45)% of valid TOF measurements with TOF-ratio value over 90% displayed in the device, nonetheless exhibiting T1 inferior to 70%.

All patients were safely extubated according to GCP guidelines with a TOF-ratio > 90% and no post-surgical complications reported. Nonetheless, this study aimed to evaluate the accuracy of TOF measurement, establishing and investigating a fair approach to define recovery criteria, for improved extubation conditions. Thus, the amount of patients that detected a TOF-ratio over 90% and T1 > 70% in the 30 min just before performing extubation were evaluated. From this analysis, it was possible to detect that from the TOF-ratio responses over 90%, in 39% of the patients the TOF measurements had type 1 error at all times (with invalid fade effect), 32% of patients had a type 2 error not ensuring the recovery of T1 > 70% and only 9 cases presented a valid TOF measurement with no errors. This corresponds to only 29% of the patients that presented an accurate TOF response before being extubated, by measuring the expected fade and TOF-ratio > 90% with significate magnitude of T1. The usage of NMB agents during surgeries occurs every day worldwide and the incidence of incomplete recovery of neuromuscular function remains significate (45%). This generally contributes to adverse respiratory events that are associated with increase in morbidity and mortality [14, 15]. Although, these error analysis show significate drawbacks in terms of the outcome generated, which may increase the unthrusting feeling by the professionals, the use of NMB monitoring during the procedures should not be rejected. Careful assessment of the NMB function must be done along the procedures, being essential to evaluate the neuromuscular relaxation after administration of NMB agents.

This study addresses the inaccuracy or defective displaying of data in TOF-Watch SX® monitor device acquired after TOF stimulations. However, it should be emphasized that the error analysis is not regarding the monitor itself or the stimuli response values, but only regarding the outcome of data transmitted to the user via screen display, namely the lack of additional information regarding the amplitude of the twitch responses. Although, this study only proved TOF-Watch SX® errors, other monitors have shown similar approaches with non-robust outcomes. Further studies should be conducted to assess the incidence of faulty decisions generated due to the inaccurate/partial data displayed.

5 Conclusion

The information provided in the monitoring device display after TOF stimuli is often inaccurate and misleading. Results showed that the data displayed in the TOF-Watch SX® screen device is often limited to TOF count and TOF-ratio regardless of fade effect or other specification concerning T1–T4 amplitude values. Moreover, clinicians rely on TOF-ratio > 90% data display to ensure conditions for extubation regardless of the significate recovery of twitch magnitude. This study proved there is a high percentage of errors and misleading TOF-ratios: non-faded TOF responses are displayed as valid monitoring results in 42.0 (17.5)% of the TOF stimuli applied; and T4 and T1 have similar but low amplitudes (<70%) and indicate TOF-ratio > 90%, which is interpretable as good extubation conditions, in 63 (45)% of these TOF measurements (in the 30 min before extubation).

The monitor should provide evidence when no fade effect is detected and/or concerning twitch amplitude information to support clinician decision making regarding dosing or residual NMB. Careful evaluation of the NMB is recommended and additional efforts should be placed in the development of improved data transmission by monitoring devices, to avoid errors that may misguide professionals conduct.

Acknowledgements. This is a retrospective analysis of Clinical trial NCT02484651 data results, in which written and informed consent was obtained from all patients included in the study. Authors acknowledge the funding of projects LAETA and FCT-MIT Portugal PhD program (PD/BD/114378/2016)

Conflict of Interest. The authors have no conflict of interest to declare.

References

1. Butterworth, J., Mackey, D., Wasnick, J.: Morgan & Mikhail's Clinical Anesthesiology, 5th edn. McGraw Hill, New York (2013)
2. Fuchs-Buder, T., Claudius, C., Skovgaard, L.T., Eriksson, L.I., Mirakhur, R.K., Viby-Mogensen, J.: Good clinical research practice in pharmacodynamic studies of neuromuscular blocking agents II: the Stockholm revision. Acta Anaesthesiol. Scand. **51**(7), 789–808 (2007)
3. Fabregat López, J., Candia Arana, C.A., Castillo Monzón, C.G.: Neuromuscular monitoring and its importance in neuromuscular blockade. Colomb. J. Anesthesiol. **40**(4), 293–303 (2012)
4. Fuchs-Buder, T.: Neuromuscular Monitoring in Clinical Practice and Research. Springer, Heidelberg (2010)
5. Miller, R.D., Cohen, N.H., Eriksson, L.I., Fleisher, L.A., WienerKronish, J.P., Young, W.L.: Miller's Anesthesia, 8th edn. Elsevier Saunders, USA (2015)
6. Hemmerling, T.M., Le, N.: Brief review: Neuromuscular monitoring: an update for the clinician. Can. J. Anesth. Can. d'anesthésie **54**(1), 58–72 (2007)
7. Fagerlund, M.J., Eriksson, L.I.: Current concepts in neuromuscular transmission. BJA Br. J. Anaesth. **103**(1), 108–114 (2009)
8. Eriksson, L.I., Lennmarken, C., Jensen, E., Viby-Mogensen, J.: Twitch tension and train-of-four ratio during prolonged neuromuscular monitoring at different peripheral temperatures. Acta Anaesthesiol. Scand. **35**(3), 247–252 (1991)
9. Han, J.U.: Train-of-Four monitoring: overestimation. Korean J. Anesthesiol. **60**(5), 311–312 (2011)
10. Brull, S.J., Kopman, A.F.: Current status of neuromuscular reversal and monitoring. Anesthesiology **126**(1), 173–190 (2017)
11. Murray, M.J., Cowen, J., DeBlock, H., Erstad, B., Gray, A.W., Tescher, A.N., McGee, W. T., Prielipp, R.C., Susla, G., Jacobi, J., Nasraway, S.A., Lumb, P.D.: Clinical practice guidelines for sustained neuromuscular blockade in the adult critically ill patient. Crit. Care Med. **30**(1), 142–156 (2002)

12. Murray, M.J., Deblock, H., Erstad, B., Gray, A., Jacobi, J., Jordan, C., McGee, W., McManus, C., Meade, M., Nix, S., Patterson, A., Sands, M.K., Pino, R., Tescher, A., Arbour, R., Rochwerg, B., Murray, C.F., Mehta, S.: Clinical practice guidelines for sustained neuromuscular blockade in the adult critically ill patient. Crit. Care Med. **44**(11), 2079–2103 (2016)
13. Grayling, M., Sweeney, B.P.: Recovery from neuromuscular blockade: a survey of practice. Anaesthesia **62**(8), 806–809 (2007)
14. Esteves, S., Martins, M., Barros, F., Barros, F., Canas, M., Vitor, P., Seabra, M., Castro, M. M., Bastardo, I.: Incidence of postoperative residual neuromuscular blockade in the postanaesthesia care unit: an observational multicentre study in Portugal. Eur. J. Anaesthesiol. **30**(5), 243–249 (2013)
15. Boos, M.C.: In the surgical patient requiring neuromuscular blockade, is there an increased incidence of postoperative adverse respiratory events with Rocuronium or Vecuronium? Doctoral Projects **8** (2015)

Device for Measuring Protection in Sunglasses Against Harmful Blue Light

Artur D. Loureiro(✉) and Liliane Ventura

Ophthalmic Instrumentation Laboratory (LIO), Department of Electrical and Computer Engineering, Sao Carlos School of Engineering (EESC), University of Sao Paulo (USP), São Carlos, SP, Brazil
{artur.loureiro,lilianeventura}@usp.br

Abstract. There is a strong association between high energy blue light exposure and development of some ocular diseases such as age-related macular degenera-tion. In recent years there has been increased people's concern about the damage caused by blue light and ways to prevent them. This study aims to develop and build a device for measuring protection against harmful blue light to human eyes in sunglasses and to determine whether this protection is adequate, adopting as criterion its blue light transmittance be less than 1.2 times its luminous transmittance. To obtain the proper spectral weighting functions, the measurement will consist of linear combinations of the responses of a TCS3472 photosensor illuminated by a high brightness white LED. The agreement between the measurements made with the proposed device and a gold standard will be analyzed by the Bland-Altman method. Sunglasses blue light protection values will be measured to analyze whether sunglasses in brazilian market are protected against harmful blue light. The developed device will be available to public to measure their own sunglasses.

Keywords: Transmittance measurement · Blue light · Sunglasses

1 Introduction

1.1 Blue Light and Sunglasses Standards

In recent years, there has been an increase in people's concern about the harmfulness of exposure to blue light and ways to prevent it. There is a strong association between exposure to high energy blue light (from 400 nm to 500 nm) and damage to ocular structures such as cornea [1, 2], conjunctiva [1] and retina [3–5]. And the use of blue filters can prevent the frequency of eye damage [5]. The blue light, in addition to the other visible wavelengths, has little contribution to visual acuity, and it is focused before the retina because of its short wavelength, which can cause ocular fatigue. Sunglasses that block harmful blue light are usually disapproved in the suitability tests for international standards, since a minimum of blue light is required for the recognition of traffic lights [6]. An old British standard (BS2724), to ensure protection against blue light, defines as the maximum permissible limit for transmittance of harmful blue light 1.2 times the luminous transmittance. Barker [7] and Dain [6] agree that there should be

J. Henriques et al. (Eds.): MEDICON 2019, IFMBE Proceedings 76, pp. 776–779, 2020.
https://doi.org/10.1007/978-3-030-31635-8_92

an upper limit for the transmittance of blue light for eye protection against high levels of blue light in relation to total light transmittance. This study aims to develop and build a device for measuring in sunglasses luminous transmittance, harmful-to-human-eye blue light transmittance and whether the protection against blue light is adequate (blue light transmittance less than 1.2 times the luminous transmittance). The definitions of luminous transmittance and blue light transmittance adopted are the ones used in brazilian standard for sun protection glasses for general use, ABNT NBR ISO 12312-1 [8].

2 Methods

2.1 Prototype Development

To obtain the appropriate weights for the measurement of luminous and blue light transmittances, linear combinations were made with the four responses of a TCS3472 photosensor illuminated by a white high brightness light emitting diode (LED). Measurements of these transmittances will be made with the device that will report whether the measured sunglasses have blue light blocking filters using as a criterion the lenses have blue light transmittance less than 1.2 times their luminous transmittance. To ensure that the device will remain calibrated, a baseline will be measured prior to the measurement of each sample. The agreement between the measurements made with the proposed device and a gold standard (Cary 5000 spectrophotometer) will be analyzed by the Bland-Altman method. In addition, sunglasses will be tested to check if the sunglasses available in brazilian market have protection against harmful blue light.

3 Results

The spectral weighting functions obtained from the four channels of the TCS3472 photosensor illuminated by a white LED are shown in Fig. 1. These functions were combined linearly to obtain functions that approximate the weighting functions for luminous and harmful blue light transmittances, shown in the graphs of Fig. 2.

4 Discussion and Conclusion

The use of linear combinations of responses from photosensors for approximations of weighting functions allows us to obtain new fixed and known weighting functions with freedom sufficient to approximate weighting functions different from those resulting from the combination of the light source and the photosensor channels. The device, once developed, will be available to the public to measure their own sunglasses. In addition to being able to be used by the general public to access information about their sunglasses, this device has potential for use in the trade to add value in sales of sunglasses and in the industry for quick conference of sunglasses produced.

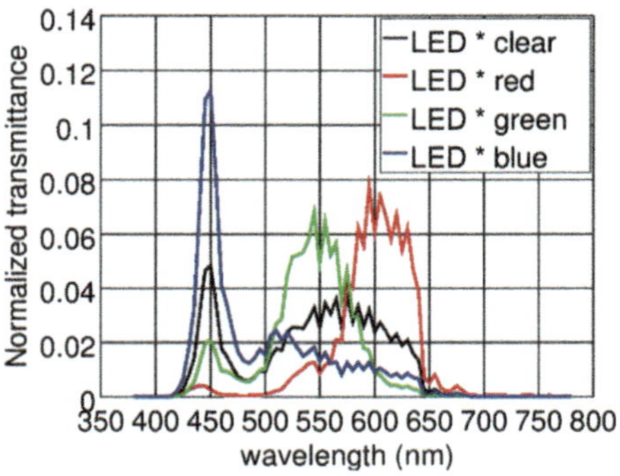

Fig. 1. Spectral weighting functions obtained by combining a high brightness white LED with a TCS3472 photosensor.

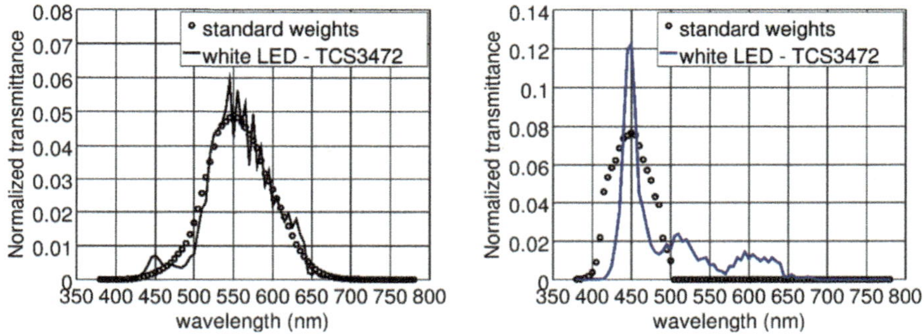

Fig. 2. Weighting functions obtained by linear combinations of the functions of Fig. 1 (continuous line) superimposed on functions defined in standard for luminous (left) and blue light (right) transmittances, respectively (in empty circles).

Acknowledgments. The authors declare that there is no conflict of interest. This study was financed in part by the Coordenação de Aperfeiçoamento de Pessoal de Nível Superior - Brasil (CAPES) - Finance Code 001 and by Fundação de Amparo à Pesquisa do Estado de São Paulo (FAPESP) 2018/16275-2.

References

1. Marek, V., Mélik-Parsadaniantz, S., Villette, T., Montoya, F., Baudouin, C., Brignole-Baudouin, F., Denoyer, A.: Blue light phototoxicity toward human corneal and conjunctival epithelial cells in basal and hyperosmolar conditions. Free Radic. Biol. Med. **126**, 27–40 (2018)
2. Niwano, Y., Kanno, T., Iwasawa, A., Ayaki, M., Tsubota, K.: Blue light injures corneal epithelial cells in the mitotic phase in vitro. Br. J. Ophthalmol. **98**(7), 990–992 (2014)

3. Narimatsu, T., Negishi, K., Miyake, S., Hirasawa, M., Osada, H., Kurihara, T., et al.: Blue light-induced inflammatory marker expression in the retinal pigment epithelium-choroid of mice and the protective effect of a yellow intraocular lens material in vivo. Exp. Eye Res. **132**, 48–51 (2015)
4. Nakamura, M., Yako, T., Kuse, Y., Inoue, Y., Nishinaka, A., Nakamura, S., et al.: Exposure to excessive blue LED light damages retinal pigment epithelium and photoreceptors of pigmented mice. Exp. Eye Res. **177**, 1–11 (2018)
5. Ueda, T., Nakanishi-Ueda, T., Yasuhara, H., Koide, R., Dawson, W.W.: Eye damage control by reduced blue illumination. Exp. Eye Res. **89**(6), 863–868 (2009)
6. Dain, S.J.: Sunglasses and sunglass standards. Clin. Exp. Optom. **86**(2), 77–90 (2003)
7. Barker, F.M.: Does the ANSI 280.3 nonprescription sunglass and fashion eyewear standard go far enough? Optom. Vis. Sci. **67**, 431–434 (1990)
8. Associação Brasileira de Normas Técnicas. ABNT NBR ISO 12312-1:2015 - Proteção dos olhos e do rosto - Óculos para proteção solar e óculos relacionados Parte 1: Óculos para proteção solar para uso geral (2015)

Backscattered Ultrasound Periodicity Characterization on Trabecular Bone-Mimicking Phantoms: A Spectral and Wavelets Approach

Christiano Bittencourt Machado[1(✉)], Mahmoud Meziri[2], Wagner Coelho de Albuquerque Pereira[4], and Guillermo Cortela[3]

[1] Laboratory of Physics in Rehabilitation, Estácio de Sá University, Rua Eduardo Luiz Gomes 134, Niterói, RJ 24020-340, Brazil
cbmfisio@gmail.com

[2] Laboratoire de Magnétisme et de Spectroscopie des Solides (LM2S), Université Badji Mokhtar, 23000 Annaba, Algeria

[3] Instituto de Fisica, Facultad de Ciencias, Universidad de La Republica, Iguá 4225 y Mataojo, 11400 Montevideo, Uruguay

[4] Biomedical Engineering Program, COPPE, UFRJ, Av. Horácio Macedo, 2030-101 – Block H, Cidade Universitária, Rio de Janeiro, RJ 21941-450, Brazil

Abstract. In an attempt to provide better diagnostic techniques using quantitative ultrasound to assess bone's microarchitecture (e.g. osteoporosis), this work aimed at characterizing the periodicity of trabecular bone-mimicking phantoms using backscattered ultrasound signals, by means of spectral methods and the continuous wavelet transform (CWT). Two transducers (3.5 and 7.5 MHz) were used on a pulse-echo configuration. Two bone-mimicking phantoms were studied, with different stiffness but similar mean scatterer spacing (MSS) distributions (varying from 1.5 to 2.5 mm). Transducers were translated parallel to the phantoms to collect signals for processing with the spectral autocorrelation (SAC), singular spectrum analysis (SSA), quadratic transformation (Simon's method) and continuous wavelet transform for MSS estimation. Some methods were able to estimate MSS values inside the range informed by the factory. Differences in trabecular thickness led to differences in MSS estimates, mainly for 7.5 MHz. There is a possibility that spectral methods find more than one periodicity pattern.

Keywords: Ultrasound · Mean scatterer spacing · Trabecular bone

1 Introduction

During ultrasonographic examination, clinicians may mislead initial stages of tissue pathologies because of an inherent high level of subjectivity. During the past decades, quantitative ultrasound (QUS) techniques have been explored as an interesting

J. Henriques et al. (Eds.): MEDICON 2019, IFMBE Proceedings 76, pp. 780–786, 2020.
https://doi.org/10.1007/978-3-030-31635-8_93

approach to overcome this limitation and supports clinical decision following medical ultrasound (US) imaging [1].

One of the QUS parameters is the mean scatterer spacing (MSS). Briefly MSS can be estimated by finding the spatial periodicity-related frequency (f) in the spectrum of backscattered US signals. Spectral analysis methods have been proposed to find f, and then use the relation MSS = $c/2f$ (c is the US longitudinal velocity of the propagating medium) [2]. Some findings have showed that MSS alone cannot discriminate healthy and pathological tissues, bringing to attention the possibility of multiparametric approaches [3]. The application of Wavelet transforms has been also applied for tissue characterization with backscattered signals [4, 5] and it can be considered as a potential tool for periodicity estimation in quasi-periodic media.

Osteoporosis and other bone diseases disrupt bone microarchitecture. Imaging techniques (high-resolution computed tomography, magnetic resonance, dual-energy X-ray absorptiometry) can be used nowadays to assess bone microstructure features. Pereira et al. [6] were able to estimate MSS values which correlated significantly to MSS measured independently from synchrotron microtomography.

In view of the above, the present work aimed at characterizing the periodicity of trabecular bone-mimicking phantoms using backscattered ultrasound radiofrequency (RF) signals, by means of spectral methods and the continuous wavelet transform (CWT).

2 Materials and Methods

The following section is divided in three topics for a better understanding: (1) a background on signal processing techniques used for MSS estimates; (2) an explanation on the experimental setup with bone phantoms; and (3) statistical analysis.

2.1 Background on Signal Processing Techniques

For MSS estimates, the following techniques were used. Firstly, the Spectral Autocorrelation (SAC). It takes the RF signal and generates an autocorrelation matrix in frequency domain. From the off-diagonal components of SAC matrix, the periodicity-related frequency f is detected, where there is no influence of diffuse scatterers. For more details, please read [2].

The Singular Spectrum Analysis (SSA) is based on the Principal Component Analysis (PCA). The N-point envelope signal is expanded with respect to a new orthonormal basis of M vectors, the eigenvectors (E_k) ($1 \leq k \leq M$). These vectors are obtained by the diagonalisation of the correlation matrix (C_x) of the N-point signal. The C_x matrix ($N \times N - M + 1$) is constructed by moving an M-point window ($M < N$), one point at a time. Pairs of eigenvalues are then selected to construct the periodic part of the signal using an heuristic criterion. The frequency f is identified from the Fourier transform (FFT) of the reconstructed signal. More details in [2].

The Simon's method (SIM) is a quadratic transformation in time-domain to explore the spectral redundancy of periodic signals. The envelope of the RF signal is squared, and an FFT of the result is computed. A window in the frequency domain is used (where f is expected to appear) to search for this frequency [2].

Finally, the continuous wavelet transform (CWT), which is a collection of multiresolution time-frequency representations, is obtained by a fully scalable modulated window shifted along the signal. A correlation of the signal $x(t)$ with the mother-wavelet $\psi_{u,s}(t)$ is computed at the scale s and position u ($Wf(u,s)$) [7]. The mother-wavelet Daubechies (db) 1 was applied here. The CWT of the signal $x(t)$ is obtained on an appropriate scale J, chosen after a visual assessment of the scale vs. time representation, $W_J x(t)$. Finally the coherent part of the RF signal can be visualized with a modulus maxima (MM) sequence with peaks representing scatterer positions. The average value of the distances between each two peaks is considered as the MSS.

2.2 Experimental Setup

Two trabecular bone-mimicking phantoms (Fig. 1) with different stiffness were used (Open Cell Rigid Foams, Sawbones®, USA): P1 (compressive strength = 0.28 MPa; compressive modulus = 18.6 MPa) and P2 (compressive strength = 0.67 MPa; compressive modulus = 53.0 MPa.). Concerning trabecular thickness, P1 presents 0.38 mm and P2 presents 0.44 mm [8]. The factory also states that both of them have a cell structure over 95% open and cell size varying from 1.5 to 2.5 mm.

Fig. 1. Open cell rigid foams (Sawbones®, Pacific Research Laboratories, Inc., USA), P1 (left) and P2 (right).

The phantoms were immersed in a water tank at 21 °C during three hours before data acquisition to extract diluted gases. Then, a 3.5 and a 7.5 MHz-ultrasound circular transducers (Olympus®) were put perpendicular to the phantom on a pulse-echo configuration, to send and receive ultrasound signals, which were digitized at 50 Msamples/s (Tektronix TDS 2024B oscilloscope). For each frequency and phantom, a total of 600 RF signals were acquired in different places, using an Arrick Robotics® XY-18 mechanical scanner (steps of 0.2 mm). Each RF-line was cut using a temporal window of 1,048 points from the beginning of first echo and then processed with SAC, SSA, SIM and CWT methods using Matlab® (Mathworks Inc., USA).

2.3 Statistical Analysis

Friedman's test (a non-parametric statistical test similar to the ANOVA for repeated measures) was applied followed by a Tukey-Kramer post-hoc test (to decide which cases are significantly different from each other) with a significance level of 0.05.

3 Results

Figure 2 shows RF signals acquired from the experimental setup, using two frequencies (3.5 and 7.5 MHz) and two phantoms (P1 and P2). Tables 1 and 2 show a descriptive analysis of the results (average, standard deviation, median).

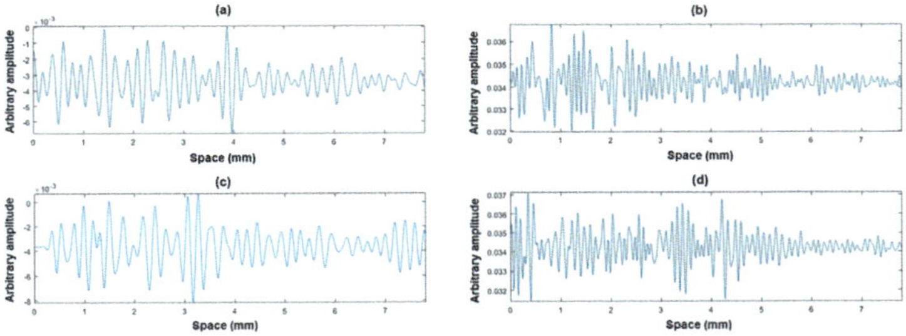

Fig. 2. Examples of RF signals acquired: (a) P1 at 3.5 MHz; (b) P1 at 7.5 MHz; (c) P2 at 3.5 MHz; (d) P2 at 7.5 MHz. The x-axis represents space (mm) considering speed of sound of water = 1,498 m/s (amplitude is in an arbitrary scale).

Friedman's test showed that some MSS distributions were different from each other ($p < 0.001$). Figure 3 shows the results from the post-hoc test for each experimental setup (total of 16 setups). It is possible to see that the lower mean ranks were obtained for CWT. Figure 4 depicts the histogram of MSS estimates for phantom P2 at 3.5 MHz using the four techniques. It can be observed that the results varied according to the applied method. CWT presented lower MSS estimates than spectral methods.

4 Discussion and Conclusions

The use of MSS to characterize tissue microstructure has been investigated for decades. However, only a few works focused on the periodicity characterization of trabecular bone using US. The present study is the first to propose the use of spectral analysis methods, together with wavelet transform, to estimate MSS in bone.

It is possible to observe in Fig. 2 that it is almost impossible to visually detect the periodicity of scatterers inside the bone phantom, even with higher frequencies. After signal processing, Tables 1 and 2 showed that some methods were able to estimate MSS values inside the range between 1.5 and 2.5 mm (as informed by the factory).

Table 1. MSS estimates for phantom P1 (SD – standard deviation).

	3.5 MHz			7.5 MHz		
Method	Average (mm)	SD (mm)	Median (mm)	Average (mm)	SD (mm)	Median (mm)
SAC	1.96	0.77	2.54	2.46	0.28	2.53
SSA	1.99	1.63	1.90	1.58	0.79	1.09
SIM	2.11	0.55	2.53	2.59	0.40	2.54
CWT	1.06	0.33	0.94	0.63	0.12	0.63

Table 2. MSS estimates for phantom P2 (SD – standard deviation).

	3.5 MHz			7.5 MHz		
Method	Average (mm)	SD (mm)	Median (mm)	Average (mm)	SD (mm)	Median (mm)
SAC	1.80	0.32	1.90	1.37	0.46	1.01
SSA	1.72	0.46	1.90	1.30	0.86	0.95
SIM	1.91	0.46	1.90	2.81	1.29	3.80
CWT	0.67	0.07	0.67	0.59	0.05	0.57

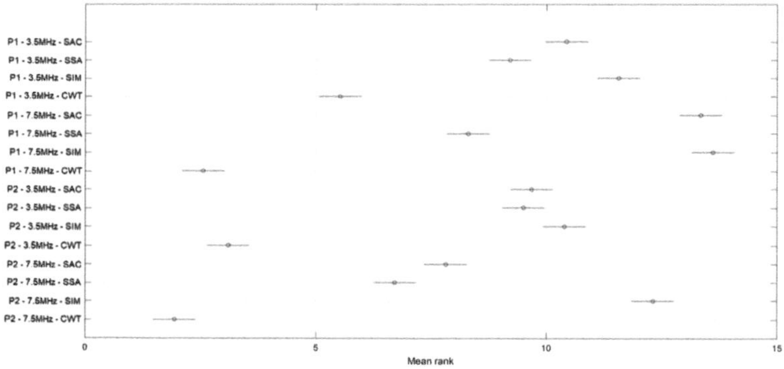

Fig. 3. Results from the post-hoc test after Friedman test (comparison of the mean rank from each experimental setup).

Some aspects must be highlighted: (1) it is not known the actual distribution of scatterers right in front of the US wave for each signal acquired; (2) CWT estimated MSS values lower than 1.06 mm in all cases, maybe because this is a technique which aims at finding the position of scatterers. For instance, the beginning and the end of the trabecula wall can be identified by ultrasound, and the average MSS is obtained with the spacing from the trabeculae and trabecular thicknesses; (3) MSS estimates are lower in phantom P2 when compared to P1. This can be explained by the fact that in P2, since it presents higher stiffness (higher compressive strength and modulus), there is a higher trabecular thickness [8], consequently leading to a decrease in scatterer spacing inside

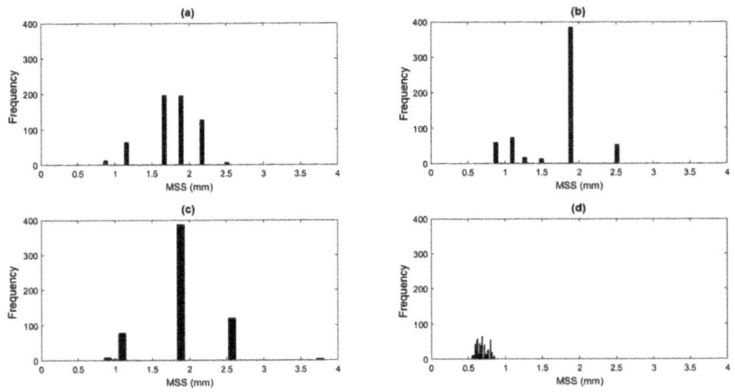

Fig. 4. Histogram of MSS estimates for phantom P2 at 3.5 MHz: (a) SAC; (b) SSA; (c) SIM; and (d) CWT.

the phantom; (4) it would be expected similar distributions of MSS estimates among the methods. However, as seen in Fig. 3, it was not the case. It can be hypothesized that CWT detected a different periodicity pattern inside the phantom (as explained above). On the other hand, the possibility of spectral methods in finding more than one periodicity pattern (as shown in the histograms in Fig. 4 and in previous works [2, 6]), for example may explain the significant difference ($p < 0.001$) among the distributions; (5) CWT seems to be more precise in MSS estimates, however far from the range of scatterer spacing of the phantom (Fig. 4).

The use of MSS as a clinical tool is promising; nonetheless further effort must be made. Adjustments in signal processing methods, the real-time implementation of MSS estimates, multiparametric QUS approaches (including imaging techniques), and *in vivo* studies are undoubtedly some of the next steps.

Conflict of Interest. We have no conflict of interest to declare.

References

1. Mamou, J., Oelze, M.L.: Quantitative Ultrasound in Soft Tissues. Springer, New York (2013)
2. Machado, C.B., Pereira, W.C.A., Meziri, M., Laugier, P.: Characterization of in vitro healthy and pathological human liver tissue periodicity using backscattered ultrasound signals. Ultrasound Med. Biol. **32**(6), 649–657 (2006)
3. Bouzitoune, R., Meziri, M., Machado, C.B., Padilla, F., Pereira, W.C.A.: Can early hepatic fibrosis stages be discriminated by combining ultrasonic parameters? Ultrasonics **68**, 120–126 (2016)
4. Tang, X., Abeyratne, U.R.: Wavelet transforms in estimating scatterer spacing from ultrasound echoes. Ultrasonics **38**, 688–692 (2000)
5. Zhou, Z., Sheng, L., Wu, S., Yang, C., Zeng, Y.: Ultrasonic evaluation of microwave-induced thermal lesions based on wavelet analysis of mean scatterer spacing. Ultrasonics **53**, 1325–1331 (2013)

6. Pereira, W.C.A., Bridal, L., Coron, A., Laugier, P.: Singular spectrum analysis applied to backscattered ultrasound signals from in vitro human cancellous bone specimens. IEEE Trans. Ultrason. Ferroelectr. Freq. Control **51**, 302–312 (2004)
7. Percival, D.B., Walden, A.T.: Wavelet Methods for Time Series Analysis. Cambridge University Press, London (2000)
8. Johnson, A.E., Keller, T.S.: Mechanical properties of open-cell foam synthetic thoracic vertebrae. J. Mater. Sci. Mater. Med. **19**, 1317–1323 (2008)

Short-Term Hemodynamic Variability in Supine and Tilted Position in Young Men

Gerard Cybulski[1(✉)], Edward Koźluk[2], Agnieszka Piątkowska[2], Ewa Michalak[3], Anna Stępniewska[4,5], Anna Gąsiorowska[5], and Wiktor Niewiadomski[5]

[1] Institute of Metrology and Biomedical Engineering, Department of Mechatronics, Warsaw University of Technology, Warsaw, Poland
G.Cybulski@mchtr.pw.edu.pl
[2] Second Chair and Department of Cardiology, Medical University of Warsaw, Warsaw, Poland
[3] Cardinal Stefan Wyszynski Institute of Cardiology, Warsaw, Poland
[4] University of Mazovia in Warsaw, Warsaw, Poland
[5] Department of Applied Physiology, Mossakowski MRC, PAS, Warsaw, Poland

Abstract. The aim of the study was to evaluate short-term changes in hemodynamic parameters observed in supine and tilted positions in six young men (age: 21–25). The cardiac inter-beat interval (RR), stroke volume (SV), ejection time (ET) and pre-ejection period (PEP) parameters were followed over two six-minute periods, in supine position and 10 min after a 60-degree head-up tilting maneuver, using continuously recorded impedance cardiography (ICG) and electrocardiography (ECG) signals. Hemodynamic variability was evaluated using coefficient of variation (CV), standard deviation (SD), and quartile deviation (QD). For the supine position, the mean (M), SD, CV and QD of the observed parameters were as follows. SV: 79 ml, 16 ml, 20%, 15 ml. RR: 905 ms, 106 ms, 12%, 163 ms. ET: 308 ms, 29 ms, 10%, 33 ms. PEP: 105 ms, 19 ms, 18%, 28 ms. In the tilted position, the following were observed. SV: 54 ml, 12 ml, 21%, 15 ml. RR: 706 ms, 68 ms, 10%, 68 ms. ET: 259 ms, 38 ms, 14%, 28 ms. PEP: 123 ms, 13 ms, 12%, 15 ms. The changes in hemodynamic variability caused by tilting were not unidirectional.

Keywords: Hemodynamic variability · Impedance cardiography · Tilt test

1 Introduction

1.1 Importance of Hemodynamic Variability

The analysis of cardiovascular system response to an orthostatic head-up tilt test may provide some diagnostic data on autonomic control [1] and help to predict the occurrence of orthostatic syncope [2]. Cardiac rhythm variability has been intensively studied in supine and other body positions in humans. However, limited data are available regarding fluctuations of hemodynamic parameters, especially during orthostatic tests.

J. Henriques et al. (Eds.): MEDICON 2019, IFMBE Proceedings 76, pp. 787–792, 2020.
https://doi.org/10.1007/978-3-030-31635-8_94

Cybulski et al. [3] analyzed central hemodynamic variability in a small group of subjects who either had atrial fibrillation or were healthy. They concluded that the variability of SV, estimated by the coefficient of variation, was mainly caused by changes in the amplitude of the signal and that variation in ET has a minor effect. Siebert et al. [4] analyzed stroke volume variability and heart rate power spectrum in relation to posture changes in healthy subjects. Beat-to-beat variability of stroke volume output velocity (SVOV) at rest has also been measured in healthy subjects and in patients 10–14 days after acute myocardial infarction [5].

1.2 Motivation and the Aim of the Study

The analysis of changes in stroke volume and systolic time intervals (including their variability) seems to be a possible source of diagnostic information on mutual coupling between the cardiovascular system and autonomic system dysfunction or modification induced by several environmental and/or pathological factors [6, 7]. Perhaps, it could be used for: monitoring the effects of neurodegenerative processes [8], analysis of sleep apnea disorders [9], monitoring the effectiveness of physical training and exercise [10] and predicting orthostatic intolerance [2]. Since we performed similar study in young women [11] we also intended to check if young men presented the diverse hemodynamic variability.

The purpose of the study was to evaluate short-term variability in hemodynamic parameters - stroke volume (SV), ejection time (ET), pre-ejection period (PEP) – observed in the supine and tilted positions in young, healthy men.

2 Materials and Methods

2.1 Subjects, Procedure, Method and Instrumentation

We used data from our impedance cardiography signal database, to estimate hemodynamic variability before and after the tilt test. These data were originally collected for a comparison of the results of hemodynamic parameters obtained from signals recorded simultaneously using two methods: ICG and pulse Doppler echocardiography. The result of that comparison was published earlier [12, 13]. In this study, we used the signals obtained from six young, healthy male subjects (ages 21–34). Impedance cardiography (ICG) and electrocardiographic (ECG) signals were recorded continuously in the supine position and following a 60-degree head-up tilt maneuver.

We used our previously described ambulatory impedance cardiography wearable system (Reomonitor) [12–14]. This device was constructed for non-invasive acquisition of central hemodynamic data during everyday activity. Stroke volume was evaluated using the Kubicek formula [15]. The validation of impedance cardiography have been performed many times [16, 17]. Reomonitor was verified using echocardiography in both the supine and tilted position [13], showing the ability of the system to distinguish between those two body position.

2.2 Statistical Analysis

Hemodynamic variability was evaluated by means of descriptive statistic parameters: standard deviation (SD), coefficient of variation (CV) and quartile deviation (QD) of the SV derived from impedance cardiography (ICG), ET and PEP. The intervals between cardiac beats (RR) were also analyzed. For each parameter in both supine and tilted position, the mean values (M) were calculated. QD was calculated as half of the difference between upper quartile (Q3) and lower quartile (Q1). Analysis was performed for two six-minute periods, recorded in the supine position and 10 min after the tilting maneuver. Artefacts were removed from the analysis based on the clearly defined criteria for each variable.

3 Results

The descriptive parameters of stroke volume in supine and tilted positions are presented in Table 1. Tilting caused a decrease in the mean value of SV. However we observed increase in CV, and decrease in QD.

Table 1. Descriptive characteristics of stroke volume (SV) in the supine and tilted positions.

Stroke volume (SV)	Supine	Tilted
Mean [ml]	79	54
Standard deviation [ml]	16	12
Coefficient off variation [%]	20	21
Quartile deviation [ml]	15	12

Table 2 contains descriptive statistical parameters for ET. Tilting caused a decrease in the mean value of ET. SD and CV increased but were not significantly modified, QD decreased.

Table 2. Descriptive characteristics of ejection time (ET) in the supine and tilted positions.

Ejection time (ET)	Supine	Tilted
Mean [ms]	308	259
Standard deviation [ml]	29	38
Coefficient off variation [%]	10	14
Quartile deviation [ms]	33	28

Table 3 contains descriptive statistical parameters for PEP. Tilting caused an increase in the mean value of PEP. SD, CV and QD decreased.

Table 3. Descriptive characteristics of pre-ejection period (PEP) in the supine and tilted positions.

Pre-ejection period (PEP)	Supine	Tilted
Mean [ms]	105	123
Standard deviation [ml]	19	13
Coefficient off variation [%]	18	12
Quartile deviation [ms]	28	15

Table 4 contains the descriptive statistics for the amplitude (AMP) of dz/dtmax. Tilting caused no significant decrease in the mean value of AMP. SD, QD and CV did not change significantly.

Table 4. Descriptive characteristics of amplitude (AMP) in the supine and tilted positions.

Amplitude (AMP)	Supine	Tilted
Mean [Ω/s]	1.85	1.76
Standard deviation [Ω/s]	0.27	0.21
Coefficient off variation [%]	14	12
Quartile deviation [Ω/s]	0.31	0.27

Table 5 contains the statistics for RR. Tilting caused the decrease of mean value of RR, significant decrease of SD, QD and no change in CV.

Table 5. Descriptive characteristics of R-R intervals (RR) in the supine and tilted positions.

R-R intervals (RR)	Supine	Tilted
Mean [ms]	905	706
Standard deviation [ml]	106	68
Coefficient off variation [%]	12	10
Quartile deviation [ms]	163	68

4 Discussion and Conclusions

The advantages/disadvantages and possible application of the ICG method have been discussed in several review publications [14, 18, 19]. Fellahi and Fisher [19] stated that, despite some disappointing results for the estimation of absolute values of cardiac output in comparison with reference methods, ICG might be "interesting in guiding the spontaneous or induced changes in cardiac output and medical decision-making at the bedside". It was pointed out that an ambulatory version of ICG could be used to record transient events, which would be difficult or even impossible to visualize using other, well established, "classical" methods [14].

We observed that SV decrease in men after tilting was deeper than in women. Also the RR interval decreased markedly in men in comparison to women. For ET and PEP the response to tilting in man and women were similar. As regards AMP in women we observed increase after tilting whereas in man AMP decreased after tilting [11].

The main research limitation of the study is the small number of examined subjects. Another one is associated with uncertainty of the method, regarding the limitations of the proper detection of the B-point on the dz/dt curve in ICG [20]. This results in uncertainty in determination of systolic time intervals (ET, PEP, and their derivatives). We could only assume that uncertainty related to B-point detection is constant despite the body position.

These preliminary results show that relative measures of hemodynamic parameter variability after tilting were different when compared to those observed in the supine position. We cannot conclude about the uniform direction of that changes. Considering mean values and the variability estimators we may conclude that the response to tilt in woman and men were different [11].

Acknowledgements. The study was supported by the research programs of institutions the authors are affiliated with.

References

1. Cybulski, G.: Influence of age on the immediate cardiovascular response to the orthostatic maneouvre. Eur. J. Appl. Physiol. **73**, 563–572 (1996)
2. Koźluk, E., Cybulski, G., Piątkowska, A., Zastawna, I., Niewiadomski, W., Strasz, A., Gąsiorowska, A., Kempa, M., Kozłowski, D., Opolski, G.: Early hemodynamic response to the tilt test in patients with syncope. Arch. Med. Sci. **10**(6), 1078–1085 (2014)
3. Cybulski, G., Ziółkowska, E., Książkiewicz, A., Łukasik, W., Niewiadomski, W., Kodrzycka, A., Pałko, T.: Application of impedance cardiography ambulatory monitoring device for analysis of central hemodynamics variability in atrial fibrillation. Comput. Cardiol. **26**, 563–566 (1999)
4. Siebert, J., Drabik, P., Lango, R., Szyndler, K.: Stroke volume variability and heart rate power spectrum in relation to posture changes in healthy subjects. Med. Sci. Monitor. **10**(2), MT31-7 (2004)
5. Zubarev, M.A., Schekotov, V.V., Parandey, O.R., Dumler, A.A.: Beat – to - beat variability of stroke volume output velocity measured by an impedance cardiographic method. In: IFMBE Proceedings, vol. 17, pp. 579–581 (2007)
6. Toska, K., Eriksen, M.: Respiration-synchronous fluctuations in stroke volume, heart rate and arterial pressure in humans. J. Physiol. **472**, 501–512 (1993)
7. Niizeki, K., Saitoh, T.: Analysis of cardiorespiratory phase coupling and cardiovascular autonomic responses during food ingestion. Physiol. Behav. **159**, 1–13 (2016)
8. Marongiu, E., Olla, S., Magnani, S., Palazzolo, G., Sanna, I., Tocco, F., Marcelli, M., Loi, A., Corona, F., Mulliri, G., Concu, A., Crisafulli, A.: Metaboreflex activity in multiple sclerosis patients. Eur. J. Appl. Physiol. **115**(12), 2481–2490 (2015)
9. Ahmad, S., Batkin, I., Kelly, O., Dajani, H.R., Bolic, M., Groza, V.: Multiparameter physiological analysis in obstructive sleep Apnea simulated with Mueller Maneuver. IEEE Trans. Instrum. Measure. **62**(10), 2751–2762 (2013)

10. Gratze, G., Mayer, H., Luft, F.C., Skrabal, F.: Determinants of fast marathon performance: low basal sympathetic drive, enhanced postcompetition vasodilatation and preserved cardiac performance after competition. Br. J. Sports Med. **42**(11), 882–888 (2008)
11. Cybulski, G., Koźluk, E., Piątkowska, A., Michalak, E., Strasz, A., Gąsiorowska, A., Niewiadomski, W.: Short-term hemodynamic variability in supine and tilted position in young women. Comput. Cardiol. **43**, 513–516 (2016). https://doi.org/10.22489/CinC.2016.146-481
12. Cybulski, G., Książkiewicz, A., Łukasik, W., Niewiadomski, W., Pałko, T.: Ambulatory monitoring device for central hemodynamics and ECG signals recording on PCMCIA flash memory cards. Comput. Cardiol. **22**, 505–507 (1995)
13. Cybulski, G., Michalak, E., Koźluk, E., Piątkowska, A., Niewiadomski, W.: Stroke volume and systolic time intervals: beat-to-beat comparison between echocardiography and ambulatory impedance cardiography in supine and tilted positions. Med. Biol. Eng. Compu. **42**, 707–711 (2004)
14. Cybulski, G.: Ambulatory impedance cardiography. In: The Systems and Their Applications. Series: Lecture Notes in Electrical Engineering. Springer, Heidelberg (2011)
15. Kubicek, W.G., Karnegis, J.N., Patterson, R.P., Witsoe, D.A., Mattson, R.H.: Development and evaluation of an impedance cardiac output system. Aerosp. Med. **37**, 1208–1212 (1966)
16. Ebert, T.J., Eckberg, D.L., Vetrvec, G.M., Cowley, M.J.: Impedance cardiograms reliably estimate beat-by-beat changes of left ventricular stroke volume in humans. Cardiovasc. Res. **18**, 354–360 (1984)
17. Scherhag, A., Kaden, J.J., Kentschke, E., Sueselbeck, T., Borggrefe, M.: Comparison of impedance cardiography and thermodilution-derived measurements of stroke volume and cardiac output at rest and during exercise testing. Cardiovasc. Drugs Ther. **19**(2), 141–147 (2005)
18. Cybulski, G., Strasz, A., Niewiadomski, W., Gąsiorowska, A.: Impedance cardiography: recent advancements. Cardiol. J. **19**(5), 550–556 (2012)
19. Fellahi, J.L., Fischer, M.O.: Electrical bioimpedance cardiography: an old technology with new hopes for the future. J. Cardiothorac. Vasc. Anesth. **28**(3), 755–760 (2014)
20. Lozano, D.L., Norman, G., Knox, D., Wood, B.L., Miller, B.D., Emery, C.F., Berntson, G. G.: Where to B in dZ/dt. Psychophysiology **44**(1), 113–119 (2007)

Efficacy of Machine Learning in Predicting the Kind of Delivery by Cardiotocography

Giovanni Improta[1(✉)], Carlo Ricciardi[2], Francesco Amato[3], Giovanni D'Addio[4], Mario Cesarelli[4], and Maria Romano[4,5]

[1] Department of Public Health, University Hospital of Naples Federico II, Naples, Italy
ing.improta@gmail.com

[2] Department of Advanced Biomedical Sciences, University Hospital of Naples Federico II, Naples, Italy

[3] Department of Electrical Engineering and Information Technology, DIETI, University of Naples Federico II, Naples, Italy

[4] Istituti Clinici Scientifici Maugeri IRCCS, Telese Terme, BN, Italy
mariar@unicz.it

[5] Department of Experimental and Clinical Medicine (DMSC), University "Magna Graecia" of Catanzaro, Catanzaro, Italy

Abstract. It is well known that the interpretation of cardiotocographic (CTG) signals is still subjective and prone to misinterpretations; as a consequence, there has been an increase of cesarean sections, often not necessary, and initial expectations of significantly reducing perinatal morbidity and mortality have been unattended. Nevertheless, in developed countries, CTG is still the most widely employed prenatal technique for monitoring fetal health; and in many countries it represents a medical report with legal value. To overcome the drawbacks of the visual interpretation, many computerized systems for automatic or semi-automatic analysis of CTG have been developed in the last years. Recently, in order to support the diagnosis process, and to increase the predictive capability of these systems, also other techniques such as artificial neural network, deep learning and machine learning have been introduced. In previous works of the authors, software for automatic CTG analysis has been developed and described in detail. Now, by employing a dataset of features extracted from CTG signals with that software, to enhance its performances, different algorithms, such as J48, Adaboosting, Random Forests and Gradient Boosted Tree, have been tested to predict whether a birth would be a caesarean section or a vaginal delivery. The RF algorithm showed the best performance, since it reached the highest accuracy (87.6%), precision (87.9%) and AUCROC (93.0%). These preliminary results are very satisfying and encouraging; they confirm that to enrich the CTG analysis software with this methodology can help to significantly improve CTG classification.

Keywords: Cardiotocography · Machine learning · Caesarean section

J. Henriques et al. (Eds.): MEDICON 2019, IFMBE Proceedings 76, pp. 793–799, 2020.
https://doi.org/10.1007/978-3-030-31635-8_95

1 Introduction

Cardiotocography (CTG), as it is known, is the method most used in prenatal diagnostic for the surveillance of fetal well-being both in antepartum and in intrapartum period. It born with the aim of early detecting fetal hypoxia and consists of the simultaneous recording of Fetal Heart Rate (FHR) and Uterine Contractions (UC).

Although the introduction of this technique in the 60 s has allowed a reduction in perinatal mortality and morbidity, its diagnostic accuracy is still far from being fully satisfactory. In fact, the evaluation of cardiotocographic traces, currently still performed, in most cases, by visual inspection, is highly subjective and qualitative. Moreover, the number of false positives is still too high.

To overcome these problems many computerized analysis systems have been introduced [1–4], which provide an objective numerical evaluation of the parameters usually evaluated from CTG signals.

However, differently from the initial expectation, it has not yet been possible to drastically reduce the number of unnecessary operative births and cesarean sections. In recent years, therefore, other computer-aided methods are being developed supporting diagnosis process, in particular artificial neural network techniques, deep learning and machine learning have been introduced [5–7].

Machine learning was born to discover hidden patterns in large datasets and exploit the amount of data that has been growing in the last years and has brought to the issue of the so called "Big Data" [8]. In healthcare, it has been applied in a variety of medical domains [9–11], such as identification of patients' needs, remote monitoring of devices, development of predictive models for helping in the decision-making process; and mostly in solving diagnostic and prognostic problems.

Aim of this paper is to test different algorithms, J48, Adaboosting (ADA-B), Random Forests (RF) and Gradient Boosted Tree (GBT), on a dataset of features extracted from CTG signals to predict whether a birth would be a caesarean section (C in the following) or a vaginal, spontaneous delivery (S in the following).

2 Materials and Methods

2.1 Dataset

For this work, 300 CTG signals were employed. They have been extracted from a database of CTG signals recorded by means of Philips devices, during routine fetal monitoring from 2000 to 2009, in clinical environments, from women, between 28th and 42nd week of gestation, who were in a rest position and did not taken any drug. Each signal lasts at least 20 min and was classified as of good quality, i.e. with a signal loss less than 20% and a total number of outliers less than 5% of FHR length (a signal with too many outliers was judged poorly reliable) [12]. Neonatal parameters such as Apgar score, new-born weight and other information such as eventual maternal pathologies had been also collected when available.

2.2 Software for Extraction of CTG Features

In previous works, a software for automated analysis of CTG was developed and presented [1, 13–15]. It implements a preprocessing to recover the so-called real FHR series, elsewhere described [16], and then an automatic procedure for measuring the most relevant parameters derivable from cardiotocographic signals, in time domain, in frequency domain, after a cubic spline interpolation, and by means of non-linear techniques. In particular, the following features were considered for this study:

- Week of gestation (estimated from the last menstruation – "W")
- Mean of FHR ("FHRm")
- Presence of tachycardia (defined as a rhythm increase over 160 bpm for at least 10 min – "tachi")
- Number of accelerations [1]
- Number of decelerations [1]
- Number of uterine contractions [1]
- CTG recorded in intrapartum period, antepartum period, antepartum but with some contractions ("I\A\C")
- Short term variability [13]
- Power in low frequency of FHR power spectrum (0.05–0.2 Hz, "LF") [14]
- Power in high frequency of FHR power spectrum (0.2–1 Hz, "HF") [14]
- Sympathovagal balance (ratio between LF and HF) [14]
- Variability index from RR series (index estimated by means of symbolic dynamics analysis "VI_{RR}") [17]
- Variability index from FHR series ("VI_{FHR}")
- Sample entropy (for quantifying the regularity/irregularity or randomness of the fluctuations of heart beats "SampEn") [18]
- Minor axis of the data map obtained with the Poincarè technique ("SD1") [19]
- Major axis of the data map obtained with the Poincarè technique ("SD2")
- Fractal dimension (estimated by means of Higouchi algorithm "Hfd")

2.3 Tool and Algorithms

Knime analytics platform was employed for this study due to its acknowledged potency [20]; it allows to implement a wide range of algorithms using workflows and without the necessity to use a programming language. Moreover, it was already employed in other biomedical studies [21].

As regards the employed algorithms, J48 consists in the Java implementation of the well-known decision tree, a structure made up of leaves and nodes that represent classes and splitting phases based on the attributes, respectively [22]. RF exploits the principle of randomization and bagging and, as the name itself explains, consists of an ensemble of trees that makes a forest [23]. The training phase is conducted separately for each tree and randomness is injected in this process to reduce the variance of the predictions. Randomness is injected by subsampling the original dataset in each iteration to get a different training set or considering different random subsets of features to split on at each tree node. To make a prediction on a new instance, RF aggregates the predictions by majority vote.

ADA-B is an ensemble method based on boosting. The ADA-B training selects only the features that improve the predictive power of the model, reducing complexity in terms of dimension [24].

GBT is the most complex algorithm because it uses randomization, bagging and boosting on decision trees [25].

Since only one dataset was available, a procedure of 10 folds cross-validation was implemented for all the algorithms: the dataset is divided in 10 mutually exclusive folds of approximately same dimensions and the model is trained k times on k-1 folds and tested on one-fold. The final results are the average of those obtained cross-validating the dataset [26].

Finally, in order to evaluate algorithms' performance in predicting a cesarean section, the following evaluation metrics were computed [27]:

- Accuracy = (TP + TN)/tot;
- Mean error = 1 − accuracy;
- Precision = TP/(TP + FP);
- Recall = Sensitivity = TP/(TP + FN);
- Specificity = TN/(TN + FP);
- AUCROC = area under the curve ROC (sensitivity-specificity) and represents the goodness of a model in distinguishing two classes;

where TP = True Positive; TN = True Negative; FN = False Negative; FP = False Positive; tot = total number of events; AUCROC = Area Under the Curve Receiving Characteristic Operator. Recall and sensitivity share the same definition but are used in different fields; for the sake of completeness, they are both cited in this paper.

3 Results

A procedure to handle missing data was chosen: each missing value was re-placed by the mean value of the features. Then, the SMOTE technique was implemented in order to balance the dataset; thus, the number of records rose from 300 to 450.

Values obtained for the evaluation metrics are summarized in Table 1.

Table 1. Algorithms' performance (all values represent percentages - %).

	Accuracy	Recall	Precision	Sensitivity	Specificity	AUCROC
J48	80.2	79.6	80.6	79.6	80.9	79.9
ADA-B	85.1	81.8	87.6	81.8	88.4	91.6
RF	87.6	87.1	87.9	87.1	88.8	93.0
GBT	86.4	88.9	84.7	88.9	84.0	92.6

From Table 1, it is possible to observe that the RF algorithm reached the highest accuracy (87.6%), precision (87.9) and AUCROC (93.0) while GBT obtained the highest recall and sensitivity (88.9%). Excluding J48, all the AUCROC overcame 90.0% proving the goodness of these models in classifying data.

The lowest results were obtained by J48 as regards all evaluation metrics.

For sake of clearness, Table 2 represents the confusion matrix of the algorithm that obtained the best accuracy and AUCROC (RF).

Table 2. Confusion matrix.

	C	S
C	196	29
S	27	198

4 Discussion

The interpretation of CTG signals is still subjective and subject to misinterpretation. Both in the clinical and in research field, for trying to solve this problem, computerized systems for automatic analysis have been developed, which can support physicians in formulating the diagnosis of fetal wellness or distress.

As part of this research, in previous works, an analysis software was developed that allows the detection of a rather high number of information from CTG signals, both in antepartum and in intrapartum period, including parameters based on the morphological analysis of the signal, still used in clinical practice, and parameters that cannot be evaluated by naked eye, based on frequency domain analysis and non-linear techniques, also inspired by results obtained on adult subjects [17–19, 28].

In this work, in accordance with the most recent literature, a machine learning technique was introduced to further improve the performance of the software and, in particular, to improve the prediction of caesarean section.

Among the characteristics in output from the analysis software, those that have been considered to have a higher discriminating power have been chosen as input to the classifier. Then, different algorithms have been tested and some evaluation metrics were estimated through the implementation of the decision tree (J48) and all its empowered implementations.

As expected, J48 obtained the worst result; indeed, it is the simplest classifier while the other algorithms exploit the principles of randomization, bagging and boosting or a combination of them that empower consistently their performance.

In literature different researchers tried to predict outcome of newborn children starting from CTG signals. Georgoulas et al. used Support Vector Machine to predict normal and abnormal cases of CTG employing time and frequency domain features getting an accuracy of 86.1% [29]. Similarly, Krupa et al. implemented Support Vector Machine to classify CTG in "normal" or "at risk" getting an accuracy over the test data of 87% [30] while Comert and Kocamaz employed a technique based on signal and image processing in order to classify CTG signals and detect hypoxic foetuses getting an accuracy of 87% [31]. However, Fergus et al. conducted the most similar study to ours focusing on different algorithms to predict a vaginal delivery or a caesarean section [7]. They implemented different ensemble methods and identified Fisher Linear Discriminant Analysis, RF, and Support Vector Machine models as the best options, employing 13 features and reaching a sensitivity of 87% and a specificity of 90%.

In our study we decided to focus on decision tree and its more powerful forms employing 17 relevant parameters as features. Moreover, as suggested by Kohavi et al. we decided to employ 10 folds cross-validation [26]. The proposed methodology allowed to reach comparable values of sensitivity and specificity (about 87% and 89% respectively), so confirming that it is a promising way for improving CTG interpretation and classification.

Obviously, the work can be further expanded, for example it would be useful to train the classifier to recognize particular patterns, which include for example so-called saltatory (characterized by an increased variability of the FHR signal > 25 bpm) or sinusoidal CTG signals, which are typical of pathological conditions; or introduce features related to the morphology of the decelerations. Furthermore, a wider range of algorithms could be tested, exploiting different combinations of algorithms to create more powerful ensemble techniques in order to improve the scores, and a wrapper method could be employed to identify the best subset of features.

Conflict of Interest Declaration. The authors declare they have no conflict of interests.

References

1. Romano, M., Bifulco, P., Ruffo, M., Improta, G., Clemente, F., Cesarelli, M.: Software for computerised analysis of cardiotocographic traces. Comput. Methods Programs Biomed. **124**, 121–137 (2016)
2. Ayres-de-Campos, D., Bernardes, J., Garrido, A., Marques-de-Sa, J., Pereira-Leite, L.: SisPorto 2.0: a program for automated analysis of cardiotocograms, J. Matern. Fetal Neonatal. Med. **9**(5), 311–318 (2000)
3. Al-Yousif, S.N., Ali, M.M.: Cardiotocography trace pattern evaluation using MATLAB program. In: Proceedings of International Conference on Biomedical Engineering and Technology - ICBET 2011 (2011)
4. Parer, J.T., Hamilton, E.F.: Comparison of 5 experts and computer analysis in rule-based fetal heart rate interpretation. Am. J. Obstet. Gynecol. **203**(5), 451–461 (2010)
5. Georgoulas, G., Stylios, C., Groumpos, P.P.: Integrated approach for classification of cardiotocograms based on independent component analysis and neural networks. In: Proceeding of 11th IEEE Mediterranean Conference on Control and Automation, 18–20 June 2003, Rodos, Greece (2003)
6. Huang, M.L., Hsu, Y.Y.: Fetal distress prediction using discriminant analysis, decision tree, and artificial neural network. J. Biomed. Sci. Eng. **5**, 526–533 (2012)
7. Fergus, P., Selvaraj, M., Chalmers, C.: Machine learning ensemble modelling to classify caesarean section and vaginal delivery types using cardiotocography traces. Comput. Biol. Med. **93**, 7–16 (2018)
8. Michie, D., Spiegelhalter, D.J., Taylor, C.C.: Machine learning. Neural Stat. Classif. **13** (1994)
9. Magoulas, G.D., Prentza, A.: Machine learning in medical applications. In: Advanced Course on Artificial Intelligence, pp. 300–307. Springer, Heidelberg, July 1999
10. Patel, H.B., Gandhi, S: A review on big data analytics in healthcare using machine learning approaches. In: 2018 2nd International Conference on Trends in Electronics and Informatics (ICOEI), pp. 84–90 (2018)
11. Collins, G.S., Mallett, S., Omar, O., Yu, L.M.: Developing risk prediction models for type 2 diabetes: a systematic review of methodology and reporting. BMC Med. **9**(1), 103 (2011)

12. Romano, M., Faiella, G., Bifulco, P., D'Addio, G., Clemente, F., Cesarelli, M.: Outliers detection and processing in CTG monitoring. In: IFMBE Proceedings Medicon 2013, 25–28 September 2013, Seville, Spain, vol. 41, pp. 651–654 (2014)
13. Cesarelli, M., Romano, M., Bifulco, P.: Comparison of short term variability indexes in cardiotocographic foetal monitoring. Comput. Biol. Med. **39**(2), 106–118 (2009)
14. Romano, M., Iuppariello, L., Ponsiglione, A.M., Improta, G., Bifulco, P., Cesarelli, M.: Frequency and time domain analysis of foetal heart rate variability with traditional indexes: a critical survey. Comput. Math. Methods Med. **2016**, 12 (2016). Article ID 9585431
15. Cesarelli, M., Romano, M., Bifulco, P., Improta, G., D'Addio, G.: An application of symbolic dynamics for FHRV assessment. Stud. Health Technol. Inform. **180**, 123–127 (2012)
16. Cesarelli, M., Romano, M., Bifulco, P., Fedele, F., Bracale, M.: An algorithm for the recovery of fetal heart rate series from CTG data. Comput. Biol. Med. **37**(5), 663–669 (2007)
17. Romano, M., D'Addio, G., Clemente, F., Ponsiglione, A.M., Improta, G., Cesarelli, M.: Symbolic dynamic and frequency analysis in foetal monitoring. In: IEEE International Symposium on Medical Measurements and Applications (MeMea) Proceedings, pp. 545–548 (2014)
18. Lake, D.K., Moorman, J.R., Hanqing, C.: Sample entropy estimation using sampan. PhysioToolkit, open source software for biomedical science and engineering
19. Fishman, M., Jacono, F.J., Park, S., Jamasebi, R., Thungtong, A., Loparo, K.L., Dick, T.E.: A method for analyzing temporal patterns of variability of a time series from poincare plots. J. Appl. Physiol. **113**(2), 297–306 (2012)
20. Sharma, N., Bansal, K.L.: Comparative study of data mining tools. J. Adv. Database Manage. Syst. **2**, 35–41 (2015)
21. Romeo, V., Ricciardi, C., Cuocolo, R., et al.: Machine learning analysis of MRI-derived texture features to predict placenta accreta spectrum in patients with placenta previa. Magn. Reson. Imaging (2019). https://doi.org/10.1016/j.mri.2019.05.017. ISSN 0730-725X
22. Quinlan, J.R.: Programs for Machine Learning. Elsevier (2014)
23. Breiman, L.: Random forests. Mach. Learn. **45**, 5–32 (2001)
24. Freund, Y., Schapire, R.E.: A decision-theoretic generalization of on-line learning and an application to boosting. J. Comput. Syst. Sci. **55**(1), 119–139 (1997)
25. Friedman, J.: Greedy function approximation: a gradient boosting machine. Ann. Stat. **29**, 1189–1232 (2001)
26. Kohavi, R.: A study of cross-validation and bootstrap for accuracy estimation and model selection. IJCAI **14**(2), 1137–1145 (1995)
27. Hossin, M., Sulaiman, M.N.: A review on evaluation metrics for data classification evaluations. Int. J. Data Min. Knowl. Manage. Process **5**(2), 1 (2015)
28. Cusenza, M., Accardo, A., D'Addio, G., Corbi, G.: Relationship between fractal dimension and power-law exponent of heart rate variability in normal and heart failure subjects. Comput. Cardiol. **37**, 935–938 (2010)
29. Georgoulas, G., Stylios, C., Bernades, J., Groumpos, P.: Classification of cardiotocograms using support vector machines. In: Proceedings 10th IFAC Symposium on Large Scale Systems: Theory and Applications – LSS 2004 (2004)
30. Krupa, N., Ali, M., Zahedi, E., Ahmed, S., Hassan, F.M.: Antepartum fetal heart rate feature extraction and classification using empirical mode decomposition and support vector machine. Biomed. Eng. Online **10**(1), 6 (2011)
31. Cömert, Z., Kocamaz, A.F.: A study based on gray level co-occurrence matrix and neural network community for determination of hypoxic fetuses. In: International Artificial Intelligence and Data Processing Symposium – IDAP, pp. 569–573 (2016)

Analysis of the Effect of Natural and Simulated Sun Exposure on Sunglasses Lenses: A Study on Materials Degradation

Leonardo Mariano Gomes(✉), Mauro Masili, and Liliane Ventura

Ophthalmic Instrumentation Laboratory, Department of Electrical Engineering, São Carlos School of Engineering, University of São Paulo, São Carlos, SP, Brazil
{leo.gomes, lilianeventura}@usp.br

Abstract. This work aimed to investigate the degradation of sunglasses lenses after a short period of exposure to solar radiation in two different ones. A set of 30 pairs of sunglasses lenses was tested, and the right lenses were exposed in an automatic sun exposure station, i.e. under natural conditions, and the left lenses were exposed in a solar simulator according to the parameters of the standard. In the solar station, the total exposure time was 115 h and in the solar simulator, 100 h. Lens spectroscopy measurements were taken before and after the exposure period for each experiment, and the UV, Visible and Infrared (IR) transmittances of the lenses were calculated by means of these measures, in addition to the accumulated erythemal UV radiation dose for both conditions. Exposure time was not sufficient to cause degradation of the lenses, but it was possible to verify some characteristics, such as differences between lenses of the same model, and differences between the conditions of the experiments that may influence the future results and the degradation of the lenses.

Keywords: Sunglasses · UV radiation · Material degradation · Sun exposure · UV index

1 Introduction

According to NBR ISO 12312-1: 2015 [1], sunglasses must be protected against ultraviolet (UV) radiation. UV protection is related to the degree of darkening of the lens, i.e. the darker the lens, the greater its protection level, and the lens blackness is related to its luminous transmittance. The hypothesis that UV protection in sunglasses lenses degrades from long periods of exposure has never been experimentally tested, although there are studies related to the degradation of polymeric materials under natural conditions and on the mechanical resistance of lenses after short periods of exposure [2]. Studies show that the time of use of the same pair of sunglasses by the Brazilian population is approximately 2 years [3]. Although there is a test in the standard that has the objective of verifying the degradation of the lenses, which is called radiation resistance test [1, 4], theoretical studies show that this test may be ineffective in determining the UV protection of the lenses [5]. Our laboratory has also

J. Henriques et al. (Eds.): MEDICON 2019, IFMBE Proceedings 76, pp. 800–810, 2020.
https://doi.org/10.1007/978-3-030-31635-8_96

analyzed the degradation of sunglasses lenses after long-term irradiation within two different solar simulators [6]. This work aims to compare the transmittance of a set of lenses before and after a period of sun exposure in two different conditions: in an automatic exposure station and in a solar simulator. This comparison is important because, even some studies using solar simulators have already been conducted, the exposition of the sunglasses to real conditions imposes other aspects that might influence the degradation of sunglasses. Also, it is known that solar simulators and test standards for sunglasses are not effective on determining if the sunglasses will resist and continue to protect after long-term exposition in real conditions [5]. Humidity and temperature variations are also important in a smaller degree than UV radiation for materials degradation, and in our setup, those variables are also measured for future analysis. Most of the studies focuses on exposing materials for short-periods of time, and in our study, the aim is to expose sunglasses lenses for a long period (1–2 years), simulating real conditions of daily using of the accessory. The importance of this study is that its results might directly benefit the ocular health of sunglasses users. After the long-term exposition of the lenses, it will be possible to determine how sunglasses degrades in real conditions, and compare this degradation with the effect caused by the solar simulator recommended by the international sunglasses standards. In summary, this study aims to evaluate if the quality of sunglasses lenses is in adequacy with the standards after long-term real exposure conditions.

Although efforts of our team has been progressively done in order to have an eye safe standard for sunglasses for our tropical country [3, 5, 6], as we did in 2013 [7], when we were part of committee for revising the Brazilian standard (NBR15111:2013), it has been repealed in 2015, for industries convenience and one of the many reasons is that there is no lab in Brazil able to perform every required test on sunglasses standards, which falls back to the issue of not having testing equipment at affordable prices. So, this work is one of the many systems that our lab is developing [8–12], to bridge this gap.

2 Materials and Methods

We have developed an automatic sun exposure station for sunglasses which consists of panels capable of holding 60 lenses, sensors (UV, rain, temperature, humidity); motor driven housing for protecting the panel when at night and from the rain, an electronic control unity for controlling the system functions and for enabling remote communication, and an online control panel, for displaying sensor information in real time and the exposure conditions.

For the experiment to be carried out, the automatic sunglasses lens exposure system [13], which was developed by the laboratory team, was used as well as the solar simulator [14]. Before and after the exposure period, the lenses were measured on a Varian Cary-5000 spectrophotometer.

2.1 Exposition in the Solar Station

A set of 60 lenses were installed in acrylic panels, which are installed in a metal frame with rails. The control system of the automatic exposure station is responsible for controlling the opening and closing of a structure that protects or exposes the lenses. Climate sensors are responsible for informing the system of current conditions (temperature, air humidity, rainfall and UV radiation), so that the system can decide, also based on the current time and times for sunrise and sunset, if the lenses should be exposed or not. The control is automatic, but can also be done manually by an online control panel. A photo of the exposure system installed on the roof of a building, and an image of its control panel, are depicted in Figs. 1 and 2.

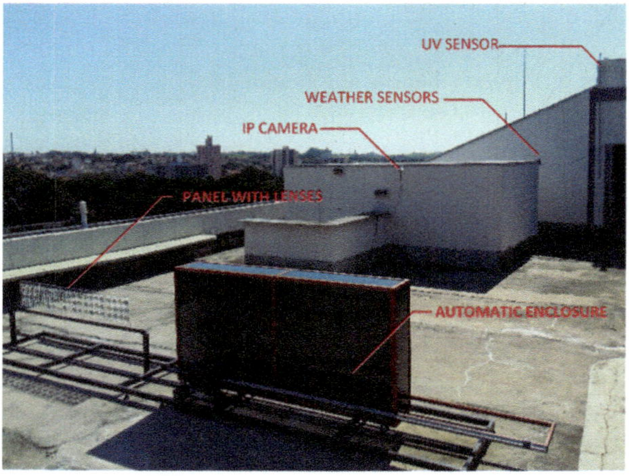

Fig. 1. Automatic exposure system with weather and UVI sensors and lenses.

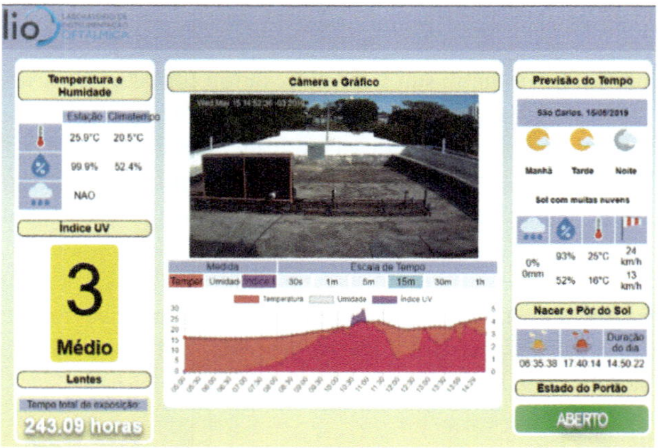

Fig. 2. Online remote monitoring system of the automatic exposure station.

The image transmitted in real time by the IP camera installed next to the system is used to assist the monitoring and manual control of the system in cases where the automatic control fails for some reason. In Fig. 3, a photo is shown in which two conditions for system closure and lens protection occur: rain and the nighty period. Under these conditions, it is not desirable for the lenses to be exposed because of rainfall, particulate materials may be deposited on the lens, thereby disrupting the absorption of solar radiation by the lens, and at night there is no solar radiation of interest.

Fig. 3. Image of the solar station of automatic exposure protecting the lenses against rain, and at night time.

The sunglasses lenses were removed from their frames, were pierced twice at each end so that, with the aid of a fabric elastic, they were installed in the acrylic panels of the station. Twelve lenses were installed per panel, which were arranged vertically to simulate the actual situation of use by a person. The lenses were exposed for a period of about two weeks during the month of February 2019 until the total sum of the exposure period, which is automatically calculated by the control system, was 115 h. Sensor data, as well as system events, were recorded in files.

2.2 Exposition in the Solar Simulator

The exposure of the lenses in the solar simulator was done using a XENON-SUNTEST equipment, from LEMA manufacturer. This simulator meets the requirements for the test of resistance to solar radiation, of the NBR ISO 12312-1: 2015 standard. The equipment has a 450 W OFR xenon arc lamp, powered by a stabilized electric current of 25 A and has a solar emission spectrum that approaches the D65 standard. In the test described in the standard, the lenses must be exposed at a distance of 300 mm from the lamp for a period of 50 h. In each exposure cycle, a total of 15 lenses can be exposed, and therefore we perform 2 exposure cycles for each of the two sets of 15 right lenses, adding a total of 100 exposure hours for each sample (Fig. 4).

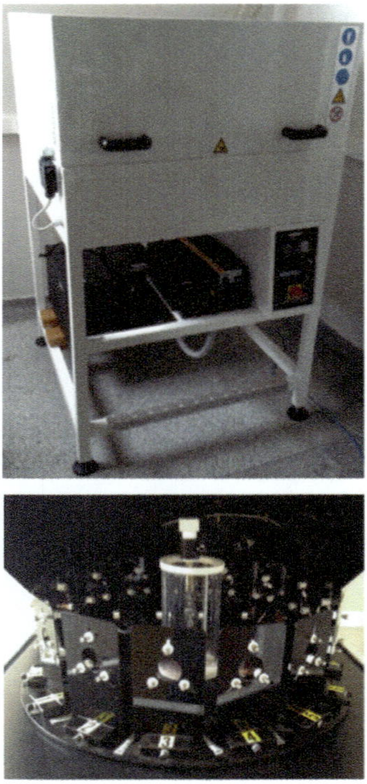

Fig. 4. Solar simulator (top) and lenses arranged around the lamp for the exposure experiment (bottom).

2.3 Spectrophotometer Lenses Measurement

After the exposure period of 115 h in the solar station and 100 h in the solar simulator, the lenses had their transmittance spectrum measured in spectrophotometer CARY 5000 - VARIAN. Measurements were taken in 280 nm to 2000 nm range, with a 5 nm step (as required by standard NBR ISO12312-1:2015. In order to minimize positioning errors in the transmittance measurements, the lenses had their geometric center determined, as well as the holders used for spectroscopy. This determination of the center of the lens is important for the gradient colored lenses, since a slight change in lens positioning may induce systematic errors in spectroscopy. Figure 5 shows the holder and lens centered set.

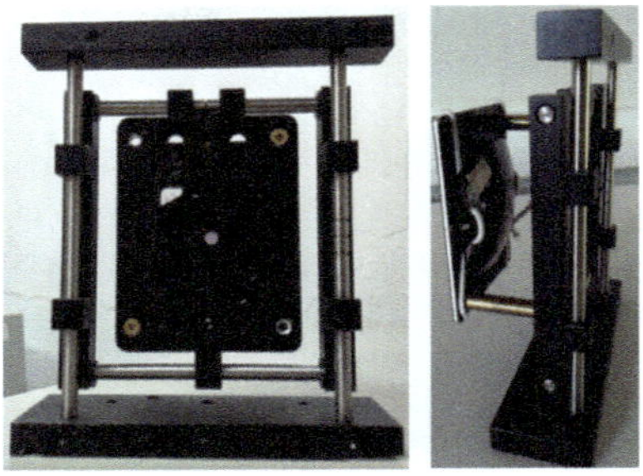

Fig. 5. Holder for positioning lens in the spectrophotometer.

3 Results

Spectroscopy was performed previous to sun and solar simulator irradiations. Then lenses were submitted to both systems exposure. After the spectrophotometric measurements of the transmittance of the exposed lenses in both systems, the following evaluations were done for the same dose.

3.1 Same Sunglasses Lenses

In Figs. 6 and 7, the transmittance spectrum for two pairs of lenses of the set of 30 pairs are shown. The spectra in the 280 nm - 2000 nm range of right and left lenses of the same model are compared in the period prior to sun exposure. For the right lens 06 132, the transmittance values were $\tau_V = 17.56\%$, $\tau_{SUV} = 1.26\%$ and $\tau_{SIR} = 88.78\%$, while for the left lens $\tau_V = 15.45\%$, $\tau_{SUV} = 1.18\%$ and $\tau_{SIR} = 89.75\%$. For the right lens 10 114, $\tau_V = 12.81\%$, $\tau_{SUV} = 0\%$ and $\tau_{SIR} = 80.52\%$, and for the left lens $\tau_V = 15.39\%$, $\tau_{SUV} = 0\%$ and $\tau_{SIR} = 92.94\%$.

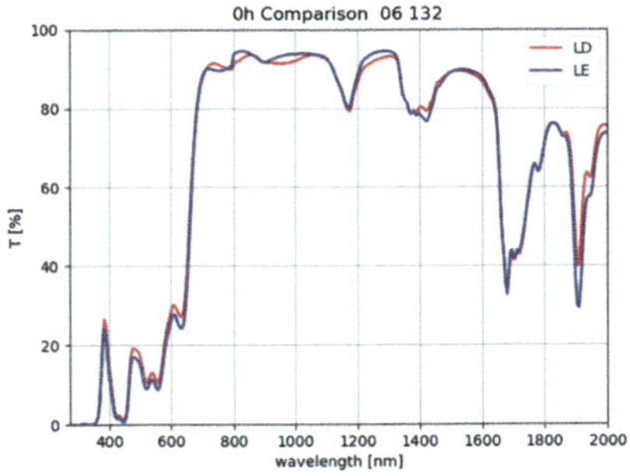

Fig. 6. Transmittance spectrum for the initial lens condition 06 132. LD (right lens) and LE (left lens).

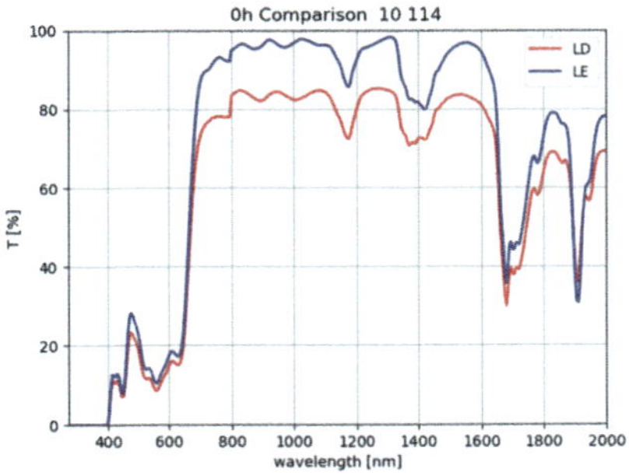

Fig. 7. Transmittance spectrum for the initial lens condition 10 114. LD (right lens) and LE (left lens).

3.2 Exposition in the Solar Station

In Fig. 8 the transmittance spectrum is compared to a lens before and after the exposure period of 115 h at the solar station (approximately 2 rainy weeks). The transmittance increase behavior for the entire range of the spectrum was not observed for all samples, and in some samples the transmittance decrease was observed. This increase was less than 2% of the absolute value of τ_V and τ_{SUV}.

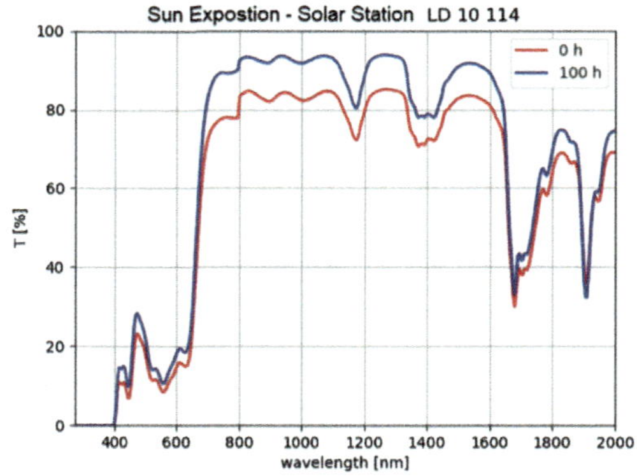

Fig. 8. Comparison of the transmittance spectrum of the right lens 10 114 for the condition before and after the exposure of 115 h in the solar station.

3.3 Exposition in the Solar Simulator

In Fig. 9 the transmittance spectrum is compared to a lens before and after the exposure period of 100 h in the solar simulator. The same observation about the exposure condition in the solar station can be made for the exposure in the simulator, although for this lens specifically, a smaller transmittance difference in the IR band was measured than in the previous case.

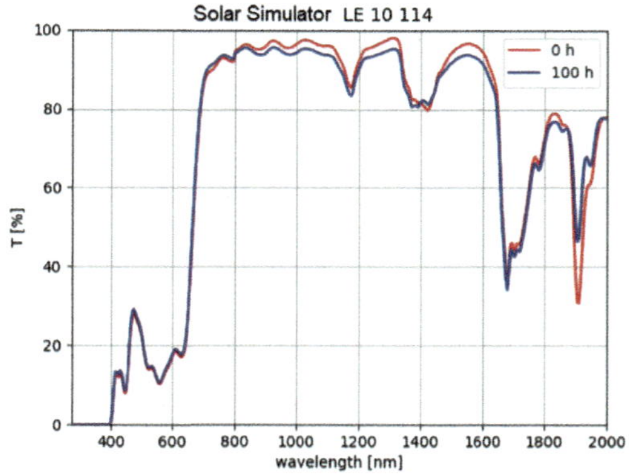

Fig. 9. Comparison of the transmittance spectrum of the left lens 10 114 for the condition before and after the 100 h exposure in the solar simulator.

3.4 Visible, UV and IR Transmittances

Visible (τ_V), UV (τ_{SUV}) and IR (τ_{SIR}) transmittances were calculated for all samples before and after exposure at the station and the solar simulator. The relative transmittance difference of the lenses after exposure with respect to their pre-exposure conditions was calculated. The transmittance variation characteristic is presented in Table 1 for the sun exposure, and in Table 2 for the exposure in the solar simulator. The calculations were done according to the equations presented in ISO 12311: 2013 [4]. Although the transmittance varied for the two exposure conditions, in both cases this difference (increase or decrease) in transmittance was not greater than 2% for τ_V and τ_{SUV}. The τ_{SIR} value has been shown to be more sensitive to the measurement process than the other transmittances, and this is related to the positioning of the samples in the measuring apparatus, not to the degradation by the radiation itself. This fact, that the degradation still does not cause perceptible effects on the transmittances, is evidenced by the fact that there is no trend of variation of the light transmittance in all the samples (only increase).

Table 1. Changing the transmittance of the lenses after 115 h in the solar station.

Condition	τV	τ_{SUV}	τ_{SIR}
Increased	11	6	5
Decreased	19	10	25
None	0	14	0

Table 2. Changing the transmittance of the lenses after 100 h in the solar simulator.

Condition	τ_V	τ_{SUV}	τ_{SIR}
Increased	20	11	14
Decreased	10	4	16
None	0	15	0

3.5 Erythemal Irradiance and UV Dose Calculation

An important parameter in the study of sunglass lens degradation, and subsequent comparison between the types of exposure (station and solar simulator) is the Erithemic UV Radiation Dose (UVR). This value is calculated by means of erythremic irradiance values, which in turn is calculated by converting the recorded UVI values into units of mW/m^2 [15, 16]. Calculation of the cumulative dose value will serve as a reference for both the exposure at the solar station and the solar simulator, since the amount of radiation that causes changes in lens transmittance can be calculated. In order to calculate the dose of UVR in the solar station, the files with UV sensor values for the exposition period and the records of the station's operation were used, which included the opening and closing times, i.e. the duration of the daily exposures. In Fig. 10, the UVI charts for the 02/22/2019, as well as the station's opening and closing periods (high level in blue) are shown. Still in Fig. 10, the graph of the erythremic irradiance calculated from the UVI is shown, as well as the graph of the Dose accumulated during the day.

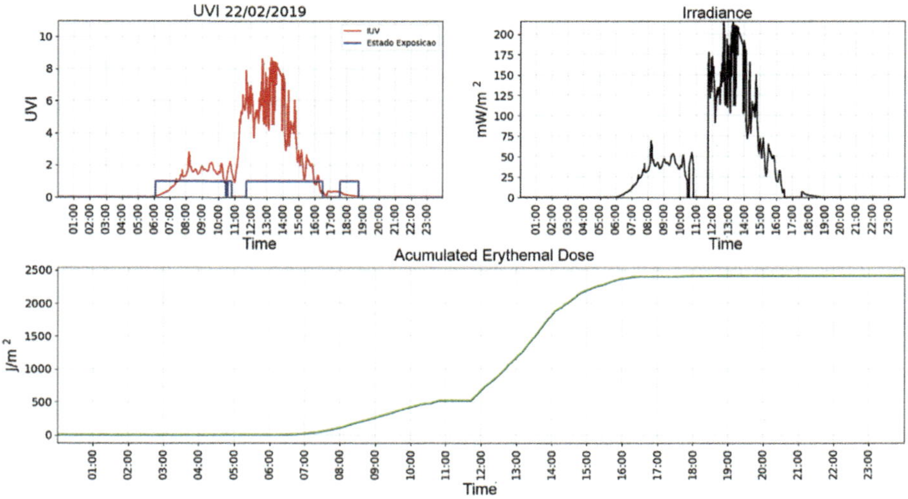

Fig. 10. Representation of the UVI for the 22/02/2019, in which the exposure of the lenses occurred in the solar station; Erythemic irradiance calculated by means of the recorded UVI values and the accumulated Erythemic URV Dose during that day.

For the case of exposure in the solar simulator, the UVI corresponding to the lamp parameters and distance between lamp and lenses was calculated. For the calculation of the UVI, the emission spectrum of the solar simulator lamp, provided by the manufacturer, was used and this spectrum was weighted with the erythemic curve. After this calculation, which is equivalent to a constant UVI during the entire exposure cycle, 100 h, the equivalent Erythmic Dose was calculated.

4 Discussion

In the comparison of the right and left lenses of the same model of sunglasses, minor or major differences in the transmittance spectrum may occur, as shown in Figs. 6 and 7, respectively. In theory, the transmittance spectrum of both lenses of the same pair of sunglasses should be as close as possible, but differences may occur in manufacturing. In the case of gradient lenses, the values of the transmittance spectrum are very sensitive to the position the lens is placed in the holder and the measurement is made, and therefore one must always try to position with reference in the geometric center of the lens.

The transmittance spectrum of right lens 10 114 (Fig. 8) showed higher values after exposure for 115 h in the solar station. The figure shows that there is no transmittance of the UV region (below 400 nm), and after calculating the transmittance values, an increase of τ_V of 12.85% to 15.86% was observed, and τ_{SIR} increased from 80.52% to 89.45%.

For the same 10 114 lens, but the left side, the transmittance spectrum is shown in Fig. 9. After exposure for 100 h in the solar simulator, there was an increase of τ_V from 15.39% to 16.02%, and a decrease of τ_{SIR} from 92.94% to 91.29%. As in the case of the right lens, the left lens did not show transmittance in the UV range.

When analyzing the variation of the transmittances of the whole set of lenses for the two solar exposition experiments (Tables 1 and 2), it was observed that after the exposure in the solar simulator, there was a small tendency to increase both τ_V and τ_{SUV}. Lenses that did not show transmittance to the region of the specified spectrum were marked as unchanged. In the case of exposure in the solar station, a tendency was observed in the decrease of the three transmittances (Visible, UV and IR).

After the processing of the files of the sensors of the station and of the events of the solar station, it was possible to calculate the accumulated UVR dose for the entire exposure period, which was 115.09 h. The calculated value of the cumulative dose considered the UVI sensor records only when the lenses were being exposed to the Sun, and for the exposure period analyzed the dose was 29.465 kJ/m^2.

In the solar simulator, the accumulated erythemal UVR dose was also calculated, since this value will be used in the future to compare the two methods of sun exposure. Considering that the UV lamp and the distance from the samples to the lamp are equivalent to a constant UVI of 2.787, for the 100 h cycle the dose value was 25.082 kJ/m^2.

5 Conclusion

The analysis of the variation of the transmittances of sunglasses before and after exposure in two different systems (station and solar simulator) is part of a work in progress. A more evident trend is expected as the exposure period increases in both systems, and this will result in degradation of sunglasses lenses.

Although the exposure dose values for the two methods were close, in this work we are comparing the effect of radiation to the before and after lenses on each exposure method, rather than between the methods. In the future, the dose value will be used as the criterion for closing the exposure of the lenses, so a direct comparison between the two methods studied can be made: the solar simulator and the solar station.

Acknowledgements. The authors are thankful for FAPESP 2018/16275-2.

References

1. ASSOCIAÇÃO BRASILEIRA DE NORMAS TÉCNICAS. NBR ISO 12312-1: Proteção dos olhos e do rosto – óculos para proteção solar e óculos relacionados – parte 1: óculos para proteção solar para uso geral, Rio de Janeiro, 26 p. (2015)
2. Chou, B.R., Dain, S.J., Cheng, B.B.: Effect of ultraviolet exposure on impact resistance of ophthalmic lenses. Optom. Vis. Sci. **92**(12), 1154–1160 (2015)
3. Masili, M., Schiabel, H., Ventura, L.: Contribution to the radiation protection. Radiat. Prot. Dosimetry, 1–9 (2014)
4. International Standard Organization ISO 12311: Personal protective equipment—Test methods for sunglasses and related eyewear (2013)
5. Masili, M., Ventura, L.: Equivalence between solar irradiance and solar simulators in aging tests of sunglasses. Biomed. Eng. Online (Online), 86–98 (2016)
6. Masili, M., Duarte, F.O., White, C.C., Ventura, L.: Degradation of sunglasses filters after long-term irradiation within solar simulator. Eng. Fail. Anal. **1**, 1 (2019)
7. Ventura, L., et al.: ABNT NBR 15111:2013 - Óculos para proteção solar, filtros para proteção solar para uso geral e filtros para observação direta do sol. ABNT **1**, 1–47 (2013)
8. Masili, M., Ventura, L.: Building a resistance to ignition testing device for sunglasses and analysing data: a continuing study for sunglasses standards. Biomed. Eng. Online **16**, 1–10 (2017)
9. Loureiro, A.D., Gomes, L., Ventura, L.: Transmittance variations analysis in sunglasses lenses post sun exposure. J. Phys. Conference Series (Print) **733**, 012028 (2016)
10. Mello, M.M., Ventura, L.: Cálculo da transmitância visível e semafórica com função de ponderação alternativa. Revista Brasileira de Física Médica (Online) **7**, 99–104 (2013)
11. Lincoln, V.A.C., Ventura, L., Faria e Sousa, S.J.: Ultraviolet analysis of donated corneas: a portable prototype. Appl. Opt. **49**, 4890–4897 (2010)
12. Mello, M.M., Lincoln, V.A.C., Ventura, L.: Self-service kiosk for testing sunglasses. Biomed. Eng. Online (Online) **13**, 45 (2014)
13. Gomes, L.M., Masili, M., Momesso, G.A., Silva, F.M., Ventura, L.: Solar exposure of sunglasses: aging test display. In: Proceedings SPIE 10474, Ophthalmic Technologies XXVIII, 104741 N (2018)
14. LEMA: Unit to determine the stability to UV radiation. Fonte: LemaParma. http://www.lemaparma.it/old/pdf/ottica.strumenti.controllo.xenon.suntest.pdf. Accessed 23 Mar 2019
15. de Paula Corrêa, M.: Solar ultraviolet radiation: properties, characteristics and amounts observed in Brazil and South America. An. Bras. Dermatol. **90**(3) (2015)
16. Masili, M., Schiabel, H., Ventura, L.: Cálculo da Irradiância e Exposição Radiante para a Radiação UV Solar: Contribuição Das Componentes Direta, Difusa e Refletida. Anais do XXIV Congresso Brasileiro de Engenharia Biomédica - CBEB 2014, Uberlândia (2014)

Eye Scan Ultrasound System for Automatic Cataract Detection: From a Preclinical to a Clinical Prototype

Lorena Petrella[1(✉)], Marco Gomes[1,2], Fernando Perdigão[1,2], Mario Santos[1], Paulo Fernandes[1,3], Carlos Pinto[1,2], Sandrina Nunes[4], Miguel Morgado[3], Miguel Caixinha[1,5], and Jaime Santos[1]

[1] Department of Electrical and Computer Engineering, University of Coimbra, 3030-790 Coimbra, Portugal
lorena.petrella@gmail.com, miguel.caixinha@gmail.com
[2] Instituto de Telecomunicações – Pole of Coimbra, University of Coimbra, 3030-790 Coimbra, Portugal
[3] Department of Physics, University of Coimbra, 3004-516 Coimbra, Portugal
[4] Association for Innovation and Biomedical Research on Light and Image, 3000-548 Coimbra, Portugal
[5] Department of Physics, University of Beira Interior, 6291-001 Covilhã, Portugal

Abstract. Cataract is a pathology associated to the loss of the normal lens transparency, and its progression can result in a total loss of vision. The gold standard diagnostic method consists on qualitative observation through a slit lamp coupled to a microscope. This method has two limitations: incipient cataract may not be detected and the cataract hardness is subjectively evaluated. It can result on a late diagnosis and/or on surgical complications when phacoemulsification surgery (the most common therapeutic approach) is conducted. On this study we present a new technology for objective cataract characterization, based on ultrasounds. The Eye Scan Ultrasound System (ESUS) allows the characterization of cataract type, severity and hardness based on features extracted from A-scan signals, working at a nominal frequency of 20 MHz. A total of 27 features were extracted and analyzed through automatic classification algorithms. Preclinical and clinical prototypes of the ESUS have been implemented. From data collected on the preclinical phase, the system precision, sensitivity and specificity values were 99.7%, with a relative absolute error of 0.4%. Clinical phase is in progress.

Keywords: Cataract · Ultrasounds · Diagnostic

1 Introduction

Cataract is a pathology associated to the loss of the normal lens transparency, and its progression can result in a total loss of vision. It is originated from abnormal protein aggregation, disrupting the protein arrangement. According to the World Health Organization, cataract is the most frequent cause of blindness, being responsible for

J. Henriques et al. (Eds.): MEDICON 2019, IFMBE Proceedings 76, pp. 811–819, 2020.
https://doi.org/10.1007/978-3-030-31635-8_97

about 51% of the total registered cases [1]. About 10% of the active population present cataract related with aging [2]. Currently, cataract is exclusively treated with surgical procedures, being the phacoemulsification surgery the most common technique on developed countries. Nevertheless, the development of pharmacological therapies has gained increasing attention in the last years [3, 4].

For clinical prognostic and therapeutic purposes, it is therefore decisive the early cataract detection and the classification of its evolution phase. The gold standard diagnostic method consists on the non-invasive observation through a slit lamp coupled to a microscope. A physician characterizes type and severity of the cataracts by qualitative analysis of the lens images. Nevertheless, this technique has some limitations. On the one hand, incipient cataract could not be identified through slit lamp observations, which can retard the diagnostic and increase therapeutic difficulties. On the other hand, the cataract hardness estimation is quite subjective by this technique, and it is directly related to the phacoemulsification energy to be applied on surgeries. Several surgical complications are associated with the use of excessive phacoemulsification energies [5], which may lead to partial or total vision loss.

The use of ultrasound techniques in ophthalmology emerged in the sixties, making use of both A-scan and B-scan modes [6]. Time and frequency domain properties of the A-scan backscattered signals convey significant diagnostic information from the tissues where it propagates. Initially, A-scan signals were used on axial length measurements, and posteriorly extend to the characterization of diverse pathologies as being the presence of intraocular foreign bodies, retinal detachment, hemorrhage, tumors, among others. Moreover, attenuation and scattering of A-scan signals have been related to different cataract severities [7], while sound velocity and attenuation coefficient have been related to cataract hardness and phacoemulsification energy on *ex vivo* animal tests [8]. On the other hand, high-frequency B-mode characterization of diverse ocular pathologies was first published in 1991 [9]. Pathologies such as conjunctiva, cornea and sclera disorders, glaucoma, anterior segment tumors, trauma, and others, were characterized.

The research team, under the research project CATARATA, conducted an innovative study in animal models of cataract, based on A-scan signals analysis (in the range of 20 MHz–25 MHz), using a custom made eye scan ultrasound system (ESUS). It was possible to detect incipient cataract, characterize their severity as well as cataract hardness evaluation; on this sense two approaches were implemented. The first approach was conducted through *ex vivo* porcine lenses [10–12], where the cataract model was obtained by immersing the lens on ethanol:2-propanol:formalin solution (3:3:4) being the cataract severity controlled by the immersion time. The second approach was based on *in vivo* rodent (Wistar rats) cataract models [13]. Cataract was induced by administration of a single subcutaneous injection of sodium selenite, being the cataract severity dosage-dependent.

In the present study we describe the main characteristics of the hardware and software constituting the ESUS, as well as their performance assessment. Moreover, a new research project CATARACTUS, being conducted by the same research team, extends the ESUS applications to clinical studies. In this sense, we also present here the main adaptations introduced to the system and discuss about the expected performance.

2 Materials and Methods

2.1 Preclinical Prototype

The Eye Scan Ultrasound System (ESUS) constructed for animal studies, consist on a 20 MHz probe (custom made by Imasonic SAS, Bourgogne-Franche-Comté, France), with a focal length of 8.9 mm and an active diameter of 3 mm (Fig. 1).

The probe is connected to a pulser/receiver (Compact Pulser, US Ultratek Inc., Los Angeles, USA), which in turn connects to an A/D converter, 14 bits per sample and a sampling frequency of 100 MHz (FCM 150, 4DSP LLC™, Texas, USA). The signal-processing unit was built up on a ZeadBoard, a low-cost and easy reprogrammable development board, based in the System-on-chip (SoC) Zynq-7000 architecture of Xilinx. The cataract classification algorithms are firstly developed in Waikato Environment for Knowledge Analysis (WEKA) software [14], and then transferred to the ZeadBoard unit for real-time operation.

This prototype is protected through the Patent PT109646 B [15].

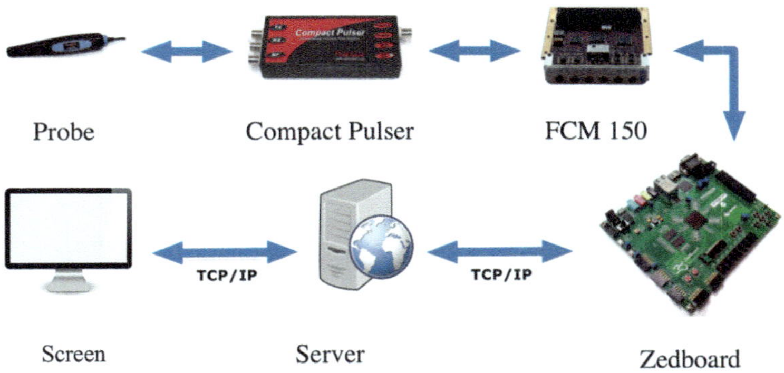

Fig. 1. ESUS design for preclinical studies.

2.2 Clinical Prototype

For human applications, some modifications were introduced in the hardware to provide a more compact and reliable system, in compliance with the international standard ISO 14155:2018 and the regulation (EU) 2017/745 [16, 17], as described below (Fig. 2). The probe (the same specified in 2.1) is connected to a data acquisition board xSCAN (custom made by Tribosonics Ltd., Sheffield, UK). xSCAN assemble diverse functions, namely, creates a tone burst for probe excitation, digitalized the received signals, implement the signal conditioning and communicates with the computer. The main xSCAN characteristics are specified in Table 1. The output of the acquisition board is connected to a computer (VM65 N, AsusTek Computer Inc., Nieuwegein, Netherlands) via Ethernet.

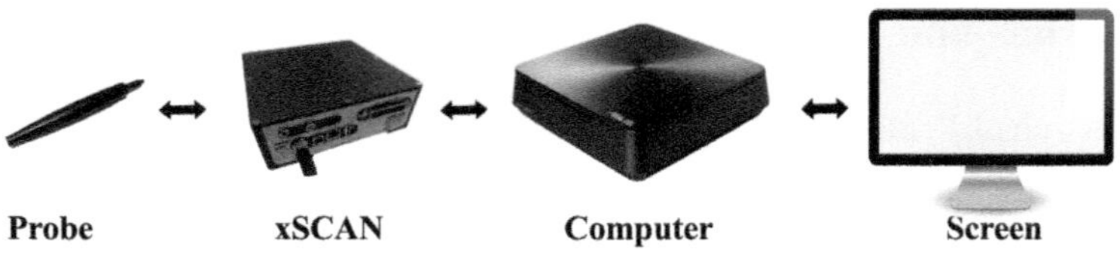

Fig. 2. ESUS design for clinical studies.

The signal analysis is implemented in C++ and MATLAB and the user's interface is constructed in MATLAB. This interface allows the acquisition configuration, the real-time visualization of A-scans, to save the acquired signals and is programmed for automatic cataract detection and classification.

Table 1. Main characteristics of the xSCAN system.

Characteristic	Value
Ultrasonic channels	2
Sampling frequency	100 MHz
Pulse voltage	±55 V
Pulse repetition frequency	Up to 10 kHz
Pulse width	Up to 5 μs
Pulse train	Up to 5 cycles
Pulse type	Variable
Amplifier gain	Up to 80 dB
Ethernet	Up to 20 MB/s

2.3 ESUS Operation

The implemented system stands out on three main aspects (Fig. 3): (1) from the A-scan backscattered signal, the system automatically recognizes ocular structures interfaces and locates cataract; (2) the system performs real-time detection, classification and hardness-characterization of cataract; (3) the system is reconfigurable and has learning capabilities.

The ESUS software operation applies for both, preclinical and clinical applications, and the steps involved on signal analysis are described below.

Database. The preclinical study included the signals collected from 14 control (healthy) rats, 12 rats with incipient cataract, 13 with moderate cataract and 11 with severe cataract. For each animal, the two eyes were studied. For clinical applications the data collection will begin after the clinical trial is approved by the regulatory authorities.

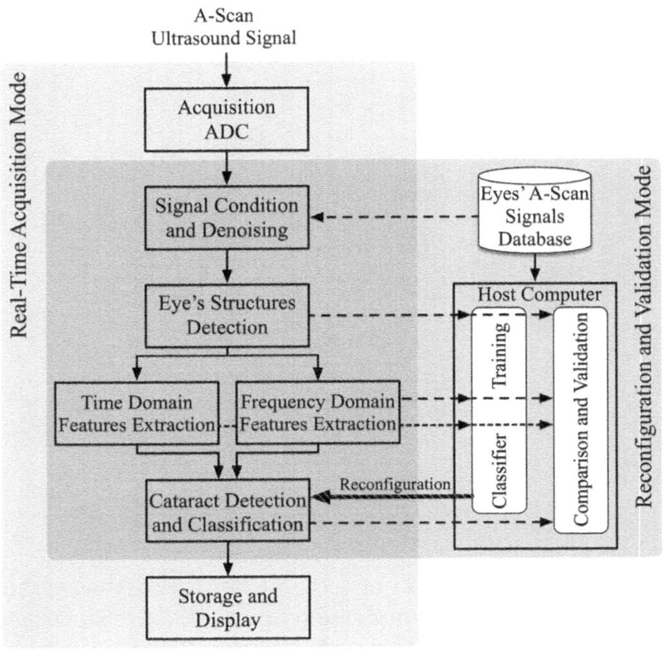

Fig. 3. ESUS operation

Signal Conditioning and Denoising. The first step consists on reducing signal noise and increase the SNR. In this sense, a band-pass filter with cutoff frequencies of 1 MHz and 50 MHz was applied, followed of a moving average of 20 A-scans.

Eye's Structures Detection. From the averaged signal, reflections from the diverse ocular interfaces can be identified (Fig. 4). The first two high-amplitude echoes correspond to the anterior and posterior surfaces of the cornea, that overlap in time. The third echo arises from to the anterior lens capsule interface. Following, under the presence of cataract, a strong echo from the cataract location is also observed, followed of a weak reflection from the posterior lens capsule interface. Retina, choroid and periorbital fat also present weak echoes.

Algorithms were developed for the automatic detection of the eye's interfaces of interest, i.e. cornea, anterior lens capsule and posterior lens capsule. The starting point consists on the identification of the global signal maximum, i.e. the emitted pulse. Then, the algorithm defines successive searching windows for the remaining interfaces, within a predefined interval. These searching intervals are estimated based on prior knowledge of the distances between successive interfaces, and the propagation time. Within these windows, local maxima are identified (more details about interfaces identification are given in [13]).

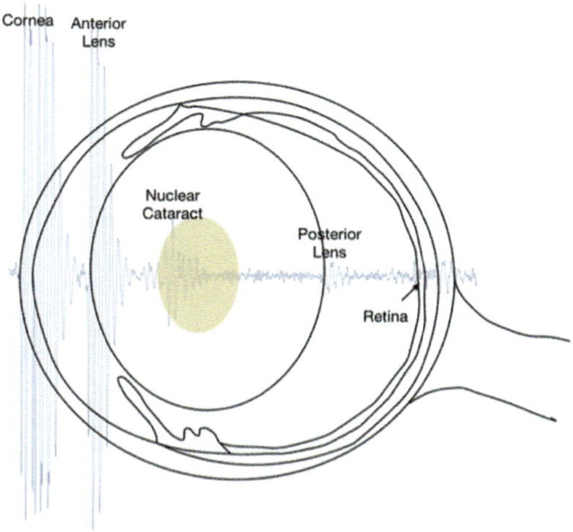

Fig. 4. A-scan signal from rat's eye and structures identifications.

Feature Extraction. Once the anterior and posterior lens capsules echoes have been identified, features were extracted from the backscattered signal contained within the lens. Twenty-seven features were extracted, including time domain features (f1–f10), statistical time domain features (f11–f19) and frequency-derived features (f20–f27), as specified in Table 2.

Table 2. Features extracted from backscattered signals.

N°	Description
f1	Propagation time in the lens path
f2	Maximum absolute amplitude of the signal inside the lens
f3	Propagation time until maximum absolute amplitude of the signal inside the lens
f4	Normalized propagation time (f1/f3)
f5	Decaying time with a threshold of 80% of maximum amplitude
f6	Maximum duration of an excursion of the absolute amplitude above the threshold
f7	The starting instant where f6 occurs
f8	The cumulative duration of all excursions above the threshold
f9	The area of the envelope above the threshold
f10	Maximum excursion of the signal inside the lens
f11	Ratio of the SD of the absolute value of the signal inside the lens by the SD of the absolute value of the anterior lens reflection
f12	Maximum signal excursion of the signal of all frames along the lens
f13	Initial position of the frame where f12 occurs
f14	Energy of the frame where f12 occurs
f15	SD of the frame where f12 occurs

(*continued*)

Table 2. (*continued*)

Nº	Description
f16	RMS value of the frame where f12 occurs
f17	The anterior to posterior lens amplitude ratio
f18	The ratio of kurtosis coefficients of the anterior and posterior lens reflections
f19	The ratio of spectral flatness measures of the posterior and anterior lens
f20	The ratio of roll-off 95% BWs of the anterior and posterior lens reflection
f21	The ratio of spectral centroids of the anterior and posterior lens reflection
f22	Frequency downshift from anterior to posterior lens reflection
f23	Central frequency deviation (f22/f1)
f24	The y-intercept value of the linear fitted transfer function
f25	The slope of the linear fitted transfer function
f26	Quadratic fitting coefficient
f27	Attenuation coefficient value in the lens for 10 MHz

For each averaged A-scan, a feature vector was produced. For every fifty feature vectors, a final vector was computed consisting on the median value for each feature.

Cataract Detection and Classification. In attempt to optimize the classification performance, the extracted features were firstly subject to principal component analysis (PCA). The number of generated components was defined such that 95% of the variance of the original data was explained.

Diverse classifiers including Bayes network, multilayer perceptron, random forest and support vector machine (SVM) were evaluated. For classifiers' training, a k-fold cross-validation was used with k = 10. The classification performance was evaluated through F-measure, accuracy, precision, sensibility and specificity measurements.

Reconfiguration and Validation. The system encompasses learning, reconfiguration and reprogramming capabilities. The classifier was designed to allow consecutive retraining based on the reference database that grows with new-collected data (after validation by the physician). This allows constant improvements of the system performance.

2.4 Cataract Hardness Assessment

The cataract hardness was also correlated with the extracted features. The hardness values used as reference were measured *ex vivo* through the NanoTest (Micro Materials Ltd, Wrexham, UK), including cortex and nucleus regions.

3 Results

The implemented algorithms successfully identified the interfaces of eye's structures (lens thickness for *ex vivo* measurement = 3.52 ± 0.49 mm vs. *in vivo* estimation = 3.54 ± 0.01 mm). The algorithms address the challenge reported by other authors [7], since the posterior lens capsule echo present low amplitude, which in some cases lies on the same order of magnitude of the noise.

All the extracted features showed a statistically significant difference between healthy lens and that with cataract, as well as between the diverse degree of severity (incipient, moderate and severe), with $p < 0.01$.

All the classifiers presented performance higher than 95% for the classification of cataract severity, while the best performance was obtained for the SVM classifier, attaining precision, sensitivity and specificity values of 99.7%, with a relative absolute error of 0.4%.

The feature reduction process using PCA led to twelve components, which accounted for 95.1% of the cumulative variance of the features. Also, in this case, the SVM classifier presented the highest performance with sensitivity and specificity of 99.2%, and with a relative absolute error of 1.1%. However, the data dimensionality reduction obtained by the PCA did not improve the classification performance.

Moreover, it was confirmed that lens hardness increases with cataract severity, and a moderate to good correlation was found between lens hardness and 23 of the studied features.

4 Discussion and Conclusion

The ESUS represents a new technological approach for cataract diagnosis and therapeutic planning, which solves the limitations found with the standard diagnostic methods. It is a minimally invasive device, with a simple clinical procedure, which can present objective cataract classification in real-time. One disadvantage of this method compared to the standard diagnostic method (slit lamps), is that the probe-eye coupling can cause a minor discomfort on patients. However, the benefits that could be obtained with this technique, overcomes this negative aspect.

So far, only preclinical data have been collected, while the beginning of clinical data collection is expected shortly. Despite algorithms should be adapted to human data, it is reasonable to expect a similar performance as in preclinical studies, since the hardware and the logistic to be used are similar in both cases.

Finally, we expect that ESUS results in a safe and reliable diagnostic technique, able to objectively classify cataract according their type and severity, as well as estimates cataract hardness and the optimal energy to be applied on phacoemulsification surgeries. The expected benefits are to avoid surgical procedures on incipient cataract cases, turn surgical procedures safer when phacoemulsification surgery must be conducted, and to reduce surgical times.

Acknowledgment. This work was supported by national funds and co-funded by POCI – COMPETE2020 and by FEDER – PT2020 partnership, through the Portuguese Foundation for Science and Technology (FCT) under the projects CATARATA (PTDC/DTP-PIC/0419/2012) and CATARACTUS (POCI-01-0145-FEDER-028758), and by UID/EEA/50008/2019 funds.

Conflict of Interest. The authors declare that there is no conflict of interest.

References

1. WHO: Global data on visual impairements, vol. 2012, p. 17 (2010)
2. Nowak, M.S., Smigielski, J.: The prevalence of age-related eye diseases and cataract surgery among older adults in the city of Lodz, Poland. J. Ophthalmol. 1–7 (2015). https://doi.org/10.1155/2015/605814
3. Abdelkader, H., Alany, R.G., Pierscionek, B.: Age-related cataract and drug therapy: opportunities and challenges for topical antioxidant delivery to the lens. J. Pharm. Pharmacol. **67**, 537–550 (2015). https://doi.org/10.1111/jphp.12355
4. Zhao, L., Chen, X.-J., Zhu, J., Xi, Y.-B., Yang, X., Hu, L.-D., et al.: Lanosterol reverses protein aggregation in cataracts. Nature **523**, 607–611 (2015). https://doi.org/10.1038/nature14650
5. Chan, E., Mahroo, O.A.R., Spalton, D.J.: Complications of cataract surgery. Clin. Exp. Optom. **93**, 379–389 (2010). https://doi.org/10.1111/j.1444-0938.2010.00516.x
6. Silverman, R.H.: Focused ultrasound in ophthalmology. Clin. Ophthalmol. **10**, 1865–1875 (2016)
7. Paunksnis, A., Kurapkiene, S., Maciulis, A., Kopustinskas, A., Paunksniene, M.: Ultrasound quantitative evaluation of human eye cataract. Informatica **18**, 267–278 (2007)
8. Huang, C.C., Ameri, H., DeBoer, C., Rowley, A.P., Xu, X., Sun, L., et al.: Evaluation of lens hardness in cataract surgery using high-frequency ultrasonic parameters in vitro. Ultrasound Med. Biol. **33**, 1609–1616 (2007). https://doi.org/10.1016/j.ultrasmedbio.2007.05.002
9. Pavlin, C.J., Harasiewicz, K., Sherar, M.D., Foster, F.S.: Clinical use of ultrasound biomicroscopy. Ophthalmology **98**, 287–295 (1991). https://doi.org/10.1016/S0161-6420(91)32298-X
10. Caixinha, M., Jesus, D.A., Velte, E., Santos, M.J., Santos, J.B.: Using ultrasound backscattering signals and Nakagami statistical distribution to assess regional cataract hardness. IEEE Trans. Biomed. Eng. **61**, 2921–2929 (2014). https://doi.org/10.1109/TBME.2014.2335739
11. Caixinha, M., Santos, M., Santos, J.: Automatic cataract hardness classification ex vivo by ultrasound techniques. Ultrasound Med. Biol. **42**, 989–998 (2016). https://doi.org/10.1016/j.ultrasmedbio.2015.11.021
12. Caxinha, M., Velte, E., Santos, M., Perdigão, F., Amaro, J., Gomes, M., et al.: Automatic cataract classification based on ultrasound technique using machine learning: a comparative study. Phys. Proc. **70**, 1221–1224 (2015). https://doi.org/10.1016/j.phpro.2015.08.263
13. Caixinha, M., Amaro, J., Santos, M., Perdigao, F., Gomes, M., Santos, J.: In-Vivo automatic nuclear cataract detection and classification in an animal model by ultrasounds. IEEE Trans. Biomed. Eng. **63**, 2326–2335 (2016). https://doi.org/10.1109/tbme.2016.2527787
14. Pal, C., Frank, E., Hall, M., Witten, I.: Data Mining: Practical Machine Learning Tools and Techniques, 4th edn. Elsevier, Cambridge (2017). https://doi.org/10.1016/b978-0-12-804291-5.00015-5
15. Santos, J.B., Gomes, M., Santos, M., Perdigão, F., Morgado, M., Caixinha, M., et al.: Sistema ultrassónico de exame ocular para deteção precoce e classificação da catarata em tempo real. PT109646B (2019)
16. International Organization for Standarization: ISO 14155:2018 Clinical investigation of medical devices for human subjects - Good clinical practice (2018). ISO/DIS 14155:2018(E)
17. Council of the European Union: European Parlament. Regulation (EU) 2017/745 of the European Parliament and of the Council (2017). JOL_2017_117_R_0001

Regular Sessions: Information Technology in Health Systems

Utilizing Incremental Learning for the Prediction of Disease Outcomes Across Distributed Clinical Data: A Framework and a Case Study

Vasileios C. Pezoulas[1], Themis P. Exarchos[1,2], Konstantina D. Kourou[1,3], Athanasios G. Tzioufas[4], Salvatore De Vita[5], and Dimitrios I. Fotiadis[1(✉)]

[1] Unit of Medical Technology and Intelligent Information Systems, University of Ioannina, 45110 Ioannina, Greece
fotiadis@cc.uoi.gr
[2] Department of Informatics, Ionian University, 49100 Corfu, Greece
[3] Institute of Molecular Biology and Biotechnology (IMBB), Foundation for Research and Technology Hellas (FORTH), 45110 Ioannina, Greece
[4] Department of Pathophysiology, National and Kapodistrian University of Athens, 15772 Athens, Greece
[5] Department of Medical and Biological Sciences, Udine University, 33100 Udine, Italy

Abstract. In this work, we highlight the need of a supervised learning framework for disease predictive modeling across distributed clinical data to overcome the privacy limitations that are introduced by centralized analysis. Towards this direction, a computational framework is proposed, consisting of six incremental learning algorithms that are based on Stochastic Gradient Descent, Naïve Bayes, and Gradient Boosting Trees, to provide new insight on the construction of supervised learning models across clinical data that are stored in multiple locations. The applicability of the proposed framework is demonstrated through a preliminary case study, where a distributed lymphoma prediction model is constructed across private cloud spaces that consist of clinical data from patients that have been diagnosed with primary Sjögren's Syndrome (pSS). Our results reveal the dominance of the Gradient Boosting Trees, yielding an average accuracy 91.6% and sensitivity 87.5% towards the correct identification of lymphoma cases.

Keywords: Predictive modeling · Supervised learning · Incremental learning · Distributed environments

1 Introduction

In healthcare, distributed environments [1–3] have gained a lot of attention due to the overwhelming need to process massive amounts of daily generated medical data from different clinical centers, in distributed databases, towards the development of prediction models for disease management (e.g., risk stratification models for chronic

J. Henriques et al. (Eds.): MEDICON 2019, IFMBE Proceedings 76, pp. 823–831, 2020.
https://doi.org/10.1007/978-3-030-31635-8_98

diseases). These large volumes of daily accumulated data, known as big data, require high computational resources, a fact that poses significant barriers towards the effective processing of these data [1–3]. An additional but crucial demand though, is the need to develop disease prediction models across clinical data that are stored in dispersed sites, without the data leaving these sites at all [4]. A centralized database model is prone to privacy breach due to the underlying legal, ethical and security threats [2, 5]. Moreover, the application of conventional machine learning algorithms for distributed learning is not always feasible due to the non-convex optimization strategy that the majority follows [6].

Towards this direction, several batch processing methods have been proposed so far to effectively deal with the need to develop distributed learning models [6–8]. Online learning [7, 9] is an example of such an approach which manages to update a training model according to a cost function that considers for additional information from newly coming data. The online learning approach uses stochastic optimization methods to update an existing model on upcoming training samples by minimizing a global cost function, where the model can be either updated on an individual upcoming sample or on a series of accumulated samples (known as mini-batch processing) [7]. Online learning can be realized using linear supervised learning algorithms, such as, Support Vector Machines (SVMs) with linear kernels [10], hybrid online learning methods based on non-linear kernels [11], and stochastic gradient descent (SGD) approaches for convex optimization [12], among others.

Another interesting approach that shares many similarities with the former one, is incremental learning [6, 7, 13], which aims to adapt an existing data model on constantly upcoming data streams. In contrast to online learning, the latter does not always need to be applied at an "online" manner. Incremental learning is a batch processing method that can be used to train a supervised learning model on an initial batch, and then adjust the existing model on a series of upcoming batches, due to its ability to solve additive cost problems [7, 13]. This makes incremental learning ideal in the case where large-scale data need to processed "offline", i.e., when the data are already stored in distributed sites (e.g., dispersed databases). The existing incremental learning methods are based on SGD for convex optimization [13, 14], stochastic gradient boosting based on ensemble classifiers for achieving high performance [15], Multinomial Naïve Bayes and variants [7, 16], etc. Meta-learning [17] techniques have been also proposed for combining individual classification outcomes (metadata) from distributed data, which, however, suffer from biases that are introduced during the assembly stage [18].

Here, we take advantage of the mathematical basis of incremental learning to deploy a computationally efficient and secure strategy for disease prediction modeling across data that are stored in multiple sites by assuming that: (i) the clinical data are stored in private cloud spaces and are harmonized through semantic interlinking methods [19], and (ii) the model for predicting disease outcomes is continuously adjusted to the data that lie across private cloud spaces. A case study is conducted to highlight the applicability of the framework by constructing a distributed binary lymphoma prediction model across private cloud spaces consisting of clinical data from patients with primary Sjögren's Syndrome (pSS); a chronic autoimmune disease that exhibits salivary gland dysfunction, with 5% of the pSS patients being prone to lymphoma development [20]. Our results demonstrate the superiority of the distributed trees towards the precise discrimination of lymphoma cases, against the conventional methods, with an average accuracy 91.6%, area under the curve score 94%, and sensitivity 87.5%.

2 Incremental Learning Strategies

2.1 Stochastic Gradient Descent

Let's assume that we have a set of training observations, i.e., clinical parameters, $\{(\boldsymbol{x}_1, y_1), (\boldsymbol{x}_2, y_2), \ldots, (\boldsymbol{x}_N, y_N)\}$, where $\boldsymbol{x}_i \in \boldsymbol{R}^N$, and y_i is the disease outcome. For the binary classification problem, we assume that $y_i \in \{-1, +1\}$. The goal of the stochastic gradient descent (SGD) approach is to seek for a linear loss function, say $\varphi(f(\boldsymbol{x}_i), \boldsymbol{y}_i)$, that minimizes the following equation [14]:

$$E(\boldsymbol{w}) = argmin\left(\frac{1}{N}\sum\nolimits_{i=1}^{N} \varphi(f(\boldsymbol{x}_i), \boldsymbol{y}_i) + ar(\boldsymbol{w})\right), \tag{1}$$

where $\boldsymbol{w}$ is a weight vector, $\varphi(.)$ is a loss function, a is a positive hyperparameter, $r(\boldsymbol{w})$ is a regularizer, $f(\boldsymbol{x}_i)$ is a linear function with an interception parameter b [14]:

$$f(\boldsymbol{x}_i) = \boldsymbol{w}^T \boldsymbol{x}_i + b. \tag{2}$$

Solving (1) yields the SGD update rule [12, 14]:

$$\tilde{w}_t = \tilde{w}_{t-1} - \eta_t \left(\frac{\partial \varphi(f(\boldsymbol{x}_i), \boldsymbol{y}_i)}{\partial w} + a\frac{\partial r(w)}{\partial w}\right), \tag{3}$$

where t is the time step, $\tilde{w}_{t-1}$ is the weight estimation at step $t-1$, η_t is a non-negative learning rate parameter, S^{-1} is a constant matrix which is used for the convergence, and $\varphi'(.)$ is the gradient of the loss function $\varphi(.)$.

By replacing the loss function in (3) we can obtain different binary classifiers. For example, if we replace the regularization term in (3): (i) with a logistic loss function: $\varphi(f(\boldsymbol{x}_i), \boldsymbol{y}_i) = ln(1 + exp(-\boldsymbol{y}_i f(\boldsymbol{x}_i)))$, we can obtain the logistic regression classifier, (ii) with the hinge loss function: $\varphi(f(\boldsymbol{x}_i), \boldsymbol{y}_i) = max(0, 1 - \boldsymbol{y}_i f(\boldsymbol{x}_i))$, we can then obtain the Support Vector Machines (SVM) classifier, and (iii) with the Perceptron loss function: $\varphi(f(\boldsymbol{x}_i), \boldsymbol{y}_i) = max(0, -\boldsymbol{y}_i f(\boldsymbol{x}_i))$, we can obtain the Perceptron classifier. The regularization function, $r(w)$, can be also replaced to derive either an $l1$- or $l2$-norm regularization, or the elastic-net regularization for a combination of both [21].

2.2 Multinomial Naïve Bayes

The Multinomial Naïve Bayes (NB) classifier is preferred in the case where the features are discrete [16]. Given an N-dimensional input vector, assume $x = (x_1, x_2, \ldots, x_N)$, where each x_i is the frequency of an event t_i, the probability that x belongs to the class, assume c_k, is given by the multinomial distribution [16]:

$$P(c_k|\boldsymbol{x}) \propto P(c_k)\prod\nolimits_{i=1}^{N} P(t_i|c_k), \tag{4}$$

where $P(t_i|c_k)$ is the conditional probability of t_i occurring in class c_k. By deriving the logarithmic expression of (4), we can calculate the class with the highest probability or the maximum a-posterior (MAP) class, as in [16]:

$$c_{MAP} = argmax_{c_k}\left[log(P(c_k)) + \sum\nolimits_{i=1}^{N} log(P(t_i|c_k))\right], \tag{5}$$

which can then be solved as a linear equation [16].

2.3 Gradient Boosting Using Regression Trees

The gradient boosting method has been widely adopted as a straightforward approach for creating additive regression models by fitting pre-defined base learners to minimize a loss function (e.g., a logistic loss) using pseudo-residuals [15]. Given a set of N-observations $\{(\boldsymbol{x}_1, y_1), (\boldsymbol{x}_2, y_2), \ldots, (\boldsymbol{x}_N, y_N)\}$, where $\boldsymbol{x}_i \in \boldsymbol{R}^N$, the goal is to obtain an estimated function, $\tilde{F}(\boldsymbol{x})$, mapping $\boldsymbol{x}$ to y, that minimizes the expected value of a loss function, assume $L(y, F(\boldsymbol{x}))$. The gradient boosting algorithm incrementally seeks for estimations of the mapper at a stage $m \in M$, assume $F_m(x)$, as follows [15]:

$$F_m(\boldsymbol{x}) = F_{m-1}(\boldsymbol{x}) + p_m \cdot h(\boldsymbol{x}; \boldsymbol{a_m}), \tag{6}$$

where p_m is the line search [15]:

$$p_m = argmin_p \sum\nolimits_{i=1}^{N} \boldsymbol{L}(y_i, F_{m-1}(\boldsymbol{x_i}) + p \cdot h(\boldsymbol{x}; \boldsymbol{a_m})), \tag{7}$$

and $h(\boldsymbol{x}; \boldsymbol{a_m})$ is a base learner with parameters $\boldsymbol{a_m}$ [15]:

$$\boldsymbol{a_m} = argmin \sum\nolimits_{i=1}^{N} \left[\tilde{y}_i - \beta h(\boldsymbol{x_i}; \boldsymbol{a})\right]^2, \tag{8}$$

with the "pseudo-residual" given by the following equation:

$$\tilde{y}_i = -\left[\frac{\partial L(y_i, F(\boldsymbol{x_i}))}{\partial F(\boldsymbol{x_i})}\right]_{F(x)=F_{m-1}(x)}. \tag{9}$$

Using regression trees as base learners, (6) becomes:

$$F_m(\boldsymbol{x}) = F_{m-1}(\boldsymbol{x}) + p_m \sum\nolimits_{l=1}^{L} b_{lm} 1(\boldsymbol{x} \in R_{lm}), \tag{10}$$

where at each m, a regression tree partitions the original space into L-disjoint regions $\{R_{lm}\}_1^L$, where R_{lm} is the L-terminal node, b_{lm} is the mean of (9) in each region R_{lm}, and

1(.) is an indicator function (1: true, 0: otherwise). Here, we work on the extreme gradient boosting framework [22] to additively update regression tree models that are constructed across distributed clinical data.

3 The Proposed Framework

The proposed framework for distributed learning consists of a sequential procedure (Fig. 1) which involves a central processing engine (CPE) that establishes the communication with each private cloud space to enable incremental learning. The prediction model is initially trained on the data that lie in private space A. The model is incrementally updated across the rest of the $M-1$ private spaces, i.e., B, and C, and evaluated on private space M, yielding the final model that is distributed across all spaces. Each private cloud space belongs to a clinical cohort and is part of the NextCloud infrastructure [23]. The framework was implemented as a representational state transfer (REST) web service in Python 3.6 [24]. Four performance evaluation methods were recruited including, the accuracy, sensitivity, and area under the curve (AUC) scores.

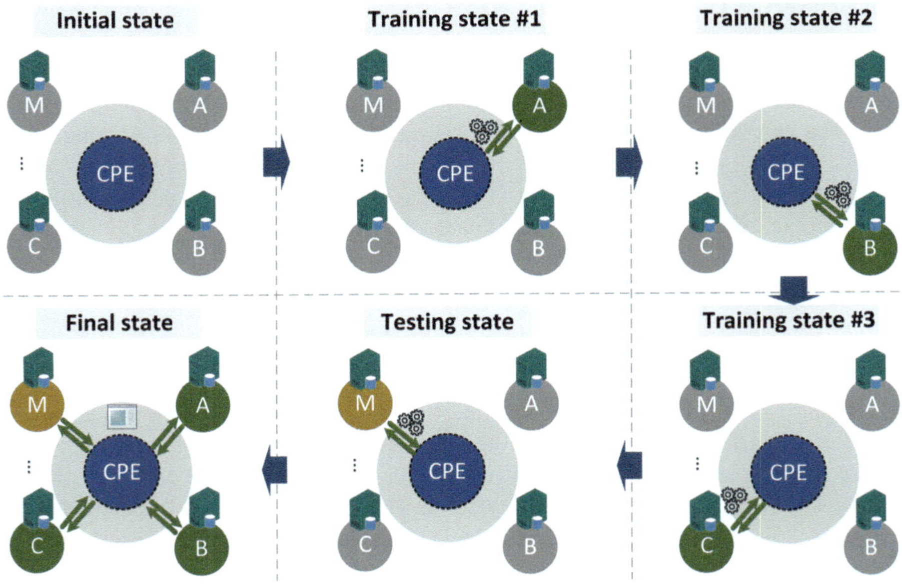

Fig. 1. Steps towards incremental learning and validation across M-sites.

4 Preliminary Case Study

4.1 Data Collection

To demonstrate the efficacy of the presented framework we acquired anonymized clinical data from the University of Athens (UoA) cohort which consists of patients that have been diagnosed with primary Sjögren's Syndrome (pSS), with the purpose of developing a binary lymphoma prediction model. Three datasets, assume, *A*, *B*, and *C*, were used for training purposes and four datasets, assume, *T*1, *T*2, *T*3, and *T*4 were used for validation. Each training dataset consists of 70 pSS patients (50 non-lymphoma; 20 lymphoma) whereas each testing dataset consists of 24 pSS patients (20 non-lymphoma; 4 lymphoma). Each dataset includes a set of 68 pSS-related features including laboratory and demographic information [20].

4.2 Distributed Model Training

For demonstration purposes and considering the imbalance between the two groups, we worked on the extreme gradient boosting [22] classifier to construct an optimized, distributed supervised learning model for predicting binary lymphoma outcomes (i.e., "0": no lymphoma, "1": lymphoma) based on tree ensembles. The simple regression models (Sect. 2.1), the Multinomial Naïve Bayes (Sect. 2.2) and the neural networks (Sect. 2.1) were also constructed for comparison purposes. Regarding the gradient boosting trees, the maximum depth was set to 6 levels using a step size value of 0.3 to prevent overfitting along with a binary logistic loss function as the learning objective [22]. The simple regression models, the neural networks, and the Multinomial Naïve Bayes, were developed based on the specifications that have been described in Sect. 2.1 and Sect. 2.2, respectively. The pipeline of Fig. 1 was executed, where the initial prediction model was trained on dataset *A*, updated on *B*, and re-updated on *C*, yielding the final one. For each supervised learning configuration, the lymphoma prediction model was evaluated on each testing dataset.

4.3 Distributed Model Evaluation and Comparison

According to Fig. 2, the gradient boosting trees exhibit the highest performance with an average AUC score 0.94 across the four testing datasets along with an average accuracy 0.916 and an average sensitivity 0.875. The performance of the SGD-based methods of Sect. 2.1 was poor (<60% accuracy), since these methods are unable to deal with the class imbalance and the associations between the features during the training stage. The same issue was observed for the neural networks (due to the small-scale datasets) and the NB model.

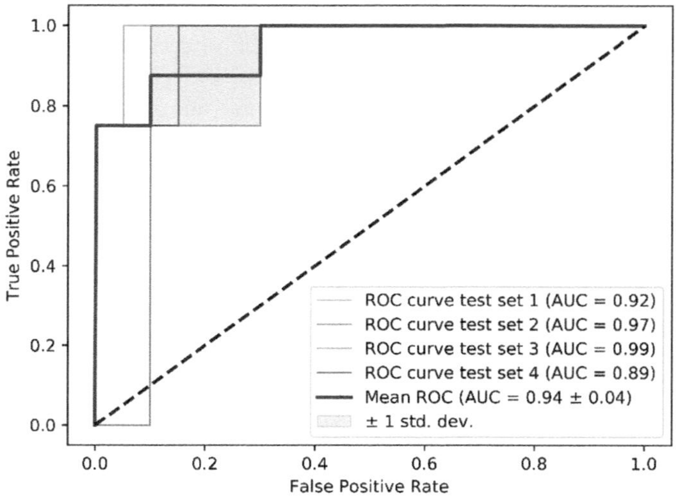

Fig. 2. Prediction performance across the four testing datasets using the gradient boosting trees.

5 Conclusions

The problem of supervised learning towards predictive modeling for effective disease management in distributed environments was examined in the context of incremental learning by considering the problem of adjusting an initial data model on data that lie across multiple sites. The stochastic gradient descent approaches along with the gradient boosting schema yield an increased prediction performance for binary classification problems that are apt to the prediction of disease-oriented outcomes across distributed clinical data against conventional methods, such as, regression models, neural networks and Naïve Bayes. The proposed framework uses additive learning to incrementally develop binary prediction models across clinical data that are stored in private spaces in contrast to the meta-learning schema, where the classification algorithm runs individually on each site and the prediction outcomes of the separate classifiers are combined, a fact that yields inaccurate results due to the biases that are introduced by the integration stage. Moreover, the framework is cloud-based, a fact that overcomes the need to install any kind of software on premises (in contrast to [4]).

The fact that the current incremental learning strategies are based on non-linear classifiers hampers the incorporation of more complex classifiers, such as, the convolutional neural networks, for deep learning, although the batch normalization technique has been recently proposed for dealing with the nonlinearities on each layer [25]. Furthermore, the data on each site need to be harmonized otherwise the distributed learning schema will fail. In addition, the overall execution time for incremental learning is affected by the number of sites, due to the large number of communication links that need to be established between the central processing node and the sites. This issue however can be reduced by the development of efficient network links within the platform.

The proposed framework overcomes significant security threats that are posed by centralized analysis, and deals with the development of accurate predictive models for disease management in distributed environments. The case study on pSS reveals promising results regarding the applicability of the framework towards the precise identification of pSS patients with lymphoma for effective patient stratification, where the class imbalance is large. Additional applications on multiple dispersed datasets are necessary to validate the applicability of the framework. New practices for large-scale convex optimization towards incremental deep learning and incremental feature selection [26] will be also taken into consideration for future work.

Acknowledgement. This project has received funding from the European Union's Horizon 2020 research and innovation programme under grant agreement No 731944 and from the Swiss State Secretariat for Education, Research and Innovation SERI under grant agreement 16.0210.

References

1. Obermeyer, Z., Emanuel, E.J.: Predicting the future—big data, machine learning, and clinical medicine. N. Engl. J. Med. **375**(13), 1216–1219 (2016)
2. Raghupathi, W., Raghupathi, V.: Big data analytics in healthcare: promise and potential. Health Inf. Sci. Syst. **2** (2014)
3. Krumholz, H.M.: Big data and new knowledge in medicine: the thinking, training, and tools needed for a learning health system. Health Aff. **33**(7), 1163–1170 (2014)
4. Deist, T.M., Jochems, A., van Soest, J., Nalbantov, G., Oberije, C., Walsh, S., et al.: Infrastructure and distributed learning methodology for privacy-preserving multi-centric rapid learning health care: euroCAT. Clin. Transl. Radiat. Oncol. **19**(4), 24–31 (2017)
5. Brown, J.S., Holmes, J.H., Shah, K., Hall, K., Lazarus, R., Platt, R.: Distributed health data networks: a practical and preferred approach to multi-institutional evaluations of comparative effectiveness, safety, and quality of care. Med. Care **48**, S45–S51 (2010)
6. Bertsekas, D.P.: Incremental gradient, subgradient, and proximal methods for convex optimization: a survey. Optim. Mach. Learn. **2010**(1–38), 3 (2011)
7. Gepperth, A., Hammer, B.: Incremental learning algorithms and applications. In: European Symposium Proceedings on Artificial Neural Networks, Bruges (2016)
8. Bekkerman, R., Bilenko, M., Langford, J.: Scaling Up Machine Learning: Parallel and Distributed Approaches. Cambridge University Press, Cambridge (2011)
9. Shalev-Shwartz, S.: Online learning and online convex optimization. Found. Trends Mach. Learn. **4**(2), 107–194 (2012)
10. Zheng, J., Shen, F., Fan, H., Zhao, J.: An online incremental learning support vector machine for large-scale data. Neural Comput. Appl. **22**(5), 1023–1035 (2013)
11. Si, S., Kumar, S., Yang, L.: Nonlinear Online Learning with Adaptive Nyström Approximation. arXiv:1802.07887 (2018)
12. Bottou, L.: Large-scale machine learning with stochastic gradient descent. In: COMPSTAT 2010 Proceedings, Physica-Verlag HD, pp. 177–186 (2010)
13. Bertsekas, D.P.: Incremental proximal methods for large scale convex optimization. Math. Program. **129**(2), 163–195 (2011)
14. Zhang, T.: Solving large scale linear prediction problems using stochastic gradient descent algorithms. In: 21st ACM Proceedings on Machine Learning, Banff, p. 116 (2004)

15. Friedman, J.H.: Greedy function approximation: a gradient boosting machine. Ann. Stat. **29**, 1189–1232 (2001)
16. Manning, C.D., Raghavan, P., Schütze, H.: Introduction to Information Retrieval. Cambridge University Press, Cambridge (2008)
17. Zeng, L., Li, L., Duan, L., Lu, K., Shi, Z., Wang, M., Luo, P.: Distributed data mining: a survey. Int. J. Inform. Tech. Manag. **13**(4), 403–409 (2012)
18. Prodromidis, A., Chan, P., Stolfo, S.: Meta-learning in distributed data mining systems: issues and approaches. Adv. Distrib. Parallel Knowl. Discov. **3**, 81–114 (2000)
19. Kourou, K.D., Pezoulas, V.C., Georga, E.I., Exarchos, T.P., Tsanakas, P., Tsiknakis, M., Varvarigou, T., De Vita, S., Tzioufas, A., Fotiadis, D.I.: Cohort harmonization and integrative analysis from a biomedical engineering perspective. IEEE Rev. Biomed. Eng. **12**, 303–318 (2018)
20. Fragkioudaki, S., Mavragani, C.P., Moutsopoulos, H.M.: Predicting the risk for lymphoma development in Sjogren syndrome: an easy tool for clinical use. Medicine **95**(25), e3766 (2016)
21. Zou, H., Hastie, T.: Regularization and variable selection via the elastic net. J. Roy. Stat. Soc. B **67**(2), 301–320 (2005)
22. Chen, T., Guestrin, C.: XGBoost: a scalable tree boosting system. In: 22nd ACM SIGKDD Proceedings on Knowledge Discovery and Data Mining, San Francisco, pp. 785–794 (2016)
23. Nextcloud – Protecting your data. https://nextcloud.com/. Accessed 19 Mar 2019
24. Pedregosa, F., Varoquaux, G., Gramfort, A., Michel, V., Thirion, B., Grisel, O., Vanderplas, J.: Scikit-learn: machine learning in Python. J. Mach. Learn. Res. **12**, 2825–2830 (2011)
25. Ioffe, S., Szegedy, C.: Batch normalization: accelerating deep network training by reducing internal covariate shift. arXiv:1502.03167 (2015)
26. Xu, Z., Huang, G., Weinberger, K., Zheng, A.X.: Gradient boosted feature selection. In: 20th ACM SIGKDD Proceedings on Knowledge Discovery and Data Mining, New York, pp. 522–531 (2014)

EmERGE Platform: A New mHealth Solution for People Living with HIV

Paloma Chausa[1(✉)], Francisco J. Gárate[1], Cesar Cáceres[1,3], Edward Wallitt[4], Jennifer Whetham[5], and Enrique J. Gómez[1,2] and the EmERGE Consortium

[1] Biomedical Engineering and Telemedicine Centre, ETSI Telecomunicación, Center for Biomedical Technology, Universidad Politécnica de Madrid, Madrid, Spain
{pchausa, fgarate, ccaceres, egomez}@gbt.tfo.upm.es
[2] Centro de Investigación Biomédica en Red, Biomateriales y Nanomedicina (CIBER-BBN), Madrid, Spain
[3] Universidad Rey Juan Carlos, Madrid, Spain
cesar.caceres@urjc.es
[4] Podmedics, London, UK
ed.w@inductionhealthcare.com
[5] Department of HIV/Genito-Urinary Medicine, Brighton and Sussex University Hospitals NHS Trust, Brighton, UK
j.whetham@nhs.net

Abstract. The EmERGE platform is a novel mHealth solution that supports a new paradigm for HIV care. Following a rigorous co-design approach, it provides users with a system that can link securely into the health records of an individual, select the data required, and once reviewed by a clinician send the data and clinical opinion confidentially to the individual's smart device. The proposed clinical pathway aims to facilitate patient empowerment and self-management of HIV while reducing face-to-face consultations for people living with stable HIV. The EmERGE platform has been deployed at 5 European clinical sites and it is being successfully used by more than 2000 individuals. A formal evaluation is currently underway using a tailored Health Technology Assessment process specifically developed for the assessment of mHealth solutions. The evaluation will assess the impact of the EmERGE pathway on patient empowerment, quality of life, patient related outcomes and experience measures, usability, quality of care and cost in diverse health systems.

Keywords: mHealth · PLWH · Patient empowerment · Self-management · HIV · Health Technology Assessment · Co-design · Healthcare pathways

1 Introduction

With the advent of effective antiretroviral therapy (ART) HIV is now regarded as a chronic illness with a normal life expectancy in individuals who have access to testing, treatment and care as in the European Union [1]. The majority of individuals have an undetectable HIV viral load (the aim of ART) on stable treatment and many are now seeking to reduce the impact of HIV on their lives.

J. Henriques et al. (Eds.): MEDICON 2019, IFMBE Proceedings 76, pp. 832–838, 2020.
https://doi.org/10.1007/978-3-030-31635-8_99

People living with HIV (PLWH) are currently seen by an HIV clinician every 3–6 months in accordance with European guidelines [2]. The number of new diagnoses has continued to increase in many parts over recent years; as people are living longer healthier lives the number of people needing clinical care continues to increase without a proportionate increase in resource. Therefore, it is imperative that new models of service provision are explored and evaluated.

Previous experiences have demonstrated that web-based telemedicine approaches can be considered as feasible and safe tools for the management of stable HIV infection [4]. Similarly, mHealth interventions have improved health outcomes and self-management when used by patients with chronic disease conditions [5]. Thus, mHealth supported pathways of care provide an opportunity to offer an alternative to traditional models care that can reduce appointments for individuals living with stable HIV and increase capacity to see those living with more complex disease.

The EmERGE project (Evaluating mHealth technology in HIV to improve Empowerment and healthcare utilisation: Research and innovation to Generate Evidence for personalised care, http://www.emergeproject.eu/) is a five year Horizon 2020 funded project which is assessing the impact of a co-designed mHealth supported pathway of care for people living with medically stable HIV in diverse settings across 5 European countries. Impact is being assessed in terms of patient empowerment, quality of life, person-centred symptoms and concerns, quality of care and costs. The EmERGE mHealth platform supporting the pathway integrates with clinic ICT systems to reduce face-to-face appointments in patients living with stable HIV and is described in more detail here.

2 General Description and EmERGE Platform Architecture

Overall, the goal of the EmERGE platform is to provide relevant, timely and secure delivery of health care data to people living with stable HIV infection through the use of novel mHealth tools. The optimal features of the platform have been determined through co-design work with PLWH and clinicians [5, 6].

The EmERGE platform consists of the Adaptor, the Mobile Application, the Clinical Web Application (CWA) and the Messaging Service. This group of software allows patient data to be gathered from the hospital network, formatted appropriately, and sent to and displayed on patient devices.

The basic purpose of the CWA is to provide clinical users with an interface to register and manage patient users and their mobile devices. In addition, the application provides executive control over delivery of information such as blood tests results, medication lists and appointment information to mobile devices. The function of the Mobile Application is to receive user specific information from the CWA. The Messaging Service is the bridge between the CWA and the patient Mobile Application. Medication information, results, and appointments for specific individuals are sent to the Messaging Service to be served to patient devices on request. The Adaptor provides a single API for the CWA to extract the data via an interface with each hospital database. It ensures that the data are in a single format for onward processing and facilitates future integration with other clinical systems.

The basic structure of the EmERGE platform and the interaction of the four key elements is shown in the schematic below (Fig. 1).

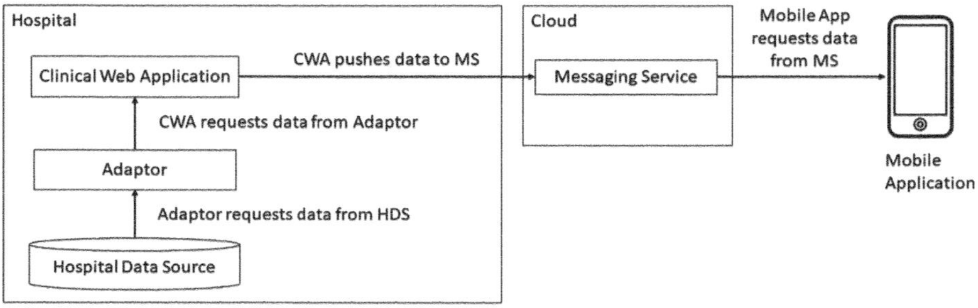

Fig. 1. EmERGE platform architecture

3 EmERGE Platform Specification

3.1 Functional Requirements

The main users of the EmERGE platform are the clinical and administrative staff of the clinical sites involved and PLWH – currently those enrolled in the clinical study designed to evaluate this new health care paradigm. Clinicians use the Clinical Web Application, a software application hosted inside the hospital network to provide the interface for the EmERGE system. The main functionalities of the CWA – as co-designed with clinicians - are described in the following table (Table 1).

Table 1. Clinical Web Application functionalities.

Functionality	Description
Authentication	Unique username and password to login to the CWA
Patient/device registration	Registration of patients and the associated mobile devices
All patients visualization	Search and filtering of all registered patients
Patient data visualization	Overview of each patient user: current medication list (name, dose/frequency and date of commencement), blood test results (graphically and in a table format) and appointment lists
'Virtual clinics' appointment management	Send data to the mobile application; re-book the patient for a future revision; pause the patient in the system
Messaging to patients	Send messages to a specific patient
Edition of FAQs	Provision of Frequently Ask Questions
Users management	Creation and elimination of users and assignment of roles

PLWH use the Mobile Application to receive their data including laboratory results, information about the treatment and appointments with the aim of empowering individuals, to self-manage their infection. The main functionalities of the Mobile Application – as co-designed with the community - are described in the following table (Table 2).

Table 2. Mobile Application functionalities.

Functionality	Description
Authentication	Patient Users are required to login to the application on each use through passcode or biometrics
Medication visualization	Visualization of medication list (name of medication, dose/frequency and date of commencement). Link to Liverpool HIV drugs interaction webpage
Laboratory results visualization	Visualization of blood test results: current and historical data; evolution graphs
Appointments	Visualization of next appointments. Integration with patient mobile calendar
Messages	Messaging from Hospital clinicians
Help desk and information	Support and access to FAQs
Application Settings	Erase application data, change passcode, change security question, ask for a new password, study withdrawal

3.2 Technology Used

The Mobile Application has been built using a hybrid approach provided by the Ionic Framework (http://ionicframework.com). A hybrid approach enables creation of native iPhone and Android applications from a single and maintainable codebase. Apps can be built with Web technologies provided by Ionic (CSS, HTML5, Sass, etc.) and then distributed through native app stores to be installed on devices. For Android, Ionic supports Android 4.1 and up. For iOS, Ionic supports iOS 7 and up.

Both the Clinical Web Application and the Adaptor are implemented using Ruby on Rails version 4.2 and Ruby version 2.3. Ruby on Rails (https://rubyonrails.org/), is an open-source, mature and well-maintained server-side web application framework written in Ruby. Ruby on Rails is a model–view–controller (MVC) framework, providing default structures for a database, a web service, and web pages. It encour-ages and facilitates the use of web standards such as JSON or XML for data transfer, and HTML, CSS and JavaScript for display and user interfacing.

Heroku (https://www.heroku.com/) is used to host the Messaging Service. Heroku is a cloud platform as a service (PaaS) supporting several programming languages. It lets the developer build, run and scale applications. Apart from the production MS, there is a full staging version system set up on Heroku for showing off the system and testing them.

3.3 User Interfaces and Graphical Design

The EmERGE platform offers two main User Interfaces: the Clinical Web Application for clinicians and the Mobile Application for PLWH. The CWA has been built using bootstrap-sass 3 for the graphical design. Bootstrap is a popular open-source framework that allows web designers of all skill levels to quickly build a site. This version of Bootstrap uses Sass (https://sass-lang.com) as the pre-processor of choice. Sass (Syntactically Awesome Style Sheets) is a pre-processor scripting language that is interpreted or compiled into Cascading Style Sheets (CSS). The Mobile Application also uses Sass for the graphical design. Figure 2 shows the CWA and Mobile Application interfaces.

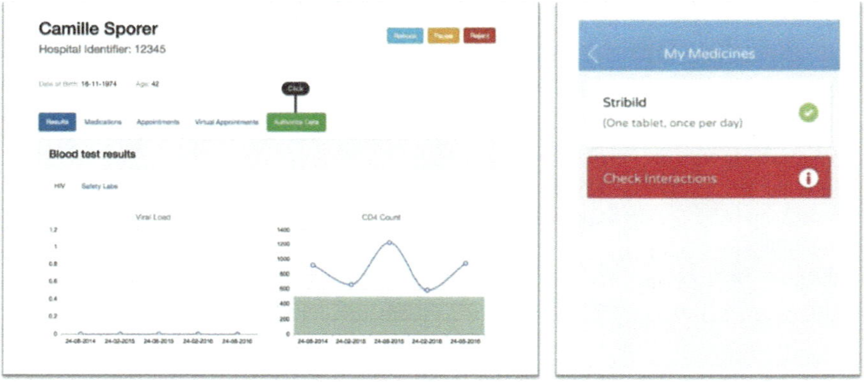

Fig. 2. CWA and Mobile Application User Interfaces

3.4 Security and Privacy

Security is a key factor for our patient group and for their clinicians. The CWA within each site is to be regarded as a 'walled-off data silo'. It exists within the existing IT infrastructure of the clinical site and should not be open for external access. As such, unless connected via a VPN, access to the CWA is not possible outside of the hospital network.

All clinical and demographic data stored locally within the CWA's local database are encrypted to a minimum of 256 bit AES. Transmission of data outside of the hospital network, from the CWA to the Mobile Application is encrypted over Secure Sockets Layer (SSL) and includes no patient identifying information.

4 EmERGE Platform Evaluation Procedure

A large prospective cohort study, undertaken in five European sites is being performed to validate the EmERGE platform supported pathway of care using a tailored Health Technology Assessment (HTA) process, Model for Assessment of Telemedicine

Applications (MAST) and will assess impact in terms of patient empowerment, quality of life, patient related outcome and experience measures; health economics, sociotechnical aspects and clinical outcomes.

The five clinical sites involved in the study are HIV clinics in Antwerp, Barcelona, Brighton, Lisbon and Zagreb. The general care of PLWH is very similar in the five clinical sites, specifically in the follow up of individuals living with stable HIV. PLWH who are participating in the study have visits at baseline, month 6, 12, 18, 24 and 30. At baseline, participants visit the hospital for a face-to-face visit and are introduced to the EmERGE mobile application. At six months the individual will attend for bloods tests as per standard of care. The results will be checked by a clinician and then be pushed through to the patient's mobile phone if the results are unremarkable. Visits at 18 and 30 months will be conducted in a similar fashion. At 12 months and 24 months the patient will be seen by their clinical team in person (or virtually if normally cared for within the Barcelona Virtual Hospital).

The recruitment of patients started on April 2017 and the platform is successfully being used by 2251 patients and more than 20 clinicians. Specifically, the evaluation includes the assessment of the impact of the EmERGE platform on the items listed in the Table 3.

Table 3. Evaluation of EmERGE platform.

Variable	Evaluation procedure
Self-management and empowerment	Patient Activation Measure (PAM-13) questionnaire at baseline, months 12 and 24
Acceptability/usability	Satisfaction System Usability Scale (SUS) at months 12 and 24
Clinical outcomes	CD4 count, viral load, percentage of patients with changes in ART) at months 6, 12, 18, 24 and 30
Quality of life	EQ-5D-5L, PROQOL-HIV questionnaires at baseline, months 12 and 24
Adherence	Modified Medication Adherence Self-Report Inventory (M-MASRI) questionnaire at baseline, months 12 and 24

An electronic Case Report Form (eCRF) has been developed to collect patient questionnaires and the clinical related data. It has been built using Ruby on Rails and PostgreSQL as Data Management System. The eCRF is hosted by Heroku and all the information is encrypted by means of bcrypt.

5 Conclusions

The EmERGE platform supports a novel pathway of care for individuals living with stable HIV infection. In this paper we describe the technology that has been developed and integrated, and is currently being used by individuals, in diverse care settings across five European countries.

This technology has been co-designed with community and clinicians [5, 6] and a formal evaluation with over 2000 individuals is currently underway using a tailored Health Technology Assessment process, specifically developed for the assessment of mHealth solutions.

The final goal aims to demonstrate the benefits to patients and simultaneous increases in cost-effectiveness for healthcare providers by reducing face-to-face consultations: assessing the intervention in terms of patient empowerment, quality of life, patient related outcomes and experience measures, usability, quality of care and cost.

Acknowledgements. This project has received funding from the European Union's Horizon 2020 research and innovation programme under grant agreement No 643736. The dissemination activities within the EmERGE project do not represent the opinion of the European Community and only reflects the opinion of the authors. The scientific members of the EmERGE consortium are: Ludwig Apers; Eduard Beck; Josip Begovac; Margarida Borges; Stephen Bremner; Paloma Chausa; Mary Darking; Maria Dutarte; Duncan Fatz; Steven Hoornaert; David García; Felipe García; Enrique J. Gomez; Flis Henwood; Agathe Leon; Sundhiya Mandalia; Benjamin Marent; Rosa Muñoz; Gonçalo Rodrigues; Eugenio Teofilo; Brian West; Jennifer Whetham; Platonas Yfantopoulos; Sime Zekan. Our thanks also to our advisory board which includes: Cesar Caceres; Richard Harding & Edward Wallitt.

References

1. May, M., et al.: AIDS [Epub ahead of print]; Impact on life expectancy of HIV-1 positive individuals of CD4 + cell count and viral load response to antiretroviral therapy: UK cohort study, 19 February 2014
2. EACS guidelines version 9.1, October 2018 http://www.eacsociety.org/files/2018_guidelines-9.1-english.pdf. Accessed 25 Apr 2019
3. León, A., Cáceres, C., Fernández, E., Chausa, P., Martin, M., Codina, C., Rousaud, A., Blanch, J., Mallolas, J., Martinez, E., Blanco, J.L., Laguno, M., Larrousse, M., Milinkovic, A., Zamora, L., Canal, N., Miró, J.M., Gatell, J.M., Gómez, E.J., García, F.: A new multidisciplinary home care telemedicine system to monitor stable chronic human immunodeficiency virus-infected patients: a randomized study. PLoS ONE 1932–6203 (2011)
4. Lee, J.A., Choi, M., Lee, S.A., Jiang, N.: Effective behavioral intervention strategies using mobile health applications for chronic disease management: a systematic review. BMC Med. Inform. Decis. Mak. **18**(1), 12 (2018). https://doi.org/10.1186/s12911-018-0591-0
5. Marent, B., Henwood, F., Darking, M., on behalf of the EmERGE Consortium: Ambivalence in digital health: co-designing an mHealth platform for HIV care. Soc. Sci. Med. **215**, 133–241 (2018)
6. Marent, B., Henwood, F., Darking, M., EmERGE Consortium: Development of an mHealth platform for HIV care: gathering user perspectives through co-design workshops and interviews. J. Med. Internet Res. (2018). https://doi.org/10.2196/preprints.9856

Machine Learning Algorithms Predict Body Mass Index Using Nonlinear Trimodal Regression Analysis from Computed Tomography Scans

Marco Recenti[1(✉)], Carlo Ricciardi[1,2], Magnus Gìslason[1], Kyle Edmunds[1], Ugo Carraro[3], and Paolo Gargiulo[1,4(✉)]

[1] Institute of Biomedical and Neural Engineering, Reykjavik University, Reykjavik, Iceland
marco18@ru.is
[2] Department of Advanced Biomedical Sciences, University Hospital of Naples 'Federico II', Naples, Italy
[3] IRRCS Fondazione Ospedale San Camillo, Venice, Italy
[4] Department of Science, Landspitali, Reykjavik, Iceland

Abstract. In this study Machine Learning supervised regression and classification algorithms are used to predict Body Mass Index (BMI), starting from Computed Tomography scans (CT). From each patient CTs, 11 parameters describing muscle, connective tissue and fat, are extracted creating a patient specific soft tissue profile called Nonlinear Trimodal Regression Analysis (NTRA). Regression and classification are applied in order to predict and classify BMI using Tree-Based algorithms. A proper Train-Test division of the dataset is applied using k_fold Cross-Validation. Various combinations of features are employed with k_fold division in order to obtain the best coefficient of determination (R^2) as evaluator of the quality of regression's prediction. Afterward, BMI is divided into 3 and 5 classes and the same methodology is used to classify it. For this analysis, the accuracy parameter is calculated to evaluate the quality of the results. The max R^2 is 0,83 and it is obtained using the 11 NTRA parameters as regressors, k_fold = 16, and the Gradient-Boosting Algorithm. The amplitude of the connective and fat tissue always covers more than 50% of all the feature importance. The best accuracy was 0,80 for 3 classes and 0,74 for 5 classes. The results prove that the 11 NTRA parameters can have a very significant predictive value and the same methodology can be applied in future works to predict other physiological parameters and comorbidities.

Keywords: Machine Learning · Regression · Classification · Computed Tomography · Body Mass Index

1 Introduction

Artificial Intelligence (AI) technologies, particularly Machine Learning (ML) algorithms, are commonly used in healthcare applications. The increasing availability of healthcare data and the development of big data analytics methods made the success of

J. Henriques et al. (Eds.): MEDICON 2019, IFMBE Proceedings 76, pp. 839–846, 2020.
https://doi.org/10.1007/978-3-030-31635-8_100

ML possible in different healthcare applications (i.e. the image processing or predictive system [1, 2]).

ML has potential to be applied in a wide range of situations like the investigation on the effectiveness of medical treatments, or the identification and tracking of chronic disease states in high risk patients that help in the design of appropriate interventions or moreover the minimization of fraud and abuse by the identification of unusual or abnormal patterns [3].

There are many examples in literature encouraging the implementation of data mining algorithms in different fields of medicine in order to discover hidden patterns or new information in large dataset [4, 5]. Age Gene/Environment Susceptibility Study (AGES) is a large dataset with more than 10 thousand features, designed to examine risk factors and gene/environment interaction, in relation to disease and disability in old age [6, 7].

Therefore, the purpose of this study is to test the potential prediction power of the Nonlinear Trimodal Regression Analysis (NTRA) parameters. This work establishes a methodology, using ML algorithms, to associate NTRA with Body Mass Index (BMI). Our final goal is to understand if NTRA parameters can also be used in cardiocirculatory or motor disease prediction. BMI is the starting point to test the effectiveness of the methodology.

2 Materials and Methods

2.1 Database

The database used in this paper is AGES: it is composed of AGES I and AGES II (same measurements on the same patients but 5 years after AGES I). Both have 3157 patients, so in total, AGES I+II is made up of 6314 patients.

In this study we used a subset of the dataset: each patient has 11 parameters extracted from a mid-femur CT scan plus the following physiological measurements:

- Body Mass Index (BMI) [kg/m2];
- Cholesterol (Chol) [mmol/L];
- 4 lower extremity function (LEF) biometric parameters:
  - Time Up-to Go (TUG) [sec];
  - Normal and Fast Gait Speed (NGait, FGait) [sec];
  - Isometric Leg Strength (ISO) [Newtons].

Some "Not a Number" (NaN) values are present in the database and must be handled. NaN values cannot be read during the ML algorithms, so some solutions were found to overcome this limit. Only 8 NaN are present in AGES I+II as regards BMI, so these are substituted with the mean value of the feature. In the case of all the other measurements, the NaN values are too many (i.e. >400 for FGait in AGES I+II) and the mean solution would compromise the results; thus, the solution adopted is to delete the patient's row. There are no NaN values in the 11 parameters extracted from the CT, the so-called NTRA data.

After some operations using the Hounsfield unit (HU), this method, described in detail in [8, 9], gives as results 11 parameters: 4 related to the Fat, 4 to the Muscle (musc) and 3 to the ConnectiveTissue (conn). They are Amplitude (N), Location (μ), Width (σ) and Skewness (α). The α is not present for the Connective Tissue.

2.2 Supervised ML Tree-Based Algorithms

In literature lots of supervised ML algorithms are present, whose effectivity depends on the uses and applications in healthcare [1]. In this paper, 4 Tree-Based algorithms are considered both for regression and classification analysis. They are denominated Tree-Based because the basic units on which they are built are regression and classification trees [10]. They were taken into consideration because they do not need normalization or standardization of the initial data and the feature importance is easy to extract.

The implemented algorithms are 4. The first is Random Forest (RF) which operate by constructing a multitude of decision trees during the training time and outputting the class that is the mode of the classes (classification) or mean prediction (regression) of the individual trees [11, 12]. The second one is Extremely Randomized Tree, EXTRA Tree (EXT) in which randomness goes one step further in the way splits are computed. It consists of randomizing both attribute and cut-point while splitting a node of the tree. It can build a totally randomized tree in which the structures are independent of the output values of the learning sample [13]. The third one is Ada Boosting (ADA-B) where the principal core is to fit a sequence of weak learners (such as small decision trees) on repeatedly modified versions of the training data. To give the final estimation, all the predictions from the small trees are combined through a weighted majority vote [14, 15]. As fourth algorithm Gradient-Boosting (GRAD-B) is used: it consists in a generalization of boosting to arbitrary differentiable loss function [16, 17].

The coding language used is Python (PY) with the relative ML library Scikit-Learn (SL). For each previous algorithm SL provides a specific function with default values, for the number of trees in the forest (n_estimators), and the maximum depth of the tree (max_depth) [18]. They can be manually modified in order to obtain better results.

2.3 Regression and Classification

Regression algorithms need a regressand (the continuous value that is predicted) and some regressors (the features used to predict the regressand). To estimate the performances of the predictions the Coefficient of Determination (R^2) is taken into consideration.

Concerning classification, the Accuracy Parameter (or Jaccard Index (JI)) is used to evaluate the results. JI compares the predicted class with the true one: if JI = 1 all the data are classified correctly. In order to apply classification algorithms, BMI needs a preliminary binning that is conducted according to the distribution of the data. BMI is divided in 3 and in 5 classes, as shown in Table 1.

Table 1. BMI [kg/m2] classes

3 classes		5 classes	
Low	BMI $\leq$ 25	Low	BMI < 20
		Medium-Low	20 < BMI $\leq$ 25
Medium	25 < BMI $\leq$ 30	Medium	25 < BMI $\leq$ 30
High	BMI > 30	Medium-High	30 < BMI $\leq$ 35
		High	BMI > 35

2.4 Cross-Validation – k_fold

In order to apply ML algorithms, the database must be divided into Train and Test set. Cross-Validation method is used to have a better vision of all the possible R^2 and JI results respectively in regression and classification. The dataset is split into k consecutive folds: k-1 folds are used to train and 1 for the test. Training is done for each combination of folds. The mean of R^2 and JI of all the folds can be calculated as well as the maximum value. With fewer number of folds, the train set is larger, and the test set is smaller and viceversa with more folds. An unbalanced dimension of folds does not allow to obtain reliable results, so k_fold division is applied using 8, 12, 16, 18 folds.

2.5 Feature Selection

Different selections of features are used as regressors:

- NTRA: the 11 parameters above-mentioned;
- NTRA+LEF: Chol and LEF were added;
- NTRAS1: a selection of the 11 NTRA based on the results of [5] (All the N, μ_fat, μ_conn, σ_fat, σ_musc, α_musc);
- NTRAS2: a selection based on the results of the feature importance obtained with only NTRA features (Fig. 3) (All the N and σ);
- NTRAS1+LEF: NTRAS1 feats plus Chol and LEF meas;
- NTRAS2+LEF: NTRAS2 feats plus Chol and LEF meas;

The same selections are used also for the classification.

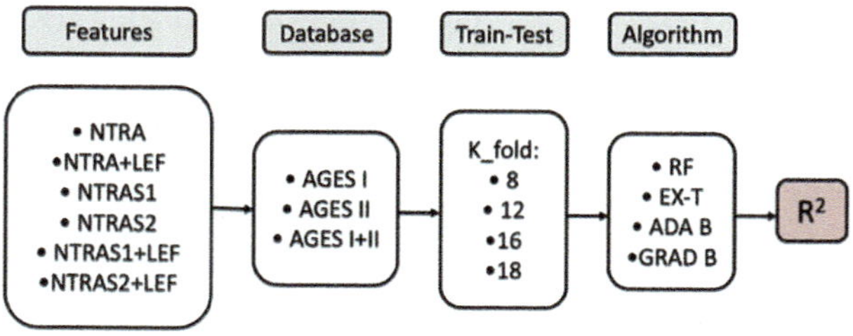

Fig. 1. Regression methodology: R^2 value is obtained for each combination of features, databases, k_fold divisions and Tree-Based algorithms.

2.6 Methodology

In order to reach the best R^2 result in regression, lots of combinations have been tried, mixing feature selections used as regressors, the 3 databases, k_fold divisions and algorithms like shown in Fig. 1. For example, using the NTRA features on the AGES I+II database with a k_fold division of 8 sets using the ADA-B algorithm, 8 different R^2 are obtained, one for each test set.

The same methodology is applied also for the classification obtaining JI as results from all the possible combinations.

3 Results and Discussion

The following results are obtained using as Database AGES I+II: employing only one database, the number of patients is about the half, and the results of R^2 and JI are worse than the ones obtained using AGES I+II.

3.1 Regression

In Fig. 2 the mean and the max R^2 obtained for each of the 4 algorithms with all the combination of features selections and k_fold divisions are showed.

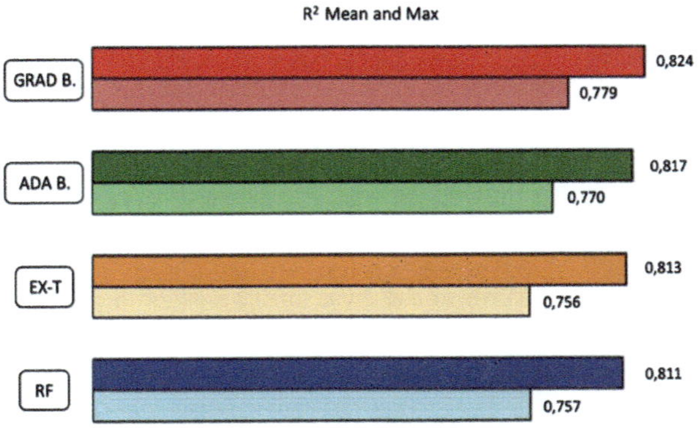

Fig. 2. Mean and Max values of R^2 for the 4 algorithms (with default values) obtained combining all feature selections and all the k_fold divisions.

Those are obtained using the default values of the functions in SL [15]. Clearly, GRAD-B gives always the best results followed by the ADA-B algorithm. The others have comparable results. The highest R^2 was 0,8305 and was obtained modifying the default values, setting on GRAD-B algorithm the n_estimators = 200 and combining it with NTRA features, and k_fold = 16. The highest result with a simple train-test division, where test is 20% of the dataset, is obtained using the GRAD-B algorithm and correspond to R^2 = 0,785 having the 11 NTRA parameters as features.

Using the NTRA selection of features, the R^2 results are the highest: selecting only some of the 11 NTRA parameters the prediction does not improve. The same happens if LEF and Chol are added to NTRA or NTRAS1 and NTRAS2. Chol, FGait, NGait, TUG and ISO are not relevant as features for the prediction. As shown in Fig. 3 the most important features are N_conn and N_fat that in each possible combination always cover more than 50% of the total importance. The importance of Chol and LEF measurements is always <0,015 (from 0 to 1). For GRAD-B and RF the most important feature is N_conn while for EX-T and ADA-B the most relevant is N_fat.

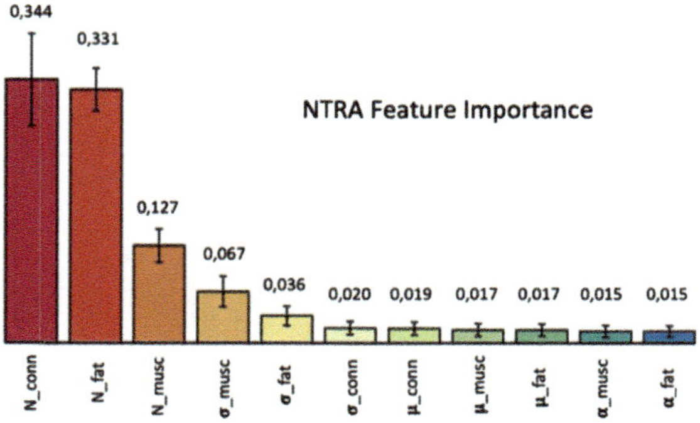

Fig. 3. Feature Importance mean (with relative std) from 0 to 1 using only NTRA as regressors. The mean of the values is obtained with the combination of all k_fold divisions, algorithms applied on AGES I+II.

3.2 Classification

Table 2 shows the results of the classification using JI as an indicator of the accuracy.

Table 2. BMI Classification - JI Mean and Max for the 4 algorithms obtained combining all feature selections and all the k_fold divisions

ML Algorithm	JI (*3* Classes) •Mean •Max	JI (*5* Classes) •Mean •Max
Random Forest	•0,701 •0,756	•0,642 •0,712
EXTRA Tree	•0,696 •0,766	•0,638 •0,715
ADA Boosting	•0,631 •0,714	•0,565 •0,618
Gradient-Boosting	•0,732 •0,797	•0,679 •0,741

The results, as expected, are better with 3 classes, but the ones with 5 classes are not far from the previous ones, so both the classifications can be considered reliable. The maximum JI (0,797) is obtained using GRAD-B with 3 classes, combined with NTRAS1 and k_fold = 18. Using the train-test division with test equal to 20% of the set of data together with the GRAD-B algorithm and the 11 NTRA parameters as features, the best JI with 3 classes is 0,755 and 0,678 with 5 classes.

4 Discussion and Conclusions

Our study provides good results in terms of prediction, using both regression and classification for the BMI starting from a CT scan and adding eventually other physiological parameters. The database we use contains more than 10 thousand of features that could be exploited to find new relationships among variables related to patients' state of health. Despite being powerful algorithms, in this paper we explored only Tree-Based ones; thus, future works could exploit a wider range of these techniques. Moreover, the predictions can be extended in future works also to other parameters present in the AGES database, presuming that the results can be equally positive. The investigation of the connections among these variables and the risk factors could provide also useful results. Finally, this paper provides a novel approach to study correlation between physiology and imaging.

Acknowledgment. We want to thank all the staff and the participants of the AGES-Reykjavik study for their important contribution: The Age, Gene/Environment Susceptibility Reykjavik Study has been funded by NIH contract N01-AG12100, the NIA Intramural Research Program, Hjartavernd (the Icelandic Heart Association), and the Althingi (the Icelandic Parliament).

Conflict of Interest Declaration. The authors declare they have no conflict of interests.

References

1. Jiang, F., Jiang, Y., Zhi, H.: Artificial intelligence in healthcare: past, present and future, Stroke and Vascular Neurology (2017)
2. Holzinger, A.: ML for health informatics. In: LNAI 9605, pp. 1–24 (2016). https://doi.org/10.1007/978-3-319-50478-0_1
3. Koh, H.C., Tan, G., et al.: Data mining applications in healthcare. J. Healthc. Inf. Manag. **19** (2), 65 (2011)
4. Romeo, V., Ricciardi, C., Cuocolo, R., et al.: Machine learning analysis of MRI-derived texture features to predict placenta accreta spectrum in patients with placenta previa. In: Magnetic Resonance Imaging (2019). ISSN 0730-725X, https://doi.org/10.1016/j.mri.2019.05.017
5. Stanzione, A., Cuocolo, R., Cocozza, S., Romeo, V., Persico, F., Fusco, F., Nicola Longo, M.D., Arturo Brunetti, M.D., Imbriaco, M.: Detection of extraprostatic extension of cancer on biparametric MRI combining texture analysis and machine learning: preliminary results. Acad. Radiol. (2019)

6. Harris, T.B., Launer, L.J., Eiriksdottir, G., Kjartansson, O., Jonsson, P.V., Sigurdsson, G., et al.: Age, gene/environment susceptibility– reykjavik study: multidisciplinary applied phenomics. Am. J. Epidemiol. **165**(9), 1076–1087 (2007). https://doi.org/10.1093/aje/kwk115. PMID: 17351290
7. Johannesdottir, F., Aspelund, T., Siggeirsdottir, K., Jonsson, B., Mogensen, B., Sigurdsson, S., et al.: Mid-thigh cortical bone structural parameters, muscle mass and strength, and association with lower limb fractures in older men and women (AGES-Reykjavik Study). Calcif. Tissue Int. **905**, 354–364 (2012)
8. Edmunds, K.J., Gislason, M., Sigurðsson, S., Guðnason, V., Harris, T.B., Carraro, U., Gargiulo, P.: Advanced quantitative methods in correlating sarcopenic muscle degeneration with lower extremity function biometrics and comorbidities. PLoS ONE **13**(3), e0193241 (2018). https://doi.org/10.1371/journal.pone.0193241
9. Edmunds, K.J., Arnadottir, I., Gislason, M., Carraro, U., Gargiulo, P.: Nonlinear trimodal regression analysis of radiodensitometric distributions to quantify sarcopenic and sequelae muscle degeneration. Comput. Math. Methods Med., 8932950 (2016). PMID: 28115982. https://doi.org/10.1155/2016/8932950
10. Breiman, L., Friedman, J., Olshen, R., Stone, C.: Classification and Regression Trees. Wadsworth, Belmont (1984)
11. Ho, T.K.: Random decision forests. In: Proceedings of the 3rd International Conference on Document Analysis and Recognition, Montreal, QC, 14–16 August 1995, pp. 278–282 (1995)
12. Ho, T.K.: The random subspace method for constructing decision forests. IEEE Trans. Pattern Anal. Mach. Intell. **20**(8) (1998). https://doi.org/10.1007/s10994-006-6226-1
13. Guerts, P., Ernst, D., Wehenkel, L.: Extremely randomized trees. Mach. Learn. **63**, 3–42 (2006). https://doi.org/10.1007/s10994-006-6226-1
14. Freund, Y., Shapire, R.E.: A decision-theoretic generalization of on-line learning and an application to boosting. J. Comput. Syst. Sci. **55**, 119–139 (1997). Article no. SS971504
15. Drucker, H.: Improving regressors using boosting techniques. In: ICML 1997 (1997)
16. Friedman, J.H.: Greedy function approximation: a gradient-boosting machine. Technical report, Department of Statistics, Stanford University (1999)
17. Friedman, J.H.: Stochastic gradient-boosting. Comput. Stat. Data Anal. **38**(4), 367–378 (2002). https://doi.org/10.1016/S0167-9473(01)00065-2
18. Pedregosa, F., Varoquaux, G., Gramfort, A., Michel, V., Thirion, B., Grisel, O., Blondel, M., Prettenhofer, P., Weiss, R., Dubourg, V., Vanderplas, J.: Scikit-learn: machine learning in Python. J. Mach. Learn. Res. **12**, 2825–2830 (2011)

Is It Possible to Predict Cardiac Death?

Carlo Ricciardi[1(✉)], Valeria Cantoni[2], Roberta Green[2], Giovanni Improta[3], and Mario Cesarelli[4,5]

[1] Department of Advanced Biomedical Sciences, University Hospital of Naples 'Federico II', Naples, Italy
carloricciardi.93@gmail.com

[2] Department of Advanced Biomedical Sciences, University Federico II, Naples, Italy

[3] Department of Public Health, University Hospital of Naples 'Federico II', Naples, Italy

[4] DIETI, University of Naples 'Federico II', Naples, Italy

[5] Istituti Clinici Scientifici Maugeri IRCCS, Telese Terme (BN), Italy

Abstract. Cardiovascular diseases are the leading cause of death in all the world; despite having the knowledge of the main risk factors, they keep on being complicated pathologies to deal with. Cardiovascular management has introduced a lot of parameters as regards patients' state of health; particularly, nuclear cardiology with Stress single-photon emission computed tomography myocardial perfusion imaging can carry out interesting parameters that have encouraged researchers to apply machine learning techniques to predict whether patients will die due to a cardiac event or not. The dataset consisted of 661 patients that were evaluated for suspected of known coronary artery disease at the Department of Advanced Biomedical Sciences of the University Hospital "Federico II" in Naples. Knime analytics platform was employed to implement a decision tree and Random forests. After a procedure of features reduction, 29 features were included, and the overall accuracy was 91.0%, while recall, precision, sensitivity and specificity overcame the value of 90.0%. This implementation shows the feasibility of machine learning combined with data coming from nuclear cardiology. Moreover, the possibility to predict cardiac death exploiting clinical data and parameters carried out from instrumental exams would help clinicians to provide patients with the best treatments and interventions.

Keywords: Cardiology, coronary artery disease · Machine learning

1 Introduction

According to the World Health Organization, cardiovascular diseases are the first cause of death in the world, including: cerebrovascular disease, coronary heart disease, peripheral arterial disease, rheumatic heart disease, congenital heart disease and deep vein thrombosis and pulmonary embolism. Many risk factors have been identified in literature [1]: hypertension, dyslipidemia, smoking, a family history of premature coronary disease, chronic kidney disease, and the presence of albuminuria are all included in previous studies [2, 3]. However, epidemiologic data show that improved

J. Henriques et al. (Eds.): MEDICON 2019, IFMBE Proceedings 76, pp. 847–854, 2020.
https://doi.org/10.1007/978-3-030-31635-8_101

control of cardiac risk factors has resulted in a temporal decrement in the incidence and the severity of coronary artery disease (CAD) as well as its related mortality [4].

Cardiovascular management introduced the issue of handling a lot of parameters and features related to patients' state of health; the increasing amount of data is making their collection more convenient in electronic medical records [5, 6]. Information and communication technology (ICT) researchers have studied many methodologies to exploit these data and help clinicians in decision-making with both automatized [7, 8] and managerial approaches such as Lean Thinking and Six Sigma methodologies [9, 10]. The quantity of data is increasing together with the power of analytics algorithms that have the capability to make accurate prediction. This relates to the feasibility of Machine learning (ML), a branch of data mining that has been applied for many years in medicine whether to predict the outcome of patients or to classify them into two or more classes. However, technology assessment, comparison between treatments, management of resources in hospitals and support in planning for health facilities are other possible applications. The advantages brought by information technology through machine learning in healthcare are widely discussed in literature [11, 12] but its power is easily recognisable in all clinical contexts because it allows clinicians to commit a fewer number of mistake and to have the possibility to make both predictions and risk stratification of patients.

Stress single-photon emission computed tomography myocardial perfusion imaging (MPI) accounts for the majority of tests currently performed for ischemia detection to evaluate patients with suspected or known coronary artery disease (CAD). It is a non-invasive approach characterized by high sensitivity, specificity and negative predictive value for both diagnostic and prognostic applications [13–16]. Moreover, MPI has taken a central role in risk stratification of patients with suspected or known CAD. Accurate risk stratification has become increasingly important to optimize patient outcomes and contains rapidly escalating medical care costs [17].

Clinicians can carry out some quantitative parameters as regards myocardial perfusion and left ventricular function thanks to the reconstruction of images [18]. Juarez-Orozco et al. enhanced the bound between nuclear cardiology techniques (such as SPECT and PET) and machine learning in cardiology to predict fatal and adverse cardiovascular events [19].

Because of this investigation through literature, the aim of this study is to apply J48 and its empowered version, Random Forests (RF), to verify their feasibility (adopting different evaluation metrics) in predicting cardiac death or the survival probability on a cohort of patients who underwent stress MPI.

2 Materials and Methods

2.1 Dataset

Our dataset consisted of a random sample of patients with suspected or known coronary artery disease underwent gated stress MPI for the assessment of myocardial ischemia at the department of Advanced Biomedical Sciences of the University Hospital "Federico II" in Naples.

As part of the baseline examination, clinical teams collected information on traditional cardiovascular risk factors, including age, gender, blood pressure, smoking history, serum cholesterol, family history of CAD, rest ECG characteristics, diabetes and its complications as well as the results of ECG stress testing. These data were verified and complemented with demographic and clinical information collected from medical records. From these clinical variables, the pre-test CAD likelihood was calculated for each patient by dedicated software based on Bayesian analysis of pre-scan patients' data (Cadenza, version 5.04.3, Advanced Heuristics, Inc., Bainbridge Island, Washington) [20]. According to pre-test likelihood of CAD, patients with a likelihood score <0.15 were considered at low risk, while those with a score between 0.15 and 0.85 were considered at intermediate risk. In patients with known CAD, the software automatically calculated the pre-test likelihood of ischemia.

The follow-up of 661 patients was obtained via phone questionnaire to all patients and general practitioners or cardiologists records. The outcome was cardiac death; the death was of cardiac origin if the primary cause was defined as acute myocardial infarction, congestive heart failure, valvular heart disease, sudden cardiac death, or cardiac interventional/surgical procedure related. The main characteristics of patients are summarized in Table 1. The mean follow-up was 4.9 ± 2.9 years.

Table 1. Patients' main characteristics. Values are expressed as mean value ± standard deviation or as number (percentage) of subjects

Variables	Values
Age [years]	63.8 ± 11.1
Male Gender, n (%)	451 (68.2)
Weight [cm]	76.6 ± 15.2
Height [cm]	165.5 ± 9.1
Diabetes, n (%)	242 (36.6)
Diabetes duration [years]	12.2 ± 10.6
Glycemia [mg/dl]	109.4 ± 45.9
Hypertension, n (%)	498 (75.3)
Hypercholesterolemia, n (%)	387 (58.5)
LDL [mg/dl]	108.4 ± 52.0
HDL [mg/dl]	49.1 ± 13.4
Known CAD, n [%]	290 (43.9)

2.2 Tool, Algorithms and Evaluation Metrics

Knime analytics platform was employed for this analysis due to its acknowledged validity [21] that made researchers use it for different applications such as radiomics [22]. It consists of a "business model" that allows users to connect different nodes for creating new workflows implementing a wide range of algorithm; moreover, it includes many extensions, such as Weka and Python. J48 is the Java implementation of decision trees, easy structures made up of nodes and leaves [23]. RF are algorithms under the ensemble learning group, included in this platform [24]. They are powerful classifiers and regressors, exploiting the principles of randomization and bagging. The training phase is conducted distinctly for each tree and randomness is injected in this process to reduce the variance of the predictions. Randomness is introduced by subsampling the main dataset on each iteration to obtain various training sets. To make a prediction on a new instance, RF aggregate the predictions by majority vote.

Furthermore, in presence of imbalanced dataset, "smote" technique (Synthetic Minority Over-sampling Technique) can be applied and is implemented within the platform; it allows users to oversample minority classes and make more correct evaluation of performances [25]. In order to assess them, some evaluation metrics were employed in this study [26]: accuracy, error, recall, precision, specificity, sensitivity, Area Under the Curve Receiver Operating Characteristic (AUCROC).

3 Results

Firstly, a procedure to handle missing data was implemented: each missing value was replaced with the mean value of the attribute, although a median or a most frequent value substitution could be used, too. Then, the reduction of features was obtained through the computation of a matrix of correlation among all the features; the most correlated attributes, overcoming a previously chosen threshold, were not included in the algorithm because they did not add information to the classifier. We started from 150 features but 29 were included in the algorithms (Table 2).

Table 2. Features included in the algorithms

Weight	Chest_Pain	Stress_summed_wall_motion
Therapy	Dyspnea	D-SPECT
Nitrate	Previous_stroke	Supine_Acquisition
Anti-arrhythmic	Left_Bundle_Branc_Block	Stress_minutes_duration
Angiotensin_Receptor Blockers	Acute_Coronary_Syndrome	Peak_Heart_Rate
Therapy_discontinuation	Resting_Electrocardiogram	Resting_Systolic_Blood_Pressure
Diabetes	Echocardiography	ST_alteration
Dyslipidemia	Ejection_Fraction (measured by Echocardiography)	Chest_pain_type
Smoking_history	Summed_difference_score	Coronary_flow_reserve
Family_history_of_CAD	Gated_Acquisition	Cardiac_Event (class)

Since the number of positives was minor than the number of negatives, the smote technique was applied in order to get a balanced dataset: the number of records rose from 661 to 834. The augmented dataset was split into training and test sets, 80% and 20% of the total number of records, respectively.

In conclusion, the evaluation metrics were computed and are reported in Table 3.

Table 3. Scores

	Accuracy [%]	Error [%]	Precision [%]	Recall [%]	Sensitivity [%]	Specificity [%]
J48	81.0	19.0	78.3	83.3	78.3	84.5
RF	91.0	9.0	91.5	90.4	90.4	91.7

Employing RF, the overall accuracy resulted in a surprising 91.0% while sensitivity and specificity were 90.4% and 91.7% respectively. Recall and precision overcame the threshold of 90%. The AUCROC for RF was 0.961 and is represented in Fig. 1 while the AUCROC of J48 was 0.763. Of course, J48 obtained lower results in all scores.

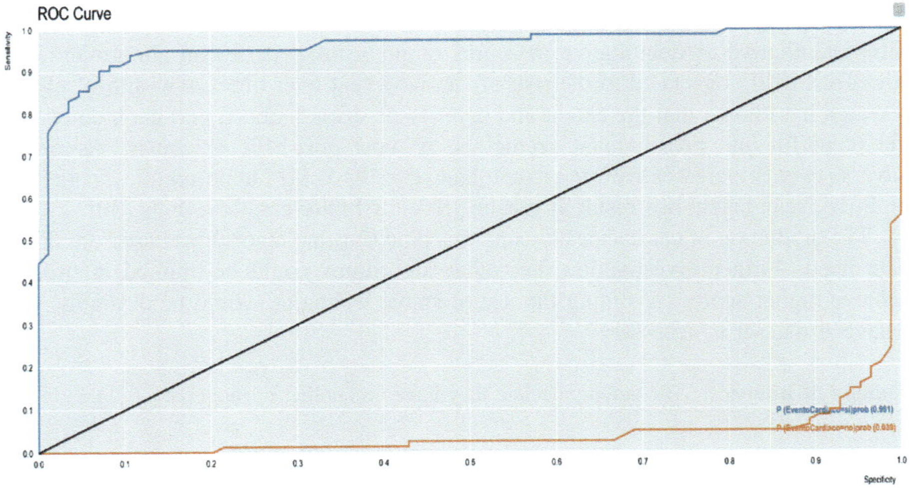

Fig. 1. ROC curve

4 Discussion and Conclusion

To sum up, clinical data and numeric parameters calculated during instrumental exams were collected as regards patients with suspected of known CAD at the department of Biomedical Advanced Sciences in the University Hospital "Federico II" of Naples. Follow-up data were also collected to record a cardiac adverse event, or a cardiac death eventually occurred. All these data were used in order to learn two models and predict whether people were alive or not.

Kwon et al. tried to predict cardiac arrest in hospitals using deep learning algorithms using four vital signs as predictor variables and got an accuracy of 91.0% and 92.6% through RF and deep learning-based systems, respectively [27]. Murukesan et al. employed 34 features (time domain, frequency domain and non-linear domain) extracted by heart rate variability signals overcoming the accuracy of 95% [28]. Alonso et al. made the same study on SPECT data reaching out an accuracy of 83.0% through support vector machine, cross-validation and employing 49 features [29].

As a result, the accuracy of 91.0% obtained through 29 features can be considered a valuable score. This model could help clinicians in the evaluation of patients' risk, giving to cardiologists an evidence-based "opinion". Of note, one of the features evaluated is coronary flow reserve (CFR) that is used to describe the amount of additional blood flow that can be supplied to the heart over baseline blood flow. The absence of CFR implies maximal vasodilatation of the resistance vessels at rest and an inability to further increase, representing a good parameter to improve clinical decision-making.

Even though diagnosis remains a primary indication for CAD testing, risk stratification has become important in patients' management. In fact, the established concept of guided care in the present era is linked to the concept of risk stratification using different imaging modalities. Accordingly, the optimal improvement in outcome may be achieved by linking high-risk imaging measures to risk-reducing therapies. Different semi-quantitative or quantitative measures of myocardial perfusion abnormalities on stress-rest MPI are related to the risk of cardiac event over time. A wealth of clinical research has shown that the extent and severity of stress-induced perfusion defects are the powerful and predominant predictors of most post-MPI outcomes. As risk of adverse events increases with the extent and severity of MPI abnormalities, patients in high-risk scan categories result at greatest risk for future cardiac events [30].

The prediction of cardiac death using machine learning algorithms based on SPECT data needs further investigation, i.e. other algorithms could be applied in order to achieve higher scores. Avoiding the use of smote technique would be desirable, yet a balanced dataset is necessary.

Conflict of Interests. The authors declare they have no conflict of interests.

References

1. American Diabetes Association: 9. Cardiovascular disease and risk management: standards of medical care in diabetes—2018. Diabetes care **41**(Supplement 1), S86–S104 (2018)
2. Anderson, K.M., Odell, P.M., Wilson, P.W., et al.: Cardiovascular disease risk profiles. Am. Heart J. **121**(1), 293–298 (1991)
3. Thayer, J.F., Yamamoto, S.S., Brosschot, J.F.: The relationship of autonomic imbalance, heart rate variability and cardiovascular disease risk factors. Int. J. Cardiol. **141**(2), 122–131 (2010)
4. Megna, R., Zampella, E., Assante, R., et al.: Temporal trends of abnormal myocardial perfusion imaging in a cohort of Italian subjects: relation with cardiovascular risk factors. J. Nucl. Cardiol. (2019). https://doi.org/10.1007/s12350-019-01630-1
5. Chen, M., Mao, S., Liu, Y.: Big data: a survey. Mobile Netw. Appl. **19**(2), 171–209 (2014)
6. Jensen, P.B., Jensen, L.J., Brunak, S.: Mining electronic health records: towards better research applications and clinical care. Nature Rev. Genet. **13**(6), 395–405 (2012)

7. Converso, G., Improta, G., Mignano, M., et al.: A simulation approach for agile production logic implementation in a hospital emergency unit. In: International Conference on Intelligent Software Methodologies, Tools, and Techniques, pp. 623–634. Springer, Cham, September 2015
8. Improta, G., Russo, M.A., Triassi, M., et al.: Use of the AHP methodology in system dynamics: Modelling and simulation for health technology assessments to determine the correct prosthesis choice for hernia diseases. Math. Biosci. **299**, 19–27 (2018)
9. Improta, G., Balato, G., Ricciardi, C., et al.: Lean Six Sigma in healthcare: fast track surgery for patients undergoing prosthetic hip replacement surgery. TQM J. (2019). https://doi.org/10.1108/TQM-10-2018-0142
10. Improta, G., Cesarelli, M., Montuori, P., et al.: Reducing the risk of healthcare-associated infections through Lean Six Sigma: the case of the medicine areas at the Federico II University Hospital in Naples (Italy). J. Eval. Clin. Pract. **24**(2), 338–346 (2018)
11. Jha, S., Topol, E.J.: Adapting to Artificial Intelligence: radiologists and pathologists as information specialists. JAMA **316**, 2353–2354 (2016)
12. Chen, M., Hao, Y., Hwang, K., et al.: Disease prediction by machine learning over big data from healthcare communities. IEEE Access **5**, 8869–8879 (2017)
13. Acampa, W., Petretta, M., Evangelista, L., et al.: Stress cardiac single-photon emission computed tomographic imaging late after coronary artery bypass surgery for risk stratification and estimation of time to cardiac events. J. Thorac. Cardiovasc. Surg. **136**, 46–51 (2008)
14. Petretta, M., Acampa, W., Evangelista, L., et al.: Reclassification of cardiovascular risk by myocardial perfusion imaging in diabetic patients with abnormal resting electrocardiogram. Nutr. Metab. Cardiovasc. Dis. **24**, 588–593 (2014)
15. Cantoni, V., Green, R., Acampa, W., et al.: Long-term prognostic value of stress myocardial perfusion imaging and coronary computed tomography angiography: a meta-analysis. J. Nucl. Cardiol. **23**, 185–197 (2016)
16. Green, R., Cantoni, V., Petretta, M., et al.: Negative predictive value of stress myocardial perfusion imaging and coronary computed tomography angiography: a meta-analysis. J. Nucl. Cardiol. **25**, 1588–1597 (2018)
17. Metz, L.D., Beattie, M., Hom, R., et al.: The prognostic value of normal exercise myocardial perfusion imaging and exercise echocardiography: a meta-analysis. J. Am. Coll. Cardiol. **49**, 227–237 (2007)
18. Cuocolo, A., Petretta, M., Acampa, W., et al.: Gated SPECT myocardial perfusion imaging: the further improvements of an excellent tool. Q. J. Nucl. Med. Mol. Imaging **54**, 129–144 (2010)
19. Juarez-Orozco, L.E., Martinez-Manzanera, O., Storti, A.E., et al.: Machine learning in the evaluation of myocardial ischemia through nuclear cardiology. Curr. Cardiovasc. Imaging Rep. **12**(2), 5 (2019)
20. Diamond, G.A., Staniloff, H.M., Forrester, J.S., et al.: Computer assisted diagnosis in the noninvasive evaluation of patients with suspected coronary artery disease. J. Am. Coll. Cardiol. **1**, 444–558 (1983)
21. Warr, W.A.: Scientific workflow systems: pipeline pilot and KNIME. J. Comput. Aided Mol. Des. **26**(7), 801–804 (2012)
22. Romeo, V., Ricciardi, C., Cuocolo, R., et al.: Machine learning analysis of MRI-derived texture features to predict placenta accreta spectrum in patients with placenta previa. Magn. Reson. Imaging (2019). https://doi.org/10.1016/j.mri.2019.05.017
23. Quinlan, J.R.: C4.5: Programs for Machine Learning. Elsevier, San Francisco (2014)
24. Breiman, L.: Random forests. Mach. Learn. **45**, 5–32 (2001)

25. Chawla, N., Bowyer, K., Hall, L., et al.: SMOTE: synthetic minority over-sampling technique. J. Artif. Intell. Res. **16**, 321–357 (2002). https://doi.org/10.1613/jair.953
26. Hossin, M., Sulaiman, M.N.: A review on evaluation metrics for data classification evaluations. Int. J. Data Min. Knowl. Manag. Process. **5**, 1–11 (2015)
27. Kwon, J.M., Lee, Y., Lee, Y., et al.: An algorithm based on deep learning for predicting in-hospital cardiac arrest. J. Am. Heart Assoc. **7**(13), e008678 (2018). https://doi.org/10.1161/JAHA.118.008678
28. Murukesan, L., Murugappan, M., Iqbal, M., et al.: Machine learning approach for sudden cardiac arrest prediction based on optimal heart rate variability features. J. Med. Imaging Health Inf. **4**(4), 521–532 (2014)
29. Alonso, D.H., Wernick, M.N., Yang, Y., et al.: Prediction of cardiac death after adenosine myocardial perfusion SPECT based on machine learning. J. Nucl. Cardiol., 1–9 (2018)
30. Acampa, W., Salvatore, M., Cuocolo, A.: Prognostication in the era of a new stressor for myocardial perfusion imaging. J. Nucl. Cardiol. **22**(6), 1222–1224 (2015). https://doi.org/10.1007/s12350-015-0075-x

On the Privacy Enhancement of In-Transit Health Data Inspection: A Preliminary Study

Jorge Sancho[1(✉)], Gert Læssøe Mikkelsen[2], Jonas Lindstrøm[2], José García[1], and Álvaro Alesanco[1]

[1] Aragón Institute of Engineering Research (I3A), University of Zaragoza, Zaragoza, Spain
jslarraz@unizar.es

[2] Alexandra Instituttet, 8200 Aarhus N, Denmark

Abstract. A new healthcare paradigm has emerged lately placing patients at the center of the healthcare scenario, providing more control over their data. Allowing patients to access data from outside healthcare organizations may suppose a security risk and should be done cautiously. Perimeter security elements, such as firewalls, are placed at organizations boundaries to inspect the in-transit data, deciding whether it should be forwarded or not. However, a disjunctive appears, if health data is kept encrypted when traversing the firewall, it could not provide its functionality properly while revealing it could be a violation of the patient privacy. In this work we propose a new method based on multi party computation to perform a secure fine-grained deep packet inspection of health data and results are presented for different number of conditions and packets sizes when using the FHIR standard. Finally, we conclude that although the performance degradation introduced by the proposed method could be too high for a general use, it could be justified by the sensitivity of the involved data in healthcare scenarios.

Keywords: HL7 FHIR · Health data inspection · Data privacy · Secure Computation

1 Introduction

In the past few years, healthcare systems have quickly evolved driven by the new social necessities that continuously appear and the Information and Communications Technologies (ICT) advancements. A new healthcare paradigm has recently emerged that moves the patient from his/her position as a mere piece in the healthcare environment to its center. This patient's empowerment implies, among other things, giving patients more control over his health-related data, allowing them to access their resources stored in healthcare institutions, creating new ones and authorizing third parties to use his/her information. However, this change of paradigm and the required system modifications should be addressed carefully to avoid incurring on security issues and facing its disastrous consequences.

Currently most systems exchange health data ciphered to guarantee its privacy. However, the use of encryption, although essential, it is not the definitive solution for

J. Henriques et al. (Eds.): MEDICON 2019, IFMBE Proceedings 76, pp. 855–860, 2020.
https://doi.org/10.1007/978-3-030-31635-8_102

all existent threats. For example, information leakages may still occur, caused by misconfigured devices or malicious behaviours from inside the organization. An information leakage in a healthcare scenario may lead to serious legal implications due to the sensibility of the involved information. To prevent this kind of concerns, firewalls, which are placed on domains boundaries, inspect the traffic and decide whether it should be forwarded or not depending on a set of policies. However, firewalls need unencrypted data to make full informed forwarding decision. This situation presents a disjunctive, if the firewall is not able to decrypt the information, it would not be able to provide full informed decisions and could not help to prevent the undesired information leakages. However, allowing the firewall to decrypt the data for its inspection could be an inadmissible privacy violation, especially when sensible data such as health related one is involved.

Fortunately, in the past few years some new approaches based on Secure Computation (SC) primitives have been proposed to tackle with this issue. These methods enable the firewall to enforce some policies directly on the encrypted traffic without compromising the privacy of the data being evaluated, at the cost of a decrease in data inspection performance. First approaches can be found on [1, 2] where Searchable Encryption is used to determine whether some patterns of interest are present in the encrypted payload. To that end, the payload is firstly tokenized and then these tokens are matched against the rule set being enforced. In [3] authors propose a new encryption scheme called PrefixMatching that allows to match an encrypted packet field against encrypted prefixes or ranges. The use of this encryption scheme enables the middlebox to perform some new functionalities like load balancing, NAT or traffic classification. All these methods, which analyse a tokenized version of the payload, present a big drawback since it is impossible for the firewall to be sure that the tokens being processed really belong to a certain packet without fully trusting the sender, who might be compromised. Finally, in [4] authors propose a different approach based on Secure Multiparty Computation (SMC) to provide range matching functionality. In this approach there is an input middlebox which splits the packet in shares, each share is sent to an instance of the functionality middlebox which jointly perform the evaluation and finally partial results obtained are reassembled in the local gateway. As pointed in [5] the security of this protocol is weak and performing the same operations in more than one instance results in a high computational redundancy. All in all, the functionality achievable with existent methods could be quite limited to inspect structured data, such as the used to exchange health information, where the same keyword means different thing depending on the context.

In this paper, we present a new inspection method to take advantage of these meaningful subtle features that could go unnoticed for other efficient but simpler methods. The rest of this paper is organized as follows: in Sect. 2, our proposed method will be explained; Sect. 3 includes all the analysis and results obtained applying the method explained in Sect. 2, while Sect. 4 provides the conclusions obtained in this work.

2 Methodology

Our new method is based on Garbled Circuits [6] and enables a firewall to inspect encrypted healthcare data while its privacy remains intact. Garbled Circuits is a Secure Multiparty Computation (SMC) protocol that involves two parties, a garbler and an evaluator, which jointly compute the output of a function of interest over their private inputs. First, both parties decide the function being evaluated and agree on a Boolean circuit representation before-hand. At the start of the protocol, the garbler "encrypts" his own input assigning a random label to each input bit and garble the circuit creating a new truth table for each gate on it. Then, the garbler sends the labels associated with his input and the new truth tables to the evaluator, who also received the encrypted version of his own input by oblivious transfer [7]. New truth tables are created in a special way so that, knowing the labels associated with a gate inputs, the evaluator is able to obtain the label associated with its output. However, anything about the real values associated with that label could be inferred without the help of the garbler. The evaluator iterates over all the gates in the circuit until he obtains all the labels related with the circuit output. Finally, both parties cooperate to learn the real values associated with that output labels. The use of this protocol guarantees that after its execution both parties learn just the function output, while anything related with the other party input or any intermediate value remains hidden.

In the proposed scenario, healthcare data would be exchanged using the last generation standard from the High Level Seven (HL7), Fast Healthcare Interoperability Resources (FHIR). Typically, FHIR flows are sent over Hypertext Transfer Protocol Secure (HTTPs) connections which relays on Transport Layer Secure (TLS) to provide confidentiality and integrity to exchanged data while the resources are sent in the body of the HTTP packets encoded using the XML or JSON formats. FHIR packets would be privately inspected using the Garbled Circuit protocol explained before. To that end, the firewall would act as evaluator providing the encrypted packet and the policy set while the FHIR server would be the garbler providing the encryption key as private input. The function to be evaluated has been designed to firstly decrypt the payload of packets and then check the rules in the policy set, while the use of garbled circuits ensures that the plain text is never revealed to neither the garbler nor the evaluator. Since Garbled Circuits allow to perform arbitrary function evaluation, different encryption algorithms could be used and almost any kind of policies could be enforced, and limitations on the achievable functionality would be imposed by the performance restrictions of the concrete use case. In the proposed scenario, policy evaluation is performed in two steps, checking the HTTP headers first and the packet content, when present, after that. For the evaluation of the packet headers, each condition in the policy would verify if a given header is present in the packet and the value of this header, by using a simple string matching. The evaluation of the content, which in this scenario would be a FHIR resource coded using the JSON format, is more complex. In this case, each condition in the policy verifies the value of a single field in the resource. For example, one condition could check if an observation is linked with a patient or the type of data represented by this observation. This evaluation requires a deep inspection through the resource structure considering the relationship between the different keywords of the JSON document. Finally, after the circuit evaluation, both parties learn whether the evaluated packet meets the policies or not without learning anything about the other party inputs.

3 Results

The proposed inspection method has been implemented using the Obliv-C framework [8] running Garbled Circuits for the honest-but-curious threat model. Several analyses have been performed to assess the viability of using this method in an operational scenario. Tests have been performed using Amazon Web Services (AWS) EC2 c5.large instances, which have 2 virtual CPUs and 4 GB RAM. One instance has been used for each party (garbler and evaluator) needed to consider the communication cost required by the Garbled Circuit protocol in a realistic way. Results have been obtained assuming that packets are encrypted using AES-128 working on Chained Block Chipper (CBC) mode which is commonly used by TLS and FHIR resources are encoded using the JSON format.

Times taken to perform the different tasks that compose the inspection function under different conditions are shown in Fig. 1. These times, which include the time associated with the data decryption, the time required to evaluate the headers and the time spent for the content evaluation, would be used to characterize the behaviour of the inspection function when the tasks input sizes and the number of conditions being enforced vary. For a given packet, tasks input sizes would correspond to the whole encrypted packet for the decryption task, the HTTP headers for headers evaluation and the FHIR resource for the content evaluation, respectively. We can derive the time needed to complete the packet inspection from the results presented in the figure: the total evaluation time can be calculated as the addition of the times shown for the three independent tasks using the correct input sizes for each of them.

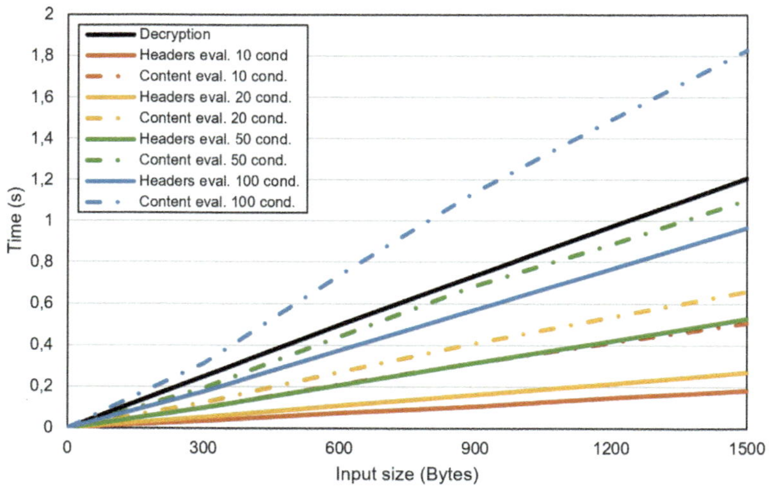

Fig. 1. Performance results.

In this figure we can observe that the decryption time (black line), which is independent of the number of conditions is linearly dependent with the input size. On the other hand, times associated with the evaluation of the headers (colorful continuous lines) and the content (colorful dotted lines) are linearly dependent with both the input size and the number of conditions. We can also observe that for the analysed encryption scheme (AES-128 CBC) the time spent for packet decryption would be much bigger than the required for policy evaluation when the number of conditions being enforced is small (less than 50 for headers and less than 20 for content).

To contextualize this information times required to perform some usual operations of patient's daily follow-up have been obtained when using the proposed method. These times correspond to the whole process, including the inspection of both the request and the response packets. To obtain these times some reasonable assumptions have been done. First, a fixed number of 20 conditions for the headers and 20 more for the content are being enforced. We have considered that the HTTP headers size would be 250 bytes in all cases, blood pressure (BP) observations (scalar values) have a size of 450 bytes while a 10 s plethysmography (PPG) observation (low frequency unidimensional signal) has a size of 1200 bytes. Given all this, times used to read and create a BP observation are 1,4 and 1,8 s respectively while times used to read and create a PPG observation are 2,3 and 3,6 s respectively. As expected, times invested to create a resource are greater than the required to read it since in a create operation the resource goes in the body of the http packets both in the request and the response while for a read operation the resource is only sent in the response, holding the request body empty.

4 Conclusions

In this work we show how Secure Computation methods could be applied to healthcare systems to satisfy the heavy security requirements of these kind of system, guaranteeing the privacy of the patients' data. Proposed inspection method could be applied to perform a secure fine-grained deep packet inspection by exploring the packet content even when it is formatted with complex high structured format such as JSON. At the best of our knowledge this is the first approach that allows to perform deep packet inspection securely in the application layer.

Although the performance degradation introduced by the proposed method could be inacceptable for a general use, it could be justified in healthcare scenarios by the sensitivity of the involved data. Besides, further research is still needed with more data exchange scenarios to determine the real performance impact analysing the achievable throughput when using the proposed method rather than just considering the delay introduced by it. Determining whether the use of other encryption schemes could be more appropriated and developing new efficient functionalities for the policy evaluation is also required.

Acknowledgements. Research funded by Ministerio de Economía, Industria y Competitividad from Gobierno de España and European Regional Development Fund (TIN2016-76770-R), Gobierno de Aragón and FEDER "Construyendo Europa desde Aragón" (T31_17R) and Ministerio de Educación, Cultura y Deporte from Gobierno de España via a doctoral grant to the first author (FPU15/04841).

Conflict of Interest. The authors declare that they have no conflict of interest.

References

1. Sherry, J., et al.: BlindBox: deep packet inspection over encrypted traffic. ACM SIGCOMM Comput. Commun. Rev. **45**(4), 213–226 (2015)
2. Yuan, X., et al.: Privacy-preserving deep packet inspection in outsourced middleboxes. In: IEEE INFOCOM 2016-The 35th Annual IEEE International Conference on Computer Communications, pp. 1–9. IEEE (2016)
3. Lan, C., et al.: Embark: securely outsourcing middleboxes to the cloud. In: NSDI, vol. 16, pp. 255–273 (2016)
4. Asghar, H., et al.: SplitBox: toward efficient private network function virtualization. In: Proceedings of the 2016 Workshop on Hot Topics in Middleboxes and Network Function Virtualization, pp. 7–13. ACM (2016)
5. Wang, C., et al.: Toward secure outsourced middlebox services: practices, challenges, and beyond. IEEE Netw. **32**(1), 166–171 (2018)
6. Yakoubov, S.: A gentle introduction to Yao's Garbled circuits. https://pdfs.semanticscholar.org/160d/ce6103b9f1b35640975cea951577c7e37096.pdf. Accessed 27 Mar 2019
7. Rabin, M.: How to exchange secrets with oblivious transfer. IACR Cryptology ePrint Archive, vol. 2005, p. 187 (2005)
8. Zahur, S., Evans, D.: Obliv-C: a language for extensible data-oblivious computation. IACR Cryptology ePrint Archive, vol. 2015, p. 1153 (2015)

rOral: Use of a Teledentistry System for Remote Images Assessment in Oral Health Education Workflows

Raquel Sebastião[1,2(✉)], Ilídio C. Oliveira[1,2], Ricardo Felgueiras[2], and Nélio J. Veiga[3,4]

[1] Institute of Electronics and Informatics Engineering of Aveiro, University of Aveiro, Campus de Santiago, 3810-193 Aveiro, Portugal
raquel.sebastiao@ua.pt
[2] Department of Electronics, Telecommunications and Informatics, University of Aveiro, Campus de Santiago, 3810-193 Aveiro, Portugal
[3] Health Sciences Institute, Universidade Católica Portuguesa, Viseu, Portugal
[4] Centre for Interdisciplinary Research in Health (CIIS) – Universidade Católica Portuguesa, Lisbon, Portugal

Abstract. Medical training in oral health includes interpreting images of the oral cavity. The use of extended datasets, real cases, and the ability to monitor the progress of students is beneficial. In this work, we investigate the feasibility of using a teledentistry solution to support medical training. The rOral integrated teledentistry solution, developed by the research team, allows the decoupling of image acquisition and image review: images of the oral cavity can be recorded using the camera of regular mobile devices, near the subject of care; the images are then uploaded to a secure storage and reviewed by experts, using a web environment. This workflow, primarily designed for remote oral health diagnosis (e.g.: population screening), can be used for dentistry education, allowing students to play the role of experts. A population of dentistry students was involved in the present study and used rOral to access anonymized cases to form a diagnosis. The results show that the existing solution, planned for remote diagnosis, is suitable for education and that smartphone-acquired images can be used in diagnosis assessment activities.

Keywords: Mobile computing · mHealth · Telemedicine · Teledentistry · Oral care · Medical training

1 Introduction

Oral diseases affect overall health, and although the significant socio-economic impacts, they are often neglected in public health policy [1], pointing out the need of oral health screening in the population, as promoting oral health will lead to wellbeing and life improvements.

Teledentistry in screening protocols has been used recently, namely in screening for new patient orthodontic referrals [2, 3], screening for Early Childhood Caries (ECC) in children at school [4] and preschool [5], screening for caries in patients at a dental

J. Henriques et al. (Eds.): MEDICON 2019, IFMBE Proceedings 76, pp. 861–868, 2020.
https://doi.org/10.1007/978-3-030-31635-8_103

clinic [6, 7] and in nursing homes. These studies found teledentistry to have at least comparable results to traditional oral examinations and, therefore, considered it to be a good alternative.

In this context, the developed rOral system, enables the remote access to oral cavity photographs, captured with mobile devices like smartphones, near the subject of care, and the posterior review by experts, using a web application.

rOral was designed to help in remote oral health assessment, with potential uses in screening, enhance the access to oral care for the underserved populations, and in dental education, to deliver alternative ways of teaching [8], like interactive video-conferencing or web-based online courses [9, 10].

In this work, we present a study involving dental students and faculty members to validate the suitability and readiness of the rOral supported workflows for medical training, focusing on image-based diagnosis.

This paper is organized as follow. The next section presents a brief overview of the current state of technology. The following section introduces the proposed rOral system and afterwards the rORAL is evaluated, presenting the main findings regarding accordance between face-to-face appointment and the evaluations made through the rOral-review. It ends with the main conclusions of the work and further improvements.

2 Related Work

The technology advances enabled the widespread use of mobile devices, with support for connectivity and camera sensors, offering an appealing instrument to capture and share images in a telehealth platform.

Teledentistry is a specific field of telemedicine for dental care, consisting on the use of technology to help screen for oral diseases and generally to help providing oral care. Although telemedicine is about the clinical uses of technology, teledentistry encompasses both remote clinical and remote non-clinical uses of technology for oral care. Teledentistry has been used both as a replacement for face-to-face appointments [11, 12], to provide care for underserved populations, for screening and on education, to provide interesting and alternative ways of teaching [8].

Regarding teledentistry in education, there are two main categories: online courses and interactive video conferencing. They both provide ways to get education remotely and have produced satisfactory results [13, 14]. Interactive video conferencing consists in several parties having a two-way communication over a live video call. Besides allowing the user (student) to receive immediate feedback [8, 15], it is also suitable for long distance learning in dentistry, reducing significantly travel time and expenses [9].

With the purpose of oral screening, a teledentistry system similar to the developed rOral was considered useful [16]. This system was designed with an Android app for image acquisition and upload, a cloud-based server that enabled data entry, data storage, data retrieval and image storage, a website to review and assess the records stored in the server, a teledental assistant to capture images of the patients and a grader (a dentist) to review the images. Within this application field the accuracy of dental images is a critical target, been gathered by photographic/videographic cameras and by smartphone cameras. Recent studies concluded that smartphone cameras could obtain results at least as good as digital cameras, and that photographic analysis had results similar to face-to-face appointment [14, 17, 18].

3 rOral Platform

The rOral platform supports asynchronous image-based diagnosis workflows, splitting the tasks of image acquisition and image review. This system is composed by four main components: rOral-mob, rOral-review, rOral-admin and the backend services. Each of these subsystems has a specific role in the platform:

1. rOral-mob: application for mobile devices, for practical recording of oral care images and insertion of patient data, to be used by Operators. A native Android application was implemented.
2. rOral-review: a web-based application for experts, allowing to manage worklists and clinical reporting, to be used by Evaluators. In this component, the evaluator also has the possibility to ask for a second opinion of a colleague.
3. rOral-admin: web application to create and manage oral health studies and their participants, to be used by the head of each study.

In the rOral system, there are different types of actors. Each type of actor has its own role and set of actions that they can perform:

1. **Operator**
   An Operator is the user of rORAL-mob. They are the only ones in the system that have direct contact with the Patients. They can be, for example, dentistry students, nurses or caregivers. Their role is to collect information and photographs from the patients, create a clinical case and submit them for evaluation, providing the data for the Evaluators to review. For that they must enrol in a Study using the credentials given by the Point of Contact, create a Patient and insert their personal data (name, gender and date of birth), create an Episode for them and then capture the Photographs and submit them for evaluation.
2. **Evaluator**
   The Evaluator is the person who submits Evaluations for the Episodes. Their goal is to make an Evaluation based on the Photographs and other information from the Patient. They can be students in a university or dentists, for example. An Evaluator is always linked to a Domain. This means that they can only review Episodes from that Domain, making the process more easily coordinated by the Domain's Point of Contact.

With respect to the quality of images taken by smartphone cameras and images taken by autonomous digital cameras, a recent study concluded that smartphone cameras have performances similar to the digital cameras and can replace these in image-based teleconsultation [19]. Despite that conclusion, the rOral system has basic support to enhance an image through transformations of brightness and contrast, as well as crop the image to better evidence a special area of interest. rOral allows, at each case evaluation, to ask for a second opinion/assessment from another oral health professional; this workflow is also suitable to promote cases discussion. Figure 1 provides overview the rOral-mob and rOral-review user interfaces. The left image shows the information of an episode, the photographs of a patient's oral cavity, and the possibility of image transformations. The right image shows the review environment seen by evaluators: the patient and episode information (such as gender, age and date), the photographs taken and the space to write the review of the clinical case.

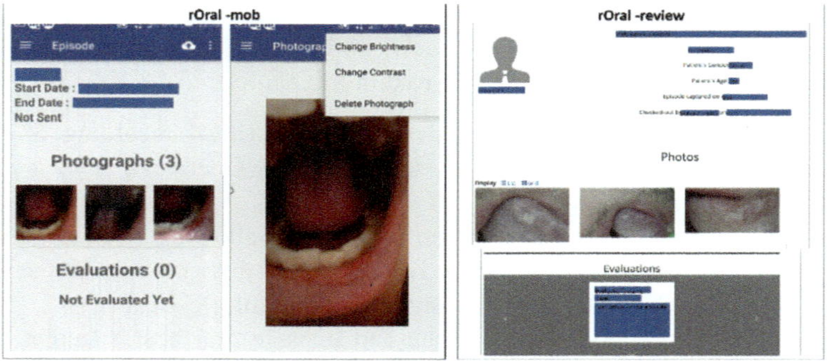

Fig. 1. Overview of an episode of a patient in rOral-mob (left) and in rOral-review (right).

3.1 Implementation

For the mobile clinical cases acquisition subsystem, the rOral-mob was developed as an Android native application. Operators take photographs (common Android Intents[1] resolution mechanisms pick the photo capture application) that are saved to the application encrypted storage. Operators can also perform a minimal image enhancing using the toolbox available (e.g.: crop, contrast and bright adjusting). Zooming an image is also possible (supported in chrisbanes/PhotoView[2] library).

rORAL-PoC, rORAL-review and the backend are deployed in a server environment (Heroku[3] cloud infrastructure), including the studies management, review and evaluation, integration API and file storage components. These backend services were implemented using the Python-based Django[4] framework, which offers a convenient framework for web applications development.

The integration API is the system component responsible for handling the requests made by rORAL-mob. The API requires that users authenticate the sessions to retrieve a secure token, required to each subsequent API call. The authentication credentials are kept at the server side, in which the only the hash of the password is stored.

For data security measures, PINs and QR codes were used to allow the Operators to use the app and an ID and PIN combination or a QR code was necessary for the Operators to join Studies. To enforce the privacy of the patient, their names and other

[1] https://developer.android.com/reference/android/content/Intent.html.
[2] https://github.com/chrisbanes/PhotoView.
[3] https://www.heroku.com/.
[4] https://www.djangoproject.com/.

personal identifiers (unique) aren't stored in the backend, nor shown to the Evaluators. What is stored is a unique key that each Domain, represented by a Point of Contact, can define what it is and the associated generation rules. Data stored in the server side uses surrogate identifiers to link the uploaded cases to the patient cases in the mobile acquisition system. This means that even if an unauthorized entity could access the cloud data, no direct patient identifiers are available (pseudonymized).

3.2 Example of Medical Training Usage

rOral can be used for remote assessment of images in dental education. As a proof of concept, rOral was evaluated at the Health Sciences Institute of the Portuguese Catholic University (ICS-Viseu).

The proposed protocol for the study was the following:

1. A manager, registered as Point of Contact, created a study in rORAL-PoC, preparing it for the students;
2. A teacher performed a face-to-face examination of the patients that participated in the study; in addition, the teacher would then use rORAL-mob and, as an Operator, registered the patient and took photographs of their oral cavity, creating a clinical case;
3. The teacher then submitted the cases for evaluation, uploading to the remote server using rORAL-mob (the cases may also be submitted automatically, if that option in the rORAL-mob settings was selected);
4. Students logged in rORAL-review, as an Evaluator, and browsing through the cases needing evaluation, assessed each case;
5. The teacher then went on rORAL-PoC to access the evaluations made by the students and compared them to his face-to-face examination;

4 Evaluation

This section presents the assessment of the proposed rOral in medical training of oral disease screening. The proposed rORAL was evaluated by assessing the accordance between diagnosis performed by dentists, through direct examination, and diagnosis performed remotely, through the rORAL-review, by groups of students.

The participation in the study was voluntary. All the participants signed an informed consent, and their participation did not bring and will not bring any risk or do any harm to their physical and moral integrity. The data collected during the study was pseudonymized and the study was approved by the Ethical Committee of the University of Aveiro.

The protocol of the study was the following. After registering the patient in the rORAL-mob, the dental assistants created clinical cases and collected several photographs of the oral cavity of the patient. Those cases were then uploaded to the remote server for later evaluation through the rORAL-review. Afterwards, the dentist performed a direct examination and reported the respective diagnosis, which was considered as gold standard.

The remote evaluation of cases was performed in a teaching environment with students of the Integrated Master in Dental Medicine at the Health Sciences Institute of the Portuguese Catholic University (ICS-Viseu), in the context of the subject of gerodontology, under the supervision of a teacher. In these classes, 5 groups of students were presented through the rOral-review with clinical cases and each group performed their evaluation. Eight patients, predominantly males (5), were enrolled in this study, with ages ranging from 21 to 89 years old (61 ± 23).

Table 1. Results of the assessments made in the medical training of rORAL.

P #	Gold Standard	Group 1	Group 2	Group 3	Group 4	Group 5
1	Gengivitis	match	match	match	match	match
2	NUP	match	match	match	different	match
3	Periodontal disease	match	different	match	different	match
4	Cervical carie	match	match	match	match	different
5	Leukoplakia	match	match	match	match	match
6	Radicular carie	different	match	different	match	match
7	Dental erosion	match	match	match	different	match
8	Fistula	match	match	match	match	match

The diagnoses made during the clinical oral examination and the comparison with the diagnoses of each group of students are presented in Table 1. The first column states the patient, the second column states the diagnosis made by the dentist during the face-to-face examination and the remain columns state the diagnosis made by the 5 groups of students, that only assessed the patient by looking at their clinical case in rORAL-review. Each of these cells are colored either in green, if the diagnosis matches the direct examination, or in yellow, if the diagnosis does not match the direct examination.

From the above results, one can observe, that except for the fourth group, the remain four groups agree with the gold standard in 7 of the 8 cases, obtaining an agreement percentage of 87.5%, while the group number four agreed only 5 of the time, with a percentage of 62.5%.

These results show that the students were able to perform their assessment on clinical cases accurately through the images presented at the rOral-review, which sustain the feasibility of the proposed system in medical training classes and its use as a mean to present the students with more clinical cases of each diagnose.

5 Conclusions and Further Research

The main objective of this work was to establish whether a teledentistry platform, engineered for remote image assessment by experts, could be beneficial for medical training settings. The developed rOral is a system capable of taking photographs of the oral cavity and share them, securely, in a platform where they could be seen and reviewed by experts, supporting teledentistry workflows for medical training and education and for population screening.

The feasibility of using rOral in the medical training and education context was proven. With the access to a remote area to review cases, the students have access to more real clinical cases for each diagnosis and can discuss with colleagues on several issues related to a disease. Therefore, the usage of such a system is considered advantageous not only by the point of view of students but also by teachers. The obtained results sustain this conclusion and the students have endorsed the use of rOral to access more clinical cases. Moreover, both students and dentists were also given two different questionnaires to fill regarding the usability of the system. The grades of those questionnaires revealed that they found rORAL, in general, to be useful and easy to use.

Besides fulfilling this purpose, dentists agreed that rORAL could help in screening and early diagnosis of oral diseases in the elderly at day care centers. The use of rORAL in such scenario helps with oral diseases prevention, which has high impact upon the society and the individual. This demand is even more important considering the reduced mobility that commonly affects the elderly. The outcomes support the usability of the rORAL platform, and the students and dentists enrolled in this study agreed that it could be a feasible tool to use in screening processes.

Although the positive obtained results, during these studies, the evaluators used different terminology to describe the same diagnosis. So, to prevent redundancy, it is advisable to restrict each diagnosis to a unique terminology. Therefore, the adoption of International Classification of Diseases, 10th Revision, Clinical Modification (ICD-10-CM)[5] is suggested as a further improvement of rORAL-review to restrict the terminology in the evaluations.

Since the quality of the photographs taken is crucial for the accuracy of the evaluations, it is also worthy of note that rOral can automatically perform image quality improvements on the photographs taken, for example, the enhancement of specific parts of the photographs relevant for oral care, and brightness correction.

Acknowledgments. This work is funded by national funds (OE), through FCT – Fundação para a Ciência e a Tecnologia, I.P., in the scope of the framework contract foreseen in the numbers 4, 5 and 6 of the article 23, of the Decree-Law 57/2016, of August 29, changed by Law 57/2017, of July 19. This work was also supported by the FCT through national funds, and co-funded by the FEDER, within the PT2020 Partnership Agreement and COMPETE2020 under projects IEETA (UID/CEC/00127/2013) and SOCA (CENTRO-01-0145-FEDER-000010). Raquel Sebastião also acknowledges her Post-Doc grant (BPD/UI62/6777/2018).

[5] http://www.who.int/classifications/icd/icdonlineversions/en/.

References

1. Jin, L.J., et al.: Global burden of oral diseases: emerging concepts, management and interplay with systemic health. Oral Dis. **22**, 609–619 (2016)
2. Mandall, N.A., et al.: Teledentistry for screening new patient orthodontic referrals. Part 1: a randomised controlled trial. Br. Dent. J. **199**(10), 659–662 (2005)
3. Mandall, N.A., Qureshi, U., Harvey, L.: Teledentistry for screening new patient orthodontic referrals. Part 2: GDP perception of the referral system. Br. Dent. J. **199**(11), 727–729 (2005)
4. Subbalekshmi, T., Anandan, V., Apathsakayan, R.: Use of a teledentistry-based program for screening of early childhood caries in a school setting. Cureus **9**(7), e1416 (2017)
5. Kopycka-Kedzierawski, D.T., Billings, R.J., McConnochie, K.M.: Dental screening of preschool children using teledentistry: a feasibility study. Pediatr. Dent. **29**(3), 209–213 (2007)
6. Estai, M., et al.: Comparison of a smartphone-based photographic method with face-to-face caries assessment: a mobile teledentistry model. Telemed. e-Health **23**(5), 435–440 (2017)
7. Estai, M., et al.: The efficacy of remote screening for dental caries by mid-level dental providers using a mobile teledentistry model. Commun. Dent. Oral Epidemiol. **44**(5), 435–441 (2016)
8. Chen, J.W., et al.: Teledentistry and its use in dental education. J. Am. Dent. Assoc. **134**(3), 342–346 (2003)
9. Ignatius, E., et al.: Teledentistry in dental specialist education in Finland. J. Telemed. Telec. **12**(7), 46–49 (2006)
10. Chhabra, N., et al.: Role of teledentistry in dental education: need of the Era. J. Clin. Diagnostic Res. **5**(7), 1486–1488 (2012)
11. Fricton, J., Chen, H.: Using teledentistry to improve access to dental care for the underserved. Dent. Clin. North Am. **53**(3), 537–548 (2009)
12. Alipour-Rocca, L., Kudryk, V., Morris, T.: TME3/347: a teledentistry consultation system and continuing dental education via internet. J. Med. Internet Res **1**(suppl. 1), e110 (1999)
13. Linjawi, A.L.: Students' attitudes towards an on-line orthodontic learning resource. Eur. J. Dent. Educ. **13**(2), 87–92 (2009)
14. Pilcher, E.S.: Students' evaluation of online course materials in fixed prosthodontics: a case study. Eur. J. Dent. Educ.: off. J. Assoc. Dent. Educ. Eur. **5**(2), 53–59 (2001)
15. Corrêa, L., et al.: Teaching oral surgery to undergraduate students: a pilot study using a web-based practical course. Eur. J. Dent. Educ. **7**(3), 111–115 (2003)
16. Estai, M., et al.: A proof-of-concept evaluation of a cloud-based store-and-forward telemedicine app for screening for oral diseases. J. Telemed. Telec. **22**(6), 319–325 (2015)
17. Meurer, M.I., et al.: Accuracy of dental images for the diagnosis of dental caries and enamel defects in children and adolescents: a systematic review. J. Telemed. Telecare **21**(8), 449–458 (2015)
18. Queyroux, A., et al.: Accuracy of teledentistry for diagnosing dental pathology using direct examination as a gold standard: results of the tel-e-dent study of older adults living in nursing homes. J. Am. Med. Dir. Assoc. **18**(6), 528–532 (2017)
19. Boissin, C., et al.: Can we trust the use of smartphone cameras in clinical practice? laypeople assessment of their image quality. Telemed. e-Health **21**(11), 887–892 (2015)

Investigations on a Computer Application for Tracking the Mean Glandular Breast Dose Profile in Mammography

Homero Schiabel[1(✉)], Bruno Barufaldi[2], and Eny M. Ruberti Filha[3]

[1] S. Carlos School of Engineering, University of S. Paulo, Av. Trabalhador Saocarlense 400, Sao Carlos, SP 13566-590, Brazil
homero@sc.usp.br
[2] Department of Radiology, University of Pennsylvania, 3620 Hamilton Walk, Philadelphia, PA 19104, USA
[3] Ruberti & Ruberti Consulting, Rua Girassol 1585-12, Sao Paulo, SP 05433-002, Brazil

Abstract. As quality assurance standards determine the use of single phantoms to be used in evaluation tests in mammography, checking the profile of the patients submitted to exams in each equipment is important to drive these tests. In addition, this profile is also relevant in investigations on the actual doses of radiation received by the patients during the exams, mainly in the current digital mammography systems. Therefore this work investigates these aspects regarding a single DR mammography unit, working in a public radiological service, by applying a novel software designed to extract data needed to the quality assurance evaluation from the DICOM files corresponding to the outcomes of the image exam. This software refers to a computational tracking scheme to investigate characteristics of the population submitted to the mammography exams, providing information on the dose profile yielded by the digital equipment. Data obtained by this application are compared to those provided by classical experimental tests and they were verified in agreement for such unit. They show that the most of patients relative to this study has a breast thickness profile above of that commonly used in test phantoms, being in the range of 52 to 70 mm. Furthermore the dose profile of this population was verified below the standard limits, which evidences the proposed software application as a useful tool to aid quality assurance programs in digital mammography.

Keywords: Digital mammography · DICOM files · Radiation dose

1 Introduction

The need of diagnostic standards implementation in order to assure the quality of radiological services is adequately established in the IEC 61223-1 standard [1]. This standard also addresses the quality of each mammography equipment, including operational techniques and acceptance tests. In Brazil, as effect of standards regarding this issue, the Brazilian Quality Program in Mammography was created as a country model for mammography image assurance. And digital mammography quality guidelines – which are

J. Henriques et al. (Eds.): MEDICON 2019, IFMBE Proceedings 76, pp. 869–873, 2020.
https://doi.org/10.1007/978-3-030-31635-8_104

based on test results, especially with phantoms defined in national and international standards – aims on breasts with standard thicknesses and glandular composition. However breast composition is well known as variable according to the population involved [2]. Therefore determining each typical population characteristics is useful in helping to define reference levels of radiation dose absorbed by the breast tissue. In current digital mammography procedures, all images are recorded according the DICOM protocol [3] and the corresponding files register important information about the exam and exposures conditions, such as breast thickness, and dose received by the breast during the exam, among others. Hence investigating such technical data from the DICOM header can lead to the improvement of the image acquisition process. In addition, a computational tool that provides information on the doses profile from data extracted from these files could manage the image acquisition process, together with the characterization of the population profile, which could complement the usual quality assurance tests in digital mammography, following the quality criteria required by medical staff and international standards.

A previous work developed by our team [4, 5] attempted in obtaining such data from the DICOM headers of a set of images corresponding to 2D digital mammography and digital breast tomosynthesis exams. Changes were implemented in such approach so that a novel software in JAVA was developed and tested with a database formed by cases of patients who underwent mammography exams on a GE Senographe DS equipment at a Hospital in S. Paulo (Brazil). The present work describes the application of this computational scheme in such database, also showing a comparison of the results obtained from a set of exposures performed in that mammography unit.

2 Methods

In addition to each digital breast image of the patient, there is text-format data in each file resulting from the exam which is transferred and stored together in image databases in the PACS (Picture Archiving and Communication System) of the radiological diagnosis centers. This information is stored according to the DICOM protocol, whose header of the corresponding file allows identifying and characterizing the equipment, also informing the patient's age, technical factors used in image generation, dose of radiation incident on the surface of the skin and absorbed by the mammary tissue, and thickness and compression of the breast. Both the image and this information can be retrieved – from the PACS, for example – and used to evaluate the image quality. The recent modification introduced in the previous procedure [4, 5] has generated an application called *ReadDICOM* which, once in execution, makes it possible to select a folder in the computer on which the complete DICOM files from the set of exams to be evaluated are stored. The application allows thus to produce a file with a sheet where data selected from the headers are registered.

A classical experimental procedure was performed in addition with the mammography equipment under evaluation. This procedure has the purpose of estimating the actual mean glandular dose (MGD) from the values of *kerma* measured with PMMA and those calculated for the breast tissue for a GE Senographe DS digital mammography system. The MGD calculation is based on the description of the International Code for Dosimetry in Radiological Diagnosis [6] and the calculation demonstrated by Dance et al. [7], deriving from a product of parameters as the air *kerma*, two conversion factors depending on the standard density, standard thickness, half value layer (HVL) and the target-filter combination [7, 8]. The relationship between air *kerma* and skin dose of the study patients was evaluated. A calibrated dosimeter (Accu-Gold AGMS-M +, Radcal, Monrovia, CA) was also used to obtain HVL values and to accurately determine the air *kerma* of the X-ray tube in order to validate the information provided by the manufacturer.

3 Results

Table 1 shows the air *kerma* values measured with PMMA in the thicknesses between [10–70] mm and those calculated for the tissue [9] at the corresponding ranges for the GE Senographe DS equipment. For the calculation of MGD, the technical data and the characteristics of the breast tissues of about 250 patients were considered.

Table 1. Measured and calculated air *kerma* values for PMMA and measured MGD [9].

PMMA thickness (mm)	Breast tissue thickness (mm)	Filter/target	*Kerma/* air range (mGy)	MGD (PMMA) (mGy)	MGD (breast tissue) (mGy)	Limit MGD (mGy)
20	[20–30]	Mo/Mo	[2.12–5.10]	0.88	1.06	1.0
30	[31–40]	Mo/Rh	[2.76–7.01]	1.30	1.23	1.5
40	[41–45]	Mo/Rh	[3.50–7.98]	1.48	1.34	2.0
50	[51–60]	Rh/Rh	[4.47–14.26]	1.75	1.63	3.0
60	[61–70]	Rh/Rh	[4.81–14.78]	1.85	1.76	4.5
70	>71	Rh/Rh	[7.30–15.68]	2.13	1.94	6.5

When applying the software to extract data from the DICOM headers of this set of images, there is a large amount of correlational data possible to determine, as entrance or glandular dose as a function of breast thicknesses, individual doses per patient, breasts thicknesses range according to patients' ages. Two graphical relations are shown in this current approach to illustrate these data: the registered MGD as a function of the breasts thicknesses for all the patients in the study, and the histogram relative to the number of cases for all the range of breast thicknesses (Figs. 1 and 2).

Considering just over 250 DICOM files (approximately 1000 images) from exams performed in this DR system, the time of getting organized all data was only 8 s.

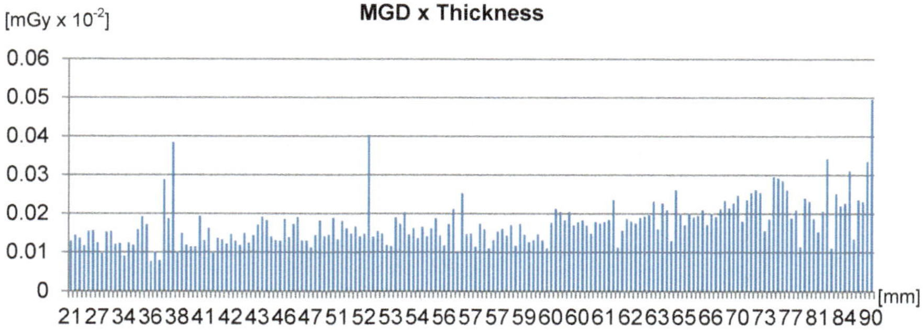

Fig. 1. Graph relative to the mean glandular dose registered from the DICOM header for all the 250 patients in the dataset as a function of the breasts thicknesses.

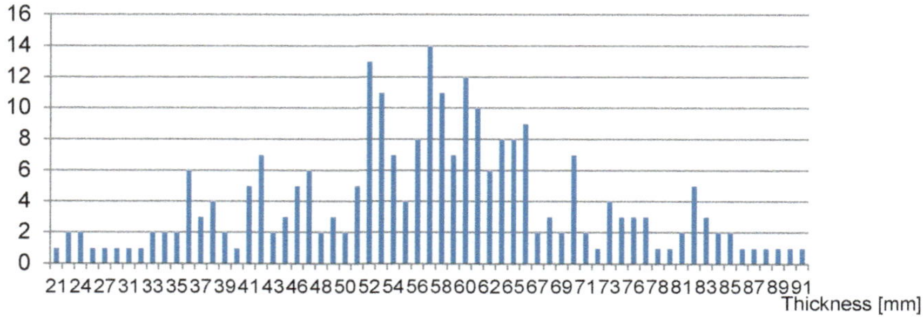

Fig. 2. Number of cases relatively to the breast thicknesses registered in the dataset

4 Discussions and Conclusion

The comparison between data from the experimental measurements and from the DICOM files shows consistency. Taking individual values for each breast thickness registered in the dataset, about 20% of them in average showed doses slightly above the experimental values given in Table 1. However, only two cases from the DICOM files headers have been overcome the MGD limit standard for that thickness range while the experimental measured values have all been below those limits (mainly for the thicker breasts).

The main age range of patients from this study is 50–70, but unlike what is usually anticipated by the quality assurance standards, the breast thickness profile of such a population is in the range between 50 and 70 mm – thickness interval in the histogram of Fig. 2 where the most cases are, with a peak between 57 and 60 mm. Thus this characteristic should be taken into account when running quality assurance tests in the equipment serving this population – the breast phantoms used in the most tests are often 45–50 mm thick. Therefore, it is important to use the data obtained in the quality control tests that evaluate the calculated dose and the contrast-to-noise ratio aiming at the appropriate use of the equipment and the decrease of the doses without impairing the image quality. Furthermore, as the comparative tests also indicated a good agreement between the measured dose values and those obtained from the mammography equipment outcome files, the use of our software can then mean a more practical and fast way to assist the quality assurance procedures for digital mammography systems.

As a conclusion, these tests allowed to determine a general compliance between measured data, obtained by classical experimental procedures for dose evaluation in mammography systems, and those extracted from the outcome DICOM files for this particular equipment. As a consequence, our proposed software application can be a useful tool to achieve in a simpler and faster way all the data needed to many of the quality assurance evaluations stated by the standards – here with focus mainly on the dose rates during the exams – and then also to infer important features relative to the population profile involved with such a digital mammography service.

Acknowledgments. Authors wish to thank Dr. Fatima Faloppa Rodrigues Alves, and Prof. Simone Elias, MD, by technical data, discussion and licensing provided, and to Dr. Bruno Barufaldi and Dr. Bruno R. N. Matheus by the technical assistance. They are also thankful to FAPESP by the financial support to this research.

References

1. IEC 61223-3: Evaluation and routine testing in medical imaging departments - Part 1: General Aspects, [s.l: s.n.] (1993)
2. Yaffe, M.J., et al.: The myth of the 50-50 breast. Med. Phys. **36**(12), 5437–5443 (2009)
3. NEMA: DICOM PS3.6 2014b - Data Dictionary. Rosslyn, VA, [s.n.] (2014). http://medical.nema.org/standard.html
4. Barufaldi, B., et al.: A radiation dose reporting system for mammography and digital breast tomosynthesis. In: Radiological Society of North America (RSNA), Chicago, USA (2014)
5. Barufaldi, B., et al.: Design and implementation of a radiation dose tracking and reporting system for mammography and digital breast tomosynthesis. Physica Medica **58**, 131–140 (2019)
6. IAEA: Dosimetry in Diagnostic radiology: An International Code of Practice, Vienna, Áustria, [s.n.] (2007)
7. Dance, D.R., Young, K.C., Van Engen, R.E.: Further factors for the estimation of mean glandular dose using the United Kingdom, European and IAEA breast dosimetry protocols. Phys. Med. Biol. **54**(14), 4361–4372 (2009)
8. IAEA: Quality Assurance Programme for Digital Mammography, Vienna, Austria, [s.n.] (2011)
9. Alves, F.F.R.: Estudo Longitudinal da Qualidade da Imagem Mamográfica em Sistemas Digitais Associado ao Processo de Otimização da Dose Média Glandular [tese]. Universidade Federal de São Paulo, São Paulo (SP) (2014)

Cuffless Blood Pressure Estimation Only an iPhone: Investigation on Cold Pressor Tests

Ippei Harada[1(✉)], Noriyuki Mochizuki[1], Peter Rolfe[2,3], Masahiro Shibata[4], and Takehiro Yamakoshi[1]

[1] Information Systems and Engineering, Graduate School of Engineering, Fukuoka Institute of Technology, Fukuoka-shi, Ishikawa, Japan
mhm18105@bene.fit.ac.jp, yamakoshi@fit.ac.jp

[2] Department of Automatic Measurement and Control, Harbin Institute of Technology, Harbin, China

[3] Oxford BioHorizons Ltd., Maidstone, UK

[4] Department of Bioscience and Engineering, Shibaura Institute of Technology, Saitama-shi, Saitama, Japan

Abstract. Cuffless blood pressure (BP) measurement is an all-inclusive term for a method that aims to measure BP without using a cuff. In our previous article, we reported the development of a simple cuffless technique based on the haemodynamic expression, using only pulse rate (PR) and modified normalized pulse volume (mNPV) that can be measured using only a smartphone. The stressor used in our previous article was the cardiac-dominant pattern (mental arithmetic); however, here we have experimented with another stressor, namely, a vascular-dominant pattern in a cold pressor test, to verify this theory in 7 healthy young males and 6 healthy young females (N = 22). As a result, the mean arterial pressure accuracy was improved (r = 0.727) compared with that achieved in the previous study.

Keywords: Cuffless blood pressure · Heamodynamic · Circulatory equation

1 Introduction

High blood pressure (BP) is a well-known risk factor for the global burden of disease and mortality, as well as for cardiovascular diseases. BP varies dramatically from beat to beat, minute to minute, and year to year. Thus, frequent BP checks are desirable. The direct measurement of BP requires an invasive technique, which is limited to particular clinical situations; therefore, non-invasive methods are preferred in most situations.

Cuffless BP is now an all-inclusive term for methods that measure or estimate BP without the use of a cuff. The advantages of cuffless BP include being free from the inconvenience or pain associated with inflation of a cuff and avoiding the need for a pressurisation mechanism, which adds to the complexity and cost of the instrument. We have therefore developed a simple cuffless method that uses pulse rate (PR) and modified normalized pulse volume (mNPV) as an indicator of finger vascular tone [1] that can be measured using only a smartphone [2], based on the knowledge that mean arterial pressure (MAP) = cardiac output (CO) × total peripheral resistance (TPR), where CO and TPR are correlated with PR and mNPV, respectively. This is a simple

J. Henriques et al. (Eds.): MEDICON 2019, IFMBE Proceedings 76, pp. 874–877, 2020.
https://doi.org/10.1007/978-3-030-31635-8_105

linear polynomial equation obtained by multiplying both sides with a logarithmic transformation: ln MAP = a × ln PR + b × ln mNPV + constant [3].

2 Methods

2.1 Participants

A total of 13 volunteers (female: 2 people, man: 9 people) participated in this study. The criteria for inclusion in the study were having no current cardiovascular disease except for high BP, and not taking any prescription medication. Written informed consent was obtained from participants after they were provided with a complete description of the study. This study was approved by the ethics committee of Fukuoka Institute of Technology.

2.2 Materials and Measurements

Finger photo-plethysmogram was measured from the left index finger using a smartphone (iPhone 7, Apple Inc.), within which the iPhysioMeter^{SM} app was installed [2]. The iPhysioMeter^{SM} is a software program that was designed to run on iPhone 5 s and later smartphones running iOS 9.2 or later (Apple Inc.) and to allow a smartphone to be a reflectance mode photo-plethysmography using the built-in LED flash light and built-in CMOS camera as a light source and photodetector, respectively. The iPhysioMeter^{SM} has a 60 Hz sampling speed. SBP (systolic BP), MAP, and DBP (diastolic BP) were measured using a brachial cuff sphygmomanometer (Omron, HEM-907) attached to the right arm. MAP was calculated using the following formula: MAP = DBP + (SBP − DBP)/3.

2.3 Measurement Procedure

The experiment was performed in a 4 × 5 m soundproof room maintained at a temperature of approximately 25 °C. After the brachial cuff of the sphygmomanometer was attached, the participants sat in a chair with both of their hands on a desk and holding a smartphone in their left hand while keeping their index finger on the CMOS camera. The brachial cuff and fingers were supported at heart level. Each participant was instructed to keep as still as possible until the end of the experiment in order to minimize movement artifacts.

2.4 Data Analysis

The beat-by-beat PR and ln mNPV were averaged for each quarter period (*i.e.*, 45 s), except for the section including noise. The averages derived from the smartphone data, including over 30% outliers, were treated as missing values. BP was mainly measured within the first and third quarters of the baseline and cold pressor test; therefore, only these averages were forwarded for further analysis. To examine the agreement of the PR and mNPV measurements derived from the smartphone, geometric mean regression and *Bland–Altman* analyses were conducted as shown in Fig. 1. Multiple linear regression analyses, where brachial ln MAP was applied as each dependent variable, and ln PR and ln mNPV as dependent variables, was conducted.

3 Results and Discussion

The correlation between proposed method and cuff oscillometry are the accuracy represented by *r* values = 0.727 for MAP, together with their *Bland–Altman* plots (see Fig. 1) was superior to our previous study [3]. This results also suggest that the proposed method can estimate blood pressure with accuracy comparable/superior to pulse transit time (PTT)-based method [4]. The present method could potentially convert a large number of smartphones into simplified sphygmomanometers. In the future, we will increase the number of subjects for wide range of people and improve accuracy.

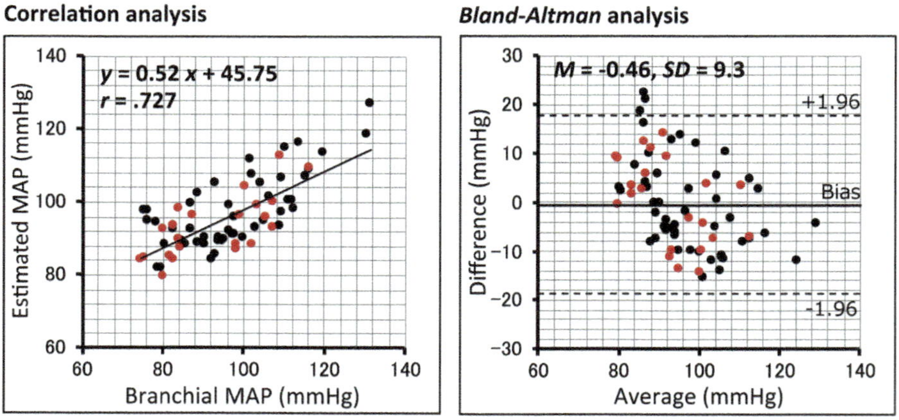

Fig. 1. The accuracy of blood pressure estimation attained by the proposed method. Scatterplots of mean arterial pressure (MAP) estimated using only the smartphone and measured using a brachial cuff sphygmomanometer (N = 49, black plots was previous data [2] + 22, red plots is new data). Solid line represents the regression line and its formula, together with *r* value, is shown in each scatterplot. Corresponding *Bland–Altman* plots. Solid line and dashed lines represent fixed bias ± 1.96 standard deviation (SD) range, respectively. Average = (estimate + brachial)/2, Difference = estimate − brachial.

Conflicts of Interests. The authors declare that there are no conflicts of interests.

References

1. Lee, J.H., Matsumura, K., Yamakoshi, T., Rolfe, P., Tanaka, N., Kim, K.H., Yamakoshi, K.: Validation of normalized pulse volume in the outer ear as a simple measure of sympathetic activity using warm and cold pressor tests: towards applications in ambulatory monitoring. Physiol. Measur. **34**(3), 359–375 (2013)

2. Matsumura, K., Yamakoshi, T.: iPhysioMeter: a new approach for measuring heart rate and normalized pulse volume using only a smartphone. Behav. Res. Methods **45**(4), 1272–1278 (2013)
3. Matsumura, K., Toda, S., Rolfe, P., Yamakoshi, T.: Cuffless blood pressure estimation using only a smartphone. Sci. Rep. **8**, Article No. 7298 (2018)
4. Ding, X. Yan, B.P., Zhang, Y.T., Liu, J., Zhao, N., Tsang, H.K.: Pulse transit time based continuous cuffless blood pressure estimation: a new extension and a comprehensive evaluation. Sci. Rep. **7**, Article No. 11554 (2017)

The UBORA E-Infrastructure for Open Source Innovation in Medical Technology

Carmelo De Maria[1(✉)], Licia Di Pietro[1], Andres Diaz Lantada[2], Alice Ravizza[3], Mannan Mridha[4], Janno Torop[5], June Madete[6], Philippa Makobore[7], and Arti Ahluwalia[1]

[1] University of Pisa, Pisa, Italy
carmelo.demaria@unipi.it
[2] Universidad Politecnica de Madrid, Madrid, Spain
[3] Politecnico di Torino, Turin, Italy
[4] Royal Institute of Technology, Stockholm, Sweden
[5] University of Tartu, Tartu, Estonia
[6] Kenyatta University, Nairobi, Kenya
[7] Uganda Industrial Research Institute, Kampala, Uganda

Abstract. The development of medical devices with open source and collaborative design methodologies has the potential to increase the access to medical technologies, thanks to a feasible reduction of design, management, maintenance, and repairing costs linked to the open access of device blueprints. UBORA is an e-infrastructure for the co-design of open source medical devices, which promotes the compliance with internationally recognized quality standards and regulations for safety and efficacy of devices, taking the EN ISO 13485:2016 and the EU MDR 2017/745 as inspiration. UBORA guides the user through a systematic design process, from the identification of clinical needs, of risks class and relevant standards for the device, and provides project management tools, including a repository, finalized to the preparation of the pre-production device dossier. The process is supervised by expert mentors, which ensure that safety and efficacy criteria are fulfilled. The UBORA e-infrastructure is in line with the 2030 Agenda for the Sustainable Development Goals, promoting and strengthening the initiatives of an international community of designers, healthcare providers and policy-makers, toward the reduction of inequalities in the access to medical devices.

Keywords: Open source medical devices · E-infrastructure · Safety · Regulation

1 Introduction

Medical technology is one of the pillars of an effective healthcare system, as recognized by the United Nations Member States in the 2030 Agenda for Sustainable Development Goals (SDGs) [1]. The overall high costs of medical devices (MDs) can be linked to their long life-cycle (design, prototyping, manufacturing, labelling and packaging, provision, installation, operation, maintenance, repair and disposal), during which each

J. Henriques et al. (Eds.): MEDICON 2019, IFMBE Proceedings 76, pp. 878–882, 2020.
https://doi.org/10.1007/978-3-030-31635-8_106

step is strictly normed, to guarantee their efficacy and the safety of patients, healthcare providers and bystanders. These costs create a barrier for the accessibility of medical technologies: even removing the charge on a single step could not make the difference. For example, the World Health Organization (WHO) estimates that in low income countries more than 80% of medical equipment is donated, but only 10–30% of these become operational, given the high operating cost, the lack of personnel and the frequent failures due to harsh environment, extreme climate conditions, humidity, dust and power instability [2–6]. These conditions, not foreseen in the design phase, cause more frequent failures, and determine a higher request for spare parts, which are expensive and difficult to find, making maintenance and repairing as problematic as acquisition [7]. Developing sustainable medical technologies to make healthcare affordable to a larger population, and thus reducing global inequalities, can only be performed taking into account the cultural and socio-economic context in which these will be applied [8].

Compared to the traditional methods of design, the open-source (OS) and collaborative approach can be a possible alternative. By definition, OS means making the design, documentation, source-code, blueprints, ideas and results available for the public. Having the software, electronic and hardware design available under an OS license and in the most suitable file format to study, modify, improve and contribute to the design, potentially leads to very rapid and more reliable innovation [9]. The application of the OS approach to MD design has been proved to offer a unique combination of advantages, such as increasing safety, security, reliability and reducing costs [10]. Until a few years ago, the development of MDs was essentially linked to companies and universities, but recently several examples of open-source MDs (OSMDs) have appeared on the web [11]; however, only some of them have been designed to be compliant with MD legislation [12–14]. It is crucial to ensure the safety and efficacy requirements of medical technology and for this reason the adoption of open resources must follow the standards and the current regulations [15]. To this end, the UBORA e-infrastructure, which merges the OS concepts with the safety and efficacy requirements enforced by the EU Regulation on MDs (MDR) 2017/745, has recently been established [16, 17].

2 The UBORA E-Infrastructure

UBORA provides a structured framework, inspired by the EN ISO 13485:2016 [18] for the identification of needs, risk class, relevant standards, management of computer-aided modeling files and preparation of the pre-production device dossier. Each stage is vetted and monitored by expert mentors to ensure that safety criteria are met during the design process. The e-infrastructure includes 3 main sections devoted to the development of OSMDs ("Clinical needs", "Project management", "Resource section"), a part for creating and strengthening a "Community" of developers, mentors and stakeholders, and software tools for the management of the e-infrastructure (Fig. 1).

Clinical Needs Section
The clinical needs section is aimed at identifying bioengineering design specifications linked to a particular clinical need. The section provides a unique environment for healthy discussion between patients, healthcare providers and engineers, to ensure that clinical needs are turned into safe and affordable projects.

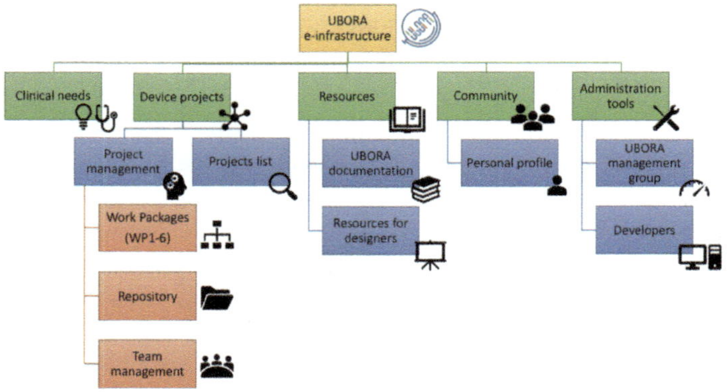

Fig. 1. Overall organization of UBORA e-infrastructure

Project Management Section
This section is a MD oriented project management environment, including a repository, available for free download of project blueprints which have completed the development and compliance assessment phases, under the coordination of expert mentors. This section provides a meta-structure composed of 6 Work Packages (WPs). The medical need and product specification section (WP1) is designed to specify the clinical need and the scope of the MD and to identify the product requirements, as well as clinical areas related to the technology. A list of suggested keywords for indexing the projects are also provided, on the basis of clinical need (e.g. rehabilitation, prevention, etc.), clinical area (e.g. cardiology, neurosurgery, etc.) and technology (e.g. mobile-based, ergonomic support, etc.): this scheme will promote a FAIR data management, ensuring Findability, Accessibility, Interoperability, and Reusability of data [19]. In WP1, the risk class and applicable standards for the specific MD under development are identified. The device classification follows the rules described in Annex VIII of MDR 2017/745. In order to improve usability and minimize human error, the rules are arranged as a decision tree, which guides the designer(s) through a flow of simple (mainly yes/no) questions. Using the same approach, UBORA allows the identification of safety standards, which are usually difficult to identify through keywords in a search engine. WP2 leads the designer in the development of different embodiments/ideas of the same MD, which will be analysed and ranked in a polling section, on the basis of feasibility, performance, usability and safety aspects. In WP3, focused on design and prototyping, the designers collaboratively work on the MD idea selected in WP2: the project management system suggests the organization in subsections related to mechanical components, electronics/firmware, software, and their integration. Once WP3 is concluded, WP4 leads the developers in the

testing phase to demonstrate the compliance with relevant standards. WP5 is aimed at completing the technical documentation of the device in order to publish the project for downloading, and provides a lean business model canvas [20], customized for OSMDs, to help the potential transfer of the technology to industry. Finally, WP6 guides the developers in creating promotional videos and brochures, toward a possible crowd-funding campaign.

Resource Section

The resource section contains a selected teaching/learning material on biomedical engineering, as well as the documentation and video tutorials to use and further develop the e-infrastructure, whose source code will be released as open source by the end of the UBORA project. In particular UBORA is built on Microsoft's ASP.NET Core web framework, using C# as programming language for the back-end, while the front- end is a multi-page application based on JavaScript, rendered on the server side and then sent to the client (browser).

UBORA was officially launched to the public during the conference of Collaborative Biomedical Engineering for Open Source Medical Technologies and the successive UBORA Design School 2018, held in Pisa from the 1st to the 7th September. To date the e-infrastructure, available at https://platform.ubora-biomedical.org, has 485 users and 286 projects, at different stage of development (Fig. 2), both on high-tech innovative MDs, such as a wearable device for the breast self-examination [14] or 3D printed splints, and smart accessories to existing MDs, as a timer for the AMBU bag.

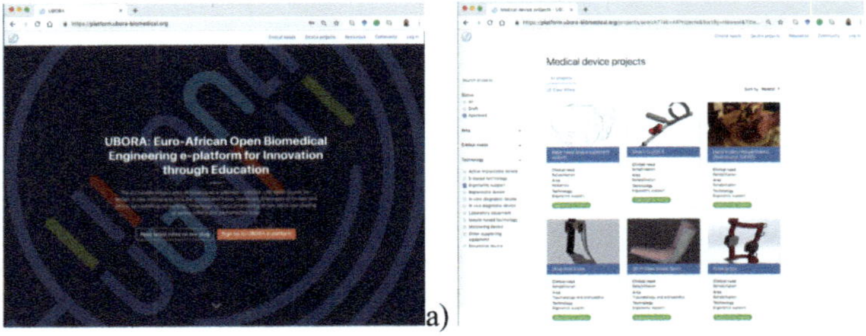

a) b)

Fig. 2. UBORA e-infrastructure. Landing page (a) and some medical device projects, selected through specific keywords (b)

3 Conclusion

The UBORA e-infrastructure provides a framework to develop ISO compliant MDs starting from clinical needs by sharing idea, blueprints and data. If properly implemented, these devices can be considered in line with the MDR 2017/745 from the design point of view. In a nutshell, its final aim is to promote well-being for all, increasing access to MDs and moving towards health equity in accordance with the UN SDGs.

Acknowledgements. This project has received funding from the European Union's Horizon 2020 research and innovation programme under grant agreement N° 731053. The authors acknowledge all students and mentors who daily animate the UBORA e-infrastructure.

Conflict of Interest. The authors have no conflict of interest to disclose.

References

1. Sustainable Development Goals. https://www.un.org/sustainabledevelopment/sustainable-development-goals/. Accessed May 2019
2. World Health Organization. Medical devices: managing the mismatch: an outcome of the priority medical devices project. World Health Organization (2010)
3. Iadanza, E., Dyro, J.: Clinical Engineering Handbook. Elsevier (2004)
4. World Health Organization. Barriers to innovation in the field of medical devices. Background paper 6. World Health Organization, Geneva (2010)
5. Malkin, R.A.: Design of health care technologies for the developing world. Ann. Rev. Biomed. Eng. **9**, 567–587 (2007)
6. Lustick, D.R., Zaman, M.H.: Biomedical engineering education and practice challenges and opportunities in improving health in developing countries. In: Atlanta Conference on Science and Innovation Policy 2011. IEEE (2011)
7. Malkin, R.A.: Barriers for medical devices for the developing world. Expert Rev. Med. Dev. **4**(6), 759–763 (2007)
8. Douglas, T.S.: Biomedical engineering education in developing countries: research synthesis. EMBC, IEEE (2011)
9. Perens, B.: The open source definition. In: Open Sources: Voices from the Open Source Revolution, vol. 1, pp. 171–188 (1999)
10. Williams, A., Gibb, A., Weekly, D.: Research with a hacker ethos: what DIY means for tangible interaction research. Interactions **19**(2), 14–19 (2012)
11. Niezen, G., Eslambolchilar, P., Thimbleby, H.: Open-source hardware for medical devices. BMJ innovations bmjinnov-2015 (2016)
12. Gamma Cardio Soft S.r.l. http://www.gammacardiosoft.it/openecg/. Accessed May 2019
13. Ferretti, J., Di Pietro, L., De Maria, C.: Open-source automated external defibrillator. HardwareX **2**, 61–70 (2017)
14. Arcarisi, L., Di Pietro, L., Carbonaro, N., Tognetti, A., Ahluwalia, A., De Maria, C.: Palpreast—a new wearable device for breast self-examination. Appl. Sci. **9**, 381 (2019)
15. De Maria, C., Mazzei, D., Ahluwalia, A.: Improving African health care through open source biomedical engineering. Int. J. Adv. Life Sci. **7**(1), 10–19 (2015)
16. UBORA: Euro-Africa Open Biomedical Engineering Innovation e-platform for Innovation through Education. https://platform.ubora-biomedical.org/. Accessed May 2019
17. European Medical Device Regulation 2017/745. https://eur-lex.europa.eu/legal-content/EN/TXT/PDF/?uri=OJ:L:2017:117:FULL&from=EN. Accessed May 2019
18. International Organization of Standardization EN ISO 13485:2016 Medical devices – Quality management system- Requirements for regulatory purpose certification. https://www.iso.org/standard/59752.html. Accessed May 2019
19. Wilkinson, M.D., Dumontier, M., Aalbersberg, I.J., Appleton, G., Axton, M., Baak, A., Bouwman, J., et al.: The FAIR guiding principles for scientific data management and stewardship. Sci. Data **3**, 160018 (2016)
20. Trimi, S., Berbegal-Mirabent, J.: Business model innovation in entrepreneurship. Int. Entrepreneurship Manage. J. **8**(4), 449–465 (2012)

Design and Implementation of a Web-Based Platform to Support Research in X-Ray Breast Imaging

Adelina Doycheva(✉), Nikolay Dukov, and Kristina Bliznakova

Laboratory of Computer Simulations in Medicine,
Technical University of Varna, 9010 Varna, Bulgaria
csmlab@tu-varna.bg

Abstract. This paper presents a web-based platform dedicated to evaluation studies in x-ray breast imaging. The backend development is based on ASP. NET and C#, while the frontend is implemented under HTML, CSS and JavaScript. The platform is designed to own simple and intuitive graphical user-friendly interface. Questions to be evaluated are entered by the users and stored in the application database. In the same database, two other features are kept as well: the results of the evaluation and the regions of interest taken from patient and simulated mammography images. The content of the database is subject to modification, as well as further addition of data. The designed web-based platform was tested in a particular evaluation study, which concerned the realism of the generated simulated mammograms from computational breast models. The new approach in evaluating images was estimated to be more convenient and faster compared to our previous experience. The system will be further updated and will be exploited in the routine research work of the laboratory as well as will be shared with other research groups.

Keywords: Evaluation of mammography images · Web-based platform · Data analysis

1 Introduction

Breast cancer is by far the most common incident form of cancer for women below seventy years of age. The statistics show that 1 in 8 women in EU-28 develops breast cancer [1] as 20% of this kind of cancers occur in women younger than 50 years [2]. Despite technological advances, such as the digital mammography, the national screening programs, the introduction of the CAD systems in clinical routine, screening and diagnosing of cancers hidden in breast dense parenchyma still remain a challenging task. Furthermore, contrary to expectations, the mortality continues to rise in many EU countries, including our home country Bulgaria [3].

Amongst the most important factors related to the challenging task is relevant to the technical limitations of the current technology to screen dense breasts [4–6]. Efforts of research and industry communities have been directed towards the development of new techniques for screening and diagnosing the breast: breast CT, tomosynthesis, contrast-enhanced mammography. These are evaluated for their ability to improve breast cancer

J. Henriques et al. (Eds.): MEDICON 2019, IFMBE Proceedings 76, pp. 883–890, 2020.
https://doi.org/10.1007/978-3-030-31635-8_107

margins, which is usually done based on software and physical models of the breast and tumors.

Within the frame of the EU project Maxima (http://maxima-tuv.eu/), scientists prepared models of tumors, which were both mathematically modelled and segmented from patient images [7, 8]. These models are now used in education activities such as the European course [9, 10] as well as in feasibility studies related to development and optimization of new breast imaging technologies [11, 12]. For subjective evaluation of results, a dedicated software may be exploited. Such software is usually in-house developed by researchers and includes a graphical user interface with possibilities to display several ROIs from mammography images for evaluation purposes. One such software is the ViewDEX [13, 14], which is Java-based and can be run from any disc area connected to a PC. Further developments of such tools may benefit of a more universal platform, which can be access through internet from anywhere. This project aims at developing of a web-based platform dedicated to evaluation purposes in x-ray breast imaging.

2 Materials and Methods

The designed web-based system is based on a client/server model and can be run from any PC connected to internet.

2.1 Overview of the System Architecture

The WEB interface is created by using the standard markup language for web pages HTML as well as CSS and JavaScript. This translates to the front-end part being a combination of three different programming languages which visualize the user's interface. The back-end part of the project is created on .NET platform using C#.

The application is built with MVC architecture. This architectural pattern separates the application into three interconnected components – *Model*, *View* and *Controller* (shown in Fig. 1). The *Model* is the main component of the MVC architecture. It maintains the data, manages the logic and the rules of the application. The *View* concerns the user interface. It is a presentation of the model in a particular format. *View* displays information such as tables and diagrams, data from the model to the users and also enables them to modify the data. The *Controller* in MVC architecture receives the user input and converts it to commands for the *Model* or *View* modules.

The project includes an existing database created on SQL Server Management Studio 2018. The later software application is used to configure and manage all components of SQL Server. It also provides tools for project development and management. The file system is created in order to serve the aim of the application: to store images, results and questions.

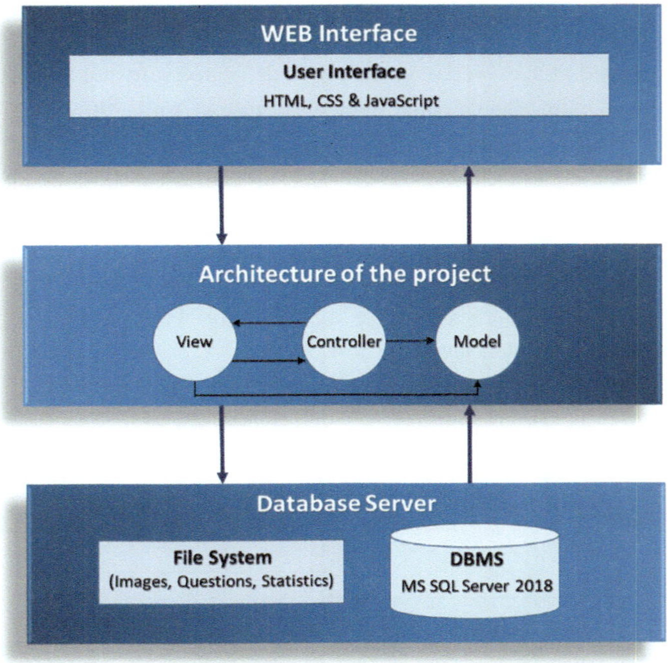

Fig. 1. Block-diagram of the application's architecture.

2.2 Database Organisation

The database is comprised of images, evaluation questions and results from the evaluation of each user. In the later case, the results are exported into excel files. The selected images were with a size of 250 × 250 mm from digital mammography. The images were obtained by selecting regions of interest (ROIs) with size of 250 × 250 mm from the mammograms, which contain healthy tissue and tumor tissue and are separated in three

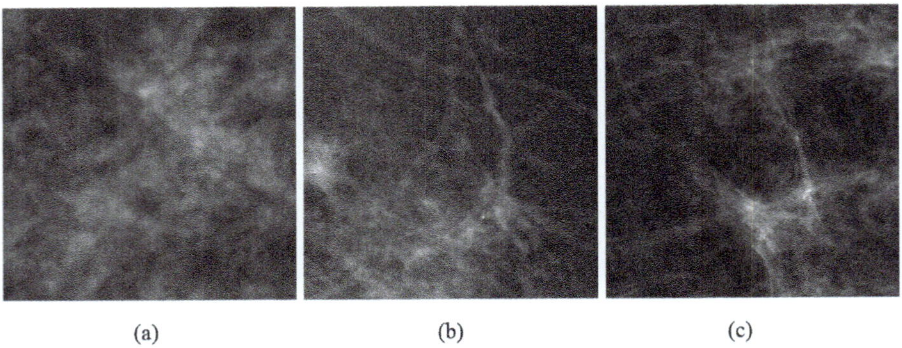

Fig. 2. Images in the database: (a) simulated, (b) patient-based without lesions and (c) patient-based with lesion.

categories – (i) healthy, (ii) with lesions and (iii) simulated. The real images are patient images acquired in a digital mammography unit using common acquisition protocol, while the simulated images were created with a dedicated software tool [15, 16], which takes into account the specific mammography unit as well. Simulated images were generated for the geometry acquisition similar to the Siemens Mammomat mammography unit at which the patient images were obtained.

Example images used in the local database of the application are shown in Fig. 2.

2.3 Use Case Diagram

The main steps of using this platform are shown in Fig. 3. Initially, the user logs in the system by using a dedicated menu for this.

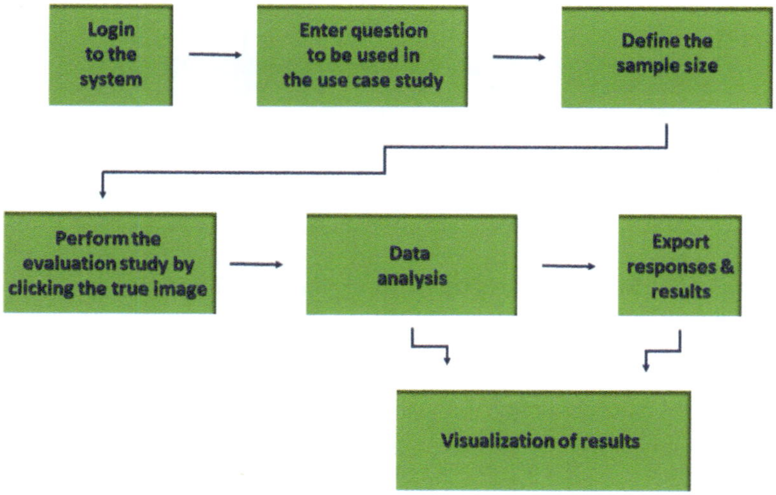

Fig. 3. Workflow of the process of creation of an evaluation study.

Then, the user may choose either a question that is already stored in the database or to define a new one. Further, the sample size and number of images to be used per screen are defined. These are the necessary requirements to start the evaluation, which is completed by clicking the *Results* button (Fig. 4). In the *Results* panel an overview of the answers is available. The user may change its answer on a given question before submitting its final evaluation. Presently, the results are exported in excel format and processing may be performed in Excel, as long as the full implementation and successful testing of the visualization of the result in the application is done.

3 Results and Discussion

Figure 4 depicts the main window of the implemented web-based system. In this particular case, the evaluators were answering the question "Which ROI contains a lesion". For this purpose, we specified four images to be loaded: three from the folder with normal tissue and one from the folder with ROIs with diagnosed lesions.

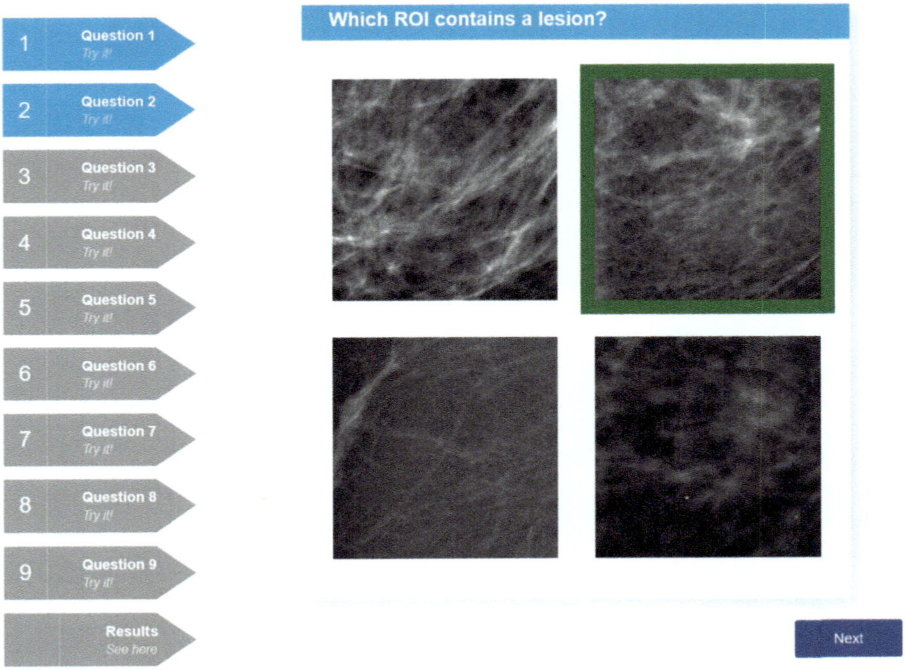

Fig. 4. Screen shot of the application, depicting the "answering" process of the question "Which ROI contains a lesion".

Similarly, the web-based system was used in a current study to evaluate the realism of simulated synthetic mammograms in comparison to real (patient) mammograms. For this purpose, we used two ROIs for evaluation, as one of these two was taken from a patient image, while the other one was taken from simulated projection images (Fig. 5).

The results of this short study are shown in Fig. 6. Similar window with results is generated for the study, shown in Fig. 4.

The presented platform is under continuous improvement and testing phases. Presently, algorithms for image selection from different categories are being optimized for efficiency. Furthermore, improvements are currently made towards better presentation and storage of the results of the evaluations. This will allow the application to operate with less memory and be more power efficient. In addition, an option will be added to the application, which allows the user to write an opinion about the application in a separate

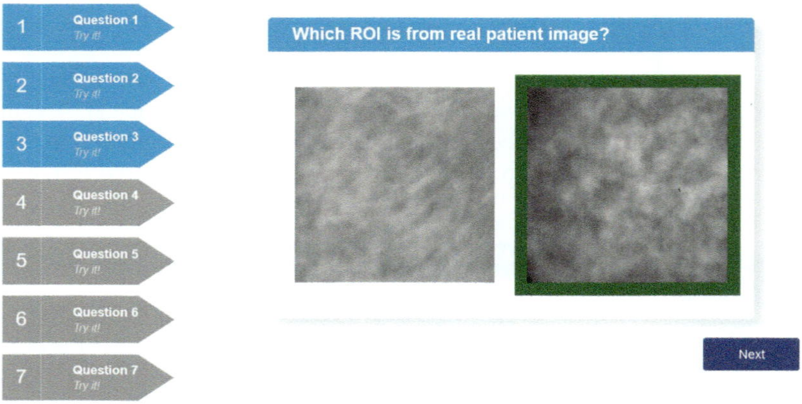

Fig. 5. Design of the study for evaluation of the realism of generated simulated images. The user is asked to select the real ROI.

Fig. 6. Results of the evaluation study, demonstrated in Fig. 5.

field. The feedback will be stored in a different page accessible by the admin. Moreover, a responsive view of the form is to be applied so users may use the application from their mobile devices. All these steps are directed towards the main goal of creating a powerful, accessible and easy to use application for evaluation of x-ray imaging studies.

4 Conclusions

This paper presented the initial design and implementation of a web-based platform dedicated to evaluation studies in x-ray breast imaging. It is based on MVC architecture and the backend development is implemented on ASP.NET and C#, while the frontend development is based on HTML, CSS and JavaScript, respectively. The platform was tested in a short study aiming to evaluate the realism of the simulated mammograms. The platform was evaluated by our radiologists as intuitive, easy to operate and user friendly compared to the previous experience which was related to sending the images for evaluation and asked them to fill a separate template for their evaluation.

Acknowledgements. This research is supported by the Bulgarian National Science Fund under grant agreement DN17/2.

References

1. Ferlay, J., Soerjomataram, I., Ervik, M., Dikshit, R., Eser, S., Mathers, C., Rebelo, M., Parkin, D., Forman, D., Bray, F.: GLOBOCAN 2012 v1.0, Cancer Incidence and Mortality Worldwide: IARC CancerBase 2013. International Agency for Research on Cancer, No. 11, Lyon, France (2015)
2. Curado, M.P.: Breast cancer in the world: Incidence and mortality. Salud Publica Mexico **53**, 372–384 (2011)
3. BNCR, Bulgarian National Cancer Registry, Cancer incidence in Bulgaria (2013)
4. Schulz-Wendtland, R., Fuchsjager, M., Wacker, T., Hermann, K.P.: Digital mammography: an update. Eur. J. Radiol. **72**, 258–265 (2009)
5. Evans, W.P., Burhenne, L.J.W., Laurie, L., O'Shaughnessy, K.F., Castellino, R.A.: Invasive lobular carcinoma of the breast: mammographic characteristics and computer-aided detection. Radiology **225**, 182–189 (2002)
6. Birdwell, R.L., Ikeda, D.M., O'Shaughnessy, K.F., Sickles, E.A.: Mammographic characteristics of 115 missed cancers later detected with screening mammography and the potential utility of computer-aided detection. Radiology **219**, 192–202 (2001)
7. Dukov, N., Bliznakova, K., Feradov, F., Ridlev, I., Bosmans, H., Mettivier, G., Russo, P., Cockmartin, L., Bliznakov, Z.: Models of breast lesions based on three-dimensional X-ray breast images. Physica Med. Eur. J. Med. Phys. **57**, 80–87 (2019)
8. Dukov, N., Bliznakov, Z., Buliev, I., Bliznakova, K.: Creation of computational breast phantoms with extracted abnormalities from real patient images. In: World Congress on Medical Physics and Biomedical Engineering 2018, vol. 1, no. 68, pp. 213–217 (2019)
9. Bliznakova, K., Buliev, I., Bosmans, H., Bliznakov, Z.B., Vassileva, J., Padovani, R., Christofides, S., Van Peteghem, N., Caruana, C.: How can biomedical engineers benefit from the new expert level course of the EUTEMPE-RX project. In: 6th European Conference of the International Federation for Medical and Biological Engineering, vol. 45, pp. 765–768 (2015)
10. Bliznakova, K., Buliev, I., Bliznakov, Z.: Implementation of the "Anthropomorphic Phantoms" educational module from the European EUTEMPE-RX course. Phys. Medica **32**, 270 (2016)

11. Mettivier, G., Bliznakova, K., Di Lillo, F., Sarno, A., Russo, P.: Evaluation of the BreastSimulator software platform for breast tomography: preliminary results. In: Lecture Notes in Computer Science (including subseries Lecture Notes in Artificial Intelligence and Lecture Notes in Bioinformatics), pp. 145–151 (2016)
12. Bliznakova, K., Russo, P., Kamarianakis, Z., Mettivier, G., Requardt, H., Bravin, A., Buliev, I.: In-line phase-contrast breast tomosynthesis: a phantom feasibility study at a synchrotron radiation facility. Phys. Med. Biol. **61**, 6243–6263 (2016)
13. Hakansson, M., Svensson, S., Zachrisson, S., Svalkvist, A., Bath, M., Mansson, L.G.: VIEWDEX: an efficient and easy-to-use software for observer performance studies. Radiat. Prot. Dosimetry. **139**, 42–51 (2010)
14. Hakansson, M., Svensson, S., Zachrisson, S., Svalkvist, A., Bath, M., Mansson, L.G.: ViewDEX 2.0: a Java-based DICOM-compatible software for observer performance studies. In: Medical Imaging 2009: Image Perception, Observer Performance, and Technology Assessment, vol. 7263 (2009)
15. Bliznakova, K., Suryanarayanan, S., Karellas, A., Pallikarakis, N.: Evaluation of an improved algorithm for producing realistic 3D breast software phantoms: application for mammography. Med. Phys. **37**, 5604–5617 (2010)
16. Bliznakova, K., Sechopoulos, I., Buliev, I., Pallikarakis, N.: BreastSimulator: a software platform for breast x-ray imaging research. J. Biomed. Graph. Comput. **2**, 1–14 (2012)

Empowering Diabetic Patients Using Gadgets and Mobile App

Sara Zulj(✉), Goran Seketa, Dominik Dzaja, Luka Celic, Igor Lackovic, and Ratko Magjarevic

Faculty of Electrical Engineering and Computing, University of Zagreb, Zagreb, Croatia
sara.zulj@fer.hr

Abstract. In this paper, we present our efforts towards more personalized diabetes treatment using the mobile application. The application helps to overcome the problem of missing and inaccurate data on blood glucose level, therapy, physical activity and other parameters, and allows development of the algorithms based on daily data of diabetic patients. For the purpose of our research, we developed two devices: myGluco - a small, portable device to be included into the glucose measurement kit patients carry with them on daily basis intended to transfer the saved blood glucose data from the glucose meter to the diabetic patients monitoring platform, and myWrist – the second small, portable device to track patients' physical activity level and to serve in guiding the patients in performing strength exercises and to enable recording of their performance during exercising.

Keywords: Diabetes · Self-management · Glucose monitoring · Physical activity monitoring

1 Introduction

World Health Organization (WHO) estimated in 2014 approximately 9% of adults have diabetes and diabetes was the direct cause of 1.5 million deaths worldwide in 2012. Diabetes Melitus type 1 is considered not to be preventable, however type 2 is mainly the result of insufficient physical activity and excessive body weight and therefore manageable [1].

There is evidence suggesting that a chronic disease self-management program can improve health status and reduce the number of hospitalizations and days in the hospital [2, 3]. Important part of the treatment of the disease is to train patients to self-manage the disease by keeping their blood glucose levels within the normal range, and also to control other symptoms in order to prevent health problems which develop later in life due to diabetes.

One of the main obstacles in the management of diabetes found among physicians is often missing and inaccurate data on blood glucose level, therapy, physical activity and other parameters that influence treatment [4]. We removed the need for handwritten diabetes diaries, which often led to incorrect data, by introducing the system for physicians to download the data from glucometers into the database [5]. The system,

J. Henriques et al. (Eds.): MEDICON 2019, IFMBE Proceedings 76, pp. 891–897, 2020.
https://doi.org/10.1007/978-3-030-31635-8_108

introduced into the Croatian National Health System, improves the trust between the patient and physician, and creates a better platform to make decisions on the further treatment. Although the data on blood glucose levels downloaded from the glucose meters are already trusted and authorized by physicians, the data on physical activity, meals and other daily parameters are not digitally obtained by the physician upon patient's visit.

In this paper, we describe further step in providing better treatment of diabetic patients, with more accurate data obtained by the patients – a mobile application with two custom designed devices for use by the patient.

2 Devices to Support Self-management of Diabetes

We developed two devices to support the patients' need to self-manage the disease. To overcome the problem of data entry, we developed miniaturized interface device myGluco to read data from the memory of most glucometer models available on Croatian market to patient's smartphone. myGluco connects to any smartphone using headphone audio jack and any supported glucose meter using custom adapter. Monitoring of basic daily physical activity of the patient is enabled on the myGluco device for patients' convenience.

Second device, myWrist is developed to replace inaccurate descriptive information on physical activity and exercising. It has integrated activity sensors (accelerometer, gyroscope and magnetometer) for qualitative and quantitative tracking of motion during exercising. More detailed structure of the sensor node and processing of the activity data have been described earlier [6, 7].

3 Mobile Application to Support Self-management of Diabetes

In order to support patients in daily self-management, we developed the mobile application. Access to patient's data on mobile app is available to registered users – patients with diabetes and/or those who care of them (e.g. parents of the children patients) after identification and authorization. The application is developed for Android and iOS in English and Croatian language. Although the app can be partially used without above described devices, patients are encouraged to use them in order to provide more accurate, unchanged data on their blood glucose values and their physical activity.

All data are stored in the database and can also be accessed by other users (e.g. physicians, trainers) on different platforms after the authorization has been given. That way, any necessary interventions between visits to the physician are made possible remotely.

Figure 1 shows the *diabICT* mobile app start screen in the d*iabICT* app. The screen shows an overview of the latest entered data per category. Mobile application consists of several categories to cover the important requirements in diabetes self-management and care: blood glucose management, meal management, exercising and therapy.

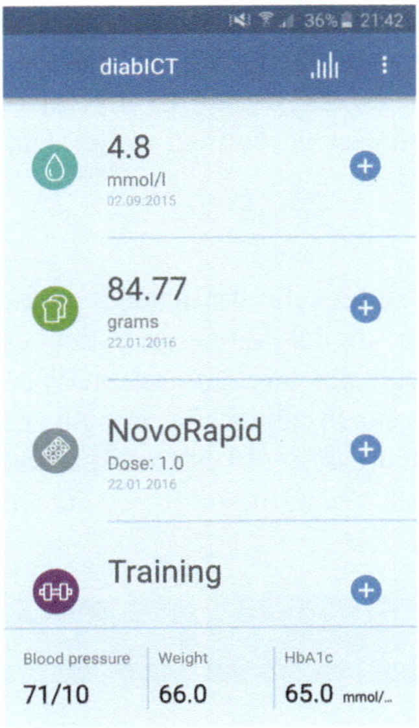

Fig. 1. diabICT start screen with an overview

3.1 Blood Glucose

Patients are encouraged to log in the data using the above described device myGluco. Each glucose meter is linked to one patient so that the data can be more trusted in

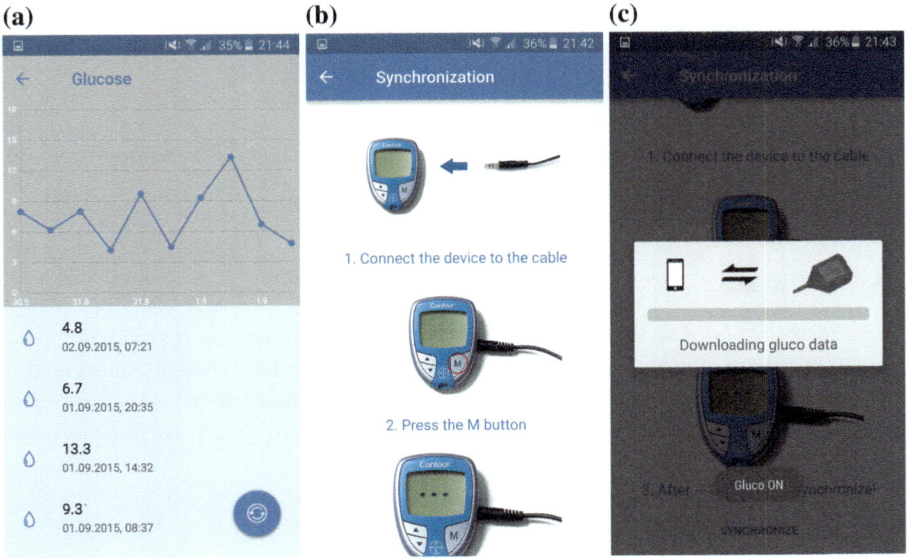

Fig. 2. Blood glucose category in diabICT app: (a) trend overview, (b) download instructions, (c) process of data download

further use by physicians. The patient can see previously logged data and the graphical trend (Fig. 2a). When downloading new data, patient is given visual and textual instructions for the correct setup and use of the myGluco device (Fig. 2b and c). After the download is finished, the user is returned to the data trend screen (Fig. 2a).

3.2 Therapy

The menu with medications and insulin for managing diabetes (Fig. 3a) allows patients to track therapy in the form of digital diary. The patient can see the list and bar chart of previously taken therapy with the amounts taken. When entering new therapy, the patient is given a simple menu to adjust the amount, mark before or after the meal, personal remark and timestamp (Fig. 3b) of the data that is being collected.

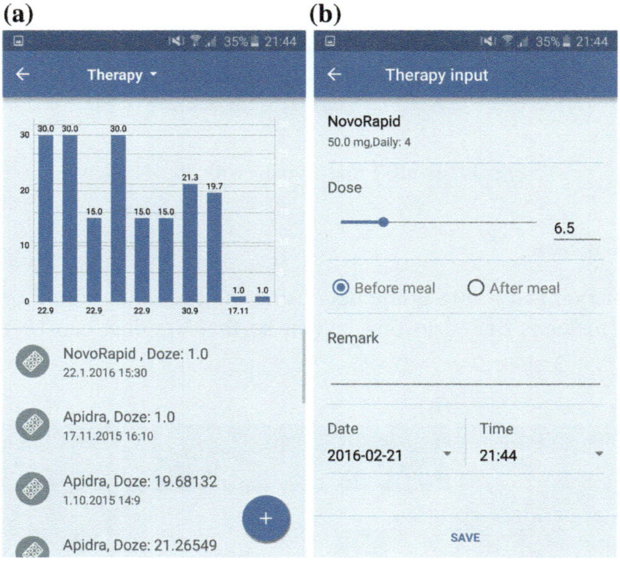

Fig. 3. Therapy category in diabICT app: (a) trend overview, (b) new input entry

3.3 Nutrition

Daily meals are stored for better management of nutritional macros and immediate insight in the amount of carbohydrates in patient's food intake. Patient can search for the individual ingredients and meals (Fig. 4a) or create saved user meals (Fig. 4b) to log it to the daily nutrition diary. Figure 4c shows the daily overview of food intake.

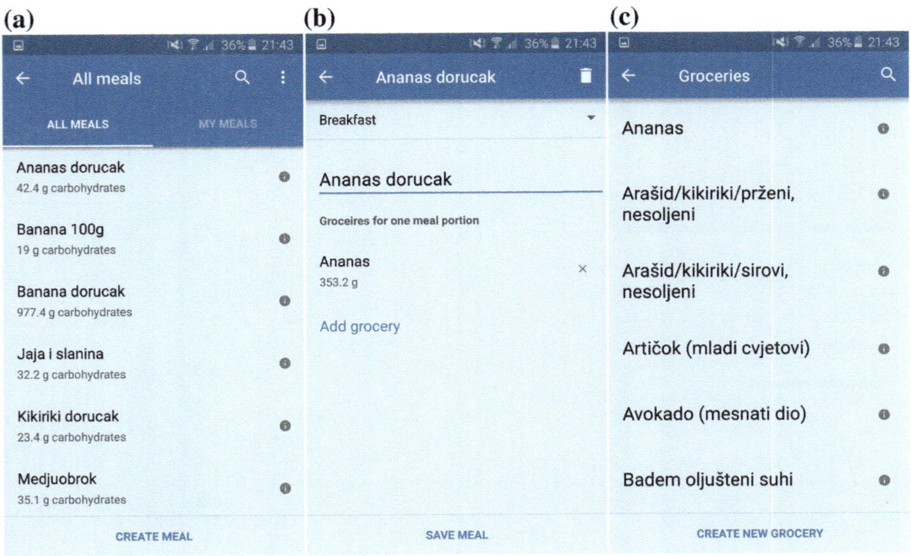

Fig. 4. Nutrition category in diabICT app: (a) meal list, (b) creation of new meal, (c) ingredient list

3.4 Assisted Exercise and Physical Activity

Mobile application enables patients to spontaneously track their physical activity levels if myGluco device is carried in the pocket throughout the day. On top of estimating basic physical activity level, patient can choose to perform different exercises using a myWrist sensor node. Application stores schedules within exercise menu with common strength exercises that patient can choose from given the level of patient's ability. Application guides patient to connect the device first (Fig. 5a) and starts Virtual Trainer mode (Fig. 5b). In Virtual Trainer mode, the patient can see how the exercise should be performed, and see their own movement for comparison. The menu also shows overall progress in scheduled exercises made by patient daily (Fig. 5c).

3.5 Other

The patient can enter other data to track progress of other important parameters as advised by the physician (blood pressure, weight, as well as laboratory data, HbA1c).

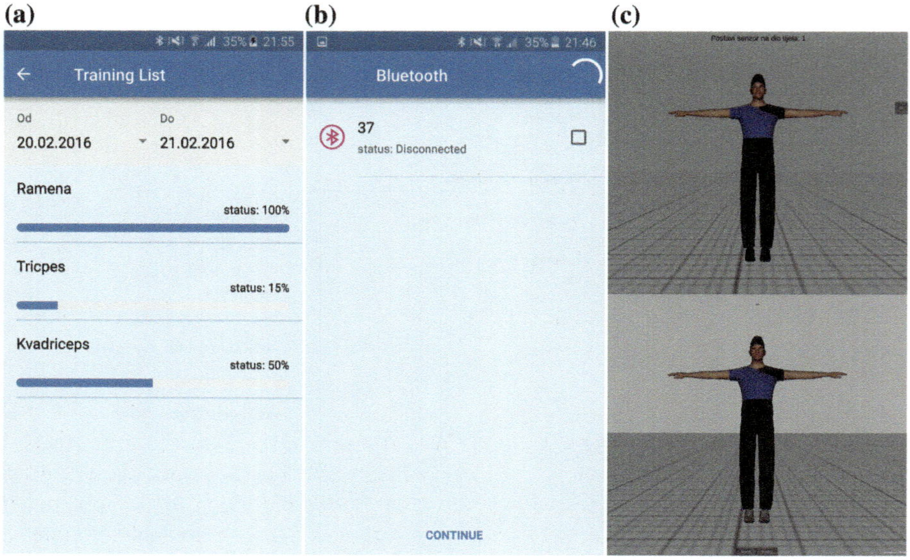

Fig. 5. Assisted exercise and physical activity category: (a) schedule exercises with progress, (b) device connection status, (c) virtual trainer

4 Further Steps in Personalized Treatment

Two main goals are considered for further steps in more personalized treatment of diabetic patients: (a) learning from patient patterns to develop new and innovative algorithms for personalized alerts and guidance, and (b) introducing the data provided by patient out of the hospital environment into personal/electronic health record.

First goal can be achieved when enough data are collected. The data are then to be used to develop new insights and algorithms with the aim to empower patients to make better choices on immediate decisions, and to provide well timed alerts when the trends are suggesting the need for change in the patient's behavior or in therapy.

Data on measured values may be entered into the diabICT e-platform on a daily basis so that the physicians can approach the data and intervene without delay. However, mHealth applications are rarely accepted into NHS due to inability to properly identify and authentify the user. This could be overcome by using eID scheme [8]. This would allow some of the data collected by the patient to be part of national personal/electronic health record. This would be an improvement for physicians as it would allow them to have all the necessary data (patient input, lab results, therapy prescribed) in one system.

5 Conclusion

Our experience showed that there is a group of patients, especially among younger diabetics, who show interest in using mobile ICT-based tools for self-management of their condition and who find they make them self-management easier. Their feedback is valuable in improving the functionalities of the system and the platform. In spite of the presence of some globally accessible platforms for diabetes self-management, which in principle could be applied locally, some national regulations, needs and customs have to be included in development to achieve user acceptability of the self-management tools.

Acknowledgment. The research leading to these results has partially received funding from the Central Finance and Contracting Agency of the Croatian Ministry of Science, Education and Sports under agreement n°RC.2.2.08-0010.

References

1. World Health Organization: Definition, diagnosis and classification of diabetes mellitus and its complications: report of a WHO consultation. Part 1, Diagnosis and classification of diabetes mellitus. No. WHO/NCD/NCS/99.2. World Health Organization, Geneva (1999)
2. Lorig, K.R., Sobel, D.S., Stewart, A.L., Brown Jr., B.W., Bandura, A., Ritter, P., Gonzalez, V.M., Laurent, D.D., Holman, H.R.: Evidence suggesting that a chronic disease self-management program can improve health status while reducing hospitalization: a randomized trial. Med. Care, 5–14 (1999)
3. Sarwar, N., Gao, P., Seshasai, S.R., Gobin, R., Kaptoge, S., Di Angelantonio, E., et al., Emerging Risk Factors Collaboration: Diabetes mellitus, fasting blood glucose concentration, and risk of vascular disease: a collaborative meta-analysis of 102 prospective studies. Lancet **375**, 2215–2222 (2010)
4. Prašek, M.: Self control diary–challenges of new technological possibilities. In: 5th European Conference of the International Federation for Medical and Biological Engineering, pp. 972–973. Springer, Heidelberg (2011)
5. Zulj, S., Celic, L., Grgurevic, M., Prasek, M., Magjarevic, R.: Pilot project: ICT system for management and self-management of diabetes. In: International Conference on Biomedical and Health Informatics, pp. 199–203. Springer, Singapore (2015)
6. Šeketa, G., Džaja, D., Žulj, S., Celic, L., Lackovic, I., Magjarevic, R.: Real-time evaluation of repetitive physical exercise using orientation estimation from inertial and magnetic sensors. In: First European Biomedical Engineering Conference for Young Investigators, pp. 11–15. Springer, Singapore (2015)
7. Pozaic, T., Džaja, D., Varga, M., Matec, I., Celic, L., Lackovic, I., Magjarevic, R.: Quantitative and qualitative assessment of assisted strength exercising. In: The International Conference on Health Informatics, pp. 71–75. Springer, Cham (2014)
8. Overview of the Croatian eID scheme, pp. 1–16 (2018)

A New Software Tool for Analyzing Mental Health Data in a Spanish Region

Diego Calvo Barreno[1], Susel Góngora Alonso[1(✉)], Isabel de la Torre Díez[1], Miguel López Coronado[1], and Manuel Franco[2]

[1] Department of Signal Theory and Communications, and Telematics Engineering, University of Valladolid, 47011 Valladolid, Spain
susel.gongora@uva.es
[2] Psiquiatry Service, Hospital Zamora, 49004 Zamora, Spain

Abstract. Mental Health disorders such as schizophrenia, depression, and dementia have a great impact on society worldwide. Recent developments have witnessed numerous advances in telemedicine that allow remote monitoring of elderly people with diseases of this pathology. One of key factors influencing recent advances in health information systems is the use of data extraction robust techniques used to extract knowledge from medical databases. The main objective of the paper is to develop a web application that allows extracting knowledge from a database of patients with psychiatric disorders and helps health personnel to know in depth the existing patient profiles and thus improve decision making. The scenario chosen for this study consists in a government of Castile and Leon database of 53641 income records of patients with Mental Health disorders between 2005 and 2015. Applying descriptive statistics for the analysis of data, the results show different parameters regarding Hospitalization behavior such as rate, average, hospitalization days of the patient's number per hospital and outsourcing services.

Keywords: Application · Hospitalizations · Mental health · R software

1 Introduction

Mental Health's diseases are greatly influencing society [1]. Mental illnesses such as autism, bipolar disorder and schizophrenia have a particularly high impact on affected individuals and their families, and they represent a heavy economic burden for the health care system [2, 3]. These disorders are complex, with a significant degree of shared clinical presentation and may share common genetic origins [4].

Mental Health is an area with great needs for improvement, as the volume of data is growing, and its objective is to improve the classification and diagnostic analysis to better understand the accumulation of patient data and make them profitable [5, 6]. Data mining techniques and statistical analysis are the most used tools for prediction in this area.

The statistical analysis includes the concept of collection, settlement, scrutiny, representation, manipulation, clarification and presentation of significant information

J. Henriques et al. (Eds.): MEDICON 2019, IFMBE Proceedings 76, pp. 898–906, 2020.
https://doi.org/10.1007/978-3-030-31635-8_109

[7]. There are two statistical techniques: descriptive and inferential statistics, in this study we will use descriptive statistics to analyze the data in the application.

Currently, many compliance monitoring technologies have been developed for medical purposes based on smartphones, sensors, etc. [8]. In [6] focus on the development of a mobile/web application, through which patients send their symptomatic queries to doctors through a server. The proposed model is the solution for the efficient collection of data, the provision of integrated medical care with analysis. While [2] propose a context-based approach, with the objective of providing medical supervisors with a series of customized applications and services aimed at exploiting the multi-parameter contextual data collected through a long-term monitoring procedure.

In societies, the diagnosis of mental disorders in individuals is often deficient due to lack of understanding of their behavior, symptoms and inadequate knowledge. Many times, an inadequate diagnosis leads to an incorrect treatment that can cause an irreversible deterioration in the mental state of the client, including hospitalization and/or premature death [9].

In the present work we show the development and evaluation of data analysis software for its exploitation in Mental Health, in order to obtain patterns of hospitalization, diagnosis and therapies.

There are similar publications in the field of Mental Health that base their study on developing an online family health tree system that allows discarding types of patients with high risk depression and having a good risk management for health [10]. In [4] investigate how text mining based on Electronic Medical Records (EMRs) and statistical analysis could define the phenotypic limits of three major neuropsychiatric diseases: autism, bipolar disorder and schizophrenia. The results can help to define the limits between different mental illnesses, facilitate the creation of cohorts for clinical and genomic studies and reveal how clear the limits are defined by the experts.

In [9] present a semiautomatic system that helps in the preliminary diagnosis of the patient with psychological disorder. In [11] present the PRISM platform that allows patients and physicians to use continuous sources of passive data for an early and accurate diagnosis, as well as constant monitoring of patients suffering from mental disorders. While in [12] they use the RStudio tool through different R codes to extract the tweets from the different mHealth applications and analyze the feelings and emotions of the users, classifying them as positive, negative or neutral, with the objective of preventing the depression.

The methodology used in this review is described below. Afterwards, the results obtained the discussion of them and the final conclusions of this research work will be finalized.

2 Methodology

The developed system is based on a master-slave architecture, using the data analytics language par excellence R [13]. It is a free software programming language, rich in several packages of statistical analysis and computer graphics, it is also an information management installation and efficient repository [14].

The development framework used to build the graphical interface has been Shiny since it provides a series of advantages such as simple control over layout, interactivity through its functionality called reactivity and a multitude of combos to perform data filtering in a simple and fast way [15]. While the development environment used has been RStudio due to the flexibility and ease of use that it provides.

2.1 Database

It is based on an anonymous dataset of psychiatric patients from the 11 public hospitals of Castile and Leon. The dataset contains 53,641 records from 2005 to 2015. The data used follow the international standard for coding diseases CIE-9.

The initial dataset has 42 fields; once we have made a series of treatments we obtain a dataset of 21 fields, as shown in Fig. 1. Initially irrelevant variables are eliminated for the study such as: descriptions of diagnoses or therapies. Null values, double blanks, special characters are also eliminated and the capitals and lowercase letter of each value is standardized in order to avoid false categories. Subsequently the variables are recategorized: hospitalization days, patient's age, hospital unit, discharge reason and admission type, for some of them as the hospital unit a more exhaustive study had to be done since the original database did not have this categorization in a direct way.

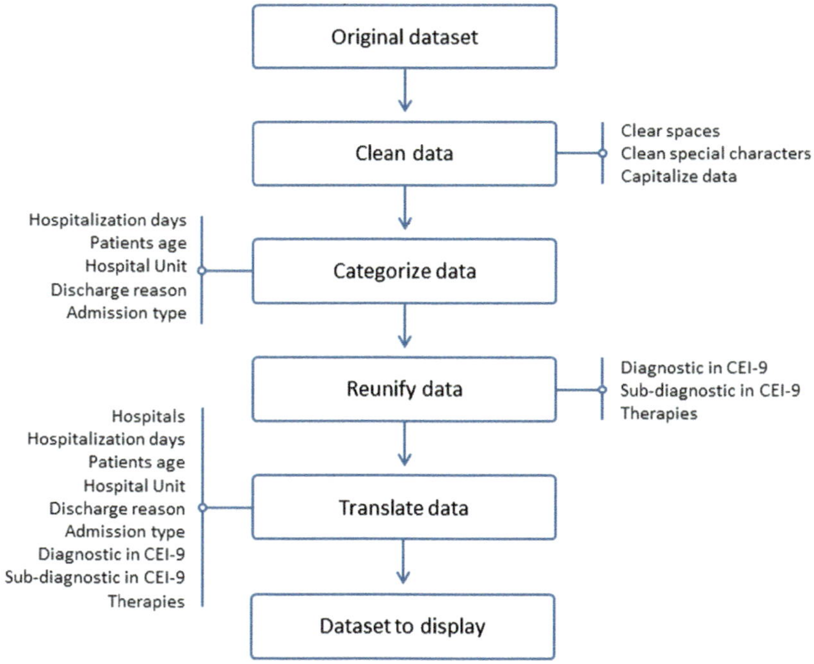

Fig. 1. Flow chart.

Once the variables are re-categorized, all the diagnoses are reunited in a single category with multiple values and all the therapies that apply to each patient in another category. Finally, all the variables are translated, for this purpose, in the majority of cases, self-generated translation files have been used and in other cases, as in the diagnoses, the official table of international classification of diseases [16].

2.2 Statistic Techniques

Several statistical techniques have been implemented using the graphic package "gg-plot2" of R that allow extracting knowledge from the dataset, these techniques can be categorized into 4 blocks:

Block 1 of Trend Evolution. Multiple line graphs, bar chart, horizontal bar chart, grouped bar chart, and Japanese box or candle charts are used.

Block 2 of Representation of a Part with Respect to Total. Circular graphs and bar graphs are used.

Block 3 of Frequency Distribution. Histograms are used.

Block 4 of Comparison. Japanese box or candle charts and grouped bar charts are used.

3 Results

3.1 Software

In this study we have developed a web application capable of extracting knowledge of each of the patient's dataset variables with psychiatric disorders in the region of Castile and Leon, Spain for the period between 2005 and 2015.

The extracted knowledge can be dynamically segmented with respect to numerous criteria such as: hospital, hospitalization days, patient age, hospital unit, discharge reason, admission type, and group of diagnostic or therapy. It can also be categorized into 4 blocks:

Block 1 Hospitalization. It shows graphs of evolution where it visualizes the number and averages of hospitalizations, the hospitalizations ratio with respect to the global population of each community and hospitalization average days of the patients.

Block 2 Patients. It shows typology graphs of the patients and categorizes them by discharge reason, age, hospitalization days, admission type or recurrency.

Block 3 Diagnoses. It shows the cases number graphs presented by each diagnostic block or sub-block according to the CIE-9standard.

Block 4 Therapies. It shows graphs of the most repeated therapies and the ratio when no therapy is applied.

The dynamic segmentation with respect to the aforementioned criteria applied to the elements of each one of the blocks provides a great potential for performing psychiatric analyzes, providing the health personnel answers to many questions that until now had no answer.

Figure 2 shows the graphic interface of the application, from which results of hospitalization (rate, average, outsourcing services), diagnosis and therapies of patients with psychiatric disorders are obtained. The main filters take into account the patient age, provinces and the hospitalization days.

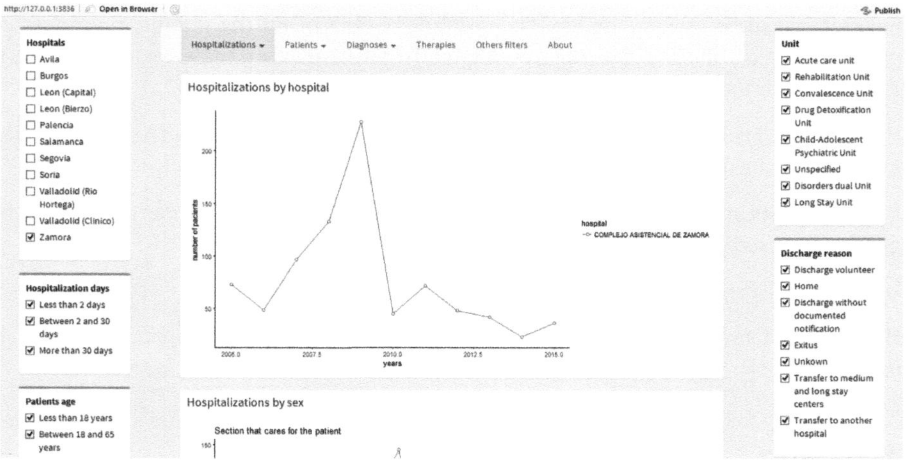

Fig. 2. Application interface.

Afterwards in the next section we show the results of evaluations carried out with the data proposed in the methodology.

3.2 Evaluation of Database with Real Data

As a result of the database analyzed, the statistics per year of the patient's number with hospitalization in each of the provinces is obtained, where the university hospital of Valladolid has the highest number reaching the highest number of 290 patients in 2013.

Regarding gender, the number of female patients in the years 2005, 2006, 2010 and 2015 exceeded the number of hospitalizations in relation to male patients (See Fig. 3), taking into account all age ranges. The results vary when filtering in patients younger than 18 years and older than 65 years, women being the most prevalent hospitalizations.

The prevalent diseases were classified in the database taking into account the CIE-9 international standard for the selected period. When screening patients aged between 18 and 65 years to determine the diagnosis percentage of the most prevalent mental illnesses in the 11 hospitals of the region, we obtained (See Fig. 4) that psychosis has the highest percentage of occurrence 60%, followed by neurotic, personality and non-psychotic disorders that make up 26% of all patients in that range.

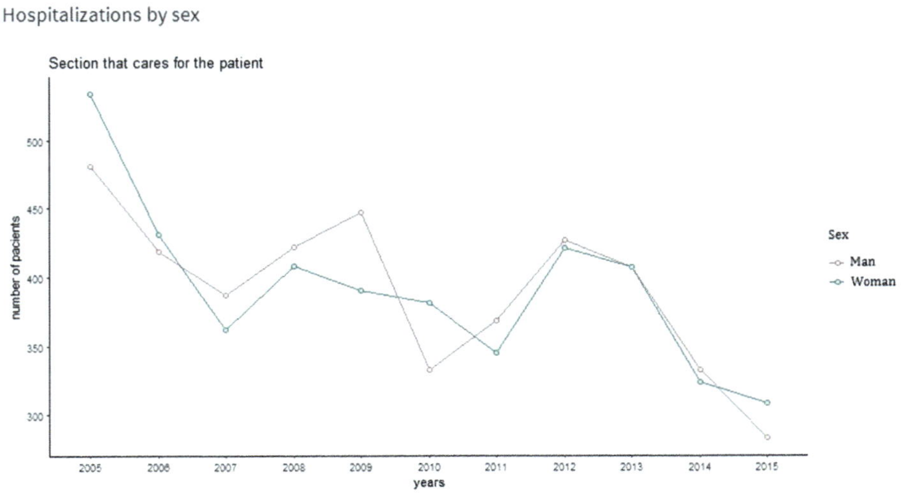

Fig. 3. Statistics hospitalization by gender.

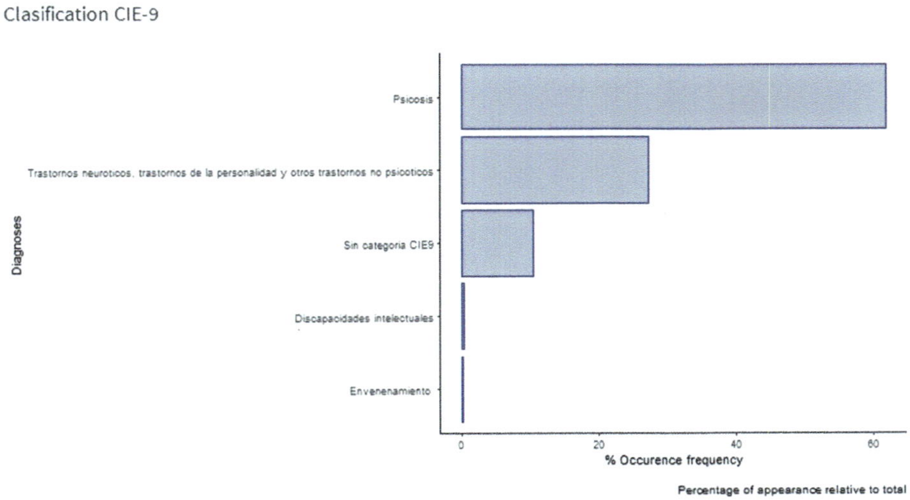

Fig. 4. Diagnostic classification according to CIE-9.

The application determines the most common therapies in patients taking into account range of ages and hospitals. The therapies with higher percentages of occurrence are: in the hospitals of Zamora, Burgos and Valladolid (Clinical) the determination of the psychiatric mental state, another vein puncture and the computerized axial tomography of the head. In Valladolid (Río Hortega), Palencia and Ávila is: another psychiatric therapy with drugs, the hospitals of Soria and Segovia is: the electrocardiogram while in León (Capital) and León (Bierzo) are: EM blood, another microscopic examination and another interview with psychiatric evaluation respectively. Figure 5 shows the most prevalent therapy taking into account the 11 hospitals.

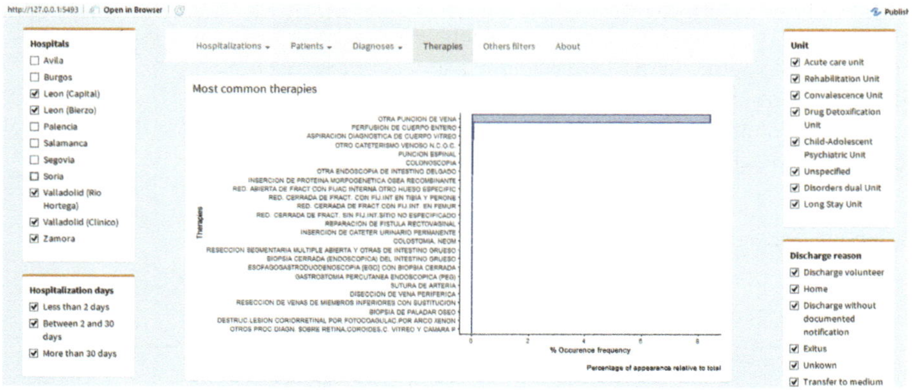

Fig. 5. Statistics of most common therapies in hospital of Zamora, Valladolid and León.

The application also contains a series of additional filters that allows more experienced users to obtain more accurate statistics regarding a more specific type of disorder such as suicide, poisoning, drug abuse, etc.

4 Discussion

Currently there are many medical applications for the exploitation of data, making use of both descriptive statistics and Data Mining techniques. How to achieve a more efficient result of medical care remains an open question for future work where medical data grow at a dizzying pace. A distributed system should be implemented to exchange data between laboratories, hospital systems, clinical centers and other interested parties. Traditional analysis tools have become friendly and transparent and can be used to improve the effectiveness of data analysis [17].

In this study, we developed a web application based on data from patients with psychiatric disorders in the region of Castile and Leon, Spain, with the aim of obtaining patterns of hospitalization, diagnosis and therapies in Mental Health.

The data were analyzed using descriptive statistics; in this case it takes into account age ranges of patients, gender, hospital where it belongs, admission days, admission type, among others. The results show different parameters regarding hospitalization behavior such as rate, average, hospitalization days of the patient's number per hospital and outsourcing services.

The application allows to categorize the number of patients by age, hospitalization days, admission and recurrence, as well as categorizes the diseases diagnosed in Mental Health following the CIE-9 standard. Finally, it allows determining the occurrence frequency percentage of the most common therapies by hospitals included in the study.

The descriptive analysis tools used in the application are sufficiently flexible to allow experienced and less experienced analysts to perform useful analyzes in other Health pathologies. An important factor to highlight in the application is its ability to update the results according to the initial parameters you want to estimate, for hospitalizations, diagnosis and therapies. In addition, the design of the interface is sufficiently generic to allow users with less experience in data analytics to handle it with ease.

The most notable limitation of this study has been that the data related to the dataset used only included the Autonomous Community of Castilla and León and therefore it has not been possible to extract comparisons of patients with other autonomous communities or countries.

5 Conclusion

The use of descriptive statistics as well as Data Mining applied to the database can increase the confidence level of the results obtained. In fact, the era of big data offers new opportunities to better understand the behavior of health data as well as the prediction of risk factors and patterns in diseases such as Mental Health. Hence, as future lines we intend to apply to the aforementioned database machine learning and deep learning techniques: (1) clustering of patients to determine pathological profiles, (2) use of neural networks for the prediction of pathologies, (3) use of machine learning for the prediction of therapies from diagnosed pathologies; and (4) comparison of supervised learning techniques for predicting diseases of Mental Health.

Acknowledgments. This research has been partially supported by European Commission and the Ministry of Industry, Energy and Tourism under the project AAL-20125036 named *"Wetake Care: ICT- based Solution for (Self-) Management of Daily Living"*.

Conflict of Interest. No author has a conflict of interest with the contents of this manuscript.

References

1. Becker, D., van Breda, W., Funk, B., Hoogendoorn, M., Ruwaard, J., Riper, H.: Predictive modeling in e-mental health: a common language framework. Internet Interv. **12**, 57–67 (2018)
2. Panagiotakopoulos, T.C., Lyras, D.P., Livaditis, M., Sgarbas, K.N., Anastassopoulos, G.C., Lymberopoulos, D.K.: A contextual data mining approach toward assisting the treatment of anxiety disorders. IEEE Trans. Inf. Technol. Biomed. **14**(3), 567–581 (2010)
3. Chen, W.-Y., Chen, L.-Y., Liu, H.-C., Wu, C.-S., Yang, S.-Y., Pan, C.-H., et al.: Antipsychotic medications and stroke in schizophrenia: a case-crossover study. PLoS One **12**(6), e0179424 (2017)
4. Lyalina, S., Percha, B., Lependu, P., Iyer, S.V., Altman, R.B., Shah, N.H.: Identifying phenotypic signatures of neuropsychiatric disorders from electronic medical records. J. Am. Med. Inform. Assoc. **20**, 297–305 (2013)

5. Frantzidis, C.A., Bratsas, C., Klados, M.A., Konstantinidis, E., Lithari, C.D., Vivas, A.B., et al.: On the classification of emotional biosignals evoked while viewing affective pictures: an integrated data-mining-based approach for healthcare applications. IEEE Trans. Inf. Technol. Biomed. **14**(2), 309–318 (2010)
6. Panda, M., Ali, S.M., Panda, S.K.: Big data in health care: a mobile based solution. In: 2017 International Conference on Big Data Analytics and Computational Intelligence, ICBDAC, pp. 149–152. (2017)
7. Das, N., Das, L., Rautaray, S.S., Pandey, M.: Big data analytics for medical applications. Int. J. Mod. Educ. Comput. Sci. **10**(2), 35–42 (2018)
8. He, T., Ogunti, R., Puppala, M., Chen, S., Yu, X., Mancuso, J.J., et al.: A smartphone app framework for segmented cancer care coordination. In: 3rd IEEE EMBS International Conference on Biomedical and Health Informatics, BHI, pp. 372–375 (2016)
9. Azar, G., Gloster, C., El-Bathy, N., Yu, S., Neela, R.H., Alothman, I.: Intelligent data mining and machine learning for mental health diagnosis using genetic algorithm. In: IEEE International Conference on Electro Information Technology, pp. 201–206 (2015)
10. Chang, C.C., Chen, R.S., Liao, K.H., Sun, P.R.: Estimation of depression risky of family health tree system - using the family functioning scale. In: Proceedings - International Conference on Management and Service Science, MASS (2009)
11. Kamdar, M.R., Wu, M.J.: PRISM: a data-driven platform for monitoring mental health. In: Pacific Symposium on Biocomputing, vol. 21, pp. 333–344 (2016)
12. Pai, R.R., Alathur, S.: Assessing mobile health applications with twitter analytics. Int. J. Med. Informatics **113**, 72–84 (2018)
13. Assink, M., Wibbelink, C.J.M.: Fitting three-level meta-analytic models in R: a step-by-step tutorial. Quant. Methods Psychol. **12**(3), 154–174 (2016)
14. Pradhananga, Y., Karande, S., Karande C.: CBA: cloud-based bigdata analytics. In: Proceedings - 1st International Conference on Computing, Communication, Control and Automation ICCUBEA, pp. 47–51 (2015)
15. Moraga, P.: SpatialEpiApp: a Shiny web application for the analysis of spatial and spatio-temporal disease data. Spat. Spatio-temporal Epidemiol. **23**, 47–57 (2017)
16. eCIE9MC Clasificación Internacional de Enfermedades. http://eciemaps.msssi.gob.es/ecieMaps/browser/index_9_mc.html. Accessed 18 Jun 2018
17. Ta, V.D., Liu, C.M., Nkabinde, G.W.: Big data stream computing in healthcare real-time analytics. In: Proceedings of the 2016 IEEE International Conference on Cloud Computing and Big Data Analysis ICCCBDA, pp. 37–42 (2016)

Feasibility of Machine Learning in Predicting Features Related to Congenital Nystagmus

Giovanni D'Addio[1(✉)], Carlo Ricciardi[1,2], Giovanni Improta[3], Paolo Bifulco[1,4], and Mario Cesarelli[1,4]

[1] Istituti Clinici Scientifici Maugeri IRCCS, Naples, Italy
gianni.daddio@icsmaugeri.it
[2] Department of Advanced Biomedical Sciences, University Hospital of Naples 'Federico II', Naples, Italy
[3] Department of Public Health, University Hospital of Naples 'Federico II', Naples, Italy
[4] DIETI, University of Naples 'Federico II', Naples, Italy

Abstract. Congenital nystagmus is an ocular-motor disease affecting people's visual acuity since their first years of life. Electrooculography is used to perform eye tracking in these patients, giving the possibility to extract a wide variety of parameters. The relationships among all these variables were analysed in the past and the aim of this paper is to perform a new analysis employing more recent techniques, those of machine learning. The electrooculography of 20 patients was recorded, signals were pre-processed, and some parameters were extracted through a custom-made software. Knime analytics platform was chosen in order to build predictive models using Random Forests and Logistic Regression Tree algorithms and some evaluation metrics were computed. The visual acuity and the variability of eye positioning were predicted employing five and six variables, respectively. In terms of coefficient of determination, visual acuity had values over 0.72 and variability of eye positioning over 0.70. Compared to the results obtained without machine learning algorithms during the past years, these values become more valuable. In conclusion, this approach showed its feasibility in detecting relationships among variables related to congenital nystagmus; it could be tested in order to find new and stronger relationships among these variables and be of support for clinicians.

Keywords: Nystagmus · Data mining · Biomedical signals

1 Introduction

During the first period of life, babies can develop congenital nystagmus (CN), an ocular-motor disease that makes visual acuity (VA) worsen [1]. CN consists of involuntary, conjugated and rhythmical horizontal to and fro movements and patients affected by it have a disrupted fixation due to rapid movements of the target image onto the retina [2]. The process of eye tracking is performed in different ways [3], particularly electrooculography (EOG) exploits measurements of skin potentials, electrodes are placed around the eyes, it is possible to catch small differences in the skin potential that are related to eye movements [4, 5]. Differently, Infrared-oculography (IROG) is a

J. Henriques et al. (Eds.): MEDICON 2019, IFMBE Proceedings 76, pp. 907–913, 2020.
https://doi.org/10.1007/978-3-030-31635-8_110

non-invasive method to define time of foveation, an indirect measure of VA. The eye is illumined by infrared light which is reflected by the sclera and the difference between the amounts of infrared light reflected from the eye gives the information about the changes of eye position. Relationships among visual acuity, baseline oscillations, variability of eye positioning (SDp), nystagmus foveation periods (Tf), nystagmus amplitude and frequency were deeply studied in the past literature. Cesarelli et al. proposed an exponential relationship between SDp and VA while Bifulco et al. investigated the relationship between the amplitude of the baseline oscillation and the SDp, both focused on foveation periods [6, 7]. Some authors focused on automatically detecting nystagmus and analysed its relationships with other parameters (i.e. VA) [8–10]. Sheth et al. proposed different associations such as VA and nystagmus features, standard deviations of eye velocities and eye-positioning [11].

Nowadays, the huge generation of data has given the chance to create new techniques that are used to discover hidden patterns in datasets: machine learning [12, 13]. The development of information technology has allowed physicians and engineers to help clinicians during both the complicated task of diagnosis and prognosis of patients [14–17], although some challenges have to be faced undoubtedly, as regard management, processing and understanding of big data [18]. The applications of data mining and machine learning in the field of ophthalmology have the tendency to tackle the thematic of brain computer interfaces, particularly through the combination of EOG and electroencephalography [19–21]. Nevertheless, some efforts were made in order to make classifications based on eye-movements recording; Zemblys et al. employed machine-learning techniques to detect events in eye tracking data and tested different state-of-art algorithms [22, 23].

The aim of this paper is to investigate the relationships among different parameters of nystagmus through two machine learning algorithms (Random Forests and Logistic Regression Tree) and carry out some evaluation metrics to make also a comparison with past results without this approach. The dataset used for this paper was the same of Cesarelli et al. and Bifulco et al. in 2000 and 2002, respectively [6, 7].

2 Materials and Methods

The EOG of 20 patients was recorded twice (both right and left eyes). Patients were 10 males and 10 females affected by CN and whose age was between 6 and 34 years. Different types of nystagmus affected each of them and binocular, horizontal eye-movements, at different gaze position, were performed.

Light stimuli were shown to patients, at a fixation distance of 1 m using an arched, horizontal LED-bar. Moreover, a head support and a chin rest were used in order to reduce head motion. The sequence of light stimuli lasted 2 min, each position was held for 10 s and the sequence of angles was the following: 0°, 5°, 10°, 20°, 30°, 0, −5°, −10°, −20°, −30° and 0°. Both EOG and IROG were employed for the eye-movement recording with a sampling frequency of 200 Hz. IROG was employed to deal with non-collaborative patients such as children.

The classic Landolt Cs technique was performed to measure VA, which was expressed in tenths. The Bioengineering Unit (Department of Electronic Engineering,

University of Naples 'Federico II') designed a specific software to process EOG signals and to extract nystagmus features.

Examples of signals acquired through this instrumentation are in Fig. 1.

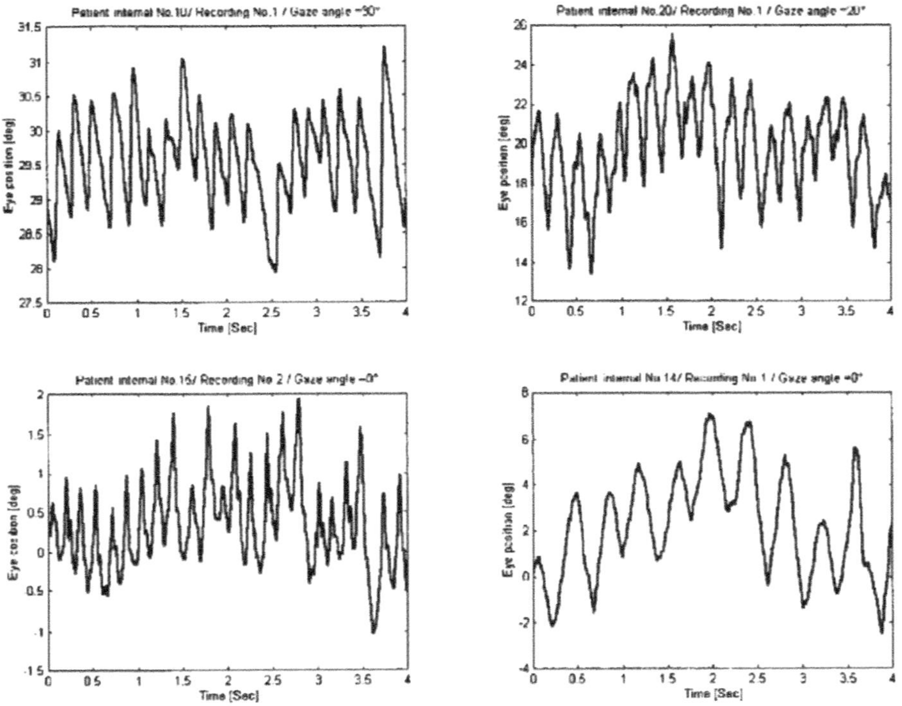

Fig. 1. Examples of 4-seconds-long horizontal eye movement recordings. Internal patient numbering, the recording number (1 or 2) and the gaze angle are indicated.

2.1 Pre-processing and Nystagmus Feature Extraction

A low-pass filter (70 Hz 3 dB cut-off frequency) was applied to all signals so as a notch filter (centred at 50 Hz) to reduce power line noise. A second pre-processing phase consisted of eliminating the DC and any linear component to avoid electrode polarisation signal components. A low-pass differentiation algorithm for biological systems was used to calculate eye velocities from eye positions [24].

Nystagmus cycles were identified and features (Frequency, amplitude, intensity, waveform shape, nystagmus foveation periods and both amplitude and frequency of baseline oscillation) extracted through a software designed by Juhola [25]. The exact definitions of features were those proposed by Cesarelli et al. and Bifulco et al. [6, 7]. The foveation window was different from the one proposed by Dell'Osso et al. [26].

2.2 Tool, Algorithm and Evaluation Metrics

Knime analytic platform, widely acknowledged as a good example of scientific workflow system, was employed to implement the algorithms. This platform was discussed in literature and exploited for healthcare applications [27, 28]. It has the possibility to install plugins and use extensions provided by R, Weka, Matlab, Python. Breiman and Landwehr et al. implemented the versions of Random Forests and Logistic Regression Tree that were used in this study [29, 30]. They are both based on the decision tree, whose basic idea is to break a complex decision into several simpler ones. It is a structure, made up of leaves and nodes, which represents a predicted value and an attribute, respectively. Random Forests consist of an ensemble of decision trees that can be both a powerful classifier and a regressor. Each tree considers a random and limited subset of features in each node having access only to a random set of the training data. This logic increases diversity in the forest conducting to more robust global predictions. The second algorithm consists of a decision tree whose leaves contain linear regression. Their performance is assessed with the following evaluation metrics: coefficient of determination (R^2), mean absolute error, mean squared error, root mean squared deviation, mean signed difference. The features considered for predicting VA were the same of Cesarelli et al.:

1. Age;
2. Amplitude of nystagmus;
3. Frequency of nystagmus;
4. Tf;
5. SDp;

The features included in the algorithms to predict SDp were the same of Bifulco et al.:

1. Age;
2. Amplitude of nystagmus;
3. Frequency of nystagmus;
4. Amplitude of baseline oscillation;
5. Frequency of baseline oscillation;
6. VA.

Moreover, a leave one out cross-validation was applied: the models were learned on all the records minus one and tested on that last one "that was left out". This procedure is repeated for a number of times equal to the number of records (in this case 20 times).

3 Results

Two models were built using the Random Forest with 50 trees and Logistic Regression Tree, and then tested by computing the above-mentioned evaluation metrics. The targets of these regressions were the VA and SDp (computed and measured as Cesarelli et al. [6]) and the scores are presented in Table 1. For both predictions, a leave one out cross-validation was employed.

Table 1. Evaluation metrics for the algorithms of the study

Algorithms	Target	R^2	Mean absolute error	Mean squared error	Root mean squared deviation	Mean signed difference
Random forests	VA	0.85	0.05	0.00	0.06	0.00
	SDp	0.73	0.25	0.12	0.35	0.01
Logistic regression tree	VA	0.72	0.06	0.01	0.08	0.01
	SDp	0.70	0.28	0.14	0.37	0.02

The coefficient of determination was between 0.70 and 0.85 for the two predictions made by Random Forests and for the VA the errors were all under the value 0.1. Random Forests showed better capability to predict VA than SDp. Logistic Regression Tree showed lower coefficient of determinations compared to those obtained by Random Forests. The biggest difference was in predicting VA (R^2 overcame 0.8 through Random Forests and was 0.72 through Logistic Regression Tree).

4 Discussion and Conclusion

Firstly, the EOG of 20 people affected by CN was recorded for both eyes. Signals were pre-processed and some features were extracted through a custom-made software, developed by the Bioengineering Unit of the University of Naples "Federico II". Then, Random Forests and Logistic Regression Tree were used to build a model through Knime analytics platform, obtaining some scores to evaluate their performance.

Cesarelli et al. [6] obtained the best coefficient of determination (R^2 = 0.85) using an exponential model based on SDp and Tf to compute VA. Bifulco et al., instead, proposed a relationship between the baseline oscillation, considered as a sinusoidal signal, in eye movement recordings during fixation and SDp, which is related to VA obtaining values under 0.70 for R^2 [7].

Compared to those results, machine learning algorithms have showed a good feasibility, considering that SDp was predicted with R^2 = 0.73 through Random Forests and R^2 = 0.70 through Logistic Regression Tree while VA was predicted with a R^2 greater than 0.70. Particularly, in this analysis Random Forests have showed greater potential than Logistic Regression Tree in predicting nystagmus features.

Zemblys et al. showed that Random Forests exhibited the best eye-movement event classification performance [22, 23] but they did not test Logistic Regression Tree; our study confirms the great results obtained by Zemblys et al. through Random Forests, enhances its strength also in regression analysis and gives insights as regards the possibility to employ new algorithm.

Indeed, this study has future perspectives: a few signals were available, and their recording and processing were conducted through old instrumentation, it means that the quality of signals and features could be better with more modern instrumentation. Only two algorithms were employed for this assessment; so, in the future, more algorithms could be implemented and a wrapper methodology could be used to identify the most relevant variables in order to optimize the scores. The results coming from this study are valuable but need further investigation.

Conflict of Interest Declaration. The authors declare they have no conflict of interests.

References

1. Dell'Osso, L.F., Van Der Steen, J., Steinman, R.M., Collewijn, H.: Foveation dynamics in congenital nystagmus. I: Fixation. DocOphthalmol **79**, 1–23 (1992)
2. Dickinson, C.M., Abadi, R.V.: The Influence of the nystagmoid oscillation on contrast sensitivity in normal observers. VisionRes **25**, 1089–1096 (1985)
3. Ciesla, M., Koziol, P.: Eye pupil location using webcam. arXiv preprint arXiv:1202.6517 (2012)
4. Bieda, R., Jaskot, K., Łazarski, J.: Nystagmus detection system. In: Advanced Technologies in Practical Applications for National Security, pp. 59–73. Springer, Cham (2018)
5. Binias, B., Palus, H., Jaskot, K.: Real-time detection and filtering of eye blink related artifacts for brain-computer interface applications. In: Man–Machine Interactions, vol. 4, pp. 281–290. Springer, Cham (2016)
6. Cesarelli, M., Bifulco, P., Loffredo, L., Bracale, M.: Relationship between visual acuity and eye position variability during foveations in congenital nystagmus. Doc. Ophthalmol. **101**(1), 59–72 (2000)
7. Bifulco, P., Cesarelli, M., Loffredo, L., Sansone, M., Bracale, M.: Eye movement baseline oscillation and variability of eye position during foveation in congenital nystagmus. Doc. Ophthalmol. **107**(2), 131–136 (2003)
8. Cesarelli, M., D'Addio, G., Loffredo, L., Daniele, A.: A system to automatically analyse nystagmus. In: Otorino Rossi Award Conference - International Workshop on Eye Movements, pp. 262–264 (1994)
9. Cesarelli, M., Loffredo, L., Bifulco, P.: Relationship between visual acuity and oculogram baseline oscillations in congenital nystagmus. In: Proceedings of the 4th European Conference on Engineering and Medicine, Warsaw, pp. 301–302 (1997)
10. Cesarelli, M., Bifulco, P., Loffredo, L.: EOG baseline oscillation in congenital nystagmus. In: VIII Mediterranean Conference on Medical Biological Engineering and Computing - MEDICON 1998, Lemesos - Cyprus, 14–17 June 1998. CD-ROM 19.3 (1998)
11. Sheth, N.V., Dell'Osso, L.F., Leigh, R.J., Van Doren, C.L., Peckham, H.P.: The effects of afferent stimulation on congenital nystagmus foveation periods. Vis. Res. **35**, 2371–2382 (1995)
12. Jensen, P.B., Jensen, L.J., Brunak, S.: Mining electronic health records: towards better research applications and clinical care. Nat. Rev. Genet. **13**(6), 395 (2012)
13. Chen, M., Mao, S., Liu, Y.: Big data: a survey. Mob. Netw. Appl. **19**(2), 171–209 (2014)
14. Guarino, F., Russo, M.A., Franzese, M., et al.: A novel shiny platform for the geo-spatial analysis of large amount of patient data (No. e3335v1). PeerJ Preprints (2017). https://doi.org/10.7287/peerj.preprints.3335v1
15. Revetria, R., Catania, A., Cassettari, L., et al.: Improving healthcare using cognitive computing-based software: an application in emergency situation. In: International Conference on Industrial, Engineering and Other Applications of Applied Intelligent Systems, pp. 477–490. Springer, Heidelberg (2012). https://doi.org/10.1007/978-3-642-31087-4_50
16. Improta, G., Russo, M.A., Triassi, M., et al.: Use of the AHP methodology in system dynamics: modelling and simulation for health technology assessments to determine the correct prosthesis choice for hernia diseases. Math. Biosci. **299**, 19–27 (2018). https://doi.org/10.1016/j.mbs.2018.03.004

17. Romano, M., D'Addio, G., Clemente, F., et al.: Symbolic dynamic and frequency analysis in foetal monitoring. In 2014 IEEE International Symposium on Medical Measurements and Applications (MeMeA), pp. 1–5. IEEE (2014). https://doi.org/10.1109/memea.2014.6860122
18. Dinov, I.D.: Methodological challenges and analytic opportunities for modeling and interpreting big healthcare data. Gigascience **5**(1), 12 (2016)
19. Fatourechi, M., Bashashati, A., Ward, R.K., Birch, G.E.: EMG and EOG artifacts in brain computer interface systems: a survey. Clin. Neurophysiol. **118**(3), 480–494 (2007)
20. Punsawad, Y., Wongsawat, Y., Parnichkun, M.: Hybrid EEG-EOG brain-computer interface system for practical machine control. In: 2010 Annual International Conference of the IEEE Engineering in Medicine and Biology, pp. 1360–1363. IEEE, August 2010
21. Witkowski, M., Cortese, M., Cempini, M., Mellinger, J., Vitiello, N., Soekadar, S.R.: Enhancing brain-machine interface (BMI) control of a hand exoskeleton using electrooculography (EOG). J. Neuroeng. Rehabil. **11**(1), 165 (2014)
22. Zemblys, R.: Eye-movement event detection meets machine learning. Biomed. Eng. **20**(1) (2016)
23. Zemblys, R., Niehorster, D.C., Komogortsev, O., et al.: Behav Res **50**, 160 (2018). https://doi.org/10.3758/s13428-017-0860-3
24. Usui, S., Amidror, I.: Digital low-pass differentiation for biological signal processing. IEEE Trans. Biomed. Eng. **29**, 686–693 (1982)
25. Juhola, M.: Detection of nystagmus eye movements using a recursive digital filter. IEEE Trans. Biomed. Eng. **35**, 389–395 (1988)
26. Dell'Osso, L.F., Darof, R.B.: Congenital nystagmus waveform and foveation strategy. Doc. Ophthalmol. **39**, 155–182 (1975)
27. Romeo, V., Ricciardi, C., Cuocolo, R., et al.: Machine learning analysis of MRI-derived texture features to predict placenta accreta spectrum in patients with placenta previa. Magn. Reson. Imaging (2019). https://doi.org/10.1016/j.mri.2019.05.017
28. Al-Khoder, A., Harmouch, H.: Evaluating four of the most popular open source and free data mining tools. Int. J. Acad. Sci. Res. **3**(1), 13–23 (2015)
29. Breiman, L.: Random forests. Mach. Learn. **45**(1), 5–32 (2001)
30. Landwehr, N., Hall, M., Frank, E.: Logistic model trees. Mach. Learn. **59**(1–2), 161–205 (2005)

A Smartphone Based Survey to Investigate the Cyber-Risk Perception on the Health-Care Professionals

Daniele Giansanti[1(✉)], Mauro Grigioni[1], Lisa Monoscalco[2], and Rosario Alfio Gulino[2]

[1] ISS, Centro TISP, via R. Elena 299, 00131 Rome, Italy
daniele.giansanti@iss.it
[2] Università di Tor Vergata, Rome, Italy

Abstract. A recent discussion has been opened on the cyber-risk in health care. Today the biomedical and the health informatics are under cyberattack. The study faces (a) from a general point of view the problem of the cybersecurity in the health care with particular attention to the medical devices of class 3 implantable active (which are vulnerable to the cyber-attacks) to the PACS, to the data and to the networks; (b) proposes and test a smartphone based survey to investigate the self-perception of the cybersecurity on the health-care actors.

Keywords: Cybersecurity · E-health · Medical devices

1 Introduction

A recent discussion has been opened on the cyber-risk in health care. Today the biomedical and the health informatics are under cyber-attack. The biomedical devices which are connected to elaborators are therefore not immune from this risk; artificial pancreas and pacemakers have in particular been found to be vulnerable [1–12]. This means that there is the need of particular attention to the cybersecurity issues during the design of these and of the similar *devices of Class 3-active implantable* connected to the elaborators. Also medical software, health care networks and patients' data are under the same risk [13–15] (Fig. 1).

2 The Cyber-Risk: Between Attacks and Opportunities

A recent discussion has been opened on the cyber-risk in health care. Today the biomedical and the health informatics are under cyber-attack. The biomedical devices which are connected to elaborators are therefore not immune from this risk; artificial pancreas and pacemakers have in particular been found to be vulnerable [1–12]. This means that there is the need of particular attention to the cybersecurity issues during the design of these and of the similar *devices of Class 3-active implantable* connected to the elaborators. Also medical software, health care networks and patients' data are under the same risk [13–15] (Fig. 1).

J. Henriques et al. (Eds.): MEDICON 2019, IFMBE Proceedings 76, pp. 914–923, 2020.
https://doi.org/10.1007/978-3-030-31635-8_111

2.1 The Pacemaker and the Cyber-Risk

According to FDA [7–12], there are more than 350 thousand cardiac devices at risk of cyber attacks in the United States and promptly, manufacturers, according to the recommendations for patients and health personnel, have started to install the safety updates for the devices. The agency's statement highlights the high danger that these pacemakers may somehow end up in the hands of cyber criminals. Being electronic devices connected on the network, each criminal is able to tamper with them remotely. The pacemakers are tiny devices powered by batteries that, inserted surgically under the chest, are able to correct, in the event that the drugs are not effective, the disturbances of the heart rhythm, intervening in the case is beating too fast, too slowly, or inordinately. Remote access to these devices is ensured by radio links that allow communications with the outside world, with physicians, to monitor the vital functions of the patient. Among the main problems identified by the FDA, therefore, there are the cybersecurity of the devices, the capacity of the batteries and delays in firmware updates. The doubts about the safety of the pacemakers in question had been raised as early as 2018. Several vulnerabilities were identified on some pacemaker models from different suppliers. At the moment, national bodies such as FDA have not registered attacks or attempts to official attacks on the sanitary equipment in question, but the volume of the devices assessed as 'vulnerable' by the survey has alarmed the national and super national institutions.

2.2 The Artificial Pancreas and the Cyber-Risk

The artificial pancreas [1–6], as it is well known is a device designed for the control of blood glucose in diabetes. It is controlled by software that runs on mobile platforms such as laptops, tablets and smartphones, and operates on wireless networks under local or remote medical supervision. Given that optimal function is fundamental to the health of the individual, safety and privacy, the risk of threats to these assumptions aimed towards the artificial pancreas is of utmost importance. In a recent article published in Diabetes Technology & Therapeutics [1] the co-authors have reviewed recent studies on artificial pancreas systems aimed at identifying safety vulnerabilities related to both internal factors such as external users that could put users at risk. These include software integrity and the risk of such threats from viruses, worms or spyware, access to personal information, or device manipulation. External concerns may include exposure to electromagnetic fields that could interfere with the proper functioning of the system. In [1] it is proposed the need for a more formal and coherent approach in reporting the technical characteristics and performance of an artificial pancreas during experimental studies. This would include the various components of these complex systems, which typically include a continuous glucose monitor, glucometer, and insulin pump.

2.3 The PACS and the Cyber-Risk

As it is well known also some software used in health care are considered medical devices. Are medical devices for example the software, allowing a diagnosis. The Picture Archiving and Communication System (PACS) [12–15] according to this are defined by the FDA as a Class II device. They are heterogeneous interoperable complex systems embedding several components such as computers, workstations, scanners, digital archives, printers, hubs, software. In particular in these systems there are several software components allowing image downloading, uploading, manipulation with important implication with the data integrity. Furthermore the PACS are a depositary of patients' data with important implication on the data privacy and protection. Therefore the PACS requires controls that provide significant integrity, availability, and confidentiality assurances. Particular attention must therefore be put in the management of each issue of these system with particular consideration to the complexity and the integration of the parts.

2.4 The Health Care Networks with the Patients' Data and the Cyber Attack

Hospital companies [13–15] today depend 100% on IT equipment. The computer security of the hospital networks, and computers of hospitals is, as recent cases in many parts of the world demonstrate, an aspect that is no longer negligible. The risk increases with the percentage of dematerialization of administrative processes, of the dependence on computerized medical devices and of the quantity and type of data stored by the hospitals. Modern attacks are no longer transmitted by simple computer viruses, but by real complex systems, managed by increasingly capable and resourceful criminal organizations. This means that any device, be it a computer, a network, a medical device, can be penetrated and therefore violated in terms of privacy, in terms of operations (with sometimes disastrous consequences) or in terms of truthfulness of information. Any organization that is directly proportional to its criticality, and the hospitals have a criticality of the highest level since the operation is linked to the health of people, must take action against the now current cyber risk. One of the most striking cases, in order of time, is certainly the attack on the Hollywood Presbyterian Medical Center which completely blocked access to any patient information for a whole week and which was only solved by the payment of a substantial redemption ($ 17,000 compared to the initially requested 3.6 million) to criminals. Another recent case is related to hackers intrusion to the University Hospital of Los Angeles, UCLA Health, violating the privacy of millions of people.

2.5 Recent Initiatives to Face the Cyber-Attacks

In USA the attention is high. The FDA after having issued the guidelines on the management of cybersecurity in medical devices before October 2014 (pre-market phase), published in December 2016 [16] also a guide for the "post-market" phase, with a view to guaranteeing an adequate level of protection of biomedical devices throughout their life cycle, from the initial development and design phases to after-sales support.

In Europe the attention is recently demonstrated by the recent General European Data Protection Regulation (GDPR), which became recently applicable in all EU countries in 2018. The GDPR and the FDA Guidelines demonstrate the regulator's attention to the security measures to be adopted to protect biomedical devices, also due to the increasing digitalization of the sector, which, on the one hand, leads to an improvement in the quality of services offered by health professionals, also due to the emergence of new care-delivery models (e.g. virtual care, mobile health, wearable and implantable patient monitoring devices), exposes the operators to more and more frequent and sophisticated cyber attacks, whose goal was for example to violate intellectual property and personal data.

3 The Proposal of an Investigation

3.1 The Idea

The general idea was to set-up a study aimed at investigating the perception of cybersecurity on the actors of the health care (physicians, doctor specialists, health care professionals, administrative, technicians, heads of Operative units, clinical engineers, clinical engineers heads of operative units, administrators, head administrators, operators, etc.).

The first step of the study was to configure a questionnaire to submit to the subjects/actors involved into the study. The second step was to collect and analyse the data. It was decided to arrange and submit the questionnaire electronically using the email or messengers tools.

3.2 The Choice of the Survey

It was decided to use an online survey creator with a large distribution in the community. For this reason it was chosen Microsoft Forms.

It allows the electronic creation of the survey. An internet link becomes available at the end of the process of the creation. This link can be sent using the *email* tool or a *messenger* tool.

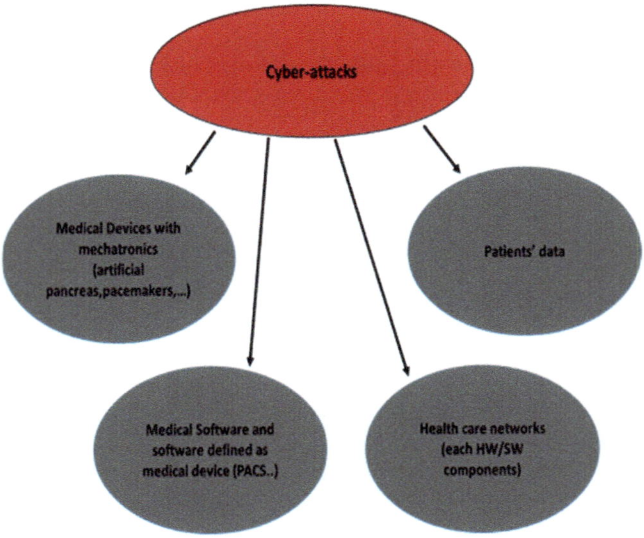

Fig. 1. Health care under cyber attack

When the remote user send the compiled form, the data is automatically stored and accessible using excel or by means of an automatically generated report.

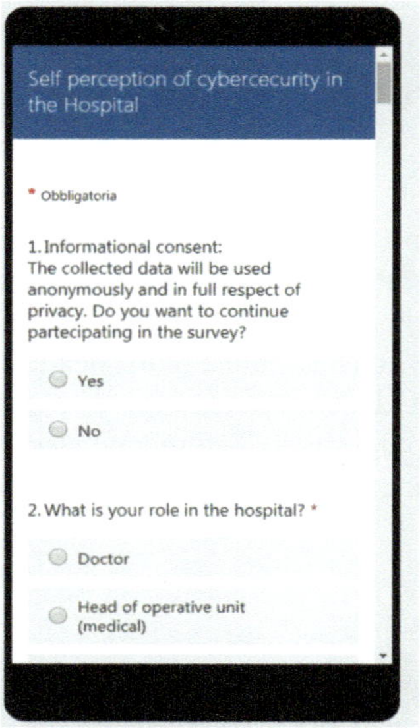

Fig. 2. The starting of the survey

4 Outcome

4.1 The Survey

The generated link is the following one https://forms.office.com/Pages/ResponsePage.aspx?id=DQSIkWdsW0yxEjajBLZtrQAAAAAAAAAAAAZ__gdk7kpUNUxSUE1WT1hLRTRBOTVIUjdDMTY2NlhIQy4u

Figure 2 shows the starting of the survey. The Figs. 3, 4 and 5 elucidate some parts of the survey. Figure 8 shows the end of the survey.

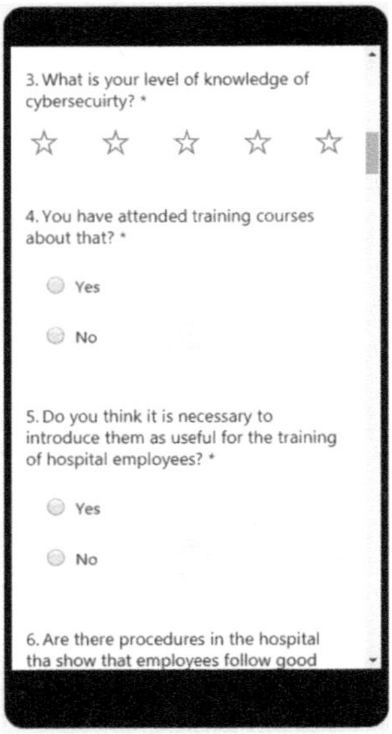

Fig. 3. The first print screen of the survey

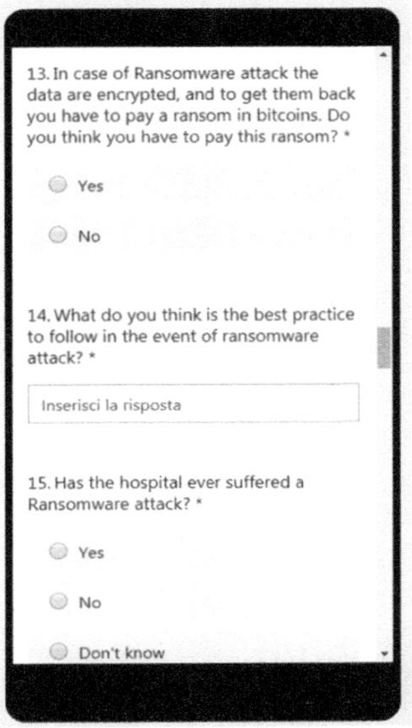

Fig. 4. The second print screen of the survey

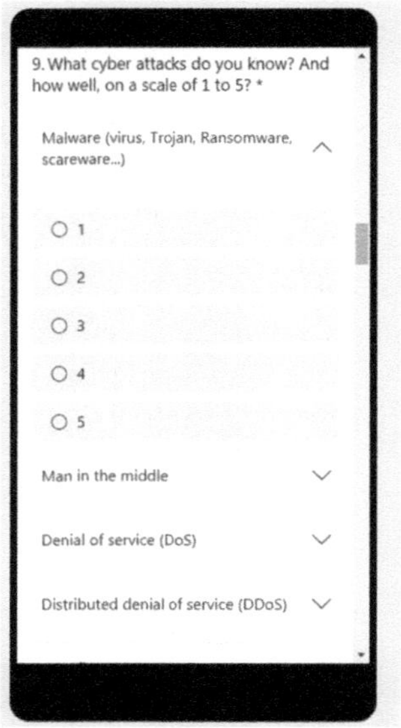

Fig. 5. The third print screen of the survey

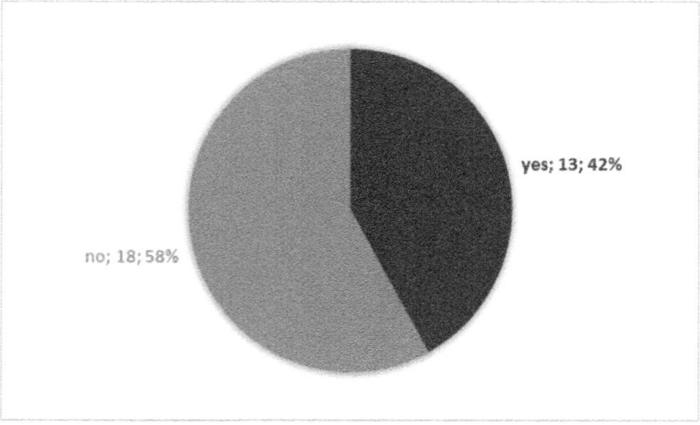

Fig. 6. Actors having had specific training on the cybersecurity

4.2 The Preliminary Data Analysis

At the moment we have just started with a first the submission of the surveys. We have collected 31 surveys. The submission is continuing and a fine data-mining will follow. The following considerations are based on a first sample used to test the methodology. The mean value of the self declared knowledge on the cyber-security was 2.74.

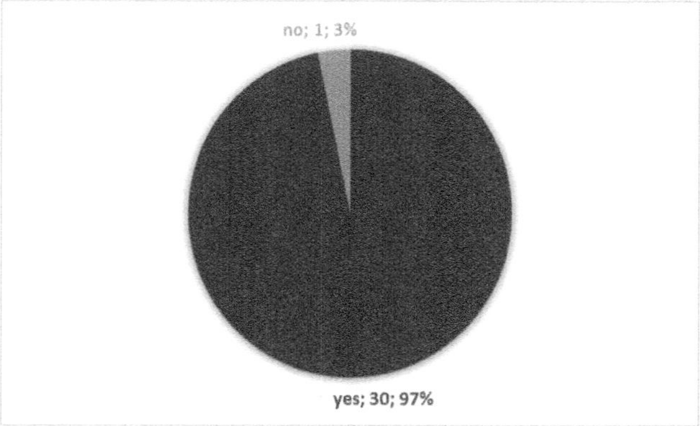

Fig. 7. Opinion on the introduction of specific training

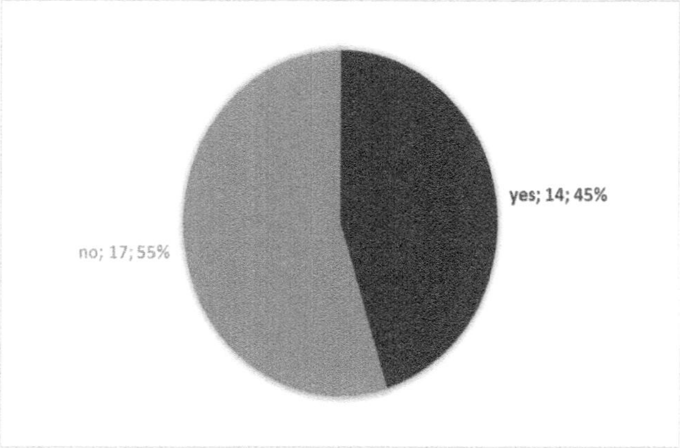

Fig. 8. Self perception of the suitability of the initiatives on the cybersecurity

The following analysis reports the outcome relevant to the received training and the diffusion of the "culture of the cybersecurity". In details Fig. 6 discriminates the actors having or not having had specific training.

Figures 7 and 8 respectively elucidate the opinion on the introduction of the training on the cybersecurity and of the self-perception of the initiatives suitability in the Hospital on the cybersecurity.

5 Conclusion and Work in Progress

The study faced the complex question of the cybersecurity in the Hospitals. According to the cyber-risks the work showed the characteristics of this problem on (a) the medical device of Class III implantable (artificial pancreas, pacemaker), (b) the medical software and software as medical device; (c) the networks in health care; (d) the patients' data.

The study also proposed a survey to collect data relevant to the self perception of cybersecurity in the Hospitals.

The survey electronically developed using the tool Microsoft Form was successfully submitted to a first sample of actors in the health care system.

The first next step will be focused to (1) the enlargement of the sample and to (2) the consequent data-mining with particular consideration to the different subjects categories.

The second next step will be to stimulate the stake-holders on the obtained outcome.

References

1. O'Keeffe, D.T., Maraka, S., Basu, A., Keith-Hynes, P., Kudva, Y.C.: Cybersecurity in artificial pancreas experiments. Diabetes Technol. Ther. **17**(9), 664–666 (2015). https://doi.org/10.1089/dia.2014.0328. Epub 29 Apr 2015
2. Doyle, F.J., Huyett, L.M., Bok Lee, J., Zisser, H.C., Dassau, E.: Closed-loop artificial pancreas systems: engineering the algorithms. Diabetes Care **37**, 1191–1197 (2014)
3. Picton, P.E., Yeung, M., Hamming, N., Desborough, L., Dassau, E., Cafazzo, J.A.: Advancement of the artificial pancreas through the development of interoperability standards. J. Diabetes Sci. Technol. **7**, 1066–1070 (2013)
4. Maisel, W.H., Kohno, T.: Improving the security and privacy of implantable medical devices. New Engl. J. Med. **362**, 1164–1166 (2010)
5. Food and Drug Administration: Content of Premarket Submissions for Management of Cybersecurity in Medical Devices—Guidance for Industry and Food and Drug Administration Staff. www.fda.gov/downloads/MedicalDevices/DeviceRegulationandGuidance/GuidanceDocuments/UCM356190.pdf. Accessed 1 Mar 2015
6. Public Workshop—Collaborative Approaches for Medical Device and Healthcare Cybersecurity, October 21–22, 2014, Arlington VA. www.fda.gov/MedicalDevices/NewsEvents/WorkshopsConferences/ucm412979.htm. Accessed 1 Mar 2015
7. Baranchuk, A., Alexander, B., Campbell, D., Haseeb, S., Redfearn, D., Simpson, C., Glover, B.: Pacemaker cybersecurity. Circulation **138**(12), 1272–1273 (2018). https://doi.org/10.1161/CIRCULATIONAHA.118.035261. PubMed PMID: 29748189
8. Muddy Waters Capital LLC. MW is short St. Jude Medical (STJ:US). Muddy Waters Research (2016). https://d.muddywatersresearch.com/research/stj/mw-is-short-stj/. Accessed 13 Feb 2018
9. Baranchuk, A., Refaat, M.M., Patton, K.K., Chung, M.K., Krishnan, K., Kutyifa, V., Upadhyay, G., Fisher, J.D., Lakkireddy, D.R.: Cybersecurity for cardiac implantable electronic devices: what should you know? J. Am. Coll. Cardiol. **71**, 1284–1288 (2018). https://doi.org/10.1016/j.jacc.2018.01.023. American college of cardiology's electrophysiology section leadership

10. U.S. Food and Drug Administration: Cybersecurity vulnerabilities identified in St. Jude Medical's implantable cardiac devices and Merlin@home transmitter: FDA safety communication. U.S. Food and Drug Administration.2017. https://www.fda.gov/medicaldevices/safety/alertsandnotices/ucm535843.htm. Accessed 13 Feb 2018
11. Kramer, D.B., Fu, K.: Cybersecurity concerns and medical devices: lessons from a pacemaker advisory. JAMA **318**, 2077–2078 (2017). https://doi.org/10.1001/jama.2017.15692
12. Ransford, B., Kramer, D.B., Foo Kune, D., de Medeiros, J.A., Yan, C., Xu, W., Crawford, T., Fu, K.: Cybersecurity and medical devices: a practical guide for cardiac electrophysiologists. Pacing Clin. Electrophysiol. **40**, 913–917 (2017). https://doi.org/10.1111/pace.13102
13. Kruse, C.S., Frederick, B., Jacobson, T., Monticone, D.K.: Cybersecurity in healthcare: a systematic review of modern threats and trends. Technol. Health Care CW Jobs [Internet]. Cyber Crime Timeline, London, UK (2016). http://www.cwjobs.co.uk/careers-advice/itglossary/cyber-crime-timeline. Accessed 09 Aug 2016. (Archived by WebCiteR at http://www.webcitation.org/6je8V9cyI)
14. Coronado, A.J., Wong, T.L.: Healthcare cybersecurity risk management: keys to an effective plan. Biomed. Instrum. Technol. **48**(Suppl), 26–30 (2014). https://doi.org/10.2345/0899-8205-48.s1.26
15. Loughlin, S., Fu, K., Gee, T., Gieras, I., Hoyme, K., Rajagopalan, S.R., et al.: A roundtable discussion: safeguarding information and resources against emerging cybersecurity threats. Biomed. Instrum. Technol. **48**, 8–17 (2014). https://doi.org/10.2345/0899-8205-48.s
16. U.S. Food & Drug Administration: Postmarket Management of Cybersecurity in Medical Devices: Guidance for Industry and FDA Staff, 28 Dec 2016. http://bit.ly/2nOYGB8

ICT4MOMs: An ICT Integrated Approach to Monitor and Manage Pregnancy Development

Maria G. Signorini[1(✉)], Nicolò Pini[1], Danilo Pani[2], and Giovanni Magenes[3]

[1] Dip. di Elettronica, Informazione e Bioingegneria (DEIB), Politecnico di Milano, Piazza Leonardo da Vinci 32, 20133 Milan, Italy
{mariagabriella.signorini,nicolo.pini}@polimi.it
[2] Dip. di Ingegneria Elettrica ed Elettronica, Università di Cagliari, Cagliari, Italy
danilo.pani@unica.it
[3] Department of Electrical, Computer and Biomedical Engineering, University of Pavia, Pavia, Italy
giovanni.magenes@unipv.it

Abstract. ICT4MOMs project aims at implementing a novel remote ICT service towards the monitoring and prediction of maternal and fetal conditions throughout pregnancy. The envisioned application is based on the integration of wearable sensors and devices connected by an ad-hoc smartphone app in communication with an ob-gyn clinical center. Advanced signal and image processing software tools will be developed for extracting information from the recorded signals, namely: fetal heart rate, uterine contractions, continuous glucose sensors and portable US probe. Once validated by the clinical partners, the collected dataset will be used for a multivariate analysis based on soft computing classifiers and machine learning techniques. Based on the growing literature providing evidence on the fact that mother-fetus system should be considered as a whole, in this proposal pregnancy is conceptualized as a continuously evolving system which needs to be investigated by means of time-varying approaches. The crucial expected outcome is the integration of the established clinical knowledge with the results of computational analysis. Such multilevel integration is expected to provide reliable and translatable clinical guidelines towards a novel pregnancy management encompassing a more inclusive monitoring framework designed on a patient-specific level. The project was recently funded by the Italian Government—Progetti di Interesse Nazionale (PRIN) under the grant number 2017RR5EW3 for the duration of three years (2019–2021).

Keywords: Bioengineering · Advanced signal processing · Remote monitoring · ICT service · Fetal medicine · Biosensors · Modelling of biological systems

J. Henriques et al. (Eds.): MEDICON 2019, IFMBE Proceedings 76, pp. 924–929, 2020.
https://doi.org/10.1007/978-3-030-31635-8_112

1 State of the Art

About 480,000 babies were born in Italy in 2016. As in the majority of industrial country, each woman undergoes at least three or more ambulatory monitoring tests throughout pregnancy. The most employed diagnostic examination is the cardiotocography (CTG). CTG provides both fetal heart rate (FHR) estimation by a Doppler US probe, and uterine contraction signal, by a pressure sensor [1]. The total amount of CTG recordings in our country is more than 1 million per year with enormous costs for the national health care system. Despite the non-negligible economic burden, since its introduction in the early 70's, CTG led to a noticeable decrease of fetal morbidities and mortality. For these reasons, it represents the almost unique technology to survey the fetal development by monitoring fetal wellbeing noninvasively and in a quite straightforward way [2]. The growing availability of long-term FHR recordings opened new diagnostic perspectives. In fact, fetal heart beat changes have been reported as a reliable sign for many pathological conditions as well as a sufficient tool to assess wellbeing in a diagnostic perspective [3, 4].

Despite the growing interest in the field of fetal care, in the last few decades, the research has limited the investigation on a single methodological approach that could be the gold standard in discriminating healthy from pathological fetuses. In parallel to signal processing advancement, a better understanding of physiological mechanism impinging fetal heart rate (FHR) control showed how autonomic nervous system (ANS) acts on a very complex way. Thus, researchers started to understand that the problem could not be solved by using a single index, but on the contrary it should make use of various indices, extracted through different methodological approaches, to provide a reliable antenatal diagnosis. A clear example of what above discussed is the work by Magenes et al. [5] focusing on the early identification of Intra Uterine Growth Restricted (IUGR) fetuses. The multivariate model makes use of a small set of parameters and clearly shows how it outperforms accuracy, sensitivity and specificity of each parameter in a univariate approach.

In the last two decades, in parallel to signal processing advancements and supported by the outstanding improvement in ICT, research has been very active in developing wearable devices and systems for medical-oriented applications, often aimed at continuously monitoring patients in their environment and during daily life activities. The rational is inspired by the vision of a sustainable health system which can manage within the same degree of reliability acute care (in hospital or emergency departments), care of outpatients (mainly chronic) but even healthy citizens during their normal life, to the ultimate aim of preventing possible diseases by tele-monitoring [6]. Nowadays, home care is considered a key issue to enforce a better control on a set of chronic diseases, delay the onset of their complications, reduce hospitalizations and save related costs. Recently, wearable systems contributed to a great impulse to remote monitoring given the introduction of low-cost wearable sensors and devices that can be embedded into clothes for tele-monitoring purposes. It was also demonstrated how pregnant women represent a subject category for which wearable tele-monitoring is particularly effective, potentially leading to high savings [7, 8].

2 Project Framework and Implementation

ICT4MOMs aims at integrating existing technologies [7] and medical knowledge [8] with novel data analysis methodologies [5] to monitor and predict mother and fetal conditions during pregnancy with a specific focus on the antepartum period.

Starting from the assumption that a continuous tracking of pregnancy evolution in time is the key characteristic to proper address maternal and fetal wellbeing, this project proposes a novel approach to redefine pregnancy monitoring within the objective to overcome the "acute and sporadic care". To achieve such an innovative design, the effort is aimed at integrating the expertise in prenatal pathophysiology, technology and advance data analysis to develop a novel personalized pregnancy control system. The monitoring architecture will be based on the following pillars: telemedicine, wearable body sensors, integrated systems, data mining, personalized medicine, and predictive analytics as summarized in Fig. 1.

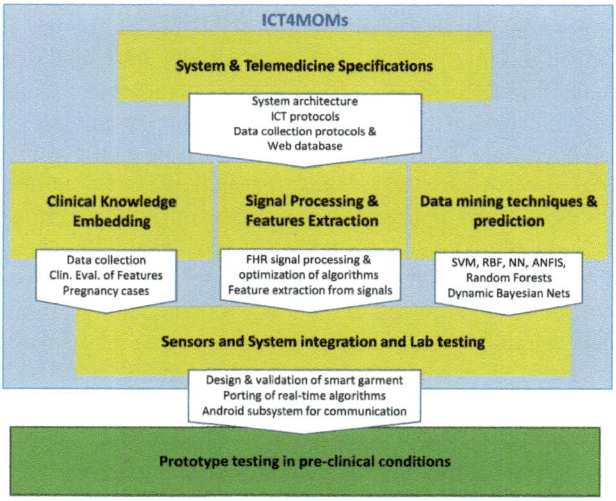

Fig. 1. ICT4MOMs project workflow encompassing data collection, data analysis, technological integration and testing towards the development of a novel framework for maternal and fetal monitoring throughout pregnancy.

Coming to the details about maternal monitoring, this project proposes to record:

- **Vital signs:** heart rate, respiratory rate, activity and position. The alteration of several indicators of maternal wellbeing is expected to provide indirect indication on fetal state and early signs of possible adverse medical condition for both subjects.
- **Lifestyles and dietary habits:** in the last decades growing attention has been focused on the mechanisms by which lifestyles and diet affect pregnancy and how they may affect the fetus later on in life. To this end, mothers are provided with

questionnaires administered by clinicians to collect a wide spectrum of qualitative information.
- **Glucose level**: a considerable number of continuous glucose monitoring (CGM) solutions are available on the market thanks to advancements in manufacturing glucose sensors. Tracking glucose level trends by CGM is expected to provide information about possible pathological conditions such as gestational diabetes.

On the fetal side this proposal aims at integrating the following:

- **Fetal heart rate and uterine contractions:** In 2010, within the framework of project Telefetalcare [7], it has been initiated the design and realization of prototype for a wearable fetal telemonitoring system able to collect FHR by means of a wearable set of sensors. This proposal aims at improving the described prototype and integrating the on-board pre-processing and wireless connectivity.
- **Portable UltraSound:** alongside the signal processing contribution, information integration encompasses image processing. Specifically, the idea is to use portable UltraSound (US) devices such as Lumifiy by Philips which is already intended for general public use.

From a technological perspective, a novel unobtrusive sensing subsystem will be developed on the experience of the TelefetalCare project and the other successful researches on real-time FECG processing and textile electrodes developments [9]. The integration of the above-described data collection and technological solutions is expected to play a crucial role in assessing both maternal and fetal well-being during pregnancy. In this context, machine learning tools such as: time-depend neural networks or recurrent neural networks trained with extended Kalman filter would be employed. The multidimensional dataset will constitute the starting point for the implementation of predictive models capable of providing reliable indications of adverse health outcomes within an acceptable accuracy.

To summarize, the innovative contribution of the described project envisions a platform with the ability to provide valuable real-time clinical-grade information about maternal and fetal wellbeing, contributing to the prevention of adverse events also by interactive coaching, supported by an Android-based m-health system integrating off-the-shelf and novel sensors for unobtrusive tele-monitoring.

3 Expected Impact and Outcome

The project aims at addressing pregnancy monitoring encompassing different disciplines, disseminating results and advancements in the scientific, technological, social and economic fields. The envisioned application is to provide mothers with an integrated telemedicine service focused on pregnancy, which makes use of a smartphone and a low cost, compact, wearable and easy-to-use device. The global market for wearables in healthcare is expected to reach a revenue of $18.9 Billion in 2020. Specifically, Regulated/Clinical-Grade Wearables is the most promising product segment within healthcare wearables. Another adjacent market that can be considered for estimating our potential global market size is Prenatal Products.

Given these considerations, the potential market in the context of this project is very broad. Furthermore, this project is addressed to pregnant women as a whole, not narrowing the focus to a specific target, thus increasing the probability to penetrate into the market. The proposed data integration has the potentialities to dramatically increase the scientific knowledge on regulatory mechanism during pregnancies being either physiological or pathological ones. The proposed design of closely monitoring mothers during pregnancy is expected to provide significant advancement in targeting early signs of pathology and reduce adverse outcomes by timely intervention. At the same time, even the understanding of specific mechanisms and scientific research per se will receive valuable drive.

The telemedicine infrastructure conceived in this proposal, is expected to produce two main results:

(1) a prototype model of a remote monitoring health service during pregnancy which includes suggestions for the lifestyle and prediction of possible adverse outcomes;
(2) a conceptual framework for developing general home care prevention services, following the increasing health care policy of prevention. Moreover, it will provide a further example that wearable devices, by themselves, are not sufficient for a correct management of patients/citizens, and an ad-hoc ICT infrastructure must be carefully designed and organized.

Lastly, an additional category of subjects which would benefit from the proposed framework is risk pregnancies. The latter are usually resulting in preterm birth usually generate unpleasant effects on both mothers and babies leading to an overall pregnancy- and childbirth-related hospital expenses. Thus, more efficient remote monitoring technologies as the ones proposed by ICT4MOMs are expected to provide solutions to overcome several pregnancy risk factors and improve pregnancy management.

References

1. ACOG Technical Bulletin: Fetal heart rate patterns: monitoring, interpretation, and management. Int. J. Gynecol. Obstet. **51**(1), 65–74 (1995)
2. Van Geijn, H.P.: Developments in CTG analysis. Baillieres Clin. Obstet. Gynaecol. **10**(2), 185–209 (1996)
3. Mantel, R., et al.: Computer analysis of antepartum fetal heart rate: 1. Baseline determination. Int. J. Biomed. Comput. **25**(4), 261–272 (1990)
4. Mantel, R., et al.: Computer analysis of antepartum fetal heart rate: 2. Detection of accelerations and decelerations. Int. J. Biomed. Comput. **25**(4), 273–286 (1990)
5. Magenes, G., et al.: Comparison of data mining techniques applied to fetal heart rate parameters for the early identification of IUGR fetuses. In: 2016 38th Annual International Conference of the IEEE Engineering in Medicine and Biology Society (EMBC), pp 916–919. IEEE (2016)
6. Gaikwad, R., et al.: The role of home-based information and communications technology interventions in chronic disease management: a systematic literature review. Health Inform. J. **15**(2), 122–146 (2009)

7. Fanelli, A., et al.: Telefetalcare: a first prototype of a wearable fetal electrocardiograph. In: Proceedings of Annual International Conference of the IEEE Engineering in Medicine and Biology Society, EMBS, pp. 6899–6902 (2011)
8. Di Lieto, A., et al.: Four years' experience with antepartum cardiotocography using telemedicine. J. Telemed. Telec. **12**(5), 228–233 (2006)
9. Pani, D., et al.: NInFEA: an embedded framework for the real-time evaluation of fetal ECG extraction algorithms. Biomed. Tech. Eng. **58**(1), 13–26 (2013)

Regular Sessions: Assistive Technologies

Smart Shirt for Uncontrolled Movement Retraining

Peteris Eizentals[1(✉)], Alexei Katashev[1], Alexander Oks[1], and Guna Semjonova[2]

[1] Riga Technical University, Riga LV1658, Latvia
peteris.eizentals@gmail.com
[2] Riga Stradins University, Riga LV1007, Latvia

Abstract. Uncontrolled movement is a major cause of musculoskeletal pain and pathology. Retraining has proven to be an effective method for movement control improvement but it is a time-consuming process that requires much attention from a physiotherapist. The efficiency of movement retraining can be increased by independent work of a patient but it can prove difficult or potentially harmful if the patient lacks understanding of the correct movement or cannot verify the accuracy of the movement. This paper introduces a smart shirt system for movement control retraining assistance. The smart shirt system consists of 11 textile stretch sensors attached to a compression shirt for movement monitoring. To verify the ability of the shirt to assist uncontrolled movement retraining, a set of 12 tests were selected from the Kinetic Control method, and measurements were obtained both for correct and incorrect movements. The acquired measurements were analyzed to determine signals from sensors that would give the most significant information for each test depending on the involved anatomy. It was confirmed that the smart shirt system could be employed for evaluation of the test performance.

Keywords: Uncontrolled movement · Smart textile · Smart shirt

1 Introduction

Uncontrolled movement (UCM) is a musculoskeletal motor control dysfunction, where a motion of the body is performed without complete control over the involved muscles. Lack of muscle control during motion can induce abnormal range or translation of the motion, which in turn leads to increased stress and strain in the involved anatomy [1]. There is a growing amount of evidence linking UCM to musculoskeletal pain and pathology [2–4]. Kinetic Control is a set of tests developed for diagnoses, assessment, and retraining of the most common UCM [1]. Retraining of specific movements has been shown to be an effective method for improvement of movement control [5, 6]. However, UCM retraining is a time-consuming task that requires considerable effort from the side of a patient, as certain tasks have to be performed repeatedly many times in a defined way. Naturally, it is not feasible for a physiotherapist to provide constant supervision thus self-monitoring from the patient must be used for most of the course after the initial movement teaching provided by the physiotherapist. The efficiency of

J. Henriques et al. (Eds.): MEDICON 2019, IFMBE Proceedings 76, pp. 933–937, 2020.
https://doi.org/10.1007/978-3-030-31635-8_113

retraining is directly dependent on the quality at which the patient is able to perform the exercises, therefore there is a necessity for a simple, user-friendly uncontrolled movement retraining assistance method that could be used under home conditions or in a clinic.

This paper proposes a novel method for UCM evaluation and retraining assistance by using a smart textile shirt. A simple experiment was performed to evaluate the feasibility of the proposed method, where 12 tests were selected by a physiotherapist and a person was instructed to performed each of the selected tests both correctly and incorrectly (according to the physiotherapist). The measured signals from all sensors were then analyzed to determine if the correct movements can be distinguished from the incorrect. For each of the performed tests at least one sensor was determined, which had a significant change of the signal when a movement was performed incorrectly.

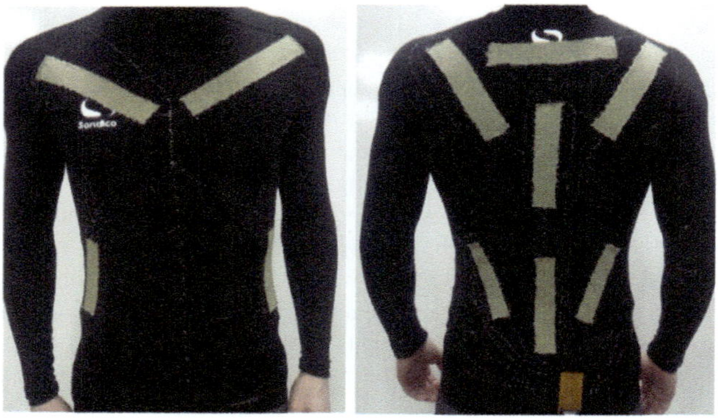

Fig. 1. Smart shirt front (left) and back (right)

2 Materials and Methods

2.1 Smart Shirt

The smart shirt was produced by attaching knitted piezoresistive textile stretch sensors [7] to a commercial long sleeve compression shirt (Fig. 1). The shirt and consequently the sensors are stretched or contracted due to movements of the body, inducing a measurable change in sensor electrical resistivity even for relatively small movements. In total 11 sensors were attached to the shirt at following anatomic sites: both sides, over each pectoral muscle, on lumbar vertebra, on thoracic vertebra, between scapulae, on the lower back, and on the upper back over scapulae. Each position was chosen with a particular movement in mind, e.g. the sensor over lumbar and thoracic vertebra should be able to monitor the spine flexion at the lumbar and thoracic areas, while pectoral sensors should monitor the scapular protraction and retraction. The resistance of each sensor was measured by a microcontroller at the speed of 10 Hz and sent to the computer, where it was displayed in real time for the user in the form of a graph.

2.2 Experimental Overview

The Kinetic Control method described in [1] consists of 82 tests for all major uncontrolled movement evaluation and retraining. An experienced physiotherapist selected 12 out of 82 tests that cover all of the available uncontrolled movement regions for the shirt evaluation test. A volunteer (male, 31 years old, 75 kg, 181 cm, no history of skeletomuscular complications) was then instructed to perform each of the selected tests several times according to the directions of the physiotherapist both in the correct and an incorrect way. The smart shirt measurement was acquired and analyzed to verify that the correct execution of the tests could be distinguished from the incorrect. The selected tests were (using the original nomenclature) T7 Ischial weight bearing test for lumbar flexion UCM; T10 Chest lift test for lumbar extension UCM; T19 Top leg turn out test for lumbopelvic rotation UCM; T41 Lower neck lean test for rotation and side-bend UCM; T45 Bilateral forward reach test for thoracic flexion UCM; T46 Bilateral overhead reach test for thoracic extension UCM; T51 Pelvic twist test and T54 One arm lift test for thoracic rotation UCM; T62 Arm flexion test for scapula and glenohumeral UCM; T65 Vertical trunk single leg ¼ squat test for hip flexion UCM; T71 Single knee lift and extension test for hip extension UCM, and T75 Single leg high knee lift test for hip lateral rotation/abduction UCM [1].

3 Results and Discussion

The performed exercises can be sorted by the target anatomy in following groups: lumbar spine, thoracic spine, scapulae and hips. The target anatomy in each test had to remain stable in a certain position while test movement was performed in full range. An anatomy was defined stable if the corresponding sensor would register no significant electrical resistance change during the movement. This was verified by comparing the measurement for both correctly and incorrectly performed exercises, which was assessed by the physiotherapist. Due to the space restrictions, only a very short result description is given for each test, and only one example is given here graphically (see Fig. 2).

During a correct movement of the *ischial weight bearing* test and the *chest lift test* no change in lumbar spine and lower back sensor measurement was observed, however, when the movement was performed incorrectly, the lumbar region movement was present in the lumbar spine and lower back sensor reading.

The correct movement of the *lower neck lean test* produced no significant change in any sensor reading as the body was stationary while peaks in the scapular and pectoral sensor reading was produced by the incorrect movement.

The *bilateral forward reach test*, when performed correctly, produced no significant change in the thoracic spine sensor measurement, while it could be seen in the incorrect movement measurement.

It was noted that during the *bilateral overhead reach test* the thoracic spine sensor itself was too sensitive to arm movement and therefore was not applicable for thoracic spine movement measurement. However, incorrect test performance was clearly visible in the lumbar spine measurement. This indicates that interpretation of sensor data during arm movements is complicated as due to any movement the shirt might be stretched in sites that are not directly related to the movement itself. Therefore, great care must be taken when interpreting such data.

The *pelvic twist test*, when performed correctly, produced the most change in signal for the lower back sensor opposite to the movement direction, while for the incorrect movement the pectoral and scapular sensors clearly showed the rotation of the thorax but the signal for lower back decreased.

A quite significant difference between correct and incorrect movement of the *one arm lift test* could be seen for opposite scapular and thoracic spine sensors while the opposite pectoral sensor shows only a slightly increased change for each arm lift.

When comparing the measurement of the *arm flexion test* for correct and incorrect movement, the most significant difference in measurement was observed for scapular and side sensors corresponding to the side of the lifted arm, and the thoracic spine sensor.

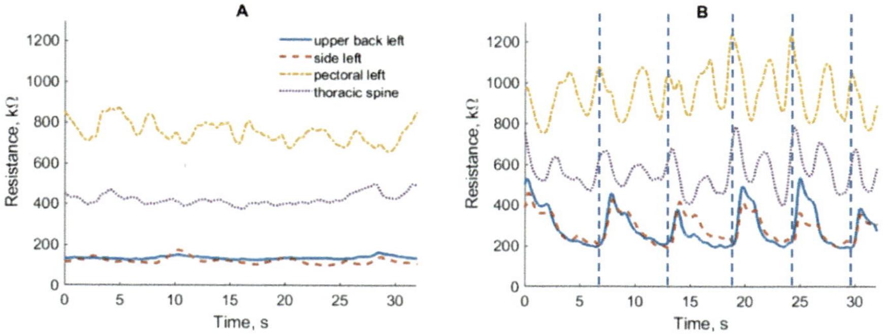

Fig. 2. T62 Arm flexion test for scapula and glenohumeral UCM, performed correctly (A) and incorrectly (B). Vertical lines in the figure B approximately correspond to the moment of lifting the left arm up. In both images lifting of the arm tugs on the pectoral sensor, but there should be no movement in the other regions, which can be seen in figure A.

The *vertical trunk single leg ¼ squat test*, when performed incorrectly, produced quite dominant spikes for the measurement of the spine and lower back sensors, while during the correct movement the resistance of these sensors had almost no change.

According to the measurement data, when the *single knee lift and extension test* was performed correctly, no significant change in any sensor was recorded.

The *single leg high knee lift test*, when the motion is performed correctly, produced almost no change in the lower back and side sensor signal, while the hip motion during the incorrect motion could be clearly seen.

4 Conclusions

This paper described a novel method for uncontrolled movement retraining support by a smart shirt. According to this method, a smart shirt system is used to monitor the movement of involved anatomy while a user is performing a test from Kinetic Control method, which enables the user to do the retraining tasks without the supervision of a physiotherapist and therefore improve the efficiency of retraining. The benefits of this system is the relative ease of use as there is no need to attach sensors to the body.

As the future work, all of the Kinetic Control tests will be analyzed in the same manner and clinical tests will be performed to assess the contribution of the system for uncontrolled movement retraining. The results from the present tests will be used to improve the sensor positions and orientations.

Acknowledgment. This work has been supported by the European Regional Development Fund within the Activity 1.1.1.2 "Post-doctoral Research Aid" of the Specific Aid Objective 1.1.1 "To increase the research and innovative capacity of scientific institutions of Latvia and the ability to attract external financing, investing in human resources and infrastructure" of the Operational Programme "Growth and Employment" (No. 1.1.1.2/VIAA/1/16/153).

Conflict of Interests. Authors declare no conflict of interests.

References

1. Comerford, M., Mottram, S.: Kinetic Control: The Management of Uncontrolled Movement. Elsevier, Amsterdam (2012)
2. Sahrmann, S.A.: Diagnosis and Treatment of Movement Impairment Syndromes. Mosby, St Louis (2002)
3. van Dillen, L.R., Maluf, K.S., Sahrmann, S.A.: Further examination of modifying patient-preferred movement and alignment strategies in patients with low back pain during symptomatic tests. Man. Ther. **14**(1), 52–60 (2009)
4. Dankaerts, W., O'Sullivan, P., Burnett, A., Straker, L., Davey, P., Gupta, R.: Discriminating healthy controls and two clinical subgroups of nonspecific chronic low back pain patients using trunk muscle activation and lumbosacral kinematics of postures and movements: a statistical classification model. Spine **34**(15), 1610–1618 (2009)
5. Jull, G.A., Falla, D., Vicenzino, B., Hodges, P.W.: The effect of therapeutic exercise on activation of the deep cervical flexor muscles in people with chronic neck pain. Man. Ther. **14**(6), 696–701 (2009)
6. Roussel, N.A., Daenen, L., Vissers, D., Lambeets, D., Schutt, A., Van Moorsel, A., et al.: Motor control and physical fitness training to prevent musculoskeletal injuries in professional dancers. In: 3rd International Conference on Movement Dysfunction, 30 October–1 November, Edinburgh, UK (2009). Man. Ther. **14**(1), S22
7. Oks, A., Katashev, A., Litvak, J.: Knitted resistive fabric: properties and applications. Mater. Sci. Text. Cloth. Technol. **9**, 28 (2015). https://doi.org/10.7250/mstct.2014.005

Computational Fluid Dynamics of Blood Flow at the Left Atrium and Left Atrium Appendage

Grigoris I. Grigoriadis[1], Antonis I. Sakellarios[2], Katerina Naka[3], Ioanna Kosmidou[4], Christopher Ellis[5], Lampros K. Michalis[3], and Dimitrios I. Fotiadis[1,2](✉)

[1] Unit of Medical Technology and Intelligent Information Systems, Department of Materials Science and Engineering, University of Ioannina, GR 45110 Ioannina, Greece
greg8grigoriadis@gmail.com, fotiadis@cc.uoi.gr
[2] Department of Biomedical Research, FORTH-IMBB, GR 45110 Ioannina, Greece
[3] Michaelideion Cardiac Center, Department of Cardiology in Medical School, University of Ioannina, GR 45110 Ioannina, Greece
{anaka, lmihalis}@cc.uoi.gr
[4] Columbia University Medical Center, New York, NY, USA
[5] Vanderbilt University Medical Center, Nashville, TN, USA

Abstract. In this study, we analyze local hemodynamics in the Left Atrium (LA) and Left Atrial Appendage (LAA) using Computational Fluid Dynamics (CFD). Our scope is to describe a novel application of CFD to analyze/describe the hemodynamic behavior at the LA and LAA according to different LAA anatomic morphologies and fluid velocities. 3D semi-automated reconstruction approach used to segment and reconstruct left atrium geometries and obtain the computational domain. Cardiac computed tomography scans (CTs) used from 3 different patients. Calculated velocities obtained from simulations varied significantly according to LAA morphology as follows: broccoli (0.1328 m/s), chicken wing (0.2468 m/s) and windsock (0.1268 m/s). We calculated as well the wall shear stress (WSS) at left atrium chamber as mean values 3 Pa, 5 Pa, 4 Pa for the patients, respectively.

Keywords: Left Atrium (LA) · Left Atrial Appendage (LAA) · Blood flow modelling · Cardiovascular modeling · CFD · Appendage morphologies

1 Introduction

The left atrium (LA) and left atrial appendage (LAA) are interlinked cardiac structures constituting the most significant source of cardioembolic events, predominantly in patients with atrial fibrillation but also mitral valve disease and left ventricular dysfunction [1]. The LA is a highly complex anatomic structure given multiple inlets at the insertion of the pulmonary veins (PV), single outlet at the mitral valve (MV) orifice and the presence of the LAA, a blind-ended structure on the anterior-lateral wall of the LA whose anatomic form is highly variable. The significance of the LA and LAA in several

J. Henriques et al. (Eds.): MEDICON 2019, IFMBE Proceedings 76, pp. 938–946, 2020.
https://doi.org/10.1007/978-3-030-31635-8_114

clinical conditions, such as atrial fibrillation, valvular disease is critical. Three-dimensional (3D) reconstruction of the LA and surrounding structures with combination of CFD modeling, it may provide further insight into the pathophysiologic process. Also, with this approach we can obtain information about the function or dysfunction of the heart.

Morphologies of the LAA are classified into four major categories: the "Chicken wing" (48%), the "Cactus" (30%), the "Windsock" (19%), and the "Cauliflower" (3%) [2]. According to a recent study [3], the risk of stroke is 54% lower in patients with chicken wing morphology compared to other anatomic variants.

Mathematical and numerical modelling of the cardiovascular system has attracted remarkable interest from the research community due to its high potential impact on physiology, risk stratification and patient outcomes. However, there is limited data on numerical hemodynamics in the LA and LAA as previous studies have shown [4–6].

In this work, employing computational fluid dynamics we are investigating the impact of anatomic morphology of local hemodynamics and wall shear stress.

2 Materials and Methods

2.1 Dataset

The dataset consists of cardiac computed tomography (CT) scans from 3 male patients (SIEMENS CT VA1 DUMMYCT). Each set consists of 300 slices at 512 × 512 pixel. The slice thickness is 0.75 mm and the pixel spacing is 0.4 mm for each volume. Transesophageal Echocardiography (TEE) data for LA, LAA and PVs were further used for setting up boundary conditions and model validation. A representative CT frame is shown in Fig. 1.

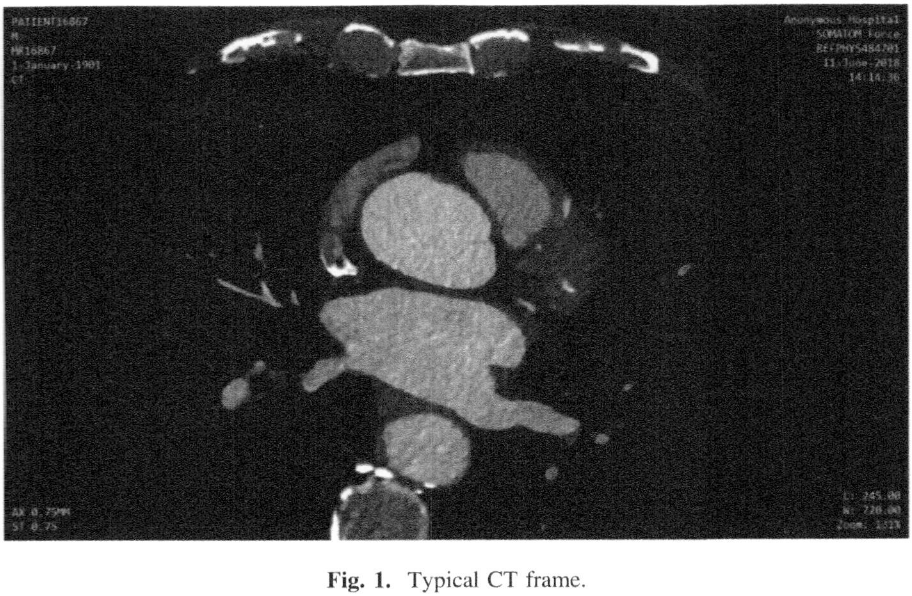

Fig. 1. Typical CT frame.

2.2 Reconstruction

The LA consists of two cavities: the LA body, with blood inflow through the pulmonary veins and outflow at the mitral valve, and the LAA. The PVs are four separate vessels with an average diameter 10 mm. Their names come from their position as is indicated in Fig. 2: right and left superior (RSPV, LSPV), right and left inferior (RIPV, LIPV) [7]. The LA is interconnected with the left ventricle by the mitral valve, through which blood flows forward from the LA to the LV. In Table 1. Geometries data. we summarize the patient's specific geometry data. This volume values indicate patients with the existence of AF conditions. The calculated volumes correspond to volumes found in literature [4]. These large volumes indicate AF conditions.

Semi-automatic segmentation is carried out, and validated by experts. A contrast enhancement method was used according to [8]. A 3D region growing method is applied to fill and interpolate the segmented 300 CT slices and obtain the 3D anatomical model [9].

The left atrium models consist of the main atrial cavity, the 4 pulmonary veins and the left atrial appendage. Particular attention was paid to the left atrial appendage, to achieve as much accuracy as it can be obtained by our approach. An example of the 3D reconstructed geometries in a posterior and anterior view is presented in Fig. 2.

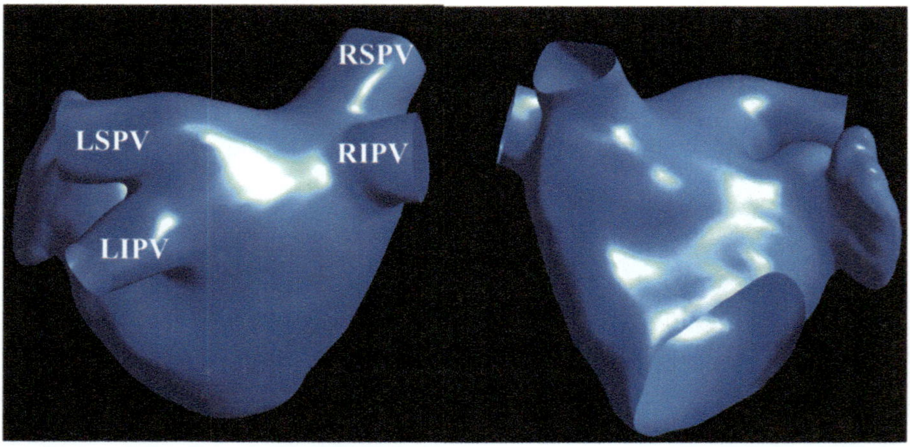

Fig. 2. Reconstructed Left Atrium (LA) 3D model. Pulmonary veins demonstrated by name.

Table 1. Geometries data.

Patients	LA volume (ml)	Heart rate (BPM)	Orifice area (cm^2)
Patient 1	164.5	85	5.0427
Patient 2	144.93	70	5.9409
Patient 3	124.29	87	3.9021

2.3 Mesh Sensitivity Analysis

The three reconstructed models were discretized into tetrahedral elements. The WSS values were used to check the mesh convergence. A sensitivity analysis and error analysis has been conducted. The final models we generated are discretized with 10,350,012 elements for patient 1, 10,226,918 for patient 2 and 8,672,982 for patient 3.

2.4 Flow Measurements

The blood flow velocity is estimated using Transesophageal Echocardiography (TEE) data Fig. 3. The extracted velocities are time dependent, so transient simulations must be performed. We obtain specific velocities at measured points such as the PVs and LAA. First, the values at the PV's are applied as inlet boundary conditions and a fully cardiac cycle profile is employed. The other values are used for validation.

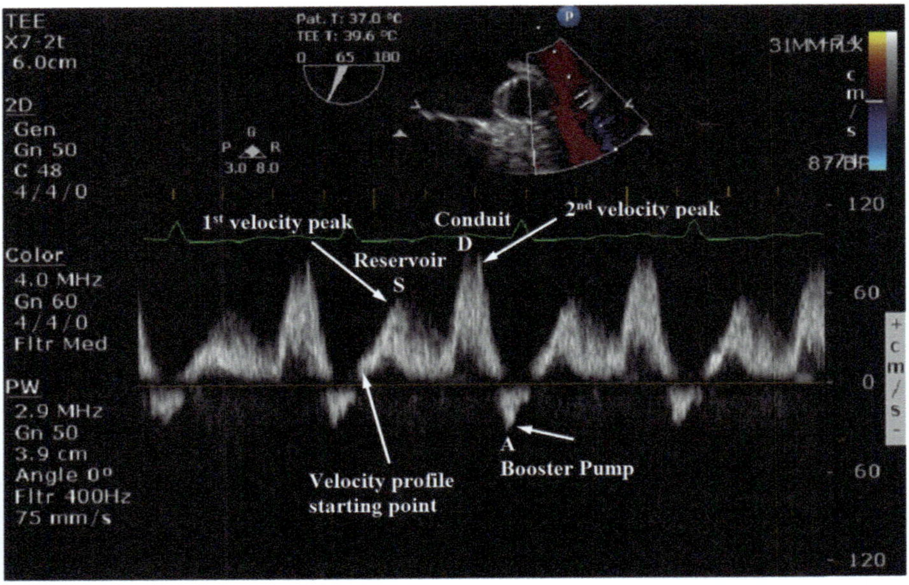

Fig. 3. Pulmonary Veins Transesophageal Echocardiography. Arrows indicating the phases of flow at pulmonary veins and its maximum values.

2.5 Blood Flow Simulations

Steady state simulations were executed to obtain initial flow conditions which are then applied in the transient simulations. We assumed that the atrium walls are rigid smooth walls. The solver we used, is based on a hybrid Finite Element (FE) - Finite Volume (FV) approach, to discretize Navier-Stokes equations. Modelling the blood flow is carried out using the Navier-Stokes and the continuity equations:

$$\rho \frac{\partial v}{\partial r} + \rho(v \times \nabla)v - \nabla \times \tau = 0, \tag{1}$$

$$\nabla \times (\rho v) = 0, \tag{2}$$

where v is the blood velocity and τ is the stress tensor, defined as:

$$\tau = -p\delta_{ij} + 2\mu\varepsilon_{ij}, \tag{3}$$

where δ_{ij} is the Kronecker delta, μ is the blood dynamic viscosity, ρ is the blood pressure and ε_{ij} is the strain tensor calculated as:

$$\varepsilon_{ij} = \frac{1}{2}(\nabla \mathrm{v} + \nabla v^T). \tag{4}$$

Blood was assumed as incompressible Newtonian fluid with density ρ = 1060 kg/m^3 and dynamic viscosity μ = 0.0035 Pa s [10].

For Newtonian fluids the shear stress is determined according to Newton's law by the equation:

$$\tau = \mu \frac{du}{dy}, \tag{5}$$

where μ is the kinetic viscosity, u is the velocity of the fluid, y is the distance from the surface and $\frac{du}{dy}$ the velocity gradient known as the shear rate $\dot{\gamma}$.

We assumed flow is transitional, i.e. the flow is at an intermediate condition between laminar and turbulent. We calculated the Reynolds number to be Re = 2683,484 based on the PVs diameter d = 0.0115 m and an average velocity value from the inlet profile, assuming that density and dynamic viscosity remain the same as we mentioned above. This Reynolds number indicates transitional flow conditions. For that purposes we used a k-epsilon model to solve the motion of the blood flow into the left atrium. The k-epsilon model is a protrusive turbulence model and it has been proven to be stable and robust. In this study CFD software *ANSYS® Academic CFX, Release 15.7, Help System, ANSYS, Inc.*, was employed to model the blood flow in left atrium.

2.6 Boundary Conditions

For the steady state simulation, 0.1 m/s used as inlet velocity value where the filling of atrium starts. A time dependent uniform velocity profile is applied at the four pulmonary veins (PVs). The total simulation time was three cardiac cycles of about 0.8 s each (depending on patient's heart rate) and every cycle was divided in 30 timesteps. An opening outlet condition with total pressure 5 kPa as a mean value of MV at AF conditions was chosen according to [11] and expert cardiologists guidance, to allow a small amount of natural blood backflow, which occurs due to atrial dilation [12]. A no-slip, no-penetration condition was imposed at the rigid walls.

3 Results

The results are produced at the peak of pulmonary venous systolic velocity (S) at the 3rd cardiac cycle where the solution is fully converged. The WSS at different areas in the LA is shown in Fig. 7. The low velocities and the strange morphology of the appendage make the conditions suitable for blood stagnation [3, 13]. The flow inside the LAA it's hard to pass through its blinded-end due to its anatomical complexity and we can observe it at the velocity streamlines at Fig. 5. This varies among unique LAA morphologies and most of them make flow conditions difficult. The calculated velocities at LAA's orifice and tip are presented at the following Table 2. At Fig. 4.

Table 2. Computed velocities and wall shear stress.

Results	Patient 1	Patient 2	Patient 3
LAA morphology	Cauliflower	Chicken wing	Windsock
TEE recordings	0.15	0.4	0.3
LAA velocity orifice (m/s)	0.1328	0.2468	0.1268
LAA velocity tip (m/s)	0.0030	0.0217	0.0012
WSS (Pa)	2.7470	5.5073	4.2487

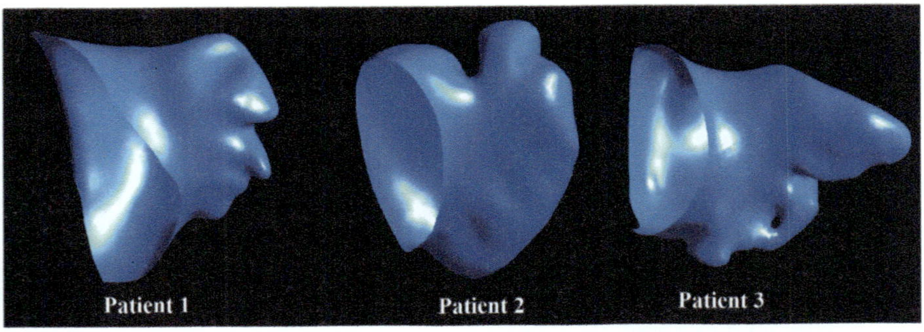

Fig. 4. Patients left atrial appendage 3D morphology.

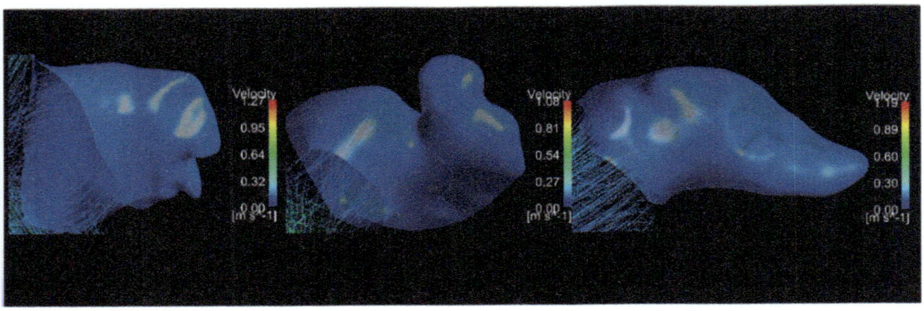

Fig. 5. Velocity stream lines at appendages. Patients 1, 2, 3 respectively.

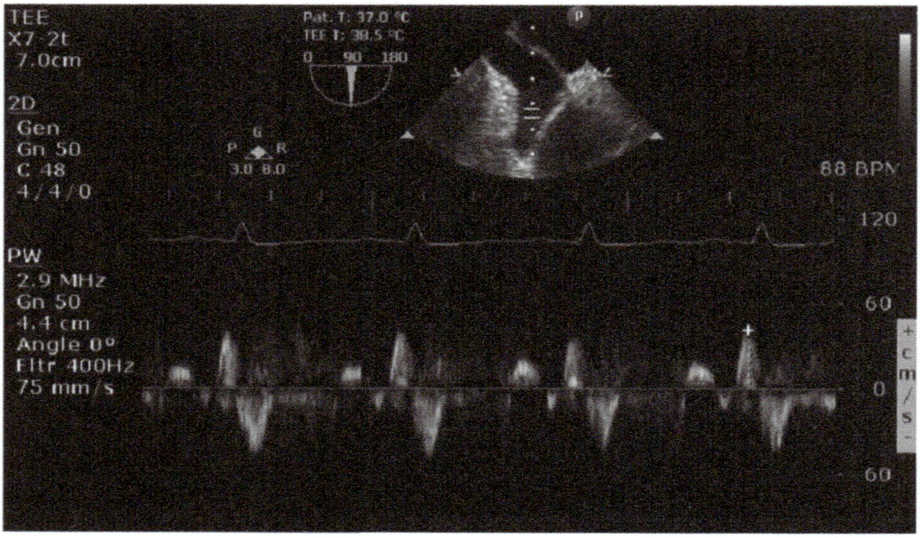

Fig. 6. Left Atrial Appendage velocity measured by PW doppler at a specific point into the LAA.

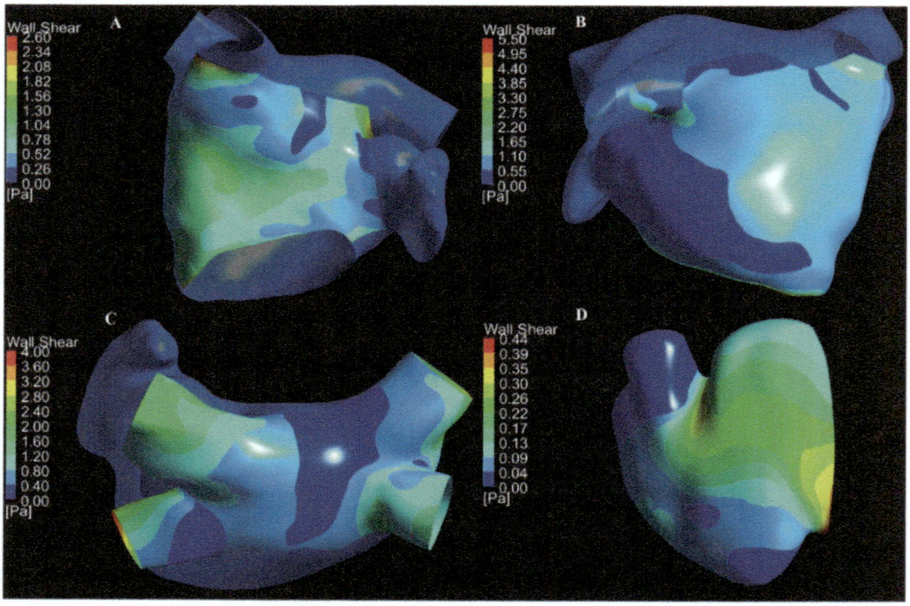

Fig. 7. Patient 2 WSS at: (A, C) Posterior LA, LAA (B, D) Anterior LA, LAA.

appendage morphologies for our three patients are presented. The blood flow has lower velocities at the tip of the LAA indicating that this area is proper for blood stagnation and thrombosis. Furthermore, this low allows the blood to pool and this can cause clots, which they may travel to brain. Moreover, the continuous changes in the blood velocities indicate intermediate changes at the wall shear stresses. This continuous changes of WSS is critical for thrombus detachment and a stroke event.

4 Discussion and Conclusions

CFD results indicated that both velocity and WSS are decreased along the LAA, from the orifice to the tip, making the LAA tip more prone to fluid stagnation and to thrombus formation. Our computational models aimed to highlight the velocity findings of LAA. The morphology of the LAA is a critical factor which is direct connected with velocity values. There is not significant difference with findings from similar clinical studies done by medical doctors [13, 14] considering the limitations of this study. Moreover, our findings are in plausible range from the TEE velocities as we can observe at Fig. 6. The wall shear stress is actually the endothelial shear stress and it is one of the most important hemodynamic factors. Low WSS is a marker as well as cause of endothelial dysfunction, and increases local thrombogenicity [15]. Thus, simulated data can be used to investigate flow paterns and clinicaly important mesurements such as wall shear stress.

In the current study, we demonstrate an approach to computation of blood flow velocities and wall shear stresses at the LA and LAA. Our approach examines the blood flow behavior at LA but also the impact of the LAA morphology on blood stagnation and flow. The ups and downs of the blood flow and its unsteady profile, have set the velocity as a critical factor for the LA and LAA functionality. Thrombus formation and detachment is a critical risk that is directly connected with the flow velocity.

5 Limitations

This study has several limitations. First, we have simulated the blood flow in such a way where the motion of fluid is caused by implementing velocity values as inlets. Instead of this approach, a more realistic way to model heart function is to determine the wall motion, as a physiological heart does. Second, we have not specified the state of the mitral valve (opened/closed). The validation process was difficult, because we only know the 2D point from TEE instead of the 3D, which makes difficult to specify the z direction at our model. These limitations make our velocity findings to be different but still in plausible ranges.

References

1. Al-Saady, N.M., Obel, O.A., Camm, A.J.: Left atrial appendage: structure, function, and role in thromboembolism. Heart **82**(5), 547–554 (1999)

2. Di Biase, L., et al.: Does the left atrial appendage morphology correlate with the risk of stroke in patients with atrial fibrillation? J. Am. Coll. Cardiol. **60**(6), 531–538 (2012)
3. Lupercio, F., et al.: Left atrial appendage morphology assessment for risk stratification of embolic stroke in patients with atrial fibrillation: a meta-analysis. Heart Rhythm **13**(7), 1402–1409 (2016)
4. Markl, M., et al.: Left atrial and left atrial appendage 4D blood flow dynamics in atrial fibrillation. J. Cardiovasc. Magn. Reson. **18**(Suppl 1), O90 (2016)
5. Bosi, G.M., et al.: Computational fluid dynamic analysis of the left atrial appendage to predict thrombosis risk. Front. Cardiovasc. Med. **5**, 34–38 (2018)
6. Otani, T., Al-Issa, A., Pourmorteza, A., McVeigh, E.R., Wada, S., Ashikaga, H.: A computational framework for personalized blood flow analysis in the human left atrium. Ann. Biomed. Eng. **44**(11), 3284–3294 (2016)
7. Kim, Y.-H., Marom, E.M., Herndon, J.E., McAdams, H.P.: Pulmonary vein diameter, cross-sectional area, and shape: CT analysis. Radiology **235**(1), 43–49 (2005)
8. Valinoti, M., Fabbri, C., Turco, D., Mantovan, R., Pasini, A., Corsi, C.: 3D patient-specific models for left atrium characterization to support ablation in atrial fibrillation patients. Magn. Reson. Imaging **45**, 51–57 (2018)
9. Jin, C., et al.: Left atrial appendage segmentation and quantitative assisted diagnosis of atrial fibrillation based on fusion of temporal-spatial information. Comput. Biol. Med. **96**, 52–68 (2018)
10. Ku, D.N.: Blood flow in arteries. Ann. Rev. Fluid Mech. **29**(1), 399–434 (1997)
11. Rim, Y., Choi, A., McPherson, D.D., Kim, H.: Personalized computational modeling of mitral valve prolapse: virtual leaflet resection. PLoS ONE **10**(6), e0130906 (2015)
12. Kim, T.-S., Youn, H.-J.: Role of echocardiography in atrial fibrillation. J. Cardiovasc. Ultrasound **19**(2), 51 (2011)
13. Lee, J.M., et al.: Why is left atrial appendage morphology related to strokes? an analysis of the flow velocity and orifice size of the left atrial appendage: why is left atrial appendage morphology related to strokes? J. Cardiovasc. Electrophysiol. **26**(9), 922–927 (2015)
14. Handke, M., Harloff, A., Hetzel, A., Olschewski, M., Bode, C., Geibel, A.: Left atrial appendage flow velocity as a quantitative surrogate parameter for thromboembolic risk: determinants and relationship to spontaneous echocontrast and thrombus formation–a transesophageal echocardiographic study in 500 patients with cerebral ischemia. J. Am. Soc. Echocardiogr. **18**(12), 1366–1372 (2005)
15. Sakariassen, K.S., Orning, L., Turitto, V.T.: The impact of blood shear rate on arterial thrombus formation. Future Sci. OA **1**(4) (2015)

Powered Wheelchair Impact – User-Centered Observational Study

Inês Domingues[1](✉), João Pinheiro[1], João Silveira[1], and Anabela Correia Martins[2]

[1] Faculty of Medicine, University of Coimbra, 3046-854 Coimbra, Portugal
inesfsdomingues@gmail.com, reabmedica@hotmail.com
[2] Physiotherapy Department, IPC ESTeSC Coimbra Health School, Coimbra, Portugal
anabelacmartins@estescoimbra.pt

Abstract. There is a growing prevalence of disability worldwide, which indicates an increasing number of persons who might benefit from assistive technologies. The purpose of this observational, descriptive, cross-sectional study is to assess the psychosocial and participation impact of powered wheelchairs (PW). From May to October 2017, 30 powered wheelchair users were interviewed using the Quebec User Evaluation of Satisfaction with Assistive Technology (QUEST) scale, the Psychosocial Impact of Assistive Devices Scale (PIADS) and the Activities and Participation Profile Related to Mobility (PAPM) scale, in addition to some demographic, clinical and powered wheelchair related questions. Descriptive and correlational statistics were performed to analyze the data. The participants were satisfied with both AT and related services, with the lowest QUEST scores belonging to those who had been using their wheelchairs for a longer period of time. PAPM scores revealed significant restrictions in participation mostly among users with longer wheelchair utilization. The most satisfied were the ones with better performance in terms of social participation. The psychosocial impact, in terms of adaptability, was higher among users who transitioned from a manual to a PW compared to those who already had a PW. There was an overall positive psychosocial impact of the powered wheelchairs, and, potentially, an increase in the quality of life of the users. There's a lack of research in this area, specifically to evaluate the impact of the environmental barriers on the social participation and on the quality of life of powered wheelchair users.

Keywords: Assistive technologies · Powered Wheelchair · Psychosocial impact · Social participation

1 Introduction

Disability is an umbrella term for impairments, activity limitations or participation restrictions [1]. More than one billion people in the world live with some form of disability and its prevalence is steadily growing [2].

Participation means engaging with the activities of daily living and it is well recognized, especially after implementation of the International Classification of Functioning,

J. Henriques et al. (Eds.): MEDICON 2019, IFMBE Proceedings 76, pp. 947–953, 2020.
https://doi.org/10.1007/978-3-030-31635-8_115

Disability and Health (ICF), that the end goal of rehabilitation is to preserve or improve participation [3].

Assistive technologies (AT) are powerful tools to increase independence and improve participation [4]. An AT device can be defined as "any item, piece of equipment, or product, whether it is acquired commercially, modified, or customized, that is used to increase, maintain, or improve the functional capabilities of individuals with disabilities" [2].

Several studies showed the benefits of the AT, particularly if they contribute to enhance participation of adults with mobility problems, as well as on psychosocial condition [5, 6]. Powered Wheelchairs (PW) in particular seem to improve users' daily routines, ability to engage in mobility related activities and social participation, and increasing independence, safety and self-esteem of their users [7]. PW will improve autonomy in activities of daily living of the persons with disability [8, 9]. Moreover, it seems that the transition from a manual wheelchair (MW) to a PW increased the occupational performance, competence, adaptability and self-esteem of persons with severe disabilities [10]. However, a large number of users are reluctant to use the PW, which makes the integration process important in promoting autonomy [11].

Satisfaction with AT, including device and services, is increasingly understood as a strong outcome measure in the client-centered rehabilitation programs [3].

Thus, it is important to understand the impact of an AT on its users' participation profile, in order to improve the quality of services, based on their needs and preferences. The aim of this study was to evaluate the psychosocial impact of the PW on the social participation of its users and to assess the relation between the users' satisfaction and his/her participation profile.

2 Methods

A cross-sectional study was designed, using a sample of 30 PW users recruited in the Physical and Rehabilitation Medicine Department of the Coimbra University Hospital Centre (CHUC) and in two other institutions whose patients were PW users, specifically, the Cerebral Palsy Association of Coimbra and the Occupational Activities Centre of São Silvestre. The study was approved by the CHUC Ethic Commission (number CHUC-054-17).

Inclusion criteria were age between 18 and 64 years old, ability to understand written and spoken Portuguese language and using the current PW for at least one year and four hours per day. All individuals voluntarily gave their informed consent to participate in the study.

Data were collected between May and October of 2017. Participants responded to the Portuguese version of Quebec User Evaluation of Satisfaction with Assistive Technology, version 2.0 (QUEST) [12, 13]; the Portuguese version of Psychosocial Impact of Assistive Devices Scale (P-PIADS) [14] and the Activities and Participation Profile Related to Mobility (PAPM) [15], in addition to demographic, clinical and PW related questions.

The QUEST 2.0 is a 12-item questionnaire whose purpose is to evaluate the users' satisfaction with his/her AT (regarding its dimensions, weight, ease of adjustment, safety, durability, ease of use, comfort and effectiveness) and the related services (specifically, delivery, repairs and servicing, professional services and follow-up services). The participant rates his/her satisfaction regarding each of the 12 items using a scale of 1 to 5 and, afterwards, chooses the 3 items he/she considers to be the most important ones [13].

The PIADS is a 26-item, self-report questionnaire that evaluates the effects of an AT on the functional independence, well-being and quality of life of its user. It assesses psychosocial factors which includes both factors within the individual and other factors attributable to the environment that affect the psychological adjustment of persons with disabilities. PIADS comprises 3 subscales, specifically, competence, adaptability and self-esteem. The competence subscale, composed of 12 items, is related to the perceived impact of the AT on the users' competence, performance and productivity. The adaptability subscale, with 6 items, assesses the users' eagerness to try new challenges and take risks and his/her ability to take advantage of opportunities, thus evaluating the enabling aspects of the AT regarding participation. Finally, the self-esteem subscale, composed of 8 items, measures the perceived impact of the AT on self-confidence and emotional well-being. For each item, a score is attributed ranging from −3 (maximum negative impact) to +3 (maximum positive impact). The midpoint, zero, indicates no perceived impact [16]. For every item, the impact of the current AT, which in this specific study is always a PW, is being compared to either the previous AT used by the participant or to not using an AT at all, depending on the participant's previous situation.

The PAPM is an 18-item scale designed to measure the social participation of community dwelling adults. It assesses the difficulties experienced by individuals in performing daily life activities that may be conditioned by mobility and related to social interactions and relations, education, employment, money management and social and community life. For each item, the participant attributes a score ranging from zero (no limitation/restriction) to four (complete limitation/restriction), except for the activities that do not apply to the individual's life (NA), which are not rated. As a result, an individual's participation profile is obtained [15].

In addition, the participants replied to a form comprising demographic, clinical and AT related questions. The demographic information consisted of age, gender and occupation (before and after starting to use the current PW). Descriptive and inferential statistics were conducted, using the software "IBM SPSS Statistics (version 23 for Apple®)". Correlations between QUEST (total and subscales), P-PIADS (total and subscales), PAPM, age, time since onset of medical condition and time since adoption of the current PW were carried out through Pearson's Coefficient. Differences between groups, specifically, between users who had a MW as the previous AT compared to those who had a different PW and between users who had received training with the PW compared to those who had not, were evaluated using Student´s t-test for independent samples. Groups were compared in terms of the scores obtained in each of the scales and subscales. Cross-tabulation was used to describe the relationship between the users previous AT and whether they had received training with the current PW. Descriptive statistics were used to describe the results of QUEST, P-PIADS and PAPM and respective subscales, as well as other demographic, clinical and AT related data.

Testing for normality was executed using a Kolmogorov-Smirnov test. In inferential statistics the reference p value used was <0.05.

3 Results

From a total of 30 participants, 18 were men (60%). The mean age of was 40.63 years (SD 13.09). Regarding occupation, 22 (73.3%) were unemployed. The number of unemployed participants increased to 28 (93.3%) after the beginning of use of the current PW, with 2 of the participants (6.7%) remaining employed (one of them changed to a different job). Cerebral palsy was the most prevalent diagnosis (37%).

As presented in Table 1, the PAPM score was negatively correlated with all QUEST scores (total, AT and services), meaning the most satisfied users with the AT and/or the related services, were the ones with the best performance in terms of social participation (lower PAPM scores correspond to a better participation profile). On other hand, there was no correlation between any of the P-PIADS scores and the PAPM score, showing no relation between the psychosocial impact of the PW and the users' participation profile.

Table 1. Relation between PAPM, P-PIADS and QUEST scores (n = 30)

	PAPM		QUEST total		QUEST AT		QUEST services	
	r	p	r	p	r	p	r	p
PAPM	–	–	−0.449	0.013	−0.363	0.049	−0.383	0.037
P-PIADS competence	−0.096	0.615	0.269	0.150	0.335	0.070	0.059	0.758
P-PIADS adaptability	−0.012	0.615	0.143	0.450	0.221	0.240	−0.031	0.872
P-PIADS self-esteem	0.000	1.000	0.203	0.283	0.191	0.313	0.134	0.481
P-PIADS total	−0.053	0.782	0.237	0.207	0.291	0.119	0.058	0.759

As demonstrated in Table 2, the PAPM score was negatively correlated with age, indicating that older users had a better participation profile compared to younger ones, contrary to what might have been expected. There was a positive correlation between the PAPM score and the amount of time the PW had been used for ($r = 0.409$, $p = 0.025$). There was no correlation between the time since onset of the disease and the PAPM score. The time since the onset of the disease was negatively correlated with PIADS total, competence and adaptability scores ($r = -0.398$, $p = 0.030$; $r = -0.378$, $p = 0.039$; $r = -0.478$, $p = 0.008$).

The QUEST total and AT scores were positively correlated with the age of the participants ($r = 0.519$, $p = 0.003$; $r = 0.481$, $p = 0.007$, respectively), which means older users were more satisfied with their AT than younger ones. The amount of time the PW had been used for was negatively correlated with all the QUEST scores ($r = -0.444$, $p = 0.014$; $r = -0.370$, $p = 0.044$; $r = -0.363$, $p = 0.048$, respectively), suggesting that participants who had been using their PW for a longer period of time were less satisfied with both the AT and the related services. There was no correlation between the time of onset of the disease and any of the QUEST scores.

Table 2. Relation between PAPM, P-PIADS, QUEST and age, time using PW and time since onset of disease (n = 30)

	Age		Time using PW		Time since onset of disease	
	r	p	r	p	r	p
PAPM	−0.584	0.001	0.409	0.025	−0.159	0.402
P-PIADS competence	0.217	0.249	0.056	0.768	−0.378	0.039
P-PIADS adaptability	0.157	0.407	0.032	0.865	−0.478	0.008
P-PIADS self-esteem	0.287	0.124	0.090	0.635	−0.225	0.233
P-PIADS total	0.238	0.206	0.064	0.739	−0.398	0.030
QUEST AT	0.481	0.007	−0.370	0.044	0.068	0.720
QUEST services	0.355	0.055	−0.363	0.048	0.219	0.244
QUEST total	0.519	0.003	−0.444	0.014	0.157	0.408

4 Discussion

According to our findings, the users' social participation does not seem to be associated to the psychosocial impact of the PW, similarly to what was demonstrated by Buning, Angelo and Schmeler, who found no significant relationship between the psychosocial impact of Powered Mobility Devices (PMD) and the users' occupational performance [10]. Contrarily, a study by Martins and collaborators demonstrated a relation between higher psychosocial impact scores and a better performance in social participation, regarding different types of AT [5]. The fact that there was no correlation between the psychosocial impact and the users' participation profile might mean that they had such severe physical limitations that, despite a positive psychosocial impact of the PW, it is not enough to attenuate these limitations and improve the users' participation performance. Furthermore, a higher satisfaction was linked to a better participation profile, which had also been previously shown by de Groot and collaborators concerning MW users [17]. On the other hand, another study did not find any correlation between wheelchair user's satisfaction and participation [18].

The fact that, in this study, the younger individuals had the worst participation profile, can be explained by the particular characteristics of the participants, which included several young PW users diagnosed with cerebral palsy and that had severe limitations since birth, which also explains why younger users had been using their PW for a longer period of time compared to older ones, whose diseases in general had a later onset. Regarding to the users' age, in Martins and collaborators' study, no relation was found between age and participation profile [5]. It is important to take into consideration that the PAPM scores may be influenced by the fact that the participants only attribute a score to the activities they consider to be significant to them, choosing "not applicable" for the others [15]. These activities that were considered unimportant may, in fact, be the ones in which the users have more difficulties. However, once this questionnaire is a self-reported measure, we interpret the results as the perception of these persons with disability.

Moreover, our findings suggest a positive psychosocial impact of PW in competence, adaptability and self-esteem, as it was also previously shown by Buning, Angelo and Schmeler regarding PMD [10, 19]. Likewise, other studies showed benefits associated with PW use, namely increased independence [8, 20], well-being [8] and social participation [6, 19]. There was no evidence of negative impact of the AT due to stigmatization, as suggested in other studies [21, 22], since the self-esteem level is about the same as competency and adaptability.

There was no relationship between the psychosocial impact of the PW and the users' age, which indicates that the AT could be beneficial at any age, making sense to be prescribed through the lifespan, also shown by Martins and collaborators [5].

Considering this positive psychosocial impact of the PW, and according to the World Health Organization's definition of Quality of Life (QoL) – "[a]n individual's perception of their position in life in the context of the culture and value systems in which they live and in relation to their goals, expectations, standards and concerns. It is a broad ranging concept affected in a complex way by the person's physical health, psychological state, personal beliefs, social relationships and their relationship to salient features of their environment" [23] – we can infer that the PW increase the QoL of their users, as was also shown by previous studies [9, 24]. Accordingly, Chan and Chan demonstrated a relation between wheelchair users' social participation and QoL [18].

5 Conclusions

Taking into consideration the main aims of this study, we conclude that:

1. There was an overall positive psychosocial impact of the PW in all three PIADS areas (competence, adaptability and self-esteem) and in the PIADS total score, regardless the user's age, with a potential increase in the QoL.
2. The best participation profiles were noted among the most satisfied users.

More studies are necessary to evaluate the impact of the environmental barriers on the PW users' social participation and QoL.

References

1. WHO (World Health Organization): International Classification of Functioning, Disability and Health. In: International Classification of Functioning, Disability and Health, vol. 26, pp. 1–3, 123–134, 168–170 (2001)
2. WHO: World report on disability 2011. Am. J. Phys. Med. Rehabil. Assoc. Acad. Physiatr, vol. 91, p. 549 (2011)
3. Brandt, Å., Kreiner, S., Iwarsson, S.: Mobility-related participation and user satisfaction: construct validity in the context of powered wheelchair use. Disabil. Rehabil. Assist. Technol. **5**(5), 305–313 (2010)
4. WHO: The World Bank. Disability – a global picture. http://www.who.int/disabilities/world_report/2011/report/en/index.html. Accessed 25 Apr 2019
5. Martins, A., Pinheiro, J., Farias, B., Jutai, J.: Psychosocial impact of assistive technologies for mobility and their implications for active ageing. Technologies **4**(3), 28 (2016)

6. Salminen, A.L., Brandt, Å., Samuelsson, K., Töytäri, O., Malmivaara, A.: Mobility devices to promote activity and participation: a systematic review. J. Rehabil. Med. **41**, 697–706 (2009)
7. Samuelsson, K., Wressle, E.: Powered wheelchairs and scooters for outdoor mobility: a pilot study on costs and benefits. Disabil. Rehabil. Assist. Technol. **9**(4), 330–334 (2014)
8. Evans, S., Frank, A.O., Neophytou, C., de Souza, L.: Older adults' use of, and satisfaction with, electric powered indoor/outdoor wheelchairs. Age Ageing **36**(4), 431–435 (2007)
9. Pettersson, I., Ahlström, G., Törnquist, K.: The value of an outdoor powered wheelchair with regard to the quality of life of persons with stroke: a follow-up study. Assist. Technol. **19**(3), 143–153 (2007)
10. Buning, M.E., Angelo, J.A., Schmeler, M.R.: Occupational performance and the transition to powered mobility: a pilot study. Am. J. Occup. Ther. **55**(3), 339–344 (2001)
11. Fehr, L., Langbein, W., Skaar, S.: Adequacy of power wheelchair control interfaces for persons with severe disabilities: a clinical survey. J. Rehabil. Res. Dev. **37**(3), 353–360 (2000)
12. Rodrigues, S.: Validação intercultural do instrumento de medição - Evaluation de la Satisfaction envers une Aide Technique, versão 2.0. Escola Superior de Tecnologia da Saúde de Coimbra (2013)
13. Demers, L.: An international content validation of the Quebec User Evaluation of Satisfaction with assistive Technology (QUEST). Occup. Ther. Int. **6**(3), 159–175 (1999)
14. Martins, A.: Adaptação cultural e linguística da versão portuguesa do PIADS: Escala do impacto psicossocial das tecnologias de apoio (P-PIADS). Integrar **21–22**, 120–124 (2004)
15. Martins, A.: Development and initial validation of the activities and participation profile related to mobility (PAPM). BMC Health Serv. Res. **16**, 78–79 (2016)
16. Jutai, J., Day, H.: Psychosocial impact of assistive devices scale (PIADS). Technol. Disabil. **14**, 107–111 (2002)
17. de Groot, S., Post, M.W., Bongers-Janssen, H.M., Bloemen-Vrencken, J.H., van der Woude, L.H.: Is manual wheelchair satisfaction related to active lifestyle and participation in people with a spinal cord injury? Spinal Cord **49**(4), 560–565 (2011)
18. Chan, S.C., Chan, A.P.: User satisfaction, community participation and quality of life among Chinese wheelchair users with spinal cord injury: a preliminary study. Occup. Ther. Int. **14**(3), 123–143 (2007)
19. Devitt, R., Chau, B., Jutai, J.: The effect of wheelchair use on the quality of life of persons with multiple sclerosis. Occup. Ther. Health Care **17**(3/4), 63–79 (2003)
20. Löfqvist, C., Pettersson, C., Iwarsson, S., Brandt, A.: Mobility and mobility-related participation outcomes of powered wheelchair and scooter interventions after 4-months and 1-year use. Disabil. Rehabil. Assist. Technol. **7**(3), 211–218 (2012)
21. Hemmingsson, H., Lidstrom, H., Nygard, L.: Use of assistive technology devices in mainstream schools: students' perspective. Am. J. Occup. Ther **63**(4), 463–472 (2009)
22. Parette, P., Scherer, M.: Assistive technology use and stigma. Educ. Train. Dev. Disabil. **39**, 217–226 (2004)
23. World Health Organization. WHOQOL - Measuring Quality of Life. http://www.who.int/healthinfo/survey/whoqolqualityoflife/en/. Accessed 25 Apr 2019
24. Davies, A., De Souza, L.H., Frank, A.O.: Changes in the quality of life in severely disabled people following provision of powered indoor/outdoor chairs. Disabil. Rehabil. **25**(6), 286–290 (2003)

Virtual Assistant Prototype for Managing Medication Using Messaging Platforms

Surya Roca(✉), Manuel Hernández, Jorge Sancho, José García, and Álvaro Alesanco

Aragón Institute of Engineering Research (I3A), University of Zaragoza, Zaragoza, Spain
{surya,manuelhr,jslarraz,jogarmo,alesanco}@unizar.es

Abstract. Healthcare chatbots (a.k.a. virtual assistants) offer services, tools, and interactions to the users, providing advice, help, support, and management to different diseases. The users of this type of chatbots can be, for example, patients, caregivers, and healthcare professionals, with different needs and requirements. Sometimes, patients with chronic diseases needing to take different medications several times per day, can have difficulties to remember their medication intakes. This situation can lead to negative effects on treatment effectiveness. In this paper, we present a virtual assistant prototype, aimed to help and manage patient's medication through the use of Signal, a secure messaging platform that uses end-to-end encryption for better user privacy. The objective of this work is to create a handy tool to remind medication intakes through the creation of a supportive community where patients, caregivers, and healthcare professionals interact with helpful tools and services offered by the virtual assistant. Moreover, the developed prototype makes use of medical standards for data management and Artificial Intelligence Markup Language (AIML) for conversation support.

Keywords: Chatbot · HL7 FHIR · Medication management · Messaging platforms · mHealth scenarios · Patient support · Virtual assistant

1 Introduction

Messaging platforms have redefined the way we people communicate and interact every day. The usage of messaging platforms is a must in our interaction with other people. The number of yearly active users in chatting applications has increased by 10.4% the last year, with the usage reaching to two thousand million people worldwide by the end of 2018 [1]. Recently, the combination of Artificial Intelligence (AI) with messaging platforms has generated a new revolution in the user's care: the emergence of chatbots. Chatbots are autonomous intelligent software that provides services of conversing with end users via diverse communication channels (e.g. messaging platforms, smartphone apps, etc.) [2].

J. Henriques et al. (Eds.): MEDICON 2019, IFMBE Proceedings 76, pp. 954–961, 2020.
https://doi.org/10.1007/978-3-030-31635-8_116

Chatbots, or virtual assistants, can help in eHealth scenarios in a wide variety of situations, such as those of patients in remote areas with limitations in healthcare facilities [3].

Nowadays, there is a wide range of chatbots offering help to patients in eHealth scenarios. Ana [4], for example, is a self-anamnesis mobile chatbot application for patients, where a conversational user interface is used to simulate the patient-therapist conversation. Another example is MamaBot [5] which is an AI-chatbot for providing support to mothers, pregnant women, and families with young children. MamaBot provides help and instructions in relevant situations through Telegram [6] (a messaging platform), that does not have end-to-end encryption and, therefore, it does not provide a secure communication channel. SWITCHes [7] is a system which has an AI-powered health chatbot for interaction with users and it provides real-time data reception and transmission to their web server through a smartphone app and a web-based dashboard. CustomRXBot [2] is a chatbot developed to help healthcare professionals with custom prescriptions for dermatology through Slack [8].

An interesting approach of chatbots is the role of medication reminders because they can help in adherence to medication. eMMA [9] is a chatbot-like smartphone app to manage patient medication through a chatbot focused on supporting a consistent flow of medication data to increase medication adherence. Roborto [10] is a chatbot model system that tries to improve patients' medication adherence and inform healthcare providers about patients' conditions and improvements, using Telegram as a communication channel.

As shown in the examples explained before, most of the developed chatbots use a smartphone app to allow the communication between the chatbot and the users. However, a quarter of all downloaded smartphone applications are abandoned after the first usage [5]. Due to this fact, an alternative approach, without the usage of exclusively dedicated applications, should be used (such as messaging platforms). Developing chatbots that use messaging platforms as communication channels helps to provide new tools and functionalities without any update from the user side: chatbots are programs running in servers completely independent from the user application. Thus, any update in the chatbot is not going to affect the user smart devices. Furthermore, any chatbot system needs to provide very strict security measures to protect personal user information during all the message exchanges in healthcare scenarios, to protect the users' privacy.

In this paper, we introduce a healthcare chatbot prototype and its integration in the messaging platform Signal [11] using end-to-end encryption to provide added security and privacy to users' medical information. The chatbot aims to support and provide different functionalities such as medication reminder, summaries, notifications to caregivers, all within a scenario using standardization of data sharing models through HL7 FHIR [12] and conversation modeling using AIML [13].

The rest of the paper is structured as follows. Section 2 provides a description of the system overview. Section 3 describes the developed prototype, explains the services developed for the medication management scenario and shows some examples of real interactions with the virtual assistant. Section 4 discusses the virtual assistant. Finally, Sect. 5 concludes the paper with possible future works.

2 System Overview

The system, as shown in Fig. 1, uses the messaging platform Signal as a communication channel between the virtual assistant and the users. Signal, with its end-to-end encryption, provides privacy in our healthcare scenario.

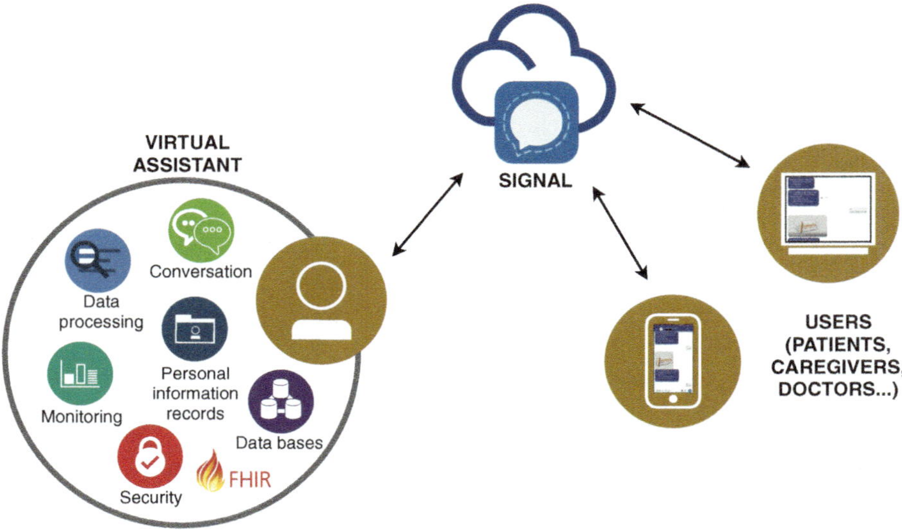

Fig. 1. Overview of the scenario.

The virtual assistant has been programmed using a modular software architecture approach which provides a flexible and adaptable framework to increase its functionalities in an easy way if required. To represent the chatbot's brain, we use Artificial Intelligence Markup Language (AIML). AIML is a recursive language based on XML that defines the stimulus-response message of the interaction and is used to create artificial intelligent agents. AIML matches the user text input into the chatbot knowledge to obtain the response to send to the user. We decided to use AIML due to it is lightweight and easy to configure [14].

The library of the Program AB [15] has been used in this prototype as our AIML 2.0 interpreter that obtains the best match between all AIML categories stored in the AIML files of the project. The conversation patterns in the current version of the virtual assistant have been developed in Spanish, but examples are in English for the sake of understandability.

3 Prototype Development

The development of our virtual assistant was performed using Android, Java, and Python as programming languages. The virtual assistant allows managing medication, setting reminders and data visualization to its users.

In order to adapt and integrate the text messaging to different scenarios where the virtual assistant interacts with the users, the text messages have been modeled using sequence diagrams and discussed with a group of healthcare professionals.

3.1 Services

At this moment, the virtual assistant offers the following services:

- **Role assignment:** users are arranged into groups. There are three different groups: patients, healthcare professionals and informal caregivers (e.g. patient's family).
- **Care group creation:** patients are the center of a care group. In order to form care groups, patients can guarantee access to their data to healthcare professionals and informal caregivers.
- **Medication management:** users are able to add, list and delete medication through interactions with the virtual assistant. They can also obtain information about their prescribed medication (images, drugs intake advice, etc.).
- **Medication reminders:** users are able to program and receive custom medication reminders.
- **Reporting facility:** users can tell the assistant to generate periodic reports about medication adherence and these records can be received by the patients' care group, including themselves.

According to the role that each user has, they can use different specific services. Healthcare professionals and informal caregivers can use medication management, medication reminders and reporting facilities to customize the patient's medication and follow the progress of the medication intake. The patient can use all the services provided and, in addition, create a care group to provide access to their data to other users such as healthcare professionals or informal caregivers.

3.2 Data Input and Storage

The input of the medication data can be done in different ways. One possible option is to do it manually by entering the medication name or national code. With this code, it is possible to search for detailed information about this drug in the Medicine Online Information Center website from the Spanish Medication Agency. After that, the virtual assistant asks further information related to that drug, such as dosage or intake duration. Another interesting possible option is to use a photo of the medical prescription. We have developed an OCR module trained with the Aragón health department prescription template, able to extract the medication details (including dosage, timespan, etc.) from the photo. Although the procedure is very accurate, possible errors may occur due to the luminosity or the angle of the page in the photograph. To guarantee that only correct data is introduced, the users are requested to check the accuracy of the extracted data in a step by step process, verify the data and change the data if it is not accurate.

Once the medication data is introduced by the user, the data is stored in the system following the medical standard HL7 FHIR to provide interoperability. This standard stores the medical information into resources. The FHIR resources used in this scenario are:

- Person, Patient, Practitioner, and Related Person to store the actors of the system.
- Medication, Medication Request, and Medication Administration to store the medical data.

Because of the sensitivity of the data stored in the system, we follow both national data protection law LO 03/2018 [16] and European GPDR [17] which dictates how personal data have to be managed to protect the user's rights and privacy.

3.3 Reminders and Notifications

Reminders are messages sent to the users at the scheduled time when they need to take a medication. The usage of reminders can help patients improve their adherence to medication. Reminders can be set by patients, caregivers or healthcare professionals (patients care group) when a new medication is added. These reminders are specific for reminding the drug dosage and the quantity. In response to these reminders, the user can specify if the medication has been taken or not. Thanks to this, the virtual assistant is able to warn any person of the patient's care group if the patient is not taking the medication using notifications.

The conversation that takes place when a user receives a new reminder of a drug administration is shown in Fig. 2a. The user can ask, as shown in the figure, the image of the medication, to help her/him to take the correct medication.

3.4 Summaries

The response (or absence of) to the reminders is used to estimate the adherence to medication and to create a summary of drugs intake. Furthermore, the proposed system allows the flexibility to the caregivers to receive the summaries of medication periodically (e.g. once every three days) or to receive a summary of the last days whenever the user asks for it (see in Fig. 2b for an example).

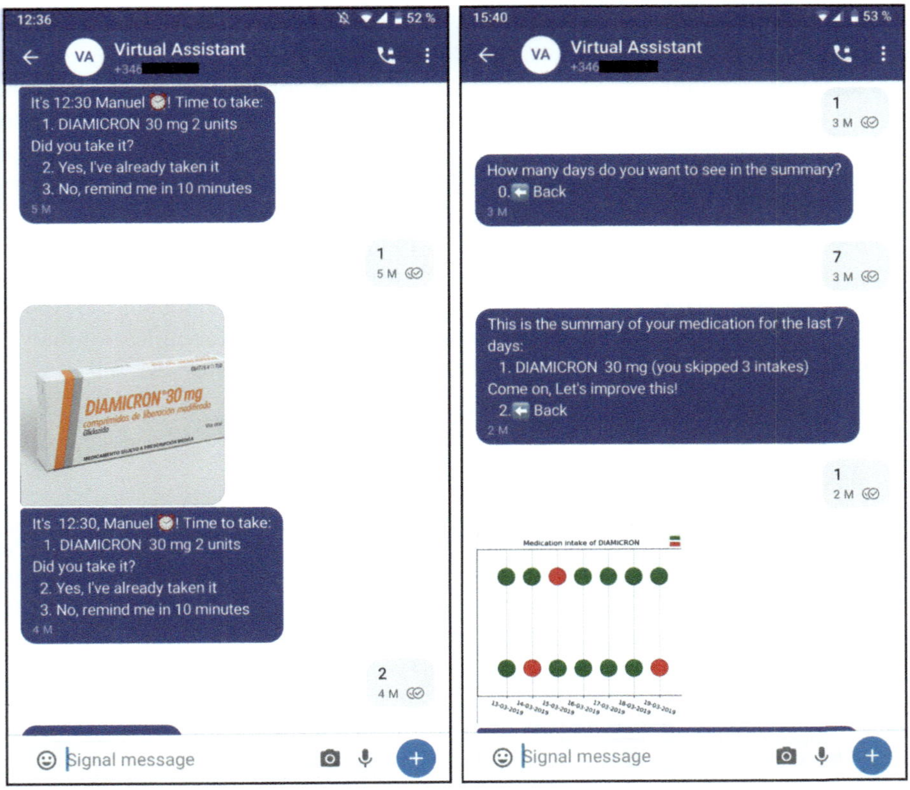

(a) Reminder example. (b) Summary example.

Fig. 2. Real interaction with the virtual assistant developed.

4 Discussion

In order to use the virtual assistant, the user needs to be connected to the Internet and be familiar with the chatting interaction (not a difficult requirement taking into account the current penetration of messaging platforms). Obviously, being connected is also a necessary condition to receive reminders and notifications.

At this moment, we have developed the virtual assistant to work with fixed keywords i.e. our virtual assistant is based on a visual-centric approach [18]. The keywords are shown in menus, where the option or the number associated with that option is understood. Otherwise, the virtual assistant is not able to interpret free text interactions (conversation-centric approach [18]).

There is an inherent limitation with OCR recognition. The OCR that we have developed only works with the prescription from our Spanish region (Aragón) due to the customization needed in the OCR module. If the system is used with other prescriptions models, the OCR recognition needs to be adapted accordingly.

Privacy of the messaging platform is paramount. We have chosen Signal because of its end-to-end encryption capabilities and its open source code approach. If the communication is not end-to-end encrypted, medical information can be eavesdropped by the messaging platform servers. It is interesting to note that any messaging platform using end-to-end encryption for user communication can be used for the deployment. Nowadays, many platforms have developed specific APIs for an easy chatbot development (e.g. Telegram) but these APIs do not use end-to-end encryption at this moment (messages are decrypted at messaging platform servers and then encrypted again to be forwarded to the destination user). Hence, such APIs should not be used for secure chatbot development. This makes the choice of platforms for chatbot communication harder since not many platforms provide with an open way to use their end-to-end encryption mode. Signal is one of the few that provides end-to-end encryption.

5 Conclusions and Future Work

Virtual assistants specifically developed to support and manage user's needs in health scenarios provide an ideal tool for building communities where healthcare professionals, patients, and informal caregivers can be involved. In this paper, we presented and discussed a virtual assistant prototype for medication management using the messaging platform Signal as a secure communication channel. The prototype is aimed at helping users in the management of medication administration and intake. The text-based interaction of the virtual assistant has been modeled with AIML for better flexibility and proactiveness in the interaction with the users. This interaction, along with the usage of the messaging platforms, provides new ways for sending reminders, notifications, and summary of medication intakes to the system users. HL7 FHIR has been used as a standard for medical data storage to allow interoperability between different healthcare systems. At this moment, we are working on performing a usability test and an adherence study with a small group of selected patients. We also aim to improve the extraction of the relevant information from the user-virtual assistant interaction so as to be able to store the relevant medical information for more personalized healthcare. This virtual assistant prototype is focused on managing medication, but also it opens the possibility to expand its functionalities, adding further dimensions for future research, such as relevant data extraction or expanding its features with the incorporation of natural language understanding techniques.

Acknowledgements. Research funded by Ministerio de Economía, Industria y Competitividad from Gobierno de España and European Regional Development Fund (TIN2016-76770-R and BES-2017-082017) and Gobierno de Aragón and FEDER "Construyendo Europa desde Aragón" (T31_17R).

Conflict of Interest. The authors declare that they have no conflict of interest.

References

1. Number of mobile phone messaging app users worldwide from 2016 to 2021 (in billions). https://www.statista.com/statistics/483255/number-of-mobile-messaging-users-worldwide/. Accessed 17 June 2019
2. Alesanco, Á., Sancho, J., Gilaberte, Y., Abarca, E., García, J.: Bots in messaging platforms, a new paradigm in healthcare delivery: application to custom prescription in dermatology. In: EMBEC & NBC 2017, pp. 185–188. Springer (2017)
3. Fadhil, A.: Beyond patient monitoring: conversational agents role in telemedicine & healthcare support for home-living elderly individuals. CoRR abs/1803.06000 (2018). http://arxiv.org/abs/1803.06000
4. Denecke, K., Lutz Hochreutener, S., Pöpel, A., May, R.: Talking to ANA: a mobile self-anamnesis application with conversational user interface. In: Proceedings of the 2018 International Conference on Digital Health, pp. 85–89. ACM (2018)
5. Vaira, L., Bochicchio, M.A., Conte, M., Casaluci, F.M., Melpignano, A.: MamaBot: a system based on ML and NLP for supporting women and families during pregnancy. In: Proceedings of the 22nd International Database Engineering & Applications Symposium, pp. 273–277. ACM (2018)
6. Telegram homepage. https://telegram.org/. Accessed 18 Mar 2019
7. Huang, C.Y., Yang, M.C., Huang, C.Y., Chen, Y.J., Wu, M.L., Chen, K.W.: A chatbot-supported smart wireless interactive healthcare system for weight control and health promotion. In: 2018 IEEE International Conference on Industrial Engineering and Engineering Management (IEEM), pp. 1791–1795. IEEE (2018)
8. Slack homepage. https://slack.com. Accessed 27 Mar 2019
9. Tschanz, M., Dorner, T.L., Holm, J., Denecke, K.: Using emma to manage medication. Computer **51**(8), 18–25 (2018)
10. Fadhil, A.: A conversational interface to improve medication adherence: towards AI support in patient's treatment. arXiv preprint arXiv:1803.09844 (2018)
11. Signal homepage. https://signal.org/. Accessed 18 Mar 2019
12. Hl7 fhir. http://www.hl7.org/FHIR/. Accessed 18 Mar 2019
13. Wallace, R.: The elements of AIML style. Alice AI Foundation (2003)
14. Satu, M.S., Parvez, M.H.: Shamim-Al-Mamun: review of integrated applications with AIML based chatbot. In: 2015 International Conference on Computer and Information Engineering (ICCIE), pp. 87–90, November 2015
15. Program AB. ALICE A.I. foundation. https://code.google.com/archive/p/program-ab/. Accessed 17 Mar 2019
16. Constitutional law 3/2018 of the 5th December concerning protection of personal data and guarantee of digital rights. https://www.boe.es/buscar/doc.php?id=BOE-A-2018-16673. Accessed 20 Mar 2019
17. Regulation (EU) 2016/679 of the European parliament and of the council of 27 April 2016. https://eur-lex.europa.eu/eli/reg/2016/679/oj. Accessed 20 Mar 2019
18. Conversational UX design - IBM. https://researcher.watson.ibm.com/researcher/view_group.php?id=8426. Accessed 20 Mar 2019

GPSR Compliance

The European Union's (EU) General Product Safety Regulation (GPSR) is a set of rules that requires consumer products to be safe and our obligations to ensure this.

If you have any concerns about our products, you can contact us on ProductSafety@springernature.com

In case Publisher is established outside the EU, the EU authorized representative is:

Springer Nature Customer Service Center GmbH
Europaplatz 3
69115 Heidelberg, Germany

Batch number: 10755090

Printed by Printforce, the Netherlands